ANNOTATED TEACHER'S EDITION

Eastern Hemisphere

HOLT

PEOPLE, PLACES, AND CHANGE

An Introduction to World Studies

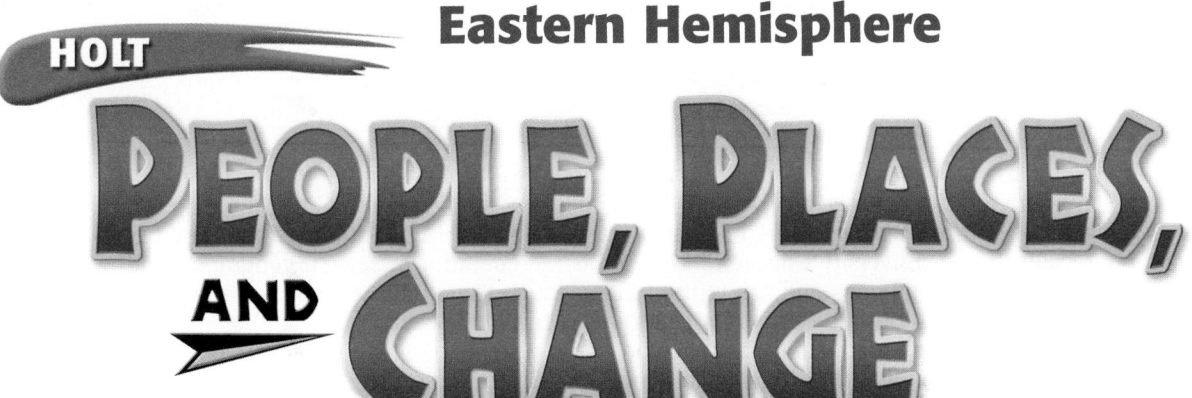

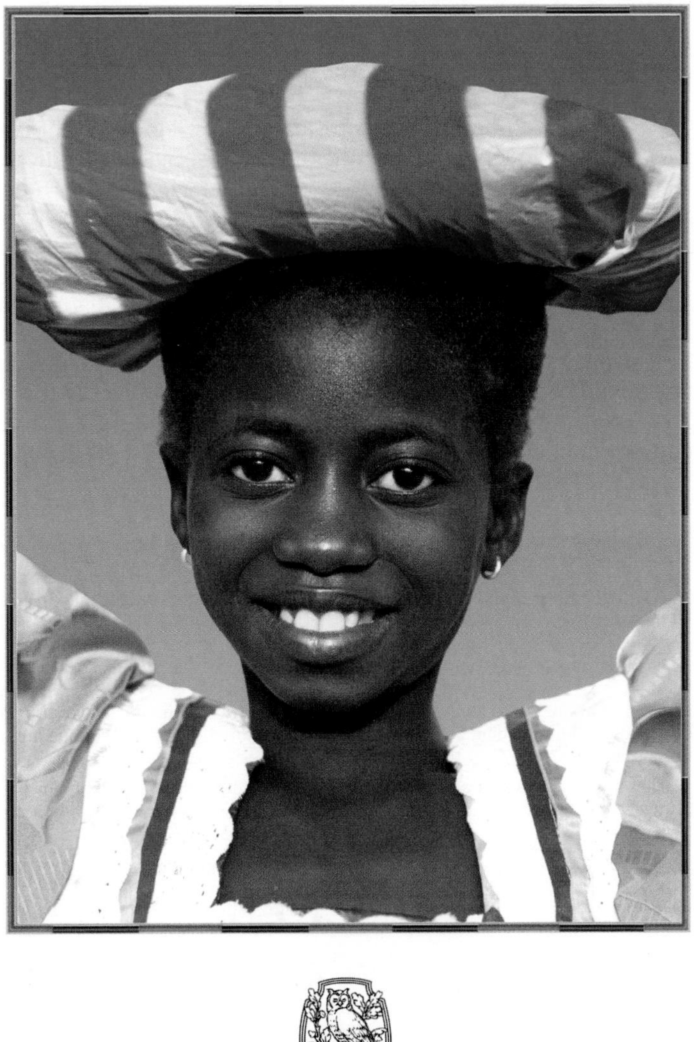

HOLT, RINEHART AND WINSTON

A Harcourt Education Company

Austin • Orlando • Chicago • New York • Toronto • London • San Diego

THE AUTHORS

Prof. Robert J. Sager is Chair of Earth Sciences at Pierce College in Lakewood, Washington. Prof. Sager received his B.S. in geology and geography and M.S. in geography from the University of Wisconsin and holds a J.D. in international law from Western State University College of Law. He is the coauthor of several geography and earth science textbooks and has written many articles and educational media programs on the geography of the Pacific. Prof. Sager has received several National Science Foundation study grants and has twice been a recipient of the University of Texas NISOD National Teaching Excellence Award. He is a founding member of the Southern California Geographic Alliance and former president of the Association of Washington Geographers.

Prof. David M. Helgren is Director of the Center for Geographic Education at San Jose State University in California, where he is also Chair of the Department of Geography. Prof. Helgren received his Ph.D. in geography from the University of Chicago. He is the coauthor of several geography textbooks and has written many articles on the geography of Africa. Awards from the National Geographic Society, the National Science Foundation, and the L. S. B. Leakey Foundation have supported his many field research projects. Prof. Helgren is a former president of the California Geographical Society and a founder of the Northern California Geographic Alliance.

Prof. Alison S. Brooks is Professor of Anthropology at George Washington University and a Research Associate in Anthropology at the Smithsonian Institution. She received her A.B., M.A., and Ph.D. in Anthropology from Harvard University. Since 1964, she has carried out ethnological and archaeological research in Africa, Europe, and Asia and is the author of more than 300 scholarly and popular publications. She has served as a consultant to Smithsonian exhibits and to National Geographic, Public Broadcasting, the Discovery Channel, and other public media. In addition, she is a founder and editor of *Anthro Notes: The National Museum of Natural History's Bulletin for Teachers* and has received numerous grants and awards to develop and lead in-service training institutes for teachers in grades 5–12. She served as the American Anthropological Association's representative to the NCSS task force on developing Scope and Sequence guidelines for Social Studies Education in grades K–12.

While the details of the young people's stories in the chapter openers are real, their identities have been changed to protect their privacy.

Cover and Title Page Photo Credits: (child image) Steve Vidler/Nawrocki Stock Photo; (bkgd) Image Copyright © 2003 PhotoDisc, Inc./HRW

For acknowledgments, see page R43, which is an extension of the copyright page.

Printed in the United States of America

ISBN 0-03-053652-9

1 2 3 4 5 6 7 8 9 032 05 04 03 02

CONTENT REVIEWERS

Robin Datel
Instructor in Geography
California State University, Sacramento

David Dickason
Professor of Geography
Western Michigan University

Dennis Dingemans
Professor of Geography
University of California, Davis

Robert Gabler
Professor of Geography
Western Illinois University

Jeffrey Gritzner
Professor of Geography
University of Montana

W. A. Douglas Jackson
Professor of Geography, Emeritus
University of Washington

Robert B. Kent
Professor of Geography and Planning
University of Akron

Kwadwo Konadu-Agyemang
Professor of Geography and Planning
University of Akron

Nancy Lewis
Professor of Geography
University of Hawaii

Bill Takizawa
Professor of Geography
San Jose State University

EDUCATIONAL REVIEWERS

Patricia Britt
Durant Middle School
Durant, Oklahoma

Marcia Caldwell
Lamar Middle School
Austin, Texas

Marcia Clevenger
Roosevelt Junior High School
Charleston, West Virginia

James Corley
Durant Middle School
Durant, Oklahoma

Maureen Dempsey
Spring Creek Middle School
Spring Creek, Nevada

Jean Eldredge
Teague Middle School
Altamonte, Florida

Cindy Herring
Old Town Elementary School
Round Rock, Texas

Lois Jordan
Pearl/Cohn Comprehensive High School
Nashville, Tennessee

Kay A. Knowles
Montross Middle School
Montross, Virginia

Wendy Mason
Corbett Junior High School
Schertz, Texas

Rebecca Minnear
Burkholder Middle School
Las Vegas, Nevada

Jane Palmer
District Supervisor for Social Studies
Sanford, Florida

Sandra Rojas
Adams City Middle School
Commerce City, Colorado

JoAnn Sadler
Curriculum Supervisor
Buffalo City Schools
Buffalo, New York

Celeste Smith
Crockett High School
Austin, Texas

Frank Thomas
Crockett High School
Austin, Texas

Susan Walker
Beaufort County School District
Beaufort, South Carolina

Field Test Teachers

Ricky A. Blackman
Rawlinson Road Middle School
Rock Hill, South Carolina

Lisa Klien
Daniels Middle School
Raleigh, North Carolina

Deborah D. Larry
Garland V. Stewart Middle School
Tampa, Florida

Linda P. Moore
Cramerton Middle School
Cramerton, North Carolina

Earl F. Sease
Portage Area School District
Portage, Pennsylvania

Christi Sherrill
Grier Middle School
Gastonia, North Carolina

John W. Watkins, Jr.
Clark Middle School
East St. Louis, Illinois

It's All About

INTERVIEWS with young people from around the world begin each chapter, making real world peer connections for your students. Plus students can read the interviews in their entirety on our Web site.

The first step to success in the social studies classroom is capturing and sustaining the interest of your students. *HOLT PEOPLE, PLACES, AND CHANGE* is designed to be open and friendly to all students, so that they develop an enthusiasm for learning and an appreciation for their world.

***HOLT PEOPLE, PLACES, AND CHANGE* offers**

- **Built-in Reading Support**
- **Technology with Instructional Value**
- **Standardized Testing Strategies and Skill Building**
- **The Best Teacher's Management System in the Industry**

RELEVANCE

CNNfyi.com™ is designed to give students in grades 6–12 access to the news about people, places, and environments around the globe while offering "real-world" articles, career and college resources, and online activities.

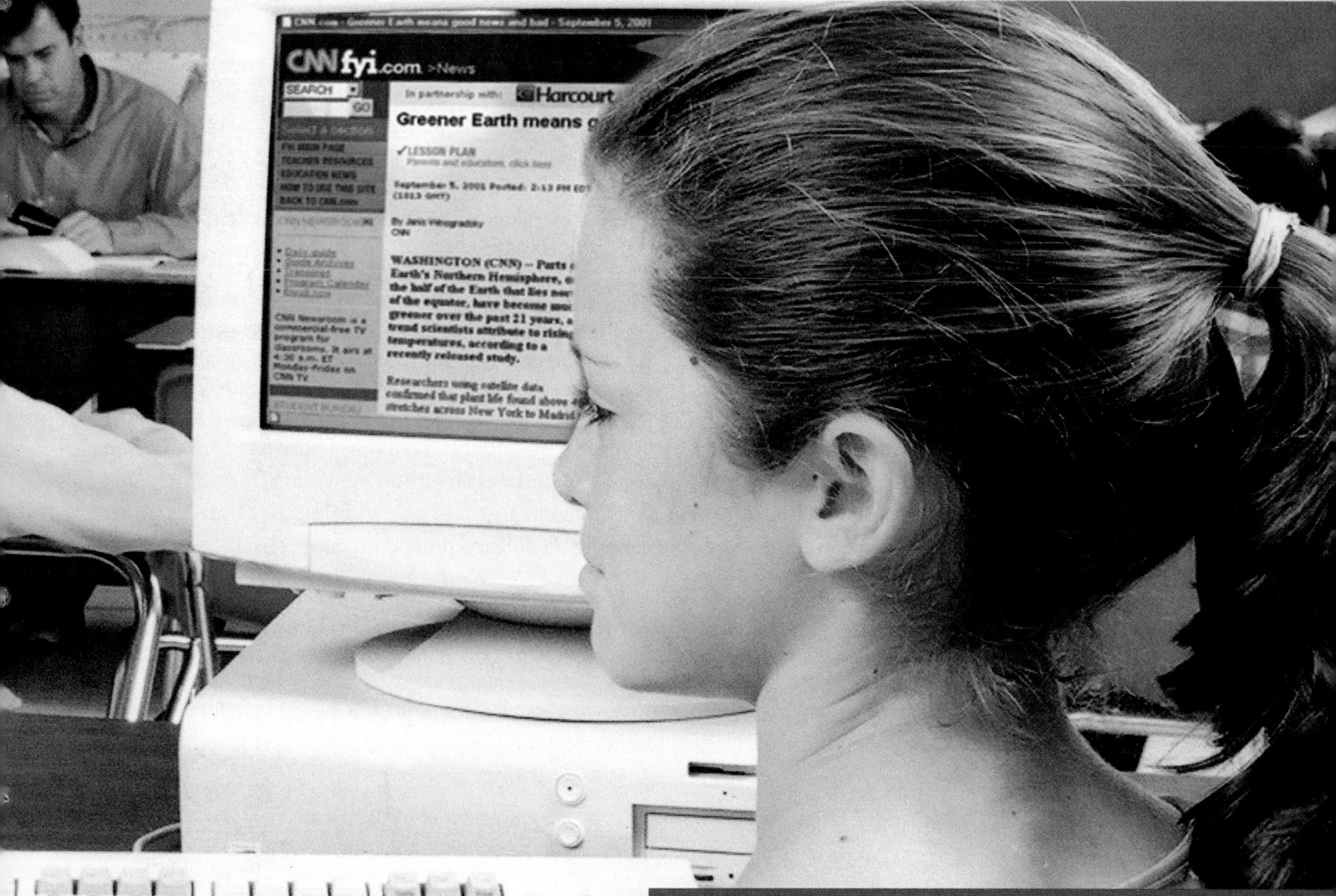

In-Text Features that Put Geography into Perspective

- Case Study
- Connecting to Art
- Connecting to History
- Connecting to Literature
- Connecting to Math
- Connecting to Science
- Connecting to Technology
- Daily Life
- Focus on Culture
- Focus on Economy
- Focus on Environment
- Focus on Government
- Focus on Regions
- Geo Skills
- Hands On Geography
- Our Amazing Planet
- Why It Matters

Reading for

At Holt, we don't assume that students know how or have any desire to make sense of what they're reading, and we develop our programs based on that assumption. We don't just ask students questions about content, we give them strategies to get to that content. Through design, research, and the help of experts like Dr. Judith Irvin, we make sure students' reading needs are covered with our programs.

Helping Students Make Sense of What They're Reading

An Essay by Dr. Judith Irvin, Ph.D.

Who in middle and high schools helps students become more successful at reading and writing informational text? When I ask this question of a school faculty, the Language Arts/English teachers point to the social studies and science teachers because they are the ones with this type of textbook. The social studies and science teachers point to the Language Arts/English teachers because they are the ones that "do" words.

I advocate teachers taking an active role in helping students learn how to use text structure and context to understand what they read. Through consistent and systematic instruction that includes modeling of effective reading behavior, teachers can assist students in becoming better readers while at the same time helping them learn more content material.

The strategies in this book are designed to assist students with getting started, maintaining focus with reading, and organizing information for later retrieval. They engage students in learning material, provide the vehicle for them to organize and reorganize concepts, and extend their understanding through writing.

When teachers combine the teaching of reading and the teaching of content together into meaningful, systematic, and corrected instruction, students can apply what they have learned to understanding increasingly more difficult and complex texts as they progress through the school years.

READING STRATEGIES FOR THE SOCIAL STUDIES CLASSROOM

by Dr. Judith Irvin,
Ph.D., Reading Education

Additional Reading Support

- **Graphic Organizer Activities**
- **Guided Reading Strategies**
- **Main Idea Activities for English Language Learners and Special-Needs Students**
- **Audio CD Program**

MEANING

Section 2 The Countries of Northeastern Europe

Read to Discover

1. What peoples contributed to the early history of northeastern Europe?
2. How was northeastern Europe's culture influenced by other cultures?
3. How has the political organization of this region changed since World War II?

Define

Indo-European

Locate

Estonia, Poland, Czech Republic, Slovakia, Hungary, Lithuania, Latvia, Prague, Tallinn, Riga, Warsaw, Vistula River, Bratislava, Budapest

Why It Matters

Jerzy Giedroye, a Polish editor living in France, helped keep the free exchange of ideas alive in Eastern Europe during Communist rule. Use CNNfyi.com or other current events sources to discover the efforts of people like Giedroye. Record your findings in your journal.

Musical score by Hungarian Béla Bartók

The Teutonic knights, a German order of soldier monks, brought Christianity and feudalism to northeastern Europe. They built this castle at Malbork, Poland, in the 1200s.

History

Migrants and warring armies have swept across Eastern Europe over the centuries. Each group of people brought its own language, religion, and customs. Together these groups contributed to the mosaic of cultures we see in Eastern Europe today.

Early History Among the region's early peoples were the Balts. The Balts lived on the eastern coast of the Baltic Sea. They spoke **Indo-European** languages. The Indo-European language family includes many languages spoken in Europe. These include Germanic, Baltic, and Slavic languages. More than 3,500 years ago, hunters from the Ural Mountains moved into what is now Estonia. They spoke a very different, non-Indo-European language. The language they spoke provided the early roots of today's Estonian and Finnish languages. Beginning around A.D. 400, a warrior people called the Huns invaded the region from Asia. Later, the Slavs came to the region from the plains north of the Black Sea.

292 Chapter 13

In the 800s the Magyars moved into the Great Hungarian Plain. They spoke a language related to Turkish. In the 1200s the Mongols rode out of Central Asia into Hungary. At the same time German settlers pushed eastward, colonizing Poland and Bohemia—the western region of the present-day Czech Republic.

Emerging Nations Since the Middle Ages, Austria, Russia, Sweden, and the German state of Prussia have all ruled parts of Eastern Europe. After World War I ended in 1918, a new map of Eastern Europe was drawn. The peace treaty created two new countries: Yugoslavia and Czechoslovakia. Czechoslovakia included the old regions of Bohemia, Moravia, and Slovakia. At about the same time, Poland, Lithuania, Latvia, and Estonia also became independent countries.

✓ READING CHECK: *Human Systems* What peoples contributed to the region's early history? Balts, hunters from Ural Mountains, Huns, Slavs, Magyars, Mongols, Germans

Culture

The culture and festivals of this region show the influence of the many peoples who contributed to its history. As in Scandinavia, Latvians celebrate a midsummer festival. The festival marks the summer solstice, the year's longest day. Poles celebrate major Roman Catholic festivals. Many of these have become symbols of the Polish nation. The annual pilgrimage, or journey, to the shrine of the Black Madonna of Częstochowa (chen-stuh-KOH-vuh) is an example.

Traditional Foods The food of the region reflects German, Russian, and Scandinavian influences. As in northern Europe, potatoes and sausages are important in the diets of Poland and the Baltic countries. Although the region has only limited access to the sea, the fish of lakes and rivers are often the center of a meal. These fish often include trout and carp. Many foods are preserved to last through the long winter. These include pickles, fruits in syrup, dried or smoked hams and sausages, and cured fish.

The Arts, Literature, and Science Northeastern Europe has made major contributions to the arts, literature, and sciences. For example, Frédéric Chopin (1810–1849) was a famous Polish pianist and composer. Marie Curie (1867–1934), one of the first female physicists, was also born in Poland. The writer Franz Kafka (1883–1924) was born to Jewish parents in Prague (PRAHG), the

Hungarian dancers perform in traditional dress.

Interpreting the Visual Record **How does this Hungarian costume compare to those you have seen from other countries?**

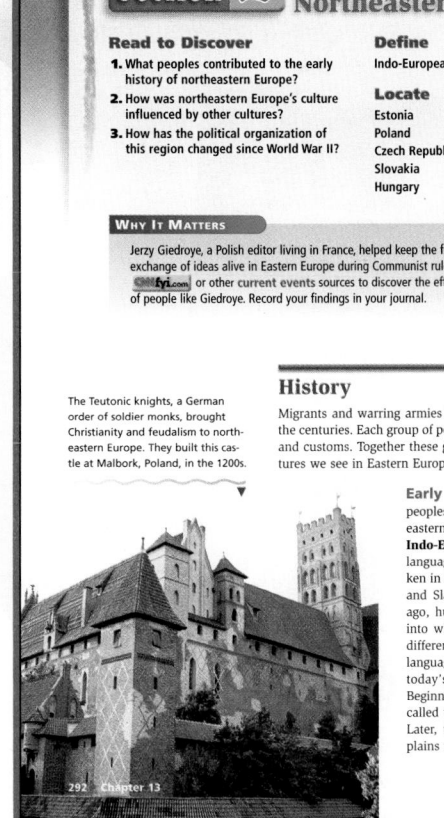

Eastern Europe • 293

Successful Readers must have:

1 AN ENGAGING NARRATIVE

Great care is taken in selecting and presenting content in a way that students will find motivating and engaging. Features such as **Youth Interviews** help students connect their own lives to the lives and cultures of other students around the world.

2 A FORECAST OF WHAT THEY WILL LEARN

Read to Discover questions give students insight into the content they will cover in the chapter to come. In features such as **Why It Matters,** students gain insight into regional issues.

3 VOCABULARY DEFINED IN CONTEXT

Important new terms are identified at the beginning of every section and are defined in context so students will develop an understanding of the contextual meaning of all terms.

4 STRATEGIES FOR UNDERSTANDING WHAT THEY READ

Through the design of the text, students are led through the content using built-in reading strategies. For example, **Reading Checks** in the text are used as a comprehension tool. The checks remind students to stop and engage with what they have read, functioning as a "Tutor in the Text."

Get Your Students

Your students love activities that get them involved with the content. That's why Holt offers active-learning resources that link directly to program content and provide a multitude of different lessons for large-group, small-group, and individual projects.

CREATIVE TEACHING STRATEGIES

These innovative teaching strategies can be utilized at various points in your lesson. The wide range of cooperative-learning activities, including learning stations and simulations, motivate your students and help them develop critical-thinking skills.

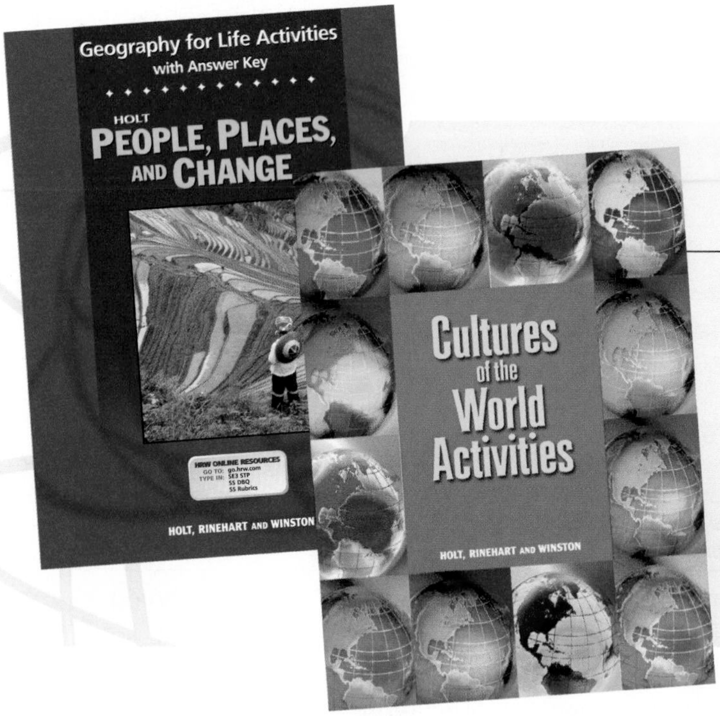

HANDS-ON GEOGRAPHY ACTIVITIES

From hands-on study of world cultures to hands-on practice with skill building, the following booklets cover it all. ***Cultures of the World Activities*** is a stand-alone booklet containing recipes, games, and craft activities. ***Geography for Life Activities with Answer Key*** contains a a problem-solving activity for each chapter, reflecting the skills and knowledge called for in the **National Geography Standards.**

GEOGRAPHY APPLICATIONS

For use in geography as well as earth science courses, here are two stand-alone booklets that organize applications with special relevance. ***Environmental and Global Issues Activities*** contains activities related to current environmental and global issues. ***Lab Activities for Geography and Earth Science*** contains laboratory activities related to physical geography and earth science.

Involved in Learning

Resources for Active Lessons

- Block Scheduling Handbook
- Creative Strategies for Teaching World Geography
- Critical Thinking Activities
- Cultures of the World Activities
- Environmental and Global Issues Activities
- Geography and Cultures Visual Resources with Teaching Activities
- Geography for Life Activities
- Interdisciplinary Activities for the Middle Grades
- Lab Activities for Geography and Earth Science
- Map Activities
- Regions of the World Map Posters
- World and Regional Outline Maps
- World History and Geography Document-Based Questions Activities

Joining Forces

CNN® Presents

Geography: Yesterday and Today

- Mapping Change
- Birth of an Island
- Global Warming
- Farming with Saltwater
- World Population—Hitting Six Billion
- What's Cooking for the Holidays?
- Do Aspen Trees Hold the Key?
- Tradition and Change—The James Bay Project
- Baseball, Mexican Style
- Preserving Heritage
- The Power of Poetry
- Destination Brazil
- Harnessing Nature
- The Fishing Life
- Europe's Immigration Challenge
- The Mountain People of Yo
- Castles for Sale
- Lake Baykal
- Cossacks Return to Power
- Dateline Kazakhstan
- Jordan's Water Crisis
- Kuwait—Restoring the Environment
- Cairo—Selling the Suburbs
- The Women of Nigeria
- The Nairobi National Park
- The Medicine Tree
- Mining in South Africa the Traditional Way
- The Yangtze (Chang) River Dam
- The Tokyo Grave Crisis
- The Highway to Prosperity
- The Borobudur Temple of Indonesia
- Indian Pop Music
- The Tengboche Monastery
- The Great Barrier Reef
- Dateline Antarctica

World Cultures: Yesterday and Today

- The Kennewick Man
- Restoring the Sphinx
- The Sacred Ganges
- Treasures from China
- Crete: Past and Present
- The Ancient Theaters of Epidaurus
- In the Shadow of Mt. Vesuvius
- The Silk Road
- Pilgrimage to Mecca
- Mother Teresa: A Devoted Life
- The Bells of Agnone
- Buddhism in Mongolia
- The Inca's Frozen Past
- Divinci for Sale
- Worlds Meet in the Americas
- Restoring St. Petersburg
- The New Globe Theater
- A Key to the Bastille
- Hong Kong: Past and Present
- Riverboats Return
- Endangered Edison
- Honoring Women's Suffrage
- Fabergé: The Czar's Jeweler
- The Farms of Zimbabwe
- Recovering Russia's Royalty
- The Killer Flu of 1918
- The Great Depression
- Japan's Royal Family
- Rosie the Riveter
- NATO at 50
- North Korea Turns 50
- A Kenyan Reggae Festival
- Dateline Iraq
- Mexico's Jewish Community
- The History of Haiti
- The Walls That Divide
- Mission to Mars

to Enrich Your Classroom

CNNfyi.com

At **CNNfyi.com**, students will love exploring news stories written by experienced journalists as well as student bureau reporters. Stories link to homework help and lesson plans.

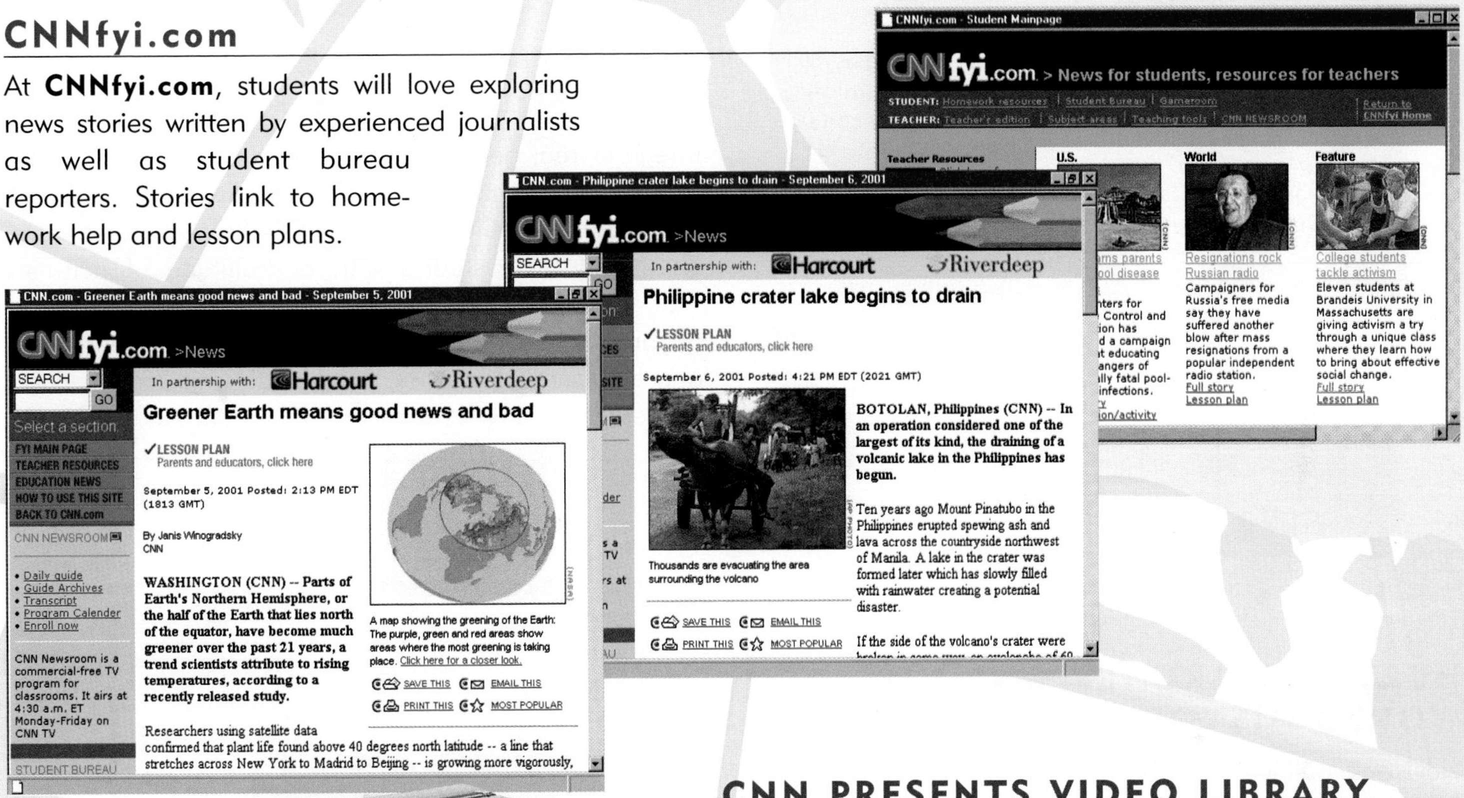

CNN PRESENTS VIDEO LIBRARY

The **CNN PRESENTS** video collection tackles the issue of making content relevant to students head on. Real-world news stories enable students to see the connections between classroom curriculum and today's issues and events around the nation and the world.

CNN PRESENTS...

- **America: Yesterday and Today, Beginnings to 1914**
- **America: Yesterday and Today, 1850 to Present**
- **America: Yesterday and Today, Modern Times**
- **Geography: Yesterday and Today**
- **World Cultures: Yesterday and Today**
- **American Government**
- **Economics**
- **September 11, 2001, Part One**
- **September 11, 2001, Part Two**

Holt is proud to team up with CNN/TURNER LEARNING® to provide you and your students with exceptional current and historical news videos and online resources that add depth and relevance to your daily instruction. This information collection takes your classroom to the far corners of the globe without students ever leaving their desks!

Your Multitalented Classroom

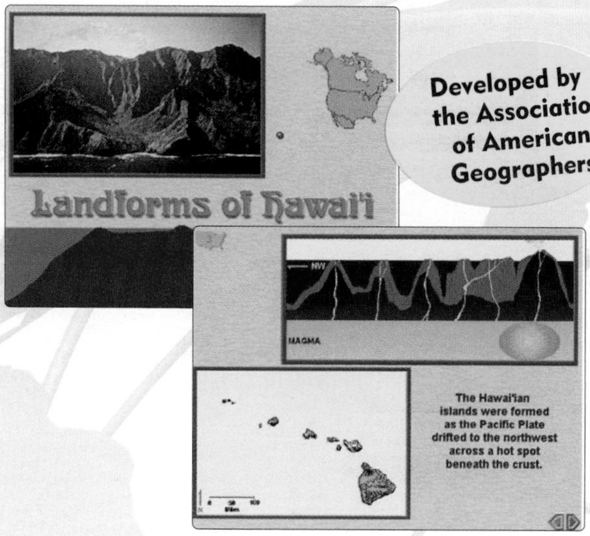

Developed by the Association of American Geographers

ACTIVITIES AND READINGS IN THE GEOGRAPHY OF THE WORLD

Integrate real geography into the topic you're studying with ***Activities and Readings in the Geography of the World (ARGWorld).*** This CD–ROM features world geography case studies with a multitude of activities that focus around geographical themes, population geography, economic geography, political geography, and environmental issues. Case studies will help teachers address the National Geography Standards.

HOLT RESEARCHER ONLINE: WORLD HISTORY AND CULTURES

New and online—students can access this outstanding research tool at **www.hrw.com**. A fully searchable database provides biographies, nation profiles and statistics, a glossary, and powerful graphic capabilities.

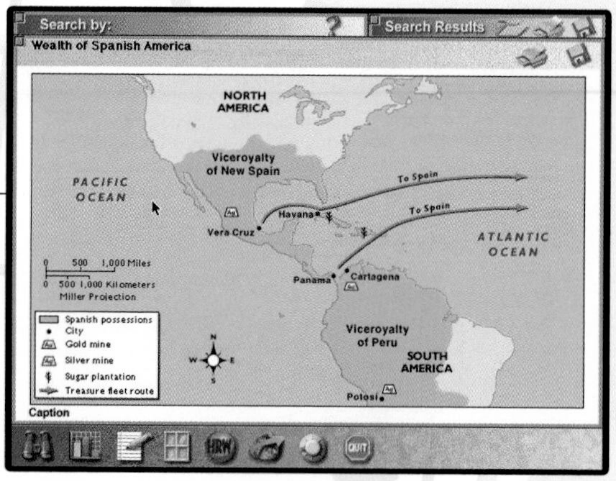

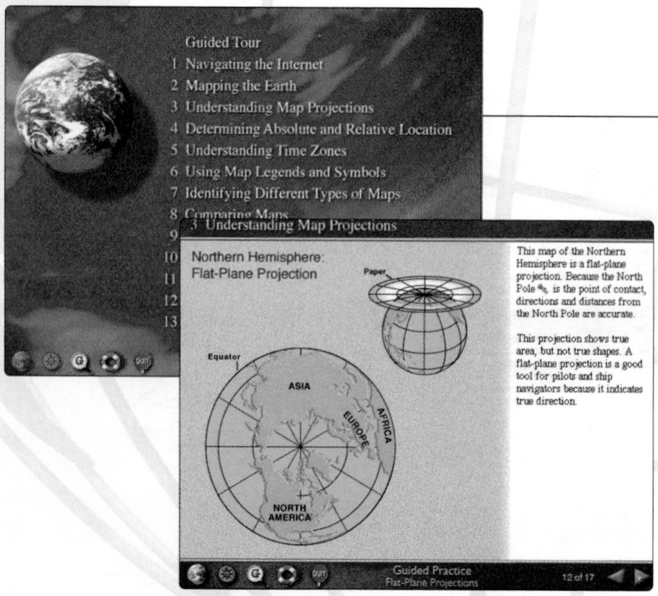

GLOBAL SKILL BUILDER CD–ROM

This CD-ROM is a comprehensive program containing interactive lessons that motivate your students to strengthen their map, graph, and computer skills. A handy ***User's Guide and Teacher's Manual*** provides student project sheets for each lesson along with optional suggestions for using the Internet to help complete the activity.

needs Multimedia Tools

THE WORLD TODAY VIDEODISC PROGRAM

This unique resource offers a stimulating outlook on world geography by showing your students the different ways geographers organize the world, and challenging them to contemplate and discuss significant world issues. Compelling video segments with in-depth content cover contemporary culture in every major world region.

PEOPLE, PLACES, AND CHANGE AUDIO CD PROGRAM

The ***Audio CD Program*** provides in-depth audio section summaries and self-check activity sheets to help those students who respond to auditory learning. Available in English and Spanish.

Other Multimedia Products

- **CNN Presents Geography: Modern Times**
- **CNN Presents Geography: Yesterday and Today**
- **CNN Presents World Cultures: Yesterday and Today**
- **Holt Researcher Online: World History and Cultures**

Technology with

go.hrw.com FOR TEACHERS

Throughout the *Annotated Teacher's Edition*, you'll find **Internet Connect** boxes that take you to specific chapter activities, links, current events, and more that correlate directly to the section you are teaching. Through **go.hrw.com** you'll find a wealth of teaching resources at your fingertips for fun, interactive lessons.

print out materials
Planner CD–ROM

nforcement,
sment Resources

Activity 13.1
eview, p. 291
.1
io Summary 13.1
Audio Summary 13.1

ctivity 13.2
iew, p. 297
io Summary 13.2
io Summary 13.2
3.2

tivity 13.3
iew, p. 303
13.3
o Summary 13.3
o Summary 13.3

needs

internet connect

HRW ONLINE RESOURCES

GO TO: go.hrw.com
Then type in a keyword.

TEACHER HOME PAGE
KEYWORD: SK3 TEACHER

CHAPTER INTERNET ACTIVITIES
KEYWORD: SK3 GT13

Choose an activity to:
• investigate the conflicts in the Balkans.
• take a virtual tour of Eastern Europe.
• learn about Baltic amber.

CHAPTER ENRICHMENT LINKS
KEYWORD: SK3 CH13

CHAPTER MAPS
KEYWORD: SK3 MAPS13

ONLINE ASSESSMENT
Homework Practice
KEYWORD: SK3 HP13
Standardized Test Prep Online
KEYWORD: SK3 STP13
Rubrics
KEYWORD: SS Rubrics

COUNTRY INFORMATION
KEYWORD: SK3 Almanac

CONTENT UPDATES
KEYWORD: SS Content Updates

HOLT PRESENTATION MAKER
KEYWORD: SK3 PPT13

ONLINE READING SUPPORT
KEYWORD: SS Strategies

CURRENT EVENTS
KEYWORD: S3 Current Events

DIRECT LAUNCH TO CHAPTER ACTIVITIES

GUIDED ONLINE ACTIVITIES

LINKS FOR EVERY SECTION

MAPS AND CHARTS

INTERACTIVE PRACTICE AND REVIEW

RUBRICS FOR SUBJECTIVE GRADING

UP-TO-DATE INFORMATION

CLASSROOM PRESENTATION SUPPORT

PRACTICE FOR READING SUCCESS

WEB RESOURCES FOR CURRENT ISSUES

Chapter Review and Assessment

IC	Interdisciplinary Activities for the Middle Grades 13, 14, 15	**A**	Chapter 13 Test
		A	Unit 3 Test
E	Readings in World Geography, History, and Culture 37, 38, and 39		Chapter 13 Test Generator (on the One-Stop Planner)
SM	Critical Thinking Activity 13		Audio CD Program, Chapter 13
REV	Chapter 13 Review, pp. 334–35	**A**	Chapter 13 Test for English Language Learners and Special-Needs Students
REV	Chapter 13 Tutorial for Students, Parents, Mentors, and Peers	**A**	Unit 3 Test for English Language Learners and Special-Needs Students
ELL	Vocabulary Activity 13		

287B

GO.HRW.COM

Online Teaching Materials

Select a Resource
HRW Online Resources

Reading Strategies written by Dr. Judith Irvin are targeted at US History and World Studies.

HRW's State Handbooks provide state-specific links and activities at the click of a mouse button.

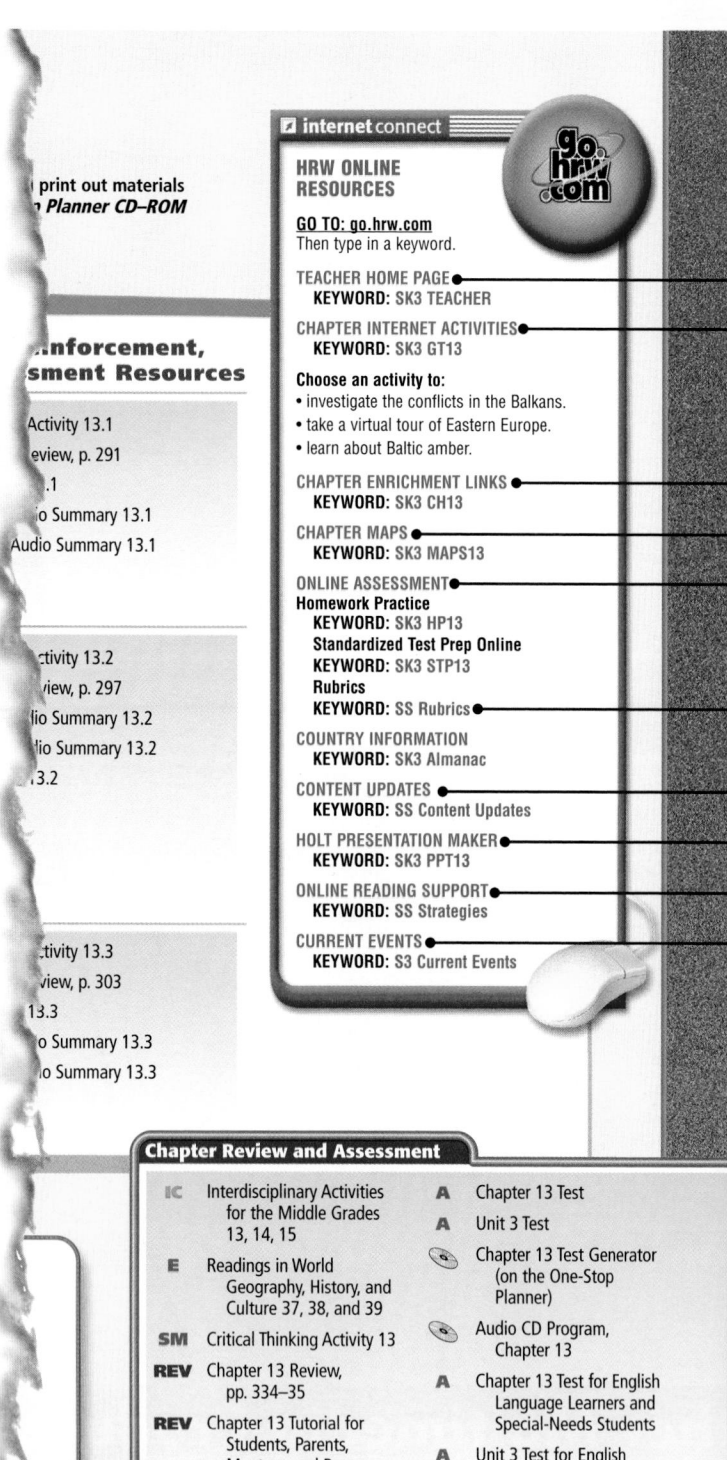

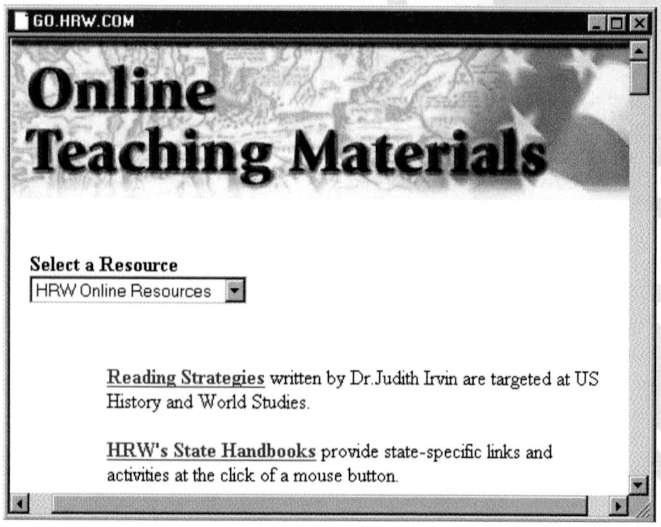

Instructional Value

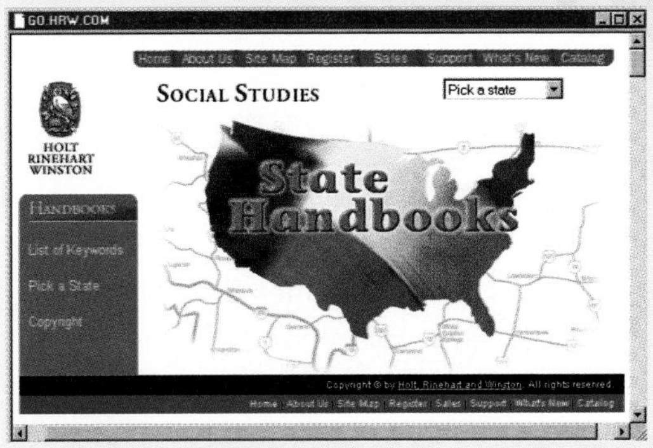

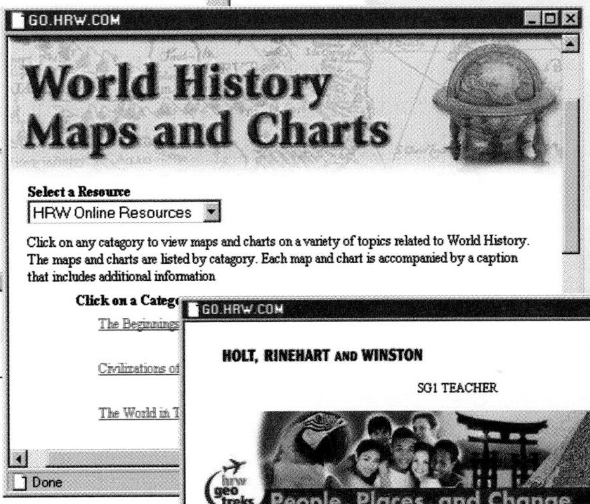

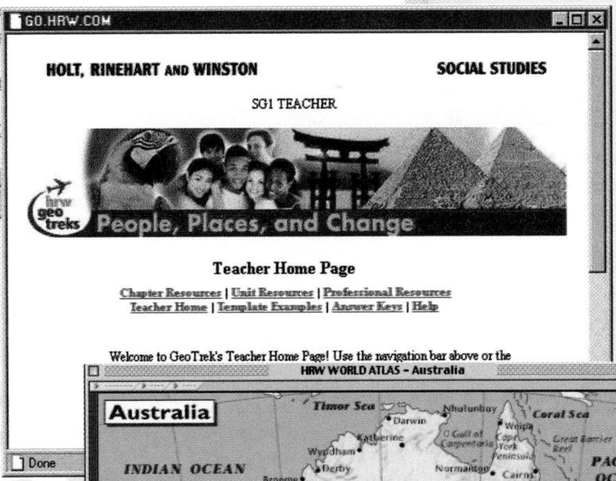

ONLINE TEACHING SUPPORT

Teacher materials on **go.hrw.com** offer you multiple resources for keeping content current. From **World History Maps and Charts** to **State Handbooks,** we've got it all.

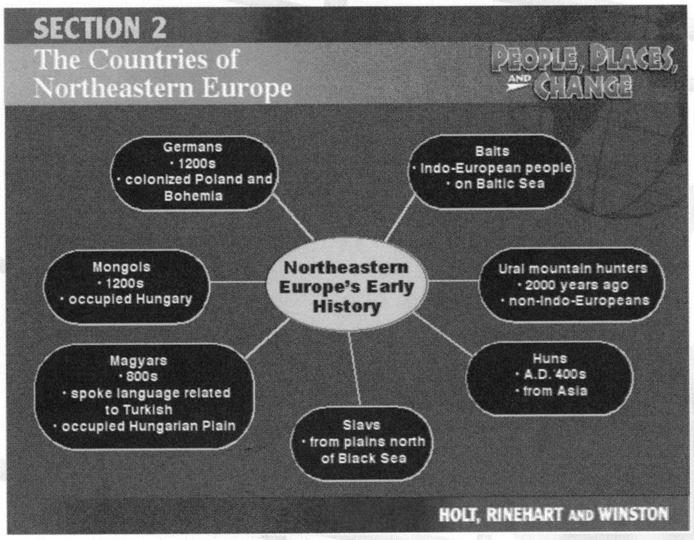

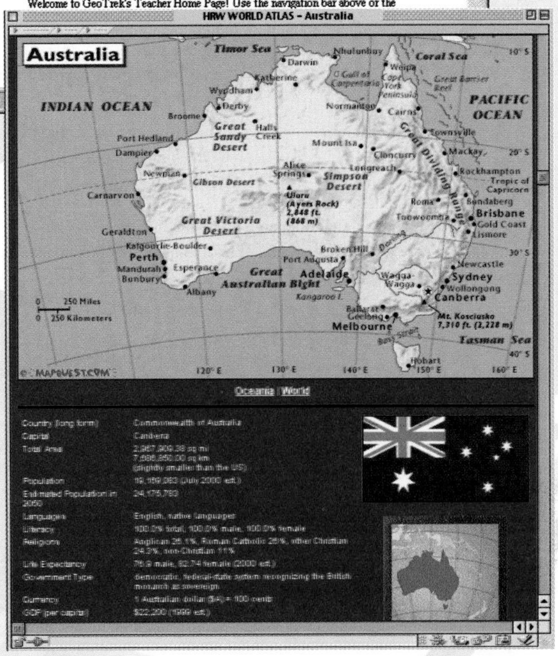

CLASSROOM PRESENTATION SUPPORT

Lecture notes and animated graphic organizers help add visual support to your classroom presentations.

Technology that

go.hrw.com FOR STUDENTS

Your students can access interactive activities, homework help, up-to-date maps, and more when they visit **go.hrw.com** and type in the keywords they find in their text.

ONLINE WORLD TRAVEL

When you log on to **go.hrw.com**, you and your students gain passage to **GeoTreks**—a site with guided Internet activities that integrate program content, spark imaginations, and promote online research skills. You'll find:

- **Interactive templates** for creating newspapers, postcards, travel brochures, guided research reports, and more
- **GeoMaps**—Interactive satellite maps of the world's regions for content review
- **Drag-and-drop exercises** to review chapter content in short, fun activities
- **Chapter Web Links** for prescreened, age-appropriate Web sites and current events

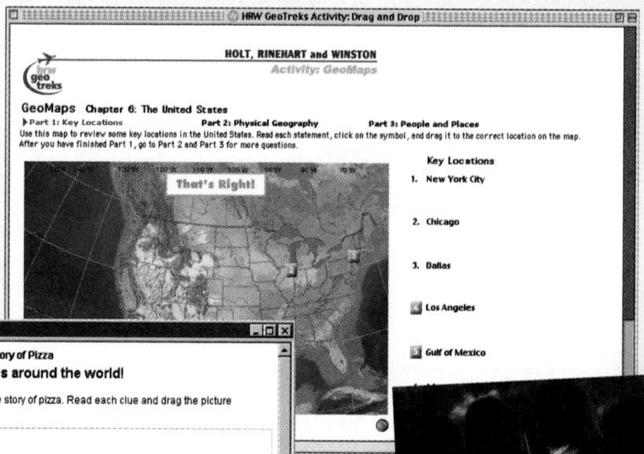

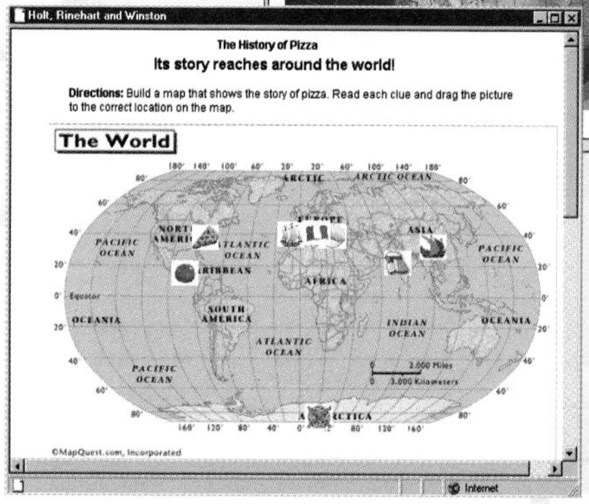

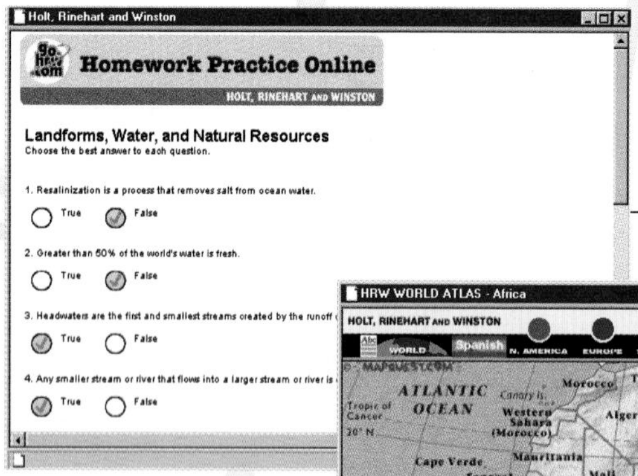

HOMEWORK PRACTICE

This helpful tool allows students to practice and review content by chapter anywhere there is a computer.

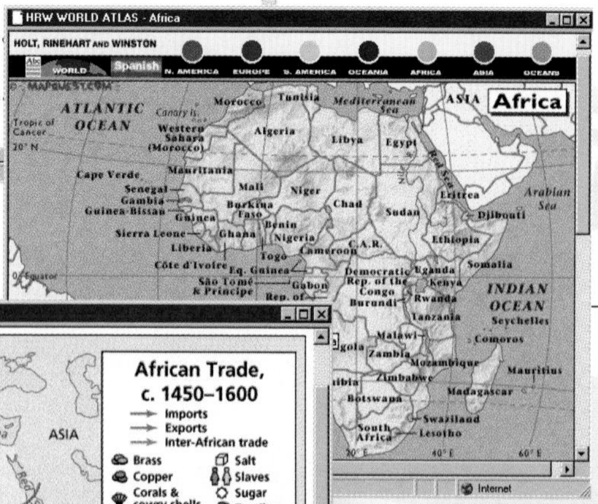

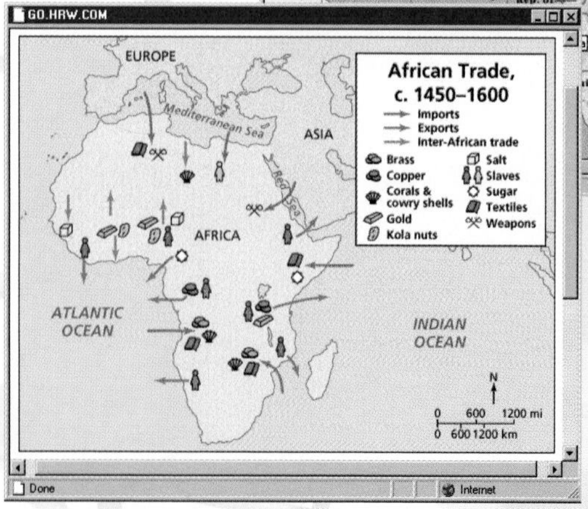

HRW ONLINE ATLAS AND HISTORICAL MAPS

The helpful online atlas contains over 300 well-rendered and clearly labeled country and state maps. Available in English and Spanish, these maps are continually updated so you can rest assured that you and your students have the latest and most accurate geographical content.

Online historical maps provide fascinating visual "snapshots" of the past. Students will relish the chance to explore medieval European trade routes, explorers' routes, ancient African kingdoms, and more.

Delivers Content

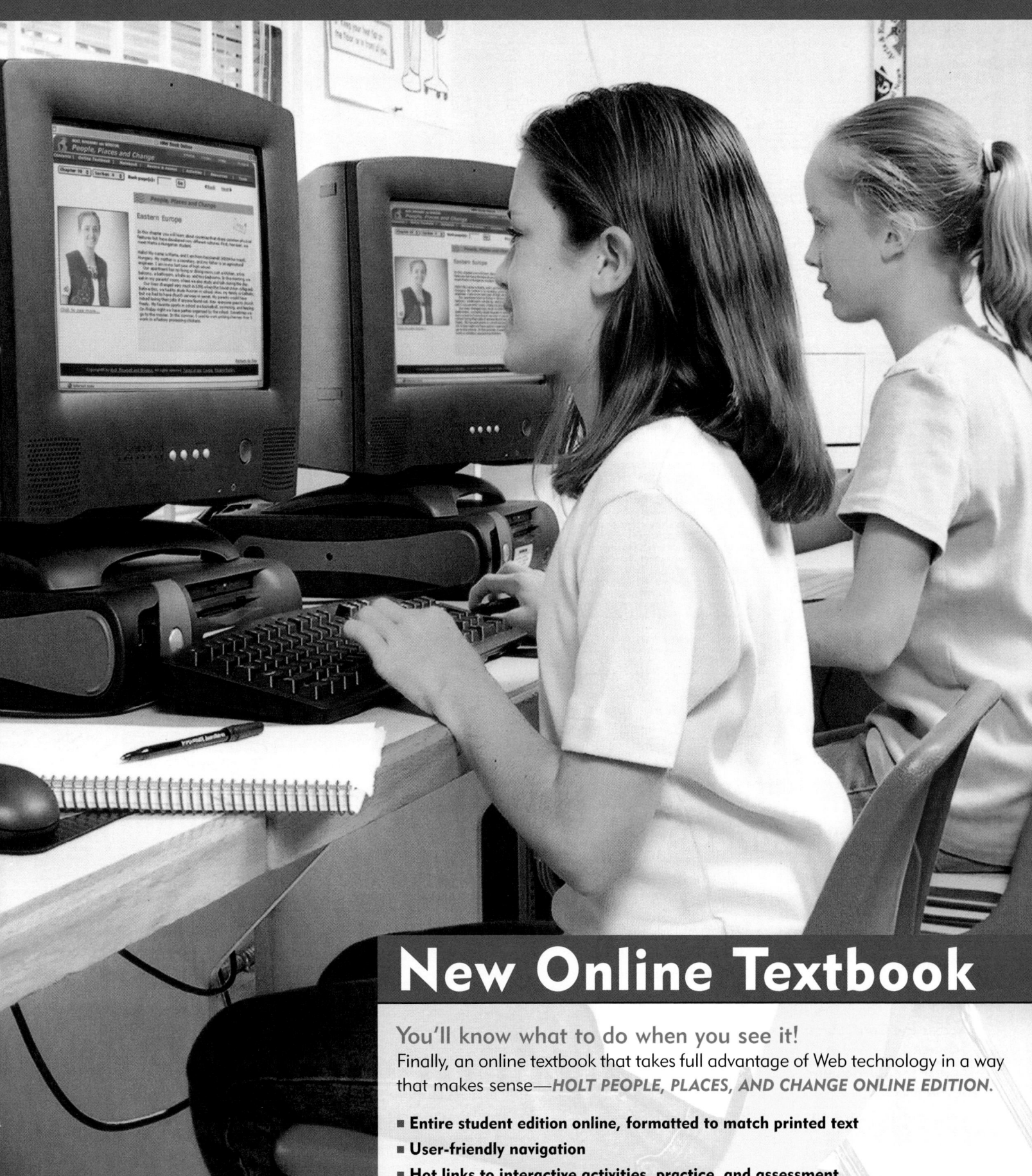

New Online Textbook

You'll know what to do when you see it!

Finally, an online textbook that takes full advantage of Web technology in a way that makes sense—***HOLT PEOPLE, PLACES, AND CHANGE ONLINE EDITION.***

- **Entire student edition online, formatted to match printed text**
- **User-friendly navigation**
- **Hot links to interactive activities, practice, and assessment**
- **Student Notebook for online responses**

Unique Teacher's

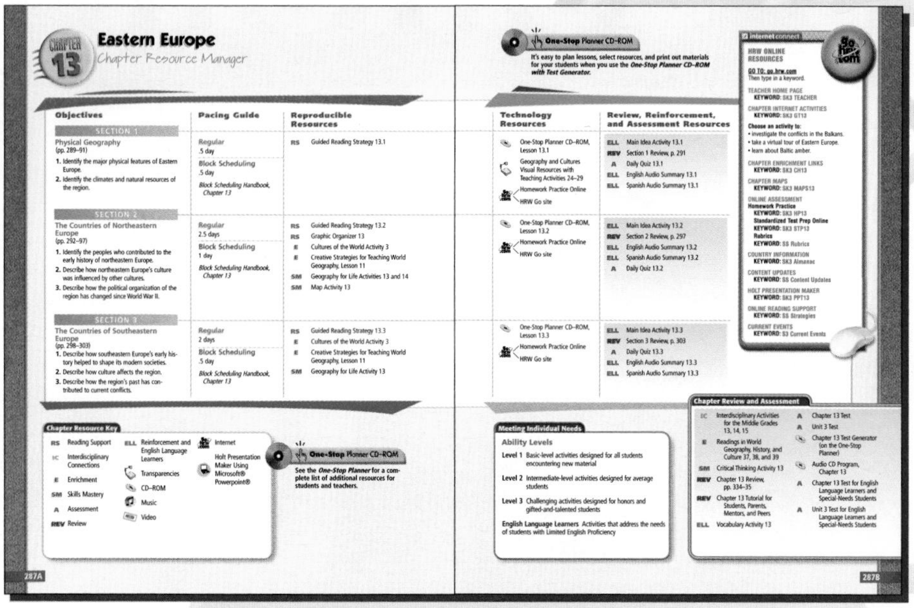

Eastern Europe

CHAPTER 13 Chapter Resource Manager

Objectives	Pacing Guide	Reproducible Resources	Technology Resources	Review, Reinforcement, and Assessment Resources
SECTION 1 Physical Geography (pp. 289–91) 1. Identify the major physical features of Eastern Europe. 2. Identify the climates and natural resources of the region.	**Regular** .5 day **Block Scheduling** .5 day *Block Scheduling Handbook, Chapter 13*	RS Guided Reading Strategy 13.1	One-Stop Planner CD–ROM, Lesson 13.1 Geography and Cultures Visual Resources with Teaching Activities 24–29 Homework Practice Online HRW Go site	ELL Main Idea Activity 13.1 REV Section 1 Review, p. 291 A Daily Quiz 13.1 ELL English Audio Summary 13.1 ELL Spanish Audio Summary 13.1
SECTION 2 The Countries of Northeastern Europe (pp. 292–97) 1. Identify the peoples who contributed to the early history of northeastern Europe. 2. Describe how northeastern Europe's culture was influenced by other cultures. 3. Describe how the political organization of the region has changed since World War II.	**Regular** 2.5 days **Block Scheduling** 1 day *Block Scheduling Handbook, Chapter 13*	RS Guided Reading Strategy 13.2 RS Graphic Organizer 13 E Cultures of the World Activity 3 E Creative Strategies for Teaching World Geography, Lesson 11 SM Geography for Life Activities 13 and 14 SM Map Activity 13	One-Stop Planner CD–ROM, Lesson 13.2 Homework Practice Online HRW Go site	ELL Main Idea Activity 13.2 REV Section 2 Review, p. 297 ELL English Audio Summary 13.2 ELL Spanish Audio Summary 13.2 A Daily Quiz 13.2
SECTION 3 The Countries of Southeastern Europe (pp. 298–303) 1. Describe how southeastern Europe's early history helped to shape its modern societies. 2. Describe how culture affects the region. 3. Describe how the region's past has contributed to current conflicts.	**Regular** 2 days **Block Scheduling** .5 day *Block Scheduling Handbook, Chapter 13*	RS Guided Reading Strategy 13.3 E Cultures of the World Activity 3 E Creative Strategies for Teaching World Geography, Lesson 11 SM Geography for Life Activity 13	One-Stop Planner CD–ROM, Lesson 13.3 Homework Practice Online HRW Go site	ELL Main Idea Activity 13.3 REV Section 3 Review, p. 303 A Daily Quiz 13.3 ELL English Audio Summary 13.3 ELL Spanish Audio Summary 13.3

Chapter Resource Key

RS Reading Support
IC Interdisciplinary Connections
E Enrichment
SM Skills Mastery
A Assessment
REV Review
ELL Reinforcement and English Language Learners
Transparencies
CD–ROM
Music
Video
Internet
Holt Presentation Maker Using Microsoft® Powerpoint®

One-Stop Planner CD–ROM
See the ***One-Stop Planner*** for a complete list of additional resources for students and teachers.

One-Stop Planner CD–ROM
It's easy to plan lessons, select resources, and print out materials for your students when you use the ***One-Stop Planner CD–ROM with Test Generator.***

internet connect
HRW ONLINE RESOURCES
GO TO: go.hrw.com
Then type in a keyword.
TEACHER HOME PAGE
KEYWORD: SK3 TEACHER
CHAPTER INTERNET ACTIVITIES
KEYWORD: SK3 GT13
Choose an activity to:
- investigate the conflicts in the Balkans.
- take a virtual tour of Eastern Europe.
- learn about Baltic amber.

CHAPTER ENRICHMENT LINKS
KEYWORD: SK3 CH13
CHAPTER MAPS
KEYWORD: SK3 MAPS13
ONLINE ASSESSMENT
Homework Practice
KEYWORD: SK3 HP13
Standardized Test Prep Online
KEYWORD: SK3 STP13
Rubrics
KEYWORD: SS Rubrics
COUNTRY INFORMATION
KEYWORD: SK3 Almanac
CONTENT UPDATES
KEYWORD: SS Content Updates
HOLT PRESENTATION MAKER
KEYWORD: SK3 PPT13
ONLINE READING SUPPORT
KEYWORD: SS Strategies
CURRENT EVENTS
KEYWORD: S3 Current Events

Meeting Individual Needs

Ability Levels
Level 1 Basic-level activities designed for all students encountering new material
Level 2 Intermediate-level activities designed for average students
Level 3 Challenging activities designed for honors and gifted-and-talented students
English Language Learners Activities that address the needs of students with Limited English Proficiency

Chapter Review and Assessment

IC Interdisciplinary Activities for the Middle Grades 13, 14, 15
E Readings in World Geography, History, and Culture 37, 38, and 39
SM Critical Thinking Activity 13
REV Chapter 13 Review, pp. 334–35
REV Chapter 13 Tutorial for Students, Parents, Mentors, and Peers
ELL Vocabulary Activity 13
A Chapter 13 Test
A Unit 3 Test
Chapter 13 Test Generator (on the One-Stop Planner)
Audio CD Program, Chapter 13
A Chapter 13 Test for English Language Learners and Special-Needs Students
A Unit 3 Test for English Language Learners and Special-Needs Students

287A

287B

In-Text Chapter Planning

TEACHER TO TEACHER

These strategies are offered in the columns of your *Annotated Teacher's Edition* and provide you with valuable, classroom-tested ideas and activities that have been developed and successfully applied by your peers.

FOOD FESTIVAL

Kolaches are popular Czech or Polish pastries. For a shortcut version, use frozen bread dough. Or, make a sweetened yeast dough from scratch. After the dough has risen, roll it out, cut it into circles 2–3 inches across, and indent the center of each. Add a filling of fruit preserves, cottage cheese with egg and sugar, or a sweetened poppyseed paste. Let rise again. Sprinkle with a streusel topping of sugar, cinnamon, butter, and a little flour. Bake at 375° for 15–20 minutes. There are many *kolache* recipes on the Internet. *Kolacky* and *kolachke* are alternate spellings.

Side-Column Annotations that Spark Curiosity

- Across the Curriculum: Art
- Across the Curriculum: History
- Across the Curriculum: Literature
- Across the Curriculum: Math
- Across the Curriculum: Science
- Across the Curriculum: Technology
- Cooperative Learning
- Cultural Kaleidoscope
- Daily Life
- Eye on Earth
- Geography sidelight
- Global Perspectives
- Historical Geography
- Linking Past to Present
- National Geography Standards
- People in the Profile
- Using Illustrations

OBJECTIVE-BASED LESSON CYCLE

With lively activities and presentation strategies such as **Let's Get Started, Building Vocabulary,** and **Graphic Organizers,** your step-by-step lesson cycle makes planning your lessons easy and productive.

Section 2

Objectives

1. Identify what peoples contributed to the early history of northeastern Europe.
2. Analyze how northeastern Europe's culture was influenced by other cultures.
3. Describe how the political organization of the region has changed since World War II.

FOCUS

LET'S GET STARTED
Copy the following question onto the chalkboard: *What does a country need to do or have to attract tourists? Possible answers: good hotels and restaurants, historical sites, beautiful scenery, entertainment, reliable transportation)* Tell students that many of the countries of northeastern Europe, particularly the Czech Republic, became popular tourist destinations during the 1990s. Before that time, few people traveled to the region because the governments were communist and tourist facilities were undeveloped. Tell students that in Section 2 they will learn more about the region's ancient and modern history.

Building Vocabulary
Write **Indo-European** on the chalkboard and call on a volunteer to read the text's definition aloud. Point out that the definition only tells about the languages the Indo-European peoples spoke—not what they looked like or what their customs were. Ask what two large regions are represented in the term. *(India and Europe)* Many languages of India are related to European languages.

TEACH

Teaching Objective 1
ALL LEVELS: (Suggested time: 10 min.) Copy the following graphic organizer onto the chalkboard, omitting the italicized answers. Call on students to fill in the outer circles with the names of peoples who contributed to northeastern Europe's early history. Then call on others to supply details about each group to fill in the circles.

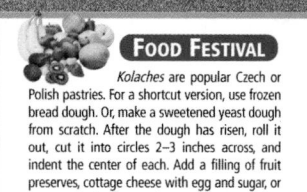

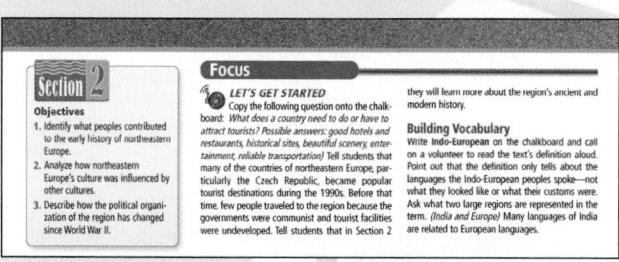

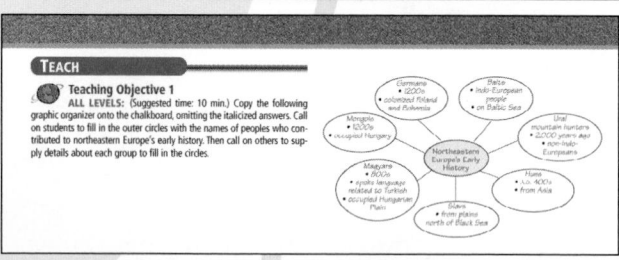

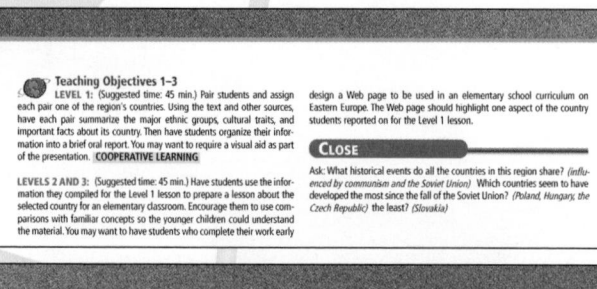

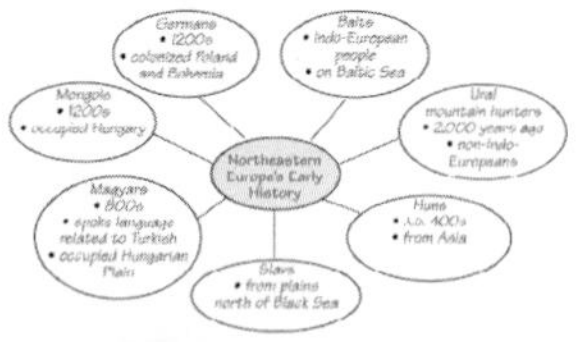

Teaching Objectives 1–3
LEVEL 1: (Suggested time: 45 min.) Pair students and assign each pair one of the region's countries. Using the text and other sources, have each pair summarize the major ethnic groups, cultural traits, and important facts about its country. Then have students organize their information into a brief oral report. You may want to require a visual aid as part of the presentation. **COOPERATIVE LEARNING**

LEVELS 2 AND 3: (Suggested time: 45 min.) Have students use the information they compiled for the Level 1 lesson to prepare a lesson about the selected country for an elementary classroom. Encourage them to use comparisons with familiar concepts so the younger children could understand the material. You may want to have students who complete their work early design a Web page to be used in an elementary school curriculum on Eastern Europe. The Web page should highlight one aspect of the country students reported on for the Level 1 lesson.

CLOSE

Ask: What historical events do all the countries in this region share? *(influenced by communism and the Soviet Union)* Which countries seem to have developed the most since the fall of the Soviet Union? *(Poland, Hungary, the Czech Republic)* the least? *(Slovakia)*

REVIEW AND ASSESS

Have students complete the Section Review. Then organize the class into small groups and have students write one quiz question each on the history, culture, and characteristics of the region. Each question should be written on a slip of paper, folded, and placed in a pile. Students should take turns reading questions to the group while the other members answer. Then have students complete Daily Quiz 16.2. **COOPERATIVE LEARNING**

RETEACH

Have students complete Main Idea Activity 16.2. Then pair students and have pairs create a page for a travel magazine featuring a country in northeastern Europe. Students may divide up tasks according to their abilities. The page should include a descriptive title, a map or flag, and a short para graph summarizing the history and culture of the country. Display and discuss the magazine pages.
ENGLISH LANGUAGE LEARNERS, COOPERATIVE LEARNING

EXTEND

The czardas (CHAHR-dash) is the national dance of Hungary. This couples dance begins with a slow section but becomes fast and lively. Many versions of the czardas have been developed. Have interested students conduct research on how to perform a basic version of the dance. You may wish to have volunteers perform the dance in class. **BLOCK SCHEDULING**

Management System

One-Stop Planner
with Test Generator

ONE-STOP PLANNER® CD–ROM WITH TEST GENERATOR

Holt brings you the most user-friendly management system in the industry with the ***One-Stop Planner CD–ROM with Test Generator.*** Plan and manage your lessons from this single disc containing all the teaching resources for ***Holt People, Places, and Change,*** valuable planning and assessment tools, and more.

- **Editable lesson plans**
- **Classroom Lecture Notes and Animated Graphic Organizers**
- **Easy-to-use test generator**
- **Previews of all teaching and video resources**
- **Easy printing feature**
- **Direct launch to go.hrw.com**

BLOCK SCHEDULING HANDBOOK

This is more than a pacing guide—it provides daily lesson plans that suggest practical ways to cover more than one textbook section in an extended class period and ways to make interdisciplinary connections.

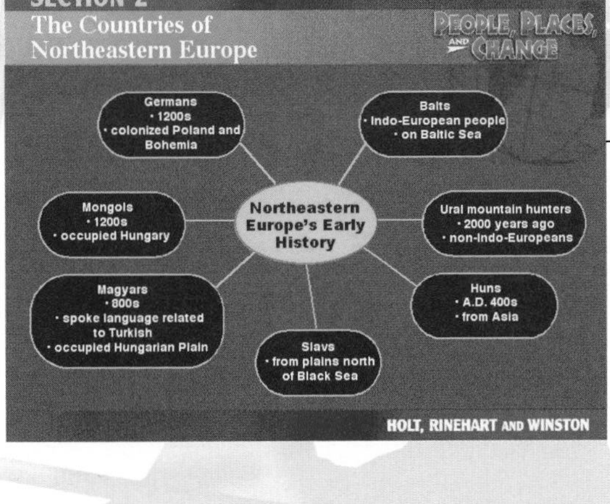

PRESENTATIONS THAT BENEFIT LEARNING

Classroom presentations and lecture notes can be accessed with ease when you use Holt's **Presentation** tool found on the ***One-Stop Planner CD–ROM***. This resource helps you spice up your presentations and gives you ideas to build on. You'll find Microsoft® PowerPoint® presentations that include lecture notes and animated graphic organizers for each chapter and section of your text.

Assessment for

REAL-WORLD APPLICATIONS

INTERACTIVE PRACTICE ACTIVITIES FOR EACH SECTION

Section 1 Review

go.hrw.com Homework Practice Online
Keyword: SK3 HP13

Define and explain: oil shale, lignite, amber

Working with Sketch Maps On a map of Eastern Europe that you draw or that your teacher provides, label the following: Baltic Sea, Adriatic Sea, Black Sea, Danube River, Dinaric Alps, Balkan Mountains, and Carpathian Mountains.

Reading for the Main Idea

1. *Places and Regions* On which three major seas do the countries of Eastern Europe have coasts?
2. *Environment and Society* What types of mineral and energy resources are available in this region? How does this influence individual economies?

Critical Thinking

3. **Making Generalizations and Predictions** Would this region be suitable for agriculture? Why?
4. **Identifying Cause and Effect** How did Communist rule contribute to the pollution problems of this region?

Organizing What You Know

5. **Summarizing** Copy the following graphic organizer. Use it to summarize the physical features, climate, and resources of the heartland, the Baltics, and the Balkans. Then write and answer one question about the region's geography based on the chart.

Region	Physical features	Climate	Resources

THE SUPERIOR TEST GENERATOR THAT REALLY WORKS!

CHAPTER 13 Reviewing What You Know

Building Vocabulary

On a separate sheet of paper, write sentences to define each of the following words:

1. oil shale
2. lignite
3. amber
4. Indo-European
5. Roma

Reviewing the Main Ideas

1. *Places and Regions* What are the major landforms of Eastern Europe?
2. *Human Systems* What groups influenced the culture of Eastern Europe? How can these influences be seen in modern society?
3. *Environment and Society* What were some environmental effects of Communist economic policies in Eastern Europe?
4. *Human Systems* What political and economic systems were most common in Eastern Europe before the 1990s? How and why have these systems changed?
5. *Human Systems* List the countries that broke away from Yugoslavia in the early 1990s. What are the greatest sources of tension in these countries?

Understanding Environment and Society

Economic Geography

During the Communist era, the Council for Economic Assistance, or COMECON, played an important role in planning the economies of countries in the region. Create a presentation comparing the practices of this organization to the practices now in place in the region. Consider the following:

- How COMECON organized the distribution of goods and services.
- How countries in the region now organize distribution.

Thinking Critically

1. **Drawing Inferences and Conclusions** How has Eastern Europe's location influenced the diets of the region's people?
2. **Analyzing Information** What is Eastern Europe's most important river for transportation and trade? How can you tell that the river is important to economic development?
3. **Identifying Cause and Effect** What geographic factors help make Warsaw the transportation and communication center of Poland? Imagine that Warsaw is located along the Baltic coast of Poland or near the German border. How might Warsaw have developed differently?
4. **Comparing/Contrasting** Compare and contrast the breakups of Yugoslavia and Czechoslovakia.
5. **Summarizing** How has political change affected the economies of Eastern European countries?

304 • Chapter 13

CRITICAL-THINKING REINFORCEMENT

Every Student

SOCIAL STUDIES SKILLS REVIEW

Building Social Studies Skills

Map ACTIVITY On a separate sheet of paper, match the letters on the map with their correct labels.

Baltic Sea	**Balkan Mountains**
Adriatic Sea	**Carpathian Mountains**
Black Sea	
Danube River	
Dinaric Alps	

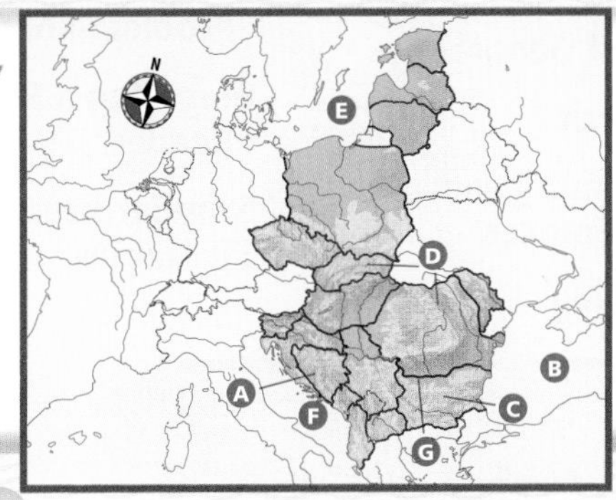

Mental Mapping Skills ACTIVITY Using the chapter map or a globe as a guide, draw a freehand map of Eastern Europe and label the following:

Bosnia and Herzegovina	**Hungary**
Croatia	**Macedonia**
Czech Republic	**Poland**
Estonia	**Yugoslavia**

WRITING ACTIVITY Imagine that you are a teenager living in Romania and want to write a family memoir of life in Romania. Include accounts of life for your grandparents under strict Soviet rule and life for your parents during the Soviet Union's breakup. Also describe your life in free Romania. Be sure to use standard grammar, spelling, sentence structure, and punctuation.

Alternative Assessment

Portfolio ACTIVITY

Learning About Your Local Geography

Cooperative Project Ask international agencies or search the Internet for help in contacting a boy or girl in Eastern Europe. As a group, write a letter to the teen, telling about your daily lives.

internet connect

Internet Activity: **go.hrw.com**
KEYWORD: **SK3 GT13**

Choose a topic to explore Eastern Europe:
- Investigate the conflicts in the Balkans.
- Take a virtual tour of Eastern Europe.
- Learn about Baltic amber.

Eastern Europe • 305

ACCESS ONLINE RUBRICS FOR GRADING PROJECTS AND PORTFOLIO ASSIGNMENTS

WORLD HISTORY AND GEOGRAPHY DOCUMENT-BASED QUESTIONS ACTIVITIES

This resource provides a wide variety of primary sources and thought-provoking questions to help students develop intelligent, well-formed opinions. Important historical and geographical themes are grouped together, allowing for scaffolded instruction.

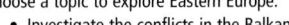

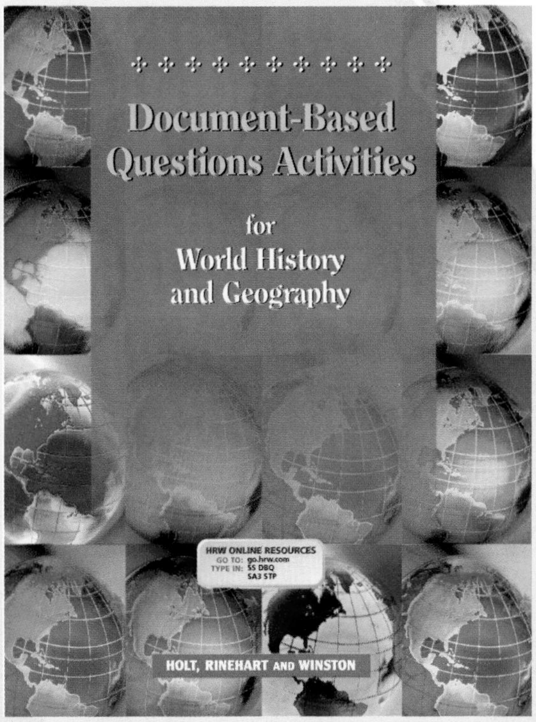

Additional Print and Technology Assessment Resources

- **Daily Quizzes**
- **Chapter Tutorials for Students, Parents, Mentors, and Peers**
- **Chapter and Unit Tests**
- **Chapter and Unit Tests for English Language Learners and Special-Needs Students**
- **Alternative Assessment Handbook**
- **Test Generator (located on the One-Stop Planner)**

HOLT

PEOPLE, PLACES, AND CHANGE

CONTENTS

UNIT 4 Russia and Northern Eurasia 310

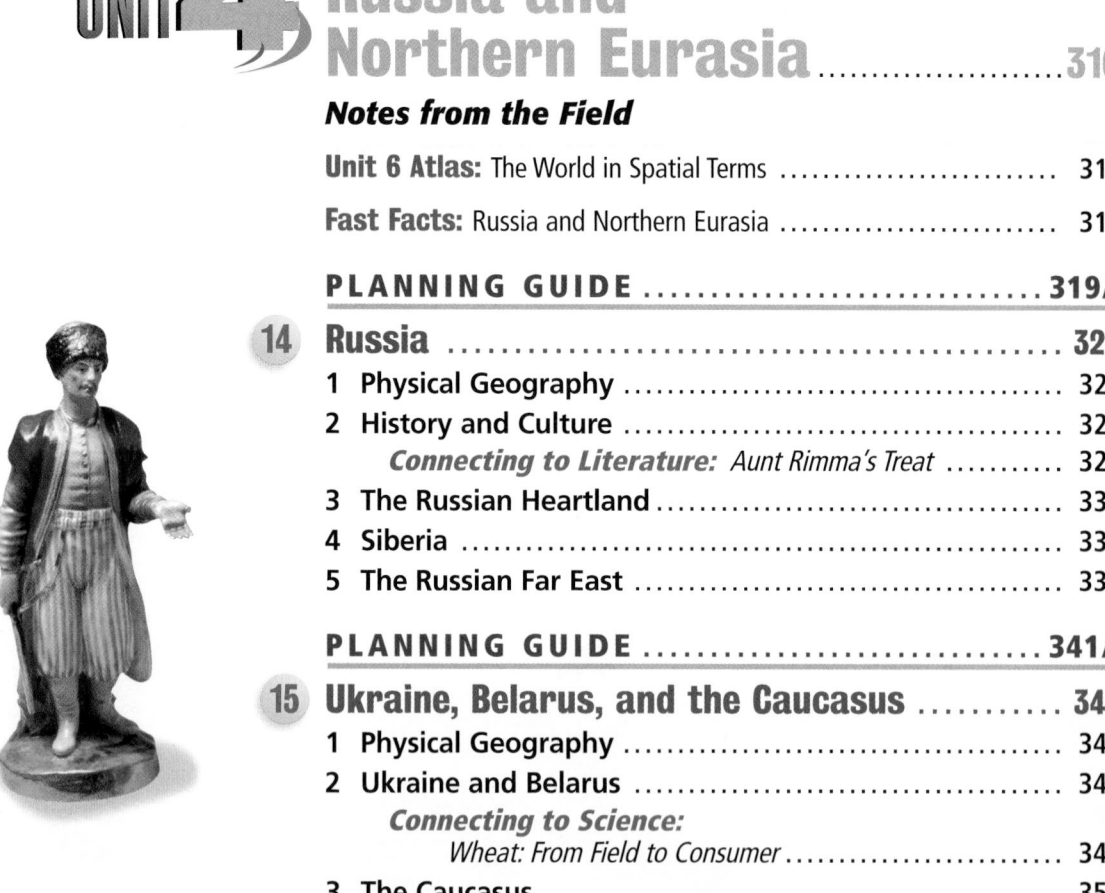

UNIT 7 East and Southeast Asia 532

Notes from the Field

UNIT 8 South Asia

UNIT 9 The Pacific World and Antarctica

REFERENCE SECTION

FEATURES

FEATURES

GeoSKILLS

HANDS ON GEOGRAPHY

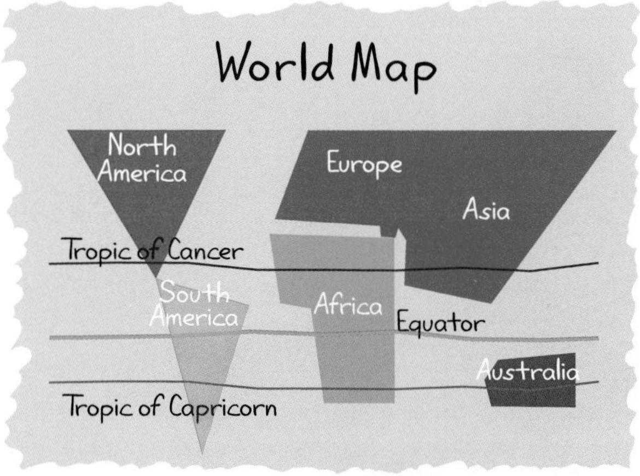

MAPS

FEATURES

DIAGRAMS, CHARTS, and TABLES

FEATURES

DIAGRAMS, CHARTS, and TABLES *continued*

How To Use Your Textbook

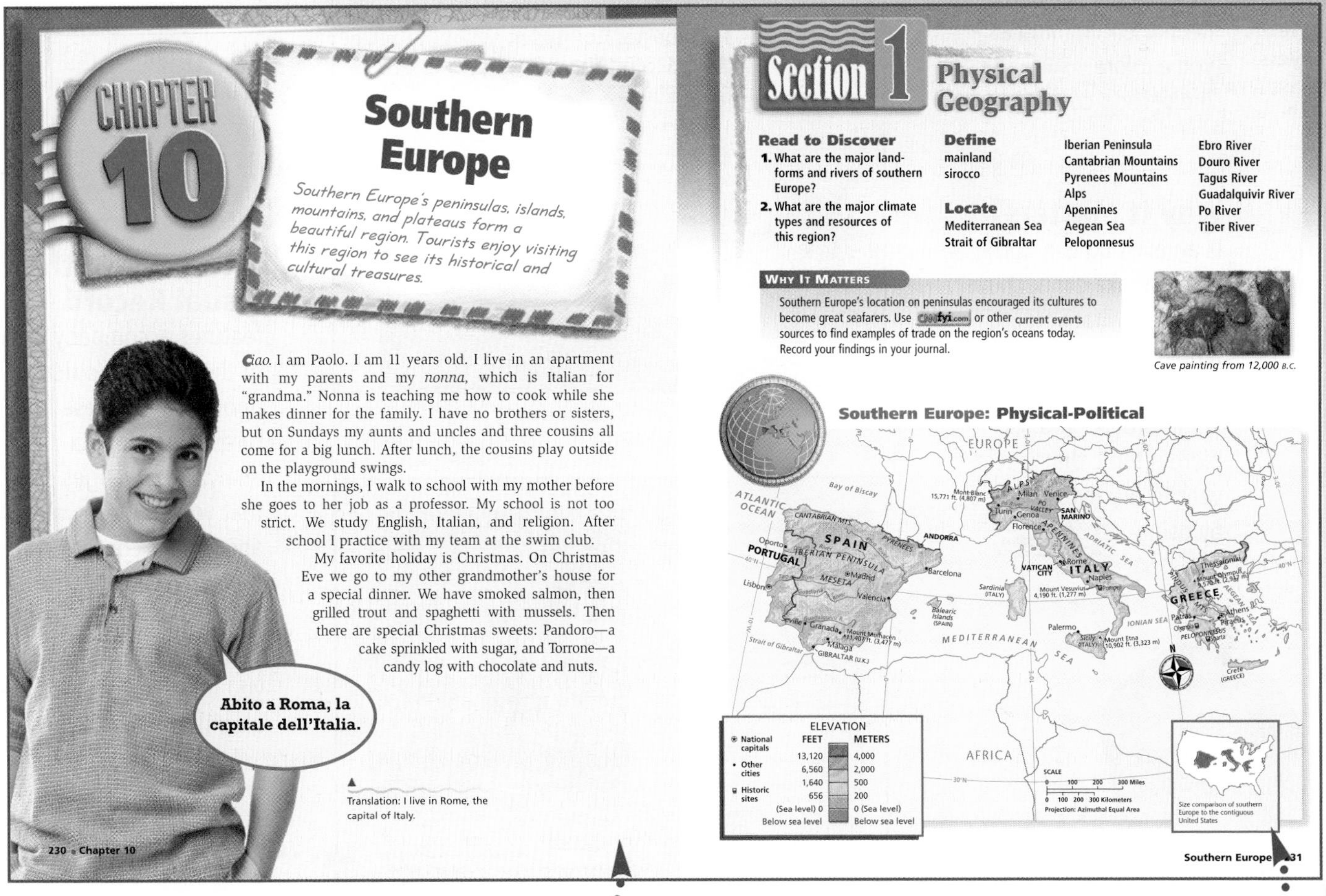
CHAPTER 10

Southern Europe

Southern Europe's peninsulas, islands, mountains, and plateaus form a beautiful region. Tourists enjoy visiting this region to see its historical and cultural treasures.

Ciao. I am Paolo. I am 11 years old. I live in an apartment with my parents and my *nonna,* which is Italian for "grandma." Nonna is teaching me how to cook while she makes dinner for the family. I have no brothers or sisters, but on Sundays my aunts and uncles and three cousins all come for a big lunch. After lunch, the cousins play outside on the playground swings.

In the mornings, I walk to school with my mother before she goes to her job as a professor. My school is not too strict. We study English, Italian, and religion. After school I practice with my team at the swim club.

My favorite holiday is Christmas. On Christmas Eve we go to my other grandmother's house for a special dinner. We have smoked salmon, then grilled trout and spaghetti with mussels. Then there are special Christmas sweets: Pandoro—a cake sprinkled with sugar, and Torrone—a candy log with chocolate and nuts.

Abito a Roma, la capitale dell'Italia.

Translation: I live in Rome, the capital of Italy.

230 Chapter 10

Section 1 Physical Geography

Read to Discover
1. What are the major landforms and rivers of southern Europe?
2. What are the major climate types and resources of this region?

Define
mainland
sirocco

Locate
Mediterranean Sea
Strait of Gibraltar
Iberian Peninsula
Cantabrian Mountains
Pyrenees Mountains
Alps
Apennines
Aegean Sea
Peloponnesus
Ebro River
Douro River
Tagus River
Guadalquivir River
Po River
Tiber River

Why It Matters
Southern Europe's location on peninsulas encouraged its cultures to become great seafarers. Use CNN fyi.com or other current events sources to find examples of trade on the region's oceans today. Record your findings in your journal.

Cave painting from 12,000 B.C.

Southern Europe: Physical-Political

Southern Europe 231

An interview with a student begins each regional chapter. These interviews give you a glimpse of what life is like for some people in the region you are about to study.

Chapter Map The map at the beginning of Section 1 in regional chapters shows you the countries you will read about. You can use this map to identify country names and capitals and to locate physical features. These chapter maps will also help you create sketch maps in section reviews.

Use these built-in tools to read for understanding.

Read to Discover questions begin each section of *Holt People, Places, and Change.* These questions serve as your guide as you read through the section. Keep them in mind as you explore the section content.

Why It Matters is an exciting way for you to make connections between what you are reading in your geography textbook and the world around you. Explore a topic that is relevant to our lives today by using CNNfyi.com.

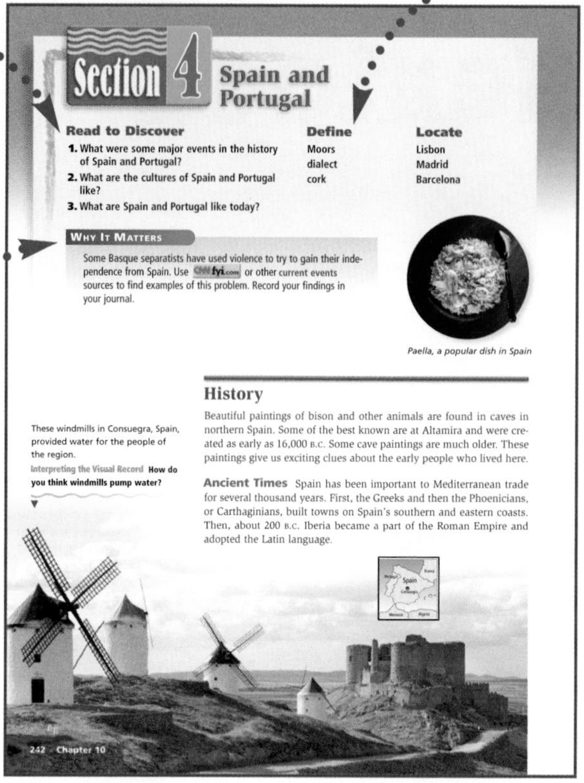

Section 4 Spain and Portugal

Read to Discover

1. What were some major events in the history of Spain and Portugal?
2. What are the cultures of Spain and Portugal like?
3. What are Spain and Portugal like today?

Define
Moors
dialect
cork

Locate
Lisbon
Madrid
Barcelona

Why It Matters

Some Basque separatists have used violence to try to gain their independence from Spain. Use CNNfyi.com or other current events sources to find examples of this problem. Record your findings in your journal.

Paella, a popular dish in Spain

History

Beautiful paintings of bison and other animals are found in caves in northern Spain. Some of the best known are at Altamira and were created as early as 16,000 B.C. Some cave paintings are much older. These paintings give us exciting clues about the early people who lived here.

Ancient Times Spain has been important to Mediterranean trade for several thousand years. First, the Greeks and then the Phoenicians, or Carthaginians, built towns on Spain's southern and eastern coasts. Then, about 200 B.C. Iberia became a part of the Roman Empire and adopted the Latin language.

These windmills in Consuegra, Spain, provided water for the people of the region.
Interpreting the Visual Record **How do you think windmills pump water?**

242 Chapter 10

Define and Locate terms are introduced at the beginning of each section. The Define terms include terms important to the study of geography and to the region you are studying. The Locate terms are important physical features or places from the region you are studying.

Interpreting the Visual Record features accompany many of the textbook's rich photographs. These features invite you to analyze the images so that you can learn more about their content and their links to what you are studying in the section. Other captions ask you to interpret maps, graphs, and charts.

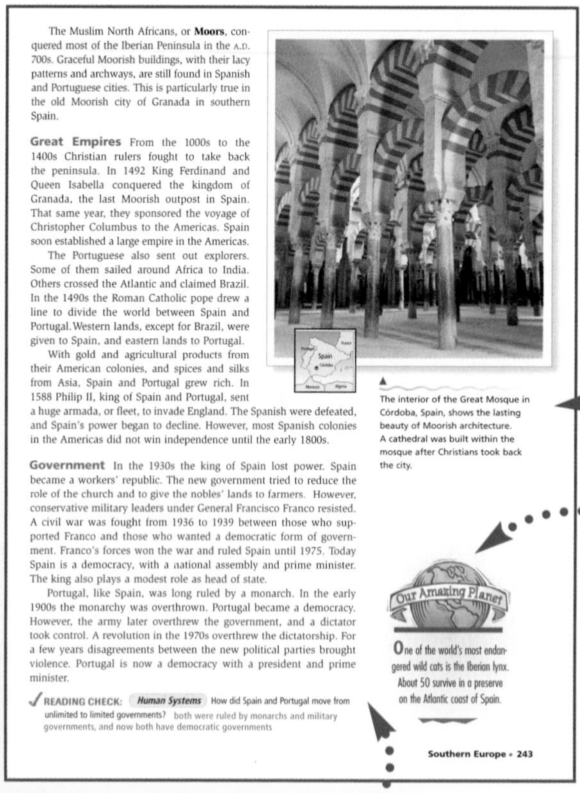

The Muslim North Africans, or **Moors**, conquered most of the Iberian Peninsula in the A.D. 700s. Graceful Moorish buildings, with their lacy patterns and archways, are still found in Spanish and Portuguese cities. This is particularly true in the old Moorish city of Granada in southern Spain.

Great Empires From the 1000s to the 1400s Christian rulers fought to take back the peninsula. In 1492 King Ferdinand and Queen Isabella conquered the kingdom of Granada, the last Moorish outpost in Spain. That same year, they sponsored the voyage of Christopher Columbus to the Americas. Spain soon established a large empire in the Americas.

The Portuguese also sent out explorers. Some of them sailed around Africa to India. Others crossed the Atlantic and claimed Brazil. In the 1490s the Roman Catholic pope drew a line to divide the world between Spain and Portugal. Western lands, except for Brazil, were given to Spain, and eastern lands to Portugal.

With gold and agricultural products from their American colonies, and spices and silks from Asia, Spain and Portugal grew rich. In 1588 Philip II, king of Spain and Portugal, sent a huge armada, or fleet, to invade England. The Spanish were defeated, and Spain's power began to decline. However, most Spanish colonies in the Americas did not win independence until the early 1800s.

Government In the 1930s the king of Spain lost power. Spain became a workers' republic. The new government tried to reduce the role of the church and to give the nobles' lands to farmers. However, conservative military leaders under General Francisco Franco resisted. A civil war was fought from 1936 to 1939 between those who supported Franco and those who wanted a democratic form of government. Franco's forces won the war and ruled Spain until 1975. Today Spain is a democracy, with a national assembly and prime minister. The king also plays a modest role as head of state.

Portugal, like Spain, was long ruled by a monarch. In the early 1900s the monarchy was overthrown. Portugal became a democracy. However, the army later overthrew the government, and a dictator took control. A revolution in the 1970s overthrew the dictatorship. For a few years disagreements between the new political parties brought violence. Portugal is now a democracy with a president and prime minister.

✓ READING CHECK: *Human Systems* How did Spain and Portugal move from unlimited to limited governments? both were ruled by monarchs and military governments, and now both have democratic governments

The interior of the Great Mosque in Córdoba, Spain, shows the lasting beauty of Moorish architecture. A cathedral was built within the mosque after Christians took back the city.

Our Amazing Planet

One of the world's most endangered wild cats is the Iberian lynx. About 50 survive in a preserve on the Atlantic coast of Spain.

Southern Europe • 243

Our Amazing Planet features provide interesting facts about the region you are studying. Here you will learn about the origins of place-names and fascinating tidbits like the size of South America's rain forests.

Reading Check questions appear often throughout the textbook to allow you to check your comprehension. As you read, pause for a moment to consider each Reading Check. If you have trouble answering the question, review the material that you just read.

Use these tools to pull together all of the information you have learned.

Critical Thinking activities in section and chapter reviews allow you to explore a topic in greater depth and to build your skills.

Homework Practice Online lets you log on to the HRW Go site to complete an interactive self-check of the material covered.

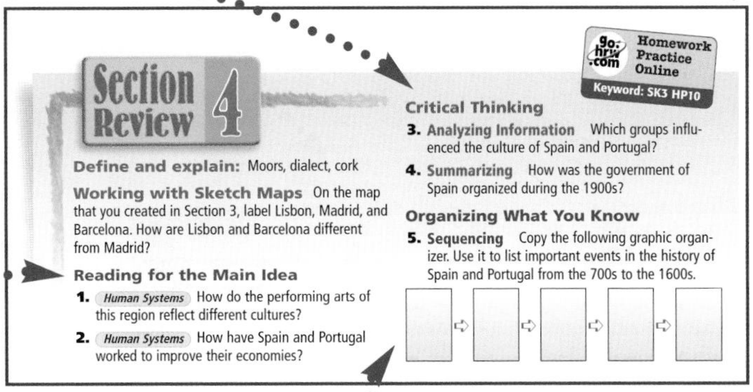
Section Review 4

go.hrw.com Homework Practice Online
Keyword: SK3 HP10

Define and explain: Moors, dialect, cork

Working with Sketch Maps On the map that you created in Section 3, label Lisbon, Madrid, and Barcelona. How are Lisbon and Barcelona different from Madrid?

Reading for the Main Idea

1. Human Systems How do the performing arts of this region reflect different cultures?
2. Human Systems How have Spain and Portugal worked to improve their economies?

Critical Thinking

3. Analyzing Information Which groups influenced the culture of Spain and Portugal?
4. Summarizing How was the government of Spain organized during the 1900s?

Organizing What You Know

5. Sequencing Copy the following graphic organizer. Use it to list important events in the history of Spain and Portugal from the 700s to the 1600s.

Reading for the Main Idea questions help review the main points you have studied in the section.

Graphic Organizers will help you pull together important information from the section.

Building Social Studies Skills activities help you develop the mapping and writing skills you need to study geography.

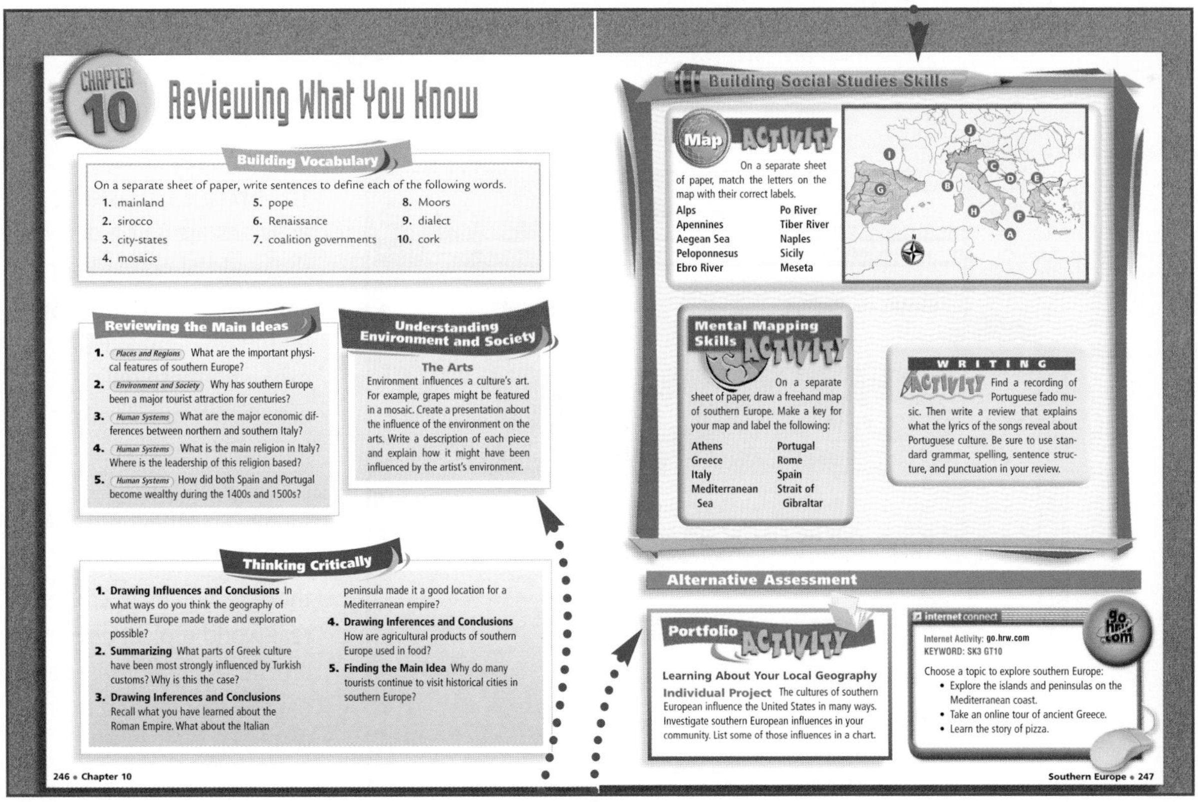
Chapter 10 Reviewing What You Know

Building Vocabulary

On a separate sheet of paper, write sentences to define each of the following words.

1. mainland
2. sirocco
3. city-states
4. mosaics
5. pope
6. Renaissance
7. coalition governments
8. Moors
9. dialect
10. cork

Reviewing the Main Ideas

1. Places and Regions What are the important physical features of southern Europe?
2. Environment and Society Why has southern Europe been a major tourist attraction for centuries?
3. Human Systems What are the major economic differences between northern and southern Italy?
4. Human Systems What is the main religion in Italy? Where is the leadership of this religion based?
5. Human Systems How did both Spain and Portugal become wealthy during the 1400s and 1500s?

Understanding Environment and Society

The Arts

Environment influences a culture's art. For example, grapes might be featured in a mosaic. Create a presentation about the influence of the environment on the arts. Write a description of each piece and explain how it might have been influenced by the artist's environment.

Thinking Critically

1. Drawing Influences and Conclusions In what ways do you think the geography of southern Europe made trade and exploration possible?
2. Summarizing What parts of Greek culture have been most strongly influenced by Turkish customs? Why is this the case?
3. Drawing Inferences and Conclusions Recall what you have learned about the Roman Empire. What about the Italian peninsula made it a good location for a Mediterranean empire?
4. Drawing Inferences and Conclusions How are agricultural products of southern Europe used in food?
5. Finding the Main Idea Why do many tourists continue to visit historical cities in southern Europe?

246 • Chapter 10

Building Social Studies Skills

Map Activity

On a separate sheet of paper, match the letters on the map with their correct labels.

Alps
Apennines
Aegean Sea
Peloponnesus
Ebro River
Po River
Tiber River
Naples
Sicily
Meseta

Mental Mapping Skills Activity

On a separate sheet of paper, draw a freehand map of southern Europe. Make a key for your map and label the following:

Athens
Greece
Italy
Mediterranean Sea
Portugal
Rome
Spain
Strait of Gibraltar

Writing Activity

Find a recording of Portuguese fado music. Then write a review that explains what the lyrics of the songs reveal about Portuguese culture. Be sure to use standard grammar, spelling, sentence structure, and punctuation in your review.

Alternative Assessment

Portfolio Activity

Learning About Your Local Geography

Individual Project The cultures of southern European influence the United States in many ways. Investigate southern European influences in your community. List some of those influences in a chart.

internet connect

Internet Activity: go.hrw.com
KEYWORD: SK3 GT10

Choose a topic to explore southern Europe:
• Explore the islands and peninsulas on the Mediterranean coast.
• Take an online tour of ancient Greece.
• Learn the story of pizza.

Southern Europe • 247

Understanding Environment and Society activities ask you to research and create a presentation expanding on an issue you have read about in the chapter.

Portfolio Activities are exciting and creative ways to explore your local geography and to make connections to the region you are studying.

Use these online tools to review and complete online activities.

go.hrw.com Homework Practice Online

Keyword: SK3 HP10

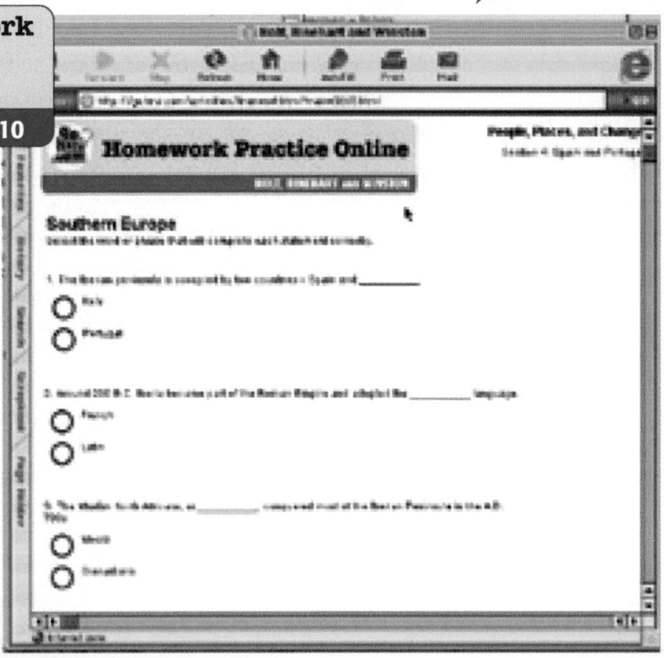

Homework Practice Online lets you log on for review anytime. You will find interactive activities for each section of the text.

internet connect

Internet Activity: go.hrw.com

KEYWORD: SK3 GT10

Choose a topic to explore southern Europe:

- Explore the islands and peninsulas on the Mediterranean coast.
- Take an online tour of ancient Greece.
- Learn the story of pizza.

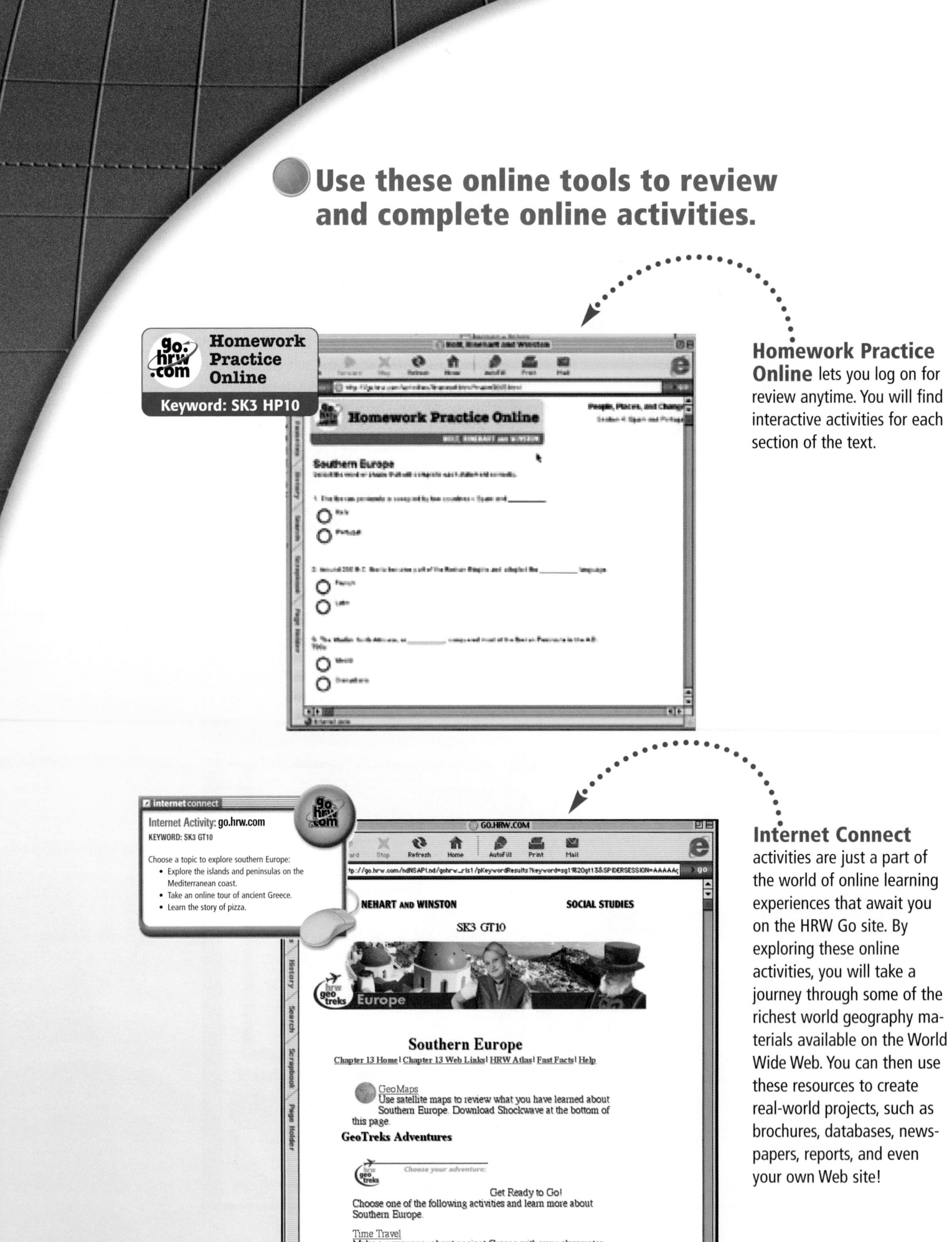

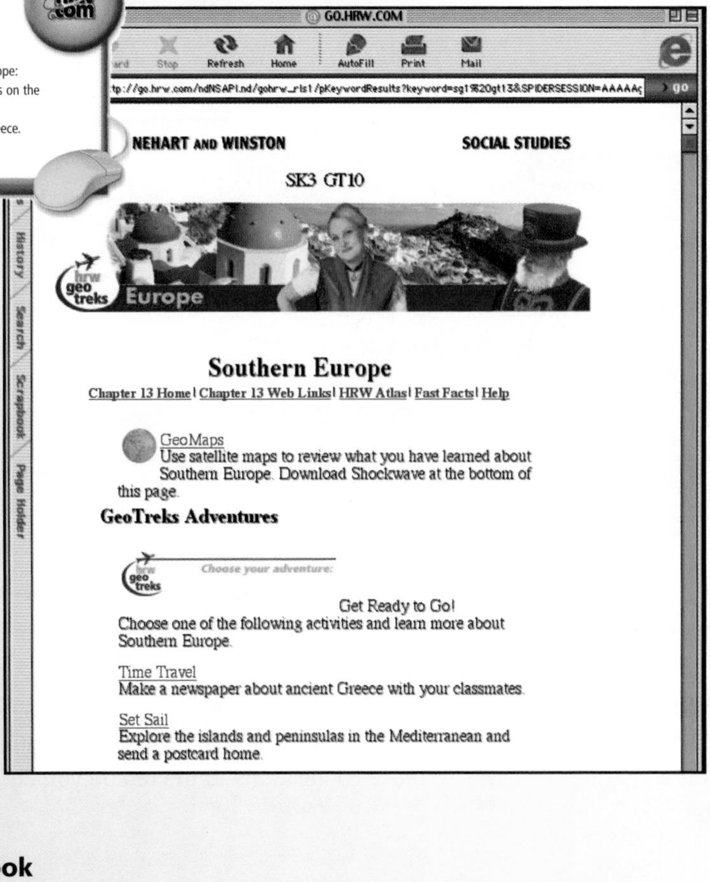

Internet Connect activities are just a part of the world of online learning experiences that await you on the HRW Go site. By exploring these online activities, you will take a journey through some of the richest world geography materials available on the World Wide Web. You can then use these resources to create real-world projects, such as brochures, databases, newspapers, reports, and even your own Web site!

Why Geography Matters

Have you ever wondered. . .

why some places are deserts while other places get so much rain? What makes certain times of the year cooler than others? Why do some rivers run dry?

The CNNfyi.com Web site

Maybe you live near mountains and wonder what processes created them. Do you know why the loss of huge forest areas in one part of the world can affect areas far away? Why does the United States have many different kinds of churches and other places of worship? Perhaps you are curious why Americans and people from other countries have such different points of view on many issues. The key to understanding questions and issues like these lies in the study of geography.

Geography and Your World

All you need to do is watch or read the news to see the importance of geography. You have probably seen news stories about the effects of floods, volcanic eruptions, and other natural events on people and places. You likely have also seen how conflict and cooperation shape the relations between peoples and countries around the world. The Why It Matters feature beginning every section of *Holt People, Places, and Change* uses the vast resources of CNNfyi.com or other current events sources to examine the importance of geography. Through this feature you will be able to draw connections between what you are studying in your geography textbook and events and conditions found around the world today.

My fall semester project, growing a garden

Geography and Making Connections

When you think of the word *geography,* what comes to mind? Perhaps you simply picture people memorizing names of countries and capitals. Maybe you think of people studying maps to identify features like deserts, mountains, oceans, and rivers. These things are important, but the study of geography includes much more. Geography involves asking questions and solving problems. It focuses on looking at people and their ways of life as well as studying physical features like mountains, oceans, and rivers. Studying geography also means looking at why things are where they are and at the relationships between human and physical features of Earth.

The study of geography helps us make connections between what was, what is, and what may be. It helps us understand the processes that have shaped the features we observe around us today, as well as the ways those features may be different tomorrow. In short, geography helps us understand the processes that have created a world that is home to more than 6 billion people and countless billions of other creatures.

Geography and You

Anyone can influence the geography of our world. For example, the actions of individuals affect local environments. Some individual actions might pollute the environment. Other actions might contribute to efforts to keep the environment clean and healthy. Various other things also influence geography. For example, governments create political divisions, such as countries and states. The borders between these divisions influence the human geography of regions by separating peoples, legal systems, and human activities.

Governments and businesses also plan and build structures like dams, railroads, and airports, which change the physical characteristics of places. As you might expect, some actions influence Earth's geography in negative ways, others in positive ways. Understanding geography helps us evaluate the consequences of our actions.

ATLAS

CONTENTS

WORLD: PHYSICAL

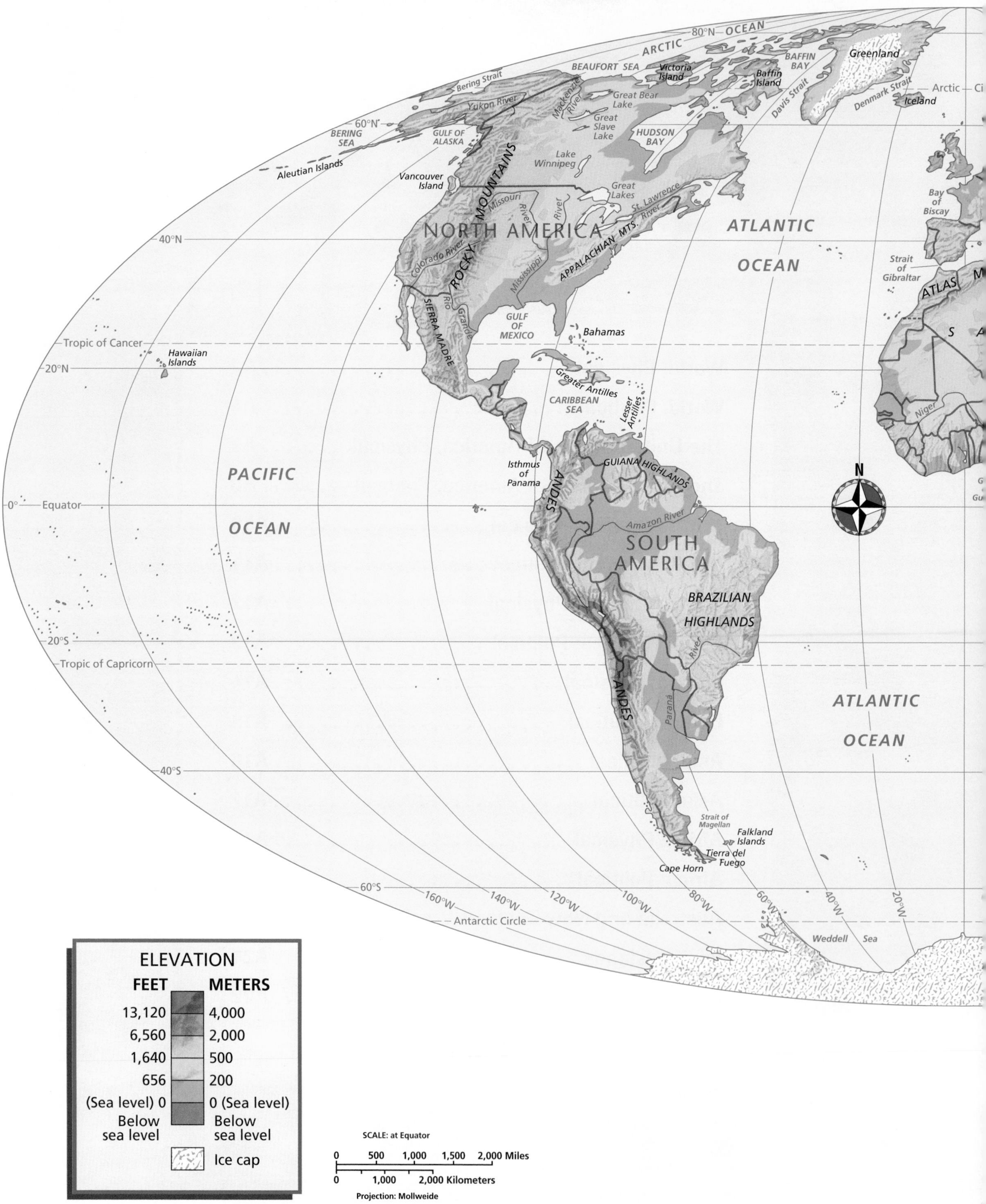

ARCTIC OCEAN
80°N
BEAUFORT SEA
Victoria Island
BAFFIN BAY
Baffin Island
Greenland
Davis Strait
Denmark Strait
Arctic Circle
Iceland
Bering Strait
Yukon River
Mackenzie River
Great Bear Lake
Great Slave Lake
HUDSON BAY
60°N
BERING SEA
GULF OF ALASKA
Aleutian Islands
Vancouver Island
ROCKY MOUNTAINS
Lake Winnipeg
Great Lakes
St. Lawrence River
Missouri River
River
NORTH AMERICA
APPALACHIAN MTS.
Colorado River
Mississippi
ATLANTIC OCEAN
Bay of Biscay
Strait of Gibraltar
ATLAS MTS.
40°N
SIERRA MADRE
Rio Grande
GULF OF MEXICO
Bahamas
Tropic of Cancer
Hawaiian Islands
20°N
Greater Antilles
CARIBBEAN SEA
Lesser Antilles
Niger
PACIFIC OCEAN
Isthmus of Panama
ANDES
GUIANA HIGHLANDS
N
0°
Equator
Amazon River
SOUTH AMERICA
BRAZILIAN HIGHLANDS
20°S
Tropic of Capricorn
River
Paraná
ATLANTIC OCEAN
40°S
Strait of Magellan
Falkland Islands
Tierra del Fuego
Cape Horn
60°S
160°W
140°W
120°W
100°W
80°W
60°W
40°W
20°W
Antarctic Circle
Weddell Sea
ELEVATION
FEET
METERS
13,120
4,000
6,560
2,000
1,640
500
656
200
(Sea level) 0
0 (Sea level)
Below sea level
Below sea level
Ice cap
SCALE: at Equator
0 500 1,000 1,500 2,000 Miles
0 1,000 2,000 Kilometers
Projection: Mollweide

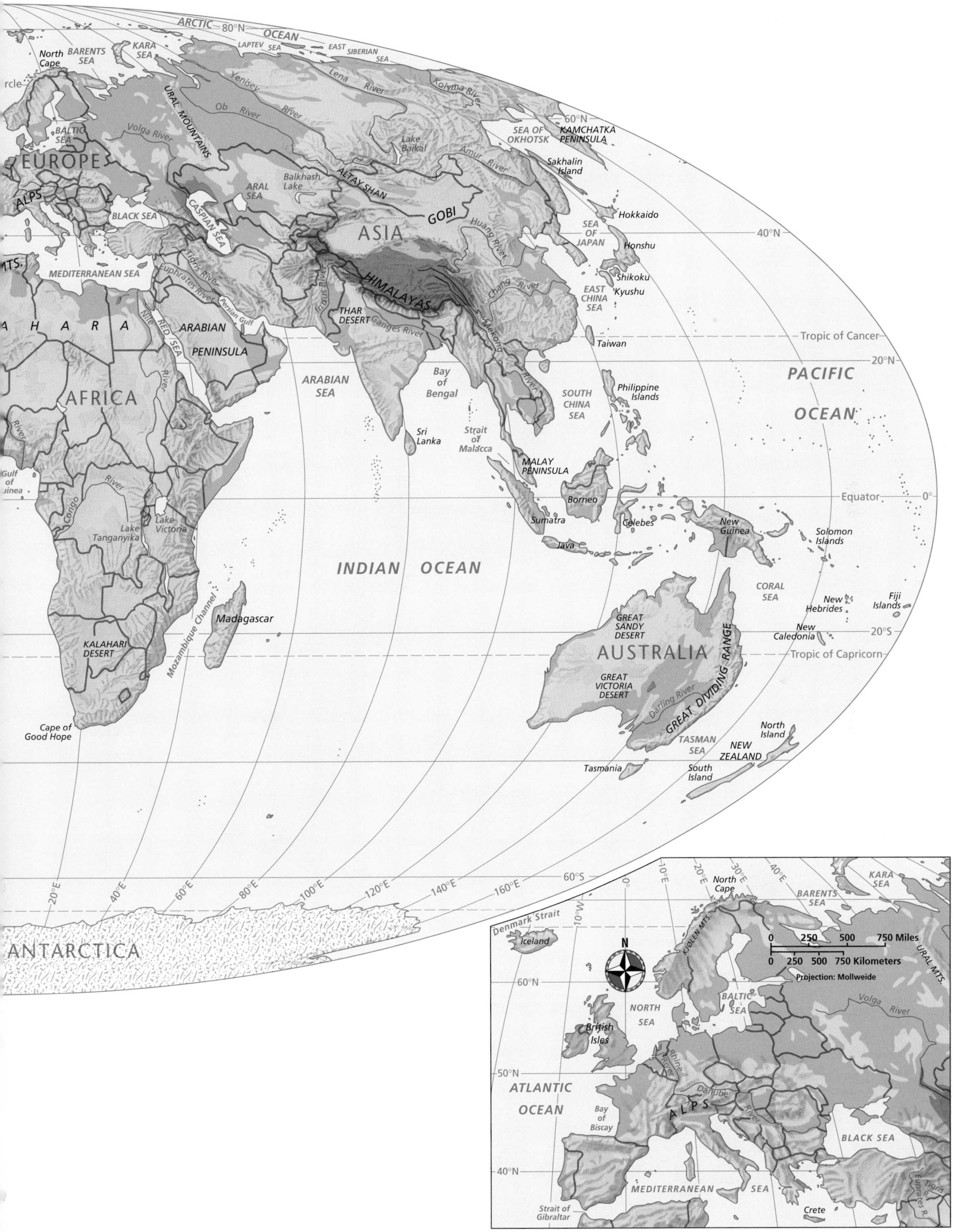
ARCTIC
80°N
OCEAN
LAPTEV SEA
EAST SIBERIAN SEA
North Cape
BARENTS SEA
KARA SEA
Yenisey River
Lena River
Kolyma River
Ob River
URAL MOUNTAINS
BALTIC SEA
Volga River
60°N
SEA OF OKHOTSK
KAMCHATKA PENINSULA
Lake Baikal
Amur River
Sakhalin Island
EUROPE
ALPS
ARAL SEA
Balkhash Lake
ALTAY SHAN
CASPIAN SEA
BLACK SEA
GOBI
Hokkaido
ASIA
Huang River
SEA OF JAPAN
40°N
Honshu
MTS.
MEDITERRANEAN SEA
Tigris River
Euphrates River
Indus River
HIMALAYAS
Chang River
Shikoku
Kyushu
EAST CHINA SEA
Persian Gulf
THAR DESERT
Ganges River
Mekong
Nile
RED SEA
ARABIAN
PENINSULA
Taiwan
Tropic of Cancer
20°N
PACIFIC
OCEAN
A H A R A
AFRICA
River
ARABIAN SEA
Bay of Bengal
SOUTH CHINA SEA
Philippine Islands
River
Sri Lanka
Strait of Malacca
MALAY PENINSULA
Gulf of Guinea
Congo
River
Borneo
Equator
0°
Lake Victoria
Lake Tanganyika
Sumatra
Celebes
New Guinea
Solomon Islands
Java
INDIAN OCEAN
CORAL SEA
New Hebrides
Fiji Islands
Mozambique Channel
Madagascar
GREAT SANDY DESERT
AUSTRALIA
GREAT DIVIDING RANGE
New Caledonia
20°S
KALAHARI DESERT
Tropic of Capricorn
GREAT VICTORIA DESERT
Darling River
Cape of Good Hope
North Island
TASMAN SEA
NEW ZEALAND
Tasmania
South Island
20°E
40°E
60°E
80°E
100°E
120°E
140°E
160°E
60°S
ANTARCTICA
10°W
0°
10°E
20°E
30°E
40°E
North Cape
BARENTS SEA
KARA SEA
Denmark Strait
Iceland
KJOLEN MTS.
N
0 250 500 750 Miles
0 250 500 750 Kilometers
Projection: Mollweide
URAL MTS.
60°N
NORTH SEA
BALTIC SEA
Volga River
British Isles
Rhine
50°N
ATLANTIC OCEAN
Danube
ALPS
River
Bay of Biscay
BLACK SEA
40°N
MEDITERRANEAN SEA
Tigris R.
Euphrates R.
Strait of Gibraltar
Crete

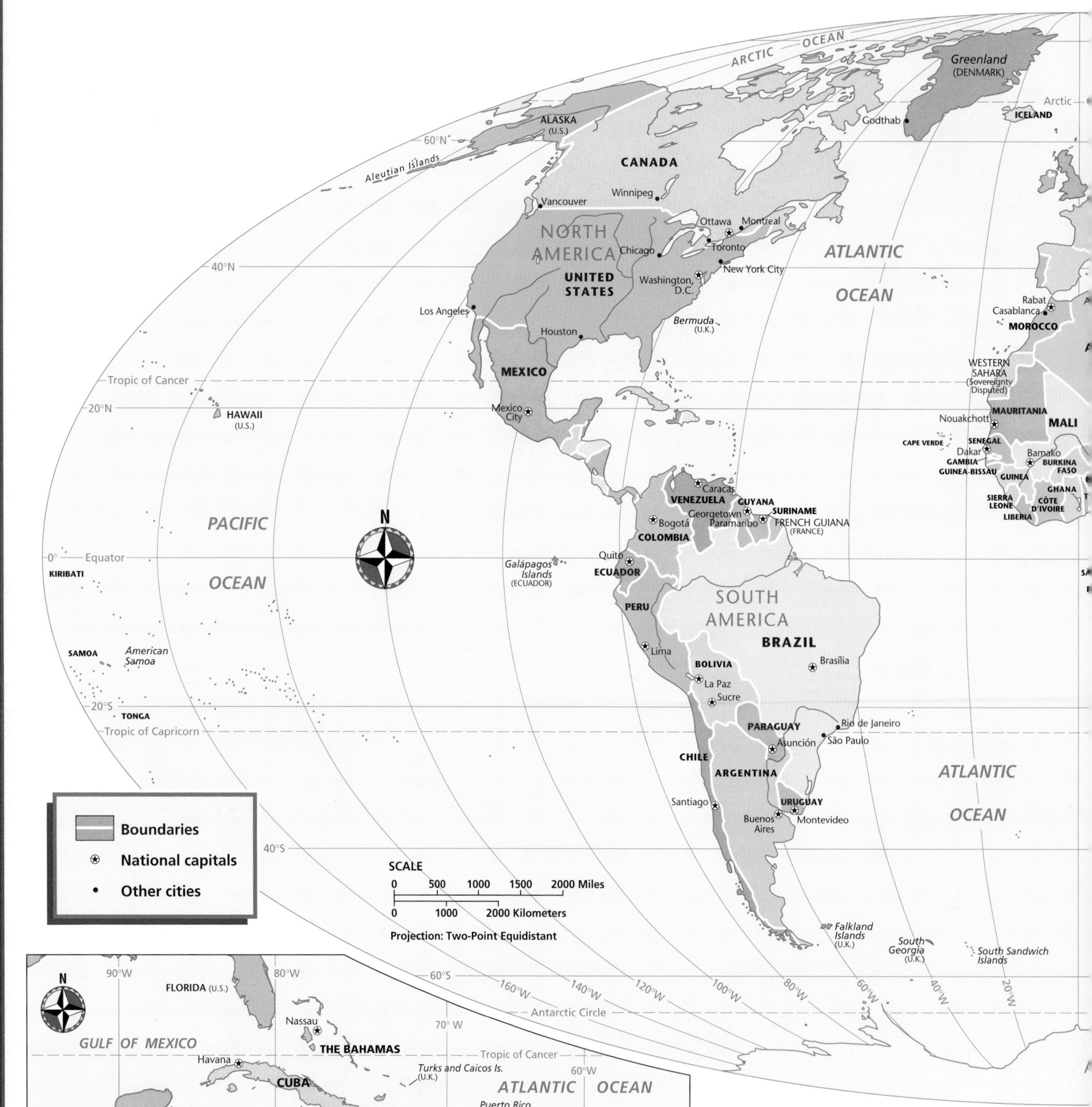
ARCTIC OCEAN
Greenland (DENMARK)
Godthab
Arctic
ICELAND
ALASKA (U.S.)
60°N
Aleutian Islands
CANADA
Winnipeg
Vancouver
NORTH AMERICA
Ottawa
Montreal
Toronto
Chicago
UNITED STATES
Washington, D.C.
New York City
ATLANTIC OCEAN
40°N
Los Angeles
Houston
Bermuda (U.K.)
Rabat
Casablanca
MOROCCO
WESTERN SAHARA (Sovereignty Disputed)
MEXICO
Tropic of Cancer
20°N
HAWAII (U.S.)
Mexico City
MAURITANIA
MALI
Nouakchott
CAPE VERDE
SENEGAL
Dakar
GAMBIA
GUINEA-BISSAU
GUINEA
Bamako
BURKINA FASO
GHANA
SIERRA LEONE
LIBERIA
CÔTE D'IVOIRE
Caracas
VENEZUELA
GUYANA
Georgetown
SURINAME
Paramaribo
FRENCH GUIANA (FRANCE)
Bogotá
COLOMBIA
PACIFIC OCEAN
N
0°
Equator
KIRIBATI
Galápagos Islands (ECUADOR)
Quito
ECUADOR
SOUTH AMERICA
PERU
BRAZIL
Lima
SAMOA
American Samoa
BOLIVIA
Brasília
La Paz
Sucre
20°S
TONGA
Tropic of Capricorn
PARAGUAY
Rio de Janeiro
Asunción
São Paulo
CHILE
ARGENTINA
ATLANTIC OCEAN
Santiago
URUGUAY
Buenos Aires
Montevideo
Boundaries
National capitals
Other cities
40°S
SCALE
0 500 1000 1500 2000 Miles
0 1000 2000 Kilometers
Projection: Two-Point Equidistant
Falkland Islands (U.K.)
South Georgia (U.K.)
South Sandwich Islands
60°S
160°W
140°W
120°W
100°W
80°W
60°W
40°W
20°W
Antarctic Circle

N
90°W
80°W
FLORIDA (U.S.)
Nassau
70° W
GULF OF MEXICO
THE BAHAMAS
Havana
Tropic of Cancer
Turks and Caicos Is. (U.K.)
CUBA
60°W
ATLANTIC OCEAN
Puerto Rico (U.S.)
Virgin Islands (U.S. and U.K.)
20°N
Cayman Is. (U.K.)
HAITI
DOMINICAN REPUBLIC
Port-au-Prince
Santo Domingo
MEXICO
BELIZE
Belmopan
JAMAICA
Kingston
1
2
Guadeloupe (FRANCE)
3
CARIBBEAN SEA
GUATEMALA
HONDURAS
Martinique (FRANCE)
Guatemala City
Tegucigalpa
Netherlands Antilles (NETHERLANDS)
4
Aruba (NETHERLANDS)
5
6
San Salvador
NICARAGUA
Managua
7
EL SALVADOR
Port-of-Spain
TRINIDAD & TOBAGO
PACIFIC OCEAN
COSTA RICA
Panama City
30° N
SCALE
San José
0 200 400 Miles
PANAMA
VENEZUELA
0 200 400 Kilometers
COLOMBIA
GUYANA
Projection: Mercator

COUNTRY	CAPITAL
1 Antigua and Barbuda	St. Johns
2 St. Kitts and Nevis	Basseterre
3 Dominica	Roseau
4 St. Lucia	Castries
5 St. Vincent and the Grenadines	Kingstown
6 Barbados	Bridgetown
7 Grenada	St. George's

	COUNTRY	CAPITAL
1	Czech Republic	Prague
2	Slovakia	Bratislava
3	Slovenia	Ljubljana
4	Croatia	Zagreb
5	Bosnia and Herzegovina	Sarajevo
6	Macedonia	Skopje
7	Yugoslavia (Serbia and Montenegro)	Belgrade
8	Lithuania	Vilnius
9	Latvia	Riga
10	Estonia	Tallinn

THE UNITED STATES OF AMERICA: PHYSICAL

PACIFIC OCEAN
Strait of Juan De Fuca
Puget Sound
Mount Rainier 14,410 ft. (4,392 m)
Columbia River
Willamette River
CASCADE RANGE
COAST RANGES
Franklin D. Roosevelt Lake
Pend Oreille
Flathead River
Clark Fork
Flathead Lake
LEWIS RANGE
BITTERROOT RANGE
Salmon River
SALMON RIVER MTS.
SAWTOOTH MTS.
COLUMBIA PLATEAU
Snake River
ROCKY MOUNTAINS
Milk River
Missouri River
Fort Peck Lake
Lake Sakakawea
Yellowstone River
Yellowstone Lake
GRAND TETONS
WIND RIVER RANGE
Wind River
Bighorn River
BIGHORN MTS.
Powder River
Gannett Peak 13,804 ft. (4,207 m)
CONTINENTAL DIVIDE
BLACK HILLS
Cheyenne River
White River
Lake Oahe
GREAT INTERIOR PLAINS
Red River
James River
Minnesota
Niobrara River
North Platte River
Missouri River
Platte River
South Platte River
Republican River
Kansas River
Smoky Hill River
FRONT RANGE
Pikes Peak 14,110 ft. (4,301 m)
Mount Elbert 14,433 ft. (4,400 m)
Cape Mendocino
Klamath River
Goose Lake
Shasta Lake
Sacramento River
CENTRAL VALLEY
SIERRA NEVADA
Pyramid Lake
Lake Tahoe
San Francisco Bay
Monterey Bay
San Joaquin River
COAST RANGES
Mount Whitney 14,494 ft. (4,419 m)
DEATH VALLEY
GREAT BASIN
Great Salt Lake
Utah Lake
WASATCH RANGE
UINTA MTS.
Green River
Colorado River
COLORADO PLATEAU
Lake Powell
San Juan River
GRAND CANYON
Lake Mead
PAINTED DESERT
SAN LUIS VALLEY
SANGRE DE CRISTO MTS.
MOJAVE DESERT
Channel Islands
Salton Sea
IMPERIAL VALLEY
Gila River
SONORA DESERT
Gulf of California
Canadian River
Keystone Lake
Arkansas
Eufaula Lake
Lake Texoma
Pecos River
Brazos
Trinity River
Colorado River
Rio Grande
Amistad Reservoir
Nueces River
Falcon Lake
Padre Island
MEXICO
GULF
50°N
45°N
40°N
35°N
30°N
25°N
130°W
125°W
120°
115°
110°
105°
100°W
Kauai
Niihau
Oahu
Molokai
Maui
Lanai
Kahoolawe
Mauna Kea 13,796 ft. (4,206 m)
Hawaii
PACIFIC OCEAN
22°N
19°N
160°W
155°W
SCALE
0 75 150 Miles
0 75 150 Kilometers
ARCTIC OCEAN
Arctic Circle
RUSSIA
Bering Strait
BROOKS RANGE
Yukon River
Tanana River
St. Lawrence Island
St. Matthew Island
Nunivak Island
Kuskokwim River
ALASKA RANGE
Mount McKinley 20,320 ft. (6,194 m)
CANADA
BERING SEA
Attu Island
Aleutian Islands
Kodiak Island
GULF OF ALASKA
Alexander Archipelago
PACIFIC OCEAN
SCALE
0 250 500 Miles
0 250 500 Kilometers
Projection: Albers Equal Area
65°N
60°N
55°N
50°N
170°E
180°
170°W
160°
150°W
140°W
130°W

To understand the relative locations of Alaska and Hawaii, as well as the vast distances separating them from the rest of the United States, see the world map.

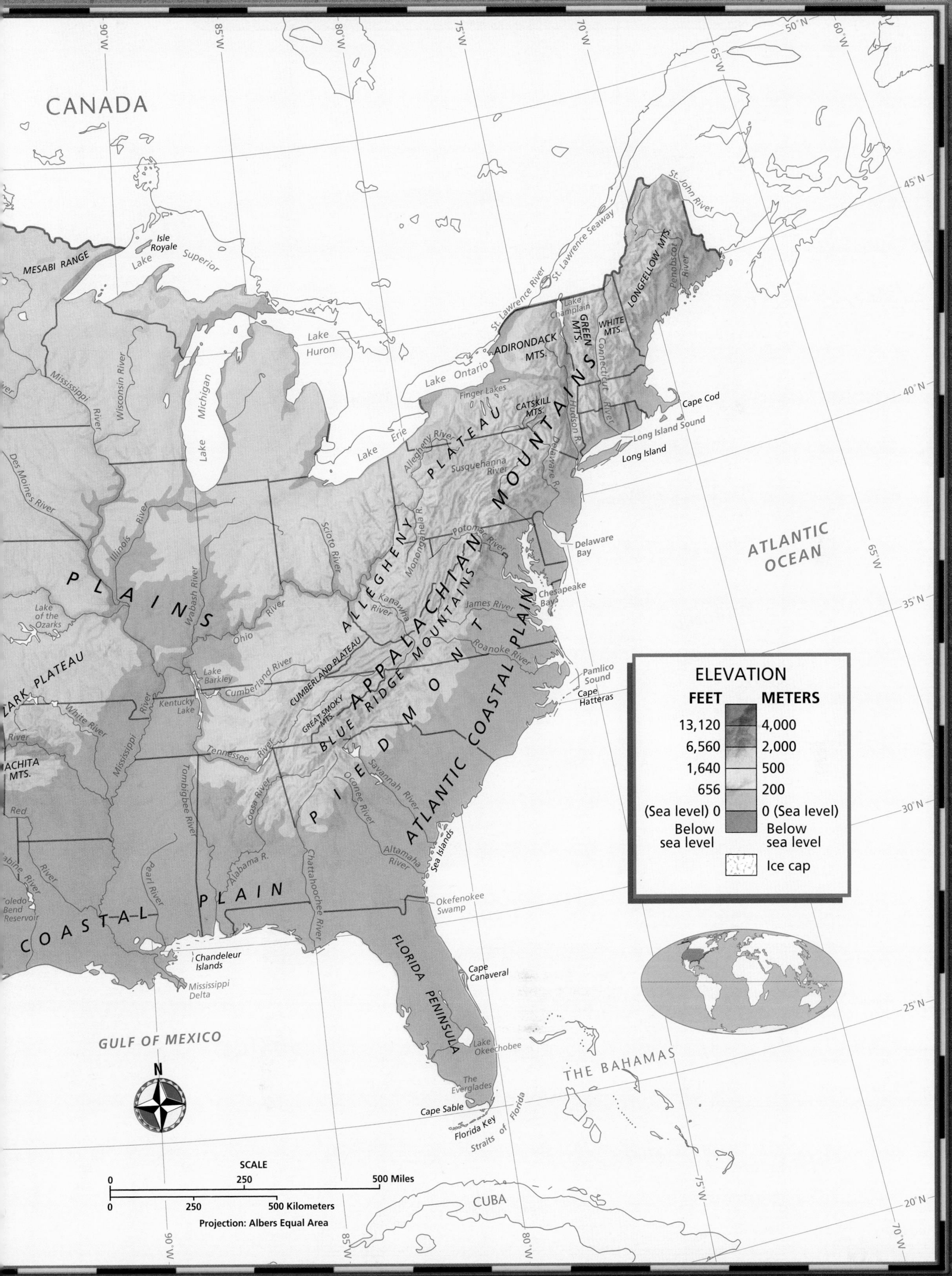

CANADA
MESABI RANGE
Isle Royale
Lake Superior
Lake Huron
Lake Michigan
Lake Ontario
Lake Erie
St. Lawrence River
St. Lawrence Seaway
St. John River
Penobscot River
LONGFELLOW MTS.
Lake Champlain
GREEN MTS.
WHITE MTS.
ADIRONDACK MTS.
Finger Lakes
CATSKILL MTS.
Connecticut River
Hudson R.
Cape Cod
Long Island Sound
Long Island
Delaware R.
Allegheny River
Susquehanna River
PLATEAU
APPALACHIAN MOUNTAINS
ALLEGHENY
Monongahela R.
Potomac River
Delaware Bay
Chesapeake Bay
James River
Kanawha River
BLUE RIDGE MOUNTAINS
CUMBERLAND PLATEAU
GREAT SMOKY MTS.
PIEDMONT
ATLANTIC COASTAL PLAIN
Roanoke River
Pamlico Sound
Cape Hatteras
ATLANTIC OCEAN
Mississippi River
Wisconsin River
Des Moines River
Illinois River
PLAINS
Wabash River
Scioto River
Ohio River
Lake of the Ozarks
OZARK PLATEAU
Lake Barkley
Cumberland River
Kentucky Lake
White River
Mississippi River
ACHITA MTS.
Red River
Tennessee River
Tombigbee River
Coosa River
Savannah River
Oconee River
Altamaha River
Sea Islands
Alabama R.
Pearl River
Chattahoochee River
Sabine River
Toledo Bend Reservoir
COASTAL PLAIN
Okefenokee Swamp
Chandeleur Islands
Mississippi Delta
FLORIDA PENINSULA
Cape Canaveral
Lake Okeechobee
The Everglades
Cape Sable
Florida Key
Straits of Florida
GULF OF MEXICO
THE BAHAMAS
CUBA
ELEVATION
FEET
METERS
13,120
4,000
6,560
2,000
1,640
500
656
200
(Sea level) 0
0 (Sea level)
Below sea level
Below sea level
Ice cap
N
SCALE
0
250
500 Miles
0
250
500 Kilometers
Projection: Albers Equal Area
90°W
85°W
80°W
75°W
70°W
65°W
60°W
50°N
45°N
40°N
35°N
30°N
25°N
20°N

THE UNITED STATES OF AMERICA: POLITICAL

PACIFIC OCEAN
Strait of Juan de Fuca
Puget Sound
Franklin D. Roosevelt Lake
Seattle
Tacoma
Olympia
Spokane
WASHINGTON
Portland
Columbia River
Salem
Eugene
OREGON
Pend Oreille
Flathead Lake
Helena
MONTANA
Missouri River
Fort Peck Lake
Yellowstone River
Billings
Lake Sakakawea
NORTH DAKOTA
Bismarck
Fargo
Red River
Lake Oahe
Minnesota
IDAHO
Boise
Snake River
Pocatello
Yellowstone Lake
WYOMING
Casper
Cheyenne
SOUTH DAKOTA
Pierre
Sioux Falls
Cape Mendocino
Goose Lake
Shasta Lake
Sacramento River
Pyramid Lake
Reno
Carson City
Lake Tahoe
NEVADA
Great Salt Lake
Salt Lake City
Utah Lake
Provo
UTAH
Green River
NEBRASKA
Platte River
Omaha
Lincoln
Missouri River
Oakland
San Francisco
San Francisco Bay
Sacramento
Stockton
Modesto
San Jose
Monterey Bay
San Joaquin R.
Fresno
CALIFORNIA
Bakersfield
Las Vegas
Lake Mead
Lake Powell
Colorado River
COLORADO
Denver
Colorado Springs
KANSAS
Topeka
Wichita
Arkansas River
Los Angeles
Long Beach
Channel Islands
Anaheim
Santa Ana
Salton Sea
San Diego
ARIZONA
Phoenix
Gila River
Tucson
Santa Fe
Albuquerque
NEW MEXICO
Canadian River
Keystone Lake
Tulsa
OKLAHOMA
Oklahoma City
Eufaula Lake
Amarillo
Lake Texoma
Lubbock
Fort Worth
Dallas
Abilene
El Paso
Odessa
TEXAS
Brazos River
Colorado River
Waco
Pecos River
Austin
Rio Grande
Amistad Reservoir
San Antonio
Houston
Corpus Christi
Laredo
Padre Island
MEXICO
Gulf of California
50°N
45°N
40°N
35°N
30°N
25°N
130°W
125°W
120°W
115°W
110°W
105°W
100°W
To understand the relative locations of Alaska and Hawaii as well as the vast distances separating them from the rest of the United States, see the world map.
Kauai
Niihau
Oahu
Honolulu
Molokai
HAWAII
Lanai
Kahoolawe
Maui
Hawaii
PACIFIC OCEAN
22°N
19°N
160°W
155°W
SCALE
0 75 150 Miles
0 75 150 Kilometers
ARCTIC OCEAN
Arctic Circle
RUSSIA
Bering Strait
Nome
Yukon River
Fairbanks
ALASKA
CANADA
St. Lawrence Island
St. Matthew Island
Nunivak Island
BERING SEA
Anchorage
Attu Island
Aleutian Islands
Kodiak Island
GULF OF ALASKA
Juneau
Alexander Archipelago
PACIFIC OCEAN
SCALE
0 250 500 Miles
0 250 500 Kilometers
Projection: Albers Equal Area
180°
170°E
170°W
160°W
150°W
140°W
130°W
65°N
60°N
55°N
50°N

CANADA
ATLANTIC OCEAN
GULF OF MEXICO
THE BAHAMAS
CUBA
MINNESOTA
WISCONSIN
MICHIGAN
IOWA
ILLINOIS
INDIANA
OHIO
PENNSYLVANIA
NEW YORK
MAINE
VT.
N.H.
MASS.
CONN.
R.I.
N.J.
DELAWARE
MD.
WEST VIRGINIA
VIRGINIA
KENTUCKY
MISSOURI
TENNESSEE
NORTH CAROLINA
SOUTH CAROLINA
ARKANSAS
MISSISSIPPI
ALABAMA
GEORGIA
LOUISIANA
FLORIDA
Boundaries
National capitals
State capitals
Other cities
SCALE
0
250
500 Miles
0
500
500 Kilometers
Projection: Albers Equal Area
N

NORTH AMERICA: PHYSICAL

ARCTIC OCEAN
North Pole
POLAR ICE PACK
ASIA
EUROPE
St. Lawrence Island
BERING SEA
Bering Strait
Nunivak Island
BROOKS RANGE
Mt. McKinley 20,320 ft. (6,194 m)
ALASKA RANGE
Yukon River
Kodiak Island
GULF OF ALASKA
YUKON PLATEAU
BEAUFORT SEA
Banks Island
Victoria Island
Queen Elizabeth Islands
Ellesmere Island
Greenland
Baffin Bay
Baffin Island
Davis Strait
Denmark Strait
Arctic Circle
Cape Farewell
Mackenzie River
Great Bear Lake
Great Slave Lake
Southampton Island
Hudson Strait
Coats Island
Mansel Island
Hudson Bay
LABRADOR SEA
Lake Athabasca
Peace River
Athabasca River
Alexander Archipelago
Queen Charlotte Islands
PACIFIC OCEAN
Vancouver Island
ROCKY MOUNTAINS
CANADIAN SHIELD
Saskatchewan River
Nelson River
Lake Winnipeg
Fraser River
Mount Rainier 14,410 ft. (4,392 m)
CASCADE RANGE
COAST RANGES
Columbia River
Anticosti Island
Newfoundland
GULF OF ST. LAWRENCE
Prince Edward Island
Cape Breton Island
St. Lawrence River
GREAT PLAINS
Missouri River
Lake Superior
Lake Michigan
Lake Huron
Lake Ontario
Lake Erie
Mississippi River
Cape Mendocino
Snake River
BLACK HILLS
Great Salt Lake
GREAT BASIN
SIERRA NEVADA
CENTRAL VALLEY
DEATH VALLEY
Mount Whitney 14,494 ft. (4,419 m)
Colorado River
COLORADO PLATEAU
Platte River
INTERIOR PLAINS
OZARK PLATEAU
Ohio River
Cumberland R.
Tennessee River
Arkansas River
Red River
Brazos River
Rio Grande
APPALACHIAN MOUNTAINS
PIEDMONT
ATLANTIC COASTAL PLAIN
Cape Cod
Long Island
ATLANTIC OCEAN
Cape Hatteras
Bermuda
Guadalupe Island
BAJA CALIFORNIA
GULF OF CALIFORNIA
SIERRA MADRE OCCIDENTAL
SIERRA MADRE ORIENTAL
GULF COASTAL PLAIN
FLORIDA PENINSULA
Cape Canaveral
Tropic of Cancer
Bahamas
GULF OF MEXICO
Florida Keys
Straits of Florida
Cuba
Greater Antilles
Hispaniola
Puerto Rico
Lesser Antilles
Jamaica
Popocatépetl 17,887 ft. (5,452 m)
YUCATÁN PENINSULA
SIERRA MADRE DEL SUR
CARIBBEAN SEA
Trinidad
Lake Nicaragua
CENTRAL AMERICA
ISTHMUS OF PANAMA
SOUTH AMERICA
Equator 0°
ELEVATION
FEET
METERS
13,120
4,000
6,560
2,000
1,640
500
656
200
(Sea level) 0
0 (Sea level)
Below sea level
Below sea level
Ice cap
N
SCALE
0 250 500 750 1,000 Miles
0 250 500 750 1,000 Kilometers
Projection: Azimuthal Equal Area

NORTH AMERICA: POLITICAL

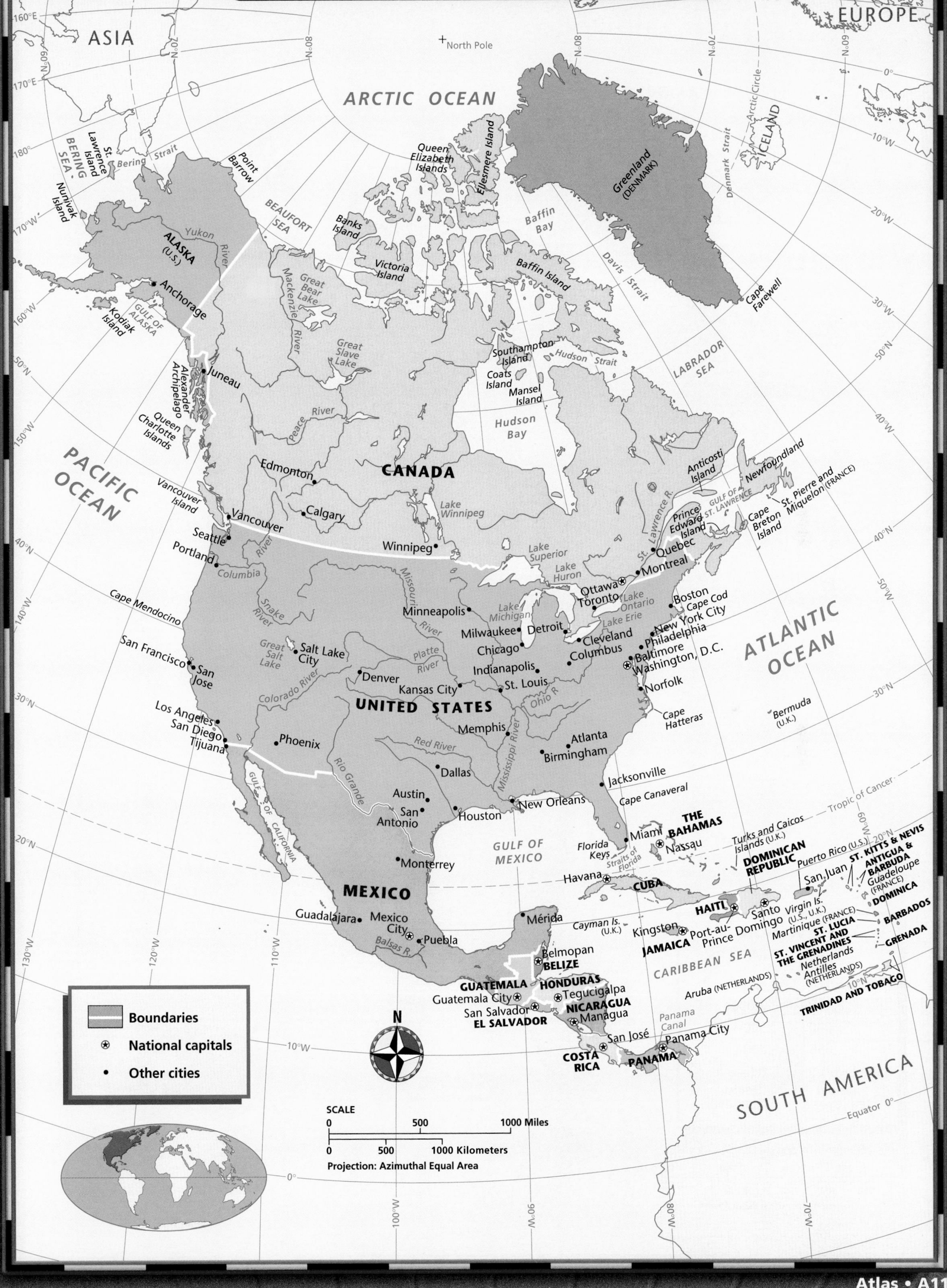

ASIA
EUROPE
North Pole
ARCTIC OCEAN
PACIFIC OCEAN
ATLANTIC OCEAN
SOUTH AMERICA
BERING SEA
Bering Strait
St. Lawrence Island
Nunivak Island
Point Barrow
BEAUFORT SEA
ALASKA (U.S.)
Yukon River
Anchorage
GULF OF ALASKA
Kodiak Island
Juneau
Alexander Archipelago
Queen Charlotte Islands
Vancouver Island
Queen Elizabeth Islands
Ellesmere Island
Banks Island
Victoria Island
Baffin Bay
Baffin Island
Greenland (DENMARK)
ICELAND
Denmark Strait
Arctic Circle
Davis Strait
Cape Farewell
LABRADOR SEA
Great Bear Lake
Mackenzie River
Great Slave Lake
Southampton Island
Coats Island
Mansel Island
Hudson Strait
Hudson Bay
Peace River
CANADA
Edmonton
Calgary
Vancouver
Lake Winnipeg
Winnipeg
Anticosti Island
Newfoundland
GULF OF ST. LAWRENCE
St. Pierre and Miquelon (FRANCE)
Prince Edward Island
Cape Breton Island
St. Lawrence R.
Quebec
Montreal
Ottawa
Toronto
Lake Superior
Lake Huron
Lake Michigan
Lake Ontario
Lake Erie
Seattle
Portland
Columbia River
Snake River
Missouri River
Cape Mendocino
Minneapolis
Milwaukee
Detroit
Chicago
Cleveland
Columbus
Boston
Cape Cod
New York City
Philadelphia
Baltimore
Washington, D.C.
Norfolk
San Francisco
San Jose
Great Salt Lake
Salt Lake City
Platte River
Denver
Kansas City
Indianapolis
St. Louis
Ohio R.
Colorado River
UNITED STATES
Cape Hatteras
Bermuda (U.K.)
Los Angeles
San Diego
Tijuana
Phoenix
Memphis
Red River
Mississippi River
Atlanta
Birmingham
Dallas
Rio Grande
Jacksonville
Cape Canaveral
Austin
San Antonio
Houston
New Orleans
Tropic of Cancer
GULF OF CALIFORNIA
GULF OF MEXICO
Florida Keys
Straits of Florida
Miami
THE BAHAMAS
Nassau
Turks and Caicos Islands (U.K.)
Monterrey
DOMINICAN REPUBLIC
Puerto Rico (U.S.)
San Juan
ST. KITTS & NEVIS
ANTIGUA & BARBUDA
Guadeloupe (FRANCE)
DOMINICA
Havana
CUBA
MEXICO
HAITI
Santo Domingo
Virgin Is. (U.S., U.K.)
Martinique (FRANCE)
BARBADOS
ST. LUCIA
ST. VINCENT AND THE GRENADINES
GRENADA
Guadalajara
Mexico City
Puebla
Balsas R.
Mérida
Cayman Is. (U.K.)
Kingston
JAMAICA
Port-au-Prince
CARIBBEAN SEA
Netherlands Antilles (NETHERLANDS)
Aruba (NETHERLANDS)
TRINIDAD AND TOBAGO
Belmopan
BELIZE
GUATEMALA
Guatemala City
HONDURAS
Tegucigalpa
San Salvador
EL SALVADOR
NICARAGUA
Managua
Panama Canal
San José
Panama City
COSTA RICA
PANAMA
Equator 0°
Boundaries
National capitals
Other cities
N
SCALE
0
500
1000 Miles
0
500
1000 Kilometers
Projection: Azimuthal Equal Area

SOUTH AMERICA: PHYSICAL

CENTRAL AMERICA
CARIBBEAN SEA
Panama Canal
GULF OF PANAMA
Margarita Island
Tobago
Trinidad
Lake Maracaibo
Orinoco River Delta
Orinoco River
LLANOS
Meta River
Angel Falls
GUIANA HIGHLANDS
Devil's Island
Cape Orange
ATLANTIC OCEAN
Cauca River
Magdalena River
Mount Tolima 18,425 ft. (5,616 m)
Malpelo Island
Amazon River Delta
Caquetá River
Rio Negro
Japurá River
AMAZON BASIN
Amazon River
Galápagos Islands
Mount Chimborazo 20,561 ft. (6,267 m)
GULF OF GUAYAQUIL
ANDES
Marañón River
Ucayali
Juruá River
Purus
Madeira River
Tapajós River
Xingu River
Tocantins River
Parnaíba River
Araguaia River
Mount Huascarán 22,205 ft. (6,768 m)
BRAZILIAN HIGHLANDS
MATO GROSSO PLATEAU
Beni River
Mamoré River
São Francisco River
PACIFIC OCEAN
Lake Titicaca
Ancohuma Peak 20,958 ft. (6,388 m)
Lake Poopó
Pilcomayo River
ATACAMA DESERT
CHACO
BRAZILIAN PLATEAU
Paraguay River
Paraná River
Salado River
Uruguay River
Tropic of Capricorn
San Ambrosio Island
San Félix Island
Mount Aconcagua 22,834 ft. (6,960 m)
Juan Fernández Islands
Río de la Plata
PAMPAS
Colorado River
GULF OF SAN MATÍAS
Chiloé Island
CHONOS ARCHIPELAGO
PATAGONIA
GULF OF SAN JORGE
Cape Tres Puntas
Bahía Grande
Strait of Magellan
Falkland Islands
TIERRA DEL FUEGO
CAPE HORN
South Georgia Islands
ELEVATION
FEET METERS
13,120 4,000
6,560 2,000
1,640 500
656 200
(Sea level) 0 0 (Sea level)
Below sea level Below sea level
SCALE
0 250 500 750 1,000 Miles
0 250 500 750 1,000 Kilometers
Projection: Azimuthal Equal Area
Equator 0°
20°N
10°N
10°S
20°S
30°S
40°S
50°S
90°W
80°W
70°W
60°W
50°W
40°W
30°W
20°W
100°W

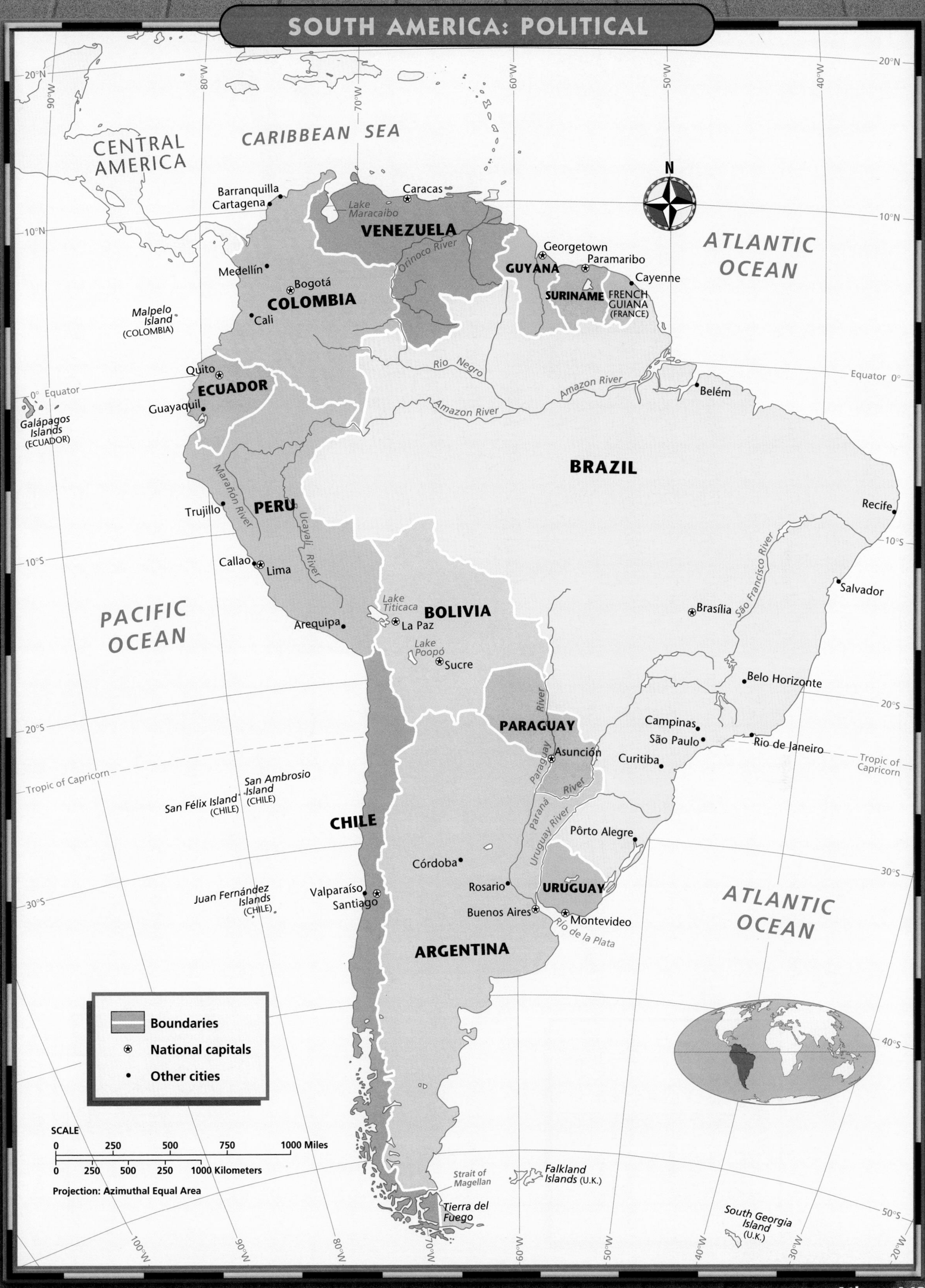
SOUTH AMERICA: POLITICAL
CARIBBEAN SEA
CENTRAL AMERICA
ATLANTIC OCEAN
PACIFIC OCEAN
N
Barranquilla
Cartagena
Caracas
Lake Maracaibo
VENEZUELA
Orinoco River
Georgetown
Paramaribo
GUYANA
Cayenne
SURINAME
FRENCH GUIANA (FRANCE)
Medellín
Bogotá
COLOMBIA
Cali
Malpelo Island (COLOMBIA)
Quito
ECUADOR
Guayaquil
Galápagos Islands (ECUADOR)
Rio Negro
Amazon River
Belém
BRAZIL
Recife
Marañon River
PERU
Ucayali River
Trujillo
Callao
Lima
São Francisco River
Salvador
Lake Titicaca
BOLIVIA
La Paz
Arequipa
Lake Poopó
Sucre
Brasília
Belo Horizonte
PARAGUAY
Paraguay River
Asunción
Campinas
São Paulo
Rio de Janeiro
Curitiba
Tropic of Capricorn
San Ambrosio Island (CHILE)
San Félix Island (CHILE)
CHILE
Paraná River
Uruguay River
Pôrto Alegre
Córdoba
Rosario
URUGUAY
Juan Fernández Islands (CHILE)
Valparaíso
Santiago
Buenos Aires
Montevideo
Río de la Plata
ATLANTIC OCEAN
ARGENTINA
Boundaries
National capitals
Other cities
SCALE
0 250 500 750 1000 Miles
0 250 500 250 1000 Kilometers
Projection: Azimuthal Equal Area
Strait of Magellan
Falkland Islands (U.K.)
Tierra del Fuego
South Georgia Island (U.K.)
20°N
10°N
0° Equator
10°S
20°S
30°S
40°S
50°S
90°W
80°W
70°W
60°W
50°W
40°W
30°W
20°W
100°W

EUROPE: PHYSICAL

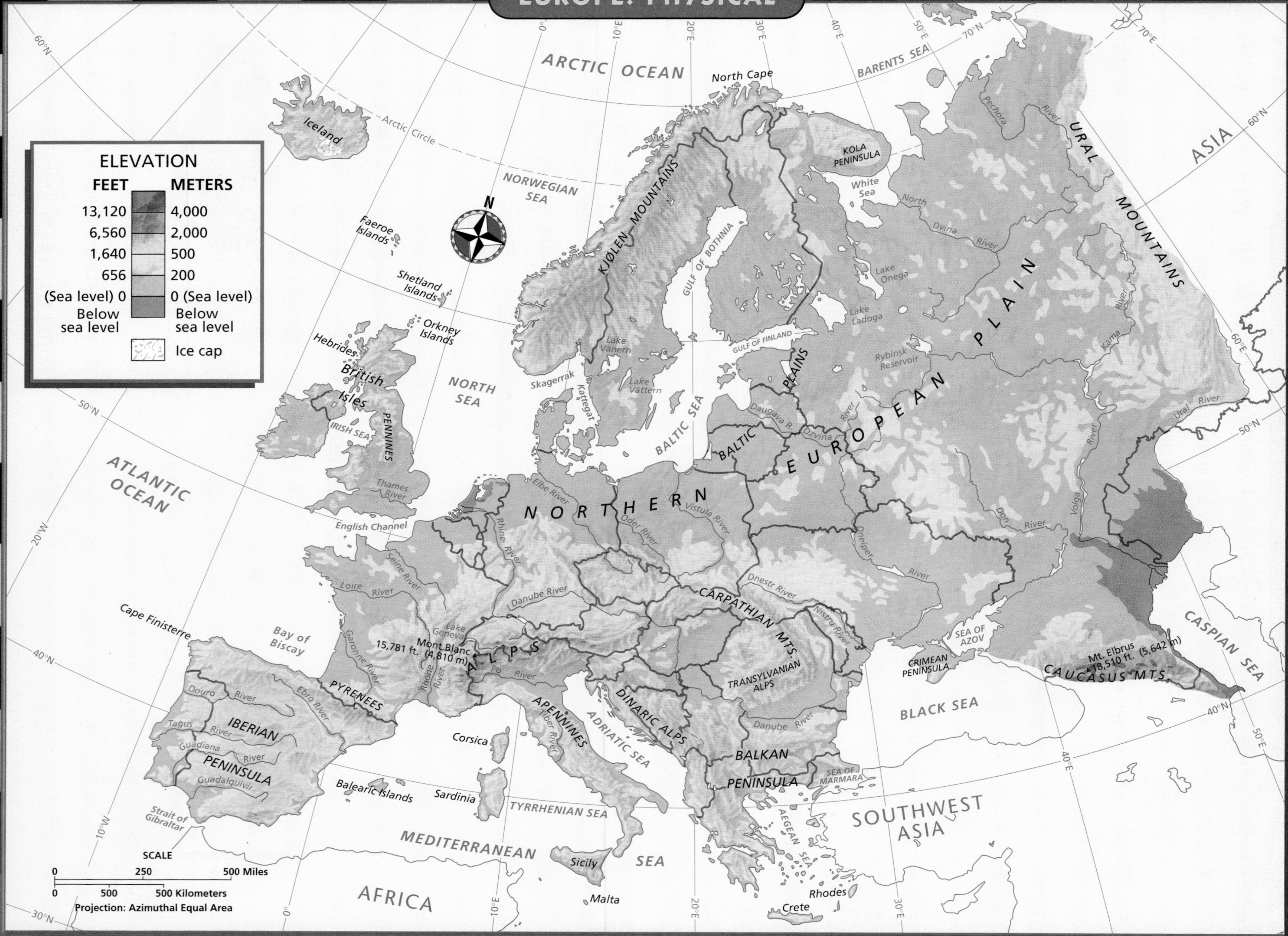
ELEVATION
FEET
METERS
13,120
4,000
6,560
2,000
1,640
500
656
200
(Sea level) 0
0 (Sea level)
Below sea level
Below sea level
Ice cap
ARCTIC OCEAN
North Cape
BARENTS SEA
Iceland
Arctic Circle
NORWEGIAN SEA
KOLA PENINSULA
White Sea
URAL MOUNTAINS
ASIA
KJØLEN MOUNTAINS
GULF OF BOTHNIA
Faeroe Islands
Shetland Islands
Orkney Islands
Hebrides
British Isles
NORTH SEA
PENNINES
IRISH SEA
Thames River
Skagerrak
Kattegat
Lake Vanern
Lake Vattern
GULF OF FINLAND
BALTIC SEA
PLAINS
BALTIC
Daugava R.
Dzvina
Lake Onega
Lake Ladoga
Rybinsk Reservoir
North Dvina River
Pechora River
Kama River
Ural River
Volga River
Don River
Dneiper River
EUROPEAN PLAIN
NORTHERN
Elbe River
Oder River
Vistula River
Rhine River
English Channel
Seine River
Loire River
Danube River
Dnestr River
Nistru River
CARPATHIAN MTS.
TRANSYLVANIAN ALPS
ATLANTIC OCEAN
Cape Finisterre
Bay of Biscay
Garonne River
Lake Geneva
Mont Blanc 15,781 ft. (4,810 m)
ALPS
Po River
Rhone River
PYRENEES
Douro River
Ebro River
Tagus River
IBERIAN PENINSULA
Guadiana River
Guadalquivir
APENNINES
Tiber River
ADRIATIC SEA
DINARIC ALPS
Corsica
Sardinia
Balearic Islands
TYRRHENIAN SEA
MEDITERRANEAN SEA
Strait of Gibraltar
Sicily
Malta
AFRICA
BALKAN PENINSULA
Danube River
SEA OF MARMARA
AEGEAN SEA
Rhodes
Crete
SOUTHWEST ASIA
BLACK SEA
CRIMEAN PENINSULA
SEA OF AZOV
Mt. Elbrus 18,510 ft. (5,642 m)
CAUCASUS MTS.
CASPIAN SEA
SCALE
0
250
500 Miles
0
500
500 Kilometers
Projection: Azimuthal Equal Area
N
60°N
50°N
40°N
30°N
70°N
20°W
10°W
0°
10°E
20°E
30°E
40°E
50°E
60°E
70°E

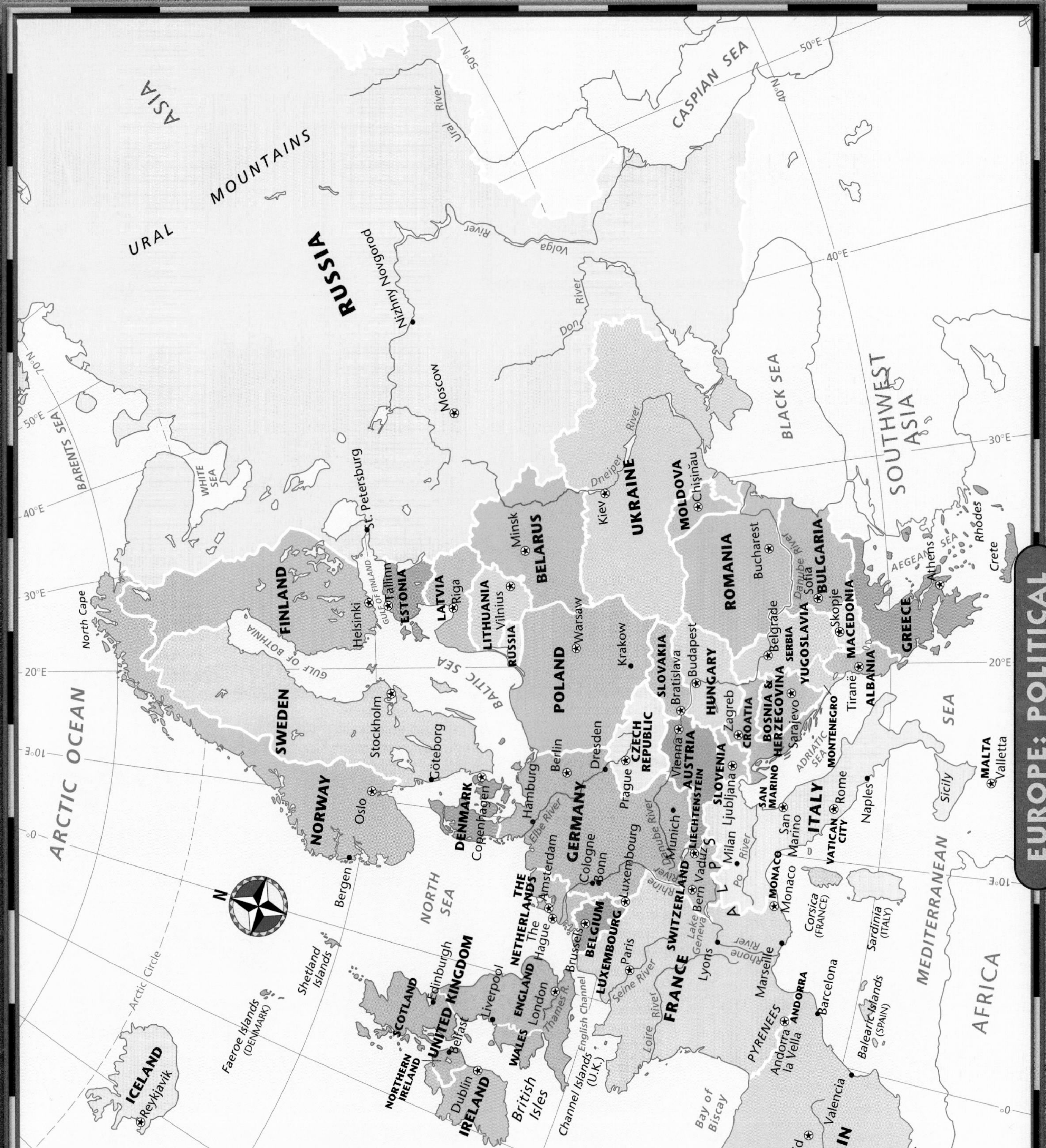
EUROPE: POLITICAL
Boundaries
National capitals
Other cities
SCALE
0 250 500 Miles
0 250 500 Kilometers
Projection: Azimuthal Equal Area
ARCTIC OCEAN
ATLANTIC OCEAN
ASIA
AFRICA
SOUTHWEST ASIA
URAL MOUNTAINS
BARENTS SEA
WHITE SEA
NORTH SEA
BALTIC SEA
GULF OF BOTHNIA
GULF OF FINLAND
CASPIAN SEA
BLACK SEA
AEGEAN SEA
ADRIATIC SEA
MEDITERRANEAN SEA
Bay of Biscay
English Channel
Strait of Gibraltar
Arctic Circle
North Cape
ICELAND
Reykjavík
Faeroe Islands (DENMARK)
Shetland Islands
British Isles
UNITED KINGDOM
SCOTLAND
ENGLAND
WALES
NORTHERN IRELAND
IRELAND
Dublin
Belfast
Edinburgh
Liverpool
London
Thames R.
Channel Islands (U.K.)
NORWAY
Oslo
Bergen
SWEDEN
Stockholm
Göteborg
FINLAND
Helsinki
DENMARK
Copenhagen
ESTONIA
Tallinn
LATVIA
Riga
LITHUANIA
Vilnius
RUSSIA
St. Petersburg
Moscow
Nizhny Novgorod
Volga River
Ural River
Don River
BELARUS
Minsk
UKRAINE
Kiev
Dnieper River
MOLDOVA
Chişinău
POLAND
Warsaw
Krakow
GERMANY
Berlin
Hamburg
Dresden
Cologne
Bonn
Munich
Elbe River
Rhine River
Danube River
THE NETHERLANDS
Amsterdam
The Hague
BELGIUM
Brussels
LUXEMBOURG
Luxembourg
FRANCE
Paris
Lyons
Marseille
Seine River
Loire River
Rhone River
SWITZERLAND
Bern
Lake Geneva
LIECHTENSTEIN
Vaduz
A L P S
AUSTRIA
Vienna
CZECH REPUBLIC
Prague
SLOVAKIA
Bratislava
HUNGARY
Budapest
SLOVENIA
Ljubljana
CROATIA
Zagreb
BOSNIA & HERZEGOVINA
Sarajevo
YUGOSLAVIA
SERBIA
MONTENEGRO
Belgrade
MACEDONIA
Skopje
ALBANIA
Tiranë
GREECE
Athens
Crete
Rhodes
ROMANIA
Bucharest
BULGARIA
Sofia
ITALY
Rome
Milan
Naples
Po River
SAN MARINO
San Marino
VATICAN CITY
MONACO
Monaco
Corsica (FRANCE)
Sardinia (ITALY)
Sicily
MALTA
Valletta
SPAIN
Madrid
Barcelona
Valencia
Seville
Balearic Islands (SPAIN)
Gibraltar (U.K.)
PYRENEES
ANDORRA
Andorra la Vella
PORTUGAL
Lisbon
Tagus River
70°N
50°N
40°N
50°E
40°E
30°E
20°E
10°E
0°
10°W
20°W
N

ASIA: PHYSICAL

ELEVATION
FEET
METERS
13,120
4,000
6,560
2,000
1,640
500
656
200
(Sea level) 0
0 (Sea level)
Below sea level
Below sea level
Ice cap
SCALE
0 500 1,000 Miles
0 500 1,000 Kilometers
Projection: Modified Oblique Conic
NORTH AMERICA
EUROPE
AFRICA
AUSTRALIA
PACIFIC OCEAN
INDIAN OCEAN
MEDITERRANEAN SEA
BLACK SEA
SEA OF AZOV
Bosporus
CASPIAN SEA
ARAL SEA
RED SEA
PERSIAN GULF
Strait of Hormuz
GULF OF OMAN
GULF OF ADEN
ARABIAN SEA
Bay of Bengal
ANDAMAN SEA
GULF OF THAILAND
SOUTH CHINA SEA
GULF OF TONKIN
EAST CHINA SEA
YELLOW SEA
SEA OF JAPAN
SEA OF OKHOTSK
BERING SEA
BARENTS SEA
KARA SEA
LAPTEV SEA
CELEBES SEA
BANDA SEA
ARAFURA SEA
JAVA SEA
Strait of Malacca
Luzon Strait
Korea Strait
Arctic Circle
Tropic of Cancer
Equator
Franz Josef Land
Novaya Zemlya
North Land
TAYMYR PENINSULA
New Siberian Islands
Wrangel Island
Aleutian Islands
KAMCHATKA PENINSULA
CENTRAL RANGE
KOLYMA MTS.
CHERSKIY RANGE
VERKHOYANSKIY RANGE
Lena River
Aldan River
STANOVOY MOUNTAINS
Sakhalin Island
Kuril Islands
Hokkaido
Honshu
Shikoku
Kyushu
Ryukyu Islands
Okinawa
Taiwan
Luzon
Philippines
Mindanao
Hainan
Borneo
Celebes
Moluccas
MAOKE MOUNTAIN
New Guinea
Sumatra
Java
Bangka
Mentawai Islands
Nicobar Islands
Andaman Islands
Sri Lanka
Maldives
Lakshadweep Islands
Socotra Island
Cyprus
SINAI PENINSULA
ANATOLIAN PLATEAU
Mount Ararat 16,945 ft. (5,165 m)
CAUCASUS MTS.
SYRIAN DESERT
AN-NAFUD
RUB' AL-KHALI
Tigris
Euphrates River
River
ZAGROS MTS.
GREAT SALT DESERT
USTYURT PLATEAU
Ural
River
URAL MOUNTAINS
TURAN LOWLAND
KYZYL KUM
KARA-KUM
Amu Darya
Syr Darya
Ishim River
Irtysh
River
Ob
WEST SIBERIAN PLAIN
S I B E R I A
Yenisey River
Lower Tunguska River
Angara River
CENTRAL SIBERIAN PLATEAU
KAZAKH UPLANDS
Lake Balkhash
ALTAY SHAN
SAYAN MOUNTAINS
Lake Baikal
YABLONOVY RANGE
Shilka River
Amur River
MONGOLIAN PLATEAU
GREATER KHINGAN RANGE
GOBI
TIAN SHAN
TARIM BASIN
TAKLIMAKAN DESERT
HINDU KUSH
KUNLUN MOUNTAINS
PLATEAU OF TIBET
Nu River
Mount Everest 29,035 ft. (8,850 m)
Brahmaputra River
H I M A L A Y A S
INDO-GANGETIC PLAIN
Ganges River
Indus River
Sutlej
THAR DESERT
DECCAN PLATEAU
Godavari River
EASTERN GHATS
WESTERN GHATS
Irrawaddy River
GREAT WALL OF CHINA
NORTH CHINA PLAIN
Huang River
QIN LING
Chang River
BOHEA HILLS
Xi River
Hong River
INDOCHINA PENINSULA
Mekong River
Chao Phraya River
MALAY PENINSULA
N
0°
10°E
20°E
30°E
40°E
50°E
60°E
70°E
80°E
90°E
100°E
110°E
120°E
130°E
140°E
150°E
160°E
170°E
180°
170°W
10°S
0°
10°N
20°N
30°N
40°N
50°N
60°N
70°N
80°N

ASIA: POLITICAL

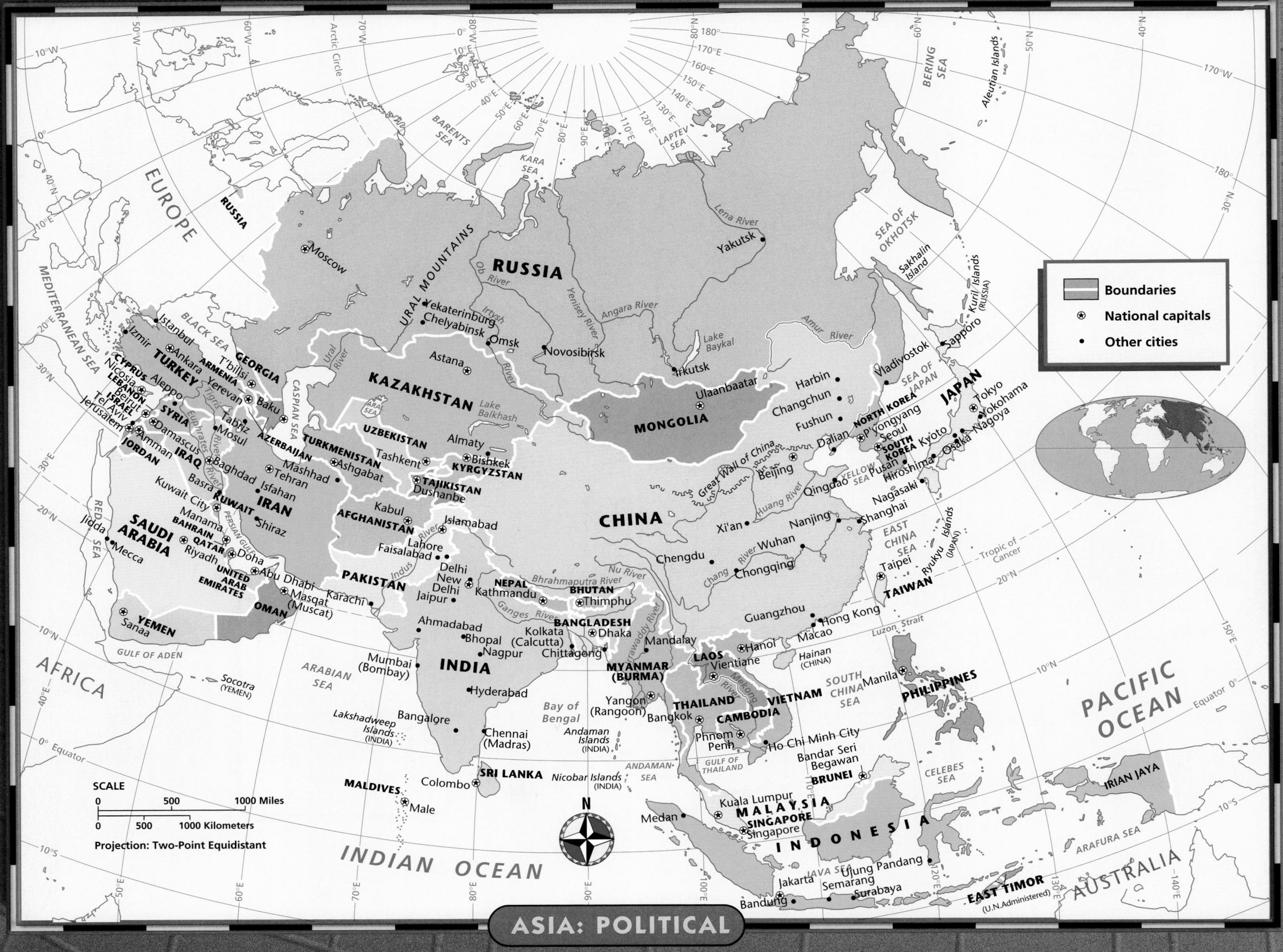

AFRICA: PHYSICAL

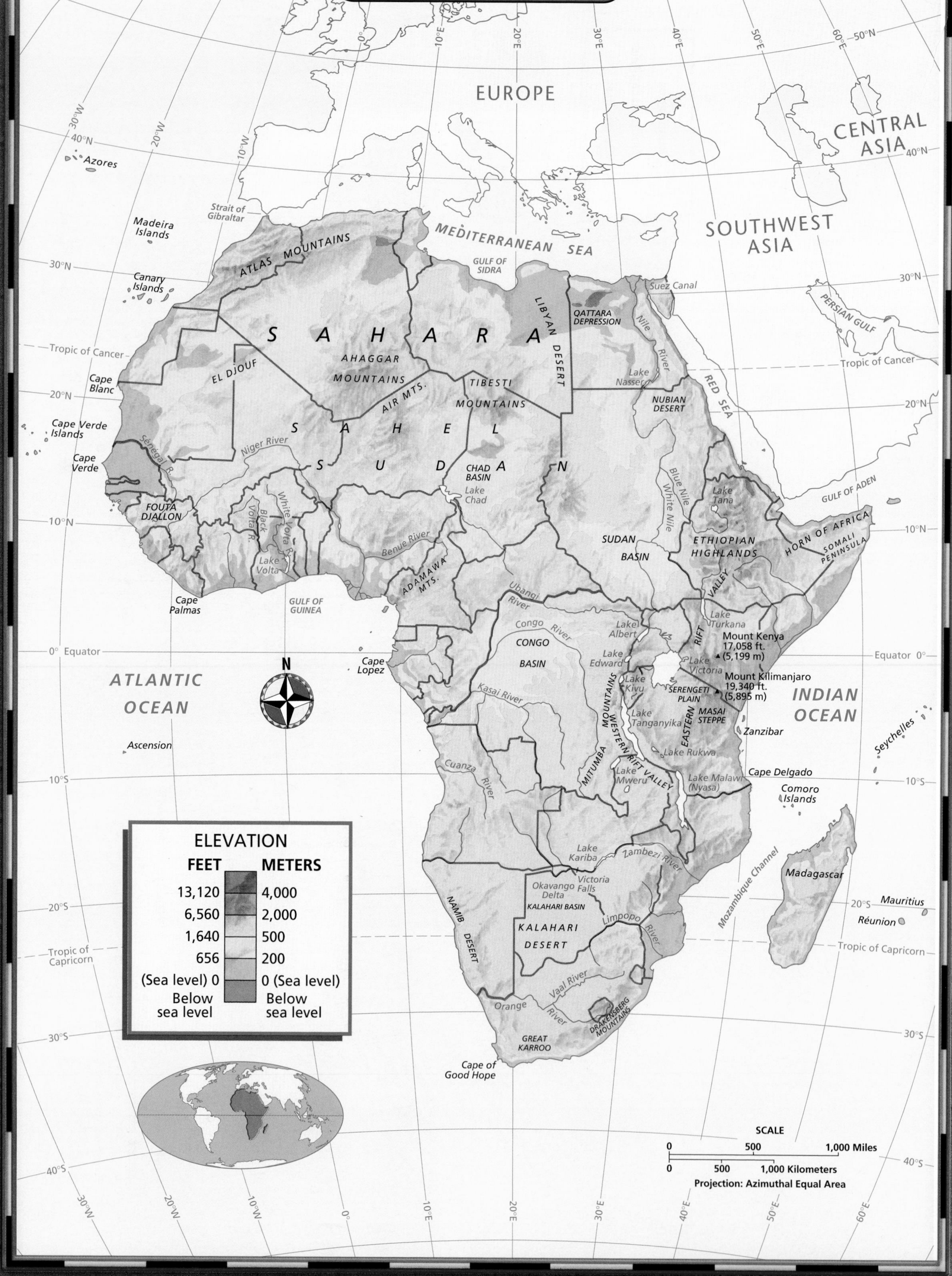
EUROPE
CENTRAL ASIA
SOUTHWEST ASIA
Azores
Madeira Islands
Strait of Gibraltar
Canary Islands
ATLAS MOUNTAINS
MEDITERRANEAN SEA
GULF OF SIDRA
Suez Canal
PERSIAN GULF
QATTARA DEPRESSION
LIBYAN DESERT
S A H A R A
AHAGGAR MOUNTAINS
EL DJOUF
TIBESTI MOUNTAINS
AIR MTS.
Nile River
Lake Nasser
RED SEA
NUBIAN DESERT
Cape Blanc
Cape Verde Islands
Cape Verde
Sénégal R.
Niger River
S A H E L
S U D A N
CHAD BASIN
Lake Chad
Blue Nile
White Nile
Lake Tana
GULF OF ADEN
FOUTA DJALLON
Black Volta R.
White Volta R.
Lake Volta
Benue River
ADAMAWA MTS.
SUDAN BASIN
ETHIOPIAN HIGHLANDS
HORN OF AFRICA
SOMALI PENINSULA
Cape Palmas
GULF OF GUINEA
Ubangi River
Congo River
CONGO BASIN
Lake Albert
Lake Edward
Lake Turkana
RIFT VALLEY
Mount Kenya 17,058 ft. (5,199 m)
Lake Victoria
Mount Kilimanjaro 19,340 ft. (5,895 m)
SERENGETI PLAIN
MASAI STEPPE
Lake Kivu
Lake Tanganyika
EASTERN
WESTERN RIFT VALLEY
MITUMBA MOUNTAINS
Lake Rukwa
Lake Mweru
Lake Malawi (Nyasa)
Zanzibar
Cape Delgado
Comoro Islands
Seychelles
INDIAN OCEAN
ATLANTIC OCEAN
Cape Lopez
Kasai River
Ascension
Cuanza River
Lake Kariba
Zambezi River
Victoria Falls
Okavango Delta
KALAHARI BASIN
KALAHARI DESERT
NAMIB DESERT
Limpopo River
Mozambique Channel
Madagascar
Mauritius
Réunion
Vaal River
Orange River
DRAKENSBERG MOUNTAINS
GREAT KARROO
Cape of Good Hope
Tropic of Cancer
Tropic of Capricorn
Equator
N
ELEVATION
FEET
METERS
13,120
4,000
6,560
2,000
1,640
500
656
200
(Sea level) 0
0 (Sea level)
Below sea level
Below sea level
SCALE
0
500
1,000 Miles
0
500
1,000 Kilometers
Projection: Azimuthal Equal Area

AFRICA: POLITICAL

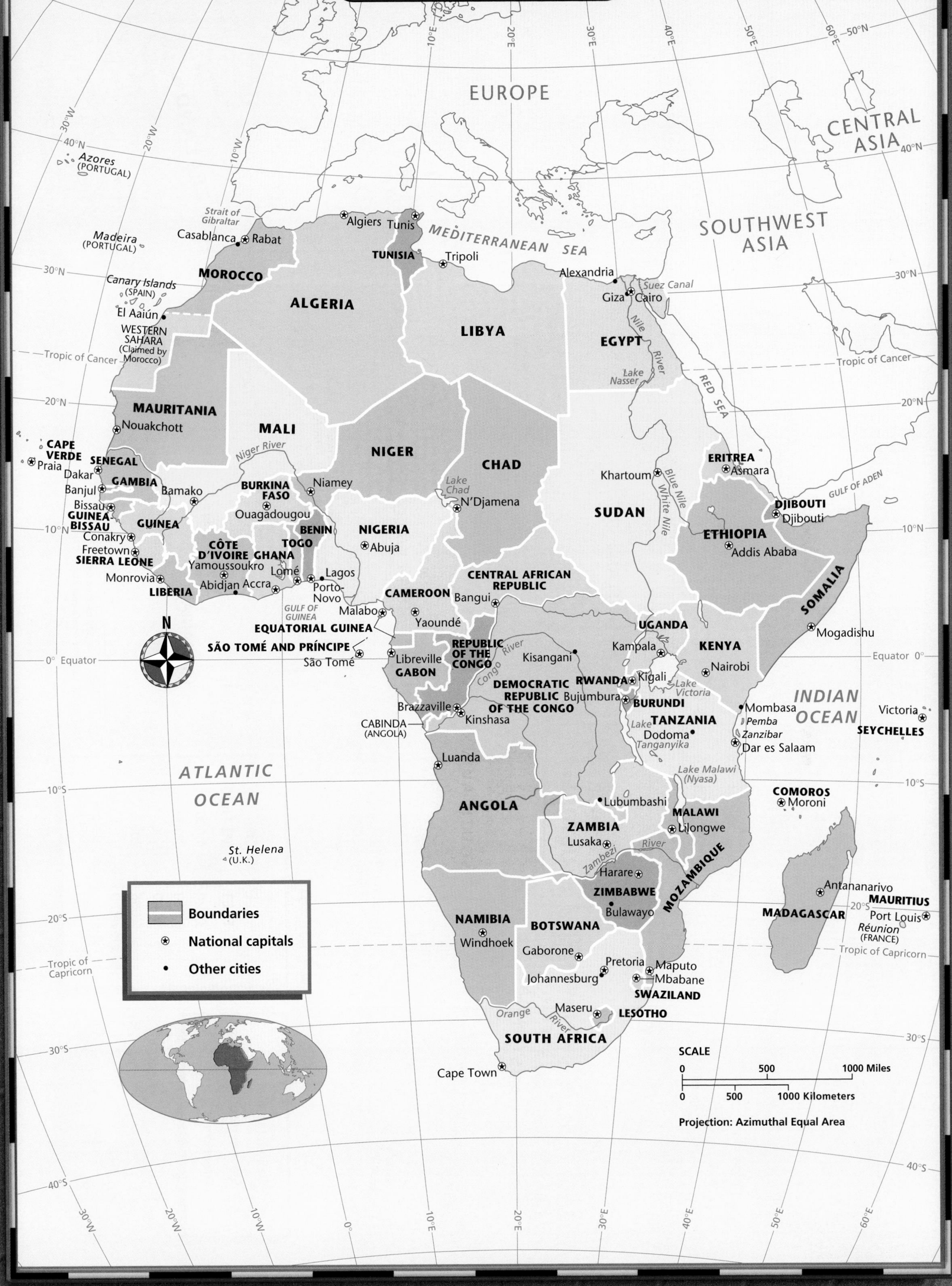

EUROPE
CENTRAL ASIA
SOUTHWEST ASIA
Azores (PORTUGAL)
Madeira (PORTUGAL)
Canary Islands (SPAIN)
Strait of Gibraltar
MEDITERRANEAN SEA
RED SEA
GULF OF ADEN
INDIAN OCEAN
ATLANTIC OCEAN
GULF OF GUINEA
St. Helena (U.K.)
Suez Canal
Nile River
Lake Nasser
Niger River
Lake Chad
Blue Nile
White Nile
Congo River
Lake Victoria
Lake Tanganyika
Lake Malawi (Nyasa)
Zambezi River
Orange River
MOROCCO
Casablanca
Rabat
WESTERN SAHARA (Claimed by Morocco)
El Aaiún
ALGERIA
Algiers
TUNISIA
Tunis
LIBYA
Tripoli
EGYPT
Alexandria
Giza
Cairo
MAURITANIA
Nouakchott
MALI
Bamako
NIGER
Niamey
CHAD
N'Djamena
SUDAN
Khartoum
ERITREA
Asmara
DJIBOUTI
Djibouti
ETHIOPIA
Addis Ababa
SOMALIA
Mogadishu
CAPE VERDE
Praia
SENEGAL
Dakar
GAMBIA
Banjul
GUINEA-BISSAU
Bissau
GUINEA
Conakry
SIERRA LEONE
Freetown
LIBERIA
Monrovia
CÔTE D'IVOIRE
Yamoussoukro
Abidjan
BURKINA FASO
Ouagadougou
GHANA
Accra
TOGO
Lomé
BENIN
Porto-Novo
NIGERIA
Abuja
Lagos
CAMEROON
Yaoundé
CENTRAL AFRICAN REPUBLIC
Bangui
EQUATORIAL GUINEA
Malabo
SÃO TOMÉ AND PRÍNCIPE
São Tomé
GABON
Libreville
REPUBLIC OF THE CONGO
Brazzaville
DEMOCRATIC REPUBLIC OF THE CONGO
Kinshasa
Kisangani
Lubumbashi
CABINDA (ANGOLA)
UGANDA
Kampala
KENYA
Nairobi
Mombasa
RWANDA
Kigali
BURUNDI
Bujumbura
TANZANIA
Dodoma
Dar es Salaam
Pemba
Zanzibar
ANGOLA
Luanda
ZAMBIA
Lusaka
MALAWI
Lilongwe
MOZAMBIQUE
Maputo
ZIMBABWE
Harare
Bulawayo
NAMIBIA
Windhoek
BOTSWANA
Gaborone
SOUTH AFRICA
Pretoria
Johannesburg
Cape Town
SWAZILAND
Mbabane
LESOTHO
Maseru
MADAGASCAR
Antananarivo
COMOROS
Moroni
SEYCHELLES
Victoria
MAURITIUS
Port Louis
Réunion (FRANCE)
40°N
30°N
Tropic of Cancer
20°N
10°N
0° Equator
10°S
20°S
Tropic of Capricorn
30°S
40°S
50°N
30°W
20°W
10°W
0°
10°E
20°E
30°E
40°E
50°E
60°E
N
Boundaries
National capitals
Other cities
SCALE
0
500
1000 Miles
0
500
1000 Kilometers
Projection: Azimuthal Equal Area

AUSTRALIA AND NEW ZEALAND

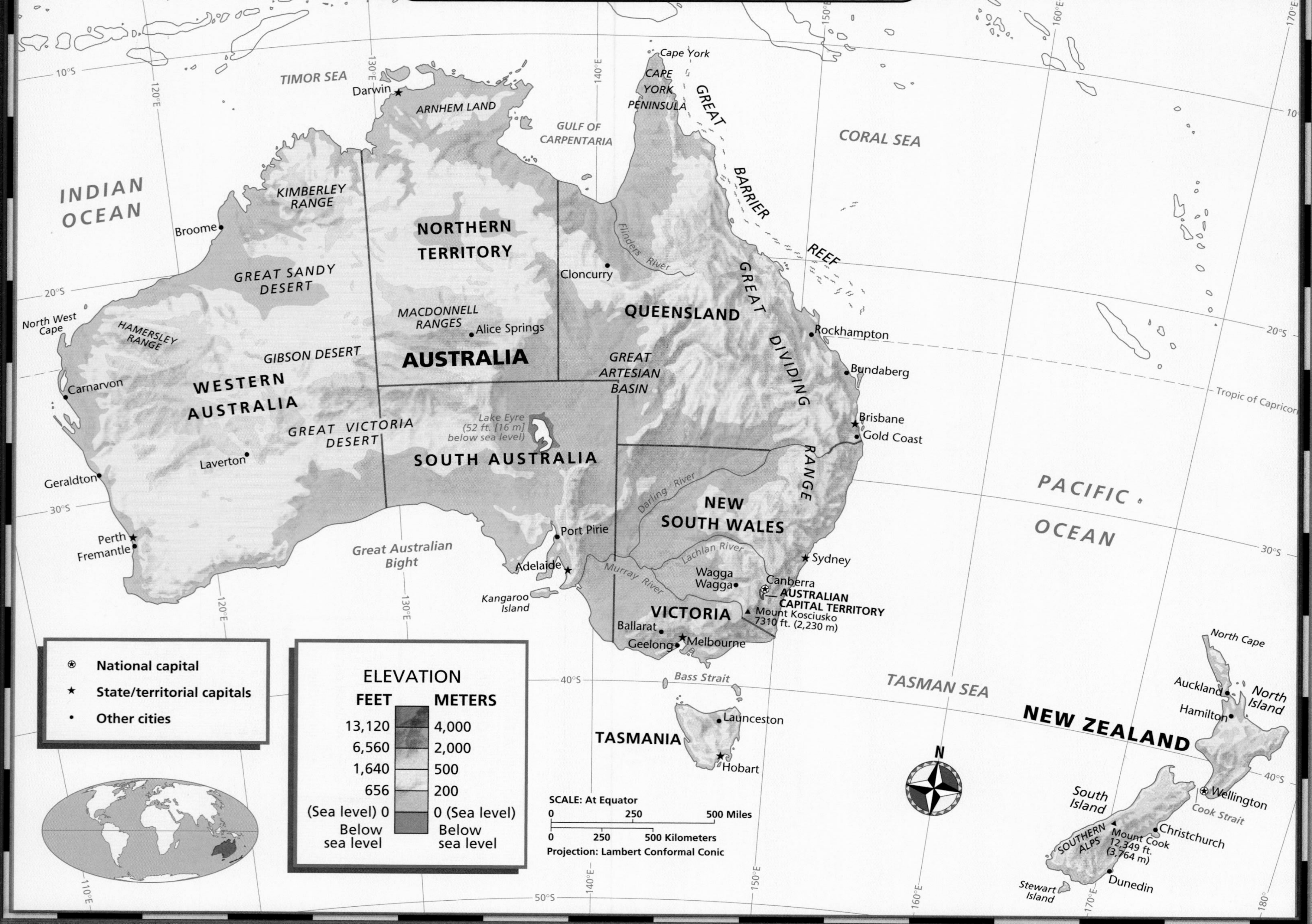

INDIAN OCEAN
TIMOR SEA
Darwin
ARNHEM LAND
GULF OF CARPENTARIA
Cape York
CAPE YORK PENINSULA
GREAT BARRIER REEF
CORAL SEA
KIMBERLEY RANGE
Broome
NORTHERN TERRITORY
Cloncurry
Flinders River
QUEENSLAND
GREAT SANDY DESERT
North West Cape
HAMERSLEY RANGE
GIBSON DESERT
MACDONNELL RANGES
Alice Springs
AUSTRALIA
GREAT ARTESIAN BASIN
GREAT DIVIDING RANGE
Rockhampton
Bundaberg
Tropic of Capricorn
Carnarvon
WESTERN AUSTRALIA
GREAT VICTORIA DESERT
Lake Eyre (52 ft. [16 m] below sea level)
Brisbane
Gold Coast
SOUTH AUSTRALIA
Laverton
Geraldton
Darling River
NEW SOUTH WALES
PACIFIC OCEAN
Perth
Fremantle
Great Australian Bight
Port Pirie
Adelaide
Kangaroo Island
Lachlan River
Sydney
Murray River
Wagga Wagga
Canberra
AUSTRALIAN CAPITAL TERRITORY
VICTORIA
Mount Kosciusko 7310 ft. (2,230 m)
Ballarat
Geelong
Melbourne
Bass Strait
TASMAN SEA
Launceston
TASMANIA
Hobart
NEW ZEALAND
North Cape
Auckland
North Island
Hamilton
Wellington
Cook Strait
South Island
SOUTHERN ALPS
Mount Cook 12,349 ft. (3,764 m)
Christchurch
Dunedin
Stewart Island
N
National capital
State/territorial capitals
Other cities
ELEVATION
FEET
METERS
13,120
4,000
6,560
2,000
1,640
500
656
200
(Sea level) 0
0 (Sea level)
Below sea level
Below sea level
SCALE: At Equator
0 250 500 Miles
0 250 500 Kilometers
Projection: Lambert Conformal Conic
10°S
20°S
30°S
40°S
50°S
110°E
120°E
130°E
140°E
150°E
160°E
170°E
180°

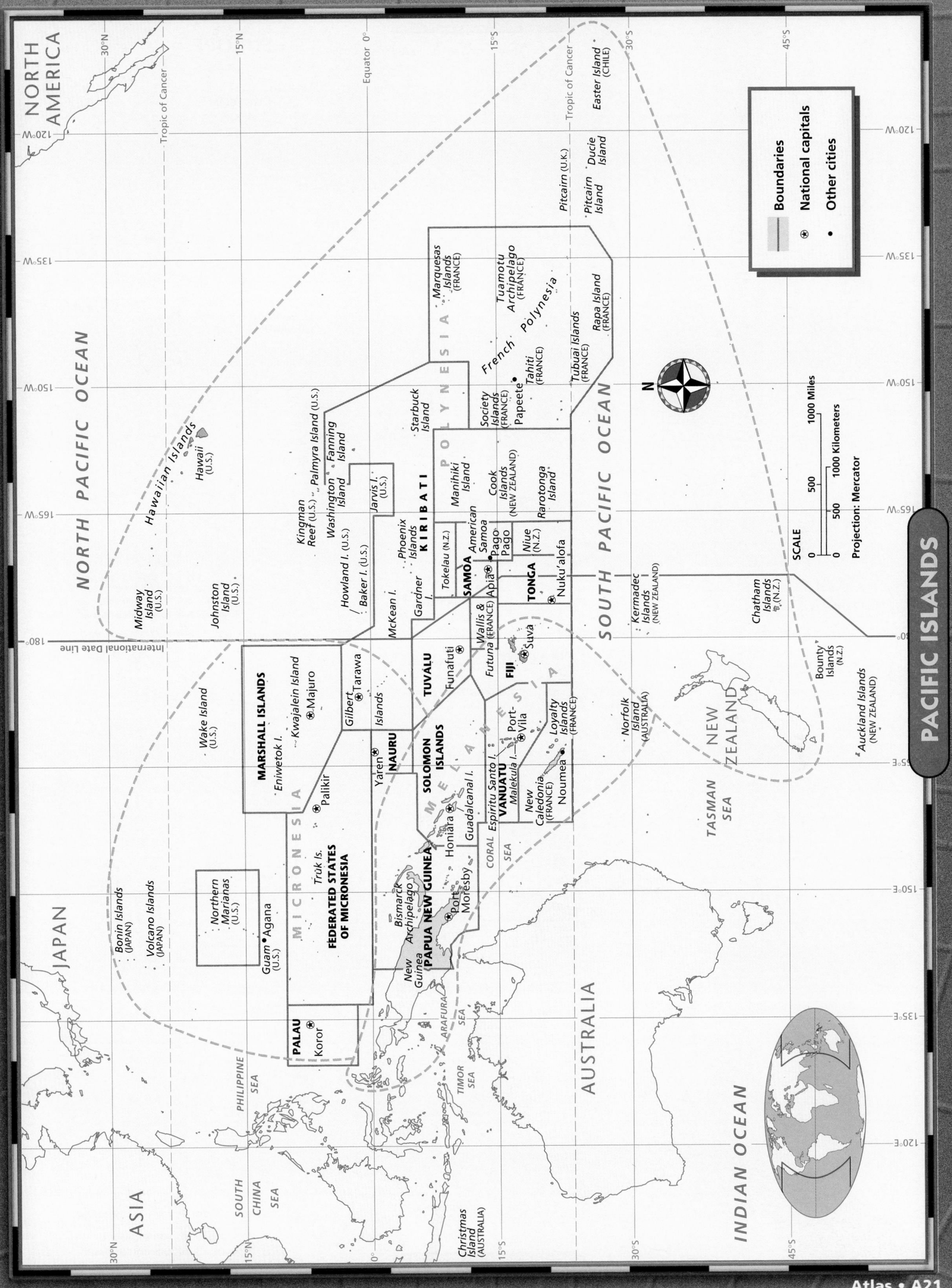
PACIFIC ISLANDS
NORTH AMERICA
NORTH PACIFIC OCEAN
SOUTH PACIFIC OCEAN
INDIAN OCEAN
ASIA
JAPAN
AUSTRALIA
NEW ZEALAND
TASMAN SEA
CORAL SEA
ARAFURA SEA
TIMOR SEA
PHILIPPINE SEA
SOUTH CHINA SEA
MICRONESIA
MELANESIA
POLYNESIA
French Polynesia
Hawaiian Islands
Hawaii (U.S.)
Midway Island (U.S.)
Johnston Island (U.S.)
Wake Island (U.S.)
Bonin Islands (JAPAN)
Volcano Islands (JAPAN)
Northern Marianas (U.S.)
Guam (U.S.)
Agana
PALAU
Koror
FEDERATED STATES OF MICRONESIA
Truk Is.
Palikir
MARSHALL ISLANDS
Eniwetok I.
Kwajalein Island
Majuro
Gilbert Islands
Tarawa
NAURU
Yaren
KIRIBATI
TUVALU
Funafuti
SOLOMON ISLANDS
Honiara
Guadalcanal I.
PAPUA NEW GUINEA
Port Moresby
New Guinea
Bismarck Archipelago
VANUATU
Espiritu Santo I.
Malekula I.
Port-Vila
New Caledonia (FRANCE)
Noumea
Loyalty Islands (FRANCE)
FIJI
Suva
Wallis & Futuna (FRANCE)
SAMOA
Apia
American Samoa
Pago Pago
Tokelau (N.Z.)
TONGA
Nuku'alofa
Niue (N.Z.)
Cook Islands (NEW ZEALAND)
Manihiki Island
Rarotonga Island
Kingman Reef (U.S.)
Palmyra Island (U.S.)
Washington Island
Fanning Island
Jarvis I. (U.S.)
Howland I. (U.S.)
Baker I. (U.S.)
Phoenix Islands
McKean I.
Gardner I.
Starbuck Island
Marquesas Islands (FRANCE)
Tuamotu Archipelago (FRANCE)
Society Islands (FRANCE)
Papeete
Tahiti (FRANCE)
Tubuai Islands (FRANCE)
Rapa Island (FRANCE)
Pitcairn (U.K.)
Pitcairn Island
Ducie Island
Easter Island (CHILE)
Kermadec Islands (NEW ZEALAND)
Norfolk Island (AUSTRALIA)
Chatham Islands (N.Z.)
Bounty Islands (N.Z.)
Auckland Islands (NEW ZEALAND)
Christmas Island (AUSTRALIA)
International Date Line
Tropic of Cancer
Equator 0°
Tropic of Capricorn
30°N
15°N
0°
15°S
30°S
45°S
120°W
135°W
150°W
165°W
180°
165°E
150°E
135°E
120°E
N
Boundaries
National capitals
Other cities
SCALE
0 500 1000 Miles
0 500 1000 Kilometers
Projection: Mercator

NORTH POLE

EUROPE
BARENTS SEA
KARA SEA
NORWEGIAN SEA
ASIA
LAPTEV SEA
GREENLAND SEA
Iceland
ATLANTIC OCEAN
ARCTIC OCEAN
NORTH POLE
Greenland (DENMARK)
International Date Line
Arctic Circle
PACK ICE
Queen Elizabeth Islands
NORTH MAGNETIC POLE
Baffin Bay
Baffin Island
BERING SEA
BEAUFORT SEA
NORTH AMERICA
PACIFIC OCEAN
90°E
60°E
30°E
150°E
30°W
80°N
70°N
60°N
50°N
180°
150°W
120°W
90°W
60°W
SCALE
0 250 500 750 Miles
0 250 500 750 Kilometers
Projection: Polar Azimuthal Equidistant

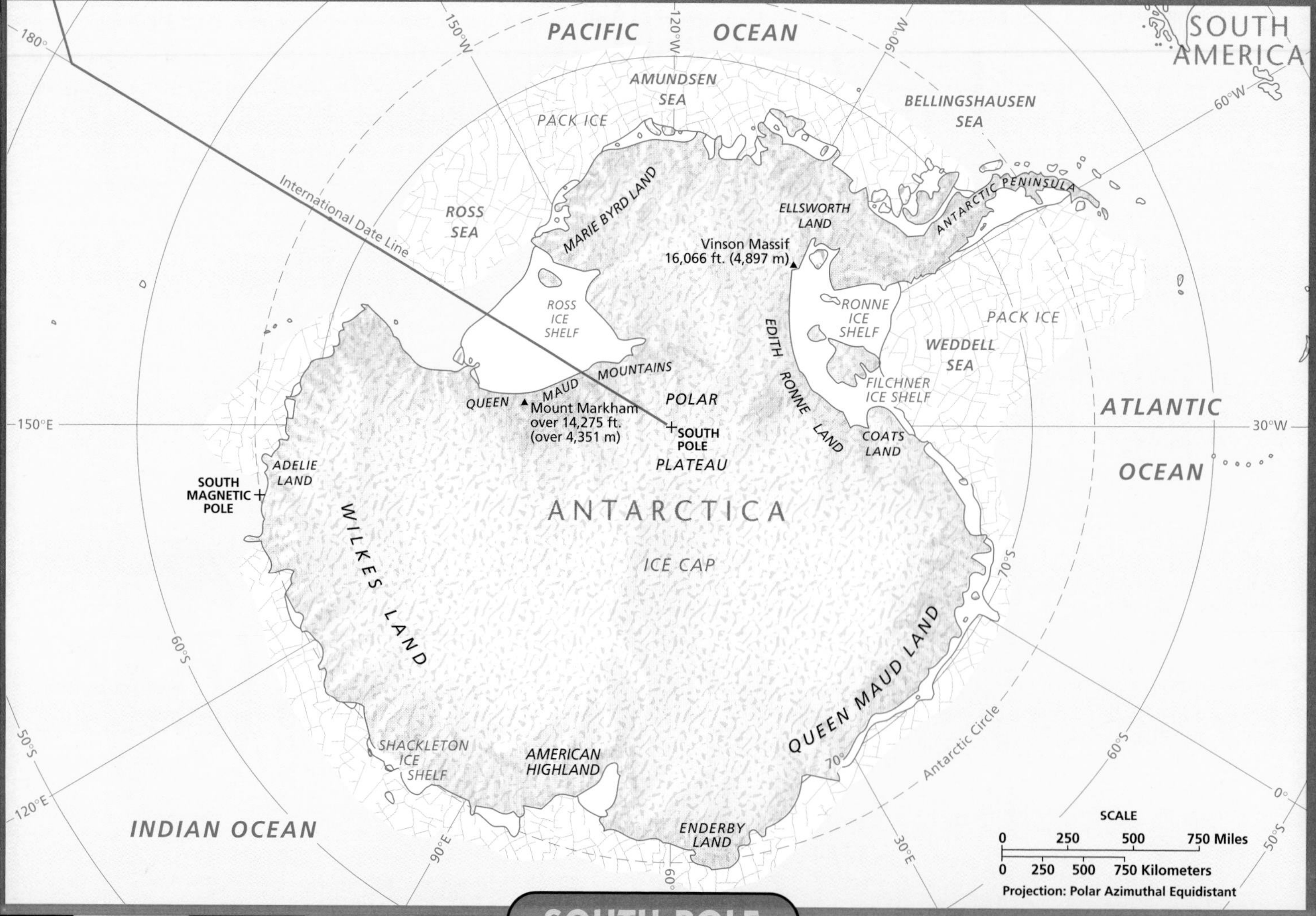

PACIFIC OCEAN
SOUTH AMERICA
AMUNDSEN SEA
BELLINGSHAUSEN SEA
PACK ICE
ROSS SEA
MARIE BYRD LAND
ELLSWORTH LAND
ANTARCTIC PENINSULA
International Date Line
Vinson Massif 16,066 ft. (4,897 m)
ROSS ICE SHELF
RONNE ICE SHELF
EDITH RONNE LAND
WEDDELL SEA
FILCHNER ICE SHELF
QUEEN MAUD MOUNTAINS
Mount Markham over 14,275 ft. (over 4,351 m)
POLAR PLATEAU
SOUTH POLE
COATS LAND
ATLANTIC OCEAN
ADELIE LAND
SOUTH MAGNETIC POLE
ANTARCTICA
ICE CAP
WILKES LAND
QUEEN MAUD LAND
SHACKLETON ICE SHELF
AMERICAN HIGHLAND
ENDERBY LAND
INDIAN OCEAN
Antarctic Circle
180°
150°W
120°W
90°W
60°W
150°E
30°W
120°E
90°E
60°E
30°E
0°
70°S
60°S
50°S
SCALE
0 250 500 750 Miles
0 250 500 750 Kilometers
Projection: Polar Azimuthal Equidistant

SOUTH POLE

SKILLS HANDBOOK

Studying geography requires the ability to understand and use various tools. This Skills Handbook explains how to use maps, charts, and other graphics to help you learn about geography and the various regions of the world. Throughout this textbook, you will have the opportunity to improve these skills and build upon them.

CONTENTS

GEOGRAPHIC Dictionary

- globe
- grid
- latitude
- equator
- parallels
- degrees
- minutes
- longitude
- prime meridian
- meridians
- hemispheres
- continents
- islands
- ocean
- map
- map projections
- compass rose
- scale
- legend

Global Skill Builder CD–ROM

You might wish to use **Mapping the Earth** from the interactive Global Skill Builder CD–ROM to reinforce students' understanding of maps and globes.

MAPPING THE EARTH

The Globe

A **globe** is a scale model of Earth. It is useful for looking at the entire Earth or at large areas of Earth's surface.

The pattern of lines that circle the globe in east-west and north-south directions is called a **grid**. The intersection of these imaginary lines helps us find places on Earth.

The east-west lines in the grid are lines of **latitude**. These imaginary lines measure distance north and south of the **equator**. The equator is an imaginary line that circles the globe halfway between the North and South Poles. Lines of latitude are called **parallels** because they are always parallel to the equator. Parallels measure distance from the equator in **degrees**. The symbol for degrees is °. Degrees are further divided into **minutes**. The symbol for minutes is ′. There are 60 minutes in a degree. Parallels north of the equator are labeled with an *N*. Those south of the equator are labeled with an *S*.

The north-south lines are lines of **longitude**. These imaginary lines pass through the Poles. They measure distance east and west of the **prime meridian**. The prime meridian is an imaginary line that runs through Greenwich, England. It represents 0° longitude. Lines of longitude are called **meridians**.

Lines of latitude range from 0°, for locations on the equator, to 90°N or 90°S, for locations at the Poles. See **Figure 1**. Lines of longitude range from 0° on the prime meridian to 180° on a meridian in the mid-Pacific Ocean. Meridians west of the prime meridian to 180° are labeled with a *W*. Those east of the prime meridian to 180° are labeled with an *E*. See **Figure 2**.

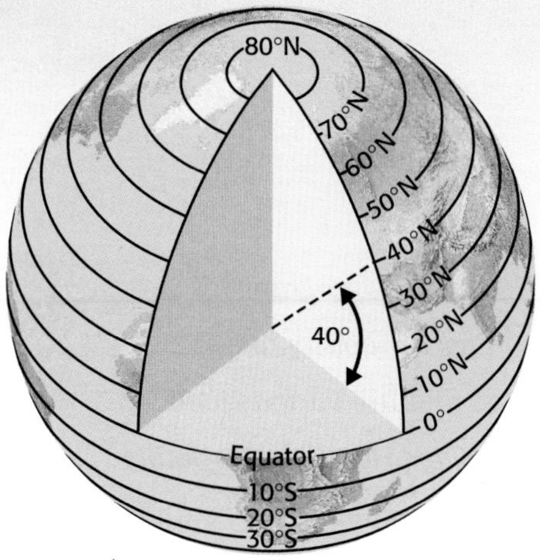

Figure 1: The east-west lines in the grid are lines of latitude.

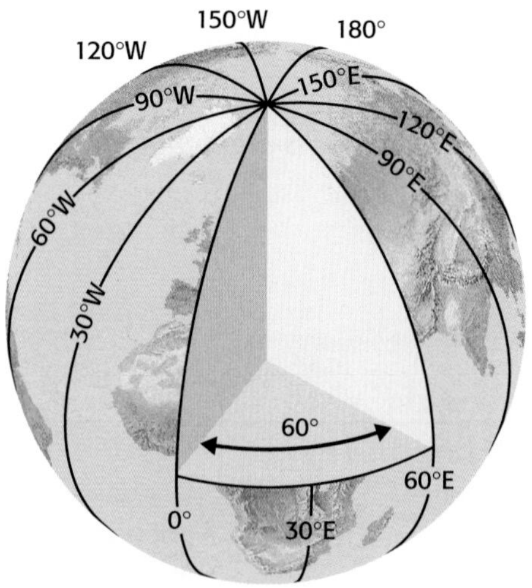

Figure 2: The north-south lines are lines of longitude.

NORTHERN HEMISPHERE

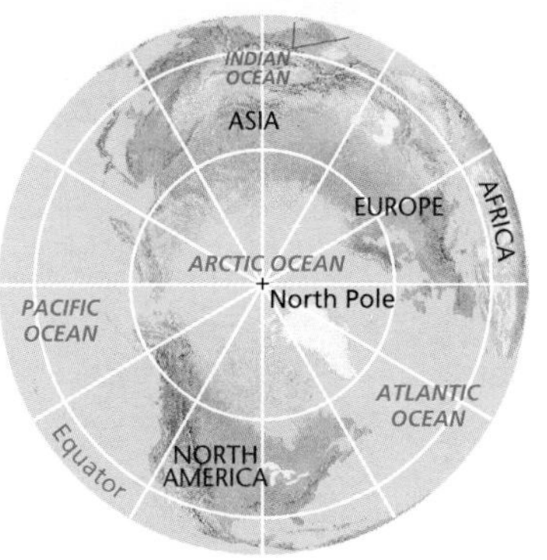

Figure 3: The hemispheres

SOUTHERN HEMISPHERE

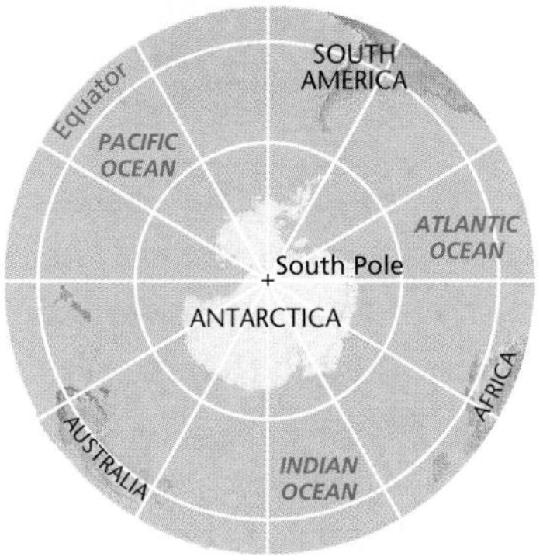

The equator divides the globe into two halves, called **hemispheres**. See Figure 3. The half north of the equator is the Northern Hemisphere. The southern half is the Southern Hemisphere. The prime meridian and the 180° meridian divide the world into the Eastern Hemisphere and the Western Hemisphere. The prime meridian separates parts of Europe and Africa into two different hemispheres. To prevent this, some mapmakers divide the Eastern and Western hemispheres at 20° W. This places all of Europe and Africa in the Eastern Hemisphere.

Our planet's land surface is organized into seven large landmasses, called **continents**. They are identified in Figure 3. Landmasses smaller than continents and completely surrounded by water are called **islands**. Geographers also organize Earth's water surface into parts. The largest is the world **ocean**. Geographers divide the world ocean into the Pacific Ocean, the Atlantic Ocean, the Indian Ocean, and the Arctic Ocean. Lakes and seas are smaller bodies of water.

EASTERN HEMISPHERE

North Pole

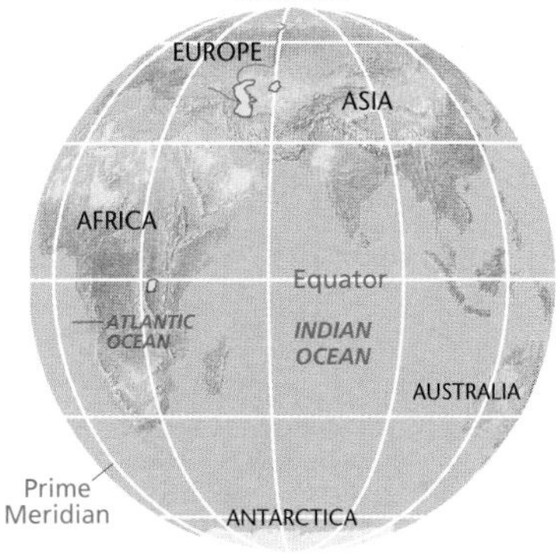

South Pole

WESTERN HEMISPHERE

North Pole

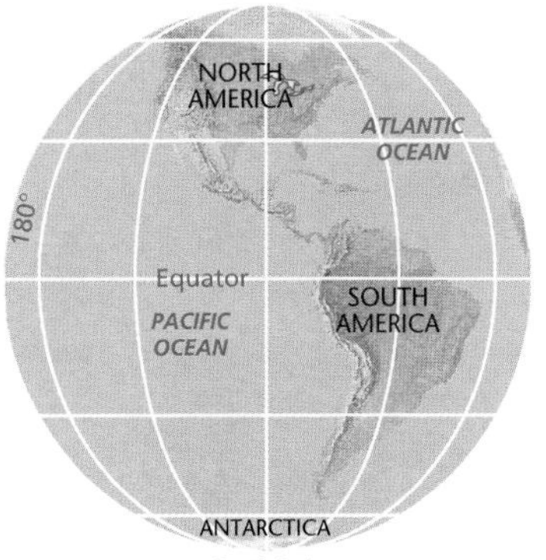

South Pole

YOUR TURN

1. Look at the Student Atlas map on page A4. What islands are located near the intersection of latitude 20° N and longitude 160° W?
2. Name the four hemispheres. In which hemispheres is the United States located?
3. Name the continents of the world.
4. Name the oceans of the world.

SKILL SIDELIGHT

Point out to students that phrases such as "Western influences," "Western world," and "Western society" do not refer to the Western Hemisphere. Rather, these phrases refer to ideas and ways of life associated with western Europe and North America.

Your Turn

Answers

1. The Hawaiian Islands
2. The United States is located in the Northern and Western Hemispheres. The other two hemispheres are the Southern and Eastern.
3. North America, South America, Africa, Europe, Asia, Australia, Antarctica
4. Pacific, Atlantic, Indian, Arctic

SKILL SIDELIGHT

Although most maps are oriented with north being "up," maps can also be oriented with the South Pole at the top. Lead a discussion about how the two different map orientations might give different impressions of the world.

Global Skill Builder CD–ROM

You might wish to use **Understanding Map Projections** from the interactive Global Skill Builder CD–ROM to reinforce students' understanding of mapmaking.

MAPMAKING

A **map** is a flat diagram of all or part of Earth's surface. Mapmakers have different ways of showing our round Earth on flat maps. These different ways are called **map projections**. Because our planet is round, all flat maps lose some accuracy. Mapmakers must choose the type of map projection that is best for their purposes. Many map projections are one of three kinds: cylindrical, conic, or flat-plane.

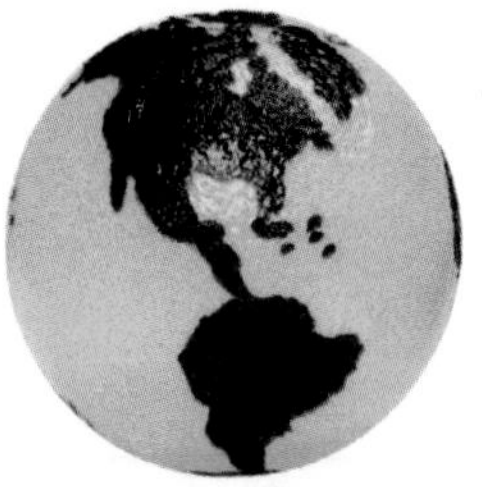

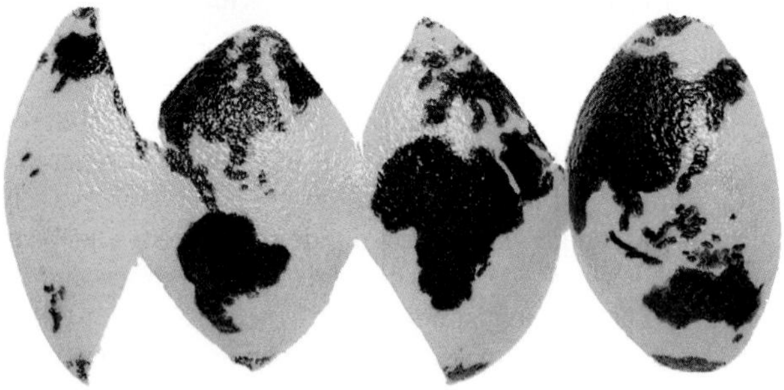

Figure 4: If you remove the peel from the orange and flatten the peel, it will stretch and tear. The larger the piece of peel, the more its shape is distorted as it is flattened. Also distorted are the distances between points on the peel.

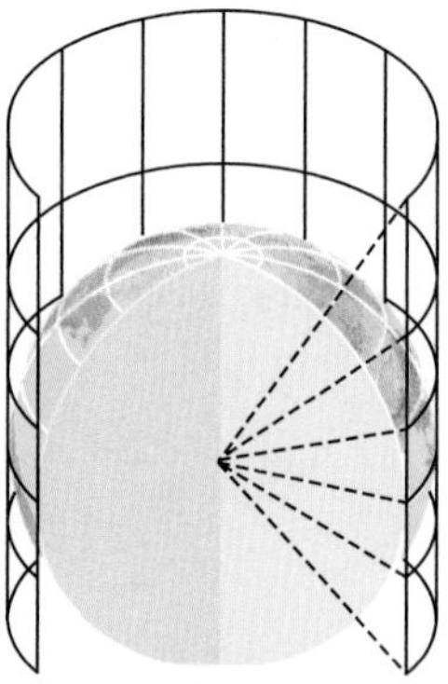

Figure 5A: Paper cylinder

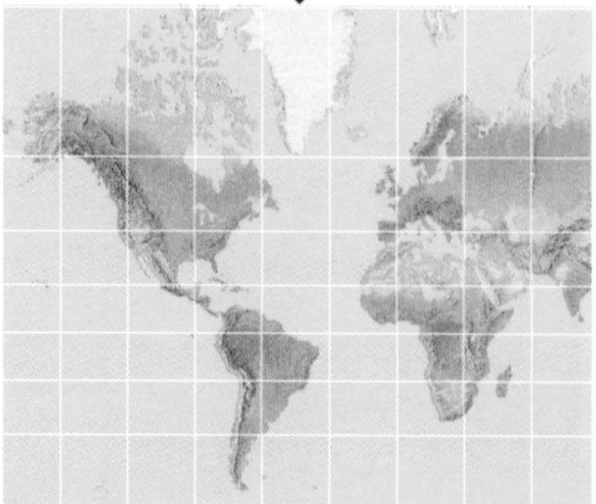

Cylindrical projections are designed from a cylinder wrapped around the globe. See **Figure 5A**. The cylinder touches the globe only at the equator. The meridians are pulled apart and are parallel to each other instead of meeting at the Poles. This causes landmasses near the Poles to appear larger than they really are. **Figure 5B** is a Mercator projection, one type of cylindrical projection. The Mercator projection is useful for navigators because it shows true direction and shape. The Mercator projection for world maps, however, emphasizes the Northern Hemisphere. Africa and South America appear smaller than they really are.

Figure 5B: A Mercator projection, although accurate near the equator, distorts distances between regions of land. This projection also distorts the sizes of areas near the poles.

Conic projections are designed from a cone placed over the globe. See **Figure 6A**. A conic projection is most accurate along the lines of latitude where it touches the globe. It retains almost true shape and size. Conic projections are most useful for areas that have long east-west dimensions, such as the United States. See the map in **Figure 6B**.

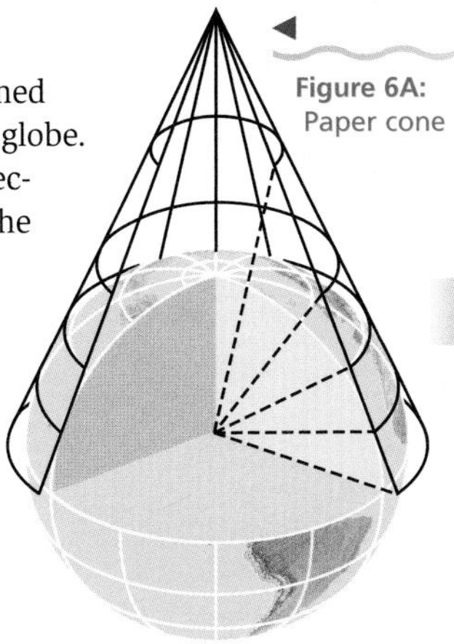

Figure 6A: Paper cone

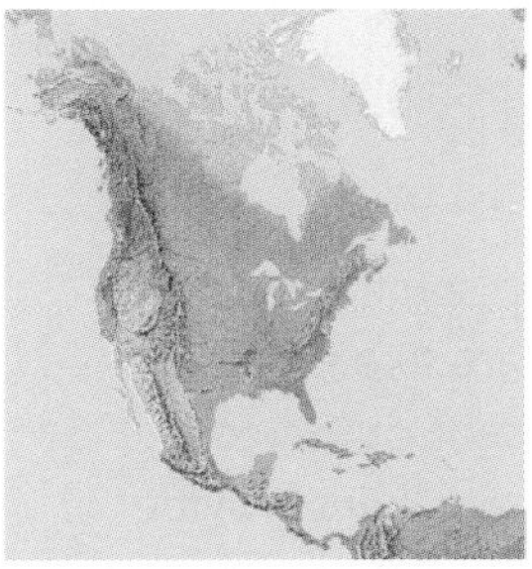

Figure 6B: Conic projection

Flat-plane projections are designed from a plane touching the globe at one point, such as at the North Pole or South Pole. See **Figures 7A** and **7B**. A flat-plane projection is useful for showing true direction for airplane pilots and ship navigators. It also shows true area. However, it distorts true shape.

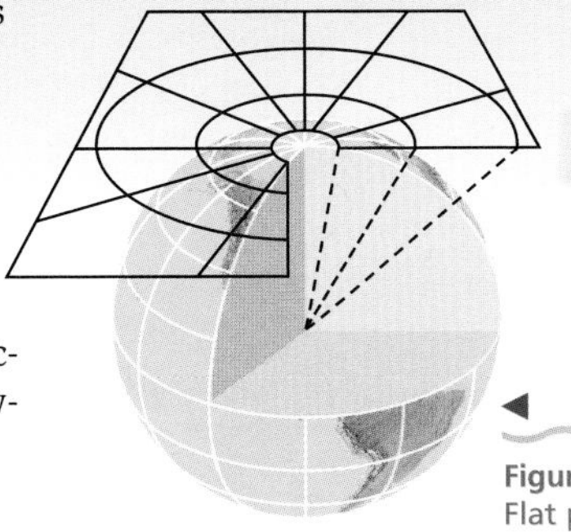

Figure 7A: Flat plane

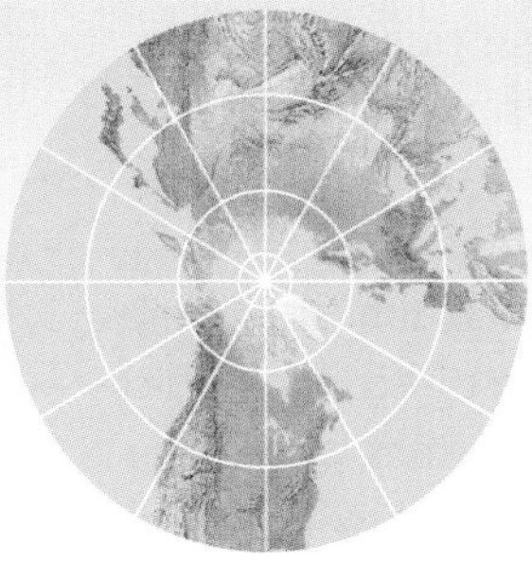

Figure 7B: Flat-plane projection

The Robinson projection is a compromise between size and shape distortions. It often is used for world maps, such as the map on page 76. The minor distortions in size at high latitudes on Robinson projections are balanced by realistic shapes at the middle and low latitudes.

YOUR TURN

1. What are three major kinds of map projections?
2. Why is a Robinson projection often used for world maps?
3. What kind of projection is a Mercator map?
4. When would a mapmaker choose to use a conic projection?

Your Turn

Answers

1. cylindrical, conic, flat-plane
2. The minor distortions in size at high latitudes are balanced by realistic shapes at the middle and low latitudes.
3. cylindrical
4. when mapping areas that have long east-west dimensions, such as the United States

SKILL SIDELIGHT

Most compass roses in this textbook are placed along lines of longitude. Directions that are true north and true south follow the longitude lines, which are curved on many map projections.

Global Skill Builder CD–ROM

You might wish to use **Map Legends and Symbols** from the interactive Global Skill Builder CD–ROM to reinforce students' understanding of map essentials.

MAP ESSENTIALS

In some ways, maps are like messages sent out in code. Mapmakers provide certain elements that help us translate these codes. These elements help us understand the message they are presenting about a particular part of the world. Of these elements, almost all maps have directional indicators, scales, and legends, or keys. Figure 8, a map of East Asia, has all three elements.

A directional indicator shows which directions are north, south, east, and west. Some mapmakers use a "north arrow," which points toward the North Pole. Remember, "north" is not always at the top of a map. The way a map is drawn and the location of directions on that map depend on the perspective of the mapmaker. Maps in this textbook indicate direction by using a **compass rose** (1). A compass rose has arrows that point to all four principal directions, as shown in Figure 8.

Figure 8: East and Southeast Asia—Physical

Mapmakers use scales to represent distances between points on a map. Scales may appear on maps in several different forms. The maps in this textbook provide a line **scale** (2). Scales give distances in miles and kilometers (km).

To find the distance between two points on the map in Figure 8, place a piece of paper so that the edge connects the two points. Mark the location of each point on the paper with a line or dot. Then, compare the distance between the two dots with the map's line scale. The number on the top of the scale gives the distance in miles. The number on the bottom gives the distance in kilometers. Because the distances are given in intervals, you will have to approximate the actual distance on the scale.

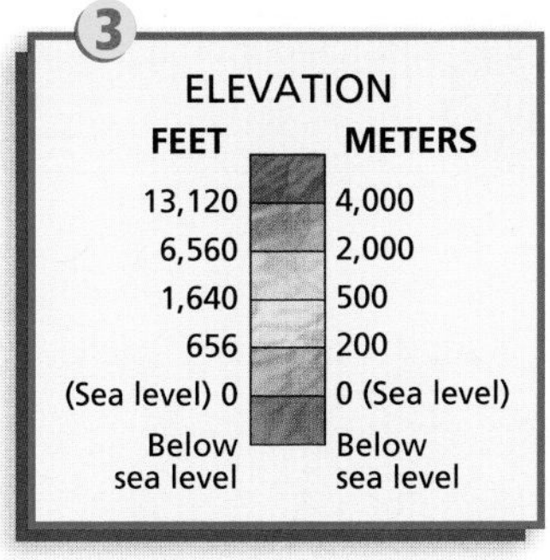

The **legend** (3), or key, explains what the symbols on the map represent. Point symbols are used to specify the location of things, such as cities, that do not take up much space on a large-scale map. Some legends, such as the one in **Figure 8**, show which colors represent certain elevations. Other maps might have legends with symbols or colors that represent things such as roads. Legends can also show economic resources, land use, population density, and climate.

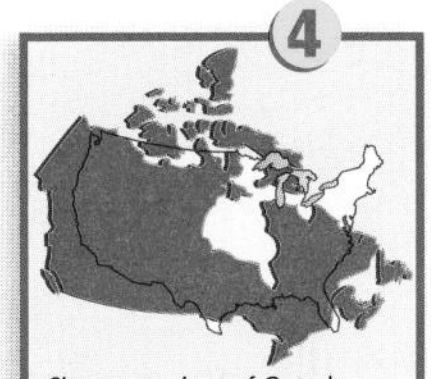

Size comparison of Canada to the contiguous United States

Physical maps at the beginning of each unit have size comparison maps (4). An outline of the mainland United States (not including Alaska and Hawaii) is compared to the area under study in that chapter. These size comparison maps help you understand the size of the areas you are studying in relation to the size of the United States.

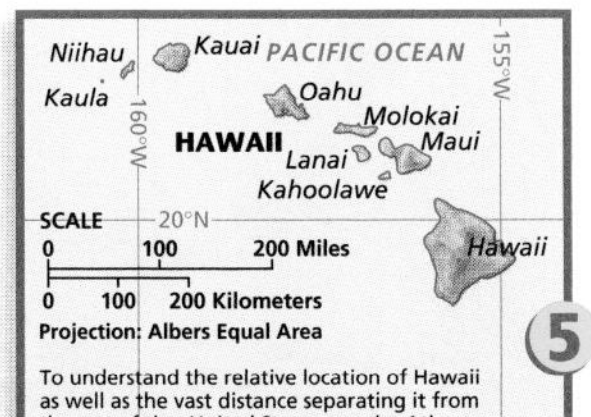

To understand the relative location of Hawaii as well as the vast distance separating it from the rest of the United States, see the Atlas map in the front of this textbook.

Inset maps are sometimes used to show a small part of a larger map. Mapmakers also use inset maps to show areas that are far away from the areas shown on the main map. Maps of the United States, for example, often include inset maps of Alaska and Hawaii (5). Those two states are too far from the other 48 states to accurately represent the true distance on the main map. Subject areas in inset maps can be drawn to a scale different from the scale used on the main map.

YOUR TURN

Look at the Student Atlas map on pages A4 and A5.

1. Locate the compass rose. What country is directly west of Madagascar in Africa?
2. What island country is located southeast of India?
3. Locate the distance scale. Using the inset map, find the approximate distance in miles and kilometers from Oslo, Norway, to Stockholm, Sweden.
4. What is the capital of Brazil? What other cities are shown in Brazil?

Your Turn

Answers

1. Mozambique
2. Sri Lanka
3. less than 500 miles (800 km)
4. Brasília is the capital. Rio de Janeiro and São Paulo are also shown.

Global Skill Builder CD–ROM

You might wish to use **Using Different Types of Maps** and **Comparing Maps** from the interactive Global Skill Builder CD–ROM to reinforce students' work with different kinds of maps.

WORKING WITH MAPS

The Atlas at the front of this textbook includes two kinds of maps: physical and political. At the beginning of most units in this textbook, you will find five kinds of maps. These physical, political, climate, population, and land use and resources maps provide different kinds of information about the region you will study in that unit. These maps are accompanied by questions. Some questions ask you to show how the information on each of the maps might be related.

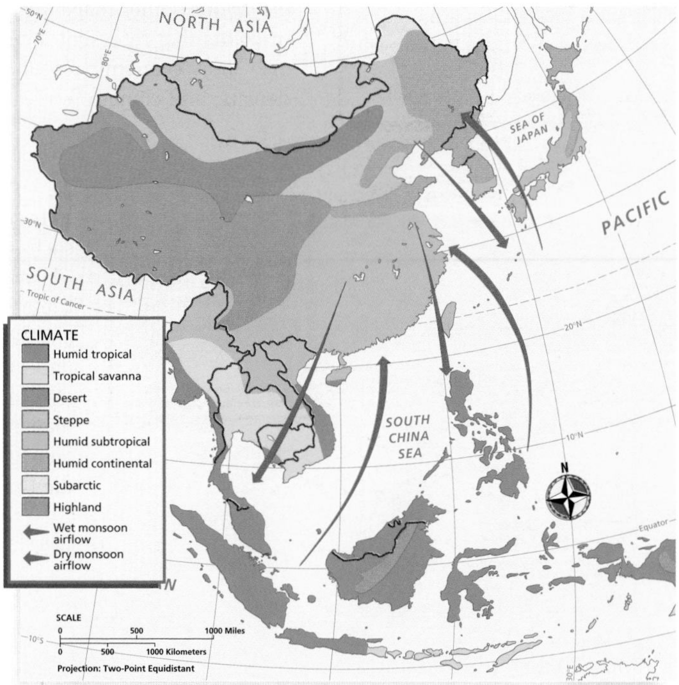

Figure 9: East and Southeast Asia—Climate

Mapmakers often combine physical and political features into one map. Physical maps, such as the one in **Figure 8** on page S6, show important physical features in a region, including major mountains and mountain ranges, rivers, oceans and other bodies of water, deserts, and plains. Physical-political maps also show important political features, such as national borders, state and provincial boundaries, and capitals and other important cities. You will find a physical-political map at the beginning of most chapters.

Mapmakers use climate maps to show the most important weather patterns in certain areas. Climate maps throughout this textbook use color to show the various climate regions of the world. See **Figure 9**. Colors that identify climate types are found in a legend with each map. Boundaries between climate regions do not indicate an immediate change in the main weather conditions between two climate regions. Instead, boundaries show the general areas of gradual change between climate regions.

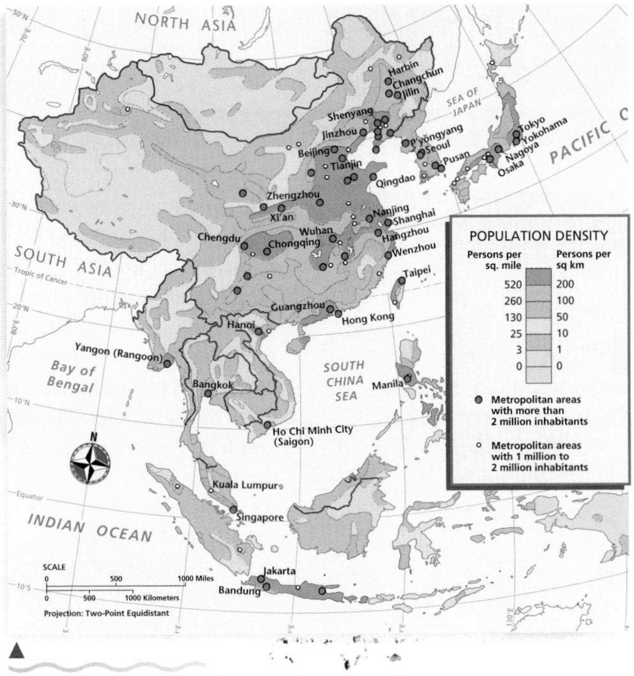

Figure 10: East and Southeast Asia—Population

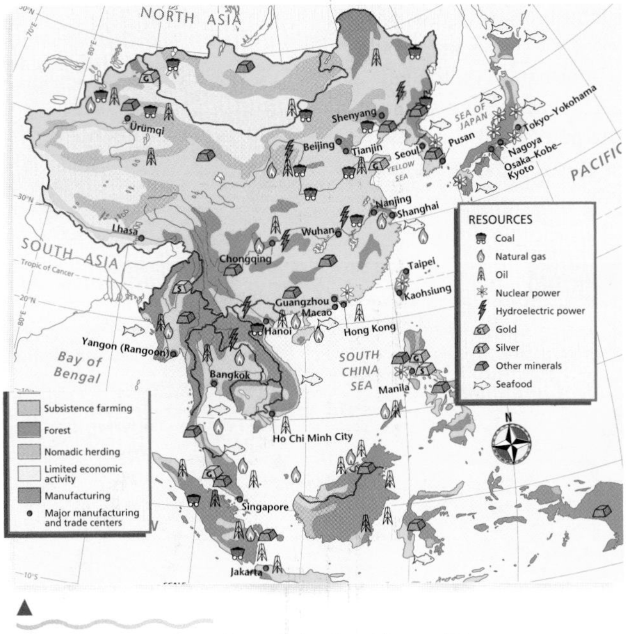

Figure 11: East and Southeast Asia—Land Use and Resources

Population maps show where people live in a particular region. They also show how crowded, or densely populated, regions are. Population maps throughout this textbook use color to show population density. See **Figure 10**. Each color represents a certain number of people living within a square mile or square kilometer. Population maps also use symbols to show metropolitan areas with populations of a particular size. These symbols and colors are shown in a legend.

Land Use and Resources maps show the important resources of a region. See **Figure 11**. Symbols and colors are used to show information about economic development, such as where industry is located or where farming is most common. The meanings of each symbol and color are shown in a legend.

YOUR TURN

1. What is the purpose of a climate map?
2. Look at the population map. What is the population density of the area around Qingdao in northern China?
3. What energy resource is found near Ho Chi Minh City?

Your Turn

Answers

1. to show the most important weather patterns in certain areas
2. more than 520 persons per square mile (200 per sq km)
3. oil

Global Skill Builder CD–ROM

You might wish to use **Reading Graphs** and **Presenting Data Graphically** from the interactive Global Skill Builder CD–ROM to reinforce students' understanding of graphics and their uses.

USING GRAPHS, DIAGRAMS, CHARTS, AND TABLES

Bar graphs are a visual way to present information. The bar graph in **Figure 12** shows the imports and exports of the countries of southern Europe. The amount of imports and exports in billions of dollars is listed on the left side of the graph. Along the bottom of the graph are the names of the countries of southern Europe. Above each country or group of countries is a vertical bar. The top of the bar corresponds to a number along the left side of the graph. For example, Italy imports $200 billion worth of goods.

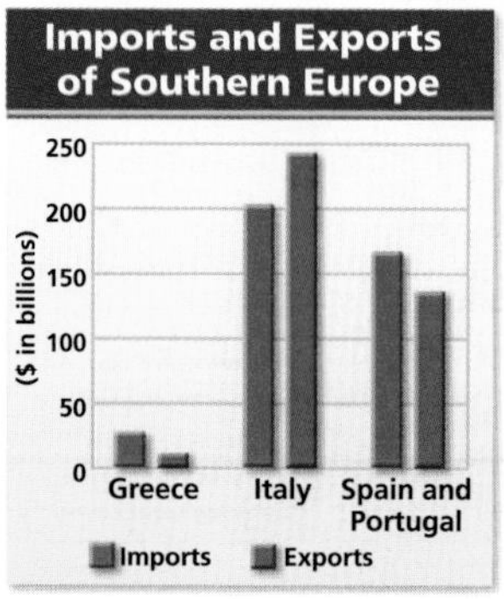

Figure 12: Reading a bar graph

Often, line graphs are used to show such things as trends, comparisons, and size. The line graph in **Figure 13** shows the population growth of the world over time. The information on the left shows the number of people in billions. The years being studied are listed along the bottom. Lines connect points that show the population in billions at each year under study. This line graph projects population growth into the future.

Figure 13: Reading a line graph

A pie graph shows how a whole is divided into parts. In this kind of graph, a circle represents the whole. The wedges represent the parts. Bigger wedges represent larger parts of the whole. The pie graph in **Figure 14** shows the percentages of the world's coffee beans produced by various groups of countries. Brazil is the largest grower. It grows 25 percent of the world's coffee beans.

Major Producers of Coffee

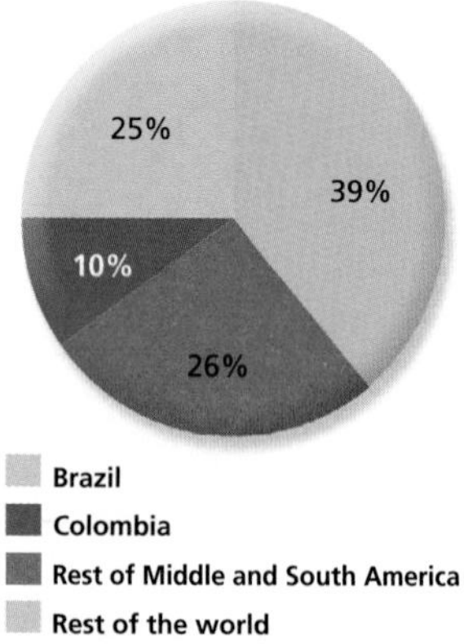

Source: United Nations, Food and Agriculture Organization

Figure 14: Reading a pie graph

Age structure diagrams show the number of males and females by age group. These diagrams are split into two sides, one for male and one for female. Along the bottom are numbers that show the number of males or females in the age groups. The age groups are listed on the side of the diagram. The wider the base of a country's diagram, the younger the population of that country. The wider the top of a country's diagram, the older the population.

Some countries have so many younger people that their age structure diagrams are shaped like pyramids. For this reason, these diagrams are sometimes called population pyramids. However, in some countries the population is more evenly distributed by age group. For example, see the age structure diagram for Germany in **Figure 15**. Germany's population is older. It is not growing as fast as countries with younger populations.

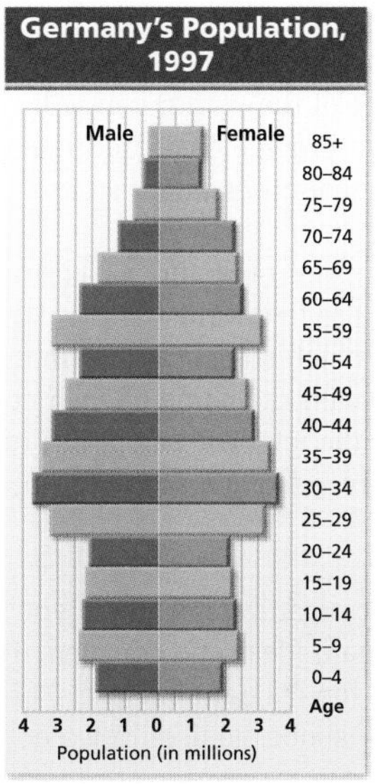

Source: U.S. Census Bureau

Figure 15: Reading an age structure diagram

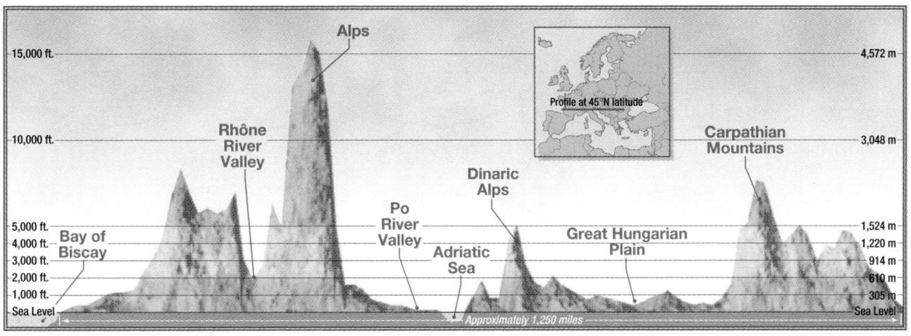

Figure 16: Reading an elevation profile

Each unit atlas includes an elevation profile. See **Figure 16**. It is a side view, or profile, of a region along a line drawn between two points.

Vertical and horizontal distances are figured differently on elevation profiles. The vertical distance (the height of a mountain, for example) is exaggerated when compared to the horizontal distance between the two points. This technique is called vertical exaggeration. If the vertical scale were not exaggerated, even tall mountains would appear as small bumps on an elevation profile.

Skill Sidelight

The population pyramid on this page also appears in the chapter about west-central Europe.

Skill Sidelight

Direct students' attention to the statistical chart on this page. Point out that most chapters contain statistical charts that provide information about the populations of countries within a region. Each statistical chart also includes information for the United States. The U.S. statistical information is provided for comparison.

Global Skill Builder CD–ROM

You might wish to use **Using Time Lines, Reading Graphs, Presenting Data Graphically,** and **Creating Graphic Organizers** from the interactive Global Skill Builder CD–ROM to reinforce students' understanding of charts, tables, and diagrams.

In each unit and chapter on the various regions of the world, you will find tables that provide basic information about the countries under study.

The countries of Spain and Portugal are listed on the left in the table in **Figure 17**. You can match statistical information on the right with the name of each country listed on the left. The categories of information are listed across the top of the table.

Graphic organizers can help you understand certain ideas and concepts. For example, the diagram in **Figure 18** helps you think about the uses of water. In this diagram, one water use goes in each oval. Graphic organizers can help you focus on key facts in your study of geography.

Time lines provide highlights of important events over a period of time. The time line in **Figure 19** begins at the left with 5000 B.C., when rice was first cultivated in present-day China. The time line highlights important events that have shaped the human and political geography of China.

Spain and Portugal

Country	Population/ Growth Rate	Life Expectancy	Literacy Rate	Per Capita GDP
Portugal	9,918,040 / 0.1%	73, male / 79, female	90% (1995)	$14,600 (1998)
Spain	39,167,744 / 0.1%	74, male / 82, female	97% (1995)	$16,500 (1998)
United States	272,639,608 / 0.9%	73, male / 80, female	97% (1994)	$31,500 (1998)

Sources: Central Intelligence Agency, *The World Factbook 1999; The World Almanac and Book of Facts 1999*

Figure 17: Reading a table

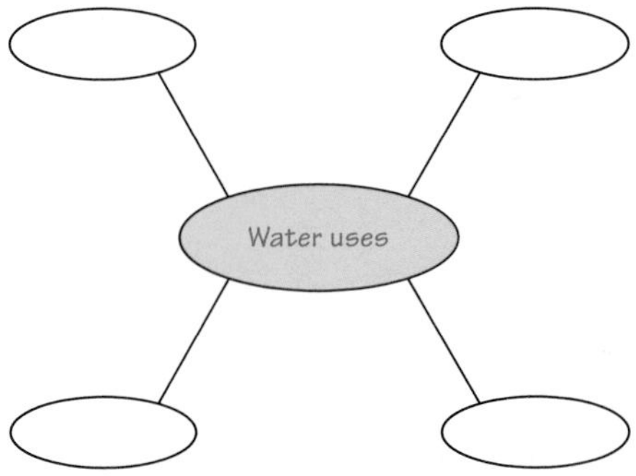

Figure 18: Graphic organizer

Historic China: A Time Line

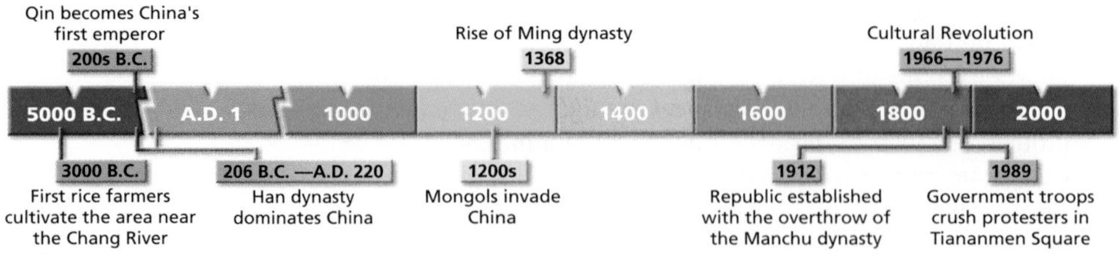

Figure 19: Reading a time line

Corn: From Field to Consumer

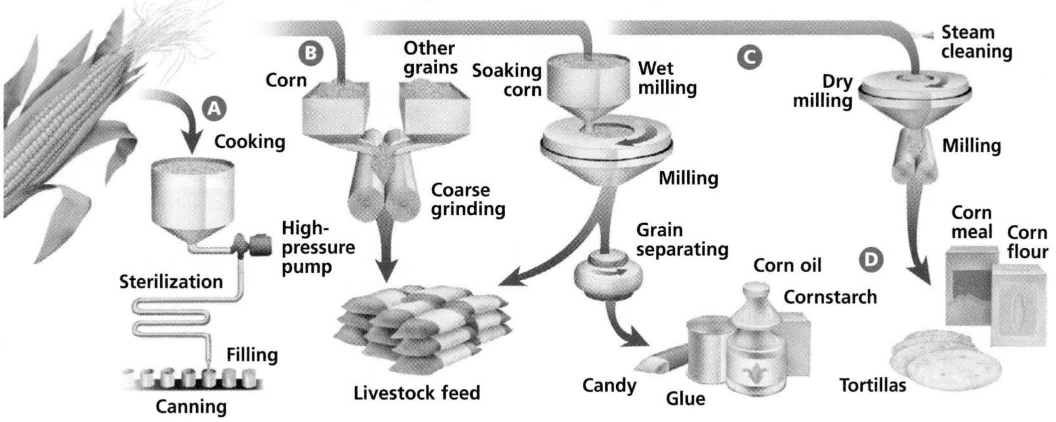

A Corn can be processed in a variety of ways. Some corn is cooked and then canned.

B Corn is ground and used for livestock feed.

C Corn also might be wet-milled or dry-milled. Then grain parts are used to make different products.

D Corn by-products, such as cornstarch and corn syrup, are used to make breads, breakfast cereals, puddings, and snack foods. Corn oil is used for cooking.

Figure 20: Reading a flowchart

Flowcharts are visual guides that explain different processes. They lead the reader from one step to the next, sometimes providing both illustrations and text. The flowchart in **Figure 20** shows the different steps involved in harvesting corn and preparing it for use by consumers. The flowchart takes you through the steps of harvesting and processing corn. Captions guide you through flowcharts.

YOUR TURN

1. Look at the statistical table for Spain and Portugal in Figure 17. Which countries have the highest literacy rate?
2. Look at the China time line in Figure 19. Name two important events in China's history between 1200 and 1400.
3. Look at Figure 20. What are three corn products?

Your Turn

Answers

1. Spain and the United States
2. Mongols invade China, rise of Ming dynasty
3. Students should name three of the following products: livestock feed, candy, glue, corn oil, cornstarch, tortillas, cornmeal, corn flour

Skill Sidelight

Point out that worldwide time zones were established in 1884. That same year, the United States was divided into four time zones. Prior to 1884, time was set locally. In addition, railroad companies set railroad times along their routes. About 100 such railroad times existed by 1883. There are six time zones in the United States today.

Global Skill Builder CD–ROM

You might wish to use **Understanding Time Zones** from the interactive Global Skill Builder CD–ROM to reinforce students' understanding of time-zone maps.

READING A TIME-ZONE MAP

The sun is not directly overhead everywhere on Earth at the same time. Clocks are set to reflect the difference in the sun's position. Our planet rotates on its axis once every 24 hours. In other words, in one hour, it makes one twenty-fourth of a complete revolution. Since there are 360 degrees in a circle, we know that the planet turns 15 degrees of longitude each hour. (360° ÷ 24 = 15°) We also know that the planet turns in a west-to-east direction. Therefore, if a place on Earth has the sun directly overhead at this moment (noon), then a place 15 degrees to the west will have the sun directly overhead one hour from now. During that hour the planet will have rotated 15 degrees. As a result, Earth is divided into 24 time zones. Thus, time is an hour earlier for each 15 degrees you move westward on Earth. Time is an hour later for each 15 degrees you move eastward on Earth.

By international agreement, longitude is measured from the prime meridian. This meridian passes through the Royal Observatory in Greenwich, England. Time also is measured from Greenwich and is called Greenwich mean time (GMT). For each time zone east of the prime meridian, clocks must be set one hour ahead of GMT. For each time zone west of Greenwich, clocks are set back one hour from GMT. When it is noon in London, it is 1:00 P.M. in Oslo, Norway, one time zone east. However, it is 7 A.M. in New York City, five time zones west.

WORLD TIME ZONES

As you can see by looking at the map below, time zones do not follow meridians exactly. Political boundaries are often used to draw time-zone lines. In Europe and Africa, for example, time zones follow national boundaries. The mainland United States, meanwhile, is divided into four major time zones: Eastern, Central, Mountain, and Pacific. Alaska and Hawaii are in separate time zones to the west of the mainland.

Some countries have made changes in their time zones. For example, most of the United States has daylight savings time in the summer in order to have more evening hours of daylight.

The international date line is a north-south line that runs through the Pacific Ocean. It is located at 180°, although it sometimes varies from that meridian to avoid dividing countries.

At 180°, the time is 12 hours from Greenwich time. There is a time difference of 24 hours between the two sides of the 180° meridian. The 180° meridian is called the international date line because when you cross it, the date and day change. As you cross the date line from the west to the east, you gain a day. If you travel from east to west, you lose a day.

YOUR TURN

1. In which time zone do you live? Check your time now. What time is it in New York?
2. How many hours behind New York is Anchorage, Alaska?
3. How many time zones are there in Africa?
4. If it is 9 A.M. in the middle of Greenland, what time is it in São Paulo?

Your Turn

Answers

1. Answers will vary. If students are in the Eastern time zone, the time is the same as New York's. The time is one hour behind New York for each time zone moving westward.
2. five
3. five (includes Seychelles and Mauritius)
4. 9 A.M. The two locations are in the same time zone. The wide north-south distance between the locations makes no difference in time.

WRITING
ABOUT GEOGRAPHY

Writers have many different reasons for writing. In your study of geography, you might write to accomplish many different tasks. You might write a paragraph or short paper to express your own personal feelings or thoughts about a topic or event. You might also write a paper to tell your class about an event, person, place, or thing. Sometimes you may want to write in order to persuade or convince readers to agree with a certain statement or to act in a particular way.

You will find different kinds of questions at the end of each section, chapter, and unit throughout this textbook. Some questions will require in-depth answers. The following guidelines for writing will help you structure your answers so that they clearly express your thoughts.

The Writing Process

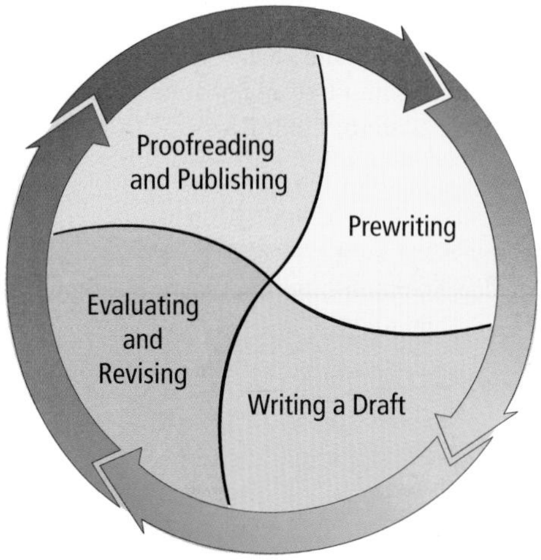

Prewriting

Prewriting is the process of thinking about and planning what to write. It includes gathering and organizing information into a clear plan. Writers use the prewriting stage to identify their audience and purpose for what is to be written.

Often, writers do research to get the information they need. Research can include finding primary and secondary sources. You will read about primary and secondary sources later in this handbook.

Writing a Draft

After you have gathered and arranged your information, you are ready to begin writing. Many paragraphs are structured in the following way:

- **Topic Sentence:** The topic sentence states the main idea of the paragraph. Putting the main idea into the form of a topic sentence helps keep the paragraph focused.
- **Body:** The body of a paragraph develops and supports the main idea. Writers use a variety of information, including facts, opinions, and examples, to support the main idea.
- **Conclusion:** The conclusion summarizes the writer's main points or restates the main idea.

Evaluating and Revising

Evaluating and Revising Read over your paragraphs and make sure you have clearly expressed what you wanted to say. Sometimes it helps to read your paragraphs aloud or to ask someone else to read them. Such methods help you identify rough or unclear sentences and passages. Revise the parts of your paragraph that are not clear or that stray from your main idea. You might want to add, cut, reorder, or replace sentences to make your paragraph as clear as possible.

Proofreading and Publishing
Before you write your final draft, read over your paragraphs and correct any errors in grammar, spelling, sentence structure, or punctuation. Common mistakes include misspelled place-names, incomplete sentences, and improper use of punctuation, such as commas. You should use a dictionary and standard grammar guides to help you proofread your work.

After you have revised and corrected your draft, neatly rewrite your paper. Make sure your final version is clean and free of mistakes. The appearance of your final version can affect how your audience perceives and understands your writing.

Practicing the Skill

1. What are the steps in the writing process?
2. How are most paragraphs formed?
3. Write a paragraph or short paper about your community for a visitor. When you have finished your draft, review it and then mark and correct any errors in grammar, spelling, sentence structure, or punctuation. At the bottom of your draft, list key resources—such as a dictionary—that you used to check and correct your work. Then write your final draft. When you are finished with your work, use pencils or pens of different colors to underline and identify the topic sentence, body, and conclusion of your paragraph.

DOING RESEARCH

Research is at the heart of geographic inquiry. To complete a research project, you may need to use resources other than this textbook. For example, you may want to research specific places or issues not discussed in this textbook. You may also want to learn more about a certain topic that you have studied in a chapter. Following the guidelines below will help you plan and complete research projects for your class.

Planning The first step in approaching a research project is planning. Planning involves deciding on a topic and finding information about that topic.

- **Decide on a Topic.** Before starting any research project, you should decide on one topic. If you are working with a group, all group members should participate in choosing a topic. Sometimes a topic will be assigned to you, but at other times you may have to choose your own. Once you have settled on a topic, make sure you can find resources to help you research it.
- **Find Information.** In order to find a particular book, you need to know how libraries organize their materials. Libraries classify their books by assigning each book a call number that tells you its location. To find the call number, look in the library's card catalog. The card catalog lists books by author, by title, and by subject. Many libraries today have computerized card catalogs. Libraries often provide instructions on how to use their computerized card catalogs. If no instructions are available, ask a library staff member for help.

Practicing the Skill Answers

1. prewriting, writing a draft, evaluating and revising, proofreading and publishing
2. with a topic sentence, body, and conclusion
3. Paragraphs will vary but should use correct grammar, spelling, sentence structure, or punctuation. Students should include a list of resources.

Practicing the Skill Answers

1. almanacs, yearbooks, and periodical indexes
2. Research projects will vary but should include a variety of sources.

Most libraries have encyclopedias, gazetteers, atlases, almanacs, and periodical indexes. Encyclopedias contain geographic, economic, and political data on individual countries, states, and cities. They also include discussions of historical events, religion, social and cultural issues, and much more. A gazetteer is a geographical dictionary that lists significant natural physical features and other places. An atlas contains maps and visual representations of geographic data. To find up-to-date facts, you can use almanacs, yearbooks, and periodical indexes.

References like *The World Almanac and Book of Facts* include historical information and a variety of statistics. Periodical indexes, particularly *The Reader's Guide to Periodical Literature*, can help you locate informative articles published in magazines. *The New York Times Index* catalogs the newspaper articles published in the *New York Times*.

You may also want to find information on the World Wide Web. The World Wide Web is the part of the Internet where people put files called Web sites for other people to access. To search the World Wide Web, you must use a search engine. A search engine will provide you with a list of Web sites that contain keywords relating to your topic. Search engines also provide Web directories, which allow you to browse Web sites by subject.

Organizing Organization is key to completing research projects of any size. If you are working with a group, every group member should have an assigned task in researching, writing, and completing your project. You and all the group members should keep track of the materials that you used to conduct your research. Then compile those sources into a bibliography and turn it in with your research project.

In addition, information collected during research should be organized in an efficient way. A common method of organizing research information is to use index cards. If you have used an outline to organize your research, you can code each index card with the appropriate main idea number and supporting detail letter from the outline. Then write the relevant information on that card. You might also use computer files in the same way. These methods will help you keep track of what information you have collected and what information you still need to gather.

Some projects will require you to conduct original research. This original research might require you to interview people, conduct surveys, collect unpublished information about your community, or draw a map of a local place. Before you do your original research, make sure you have all the necessary background information. Also, create a pre-research plan so that you can make sure all the necessary tools, such as research sources, are available.

Completing and Presenting Your Project Once you have completed your research project, you will need to present the information you have gathered in some fashion. Many times, you or your group will simply need to write a paper about your research. Research can also be presented in many other ways, however. For example, you could make an audiotape, a drawing, a poster board, a video, or a Web page to explain your research.

Practicing the Skill

1. What kinds of references would you need to research specific current events around the world?
2. Work with a group of four other students to plan, organize, and complete a research project on a topic of interest in your local community. For example, you might want to learn more about a particular individual or event that influenced your community's history. Other topics might include the economic features, physical features, and political features of your community.

ANALYZING

PRIMARY and SECONDARY SOURCES

When conducting research, it is important to use a variety of primary and secondary sources of information. There are many sources of firsthand geographical information, including diaries, letters, editorials, and legal documents such as land titles. All of these are primary sources. Newspaper articles are also considered primary sources, although they generally are written after the fact. Other primary sources include personal memoirs and autobiographies, which people usually write late in life. Paintings and photographs of particular events, persons, places, or things make up a visual record and are also considered primary sources. Because they allow us to take a close-up look at a topic, primary sources are valuable geographic tools.

Secondary sources are descriptions or interpretations of events written after the events have occurred by persons who did not participate in the events they describe. Geography textbooks such as this one, as well as biographies, encyclopedias, and other reference works, are examples of secondary sources. Writers of secondary sources have the advantage of seeing what happened beyond the moment or place that is being studied. They can provide a perspective wider than that available to one person at a specific time.

How to Study Primary and Secondary Sources

1. **Study the Material Carefully.** Consider the nature of the material. Is it verbal or visual? Is it based on firsthand information or on the accounts of others? Note the major ideas and supporting details.
2. **Consider the Audience.** Ask yourself, "For whom was this message originally meant?" Whether a message was intended for the general public or for a specific private audience may have shaped its style or content.
3. **Check for Bias.** Watch for words or phrases that present a one-sided view of a person or situation.
4. **Compare Sources.** Study more than one source on a topic. Comparing sources gives you a more complete and balanced account of geographical events and their relationships to one another.

Practicing the Skill

1. What distinguishes secondary sources from primary sources?
2. What advantages do secondary sources have over primary sources?
3. Why should you consider the intended audience of a source?
4. Of the following, identify which are primary sources and which are secondary sources: a newspaper, a private journal, a biography, an editorial cartoon, a deed to property, a snapshot of a family vacation, a magazine article about the history of Thailand, an autobiography. How might some of these sources prove to be both primary and secondary sources?

Practicing the Skill Answers

1. Primary sources are firsthand information; secondary sources are descriptions or interpretations of events written after the fact by persons who did not participate in the events they describe.
2. They provide a wider perspective.
3. Answers will vary but might mention that the intended audience indicates whether the source is biased or objective.
4. primary, primary, secondary, secondary, primary, primary, secondary, primary; answers will vary but might mention that a magazine article contains photographs, which are primary sources, while the text of the article would be a secondary source.

CRITICAL THINKING

The study of geography requires more than analyzing and understanding tools like graphs and maps. Throughout Holt People, Places, and Change, *you are asked to think critically about some of the information you are studying. Critical thinking is the reasoned judgment of information and ideas. The development of critical thinking skills is essential to learning more about the world around you. Helping you develop critical thinking skills is an important goal of* Holt People, Places, and Change. *The following critical thinking skills appear in the section reviews and chapter reviews of the textbook.*

Summarizing involves briefly restating information gathered from a larger body of information. Much of the writing in this textbook is summarizing. The geographical data in this textbook has been collected from many sources. Summarizing all the qualities of a region or country involves studying a large body of cultural, economic, geological, and historical information.

Finding the main idea is the ability to identify the main point in a set of information. This textbook is designed to help you focus on the main ideas in geography. The Read to Discover questions in each chapter help you identify the main ideas in each section. To find the main idea in any piece of writing, first read the title and introduction. These two elements may point to the main ideas covered in the text.

Also, write down questions about the subject that you think might be answered in the text. Having such questions in mind will focus your reading. Pay attention to any headings or subheadings, which may provide a basic outline of the major ideas. Finally, as you read, note sentences that provide additional details from the general statements that those details support. For example, a trail of facts may lead to a conclusion that expresses the main idea.

Comparing and contrasting involve examining events, points of view, situations, or styles to identify their similarities and differences. Comparing focuses on both the similarities and the differences. Contrasting focuses only on the differences. Studying similarities and differences between people and things can give you clues about the human and physical geography of a region.

Buddhist shrine, Myanmar

Stave church, Norway

Supporting a point of view involves identifying an issue, deciding what you think about it, and persuasively expressing your position. Your stand should be based on specific information. When taking a stand, state your position clearly and give reasons that support it.

Identifying points of view involves noting the factors that influence the opinions of an individual or group. A person's point of view includes beliefs and attitudes that are shaped by factors such as age, gender, race, and economic status. Identifying points of view helps us examine why people see things as they do. It also reinforces the realization that people's views may change over time or with a change in circumstances.

Identifying bias is an important critical thinking skill in the study of any subject. When a point of view is highly personal or based on unreasoned judgment, it is considered biased. Sometimes, a person's actions reflect bias. At its most extreme, bias can be expressed in violent actions against members of a particular culture or group. A less obvious form of bias is a stereotype, or a generalization about a group of people. Stereotypes tend to ignore differences within groups.

Political protest, India

Probably the hardest form of cultural bias to detect has to do with perspective, or point of view. When we use our own culture and experiences as a point of reference from which to make statements about other cultures, we are showing a form of bias called ethnocentrism.

Analyzing is the process of breaking something down into parts and examining the relationships between those parts. For example, to understand the processes behind forest loss, you might study issues involving economic development, the overuse of resources, and pollution.

Evaluating involves assessing the significance or overall importance of something. For example, you might evaluate the success of certain environmental protection laws or the effect of foreign trade on a society. You should base your evaluation on standards that others will understand and are likely to consider valid. For example, an evaluation of international relations after World War II might look at the political and economic tensions between the United States and the Soviet Union. Such an evaluation would also consider the ways those tensions affected other countries around the world.

Identifying cause and effect is part of interpreting the relationships between geographical events. A cause is any action that leads to an event; the outcome of that action is an effect. To explain geographical developments, geographers may point out multiple causes and effects. For example, geographers studying pollution in a region might note a number of causes.

Ecuador rain forest

Cleared forest, Kenya

Drawing inferences and drawing conclusions are two methods of critical thinking that require you to use evidence to explain events or information in a logical way. Inferences and conclusions are opinions, but these opinions are based on facts and reasonable deductions.

For example, suppose you know that people are moving in greater and greater numbers to cities in a particular country. You also know that poor weather has hurt farming in rural areas while industry has been expanding in cities. You might be able to understand from this information some of the reasons for the increased migration to cities. You could conclude that poor harvests have pushed people to leave rural areas. You might also conclude that the possibility of finding work in new industries may be pulling people to cities.

Drought in West Texas

Dallas, Texas

Making generalizations and making predictions are two critical thinking skills that require you to form specific ideas from a large body of information. When you are asked to generalize, you must take into account many different pieces of information. You then form a main concept that can be applied to all of the pieces of information. Many times making generalizations can help you see trends. Looking at trends can help you form a prediction. Making a prediction involves looking at trends in the past and present and making an educated guess about how these trends will affect the future.

Communications technology, rural Brazil

Practicing the Skill Answers

1. Plans will vary but should be present solutions to the ethnic conflict in Rwanda.
2. Problem solving will vary according to the chapter chosen.

DECISION-MAKING AND PROBLEM-SOLVING

SKILLS

Like you, many people around the world have faced difficult problems and decisions. By using appropriate skills such as problem solving and decision making, you will be better able to choose a solution or make a decision on important issues. The following activities will help you develop and practice these skills.

Decision Making

Decision making involves choosing between two or more options. Listed below are guidelines to help you with making decisions.

1. **Identify a situation that requires a decision.** Think about your current situation. What issue are you faced with that requires you to take some sort of action?
2. **Gather information.** Think about the issue. Examine the causes of the issue or problem and consider how it affects you and others.
3. **Identify your options.** Consider the actions that you could take to address the issue. List these options so that you can compare them.
4. **Make predictions about consequences.** Predict the consequences of taking the actions listed for each of your options. Compare these possible consequences. Be sure the option you choose produces the results you want.
5. **Take action to implement a decision.** Choose a course of action from your available options, and put it into effect.

Problem Solving

Problem solving involves many of the steps of decision making. Listed below are guidelines to help you solve problems.

1. **Identify the problem.** Identify just what the problem or difficulty is that you are facing. Sometimes you face a difficult situation made up of several different problems. Each problem may require its own solution.
2. **Gather information.** Conduct research on any important issues related to the problem. Try to find the answers to questions like the following: What caused this problem? Who or what does it affect? When did it start?
3. **List and consider options.** Look at the problem and the answers to the questions you asked in Step 2. List and then think about all the possible ways in which the problem could be solved. These are your options—possible solutions to the problem.
4. **Examine advantages and disadvantages.** Consider the advantages and disadvantages of all the options that you have listed. Make sure that you consider the possible long-term effects of each possible solution. You should also determine what steps you will need to take to achieve each possible solution. Some suggestions may sound good at first but may turn out to be impractical or hard to achieve.
5. **Choose and implement a solution.** Select the best solution from your list and take the steps to achieve it.
6. **Evaluate the effectiveness of the solution.** When you have completed the steps needed to put your plan into action, evaluate its effectiveness. Is the problem solved? Were the results worth the effort required? Has the solution itself created any other problems?

Practicing the Skill

1. Chapter 24, Section 2: East Africa's History and Culture, describes the challenges of religious and ethnic conflict occurring in the region. Imagine that you are an ambassador to Rwanda. Use the decision-making guidelines to help you come up with a plan to help resolve the problems there. Be prepared to defend your decision.
2. Identify a similar problem discussed in another chapter and apply the problem-solving process to come up with a solution.

Becoming a Strategic Reader

by Dr. Judith Irvin

Everywhere you look, print is all around us. In fact, you would have a hard time stopping yourself from reading. In a normal day, you might read cereal boxes, movie posters, notes from friends, T-shirts, instructions for video games, song lyrics, catalogs, billboards, information on the Internet, magazines, the newspaper, and much, much more. Each form of print is read differently depending on your purpose for reading. You read a menu differently from poetry, and a motorcycle magazine is read differently than a letter from a friend. Good readers switch easily from one type of text to another. In fact, they probably do not even think about it, they just do it.

When you read, it is helpful to use a strategy to remember the most important ideas. You can use a strategy before you read to help connect information you already know to the new information you will encounter. Before you read, you can also predict what a text will be about by using a previewing strategy. During the reading you can use a strategy to help you focus on main ideas, and after reading you can use a strategy to help you organize what you learned so that you can remember it later. *Holt People, Places, and Change* was designed to help you more easily understand the ideas you read. Important reading strategies employed in *Holt People, Places, and Change* include:

A Tools to help you **preview and predict** what the text will be about

B Ways to help you **use and analyze visual information**

C Ideas to help you **organize the information** you have learned

A. Previewing and Predicting

How can I figure out what the text is about before I even start reading a section?

Previewing and **predicting** are good methods to help you understand the text. If you take the time to preview and predict before you read, the text will make more sense to you during your reading.

1 Usually, your teacher will set the purpose for reading. After reading some new information, you may be asked to write a summary, take a test, or complete some other type of activity.

"After reading about Spain and Portugal, you will work with a partner to present a history of the countries to a travel group..."

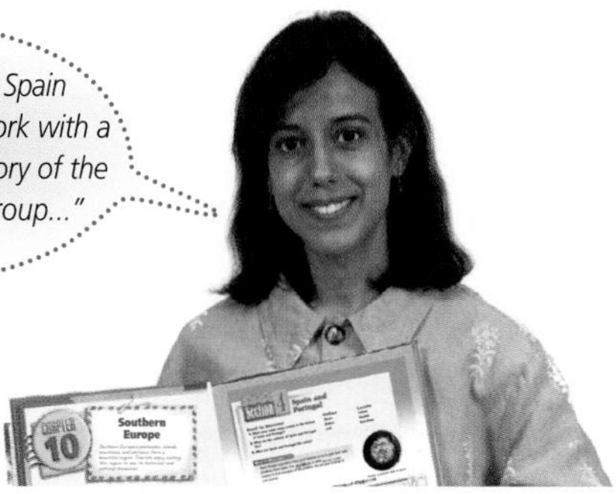

Previewing and Predicting

step 1 Identify your purpose for reading. Ask yourself what you will do with this information once you have finished reading.

▼

step 2 Ask yourself what is the main idea of the text and what are the key vocabulary words you need to know.

▼

step 3 Use signal words to help identify the structure of the text.

▼

step 4 Connect the information to what you already know.

2 As you preview the text, use **graphic signals** such as headings, subheadings, and boldface type to help you determine what is important in the text. Each section of *Holt People, Places, and Change* opens by giving you important clues to help you preview the material.

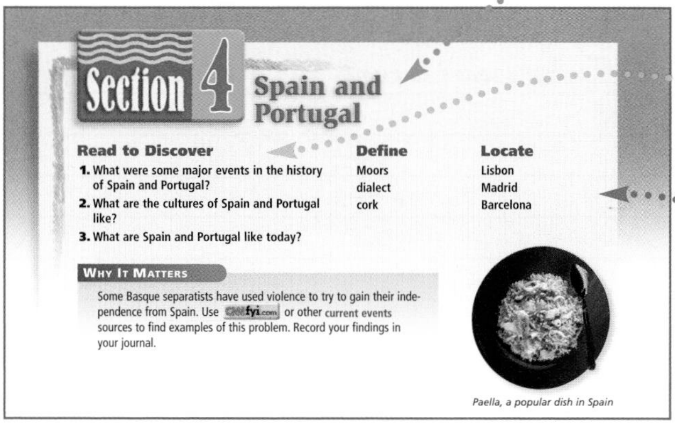

Section 4 Spain and Portugal

Read to Discover
1. What were some major events in the history of Spain and Portugal?
2. What are the cultures of Spain and Portugal like?
3. What are Spain and Portugal like today?

Define
Moors
dialect
cork

Locate
Lisbon
Madrid
Barcelona

Why It Matters
Some Basque separatists have used violence to try to gain their independence from Spain. Use CNNfyi.com or other current events sources to find examples of this problem. Record your findings in your journal.

Paella, a popular dish in Spain

Looking at the section's **main heading** and subheadings can give you an idea of what is to come.

Read to Discover questions give you clues as to the section's main ideas.

Define and **Locate** terms let you know the key vocabulary and places you will encounter in the section.

3 Other tools that can help you in previewing are **signal words**. These words prepare you to think in a certain way. For example, when you see words such as *similar to, same as,* or *different from,* you know that the text will probably compare and contrast two or more ideas. Signal words indicate how the ideas in the text relate to each other. Look at the list below for some of the most common signal words grouped by the type of text structures they include.

SIGNAL WORDS

Cause and Effect	Compare and Contrast	Description	Problem and Solution	Sequence or Chronological Order
because	different from	for instance	the question is	not long after
since	same as	for example	a solution	next
consequently	similar to	such as	one answer is	then
this led to...so	as opposed to	to illustrate		initially
if...then	instead of	in addition		before
nevertheless	although	most importantly		after
accordingly	however	another		finally
because of	compared with	furthermore		preceding
as a result of	as well as	first, second ...		following
in order to	either...or			on (date)
may be due to	but			over the years
for this reason	on the other hand			today
not only...but	unless			when

4 Learning something new requires that you connect it in some way with something you already know. This means you have to think before you read and while you read. You may want to use a chart like this one to remind yourself of the information already familiar to you and to come up with questions you want answered in your reading. The chart will also help you organize your ideas after you have finished reading.

What I know	*What I want to know*	*What I learned*

B. Use and Analyze Visual Information

How can all the pictures, maps, graphs, and time lines with the text help me be a stronger reader?

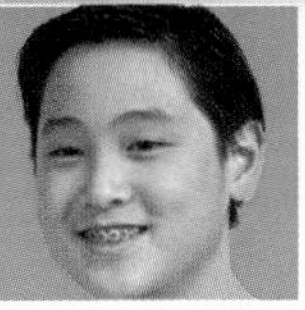

Using visual information can help you understand and remember the information presented in *Holt People, Places, and Change.* Good readers make a picture in their mind when they read. The pictures, charts, graphs, and diagrams that occur throughout *Holt People, Places, and Change* are placed strategically to increase your understanding.

1 You might ask yourself questions like these:

Why did the writer include this image with the text? What details about this image are mentioned in the text?

Analyzing Visual Information

step 1 As you preview the text, ask yourself how the visual information relates to the text.

step 2 Generate questions based on the visual information.

step 3 After reading the text, go back and review the visual information again.

step 4 Connect the information to what you already know.

2 After you have read the text, see if you can answer your own questions.

→ *Why are windmills important?*

→ *What technology do windmills use to pump water?*

→ *How might environment affect the use of windmills?*

B. Use and Analyze Visual Information *(continued)*

2 Maps, graphs, and charts help you organize information about a place. You might ask questions like these:

How does this map support what I have read in the text?

What does the information in this bar graph add to the text discussion?

→ *What is the purpose of this map?*

→ *What special features does the map show?*

→ *What do the colors, lines, and symbols on the map represent?*

Land Use and Resources

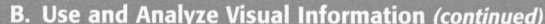

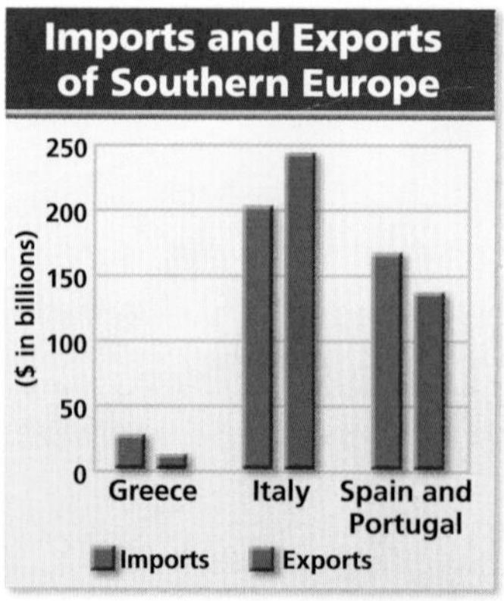

→ *What information is the writer trying to present with this graph?*

→ *Why did the writer use a bar graph to organize this information?*

3 After reading the text, go back and review the visual information again.

4 Connect the information to what you already know.

C. Organize Information

Once I learn new information, how do I keep it all straight so that I will remember it?

To help you remember what you have read, you need to find a way of **organizing information**. Two good ways of doing this are by using graphic organizers and concept maps. **Graphic organizers** help you understand important relationships—such as cause and effect, compare/contrast, sequence of events, and problem/solution—within the text. **Concept maps** provide a useful tool to help you focus on the text's main ideas and organize supporting details.

Identifying Relationships

Using graphic organizers will help you recall important ideas from the section and give you a study tool you can use to prepare for a quiz or test or to help with a writing assignment. Some of the most common types of graphic organizers are shown below.

Cause and Effect

Events in history cause people to react in a certain way. Cause-and-effect patterns show the relationship between results and the ideas or events that made the results occur. You may want to represent cause-and-effect relationships as one cause leading to multiple effects,

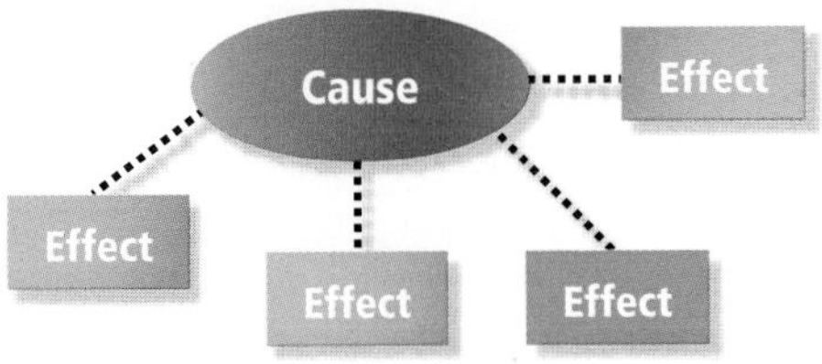

or as a chain of cause-and-effect relationships.

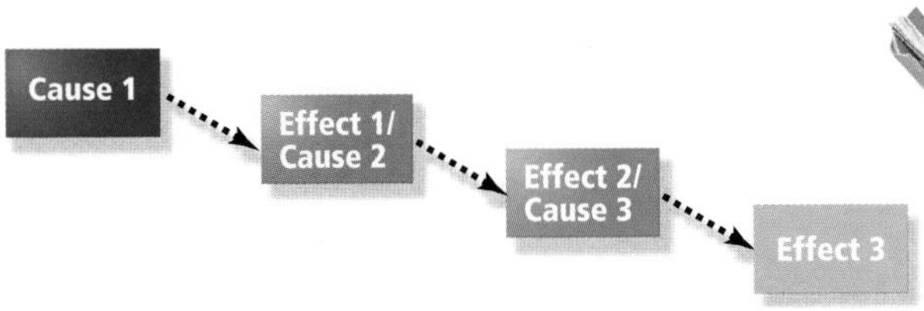

Constructing Graphic Organizers

step 1 Preview the text, looking for signal words and the main idea.

step 2 Form a hypothesis as to which type of graphic organizer would work best to display the information presented.

step 3 Work individually or with your classmates to create a visual representation of what you read.

C. Organize Information *(continued)*

Comparing and Contrasting

Graphic organizers are often useful when you are comparing or contrasting information. Compare-and-contrast diagrams point out similarities and differences between two concepts or ideas.

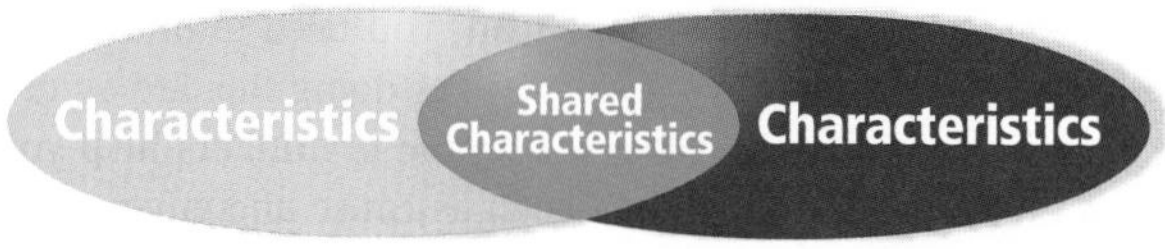

Sequencing

Keeping track of dates and the order in which events took place is essential to understanding the history and geography of a place. Sequence or chronological-order diagrams show events or ideas in the order in which they happened.

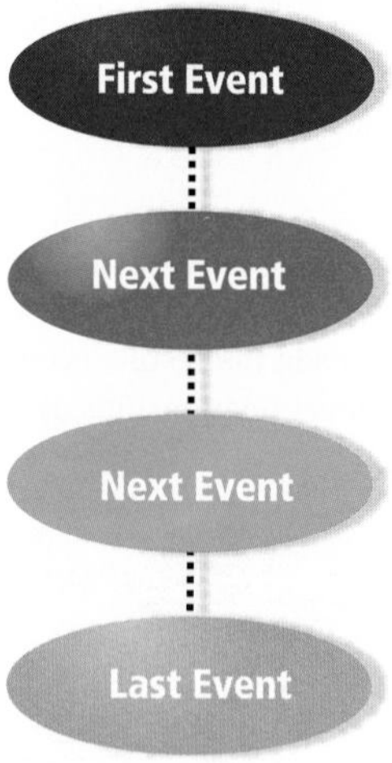

Problem and Solution

Problem-solution patterns identify at least one problem, offer one or more solutions to the problem, and explain or predict outcomes of the solutions.

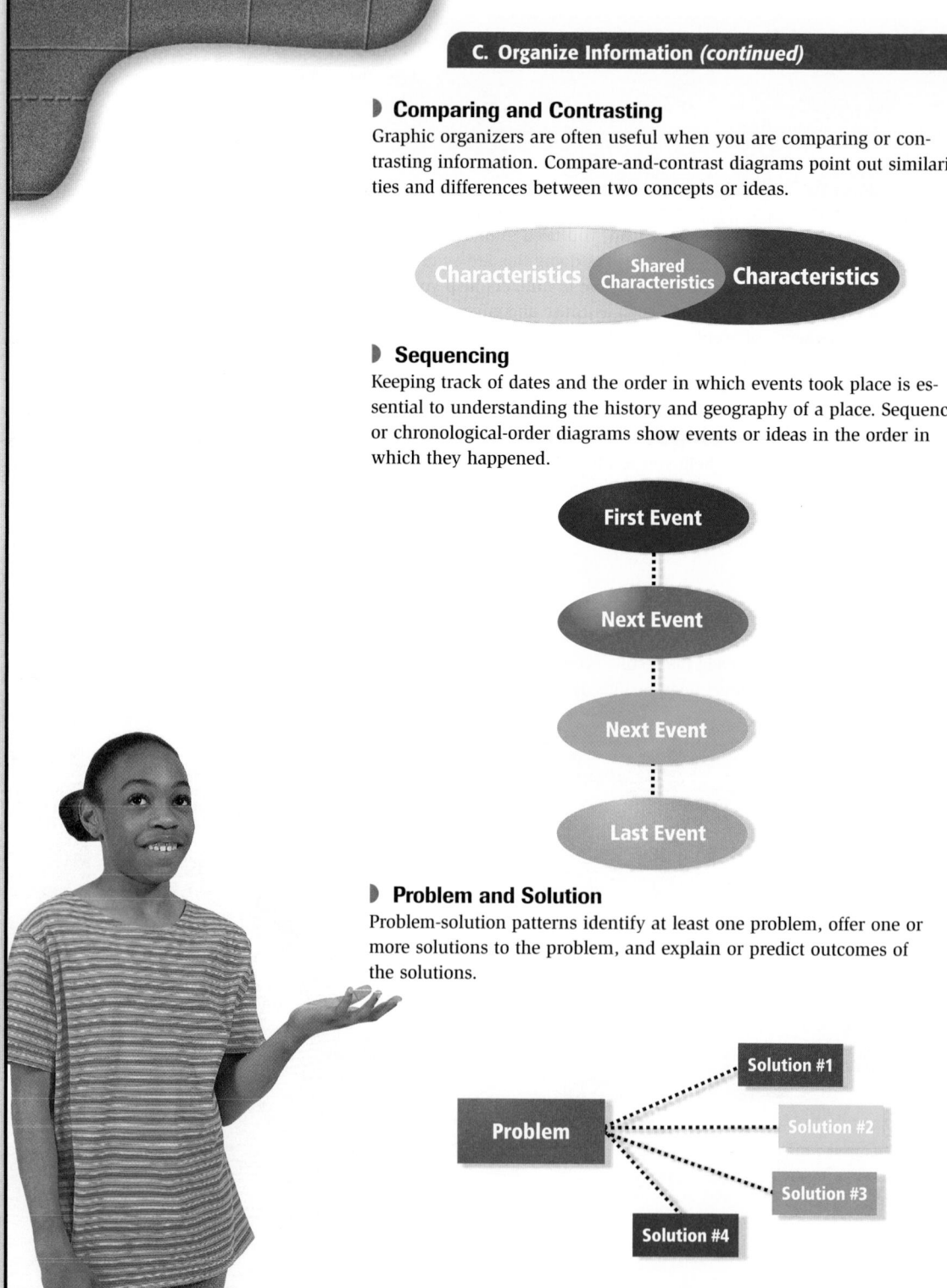

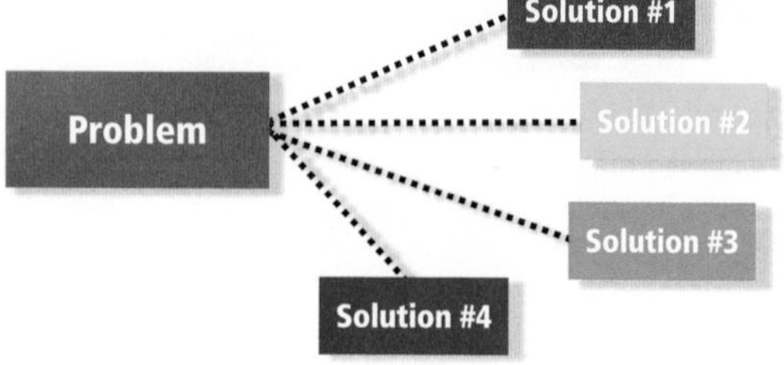

C. Organize Information *(continued)*

Identifying Main Ideas and Supporting Details

One special type of graphic organizer is the concept map. A concept map allows you to zero in on the most important points of the text. The map is made up of lines, boxes, circles, and/or arrows. It can be as simple or as complex as you need it to be to accurately represent the text. Here are a few examples of concept maps you might use.

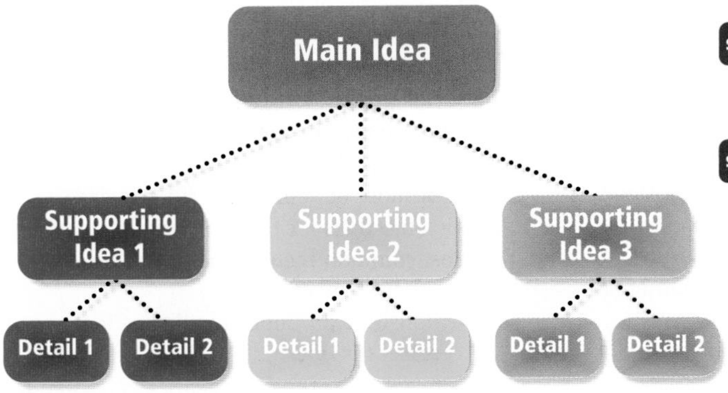

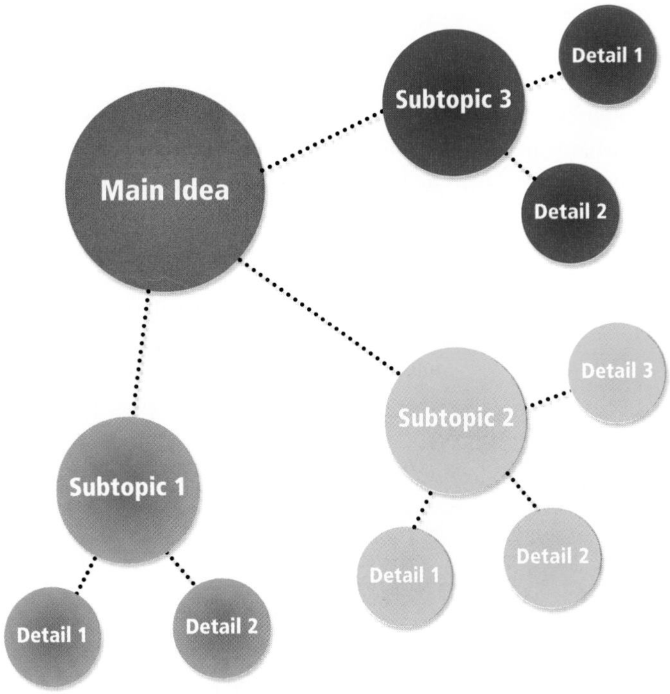

Constructing Concept Maps

step 1 Preview the text, looking for what type of structure might be appropriate to display as a concept map.

▼

step 2 Taking note of the headings, boldface type, and text structure, sketch a concept map you think could best illustrate the text.

▼

step 3 Using boxes, lines, arrows, circles, or any shapes you like, display the ideas of the text in the concept map.

Standardized Test-Taking Strategies

A number of times throughout your school career, you may be asked to take standardized tests. These tests are designed to demonstrate the content and skills you have learned. It is important to keep in mind that in most cases the best way to prepare for these tests is to pay close attention in class and take every opportunity to improve your general social studies, reading, writing, and mathematical skills.

Tips for Taking the Test

1. Be sure that you are well rested.
2. Be on time, and be sure that you have the necessary materials.
3. Listen to the teacher's instructions.
4. Read directions and questions carefully.
5. **DON'T STRESS!** Just remember what you have learned in class, and you should do well.

Practice the strategies at go.hrw.com

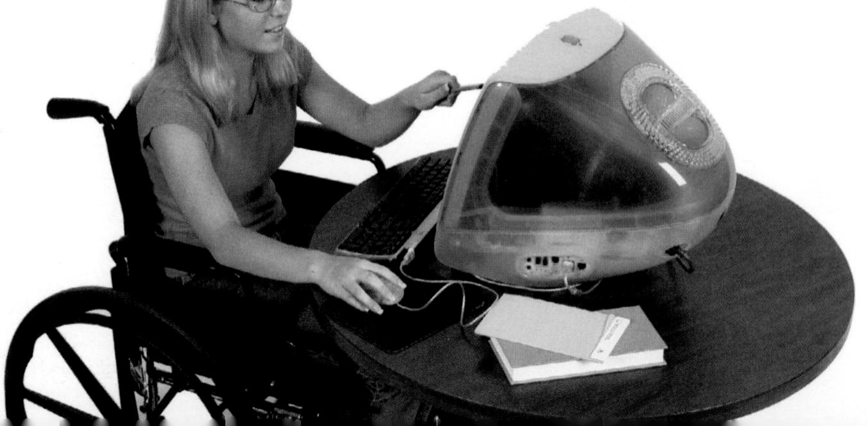

Tackling Social Studies

The social studies portions of many standardized tests are designed to test your knowledge of the content and skills that you have been studying in one or more of your social studies classes. Specific objectives for the test vary, but some of the most common include the following:

1. Demonstrate an understanding of issues and events in history.
2. Demonstrate an understanding of geographic influences on historical issues and events.
3. Demonstrate an understanding of economic and social influences on historical issues and events.
4. Demonstrate an understanding of political influences on historical issues and events.
5. Use critical thinking skills to analyze social studies information.

Standardized tests usually contain multiple-choice and, sometimes, open-ended questions. The multiple-choice items will often be based on maps, tables, charts, graphs, pictures, cartoons, and/or reading passages and documents.

Tips for Answering Multiple-Choice Questions

1. If there is a written or visual piece accompanying the multiple-choice question, pay careful attention to the title, author, and date.
2. Then read through or glance over the content of the written or visual piece accompanying the question to familiarize yourself with it.
3. Next, read the multiple-choice question first for its general intent. Then reread it carefully, looking for words that give clues or can limit possible answers to the question. For example, words such as *most* or *best* tell you that there may be several correct answers to a question, but you should look for the most appropriate answer.
4. Read through the answer choices. Always read all of the possible answer choices even if the first one seems like the correct answer. There may be a better choice farther down in the list.
5. Reread the accompanying information (if any is included) carefully to determine the answer to the question. Again, note the title, author, and date of primary-source selections. The answer will rarely be stated exactly as it appears in the primary source, so you will need to use your critical thinking skills to read between the lines.
6. Think of what you already know about the time in history or person involved and use that to help limit the answer choices.
7. Finally, reread the question and selected answer to be sure that you made the best choice and that you marked it correctly on the answer sheet.

Strategies for Success

There are a variety of strategies you can prepare ahead of time to help you feel more confident about answering questions on social studies standardized tests. Here are a few suggestions:

1. Adopt an acronym—a word formed from the first letters of other words—that you will use for analyzing a document or visual piece that accompanies a question.

Helpful Acronyms

For a document, use **SOAPS**, which stands for

S Subject
O Overview
A Audience
P Purpose
S Speaker/author

For a picture, cartoon, map, or other visual piece of information, use **OPTIC**, which stands for

O Occasion (or time)
P Parts (labels or details of the visual)
T Title
I Interrelations (how the different parts of the visual work together)
C Conclusion (what the visual means)

2. Form visual images of maps and try to draw them from memory. Standardized tests will most likely include maps showing many features, such as states, countries, continents, and oceans. Those maps may also show patterns in settlement and the size and distribution of cities. For example, in studying the United States, be able to see in your mind's eye such things as where the states and major cities are located. Know major physical features, such as the Mississippi River, the Appalachian and Rocky Mountains, the Great Plains, and the various regions of the United States, and be able to place them on a map. Such features may help you understand patterns in the distribution of population and the size of settlements.
3. When you have finished studying a geographic region or period in history, try to think of who or what might be important enough for a standardized test. You may want to keep your ideas in a notebook to refer to when it is almost time for the test.
4. Standardized tests will likely test your understanding of the political, economic, and social processes that

shape a region's history, culture, and geography. Questions may also ask you to understand the impact of geographic factors on major events. For example, some may ask about the effects of migration and immigration on various societies and population change. In addition, questions may test your understanding of the ways humans interact with their environment.

5. For the skills area of the tests, practice putting major events and personalities in order in your mind. Sequencing people and events by dates can become a game you play with a friend who also has to take the test. Always ask yourself "why" this event is important.
6. Follow the tips under "Ready for Reading" below when you encounter a reading passage in social studies, but remember that what you have learned about history can help you in answering reading-comprehension questions.

Ready for Reading

The main goal of the reading sections of most standardized tests is to determine your understanding of different aspects of a piece of writing. Basically, if you can grasp the main idea and the writer's purpose and then pay attention to the details and vocabulary so that you are able to draw inferences and conclusions, you will do well on the test.

Tips for Answering Multiple-Choice Questions

1. Read the passage as if you were not taking a test.
2. Look at the big picture. Ask yourself questions like, "What is the title?", "What do the illustrations or pictures tell me?", and "What is the writer's purpose?"
3. Read the questions. This will help you know what information to look for.
4. Reread the passage, underlining information related to the questions.

Types of Multiple-Choice Questions

1. **Main Idea** This is the most important point of the passage. After reading the passage, locate and underline the main idea.
2. **Significant Details** You will often be asked to recall details from the passage. Read the question and underline the details as you read, but remember that the correct answers do not always match the wording of the passage precisely.
3. **Vocabulary** You will often need to define a word within the context of the passage. Read the answer choices and plug them into the sentence to see what fits best.
4. **Conclusion and Inference** There are often important ideas in the passage that the writer does not state directly. Sometimes you must consider multiple parts of the passage to answer the question. If answers refer to only one or two sentences or details in the passage, they are probably incorrect.

5. Go back to the questions and try to answer each one in your mind before looking at the answers.
6. Read all the answer choices and eliminate the ones that are obviously incorrect.

Tips for Answering Short-Answer Questions

1. Read the passage in its entirety, paying close attention to the main events and characters. Jot down information you think is important.
2. If you cannot answer a question, skip it and come back later.
3. Words such as *compare, contrast, interpret, discuss,* and *summarize* appear often in short-answer questions. Be sure you have a complete understanding of each of these words.
4. To help support your answer, return to the passage and skim the parts you underlined.

5. Organize your thoughts on a separate sheet of paper. Write a general statement with which to begin. This will be your topic statement.
6. When writing your answer, be precise but brief. Be sure to refer to details in the passage in your answer.

Targeting Writing

On many standardized tests, you will occasionally be asked to write an essay. In order to write a concise essay, you must learn to organize your thoughts before you begin writing the actual piece. This keeps you from straying too far from the essay's topic.

Tips for Answering Composition Questions

1. Read the question carefully.
2. Decide what kind of essay you are being asked to write. Essays usually fall into one of the following types: persuasive, classificatory, compare/contrast, or "how to." To determine the type of essay, ask yourself questions like, "Am I trying to persuade my audience?", "Am I comparing or contrasting ideas?", or "Am I trying to show the reader how to do something?"
3. Pay attention to key words, such as *compare, contrast, describe, advantages, disadvantages, classify,* or *speculate.* They will give you clues as to the structure that your essay should follow.
4. Organize your thoughts on a sheet of paper. You will want to come up with a general topic sentence that expresses your main idea. Make sure this sentence addresses the question. You should then create an outline or some type of graphic organizer to help you organize the points that support your topic sentence.
5. Write your composition using complete sentences. Also, be sure to use correct grammar, spelling, punctuation, and sentence structure.
6. Be sure to proofread your essay once you have finished writing.

Gearing Up for Math

On most standardized tests you will be asked to solve a variety of mathematical problems that draw on the skills and information you have learned in class. If math problems sometimes give you difficulty, have a look at the tips below to help you work through the problems.

Tips for Solving Math Problems

1. Decide what is the goal of the question. Read or study the problem carefully and determine what information must be found.
2. Locate the factual information. Decide what information represents key facts—the ones you must have to solve the problem. You may also find facts you do not need to reach your solution. In some cases, you may determine that more information is needed to solve the problem. If so, ask yourself, "What assumptions can I make about this problem?" or "Do I need a formula to help solve this problem?"
3. Decide what strategies you might use to solve the problem, how you might use them, and what form your solution will be in. For example, will you need to create a graph or chart? Will you need to solve an equation? Will your answer be in words or numbers? By knowing what type of solution you should reach, you may be able to eliminate some of the choices.
4. Apply your strategy to solve the problem and compare your answer to the choices.
5. If the answer is still not clear, read the problem again. If you had to make calculations to reach your answer, use estimation to see if your answer makes sense.

UNIT 1

EXPLORING OUR WORLD

USING THE ILLUSTRATIONS

Direct students' attention to the photographs on these pages. Point out that in this unit students will learn there are two major branches of geography. One is physical geography, which deals with land, water, climate, and similar topics; the other is human geography, which involves people. Ask students which photos may relate more closely to physical geography *(rosettes, tropical landscape, iceberg)* and which relate more to human geography *(Carnival scene).*

Ask which photo shows a cold climate *(iceberg),* and which shows a warm climate *(La Digue Island).* On what familiar images or clues do we depend for the answers? *(Possible answers: ice, lush vegetation, palm trees)* Ask why the photo of the rosette plants does not give us much information about how warm or cold the climate may be. *(Because the plant is not familiar to most of us, we do not know where it grows.)*

You may want to invite students to speculate about the construction or meaning of the costumes in the Carnival photo. Lead a discussion about what kinds of parades are held in your community and what costumes the participants wear.

UNIT OBJECTIVES

1. Introduce geography as a field of study.
2. Explain Earth's position in space and the forces acting on Earth's land and water.
3. Analyze the interrelationships of wind, climate, and natural environments.
4. Identify major resources and how people use them.
5. Describe the development of cultures and the results of population expansion.
6. Learn to draw sketch maps and use them as geographic tools.

Your Classroom Time Line

To help you create a time line to display in your classroom, the most important dates and time periods discussed in each unit's chapters are compiled for you. Some additional dates have been inserted for clarity and continuity. Note that many dates, particularly those in the distant past, are approximate. In each unit, the lists begin in the sidebar on the page with the political map. You may want to have students use colored markers to differentiate among political, scientific, religious, and artistic events or achievements. You might also want to create your own categories.

UNIT 1

Exploring Our World

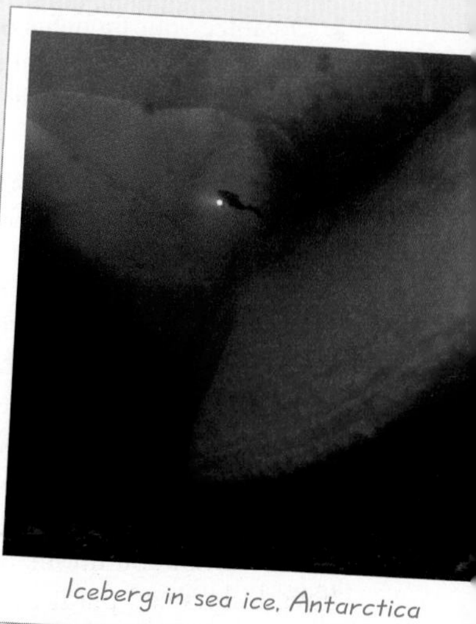

Iceberg in sea ice, Antarctica

Carnival parade in Valletta, Malta

Chapters

1 A Geographer's World
introduces the study of geography, its six essential elements, and its two main branches: human geography and physical geography.

2 Planet Earth
explores Earth's place in the solar system, the importance of water, landforms, and the geological forces that shape the land.

3 Wind, Climate, and Natural Environments
introduces wind and ocean currents, climate types, vegetation, and soils.

4 Earth's Resources
describes the distribution and use of Earth's soil, forests, water, air, minerals, and energy sources.

5 The World's People
introduces human geography, the development of cultures, and population issues.

Notes from the Field

A Physical Geographer in Mountain Environments

Professor Francisco Pérez studies tropical mountain environments. He is interested in the natural processes, plants, and environments of mountains. **WHAT DO YOU THINK?** ***What faraway places would you like to study?***

I became attracted to mountains when I was a child. While crossing the Atlantic Ocean in a ship, I saw snow-capped Teide Peak in the Canary Islands rising from the water. It was an amazing sight.

As a physical geographer, I am interested in the unique environments of high mountain areas. This includes geological history, climate, and soils. The unusual conditions of high mountain environments have influenced plant evolution. Plants and animals that live on separate mountains sometimes end up looking similar. This happens because they react to their environments in similar ways. For example, several types of tall, weird-looking plants called giant rosettes grow in the Andes, Hawaii, East Africa, and the Canary Islands. Giant rosettes look like the top of a pineapple at the end of a tall stem.

Rosette plants, Ecuador

I have found other strange plants, such as rolling mosses. Mosses normally grow on rocks. However, if a moss plant falls to the ground, ice crystals on the soil surface lift the moss. This allows it to "roll" downhill while it continues to grow in a ball shape!

I like doing research in mountains. They are some of the least explored regions of our planet. Like most geographers, I cannot resist the attraction of strange landscapes in remote places.

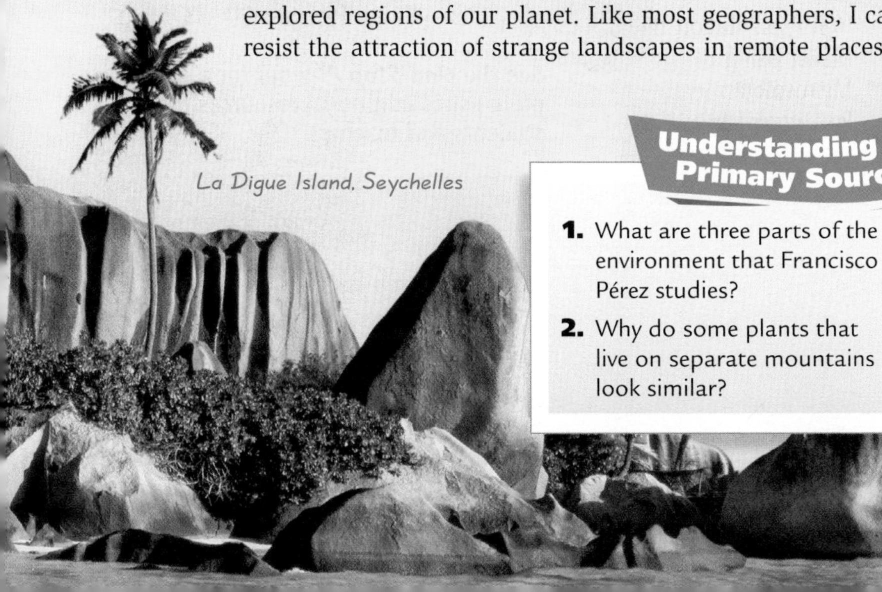
La Digue Island, Seychelles

Understanding Primary Sources

1. What are three parts of the environment that Francisco Pérez studies?
2. Why do some plants that live on separate mountains look similar?

Sturgeonfish

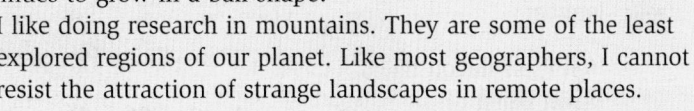

More from the Field

Living things that are not related sometimes develop similar physical traits because they live in similar environments. This process is called convergent evolution. For example, tuna (fish) and dolphins (mammals) both have streamlined bodies and fins for living in the water.

For a land-based example, compare the serval of Africa, a cat, and the maned wolf of South America, a dog. Both have long necks, long legs, and large ears. They hunt small animals in grassy plains areas. Their long legs and necks elevate their ears above the grass. As a result, they can hear the slightest sound made by their prey.

Discussion: Refer students to the food web illustration in Section 3 of Chapter 3. Lead a discussion on how different plants or animals with similar characteristics might fit into the web.

Understanding Primary Sources
Answers
1. geological history, climate, soils
2. because they react to their environments in similar ways

CHAPTER 1 A Geographer's World

Chapter Resource Manager

Objectives	Pacing Guide	Reproducible Resources
SECTION 1 **Developing a Geographic Eye** (pp. 3–5) **1.** Identify the role perspective plays in the study of geography. **2.** Describe some of the issues or topics that geographers study. **3.** Name three levels geographers can use to view the world.	**Regular** .5 day **Block Scheduling** .5 day *Block Scheduling Handbook, Chapter 1*	**RS** Guided Reading Strategy 1.1 **SM** Geography for Life Activity 1
SECTION 2 **Themes and Essential Elements** (pp. 6–10) **1.** Identify tools geographers use to study the world. **2.** Identify what shapes Earth's features. **3.** Describe how humans shape the world. **4.** Describe how studying geography helps us understand the world.	**Regular** 1.5 days **Block Scheduling** .5 day *Block Scheduling Handbook, Chapter 1*	**RS** Guided Reading Strategy 1.2 **RS** Graphic Organizer 1 **E** Creative Strategies for Teaching World Geography, Lessons 1–3 **E** Lab Activities for Geography and Earth Science, Hands-On 2 **SM** Map Activity 1
SECTION 3 **The Branches of Geography** (pp. 11–13) **1.** Describe what is included in the study of human geography. **2.** Describe what is included in the study of physical geography. **3.** Identify the types of work geographers do.	**Regular** 1 day **Block Scheduling** .5 day *Block Scheduling Handbook, Chapter 1*	**RS** Guided Reading Strategy 1.3 **E** Lab Activities for Geography and Earth Science, Hands-On 1

Chapter Resource Key

RS Reading Support

IC Interdisciplinary Connections

E Enrichment

SM Skills Mastery

A Assessment

REV Review

ELL Reinforcement and English Language Learners

Transparencies

CD–ROM

Music

Video

go.hrw.com Internet

Holt Presentation Maker Using Microsoft® Powerpoint®

One-Stop Planner CD–ROM

See the *One-Stop Planner* for a complete list of additional resources for students and teachers.

One-Stop Planner CD–ROM

It's easy to plan lessons, select resources, and print out materials for your students when you use the ***One-Stop Planner CD–ROM with Test Generator.***

Technology Resources	Review, Reinforcement, and Assessment Resources
One-Stop Planner CD–ROM, Lesson 1.1 Global Skill Builder CD–ROM, Project 1 go.hrw.com: Homework Practice Online; HRW Go site	**ELL** Main Idea Activity 1.1 **REV** Section 1 Review, p. 5 **A** Daily Quiz 1.1 **ELL** English Audio Summary 1.1 **ELL** Spanish Audio Summary 1.1
One-Stop Planner CD–ROM, Lesson 1.2 go.hrw.com: Homework Practice Online; HRW Go site	**ELL** Main Idea Activity 1.2 **REV** Section 2 Review, p. 10 **A** Daily Quiz 1.2 **ELL** English Audio Summary 1.2 **ELL** Spanish Audio Summary 1.2
One-Stop Planner CD–ROM, Lesson 1.3 Yourtown CD–ROM go.hrw.com: Homework Practice Online; HRW Go site	**ELL** Main Idea Activity 1.3 **REV** Section 3 Review, p. 13 **A** Daily Quiz 1.3 **ELL** English Audio Summary 1.3 **ELL** Spanish Audio Summary 1.3

internet connect

HRW ONLINE RESOURCES

GO TO: go.hrw.com
Then type in a keyword.

TEACHER HOME PAGE
KEYWORD: SK3 TEACHER

CHAPTER INTERNET ACTIVITIES
KEYWORD: SK3 GT1

Choose an activity to:
- learn to use online maps.
- be a virtual geographer for a day.
- compare regions around the world.

CHAPTER ENRICHMENT LINKS
KEYWORD: SK3 CH1

CHAPTER MAPS
KEYWORD: SK3 MAPS1

ONLINE ASSESSMENT
Homework Practice
KEYWORD: SK3 HP1
Standardized Test Prep Online
KEYWORD: SK3 STP1
Rubrics
KEYWORD: SS Rubrics

COUNTRY INFORMATION
KEYWORD: SK3 Almanac

CONTENT UPDATES
KEYWORD: SS Content Updates

HOLT PRESENTATION MAKER
KEYWORD: SK3 PPT1

ONLINE READING SUPPORT
KEYWORD: SS Strategies

CURRENT EVENTS
KEYWORD: S3 Current Events

Chapter Review and Assessment

- **E** Readings in World Geography, History, and Culture 1, 2
- **SM** Critical Thinking Activity 1
- **REV** Chapter 1 Review, pp. 14–15
- **REV** Chapter 1 Tutorial for Students, Parents, Mentors, and Peers
- **ELL** Vocabulary Activity 1
- **A** Chapter 1 Test
- Chapter 1 Test Generator (on the One-Stop Planner)
- Audio CD Program, Chapter 1
- **A** Chapter 1 Test for English Language Learners and Special-Needs Students

Meeting Individual Needs

Ability Levels

Level 1 Basic-level activities designed for all students encountering new material

Level 2 Intermediate-level activities designed for average students

Level 3 Challenging activities designed for honors and gifted-and-talented students

English Language Learners Activities that address the needs of students with Limited English Proficiency

CHAPTER 1

Launch into Learning

Ask students to name a country they would like to visit and to give a reason why they want to travel to that particular country. *(Examples: France, for the food; China, to see the Great Wall)* Point out that their interests could probably be the subject of serious study by a geographer. *(Examples: A geographer may study patterns in food preferences among French people or the regional use of ingredients and cooking techniques. Another may use satellite technology to find forgotten sections of the Great Wall.)* Use several of the students' suggestions to show geography's wide range. Then ask students to create "geographic studies" based on their classmates' chosen destinations.

Section 1

Objectives

1. Explain the role perspective plays in the study of geography.
2. Describe some issues or topics that geographers study.
3. Identify the three levels geographers use to view the world.

Links to Our Lives

You may want to share with your students the following reasons for gaining a basic understanding of geography as a field of study:

- Knowing the fundamentals of geography will help students learn more about all aspects of their world.
- Getting an overview from Chapter 1 will make it easier to grasp details in later chapters.
- Throughout the book, connections are made to the six essential elements of geography. These elements are explained fully in Chapter 1.
- Geography is an expanding field that includes a wide range of specializations. Students might want to consider geography as a career.

CHAPTER 1 A Geographer's World

Chart of the Mediterranean and Europe, 1559

Hand-held compass

GPS (global positioning satellite) receiver

FOCUS

LET'S GET STARTED

Select several photographs of scenes from around the world and display them in the classroom. Copy the following instructions on the chalkboard: *Choose one of the photographs and write down three questions you would like to ask about the place in the picture.* Call on students to read their questions aloud, and use their questions as the basis for a discussion about the issues that professional geographers study. Tell students that in Section 1 they will learn more about developing a geographic eye.

Building Vocabulary

Write the key terms on the chalkboard. Tell students that **perspective** is based on a word meaning "to look" and that **spatial** is based on a word meaning "space." Then, as a class, decide on a definition for **spatial perspective**. Compare this definition to the one in Section 1. Then, point out that **geography** is based on two Greek roots: *geō-*, which means "Earth," and *graphein,* which means "to write." Ask students to compare the meaning of the root words to the textbook's definition and explain the relationship. Finally, have students read the definitions for **urban** and **rural** and then provide examples of urban and rural areas in their region.

Section 1 Developing a Geographic Eye

Read to Discover

1. What role does perspective play in the study of geography?
2. What are some issues or topics that geographers study?
3. At what three levels can geographers view the world?

Define

perspective
spatial perspective
geography
urban
rural

WHY IT MATTERS

What factors would you consider if you were moving to a new town or city? You would probably want to know about its geography. Use CNNfyi.com or other **current events** sources to investigate a place you might like to live. Record your findings in your journal.

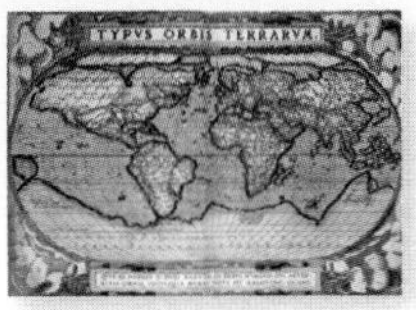

World map, 1598

Perspectives

People look at the world in different ways. Their experiences shape the way they understand the world. This personal understanding is called **perspective**. Your perspective is your point of view. A geographer's point of view looks at where something is and why it is there. This point of view is known as **spatial perspective**. Geographers apply this perspective when they study the arrangement of towns in a state. They might also use this perspective to examine the movement of cars and trucks on busy roads.

Geographers also work to understand how things are connected. Some connections are easy to see, like highways that link cities. Other connections are harder to see. For example, a dry winter in Colorado could mean that farms as far away as northern Mexico will not have enough water.

Geography is a science. It describes the physical and cultural features of Earth. Studying geography is important. Geographically informed people can see meaning in the arrangement of things on Earth. They know how people and places are related. Above all, they can apply a spatial perspective to real life. In other words, people familiar with geography can understand the world around them.

This fish-eye view of a large city shows highway patterns.

✓ **READING CHECK:** *The World in Spatial Terms* What role does perspective play in the study of geography? Geographers use perspective when they study where something is and why it is there.

Section 1 RESOURCES

Reproducible

- Block Scheduling Handbook, Chapter 1
- Guided Reading Strategy 1.1
- Geography for Life Activity 1

Technology

- One-Stop Planner CD–ROM, Lesson 1.1
- Homework Practice Online
- Global Skill Builder CD–ROM, Project 1
- HRW Go site

Reinforcement, Review, and Assessment

- Section 1 Review, p. 5
- Daily Quiz 1.1
- Main Idea Activity 1.1
- English Audio Summary 1.1
- Spanish Audio Summary 1.1

TEACH

Teaching Objective 1

ALL LEVELS: (Suggested time: 10 min.) Discuss geographers' use of spatial perspective. Then have students examine the aerial photograph on the previous page and suggest why the highways are located where they are. **ENGLISH LANGUAGE LEARNERS**

Teaching Objectives 2–3

ALL LEVELS: (Suggested time: 20 min.) Copy the following graphic organizer onto the chalkboard, omitting the italicized answers. Use it to help students understand the issues geographers study and the level at which they view the world. **ENGLISH LANGUAGE LEARNERS**

STUDY OF GEOGRAPHY

Issues/Topics	Levels
Earth's processes	*local*
relationships between people and environment	*regional*
governments	*global*
religion and food	*local, regional*
urban and rural areas	*local, regional*

PHYSICAL SYSTEMS

Geographers study how people all around the world react to Earth's processes. Tristan da Cunha is one of a group of small islands in the South Atlantic Ocean about midway between Africa and South America. It is a British territory.

A volcano 6,760 feet (2,060 m) high dominates the island. Its peak is often shrouded in clouds. Lava flows have continually shaped the island's landscape. A volcanic eruption in 1961 forced the evacuation of the island's residents. After the danger passed, most of the Tristanians returned to their isolated island.

Critical Thinking: How have Earth's physical processes affected Tristanians?

Answer: They were forced to evacuate their homeland because of a volcano.

The movement of people is one issue that geographers study. For example, political and economic troubles led many Albanians to leave their country in 1991. Many packed onto freighters like this one for the trip. Geographers want to know how this movement affects the environment and other people.

Interpreting the Visual Record **How do you think Albania has been affected by so many people leaving the country?**

Geographic Issues

Issues geographers study include Earth's processes and their impact on people. Geographers study the relationship between people and environment in different places. For example, geographers study tornadoes to find ways to reduce loss of life and property damage. They ask how people prepare for tornadoes. Do they prepare differently in different places? When a tornado strikes, how do people react?

Geographers also study how governments change and how those changes affect people. Czechoslovakia, for example, split into Slovakia and the Czech Republic in 1993. These types of political events affect geographic boundaries. People react differently to these changes. Some people are forced to move. Others welcome the change.

Other issues geographers study include religions, diet (or food), **urban** areas, and **rural** areas. Urban areas contain cities. Rural areas contain open land that is often used for farming.

✓ **READING CHECK:** *The Uses of Geography* What issues or topics do geographers study? Earth's processes, the relationship between people and environment, changes of government, religions, diet, urban areas, and rural areas

Local, Regional, and Global Geographic Studies

With any topic, geographers must decide how large an area to study. They can focus their study at a local, regional, or global level.

Local Studying your community at the local, or close-up, level will help you learn geography. You know where homes and stores are located. You know how to find parks, ball fields, and other fun places. Over time, you see your community change. New buildings are constructed. People move in and out of your neighborhood. New stores open their doors, and others go out of business.

Visual Record Answer

Answers will vary but students may mention possible political volatility of the area, a negative or fearful atmosphere, or fewer people for jobs.

internet connect
GO TO: go.hrw.com
KEYWORD: SK3 CH1
FOR: Web sites about the geographer's world

CLOSE

Ask students to imagine that they are geographers from the planet Geog and they have landed on Earth. Have students list what human activities they would study first and what sources they would use in their research.

REVIEW AND ASSESS

Have students complete the Section Review. Then have students work in groups to create short quizzes based on the section's material. Have groups exchange quizzes and complete another group's quiz. Then have students complete Daily Quiz 1.1. **COOPERATIVE LEARNING**

RETEACH

Have students complete Main Idea Activity 1.1. Then have them illustrate one of the section's topics. Ask students to explain their illustrations. **ENGLISH LANGUAGE LEARNERS**

EXTEND

Have interested students conduct research on the history of the field of geography and its influence on society. They may want to concentrate on ancient Greek or Arabic achievements. Ask them to create illustrated charts showing their research. **BLOCK SCHEDULING**

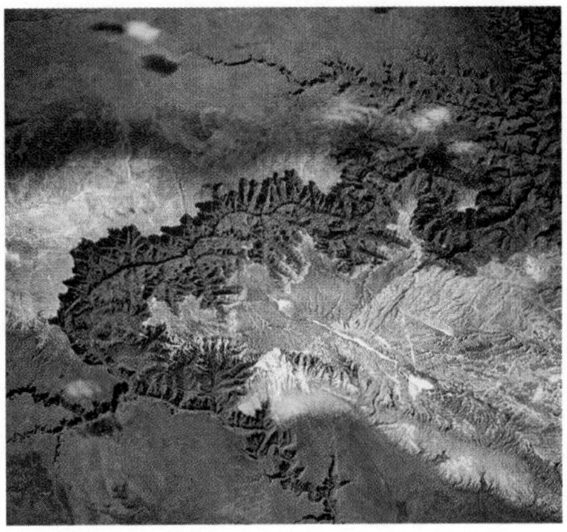

The southwest is a region within the United States. One well-known place that characterizes the landscape of the southwest is the Grand Canyon. The Grand Canyon is shown in the photo at left and in the satellite image at right.

Regional Regional geographers organize the world into convenient parts for study. For example, this book separates the world into big areas like Africa and Europe. Regional studies cover larger areas than local studies. Some regional studies might look at connections like highways and rivers. Others might examine the regional customs.

Global Geographers also work to understand global issues and the connections between events. For example, many countries depend on oil from Southwest Asia. If those oil supplies are threatened, some countries might rush to secure oil from other areas. Oil all over the world could then become much more expensive.

✓ **READING CHECK:** *The World in Spatial Terms* What levels do geographers use to focus their study of an issue or topic? local, regional, or global

go.hrw.com Homework Practice Online
Keyword: SK3 HP1

Section Review 1

Define and explain: perspective, spatial perspective, geography, urban, rural

Reading for the Main Idea

1. *The Uses of Geography* How can a spatial perspective be used to study the world?
2. *The Uses of Geography* Why is it important to study geography?

Critical Thinking

3. **Drawing Inferences and Conclusions** How do threatening weather patterns affect people, and why do geographers study these patterns?
4. **Drawing Inferences and Conclusions** Why is it important to view geography on a global level?

Organizing What You Know

5. **Finding the Main Idea** Copy the following graphic organizer. Use it to examine the issues geographers study. Write a paragraph on one of these issues.

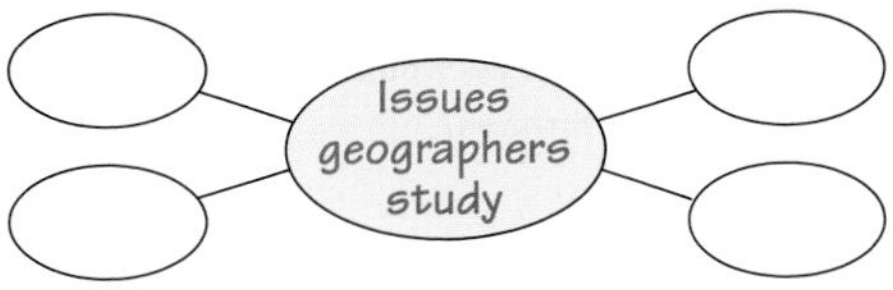

Section Review 1

Answers

Define For definitions, see: perspective, p. 3; spatial perspective, p. 3; geography, p. 3; urban, p. 4; rural, p. 4

Reading for the Main Idea

1. to understand how things are connected (NGS 3)
2. to see meaning in the arrangement of things on Earth and to understand the world (NGS 17, 18)

Critical Thinking

3. They can cause loss of life or property damage; to help people protect themselves from dangerous weather situations
4. Answers will vary but might include to gain an understanding of how events in one region can affect other regions.

Organizing What You Know

5. Answers will vary but should be issues geographers study.

Section 2

Objectives

1. Identify tools geographers use to study the world.
2. Identify what shapes Earth's features.
3. Examine how humans shape the world.
4. Explain how studying geography helps us understand the world.

Focus

LET'S GET STARTED

Write the following question on the chalkboard: *Where is your favorite shopping mall or movie theater located?* Have students respond to the question. If a student names the actual address for the building, explain that he or she has provided its absolute location. Tell students that in Section 1 they will learn about the difference between absolute and relative location and other topics of geography.

Building Vocabulary

Write the key terms on the chalkboard. Ask what we mean when we say "It's all relative" and "Absolutely!" Ask students for suggestions on how those phrases could relate to **absolute location** and **relative location**. Have students look up the remaining key terms in the text or glossary and write sentences using them.

Section 2 Resources

Reproducible

- Guided Reading Strategy 1.2
- Map Activity 1
- Creative Strategies for Teaching World Geography, Lessons 1–3
- Lab Activities for Geography and Earth Science, Hands-On 2

Technology

- One-Stop Planner CD–ROM, Lesson 1.2
- Homework Practice Online
- HRW Go site

Reinforcement, Review, and Assessment

- Section 2 Review, p. 10
- Daily Quiz 1.2
- Main Idea Activity 1.2
- English Audio Summary 1.2
- Spanish Audio Summary 1.2

Visual Record Answer

near Cairo, near the pyramids

Section 2 Themes and Essential Elements

Read to Discover

1. What tools do geographers use to study the world?
2. What shapes Earth's features?
3. How do humans shape the world?
4. How does studying geography help us understand the world?

Define

absolute location
relative location
subregions
diffusion
levees

Tombs carved out of a mountain in Turkey

WHY IT MATTERS

Geographers often study the effect that new people have on a place. Use CNNfyi.com or other current events sources to find out how the arrival of new people has changed the United States or another country. Record your findings in your journal.

▲ The location of a place can be described in many ways.

Interpreting the Visual Record **Looking at the photo of this hotel in Giza, Egypt, and at the map, how would you describe Giza's location?**

Themes

The study of geography has long been organized according to five important themes, or topics of study. One theme, *location*, deals with the exact or relative spot of something on Earth. *Place* includes the physical and human features of a location. *Human-environment interaction* covers the ways people and environments affect each other. *Movement* involves how people change locations and how goods are traded, as well as the effects of these movements. *Region* organizes Earth into geographic areas with one or more shared characteristics.

✓ **READING CHECK:** *The Uses of Geography* What are the five themes of geography? location, place, human-environment interaction, movement, region

Six Essential Elements

Another way to look at geography is to study its essential elements, or most important parts. The six essential elements used to study geography are The World in Spatial Terms, Places and Regions, Physical Systems, Human Systems, Environment and Society, and The Uses of Geography. These six essential elements will be used throughout this textbook. They share many properties with the five themes of geography.

TEACH

Teaching Objectives 1–2

ALL LEVELS: (Suggested time: 20 min.) Pair students and have each pair create a geography fact sheet for the school. Fact sheets should include the school's absolute and relative location as well as several of its physical or human characteristics. Ask volunteers to read their fact sheets to the class. Then discuss why pairs may have chosen different identifying features. **ENGLISH LANGUAGE LEARNERS, COOPERATIVE LEARNING**

Teaching Objectives 3–4

ALL LEVELS: (Suggested time: 30 min.) Have students draw maps of their neighborhoods, including homes, stores, streets, and other landmarks. Then have them locate their neighborhoods on a map of their town, city, or county. Tell students to label items on their maps that represent the relationship between environment and society. *(such as dams, recycling plants, airports, train stations, and highways)* Ask volunteers to share their maps with the class. **ENGLISH LANGUAGE LEARNERS**

Location **Every place on Earth has a location. Location is defined by absolute and relative location.**

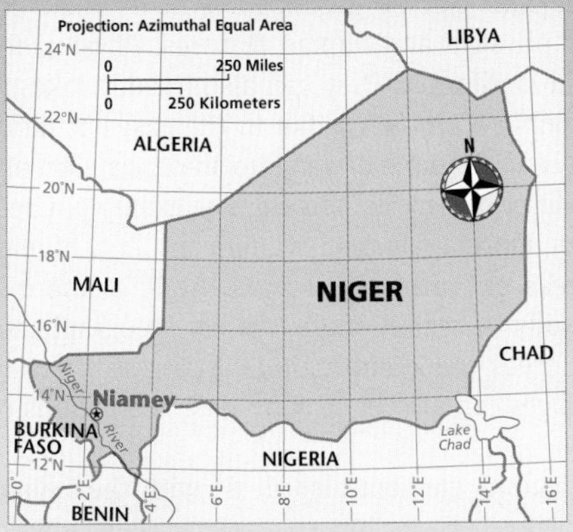

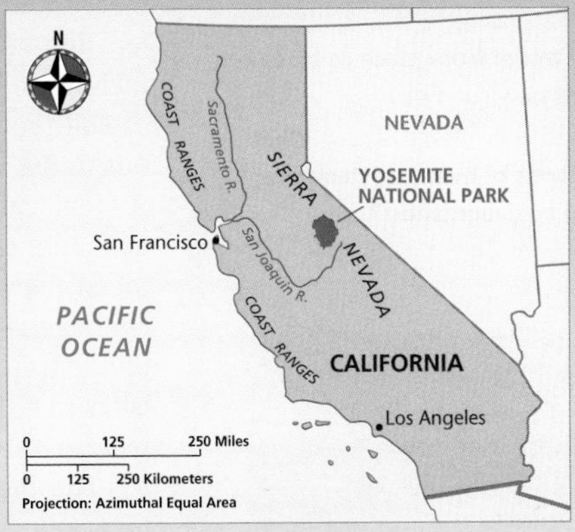

Absolute Location: the exact spot on Earth where something is found

Example: Niamey, the capital of Niger, is located at 13°31′ north latitude and 2°07′ east longitude.

Relative Location: the position of a place in relation to other places

Example: Yosemite National Park is north of Los Angeles, California, and east of San Francisco, California.

You Be the Geographer
1. Use an atlas to find the absolute location of your city or town.
2. Write a sentence describing the relative location of your home.

The World in Spatial Terms This element focuses on geography's spatial perspective. As you learned in Section 1, geographers apply spatial perspective when they look at the location of something and why it is there. The term *location* can be used in two ways. **Absolute location** defines an exact spot on Earth. For example, the address of the Smithsonian American Art Museum is an absolute location. The address is at 8th and G Streets, N.W., in Washington, D.C. City streets often form a grid. This system tells anyone looking for an address where to go. The grid formed by latitude and longitude lines also pinpoints absolute location. Suppose you asked a pilot to take you to 52° north latitude by 175° west longitude. You would land at a location on Alaska's Aleutian Islands.

Relative location describes the position of a place in relation to another place. Measurements of direction, distance, or time can define relative location. For example, the following sentences give relative location. "The hospital is one mile north of our school." "Canada's border is about an hour's drive from Great Falls, Montana."

A geographer must be able to use maps and other geographic tools and technologies to determine spatial perspective. A geographer must

▲ Places can be described by what they do not have. This photo shows the result of a long period without rain.

Linking Past to Present

Channeled Scablands Geographers think that large Ice Age floods originating in western Montana created the Channeled Scablands in eastern Washington state. A glacier blocked a river and created a glacial lake near Missoula in present-day Montana. When this ice dam broke, a wall of water perhaps 2,000 feet (610 m) high crashed through the region, carving out unusual landforms such as the Channeled Scablands—an area marked by channels, cliffs, and steep-sided canyons. Scientists suspect that water poured from the lake at 60 or more miles per hour and that the glacial lake near Missoula may have filled and emptied dozens of times.

Activity: Pair students and have them conduct research on the Channeled Scablands. Ask students to create a travel guide to the area that explains how the area was created and what tourists would see there. Guides should include photographs if possible.

You Be the Geographer Answers
1. Locations should be as accurate as the atlas used allows.
2. Sentences might refer to bodies of water, landforms, streets, or other features.

Teaching Objective 4

ALL LEVELS: (Suggested time: 30 min.) Copy the following graphic organizer onto the chalkboard, omitting the italicized answers. Use it to help students illustrate how regions and subregions help geographers understand our world. Using the United States as an example, tell students to identify regions *(such as the East or the Midwest)* and subregions *(such as their state)*. Remind students that regions and subregions may vary in size and can be categorized as cultural, economic, or political. Then have students identify subregions of one subregion and classify each into one of these three categories. **ENGLISH LANGUAGE LEARNERS**

REGIONS AND SUBREGIONS

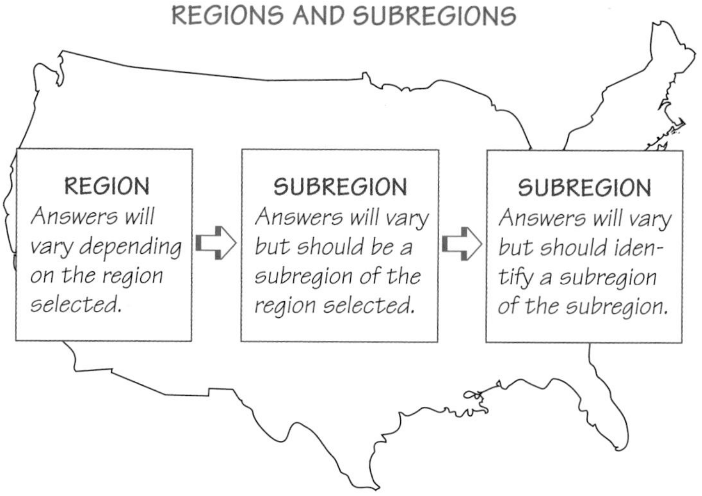

REGION
Answers will vary depending on the region selected.
SUBREGION
Answers will vary but should be a subregion of the region selected.
SUBREGION
Answers will vary but should identify a subregion of the subregion.

Environment and Society

Recycling is another example of the relationship between society and the environment. Its success depends on developing new technology to reuse collected materials and protect the environment. Products made from recycled materials include boards made of sawdust mixed with scrap wood and shrinkwrap.

Recycled materials are even used in the manufacture of shoes and clothing. Three different recycled materials are used in some hiking boots. The soles are made from used tires, the innersole padding from white paper, and the fabric uppers from used plastic bottles. Plastic bottles are also recycled to create fabric for sweaters.

Critical Thinking: How do new recycling technologies affect the environment?

Answer: They help to protect it.

internet connect

GO TO: **go.hrw.com**
KEYWORD: **SK3 CH1**
FOR: **Web sites about recycling**

Visual Record Answers ▲

the communication of ideas, the production of goods, trade, conflict, governments ►

It represents a human adaptation to Egypt's environment.

People travel from place to place on miles of new roadway.

Interpreting the Visual Record **What other forms of human systems are studied by geographers?**

▼

also know how to organize and analyze information about people, places, and environments using geographic tools.

Places and Regions Our world has a vast number of unique places and regions. Places can be described both by their physical location and by their physical and human features. Physical features include coastlines and landforms. They can also include lakes, rivers, or soil types. For example, Colorado is flat in the east but mountainous in the west. This is an example of a landform description of place. A place can also be described by its climate. For example, Greenland has long, cold winters. Florida has mild winters and hot, humid summers. Regions are areas of Earth's surface with one or more shared characteristics. To study a region more closely, geographers often divide it into smaller areas called **subregions**. Many of the characteristics that describe places can also be used to describe regions or subregions.

The Places and Regions element also deals with the human features of places and regions. Geographers want to know how people have created regions based on Earth's features and how culture and other factors affect how we see places and regions on Earth.

Physical Systems Physical systems shape Earth's features. Geographers study earthquakes, mountains, rivers, volcanoes, weather patterns, and similar topics and how these physical systems have affected Earth's characteristics. For example, geographers might study how volcanic eruptions in the Hawaiian Islands spread lava, causing landforms to change. They might note that southern California's shoreline changes yearly, as winter and summer waves move beach sand.

Geographers also study how plants and animals relate to these nonliving physical systems. For example, deserts are places with cactus and other plants as well as rattlesnakes and other reptiles, that can

Men in rural Egypt wear a long shirt called a *galabia*. This loose-fitting garment is ideal for people living in Egypt's hot desert climate. In addition, the galabia is made from cotton, an important agricultural product of Egypt.

Interpreting the Visual Record **How does the *galabia* show how people have adapted to their environment?**

►

➤ASSIGNMENT: Have students recall the most beautiful, interesting, or exciting place they have ever visited. Then have them write words or phrases that describe that place in terms of landforms, climate, animal life, plant life, language spoken, common religion, history, customs, or other physical or human characteristics. Ask students to consult primary or secondary sources for additional information. Then have students write a description of their chosen locale's relative location and find its absolute location by calculating latitude and longitude.

Teacher to Teacher

Rebecca Minnear of Las Vegas, Nevada, suggests the following activity to help students understand the six essential elements of geography. Prior to class, draw an outline map of your community on six transparencies. Organize the class into six groups and assign one element to each group. Give each group a transparency sheet and a marker. Each group should draw on the transparency ways its element relates to the community. For example, the "places and regions" group might draw features of the local landscape. The "environment and society" group might draw waterways, streets, airports, and so on. Groups should draw the parts of the community in their proper place so that when the transparencies are placed on top of each other there will be an overlap.

▲ A satellite dish brings different images and ideas to people in a remote area of Brazil.

Interpreting the Visual Record How might resources have affected the use of technology here?

live in very dry conditions. Geographers also study how different types of plants, animals, and physical systems are distributed on Earth.

Human Systems People are central to geography. Our activities, movements, and settlements shape Earth's surface. Geographers study peoples' customs, history, languages, and religions. They study how people migrate, or move, and how ideas are communicated. When people move, they may go to live in other countries or move within a country. Geographers want to know how and why people move from place to place.

People move for many reasons. Some move to start a new job. Some move to attend special schools. Others might move to be closer to family. People move either when they are pushed out of a place or when they are pulled toward another place. In the Dust Bowl, for example, crop failures pushed people out of Oklahoma in the 1930s. Many were pulled to California by their belief that they would find work there. Geographers also want to know how ideas or behaviors move from one region to another. The movement of ideas occurs through communication. There are many ways to communicate. People visit with each other in person or on the phone. New technology allows people to communicate by e-mail. Ideas are also spread through films, magazines, newspapers, radio, and television. The movement of ideas or behaviors from one region to another is known as **diffusion**.

The things we produce and trade are also part of the study of human systems. Geographers study trading patterns and how countries depend on each other for certain goods. In addition, geographers look at the causes and results of conflicts between peoples. The study of governments we set up and the features of cities and other settlements we live in are also part of this study.

Environment and Society Geographers study how people and their surroundings affect each other. Their relationship can be examined in three ways. First, geographers study how humans depend on

▲ This woman at a railway station in Russian Siberia sells some goods that were once unavailable in her country.

Interpreting the Visual Record Which essential element is illustrated in this photo?

Human Systems

A new development in the movement of goods between countries and regions has been the creation of the "megamall." At more than 5 million square feet (465,000 sq m), West Edmonton Mall in Alberta, Canada, is the largest indoor shopping and entertainment complex in the world. The Mall of America, near Minneapolis, Minnesota, is the largest mall in the United States. These malls combine hundreds of stores, full-scale amusement parks, and other attractions. They have become the top tourist destination in their regions and attract people from as far away as Asia and Europe.

Activity: Have students conduct research on sales at West Edmonton Mall and the Mall of America. Ask students to compare the sales at these malls, explain their differences, and identify what role these malls play in the movement of goods.

▲ Visual Record Answers

People in the region value the ability to obtain information from other parts of the world and have invested in costly technology.

◄ Human Systems

CLOSE

Display a picture of a well-known local landmark. Call on students to suggest how the six essential elements of geography relate to the landmark.

REVIEW AND ASSESS

Have students complete the Section Review. Then organize students into groups of four or five. Assign each group a city that appears on one of the Atlas maps in the textbook. Have students create a travel guide that describes the region in which the city is located. Then have students complete Daily Quiz 1.2. **COOPERATIVE LEARNING**

RETEACH

Have students complete Main Idea Activity 1.2. Then organize students into six groups and assign each group one of the six essential elements of geography. Ask members of each group to write a paragraph describing their element in relation to your school. **ENGLISH LANGUAGE LEARNERS**

EXTEND

Ask interested students to imagine that they have been hired to submit a building plan for a recreation center in their community. Tell them to use the six essential elements of geography to determine the center's location and construction features. Ask students to include a drawing of the building and a map showing its location within the community. **BLOCK SCHEDULING**

Section 2 Review

Answers

Define For definitions, see: absolute location, p. 7, relative location, p. 7, subregions, p. 8, diffusion, p. 9, levees, p. 10

Reading for the Main Idea

1. with maps and other geographic tools (NGS 1)
2. physical systems such as earthquakes, mountains, rivers, volcanoes, weather patterns (NGS 7)

Critical Thinking

3. through their activities, movements, settlements, modifications
4. provides clues to the past and helps geographers plan for the future
5. The World in Spatial Terms—using maps and other geographic tools to look at the world with a spatial perspective; Places and Regions—studying the physical and human features of a place; Physical Systems—systems that have shaped Earth's features; Human Systems—how people have shaped Earth's surface; Environment and Society—how people and their surroundings affect each other; The Uses of Geography—how geography helps us understand relationships among people, places, and the environment over time

▲ Open-air markets like this one in Mali provide opportunities for farmers to sell their goods.

their physical environment to survive. Human life requires certain living and nonliving resources, such as freshwater and fertile soil for farming.

Geographers also study how humans change their behavior to be better suited to an environment. These changes or adaptations include the kinds of clothing, food, and shelter that people create. These changes help people live in harsh climates.

Finally, humans change the environment. For example, farmers who irrigate their fields can grow fruit in Arizona's dry climate. People in Louisiana have built **levees**, or large walls, to protect themselves when the Mississippi River floods.

The Uses of Geography Geography helps us understand the relationships among people, places, and the environment over time. Understanding how a relationship has developed can help in making plans for the future. For example, geographers can study how human use of the soil in a farming region has affected that region over time. Such knowledge can help them determine what changes have been made to the soil and whether any corrective measures need to be taken.

✓ **READING CHECK:** *The Uses of Geography* What are the six essential elements in studying geography? The World in Spatial Terms, Places and Regions, Physical Systems, Human Systems, Environment and Society, and The Uses of Geography

Section 2 Review

go.hrw.com Homework Practice Online
Keyword: SK3 HP1

Define and explain: absolute location, relative location, subregions, diffusion, levees

Reading for the Main Idea

1. *The World in Spatial Terms* How do geographers study the world?
2. *Physical Systems* What shapes Earth's features? Give examples.

Critical Thinking

3. **Finding the Main Idea** How do humans shape the world in which they live?
4. **Analyzing Information** What benefits can studying geography provide?

Organizing What You Know

5. **Summarizing** Copy the following graphic organizer. Use it to identify and describe all aspects of each of the six essential elements.

Element	Description

Section 3

Objectives

1. Explain the study of human geography.
2. Describe the study of physical geography.
3. Investigate the types of work that geographers do.

FOCUS

LET'S GET STARTED

Ask students what they would do if they were lost in an unfamiliar part of town. Then have them work in pairs to list ways to find their way home. *(asking for directions or consulting a map)* Ask them to describe the advantages or disadvantages of these options. Have them choose one of these solutions and create a map or written directions for a place with which they are both familiar. Tell them to evaluate the effectiveness of the solution by presenting their work to another pair.

Building Vocabulary

Tell students that the word **cartography** contains the common suffix *-graphy,* which means "writing or representation." Ask students what other words they know that contain the suffix. Point out that the prefix *cart-* indicates maps, so the complete word means writing or representing maps. Divide **meteorology** also: *meteor-,* from a Greek word meaning "high in the air"; and *-logy,* meaning "a branch of learning." Then ask students to use what they just learned to define **climatology**.

Section 3 The Branches of Geography

Read to Discover

1. What is included in the study of human geography?
2. What is included in the study of physical geography?
3. What types of work do geographers do?

Define

human geography
physical geography
cartography
meteorology
climatology

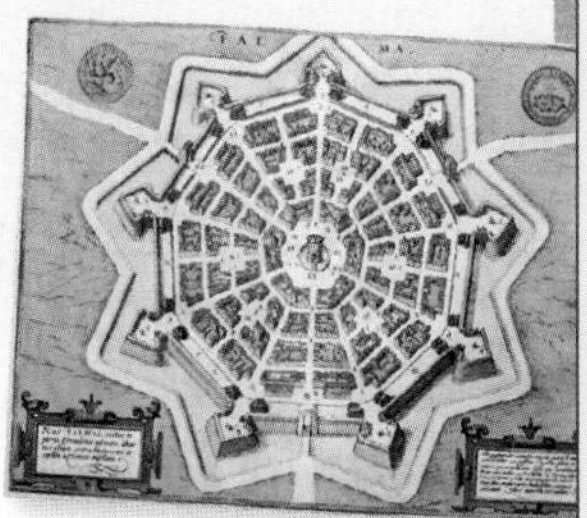

Map of an ancient fortress

WHY IT MATTERS

Nearly every year, hurricanes hit the Atlantic or Gulf coasts of the United States. Predicting weather is one of the special fields of geography. Use CNNfyi.com or other **current events** sources to find out about hurricanes. Record your findings in your journal.

Section 3 RESOURCES

Reproducible

- Guided Reading Strategy 1.3
- Lab Activities for Geography and Earth Science, Hands-On 1

Technology

- One-Stop Planner CD–ROM, Lesson 1.3
- Homework Practice Online
- Yourtown CD–ROM
- HRW Go site

Reinforcement, Review, and Assessment

- Section 3 Review, p. 13
- Daily Quiz 1.3
- Main Idea Activity 1.3
- English Audio Summary 1.3
- Spanish Audio Summary 1.3

Human Geography

The study of people, past or present, is the focus of **human geography**. People's location and distribution over Earth, their activities, and their differences are studied. For example, people living in different countries create different kinds of governments. Political geographers study those differences. Economic geographers study the exchange of goods and services across Earth. Cultural geography, population geography, and urban geography are some other examples of human geography. A professional geographer might specialize in any of these branches.

✓ **READING CHECK:** *Human Systems* How is human geography defined? as the study of people, past or present

▲ A volunteer visits a poor area of Bangladesh. Geographers study economic conditions in regions to help them understand human geography.

Physical Geography

The study of Earth's natural landscapes and physical systems, including the atmosphere, is the focus of **physical geography**. The world is full of different landforms such as deserts, mountains, and plains. Climates affect these landscapes. Knowledge of physical systems helps geographers understand how a landscape developed and how it might change.

TEACH

Teaching Objectives 1–3

ALL LEVELS: (Suggested time: 20 min.) Copy the following graphic organizer onto the chalkboard, omitting the italicized answers. Use it to help students distinguish between the study of human geography and the study of physical geography to identify the types of work geographers do. **ENGLISH LANGUAGE LEARNERS**

Geography helps us understand the world.

Physical Geography
the study of
- *Earth's natural landscapes and physical systems*
- *different landforms*

Types of work
include
- *cartography* • *meteorology* • *climatology*

Human Geography
the study of
- *people, past or present*
- *politics, economy, and culture*

Section 3 Review

Answers

Define For definitions, see: human geography, p. 11; physical geography, p. 11; cartography, p. 13, meteorology, p. 13, climatology, p. 13

Reading for the Main Idea

1. Topics are the study of people, their location and distribution, their activities, and their differences. (NGS 9)
2. by tracking Earth's larger atmospheric systems (NGS 18)

Critical Thinking

3. to learn how a landscape developed and how it might change
4. issues such as population, pollution, endangered plants and animals, or decreased or increased economic activities

Organizing What You Know

5. meteorologist—tracks weather and atmospheric conditions; climatologist—tracks atmospheric systems

Connecting to Technology Answers

1. It can help planners build roads, dams, or other structures.
2. consequences, such as greater knowledge about population, profitable economic activities, or change to the environment

CONNECTING TO Technology

A mapmaker creates a digital map.

Computer Mapping

Maps are tools that can display a wide range of information. Traditionally, maps were drawn on paper and could not be changed to suit the user. However, computers have revolutionized the art of mapmaking.

Today, mapmakers use computers to create and modify maps for different uses. They do this by using a geographic information system, or GIS. A GIS is a computer system that combines maps and satellite photographs with other kinds of spatial data—information about places on the planet. This information might include soil types, population figures, or voting patterns.

Using a GIS, mapmakers can create maps that show geographic features and relationships. For example, a map showing rainfall patterns in a particular region might be combined with data on soil types or human settlement to show areas of possible soil erosion.

The flexibility of a GIS allows people to seek answers to specific questions. Where should a new road be built to ease traffic congestion? How are changes in natural habitat affecting wildlife? These and many other questions can be answered with the help of computer mapping.

Understanding What You Read

1. How could a GIS help people change their environment?
2. What social, environmental, or economic consequences might future advances in GIS technology have?

Knowledge of physical and human geography will help you understand the world's different regions and peoples. In your study of the major world regions, you will see how physical and human geography connect to each other.

✓ **READING CHECK:** *Physical Systems* What is included in the study of physical geography? Earth's natural landscapes and physical systems, including the atmosphere

Working as a Geographer

Geography plays a role in almost every occupation. Wherever you live and work, you should know local geography. School board members know where children live. Taxi drivers are familiar with city streets. Grocery store managers know which foods sell well in certain areas.

CLOSE

Write the following statement on the chalkboard: *A cartographer's work is never done.* Ask students why might this be true. *(Possible answers: changes required by physical processes, new roads and suburbs, and political boundary changes)*

REVIEW AND ASSESS

Have students complete the Section Review. Then pair students and have each pair locate newspaper or magazine articles that relate to some aspect of human or physical geography. Have pairs write a few sentences explaining the connection between the articles and human or physical geography. Then have students complete Daily Quiz 1.3. **COOPERATIVE LEARNING**

RETEACH

Have students complete Main Idea Activity 1.3. Then have them complete the following sentence: "I used my knowledge of geography today when I . . . " Ask volunteers to read their sentences to the class. **ENGLISH LANGUAGE LEARNERS**

EXTEND

Have interested students conduct research on the history of cartography in an area that has been mapped since antiquity. Have them use copies of ancient maps to investigate how maps of that region have evolved over time and then present their findings to the class. **BLOCK SCHEDULING**

They also know where they can obtain these products throughout the year. Local newspaper reporters are familiar with town meetings and local politicians. Reporters also know how faraway places can affect their communities. Doctors must know if their towns have poisonous snakes or plants. City managers know whether nearby rivers might flood. Emergency workers in mountain towns check snow depth so they can give avalanche warnings. Local weather forecasters watch for powerful storms and track their routes on special maps.

Some specially trained geographers practice in the field of **cartography**. Cartography is the art and science of mapmaking. Today, most mapmakers do their work on computers. Geographers also work as weather forecasters. The field of forecasting and reporting rainfall, temperature, and other atmospheric conditions is called **meteorology**. A related field is **climatology**. These geographers, known as climatologists, track Earth's larger atmospheric systems. Climatologists want to know how these systems change over long periods of time. They also study how people might be affected by changes in climate.

Governments and a variety of organizations hire geographers to study the environment. These geographers might explore such topics as pollution, endangered plants and animals, or rain forests. Some geographers who are interested in education become teachers and writers. They help people of all ages learn more about the world. Modern technology allows people all over the world to communicate instantly. Therefore, it is more important than ever to be familiar with the geographer's world.

▲ Experts examine snow to help forecast avalanches. They study the type of snow, weather conditions, and landforms. For example, wet snow avalanches can occur because of the formation of a particular type of ice crystal, called depth hoar, near the ground.

✓ **READING CHECK:** *The Uses of Geography* What types of work do geographers perform? They make maps, work as weather forecasters, track atmospheric systems, or work as teachers or writers.

go.hrw.com Homework Practice Online
Keyword: SK3 HP1

Section 3 Review

Define and explain: human geography, physical geography, cartography, meteorology, climatology

Reading for the Main Idea

1. *Human Systems* What topics are included in the study of human geography?
2. *The Uses of Geography* How do people who study the weather use geography?

Critical Thinking

3. **Finding the Main Idea** Why is it important to study physical geography?
4. **Making Generalizations and Predictions** How might future discoveries in the field of geography affect societies, world economies, or the environment?

Organizing What You Know

5. **Categorizing** Copy the following graphic organizer. Use it to list geographers' professions and their job responsibilities.

CHAPTER 1 Review ANSWERS

Building Vocabulary

For definitions, see: perspective, p. 3; spatial perspective, p. 3; geography, p. 3; urban, p. 4; rural, p. 4; absolute location, p. 7; relative location, p. 7; subregions, p. 8; diffusion, p. 9; levees, p. 10; human geography, p. 11; physical geography, p. 11; cartography, p. 13; metereology, p. 13; climatology, p. 13

Reviewing the Main Ideas

1. local, regional, and global; examples will vary but should reflect the appropriate level (NGS 3)
2. absolute location—directions would define an exact spot on Earth; relative location—directions would describe the position of a place in relation to another place (NGS 1)
3. the movement of ideas or behaviors from one cultural region to another; allows people to learn new things (NGS 9)
4. to study a region more closely (NGS 5)
5. because maps are useful; answers will vary but might include meteorology and climatology (NGS 1)

ASSESS

Have students complete the Chapter 1 Test.

RETEACH

Organize students into three groups—one group for each section. Have groups summarize their section in a paragraph. Call on volunteers to read their group's paragraph to the class. Ask members of the other groups to evaluate the paragraphs. **ENGLISH LANGUAGE LEARNERS, COOPERATIVE LEARNING**

PORTFOLIO

PORTFOLIO EXTENSIONS

1. Different cultures use different methods for showing directions and locations. For example, long ago Polynesians developed shell maps to help them navigate in the vast Pacific Ocean. Have students work in groups to devise new ways to record information about a region familiar to them. Ask them to include a standard map of the area along with the new map. They should also write a legend or key for the new map.
2. Ask students to imagine that they are directing a documentary entitled The Six Essential Elements of Geography and You. Pair students and have each pair create six storyboards—one for each element. Storyboards should include a paragraph that describes how the scene represents the element.

CHAPTER 1

Review ANSWERS

Understanding Environment and Society

Information included in reports should be consistent with text material. Students should discuss the positive and negative aspects of the after-school program and of protecting endangered species. Use Rubric 12, Drawing Conclusions, to evaluate student work.

Thinking Critically

1. Answers will vary, but students might mention that a geographer identifies where things are so that connections can be made.
2. when—daily; how—answers will vary; examples might include building dams and irrigating fields
3. Students might mention the movement of people, trade networks, or the diffusion of ideas between groups.
4. both human and physical characteristics
5. by helping us see meaning in the arrangement of things on Earth

CHAPTER 1

Reviewing What You Know

Building Vocabulary

On a separate sheet of paper, write sentences to define each of the following words.

1. perspective
2. spatial perspective
3. geography
4. urban
5. rural
6. absolute location
7. relative location
8. levees
9. diffusion
10. subregions
11. human geography
12. physical geography
13. cartography
14. meteorology
15. climatology

Reviewing the Main Ideas

1. *The World in Spatial Terms* What are three ways to view geography? Give an example of when each type could be used.
2. *The World in Spatial Terms* What kind of directions would you give to indicate a place's absolute location? Its relative location?
3. *Human Systems* What is diffusion, and why is it important?
4. *Places and Regions* Why do geographers create subregions?
5. *The World in Spatial Terms* Why is cartography important? What types of jobs do geographers do?

Understanding Environment and Society

Land Use

You are on a committee that will decide whether to close a park near your school. One proposed use for the land is a building where after-school activities could be held. However, the park is the habitat of an endangered bird. Write a report describing consequences of the park closing. Then organize information from your report to create a proposal on what decision should be made.

Thinking Critically

1. **Analyzing Information** How can a geographer use spatial perspective to explain how things in our world are connected?
2. **Drawing Inferences and Conclusions** When and how do humans relate to the environment? Provide some examples of this relationship.
3. **Summarizing** How are patterns created by the movement of goods, ideas, and people?
4. **Finding the Main Idea** How are places and regions defined?
5. **Finding the Main Idea** How does studying geography help us understand the world?

Food Festival

Have students bring food items to class and use them to discuss human and physical geography. For example, a student may bring a can of green beans and note that certain soil, sunlight, and climate conditions must be present to grow the beans. Or a student may use the can to discuss how people in different parts of the country prepare green beans, or how farms and canneries affect local economies.

Chapter 1 Review and Assessment Resources

Reproducible
- Readings in World Geography, History, and Culture 1, 2
- Critical Thinking Activity 1
- Vocabulary Activity 1

Technology
- Chapter 1 Test Generator (on the One-Stop Planner)
- Audio CD Program, Chapter 1 (English and Spanish)
- HRW Go site

Reinforcement, Review, and Assessment
- Chapter 1 Review, pp. 14–15
- Chapter 1 Tutorial for Students, Parents, Mentors, and Peers
- Chapter 1 Test
- Chapter 1 Test for English Language Learners and Special-Needs Students

Building Social Studies Skills

Map Activity

On a separate sheet of paper, match the letters on the map with their correct labels.

Africa	Europe
Antarctica	North America
Asia	South America
Australia	

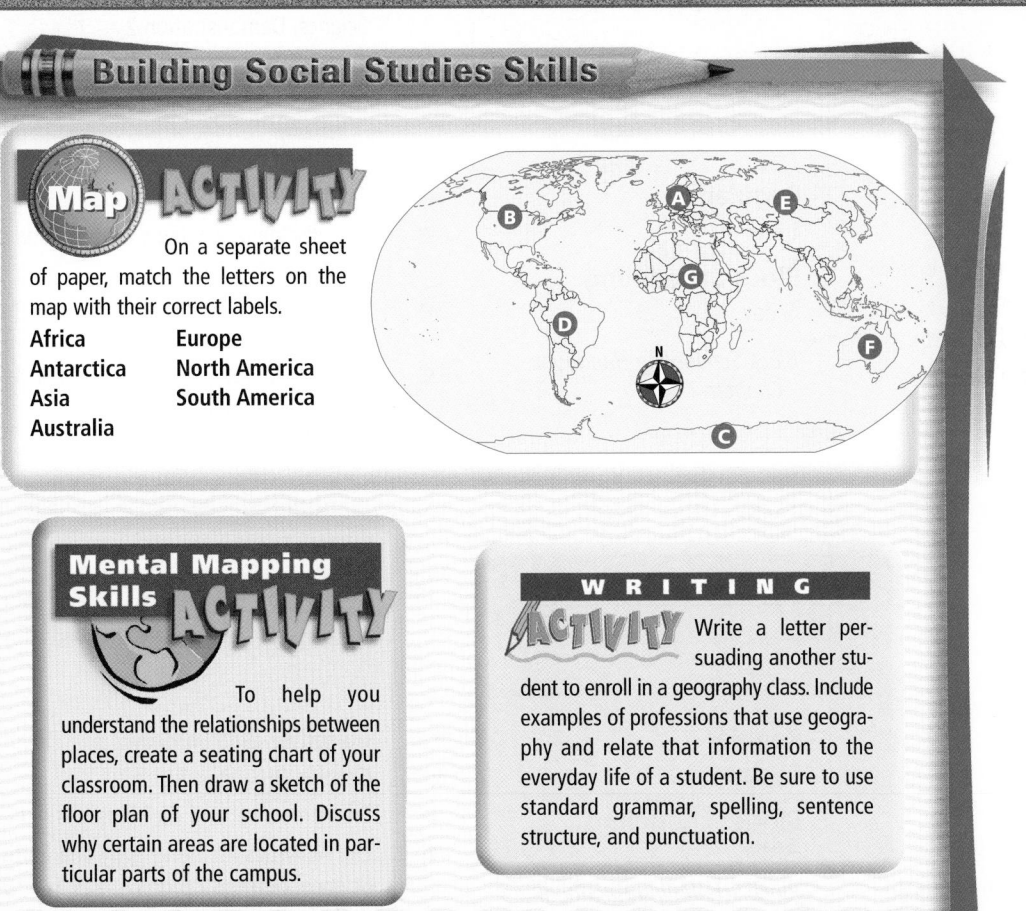

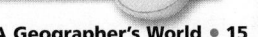

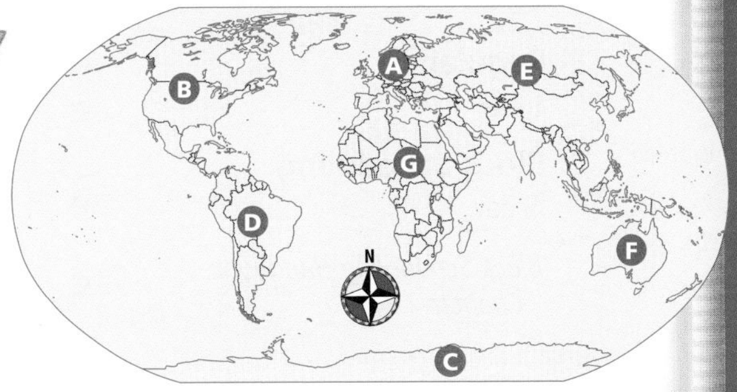

Mental Mapping Skills Activity

To help you understand the relationships between places, create a seating chart of your classroom. Then draw a sketch of the floor plan of your school. Discuss why certain areas are located in particular parts of the campus.

Writing Activity

Write a letter persuading another student to enroll in a geography class. Include examples of professions that use geography and relate that information to the everyday life of a student. Be sure to use standard grammar, spelling, sentence structure, and punctuation.

Alternative Assessment

Portfolio Activity

Learning About Your Local Geography

Individual Project How do you define your community geographically? Is your community the area around your home or school? Write two or three sentences defining your community to share with the class.

internet connect

Internet Activity: go.hrw.com
KEYWORD: SK3 GT1

Choose a topic to explore online:
- Learn to use online maps.
- Be a virtual geographer for a day.
- Compare regions around the world.

Map Activity

A. Europe
B. North America
C. Antarctica
D. South America
E. Asia
F. Australia
G. Africa

Mental Mapping Skills Activity

Seating charts, sketches, and answers will vary but should be logical for your classroom and school.

Writing Activity

Letters will vary, but should include various professions that use geography. Letters should also relate the use of geography to the everyday life of a student. Use Rubric 25, Personal Letters, to evaluate student work.

Portfolio Activity

Answers will vary but should demonstrate the understanding that communities can be defined narrowly or broadly. Use Rubric 38, Writing to Classify, to evaluate student work. Call on students to share their sentences with the class.

internet connect

GO TO: go.hrw.com
KEYWORD: SK3 Teacher
FOR: a guide to using the Internet in your classroom

CHAPTER 2

Planet Earth

Chapter Resource Manager

Objectives	Pacing Guide	Reproducible Resources
SECTION 1 **The Sun, Earth, and Moon** (pp. 17–22) **1.** Name the objects that make up the solar system. **2.** Describe what causes the seasons. **3.** Identify the four parts of the Earth system.	**Regular** 1 day **Block Scheduling** .5 day *Block Scheduling Handbook, Chapter 2*	**RS** Guided Reading Strategy 2.1 **E** Environmental and Global Issues Activity 2 **SM** Geography for Life Activity 2 **IC** Interdisciplinary Activity for the Middle Grades 2 **E** Lab Activity for Geography and Earth Science, Demonstration 2
SECTION 2 **Water on Earth** (pp. 23–27) **1.** Identify the processes that make up the water cycle and how they are connected. **2.** Describe how water is distributed on Earth. **3.** Explain how water affects people's lives.	**Regular** 1 day **Block Scheduling** .5 day *Block Scheduling Handbook, Chapter 2*	**RS** Guided Reading Strategy 2.2 **E** Interdisciplinary Activity for the Middle Grades 3 **E** Lab Activity for Geography and Earth Science, Hands-On 5 **E** Lab Activity for Geography and Earth Science, Demonstration 1
SECTION 3 **The Land** (pp. 28–33) **1.** Describe primary landforms. **2.** Describe secondary landforms. **3.** Describe how humans interact with landforms.	**Regular** 1.5 days **Block Scheduling** .5 day *Block Scheduling Handbook, Chapter 2*	**RS** Guided Reading Strategy 2.3 **RS** Graphic Organizer 2 **E** Environmental and Global Issues Activity 3 **E** Lab Activities for Geography and Earth Science, Demonstrations 3, 4, 5, 6, 7, 9, and 10 **SM** Map Activity 2

Chapter Resource Key

RS Reading Support
IC Interdisciplinary Connections
E Enrichment
SM Skills Mastery
A Assessment
REV Review
ELL Reinforcement and English Language Learners
Transparencies
CD–ROM
Music
Video
go.hrw.com Internet
Holt Presentation Maker Using Microsoft® Powerpoint®

One-Stop Planner CD–ROM

See the ***One-Stop Planner*** for a complete list of additional resources for students and teachers.

One-Stop Planner CD–ROM

It's easy to plan lessons, select resources, and print out materials for your students when you use the ***One-Stop Planner CD–ROM with Test Generator.***

Technology Resources	Review, Reinforcement, and Assessment Resources
One-Stop Planner CD–ROM, Lesson 2.1 Geography and Cultures Visual Resources with Teaching Activity 1 Homework Practice Online HRW Go site	**ELL** Main Idea Activity 2.1 **REV** Section 1 Review, p. 22 **A** Daily Quiz 2.1 **ELL** English Audio Summary 2.1 **ELL** Spanish Audio Summary 2.1
One-Stop Planner CD–ROM, Lesson 2.2 Geography and Cultures Visual Resources with Teaching Activity 7 Earth: Forces and Formations CD–ROM/Seek and Tell/ Forces and Processes Earth: Forces and Formations CD–ROM/Seek and Tell/The Earth's Surface Our Environment CD–ROM/ Seek and Tell/Natural Resources Homework Practice Online HRW Go site	**ELL** Main Idea Activity 2.2 **REV** Section 2 Review, p. 27 **A** Daily Quiz 2.2 **ELL** English Audio Summary 2.2 **ELL** Spanish Audio Summary 2.2
One-Stop Planner CD–ROM, Lesson 2.3 Earth: Forces and Formations CD–ROM/Seek and Tell/Forces and Processes Earth: Forces and Formations CD–ROM/Seek and Tell/The Earth's Surface Homework Practice Online HRW Go site	**ELL** ELL Main Idea Activity 2.3 **REV** Section 3 Review, p. 33 **ELL** English Audio Summary 2.3 **ELL** Spanish Audio Summary 2.3 **A** Daily Quiz 2.3

internet connect

HRW ONLINE RESOURCES

GO TO: go.hrw.com
Then type in a keyword.

TEACHER HOME PAGE
KEYWORD: SK3 TEACHER

CHAPTER INTERNET ACTIVITIES
KEYWORD: SK3 GT2

Choose an activity to:
- learn more about Earth's seasons.
- discover facts about Earth's water.
- investigate earthquakes.

CHAPTER ENRICHMENT LINKS
KEYWORD: SK3 CH2

CHAPTER MAPS
KEYWORD: SK3 MAPS2

ONLINE ASSESSMENT
Homework Practice
KEYWORD: SK3 HP2
Standardized Test Prep Online
KEYWORD: SK3 STP2
Rubrics
KEYWORD: SS Rubrics

COUNTRY INFORMATION
KEYWORD: SK3 Almanac

CONTENT UPDATES
KEYWORD: SS Content Updates

HOLT PRESENTATION MAKER
KEYWORD: SK3 PPT2

ONLINE READING SUPPORT
KEYWORD: SS Strategies

CURRENT EVENTS
KEYWORD: S3 Current Events

Meeting Individual Needs

Ability Levels

Level 1 Basic-level activities designed for all students encountering new material

Level 2 Intermediate-level activities designed for average students

Level 3 Challenging activities designed for honors and gifted-and-talented students

English Language Learners Activities that address the needs of students with Limited English Proficiency

Chapter Review and Assessment

E Readings in World Geography, History, and Culture 3 and 4
SM Critical Thinking Activity 2
REV Chapter 2 Review, pp. 34–35
REV Chapter 2 Tutorial for Students, Parents, Mentors, and Peers
ELL Vocabulary Activity 2
A Chapter 2 Test
Chapter 2 Test Generator (on the One-Stop Planner)
Audio CD Program, Chapter 2
A Chapter 2 Test for English Language Learners and Special-Needs Students

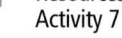

CHAPTER 2

Launch into Learning

Write *air, earth, fire,* and *water* on the chalkboard. Tell students that long ago, people thought everything was made up of these four elements. Ask students to identify different forms of these elements. Write their responses under the appropriate categories on the chalkboard. *(Examples: air—wind, tornadoes, ozone; earth—garden soil, landslides, mountains; fire—volcanoes, forest fires; water—rain, oceans, rivers, water from pipes and faucets)* Tell students that although we now know that air, earth, fire, and water are not elements, understanding their characteristics and relationships helps us understand geography and life on Earth.

Section 1

Objectives

1. Identify what objects make up the solar system.
2. Explain what causes the seasons.
3. Describe the four parts of the Earth system.

Links to Our Lives

These are among the reasons why students should take an interest in this chapter's topics:

- It is easier to understand physical processes here on Earth if we first understand our planet's relationship to the solar system.
- New discoveries about the solar system are made almost every day. We need to have background knowledge if we are to understand those findings.
- To protect our supplies of freshwater and clean air, we should know more about these precious resources.
- By learning more about how land is formed and changed, we can save lives threatened by earthquakes, volcanoes, and other hazards.

CHAPTER 2

Planet Earth

Erupting volcano

Earth as seen from space

Fossilized shell

Galileo's telescope

Space observatory, Mauna Kea, Hawaii

Focus

LET'S GET STARTED

Write the following scenario on the chalkboard: *Imagine that you were born and raised in a dark cave. When you finally come out of the cave, you see the night sky, with the Moon and stars shining. How would you explain these bright objects?* Ask students to respond to the scenario. Have volunteers share their responses. *(Students will probably say that the Moon and the stars are close to or attached to a ceiling.)* Then ask students what they already know about the relationships involving Earth, the Sun, the Moon, the planets, and the stars. Tell students that in Section 1 they will learn more about the solar system.

Building Vocabulary

Write the key terms on the chalkboard in random order. Have students group the terms by using information they already know or from clues in the words themselves. *(Students are likely to group terms relating to the* ***solar system****, nouns that relate to Earth, and terms that have a common suffix.)* Ask students to find the definitions of the terms in Section 1 or the glossary. Then have them write sentences using the terms and explaining the connections between them.

Section 1 The Sun, Earth, and Moon

Read to Discover

1. What objects make up the solar system?
2. What causes the seasons?
3. What are the four parts of the Earth system?

Define

solar system
orbit
satellite
axis
rotation
revolution
Arctic Circle
Antarctic Circle
solstice
Tropic of Cancer
Tropic of Capricorn
equinoxes
atmosphere
ozone

Why It Matters

In 2001 scientists labeled a rocky object beyond Pluto as the new largest minor planet. Use CNNfyi.com or other **current events** sources to discover more about this huge frozen rock, called 2001 KX76. Record your findings in your journal.

Mechanical model of the solar system

The Solar System

The **solar system** consists of the Sun and the objects that move around it. The most important of those objects are the planets, their moons, and relatively small rocky bodies called asteroids. Our Sun is a star at the center of our solar system. Every object in the system travels around the Sun in an **orbit**, or path. These orbits are usually elliptical, or oval shaped.

The Solar System

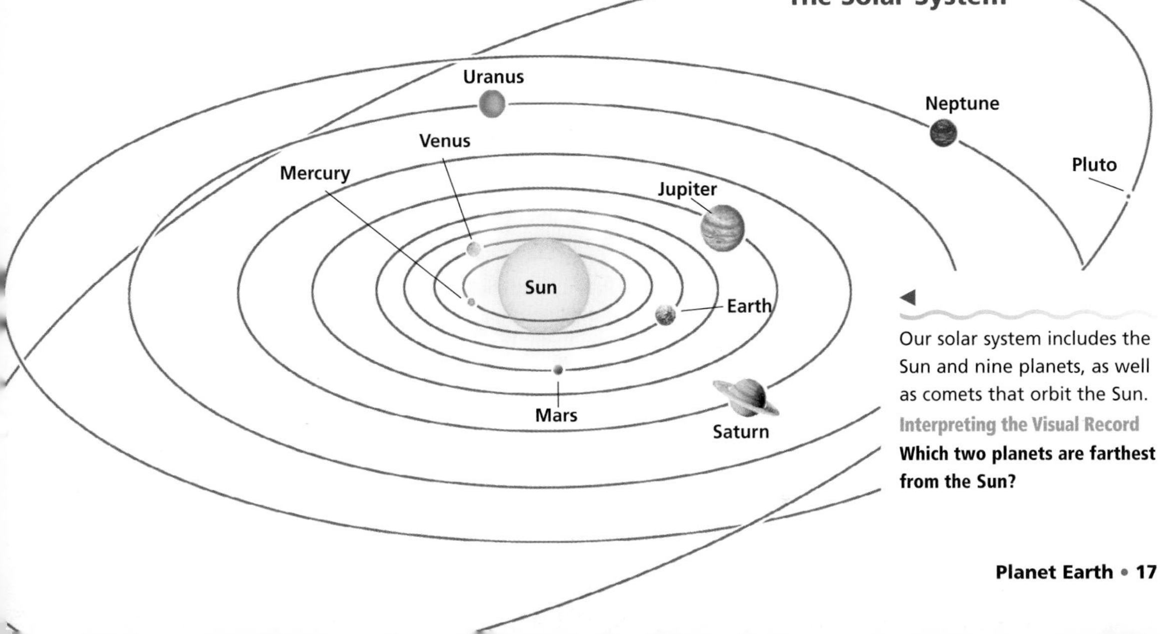

Our solar system includes the Sun and nine planets, as well as comets that orbit the Sun.

Interpreting the Visual Record
Which two planets are farthest from the Sun?

Section 1 Resources

Reproducible

- Block Scheduling Handbook, Ch. 2
- Guided Reading Strategy 2.1
- Environmental and Global Issues Activity 2
- Geography for Life Activity 2
- Interdisciplinary Activity for the Middle Grades 2
- Lab Activity for Geography and Earth Science, Demonstration 2

Technology

- One-Stop Planner CD–ROM, Lesson 2.1
- Homework Practice Online
- Geography and Cultures Visual Resources with Teaching Activity 1
- HRW Go site

Reinforcement, Review, and Assessment

- Section 1 Review, p. 22
- Daily Quiz 2.1
- Main Idea Activity 2.1
- English Audio Summary 2.1
- Spanish Audio Summary 2.1

Visual Record Answer

Neptune and Pluto

TEACH

Teaching Objective 1

LEVEL 1: (Suggested time: 15 min.) Organize students into groups. Have each group prepare a set of paper circles and label them with the names of the nine planets, the Sun, and Earth's moon. Have students arrange the elements of the solar system in the correct order on a desktop. Instruct groups to also create their own depiction of the asteroid belt and locate it correctly. **ENGLISH LANGUAGE LEARNERS, COOPERATIVE LEARNING**

LEVEL 2: (Suggested time: 40 min.) To show the relationships within our solar system, assign students to represent the Sun and each of the planets. Other students could represent major moons, including Earth's, or asteroids between Mars and Jupiter. Each student should wear a label identifying what he or she represents. Then take your class to the school's sports field for a simulation. Have the student representing the Sun stand at the goal line. Then ask students representing the planets to stand the correct distance from the "Sun," according to the chart on the following page. Later, lead a discussion on what students learned from the demonstration. *(Possible answer: The distances between outer planets are greater than between inner planets.)* **COOPERATIVE LEARNING**

National Geography Standard 15

Environment and Society

During a solar storm called a coronal mass ejection (CME), the Sun spews up to 10 billion tons (9 billion metric tons) of hot, electrically charged gas into space. The gas cloud travels at speeds reaching 1,250 miles per second (2,000 kilometers per second). If a really strong CME hits Earth, it can damage satellites, interrupt communication systems, and cause power blackouts. For example, in 1989 a CME left about 6 million people in Canada's Quebec province without electricity.

Technicians can prevent potential damage if they know when a CME is about to occur. Scientists know that periodic shifts in the Sun's magnetic field cause these ejections. They have also found that a gigantic "S" shape, seen in X-ray images taken of the Sun's surface, seems to indicate that the Sun may launch a CME within days. The "S" shapes, dubbed sigmoids, are areas about 50,000 miles (80,000 km) across, where the Sun's magnetic field has twisted back on itself.

Critical Thinking: How can scientists' knowledge about CMEs potentially affect the world?

Answer: By predicting CMEs, scientists can help people prepare for the consequences.

Our Amazing Planet

Mexico's Yucatán Peninsula has a crater that stretches 200 miles (322 km) wide. Scientists believe that it is the site where a giant asteroid crashed into Earth some 65 million years ago. They think the collision caused more than half of all species, including dinosaurs, to become extinct.

The planet nearest the Sun is Mercury, followed by Venus, Earth, and Mars. Located beyond the orbit of Mars is a belt of asteroids. Beyond this asteroid belt are the planets Jupiter and Saturn. Even farther from the Sun are the planets Uranus, Neptune, and Pluto.

The Moon Some of the planets in the solar system have more than one moon. Saturn, for example, has 18. Other planets have none. A moon is a **satellite**—a body that orbits a larger body. Earth has one moon, which is about one fourth the size of Earth. Our planet is also circled by artificial satellites that transmit signals for television, telephone, and computer communications. The Moon takes about 29½ days—roughly a month—to orbit Earth.

The Moon and Sun influence physical processes on Earth. This is because any two objects in space are affected by gravitational forces pulling them together. The gravitational effects of the Sun and the Moon cause tides in the oceans here on Earth.

The Sun Compared to some other stars, our Sun is small. It is huge, however, when compared to Earth. Its diameter is about 100 times the diameter of our planet. The Sun appears larger to us than other stars. This is because it is much closer to us than other stars. The Sun is about 93 million miles (150 million km) from Earth. The next nearest star is about 25 trillion miles (40 trillion km) away.

Scientists are trying to learn if other planets in our solar system could support life. Mars seems to offer the best possibility. It is not clear, however, if life can, or ever did, exist on Mars.

✓ **READING CHECK:** *Physical Systems* What are the main objects that make up the solar system? the Sun, the planets and their moons, asteroids

Tides are higher than normal when the gravitational pull of the Moon and the Sun combine. These tides, called spring tides, occur twice a month. Tides are lower than normal during neap tides, when the Sun and the Moon are at right angles.

Effects of the Moon and Sun on Tides

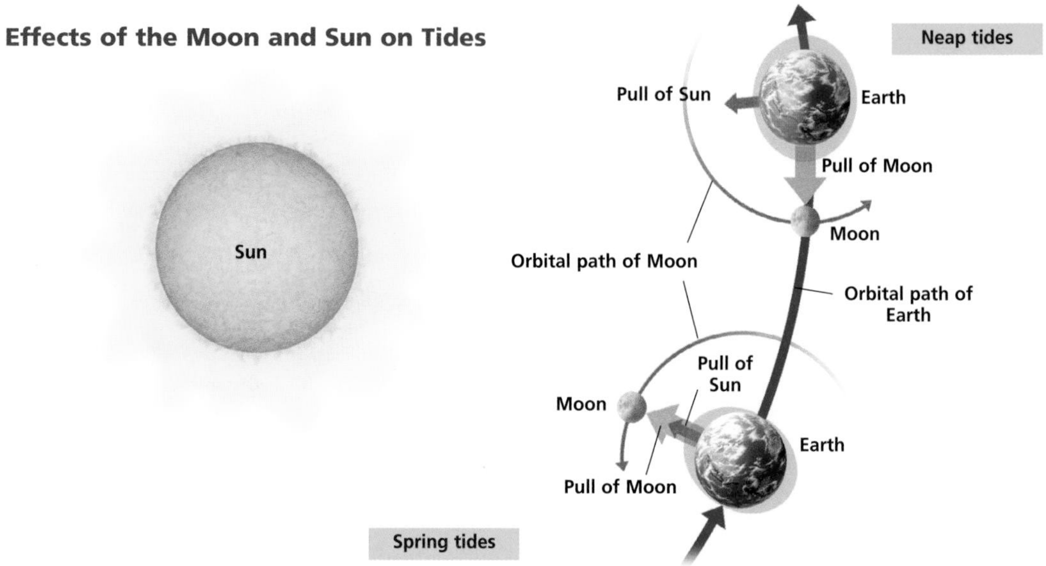

Planet	Distance from the Sun in AU*	Scaled distance in yards
Mercury	0.39	1
Venus	0.72	1.8
Earth	1	2.5
Mars	1.52	4
Jupiter	5.20	13
Saturn	9.54	24
Uranus	19.19	49
Neptune	30.06	76
Pluto	39.53	100

* AU stands for an *astronomical unit*, the average distance between Earth and the Sun.

LEVEL 3: Organize students into groups. Assign one planet to each group and have students conduct research on their planet's orbit. Students should find out how fast the planet travels and its distance from the Sun and the other planets as it travels. For example, Pluto's orbit swings inside Neptune's for part of its year. Have students use the information from their research to create a report. Make sure they create a bibliography with their report. Then have them set in motion the solar system from the Level 2 activity. Later, have each group present a brief summary of its research. **COOPERATIVE LEARNING**

Earth

Geographers are interested in how different places on Earth receive different amounts of energy from the Sun. Differences in solar energy help explain why the tropics are warm, why the Arctic region is cold, and why day is warmer than night. To understand these differences, geographers study Earth's rotation, revolution, and the tilt of its **axis**. The axis is an imaginary line that runs from the North Pole through Earth's center to the South Pole. Rotation, revolution, and tilt control the amount of solar energy reaching Earth.

Rotation One complete spin of Earth on its axis is called a **rotation**. Each rotation takes 24 hours, or one day. Earth turns on its axis, but to us it appears that the Sun is moving. The Sun seems to "rise" in the east and "set" in the west. Before scientists learned that Earth revolves around the Sun, people thought that the Sun revolved around Earth. They thought Earth was at the center of the heavens.

Revolution It takes a year for Earth to orbit the Sun, or to complete one **revolution**. More precisely, it takes 365¼ days. To allow for this fraction of a day and keep the calendar accurate, every fourth year becomes a leap year. An extra day—February 29—is added to the calendar.

Tilt The amount of the Sun's energy reaching different parts of Earth varies. This is because Earth's axis is not straight up and down. It is actually tilted, or slanted, at an angle of 23.5° from vertical to the plane of Earth's orbit. Because of Earth's tilt, the angle at which the Sun's rays strike the planet is constantly changing as Earth revolves around the Sun. For this reason, the point where the vertical rays of

▲ Photographs taken from space can tell us about Earth.

Interpreting the Visual Record Where can you see the presence of water in this view of Earth?

internet connect
GO TO: go.hrw.com
KEYWORD: SK3 CH2
FOR: Web sites about planet Earth

Angle of Sun's Rays Hitting Earth

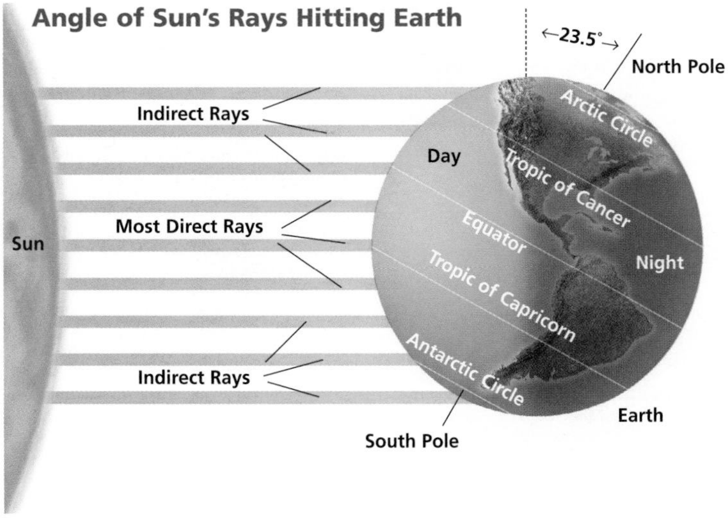

◄ The tilt of Earth's axis and the position of the planet in its orbit determine where the Sun's rays will most directly strike the planet.

Interpreting the Visual Record Which areas of Earth receive only indirect rays from the Sun?

DAILY LIFE

Near the North and South Poles bands of light called the aurora can be seen in the night sky. These great flashes and streaks of color are called the Aurora Borealis, or the Northern Lights, in the Northern Hemisphere. In the Southern Hemisphere they are the Aurora Australis, or Southern Lights. Auroras take place when the electrically charged particles—electrons and protons—of the solar wind strike gas molecules in the upper atmosphere. Different gases create different colors. At high altitudes oxygen produces red. At lower altitudes it produces a yellow-green glow. Nitrogen glows blue, red, and purple. The lights are seen most often near the Poles because the magnetic fields there attract the charged particles.

Discussion: What physical phenomena or events are part of daily life in your area? *(Possible answers: storms, droughts, and high tides)* How do residents respond to these events?

▲ **Visual Record Answers**

clouds, lakes, and oceans

◄ most areas except those closest to the equator

Teaching Objective 2

LEVELS 1 AND 2: (Suggested time: 15 min.) To help students understand how the seasons relate to the Sun's energy, ask two volunteers to perform this demonstration. Have one student act as Earth and give him or her a globe. Have the other student act as the Sun and give him or her a flashlight. Turn off the lights and have the volunteers sit on the floor. Ask the "Sun" to shine the flashlight on the globe, and ask "Earth" to slowly spin the globe. Have students notice which parts of the globe are most exposed to the Sun. Then have "Earth" make a complete revolution around the "Sun" while at the same time rotating the globe. As "Earth" revolves, have students identify the seasons in various parts of the world.

LEVEL 3: (Suggested time: 30 min.) On slips of paper write the names of cities around the world and various dates. Have each student select a city and find its location in the textbook's Atlas. Tell students to draw on a separate sheet of paper a diagram of Earth's position in relation to the Sun on that date. Then ask students to mark the location of their city and to write a sentence or two describing the season there on the assigned date.

COOPERATIVE LEARNING

Organize the class into small groups. Tell students that different cultures celebrate the beginning of the new year at different times. The starting point of a new year may be determined by seasonal occurrences such as the winter solstice or by a religious event. Ask each group to research the topic and prepare a presentation on the new year's observances of a particular culture or region. Tell them to be sure to include information on the historical origins of the observance.

Visual Record Answer

June solstice

the Sun strike Earth shifts north and south of the equator. These vertical rays provide more energy than rays that strike at an angle.

✓ **READING CHECK:** *Physical Systems* How do rotation, revolution, and tilt affect solar energy reaching Earth? They control the amount of solar energy because they determine the position of the Earth in relation to the Sun.

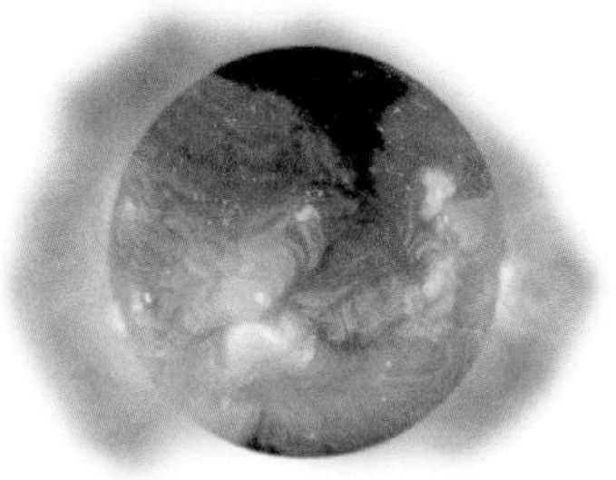

▲ The Sun's surface is always violently churning as heat flows outward from the interior.

Solar Energy and Latitude

The angle at which the Sun's rays reach Earth affects temperature. In the tropics—areas in the low latitudes near the equator—the Sun's rays are nearly vertical throughout the year. In the polar regions—the areas near the North and South Poles—the Sun's rays are always at a low angle. As a result, the poles are generally the coldest places on Earth. The **Arctic Circle** is the line of latitude located 66.5° north of the equator. It circles the North Pole. The **Antarctic Circle** is the line of latitude located 66.5° south of the equator. It circles the South Pole.

The Seasons

Each year is divided into periods of time called seasons. Each season is known for a certain type of weather, based on temperature and amount of precipitation. Winter, spring, summer, and fall are examples of seasons that are described by their average temperature. "Wet" and "dry" seasons are described by their precipitation. The seasons change as Earth orbits the Sun. As this happens, the amount of solar energy received in any given location changes.

As Earth revolves around the Sun, the tilt of the poles toward and away from the Sun causes the seasons to change.

Interpreting the Visual Record **At what point is the North Pole tilted toward the Sun?** ▼

The Seasons

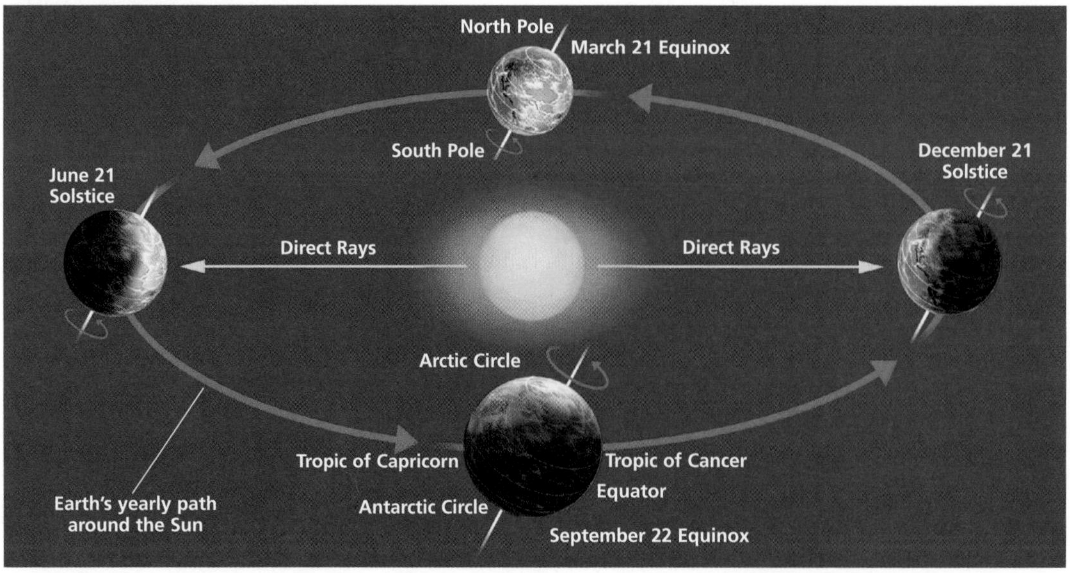

Teaching Objective 3

ALL LEVELS: (Suggested time: 20 min.) Copy the following graphic organizer onto the chalkboard, omitting the italicized answers. Use it to help students describe the four parts of the Earth system.

ENGLISH LANGUAGE LEARNERS

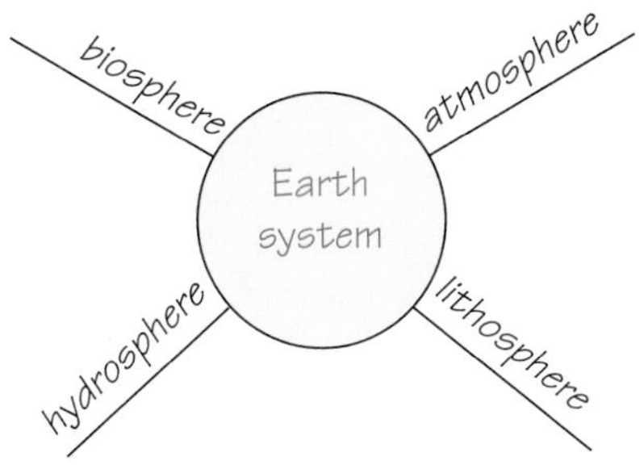

CLOSE

Call on students to suggest recent movies or television programs that depict the solar system in some way. Ask them to compare the portrayals of the scientific topics with what they have learned. Then ask them to discuss the accuracy of the filmed versions.

REVIEW AND ASSESS

Have students complete the Section Review. Then have them imagine that they are astronauts traveling through the solar system. Ask each student to write an entry in the ship's log that makes observations related to the key terms. Call on volunteers to read their entries to the class. Then have students complete Daily Quiz 2.1.

Solstice The day when the Sun's vertical rays are farthest from the equator is called a **solstice**. Solstices occur twice a year—about June 21 and about December 22. In the Northern Hemisphere the June solstice is known as the summer solstice. This is the longest day of the year and the beginning of summer. On this date the Sun's vertical rays strike Earth at the **Tropic of Cancer**. This is the line of latitude that is about 23.5° north of the equator. Six months later, about December 22, another solstice takes place. This is the winter solstice for the Northern Hemisphere. On this date the North Pole is pointed away from the Sun. The Northern Hemisphere experiences the shortest day of the year. On this date the Sun's rays strike Earth most directly at the **Tropic of Capricorn**. This line of latitude is about 23.5° south of the equator. In the Southern Hemisphere the seasons are reversed. June 21 is the winter solstice and December 22 is the summer solstice. The middle-latitude regions lie between the Tropic of Cancer and the Arctic Circle and between the Tropic of Capricorn and the Antarctic Circle.

Equinox Twice a year, halfway between summer and winter, Earth's poles are at right angles to the Sun. The Sun's rays strike the equator directly. On these days, called **equinoxes**, every place on Earth has 12 hours of day and 12 hours of night. Equinoxes mark the beginning of spring and fall. In the Northern Hemisphere the spring equinox occurs about March 21. The fall equinox occurs there about September 22. In the Southern Hemisphere the March equinox signals the beginning of fall, and the September equinox marks the beginning of spring.

Some regions on Earth, particularly in the tropics, have seasons tied to precipitation rather than temperature. Shifting wind patterns are one cause of seasonal change. For example, in January winds from the north bring dry air to India. By June the winds have shifted, coming from the southwest and bringing moisture from the Indian Ocean. These winds bring heavy rain to India. Some places in the United States also have seasons tied to moisture. East Coast states south of Virginia have a wet season in summer. These areas also have a hurricane season, lasting roughly from June to November. Some areas of the West Coast have a dry season in summer.

The seasons affect human activities. For example, in Minnesota, people shovel snow in winter to keep the walkways clear. Students waiting for a bus must wear warm clothes. The Sun rises late and sets early. As a result, people go to work and return home in darkness.

READING CHECK: *Physical Systems* How do the seasons relate to the Sun's energy? **They are determined by the position of the Earth in its orbit around the Sun and how much solar energy is being received.**

This map shows Earth's temperatures on January 28, 1997.

Interpreting the Visual Record Which hemisphere is warmer in January?

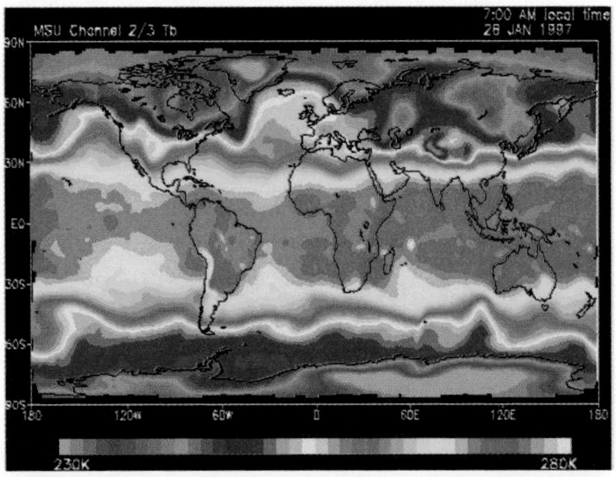

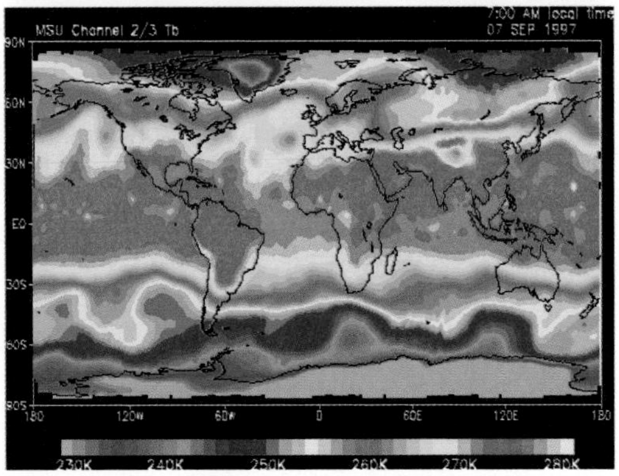

This temperature map for September 7, 1997, shows summer in the Northern Hemisphere.

Across the Curriculum

SCIENCE

Life in the Hydrosphere

Deep in Earth's oceans, scientists have found organisms, such as bacteria and tube worms, that thrive in total darkness in superheated mineral-rich water. That organisms survive under these extreme conditions supports the theory that simple forms of life may exist on Europa, one of Jupiter's moons. Europa is covered by ice several miles thick. Images from NASA's *Galileo* spacecraft indicate that liquid water may have existed under the icy crust. This "ocean" may still exist, but scientists have no definitive proof that life-forms exist on Europa.

Activity: Have students conduct research on deep-sea vents on Earth called black smokers. Call on students to suggest plots for science-fiction stories based on their findings.

internet connect

GO TO: go.hrw.com
KEYWORD: SK3 CH2
FOR: Web sites about Europa

Visual Record Answer

the Southern Hemisphere

RETEACH

Have students complete Main Idea Activity 2.1. Then organize students into three groups to create mobiles. One group's mobile should depict the objects of the solar system, the second the cause of seasons, and the third the relationships of the Earth system. **ENGLISH LANGUAGE LEARNERS**

EXTEND

Organize interested students into groups to research the discoveries of major figures in the history of astronomy, such as Nicolaus Copernicus, Galileo Galilei, Johannes Kepler, or Isaac Newton. Then have the groups work together to write a script for an imaginary conference call in which the scientists discuss their work. Ask the students to perform their scripts for the class. **BLOCK SCHEDULING**

Section 1 Review

Answers

Define For definitions, see: solar system, p. 17; orbit, p. 17; satellite, p. 18; axis, p. 19; rotation, p. 19; revolution, p. 19; Arctic Circle, p. 20; Antarctic Circle, p. 20; solstice, p. 21; Tropic of Cancer, p. 21; Tropic of Capricorn, p. 21; equinoxes, p. 21; atmosphere, p. 22; ozone, p. 22

Reading for the Main Idea

1. the Sun, the planets, their moons, and asteroids (NGS 7)
2. atmosphere, lithosphere, hydrosphere, biosphere (NGS 3)

Critical Thinking

3. rotation, revolution, and tilt
4. Students might suggest that the seasons are reversed because the tilt of Earth's axis and Earth's rotation around the Sun affect the amount of solar energy received.

Organizing What You Know

5. Answers will vary but should include that a solstice occurs when the Sun's vertical rays are farthest from the equator and that an equinox occurs when the Sun's rays strike the equator directly.

Visual Record Answer ►

the lake, rain, and snow

The Earth System

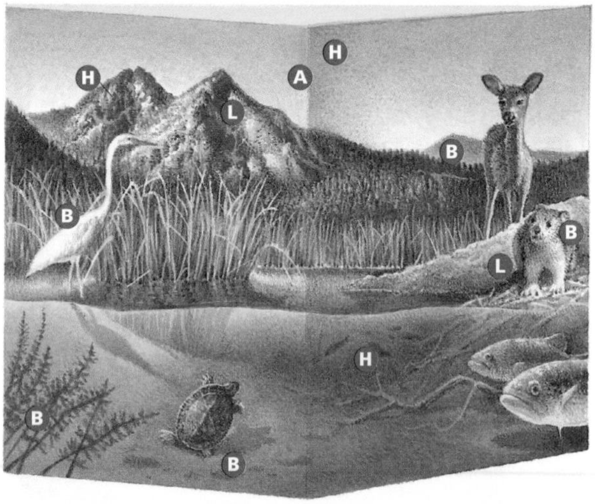

Ⓐ Atmosphere Ⓑ Biosphere
Ⓛ Lithosphere Ⓗ Hydrosphere

▲ The interactions of the atmosphere, lithosphere, hydrosphere, and biosphere make up the Earth system.

Interpreting the Visual Record **Which items in this image are part of the hydrosphere?**

The Earth System

Geographers need to be able to explain how and why places on Earth differ from each other. One way they do this is to study the interactions of forces and materials on the planet. Together, these forces and materials are known as the Earth system.

The Earth system has four parts: the **atmosphere**, the lithosphere, the hydrosphere, and the biosphere. The atmosphere is the layer of gases—the air—that surrounds Earth. These gases include nitrogen, oxygen, and carbon dioxide. The atmosphere also contains a form of oxygen called **ozone**. A layer of this gas helps protect Earth from harmful solar radiation. Another part of the Earth system is the lithosphere. The prefix *litho* means rock. The lithosphere is the solid, rocky outer layer of Earth, including the sea floor. The hydrosphere—*hydro* means water—consists of all of Earth's water, found in lakes, oceans, and glaciers. It also includes the moisture in the atmosphere. Finally, the biosphere—*bio* means life—is the part of the Earth system that includes all plant and animal life. It extends from high in the air to deep in the oceans.

By dividing Earth into these four spheres, geographers can better understand each part and how each affects the others. The different parts of the Earth system are constantly interacting in many ways. For example, a tree is part of the biosphere. However, to grow it needs to take in water, chemicals from the soil, and gases from the air.

✓ **READING CHECK:** *The World in Spatial Terms* What are the four parts of the Earth system? atmosphere, lithosphere, hydrosphere, biosphere

go.hrw.com Homework Practice Online
Keyword: SK3 HP2

Section 1 Review

Define and explain: solar system, orbit, satellite, axis, rotation, revolution, Arctic Circle, Antarctic Circle, solstice, Tropic of Cancer, Tropic of Capricorn, equinoxes, atmosphere, ozone

Reading for the Main Idea

1. *Physical Systems* What are the major objects in the solar system?
2. *The World in Spatial Terms* What are the four parts of the Earth system?

Critical Thinking

3. **Summarizing** Which three things determine the amount of solar energy reaching places on Earth?
4. **Drawing Inferences and Conclusions** Why are the seasons reversed in the Northern and Southern Hemispheres?

Organizing What You Know

5. **Finding the Main Idea** Use this graphic organizer to explain solstice and equinox.

Solstice	⇔	Equinox

Section 2

Objectives

1. Examine the processes that make up the water cycle and how they are connected.
2. Identify how water is distributed on Earth.
3. Explain how water affects people's lives.

Focus

LET'S GET STARTED

Write the following instructions on the chalkboard: *Draw as many quick sketches as you can, in two minutes, of the ways you use water throughout the day.* Discuss completed drawings as a class. Then ask students to write a summary sentence beneath their drawings. *(Possible answers: We can't live without water. We use water every day in many ways.)* Display students' sketches around the classroom. Tell students that in Section 2 they will learn more about the role of water on Earth.

Building Vocabulary

Explain that adding *-tion* to a root verb usually turns it into a noun. Point out the verbs from which the terms *evaporation, condensation*, and *precipitation* are formed *(evaporate, condense, precipitate).* Have students use these key terms in sentences relating the terms to bodies of water in their region. Have students use the U.S. map in the textbook's Atlas for help. *(Example: Precipitation that falls in Illinois may end up in the Mississippi River.)*

Section 2 Water on Earth

Read to Discover

1. Which processes make up the water cycle and how are they connected?
2. How is water distributed on Earth?
3. How does water affect people's lives?

Define

water vapor
water cycle
evaporation
condensation
precipitation
tributary
groundwater
continental shelf

Why It Matters

Scientists study other parts of our solar system to find out if water exists or might have existed elsewhere. Use CNNfyi.com or other **current events** sources to learn more about space agencies and their searches to detect water. Record your findings in your journal.

A limestone cavern

Characteristics of Water

Water has certain physical characteristics that influence Earth's geography. Water is the only substance on Earth that occurs naturally as a solid, a liquid, and a gas. We see water as a solid in snow and ice and as a liquid in lakes, oceans, and rivers. Water also occurs in the air as an invisible gas called **water vapor**.

Another characteristic of water is that it heats and cools slowly compared to land. Even on a very hot day, the ocean stays cool. A breeze blowing over the ocean brings cooler temperatures to shore. This keeps temperatures near the coast from getting as hot as they do farther inland. In winter the oceans cool more slowly than land. This generally keeps winters milder in coastal areas.

✓ **READING CHECK:** *Physical Systems* What are some important characteristics of water? occurs naturally as a solid, liquid, and gas; heats and cools slowly

The Water Cycle

The circulation of water from Earth's surface to the atmosphere and back is called the **water cycle**. The total amount of water on the planet does not change. Water, however, does change its form and its location.

Water rushes through the Stewart Mountain Dam in Arizona.

Section 2 Resources

Reproducible

- Guided Reading Strategy 2.2
- Interdisciplinary Activity for the Middle Grades 3
- Lab Activity for Geography and Earth Science, Hands-On 5
- Lab Activity for Geography and Earth Science, Demonstration 1

Technology

- One-Stop Planner CD–ROM, Lesson 2.2
- Homework Practice Online
- Geography and Cultures Visual Resources with Teaching Activity 7
- Earth: Forces and Formations, CD–ROM/Seek and Tell/Forces and Processes
- Earth: Forces and Formations, CD–ROM/Seek and Tell/The Earth's Surface
- Our Environment CD–ROM/Seek and Tell/Natural Resources
- HRW Go site

Reinforcement, Review, and Assessment

- Section 2 Review, p. 27
- Daily Quiz 2.2
- Main Idea Activity 2.2
- English Audio Summary 2.2
- Spanish Audio Summary 2.2

TEACH

Teaching Objective 1

ALL LEVELS: (Suggested time: 20 min.) Pair students. Then, using the diagram of the water cycle as a guide, have each pair create a graphic organizer showing the role of evaporation, condensation, and precipitation in the cycle. **ENGLISH LANGUAGE LEARNERS, COOPERATIVE LEARNING**

Teaching Objective 2

LEVEL 1: (Suggested time: 15 min.) Pair students and tell each pair to create flash cards for the key terms that relate to how water is distributed on Earth. Students should write an appropriate key term on one side of a flash card and its definition on the reverse. Ask volunteers to share their flash cards with the class. **ENGLISH LANGUAGE LEARNERS**

LEVELS 2 AND 3: (Suggested time: 20 min.) Give each student a map of a river that flows through your region. Be sure that the map shows the river's full course. Have students label the river's headwaters, tributaries, and any reservoirs that have been created. To conclude, lead a discussion on how water is distributed on Earth.

PHYSICAL SYSTEMS

Each year as many as 40,000 icebergs break off from the glaciers of Greenland. One of Greenland's largest reported icebergs towered 550 feet (168 m) above the Atlantic Ocean's surface. In 1987 a huge tablelike berg broke away from an Antarctic ice sheet. This iceberg stretched some 100 miles (161 km). At high latitudes these gigantic icebergs can last as long as 10 years before melting.

Could we use the freshwater locked in icebergs? Experts have tried to design ways to tow icebergs to arid regions, but expense and other problems have prevented their success.

Discussion: Have students determine why towing icebergs has not been successful. Ask them to gather information about the subject and then brainstorm possible solutions to these obstacles. Ask them what the advantages and disadvantages of the solutions might be. Then have them choose a solution and demonstrate it with tubs of water and blocks of ice.

Visual Record Answer

It might increase evaporation.

The Water Cycle

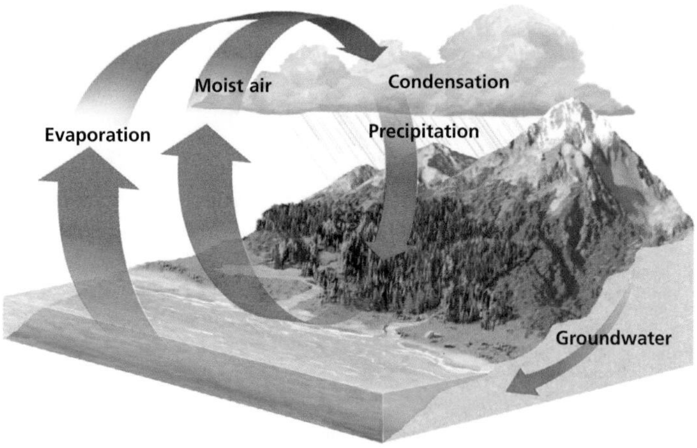

The circulation of water from one part of the hydrosphere to another depends on energy from the Sun. Water evaporates, condenses, and falls to Earth as precipitation.

Interpreting the Visual Record **How would a seasonal increase in the amount of the Sun's energy received by an area change the water cycle in that area?**

The Sun's energy drives the water cycle. **Evaporation** occurs when the Sun heats water on Earth's surface. The heated water evaporates, becoming water vapor and rising into the air. Energy from the Sun also causes winds to carry the water vapor to new locations. As the water vapor rises, it cools, causing **condensation**. This is the process by which water changes from a gas into tiny liquid droplets. These droplets join together to form clouds. If the droplets become heavy enough, **precipitation** occurs—that is, the water falls back to Earth. This water can be in the form of rain, hail, sleet, or snow. The entire cycle of evaporation, condensation, and precipitation repeats itself endlessly.

✓ **READING CHECK:** *Physical Systems* What is the water cycle? the circulation of water from Earth's surface to the atmosphere and back

This glacier is "calving"—a mass of ice is breaking off, forming an iceberg. Most of Earth's freshwater exists as ice.

Geographic Distribution of Water

The oceans contain about 97 percent of Earth's water. About another 2 percent is found in the ice sheets of Antarctica, the Arctic, Greenland, and mountain glaciers. Approximately 1 percent is found in lakes, streams, rivers, and under the ground.

Earth's freshwater resources are not evenly distributed. There are very dry places with no signs of water. Other places have many lakes and rivers. In the United States, for example, Minnesota is dotted with more than 11,000 lakes. Dry states such as Nevada have few natural lakes. In many dry places rivers have been dammed to create artificial lakes called reservoirs.

Teaching Objective 3

LEVEL 1: (Suggested time: 20 min.) Copy the following graphic organizer onto the chalkboard, omitting the italicized answers. Use it to help students explore how water affects people's lives. Have students copy the organizer into their notebooks and complete it.

ENGLISH LANGUAGE LEARNERS

WATER ISSUES

1. *floods*
2. *flood control/dams*
3. *availability of clean water*
4. *availability of adequate water supply*

Surface Water Water sometimes collects at high elevations, where rivers begin their flow down toward the lowlands and coasts. The first and smallest streams that form from this runoff are called headwaters. When these headwaters meet and join, they form larger streams. In turn, these streams join with other streams to form rivers. Any smaller stream or river that flows into a larger stream or river is a **tributary**. For example, the Missouri River is an important tributary of the Mississippi River, into which it flows near St. Louis, Missouri.

Lakes are usually formed when rivers flow into basins and fill them with water. Most lakes are freshwater, but some are salty. For example, the Great Salt Lake in Utah receives water from the Bear, Jordan, and Weber Rivers but has no outlet. Because the air is dry here, the rate of evaporation is very high. When the lake water evaporates, it leaves behind salts and minerals, making the lake salty.

Groundwater

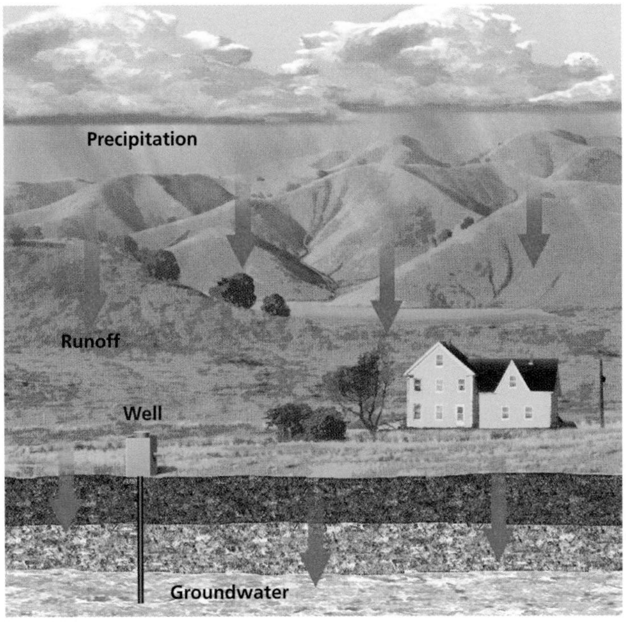

▲ In some areas where rainfall is scarce, enough groundwater exists to support agriculture.

Interpreting the Visual Record **How do people gain access to groundwater?**

Groundwater Not all surface water immediately returns to the atmosphere through evaporation. Some water from rainfall, rivers, lakes, and melting snow seeps into the ground. This **groundwater** seeps down until all the spaces between soil and grains of rock are filled. In some places, groundwater bubbles out of the ground as a spring. Many towns in the United States get their water from wells—deep holes dug down to reach the groundwater. Motorized pumps allow people to draw water from very deep underground.

Oceans Most of Earth's water is found in the oceans. The Pacific, Atlantic, Indian, and Arctic Oceans connect with each other. This giant body of water covers some 71 percent of Earth's surface. These oceans also include smaller regions called seas and gulfs. The Gulf of Mexico and the Gulf of Alaska are two examples of smaller ocean areas.

Surrounding each continent is a zone of shallow ocean water. This gently sloping underwater land—called the **continental shelf**—is important to marine life. Although the oceans are huge, marine life is concentrated in these shallow areas. Deeper ocean water is home to fewer organisms. Overall, the oceans average about 12,000 feet (about 3,700 m) in depth. The deepest place is the Mariana Trench in the Pacific Ocean, at about 36,000 feet (about 11,000 m) deep.

✓ **READING CHECK:** *The World in Spatial Terms* How is water distributed on Earth? in the oceans; ice sheets of Antarctica, the Arctic, and Greenland; and in lakes, streams, rivers, and groundwater

The Continental Shelf

▲ The continental shelf slopes gently away from the continents. The ocean floor drops steeply at the edge of the shelf.

EYE ON EARTH

We think of ocean waves as racing across miles of open water within hours. However, Rossby waves can take years to travel across the ocean. These large-scale waves may be only a few inches high, but they can be many miles long. Because they move slowly, Rossby waves can carry a "memory" of storms or other oceanic events that occurred years earlier. For example, oceanographers mapped a Rossby wave in 1994 that seemed to show evidence of a 1982–83 El Niño. Understanding Rossby waves helps meteorologists predict the weather because the waves may push powerful ocean currents away from their usual paths.

Activity: Have students conduct research on connections between ocean currents and weather prediction. Have students create graphic organizers of their findings.

▲ **Visual Record Answer**

by digging wells

LEVELS 2 and 3: (Suggested time: 10 min.) Give students a topographic map of their community or area. Ask them to identify and shade low-lying areas that might be prone to flooding.

➤ASSIGNMENT: Have students use their sketches from Let's Get Started to recall times they use water throughout the day. Have them write a list of those uses. Then have students propose how they would cope if the water supply were suddenly unavailable or polluted.

CLOSE

Tell students that in 1991, during the Persian Gulf War, a massive oil spill in the Persian Gulf threatened desalinization factories in the region. These plants remove salt from seawater and supply freshwater to countries in the region. Ask students to suggest how other natural and human-made disasters might affect water supplies.

National Geography Standard 8

Physical Systems Wetlands are vital to Earth's water supply. They are areas where the soil absorbs a great amount of water. Depending on the soil and plant life found in them, wetlands can also be called tidal flats, swamps, or bogs.

Wetlands are some of the most productive environmental systems on Earth. By absorbing a lot of moisture, they help prevent floods. The plants that grow in wetlands can remove pollution from sewage. Wetlands also serve as nurseries for fish, shrimp, and shellfish. About one third of the rare and endangered animals in the United States live in wetlands. Migrating birds depend on wetlands for rest and food. Coastal wetlands can help prevent erosion.

By the mid-1970s these vital elements in the biosphere were threatened. The continental United States had lost more than 50 percent of its original wetlands. Wetlands remain threatened worldwide.

Activity: Have students design a public-awareness campaign to focus attention on the importance of preserving the wetlands.

Visual Record Answer ▶

Answers will vary, but students might mention access to transportation, trade, and a water supply as reasons.

▲ This kelp forest is off the coast of southern California. The shallower parts of the oceans are home to many plants and animals.

Water Issues

Water plays an important part in our survival. As a result, water issues frequently show up in the news. Thunderstorms, particularly when accompanied by hail or tornadoes, can damage buildings and ruin crops. Droughts also can be deadly. In the mountains, heavy snowfalls sometimes cause deadly snow slides called avalanches. Heavy fog can make driving or flying dangerous. Geographers are concerned with these issues. They work on ways to better prepare for natural hazards.

Floods Water can both support and threaten life. Heavy rains can cause floods, which are the world's deadliest natural hazard. Floods kill four out of every ten people who die from natural disasters, including hurricanes, earthquakes, tornadoes, and thunderstorms.

Some floods occur in dry places when strong thunderstorms drop a large amount of rain very quickly. The water races along on the hard, dry surface instead of soaking into the ground. This water can quickly gather in low places. Creekbeds that are normally dry can suddenly surge with rushing water. People and livestock are sometimes caught in these flash floods.

Floods also happen in low-lying places next to rivers and on coastlines. Too much rain or snowmelt entering a river can cause the water to overflow the banks. Powerful storms, particularly hurricanes, can sometimes cause ocean waters to surge into coastal areas. Look at a

Davenport, Iowa, suffered severe flooding in 1993. Many cities are located along rivers even though there is a danger of floods.

Interpreting the Visual Record **Why are many cities located next to rivers?** ▶

REVIEW AND ASSESS

Have students complete the Section Review. Then prepare a set of flash cards, each with a key term or definition on it. Hold up the cards and call on students to supply the correct term or definition. Then have students complete Daily Quiz 2.2.

RETEACH

Have students complete Main Idea Activity 2.2. Then call on students to relate each of the section's main concepts to specific locations on a map of your state. For example, students should locate tributaries, reservoirs, and areas that may be prone to flooding and explain why. **ENGLISH LANGUAGE LEARNERS**

EXTEND

Have interested students conduct research on a significant flood that occurred in the past 10 years. Tell students to provide a map of the areas hurt by the flood as well as information on how the flood affected the area's people, plants, animals, and economy. Students should also note the long-term effects of the flood. **BLOCK SCHEDULING**

map of the United States. You will see that most major cities are located either next to a river or along a coast. For this reason, floods can be a threat to lives and property.

Flood Control Dams are a means people use to control floods. The huge Hoover Dam on the Colorado River in the United States is one example. The Aswān Dam on the Nile River in Egypt and those on the Murray River in Australia are others. Dams help protect people from floods. They also store water for use during dry periods. However, dams prevent rivers from bringing soil nutrients to areas downstream. Sometimes farming is not as productive as it was before the dams were built.

▲ The Atatürk Dam in Turkey is 604 feet (184 m) high.

Interpreting the Visual Record **How have people used dams to change their physical environment?**

Clean Water The availability of clean water is another issue affecting the world's people. Not every country is able to provide clean water for drinking and bathing. Pollution also threatens the health of the world's oceans—particularly in the shallower seabeds where many fish live and reproduce.

Water Supply Finding enough water to meet basic needs is a concern in regions that are naturally dry. The water supply is the amount of water available for use in a region. It limits the number of living things that can survive in a place. People in some countries have to struggle each day to find enough water.

✓ **READING CHECK:** *Environment and Society* What are some ways in which water affects people? through the weather, the availability of clean water for drinking and bathing

Section 2 Review

go.hrw.com Homework Practice Online
Keyword: SK3 HP2

Define and explain: water vapor, water cycle, evaporation, condensation, precipitation, tributary, groundwater, continental shelf

Reading for the Main Idea

1. *Physical Systems* What are the steps in the water cycle?
2. *Environment and Society* What are some major issues related to water?

Critical Thinking

3. **Analyzing Information** What are some important characteristics of water?
4. **Finding the Main Idea** Why is water supply a problem in some areas?

Organizing What You Know

5. **Sequencing** Copy the following graphic organizer. Use it to explain the water cycle.

	⇨		⇨		⇨		⇨	

Section 2 Review

Answers

Define For definitions, see: water vapor, p. 23; water cycle, p. 23; evaporation, p. 24; condensation, p. 24; precipitation, p. 24; tributary, p. 25; groundwater, p. 25; continental shelf, p. 25

Reading for the Main Idea

1. Evaporation leads to condensation, which leads to precipitation, and then the cycle repeats itself. (NGS 7)
2. floods, flood control, availability of clean water, and the water supply (NGS 15)

Critical Thinking

3. Water is the only substance on Earth that occurs naturally as a solid, a liquid, and a gas. Water also heats and cools slowly.
4. because these areas are naturally dry and do not have enough water to meet people's needs

Organizing What You Know

5. Answers will vary but should include evaporation, condensation, and precipitation as steps.

▲ **Visual Record Answer**

Dams provide people with the ability to control flooding and with water storage for dry periods. They can also provide electrical power.

Section 3

Objectives

1. Identify primary landforms.
2. Describe secondary landforms.
3. Explain how humans interact with landforms.

Focus

LET'S GET STARTED

Write the following question on the chalkboard: *What do we mean when we say "solid as a rock," "mountain of strength," or "older than dirt"?* Conduct a class discussion. Then ask students what the phrases suggest about Earth. *(Possible answers: that it is unchangeable, permanent)* Tell students that in Section 3 they will learn that Earth is actually in motion and can change dramatically.

Building Vocabulary

Write the key terms on slips of paper and have each student draw one from a hat. You may need to have duplicates of some terms. Ask students to find the definitions by skimming the text and read the definitions to the class. Then write *shapes on Earth's surface, slow movement,* and *fast change* on the chalkboard. Have the students determine the appropriate category for each term.

Section 3 Resources

Reproducible

- Guided Reading Strategy 2.3
- Graphic Organizer 2
- Environmental and Global Issues Activity 3
- Lab Activities for Geography and Earth Science, Demonstrations 3, 4, 5, 6, 7, 9, and 10
- Map Activity 2

Technology

- One-Stop Planner CD–ROM, Lesson 2.3
- Homework Practice Online
- Earth: Forces and Formations CD–ROM/Seek and Tell/Forces and Processes
- Earth: Forces and Formations CD–ROM/Seek and Tell/The Earth's Surface
- HRW Go site

Reinforcement, Review, and Assessment

- Section 3 Review, p. 33
- Daily Quiz 2.3
- Main Idea Activity 2.3
- English Audio Summary 2.3
- Spanish Audio Summary 2.3

Section 3 The Land

Read to Discover

1. What are primary landforms?
2. What are secondary landforms?
3. How do humans interact with landforms?

Define

landforms	core	subduction	alluvial fan
plain	mantle	earthquakes	floodplain
plateau	crust	fault	deltas
isthmus	magma	Pangaea	glaciers
peninsula	lava	weathering	
plate tectonics	continents	erosion	

Why It Matters

Earth's surface has been the focus of scientific research for centuries. Use CNNfyi.com logo or other **current events** sources to explore how scientists use maps to study the surface of Earth and other planets. Record your findings in your journal.

Mount Saint Helens, in southern Washington State

Landforms

Landforms are shapes on Earth's surface. One common landform is a **plain**—a nearly flat area. A **plateau** is an elevated flatland. An **isthmus** is a neck of land connecting two larger land areas. A **peninsula** is land bordered by water on three sides.

Primary Landforms

The theory of **plate tectonics** helps explain how forces raise, lower, and roughen Earth's surface. According to this theory, Earth's surface is divided into several large plates, or pieces. There are also many smaller plates. The plates slowly move. Some plates are colliding. Some are moving apart. Others are sliding by each other. Landforms created by tectonic processes are called primary landforms. These include masses of rock raised by volcanic eruptions and deep ocean trenches. The energy that moves the tectonic plates comes from inside Earth. The inner, solid **core** of the planet is surrounded by a liquid layer called the **mantle**. The outer, solid layer of Earth is called the **crust**. Currents of heat from the core travel outward through the mantle. When the currents reach the upper mantle, rocks can melt to form **magma**.

Earth's thin crust floats on top of the liquid mantle.

The Interior of Earth

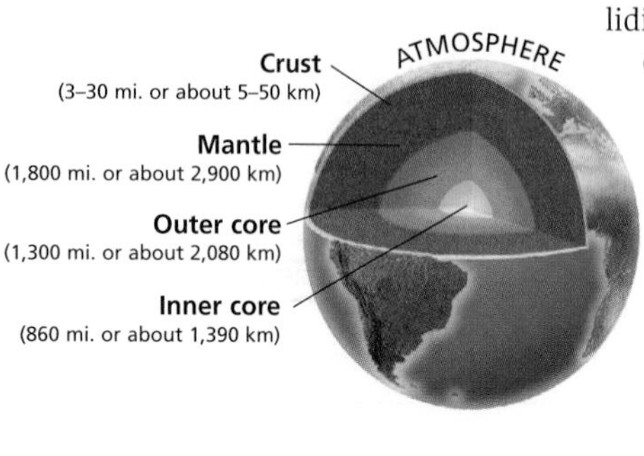

TEACH

Teaching Objective 1

LEVEL 1: (Suggested time: 45 min.) Provide students with nature and tourism magazines. Have them find and cut out photographs that show various primary landforms. Have each student create a mural with his or her photos. Call on students to tell why they identified these landforms as primary landforms. **ENGLISH LANGUAGE LEARNERS**

LEVELS 2 AND 3: (Suggested time: 30 min.) Have each student create a three-panel brochure titled "When Plates Collide." Each panel should contain a description of a primary landform created when tectonic plates collide, a diagram of the process involved in creating the landform, and an example of a place in the world where that process is occurring. Ask volunteers to share their brochures with the class.

Plate Tectonics

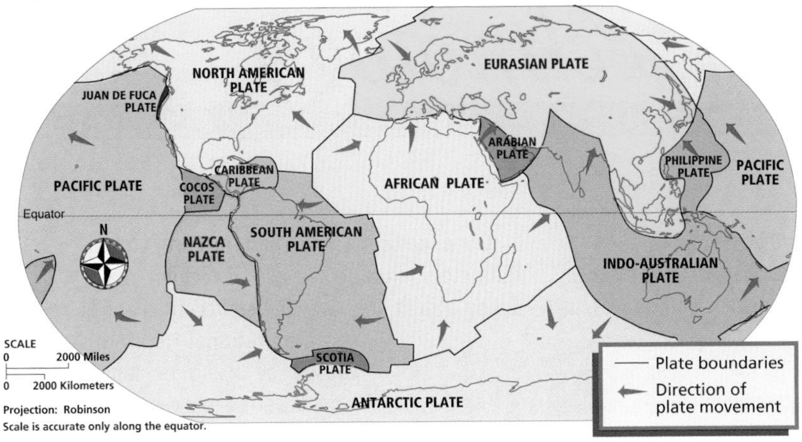

◄ The plates that make up Earth's crust are moving, usually a few inches per year. This map shows the plates and the direction of their movement.

Magma sometimes breaks through the crust to form a volcano. After reaching Earth's surface, magma is called **lava**.

Plates cover Earth's entire surface, both the land and the ocean. In general, the plates under the oceans are made of dense rock. The plates on the **continents**—Earth's large landmasses—are made of lighter rock.

Subduction and Spreading

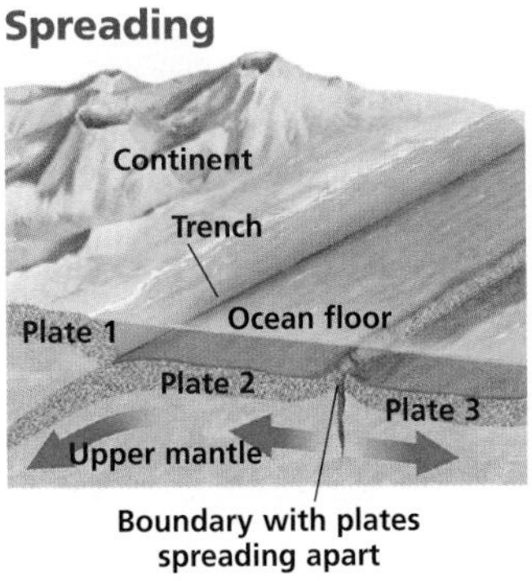

◄ Where Plate 2 pushes under Plate 1, a deep trench forms. This process is called subduction. Where Plates 2 and 3 move apart, lava creates a mid-ocean ridge.

Plates Colliding When two plates collide, one plate can be pushed under another. When this occurs under the ocean, a very deep trench is sometimes created. This is happening near Japan, where the Pacific plate is slowly moving under the Eurasian and Philippine plates. Any time a heavier plate moves under a lighter one, trenches can form. This process is called **subduction**. **Earthquakes** are common in subduction zones. An earthquake is a sudden, violent movement along a fracture within Earth's crust. A series of shocks usually results from such a movement within the crust.

The borders of the Pacific plate move against neighboring plates. This causes volcanoes to erupt and earthquakes to strike in that area. The Pacific plate's edge has been called the Ring of Fire because it is rimmed by active volcanoes. Thousands of people have died and terrible destruction has resulted from the earthquakes and volcanoes there.

When a continental plate and an ocean plate collide, the lighter rocks of the continent do not sink. Instead, they crumple and form a mountain range. The Andes in South America were formed this way. When two continental plates collide, land is lifted, sometimes to great heights. The Himalayas, the world's highest mountain range, were created by the Indo-Australian plate pushing into the Eurasian plate.

Plates Colliding

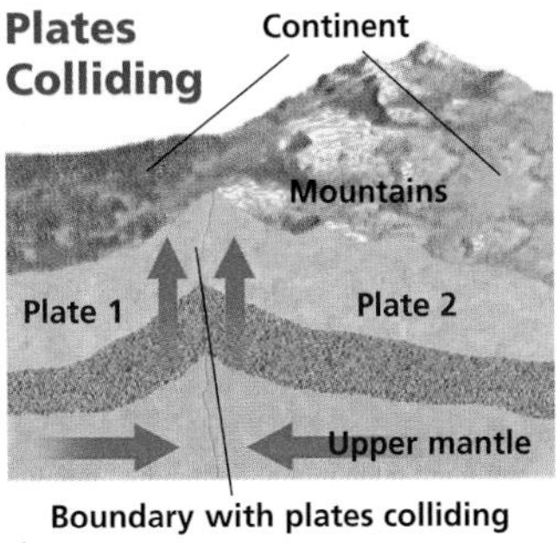

▲ Where two continental plates collide, Earth's crust is pushed upward, forming a mountain range.

Cultural Kaleidoscope

Primary Landforms and Legends

According to Irish legend, a giant named Finn Mac Cool built the Giant's Causeway, a formation of about 40,000 stone columns on the coast of Northern Ireland. By one account, Mac Cool drove the columns into place so he could walk to Scotland to fight fellow giant Benandonner. After completing the causeway, Mac Cool returned to Ireland. When Benandonner saw the causeway, he walked across it to Ireland, where he saw Mac Cool asleep. Mac Cool's wife told Benandonner that the sleeping giant was her child. Thinking that if this was the child the father must surely be unbeatable, Benandonner fled back to Scotland, destroying the causeway as he went. Similar column formations exist across the North Channel where the "causeway" would have reached Scotland.

The basalt columns were actually formed 50–60 million years ago when lava flows cooled as they reached the sea. Pressure shaped the rock into columns with three to seven sides. The columns average 330 feet (100 m) high.

Activity: Have students identify an unusual landform in your state and find out how it was formed. Then have them write and act out legends that describe the feature's origin.

Teaching Objectives 1–2

LEVEL 1: (Suggested time: 20 min.) Copy the following graphic organizer onto the chalkboard, omitting the italicized answers. Use it to help students distinguish between primary landforms and secondary landforms. **ENGLISH LANGUAGE LEARNERS**

Primary Landforms	Secondary Landforms
created by tectonic processes	*landforms that result when primary landforms break down*
masses of rock raised by volcanic eruptions and deep ocean trenches	*caused by weathering and erosion*
include mountain ranges and mid-ocean ranges	*include floodplains and deltas*

Teaching Objective 2

LEVELS 2 AND 3: (Suggested time: 45 min.) Organize students into teams and have them explore the school grounds or a nearby park to find examples of erosion. Ask them to collect rocks and classify them according to which force seems to have eroded them. Also have students determine if and why any of the three forces of erosion are not present in their region. Have students present their findings to the class. *(Students might mention that glaciers are not a force of erosion in their regions.)* **COOPERATIVE LEARNING**

EYE ON EARTH

Weathering and erosion are generally slow processes that lead to the creation of secondary landforms. Sometimes, however, the breaking down of primary landforms can be quite sudden.

For example, miners in Switzerland carved slate from the base of a mountain cliff, creating a huge overhang. Great cracks appeared in the cliff, and finally on September 11, 1881, millions of cubic yards of rock fell. The avalanche of rocks did not stop when it hit the valley floor. Instead, the broken rock continued up the valley's other side. More than 100 people were killed in this debris avalanche.

Critical Thinking: In what other ways do humans change their environment that might create conditions for debris avalanches?

Answer: cutting through mountains to build highways, constructing buildings on hillsides, cutting down trees, and other disturbances of steep terrains

Connecting to Technology Answers

1. could save lives and property
2. seismographs, tiltmeters, gravimeters, laser beams, and satellites

CONNECTING TO *Technology*

forecasting earthquakes

Since ancient times, people have tried to forecast earthquakes. A Chinese inventor even created a device to register earthquakes as early as A.D. 132.

The theory of plate tectonics gives modern-day scientists a better understanding of how and why earthquakes occur. Earthquake scientists, known as seismologists, have many tools to help them monitor movements in Earth's crust. They try to understand when and where earthquakes will occur.

The most common of these devices is the seismograph. It measures seismic waves—vibrations produced when two tectonic plates grind against each other. Scientists believe that an increase in seismic activity may signal a coming earthquake.

A scientist with a seismograph

Other devices show shifts in Earth's crust. Tiltmeters measure the rise of tectonic plates along a fault line. Gravimeters record changes in gravitational strength caused by rising or falling land. Laser beams can detect lateral movements along a fault line. Satellites can note the movement of entire tectonic plates.

Scientists have yet to learn how to forecast earthquakes with accuracy. Nevertheless, their ongoing work may one day provide important breakthroughs in the science of earthquake forecasting.

You Be the Geographer

1. What social and economic consequences might earthquake forecasting have?
2. What technology has helped with earthquake forecasting?

Plates Moving Apart When two plates move away from each other, hot lava emerges from the gap that has been formed. The lava builds a mid-ocean ridge—a landform that is similar to an underwater mountain range. This process is currently occurring in the Atlantic Ocean where the Eurasian plate and the North American plate are moving away from each other.

Plates Sliding Tectonic plates can also slide past each other. Earthquakes occur from sudden adjustments in Earth's crust. In California the Pacific plate is sliding northwestward along the edge of the North American plate. This has created the San Andreas Fault

TEACHER TO TEACHER

Kay A. Knowles, of Montross, Virginia, suggests the following activity to help students identify landforms and predict where other landforms might be found. Display a wall map and have students point out examples of these landforms: continent, isthmus, peninsula, plain, and plateau. Also, ask students to identify places where they would expect to find these landforms: primary landform, secondary landform, alluvial fan, floodplain, and delta.

Teaching Objective 3

ALL LEVELS: (Suggested time: 20 min.) Pair students and ask each pair to list examples of primary and secondary landforms. Then have students write next to each example how humans interact with these landforms. Ask volunteers to share their lists with the class.
ENGLISH LANGUAGE LEARNERS, COOPERATIVE LEARNING

zone. A **fault** is a fractured surface in Earth's crust where a mass of rocks is in motion.

Tectonic plates move slowly—just inches a year. If, however, we could look back 200 million years, we would see that the continents have moved a long way. From their understanding of plate tectonics, scientists proposed the theory of continental drift. This theory states that the continents were once united in a single supercontinent. They then separated and moved to the positions they are in today. Scientists call this original landmass **Pangaea** (pan-GEE-uh).

Continental Drift

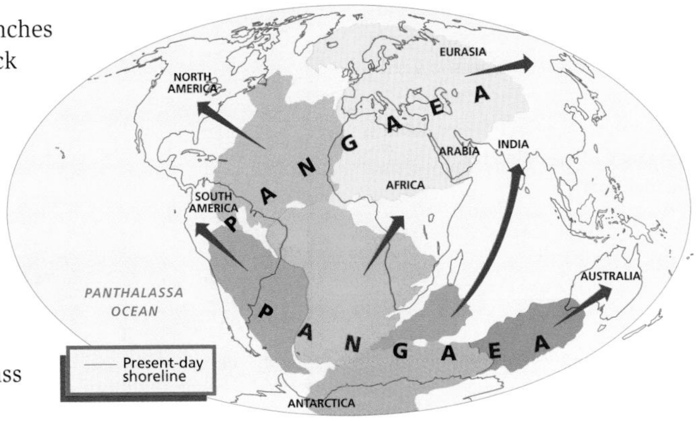

▲ About 200 million years ago it is believed there was only one continent, Pangaea, and one ocean, Panthalassa. The continental plates slowly drifted into their present-day positions.

READING CHECK: *Physical Systems* How are primary landforms formed? by the movement of the tectonic plates

Secondary Landforms

The forces of plate tectonics build up primary landforms. At the same time, water, wind, and ice constantly break down rocks and cause rocky material to move. The landforms that result when primary landforms are broken down are called secondary landforms.

One process of breaking down and changing primary landforms into secondary landforms is called **weathering**. It is the process of breaking rocks into smaller pieces. Weathering occurs in several ways. Heat can cause rocks to crack. Water may get into cracks in rocks and freeze. This ice then expands with a force great enough to break the rock. Water can also work its way underground and slowly dissolve minerals such as limestone. This process sometimes creates caves. In

These steep peaks in Chile are part of the Andes.

Interpreting the Visual Record **How do these mountains show the effects of weathering and erosion?** ▼

Linking Past to Present

Floods in Iceland According to some historical accounts, Iceland has experienced about 60 gigantic floods since the Vikings arrived in the A.D. 800s. Some scientists had doubted the accuracy of these accounts. However, events in November 1996 showed how these earlier floods could have occurred.

Water that had been melted by a volcanic eruption broke out from under the Vatnajökull (VAHT-nah-yuh-koot-uhl) glacier. Ash and steam gushed up through the glacier, melting a huge hole in the ice. The runoff carved a canyon in the ice. This canyon was 500 feet (150 m) deep and more than two miles (3.2 km) long. Magma melted more of the glacier's bottom layers. Billions of gallons of water drained into an ice-covered crater. The water raised the crater's thick ice cap and then flowed downhill underneath the glacier. Finally, the water burst out from under the ice. Blocks of ice as large as buildings ripped loose. When the flood crashed into a bridge it was flowing at 1.6 million cubic feet per second.

◄ **Model Answer**

sea level to 8,028 ft.

◄ **Visual Record Answer**

Their crags and peaks show erosion.

Close

Have students study the map on p. 29 showing continental drift. Point out that the shapes of South America and Africa were an early clue in the development of the continental-drift theory. Ask students what other kinds of evidence might indicate that now-distant continents were once joined. *(Possible answers: corresponding fossils and mineral deposits)*

Review and Assess

Have students complete the Section Review. Then pair students and instruct pairs to write sentences that illustrate the connection between two of the key terms. *(Example: A plateau is a plain that lies at a higher elevation.)* Call on volunteers to read their sentences to the class. Continue until all the major points have been covered. Then have students complete Daily Quiz 2.3.

Section 3 Review

Answers

Define For definitions, see: landforms, p. 28; plain, p. 28; plateau, p. 28; isthmus, p. 28; peninsula, p. 28; plate tectonics, p. 28; core, p. 28; mantle, p. 28; crust, p. 28; magma, p. 28; lava, p. 29; continents, p. 29; subduction, p. 29; earthquakes, p. 29; fault, p. 31; Pangaea, p. 31; weathering, p. 31; erosion, p. 32; alluvial fan, p. 32; floodplain, p. 32; deltas, p. 32; glaciers, p. 33

Reading for the Main Idea

1. by tectonic processes (NGS 7)
2. weathering—heat, cold, water, and plants; erosion—water, glaciers, and wind (NGS 7)

Critical Thinking

3. Earth's surface is divided into seven plates. These plates move and create landforms.
4. People change landforms and adapt them as needed, building dams, canals, highways through mountains, etc.

Organizing What You Know

5. colliding—ocean trenches and mountain ranges created; earthquakes and volcanic eruptions common; moving apart—hot lava emerges and mid-ocean ridges are created; sliding—faults created

Elevation is the height of the land above sea level. An elevation profile is a cross-section used to show the elevation of a specific area.

Interpreting the Visual Record **What is the range of elevation for Guadalcanal?**

Elevation Profile: Guadalcanal

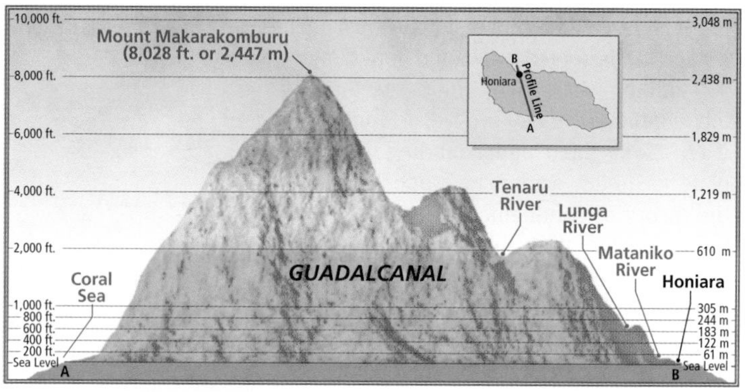

some areas small plants called lichens attach to bare rock. Chemicals in the lichens gradually break down the stone. Some places in the world experience large swings in temperature. In the Arctic the ground freezes and thaws, which tends to lift stones to the surface in unusual patterns. Regardless of which weathering process is at work, rocks eventually break down into sediment. These smaller pieces of rock are called gravel, sand, silt, or clay, depending on particle size. Once weathering has taken place, water, ice, or wind can move the material and create new landforms.

✓ **READING CHECK:** *Physical Systems* What is one way in which secondary landforms are created? through weathering

Erosion

Another process of changing primary landforms into secondary landforms is **erosion**. Erosion is the movement of rocky materials to another location. Moving water is the most common force that erodes and shapes the land.

Water Flowing water carries sediment. This sediment forms different kinds of landforms depending on where it is deposited. For example, a river flowing from a mountain range onto a flat area, or plain, may deposit some of its sediment there. The sediment sometimes builds up into a fan-shaped form called an **alluvial fan**. A **floodplain** is created when rivers flood their banks and deposit sediment. A **delta** is formed when rivers carry some of their sediment all the way to the ocean. The sediment settles to the bottom where the river meets the ocean. The Nile and Mississippi Rivers have two of the world's largest deltas.

Waves in the ocean and in lakes also shape the land they touch. Waves can shape beaches into great dunes, such as on the shore of Long Island. The jagged coastline of Oregon also shows the erosive power of waves.

River water, brown with sediment, enters the ocean.

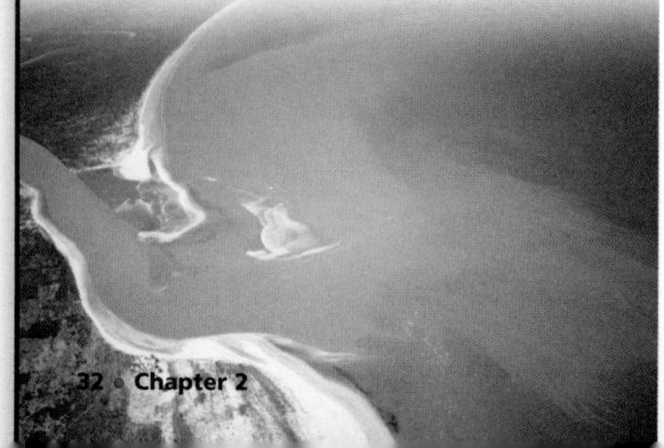

RETEACH

Have students complete Main Idea Activity 2.3. Then pair students and have them create brief descriptions of each main idea in the section. Have students consult atlases, globes, or encyclopedias to locate two specific examples of each feature or main idea. **ENGLISH LANGUAGE LEARNERS**

EXTEND

Have interested students conduct research on plate tectonics and prepare a report that focuses on one of the following topics: (1) how the theory was developed, and what evidence supports it; (2) the causes of plate movement and continental drift; and (3) practical applications of the theory, such as earthquake prediction and mineral exploration. Then, have students work together to create a script and several storyboards for a documentary film on the topic. Remind them to use standard grammar, spelling, sentence structure, and punctuation. **BLOCK SCHEDULING**

Glaciers In high mountain settings and in the coldest places on Earth are **glaciers**. These large, slow-moving rivers of ice have the power to move tons of rock.

Giant sheets of thick ice called continental glaciers cover Greenland and Antarctica. Over the past 2 million years Earth has experienced several ice ages—periods of extremely cold conditions. During each ice age continental glaciers covered most of Canada and the northern United States. The Great Lakes were carved out by the movement of a continental glacier.

Wind Wind also shapes the land. Strong winds can lift soils into the air and carry them across great distances. On beaches and in deserts wind can deposit large amounts of sand to form dunes.

Blowing sand can wear down rock. The sand acts like sandpaper to polish jagged edges. An example of rocks worn down by blowing sand can be seen in Utah's Canyonlands National Park.

✓ **READING CHECK:** *Physical Systems* What forces cause erosion? water, glaciers, wind

Our Amazing Planet

Waves 50 to 60 feet high, which the local people call Jaws, sometimes occur off the coast of Maui, Hawaii. They are caused by storms in the north Pacific and a high offshore ridge that focuses the waves' energy.

People and Landforms

Geographers study how people adapt their lives to different landforms. Deltas and floodplains, for example, are usually fertile places to grow food. People also change landforms. Engineers build dams to control river flooding. They drill tunnels through mountains instead of making roads over mountaintops. People have used modern technology to build structures that are better able to survive disasters like floods and earthquakes.

✓ **READING CHECK:** *Environment and Society* What are some examples of humans adjusting to and changing landforms? using fertile deltas for farming, drilling through mountains to make roads

go.hrw.com **Homework Practice Online** Keyword: SK3 HP2

Section 3 Review

Define and explain: landforms, plain, plateau, isthmus, peninsula, plate tectonics, core, mantle, crust, magma, lava, continents, earthquakes, fault, Pangaea, weathering, erosion, alluvial fan, floodplain, deltas, glaciers

Reading for the Main Idea

1. *Physical Systems* How are primary landforms created?
2. *Physical Systems* What forces cause weathering and erosion?

Critical Thinking

3. **Summarizing** What is plate tectonics?
4. **Finding the Main Idea** How do people affect landforms? Give examples.

Organizing What You Know

5. **Identifying Cause and Effect** Copy the following graphic organizer. Use it to describe the movement of plates and the movement's effects.

movement		resulting landforms and changes
	⇨	
	⇨	
	⇨	

CHAPTER 2 Review ANSWERS

Building Vocabulary

For definitions, see: solar system, p. 17; orbit, p. 17; solstice, p. 21; equinoxes, p. 21; atmosphere, p. 22; water cycle, p. 23; evaporation, p. 24; condensation, p. 24; precipitation, p. 24; landforms, p. 28; plate tectonics, p. 28; continents, p. 29; earthquakes, p. 29; weathering, p. 31; erosion, p. 32

Reviewing the Main Ideas

1. The rotation, revolution, and tilt of Earth limit the areas that sunlight can reach. (NGS 7)
2. The Sun's energy creates water vapor and causes winds, which carry water vapor to new locations. Water vapor rises, cools, condenses, forms clouds; precipitation occurs. (NGS 7)
3. Tectonic plates move and collide, raising, lowering, and roughening Earth's surface, and thereby shaping continents and primary landforms. (NGS 7)
4. A heavier plate is pushed under a lighter one, and sometimes a trench is formed. (NGS 7)
5. by weathering and erosion (NGS 7)

ASSESS

Have students complete the Chapter 2 Test.

RETEACH

Organize students into groups and have each group create one of three posters: a depiction of Earth's location in the solar system and how its tilt determines seasons, a cutaway view of Earth showing water distribution, or a cutaway view of Earth showing landforms and the processes that affect them, including ocean circulation. Compare the posters as a class to ensure that all major points have been covered.
ENGLISH LANGUAGE LEARNERS

PORTFOLIO EXTENSIONS

1. To demonstrate the importance of water, have students conduct research on the location of their school's drinking water. *(Possible water sources include rivers, lakes, reservoirs, and groundwater.)* Students should also find out how the school disposes of its wastewater. Have students create diagrams of their school's water-supply system.
2. Have students research the origin of local or nearby landforms. *(Possible origins include glacial action, volcanic action, and sedimentation.)* To present their findings, have students label landforms on a map according to how they were formed. You may want to have students design models that show the formation processes. Photograph the models and place the pictures in student portfolios.

CHAPTER 2 Review ANSWERS

Understanding Environment and Society

- Sources may include lakes, rivers, reservoirs, springs, and wells.
- Arid regions are more likely to have problems, but almost all areas have suffered occasional water shortages due to drought, broken pipes, flood, or other causes.
- Students' answers may include laws or ordinances regarding commercial and residential use.

Thinking Critically

1. because the atmosphere, lithosphere, hydrosphere, and biosphere are constantly interacting
2. Valleys, ravines, and canyons are cut by streams; deposits of river sediment form alluvial fans, floodplains, and deltas; glaciers move rock, carving landforms such as lakes; wind creates dunes and wears down rock.
3. These areas are usually good places for people to grow food.
4. (a) Land is lifted, creating mountains. (b) Lava emerges from the gap, creating a mid-ocean ridge.

CHAPTER 2 Reviewing What You Know

Building Vocabulary

On a separate sheet of paper, write sentences to define each of the following words.

1. solar system
2. orbit
3. solstice
4. equinoxes
5. atmosphere
6. water cycle
7. evaporation
8. condensation
9. precipitation
10. landforms
11. plate tectonics
12. continents
13. earthquakes
14. weathering
15. erosion

Reviewing the Main Ideas

1. *Physical Systems* In what ways do Earth's rotation, revolution, and tilt help determine how much of the Sun's energy reaches Earth?
2. *Physical Systems* Explain how the Sun's energy drives the water cycle. Be sure to include a discussion of the three elements of the cycle.
3. *Physical Systems* How does plate tectonics relate to the continents and their landforms?
4. *Physical Systems* Describe how subduction zones are created.
5. *Physical Systems* Describe the different ways secondary landforms are created.

Understanding Environment and Society

Water Use

The availability and purity of water are important for everyone. Research the water supply in your own city or community and prepare a presentation on it. You may want to think about the following:

- Where your city or community gets its drinking water,
- Drought or water shortages that have happened in the past,
- Actions your community takes to protect the water supply.

Thinking Critically

1. **Drawing Inferences and Conclusions** How do the four parts of the Earth system help explain why places on Earth differ?
2. **Finding the Main Idea** Describe landforms that are shaped by water and wind.
3. **Drawing Inferences and Conclusions** Why do people continue to live in areas where floods are likely to occur?
4. **Identifying Cause and Effect** Describe the landforms that result (a) when two tectonic plates collide and (b) when two plates move away from each other.
5. **Summarizing** In what ways do people interact with landforms?
6. **Analyzing Information** How might mountains be primary and secondary landforms?

Food Festival

The Sun makes life on Earth possible. It can also be used to process foods. Have students make sun tea by filling a jar with water, adding tea bags, covering the jar, and then placing it in sunlight to steep. After a few hours, have students add ice to the tea and enjoy. You might also have students locate and bring to class sun-dried food items, such as beef jerky, raisins, or tomatoes. Challenge the class to research regions where fuel is scarce and devise ways people could use solar energy to cook food. Have them report their information to the class.

Chapter 2 Review and Assessment Resources

Reproducible
- Readings in World Geography, History, and Culture 3 and 4
- Critical Thinking Activity 2
- Vocabulary Activity 2

Technology
- Chapter 2 Test Generator (on the One-Stop Planner)
- Audio CD Program, Chapter 2
- HRW Go site

Reinforcement, Review, and Assessment
- Chapter 2 Review, pp. 34–35
- Chapter 2 Tutorial for Students, Parents, Mentors, and Peers
- Chapter 2 Test
- Chapter 2 Test for English Language Learners and Special-Needs Students

Building Social Studies Skills

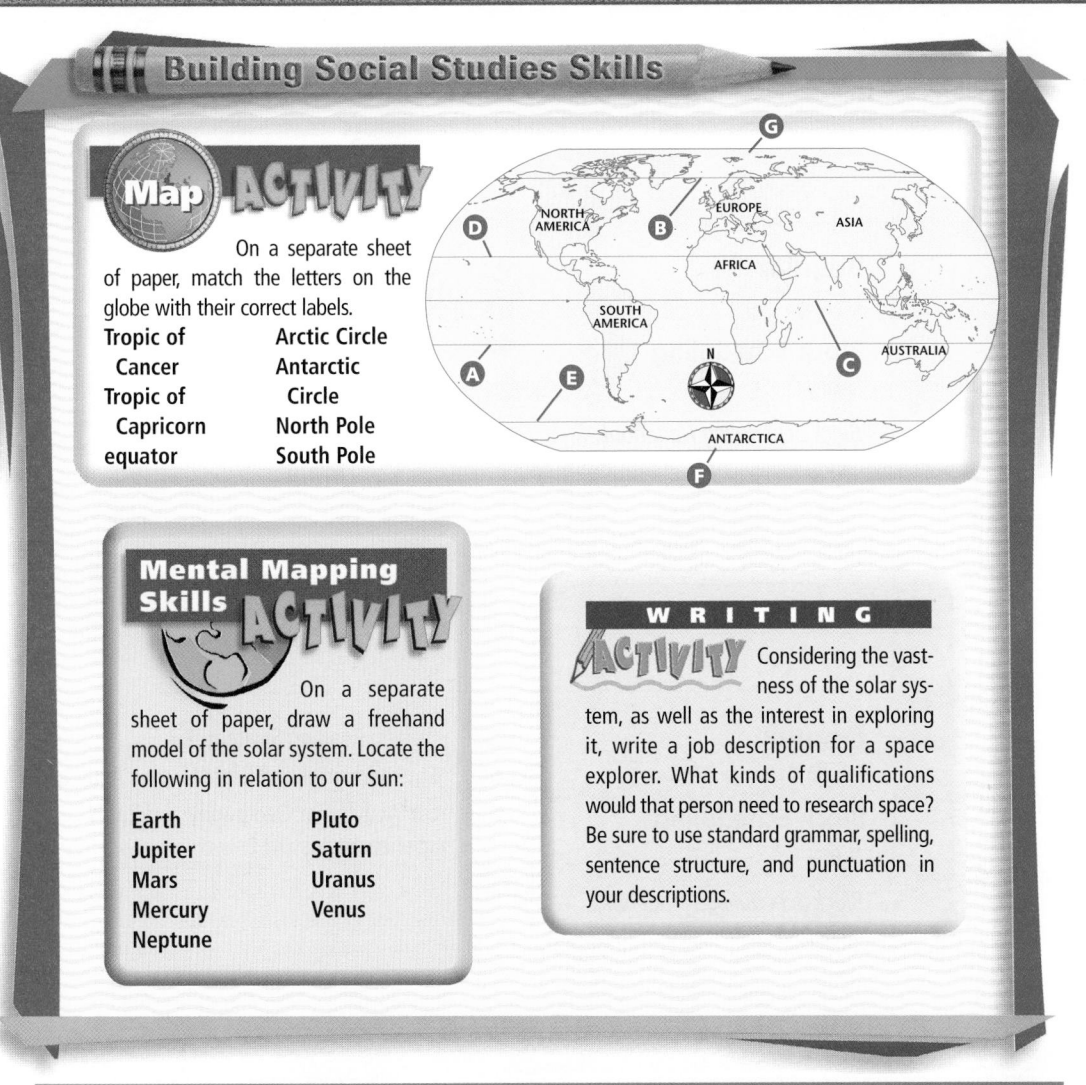

Map Activity

On a separate sheet of paper, match the letters on the globe with their correct labels.

Tropic of Cancer	Arctic Circle
Tropic of Capricorn	Antarctic Circle
equator	North Pole
	South Pole

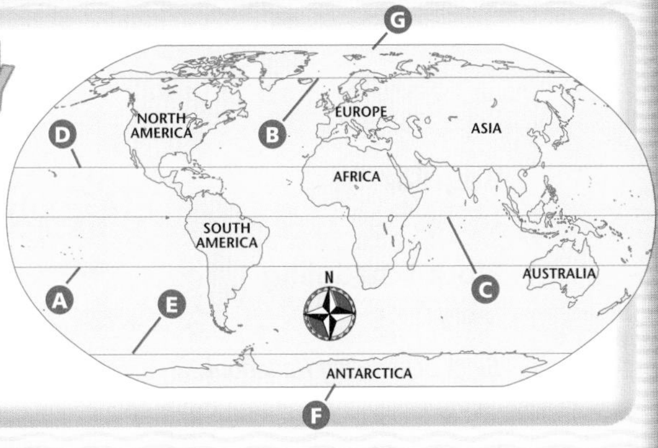

Mental Mapping Skills Activity

On a separate sheet of paper, draw a freehand model of the solar system. Locate the following in relation to our Sun:

Earth	Pluto
Jupiter	Saturn
Mars	Uranus
Mercury	Venus
Neptune	

Writing Activity

Considering the vastness of the solar system, as well as the interest in exploring it, write a job description for a space explorer. What kinds of qualifications would that person need to research space? Be sure to use standard grammar, spelling, sentence structure, and punctuation in your descriptions.

Alternative Assessment

Portfolio Activity

Learning About Your Local Geography

Individual Project Compare the latitude and longitude of your state's capital city with that of the capitals of three countries. How might the seasons be similar or different in each city?

internet connect

Internet Activity: go.hrw.com
KEYWORD: SK3 GT2

Choose a topic to explore online:
- Learn more about Earth's seasons.
- Discover facts about Earth's water.
- Investigate earthquakes.

5. People can adapt to landforms, such as by growing food in river floodplains; change landforms to fit their needs, such as by building dams.
6. primary landform if created when two tectonic plates collide; ravines and contours might be created by erosion and weathering

Map Activity

A. Tropic of Capricorn
B. Arctic Circle
C. equator
D. Tropic of Cancer
E. Antarctic Circle
F. South Pole
G. North Pole

Mental Mapping Skills Activity

Use the diagram on p. 17 to check students' models.

Writing Activity

Answers will vary, but the information included should be consistent with text material. Use Rubric 31, Resumés, to evaluate student work.

Portfolio Activity

Cities in the Southern Hemisphere have winter while those in the Northern Hemisphere have summer.

internet connect

GO TO: go.hrw.com
KEYWORD: SK3 Teacher
FOR: a guide to using the Internet in your classoom

CHAPTER 3

Wind, Climate, and Natural Environments

Chapter Resource Manager

Objectives	Pacing Guide	Reproducible Resources
SECTION 1 **Winds and Ocean Currents** (pp. 37–41) **1.** Describe how the Sun's energy changes the Earth. **2.** Explain why wind and ocean currents are important.	**Regular** 1.5 days **Block Scheduling** .5 day *Block Scheduling Handbook, Chapter 3*	**RS** Guided Reading Strategy 3.1 **SM** Geography for Life Activity 3 **E** Lab Activities for Geography and Earth Science, Demonstrations 11 and 12
SECTION 2 **Earth's Climate and Vegetation** (pp. 44–50) **1.** Describe what is included in the study of weather. **2.** Identify the major climate types, and the types of plants that live in each.	**Regular** 1.5 days **Block Scheduling** .5 day *Block Scheduling Handbook, Chapter 3*	**RS** Guided Reading Strategy 3.2 **RS** Graphic Organizer 3 **SM** Map Activity 3
SECTION 3 **Natural Environments** (pp. 51–55) **1.** Describe how environments affect life, and how they change. **2.** Identify the substances that make up the different layers of soil.	**Regular** 1.5 days **Block Scheduling** .5 day *Block Scheduling Handbook, Chapter 3*	**RS** Guided Reading Strategy 3.3 **E** Environmental and Global Issues Activity 4 **E** Lab Activity for Geography and Earth Science, Hands-On 1

Chapter Resource Key

- **RS** Reading Support
- **IC** Interdisciplinary Connections
- **E** Enrichment
- **SM** Skills Mastery
- **A** Assessment
- **REV** Review
- **ELL** Reinforcement and English Language Learners
- Transparencies
- CD–ROM
- Music
- Video
- go.hrw.com Internet
- Holt Presentation Maker Using Microsoft® Powerpoint®

One-Stop Planner CD–ROM

See the *One-Stop Planner* for a complete list of additional resources for students and teachers.

One-Stop Planner CD–ROM

It's easy to plan lessons, select resources, and print out materials for your students when you use the *One-Stop Planner CD–ROM with Test Generator.*

Technology Resources	Review, Reinforcement, and Assessment Resources
One-Stop Planner CD–ROM, Lesson 3.1 Earth: Forces and Formations CD–ROM/Seek and Tell/Forces and Processes Geography and Cultures Visual Resources with Teaching Activities 1–6 and 8 Homework Practice Online HRW Go site	**ELL** Main Idea Activity 3.1 **REV** Section 1 Review, p. 41 **A** Daily Quiz 3.1 **ELL** English Audio Summary 3.1 **ELL** Spanish Audio Summary 3.1
One-Stop Planner CD–ROM, Lesson 3.2 Earth: Forces and Formations CD–ROM/Seek and Tell/Forces and Processes *ARGWorld* CD–ROM: Using Climagraphs to Interpret Seasons Homework Practice Online HRW Go site	**ELL** Main Idea Activity 3.2 **REV** Section 2 Review, p. 50 **A** Daily Quiz 3.2 **ELL** English Audio Summary 3.2 **ELL** Spanish Audio Summary 3.2
One-Stop Planner CD–ROM, Lesson 3.3 Our Environment CD–ROM/Seek and Tell/Natural Resources Homework Practice Online HRW Go site	**ELL** Main Idea Activity 3.3 **REV** Section 3 Review, p. 55 **A** Daily Quiz 3.3 **ELL** English Audio Summary 3.3 **ELL** Spanish Audio Summary 3.3

internet connect

HRW ONLINE RESOURCES

GO TO: go.hrw.com
Then type in a keyword.

TEACHER HOME PAGE
KEYWORD: SK3 TEACHER

CHAPTER INTERNET ACTIVITIES
KEYWORD: SK3 GT3

Choose an activity to:
- learn more about using weather maps.
- follow El Niño, an ocean phenomenon that affects weather.
- build a food web.

CHAPTER ENRICHMENT LINKS
KEYWORD: SK3 CH3

CHAPTER MAPS
KEYWORD: SK3 MAPS3

ONLINE ASSESSMENT
Homework Practice
KEYWORD: SK3 HP3
Standardized Test Prep Online
KEYWORD: SK3 STP3
Rubrics
KEYWORD: SS Rubrics

COUNTRY INFORMATION
KEYWORD: SK3 Almanac

CONTENT UPDATES
KEYWORD: SS Content Updates

HOLT PRESENTATION MAKER
KEYWORD: SK3 PPT3

ONLINE READING SUPPORT
KEYWORD: SS Strategies

CURRENT EVENTS
KEYWORD: S3 Current Events

Meeting Individual Needs

Ability Levels

Level 1 Basic-level activities designed for all students encountering new material

Level 2 Intermediate-level activities designed for average students

Level 3 Challenging activities designed for honors and gifted-and-talented students

English Language Learners Activities that address the needs of students with Limited English Proficiency

Chapter Review and Assessment

E Readings in World Geography, History, and Culture 5 and 6
SM Critical Thinking Activity 3
REV Chapter 3 Review, pp. 56–57
REV Chapter 3 Tutorial for Students, Parents, Mentors, and Peers
ELL Vocabulary Activity 3
A Chapter 3 Test
Chapter 3 Test Generator (on the One-Stop Planner)
Audio CD Program, Chapter 3
A Chapter 3 Test for English Language Learners and Special-Needs Students

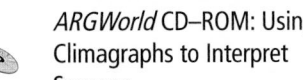

Chapter 3

Launch into Learning

Give students two minutes to complete a quick sketch of a wild animal common in your region. Remind them that even crowded cities are home to wild animals. Have the students hold up their drawings. Then call on volunteers to suggest how the animal survives: what does it eat, how does it adapt to the heat or cold, where does it get water? Tell students that weather conditions and plants play a large role in determining where animals—and people—can live.

Section 1

Objectives

1. Analyze how the Sun's energy changes Earth.
2. Explain the importance of wind and ocean currents.

Links to Our Lives

There are many reasons why studying climate and environmental issues is important. Here are some to share with your students:

- Ocean and wind currents can influence everything from vacation plans to our food supply.
- Every day, weather affects us all. We can predict it or accommodate it better if we understand it.
- We can prevent damage to the environment only if we understand the connections among climate, plants, animals, and humans.
- World events are often determined by environmental changes. We can predict events better if we recognize those cause-and-effect relationships.

Chapter 3

Wind, Climate, and Natural Environments

Tornado in Saskatoon, Canada

Diver, coral, and fish, Fiji Islands

Igloo at night, Alaska Range

FOCUS

LET'S GET STARTED

Copy the following instructions onto the chalkboard: *What would happen if, for one week, all the winds in the world were still? Write down your ideas.* Discuss student responses. *(Students might suggest that temperature differences would become extreme, that pollution would get worse, and that ocean waves would calm.)* Point out that although one cannot see it, smell it, or taste it, wind makes life as we know it possible on Earth. Tell students that in Section 1 they will learn more about winds and ocean currents.

Building Vocabulary

Write the key terms on the chalkboard and ask students if they are familiar with any of the words that make up the terms. Write students' responses beside the terms. *(Students might suggest that greenhouses are where flowers are grown, that **air pressure** measures the amount of air in tires, and that **currents** transmit electric power.)* Call on volunteers to find and read aloud the definitions in the text. Then ask volunteers to compare the text definitions to those already suggested. *(Example: Plants that would normally be killed by cold weather can survive inside a greenhouse because the glass traps heat. The **greenhouse effect** traps the Sun's energy.)*

Section 1 Winds and Ocean Currents

Read to Discover

1. How does the Sun's energy change Earth?
2. Why are wind and ocean currents important?

Define

- weather
- climate
- greenhouse effect
- air pressure
- front
- currents

WHY IT MATTERS

Changes in ocean temperatures create currents that affect Earth's landmasses. Use CNNfyi.com or other **current events** sources to find examples of the effects of changes in ocean temperatures, such as the La Niña weather pattern. Record your findings in your journal.

Earth from space

The Sun's Energy

All planets in our solar system receive energy from the Sun. This energy has important effects. Among the most obvious effects we see on our planet are those on **weather** and **climate**. Weather is the condition of the atmosphere at a given place and time. Climate refers to the weather conditions in an area over a long period of time. How do you think the Sun's energy affects weather and climate?

Energy Balance Although Earth keeps receiving energy from the Sun, it also loses energy. Energy that is lost goes into space. As a result, Earth—as a whole—loses as much energy as it gets. Thus, Earth's overall temperature stays about the same.

As you learned in Chapter 2, the Sun does not warm Earth evenly. The part of Earth in daylight takes in more energy than it loses. Temperatures rise. However, the rest of Earth is in darkness. That part of Earth loses more energy than it gets. Temperatures drop. In addition, when the direct rays of the Sun strike Earth at the Tropic of Cancer, it is summer in the Northern Hemisphere. Temperatures are warm. The Southern Hemisphere, on the other hand, is having winter. Temperatures are lower. The seasons reverse when the Sun's direct rays move above the Tropic of Capricorn. We can see that in any one place temperatures vary from day to day. From year to year, however, they usually stay about the same.

Stored Energy Some of the heat energy that reaches Earth is stored. One place Earth stores heat is in the air. This keeps Earth's surface warmer than if there were no air around it. The process by which

Plants—like this fossil palm found in a coal bed—store the Sun's energy. When we burn coal—the product of long-dead plants—we release the energy the plants had stored.

internet connect

GO TO: go.hrw.com
KEYWORD: SK3 CH3
FOR: Web sites about wind, climates, and environments

Section 1 RESOURCES

Reproducible
- Block Scheduling Handbook, Ch. 3
- Guided Reading Strategy 3.1
- Geography for Life Activity 3
- Lab Activities for Geography and Earth Science, Demonstrations 11 and 12

Technology
- One-Stop Planner CD–ROM, Lesson 3.1
- Homework Practice Online
- Earth: Forces and Formations CD–ROM/Seek and Tell/Forces and Processes
- Geography and Cultures Visual Resources with Teaching Activities 1–6 and 8
- HRW Go site

Reinforcement, Review, and Assessment
- Section 1 Review, p. 41
- Daily Quiz 3.1
- Main Idea Activity 3.1
- English Audio Summary 3.1
- Spanish Audio Summary 3.1

TEACH

Teaching Objectives 1–2

ALL LEVELS: (Suggested time: 10 min.) Copy the following graphic organizer onto the chalkboard, omitting the dashed arrows. Have each student complete the organizer by showing energy coming from the Sun *(large arrow)*, and being distributed by wind and ocean currents and by escaping into space *(small arrows)*. **ENGLISH LANGUAGE LEARNERS**

ENERGY FROM THE SUN

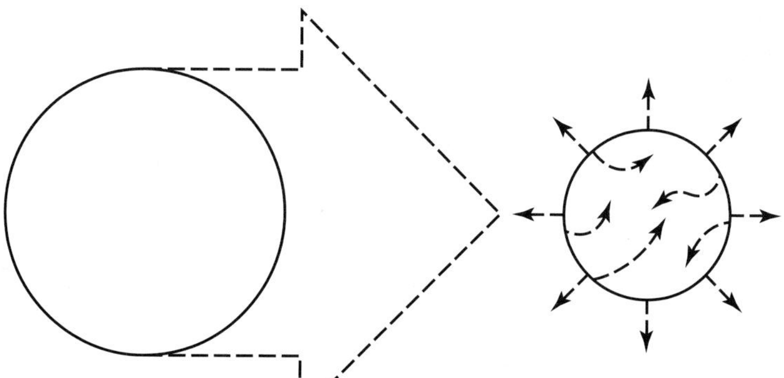

Linking Past to Present

The Little Ice Age Many scientists believe that the greenhouse effect is responsible for global temperature increases of recent decades. Temperature changes within recorded history have not always resulted in global warming, however.

During the Little Ice Age, which lasted roughly from the A.D. 1500s to the 1800s, the average global temperature dropped about 1°C.

Ice sheets advanced over farms and villages on Greenland. The Baltic Sea and Thames River, which now seldom freeze, then froze regularly. Crop failure, famine, and disease were common throughout Europe. Colonists in North America suffered through harsh winters also. Although personal accounts reveal that people endured hard times, they did not realize they were in the Little Ice Age. Climate changes are usually gradual and seen as normal within a person's lifetime.

Activity: Have students conduct research on the advance and retreat of glaciers since the Ice Age. Have them report their findings in a bar graph.

You Be the Geographer Answer

Heat would build up and global average temperatures would rise.

The Greenhouse Effect

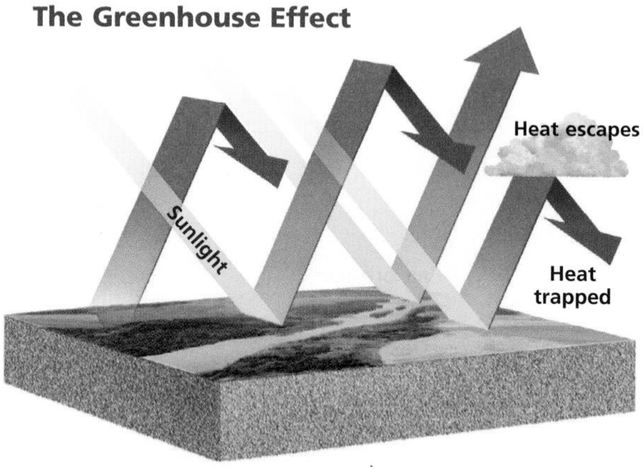

Light from the Sun passes through the atmosphere and heats Earth's surface. Most heat energy later escapes into space.

You Be the Geographer **What would happen if too much heat energy remained trapped in the atmosphere?**

Earth's atmosphere traps heat is called the **greenhouse effect**. In a greenhouse the Sun's energy passes through the glass and heats everything inside. The glass traps the heat, keeping the greenhouse warm.

Water and land store heat, too. As we learned in Chapter 2, water warms and cools slowly. This explains why in fall, long after temperatures have dropped, the ocean's water is only a little cooler than in summer. Land and buildings also store heat energy. For example, a brick building that has heated up all day stays warm after the Sun sets.

✓ **READING CHECK:** *Physical Systems* How does the Sun's energy affect Earth? It affects temperature and climate.

Wind and Currents

Air and water both store heat. When they move from place to place, they keep different parts of the world from becoming too hot or too cold. By moving air and water, winds and ocean currents move heat energy between warmer and cooler places. Different parts of the world are kept from becoming too hot or too cold.

When the wind is blowing, air is moving from one place to another. Everyone has experienced these local winds. Global winds also exist. They move air and heat energy around Earth. Ocean currents, which are caused by wind, also move heat energy.

This snow-covered waterfront town is located on Mackinac Island, Michigan.

Teaching Objective 2

ALL LEVELS: (Suggested time: 20 min.) Using the text atlas as a guide, have students draw a rough freehand map of the world and then add lines indicating the patterns of major ocean currents and wind currents described in the text. Students should label the continents and currents. **ENGLISH LANGUAGE LEARNERS**

LEVEL 2: (Suggested time: 40 min.) Have students write a children's book on wind and ocean currents. Topics discussed in the book should include air pressure and how the wind and ocean currents move energy around Earth. Students should also provide illustrations to further explain these subjects.

LEVEL 3: (Suggested time: 35 min.) Organize the students into groups of three. Assign each student air pressure, wind currents, or ocean currents as a topic. Have each student prepare a brief lesson on his or her topic to teach to the other group members. Have students teach the lesson. **COOPERATIVE LEARNING**

Air Pressure To understand why there are winds, we must understand **air pressure**. Air pressure is the weight of the air. Air is a mixture of gases. At sea level, a cubic foot of air weighs about 1.25 ounces (35 grams). We do not feel this weight because air pushes on us from all sides equally. The weight of air, however, changes with the weather. Cold air weighs more than warmer air. An instrument called a barometer measures air pressure.

When air warms, it gets lighter and rises. Colder air then moves in to replace the rising air. The result is wind. Wind travels from areas of high pressure to areas of low pressure. During the day land heats up faster than water. The air over the land heats up faster as well. Along the coast lower air pressure is located over land and higher air pressure is located over water. The air above land rises, and cool air flows in to shore to take its place. At night the land cools more quickly than the water. Air pressure over the land increases, and the wind changes direction.

Earth has several major areas where air pressure stays about the same throughout the year. Along the equator is an area of low air pressure. The pressure is low because the Sun is always warming this area.

▲ Wind shapes Earth and the life that thrives here. For example, this tree has grown in the direction blown by the area's prevailing winds.

Interpreting the Visual Record How do you think wind shapes Earth's landscape?

Reading a Weather Map

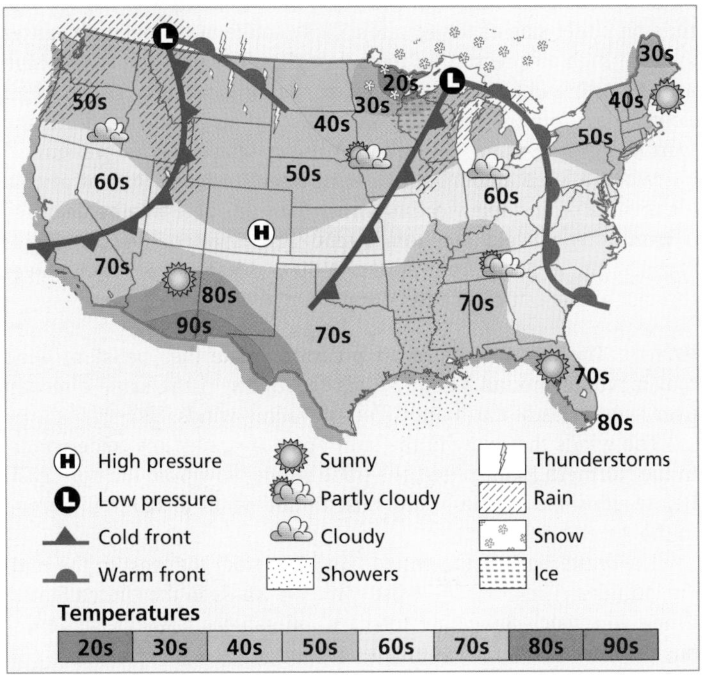

◄ Weather maps show atmospheric conditions as they currently exist or as they are forecast for a particular time period. Most weather maps have legends that explain what the symbols on the map mean. This map shows a cold front sweeping through the central United States. A low-pressure system is at the center of a storm bringing rain and snow to the Midwest. Notice that temperatures behind the cold front are considerably cooler than those ahead of the front.

You Be the Geographer What is the average temperature range for the southwestern states?

Across the Curriculum

SCIENCE

The North Atlantic Oscillation Missionaries in Greenland in the 1700s first noticed the weather cycle known as the North Atlantic Oscillation (NAO). They noted that if Greenland had a severe winter, Denmark had a mild one. The trend also worked the other way around. The NAO's effects stretch from North America's eastern seaboard to much of Europe. Its results can include higher or lower temperatures, heavier or lighter rainfall, changed ocean salinity levels, and longer growing seasons.

The NAO is measured by atmospheric pressure around Iceland and the Azores, which are islands west of Portugal. When atmospheric pressure is unusually high around Iceland it is unusually low over the Azores and vice versa. What triggers these changes in air pressure is still under study.

Activity: Have students work in groups to conduct research on the NAO and how it affects people around the world. Have students create maps to display their findings.

▲ **Visual Record Answer**

by affecting patterns of erosion

◄ **You Be the Geographer Answer**

in the 80s

CLOSE

Tell students that during the 1700s sailors sometimes called the northern edge of the northeastern trade winds belt the "horse latitudes." Ask students to speculate how this region of wind current got its name. *(Ships caught in this calm region would often run low on supplies. To save water, the horses on board would be destroyed so the people could survive.)* Ask students to suggest nicknames for the other wind belts.

REVIEW AND ASSESS

Have students complete the Section Review. Then pair students and assign one student the first Read to Discover question and assign the second student the other question. Have each student write three statements that address his or her assigned question. Have students within each pair exchange their statements. Then have students complete Daily Quiz 3.1.
COOPERATIVE LEARNING

EYE ON EARTH

The stormbeaches of southern England are known for their pebbles, which were washed up by rising sea levels and large waves some 10,000 years ago. The name of Chesil Beach comes from an old word for pebbles. Because of the prevailing winds and currents, the beach's pebbles vary in size. At the western end of this 16-mile (26 km) beach the pebbles are small, but they are larger farther east.

Fishermen coming ashore in a fog claim that they can determine their location along Chesil Beach by the size of the pebbles. Long ago, smugglers landing at night also checked the pebbles' size to determine their location.

Activity: Ask students to choose a coastal area of the United States to study. Then organize students into small groups and have each group conduct research on how winds and ocean currents have affected the dunes, sands, gravels, and other features of the area's beaches. Have each group present its findings to the class.

You Be the Geographer Answer

An area of unstable weather forms.

Pressure and Wind Systems

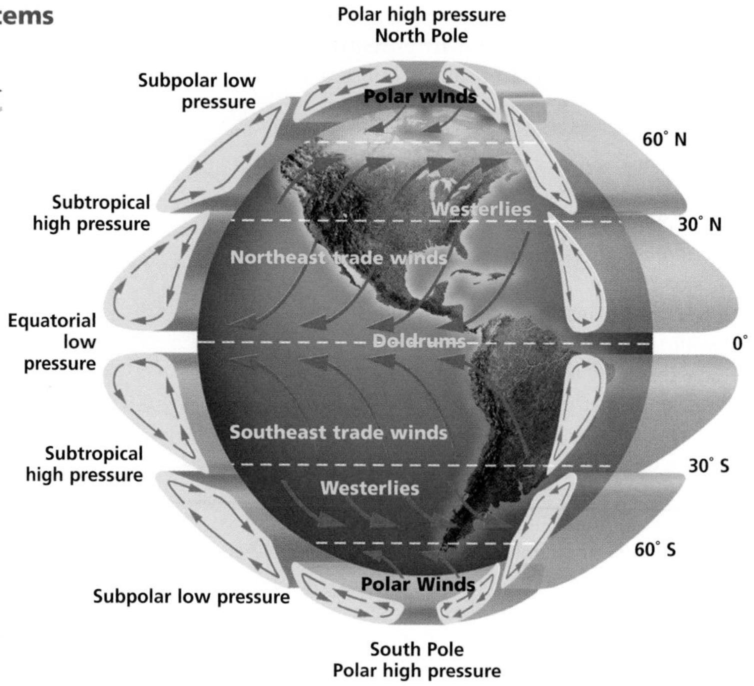

Winds between Earth's high- and low-pressure zones help regulate the globe's energy balance.

You Be the Geographer **What happens when warm westerlies come into contact with cold polar winds?**

This ship uses wind to travel. As more wind catches the sails, the ship moves faster.

This warm air rises along the equator and moves north and south. Some of the warm air cools and sinks once it reaches about 30° latitude on either side of the equator. This change in temperature causes areas of high air pressure in the subtropics. The pressure is also high at the North and South Poles because the air is so cold. This cold air flows away from the Poles. Because the cold air is heavier, it lifts the warmer air in its path. The subpolar regions have low air pressure.

When a large amount of warm air meets a large amount of cold air, an area of unstable weather forms. This unstable weather is called a **front**. When cold air from Arctic and Antarctic regions meets warmer air, a polar front forms. When this type of front moves through an area, it can cause storms.

Winds The major areas of air pressure create the "belts" of wind that move air around Earth. Winds that blow in the same direction over large areas of Earth are called prevailing winds.

The winds that blow in the subtropics are called the trade winds. In the Northern Hemisphere, the trade winds blow from the northeast. Before ships had engines, sailors used trade winds to sail from Europe to the Americas.

The westerlies are the winds that blow from the west in the middle latitudes. These are the most common winds in the United States. When you watch a weather forecast you can see how the westerlies push storms across the country from west to east.

RETEACH

Have students complete Main Idea Activity 3.1. Then organize the class into three groups and assign each group one of the Read to Discover questions. Have each group write, and then present to the class, three true-false statements related to its question. If the statement is false, call on a group member to explain why it is false. **ENGLISH LANGUAGE LEARNERS, COOPERATIVE LEARNING**

EXTEND

Have interested students conduct research on buildings designed to take advantage of stored energy. You might have them collect samples of materials from local building supply stores and organize a display. **BLOCK SCHEDULING**

Few winds blow near the equator. This area is called the doldrums. Here, warm air rises rather than blowing east or west.

Ocean Currents Winds make ocean water move in the same general directions as the air above it moves. Warm ocean water from the tropics moves in giant streams, or **currents**, to colder areas. Cold water moves in streams from the polar areas to the tropics. This moves energy between different places. Warm air and warm ocean currents raise temperatures. On the other hand, cold winds and cold ocean currents lower temperatures.

The Gulf Stream is an ocean current. It moves warm water north along the east coast of the United States. The Gulf Stream then moves across the Atlantic Ocean toward western Europe. The warm air that moves with it keeps winters mild. As a result, areas such as Ireland have warmer winters than areas in Canada that are just as far north.

Ocean currents also bring heat energy into the Arctic Ocean. In the winter, warm currents create openings in the ice. These openings give arctic whales a place to breathe. They also give people a place to catch fish. Cold water also flows out of the Arctic Ocean into warmer waters of the Pacific and Atlantic Oceans. The cold water sinks below warmer water, causing mixing. This mixing brings food to sea life.

▲ Plants and animals adapt to their particular environments on Earth.

Interpreting the Visual Record **How do you think the bearded seal is able to live in Norway's cold environment?**

✓ **READING CHECK:** *Physical Systems* How do winds and currents create patterns on Earth's surface? **They move warm air and warm water from one place to another.**

go.hrw.com Homework Practice Online
Keyword: SK3 HP3

Section 1 Review

Define and explain: weather, climate, greenhouse effect, air pressure, front, currents

Reading for the Main Idea

1. *Physical Systems* How does Earth's temperature stay balanced?
2. *Physical Systems* How does the greenhouse effect allow Earth to store the Sun's energy?

Critical Thinking

3. **Drawing Inferences and Conclusions** How would weather in western Europe be different if there were no Gulf Stream?
4. **Analyzing Information** Why were the trade winds important to early sailors?

Organizing What You Know

5. **Categorizing** Copy the following graphic organizer. Use it to show the names, locations, and directions of wind and air pressure belts.

Name	Location	Direction

Section 1 Review Answers

Define For definitions, see: weather, p. 37; climate, p. 37; greenhouse effect, p. 38; air pressure, p. 39; front, p. 40; currents, p. 41

Reading for the Main Idea

1. As Earth receives energy from the Sun, it also loses energy that escapes into space. (NGS 7)
2. Earth's atmosphere, water, and land all trap heat. (NGS 7)

Critical Thinking

3. Europe's winters would be colder.
4. Sailors used the trade winds to sail from Europe to the Americas.

Organizing What You Know

5. trade winds—subtropics, from the northeast; westerlies—middle latitudes, west to east; doldrums—equator, warm air rises but does not blow in any direction

◀ **Visual Record Answer**

The bearded seal has a layer of fat that insulates it from the cold.

CASE STUDY

Setting the Scene

Every year, hurricanes torment residents of the Caribbean islands, the coastlands bordering the Gulf of Mexico, and the East Coast of the United States. Because hurricanes get their strength from warm water, hurricane season lasts through summer and into fall. These storms carry tremendous energy. In one day an average hurricane releases at least 8,000 times the daily electrical power output of the United States. Severe hurricanes can cause billions of dollars of damage. Fewer lives are lost now than in years past, however, because early warning systems help predict the storms' paths and power. Coastal towns and cities evacuate people before the storms arrive. Satellites provide much of the information used to make storm predictions.

Building a Case

Have students read "Hurricane: Tracking a Natural Hazard" and follow the instructions in You Be the Geographer. Ask on what date the atmospheric pressure was lowest. *(10/27)* How did Mitch register on the Saffir-Simpson Scale that day? *(category 5)* Compare Hurricane Mitch with the hurricane that struck Galveston on September 8, 1900.

The storm headed for Galveston was first observed on August 30. The Weather Bureau placed Galveston under a storm warning on September 7. September 8 dawned rainy and gusty. Though the storm worsened, few residents left the city. At 6:30 P.M. a storm surge flooded the city. The lowest barometer reading was 27.91. Windspeed was estimated at more than 120 mph. By 10:00 P.M. much of the city was wrecked. As many as 8,000 city residents died.

HISTORICAL GEOGRAPHY

Hurricanes and typhoons—as these large storms are called when they occur in the Pacific Ocean—have changed history. Here are just three examples.

In 1281 the Mongol ruler Kublai Khan was ready to invade Japan, but a typhoon scattered his huge fleet of ships. A second storm, dubbed the Great Hurricane, ravaged the Caribbean in October 1780. It killed approximately 22,000 people and may be the deadliest hurricane on record. British and French fleets involved in the American Revolutionary War were both ravaged. Finally, in December 1944, during World War II, a sudden typhoon east of the Philippines caught the U.S. Third Fleet by surprise. Three destroyers, 146 aircraft, and several hundred men were lost.

Critical Thinking: How could early storm warning technology have changed world history?

Answer: Answers will vary but students might mention a successful invasion by Kublai Khan, fewer deaths in the Caribbean, and fewer ships and lives lost during the Revolutionary War and World War II.

➤ **This Case Study feature addresses National Geography Standards 4, 15, and 17.**

CASE STUDY

HURRICANE: TRACKING A NATURAL HAZARD

Hurricanes are large circulating storms that begin in tropical oceans. Hurricanes often move over land and into populated areas. When a hurricane approaches land, it brings strong winds, heavy rains, and large ocean waves.

The map below shows the path of Hurricane Fran in 1996. Notice how Fran moved to the west and became stronger until it reached land. It began as a tropical depression and became a powerful hurricane as it passed over warm ocean waters.

Scientists who study hurricanes try to predict where these storms will travel. They want to be able to warn people in the hurricane's path. Early warnings can help people be better prepared for the deadly winds and rain. It is a difficult job because hurricanes can change course suddenly. Hurricanes are one of the most dangerous natural hazards.

One way of determining a hurricane's strength is by measuring the atmospheric pressure inside it. The lower the pressure, the stronger the storm. Hurricanes are rated on a scale of one to five. Study Table 1 to see how wind speed and air pressure are used to help determine the strength of a hurricane.

Hurricane Mitch formed in October 1998. The National Weather Service (NWS) recorded Mitch's position and strength. They learned that Mitch's pressure was one of the lowest ever recorded. The NWS estimated that Mitch's maximum sustained surface winds reached 180 miles per hour.

Table 1: Saffir-Simpson Scale

Hurricane Type	Wind Speed MPH	Air Pressure MB (Inches)
Category 1	74–95	more than 980 (28.94)
Category 2	96–110	965–979 (28.50–28.91)
Category 3	111–130	945–964 (27.91–28.47)
Category 4	131–155	920–944 (27.17–27.88)
Category 5	more than 155	919 (27.16)

Source: Florida State University, <http://www.met.fsu.edu/explores/tropical.html>

Path of Hurricane Fran, 1996

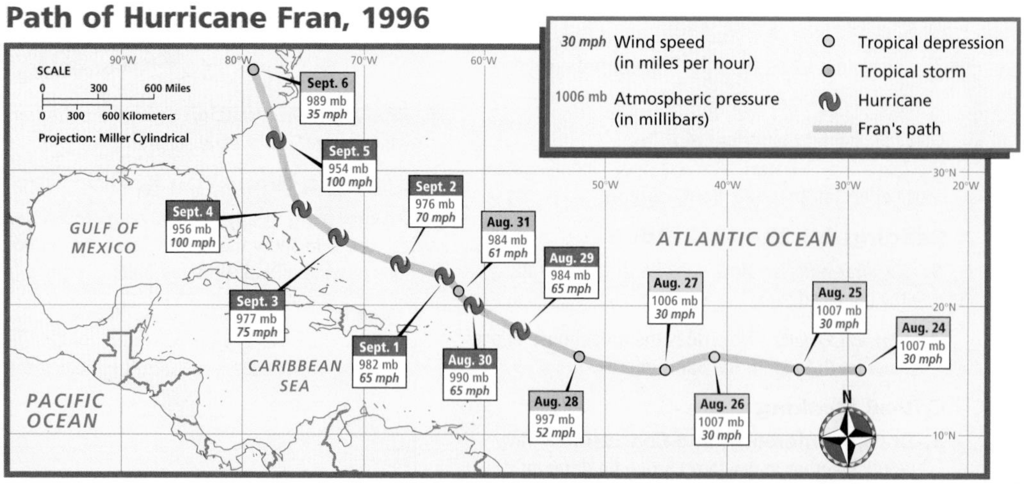

Drawing Conclusions

Lead a discussion comparing the two storms. According to wind speed and air pressure, what level storm was the Galveston hurricane? *(3)* Which was the stronger storm? *(Mitch)* Which hurricane lasted longer? *(Mitch)* Why did the 1900 storm kill so many people in such a short time? *(They had not evacuated the city.)* If there had been no warning system, how might Hurricane Mitch have affected the Caribbean region? *(It might have killed even more people.)*

What might have happened if Galveston had been warned earlier? Have students prepare and present an alternate newscast for the morning of September 9, 1900, based on this possibility.

Going Further: Thinking Critically

Locate detailed maps of the Gulf of Mexico or Atlantic coasts of the United States. Use maps of different areas or concentrate on one region. Have students work in groups to answer some or all of these questions:

- What cities and towns might be threatened by a hurricane? Can students estimate how many people live in the area?
- What routes could residents use to evacuate? What factors might slow evacuation? If they could travel about 30 mph (48 km/h), how far could people travel in one day? two days?
- What would happen if residents were warned just a few hours before a hurricane? What effect might an early warning system have on this region?

Table 2: Hurricane Mitch, 1998 Position and Strength

Date	Latitude (degrees)	Longitude (degrees)	Wind Speed (mph)	Pressure (millibars)	Storm Type
10/22	12 N	78 W	30	1002	Tropical depression
10/24	15 N	78 W	90	980	Category 2
10/26	16 N	81 W	130	923	Category 4
10/27	17 N	84 W	150	910	Category 5
10/31	15 N	88 W	40	1000	Tropical storm
11/01	15 N	90 W	30	1002	Tropical depression
11/03	20 N	91 W	40	997	Tropical storm
11/05	26 N	83 W	50	990	Tropical storm

Source: <http://www.met.fsu.edu/explores/tropical.html>

Hurricanes like Mitch cause very heavy rains in short periods of time. These heavy rains are particularly dangerous. The ground becomes saturated, and mud can flow almost like water. The flooding and mudslides caused by Mitch killed an estimated 10,000 people in four countries. Many people predicted that the region would not recover without help from other countries.

In the southeastern United States, many places have emergency preparedness units. The people assigned to these groups organize their communities. They provide food, shelter, and clothing for those who must evacuate their homes.

▲ This satellite image shows the intensity of Hurricane Mitch. With advanced technology, hurricane tracking is helping to save lives.

You Be the Geographer

1. Trace a map of the Caribbean. Be sure to include latitude and longitude lines.
2. Use the data about Hurricane Mitch in Table 2 to plot its path. Make a key with symbols to show Mitch's strength at each location.
3. What happened to Mitch when it reached land?

You Be the Geographer

1. Students may trace the map on the previous page. Or, provide an outline map to students.
2. On student maps, from its first position Hurricane Mitch should progress north-northwest toward Cuba, swing southwest toward Nicaragua, then back northeast across the Yucatán Peninsula on its way to the open Atlantic Ocean.
3. When it reached land, Mitch's strength weakened.

internet connect

GO TO: **go.hrw.com**
KEYWORD: SK3 CH3
FOR: Web sites about hurricanes

Section 2

Objectives

1. Describe what is included in the study of weather.
2. Identify the major climate types, and describe the types of plants that live in each.

Focus

LET'S GET STARTED

Copy the following instructions onto the chalkboard: *Write down three words or phrases you use to describe the weather in our area. (Possible answers: muggy, bone-chilling, gully-washer, nor'easter, raining cats and dogs)* Point out that having so many terms for our weather indicates that weather is very important to us. Tell students that in Section 2 they will learn that climate and weather affect plants and practically all other forms of life.

Building Vocabulary

Write the key terms on the chalkboard. Call on volunteers to find and read aloud the definitions in the text or glossary. Point out that **monsoon** actually has two meanings. Originally, it referred to the wind systems that bring wet and dry seasons to tropical areas. Its more common usage refers to heavy rains brought in by the winds during summer. You may also want to introduce *orographic effect* as the term for what causes a **rain shadow**.

Section 2 Resources

Reproducible
- Guided Reading Strategy 3.2
- Graphic Organizer 3
- Map Activity 3

Technology
- One-Stop Planner CD–ROM, Lesson 3.2
- Homework Practice Online
- Earth: Forces and Formations CD–ROM/Seek and Tell/Forces and Processes
- *ARGWorld* CD–ROM: Using Climagraphs to Interpret Seasons
- HRW Go site

Reinforcement, Review, and Assessment
- Section 2 Review, p. 50
- Daily Quiz 3.2
- Main Idea Activity 3.2
- English Audio Summary 3.2
- Spanish Audio Summary 3.2

Section 2 Earth's Climate and Vegetation

Read to Discover

1. What is included in the study of weather?
2. What are the major climate types, and what types of plants live in each?

Define

rain shadow
monsoon
arid
steppe climate
hurricanes
typhoons
tundra climate
permafrost

Why It Matters

Flooding often becomes a problem during severe weather. Use CNNfyi.com or other **current events** sources to find examples of the effects of flooding on nations around the world. Record your findings in your journal.

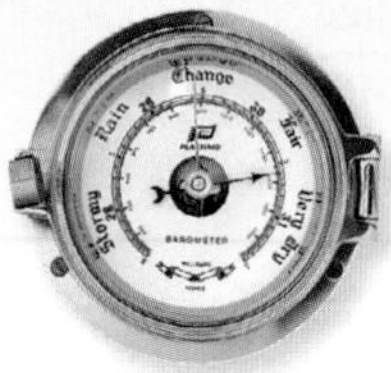

A barometer

Nature has many incredible sights. Lightning is one of the most spectacular occurrences. These flashes of light are produced by a discharge of atmospheric electricity.

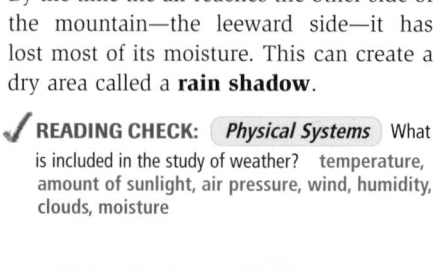

Weather

As you have read, the condition of the atmosphere in a local area for a short period of time is called weather. Weather is a very general term. It can describe temperature, amount of sunlight, air pressure, wind, humidity, clouds, and moisture.

When warm and cool air masses come together, they form a front. Cold air lifts the warm air mass along the front. The air that is moved to higher elevations is cooled. If moisture is present in the lifted air, this cooling causes clouds to form. The moisture may fall to Earth as rain, snow, sleet, or hail. Moisture that falls in any form is called precipitation.

Another type of lifting occurs when warm, moist air is blown up against a mountain and forced to rise. The air cools as it is lifted, just as when air masses collide. Clouds form, and precipitation falls. The side of the mountain facing the wind—the windward side—often gets heavy rain. By the time the air reaches the other side of the mountain—the leeward side—it has lost most of its moisture. This can create a dry area called a **rain shadow**.

✓ **READING CHECK:** *Physical Systems* What is included in the study of weather? **temperature, amount of sunlight, air pressure, wind, humidity, clouds, moisture**

TEACH

Teaching Objective 1

LEVEL 1: (Suggested time: 30 min.) Have students read the section. Then pair students and have each pair create an informational brochure titled "The Study of Weather." In their brochures, students should identify and create a symbol for each of the elements of weather that can be studied. *(Brochures should identify temperature, amount of sunlight, air pressure, wind, humidity, clouds, and moisture.)* Ask volunteers to present and explain their brochures to the class.
ENGLISH LANGUAGE LEARNERS, COOPERATIVE LEARNING

LEVEL 2: (Suggested time: 30 min.) Prepare photocopies of the local weather map from your daily newspaper or the U.S. weather map in a national paper. Review the "Reading a Weather Map" diagram in Section 1. Have students work in pairs to circle and label the information on their map that illustrates the conditions that affect weather. *(See the Level 1 lesson for the correct elements of weather.)*
COOPERATIVE LEARNING

Landforms and Precipitation

As moist air from the ocean moves up the windward side of a mountain, it cools. The water vapor in the air condenses and falls in the form of rain or snow. Descending, the drier air then moves down the leeward side of the mountain. This drier air brings very little precipitation to areas in the rain shadow.

Low-Latitude Climates

If you charted the average weather in your community over a long period of time, you would be describing the climate for your area. Geographers have devised ways of describing climates based mainly on temperature, precipitation, and natural vegetation. In general, an area's climate is related to its latitude. In the low latitudes—the region close to the equator—there are two main types of climates: humid tropical and tropical savanna.

Humid Tropical Climate A humid tropical climate is warm and rainy all year. People living in this climate do not see a change from summer to winter. This is because the Sun heats the region throughout the year. The heat in the tropics causes a great deal of evaporation and almost daily rainstorms. One of the most complex vegetation systems in the world—the tropical rain forest—exists in this climate. The great rain forests of Brazil, Indonesia, and Central Africa are in the equatorial zone.

Some regions at higher latitudes have warm temperatures all year but have strong wet and dry seasons. Bangladesh and coastal India, for example, have an extreme wet season during the summer. Warm, moist air from the Indian Ocean reaches land. The air rises and cools, causing heavy rains. The rains continue until the wind changes direction in the fall. This seasonal shift of air flow and rainfall is known as a **monsoon**. A monsoon may be wet or dry. The monsoon system is particularly important in Asia.

The people of Tamil Nadu, India, adjust to the monsoon season.

Interpreting the Visual Record

What problems might people face during the wet monsoon?

ENVIRONMENT AND SOCIETY

The monsoon is a seasonal wind that shifts direction twice a year. Summer monsoon winds bring heavy rains from the oceans to Asia. Beginning in October, the winter monsoon brings cool, dry air to India and China and rain to Indonesia and Australia.

The monsoon is vital to Asian agriculture and to the survival of billions of people, particularly those who rely on the success of a single harvest. The weather system formed by monsoon winds transfers heat to and from Asia and helps balance temperatures. Without the monsoons, regions that receive the Sun's direct rays would be scorched and other areas would be severely cold.

Critical Thinking: How might Asian farmers reduce their dependence on monsoons?

Answer: Farmers could dig deep wells or develop irrigation systems.

internet connect

GO TO: go.hrw.com
KEYWORD: SK3 CH3
FOR: Web sites about sprites

Visual Record Answer

flooded streets and buildings, contaminated water supplies, drowned livestock, and others

Teaching Objective 2

ALL LEVELS: (Suggested time: 10 min.) To help students understand the relationship between climate and latitude, copy the following graphic organizer onto the chalkboard, omitting the italicized answers. Pair students and have each pair complete the organizer by filling in the correct latitudinal groupings. Ask students what climate types are not represented on the organizer. *(Students should mention desert, steppe, and highland climates.)* Call on volunteers to describe those climates and their locations. Then have students describe the types of plants that live in each climate. **ENGLISH LANGUAGE LEARNERS, COOPERATIVE LEARNING**

CLIMATE TYPES AND LATITUDE

high	*subarctic, tundra, and ice cap*
middle	*Mediterranean, humid subtropical, marine west coast, and humid continental*
low	*humid tropical and tropical savanna*
middle	*Mediterranean, humid subtropical, marine west coast, and humid continental*
high	*subarctic, tundra, and ice cap*

GLOBAL PERSPECTIVES

The Sahara was once a vast grassland. Researchers recently concluded that the Sahara's transformation into a desert was triggered by changes in Earth's orbit and the tilt of Earth's axis. These changes occurred in two phases, with the first occurring some 6,700 to 5,500 years ago and the second 4,000 to 3,600 years ago.

The orbital changes caused fairly sudden changes in North Africa's climate. Rains stopped coming to the Sahara and regional temperatures rose. Within a few hundred years the moist Sahara became a desert shrubland. Researchers think that ancient civilizations in the Sahara may have moved to the Nile River valley in response to the climate changes.

Critical Thinking: What effect did the change that occurred in Earth's orbit have on African civilizations?

Answer: Researchers believe people may have moved from the Sahara to the Nile River valley because of climate changes.

Visual Record Answer ►

lack of vegetation

Our Amazing Planet

In terms of loss of life, the worst natural disaster in U.S. history is still the hurricane that hit Galveston, Texas, on September 8, 1900. As many as 8,000 people died in Galveston.

Tropical Savanna Climate There is another type of tropical climate that has wet and dry seasons. However, this climate does not have the extreme shifts found in a monsoon climate. It is called a tropical savanna climate. The tropical savanna climate has a wet season soon after the warmest months. It has a dry season soon after the coolest months. Total rainfall, however, is fairly low. Vegetation in a tropical savanna climate is grass with scattered trees and shrubs.

✓ **READING CHECK:** *Physical Systems* Which climate regions are in the low latitudes? humid tropical and tropical savanna

Dry Climates

Temperature and precipitation are the most important parts of climate. Some regions, for example, experience strong hot and cold seasons. Other regions have wet and dry seasons. Some places are wet or dry all year long. If an area is **arid** (dry) it receives little rain. Arid regions usually have few streams and plants.

Desert Climate Most of the world's deserts lie near the tropics. The high air pressure and settling air keep these desert climate regions dry most of the time. Other deserts are located in the interiors of continents and in the rain shadows of mountains. Few plants can survive in the driest deserts, so there are many barren, rocky, or sandy areas. Dry air and clear skies permit hot daytime temperatures and rapid cooling at night.

Steppe Climate Another dry climate—the **steppe climate**—is found between desert and wet climate regions. A steppe receives more rainfall than a desert climate. However, the total amount of precipitation is still low. Grasses are the most common plants, but trees can grow along creeks and rivers. Farmers can grow crops but usually need to irrigate. Steppe climates occur in Africa, Australia, Central Asia, eastern Europe, in the Great Plains of the United States and Canada, and in South America.

✓ **READING CHECK:** *The World in Spatial Terms* What are the dry climates? desert and steppe

Palo Duro Canyon State Park in Texas attracts thousands of visitors each year.

Interpreting the Visual Record How can you tell that this canyon is located in a dry climate?

ALL LEVELS: (Suggested time: 10 min.) Tell students to use the world climate regions map on this page to locate the region where they live. Have them refer to the chart below the map to identify the state's climate region. Then tell them to read the description of the weather patterns for their climate region. Ask whether the description corresponds to their observations of weather patterns in the area. **ENGLISH LANGUAGE LEARNERS**

LEVEL 3: (Suggested time: 30 min.) Have students write trivia questions about each of the climate types. Tell students to refer to the world climate regions map on this page and the text atlas to help them write questions. *(Possible questions: What is the coldest temperature ever recorded at the South Pole? How hard can winds blow in the tropics? What is the hottest place in the middle latitudes?)* Then have students conduct research to find the answers to the questions.

World Climate Regions

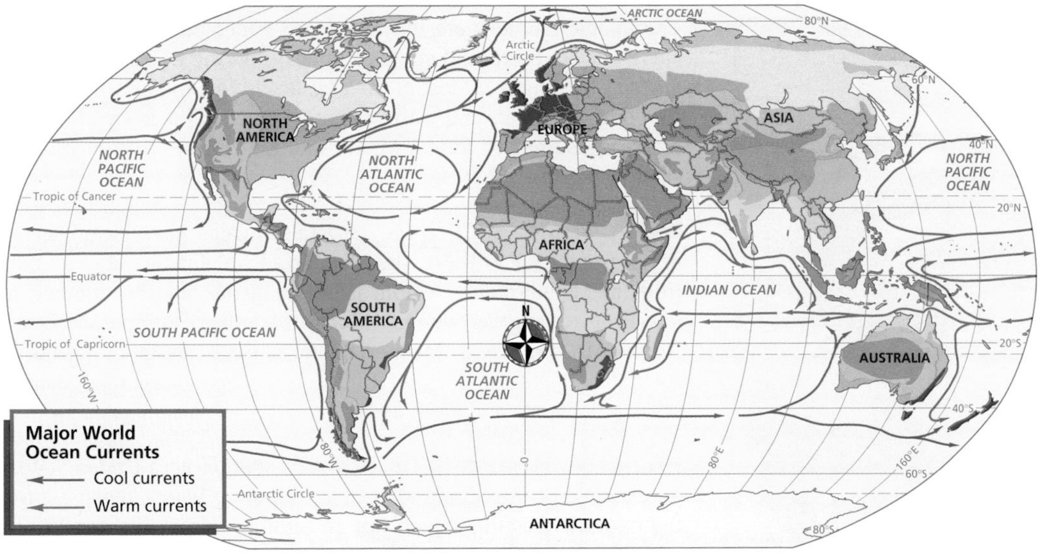

	Climate	Geographic Distribution	Major Weather Patterns	Vegetation
Low Latitudes	HUMID TROPICAL	along the equator	warm and rainy year-round, with rain totaling anywhere from 65 to more than 450 in. (165–1,143 cm) a year	tropical rain forest
	TROPICAL SAVANNA	between the humid tropics and the deserts	warm all year; distinct rainy and dry seasons; at least 20 in. (51 cm) of rain during the summer	tropical grassland with scattered trees
Dry	DESERT	centered along 30° latitude; some middle-latitude deserts are in the interior of large continents and along their western coasts	arid; less than 10 in. (25 cm) of rain a year; sunny and hot in the tropics and sunny with wide temperature ranges during the day in middle latitudes	a few drought-resistant plants
	STEPPE	generally bordering deserts and interiors of large continents	semiarid; about 10–20 in. (25–51 cm) of precipitation a year; hot summers and cooler winters with wide temperature ranges during a day	grassland; few trees
Middle Latitudes	MEDITERRANEAN	west coasts in middle latitudes	dry, sunny, warm summers and mild, wetter winters; rain averages 15–20 in. (38–51 cm) a year	scrub woodland and grassland
	HUMID SUBTROPICAL	east coasts in the middle latitudes	hot, humid summers and mild, humid winters; rain year-round; coastal areas are in the paths of hurricanes and typhoons	mixed forest
	MARINE WEST COAST	west coasts in the upper-middle latitudes	cloudy, mild summers and cool, rainy winters; strong ocean influence; rain averages 20–60 in. (51–152 cm) a year	temperate evergreen forest
	HUMID CONTINENTAL	east coasts and interiors of upper-middle latitude continents	four distinct seasons; long, cold winters and short, warm summers; amounts of precipitation a year vary	mixed forest
High Latitudes	SUBARCTIC	higher latitudes of the interior and east coasts of continents	extremes of temperature; long, cold winters and short, warm summers; little precipitation all year	northern evergreen forest
	TUNDRA	high-latitude coasts	cold all year; very long, cold winters and very short, cool summers; little precipitation	moss, lichens, low shrubs; permafrost marshes
	ICE CAP	polar regions	freezing cold; snow and ice year-round; little precipitation	no vegetation
	HIGHLAND	high mountain regions	temperatures and amounts of precipitation vary greatly as elevation changes	forest to tundra vegetation, depending on elevation

Cultural Kaleidoscope

Religion and Weather Cultural and religious explanations of climate and weather have influenced people's understanding of the physical world. The Hopi's rain dance reveals that culture's attitude toward nature. By dancing and making offerings, the Hopi show respect to the kachinas, or spirits of ancestors, that they believe control the natural world. The Hopi believe the kachinas will return the favor by sending rain. Many other cultures have ceremonies designed to affect the weather.

Drought conditions in Texas during 1999 prompted community and church leaders there to pray for rain. The mayor of one city proclaimed a day of prayer for rain. One church pastor noted that although farmers can use the most modern equipment and seed, the success of a harvest is often beyond human control.

Activity: Have students conduct research on ways that weather has been incorporated into other cultures' religious beliefs or folklore and analyze the similarities and differences. Have students present their findings to the class.

➤ASSIGNMENT: Prepare a list of 10 to 20 major world cities. Include cities in all climate zones (except for the ice cap zone). Give students copies of the list and have them locate the cities in the text atlas or on a classroom globe. Then have them write down each city's climate type and the temperatures, moisture, vegetation, sunlight levels, and storms that would be common there.

TEACHER TO TEACHER

Maureen Dempsey of Spring Creek, Nevada, suggests the following activity to help students summarize climate types. Organize students into small groups. Have each group fold and staple 12 sheets of white paper together to create booklets titled *Climates of the World.* Students should describe each climate type on one page and illustrate it on the opposite page. They may draw pictures, cut pictures from magazines, or download images from the Internet. Each book should include a table of contents and a cover illustration.

National Geography Standard 14

Environment and Society. Native to the Mediterranean region, the olive tree is important to the culture, history, and economy of that region. According to myth, the goddess Athena gave the ancient Greeks the olive tree to use for food, fuel, and healing wounds.

Wild olive trees are believed to have originated in Asia Minor and spread westward. The Romans then introduced the trees into regions they conquered. Later, Arabs began cultivating the olive. By the 1700s olive groves had been planted by the Spanish in places such as California, Mexico, and Chile.

Activity: Have students conduct research on other food plants that have been introduced into various geographical regions that share the same climate. Then have students create databases that show the means and results of the plants' introduction into different regions.

Middle-Latitude Climates

The middle latitudes are the two broad zones between Earth's polar circles (66.5° north and south latitudes) and the tropics (23.5° north and south latitudes). Most of the climates in the middle latitudes have cool or cold winters and warm or hot summers. Climates with wet and dry seasons are also found. Middle-latitude climates may include rain shadow deserts.

Mediterranean Climate Several climates have clear wet and dry seasons. One of these, the Mediterranean climate, takes its name from the Mediterranean region. This climate has hot, dry summers followed by cooler, wet winters. Much of southern Europe and coastal North Africa have a Mediterranean climate. Parts of California, Australia, South Africa, and Chile do as well. Vegetation includes scrub woodlands and grasslands.

▲ The Mediterranean coast of France is part of the Riviera. The Riviera is located in a Mediterranean climate region.

Interpreting the Visual Record **What features would make this area popular with tourists?**

Humid Subtropical Climate The southeastern United States is an example of the humid subtropical climate. Warm, moist air from the ocean makes this region hot and humid in the summer. Winters are mild, but snow falls occasionally. People in a humid subtropical climate experience **hurricanes** and **typhoons**. These are tropical storms that bring violent winds, heavy rain, and high seas. The humid subtropical climate supports areas of mixed forests where deciduous and coniferous forests blend. Deciduous trees lose their leaves during the fall each year. Coniferous trees have needle-shaped leaves that remain green year-round.

Marine West Coast Climate Some coastal areas of North America and much of western Europe have a marine west coast climate. Westerly winds carry moisture from the ocean across the land, causing winter rainfall. Evergreen forests can grow in these regions because of regular rain.

The Mediterranean has a variety of vegetation. This Mediterranean scrub forest is in Corfu, Greece.

Interpreting the Visual Record **What kinds of vegetation do you see in this scrub forest?** ►

Visual Record Answers

▲ Students might suggest hills, beaches, warm water, and the region's mild climate.

► scrub bushes and trees

CLOSE

Point out to students that people live in almost every climate on Earth, including some of the most extreme. Have students describe those extreme climates and make suggestions on ways that people have adapted to them.

REVIEW AND ASSESS

Have students complete the Section Review. Then have each student choose a climate type and write eight adjectives describing that climate. Have students read their descriptions to the class, and have other students guess which climate type is being described. Then have students complete Daily Quiz 3.2.

Humid Continental Climate Farther inland are regions with a humid continental climate. Winters in this region bring snowfall and cold temperatures, but there are some mild periods too. Summers are warm and sometimes hot. Most of the shifting weather in this climate region is the result of cold and warm air coming together along a polar front. Humid continental climates have four distinct seasons. Much of the midwestern and northeastern United States and southeastern Canada have a humid continental climate. This climate supports mixed forest vegetation.

✓ **READING CHECK:** *The World in Spatial Terms* What are the middle-latitude climates? Mediterranean, humid subtropical, marine west coast, humid continental

High-Latitude Climates

Closer to the poles we find another set of climates. They are the high-latitude climates. They have cold temperatures and little precipitation.

Subarctic Climate The subarctic climate has long, cold winters, short summers, and little rain. In the inland areas of North America, Europe, and Asia, far from the moderating influence of oceans, subarctic climates experience extreme temperatures. However, summers in these regions can be warm. In the Southern Hemisphere there is no land in the subarctic climate zone. As a result, boreal (BOHR-ee-uhl) forests are found only in the Northern Hemisphere. Trees in boreal forests are coniferous and cover vast areas in North America, Europe, and northern Asia.

Tundra Climate Farther north lies the **tundra climate**. Temperatures are cold, and rainfall is low. Usually just hardy plants, including mosses, lichens, and shrubs, survive here. Tundra summers are so short and cool that a layer of soil stays frozen all year. This frozen layer is called **permafrost**. It prevents water from draining into the soil. As a result, many ponds and marshes appear in summer.

Ice Cap Climate The polar regions of Earth have an ice cap climate. This climate

Wildlife eat the summer vegetation in the Alaskan tundra.

Interpreting the Visual Record **Why is there snow on the mountain peaks during summer?**

PHYSICAL SYSTEMS

Permafrost lies under some 20 to 25 percent of the world's land surface. It occurs in more than 50 percent of Russia and Canada and more than 80 percent of Alaska.

Permafrost gives scientists a window into the plant and animal life of the past. The various layers of permafrost contain plant and animal remains from different periods of Earth's history. Some of these layers are more than 30,000 years old.

Scientists also use permafrost to assess the rate of global warming. By studying ground temperatures and preserved plant and animal life, scientists can understand past climatic conditions and current temperature changes.

Critical Thinking: What might the presence of oak tree remains in a layer of permafrost indicate about past temperatures in a tundra region?

Answer: Students might suggest that the tundra region's temperatures were warmer in a previous era.

◄ **Visual Record Answer**

High elevations keep temperatures low.

RETEACH

Have students complete Main Idea Activity 3.2. Then have them choose a certain locale and create postcards they would send from there. On the picture side they should illustrate the climate for that area and on the other write a description of the weather on a typical day of their "vacation." Call on volunteers to read their postcards, until all the major climate types have been reviewed. Display the postcards around the classroom.
ENGLISH LANGUAGE LEARNERS

EXTEND

Have interested students perform a simple experiment to illustrate dew point—the temperature at which water vapor begins to condense. Put water into a coffee can and gradually add ice cubes while swirling a thermometer carefully in the icy water. Record the temperature at the moment when condensation forms on the outside of the coffee can. This temperature is the dew point. Have students conduct research on the relationship between relative humidity and dew point to explain the results of their experiment. Students may also want to investigate places in the world where very little rain falls and where dew sustains plants and animals. Have them briefly report their findings to the class.
BLOCK SCHEDULING

Section 2 Review

Answers

Define For definitions, see: rain shadow, p. 44; monsoon, p. 45; arid, p. 46; steppe climate, p. 46; hurricanes, p. 48; typhoons, p. 48; tundra climate, p. 49; permafrost, p. 49

Reading for the Main Idea

1. Weather is the condition of the atmosphere at a given place and time. Climate refers to regional weather conditions over a long period of time. (NGS 7)
2. humid tropical, tropical savanna, desert, steppe, Mediterranean, humid subtropical, marine west coast, humid continental, subarctic, tundra, ice cap, and highland (NGS 3)

Critical Thinking

3. to be prepared for regular seasonal changes as well as potentially dangerous weather conditions
4. the closer a latitude is to the equator, the warmer the climate, generally

Organizing What You Know

5. For climate types, latitudes, and characteristics, see the table on p. 47.

Visual Record Answer ►

The building is located underground and appears to be well insulated.

▲ The Amundsen-Scott South Pole Station is a research center in Antarctica.

Interpreting the Visual Record **How does the design of this building show ways people have adapted to the ice cap climate?**

is cold—the monthly average temperature is below freezing. Precipitation averages less than 10 inches (25 cm) annually. Animals adapted to the cold, like walruses, penguins, and whales, are found here. No vegetation grows in this climate.

✓ **READING CHECK:** *The World in Spatial Terms* How are tundra and ice cap climates different? Ice cap—snow and ice year-round, no vegetation; tundra—short, cool summer, hardy plants

Highland Climates

Mountains usually have several different climates in a small area. These climate types are known as highland climates. If you went from the base of a high mountain to the top, you might experience changes similar to going from the tropics to the Poles! The vegetation also changes with the elevation. It varies from thick forests or desert to tundra. Lower mountain elevations tend to be similar in temperature to the surrounding area. On the windward side, however, are zones of heavier rainfall or snowfall. As you go uphill, the temperatures drop. High mountains have a tundra zone and an icy summit.

✓ **READING CHECK:** *The World in Spatial Terms* What are highland climate regions? the climates around mountains; they differ greatly in a relatively small area

Section 2 Review

go.hrw.com Homework Practice Online
Keyword: SK3 HP3

Define and explain: rain shadow, monsoon, arid, steppe climate, hurricanes, typhoons, tundra climate, permafrost

Reading for the Main Idea

1. *Physical Systems* How is weather different from climate?
2. *The World in Spatial Terms* What are the major climate regions of the world?

Critical Thinking

3. **Drawing Inferences and Conclusions** Why is it important to understand weather and climate patterns?
4. **Analyzing Information** How does latitude influence climate?

Organizing What You Know

5. **Categorizing** Copy the following graphic organizer. Use it to describe Earth's climate types.

Climate	Latitudes	Characteristics

Section 3

Objectives

1. Explain how environments affect life, and describe how they change.
2. Identify the substances that make up the different layers of soil.

FOCUS

LET'S GET STARTED

Copy the following question onto the chalkboard: *What are three words or phrases that come to mind when you think of the plant life in this area?* If you are in a large city, you may want to specify a nearby park or wilderness area. *(Examples: forest, woods, prairie, cactus, scrub oak, wildflowers)* Write student responses on the chalkboard. As you teach the lesson, refer to the students' descriptions. Tell students that in Section 3 they will learn more about natural environments.

Building Vocabulary

Write **photosynthesis** on the chalkboard. Point out that *photo-* means "light." Have students find the definition in the text and relate the prefix's meaning to the term. What other words with this prefix do students know? *(Examples: photograph, photocopy)* Then have students create definitions for the Define terms that are compound words—**food chain, plant communities, plant succession**—based on what they know about the separate words. Check all definitions against the text. Have volunteers find and read aloud the remaining Define terms' definitions in the section.

Section 3 Natural Environments

Read to Discover

1. How do environments affect life, and how do they change?
2. What substances make up the different layers of soil?

Define

- extinct
- ecology
- photosynthesis
- food chain
- nutrients
- plant communities
- ecosystem
- plant succession
- humus

WHY IT MATTERS

Plants and animals depend on their environment for survival. Use CNNfyi.com or other **current events** sources to find out how conservationists work to save animals whose environments are threatened. Record your findings in your journal.

Acorns, the nut of an oak tree

Environmental Change

Geographers examine the distribution of plants and animals and the environments they occupy. They also study how people change natural environments. Changes in an environment affect the plants and animals that live in that environment. If environmental changes are extreme, some types of plants and animals may become **extinct**. This means they die out completely.

▲ Plants use a particular environment's sunlight, water, gases, and minerals to survive.

Ecology and Plant Life The study of the connections among different forms of life is called **ecology**. One process connecting life forms is **photosynthesis**—the process by which plants convert sunlight into chemical energy. Roots take in minerals, water, and gases from the soil. Leaves take in sunlight and carbon dioxide from the air. Plant cells take in these elements and combine them to produce special chemical compounds. Plants use some of these chemicals to live and grow.

Plant growth is the basis for all the food that animals eat. Some animals, like deer, eat only plants. When deer eat plants, they store some of the plant food energy in their bodies. Other animals, like wolves, eat deer and indirectly get the plant food the deer ate. The plants, deer, and wolves together make up a **food chain**. A food chain is a series of organisms in which energy is passed along.

Section 3 RESOURCES

Reproducible

- Guided Reading Strategy 3.3
- Environmental and Global Issues Activity 4
- Lab Activity for Geography and Earth Science, Hands-On 1

Technology

- One-Stop Planner CD–ROM, Lesson 3.3
- Homework Practice Online
- Our Environment CD–ROM/Seek and Tell/Natural Resources
- HRW Go site

Reinforcement, Review, and Assessment

- Section 3 Review, p. 55
- Daily Quiz 3.3
- Main Idea Activity 3.3
- English Audio Summary 3.3
- Spanish Audio Summary 3.3

TEACH

Teaching Objective 1

LEVELS 1 AND 2: (Suggested time: 30 min.) Pair students and have each pair write a paragraph to explain how environments affect life and how environments can change. *(Paragraphs should explain that different environments sustain different plants and animals and that changes to an ecosystem can result in the extinction of plants and animals. Paragraphs should also explain that environmental changes can be a result of natural disasters or human actions.)* Have volunteers read their paragraphs to the class. **COOPERATIVE LEARNING**

LEVEL 3: (Suggested time: 30 min.) Call on students to name a popular nearby natural area that includes a plant community, such as a park, forest, or riverbank. Ask students to identify potential natural and human events that could damage the ecosystem. *(Possible answers: fire, flood, drought, hurricane, clearing for new development)* Then have them gather information about the area and write a letter to the editor of the local newspaper listing the consequences of these damaging events. *(Possible answers: extinction of plants and animals)* Have students propose strategies in their letters for protecting the ecosystem. Then ask them to list the advantages and disadvantages of these options and choose one strategy to implement. Have them design a proposal for the city council that shows how to implement the solution and explains why it would be effective.

ENVIRONMENT AND SOCIETY

England's peppered moths are either light or dark. Scientist H. B. D. Kettlewell noticed that before 1848 there were few of the dark variety. Yet 50 years later, most were dark.

What had happened? Soot from the area's new factories had darkened the white birch trees, making it easier for moth-eating birds to spot the light moths that landed on them. As a result, fewer of the light moths survived to reproduce and pass on the gene for light coloring to later generations. The change in color was due to the process known as natural selection.

Discussion: What changes occurring in your area might give advantages or disadvantages to certain animals? *(Possible answers: Habitat destruction may favor animals that can live near humans. Drought may threaten species that require a steady supply of water.)*

Food Web

Food chains rarely occur alone in nature. More common are food webs—interlocking networks of food chains. This food web includes tiny organisms, such as parasites and bacteria, as well as human beings and other large mammals.

Not all predators are big. In Antarctica's dry areas a microscopic bacteria-eating worm that can survive years of being freeze-dried is at the top of the food chain.

Limits for Life Plants and animals cannot live everywhere. They are limited by environmental conditions. Any type of plant or animal tends to be most common in areas where it is best able to live, grow, and reproduce. A simple example shows why you do not find every kind of plant and animal in all regions on Earth. Trees do not survive in the tundra because it is too cold and there is not enough moisture. At the edge of the tundra, however, there are large boreal forests. Once in a while, wind carries tree seeds into the tundra. Some of the seeds sprout and grow. These young trees are generally small and weak. Eventually they die.

In this example life is limited by conditions of temperature and moisture. Other factors that can limit plant life include the amount of light, water, and soil **nutrients**. Nutrients are substances promoting growth. Some plant and animal life is limited by other plants and animals that compete for the same resources.

Teaching Objective 2

ALL LEVELS: (Suggested time: 10 min.) To help students understand the composition of the three soil layers, copy the following graphic organizer onto the chalkboard, omitting the italicized answers. Call on volunteers to supply the names of the three layers of soil and descriptions for each layer. Use students' answers to fill in the graphic organizer on the chalkboard. **ENGLISH LANGUAGE LEARNERS**

SOIL LAYERS

topsoil	*contains humus, insects, and plants*
subsoil	*contains deep roots*
broken rock	*contains rocks that eventually break down into more soil*

CONNECTING TO *Science*

There are more than 2,000 species of harmless mushrooms.

Soil Factory

The next time you see a fallen tree in the forest, do not think of it as a dead log. Think of it as a soil factory. As the tree decays and crumbles, it adds valuable nutrients to the forest soil. These nutrients enrich the soil. They also make it possible for new trees and plants to grow.

The fallen tree does not do its work alone, however. It is aided by many different living organisms that break down the wood and turn it into humus. Humus is a rich blend of organic material that mixes with the soil. A downed tree that lies on the forest floor is buzzing with the activity of hundreds of species of insects and plants that live and work inside it.

When a tree falls, insects, bacteria and other microorganisms invade the wood and start the process of decay. Insects like weevils, bark beetles, carpenter ants, and termites, among others, bore into the wood and break it down further. These insects, in turn, attract birds, spiders, lizards, and other predators who feed on insect life. Before long, the fallen tree is brimming with life. This happens even as the tree breaks apart on the forest floor.

Fallen trees provide as much as one third of the organic matter in forest soil. As forest ecologist Chris Maser notes, "Dead wood is no wasted resource. It is nature's reinvestment in biological capital."

You Be the Geographer

1. How do fallen trees decompose?
2. How is fertile soil produced by downed trees?

Plant Communities Groups of plants that live in the same area are called **plant communities**. In harsh environments, such as tundra, plant communities tend to be simple. They may be made up of just a few different types of plants. Regions that receive more rainfall and have more moderate temperatures tend to have more complex plant communities. The greatest variety of plants and animals can be found in the rain forests of the tropics. This environment is warm and moist all year.

Each plant community has plants adapted to the environment of the region. Near the Arctic, for example, the shortage of sunlight limits plant growth. Some of the flowers that grow here turn to follow the Sun as it moves across the sky. Thus the flowers collect as much sunlight as they can. In other regions there are plants adapted to survive in poor soils or with limited moisture. Some large trees, for example,

National Geography Standard 7

Physical Systems In the 1980s scientists noticed that spotted owls in the Pacific Northwest were declining. The old-growth forests where the owls lived were being cut down.

The owls became the subject of conflict between environmentalists who wanted to protect the owl's habitat and the timber industry and others who worried about the economic effects of ending lumbering. In the late 1980s the federal government put the owl on the Threatened Species List and set out to create a plan to conserve old-growth forests. Scientists are still working on a plan to track the owls to learn whether conservation plans have helped.

Activity: Have students create dioramas that show old-growth and second-growth forests. Students may need to do additional research.

Connecting to Science

Answers

1. aided by different organisms that break down the wood
2. They add organic matter.

internet connect

GO TO: go.hrw.com
KEYWORD: SK3 CH3
FOR: Web sites about forests

CLOSE

Have students look through the text's later chapters for photographs showing various plant communities or natural environments. Ask volunteers to choose a photograph and compare what they have learned in Section 3 with what they see in the photograph. Encourage the class to brainstorm how the plants and animals pictured fit into a food chain or ecosystem.

REVIEW AND ASSESS

Have students complete the Section Review. Then have them create flash cards with important terms and phrases on one side and full descriptions on the back. Students should work in pairs to review the flash cards. Then have students complete Daily Quiz 3.3. COOPERATIVE LEARNING

Section 3 Review

Answers

Define For definitions, see: extinct, p. 51; ecology, p. 51; photosynthesis, p. 51; food chain, p. 51; nutrients, p. 52; plant communities, p. 53; ecosystem, p. 54; plant succession, p. 54; humus, p. 55

Reading for the Main Idea

1. Plants and animals cannot adapt to every climate. A rain forest plant would not get enough moisture in a desert. (NGS 8)
2. layers of topsoil, subsoil, and broken rock; insects and bacteria live in the soil and produce humus (NGS 7)

Critical Thinking

3. the cold climate and predominance of ice
4. When natural or human forces disturb a plant community, the community may be replaced by a different group of plants suited to the new conditions.

Organizing What You Know

5. plants convert sunlight into energy through photosynthesis; an insect eats a plant; a fish eats the insect; a human eats the fish

You Be the Geographer Answer

grasses and wildflowers

have deep roots that can reach water and soil nutrients far below the surface. Their spreading branches also collect large amounts of sunlight for photosynthesis.

Plants adapt to the sunlight, soil, and temperature of their region. They may also be suited to other plants found in their communities. For example, many ferns grow well in the shade found underneath trees. Some vines grow up tree trunks to reach sunlight.

Forest Succession After a Fire

Ⓐ Forest fire in progress
Ⓑ Early plant growth
Ⓒ Middle stage
Ⓓ Forest recovered

▲ Difficult conditions after a forest fire mean that the first plants to grow back must be very hardy.

You Be the Geographer **In this series of photos, which plants are the first to grow back?**

All of the plants and animals in an area together with the nonliving parts of their environment—like climate and soil—form what is called an **ecosystem**. The size of an ecosystem varies, depending on how it is defined. A small pond, for example, can be considered an ecosystem. The entire Earth can also be considered an ecosystem.

Ecosystems can be affected by natural events like droughts, fires, floods, severe frosts, and windstorms. Human activities can also disturb ecosystems. This happens when land is cleared for development, new kinds of plants and animals are brought into an area, or pollution is released into air and water.

Plant Succession When natural or human forces disturb a plant community, the community may be replaced by a different group of plants suited to the new conditions. The gradual process by which one group of plants replaces another is called **plant succession**.

To better understand plant succession, imagine an area just after a forest fire. The first plants to return to the area need plenty of sunshine. These plants hold the soil in place. They also provide shade for the seeds of other plants. Gradually, seeds from small trees and shrubs grow under the protection of the first plants. These new plants grow taller and begin to take more and more of the sunlight. Many of the smaller plants die. Later, taller trees in the area replace the shorter trees and shrubs that grew at first.

It is important to remember that plant communities are not permanent. The conditions they experience change over time. Some changes affect a whole region. For example, a region's climate may gradually become colder, drier, warmer, or wetter. Additional changes may occur if new plants are introduced to the community.

✓ **READING CHECK:** *Physical Systems* How do environments affect life, and how do they change? They determine the plants and animals that live in the environment; natural or human forces can change them.

RETEACH

Have students complete Main Idea Activity 3.3. Then pair students and have each pair choose two key terms at random. Ask pairs to write a sentence that relates the terms to each other. *(Example: If they do not receive enough nutrients, plants can become extinct.)* Tell students to be sure to use standard grammar, spelling, sentence structure, and punctuation. Call on volunteers to read their sentences. Discuss the sentences until all of the section's major points have been covered.
ENGLISH LANGUAGE LEARNERS, COOPERATIVE LEARNING

EXTEND

Have interested students contact a local or state wildlife agency for information on predator species in your region. Have students use databases to conduct research on the predator's natural environment and to create a food chain that shows the connections of the species to plants, other animals, and people. **BLOCK SCHEDULING**

Soils

In any discussion of plants, plant communities, or plant succession, it is important to know about the soils that support plant life. All soils are not the same. The type of soil in an area can contribute to the kinds of plants that can be grown there. It can also affect how well a plant grows. Plants need soil with minerals, water, and small air spaces if they are to survive and grow.

Soils contain decayed plant and animal matter, called **humus**. Soils rich in humus are fertile. This means they can support an abundance of plant life. Humus is formed by insects and bacteria that live in soil. They break down dead plants and animals and make the nutrients available to plant roots. Insects also make small air spaces as they move through soils. These air pockets contain moisture and gases that plant roots need for growth.

The processes that break down rocks to form soil take hundreds or even thousands of years. Over this long period of time soil tends to form layers. If you dig a deep hole, you can see these layers. Soils typically have three layers. The thickness of each layer depends on the conditions in a specific location. The top layer is called topsoil. It includes humus, insects, and plants. The layer beneath the surface soil is called the subsoil. Only the deep roots of some plants, mostly trees, reach the subsoil. Underneath this layer is broken rock that eventually breaks down into more soil. As the rock breaks down it adds minerals to the soil.

Soils can lose their fertility in several ways. Erosion by water or wind can sweep topsoil away. Nutrients can also be removed from soils by leaching. This occurs when rainfall dissolves nutrients in topsoil and washes them down into lower soil layers, out of reach of most plant roots.

✓ **READING CHECK:** *Physical Systems* What are the physical processes that produce fertile soil? Insects and bacteria live in the soil and form humus.

Soil Layers

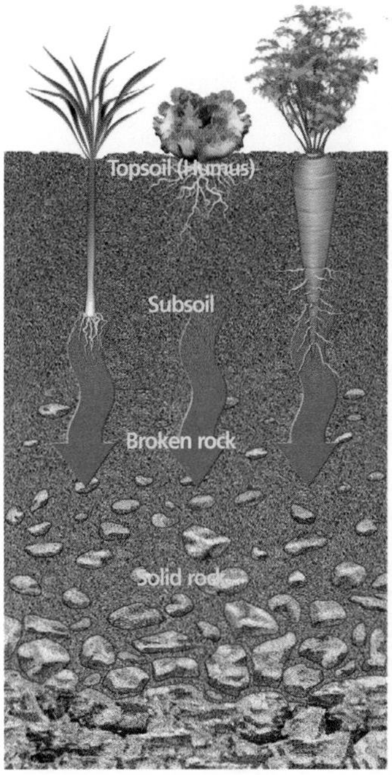

▲ The three layers of soil are the topsoil, subsoil, and broken rock.
Interpreting the Visual Record What do you think has created the cracks in the rocky layer below the broken rock?

go.hrw.com **Homework Practice Online**
Keyword: SK3 HP3

Section 3 Review

Define and explain: extinct, ecology, photosynthesis, food chain, nutrients, plant communities, ecosystem, plant succession, humus

Reading for the Main Idea

1. *Physical Systems* Why cannot all plants and animals live everywhere? Provide an example to illustrate your answer.
2. *Physical Systems* What makes up soil, and what produces fertile soil?

Critical Thinking

3. **Analyzing Information** What keeps tundra plant communities simple?
4. **Summarizing** How does plant succession occur?

Organizing What You Know

5. **Sequencing** Copy the following graphic organizer. Use it to describe a food chain.

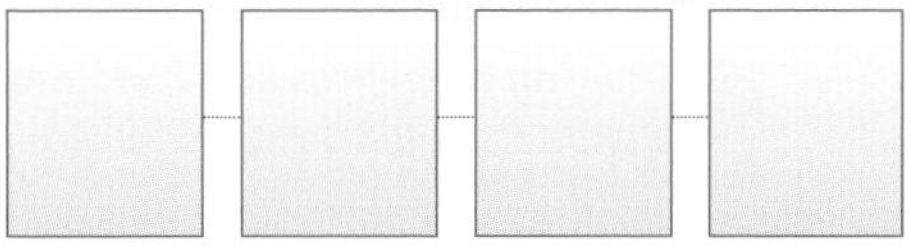

CHAPTER 3 Review ANSWERS

Building Vocabulary

For definitions, see weather, p. 37; climate, p. 37; greenhouse effect, p. 38; air pressure, p. 39; front, p. 40; currents, p. 41; rain shadow, p. 44; monsoon, p. 45; arid, p. 46; permafrost, p. 49; extinct, p. 51; ecology, p. 51; nutrients, p. 52; ecosystem, p. 54; humus p. 55

Reviewing the Main Ideas

1. trade winds—blow from northeast in the Northern Hemisphere; westerlies—blow from west in the middle latitudes; doldrums—warm air rises rather than blowing east or west (NGS 3)
2. The side of the mountain facing the wind might get heavy rain, and the other side of the mountain might not get much rain. (NGS 7)
3. low latitudes, dry, middle latitudes, high latitudes, and highland (NGS 3)
4. clear land; introduce new plants, animals; create pollution (NGS 14)
5. topsoil, subsoil, and broken rock; insects and bacteria that live in the soil form humus (NGS 7)

◄ **Visual Record Answer**

rainfall that has penetrated the layer of broken rock

ASSESS

Have students complete the Chapter 3 Test.

RETEACH

Organize the class into three groups—one for each section. Have the groups discuss the text material among themselves, focusing on how factors discussed in the other sections affect, or are affected by, the concepts in their assigned sections. Then have them record these connections on a butcher-paper mural titled "Wind, Climate, and Natural Environments." **ENGLISH LANGUAGE LEARNERS, COOPERATIVE LEARNING**

PORTFOLIO EXTENSIONS

1. Have students conduct research on ways sailors from the 1400s to 1700s used global wind currents to help them navigate. Students might compare the early sea voyages with the first nonstop balloon trip around the world in 1999. Using a world map students can study global wind currents and routes taken by adventurers. Have students write a paragraph on the results.
2. Have students obtain instructions for building a barometer, rain gauge, or home weather station. Ask them to build the device and then collect data for a certain length of time. Have them record their data in a chart and compare it with the official local weather reports. Take photographs of the measuring device(s), attach them to the reports, and place them with the reports in students' portfolios.

CHAPTER 3 Review ANSWERS

Understanding Environment and Society

Outlines and presentations will vary. Students should present information on the formation of tornadoes as well as forecast changes and precautions. Use Rubric 29, Presentations, to evaluate student work.

Thinking Critically

1. Earth's atmosphere, water, and land trap heat.
2. They create warming and cooling currents that move according to the winds' directions.
3. to be prepared for regular seasonal changes as well as potentially dangerous weather conditions
4. to be able to correlate climates to geographic regions
5. to know which soils can support plant life and to understand how to maintain soil fertility

Map Activity

A. hurricanes on the East Coast
B. forest fires on the West Coast
C. hurricanes on the Gulf Coast
D. tornadoes in Texas, Oklahoma, Kansas, and Nebraska

CHAPTER 3 Reviewing What You Know

Building Vocabulary

On a separate sheet of paper, write sentences to define each of the following words.

1. weather
2. climate
3. greenhouse effect
4. air pressure
5. front
6. currents
7. rain shadow
8. monsoon
9. arid
10. permafrost
11. extinct
12. ecology
13. nutrients
14. ecosystem
15. humus

Reviewing the Main Ideas

1. *The World in Spatial Terms* What are the major wind belts? Describe each.
2. *Physical Systems* How is the area around mountains affected by precipitation?
3. *The World in Spatial Terms* Into what main divisions can climate be grouped?
4. *Environment and Society* What do people do to change ecosystems?
5. *Physical Systems* What elements make up soil? How is fertile soil created?

Understanding Environment and Society

Tornadoes

Prepare an outline for a presentation on tornadoes. As you prepare your presentation from the outline, think about the following:

- How tornadoes are formed.
- How experts are able to forecast their occurrence more accurately.
- The safety precautions taken by people in tornado-prone areas.

Thinking Critically

1. **Finding the Main Idea** How does Earth store the Sun's energy?
2. **Analyzing Information** What effect do wind patterns have on ocean currents?
3. **Drawing Inferences and Conclusions** Why is it important to study the weather?
4. **Drawing Inferences and Conclusions** Why is it important to understand the concept of latitude when learning about Earth's many climates?
5. **Drawing Inferences and Conclusions** Why is it important for scientists to study soils?

Food Festival

Fruits and vegetables in the supermarket come from a range of environments. Have students interview market managers to learn where produce items were grown. They could organize a display linking the fruits and vegetables with descriptions of the climate and soil conditions the plants require. Then wash, peel, and eat those items that do not require cooking.

Chapter 3 Review and Assessment Resources

Reproducible
- Readings in World Geography, History, and Culture 5 and 6
- Critical Thinking Activity 3
- Vocabulary Activity 3

Technology
- Chapter 3 Test Generator (on the One-Stop Planner)
- Audio CD Program, Chapter 3
- HRW Go site

Reinforcement, Review, and Assessment
- Chapter 3 Review, pp. 56–57
- Chapter 3 Tutorial for Students, Parents, Mentors, and Peers
- Chapter 3 Test
- Chapter 3 Test for English Language Learners and Special-Needs Students

Building Social Studies Skills

Map Activity

On a separate sheet of paper, match the letters on the map with their correct labels.

The following natural disasters are often experienced in the United States:

- **hurricanes on the East Coast**
- **hurricanes on the Gulf Coast**
- **forest fires on the West Coast**
- **tornadoes in Texas, Oklahoma, Kansas, and Nebraska**

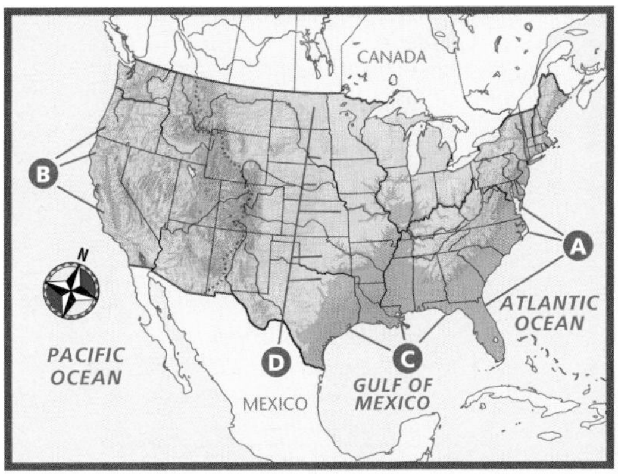

Mental Mapping Skills Activity

Draw a freehand map of the globe. Draw lines to show the equator and low, middle, and high latitudes. Draw the continents in the appropriate areas.

Writing Activity

After studying different climate regions, decide in which of the regions you would like to live. Write a journal entry describing what your life would be like in this particular area. Be sure to use standard grammar, spelling, sentence structure, and punctuation in your story.

Alternative Assessment

Portfolio Activity

Learning About Your Local Geography

Understanding Cause and Effect

Research your local weather patterns. Create a graph that shows how recent weather conditions have affected humans, plants, and animals in your part of the state.

internet connect

Internet Activity: go.hrw.com
KEYWORD: SK3 GT3

Choose a topic to explore online:
- Learn more about using weather maps.
- Follow El Niño, an ocean phenomenon that affects weather.
- Build a food web.

Mental Mapping Skills Activity

Maps will vary; students should accurately indicate the locations of the equator and the low, middle, and high latitudes. The continents should be drawn in the appropriate areas.

Writing Activity

Journal entries will vary; each story should accurately describe the chosen region. Use Rubric 40, Writing to Describe, to evaluate student work.

Portfolio Activity

Flow charts will vary; however, they should show the relationship between recent conditions and humans, plants, and animals. Use Rubric 7, Charts, to evaluate student work.

internet connect

GO TO: go.hrw.com
KEYWORD: SK3 Teacher
FOR: a guide to using the Internet in your classroom

CHAPTER 4

Earth's Resources

Chapter Resource Manager

Objectives	Pacing Guide	Reproducible Resources
SECTION 1 **Soil and Forests** (pp. 59–61) **1.** Identify the processes that threaten soil fertility. **2.** Describe why forests are valuable resources. **3.** Identify the human activities that can help and hurt forests.	**Regular** .5 day **Block Scheduling** .5 day *Block Scheduling Handbook, Chapter 4*	**RS** Guided Reading Strategy 4.1 **SM** Geography for Life Activity 4 **SM** Map Activity 4
SECTION 2 **Water and Air** (pp. 62–64) **1.** Describe why water is an important resource. **2.** Identify what threatens our supply of freshwater and how we can protect these supplies. **3.** Identify some problems caused by air pollution.	**Regular** .5 day **Block Scheduling** .5 day *Block Scheduling Handbook, Chapter 4*	**RS** Guided Reading Strategy 4.2 **E** Environmental and Global Issues Activities 1, 2, 5 **E** Lab Activities for Geography and Earth Science, Demonstration 8; Hands-On 3
SECTION 3 **Minerals** (pp. 65–67) **1.** Explain what minerals are. **2.** Identify the two types of minerals.	**Regular** .5 day **Block Scheduling** .5 day *Block Scheduling Handbook, Chapter 4*	**RS** Guided Reading Strategy 4.3
SECTION 4 **Energy Resources** (pp. 68–71) **1.** Identify the three main fossil fuels. **2.** Identify the four renewable energy sources. **3.** Identify the issues that surround the use of nuclear power.	**Regular** .5 day **Block Scheduling** .5 day *Block Scheduling Handbook, Chapter 4*	**RS** Guided Reading Strategy 4.4 **RS** Graphic Organizer 4 **E** Environmental and Global Issues Activities 6, 7 **IC** Interdisciplinary Activity for the Middle Grades 4

Chapter Resource Key

- **RS** Reading Support
- **IC** Interdisciplinary Connections
- **E** Enrichment
- **SM** Skills Mastery
- **A** Assessment
- **REV** Review
- **ELL** Reinforcement and English Language Learners
- Transparencies
- CD–ROM
- Music
- Video
- go.hrw.com Internet
- Holt Presentation Maker Using Microsoft® Powerpoint®

One-Stop Planner CD–ROM

See the *One-Stop Planner* for a complete list of additional resources for students and teachers.

One-Stop Planner CD–ROM

It's easy to plan lessons, select resources, and print out materials for your students when you use the ***One-Stop Planner CD–ROM with Test Generator.***

Technology Resources	Review, Reinforcement, and Assessment Resources
One-Stop Planner CD–ROM, Lesson 4.1 Homework Practice Online HRW Go site	ELL Main Idea Activity 4.1 REV Section 1 Review, p. 61 A Daily Quiz 4.1 ELL English Audio Summary 4.1 ELL Spanish Audio Summary 4.1
One-Stop Planner CD–ROM, Lesson 4.2 Homework Practice Online HRW Go site	ELL Main Idea Activity 4.2 REV Section 2 Review, p. 64 A Daily Quiz 4.2 ELL English Audio Summary 4.2 ELL Spanish Audio Summary 4.2
One-Stop Planner CD–ROM, Lesson 4.3 Homework Practice Online HRW Go site	ELL Main Idea Activity 4.3 REV Section 3 Review, p. 67 A Daily Quiz 4.3 ELL English Audio Summary 4.3 ELL Spanish Audio Summary 4.3
One-Stop Planner CD–ROM, Lesson 4.4 Homework Practice Online HRW Go site	ELL Main Idea Activity 4.4 REV Section 4 Review, p. 71 A Daily Quiz 4.4 ELL English Audio Summary 4.4 ELL Spanish Audio Summary 4.4

internet connect

HRW ONLINE RESOURCES

GO TO: go.hrw.com
Then type in a keyword.

TEACHER HOME PAGE
KEYWORD: SK3 TEACHER

CHAPTER INTERNET ACTIVITIES
KEYWORD: SK3 GT4

Choose an activity to:
- trek through different kinds of forests.
- investigate global warming.
- make a recycling plan.

CHAPTER ENRICHMENT LINKS
KEYWORD: SK3 CH4

CHAPTER MAPS
KEYWORD: SK3 MAPS4

ONLINE ASSESSMENT
Homework Practice
KEYWORD: SK3 HP4
Standardized Test Prep Online
KEYWORD: SK3 STP4
Rubrics
KEYWORD: SS Rubrics

COUNTRY INFORMATION
KEYWORD: SK3 Almanac

CONTENT UPDATES
KEYWORD: SS Content Updates

HOLT PRESENTATION MAKER
KEYWORD: SK3 PPT4

ONLINE READING SUPPORT
KEYWORD: SS Strategies

CURRENT EVENTS
KEYWORD: S3 Current Events

Meeting Individual Needs

Ability Levels

Level 1 Basic-level activities designed for all students encountering new material

Level 2 Intermediate-level activities designed for average students

Level 3 Challenging activities designed for honors and gifted-and-talented students

English Language Learners Activities that address the needs of students with Limited English Proficiency

Chapter Review and Assessment

- E Readings in World Geography, History, and Culture 7 and 8
- SM Critical Thinking Activity 4
- REV Chapter 4 Review, pp. 72–73
- REV Chapter 4 Tutorial for Students, Parents, Mentors, and Peers
- ELL Vocabulary Activity 4
- A Chapter 4 Test
- Chapter 4 Test Generator (on the One-Stop Planner)
- Audio CD Program, Chapter 4
- A Chapter 4 Test for English Language Learners and Special-Needs Students

CHAPTER 4

Launch into Learning

Call on a student to select a classroom object at random. *(Examples: book, jacket, globe)* Ask students to name the various materials from which the object is made. *(Example: jacket—cotton, synthetic fibers, metal)* Challenge them to identify the origins of these materials *(cotton needs soil and water to grow, synthetic fibers from petroleum, metal zipper from ore).* Tell students they are naming resources—supplies that can be used to create something else. Then tell students they will learn more about types and uses of resources in this chapter.

Section 1

Objectives

1. Describe the physical processes that produce fertile soil.
2. Describe the processes that threaten soil fertility.
3. Explain why forests are valuable resources and what efforts are being taken to preserve them.

Links to Our Lives

You may want to emphasize the importance of studying the materials and substances on which we depend by discussing with your students the following points:

- Our quality of life—and life itself—can depend on how wisely we use our resources.
- Some resources may run out in the foreseeable future. We need to know how to make those resources last as long as possible.
- We make decisions about using resources every day, even if we do not realize it. We need to be well informed if we are to make good decisions.
- We must explore new sources of energy that can replace those, such as oil and natural gas, that cannot be renewed.

CHAPTER 4

Earth's Resources

Precious gems

Wind turbines

Valley in the Andes of Ecuador

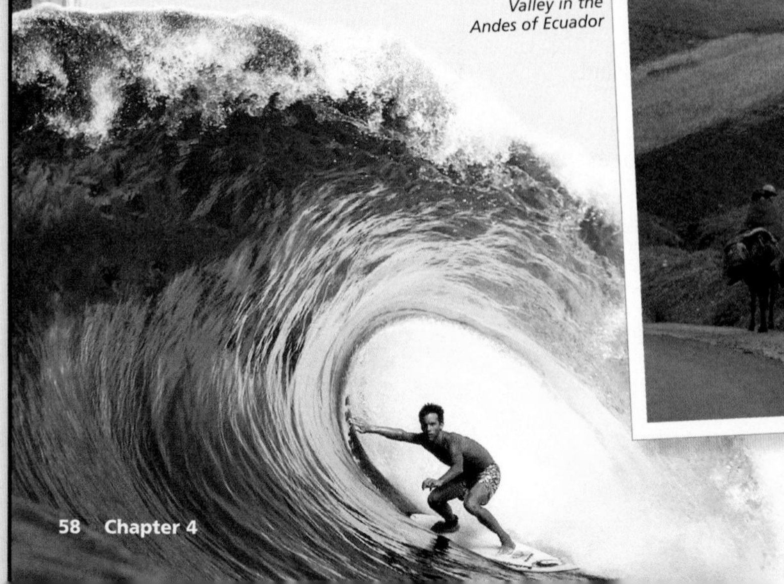

FOCUS

LET'S GET STARTED

Copy the following riddle onto the chalkboard and ask students to solve it.

It lies all around us. We don't give it a thought,
But it puts food on the table. What it gives can't be bought.

(The answer is soil.) Tell students that in Section 1 they will learn more about the importance of soil and forests.

Building Vocabulary

Write the key terms on the chalkboard. Point out that the prefix *re-* often means "again." So, **renewable resources** are those that can be new again. **Reforestation** indicates that a forest is growing again in the same place. To end a word in *-ation* indicates a process or action. Ask students to relate the suffix to the definitions of **crop rotation, desertification, deforestation,** and reforestation that they find in Section 1 or in the text's glossary. Then ask them to write sentences using each of the terms.

Section 1 Soil and Forests

Read to Discover

1. What physical processes produce fertile soil?
2. What processes threaten soil fertility?
3. Why are forests valuable resources, and how are they being protected?

Define

renewable resources
crop rotation
terraces
desertification
deforestation
reforestation

WHY IT MATTERS

Many people work to balance concerns about saving the rain forests with local economic needs. Use CNNfyi.com or other **current events** sources to learn more about these efforts. Record your findings in your journal.

Corn, a major food crop

Section 1 RESOURCES

Reproducible
- Block Scheduling Handbook, Chapter 4
- Guided Reading Strategy 4.1
- Map Activity 4

Technology
- One-Stop Planner CD–ROM, Lesson 4.1
- Homework Practice Online
- HRW Go site

Reinforcement, Review, and Assessment
- Section 1 Review, p. 61
- Daily Quiz 4.1
- Main Idea Activity 4.1
- English Audio Summary 4.1
- Spanish Audio Summary 4.1

Soil

Soil is one of the most important **renewable resources** on Earth. Renewable resources are those that can be replaced by Earth's natural processes.

Soil types vary depending on geographic factors. As you learned in Chapter 3, soil contains rock particles and humus. It also contains water and gases. Because soil types vary, some are better able to support plant life than others.

Soil Fertility Soil conservation—protecting the soil's ability to nourish plants—is one challenge facing farmers. Plants must take up nutrients like calcium, nitrogen, phosphorus, and potassium in order to grow. These essential nutrients may become used up if fields are always planted with the same crops. Farmers can add these nutrients to soils in the form of fertilizers, sometimes called plant food. The first fertilizers used were manures. Later, chemical fertilizers were used to increase yields.

Some farmers choose not to use chemical fertilizers. Others cannot afford to use them. They rely on other ways to keep up the soil's ability to produce. One such method is **crop rotation**. This is a system of growing different crops on the same land over a period of years.

On this farm, fields are planted with corn and alfalfa. Alfalfa is a valued crop because it replaces nutrients in the soil.

TEACH

Teaching Objective 1

ALL LEVELS: (Suggested time: 45 min.) Lead a discussion on the physical processes that produce fertile soil. Then have students find pictures in books and magazines that depict erosion, failing crops, desertification, or development onto farmland. Have them cut out or reproduce the pictures, paste them on posters, and label their posters with a paragraph that describes the problem illustrated and suggests a solution.
ENGLISH LANGUAGE LEARNERS

Teaching Objective 2

ALL LEVELS: (Suggested time: 25 min.) Copy the following graphic organizer onto the chalkboard, omitting the italicized answers. Use it to help students list ways people use forests. After discussing the completed organizer, point out that timber is an important natural resource. Ask them to describe how more timber can be produced. *(reforestation).*
ENGLISH LANGUAGE LEARNERS

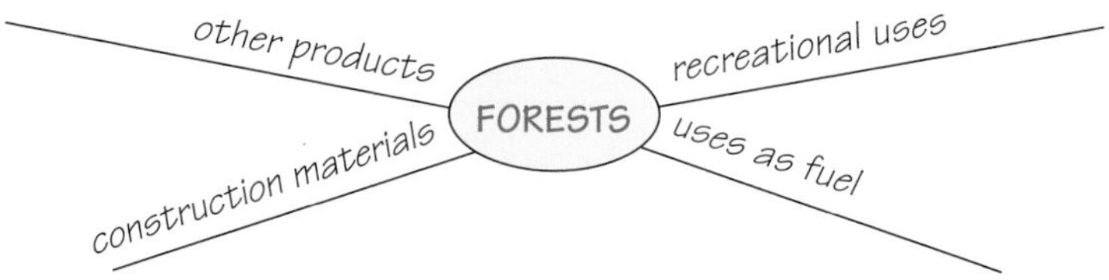

Linking Past to Present
Deforestation

Although they did not have access to modern technology, ancient peoples often dramatically affected natural landscapes. Some archaeologists believe that during the Neolithic era—beginning around 6000 B.C.—people may have deforested large areas of central and western Europe. They felled trees to clear land for farming and used the lumber for building and fuel.

Activity: Have students conduct research on deforestation that took place during ancient and modern times. Then have them compare the deforestation that took place during the two time periods in terms of causes, methods, and results. Ask students to draw conclusions about the long-term effects of deforestation.

internet connect
GO TO: go.hrw.com
KEYWORD: SK3 CH4
FOR: Web sites about deforestation

A row of poplar trees divides farmland near Aix-en-Provence, France.
Interpreting the Visual Record What effect will these trees have on erosion?

Salty Soil Salt buildup is another threat to soil fertility. In dry climates farmers must irrigate their crops. They use well water or water brought by canals and ditches. Much of this water evaporates in the dry air. When the water evaporates, it leaves behind small amounts of salt. If too much salt builds up in the soil, crops cannot grow.

Erosion The problem of soil erosion is faced by farmers all over the world. Farmers have to work to keep soil from being washed away by rainfall. Soil can also be blown away by strong winds. To prevent soil loss, some farmers plant rows of trees to block the wind. Others who farm on steep hillsides build **terraces** into the slope. Terraces are horizontal ridges like stair steps. By slowing water movement the terraces stop the soil from being washed away. They also provide more space for farming.

Loss of Farmland The loss of farmland is a serious problem in many parts of the world. In some places farming has worn out the soil, and it can no longer grow crops. Livestock may then eat what few plants remain. Without plants to hold the soil in place, it may blow away. The long-term process of losing soil fertility and plant life is called **desertification**. Once this process begins, the desert can expand as people move on to better soils. They often repeat the destructive practices, damaging ever-larger areas.

Farmland is also lost when cities and suburbs expand into rural areas. Nearby farmers sell their land. The land is then used for housing or businesses, rather than for agriculture. This is happening in many poorer countries. It also happens in richer nations like the United States.

✓ **READING CHECK:** *Environment and Society* What physical processes help soil fertility, and what threatens it? fertilizers, crop rotation; salt buildup, erosion, too much farming, desertification, urbanization

Our Amazing Planet

About 47,000 aspen trees in Utah share a root system and a set of genes. These trees are actually a single organism. These aspens cover 106 acres (43 hectares) and weigh at least 13 million pounds (5.9 million kg). Together they are probably the world's heaviest living thing.

Visual Record Answer

reduce erosion by blocking wind

Close

Have students suggest ways they might help preserve forests and protect fertile soil.

Review and Assess

Have students complete the Section Review. Then have each student write five true or false statements based on the text. Call on volunteers to read their statements aloud and have the class determine if the statements are true or false. Require that statements judged false be converted to true statements. Then have students complete Daily Quiz 4.1.

Reteach

Have students complete Main Idea Activity 4.1. Then create a two-column chart on the chalkboard. Label the columns *Soil* and *Forests* and label the rows *Importance, Problems,* and *Solutions.* Fill in the chart as a class. **ENGLISH LANGUAGE LEARNERS**

Extend

Assign groups of students a country. Have them identify and research a problem in their country's forests. Tell them to list and consider options to solve the problem and to discuss the merits of the options. Then have them draw up action plans for preserving their forests while providing for local needs. Invite the class to evaluate the plans. **COOPERATIVE LEARNING, BLOCK SCHEDULING**

Forests

Forests are renewable resources because new trees can be planted in a forest. If cared for properly, they will be available for future generations. Forests are important because they provide both people and wildlife with food and shelter. People depend on forests for a wide variety of products. Wood products include lumber, plywood, and shingles for building houses. Other wood-based manufactured products include cellophane, furniture, some plastics, and fibers such as rayon. Trees also supply fats, gums, medicines, nuts, oils, turpentine, waxes, and rubber. The forests are valuable not only for their products. People also use forests for recreational activities, such as camping and hiking.

Deforestation The destruction or loss of forest area is called **deforestation**. It is happening in the rain forests of Africa, Asia, and Central and South America. People clear the land and use it for farming, industry, and housing. Pollution also causes deforestation.

Protecting Forests Many countries, including the United States, are trying to balance their economic needs regarding forests with conservation efforts. Since the late 1800s Congress has passed laws to protect and manage forest and wilderness areas. In addition to protecting forests, people can also plant trees in places where forests have been cut down. This replanting is called **reforestation**. In some cases the newly planted trees can be "harvested" again in a few years.

▲ Villagers work on a reforestation project in Cameroon.

✓ **READING CHECK:** *Environment and Society* Why is preserving forests important, and how can people contribute to it? **They provide food, shelter, and materials for recreation, many products; through laws and reforestation.**

go.hrw.com Homework Practice Online
Keyword: SK3 HP4

Section 1 Review

Define and explain: renewable resources, crop rotation, terraces, desertification, deforestation, reforestation

Reading for the Main Idea

1. *Environment and Society* What physical processes produce fertile soil?
2. *Environment and Society* What do forests provide, and how can they be protected?

Critical Thinking

3. **Finding the Main Idea** Which two natural forces contribute to soil erosion? How can erosion be prevented?
4. **Drawing Inferences and Conclusions** Why is it important to manage forests?

Organizing What You Know

5. **Categorizing** Copy the following graphic organizer. Use it to discuss whether or not rain forests should be protected.

Reasons for . . .	Reasons against . . .

Section 1 Review Answers

Define For definitions, see: renewable resources, p. 59; crop rotation, p. 59; terraces, p. 60; desertification, p. 60; deforestation, p. 61; reforestation, p. 61

Reading for the Main Idea

1. using fertilizers and crop rotation, eliminating salt buildup, preventing erosion, preventing desertification (NGS 14, 16)
2. possible answers: lumber, plywood, shingles, cellophane, furniture, plastics, rayon, fats, gums, medicines, nuts, oil, turpentine, waxes, and rubber; protection through laws and reforestation (NGS 14, 16)

Critical Thinking

3. wind and water; through planting rows of trees, building terraces
4. possible answer: so they will continue to produce the goods people use, provide recreation, and shelter wildlife

Organizing What You Know

5. possible reasons for—products that forests supply, wildlife protection, oxygen supply; possible reasons against—land needed for farms, wood needed for industry and housing

Section 2

Objectives

1. Explain why water is an important resource.
2. Identify threats to our water supply, and describe ways to conserve it.
3. Analyze some of the problems caused by air pollution.

FOCUS

LET'S GET STARTED

Copy the following statement and question onto the chalkboard: *The average American uses 168 gallons (636 l) of water daily. What are some ways you and your family can save water?* (Students' answers might include turning off the tap while brushing teeth, quick showers, installing water-saving toilets, and running the dishwasher only when full.) Tell students that in Section 2 they will learn more about water and air.

Building Vocabulary

Write the key terms on the chalkboard. Call on volunteers to underline the parts of the words that resemble elements of familiar words *(semi-, aqu-)* and to tell what words they resemble. *(Possible answers: semiweekly, aquatic)* Have other students look up the key terms and infer how they relate to the familiar words. Ask students to find definitions for the remaining key terms and use the terms in sentences.

Section 2 RESOURCES

Reproducible
- Guided Reading Strategy 4.2
- Environmental and Global Issues Activities 1, 2, 5
- Lab Activities for Geography and Earth Science, Demonstration 8; Hands-On 3

Technology
- One-Stop Planner CD–ROM, Lesson 4.2
- Homework Practice Online
- HRW Go site

Reinforcement, Review, and Assessment
- Section 2 Review, p. 64
- Daily Quiz 4.2
- Main Idea Activity 4.2
- English Audio Summary 4.2
- Spanish Audio Summary 4.2

Section 2 Water and Air

Read to Discover

1. Why is water an important resource?
2. What threatens our supplies of freshwater, and how can we protect these supplies?
3. What are some problems caused by air pollution?

Define

semiarid
aqueducts
aquifers
desalinization
acid rain
global warming

A water tank

WHY IT MATTERS

Countries around the world have worked together to resolve problems regarding Earth's atmosphere. Use CNNfyi.com or other **current events** sources to learn about these efforts. Record your findings in your journal.

Water

Dry regions are found in many parts of the world, including the western United States. There are also **semiarid** regions—regions that receive a small amount of rain. Semiarid places are usually too dry for farming. However, these areas may be suitable for grazing animals.

Water Supply Many areas of high mountains receive heavy snowfall in the winter. When that snow melts, it forms rivers that flow from the mountains to neighboring regions. People in dry regions use various means to bring the water where it is needed for agricultural and other uses. They build canals, reservoirs, and **aqueducts**—artificial channels for carrying water.

Some places have water deep underground in **aquifers**. These are water-bearing layers of rock, sand, or gravel. Some are quite large. For example, the Ogallala Aquifer stretches across the Great Plains from Texas to South Dakota. People drill wells to reach the water in the aquifer.

People in dry coastal areas have access to plenty of salt water. However, they typically do not have enough freshwater. In Southwest Asia this situation is common. To create a supply of

As people make their homes in dry areas, they add to the demand for water.

TEACH

Teaching Objective 1

ALL LEVELS: (Suggested time: 30 min.) Have students design covers for a magazine titled *Water Weekly.* As part of their covers, have students include article titles that suggest why water is an important resource. Ask volunteers to present their covers to the class. Then display the covers around the classroom. **ENGLISH LANGUAGE LEARNERS**

Teaching Objectives 2–3

ALL LEVELS: (Suggested time: 30 min.) Copy the following graphic organizer onto the chalkboard, omitting the italicized answers, and have students copy it into their notebooks. Ask half of the class to fill in the water side and the other half to complete the air material. When they are finished, have students compare their charts and fill in the gaps. **ENGLISH LANGUAGE LEARNERS**

	Water Pollution	Air Pollution
Causes	*chemical fertilizers, pesticides, industrial wastes*	*burning fuel*
Results	*reduced water supply, harm to other living things*	*threatens health, causes acid rain, may damage ozone layer*
Solutions	*limit pollutants*	*find cleaner energy sources*

freshwater, people in these places have built machines that take the salt out of seawater. This process, known as **desalinization**, is expensive and takes a lot of energy. However, in some places, it is necessary.

Water Conservation In recent decades people have developed new ways to save water. Many factories now recycle water. Farmers are able to irrigate their crops more efficiently. Cities build water treatment plants to purify water that might otherwise be wasted. Some people in dry climates are using desert plants instead of grass for landscaping. This means that they do not need to water as often.

It is important for people in all climates to conserve water. Wasting water in one location could mean that less water is available for use in other places.

Pivoting sprinklers irrigate these circular cornfields in Kansas.

Interpreting the Visual Record Why might irrigation be necessary in these fields?

Water Quality Industries and agriculture also affect the water supply. In many countries there are still places that cannot afford to build closed sewer systems. Some factories also operate without pollution controls because such controls would add a great deal of cost to operation.

Industrialized countries like the United States have water treatment plants and closed sewer systems. However, water can still be polluted when farmers use too much chemical fertilizer and pesticides. These chemicals can get into local streams. Waste from industries may also contain chemicals, metals, or oils that can pollute streams and rivers.

Rivers carry pollution to the oceans. The pollution can harm marine life such as fish and shellfish. Eating marine life from polluted waters can make people sick. So can drinking polluted water. Balancing industrial and agricultural needs with the need for clean water continues to be a challenge faced by many countries.

✓ **READING CHECK:** *Environment and Society* Why is water an important resource, and how does its availability affect people? People need it for drinking and farming; if water is polluted or wasted, people will not have enough for everyday life.

Air

Air is essential to life. Plants and animals need the gases in the air to live and grow.

Human activities can pollute the air and threaten the health of life on the planet. Burning fuels for heating, for transportation, and to power factories releases chemicals into the air. Particularly in large cities, these chemicals build up in the air. The chemicals create a mixture called smog.

Some cities have special problems with air pollution. Denver, Los Angeles, and Mexico City, for example, are located in bowl-shaped valleys that can trap air pollution. This pollution sometimes builds up to levels that are dangerous to people's health.

Heavy smog clouds the Los Angeles skyline.

GLOBAL PERSPECTIVES

People have been building dams for at least 5,000 years. More than 800,000 dams have been constructed around the world.

Dams and the reservoirs created by them can provide a steady water supply, hydroelectricity, protection from floods, and increased fish yields. However, they can also displace people, submerge farmland and cultural sites, disrupt animal migration patterns, and even increase the risk of certain diseases. Malaria, for example, is spread by mosquitoes that can breed in the standing water of reservoirs. For these and other reasons, dam construction has slowed in recent years.

Activity: Have students search news media for information on dams in your state. What controversies have arisen? What possible solutions have been suggested to address these controversies? Do you think that future scientific discoveries will lessen the negative effects of dams on the environment? Why or why not?

◄ **Visual Record Answer**

lack of adequate rainfall

CLOSE

Lead a discussion on water and air issues your community may face in coming years. *(Possible issues: reduced water supply from falling aquifer level, more pollution from increased traffic or industry, hurricanes threatening water supplies)* What might happen if new technology makes more water available in dry locations?

REVIEW AND ASSESS

Have students complete the Section Review. Then pair students. Have pairs find important terms in each subsection and take turns relating the terms to the main points. Then have students complete Daily Quiz 4.2.
COOPERATIVE LEARNING

RETEACH

Have students complete Main Idea Activity 4.2. Then have them design posters that illustrate the main points of the section. Ask volunteers to present their posters to the class. Then display the posters around the classroom. **ENGLISH LANGUAGE LEARNERS**

EXTEND

Have students conduct research on aqueducts built by the ancient Romans to supply freshwater to their cities. Students who like to work with their hands may want to construct a scale model. **BLOCK SCHEDULING**

Section 2 Review

Answers

Define For definitions, see: semiarid, p. 62; aqueducts, p. 62; aquifers, p. 62; desalinization, p. 63; acid rain, p. 64; global warming, p. 64

Reading for the Main Idea

1. They build canals, reservoirs, and aqueducts. (NGS 14)
2. allow chemical fertilizers, pesticides, and industrial waste into water supply (NGS 14)
3. People burn fuel for heating, for transportation, and to power factories. (NGS 14)

Critical Thinking

4. The process of desalinization is expensive.

Organizing What You Know

5. possible answers—threatens health, creates acid rain, damages ozone layer, may cause global warming

This satellite image of the Southern Hemisphere shows a thinning in the ozone layer in October 1979.

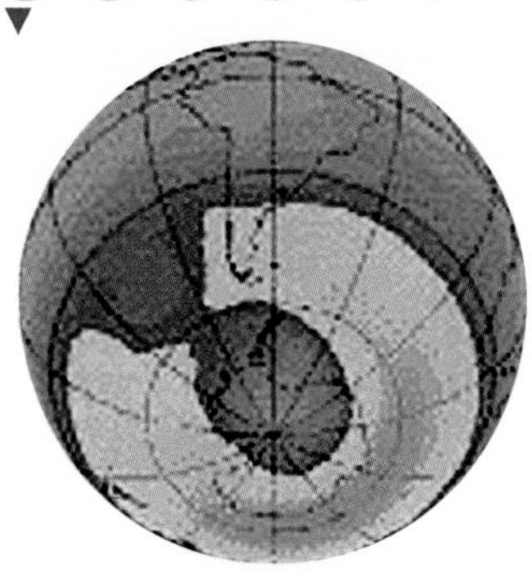

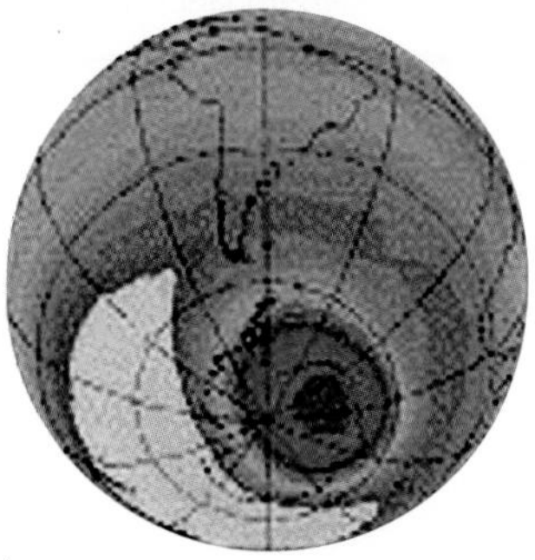

By October 1992 the thinning area, shown in purple, had grown much larger.

Acid Rain When air pollution combines with moisture in the air, it can form a mild acid. This can be similar in strength to vinegar. When it falls to the ground, this moisture is called **acid rain**. It can damage or kill trees. Acid rain can also kill fish.

Many countries have laws to limit pollution. However, pollution is an international problem. Winds can blow away air pollution, but the wind is only moving the pollution to another place. Pollution can pass from one country to another. It can even pass from one continent to another. Countries that limit their own pollution can still be affected by pollution from other countries.

Pollution and Climate Change Smog and acid rain are short-term effects of air pollution. Air pollution may also have long-term effects by changing conditions in Earth's atmosphere. Certain kinds of pollution damage the ozone in the upper atmosphere. This ozone layer protects living things by absorbing harmful ultraviolet light from the Sun. Damage to Earth's ozone layer may cause health problems in people. For example, it could lead to an increase in skin cancer.

Another concern is **global warming**—a slow increase in Earth's average temperature. The Sun constantly warms Earth's surface. The gases and water vapor in the atmosphere trap some of this heat. This helps keep Earth warm. Without the atmosphere, this heat would return to space. Evidence suggests that pollution causes the atmosphere to trap more heat. Over time this would make Earth warmer.

Scientists agree that Earth's climate has warmed during the last century. However, they disagree about exactly why. Some scientists say that temperatures have warmed because of air pollution caused by human activities, particularly the burning of fossil fuels. Others think warmer temperatures have resulted from natural causes. Scientists also disagree about what has caused the thinning of the ozone layer.

✓ **READING CHECK:** *Environment and Society* What are the different points of view about air pollution and climate change? Some scientists think that air pollution has created warmer temperatures; others think that warmer temperatures are the result of natural changes.

go.hrw.com Homework Practice Online
Keyword: SK3 HP4

Section 2 Review

Define and explain: semiarid, aqueducts, aquifers, desalinization, acid rain, global warming

Reading for the Main Idea

1. *Environment and Society* How have people changed the environment to increase the water supply in drier areas?
2. *Environment and Society* How do people cause water pollution?
3. *Environment and Society* What kinds of human activities have polluted the air?

Critical Thinking

4. **Drawing Inferences and Conclusions** Why is desalinization rarely practiced?

Organizing What You Know

5. **Identifying Cause and Effect** Use this graphic organizer to explain air pollution.

Causes: — Air Pollution — Effects:

Section 3

Objectives

1. Define minerals.
2. Identify the two types of minerals.

Focus

LET'S GET STARTED

Copy these questions onto the chalkboard: *Have you played the guessing game that begins by asking if something is animal, vegetable, or mineral? What does mineral mean in this context?* Students might suggest that a mineral is something that is not alive, rocks, or materials dug from the ground. Call on volunteers to list minerals with which they may already be familiar *(aluminum, building stone, precious and semiprecious stones, salt, and so on).* Tell students that in Section 3 they will learn more about minerals.

Building Vocabulary

Write the key terms on the chalkboard. Point out that the suffix *-ic* means "of or relating to" and that the prefix *non-* means "not." Have students relate these meanings to the word "metal" to arrive at definitions for **metallic** and **nonmetallic minerals**. Then ask students to use the glossary to define **minerals**. Call on a volunteer to use the definition of "renewable" in Section 1 to define **nonrenewable resources**.

Section 3 Minerals

Read to Discover

1. What are minerals?
2. What are the two types of minerals?

Define

nonrenewable resources
minerals
metallic minerals
nonmetallic minerals

WHY IT MATTERS

Most minerals are dug from deep in the ground by miners. Use CNNfyi.com or other **current events** sources to learn about life as a miner. Record your findings in your journal.

Quartz crystals

Minerals

You have learned that renewable resources such as trees are always being produced. **Nonrenewable resources** are those that cannot be replaced by natural processes or are replaced very slowly.

Earth's crust is made up of substances called **minerals**. Minerals are an example of a nonrenewable resource. They provide us with many of the materials we need. More than 3,000 minerals have been identified, but fewer than 20 are common. Around 20 minerals make up most of Earth's crust. Minerals have four basic properties. First, they are inorganic. Inorganic substances are not made from living things or the remains of living things. Second, they occur naturally, rather than being manufactured like steel or brass. Third, minerals are solids in crystalline form, unlike petroleum or natural gas. Finally, minerals have a definite chemical composition or combination of elements. Although all minerals share these four properties, they can be very different from one another. Minerals are divided into two basic types: metallic and nonmetallic.

Metallic Minerals Metals, or **metallic minerals**, are shiny and can conduct heat and electricity. Metals are solids at normal room temperature. An exception is mercury, a metal that is liquid at room temperature.

Gold is one of the heaviest of all metals and is easily worked. For thousands of years people have highly valued gold. Precious metals are commonly made into jewelry and coins. Silver and platinum are other precious metals.

The Most-Common Elements in Earth's Crust

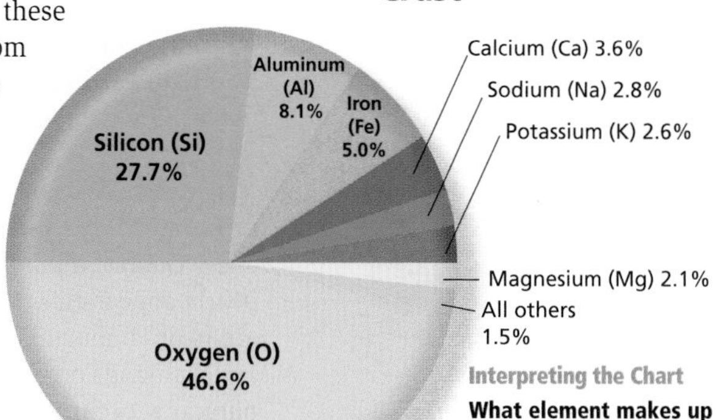

Interpreting the Chart What element makes up almost 47 percent of Earth's crust?

Section 3 Resources

Reproducible
- Guided Reading Strategy 4.3

Technology
- One-Stop Planner CD–ROM, Lesson 4.3
- Homework Practice Online
- HRW Go site

Reinforcement, Review, and Assessment
- Section 3 Review, p. 67
- Daily Quiz 4.3
- Main Idea Activity 4.3
- English Audio Summary 4.3
- Spanish Audio Summary 4.3

◄ **Chart Answer**

oxygen

TEACH

Teaching Objectives 1–2

ALL LEVELS: (Suggested time: 30 min.) Copy the following graphic organizer onto the chalkboard, omitting the italicized answers. Use it to help students learn more about minerals.
ENGLISH LANGUAGE LEARNERS

MINERALS		
Definition: *substances that make up Earth's crust*		
Types	*Metallic*	*Nonmetallic*
Characteristics	*shiny, good conductors of heat and electricity*	*varied appearance, not good conductors of heat and electricity*
Examples	*aluminum, iron, gold, silver, and other precious metals*	*quartz, talc, diamonds, gemstones, salt, and sulfur*

DAILY LIFE

One of the most widely used nonmetallic minerals is salt. Salt is plentiful, inexpensive, and essential to the health of people and animals. We are most familiar with salt as a cooking ingredient, but it is also used in many other products and processes.

Salt removes unwanted minerals from the water supply. Farm animals and poultry consume salt. Food processors, such as pickle makers, use tons of the mineral. Salt makes dyes colorfast in fabric. Film companies use salt in chemical solutions. Health spas offer salt baths and rubs. Salt producers have claimed that the precious mineral has 14,000 uses.

Activity: Have students examine labels of common household products and foods to find salt as an ingredient. You might also have them conduct research on the industrial uses for salt.

Connecting to Art

Answers

1. They use recycled objects to create art.
2. less-developed countries

internet connect
GO TO: go.hrw.com
KEYWORD: SK3 CH4
FOR: Web sites about mining

CONNECTING TO Art

Art made from recycled products

Junk Art

Many Americans now make a habit of recycling. They put their cans, bottles, and newspapers by the curb for pickup or take them to recycling centers. However, some people do their recycling in a different way. They use their junk to create art.

Much recycled art is folk art. These objects have a practical purpose but are made with creativity and a sense of style. The making of folk art objects from junk is common in the world's poorer countries, where resources are scarce.

Some examples of recycled folk art include dust pans made from license plates (Mexico), jugs made from old tires (Morocco), briefcases made from flattened tin cans (Senegal), and a toy helicopter made from plastic containers and film canisters (Haiti). As scientist Stephen J. Gould has written, "In our world of material wealth, where so many broken items are thrown away rather than mended . . . we forget that most of the world fixes everything and discards nothing."

Americans do have a tradition of making recycled art, however. The Amish make quilts from old scraps of cloth. Other folk artists build whimsical figures out of bottle caps and wire. Some modern artists create sculptures from "found objects" like machine parts, bicycle wheels, and old signs. Junk art can even be fashion. One movie costume designer went to the Academy Awards ceremony wearing a dress made of credit cards!

Understanding What You Read

1. How do some societies turn junk into art?
2. Where is much recycled folk art made?

Iron is the cheapest metal. Iron can be combined with certain other minerals to make steel. Aluminum is another common metal. This lightweight metal is used in such items as soft drink cans and airplanes. We handle copper every time we pick up a penny.

▲ Gold is a metallic mineral.

Nonmetallic Minerals Minerals that lack the characteristics of a metal are called **nonmetallic minerals**. These vary in their appearance. Quartz, a mineral often found in sand, looks glassy. Talc has a pearly appearance. Most nonmetallic minerals have a dull surface and are poor conductors of heat or electricity.

Diamonds are minerals made of pure carbon. They are the hardest naturally occurring substance. The brilliant look of diamonds has made them popular gems. Their hardness makes them valuable for industrial use. Other gemstones, like rubies, sapphires, and emeralds, are also nonmetallic minerals.

Close

Ask students to write letters to your local newspaper urging the community to begin, continue, or expand a recycling program.

Review and Assess

Have students complete the Section Review. Then call on a volunteer to name a common mineral. Instruct the student to call on another to suggest a use for that mineral. Continue until all the minerals discussed in the section have been reviewed. Then have students complete Daily Quiz 4.3.

Reteach

Have students complete Main Idea Activity 4.3. Then prepare flash cards of the section's Define terms and names of minerals. As you display a card, call on a student to find the sentences that discuss the concept and explain the term to the class. **ENGLISH LANGUAGE LEARNERS**

Extend

Interested students may want to conduct research on the many ways that cultures have used precious and semiprecious stones throughout history. *(Examples: in paint, as medicine, as ritual objects, to decorate bookcovers)* Have students use their research to create entries for a pamphlet to accompany a museum exhibit displaying these uses. **BLOCK SCHEDULING**

Mineral and Energy Resources in the United States

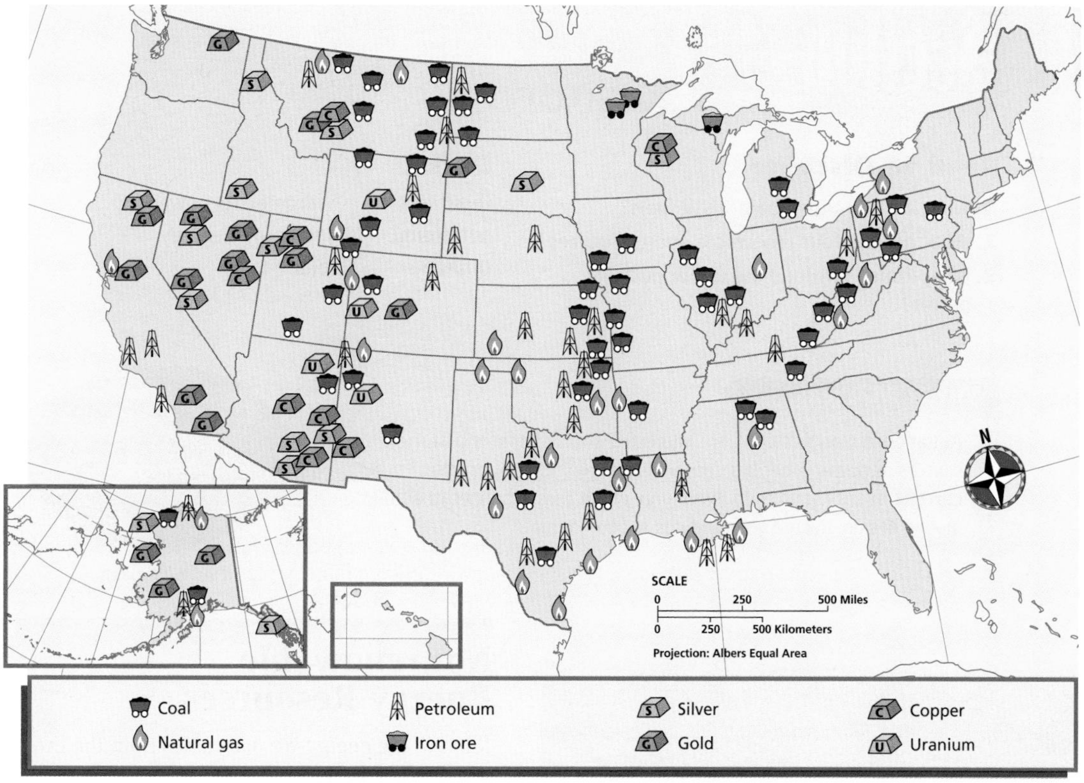

Other mineral substances also have important uses. For example, people need salt to stay healthy. Sulfur is used in many ways, from making batteries to bleaching dried fruits. Graphite, another form of carbon, is used in making pencils.

✓ **READING CHECK:** *Physical Systems* What are the two types of minerals?
metallic and nonmetallic

Section 3 Review

go.hrw.com Homework Practice Online
Keyword: SK3 HP4

Define and explain: nonrenewable resources, minerals, metallic minerals, nonmetallic minerals

Reading for the Main Idea

1. *Physical Systems* What is a nonrenewable resource?
2. *Environment and Society* Why are minerals important? What is the difference between a metallic mineral and a nonmetallic mineral?

Critical Thinking

3. **Summarizing** How can minerals be used?
4. **Drawing Inferences and Conclusions** What makes gold such an important mineral?

Organizing What You Know

5. **Contrasting** Copy the following graphic organizer. Use it to contrast types of minerals.

Metallic minerals	Nonmetallic minerals

Section 3 Review Answers

Define For definitions, see: nonrenewable resources, p. 65; minerals, p. 65; metallic minerals, p. 65; nonmetallic minerals, p. 66

Reading for the Main Idea

1. cannot be replaced by natural processes or are replaced very slowly (NGS 7)
2. metallic—shiny, can conduct heat and electricity, most are solid at room temperature; nonmetallic—varied in appearance, poor conductors of heat and electricity (NGS 15)

Critical Thinking

3. Answer may include as jewelry and coins, to make steel, in soft drink cans and airplanes, for industrial use, to stay healthy, to make batteries, to bleach dried fruits, or in making pencils.
4. can be used as money and made into jewelry

Organizing What You Know

5. metallic—shiny, can conduct heat and electricity, most are solid at room temperature; nonmetallic—varied in appearance, poor conductors of heat and electricity

Section 4

Objectives

1. Describe the three main fossil fuels.
2. Identify four renewable energy sources.
3. Analyze the issues surrounding nuclear power.

Focus

LET'S GET STARTED

Copy the following instructions onto the chalkboard: *What are three ways you use energy? What are the sources of that energy?* Students may mention driving, cooking, heating, air conditioning, or operating appliances as ways to use energy. Possible answers for sources of energy might be natural gas, heating oil, gasoline, or electricity. Discuss responses. Then ask students how they would be affected if those energy sources were no longer available. Tell students that in Section 4 they will learn more about Earth's energy resources.

Building Vocabulary

Write the key terms on the chalkboard and underline *petro, hydro, geo,* and *sol.* Tell students that these Greek and Latin word fragments mean "rock," "water," "earth," and "sun" respectively. Have students infer the definitions of the Define terms that contain the word fragments and check the glossary to confirm. Ask students what other words they know that contain those prefixes. *(Possible answers: petrochemical, hydraulic, geology, solstice)* Call on volunteers to read the definitions for **fossil fuels** and **refineries**.

Section 4 Resources

Reproducible

- Guided Reading Strategy 4.4
- Graphic Organizer 4
- Environmental and Global Issues Activities 6, 7
- Interdisciplinary Activity for the Middle Grades 4

Technology

- One-Stop Planner CD–ROM, Lesson 4.4
- Homework Practice Online
- HRW Go site

Reinforcement, Review, and Assessment

- Section 4 Review, p. 71
- Daily Quiz 4.4
- Main Idea Activity 4.4
- English Audio Summary 4.4
- Spanish Audio Summary 4.4

Section 4 Energy Resources

Read to Discover

1. What are the three main fossil fuels?
2. What are the four renewable energy sources?
3. What issues surround the use of nuclear power?

Define

fossil fuels
petroleum
refineries
hydroelectric power
geothermal energy
solar energy

Why It Matters

Automobile manufacturers are interested in building cars that use sources of power other than gasoline. Use CNNfyi.com or other **current events** sources to learn more about these changes in automobile design. Record your findings in your journal.

An oil tanker

A miner digs coal in a narrow tunnel.

Nonrenewable Energy Resources

Most of the energy we use comes from the three main **fossil fuels**: coal, petroleum, and natural gas. Fossil fuels were formed from the remains of ancient plants and animals. These remains gradually decayed and were covered with sediment. Over long periods of time, pressure and heat changed these materials into fossil fuels. All fossil fuels are nonrenewable resources.

Coal Until the 1900s people mostly used wood and coal as sources of energy. Coal was used to make steel in giant furnaces and to run factories. However, the burning of coal polluted the air. The way coal is burned today has improved so that it releases less pollution. This new technology is more expensive, however. Some of the largest coal deposits are located in Australia, China, Russia, India, and the United States.

Petroleum Industrialized societies now use **petroleum**—an oily liquid—for a variety of purposes. When it is first pumped out of the ground, petroleum is called crude oil. It is then shipped or piped to **refineries**—factories where crude oil is processed, or refined. Petroleum is made into gasoline, diesel and jet fuels, and heating oil.

TEACH

Teaching Objectives 1–3

ALL LEVELS: (Suggested time: 30 min.) Copy the following graphic organizer onto the chalkboard, omitting the italicized answers. Pair students and have the pairs complete the organizer. Discuss any gaps in the students' charts. **ENGLISH LANGUAGE LEARNERS, COOPERATIVE LEARNING**

FUELS

		Advantage	Disadvantage
FOSSIL	Coal	*plentiful*	*pollutes, nonrenewable*
	Petroleum	*plentiful*	*pollutes, not evenly distributed, nonrenewable*
	Natural gas	*plentiful, burns clean*	*hard to transport, nonrenewable*
RENEWABLE	Hydroelectric Power	*clean, renewable*	*affects habitats, drowns farms and forests*
	Wind Power	*clean, renewable*	*needs steady wind*
	Geothermal Energy	*clean, renewable*	*only in limited areas*
	Solar Energy	*clean, renewable*	*expensive*
	Nuclear Energy	*reduces need to burn fossil fuels*	*possible accidents, waste*

Who Has the Oil?

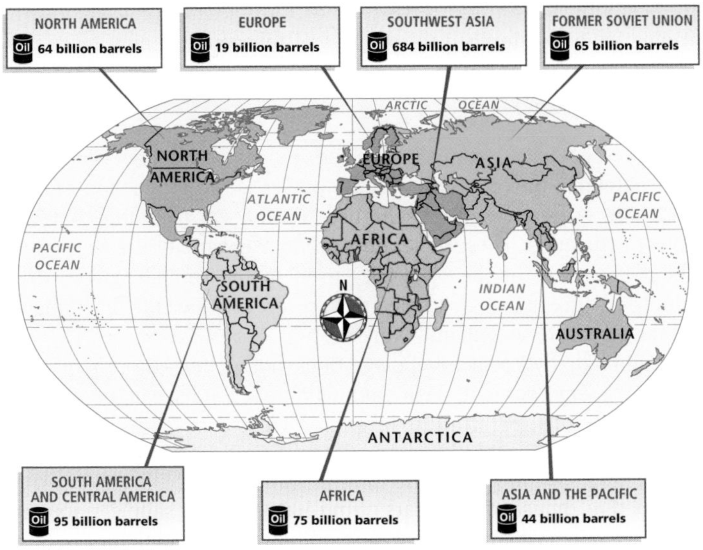

Source: *BP Amoco Statistical Review of World Energy 2001*

◄ More than half of all known oil deposits are located in Southwest Asia. New deposits are still being found, but Southwest Asia will probably continue to hold the largest share.

Petroleum is not evenly distributed on Earth. Of the oil reserves that have been discovered, more than 65 percent are found in Southwest Asia. Most of that is in Saudi Arabia. North America has around 7 percent of the world's known oil, while South America has around 10 percent. Other regions have oil in small amounts.

Natural Gas The use of natural gas is growing rapidly. Natural gas is gas that comes from Earth's crust through natural openings or drilled wells. Large natural gas fields are found in Russia and Southwest Asia. Northern Canada also has large amounts of natural gas. However, the fields are located in the far north. Frozen seas and low temperatures make it difficult to pump and ship the gas safely.

Natural gas is the cleanest-burning fossil fuel. It produces much less air pollution than gasoline or diesel fuel. It is usually transported by pipeline, making it most useful for factories and electrical plants. It is also used for heating and cooking. Vehicles that run on natural gas have to carry the fuel in special, bulky containers. However, some large cities now have buses and taxis that use natural gas to help cut down on air pollution.

The Alaska pipeline carries oil from north to south across Alaska.

Interpreting the Visual Record **Why do you think this pipeline is above ground?** ▼

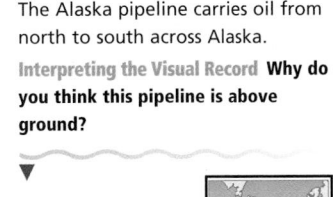

✓ READING CHECK: *Physical Systems* What physical processes produced fossil fuels? The remains of ancient plants and animals decayed and were covered with sediment. Pressure and heat changed the materials to fossil fuels.

Across the Curriculum

SCIENCE

Methane Hydrates A methane hydrate crystal consists of a natural gas molecule surrounded by water molecules. Methane hydrates resemble regular ice, but unlike ice they can burn! The compounds are abundant in nature, particularly on the ocean floor, and might provide a new source of energy. In fact, the amount of carbon that is found in the global stores of hydrates may be twice as large as the total amount of carbon found in all other fossil fuels.

However, because they are formed under pressure, hydrates can disintegrate when removed from the ocean. Rockslides could make their removal from the ocean floor hazardous. Furthermore, in terms of its potential impact on global warming, methane as a greenhouse gas is 10 times stronger than carbon dioxide, which is released in the burning of traditional fossil fuels.

Discussion: Have students make predictions about the economic and environmental consequences of a new technology that would eliminate the hazards of using methane hydrates.

◄ **Visual Record Answer**

for easier access for repairs

Teaching Objective 1

LEVEL 2: (Suggested time: 20 min.) Have students create a flowchart showing how fossil fuels are formed. Invite volunteers to present their flowcharts to the class. **ENGLISH LANGUAGE LEARNERS**

➤ASSIGNMENT: Have students design yellow-pages ads for companies that harness and sell hydroelectric power, wind energy, geothermal energy, and solar energy. Challenge students to review previous chapters for hints about other potential sources of renewable energy *(tidal energy, wave energy)* and to design ads for them as well. **ENGLISH LANGUAGE LEARNERS**

TEACHER TO TEACHER

Jane Palmer of Sanford, Florida, suggests the following activity. Organize the class into six groups. Have two groups represent countries with many resources, another two groups represent countries with some resources, and the final two groups represent countries with few resources. Distribute building supplies (craft sticks, wood blocks, glue, and so on) to each group to represent that country's resources. For example, the countries that receive many materials would have more resources. Instruct each group to construct a small building using their supplies. Then ask students how they felt about using limited resources. Discuss challenges they faced and what could have been done to meet those challenges.

Section 4 Review

Answers

Define For definitions, see: fossil fuels, p. 68; petroleum, p. 68; refineries, p. 68; hydroelectric power, p. 70; geothermal energy, p. 70; solar energy, p. 71

Reading for the Main Idea

1. coal, petroleum, natural gas; plant and animal remains decayed, were covered with sediment, and became fuel through heat and pressure. (NGS 7)
2. water, wind, geothermal energy, Sun; water—hydroelectric power, wind—wind turbines, geothermal energy—captures Earth's heat, Sun—solar panels absorb heat (NGS 7)

Critical Thinking

3. affect fish and wildlife habitats
4. nuclear accidents and waste

Organizing What You Know

5. renewable—can be replaced by Earth's natural processes; nonrenewable—cannot

Visual Record Answer ▶

People have taken advantage of a windy location through using turbines for power.

internet connect

GO TO: go.hrw.com
KEYWORD: SK3 CH4
FOR: Web sites about Earth's resources

Renewable Energy Resources

People have also learned to use several renewable energy resources. They include water power, wind power, heat from within Earth, and the Sun's energy.

Water The most commonly used renewable energy source is **hydroelectric power**. Hydroelectric power is the production of electricity by waterpower. Dams harness the energy of falling water to power generators. These generators produce electricity. Dams produce about 9 percent of the electricity used in the United States. Other countries producing hydroelectric power include Brazil, Canada, China, Egypt, New Zealand, Norway, and Russia.

Although hydroelectric power does not pollute the air, it does affect the environment. Fish and wildlife habitats can be affected when reservoirs, or artificial lakes, are created by dams. Reservoirs may also cover farmland and forests.

Wind For thousands of years, wind has powered sailing ships and boats. People have also long used wind power to turn windmills. Such mills were used to pump water out of wells or to grind grain into flour. In some places windmills are still used for these purposes.

Today, wind has a new use. It can create electricity by turning a system of fan blades called a turbine. "Wind farms" with hundreds of wind turbines have been built in some windy places.

Geothermal Energy The heat of Earth's interior—**geothermal energy**—can also be used to generate electricity. This internal heat escapes through hot springs and steam vents on Earth's surface. This geothermal energy can be captured to power electric generators.

Wind turbines are just one source of electricity.

Interpreting the Visual Record How does this photo show human adaptation to the environment? ▼

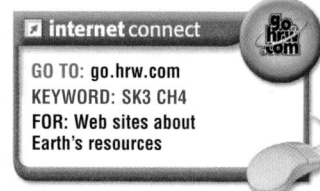

CLOSE

Have students imagine that they live in a community that is considering building a nuclear power plant. Ask each student to jot down two points—either for or against nuclear energy—that he or she would include in a speech at a public information meeting. Call on volunteers to make their points for both sides of the issue.

REVIEW AND ASSESS

Have students complete the Section Review. You might also have them prepare an outline of the section. Then have students complete Daily Quiz 4.4.

RETEACH

Have students complete Main Idea Activity 4.4. Then have them create crossword puzzles using the section's key terms and concepts. Ask students to exhange and solve each others' puzzles.
ENGLISH LANGUAGE LEARNERS

EXTEND

Have interested students conduct research on the Alaska pipeline—the controversies surrounding its construction, how those controversies were addressed, and the impact the pipeline has had on Alaska's economy.
BLOCK SCHEDULING

The Sun The Sun's heat and light are known as **solar energy**. This energy can be used to heat water or homes. Special solar panels also absorb solar energy to make electricity. In the past, converting solar energy into electricity was expensive. New research, however, may make solar energy cheaper in the future.

✓ **READING CHECK:** *Physical Systems* What are four renewable energy sources? water, wind, geothermal energy, the Sun

▲ Experimental cars like this one run on solar energy.

Nuclear Energy

In the late 1930s scientists discovered nuclear energy. They learned that energy could be released by changes in the nucleus, or core, of atoms. Nuclear energy was first used to make powerful bombs. Scientists found out that nuclear energy can also be used to produce electricity. Lithuania, France, Belgium, Ukraine, and Sweden rely on nuclear energy for much of their power. This reduces their dependence on imported oil.

There are serious concerns about the use of nuclear power. Several nuclear power plants have had accidents. In 1986 an accident at a nuclear reactor in Chernobyl in Ukraine killed dozens of people. It also caused cancer in thousands more. Homes and farms in the area of the reactor had to be abandoned.

Nuclear energy produces waste that remains dangerous for thousands of years. People do not agree on how nuclear waste can be stored or transported safely. Because of this some countries are reducing their dependence on nuclear energy. Denmark and New Zealand avoid it altogether. The United States does not intend to expand its use of nuclear energy. As a result, scientists continue to search for other renewable energy sources.

✓ **READING CHECK:** *Environment and Society* What are the concerns associated with nuclear power? accidents and nuclear waste

Section 4 Review

go.hrw.com Homework Practice Online
Keyword: SK3 HP4

Define and explain: fossil fuels, petroleum, refineries, hydroelectric power, geothermal energy, solar energy

Reading for the Main Idea

1. *Physical Systems* What are the three main fossil fuels, and how were they formed?
2. *Physical Systems* What are the four most common renewable energy sources? Describe how each of these harnesses energy.

Critical Thinking

3. **Finding the Main Idea** How do dams affect the environment?
4. **Analyzing Information** What are some of the problems associated with nuclear energy?

Organizing What You Know

5. **Contrasting** Copy the following graphic organizer. Use it to contrast the characteristics of renewable and nonrenewable resources.

Renewable resources	↔	Nonrenewable resources

CHAPTER 4 Review ANSWERS

Building Vocabulary

For definitions, see: renewable resources, p. 59; desertification, p. 60; deforestation, p. 61; reforestation, p. 61; aquifers, p. 62; acid rain, p. 64; nonrenewable resources, p. 65; fossil fuels, p. 68; hydroelectric power, p. 70; solar energy, p. 71

Reviewing the Main Ideas

1. plants need soil to grow; by using fertilizers responsibly, rotating crops, and preventing salt buildup and erosion (NGS 15)
2. canals, reservoirs, aqueducts; removed salt from seawater (NGS 14)
3. substances that are part of Earth's crust; provide us with many materials we need (NGS 15)
4. to run factories, power plants, and vehicles; heat buildings (NGS 15)
5. Water creates hydroelectric power; wind creates electricity; geothermal powers electric generators; Sun heats water or homes and can be used to generate electricity. (NGS 15)

ASSESS

Have students complete the Chapter 4 Test.

RETEACH

Organize the class into nine groups—for water, air, soil, forests, metallic minerals, nonmetallic minerals, nonrenewable energy sources, renewable energy sources, and nuclear energy. Have each group create an idea web of the material related to its topic. Compare, discuss, and display the webs around the classroom. **ENGLISH LANGUAGE LEARNERS, COOPERATIVE LEARNING**

PORTFOLIO

PORTFOLIO EXTENSIONS

1. Have students collect useful minerals, label the samples, and construct flowcharts that trace the raw minerals through processing to their final uses. They may want to construct models of expensive or unavailable minerals. Photograph the project results for placement in student portfolios.
2. Ask students to name natural phenomena that link countries without respect to boundaries. *(winds, rain, waterways, wildlife migration)* Then have them explain ways that pollution created in one state or country can contaminate the air, water, or land of another. Have students create maps showing the spread of contaminants. Then have them explain why pollution control might be a source of conflict among countries. *(economic reasons)*

CHAPTER 4

Review ANSWERS

Understanding Environment and Society

Answers will vary, but the information included should be consistent with text material. Verify that students have compared environmental conditions and included a description of preventive measures. Use Rubric 29, Presentations, to evaluate student work.

Thinking Critically

1. Some people want to clear the land for farms, livestock, or industry. Others believe the rain forests should be protected and preserved. The rain forests are being deforested.
2. water—chemical fertilizers, pesticides, and industrial wastes in water supply; air—burning fuel for heating, transportation, and to power factories; pollution threatens health of people and animals, damages crops and forests, causes thinning in ozone layer, may cause global warming
3. provide materials we want and need
4. metallic—shiny, can conduct heat and electricity, most are solid at room temperature; nonmetallic—varied in

CHAPTER 4

Reviewing What You Know

Building Vocabulary

On a separate sheet of paper, write sentences to define each of the following words.

1. renewable resources
2. desertification
3. deforestation
4. reforestation
5. aquifers
6. acid rain
7. nonrenewable resources
8. fossil fuels
9. hydroelectric power
10. solar energy

Reviewing the Main Ideas

1. *Environment and Society* Why is fertile soil important, and how is it created?
2. *Environment and Society* What actions have people taken to increase their supply of water?
3. *Environment and Society* What are minerals, and why are they important to us?
4. *Environment and Society* For what are fossil fuels used?
5. *Environment and Society* What are the four most common renewable energy sources, and how is each used?

Understanding Environment and Society

Chernobyl

On April 26, 1986, a nuclear reactor exploded in Chernobyl. Radiation caused serious problems in Ukraine and Belarus, as well as in Eastern and Western Europe. Create an outline for a presentation based on information about the following:

- Conditions of the environment then and now.
- Preventive measures taken to prevent future accidents.

Thinking Critically

1. **Drawing Inferences and Conclusions** Why does the issue of use of rain forests cause such disagreement? What is happening to Earth's rain forests?
2. **Summarizing** What are the major causes of water and air pollution, and how does pollution affect life on this planet?
3. **Drawing Inferences and Conclusions** Why are minerals valued by society?
4. **Contrasting** How are metallic minerals different from nonmetallic minerals?
5. **Drawing Inferences and Conclusions** Why does the use of nuclear energy continue to be a debated topic?

Food Festival

In countries where farmland is scarce, farmers are increasingly using hydroponic agriculture. Plants grown hydroponically are raised in a water-based nutrient solution. Ask students to visit a local grocery and to locate items in the produce department that are grown hydroponically. Students may need to ask workers at the grocery for help. The class may also want to try to grow plants, such as tomatoes, hydroponically.

Chapter 4 Review and Assessment Resources

Reproducible
- Readings in World Geography, History, and Culture 7, 8
- Critical Thinking Activity 4
- Vocabulary Activity 4

Technology
- Chapter 4 Test Generator (on the One-Stop Planner)
- Audio CD Program, Chapter 4
- HRW Go site

Reinforcement, Review, and Assessment
- Chapter 4 Review, pp. 72–73
- Chapter 4 Tutorial for Students, Parents, Mentors, and Peers
- Chapter 4 Test
- Chapter 4 Test for English Language Learners and Special-Needs Students

Building Social Studies Skills

Map Activity

On a separate sheet of paper, match the letters on the map with their correct region. Then write the amount of oil known to be located in each region.

Africa	South and Central America
Europe	Southwest Asia
former Soviet Union	

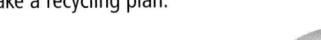

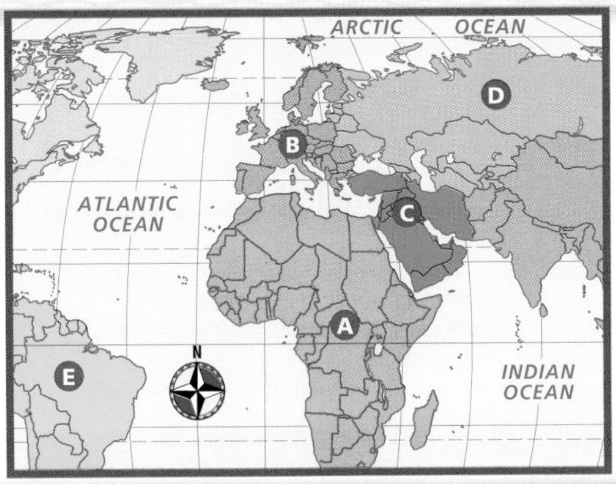

Mental Mapping Skills Activity

On a separate sheet of paper, draw a freehand map of the United States. Label areas that have copper. Compare this map with a physical map of the United States. What, if any, physical features are located in the same regions as copper deposits?

Writing Activity

Write a short paper explaining which mineral is most important to you. Justify your selection with facts you have learned. Be sure to use standard grammar, spelling, sentence structure, and punctuation.

Alternative Assessment

Portfolio Activity

Learning About Your Local Geography

Environmental Issues Study your local environment. What issues of preservation and use are important to the people of your community? How does your government handle these issues?

internet connect

Internet Activity: go.hrw.com
KEYWORD: SK3 GT4

Choose a topic to explore Earth's resources:
- Trek through different kinds of forests.
- Investigate global warming.
- Make a recycling plan.

appearance, are poor conductors of heat and electricity

5. Nuclear energy offers great benefits but can create serious problems.

Map Activity

A. Africa
B. Europe
C. Southwest Asia
D. former Soviet Union
E. South and Central America

Mental Mapping Skills Activity

Maps will vary, but should show areas that have copper.

Writing Activity

The information included should be consistent with text material. Use Rubric 38, Writing to Classify, to evaluate student work.

Portfolio Activity

Answers will vary. Verify that students discuss both the nature of the local issues and how the local government has dealt with them. Use Rubric 30, Research, to evaluate student work.

internet connect

GO TO: go.hrw.com
KEYWORD: SK3 Teacher
FOR: a guide to using the Internet in your classroom

CHAPTER 5

The World's People

Chapter Resource Manager

Objectives	Pacing Guide	Reproducible Resources
SECTION 1 **Culture** (pp. 75–80) **1.** Explain the term *culture.* **2.** Identify the influences that help cultures develop. **3.** Describe how agriculture has affected the development of culture.	**Regular** 2.5 days **Block Scheduling** 1 day *Block Scheduling Handbook, Chapter 5*	**RS** Guided Reading Strategy 5.1 **E** Creative Strategies for Teaching World Geography, Lessons 4, 5 **SM** Geography for Life Activity 5 **E** Lab Activity for Geography and Earth Science, Hands-On 4 **SM** Map Activity 5
SECTION 2 **Population, Economy, and Government** (pp. 81–86) **1.** Explain why population density varies, and identify how the world's population has changed. **2.** Explain how geographers describe and measure economics. **3.** Identify the connections between economics and politics. **4.** Explain how governments differ.	**Regular** 2.5 days **Block Scheduling** .5 day *Block Scheduling Handbook, Chapter 5*	**RS** Guided Reading Strategy 5.2 **RS** Graphic Organizer 5
SECTION 3 **Population Growth Issues** (pp. 87–89) **1.** Identify the problems associated with high and low population growth rates. **2.** Identify two different views of population growth and resources.	**Regular** .5 day **Block Scheduling** .5 day *Block Scheduling Handbook, Chapter 5*	**RS** Guided Reading Strategy 5.3 **E** Environmental and Global Issues Activity 8

Chapter Resource Key

- **RS** Reading Support
- **IC** Interdisciplinary Connections
- **E** Enrichment
- **SM** Skills Mastery
- **A** Assessment
- **REV** Review
- **ELL** Reinforcement and English Language Learners
- Transparencies
- CD–ROM
- Music
- Video
- go.hrw.com Internet
- Holt Presentation Maker Using Microsoft® Powerpoint®

One-Stop Planner CD–ROM

See the ***One-Stop Planner*** for a complete list of additional resources for students and teachers.

One-Stop Planner CD-ROM

It's easy to plan lessons, select resources, and print out materials for your students when you use the *One-Stop Planner CD-ROM with Test Generator.*

Technology Resources	Review, Reinforcement, and Assessment Resources
One-Stop Planner CD-ROM, Lesson 5.1 Geography and Cultures Visual Resources with Teaching Activities 9 and 10 Homework Practice Online HRW Go site	ELL Main Idea Activity 5.1 REV Section 1 Review, p. 80 A Daily Quiz 5.1 ELL English Audio Summary 5.1 ELL Spanish Audio Summary 5.1
One-Stop Planner CD-ROM, Lesson 5.2 *ARGWorld* CD-ROM: Mapping Urban Growth *ARGWorld* CD-ROM: Strategic Straits and Choke Points *ARGWorld* CD-ROM: Technological Change and Factory Locations *ARGWorld* CD-ROM: Representation in the United Nations Homework Practice Online HRW Go site	ELL Main Idea Activity 5.2 REV Section 2 Review, p. 86 A Daily Quiz 5.2 ELL English Audio Summary 5.2 ELL Spanish Audio Summary 5.2
One-Stop Planner CD-ROM, Lesson 5.3 Yourtown CD-ROM Homework Practice Online HRW Go site	ELL Main Idea Activity 5.3 REV Section 3 Review, p. 89 A Daily Quiz 5.3 ELL English Audio Summary 5.3 ELL Spanish Audio Summary 5.3

internet connect

HRW ONLINE RESOURCES

GO TO: go.hrw.com
Then type in a keyword.

TEACHER HOME PAGE
KEYWORD: SK3 TEACHER

CHAPTER INTERNET ACTIVITIES
KEYWORD: SK3 GT5

Choose an activity to:
- visit famous buildings and monuments around the world.
- compare facts about life in different countries.
- examine world population growth.

CHAPTER ENRICHMENT LINKS
KEYWORD: SK3 CH5

CHAPTER MAPS
KEYWORD: SK3 MAPS5

ONLINE ASSESSMENT
Homework Practice
KEYWORD: SK3 HP5
Standardized Test Prep Online
KEYWORD: SK3 STP5
Rubrics
KEYWORD: SS Rubrics

COUNTRY INFORMATION
KEYWORD: SK3 Almanac

CONTENT UPDATES
KEYWORD: SS Content Updates

HOLT PRESENTATION MAKER
KEYWORD: SK3 PPT5

ONLINE READING SUPPORT
KEYWORD: SS Strategies

CURRENT EVENTS
KEYWORD: S3 Current Events

Chapter Review and Assessment

- E Readings in World Geography, History, and Culture 9 and 10
- SM Critical Thinking Activity 5
- REV Chapter 5 Review, pp. 90–91
- REV Chapter 5 Tutorial for Students, Parents, Mentors, and Peers
- ELL Vocabulary Activity 5
- A Chapter 5 Test
- A Unit 1 Test
- Chapter 5 Test Generator (on the One-Stop Planner)
- Audio CD Program, Chapter 5
- A Chapter 5 Test for English Language Learners and Special-Needs Students
- A Unit 1 Test for English Language Learners and Special-Needs Students

Meeting Individual Needs

Ability Levels

Level 1 Basic-level activities designed for all students encountering new material

Level 2 Intermediate-level activities designed for average students

Level 3 Challenging activities designed for honors and gifted-and-talented students

English Language Learners Activities that address the needs of students with Limited English Proficiency

CHAPTER 5

Launch into Learning

Display a news-story photo that shows elements of physical and human geography. *(Examples: natural disaster, new medicine discovered in rain forest)* Ask students to list the physical and human elements shown or implied in the photo. *(For example, for a hurricane: physical—rain, mud slides, beach erosion, high tides; human—people forced from homes, businesses hurt)* Point out to students that their interpretation of the photo would be incomplete if they left out either the physical or human issues. Similarly, the study of geography must include the human element. Tell students they will learn more about the human element of geography in this chapter.

Section 1

Objectives

1. Define culture and culture region.
2. Identify the importance of cultural symbols.
3. Trace how cultures develop.
4. Explain how agriculture affected culture.

Links to Our Lives

You may want to emphasize the importance of learning the basics of human geography by sharing these points with your students:

- In order to understand individual cultures, we need to understand basic concepts of culture.
- Cultural differences affect world events and our daily lives. We should understand why cultures are different.
- We will participate more effectively in our own culture if we know how it developed.
- Population issues affect food production, international conflicts, and environmental questions—all of which can affect Americans directly.
- We must interpret population issues to make wise decisions about how we use resources.

CHAPTER 5

The World's People

The Colosseum, Rome, Italy

1998 Olympic opening ceremony, Nagano, Japan

Easter Island, Chile

FOCUS

LET'S GET STARTED

Copy the following instructions onto the chalkboard: *What are some organized events that occur regularly in our community? Work with a partner to list as many as you can.* Have students write down their answers. Discuss responses. *(Possible answers: parades, festivals, garage sales, sports tournaments, concerts)* Point out to students that the events they listed are part of the community's culture. You may want to ask if students know of similar events elsewhere. If so, compare those events with the students' lists to note ways in which other communities have different cultures. Tell students that in Section 1 they will learn more about culture.

Building Vocabulary

Write the Define terms on the chalkboard. Underline the prefixes *multi-* and *sub-*. Explain that they mean, respectively, "many" and "below" or "almost." Ask students to find the terms' definitions in the text or glossary and to relate the prefix meanings to **multicultural** and **subsistence agriculture**. Call on students to read the remaining definitions. Ask each student to choose a Define term and write a sentence to define it. Call on students to read their sentences. Continue until all of the Define terms have been covered.

Section 1 Culture

Read to Discover

1. What is culture?
2. Why are cultural symbols important?
3. What influences how cultures develop?
4. How did agriculture affect the development of culture?

Define

culture
culture region
culture traits
ethnic groups
multicultural
race
acculturation
symbol
domestication
subsistence agriculture
commercial agriculture
civilization

WHY IT MATTERS

Throughout history, culture has both brought people together and created conflict among groups. Use CNNfyi.com or other **current events** sources to learn about cultural conflicts around the globe. Record your findings in your journal.

Flags at the South Pole

Section 1 RESOURCES

Reproducible
- Block Scheduling Handbook, Chapter 5
- Guided Reading Strategy 5.1
- Creative Strategies for Teaching World Geography, Lessons 4, 5
- Map Activity 5
- Geography for Life Activity 5
- Lab Activity for Geography and Earth Sciences, Hands-On 4

Technology
- One-Stop Planner CD–ROM, Lesson 5.1
- Homework Practice Online
- Geography and Cultures Visual Resources with Teaching Activities 9 and 10
- HRW Go site

Reinforcement, Review, and Assessment
- Section 1 Review, p. 80
- Daily Quiz 5.1
- Main Idea Activity 5.1
- English Audio Summary 5.1
- Spanish Audio Summary 5.1

Aspects of Culture

The people of the world's approximately 200 countries speak hundreds of different languages. They may dress in different ways and eat different foods. However, all societies share certain basic institutions, including a government, an educational system, an economic system, and religious institutions. These vary from society to society and are often based on that society's **culture**. Culture is a learned system of shared beliefs and ways of doing things that guides a person's daily behavior. Most people around the world have a national culture shared with people of their own country. They may also have religious practices, beliefs, and language in common with people from other countries. Sometimes a culture dominates a particular region. This is known as a **culture region**. In a culture region, people may share certain **culture traits**, or elements of culture, such as dress, food, or religious beliefs. West Africa is an example of a culture region. Culture can also be based on a person's job or age. People can belong to more than one culture and can choose which to emphasize.

Race and Ethnic Groups Cultural groups share beliefs and practices learned from parents, grandparents, and ancestors. These groups are sometimes called **ethnic groups**. An ethnic group's shared culture may include its religion, history, language, holiday traditions, and special foods.

When people from different cultures live in the same country, the country is described as **multicultural** or multiethnic. Many countries

Thousands of Czechs and Germans settled in Texas in the mid-1800s. Dancers from central Texas perform a traditional Czech dance. ▼

TEACH

Teaching Objectives 1 and 3

ALL LEVELS: (Suggested time: 30 min.) Ask students to write on a slip of paper three of the cultures or culture regions to which they belong. Collect the suggestions. On the chalkboard, draw three columns. Label them *Culture, History,* and *Environment.* Choose a range of the student suggestions. As you record each group on the chart, lead a discussion on how it fulfills the definition of culture and how history and environment helped shape that culture. **ENGLISH LANGUAGE LEARNERS**

Teaching Objective 2

ALL LEVELS: (Suggested time: 45 min.) Point out that gestures are symbols and that they vary among cultures. Call on volunteers to demonstrate how they would show appreciation after a school play or choir concert. *(polite applause)* and after the home team's winning goal at a sports event. *(cheers, high fives).* Then have students conduct research on how people in other countries display approval in similar situations. Ask them to speculate how history and the environment have shaped the customs. Tell them to consider the effect of having a culture region divided by political boundaries. How might gestures or symbols in general change? You may need to extend this activity to another class period so that students can use other sources. **ENGLISH LANGUAGE LEARNERS**

DAILY LIFE

Peoples of many races and ethnic groups often enjoy the same entertainments. For example, a board game called *mancala* is popular in many parts of the world. *Mancala* is possibly the oldest board game in the world. Egyptians played this counting and strategy game before 1400 B.C. It is popular across Africa, among all ages and social classes.

Mancala is played on boards of different sizes and shapes or simply with shallow holes in the ground. Counters can be anything from seashells to seeds. Versions of *mancala* can be found thousands of miles from Africa, in Southeast Asia. The game's actual origin is not known, however.

Activity: Have students conduct research on the rules for playing *mancala.* Ask them to determine similarities and differences in rules from different countries. Then have them construct *mancala* boards from egg cartons. Hold a *mancala* tournament in your classroom.

internet connect

GO TO: **go.hrw.com**
KEYWORD: **SK3 CH5**
FOR: **Web sites about *mancala***

World Religions

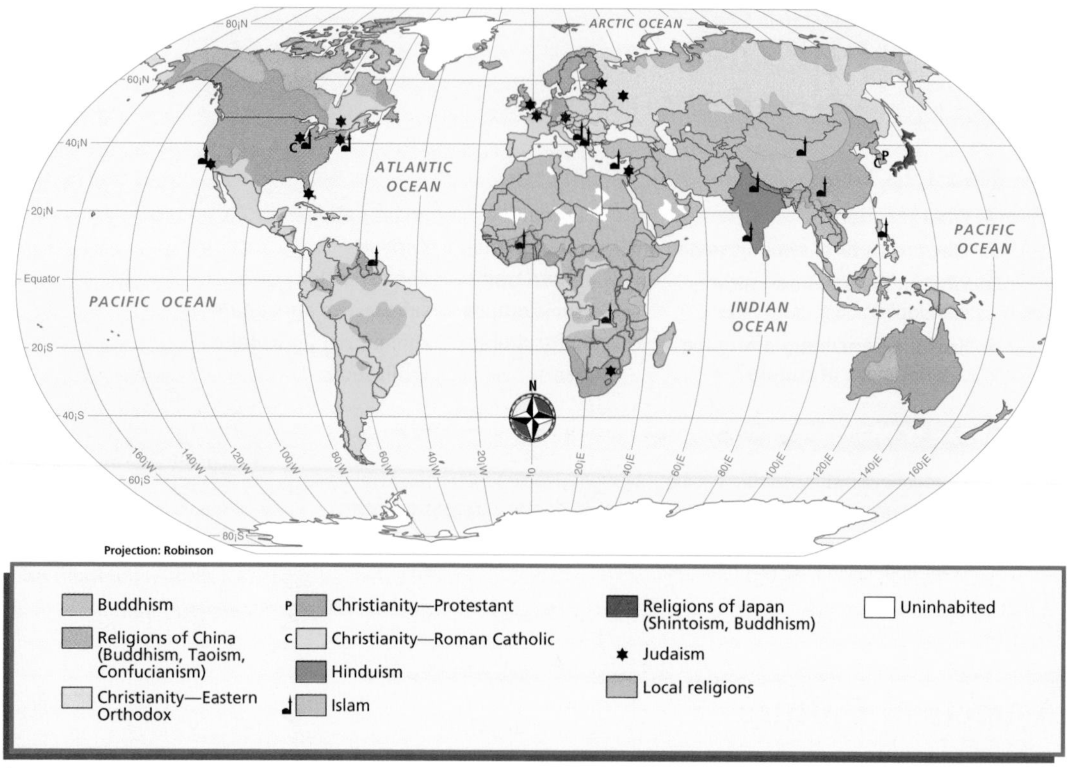

Religion is one aspect of culture.

A disc jockey sits at the control board of a Miami radio station that plays Cuban music.

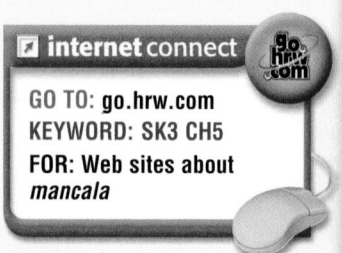

are multicultural. In some countries, such as Belgium, different ethnic groups have cooperated to form a united country. In other cases, such as in French-speaking Quebec, Canada, ethnic groups have frequently been in conflict. Sometimes, people from one ethnic group are spread over two or more countries. For example, Germans live in different European countries: Germany, Austria, and Czechoslovakia. The Kurds, who are a people with no country of their own, live mostly in Syria, Iran, Iraq, and Turkey.

Race is based on inherited physical or biological traits. It is sometimes confused with ethnic group. For example, the Hispanic ethnic group in the United States includes people who look quite different from each other. However, they share a common Spanish or Latin American heritage. As you know, people vary in physical appearance. Some of these differences have developed in response to climate factors like cold and sunlight. Because people have moved from region to region throughout history, these differences are not clear-cut. Each culture defines race in its own way, emphasizing particular biological and ethnic characteristics. An example can be seen in Rwanda, a country in East Africa. In this country, the Hutu and the Tutsi have carried on a bitter civil war. Although both are East African, each one

Teaching Objective 4

LEVEL 1: (Suggested time: 30 min.) Copy the following graphic organizer onto the chalkboard, omitting the italicized answers. Use it to help students describe how agriculture affected culture. Call on students to fill in the boxes. Then have them choose a step and illustrate it. Display the illustrated diagrams. **ENGLISH LANGUAGE LEARNERS**

AGRICULTURE AND CIVILIZATION

Humans eat wild foods and migrate. → *Humans domesticate plants and animals.* → *People develop agriculture and change the environment.* → *Agriculture provides surplus food.* → *Towns and cities lead to civilization.*

LEVELS 2 AND 3: (Suggested time: 45 min.) Organize students into groups. Have each group create a scenario that describes what human life would be like today if the progression from hunting and gathering to civilization had been interrupted. Instruct students to create feasible scenarios. You may want to have students act out their ideas. Have group members summarize in a paragraph the reasoning they used. **COOPERATIVE LEARNING**

considers itself different from the other. Their definition of race involves height and facial features. Around the world, people tend to identify races based on obvious physical traits. However, these definitions of race are based primarily on attitudes, not actual biological differences.

Cultural Change Cultures change over time. Humans invent new ways of doing things and spread these new ways to others. The spread of one culture's ways or beliefs to another culture is called diffusion. Diffusion may occur when people move from one place to another. The English language was once confined to England and parts of Scotland. It is now one of the world's most widely spoken languages. English originally spread because people from England founded colonies in other regions. More recently, as communication among cultures has increased, English has spread through English-language films and television programs. English has also become an international language of science and technology.

People sometimes may borrow aspects of another culture as the result of long-term contact with another society. This process is called **acculturation**. For example, people in one culture may adopt the religion of another. As a result, they might change other cultural practices to conform to the new religion. For example, farmers who become Muslim may quit raising pigs because Islam forbids eating pork.

✓ **READING CHECK:** *Human Systems* What is the definition of culture? a learned system of shared beliefs and ways of doing things that guides a person's daily behavior

Cultural Differences

A **symbol** is a sign that stands for something else. A symbol can be a word, a shape, a color, or a flag. People learn symbols from their culture. The sets of sounds of a language are symbols. These symbols have meaning for the people who speak that language. The same sound may mean something different to people who speak another language. The word *bad* means "evil" in English, "cool" to teenagers, and "bath" in German.

If you traveled to another country, you might notice immediately that people behave differently. For instance, they may speak a different language or wear different clothes. They might celebrate different holidays or salute a different flag. Symbols reflect the artistic, literary, and religious expressions of a society or culture. They also reflect that society's belief systems. Language, clothing, holidays, and flags are all symbols. Symbols help people communicate with each other and create a sense of belonging to a group.

✓ **READING CHECK:** *Human Systems* How do symbols reflect differences among societies and cultures? They show the societies' particular cultural expressions or belief systems.

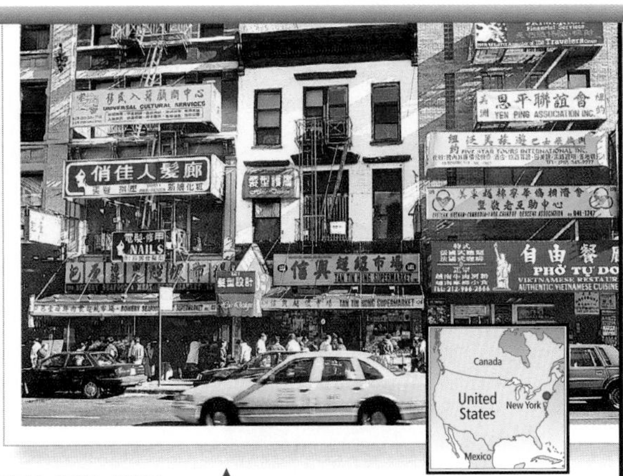

▲ Cultures mix in New York City's Chinatown.

Interpreting the Visual Record **Why might immigrants to a new country settle in the same neighborhood?**

internet connect

GO TO: go.hrw.com
KEYWORD: SK3 CH5
FOR: Web sites about the world's people

Fans cheer for the U.S. Olympic soccer team.

Interpreting the Visual Record **Why do you think symbols such as flags create strong emotions?**

▼

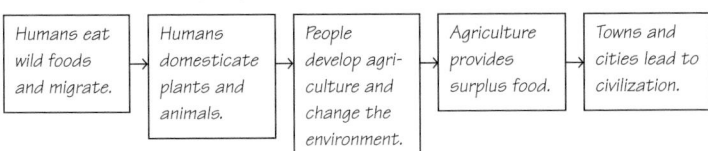

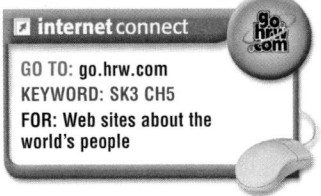

National Geography Standard 10

Human Systems Diffusion has carried elements of modern cultures into the far corners of the world, from vast deserts to rain forests.

In 1972, newspapers reported that a "lost" culture had been found in the Philippines. The Tasaday supposedly led an isolated existence and had never before encountered other cultures. However, in 1986 the story was revealed as a hoax. The Tasaday were well aware of modern cultures.

Discussion: Lead a discussion on these questions: Do isolated peoples have the right to be left alone? What might be the consequences of improved communication between these cultures and modern cultures?

▲ **Visual Record Answers**

possible answers: mutual support, common language, similar lifestyles

◄ because they can evoke strong feelings of national pride

➤ASSIGNMENT: Tell students that the cultures to which we belong affect practically all aspects of our daily lives—what we wear, how we talk, what we do at school, our choices in friends, how we relate to our families, what we do for fun, and so on. Have students write either a paragraph or a list of ways in which culture influences what they do throughout the day. For each event, ask students to speculate how someone in another culture might perform that action. You may want to have students discuss what factors would affect that choice. For example, we may eat store-bought cereal for breakfast. In another culture, a Chinese teen might eat boiled rice instead because his or her family members raise rice on their farm.

Teacher to Teacher

Lois Jordan, of Nashville, Tennessee, suggests this research project to help students explore how minority groups can affect a country's development and culture. Organize the class into teams. Assign one country per team. Have each team use primary and secondary sources to research the country's minority groups and the geographical or historical reasons why the groups are there. Students should try to answer questions such as these: Are members of the minority groups spread throughout the country or concentrated in one area? In what ways are the minority cultures different from the majority culture? How do the majority and minority cultures relate to each other politically and socially? Have the students share their findings in a panel discussion.

Linking Past to Present

Nile River Cultures Without the Nile River, the civilization whose people built the pyramids and the Sphinx could never have existed. The thousands of workers who labored on these and other monumental projects could be fed only because the river made intensive agriculture possible. The Nile Valley and Delta have been densely populated for thousands of years. Hunter-gatherers may have started to move into the Nile Valley by 12,000 B.C. By about 3,000 B.C. a great civilization flourished.

Alexandria, Egypt, was built next to the Nile Delta. During the 200s B.C. Alexandria may have been the largest metropolis in the world. Cleopatra's palace was there. During the 1980s, archaeologists began excavating the ruins of the city, which is now under water. They mapped the Royal Quarter, where Cleopatra had lived. Today, Alexandria is Egypt's second-largest city, with more than 4 million inhabitants.

Visual Record Answers ▲

possible answers: embroidery, fancy headdress

►

possible answers: streets narrower and appear to be laid out in a random pattern, houses closer together and with flat roofs

▲ A couple prepares for a wedding ceremony in Kazakhstan.

Interpreting the Visual Record **What aspects of these people's clothing indicate that they are dressed for a special event?**

The layout of Marrakech, Morocco, is typical of many North African cities.

Interpreting the Visual Record **How are the streets and houses of Marrakech different from those in your community?** ▼

Development of a Culture

All people have the same basic needs for food, water, clothing, and shelter. People everywhere live in families and mark important family changes together. They usually have rituals or traditions that go with the birth of a baby, the wedding of a couple, or the death of a grandparent. All human societies need to deal with natural disasters. They must also deal with people who break the rules of behavior. However, people in different places meet these needs in unique ways. They eat different foods, build different kinds of houses, and form families in different ways. They have different rules of behavior. Two important factors that influence the way people meet basic needs are their history and environment.

History Culture is shaped by history. A region's people may have been conquered by the same outsiders. They may have adopted the same religion. They may have come from the same area and may share a common language. However, historical events may have affected some parts of a region but not others. For example, in North America French colonists brought their culture to Louisiana and Canada. However, they did not have a major influence on the Middle Atlantic region of the United States.

Cultures also shape history by influencing the way people respond to the same historical forces. Nigeria, India, and Australia were all colonized by the British. Today each nation still uses elements of the British legal system, but with important differences.

Environment The environment of a region can influence the development of culture. For example, in Egypt the Nile River is central to people's lives. The ancient Egyptians saw the fertile soils brought by the flooding of the Nile as the work of the gods. Beliefs in mountain spirits were important in many mountainous regions of the world. These areas include Tibet, Japan, and the Andes of South America.

CLOSE

Call on students to suggest foods that are typical of a local culture. *(Possible answers: pierogi, quesadillas, dim sum, gumbo, grits)* Ask them to speculate how the popularity of that dish spread to the region. *(Possible answers: television ads, families moving to new regions, national restaurant franchises)* Point out how food is related to the main topics of the section.

REVIEW AND ASSESS

Have students complete the Section Review. Then have them work in pairs to draw flow charts or other graphic organizers of the section material. Then have students complete Daily Quiz 5.1. COOPERATIVE LEARNING

Culture also determines how people use and shape their landscape. For example, city plans are cultural. Cities in Spain and its former colonies are organized around a central plaza, or square, with a church and a courthouse. On the other hand, Chinese cities are oriented to the four compass points. American cities often follow a rectangular grid plan. Many French city streets radiate out from a central core.

✓ **READING CHECK:** *Human Systems* What are some ways in which culture traits spread? through historical events such as conquest by outsiders or colonization

Development of Agriculture

For most of human history people ate only wild plants and animals. When the food ran out in one place, they migrated, or moved to another place. Very few people still live this way today. Thousands of years ago, humans began to help their favorite wild plant foods to grow. They probably cleared the land around their campsites and dumped seeds or fruits in piles of refuse. Plants took root and grew. People may also have dug water holes to encourage wild cattle to come and drink. People began cultivating the largest plants and breeding the tamest animals. Gradually, the wild plants and animals changed. They became dependent on people. This process is called **domestication**. A domesticated species has changed its form and behavior so much that it depends on people to survive. Domestic sheep can no longer leap from rock to rock like their wild ancestors. However, the wool of domestic sheep is more useful to humans. It can be combed and twisted into yarn.

▲ This ancient Egyptian wall painting shows domesticated cattle.

Interpreting the Visual Record **Can you name other kinds of domesticated animals?**

Domestication happened in many parts of the world. In Peru llamas and potatoes were domesticated. People in ancient Mexico and Central America domesticated corn, beans, squash, tomatoes, and hot peppers. None of these foods was grown in Europe, Asia, or Africa before the time of Christopher Columbus's voyages to the Americas. Meanwhile, Africans had domesticated sorghum and a kind of rice. Cattle, sheep, and goats were probably first raised in Southwest Asia. Wheat and rye were first domesticated in Central Asia. The horse was also domesticated there. These domesticated plants and animals were unknown in the Americas before the time of Columbus.

ENVIRONMENT AND SOCIETY

We know cats, dogs, cattle, horses, sheep, and other familiar animals have served or been dependent on humans for many centuries. Creatures we think of as living in the wild have also been domesticated.

Cheetahs were tamed perhaps 5,000 years ago. Ancient Egyptians and early rulers of India kept them as pets and trained them to hunt. Hunting with falcons and hawks has been known since before 700 B.C. Elephants have served as beasts of burden and have been used to drag heavy equipment over difficult terrain. As early as the 200s B.C. and as late as the 1940s elephants have been used in warfare.

Activity: What other examples of domesticated animals can students find in books and nature television programs? To depict their findings, have them paint a mural in the style of a cave painting.

◄ **Visual Record Answer**

possible answers: cats, chickens, dogs, ducks, geese, goats, horses, pigs, sheep, turkeys

RETEACH

Have students complete Main Idea Activity 5.1. Then have them work in groups to invent a previously unknown culture. Have them write sentences describing the features and development of their culture by using the Define terms. Discuss the invented cultures to check on students' understanding of key concepts. **ENGLISH LANGUAGE LEARNERS, COOPERATIVE LEARNING**

EXTEND

Have interested students use primary and secondary sources to research how archaeology has shed light on when and where various plants and animals were domesticated for use as food. Then challenge them to choose a certain time period and region and write recipes appropriate to the available foods. Students may want to prepare an "ancient" meal for the class. Substitutions will be necessary. **BLOCK SCHEDULING**

Section 1 Review

Answers

Define For definitions, see: culture, p. 75; culture region, p. 75; culture trait, p. 75; ethnic groups, p. 75; multicultural, p. 75; race, p. 76; acculturation, p. 77; symbol, p. 77; domestication, p. 79; subsistence agriculture, p. 80; commercial agriculture, p. 80; civilization, p. 80

Reading for the Main Idea

1. People may belong to cultures based on where they live, their job, religious practices, beliefs, or age. (NGS 10)
2. Government, education, an economic system, religious institutions are basic to all. (NGS 10)

Critical Thinking

3. history—conquered by the same outsiders, have same religion or language; environment—affects religion, land use and planning; French colonists brought their culture to Louisiana; Nile River's influence on Egyptians
4. People built permanent settlements; surplus of food developed; population grew; civilizations formed.

Organizing What You Know

5. possible answers: religion, age, job, race, where we live, language and other cultural symbols

Our Amazing Planet

Thousands of years ago, domesticated dogs came with humans across the Bering Strait into North America. A breed called the Carolina dog may be descended almost unchanged from those dogs. The reddish yellow, short-haired breed also appears to be closely related to Australian dingoes.

Agriculture and Environment Agriculture changed the landscape. To make room for growing food, people cut down forests. They also built fences, dug irrigation canals, and terraced hillsides. Governments were created to direct the labor needed for these large projects. Governments also defended against outsiders and helped people resolve problems. People could now grow enough food for a whole year. Therefore, they stopped migrating and built permanent settlements.

Types of Agriculture Some farmers grow just enough food to provide for themselves and their own families. This type of farming is called **subsistence agriculture**. In the wealthier countries of the world, a small number of farmers can produce food for everyone. Each farm is large and may grow only one product. This type of farming is called **commercial agriculture**. In this system companies rather than individuals or families may own the farms.

Agriculture and Civilization Agriculture enabled farmers to produce a surplus of food—more than they could eat themselves. A few people could make things like pottery jars instead of farming. They traded or sold their products for food. With more food a family could feed more children. As a result, populations began to grow. More people became involved in trading and manufacturing. Traders and craftspeople began to live in central market towns. Some towns grew into cities, where many people lived and carried out even more specialized tasks. For example, cities often supported priests and religious officials. They were responsible for organizing and carrying out religious ceremonies. When a culture becomes highly complex, we sometimes call it a **civilization**.

✓ **READING CHECK:** *Environment and Society* In what ways did agriculture affect culture? Permanent settlements developed; a surplus of food developed; the population grew; civilizations developed.

go.hrw.com Homework Practice Online
Keyword: SK3 HP5

Section 1 Review

Define and explain: culture, culture region, culture trait, ethnic groups, multicultural, race, acculturation, symbol, domestication, subsistence agriculture, commercial agriculture, civilization

Reading for the Main Idea

1. *Human Systems* How can an individual belong to more than one cultural group?
2. *Human Systems* What institutions are basic to all societies?

Critical Thinking

3. **Drawing Inferences and Conclusions** In what ways do history and environment influence or shape a culture? What examples can you find in the text that explain this relationship?
4. **Analyzing Information** What is the relationship between the development of agriculture and culture?

Organizing What You Know

5. **Summarizing** Copy the following graphic organizer. Use it to describe culture by listing shared beliefs and practices.

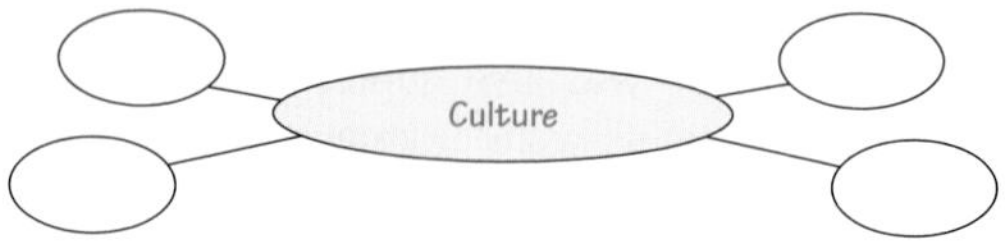

Section 2

Objectives

1. Explain why population density varies, and describe how the world's population has changed.
2. Identify ways that geographers describe and measure economies.
3. Identify the different economic systems.
4. Explain how governments differ.

FOCUS

LET'S GET STARTED

Write these instructions on the chalkboard: *Everyone except those on the first row should move to the back third of the room. I will explain soon.* Then tell students the crowded area represents a densely populated country, such as India, and the front of the room a thinly populated one, such as Mongolia. Discuss with students the advantages and disadvantages they would experience as citizens of these countries, based on their population density. Tell students that in Section 2 they will learn more about population issues.

Building Vocabulary

Write the Define terms on the chalkboard. Call on volunteers to read the definitions aloud. Then label some of them according to these categories: Economic Activities (**primary, secondary, tertiary, quaternary industries**); Ways to Measure Economic Development (**gross national product, gross domestic product, developed countries,** and **developing countries**); Economic Systems (**free enterprise; market, command, traditional economies**).

Section 2 Population, Economy, and Government

Read to Discover

1. Why does population density vary, and how has the world's population changed?
2. How do geographers describe and measure economies?
3. What are the different types of economic systems?
4. How do governments differ?

Define

primary industries
secondary industries
tertiary industries
quaternary industries
gross national product
gross domestic product
developed countries
developing countries
free enterprise
factors of production
entrepreneurs
market economy
command economy
traditional economy
democracy
unlimited government
limited government

WHY IT MATTERS

Human population has increased dramatically in the past 200 years. Use CNNfyi.com or other **current events** sources to find current projections for global or U.S. populations. Record your findings in your journal.

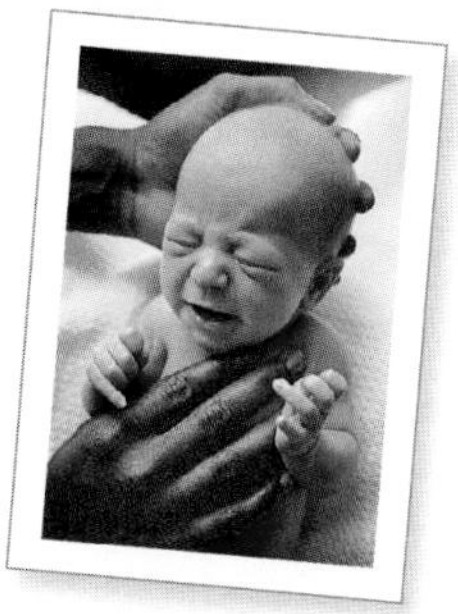

Newborn baby

Calculating Population Density

The branch of geography that studies human populations is called demography. Geographers who study it are called demographers. They look at such things as population size, density, and age trends. Some countries are very crowded. Others are only thinly populated. Demographers measure population density by dividing a country's population by its area. The area is stated in either square miles or square kilometers. For example, the United States has 74 people for every square mile (29/sq km). Australia has just 6 people per square mile (2.3/sq km). Japan has 869 people per square mile (336/sq km), and Argentina has 35 people per square mile (14/sq km).

These densities include all of the land in a country. However, people may not be able to live on some land. Rugged mountains, deserts, frozen lands, and other similar places usually have very few people. Instead, people tend to live in areas where the land can be farmed. Major cities tend to be located in these same regions of dense population.

✓ **READING CHECK:** *Human Systems* What is population density?

Differences in Population Density

Looking at this book's maps of world population densities allows us to make generalizations. Much of eastern and southern Asia is very

Shoppers crowd a street in Tokyo, Japan.

Section 2 **RESOURCES**

Reproducible
- Guided Reading Strategy 5.2
- Graphic Organizer 5

Technology
- One-Stop Planner CD–ROM, Lesson 5.2
- *ARGWorld* CD–ROM: Mapping Urban Growth
- *ARGWorld* CD–ROM: Strategic Straits and Choke Points
- *ARGWorld* CD–ROM: Technological Change and Factory Locations
- *ARGWorld* CD–ROM: Representation in the United Nations
- Homework Practice Online
- HRW Go site

Reinforcement, Review, and Assessment
- Section 2 Review, p. 86
- Daily Quiz 5.2
- Main Idea Activity 5.2
- English Audio Summary 5.2
- Spanish Audio Summary 5.2

TEACH

Teaching Objective 1

LEVEL 1: (Suggested time: 15 min.) Lead a class discussion on why people live where they do in your state. Encourage students to consider issues such as climate, availability of jobs, cultural and educational opportunities, and proximity to transportation facilities. Then call on volunteers to suggest how state population patterns can be compared to entire countries. *(People have the same basic needs and wants everywhere. More people live where their wants can be fulfilled most easily.)*
ENGLISH LANGUAGE LEARNERS

LEVELS 2 AND 3: (Suggested time: 30 min.) In advance, prepare names of countries on slips of paper. Include densely populated nations *(United Kingdom, Bangladesh, Japan)* and sparsely populated ones *(Mongolia, Australia, Canada).* Have students pick a country and find population data on that country in the text or in other secondary sources. They should also look for information on the country's resources and climate. Then have students write a paragraph describing the influences on that country's population density. Ask students to use the Fast Facts features or a world almanac to calculate the population density for their chosen country.

National Geography Standard 9

Human Systems Our government maintains records of population densities in the United States. Every 10 years, the U.S. Census Bureau counts the people who live here and gathers information about them. Following are some interesting facts about the 2000 census form:

- For the first time, there was a question about grandparents as caregivers.
- Respondents could mark "one or more" for their racial group.
- The mail-in census form was printed in English, Spanish, Chinese, Tagalog, Vietnamese, and Korean. Language Assistance Guides were available in 44 other languages.
- Census workers counted homeless people at campgrounds, shelters, soup kitchens, and other places.

Discussion: Lead a discussion on what these facts about the census form can tell us about changes in American society.

Chart Answer ►

1800

densely populated. There are dense populations in Western Europe and in eastern areas of North America. There are also places with very low population densities. Canada, Australia, and Siberia have large areas where few people live. The same is true for the Sahara Desert. Parts of Asia, South America, and Africa also have low population densities.

Heavily populated areas attract large numbers of people for different reasons. Some places have been densely populated for thousands of years. Examples include the Nile River valley in Egypt and the Huang River valley in China. These places have fertile soil, a steady source of water, and a good growing climate. These factors allow people to farm successfully. People are also drawn to cities. The movement of people from farms to cities is called urbanization. In many countries people move to cities when they cannot find work in rural areas. In recent years this movement has helped create huge cities. Mexico City, Mexico; São Paulo, Brazil; and Lagos, Nigeria, are among these giant cities. Some European cities experienced similar rapid growth when they industrialized in the 1800s. During that period people left farms to come to the cities to work in factories.

✓ **READING CHECK:** *Human Systems* Why does population density vary?
because of what draws people to certain areas—fertile soil, water, good climate for agriculture, employment

World Population Growth

Source: U.S. Census

Interpreting the Chart **Approximately when did world population growth begin to increase significantly?**

Population Growth

Researchers estimate that about 10,000 years ago the world's entire human population was less than 10 million. The annual number of births was roughly the same as the annual number of deaths.

After people made the shift from hunting and gathering to farming, more food was available. People began to live longer and have more children. The world's population grew. About 2,000 years ago, the world had some 200 million people. By A.D. 1650 the world's population had grown to about 500 million. By 1850 there were some 1 billion people. Better health care and food supplies helped more babies survive into adulthood and have children. By 1930 there were 2 billion people on Earth. Just 45 years later that number had doubled to 4 billion. In 1999 Earth's population reached 6 billion. By the year 2025 the world's population could grow to about 9 billion.

Births add to a country's population. Deaths subtract from it. The number of births per 1,000 people in a year is called the birthrate. Similarly, the death rate is the annual number of deaths per 1,000 people. The birthrate minus the death rate equals the rate of natural increase. This number is expressed as a percentage. A country's population changes when people enter or leave the country.

✓ **READING CHECK:** *Human Systems* What is the rate of natural increase?
a country's birthrate minus its death rate

Teaching Objective 2

ALL LEVELS: (Suggested time: 20 min.) Have students create a time line of the world's population growth, using the years stated in the text as milestones. If some students complete their time lines early, you may want to have them compile their work to create one time line, enlarge it, and illustrate it on a butcher-paper mural.
ENGLISH LANGUAGE LEARNERS

Teaching Objective 3

LEVEL 1: (Suggested time: 30 min.) Organize the class into groups. Have each group examine one common object *(dish, book, sock, eyeglasses)* and discuss among themselves what role primary, secondary, tertiary, and quaternary industries have played in the production of the object. Then have each group create a flow chart to display their ideas.
ENGLISH LANGUAGE LEARNERS, COOPERATIVE LEARNING

LEVELS 2 AND 3: (Suggested time: 20 min.) Organize the class into two groups. Using the characteristics noted in the section, have one half write a what-I-did-today journal entry for a youngster in an imaginary developed country and the other half do the same for an imaginary developing country. Call on volunteers to read their entries. Lead a discussion on how the journal entries compare to the characteristics mentioned in the section.
COOPERATIVE LEARNING

Economic Activity

All of the activities that people do to earn a living are part of a system called the economy. This includes people going to work, making things, selling things, buying things, and trading services. Economics is the study of the production, distribution, and use of goods and services.

Types of Economic Activities Geographers divide economic activities into primary, secondary, tertiary, and quaternary industries. **Primary industries** are activities that directly involve natural resources or raw materials. These industries include farming, mining, and cutting trees.

The products of primary industries often have to go through several stages before people can use them. **Secondary industries** change the raw materials created by primary activities into finished products. For example, the sawmill that turns a tree into lumber is a secondary industry.

Tertiary industries handle goods that are ready to be sold to consumers. The stores that sell products are included in this group. The trucks and trains that move products to stores are part of this group. Banks, insurance companies, and government agencies are also considered tertiary industries.

The fourth part of the economy is known as **quaternary industries**. People in these industries have specialized skills. They work mostly with information instead of goods. Researchers, managers, and administrators fall into this category.

Economic Indicators A common means of measuring a country's economy is the **gross national product** (GNP). The GNP is the value of all goods and services that a country produces in one year. It includes goods and services made by factories owned by that country's citizens but located in foreign countries. Most geographers use **gross domestic product** (GDP) instead of GNP. GDP includes only those goods and services produced within a country. GDP divided by the country's population is called per capita GDP. This figure shows individual purchasing power and is useful for comparing levels of economic development.

A **Primary industry:** A dairy farmer feeds his cows.
B **Secondary industry:** Cheese is prepared in a factory.
C **Tertiary industry:** A grocer is selling cheese to a consumer.
D **Quaternary industry:** A technician inspects dairy products in a lab.

READING CHECK: *Human Systems* What are primary, secondary, tertiary, and quaternary industries? primary—involve natural resources or raw materials; secondary—change raw materials into finished products; tertiary—handle goods for consumers; quaternary—skilled workforce working with information instead of goods

GLOBAL PERSPECTIVES

Beliefs and events can influence population issues. For example, in 1967 Iran became one of the first developing countries to institute a policy to slow population growth. The government declared family planning a human right and promoted it as a policy.

After Iran's 1979 Islamic Revolution, earlier efforts to limit family size were criticized as pro-West. The population growth rate climbed to more than 3 percent per year. However, a long war with Iraq badly damaged Iran. The leaders decided that it would be hard to rebuild their country while the population continued to expand so rapidly. As a result, the policy changed back to favoring small families.

Discussion: Lead a discussion on the relationship among religion, philosophical ideas, and Iranian culture with regards to population growth.

Teaching Objectives 3 and 4

ALL LEVELS: (Suggested time: 15 min.) Copy the following graphic organizer onto the chalkboard, omitting the italicized answers. Use it to help students understand the connection between economics and politics. Fill in the chart as a class. Point out that many countries do not follow the chart's model exactly but are a mixture of economic and political systems. Then have students discuss the concepts of limited and unlimited government. Have them work in groups to find examples of each type of government. **ENGLISH LANGUAGE LEARNERS, COOPERATIVE LEARNING**

ECONOMICS AND POLITICS

	Developed Countries	Developing Countries
Economy	*free enterprise*	*government control, communism*
Political System	*democracy*	*communism*

LEVELS 2 AND 3: (Suggested time: 20 min.) Have students formulate and write hypotheses to explain why developed countries are usually based on free enterprise and democracy and why the economies of developing countries are often controlled by the government. Tell them to be sure to describe the benefits of the U.S. free enterprise system.

Across the Curriculum
MATH

Per Capita GDP A country's gross domestic product (GDP) indicates the total size of its economy. However, it may not be a clear indication of an average citizen's wealth. For that purpose per capita GDP is probably more useful.

For example, China's GDP for 1998 was estimated at $4.4 trillion, making China a major economic power. However, China's per capita GDP was only $3,600, because the GDP figure was divided among China's immense population of more than 1 billion. In contrast, Chile's GDP the same year was about $184.6 billion. With a population of about 15 million, Chile's per capita GDP was $12,500. This comparison indicates that an average citizen in Chile is probably better off financially than an average citizen in China.

Activity: Use an almanac to provide students with GDP and population figures for several countries. Have each student choose a country and divide its gross GDP by its population to calculate its per capita GDP. Ask them to write a brief paragraph describing the information they found.

Chart Answer

France and Poland

Economic Development

Geographers use various measures including GNP, GDP, per capita GDP, life expectancy, and literacy to divide the world into two groups. Industrialized countries like the United States, Canada, Japan, and most European countries are wealthier. They are called **developed countries**. These countries have strong secondary, tertiary, and quaternary industries. They have good health care systems. Developed countries have good systems of education and a high literacy rate. The literacy rate is the percentage of people who can read and write. Most people in developed countries live in cities and have access to telecommunications systems—systems that allow long-distance communication. Geographers study the number of telephones, televisions, or computers in a country. They sometimes use these figures to estimate the country's level of technology.

Developing countries make up the second group. They are in different stages of moving toward development. About two thirds of the world's people live in developing countries. These countries are poorer. Their citizens often work in farming or other primary industries and earn low wages. Cities are often crowded with poorly educated people hoping to find work. They usually have little access to health care or telecommunications. Some developing countries have made economic progress in recent decades. South Korea and Mexico are good examples. These countries are experiencing strong growth in manufacturing and trade. However, some of the world's poorest countries are developing slowly or not at all.

✓ **READING CHECK:** *Human Systems* How do geographers distinguish between developed and developing countries? by examining GNP, GDP, per capita GDP, life expectancy, and literacy

Comparing Developed and Developing Countries

Country	Population	Rate of Natural Increase	Per Capita GDP	Life Expectancy	Literacy Rate	Telephone Lines
United States	281.4 million	0.9%	$ 36,200	77	97%	194 million
France	59.5 million	0.4%	$ 24,400	79	99%	35 million
South Korea	47.9 million	0.9%	$ 16,100	75	98%	24 million
Mexico	101.8 million	1.5%	$ 9,100	72	90%	9.6 million
Poland	38.6 million	–0.03%	$ 8,500	73	99%	8 million
Brazil	174.4 million	0.9%	$ 6,500	63	83%	17 million
Egypt	69.5 million	1.7%	$ 3,600	64	51%	3.9 million
Myanmar	42 million	0.6%	$ 1,500	55	83%	250,000
Mali	11 million	3.0%	$ 850	47	31%	23,000

Sources: Central Intelligence Agency, *The World Factbook 2001*

Interpreting the Visual Record **Which countries have the highest literacy rate?**

CLOSE

To illustrate the meaning of 6 billion, the approximate total world population, challenge students to calculate how long it would take for 6 billion seconds to go by *(approximately 190 years).*

REVIEW AND ASSESS

Have students complete the Section Review. Then refer them to the list of Define terms. Call on students to create a sentence that relates the first term to the second, then the second to the third, and so on. Then have students complete Daily Quiz 5.2.

RETEACH

Have students complete Main Idea Activity 5.2. Then have students work in pairs to create web diagrams to illustrate the section's main points.
ENGLISH LANGUAGE LEARNERS, COOPERATIVE LEARNING

Economic Systems

Countries organize their economies in different ways. Most developed countries organize the production and distribution of goods and services in a system called **free enterprise**. The United States operates under a free enterprise system. There are many benefits to this system. Companies are free to make whatever goods they wish. Employees can seek the highest wages for their work. People, rather than the government, control the **factors of production**. Factors of production are the things that determine what goods are produced in an economy. They include the natural resources that are available for making goods for sale. They also include the capital, or money, needed to pay for production and the labor needed to manufacture goods. The work of **entrepreneurs** (ahn-truh-pruh-NUHRS) makes up a fourth factor of production. Entrepreneurs are people who start businesses in a free enterprise system. Business owners in a free enterprise system sell their goods in a **market economy**. In such an economy, business owners and customers make decisions about what to make, sell, and buy.

In contrast, the governments of some countries control the factors of production. The government decides what, and how much, will be produced. It also sets the prices of goods to be sold. This is called a **command economy**. Some countries with a command economy are governed by an economic and political system called communism. Under communism, the government owns almost all the factors of production. Very few countries today are communist. Cuba is an example of a communist nation.

Finally, there are some societies around the world that operate within a **traditional economy**. A traditional economy is one that is based on custom and tradition. Economic activities are based on laws, rituals, religious beliefs, or habits developed by the society's ancestors. The Mbuti people of the Democratic Republic of the Congo practice a traditional economy, for example.

✓ **READING CHECK:** *Human Systems* What are the three types of economies? market, command, and traditional

Large shopping malls, such as this one in New York City's Trump Tower, are common in countries that have market economies and a free enterprise system. Shoppers here can find a wide range of stores and goods concentrated in one area.

COOPERATIVE LEARNING

Organize the class into small groups. Tell students to imagine that an automobile company wants to build a factory in a developing country.

Have some members of each group play the role of automobile executives, and have others be officials of the developing country's government. Assign one member of each group to conduct research on (or invent) characteristics that describe the country (location, labor force, resources, environment, type of government). Ask students to address these questions and others while they negotiate the business deal: How much control will government officials have over the factory? Who will decide which designs to use? What will happen if the factory pollutes the air or water? What will happen if car buyers have complaints? Then set negotiations into motion. Challenge students to draw conclusions about the relationships between economics and politics.

EXTEND

No longer does one have to live in a large city to be employed by a large company. Have interested students conduct research on the effects of telecommuting—using an electronic linkup with a central office to work out of one's home—on population densities in the United States. Students may want to concentrate on one of your state's major cities and the towns nearby. Have students report their findings in a bar graph. You may want to have them interview a telecommuter for additional information.
BLOCK SCHEDULING

➤ASSIGNMENT: Have students discuss how different forms of government function in society. Then have them create editorial cartoons that express the different forms. Call on volunteers to present their cartoons to the class. Discuss and display all the cartoons.

Section 2 Review Answers

Define For definitions, see: population density, p. 81; primary industries, p. 83; secondary industries, p. 83; tertiary industries, p. 84; quaternary industries, p. 84; gross national product, p. 84; gross domestic product, p. 84; developed countries, p. 84; developing countries, p. 84; free enterprise, p. 85; factors of production, p. 85; market economy, p. 85; command economy, p. 85; traditional economy, p. 85; democracy, p. 86; communism, p. 86; unlimited governments, p. 86; limited government, p. 86

Reading for the Main Idea

1. good soil, water, jobs (NGS 15)
2. free enterprise, democracy, high per capita GDP, high life expectancy and literacy (NGS 14)

Critical Thinking

3. market—business owners and customers decide what to sell and buy; command—government makes decisions; traditional—based on custom (NGS 14)
4. in ancient Greece, results of American, French Revolutions; made governments accountable

Organizing What You Know

5. Answers will vary.

World Governments

Just as countries use different economic systems, they also have different ways of organizing their governments. Some countries are controlled by one ruler, such as a monarch or a dictator. For example, Saudi Arabia is ruled by a monarch. King Fahd bin Abd al-Aziz Al Saud is both the chief of state and the head of government.

In other countries, a relatively small group of people controls the government. Many countries—including the United States, New Zealand, and Germany—have democratic governments. In a **democracy**, voters elect leaders and rule by majority. Ideas about democratic government began in ancient Greece. The American and French Revolutions established the world's first modern democratic governments in the late 1700s. Today, most developed countries are democracies with free enterprise economies. However, some countries with democratic governments, such as Russia and India, struggle with economic issues.

How a government is organized determines whether it has limited or unlimited powers. **Unlimited governments**, such as the French monarchy before the French Revolution, have total control over their citizens. They also have no legal controls placed on their actions. In a **limited government**, government leaders are held accountable by citizens through their constitutions and the democratic process. These limitations help protect citizens from abuses of power. Today, many countries around the world, including the United States, have limited governments.

✓ **READING CHECK:** *Human Systems* What is the difference between limited and unlimited government? Unlimited governments have total control over their citizens, while limited governments hold leaders accountable to their citizens and follow a democratic process.

Section 2 Review

go.hrw.com **Homework Practice Online**
Keyword: SK3 HP5

Define and explain: primary industries, secondary industries, tertiary industries, quaternary industries, gross national product, gross domestic product, developed countries, developing countries, free enterprise, factors of production, entrepreneurs, market economy, command economy, traditional economy, democracy, unlimited governments, limited government

Reading for the Main Idea

1. *Environment and Society* What geographic factors influence population density?
2. *Human Systems* What characteristics do developed countries share?

Critical Thinking

3. **Finding the Main Idea** What are the different economic systems? Describe each.
4. **Drawing Inferences and Conclusions** How did democracy develop, and why did it help create limited government?

Organizing What You Know

5. **Summarizing** Copy the following graphic organizer. Use it to study your local community and classify the businesses in your area.

Primary Industries	Secondary Industries	Tertiary Industries	Quaternary Industries
•	•	•	•
•	•	•	•

Section 3

Objectives

1. Identify problems associated with high and low population growth rates.
2. Compare two opposing views on population growth and resources.

Focus

LET'S GET STARTED

Copy the following instructions onto the chalkboard: *Look back through the Unit 1 chapters. Find a photo that you think shows a place with high population density. What made you choose that picture?* Allow students time to select a photo. Discuss student choices and what they see in the photos. Lead a discussion on how a high population density might affect daily life, as reflected in the photographs. Tell students that in Section 3 they will learn more about population growth and the issues associated with the topic.

Building Vocabulary

Write **carrying capacity** on the chalkboard. Have a student read the definition. Elicit a discussion comparing the carrying capacity of the entire Earth to that of a smaller environment. For example, have students consider the carrying capacity of your school. Ask students if they think the school could support a larger population or if it has reached its carrying capacity.

Section 3 Population Growth Issues

Read to Discover

1. What problems are associated with high and low population growth rates?
2. What are two different views of population growth and resources?

Define

scarcity
carrying capacity

Why It Matters

Some people believe that the world's limited resources cannot support a rapidly increasing human population. Use CNNfyi.com or other **current events** sources to find out what issues are raised by world population growth. Record your findings in your journal.

Population sign for Anatone, Washington

Section 3 Resources

Reproducible

- Guided Reading Strategy 5.3
- Environmental and Global Issues Activity 8

Technology

- One-Stop Planner CD–ROM, Lesson 5.3
- Yourtown CD–ROM
- Homework Practice Online
- HRW Go site

Reinforcement, Review, and Assessment

- Section 3 Review, p. 89
- Daily Quiz 5.3
- Main Idea Activity 5.3
- English Audio Summary 5.3
- Spanish Audio Summary 5.3

Population Growth Rates

With the help of technology, humans can survive in a wide range of environments. People can build houses and wear clothing to survive in cold climates. Food can be grown in one place and shipped to another. For these reasons and others the human population has grown tremendously.

Population growth rates differ from place to place. Many developed countries have populations that are growing very slowly, holding steady, or even shrinking. However, the populations of most developing countries continue to grow rapidly.

An Inuit family in northern Canada uses a snowmobile to pull a sled.

Growth Rate Issues In general, a high population growth rate will hinder a country's economic development. Countries must provide jobs, education, and medical care for their citizens. A rapidly growing population can strain a country's resources and lead to **scarcity**—when demand is greater than supply. Many of the countries with the highest growth rates today are among the world's poorest.

However, a shortage of young people entering the workforce lowers a country's ability to produce goods. Young people are needed to replace older people who retire or die. Many countries with very low growth rates or shrinking overall populations must support a growing number of older people. These people may need more health care.

TEACH

Teaching Objectives 1–2

LEVEL 1: (Suggested time: 15 min.) Copy the following graphic organizer onto the chalkboard, omitting the italicized answers. Use it to help students understand population growth issues. Fill in the advantages and disadvantages boxes as a class. Then, using the chart, lead a discussion to summarize the two sides of the population and resources issue. **ENGLISH LANGUAGE LEARNERS**

POPULATION GROWTH

	Advantages	Disadvantages
Low Population Growth Rate	*high standard of living, enough resources to go around*	*lowers country's ability to produce, large number of older people that may need financial support*
High Population Growth Rate	*increased political power, high productivity*	*hinders economic development, can strain resources*

Section 3 Review

Answers

Define For definition, see: carrying capacity, p. 89

Reading for the Main Idea

1. Rapid population growth can hinder a country's economic development. (NGS 9)
2. Earth can easily support a much larger human population; Earth has reached its carrying capacity. (NGS 9)

Critical Thinking

3. possible answers: Humans can survive in a wide range of environments; people can build houses and wear clothing to survive in cold climates; food can be grown in one place and shipped to another.
4. through economic interdependence—trading—or through conflicts with other countries

Organizing What You Know

5. ability—new fertilizers, special seeds, new energy sources, better use of existing resources; inability—limited amount of land available for farming, shortage of freshwater, oil running out, pollution

Visual Record Answer ►

small plots, animals used for power

Our Amazing Planet

About 1,000 years ago, people began carving out homes, churches, stables, and other "buildings" from the cone-shaped rock formations of Turkey's Cappadocia (ka-puh-DOH-shuh) region. Today, some of the larger spaces have been made into restaurants.

Uneven Resource Distribution Natural resources such as fresh water, minerals, and fertile land are not distributed evenly. A country's resources cannot always support its population. However, the country may be able to acquire needed resources by trading with other countries. In this system of economic interdependence, two countries can exchange resources or goods so that each gets what it needs. Japan, for example, has few energy resources but is a world leader in manufacturing. Japan sells manufactured goods to others, particularly the United States. Japan then uses the money to buy oil from Saudi Arabia.

The need for scarce resources usually leads countries to trade peacefully. However, it can also lead to military conflict. One country might try to take over a resource-rich area of a neighboring country. If we hope to avoid future wars over resources, the world's people must share resources more equally.

✓ **READING CHECK:** *Human Systems* How does scarcity of resources affect international trade and economic interdependence? It causes countries to trade peacefully but can also lead to military conflict.

World Population and Resources

Some people think Earth can easily support a much larger human population. They base this view partly on history. For example, new fertilizers and special seeds mean more food can be grown today than ever before. They also think that scientists will probably discover new energy resources. The Sun, for example, is a vast energy source. Today we can only use a fraction of its power, however. Another way to support more people would be to make better use of existing resources. For example, we can recycle materials and reduce the amount of waste we generate.

Many farmers in India still use traditional methods.

Interpreting the Visual Record What in this photo indicates a less developed agricultural society? ▼

CLOSE

Ask students how decisions they make now and in the future may affect population and resource distribution issues.

REVIEW AND ASSESS

Have students complete the Section Review. Then have students complete Daily Quiz 5.3.

RETEACH

Have students complete Main Idea Activity 5.3. Then use colored toothpicks to symbolize resources: water (blue), farmland (green), food (yellow), and mineral and energy resources (red). Each student is to represent a country. Give many toothpicks of all colors to a few students, just a few toothpicks to more students, and just one or two to the remainder of the students. Explain the colors' meaning. Ask students to decide among themselves how to distribute their resources. After the exercise, elicit a discussion based on these questions: What happened when countries with few resources tried to increase their wealth? What might have happened if there were twice as many students in the class, but the same number of resources? **ENGLISH LANGUAGE LEARNERS**

EXTEND

Have students conduct research on the role of population expansion in an era of invasion or conquest in world history, such as European colonialism. Ask them to create a cause-and-effect chart. **BLOCK SCHEDULING**

A buildup of salt has ruined this field in southern Iraq. This land is now useless for growing crops, thus contributing to the lack of land suitable for farming.

Other people hold the opposite opinion. They argue that the world is already showing signs of reaching its **carrying capacity**. Carrying capacity is the maximum number of a species that can be supported by an area's scarce resources. The amount of land available for farming is shrinking. Many areas are experiencing a shortage of fresh water. Oil, a nonrenewable resource, will eventually run out. Pollution is damaging the atmosphere and the oceans. The rich nations of the world are not always willing to share with poorer nations. In the future, these people think food and water supplies will run short in many countries. People without enough to eat will become ill more easily, leading to widespread disease. Such problems may lead a country to invade its neighbors to capture resources.

These are challenging issues that reach into all areas of life. They will become even more important in the future. A better understanding of geography will help you understand and deal with these issues.

✓ **READING CHECK:** *Environment and Society* What are two different views on population growth and resources? Some people think Earth can support a much larger population; others think Earth is already reaching its carrying capacity.

Section 3 Review

go.hrw.com Homework Practice Online
Keyword: SK3 HP5

Define and explain: scarcity, carrying capacity

Reading for the Main Idea

1. *Human Systems* What problems are associated with rapid population growth?
2. *Human Systems* What are two viewpoints about future population growth?

Critical Thinking

3. **Analyzing Information** How has technology helped the worldwide human population grow?
4. **Finding the Main Idea** How do countries deal with the uneven distribution of resources?

Organizing What You Know

5. **Contrasting** Copy the following graphic organizer. Use it to discuss two arguments about population growth.

Earth's ability to support a much larger population	⇔	Earth's inability to support a much larger population

CHAPTER 5 Review ANSWERS

Building Vocabulary

For definitions, see: culture, p. 75; culture region, p. 75; culture traits, p. 75; ethnic groups, p. 75; multicultural, p. 75; acculturation, p. 77; symbol, p. 77; domestication, p. 79; subsistence agriculture, p. 80; civilization, p. 80; entrepreneurs, p. 85; limited government, p. 86; command economy, p. 85; market economy, p. 85; factors of production, p. 85; free enterprise, p. 85; carrying capacity, p. 89

Reviewing the Main Ideas

1. learned system of shared beliefs and ways of doing things that guide a person's daily behavior; to learn more about the world and its people (NGS 10)
2. subsistence—farmers growing just enough food for their own families; commercial—farmers growing food for consumers (NGS 14)
3. Some governments are controlled by one ruler, such as a monarch. In others, groups of people control the government. (NGS 9)
4. useful to indicate the country's overall level of technology (NGS 11)
5. A rapid population growth rate can strain a country's resources. (NGS 16)

ASSESS

Have students complete the Chapter 5 Test.

RETEACH

Organize the class into three groups—one for each section. Have students work in groups or alone to concentrate on topics within the sections. Tell them to create posters with captions to summarize their assigned section's main points. Display the posters. COOPERATIVE LEARNING

PORTFOLIO EXTENSIONS

1. Have students collect symbols and logos—on signs, clothing, and food and beverage labels, for example—that are part of American culture. They may need to photograph or draw some symbols. Ask students how they would interpret the images if they lived in another culture. Have them write brief explanations of the symbols' meaning. Include descriptions in portfolios.
2. Discuss what changes ancient peoples would have experienced as the population growth rate increased and changed the relative supply of the factors of production. Explain to students that writing did not develop until after the world's population had increased dramatically. Then have students create "cave paintings" on butcher paper to express those changes caused by the increased population growth rate.

Chapter 5 Review ANSWERS

Understanding Environment and Society

Answers will vary, but the information included should be consistent with text material. Presentations should address the points listed. Use Rubric 29, Presentations, to evaluate student work.

Thinking Critically

1. Possible answer: because differences are not clear-cut; people define races differently
2. doubled four times; by 2025; possible answers: development hindered, crowding, pollution, resource depletion
3. primary, secondary, tertiary, quaternary; farming, saw mill, bank, researcher
4. traditional, market, command; allows for economic development and individual freedom
5. Answers will vary according to students' opinions.
6. unlimited—totalitarian, undemocratic, no legal controls placed on their actions, French monarchy; limited—held accountable by their citizens, United States

Chapter 5 Reviewing What You Know

Building Vocabulary

On a separate sheet of paper, write sentences to define each of the following words.

1. culture
2. ethnic groups
3. multicultural
4. acculturation
5. symbol
6. domestication
7. subsistence agriculture
8. civilization
9. limited government
10. entrepreneurs
11. command economy
12. market economy
13. factors of production
14. free enterprise
15. carrying capacity

Reviewing the Main Ideas

1. (Human Systems) What is culture, and why should people study it?
2. (Environment and Society) What is the difference between subsistence and commercial agriculture?
3. (Human Systems) What are some of the different ways countries organize governments?
4. (Human Systems) Why are telecommunications devices useful as economic indicators?
5. (Environment and Society) How do population growth rates affect resources?

Understanding Environment and Society

Domestication

Do you know how, when, and why people first domesticated dogs, cats, pigs, and hawks? What about oranges? Pick a domesticated plant or animal to research. As you prepare your presentation consider the following:

- Where the crop or animal was first domesticated.
- Differences between it and its wild ancestors.
- Humans spreading it to new areas.

Thinking Critically

1. **Drawing Inferences and Conclusions** Why are ethnic groups sometimes confused with races?
2. **Making Generalizations and Predictions** Over the past 2,000 years, how many times has world population doubled? By when is it projected to double again? What might be the effect of this increase?
3. **Finding the Main Idea** What are the four basic divisions of industry? Give examples.
4. **Summarizing** What are the three economic systems, and what are the benefits of the U.S. free-enterprise system?
5. **Making Generalizations and Predictions** Do you think Earth has a carrying capacity for its human population? Why or why not?
6. **Summarizing** Explain unlimited and limited government. Give examples of each.

Food Festival

Ask students to bring foods to class that reflect their families' culture. If most of the students are from similar backgrounds, have them concentrate on family variations of a common dish. For example, there are innumerable ways to make salsa, a spicy condiment for Mexican food. Jewish students of Eastern European heritage might compare how their families make kugel, a baked pudding of noodles or potatoes.

Chapter 5 Review and Assessment Resources

Reproducible
- Readings in World Geography, History, and Culture 9 and 10
- Critical Thinking Activity 5
- Vocabulary Activity 5

Technology
- Chapter 5 Test Generator (on the One-Stop Planner)
- HRW Go site
- Audio CD Program, Chapter 5

Reinforcement, Review, and Assessment
- Chapter 5 Review, pp. 90–91
- Chapter 5 Tutorial for Students, Parents, Mentors, and Peers
- Chapter 5 Test
- Chapter 5 Test for English Language Learners and Special-Needs Students
- Unit 1 Test
- Unit 1 Test for English Language Learners and Special-Needs Students

Building Social Studies Skills

Map Activity

On a separate sheet of paper, match the letters on the map with their correct labels.

Buddhism
Christianity—Eastern Orthodox
Christianity—Protestant
Christianity—Roman Catholic
Hinduism
Islam

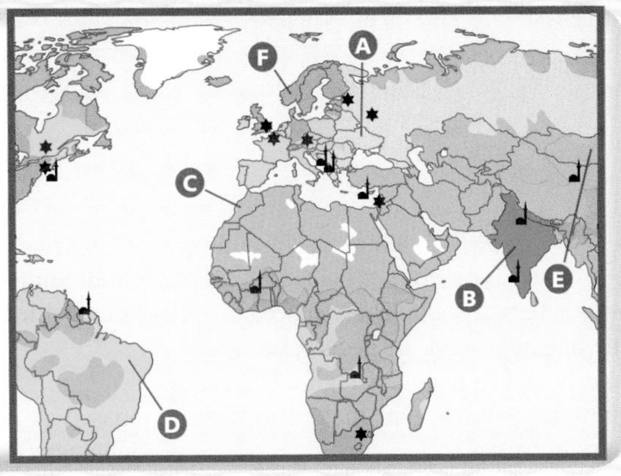

Mental Mapping Skills Activity

Draw a map of the world and label Japan, Australia, the United Kingdom, and Argentina. Based on your knowledge of climates, population density, and resources, which of these countries probably depend on imported food? Which ones probably export food? Express this information on your map.

Writing Activity

Study the economy of your local community. Has the local economy grown or declined since 1985? Why? Predict how your local area could change economically during the next 10 years. Be sure to use standard grammar, spelling, sentence structure, and punctuation.

Alternative Assessment

Portfolio Activity

Learning About Your Local Geography

Factors of Production Recall the discussion of the factors of production. Create a model showing how these factors influence the economy of your community.

internet connect

Internet Activity: go.hrw.com
KEYWORD: SK3 GT5

Choose a topic to explore online:
- Visit famous buildings and monuments around the world.
- Compare facts about life in different countries.
- Examine world population growth.

Map Activity
A. Christianity—Eastern Orthodox
B. Hinduism
C. Islam
D. Christianity—Roman Catholic
E. Buddhism
F. Christianity—Protestant

Mental Mapping Skills Activity
Maps will vary, but listed places should be labeled in their approximate locations. Students should support their answers.

Writing Activity
Answers will vary, but the information included should be consistent with text material. Students' predictions should be supported by logical arguments. Use Rubric 37, Writing Assignments, to evaluate student work.

Portfolio Activity
Responses will vary, but students should support their arguments logically. Use Rubric 28, Posters, to evaluate student work.

internet connect

GO TO: go.hrw.com
KEYWORD: SK3 Teacher
FOR: a guide to using the Internet in your classroom

Focus On Regions

Identifying Local Regions

Ask students to write down regions in which they live, go to school, shop, and pursue other daily activities. Call on volunteers for their responses. Record them on the chalkboard, placing the largest regions (continent, country) at the top and placing smaller regions (neighborhoods, boroughs, blocks) at the bottom. Note at what point descriptions of regions where students live differ from each other (probably at the town or neighborhood level). Draw a line separating these smaller divisions from the larger ones. Have students work in small groups to draw large maps showing these local regions and their relationships.

Geography Sidelight

Physical, political, economic, and cultural changes can affect the way we define regions and draw their boundaries.

Physical changes are often slow to occur, but they are important. Beaches erode. Rivers run dry. Forests grow up on abandoned farmland. Political changes also bring regional redefinitions. For example, one large country may break up into several smaller ones, or small countries may unite to form a larger country.

Economic changes occur as different resources or economic activities decline or increase in importance. For example, a community once known for its manufacturing industries may now be part of an area that focuses on services. Finally, changing cultural characteristics can reshape regions as people of various language, religious, or ethnic groups move across regional boundaries.

Critical Thinking: What factors affect how we define regions?

Answer: physical, political, economic, and cultural changes

➤ This Focus On Regions feature addresses National Geography Standards 5 and 6.

Focus On Regions

What is a Region?

Think about where you live, where you go to school, and where you shop. These places are all part of your neighborhood. In geographic terms, your neighborhood is a region. A region is an area that has common features that make it different from surrounding areas.

What regions do you live in? You live on a continent, in a country, and in a state. These are all regions that can be mapped.

Regions can be divided into smaller regions called subregions. For example, Africa is a major world region. Africa's subregions include North Africa, West Africa, East Africa, central Africa, and southern Africa. Each subregion can be divided into even smaller subregions.

Regional Characteristics Regions can be based on physical, political, economic, or cultural characteristics. Physical regions are based on Earth's natural features, such as continents, landforms, and climates. Political regions are based on countries and their subregions, such as states, provinces, and cities. Economic regions are based on money-making activities such as agriculture or industries. Cultural regions are based on features such as language, religion, or ethnicity.

▲ East Africa is a subregion of Africa. It is an area of plateaus, rolling hills, and savanna grasslands.

Going Further: Thinking Critically

Have students compare their maps. Note areas where regions overlap and ask students how they might describe these areas *(transition zones).* Also have them look at how maps of the same regions have different boundaries. Point out that these are perceived regions. The word *perceived* has to do with getting information from one's senses. Therefore, perceived regions differ according to how one "sees" the region. Ask students how they decided on the boundaries that they drew. Use their responses to illustrate the differences in perceived regions. Finally, review formal and functional regions and ask whether any regions shown on the maps fulfill the definitions.

Major World Regions

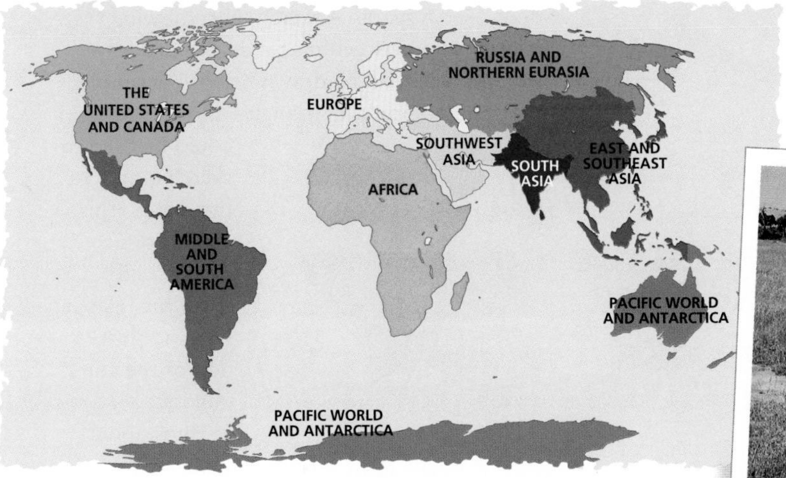

The international border between Kenya and Tanzania is a clearly defined regional boundary.

Regional Boundaries All regions have boundaries, or borders. Boundaries are where the features of one region meet the features of a different region. Some boundaries, such as coastlines or country borders, can be shown as lines on a map. Other regional boundaries are less clear.

Transition zones are areas where the features of one region change gradually to the features of a different region. For example, when a city's suburbs expand into rural areas, a transition zone forms. In the transition zone, it may be hard to find the boundary between rural and urban areas.

Types of Regions There are three basic types of regions. The first is a formal region. Formal regions are based on one or more common features. For example, Japan is a formal region. Its people share a common government, language, and culture.

The second type of region is a functional region. Functional regions are based on movement and activities that connect different places. For example, Paris, France, is a functional region. It is based on the goods, services, and people that move throughout the city. A shopping center or an airport might also be a functional region.

The third type of region is a perceived region. Perceived regions are based on people's shared feelings and beliefs. For example, the neighborhood where you live may be a perceived region.

The three basic types of regions overlap to form complex world regions. In this textbook, the world is divided into nine major world regions (see map above). Each has general features that make it different from the other major world regions. These differences include physical, cultural, economic, historical, and political features.

Understanding What You Read

1. Regions can be based on what types of characteristics?
2. What are the three basic types of regions?

Understanding What You Read

Answers

1. Regions can be based on physical, political, economic, or cultural characteristics.
2. The three basic types of regions are: formal, functional, and perceived.

Going Further: Thinking Critically

Prepare two sets of words on slips of paper. One set should consist of action words, such as *go, fly, run, skip, dig, walk, swim, climb,* and *drive.* The other set should consist of nouns having to do with places, such as *tree, house, cliff, beach, road, river, mountain, church, rock, cave, highway,* and *ocean.* Students will draw at least one word of each type from a hat. Using their chosen words at least once, students should write travel or adventure stories set in familiar places. You may want to require that the stories be a certain length or that the students mention a certain number of places in their stories. Tell students that when they finish they should be able to draw maps of the settings through which the characters travel. When all the students have completed their stories, have them exchange papers to draw maps of each other's story settings.

PRACTICING THE SKILL

1. The prime meridian extends through western Europe (England, western France, northeastern Spain) and western Africa (Algeria, Mali, Burkina Faso, Togo, Ghana).
2. Students' sketch maps should show the equator, Tropic of Cancer, Tropic of Capricorn, prime meridian, and continents in their approximate locations.
3. Answers will vary. Students might notice that the international date line does not cross any major landmasses, the Southern Hemisphere has much more water than land, South America extends farther south than Africa, or other similar facts.

➤ **This GeoSkills feature addresses National Geography Standards 1, 2, and 3.**

GeoSkills

Building Skills for Life: Drawing Mental Maps

We create maps in our heads of all kinds of places—our homes, schools, communities, country, and the world. Some of these places we know well. Others we have only heard about. These images we carry in our heads are shaped by what we see and experience. They are also influenced by what we learn from news reports or other sources. Geographers call the maps that we carry around in our heads mental maps.

We use mental maps to organize spatial information about people and places. For example, our mental maps help us move from classroom to classroom at school or get to a friend's home. A mental map of the United States helps us list the states we would pass through driving from New York City to Miami.

We use our mental maps of places when we draw sketch maps. A sketch map showing the relationship between places and the relative size of places can be drawn using very simple shapes. For example, triangles and rectangles could be used to sketch a map of the world. This quickly drawn map would show the relative size and position of the continents.

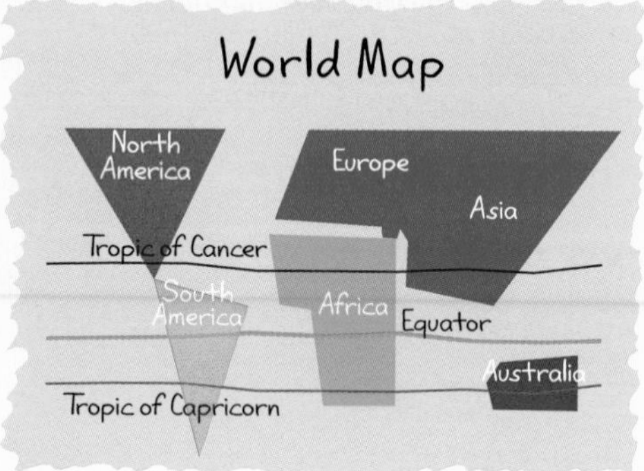

Think about some simple ways we could make our map of the world more detailed. Adding the equator, Tropic of Cancer, and Tropic of Capricorn would be one way. Look at a map of the world in your textbook's Atlas. Note that the bulge in the continent of Africa is north of the equator. Also note that all of Asia is north of the equator. Next note that the Indian subcontinent extends south from the Tropic of Cancer. About half of Australia is located north of the Tropic of Capricorn. As your knowledge of the world increases, your mental map will become even more detailed.

PRACTICING THE SKILL

1. Look at the maps in your textbook's Atlas. Where does the prime meridian fall in relation to the continents?
2. On a separate sheet of paper, sketch a simple map of the world from memory. First draw the equator, Tropic of Cancer, Tropic of Capricorn, and prime meridian. Then sketch in the continents. You can use circles, rectangles, and triangles.
3. Draw a second map of the world from memory. This time, draw the international date line in the center of your map. Add the equator, Tropic of Cancer, and Tropic of Capricorn. Now sketch in the continents. What do you notice?

Lab Report

Answers

1. Important features on student maps will probably be their own homes, friends' homes, major streets, schools, restaurants, stores or malls, and recreation facilities.
2. Students' maps will probably omit secondary streets, boundaries of political divisions, elevation figures, contour lines, and similar details. They may record more labeled individual buildings or natural features than the published maps.
3. Answers will vary. Students might suggest that their maps include more streets, features be more to scale, or other changes be made.

HANDS ON GEOGRAPHY

Mental maps are personal. They change as we learn more about the world and the places in it. For example, they can include details about places that are of interest only to you.

What is your mental map of your neighborhood like? Sketch your mental map of your neighborhood. Include the features that you think are important and that help you find your way around. These guidelines will help you get started.

1. Decide what your map will show. Choose boundaries so that you do not sketch more than you need to.
2. Determine how much space you will need for your map. Things that are the same size in reality should be about the same size on your map.
3. Decide on and note the orientation of your map. Most maps use a directional indicator. On most maps, north is at the top.
4. Label reference points so that others who look at your map can quickly and easily figure out what they are looking at. For example, a major street or your school might be a reference point.
5. Decide how much detail your map will show. The larger the area you want to represent, the less detail you will need.
6. Use circles, rectangles, and triangles if you do not know the exact shape of an area.
7. As you think of them, fill in more details, such as names of places or major land features.

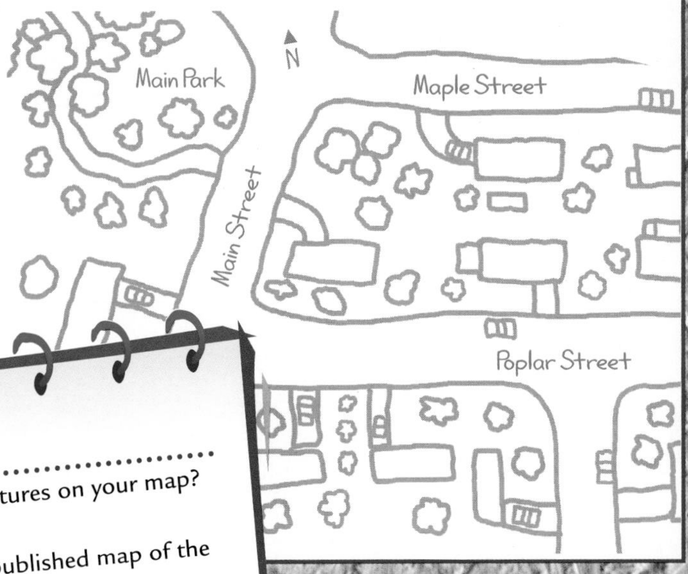

Lab Report

1. What are the most important features on your map? Why did you include them?
2. Compare your sketch map to a published map of the area. How does it differ?
3. At the bottom, list three ways that you could make your sketch map more complete.

UNIT 2

GAINING A HISTORICAL PERSPECTIVE

Using the Illustrations

Direct students' attention first to the large photograph and then to the smaller photographs on this page and the next. Ask students how they think all of the photographs are connected. *(The large photograph shows people digging at a site where there are ancient ruins and objects. The smaller photographs are some of the artifacts that have been found there.)* Have students read the captions and identify the objects. Point out that people must work very carefully at a site where there are building ruins and objects. Ask students why this is necessary *(in order not to destroy or damage anything; to preserve the historical record)*. Also ask what kinds of tools students think people use at an archaeological site. *(Possible answers: hands, brushes, small picks, rakes, shovels.)*

Tell students that once an object is dug up, it is cataloged—numbered and listed in a book. Information about the type of object, where it was found, the date, and other appropriate information is recorded. Ask students why they think this might be important *(to keep track of objects; to let others who study the objects know about them)*.

Unit Objectives

1. Analyze life in prehistoric times and the development of early civilizations.
2. Examine changes in the political, social, and religious systems of the ancient world.
3. Identify the significance of the Middle Ages, Renaissance, Reformation, and Scientific and Industrial Revolutions.
4. Interpret the effects of European exploration, expansion, and colonization on other parts of the world.
5. Describe the political, social, religious, and military events that gave birth to the modern world.
6. List the political, social, religious, and military events that shaped the modern world we live in today.

UNIT 2

Gaining a Historical Perspective

Young people working at one of York Archaeological Trust's sites

Viking pot found at York

CHAPTERS

An Archaeologist at Work

Dr. Ailsa Mainman is an archaeologist. She studies the remains and ruins of past cultures. She lives in York, in the northeastern part of England. She works for the York Archaeological Trust. Here she describes her work. **WHAT DO YOU THINK?** ***What part of Dr. Mainman's work would you enjoy the most?***

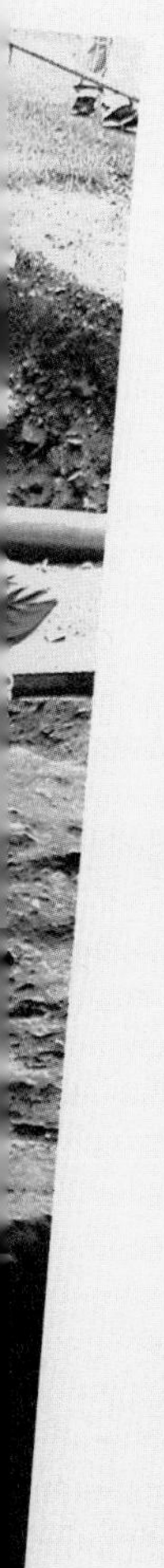

I chose archaeology because I loved history but I wanted to be in touch with real things, not just books. It still gives me a real thrill to hold something that was found in one of our digs. The object might be hundreds, sometimes thousands of years old. I like to think about what the people and their lives were like.

We have schoolchildren who come in the summer to help out on the digs, including some from the United States. They wash, sort, and draw these finds (valuable discoveries). Then the children try to work out what the objects are. Even broken bits of pot or bone have a lot to tell us about how people used to live.

Leather boot found at York

Everyone thinks Vikings were just fierce warriors and raiders. They were, of course, but they were so much more. We have recreated the Viking city of Jorvik, which thrived in York beginning in A.D. 866. You can travel through the streets and houses and see the artisans and craftsmen. The Vikings were skilled at many crafts. They worked with gold, silver, iron, antler, bone, glass, and many other materials.

The Vikings were also tremendous shipbuilders. They explored and then settled in Iceland, Greenland, and Canada. Their trade routes linked them with Eastern Europe and even China!

Viking artifacts found at York

Understanding Primary Sources

1. What has Dr. Mainman and others learned about the Vikings from the finds at their archaeological digs?
2. How does Dr. Mainman describe the Vikings?

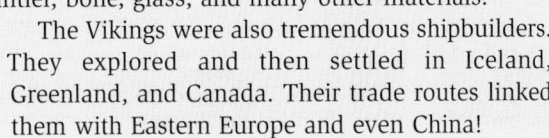

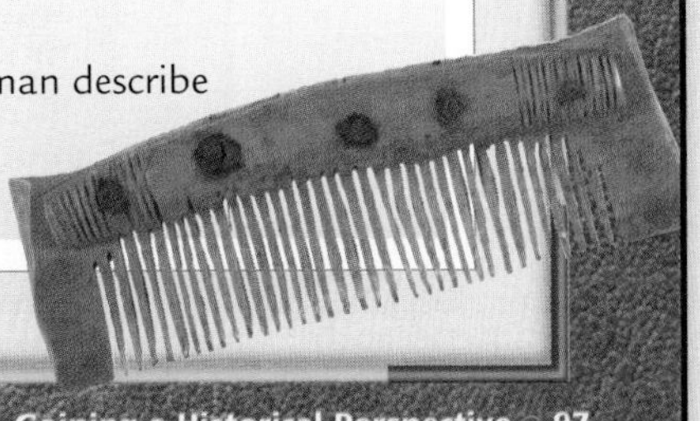

Comb found at York

MORE FROM THE FIELD

Over 35,000 objects have been recovered from the Jorvik site pictured on the opposite page. Most were in good condition because of the waterlogged soils in which they had been buried.

Once artifacts are out of the soil, the York Archaeological Trust's laboratories use various techniques to prevent them from decaying. The methods depend on the type of material. For example, wood can impregnated with wax and then freeze-dried. Once the objects have been treated they have to be kept, whether in storage or on display, in conditions which suit them—neither too hot nor too cold, not too damp or too dry.

Activity: Have students conduct research on an archaeological dig, either in your area or overseas. Ask them to speculate on what day-to-day life might be like on such a dig.

Understanding Primary Sources

Answers

1. what the Vikings were like, how they lived, tools and other objects they made and used
2. fierce warriors and raiders, skilled craftworkers, shipbuilders, explorers, traders

OVERVIEW

In this unit, students will explore the story of human life from earliest times to the present in order to gain a historical perspective of world cultures. This story encompasses politics and government, science and technology, religion and ideas, the arts, and daily life. It describes events and patters that have influenced the course of world history from the Stone Age to the age of computers and space exploration, from the glories of ancient Rome and Greece to the brutality of the two world wars.

In prehistoric times, humans moved from hunting and gathering to using tools and practicing agriculture. Ancient civilizations developed complex societies and produced great and lasting works of art. Over centuries, curiosity and exploration led human beings to every corner of the world and new ways of thinking led to innovative governments with more individual freedoms. In the modern world—as in ancient times—people continue to wage wars, but they have tempered these violent periods with quests for peace. Students will end the unit by considering what contributions their own generation might make to the world's ongoing history.

Your Classroom Time Line

These are the major dates and time periods for this unit. You may want to have students watch for them as they progress through the unit.

- c.* 3,700,000 B.C. Wandering hominids leave footprints in volcanic ash.
- 2,500,000 B.C. The first known stone tools are made.
- c. 1,800,000 B.C. Hominids migrate from Africa to Asia.
- c. 400,000 B.C.–100,000 B.C. The first *Homo sapiens* appear.
- c. 33,000 B.C. Cro-Magnon people create cave paintings.
- c. 15,000 B.C. Humans now inhabit Africa, Europe, Asia, North America, and Australia.
- c. 8000 B.C. Agricultural societies are developed in Mesopotamia.
- c. 3200 B.C. Upper and Lower Egypt are united.
- c. 2500 B.C. The Harappan civilization appears in the Indus River valley.
- c. 2300 B.C. Indus River valley people trade with people of the Tigris and Euphrates River valleys.
- c. Late 1000s B.C. The Zhou dynasty begins in China.

*c. stands for *circa* and means "about."

UNIT 2

Chapter 6
3,700,000 B.C–A.D. 476
The Ancient World

Gold funeral mask of Pharaoh Tutankhamen

Greek vase showing potters at work

Ancient Chinese art

2,500,000 B.C. | 500,000 B.C. | 8000 B.C. | 4000 B.C. | 1 B.C. | A.D. 500

2,500,000 B.C.
Science and Technology
The first stone tools appear.

c. 400,000 B.C.–100,000 B.C.
Global Events
The first *Homo sapiens* appear.

c. 8000 B.C.
Science and Technology
Agricultural societies develop in Mesopotamia.

c. 3200 B.C.
Politics
Upper and Lower Egypt are united.

c. 2500 B.C.
Global Events
The Harappan civilization appears in the Indus River valley.

c. Late 1000s B.C.
Politics
The Zhou dynasty begins in China.

c. 800s B.C.–700s B.C.
Politics
Sparta and Athens develop into powerful city-states.

A.D. 476
Global Events
The Western Roman Empire falls.

The Acropolis, Athens

Using the Illustrations

Direct students' attention to the image of the Egyptian funeral mask on the opposite page and ask what they think the function of a funeral mask might have been *(Possible answer: showed how the dead person had looked in life).* Tell students to examine the picture of the Acropolis. Call on volunteers to suggest how the structure is still standing after more than 2,000 years *(Possible answers: sound structure, good engineering).* What does this building suggest about the ancient Greeks? *(Possible answer: advanced society that appreciated beauty and symmetry)* Ask students what kind of ship might have borne a carving like the one on the facing page. *(Possible answer: a war ship that sought to frighten enemies)*

Direct students' attention to the picture of the crusaders on this page and have them find the first Crusade on the time line. Ask if they know who the crusaders were. Have students read the caption that accompanies the Gutenberg Bible and ask them to explain why they think the invention of movable type was an important historical development. Finally, ask students to compare the two portraits of people on this page and speculate as to why they were important. *(Catherine politically, Galileo as a scientist)*

Chapter 7
A.D. 432–1800

The World in Transition

Viking carving of a lion's head

Crusaders at the gates of Jerusalem

Catherine the Great, empress of Russia

800
Politics
Charlemagne is crowned Emperor of the Romans by Pope Leo III.

800–900s
Politics
The Vikings invade Western Europe.

1096
Global Events
The first Crusade begins.

1271
Global Events
Marco Polo begins his trip to China.

1347–1351
Global Events
The Black Death sweeps through Europe.

c. 1450
Science and Technology
Johannes Gutenberg invents the movable type printing press.

1492
Global Events
Christopher Columbus makes his first voyage to America.

1517
Daily Life
Martin Luther posts his 95 theses.

1632
Science and Technology
Galileo proves that Earth revolves around the Sun.

1762
Politics
Catherine the Great becomes Empress of Russia.

1000 — 1200 — 1400 — 1600 — 1800

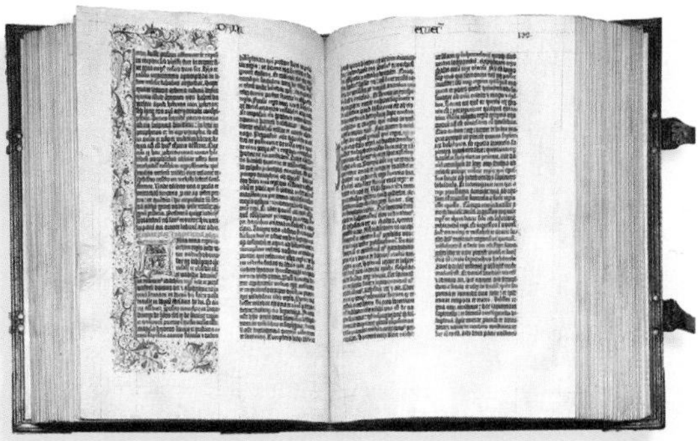

A Gutenberg Bible

Galileo Galilei

Your Classroom Time Line (continued)

- c. 800s B.C.–700s B.C. Sparta and Athens develop into powerful city-states.
- 27 B.C. The Roman Republic becomes the Roman Empire.
- A.D. 476 The Western Roman Empire falls.
- 800 Charlemagne is crowned Emperor of the Romans by Pope Leo III.
- 800–900s The Vikings invade Western Europe.
- 1096 The first Crusade begins.
- 1271 Marco Polo begins his trip to China.
- c. 1345 Notre Dame Cathedral in Paris is completed.
- 1347–1351 The Black Death sweeps through Europe.
- c. 1450 Johannes Gutenberg invents the movable type printing press.
- 1492 Christopher Columbus makes his first voyage to America.
- 1503–1506 Leonardo da Vinci paints the *Mona Lisa.*
- c. 1514 Portuguese traders reach China.
- 1517 Martin Luther posts his 95 theses.
- 1558 Elizabeth I becomes queen of England.
- 1563 The first printing press is built in Europe.

Using the Illustrations

Ask students to examine the picture of the new Globe Theatre. Explain that the theater was built to resemble the one where Shakespeare's plays were originally performed. Ask students to speculate why this theatre was built and what could be inferred about Shakespeare's importance in Britain *(built to honor Shakespeare; a very great writer)*.

Point out that the plate celebrating the coronation of William and Mary also honors figures in British history. Why might people have celebrated this coronation? *(popular monarchs)* Tell students that the coronation was celebrated in England in 1691 after the Glorious Revolution.

Direct students' attention to the picture of women marching on Versailles. Explain that Versailles was the palace of the French king and queen before the French Revolution. Ask students what the women are carrying and why *(weapons, to attack the king and queen during the Revolution)*.

Your Classroom Time Line (continued)

- **1594–1595** William Shakespeare writes *Romeo and Juliet.*
- **1600** The British East India Company is created to control trade with Asia.
- **1607** Jamestown, Virginia, the first permanent English settlement in America, is founded.
- **1616** William Shakespeare dies.
- **1632** Galileo discovers that Earth revolves around the Sun.
- **1687** Isaac Newton publishes his most famous work, *Principia.*
- **1688** The Glorious Revolution occurs in England.
- **1690** John Locke publishes his ideas about government.
- **1708** British traders dominate trade between Europe and India.
- **1737** Samuel F.B. Morse invents the telegraph.
- **1747** The Black Death sweeps through Europe.
- **1762** Catherine the Great becomes Empress of Russia.
- **1763** The Seven Years' War ends.
- **1769** James Watt builds the first steam engine.
- **1776** The American colonies declare independence from Great Britain.

UNIT 2

Chapters 8 & 9
1550–Present
The Modern World

Plate celebrating the coronation of William III and Mary II

Women marching on Versailles during the French Revolution

1550 | 1650 | 1700 | 1750 | 1800

1558 Politics
Elizabeth I becomes queen of England.

1594-95 The Arts
William Shakespeare writes *Romeo and Juliet.*

1687 Science and Technology
Isaac Newton publishes his most famous work, *Principia.*

1688 Politics
The Glorious Revolution occurs in England.

1737 Science and Technology
Samuel F. B. Morse invents the telegraph.

1763 Global Events
The Seven Years' War ends.

1769 Science and Technology
James Watt builds the first steam engine.

1776 Politics
The American colonies declare independence from Great Britain.

1789 Politics
The United States Constitution is ratified.

1789 Politics
The French Revolution begins.

Re-creation of Shakespeare's Globe Theatre

William Shakespeare

Early steam locomotive

Direct students' attention to the photograph of the Wright Brothers' plane. Have them contrast this plane with modern jets. Then encourage students to think about how the invention of the airplane changed the world.

Have students find the image on this page that shows the destruction caused by World War I. What was once in the place shown in the photograph? How can students tell? *(remains of buildings, paving stones on road)* Ask students to contemplate what it might feel like to be a person in a town destroyed by war.

Direct students' attention to the photo of Boris Yeltsin on this page and ask if they know he is. Ask students to describe Yeltsin's attitude in the picture *(Possible answers: happy, victorious)* Tell students that Yeltsin has been a major figure in recent Russian history and that they will learn about his accomplishments in this unit.

Finally, ask students to focus on the photograph of the September 11 interfaith memorial. Call on volunteers to speculate what the people in the photo were feeling and why have they lit candles.

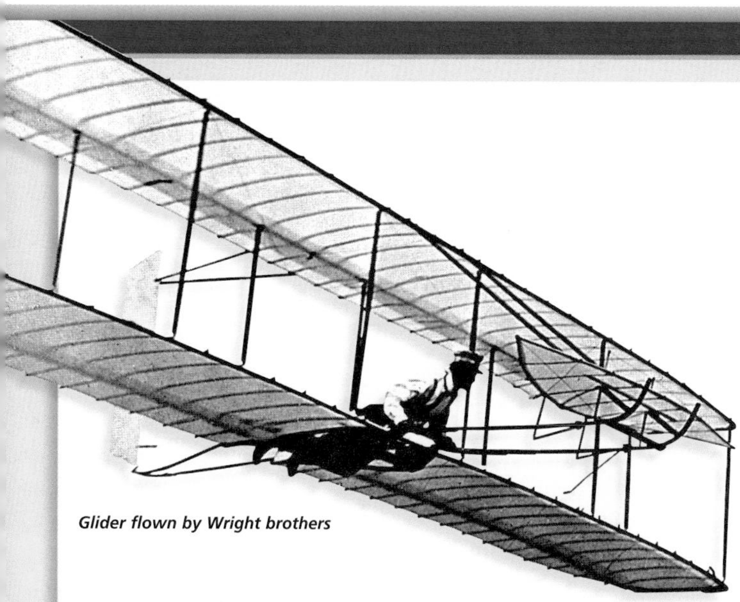

Glider flown by Wright brothers

Russian President Boris Yeltsin (center)

1815 Global Events Napoleon is defeated at Waterloo.

1850

1865 Daily Life Slavery is abolished in the United States at the end of the Civil War

1900

1903 Science and Technology The Wright brothers build the first working airplane.

1917 Global Events The United States enters World War I.

1917 Politics The Russian Revolution is fought.

1929 Daily Life The Great Depression begins.

1945 Global Events World War II ends.

1950

1957 Science and Technology The first satellite in space, Sputnik, is launched.

1989 Global Events The Soviet Union collapses.

2000

2001 Global Events Terrorists attack the World Trade Center in New York and the Pentagon in Washington, D.C.

World War I

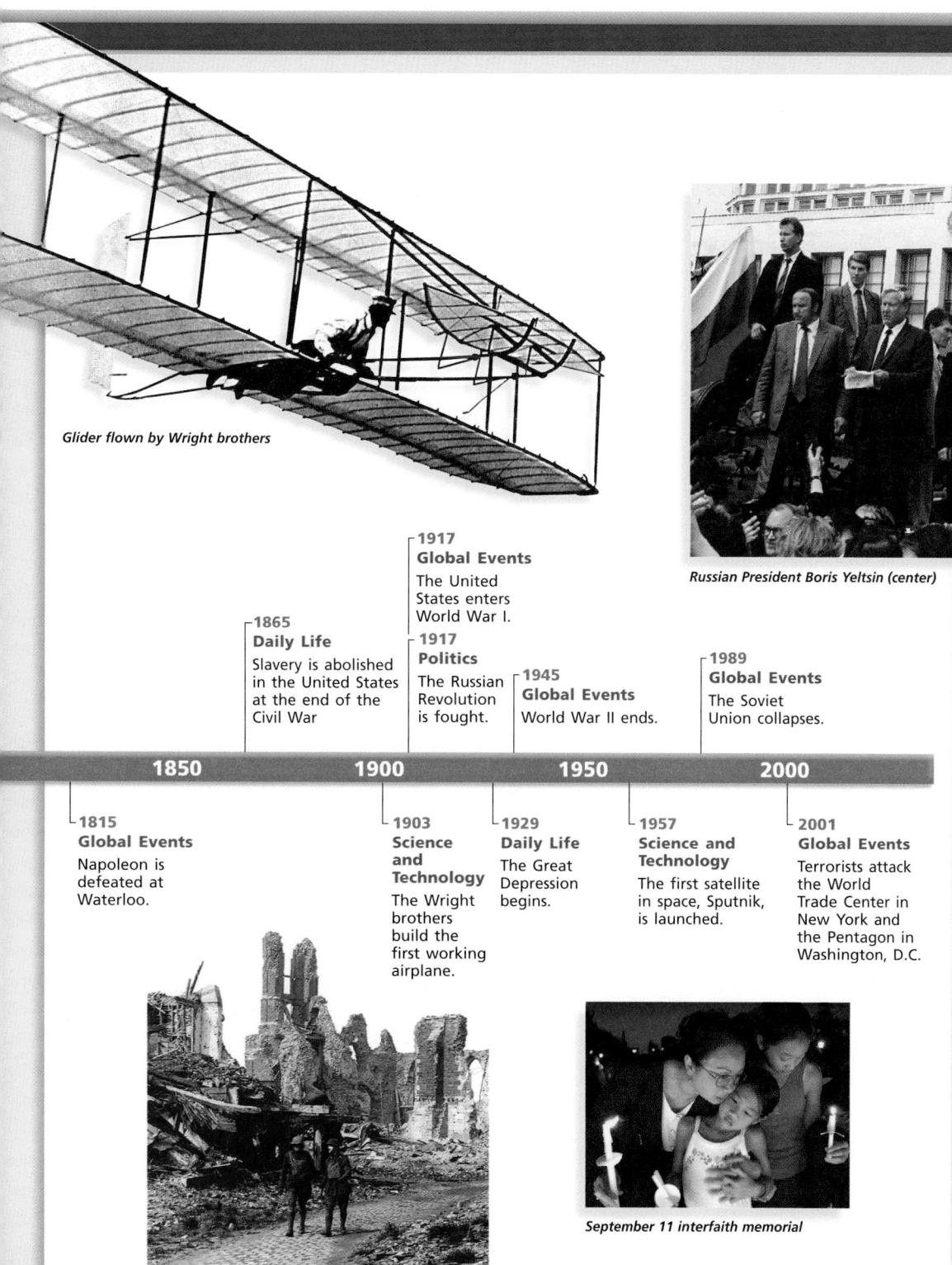

September 11 interfaith memorial

Your Classroom Time Line (continued)

- 1780s The first factories open in England.
- 1789 The French Revolution begins.
- 1789 The United States Constitution is ratified.
- 1815 Napoleon is defeated at Waterloo.
- 1848 Revolutions break out across Europe.
- 1865 Slavery is abolished in the United States at the end of the Civil War.
- 1876 Alexander Graham Bell invents the telephone.
- 1903 The Wright brothers build the first working airplane.
- 1914 The Panama Canal opens.
- 1917 The Russian Revolution is fought.
- 1917 The United States enters World War I.
- 1918 World War I ends.
- 1929 The Great Depression begins.
- 1945 World War II ends.
- 1957 The first satellite in space, Sputnik, is launched.
- 1973 American forces pull out of the Vietnam War.
- 1989 The Soviet Union collapses.
- 2001 Terrorists attack the World Trade Center in New York and the Pentagon in Washington, D.C.

CHAPTER 6 The Ancient World

Chapter Resource Manager

Objectives	Pacing Guide	Reproducible Resources
SECTION 1 **The Birth of Civilization** (pp. 103–108) **1.** Examine the discoveries scientists have made about life in prehistoric times **2.** Describe the four characteristics of civilization. **3.** Identify the locations of the first civilizations.	**Regular** 1 day **Block Scheduling** .5 day *Block Scheduling Handbook, Chapter 6*	**RS** Guided Reading Strategy 6. **RS** Graphic Organizer 6.1
SECTION 2 **The First Civilizations** (pp. 109–116) **1.** Explain how physical geography shaped the civilizations of the Fertile Crescent. **2.** Describe life in ancient Egypt. **3.** Analyze how various peoples contributed to ancient Indian civilization. **4.** Identify some achievements of ancient China.	**Regular** 1 day **Block Scheduling** .5 day *Block Scheduling Handbook, Chapter 6*	**RS** Guided Reading Strategy 6.2 **E** Creative Strategies for Teaching World Geography, Lesson 21
SECTION 3 **Early Sub-Saharan Africa** (pp. 122–129) **1.** Explain how historians study cultures that left no written records. **2.** Describe Kush and Aksum. **3.** Explore how trade affected the growth of kingdoms in Sub-Saharan Africa.	**Regular** 1 day **Block Scheduling** .5 day *Block Scheduling Handbook, Chapter 6*	**RS** Guided Reading Strategy 6.3
SECTION 4 **Greece and Rome** (pp. 122–129) **1.** Describe how ancient Greek civilization developed. **2.** Describe the events that led to the birth and decline of the Roman Empire. **3.** Explain how Christianity began.	**Regular** 2 days **Block Scheduling** .5 day *Block Scheduling Handbook, Chapter 6*	**RS** Guided Reading Strategy 6.4

Chapter Resource Key

RS Reading Support

IC Interdisciplinary Connections

E Enrichment

SM Skills Mastery

A Assessment

REV Review

ELL Reinforcement and English Language Learners

Transparencies

CD–ROM

Music

Video

go.hrw.com Internet

Holt Presentation Maker Using Microsoft® Powerpoint®

One-Stop Planner CD–ROM

See the *One-Stop Planner* for a complete list of additional resources for students and teachers.

One-Stop Planner CD–ROM

It's easy to plan lessons, select resources, and print out materials for your students when you use the ***One-Stop Planner CD–ROM with Test Generator.***

Technology Resources	Review, Reinforcement, and Assessment Resources
One-Stop Planner CD-ROM Lesson 6.1 *ARGWorld* CD–ROM Homework Practice Online HRW Go site	**ELL** Main Idea Activity 6.1 **REV** Section 1 Review, p. 108 **A** Daily Quiz 6.1 **ELL** English Audio Summary 6.1 **ELL** Spanish Audio Summary 6.1
One-Stop Planner CD-ROM Lesson 6.2 *ARGWorld* CD–ROM Homework Practice Online HRW Go site	**ELL** Main Idea Activity 6.2 **REV** Section 2 Review, p. 116 **A** Daily Quiz 6.2 **ELL** English Audio Summary 6.2 **ELL** Spanish Audio Summary 6.2
One-Stop Planner CD-ROM Lesson 6.3 *ARGWorld* CD–ROM Homework Practice Online HRW Go site	**ELL** Main Idea Activity 6.3 **REV** Section 3 Review, p. 121 **A** Daily Quiz 6.3 **ELL** English Audio Summary 6.3 **ELL** Spanish Audio Summary 6.3
One-Stop Planner CD-ROM Lesson 6.4 *ARGWorld* CD–ROM Homework Practice Online HRW Go site	**ELL** Main Idea Activity 6.4 **REV** Section 4 Review, p. 129 **A** Daily Quiz 6.4 **ELL** English Audio Summary 6.4 **ELL** Spanish Audio Summary 6.4

internet connect

HRW ONLINE RESOURCES

GO TO: go.hrw.com
Then type in a keyword.

TEACHER HOME PAGE
KEYWORD: SK3 TEACHER

CHAPTER INTERNET ACTIVITIES
KEYWORD: SK3 GT6

Choose an activity to:
- list different divisions of labor in modern society.
- report on Minoan civilization.
- create a newspaper article about ancient Athens.

CHAPTER ENRICHMENT LINKS
KEYWORD: SK3 CH6

CHAPTER MAPS
KEYWORD: SK3 MAPS6

ONLINE ASSESSMENT
Homework Practice
KEYWORD: SK3 HP6
Standardized Test Prep Online
KEYWORD: SK3 STP6
Rubrics
KEYWORD: SS Rubrics

COUNTRY INFORMATION
KEYWORD: SK3 Almanac

CONTENT UPDATES
KEYWORD: SS Content Updates

HOLT PRESENTATION MAKER
KEYWORD: SK3 PPT6

ONLINE READING SUPPORT
KEYWORD: SS Strategies

CURRENT EVENTS
KEYWORD: S3 Current Events

Meeting Individual Needs

Ability Levels

Level 1 Basic-level activities designed for all students encountering new material

Level 2 Intermediate-level activities designed for average students

Level 3 Challenging activities designed for honors and gifted-and-talented students

English Language Learners Activities that address the needs of students with Limited English Proficiency

Chapter Review and Assessment

E Readings in World Geography, History, and Culture 29, 50, 57
SM Critical Thinking Activity 6
REV Chapter 6 Review, pp. 130–131
REV Chapter 6 Tutorial for Students, Parents, Mentors, and Peers
ELL Vocabulary Activity 6
A Chapter 6 Test
A Unit 2 Test
Chapter 6 Test Generator (on the One-Stop Planner)
Audio CD Program, Chapter 6
A Chapter 6 Test for English Language Learners and Special-Needs Students
A Unit 2 Test for English Language Learners and Special-Needs Students

CHAPTER 6

Launch Into Learning

Ask students what they think an archaeologist does and what a historian does. Discuss the differences. *(An archaeologist studies objects such as pottery, clothing, jewelry, and tools, while a historian studies written records such as documents, letters, and journals.)* Point out that the word history generally refers to anything that happened in the past. However, history specifically refers to events since people developed writing, about 5,000 years ago. Civilizations thrived long before people invented writing. Events that occurred before writing was developed make up the period referred to as prehistory. Tell students that they will learn about many cultures, both prehistoric and historic, in this chapter.

Section 1

Objectives

1 Examine the discoveries scientists have made about life in prehistoric times.

2. Describe the four characteristics of civilization.

3. Identify the locations of the first civilizations.

Links to Our Lives

You might want to share with your students the following reasons for learning about early cultures:

- The study of ancient remains and artifacts affects our understanding of humans and of our place on Earth.
- Civilizations today share many characteristics with the first civilizations.
- The development of agriculture was key to the development of civilization. Agriculture is still an important issue, since the world still needs to be fed.
- Some parts of the world whose roots go back thousands of years are facing conflicts today.
- Members of some ethnic groups are proud to trace their heritage back to ancient times.
- Much of history is based on the exchange of ideas, goods, and technology, all of which are still important issues today.

CHAPTER 6 The Ancient World

This period lasted for thousands of years and saw the rise and fall of countless cultures. Even though these civilizations existed many years ago, you may find that you have something in common with students from that time.

Statue of mythical Sumerian king Gilgamesh

Cuneiform tablet

The city of Sumer was built more than 4,000 years ago. It was located in a river valley in what is now Iraq. Sumerian boys who showed intelligence and skill were trained to be scribes. They learned to read and write in cuneiform, a type of picture writing. They also learned basic mathematics. Teachers often punished poor performance with beatings.

Schoolboys who were late for class were also punished harshly. One Sumerian boy wrote about how afraid he was to explain his lateness to his teacher. Too scared to speak, he entered the room and bowed deeply to the teacher. History does not record the outcome of the boy's tardiness.

Sumerian mosaic, c. 2500 B.C.

FOCUS

LET'S GET STARTED

Copy the following questions onto the chalkboard. *How would you survive if you were stranded on a remote island? How would you get food? What would you use for tools and shelter?* Discuss responses. *(Possible answers: food—hunt fish and small animals, gather seeds, nuts, berries; tools—sticks, rocks; shelter—branches, caves)* Point out to students that early humans had to survive in much the same way. Tell students that in Section 1 they will learn about early humans and how they lived.

Building Vocabulary

Write the key terms on the chalkboard. Remind students that the prefix *pre-* means "before." Ask students to infer the meaning of **prehistoric** *("before history")*. Ask students to identify other words that begin with the prefix *pre-*. Point out that the suffix *-tion* at the end of a word indicates a process or an action. Have students infer the meanings of the words **civilization** and **irrigation** using what they know about each base word and the suffix. Call on volunteers to read the definitions for all the terms from the text.

Section 1 The Birth of Civilization

Read to Discover

1. What discoveries have scientists made about life in prehistoric times?
2. What are the four characteristics of civilization?
3. Where were the first civilizations located?

Define

- hominid
- prehistory
- nomads
- land bridges
- irrigation
- division of labor
- history

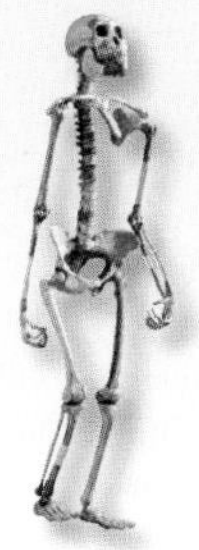

Hominid skeleton from about 3 million years ago

WHY IT MATTERS

Scientists continue to uncover clues about how ancient people lived. Use CNNfyi.com or other **current events** sources to learn about recent discoveries. Record your findings in your journal.

Section 1 RESOURCES

Reproducible
- Block Scheduling Handbook, Chapter 6
- Guided Reading Strategy 6.1
- Graphic Organizer 6

Technology
- One-Stop Planner CD–ROM, Lesson 6.1
- Homework Practice Online
- HRW Go site

Reinforcement, Review, and Assessment
- Main Idea Activity 6.1
- Section 1 Review, p. 108
- Daily Quiz 6.1
- English Audio Summary 6.1
- Spanish Audio Summary 6.1

Prehistory

In the late 1970s in Tanzania, a country in East Africa, scientist Mary Leakey discovered parts of a skeleton dating back millions of years. She believed the bones were those of a **hominid**, an early human-like creature. Scientists use the remains of bodies and other objects they have found to make educated guesses about hominid life. For example, scientists can tell that hominids stood upright and used primitive tools made of stone.

The period during which hominids and even early humans lived is called **prehistory**. This means that no written records were made for historians to examine. The period of prehistory in which stone tools were used is called the Stone Age. It began about 2.5 million years ago and lasted for more than 2 million years.

Scientists carefully unearth objects in "digs" such as this one.

Interpreting the Visual Record What kinds of objects might these scientists be looking for?

◄ **Visual Record Answer**

skeletal remains and stone tools

TEACH

Teaching Objective 1

ALL LEVELS: (Suggested time: 15 min.) Ask students to look in the text for examples of the types of information scientists have gathered about prehistoric times. Then ask students to provide specific examples for each one. *(Example: type of information—how early humans got food; specific examples—gathered plants, hunted, developed agriculture)*

LEVEL 1: (Suggested time: 30 min.) Have students complete the All Levels activity. Then copy the following graphic organizer onto the chalkboard, omitting the italicized answers. Use the organizer to help students examine the kinds of resources scientists believe early people used to survive. Have students work in pairs to complete it.

ENGLISH LANGUAGE LEARNERS, COOPERATIVE LEARNING

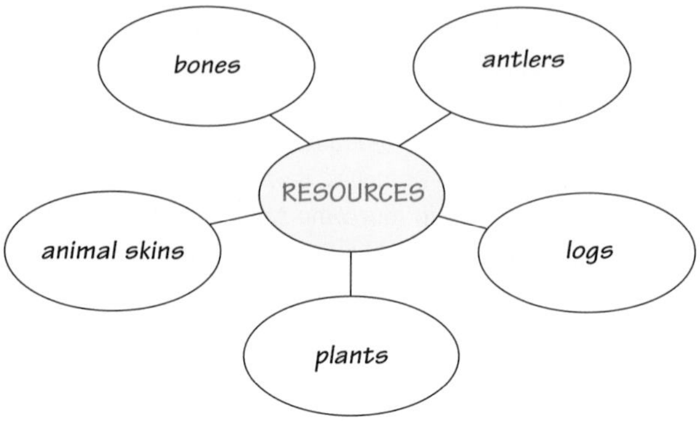

Across the Curriculum
TECHNOLOGY

Fire Archaeologists have discovered hearths, or fireplaces, along with the remains of hominids who lived approximately 500,000 years ago. With the ability to control fire, they apparently started to cook their meat, making it easier to chew. As a result, the need for powerful jaws and large teeth eventually declined. The mastery of fire also allowed people to live in colder regions of the world.

Critical Thinking: Ask students how hominids may have learned to master fire. *(Answer: Lightning may have started a fire, and hominids learned how to keep it burning or how to use the fire to start a new fire.)*

internet connect

GO TO: go.hrw.com
KEYWORD: SK3 CH6
FOR: Web sites about archaeology

Visual Record Answer

because other materials were not available in sufficient quantities

This cave painting in Lascaux, France, was made before the invention of writing. People may have drawn on cave walls to express ideas.

internet connect

GO TO: go.hrw.com
KEYWORD: SK3 CH6
FOR: Web sites about the ancient world

The First Humans Early humans that looked like modern people probably appeared during the Stone Age between 100,000 and 400,000 years ago. These first humans, called *Homo sapiens*, may have first lived in Africa. They were **nomads** who moved from place to place in search of food. They lived on seeds, fruits, nuts, and other plants that they gathered. In time they also began to hunt small animals.

Migration Within the last 1.7 million years, Earth has gone through several periods of very cold weather. Together these periods are known as the Ice Age. During each period, large parts of Earth's surface were covered with ice. Sea levels dropped, leaving strips of dry land called **land bridges** between continents. One such land bridge connected the eastern part of Asia with what is now Alaska. Scientists think that early humans and animals migrated from Africa, into Asia, and across the land bridge onto the North American continent. Over time, humans spread to all parts of the world.

Later Developments In time *Homo sapiens* began to make more advanced tools. They were able to hunt larger animals with spears. They made clothes from animal skins. They also learned how to control fire and how to use it for warmth and cooking. Between 37,000 and 27,000 years ago, people began to create art to express their ideas. Carved ivory figures show that some groups had time for activities besides hunting and tool making. Beautiful paintings found on the walls of caves in France and Spain show graceful, elegant animals such as bison, bulls, and horses.

In northeastern Europe, early humans used the bones of giant mammoths to build shelters similar to this museum model.

Interpreting the Visual Record **Why did people use bones to build a shelter instead of materials such as wood or stone?**

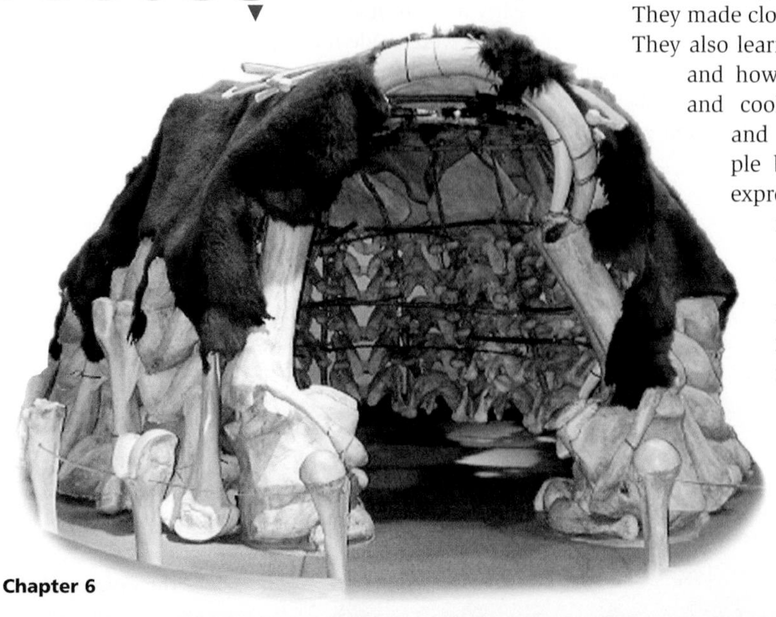

LEVEL 2: (Suggested time: 30 min.) Have students work in pairs to create time lines that show what scientists believe to be the progress of humans from the appearance of *Homo sapiens* between 400,000 and 410,000 years ago up to the time that the first cities appeared about 5,000 years ago. On their time lines, have students list the developments and achievements of early humans. *(Example: 10,000 to 5,500 years ago.—made better tools)*

LEVEL 3: (Suggested time: 40 min.) Have students complete the Level 2 activity. Then have them combine their time lines into one and reproduce it on a long piece of butcher paper. Ask students to illustrate the time line as if it were a cave painting. Refer them to the illustration of the cave paintings of Lascaux on the previous page. You may want to provide additional resources and allocate extra time for this activity.

➤ASSIGNMENT: Tell students that agriculture, one of the most important achievements of early humans, affects all aspects of our daily lives. Have each student write a paragraph that tells how agriculture affects him or her personally each day. *(Example: Our breakfast cereal is made from grains grown by farmers. Many of the clothes we wear are made of cotton.)*

Later, between 10,000 and 5,500 years ago, people learned to make sharper tools by grinding and polishing stone. With better tools, people developed better methods of hunting. They made bows and arrows, which made hunting easier. They shaped fishhooks and harpoons from bones and antlers. People hollowed out logs to make canoes to fish in deep water and to cross rivers. Also around this time, people tamed the dog. Dogs helped people hunt. They may also have warned people if wild animals or strangers were approaching.

In the late Stone Age people learned to practice agriculture. We do not know why people made the change from gathering grains and other plants to growing them, but life changed drastically when they did. Instead of moving from place to place to hunt animals and gather food that grew wild, people began to stay in one place. They became farmers. People also domesticated animals such as cattle and sheep. That means people tamed animals that had been living wild.

The Importance of Agriculture Agriculture changed the ways in which people interacted with their environment. To grow food, people had to find ways to control and change their environment. They cleared forested areas to make room for fields. They invented **irrigation** systems, digging ditches and canals to move water from rivers to fields where crops grew.

Agriculture also changed the ways in which people interacted with each other. Because people who farmed stayed in one place, they began to live in larger groups and form societies. By about 9000 B.C., people began to live in permanent settlements and villages. Because farming made food more plentiful, populations increased. Small villages eventually grew into cities. In towns and cities, people shared new ideas and methods of doing things. Historians think that the first cities may have been founded more than 10,000 years ago. Jericho, the world's oldest known city, was founded at that time on the west bank of the Jordan River.

Our Amazing Planet

The dog was the first animal that humans tamed. Dogs and people have been companions for thousands of years. Some Stone Age people were even buried with their dogs.

Before people developed irrigation they depended on yearly floods of rivers such as the Nile to water fields.

▼

Cultural Kaleidoscope

Even in today's modern world, some people live in ways similar to those of Late Stone Age people. For example, the Yanomamo of Brazil's rain forest live as subsistence farmers and hunters. Yanomamo farmers grow plantains and bananas. They supplement their diets with nuts, seed pods, and mushrooms. Their meat comes from a variety of game animals, including the anaconda, a huge snake, which they hunt with bows and poisoned arrows. Most Yanomamo live in small villages. The Yanomamo had no contact with outsiders until the late 1940s.

Activity: Have students conduct research on nonindustrial societies in today's world. Students could choose among indigenous peoples of New Guinea, the Amazon rain forest, parts of Africa, or other places. Ask students to concentrate on the skills and knowledge that help the people live in their particular environments.

Teaching Objective 2

LEVEL 1: (Suggested time: 30 min) Have students list the four main characteristics shared by civilizations *(organized society, produce extra food, live in towns and cities with governments, practice division of labor).* Ask each student to write a short paragraph explaining how each characteristic relates to the development of civilization. *(Answers will vary, but should note that organized society allows for cooperation and the development of government, laws, and customs; surplus food allows time for activities besides survival; government helps ensure order; and division of labor allows people to do different jobs.)*

LEVELS 2 AND 3: (Suggested time: 10 min) Have students complete the Level 1 activity. Then ask students to identify two more accomplishments characteristic of civilizations *(a calendar and some form of writing).* Have students work in pairs to draw two-column charts with one of these characteristics at the top of each column. Then ask students to list the reasons why each characteristic contributes to a civilization. *(Examples: A calendar told people when to plant crops and when to expect rain; writing allowed people to keep records and communicate more easily.)* Lead a discussion about the significance of calendars and writing in our civilization. **COOPERATIVE LEARNING**

Across the Curriculum

SCIENCE

Inca Farming Practices
Agriculture in South America began long ago. Scientific examination of organic material in the soil shows that people farmed the Andes region of Peru intensively from about 2000 B.C. to A.D. 100. When the region's climate cooled, farming declined, and soil erosion increased.

About 1230 A.D., the climate warmed, and the Inca took over the area. Archaeological evidence suggests that the Inca may have used sophisticated soil conservation techniques. They built a long canal to irrigate fields, terraced the hillsides to grow crops, and planted alder trees that helped curb soil erosion. After the Spanish conquest of Peru, the terraces deteriorated. Today, the alder trees are found only in a few remote ravines.

Activity: Have students conduct research on ancient farming practices developed at various sites around the world and report their findings to the class.

Interpreting the Map Answer

in parts of Southwest Asia and China

Emergence of Agriculture

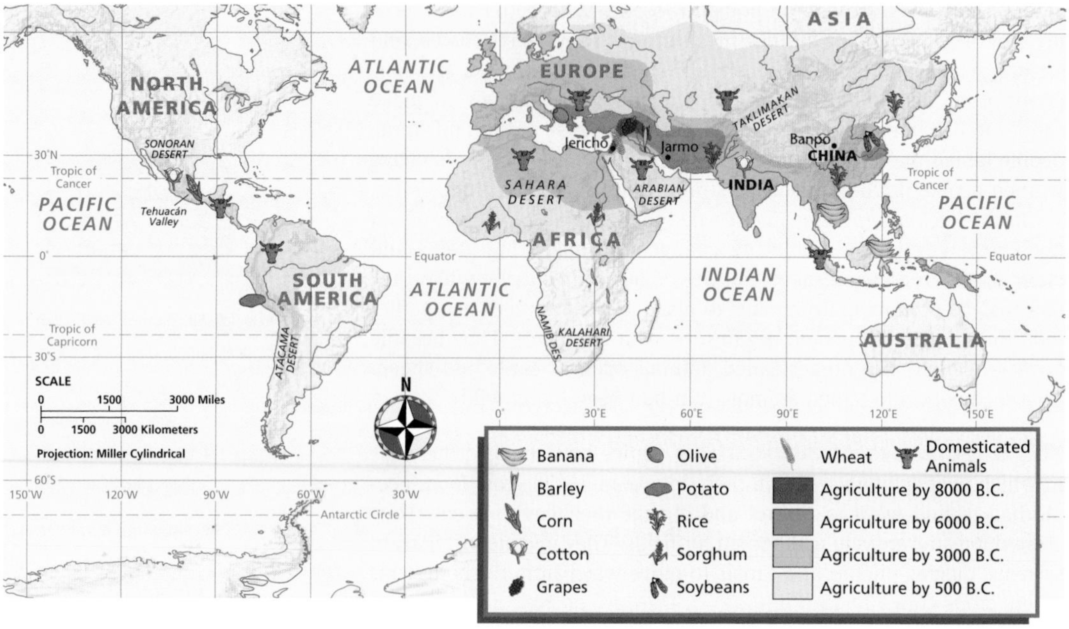

▲ The practice of agriculture spread over a period of thousands of years.

Interpreting the Map **Where was agriculture first developed?**

Because of the ways it changed people's lives, the development of agriculture was enormously important. In fact, learning how to grow food prepared the way for a new chapter in the story of human life—the story of civilization.

✓ **READING CHECK:** *Summarizing* How did agriculture change the ways in which people lived? As farmers they didn't have to migrate to hunt for food.

▲ People made stone and ivory tools during the Stone Age.

The Beginnings of Civilization

Historians describe civilization as having four basic characteristics. First, a civilization is made up of people who live in an organized society, not simply as a loosely connected group. Second, people are able to produce more food than they need to survive. Third, they live in towns or cities with some form of government. And fourth, they practice **division of labor**. This means that each person performs a specific job.

Agriculture and Civilization How did the development of agriculture affect the growth of civilization? Before agriculture, people spent almost all of their time simply finding food. When people were able to grow their own food, they could produce more than they needed to survive. This meant that some people did not have to grow food at all. They had time to develop other skills, such as making pottery, cloth, and other goods. These people could trade the goods they produced and the services they offered for food or other needs.

Teaching Objective 3

ALL LEVELS: (Suggested time: 20 min) Ask students to imagine that they are members of a farming community searching for a new home. Ask students what two features they would require in a new location *(rich, fertile soil and water for their crops).* Then call on volunteers to explain briefly why the earliest civilizations developed in river valleys. Have them locate the four river valleys where civilization began on the world map in the textbook's atlas. You may want to extend the activity by leading a discussion about what hazards these river valleys may have posed, in contrast to their advantages. *(Possible answers: exposure to invasion, floods)*

Close

Call on volunteers to discuss how the illustrations in this section reflect the development and achievements of early humans.

Review and Assess

Have students complete the Section Review. Then have pairs of students create a set of 10 question-and-answer flash cards about early humans and the birth of civilization. Pairs of students can then quiz each other with their cards. Then have students complete Daily Quiz 6.1.

ENGLISH LANGUAGE LEARNERS, COOPERATIVE LEARNING

Trade Once people began to trade, they had to deal with each other in more complex ways than before. Disagreements arose, creating a need for laws. Governments and priesthoods developed to fill that need. Governments made laws and saw that they were obeyed. Religion taught people what they should and should not do.

When people traded, they traveled to places where their goods were wanted and where they could get the things they wanted and needed. Some places where people exchanged goods grew into cities. In these cities, people traded not only goods but also ideas. Over time people built palaces, temples, and other public buildings in their cities.

The Development of Writing Trade, like business today, required people to keep records. Written languages may have developed from this need. The invention of written language began about 3,000 B.C. Farmers also needed a method to keep track of seasonal cycles. They had to know when it was time to plant new crops and when they could expect rain. Over time, they developed calendars.

Once they had writing and calendars, people began to keep written records of events. **History**, which is the written record of human civilization, had begun.

✓ **READING CHECK:** *Identifying Cause and Effect* How did trade lead to the development of writing? Written records were needed to produce and exchange goods.

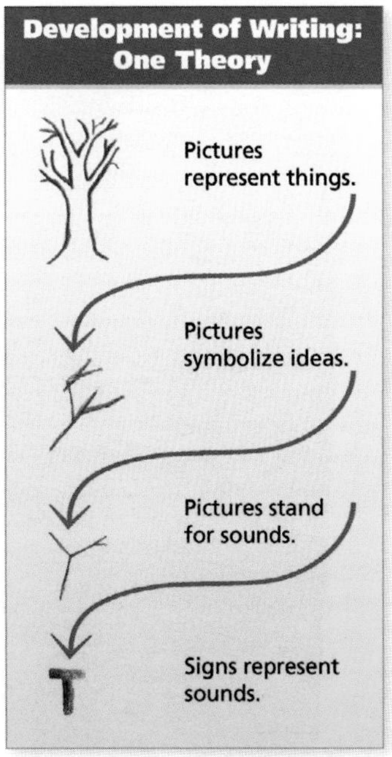

The invention of the alphabet may have begun from pictures. This flowchart shows the possible development of the letter T.

River Valley Civilizations

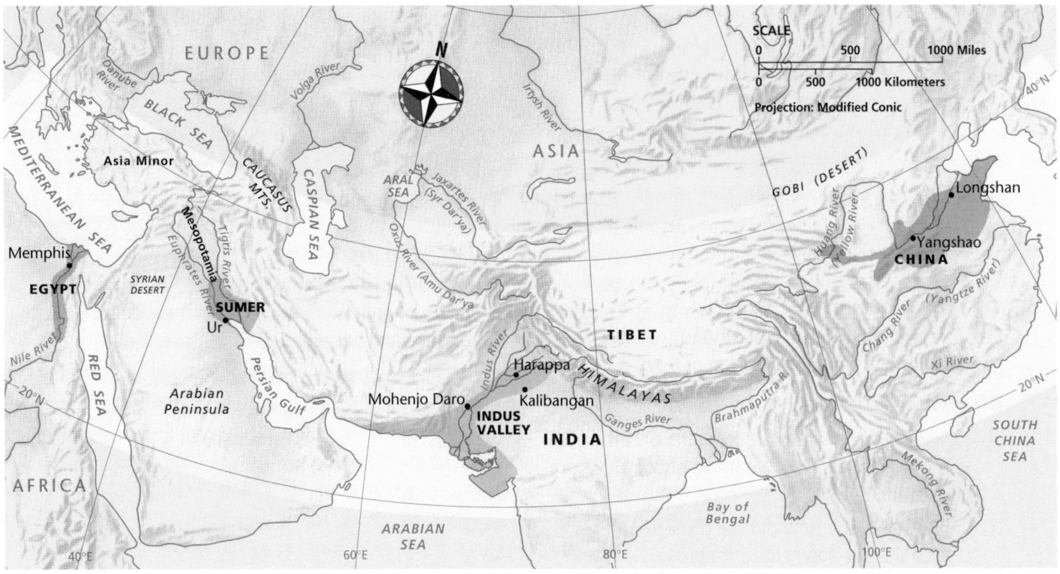

Linking Past to Present

Trade and Travel in Egypt The development of travel by water has stimulated the growth of trade. Throughout history, water travel has usually been cheaper and faster than travel overland. As a result, goods transported by water are usually cheaper than those that must come by land from far away.

Living along the Nile River, the ancient Egyptians were some of the earliest developers of water transportation. An image on a pot dated from about 3200 B.C. shows that Egyptians were already using sails to travel on the Nile. Early boats floated north downriver toward the Mediterranean Sea. Then their captains could raise sails and be carried back upstream by the wind, which most of the time blows from the north. Rafts and barges carried goods up and down the Nile, and ferries traveled across the river. Egypt's rulers, the pharaohs, also used Nile River boats to send messages throughout the land.

RETEACH

Have students complete Main Idea Activity English 6.1. Then have students copy the headings and subheadings in the section. Ask them to write sentences stating the main ideas for each heading and subheading.
ENGLISH LANGUAGE LEARNERS

EXTEND

Invite interested students to conduct research about scientists who study ancient hominid and human remains, the methods they use, and the information they gather. Students might research "Lucy"—the female hominid that may have lived 3 million years ago—whose skeletal remains were discovered by Donald Johanson in Ethiopia in 1974. Others might conduct research on the findings of Mary and Louis Leakey in Tanzania in the late 1970s. Encourage students to supplement their reports with graphics and illustrations. Have students present their findings to the class.
BLOCK SCHEDULING

Section 1 Review

Answers

Define For definitions, see: hominid, p. 103; prehistory, p. 103; nomads, p. 104; land bridges, p. 104; irrigation, p. 105; division of labor, p.106; history, p. 107

Reading for the Main Idea

1. Possible answers: Hominids stood upright 2.5 million years ago; humans created art 37,000 years ago. (NGS 17)
2. People traded the goods they made and the services they offered for food or other goods they needed. (NGS 11)
3. Water and fertile soil were necessary for farming. (NGS 5)

Critical Thinking

4. Disagreements arising from trade created the need for law and order and for teachings about right and wrong behavior.

Organizing What You Know

5. Outer ovals—people live in an organized society; able to produce more food than they need to survive; live in towns or cities with a government; practice division of labor

▲ Scientists discovered the remains of mud-brick houses at Catalhüyük (chah-TUHL-hoo-YOOKH) in Turkey. This was one of the world's first cities.

The First Civilizations

The world's first civilizations developed around four great river valleys. The earliest arose along the valley of the Tigris and Euphrates Rivers in Southwest Asia. Called Mesopotamia, this area was located in what is now Iraq. The ancient Egyptian civilization grew up around the valley of the Nile River. The first civilizations of India were centered on the Indus Valley. Early Chinese civilization began in the valley of the Huang, or Yellow River. (See the map on the previous page.)

These four great river valleys provided fertile soil for growing crops and rich water sources to irrigate crops. The people who lived in these valleys developed advanced civilizations. They learned how to make tools and weapons out of metal, first bronze and then iron. This was the end of the Stone Age. The civilizations that formed in each of these areas developed and declined for different reasons. All of them created written records of their cultures and societies. Thus, they mark the beginning of human history.

✓ **READING CHECK:** *Drawing Inferences* How did geography influence the beginnings of history? River valleys promoted irrigation and fertile soil for growing food.

Section 1 Review

go.hrw.com Homework Practice Online
Keyword: SK3 HP6

Define and explain: hominid, prehistory, nomads, history, irrigation, division of labor.

Reading for the Main Idea

1. *Human Systems* Describe some important discoveries which scientists have made about life in prehistoric times.
2. *Human Systems* How did the division of labor lead to the development of trade?
3. *Environment and Society* Why were early civilizations located around river valleys?

Critical Thinking

4. **Identifying Cause and Effect** How did the needs of early civilizations lead to the development of government and priesthoods?

Organizing What You Know

5. **Summarizing** Copy the following graphic organizer. Use it to describe the traits of civilization.

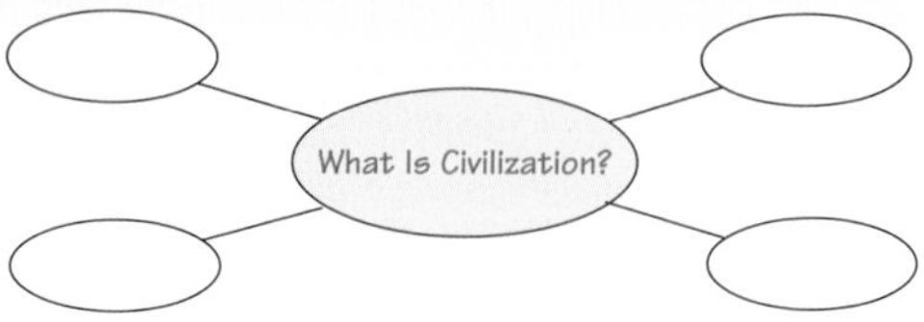

Section 2

Objectives

1. Explain how physical geography shaped the civilizations of the Fertile Crescent.
2. Describe life in ancient Egypt.
3. Analyze the ways in which various peoples contributed to ancient Indian civilization.
4. Identify some achievements of ancient China.

Focus

LETS GET STARTED

Copy the following question onto the chalkboard: *What are some images you associate with ancient Egypt?* Allow students time to record their responses. *(Possible answers: pyramids, the Sphinx, pharoahs, Cleopatra)* Point out that ancient Egypt was one of the world's first civilizations. Tell students that they will learn more about Egypt and other early civilizations in this section.

Building Vocabulary

Write the key terms on the chalkboard. Call on volunteers to locate and read their definitions. Point out that **dynasty** is derived from a Greek word meaning "lordship" or "rule." **Pharaoh** comes from an Egyptian word for "great house." **Hieroglyphics** comes from two Greek words—*hiero* means "sacred," and *glyphein* means "to carve." **Dialect** also comes from Greek—*dia-* means "between" and *legein* means "to talk." Ask students how these roots are reflected in the meaning of each word.

Section 2 The First Civilizations

Read to Discover

1. How did physical geography shape the civilizations of the Fertile Crescent?
2. What was life like in ancient Egypt?
3. How did various peoples contribute to ancient Indian civilization?
4. What were some achievements of ancient China?

Define

dynasty
pharaohs
hieroglyphics
subcontinent
dialects

Locate

Fertile Crescent
Tigris River
Euphrates River
Jordan River
Nile Valley
Indus River
Huang River

Sumerian sheep's head sculpture

Why It Matters

Scientists continue to learn more about how ancient people lived and worked. Use CNNfyi.com or other **current events** sources to learn more about recent discoveries. Record your findings in your journal.

Section 2 Resources

Reproducible

- Guided Reading Strategy 6.2
- Creative Strategies for Teaching World Geography, Lesson 21

Technology

- One-Stop Planner CD-ROM Lesson 6.2
- Homework Practice Online
- HRW Go site

Reinforcement, Review, and Assessment

- Main Idea Activity 6.2
- Section 2 Review, p. 116
- Daily Quiz 6.2
- English Audio Summary 6.2
- Spanish Audio Summary 6.2

The Fertile Crescent

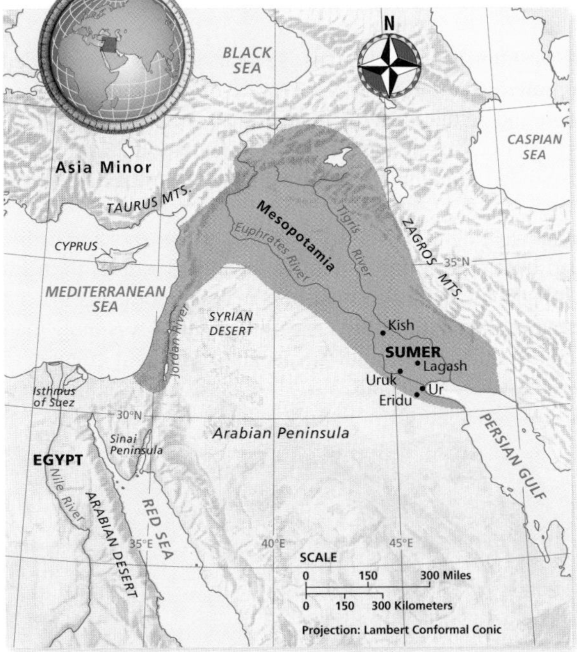

The Tigris, Euphrates, and Jordan Rivers are three of the main rivers in the Fertile Crescent.

The Fertile Crescent

A strip of fertile land begins at the Isthmus of Suez and arcs through Southwest Asia to the Persian Gulf. This rich farmland region is known as the Fertile Crescent. By 8000 B.C., farmers in this region had begun to grow crops. In time they learned to work together to control flooding and to irrigate their fields. A new civilization developed as a result.

The Land The Tigris and the Euphrates (yoo-FRAY-teez) Rivers flow through the Fertile Crescent. These rivers start in what is now known as Turkey, and flow southeast, joining before they reach the Persian Gulf. In the past, the Tigris and Euphrates Rivers frequently flooded. Ancient people built a system of canals and dikes to bring water to their fields and to return water to the rivers after floods. Several civilizations grew up in the Fertile Crescent. Each of them eventually declined and disappeared. Internal quarreling and poor leadership weakened them. No natural barriers protected them from invasion, and conquerors took over many of these kingdoms.

TEACH

Teaching Objective 1

LEVELS 1 AND 2: (Suggested time: 15 min.) Lead a class discussion about the influences of physical geography on the development of civilization in the Fertile Crescent. Ask students how the region's landscape contributed to the rise of early civilizations. *(The region's rich soil was excellent for agriculture and allowed farmers to produce excess crops. In addition, the rivers of the area provided enough water to irrigate the fields.)* Then ask students how the region's physical geography helped lead to the collapse of societies. *(Because there were no natural boundaries surrounding the Fertile Crescent, foreign peoples could and did invade and conquer its residents.)* **ENGLISH LANGUAGE LEARNERS**

LEVEL 3: (Suggested time: 45 min.) Have students use library or Internet resources to locate the homeland of either the Babylonians or the Persians. Have students note the location of this homeland on a physical map of Southwest Asia. Then ask students to trace a path these invaders may have taken to reach the Fertile Crescent. Ask students what they notice about the landscape along this route. *(Students will probably notice that the land is flat. There are no barriers to prevent invaders from reaching the Fertile Crescent.)* Lead a discussion about how the flat terrain of Southwest Asia influenced the history of conquest in the region.

EYE ON EARTH

Geographically, the Fertile Crescent is readily accessible from almost all directions. The mountains to the north and east have many valleys and natural passes. The desert to the west and south is easy to cross, as are the many rivers in the area. Thus, the Sumerians had few natural boundaries to protect their cities and farmland from invasion by nomadic peoples.

Critical Thinking: Ask students to suggest how Mesopotamia's lack of natural boundaries might also have been an advantage.

Answer: Although Mesopotamia was vulnerable to invasion, the Sumerians were able to reach the areas around it to engage in trade, exchange ideas, or invade their neighbors.

Our Amazing Planet

The Sumerians created a number system based on 60. They divided a circle into 360°, or six 60s. Modern clocks and compasses use the same system.

As ruler, Hammurabi made great contributions to Babylonian society. He is best remembered for his code of laws.

The Sumerians Historians believe that the Sumerians were the first civilization in the Fertile Crescent. They settled in the lower part of the Tigris-Euphrates valley, in an area called Sumer. There, they created what became known as the Sumerian civilization. Most Sumerians were farmers. After a time they were able to grow extra food. This allowed some people to become artisans and traders.

The Sumerians may have been the first people to use the wheel. Sumerian builders were the first to use the column, the vault, and the dome. The Sumerians also invented the world's first system of writing. Eventually, the Sumerians were conquered. Hundreds of years later, a new civilization, the Babylonians, grew up in the area.

The Babylonians In some ways the Babylonian and Sumerian civilizations were similar. The people were farmers, artisans, and traders. Like the Sumerians, the Babylonian merchants traded goods with distant parts of the Fertile Crescent and with Egypt and India.

Babylonian women had some legal and economic rights, including property rights. Women could be merchants, traders, or even scribes.

The greatest of the Babylonian kings was Hammurabi (ham-uh-RHAB-ee). In the 1790s B.C. he conquered most of the Tigris-Euphrates Valley. Hammurabi was not only a powerful military leader but also an outstanding political leader. However, he is best remembered as a lawmaker. He put together a group of several hundred laws governing all aspects of life. The collection is known as the Code of Hammurabi. Ideas from the Code are still found in laws today.

The Persians After the Babylonians, several other peoples settled in Tigris-Euphrates valley. By about 550 B.C., the Persians had conquered Babylon and settled in what is now known as Iran. The kings of Persia waged many wars to add territory to their empire. Eventually, the empire expanded to include everything between India and southeastern Europe.

The Persians were more tolerant of local customs than some earlier conquerors had been. For example, they allowed local peoples to maintain their own religions. They also built a huge system of roads to hold the empire together. At one time, the Persian Royal Road was more than 1,500 miles (2,410 km) long. It allowed people of different cultures to exchange ideas, customs, and goods.

The ruins of Persepolis reflect the former glory of the Persian Empire. King Darius I built Persepolis in about 500 B. C. as the capital of his empire.

Teaching Objective 2

LEVEL 1: (Suggested time: 30 min.) Copy the following graphic organizer on the chalkboard, omitting the italicized answers. Have students work in pairs to fill in the organizer with a description of the class structure of ancient Egypt. Tell each pair of students to fill in each section by identifying the people who belonged to each class and describing their functions in society. Invite pairs to share and discuss their completed charts with the class. **ENGLISH LANGUAGE LEARNERS, COOPERATIVE LEARNING**

Ruler
pharaoh—controlled government, served as judge, high priest, general of the army

Upper Classes
priests—religious leaders
scribes—kept records
government officials—helped run the government

Lower Classes
peasant farmers—farmed the land; grew food for the pharaoh, served in military, built canals and pyramids

Persian Empire about 500 B.C.

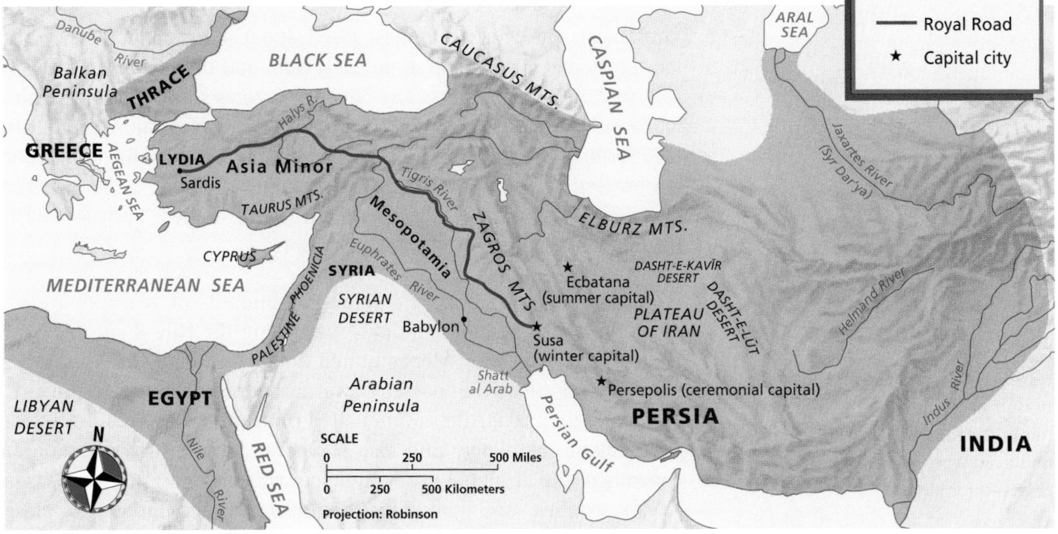

▲ At its height, the Persian Empire stretched from southeastern Europe to the Indus River in southwestern Asia.

The Phoenicians The Phoenician (fi-NEE-shuhn) civilization developed around 1200 B.C. It had its origins in what is now Israel, Lebanon, and Syria. The land to the east was not good for farming, so the Phoenicians turned to the sea. They may even have sailed as far as Britain and Western Africa. Phoenician traders set up colonies around the Mediterranean.

Their most important advance was the invention of the Phoenician alphabet. Earlier writing systems had thousands of symbols, but the Phoenician alphabet had just 22 letters. Other western cultures adopted and modified this alphabet for their own use.

The Hebrews and Judaism South of Phoenicia was a strip of land known as Canaan. Many peoples lived in this region at different times. Among them were the Hebrews, ancestors of the modern Jews. Early on, the Hebrews moved to Egypt, probably to escape drought and famine. Eventually the Egyptians made the Hebrews slaves. They remained enslaved for 400 years, until the 1200s or 1300s B.C.

At that time, a great Hebrew leader arose named Moses. Moses said the Hebrew god, Yahweh, had sent him to form a nation in Canaan. Moses led the Hebrews out of Egypt. They settled once again in Canaan, but only after they had wandered in the desert for many years. They were also forced to battle other peoples who had settled there. The Hebrews followed a new code of laws, the Ten Commandments. They are a key part of Judaism, and deal with the Hebrews' relationship to Yahweh. They also emphasize the importance of family and human life, as well as exercising self-control.

✓ **READING CHECK:** *Summarizing* What are some important contributions of ancient civilizations in the Fertile Crescent? writing system, codes of law, setting up colonies along the Meditterranean

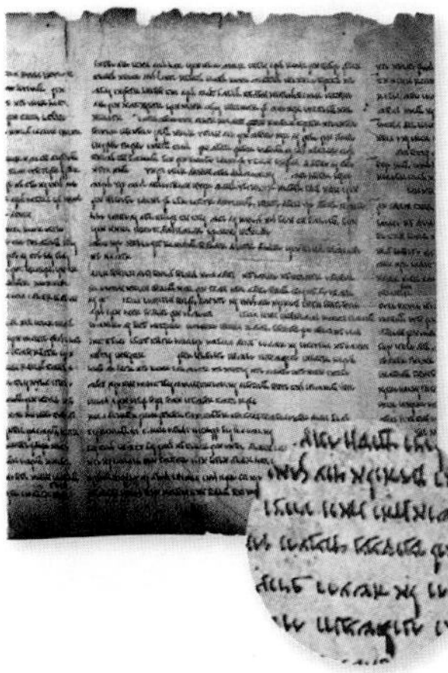

▲ The Dead Sea Scrolls contain details about the history and principles of Judaism.

Cultural Kaleidoscope

The Alphabet The spread of the alphabet is a good example of how commerce can speed cultural diffusion. The Phoenicians used writing in their businesses to draw up contracts and record bills. Their trading partners saw these written records. They probably also saw the advantages of them. The Phoenician traders spread the knowledge of alphabetical writing throughout the Mediterranean world. The Phoenician alphabet was adopted by the Greeks, who improved upon it by adding signs for vowel sounds. Later, the Romans copied the Greek alphabet and eventually developed the alphabet we use now.

Critical Thinking: How might the history of writing have been different if the Phoenicians had not been traders?

Answer: Europeans might have developed an alphabet later in history; their alphabets might have been influenced by other cultures, such as Arabic or Chinese cultures.

LEVEL 2: (Suggested time: 20 min.) Tell students to imagine that they are peasant farmers in ancient Egypt. Have students write letters they might dictate to a scribe describing what they have done over the past several days. Encourage students to note their feelings about the things they have seen and done and their thoughts about the upper classes.

LEVEL 3: (Suggested time: 30 min.) Point out to students that some of the symbols used in Egyptian hieroglyphics resembled the objects they represented. For example, the symbol for *eat* might show a person with a hand held to his mouth. Challenge students to create hieroglyphic characters for some common English words. Then have each student write a story or paragraph using these characters in place of the words they represent. Have students exchange paper and attempt to decipher the hieroglyphics they have received. Then lead a discussion about Egyptian hieroglyphics and the attempts by archaeologists to decipher them.

Environment and Society

Archaeologists have discovered settlements near the Egypt-Sudan border that were built before the Sahara became a desert about 6,000 years ago. Evidence suggests that the inhabitants of these villages had developed agriculture and domesticated various animals, long before similar activities were begun in the Nile Valley. These villagers also cut huge blocks of stone that some scholars believe might be the forerunners of Egypt's pyramids. Archaeologists believe that increasingly harsh conditions drove these people into the Nile Valley, where within a few centuries an advanced civilization had developed.

Discussion: Lead a discussion about the possible effects of climate or vegetation changes on societies.

Interpreting the Map Answer ►

the Mediterranean Sea

Visual Record Answer ►

Students might suggest that she wanted to command the same respect as her male predecessors.

▲ Hatshepsut (hat-SHEP-soot) declared herself pharaoh after the death of her husband. She dressed the part, and even wore the false beard reserved for kings.

Interpreting the Visual Record **Why do you think Hatshepsut dressed like a male pharaoh?**

The Kingdom of Egypt around 1450 B.C.

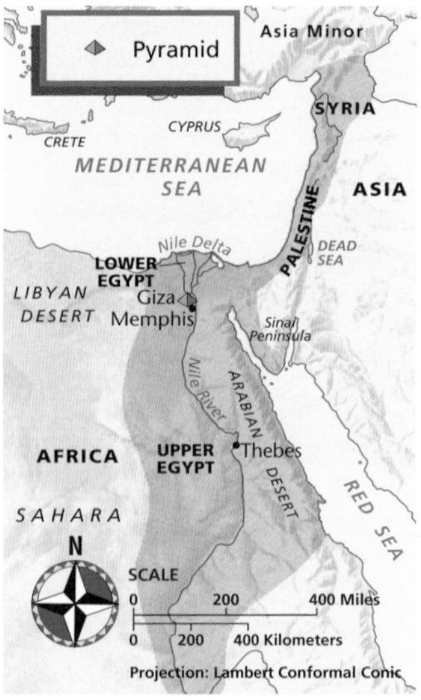

▲ Interpreting the Map **Into which body of water does the Nile River flow?**

The Nile Valley

The Nile is the world's longest river, about 4,160 miles (6,693 km) long. It floods at the same time each year, and the waters spread fine soil over the river's banks. The Nile River flows south to north, which made it possible for early peoples to travel upland to the Mediterranean Sea. Deserts and seas afforded early civilizations natural protection from invaders. The Isthmus of Suez, a land bridge between Africa and Asia, provided trade routes between early Egyptian civilizations and their neighbors.

Government and Society Over hundreds of years, two kingdoms with distinct cultures developed along the Nile River in Egypt. Around 3200 B.C., King Menes united them and founded a **dynasty**, a family line of rulers that passes power from one generation to the next. In later years, these rulers were called **pharaohs**. They controlled the government completely and also served as judges, high priests, and generals of the armies. Egyptian society was divided into two classes. Priests, scribes, and government officials formed the upper class. Peasants and farmers formed the lower class. They had to grow food for the pharaoh. Some were also forced to serve in the military or work on building projects such as canals or the pyramids.

During the time of the pharaohs, Egypt's contact with other parts of the world grew either through conquest or trade. By about 1085 B.C., Egypt had expanded into what is now Syria, Israel, and Libya. The Egyptians traded with peoples throughout Southwest Asia and North Africa.

Achievements Egyptian architects and engineers were among the best in the ancient world. They built the pyramids, the Great Sphinx, and other monuments that still stand today. They developed a writing system that used pictures and symbols called **hieroglyphics**.

Teaching Objective 3

ALL LEVELS: (Suggested time: 20 min.) Copy the graphic organizer onto the chalkboard, omitting the italicized answers. Ask students to copy it into their notebooks. Then have students work in pairs to complete the organizer by noting the ways in which various peoples contributed to ancient Indian civilization. Call on volunteers to share their webs with the class. **ENGLISH LANGUAGE LEARNERS, COOPERATIVE LEARNING**

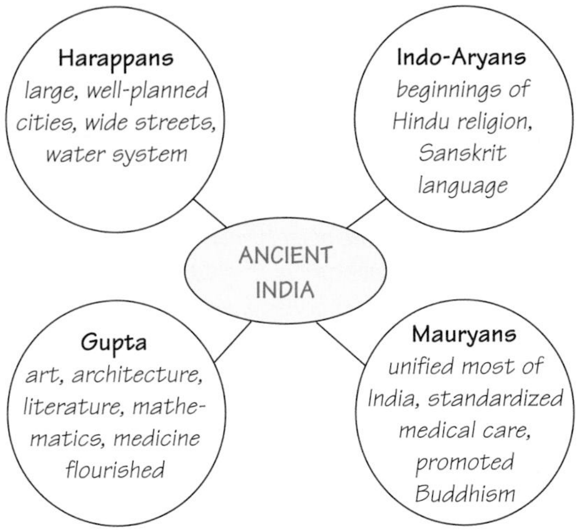

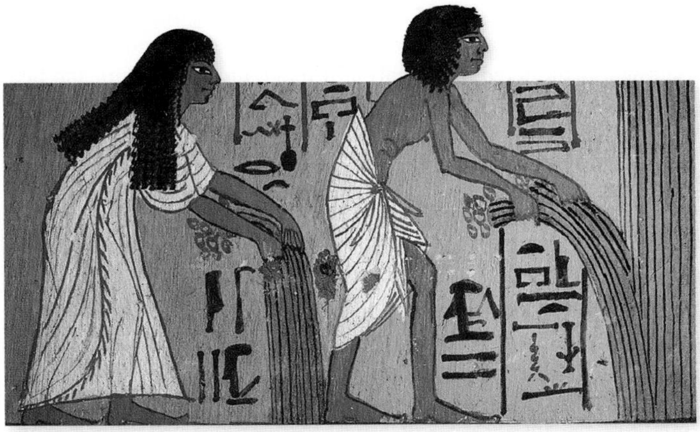

This mural shows Egyptian farmers processing grain.

Egyptians used a number system based on 10, and they understood both fractions and geometry. They had an accurate 365-day calendar, and they made important discoveries in medicine.

After Ramses the Great, who ruled in the 1200s B.C., Egypt had no great leaders. Attacks from foreign peoples, including the Phoenicians, the Persians, and the Greeks, weakened Egypt. By the 500s B.C., Egypt was no longer ruled by Egyptians.

✓ **READING CHECK:** *Analyzing* How was ancient Egypt protected from invaders? surrounded by deserts and seas

The Indus River Valley

East of the Fertile Crescent is the Indus River. It flows through what are India and Pakistan today. Several civilizations developed in the Indus River Valley.

The Harappans The first great civilization on the Indian **subcontinent** developed in the Indus River Valley around 2500 B.C. A subcontinent is a very large land mass that is smaller than a continent. These people are called Harappans, after one of their most important cities, Harappa. The civilization was located in what is known today as Pakistan, but it also extended into India.

Historians know very little about the Harappan people. They lived in large, well-planned cities with wide streets and a water system complete with public baths and brick sewers. Harappan civilization lasted a thousand years. Historians are not sure how it ended.

This baked-clay sculpture was made by the Harappan civilization.

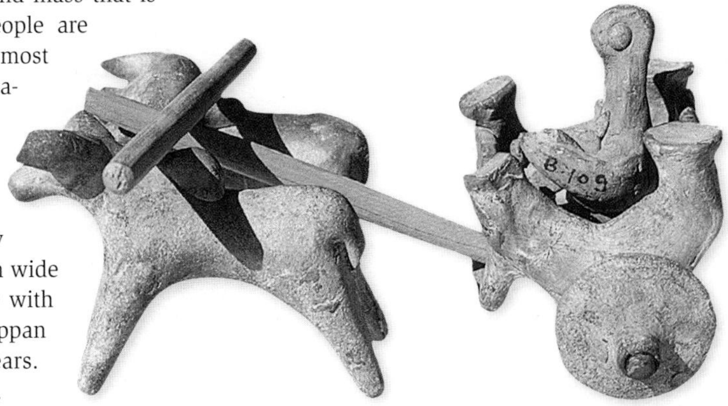

Human Systems

In addition to Harappa, archaeologists have uncovered many other sites along the Indus River valley that were a part of the Harappan civilization. Among these sites are seaports that give evidence of Harappan trading and commerce. The port of Lothal, for example, had an enclosed brick shipping dock that was more than 700 feet long. There was also a sluice gate that made it possible to regulate the water level so that ships could be loaded at high or low tide.

Activity: Have students conduct research on modern devices such as dry docks and canal locks that perform functions similar to that of the sluice gate. Interested students may want to create a diagram or build a model of one of these devices and use it in a presentation to show how the device works.

Teaching Objective 4

LEVEL 1: (Suggested time: 20 min.) Call on students to list some of the achievements of ancient China. As students name achievements, write them on the chalkboard. *(Possible answers: calculated precise length of year, tracked movement of the planets, invented sun dial, invented water clock, invented machine to measure earthquakes, developed herbal medicines, learned how to set broken bones, invented paper, invented printing press.)* Ask each student to choose one achievement and to write a few sentences explaining why he or she thinks it was important. Call on volunteers to share their sentences, have the class vote on ancient China's most important achievement.
ENGLISH LANGUAGE LEARNERS, COOPERATIVE LEARNING

LEVEL 2: (Suggested time: 30 min.) Provide students with art supplies and have each student create a poster illustrating one of ancient China's contributions to the world. Encourage students to use library or Internet resources to learn more about the contributions they have chosen. Call on volunteers to share their work with the class. Display the posters around the classroom. **ENGLISH LANGUAGE LEARNERS**

Across the Curriculum

LITERATURE

Folk Tales The *Panchatantra* is an anonymous collection of Buddhist fables most likely compiled sometime before A.D. 500. The framework of the *Panchatantra* consists of one narrative thread that holds together several different stories. The stories, which combine both prose and verse, were originally intended for the instruction of sons of royalty.

The Jakata tales are popular Indian stories, also taken from Buddhist writings. Some of them use animals as characters to teach lessons about kindness. Their theme is good versus evil.

Discussion: Find and read to the class stories from the *Panchatantra* or the Jakata tales. You may wish to consider the story of the mice and the elephants from the *Panchatantra* or a version of the Cinderella story from the Jakata tales. After you have read the story, lead a class discussion about its moral.

Interpreting the Map ►

They had river highways from deep inland to the sea.

▲ This Indian sculpture shows the Hindu god Vishnu.

The Indo-Aryans By about 1500 B.C. a new group of people had come into northern India. They were animal herders and skilled warriors who conquered all of northern India and mixed with the people already living there. Scholars call these people Indo-Aryans. Their looks and their culture were different from the people they conquered. They introduced many new ideas and a strict class system to India.

Most of what we know about the Indo-Aryans comes from the Vedas. These are the Indo-Aryans' great works of religious literature. For centuries people memorized the Vedas and retold them to their children. Later, the Indo-Aryans developed writing. Scholars wrote the Vedas down. This period of Indian history is sometimes called the Vedic Age. It lasted until about 1000 B.C.

The effect of Indo-Aryan culture is still felt in India today. Indian and Indo-Aryan religious beliefs and customs mixed as well, forming the beginnings of the Hindu religion. Their language, Sanskrit, became the basis for the main language of modern India.

The Mauryans In the early 500s B.C., 16 different kingdoms existed in northern India. By about 320 B.C. much of India was united under one able ruler, Candragupta Maurya. It became known as the Mauryan Empire. Candragupta made many advances. He established a system of efficient administrators to help govern the empire. He also set standards for doctors.

His grandson Aśoka extended the empire to include almost all of India, except for the southern tip. He tried to unite the many different peoples who came under his rule. Eventually he gave up war because of the pain and suffering it caused. Aśoka took up Buddhism and furthered its spread through India. He even sent missionaries to other countries in an effort to encourage the growth of Buddhism. Aśoka worked to improve living conditions in his empire. He dug wells and built rest houses along trade routes. He also planted trees along the roads to make travelers more comfortable. After his death the Mauryan Empire slowly lost strength. It collapsed about 140 years later.

Indus River Valley

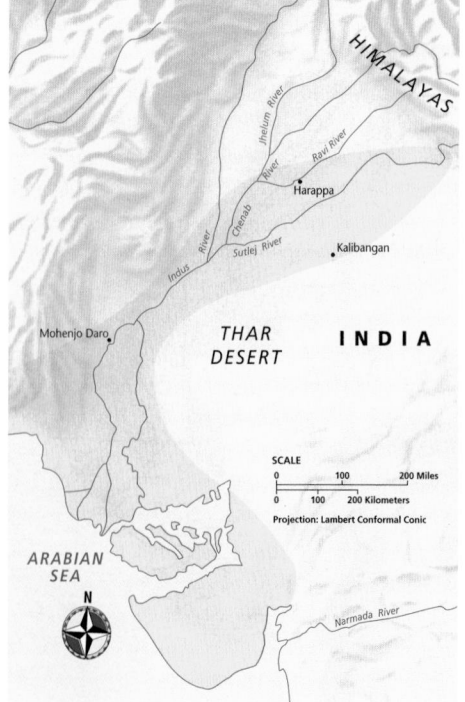

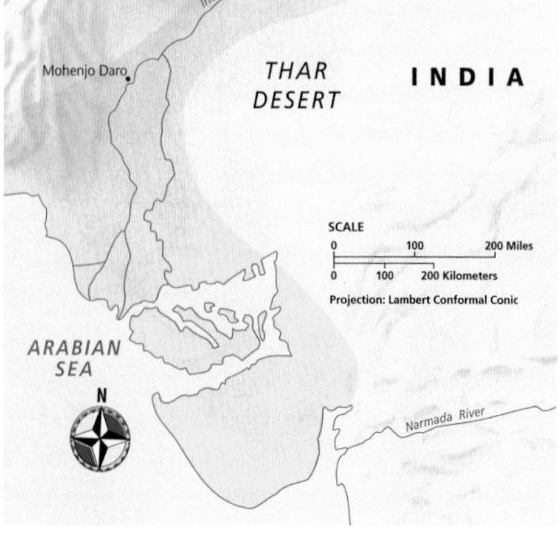

The Gupta The Gupta Empire ruled India beginning around A.D. 320. By A.D. 400, the Gupta controlled the entire northern part of India. During this time, both Buddhism and Hinduism attracted many followers. Art, architecture, literature, mathematics and medicine developed under the Guptas. People came from all over Asia to learn at Indian universities. However, in the late 400s, invaders from Central Asia attacked India. The Gupta Empire fell about 550.

✓ **READING CHECK:** *Summarizing* What are the two main religions practiced in the Indus River valley? **Hinduism and Buddhism**

LEVEL 3: (Suggested time: 40 min.) Have students work in groups to create illustrated time lines showing some of the achievements of ancient Chinese civilization. Once students have completed their time lines, ask them to draw conclusions about the relationships between some of these achievements. *(For example, the printing press would probably not have been invented had paper not been invented first. Both determining the length of the year and calculating the movements of the planets involved observing the heavens.)* Call on students to share their conclusions with the class. **COOPERATIVE LEARNING**

➤ASSIGNMENT: Provide each student with an outline map of Asia and North Africa. Have each student color the map to indicate the location of the world's first four civilizations. Then have students mark their maps with symbols that represent some of the achievements of each civilization. Remind students to include legends with their maps.

Early China, c. 5000 B.C.—c. 10000 B.C.

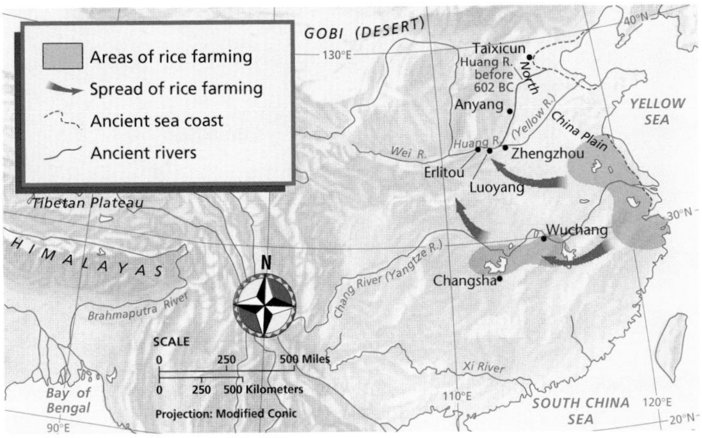

The Huang River Valley

Ancient China was cut off from other civilizations by mountains and the Gobi Desert. Its isolation led to the development of a unique culture. The ancient northern Chinese people began forming dynasties in the Huang River Valley around 2000 B.C.

The Shang People have lived in the Huang Valley since prehistoric times. The first we know of were the Shang. The Shang people invaded the Huang Valley and established China's first dynasty. They brought irrigation and flood control to the Huang Valley.

The Shang developed a written language that could be used to write all the **dialects**—or different forms—of Chinese. Over more than 100 years, many different invaders attacked the Shang. In about 1050 B.C., the Zhou (JOH) overthrew the Shang and established their own dynasty.

The Zhou Under the Zhou and the two dynasties after it, China became a large and powerful state. Zhou rulers allowed each territory to have its own leader. These leaders fought each other as well as outside invaders. Philosophers tried to bring peace and harmony back into daily life. Some of their ideas have been carried into modern times.

The Qin The Qin (CHIN) dynasty lasted only 15 years, but it made many important changes in Chinese life. The Western name *China* comes from *Qin*. The Qin built huge walls to protect their borders. Later dynasties added to and connected these walls. Today, we call this structure the Great Wall of China.

In 206 B.C., the Qin people grew unhappy with their ruler. A commoner led a revolt and seized power. He established the Han dynasty. It ruled China for about 400 years.

These are life-size clay figures from the tomb of China's first emperor, Cheng also called Shih Huang Ti. He unified China under Qin rule.

Across the Curriculum
SCIENCE

Silk During the Shang dynasty, the Chinese learned to raise silkworms. Silkworms are not worms at all, but rather caterpillars. The most commonly raised silkworms eat only the leaves of mulberry trees. The silk comes from the cocoons the caterpillars spin and live in while they are changing into moths.

In the silk-making process, heat is used to kill the caterpillar inside the cocoon. Next the cocoon is carefully unwound, producing a continuous silk thread. These threads are then spun into silk cloth. About 100 cocoons are needed to make a silk tie, and more than 600 for a blouse. A silk kimono can require 3,000 silkworms, which will eat about 135 pounds of mulberry leaves before they spin their cocoons.

Activity: Have students make posters illustrating the silk-making process.

CLOSE

Display a world map for the class. As you name the peoples described in this section, have volunteers go to the map and point to the area where that people lived.

REVIEW AND ASSESS

Have students complete the Section Review. Then have students choose one of the civilizations discussed in the section and write three statements about it. These statements should begin "I come from a civilization that....," or "In my civilization...." Have students play a who-am-I game using the statements. Then have students complete Daily Quiz 6.2.
ENGLISH LANGUAGE LEARNERS, COOPERATIVE LEARNING

RETEACH

Have students complete Main Idea Activity 6.2. Have students design graphic organizers using the material presented in this section. Then have students exchange and complete the organizers.
ENGLISH LANGUAGE LEARNERS

EXTEND

Have interested students conduct research on another ancient civilization of the Fertile Crescent — the Akkadians, Hittites, Assyrians, Chaldeans, or Lydians. Have them note similarities and difference between the selected civilization and the ones discussed in this section. Call on students to share their findings with the class. **BLOCK SCHEDULING**

Section 2 Review

Answers

Define dynasty, p. 112; pharaohs, p. 112; hieroglyphics, p. 113; subcontinent, p. 113; dialects, p. 115

Reading for the Main Idea

1. ability to grow surplus food (NGS 11)
2. Possible answers: wheel, dome, written language, law code (NGS 13)

Critical Thinking

3. All developed along rivers, making trade and communication easier. Some were protected and/or isolated by mountains, deserts.
4. Possible answers: weak leadership, wars, invasion

Organizing What You Know

5. 3200 B.C.—First Egyptian dynasty; 2500 B.C.–2000 B.C.—Harappan culture, dynasties in Huang valley; 2000 B.C.–1500 B.C.—Hammurabi, Indo-Aryans; 1500 B.C.–1000 B.C.—Ramses the Great, Phoenician civilization, Zhou overthrow Shang, Egypt expands; 1000 B.C.–500 B.C.—Moses, Persian Empire; 500 B.C. – A.D. 500—India united, Qin dynasty, Han dynasty, Guptas

In ancient China people who wanted to be government officials had to study hard in order to take civil service tests. Failing the test was considered a disgrace.

The Han Han rulers expanded the empire until it was larger than the Roman Empire, and extended the Great Wall. They introduced a civil service system to run the daily business of government. It was designed to be fair and to reward ability. People had to pass an exam in order to hold civil service jobs. Traders used the Silk Road, which stretched from China all the way to the Mediterranean region, to trade with Greece and Rome.

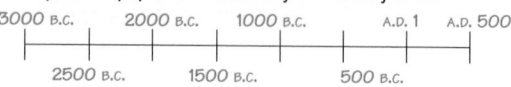

This device, developed around A.D. 132, warned of earthquakes. Ground tremors would cause metal balls to drop from the dragons' mouths to the frogs below.

Advances of the Huang Valley Civilizations These people of ancient China figured out the precise length of a year, and tracked the movement of the planets. They invented the sundial, the water clock, and a machine to measure earthquakes. Chinese doctors developed herbal medicines and learned how to set bones. By A.D. 105, the Chinese had invented paper, which would not be used in other parts of the world until 500 years later. They also invented a printing press. Use of printed books helped to establish a common culture across China.

✓ **READING CHECK:** *Main Idea* What were some of the achievements of the ancient Chinese? calendar, water clock, sundial, machine to measure earthquakes, medicines, paper, printing press

Section 2 Review

Define and explain: dynasty, pharaohs, hieroglyphics, subcontinent, dialects

Reading for the Main Idea

1. *Environment and Society* What made it possible for people in the Fertile Crescent to stop farming and become artisans and traders?
2. *Analyzing* What are some of the lasting contributions made by the peoples of the Fertile Crescent?

Critical Thinking

3. **Making Generalizations** How did geography affect the ancient civilizations in this section?
4. **Human Systems** What caused many of the ancient civilizations to disappear?

Organizing What You Know

5. **Timeline** Copy the following timeline onto a large piece of paper. Fill in as many events as you can.

3000 B.C. | 2500 B.C. | 2000 B.C. | 1500 B.C. | 1000 B.C. | 500 B.C. | A.D. 1 | A.D. 500

Section 3

Objectives

1. Explain how historians study cultures that left no written records.
2. Describe the kingdoms of Kush and Aksum.
3. Explore how trade affected the growth of kingdoms in Sub-Saharan Africa.

Focus

LET'S GET STARTED

Copy the following question onto the board: *What geographical features might help trade and travel flourish in a region?* Allow students time to record their thoughts. *(Possible answers: rivers, seas, lack of natural barriers such as mountains, natural resources.)* Discuss responses. Point out that in Africa, kingdoms arose in places that had these very features. Tell students that in this section they will learn more about the kingdoms of ancient Africa.

Building Vocabulary

Write **oral history** on the chalkboard and have students read its definition from the glossary. Remind students that many civilizations existed for centuries before they developed written languages. The histories of these peoples were passed on from one generation to the next in spoken form. Challenge students to think of oral traditions that are passed along in their own cultures.

Section 3 Early Sub-Saharan Africa

Read to Discover

1. How do historians study cultures that left no written records?
2. What were the kingdoms of Kush and Aksum like?
3. How did trade affect the growth of kingdoms in Sub-Saharan Africa?

Define

oral history

Locate

Nile River
Red Sea
Zimbabwe
Mauritania
Niger River
Timbuktu

Why It Matters

Sub-Saharan Africa was a key player in ancient trade between cultures. Use CNNfyi.com or other **current events** sources to find out about international trade today. Record your findings in your journal.

Bronze statue from Benin

Ancient Africa

Egypt was the first kingdom to develop in Africa, but it was not the only one. South of the vast desert called the Sahara, many different cultures developed. In time some of these groups established city-states, kingdoms, and even empires. It is sometimes difficult, however, for historians to study the earliest African civilizations.

Language One reason for this difficulty is the absence of written records. Spoken language developed in Sub-Saharan Africa long before writing did. Historians therefore must find other ways to study early cultures. One common method they use is the study of **oral history**, spoken information passed from one generation to the next. This information is often contained in stories or songs. Many of the stories tell about great kings and heroes from the past. In some parts of Africa these stories are still told today. Scholars today study these tales to learn about Africa's past.

The Lion Temple was built by the people of Kush, one of the largest kingdoms of Sub-Saharan Africa.

Section 3 Resources

Reproducible

- Guiding Reading Strategy 6.3

Technology

- One-Stop Planner CD-ROM Lesson 6.3
- Homework Practice Online
- HRW Go site

Reinforcement, Review, and Assessment

- Main Idea Activity 6.3
- Section 3 Review, p. 121
- Daily Quiz 6.3
- English Audio Summary 6.3
- Spanish Audio Summary 6.3

TEACH

Teaching Objective 1

LEVELS 1 AND 2: (Suggested time: 15 min.) Copy the following graphic organizer onto the chalkboard, omitting the italicized answers. Ask students to copy it into their notebooks. Then pair students and have the pairs fill in the chart with information about how historians study early African civilizations. Lead a discussion about how these methods are used to study cultures and periods in which there was no written language. **ENGLISH LANGUAGE LEARNERS, COOPERATIVE LEARNING**

STUDYING EARLY AFRICAN SOCIETIES

Historians study...	to learn about...
oral histories	*tales of kings and heroes from the past*
modern languages	*other cultures with which the society had contact*
trade goods	*parts of the world with which the society traded*

Linking Past to Present

The Great Enclosure of Musawwarat is a Kushite ruin in what is today the country of Sudan. It consists of sandstone structures surrounding a temple. The temple is on a raised platform and can be entered only by climbing a ramp. Recent work by archaeologists shows that underground pipes for irrigation supplied water to gardens within the enclosure.

Irrigation is still a necessary part of farming in North Africa. In Egypt, for example, a great deal of money has been invested in building canals, drains, dams, and water pumps.

Discussion: Lead a discussion on why irrigation systems were probably very important in Kush. Point out to students that little rain falls in this area. Water from the Nile must be carried to fields.

Visual Record Answer ►

wood, grasses, other natural materials

▲ Many houses built in Africa today are similar to those built in ancient times. **Interpreting the Visual Record** **From what materials are traditional houses like this made?**

The modern languages of Africa can also provide clues about the continent's history. Historians have tried to figure out which African languages are related. This is often a clue that cultures have had contact with each other in the past. Some African languages are related to languages spoken in other parts of the world. Scholars have learned from this that early Africans had contact with people from distant lands. Other evidence suggests the same thing. Bananas, for example, have been popular in Africa for a long time, but the banana tree is not native to Africa. Early traders must have brought the plant from Asia.

Life in Early Africa Historians have been able to piece together a picture of life in ancient Sub-Saharan Africa. There is still, however, much that they do not know. Across most of the continent, life was centered around small villages. Most of the time, all of the people in a village were related to each other. This helped to tie the village together. Older members of a community were often its leaders. Everyone in the village respected and obeyed them. Women played many roles in early African societies. They were responsible for all farming. In addition, people in some societies traced their family lines through their mothers.

Life was very similar in many villages across Africa. Most people were involved in farming, herding, or fishing. Basic agricultural practices did not vary much from place to place. Most people in early Africa also had similar religious beliefs. They believed that spirits, including the spirits of their ancestors, were all around. They also believed in many gods who controlled nature and human activities.

✓ **READING CHECK:** *Summarizing* What do historians study to learn about ancient Africa? oral histories, languages, plants

This gold plaque and tube show the influence of Egypt on Kushite art. ▼

Kush and Aksum

Two great ancient kingdoms developed along the Nile south of Egypt. The first was Kush. It was eventually conquered by Aksum, its neighbor to the south.

Kush Kush was probably founded around 2000 B.C. In about 1500 B.C., however, it was conquered by Egypt. Egypt ruled Kush for about 500 years. The two countries were never completely unified, however, and Kush remained largely free from Egyptian control. In about 720 B.C., Kush invaded and conquered Egypt. The kings of Kush ruled Egypt for about 50 years. In the 600s B.C., invasions by groups from Southwest Asia weakened the kingdom. Kush survived, however, and became a powerful kingdom again.

LEVEL 3: (Suggested time: 15 min.) On the chalkboard, write the word *good.* Ask students if they know the words for water in Spanish, French, or Italian. Then write the translated words on the chalkboard. *(Spanish: bueno; Italian: buono; French: bon)* Help students find similarities among the words and discuss why the words are similar. *(They all come from the Latin word* bonus. *Spain, Italy, and France were once part of the Roman Empire, where Latin was spoken.)* Explain that analyzing modern language can sometimes reveal clues about relationships between ancient cultures. Point out that historians have studied African languages in this way to learn about early cultures in Africa. **ENGLISH LANGUAGE LEARNERS**

Teaching Objective 2

LEVEL 1: (Suggested time: 10 minutes) Lead a class discussion about what life might have been like in Kush or Aksum. Ask students to name some careers that people might have had in these kingdoms. *(Possible answers: farmer, trader, soldier, artisan)* Then ask students to find examples in the text of some of the achievements of each kingdom. **ENGLISH LANGUAGE LEARNERS**

The people of Kush were mostly traders. They traveled along the Nile to trade goods and ideas with Egypt. Because of this trade, Kush's culture was in many ways similar to Egypt's. Kush also lay along the trade routes between the Red Sea and the Nile. Caravans traveling between these two areas passed through Kush. This allowed the kingdom to become a rich trading center. The Kushites built huge pyramids and temples. They made beautiful pottery and jewelry. They also developed a written form of their language.

Aksum Another powerful kingdom arose to the southeast of Kush. This was Aksum. It lay in a hilly area called the Ethiopian Highlands. Like Kush, Aksum controlled trade routes between the Red Sea and the Nile. In addition, Aksum lay on trade routes between the Red Sea and central Africa. By the A.D. 100s, Aksum had grown into a major trading kingdom. Two hundred years later, it had also become a strong military power. In about A.D. 350 Aksum conquered Kush.

At the time of its conquest of Kush, Aksum was ruled by King 'Ēzānā. He was a strong leader who had control over other rulers in the area. While he was king, 'Ēzānā became a Christian. He made Christianity the official religion of Aksum. The form of Christianity that 'Ēzānā practiced included some local traditions, customs, and beliefs. This made it popular with other people throughout the region. Christianity slowly became a powerful influence in East Africa.

From the A.D. 300s to the 600s, Aksum continued to grow rich from trade. By this time it controlled nearly all the trade on the western shore of the Red Sea. By the late A.D. 500s, however, Aksum had begun to decline for several reasons. One reason may have been soil exhaustion. Farmers had been growing crops in this region for hundreds of years. Over time, it probably became more difficult for farmers to produce enough food to support the population.

Increased competition for trade also hurt Aksum. By the 700s the Persian Empire had become a powerful trading state. The Persians took over much of the trade along the Red Sea that Aksum had once controlled. Muslim traders from the north also took trade away from Aksum. Faced with these new trading empires, Aksum slowly lost most of its economic and political power.

READING CHECK: ***Contrasting*** How was the form of Christianity practiced in Aksum different from the Christianity practiced elsewhere? included local customs and beliefs

Kush and Aksum

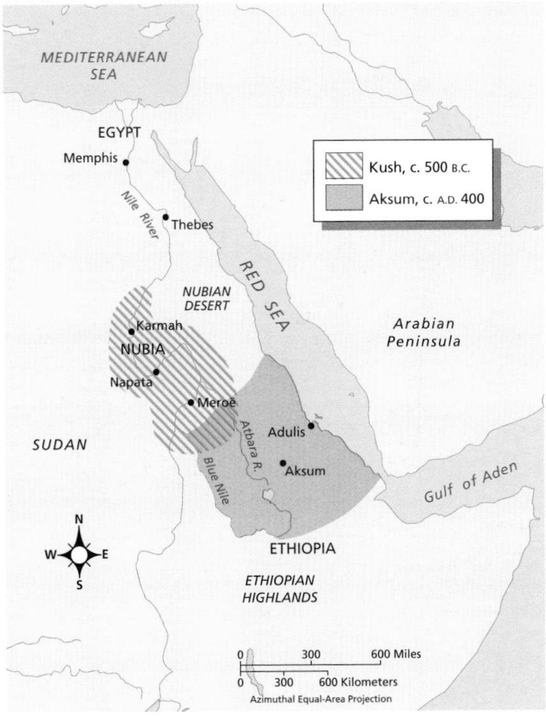

Aksum conquered Kush by about A.D. 350.

Interpreting the Map On what river would goods from Kush be sent to Egypt?

This coin pictures King 'Ēzānā, who made Aksum the first Christian kingdom in Africa.

Across the Curriculum

ART

Sculpture The artisans of Aksum were known for their work with stone. They are perhaps best known for making massive stone obelisks called stelae. Some of these monumental pillars reached nearly 100 feet in height and were carved from single blocks of stone. Historians think that the carvings on these stelae commemorated events in the lives of rulers or the community.

Activity: Have students make their own stelae using clay or cardboard and marker pen. Then have students explain the markings on their stelae to the class.

Interpreting the Map

the Nile

LEVELS 2 AND 3: (Suggested time: 45 min.) Organize students into groups and have each group design an illustrated cover for a magazine called "Ancient Africa" that will feature Kush and Aksum. Then have the groups write short summaries of articles that could be included in an issue of the magazine. Topics might include history, trade, arts, architecture, geography. **COOPERATIVE LEARNING**

Teaching Objective 3

ALL LEVELS: (Suggested time: 40 min.) On the chalkboard, write the names of the kingdoms discussed in this section: Kush, Aksum, Zimbabwe, Ghana, Mali, and Songhay. Then organize the class into pairs. Have the students in each pair discuss how trade affected the growth each kingdom. Have each pair of students discuss each of the listed kingdoms for five minutes. Then direct students to move on to the next kingdom. Then lead a class discussion about the importance of trade in early African history. **ENGLISH LANGUAGE LEARNERS, COOPERATIVE LEARNING**

Linking Past to Present

Ghana From about A.D. 1000 to 1400, the African gold trade was centered on the southeastern coast. By 1500, it had moved to the southern coast of West Africa. There, Africans traded gold to Europeans in exchange for weapons. This part of West Africa came to be called the Gold Coast. For a time it was a British colony. In 1957, it became the first African colony to gain independence from European control. Kwame Nkrumah, the first president of the new country, chose to name it Ghana after the ancient West African trading kingdom.

Critical Thinking: Ask students speculate about why Nkhruma might have chosen the name Ghana for the new country.

Answer: Perhaps he thought of the name as a link to the country's past, when the region was not ruled by others.

African Trading States, c. A.D. 1230–1591

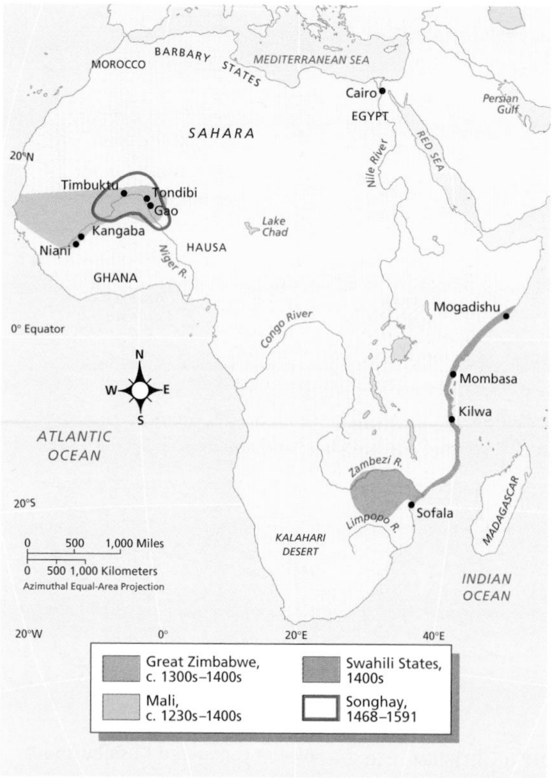

The Trading States

Kush and Aksum were the largest kingdoms in ancient Africa, but they were not the only ones. Like these two kingdoms, most of the civilizations of Sub-Saharan Africa were great traders. Across the continent, trade connected Africa with the rest of the ancient world.

East Africa Along the coast of East Africa, dozens of small city-states appeared. The earliest such city-states, like Mogadishu (moh-guh-DEE-shoo) and Mombasa, were located in the northern part of the region. Eventually city-states like these lined most of the eastern coast of Africa. They controlled trade in the Indian Ocean. They sold gold, ivory, hides, and tortoise shells to traders from around the world. They also sold slaves. In exchange, these traders brought weapons and porcelain to Africa.

Eventually a new culture arose out of this trade. A new language, called Swahili, developed. It was an African language but it also included many words from Arabic. For this reason, the city-states of East Africa are sometimes called the Swahili States.

Zimbabwe Later traders established settlements farther south. By A.D. 900, gold had been discovered in the interior of southeastern Africa. This gold was shipped to the coast on the Zambezi River. The discovery of gold increased trade along the Indian Ocean. Many peoples wanted to trade for it. African kingdoms fought to control the gold trade.

Ruins of the Great Zimbabwe fortress near Masvingo, Zimbabwe

A group of people called the Shona gained control of the region in the 1200s. Their kingdom was centered around Great Zimbabwe, a huge stone city of more than 10,000 people. *Zimbabwe* is a Shona word that means "stone houses." From Great Zimbabwe, the Shona controlled the gold trade out of southeastern Africa. They also held great political power. Their kingdom lasted until the 1400s.

West Africa In West Africa, trade centered on the Atlantic Ocean. Many trading societies developed between the Atlantic and Lake Chad. The most powerful trading societies in these region were those who controlled trade routes across the Sahara. Traders from the north brought salt across the desert in exchange for gold from West Africa. Cities grew along these trade routes.

Close

On the chalkboard, write the heading "Steps in the History of Sub-Saharan Africa." Below this draw outlines of seven footprints in a horizontal line. Have volunteers fill in the footsteps with facts from the text in chronological order.

Review and Assess

Have students complete the Section Review. Then ask each student to write three trivia questions based on the text. Have students answer each other's questions. Then have students complete Daily Quiz 6.3.

Reteach

Have each student choose one of the kingdoms studied in this section. Have students imagine that they are staff members on an educational television program. Have them write brief scripts that introduce viewers to their chosen kingdoms. Have volunteers read their scripts.

ENGLISH LANGUAGE LEARNERS, COOPERATIVE LEARNING

Extend

Have interested students conduct research on the folklore of the regions studied in this section and write a retelling of one of the tales they discover. Encourage students to include illustrations and to present their stories to the class. **BLOCK SCHEDULING**

The first of the great West African trading kingdoms was Ghana. It grew up around a trading village in what is now Mauritania. Gold trade made the kings of Ghana rich and powerful. They built strong armies and conquered new lands. Eventually, however, Ghana began to decline. In about A.D. 1235, a neighboring people conquered Ghana. They set up a new empire called Mali.

Mali covered all of the area that had been Ghana. It also included new lands to the north and west. Mali reached the height of its power in the early 1300s under a ruler named Mansa Mūsā. He was a strong supporter of education and the arts. During his reign, the city of Timbuktu became an important center of learning and trade. People came from as far away as Egypt and Arabia to study at the university there. After Mansa Mūsā died, Mali began to grow weak. In 1468, rebels captured Timbuktu. They set up a new kingdom called Songhay.

Songhay was a powerful trading kingdom. It was centered on the city of Gao, a key trade city on the Niger River. By this time the Niger had become an important trade route. The kings of Songhay encouraged the growth of Islamic teachings at the university in Timbuktu. This helped Timbuktu remain an important cultural and trading center. Goods from Europe, India, and China were exchanged there. Timbuktu also remained a center of learning. Books sold for very high prices in the markets there. Songhay remained a powerful state until 1591. In that year it was conquered by Moroccan troops.

▲ Mansa Mūsā was a devout Muslim who encouraged the building of mosques, or Islamic centers for worship. This one, in Mali, was built in the 1300s.

✓ **READING CHECK:** *Identifying Cause and Effect* What led to the development of strong kingdoms in West Africa? trade with merchants from across the Sahara

Section 3 Review

go.hrw.com Homework Practice Online Keyword: SK3 HP6

Define and explain: oral history

Reading for the Main Idea

1. *Human Systems* What do scholars study to learn about early African civilizations?
2. *Places and Regions* What were the two most powerful kingdoms of Sub-Saharan Africa?
3. *Human Systems* What were some items traded by early African civilizations?

Critical Thinking

4. **Drawing Inferences and Conclusions** Why were the most powerful kingdoms of Sub-Saharan Africa located on coasts or rivers?

Organizing What You Know

5. **Comparing and Contrasting** Copy the following chart. Use it to name and describe the powerful trading kingdoms of Sub-Saharan Africa.

State	Location in Africa	Features

Section 3 Review Answers

Define For definition, see: oral history, p. 117

Reading for the Main Idea

1. oral history, languages, trade goods (NGS 17)
2. Kush and Aksum (NGS 12)
3. Possible answers: gold, ivory, hides, tortoise shells, slaves (NGS 10)

Critical Thinking

4. trade routes generally over water instead of over land; easier to ship goods

Organizing What You Know

5. Swahili states—east coast; traded across Indian Ocean; Great Zimbabwe—interior southeast Africa; created by Shona, traded gold, built fortresses; Ghana—West Africa; traded with merchants from across Sahara; powerful army; Mali—West Africa; conquered Ghana, built Timbuktu; Songhay—on Niger River; powerful traders, made Timbuktu a center of learning

Section 4

Objectives

1. Describe how ancient Greek civilization developed.
2. Discuss the events that led to the birth and decline of the Roman Empire.
3. Explain how Christianity began.

Focus

LETS GET STARTED

Copy the following instructions onto the chalkboard: *Examine the pictures in the section titled "The Greek and Roman Worlds." Write a sentence about what you can tell about the Greek and Roman civilizations by looking at these pictures.* Discuss responses. *(Possible answer: These civilizations had powerful leaders, beautiful art, and impressive buildings and other structures.)* Tell students that in this section they will learn about these great ancient civilizations.

Building Vocabulary

Write the key terms on the chalkboard. Call on individual students to locate the definitions in the text and to read them aloud. Then challenge students to write sentences that connect two of the terms. *(Example: A **direct democracy** and a **republic** are two forms of government.)*

Section 4 Resources

Reproducible
- Guided Reading Strategy 6.4
- Map Activity 6

Technology
- One-Stop Planner CD-ROM, Lesson 6.4
- Homework Practice Online
- HRW Go Site

Reinforcement, Review, and Assessment
- Section 4 Review, p. 129
- Daily Quiz 6.4
- Main Idea Activity 6.3
- English Audio Summary 6.4
- Spanish Audio Summary 6.4

Section 4 The Greek and Roman Worlds

Read to Discover

1. How did ancient Greek civilization develop?
2. What events led to the birth and decline of the Roman Empire?
3. How did Christianity begin?

Define

city-states
direct democracy
republic
aqueduct

Locate

Aegean Sea
Greece
Crete
Athens
Italy
Rome
Adriatic Sea
Mediterranean Sea
Alps
Jerusalem

WHY IT MATTERS

Both the Greek and Roman governments were based on the idea that people can govern themselves. Use CNNfyi.com or other **current events** sources to find out about how citizens of the United States and other countries take part in government.

Greek coins, c. 500s B.C.

Aegean Civilization, c. 1450 B.C.-700 B.C.

The Early Greeks

By 2000 B.C. civilizations were developing in the Nile River valley and the Fertile Crescent. At the same time another civilization was forming near the Balkan peninsula and the Aegean Sea. The people who settled this area later became known as the Greeks. From the Greeks came many of the ideas that formed the foundation for modern western civilization.

Geography and Greek Civilization

Geography has much to do with the way the early Greeks lived. Greece is a rugged country that is made up of many peninsulas and islands separated by narrow waters. The land is covered with high mountains. They separated groups of people who lived in the valleys. These landforms contributed to the development of separate communities rather than one large and united kingdom. Because it was difficult to travel through the mountains, some Greeks preferred to travel by sea. Many became fighters, sailors, and traders.

TEACH

Teaching Objective 1

LEVEL 1: (Suggested time: 20 min.) Copy the following graphic organizer onto the chalkboard, omitting the italicized answers. Have students use it to compare and contrast the cultures of the Minoans and the Mycenaeans. Then ask students to create their own graphic organizers to compare and contrast the cultures of Sparta and Athens. *(Organizer designs will vary. Sparta—strong government, little personal freedom, powerful army, little development of arts and sciences; Athens—direct democracy, center of learning, literature and the arts, philosophy and science)*

ENGLISH LANGUAGE LEARNERS

	Minoans	Mycenaeans
Location	*island of Crete*	*Greek mainland*
Achievements	*palaces, writing system, beautiful art*	*fortlike walled cities*
Reasons for decline	*conquered by Mycenaeans*	*earthquakes, wars*

CONNECTING TO Art

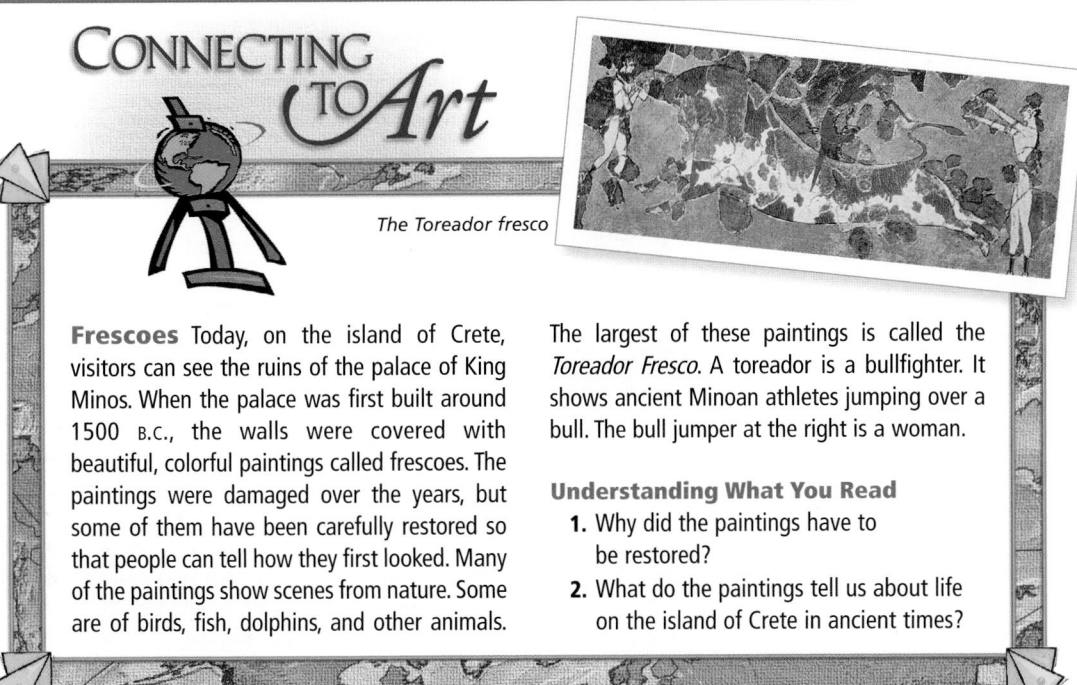

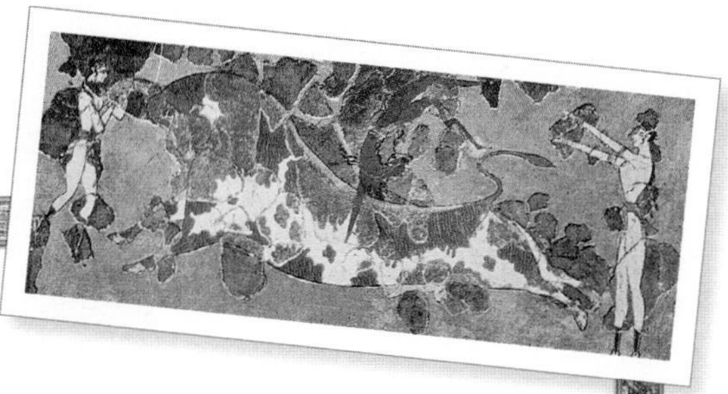

The Toreador fresco

Frescoes Today, on the island of Crete, visitors can see the ruins of the palace of King Minos. When the palace was first built around 1500 B.C., the walls were covered with beautiful, colorful paintings called frescoes. The paintings were damaged over the years, but some of them have been carefully restored so that people can tell how they first looked. Many of the paintings show scenes from nature. Some are of birds, fish, dolphins, and other animals. The largest of these paintings is called the *Toreador Fresco.* A toreador is a bullfighter. It shows ancient Minoan athletes jumping over a bull. The bull jumper at the right is a woman.

Understanding What You Read

1. Why did the paintings have to be restored?
2. What do the paintings tell us about life on the island of Crete in ancient times?

The Minoans About 100 years ago on the island of Crete, scientists found the remains of the earliest Greek civilization. By 2000 B.C., the Minoan people had developed a great civilization. From the evidence scientists found, we know that the Minoans built cities and grand palaces that even had running water. We also know that they developed a system of writing and that Minoan artists carved beautiful statues from gold, ivory, and stone. Because Crete's soil was very poor, farming was not very productive. Many people became sailors and fishers. By 1400 B.C., the Minoan civilization began to decline. It was conquered by a group that lived on the Greek mainland, the Mycenaeans (my-suh-NEE-unhz).

The Mycenaeans The Mycenaeans controlled the Greek mainland from about 1600 B.C. to about 1200 B.C. They were a warlike people who lived in tribes. Each tribe had its own chief. The Mycenaean tribes built fort-like cities surrounded by stone walls. They carried out raids on other peoples throughout the eastern Mediterranean. Once they conquered the Minoans, they adopted many aspects of their civilization. For example, they used the Minoan system of writing. By 1200 B.C. earthquakes and wars had destroyed most of the Mycenaean cities. A later Greek poet named Homer wrote a long poem called the *Iliad,* which pulls together about 400 years of historical events, legends, and folk tales. It tells the story of the Trojan War. The Mycenaeans were the Greeks Homer wrote about in that story.

✓ **READING CHECK:** *Analyzing* How did the geography of the land affect the way the Minoans lived? **did not become farmers because the land was poor; earned living as fishers and sailors because of location close to large bodies of water.**

Across the Curriculum
ART

Architecture Although little is known about the earliest palaces, the later palaces of the Minoans are distinctive for the quality of their workmanship and the sophistication of their designs. The carefully planned palaces had wide staircases and multiple stories, vistas, and storage rooms. Artisans painted the interior walls with scenes of daily life. Even built-in bathtubs and running water could be found inside the palaces.

Critical Thinking: What do the remains of the palaces show about the society that built them?

Answer: The palaces show that the Minoans were quite advanced and that they were concerned with practical function as well as beauty.

Connecting to Art

Answers

1. The paintings became damaged over the years.
2. The paintings show that the Minoans saw beauty in nature, that they liked sports, and that women participated in sports.

LEVEL 2: (Suggested time: 30 min.) Have students complete the Level 1 activity. Then use the text to review the physical features of Greece's geography. Ask students to write a few sentences about ways in which the development of Greek civilization was influenced by these features. *(Possible answers: many islands and high mountains—separate communities instead of one united kingdom; rugged land and nearness of sea—fighters, sailors, traders rather than farmers)*

LEVEL 3: (Suggested time: 30 min.) Point out to the class that the Greeks wrote many plays—some funny, some sad—that are still performed today. Organize the class into groups. Ask students to imagine that they are Greek playwrights. Have each group determine what events in Greek history the members would highlight in a play titled The Greatness of Greece and report the selections to the class. Then ask each group to outline a script for a brief scene from its play. You may want to extend the activity by having students perform their scenes. **COOPERATIVE LEARNING**

DAILY LIFE

In ancient Athens, men held all legal rights, and a woman who wished to be thought of as respectable would not even leave her house. Marriage was largely a financial agreement. Women played such a minor role in Greek society that they were often not named publicly until their death. As a result, historians have very little evidence about Athenian women.

Critical Thinking: Ask students to suggest ways that modern scholars have learned about the Greeks' attitudes toward women.

Answer: Student should suggest that scholars have probably learned about ancient Greek attitudes from written accounts of the time.

internet connect

GO TO: go.hrw.com
KEYWORD: SK3 CH6
FOR: Web sites about Ancient Greece

Visual Record Answers

It shows that Spartan soldiers were strong and wore elaborate helmets.

They valued both wisdom and military skill.

Sparta's men were expected to serve in the military until they were 60 years old.

Interpreting the Visual Record **What does this bronze figure show about Spartan soldiers?**

The Greek City-States

Historians do not know much about how the Greeks lived for the next 400 years. However, by the 800s B.C. the Greeks lived in separate **city-states**. A city-state was a city or town that had its own government and laws and controls the land surrounding it. Each city-state had its own calendar, money, and system of weights and measures. Two of the largest city-states, Sparta and Athens, developed in different ways.

Sparta Spartans were very loyal to their city-state. It had a strong government that allowed little personal freedom. Sparta also had a powerful army. At the age of seven, boys left home to be trained as soldiers. When men grew older and left the army, they were expected to work for the public good. Girls received strict training at home. Both boys and girls were educated and encouraged to study music. However, Spartan culture left little time for the development of the arts, literature, philosophy, and science.

Athens The Athenians developed a form of government called **direct democracy**. In a direct democracy, citizens take part in making all decisions. Athenians also had courts where cases were decided by juries. Jurors were chosen by lot. The jury voted on each case by secret ballot. While the Athenians had much more freedom than the Spartans, Athenian women could not participate in government. Slavery was also permitted.

Athenians made great contributions to the arts, literature, philosophy, and science. Builders, artists, and sculptors created beautiful temples and other works. Athenian writers produced great works of literature. They were the first people to write drama, or plays. Scientists discovered new laws of mathematics and developed a system to classify animals and plants. The Greek physician Hippocrates is considered by many to be the father of medical science. Philosophers such as Socrates, Plato, and Aristotle studied questions of reality and human existence. Their thoughts formed the basis for many later ideas at the heart of Western culture.The 400s B.C. are known as the Golden Age of Athens.

This photo shows part of a temple in honor of Athena, the Greek goddess of warfare and wisdom. It was built between about 421 B.C. and 415 B.C. Athens was named for Athena.

Interpreting the Visual Record **What does this tell you about Athenian values?**

READING CHECK: *Comparing and Contrasting* How did Athenian way of life differ from the Spartan way of life? **Spartans: a strict military life; Athenians: developed in the arts, in education, and in the sciences.**

Teaching Objective 2

LEVEL 1: (Suggested time: 20 min.) Copy the following graphic organizer onto the chalkboard, omitting the italicized answers. Have pairs of students work together to complete and to understand cause-and-effect relationships as they read about Rome and its eventual decline. Then lead a discussion about the events in the chart, asking students to fill in more details. **ENGLISH LANGUAGE LEARNERS, COOPERATIVE LEARNING**

Cause	Effect
Romans build bridges and roads.	*The Roman Republic grows.*
Julius Caesar comes to power.	*Romans conquer more territory.*
Roman Empire controls vast lands.	*Rome can no longer be ruled by one person.*
Dishonest leaders neglect the empire.	*Invaders threaten Rome's borders, Rome is drained of its resources, civil wars begin, and rising prices and taxes make daily life hard.*

Alexander the Great

While Athens and other city-states fought, weakened, and declined, Macedonia, a kingdom to the north, gained strength. Macedonia's king, Philip II, took control of Greece and united it under his rule. When he was killed in 336 B.C., his son, who would become known as Alexander the Great, took his place. Alexander conquered much of the known world—including the Greek city-states—in a short period of time. He built cities that grew to be great centers of learning. Alexander admired the Greeks. Travel and trade along the routes that linked Alexander's empire to Asia spread Greek culture and ideas throughout the world. After Alexander died in 323 B.C., his empire was split into smaller kingdoms. In time, the Romans conquered Alexander's empire.

READING CHECK: ***Drawing Conclusions*** How did Alexander help spread Greek culture throughout his empire? had widespread trade routes built; encouraged trade and travel; respected Greek ideas and culture.

Alexander's education prepared him to be a great leader. He got his military training from his father and his education from the great Greek philosopher Aristotle.

The Early Romans

In about 750 B.C., while the city-states were growing in Greece, people called Latins were settling in villages along the Tiber River in Italy. In time, they united to form the city of Rome. During the late 600s B.C. they were conquered by a people called the Etruscans from the north. The Etruscans brought written language to Rome. Skilled and clever craftspeople, they built paved roads and sewers. Rome grew into a large and successful city. In time, the Greeks also settled in Italy. Their ideas and culture strongly influenced the Romans. Even the Roman religion was partly based on Greek beliefs.

Geography and Roman Civilization In some ways, the geography of Italy helped a great civilization to develop there. Italy is protected by the Alps to the north. The Mediterranean Sea to the west and the Adriatic Sea to the east of this boot-shaped peninsula made trade and travel easy. Not everything about Rome's location, however, was good. Geography also posed some problems. Passages through the Alps left Italy open to invasions. The peninsula's long coastlines left it open to attack from the sea.

READING CHECK: ***Identifying Cause and Effect*** How did the geography of Italy help the Roman civilization grow? location by the sea made trade and travel easy; protected by the Alps against northern invaders

Ancient Italy. c 600 B.C.

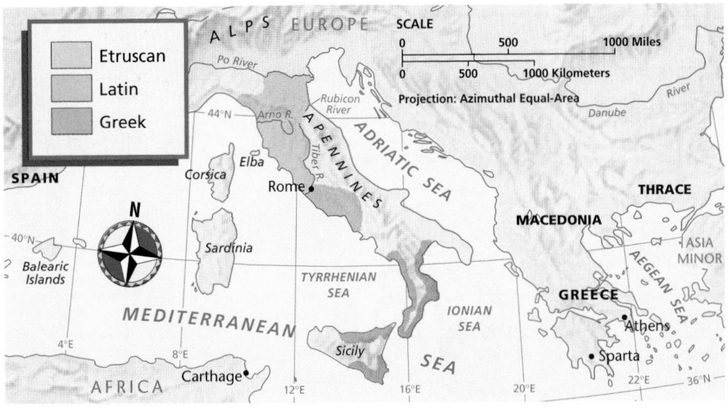

Linking Past to Present

Alexander's Cities Many legends surround Alexander the Great. One of the best-known involves his horse, Bucephalus. According to legend, Bucephalus was a gift to Philip, but no one could ride him. The horse shied and reared whenever someone tried to mount him. Alexander noticed that the horse was merely afraid of its own shadow. He turned the horse to face the sun and mounted him easily. Acknowledging his son's wisdom, Philip gave Bucephalus to Alexander, who rode him for many years. When the horse died on a campaign, Alexander founded a city at the spot and named it Alexandria Bucephala.

Activity: Have students conduct research to find maps showing Alexander's empire and the cities he founded, including Alexandria Bucephala. Then ask students to investigate the origins of the other cities whose names begin with Alexandria.

LEVEL 2: (Suggested time: 15 minutes) Lead a class discussion comparing and contrasting the government of Greece—a direct democracy—with the government of Rome—a republic. Students should note that the main difference between the two is that in a direct democracy the people themselves vote on new laws, whereas in a republic the people elect officials to make laws for them. Ask students which type of government is more like the one we have in the United States today. *(The government of the United States is more like the government of Rome because we elect our lawmakers.)* Then have each student develop a time line of Rome's growth and decline. Compare and discuss time lines.

LEVEL 3: (Suggested time: 20 min.) Ask students how Italy's location and geographic features created both advantages and disadvantages for Rome. *(The Alps protected Rome from invasion, and the sea made travel easy. Both the sea and the natural passes through the Alps left Rome open to invasion.)* Have students work in pairs to create ways to present this information graphically. Ask students to relate these factors to Rome's growth and decline. *(Students might create charts with two columns headed by plus and minus signs, listing advantages and disadvantages in the appropriate columns. They might draw maps with labeled arrows pointing to the appropriate features.)* **COOPERATIVE LEARNING**

COOPERATIVE LEARNING

Throughout history, communities have erected statues of people who contributed to the public good. Because many Romans viewed Julius Caesar as this kind of person, numerous statues of Caesar stood in Rome or in other parts of the Roman Empire. Today many of those statues are in museums.

Organize students into groups. Have each group select a person in your community who students believe deserves to be honored by a statue. Have each group decide on the statue's materials, size, and location. Then ask students to create a plaque that might appear at the base of the statue describing the person and his or her contributions to the community.

Visual Record Answer

Some Roman leaders may have felt he had too much power.

Two of the men who killed Caesar were his friends.

Interpreting the Visual Record **Why do you think some Roman leaders feared Caesar?**

The Roman Republic

In 509 B.C. a group of wealthy Romans overthrew the Etruscan king. They replaced the Etruscan rule with a **republic**. A republic is a government in which voters elect leaders to run the state. For the next 200 years, the Romans fought many wars. Their well-trained armies built bridges and roads. The republic grew as the Romans gained new territories. In time, however, controlling such a vast area became a problem for Rome's leaders.

Julius Caesar By 60 B.C. a popular public speaker named Julius Caesar began to win support among Rome's poor. Caesar soon became a powerful general who conquered more territory for the republic. He extended the rule of Rome to present-day France. He even marched into Egypt. He put Cleopatra, a daughter of the ruling family, on the throne as a Roman ally. Victorius, Caesar returned to Rome in 46 B.C. Two years later Roman officials made him ruler for life. However, there were some Roman leaders who feared Caesar's power. On March 15, 44 B.C. Caesar was assassinated.

✓ **READING CHECK:** *Analyzing* What events led to Caesar's rise to power? developed a following among the poor; became a great general and conquered lands so that the Roman empire greatly expanded

The Roman forum was the center of Rome's government.

ALL LEVELS: (Suggested time: 40 min.) Organize students into small groups. Have each group create a chart with two headings—"Rome's Greatness" and "Rome's Decline." As students read about the Roman Empire, have them record details that contributed to Rome's greatness and details that contributed to Rome's decline on the chart. *(Examples: Rome's Greatness—well-trained armies, excellent roads and bridges, the Roman empire grew; Rome's Decline—dishonest leaders, invasions, civil wars, high taxes and prices)*

ENGLISH LANGUAGE LEARNERS, COOPERATIVE LEARNING

Teaching Objective 3

ALL LEVELS: (Suggested time: 30 min.) Organize students into groups and ask each group to create a children's picture book to be titled The Rise of Christianity. Have each group create an outline for its book. You may want to have each student create text and illustrations for one or more pages. The book should summarize the key points that students have read in the text. You may want to provide additional resources.

COOPERATIVE LEARNING

The Roman Empire

After Caesar's death, his grandnephew Octavian became the leader of the Roman world. The Romans called Octavian *Augustus*, which meant the "honored one." He became known as Augustus Caesar. Historians refer to Augustus as the first emperor of Rome. Augustus, however, never actually used the title of emperor.

Under Augustus, the Roman republic became a great empire. Augustus sent his armies out to conquer new lands. Soon the empire extended from Spain to Syria and from Egypt to the Sahara Desert. In the north, the empire reached all the way to the Rhine and Danube Rivers.

Pax Romana The reign of Augustus began a 200-year period called the Pax Romana, which means "Roman Peace." During this time Rome was a stable and peaceful empire. Laws became more fair. Widespread trade created a strong economy. In order to trade, people had to travel, so the Roman army built roads and bridges that helped unite the vast empire. The Roman army also helped keep the peace by defending Rome's borders against outside invaders.

✓ **READING CHECK:** ***Summarizing*** What was the Roman Empire like during the Pax Romana? a stable and peaceful empire with fair laws; feeling of security against foreign invasion

Our Amazing Planet

The Roman army built many great roads and bridges. If you travel in Europe today, you may use the same roads the Romans built some 2000 years ago.

Trade in the Roman Empire A.D. 117

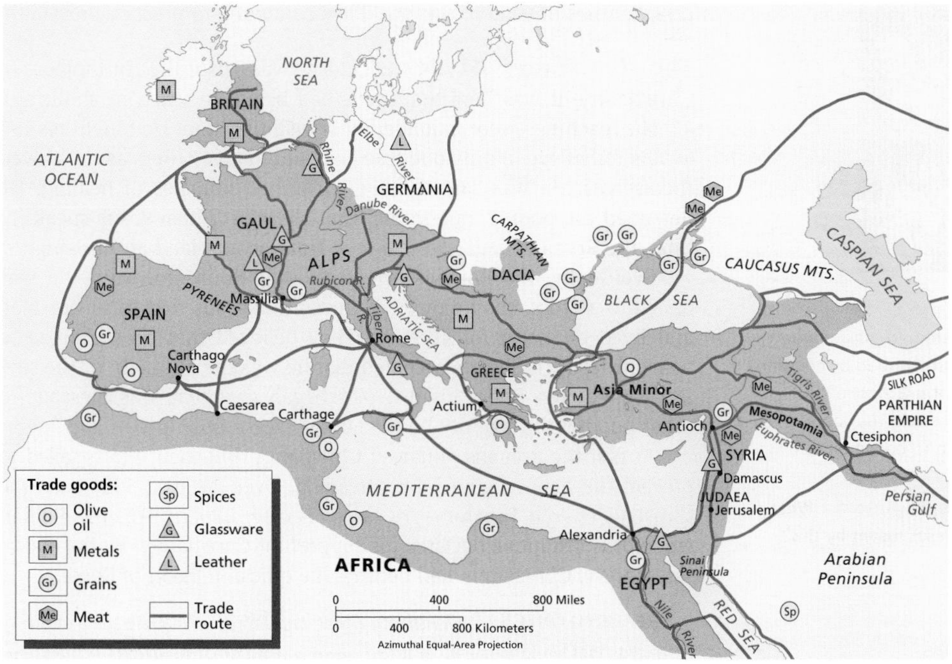

Linking Past to Present

Latin The Latin alphabet was derived from the Etruscan alphabet. The earliest inscription in Latin was found on a cloak pin dating back to the 700s B.C. Today the Latin alphabet is the most widely used alphabetic system in the world. It is the alphabet of the English language and of most European languages.

Although today few people actually speak Latin, it has influenced many modern languages, including English. Numerous words from the medical, scientific, and legal professions are made up of Latin root words, prefixes, and suffixes. For example, the word *legal* comes from the Latin root, *legis*, meaning "law." Other words such as legislature and legitimate are also derived from this same root. The scientific classification system for plant and animal names is based on Latin words. For example, *felis domesticus* is the scientific term for "house cat."

Activity: Ask students to use a dictionary to find familiar words that are based on Latin words or roots.

TEACHER TO TEACHER

Peggy Altoff of Baltimore, Maryland, suggests the following activity to help students learn about the influence of Greece and Rome on the modern world: Have students compare the photographs of Greek and Roman buildings in this section and other sources to public buildings in your community. Have them describe the similarities and differences.

CLOSE

Lead a discussion on how the histories of Greece and Rome are different. *(Examples: Greece consisted of separate city-states, while Rome was more unified; Greece did not became an empire until Philip II, a Macedonian, took control; Greece contained direct democracies, while Rome was a republic; Roman rulers governed huge territories, but Greek rulers did not; Greece was conquered by the Roman Empire, while Rome fell to invading tribes.)* You may want to extend the activity by asking students in which civilization they would rather have lived and why.

Section 4 Review

Answers

Define For definitions, see: city-states, p. 116; direct democracy, p. 116; republic, p. 118; aqueduct, p. 120

Reading for the Main Idea

1. Mountains separated settlements of people, allowing the growth of separate states. (NGS 5)
2. Augustus expanded the empire by conquering new lands. (NGS 10)
3. Jews believed Jesus to be their savior. His teachings became the foundation of Christianity. (NGS 6)

Critical Thinking

4. Direct democracy involves all citizens in making decisions. In the United States, elected leaders make decisions.

Organizing What You Know

5. Ancient Greece: direct democracy; Mycenaeans, Minoans; rugged mountains, many islands; Ancient Rome: Republic, then empire; Latins, Etruscans; boot-shaped peninsula, long coastlines

Visual Record Answer ▶

Jesus may have been referring to children's faithfulness, trust, and innocence.

▲ Aqueducts were stone canals that carried water from mountains to the city.

▲ This modern stained glass window shows Jesus surrounded by children. According to the Gospel of Matthew, Jesus said that people should change to become more like children.

Interpreting the Visual Record **What do you think Jesus meant by this?**

Rome's Achievements

The Romans made many important advances in science and engineering. They built temples, palaces, arenas, bridges, and roads. They figured out a way to transport water from one place to another by a system of aqueducts. An **aqueduct** is a sloped bridge-like structure that carries water. Skilled Roman architects designed buildings with domes and arches.

Some Romans were also very talented writers. The verses of ancient Roman poets such as Virgil, Horace, and Ovid are still read today. The biographer Plutarch wrote of the lives of famous Greeks and Romans. The Roman language, Latin, was used for many centuries. It became the basis for many modern languages including French, Spanish, Italian, Portuguese, and Romanian.

✓ **READING CHECK:** *Summarizing* What did Roman engineers contribute to their society? built great temples with domes and arches; created aqueducts to transport water.

The Rise of Christianity

In A.D. 6, the land called Judea came under Roman control. Judea was the home of the Jews. Because the Jews wanted a separate state, they rebelled against Roman rule. After a series of uprisings, the Romans sacked the city of Jerusalem. They destroyed all but the western wall of the city. In A.D. 135, the Roman emperor Hadrian defeated the Jews and banned them from the city of Jerusalem. The Jews continued to build communities outside of the city. They continued to practice their faith.

The Teachings of Jesus Judea was also the birthplace of Christianity. It was here that Jesus had begun to teach in about A.D. 27. His teachings were grounded in Jewish tradition. He taught his followers to believe in only one true God and to love others as they loved themselves. He was said to have performed miracles of healing. He defended the poor. People came from all over to hear Jesus speak. In time, the Romans began to fear that Jesus would lead an uprising.

Eventually, the Romans arrested Jesus. Soon afterward, he was nailed to a cross. His followers believed that Jesus rose from the dead and lived on Earth for 40 days. They believed that he then rose to heaven. Jesus' followers accepted that he was the Messiah, or the savior of the Jews. Soon, his disciples began to spread this message to other people. Christianity had begun to take root.

At first the Romans outlawed Christianity, but their efforts failed to prevent the new religion from spreading. Over the next 300 years, the Christian church became very large. Finally, in A.D. 312, the Roman emperor Constantine declared his support for Christianity. By the end of the century, Christianity had become the official religion of Rome.

✓ **READING CHECK:** *Identifying Cause and Effect* What are some of the events that led to the death of Jesus? developed a great following; Roman fear that he would lead an uprising

REVIEW AND ASSESS

Have students complete the Section Review. Then have each student write a question about the content under each subhead. Call on students to read their questions and on others to provide answers. Continue until all questions have been answered correctly. Then have students complete Daily Quiz 6.4.

RETEACH

Have students complete Main Idea Activity 6.4. Then organize students into three groups—one for each Read to Discover question. Have group members write three true-false statements that relate to their question. Ask class members to decide if statements are true or false and then change false statements to true statements. **ENGLISH LANGUAGE LEARNERS**

EXTEND

Ask interested students to conduct research about daily life in ancient Rome. Have students create displays with diagrams and illustrations depicting the food people ate, the homes they lived in, the kinds of work they did, and the entertainment they enjoyed . **BLOCK SCHEDULING**

The Decline of the Roman Empire

In the early a.d. 200s, Rome faced troubled times. Ambitious generals frequently decided to seize power for themselves. Some of them even assassinated emperors and took their places. Over time, the army lost its loyalty to Rome. Soldiers became more interested in becoming wealthy than in defending the empire. Dishonest leaders fought for power. Their neglect of the empire made it possible for invaders to threaten the borders. Invasions were costly and drained the empire of its resources. Inside the empire, civil wars had begun. Daily life became more difficult as taxes and the cost of goods rose higher.

A Split in the Empire By A.D. 284 the Roman Empire could no longer be ruled well by one person. The emperor Diocletian selected a co-emperor to help rule. About 20 years later, the emperor Constantine, who had accepted Christianity, took over the eastern part of the empire. Constantine was a strong ruler. The empire in the east fared much better than the crumbling and weakened empire in the west.

The Fall of Rome As the years passed, invasions from the north continued. Groups such as the Vandals, Visigoths, and Huns set up tribal kingdoms within the empire. In 476, the last emperor of Rome was overthrown by invaders. This marked the end of the empire in the west. The empire of the east was able to fight off invaders. This part of the empire became known as the Byzantine Empire. It lasted until 1453 when it fell to the Ottoman Turks.

✓ **READING CHECK:** *Finding the Main Idea* How did weak leadership lead to the fall of Rome? neglected the empire; allowed invaders to set up tribal kingdoms and overthrow the last emperor in the west

Section 4 Review

go.hrw.com Homework Practice Online
Keyword: SK3 HP6

Define and explain: city-states, direct democracy, republic, aqueduct

Reading for the Main Idea

1. *Environment and Society* How did the geography of Greece lead to the development of its city-states?
2. *Human Systems* What events under the rule of Augustus helped the Roman republic become the Roman empire?
3. *Human Systems* How did the problems in Judea lead to the rise of Christianity?

Critical Thinking

4. **Comparing and Contrasting** How did the direct democracy of the Athenians differ from government in the United States today?

Organizing What You Know

5. Copy the following graphic organizer. Use it to compare the governments, people, and geography of ancient Greece and ancient Rome.

	Ancient Greece	Ancient Rome
Government		
Early People		
Geography		

CHAPTER 6 Review ANSWERS

Building Vocabulary

For definitions, see: hominid, p. 103; prehistory, p. 103; nomads, p. 104; land bridges, p. 104; irrigation, p. 105; civilization, p. 106; division of labor, 106; history, p. 107; dynasty, p. 112; pharaohs, p. 112; hieroglyphics, p. 113; city-states, p. 124; direct democracy, p. 124; republic, p. 126; aqueduct, p. 128

Reviewing the Main Ideas

1. People live in an organized society, produce more food than they need, live in towns or cities with governments, and practice division of labor. (NGS 12)
2. When people began to farm, they stayed in one place instead of living as wandering nomads. (NGS 15)
3. The Nile Valley is surrounded by deserts and seas. (NGS 4)
4. City-states were separated by rugged landforms. They developed their own governments, laws, and cultures. (NGS 4)
5. Possible answers: architecture, literature, engineering, government (NGS 12)

Assess

Have students complete the Chapter 6 Test.

Reteach

Organize the class into six teams. Have each time write ten short-answer questions about the content of the chapter. Be sure that students' questions cover the content of all sections. Then pair up the teams and have them quiz each other with their questions.
ENGLISH LANGUAGE LEARNERS, COOPERATIVE LEARNING

Portfolio Extensions

1. Ask student to imagine they are archaeologists excavating a site once inhabited by one of the ancient civilizations discussed in this chapter. Have them create a list of objects they think they might find that would indicate what life was like in that civilization. Then have students write explanations or draw pictures of each item.
2. Have students conduct research on the Olympic Games in ancient Greece. Students should include information on why the games were held, what events took place at the games, and who was allowed to participate and view the games. Then have students design a commemorative stamp to celebrate the ancient games, or design an original medal that may be awarded to winning participants.

Chapter 6 Review Answers

Understanding History and Society

Presentations will vary but should include the effects of tool-making, agriculture, and trade on human civilization. Use Rubric 29, Presentations, to evaluate student work.

Thinking Critically

1. use of tools marked the beginning of technology necessary for development of civilization
2. All of these allowed people to keep records.
3. Possible answers: people will develop more advanced technology, further explore the universe, live peacefully
4. Similarities should include religion and achievements in the arts. Differences should include Rome's imperial government and conquests, and Greece's ideas and lack of political unity.
5. Possible answers: weak or corrupt rulers, war, economic problems, being conquered by other peoples

Chapter 6 Reviewing What You Know

Building Vocabulary

On a separate sheet of paper, write sentences to define each of the following words.

1. hominid
2. prehistory
3. nomads
4. land bridges
5. irrigation
6. civilization
7. division of labor
8. history
9. dynasty
10. pharaohs
11. hieroglyphics
12. city-states
13. direct democracy
14. republic
15. aqueduct

Reviewing the Main Ideas

1. *Human Systems* What are the four characteristics of civilization?
2. *Environment and Society* How did the development of agriculture lead to the growth of villages and towns?
3. *Environment and Society* How did the land protect Egypt from invaders?
4. *Environment and Society* Why did the city-states in ancient Greece develop differently from each other?
5. *Human Systems* What were some of the ancient Romans' achievements?

Understanding History and Society

The Road to Civilization

People of the early civilizations lived very differently from the first humans. Make a flow chart that shows the development of civilization from nomadic hunters and gatherers to city dwellers. As you prepare your presentation about human development, consider the following:

- How people made better tools.
- How agriculture changed society.
- How trade affected the way people lived.

Thinking Critically

1. **Drawing Conclusions** Why is the ability to make and use tools an important step in human development?
2. **Drawing Inferences** Why are the developments of a calendar, a system of counting, and a system of writing so important to a civilization?
3. **Predicting** How do you think civilization might change in the future?
4. **Comparing and Contrasting** How were the cultures of the ancient Greeks and the ancient Romans similar? How were they different?
5. **Identifying Cause and Effect** Give at least two reasons why civilizations decline.

Food Festival

Barley is one of the world's oldest grain crops. It was grown in ancient times and was a staple in both Roman and Greek diets. Today, pearl barley, which has had its hull and outer bran removed is popular in soups, porridges, and for making flour for flat breads. Have students find recipes for barley soup. Ask students to bring their soups to class to sample. You may wish to provide figs, dates, and raisins such as those that would have been found at a Roman banquet.

Chapter 6 Review and Assessment Resources

Reproducible
- Readings in World Geography, History, and Culture 29, 50, 57
- Critical Thinking Activity 6
- Vocabulary Activity 6

Technology
- Chapter 6 Test Generator (on the One-Stop Planner)
- Audio CD Program, Chapter 6
- HRW Go site

Reinforcement, Review, and Assessment
- Chapter 6 Review, pp. 130–131
- Chapter 6 Tutorial for Students, Parents, Mentors, and Peers
- Chapter 6 Test
- Chapter 6 Test for English Language Learners and Special-Needs Students

Building Social Studies Skills

Map Activity

On a separate sheet of paper, match the letters on the map with their correct labels.

Macedon	**Asia Minor**
Ionian Sea	**Athens**
Aegean Sea	**Crete**
Knossos	

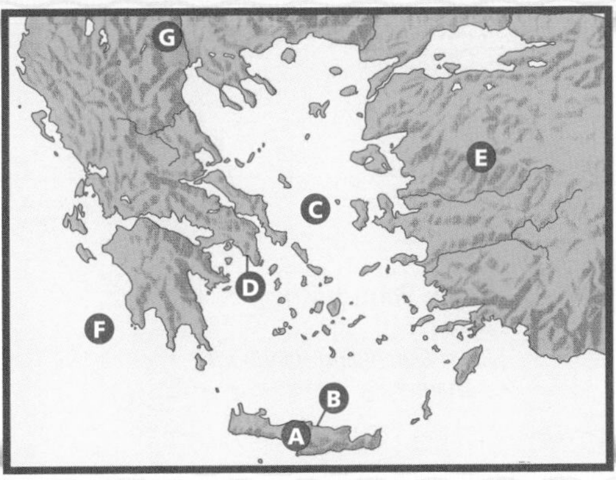

Mental Mapping Skills Activity

On a separate sheet of paper, draw a freehand map of the Fertile Crescent. Make a key for the map and label the following:

Tigris River	**Jordan River**
Euphrates River	

Writing Activity

The *Iliad,* a long poem by the ancient Greek poet Homer, tells the story of a great war. Write a short poem that tells the story of another event from ancient history. Your poem does not have to rhyme.

Alternative Assessment

Portfolio Activity

Learning About Your Local History

Early Settlers Research how your town or city was founded. On poster board, create a display that shows who the first people were to settle in your town.

internet connect

Internet Activity: **go.hrw.com**
KEYWORD: SK3 GT6

Choose a topic to explore about the ancient world:
- List different divisions of labor in early civilization.
- Report on Minoan civilization.
- Create a newspaper article about ancient athens.

Map Activity

A. Crete
B. Knossos
C. Aegean Sea
D. Athens
E. Asia Minor
F. Ionian Sea
G. Macedon

Mental Mapping Skills Activity

Maps will vary, but listed rivers should appear in their approximate locations.

Writing Activity

Poems will vary, but should accurately describe a historical event. Use Rubric 29, Poems and Songs, to evaluate student work.

Portfolio Activity

Posters will vary, but students should identify the first people who settled in your town or community. Use Rubric 28, Posters, to evaluate student work.

internet connect

GO TO: **go.hrw.com**
KEYWORD: **SK3 Teacher**
FOR: **a guide to using the Internet in your classroom**

CHAPTER 7

The World in Transition

Chapter Resource Manager

Objectives	Pacing Guide	Reproducible Resources
SECTION 1 **Empires in Asia** (pp. 133–36) **1.** Explain how the growth of Islam influenced Asian history. **2.** Name the dynasties that helped to shape China. **3.** Identify key events in Japan's history.	**Regular** 1 day **Block Scheduling** .5 day *Block Scheduling Handbook, Chapter 7*	**RS** Guided Reading Strategy 7.1 **RS** Graphic Organizer 7
SECTION 2 **The Middle Ages** (pp. 137–42) **1.** Define the Middle Ages. **2.** Describe society during the Middle Ages. **3.** Explain how the Middle Ages came to an end.	**Regular** 1 day **Block Scheduling** .5 day *Block Scheduling Handbook, Chapter 7*	**RS** Guided Reading Strategy 7.2 **SM** Map Activity 7
SECTION 3 **Renaissance and Reformation** (pp. 143–48) **1.** Identify the main interests of Renaissance scholars. **2.** Explain how people's lives changed during the Renaissance. **3.** Describe the changes that took place during the Reformation and Counter-Reformation.	**Regular** 1.5 days **Block Scheduling** .5 day *Block Scheduling Handbook, Chapter 7*	**RS** Guided Reading Strategy 7.3
SECTION 4 **The Age of Exploration and Conquest** (pp. 149–55) **1.** Describe the Scientific Revolution. **2.** Identify factors that triggered the Age of Exploration. **3.** Explain how the English monarchy differed from others in Europe. **4.** Describe the relationship between England and its American colonies.	**Regular** 1.5 days **Block Scheduling** .5 day *Block Scheduling Handbook, Chapter 7*	**RS** Guided Reading Strategy 7.4

Chapter Resource Key

- **RS** Reading Support
- **IC** Interdisciplinary Connections
- **E** Enrichment
- **SM** Skills Mastery
- **A** Assessment
- **REV** Review
- **ELL** Reinforcement and English Language Learners
- Transparencies
- CD–ROM
- Music
- Video
- go.hrw.com Internet
- Holt Presentation Maker Using Microsoft® Powerpoint®

One-Stop Planner CD–ROM

See the *One-Stop Planner* for a complete list of additional resources for students and teachers.

One-Stop Planner CD–ROM

It's easy to plan lessons, select resources, and print out materials for your students when you use the ***One-Stop Planner CD–ROM with Test Generator.***

Technology Resources	Review, Reinforcement, and Assessment Resources
One-Stop Planner CD–ROM, Lesson 7.1 *ARGWorld* CD–ROM Homework Practice Online HRW Go site	**ELL** Main Idea Activity 7.1 **REV** Section 1 Review, p. 136 **A** Daily Quiz 7.1 **ELL** English Audio Summary 7.1 **ELL** Spanish Audio Summary 7.1
One-Stop Planner CD–ROM, Lesson 7.2 *ARGWorld* CD–ROM Homework Practice Online HRW Go site	**ELL** Main Idea Activity 7.2 **REV** Section 2 Review, p. 142 **A** Daily Quiz 7.2 **ELL** English Audio Summary 7.2 **ELL** Spanish Audio Summary 7.2
One-Stop Planner CD–ROM, Lesson 7.3 *ARGWorld* CD–ROM Homework Practice Online HRW Go site	**ELL** Main Idea Activity 7.3 **REV** Section 3 Review, p. 148 **A** Daily Quiz 7.3 **ELL** English Audio Summary 7.3 **ELL** Spanish Audio Summary 7.3
One-Stop Planner CD–ROM, Lesson 7.3 *ARGWorld* CD–ROM Homework Practice Online HRW Go site	**ELL** Main Idea Activity 7.4 **REV** Section 3 Review, p. 155 **A** Daily Quiz 7.4 **ELL** English Audio Summary 7.4 **ELL** Spanish Audio Summary 7.4

internet connect

HRW ONLINE RESOURCES

GO TO: go.hrw.com
Then type in a keyword.

TEACHER HOME PAGE
KEYWORD: SK3 TEACHER

CHAPTER INTERNET ACTIVITIES
KEYWORD: SK3 GT7

Choose an activity to:
- write a report on daily life in the Middle Ages.
- create a biography of a Renaissance artist or writer.
- learn more about an explorer described in this chapter.

CHAPTER ENRICHMENT LINKS
KEYWORD: SK3 CH7

CHAPTER MAPS
KEYWORD: SK3 MAPS7

ONLINE ASSESSMENT
Homework Practice
KEYWORD: SK3 HP7
Standardized Test Prep Online
KEYWORD: SK3 STP7
Rubrics
KEYWORD: SS Rubrics

COUNTRY INFORMATION
KEYWORD: SK3 Almanac

CONTENT UPDATES
KEYWORD: SS Content Updates

HOLT PRESENTATION MAKER
KEYWORD: SK3 PPT7

ONLINE READING SUPPORT
KEYWORD: SS Strategies

CURRENT EVENTS
KEYWORD: S3 Current Events

Meeting Individual Needs

Ability Levels

Level 1 Basic-level activities designed for all students encountering new material

Level 2 Intermediate-level activities designed for average students

Level 3 Challenging activities designed for honors and gifted-and-talented students

English Language Learners Activities that address the needs of students with Limited English Proficiency

Chapter Review and Assessment

- **SM** Readings in World Geography, History, and Culture 66
- **SM** Critical Thinking Activity 7
- **REV** Chapter 7 Review, pp. 156–57
- **REV** Chapter 7 Tutorial for Students, Parents, Mentors, and Peers
- **ELL** Vocabulary Activity 7
- **A** Chapter 7 Test
- **A** Unit 2 Test
- Chapter 7 Test Generator (on the One-Stop Planner)
- Audio CD Program, Chapter 7
- Chapter 7 Test for English Language Learners and Special-Needs Students
- **A** Unit 2 Test for English Language Learners and Special-Needs Students

CHAPTER 7

LAUNCH INTO LEARNING

Ask students to suggest reasons why people travel today. *(Possible answers: for pleasure, for business, for religious purposes, to explore a new place)* Tell students that throughout history people have had similar reasons for travel. Some people were involved in the business of trading goods. Others went on pilgrimages to religious shrines and holy places. Still others went to search for and explore new lands in order to gain political and economic power. Point out that while all of these people were experiencing change away from home, the people, places, ideas, and institutions of the time were also changing. Tell students that they will learn about several periods of great change in Chapter 7.

Section 1

Objectives

1. Explain how the growth of Islam influenced Asian history.
2. Identify the dynasties that helped to shape Chinese history.
3. Describe some key events in Japan's history.

LINKS TO OUR LIVES

You might want to share with your students the following reasons for learning about these periods of change:

- Some challenges facing governments today are similar to those faced in the Middle Ages and the Renaissance.
- Conflicts that began during this period are still flaring up in parts of the world today.
- Certain patterns of change that began in this period, such as the shift of population from rural areas to urban, have continued to the present.
- The works of many Renaissance artists are still admired today.
- In some parts of the world, the national boundaries recognized today were drawn by colonizing powers during this period.

CHAPTER 7

The World in Transition

The next 2000 years in human history were a time of great change. In this chapter you will read about the growth of new empires, the development of new political systems, and the search for new ideas.

During the Middle Ages about 1,000 years ago, society was separated into distinct classes. Certain young women of the time who were born into noble families had to undergo training on how to behave.

Young girls from the families of lesser nobles often went to live in the households of higher-ranking noblewomen. There they would be trained in the skills and responsibilities that were expected of women in their rank.

Generally a young noblewoman was taught to sew, to weave, to cook, to play musical instruments, and to sing. She also learned the social conduct that was proper for women of the nobility. In some cases girls and young women were also instructed in the skills of household supervision.

Young noblewoman from the Middle Ages

Knight with female admirers

Papal palace in Avignon, France

FOCUS

LET'S GET STARTED

Write the following passage on the chalkboard: *What do you think the following statement means? Do unto all men as you would that they should do unto you.* Discuss responses. Point out that this passage comes from the Qur'an, the holy book of Islam, but that similar statements are found in the Bible and the writings of the Chinese philosopher Confucius. Tell students that in this section they will learn about the rise and history of Islam and about the great empires of China and Japan.

Building Vocabulary

Write the key terms on the board. Call on volunteers to find the definitions of the terms in the glossary and read them aloud. Ask students to identify those words that refer to people **(caliph, shah, daimyo, shogun,** and **samurai)**. You may wish to point out that *caliph* and ***mosques*** are Arabic words, *daimyo*, *shogun*, and *samurai* are Japanese words, and *shah* is a Persian word.

Section 1 Empires in Asia

Read to Discover

1. How did the growth of Islam influence Asian history?
2. What dynasties helped to shape Chinese history?
3. What were some key events in Japan's history?

Define

caliph
mosques
shah
daimyo
shogun
samurai

Locate

Arabian Peninsula
Mecca
Medina
Constantinople
China
Mongolia
Japan

WHY IT MATTERS

Asia is a continent of many different peoples and many different cultures. Use CNNfyi.com or other **current events** sources to find out about Asian cultures today. Record your findings in your journal.

A page from the Qur'an

Section 1 RESOURCES

Reproducible
- Block Scheduling Handbook, Chapter 7
- Guided Reading Strategy 7.1
- Graphic Organizer 7

Technology
- One-Stop Planner CD–ROM, Lesson 7.1
- Homework Practice Online
- HRW Go site

Reinforcement, Review, and Assessment
- Section 1 Review, p. 136
- Daily Quiz 7.1
- Main Idea Activity 7.1
- English Audio Summary 7.1
- Spanish Audio Summary 7.1

The Birth of Islam

By the year 600, the western Roman Empire had fallen apart. In Southwest Asia, however, a new empire was growing. It was based on a new religion called Islam. Muhammad, the founder of Islam, lived from about 570 to 632. He was born in Mecca, a city in the western part of the Arabian Peninsula. When Muhammad was about 40, he announced that he had been visited by an angel who told him to spread the word of God. The word *God* in Arabic is *Allah*. At first Muhammad was opposed by the leaders of Mecca. He left the city and traveled north to the city now called Medina. There Muhammad quickly gained many followers. People who follow the teachings of Islam are called Muslims. In 630 Muhammad and an army of his followers took control of Mecca. More people began to follow his teaching, and Islam began to spread through Southwest Asia.

The Teachings of Islam Muslims believe that Allah's message to Muhammad is contained in the Qur'an, Islam's holy book. It contains rules and instructions on how to lead a good life. Among other things, these rules require all Muslims to pray five times every day and to give money to the poor. The Qur'an also instructs Muslims to live humble lives, be tolerant, and avoid pork and alcoholic beverages.

The Kaaba in the Great Mosque at Mecca—one of Islam's holiest sites. Millions of Muslims visit it every year.

TEACH

Teaching Objective 1

ALL LEVELS: (Suggested time: 15 min.) Display or draw a map of Asia on the chalkboard. Call on volunteers to draw arrows on the map to show how Islam spread from Arabia to other parts of Asia, Africa, and Europe. Then lead a discussion about the influences of Islamic culture on societies that developed in these areas.

ENGLISH LANGUAGE LEARNERS

Teaching Objective 2

ALL LEVELS: (Suggested time: 45 min.) Organize the class into three groups and have each group prepare an argument for a debate about which early dynasty made the greatest contributions to Chinese history. One group should prepare arguments for the Sui dynasty, another for the T'ang dynasty, and the third for the Yuan dynasty. Encourage students to consult encyclopedias or other reference materials for additional information. Stage the debate in your classroom.

ENGLISH LANGUAGE LEARNERS, COOPERATIVE LEARNING

PLACES AND REGIONS

Constantinople lies between eastern Europe and Asia at the point that connects the Black Sea with the Mediterranean Sea. This has made it useful as both a barrier and as a bridge. It was an important center for religion, culture, and power for centuries. After the Ottoman Turks conquered the city in 1453 they changed its name to Istanbul.

Critical Thinking: Why was Constantinople such an important conquest for the Ottomans?

Answer: The city's position gave the Ottomans a foothold in Europe and a seaport.

Visual Record Answer

treating the patient's arm

This piece of art shows a Muslim doctor treating a patient.

Interpreting the Visual Record **What is the doctor in this image doing?**

The Spread of Islam By Muhammad's death in 632, his followers controlled most of the Arabian Peninsula. Soon after he died, however, his followers broke into two groups. They could not agree who should be **caliph**, or leader. The two groups eventually became known as the Sunni and the Shia.

This division did not slow the growth of Islam. Within about 100 years of Muhammad's death, Muslim armies had conquered the Arabian Peninsula, North Africa, Spain, and Persia. Many of the people who lived in these areas converted to Islam.

Muslim Civilization The Muslim Empire grew rich from trade with other parts of the world. Goods from Muslim lands were in demand across Europe, Africa, India, and China. Centers of learning like Córdoba and Toledo in Spain attracted scholars from around the world. Muslim doctors were skilled with surgery and medicines. Geographers drew beautiful maps, and mathematicians created the number system we use today. Throughout the Islamic world, architects built beautiful places of worship called **mosques**.

✓ **READING CHECK:** *Summarizing* What were some achievements of the Islamic Empire? extensive trade, centers of learning, medicine, maps, number system, mosques

internet connect

GO TO: go.hrw.com
KEYWORD: SK3 CH7
FOR: Web sites about the world in transition

Islamic Empires

During the 1300s a group of Muslim warriors captured most of what is now Turkey. They were called the Ottoman Turks. In 1453 they took the Christian city of Constantinople and made it their capital. The Ottoman Empire continued to grow in the 1500s under the leadership of Süleyman. Under his rule, the Ottomans ruled much of eastern Europe, western Asia, and northern Africa. Although it slowly lost power, the empire lasted until 1922.

Another Muslim kingdom, the Safavid Empire, included much of what is now Iran. It was founded about 1500 when a Shia Muslim ruler conquered Persia. He took the title **shah**, an ancient Persian word for king. The Safavids built their wealth on trade. Their capital, Esfahan, was one of the world's most beautiful cities. The Safavid Empire collapsed by 1736.

A third Muslim empire, the Mughal Empire, was founded in northern India in the 1500s. Its first leader was a Mongol leader called Babur the Tiger. One of the most famous Mughal rulers was Shah Jahan, who built the famous Taj Mahal in the mid-1600s. Soon after his reign, however, the Mughal Empire began to crumble. In the 1700s, the British began to take over Mughal lands.

✓ **READING CHECK:** *Finding the Main Idea* What were three major Islamic empires? Ottoman, Safavid, Mughal

The Safavids built this mosque in Persia. Many mosques are decorated with geometric designs.

Teaching Objective 3

All LEVELS: (Suggested time: 30 min.) Copy the following graphic organizer onto the chalkboard, omitting the italicized answers. Tell students it represents a time line of Japanese history. Organize the class into groups and have each group fill in the time line with key events that took place on each date. Then have the groups draw pictures to illustrate each event. **ENGLISH LANGUAGE LEARNERS, COOPERATIVE LEARNING**

1853	1603	1868
Japan is opened to the West.	*Tokygawa shogunate comes to power.*	*The Meiji Restoration brings emperor back to power.*

CLOSE

Have students look through the images in this section and choose one or two that they think best represent the history of Asian empires. Then call on volunteers to present their choices to the class and to explain the reasoning behind each choice. Encourage the class to discuss each of the pictures presented.

Empires in East Asia

Around the same time that Islamic leaders building empires in western Asia, powerful rulers were doing the same farther east. Beginning in about 500, great empires began to develop in China, Mongolia, and Japan.

China When the Han dynasty collapsed in 220, China broke up into several small kingdoms. The political system broke apart, and Chinese society was in disorder for many years. Finally the Sui dynasty reunited China in the late 500s. They ruled China for only a short time but achieved a great deal. Their greatest accomplishment was the Grand Canal, which linked north and south China. It is the oldest and longest canal system in the world.

The T'ang dynasty followed the Sui. They ruled China for about 300 years. The T'ang were powerful military leaders who saved China from invading Turks. Under the T'ang, China grew larger and more powerful. It became the world's most advanced country. Great works of art and literature were produced during this period.

When the T'ang dynasty collapsed, the Sung came to power. They were great inventors. During their rule, the Chinese developed gunpowder and movable type, which allowed them to mass produce books. They were also experts in irrigating farmlands.

The Mongols were known as skilled warriors. They fought mostly from horseback.

The Mongol Empire An army of Mongols under Genghis Khan invaded China from the north in the early 1200s. They swept into China from what is now called Mongolia. They were skilled warriors who fought on horseback. Under Genghis Khan, the Mongols captured the city now called Beijing.

Two grandsons of Genghis Khan added to the Mongol Empire. One, named Batu, led an army into Europe around 1240. His army was called the Golden Horde. They reached as far as Vienna before they were turned back. The other grandson, Kublai Khan, completed the conquest of China and founded the Yuan dynasty. Kublai Khan was a wise and powerful ruler. He improved communications by creating a system of fast-moving **couriers**, or messengers on horseback. During his reign, contact between China and Europe increased. Kublai Khan even allowed Italian trader Marco Polo to become his special representative within the empire. When Kublai Khan died in 1294, he was followed by weak rulers. The Chinese rebelled. The Yuan dynasty fell in 1368.

✓ **READING CHECK:** *Analyzing Information* What were some achievements of the Mongol Empire? **conquest of China and parts of Europe, courier system, contact with Europe**

The Grand Canal, the great project of the Sui Dynastry, was actually completed by the Mongols in the 1200s. The canal, which has been extended since the 1950s, is still used today.

Across the Curriculum
LITERATURE
Japanese Haiku

Many forms of poetry have developed in various societies around the world. Each culture's poetry has distinctive sounds and styles. Japanese haiku poems, for example, follow a very rigid form. Consisting of only three lines, the first line in the haiku contains five syllables, the second contains seven, and the third has five.

Traditional Japanese haiku deals with nature and contains strong, suggestive imagery. You may wish to locate examples of traditional haiku and read them to your class.

Activity: Have students write their own haiku poems in English on subjects of their choice. Call on volunteers to read their poems aloud to the class.

REVIEW AND ASSESS

Have students complete the Section Review. Then have each student write four fill-in-the-blank questions about people or events in the section. Have students exchange papers with a partner and answer the questions they receive. Then have students complete Daily Quiz 7.3.

RETEACH

Have students complete Main Idea Activity 7.1. Then organize students into groups and assign each group an Asian empire. Have each group prepare an oral report to give to the class describing life in its assigned empire. **ENGLISH LANGUAGE LEARNERS, COOPERATIVE LEARNING**

EXTEND

Have interested students conduct research on the opening of Japan by Commodore Perry. Ask them to interpret how the Japanese responded to the Americans and how the Americans reacted to the Japanese. Have students report their findings in a news article written for either an American or a Japanese newspaper of the time. **BLOCK SCHEDULING**

Section 1 Review

Answers

Define For definitions, see: caliph, p. 133; mosques, p. 134; shah, p. 134; daimyo, p. 136; shogun, p. 136; samurai, p. 136

Reading for the Main Idea

1. Many people in areas conquered by Muslim armies converted to Islam. (NGS 13)
2. developed movable type that allowed for mass production of books; developed gunpowder; irrigated farmland (NGS 13)
3. The emperor had little power; real control was in the hands of shogun and daimyo. (NGS 10)

Critical Thinking

4. Japanese leaders thought it needed to be protected from outside influences. Americans saw it as a potential trading partner.

Organizing What You Know

5. Sui—reunited China; built Grand Canal; T'ang—enlarged China; made it strong and powerful; great art and literature; Sung—new inventions, irrigation; Yuan—improved communications; contact with Europe

Samurai warriors wore elaborate suits of armor.

Japan Until the early 300s Japan had no central government. At that time, however, the first Japanese emperor came to power. The first emperor was a member of the Yamato clan. Since the 300s, every Japanese emperor has been a member of this same imperial family. For many years, Japan's government was very much like China's. Beginning in the 800s, however, a new system began to develop. Although the emperor remained the head of government, real power was held by local lords called **daimyo**. The most powerful daimyo was sometimes given the title **shogun**. The daimyo and the shogun were protected by hired warriors called **samurai**. Under this system Japan had no strong central government. Local wars were common.

This changed in 1603 when the Tokugawa shogunate, or dynasty, came to power. Tokugawa leaders built a strong central government by limiting the power of other daimyo. Because they feared that Europeans would threaten their society, the Tokugawa did not allow European traders into Japan. Japanese people were not allowed to travel outside the country.

In 1853 American warships under Commodore Matthew Perry arrived in Japan. Perry had been sent by President Millard Fillmore to trade with Japan. The shogun agreed to begin trading the next year. Soon Japan opened its ports to European countries as well.

In 1868 a group of samurai overthrew the last shogun. They restored political power to the emperor. This political revolution is called the Meiji Restoration. *Meiji* means "enlightened rule." After this revolution, Japan entered a new modern age. Japan's old class system was abandoned. A new constitution passed in 1899 gave more Japanese a say in the government. The economy also began to modernize. By 1900 Japan had become the first industrialized Asian country.

✓ **READING CHECK:** *Summarizing* What was the Meiji Restoration?
revolution by samurai to overthrow shogun and restore emperor

go.hrw.com Homework Practice Online
Keyword: SK3 HP7

Section 1 Review

Define and explain: caliph, mosques, shah, daimyo, shogun, samurai

Reading for the Main Idea

1. *Human Systems* How did Islam spread through Southwest Asia?
2. *Human Systems* How did the Sung dynasty improve life in China?
3. *Human Systems* What was the government of Japan like after 800?

Critical Thinking

4. **Identifying Points of View** How do you think Japanese leaders and American traders viewed Japan differently?

Organizing What You Know

5. **Categorizing** Copy the following graphic organizer. Use it to describe the major dynasties of China.

Dynasty	Achievements
Sui	
Tang	
Sung	
Yuan	

Section 2

Objectives

1. Define the Middle Ages.
2. Describe society in the Middle Ages.
3. Explain how the Middle Ages came to an end.

Focus

LET'S GET STARTED

Copy the following question onto the board: *Why do you think the period of European history following the fall of the Roman Empire is called the Middle Ages?* Call on volunteers to share their answers. Point out that the name stems from the position of the period between the ancient and modern worlds. Tell students that they will learn more about the Middle Ages in this section.

Building Vocabulary

Write the key terms on the board. Call on volunteers to find the definitions of the terms in the glossary and to read them to the class. Then ask students write sentences that link pairs of related words. *(For example,* ***vassals*** *received* ***fiefs*** *from their lords.* ***Knights*** *practiced* ***chivalry****.* ***Clergy*** *and* ***cathedrals*** *were both parts of the Catholic Church.)*

Section 2 The Middle Ages

Read to Discover

1. What were the Middle Ages?
2. What was society like during the Middle Ages?
3. How did the Middle Ages come to an end?

Define

feudalism, nobles, fief, vassals, knight, chivalry, manors, serfs, clergy, cathedrals, Crusades, middle class, vernacular

Locate

France, England, Normandy, Norway, Denmark, Sweden, Jerusalem, Spain

A medieval knight

Why It Matters

The desire to control Jerusalem, which led to the Crusades, is still a cause of conflict. Use CNNfyi.com or other **current events** sources to learn what is happening in Jerusalem today. Record your findings in your journal.

Section 2 Resources

Reproducible

- Block Scheduling Handbook, Chapter 7
- Guided Reading Strategy 7.2
- Graphic Organizer 7
- Map Activity 7

Technology

- One-Stop Planner CD–ROM, Lesson 7.2
- Homework Practice Online
- HRW Go site

Reinforcement, Review, and Assessment

- Main Idea Activity 7.2
- Section 2 Review, p. 142
- Daily Quiz 7.2
- English Audio Summary 7.2
- Spanish Audio Summary 7.2

The Rise of the Middle Ages

In the year A.D. 476, the last of the western Roman emperors was defeated by invading Germanic tribes. These invaders from the north brought new ideas and traditions that gradually developed into new ways of life for people in Europe. Historians see the years between the last of the Roman emperors and the beginnings of the modern world in about 1500 as a period of change. Because it falls between the ancient and modern worlds, this time in history is called the Middle Ages or the medieval period. *Medieval* comes from the Latin for "middle age."

The time from the 400s to around 1000 is known as the Early Middle Ages. As this period began, the Roman system of laws and government had broken down. Western Europe was in a state of disorder. It was divided into many kingdoms ruled by kings who had little authority. For example, Britain was largely controlled by two Germanic tribes, the Angles and the Saxons. These groups had established several independent kingdoms.

Nobles in the Middle Ages built their own castles for protection.

TEACH

Teaching Objective 1

ALL LEVELS: (Suggested time: 10 min.) Have each student write a brief, dictionary-style definition of the term *Middle Ages.* Remind students that their definitions should note both the appropriate time period and the geographic area covered by the Middle Ages. When they have finished, have students compare their definitions with the definition in a classroom dictionary. Then call on volunteers to name characteristics that define the European Middle Ages. **ENGLISH LANGUAGE LEARNERS**

TEACHER TO TEACHER

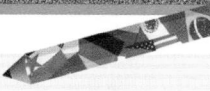

Scott Whitlow of Round Rock, Texas, suggests the following activity to help students better understand the Middle Ages. Ask students to name some images that come to mind when they think of the Middle Ages. Copy their list of images onto the chalkboard. *(Students will likely mention castles, knights in armor, princesses, swords, and so on.)* Refer to this list as you go through the main ideas in this section. At the end of the lesson, ask students if and how their ideas about the Middle Ages have changed.

Across the Curriculum

Weaponry

The Franks were famous for their swords, which were in great demand across Europe. These swords were known for their balance and strength. Frankish foot soldiers used short swords and carried spears, bows and arrows, and shields. Soldiers who rode on horseback carried both long and short swords, spears, and round shields. Charlemagne's mounted soldiers also wore simple helmets and other basic forms of armor, representing what may be the early stages of the armored knight of later centuries.

Activity: Have students conduct research on the weapons of the Middle Ages. Ask them to make sketches of some of the weapons used during the period.

Visual Record Answer

strength, intelligence, dedication

Map Answer

the Vikings

Charlemagne (742-814) united most of the Christian lands in western Europe. He built libraries and supported the collection and copying of Roman books. His rule became a model for later kings in medieval Europe.

Interpreting the Visual Record
What qualities made Charlemagne a good king?

internet connect
GO TO: go.hrw.com
KEYWORD: SK3 CH7
FOR: Web sites about the world in transition

The Franks The Franks were one of the Germanic tribes that moved into western Europe. In the 490s, Clovis, the king of the Frankish tribes, became a Christian and gained the support of the church. He conquered other Frankish tribes and won control of the territory of Gaul. Today this area is called France after the Franks. In 732 a Frankish army under King Charles Martel held off an army of Spanish Moors who had invaded his kingdom. This conflict is called the Battle of Tours. This defeat drove the Muslim Moors back into Spain, creating a border between the Christian and Muslim worlds.

The greatest Frankish king, Charlemagne (SHAR-luh-mayn), ruled from 768 to 814. A strong, smart leader, Charlemagne established schools and encouraged people to learn to read and write. His greatest accomplishment was to unite most of western Europe under his rule. Charlemagne's empire included most of the old Roman Empire plus some new additional territory. It later became known as the Holy Roman Empire because the pope declared Charlemagne "Emperor of the Romans."

After Charlemagne died, his grandsons weakened the empire by dividing it among themselves. Muslims invaded from the south. Slavs invaded from the east. From the north came the dreaded Vikings.

The Vikings During the 800s and 900s the Vikings were feared throughout western Europe. They came from what are now the countries of Denmark, Norway, and Sweden. The Vikings were not only farmers but also skilled sailors and fierce warriors who raided towns along the coasts of Europe. Eventually these invaders settled in England, Ireland, and other parts of Europe. A large Viking settlement in northwestern France gave that region its name. It is called Normandy, from the French word for "Northmen." The Vikings there came to be known as the Normans.

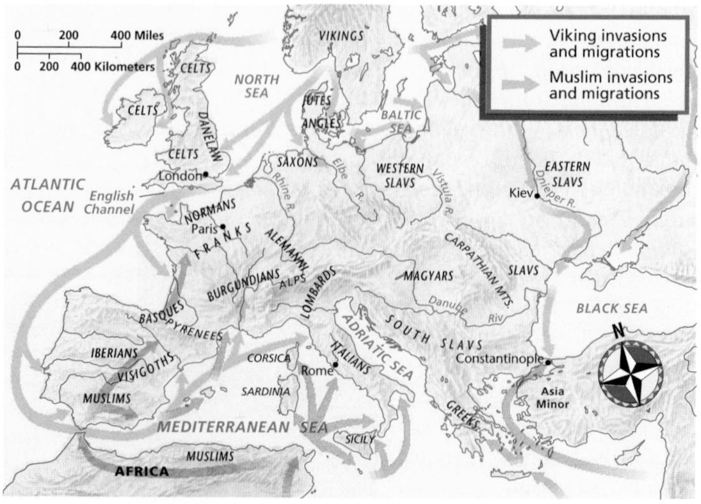

Between A.D. 600 and 1000, many people invaded western Europe.

Interpreting the Map **What group of invaders were most active in the Mediterranean area?**

Teaching Objective 2

ALL LEVELS: (Suggested time: 20 min.) Copy the following graphic organizer onto the board, omitting the italicized answers. Have each student complete the organizer with factors that led to the Great Depression. Ask volunteers to share their answers with the class. Use their completed organizers to lead a discussion about the beginnings of the Great Depression. **ENGLISH LANGUAGE LEARNERS**

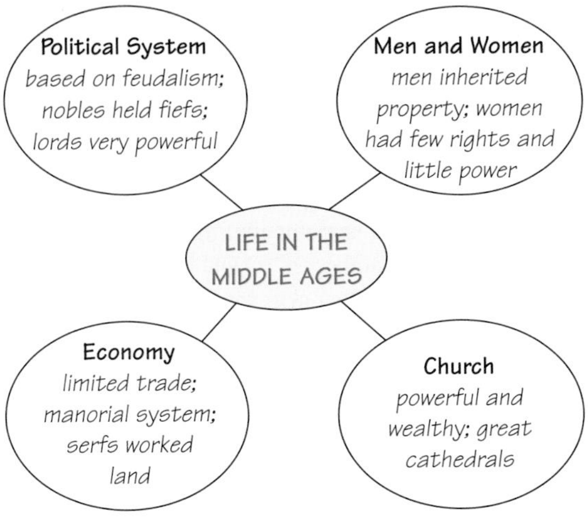

Political System
based on feudalism; nobles held fiefs; lords very powerful
Men and Women
men inherited property; women had few rights and little power
LIFE IN THE MIDDLE AGES
Economy
limited trade; manorial system; serfs worked land
Church
powerful and wealthy; great cathedrals

Life in the Middle Ages

Feudalism Within 100 years of Charlemagne's death, the organized central government he had put in place was gone. By the 900s most of Europe was governed by local leaders under a system known as **feudalism**. It was a way of organizing and governing people based on land and service. In most feudal societies, the king, who owned all the land in his kingdom, granted some lands to **nobles**—people who were born into wealthy, powerful families. The grant of land was called a **fief**. Nobles had complete power over their land—power to collect taxes, enforce laws, and maintain armies. In return for land, they became **vassals** of the king. This means that they promised to serve the king, especially in battle. A noble could, in turn, grant fiefs to lesser nobles. In so doing, that noble would become a lord, and the lesser nobles would become his vassals. A vassal owed service—especially military service—to his lord.

This stained-glass window shows a lord and his vassals.

Interpreting the Visual Record
What details in the picture show that the lord is more powerful than the vassals?

Feudalism was a very complex system. Its rules varied from kingdom to kingdom. Feudal relationships in France, for example, were not the same as those in Germany. The relationships between kings and nobles in England were very different from those in either France or Germany. In addition, the nature of feudal relationships were constantly changing. Laws that governed a king's or a vassal's behavior one year might not apply just a few years later. It was sometimes very difficult, even during the Middle Ages, for people to keep track of their feudal obligations.

Nevertheless, powerful lords were the ruling class in Europe for more than 400 years. Some lords were so powerful that the king remained on the throne only with their support. Over time it became the custom that the owner of a fief would pass his land on to his son. By about 1100 the custom was that the eldest son inherited his father's land. Women had few rights when it came to owning property. If a woman who owned land married, her husband gained control of her land.

Knights The most common type of nobleman was the **knight**, or warrior, who received land from a lord in return for military service. Knights lived by a code of behavior called **chivalry**. This code said that a knight had to be brave, fight fairly, be loyal, and keep his word. He had to treat defeated enemies with respect and be polite to women. In battle, a knight wore heavy metal armor and a metal helmet. He carried a sword, a shield, a lance, and other weapons. Knights had plenty of opportunities to fight. In addition to large-scale wars that occurred during the Middle Ages, frequent smaller battles took place between lords who tried to seize each other's lands.

To become a knight, a boy usually had to come from a noble family. Boys began their training at the age of seven.

Interpreting the Visual Record **How do the knights in this picture look, and how do you think they feel?**

Across the Curriculum
LITERATURE
King Arthur

Legends King Arthur is a mythical figure from the Middle Ages. Legend tells that Arthur was the son of a king named Uther but was raised by one of the king's vassals, Sir Ector, who never told the boy of his ancestry. Arthur became king by pulling a magic sword out of a stone. At his favorite castle, Camelot, he presided over his loyal knights of the Round Table.

Sir Thomas Malory wrote *Le Morte d'Arthur (The Death of Arthur)* around 1469. Since then, many poems and stories have been written about the adventures of King Arthur and his knights. These include T.H. White's *The Once and Future King,* the movie *The Sword in the Stone,* and the Broadway musical *Camelot.*

Discussion: Ask students to share stories about King Arthur with which they are familiar. Lead a discussion about historical truths about the Middle Ages that can be seen in these tales.

Visual Record Answer

king sitting on a throne and wearing a crown; vassals at a lower level

Visual Record Answer

look strong and brave; possibly feel proud, excited, a little scared

LEVELS 2 AND 3: (Suggested time: 45 min.) (Suggested time: 45 min.) Organize the class into groups. Ask each group to write a skit in which group members play the parts of figures in medieval society, such as kings, nobles, knights, and serfs. Tell students that their skits should demonstrate the ways in which various members of society interacted with one another. Call on volunteers to perform their skits for the class. COOPERATIVE LEARNING

Teaching Objective 3

ALL LEVELS: (Suggested time: 25 min.) Call on students to find in the text some reasons the Middle Ages ended. Ask volunteers to come forward and write these reasons on the board. *(Reasons include the growth of cities, increased interest in education and trade, a rise in the power of kings, and the weakening of the Church.)* Then lead a class discussion about how each of the given reasons contributed to the decline of the Middle Ages. ENGLISH LANGUAGE LEARNERS

Human Systems

The Crusades brought Europeans and their customs to the Holy Land. After the capture of Jerusalem, the crusaders set up four small states. They introduced feudalism and subdivided the land among feudal lords. Trade between Europe and the Holy Land sprang up. Italian ships carried goods back and forth.

The European occupiers were affected as well. Christians and Muslims lived alongside each other and grew to respect each other. Many Europeans adopted Eastern customs and began to wear Eastern clothes and eat Eastern foods.

Activity: Ask students to imagine that they are Europeans living in one of the Crusader States. Have them write letters to family members back home comparing and contrasting life in the Holy Land with life in Europe.

Visual Record Answer

hard work, simple tools, no machinery

This picture, made during the 1400s, shows peasants at work during a harvest.

Interpreting the Visual Record **What do you think farming was like in the Middle Ages?**

The Manorial System Trade declined after the end of the Roman Empire. Most people took up farming for a living. The large farm estates which some nobles developed were called **manors**. Such manors included large houses, farmed lands, wooded land, pastures, fields, and villages. The lord of a manor ruled over peasants called **serfs** who lived on his land. Serfs were poor and had no rights. They had to work the lord's land and give him part of their crops. They could not leave the manor without the lord's permission. A manor was usually self-sufficient. Almost everything people needed, including food and clothing, was produced right there.

The Church One of the largest and wealthiest landowners during the Middle Ages was the Catholic Church. Headed by the pope, the church was enormously powerful, with its own laws and courts.

Officials of the church were known as the **clergy**. Beneath the pope were bishops and priests. Other members of the clergy were monks, who lived in monasteries, and nuns, who lived in convents. While most ordinary people could not read or write, many members of the clergy were educated. In monasteries, monks prayed, studied, and copied ancient books.

Eventually, huge churches, called **cathedrals**, were built. Cathedrals cost a great deal of money and were beautifully decorated. Some of the most common decorations were elaborate stained-glass windows.

The Crusades In the late 1000s, the pope asked the lords of Europe to join in a great war against the Turks, who had gained control of Palestine, which the Christians called the Holy Land. This war turned into a long series of battles called the **Crusades**. The First Crusade lasted from 1095 to 1099. Crusaders captured Jerusalem and killed many of the Muslims and Jews who lived there. However, over the next 100 years, the Turks won back the land they had lost. Three more major Crusades were launched. Although the Holy Land was not recaptured, the Crusades led to important changes in Europe.

✓ **READING CHECK:** *Summarizing* How did feudalism make the nobles and their vassals depend on each other? Lords needed soldiers; vassals needed a means of support.

Knights in the Middle Ages often fought from horseback.

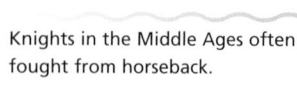

Teaching Objectives 1–3

ALL LEVELS: (Suggested time: 35 min.) Organize students into groups and have each group construct a time line that illustrates one aspect of the Middle Ages. For example, one group might examine political developments while another looks at military history. Encourage group members to conduct research to find additional material for their time lines. Display the completed time lines in the classroom.

ENGLISH LANGUAGE LEARNERS, COOPERATIVE LEARNING

CLOSE

Have students write newspaper headlines about events that occurred or might have occurred—based on students' knowledge of daily life—during the Middle Ages. You may wish to show students examples of well-written headlines from current newspapers. Call on volunteers to share their headlines with the class.

The Bayeaux Tapestry, part of which is shown here, tells the story of the Norman Conquest of England.

The High Middle Ages

The Crusades brought about major economic and political changes in Europe. The period following the Crusades to about 1300 is known as the High Middle Ages.

Stronger Nations Many lords sold their lands to raise money in order to join the Crusades. Without land, they had no power. In addition, many lords died in the Crusades. With fewer powerful lords, kings grew stronger. By the end of the Middle Ages, England, France, and Spain had become powerful nations. Strong central governments and the decline of the nobility's power helped to bring about the end of feudalism in Europe.

In 1066 William, Duke of Normandy in northwestern France, claimed the English throne. He landed in England, defeated the Anglo-Saxon army, and was crowned King William I of England. He became known as William the Conqueror. William built a strong central government in England. When William's great-grandson John took the throne, however, he pushed the nobles too far by raising taxes. In 1215 a group of nobles forced King John to sign Magna Carta, one of the most important documents in European history.

People exchanged goods at trade fairs in the Middle Ages.

Magna Carta stated that the king could not collect new taxes without the consent of the Great Council, a body of nobles and church leaders. The king could not take property without paying for it. Any person accused of a crime had the right to a trial by jury. The most important provision of Magna Carta was that the law, not the king, is the supreme power in England. The king had to obey the law. The Great Council was the forerunner of England's Parliament, which governs Great Britain today.

The Growth of Trade The Crusades increased Europeans' demand for Asian dyes, medicines, silks, and spices. People also began buying lemons, apricots, melons, rice, and sugar from Asia. In exchange, Europeans traded timber, leather, wine, glassware, and woolen cloth.

Increased trade led to the growth of manufacturing and banking. A **middle class** of merchants and craftsmen arose between the nobility and the peasants. The late medieval economy formed the basis for our modern economic system.

Across the Curriculum

MATH

Trade Fairs Fairs in the Middle Ages were colorful events. They were also the main method of carrying out trade. Because of fairs, easier ways to trade developed.

For example, some goods were sold by length or weight. Because people came from all over Europe to trade fairs, a standard system of weights and measures was needed. The troy weight was set to weigh gold and silver. This weight, named for the town of Troyes, France, is still used today.

Trade fairs also helped develop the bill of exchange. This note was a written promise to pay a sum of money at a later time.

Activity: Have students imagine that they are cloth traders at a fair in the Middle Ages. One meter of cloth is worth 3 ducats (a currency of the time). The spice trader nearby sells pepper at 12 ducats per ounce. How much cloth would they need to sell to buy 4 ounces of pepper? *(16 meters)*

REVIEW AND ASSESS

Have students complete the Section Review. Then have students work in pairs to review the section material. Have each student write one question about each of the subjects discussed in the text to ask his or her partner. Then have students complete Daily Quiz 7.2.

RETEACH

Have students complete Main Idea Activity 7.2. Then organize the class into groups. Tell each group to create a two-column chart with the headings "Early Middle Ages" and "High Middle Ages." Have students fill in the space beneath each heading with information about each period. Ask volunteers to write their charts on the board.

EXTEND

Have interested students conduct research on the art, music, or literature of the Middle Ages. Encourage students to choose a specific topic to research, such as stained glass windows or medieval instruments. Have each student present his or her findings in a report or poster.

Section 2 Review

Answers

Define For definitions, see: feudalism, 139; nobles, 139; fief, 139; vassals, 139; knight, 139; chivalry, 139; manor, 140; serf, 140; clergy, 140; cathedrals, 140; Crusades, 140; middle class, 141; vernacular, 142

Reading for the Main Idea

1. refers to time between classical age and modern world (NGS 10)
2. feudalism; powerful lords sometimes more powerful than kings; women had few rights; little travel or trade; manorialism; lords owned self-sufficient manors; serfs worked land; most people not educated; church very powerful (NGS 11
3. breakdown of feudalism and the manorial system, weakening of church power; growth of trade, growth of cities; growth of universities; stronger kings and nations (NGS 13)

Critical Thinking

4. took supreme power away from a king, and made the law the supreme power; idea has influenced national governments in modern times

Organizing What You Know

5. noble: collect taxes, enforce laws, maintain army
   vassal: serve king or lord, especially in battle

Between 1347 and 1351 the Black Death killed about one third of Europe's entire population.

A lecturer teaches at a university during the Middle Ages. Medieval universities taught religion, the liberal arts, medicine, and law.

The Growth of Cities As Europe's economy got stronger, cities grew. Centers for trade and industry, cities attracted merchants and craftsmen as well as peasants, who hoped to find opportunities for better lives and more freedom. Both the manorial system and the feudal system began to fall apart.

During the Middle Ages, cities were crowded and dirty. When disease struck, it spread rapidly. In 1347 a deadly disease called the Black Death swept through Europe. Even this disastrous plague, however, had some positive effects. With the decrease in population came a shortage of labor. This meant that people could begin to demand higher wages for their work.

Education and Literature Most people could not read or write. As cities grew and trade increased, so did the demand and need for education. Between the late 1000s and the late 1200s, four important universities developed in England, France, and Italy. By the end of the 1400s, many more universities had opened throughout Europe.

Most people during the Middle Ages did not speak, read, or write Latin, the language of the Church. They spoke **vernacular** languages—everyday speech that varied from place to place. Writers such as Dante Alighieri in Italy and Geoffrey Chaucer in England began writing literature in vernacular languages. Dante is best known for *The Divine Comedy.* Chaucer's most famous work is *The Canterbury Tales.*

The End of the Middle Ages The decline of feudalism and the manorial system, the growth of stronger central governments, the growth of cities, and a renewed interest in education and trade brought an end to the Middle Ages. In addition, stronger kings challenged the power of the Catholic church. By the end of the 1400s, a new age had begun.

✓ **READING CHECK:** *Finding the Main Idea* Why did kings become more powerful after the Crusades? Lords sold land to raise money for the Crusades; many died in battle.

Section 2 Review

go.hrw.com Homework Practice Online
Keyword: SK3 HP7

Define and explain: feudalism, nobles, fief, vassals, knight, chivalry, manors, serfs, clergy, cathedrals, Crusades, middle class, vernacular

Reading for the Main Idea

1. *Human Systems* Why is the time between the A.D. 400s and about 1500 called the Middle Ages?
2. *Human Systems* What was life like in the Middle Ages?
3. *Human Systems* What led to the end of the Middle Ages?

Critical Thinking

4. **Evaluating** Why is Magna Carta considered one of the most important documents in European history?

Organizing What You Know

5. **Analyzing** Copy the following graphic organizer. Use the right-hand column to describe briefly each person's responsibility.

Person	Responsibility
noble	
vassal	

Section 3

Objectives

1. Identify the main interests of Renaissance scholars.
2. Explain how people's lives changed during the Renaissance.
3. Explain how the English monarchy differed from others in Europe.
4. Describe changes that occurred during the Reformation and Counter-Reformation.

Focus

LET'S GET STARTED

Copy the following instructions onto the chalkboard: *Look at the paintings from the Middle Ages in Section 2. Then look at the artwork from the Renaissance in Section 3. How are they different?* Discuss responses. Explain that art in the Renaissance became more sophisticated and realistic. Tell students that they will learn more about the Renaissance in this section.

Building Vocabulary

Write the key terms on the chalkboard. Underline *human* in **humanists.** Point out that humanist thinkers in the Renaissance stressed the importance of human reason. Then underline *protest* and *reform* in **Protestants** and **Reformation.** Explain that the Protestants of the Reformation protested against certain teachings and wanted to reform the Catholic Church. Point out that *counter-* in **Counter-Reformation** means "against." Tell students that this movement was staged against the effects of the Reformation. Call on a volunteer to read the definitions of all the key terms.

Section 3 The Renaissance and Reformation

Read to Discover

1. What were the main interests of Renaissance scholars?
2. How did people's lives change during the Renaissance?
3. What changes took place during the Reformation and Counter-Reformation?

Define

Renaissance
humanists
Reformation
Protestants
Counter-Reformation

Locate

Rome
Florence

Why It Matters

Works of Renaissance art have remained popular for hundreds of years. Use CNNfyi.com or other **current events** sources to learn about Renaissance paintings displayed in museums around the world today. Record your findings in your journal.

Leonardo da Vinci's sketch of a flying machine

Section 3 Resources

Reproducible

- Guided Reading Strategy 7.3

Technology

- One-Stop Planner CD–ROM, Lesson 7.3
- Homework Practice Online
- HRW Go site

Reinforcement, Review, and Assessment

- Main Idea Activity 7.3
- Section 3 Review, p. 148
- Daily Quiz 7.3
- English Audio Summary 7.3
- Spanish Audio Summary 7.3

New Interests and New Ideas

The Crusades and trade in distant lands caused great changes in Europe. During their travels, traders and Crusaders discovered scholars who had studied and preserved Greek and Roman learning. While trading in Southwest Asia and Africa, people learned about achievements in science and medicine. Such discoveries encouraged more curiosity. During the 1300s, this new creative spirit developed and sparked a movement known as the **Renaissance** (re-nuh-SAHNS). This term comes from the French word for "rebirth." The Renaissance brought fresh interest in exploring the achievements of the ancient world, its ideas, and its art.

The powerful Medici family ruled Florence for most of the Renaissance. Banker Cosimo de' Medici, seen here, was a great supporter of the arts.

Beginning of the Renaissance The Renaissance started in Italy. Italian cities such as Florence and Venice had become rich through industry and trade. Among the population was a powerful middle class. Many members of this class were wealthy and well-educated. They had many interests beside their work. Many studied ancient history, the arts, and education. They used their fortunes to support painters, sculptors, and architects, and to encourage learning. Scholars revived the learning of ancient Greece and Rome. Enthusiasm for art and literature increased. Over time the ideas of the Renaissance spread from Italy into other parts of Europe.

The Humanities As a result of increased interest in ancient Greece and Rome, scholars encouraged the study of subjects that had been taught in ancient Greek and Roman schools. These subjects, including history, poetry, and grammar, are called the humanities.

TEACH

Teaching Objective 1

LEVELS 1 AND 2: (Suggested time: 15 min.) Write the following questions on the chalkboard: *What events and discoveries sparked the Renaissance? What subjects were scholars interested in during this period?* Tell students to write down their responses. Then call on volunteers to share their answers and discuss them in class. Lead a class discussion about the ideas and beliefs that inspired Renaissance artists.
ENGLISH LANGUAGE LEARNERS

LEVEL 3: (Suggested time: 30 minutes) Have pairs of students work together to write fictional interviews with Renaissance figures like Michelangelo or Leonardo da Vinci. Tell students that their interviews should note the ideas that inspired their subjects to create the works for which they are famous. Call on volunteers to present their interviews in class and to discuss how their subjects represented the Renaissance spirit.
COOPERATIVE LEARNING

USING ILLUSTRATIONS

Have students examine the painting by Peter Breughel on this page. Ask them why the subject matter of the painting is unusual. *(Possible answer: It is unusual for a painting to focus on children at play.)* Also ask how the painting reflects Renaissance ideas. *(It shows normal people in their daily lives; it is realistic; it shows people enjoying themselves.)*

Activity: Have students find other examples of paintings by Breughel or other Renaissance artists. Ask them to write short descriptions of the works and how they reflects Renaissance ideas. Have students display the paintings and read their descriptions in class.

Humanists, the people who studied these subjects, were practical. They wanted to learn more about the world and how things worked. Reading ancient texts helped them recover knowledge that had been forgotten or even lost. They believed that people should support the arts. They also thought that education was the only way to become a well-rounded person. People were urged to focus on what they could achieve in this life.

✓ **READING CHECK:** *Cause and Effect* How did the Renaissance begin?
Trade and Crusades bring people into contact with ideas; curiosity about the world awakened.

Isabella d'Este was a very intelligent and powerful member of a wealthy Italian noble family. She was educated in languages and poetry. She supported the arts and hired noted architects to design parts of her palace.

The Creative Spirit

During the Renaissance, interest in painting, sculpture, architecture, and writing was renewed. Inspired by Greek and Roman works, artists produced some of the world's greatest masterpieces for private buyers as well as for churches and other public places.

Art Leonardo da Vinci and Michelangelo truly represented the Renaissance. Leonardo achieved the Renaissance ideal of excelling in many things. He was not only a painter but also an architect, engineer, sculptor, and scientist. He sketched plants and animals. He made detailed drawings of a flying machine and a submarine. He used mathematics to organize space in his paintings and knowledge about the human body to make figures more realistic. Michelangelo was not only a brilliant sculptor, but also an accomplished painter, musician, poet, and architect.

Northern European merchants carried Italian paintings home, and painters went from northern Europe to study with Italian masters. In time, Renaissance ideas spread into northern and western Europe.

This painting by Pieter Brueghel shows children's games in the 1500s. Many Renaissance painters chose to focus their attention on daily life.
Interpreting the Visual Record **What activities do you see that children still do today?**

Visual Record Answer

Possible answers: somersaults, tag, hoop spinning, tug-of-war, hanging from a crossbar, leap-frog

Teaching Objective 2

LEVEL 1: (Suggested time: 20 min.) Organize the class into groups and assign each group one aspect of life in the Renaissance. *(Possible topics include cities, foods, trade, and so on.)* Have each group write a paragraph that explains how their assigned aspect changed during the period. Call on a volunteer from each group to share its findings with the class.

ENGLISH LANGUAGE LEARNERS, COOPERATIVE LEARNING

LEVEL 2: (Suggested time: 25 min.) Ask students to think about how daily life changed during the Renaissance. Then have each student write two journal entries. One entry should describe a typical day in the life of a teenager in the Middle Ages, while the other should describe a teenager's life in the Renaissance. Call on volunteers to read their entries and discuss how life changed during the Renaissance.

ENGLISH LANGUAGE LEARNERS

CONNECTING TO *Technology*

The Printing Press

During the Middle Ages books were written and copied by hand. It took a long time and great expense to produce a book. A German inventor, Johannes Gutenberg, developed a printing press in the 1400s. It could print much faster than a human could write. Gutenberg used his printing press to print copies of the Bible. This began the era of the printed book, which had a huge impact on the world of learning. Books printed on the printing press helped to spread the ideas of the Renaissance, and later of the Reformation.

Understanding What You Read

1. How was the printing press an improvement over the old ways of making books?
2. Why do you think the printing press helped to spread ideas?

Writing Writers of the time expressed the attitudes of the Renaissance. Popular literature was written in the vernacular, the people's language, instead of in Latin. Dutch writer Desiderius Erasmus criticized ignorance and superstition is his work *In Praise of Folly*. In *Gargantua*, French writer François Rabelais promoted the study of the arts and sciences. Spanish writer Miguel de Cervantes wrote *Don Quixote* in which he mocked the ideals of the Middle Ages. Italian writers such as Machiavelli and Baldassare Castiglione wrote handbooks of proper behavior for rulers and nobles.

Of all Renaissance writers, William Shakespeare is probably the most widely known. He was talented at turning popular stories into great drama. His plays and poetry show a great understanding of human nature. He used the popular English language of his time to skillfully express the thoughts and human feelings of his characters. Many of Shakespeare's subjects and ideas are still important to people today.

William Shakespeare wrote such famous plays as *Romeo and Juliet*, *Hamlet*, and *Macbeth*.

✓ **READING CHECK:** *Summarizing* What was the Renaissance attitude?

Curiosity about world; focus on achievement in this world; involvement in the arts; ed cation important.

Across the Curriculum

LITERATURE

Thomas More's *Utopia* One important Renaissance figure was the English writer Thomas More. In 1516 More published *Utopia*, a book that became very popular throughout Europe. In the book, More criticized corrupt and harsh governments in Europe and the inequality in European society. He imagined a world in which all male citizens were equal and everyone worked to support society. As a result of More's book, the word *utopia* has come to mean "an ideal place or society."

Activity: Have students conduct research on great writers or thinkers of the Renaissance. Have each student create a collector's card (much like a baseball card) with a picture, short biography and list of the person's achievements.

Connecting to Technology

Answers

1. Copying by hand took a long time and made books expensive. Printing was much quicker and reduced the costs of books.
2. Printed material can be passed around a wider area and more quickly than word of mouth.

LEVEL 3: (Suggested time: 45 min.) Provide students with encyclopedias or other reference books. Organize students into groups and have them conduct additional research on the Renaissance. Direct them to focus on changes that took place in the cities, in work and daily life, and in the arts and education. Then have each group create a poster based on its findings. Invite groups to display their posters and discuss them in class.
COOPERATIVE LEARNING

Teaching Objective 3

ALL LEVELS: (Suggested time: 25 min.) Copy the graphic organizer on the next page onto the board, omitting the italicized answers. Have students complete it with descriptions of some of the changes that took place during the Reformation and Counter-Reformation. Then lead a discussion about the information presented on the graphic organizer.
ENGLISH LANGUAGE LEARNERS

Environment and Society

The depopulation caused by the Black Death had major consequences for both the economy and environment of Europe. A shortage of farm workers caused many landowners to stop farming a portion of their land. This resulted in fewer farm goods and reduced trade. It also meant that forests that had once been cut down to create farmlands could grow back again. Across Europe, many cleared fields were allowed to return to their natural, forested condition.

Discussion: Ask students how reforestation might have been beneficial to Europeans *(provided wood for building and heating; created a healthier environment).* Then lead a discussion about other environmental effects that might result from a shrinking or growing population.

Visual Record Answer

Women were becoming educated and could take over more of the work once reserved for men.

During the Renaissance, more people learned to read and write. This painting shows a couple working together in their banking business. **Interpreting the Visual Record** **What does this painting suggest about the changing role of women during the Renaissance?**

The Renaissance and Daily Life

The Renaissance was not only a time of learning, art, and invention. It was also a time of change in people's daily lives. As the manorial system of the Middle Ages fell apart, many peasants left the manors on which they had lived. Because there were fewer people to work the land, many of these peasants could now demand wages for their labor. For the first time, they had money to spend. As Europe's population began to increase again after the Black Death, however, prices rose very quickly. Only wealthy people could afford many goods.

Although they now had some money, most peasants were still poor. Some migrated to cities in search of work. Instead of raising their food, they bought it in shops. In the 1500s traders brought to Europe new vegetables such as beans, lettuce, melons, spinach, and tomatoes. Traders also brought new luxury items such as coffee and tea. As the idea of the printing press caught on in Europe, books became more common. More and more people learned how to read. Gradually a new way of life developed, and the quality of life slowly began to change.

✓ **READING CHECK:** *Comparing and Contrasting* In what ways did life in Europe change during the Renaissance? Migration to cities; less dependence on farming as a means of support; work for wages; new foods, more varied diets.

The city of Florence was the center of the early Renaissance. The large domed building is the Duomo, the cathedral of Florence

Changes of the Reformation and Counter-Reformation

- more churches
- emphasis on education and literacy
- less power for Catholic Church
- more power for monarchs and national governments

Close

Have students look through the images in this section. Call on volunteers to explain how each of the images reflects the changes that occurred in art or society during the Renaissance and Reformation.

The Reformation

As humanism became more popular, people began to question their religious beliefs. Northern humanists thought the Roman Catholic Church had become too powerful and too worldly. They thought that it was too rich and owned too much land and that it had lost the true message of Jesus. Some people began to question the pope's authority. The humanists' claims sparked a movement that split the church in western Europe during the 1500s. This movement is called the **Reformation**.

Martin Luther A German monk named Martin Luther disagreed with the Catholic Church about how people should act. The Church taught that the way to heaven lay in attending church, giving money to the church, and doing good deeds. Luther said that the way to heaven was simply to have faith in God. He argued that the Bible was the only authority for Christians. The printing press helped Luther's ideas spread. He gained followers who became known as **Protestants** because they protested against the Catholic Church's teachings and practices. Luther eventually broke with the Church and founded the Lutheran Church.

Martin Luther believed that God viewed all people of faith equally.

John Calvin Another important thinker of the Reformation was John Calvin. Many of his ideas are similar to those of Martin Luther. Like Luther, Calvin taught that the Bible was the most important element of Christianity. Priests and other clergy were not necessary. Unlike Luther, however, Calvin believed that God had already decided who was going to go to heaven, even before these people were born. He encouraged his followers to dedicate themselves completely to God and to live lives of self-restraint.

Calvin's teachings were very popular, particularly in Switzerland. In 1536 he and his followers took over the city of Geneva. There they passed laws requiring that everyone live according to Calvinist teachings.

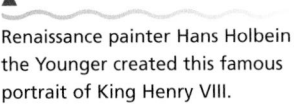

Renaissance painter Hans Holbein the Younger created this famous portrait of King Henry VIII.

Henry VIII of England Henry brought major religious change to England. At first he was a great defender of the Roman Catholic Church, but this changed after a conflict with the pope. Henry wanted a son to inherit his throne, but his wife could not have more children after their daughter was born. Henry asked the pope for permission to divorce her, but the pope refused. Henry then claimed that the pope did not have authority over the powerful English monarchy. He broke away from the Roman Catholic Church and had laws passed that created the Church of England. The new Church granted Henry VIII a divorce.

Linking Past to Present

Protestant Geneva In the mid-1500s John Calvin established a religious government in Geneva, Switzerland. The city became a Protestant stronghold. Legend has it that Geneva's Protestant leaders owed their success to a pot of soup.

In 1602 the Catholic Duke of Savoy in France sent an army to drive the Protestants out of Geneva. French soldiers tried to slip quietly over the city walls, but an alert cook heard them and emptied a cauldron of hot soup onto their heads. She then inspired her fellow townspeople to fight while the Genevan army readied itself for battle. The city was saved and the Protestant leaders remained in power. These same leaders later urged the city's goldsmiths to stop making jewelry, because wearing jewelry had been forbidden. The jewelers then focused their talents on what is now one of Switzerland's most famous industries—watchmaking.

Critical Thinking: How did the religious government of Geneva affect its economy?

Answer: By forbidding the wearing of jewelry, the government encouraged the growth of watchmaking.

REVIEW AND ASSESS

Have students complete the Section Review. Then hand out index cards and have students write questions for a quiz game about the Renaissance, the Reformation, and the Counter-Reformation. Have students use their cards to quiz each other. Then have students complete Daily Quiz 7.3.

RETEACH

Have students complete Main Idea Activity 7.3. Then have students create charts with three column headings—"Renaissance," "Reformation," and "Counter-Reformation"—and two row labels, "Origins" and "Features." Have them fill out the chart with the appropriate information.
ENGLISH LANGUAGE LEARNERS

EXTEND

Have interested students conduct research on the Renaissance in northern Europe, where art developed quite differently than it did in Italy. Encourage students to select one figure or work from the Northern Renaissance and compare it to the art of southern Europe in the same time period.
BLOCK SCHEDULING

Section 3 Review

Answers

Define For definitions, see: Renaissance, p. 143; humanists, p. 143; Reformation, p. 147; Protestants, p. 147; Counter-Reformation, p. 148

Reading for the Main Idea

1. Travel brought people into contact with ideas; curiosity about the world awakened. (NGS 6)
2. developing cities; new foods; new sources of income; more freedom; work for hire (NGS 15)
3. many different churches; more emphasis on education; increased power of national governments and monarchs (NGS 12)

Critical Thinking

4. The Pope and the Church during the Middle Ages had tremendous power over nations and monarchs. The less powerful the Church became, the more powerful nations and governments became.

Organizing What You Know

5. d'Este: patron of the arts; Luther: started Protestantism; Gutenberg: developed a printing press; Leonardo: great artist and inventor; Shakespeare: popular plays and poems

Visual Record Answer ►

education, literacy

The Counter-Reformation In response to the rise of Protestantism, the Catholic Church attempted to reform itself. This movement is called the **Counter-Reformation**. Church leaders began to focus more on spiritual matters and on making Church teachings easier for people to understand. They also attempted to stop the spread of Protestantism. Since about 1478, Spanish leaders had put on trial and severely punished people who questioned Catholic teachings. Leaders of what was called the Spanish Inquisition saw their fierce methods as a way to protect the Catholic Church from its enemies. During the Counter-Reformation, the pope brought the Spanish Inquisition to Rome.

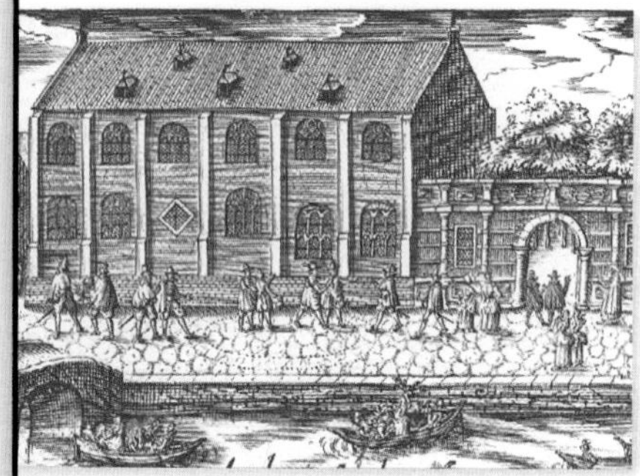

▲ Many schools, including the Dutch University of Leiden shown here, were established during the Reformation.

Interpreting the Visual Record **What goal of the Renaissance humanists was shared by both Catholics and Protestants?**

Results of Religious Struggle Terrible religious wars broke out in France, Germany, the Netherlands, and Switzerland after the Reformation. By the time these wars ended, important social and political changes had occurred in Europe. Many different churches arose in Europe.

A stronger interest in education arose. Catholics saw education as a tool to strengthen people's belief in the teachings of the Church. Protestants believed that people could find their own way to Christian faith by studying the Bible. Although both Catholics and Protestants placed importance on literacy, the ability to read, education did not make people more tolerant. Both Catholic and Protestant leaders opposed views that differed from their own.

As Protestantism became more popular, the Catholic Church lost some of its power. It was no longer the only church in Europe. As a result, it lost some of the tremendous political power it had held there. As the power of the Church and the pope decreased, the power of monarchs and national governments increased.

✓ **READING CHECK:** *Identifying Cause and Effect* How did the religious conflicts of the 1500s change life in Europe? **Many different churches appeared; education more important; national governments gained power.**

Section 3 Review

go.hrw.com Homework Practice Online
Keyword: SK3 HP7

Define and explain: Renaissance, humanists, Reformation, Protestants, Counter-Reformation

Reading for the Main Idea

1. *Human Systems* What brought about the Renaissance?
2. *Human Systems* What were some important changes in daily life during the Renaissance?
3. *Human Systems* After the Reformation and Counter-Reformation, how was life in Europe different?

Critical Thinking

4. **Drawing Inferences** Why did national governments gain strength as the power of the Catholic Church declined?

Organizing What You Know

5. **Categorizing** Copy and complete the following graphic organizer with the achievements of some key people of the Renaissance and the Reformation.

Person	Achievement

Section 4

Objectives

1. Describe the Scientific Revolution.
2. Identify factors that triggered the Age of Exploration.
3. Explain how the English monarchy differed from others in Europe.
4. Describe the relationship between England and its American colonies.

Focus

LET'S GET STARTED

Copy the following question onto the chalkboard: *How has science changed the world? Write down your ideas.* Discuss student responses. Point out that many of the scientific advances of the modern age have their roots in the Scientific Revolution. Tell students that they will learn more about scientific advances and other developments of the 1500s and 1600s in this section.

Building Vocabulary

Write the key terms on the board. Ask students to read the definitions from the glossary aloud. Have students discuss what kind of government comes to mind when they hear **monarchy**. Point out that in addition to a monarchy, the United Kingdom has a **Parliament** that is responsible for making laws. This makes their government a **limited monarchy**. Have students read the definitions for the other key terms from the glossary.

Section 4 Exploration and Conquest

Read to Discover

1. What was the Scientific Revolution?
2. What started the Age of Exploration?
3. How was the English monarchy different from others in Europe?
4. What was the relationship between England and its American colonies?

Define

Scientific Revolution
Age of Exploration
colony
mercantilism
absolute authority
monarchy
limited monarchy
Parliament
Puritan
constitution
Restoration

Locate

India
China
Spain
Portugal
Mexico
Bahamas
France
Russia
Austria
England

Why It Matters

Countries often try to increase their power through conquest. Use CNNfyi.com or other **current events** sources to find examples of conquest going on today. Record your findings in your journal.

Galileo's telescope

Section 4 Resources

Reproducible
- Guided Reading Strategy 7.4

Technology
- One-Stop Planner CD–ROM, Lesson 7.4
- Homework Practice Online
- HRW Go site

Reinforcement, Review, and Assessment
- Main Idea Activity 7.4
- Section 4 Review, p. 155
- Daily Quiz 7.4
- English Audio Summary 7.4
- Spanish Audio Summary 7.4

The Scientific Revolution

You have read that Renaissance humanists encouraged learning, curiosity, and discovery. The spirit of the Renaissance paved the way for a development during the 1500s and 1600s known as the **Scientific Revolution**. During this period, Europeans began looking at the world in a different way. Using new instruments such as the microscope and the telescope, they made more accurate observations than were possible before. They set up scientific experiments and used mathematics to learn about the natural world.

This scientific approach produced new knowledge in the fields of astronomy, physics, and biology. For example, in 1609 Galileo Galilei built a telescope and observed the sky. He eventually proved that an earlier scientist, Copernicus, had been correct in saying that the planets circle the sun. Earlier, people had believed that the planets moved around the Earth. In 1687 Sir Isaac Newton explained the law of gravity. In the 1620s William Harvey discovered the circulation of blood.

Other discoveries and advances such as better ships, improved maps, compasses and other sailing equipment allowed explorers to venture farther over the seas than before. These discoveries paved the way for the Age of Exploration.

✓ **READING CHECK:** *Identifying Cause and Effect* What brought about the Scientific Revolution? Renaissance attitude of curiosity, a desire to learn and discover.

The model pictured below is of an English explorer's ship. Although they appear tiny and fragile by today's standards, ships like this carried European explorers to new lands around the world.

TEACH

Teaching Objective 1

LEVEL 1: (Suggested time: 20 min.) Ask students to name some of the defining characteristics of the Scientific Revolution. Then have each student write a paragraph summarizing those aspects. Tell them that their paragraphs should describe the Scientific Revolution by answering the five "W" questions: Who? What? When? Where? Why?

ENGLISH LANGUAGE LEARNERS

LEVELS 2 AND 3: (Suggested time: 45 min.) Provide students with encyclopedias or other reference books and have them conduct research about key figures of the Scientific Revolution. Then have each student prepare a brief report about his or her chosen person's life and achievements. Challenge students to evaluate the historic significance of the scientist's work and its influence on the modern world.

Across the Curriculum

SCIENCE

Early Mapmaking The interest of Renaissance scholars in the work of ancient geographers like Ptolemy led to improvements in mapmaking. Most scholars of the time knew the world was round. During the Renaissance, new information about Africa and Asia was added to ancient maps, but the Americas were not yet known.

With the Age of Exploration, mapmaking made great advances. The discovery of new lands and the use of new navigational tools like the compass improved the accuracy of maps. The most famous mapmaker of the era was Gerardus Mercator, who published his map of the world in 1569. Mercator developed a technique, called the Mercator Projection, which allowed him to represent the curved surface of Earth on a flat map.

Discussion: Show students examples of old maps from reference books or the Internet. Lead a class discussion comparing and contrasting those maps with recent ones.

Map Answer

Sailed around the tip of South America through what is now called the Strait of Magellan

This map shows the routes taken by Portuguese, Spanish, French, English, and Dutch explorers. They sailed both east and west to discover new lands.

Interpreting the Map **Find Magellan's course on the map. Describe the route he found from the Atlantic Ocean to the Pacific Ocean.**

European Explorations, 1492–1535

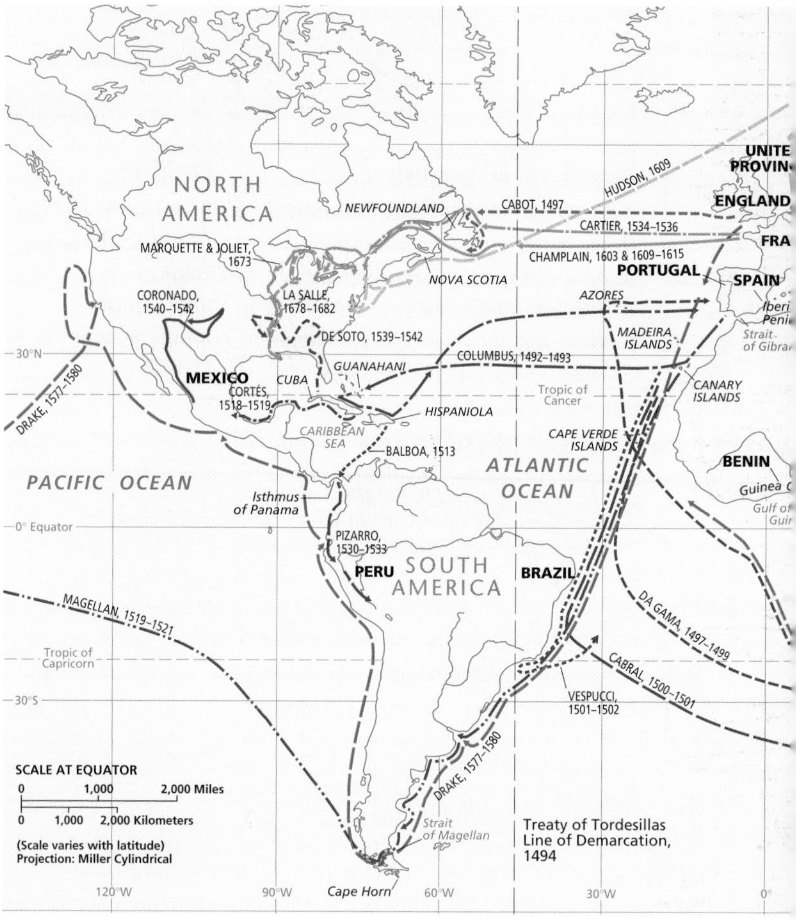

The Dutch artist Jan Vermeer, who painted during the 1600s, often showed his subjects in the midst of work activities. Vermeer's *The Astronomer* shows research and work connected to mapmaking.

The Age of Exploration

Europeans were eager to find new and shorter sea routes so that they could trade with India and China for spices, silks, and jewels. The combination of curiosity, technology, and the demand for new and highly valued products launched a period known as the **Age of Exploration**.

A member of the Portuguese royal family named Prince Henry encouraged Portugal to become a leader in exploration. He wanted to find a route to the rich spice trade of India. Portuguese explorers did eventually succeed in finding a way. They reached India by sailing around Africa.

Hoping to find another route to India, Spain sponsored the voyage of the Italian navigator Christopher Columbus. He hoped to find a direct route to India by sailing westward across the Atlantic Ocean. In

Teaching Objective 2

LEVELS 1 AND 2: (Suggested time: 30 min.) Copy the following graphic organizer onto the board, omitting the italicized answers. Ask students to copy it into their notebooks. Then pair students and have each pair fill in the chart with factors that led to the beginning of the Age of Exploration. Call on volunteers to share their answers with the class. Discuss the information in students' charts.

ENGLISH LANGUAGE LEARNERS, COOPERATIVE LEARNING

Scientific factors	Economic factors	Early events
• *curiosity about the world* • *better ships, maps, and equipment*	• *desire for trade with Asia* • *search for shorter trade routes by sea*	• *Portuguese discovered sea route around Africa* • *Columbus landed in Americas*

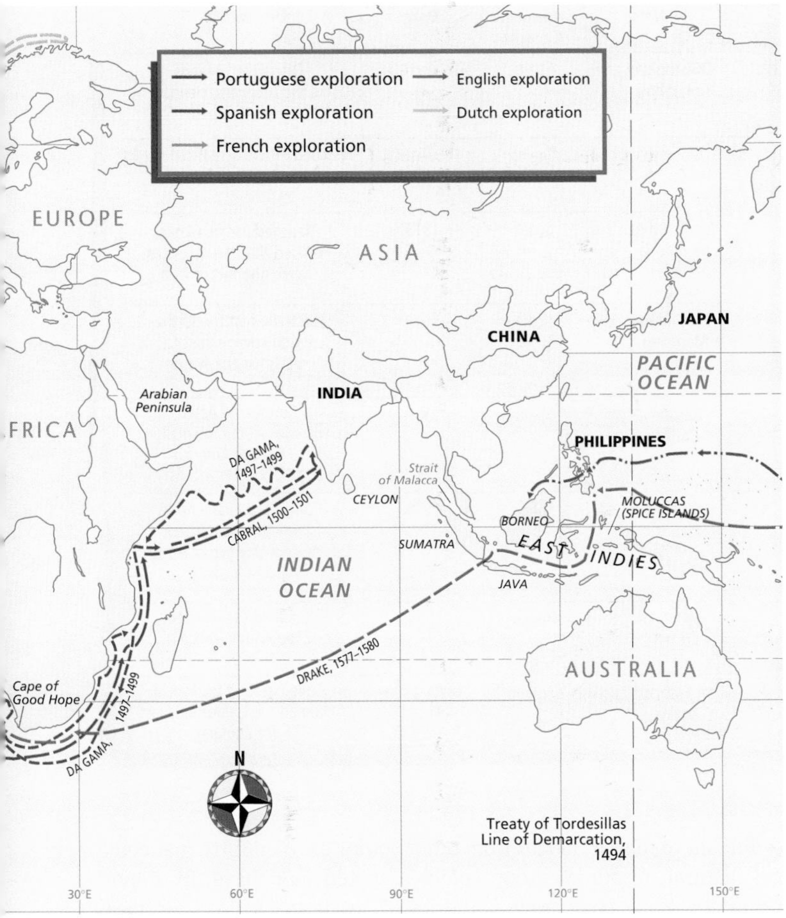

▲ Sailors of the 1500s had many new tools such as this astrolabe to hold a course and measure their progress. A sailor would sight a star along the bar of the astrolabe. By lining the bar with markings on the disk, he could figure out the latitude of the ship's position.

1492 Columbus reached an island in what is now called the Bahamas. Because he had no idea that the Americas lay between Europe and Asia, Columbus believed he had reached the east coast of India.

Later Spanish explorers who knew of the Americas were motivated more by the promise of conquest and riches than by curiosity and the opening of trade routes. The chart on the following page provides an overview of the major explorers from the late 1400s through the 1500s, their voyages, and their accomplishments.

Conquest and Colonization Over time the Spanish, French, English, Dutch, and others established American colonies. A colony is a territory controlled by people from a foreign land. As they expanded overseas, Europeans developed an economic theory called **mercantilism**. This theory said that a government should do everything it could to increase its wealth. One way it could do so was by

Global Perspectives

One way that colonial powers kept economic control over their colonies was to erect trade barriers, such as tariffs. These barriers helped prevent colonies from trading with other countries. Similar barriers still exist in varying degrees in modern nations, mainly as a way to protect local industries from foreign competition.

Today, however, many countries support free trade policies that lower barriers to foreign trade. Free trade agreements like the North American Free Trade Agreement (NAFTA) between the United States, Canada, and Mexico have reduced tariffs and stimulated foreign trade. Many economists believe that free trade benefits all nations in the long run. However, critics of free trade note that it can also cause job losses and result in the movement of factories from high-wage countries to low-wage countries.

Discussion: Lead a class discussion about the benefits and drawbacks of free trade. Ask students to evaluate the arguments, pro and con, and decide whether free trade is a good idea or not.

LEVEL 3: (Suggested time: 50 min.) Write the following statement on the board: *Without the scientific discoveries of the 1400s and 1500s, the Age of Exploration would not have been possible.* Challenge students to write essays that either support or dispute this statement. Encourage students to conduct additional research on the Age of Exploration if necessary. Call on volunteers to read and discuss their essays in class.

Teaching Objective 3

LEVEL 1: (Suggested time:15 min.) Copy the graphic organizer on the next page onto the board, omitting the italicized answers. Have students copy it into their notebooks and fill it in with descriptions of continental European monarchies and the British monarchy. Tell students that the space where the circles overlap should be filled in with features common to both types of monarchies. When students have finished, ask them how the English monarchy was different from others in Europe. *(Its power was limited by laws.)* **ENGLISH LANGUAGE LEARNERS**

National Geography Standard 10

Human Systems The conquest and colonization of the Americas brought a diffusion of culture and products between the Old and New Worlds. One result was the transfer of foods and animals, known as the Columbian Exchange.

The effects of this exchange are still felt today. For example, diets on both sides of the Atlantic changed. Potatoes from the Andes now feed millions of people in Europe. Corn, first grown by Native Americans, is a staple around the world. Wheat was not grown in the Americas until the Europeans arrived. The tomato first grew not in Italy, but in the Americas. Horses, cattle, goats, sheep, and other animals came from Europe, while North America contributed the turkey, the gray squirrel, and the muskrat.

Activity: Have students conduct research on the Columbian Exchange and create charts or posters detailing its features and effects.

Interpreting the Chart **Which explorer do you think made the most important discovery? Give reasons for your answer.**

European Explorers

Name	Sponsoring Country	Date	Accomplishment
Christopher Columbus	Spain	1492–1504	Discovered islands in the Americas, claimed them for Spain
Amerigo Vespucci	Spain, Portugal	1497–1504	Reached America, realized it was not part of Asia
Vasco Núñez de Balboa	Spain	1513	Reached Pacific Ocean, proved that the Americas were not part of Asia
Ferdinand Magellan	Spain	1519	Made the first round-the-world voyage, first to reach Asia by water
Hernán Cortés	Spain	1519	Conquered Aztec civilization, brought smallpox to Central America
Francisco Pizarro	Spain	1530	Conquered Inca empire, claimed the land from present-day Ecuador to Chile for Spain
Jacques Cartier	France	1535	Claimed the Quebec region for France
Sir Francis Drake	England	1577–1580	Sailed around the World, claimed the California coast for England

European explorers sailed in search of goods like gold and cinnamon. In addition, some found new foods, like tomatoes, to bring back to Europe.

selling more than it bought from other countries. A country that could get natural resources from colonies would not have to import resources from competing countries. The desire to win overseas sources of materials helped fuel the race for colonies. The Age of Exploration changed both Europe and the lands it colonized. Colonized lands did benefit from these changes. However, in general, Europe gained the most. During this time, goods, plants, animals, and even diseases were exchanged between Europe and the Americas.

The Slave Trade A tragic result of exploration and colonization was the spread of slavery. During the 1500s Europeans began to use enslaved Africans to work in their colonies overseas. In exchange for slaves, European merchants shipped cotton goods, weapons, and liquor to Africa. These slaves were sent across the Atlantic to the Americas, where they were traded for goods such as sugar and cotton. These goods were then sent to Europe in exchange for manufactured products to be sold in the Americas. Conditions aboard the slave ships were horrific, and slaves were treated brutally. Many died crossing the Atlantic.

✓ **READING CHECK:** ***Identifying Cause and Effect*** What were two results of European exploration? increased trade and the slave trade

Features of European Monarchies

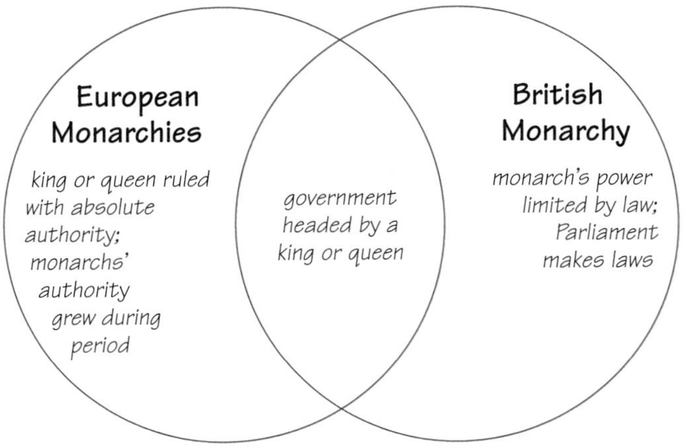
European Monarchies
king or queen ruled with absolute authority; monarchs' authority grew during period
government headed by a king or queen
British Monarchy
monarch's power limited by law; Parliament makes laws

LEVELS 2 AND 3: (Suggested time: 30 min.) Organize students into groups and have each group create a chart showing how Parliament exercised power over English rulers in the 1500s and 1600s. Charts should describe Parliament and its functions and trace its influence through events like the English Civil War, the Restoration, and the Glorious Revolution. Have students discuss their charts in class. **COOPERATIVE LEARNING**

Monarchies in Europe

Wealth flowed into European nations from their colonies. At the same time the Church's power over rulers and governments lost strength. The power of monarchs increased. In France, Russia, and Central Europe, monarchs ruled with **absolute authority**, meaning they alone had the power to make all the decisions about governing their nation. This situation would not change much until the 1700s.

France was ruled by a royal family called the Bourbons. Its most powerful member was Louis XIV, who ruled France from 1643 to 1715. Like many European monarchs, Louis believed that he had been chosen by God to rule. He had absolute control of the government and made all important decisions himself. Under Louis, France became a very powerful nation. In Russia, the Romanov dynasty came to power in the early 1600s. The most powerful of the Romanov czars was Peter the Great, who took the throne in 1682. He wanted to make Russia more like countries in Western Europe. Like Louis XIV, Peter the Great was an absolute monarch who strengthened his country. In Central Europe, two great families competed for power. The Habsburgs ruled the Austrian Empire, while the Hohenzollerns controlled Prussia to the north.

England's situation was different. When King John signed Magna Carta during the Middle Ages, he set a change in motion for England's government. England became a **limited monarchy**. This meant that the powers of the king were limited by law. By the 1500s, **Parliament**, an assembly made up of nobles, clergy, and common people, had gained the power to pass laws and make sure they were upheld.

English Civil War English monarchs such as Henry VIII and Elizabeth I had to work with or around Parliament to achieve their political goals. Later English monarchs fought with Parliament for power. Some even went to war over this issue. The struggle between king and Parliament reached its peak in the mid-1600s. Armies of Parliament supporters under Oliver Cromwell defeated King Charles I, ended the monarchy, and proclaimed England a commonwealth, a nation in which the people held most of the authority.

A special court tried Charles I for crimes against the people. Oliver Cromwell, a **Puritan**, took control of England. Puritans were a group of Protestants who thought that the Church of England was too much like the Catholic Church. The Puritans were a powerful group in Parliament at the time, and Cromwell was their leader.

Empress Maria Theresa of Austria was a member of the Habsburg family.

Oliver Cromwell led the Puritan forces that overthrew the English monarchy. He ruled England from 1653 to 1658.

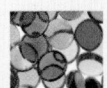

Cultural Kaleidoscope

African Culture in Latin America The slave trade brought Africans and African culture to many parts of the Americas. Large populations of African ancestry live in Brazil, on the Caribbean islands and Caribbean coast of South and Central America, and along the Pacific coast of South America as far south as Lima, Peru.

African influences are a key feature of the cultural mosaic that characterizes Latin America today. Music is one notable example. Reggae, calypso, samba, salsa, cumbia, and merengue are just a few of the musical styles that bear the strong imprint of African rhythms. Music is not the only area of influence. African cultural influences affect everything from dance, art, and literature to food, language, and religion.

Activity: Have students research a cultural feature of the Americas that reflects African influence. They might choose a style of music, a type of cooking, an art form or dance, or a particular writer or musician. Ask them to present brief oral reports based on their research.

Teaching Objective 4

ALL LEVELS: (Suggested time: 15 min.) Call on a volunteer to identify one policy passed by the British government with regard to its American colonies. Have the student write his or her answer on the chalkboard. Then call on another student to identify the response of the American colonists to this policy. Have that student write the response next to the first answer on the board. Repeat this process until students have names several policies that defined the relationship between Great Britain and the American colonies. Then lead a class discussion about the nature of this relationship. **ENGLISH LANGUAGE LEARNERS**

➤ASSIGNMENT: Have each student design a movie poster for a film set in or about the Scientific Revolution, the Age of Exploration, or another aspect of the 1500s and 1600s. Direct students to create names for their movies and a few characters that might appear in them. Students may also wish to suggest contemporary actors who could appear in the film. Call on volunteers to share their posters with the class. **COOPERATIVE LEARNING**

Section 4 Review

Answers

Define For definitions, see: Scientific Revolution, p. 149; Age of Exploration, p. 150; colony, p. 151; mercantilism, p. 152; absolute authority, p. 153; monarchy, p. 153; limited monarchy, p. 153; Parliament, p. 153; Puritan, p. 153; constitution, p. 154; Restoration, p. 154

Reading for the Main Idea

1. Improvements in science and technology allowed sailors to explore distant lands. (NGS 12)
2. curiosity, a desire for new products, and desire for wealth (NGS 12)
3. others absolute; England became limited monarchy (NGS 13)

Critical Thinking

4. Policy of mercantilism antagonized the colonists by disregarding their rights.

Organizing What You Know

5. Europe to Africa—cotton goods, weapons, liquor; Americas to Europe—sugar and cotton; Europe to the Americas—manufactured goods; Africa to the Americas—slaves

Visual Record Answer ►

to make a public example and public statement; a warning to any monarch who might try to gain power

The death warrant of Charles I was signed and sealed by members of Parliament. Parliament chose to behead King Charles I in public.

Interpreting the Visual Record

Why might Parliament have decided to have Charles I beheaded where everyone could see?

Cromwell's Commonwealth Cromwell controlled England for about five years. He used harsh methods to create a government that represented the people. Twice he tried to establish a **constitution**, a document that outlined the country's basic laws, but his policies were unpopular. Discontent became widespread. In 1660, two years after Cromwell died, Parliament invited the son of Charles I to rule England. Thus the English monarchy was restored under Charles II. This period of English history was called the **Restoration**.

Charles II, shown here as a boy, became king following the fall of Cromwell's commonwealth in 1660.

Last Change in Government Cheering crowds greeted Charles II when he reached London. One observer recalled that great celebrations were held in the streets, which were decorated with flowers and tapestries. People hoped that the Restoration would bring peace and progress to England.

Although England had a king again, the Civil War and Cromwell's commonwealth had made lasting changes in the government. Parliament strictly limited the king's power.

The Glorious Revolution When Charles II died, his brother became King James II. James's belief in absolute rule angered Parliament. They demanded that he give up the throne and invited his daughter, Mary, and her Dutch husband, William of Orange, to replace him. This transfer of power, which was accomplished without bloodshed, was called the Glorious Revolution. The day before William and Mary took the throne in 1689, they had to agree to a document called the Declaration of Rights. It stated that Parliament would choose who ruled the country. It also said that the ruler could not make laws, impose taxes, or maintain an army without Parliament's approval. By 1700 Parliament had replaced the monarchy as the major source of political power in England.

✓ **READING CHECK:** *Drawing Inferences* How did the English Civil War and events that followed affect the English government? Parliament strictly limited monarch's power.

REVIEW AND ASSESS

Have students complete the Section Review. Then ask each student to write a question about a person or event discussed in this section. Have students read their questions to the class and allow the class to answer. Then have students complete Daily Quiz 7.4.

RETEACH

Have students complete Main Idea Activity 7.4. Then organize students into groups and have each group write a skit based on the information in Section 3. Each skit should cover one of the three main topics discussed in this section. Call on volunteer groups to perform their skits for the class.
ENGLISH LANGUAGE LEARNERS, COOPERATIVE LEARNING

EXTEND

Have students conduct additional research on one of the explorers named on the chart in this section. Direct students to find information about the explorer's life and accomplishments. Then have each student prepare a brief biography of his or her chosen explorer and present it to the class.
BLOCK SCHEDULING

English Colonial Expansion

During the 1600s, English explorers began claiming and conquering lands overseas. In 1607 the British established Jamestown in what is now the state of Virginia. Jamestown was the first permanent English settlement in North America. In 1620, settlers founded Plymouth in what is now Massachusetts.

Mercantilism and the British Colonies The British government, with its policy of mercantilism, thought that the colonies should exist only for the benefit of England. Parliament passed laws that required colonists to sell certain products only to Britain, even if another country would pay a higher price. Other trade laws imposed taxes on sugar and other goods that the colonies bought from non-British colonies.

Resistance in the Colonies The American colonists saw these trade laws as a threat to their liberties. They found many ways to break the laws. For example, they avoided paying taxes whenever and however they could. Parliament, however, continued to impose new taxes. With each new tax, colonial resistance increased. Relations between England and the colonies grew steadily worse. The stage was set for revolution.

The settlers at Jamestown settled close by the James River.

Interpreting the Visual Record

Why do you think the colonists built their settlement in the manner shown here?

The Granger Collection, New York

✓ **READING CHECK:** *Finding the Main Idea* How did England regard the American colonies? as existing only for the benefit of England

go.hrw.com Homework Practice Online
Keyword: SK3 HP7

Section 4 Review

Define and explain: Scientific Revolution, Age of Exploration, colony, mercantilism, absolute authority, monarchy, limited monarchy, Parliament, Puritan, constitution, Restoration

Reading for the Main Idea

1. *Human Systems* How did the Scientific Revolution aid European exploration?
2. *Places and Regions* What prompted Europeans to explore and colonize land overseas?
3. *Human Systems* How was England different from other monarchies in Europe?

Critical Thinking

4. **Drawing Inferences** How did England's treatment of the American colonies set the stage for revolution?

Organizing What You Know

5. **Summarizing** Copy the following graphic organizer. Use it to show how the slave trade worked between Europe, Africa, and the Americas. Alongside each arrow, list the items that were traded along that route.

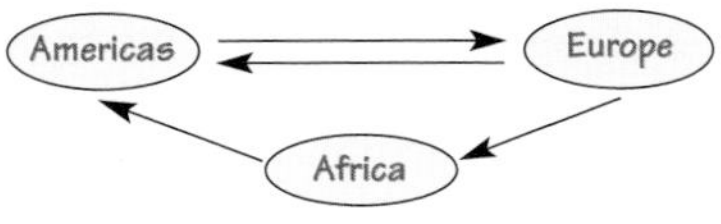

CHAPTER 7 Review ANSWERS

Building Vocabulary

For definitions, see: : mosques, p. 134; shah, p. 134; samurai, p. 136; feudalism, p. 139; vassals, p. 139; chivalry, p. 139; serfs, p. 140; clergy, p. 140; Renaissance, p. 143; humanists, p. 144; Reformation, p. 147; Scientific Revolution, p. 149; colony, p. 150; mercantilism, p. 150; constitution, p. 153

Reviewing the Main Ideas

1. trade, centers of learning, medicine, maps, number system, mosques (NGS 12)
2. philosophy that focused on life here and now and on the importance on individual achievement (NGS 10)
3. new churches, increased interest in education, reforms in Catholic Church; national governments gained power (NGS 12)
4. fostered curiosity and through scientific and technological improvements (NGS 15)
5. Colonies could help a nation increase its wealth by gaining access to labor and natural resources. Colony would give the home country a market for its products. (NGS 11)

◄ **Visual Record Answer**

for defense against attack

ASSESS

Have students complete the Chapter 7 Test.

RETEACH

Pair students and assign each pair one of the chapter's topics. Have each pair write the main points of the assigned topic on a transparency. Then have pairs teach the material to the class, using the transparency and an overhead projector. Students may write the main points on the chalkboard if no overhead projector is available. **ENGLISH LANGUAGE LEARNERS, COOPERATIVE LEARNING**

PORTFOLIO EXTENSIONS

1. Castles were a notable feature of the Middle Ages. Have students research medieval castles and create a guidebook for tourists exploring a castle. The guidebook should explain the different parts of the castle and what they were used for. Students should include illustrations to accompany their explanations.
2. Ask students to imagine that they are explorers of the early 1500s who wish to make a voyage to the New World. Have students work in groups to write a skit in which they petition the monarch of their country to finance the trip. In the skit, the explorer should provide reasons to convince the monarch that the trip will be worth while. Have each group perform its skit. Place scripts in student portfolios.

Chapter 7 Review ANSWERS

Understanding History and Society

Presentations will vary, but information included should be consistent with text. Students should demonstrate understanding that land was granted in exchange for military service.

Thinking Critically

1. Possible answers: united society, generally peaceful times, powerful military protection from invaders, wise and strong leadership.
2. People who live in big cities and make their living through trade need to communicate with each other and keep written records.
3. Catholic Church lost much political power; monarchs and national governments gained power.
4. increased Great Britain's wealth, but pushed the colonists into rebellion
5. Parliament replaced the monarchy as the major political power in Great Britain.

Chapter 7 Reviewing What You Know

Building Vocabulary

On a separate sheet of paper, write sentences to define each of the following words.

1. mosques
2. shah
3. samurai
4. feudalism
5. vassals
6. chivalry
7. serfs
8. clergy
9. Renaissance
10. humanists
11. Reformation
12. Scientific Revolution
13. colony
14. mercantilism
15. constitution

Reviewing the Main Ideas

1. (Human Systems) What were some achievements of the Islamic Empire?
2. (Human Systems) What was humanism?
3. (Human Systems) What were the results of the Reformation and Counter-Reformation?
4. (Environment and Society) How did the Scientific Revolution pave the way for European exploration?
5. (Human Systems) What was mercantilism?

Understanding History and Society

The Feudal System

Throughout Europe in the Middle Ages, the feudal system was the most important political structure. Create a chart to explain the relationships among kings, lords, vassals, and knights. Consider the following:

- Who granted lands and who received the grants.
- How a person became a knight.
- What a king or lord expected from his vassals.

Thinking Critically

1. **Analyzing** Under what kind of social conditions did people of the early Chinese and Mongol Empires produce their achievements?
2. **Identifying Cause and Effect** Why did the growth of cities and trade during the High Middle Ages increase the need for education?
3. **Contrasting** How did political power in Europe change after the Counter-Reformation?
4. **Supporting a Point of View** How did mercantilism both help and harm Great Britain as a colonial power?
5. **Identifying Cause and Effect** How did the English Civil War change the politics in Great Britain?

FOOD FESTIVAL

Many food combinations we take for granted today exist only because of the exchange of plants between the Americas and Europe. Salsa is a good example. Peppers and tomatoes came from the Americas, while onions and coriander—also called cilantro—were known in Europe. To make salsa combine 3 cups of chopped tomatoes, a half cup of chopped onion, 2 tablespoons of chopped coriander, and 1 or 2 chopped jalapeño peppers. Serve with tortilla chips. They are made from corn—another plant from the Americas.

CHAPTER 7 REVIEW AND ASSESSMENT RESOURCES

Reproducible
- Readings in World Geography, History, and Culture 66
- Critical Thinking Activity 7
- Vocabulary Activity 7

Technology
- Chapter 7 Test Generator (on the One-Stop Planner)
- HRW Go site
- Audio CD Program, Chapter 7

Reinforcement, Review, and Assessment
- Chapter 7 Review, pp. 156–157
- Chapter 7 Tutorial for Students, Parents, Mentors, and Peers
- Chapter 7 Test
- Chapter 7 Test for English Language Learners and Special-Needs Students

Building Social Studies Skills

Map ACTIVITY

On a separate sheet of paper, match the letters on the map with their correct labels.

Spain	**Mexico**
Portugal	**Europe**
South America	
North America	

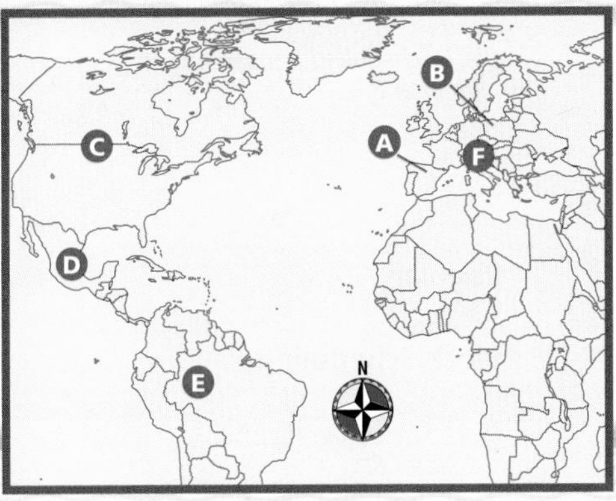

Mental Mapping Skills ACTIVITY

On a separate sheet of paper, draw a freehand map of the world. On your map, sketch and label each of the following explorers' routes:

Columbus **Drake**
Magellan

WRITING ACTIVITY

Imagine that you are a peasant during the Middle Ages. Write a short dialogue in which you talk with another person about your lives on a manor. In your dialogue, discuss both positive and negative aspects of that life. Be sure to use standard grammar, spelling, sentence structure, and punctuation.

Alternative Assessment

Portfolio ACTIVITY

Learning About Your Local History

Local Artists and Writers The Renaissance produced many great artists and writers. Conduct research on the Internet or in your local library to find out about artists and writers in your state or region. Write a brief report on one of these artists or writers.

internet connect

Internet Activity: **go.hrw.com**
KEYWORD: **SK3 GT7**

Choose a topic to explore about the world in transition:
- Write a report on daily life in the Middle Ages.
- Create a biography of a Renaissance artist or writer.
- Learn more about an explorer described in this chapter.

Map Activity
A. Spain
B. Portugal
C. North America
D. Mexico
E. South America
F. Europe

Mental Mapping Skills Activity
Maps will vary, but should show the approximate routes taken by each explorer.

Writing Activity
Students' dialogues will vary but should accurately reflect aspects of the manorial system. Use Rubric 40, Writing to Describe, to evaluate student work.

Portfolio Activity
Reports will vary, but should accurately describe a local artist or writer. Use Rubric 30, Research, to evaluate student work.

internet connect

GO TO: **go.hrw.com**
KEYWORD: **SK3 Teacher**
FOR: **a guide to using the Internet in your classroom**

CHAPTER 8 The Birth of the Modern World

Chapter Resource Manager

Objectives	Pacing Guide	Reproducible Resources
SECTION 1 **The Enlightenment** (pp. 159-62) **1.** Define the Enlightenment. **2.** Describe the ideas of government suggested by Enlightenment thinkers. **3.** Explain how the Enlightenment led to changes in society.	**Regular** 1 day **Block Scheduling** .5 day *Block Scheduling Handbook, Chapter 8*	**RS** Guided Reading Strategy 8.1 **RS** Graphic Organizer 8
SECTION 2 **The Age of Revolution** (pp. 163–170) **1.** Describe the start and results of the American Revolution. **2.** Explain how the French Revolution changed France. **3.** Explain how Europe changed during and after the Napoléonic era.	**Regular** 1.5 days **Block Scheduling** .5 day *Block Scheduling Handbook, Chapter 8*	**RS** Guided Reading Strategy 8.2 **SM** Map Activity 8
SECTION 3 **The Industrial Revolution** (pp. 171–75) **1.** Describe the beginnings of the Industrial Revolution. **2.** Explain how developments in transportation and communications spread industrial development. **3.** Identify business features that affected life in the Industrial Age.	**Regular** 1 day **Block Scheduling** .5 day *Block Scheduling Handbook, Chapter 8*	**RS** Guided Reading Strategy 8.3
SECTION 4 **Expansion and Reform** (pp. 176–81) **1.** Describe changes that occurred in Europe and the United States after 1850. **2.** Identify factors that led to reform in the later 1800s. **3.** Explain how nationalism changed the map of Europe in the mid-1800s.	**Regular** 1 day **Block Scheduling** .5 day *Block Scheduling Handbook, Chapter 8*	**RS** Guided Reading Strategy 8.4

Chapter Resource Key

RS Reading Support

IC Interdisciplinary Connections

E Enrichment

SM Skills Mastery

A Assessment

REV Review

ELL Reinforcement and English Language Learners

Transparencies

CD–ROM

Music

Video

go.hrw.com Internet

Holt Presentation Maker Using Microsoft® Powerpoint®

One-Stop Planner CD–ROM

See the ***One-Stop Planner*** for a complete list of additional resources for students and teachers.

One-Stop Planner CD–ROM

It's easy to plan lessons, select resources, and print out materials for your students when you use the ***One-Stop Planner CD–ROM with Test Generator.***

Technology Resources	Review, Reinforcement, and Assessment Resources
One-Stop Planner CD–ROM, Lesson 8.1 *ARGWorld* CD–ROM Homework Practice Online HRW Go site	**ELL** Main Idea Activity 8.1 **REV** Section 1 Review, p. 162 **A** Daily Quiz 8.1 **ELL** English Audio Summary 8.1 **ELL** Spanish Audio Summary 8.1
One-Stop Planner CD–ROM, Lesson 8.2 *ARGWorld* CD–ROM Homework Practice Online HRW Go site	**ELL** Main Idea Activity 8.2 **REV** Section 2 Review, p. 170 **A** Daily Quiz 8.2 **ELL** English Audio Summary 8.2 **ELL** Spanish Audio Summary 8.2
One-Stop Planner CD–ROM, Lesson 8.3 *ARGWorld* CD–ROM Homework Practice Online HRW Go site	**ELL** Main Idea Activity 8.3 **REV** Section 3 Review, p. 175 **A** Daily Quiz 8.3 **ELL** English Audio Summary 8.3 **ELL** Spanish Audio Summary 8.3
One-Stop Planner CD–ROM, Lesson 8.3 *ARGWorld* CD–ROM Homework Practice Online HRW Go site	**ELL** Main Idea Activity 8.4 **REV** Section 3 Review, p. 181 **A** Daily Quiz 8.4 **ELL** English Audio Summary 8.4 **ELL** Spanish Audio Summary 8.4

internet connect

HRW ONLINE RESOURCES

GO TO: go.hrw.com
Then type in a keyword.

TEACHER HOME PAGE
KEYWORD: SK3 TEACHER

CHAPTER INTERNET ACTIVITIES
KEYWORD: SK3 GT8

Choose an activity to:
- explore the ideas of the Enlightenment.
- investigate the causes of the French Revolution.
- understand capitalism.

CHAPTER ENRICHMENT LINKS
KEYWORD: SK3 CH8

CHAPTER MAPS
KEYWORD: SK3 MAPS8

ONLINE ASSESSMENT
Homework Practice
KEYWORD: SK3 HP8
Standardized Test Prep Online
KEYWORD: SK3 STP8
Rubrics
KEYWORD: SS Rubrics

COUNTRY INFORMATION
KEYWORD: SK3 Almanac

CONTENT UPDATES
KEYWORD: SS Content Updates

HOLT PRESENTATION MAKER
KEYWORD: SK3 PPT8

ONLINE READING SUPPORT
KEYWORD: SS Strategies

CURRENT EVENTS
KEYWORD: S3 Current Events

Meeting Individual Needs

Ability Levels

Level 1 Basic-level activities designed for all students encountering new material

Level 2 Intermediate-level activities designed for average students

Level 3 Challenging activities designed for honors and gifted-and-talented students

English Language Learners Activities that address the needs of students with Limited English Proficiency

Chapter Review and Assessment

SM Critical Thinking Activity 8
REV Chapter 8 Review, pp. 1582–83
REV Chapter 8 Tutorial for Students, Parents, Mentors, and Peers
ELL Vocabulary Activity 8
A Chapter 8 Test
A Unit 2 Test
Chapter 8 Test Generator (on the One-Stop Planner)
Audio CD Program, Chapter 8
A Chapter 8 Test for English Language Learners and Special-Needs Students
A Unit 2 Test for English Language Learners and Special-Needs Students

CHAPTER 8

Launch Into Learning

Challenge students to write down the century in which the following concepts were introduced or became important: 1. Religious institutions should not control governments. 2. People have the right to decide on their country's form of government and to participate in it. 3. People are born equal and remain equal before the law. 4. Machines can produce more goods more quickly than can individual craftworkers. 5. A government should protect the rights of its citizens. Then point out that all of these now-common ideas either emerged or became widely accepted during the 1700s. Tell students they will learn more about how our modern world developed in this chapter.

Section 1

Objectives

1. Define the Enlightenment.
2. Describe the ideas about government suggested by Enlightenment thinkers.
3. Explain how the Enlightenment led to changes in society.

Links to Our Lives

You might want to share with your students the following reasons for learning more about the 1700s and 1800s:

- Debates involving science and religion are still in the news and illustrate deep divisions between some segments of society.
- Monarchies still exist in the world. They differ widely in how well they meet the needs of their people.
- Revolutions that change governments still take place today. Some depose tyrants, while others bring tyrants to power.
- In much of the world, economies are based on industry and technology.

CHAPTER 8

The Birth of the Modern World

The basis for the democratic and modern world we live in today dates back to the 1700s and 1800s. When you read this chapter, you might just wish you had been part of this stirring time of upheaval and great accomplishment.

Drummer boy from the American Revolution

From the mid-1760s through the mid-1770s, British policies and actions toward the 13 colonies in North America became intolerable. Many colonists called Patriots believed that independence from Great Britain was necessary to guarantee their freedom and right to a government of, for, and by the people.

When the American Revolution started in April 1775, patriots of all ages joined the fight for independence. Many young boys between the ages of 14 and 16 enlisted in the American army. Other boys, some as young as six, served as drummers for the troops, like the one pictured on this page. Their job was to signal commands, which sometimes put them in the midst of battle.

The American navy had its share of young sailors as well. Small boys served as deckhands or "powder monkeys." They carried ammunition to the gunners during battle.

Revolutionary War cannon

Harbor at Charleston, South Carolina

FOCUS

LET'S GET STARTED

Write the word *Enlightenment* on the chalkboard. Ask students to copy it and then circle a familiar five-letter word they see within the word (light). Next, ask students what it means when a cartoonist shows a light bulb shining over a character's head. Discuss responses. *(Possible answer: The person has had a bright idea or "seen the light.")* Tell students that in Section 1 they will read about thinkers who had bright new ideas about government and society.

Building Vocabulary

Call on volunteers to locate the definitions of the words **secularism** and **individualism** in the text and read them aloud. Ask students what the words have in common. *(Possible answer: They are both about ideas.)* Next have students identify the two terms that are compound words (**popular sovereignty** *and* **social contract**). Call on volunteers to discuss the meanings of the individual words that make up each term. Then have students locate and read the definitions for the remaining terms. Ask students to write sentences for the terms.

Section 1 The Enlightenment

Read to Discover

1. What was the Enlightenment?
2. What ideas about government did Enlightenment thinkers suggest?
3. How did the Enlightenment lead to changes in society?

Define

- Enlightenment
- reason
- secularism
- individualism
- popular sovereignty
- social contract

Section 1 RESOURCES

Reproducible
- Block Scheduling Handbook, Chapter 8
- Guided Reading Strategy 8.1
- Graphic Organizer 8

Technology
- One-Stop Planner CD–ROM, Lesson 8.1
- Homework Practice Online
- HRW Go site

Reinforcement, Review, and Assessment
- Section 1 Review, p.150
- Daily Quiz 8.1
- Main Idea Activity 8.1
- English Audio Summary 8.1
- Spanish Audio Summary 8.1

WHY IT MATTERS

Our American government and way of life are largely based on the ideas of the Enlightenment. Use CNNfyi.com or other **current events** sources to find out how these ideas continue to cause changes today all around the world. Record your findings in your journal.

Monticello, home of American Enlightenment thinker, Thomas Jefferson

The Birth of a New Age

You have read about how greatly European society changed during the Middle Ages, the Renaissance, the Reformation, and the Scientific Revolution. You have also learned how advances in science and technology paved the way for exploration and expansion overseas. As time went on, many changes in the world continued to take place, especially in people's thinking about their relationship with their nation's government.

From the mid 1600s through the 1700s, most European countries were ruled by monarchs. These rulers increasingly wanted absolute control over their governments and their subjects. Many, although not all, literary people, scientists, and philosophers, or thinkers, in these countries saw the need for change. They believed it was necessary to combat the political and social injustices people suffered every day. These critics claimed that the monarchs and nobility took too much from the common people and gave back too little. They wanted a new social order that was fairer for all people. These ideas were published in books, pamphlets, plays, and newspapers. Historians call this era of new ideas the **Enlightenment**.

✓ **READING CHECK:** *Identifying Cause and Effect* What about the political and social order of European nations caused many people to want change? monarchs' desire for absolute control over their governments and their subjects; people suffering from inequality and injustice in daily life.

▲ John Locke (1632-1704) was an important English philosopher. He is considered the founder of the Enlightenment in England.

TEACH

Teaching Objectives 1–2

LEVELS 1 AND 2: (Suggested time: 10 min.) Have students list some of the Enlightenment's beliefs about human beings and nature. *(Possible responses: The laws of nature govern all creatures; people can discover natural laws through reason; following natural law improves people's lives.)* Ask students what bigger idea these beliefs led to during the Enlightenment. *(Possible responses: They all led to the idea that people should have a part in governing themselves; people should have freedom.)*

LEVEL 3: (Suggested time: 15 min.) Pair students and tell each pair to create flash cards for the key terms *reason, secularism,* and *individualism.* Students should write each key term on one side of a flash card and its definition and details about how it affected the Enlightenment on the other side. **ENGLISH LANGUAGE LEARNERS, COOPERATIVE LEARNING**

Teaching Objective 3

ALL LEVELS: (Suggested time: 30 min.) Copy the graphic organizer on the following page onto the chalkboard, omitting the italicized answers. Have students work in pairs to complete the chart by listing three problems Enlightenment thinkers identified and their solutions to these problems. **COOPERATIVE LEARNING**

DAILY LIFE

The revolutionary ideas of Enlightenment thinkers did not spread rapidly to the general population of England and France. During the 1700s, most people in these countries could not read well, if they could read at all. Few common people had money or time to buy and read books. So in general, only privileged members of society met to discuss Enlightenment ideas. The daily lives of these wealthy and cultured people had little in common with the daily lives of most people. Nevertheless, the writings and conversations of Enlightenment thinkers gradually spread the new ideas far and wide.

Critical Thinking: How does requiring everyone to go to school and providing free education affect society?

Answer: If everyone can read, everyone has access to new ideas about government, and new ideas often lead to positive changes.

▲ Mary Wollstonecraft was a British writer of the Enlightenment. She argued that women should have the same rights as men, including the right to an equal education.

internet connect

GO TO: go.hrw.com
KEYWORD: SK3 CH8
FOR: Web sites about the birth of the modern world

Enlightenment Thinking

The Enlightenment is also called the Age of Reason. At this time, scientists began to use **reason**, or logical thinking, to discover the laws of nature. They believed that the laws of nature governed the universe and all its creatures. Some also thought there was a natural law that governed society and human behavior. They tried to use their powers of reasoning to discover this natural law. By following natural law, they hoped to solve society's problems and improve people's lives.

While religion was important to some thinkers, other thinkers played down its importance. Playing down the importance of religion became known as **secularism**. The ideas of secularism and **individualism**—a belief in the political and economic independence of individuals—would later influence some ideas about the separation of church and state in government. These ideas led to more rights for all people, individual freedoms, and government by the people.

The Enlightenment in England The English philosopher John Locke believed that natural law gave individuals the right to govern themselves. Locke wrote that freedom was people's natural state. He thought individuals possessed natural rights to life, liberty, and property. Locke also claimed people should have equality under the law.

Much of Locke's writing focused on government. Locke argued that government should be based on an agreement between the people and their leaders. According to Locke, people give their rulers the power to rule. If the ruler does not work for the public good, the people have the right to change the government. Locke's writings greatly influenced other thinkers of the Enlightenment. They also influenced the Americans who shaped and wrote the Declaration of Independence and the Constitution.

The Enlightenment in France In France, the thinkers of the Enlightenment believed that science and reason could work together to improve people's lives. They spoke out strongly for individual rights, such as freedom of speech and freedom of worship.

▶ *The Encyclopedia*, published by Enlightenment philosophers, became the most famous publication of the period.

ENLIGHTENMENT IDEAS ABOUT SOCIETY	
Problems with Government or Society	**Enlightenment Solutions to These Problems**
Everyone has to belong to the same official faith.	*People should have freedom of speech.*
People should have freedom of religion.	*People feel as though they have no political freedom.*
People are being jailed for criticizing the king and nobility.	*People must choose own government.*

Close

Ask students to study the illustrations and read the captions in Section 1. For each one, ask: *What does this illustration and caption tell you about the Enlightenment?* Discuss responses.

Review and Assess

Have students complete the Section Review. Then ask each student to choose a person, event, or idea from the section and write a few sentences describing the topic. Ask volunteers to read their sentences aloud while other students identify who or what is being described. Continue until all the major topics have been discussed. Then have students complete Daily Quiz 8.1.

Voltaire The French writer Voltaire was a leading voice of the Enlightenment. As a young man, Voltaire became a famous poet and playwright. He used his wit to criticize the French monarchy, the nobility, and the religious controls of the church. His criticisms got him into trouble. He eventually went to England after being imprisoned twice.

In England, Voltaire was delighted by the freedom of speech he found. In defense of this freedom, he wrote, "I may disapprove of what you say, but I will defend to the death your right to say it."

Voltaire also studied the writings of John Locke. When Voltaire returned to France, he published many essays and tales. These writings explored Enlightenment ideas, such as justice, good government, and human rights.

Rousseau Jean Jacques Rousseau (roo-SOH) was another French thinker of the Enlightenment. He believed that people could only preserve their freedom if they chose their own government, and that good government must be controlled by the people. This belief is called **popular sovereignty**.

Rousseau's most famous book, *The Social Contract*, published in 1762, expressed his views. "Man was born free, and everywhere he is in chains," Rousseau wrote. He meant that people in society lose the freedom they have in nature. Like Locke, Rousseau believed that government should be based on an agreement made by the people. He called this agreement the **social contract**.

The Encyclopedia *The Encyclopedia* was the most famous publication of the Enlightenment. It brought together the writings of Voltaire, Rousseau, and other philosophers. The articles in *The Encyclopedia* covered science, religion, government, and the arts. Many articles criticized the French government and the Catholic church. Some philosophers went to jail for writing these articles. Nevertheless, the *Encyclopedia* helped spread Enlightenment ideas.

✓ **READING CHECK:** *Summarizing* What did the thinkers of the Enlightenment believe? A natural law governed human life and society; truth about natural law could be learned; human problems solved through reason.

Our Amazing Planet

In 1717, when Voltaire was 23, he spent eleven months in prison for making fun of the government. During that time, he wrote his first play. Its success made him the greatest playwright in France.

▲ Jean-Jacques Rousseau

◄ In France, writers and artists gathered each week at meetings like the one shown in this painting. Their purpose was to discuss the new ideas of the Enlightenment.

Interpreting the Visual Record
How might the group pictured here encourage the free sharing of ideas?

Linking Past to Present

The Enlightenment and the U.S. Constitution Point out the illustration of the U.S. Constitution on page 149 and have students read the caption. Ask them how Enlightenment ideas affected the kind of government we have in the United States today. *(The U.S. government is divided into three branches, as the philosopher Baron de Montesquieu thought governments should be.)*

Activity: Organize the class into three groups. Ask each group to conduct research on one of the three branches of the U.S. government—the executive, the legislative, or the judicial. Have the groups investigate the powers of its branch and the individuals and groups that are part of it. After each group has done its research, have students meet together to draw a chart that presents the three branches and their powers. Students might also create a chart that shows how the branches interact with each other.

◄ **Visual Record Answer**

Sitting in a semicircle in comfortable surroundings, people would feel free to exchange ideas and information.

Reteach

Have students complete Main Idea Activity 8.1. Then organize the class into two groups. Ask the first group to pretend they are common people living in England or France before the Enlightenment. Have them list some of the problems they have with their society and government. Ask the second group to pretend they are typical Enlightenment thinkers. Ask them to explain the changes they would make in government and society and how these changes would improve people's lives. Have the groups compile their ideas on a Before and After chart. **ENGLISH LANGUAGE LEARNERS, COOPERATIVE LEARNING**

Extend

Using *Bartlett's Book of Familiar Quotations* or other reference sources, have interested students compile a list of famous quotations from Enlightenment thinkers. Students can begin with the Voltaire and Rousseau quotes given in the section. To assess students' understanding, ask them to paraphrase each quotation and explain why they think it is memorable or significant. **BLOCK SCHEDULING**

Section 1 Review

Answers

Define For definitions, See: Enlightenment, p. 147; reason, p. 148; secularism, p. 148; individualism, p. 148; popular sovereignty p. 149; social contract, p. 149

Reading for the Main Idea

1. Thinkers tried to use reason to discover natural laws that governed society. (NGS 13)
2. equality under the law, basic rights, popular sovereignty. (NGS 13)

Critical Thinking

3. It included criticism of the government and the church.
4. Ideas such as equality, rights to life and liberty, and popular sovereignty appear in the Declaration of Independence. Three branches of government are in the U.S. Constitution.

Organizing What You Know

5. Locke—England; natural rights; equality; government based on agreement between people and leaders; Voltaire—France; freedom of speech; government good and just; Rousseau—France; social contract.

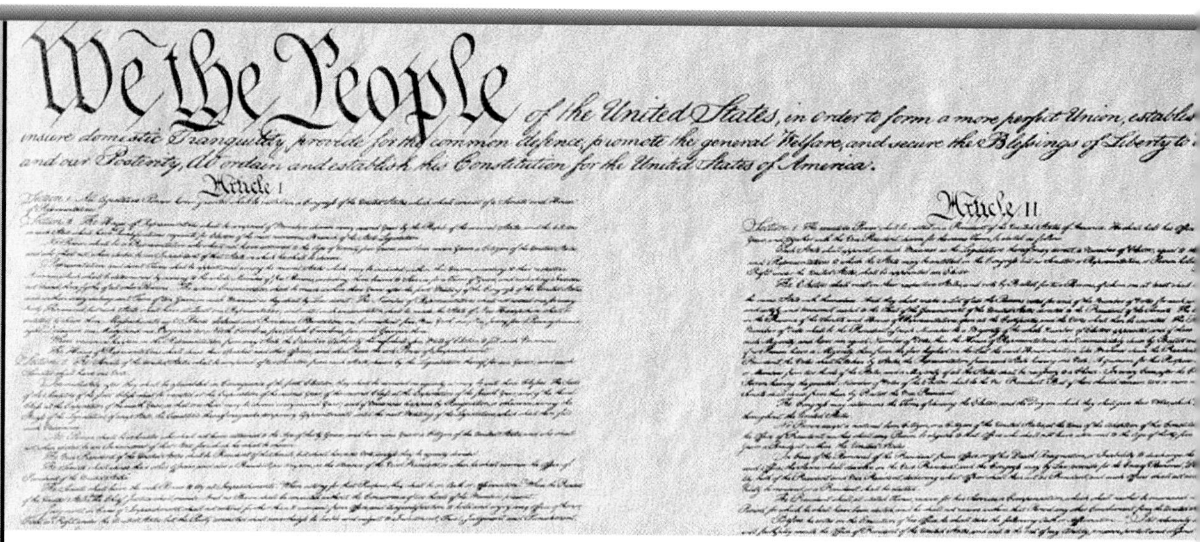

▲ The philosopher Baron de Montesquieu (MOHN-tes-kyoo) thought governments should be divided into three branches. His ideas helped the writers of the U.S. Constitution form our government.

The Enlightenment and Society

When the philosophers began to publish their ideas, there was little freedom of expression in Europe. Most countries were ruled by absolute monarchs. Few people dared to criticize the court or the nobility. Most nations had official religions, and there was little toleration of other faiths.

As time passed, Enlightenment ideas about freedom, equality, and government became more influential. Eventually, they inspired the American and French revolutions. In that way, the Enlightenment led to more freedom for individuals and to government by the people.

Section 1 Review

go.hrw.com Homework Practice Online
Keyword: SK3 HP8

Define and explain Enlightenment, reason, secularism, individualism, popular sovereignty, social contract

Reading for the Main Idea

1. *Human Systems* Why was the Enlightenment also called the Age of Reason?
2. *Human Systems* What important ideas about government came from Enlightenment thinkers?

Critical Thinking

3. **Drawing Conclusions** Why might the French nobility and the church dislike *The Encyclopedia*?
4. **Analyzing** In what ways did John Locke and other philosophers of the Enlightenment help pave the way for democracy in the United States?

Organizing What You Know

5. **Categorizing** Copy the following graphic organizer. Use details from the chapter to fill it in. Then write a title for the chart.

Writer	Country	Important Ideas
Locke		
Voltaire		
Rousseau		

Section 2

Objectives

1. Describe the start and results of the American Revolution.
2. Examine how the French Revolution changed France.
3. Explain how Europe changed during and after the Napoléonic era.

FOCUS

LETS GET STARTED

Copy the following question onto the chalkboard: *What are some ways in which you might challenge governmental rules or laws that seem unjust?* Discuss responses and write a summary of students' suggestions on the chalkboard. Tell students that in Section 2 they will learn how both American and French citizens responded to their governments' injustices.

Building Vocabulary

Write the key terms on the chalkboard. Ask students to identify base words within some of the terms. *(Examples:* **Loyalists**—*loyal,* **alliance**—*ally,* **oppression**—*oppress,* **reactionaries**—*react)* Have students use what they know about the base words to speculate on the meanings of these key terms. Then have students locate and read the definitions for all the key terms, checking the inferences they have made about the meanings.

Section 2 The Age of Revolution

Read to Discover

1. What started the American Revolution and what were its results?
2. How did the French Revolution change France?
3. How did Europe change during and after the Napoléonic Era?

Define

- Patriots
- Loyalists
- alliance
- oppression
- Reign of Terror
- balance of power
- reactionaries

WHY IT MATTERS

The revolutions of the 1700s in the United States and France gave many new rights and freedoms to ordinary citizens. Use CNNfyi.com or other **current events** sources to find examples of recent revolutions that have occurred in countries around the world.

American teapot with anti-Stamp Act slogan

The American Revolution

Enlightenment philosophers' ideas about freedom, equality, and government were not confined to Europe in the 1700s. By the 1750s, the British had established 13 colonies along the Atlantic Coast in North America. These British colonists had developed a new way of life and a new relationship with their home country. The colonists held their own elections and made their own laws. However, the colonists were still British subjects, and they had no representation in the British Parliament.

The Growing Conflict While the British had colonies along the Atlantic Coast in North America, the French colonies—New France—lay to the north and west. As British colonists pushed westward into French-controlled territory, tensions mounted.

France and Great Britain had long been enemies in Europe. In 1754, their conflict spilled over into North America, sparking the French and Indian War. In Europe this war was called the Seven Years' War. It began in 1756 and ended in 1763. As the victor in this war, the British gained control of most of North America.

To help pay for the war, the British taxed goods that their colonists in North America needed. Many colonists thought these new taxes were unfair, since they had no representatives in Parliament to express their views. Americans resisted the new taxes by refusing to buy British goods.

▲ The Stamp Act required Americans to purchase stamps like this one and to place them on many types of public documents.

Section 2 RESOURCES

Reproducible

- Guided Reading Strategy 8.2
- Map Activity 8

Technology

- One-Stop Planner CD–ROM, Lesson 8.2
- Homework Practice Online
- HRW Go site

Reinforcement, Review, and Assessment

- Section 2 Review, p. 158
- Daily Quiz 8.2
- Main Idea Activity 8.2
- English Audio Summary 8.2
- Spanish Audio Summary 8.2

TEACH

Teaching Objective 1

LEVEL 1: (Suggested time: 30 min.) Copy the following graphic organizer onto the chalkboard, omitting the italicized answers. Pair students and have the pairs record causes of the American Revolution in the first column and results in the second column. Discuss the charts.
ENGLISH LANGUAGE LEARNERS, COOPERATIVE LEARNING

CAUSES AND RESULTS OF THE AMERICAN REVOLUTION

Causes of the American Revolution	Results of the American Revolution
Americans were influenced by Enlightenment ideas about freedom and popular sovereignty.	*The British recognized the independence of the United States.*
Americans were angered by taxation without representation.	*All land east of the Mississippi belonged to the United States.*
Colonies united against Great Britain by setting up and sending delegates to the Continental Congresses.	*Americans wrote the Articles of Confederation and the later the Constitution and set up a democracy.*

National Geography Standard 6

Places and Regions The 13 original colonies are usually grouped according to region. The Northern, or New England, Colonies were Connecticut, Massachusetts, New Hampshire, and Rhode Island. The Middle Colonies were Delaware, New Jersey, New York, and Pennsylvania. The Southern Colonies were Georgia, Maryland, North Carolina, South Carolina, and Virginia.

Activity: Have students label the 13 colonies on maps of the United States. Ask them to color code the colonies by region. You may also want students to conduct research on battles of the American Revolution and mark their locations on their maps.

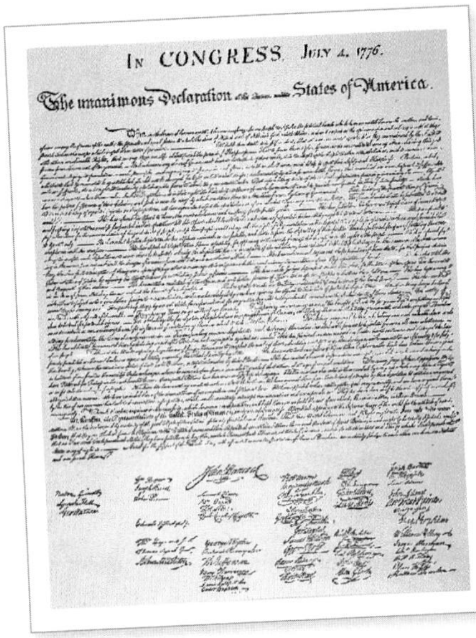
IN CONGRESS. JULY 4, 1776.
The unanimous Declaration of the thirteen united States of America.

The Declaration of Independence declared the American colonies free from British control. It was adopted on July 4, 1776—now celebrated as Independence Day.

As their unhappiness increased, the colonies united against the British. In 1774, 12 colonies sent representatives to the First Continental Congress. The Congress pledged to stop trade with Britain until the colonies had representation in Parliament.

Some American colonists believed the best way to guarantee their rights was to break away from British rule. Colonists called **Patriots** wanted independence. They made up one third of the population. Another third, the **Loyalists**, wanted to remain loyal to Great Britain. The rest of the colonists were undecided.

The Declaration of Independence In 1776 the Continental Congress adopted the Declaration of Independence. Thomas Jefferson was the Declaration's main author. The Declaration clearly showed the influence of Enlightenment thinkers, especially Locke and Rousseau. The Declaration stated that that "all men are created equal" and have the right to "life, liberty, and the pursuit of happiness." The ideal of individual liberty was only applied in a limited way. Women and slaves were not included. Nevertheless, the Declaration was still a great step forward toward equality and justice.

Locke's and Rousseau's ideas about popular sovereignty were clearly seen in the Declaration. It stated that all powers of government come from the people. It said that no government can exist without the consent of its citizens and that government is created to protect individual rights. In addition, it stated that if a government fails to protect these rights, the people may change it and set up a new government.

The first battle of the American Revolution was fought in Lexington, Massachusetts, on April 19, 1775.

War and Peace By the time the Declaration of Independence was written, the colonies were already at war with Great Britain. At first, the British seemed unbeatable. Then in late 1777, France formed an **alliance** with the Americans. An alliance is an agreement formed to help both sides. By helping the Americans, France hoped to weaken the British Empire.

In 1781 the American forces—commanded by George Washington—and their French allies defeated the main British army in Virginia. The Americans had won the Revolutionary War. The final peace terms were settled in the Treaty of Paris in 1783. The British recognized the independence of the United States. All land east of the Mississippi now belonged to the new country.

✓ **READING CHECK:** *Finding the Main Idea* How did the ideas of the Enlightenment influence the American Revolution? **The ideas liberty, equality, and popular sovereignty inspired the move toward self-government.**

LEVELS 2 AND 3: (Suggested time: 30 min.) Organize the students into small groups representing Patriots and Loyalists. Ask each group to give an account of the events leading up to the Revolution, as seen from its particular point of view. Similarly, the two groups can give their views of the Declaration of Independence and the events of the war and postwar period. You may want to have students present their views to the class in the form of skits or short plays. **COOPERATIVE LEARNING**

➤ASSIGNMENT: Have students write newspaper editorials dated July 4, 1776, about the signing of the Declaration of Independence. Tell students to write the editorial from the point of view of a Patriot newspaper editor. In addition to giving factual details about this important document, the editorial should include a vision of the country that might one day be formed on the principles in the Declaration. You may want to extend the assignment by asking students what arguments Loyalists would raise against the Patriot editorials.

Effects of American Independence

In 1777 the Americans adopted a plan of government called the Articles of Confederation. The Articles set up a central government, but it was purposely weak. Many Americans did not trust that a central government would always protect the individual rights and liberties they had fought for in the Revolution. Thus Congress could not levy taxes or coin money. It could not regulate trade. Within 10 years, however, it became clear that a weak central government was not helping the country to work as a whole.

In May 1787 delegates from all the states met at a convention in Philadelphia to revise the Articles. The delegates soon realized that making changes in the Articles would not be enough. They decided instead to write a new constitution.

After choosing George Washington to preside over the convention, the delegates went to work. They wanted a strong central government. They also wanted some powers kept for the states. As a result, the new Constitution they wrote set up a federal system of government. This is a system of government in which power is divided between a central government and individual states. The central government was given several important powers. It could declare war, raise armies, and make treaties. It could coin money and regulate trade with foreign countries. The states and the people kept all other powers. The Constitution was approved in 1789. The federal government had three branches. Each branch acted as a check on the power of the others. The executive branch enforced the laws. The legislative branch made the laws. The judicial branch interpreted the laws.

The American Revolution and the writing of the U.S. Constitution were major events in world history. Enlightenment ideas were finally put into practice. The success of the American democracy also encouraged people around the world. They realized they could fight for political freedoms, too.

Of course, American democracy in 1789 was not perfect. Women had few rights, and slaves had no rights at all. Still, the world now had a democratic country that inspired the loyalty of most of its citizens.

✓ **READING CHECK:** *Comparing and Contrasting* How was the new Constitution different from the old Articles of Confederation? **The Constitution gave the central government important powers, but allowed states to retain most of their powers to govern themselves.**

The United States in 1783

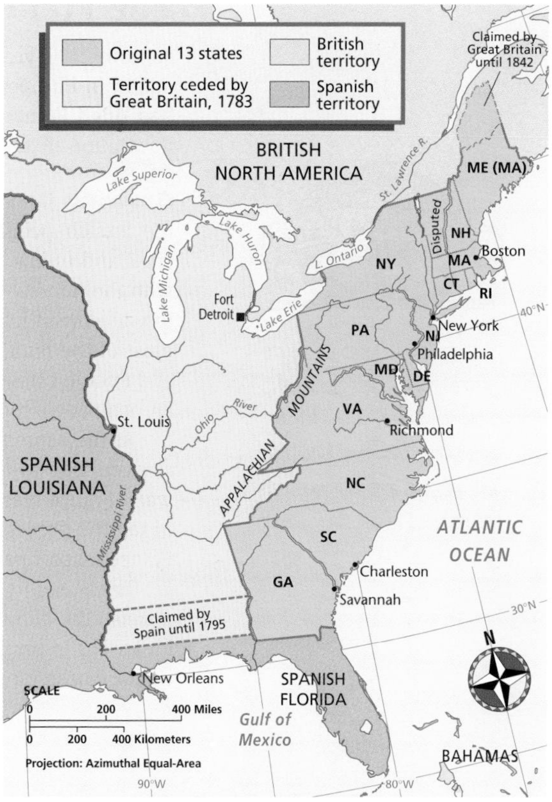

▲ The Treaty of Paris doubled the size of the United States.

Interpreting the Map **What nation controlled the region to the south of the United States? To the north? To the west?**

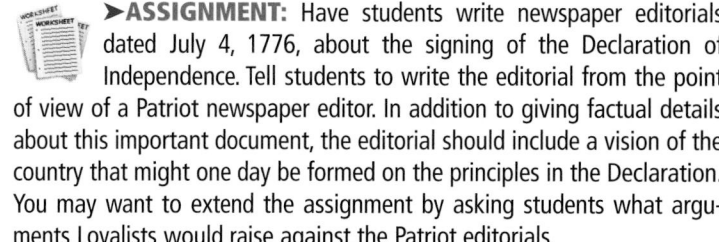

▲ George Washington (1732–1799) led the American troops to victory in the Revolution and was elected the first president of the United States.

Linking Past to Present

The U.S.

Congress At the 1787 Constitutional Convention, delegates from the large states wanted representation in Congress to be based on population. The small states objected, fearing they would have too little power in Congress. They wanted all states to have the same representation. Eventually the delegates reached an agreement called the Great Compromise. It provided for a Congress with two parts, or houses. In one of them—the House of Representatives—the number of representatives would depend on population. In the other—the Senate—each state would have the same number of representatives.

Activity: Have students find the names of the U.S. senators and representatives from your state. Ask them also to contrast the number of representatives from your state with those from larger and smaller states.

◄ **Map Answer**

Spain; Great Britain; Spain

Teaching Objective 2

LEVEL 1: (Suggested time: 15 min.) Ask students to use the following as the main headings of an outline: Causes of the French Revolution, The Outbreak of the French Revolution,. The End of the Monarchy, and The French Republic. Beneath each main heading, ask students to enter at least two or three subheadings that provide important ideas about the heading.

LEVEL 2: (Suggested time: 30 min.) Have pairs of students turn the following headings and subheadings in the text into questions: Growing Discontent, The Outbreak of Revolution, The End of the Monarchy, The French Republic. For example, the heading Growing Discontent might be rephrased as the question, *"Why was there growing discontent in France in the 1770s"?* Have students work together to write questions and then answer them. **COOPERATIVE LEARNING**

LEVEL 3: (Suggested time: 30 min.) Ask students to pretend they are Parisian citizens in July of 1789. Have them write diary entries about the events leading up to the storming of the Bastille. Each student's entry should include a description of the suffering that had taken place in Paris as well as an explanation for why the Bastille was chosen as a target.

USING ILLUSTRATIONS

Have students examine the painting of the Bastille on this page. Ask students why freeing the prisoners was a major accomplishment for the mob of angry citizens. *(Possible answer: The prison was a symbol of the king's power.)* Ask students to note other details about this historic event based on the painting. (*Possible answers: The mob seems to have set fire to part of the prison; the mob was armed with at least one cannon; some soldiers in uniform seem to have taken part in the attack.)* Invite students to draw conclusions about how the storming of the Bastille might have affected the revolutionaries. *(Possible answer: They would have been emboldened by this success to take further steps against the king and nobility.)*

You may also want to have students conduct further research on the storming of the Bastille and compare these facts with what they see in the illustration.

The French Revolution

For over 100 years France had been the largest and most powerful nation in Europe. For all of this time, a monarch with absolute power had ruled France. Yet within months of the beginning of the French Revolution in 1789, the king lost all power.

Growing Discontent As the United States won its independence, the French people struggled against **oppression**. Oppression is the cruel and unjust use of power against others. By the 1770s discontent with the nobility was widespread. Food shortages and rising prices led to widespread hunger. To make matter worse, the nobles, who owned most of the land, raised rents. Taxes were also raised on the peasants and middle classes while the nobles and the clergy paid no taxes. Some French people took to the streets, rioting against high prices and taxes.

At the same time, the French monarchy was losing authority and respect. Due to the king's expensive habits and spending on foreign wars, France was in deep debt. To pay the debts, King Louis XVI tried to tax the nobles and the clergy. When they refused to pay the taxes, France faced financial collapse.

The French peasants and middle classes had different complaints against the king. They did, however, share certain Enlightenment ideas. For example, they spoke of liberty and equality as their natural rights. These ideas united them against the king and nobles.

Our Amazing Planet

The nobles of France seemed to care little for the suffering of ordinary people. When Marie-Antoinette, the wife of King Louis XVI, was told that many peasants had no bread to eat, she is said to have replied, "Let them eat cake."

The Outbreak of Revolution In 1789 a group representing the majority of the people declared itself to be the National Assembly. It was determined to change the existing government. This action marked the beginning of the French Revolution.

When Louis XVI moved troops into Paris and Versailles, there was fear that the soldiers would drive out the National Assembly by force. In Paris, the people took action. Angry city dwellers destroyed the Bastille prison, which they called a symbol of royal oppression. The violence spread as peasants attacked manor houses and monasteries throughout France.

On July 14, 1789, a crowd destroyed the Bastille, freeing its prisoners. Bastille Day marked the spread of the Revolution and is celebrated in France every July 14.

ALL LEVELS: (Suggested time: 30 min.) Copy the following Venn diagram onto the chalkboard, omitting the italicized answers. Have students use the diagram to compare and contrast the causes of the French Revolution with the causes of the American Revolution. Then lead a discussion about how the French Revolution changed France.
ENGLISH LANGUAGE LEARNERS

CAUSES OF THE REVOLUTIONS

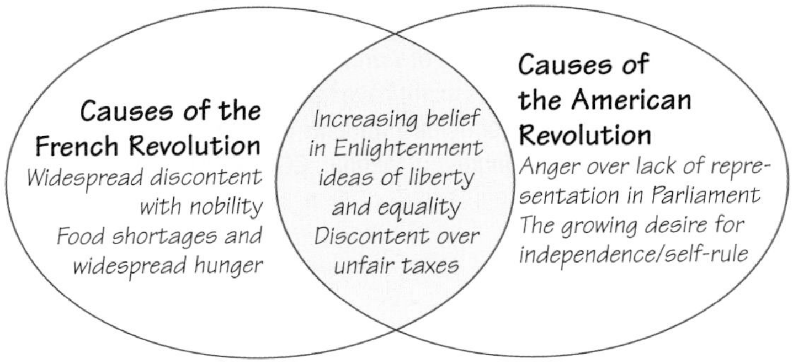
Causes of the French Revolution
Widespread discontent with nobility
Food shortages and widespread hunger
Increasing belief in Enlightenment ideas of liberty and equality
Discontent over unfair taxes
Causes of the American Revolution
Anger over lack of representation in Parliament
The growing desire for independence/self-rule

The National Assembly quickly took away the privileges of the clergy and nobles. Feudalism was ended and peasants were freed from their old duties. The National Assembly also adopted the *Declaration of the Rights of Man and Citizen.* This document stated that men are born equal and remain equal under the law. It guaranteed basic rights and also defined the principles of the French Revolution—liberty, equality, and fraternity.

The End of the Monarchy In 1791 the National Assembly completed a constitution for France. This constitution allowed for the king to be the head of the government, but limited his authority. The constitution divided the government into three branches—executive, legislative, and judicial. Louis XVI pretended to agree to this new government. In secret, he tried to overthrow it. When he and his family tried to escape France in 1791, they were arrested and sent back to Paris.

The French Republic In 1792 a new group of people, the National Convention, gathered and declared France a republic. The National Convention also put Louis XVI on trial as an enemy of the state. Urged on by a lawyer named Maximilien Robespierre, the members found the king guilty. In 1793, the king was sent to the guillotine, a machine that dropped a huge blade to cut off a person's head.

Robespierre was the leader of a political group known as the Jacobins. Many of the Jacobins wanted to bring about sweeping reforms that would benefit all classes of French society. As the French Revolution went on, the Jacobins gained more and more power. By the time Louis XVI was executed, Robespierre was probably the most powerful man in France. He and his allies controlled the actions of the National Convention.

LIBERTÉ,
EGALITÉ,
FRATERNI
TÉ OU LA
MORT

This poster summarizes the main goals of the French Revolution: liberty, equality, and fraternity. Fraternity means "brotherhood."

The frequent use of the guillotine shocked people in France, Europe, and the United States. As a result, the French Revolution lost many supporters.

Linking Past to Present

Louis XVI's lavish lifestyle contributed to his unpopularity. Maintaining his immense palace, Versailles, was a drain on the treasury. The palace was also a striking example of extravagance at the taxpayers' expense. While Louis XVI ruled, 10,000 people lived or worked in his household. There were special rooms for a clock, the king's throne, and the royal billiard table. The grounds at Versailles covered 2,000 acres and included miles of wide pathways and thousands of exotic plants. Versailles was sacked during the Revolution. Only a portion of its grandeur has been restored.

Activity: In 1999 a powerful storm swept through France and downed some 10,000 of the trees at Versailles. Have some students conduct research on the restoration of the palace's gardens. Ask a second group to investigate ongoing restoration of Versailles' interior. Have a third group research Versailles' role as a major tourist attraction.

Teaching Objective 3

LEVEL 1: (Suggested time: 30 minutes) Have pairs of students play the roles of ordinary French citizens who have lived through the Napoléonic era and the Congress of Vienna. Ask students to create conversations that the French citizens might have had about the accomplishments of their emperor and his later defeat. Students can also share their feelings about the events at the Congress of Vienna. **COOPERATIVE LEARNING**

LEVELS 2 AND 3: (Suggested time: 30 minutes) Have students work in groups to write newspaper headlines that tell about the accomplishments of Napoléon and the work of the Congress of Vienna. Ask each student to choose one of the group's headlines and write the opening, or lead, paragraph for an article about the event. You may want to provide additional resources. Have students read their paragraphs aloud to the class. You may want to extend the activity by having students complete the newspaper, with comics, an advice column, classified ads, and other features appropriate to the place and time. **COOPERATIVE LEARNING**

Across the Curriculum
MUSIC

National Anthems Play a recording of the French national anthem, *La Marseillaise*, for the class. Ask students to share their reactions to the music. Tell students that the anthem was written in 1792 by a captain in the French army, Rouget de Lisle. Originally a marching song, the words were sung by soldiers from Marseilles, a city in southern France, who were going to defend their homeland against invaders from Austria. Play the music again and ask students why a marching song is appropriate for a national anthem. *(Possible answer: The stirring rhythm inspires patriotism and feelings of pride among citizens.)*

Activity: Have students research the history of the American national anthem. Then have students compare and contrast *The Star-Spangled Banner* and *The Marseillaise*.

A group called the Jacobins controlled the National Convention. Robespierre was a powerful leader of the Jacobins.

Under Robespierre, the National Convention looked for other enemies. Anyone who had supported the king or criticized the revolution was a suspect. Thousands of people—nobles and peasants alike—died at the guillotine. This period, called the **Reign of Terror**, ended in 1794 when Robespierre himself was put to death. Despite the terror, the revolutionaries did achieve some goals. They replaced the monarchy with a republic. They also gave peasants and workers new political rights. They opened new schools and supported the idea of universal elementary education. They established wage and price controls in an effort to stop inflation. They abolished slavery in France's colonies. They encouraged religious tolerance.

Between 1795 and 1799, a government called the Directory tried to govern France. A new two-house legislature was created to make laws. This legislature also elected five officials called directors to run the government. The people selected to be directors, however, could not agree on many issues. They were corrupt and quarreled about many issues. They quickly became unpopular with the French people. In addition, by 1799 enemy armies were again threatening France. Food shortages were causing panic in the cities. Many French people concluded their country needed one strong leader to restore law and order.

✓ **READING CHECK:** *Summarizing* Why did many French peasants and poor workers support the Revolution? believed in liberty and equality, tired of being oppressed

Napoléon had a way of getting the public's attention. He was very popular with the French people.

The Napoléonic Era

In 1799 a young general named Napoléon Bonaparte overthrew the Directory and took control of the French government. Most people in France accepted Napoléon. In turn he supported the changes brought about by the Revolution. In 1804 France was declared an empire, and Napoléon was crowned emperor.

Napoléon as Emperor In France Napoléon used his unlimited power to restore order. He organized French law into one system—the Napoléonic Code. He set up the Bank of France to run the country's finances. Influenced by the Enlightenment, he built schools and universities.

A brilliant general, Napoléon won many land battles in Europe. By 1809 he ruled the Netherlands and Spain. He forced Austria and Prussia to be France's allies. He abolished the Holy Roman Empire. He also unified the northern Italian states into the Kingdom of Italy, under his control. Within five years of becoming emperor, Napoléon had reorganized and dominated Europe. Because of the important role that he played, the wars that France fought from 1796 until 1815 are called the Napoléonic Wars.

ALL LEVELS: (Suggested time: 30 min.) Copy the following graphic organizer onto the chalkboard, omitting the italicized answers. Invite students to summarize some of Napoléon's achievements by describing conditions in France and Europe before and after Napoléon's rise to power.

NAPOLÉON'S FRANCE

Before Napoléon	After Napoléon
enemy armies threatening France	*France had reorganized and dominated Europe.*
law and order breaking down	*French law organized into one system—the Napoléonic Code*
food shortages in cities	*France's finances run by Bank of France*
feudalism and serfdom common in countries around France	*Feudalism and serfdom abolished by Napoléon*

Napoléon also made changes in the lands he controlled. He put the Napoléonic Code into effect in the countries he conquered and abolished feudalism and serfdom. He also introduced new military techniques throughout Europe. Without intending to, the French increased feelings of loyalty and patriotism among the people Napoléon had conquered. In some places this increased opposition to French rule. Over time, the armies of Napoléon's enemies grew stronger.

In 1812 Napoléon invaded Russia with more than 500,000 soldiers. The invasion was a disaster. The cold Russian winter, hunger, and disease claimed the lives of most of the French soldiers. Napoléon finally ordered his soldiers to retreat.

This painting captures the glory of Napoléon as a military leader.

Napoléon's Defeat The monarchs of Europe took advantage of Napoléon's weakened state. Prussia, Austria, and Great Britain joined together to invade France. These allies captured Paris in 1814. Napoléon gave up the throne and went into exile on the island of Elba, near Italy. Louis XVIII, the brother of the executed king, was made the new king of France.

The following year Napoléon made a short-lived attempt to retake his empire. This period is known as the Hundred Days. Between March and June of 1815, Napoléon regained control of France. The king fled into exile. Soon, however, Napoléon's enemies sent armies against him. The other European nations defeated him at Waterloo in Belgium. Napoléon was sent to St. Helena, a small island in the South Atlantic. He lived there under guard and died in 1821. In 1840 the British allowed the French to bring Napoléon's remains back to Paris, where they lie to this day.

Governments of France, 1774–1814

1774	Louis XVI becomes king.
1789	Third Estate, as the National Assembly, assumes power.
1791	Legislative Assembly, with Louis XVI as constitutional monarch, begins rule.
1799	Napoléon establishes himself as First Consul.
1804	Napoléon is crowned emperor.
1814	Napoléon is defeated and the monarchy is restored.

After being ruled as a republic and then an empire, France became a monarchy once again in 1814.

Europe After Napoléon During the years of Napoléon's rule, France had become bigger and stronger than the other countries in Europe. After Napoléon's defeat, delegates from all over Europe met at the Congress of Vienna. Their goal was to bring back the **balance of power** in Europe. Having a balance of power is a way to keep peace by making sure no one nation or group of nations becomes too powerful.

Napoléon had not always upheld the ideals of the French Revolution, but he did extend their influence throughout Europe. This led other governments to fear that rebellions against monarchy might spread. Having defeated Napoléon, the major European powers wanted to restore order, keep the peace, and suppress the ideas of the revolution.

USING ILLUSTRATIONS

Jacques-Louis David (1748–1825) was the leading French painter during the French Revolution and the Napoléonic era. His popular paintings helped build support for the Revolution. In fact, as a member of the Jacobin political party, David himself voted to send King Louis XVI to the guillotine. Later the artist recorded the big events of Napoléon's life. Have students study David's portrait of Napoléon on this page. Ask students to note some techniques the painter has used to portray Napoléon in a positive way. *(Possible responses: Napoléon is shown leading a dramatic military charge; he seems larger than life and is pointing the way with his right hand; high in the mountains and close to the sky, Napoléon is shown as someone to look up to.)*

CLOSE

Play the *1812 Overture*, by Pyotr Ilich Tchaikovsky, for the class. This famous symphony commemorates Napoléon's ultimate defeat by the brutal Russian winter. Challenge students to listen for the Russian church music and folk music that the composer worked into the score. Tchaikovsky even incorporated the French national anthem into the music.

REVIEW AND ASSESS

Have students complete the Section Review. Then call on volunteers to read aloud terms, phrases, dates, and names from the section. Ask other students to explain the words' significance. Have students complete Daily Quiz 8.2.

RETEACH

Have students complete Main Idea Activity 8.2. Then have pairs of students portray TV news reporters and anchorpersons and report on events of the American or French Revolutions. Have pairs deliver their oral reports of the events in chronological order. **ENGLISH LANGUAGE LEARNERS**

EXTEND

Have interested students conduct research on one of the following historical American gatherings: the First Continental Congress (1775), the Second Continental Congress (1776), the Constitutional Convention (1787). Or, students can research these important French gatherings: the National Assembly (1789); the National Convention (1792), the Committee of Public Safety (1793). **BLOCK SCHEDULING**

Section 2 Review

Answers

Define For definitions, see Patriots, p. 152; Loyalists, p. 152; alliance, p. 152; oppression, p. 154; Reign of Terror, p. 156; balance of power, p. 157; reactionaries, p. 158

Reading for the Main Idea

1. Colonists thought it was unfair that they had no representation in Parliament. (NGS 13)
2. Causes—rising costs of food and taxes; resentment toward the nobility and clergy; Effects—new ruling bodies, end of the monarchy; improvements in some areas of life and new political rights (NGS 11)
3. restored order to France, reorganized and dominated Europe, established reforms (NGS 13)

Critical Thinking

4. It was not until 1789 that the Constitution went into effect and the modern American political system was created.

Organizing What You Know

5. 1776—Declaration of Independence; 1783—Treaty of Paris; 1789—U.S. Constitution approved; 1789—Mob storms Bastille; 1799—Napoléon seizes control of France; 1812—Napoléon invades Russia; 1814—Congress of Vienna; 1815—Napoléon defeated

Many delegates to the Congress of Vienna were **reactionaries**. Reactionaries not only oppose change. They would like to actually undo certain changes. In this case they wanted to return to an earlier political system. These delegates were not comfortable with the ideals of the French Revolution, such as liberty and equality. They worried that these ideals would overturn the monarchies in their own countries.

One of the most influential leaders at the Congress of Vienna was Prince Metternich of Austria. To protect his absolute power in Austria, Metternich suppressed ideas such as freedom of speech and of the press. He encouraged other leaders to censor newspapers and to spy on individuals they suspected of revolutionary activity.

The Congress of Vienna redrew the map of Europe. Lands that Napoléon had conquered were taken away from France. In the end, France's boundaries were returned to where they had been in 1790. Small countries around France were combined into bigger, stronger ones. This was done to prevent France from ever again threatening the peace of Europe. France also had to pay other countries for the damages it had caused. Ruling families were returned to their thrones in Spain, Portugal, and parts of Italy. Switzerland alone kept its constitutional government but had to promise to remain neutral in European wars.

The Congress of Vienna also led to an alliance between Great Britain, Russia, Prussia, and Austria. The governments of these countries agreed to work together to keep order in Europe. For 30 years the alliance successfully prevented new revolutions in Europe.

Many of Europe's royal families came to Vienna during the winter of 1814–15. They attended balls while diplomats and rulers discussed the situation of Europe after Napoléon.

✓ **READING CHECK:** *Drawing Conclusions* Why did the other countries of Europe want to defeat Napoléon? to restore a balance of power in Europe

Section 2 Review

go.hrw.com Homework Practice Online
Keyword: SK3 HP8

Define and explain Patriots, Loyalists, alliance, oppression, Reign of Terror, balance of power, reactionaries

Reading for the Main Idea

1. *Human Systems* Why did the 13 American colonies rebel against Great Britain?
2. *Human Systems* What were the causes and effects of the French Revolution?
3. *Human Systems* How did Napoléon change France and the rest of Europe?

Critical Thinking

4. **Supporting a Point of View** Many people argue that the United States was not really created until 1789. Why do you think this is so? Explain your answer.

Organizing What You Know

5. **Identifying Time Order** Copy the following time line. Use it to list some important events of both the American Revolution and the French Revolution.

1775 — 1780 — 1785 — 1790 — 1795 — 1800

Section 3

Objectives

1. Describe the beginnings of the Industrial Revolution.
2. Explain how developments in transportation and communications spread industrial development.
3. Identify business features that affected life in the Industrial Age.

Focus

LET'S GET STARTED

Write the word *revolution* on the chalkboard and ask students to think about its various meanings. Remind students that the previous section described political revolutions brought about by military means. Point out that a second definition for revolution is "a complete change." Have students name some major changes in society that have been or might be called revolutions. Tell students that they will learn about one of these, the Industrial Revolution, in this section.

Building Vocabulary

Write the key terms on the chalkboard. Circle the word *production* in **factors of production** and **mass production**. Ask students to provide a definition for *production (the act of making things).* Based on this definition, have students infer what the key terms might mean *(mass production—making large numbers of things; factors of production—items needed to produce things).* Then have students locate and read the definitions of the other key terms.

Section 3 The Industrial Revolution

Read to Discover

1. How did the Industrial Revolution begin?
2. What developments in transportation and communications helped spread industrial development?
3. What features of business affected life in the Industrial Age?

Define

Industrial Revolution
factors of production
capital
factories
capitalism
mass production

Why It Matters

The high standards of living that most American enjoy today were made possible by the Industrial Revolution. Use CNNfyi.com or other **current events** sources to find out more about industry and industrialized nations.

An early steam locomotive

The Origins of the Industrial Revolution

In the early 1700s inventors began putting the ideas of the Scientific Revolution to work by creating many new machines. Advances in industry, business, transportation, and communications changed people's lives around the world in almost every way. This period, which lasted through the 1700s and 1800s, was called the **Industrial Revolution**.

New Needs in Agriculture The first stages of the Industrial Revolution took place in agricultural communities in Great Britain. Ways of dividing, managing, and using the land had changed greatly since the Middle Ages. People had begun to think about land in new ways. Wealthy farmers began to buy more land to increase the size of their farms. Small farmers, unable to compete with these large operations, sometimes lost their land. At the same time, Europe's population continued to grow, which meant that the demand for food grew as well. Farmers recognized the need to improve farming methods and increase production.

One such farmer was Jethro Tull. He invented a new farm machine, called a seed drill, for planting seeds in straight rows. More inventors soon followed with other new farm machines. The machinery made farms more productive, and farmers were able to grow more food with fewer workers. As a result, many farm workers lost their jobs. Many of these people moved to cities to look for other kinds of work.

This painting shows the original McCormick reaper, used to cut grain. It was invented by Cyrus H. McCormick in 1831.

Section 3 Resources

Reproducible
- Guided Reading Strategy 8.3

Technology
- One-Stop Planner CD–ROM, Lesson 8.3
- Homework Practice Online
- HRW Go site

Reinforcement, Review, and Assessment
- Main Idea Activity 8.3
- Section 3 Review, p. 163
- Daily Quiz 8.3
- English Audio Summary 8.3
- Spanish Audio Summary 8.3

TEACH

Teaching Objective 1

ALL LEVELS: (Suggested time: 30 min.) Copy the following graphic organizer onto the chalkboard, omitting the italicized answers. Ask students to copy it into their notebooks. Pair students and have each complete the chart with descriptions of how Great Britain was able to begin to industrialize. **ENGLISH LANGUAGE LEARNERS**

THE BEGINNINGS OF THE INDUSTRIAL REVOLUTION	
Factor of Production	How It Was Supplied in Britain
Land	*Britain had rich deposits of coal and iron ore and fast-moving rivers for water power.*
Labor	*Many workers left farms due to changes in agriculture. They looked for jobs in factories.*
Capital	*British had money and tools to buy and outfit factories.*

Linking Past to Present

The Textile Industry Changes in the factors of production, especially labor, have played an important role in the American textile industry over the last 200 years. In the early 1800s most textile mills were established in New England near fast-moving rivers that provided power. Displaced farm workers and immigrants provided much of the labor.

By the early 1900s the old textile mills of New England were shutting down as capitalists moved their operations to the South to take advantage of lower labor costs. Then, in the late 1900s, many southern mills were closed as textile companies again looked for cheaper labor in foreign countries.

Activity: Have students conduct research to create time lines illustrating key events in the history of the American textile industry.

Our Amazing Planet

One early water-powered machine in an English mill was said to spin more than 300 million yards of silk thread every day!

Factors of Production The Industrial Revolution began in Great Britain because the country had the right **factors of production**. These are items necessary for industry to grow. They include land, natural resources, workers, and **capital**. Capital refers to the money, and tools needed to make a product.

Great Britain had rich deposits of coal and iron ore. It also had many rivers to provide water power for **factories** and transportation. Money was available, since many British people had grown wealthy during the 1700s. They were willing to invest their money in new businesses. The British government allowed people to start businesses and protected their property. Labor was available since many ex-farm workers needed jobs.

✓ **READING CHECK:** *Summarizing* What factors of production helped Great Britain to develop early industries? land; natural resources including coal, iron ore, and many rivers; investment capital; many available workers; and political stability

The Growth of Industry

As mentioned earlier, agricultural needs led to new machines and methods for farming. People in other industries began to wonder how machines could help them as well. For example, before the early 1700s, British people had spun thread and woven their own cloth at home on simple spinning wheels and looms. It was a slow process, and the demand for cloth was always greater than the supply.

This painting shows one artist's view of a factory. By 1800, textiles made in English factories were shipped all over the world.

The Textile Industry To speed up cloth making in the early 1700s, English inventors built new types of spinning machines and looms. In 1769 Richard Arkwright invented a water-powered spinning machine. He eventually set up his spinning machines in mills and hired workers to run them. Workers earned a fixed rate of pay for a set number of hours of work. Arkwright brought his workers and machinery together in a large building called a factory. Arkwright's arrangement with his workers was the beginning of the factory system.

In 1785 Edmund Cartwright built a water-powered loom. It could weave cloth much faster than could a hand loom. In fact, one worker with a powered loom could produce as much cloth as several people with traditional ones. Each new invention that improved the spinning and weaving process led to more inventions and improvements.

Teaching Objective 2

LEVEL 1: (Suggested time: 25 min.) Provide students with blank index cards. Have each student make a flash card for each of the following items: *water-powered spinning machines and looms, factories, steam engine, canals, paved roads, locomotive, telegraph.* On the back of each card, have students write a sentence or two that describes how each item helped industry develop and spread. **ENGLISH LANGUAGE LEARNERS, COOPERATIVE LEARNING**

LEVELS 2 AND 3: (Suggested time: 45 min.) Tell students to imagine they are the inventors of one of the machines described in this section. Ask each student to write a short speech in which he or she describes his invention and how it helped the Industrial Revolution to begin or spread. Encourage students to conduct additional research to find more information for their speeches. Call on volunteers to deliver their speeches.

The factory system soon spread to other industries. Machines were invented to make shoes, clothing, furniture, and other goods. Machines were also used for printing, papermaking, lumber and food processing, and for making other machines. More and more British people went to work in factories and mills.

The Steam Engine Early machines in factories were driven by water power. This system, however, had drawbacks. It meant that a factory had to be located on a stream or river, preferably next to a waterfall or dam. In many cases these streams and rivers were far from raw materials and overland transportation routes. The water flow in rivers can change from season to season, and sometimes rivers run dry. People recognized that a lighter, movable, and more dependable power source was needed. Many inventors thought using steam power to run machines was the answer.

Steam engines boil water and use the steam to do work. Early steam engines were not efficient though. In 1769 James Watt, a Scottish inventor, built a modern steam engine that did work well. With Watt's invention, steam power largely replaced water power. This meant that factories could be built anywhere.

The factory system changed the lives of workers. In the past, workers had taken years to learn their trades. In a factory, however, a worker could learn to run a machine in just a day or two. Factory owners hired unskilled workers—often young men, women, and children—and paid them as little as possible. As a result, the older skilled workers were often out of work.

▲ New industries needed much steel for machinery. The Bessemer converter, invented in the 1850s, was a cheaper, better way to make steel.

✓ **READING CHECK:** *Cause and Effect* How was the textile industry created in Great Britain? the invention of machines to spin and weave produced enough cloth to sell at home and abroad

The Spread of the Industrial Revolution

Great Britain quickly became the world's leading industrial power. British laws encouraged people to use capital to set up factories. Great Britain's stable government was good for industry too.

The rest of Europe did not develop industry as quickly. For one thing, the French Revolution and Napoléon's wars had disrupted Europe's economies. That made it difficult to put the factors of production to work. Many countries also lacked the resources needed to industrialize.

HUMAN SYSTEMS

During the early decades of the Industrial Revolution, many serious problems arose in Great Britain. As more and more people migrated to industrial cities, severe overcrowding occurred. Unsanitary conditions led to outbreaks of disease. Workers, even young children, labored 10 to 14 hours a day, six days a week. Their factory jobs were monotonous and often dangerous. Wages were kept as low as possible, and workers often went hungry. Conditions were so bad that factory employees occasionally rioted, destroying the machines that they blamed for their misery.

Critical Thinking: Why do you think the conditions of industrial workers gradually improved during the 1800s?

Answer: Politicians began to act in the interests of the working class; laws were passed to keep workers safer and healthier.

Teaching Objective 3

LEVEL 1: (Suggested time: 15 min.) Lead a class discussion on inventions and business practices that made mass production possible. *(Possible answers: Capitalists divided manufacturing into a series of steps; interchangeable parts were used to make products; assembly lines carried parts to workers.)* Then have students describe some of the effects of mass production for society in general. *(Possible answer: The cost of goods was lowered; more people could buy goods and enjoy a higher standard of living.)*

LEVELS 2 AND 3: (Suggested time: 20 min.) Ask students to imagine they have just gotten jobs in one of the large new factories of the late 1800s. In a letter to a relative or friend, have each student describe the factory in which he or she works and describe how his or her work is organized. Direct students to note in their letters which aspects of their jobs they like and which they dislike. Remind students that their letters should mention what type of product is made in the factory.

Across the Curriculum
TECHNOLOGY

Early Inventions Early steam engines lost so much steam and power that they were too costly to operate. By studying the properties of steam, James Watt was able to design a condensing chamber that made the engine more efficient and less likely to lose steam.

Samuel Morse's early telegraph worked much like a doorbell. The sending device was a switch, or key, that acted like a doorbell button. When pressed, the key completed the telegraph circuit, allowing current to flow to a receiving sounder. When the key was released, the circuit was broken and no current flowed.

Activity: Invite interested students to research the scientific principles on which the steam engine or telegraph are based. Ask these students to prepare short talks about these inventions, using diagrams that can be shown on the board or presented on paper as handouts.

The Industrial Revolution did spread quickly to the United States though. The United States had a stable government, rich natural resources, and a growing labor force. Americans were quick to adapt British inventions and methods to their own industries.

These steamboats from the 1850s carried people and goods on the Mississippi River.

Transportation Since the Middle Ages, horse-drawn wagons had been the main form of transportation in Europe. Factory owners needed better transportation to get raw materials and send goods to market. To move goods faster, stone-topped roads were built in Europe and the United States. Canals were dug to link rivers. The steam engine was also put to work in transportation. In 1808 American inventor Robert Fulton built the first steamboat. Within a few decades, steamships were crossing the Atlantic.

Steam also powered the first railroads. An English engineer, George Stephenson, perfected a steam locomotive that ran on rails. By the 1830s, railways were being built across Great Britain, mainland Europe, and the United States.

Communication Even before 1800, scientists had known that electricity and magnetism were related. American inventor Samuel F. B. Morse put this knowledge to practical use. Morse sent an electrical current through a wire. The current made a machine at the other end click. Morse also invented a code of clicking dots and dashes to send messages this way.

Morse's inventions—the telegraph and the Morse code—brought about a major change in communications. Telegraph wires soon stretched across continents and under oceans. Suddenly information and ideas could travel at the speed of electricity.

The telegraph revolutionized communications in the 1850s. This device is a telegraph receiver.

Life in the Industrial Age

The 1800s are sometimes called the Industrial Age. This was an age of new inventions. It was a time when businesses found new ways to produce and distribute goods. The owners of factories in the Industrial Age often became very wealthy. Low factory wages, however, meant many workers faced poverty.

CLOSE

Call on students to summarize some of the ways in which the Industrial Revolution changed everyday life and work. Then ask students if they would have preferred to live before, during, or after the Industrial Revolution.

REVIEW AND ASSESS

Have students complete the Section Review. Call on students to choose one of the key terms from this section and summarize its significance with regard to the Industrial Revolution. Then have students complete Daily Quiz 8.3.

RETEACH

Have students complete Main Idea Activity 8.3. Ask students to imagine that they are a group of American capitalists from the late 1800s trying to decide what type of factory to build and where to build it. Have students discuss how the factors of production, as well as transportation and communication, might affect their decision. **ENGLISH LANGUAGE LEARNERS**

EXTEND

Invite students to conduct research on changes in the factory system since the Industrial Revolution. Encourage students to focus their attention on laws and policies that were passed to protect the rights and safety of workers. Call on volunteers to share their findings with the class. **BLOCK SCHEDULING**

The Rise of Capitalism In the late 1800s European and American individuals owned and operated factories. This economic system is called **capitalism**. In a capitalist system, individuals or companies, not the government, control the factors of production.

The early capitalists wanted to make as much profit as possible from their factories. They divided each manufacturing process into a series of steps. Each worker performed just one of the steps, over and over again. This division of labor meant workers could produce more goods in less time.

Factory owners used machines to make the parts for their products. These parts were identical and interchangeable. To speed up production, the parts were carried to the workers in the factory. Each worker added one part, and the product moved on to the next worker. This method of production is called an assembly line.

▲ In the early 1900s, Henry Ford used an assembly line to build cars.

Interpreting the Visual Record **How do you think the assembly line might have made work easier for these people?**

Mass Production The division of labor, interchangeable parts, and the assembly line made mass production possible. **Mass production** is a system of producing large numbers of identical items. Mass production lowered the cost of clothing, furniture, and other goods. It allowed more people to buy manufactured products and to enjoy a higher standard of living.

✓ READING CHECK: *Finding the Main Idea* What is capitalism and how did it affect the Industrial Revolution? private ownership of business; more capital available to invest in factories and inventions

go.hrw.com Homework Practice Online
Keyword: SK3 HP8

Section 3 Review

Define and explain Industrial Revolution, factors of production, capital, factories, capitalism, mass production

Reading for the Main Idea

1. *Environmental and Society* Where did the Industrial Revolution begin and why?
2. *Human Systems* What advances in transportation and communications helped to spread the Industrial Revolution?
3. *Human Systems* How did capitalism and mass production affect people's standard of living in the late 1800s?

Critical Thinking

4. **Drawing Conclusions** Why do you think the steam engine was such an important invention of the Industrial Revolution?

Organizing What You Know

5. **Categorizing** Copy the following graphic organizer. Use it to describe some important inventions of the Industrial Revolution.

Invention	Inventor	Importance
seed drill		
spinning machine		
water-powered loom		
steam engine		
steam locomotive		
telegraph		

Section 3 Review Answers

Define For definitions, see: Industrial Revolution, p. 159; factors of production, p. 160; capital, p. 160; factory system, p. 160; capitalism, p. 163; mass production, p. 163

Reading for the Main Idea

1. Great Britain; sufficient land, resources, capital, labor; stable government (NGS 16)
2. roads, canals, steamship, locomotive, telegraph (NGS 11)
3. lowered the cost of goods, which improved standard of living (NGS 11)

Critical Thinking

4. allowed the development of mechanized industries and transportation systems

Organizing What You Know

5. seed drill—Jethro Tull, mechanized farming; spinning machine—Richard Arkwright, made thread easier to spin; water-powered loom—Edmund Cartwright, produced cloth faster; steam engine—James Watt, new source of power; steam locomotive—George Stephenson, transport people and goods quickly; telegraph—Samuel F.B. Morse, allowed long-distance communication

◄ **Visual Record Answer**

They only had to learn how to perform one task and did not have to move around much.

Section 4

Objectives

1. Describe changes that occurred in Europe and the United States after 1850.
2. Identify factors that led to reform in the later 1800s.
3. Explain how nationalism changed the map of Europe in the mid-1800s.

Focus

LET'S GET STARTED

Copy the following question onto the chalkboard: *What are some forms of new technology that affect your lives every day?* Discuss responses. *(Possible answers: cell phones, e-mail, personal computers, video games)* Ask students to name some older inventions that have made this newer technology possible. *(Possible answers: electricity, telephones, television)* Tell students that they will learn about the development of some of these inventions and how they affected peoples lives in Section 4.

Building Vocabulary

Write the key terms on the chalkboard. Explain that a number of these terms have Latin roots. **Literacy** comes from the Latin word *littera*, meaning "letter." **Emigrate** is derived from the verb *migrare*, meaning "to wander," and the prefix *e-*, meaning "out." **Suburbs** is a combination of the word *urbs*, meaning "city," and the prefix *sub-*, "below." Ask students to think of other words based on the same roots. *(Possible answers: literature, migration, urban, submarine)* Then have students find and read the definitions of all the key terms.

Section 4 Resources

Reproducible
- Guided Reading Strategy 8.4

Technology
- One-Stop Planner CD–ROM, Lesson 8.4
- Homework Practice Online
- HRW Go site

Reinforcement, Review, and Assessment
- Main Idea Activity 8.4
- Section 3 Review, p. 169
- Daily Quiz 8.4
- English Audio Summary 8.4
- Spanish Audio Summary 8.4

Section 4 Expansion and Reform

Read to Discover

1. How did life in Europe and America change after 1850?
2. What led to reforms in the later 1800s?
3. How did nationalism change the map of Europe in the mid-1800s?

Define

working class
literacy
emigrate
suburbs
reform
suffragettes
nationalism

Why It Matters

By the later 1800s, new ideas and technology began to improve city life. Use CNNfyi.com or other **current events** sources to find out about solutions to today's urban problems.

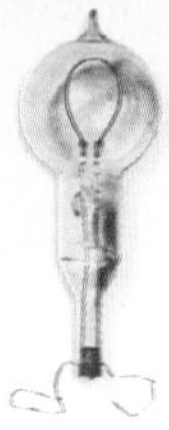

Thomas Edison's electric light bulb

The Rise of the Middle Class

The Industrial Revolution changed how people in Europe and America worked and lived. Industries and cities grew, and new inventions made life easier.

During the later 1800s many people became better educated. Some became wealthy. This group included bankers, doctors, lawyers, professors, engineers, factory owners, and merchants. Also in this group were the managers who helped keep industries running. Together these people and their families were known as the middle class. Membership in the middle class was based upon economic standing rather than upon birth.

Doctors were one of the groups who made up the middle class of the late 1800s. Medical advances made during this period made their jobs safer and more efficient.

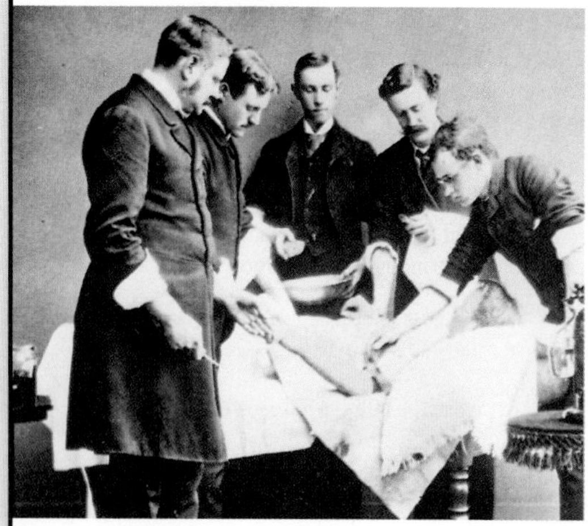

The ideas of the middle class influenced many areas of life in Western Europe and in the United States. Over time, the middle class's wealth, social position, lifestyle, and political power grew. Government leaders began turning to some middle-class individuals for advice, particularly about business and industry.

Many middle-class families had enough money that women did not need to work outside the home. They cleaned, cooked, and took care of the children, often with hired help. In the mid-1800s, however, many middle-class women started to express a desire for roles outside the home.

TEACH

Teaching Objective 1

ALL LEVELS: (Suggested time: 30 minutes) Copy the following graphic organizer onto the chalkboard, omitting the italicized answers. Ask students to copy it into their notebooks. Pair students and have each pair fill in the chart, identifying one or more effects for each listed cause. Then invite students to add additional examples of causes and effects that they read about in this section. Call on volunteers to share their answers with the class. **ENGLISH LANGUAGE LEARNERS, COOPERATIVE LEARNING**

CHANGES IN THE LATE 1800S

Cause: Effect:	**Scientists learned more about food, health, and disease.** *People were able to live longer and healthier lives.*
Cause: Effect:	**Governments required all children to go to school.** *Literacy became widespread. More books and magazines were published. People became more informed.*
Cause: Effect:	**City dwellers had more time and money for entertainment.** *Theaters and libraries opened. Sports teams became more organized. Parks were built*

For some women, doing something outside the home meant independence. It was also a way to earn a living. During the late 1800s more jobs opened up to women. They became nurses, secretaries, telephone operators, and teachers.

✓ **READING CHECK:** *Finding the Main Idea* What role did the middle class play in the society of the late 1800s? became the professional and management people of society; gave advice to the government about business and industry

▲ After 1870, more and more women went to high school. For the first time they began to study the same subjects as men did.

The Growth of Society

The middle class was not the only group of people to enjoy the benefits of the Industrial Age. By the 1870s life was improving in some ways for both the middle class and the **working class,** people who worked in factories and mines.

Technology and Communication In the 1870s a tremendous new power source was developed. That power source was electricity. This led to a new wave of inventions in Western Europe and the United States. The electric generator produced the power needed to run all kinds of machines and engines. Thomas Edison's electric light bulb created a new way of lighting rooms, streets, and cities. Alexander Graham Bell's telephone made it possible to transmit the human voice over long distance.

In the late 1800s the first successful gasoline-driven automobile was built. In 1908 American inventor Henry Ford produced the Model T. This was the first automobile to become popular with American buyers.

Other Advances Advances in science and medicine also transformed people's lives. Scientists discovered more about the connection between food and health. This new knowledge, plus new information about diseases, made it possible for people to live healthier, longer lives.

Scientists of this time also made great advances in the fields of chemistry and physics. For example, they formed new theories about the structure of the atom and organized all known elements into the periodic table. It was also at this time that X-rays were discovered and first used in medicine. Scientists like Max Planck and Albert Einstein developed new ideas that changed the study of physics. Their ideas were the basis for the work of many later experiments.

Several new fields of study, together called the social sciences, gained popularity during this period. Scholars saw these fields as a way to study people as members of society. The social sciences include such fields as economics, politics, anthropology, and psychology. The study of history also changed. Historians searched for evidence of the past in documents, diaries, letters, and other written sources. As a result, new views of history began to emerge from their research.

◄ Scientists of the 1800s used microscopes like this one to study cells.

Across the Curriculum

SCIENCE

The Fight Against Disease

In the struggle against disease, the discoveries of Louis Pasteur, Joseph Lister, and Robert Koch mark an important turning point. The French chemist Louis Pasteur identified bacteria and explained how they reproduce and cause disease. Joseph Lister, an English surgeon, used Pasteur's findings to develop *antisepsis*—the use of chemicals to kill disease-causing bacteria. Robert Koch, a German doctor, further confirmed Pasteur's findings by identifying the germs that caused tuberculosis and cholera.

Activity: Ask interested students to conduct research on one of these three scientists and to describe the experiments he used to make his discoveries.

Teaching Objective 2

LEVEL 1: (Suggested time: 15 min.) Have students draw three columns in their notebooks labeled *Reforms in Great Britain, Reforms in France*, and *Reforms in the United States.* Then have students look through the text to find examples of reforms to list under each heading. Call on volunteers to write their answers on the chalkboard. **ENGLISH LANGUAGE LEARNERS**

LEVELS 2 AND 3: (Suggested time: 30 min.) Organize the students into small groups and have the members of each group imagine that they are living in Great Britain, France, or the United States in the mid-1800s. Have the groups discuss the reforms that are occurring in their respective countries and how these reforms might affect them. Then have each student write a letter to a member of another group describing the reforms and his or her thoughts about them. **COOPERATIVE LEARNING**

Linking Past to Present

Colleges for Women By the end of the 1800s many countries offered elementary education for girls, but secondary education was limited. Some people argued that many subjects were not necessary or proper for women. In the United States, Great Britain, and France, secondary education for girls focused on languages, literature, and home economics—not the sciences or mathematics. Some people objected to these differences.

A British woman named Emily Davies urged her government to prepare women to attend universities. However, few colleges admitted women as students during the 1800s. Thus, colleges just for women began to appear in Great Britain and the United States. Some colleges today still admit only women.

Discussion: Lead a discussion based on the following questions: Is there still a need for women's colleges? Most colleges that once admitted only men are now co-educational. Should all colleges be co-educational?

Visual Record Answer

Possible answers: People dressed up to play sport. Sports were popular with people who had money.

Public Education After 1870, governments in Europe and the United States required all children to attend school. The spread of education had many benefits. As **literacy**—the ability to read—became widespread, more books and magazines were published. Newspapers that carried stories from all over the world also became very popular. By reading them, citizens became more informed about their governments.

As greater numbers of people learned to read, newspapers competed for their attention. They published eye-catching stories and cartoons.

Arts and Entertainment City dwellers of the late 1800s found themselves with more time and money for entertainment. Theaters opened to meet a demand for concerts, plays, and vaudeville shows. Art collections were made available to the public by displaying them in museums. Free public libraries opened in many cities. Sports became more organized, and cities began to sponsor teams with official rules and national competitions. Many cities began to construct public parks. These parks allowed people who lived in the cities to enjoy outdoor activities. By the end of the 1800s many of these parks had begun to include playgrounds for children.

A Growing Population One of the greatest changes of the later 1800s was the rapid growth of cities in the United States. Faced with crowded, dirty cities and seeking new opportunities, many Europeans chose to emigrate to the United States. To **emigrate** means to leave one country to live in another. The United States was not the only destination for these emigrants. Many also chose to seek new lives in South America, Africa, Australia, and New Zealand.

Between 1870 and 1900 more than 10 million people left Europe for the United States. These newcomers hoped to find economic opportunities. Some sought political and religious freedom as well.

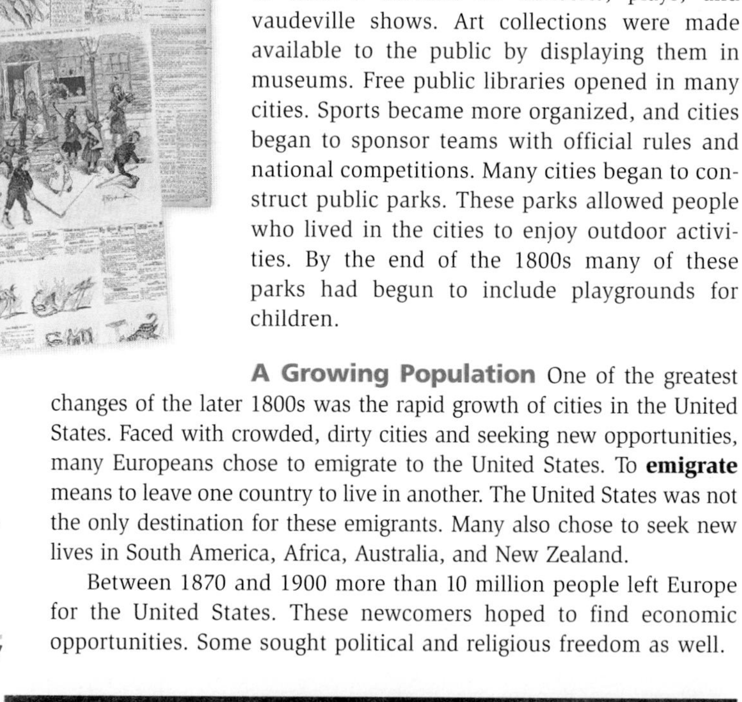

This is a painting of a croquet game in a public park. It reflects the increased participation in free-time activities during the late 1800s.

Interpreting the Visual Record **What does this painting suggest about sports in the later 1800s?**

TEACHER TO TEACHER

Madeleine Schmitt of St. Louis, Missouri, suggests the following activity to help students understand the rise of realism in music and art. Have students listen to a piece of music written by a composer of the romantic period such as Brahms, Liszt, or Schubert. Ask students to name elements of the music that would define it as "romantic." Then have them compare how the "romantic" mood of such music differed from the real lives of most people. Lead a discussion about how this difference led to the rise of the realism movement.

Teaching Objective 3

ALL LEVELS: (Suggested time: 30 min.) Organize the class into six groups and assign each group one of the following topics: Italy before unification, Italy after unification, Germany before unification, Germany after unification, Russia before Alexander II, or Russia after Alexander II. Have each group design a poster that describes its assigned country during the appropriate time period. Encourage students to conduct additional research to find a map of the country to include on the poster. Display the posters in pairs grouped by the countries they describe.

ENGLISH LANGUAGE LEARNERS, COOPERATIVE LEARNING

CONNECTING TO Art

A New Art Period

Works of art not only show the values of an artist but also the values of the society in which the artist lives. The American and French Revolutions, for example, changed society deeply. Many artists were inspired to paint stirring scenes from history and nature. These artists were called *romantics*. Their scenes showed life to be more exciting and satisfying than it normally is.

By the mid-1800s, however, many artists had rejected romanticism. These artists instead wanted to portray life as it really was. A style called *realism* developed. The realists painted ordinary living conditions and familiar settings. They tried to re-create what they saw around them, accurately and honestly.

Honore Daumier was a French realist. He painted *The Washerwoman* during the Industrial Revolution in France, a time when many city workers were struggling to survive. The subject in the painting is one of these workers.

Understanding What You Read

1. How was realism different from romanticism?
2. What types of subjects and scenes might a realist paint?

City Improvements Faced with rapidly growing populations and changes in society, many cities needed civic improvements. Local governments began to provide water and sewer service to city dwellers. Many streets were also paved. In 1829 London organized a police force. Many other cities soon had police forces, too.

Cities around the world grew rapidly in the 1800s. By the early 1900s, more people lived in cities than in the country. Many cities—including New York, London, Paris, and Berlin—had populations of more than 1 million.

Cities also created public transportation systems. Horse-drawn streetcars and buses were used mainly within cities. Trains, however, could take people far outside a city. As a result, some people began to move outside cities to areas called **suburbs**. These people usually took trains into cities each morning and returned to their homes in the suburbs at night.

✓ **READING CHECK:** *Summarizing* What were some advances of the later 1800s? Some possible answers: electricity, electric light, telephone, new knowledge in science and medicine, automobile, new city transportation, public education

Our Amazing Planet

Between 1865 and 1900 most American cities doubled or tripled in size. Much of this growth was due to immigration. Many immigrants to the United States moved to New York City. By 1900 the city's population was nearly five times larger than it had been in 1850.

DAILY LIFE

The development of public transportation encouraged city dwellers to move to outlying suburbs. The first suburbs tended to grow along train and trolley lines and close to the city center. However, as time passed and land costs increased, suburbs were built farther from cities and public transportation. The automobile eventually provided more mobility for people who lived in the suburbs.

Activity: Invite students to conduct research on how the areas surrounding a local city were suburbanized. Ask students how the locations of public transportation routes influenced the development of suburbs.

Connecting to Art

Answers

1. realism—ordinary scenes from life just as they are; romanticism—showed life to be more exciting than it really is
2. familiar settings; crowded slums, factories, nightclubs

CLOSE

Ask students to name and discuss some of the changes that occurred during the second half of the 1800s that were important to people's well-being. As students name reforms, write them on the chalkboard. Then have the class vote on which reforms they consider to be the most important. Call on volunteers to explain their reasoning.

REVIEW AND ASSESS

Have students complete the Section Review. Then name a person, event, or object described in this section. Ask students to explain its significance using the "W" questions: *who* or *what* was the person, event or object; *where* and *when* did it happen; and *why* is it significant? Then have students complete Daily Quiz 8.4.

Section 4 Review

Answers

Define For definitions, see: working class, p. 165; literacy, p. 166; suburbs, p. 167; emigrate, p.167; reform, p. 168; suffragettes, p. 168; nationalism, p. 169

Reading for the Main Idea

1. advances in technology, communication, and other areas; water and sewer systems; streets paved and lit; police forces established (NGS 13)
2. improved working conditions and benefits; abolished slavery; men got the right to vote (NGS 13)
3. The love of country stirred some people to fight to unify the independent states in Italy and in Germany. (NGS 13)

Critical Thinking

4. rapid growth of cities; many civic improvements

Organizing What You Know

5. Giuseppe Garibaldi—Italy; led an army to unify the states of Italy; Otto von Bismarck—Prussia; Waged war with Austria and France to free German states under their control; united German states into the German Empire; Alexander II—Russia; freed the serfs and tried to modernize Russia

Political and Social Reform

In addition to great advances in technology, communications, science, and medicine, the mid-1800s and early 1900s saw many political and social reforms. To **reform** something is to remove its faults. Around the world, citizens worked to improve their governments and societies.

Emmeline Pankhurst (1858–1928) led many demonstrations and marches on behalf of the women's suffrage movement in Great Britain.

Great Britain In Great Britain reformers passed laws that allowed male factory workers in cities to vote. Laws were also passed to improve conditions in factories, and slavery was abolished. New laws were passed to provide health insurance, unemployment insurance, and money for the elderly.

Beginning in the late 1800s, many women in Great Britain became **suffragettes**. These women campaigned for their right to vote. They were led by outspoken women like Emmeline Pankhurst. British women gained this right in 1928.

France A revolution in France in 1848 forced the king from the throne. A new government called the Second Republic was established. It guaranteed free speech and gave the vote to all men. In 1875, a new French constitution established the Third Republic. This constitution lasted for nearly 70 years.

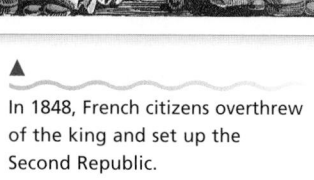

RÉPUBLIQUE FRANÇAISE.
Combat du peuple parisien dans les journées des 22, 23 et 24 Février 1848.

In 1848, French citizens overthrew of the king and set up the Second Republic.

United States The issue of slavery divided the United States in the mid-1800s. In late 1860 and early 1861 several southern states broke from the Union to form the Confederate States of America. The Civil War followed and raged until 1865 when the Confederacy surrendered. Congress then amended the Constitution to abolish slavery and grant citizenship to former slaves. The vote was given to all men, regardless of race or color.

Many reforms were the results of efforts by women such as Elizabeth Cady Stanton and Lucretia Mott. As early as 1848 they had campaigned for the abolition of slavery, equality for women, and the right to vote. In the 1890s and early 1900s, the movement for women's right to vote grew stronger. The Nineteenth Amendment to the Constitution, ratified in 1920, finally gave women this right.

✓ **READING CHECK:** *Comparing and Contrasting* How were reforms in Great Britain, France, and the United States in the mid-1800s and early 1900s similar? They all gave men the right to vote. Great Britain and the United States abolished slavery.

RETEACH

Have students complete Main Idea Activity 8.4. Then tell them to imagine that, like Rip Van Winkle, they had fallen asleep in 1840 and not woken up again until 1880. Ask: What changes might they spot walking around the streets of a major American city? What differences might they notice traveling through countries in Europe, such as England, France, Italy, Germany, and Russia? Call on students to share their answers.
ENGLISH LANGUAGE LEARNERS

EXTEND

Have interested students conduct research on the contributions of immigrants to American society. Students might wish to select one individual to study or may choose to look at a broader subject. Have students design collages that illustrate their findings. **BLOCK SCHEDULING**

Nationalism in Europe

Nationalism is the love of one's country more than the love of one's native region or state. In the 1800s, nationalism led to the unification of Italy and of Germany. It was also a driving force for change in Russia.

In the early 1800s the Congress of Vienna had divided Italy into several states, some of which were ruled by Austria. In the 1850s and 1860s, a nationalist named Giuseppe Garibaldi led a movement to unify these states. He and his army defeated the Austrians and their French allies and drove them out of Italy. Largely because of his efforts, most of present-day Italy had been unified by 1861. In that year Victor Emmanuel II was made the king of Italy.

Germany in the mid-1800s was a patchwork of 39 independent states. The largest was Prussia, ruled by William I. In 1862 he appointed Otto von Bismarck one of his advisers. Both men wanted to make Germany into a powerful unified country. Bismarck convinced the other German states to join in this effort and to declare war first on Austria, Prussia's chief rival, and then on France. After the war, the German states were joined together into the German Empire. William I became the first kaiser, or emperor.

In the 1800s Russia had more territory and people than any other country in Europe. Its economy, however, was not as developed as those of other countries. People from Russia's many ethnic groups felt very little unity with each other. In the 1850s Czar Alexander II tried to introduce major reforms. He freed all the serfs in Russia and introduced political changes. Later czars, however, tried to undo these reforms. Censorship and discrimination against minorities became widespread. This repression created an explosive situation in Russia. In 1905, a group of revolutionaries tried to overthrow the czar but failed.

Otto von Bismarck (1815–1898) was known for his strong will and determination.

✓ **READING CHECK:** *Drawing Inferences* How did nationalism help reshape nations? led to the unification of Italy and Germany; led to social and political changes that brought more unity and equality to people in Russia

Section 4 Review

go.hrw.com Homework Practice Online
Keyword: SK3 HP8

Define and explain working class, literacy, emigrate, suburbs, reform, suffragettes, nationalism

Reading for the Main Idea

1. *Human Systems* What allowed people's lives to improve during the last half of the 1800s?
2. *Human Systems* How did reforms of the later 1800s and early 1900s affect people's lives?
3. *Human Systems* How did nationalism lead to the unification of Italy and Germany in the mid-1800s?

Critical Thinking

4. **Analyzing** What effect did immigration of the later 1800s have on the United States?

Organizing What You Know

5. **Categorizing** Copy the following chart. List the home country of each leader. Then give details about his accomplishments.

Leader	Country	Accomplishments
Giuseppe Garibaldi		
Otto von Bismarck		
Czar Alexander II		

CHAPTER 8 Review ANSWERS

Building Vocabulary

For definitions, see: Enlightenment, p.147; reason, p.148; individualism, p.148; popular sovereignty, p.149; Patriots, p. 152; alliance, p. 152; oppression, p. 154; factors of production, p.160; capital, p.160; capitalism, p163; mass production, p.163; literacy, p. 166; emigrate, p. 167; reform, p.168; nationalism, p. 169

Reviewing the Main Ideas

1. a natural law that would help them solve society's problems and improve people's lives (NGS 10)
2. alliance with France (NGS 13)
3. land, natural resources, workers, and capital (NGS 14)
4. Governments required all children to go to school, so more public schools were built; spread of literacy led to more books, magazines, and newspapers read, so citizens became more informed (NGS 16)
5. led to unification of Italy and Germany and inspired reforms (NGS 10)

Assess

Have students complete the Chapter 8 Test.

Reteach

Organize students into groups and assign each group one of the following time spans: 1700–1750, 1751–1800, 1801–1850, 1851–1900. Have the members of each group list and describe some events and developments that occurred during its assigned time span. Working in chronological order, have members of each group present these events to the whole class, explaining why they are significant. **ENGLISH LANGUAGE LEARNERS, COOPERATIVE LEARNING**

Portfolio Extensions

1. Organize the class into small groups and ask each group to write a short script for a historical skit about one of the events described in this chapter. After students write and rehearse their scenes, ask them to perform their skits for the class. Take pictures of the performances and include these with the scripts in student portfolios.
2. Have students build dioramas that show memorable events from the American or French Revolution or scenes from everyday life during the Industrial Revolution or Age of Reform. Ask each student to write a short description of the scene he or she has depicted. Place these descriptions and photos of the dioramas in student portfolios.

Chapter 8 Review Answers

Understanding History and Society

Presentations will vary, but should mention such things as sanitation, a police force, and public transportation. Use Rubric 7, Charts, to evaluate student work.

Thinking Critically

1. Enlightenment ideas of equality and popular sovereignty were the basis for the Declaration of Independence and Constitution.
2. to achieve a basic change in their political and economic situations
3. restored order to France; brought legal, financial, and educational changes to much of Europe; reorganized Europe's boundaries and governments
4. Answers will vary.
5. Members of the middle class owned most businesses and factories. Their wealth, social position, lifestyle, and political power affected all areas of society.

Chapter 8 Reviewing What You Know

Building Vocabulary

On a separate sheet of paper, write sentences to define each of the following.

1. Enlightenment
2. reason
3. individualism
4. popular sovereignty
5. Patriots
6. alliance
7. oppression
8. factors of production
9. capital
10. capitalism
11. mass production
12. literacy
13. emigrate
14. reform
15. nationalism

Reviewing the Main Ideas

1. *Human Systems* What did Enlightenment thinkers hope to discover?
2. *Human Systems* What led to the defeat of the British in the American Revolution?
3. *Environment and Society* What conditions in Great Britain gave rise to the Industrial Revolution?
4. *Environment and Society* How did education change greatly in Europe and the United States during the later 1800s and what were the benefits?
5. *Human Systems* What were the effects of nationalism on European nations in the late 1800s?

Understanding History and Society

Plans for Reform

Imagine you are the mayor of a large European city in the mid-1800s. Using a chart, make a presentation that lists and describes the reforms and changes you think should be made in your city. Consider the following:

- Healthy and safety issues.
- Communication and transportation needs.
- Education needs.

Thinking Critically

1. **Analyzing** How did Enlightenment ideas influence American democracy?
2. **Drawing Conclusions** Why did revolutions break out in America and France in the late 1700s?
3. **Analyzing** How did Napoléon Bonaparte change France and the rest of Europe in the early 1800s?
4. **Supporting a Point of View** Which three advances in the later 1800s do you think changed people's lives the most? Explain your answer.
5. **Evaluating** Why was the middle class so important to the Industrial Revolution?

Food Festival

American eating habits and diets have undergone a revolution since colonial days. Encourage students to conduct research on foods that were popular among people of the New England, Middle, and Southern colonies and to prepare a menu that might have been used in a colonial inn around 1750. If possible, invite students to prepare samples of some of these colonial dishes.

Chapter 8 Review and Assessment Resources

Reproducible
- Readings in World Geography, History, and Culture 25, 26, and 27
- Critical Thinking Activity 8
- Vocabulary Activity 8

Technology
- Chapter 8 Test Generator (on the One-Stop Planner)
- Audio CD Program, Chapter 8
- HRW Go site

Reinforcement, Review, and Assessment
- Chapter 8 Review, pp. 170–171
- Chapter 8 Tutorial for Students, Parents, Mentors, and Peers
- Chapter 8 Test
- Chapter 8 Test for English Language Learners and Special-Needs Students
- Unit 2 Test
- Unit 2 Test for English Language Learners and Special-Needs Students

Building Social Studies Skills

Map Activity

On a separate sheet of paper, match the letters on the map with their correct labels.

Maine	Pennsylvania
New Hampshire	Maryland
Massachusetts	Virginia
Connecticut	North Carolina
New York	Georgia

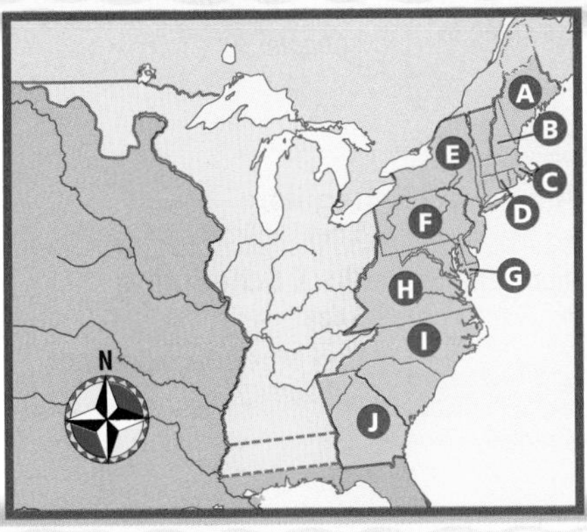

Mental Mapping Skills Activity

On a separate sheet of paper, draw a freehand map of Europe in the late 1800s. Make a key for your map and label the following:

Great Britain	Italy
France	Russia
Germany	Austria

Writing Activity

Imagine you are a news reporter in France in 1789, just as the French Revolution is breaking out. Write a news story about why the French people are in revolt and what they hope to achieve. Be sure to use standard grammar, spelling, sentence structure, and punctuation.

Alternative Assessment

Portfolio Activity

Learning About Your Local History

Industries in Your Town Research one or two of the main industries of your town, region, or state. Create a chart to explain when, where, why, and how the industry or industries were developed.

internet connect

Internet Activity: go.hrw.com
KEYWORD: SK3 GT8

Choose a topic to explore about the birth of the modern world:
- Explore the ideas of the Enlightenment.
- Investigate the causes of the French Revolution.
- Understand capitalism.

Map Activity
A. Maine
B. New Hampshire
C. Massachusetts
D. Connecticut
E. New York
F. Pennsylvania
G. Maryland
H. Virginia
I. North Carolina
J. Georgia

Mental Mapping Skills Activity
Maps will vary but listed places should be labeled in their approximate locations.

Writing Activity
News stories will vary but should include information about the French Revolution that is consistent with details in the text. Use Rubric 23, Newspapers, to evaluate student work.

Portfolio Activity
Answers will vary but should accurately describe local industries. Use Rubrics 7, Charts, and 30, Research, to evaluate student work.

internet connect

GO TO: go.hrw.com
KEYWORD: SK3 Teacher
FOR: a guide to using the Internet in your classroom

CHAPTER 9

The Modern World

Chapter Resource Manager

Objectives	Pacing Guide	Reproducible Resources
SECTION 1 **World War I** (pp. 185–89) **1.** Identify the causes of World War I. **2.** Explain how science and technology made this war different from earlier wars. **3.** Investigate how the world changed because of World War I.	**Regular** 1 day **Block Scheduling** .5 day *Block Scheduling Handbook, Chapter 9*	**RS** Guided Reading Strategy 9.1 **RS** Graphic Organizer 9
SECTION 2 **The Great Depression and the Rise of Dictators** (pp. 190–93) **1.** Identify the causes of the Great Depression. **2.** Describe a dictatorship. **3.** Analyze how the Great Depression helped dictators come to power in Europe.	**Regular** 1 day **Block Scheduling** .5 day *Block Scheduling Handbook, Chapter 9*	**RS** Guided Reading Strategy 9.2
SECTION 3 **Nationalist Movements in Africa and Asia** (pp. 194–97) **1.** Explain how World War I increased feelings of nationalism in Africa and Asia. **2.** Identify when and how African colonies became independent. **3.** Describe how the communists came to power in China.	**Regular** .5 day **Block Scheduling** .5 day *Block Scheduling Handbook, Chapter 9*	**RS** Guided Reading Strategy 9.3
SECTION 4 **World War II** (pp. 198–203) **1.** Identify the causes of World War II. **2.** Describe the Holocaust. **3.** Analyze how World War II came to an end.	**Regular** 1 day **Block Scheduling** .5 day *Block Scheduling Handbook, Chapter 9*	**RS** Guided Reading Strategy 9.4
SECTION 5 **The World Since 1945** (pp. 204–11) **1.** Describe the Cold War. **2.** Identify scenes of conflict in the world since World War II. **3.** Investigate important events that happened at the end of the 1900s.	**Regular** 1 day **Block Scheduling** .5 day *Block Scheduling Handbook, Chapter 9*	**RS** Guided Reading Strategy 9.5 **E** Creative Strategies for Teaching World Geography, Lesson 10 **SM** Map Activity 9

Chapter Resource Key

RS Reading Support
IC Interdisciplinary Connections
E Enrichment
SM Skills Mastery
A Assessment
REV Review
ELL Reinforcement and English Language Learners
Transparencies
CD–ROM
Music
Video
Internet
Holt Presentation Maker Using Microsoft® Powerpoint®

One-Stop Planner CD–ROM

See the ***One-Stop Planner*** for a complete list of additional resources for students and teachers.

One-Stop Planner CD–ROM

It's easy to plan lessons, select resources, and print out materials for your students when you use the ***One-Stop Planner CD–ROM with Test Generator.***

Technology Resources	Review, Reinforcement, and Assessment Resources
One-Stop Planner CD–ROM, Lesson 9.1 *ARGWorld* CD–ROM Homework Practice Online HRW Go site	**ELL** Main Idea Activity 9.1 **REV** Section 1 Review, p. 189 **A** Daily Quiz 9.1 **ELL** English Audio Summary 9.1 **ELL** Spanish Audio Summary 9.1
One-Stop Planner CD–ROM, Lesson 9.2 *ARGWorld* CD–ROM Homework Practice Online HRW Go site	**ELL** Main Idea Activity 9.2 **REV** Section 2 Review, p. 193 **A** Daily Quiz 9.2 **ELL** English Audio Summary 9.2 **ELL** Spanish Audio Summary 9.2
One-Stop Planner CD–ROM, Lesson 9.3 *ARGWorld* CD–ROM Homework Practice Online HRW Go site	**ELL** Main Idea Activity 9.3 **REV** Section 3 Review, p. 197 **A** Daily Quiz 9.3 **ELL** English Audio Summary 9.3 **ELL** Spanish Audio Summary 9.3
One-Stop Planner CD–ROM, Lesson 9.4 *ARGWorld* CD–ROM Homework Practice Online HRW Go site	**ELL** Main Idea Activity 9.4 **REV** Section 4 Review, p. 203 **A** Daily Quiz 9.4 **ELL** English Audio Summary 9.4 **ELL** Spanish Audio Summary 9.4
One-Stop Planner CD–ROM, Lesson 9.5 *ARGWorld* CD–ROM Homework Practice Online HRW Go site	**ELL** Main Idea Activity 9.5 **REV** Section 5 Review, p. 211 **A** Daily Quiz 9.5 **ELL** English Audio Summary 9.5 **ELL** Spanish Audio Summary 9.5

internet connect

HRW ONLINE RESOURCES

GO TO: go.hrw.com
Then type in a keyword.

TEACHER HOME PAGE
KEYWORD: SK3 TEACHER

CHAPTER INTERNET ACTIVITIES
KEYWORD: SK3 GT9

Choose an activity to:
- learn about the effects of the Treaty of Versailles.
- write a report about Anne Frank.
- create a poster about the causes and effects of global warming.

CHAPTER ENRICHMENT LINKS
KEYWORD: SK3 CH9

CHAPTER MAPS
KEYWORD: SK3 MAPS9

ONLINE ASSESSMENT
Homework Practice
KEYWORD: SK3 HP9
Standardized Test Prep Online
KEYWORD: SK3 STP9
Rubrics
KEYWORD: SS Rubrics

COUNTRY INFORMATION
KEYWORD: SK3 Almanac

CONTENT UPDATES
KEYWORD: SS Content Updates

HOLT PRESENTATION MAKER
KEYWORD: SK3 PPT9

ONLINE READING SUPPORT
KEYWORD: SS Strategies

CURRENT EVENTS
KEYWORD: S3 Current Events

Meeting Individual Needs

Ability Levels

Level 1 Basic-level activities designed for all students encountering new material

Level 2 Intermediate-level activities designed for average students

Level 3 Challenging activities designed for honors and gifted-and-talented students

English Language Learners Activities that address the needs of students with Limited English Proficiency

Chapter Review and Assessment

E Readings in World Geography, History, and Culture 32, 34, 38, 60, 65, 74

SM Critical Thinking Activity 9

REV Chapter 9 Review, pp. 212–13

REV Chapter 9 Tutorial for Students, Parents, Mentors, and Peers

ELL Vocabulary Activity 9

A Chapter 9 Test

A Unit 2 Test

Chapter 9 Test Generator (on the One-Stop Planner)

Audio CD Program, Chapter 9

A Chapter 9 Test for English Language Learners and Special-Needs Students

A Unit 2 Test for English Language Learners and Special-Needs Students

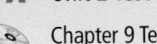

Launch Into Learning

Ask students to think about some things that have been invented or improved during their lifetimes. Then ask if any students have heard stories about their parents' lives when they were in school. Ask if any students know how their grandparents or great-grandparents lived when they were young. Encourage students to share these stories with the class. If no students have any family accounts, ask if they have seen pictures or read stories about life in the early to mid-1900s. Point out that nearly every aspect of society has undergone drastic changes since 1900. Tell students that they will learn about these changes—and the major events that caused them—in this chapter.

Section 1

Objectives

1. Identify the causes of World War I.
2. Examine how science and technology made this war different from earlier wars.
3. Investigate how the world changed because of World War I.

Links to Our Lives

You might want to share with your students the following reasons for learning about the people and events of the 1900s:

- World War I and World War II led to the redrawing of national boundaries around the world. Some of the boundaries created after these wars are still a source of conflict.
- Countries that underwent great changes in the 1900s, such as Russia and other previously communist countries, are still affected by these changes.
- Organizations like the United Nations and NATO founded after World War II remain active in political affairs today.
- Science and technology continue to change our lives.
- Entertainment media developed in the 1900s—radio, movies, and television—are enjoyed by billions of people around the world.

Chapter 9
The Modern World

The 1900s were filled with change. Great wars, economic depressions, horrible injustices, and tremendous technological advances have all taken place. You will find out how it all happened in this chapter.

One of the events that affected the world in the 1900s was the Great Depression. In the early 1930s millions of workers throughout the world could not find jobs and people had no money to buy goods.

During the Great Depression, young people faced special problems. In some cases, parents expected children to work when the parents themselves could not. Children were often a burden in poor families. For many youngsters, running away seems the only solution. At one point, almost 250,000 teenaged "hoboes" were roaming the United States. Many of these young people searched for any kind of work or odd job that they could find.

Poster from the Great Depression

YEARS OF DUST
RESETTLEMENT ADMINISTRATION
Rescues Victims
Restores Land to Proper Use

Migrant child

TEXAS 1939

Depression-era farmhouse

Focus

LET'S GET STARTED

Copy this question and instructions onto the chalkboard: *What do you think the expression "world war" means?* Discuss student responses. *(Possible answer: a war involving many countries or affecting a large area)* Call on students to suggest some reasons a small or local conflict between two countries might turn into a world war. *(Students might say that a country could have many enemies or that other countries could take sides in an existing conflict.)* Tell students that in Section 1 they will learn about the causes and results of the First World War.

Building Vocabulary

Write the key terms on the chalkboard. Call on volunteers to find the definitions of the terms in the glossary and read them to the class. Point out the words **nationalism** and **militarism**, and circle the suffix *-ism.* Explain that this suffix indicates an idea, concept, or behavior. Ask students to use the suffix to explain the meaning of these two words. *(Possible answer: Nationalism is the idea that one should honor one's nation; militarists act in a way that glorifies the military.)* Ask students to name some other words ending in this suffix. *(Possible answers: terrorism, heroism, athleticism)*

Section 1 World War I

Read to Discover

1. What were the causes of World War I?
2. How did science and technology make this war different from earlier wars?
3. How was the world changed because of World War I?

Define

militarism
U-boats
armistice

Locate

England
France
Germany
Russia
Austria-Hungary
Ottoman Empire
Serbia
Sarajevo

Why It Matters

World War I started in the Balkans—a region that is still the scene of much conflict. Use CNNfyi.com or other **current events** sources to learn what is happening in this region today. Record your findings in your journal.

German Poster from World War I

Section 1 Resources

Reproducible
- Block Scheduling Handbook, Chapter 9
- Guided Reading Strategy 9.1
- Graphic Organizer 9

Technology
- One-Stop Planner CD–ROM, Lesson 9.1
- Homework Practice Online
- HRW Go site

Reinforcement, Review, and Assessment
- Main Idea Activity 9.1
- Section 1 Review, p. 189
- Daily Quiz 9.1
- English Audio Summary 9.1
- Spanish Audio Summary 9.1

Beginning of World War I

By the early 1900s, countries across Europe were competing for power. They built up strong armies to protect themselves and their interests. Powerful nations feared each other. Tensions were high. The stage was set for war.

The spirit of nationalism was still strong in Europe in the early 1900s. Nationalism is a fierce pride in one's country. Many European countries wanted more power and more land. They built strong armies, and threatened to use force to get what they wanted. The use of strong armies and the threat of force to gain power is called **militarism**.

Europe's leaders did not trust one another. To protect their nations against strong enemies, they formed alliances. An alliance is an agreement between countries. If a country is attacked, its allies—the members of the alliance—help it fight.

By 1907 Europe was divided into two opposing sides. Germany, Austria-Hungary, and Italy had formed one alliance. England, France, and Russia had formed another.

The attention of both alliances was soon drawn to the Balkans, a region in southeastern Europe. In 1878, Serbia, part of this region had become an independent country. Serbian nationalists now wanted control of Bosnia and Herzogovina, which belonged to Austria-Hungary.

On June 28, 1914, a Serbian nationalist shot and killed the heir to the Austro-Hungarian throne, Archduke Francis Ferdinand. As a result, Austria-Hungary declared war on Serbia. Russia supported Serbia; Germany supported Austria-Hungary. With Russia and its allies on one side and Germany and its allies on the other, conflict quickly spread.

This drawing of the killing of Archduke Francis Ferdinand in Sarajevo was published in French newspapers.

TEACH

Teaching Objective 1

ALL LEVELS: (Suggested time: 20 min.) Copy the following graphic organizer onto the board, omitting the italicized answers. Have students fill in the chart with descriptions and examples of the factors that led to World War I. After students have completed their charts, lead a class discussion about how each of the listed factors contributed to the outbreak of war.

THE CAUSES OF WORLD WAR I

Nationalism	Militarism
People began to feel fierce pride in their countries. *Countries wanted more land.*	*Countries built strong armies.* *Strong countries used the threat of force to gain power.* *Countries formed alliances to maintain the balance of power.*

DAILY LIFE

In addition to working at important jobs and serving in the armed forces in Europe, American women contributed to the Allies' war effort by saving food. At the time of World War I, most homemakers were women. Some 20 million homemakers signed pledges promising not to serve meat on Mondays or bread on Wednesdays and to grow their own vegetables in "Victory Gardens."

In return, they received stickers to display in the windows of their homes, showing that they were helping with the war effort. Their voluntary rationing efforts helped the United States double its shipments of food to the Allies in Europe, who badly needed it.

Discussion: During times of crisis, why are people willing to make sacrifices that they might ordinarily complain about?

Possible answers: They feel patriotic; they want to help others; they want to join in with what others are doing.

internet connect
GO TO: go.hrw.com
KEYWORD: SK3 CH9
FOR: Web sites about the modern world

In August 1914 Germany declared war on Russia. Russia was allied with France, so Germany declared war on France, too. England declared war on Germany. Japan also declared war on Germany. England, France, Russia, and Japan became known as the Allied Powers. The alliance of Germany, Austria-Hungary, the Ottoman Empire, and Bulgaria was called the Central Powers. Later in the war, Italy left the Central Powers and joined the Allied Powers. Eventually, the Allied Powers included 32 countries.

✓ **READING CHECK:** *Identifying Cause and Effect* How did militarism and alliances help set the stage for war in Europe? **Militarism: countries had standing armies ready to go to war. alliances: countries had to go to war to help their allies.**

A New Kind of War

New weapons played a major role in World War I. Germany introduced submarines, which were called **U-boats**. This name is short for "underwater boats." Germany also introduced poison gas, which was later used by both sides and caused great loss of life. Other new weapons included large, long-range cannons and the machine gun. Machine guns could kill hundreds of people in a few minutes.

World War I was also the first war to use the airplane. At first airplanes were used mainly to observe enemy troops. Later, machine guns were placed on airplanes, so they could fire on troops and shoot at each other in the sky. England also introduced the tank during the war. This huge, heavy vehicle could not be easily stopped. With machine guns mounted on them, tanks could kill large numbers of soldiers.

The Allied Powers and the Central Powers divided Europe into two opposing sides.

Europe at the Beginning of World War I

Teaching Objective 2

All LEVELS: (Suggested time: 30 min.) Organize the class into four groups. Have each group prepare a chart, diagram, or report that shows how new weapons and technologies affected the way World War I was fought. Tell students that their projects should clearly demonstrate the differences between World War I and early wars in which the United States was involved. Encourage interested students to note developments from World War I that are still employed in wars today.
ENGLISH LANGUAGE LEARNERS, COOPERATIVE LEARNING

Teaching Objective 3

LEVELS 1 and 2: (Suggested time: 10 min.) Ask students to find examples in the text of ways in which society changed after World War I. As students name changes, write them on the chalkboard. Then ask students to name the common theme that links all these things. *(People thought the world no longer made sense and felt the need to experiment with new ideas.)*

LEVEL 3: (Suggested time: 30 min.) Obtain samples of art and music from the era immediately following World War I. Play or show these materials to the class. Lead a discussion of how this music and art represent changes brought about by the war.

Connecting to Science

U-Boats

Cutaway view of a German U-Boat

A U-boat had two hulls, or walls. Between the hulls was an open space. When the U-boat took water into this space, the extra weight made the boat sink below the surface of the ocean. When water was pumped out of the space, the boat became lighter, and the U-boat rose.

When the U-boat was under water, the men on the boat could look at the surface through a tube called a periscope. U-boats could also fire torpedoes while underwater. This allowed them to make devastating surprise attacks on enemy ships.

Understanding What You Read

1. How did a U-boat submerge and surface?
2. Why were U-boats such a threat to the Allies' ships?

The Early Years of the War Early in the war, Germany attacked France. The German army almost reached Paris, the French capital. However, Russia attacked Germany and Austria-Hungary, forcing Germany's attention east. At sea, England used its powerful navy to stop supplies from reaching Germany by ship. Germany used its deadly U-boats to sink ships carrying supplies to Great Britain.

At first, both sides thought they would win a quick victory. They were wrong. Armies dug in for a long and costly fight. World War I would go on for four years.

The United States and World War I At first, the United States stayed out of World War I. In 1917, however, Germany tried to persuade Mexico to join the Central Powers. The Germans promised to help Mexico retake Arizona, New Mexico, and Texas from the United States after the war. This angered many Americans.

At the same time, German U-boats were attacking American ships carrying supplies to the Allies. Many ships were sunk and many Americans died.

The United States had another motivation for joining the war. The major Allied countries had moved toward democracy, but the Central Powers had not. President Woodrow Wilson told Congress that "the world must be made safe for democracy." On April 6, 1917, the United States declared war on Germany.

American soldiers march through Paris during World War I.

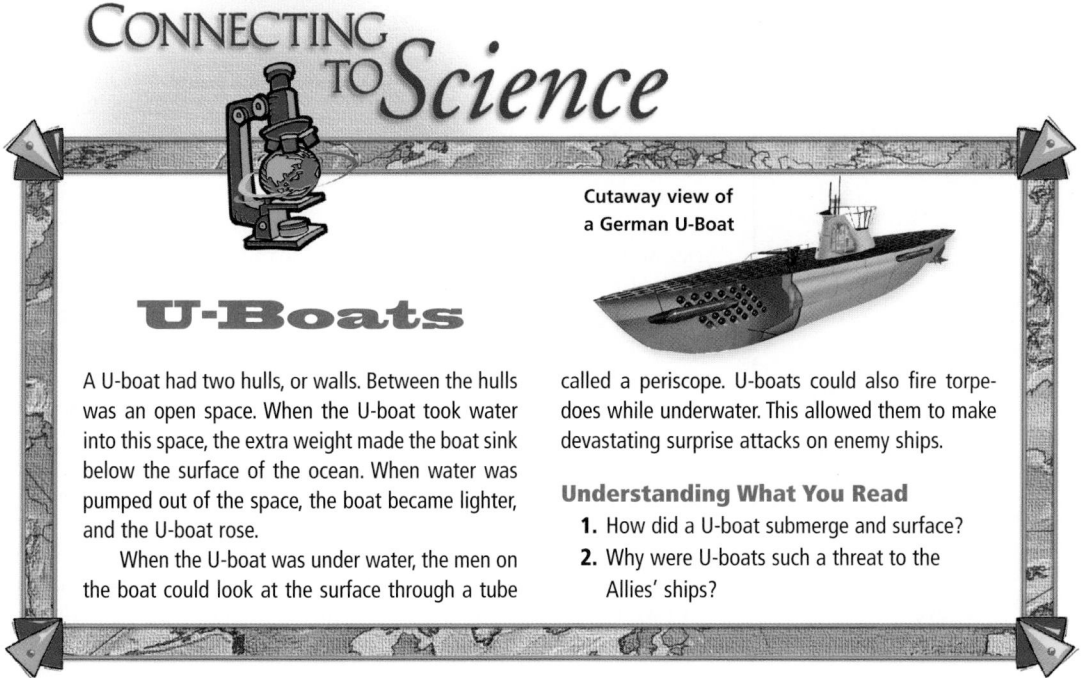

✓ **READING CHECK:** *Summarizing* Why did the United States enter World War I? Central Powers tried to get Mexico to join them; German U-boats attacked U.S. ships; many Americans died; the U.S. wanted to make the world safe for democracy.

Global Perspectives

When Germany attacked France in 1914, the German army got as far as the Marne River, near Paris. But then their advance was stopped by the French and British armies. Both sides dug long deep trenches to protect their troops. These trenches extended across France and Belgium, from Switzerland to the North Sea. This broad band become known as the western front. The troops stayed in the trenches and attacked each other's positions, but neither side could gain much ground. Territory that was captured in one attack was quickly lost, sometimes in the very next attack. This kind of fighting became known as trench warfare.

Discussion: Lead a discussion on the following question: What do you think daily life was like for the soldiers in the trenches?

Understanding What You Read

Answers

1. by controlling the amount of water contained between its hulls
2. Possible answer: They could attack ships from underwater.

CLOSE

Point out to students that the term *World War I* was not actually used during the war itself. People who lived through the war often referred to it as the "war to end all wars." Ask students what they think people meant by this name. *(Possible answer: The war was so long and destructive that people thought it would discourage countries from ever going to war again.)* Then ask students why we no longer use this name. *(World War I did not, obviously, end warfare. World War II was fought later.)*

REVIEW AND ASSESS

Have students complete the Section Review. Then organize students into teams. Give each team an index card. Have the members of each team choose an event or person discussed in the chapter and write a question about that event or person on their cards. Pass the cards from team to team. Ask each team to copy each question into their notebooks and to answer them. Continue until all each team has seen every question. Then have students complete Daily Quiz 9.1. **COOPERATIVE LEARNING**

HUMAN SYSTEMS

President Woodrow Wilson felt that World War I was a European affair. He wanted the United States to stay out of the war, and many Americans agreed. Even before entering the war on the Allied side in 1917, however, the United States sent food and weapons to both the Allies and the Central Powers.

Critical Thinking: What part did geography play in making Americans think that the war was a European affair?

Answer: The Atlantic Ocean separated the United States from Europe; the vast distance made Americans feel they were not involved in European matters.

Graph Answer

Russia; Germany

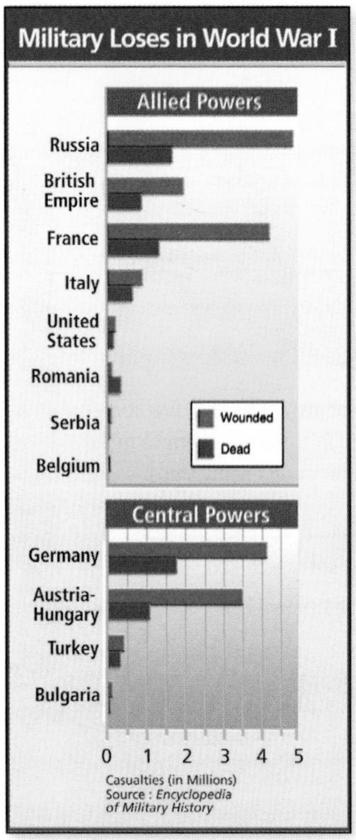

Interpreting the Graph **Which of the Allied Powers had the highest number of total casualties? Which of the Central Powers had the highest casualties?**

The War Ends

During World War I, Russian citizens held protests and demonstrations because they did not have enough food and because so many Russians were dying in the war. The Russian army joined the people in their protests. In March 1917 the czar, or king, was overthrown and put in prison.

A new government was set up. Political groups called soviets, or councils, were also formed. The most powerful soviet leader was Vladimir Lenin. He offered the Russian people peace, food, and land. Lenin's ideas were part of an economic and political system known as communism. On November 7, 1917, Lenin's followers took control of Russia.

Lenin's government signed a peace treaty with the Central Powers, and Russia withdrew from the war. In 1918 Lenin's followers established a communist party. Some Russians wanted the czar to return, and Civil War broke out. The Communists won. In 1922 they renamed their country the Union of Soviet Socialist Republics, or the Soviet Union.

With Russia out of the war, the tide began to turn in favor of Germany. The German army advanced on Paris. However, when the United States entered the war, the German army was pushed back to its own border. Germany's allies began to surrender. At last, Germany itself surrendered. An **armistice** was signed. An armistice is an agreement to stop fighting.

The fighting stopped on November 11, 1918. More than 8.5 million soldiers had been killed, and 21 million more wounded. Millions who did not fight died from starvation, disease, and bombs.

Making Peace In January 1919 the Allied nations met near Paris to decide what would happen now that the war was over. This meeting came to be known as the Paris Peace Conference.

President Wilson wanted fair peace terms to end the war. He felt that harsh terms might lead to future wars. His ideas were called the Fourteen Points. These ideas called for no secret treaties, freedom of the seas for everyone, and the establishment of an association of nations to promote peace and international cooperation. That association, the League of Nations, was formed later but the United States never joined.

Other Allied leaders wanted to punish Germany. They felt that Germany had started the war and should pay for it. They believed that the way to prevent future wars was to make sure that Germany could never become powerful again.

The agreement these leaders finally reached became known as the Treaty of Versailles. Germany was forced to admit it had started the war and to pay money to the Allies. Germany also lost territory. The treaty stated that Germany could not make tanks, military planes, large weapons, or submarines. The United States never agreed to the Treaty of Versailles. It eventually signed a separate peace treaty with Germany.

▲ Allied Leaders at the Paris Peace Conference. President Woodrow Wilson is at the far right.

RETEACH

Have students complete Main Idea Activity 9.1. Then organize the class into four groups. Assign each group one of the topics discussed in this section. Have each group write a list of facts about its topic. Encourage students to illustrate their lists. Then have each group present and explain its list to the class.

ENGLISH LANGUAGE LEARNERS, COOPERATIVE LEARNING

EXTEND

Have interested students conduct research on important battles or other major events of World War I. Ask them to prepare brief presentations on their findings. Have students give their presentations to the class, explaining the significance of the events they have researched. Encourage students to prepare maps or other visual aids to accompany their presentations. **BLOCK SCHEDULING**

A New Europe World War I changed the map of Europe. France and Belgium gained territory that had belonged to Germany. Austria, and Hungary became separate countries. Poland and Czechoslovakia gained their independence. Bosnia and Herzegovina, Croatia, Montenegro, Serbia, and Slovenia were united as Yugoslavia. Finland, Estonia, Latvia, and Lithuania, all of which had been part of Russia, also became independent nations. Bulgaria and the Ottoman Empire likewise lost territory.

✓ **READING CHECK:** *Comparing and Contrasting* How did the peace terms Woodrow Wilson wanted compare to those in the Treaty of Versailles? Wilson : fair peace terms that would not contribute to future wars. Treaty had the terms Allied leaders wanted: punish Germany and make sure it could not become powerful again.

A New World

After World War I, the world was very different. New ideas, new art, new music, and new kinds of books reflected the feeling that the world no longer made sense. Some writers called the people who had been through the war "the lost generation." Composers wrote music that sounded different from the music people were used to hearing. Many people thought it didn't sound pretty. Artists like Pablo Picasso and Salvador Dali created paintings that looked more like scenes from dreams than from the real world. People were tired of war. They wanted to have fun, and not worry so much about what might happen tomorrow. Jazz music, which gave musicians more freedom, became popular. Women wanted more freedom. They began to wear their hair and their skirts short. In the United States, women demanded and won the right to vote.

▲ Women vote in the United States, c. 1920.

✓ **READING CHECK:** *Drawing Inferences* Why were there so many new ideas and new kinds of art after World War I? World no longer seemed to make sense; old ideas didn't work anymore; people wanted new ideas, more freedom, more fun, and not to worry about the future.

go.hrw.com Homework Practice Online
Keyword: SK3 HP9

Section 1 Review

Define and explain: militarism, U-boats, armistice

Reading for the Main Idea

1. *Human Systems* What was Europe like just before World War I?
2. *Human Systems* What role did science and technology play in the war?
3. *Human Systems* How did World War I change the world?

Critical Thinking

4. **Drawing Inferences and Conclusions** How might World War I have been different if there had not been alliances in Europe?

Organizing What You Know

5. **Categorizing** Copy the following chart. Use it to list the members of each alliance.

Allied Powers	Central Powers

Section 1 Review

Answers

Define For definitions, see: nationalism, p.185; militarism, p.185; U-boats, p. 186; armistice, p.188

Reading for the Main Idea

1. Militarism encouraged nations to build strong armies; nations did not trust one another; alliances meant countries had to go to war to help their allies. (NGS 13)
2. New weapons were a big reason for the war's high death toll. (NGS 14)
3. New weapons led to higher death tolls in future wars; the map of Europe changed; new nations created; League of Nations created; new ideas and new kinds of art, music, and books developed. (NGS 10)

Critical Thinking

4. The war would not have been as widespread; it might have been a war between only Serbia and Austria-Hungary.

Organizing What You Know

5. Allied Powers—England, France, Russia, Japan, Italy, United States; Central Powers—Germany, Austria-Hungary, Ottoman Empire, Bulgaria

Section 2

Objectives

1. Identify the causes of the Great Depression.
2. Describe a dictatorship.
3. Analyze how the Great Depression helped dictators come to power in Europe.

FOCUS

LET'S GET STARTED

Copy the following question onto the board: *What do you think would happen if, suddenly, most people had no money?* Discuss student responses. *(Possible answers: People might go hungry; businesses might close as people stopped spending; many might panic or come to depend on government assistance.)* Point out that these things happened in the 1930s during the Great Depression. Tell students that in Section 2 they will learn more about the Great Depression and how it affected the world.

Building Vocabulary

Write the key terms on the board. Call on volunteers to find the definitions of the terms in the textbook, and read them to the class. Point out the suffix *–or* in the word **dictator**, meaning "one who." Ask students if they are familiar with the word *dictate*. Call on a volunteer to explain how its meaning is reflected in the definition of *dictator*. *(A dictator's rules dictate the actions of a country's people.)* Then ask what *collective* means *(working together)*. Ask how this meaning relates to the definition of **collective farms**.

Section 2 RESOURCES

Reproducible
- Guided Reading Strategy 9.2

Technology
- One-Stop Planner CD–ROM, Lesson 9.2
- Homework Practice Online
- HRW Go site

Reinforcement, Review, and Assessment
- Main Idea Activity 9.2
- Section 2 Review, p. 193
- Daily Quiz 9.2
- English Audio Summary 9.2
- Spanish Audio Summary 9.2

Section 2 The Great Depression and the Rise of Dictators

Read to Discover

1. What led to the Great Depression?
2. What is a dictatorship?
3. How did the Great Depression help dictators come to power in Europe?

Define

stock market
bankrupt
The Great Depression
New Deal
dictator
fascism
communism
police state
collective farms

Locate

New York

WHY IT MATTERS

Some countries in today's world are ruled by dictators. Use CNNfyi.com or other **current events** sources to find examples of dictators and learn how they came to power. Record your findings in your journal.

Library of Congress

Dorothea Lange's photograph
Migrant Mother

▲ During the stock market crash, people rushed to get money out of banks.

The Great Depression

During the 1920s industrialized countries such as the United States and Great Britain experienced great economic growth. However, less than 10 years later, many of these countries struggled with high rates of unemployment and poverty.

Causes of the Depression During World War I, much farmland in Europe was destroyed. Farmers all over the world planted more crops to sell food to European countries. Many American farmers borrowed money to buy farm machinery and more land. When the war ended, there was less demand for food in Europe. Prices went down. Farmers could not pay back the money they had borrowed. Many lost their land.

During the 1920s, the **stock market** did very well. The stock market is an organization through which shares of stock in companies are bought and sold. People who buy stock are buying shares in a company. If people sell their stock when the price per share has risen above the original price, they make a profit.

In the 1920s, stock prices rose very high and many people invested their money in the stock market. People thought stock prices would stay high, so they borrowed money to buy more stocks. Unfortunately, stock prices fell. Low stock prices made people rush to sell their shares before they lost any more money. So many people all selling stock at once drove prices down even more quickly. Finally the stock marked crashed, or hit bottom, on October 29, 1929.

TEACH

Teaching Objective 1

ALL LEVELS: (Suggested time: 20 min.) Copy the following graphic organizer onto the board, omitting the italicized answers. Have each student complete the organizer with factors that led to the Great Depression. Ask volunteers to share their answers with the class. Use their completed organizers to lead a discussion about the beginnings of the Great Depression. **ENGLISH LANGUAGE LEARNERS**

People who had borrowed to buy stocks suddenly had to pay back the money. They rushed to the banks to take money out. But the banks did not have enough money to give everyone their savings all at once. In a very short time, banks, factories, farms, and people went **bankrupt**. This meant they had no more money.

This is a breadline in New York City, during the Great Depression. People who could not find jobs stood in these lines to receive free food from the government.

This was the beginning of the **Great Depression**. All over the world, prices and wages fell, banks closed, business slowed or stopped, and people could not find jobs. Many people were poor. Many did not even have enough money to buy food. Some people sold apples on street corners to make a little money.

Governments around the world tried to lessen the effects of the Great Depression. Some limited the number of imports they allowed into their countries. They thought this would encourage citizens to buy products made by businesses in their own countries. This plan did not work. In fact, the loss of foreign markets for their products drove many countries even further into debt.

The New Deal In 1932 Franklin D. Roosevelt became president of the United States. He created a program to help end the Great Depression. This program was called the **New Deal**. The federal government gave money to each state to help people. The government also created jobs. It hired people to construct buildings and roads and work on other projects.

Laws were passed to regulate banks and stock exchanges better. In 1938, Congress passed the Fair Labor Standards Act. It established the lowest amount of money a worker could be paid to keep a healthy standard of living. Many people today refer to it as the minimum wage law. Congress also guaranteed workers the right to form unions so they could demand better pay and better working conditions. The Social Security Act, passed in 1935, created benefits for people who were unemployed or elderly.

Under the New Deal, the United States became deeply involved in the well-being of its citizens. For the first time, the government created large-scale social programs to better the lives of American citizens. The New Deal did not, however, completely end the Great Depression in the United States. Government programs helped the economy grow somewhat, but they were not enough to solve the economic crisis completely. The Great Depression would not end in the United States until World War II.

✓ **READING CHECK:** *Identifying Cause and Effect* How did the Great Depression start? Stock prices dropped; then rush to sell before the loss was great; selling panic drove stock values down more; many people lost all their money.

Our Amazing Planet

The Great Depression affected the entire world. By 1932, more than 30 million people throughout the world could not find jobs.

USING ILLUSTRATIONS

Direct students' attention to the picture of the New York City breadline on this page. Ask students what details they see in the image that reflect life during the Great Depression. List students' responses on the board and discuss each one. *(Possible answers: the line is long; many people needed free food; many people in the line are well dressed, showing that even people who once had money needed free food; all types of people needed free food; people seem to be getting only bread, but are willing to stand in line for it; people seem to be orderly; breadlines were common.)*

➤**ASSIGNMENT:** Have each student write a short story that describes what they think daily life might have been like in the Great Depression. Tell students that their stories should note some of the hardships that people faced in the 1930s. Encourage them also to note the effects of government programs passed to alleviate some of this hardship. Call on volunteers to share their stories with the class.

Teaching Objective 2–3

ALL LEVELS: (Suggested time: 30 min.) Ask students to name some rights that people have in a democracy. *(Freedom of speech; freedom of the press; freedom of religion; right to a fair trial; free elections; different political parties)* List each response on the board and ask students why this freedom is important to them. Explain that people who live in dictatorships often do not enjoy any of these freedoms. Ask students to find examples in the text of what life was like in Italy and Germany under Mussolini and Hitler. Then lead a discussion about factors that led to the rise of each of these dictators. **ENGLISH LANGUAGE LEARNERS** **COOPERATIVE LEARNING**

HUMAN SYSTEMS

When Mussolini's Fascist party first tried to gain power, it failed. The Fascists then turned to violence. They fought with communists in the streets of Italy's cities. They beat up people who tried to organize unions. They broke strikes by using force against the workers who were striking. These actions gained Mussolini's Fascists the support of the police, the military, and many Italians.

Discussion: Ask students how they would feel if a political party in the United States tried to gain power by using the same methods the Fascists used in Italy. Lead a discussion on how political groups in the United States today gain support for their ideas.

When he became dictator of Italy, Benito Mussolini took the title *il Duce* (il DOO-chay), Italian for "the leader."

Hitler was a very powerful speaker. He often twisted the truth in his speeches. He claimed that the bigger a lie was, the more likely people would be to believe it.

The Rise of Dictators in Europe

As the Great Depression continued in Europe, life got harder. Many people became unhappy with their governments, which were not able to help them. In some countries, people were willing to give up democracy to have strong leaders who promised them more money and better lives. It was easy for dictators to take control of these governments. A **dictator** is an absolute, or total, ruler. A government ruled by a dictator is called a dictatorship. Powerful dictators seized control of Italy and Germany.

Italy Becomes a Dictatorship Benito Mussolini told the Italians that he had the answers to their problems. Mussolini called his ideas **fascism** and started the Fascist Party. Fascism was a political movement that put the needs of the nation above the needs of the individual. The nation's leader was supposed to represent the will of the nation. The leader had total control over the people and the economy.

Fascist nations became strong through militarism. Their leaders feared communism for a good reason. **Communism** promised a society in which property would be shared by everyone. Mussolini promised that he would not let communists take over Italy. He also promised he would bring Italy out of the Great Depression and return Italy to the glory of the Roman Empire.

In 1924, the Fascist Party won Italy's national election. Mussolini took control of the government and became a dictator. He turned Italy into a police state. A **police state** is a country in which the government has total control over people and uses secret police to find and punish people who rebel or protest.

Germany After World War I many Germans felt that their government had betrayed them by signing the Treaty of Versailles. Many also blamed the German government for the unemployment and inflation brought by the Great Depression. Several groups attempted to overthrow and replace the old government. Eventually a new party called the Nazi Party gained power. It too was a fascist party. Adolf Hitler was its leader.

Hitler promised to break the treaty of Versailles. He said he would restore Germany's economy, rebuild Germany's military power, and take back territory that Germany had lost after the war. Hitler told the Germans that they were superior to other people. Many Germans eagerly listened to Hitler's message. They thought he would restore Germany to its former power.

CLOSE

Ask students to suggest images they think best represent the Great Depression in the United States. Discuss student responses.

REVIEW AND ASSESS

Have students complete the Section Review. Then organize students into groups and have each group prepare five questions about a person from this section for a quiz game. Then have students complete Daily Quiz 9.2.
ENGLISH LANGUAGE LEARNERS, COOPERATIVE LEARNING

RETEACH

Have students complete Main Idea Activity 9.2. Then organize the class into four groups. Assign each group one of the following countries: United States; Italy; Germany; Soviet Union. Have each group create a list of newspaper headlines that might have appeared in their country during the 1930s. Then have each share its headlines with the class.
ENGLISH LANGUAGE LEARNERS, COOPERATIVE LEARNING

EXTEND

Have interested students conduct research on programs of the New Deal. Have each of them write a brief report on one New Deal program and how it helped the country to recover from the Depression.
BLOCK SCHEDULING

The Nazis quickly gained power in Germany. In 1933 Hitler took control of the German government. He made himself dictator and used the title *der Führer* (FYOOR-ur), which is German for "the leader." He turned Germany into a police state. Newspapers and political parties that opposed the Nazis were outlawed. Groups of people that Hitler claimed were inferior, especially Jews, lost their civil liberties.

Hitler began to secretly rebuild Germany's army and navy. He was going to make Germany a mighty nation again. He called his rule the Third Reich. *Reich* is the German word for *empire*. In 1936 Hitler formed a partnership with Mussolini called the Rome-Berlin Axis.

The Soviet Union Russia had suffered terribly during World War I. Lenin's Communist government had promised an ideal society in which people would share things and live well. However, most Russians remained poor.

After Lenin died, Joseph Stalin gained control of the Communist Party. Stalin's government took land from farmers and forced farmers to work on large **collective farms** owned and controlled by the central government. Stalin also tried to industrialize the Soviet Union. However, for ordinary Russians Food and manufactured goods remained scarce.

Religious worship was forbidden. Artists were even told what kind of pictures to make. Secret police spied on people. If people did not obey Stalin's policies they were arrested and put in jail or killed. Scholars think that by 1939 more than 5 million people had been arrested, deported, sent to forced labor camps, or killed.

For many years, many Soviet people thought Joseph Stalin was a great hero. Later, people became more aware of his responsibility for the deaths of millions of Russians for "crimes against the state."

✓ **READING CHECK:** *Analyzing* How did Hitler use the Treaty of Versailles to help him gain power in Germany? Germans' problems blamed on Treaty of Versaille; Hitler promised to ignore the treaty and rebuild Germany's economy and military power.

go.hrw.com Homework Practice Online Keyword: SK3 HP9

Section 2 Review

Define and explain: bankrupt, The Great Depression, New Deal, dictator, fascism, communism, police state, collective farms

Reading for the Main Idea

1. *Human Systems* What happened during the Great Depression?
2. *Human Systems* What is life like in a dictatorship?
3. *Human Systems* How did European dictators take advantage of the Great Depression to gain power?

Critical Thinking

4. **Making Inferences and Conclusions** How were Woodrow Wilson's concerns about the Treaty of Versailles proven correct by the rise of Adolf Hitler?

Organizing What You Know

5. **Identifying Cause and Effect** Copy the following graphic organizer. Fill it in to summarize what happened in the Great Depression.

Cause	Effect
Europe needs food during the war.	
The war ends and crop prices fall.	
Stock prices rise very high.	
People rush to sell their stocks.	
People rush to take money out of the banks.	

Section 2 Review Answers

Define For definitions see: stock market, p.191; bankrupt, p.191; the Great Depression, p. 191; New Deal, p. 191; dictator, p.192; fascism, p.192; police state, p. 192; collective farms, p. 193

Reading for the Main Idea

1. prices and wages fell; banks closed; business slowed or stopped; people could not find jobs; many people were poor; dictators came to power in some countries (NGS 11)
2. no freedom of speech; no freedom of the press; no opposing political parties; no free elections; police state; secret police (NGS 13)
3. Great Depression caused economic problems; dictators said they would solve the problems (NGS 13)

Critical Thinking

4. Wilson felt harsh peace terms might lead to another war; Hitler used German anger about the treaty to gain power.

Organizing What You Know

5. Farmers plant more crops, borrow money for land and machinery; farmers cannot pay debts, lose their land; people borrow money to buy stocks; stock prices fall quickly; banks do not have enough money, people cannot get money.

Section 3

Objectives

1. Explain how World War I increased feelings of nationalism in Africa and Asia.
2. Identify when and how African colonies became independent.
3. Describe how the communists came to power in China.

Focus

LET'S GET STARTED

Copy the following question and instructions onto the chalkboard: *What do you know about recent events in Africa and Asia? Write a few sentences to describe some of these events.* Discuss student responses. Point out that many countries on these two continents were controlled by European powers until the mid-1900s. In some places this colonial history continues to influence events. Tell students they will learn more about Africa and Asia in the 1900s in this section.

Building Vocabulary

Write the key terms on the chalkboard and have students read their definitions. Point out that **boycotted** is derived from the name of Charles Boycott, an English landowner in Ireland whose Irish tenants took nonviolent action against him in the 1880s. Ask them how this relates to the meaning of *boycotted.* Then point out that **apartheid** is from the Afrikaans language spoken in South Africa. It means "apartness." Have a student explain what apartheid meant for Africans and other minorities in South Africa.

Section 3 Resources

Reproducible
- Block Scheduling Handbook, Chapter 9
- Guided Reading Strategy 9.3
- Graphic Organizer 9

Technology
- One-Stop Planner CD–ROM, Lesson 9.3
- Homework Practice Online
- HRW Go site

Reinforcement, Review, and Assessment
- Section 3 Review, p. 198
- Daily Quiz 9.3
- Main Idea Activity 9.3
- English Audio Summary 9.3
- Spanish Audio Summary 9.3

Section 3 Nationalist Movements in Africa and Asia

Read to Discover

1. How did World War I increase feelings of nationalism in Africa and Asia?
2. When and how did African colonies become independent?
3. How did the communists come to power in China?

Define

boycotted
apartheid

Locate

Ghana
South Africa
India
Pakistan
Japan
China
Egypt
Iran
Turkey

Why It Matters

Many countries in Africa and Asia are still working to strengthen their economies and governments. Use CNNfyi.com or other **current events** sources to find out how these countries participate in world affairs today. Record your findings in your journal.

A political campaign button from South Africa

Beginnings of Nationalism in Africa and Asia

Until the years following World War I, nearly all of Africa and many parts of Asia were controlled by European countries. As European colonies these places gained access to better health care, more effective farming methods, and improved roads and railroads. The people who lived in these colonies, however, had few rights. They played little part in running their countries, and their cultures were usually not nurtured or respected.

Around the end of World War I, the people in these colonies began to express resentment over this European control. They developed more pride in their own cultures and national identities. These feelings of nationalism led to a demand for self-rule. By the 1930s many colonies in Africa and Asia were calling for independence.

Indian citizens protest against British rule in India.

Interpreting the Visual Record Why do you think they are protesting?

Visual Record Answer

They want more of a say in governing India

READING CHECK: *Summarizing* Why did many colonies in Africa and Asia want to become independent? people had few rights, no part in running countries; cultures not respected

TEACH

Teaching Objective 1

LEVELS 1 AND 2: (Suggested time: 15 min.) Call on volunteers to recall some of the ideas that grew out of World War I and its aftermath. *(Possible answers: freedom, international cooperation, changes in international boundaries)* Then ask students to describe how these ideas influenced the development of nationalist movements in Africa and Asia. *(Possible answers: Africans and Asians were inspired to try to achieve freedom. They may have hoped European powers would cooperate to grant them increased freedom.)*. **ENGLISH LANGUAGE LEARNERS**

LEVEL 3: (Suggested time: 20 min.) Have students imagine they are African soldiers fighting in the British army during World War I. Have them write a short letter home describing how life among Europeans is different from life in a European colony. Direct students to mention particular ideas they might have picked up from their fellow soldiers. Call on volunteers to share their letters with the class.

African Nationalism

The growth of nationalism in Africa was in part caused by World War I. However, most countries did not actually become independent until after World War II. African soldiers who had fought in British and French armies during the war had been exposed to new ideas. They probably learned about European political systems from their fellow soldiers. When they returned home, they brought some new ideas with them. Some began to protest against racism and political oppression.

These protests were linked to a worldwide movement called Pan-Africanism. It was begun in England, the United States, and the West Indies by people who wanted cultural unity and equality for everyone of African heritage. Members of the Pan-African movement wanted two things. First, they wanted to end European control of Africa. They also wanted Africa to become a unified homeland for all people of African descent. Despite these efforts, however, it took many years for most African colonies to gain independence. In some countries the people were able to achieve their goals peacefully. In others, however, the fight for independence became violent.

▲ The Organization of African Unity (OAU), whose flag is seen here, grew out of the Pan-African movement. It was formed in 1963. More than 50 countries belong to this organization today.

British Colonies The first British African colony to gain independence was the Gold Coast. Protestors staged demonstrations and **boycotted**, or refused to buy, British goods. Finally in 1957 the British agreed to let the people of the Gold Coast choose their own government. The people voted for independence. The Gold Coast became the independent nation of Ghana. Other British colonies followed this example. The colonies that became Kenya, Malawi, Nigeria, and Zambia all won their independence by 1970. Zimbabwe became independent in 1980.

Other Colonies By 1962 France had granted independence to nearly all of its African colonies. Other European countries were not as quick to give up their colonies. Belgium gave up any claim to its Congo colony—which later became the Democratic republic of the Congo—after violence broke out in 1959. Portugal also refused to free its colonies until after bloody civil wars. Angola, the last Portuguese colony in Africa, won its independence in 1975.

◄ Kwame Nkrumah (at left) led the struggle to win independence for Ghana. He became the country's first leader.

DAILY LIFE

One reason the Portuguese did not want to give up their African colonies was because they were so profitable. The soils of Angola and Mozambique were perfect for growing coffee, cotton, and sugar. Large plantations growing these products became the backbone of these colonies' economies.

In the 1960s, 160,000 white immigrants arrived to compete for jobs and land with Africans. In just a decade, white farmers, who made up only 2 percent of the farming population, owned 40 percent of the best land. This injustice led to rebellions in Angola in 1961 and in Mozambique three years later.

Critical Thinking: What economic theory led the Portuguese to establish colonies in Africa?

Answer: mercantilism

Teaching Objective 2

LEVEL 1: (Suggested time: 25 min.) Organize the class into groups and assign each group one of the African colonies whose independence movement is described in this chapter. Then have each group create a poster than illustrates one aspect of that country's experience. Display the posters around the classroom.
ENGLISH LANGUAGE LEARNERS, COOPERATIVE LEARNING

LEVELS 2 AND 3: (Suggested time: 45 min.) Tell students to imagine that they have just been elected to lead newly independent African countries. Ask them to think about what changes they might make to their countries and their governments to assure better futures for their people. Have each student write a short campaign speech expressing his or her ideas. Encourage students to note problems that their countries might have faced as colonies before they achieved independence.
ENGLISH LANGUAGE LEARNERS

HUMAN SYSTEMS

Although South Africa witnessed much violence and protest under apartheid, a number of individuals stood up for peace and made a difference. Four South Africans have won the Nobel Peace Prize since 1960. The first was Zulu chief Albert Luthuli. He was honored in 1960 for his nonviolent sturggle against racial discrimination. In 1984, Anglican bishop Desmond Tutu won the prize for his anti-apartheid work. He is the first black Anglican bishop in his country and the first black archbishop. In 1993 Nelson Mandela and South African president F.W. de Klerk shared the Nobel Peace Prize for their efforts to transform South Africa into a multiracial democracy. Mandela was elected president the following year.

Activity: Have students conduct research on one of these four Noel Peace Prize winners and how their actions promoted peace in their country. Than have students write short reports about their findings.

Visual Record Answer

Possible answer: Mandela wanted to undo some of the lingering effects of apartheid.

In 1986 Nigerian Wole Soyinka became the first African writer to win the Nobel Prize in literature.

Nelson Mandela called on the people of South Africa to "heal the wounds of the past."

Interpreting the Visual Record What do you think the phrase "heal the wounds of the past" means?

South Africa The Independent Union of South Africa was created in 1910. However, from the beginning, its government had been controlled by the descendents of British and Dutch settlers. White South Africans enforced a policy of **apartheid**. This was a system of laws that denied black South Africans any political rights. Many South Africans—both black and white—protested apartheid laws. Some protests turned violent, and others were violently put down. Many people were killed. Black leaders like Nelson Mandela were thrown into prison. Other countries around the world also disapproved of apartheid policies. Some of these countries refused to trade or have any dealings with South Africa.

Things changed in the 1990s. South Africa banned apartheid. Nelson Mandela was freed from prison and was elected president in 1994. Mandela and his government worked to establish a new government based on equality for everyone.

✓ **READING CHECK:** *Contrasting* How was South Africa different from other African countries? not a European colony

Life in Independent Africa

Independence did not solve all of the problems of the former African colonies. In addition, new challenges faced the newly independent countries. At the same time, however, people around the world developed a new interest in Africa and its cultures.

Challenges Many political leaders of the new African countries were inexperienced. If they were unable to improve conditions in their countries, military leaders sometimes took control. Many African countries were run by military dictatorships through the late 1900s. Civil wars broke out between ethnic groups in some countries. Thousands of people died. Many countries also fell into debt. Crops failed due to droughts and the overuse of land, and millions more people died. Outbreaks of diseases like malaria and AIDS have also killed many Africans.

Cultural Revival As the demand for independence in Africa grew, people around the world developed in interest in African culture. Authors like Nigeria's Wole Soyinka and Chinua Achebe won awards for their stories about African life. African music and art became more popular around the world. Also, African directors made movies that were appreciated in many countries.

✓ **READING CHECK:** *Summarizing* How did life in Africa change after independence? new challenges faced independent countries; worldwide interest in African culture

Teaching Objective 3

ALL LEVELS: (Suggested time: 20 min.) Copy the following graphic organizer onto the chalkboard, omitting the italicized answers. Ask students to fill in the organizer with details that describe the development of communism in China. Lead a discussion on the development of Chinese communism and its effects on Chinese culture.

ENGLISH LANGUAGE LEARNERS, COOPERATIVE LEARNING

THE DEVELOPMENT OF COMMUNISM IN CHINA

Nationalist Party is formed by Chinese who resent foreign control. → *Chiang Kai-shek takes control of government.* → *A group breaks away from Nationalist Party to form Chinese Communist Party.* → *Mao Zedong leads the Long March.* → *Mao and the communists establish the People's Republic of China.*

Nationalism in Asia

Africa was not the only region in which people wanted to break free of European control. A spirit of nationalism similar to the one that affected Africa swept through Asia in the mid-1900s.

India Great Britain had promised India more self-government in return for troops and money during World War I. People in both countries, however, were divided on the question of Indian independence. Some wanted India to remain part of the British Empire. Others wanted the country to be completely free from European influence. An Indian lawyer named Mohandas Gandhi led nonviolent protests against British control. Largely due to Gandhi's courageous and inspiring leadership, India won its independence in 1947.

Within India, however, Muslims and Hindus did not get along. When the British withdrew in 1947, they created two new countries. India was mainly Hindu. Pakistan was mostly Muslim. In 1948 Gandhi was assassinated by a Hindu who thought the leader was too kind to Muslims. Jawaharlal Nehru became the first prime minister of independent India. His daughter, Indira Gandhi, became prime minister in 1966.

▲ Mohandas Gandhi led India's independence movement.

Japan By the early 1900s Japan had emerged as a world power. It was quickly becoming a major industrial power as well. When the Great Depression began in 1929, however, many Japanese felt that the country should turn away from Western ideas. They wanted the country to return to its own traditions. The country became less democratic and more militaristic. After it was defeated in World War II, Japan went through difficult times politically and economically. In time the Japanese rebuilt their economy. Since the 1950s, Japan has become a modern, technologically advanced, and democratic nation. It has strong ties to other countries in Asia and the West.

China A new dynasty, the Qing, rose to power in China in the 1600s. Under the Qing, trade between China and the West increased. Over time, other countries came to dominate the Chinese economy and government. By 1900 China was completely under foreign control. That year a group of Chinese tried to force all foreigners out of China but were defeated in the Boxer Rebellion.

By 1912 nationalist feelings had begun to grow among Chinese who resented foreign control. The Nationalist Party, led by Sun Yat-sen, wanted China to become more industrial and democratic. The Nationalists overthrew the last Qing emperor and took control of China in 1912. When Sun died, the military leader Chiang Kai-shek took over the Nationalist Party. He set up a one-party government in China and became a military dictator.

When Chiang Kai-shek took over the Nationalist Party, powerful warlords still ruled most of the country through their personal armies. Under Chiang, the Nationalist army broke the power of the warlords and unified China.

Interpreting the Visual Record What does this photograph suggest about Chiang Kai-shek? ▼

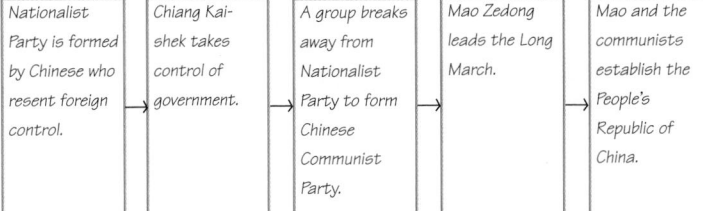

Across the Curriculum
ART
The Cultural Revolution The Chinese Communist Party viewed all art as simply another tool to be used by the revolution. During the Cultural Revolution (1966-1967), nearly all forms of art were examined carefully for their "socialist content."

Traditional art forms, such as landscape painting, had to portray some aspect of socialism. A waterfall, for example, may have been painted as running a hydroelectric plant or serving some other useful function for the revolution. Operas written during the Cultural Revolution often portrayed local communists in action, waving huge red banners and quoting Mao Zedong while attacking capitalist "imperialism."

Critical Thinking: Why might Chinese artists have thought of this system?

Answer: Students may respond that some artists probably disliked it because it restricted creative freedom, but that others might have been glad to have state support.

◄ **Visual Record Answer**

He looks powerful, like a person who is in command.

CLOSE

Call on a volunteer to point out on a wall map one of the countries discussed in this section. Then have him or her call on another student to summarize that country's path to independence. Repeat this process.

REVIEW AND ASSESS

Have students complete the Section Review. Then have each student write five statements about a person or group discussed in this section. Pair students and have them read each statement to their partners. Have the partners identify the person or group described by each statement. Then have students complete Daily Quiz 9.3. **COOPERATIVE LEARNING**

RETEACH

Have students complete Main Idea Activity 9.3. Then have students work in pairs to create outlines of the section material. Discuss their outlines. **ENGLISH LANGUAGE LEARNERS, COOPERATIVE LEARNING**

EXTEND

Have interested students conduct research on the current status of one African country described in this section. Ask them to examine how the problems that faced the country at the time of independence have been addressed or solved. Have students share their findings with the class. **BLOCK SCHEDULING**

Section 3 Review Answers

Define For definitions, see: boycotted, p. 195; apartheid, p. 196

Reading for the Main Idea

1. African soldiers who fought in the war were exposed to new ideas. (NGS 13)
2. Muslims and Hindus did not get along. Pakistan was created as a home for the Muslims. (NGS 12)
3. They wanted fair treatment for peasants and workers. (NGS 11)

Critical Thinking

4. Other countries disapproved of apartheid policies and put pressure on the government by refusing to trade with South Africa.

Organizing What You Know

5. Answers will vary but should mention African soldiers returning from World War I, the beginnings of nationalist movements, the demand for self-rule, and the granting of independence.

Mao Zedong wanted to modernize China. He and his followers destroyed thousands of books and works of art that reflected China's history.

In 1921 a group broke away from the Nationalist Party to form the Chinese Communist Party. They demanded fair treatment for peasants and workers. In the early 1930s Chiang's forces drove the communists into northwestern China. Chased by Nationalist troops, many died on the 6,000-mile trip, known as the Long March.

On this trip, a man named Mao Zedong established himself as the leader of the communists. Mao and his followers encouraged the peasants and workers to support a revolution. In 1949, Communists led by Mao took over China and established the People's Republic of China. Mao soon became a dictator and instituted sweeping changes in China. The Nationalists, opposed to communism, fled to the island of Taiwan.

Other Asian Countries Other countries in Asia also sought independence after World War I. In 1921 Reza Khan, a Persian army officer, seized control of the Persian government. He took the title *shah*, the ancient Persian word for king, and changed his country's name to Iran. Reza Shah Pahlavi did much to modernize Iran, but he ruled the country as a dictator.

Turkey was occupied by Greek troops after World War I. In 1922, Turkish nationalists led by Mustafa Kemal drove the Greeks out of their country. Kemal changed his name to Kemal Atatürk and established the Republic of Turkey. Atatürk modernized Turkey, improving education and giving women the right to vote.

✓ **READING CHECK:** *Summarizing* What was the Long March? **6,000-mile trip by Chinese Communists to escape Nationalists**

Section 3 Review

go.hrw.com Homework Practice Online Keyword: SK3 HP9

Define and explain: boycotted, apartheid

Reading for the Main Idea

1. *Human Systems* How did World War I lead to the growth of nationalism in Africa?
2. *Human Systems* What led to the creation of Pakistan as an independent country?
3. *Human Systems* What were the goals of the Chinese Communist Party?

Critical Thinking

4. **Drawing Inferences and Conclusions** How did other countries influence South Africa's decision to end apartheid?

Organizing What You Know

5. **Identifying Cause and Effect** Copy the following graphic organizer. Use it describe the process that led to the growth of nationalism and independence for African colonies.

	⇨	⇨

Section 4

Objectives

1. Identify the causes of World War II.
2. Describe the Holocaust.
3. Analyze how World War II came to an end.

Focus

LET'S GET STARTED

Copy these instructions onto the board: *Think about television programs, movies, and books with which you are familiar. What images do you associate with World War II?* Discuss student responses. *(Students might mention Nazis, tanks, fighter planes, aircraft carriers, the Holocaust, or similar images.)* Tell students that they will learn more about the history of World War II in this section.

Building Vocabulary

Write the key terms on the chalkboard. Call on volunteers to locate and read the definitions in the glossary or text. Point out the prefix *anti-* in **anti-Semitism** means "against." *Semitism* comes from *Semite*, a word that refers to several peoples from Southwest Asia, including the Hebrews. **Genocide** is derived from the Latin words *gens*, which means "race" or "people," and *caedere*, meaning "to kill." Have students write a sentence using the word *aggression.*

Section 4 World War II

Read to Discover

1. What were the causes of World War II?
2. What was the Holocaust?
3. How did World War II end?

Define

aggression
anti-Semitism
genocide
Holocaust

Locate

Japan
Italy
Germany
Hawaii
Ethiopia
Poland
Sicily
Pearl Harbor
Normandy

Why It Matters

Acts of aggression still happen. Use CNNfyi.com or other **current events** sources to find examples of aggression in today's world. Record your findings in your journal.

National Iwo Jima Memorial Monument

Threats to World Peace

During the 1930s, Japan, Italy, and Germany committed acts of aggression against other countries. **Aggression** is warlike action, such as an invasion or an attack. At first, little was done to stop them. Eventually, their actions led to a full-scale war that involved much of the world.

In 1931, Japanese forces took control of Manchuria, a part of China. The League of Nations protested, but took no military action to stop Japan. Continuing its aggressive actions, Japan succeeded in controlling about one fourth of China by 1939. At about the same time, Italy invaded Ethiopia, a country in East Africa. Many countries protested, but they did not want to go to war again. Like Japan, Italy saw that the rest of the world would not try hard to stop its aggression.

Pablo Picasso painted *Guernica* after the Spanish town of the same name was bombed.

Interpreting the Visual Record What human feelings about war does Picasso express?

Section 4 Resources

Reproducible

- Guided Reading Strategy 9.4

Technology

- One-Stop Planner CD–ROM, Lesson 9.4
- Homework Practice Online
- HRW Go site

Reinforcement, Review, and Assessment

- Main Idea Activity 9.4
- Section 4 Review, p. 203
- Daily Quiz 9.4
- English Audio Summary 9.4
- Spanish Audio Summary 9.4

Visual Record Answer

Picasso's broken figures show war victims' fear, suffering, horror, and despair.

TEACH

Teaching Objective 1

ALL LEVELS: (Suggested time: 20 min.) Copy the following graphic organizer onto the board, omitting the italicized answers. Call on students to fill in the boxes on the chart with actions taken by each country that led to the outbreak of World War II. Then lead a discussion about the underlying causes that led to these actions.

ENGLISH LANGUAGE LEARNERS

ACTIONS AND EVENTS THAT LED TO WORLD WAR II	
Japan	*1. took control of Manchuria* *2. conquered more of China*
Italy	*1. invaded Ethiopia*
Germany	*1. made Austria part of Germany* *2. took over the Sudetenland* *3. conquered Czechoslovakia* *4. invaded Poland*

DAILY LIFE

Even before World War II began, the leaders of Great Britain began making plans to protect its people from German air attacks. The British government sent thousands of children away from London to the countryside, where German bombers were less likely to attack. Moving children separated families, but it saved many lives.

When the war began, London was bombed night and day by German planes. People took shelter in subways and the basements of buildings. These bombing raids became known as the Blitz. The British people were famous for bravely carrying on with their normal lives as much as possible.

Discussion: Lead a discussion about what life was like during the Blitz. Call on students to share their ideas and impressions.

Visual Record Answer

Possible answer: They are unhappy that the Germans have occupied their homeland.

Women cry as they give the Nazi salute to German troops in Sudentenland.

Interpreting the Visual Record **Why do you think these women are showing strong feelings?**

Spanish Civil War In 1936, civil war broke out in Spain. On one side were fascists led by General Francisco Franco. Both Italy and Germany sent troops and supplies to help Franco's forces. On the other side were Loyalists, people loyal to the elected Spanish government. The Soviet Union sent aid to the Loyalists. Volunteers from France, Great Britain, and the United States also fought on their side, but their help was not enough. In 1939, the fascists defeated the Loyalists. Franco set up a dictatorship. He ended free elections and most civil rights. By the end of the 1930s, it was clear that fascism was growing in Europe.

Hitler's Aggressions In the late 1930s many Germans lived in Austria, Czechoslovakia, and Poland. Hitler wanted to unite these countries to bring all Germans together. In 1938 German soldiers marched into Austria, and Hitler declared Austria to be part of the Third Reich. Great Britain and France protested but did not attack Germany. Later that year Hitler took over the Sudentenland, a region of western Czechoslovakia. Other European countries were worried, but they still did not want a war. Hitler soon conquered the rest of Czechoslovakia.

Eventually Britain and France realized they could not ignore Hitler. They asked the Soviet Union to be their ally in a war against Germany. However, Soviet leader Joseph Stalin had made a secret plan with Hitler. They decided that their countries would never attack each other. This deal was called the German-Soviet nonaggression pact.

In September 1939 Hitler invaded Poland. Two days later, Great Britain and France declared war on Germany. World War II had begun. On one side were Germany, Italy, and Japan. They called themselves the Axis Powers, or the Axis. Great Britain, France, and other countries that fought against the Axis called themselves the Allies.

✓ **READING CHECK:** ***Drawing Inferences*** How might British and French leaders have prevented World War II? **stopped aggression of Japan, Italy, and Germany; stopped Hitler before he became too powerful**

Joseph Stalin (second from right) made an agreement that the Soviet Union and Germany would not attack each other.

Teaching Objective 2

ALL LEVELS: (Suggested time: 40 min.) Organize the class into groups and have each group design a museum exhibit about the Holocaust. Tell students that their designs should list the causes of the Holocaust and describe events associated with it. Call on students to share their ideas with the class. **ENGLISH LANGUAGE LEARNERS** **COOPERATIVE LEARNING**

Teaching Objective 3

ALL LEVELS: (Suggested time: 30 min.) Have students work in pairs to create time lines of the last years of World War II. Direct students to fill their time lines with events from the text. Next to each entry on their time lines, have students note how each event contributed to ending the war. Display the completed time lines around the classroom. **ENGLISH LANGUAGE LEARNERS, COOPERATIVE LEARNING**

◀ Carrying whatever belongings they can, people in northern France try to escape from attacks.

War

At the beginning of the war, the Germans won many victories. Poland fell in one month. In 1940 Germany conquered Denmark, Norway, the Netherlands, Belgium, and Luxembourg. In June 1940 Germany invaded and quickly defeated France. In less than one year, Hitler had gained control of almost all of western Europe. Next he sent German planes to bomb Great Britain. The British fought back with their own air force. This struggle became known as the Battle of Britain.

In June 1941, Hitler turned on his ally and invaded the Soviet Union. As winter set in, however, the Germans found themselves vulnerable to Soviet attacks. Without enough supplies, Hitler's troops were defeated by a combination of the freezing Russian winter and the Soviet Red Army. For the first time in the war, German soldiers were forced to retreat.

The United States Many people in the United States did not want their country to go to war. The United States sent supplies, food, and weapons to the British but did not actually enter the war until 1941. In that year Japan was taking control of Southeast Asia and the Pacific. Seeing the United States as a possible enemy, Japanese military leaders attempted to destroy the U.S. naval fleet in the Pacific. On December 7, 1941, Japan launched a surprise air attack on the naval base at Pearl Harbor, Hawaii. The attack sank or damaged U.S. battleships and killed more than 2,300 American soldiers. The next day President Franklin D. Roosevelt announced that the United States was at war with Japan. Great Britain also declared war on Japan. Three days later, Germany and Italy—both allies of Japan—declared war on the United States. In response, Congress declared war on both countries.

✓ **READING CHECK:** *Evaluating* How were Hitler's invasion of the Soviet Union and Japan's attack on Pearl Harbor turning points in the war? invasion of Soviet Union led to Germany's retreat; attack on Pearl Harbor brought U.S. into the war

Our Amazing Planet

After France was forced to sign a peace agreement with Germany and Italy in 1940, some French groups escaped to North Africa or to Great Britain and formed the Free French army. Others remained in France and formed an "underground" movement to resist the Germans secretly.

EXTRA
Los Angeles Times
NIGHT Pictorial
IT'S WAR!
Hostilities Declared by Japanese; 350 Reported Killed in Hawaii Raid
U.S. Battleships Hit; 7 Die in Honolulu
Air Bombs Rained on Pacific Bases

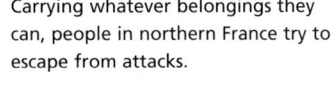

▲ Newspapers around the country ran headlines similar to this one after the Japanese attack on Pearl Harbor.

Interpreting the Visual Record **How do you think Americans felt when they saw headlines like this one?**

National Geography Standard 9

Human Systems In many occupied cities, the Nazis crowded Jews into small areas called *ghettos*. The ghetto in Warsaw, Poland, was surrounded by a wall topped with barbed wire and held 500,000 people in a small, cramped area. As the Germans began to deport Jews from the Warsaw ghetto to the death camps, thousands hid in underground rooms and bunkers. German troops tried to enter the ghetto, only to be driven out by Jewish resistance forces. Starting in April 1943, about 700 Jews armed with little more than pistols battled 2,000 German soldiers and tanks for 27 days.

Discussion: Lead a class discussion about the meaning of the word *ghetto* today. Ask students how this meaning is related to the historical one.

◀ **Visual Record Answer**

Americans probably felt angry, sad, upset, frightened, and betrayed. Many probably wanted to fight back.

➤**ASSIGNMENT:** Tell students to imagine that they are American newspaper editors in 1941 immediately following the bombing of Pearl Harbor. Have each student write an editorial for his or her newspaper arguing whether or not the United States should enter the war.

TEACHER TO TEACHER

Alfred J. Hamel, of Worcester, Massachusetts, suggests the following activity to help students understand the war in the Pacific. The American-led strategy called island hopping was eventually successful in defeating the Japanese. Have students conduct research on the Pacific war and draw maps that show which Pacific islands were taken by the Allies. Ask students to explain how the taking of these islands helped lead to the defeat of the Japanese.

National Geography Standard 17

The Uses of Geography
The war against Japan stretched across the Pacific Ocean. Therefore ships played a very important role in this part of World War II. U.S. submarines attacked Japanese shipping in an effort to cut off Japan's oil supply. Aircraft carriers transported planes to scenes of battle.

Early in the war, Japan advanced eastward across the Pacific Ocean by capturing many Pacific islands. To take back this territory, Allied forces used a strategy called "island hopping." They attacked only certain Japanese-held islands, skipping others, but leaving them without supplies. The plan was successful in stopping Japan and pushing its forces back across the Pacific.

Discussion: Lead a class discussion on this topic: How was the war in the Pacific different from the war in Europe?

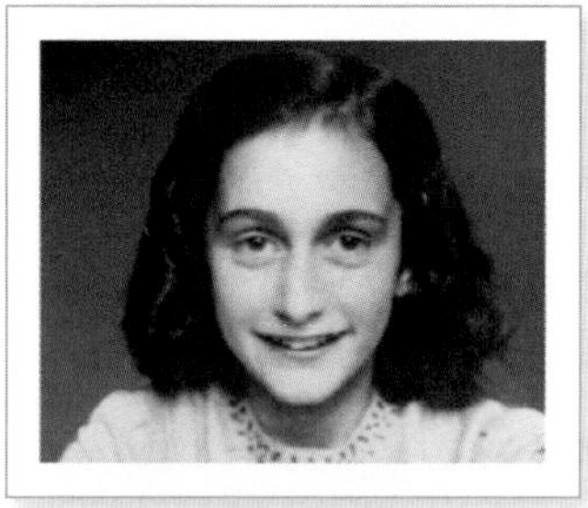

Anne Frank (1930-1945) was a Jewish teenager. During the Holocaust her family hid in an attic for two years to escape the Nazis. Anne kept a diary in which she wrote her thoughts and feelings.

Peace Memorial Park marks the spot where the first atomic bomb was dropped August 6, 1945, in Hiroshima, Japan

The Holocaust

Hitler believed that Germans were a superior people, and planned to destroy or enslave people whom he believed were inferior. Hitler hated many peoples, but he particularly hated the Jews. Hatred of Jews is called **anti-Semitism**. The Nazis rounded up Europe's Jews and imprisoned them in concentration camps.

Death Camps In 1941, Hitler ordered the destruction of Europe's entire Jewish population. The Nazis built death camps in Poland to carry out this plan. People who could work were forced into slave labor. Those who could not work were sent to gas chambers where they were killed. Some Jews were shot in large groups. Thousands of other people died from conditions in the camps. The dead were buried in mass graves or burned in large ovens.

By the time the Nazi government fell, its leaders and followers had murdered an estimated 6 million European Jews. The Nazi **genocide**, the planned killing of a race of people, is called the **Holocaust**. Millions of non-Jews were also killed.

Resisting the Nazis Some Jews tried to fight back. Others hid. Most, however, were unable to escape. Many Europeans ignored what was happening to the Jews, but some tried to save people from the Holocaust. The Danes helped about 7,000 Jews escape to Sweden. In Poland and Czechoslovakia, the German businessman Oskar Schindler saved many Jews by employing them in his factories.

✓ **READING CHECK:** *Summarizing* What were Nazi concentration camps like? places for slave labor and for death in the gas chambers, or by being shot, or through starvation and disease

The End of the War

In 1942 the Germans tried to capture the Soviet city of Stalingrad. The battle lasted six months, but the Soviet defenders held out. The Germans were never able to take the city. This was a major blow to the Germans, who never fully recovered from this defeat. At the same time, American and British forces defeated the Germans in Africa. The war began to turn in favor of the Allies. That same year in the Pacific Japan lost several important battles. Led by the United States, Allied forces—including troops from Australia and New Zealand—began a campaign to regain some of the Pacific islands Japan had taken. Slowly, the Allies pushed the Japanese forces back across the Pacific Ocean.

In the summer of 1943 the Allies captured the island of Sicily in Italy. Italians forced Mussolini to resign, and Italy's new leader dissolved the Fascist Party. In September, Italy agreed to stop fighting the Allies.

Victory in Europe On June 6, 1944, Allied forces landed on the beaches of Normandy in northern France. This was the D-Day invasion. The invasion was a success. In August, Allied troops entered Paris. By September they were at Germany's western border. With the

Close

Call on students to name individuals who played major roles in World War II. As people are named, call on other students to describe the actions for which they are known.

Review and Assess

Have students complete the Section Review. Then ask students to write one sentence about the role each of the following countries played in World War II: Italy, Japan, Germany, England, Soviet Union, United States. Then have students complete Daily Quiz 9.4.
ENGLISH LANGUAGE LEARNERS

Reteach

Have students complete Main Idea Activity 9.3. Then copy onto the chalkboard each of the major topics discussed in this section. Call on students to provide facts that apply to each topic.
ENGLISH LANGUAGE LEARNERS, COOPERATIVE LEARNING

Extend

Have interested students conduct research on the American home front during World War II. Direct students to pay particular attention to the roles of women during the war. Call on volunteers to share their findings with the class. **BLOCK SCHEDULING**

Soviets attacking Germany from the east, the Nazis' defenses fell apart. On April 30 Hitler killed himself, and within a week, Germany surrendered.

Victory over Japan Fighting continued in the Pacific. The Allies bombed Japan, but the Japanese would not surrender. Finally President Harry Truman decided to use the atomic bomb against Japan. On August 6, 1945, the most powerful weapon the world had ever seen was dropped on the city of Hiroshima. The bomb reduced the city to ashes and destroyed the surrounding area. About 130,000 people were killed and many more were injured. Countless more people died later. On August 9 another atomic bomb was dropped on the Japanese city of Nagasaki. Five days later Japan surrendered.

A New Age World War II resulted in more destruction than any other war in history. More than 50 million people were killed, and millions more were wounded. Unlike in most earlier wars, many of the people killed were civilians. **Civilians** are people who are not in the military. Millions were killed in the Holocaust. Thousands were killed by bombs dropped on cities in Europe and Japan. Thousands more died in prison camps in Japan and the Soviet Union. In time, people began to question how such cruel acts against human life and human rights were allowed to happen, and how they could be prevented in the future.

The American use of the atomic bomb began the atomic age. With it, came many questions and fears. How would this new weapon be used? What effect would it have on future wars? After World War II, world leaders would struggle with these questions.

✓ **READING CHECK:** *Analyzing* How was World War II unlike any war that came before it? most destructive war in history; human rights ignored in new ways; atomic bomb introduced

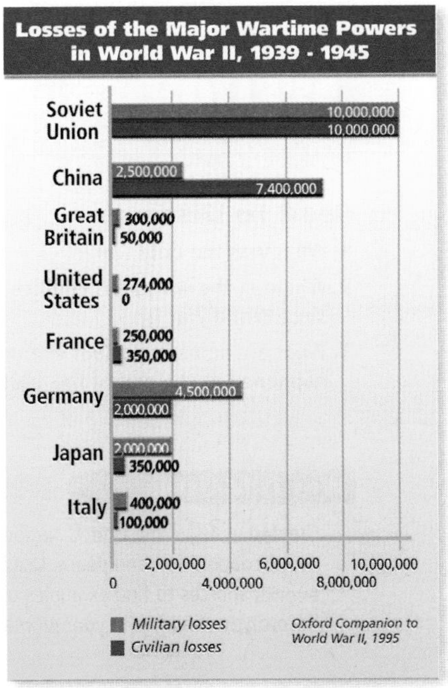

Interpreting the Graph What three countries had the highest civilian losses? What do you think caused these losses?

go.hrw.com Homework Practice Online
Keyword: SK3 HP9

Section 4 Review

Define and explain: aggression, anti-Semitism, genocide, Holocaust

Reading for the Main Idea

1. *Human Systems* What events led to World War II?
2. *Human Systems* What happened during the Holocaust?
3. *Human Systems* What were the results of World War II?

Critical Thinking

4. **Cause and Effect** How did the rise of fascism in Europe lead to World War II?

Organizing What You Know

5. **Drawing Inferences and Conclusions** Copy the following graphic organizer. Fill it in, telling why each event was important in the war.

Event	Why Important?
Hitler invades Poland.	
Hitler gains control of western Europe.	
Germany invades the Soviet Union.	
Japan attacks Pearl Harbor.	
The Allies invade Europe on D-Day.	
The United States drops the atomic bomb on Japan.	

Section 4 Review

Answers

Define For definitions, see: aggression, p.199; anti-Semitism, p.199; genocide, p.202

Reading for the Main Idea

1. acts of aggression by Japan, Italy, and Germany (NGS 13)
2. Jews were put in concentration camps where millions were murdered; millions of others who the Nazis thought were not pure were murdered. (NGS 9)
3. Axis defeated; millions of soldiers and civilians killed; enormous destruction; atomic age began (NGS 13)

Critical Thinking

4. Fascist nations became strong through militarism; Mussolini and Hitler set out to restore their nations' glory and power by invading other countries.

Organizing What You Know

5. World War II begins; Germany begins bombing England; Germans retreat for first time in war; United States enters war; Allies begin to retake Europe; Japan surrenders, war ends.

◄ **Graph Answer**

Soviet Union, China, Germany; locations of much of the fighting

Section 5

Objectives

1. Describe the Cold War.
2. Identify scenes of conflict in the world since World War II.
3. Investigate important events that happened at the end of the 1900s.

FOCUS

LET'S GET STARTED

Copy the following instructions onto the board: *What are some recent events that have affected the entire world. Write down a few of these events.* Allow students time to write their answers. Discuss the responses and list them on the board. Point out that our world is constantly changing. Tell students that in Section 5 they will learn some changes that have happened since 1945 and how they have shaped our world.

Building Vocabulary

Write **globalization** and **arms race** on the chalkboard. Underline the word *global* in *globalization.* Ask students to suggest possible meanings for the word based on the definition of the root word. Then point out that the word *arms* in *arms race* refers to weapons. Call on a volunteer to suggest a definition for the phrase. Then have a student locate and read the definitions for all the key terms.

Section 5 RESOURCES

Reproducible
- Guided Reading Strategy 9.5
- Creative Strategies for Teaching World Geography, Lesson 10
- Map Activity 9

Technology
- One-Stop Planner CD–ROM, Lesson 9.5
- Homework Practice Online
- HRW Go site

Reinforcement, Review, and Assessment
- Main Idea Activity 9.5
- Section 5 Review, p. 211
- Daily Quiz 9.5
- English Audio Summary 9.5
- Spanish Audio Summary 9.5

Section 5 The World Since 1945

Read to Discover

1. What was the Cold War?
2. Where in the world has conflict arisen since World War II?
3. What are some important events that happened at the end of the 1900s?

Define

bloc
arms race
partition
globalization

Locate

India
Israel
China
Taiwan
Korea
Vietnam
Cuba

WHY IT MATTERS

Created in 1945, the United Nations continues to promote international cooperation and peace. Use CNNfyi.com or other **current events** sources to find examples of UN action to stop violence. Record your findings in your journal.

compact discs

The Cold War

Although the Soviet Union and the United States were allies in World War II, the alliance fell apart after the war. The former allies clashed over ideas about freedom, government, and economics. Because this struggle did not turn into a shooting or "hot" war, it is known as the Cold War.

A Struggle of Ideas The struggle that started the Cold War was between two ideas—communism and capitalism. Communism is an economic system in which a central authority controls the government and the economy. Capitalism is a system in which businesses are privately owned. During the Cold War, the Soviet government functioned as a dictatorship which controlled the economy. However, the United States and other democratic nations practiced some form of capitalism.

In June 1948 the Soviets set up a blockade along the East German border to prevent supplies from getting into West Berlin. The people of West Berlin faced starvation. The United States and Great Britain organized an airlift to supply West Berlin. Food and supplies were flown in daily to the people.

TEACH

Teaching Objective 1

All LEVELS: (Suggested time: 20 min.) Copy the following graphic organizer onto the board, omitting the italicized answers. Ask students to reread the discussion of the Cold War in their textbooks and to provide a definition for the term. Remind students that the Cold War was largely a conflict between the United States and its allies and the Soviet Union and its allies. Have students complete the graphic organizer with descriptions of these two sides. Lead a discussion about the completed organizer.

ENGLISH LANGUAGE LEARNERS, COOPERATIVE LEARNING

THE COLD WAR

Western Bloc	Eastern Bloc
•**Strongest Country** *United States*	•**Strongest Country** *Soviet Union*
•**Economic System** *capitalism*	•**Economic System** *communism*
•**Political System** *democracy*	•**Political System** *not democratic*
•**Alliances** *NATO*	•**Alliances** *Warsaw Pact*
•**Economy** *rapid economic growth*	•**Economy** *shortages of goods, food, money*

Europe Divided After World War II Joseph Stalin, leader of the Soviet Union, brought most countries in Eastern Europe under communist control. The Soviet Union and those communist-controlled countries were known as the Eastern bloc. A **bloc** is a group of nations united under a common idea or for a common purpose. The United States and the democracies in Western Europe were known as the Western bloc. While Western countries experienced periods of great economic growth, industries in most communist countries did not develop. People in these countries suffered from shortages of goods, food, and money.

Two Germanies After World War II the Allies divided Germany into four zones to keep it from becoming powerful again. Britain, France, the United States, and the Soviet Union each controlled a zone. Germany could no longer have an army, and the Nazi Party was outlawed. By 1948, the Western Allies were ready to unite their zones, but the Soviets did not want Germany united as a democratic nation. The next year the American, British, and French zones became the Federal Republic of Germany, or West Germany. The Soviets established the German Democratic Republic, or East Germany. The city of Berlin, although part of East Germany, was divided into East and West Berlin, with West Berlin under Allied control. The Berlin Wall, which became a famous symbol of the Cold War, separated the two parts of the city.

The Soviet Union After Stalin Joseph Stalin, who had led the Soviet Union through World War II, died in 1953. The next Soviet leader was Nikita Khrushchev. He criticized Stalin's policies and reduced the government's control over the economy.

During the 1950s and 1960s some Eastern bloc nations tried to break free of communism. In East Germany, Czechoslovakia, and Hungary, for example, people rebelled against Soviet control, but the Soviets crushed these revolts.

The Berlin Wall did not stop people from trying to reach West Berlin. More than 130 people died trying to escape over the heavily guarded wall.

Interpreting the Visual Record **Why do you think people risked death to escape communism?**

HUMAN SYSTEMS

When the Berlin Wall was built in 1961, only a few border crossings were left open. The most famous of these was named Checkpoint C, but it was generally referred to as Checkpoint Charlie. Over time, it became the main crossing point between the two parts of the city. By the time the wall came down, thousands of East Germans had tried to enter West Berlin, often through Checkpoint Charlie. Some tried simply to elude the checkpoint guards, but others sought more unusual passage. One man even reportedly smuggled his girlfriend through the checkpoint by fitting her inside the seat of his car. Many similarly amazing stories of daring escapes are told at the Checkpoint Charlie Museum in Berlin.

Activity: Have students draw pictures or construct models of the Berlin Wall before it fell. Direct students to mark the location of Checkpoint Charlie.

Visual Record Answer

They wanted to live in a democracy, where they would have more freedom and economic opportunity.

TEACH

Teaching Objective 2

ALL LEVELS: (Suggested time: 30 min.) Organize the class into six groups and assign each group one of the following areas of conflict: Southwest Asia, Korea, Cuba, Vietnam, Northern Ireland, and Yugoslavia. Have the members of each group discuss the conflicts that have arisen in their assigned area and prepare a chart that includes the approximate date of the conflict, reason for conflict, and the outcome. Have each group present its chart to the class and lead a discussion about the information in the chart. **ENGLISH LANGUAGE LEARNERS, COOPERATIVE LEARNING**

➤ASSIGNMENT: Have each student consult newspapers, television news programs, Internet news sites, or other sources to find information about a conflict raging in the world today. Have each student write a brief paragraph that tell something about this conflict. Students might note the cause of their chosen conflicts or efforts that have been undertaken to resolve them.

GLOBAL PERSPECTIVES

The United Nations is composed of six main bodies. The General Assembly investigates disputes and recommends action to resolve them. The Security Council these actions and approves all military actions. The International Court of Justice decides questions of international law. The Economic and Social Council sponsors trade and human rights organizations. The Trusteeship Council controls territories that are under UN supervision. Finally, the Secretariat runs the UN itself.

Critical Thinking: What are some things the UN can do to settle disputes?

Answer: encourage international communication, take military action, sponsor economic actions

Visual Record Answer ►

The flags are the flags of the NATO member nations; the compass stands for the four corners of the globe. The emblem means that NATO will protect its members wherever they are.

▲ In this picture U.S. President Harry Truman signs the North Atlantic Pact, which created NATO. Shown above is the NATO emblem.

Interpreting the Visual Record How would you explain the NATO emblem?

The United Nations World leaders did not want the Cold War to turn "hot." Although the League of Nations had failed to prevent World War II, people still wanted an international organization that could settle problems peacefully. In April, 1945, the United Nations (UN) was created. Its purpose was to solve economic and social problems as well as to promote international cooperation and maintain peace. Representatives of 50 countries formed the original United Nations. Today there are nearly 200 member nations. The six official languages of the United Nations are Arabic, Chinese, English, French, Russian, and Spanish. The headquarters of the United Nations are in New York City. It also has offices in Geneva, Switzerland, and Vienna, Austria.

New Alliances Fearing war but hoping to preserve peace, nations around the world formed new alliances. In 1949, 12 Western nations, including the United States, created the North Atlantic Treaty Organization (NATO). In 1954 the Southeast Asia Treaty Organization (SEATO) was created in an attempt to halt the spread of communism in Southeast Asia. Many Eastern bloc countries, including the Soviet Union, signed the Warsaw Pact in 1955. The Warsaw Pact countries had more total troops than the NATO members. This difference in the number of troops encouraged the Western powers to rely on nuclear weapons to establish a balance of power.

▲ Mikhail Gorbachev, with his wife, Raisa. In 1990, Gorbachev won the Nobel Peace Prize for his reform work in the Soviet Union.

The End of the Cold War Throughout the Cold War, the Soviet Union and the United States had been a waging an **arms race**. The countries competed to create more advanced weapons and to have more nuclear missiles than each other. The arms race was expensive and took its toll on the already shaky Soviet economy.

In 1985, Mikhail Gorbachev became head of the Soviet Union. He reduced government control of the economy and increased individual liberties, such as freedom of speech and the press. He also improved relations with the United States. These reforms in the Soviet Union encouraged democratic movements in Eastern bloc countries. In 1989, Poland and Czechoslovakia threw off communist rule. In November, the Berlin Wall came down. In October 1990, East and West Germany became one democratic nation. Soviet republics also began to seek freedom and independence. By the end of 1991, the Soviet Union no longer existed. The Cold War was over. The arms race could stop.

Teaching Objective 3

ALL LEVELS: (Suggested time: 20 min.) Copy the following graphic organizer onto the board, omitting the italicized answers. Ask students to copy it into their notebooks. Have student fill in the organizer with important political events that occurred at the end of the 1900s. Ask volunteers to present their completed charts to the class. Lead a class discussion about the events that appear on the graphic organizers.
ENGLISH LANGUAGE LEARNERS, COOPERATIVE LEARNING

In October 1990 young people in Berlin wave German flags to celebrate the reunification of Germany.

The breakup of the Soviet Union created several independent countries. Russia was the largest of these new nations. Its new leader was Boris Yeltsin. Under Yeltsin, Russia moved toward democracy. Yeltsin also improved Russia's relations with the West. In 2000, Vladimir Putin became leader of Russia. Under Putin, relations with the United States improved further.

Some tension arose, however, between Russia and the former Soviet republics. For example, Russia and the Ukraine clashed over military issues in the 1990s.

✓ **READING CHECK:** *Evaluating* How did the end of the Soviet Union affect the world? ended Cold War; stopped arms race; new countries created; improved relations between United States and Russia

Other World Conflicts

Since the end of World War II, many conflicts have shaken the world. Some have been resolved, while others continue to threaten world peace. The lessons of two world wars and the threat of mass destruction that would result from a nuclear war has kept these conflicts contained. World War III has not occurred, and the hope of people everywhere is that it never will.

Southwest Asia After World War I Britain said it would help create a Jewish homeland in Palestine, a region of Southwest Asia. Many Arab nations, however, wanted an Arab state in Palestine. In 1947 the UN voted to **partition**, or divide, Palestine, creating both a Jewish state and an Arab state. While the Arabs rejected this plan, Jewish leaders in Palestine accepted it. In May 1948, Israel was established as a Jewish state.

Israel's Knesset, or parliament, meets here. The Knesset is the supreme power in Israel.

USING VISUAL RESOURCES

Have students study the picture on this page of people celebrating the reunification of Germany. Ask students what details they see that tell them what this celebration was like and what German reunification meant to the people in the picture. *(Possible answers: Many of the people in the picture are young, which suggests that reunification was very important to young Germans. The German flag is a symbol of one united Germany. The photograph was taken at night, which suggests that the celebration went on for a long time.)*

Activity: Have students write a journal entry that might have been written by one of the young people in the photograph. Invite volunteers to share their entries with the class.

LEVEL 1: (Suggested time: 10 min.) Call on students to name some of the advancements made at the end of the 1900s. As students name development, write them on the chalkboard. Discuss student responses. After the discussion, have students vote on which of the listed achievements they think will have the greatest influence on the future.

ENGLISH LANGUAGE LEARNERS

LEVEL 2: (Suggested time: 30 min.) Tell students to imagine that they are historians from the year 2250 who have been asked to write descriptions of the major events of the late 1990s and early 2000s for middle school students. Have each student write a short description of one event that explains its significance to the world. Encourage students to share their writing with the class.

GLOBAL PERSPECTIVES

In 2000, 50 years after the beginning of the Korean War, the leaders of North and South Korea met for a historic three-day summit at P'yŏngyang. The leaders agreed to reunited families separated by political boundaries since the Korean War and to promote economic development in both countries. They also pledged to work toward reunification through increased cultural, athletic, medical, and environmental cooperation and exchanges.

Part of the agreement went into effect immediately. Two hundred families were brought together later that year. In addition, members of the North and South Korean Olympic teams marched together at the 2000 Olympics.

Activity: Have students use the Internet, almanacs, and other current events sources to find recent information about reunification efforts.

At the end of the Korean War, the two sides set up a neutral area, called the demilitarized zone, or DMZ. It is a buffer zone, and no military forces from either side may enter the area.

The establishment of Israel enraged many Palestinian Arabs and Arab nations. Attacked by neighboring Arab countries, Israel fought back. By early 1949 a cease-fire was reached. Israel survived, but Palestinian Arabs had no homeland. In 1967, tensions between Israel and its Arab neighbors exploded into war again. In what became known as the Six-Day War, Israel captured territory from Egypt, Syria, and Jordan. After the Six-Day War, the Palestine Liberation Organization (PLO), led by Yasir Arafat, launched many attacks on Israel.

The many attempts to bring peace to the Middle East have failed. The Israelis and the Arabs do not trust each other. Both sides make demands that the other side will not meet. When one side commits acts of violence, the other side strikes back. The failure to achieve peace in the Middle East continues to be one of the most disturbing issues facing the world.

Korea At the end of world War II the Soviet Union controlled northern Korea. U.S. troops controlled southern Korea. A Communist government took power, and in 1950, North Korea invaded South Korea. The United Nations sent troops to stop the invasion. The Korean War lasted until 1953, when a cease-fire was signed. Korea remains divided.

Cuba In 1959 Fidel Castro established a Communist government in Cuba. In 1961 President John F. Kennedy approved an invasion of Cuba by anti-Castro forces. The invasion failed, and Castro turned to the Soviet Union for support. The Soviet Union, which by this time possessed nuclear weapons, sent nuclear missiles to Cuba. Kennedy demanded that the missiles be withdrawn, but Khruschev refused. NATO and Warsaw Pact military forces prepared for combat. For several days the Cuban missile crisis held the world on the brink of nuclear war. Finally, the Soviet Union agreed to remove its missiles, and the United States promised it would not invade Cuba.

Vietnam Vietnam had been a colony of France for more than 60 years. In 1945 Ho Chi Minh, a Communist leader, declared Vietnam independent. In 1954 Vietnam became divided. North Vietnam was communist. South Vietnam was not. In the late 1950s, when North Vietnam invaded South Vietnam, the United States sent troops to South Vietnam to fight the communists. Many Americans were unhappy that the country had become involved in this war, and American troops pulled out of Vietnam in 1973. South Vietnam surrendered in 1975. Nearly 1.7 million Vietnamese and about 58,000 Americans lost their lives in the Vietnam War. In 1976 Vietnam was united under a Communist government.

LEVEL 3: (Suggested time: 50 min.) Organize the class into small groups. Have each group create a cause-and-effect diagram that illustrates the spread of communism in the mid-1900s and the events that led to its decline in the late 1900s. Tell students that their diagrams should clearly identify the ideas and actions that brought an end to communist governments in the Soviet Union and other countries.

Then have each group select a particular country in which to study the effects of the decline of communism. Have each group consult library and Internet resources to prepare a report about its chosen country. These reports should note when and how the chosen countries became free of communism, what changes have taken place in the countries, and what is happening in the countries today. Call on volunteers from each group to share the reports with the class. **BLOCK SCHEDULING**

Northern Ireland When the Republic of Ireland gained independence from Great Britain in 1922, the territory of Northern Ireland remained part of Britain. The Protestant majority in Northern Ireland controlled both the government and the economy. This caused resentment among Northern Irish Catholics. During the late 1960s Catholic protests began to turn violent. The British have tried to resolve the conflict both through political means and military force. While the situation has improved, a permanent peaceful solution has not yet been found.

The Breakup of Yugoslavia Yugoslavia was created after World War I by uniting several formerly independent countries. These included Bosnia, Croatia, Slovenia, and Serbia. After the fall of communism in eastern Europe, Eastern Orthodox Serbs tried to dominate parts of Yugoslavia where people were mainly Roman Catholic or Muslim. Fighting broke out between Serbia and Croatia, which was mainly Roman Catholic. Yugoslavia was once again divided into several countries in the early 1990s, but this did not end the violence. In 1992 Bosnian Serbs began a campaign of terror and murder intended to drive the Muslims out of Bosnia. Finally, NATO bombed Serbian targets in 1995, and the fighting stopped. Several Serb leaders were tried as war criminals.

The War on Terrorism On September 11, 2001, terrorists attacked the World Trade Center in New York City and the Pentagon in Washington, D.C. Following these attacks, the United States asked for support of nations around the world. Many nations, including Russia, China, Cuba, Pakistan, and Saudi Arabia, supported the United States in what President George W. Bush called a "war on terrorism." That show of support indicated that nations of the world might be beginning to leave behind some of the struggles of the last century.

✓ **READING CHECK:** *Evaluating* What has prevented conflicts in different parts of the world from becoming world wars? fear of another world war; fear of the destruction from a nuclear war

▲ After the Vietnam War more than a million people fled Vietnam by boat.

Interpreting the Visual Record What kinds of conditions might make people willing to leave their homes?

◄ Mostar, the unofficial capital of Bosnia and Herzegovina, was bombed heavily in the 1990s.

National Geography Standard 6

Places and Regions U.S. actions in the Vietnam war left Americans divided over what the country's basic values and role in the world should be. It also left some Americans with an attitude of distrust toward their government and political leaders that continues to be felt today.

The effects of the Vietnam War on U.S. foreign policy have been equally significant. The level of American involvement in Southwest Asia, Africa, and the Balkans has been increased by the Vietnam experience. Because of its broad global and domestic effects, many historians consider the Vietnam War to be the defining event in American history since World War II.

Activity: Have students conduct research in primary sources such as letters, newspapers, and magazines from the Vietnam War era and present their findings to the class. Have students compare and discuss the different feelings and opinions expressed in the sources.

◄ **Visual Record Answer**

war, dictatorship, poverty, terrorism, lack of economic opportunity

CLOSE

Ask students to list the three most significant challenges they think the world faces today. Lead a brief discussion about these issues and steps that have been taken to address them.

REVIEW AND ASSESS

Have students complete the Section Review. Then have each student write four multiple choice questions about one of the topics discussed in this section. Collect the questions and redistribute them to the class so that each student receives questions prepared by another student to answer. Then have students complete Daily Quiz 9.5.
ENGLISH LANGUAGE LEARNERS

Section 5 Review

Answers

Define For definitions, see: bloc, p. 204; arms race, p. 206; partition, p. 207; globalization, p. 211

Reading for the Main Idea

1. struggle between communism and capitalism; arms race; new alliances among countries (NGS 13)
2. Israeli Jews and Palestinian Arabs disagree over who should control Palestine. Many Arabs want to eliminate the country of Israel. (NGS 13)
3. space travel; computers; the Internet; genetics; cloning; global cooperation (NGS 10)

Critical Thinking

4. Answers will vary, but should be supported.

Organizing What You Know

5. Stalin—brought eastern Europe under communist control; Gorbachev—began reform in Soviet Union, improved relations with United States; Yeltsin—led Russia toward democracy; Bush—brought nations together to fight terrorism; Arafat—created PLO

Visual Record Answer

courage, generosity, willingness to take risks in order to help others

▲ After the terrorist attacks, members of New York City's fire and police departments risked their lives to help people.

Interpreting the Visual Record **What qualities can a terrible disaster bring out in people?**

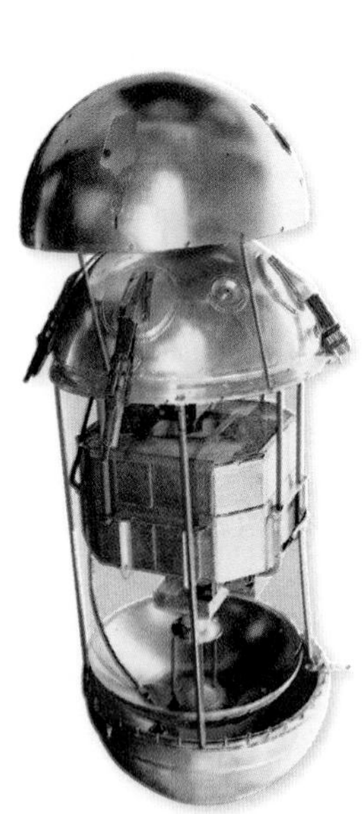

▲ The launch of the *Sputnik* satellite, shown above, began a space race between the U.S. and the Soviet Union to see who would reach the moon first.

Progress and Problems

During the late 1900s enormous advances of all kinds occurred in science, medicine, space travel, communication, and technologies. Also during this period, however, problems developed that must be addressed in the new century.

Space On October 4, 1957, the Soviet Union launched *Sputnik*, the world's first space satellite. The space age had begun. In 1961 Soviet Yury Gagarin became the first person to travel into space. In 1969 Neil Armstrong, an American, became the first person to walk on the moon. Beginning in the late 1990s, the United States, Russia, and 14 other nations worked together to build an International Space Station (ISS) The first crew members reached the ISS late in 2000. Researchers in space have conducted many useful experiments in geography, engineering, medicine, and other fields of study.

Genetics Genes are small units in the body that determine all our physical characteristics. Knowledge of genetics is leading to new cures for diseases. It has also given scientists the ability to clone, or make an identical copy of an animal. In 1997, for example, scientists cloned a sheep. Although cloning has raised some ethical debates that have not been resolved, it is still a remarkable scientific advance.

The Computer Age The first modern computer, called ENIAC, was built in 1946. It was so large that it filled an entire room. By the end of the 1950s, however, the manufacturing of computers was a growing industry. Over time, computers became smaller and faster. Now many people around the world use computers every day at home, at work, and at school. They are important parts of many products, from cars to medical equipment to rockets. New technologies like the Internet and the World Wide Web have made computers even more useful. Today computers can send messages around the world in just a few seconds.

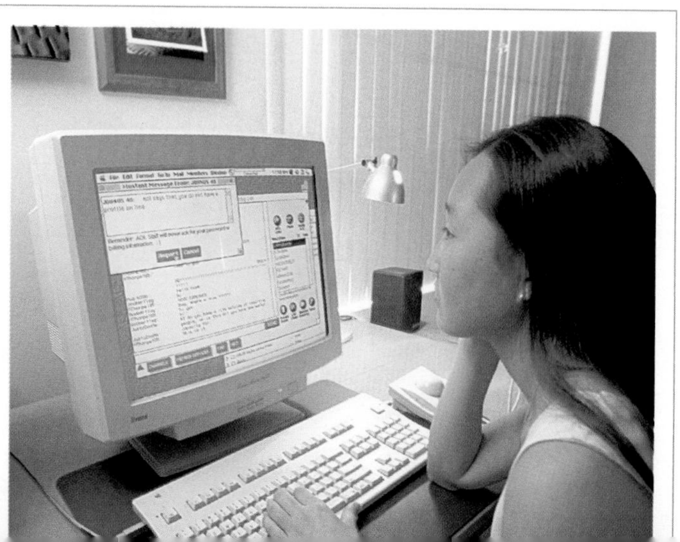

Reteach

Have students complete Main Idea Activity 9.5. Then pair students and have each pair write newspaper headlines that describe five of the events described in this section. Encourage the pairs to choose the five events they think are most significant in world affairs. Call on volunteers to share their headlines with the class. Display students' headlines around the classroom.

ENGLISH LANGUAGE LEARNERS, COOPERATIVE LEARNING

Extend

Have interested students conduct research on an environmental issue facing the United States and prepare a presentation on this issue. Presentations should include a description of the issue addressed as well as recommendations for solving problems related to the issue. Have students give their presentations to the class. **BLOCK SCHEDULING**

Globalization Faster ways of traveling and communicating have brought countries and cultures around the world closer together. As a result, people's lifestyles are becoming more similar. For example, people in different parts of the world now eat the same kinds food, wear the same types of jeans, and listen to the same music. This process, in which connections around the world increase and cultures share similar practices, is called **globalization**. One example of globalism is the European Union (EU). The EU includes twelve European nations that agreed to drop trade barriers and to adopt a common currency—the Euro.

An Endangered Environment Modern technology has improved the world in many ways, but it has also caused problems. For example, burning certain fuels causes pollution. As rain falls through the polluted air, it becomes acid rain, which kills trees and plants. Burning these same fuels also releases excess carbon dioxide into the atmosphere. Some scientists say this causes a greenhouse effect, trapping heat and causing temperatures on Earth to rise. Some scientists also say that Earth's ozone layer is thinning out. Ozone is a gas that shields our planet from some of the sun's rays which are harmful to plants and animals.

Looking Ahead The world faces many challenges, but the prospect for the future is hopeful if we can meet them. If we can live in peace, preserve the environment, and use science and technology for the benefit of all, the world of tomorrow can be a truly wonderful place.

These coins and bills are part of the European Union's currency system, the Euro.

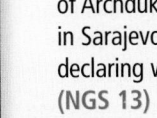

✓ **READING CHECK:** *Summarizing* How have some modern scientific advances changed the world? space travel, computers, the Internet, knowledge of genetics, identifying the greenhouse effect

Section 5 Review

go.hrw.com Homework Practice Online
Keyword: SK3 HP9

Define and explain: bloc, arms race, partition, globalization

Reading for the Main Idea

1. *Human Systems* What happened during the Cold War?
2. *Human Systems* What are the reasons for conflict in Southwest Asia?
3. *Human Systems* What were some important advances made in 1900s?

Critical Thinking

4. **Supporting a Point of View** Do you think most countries believe it is in their best interest to keep peace? Why?

Organizing What You Know

5. **Lead in to come?** Copy the following graphic organizer. Fill it in, telling how each person shaped events in the late 20th century.

Joseph Stalin	
Eleanor Roosevelt	
Mao Zedong	
Mikhail Gorbachev	
George W. Bush	

Chapter 9 Review Answers

Building Vocabulary

For definitions, see: militarism, p. 185; armistice, p. 188; stock market, p. 190; bankrupt, p. 191; dictator, p. 192; fascism, p. 192; police state, p. 192; boycotted, p. 195; apartheid, p. 196; aggression, p. 199; anti-Semitism, p. 201; genocide, p. 202; bloc, p. 204; arms race, p. 206; globalization, p. 211

Reviewing the Main Ideas

1. nationalism, militarism; killing of Archduke Francis Ferdinand in Sarajevo; Austria-Hungary declaring war on Serbia. (NGS 13)
2. promised to solve economic woes of the Great Depression, and to restore countries' former glory (NGS 11)
3. gained control of their land and lives, they cultural revival; inexperienced leaders and dictators; civil wars; failed economies (NGS 12)
4. Japan attacked Pearl Harbor. (NGS 13)
5. new alliances; world divided into Western bloc and Eastern bloc; Germany divided into communist and noncommunist sections; West's attempts to stop spread of communism; arms race; space race (NGS 12)

ASSESS

Have students complete the Chapter 9 Test.

RETEACH

Organize the class into five groups and assign each group one of the sections in this chapter. Have each group create the front page of a newspaper that describes the events discussed in their assigned sections. The page should include a few short articles with headlines addressing the major topics in the section.

ENGLISH LANGUAGE LEARNERS, COOPERATIVE LEARNING

PORTFOLIO

PORTFOLIO EXTENSIONS

1. Organize the class into groups and have each group create a short play that depicts a major event from Chapter 9. Have each group present its play to the class. Remind students that they should create programs to accompany their plays, listing the cast, describing the event depicted, and explaining the plot of the play. Place these programs in student portfolios.
2. Have students write paragraphs or create models that represent what they consider to be the most significant development of the late 1900s. Encourage students to be creative in their work. Place the paragraphs or photographs of the models in student portfolios.

CHAPTER 9

Review ANSWERS

Understanding History and Society

Charts will vary but should note countries created after World War I, World War II, and the breakup of the Soviet Union. Use Rubric 7, Charts, to evaluate student work.

Thinking Critically

1. Germans were angry about Treaty of Versailles; Hitler used this anger to gain support and power.
2. created more job opportunities, helped set up minimum wage, and supported workers' rights to form unions
3. In both countries protests and demonstrations led to the downfall of white power. India was fighting for its independence, while in South Africa people were fighting for racial equality.
4. Aggression helped Germany, Italy, and Japan gain power and territory; war was finally declared to stop the aggression
5. Possible answers: lessons of two world wars, UN intervention, the threat of mass destruction posed by a nuclear war

CHAPTER 9

Reviewing What You Know

Building Vocabulary

On a separate sheet of paper, write sentences to define each of the following words.

1. militarism
2. armistice
3. stock market
4. bankrupt
5. dictator
6. fascism
7. police state
8. boycotted
9. apartheid
10. aggression
11. anti-Semitism
12. genocide
13. bloc
14. arms race
15. globalization

Reviewing the Main Ideas

1. *Human Systems* What ideas, actions, and incidents were major causes of World War I?
2. *Human Systems* How did dictators come to power in Europe during the 1930s?
3. *Human Systems* What problems were solved and which were created by African independence?
4. *Human Systems* What single event finally caused the United States to enter World War II?
5. *Human Systems* How did the Cold War shape the second half of the 1900s?

Understanding Environment and Society

The Changing World Map

The world map has changed a great deal during the 1900s. Create a presentation that shows some of these changes. Your presentation should include a chart that describes the changes and shows where and when they happened. Consider the changes that occurred:

- After World War I.
- After World War II.
- After the Cold War.

Thinking Critically

1. **Drawing Conclusions** How did the Treaty of Versailles help set the stage for World War II?
2. **Evaluating** How did President Roosevelt's program, the New Deal, help workers?
3. **Comparing and Contrasting** How were the struggles against European domination in South Africa and in India similar? How were they different?
4. **Analyzing** What role did aggression, and the response of world leaders to aggression, play in World War II?
5. **Evaluating** What prevented conflicts from becoming major world wars in the late 1900s?

Food Festival

The expression "An army marches on its stomach" suggests how important food is to soldiers. In World War II, the U.S. military developed K rations so soldiers would have something to eat when no other food was available. Soldiers carried K rations in their packs. K Rations contained biscuits made from dried beef, malted milk tablets, canned veal loaf, coffee, sugar, canned ham spread, soup cubes, chocolate, sausage, and lemon powder. Sometimes K rations were the only food soldiers had to eat.

Chapter 9 Review and Assessment Resources

Reproducible
- Readings in World Geography, History, and Culture 32, 34, 38, 60, 65, 74
- Critical Thinking Activity 9
- Vocabulary Activity 9

Technology
- Chapter 9 Test Generator (on the One-Stop Planner)
- Audio CD Program, Chapter 9
- HRW Go site

Reinforcement, Review, and Assessment
- Chapter 9 Review, pp. 212–213
- Chapter 9 Tutorial for Students, Parents, Mentors, and Peers
- Chapter 9 Test
- Chapter 9 Test for English Language Learners and Special-Needs Students
- Unit 2 Test
- Unit 2 Test for English Language Learners and Special-Needs Students

Building Social Studies Skills

Map Activity

On a separate sheet of paper, match the letters on the map with their correct labels.

Austria	**Italy**
France	**Poland**
Germany	**Russia**
Great Britain	

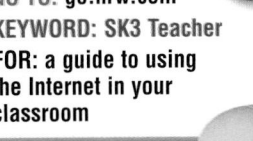

Mental Mapping Skills Activity

On a separate sheet of paper, draw a freehand map of Western Europe. Label Germany, Italy, France, and Great Britain. Shade the countries that sided with the Allies in World War II one color. Shade the countries that were Axis Powers a different color. Make a key for your map.

Writing Activity

Imagine that you are an American journalist in India just before it gained independence in 1947. Write a news article about the events that have led up to independence, and about problems the new nation may face. Be sure to use standard grammar, spelling, sentence structure, and punctuation.

Alternative Assessment

Portfolio Activity

Learning About Your Local History

Historical People and Places

Research how your town is connected to an event from the 1900s. Write a report on a key person, place, or object and explain its historical importance. Include photographs or drawings in your report.

internet connect

Internet Activity: go.hrw.com
KEYWORD: SK3 GT9

Choose a topic about the modern world.
- Learn about the Treaty of Versailles.
- Write a report about Anne Frank.
- Create a poster about the causes and effects of global warming.

Map Activity

A. Great Britain
B. France
C. Germany
D. Russia
E. Austria
F. Italy
G. Poland

Mental Mapping Skills Activity

Maps should show correct locations of Germany, Italy, France, and Great Britain. They should show that Great Britain and France were on the side of the Allies and that Germany and Italy were on the side of the Axis.

Writing Activity

Articles will vary but should be consistent with text material. Use Rubric 40, Writing to Describe, to evaluate student work.

Portfolio Activity

Reports will vary. Students should identify a local person, place, or thing that is connected to a local historical event and write about this connection. Use Rubric 42, Writing to Inform, to evaluate student work.

internet connect

GO TO: go.hrw.com
KEYWORD: SK3 Teacher
FOR: a guide to using the Internet in your classroom

FOCUS ON GOVERNMENT

Studying Government Agencies

Tell students that many different agencies of the U.S. government are responsible for guaranteeing the safety and security of American citizens from terrorists and terrorism. These include the Federal Bureau of Investigation (FBI), the Central Intelligence Agency (CIA), the Immigration and Naturalization Service (INS), the Federal Aviation Agency (FAA), and the recently formed Office of Homeland Security. Organize the class into groups and assign one of the federal agencies to each group. Ask each group to conduct research on its assigned agency. Direct the groups to create charts about these agencies. In one column of the chart, have students list the agency's main responsibilities and tasks. In the other, have them list ways in which the agency might combat terrorism either at home or around the world. Have each group present its chart to the class. Then stage a roundtable discussion about these federal agencies and the fight against terrorism.

GEOGRAPHY SIDELIGHT

One reason for the initial success of the war on terrorism was that Afghanistan had become a friendless regime. Isolated geographically, Afghanistan had cut cultural ties with the rest of the world. Few countries recognized the Taliban as the legitimate government.

The Taliban, an Islamic fundamentalist group, rose to power in the late 1980s. By the late 1990s they controlled more than 90 percent of the country. The United States imposed sanctions against Afghanistan in 1999 when the Taliban refused to turn over Osama bin Laden, who had ordered the bombing of American embassies in Kenya and Tanzania the previous year. In early 2001, the Taliban angered countries around the world when it destroyed two ancient Buddha statues in Bamiyan, near Kabul.

Critical Thinking: How might the war on terrorism have been different had Afghanistan had more allies?

Answer: The coalition might have had to fight Afghanistan's allies and trading partners.

➤ **This Focus on Government feature addresses National Geography Standard 13.**

FOCUS ON GOVERNMENT

Combating Global Terrorism

Early in the morning of September 11, 2001, two passenger jets crashed into the two towers of the World Trade Center. Another jet hit the Pentagon outside of Washington, D.C., and a fourth crashed in southern Pennsylvania. All of these planes were piloted by terrorists, people who use violence to achieve political goals. An investigation by the U.S. government led officials to conclude that al Qaeda, a terrorist organization led by Osama bin Laden, was responsible for these attacks. At the time of the attacks, bin Laden was living in Afghanistan, a landlocked country in Southwest Asia.

▲ President George W. Bush addressed a joint session of Congress on September 20. During the session he pledged to use the full might of the nation in a war on international terrorism

The War on Terror The terrorist attacks of September 11, 2001, were not the first on American soil or American interests. Terrorists tried to destroy the World Trade Center in New York in 1993. American citizens and embassies abroad have also been targeted. U.S. allies have also been victims of terrorists. For years, the U.S. government and its allies went to great lengths to combat terrorism. The terrible attacks of September 11, however, demanded the strongest possible retaliation.

The New York City sky filled with smoke after the attack. ▼

Calling the attacks an act of war, President George W. Bush declared war on terrorism. "Either you are with us or you are with the terrorists," President Bush told the world.

Forming an International Coalition The United States wanted to combat terrorists on a global scale. To do that, President Bush formed an international coalition. Great Britain was an early partner in the coalition, and Prime Minister Tony Blair enlisted the support of many nations. A new era in global relations began as China, Russia, and more than 50 other countries joined the coalition.

Going Further: Thinking Critically

Ask students to discuss why it is essential for the different agencies of the United States to share information and cooperate among themselves to fight terrorism. Have students speculate on how the Office of Homeland Security might make this kind of cooperation possible. Ask students to use the results of their research in the previous activity to suggest kinds of information that each agency might be able to provide to other organizations. Finally, using what they have learned, have students work together to design a diagram that shows how the various government agencies they have studied work together to combat terrorism.

Asia Political

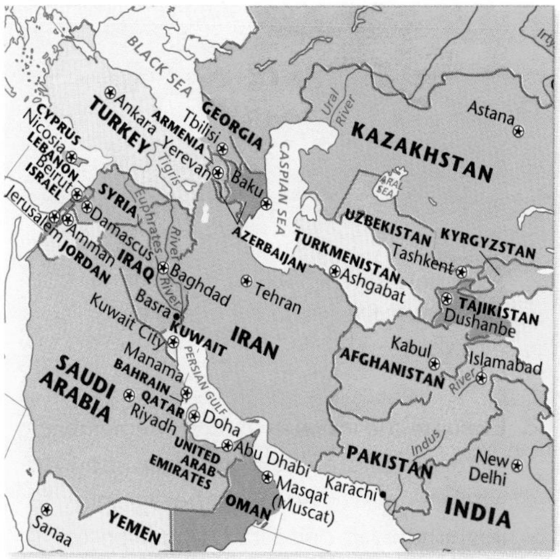

Afghanistan is a landlocked country. It shares borders with Pakistan, Iran, Turkmenistan, Uzbekistan, and Tajikistan.

Much of Afghanistan is covered by rugged mountainous terrain. This landscape provided hiding places for al Qaeda terrorist camps and cells.

Some countries supplied military support for the fight against terrorists in Afghanistan. Others, including Pakistan and Uzbekistan, allowed coalition forces to use their air bases near Afghanistan. The coalition included many Muslim and Arab nations.

Coalition forces achieved victory quickly in Afghanistan. By the end of 2001, the Taliban, Afghanistan's ruling government, was forced from power and many al Qaeda terrorists were captured or killed. Despite this success, the war on terrorism is not over. Terrorist cells are still active in other countries. President Bush warned that new battles will have to be fought to eliminate them.

A New Government for Afghanistan

After the fall of the Taliban, representatives from all over Afghanistan formed a new government. Their country faced difficult problems. Years of war had left the country in ruins. As part of the war on terrorism, coalition forces provided food and medical care and also promised to rebuild Afghanistan.

Homeland Security While fighting terrorism in Afghanistan, the U. S. government also acted to block future terrorist attacks at home. The nation's borders and coastline are patrolled more carefully. New procedures make air travel more secure.

President Bush also created the Office of Homeland Security. Its job is to coordinate different security and intelligence agencies. As a result, officials can better identify people with links to terrorists.

Understanding What You Read

1. How are the United States and other nations fighting global terrorism?
2. In what ways are Americans defending their homeland against terror?

Understanding What You Read

Answers

1. The United States and more than 50 other countries have formed a coalition to fight terrorism. Some countries supply military support. Others allow the coalition to use their air bases.
2. The United States is patrolling its borders more carefully and working to make air travel safer. The United States has also established an Office of Homeland Security to coordinate information about terrorists between security and intelligence agencies.

Going Further: Thinking Critically

Explain that the development of photography has contributed a great deal to what we know about the past leaders of our country. Gather images of painted portraits of early presidents, such as George Washington and Thomas Jefferson, and show them to the class. Point out that we have no photographs of these men. Next show a few photographs of Abraham Lincoln. Explain to students that these photographs were taken at a time when the camera was still a relatively new invention. Finally, show some candid photographs of recent presidents, such as John F. Kennedy playing with his children or Bill Clinton playing golf. Explain that our modern presidents usually face dozens of photographers almost every day.

After students study the visual evidence you have presented, ask them to speculate about how photography may have changed our attitudes and ideas about our leaders and other public figures. How might general attitudes toward leaders been different in the past before photography and video? *(Students may note that images and posed photographs present a more formal view of individuals, while candid shots make even powerful figures seem more approachable and human.)*

PRACTICING THE SKILL

1. Students should identify the approximate year or decade when each photograph was taken. Depending on the photos, students might list details that describe clothing and home furnishing styles, buildings, automobiles, toys, tools, and other everyday objects.
2. Suggest that students cover the caption before they begin to study the photo. In creating the "story" that a photo tells, ask students to think about the values that the photographer or subjects in the photo seem to advance.
3. Ask students to choose photos that tell about different areas and aspects of American life. These areas, for example, might include business, sports, technology, art and culture, education, and so on.

➤ **This GeoSkills feature addresses National Geography Standards 2, 4, 6, 15, and 18.**

GeoSKILLS

Building Skills for Life: Using Visual Evidence

Pieter Brueghel the Younger painted this scene in the 1500s.

We learn about the past from many different sources. One important source of information is visual evidence. Paintings, statues, drawings, and photographs are all examples of visual evidence. Pictures and paintings provide many clues about life in the past. Sometimes they show the clothing people wore, the houses they lived in, or the games they played. Often they document, or prove, that certain events were an important part of a culture.

As with all historical sources, you need to study visual evidence carefully. Often, artists and photographers only show what they want you to see. To use visual evidence effectively, follow these steps:

1. **Identify the visual evidence.** Study a picture or painting carefully. Pay attention to details. Ask yourself who the subjects of the picture are and what they might be doing. Look at the painting on this page. How would you describe the people you see? What are they doing?
2. **Evaluate the visual evidence.** Remember, a picture or photo does not always show the whole story. The artist or photographer might have left out details on purpose. Thinking about the artist's purpose will help your decide whether visual evidence is reliable. What purpose do you think Brueghel had for painting his picture? Do you think it is a reliable record of how some people lived?
3. **Learn from the visual evidence.** After studying a picture carefully, use the details that you note to draw conclusions. Your conclusions should help you understand how some people lived at a particular time and place in the past. What conclusions can you draw from the details in Brueghel's painting?

PRACTICING THE SKILL

1. Study an old family album or library book that contains photographs taken long ago in your city or state. What do the pictures tell you about life in the time in which they were taken?
2. Study a picture in a newspaper or magazine without reading the caption. What story does the photo tell about our culture? Why do you think the photographer has chosen to document this event in this way?
3. What visual evidence would best capture our way of life today? Choose five pictures. Explain why each one might help people of the future understand American life today.

HANDS ON GEOGRAPHY

Paintings, drawings, statues, sculpture, and photographs are all types of visual evidence. Historians study the details in visual evidence to draw conclusions about the past.

Look at the two examples of visual evidence below. The first is a sculpture. It shows a battle that took place in ancient Greece about 2,500 years ago. The second is a painting from 1862 of a British railway station. What details do you notice in each example? What information do they help you to know?

This relief shows the Battle of Marathon that took place in 490 B.C.

Artist William Frith painted this scene in 1862.

Lab Report

1. Look at the sculpture of the Battle of Marathon. What conclusions can you draw about warfare in ancient Greece?
2. Look at the painting. What do you think the artist's purpose was painting this scene?

Lab Report

Answers

1. Possible conclusions: The ancient Greeks fought with heavy shields and swords. Aside from heavy battle helmets, they wore little armor in war. They used horses in battle, either as mounts for warriors or to pull chariots and equipment.
2. Possible conclusions: The artist wanted to portray England's modern transportation system in a glowing light; he wanted to capture the noise and bustle of the large and ornate terminal; he wanted to show that crowds of well-dressed and probably wealthy people took advantage of the new mode of transportation.

UNIT 3

EUROPE

Using the Illustrations

Direct students' attention to the photographs on these pages. Tell them that instead of streets Venice, Italy, has canals. The city's only paved streets are narrow and for pedestrians only. Venice was built on a group of islands in a lagoon. You may want to read the sidebar feature about Venice's gondolas in the chapter on southern Europe. Ask students to speculate about all the ways their lives would be different if the major thoroughfares of their community were canals instead of streets.

Remind students that various forces of erosion shape the land. Ask how the valley in the small photo may have been created *(by a glacier).* Tell students that in the chapter on northern Europe they will learn more about fjords, like the one pictured on this page, and how they were formed.

Direct students' attention to the photograph of the swan. Point out that Europe is densely populated and that very few large wild animals live there except in nature preserves.

Finally, refer to the photo taken in Prague. The writer Franz Kafka lived on the street pictured. The house in the center is tiny, and visitors must stoop to enter it.

Unit Objectives

1. Describe the major landforms, bodies of water, climates, and resources of Europe.
2. Explain the historical development of the European nations and the influence of European cultures on other parts of the world.
3. Identify different cultural groups in Europe and understand how they interact with one another.
4. Discuss Europe's influence in international affairs.
5. Interpret special-purpose maps to analyze relationships among climates, population patterns, and economic resources in Europe.

UNIT 3

Europe

CHAPTER 10
Southern Europe

CHAPTER 11
West-Central Europe

CHAPTER 12
Northern Europe

CHAPTER 13
Eastern Europe

Geirangerfjord, Norway

Regatta on the Grand Canal, Venice, Italy

CHAPTERS

10 Southern Europe explores Portugal and the Mediterranean countries—Greece, Italy, and Spain.

11 West-Central Europe focuses on France, Germany, Belgium, the Netherlands, Switzerland, and Austria.

12 Northern Europe describes the United Kingdom, the Republic of Ireland, and the Scandinavian countries.

13 Eastern Europe describes the political and economic changes and ethnic conflicts that have characterized Eastern Europe in recent years.

Notes from the Field

A Professor in the Czech Republic

Meredith Walker teaches English at Clemson University. She also teaches English As a Second Language courses. Here she describes a visit to Prague, the capital of the Czech Republic. **WHAT DO YOU THINK?** ***Does Prague sound like a city you would like to see?***

The castle in Prague sits on a hill high above the Vltava River. This river divides the city. A castle has stood on that hillside for 1,000 years. Rising from the inner courtyard of the castle is a huge medieval building, St. Vitus' Cathedral. Together, the castle and cathedral look almost magical, especially at night when spotlights shine on them. The castle is the most important symbol of the city. A great Czech writer, Franz Kafka, believed that the castle influenced everything and everyone in the city.

Another important landmark in Prague is Charles Bridge, one of eight bridges that cross the Vltava River. The bridge is part of the route that kings once traveled on their way to the castle to be crowned. Today, large crowds of tourists walk there. Many stop to photograph some of the 22 statues that line the bridge. No cars are allowed on the bridge today.

The first time I walked across the bridge, heading up toward the castle, it was at night. Fireworks were lighting the sky all around me. It was a colorful welcome.

Street scene in Prague, Czech Republic

Portuguese girls in native costume

Understanding Primary Sources

1. How does the city's physical geography help make the castle a symbol of Prague?
2. How did the fireworks display affect Meredith Walker's feelings about Prague?

Whooper swan

MORE FROM THE FIELD

Events in Prague have given the world an unusual word: *defenestration*, which means "a throwing of a person or thing out of a window." The word is taken from the Latin word *fenestra*, which means "window." In 1419, religious reformers defenestrated the entire town council. A second case was part of a Protestant rebellion. Three Catholic officials were thrown from a window in Prague's Hradčany Castle. The Second Defenestration of Prague, as the episode was called, led directly to the Thirty Years' War (1618–48).

In the aftermath of the Communist takeover of 1948, Jan Masaryk, the non-Communist foreign minister, committed suicide or was assassinated in a fall from his office window. Controversy arose after poet and screenwriter Bohumil Hrabal fell to his death from the window of his fifth-floor hospital room in 1997.

Activity: Have students conduct research on how certain buildings in Eastern Europe relate to the local or regional history of the area in which they are located.

Understanding Primary Sources

Answers

1. It is on a hill.
2. It made her feel more welcome.

OVERVIEW

In this unit students will learn about the landscape and cultures of one of the world's most densely settled continents.

A wide range of climates, from steppe to ice-cap, has allowed Europeans to use the land in many ways. In general, the farms are so productive that the majority of the population can live in cities and towns and work in manufacturing and service jobs.

Europe's past is dynamic and turbulent. Countless leaders have gone into battle to control its resources. Explorers, conquerors, and colonists spread out from Europe and have influenced most parts of the world. The continent was at the center of two world wars. Ethnic conflicts continue to trouble postwar Europe. Recently created republics struggle to find their place among the other countries.

Over the centuries, European art, architecture, music, literature, and drama have been valued by people around the world. The region's international influence has grown since many of the countries have joined the European Union.

PEOPLE IN THE PROFILE

Note that the elevation profile crosses the Great Hungarian Plain. The nation of Hungary dates back to when the nomadic Magyar people, whose ancestors came from central Russia, moved into the sparsely populated plain during the A.D. 890s.

The Magyars spent much of their time on horseback. They lived mainly on meat, mare's milk, and fish. The Magyars were organized into clans, which in turn banded together into tribes.

Until the mid-900s Magyars raided far into western Europe. They were skilled with their favorite weapon, the bow and arrow, and their horses were fast. The Magyars captured slaves and stole treasure until Otto the Great of Germany stopped them. According to tradition, they adopted Christianity on Christmas Day, 1000, when, with the pope's approval, the Magyar king, Stephen I, was crowned.

Today, most Hungarians still call themselves Magyars. There are large Magyar minorities in other countries, such as Romania. In fact, perhaps half of the world's Magyars live in countries other than Hungary.

Critical Thinking: Where did the Magyar people originate?

Answer: central Russia

UNIT 3 ATLAS The World in Spatial Terms

Europe

Elevation Profile

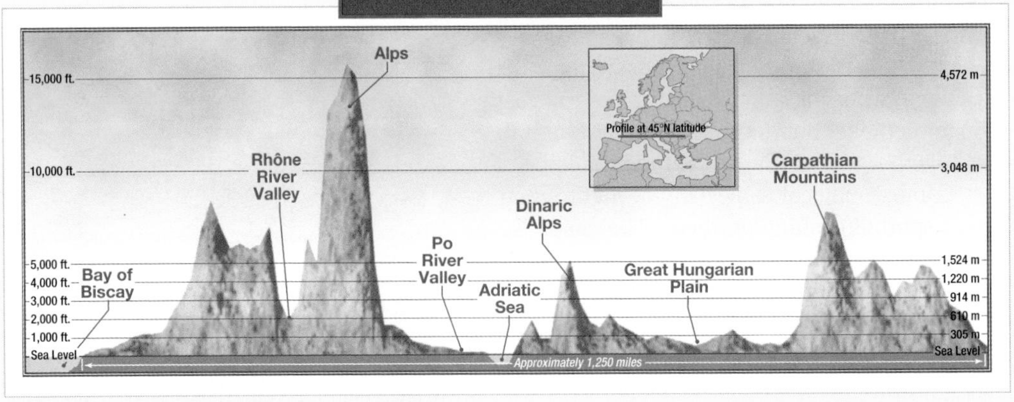

GEOSTATS:

- World's largest island: Greenland–839,999 sq. mi. (2,175,597 sq km)
- World's smallest independent country: Vatican City–0.17 sq. mi. (0.44 sq km)
- World's northernmost town: Longyearbyen, Spitsbergen Island, Norway–about 77.5°N

The United States and Europe: Comparing Sizes

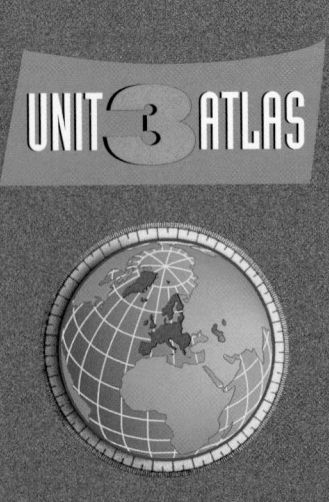

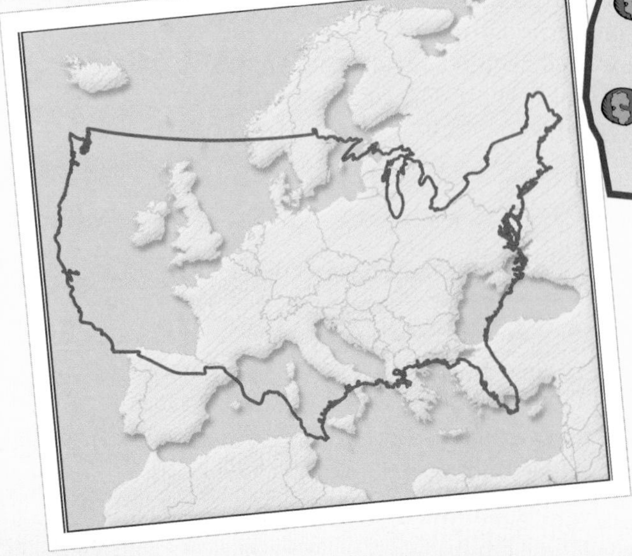

Using the Physical Map

Direct students' attention to the **physical map** on this page. Ask a volunteer to name the highest physical feature on the map and its location *(Mont Blanc, on France's border with Italy and Switzerland).* Then ask them to identify the area that is below sea level *(large area in the Netherlands).* Have students compare this map to the **population map** and have them suggest a problem that people of the Netherlands had to solve before the country could support a large population *(how to keep the North Sea from flooding the land).*

Ask students to suggest areas where no people live *(southeast Iceland, most of Greenland).* Then ask how they could guess this without looking at the population map *(presence of ice caps).*

Europe: Physical

UNIT 3 ATLAS

1. *Places and Regions* What are two mountain ranges that occupy peninsulas?
2. *Places and Regions* What are the two major plains of Europe? Which is larger?

Critical Thinking

3. **Analyzing Information** In the days before air travel, which physical feature would have made it difficult to travel between Italy and the countries to its north?
4. **Analyzing Information** Which physical feature would have made travel between Greece, Italy, and Spain fairly easy?

internet connect

ONLINE ATLAS
GO TO: go.hrw.com
KEYWORD: SK3 MapsU3
Web links to online maps of the region.

Physical Map

Answers

1. Apennines and Kjølen Mountains
2. Northern European Plain and Great Hungarian Plain; Northern European Plain

Critical Thinking

3. Alps
4. Mediterranean Sea

Using the Political Map

Tell the class that Europe is often called a peninsula of peninsulas. On a wall map, trace the outline of Europe so that students can observe how the entire continent is a peninsula of the Eurasian land mass. Then ask them to look at the **political map** on this page and to identify countries or pairs of countries that are peninsulas. *(Examples: Spain and Portugal, Italy, Denmark, Norway and Sweden, Greece)*

Also ask students to identify the largest country in Europe and those that are so small that their area is not apparent on the map *(France; Andorra, Liechtenstein, Malta, Monaco, San Marino, Vatican City)*. Then ask which country would be the largest if territory under its control were included in its area *(Denmark, because it controls Greenland)*. Point out that most of the large islands and island groups in the Mediterranean Sea are not independent countries. Call on volunteers to identify the countries to which the islands belong *(Balearic Islands, Spain; Corsica, France; Sardinia and Sicily, Italy; Crete, Greece)*.

internet connect

ONLINE ATLAS
GO TO: **go.hrw.com**
KEYWORD: **SK3 MapsU3**

Your Classroom Time Line

These are the major dates and time periods for this unit. Have students enter them on the time line you created earlier. You may want to watch for these dates as students progress through the unit.

- c.* 16,000 B.C. Early peoples create cave paintings at Altamira, Spain.
- c. 2000 B.C. Complex civilization exists on Crete.
- c. 800 B.C. City-states organize on Greek mainland.
- c. 750 B.C. The Latins establish Rome.
- c. 750–450 B.C. The Celts come to the British Isles.
- c. 400s B.C. Construction of the Parthenon begins.
- c. 300 B.C. Euclid writes *Elements*.

*c. stands for *circa* and means "about."

Political Map

Answers

1. Spain, France
2. Pyrenees

Critical Thinking

3. Bosnia and Herzegovina, Slovenia
4. It is separated from other countries by the English Channel, the Atlantic Ocean, and the North Sea.

Europe: Political

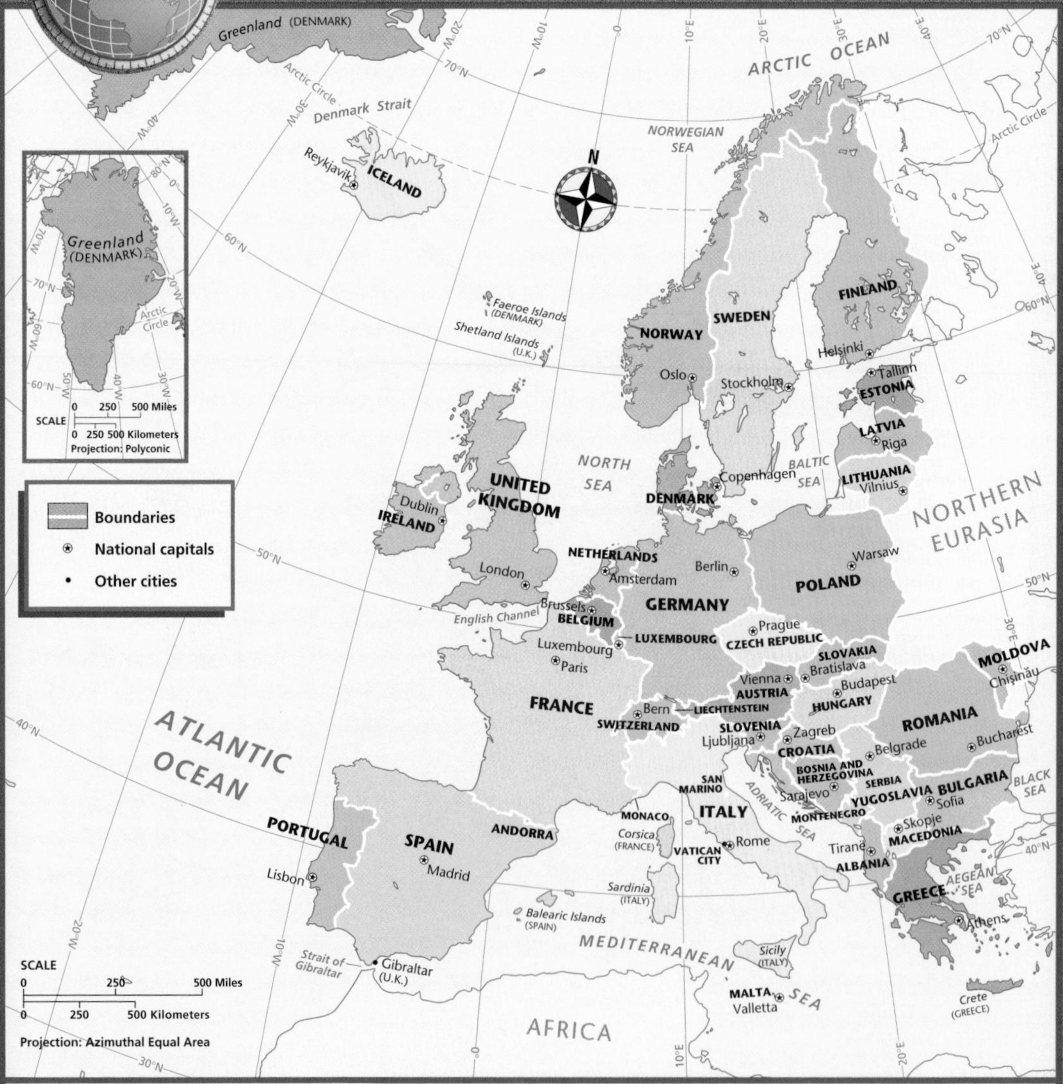

1. *Places and Regions* Which countries border both the Atlantic and the Mediterranean?

2. *Places and Regions* Compare this map to the **physical map** of the region. Which physical feature helps form the boundary between Spain and France?

Critical Thinking

3. **Comparing** Which countries have the shortest coastlines on the Adriatic Sea?

4. **Drawing Inferences and Conclusions** The United Kingdom has not been invaded successfully since A.D. 1066. Why?

USING THE CLIMATE MAP

Direct students' attention to the **climate map** on this page. Ask one student to call out the names of the individual countries that make up Europe and have other students name the main climate types in each country. Have students compare this map to the **political map** to answer these questions: Which island countries have only a marine west coast climate? *(Ireland, United Kingdom)* Which Atlantic coast country has only a Mediterranean climate? *(Portugal)* Which large Eastern European country is divided about equally between marine west coast and humid continental climate regions? *(Poland)*

UNIT 3 ATLAS

Europe: Climate

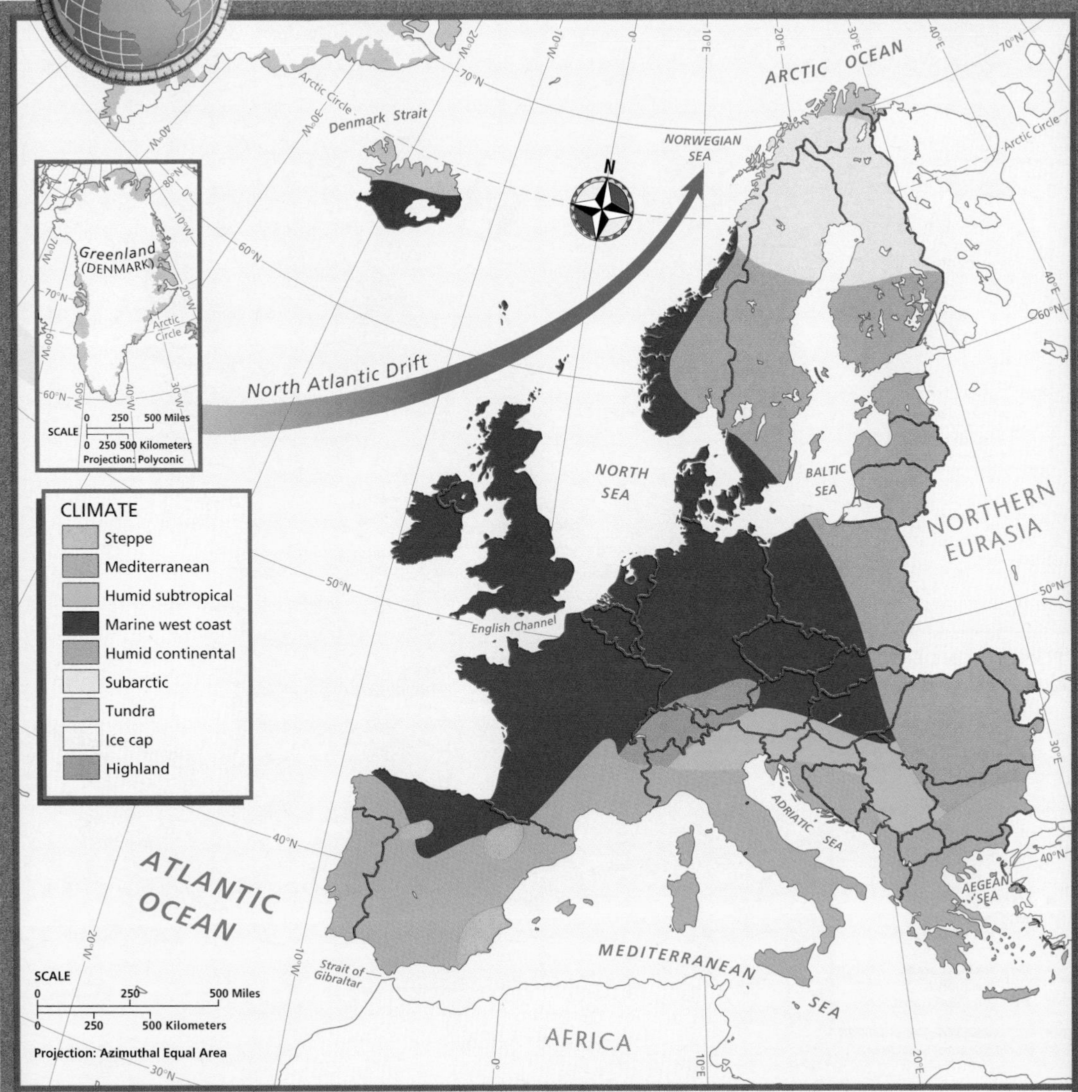

1. *Physical Systems* Which climate type takes its name from a sea in the region?
2. *Places and Regions* Which two independent countries have climate types that are not found in any other European country? Which climate types do these two countries have?

Critical Thinking

3. **Comparing** Compare this map to the **physical** and **population maps**. Which physical feature in central Europe has a highland climate and relatively few people? This physical feature is in which countries?

Your Classroom Time Line (continued)

- **200s–100s B.C.** Rome begins to expand.
- **A.D. 100s** The Roman Empire is at its greatest extent.
- **early 300s** Constantine adopts Christianity.
- **400s** The Roman Empire is divided into two parts.
- **400s** Angles and Saxons migrate to the British Isles.
- **400s** Huns invade eastern Europe.
- **476** Rome falls to invaders.
- **early 700s** Moors conquer the Iberian Peninsula.
- **800** Charlemagne is crowned Emperor of the Romans.
- **1066** The duke of Normandy invades the British Isles.
- **1100s** England conquers Ireland.
- **1200s** The Mongols invade Hungary.
- **1300s** The Renaissance begins in Italy.

Climate Map

Answers

1. Mediterranean
2. Iceland and Spain; tundra, marine west coast, ice cap; marine west coast, Mediterranean, steppe

Critical Thinking

3. Alps; France, Italy, Germany, Austria, Switzerland

UNIT 3 ATLAS

internet connect

ONLINE ATLAS
GO TO: **go.hrw.com**
KEYWORD: **SK3 MapsU3**

USING THE POPULATION MAP

As students examine the **population map** on this page, point out that many European cities are located on or near rivers. Have students compare this map to the **physical map** to identify some of these cities and rivers. Students may need to use the chapter maps to confirm detail. *(Possible answers: London—Thames, Paris—Seine, Rome—Tiber, Rotterdam—Rhine, Budapest—Danube)*

Ask students to identify the countries with the largest areas of high population density *(United Kingdom, Netherlands, Belgium, Germany, Italy).*

Your Classroom Time Line (continued)

- **1337–1453** England and France fight the Hundred Years' War.
- **1453** Ottoman Turks conquer Constantinople.
- **1492** King Ferdinand and Queen Isabella take Granada, Spain, from Moors. Christopher Columbus sails to America.
- **1500s** Germany becomes the center of the Reformation.
- **1503–06** Leonardo da Vinci paints *Mona Lisa.*
- **1588** The English defeat the Spanish Armada.
- **1616** William Shakespeare dies.
- **1789** The French Revolution begins.
- **1815** Napoléon Bonaparte is defeated.
- **1840s** The Irish potato famine occurs.
- **1867** The Austro-Hungarian Empire is created.
- **1871** Prussia unites Germany.

Population Map

Answers

1. cold subarctic, tundra climates
2. Latvia, Lithuania

Critical Thinking

3. Charts, graphs, databases, and models should accurately reflect population figures for cities on the maps.

Europe: Population

UNIT 3 ATLAS

1. *Places and Regions* Examine the **climate map**. Why are northern Norway, Sweden, and Finland so thinly populated?
2. *Places and Regions* Compare this map to the **political map**. Which countries have between 25 and 130 persons per square mile in all areas?

Critical Thinking

3. **Analyzing Information** Use the map to create a chart, graph, database, or model of population centers in Europe.

Using the Land Use and Resources Map

Focus students' attention on the **land use and resources map** on this page. Ask students to name countries that do not have major manufacturing and trade centers *(Iceland, Estonia, Latvia, Lithuania, Yugoslavia, Albania, Macedonia, Croatia, Bosnia and Herzegovina)* and to identify other economic activities that are important to these countries *(commercial farming, livestock raising, some mineral production).* Then ask which countries appear to have the largest number of major manufacturing and trade centers. *(Germany and Italy)*

Europe: Land Use and Resources

UNIT 3 ATLAS

1. *Places and Regions* What is the only country in the region that uses geothermal power?
2. *Places and Regions* Which country in the region mines uranium?
3. *Places and Regions* In which body of water is oil and gas production concentrated?

Critical Thinking

4. **Analyzing Information** Use the map on this page to create a chart, graph, database, or model of economic activities in Europe.

Your Classroom Time Line (continued)

- **1914–18** World War I is fought.
- **1921** Most of Ireland becomes independent.
- **1933** The Nazi Party seizes power in Germany.
- **1936–39** The Spanish Civil War is fought.
- **1939** World War II begins.
- **1944** Allies invade Normandy.
- **1945** Germany is defeated.
- **1956** Revolt in Hungary against the Soviet Union fails.
- **1961** The Berlin Wall is built.
- **1974** Greece becomes a republic.
- **1975** Francisco Franco's rule of Spain ends.
- **1989** Soviet control of Eastern Europe ends.
- **1989** The Berlin Wall is opened.
- **1993** Czechoslovakia divides into two countries.
- **1990s** Civil war and fighting in former Yugoslavia begins.

Land Use and Resources Map

Answers

1. Iceland
2. France
3. North Sea

Critical Thinking

4. Charts, graphs, databases, and models should accurately reflect information from the map.

Fast Facts

EUROPE

FAST FACTS ACTIVITIES

LEVEL 1: (Suggested time: 30 min.) Ask students to look at the population and area figures on these pages. Tell them that the United States has a population of more than 281.4 million people living in an area of 3,717,792 square miles. Have students work with a partner to compare the U.S. population per square mile with that of the European countries. Ask the students to decide if the population in the United States is more or less dense than the population of the European countries. *(Population density in Europe is higher than it is in the United States.)*

Then ask each pair of students to share their answer with another pair of students. Encourage the students to define the concept of population density and to demonstrate how they determined whether the United States or European countries tended to have a higher population density. *(Divide population by square miles. A larger number indicates a higher density—more people per square mile.)*

LEVEL 2: (Suggested time: 25 min.) Have students examine the figures for population and cars for the European countries. Mention that the U.S. population of approximately 281 million owns more than 130 million cars. Those figures translate roughly to one car for every two people in the United States, which

UNITED STATES OF AMERICA

CAPITAL: Washington, D.C.

AREA: 3,717,792 sq. mi. (9,629,091 sq km)

POPULATION: 281,421,906

MONEY: U.S. dollar (US$)

LANGUAGES: English, Spanish (spoken by a large minority)

CARS: 131,838,538

Fast Facts Europe

ALBANIA

CAPITAL: Tiranë
AREA: 11,100 sq. mi. (28,748 sq km)
POPULATION: 3,510,484
MONEY: lek (ALL)
LANGUAGES: Albanian, Greek
CARS: data not available

ANDORRA

CAPITAL: Andorra la Vella
AREA: 181 sq. mi. (468 sq km)
POPULATION: 67,627
MONEY: euro (€), 1-01-2002
LANGUAGES: Catalan (official), French Castilian
CARS: 35,358

AUSTRIA

CAPITAL: Vienna
AREA: 32,378 sq. mi. (83,858 sq km)
POPULATION: 8,150,835
MONEY: euro (€), 1-01-2002
LANGUAGES: German
CARS: 3,780,000

BELGIUM

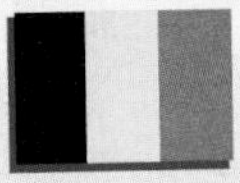

CAPITAL: Brussels
AREA: 11,780 sq. mi. (30,510 sq km)
POPULATION: 10,258,762
MONEY: euro (€), 1-01-2002
LANGUAGES: Dutch, French, German
CARS: 4,420,000

BOSNIA AND HERZEGOVINA

CAPITAL: Sarajevo
AREA: 19,741 sq. mi. (51,129 sq km)
POPULATION: 3,922,205
MONEY: marka (BAM)
LANGUAGES: Croatian, Serbian, Bosnian
CARS: data not available

BULGARIA

CAPITAL: Sofia
AREA: 42,822 sq. mi. (110,910 sq km)
POPULATION: 7,707,495
MONEY: lev (BGL)
LANGUAGES: Bulgarian
CARS: 1,650,000

CROATIA

CAPITAL: Zagreb
AREA: 21,831 sq. mi. (56,542 sq km)
POPULATION: 4,334,142
MONEY: Croatian kuna (HRK)
LANGUAGES: Croatian
CARS: 698,000

CZECH REPUBLIC

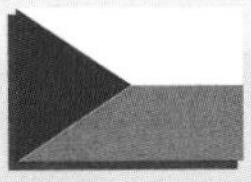

CAPITAL: Prague
AREA: 30,450 sq. mi. (78,866 sq km)
POPULATION: 10,264,212
MONEY: Czech koruna (CZK)
LANGUAGES: Czech
CARS: 4,410,000

DENMARK

CAPITAL: Copenhagen
AREA: 16,639 sq. mi. (43,094 sq km)
POPULATION: 5,352,815
MONEY: Danish krone (DKK)
LANGUAGES: Danish, Faroese, Greenlandic (an Inuit dialect), German
CARS: 1,790,000

ESTONIA

CAPITAL: Tallinn
AREA: 17,462 sq. mi. (45,226 sq km)
POPULATION: 1,423,316
MONEY: Estonian kroon (EEK)
LANGUAGES: Estonian (official), Russian, Ukrainian, English, Finnish
CARS: 338,000

Countries not drawn to scale.

is far more than in most European countries. You may want to invite students to verify this statement by making rough estimates of the European figures.

Have the students refer to the accompanying table listing the miles of railroads in selected European countries. Point out the figure for the United States, and mention that the United States is larger than all of the European countries combined. (Students may verify this statement by totaling the areas of the European countries and comparing the result to the area of the United States.) Ask students to write a sentence or two explaining why they think Europe has fewer cars. *(Possible answers: Europe's population density may indicate that many people live in apartments and do not have a place to park a car. Because Europe has an extensive rail system, people do not need to rely as much on cars.)*

UNIT 3 ATLAS

FINLAND

CAPITAL: Helsinki
AREA: 130,127 sq. mi. (337,030 sq km)
POPULATION: 5,175,783
MONEY: euro (€), 1-01-2002
LANGUAGES: Finnish, Swedish
CARS: 1,940,000

ICELAND

CAPITAL: Reykjavik
AREA: 39,768 sq. mi. (103,000 sq km)
POPULATION: 277,906
MONEY: Icelandic krona (ISK)
LANGUAGES: Icelandic
CARS: 132,468

FRANCE

CAPITAL: Paris
AREA: 211,208 sq. mi. (547,030 sq km)
POPULATION: 59,551,227
MONEY: euro (€), 1-01-2002
LANGUAGES: French
CARS: 25,500,000

IRELAND

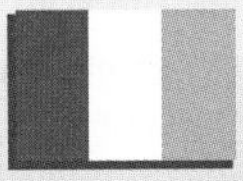
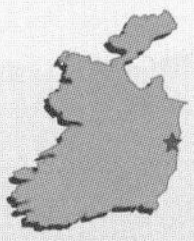

CAPITAL: Dublin
AREA: 27,135 sq. mi. (70,280 sq km)
POPULATION: 3,840,838
MONEY: euro (€), 1-01-2002
LANGUAGES: English, Irish (Gaelic)
CARS: 1,060,000

GERMANY

CAPITAL: Berlin
AREA: 137,846 sq. mi. (357,021 sq km)
POPULATION: 83,029,536
MONEY: euro (€), deutsche mark (DM)
LANGUAGES: German
CARS: 41,330,000

ITALY

CAPITAL: Rome
AREA: 116,305 sq. mi. (301,230 sq km)
POPULATION: 57,679,825
MONEY: euro (€), 1-01-2002
LANGUAGES: Italian, German, French, Slovene
CARS: 31,000,000

GREECE

CAPITAL: Athens
AREA: 50,942 sq. mi. (131,940 sq km)
POPULATION: 10,623,835
MONEY: euro (€), 1-01-2002
LANGUAGES: Greek (official), English, French
CARS: 2,340,000

LATVIA

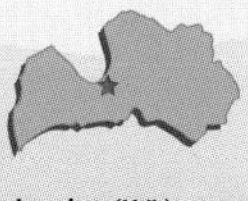

CAPITAL: Riga
AREA: 24,938 sq. mi. (64,589 sq km)
POPULATION: 2,385,231
MONEY: Latvian lat (LVL)
LANGUAGES: Lettish (official), Lithuanian, Russian
CARS: 252,000

HUNGARY

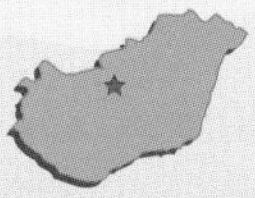

CAPITAL: Budapest
AREA: 35,919 sq. mi. (93,030 sq km)
POPULATION: 10,106,017
MONEY: forint (HUF)
LANGUAGES: Hungarian
CARS: 2,280,000

LIECHTENSTEIN

CAPITAL: Vaduz
AREA: 62 sq. mi. (160 sq km)
POPULATION: 32,528
MONEY: Swiss franc (CHF)
LANGUAGES: German
CARS: data not available

Sources: Central Intelligence Agency, *The World Factbook 2001; The World Almanac and Book of Facts 2001;* population figures are 2001 estimates.

UNIT 3 ASSESSMENT RESOURCES

Reproducible
- Unit 3 Test
- Unit 3 Test for English Language Learners and Special-Needs Students

internet connect

COUNTRY STATISTICS
GO TO: go.hrw.com
KEYWORD: SK3 FactsU3

Highlights of Country Statistics
- *CIA World Factbook*
- Library of Congress country studies
- Flags of the world

Miles of Railroad in Selected European Countries and the United States

Austria	3,524	Norway	2,485
Belgium	2,093	Poland	14,904
Denmark	1,780	Romania	7,062
Finland	3,641	Slovakia	2,277
France	19,847	Spain	8,252
Germany	54,994	Sweden	6,756
Greece	1,537	Switzerland	3,132
Hungary	8,190	United Kingdom	23,518
Italy	9,944	United States	137,900
Netherlands	1,702		

Source: *The World Almanac and Book of Facts 2001*

Lithuania

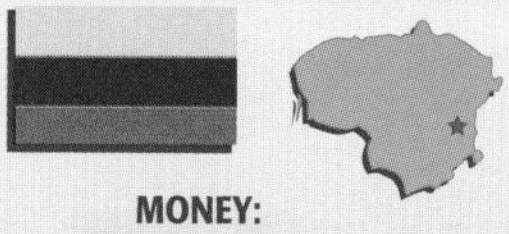

CAPITAL: Vilnius
AREA: 25,174 sq. mi. (65,200 sq km)
POPULATION: 3,610,535
MONEY: litas (LTL)
LANGUAGES: Lithuanian (official), Polish, Russian
CARS: 653,000

Monaco

CAPITAL: Monaco
AREA: 0.75 sq. mi. (1.95 sq km)
POPULATION: 31,842
MONEY: euro (€), 1-01-2002
LANGUAGES: French (official), English, Italian, Monegasque
CARS: 17,000

Luxembourg

CAPITAL: Luxembourg
AREA: 998 sq. mi. (2,586 sq km)
POPULATION: 429,080
MONEY: euro (€), Luxembourg franc (LuxF)
LANGUAGES: Luxembourgish, German, French
CARS: 231,666

Netherlands

CAPITAL: Amsterdam
AREA: 16,033 sq. mi. (41,526 sq km)
POPULATION: 15,981,472
MONEY: euro (€), 1-01-2002
LANGUAGES: Dutch
CARS: 5,810,000

Macedonia

CAPITAL: Skopje
AREA: 9,781 sq. mi. (25,333 sq km)
POPULATION: 2,046,209
MONEY: Macedonian denar (MKD)
LANGUAGES: Macedonian, Albanian, Turkish, Serbo-Croatian
CARS: 263,000

Norway

CAPITAL: Oslo
AREA: 125,181 sq. mi. (324,220 sq km)
POPULATION: 4,503,440
MONEY: Norwegian krone (NOK)
LANGUAGES: Norwegian
CARS: 1,760,000

Malta

CAPITAL: Valletta
AREA: 122 sq. mi. (316 sq km)
POPULATION: 394,583
MONEY: Maltese lira (MTL)
LANGUAGES: Maltese (official), English (official)
CARS: 122,100

Poland

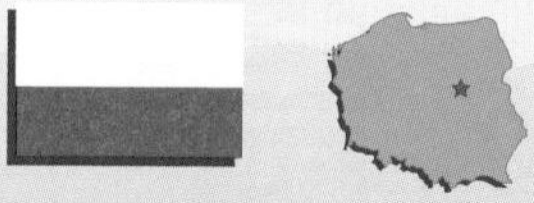

CAPITAL: Warsaw
AREA: 120,728 sq. mi. (312,685 sq km)
POPULATION: 38,633,912
MONEY: zloty (PLN)
LANGUAGES: Polish
CARS: 7,520,000

Moldova

CAPITAL: Chişinău
AREA: 13,067 sq. mi. (33,843 sq km)
POPULATION: 4,431,570
MONEY: Moldovan leu (MDL)
LANGUAGES: Moldovan (official), Russian, Gagauz
CARS: 169,000

Portugal

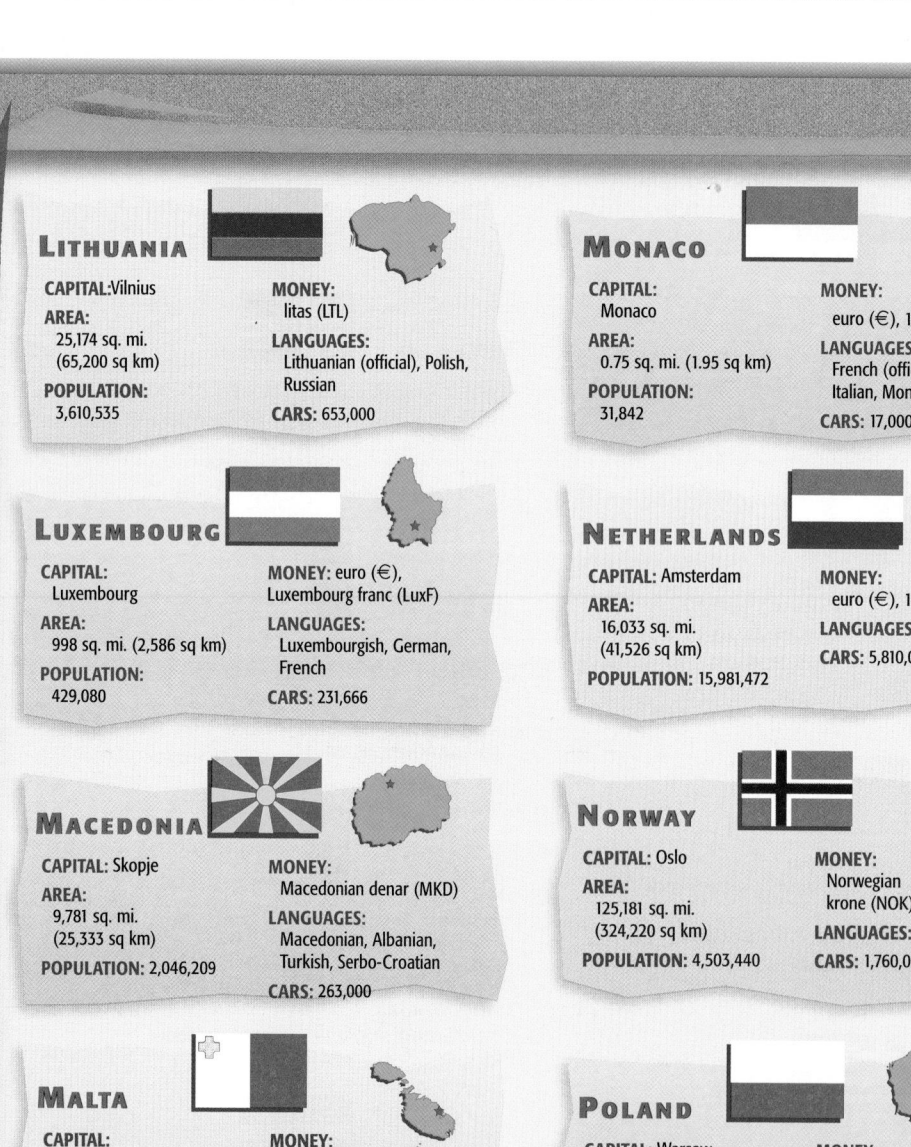

CAPITAL: Lisbon
AREA: 35,672 sq. mi. (92,391 sq km)
POPULATION: 10,066,253
MONEY: euro (€), 1-01-2002
LANGUAGES: Portuguese
CARS: 2,950,000

Countries not drawn to scale.

LEVEL 3: (Suggested time: 30 min.) Tell students that Europe has an economic organization called the European Union (EU). It was formed to strengthen the economies of European nations. In 1991, countries of the EU began to discuss adopting a common currency, to be called the euro. Point out the euro symbol in the Fast Facts data. By May of 1998, nearly every interested EU country had met the requirements for membership, which include maintaining a low inflation rate, low budget deficit, and low interest rate. Soon after, the European National Bank was formed. Then, on January 1, 2002, the countries adopted the euro as their standard currency.

Ask students to consider the effect that changing to a common currency might have on a nation's culture, economy, and relationships with neighboring countries. Have them imagine that they live in Europe in 1991 and are weighing the pros and cons of changing to the euro. In one paragraph, students should give reasons why they support their country's adoption of a common currency. In another paragraph, they should defend a position against adopting the euro. Invite them to refer back to their arguments as they learn more about the culture, economies, and relationships among European countries.

UNIT 3 ATLAS

Romania

CAPITAL: Bucharest
AREA: 91,699 sq. mi. (237,500 sq km)
POPULATION: 22,364,022
MONEY: leu (ROL)
LANGUAGES: Romanian, Hungarian, German
CARS: 2,390,000

San Marino

CAPITAL: San Marino
AREA: 23.6 sq. mi. (61.2 sq km)
POPULATION: 27,336
MONEY: euro (€), 1-01-2002
LANGUAGES: Italian
CARS: 24,825

Slovakia

CAPITAL: Bratislava
AREA: 18,859 sq. mi. (48,845 sq km)
POPULATION: 5,414,937
MONEY: Slovak koruna (SKK)
LANGUAGES: Slovak (official), Hungarian
CARS: 994,000

Slovenia

CAPITAL: Ljubljana
AREA: 7,820 sq. mi. (20,253 sq km)
POPULATION: 1,930,132
MONEY: tolar (SIT)
LANGUAGES: Slovenian, Serbo-Croatian
CARS: 657,000

Spain

CAPITAL: Madrid
AREA: 194,896 sq. mi. (504,782 sq km)
POPULATION: 40,037,995
MONEY: euro (€), peseta (Pta)
LANGUAGES: Castilian Spanish, Catalan, Galician
CARS: 15,300,000

Sweden

CAPITAL: Stockholm
AREA: 173,731 sq. mi. (449,964 sq km)
POPULATION: 8,875,053
MONEY: Swedish krona (SEK)
LANGUAGES: Swedish
CARS: 3,700,000

Switzerland

CAPITAL: Bern
AREA: 15,942 sq. mi. (41,290 sq km)
POPULATION: 7,283,274
MONEY: Swiss franc (CHF)
LANGUAGES: German, French, Italian
CARS: 3,320,000

United Kingdom

CAPITAL: London
AREA: 94,525 sq. mi. (244,820 sq km)
POPULATION: 59,647,790
MONEY: British pound (GBP)
LANGUAGES: English, Welsh, Scottish form of Gaelic
CARS: 25,590,000

Vatican City

CAPITAL: Vatican City
AREA: 0.17 sq. mi. (0.44 sq km)
POPULATION: 890
MONEY: euro (€), 1-01-2002
LANGUAGES: Italian, Latin, French
CARS: data not available

Yugoslavia

CAPITAL: Belgrade
AREA: 39,517 sq. mi. (102,350 sq km)
POPULATION: 10,677,290
MONEY: New Yugoslav dinar (YUM)
CARS: 1,000,000
LANGUAGES: Serbian, Albanian

internet connect
COUNTRY STATISTICS
GO TO: go.hrw.com
KEYWORD: SK3 FactsU3
FOR: more facts about Europe

Sources: Central Intelligence Agency, *The World Factbook 2001; The World Almanac and Book of Facts 2001;* population figures are 2001 estimates.

CHAPTER 10

Southern Europe

Chapter Resource Manager

Objectives	Pacing Guide	Reproducible Resources
SECTION 1 **Physical Geography** (pp. 231–33) **1.** Identify the major landforms and rivers of southern Europe. **2.** Identify the major climate types and resources of this region.	**Regular** .5 day **Block Scheduling** .5 day *Block Scheduling Handbook, Chapter 10*	**RS** Guided Reading Strategy 10.1 **RS** Graphic Organizer 10 **E** Creative Strategies for Teaching World Geography, Lessons 10 and 11 **SM** Geography for Life Activity 10 **IC** Interdisciplinary Activity for Middle Grades 4
SECTION 2 **Greece** (pp. 234–37) **1.** Identify some achievements of the ancient Greeks. **2.** Identify two features of Greek culture. **3.** Describe what Greece is like today.	**Regular** 1 day **Block Scheduling** .5 day *Block Scheduling Handbook, Chapter 10*	**RS** Guided Reading Strategy 10.2 **E** Cultures of the World Activity 3 **IC** Interdisciplinary Activity for Middle Grades 16
SECTION 3 **Italy** (pp. 238–41) **1.** Describe the early history of Italy. **2.** Describe how Italy has added to world culture. **3.** Describe what Italy is like today.	**Regular** 1 day **Block Scheduling** .5 day *Block Scheduling Handbook, Chapter 10*	**RS** Guided Reading Strategy 10.3 **E** Cultures of the World Activity 3 **SM** Map Activity 10
SECTION 4 **Spain and Portugal** (pp. 242–45) **1.** Identify some major events in the history of Spain and Portugal. **2.** Describe the cultures of Spain and Portugal. **3.** Describe what Spain and Portugal are like today.	**Regular** 1 day **Block Scheduling** .5 day *Block Scheduling Handbook, Chapter 10*	**RS** Guided Reading Strategy 10.4 **E** Cultures of the World Activity 3

Chapter Resource Key

RS Reading Support

IC Interdisciplinary Connections

E Enrichment

SM Skills Mastery

A Assessment

REV Review

ELL Reinforcement and English Language Learners

Transparencies

CD–ROM

Music

Video

go.hrw.com Internet

Holt Presentation Maker Using Microsoft® Powerpoint®

One-Stop Planner CD–ROM

See the *One-Stop Planner* for a complete list of additional resources for students and teachers.

One-Stop Planner CD–ROM

It's easy to plan lessons, select resources, and print out materials for your students when you use the ***One-Stop Planner CD–ROM with Test Generator.***

Technology Resources	Review, Reinforcement, and Assessment Resources
One-Stop Planner CD–ROM, Lesson 10.1 Geography and Cultures Visual Resources with Teaching Activities 24–29 Homework Practice Online HRW Go site	**ELL** Main Idea Activity 10.1 **REV** Section 1 Review, p. 233 **A** Daily Quiz 10.1 **ELL** English Audio Summary 10.1 **ELL** Spanish Audio Summary 10.1
One-Stop Planner CD–ROM, Lesson 10.2 Homework Practice Online HRW Go site	**ELL** Main Idea Activity 10.2 **REV** Section 2 Review, p. 237 **A** Daily Quiz 10.2 **ELL** English Audio Summary 10.2 **ELL** Spanish Audio Summary 10.2
One-Stop Planner CD–ROM, Lesson 10.3 *ARGWorld* CD–ROM: Conditions and Connections in Renaissance Europe Homework Practice Online HRW Go site	**ELL** Main Idea Activity 10.3 **REV** Section 3 Review, p. 241 **A** Daily Quiz 10.3 **ELL** English Audio Summary 10.3 **ELL** Spanish Audio Summary 10.3
One-Stop Planner CD–ROM, Lesson 10.4 Homework Practice Online HRW Go site	**ELL** Main Idea Activity 10.4 **REV** Section 4 Review, p. 245 **A** Daily Quiz 10.4 **ELL** English Audio Summary 10.4 **ELL** Spanish Audio Summary 10.4

internet connect

HRW ONLINE RESOURCES

GO TO: go.hrw.com
Then type in a keyword.

TEACHER HOME PAGE
KEYWORD: SK3 TEACHER

CHAPTER INTERNET ACTIVITIES
KEYWORD: SK3 GT10

Choose an activity to:
- explore the islands and peninsulas on the Mediterranean coast.
- take an online tour of ancient Greece.
- learn the story of pizza.

CHAPTER ENRICHMENT LINKS
KEYWORD: SK3 CH10

CHAPTER MAPS
KEYWORD: SK3 MAPS10

ONLINE ASSESSMENT
Homework Practice
KEYWORD: SK3 HP10
Standardized Test Prep Online
KEYWORD: SK3 STP10
Rubrics
KEYWORD: SS Rubrics

COUNTRY INFORMATION
KEYWORD: SK3 Almanac

CONTENT UPDATES
KEYWORD: SS Content Updates

HOLT PRESENTATION MAKER
KEYWORD: SK3 PPT10

ONLINE READING SUPPORT
KEYWORD: SS Strategies

CURRENT EVENTS
KEYWORD: S3 Current Events

Meeting Individual Needs

Ability Levels

Level 1 Basic-level activities designed for all students encountering new material

Level 2 Intermediate-level activities designed for average students

Level 3 Challenging activities designed for honors and gifted-and-talented students

English Language Learners Activities that address the needs of students with Limited English Proficiency

Chapter Review and Assessment

- **IC** Interdisciplinary Activity for the Middle Grades 13, 14, 15
- **E** Readings in World Geography, History, and Culture 28, 29, and 30
- **SM** Critical Thinking Activity 10
- **REV** Chapter 10 Review, pp. 276–77
- **REV** Chapter 10 Tutorial for Students, Parents, Mentors, and Peers
- **ELL** Vocabulary Activity 10
- **A** Chapter 10 Test
- Chapter 10 Test Generator (on the One-Stop Planner)
- Audio CD Program, Chapter 10
- **A** Chapter 10 Test for English Language Learners and Special-Needs Students

CHAPTER 10

Launch into Learning

In a democracy such as the United States, people participate in the government in a variety of ways. Citizens vote, hold office, write letters, and demonstrate peacefully. Ask students where they think Americans got these ideas. Have them locate Greece and Italy on the map on the following page. Tell students that in the 400s B.C. citizens of the Greek city-state of Athens voted on their government's decisions. Romans began to elect people to represent them in their government at about the same time. Tell students they will learn more about these cultures, what came after them, and the nearby countries of Spain and Portugal in this chapter.

Section 1

Objectives

1. Identify the major landforms and rivers of southern Europe.
2. Examine the major climate types and resources of this region.

Links to Our Lives

These are reasons why studying southern Europe might interest students:

- Many ideas that are basic to the Western world's way of life originated with the ancient Greeks and Romans.
- Spanish and Portuguese sailors opened the New World to European exploration and settlement.
- Artists, musicians, writers, scientists, and scholars from southern Europe have made significant contributions to the world's cultural heritage.
- The U.S. economy is closely linked with the economies of these countries.

CHAPTER 10

Southern Europe

Southern Europe's peninsulas, islands, mountains, and plateaus form a beautiful region. Tourists enjoy visiting this region to see its historical and cultural treasures.

Ciao. I am Paolo. I am 11 years old. I live in an apartment with my parents and my *nonna,* which is Italian for "grandma." Nonna is teaching me how to cook while she makes dinner for the family. I have no brothers or sisters, but on Sundays my aunts and uncles and three cousins all come for a big lunch. After lunch, the cousins play outside on the playground swings.

In the mornings, I walk to school with my mother before she goes to her job as a professor. My school is not too strict. We study English, Italian, and religion. After school I practice with my team at the swim club.

My favorite holiday is Christmas. On Christmas Eve we go to my other grandmother's house for a special dinner. We have smoked salmon, then grilled trout and spaghetti with mussels. Then there are special Christmas sweets: Pandoro—a cake sprinkled with sugar, and Torrone—a candy log with chocolate and nuts.

▲ Translation: I live in Rome, the capital of Italy.

FOCUS

LET'S GET STARTED

Copy the following instructions onto the chalkboard: *Look through Section 1 and find a picture you like. What do you want to know about that picture? Write down a question.* Ask volunteers to share their questions with the class. Discuss some of the questions. Invite students to suggest answers to each other's questions. Tell students that in Section 1 they will learn more about the physical geography of southern Europe.

Using the Physical-Political Map

Have students examine the map on this page. Remind students that Europe is often called a peninsula of peninsulas. Point out the Iberian, Italian, and Greek peninsulas and call on students to name the bodies of water around them. Ask students how the countries' locations may have influenced the region's history and economy. *(Access to the sea would help trade and travel.)*

Section 1 Physical Geography

Read to Discover

1. What are the major landforms and rivers of southern Europe?
2. What are the major climate types and resources of this region?

Define

mainland
sirocco

Locate

Mediterranean Sea
Strait of Gibraltar
Iberian Peninsula
Cantabrian Mountains
Pyrenees Mountains
Alps
Apennines
Aegean Sea
Peloponnesus
Ebro River
Douro River
Tagus River
Guadalquivir River
Po River
Tiber River

WHY IT MATTERS

Southern Europe's location on peninsulas encouraged its cultures to become great seafarers. Use CNNfyi.com or other **current events** sources to find examples of trade on the region's oceans today. Record your findings in your journal.

Cave painting from 12,000 B.C.

Southern Europe: Physical-Political

Section 1 RESOURCES

Reproducible

- Block Scheduling Handbook, Chapter 10
- Graphic Organizer 10
- Guided Reading Strategy 10.1
- Geography for Life Activity 10
- Interdisciplinary Activity for the Middle Grades 4
- Creative Strategies for Teaching World Geography, Lessons 10 and 11

Technology

- One-Stop Planner CD–ROM, Lesson 10.1
- Homework Practice Online
- Geography and Cultures Visual Resources with Teaching Activities 24–29
- HRW Go site

Reinforcement, Review, and Assessment

- Section 1 Review, p. 263
- Daily Quiz 10.1
- Main Idea Activity 10.1
- English Audio Summary 10.1
- Spanish Audio Summary 10.1

Teach

Teaching Objectives 1–2

ALL LEVELS: (Suggested time: 30 min.) Copy the following graphic organizer onto the chalkboard, omitting the italicized answers. Use it to help students understand the physical geography of southern Europe. Have students complete the organizer. **ENGLISH LANGUAGE LEARNERS**

The Physical Geography of SOUTHERN EUROPE

Shared Characteristics:
1. peninsulas 2. mountains 3. rivers 4. climate 5. resources

Spain and Portugal	Italy	Greece
1. *Iberian Peninsula*	1. *shaped like a boot*	1. *largest peninsula is Peloponnesus*
2. *Cantabrian and Pyrenees*	2. *southern Alps, Apennines*	2. *very mountainous*
3. *several east-west*	3. *Po and Tiber*	3. *most rivers are short*
4. *some semiarid climates; northern Spain is cool and humid*	4. *sirocco*	4. *warm and sunny*
5. *trade, fishing, iron ore, beaches*	5. *trade, marble*	5. *bauxite, chromium, lead, marble, and zinc*

Eye on Earth

Sicily's Mount Etna is the tallest and most active volcano in Europe. Its name comes from a Greek word meaning "I burn." The mountain has three ecological zones, each with its own vegetation. The lowest zone is fertile and rich in citrus fields, olive groves, and vineyards. Catania, a city of about 330,000 residents, is located on the mountain's lowest slopes. Forests are found farther up the mountain in the second ecological zone. Ash, sand, and lava fragments cover the mountain at heights more than 6,500 feet (1,980 m) in the final zone.

Geologists estimate that Mount Etna has been active for more than 2.5 million years. The mountain has had more than 100 serious eruptions in the past 2,500 years. The resulting lava flows have repeatedly destroyed villages, fields, and vineyards.

Discussion: Point out Catania and Mount Etna on a map of Italy. Ask students to identify any advantages that might counterbalance the city's dangerous location. *(possible answers: seaport, fertile fields from lava flows)*

internet connect

GO TO: go.hrw.com
KEYWORD: SK3 CH10
FOR: Web sites about southern Europe

Imports and Exports of Southern Europe

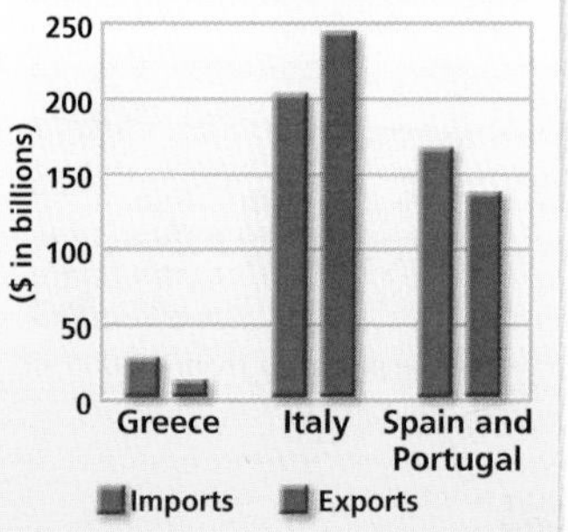

Source: Central Intelligence Agency, *The World Factbook, 2001*

Interpreting the Graph **Which country has the fewest imports and exports?**

Physical Features

Southern Europe is also known as Mediterranean Europe because most of its countries are on the sea's shores. The Mediterranean Sea stretches some 2,300 miles (3,700 km) from east to west. *Mediterranean* means "middle of the land" in Latin. In ancient times, the Mediterranean was considered the center of the Western world, since it is surrounded by Europe, Africa, and Asia. The narrow Strait of Gibraltar (juh-BRAWL-tuhr) links the Mediterranean to the Atlantic Ocean.

The Land Southern Europe is made up of three peninsulas. Portugal and Spain occupy one, Italy occupies another, and Greece is located on a third peninsula. Portugal and Spain are on the Iberian (eye-BIR-ee-uhn) Peninsula. Much of the peninsula is a high, rocky plateau. The Cantabrian (kan-TAY-bree-uhn) and the Pyrenees (PIR-uh-neez) Mountains form the plateau's northern edge. Italy's peninsula includes the southern Alps. A lower mountain range, the Apennines (A-puh-nynz), runs like a spine down the country's back. Islands in the central and western Mediterranean include Italy's Sicily and Sardinia (sahr-DI-nee-uh), as well as Spain's Balearic (ba-lee-AR-ik) Islands.

Greece's **mainland**, or the country's main landmass, extends into the Aegean (ee-JEE-uhn) Sea in many jagged little peninsulas. The largest one is the Peloponnesus (pe-luh-puh-NEE-suhs). Greece is mountainous and includes more than 2,000 islands. The largest island is Crete (KREET).

On all three peninsulas, coastal lowlands and river valleys provide excellent areas for growing crops and building cities. Soils on the region's uplands are thin and stony. They are also easily eroded. In this area of young mountains, earthquakes are common. They are particularly common in Greece and Italy.

The Rivers Several east-west rivers cut through the Iberian Peninsula. The Ebro River drains into the Mediterranean. The Douro, Tagus, and Guadalquivir (gwah-thahl-kee-VEER) Rivers, however, flow to the Atlantic Ocean. The Po (POH) is Italy's largest river. It creates a fertile agricultural region in northern Italy. Farther south, along the banks of the much smaller Tiber River, is the city of Rome.

✓ **READING CHECK:** *Physical Systems* What physical processes cause problems in parts of southern Europe? erosion, earthquakes

Greece, a land of mountains and sea, is home to the ancient city of Lindos on the island of Rhodes.

Graph Answer

Greece

CLOSE

Have students review the questions they wrote for the Let's Get Started Activity. Ask students if they can answer more of them now. What else do they need to learn to find the answers?

REVIEW AND ASSESS

Have students complete the Section Review. Then organize students into teams. Give each team an index card with a Define/Locate term on one side. Have teams create a question about the word or term on their card. Pass the cards from team to team. Ask each team to read the question aloud and to answer it. Continue until all the questions have been answered. Then have students complete Daily Quiz 10.1.
COOPERATIVE LEARNING

RETEACH

Have students complete Main Idea Activity 10.1. Then have volunteers come to the chalkboard and write explanations of the main physical features discussed in Section 1. **ENGLISH LANGUAGE LEARNERS**

EXTEND

Have interested students conduct research on the connections between the physical geography of southern Europe and economic activities relating to agriculture. Have students consider climate, soil, and vegetation in their study. Ask them to report their findings in the form of a graph or a map. **BLOCK SCHEDULING**

Climate and Resources

Much of southern Europe enjoys a warm, sunny climate. Most of the rain falls during the mild winter. Rainfall sometimes causes floods and mudslides due to erosion from overgrazing and deforestation. A hot, dry wind from North Africa called a **sirocco** (suh-RAH-koh) picks up some moisture over the Mediterranean Sea. It blows over Italy during spring and summer. The Po Valley is humid. Northern Italy's Alps have a highland climate. In Spain, semiarid climates are found in pockets. Northern Spain is cool and humid.

Southern Europeans have often looked to the sea for trade. Important Mediterranean ports include Barcelona, Genoa, Naples, Piraeus (py-REE-uhs)—the port of Athens—and Valencia. Lisbon, the capital of Portugal, is an important Atlantic port. The Atlantic Ocean supports Portugal's fishing industry. Although the Mediterranean suffers from pollution, it has a wealth of seafood.

The region's resources vary. Northern Spain has iron ore mines. Greece mines bauxite, chromium, lead, and zinc. Italy and Greece quarry marble. Falling water generates hydroelectricity throughout the region's uplands. Otherwise, resources are scarce.

The region's sunny climate and natural beauty have long attracted visitors. Millions of people explore castles, museums, ruins, and other cultural sites each year. Spain's beaches help make that country one of Europe's top tourist destinations.

▲ Workers prepare to separate a giant block of marble from a wall in a quarry in Carrara, Italy.

✓ **READING CHECK:** *Places and Regions* What climate types and natural resources are found in the region? mostly warm, sunny; also highland, semiarid, cool, humid; seafood; iron ore, bauxite, chromium, lead, zinc, marble

Section 1 Review

go.hrw.com Homework Practice Online
Keyword: SK3 HP10

Define and explain: mainland, sirocco

Working with Sketch Maps On a map of southern Europe that you draw or that your teacher provides, label Greece, Italy, Spain, and Portugal. Also label the Mediterranean Sea and the Strait of Gibraltar.

Reading for the Main Idea

1. *Places and Regions* Why is southern Europe known as Mediterranean Europe?
2. *Places and Regions* What countries occupy the region's three main peninsulas?
3. *Environment and Society* Why might people settle in river valleys and in coastal Southern Europe?

Critical Thinking

4. **Drawing Inferences and Conclusions** In what ways could the region's physical geography aid the development of trade?

Organizing What You Know

5. **Summarizing** Copy the following graphic organizer. Use it to list the region's physical features, climates, and resources.

	Physical Features	Climate	Resources
Spain and Portugal			
Italy			
Greece			

Section 1 Review

Answers

Define For definitions, see: mainland, p. 232; sirocco, p. 233

Working with Sketch Maps Maps will vary, but listed places should be in their approximate locations.

Reading for the Main Idea

1. because most of its countries are on the sea's shores (NGS 4)
2. Portugal, Spain, Italy, and Greece (NGS 4)
3. access to water transportation, trade, natural resources like fertile soil, moderate climate (NGS 15)

Critical Thinking

4. access to the sea, many good harbors

Organizing What You Know

5. Spain and Portugal—on the Iberian Peninsula rocky plateau, Cantabrian and Pyrenees Mountains, Balearic Islands; east-west rivers; warm, sunny; seafood, iron ore, beaches; Italy—southern Alps, Apennines, islands of Sicily and Sardinia; Po and Tiber Rivers; warm, sunny, sirocco; seafood, marble; Greece—large and small peninsulas, mountains, Crete; warm, sunny; seafood, bauxite, chromium, lead, zinc, marble

Section 2

Objectives

1. Describe some of the achievements of the ancient Greeks.
2. Identify two features of Greek culture.
3. Examine what Greece is like today.

Focus

LET'S GET STARTED

Copy the following questions onto the chalkboard: *What is one way your life would be different if U.S. citizens had no say in the government? What if you did not know about atoms or the true size of Earth?* Ask students to respond to the questions in writing. Ask volunteers to read their answers. Tell students that our form of government, physics, and geography are part of what we have learned from the ancient Greeks. Tell students that they will learn more about Greece in Section 2.

Building Vocabulary

Write the key terms on the chalkboard. Have students suggest a meaning for **city-states**, based on the component words, and check their definitions against the text. Have students find the definition of **mosaic**. Tell them that the word comes from *muse*. The muses were nine Greek goddesses of the arts. Ask students how mosaic might relate to goddesses of the arts.

Section 2 Resources

Reproducible

- Guided Reading Strategy 10.2
- Cultures of the World Activity 3
- Interdisciplinary Activity for the Middle Grades 16

Technology

- One-Stop Planner CD–ROM, Lesson 10.2
- Homework Practice Online
- HRW Go site

Reinforcement, Review, and Assessment

- Section 2 Review, p. 237
- Daily Quiz 10.2
- Main Idea Activity 10.2
- English Audio Summary 10.2
- Spanish Audio Summary 10.2

Section 2 Greece

Read to Discover

1. What were some of the achievements of the ancient Greeks?
2. What are two features of Greek culture?
3. What is Greece like today?

Define

city-states
mosaics

Locate

Athens
Thessaloníki

Why It Matters

The Olympic Games first started in Greece. Today we continue the tradition and hold both Summer and Winter Games. Use CNNfyi.com or other current events sources to find out more about the modern Olympic Games. Record your findings in your journal.

Ancient coin of Alexander the Great

History

The Greek islands took an early lead in the development of trade and shipping between Asia, Africa, and Europe. By about 2000 B.C. large towns and a complex civilization existed on Crete.

Ancient Greece About 800 B.C. Greek civilization arose on the mainland. The mountainous landscape there favored small, independent **city-states**. Each Greek city-state, or *polis,* was made up of a city and the land around it. Each had its own gods, laws, and form of government. The government of the city-state of Athens was the first known democracy. Democracy is the form of government in which all citizens take part. Greek philosophers, artists, architects, and writers made important contributions to Western civilization. For example, the Greeks are credited with inventing theater. Students still study ancient Greek literature and plays.

Eventually, Greece was conquered by King Philip. Philip ruled Macedonia, an area north of Greece. About 330 B.C. Philip's son, Alexander the Great, conquered Asia Minor, Egypt, Persia, and part of India. His empire combined Greek culture with influences from Asia and Africa. In the 140s B.C. Greece and Macedonia were conquered by the Roman Empire.

The Greeks believed that the Temple of Delphi—shown below—was the center of the world.

Interpreting the Visual Record **Why do you think remains of this temple have lasted for so many years?**

Visual Record Answer

Students might suggest that it was considered a sacred place and was made of durable materials.

TEACH

Teaching Objective 1

ALL LEVELS: (Suggested time: 20 min.) Copy the following graphic organizer onto the chalkboard, omitting the italicized answers. Use it to help students understand the achievements of the ancient Greeks. **ENGLISH LANGUAGE LEARNERS**

Ancient Greek Achievements

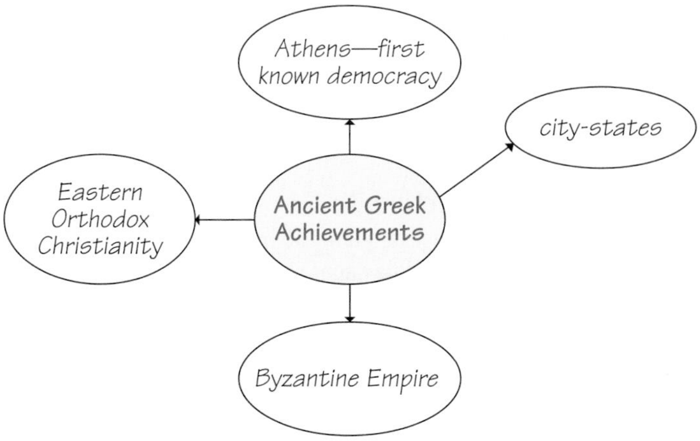

The Byzantine Empire About A.D. 400 the Roman Empire was divided into two parts. The western half was ruled from Rome. It soon fell to Germanic peoples the Romans called barbarians. Barbarian means both *illiterate* and *wanderer*. The eastern half of the Roman Empire was known as the Byzantine Empire. It was ruled from Constantinople. Constantinople was located on the shore of the Bosporus in what is now Turkey. This city—today known as Istanbul—served as a gathering place for people from Europe and Asia. The Byzantine Empire carried on the traditions of the Roman Empire for another 1,000 years. Gradually, an eastern form of Christianity developed. It was influenced by Greek language and culture. It became known as Eastern Orthodox Christianity. It is the leading form of Christianity in Greece, parts of eastern Europe, and Russia.

The Acropolis in Athens was built in the 400s B.C. The word *acropolis* is Greek for "city at the top."

Interpreting the Visual Record **Why would it be important to build a city on a hill?**

Turkish Rule In 1453 Constantinople was conquered by the Ottoman Turks, a people from Central Asia. Greece and most of the rest of the region came under the rule of the Ottoman Empire. It remained part of this empire for nearly 400 years. In 1821 the Greeks revolted against the Turks, and in the early 1830s Greece became independent.

Government In World War II Greece was occupied by Germany. After the war Greek communists and those who wanted a king and constitution fought a civil war. When the communists lost, the military took control. Finally, in the 1970s the Greek people voted to make their country a republic. They adopted a new constitution that created a government with a president and a prime minister.

✓ **READING CHECK:** *Human Systems* What were some of the achievements of ancient Greece? **democracy, contributions of Greek philosophers, artists, architects, and writers**

Culture

Turkish influences on Greek art, food, and music can still be seen. However, Turkey and Greece disagree over control of the islands and shipping lanes of the Aegean Sea.

Religion Some 98 percent of Greeks are Eastern Orthodox Christians, commonly known as Greek Orthodox. Easter is a major holiday and cause for much celebration. The traditional Easter meal is eaten on Sunday—roasted lamb, various vegetables, Easter bread, and many desserts. Because the Greek Orthodox Church has its own calendar, Christmas and Easter are usually celebrated one to two weeks later than in the West.

The people of Karpathos and their religious leaders are participating in an Easter celebration.

Linking Past to Present

The Seafaring Greeks Greece has long been closely linked to the sea. Many Greek cities were established where ships could drop anchor. Naval power allowed Athens to dominate the Aegean area in the 400s B.C.

Modern Greece also depends on the sea. Shipping and tourism are its major sources of income. The Greek merchant fleet—more than 2,000 ships—is one of the world's largest.

Critical Thinking: Point out to students that ancient Romans developed their navy more slowly than did the Greeks. Have students consult the physical-political map in this chapter and then ask them why might this be so.

Answer: Italy has fewer harbors, and therefore travel by sea was more difficult there. Also, because Romans could easily cross the Apennines and trade inland, they did not depend as heavily on the sea.

Visual Record Answer

Students might suggest that a city on a hill is easier to defend against attack.

Teaching Objective 2

ALL LEVELS: (Suggested time: 20 min.) Pair students and have one student in each pair write a few sentences about religion in Greece. Have the other student write about the arts. Then have pairs explain their topics to each other. **ENGLISH LANGUAGE LEARNERS, COOPERATIVE LEARNING**

Teaching Objective 3

ALL LEVELS: (Suggested time: 30 min.) Pair students and give each pair two sheets of heavy white paper, several pieces of colored paper, scissors, and glue. Have students cut the colored paper into small squares and then arrange the squares to create mosaics depicting some of the agricultural products of Greece. **ENGLISH LANGUAGE LEARNERS, COOPERATIVE LEARNING**

➤ASSIGNMENT: Have students write about some of the objects in their everyday lives that are related to ideas from ancient Greece. *(examples: Newspapers reflect freedom of speech, which is important to our democracy. Televisions broadcast dramas, which the Greeks developed.)*

CLOSE

Tell students that the first Olympic Games were held in Greece in 776 B.C. Stage a Class Olympics based on what students have learned about Greece. Events might be organized around physical features, history, or culture. Organize team or individual events.

ENVIRONMENT AND SOCIETY

Greece became an industrial nation in the 1970s. Since then heavy air and water pollution have made some people ill and eroded the marble of many ancient monuments and statues. Recent efforts have reduced some air pollution, but heavy traffic still dirties the air. In some places untreated sewage and industrial wastes have polluted the Mediterranean Sea.

Connecting to Math

Answers

1. using deduction in mathematical proofs, developing the Pythagorean Theorem, stating the basic principles of geometry, and calculating the value of *pi*
2. the creation of a model of the solar system, the estimation of the circumference of Earth, the treatment of medicine as a science, the gathering of information on plants and animals, and an understanding of the importance of observation and classification

CONNECTING TO *Math*

Greek postage stamp of the Pythagorean Theorem

Greek Math and Science

Greek civilization made many contributions to world culture. We still admire Greek art and literature. Greek scholars also paved the way for modern mathematics and science.

More than 2,000 years ago Thales (THAY-leez), a philosopher, began the use of deduction in mathematical proofs. Pythagoras (puh-THAG-uh-ruhs) worked out an equation to calculate the dimensions of a right triangle. The equation became known as the Pythagorean Theorem. By 300 B.C. Euclid (YOO-kluhd) had stated the basic principles of geometry in his book *Elements*. Soon after, Archimedes (ahr-kuh-MEED-eez) calculated the value of *pi*. This value is used to measure circles and spheres. He also explained how and why the lever, a basic tool, works.

Aristarchus (ar-uh-STAHR-kuhs), an astronomer, worked out a model of the solar system. His model placed the Sun at the center of the universe. Eratosthenes (er-uh-TAHS-thuh-neez) estimated the circumference of Earth with great accuracy.

Two important figures in the life sciences were Hippocrates (hip-AHK-ruh-teez) and Aristotle (AR-uh-staht-uhl). Hippocrates was a doctor who treated medicine as a science. He understood that diseases have natural causes. Aristotle gathered information on a wide variety of plants and animals. He helped establish the importance of observation and classification in the study of nature. In many ways, the Greeks began the process of separating scientific fact from superstition.

Understanding What You Read

1. How did Greeks further the study of mathematics?
2. What were some of the Greeks' scientific achievements?

This Greek vase shows a warrior holding a shield. Kleophrades is thought to have made this vase, which dates to 500 B.C.

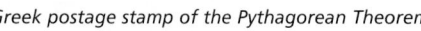

This mosaic of a dog bears the inscription "Good Hunting."

The Arts The ancient Greeks produced beautiful buildings, sculpture, poetry, plays, pottery, and gold jewelry. They also made **mosaics** (moh-ZAY-iks)—pictures created from tiny pieces of colored stone—that were copied throughout Europe. The folk music of Greece shares many features with the music of Turkey and Southwest Asia. In 1963 the Greek writer George Seferis won the Nobel Prize in literature.

✓ **READING CHECK:** Why is there a Turkish influence in Greek culture? **because the Ottoman Turks conquered Greece and controlled it for nearly 400 years**

REVIEW AND ASSESS

Have students complete the Section Review. Then have them reread the subsection on Greek history in Section 2. Pair students and tell one student to close his or her textbook and to summarize the main points. The other student in each pair should keep the book open and ask questions to encourage responses. Have students reverse the roles for the next subsection. Continue until the entire section is completed. Then have students complete Daily Quiz 10.2. **COOPERATIVE LEARNING**

RETEACH

Have students complete Main Idea Activity 10.2. Then tell students about the battle in 490 B.C. at Marathon, where Greek soldiers defeated the Persians. According to legend, a messenger ran more than 26 miles to report the news. Modern marathons commemorate the event. Hold a marathon in class. Organize students into teams, mark off 26 spaces on the floor, and allow "runners" to advance one "mile" for each fact from Section 2 that the team can report to the class.
ENGLISH LANGUAGE LEARNERS

EXTEND

Have interested students conduct research on Greek mythology and how it explained natural phenomena. Each student should focus on one myth. Then have students compare the Greek explanation with their culture's explanation. Ask volunteers to present their findings to the class.
BLOCK SCHEDULING

Greece Today

When people think of Greece now, they often recall the past. For example, many have seen pictures of the Parthenon, a temple built in the 400s B.C. in Athens. It is one of the world's most photographed buildings.

Economy Greece today lags behind other European nations in economic growth. More people work in agriculture than in any other industry. However, only about 19 percent of the land can be farmed because of the mountains. For this reason old methods of farming are used rather than modern equipment. Farmers raise cotton, tobacco, vegetables, wheat, lemons, olives, and raisins.

Service and manufacturing industries are growing in Greece. However, the lack of natural resources limits industry. Tourism and shipping are key to the Greek economy.

Cities About 20 percent of the Greek labor force works in agriculture. About 40 percent of Greeks live in rural areas. In the past few years, people have begun to move to the cities to find better jobs.

Athens, in central Greece, is the capital and by far the largest city. About one third of Greece's population lives in the area in and around Athens. Athens and its seaport, Piraeus, have attracted both people and industries. Most of the country's economic growth is centered there. However, the city suffers from air pollution, which causes health problems. Air pollution also damages historical sites, such as the Parthenon. Greece's second-largest city is Thessaloníki. It is the major seaport for northern Greece.

✓ **READING CHECK:** *Environment and Society* How does scarcity of natural resources affect Greece's economy? **It limits industry and forces Greece to rely on tourism and shipping.**

Greece

Country	Population/ Growth Rate	Life Expectancy	Literacy Rate	Per Capita GDP
Greece	10,623,835 / 0.2%	76, male / 81, female	95%	$17,200
United States	281,421,906 / 0.9%	74, male / 80, female	97%	$36,200

Sources: Central Intelligence Agency, *The World Factbook 2001;* U.S. Census Bureau

Interpreting the Chart **Why might life expectancy be higher in Greece than in the United States?**

Section 2 Review

go.hrw.com Homework Practice Online
Keyword: SK3 HP10

Define and explain: city-states, mosaics

Working with Sketch Maps On the map you created in Section 1, label Athens and Thessaloníki. What physical features do these cities have in common? What economic activities might they share?

Reading for the Main Idea

1. *Human Systems* What groups influenced Greek culture?
2. *Human Systems* For what art forms is Greece famous?

Critical Thinking

3. **Drawing Inferences and Conclusions** How did the physical geography of this region influence the growth of major cities?
4. **Finding the Main Idea** On what does Greece rely to keep its economy strong?

Organizing What You Know

5. **Sequencing** Create a time line that documents the history of ancient Greece from 2000 B.C. to A.D. 1453.

2000 B.C. ——— A.D. 1453

Section 2 Review

Answers

Define For definitions, see: city-states, p. 234; mosaics, p. 236

Working with Sketch Maps Maps will vary, but listed places should be labeled in their approximate locations. Athens and Thessaloníki are both near the sea; shipping, fishing

Reading for the Main Idea

1. Macedonians, Romans, and Turks (NGS 10)
2. buildings, sculpture, poetry, plays, pottery, gold jewelry, mosaics, and folk music (NGS 10)

Critical Thinking

3. The mountainous landscape favored the creation of small city-states.
4. tourism and shipping

Organizing What You Know

5. Answers will vary but might include: 2000 B.C.—civilization on Crete; 800 B.C.—civilization on mainland; about 330 B.C.—Alexander the Great creates empire; 140s B.C.—conquered by Rome; A.D. 400s—Byzantine Empire begins; 1453—Ottoman Turks conquer Constantinople.

▲ **Chart Answer**

Answers will vary, but students might mention diet or lifestyle.

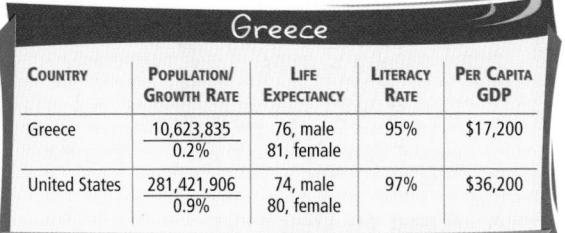

Section 3

Objectives

1. Explore the early history of Italy.
2. Describe how Italy has added to world culture.
3. Explain what Italy is like today.

FOCUS

LET'S GET STARTED

Copy the following question onto the chalkboard: *What do you already know about Rome or the Roman Empire?* Discuss responses. *(Possible answers: chariot races, gladiators, the persecution of Christians)* Tell students that although Rome spread and maintained its power by force, it also made lasting contributions to literature, language, law, city planning, architecture, engineering, and other fields. Tell students that in Section 3 they will learn more about ancient Rome and modern Italy.

Building Vocabulary

Write the key terms on the chalkboard. Explain that **pope** comes from the Latin word *papa,* meaning "father," and that the pope is considered the father of the Catholic church. Tell students that the word **Renaissance** combines *re-,* which means "again," with a Latin word, *nasci,* which means "to be born." Renaissance means "to be born again" or "rebirth." Point out that the Renaissance was a rebirth of learning. Call on a volunteer to find and read the definition of **coalition governments**.

Section 3 RESOURCES

Reproducible

- Guided Reading Strategy 10.3
- Cultures of the World Activity 3
- Map Activity 10

Technology

- One-Stop Planner CD–ROM, Lesson 10.3
- Homework Practice Online
- *ARGWorld* CD–ROM: Conditions and Connections in Renaissance Italy
- HRW Go site

Reinforcement, Review, and Assessment

- Map Activity 10
- Section 3 Review, p. 241
- Daily Quiz 10.3
- Main Idea Activity 10.3
- English Audio Summary 10.3
- Spanish Audio Summary 10.3

Visual Record Answer ►

Students might suggest that the contruction is of high quality.

Section 3 Italy

Read to Discover

1. What was the early history of Italy like?
2. How has Italy added to world culture?
3. What is Italy like today?

Define

pope
Renaissance
coalition governments

Locate

Rome
Genoa
Naples
Milan
Turin
Florence

WHY IT MATTERS

Leonardo da Vinci and Galileo are just two famous Italians who have made significant contributions to science and art. Use CNNfyi.com or other current events sources to find examples of recent Italian scientists and artists. Record your findings in your journal.

Artifact from a warrior's armor, 400s B.C.

▲ The Appian Way was a road from Rome to Brindisi. It was started in 312 B.C. by the emperor Claudius.

Interpreting the Visual Record How do you think this road has withstood more than 2,000 years of use?

History

About 750 B.C. a tribe known as the Latins established the city of Rome on the Tiber River. Over time, these Romans conquered the rest of Italy. They then began to expand their rule to lands outside Italy.

Roman Empire At its height about A.D. 100, the Roman Empire stretched westward to what is now Spain and Portugal and northward to England and Germany. The Balkans, Turkey, parts of Southwest Asia, and coastal North Africa were all part of the empire. Roman laws, roads, engineering, and the Latin language could be found throughout this huge area. The Roman army kept order, and people could travel safely throughout the empire. Trade prospered. The Romans made advances in engineering, including roads and aqueducts—canals that transported water. They also learned how to build domes and arches. Romans also produced great works of art and literature.

About A.D. 200, however, the Roman Empire began to weaken. The western part, with its capital in Rome, fell in A.D. 476. The eastern part, the Byzantine Empire, lasted until 1453.

Roman influences in the world can still be seen today. Latin developed into the modern languages of French, Italian, Portuguese, Romanian, and Spanish. Many English words have Latin origins as well. Roman laws and political ideas have influenced the governments and legal systems of many modern countries.

TEACH

Teaching Objectives 1–2

ALL LEVELS: (Suggested time: 40 min.) Copy the following graphic organizer onto the chalkboard, omitting the italicized answers. Use it to help students explore the early history of Italy and how it has added to world culture. Complete the organizer as a class. Then pair students and have each pair write one sentence about each entry. **ENGLISH LANGUAGE LEARNERS, COOPERATIVE LEARNING**

The History and Culture of ITALY

History	Culture
• *750 B.C. Rome established*	• *Latin*
• *Roman Empire*	• *Roman Catholic Church*
• *Christianity*	• *Mediterranean diet*
• *Renaissance*	• *glassware*
• *coalition governments*	• *jewelry*
	• *painting*
	• *sculpture*

The Colosseum is a giant amphitheater. It was built in Rome between A.D. 70 and 80 and could seat 50,000 people.

Interpreting the Visual Record **For what events do you think the Colosseum was used? What type of modern buildings look like this?**

Christianity began in the Roman province of Judaea (modern Israel and the West Bank). It then spread through the Roman Empire. Some early Christians were persecuted for refusing to worship the traditional Roman gods. However, in the early A.D. 300s the Roman emperor, Constantine, adopted Christianity. It quickly became the main religion of the empire. The **pope**—the bishop of Rome—is the head of the Roman Catholic Church.

The Renaissance Beginning in the 1300s a new era of learning began in Italy. It was known as the **Renaissance** (re-nuh-SAHNS). In French this word means "rebirth." During the Renaissance, Italians rediscovered the work of ancient Roman and Greek writers. Scholars applied reason and experimented to advance the sciences. Artists pioneered new techniques. Leonardo da Vinci, painter of the *Mona Lisa,* was also a sculptor, engineer, architect, and scientist. Another Italian, Galileo Galilei, perfected the telescope and experimented with gravity.

Christopher Columbus opened up the Americas to European colonization. Although Spain paid for his voyages, Columbus was an Italian from the city of Genoa. The name *America* comes from another Italian explorer, Amerigo Vespucci.

Government Italy was divided into many small states until the late 1800s. Today Italy's central government is a democracy with an elected parliament. Italy has had many changes in leadership in recent years. This has happened because no political party has won a majority of votes in Italian elections. As a result, political parties must form **coalition governments**. A coalition government is one in which several parties join together to run the country. Unfortunately, these coalitions usually do not last long.

READING CHECK: *Human Systems* How is the Italian government different from that of the United States? many changes in leadership, coalition governments

Leonardo da Vinci painted the *Mona Lisa* about 1503–06.

Interpreting the Visual Record **Why do you think Leonardo's painting became famous?**

Across the Curriculum
TECHNOLOGY

Roman Engineering The Romans were the best civil engineers of the ancient world. They used arches, concrete, and tunnel-like vaults to create bridges and buildings.

The Romans also built more than 50,000 miles of roads, which crossed an area that now includes 30 countries. The roads were built in layers of rough stone, gravel, and sand. Smooth paving stones covered the surface, which was higher in the center so that water would drain to the sides.

The Romans were also famous for their aqueducts, or covered stone water channels. Aqueducts supplied the city of Rome with more than 200 million gallons of water each day.

Activity: Organize students into groups to research, build, and label models of aqueducts, baths, or arenas.

internet connect
GO TO: **go.hrw.com**
KEYWORD: **SK3 CH10**
FOR: **Web sites about Rome**

▲ Visual Record Answer

sporting events and public displays; stadiums

◄

Students might mention the subject's mysterious smile.

Teaching Objective 3

ALL LEVELS: (Suggested time: 20 min.) Organize students into triads. Have each triad write a brief magazine article titled Italy Today. Articles should focus on Italy's economy, its cities, and the differences between northern and southern Italy.

ENGLISH LANGUAGE LEARNERS, COOPERATIVE LEARNING

CLOSE

Have volunteers write on the chalkboard one fact about Roman and Italian contributions to culture. To prompt students, you might first write categories such as architecture, politics, religion, art, food, and fashion on the chalkboard.

REVIEW AND ASSESS

Have students complete the Section Review. Then organize the class into groups of four, to represent the four horses of a Roman chariot team. Have each group review the sections. Then stage a chariot race by clearing a space in the classroom, asking questions of each group, and allowing the groups to take one step forward with each correct answer. Then have students complete Daily Quiz 10.3.

RETEACH

Have students complete Main Idea Activity 10.3. Then create sets of index cards with the names of major historical periods and famous Italians written on them. Organize the class into groups. Give each group a set of cards.

USING ILLUSTRATIONS

Focus students' attention on the *Mona Lisa* on the previous page. Remind students that Italian artist Leonardo da Vinci painted the *Mona Lisa* about 500 years ago. Ask them to identify other places they may have seen the image. *(possible answers: T-shirts, cartoons, and advertisements)* Point out that images of the *Mona Lisa* have been popular for many years and that the painting is recognized around the world. How have modern media helped spread the image to people around the world? You may want to challenge students to search the media for more images that began as part of another country's culture but are now encountered worldwide.

Visual Record Answers ▲

because they had no electricity or gas to create artificial light

►

by allowing people to come together

This view of the Pantheon in Rome shows the oculus—the opening at the top. The Pantheon was built as a temple to all Roman gods.

Interpreting the Visual Record Why did the Romans need to design buildings that let in light?

▼

In Rome people attend mass in St. Peter's Square, Vatican City. Vatican City is an independent state within Rome.

Interpreting the Visual Record How does a plaza, or open area, help create a sense of community?

▼

Culture

People from other places have influenced Italian culture. During the Renaissance, many Jews who had been expelled from Spain moved to Italian cities. Jews often had to live in segregated areas called ghettoes. Today immigrants have arrived from former Italian colonies in Africa. Others have come from the eastern Mediterranean and the Balkans.

Religion and Food Some 98 percent of Italians belong to the Roman Catholic Church. The leadership of the church is still based in the Vatican in Rome. Christmas and Easter are major holidays in Italy. Italians also celebrate All Saints Day on November 1 by cleaning and decorating their relatives' graves.

Italians enjoy a range of regional foods. Recipes are influenced by the history and crops of each area. In the south, Italians eat a Mediterranean diet of olives, bread, and fish. Dishes are flavored with lemons from Greece and spices from Africa. Tomatoes, originally from the Americas, have become an important part of the diet. Some Italian foods, such as pizza, are popular in the United States. Modern pizza originated in Naples. Northern Italians eat more rice, butter, cheeses, and mushrooms than southern Italians.

The Arts The ancient Romans created beautiful glassware and jewelry as well as marble and bronze sculptures. During the Renaissance, Italy again became a center for art, particularly painting and sculpture. Italian artists discovered ways to make their paintings more lifelike. They did this by creating the illusion of three dimensions. Italian writers like Francesco Petrarch and Giovanni Boccaccio wrote some of the most important literature of the Renaissance. More recently, Italian composers have written great operas. Today, Italian designers, actors, and filmmakers are celebrated worldwide.

✓ **READING CHECK:** *Human Systems* What are some examples of Italian culture? Italian food, such as pizza; glassware, jewelry, painting, sculpture, literature, opera, film

Have students place the cards in correct chronological order as quickly as possible. Have each group select one card and use the text to tell the class about the time period or person on that card.
ENGLISH LANGUAGE LEARNERS, COOPERATIVE LEARNING

EXTEND

Have interested students conduct research on the major languages derived from Latin—French, Italian, Portuguese, and Spanish. Challenge them to find and compare common words in those languages. Students might create web diagrams to share their findings with the class. **BLOCK SCHEDULING**

TEACHER TO TEACHER

Lois Jordan, of Nashville, Tennessee, suggests the following activity to help students learn more about Italian culture. Have students conduct research on what typical Italians might eat. Then have students compare the Italian diet with the diet in a region or country with different physical and climatic characteristics, such as northern Canada. Students might investigate how climate and natural resources affect diet, how much of the average person's income is spent on food, and whether most people eat their meals at home or elsewhere.

Italy Today

Italy is slightly smaller than Florida and Georgia combined, with a population of about 57 million. A shared language, the Roman Catholic Church, and strong family ties continue to bind Italians together.

Economy After its defeat in World War II, Italy rebuilt its industries in the north. Rich soil and plenty of water make the north Italy's "breadbasket," or wheat growing area. Italy's most valuable crop is grapes. Although grapes are grown throughout the country, northern Italy produces the best crops. These grapes help make Italy the world's largest producer of wine. Tourists are also important to Italy's economy. They visit northern and central Italy to see ancient ruins and Renaissance art. Southern Italy remains poorer with lower crop yields. Industrialization there also lags behind the north. Tourist resorts, however, are growing in the south and promise to help the economy.

Italy

Country	Population/ Growth Rate	Life Expectancy	Literacy Rate	Per Capita GDP
Italy	57,679,825 .07%	76, male 83, female	98%	$22,100
United States	281,421,906 0.9%	74, male 80, female	97%	$36,200

Sources: Central Intelligence Agency, *The World Factbook 2001;* U.S. Census Bureau

Interpreting the Chart **What is the difference in the growth rate of Italy and the United States?**

Cities The northern cities of Milan, Turin, and Genoa are important industrial centers. Their location near the center of Europe helps companies sell products to foreign customers. Also in the north are two popular tourist sites. One is Venice, which is famous for its romantic canals and beautiful buildings. The other is Florence, a center of art and culture. Rome, the capital, is located in central Italy. Naples, the largest city in southern Italy, is a major manufacturing center and port.

✓ **READING CHECK:** *Environment and Society* What geographic factors influence Italy's economy? Rich soil and plenty of water are good for growing crops, particularly grapes.

Section 3 Review

go.hrw.com Homework Practice Online
Keyword: SK3 HP10

Define and explain: pope, Renaissance, coalition governments

Working with Sketch Maps On the map you created in Section 2, label Florence, Genoa, Milan, Naples, Rome, and Turin. Why are they important?

Reading for the Main Idea

1. *Human Systems* What were some of the important contributions of the Romans?
2. *Human Systems* What are some art forms for which Italy is well known?

Critical Thinking

3. **Finding the Main Idea** Which of Italy's physical features encourage trade? Which geographical features make trading difficult?
4. **Analyzing Information** Why is the northern part of Italy known as the country's "breadbasket"?

Organizing What You Know

5. **Finding the Main Idea** Copy the following graphic organizer. Use it to describe the movement of goods and ideas to and from Italy during the early days of trade and exploration.

[ ] ⇨ [Italy] ⇨ [ ]

Section 3 Review Answers

Define For definitions, see: pope, p. 239; Renaissance, p. 239; coalition governments, p. 239

Working with Sketch Maps Maps will vary, but listed places should be labeled in their approximate locations. These cities serve as centers of culture, industry, and government.

Reading for the Main Idea

1. Latin, art, literature, roads, aqueducts, domes, arches, laws (NGS 10)
2. architecture, glassware, jewelry, sculpture, painting, literature, opera, drama, and film (NGS 10)

Critical Thinking

3. access to the sea and its location at the center of the Mediterranean; few major rivers or harbors and its central mountains
4. its plentiful wheat harvests

Organizing What You Know

5. to—Jews expelled from Spain, African and Balkan immigrants; from—laws, roads, engineering, language, art, literature, architecture, Christianity, Renaissance ideas

▲ **Chart Answer**

Italy's growth is declining, while that of the United States is increasing.

Section 4

Objectives

1. Identify some major events in the history of Spain and Portugal.
2. Describe the cultures of Spain and Portugal.
3. Investigate what Spain and Portugal are like today.

Focus

LET'S GET STARTED

Copy the following instructions onto the chalkboard: *Use the Fast Facts features in your textbook to find countries other than Spain or Portugal where Spanish or Portuguese is spoken. How many can you list?* Discuss students' lists. Ask students how they think these languages spread so far. Tell them that in Section 4 they will learn more about the history and culture of Spain and Portugal.

Building Vocabulary

Write the key terms on the chalkboard. Write *moor* next to **Moors**. Explain that the capitalized term refers to Muslims from North Africa who conquered much of Spain in the A.D. 700s. Point out that **dialect** comes from Greek, Latin, and French roots that mean "between" or "over" (dia-) and "to speak" (-lect). Have students relate the roots to the term. Conclude by having a volunteer find and read aloud the definition of the term **cork.**

Section 4 RESOURCES

Reproducible

- Guided Reading Strategy 10.4
- Cultures of the World Activity 3

Technology

- One-Stop Planner CD–ROM, Lesson 10.4
- HRW Go site

Reinforcement, Review, and Assessment

- Section 4 Review, p. 245
- Daily Quiz 10.4
- Main Idea Activity 10.4
- English Audio Summary 10.4
- Spanish Audio Summary 10.4

Visual Record Answer

Blades catch the wind and allow a mechanical system to draw water.

Section 4 Spain and Portugal

Read to Discover

1. What were some major events in the history of Spain and Portugal?
2. What are the cultures of Spain and Portugal like?
3. What are Spain and Portugal like today?

Define

Moors
dialect
cork

Locate

Lisbon
Madrid
Barcelona

WHY IT MATTERS

Some Basque separatists have used violence to try to gain their independence from Spain. Use CNNfyi.com or other current events sources to find examples of this problem. Record your findings in your journal.

Paella, a popular dish in Spain

History

Beautiful paintings of bison and other animals are found in caves in northern Spain. Some of the best known are at Altamira and were created as early as 16,000 B.C. Some cave paintings are much older. These paintings give us exciting clues about the early people who lived here.

Ancient Times Spain has been important to Mediterranean trade for several thousand years. First, the Greeks and then the Phoenicians, or Carthaginians, built towns on Spain's southern and eastern coasts. Then, about 200 B.C. Iberia became a part of the Roman Empire and adopted the Latin language.

These windmills in Consuegra, Spain, provided water for the people of the region.

Interpreting the Visual Record How do you think windmills pump water?

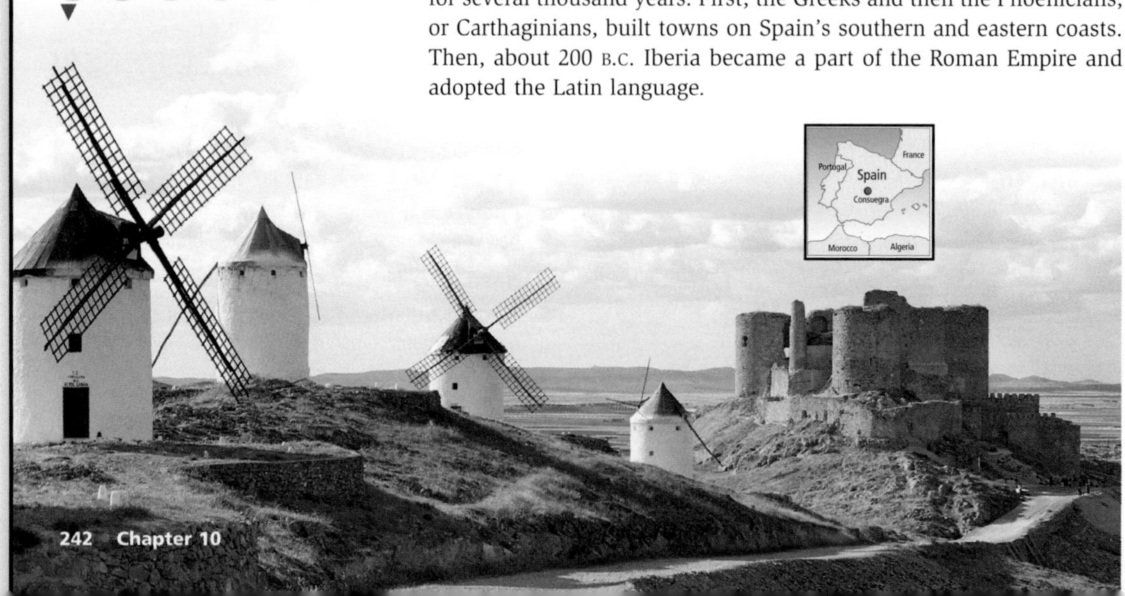

TEACH

Teaching Objective 1

ALL LEVELS: (Suggested time: 30 min.) Tell students to imagine that they are tour guides in charge of a trip to Spain and Portugal. Before they leave on the trip they must present a brief history of Spain and Portugal to their clients. Pair students and have each pair write a presentation. Ask volunteers to present their histories to the class. **ENGLISH LANGUAGE LEARNERS, COOPERATIVE LEARNING**

Teaching Objectives 2–3

ALL LEVELS: (Suggested time: 20 min.) Copy the graphic organizer on the following page onto the chalkboard, omitting the italicized answers. Use it to help students compare the cultures of Spain and Portugal and what the countries are like today. Have each student complete it. **ENGLISH LANGUAGE LEARNERS**

The Muslim North Africans, or **Moors**, conquered most of the Iberian Peninsula in the A.D. 700s. Graceful Moorish buildings, with their lacy patterns and archways, are still found in Spanish and Portuguese cities. This is particularly true in the old Moorish city of Granada in southern Spain.

The interior of the Great Mosque in Córdoba, Spain, shows the lasting beauty of Moorish architecture. A cathedral was built within the mosque after Christians took back the city.

Great Empires From the 1000s to the 1400s Christian rulers fought to take back the peninsula. In 1492 King Ferdinand and Queen Isabella conquered the kingdom of Granada, the last Moorish outpost in Spain. That same year, they sponsored the voyage of Christopher Columbus to the Americas. Spain soon established a large empire in the Americas.

The Portuguese also sent out explorers. Some of them sailed around Africa to India. Others crossed the Atlantic and claimed Brazil. In the 1490s the Roman Catholic pope drew a line to divide the world between Spain and Portugal. Western lands, except for Brazil, were given to Spain, and eastern lands to Portugal.

With gold and agricultural products from their American colonies, and spices and silks from Asia, Spain and Portugal grew rich. In 1588 Philip II, king of Spain and Portugal, sent a huge armada, or fleet, to invade England. The Spanish were defeated, and Spain's power began to decline. However, most Spanish colonies in the Americas did not win independence until the early 1800s.

Government In the 1930s the king of Spain lost power. Spain became a workers' republic. The new government tried to reduce the role of the church and to give the nobles' lands to farmers. However, conservative military leaders under General Francisco Franco resisted. A civil war was fought from 1936 to 1939 between those who supported Franco and those who wanted a democratic form of government. Franco's forces won the war and ruled Spain until 1975. Today Spain is a democracy, with a national assembly and prime minister. The king also plays a modest role as head of state.

Portugal, like Spain, was long ruled by a monarch. In the early 1900s the monarchy was overthrown. Portugal became a democracy. However, the army later overthrew the government, and a dictator took control. A revolution in the 1970s overthrew the dictatorship. For a few years disagreements between the new political parties brought violence. Portugal is now a democracy with a president and prime minister.

✓ **READING CHECK:** *Human Systems* How did Spain and Portugal move from unlimited to limited governments? both were ruled by monarchs and military governments, and now both have democratic governments

Our Amazing Planet

One of the world's most endangered wild cats is the Iberian lynx. About 50 survive in a preserve on the Atlantic coast of Spain.

Across the Curriculum
MUSIC

Flamenco The musical and dance performance known as flamenco developed in southern Spain hundreds of years ago. Flamenco reflects Gypsy, Andalusian, and Arabic influences, among others. Early flamenco combined *cante* (song) and *baile* (dance), accompanied by rhythmic handclapping.

The golden age of flamenco lasted from 1869 to 1910. It became popular in cafés, and many performers added *guitarra,* or guitar playing. In the 1900s jazz, salsa, and bossa nova influenced flamenco. Flamenco has gained popularity recently, and some of its spontaneity has been replaced by rehearsed routines.

Activity: Have interested students conduct research on the popularity of flamenco guitar music and the rhythmic handclapping, called *palmas,* that accompanies it. Play a recording of a flamenco song in class. Ask students who have researched *palmas* to lead the class in clapping.

internet connect go.hrw.com
GO TO: go.hrw.com
KEYWORD: SK3 CH10
FOR: Web sites about flamenco

Chart Answer (p. 244)

They are almost the same.

The Culture of SPAIN and PORTUGAL

Food and Festivals	The Arts	Today
• olives and olive oil • limes • wine • fish • wheat • foods from the Americas • Roman Catholic holidays • bullfights	• porcelain • fado singers • flamenco dancers • Picasso	• European Union • agricultural products: wine, fruit, olive oil, olives, and cork • clothing • timber products • cars and trucks • tourism

CLOSE

Tell students about *castell* building—a sport popular in the Spanish region of Catalonia. *Castellers* build castles out of people. Dressed in traditional clothing, a group pushes together tightly to form the foundation of the castle. Other people press in against them. Lighter, barefoot participants then scramble over their backs to stand on their shoulders. Finally a child climbs to the top. In this way, Catalonians build human towers about 35 feet (10.6 m) tall. Ask students to speculate on how this custom began.

Section 4 Review

Answers

Define For definitions see: Moors, p. 243; dialect, p. 244; cork, p. 245

Working with Sketch Maps Maps will vary, but should correctly depict the locations of the three cities. Lisbon and Barcelona are seaports while Madrid is located inland.

Reading for the Main Idea

1. Portuguese fado music and Spanish flamenco dancing reflect African influence. (NGS 10)
2. by joining the European Union, which provides free trade, travel, exchange of workers, and rapid growth (NGS 11)

Critical Thinking

3. Greeks, Phoenicians, Romans, and Moors
4. 1930s—king lost power and Spain became a workers' republic; 1936–1939—civil war, which led to conservative military rule under Francisco Franco until 1975; since 1975—democracy

Organizing What You Know

5. Graphic organizers will vary but should include the following information: A.D. 700s—Moorish advance on area; 1492—kingdom of Granada conquered and Columbus sails to the Americas; 1588—defeat of Spanish Armada.

Spain and Portugal

Country	Population/ Growth Rate	Life Expectancy	Literacy Rate	Per Capita GDP
Portugal	10,066,253 0.2%	72, male 80, female	87%	$15,800
Spain	40,037,995 0.1%	75, male 83, female	97%	$18,000
United States	281,421,906 0.9%	74, male 80, female	97%	$36,200

Sources: Central Intelligence Agency, *The World Factbook 2001;* U.S. Census Bureau

Interpreting the Chart **How do the growth rates of these countries compare?**

Culture

The most widely understood Spanish **dialect** (DY-uh-lekt), or variation of a language, is Castilian. This is the form spoken in central Spain. Spanish and Portuguese are not the only languages spoken on the Iberian Peninsula, however. Catalan is spoken in northeastern Spain (Catalonia). Basque is spoken by an ethnic group living in the Pyrenees.

Spain faces a problem of unrest among the Basque people. The government has given the Basque area limited self-rule. However, a small group of Basque separatists continue to use violence to protest Spanish control.

Food and Festivals Spanish and Portuguese foods are typical of the Mediterranean region. Many recipes use olives and olive oil, lemons, wheat, wine, and fish. Foods the explorers brought back from the Americas—such as tomatoes and peppers—are also important.

Both Spain and Portugal remain strongly Roman Catholic. The two countries celebrate major Christian holidays like Christmas and Easter. As in Italy, each village has a patron saint whose special day is the occasion for a fiesta, or festival. A bull fight, or *corrida,* may take place during the festival.

The Arts Spanish and Portuguese art reflects the many peoples who have lived in the region. The decoration of Spanish porcelain recalls Islamic art from North Africa. The sad melodies of the Portuguese fado singers and the intense beat of Spanish flamenco dancing also show African influences. In the 1900s the Spanish painter Pablo Picasso boldly experimented with shape and perspective. He became one of the most famous artists of modern times.

✓ **READING CHECK:** How has the mixture of different ethnic groups created some conflict in Spanish society? **Some Basque people want to be separate from Spain and are using violence to achieve that goal.**

Flamenco dancers perform at a fair in Málaga, Spain.

Review and Assess

Have students complete the Section Review. Then organize the class into triads. Have each triad work together to create an outline of Section 4. Call on volunteers to write their outlines on the chalkboard. Compare the outlines and fill in any blanks. To conclude have students complete Daily Quiz 10.4. **COOPERATIVE LEARNING**

Reteach

Have students complete Main Idea Activity 10.4. Then have them suggest words that describe Spain or Portugal. Have the class decide whether the word relates to the physical or the cultural geography of the region. **ENGLISH LANGUAGE LEARNERS**

Extend

Have interested students conduct research on a Spanish artist, such as El Greco, Goya, Murillo, or Velázquez. Ask students to prepare posters illustrating a work by their chosen artist. Have students include a statement about how the artist's work reflects Spanish cultural traditions. **BLOCK SCHEDULING**

Porto, Portugal, combines modern industry with the historical sea trade.

Spain and Portugal Today

Like Greece and Italy, both Spain and Portugal belong to the European Union (EU). The EU allows free trade, travel, and exchange of workers among its members. The economies of Spain and Portugal have been growing rapidly. However, they remain poorer than the leading EU countries.

Agricultural products of Spain and Portugal include wine, fruit, olives, olive oil, and **cork**. Cork is the bark stripped from a certain type of oak tree. Spain exports oranges from the east, beef from the north, and lamb from ranches on the Meseta. Portugal also makes and exports clothing and timber products. Spain makes cars and trucks, and most of its industry is located in the north. Tourism is also an important part of the Spanish economy. This is particularly true along Spain's coasts and on the Balearic Islands.

Portugal's capital and largest city is Lisbon. It is located on the Atlantic coast at the mouth of the Tagus River. Madrid, Spain's capital and largest city, is located inland on the Meseta. Spain's second-largest city is the Mediterranean port of Barcelona.

✓ **READING CHECK:** *Places and Regions* What are Spain and Portugal like today? rapidly growing economies with agricultural exports and industry

A worker uses an ax to strip the bark from a cork oak.

go.hrw.com Homework Practice Online — Keyword: SK3 HP10

Section 4 Review

Define and explain: Moors, dialect, cork

Working with Sketch Maps On the map that you created in Section 3, label Lisbon, Madrid, and Barcelona. How are Lisbon and Barcelona different from Madrid?

Reading for the Main Idea

1. *Human Systems* How do the performing arts of this region reflect different cultures?
2. *Human Systems* How have Spain and Portugal worked to improve their economies?

Critical Thinking

3. **Analyzing Information** Which groups influenced the culture of Spain and Portugal?
4. **Summarizing** How was the government of Spain organized during the 1900s?

Organizing What You Know

5. **Sequencing** Copy the following graphic organizer. Use it to list important events in the history of Spain and Portugal from the 700s to the 1600s.

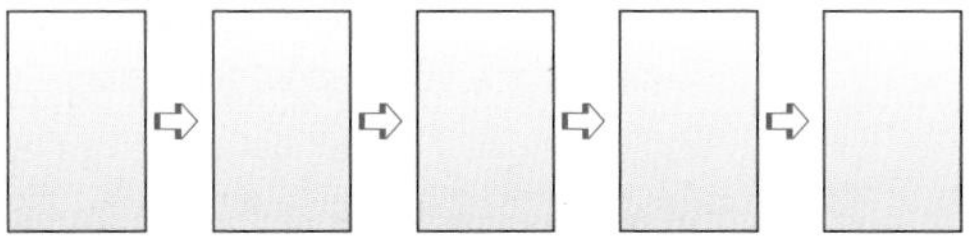

Chapter 10 Review Answers

Building Vocabulary

For definitions, see: mainland, p. 232; sirocco, p. 233; city-states, p. 234; mosaics, p. 236; pope, p. 239; Renaissance, p. 239; coalition governments, p. 239; Moors, p. 243; dialect, p. 244; cork, p. 245

Reviewing the Main Ideas

1. peninsulas (Iberian, Italian, Greek); mountains (Cantabrian, Pyrenees, Alps, Apennines); a volcano (Vesuvius); and thousands of islands (NGS 4)
2. warm, sunny climate; natural beauty; numerous historical and cultural sites (NGS 15)
3. northern—importance of wine industry, more industrialization, prosperous; southern—less industrialized, poorer (NGS 12)
4. Roman Catholicism; in the Vatican in Rome (NGS 10)
5. by acquiring and trading gold and agricultural products from their New World colonies, and spices and silks from Asia (NGS 11)

ASSESS

Have students complete the Chapter 10 Test.

RETEACH

Have students prepare segments for a television broadcast titled Focus on Southern Europe. Organize the class into five groups. Assign each group one of the following topics: physical features, major historical events, contributions to world culture, economies, and cities. Have groups write scripts and present their newscasts to the class.

ENGLISH LANGUAGE LEARNERS, COOPERATIVE LEARNING

PORTFOLIO EXTENSIONS

COOPERATIVE LEARNING

1. Organize students into five groups and have each group do research and create an illustrated time line showing the development of Greek contributions to one of the following topics: art, architecture, literature, science, and philosophy. Have the groups collaborate to make connections between the events on the time lines. *(Example: a scene from the Iliad might be depicted on artwork)*
2. Not all citizens of modern Spain consider themselves Spaniards. Many people living in the country's northeastern Basque region want an independent Basque nation. Have students conduct research on the Basque conflict and debate whether the region should be allowed to break away from Spain. Place research notes and debate outlines in student portfolios.

CHAPTER 10 Review ANSWERS

Understanding Environment and Society

Presentations will vary, but the information included should be consistent with text material. Use Rubric 40, Writing to Describe, to evaluate student work.

Thinking Critically

1. Answers will vary but should refer to the region's access to the Mediterranean Sea and the Atlantic Ocean.
2. art, food, and music; the proximity of the two countries and former Turkish rule of Greece
3. its location in the center of the Mediterranean world
4. Olives are used to make olive oil; grapes are used to make wine; and other staple crops such as lemons, mushrooms, tomatoes, and wheat are used in many dishes.
5. the region's cultural events, historical sites, and pleasant climate

CHAPTER 10 Reviewing What You Know

Building Vocabulary

On a separate sheet of paper, write sentences to define each of the following words.

1. mainland
2. sirocco
3. city-states
4. mosaics
5. pope
6. Renaissance
7. coalition governments
8. Moors
9. dialect
10. cork

Reviewing the Main Ideas

1. *Places and Regions* What are the important physical features of southern Europe?
2. *Environment and Society* Why has southern Europe been a major tourist attraction for centuries?
3. *Human Systems* What are the major economic differences between northern and southern Italy?
4. *Human Systems* What is the main religion in Italy? Where is the leadership of this religion based?
5. *Human Systems* How did both Spain and Portugal become wealthy during the 1400s and 1500s?

Understanding Environment and Society

The Arts

Environment influences a culture's art. For example, grapes might be featured in a mosaic. Create a presentation about the influence of the environment on the arts. Write a description of each piece and explain how it might have been influenced by the artist's environment.

Thinking Critically

1. **Drawing Influences and Conclusions** In what ways do you think the geography of southern Europe made trade and exploration possible?
2. **Summarizing** What parts of Greek culture have been most strongly influenced by Turkish customs? Why is this the case?
3. **Drawing Inferences and Conclusions** Recall what you have learned about the Roman Empire. What about the Italian peninsula made it a good location for a Mediterranean empire?
4. **Drawing Inferences and Conclusions** How are agricultural products of southern Europe used in food?
5. **Finding the Main Idea** Why do many tourists continue to visit historical cities in southern Europe?

FOOD FESTIVAL

This is the perfect opportunity for a pizza party. You may want to have students bring different toppings for purchased crusts. Or, have students conduct research on the history of pizza and make one as historically accurate as possible. For example, in Italy, pizza is sometimes made without tomato sauce. To celebrate Spain, enjoy tapas. These appetizers can range from olives, cubes of cheese, or ham to fancier cold omelets, stuffed peppers, or small sandwiches.

CHAPTER 10 REVIEW AND ASSESSMENT RESOURCES

Reproducible
- Readings in World Geography, History, and Culture 28, 29, and 30
- Critical Thinking Activity 10
- Vocabulary Activity 10

Technology
- Chapter 10 Test Generator (on the One-Stop Planner)
- HRW Go site
- Audio CD Program, Chapter 10

Reinforcement, Review, and Assessment
- Chapter 10 Review, pp. 246–47
- Chapter 10 Tutorial for Students, Parents, Mentors, and Peers
- Chapter 10 Test
- Chapter 10 Test for English Language Learners and Special-Needs Students

Building Social Studies Skills

Map ACTIVITY

On a separate sheet of paper, match the letters on the map with their correct labels.

Alps	Po River
Apennines	Tiber River
Aegean Sea	Naples
Peloponnesus	Sicily
Ebro River	Meseta

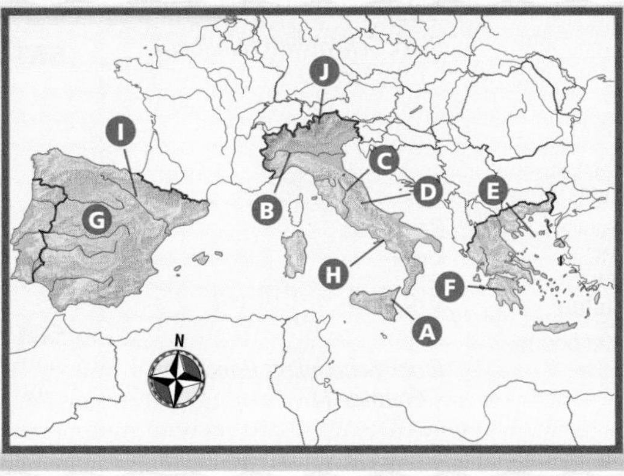

Mental Mapping Skills ACTIVITY

On a separate sheet of paper, draw a freehand map of southern Europe. Make a key for your map and label the following:

Athens	Portugal
Greece	Rome
Italy	Spain
Mediterranean Sea	Strait of Gibraltar

WRITING ACTIVITY

Find a recording of Portuguese fado music. Then write a review that explains what the lyrics of the songs reveal about Portuguese culture. Be sure to use standard grammar, spelling, sentence structure, and punctuation in your review.

Alternative Assessment

Portfolio ACTIVITY

Learning About Your Local Geography

Individual Project The cultures of southern European influence the United States in many ways. Investigate southern European influences in your community. List some of those influences in a chart.

internet connect

Internet Activity: **go.hrw.com**
KEYWORD: **SK3 GT10**

Choose a topic to explore southern Europe:
- Explore the islands and peninsulas on the Mediterranean coast.
- Take an online tour of ancient Greece.
- Learn the story of pizza.

Map Activity

A. Sicily	F. Peloponnesus
B. Po River	G. Meseta
C. Tiber River	H. Naples
D. Apennines	I. Ebro River
E. Aegean Sea	J. Alps

Mental Mapping Skills Activity

Maps will vary but listed places should be labeled in their approximate locations.

Writing Activity

Reviews will vary but should include information about Portuguese culture. Use Rubric 37, Writing Assignments, to evaluate student work.

Portfolio Activity

Answers will vary, but the information included should be consistent with text material. Use Rubric 7, Charts, to evaluate student work.

internet connect

GO TO: **go.hrw.com**
KEYWORD: **SK3 Teacher**
FOR: a guide to using the Internet in your classroom

CHAPTER 11

West–Central Europe

Chapter Resource Manager

Objectives	Pacing Guide	Reproducible Resources
SECTION 1 **Physical Geography** (pp. 249–51) **1.** Identify the area's major landform regions. **2.** Describe the role rivers, canals, and harbors play in the region. **3.** Identify west-central Europe's major resources.	**Regular** .5 day **Block Scheduling** .5 day *Block Scheduling Handbook, Chapter 11*	**RS** Guided Reading Strategy 11.1 **RS** Graphic Organizer 11
SECTION 2 **France** (pp. 252–55) **1.** Identify which foreign groups affected the historical development of France. **2.** Identify the main features of French culture. **3.** Identify the products France exports.	**Regular** 1 day **Block Scheduling** .5 day *Block Scheduling Handbook, Chapter 11*	**RS** Guided Reading Strategy 11.2 **E** Cultures of the World Activity 3 **E** Creative Strategies for Teaching World Geography, Lessons 10 and 11 **SM** Geography for Life Activity 11
SECTION 3 **Germany** (pp. 256–59) **1.** Describe the effects wars have had on Germany. **2.** Identify Germany's major contributions to world culture. **3.** Describe how the division of Germany affected its economy.	**Regular** 1 day **Block Scheduling** .5 day *Block Scheduling Handbook, Chapter 11*	**RS** Guided Reading Strategy 11.3 **E** Creative Strategies for Teaching World Geography, Lessons 10 and 11 **SM** Map Activity 11
SECTION 4 **The Benelux Countries** (pp. 260–62) **1.** Describe how larger countries influenced the Benelux countries. **2.** Describe what this region's culture is like. **3.** Describe what the Benelux countries are like today.	**Regular** 1 day **Block Scheduling** .5 day *Block Scheduling Handbook, Chapter 11*	**RS** Guided Reading Strategy 11.4 **E** Creative Strategies for Teaching World Geography, Lessons 10 and 11
SECTION 5 **The Alpine Countries** (pp. 263–65) **1.** Identify some of the major events in the history of the Alpine countries. **2.** Identify some of the cultural features of the region. **3.** Describe how the economies of Switzerland and Austria are similar.	**Regular** 1 day **Block Scheduling** .5 day *Block Scheduling Handbook, Chapter 11*	**RS** Guided Reading Strategy 11.5 **E** Creative Strategies for Teaching World Geography, Lessons 10 and 11

Chapter Resource Key

RS Reading Support
IC Interdisciplinary Connections
E Enrichment
SM Skills Mastery
A Assessment
REV Review
ELL Reinforcement and English Language Learners
Transparencies
CD–ROM
Music
Video
go.hrw.com Internet
Holt Presentation Maker Using Microsoft® Powerpoint®

One-Stop Planner CD–ROM

See the ***One-Stop Planner*** for a complete list of additional resources for students and teachers.

One-Stop Planner CD-ROM

It's easy to plan lessons, select resources, and print out materials for your students when you use the ***One-Stop Planner CD-ROM with Test Generator.***

Technology Resources	Review, Reinforcement, and Assessment Resources
One-Stop Planner CD-ROM, Lesson 11.1 Geography and Cultures Visual Resources with Teaching Activities 24–30 Homework Practice Online HRW Go site	**ELL** Main Idea Activity 11.1 **REV** Section 1 Review, p. 251 **A** Daily Quiz 11.1 **ELL** English Audio Summary 11.1 **ELL** Spanish Audio Summary 11.1
One-Stop Planner CD-ROM, Lesson 11.2 Homework Practice Online HRW Go site	**ELL** Main Idea Activity 11.2 **REV** Section 2 Review, p. 255 **A** Daily Quiz 11.2 **ELL** English Audio Summary 11.2 **ELL** Spanish Audio Summary 11.2
One-Stop Planner CD-ROM, Lesson 11.3 Homework Practice Online HRW Go site	**ELL** Main Idea Activity 11.3 **REV** Section 3 Review, p. 259 **A** Daily Quiz 11.3 **ELL** English Audio Summary 11.3 **ELL** Spanish Audio Summary 11.3
One-Stop Planner CD-ROM, Lesson 11.4 Homework Practice Online HRW Go site	**ELL** Main Idea Activity 11.4 **REV** Section 4 Review, p. 262 **A** Daily Quiz 11.4 **ELL** English Audio Summary 11.4 **ELL** Spanish Audio Summary 11.4
One-Stop Planner CD-ROM, Lesson 11.5 *ARGWorld* CD-ROM: Ski Resorts in Switzerland Homework Practice Online HRW Go site	**ELL** Main Idea Activity 11.5 **REV** Section 5 Review, p. 265 **A** Daily Quiz 11.5 **ELL** English Audio Summary 11.5 **ELL** Spanish Audio Summary 11.5

internet connect

HRW ONLINE RESOURCES

GO TO: go.hrw.com
Then type in a keyword.

TEACHER HOME PAGE
KEYWORD: SK3 TEACHER

CHAPTER INTERNET ACTIVITIES
KEYWORD: SK3 GT11

Choose an activity to:

- tour the land and rivers of Europe.
- travel back in time to the Middle Ages.
- visit schools in Belgium and the Netherlands.

CHAPTER ENRICHMENT LINKS
KEYWORD: SK3 CH11

CHAPTER MAPS
KEYWORD: SK3 MAPS11

ONLINE ASSESSMENT
Homework Practice
KEYWORD: SK3 HP11
Standardized Test Prep Online
KEYWORD: SK3 STP11
Rubrics
KEYWORD: SS Rubrics

COUNTRY INFORMATION
KEYWORD: SK3 Almanac

CONTENT UPDATES
KEYWORD: SS Content Updates

HOLT PRESENTATION MAKER
KEYWORD: SK3 PPT11

ONLINE READING SUPPORT
KEYWORD: SS Strategies

CURRENT EVENTS
KEYWORD: S3 Current Events

Meeting Individual Needs

Ability Levels

Level 1 Basic-level activities designed for all students encountering new material

Level 2 Intermediate-level activities designed for average students

Level 3 Challenging activities designed for honors and gifted-and-talented students

English Language Learners Activities that address the needs of students with Limited English Proficiency

Chapter Review and Assessment

IC Interdisciplinary Activities for the Middle Grades 13, 14, 15

E Readings in World Geography, History, and Culture 31–33

SM Critical Thinking Activity 11

REV Chapter 11 Review, pp. 266–67

REV Chapter 11 Tutorial for Students, Parents, Mentors, and Peers

ELL Vocabulary Activity 11

A Chapter 11 Test

Chapter 11 Test Generator (on the One-Stop Planner)

Audio CD Program, Chapter 11

A Chapter 11 Test for English Language Learners and Special-Needs Students

CHAPTER 11

Launch into Learning

Write the names of the countries of west-central Europe on the chalkboard. Then ask students leading questions about the countries. *(Examples: What is the Eiffel Tower, and where is it located? What country used to be two separate countries but has been reunited? What kind of cheese is full of holes? What country do you associate with tulips and windmills?)* Students will probably know the answers to some of these questions. Encourage them to ask other questions based on yours. *(Examples: What do windmills actually do? Why are they important?)* Create a web of student responses and questions on the chalkboard. Have a volunteer copy the web into his or her notes to allow class members to compare their initial questions to what they learn in this chapter.

Section 1

Objectives

1. Describe the area's major landform regions.
2. Analyze the role that rivers, canals, and harbors play in the region.
3. Identify west-central Europe's major resources.

Links to Our Lives

There are many reasons why American students should know more about the countries of west-central Europe. Here are a few of them:

- All of the countries of west-central Europe are friendly to the United States and have strong economic ties to our country.
- Although Germany was our enemy in both world wars, it is now a major ally. The United States has military bases in Germany.
- A large percentage of Americans trace their ancestry to the region.
- The region has produced some of the world's greatest composers, writers, and artists.
- Millions of American tourists enjoy the region's beauty, cuisine, culture, and sports.

CHAPTER 11

West-Central Europe

West-central Europe is an important agricultural, industrial, and manufacturing area. The countries of this region export many different products. They are some of the richest countries in the world.

Gruss dich (Hello). My name is Lizzi (LEE-zee). I live in southern Germany in the village of Deutenhausen. Lizzi is short for Felicitas—my grandmother's name—which means "happiness." I am in the eighth grade at the gymnasium, or high school. I live in a big house on a farm with my three older sisters, my parents, and my grandmother. My parents are farmers and also own a restaurant. In summer, I make sure the cows have enough water and I chase the geese home. I also help my parents chop vegetables in the restaurant. I don't want to be a farmer when I grow up! I hope to become a doctor and work in an emergency room.

At about 7:30, I take the bus to school in Weilheim, about 2 miles (3 km) away. My favorite subject is art. My school is not very strict, and we do not wear uniforms. I study German, geography, English, and Latin. Next year I will start classical Greek.

After school is over at 12:30, I go home to have lunch with my grandmother. Then, I play with my friends outdoors, even when it rains.

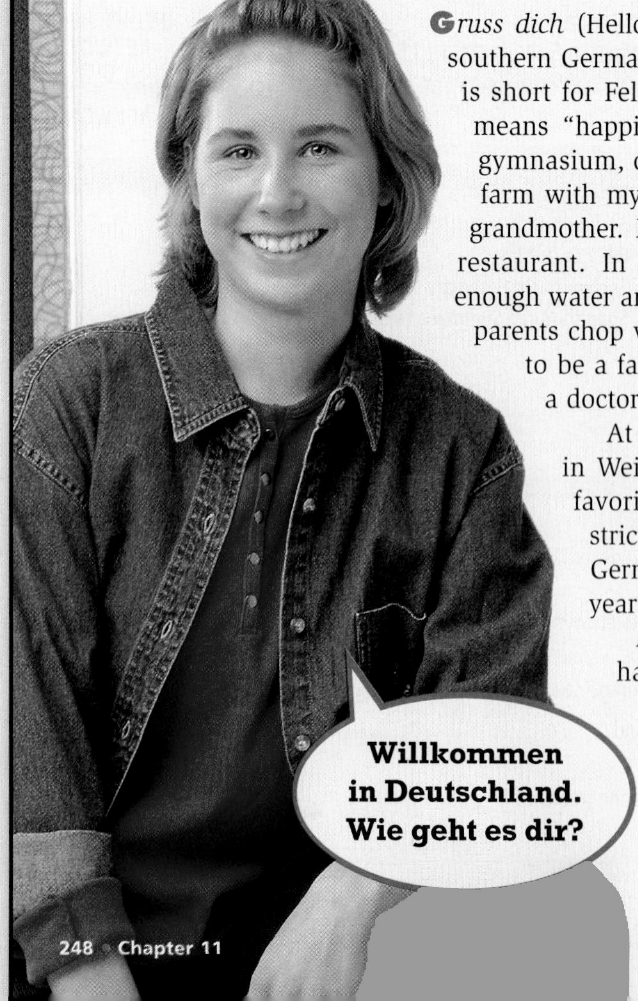
Willkommen in Deutschland. Wie geht es dir?

Translation: Welcome to Germany! How are you?

Focus

LET'S GET STARTED

Copy the following question onto the chalkboard: *What are some of the sports at which European athletes have excelled?* Discuss responses. *(Possible answers: bicycling, skiing, ice skating, sailing, or mountain climbing)* Then ask students to suggest what these sports might indicate about the landforms of west-central Europe. *(The region includes mountains, water, plains, cold and warm climates.)* Tell students that they will learn about the physical geography of west-central Europe in Section 1.

Using the Physical-Political Map

Have students examine the map on this page. Then have them name countries that fit these categories: countries on the North Sea *(Germany, Netherlands, Belgium, France)*, countries with elevations over 6,560 feet (1,999 m) *(France, Germany, Switzerland, Austria, Liechtenstein)*, landlocked countries *(Switzerland, Austria, Liechtenstein, Luxembourg)*, country on the Bay of Biscay and the English Channel *(France)*. Have students list the countries through which the Loire, Rhine, and Danube Rivers flow.

Section 1 Physical Geography

Read to Discover

1. Where are the area's major landform regions?
2. What role do rivers, canals, and harbors play in the region?
3. What are west-central Europe's major resources?

Define

navigable
loess

Locate

Northern European Plain
Pyrenees
Alps
Seine River
Rhine River
Danube River
North Sea
Mediterranean Sea
English Channel
Bay of Biscay

Why It Matters

Nuclear power is important in west-central Europe. However, nuclear reactors can pose problems for the environment. Use CNNfyi.com or other **current events** sources to find out more about alternative sources of power. Record your findings in your journal.

Neuschwanstein Castle, Germany

West-Central Europe: Physical-Political

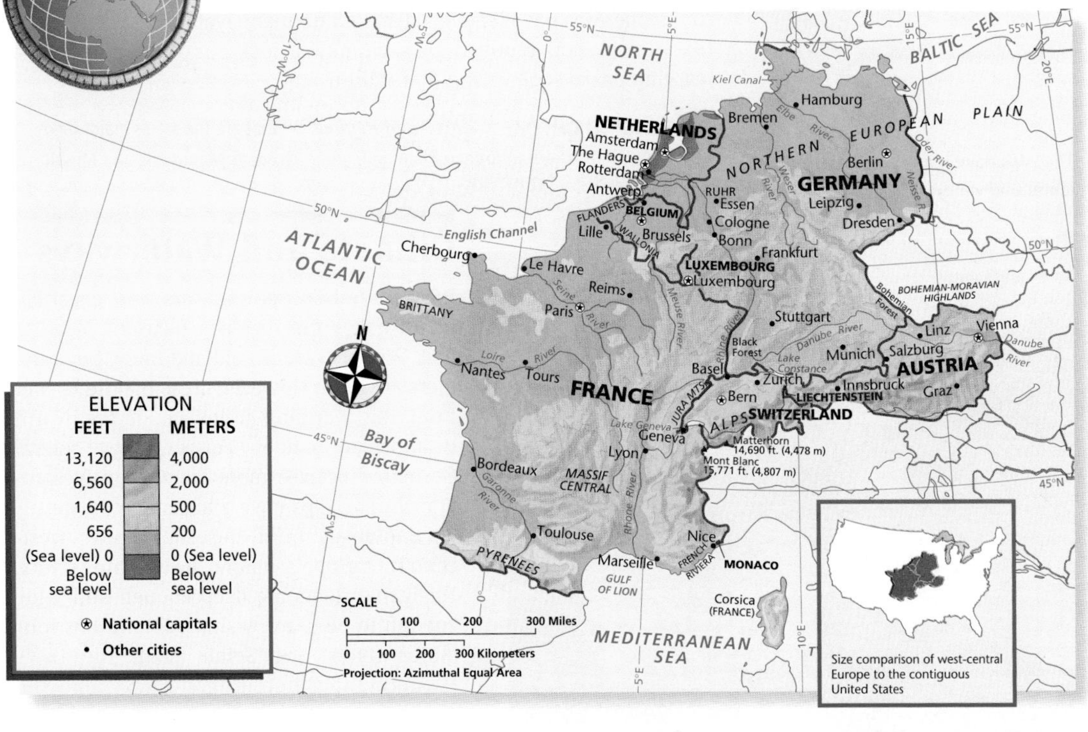

Section 1 Resources

Reproducible

- Block Scheduling Handbook, Chapter 11
- Graphic Organizer 11
- Guided Reading Strategy 11.1

Technology

- One-Stop Planner CD–ROM, Lesson 11.1
- Homework Practice Online
- Geography and Visual Resources with Teaching Activity 24–30
- HRW Go site

Reinforcement, Review, and Assessment

- Section 1 Review, p. 251
- Daily Quiz 11.1
- Main Idea Activity 11.1
- English Audio Summary 11.1
- Spanish Audio Summary 11.1

TEACH

Teaching Objectives 1–3

ALL LEVELS: (Suggested time: 20 min.) Copy the following graphic organizer onto the chalkboard, omitting the italicized answers. Have each student complete the organizer by filling in west-central Europe's major landform regions; the role of canals, rivers, and harbors in the region; and the region's major resources. Ask volunteers to share their answers with the class. **ENGLISH LANGUAGE LEARNERS**

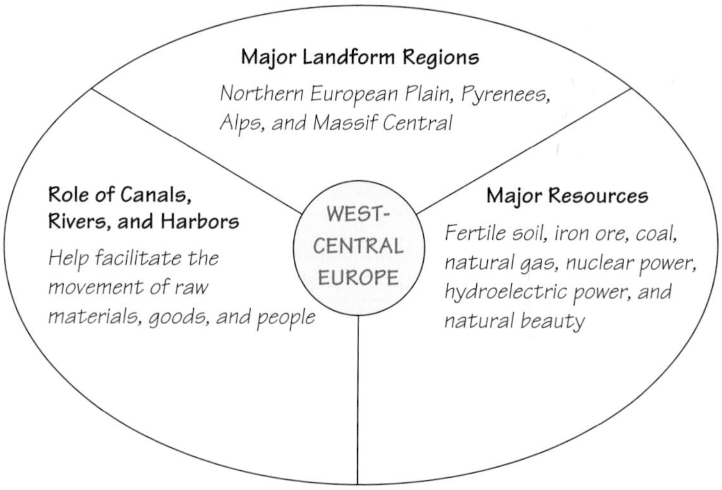

EYE ON EARTH

During the 1880s commercial fishers would take about 250,000 Atlantic salmon each year from the Rhine River. However, in 1958 the last known salmon was pulled from the river. What had happened?

For decades, the mighty Rhine had been dredged and straightened. This process altered the water's flow, clarity, and temperature. Overfishing and dumping of toxic chemicals from factories depleted the salmon and other species. Then, in 1986, following a fire at a chemical plant, over 30 tons (27 metric tons) of dyes, herbicides, pesticides, fungicides, and mercury poured into the river. The Rhine seemed to be poisoned forever.

European governments and citizens responded by banding together to clean up the river. Finally, in 1990, an Atlantic salmon was pulled from the Sieg, a tributary of the Rhine. The "king of fish" had returned to live and breed.

Discussion: Lead a discussion on European citizens' working to clean up the Rhine River and the importance of civic participation in democratic societies.

Visual Record Answer

Many industries might locate near the Rhine to facilitate the transportation of goods.

The Rhine River has been an important transportation route since Roman times.

Interpreting the Visual Record **How might the Rhine influence the location of German industry?**

internet connect
GO TO: go.hrw.com
KEYWORD: SK3 CH11
FOR: Web sites about west-central Europe

The Alps have many large glaciers, lakes, and valleys.

Physical Features

West-central Europe includes France, Germany, Belgium, the Netherlands, Luxembourg, Switzerland, and Austria. Belgium, the Netherlands, and Luxembourg are called the Benelux countries. The word Benelux is a combination of the first letters of each country's name. They are also sometimes called the Low Countries. Large areas of Switzerland and Austria lie in the Alps mountain range. For this reason, they are called the Alpine countries.

Lowlands The main landform regions of west-central Europe are arranged like a fan. The outer edge of the fan is the Northern European Plain. Brittany, a peninsula jutting from northern France, rises slightly above the plain. In Belgium and the Netherlands, the Northern European Plain dips below sea level.

Uplands Toward the middle of the fan a wide band of uplands begins at the Pyrenees (PIR-uh-neez) Mountains. Another important uplands region is the Massif Central (ma-SEEF sahn-TRAHL) in France. Most of the southern two thirds of Germany is hilly. The Schwarzwald (SHFAHRTS-vahlt), or Black Forest, occupies the southwestern corner of Germany's uplands region.

Mountains At the center of the fan are the Alps, Europe's highest mountain range. Many peaks in the Alps reach heights of more than 14,000 feet (4,267 m). The highest peak, France's Mont Blanc (mawn BLAHN), reaches to 15,771 feet (4,807 m). Because of their high elevations, the Alps have large glaciers and frequent avalanches. During the Ice Age, glaciers scooped great chunks of rock out of the mountains, carving peaks such as the Matterhorn.

✓ **READING CHECK:** *Places and Regions* What are the area's major landforms? Northern European Plain; uplands—Pyrenees Mountains and Massif Central, Black Forest; Alps

Climate and Waterways

West-central Europe's marine west coast climate makes the region a pleasant place to live. Winters can be cold and rainy, but summers are mild. However, areas that lie farther from the warming influence of the North Atlantic are colder. For example, central Germany receives more snow than western France. The Alps have a highland climate.

Snowmelt from the Alps feeds west-central Europe's many **navigable** rivers. Navigable rivers are deep enough and wide enough to be used by ships. France has four major rivers: the Seine (SEN), the Loire (LWAHR), the Garonne (gah-RAWN), and the

Close

Call on volunteers to use the physical-political map to name the landforms and rivers they would encounter if they were to travel eastward from Cherbourg to Berlin or southward from Hamburg to Marseille.

Review and Assess

Have students complete the Section Review. Then call on a student to choose a place on the physical-political map of west-central Europe and on another student to describe its physical geography. Continue until all major features have been covered. Then have students complete Daily Quiz 11.1.

Reteach

Have students complete Main Idea Activity 11.1. Then pair students and have each pair create a jumble puzzle of 10 words from the key terms or places on the physical-political map. Students should scramble the letters of each term and write a clue describing it.

ENGLISH LANGUAGE LEARNERS, COOPERATIVE LEARNING

Extend

Have interested students conduct research on the mistral—a dry, cold northerly wind that blows from the Alps through the Rhone Valley, or the foehn—a warm, dry wind that blows down from the Alps into Switzerland. Ask students to use their research to create a poster illustrating what creates these winds and how they affect daily life.

Rhone (ROHN). Germany has five major rivers: the Rhine (RYN), the Danube (DAN-yoob), the Elbe (EL-buh), the Oder (OH-duhr), and the Weser (VAY-zuhr). These rivers and the region's many canals are important for trade and travel. Many large harbor cities are located where rivers flow into the North Sea, Mediterranean Sea, English Channel, or Bay of Biscay. The region's heavily indented coastline has hundreds of excellent harbors.

✓ **READING CHECK:** *Environment and Society* What economic role do rivers, canals, and harbors play in west-central Europe? They are important for trade and travel.

▲ The Grindelwald Valley in Switzerland has excellent pastures.

Resources

Most of the forests that once covered west-central Europe were cut down centuries ago. The fields that remained are now some of the most productive in the world. Germany's plains are rich in **loess** (LES)—fine, wind-blown soil deposits. Germany and France produce grapes for some of the world's finest wines. Switzerland's Alpine pastures support dairy cattle.

The distribution of west-central Europe's mineral resources is uneven. Germany and France have deposits of iron ore but must import oil. Energy resources are generally in short supply in the region. However, there are deposits of coal in Germany and natural gas in the Netherlands. Nuclear power helps fill the need for energy, particularly in France and Belgium. Alpine rivers provide hydroelectric power in Switzerland and Austria. Natural beauty is perhaps the Alpine countries' most valuable natural resource, attracting millions of tourists every year.

✓ **READING CHECK:** *Environment and Society* What geographic factors contribute to the economy of the region? Rich soils are good for farming; Alpine pastures support dairy cattle; lack of energy resources leads to energy imports; natural beauty supports tourism.

BUILD ON WHAT You Know

Do you remember what you learned about hydroelectric power? See Chapter 4 to review.

go.hrw.com Homework Practice Online Keyword: SK3 HP11

Section 1 Review

Define and explain: navigable, loess

Working with Sketch Maps On a map of west-central Europe that you draw or that your teacher provides, label the following: the Northern European Plain, Alps, North Sea, Mediterranean Sea, English Channel, and Bay of Biscay.

Reading for the Main Idea

1. *Places and Regions* What are the landform regions of west-central Europe?
2. *Places and Regions* What type of climate dominates this region?

Critical Thinking

3. **Making Generalizations and Predictions** What might be the advantages of having many good harbors and navigable rivers?
4. **Drawing Inferences and Conclusions** How do you think an uneven distribution of resources has affected this region?

Organizing What You Know

5. **Categorizing** Copy the following graphic organizer. Use it to describe the major rivers of west-central Europe. Add rows as needed.

River	Country/Countries	Flows into. . .

Section 1 Review

Answers

Define For definitions, see: navigable, p. 250; loess, p. 251

Working with Sketch Maps Maps will vary, but listed places should be labeled in their approximate locations.

Reading for the Main Idea

1. the Northern European Plain, the Pyrenees, the Alps, and the Massif Central (NGS 4)
2. marine west coast (NGS 4)

Critical Thinking

3. They would help move materials, goods, and people.
4. influenced types of goods and energy the region produces, what resources it must import

Organizing What You Know

5. Seine—France, English Channel; Loire—France, Bay of Biscay; Garonne—France, Bay of Biscay; Rhone—France, Mediterranean; Rhine—Switzerland, Germany, the Netherlands, North Sea; Danube—Germany and Austria; Elbe—Germany, North Sea; Oder—Germany, the Baltic; and Weser—Germany, North Sea

Section 2

Objectives

1. Discuss which foreign groups have affected France's history.
2. Describe the main features of French culture.
3. Identify the products that France exports.

Focus

LET'S GET STARTED

Copy the following question onto the chalkboard: *What are three things that come to mind when you think of France?* Discuss responses. *(Possible answers: fashion, food, Eiffel Tower)* Then ask students where they might have received their impressions of France. *(movies, television commercials, news broadcasts)* You may want to ask what important aspects of the country are missing from the list. *(Possible answers: history, daily life, economy)* Tell students that in Section 2 they will learn more about France.

Building Vocabulary

Copy the key terms **medieval**, **NATO**, and **impressionism** onto the chalkboard. Tell students that *medieval* is one of the most frequently misspelled words in the English language. Have a volunteer read aloud the word's definition and history from the text. For **NATO**, introduce the concept of acronyms. Finally, tell students that **impressionism** comes from a painting by Claude Monet titled *Impression—Sunrise.* Ask students why Monet may have given his painting this title. *(It was his impression of the sunrise, not a faithful reproduction of every detail.)*

Section 2 Resources

Reproducible
- Guided Reading Strategy 11.2
- Cultures of the World Activity 3
- Geography for Life Activity 11
- Creative Strategies for Teaching World Geography, Lessons 10 and 11

Technology
- One-Stop Planner CD–ROM, Lesson 11.2
- Homework Practice Online
- HRW Go site

Reinforcement, Review, and Assessment
- Section 2 Review, p. 255
- Daily Quiz 11.2
- Main Idea Activity 11.2
- English Audio Summary 11.2
- Spanish Audio Summary 11.2

Section 2 France

Read to Discover

1. Which foreign groups affected the historical development of France?
2. What are the main features of French culture?
3. What products does France export?

Define

medieval
NATO
impressionism

Locate

Brittany
Normandy
Paris
Marseille
Nice

Why It Matters

France is a key member of NATO, the North Atlantic Treaty Organization. Use CNNfyi.com or other **current events** sources to find out more about NATO. Record your findings in your journal.

French croissants

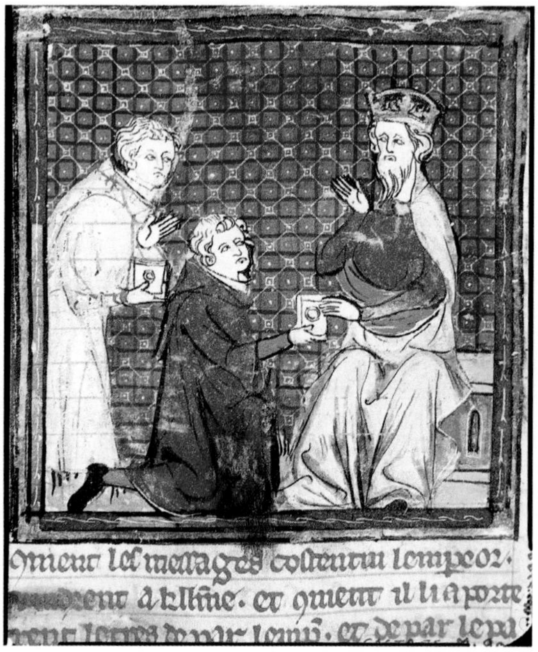

In this illustration messengers inform Charlemagne of a recent military victory.

History

France has been occupied by people from many other parts of Europe. In ancient times, France was part of a region known as Gaul. Thousands of years ago, people moved from eastern Europe into Gaul. These people spoke Celtic languages related to modern Welsh and Gaelic. Breton is a Celtic language still spoken in the region of Brittany.

Early History About 600 B.C. the Greeks set up colonies on Gaul's southern coast. Several centuries later, the Romans conquered Gaul. They introduced Roman law and government to the area. The Romans also established a Latin-based language that developed into French.

Roman rule lasted until the A.D. 400s. A group of Germanic people known as the Franks then conquered much of Gaul. It is from these people that France takes its name. Charlemagne was the Franks' greatest ruler. He dreamed of building a Christian empire that would be as great as the old Roman Empire. In honor of this, the pope crowned Charlemagne Emperor of the Romans in

TEACH

Teaching Objective 1

ALL LEVELS: (Suggested time: 15 min.) Copy the following graphic organizer onto the chalkboard, omitting the italicized answers. Have each student complete the organizer by describing how Celts, Romans, Franks, and Normans affected France's historical development. Ask volunteers to share their answers with the class. **ENGLISH LANGUAGE LEARNERS**

INFLUENCES ON FRANCE'S EARLY HISTORY

Celts	Romans	Franks	Normans
• *migrated from eastern Europe to Gaul*	• *conquered Gaul*	• *conquered Gaul*	• *migrated from Normandy and conquered England*
• *introduced Celtic languages, including Breton*	• *introduced Roman law and government and established Latin-based language that developed into French*	• *Frankish emperor Charlemagne strengthened government and improved education*	• *Norman kings of England claimed throne of France, which led to Hundred Years' War*

A.D. 800. During his rule, Charlemagne did much to strengthen government and improve education and the arts in Europe.

The Franks divided Charlemagne's empire after his death. Invading groups attacked from many directions. The Norsemen, or Normans, were one of these groups. They came from northern Europe. The area of western France where the Normans settled is known today as Normandy.

The period from the collapse of the Roman Empire to about 1500 is called the Middle Ages, or **medieval** period. The word medieval comes from the Latin words *medium,* meaning "middle," and *aevum,* meaning "age." During much of this period kings in Europe were not very powerful. They depended on cooperation from nobles, some of whom were almost as powerful as kings.

In 1066 a noble, the duke of Normandy, conquered England, becoming its king. As a result, the kings of England also ruled part of France. In the 1300s the king of England tried to claim the throne of France. This led to the Hundred Years' War, which lasted from 1337 to 1453. Eventually, French armies drove the English out of France. The French kings then slowly increased their power over the French nobles.

During the Middle Ages the Roman Catholic Church created a sense of unity among many Europeans. Many tall, impressive cathedrals were built during this time. Perhaps the most famous is the Cathedral of Notre Dame in Paris. It took almost 200 years to build.

French and English knights clash in this depiction of the Hundred Years' War.

Revolution and Napoléon's Empire From the 1500s to the 1700s France built a global empire. The French established colonies in the Americas, Asia, and Africa. During this period most French people lived in poverty and had few rights. In 1789 the French Revolution began. The French overthrew their king and established an elected government. About 10 years later a brilliant general named Napoléon Bonaparte took power. As he gained control, he took the title of emperor. Eventually, Napoléon conquered most of Europe. Napoléon built new roads throughout France, reformed the French educational system, and established the metric system of measurement. In 1815 an alliance including Austria, Great Britain, Prussia, and Russia finally defeated Napoléon. The French king regained the throne.

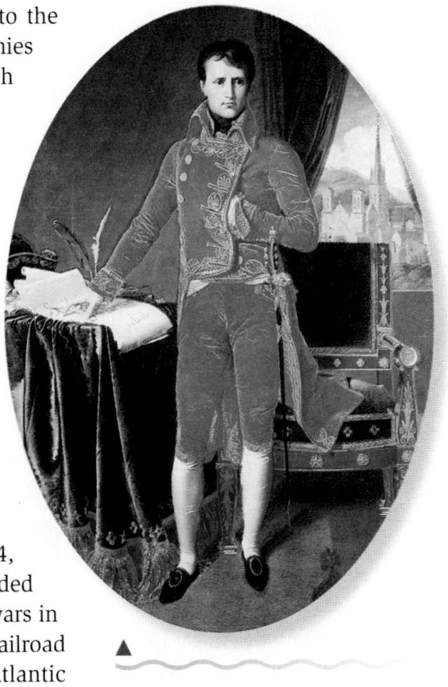

Napoléon Bonaparte became the ruler of France and conquered most of Europe.

World Wars During World War I (1914–18) the German army controlled parts of northern and eastern France. In the early years of World War II, Germany defeated France and occupied the northern and western parts of the country. In 1944, Allied armies including U.S., British, and Canadian soldiers landed in Normandy and drove the Germans out. However, after two wars in 30 years France was devastated. Cities, factories, bridges, railroad lines, and train stations had been destroyed. The North Atlantic Treaty Organization, or **NATO**, was formed in 1949 with France as a founding member. This military alliance was created to defend Western Europe against future attacks.

Linking Past to Present

Links between England and France During part of the Middle Ages, England and France were closely linked politically. In 1066 William, duke of Normandy conquered. Later, King Henry II of England also ruled Normandy and western France because his wife had inherited a huge piece of France. Only after its defeat in the Hundred Years' War did England lose most of its French territory. For the next several centuries, England and France were often in conflict.

In 1994 a new era in British-French cooperation dawned with the completion of the "Chunnel." It consists of two railroad tunnels and one service tunnel that link France and England under the English Channel. These tunnels have helped stimulate trade and economic development.

Critical Thinking: How did substantial resources affect the creation of the Chunnel?

Answer: A project like the Chunnel would be very expensive and could only be completed by wealthy countries.

internet connect

GO TO: go.hrw.com
KEYWORD: SK3 CH11
FOR: Web sites about the Chunnel

Teaching Objectives 2–3

LEVEL 1: (Suggested time: 30 min.) Pair students and have each pair create a collage depicting some of the main features of French culture as well as products that France exports. *(The collages' images and symbols should pertain to Roman Catholicism; predominance of the French language; traditions of wine and cheese making; various immigrant cultures; French literary, artistic, and philosophical traditions; and exports such as wheat, olives, cars, airplanes, shoes, clothing, machinery, and chemicals.)* Display collages around the classroom. **ENGLISH LANGUAGE LEARNERS, COOPERATIVE LEARNING**

LEVELS 2 AND 3: (Suggested time: 30 min.) Have each student write a short poem that describes aspects of France's culture and economy. Have volunteers recite their poems to the class.

TEACHER TO TEACHER

Rebecca Minnear of Las Vegas, Nevada, suggests the following activity to help students learn more about France: Have each student create two postcards to illustrate and describe aspects of France's economy, history, and culture. Students should create an illustration for one side of the postcard. On the reverse side of the card, students should write a note to a friend or family member describing the significance of the illustration. Have volunteers present and explain their postcards to the class.

DAILY LIFE

Have the class look up the word *baccalaureate* in a dictionary. *(The word refers to the bachelor's degree bestowed by a college or university, or to a religious service for graduates.)*

Explain that in France *baccalaureate* refers to a difficult examination students take when they have completed high school. Students who want to go to a university or find a good job must do well on *"le bac."* The emphasis placed on the baccalaureate exam reflects the high value the French place on education.

Activity: Have students acquire information about various placement examinations from a high school counselor. Then lead a discussion comparing the French examination system to systems found in the United States.

Visual Record Answers

▲ no need to convert currencies, eased trade

▶ modern vehicle, otherwise still traditional

Section 2 Review

Answers

Define For definitions, see: medieval, p. 253; NATO, p. 253; impressionism, p. 255

▲ The euro replaced the currencies of most of the individual EU countries.

Interpreting the Visual Record What are the advantages of a shared currency?

▲ Many French cheeses, such as Brie, Camembert, and Roquefort, are named after the places where they are made.

Government In the 1950s and 1960s most French colonies in Asia and Africa achieved independence. However, France still controls several small territories around the world. Today, France is a republic with a parliament and an elected president. France is also a founding member of the European Union (EU). France is gradually replacing its currency, the franc, with the EU currency, the euro.

✓ **READING CHECK:** *Human Systems* Which foreign groups have affected France's historical development? **Greeks, Romans, Franks, Normans, English, Germans**

Culture

About 90 percent of French people are Roman Catholic, and 3 percent practice Islam. Almost all French citizens speak French. However, small populations of Bretons in the northwest and Basques in the southwest speak other languages. In Provence-Alpes-Côte d'Azur and Languedoc-Roussillon in the south and on the island of Corsica, some people speak regional dialects along with French. Immigrants from former colonies in Africa, the Caribbean, and Southeast Asia also influence French culture through their own styles of food, clothing, music, and art.

Customs In southern France people eat Mediterranean foods like wheat, olives and olive oil, cheeses, and garlic. In the north food is more likely to be prepared with butter, herbs, and mushrooms. Wine is produced in many French regions, and France produces more than 400 different cheeses. French people celebrate many festivals, including Bastille Day on July 14. On this date in 1789 a mob stormed the Bastille, a royal prison in Paris. The French recognize this event as the beginning of the French Revolution.

Workers harvest grapes at a vineyard in the Rhone Valley near Lyon.

Interpreting the Visual Record Has modern technology changed the grape-growing process? ▶

Close

Display impressionist paintings for students and lead a discussion about how they depict the effects of light.

Review and Assess

Have students complete the Section Review. Then teach the class the phrase *"Je sais"* (ZHUH SAY)—meaning "I know" in French. Call on volunteers to say *Je sais* and follow it with an important fact pertaining to Section 2. Then have students complete Daily Quiz 11.2.

Reteach

Have students complete Main Idea Activity 11.2. Then organize students into small groups and have each group prepare an outline of Section 2. **ENGLISH LANGUAGE LEARNERS, COOPERATIVE LEARNING**

Extend

Have interested students compile a "Who's Who" of famous French musicians, writers, political leaders, and other historical figures. Examples include Joan of Arc, Napoléon, Debussy, and Hugo, among others. Have each student choose two or three figures and write a short description of his or her chosen figures' significance. Arrange the descriptions in alphabetical order. Compile the biographies into a scrapbook, and use it as a classroom reference.

The Arts and Literature France has a respected tradition of poetry, philosophy, music, and the visual arts. In the late 1800s and early 1900s France was the center of an artistic movement called **impressionism**. Impressionist artists tried to capture the rippling of light rather than an exact, realistic image. Famous impressionists include Monet, Renoir, and Degas. French painters, like Cézanne and Matisse, influenced styles of modern painting. Today, France is a world leader in the arts and film industry.

✓ **READING CHECK:** *Human Systems* How did French art affect the world? French painters are respected around the world and influenced modern painting.

France

Country	Population/ Growth Rate	Life Expectancy	Literacy Rate	Per Capita GDP
France	59,551,227 0.4%	75, male 83, female	99%	$24,400
United States	281,421,906 0.9%	74, male 80, female	97%	$36,200

Sources: Central Intelligence Agency, *The World Factbook 2001;* U.S. Census Bureau

Interpreting the Chart **How do France's life expectancy and literacy rate compare to those of the United States?**

France Today

France is a major agricultural and industrial country. Its resources, labor force, and location in the heart of Europe have helped spur economic growth. France exports wheat, olives, wine, and cheeses as well as other dairy products. French factories produce cars, airplanes, shoes, clothing, machinery, and chemicals. France's largest city is Paris, which has nearly 10 million people in its metropolitan area. Other major cities include Marseille, Nice, Lyon, and Lille. France's major cities are linked by high-speed trains and excellent highways.

✓ **READING CHECK:** *Human Systems* What are some products that France exports? wheat, olives, wine, cheeses, cars, airplanes, shoes, clothing, machinery, chemicals

Section 2 Review

go.hrw.com Homework Practice Online
Keyword: SK3 HP11

Define and explain: medieval, NATO, impressionism

Working with Sketch Maps On the map you created in Section 1, label Brittany, Normandy, Paris, Marseille, and Nice.

Reading for the Main Idea

1. *Human Systems* What were the main periods of French history?
2. *Human Systems* What are the main features of French culture?

Critical Thinking

3. **Finding the Main Idea** What were some long-lasting achievements of Charlemagne and Napoléon?
4. **Summarizing** What is the French economy like?

Organizing What You Know

5. **Identifying Cause and Effect** Copy the following graphic organizer. Use it to list the causes and effects of the Hundred Years' War.

	⇨		⇨		⇨		⇨	

Working with Sketch Maps Maps will vary, but listed places should be labeled in their approximate locations.

Reading for the Main Idea

1. Greek colonization, Romans, Franks, medieval period, colonial empire, French Revolution, Napoleonic era, world wars (NGS 13)
2. mostly Roman Catholic; French language; immigrant cultures; popularity of wine and cheese; respected cultural traditions (NGS 10)

Critical Thinking

3. Charlemagne—strengthened government, helped to improve education and arts in Europe; Napoléon—built new roads, reformed French educational system, instituted metric system in France
4. major agricultural and industrial producer and exporter; resources, location

Organizing What You Know

5. duke of Normandy conquers England; Norman kings of England continue to rule part of France; English king tries to claim French throne; Hundred Years' War; French drive English out

▲ **Chart Answer**

They are higher.

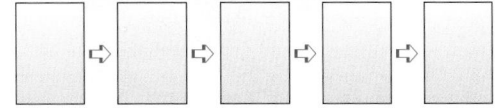

Section 3

Objectives

1. Examine the effects that wars have had on Germany.
2. Describe Germany's major contributions to world culture.
3. Analyze how the division of Germany affected its economy.

Focus

LET'S GET STARTED

Copy the following passage onto the chalkboard: *Imagine that the American Civil War had resulted in the creation of two separate countries. Imagine that the two "Americas" were separated for 40 years before they were reunified. What is one problem that the country might face after reunification?* Discuss responses. Explain that after World War II Germany was divided into two countries that have since been reunited. Tell students that in Section 3 they will learn more about Germany.

Building Vocabulary

Locate a newspaper headline that includes the word *reform.* Discuss with students how the word is used in the headline and article. Compare modern political reform with the **Reformation** of the 1500s. Then tell students that the word **Holocaust** comes from the Greek *holokauston,* which means "that which is completely burnt." Tell students that the Holocaust refers to the mass killing, or genocide, of millions of Jews and other people by the Nazis during World War II. Finally, have a volunteer read aloud the definition of the word **chancellor** from the glossary.

Section 3 Resources

Reproducible
- Guided Reading Strategy 11.3
- Map Activity 11
- Creative Strategies for Teaching World Geography, Lessons 10 and 11

Technology
- One-Stop Planner CD–ROM, Lesson 11.3
- Homework Practice Online
- HRW Go site

Reinforcement, Review, and Assessment
- Section 3 Review, p. 259
- Daily Quiz 11.3
- Main Idea Activity 11.3
- English Audio Summary 11.3
- Spanish Audio Summary 11.3

Visual Record Answer

on a hill overlooking a river

Section 3 Germany

Read to Discover

1. What effects have wars had on Germany?
2. What are Germany's major contributions to world culture?
3. How did the division of Germany affect its economy?

Define

Reformation
Holocaust
chancellor

Locate

Berlin
Bonn
Essen
Frankfurt
Munich
Hamburg
Cologne

Why It Matters

The Berlin Wall had fallen by 1990, and East and West Germany were reunited as a single nation. What have been the effects of reunification? Use CNNfyi.com or other **current events** sources to find out more about Germany. Record your findings in your journal.

A VW Turbo beetle

This medieval castle overlooks a German town.

Interpreting the Visual Record **What geographic features made this a good place to build a fortress?**

History

Many Germans are descendants of tribes that migrated from northern Europe in ancient times. The Romans conquered the western and southern fringes of the region. They called this land Germania, from the name of one of the tribes that lived there.

The Holy Roman Empire When the Roman Empire collapsed, the Franks became the most important tribe in Germany. The lands ruled by the Frankish king Charlemagne in the early 800s included most of what is now Germany. Charlemagne's empire was known as the Holy Roman Empire.

Reformation and Unification During the 1500s Germany was the center of the **Reformation**—a movement to reform Christianity. The reformers were called Protestants. Protestants rejected many practices of the Roman Catholic Church. At the time, Germany was made up of many small states. Each state was ruled by a prince who answered to the Holy Roman emperor. Many of the princes became Protestants. This angered the Holy Roman emperor, who was Catholic. He sent armies against the princes. Although the princes won the right to choose the religion of their states, conflict continued. This conflict eventually led to the Thirty Years' War (1618–48). This war was costly. Many towns were destroyed and nearly one third of the

TEACH

Teaching Objective 1

ALL LEVELS: (Suggested time: 10 min.) Copy the following graphic organizer onto the chalkboard, omitting the italicized answers. Ask each student to complete the organizer by describing the effects of the Thirty Years' War and World Wars I and II on Germany. Ask volunteers to share their answers with the class. **ENGLISH LANGUAGE LEARNERS**

EFFECTS OF WARS ON GERMANY

Thirty Years' War	World War I	World War II
• *many towns destroyed* • *nearly one third of the population died*	• *lost territory and overseas colonies* • *paid heavy fines after the war*	• *Jewish population was nearly wiped out* • *divided into two countries*

Teaching Objective 1

LEVELS 2 AND 3: (Suggested time: 30 min.) Tell students to imagine that they are Germans who oppose their country's involvement in any military actions. Then ask each student to write an antiwar speech. Students' speeches should reference the effects of previous wars on Germany. Ask volunteers to deliver their speeches to the class.

population died. Germany remained divided for more than 200 years. In the late 1800s Prussia, the strongest state, united Germany.

World Wars In 1914 national rivalries and a conflict in the Balkans led to World War I. Austria, Germany, and the Ottoman Empire, later joined by Bulgaria, fought against Britain, France, and Russia, later joined by Italy and the United States. By 1918 Germany and its allies were defeated.

During the 1920s Austrian war veteran Adolf Hitler led a new political party in Germany called the Nazis. The Nazis took power in 1933. In the late 1930s Germany invaded Austria, Czechoslovakia, and finally Poland, beginning World War II. By 1942 Germany and Italy had conquered most of Europe. The Nazis forced many people from the occupied countries into concentration camps to be enslaved or killed. About 6 million Jews and millions of other people were murdered in a mass killing called the **Holocaust**.

To defeat Germany, several countries formed an alliance. These Allies included Britain, the Soviet Union, the United States, and many others. The Allies defeated Germany in 1945. Germany and its capital, Berlin, were divided into Soviet, French, British, and U.S. occupation zones. Britain, France, and the United States later combined their zones to create a democratic West Germany with its capital at Bonn. In its zone, the Soviet Union set up the Communist country of East Germany with an unlimited totalitarian government. Its capital became East Berlin; however, West Berlin became part of West Germany. In 1961 the East German government built the Berlin Wall across the city to stop East Germans from escaping to the West.

Reunification and Modern Government West Germany's roads, cities, railroads, and industries were rebuilt after the war with U.S. financial aid. East Germany was also rebuilt, but it was not as prosperous as West Germany. Unlike the West German government, the East German government allowed people very little freedom. Also, its command economy—managed by the government—was less productive than the free enterprise, market system of West Germany. In the late 1980s East Germans and people throughout Eastern Europe demanded democratic reform. In 1989 the Berlin Wall was torn down. In 1990 East and West Germany reunited. Germany's capital again became Berlin. Today, all Germans enjoy democratic rights. A parliament elects the president and prime minister, or **chancellor**. Germany is a member of the EU and NATO.

✓ **READING CHECK:** *Human Systems* How were the economies of East and West Germany organized following World War II? **East Germany—command economy controlled by a communist government; West Germany—free enterprise, market-based economy**

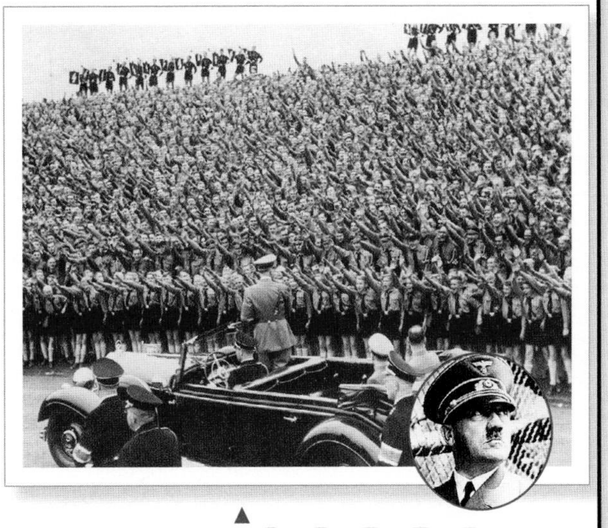

German youth salute Adolf Hitler in Nürnberg in 1938.

For nearly 30 years the Berlin Wall separated East and West Berlin. Many people in West Berlin protested by painting graffiti on the wall.

Interpreting the Visual Record Why would the government make the wall solid instead of a barrier that would allow people visual access?

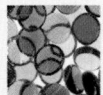

Cultural Kaleidoscope

The Jews of Berlin When Adolf Hitler came to power in 1933, more than 170,000 Jews lived in Berlin. By 1945 Berlin's Jewish population stood at some 5,000. The vast majority of Berlin's Jews had fled or been killed.

Now Berlin's Jewish community is growing again. Many of the city's new Jewish residents are emigrants from Russia. New Jewish shops, restaurants, schools, and even a chess team have been established. A new Jewish museum has opened. A synagogue that had been damaged by the Nazis and almost destroyed by Allied bombs has been restored.

Activity: Have each student conduct research on the history of Jews in Germany and create a time line to depict major events.

Visual Record Answer

Possible answer: The government did not want people to communicate through the wall or see that conditions were better on the other side.

Teaching Objective 2

ALL LEVELS: (Suggested time: 30 min.) Tell students to imagine that they have been hired to design a Web page that highlights Germany's major contributions to world culture. Then have each student design a mock Web page that includes headlines and links. *(Web pages should mention the invention of movable metal type, the development of classical music, including the work of Bach and Beethoven as well as the operas of Wagner, and German traditions in literature and the arts.)* Ask volunteers to present their Web pages to the class. **ENGLISH LANGUAGE LEARNERS**

Teaching Objective 3

LEVEL 1: (Suggested time: 15 min.) Tell students to imagine that they are journalists who are writing about how the division of Germany affected the country's economy. Then pair students and have pairs write three headlines that describe the effects of the country's division on the economy. Have volunteers read their headlines to the class. **ENGLISH LANGUAGE LEARNERS, COOPERATIVE LEARNING**

LEVELS 2 AND 3: (Suggested time: 30 min.) Have each student create an editorial cartoon that might accompany an article printed below one of the headlines from the Level 1 activity.

Across the Curriculum

ART

Landscape Painting Some of the world's greatest landscape artists have been from west-central Europe. Among these are the Dutch painter Vincent van Gogh (1853–90), French artist Antoine Watteau (1684–1721), and German painter Caspar David Friedrich (1774–1840).

Activity: Have students choose a landscape artist of the 1700s or 1800s from west-central Europe. Ask them to compare how the artist portrayed the region's landscape with photographs they find in the text and other sources. Are the physical features painted realistically? Was the painting meant to be realistic? Have students write a paragraph or two comparing and contrasting their chosen painter's depiction of the landscape with the way the landscape appears in photographs.

internet connect

GO TO: go.hrw.com
KEYWORD: SK3 CH11
FOR: Web sites about landscapes

Visual Record Answer

It is a festive, celebratory time.

Chart Answer (p. 259)

Answers will vary, but students might say that Germans value education.

Culture

About 34 percent of Germans are Roman Catholic, and 38 percent are Protestant. Most other Germans have no religious association. Many of these people are from eastern Germany, where the communist government suppressed religion from 1945 to 1990.

Diversity About 90 percent of Germany's inhabitants are ethnic Germans. However, significant numbers of Turks, Poles, and Italians have come to Germany to live and work. These "guest workers" do not have German citizenship. Germany has also taken in thousands of refugees from Eastern Europe during the last 50 years.

Customs Traditional German food emphasizes the products of the forests, farms, and seacoasts. Each region produces its own varieties of sausage, cheese, wine, and beer. German celebrations include Oktoberfest; *Sangerfast*, a singing festival; and *Fastnacht*, a religious celebration. The major German festival season is Christmas. The Germans began the custom of bringing an evergreen tree indoors at Christmas and decorating it with candles.

The Arts and Literature Germany has a great tradition of literature, music, and the arts. The first European to print books using movable metal type was a German, Johannes Gutenberg. In the 1700s and 1800s, Germany led Europe in the development of classical music. World-famous German composers include Johann Sebastian Bach and Ludwig van Beethoven. The operas of Richard Wagner revived the folktales of ancient Germany.

✓ **READING CHECK:** *Human Systems* What technology and other contributions have Germans made to world culture? movable type, classical music

Crowds gather in a German town for a Christmas market. Christmas markets have been popular in Germany for more than 400 years. From the beginning of Advent until Christmas, booths are set up on the market place in most cities. Here people can buy trees, decorations, and gifts.

Interpreting the Visual Record What does this photo suggest about the importance of Christmas in Germany?

CLOSE

Play for the class a work by one of the great German composers, such as Johann Sebastian Bach, Ludwig van Beethoven, or Johannes Brahms—sometimes called the "Three Bs."

REVIEW AND ASSESS

Have students complete the Section Review. Then pair students and have each student write five questions based on Section 3 information. Have students exchange questions with their partners and answer them. Then have students complete Daily Quiz 11.3. **COOPERATIVE LEARNING**

RETEACH

Have students complete Main Idea Activity 11.3. Then organize students into small groups and have each group create a poster to show how one of the six essential elements of geography applies to a topic in Section 3. Have volunteers present their posters to the class. **ENGLISH LANGUAGE LEARNERS, COOPERATIVE LEARNING**

EXTEND

Have interested students conduct research on how German immigrants have influenced the customs, religion, politics, language, and food of a specific U.S. region. Then have each student write a short play to share his or her findings. **BLOCK SCHEDULING**

Germany Today

Germany has a population of 83 million, more people than any other European country. Germany also has Europe's largest economy. Nearly one fourth of all goods and services produced by the EU come from Germany.

Economy Ample resources, labor, and capital have made Germany one of the world's leading industrial countries. The nation exports a wide variety of products. You may be familiar with German automakers like Volkswagen, Mercedes-Benz, and BMW. The German government provides education, medical care, and pensions for its citizens, but Germans pay high taxes. Unemployment is high. Many immigrants work at low-wage jobs. These "guest workers" are not German citizens and cannot receive many government benefits. Since reunification, Germany has struggled to modernize the industries, housing, and other facilities of the former East Germany.

Germany

Country	Population/ Growth Rate	Life Expectancy	Literacy Rate	Per Capita GDP
Germany	83,029,536 / 0.3%	74, male 81, female	99%	$23,400
United States	281,421,906 / 0.9%	74, male 80, female	97%	$36,200

Sources: Central Intelligence Agency, *The World Factbook 2001;* U.S. Census Bureau

Interpreting the Chart **What might the literacy rate of Germany suggest about its culture?**

Cities Germany's capital city, Berlin, is a large city with wide boulevards and many parks. Berlin was isolated and economically restricted during the decades after World War II. However, Germans are now rebuilding their new capital to its former splendor.

Near the Rhine River and the coal fields of Western Germany is a huge cluster of cities, including Essen and Düsseldorf. They form Germany's largest industrial district, the Ruhr. Frankfurt is a city known for banking and finance. Munich is a manufacturing center. Other important cities include Hamburg, Bremen, Cologne, and Stuttgart.

✓ **READING CHECK:** *Human Systems* How did the division of Germany affect its economy? **It delayed the modernization of East German industry, housing, and other facilities.**

go.hrw.com Homework Practice Online
Keyword: SK3 HP11

Section 3 Review

Define and explain: Reformation, Holocaust, chancellor

Working with Sketch Maps On the map you created in Section 2, label Berlin, Bonn, Essen, Frankfurt, Munich, Hamburg, and Cologne.

Reading for the Main Idea

1. *Human Systems* How did wars affect the development of Germany in the 1900s?
2. *Human Systems* What are some notable features of the German economy?

Critical Thinking

3. **Drawing Inferences and Conclusions** How has Germany's history influenced the religious makeup of the population?
4. **Summarizing** What have been some results of the unification of Germany in 1990?

Organizing What You Know

5. **Sequencing** Create a time line listing key events in the history of Germany from 1000 B.C. to 1990.

1000 B.C. ———————— A.D. 1990

Section 3 Review Answers

Define For definitions, see: Reformation, p. 286; Holocaust, p. 257; chancellor, p. 257

Working with Sketch Maps Maps will vary, but listed places should be labeled in their approximate locations.

Reading for the Main Idea

1. World War I—lost territory, had to pay heavy fines: World War II—Jewish population nearly wiped out, country divided (NGS 13)
2. leading industrial producer with many notable exports, including automobiles (NGS 11)

Critical Thinking

3. Reformation gave rise to large Protestant population; Holocaust killed all but small number of Jews.
4. capital was moved from Bonn to Berlin, and country struggling to modernize eastern zone

Organizing What You Know

5. 800s—Franks established Holy Roman Empire; 1500s—Reformation; 1618–48—Thirty Years' War; late 1800s—Germany united; 1914–18—World War I; 1933—Nazis seize power; late 1930s–1945— World War II; 1961—Berlin Wall built; 1989—wall torn down; 1990—Germany reunited

Section 4

Objectives

1. Investigate how the Benelux countries were influenced by larger countries.
2. Describe what the region's culture is like.
3. Discuss what the Benelux countries are like today.

Focus

LET'S GET STARTED

Copy the following question onto the chalkboard: *If you lived at the seashore and wanted to recover some of the land under the water, how might you do it?* Ask volunteers to explain their ideas or draw them on the chalkboard. Tell students that the people of the Netherlands have succeeded in reclaiming land from the sea by building earthen walls to keep out the seawater and then pumping water from reclaimed lands. Tell students that in Section 4 they will learn more about the Netherlands and the other Benelux countries.

Building Vocabulary

Write the word **cosmopolitan** on the chalkboard. Remind students that the Greek word *polis* referred to city-states. *Cosmos* means universe. Ask students to infer the term's meaning. Then read the definition aloud.

Section 4 Resources

Reproducible
- Guided Reading Strategy 11.4
- Creative Strategies for Teaching World Geography, Lessons 10 and 11

Technology
- One-Stop Planner CD–ROM, Lesson 11.4
- Homework Practice Online
- HRW Go site

Reinforcement, Review, and Assessment
- Section 4 Review, p. 262
- Daily Quiz 11.4
- Main Idea Activity 11.4
- English Audio Summary 11.4
- Spanish Audio Summary 11.4

Visual Record Answer

It is located at the mouth of a major river and is on the North Sea.

Section 4 The Benelux Countries

Read to Discover

1. How were the Benelux countries influenced by larger countries?
2. What is this region's culture like?
3. What are the Benelux countries like today?

Define

cosmopolitan

Locate

Flanders
Wallonia
Amsterdam
Antwerp
Brussels

WHY IT MATTERS

The Benelux countries are key members of the European Union (EU). Use CNNfyi.com or other **current events** sources to find out more about the membership, functions, and goals of the EU. Record your findings in your journal.

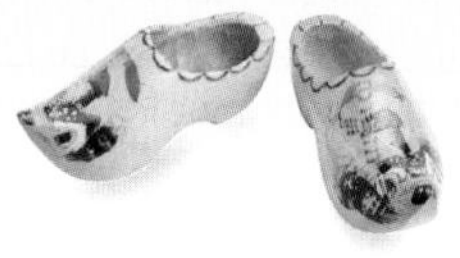

Dutch wooden shoes

The Dutch city of Rotterdam is one of the world's busiest ports.

Interpreting the Visual Record **Why might this city be an important transportation center?**

History

Celtic and Germanic tribes once lived in this region, as in most of west-central Europe. They were conquered by the Romans. After the fall of the Roman Empire and the conquests of Charlemagne, the region was ruled alternately by French rulers and by the Holy Roman emperor.

In 1555 the Holy Roman emperor presented the Low Countries to his son, King Philip II of Spain. In the 1570s the Protestants of the Netherlands won their freedom from Spanish rule. Soon after, the Netherlands became a great naval and colonial power. Belgium had been ruled at times by France and the Netherlands. However, by 1830 Belgium had broken away to become an independent kingdom.

Both world wars scarred this region. Many of the major battles of World War I were fought in Belgium. Then in World War II Germany occupied the Low Countries. In 1949 Belgium, the Netherlands, and Luxembourg were founding members of NATO. Later they joined the EU. Today, each of the three countries is ruled by a parliament and a monarch. The monarchs' duties are mostly ceremonial. The Netherlands controls several Caribbean islands. However, its former colonies in Asia and South America are now independent.

✓ **READING CHECK:** *Human Systems* How are the governments of the Benelux countries organized? Each is ruled by a parliament and a monarch, who has mostly ceremonial duties.

TEACH

Teaching Objectives 1–3

ALL LEVELS: (Suggested time: 40 min.) Copy the graphic organizer at the right onto the chalkboard, omitting the italicized answers. Have each student complete it by filling in information about the influence of larger countries on the Benelux countries and the region's culture. Then lead a discussion based on how students might spend their day if they lived in the Netherlands, Belgium, or Luxembourg.

ENGLISH LANGUAGE LEARNERS

➤ASSIGNMENT: Have each student write a paragraph that analyzes the economies of the Benelux region and proposes a reason why the Benelux countries might have become cosmopolitan.

Benelux Countries

Influences from Larger Countries	Culture of the Benelux Countries
• *Ruled by France and the Holy Roman Empire* • *Netherlands ruled by Spain* • *Belgium ruled by France and the Netherlands* • *WWI battles fought in Belgium* • *Low Countries occupied by Germany during WWII*	• *Belgians and Luxembourgers predominantly Roman Catholic* • *Dutch evenly divided between Catholics, Protestants, and nonreligious persons* • *Dutch spoken in Netherlands* • *Flemish and French spoken in Belgium*

CONNECTING TO *Technology*

Dutch Polders

A polder in the Netherlands

Much of the Netherlands lies below sea level and was once covered with water. For at least 2,000 years, the Dutch have been holding back the sea. First they lived on raised earthen mounds. Later they built walls or dikes to keep the water out. After building dikes, the Dutch installed windmills to pump the water out of reclaimed areas, called polders.

Using this system, the Dutch have reclaimed large amounts of land. Cities like Amsterdam and Rotterdam sit on reclaimed land. The dike and polder system has become highly sophisticated. Electric pumps have largely replaced windmills, and dikes now extend along much of the country's coastline. However, this system is difficult to maintain. It requires frequent and expensive repairs. Creating polders has also produced sinking lowlands and other environmental damage. As a result, the Dutch are considering changes to the system. These changes might include restoring some of the polders to wetlands and lakes.

Understanding What You Read

1. What are polders?
2. How did the Dutch use technology to live on land previously under water?

Culture

The people of Luxembourg and Belgium are mostly Roman Catholic. The Netherlands is more evenly divided among Catholic, Protestant, and those who have no religious ties. Dutch is the language of the Netherlands. Flemish is a language related to Dutch that is spoken in Flanders, the northern part of Belgium. Belgium's coast and southern interior are called Wallonia. People in Wallonia speak mostly French and are called Walloons. In the past, cultural differences between Flemish and Walloons have produced conflict in Belgium. Today street signs and other notices are often printed in both Flemish and French. The Benelux countries are also home to immigrants from Asia and Africa.

These children in Brussels, Belgium, are wearing traditional clothing. ▼

PLACES AND REGIONS

Antwerp, Belgium, is one of the world's four leading diamond-cutting centers. (The others are New York, Tel Aviv, and Mumbai.) According to legend, the first diamond was cut in Antwerp in 1476. Since the 1500s establishments that cut and deal in diamonds have flourished in a neighborhood near the central train station.

At Antwerp's Diamond Museum, employees demonstrate the art of cutting and polishing diamonds. Priceless jewelry sparkles in the museum's treasure chamber.

Activity: Have students conduct research on the diamond industry. Tell students to create a map showing sources, processing centers, and major markets.

Connecting to Technology

Answers

1. areas reclaimed from the sea
2. They have used dikes, windmills, and electric pumps to reclaim land.

Section 4 Review

Answers

Define For definition, see: cosmopolitan, p. 262

Working with Sketch Maps Maps will vary, but listed places should be labeled in their approximate locations.

CLOSE

Tell students about the "Tulip Mania" of the 1600s. Tulips were brought to the Netherlands from Turkey in the 1550s and soon became very popular. People speculated on tulip prices—they bought bulbs planning to resell them at higher prices. At the height of the craze, a single bulb of a prized variety could cost 4,000 guilders—as valuable as a ship loaded with cargo! When the tulip market crashed in 1637, many people went bankrupt.

REVIEW AND ASSESS

Have students complete the Section Review. Then pair students and have each pair write two or three verses for a new national anthem for one of the Benelux countries. Verses should incorporate information from Section 4. Ask volunteers to share their work with the class. Have students complete Daily Quiz 11.4. **COOPERATIVE LEARNING**

RETEACH

Have students complete Main Idea Activity 11.4. Then ask each student to choose one of the Benelux countries. Have each student use the textbook to compile a list of the three most important things to know about his or her chosen country. Ask volunteers to share their lists with the class. **ENGLISH LANGUAGE LEARNERS**

EXTEND

Have interested students conduct research on the mix of cultures in Belgium or another Benelux country. Then have students write short stories based on their research. **BLOCK SCHEDULING**

Reading for the Main Idea

1. Belgium and Luxembourg are predominantly Roman Catholic; Netherlands has more equal proportion of Catholics, Protestants, and nonreligious people; Dutch spoken in Netherlands; Flemish and French spoken in Belgium; Belgium and Netherlands have famous artistic traditions. (NGS 10)
2. They speak different languages and have different religious traditions and different economies. (NGS 10)

Critical Thinking

3. The Benelux countries are small and rely on international trade for their income and to obtain resources such as oil.
4. because of cultural differences

Organizing What You Know

5. Belgium—cheese, chocolates, cocoa, and diamond cutting; Luxembourg—service industries such as banking; Netherlands—flowers, cheese, chocolates, cocoa, diamond cutting, and oil refining

Chart Answer ▲

Luxembourg's

Benelux Countries

COUNTRY	POPULATION/ GROWTH RATE	LIFE EXPECTANCY	LITERACY RATE	PER CAPITA GDP
Belgium	10,258,762 / 0.2%	75, male 81, female	98%	$25,300
Luxembourg	442,972 / 1.3%	74, male 81, female	100%	$36,400
Netherlands	15,807,641 / 0.5%	75, male 81, female	99%	$24,400
United States	281,421,906 / 0.9%	74, male 80, female	97%	$36,200

Sources: Central Intelligence Agency, *The World Factbook 2001;* U.S. Census Bureau

Interpreting the Chart **Which country's per capita GDP is closest to that of the United States?**

The region's foods include dairy products, fish, and sausage. The Dutch spice trade led to dishes flavored with spices from Southeast Asia. The Belgians claim they invented french fries, which they eat with mayonnaise.

The Netherlands and Belgium have been world leaders in fine art. In the 1400s and 1500s, Flemish artists painted realistic portraits and landscapes. Dutch painters like Rembrandt and Jan Vermeer experimented with different qualities of light. In the 1800s Dutch painter Vincent van Gogh portrayed southern France with bold brush strokes and bright colors.

✓ **READING CHECK:** *Human Systems* What is the relationship between cultures in Belgium? many different cultures; in the past, Walloons and Flemish have had conflict but now cooperate

The Benelux Countries Today

The Netherlands is famous for its flowers, particularly tulips. Belgium and the Netherlands export cheeses, chocolate, and cocoa. Amsterdam and Antwerp, Belgium, are major diamond-cutting centers. The Netherlands also imports and refines oil. Luxembourg earns much of its income from services such as banking. The region also produces steel, chemicals, and machines. Its **cosmopolitan** cities are centers of international business and government. A cosmopolitan city is one that has many foreign influences. Brussels, Belgium, is the headquarters for many international organizations such as the EU and NATO.

✓ **READING CHECK:** *Places and Regions* What are the Benelux countries like today? They are major exporters and are very cosmopolitan.

Section 4 Review

go.hrw.com Homework Practice Online Keyword: SK3 HP11

Define and explain: cosmopolitan

Working with Sketch Maps On the map you created in Section 3, label Flanders, Wallonia, Amsterdam, Antwerp, and Brussels.

Reading for the Main Idea

1. *Human Systems* What are the main cultural features of the Benelux countries?
2. *Human Systems* In what ways do the people of the Benelux countries differ?

Critical Thinking

3. **Drawing Inferences and Conclusions** Why might the economies of the Benelux countries be dependent on international trade?
4. **Analyzing Information** Why have groups in Belgium been in conflict?

Organizing What You Know

5. **Comparing** Copy the following graphic organizer. Use it to compare the Benelux countries' industries.

Belgium	Luxembourg	Netherlands

Section 5

Objectives

1. Identify some of the major events in the history of the Alpine countries.
2. Describe some of the cultural features of the region.
3. Compare the economies of Switzerland and Austria.

FOCUS

LET'S GET STARTED

Copy the following passage onto the chalkboard: *Look up* cacao *in the textbook's index. What is it? Where is it grown? Why might we think of Switzerland when we think of chocolate?* Discuss responses. Tell students that in 1876 a Swiss man added concentrated milk to chocolate for the first time, forming milk chocolate. Today Switzerland imports the raw materials for making chocolate and exports the finished product. Tell students that in Section 5 they will learn more about Switzerland and Austria.

Building Vocabulary

Copy the terms **cantons** and **nationalism** onto the chalkboard and call on volunteers to read the definitions aloud from the glossary. Ask students what other names they know that refer to political divisions of territory. *(Examples: state, province, territory, county) Canton* comes from a Latin word meaning "corner." So, students might think of a canton as a corner of Switzerland. Nationalism contains the word *nation.* People who want to form their own country or nation are nationalists.

Section 5 The Alpine Countries

Read to Discover

1. What are some of the major events in the history of the Alpine countries?
2. What are some cultural features of this region?
3. How are the economies of Switzerland and Austria similar?

Define

cantons
nationalism

Locate

Geneva	Zurich
Salzburg	Basel
Vienna	Bern

Swiss cuckoo clock

WHY IT MATTERS

When people think of chocolate they often think of Switzerland, one of the world's leading producers of chocolate. Use CNNfyi.com or other **current events** sources to find out more about chocolate. Record your findings in your journal.

History

Austria and Switzerland share a history of Celtic occupation, Roman and Germanic invasions, and rule by the Holy Roman Empire.

Switzerland Swiss **cantons**, or districts, gradually broke away from the Holy Roman Empire, and in the 1600s Switzerland became independent. Today Switzerland is a confederation of 26 cantons. Each controls its own internal affairs, and the national government handles defense and international relations. Switzerland's location in the high Alps has allowed it to remain somewhat separate from the rest of Europe. It has remained neutral in the European wars of the last two centuries. Switzerland has not joined the United Nations, EU, or NATO. Because of this neutrality, the Swiss city of Geneva is home to many international organizations.

Austria During the Middle Ages, Austria was a border region of Germany. This region was the home of the Habsburgs, a powerful family of German nobles. From the 1400s onward the Holy Roman emperor was always a Habsburg. At the height of their power the Habsburgs ruled Spain and the Netherlands, as well as large areas of Germany, eastern Europe, and Italy. This empire included different

The International Red Cross in Geneva has offices in almost every country in the world.

Interpreting the Visual Record What symbol of the Red Cross is displayed on this building?

Section 5 RESOURCES

Reproducible

- Guided Reading Strategy 11.5
- Creative Strategies for Teaching World Geography, Lessons 10 and 11

Technology

- One-Stop Planner CD–ROM, Lesson 11.5
- Homework Practice Online
- *ARGWorld* CD–ROM: Ski Resorts in Switzerland
- HRW Go site

Reinforcement, Review, and Assessment

- Section 5 Review, p. 265
- Daily Quiz 11.5
- Main Idea Activity 11.5
- English Audio Summary 11.5
- Spanish Audio Summary 11.5

◄ Visual Record Answer

the flag

TEACH

Teaching Objectives 1–3

ALL LEVELS: (Suggested time: 30 min.) Copy the following graphic organizer onto the chalkboard, omitting the italicized answers. Pair students and have each pair complete the organizer with major events in the history of the Alpine countries, cultural features of each country, and economic features of each country. Ask volunteers to share their answers with the class. **ENGLISH LANGUAGE LEARNERS, COOPERATIVE LEARNING**

COMPARING AUSTRIA AND SWITZERLAND

	Austria	Switzerland
History	*Invasion by Celts, Romans, and Germanic tribes; ruled by the Holy Roman Empire; part of Habsburg, Austrian, and Austro-Hungarian Empires; became republic; annexed by Germany; became republic*	*invasion by Celts, Romans, and Germanic tribes; ruled by the Holy Roman Empire; gained independence in the 1600s*
Culture	*Predominantly Roman Catholic and German-speaking with small minority of Slovenes and Croatians; known for classical music*	*Majority of Swiss are Roman Catholic or Protestant.*
Economy	*Dairy products, including cheese; Vienna is Austria's commercial and industrial center.*	*Dairy products, including cheese; manufacturer of watches, optical instruments, and other machinery; Zurich is the center of Swiss banking.*

Section 5 Review

Answers

Define For definitions, see: cantons, p. 263; nationalism, p. 264

Working with Sketch Maps Maps will vary, but listed places should be labeled in their approximate locations.

Reading for the Main Idea

1. Switzerland—independence in 1600s, neutral; Austria—part of larger empires, republic, annexed by Germany, and republic (NGS 13)
2. Swiss—three languages; Austrian—mainly Roman Catholic, German-speaking, Slovene and Croatian minorities, known for music (NGS 10)

Critical Thinking

3. Switzerland's mountains helped the country remain neutral.
4. Napoléon's conquest led to end of Holy Roman Empire; Germany drew Austria into World War II.

Organizing What You Know

5. Students should mention the elements of culture, language, economics, and history in the section.

Visual Record Answers

It flows through many countries.

The Danube River passes through Vienna, the capital of Austria.

Interpreting the Visual Record How does this river influence movement and trade?

ethnic groups, each with its own language, government, and system of laws. The empire was united only in its allegiance to the emperor and in its defense of the Roman Catholic religion.

With the conquests of Napoléon after 1800, the Holy Roman Empire was formally eliminated. It was replaced with the Austrian Empire, which was also under Habsburg control. When Napoléon was defeated, the Austrian Empire became the dominant power in central Europe.

Through the 1800s the diverse peoples of the empire began to develop **nationalism**, or a demand for self-rule. In 1867 the Austrians agreed to share political power with the Hungarians. The Austrian Empire became the Austro-Hungarian Empire. After World War I the empire was dissolved. Austria and Hungary became separate countries. Shortly before World War II the Germans took over Austria and made it part of Germany. After the war, the Allies occupied Austria. Today Austria is an independent member of the EU.

Austrians wearing carved wooden masks celebrate the return of spring and milder weather.

✓ **READING CHECK:** *Human Systems* What were the major events in the history of the Alpine countries? **Switzerland—breaking away from Holy Roman Empire; end of that empire, beginning of Austrian Empire, shared power with Hungarians, World Wars I and II**

Culture

About 46 percent of the population in Switzerland is Roman Catholic, and 40 percent is Protestant. Austria's population is mainly Roman Catholic. Only about 5 percent of its people are Protestant, while 17 percent follow Islam or other religions.

Languages and Diversity About 64 percent of Swiss speak German, 19 percent speak French, and 8 percent speak Italian. Small groups in the southeast speak a language called Romansh. Other European languages are also spoken in Switzerland. Austria is almost entirely German-speaking, but contains small minorities of Slovenes and Croatians.

Customs Christmas is a major festival in both countries. People make special cakes and cookies at this time. In rural parts of Switzerland people take cattle up to the high mountains in late spring

CLOSE

Refer students to the comments in the chapter introduction. Have each student write a similar self-introduction for a boy or girl from Switzerland or Austria, using information from Section 5. Ask volunteers to share their introductions with the class.

REVIEW AND ASSESS

Have students complete the Section Review questions. Then have students work in pairs to create flash cards of Define and Locate terms and key concepts from the section. Ask them to quiz each other using the flash cards. Then have students complete Daily Quiz 11.5.

RETEACH

Have students complete Main Idea Activity 11.5. Then have each student choose and illustrate a major concept from Section 5. Display and discuss the illustrations. **ENGLISH LANGUAGE LEARNERS, COOPERATIVE LEARNING**

EXTEND

Have interested students conduct research on how Switzerland has been able to maintain neutrality since the 1500s. Students might focus on whether or how the country's physical geography has helped it to maintain neutrality.

and return in the fall. Their return is celebrated by decorating homes and cows' horns with flowers. A special feast is also prepared.

The Alpine region is particularly well known for its music. In the 1700s Mozart wrote symphonies and operas in the Austrian city of Salzburg. Every year a music festival is held there in his honor. Austria's capital, Vienna, is also known as a center of music and fine art.

✓ **READING CHECK:** *Human Systems* What role have the arts played in this region? The region is well known for its music.

The Alpine Countries

Country	Population/ Growth Rate	Life Expectancy	Literacy Rate	Per Capita GDP
Austria	8,150,835 / 0.2%	75, male 81, female	98%	$25,500
Switzerland	7,283,274 / 0.3%	77, male 83, female	99%	$28,600
United States	281,421,906 / 0.9%	74, male 80, female	97%	$36,200

Sources: Central Intelligence Agency, *The World Factbook 2001;* U.S. Census Bureau

Interpreting the Chart How do the populations of the Alpine countries compare with that of the United States?

The Alpine Countries Today

Switzerland and Austria both produce dairy products, including many kinds of cheese. Switzerland is also famous for the manufacturing of watches, optical instruments, and other machinery. Swiss chemists discovered how to make chocolate bars. Switzerland is a major producer of chocolate, although it must import the cocoa beans.

Switzerland and Austria are linked to the rest of Europe by excellent highways, trains, and airports. Several long tunnels allow trains and cars to pass through mountains in the Swiss Alps. Both countries attract many tourists with their mountain scenery, lakes, and ski slopes.

Located on the Danube, Vienna is Austria's commercial and industrial center. Switzerland's two largest cities are both in the German-speaking north. Zurich is a banking center, while Basel is the starting point for travel down the Rhine to the North Sea. Switzerland's capital is Bern, and Geneva is located in the west.

✓ **READING CHECK:** *Human Systems* How are the economies of Switzerland and Austria similar? They both produce dairy products and attract tourists.

Section 5 Review

go.hrw.com Homework Practice Online Keyword: SK3 HP11

Define and explain: cantons, nationalism

Working with Sketch Maps On the map you created in Section 4, label Geneva, Salzburg, Vienna, Zurich, Basel, and Bern.

Reading for the Main Idea

1. *Human Systems* What were the main events in the history of the Alpine countries?
2. *Human Systems* What are some notable aspects of Swiss and Austrian culture?

Critical Thinking

3. **Drawing Inferences and Conclusions** How might geography have been a factor in Switzerland's historical neutrality?
4. **Drawing Inferences and Conclusions** How have foreign invasions of Austria shaped its history?

Organizing What You Know

5. **Comparing/Contrasting** Use this graphic organizer to compare and contrast the culture, language, economies, and history of Switzerland and Austria.

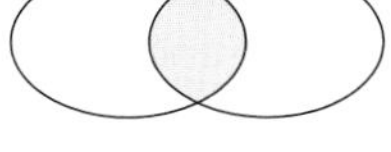

Chapter 11 Review Answers

Building Vocabulary

For definitions, see: navigable, p. 250; loess, p. 251; medieval, p. 253; impressionism, p. 255; Reformation, p. 256; Holocaust, p. 257; chancellor, p. 287; cosmopolitan, p. 262; cantons, p. 263; nationalism, p. 264

Reviewing the Main Ideas

1. marine west coast climate with cold and rainy winters but mild summers; Alps have a highland climate (NGS 4)
2. led to the Thirty Years' War in which nearly one third of the population died (NGS 13)
3. coal, natural gas, nuclear power, and hydroelectric power (NGS 4)
4. Students should name five of the following: Monet, Renoir, Degas, Cézanne, Matisse, Rembrandt, Vermeer, van Gogh, Bach, Beethoven, Wagner, or Mozart. (NGS 10)
5. NATO and the EU (NGS 11, 13)

▲ **Chart Answer**

They are much smaller.

ASSESS

Have students complete the Chapter 11 Test.

RETEACH

Copy the web created for the Launch into Learning activity. Erase or cover some of the labels and connections. Duplicate the altered web and give each student a copy. Have students work in pairs to fill in the missing information and provide additional links and labels. Display and discuss the webs. **ENGLISH LANGUAGE LEARNERS, COOPERATIVE LEARNING**

PORTFOLIO EXTENSIONS

PORTFOLIO

1. Avalanches are a constant threat in the Austrian and Swiss Alps. Have students conduct research on Alpine avalanche control measures. They might build scale models to demonstrate the most common methods. Cotton batting can substitute for snow. Place photos of the models and written descriptions of the project in portfolios.
2. Have students choose one of the countries of west-central Europe for a virtual vacation. Allot a specific amount of money for students to spend on their trip. Have them plan transportation, accommodations, food, sightseeing, and souvenirs within their budget. Have students prepare an itinerary with the costs for each major expenditure. You may want to have students find prices in each country's currency and convert the prices to dollars.

CHAPTER 11

Review ANSWERS

Understanding Environment and Society

Reports will vary, but the information included should address the topics suggested. Use Rubric 42, Writing to Inform, to evaluate student work.

Thinking Critically

1. Rivers have facilitated trade; the Alps have somewhat hindered travel and trade.
2. by building tunnels through the Alps
3. the Rhine River, Jura Mountains, Pyrenees, and Brittany; borders with Belgium, Luxembourg, and part of the border with Germany
4. The EU has its headquarters in Brussels.
5. low population growth, long life expectancy, high literacy rate, and high incomes; each of the nations is economically developed

CHAPTER 11

Reviewing What You Know

Building Vocabulary

On a separate sheet of paper, write sentences to define each of the following words.

1. navigable
2. loess
3. medieval
4. impressionism
5. Reformation
6. Holocaust
7. chancellor
8. cosmopolitan
9. cantons
10. nationalism

Reviewing the Main Ideas

1. *Places and Regions* What is the climate of west-central Europe like?
2. *Human Systems* What were the effects of religious conflicts on Germany?
3. *Places and Regions* What energy resources are available to the countries of this region?
4. *Human Systems* Name five artists from west-central Europe who have made notable contributions to culture.
5. *Human Systems* What international organizations have created ties between countries of west-central Europe since World War II?

Understanding Environment and Society

Cleaning up Pollution

Create a presentation on East German industries' pollution of the environment. Include a chart, graph, database, model, or map showing patterns of pollutants, as well as the German government's clean-up effort. Consider the following:

- Pollution in the former East Germany.
- Pollution today.
- Costs of stricter environmental laws.

Write a five-question quiz, with answers about your presentation to challenge fellow students.

Thinking Critically

1. **Drawing Inferences and Conclusions** What geographic features have encouraged travel and trade in west-central Europe? What geographic features have hindered travel and trade?
2. **Finding the Main Idea** How have the people of Switzerland altered their environment?
3. **Analyzing Information** What landform regions give France natural borders? Which French borders do not coincide with physical features?
4. **Drawing Inferences and Conclusions** Why might Brussels, Belgium, be called the capital of Europe?
5. **Comparing** What demographic factors are shared by all countries of west-central Europe today? How do they reflect levels of economic development?

Food Festival

The countries of west-central Europe are famous for their cheese. Point out that cheese comes in hundreds of varieties, based on these factors and others: the type of milk used to make the cheese (cow, goat, or sheep); fresh or ripened; soft, hard, semisoft, or semifirm; and the herbs, spices, mold spores or bacteria that have been added to the cheese. Students could search the supermarket for different types of cheese from west-central Europe, and the class could hold a cheese-tasting party. Serve French or German breads and a beverage with the cheese.

Chapter 11 Review and Assessment Resources

Reproducible
- Readings in World Geography, History, and Culture 31, 32, and 33
- Vocabulary Activity 11

Technology
- Chapter 11 Test Generator (on the One-Stop Planner)
- Audio CD Program, Chapter 11

Reinforcement, Review, and Assessment
- Chapter 14 Review, pp. 266–67
- Chapter 11 Tutorial for Students, Parents, Mentors, and Peers
- Chapter 11 Test
- Chapter 11 Test for English Language Learners and Special-Needs Students

Building Social Studies Skills

Map Activity

On a separate sheet of paper, match the letters on the map with their correct labels.

Northern European Plain	Danube River
Pyrenees	North Sea
Alps	Mediterranean Sea
Seine River	Paris
Rhine River	Berlin

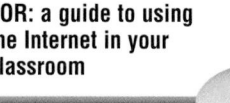

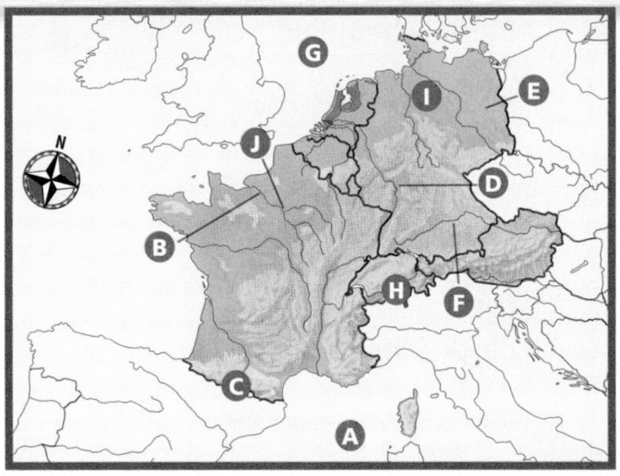

Mental Mapping Skills Activity

On a separate sheet of paper, draw a freehand map of west-central Europe. Make a key for your map and label the following:

Austria	Luxembourg
Belgium	the Netherlands
France	North Sea
Germany	Switzerland

Writing Activity

Imagine you are taking a boat tour down the Rhine River. You will travel from Basel, Switzerland, to Rotterdam, the Netherlands. Keep a journal describing the places you see and the stops you make. Be sure to use standard grammar, spelling, sentence structure, and punctuation.

Alternative Assessment

Portfolio Activity

Learning About Your Local Geography

Agricultural Products Some countries in west-central Europe are major exporters of agricultural products. Research the agricultural products of your state. Draw a map that shows important products and areas they are produced.

internet connect

Internet Activity: go.hrw.com
KEYWORD: SK3 GT11

Choose a topic to explore about west-central Europe:
- Tour the land and rivers of Europe.
- Travel back in time to the Middle Ages.
- Visit Belgian and Dutch schools.

Map Activity

A. Mediterranean Sea
B. Seine River
C. Pyrenees
D. Rhine River
E. Berlin
F. Danube River
G. North Sea
H. Alps
I. Northern European Plain
J. Paris

Mental Mapping Skills Activity

Maps will vary, but listed places should be labeled in their approximate locations.

Writing Activity

Journals will vary, but the place descriptions should be consistent with text material. Use Rubric 15, Journals, to evaluate student work.

Portfolio Activity

Maps will vary, but they should identify important products and the areas in which they are grown. Maps should be accompanied by a discussion explaining why various products are grown in different regions. Use Rubric 20, Map Creation, to evaluate student work.

internet connect

GO TO: go.hrw.com
KEYWORD: SK3 Teacher
FOR: a guide to using the Internet in your classroom

CHAPTER 12

Northern Europe

Chapter Resource Manager

Objectives	Pacing Guide	Reproducible Resources
SECTION 1 **Physical Geography** (pp. 269–271) **1.** Describe the region's major physical features. **2.** Identify the region's most important natural resources. **3.** Identify the climates that are found in northern Europe.	**Regular** .5 day **Block Scheduling** .5 day *Block Scheduling Handbook, Chapter 12*	**RS** Guided Reading Strategy 12.1 **RS** Graphic Organizer 12 **E** Creative Strategies for Teaching World Geography, Lessons 10 and 11 **E** Lab Activity for Geography and Earth Science, Demonstration 4
SECTION 2 **The United Kingdom** (pp. 272–75) **1.** Identify some important events in the history of the United Kingdom. **2.** Describe what the people and culture of the country are like. **3.** Describe the United Kingdom of today.	**Regular** 1 day **Block Scheduling** .5 day *Block Scheduling Handbook, Chapter 12*	**RS** Guided Reading Strategy 12.2 **E** Cultures of the World Activity 3 **SM** Map Activity 12
SECTION 3 **The Republic of Ireland** (pp. 278–80) **1.** Identify the key events in Ireland's history. **2.** Describe the people and culture of Ireland. **3.** Identify the kinds of economic changes Ireland has experienced in recent years.	**Regular** 1 day **Block Scheduling** .5 day *Block Scheduling Handbook, Chapter 12*	**RS** Guided Reading Strategy 12.3 **E** Cultures of the World Activity 3
SECTION 4 **Scandinavia** (pp. 281–85) **1.** Describe the people and culture of Scandinavia. **2.** Identify some important features of each of the region's countries, plus Greenland and Lapland.	**Regular** 1.5 days **Block Scheduling** .5 day *Block Scheduling Handbook, Chapter 12*	**RS** Guided Reading Strategy 12.4 **E** Cultures of the World Activity 3 **SM** Geography for Life Activity 12

Chapter Resource Key

- **RS** Reading Support
- **IC** Interdisciplinary Connections
- **E** Enrichment
- **SM** Skills Mastery
- **A** Assessment
- **REV** Review
- **ELL** Reinforcement and English Language Learners
- Transparencies
- CD–ROM
- Music
- Video
- go.hrw.com Internet
- Holt Presentation Maker Using Microsoft® Powerpoint®

One-Stop Planner CD–ROM

See the *One-Stop Planner* for a complete list of additional resources for students and teachers.

One-Stop Planner CD–ROM

It's easy to plan lessons, select resources, and print out materials for your students when you use the ***One-Stop Planner CD–ROM with Test Generator.***

Technology Resources	Review, Reinforcement, and Assessment Resources
One-Stop Planner CD–ROM, Lesson 12.1 Geography and Cultures Visual Resources with Teaching Activities 24–29 Homework Practice Online HRW Go site	ELL Main Idea Activity 12.1 REV Section 1 Review, p. 271 A Daily Quiz 12.1 ELL English Audio Summary 12.1 ELL Spanish Audio Summary 12.1
One-Stop Planner CD–ROM, Lesson 12.2 Homework Practice Online HRW Go site	ELL Main Idea Activity 12.2 REV Section 2 Review, p. 275 A Daily Quiz 12.2 ELL English Audio Summary 12.2 ELL Spanish Audio Summary 12.2
One-Stop Planner CD–ROM, Lesson 12.3 Homework Practice Online HRW Go site	ELL Main Idea Activity 12.3 REV Section 3 Review, p. 280 A Daily Quiz 12.3 ELL English Audio Summary 12.3 ELL Spanish Audio Summary 12.3
One-Stop Planner CD–ROM, Lesson 12.4 Homework Practice Online HRW Go site	ELL Main Idea Activity 12.4 REV Section 4 Review, p. 315 A Daily Quiz 12.4 ELL English Audio Summary 12.4 ELL Spanish Audio Summary 12.4

internet connect

HRW ONLINE RESOURCES

GO TO: go.hrw.com
Then type in a keyword.

TEACHER HOME PAGE
KEYWORD: SK3 TEACHER

CHAPTER INTERNET ACTIVITIES
KEYWORD: SK3 GT12

Choose an activity to:
- explore the island and fjords on the Scandinavian coast.
- visit historic palaces in the United Kingdom.
- investigate the history of skiing.

CHAPTER ENRICHMENT LINKS
KEYWORD: SK3 CH12

CHAPTER MAPS
KEYWORD: SK3 MAPS12

ONLINE ASSESSMENT
Homework Practice
KEYWORD: SK3 HP12
Standardized Test Prep Online
KEYWORD: SK3 STP12
Rubrics
KEYWORD: SS Rubrics

COUNTRY INFORMATION
KEYWORD: SK3 Almanac

CONTENT UPDATES
KEYWORD: SS Content Updates

HOLT PRESENTATION MAKER
KEYWORD: SK3 PPT12

ONLINE READING SUPPORT
KEYWORD: SS Strategies

CURRENT EVENTS
KEYWORD: S3 Current Events

Meeting Individual Needs

Ability Levels

Level 1 Basic-level activities designed for all students encountering new material

Level 2 Intermediate-level activities designed for average students

Level 3 Challenging activities designed for honors and gifted-and-talented students

English Language Learners Activities that address the needs of students with Limited English Proficiency

Chapter Review and Assessment

- IC Interdisciplinary Activities for the Middle Grades 13, 14, 15
- E Readings in World Geography, History, and Culture 34, 35, 36
- SM Critical Thinking Activity 12
- REV Chapter 12 Review, pp. 286–87
- REV Chapter 12 Tutorial for Students, Parents, Mentors, and Peers
- ELL Vocabulary Activity 12
- A Chapter 12 Test
- Chapter 12 Test Generator (on the One-Stop Planner)
- Audio CD Program, Chapter 12
- A Chapter 12 Test for English Language Learners and Special-Needs Students

CHAPTER 12

Launch into Learning

Select a brief folktale from one of the countries of northern Europe. Two possibilities are "Three Billy Goats Gruff" from Norway or one of the King Arthur stories from Great Britain. Read the story aloud to the class. You may want to invite students to sketch illustrations for the story as you read. When you have finished the story, ask students if they have heard it before. Discuss responses. Explain that Americans have inherited many cultural elements from northern Europe, such as the stories that were told around the fire on long winter nights. Tell students they will learn more about the countries of northern Europe and the connections the United States has with those countries in Chapter 12.

Section 1

Objectives

1. Describe the major physical features of northern Europe.
2. Identify the region's most important natural resources.
3. Examine the climates found in northern Europe.

Links to Our Lives

You may wish to point out to students the following reasons why we should study northern Europe:

- The United States has strong historical and economic ties to the countries of northern Europe.
- Denmark, Iceland, Norway, and the United Kingdom are members of the North Atlantic Treaty Organization (NATO). They are political allies of the United States.
- Millions of Americans have northern European ancestry.
- American culture has been influenced beyond measure by northern Europe. The language, law, commerce, government, literature, art, holidays, and food found in the United States—all these and more bear the imprint of northern Europe.

CHAPTER 12

Northern Europe

Now we will study the countries of northern Europe. First we meet Lars, a student in Norway. He lives in a place where the Sun does not rise during much of the winter.

Hi! My name is Lars. I am 13, and I live in Tromsø, one of the northernmost cities in Europe. I am in my seventh year at school. In school we study Norwegian, plus English, French or German, social studies, science, music, art, and cooking. If I do well in junior high, I will go to an academic high school and prepare for a university.

Usually I walk to school, which is about 3 km (1.9 miles) away. In the winter, everyone skis to school. The Sun never shines on many winter days because we live north of the Arctic Circle. On January 20, when the Sun appears again for just a few minutes, we celebrate Sun Day.

In the summer the Sun never sets. This is my favorite time of the year. It still can be cold then. Last summer the temperature was mostly around 6° or 7°C (about 43° or 44°F).

Jeg bor i midnattssolens land.

Translation: I live in the Land of the Midnight Sun.

FOCUS

LET'S GET STARTED

Copy the following instructions onto the chalkboard: *Look at the physical-political map of northern Europe. What outdoor sports might you enjoy if you lived in the lake region of Finland, in central Norway, in southern England, or on the western coast of Ireland? (Possible answers: Finland—ice skating, Norway—skiing, England—soccer, Ireland—sailing)* Give students time to write down their answers. Discuss responses. Point out that it is possible to participate in many sports in northern Europe because the region has a mix of landforms and climates. Tell students that in Section 1 they will learn more about the physical geography of northern Europe.

Using the Physical-Political Map

Have students examine the map on this page. Call on a volunteer to name the countries in the region. Have students classify the countries as islands or peninsulas. Ask students to name the seas that surround the region's countries and to speculate on the influence the sea has had on the economic activities of the people of northern Europe.

Section 1 Physical Geography

Read to Discover

1. What are the region's major physical features?
2. What are the region's most important natural resources?
3. What climates are found in northern Europe?

Define

fjords
lochs
North Atlantic Drift

Locate

British Isles
English Channel
North Sea
Great Britain
Ireland
Iceland
Greenland
Scandinavian Peninsula
Jutland Peninsula
Kjølen Mountains
Northwest Highlands
Shannon River
Baltic Sea

A Viking ship post

WHY IT MATTERS

An important physical feature of Northern Europe is the North Sea. Use CNNfyi.com or other **current events** sources to learn more about America's dependence on oil. Record your findings in your journal.

Northern Europe: Physical-Political

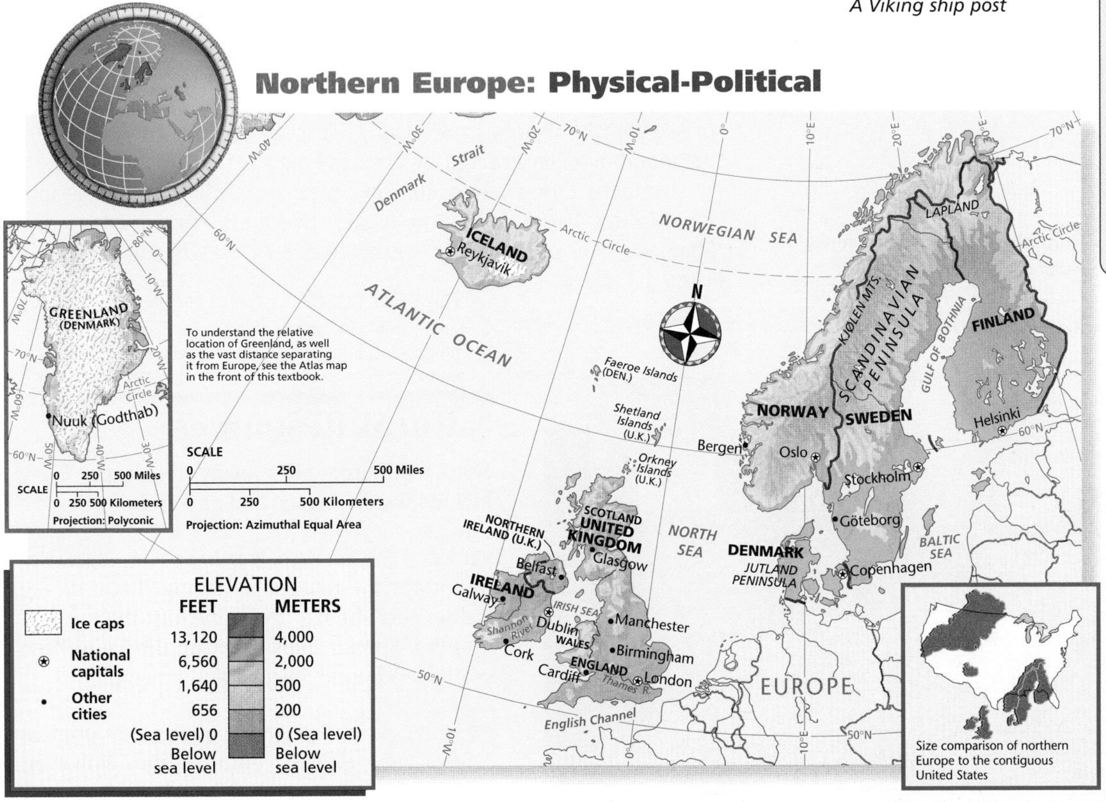

Section 1 RESOURCES

Reproducible

- Block Scheduling Handbook, Chapter 12
- Guided Reading Strategy 12.1
- Graphic Organizer 12
- Creative Strategies for Teaching World Geography, Lessons 10 and 11
- Lab Activity for Geography and Earth Science, Demonstration 4

Technology

- One-Stop Planner CD–ROM, Lesson 12.1
- Homework Practice Online
- Geography and Cultures Visual Resources with Teaching Activities 24–29
- HRW Go site

Reinforcement, Review, and Assessment

- Section 1 Review, p. 271
- Daily Quiz 12.1
- Main Idea Activity 12.1
- English Audio Summary 12.1
- Spanish Audio Summary 12.1

TEACH

Teaching Objectives 1–3

ALL LEVELS: (Suggested time: 20 min.) Copy the graphic organizer at right onto the chalkboard, omitting the italicized answers. Use it to help students classify the landforms, lakes and rivers, natural resources, and climates of northern Europe. Call on volunteers to point out the listed physical features on a wall map. Then have students brainstorm occupations in northern Europe that depend on the listed physical features and resources. *(Example: oil—oil-rig worker)* **ENGLISH LANGUAGE LEARNERS**

Physical Geography of Northern Europe

Landforms	*hills of Ireland highlands of Great Britain Kjölen Mountains lowlands of southeastern Great Britain and southern Scandinavia glaciers fjords*	*marine west coast caused by North Atlantic Drift humid continental subarctic tundra*	Climates
Lakes and rivers	*lochs of Scotland Shannon River — longest in region*	*North Sea, Baltic Sea fish, forests, soil oil and natural gas geothermal and hydroelectric power*	Natural resources

National Geography Standard 14

Environment and Society Northern Europe has many lakes, rivers, and streams. Some of these waterways are polluted.

During the 1960s increased industrial production in Europe caused airborne pollutants to drift far over the continent. These pollutants combined with water vapor in the air and fell as acid rain. Forests in Scandinavia show the damaging effects.

In winter, snow carries the pollutants to the land. When the snow melts, the acid makes its way into lakes and streams. Fish have been killed off in many lakes of Norway and Sweden. Acid rain has also corroded many important buildings.

Activity: Have students conduct research on efforts by northern European governments and organizations to solve pollution problems. Allow them to summarize their findings in a written report or in an oral presentation to the class.

internet connect

GO TO: go.hrw.com
KEYWORD: SK3 CH12
FOR: Web sites about pollution

Visual Record Answer

carved by glaciers

internet connect

GO TO: go.hrw.com
KEYWORD: SK3 CH12
FOR: Web sites about northern Europe

Our Amazing Planet

Scotland's Loch Ness contains more fresh water than all the lakes in England and Wales combined. It is deeper, on average, than the nearby North Sea.

Physical Features

Northern Europe includes several large islands and peninsulas. The British Isles lie across the English Channel and North Sea from the rest of Europe. They include the islands of Great Britain and Ireland and are divided between the United Kingdom and the Republic of Ireland. This region also includes the islands of Iceland and Greenland. Greenland is the world's largest island.

To the east are the Scandinavian and Jutland Peninsulas. Denmark occupies the Jutland Peninsula and nearby islands. The Scandinavian Peninsula is divided between Norway and Sweden. Finland lies farther east. These countries plus Iceland make up Scandinavia.

Landforms The rolling hills of Ireland, the highlands of Great Britain, and the Kjølen (CHUH-luhn) Mountains of Scandinavia are part of Europe's Northwest Highlands region. This is a region of very old, eroded hills and low mountains.

Southeastern Great Britain and southern Scandinavia are lowland regions. Much of Iceland is mountainous and volcanic. More than 10 percent of it is covered by glaciers. Greenland is mostly covered by a thick ice cap.

Coasts Northern Europe has long, jagged coastlines. The coastline of Norway includes many **fjords** (fee-AWRDS). Fjords are narrow, deep inlets of the sea set between high, rocky cliffs. Ice-age glaciers carved the fjords out of coastal mountains.

Lakes and Rivers Melting ice-age glaciers left behind thousands of lakes in the region. In Scotland, the lakes are called **lochs**. Lochs are found in valleys carved by glaciers long ago.

Northern Europe does not have long rivers like the Mississippi River in the United States. The longest river in the British Isles is the Shannon River in Ireland. It is just 240 miles (390 km) long.

✓ READING CHECK: *Places and Regions* What are the physical features of the region? large islands, peninsulas, hills, mountains, glaciers, ice cap, fjords, lakes

Fjords like this one shelter many harbors in Norway.

Interpreting the Visual Record
How are fjords created?

Natural Resources

Northern Europe has many resources. They include water, forests, and energy sources.

Water The ice-free North Sea is especially important for trade and fishing. Parts of the Baltic Sea freeze over during the winter months. Special ships break up the ice to keep sea lanes open to Sweden and Finland.

Forests and Soil Most of Europe's original forests were cleared centuries ago. However,

Close

Challenge students to suggest ways that northern Europe's physical geography may have shaped its history. *(Examples: Long coastlines and ice-free ports make long-distance trade and conquest possible. The isolation of the British Isles may have discouraged some invaders.)*

Review and Assess

Have students complete the Section Review. Then have students write a description of a country or region in northern Europe without naming it. Call on volunteers to read their descriptions and on others to determine the place described. Then have students complete Daily Quiz 12.1.

Reteach

Have students complete Main Idea Activity 12.1. Then organize them into groups to create a database of facts on the region. Each database should include physical features, resources, and climate types for each area. **ENGLISH LANGUAGE LEARNERS, COOPERATIVE LEARNING**

Extend

Have interested students conduct research on Surtsey, a volcanic island that emerged off the coast of Iceland in 1963. Ask students to include information on how the island was formed, the plants that now grow there, and the animals that either live on or visit the island. Have students write reports of these findings. **BLOCK SCHEDULING**

Sweden and Finland still have large, coniferous forests that produce timber. The region's farmers grow many kinds of cool-climate crops.

Energy Beneath the North Sea are rich oil and natural gas reserves. Nearly all of the oil reserves are controlled by the nearby United Kingdom and Norway. However, these reserves cannot satisfy all of the region's needs. Most countries import oil and natural gas from southwest Asia, Africa, and Russia. Some, such as Iceland, use geothermal and hydroelectric power.

✓ **READING CHECK:** *Environment and Society* In what way has technology allowed people in the region to keep the North Sea open during the winter? Special ships break up the ice.

Climate

Despite its northern location, much of the region has a marine west coast climate. Westerly winds blow over a warm ocean current called the **North Atlantic Drift**. These winds bring mild temperatures and rain to the British Isles and coastal areas. Atlantic storms often bring even more rain. Snow and frosts may occur in winter.

Central Sweden and southern Finland have a humid continental climate. This area has four true seasons. Far to the north are subarctic and tundra climates. In the forested subarctic regions, winters are long and cold with short days. Long days fill the short summers. In the tundra region it is cold all year. Only small plants such as grass and moss grow there.

✓ **READING CHECK:** *Places and Regions* What are the region's climates? marine west coast, humid continental, subarctic, tundra

BUILD on WHAT You Know

Do you remember what you learned about ocean currents? See Chapter 3 to review.

go.hrw.com Homework Practice Online
Keyword: SK3 HP12

Section 1 Review

Define and explain: fjords, lochs, North Atlantic Drift

Working with Sketch Maps On an outline map that you draw or that your teacher provides, label the following: British Isles, English Channel, North Sea, Great Britain, Ireland, Iceland, Greenland, Scandinavian Peninsula, Jutland Peninsula, Kjølen Mountains, Northwest Highlands, Shannon River, and Baltic Sea. In the margin, write a short caption explaining how the North Atlantic Drift affects the region's climates.

Reading for the Main Idea

1. *Places and Regions* Which parts of northern Europe are highland regions? Which parts are lowland regions?
2. *Places and Regions* What major climate types are found in northern Europe?

Critical Thinking

3. **Finding the Main Idea** How has ice shaped the region's physical geography?
4. **Making Generalizations and Predictions** Think about what you have learned about global warming. How might warmer temperatures affect the climates and people of northern Europe?

Organizing What You Know

5. **Summarizing** Copy the following graphic organizer. Use it to describe the region's important natural resources.

Water	Forests and soil	Energy

Section 1 Review
Answers

Define For definitions, see: fjords, p. 270; lochs, p. 270; North Atlantic Drift, p. 271

Working with Sketch Maps Places should be labeled in their approximate locations. The North Atlantic Drift brings mild temperatures and rain to the British Isles and coastal areas.

Reading for the Main Idea

1. hills of Ireland, highlands of Great Britain, Kjölen Mountains, much of Iceland; southeastern Great Britain, southern Scandinavia (NGS 4)
2. marine west coast, humid continental, subarctic, tundra (NGS 4)

Critical Thinking

3. carved fjords, left lakes behind
4. Possible answers: melt Iceland's glaciers and Greenland's ice cap, fill fjords, flood lowlands

Organizing What You Know

5. water—North Sea, Baltic Sea; forests and soil—coniferous forests in Sweden and Finland, soil for cool-climate crops; energy—oil, natural gas, geothermal and hydroelectric power

Section 2

Objectives

1. Identify some important events in the history of the United Kingdom.
2. Describe the people and culture of the country.
3. Explain what the United Kingdom is like today.

Focus

LET'S GET STARTED

Copy the following "formula" and question onto the chalkboard: *Celts + Romans + Angles + Saxons + Jutes + Vikings + Normans + Indians + Pakistanis + others = British people. What does this formula mean? (It refers to some of the groups of people who have come to live in what is now the United Kingdom.)* Have students write their answers. Discuss responses. Explain that the British Isles have been a "melting pot" for centuries and that its population continues to become more diverse. Tell students they will learn more about the United Kingdom in Section 2.

Building Vocabulary

Write the key terms on the chalkboard. Ask volunteers to read the definitions aloud and use each one in a sentence. Point out that the word **textile** is often used as an adjective, such as in "textile mill" or "textile market." Explain that in the United Kingdom Parliament passes laws. The United Kingdom also has a reigning king or queen; he or she is called the monarch. The combination of these two parts of the government creates a **contitutional monarchy**. Ask students if they know any place names or streets that include **glen**. The word has become common in the English language.

Section 2 Resources

Reproducible

- Guided Reading Strategy 12.2
- Map Activity 12
- Cultures of the World Activity 3

Technology

- One-Stop Planner CD–ROM, Lesson 12.2
- Homework Practice Online
- HRW Go site

Reinforcement, Review, and Assessment

- Section 2 Review, p. 275
- Daily Quiz 12.2
- Main Idea Activity 12.2
- English Audio Summary 12.2
- Spanish Audio Summary 12.2

Section 2 The United Kingdom

Read to Discover

1. What are some important events in the history of the United Kingdom?
2. What are the people and culture of the country like?
3. What is the United Kingdom like today?

Define

textiles
constitutional monarchy
glen

Locate

England
Scotland
Wales
Northern Ireland
Irish Sea
London
Birmingham
Manchester
Glasgow
Cardiff
Belfast

Why It Matters

The United States has been heavily influenced by the British people. Use CNNfyi.com or other **current events** sources to find out about present-day ties between the United States and Great Britain. Record your findings in your journal.

British crown

Early peoples of the British Isles built Stonehenge in stages from about 3100 B.C. to about 1800 B.C.

History

Most of the British are descended from people who came to the British Isles long ago. The Celts (KELTS) are thought by some scholars to have come to the islands around 450 B.C. Mountain areas of Wales, Scotland, and Ireland have remained mostly Celtic.

Later, from the A.D. 400s to 1000s, new groups of people came. The Angles and Saxons came from northern Germany and Denmark. The Vikings came from Scandinavia. Last to arrive in Britain were the Normans from northern France. They conquered England in 1066. English as spoken today reflects these migrations. It combines elements from the Anglo-Saxon and Norman French languages.

A Global Power England became a world power in the late 1500s. Surrounded by water, the country developed a powerful navy that protected trade routes. In the 1600s the English began establishing colonies around the world. By the early 1800s they had also united England, Scotland, Wales, and Ireland into one kingdom. From London the United Kingdom built a vast British Empire. By 1900 the empire covered nearly one fourth of the world's land area.

This beautiful Anglo-Saxon shoulder clasp from about A.D. 630 held together pieces of clothing.

TEACH

Teaching Objective 1

LEVEL 1: (Suggested time: 45 min.) Organize students into groups. Ask each group to make an illustrated time line of a particular 100-year period in the history of England. Be sure all significant eras are covered. Ask students to include on their time lines significant historical events and achievements that have helped make Great Britain the nation it is today. Encourage students to consult library resources for information.
ENGLISH LANGUAGE LEARNERS, COOPERATIVE LEARNING

LEVELS 2 AND 3: (Suggested time: 45 min.) Have students complete the time line in the Level 1 activity. Then ask them to choose a historical figure who lived during their 100-year period. The person chosen may be famous, such as Queen Elizabeth I or William Shakespeare, or may be a representative of now anonymous citizens, such as a worker in a textile mill of the early Industrial Revolution. Have students write monologues in which their chosen figures reflect on how they fit into or played a role in British history.

The United Kingdom also became an economic power in the 1700s and 1800s. It was the cradle of the Industrial Revolution, which began in the last half of the 1700s. Large supplies of coal and iron and a large labor force helped industries grow. The country also developed a good transportation network of rivers, canals, and railroads. Three of the early industries were **textiles**, or cloth products, shipbuilding, iron, and later steel. Coal powered these industries. Birmingham, Manchester, and other cities grew up near Britain's coal fields.

▲ Queen Elizabeth I (1533–1603) ruled England as it became a world power in the late 1500s.

Decline of Empire World wars and economic competition from other countries weakened the United Kingdom in the 1900s. All but parts of northern Ireland became independent in 1921. By the 1970s most British colonies also had gained independence. Most now make up the British Commonwealth of Nations. Members of the Commonwealth meet to discuss economic, scientific, and business matters.

The United Kingdom still plays an important role in world affairs. It is a leading member of the United Nations (UN), the European Union (EU), and the North Atlantic Treaty Organization (NATO).

The Government The United Kingdom's form of government is called a **constitutional monarchy**. That is, it has a monarch—a king or queen—but a parliament makes the country's laws. The monarch is the head of state but has largely ceremonial duties. Parliament chooses a prime minister to lead the national government.

In recent years the national government has given people in Scotland and Wales more control over local affairs. Some people think Scotland might one day seek independence.

✓ **READING CHECK:** *Human Systems* How are former British colonies linked today? Most of them make up the British Commonwealth of Nations.

The British government is seated in London, the capital. The Tower Bridge over the River Thames [TEMZ] is one of the city's many famous historical sites.

Interpreting the Visual Record
Why do you think London became a large city? ▼

USING ILLUSTRATIONS

Direct students' attention to the image of Queen Elizabeth I on this page. Have them describe her clothing and jewelry. Have students contrast Elizabeth's clothing to today's styles. Point out that the queen set fashion trends of the day. Ask students who establishes today's fads and fashions.

Then focus students' attention on the photograph of London. Lead a discussion about how the London cityscape compares to cities with which the students are familiar. Your questions might include: Do U.S. cities have more or fewer skyscrapers than London? Do U.S. cities have more or less open space in their downtown areas?

internet connect
GO TO: go.hrw.com
KEYWORD: SK3 CH12
FOR: Web sites about England

◄ **Visual Record Answer**

its location on a major river and near the sea

ALL LEVELS: (Suggested time: 10 min.) Copy the following graphic organizer onto the chalkboard, omitting the italicized answers. Call on volunteers to complete it. Add more boxes as needed. Then lead a discussion of how the United Kingdom's role in world affairs has changed over the years. Have students identify historical events and factors that have led to these changes. **ENGLISH LANGUAGE LEARNERS**

Church of England, other Protestant churches, Roman Catholicism

English, some Welsh and Gaelic

fish and chips, meats, oat porridge and cakes, potatoes

about 60 million people

Culture United Kingdom Today

religious holidays, queen's birthday

90 percent in urban areas

problems in Northern Ireland

London—capital, largest city

productive farms

stronger economy—oil and natural gas, Silicon Glen

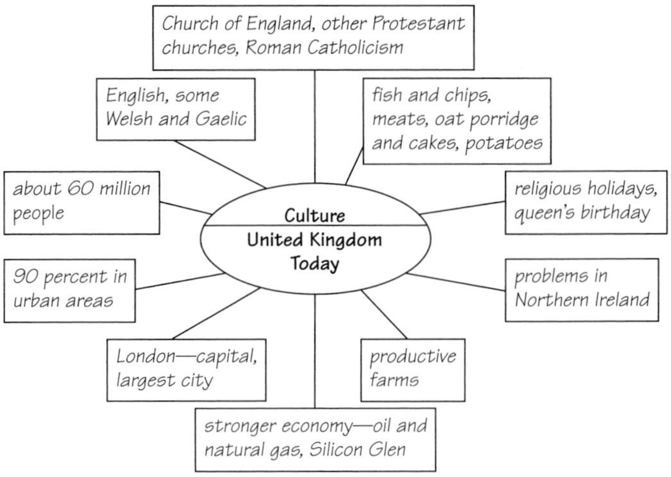

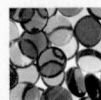

Cultural Kaleidoscope

Pakistanis in Britain

The population of Great Britain has become increasingly diverse over the years. People from around the world, particularly from countries that were once part of the British Empire, have immigrated to Great Britain. During the 1950s and 1960s many people immigrated to Britain from Pakistan. Today Britain's Pakistanis number some 500,000. Most live in large cities, where some have established restaurants and other small businesses.

Pakistanis living in Britain face many challenges, including discrimination. The unemployment rate among them is high partly because some recent immigrants speak little English. In addition, many Pakistani children in Britain receive a poor education. However, British-born Pakistanis have better job prospects. This segment of the population has grown up speaking English.

Activity: Have students research immigrant education in Britain and the United States and compare them.

Visual Record Answer ▲

Students might suggest that television exposes people around the world to each other's cultures.

Chart Answer ▶

It is higher.

▲ Millions of Americans watched the Beatles, a British rock band, perform on television in 1964. The Beatles and other British bands became popular around the world.

Interpreting the Visual Record How do you think television helps shape world cultures today?

COUNTRY	POPULATION/ GROWTH RATE	LIFE EXPECTANCY	LITERACY RATE	PER CAPITA GDP
United Kingdom	59,647,790 0.2%	75, male 81, female	99%	$22,800
United States	281,421,906 0.9%	74, male 80, female	97%	$36,200

Sources: Central Intelligence Agency, *The World Factbook 2001;* U.S. Census Bureau

Interpreting the Chart How does the literacy rate in the United Kingdom compare with that of the United States?

Culture

Nearly 60 million people live in the United Kingdom today. English is the official language. Some people in Wales and Scotland also speak the Celtic languages of Welsh and Gaelic [GAY-lik]. The Church of England is the country's official church. However, many Britons belong to other Protestant churches or are Roman Catholic.

Food and Festivals Living close to the sea, the British often eat fish. One popular meal is fish and chips—fried fish and potatoes. However, British food also includes different meats, oat porridge and cakes, and potatoes in many forms.

The British celebrate many religious holidays, such as Christmas. Other holidays include the Queen's official birthday celebration in June. In July many Protestants in Northern Ireland celebrate a battle in 1690 in which Protestants defeated Catholic forces. In recent years the day's parades have sometimes sparked protests and violence between Protestants and Catholics.

Art and Literature British literature, art, and music have been popular around the world. Perhaps the most famous British writer is William Shakespeare. He died in 1616, but his poetry and plays, such as *Romeo and Juliet*, remain popular. In the 1960s the Beatles helped make Britain a major center for modern popular music. More recently, British performers from Elton John to the Spice Girls have attracted many fans.

✓ **READING CHECK:** *Human Systems* What aspects of British culture have spread around the world? British literature, art, and music have been popular around the world.

The United Kingdom Today

Nearly 90 percent of Britons today live in urban areas. London, the capital of England, is the largest city. It is located in southeastern England. London is also the capital of the whole United Kingdom.

More than 7 million people live in London. The city is a world center for trade, industry, and services, particularly banking and insurance. London also has one of the world's busiest airports. Many tourists visit London to see its famous historical sites, theaters, and shops. Other important cities include Glasgow, Scotland; Cardiff, Wales; and Belfast, Northern Ireland.

Close

Ask students to name rock bands, television programs, fashions, cars, and fads from the United Kingdom. Ask why these influences travel to the United States so easily. (Possible answer: A common language and many cultural ties link the countries.)

Reteach

Have students complete Main Idea Activity 12.2. Then write the headings *History, Government, Culture,* and *Economy* on the chalkboard. Call on students to supply terms and phrases related to the United Kingdom for each heading. **ENGLISH LANGUAGE LEARNERS**

Review and Assess

Have students complete the Section Review. Then pair students. Ask each student to write three questions and answers related to the content of Section 2. Have students take turns reading an answer and challenging their partner to guess the corresponding question. Then have students complete Daily Quiz 12.2. **COOPERATIVE LEARNING**

Extend

Have interested students conduct research on differences between British and American word usage. Have students create a chart to present notable differences in usage. **BLOCK SCHEDULING**

The Economy Old British industries like mining and manufacturing declined after World War II. Today, however, the economy is stronger. North Sea reserves have made the country a major producer of oil and natural gas. Birmingham, Glasgow, and other cities are attracting new industries. One area of Scotland is called Silicon Glen. This is because it has many computer and electronics businesses. **Glen** is a Scottish term for a valley. Today many British work in service industries, including banking, insurance, education, and tourism.

Agriculture Britain's modern farms produce about 60 percent of the country's food. Still, only about 1 percent of the labor force works in agriculture. Important products include grains, potatoes, vegetables, and meat.

Northern Ireland One of the toughest problems facing the country has been violence in Northern Ireland. Sometimes Northern Ireland is called Ulster. The Protestant majority and the Roman Catholic minority there have bitterly fought each other. Violence on both sides has resulted in many deaths.

Many Catholics believe they have not been treated fairly by the Protestant majority. Therefore, many want Northern Ireland to join the mostly Roman Catholic Republic of Ireland. Protestants fear becoming a minority on the island. They want to remain part of the United Kingdom. Many people hope that recent agreements made by political leaders will lead to a lasting peace. In 1999, for example, Protestant and Roman Catholic parties agreed to share power in a new government. However, there have been problems putting that agreement into effect.

✓ **READING CHECK:** *Human Systems* What has been the cause of conflict in Northern Ireland? historical differences between the Protestant majority and the Roman Catholic minority

Religion: A Divided Island

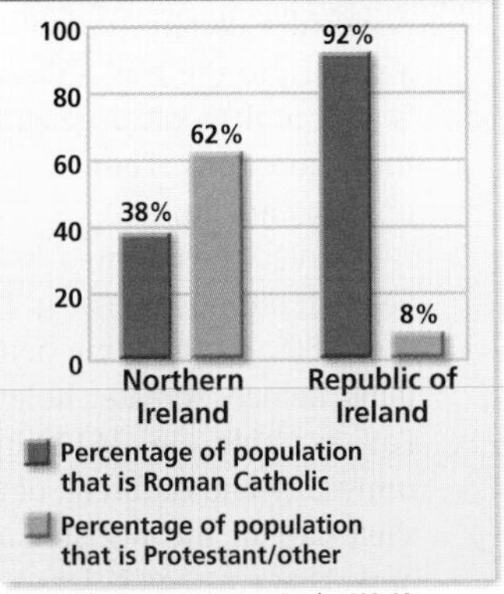

Source: *The Statesman's Year Book,* 1998–99.

Interpreting the Graph How does the number of Roman Catholics in the Republic of Ireland differ from that of Northern Ireland?

go.hrw.com Homework Practice Online
Keyword: SK3 HP12

Section 2 Review

Define and explain: textiles, constitutional monarchy, glen

Working with Sketch Maps On the map you created in Section 1, label the United Kingdom, England, Scotland, Wales, Northern Ireland, Irish Sea, London, Birmingham, Manchester, Glasgow, Cardiff, and Belfast.

Reading for the Main Idea

1. *Human Systems* What peoples came to the British Isles after the Celts? When did they come?
2. *Places and Regions* What was the British Empire?

Critical Thinking

3. **Contrasting** How is the British government different from the U.S. government?
4. **Drawing Inferences and Conclusions** Why do you think Protestants in Northern Ireland want to remain part of the United Kingdom?

Organizing What You Know

5. **Sequencing** Create a timeline that lists important events in the period.

800 B.C. —— A.D. 2000

Section 2 Review Answers

Define For definitions, see: textiles, p. 273; constitutional monarchy, p. 273; glen, p. 275

Working with Sketch Maps Places should be labeled in their approximate locations.

Reading for the Main Idea

1. Angles, Saxons, Vikings—A.D. 400s to 1000s; Normans—1066 (NGS 9)
2. colonies around the world established by England (NGS 4)

Critical Thinking

3. Britain—constitutional monarchy; U.S.—constitutional democracy
4. Possible answers: fear of losing influence or of change, desire to stay in majority

Organizing What You Know

5. 750–450 B.C.—Celts arrive; A.D. 400s–1000s—Angles, Saxons, Vikings, Normans arrive; late 1500s—England a world power; 1600s—colonies established; 1700s–1800s—Industrial Revolution; 1900—empire at its height; by 1921—United Kingdom weakened; 1970s—most colonies independent

▲ **Graph Answer**

Roman Catholics make up most of the population.

Case Study

Setting the Scene

During the mid-1800s, London was a crowded, dirty place for many of its citizens. Health care was limited by misinformation. Sanitation was poor. These factors contributed to the spread of disease. Cholera was one of the worst. Cholera victims experienced symptoms such as diarrhea and vomiting. Dehydration ultimately killed them.

Building a Case

Have students read the Case Study feature. Then have students fill in this graphic organizer on the progression of events and the effect that Dr. Snow's map had on the control of cholera.

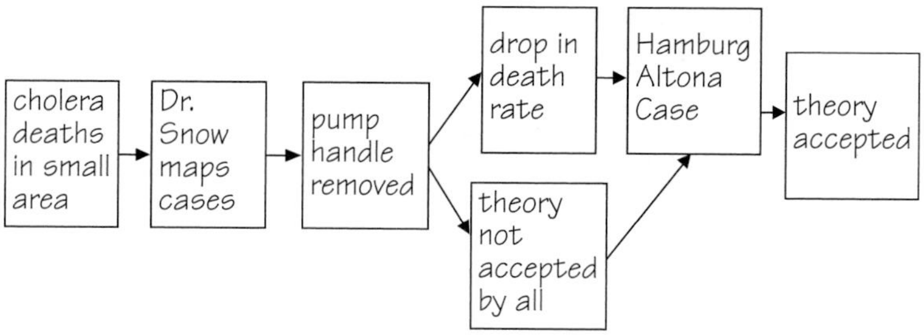

Historical Geography

The United States had its own epidemic of huge proportions early in the 1900s, but many people have never heard of it.

In March 1918, a soldier in Kansas went to an army base hospital complaining of a fever, sore throat, and headache. The soldier had influenza, more commonly called the flu. By the end of the week, the base hospital had treated about 500 ill people.

Soon, Americans were dying by the thousands. New York City reported 851 deaths in one day. San Francisco reported 5,000 new cases in December 1918. In all, more than half a million Americans died of the flu and related diseases during the epidemic.

Americans were also fighting World War I at the time U.S. soldiers accidentally spread the disease to Europe. By the time the epidemic ended in 1919, the flu had killed some 30 million people worldwide.

Activity: Have students conduct research on the spread of influenza during 1918–19 and record their findings on a world map. Can they draw any conclusions about the spread of the disease from the results?

➤ **This Case Study feature addresses National Geography Standards 1, 3, 12, 15 and 18.**

Case Study

Mapping the Spread of Cholera

Medical geographers want to discover why a disease occurs in a particular place. Does a disease occur in a certain type of environment? Is there a pattern to the way a disease spreads? Mapping is one tool medical geographers use to answer these questions. They first used maps in this way to fight cholera.

Cholera has existed in India for hundreds of years. It did not appear in Europe, however, until the 1800s. At that time better transportation systems helped spread cholera around the world. For example, in 1817 India experienced an unusually bad outbreak of the disease. India was then part of the British Empire. British soldiers and ships carried cholera to new places. By 1832 the disease had spread to the British Isles and to North America.

No one knew what caused cholera. In fact, no one knew about bacteria or how they caused disease. What was known was that sick people suffered from diarrhea and vomiting. They often died quickly. In just 10 days in 1854, more than 500 people in one London neighborhood died from cholera. Dr. John Snow thought he knew why.

Dr. Snow believed that people got cholera from dirty water. To test his theory, he mapped the location of some of London's public water pumps. (Houses at that time did not have running water.) Then he marked the location of each cholera death on his map. He found most of the deaths were scattered around the water pump on Broad Street. He persuaded officials to remove the pump's handle. After the pump was shut down, there were few new cases of cholera. Not everyone, however, believed Dr. Snow's evidence.

This illustration shows London's Regent Street in about 1850. An outbreak of cholera in this neighborhood killed hundreds of people in 1854.

Drawing Conclusions

Why was Dr. Snow's map instrumental in solving the cholera riddle? Are there other factors besides dirty water that might have caused the results recorded on the map? Challenge students to answer these questions by writing and acting out a skit of Dr. Snow's meeting with the London authorities. Conclude by asking students to suggest other issues that we might understand more fully if they were plotted on a map.

Going Further: Thinking Critically

Have students contact public health departments for information on allergies in the United States. Then have them answer these questions:

- Where are allergy cases common? Can any connections be made between the cases and natural features, such as types of vegetation?
- Do people in areas where air pollution is a problem suffer more from allergies? *(Students should provide evidence to support their conclusions.)*
- What actions should be taken next if mapping allergy sufferers and allergens (substances that cause allergic reactions) reveals clear connections? *(Students might provide small- or large-scale solutions to allergy outbreaks.)*

Cholera Deaths in London, 1854

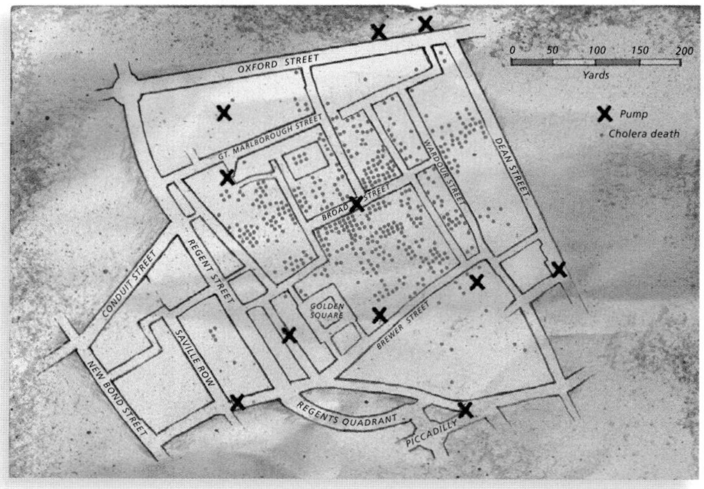

Dr. Snow plotted cholera deaths for the first 10 days of September 1854 on a map of the neighborhood. Dr. Snow believed that people were getting impure water from the Broad Street pump. A check of the pump showed that a leaking sewer had contaminated the pump's water.

Interpreting the Map **What does the map tell you about the relationship between deaths and the Broad Street pump?**

Bacteria was discovered in the 1880s. Yet many people still believed that bad air from river mud or swamps spread diseases like cholera. It took another epidemic to convince everyone that dirty water made people sick.

In 1892 an outbreak of cholera in the German port city of Hamburg ended the debate. Once again, the locations of people who became sick were mapped. This mapping was the key to proving that water could spread cholera.

At the time, a street separated Hamburg and another town, Altona. Both towns got their water from the Elbe River. There was, however, an important difference in the two towns' water supplies. Altona had a system for cleaning its water. Hamburg did not. People living on the Hamburg side of the street got sick. Those on the Altona side of the street did not. The air on both sides of the street was the same. Therefore, no one could argue that bad air caused the outbreak. Hamburg then moved quickly to install a system to clean its water.

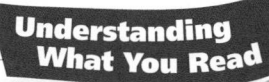

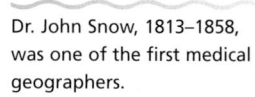

Dr. John Snow, 1813–1858, was one of the first medical geographers.

Understanding What You Read

1. Why did Dr. Snow make a map of a neighborhood's cholera deaths and the location of its water pumps?
2. Why was it important that people on the Hamburg side of the street got sick but people on the Altona side did not?

Understanding What You Read

1. Dr. Snow wanted to test his theory that people got cholera from impure water. By mapping the location of water pumps and cholera cases, he hoped to make a visual connection between the two factors.
2. It showed that bad air was not causing the cholera outbreak, since people on either side of the street breathed the same air. However, the two sides got their water from different sources. Altona had a water purification system, but Hamburg did not.

internet connect

GO TO: go.hrw.com
KEYWORD: SK3 CH12
FOR: Web sites about cholera

Interpreting the Map

Deaths were concentrated around the Broad Street pump.

Section 3

Objectives

1. Identify key events in Ireland's history.
2. Describe the people and culture of Ireland.
3. Explain the economic changes Ireland has experienced in recent years.

FOCUS

LET'S GET STARTED

Copy the following questions onto the chalkboard: *Which Irish holiday is celebrated by many Americans? How is it celebrated? (Saint Patrick's Day, wearing green, parades)* Have students write their answers. Explain that Irish immigrants have contributed much to our culture. Point out that although it is now a fairly prosperous country, Ireland has had difficult times in its past. Because of those difficulties, many people left Ireland for the United States. Tell students that in Section 3 they will learn more about the Republic of Ireland.

Building Vocabulary

Write the key terms on the chalkboard and ask students to locate their definitions in the section text. Call on volunteers to read the definitions aloud. Ask students if they recall news about **famine** in other countries. You may want to ask if they have heard of the Irish potato famine. Explain that from 1845 to 1849, a disease killed so many potato plants in Ireland that hundreds of thousands of people starved. Potatoes had been a major food source, particularly among the poor. Explain that **bog** comes from the ancient language of Ireland, Gaelic. **Peat** is found in bogs.

Section 3 RESOURCES

Reproducible
- Guided Reading Strategy 12.3
- Cultures of the World Activity 3

Technology
- One-Stop Planner CD–ROM, Lesson 12.3
- Homework Practice Online
- HRW Go site

Reinforcement, Review, and Assessment
- Section 3 Review, p. 280
- Daily Quiz 12.3
- Main Idea Activity 12.3
- English Audio Summary 12.3
- Spanish Audio Summary 12.3

Visual Record Answer

rainy temperate

Section 3 The Republic of Ireland

Read to Discover

1. What are the key events in Ireland's history?
2. What are the people and culture of Ireland like?
3. What kinds of economic changes has Ireland experienced in recent years?

Define

famine
bog
peat

Locate

Dublin
Cork
Galway

WHY IT MATTERS

Millions of Americans trace their heritage to Ireland. Use CNNfyi.com or other **current events** sources to learn about Ireland and its people today. Record your findings in your journal.

An Irish harp

History

The Irish are descendants of the Celts. Irish Gaelic, a Celtic language, and English are the official languages. Most people in Ireland speak English. Gaelic is spoken mostly in rural western areas.

English Conquest England conquered Ireland in the A.D. 1100s. By the late 1600s most of the Irish had become farmers on land owned by the British. This created problems between the two peoples. Religious differences added to these problems. Most British were Protestant, while most Irish were Roman Catholic. Then, in the 1840s, millions of Irish left for the United States and other countries because of a poor economy and a potato **famine**. A famine is a great shortage of food.

Independence The Irish rebelled against British rule. In 1916, for example, Irish rebels attacked British troops in the Easter Rising.

Stone fences divide the green fields of western Ireland, the Emerald Isle.

Interpreting the Visual Record **What kind of climate would you expect to find in a country with rich, green fields such as these?**

TEACH

Teaching Objectives 1–3

LEVEL 1: (Suggested time: 20 min.) Copy the following graphic organizer onto the chalkboard, omitting the italicized answers. Have students fill in important eras in Ireland's history. Then lead a discussion about how Ireland's history has influenced its culture and the recent economic changes in the country. **ENGLISH LANGUAGE LEARNERS**

LEVELS 2 AND 3: (Suggested time: 30 min.) Ask students to imagine that they are the prime minister of Ireland and that they are running for re-election. Have them write campaign speeches in which they display their knowledge of Ireland's past and take credit for the country's economic progress. Call on volunteers to deliver their speeches to the class.

Celtic Ireland → *English conquest* → *Immigration to United States* → *Easter Rising* → *Independence*

At the end of 1921, most of Ireland gained independence. Some counties in northern Ireland remained part of the United Kingdom. Ties between the Republic of Ireland and the British Empire were cut in 1949.

Government Ireland has an elected president and parliament. The president has mostly ceremonial duties. Irish voters in 1990 elected a woman as president for the first time.

The parliament makes the country's laws. The Irish parliament chooses a prime minister to lead the government.

✓ **READING CHECK:** *Human Systems* What are some important events in Ireland's history? **English conquest, Easter Rising, independence for most of Ireland, end of ties between Republic of Ireland and the British Empire**

Culture

Centuries of English rule have left their mark on Irish culture. For example, today nearly everyone in Ireland speaks English. Irish writers, such as George Bernard Shaw and James Joyce, have been among the world's great English-language writers.

A number of groups promote traditional Irish culture in the country today. The Gaelic League, for example, encourages the use of Irish Gaelic. Gaelic and English are taught in schools and used in official documents. Another group promotes Irish sports, such as hurling. Hurling is an outdoor game similar to field hockey and lacrosse.

Elements of Irish culture have also become popular outside the country. For example, traditional Irish folk dancing and music have attracted many fans. Music has long been important in Ireland. In fact, the Irish harp is a national symbol. Many musicians popular today are from Ireland, including members of the rock band U2.

More than 90 percent of the Irish today are Roman Catholic. St. Patrick's Day on March 17 is a national holiday. St. Patrick is believed to have brought Christianity to Ireland in the 400s.

▲ Many Irish enjoy a meal of lamb chops with mustard sauce, soda bread, carrots, and mashed potatoes.

✓ **READING CHECK:** *Human Systems* What is Irish culture like today? **English is almost universal; groups promote Irish traditions, such as the use of Gaelic and Irish sports; music is important; 90 percent of Irish are Roman Catholic.**

Ireland Today

Ireland used to be one of Europe's poorest countries. Today it is a modern, thriving country with a strong economy and growing cities.

Linking Past to Present

St. Patrick The young man who would later be called Saint Patrick was living in western England when he was captured by raiders and sold into slavery in Ireland. He is credited with converting the Irish people to Christianity in the A.D. 400s. He also introduced the Roman alphabet and Latin literature to Ireland.

According to legend, Patrick drove all the snakes in Ireland into the sea. However, evidence shows that a great freeze affected Ireland and much of the Northern Hemisphere until some 15,000 years ago. Any snakes living in Ireland would have been killed off. Because land snakes cannot migrate across water, there are no snakes living in Ireland today. The legend may refer to Patrick's eliminating ancient non-Christian beliefs from Ireland. Snakes were among many pre-Christian symbols.

Activity: Have students work in groups to conduct research on the historical background of the holidays celebrated in their communities. You may want to have them concentrate on their community's celebration of St. Patrick's Day. Have them summarize their findings in a written or oral report.

◄ **Chart Answer (p. 280)**

They are the same.

CLOSE

Play a recording of a Celtic Irish song, such as one by the Chieftains or DeDanaan, and one by a modern Irish rock band, such as U2. Ask students to compare the two songs and explain how they express characteristics of traditional and modern Irish society.

RETEACH

Have students complete Main Idea Activity 12.3. Then write *Ireland* vertically on the chalkboard. Have students suggest a phrase or sentence beginning with each letter on the board that describes Ireland. *(Example: I—Island; Ireland is an island country.)* **ENGLISH LANGUAGE LEARNERS**

REVIEW AND ASSESS

Have students complete the Section Review. Then organize the class into groups—one for each major topic in the section. Have each group write four original statements about its topic. Have students write their sentences on the chalkboard. Discuss each statement. Then have students complete Daily Quiz 12.3. **COOPERATIVE LEARNING**

EXTEND

Ask interested students to conduct research on popular sports in the Republic of Ireland. They may want to investigate which sports teens enjoy, which have professional status, how much income sports generate for the country, or the sports stars of today. Have students report their findings in a simulated television sports broadcast. **BLOCK SCHEDULING**

Section 3 Review

Answers

Define For definitions, see: famine, p. 278; bog, p. 280; peat, p. 280

Working with Sketch Maps Dublin is the capital and largest city, with many factories. It is the country's center for education, banking, and shipping.

Reading for the Main Idea

1. poor economy, potato famine (NGS 9)
2. change from agricultural to industrial economy, investment by foreign companies (NGS 11, 12)

Critical Thinking

3. possible answer: desire to remain independent and retain their own language, religion
4. more demand for housing as people moved there to work, more people able to pay higher prices

Organizing What You Know

5. Ireland: problems arose between Brit. landowners, Irish farmers; many Irish emigrated; Irish rebelled, gained independence; most Irish speak English; Irish contributed to arts, literature; most Irish are Roman Catholic; economy improved; main cities—Dublin, Cork, Galway. See Section 2 for information about UK.

Ireland

COUNTRY	POPULATION/ GROWTH RATE	LIFE EXPECTANCY	LITERACY RATE	PER CAPITA GDP
Ireland	3,840,838 / 1.1%	74, male / 80, female	98%	$21,600
United States	281,421,906 / 0.9%	74, male / 80, female	97%	$36,200

Sources: Central Intelligence Agency, *The World Factbook 2001;* U.S. Census Bureau

Interpreting the Chart **How does life expectancy in Ireland compare with that of the United States?**

Economy Until recently, Ireland was mostly an agricultural country. This was true even though much of the country is either rocky or boggy. A **bog** is soft ground that is soaked with water. For centuries, **peat** dug from bogs has been used for fuel. Peat is made up of dead plants, usually mosses.

Today Ireland is an industrial country. Irish workers produce processed foods, textiles, chemicals, machinery, crystal, and computers. Finance, tourism, and other service industries are also important.

How did this change come about? Ireland's low taxes, well-educated workers, and membership in the European Union have attracted many foreign companies. Those foreign companies include many from the United States. These companies see Ireland as a door to millions of customers throughout the EU. In fact, goods from their Irish factories are exported to markets in the rest of Europe and countries in other regions.

Cities Many factories have been built around Dublin. Dublin is Ireland's capital and largest city. It is a center for education, banking, and shipping. Nearly 1 million people live there. Housing prices rapidly increased in the 1990s as people moved there for work.

Other cities lie mainly along the coast. These cities include the seaports of Cork and Galway. They have old castles, churches, and other historical sites that are popular among tourists.

✓ **READING CHECK:** *Human Systems* What important economic changes have occurred in Ireland and why? industrial country; low taxes, well-educated workers, and EU membership have attracted foreign companies

go.hrw.com Homework Practice Online
Keyword: SK3 HP12

Section 3 Review

Define and explain: famine, bog, peat

Working with Sketch Maps On the map you created in Section 2, label Ireland, Dublin, Cork, and Galway. In the margin explain the importance of Dublin to the Republic of Ireland.

Reading for the Main Idea

1. *Human Systems* What were two of the reasons many Irish moved to the United States and other countries in the 1800s?
2. *Human Systems* What are some important reasons why the economy in Ireland has grown so much in recent years?

Critical Thinking

3. **Drawing Inferences and Conclusions** Why do you think the Irish fought against British rule?
4. **Drawing Inferences and Conclusions** Why do you suppose housing prices rapidly increased in Dublin in the 1990s?

Organizing What You Know

5. **Comparing/Contrasting** Copy the following graphic organizer. Use it to compare and contrast the history, culture, and governments of the Republic of Ireland and the United Kingdom.

Ireland	United Kingdom
Conquered by England in the 1100s	Created vast world empire

Section 4

Objectives

1. Describe the people and culture of Scandinavia.
2. Identify important features of the region's countries, Greenland, and Lapland.

FOCUS

LET'S GET STARTED

Copy the following instructions onto the chalkboard: *What do you already know about the Vikings? (Possible answers: fierce fighters, tall, blonde, wore metal helmets)* Allow students to respond and discuss responses. Explain that the Viking raiders do not represent all early Scandinavians. Many Scandinavians were farmers and merchants. Point out that most modern Scandinavians are descended from the Vikings. Tell students that in Section 4 they will learn more about ancient and modern Scandinavia.

Building Vocabulary

Write the key terms on the board and have students read the definitions. Ask a volunteer to use **neutral** in a sentence about politics or current events. For **uninhabitable**, explain that the prefix *un-* means "not." The root word *inhabit* means "to live in." The suffix *-able* means "capable of being." So, the term describes a place where no one can live. Ask students to look at a map of the region and suggest which areas may be uninhabitable. Tell students that the word **geysers** comes from the Icelandic language. It originally referred to a particular hot spring in Iceland.

Section 4 Scandinavia

Read to Discover

1. What are the people and culture of Scandinavia like?
2. What are some important features of each of the region's countries, plus Greenland, and Lapland?

Define

neutral
uninhabitable
geysers

Locate

Oslo
Bergen
Stockholm
Göteborg
Copenhagen
Nuuk (Godthab)
Reykjavik
Gulf of Bothnia
Gulf of Finland
Helsinki
Lapland

WHY IT MATTERS

Much of the fish Americans eat comes from the nations of Scandinavia. Use CNNfyi.com or other **current events** sources to learn more about the economic importance of these nations. Record your findings in your journal.

Smoked salmon, a popular food in Scandinavia

Section 4 RESOURCES

Reproducible

- Guided Reading Strategy 12.4
- Geography for Life Activity 12
- Cultures of the World Activity 3

Technology

- One-Stop Planner CD–ROM, Lesson 12.4
- Homework Practice Online
- HRW Go site

Reinforcement, Review, and Assessment

- Section 4 Review, p. 285
- Daily Quiz 12.4
- Main Idea Activity 12.4
- English Audio Summary 12.4
- Spanish Audio Summary 12.4

People and Culture

Scandinavia once was home to fierce, warlike Vikings. Today the countries of Norway, Sweden, Denmark, Iceland, and Finland are peaceful and prosperous.

The people of the region enjoy high standards of living. They have good health care and long life spans. Each government provides expensive social programs and services. These programs are paid for by high taxes.

The people and cultures in the countries of Scandinavia are similar in many ways. For example, the region's national languages, except for Finnish, are closely related. In addition, most people in Scandinavia are Lutheran Protestant. All of the Scandinavian countries have democratic governments.

✓ **READING CHECK:** *Human Systems* How are the people and cultures of Scandinavia similar? **Most languages closely related, most people are Lutheran Protestant, all governments are democratic.**

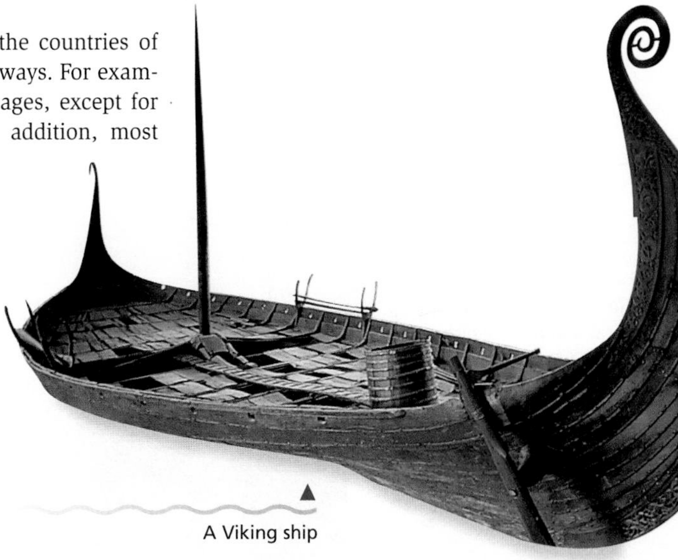

A Viking ship

TEACH

Teaching Objective 1

ALL LEVELS: (Suggested time: 10 min.) Copy the following graphic organizer onto the chalkboard, omitting the italicized answers. Call on students to fill in the lines with characteristics the Scandinavian countries share. **ENGLISH LANGUAGE LEARNERS**

Scandinavia

- *peaceful and prosperous*
- *high standards of living*
- *good health care, long life spans*
- *social programs and services government-sponsored*
- *high taxes*
- *languages closely related (except Finnish)*
- *Lutheran Protestants*
- *democratic governments*

ENVIRONMENT AND SOCIETY

Norway's long shape, severe winter climate, and sparsely populated countryside make it difficult and expensive to provide transportation services throughout the country. Winter weather limits car travel—the main road between Oslo and Bergen is closed for four to six months.

Because transportation costs are high, the Norwegian government provides or helps to pay for some forms of transportation. The government-owned railway loses money each year. There are simply too few train riders outside the large cities to make the service profitable.

Critical Thinking: Why might the Norwegian government continue to provide rail services even though it loses money?

Answer: Students might suggest that a national transportation system is vital for commerce, national defense purposes, or as a public good.

Chart Answer ▲

They are higher than those of the United States and Denmark.

Visual Record Answer ►

thwart land-based attacks; create increased access to waterways for trade

Scandinavia

Country	Population/ Growth Rate	Life Expectancy	Literacy Rate	Per Capita GDP
Denmark	5,352,815 / 0.3%	74, male 80, female	100%	$25,500
Finland	5,175,783 / 0.2%	74, male 81, female	100%	$22,900
Iceland	277,906 / 0.5%	77, male 82, female	100%	$24,800
Norway	4,503,440 / 0.5%	76, male 82, female	100%	$27,700
Sweden	8,875,053 / .02%	77, male 83, female	99%	$22,200
United States	281,421,906 / 0.9%	74, male 80, female	97%	$36,200

Sources: Central Intelligence Agency, *The World Factbook 2001;* U.S. Census Bureau

Interpreting the Chart **What is noteworthy about life expectancy in Iceland, Norway, and Sweden?**

Norway

Norway is a long, narrow, and rugged country along the western coast of the Scandinavian Peninsula. Norway once was united with Denmark and then Sweden. In 1905 Norway became independent. Today Norway is a constitutional monarchy with an elected parliament.

About 75 percent of the people live in urban areas. The largest cities are the capital, Oslo, and Bergen on the Atlantic coast. Oslo is a modern city. It lies at the end of a wide fjord on the southern coast. The city is Norway's leading seaport, as well as its industrial and cultural center.

Norway has valuable resources, especially oil and natural gas. However, Norway's North Sea oil fields are expected to run dry over the next century. A long coastline and location on the North Sea have helped make Norway a major fishing and shipping country. Fjords shelter Norway's harbors and its fishing and shipping fleets.

Sweden

Sweden is Scandinavia's largest and most populous country. It is located between Norway and Finland. Most Swedes live in cities and towns. The largest cities are Stockholm, which is Sweden's capital, and Göteborg. Stockholm is located on the Baltic Sea coast. It is a beautiful city of islands and forests. Göteborg is a major seaport.

Like Norway, Sweden is a constitutional monarchy. The country has been at peace since the early 1800s. Sweden remained **neutral** during World Wars I and II. A neutral country is one that chooses not to take sides in an international conflict.

Sweden's main sources of wealth are forestry, farming, mining, and manufacturing. Wood, iron ore, automobiles, and wireless telephones are exports. Hydroelectricity is important.

Riddarhomem—the Isle of Knights—is one of several islands on which the original city of Stockholm was built.

Interpreting the Visual Record **Why might a group of islands be a good place to build a city?** ►

Teaching Objective 2

LEVEL 1: (Suggested time: 30 min.) Organize the class into five groups. Assign one of the region's countries to each group. Have the groups design a new flag for their assigned country using colors and symbols that refer to the text material. Tell students that the flags should represent the important features of the region's countries as well as Lapland and Greenland. Ask each group to choose a spokesperson to explain its meaning to the class. Then have the spokespeople make their presentations. Finally, display students' flags around the classroom.
ENGLISH LANGUAGE LEARNERS, COOPERATIVE LEARNING

LEVELS 2 AND 3: (Suggested time: 45 min.) Have students create brochures to encourage people from other countries to invest in Scandinavian countries. Brochures should provide an introduction to the country's physical and human geography and summarize each country's economic and political characteristics as well as other important features. Instruct students to make a clear connection in their brochures between their assigned country's characteristics and its economic future. Ask students to add a catchy slogan for the country. *(example: "Iceland—Cold Name, Warm Welcome")* You may want to have students add drawings or pictures from magazines to add interest. Additional research may be necessary.

Connecting to Art

Stave Churches

A stave church in Norway

In Norway you will find some beautiful wooden churches built during the Middle Ages. They are known as stave churches because of their corner posts, or staves. The staves provide the building's basic structure. Today stave churches are a reminder of the days when Viking and Christian beliefs began to merge in Norway.

As many as 800 stave churches were built in Norway during the 1000s and 1100s. Christianity was then beginning to spread throughout the country. It was replacing the old religious beliefs of the Viking people. Still, Viking culture is clearly seen in stave buildings.

Except for a stone foundation, stave churches are made entirely of wood. Workers used methods developed by Viking boat builders. For example, wood on Viking boats was coated with tar to keep it from rotting. Church builders did the same with the wood for their churches. They also decorated the churches with carvings of dragons and other creatures. The stave church at Urnes even has a small Viking ship decorated with nine candles.

When the plague, or Black Death, arrived in Norway about 1350, many communities were abandoned. Many stave churches fell apart. Others were replaced by larger stone buildings. Today only 28 of the original buildings remain. They have been preserved for their beauty and as reminders of an earlier culture.

Understanding What You Read

1. What are staves?
2. How did stave churches reflect new belief systems in Norway?

Denmark

Denmark is the smallest and most densely populated of the region's countries. Most of Denmark lies on the Jutland Peninsula. About 500 islands make up the rest of the country.

Denmark is also a constitutional monarchy. The capital and largest city is Copenhagen. It lies on an island between the Jutland Peninsula and Sweden. Some 1.4 million people—about 25 percent of Denmark's population—live there.

About 60 percent of Denmark's land is used for farming. Farm products, especially meat and dairy products, are important exports. Denmark also has a modern industrial economy. Industries include food processing, machinery, furniture, and electronics.

Across the Curriculum

LITERATURE

Hans Christian Andersen

Among the many contributions made to the arts by Scandinavians, some of the most popular are the stories written by Danish author Hans Christian Andersen. His fairy tales are among the most frequently translated works in the world. Andersen also wrote plays and novels.

The author published his first book of stories, *Tales, Told for Children,* in 1835. It included the classic "Princess and the Pea," which was the inspiration for the Broadway musical *Once upon a Mattress.* Andersen's story "The Little Mermaid" was the basis for a popular animated movie of the same name. A bronze statue of a mermaid was placed in Copenhagen's harbor to commemorate Andersen's heroine.

Critical Thinking: In what ways have Andersen's tales spread beyond Scandinavia?

Answer: They have become popular around the world and have been used for a musical and a movie.

Connecting to Art Answers

1. corner posts of wooden churches
2. They were built as Christianity was replacing Viking religious beliefs.

Teacher to Teacher

Kay A. Knowles of Montross, Virginia, suggests the following activity to show how cultures can differ within the same region. First, have students conduct research on these topics as they relate to the British Isles and Scandinavia: origins, religions, languages, ethnic groups, conflicts, and customs. Then have students create charts to organize and present their findings.

Lead a class discussion comparing what students have learned from the activity to other countries or regions they have studied. Challenge them to draw conclusions.

Close

Lead a discussion on how a Viking might react if he or she visited Scandinavia today. Ask: Which aspects of the region might be familiar? Which would be unfamiliar?

Section 4 Review

Answers

Define For definitions, see: neutral, p. 282; uninhabitable, p. 284; geysers, p. 284

Working with Sketch Maps They are descendants of hunters from northern Asia, speak languages related to Finnish, earn money from reindeer herding or tourism.

Reading for the Main Idea

1. closely related languages (except for Finnish), Lutheran Protestant religion, democratic governments (NGS 10)
2. by catching fish, using hot water from geysers for heat (NGS 15)
3. country's original settlers not Vikings; Finnish not related to other Scandinavian languages (NGS 10)

Critical Thinking

4. possible answer: They may cause the cultures to become more separate and lead to independence for Greenland.

Organizing What You Know

5. Answers will vary but should be consistent with text.

Visual Record Answer ►

The interior is icy and uninhabitable.

▲ Greenland's capital lies on the island's southwestern shore.

Interpreting the Visual Record

Why do most people in Greenland live along the coast?

Greenland

The huge island of Greenland is part of North America, but it is a territory of Denmark. Greenland's 56,000 people have their own government. They call their island Kalaallit Nunaat. The capital is Nuuk, also called Godthab. Most of the island's people are Inuit (Eskimo). Fishing is the main economic activity. Some Inuit still hunt seals and small whales.

The island's icy interior is **uninhabitable**. An uninhabitable area is one that cannot support human settlement. Greenland's people live mostly along the southwestern coast in the tundra climate regions.

Our Amazing Planet

The Great Geysir in southwestern Iceland can spout water nearly 200 feet (61 m) into the air. Some geysers shoot steam and boiling water to a height of more than 1,600 feet (nearly 500 m)!

Iceland

Between Greenland and Scandinavia is the country of Iceland. This Atlantic island belonged to Denmark until 1944. Today it is an independent country. It has an elected president and parliament.

Unlike Greenland, Iceland is populated mostly by northern Europeans. The capital and largest city is Reykjavik (RAYK-yuh-veek). Nearly 40 percent of the country's people live there.

Icelanders make good use of their country's natural resources. For example, about 70 percent of the country's exports are fish. These fish come from the rich waters around the island. In addition, hot water from Iceland's **geysers** heats homes and greenhouses. The word *geyser* is an Icelandic term for hot springs that shoot hot water and steam into the air. Volcanic activity forces heated underground water to rise from the geyser.

Finland

Finland is the easternmost of the region's countries. It lies mostly between two arms of the Baltic Sea: the Gulf of Bothnia and the Gulf of Finland. The capital and largest city is Helsinki, which is located on the southern coast.

REVIEW AND ASSESS

Have students complete the Section Review. Then have students outline the main points of Section 4. Ask them to choose one topic from their outlines and create a simple graphic organizer to illustrate it. Call on volunteers to draw their organizers on the chalkboard and explain their use. Then have students complete Daily Quiz 12.4.

RETEACH

Have students complete Main Idea Activity 12.4. Then organize students into groups and assign each group one of the Scandinavian countries. Have each group become "experts" on their country. Give them chalk for writing on the board or materials for using the overhead projector, allow time for collaboration, and have them present a lesson on their country to the class. **ENGLISH LANGUAGE LEARNERS, COOPERATIVE LEARNING**

EXTEND

Have interested students conduct research on the important role Finnish ski troops played during the Soviet invasion of their country in 1939–40. Have students act out their findings in a "news broadcast" from the "front." **BLOCK SCHEDULING**

The original Finnish settlers probably came from northern Asia. Finnish belongs to a language family that includes Estonian and Hungarian. About 6 percent of Finns speak Swedish. Finland was part of Sweden from the 1100s to 1809. It then became part of Russia. Finland gained independence at the end of World War I.

As in the other countries of the region, trade is important to Finland. The country is a major producer of paper and other forest products as well as wireless telephones. Metal products, shipbuilding, and electronics are also important industries. Finland imports energy and many of the raw materials needed in manufacturing.

▲ Young Sami couples here are dressed in traditional clothes for the Easter reindeer races in northern Norway.

Interpreting the Visual Record **What climates would you find in the lands of the Sami?**

Lapland

Across northern Finland, Sweden, and Norway is a culture region known as Lapland. This region is populated by the Lapps, or Sami, as they call themselves.

The Sami are probably descended from hunters who moved to the region from northern Asia. The languages they speak are related to Finnish. The Sami have tried to keep their culture and traditions, such as reindeer herding. Many now earn a living from tourism.

✓ **READING CHECK:** *Human Systems* Around what activities are the economies of the countries and territories discussed in this section organized? Some are modern industrial economies, while others are organized around fishing, farming, or tourism.

go.hrw.com Homework Practice Online
Keyword: SK3 HP12

Section 4 Review

Define and explain: neutral, uninhabitable, geysers

Working with Sketch Maps On the map you created in Section 3, label the countries of Scandinavia, Oslo, Bergen, Stockholm, Göteborg, Copenhagen, Nuuk (Godthab), Reykjavik, Gulf of Bothnia, Gulf of Finland, Helsinki, and Lapland. In the margin describe the people of the Lapland region.

Reading for the Main Idea

1. *Human Systems* What are two of the cultural similarities among the peoples of Scandinavia?
2. *Environment and Society* In what ways have Icelanders adapted to their natural environment?
3. *Human Systems* How have the history and culture of Finland been different from that of other countries in Scandinavia?

Critical Thinking

4. **Making Generalizations and Predictions** How do you think the location of Greenland and the culture of its people will affect the island's future relationship with Denmark?

Organizing What You Know

5. **Summarizing** Copy the following graphic organizer. Label the center of the organizer "Scandinavia." In the ovals, write one characteristic of each country and region discussed in this section. Then do the same for the other countries, as well as for Greenland and Lapland.

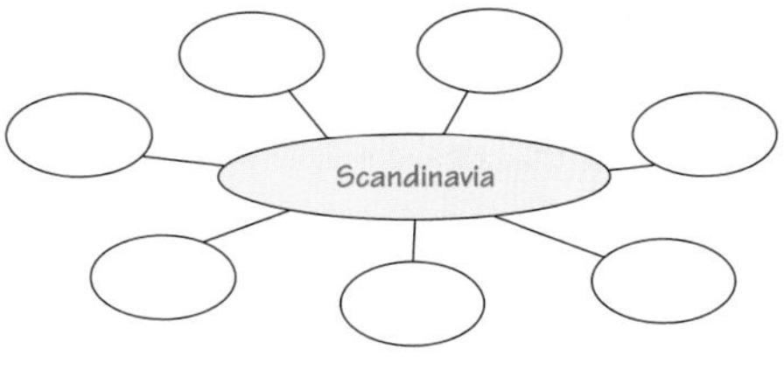

CHAPTER 15 Review ANSWERS

Building Vocabulary

For definitions, see: fjords, p. 270; lochs, p. 270; North Atlantic Drift, p. 271; constitutional monarchy, p. 273; famine, p. 278; bog, p. 280; peat, p. 280; neutral, p. 282; uninhabitable, p. 284; geysers, p. 284

Reviewing the Main Ideas

1. Great Britain, Ireland; Atlantic Ocean, North Sea, English Channel, Irish Sea (NGS 4)
2. marine west coast, humid continental, subarctic, tundra (NGS 4)
3. Greenland (NGS 4)
4. Most colonies gained their independence. The empire became the British Commonwealth of Nations. (NGS 13)
5. Roman Catholicism and Protestantism are the two major religions there; violent Catholic and Protestant groups have fought over whether or not Northern Ireland should remain part of the United Kingdom. (NGS 10)

◄ **Visual Record Answer**

subarctic or tundra

Assess

Have students complete the Chapter 12 Test.

Reteach

Organize the class into sections for the United Kingdom, Ireland, and Scandinavia. Ask students to work with a partner within their section to review the landforms, rivers, climates, resources, economy, history, governments, and cultural traits of their assigned areas. Then have the pairs meet within their section and compile lists of what they think are the 10 most important facts one should know about their region.

PORTFOLIO

Portfolio Extensions

1. Have students conduct research on John Cabot, Sir Francis Drake, Sir Walter Raleigh, or Captain James Cook and create presentations on the role the explorer played in the creation of the British Empire. Students' presentations should include an illustrated map and a version of the explorer's diary or ship's log.
2. For almost 100 years, Northern Ireland has had a tradition of political murals. Have students conduct research on the conflicts in Northern Ireland and draw murals advocating lasting peace on butcher paper. Photograph the murals and place the photograph in students' portfolios.

Chapter 12 Review Answers

Understanding Environment and Society

The North Sea is an important location for the production of oil and gas. Natural gas was discovered off the Netherlands in 1959. Norway, Denmark, Germany, Netherlands, and the United Kingdom are the main countries that drill in the North Sea. Use Rubric 29, Presentations, to evaluate student work.

Thinking Critically

1. Answers will vary, but students should mention the themes in some of these art forms, such as Shakespearean literature or Beatles music, make them universally appealing.
2. saved transportation costs because coal, which fueled factories, was available nearby
3. history of English control; English is language of commerce, government, education
4. Possible: adapted building and clothing styles; used skis, snowshoes, sleds for travel; learned to make a living from what environment makes possible, such as reindeer

Chapter 12 Reviewing What You Know

Building Vocabulary

On a separate sheet of paper, write sentences to define each of the following words.

1. fjords
2. lochs
3. North Atlantic Drift
4. constitutional monarchy
5. famine
6. bog
7. peat
8. neutral
9. uninhabitable
10. geysers

Reviewing the Main Ideas

1. *Places and Regions* What islands make up the British Isles? What bodies of water surround the British Isles?
2. *Places and Regions* What are the three main climates in northern Europe?
3. *Places and Regions* What large North Atlantic island is a territory of Denmark?
4. *Human Systems* What happened to the British Empire?
5. *Human Systems* What are the two major religions in Northern Ireland? What do these religions have to do with violence in Northern Ireland?

Understanding Environment and Society

Resource Use

Prepare a presentation, along with a map, graph, chart, model, or database, on the distribution of oil and natural gas in the North Sea. You may want to think about the following:

- How oil and natural gas are recovered there.
- How countries divided up the North Sea's oil and natural gas.

Write a five-question quiz, with answers, about your presentation to challenge fellow students.

Thinking Critically

1. **Drawing Inferences and Conclusions** Why do you think British literature, art, and music have been popular around the world?
2. **Drawing Inferences and Conclusions** Why do you think industrial cities like Birmingham and Manchester in Great Britain grew up near coal deposits?
3. **Drawing Inferences and Conclusions** Why do you think most Irish speak English rather than Gaelic?
4. **Making Generalizations and Predictions** How do you think Scandinavians have adapted to life in these very cold environments?
5. **Finding the Main Idea** Why are climates in the British Isles milder than in much of Scandinavia?

FOOD FESTIVAL

In Swedish, the word *smorgasbord* means "bread and butter table," but a smorgasbord is not just a table loaded with buttered bread. It is a complete buffet-style meal, with a variety of open-faced sandwiches, sliced meats, marinated or pickled fish, cheeses, hot or cold cooked vegetables, salads, and desserts. To create your own Swedish smorgasbord, have students bring as many of the listed food items as they can. Set the food on a large table and let students help themselves.

CHAPTER 12 Review and Assessment Resources

Reproducible
- Readings in World Geography, History, and Culture 34, 35, and 36
- Vocabulary Activity 12

Technology
- Chapter 12 Test Generator (on the One-Stop Planner)
- Audio CD Program, Chapter 12 (English and Spanish)
- HRW Go site

Reinforcement, Review, and Assessment
- Chapter Review, pp. 286–87
- Chapter 12 Tutorial for Students, Parents, Mentors, and Peers
- Chapter 12 Test
- Chapter 12 Test for English Language Learners and Special-Needs Students

Building Social Studies Skills

Map Activity

On a separate sheet of paper, match the letters on the map with their correct labels.

London	**Oslo**
Manchester	**Stockholm**
Belfast	**Copenhagen**
Dublin	**Reykjavik**
Cork	**Helsinki**

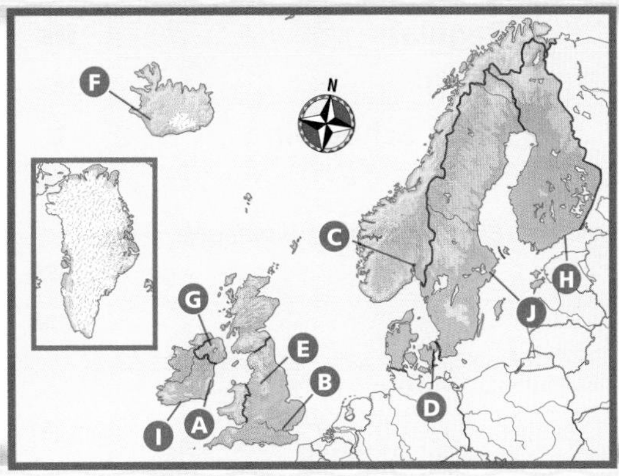

Mental Mapping Skills Activity

On a separate sheet of paper, draw a freehand map of northern Europe. Make a key for your map and label the following:

Baltic Sea	**Ireland**
Denmark	**Norway**
Finland	**Scandinavian Peninsula**
Great Britain	**Sweden**
Iceland	

WRITING ACTIVITY

Use print resources to find out more about the Vikings and how they lived in their cold climate. Write a short story set in a Viking village or on a Viking voyage. Describe daily life in the village or on the voyage. Include a bibliography showing references you used. Be sure to use standard grammar, spelling, sentence structure, and punctuation.

Alternative Assessment

Portfolio Activity

Learning About Your Local Geography

Research Project Northern Europe's social programs are supported by taxes. Research the taxes that people in your area must pay. Interview residents, asking their feelings about taxes.

internet connect go.hrw.com

Internet Activity: **go.hrw.com**
KEYWORD: **SK3 GT12**

Choose a topic to explore northern Europe:
- Explore the islands and fjords on the Scandinavian coast.
- Visit historic palaces in the United Kingdom.
- Investigate the history of skiing.

herding; developed diverse energy sources

5. Winds blowing over the North Atlantic Drift bring mild temperatures and rain.

Map Activity

A. Dublin	**F.** Reykjavik
B. London	**G.** Belfast
C. Oslo	**H.** Helsinki
D. Copenhagen	**I.** Cork
E. Manchester	**J.** Stockholm

Mental Mapping Skills Activity

Maps will vary, but listed places should be labeled in their approximate locations.

Writing Activity

Check stories to see that students have introduced several aspects of the Vikings' environment. Use Rubric 39, Writing to Create, to evaluate student work.

Portfolio Activity

Information included should be consistent with text material. Use Rubric 29, Presentations, to evaluate student work.

internet connect go.hrw.com

GO TO: **go.hrw.com**
KEYWORD: **SK3 Teacher**
FOR: a guide to using the Internet in your classroom

CHAPTER 13

Eastern Europe

Chapter Resource Manager

Objectives	Pacing Guide	Reproducible Resources
SECTION 1 **Physical Geography** (pp. 289–91) **1.** Identify the major physical features of Eastern Europe. **2.** Identify the climates and natural resources of the region.	**Regular** .5 day **Block Scheduling** .5 day *Block Scheduling Handbook, Chapter 13*	**RS** Guided Reading Strategy 13.1
SECTION 2 **The Countries of Northeastern Europe** (pp. 292–97) **1.** Identify the peoples who contributed to the early history of northeastern Europe. **2.** Describe how northeastern Europe's culture was influenced by other cultures. **3.** Describe how the political organization of the region has changed since World War II.	**Regular** 2.5 days **Block Scheduling** 1 day *Block Scheduling Handbook, Chapter 13*	**RS** Guided Reading Strategy 13.2 **RS** Graphic Organizer 13 **E** Cultures of the World Activity 3 **E** Creative Strategies for Teaching World Geography, Lesson 11 **SM** Geography for Life Activities 13 and 14 **SM** Map Activity 13
SECTION 3 **The Countries of Southeastern Europe** (pp. 298–303) **1.** Describe how southeastern Europe's early history helped to shape its modern societies. **2.** Describe how culture affects the region. **3.** Describe how the region's past has contributed to current conflicts.	**Regular** 2 days **Block Scheduling** .5 day *Block Scheduling Handbook, Chapter 13*	**RS** Guided Reading Strategy 13.3 **E** Cultures of the World Activity 3 **E** Creative Strategies for Teaching World Geography, Lesson 11 **SM** Geography for Life Activity 13

Chapter Resource Key

RS Reading Support
IC Interdisciplinary Connections
E Enrichment
SM Skills Mastery
A Assessment
REV Review
ELL Reinforcement and English Language Learners
Transparencies
CD–ROM
Music
Video
go.hrw.com Internet
Holt Presentation Maker Using Microsoft® Powerpoint®

One-Stop Planner CD–ROM

See the *One-Stop Planner* for a complete list of additional resources for students and teachers.

One-Stop Planner CD–ROM

It's easy to plan lessons, select resources, and print out materials for your students when you use the ***One-Stop Planner CD–ROM with Test Generator.***

Technology Resources	Review, Reinforcement, and Assessment Resources
One-Stop Planner CD–ROM, Lesson 13.1 Geography and Cultures Visual Resources with Teaching Activities 24–29 Homework Practice Online HRW Go site	**ELL** Main Idea Activity 13.1 **REV** Section 1 Review, p. 291 **A** Daily Quiz 13.1 **ELL** English Audio Summary 13.1 **ELL** Spanish Audio Summary 13.1
One-Stop Planner CD–ROM, Lesson 13.2 Homework Practice Online HRW Go site	**ELL** Main Idea Activity 13.2 **REV** Section 2 Review, p. 297 **ELL** English Audio Summary 13.2 **ELL** Spanish Audio Summary 13.2 **A** Daily Quiz 13.2
One-Stop Planner CD–ROM, Lesson 13.3 Homework Practice Online HRW Go site	**ELL** Main Idea Activity 13.3 **REV** Section 3 Review, p. 303 **A** Daily Quiz 13.3 **ELL** English Audio Summary 13.3 **ELL** Spanish Audio Summary 13.3

internet connect

HRW ONLINE RESOURCES

GO TO: go.hrw.com
Then type in a keyword.

TEACHER HOME PAGE
KEYWORD: SK3 TEACHER

CHAPTER INTERNET ACTIVITIES
KEYWORD: SK3 GT13

Choose an activity to:
- investigate the conflicts in the Balkans.
- take a virtual tour of Eastern Europe.
- learn about Baltic amber.

CHAPTER ENRICHMENT LINKS
KEYWORD: SK3 CH13

CHAPTER MAPS
KEYWORD: SK3 MAPS13

ONLINE ASSESSMENT
Homework Practice
KEYWORD: SK3 HP13
Standardized Test Prep Online
KEYWORD: SK3 STP13
Rubrics
KEYWORD: SS Rubrics

COUNTRY INFORMATION
KEYWORD: SK3 Almanac

CONTENT UPDATES
KEYWORD: SS Content Updates

HOLT PRESENTATION MAKER
KEYWORD: SK3 PPT13

ONLINE READING SUPPORT
KEYWORD: SS Strategies

CURRENT EVENTS
KEYWORD: S3 Current Events

Meeting Individual Needs

Ability Levels

Level 1 Basic-level activities designed for all students encountering new material

Level 2 Intermediate-level activities designed for average students

Level 3 Challenging activities designed for honors and gifted-and-talented students

English Language Learners Activities that address the needs of students with Limited English Proficiency

Chapter Review and Assessment

- **IC** Interdisciplinary Activities for the Middle Grades 13, 14, 15
- **E** Readings in World Geography, History, and Culture 37, 38, and 39
- **SM** Critical Thinking Activity 13
- **REV** Chapter 13 Review, pp. 334–35
- **REV** Chapter 13 Tutorial for Students, Parents, Mentors, and Peers
- **ELL** Vocabulary Activity 13
- **A** Chapter 13 Test
- **A** Unit 3 Test
- Chapter 13 Test Generator (on the One-Stop Planner)
- Audio CD Program, Chapter 13
- **A** Chapter 13 Test for English Language Learners and Special-Needs Students
- **A** Unit 3 Test for English Language Learners and Special-Needs Students

CHAPTER 13

Launch into Learning

Focus students' attention on the map on the following page. Call on volunteers to locate several familiar place names on the map and tell where they have heard the names before. *(Possible answers: Transylvanian Alps—Dracula movies, Bosnia or Kosovo—news broadcasts, Poland—history lessons on World War II)* Discuss what these associations indicate about the region's history. *(rich history, involved in wars)* Tell students that cultural conflicts are part, but not all, of the region's complex history. Tell students that they will learn more about the history and cultures of Eastern Europe in this chapter.

Section 1

Objectives

1. Identify the major physical features of Eastern Europe.
2. Describe the climates and natural resources of the region.

Links to Our Lives

You may wish to point out the following reasons why we should know more about Eastern Europe:

- Ethnic conflicts in Yugoslavia have resulted in military involvement by the United States, NATO, and the UN.
- Many countries in the region are establishing democratic governments after decades of communism.
- Eastern Europe is a cultural and economic crossroads between Western Europe and Asia.
- Many Americans can trace their ancestry to Eastern Europe.
- The region's foods, festivals, literature, music, and other traditions can be enjoyed by all.

CHAPTER 13

Eastern Europe

In this chapter you will learn about countries that share common physical features but have developed very different cultures. First, however, we meet Marta, a Hungarian student.

Üdvözöljük Magyarországon!

Hello! My name is Marta, and I am from Kecskemét (KECH-ke-mayt), Hungary. My mother is a secretary, and my father is an agricultural engineer. I am in my last year of high school.

Our apartment has no living or dining room, just a kitchen, a tiny balcony, a bathroom, a hallway, and two bedrooms. In the morning, we eat in my parents' room, where we also study and talk during the day.

Our lives changed very much in 1991 when the Soviet Union collapsed. Before this, we had to study Russian in school. Also, my family is Catholic, but we had to have church services in secret. My parents would have risked losing their jobs if anyone found out. Now everyone goes to church freely. My favorite sports in school are basketball, swimming, and fencing. On Friday night we have parties organized by the school. Sometimes we go to the movies. In the summer, I used to work picking cherries. Now I work in a factory processing chickens.

▲ Translation: Welcome to Hungary!

FOCUS

LET'S GET STARTED

Copy the following instructions onto the chalkboard: *Look at the map and photo in Section 1. Imagine that they illustrate magazine articles about Eastern Europe's physical geography. What might be the titles for the articles? Write down your ideas for one of the illustrations. (Example: for the photo of the Danube River—"Danube: the Lifeblood Flowing through Romania's Heart")* Discuss student responses. Tell students that in Section 1 they will learn more about the physical geography of Eastern Europe.

Using the Physical-Political Map

Tell the class that Eastern Europe has been invaded from different directions many times over the centuries. Have students examine the map on this page. Ask them to speculate how Eastern Europe's physical geography might have contributed to the region being a crossroads. *(center of the continent, many rivers for navigation, few barriers to invasion)*

Building Vocabulary

Write the key terms on the chalkboard. Have students find and read the definitions for **oil shale**, **lignite**, and **amber** in the Section 1 text or glossary. Ask: What do these materials have in common? *(They are natural resources found in Eastern Europe.)*

Section 1 Physical Geography

Read to Discover

1. What are the major physical features of Eastern Europe?
2. What climates and natural resources does this region have?

Define

oil shale
lignite
amber

Locate

Baltic Sea
Adriatic Sea
Black Sea
Danube River
Dinaric Alps
Balkan Mountains
Carpathian Mountains

WHY IT MATTERS

The Danube and its tributaries are important transportation links in Eastern Europe. The rivers also help spread pollution. Use CNNfyi.com or other **current events** sources to find examples of pollution problems facing the region's rivers. Record your findings in your journal.

Amber

Eastern Europe: Physical-Political

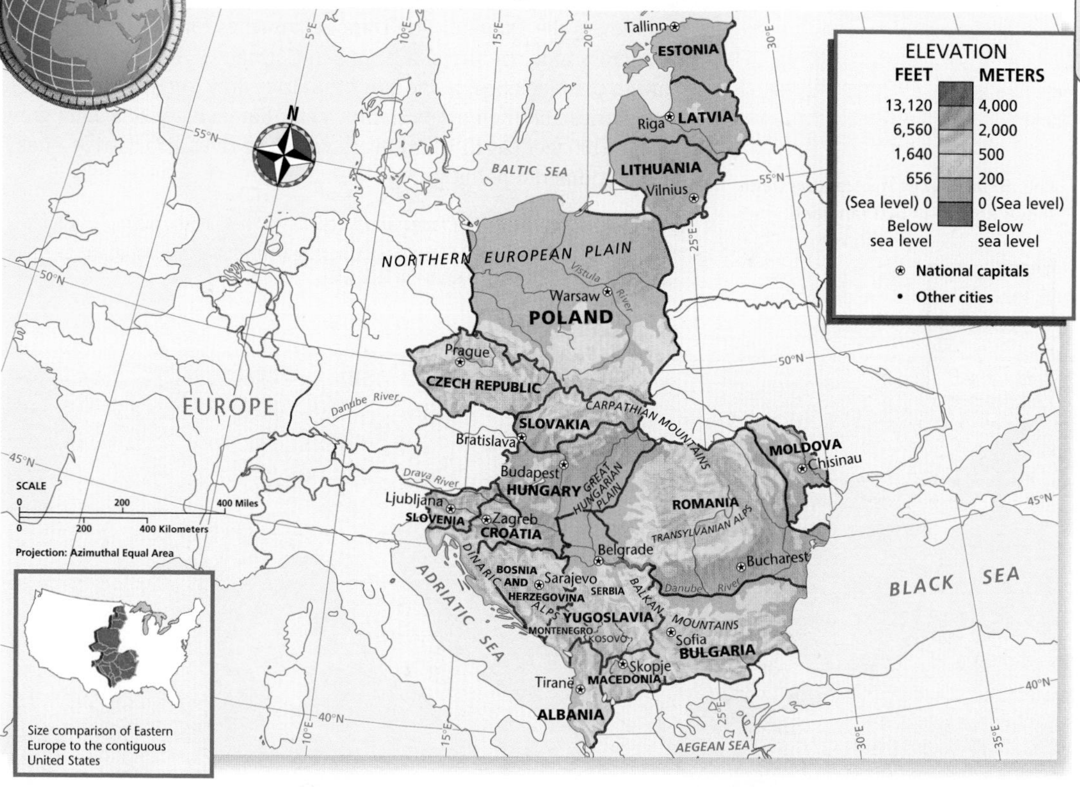

Section 1 RESOURCES

Reproducible

- Block Scheduling Handbook, Chapter 13
- Guided Reading Strategy 13.1

Technology

- One-Stop Planner CD–ROM, Lesson 13.1
- Homework Practice Online
- Geography and Cultures Visual Resources with Teaching Activities 24–29
- HRW Go site

Reinforcement, Review, and Assessment

- Section 1 Review, p. 291
- Daily Quiz 13.1
- Main Idea Activity 13.1
- English Audio Summary 13.1
- Spanish Audio Summary 13.1

TEACH

Teaching Objectives 1–2

ALL LEVELS: (Suggested time: 20 min.) Copy the following graphic organizer onto the chalkboard, omitting the italicized answers. Point out that it shows the general location of Eastern Europe's main regions. Call on students to provide words and phrases to summarize the physical features, climates, and resources of each region. Encourage students to use the unit's atlas as an additional resource. Then have students speculate how the location and availability of resources might affect economic development.

LEVELS 2 AND 3: (Suggested time: 20 min.) Have students complete the All Levels lesson. Then have each student select an Eastern European country and write a paragraph that describes its landforms, climates, rivers, and resources. Again, encourage students to use the unit's atlas as a resource. Call on volunteers to read their paragraphs to the class. Then have students speculate what effect pollution might have on the country they selected. *(Possible answer: harm rivers and air)*

The Physical Geography of Eastern Europe

Baltic Countries	Heartland	Balkan Countries
• *plains* • *cold winters* • *amber, oil shale*	• *plains, mountains* • *cold winters in east, warmer in west* • *bauxite, lignite, salt*	• *mountains* • *cold winters in east, warmer in south and west* • *bauxite, oil, lignite*

EYE ON EARTH

Refer to the section titled Climate and Resources on the opposite page. Notice that salt mines have operated in Poland since the 1200s.

Poland's Wieliczka salt mine contains more than 124 miles (200 km) of passages that connect more than 2,000 rooms. The lowest room is 1,073 feet (327 m) below the surface.

There is a tradition among the Wieliczka miners to carve the salt into churches, altars, and large statues. In recent years, increased humidity started to dissolve the carvings. An international team of scientists has made great progress in stopping the deterioration.

Critical Thinking: Once students have read the section on Poland, remind them of the salt carvings. Ask them how these carvings are an expression of Polish culture. *(Possible answer: shows importance of religion)*

internet connect
GO TO: go.hrw.com
KEYWORD: SK3 CH13
FOR: Web sites about Poland

internet connect
GO TO: go.hrw.com
KEYWORD: SK3 CH13
FOR: Web sites about Eastern Europe

Our Amazing Planet

Amber is golden, fossilized tree sap. The beaches along the eastern coast of the Baltic Sea are the world's largest and most famous source of amber. Baltic amber is approximately 40 million years old.

Physical Features

Eastern Europe stretches southward from the often cold, stormy shores of the Baltic Sea. In the south are the warmer and sunnier beaches along the Adriatic and Black Seas. We can divide the countries of this region into three groups. Poland, the Czech Republic, Slovakia, and Hungary are in the geographical heart of Europe. The Baltic countries are Estonia, Latvia, and Lithuania. Yugoslavia, Bosnia and Herzegovina, Croatia, Slovenia, Macedonia, Romania, Moldova, Bulgaria, and Albania are the Balkan countries.

Landforms Eastern Europe is a region of mountains and plains. The plains of Poland and the Baltic countries are part of the huge Northern European Plain. The Danube River flows through the Great Hungarian Plain, also called the Great Alföld.

The Alps extend from central Europe southeastward into the Balkan Peninsula. Where they run parallel to the Adriatic coast, the mountains are called the Dinaric (duh-NAR-ik) Alps. As the range continues eastward across the peninsula its name changes to the Balkan Mountains. The Carpathian (kahr-PAY-thee-uhn) Mountains stretch from the Czech Republic across southern Poland and Slovakia and into Ukraine. There they curve south and west into Romania. In Romania they are known as the Transylvanian Alps.

Rivers Eastern Europe's most important river for trade and transportation is the Danube. The Danube stretches for 1,771 miles (2,850 km) across nine countries. It begins in Germany's Alps and flows eastward to the Black Sea. Some 300 tributaries flow into the Danube. The river carries and then drops so much silt that its Black Sea delta grows by 80 to 100 feet (24 to 30 m) every year. The river also carries a heavy load of industrial pollution.

✓ **READING CHECK:** *Places and Regions* What are the main physical features in Eastern Europe? areas—Baltics, heartland, Balkans; landforms—mountains, plains; rivers—Danube, tributaries

This aerial view of the Danube Delta shows Romania's rich farmland.

CLOSE

Ask a student to describe one of the region's countries just by naming the bodies of water or countries that border it, plus one fact about its physical geography. Then have that student call on another student to name the country. Continue until all of the countries have been covered.

REVIEW AND ASSESS

Have students complete the Section Review. Then pair students. Have one student name a country in Eastern Europe and the other name a physical feature or resource found in or bordering that country. Then have students complete Daily Quiz 13.1. **COOPERATIVE LEARNING**

RETEACH

Have students complete Main Idea Activity 13.1. Then organize the class into triads and give each student an outline map of the region. Assign each group member the task of labeling either the plains, mountains, or rivers. Instruct members to exchange maps until all maps are complete.
ENGLISH LANGUAGE LEARNERS, COOPERATIVE LEARNING

EXTEND

Invite interested students to conduct research on ways that the physical geography of the Northern European Plain has affected history. They may want to create an illustrated map that shows the invasions and influences that have swept across this broad, lowland area. **BLOCK SCHEDULING**

Climate and Resources

The eastern half of the region has long, snowy winters and short, rainy summers. Farther south and west, winters are milder and summers become drier. A warm, sunny climate has drawn visitors to the Adriatic coast for centuries.

Eastern Europe's mineral and energy resources include coal, natural gas, oil, iron, lead, silver, sulfur, and zinc. The region's varied resources support many industries. Some areas of the Balkan region and Hungary are major producers of bauxite. Romania has oil. Estonia has deposits of **oil shale**, or layered rock that yields oil when heated. Estonia uses this oil to generate electricity, which is exported to other Baltic countries and Russia. Slovakia and Slovenia mine a soft form of coal called **lignite**. Nevertheless, many countries must import their energy because demand is greater than supply.

For thousands of years, people have traded **amber**, or fossilized tree sap. Amber is found along the Baltic seacoast. Salt mining, which began in Poland in the 1200s, continues in central Poland today.

During the years of Communist rule industrial production was considered more important than the environment. The region suffered serious environmental damage. Air, soil, and water pollution, deforestation, and the destruction of natural resources were widespread. Many Eastern European countries have begun the long and expensive task of cleaning up their environment.

✓ **READING CHECK:** *Environment and Society* What factors affect the location of economic activities in the region? climate, location of resources

BUILD on WHAT You Know

Do you remember what you learned about climate types? See Chapter 3 to review.

Section 1 Review

go.hrw.com Homework Practice Online
Keyword: SK3 HP13

Define and explain: oil shale, lignite, amber

Working with Sketch Maps On a map of Eastern Europe that you draw or that your teacher provides, label the following: Baltic Sea, Adriatic Sea, Black Sea, Danube River, Dinaric Alps, Balkan Mountains, and Carpathian Mountains.

Reading for the Main Idea

1. *Places and Regions* On which three major seas do the countries of Eastern Europe have coasts?
2. *Environment and Society* What types of mineral and energy resources are available in this region? How does this influence individual economies?

Critical Thinking

3. **Making Generalizations and Predictions** Would this region be suitable for agriculture? Why?
4. **Identifying Cause and Effect** How did Communist rule contribute to the pollution problems of this region?

Organizing What You Know

5. **Summarizing** Copy the following graphic organizer. Use it to summarize the physical features, climate, and resources of the heartland, the Baltics, and the Balkans. Then write and answer one question about the region's geography based on the chart.

Region	Physical features	Climate	Resources

Section 1 Review
Answers

Define For definitions, see: oil shale, p. 291; lignite, p. 291; amber, p. 291

Working with Sketch Maps Maps will vary, but listed places should be labeled in their approximate locations.

Reading for the Main Idea

1. Adriatic, Baltic, and Black Seas (NGS 4)
2. coal, iron, lead, natural gas, silver, sulfur, zinc, bauxite, oil shale, lignite, amber, salt; wide range of industries, energy shortages, need to import (NGS 4)

Critical Thinking

3. yes; wide plains, plenty of rain
4. increased production and ignored pollution

Organizing What You Know

5. heartland—Northern European Plain, Great Hungarian Plain, Carpathian Mountains; cold winters in east, milder in west; bauxite, lignite, salt; Baltics—Northern European Plain; cold winters; amber, oil shale; Balkans—Dinaric Alps, Balkan Mountains, Transylvanian Alps; cold winters in east, warm in south and west; bauxite, oil, lignite; questions will vary but should focus on geographic distributions

Section 2

Objectives

1. Identify what peoples contributed to the early history of northeastern Europe.
2. Analyze how northeastern Europe's culture was influenced by other cultures.
3. Describe how the political organization of the region has changed since World War II.

Focus

LET'S GET STARTED

Copy the following question onto the chalkboard: *What does a country need to do or have to attract tourists? (Possible answers: good hotels and restaurants, historical sites, beautiful scenery, entertainment, reliable transportation)* Tell students that many of the countries of northeastern Europe, particularly the Czech Republic, became popular tourist destinations during the 1990s. Before that time, few people traveled to the region because the governments were communist and tourist facilities were undeveloped. Tell students that in Section 2 they will learn more about the region's ancient and modern history.

Building Vocabulary

Write **Indo-European** on the chalkboard and call on a volunteer to read the text's definition aloud. Point out that the definition only tells about the languages the Indo-European peoples spoke—not what they looked like or what their customs were. Ask what two large regions are represented in the term. *(India and Europe)* Many languages of India are related to European languages.

Section 2 Resources

Reproducible

- Guided Reading Strategy 13.2
- Graphic Organizer 13
- Geography for Life Activities 13 and 14
- Map Activity 14
- Cultures of the World Activity 3
- Creative Strategies for Teaching World Geography, Lesson 11

Technology

- One-Stop Planner CD–ROM, Lesson 13.2
- Homework Practice Online
- HRW Go site

Reinforcement, Review, and Assessment

- Section 2 Review, p. 297
- Daily Quiz 13.2
- Main Idea Activity 13.2
- English Audio Summary 13.2
- Spanish Audio Summary 13.2

Section 2 The Countries of Northeastern Europe

Read to Discover

1. What peoples contributed to the early history of northeastern Europe?
2. How was northeastern Europe's culture influenced by other cultures?
3. How has the political organization of this region changed since World War II?

Define

Indo-European

Locate

Estonia
Poland
Czech Republic
Slovakia
Hungary
Lithuania
Latvia
Prague
Tallinn
Riga
Warsaw
Vistula River
Bratislava
Budapest

Why It Matters

Jerzy Giedroye, a Polish editor living in France, helped keep the free exchange of ideas alive in Eastern Europe during Communist rule. Use CNNfyi.com or other **current events** sources to discover the efforts of people like Giedroye. Record your findings in your journal.

Musical score by Hungarian Béla Bartók

History

Migrants and warring armies have swept across Eastern Europe over the centuries. Each group of people brought its own language, religion, and customs. Together these groups contributed to the mosaic of cultures we see in Eastern Europe today.

The Teutonic knights, a German order of soldier monks, brought Christianity and feudalism to northeastern Europe. They built this castle at Malbork, Poland, in the 1200s.

Early History Among the region's early peoples were the Balts. The Balts lived on the eastern coast of the Baltic Sea. They spoke **Indo-European** languages. The Indo-European language family includes many languages spoken in Europe. These include Germanic, Baltic, and Slavic languages. More than 3,500 years ago, hunters from the Ural Mountains moved into what is now Estonia. They spoke a very different, non-Indo-European language. The language they spoke provided the early roots of today's Estonian and Finnish languages. Beginning around A.D. 400, a warrior people called the Huns invaded the region from Asia. Later, the Slavs came to the region from the plains north of the Black Sea.

TEACH

Teaching Objective 1

ALL LEVELS: (Suggested time: 10 min.) Copy the following graphic organizer onto the chalkboard, omitting the italicized answers. Call on students to fill in the outer circles with the names of peoples who contributed to northeastern Europe's early history. Then call on others to supply details about each group to fill in the circles.

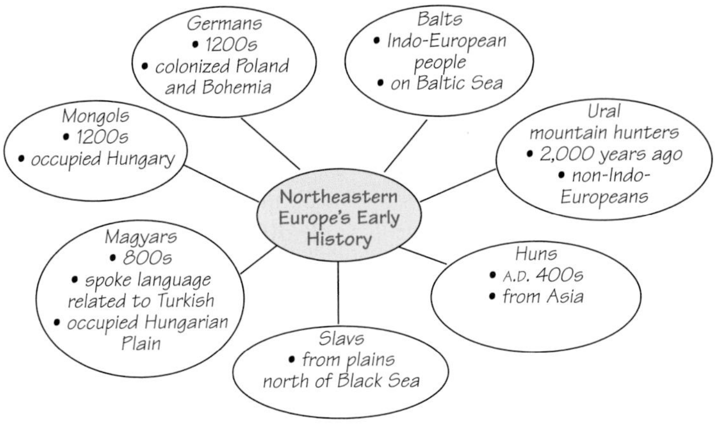

In the 800s the Magyars moved into the Great Hungarian Plain. They spoke a language related to Turkish. In the 1200s the Mongols rode out of Central Asia into Hungary. At the same time German settlers pushed eastward, colonizing Poland and Bohemia—the western region of the present-day Czech Republic.

Emerging Nations Since the Middle Ages, Austria, Russia, Sweden, and the German state of Prussia have all ruled parts of Eastern Europe. After World War I ended in 1918, a new map of Eastern Europe was drawn. The peace treaty created two new countries: Yugoslavia and Czechoslovakia. Czechoslovakia included the old regions of Bohemia, Moravia, and Slovakia. At about the same time, Poland, Lithuania, Latvia, and Estonia also became independent countries.

✓ **READING CHECK:** *Human Systems* What peoples contributed to the region's early history? **Balts, hunters from Ural Mountains, Huns, Slavs, Magyars, Mongols, Germans**

Culture

The culture and festivals of this region show the influence of the many peoples who contributed to its history. As in Scandinavia, Latvians celebrate a midsummer festival. The festival marks the summer solstice, the year's longest day. Poles celebrate major Roman Catholic festivals. Many of these have become symbols of the Polish nation. The annual pilgrimage, or journey, to the shrine of the Black Madonna of Częstochowa (chen-stuh-KOH-vuh) is an example.

Traditional Foods The food of the region reflects German, Russian, and Scandinavian influences. As in northern Europe, potatoes and sausages are important in the diets of Poland and the Baltic countries. Although the region has only limited access to the sea, the fish of lakes and rivers are often the center of a meal. These fish often include trout and carp. Many foods are preserved to last through the long winter. These include pickles, fruits in syrup, dried or smoked hams and sausages, and cured fish.

The Arts, Literature, and Science Northeastern Europe has made major contributions to the arts, literature, and sciences. For example, Frédéric Chopin (1810–1849) was a famous Polish pianist and composer. Marie Curie (1867–1934), one of the first female physicists, was also born in Poland. The writer Franz Kafka (1883–1924) was born to Jewish parents in Prague (PRAHG), the

Hungarian dancers perform in traditional dress.

Interpreting the Visual Record
How does this Hungarian costume compare to those you have seen from other countries? ▼

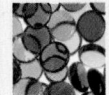

Cultural Kaleidoscope
Czech Texans

During the 1850s affordable land attracted Czech settlers to the fertile low hills of southeastern Texas. The newcomers missed the great cathedrals of their homeland. As soon as they could afford to do so, the immigrants built churches with high vaulted ceilings, stained glass windows, and religious statuary. Interior walls were elaborately painted with angels, clouds, trailing ivy, flowers, and symbols. One surviving ceiling features 66 types of flowers, vines, and shrubs.

Today, services and festivals held at the painted churches of Fayette County draw thousands of visitors, Czech and non-Czech alike. At its Veterans Day ceremony, airplanes drop flower petals on one of the churchyards to honor church members who died in World War II.

Critical Thinking: How did Czech architecture in Texas reflect architectural traditions from the homeland?

Answer: The Czechs built churches in Texas that reflected the great cathedrals of their homeland.

◄ **Visual Record Answer**

Answers will vary according to students' experiences.

Teaching Objective 2

ALL LEVELS: (Suggested time: 30 min.) Organize the class into small groups. Provide each group with colored markers and a sheet of butcher paper. Instruct the students to create a mural showing the cultural traits of the region. Ask them also to label their pictures, explaining the origins of the various contributions. Display the murals around the classroom.
ENGLISH LANGUAGE LEARNERS, COOPERATIVE LEARNING

TEACHER TO TEACHER

Jean Eldredge of Altamonte Springs, Florida, suggests the following activity to teach students about northeastern Europe: Have students choose a country in the region and write five complete sentences about it, using the material in Section 2, the unit's atlas, and any other resources available in your classroom. Collect the papers. Call on a volunteer to read one student's statements. Have the other students guess which country is being described. Limit the number of guesses. You may want to offer extra-credit points for correct identifications. Repeat the process until all the countries have been covered. This activity can be used as an initial teaching activity, to check on student progress, or to review for a test.

DAILY LIFE

Folk music is an important part of Baltic cultures. Many folk artists from the region now record for the world market.

The *kokle* is a native Latvian instrument. It is played flat on the musician's lap. The strings are plucked. Because the *kokle* can be adapted to a wide range of musical styles, musicians often include non-Latvian songs in their performances.

Modern *kokles* have 30 strings and can be several feet long. Earlier versions had between 5 and 13 strings and were hollowed from a single block of wood. A soundboard was then attached. No two instruments were exactly alike.

Critical Thinking: Why might there be an increase in folk music's popularity in Latvia and the other Baltic countries?

Answer: Students might suggest that since the countries became independent from the Soviet Union they have had more freedom to celebrate their unique cultures.

internet connect go.hrw.com

GO TO: **go.hrw.com**
KEYWORD: **SK3 CH13**
FOR: **Web sites about Baltic culture**

Chart Answer

Latvia, Lithuania

Northeastern Europe

Country	Population/ Growth Rate	Life Expectancy	Literacy Rate	Per Capita GDP
Czech Republic	10,264,212 / -0.1%	71, male / 78, female	100%	$12,900
Estonia	1,423,316 / -0.6%	64, male / 76, female	100%	$10,000
Hungary	10,106,017 / -0.3%	67, male / 76, female	99%	$11,200
Latvia	2,385,231 / -0.8%	63, male / 75, female	100%	$7,200
Lithuania	3,610,535 / -0.3%	63, male / 76, female	98%	$7,300
Poland	38,633,912 / -0.03%	69, male / 78, female	99%	$8,500
Slovakia	5,414,937 / 0.1%	70, male / 78, female	not available	$10,200
United States	281,421,906 / 0.9%	74, male / 80, female	97%	$36,200

Sources: Central Intelligence Agency, *The World Factbook 2001;* U.S. Census Bureau

Interpreting the Chart **Based on the data in the table, which two countries have the lowest levels of economic development?**

present-day capital of the Czech Republic. Astronomer Nicolaus Copernicus (1473–1543) was born in Toruń (TAWR-oon), a city in north-central Poland. He set forth the theory that the Sun—not Earth—is the center of the universe.

✓ **READING CHECK:** *Human Systems* How is the region's culture a reflection of its past and location? includes festivals and foods introduced by invaders, migrants, and neighbors

Northeastern Europe Today

Estonia, Latvia, and Lithuania lie on the flat plain by the eastern Baltic Sea. Once part of the Russian Empire, the Baltic countries gained their independence after World War I ended in 1918. However, they were taken over by the Soviet Union in 1940 and placed under Communist rule. The Soviet Union collapsed in 1991. Since then, the countries of northeastern Europe have been moving from communism to capitalism and democracy.

Estonia A long history of Russian control is reflected in Estonia today. Nearly 30 percent of Estonia's population is ethnic Russian. Russia remains one of Estonia's most important trading partners. However, Estonia is also building economic ties to other countries, particularly Finland. Ethnic Estonians have close cultural ties to Finland. In fact, the Estonian language is related to Finnish. Also, most people in both countries are Lutherans. Ferries link the Estonian capital of Tallinn (TA-luhn) with Helsinki, Finland's capital.

This suspension bridge spans the Western Dvina River in Riga, the capital of Latvia.

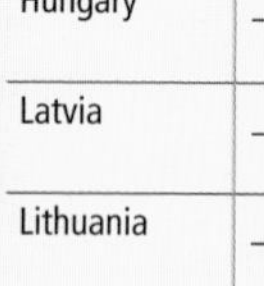

Teaching Objective 3

LEVEL 1: (Suggested time: 30 min.) Have students create covers for a special edition of a magazine titled News from the Northeast. In advance, duplicate a sheet of paper with the magazine title and date printed on it. (You may want to use the magazine cover template available on the go.hrw.com Web site at Keyword: SK3 Teacher.) Tell students that the special edition highlights the many changes that have occurred in northeastern Europe in the last 100 years or so. Have students think about what images and selected article titles would best represent the concept and re-create them for their own magazine covers.

LEVELS 2 AND 3: (Suggested time: 45 min.) Have students write an article for the News from the Northeast magazine.

➤ASSIGNMENT: Collect the articles from the Levels 2 and 3 lesson. Duplicate several, removing students' names, and distribute them to the class. Then instruct students to write a letter to the magazine's editor commenting on or disputing a point or fact in one of the articles.

Connecting to Literature

Toy robot

While he wrote many books, Czech writer Karel Capek is probably best known for his play R.U.R. *This play added the word* robot *to the English language. The Czech word* robota *means "drudgery" or forced labor. The term is given to the artificial workers that Rossum's Universal Robots factory make to free humans from drudgery. Eventually, the Robots develop feelings and revolt. The play is science fiction. Here, Harry Domin, the factory's manager, explains the origin of the Robots to visitor Helena Glory.*

ROBOT
ROBOTA

Domin: "Well, any one who has looked into human anatomy will have seen at once that man is too complicated, and that a good engineer could make him more simply. So young Rossum began to overhaul anatomy and tried to see what could be left out or simplified. . . . [He] said to himself: 'A man is something that feels happy, plays the piano, likes going for a walk, and in fact, wants to do a whole lot of things that are really unnecessary. . . .

But a working machine must not play the piano, must not feel happy, must not do a whole lot of other things. A gasoline motor must not have tassels or ornaments, Miss Glory. And to manufacture artificial workers is the same thing as to manufacture gasoline motors. The process must be of the simplest, and the product of the best from a practical point of view. . . .

Young Rossum . . . rejected everything that did not contribute directly to the progress of work—everything that makes man more expensive. In fact, he rejected man and made the Robot. My dear Miss Glory, the Robots are not people. Mechanically they are more perfect than we are, they have an enormously developed intelligence, but they have no soul."

Analyzing Primary Sources

1. Why does Rossum design the Robots without human qualities?
2. How has Karel Capel's play influenced other cultures?

Latvia Latvia is the second largest of the Baltic countries. Its population has the highest percentage of ethnic minorities. Some 57 percent of the population is Latvian. About 30 percent of the people are Russian. The capital, Riga (REE-guh), has more than 1 million people. It is the largest urban area in the three Baltic countries. Like Estonia, Latvia also has experienced strong Scandinavian and Russian influences. As well as having been part of the Russian Empire, part of the country once was ruled by Sweden. Another tie between Latvia, Estonia, and the Scandinavian countries is religion. Traditionally most people in these countries are Lutheran. In addition, Sweden and Finland are important trading partners of Latvia today.

Daily Life

As countries in northeastern Europe have gained freedom from communism, television and radio have been released from government control.

There are few comedies or dramas on television stations in the Baltic countries. Private groups have little experience creating television programs. Broadcasting is still dominated by political discussions and news programs.

Activity: Organize students into small groups and have them create a premise for a television comedy or drama set in any of the Baltic countries. The groups should share their ideas with the class.

Connecting to Literature

Answers

1. because he wished to create a machine that was simple, inexpensive, and that only contributed to the progress of work
2. by contributing the word *robot* to the English language

Teaching Objectives 1–3

LEVEL 1: (Suggested time: 45 min.) Pair students and assign each pair one of the region's countries. Using the text and other sources, have each pair summarize the major ethnic groups, cultural traits, and important facts about its country. Then have students organize their information into a brief oral report. You may want to require a visual aid as part of the presentation. **COOPERATIVE LEARNING**

LEVELS 2 AND 3: (Suggested time: 45 min.) Have students use the information they compiled for the Level 1 lesson to prepare a lesson about the selected country for an elementary classroom. Encourage them to use comparisons with familiar concepts so the younger children could understand the material. You may want to have students who complete their work early design a Web page to be used in an elementary school curriculum on Eastern Europe. The Web page should highlight one aspect of the country students reported on for the Level 1 lesson.

CLOSE

Ask: What historical events do all the countries in this region share? *(influenced by communism and the Soviet Union)* Which countries seem to have developed the most since the fall of the Soviet Union? *(Poland, Hungary, the Czech Republic)* the least? *(Slovakia)*

National Geography Standard 11

Human Systems The fall of communism in Eastern Europe has led to social improvements as well as changes in government. However, not everyone has experienced the benefits of these improvements equally. In Hungary, Slovakia, the Czech Republic, and Poland, those who have reaped the greatest benefits are generally former Communist Party officials, government workers, and private business owners. State company employees, women, retirees, and the working class have gained less.

Activity: Have students conduct research on the growth of new political parties in Eastern Europe since the fall of communism. Ask students to identify which countries are still dominated by former communists and which have developed new political parties.

Visual Record Answer

possible answers: lined with statues, has pedestrians instead of vehicular traffic

This Polish teenager wears traditional dress during a local festival.

Lithuania Lithuania is the largest and southernmost Baltic country. Its capital is Vilnius (VIL-nee-uhs). Lithuania's population has the smallest percentage of ethnic minorities. More than 80 percent of the population is Lithuanian. Nearly 9 percent is Russian, while 7 percent is Polish. Lithuania has ancient ties to Poland. For more than 200 years, until 1795, they were one country. Roman Catholicism is the main religion in both Lithuania and Poland today. As in the other Baltic countries, agriculture and production of basic consumer goods are important parts of Lithuania's economy.

Poland Poland is northeastern Europe's largest and most populous country. The total population of Poland is about the same as that of Spain. The country was divided among its neighbors in the 1700s. Poland regained its independence shortly after World War I. After World War II the Soviet Union established a Communist government to rule the country.

In 1989 the Communists finally allowed free elections. Many businesses now are owned by people in the private sector rather than by the government. The country has also strengthened its ties with Western countries. In 1999 Poland, the Czech Republic, and Hungary joined the North Atlantic Treaty Organization (NATO).

Warsaw, the capital, has long been the cultural, political, and historical center of Polish life. More than 2 million people live in the urban area. The city lies on the Vistula River in central Poland. This location has made Warsaw the center of the national transportation and communications networks as well.

Prague's Charles Bridge is lined with historical statues.

Interpreting the Visual Record How does this bridge compare to other bridges you have seen?

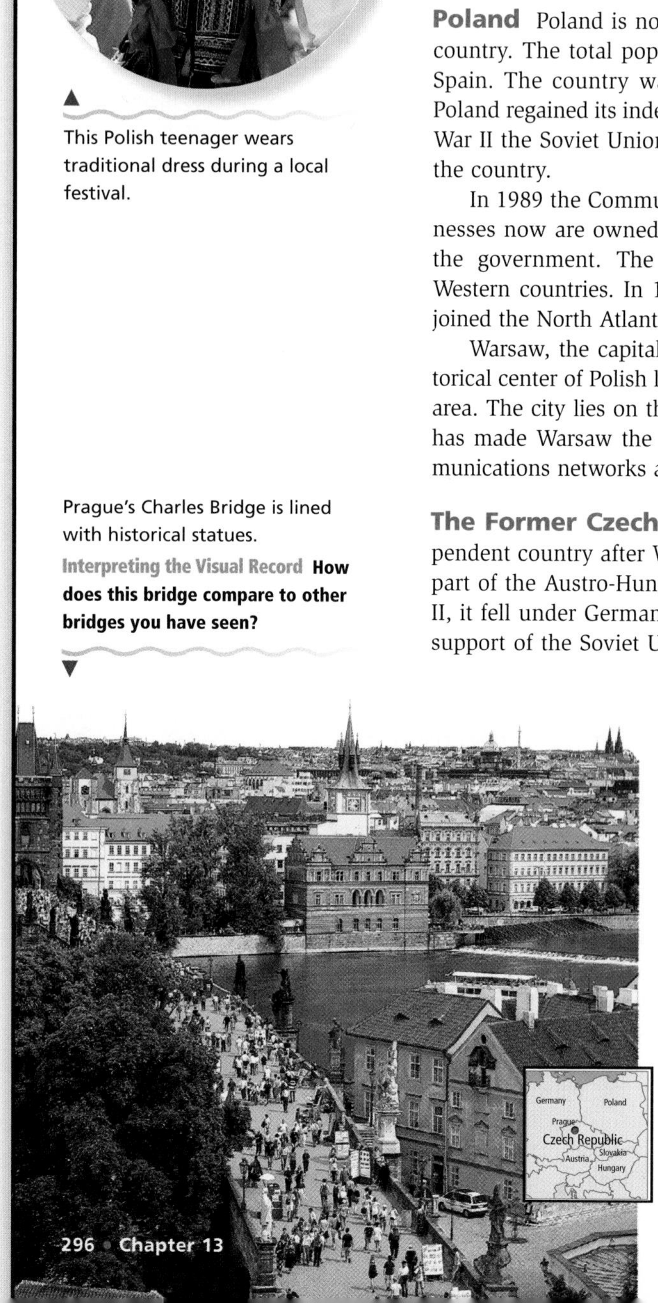

The Former Czechoslovakia Czechoslovakia became an independent country after World War I. Until that time, its lands had been part of the Austro-Hungarian Empire. Then shortly before World War II, it fell under German rule. After the war the Communists, with the support of the Soviet Union, gained control of the government. As in Poland, the Communists lost power in 1989. In 1993 Czechoslovakia peacefully split into two countries. The western part became the Czech Republic. The eastern part became Slovakia. This peaceful split helped the Czechs and Slovaks avoid the ethnic problems that have troubled other countries in the region.

The Czech Republic The Czech Republic experienced economic growth in the early 1990s. Most of the country's businesses are completely or in part privately owned. However, some Czechs worry that the government remains too involved in the economy. As in Poland, a variety of political parties compete in free elections. Czech lands have coal and other important

REVIEW AND ASSESS

Have students complete the Section Review. Then organize the class into small groups and have students write one quiz question each on the history, culture, and characteristics of the region. Each question should be written on a slip of paper, folded, and placed in a pile. Students should take turns reading questions to the group while the other members answer. Then have students complete Daily Quiz 13.2. **COOPERATIVE LEARNING**

RETEACH

Have students complete Main Idea Activity 13.2. Then pair students and have pairs create a page for a travel magazine featuring a country in northeastern Europe. Students may divide up tasks according to their abilities. The page should include a descriptive title, a map or flag, and a short paragraph summarizing the history and culture of the country. Display and discuss the magazine pages. **ENGLISH LANGUAGE LEARNERS, COOPERATIVE LEARNING**

EXTEND

The czardas (CHAHR-dash) is the national dance of Hungary. This couples dance begins with a slow section but becomes fast and lively. Many versions of the czardas have been developed. Have interested students conduct research on how to perform a basic version of the dance. You may wish to have volunteers perform the dance in class. **BLOCK SCHEDULING**

mineral resources that are used in industry. Much of the country's industry is located in and around Prague, the capital. The city is located on the Vltava River. More than 1.2 million people live there. Prague has beautiful medieval buildings. It also has one of Europe's oldest universities.

Slovakia Slovakia is more rugged and rural, with incomes lower than in the Czech Republic. The move toward a freer political system has been slow. However, progress has been made. Bratislava (BRAH-tyee-slah-vah), the capital, is located on the Danube River. The city is the country's most important industrial area and cultural center. Many rural Slovaks move to Bratislava looking for better-paying jobs. Most of the country's population is Slovak. However, ethnic Hungarians account for more than 10 percent of Slovakia's population.

Hungary Hungary separated from the Austro-Hungarian Empire at the end of World War I. Following World War II, a Communist government came to power. A revolt against the government was put down by the Soviet Union in 1956. The Communists ruled until 1989.

Today the country has close ties with the rest of Europe. In fact, most of Hungary's trade is with members of the European Union. During the Communist era, the government experimented with giving some businesses the freedom to act on their own. For example, it allowed local farm managers to make key business decisions. These managers kept farming methods modern, chose their crops, and marketed their products. Today, farm products from Hungary's fertile plains are important exports. Much of the country's manufacturing is located in and around the capital, Budapest (BOO-duh-pest). Budapest is Hungary's largest city. Nearly 20 percent of the population lives there.

✓ **READING CHECK:** *Human Systems* How have the governments and economies of the region been affected by recent history? moved from communism to democracy and private ownership following collapse of Soviet Union

▲ The Danube River flows through Budapest, Hungary.

Interpreting the Visual Record **Why might Hungary's capital have grown up along a river?**

Section 2 Review

go.hrw.com Homework Practice Online
Keyword: SK3 HP13

Define and explain: Indo-European

Working with Sketch Maps On the map you drew in Section 1, label the countries of the region, Prague, Tallinn, Riga, Warsaw, Vistula River, Bratislava, and Budapest.

Reading for the Main Idea

1. *Human Systems* How did invasions and migrations help shape the region?
2. *Places and Regions* What has the region contributed to the arts?

Critical Thinking

3. **Drawing Conclusions** What influence did the Soviet Union have on the region?
4. **Summarizing** What social changes have taken place here since the early 1990s?

Organizing What You Know

5. **Sequencing** Copy the following graphic organizer. Use it to show the history of the Baltics since 1900.

1900 —— 1945 —— 1999

Section 2 Review

Answers

Define For definition, see: Indo-European, p. 292

Working with Sketch Maps Maps will vary, but listed places should be labeled in their approximate locations. The Czech Republic, Hungary, and Slovakia do not have seacoasts.

Reading for the Main Idea

1. Balts, hunting people from Ural Mountains, Huns, Slavs, Magyars, Mongols, and Germans all moved into the area, bringing their own languages and traditions. (NGS 9)
2. music—Frédéric Chopin, Béla Bartók; literature—Franz Kafka, Karel Capek (NGS 10)

Critical Thinking

3. imposed communism
4. Change from communism to capitalism and democracy has allowed free elections, greater political and economic freedom.

Organizing What You Know

5. after World War I (1918)—became independent; during World War II (1940)—taken over by Soviet Union; after 1991—changing from communism to democracy

◄ **Visual Record Answer**

Rivers provide reliable sources of water and serve as trade routes.

Section 3

Objectives

1. Describe the early history of southeastern Europe.
2. Explain how other cultures have influenced southeastern Europe.
3. Summarize the changes in the region's governments during the 1980s and 1990s.

FOCUS

LET'S GET STARTED

Copy the following question onto the chalkboard: *How do you identify yourself in terms of your country, state, region, or ethnic group? Write down more than one term. (Examples: American, Texan, Southerner, Puerto Rican, Hispanic, African American)* Discuss responses. Then ask how students would feel if suddenly they had to define themselves as Canadians or as a member of a different ethnic group. Explain that some people in southeastern Europe have had to make similar changes. Tell students that in Section 3 they will learn more about the countries of southeastern Europe.

Building Vocabulary

Write the key terms on the chalkboard. Point out that the people who call themselves **Roma** were once known as Gypsies, when their origins were thought to be in Egypt. If your class has studied Canada, ask students if they recall another group of people who are referred to now by the name they call themselves. *(Inuit)* If they have not covered Canada, lead a discussion about the terms *Hispanic, Chicano, African American*, or others.

Section 3 RESOURCES

Reproducible

- Guided Reading Strategy 13.3
- Geography for Life Activity 13
- Cultures of the World Activity 3
- Creative Strategies for Teaching World Geography, Lesson 11

Technology

- One-Stop Planner CD–ROM, Lesson 13.3
- Homework Practice Online
- HRW Go site

Reinforcement, Review, and Assessment

- Section 3 Review, p. 303
- Daily Quiz 13.3
- Main Idea Activity 13.3
- English Audio Summary 13.3
- Spanish Audio Summary 13.3

Visual Record Answer

Greek

Section 3 The Countries of Southeastern Europe

Read to Discover

1. How did Southeastern Europe's early history help shape its modern societies?
2. How does culture both link and divide the region?
3. How has the region's past contributed to current conflicts?

Define

Roma

Locate

Bulgaria
Romania
Croatia
Slovenia
Serbia
Bosnia and Herzegovina
Albania
Yugoslavia
Kosovo
Montenegro
Macedonia
Belgrade
Podgorica
Sarajevo
Zagreb
Ljubljana
Skopje
Bucharest
Moldova
Chişinău
Sofia
Tiranë

WHY IT MATTERS

Bosnia and the other republics of the former Yugoslavia have experienced ethnic violence since independence. Use CNNfyi.com or other **current events** sources to check on current conditions in the region. Record your findings in your journal.

Dolls in traditional Croatian dress

These ancient ruins in southern Albania date to the 500s B.C.

Interpreting the Visual Record **What cultural influence does this building show?**

History

Along with neighboring Greece, this was the first region of Europe to adopt agriculture. From here farming moved up the Danube River valley into central and western Europe. Early farmers and metalworkers in the south may have spoken languages related to Albanian. Albanian is an Indo-European language.

Early History Around 750–600 B.C. the ancient Greeks founded colonies on the Black Sea coast. The area they settled is now Bulgaria and Romania. Later, the Romans conquered most of the area from the Adriatic Sea to the Danube River and across into Romania. When the Roman Empire divided into west and east, much of the Balkans and Greece became part of the Eastern Roman Empire. This eastern region eventually became known as the Byzantine Empire. Under Byzantine rule, many people of the Balkans became Orthodox Christians.

TEACH

Teaching Objective 1

LEVEL 1: (Suggested time: 20 min.) Pair students and have each pair create a flowchart to show how the ancient Greeks, the Roman Empire, the Ottoman Empire, the Austro-Hungarian Empire, and World War I affected the boundaries and religions of the countries of southeastern Europe. **ENGLISH LANGUAGE LEARNERS, COOPERATIVE LEARNING**

LEVELS 2 AND 3: (Suggested time: 30 min.) Focus students' attention on the photographs of buildings and other structures in Section 3. Ask if they have heard the phrase "if these walls could talk." Discuss the meaning of the phrase. Then have students use the information from the Level 1 lesson to write what the walls of one of those structures would say about southeastern Europe's history if they could indeed talk. You may want to invite some students to dramatize their "wall soliloquies."

Kingdoms and Empires Many of today's southeastern European countries first appear as kingdoms between A.D. 800 and 1400. The Ottoman Turks conquered the region and ruled until the 1800s. The Ottomans, who were Muslims, tolerated other religious faiths. However, many peoples, such as the Bosnians and Albanians, converted to Islam. As the Ottoman Empire began to weaken in the late 1800s, the Austro-Hungarians took control of Croatia and Slovenia. They imposed Roman Catholicism.

Slav Nationalism The Russians, meanwhile, were fighting the Turks for control of the Black Sea. The Russians encouraged Slavs in the Balkans to revolt against the Turks. The Russians appealed to Slavic nationalism—to the Slav's sense of loyalty to their country. The Serbs did revolt in 1815 and became self-governing in 1817. By 1878 Bulgaria and Romania were also self-governing.

The Austro-Hungarians responded to Slavic nationalism by occupying additional territories. Those territories included the regions of Bosnia and Herzegovina. To stop the Serbs from expanding to the Adriatic coast, European powers made Albania an independent kingdom.

In August 1914 a Serb nationalist shot and killed the heir to the Austro-Hungarian throne. Austria declared war on Serbia. Russia came to Serbia's defense. These actions sparked World War I. All of Europe's great powers became involved. The United States entered the war in 1917.

Creation of Yugoslavia At the end of World War I Austria-Hungary was broken apart. Austria was reduced to a small territory. Hungary became a separate country but lost its eastern province to Romania. Romania also gained additional lands from Russia. Albania remained independent. The peace settlement created Yugoslavia. *Yugoslavia* means "land of the southern Slavs." Yugoslavia brought the region's Serbs, Bosnians, Croatians, Macedonians, Montenegrins, and Slovenes together into one country. Each ethnic group had its own republic within Yugoslavia. Some Bosnians and other people in Serbia were Muslims. Most Serbs were Orthodox Christians, and the Slovenes and Croats were Roman Catholics. These ethnic and religious differences created problems that eventually led to civil war in the 1990s.

✓ **READING CHECK:** *Human Systems* How is southeastern Europe's religious and ethnic makeup a reflection of its past? Colonizers and conquerors introduced different religions.

This bridge at Mostar, Bosnia, was built during the 1600s. This photograph was taken in 1982.

This photograph shows Mostar after civil war in the 1990s.

Interpreting the Visual Record **What differences can you find in the two photos?**

Linking Past to Present

The Slavs The Slavs are the largest ethnic and language group of the European peoples. Long ago, the Slavs' ancestors migrated out of Asia into what is now Eastern Europe. During the A.D. 400s and 500s they moved into the Balkan Peninsula. Little unity developed among the Slavic groups, however, as they adopted the ideas and influences of their new neighbors.

In the 1800s some Slavic intellectuals and poets founded the Pan-Slav movement. (*Pan* means "all.") Its supporters hoped to unite the Slavic peoples politically and culturally. The movement did not succeed, partly because the various Slavic peoples remained separated by their own rivalries.

Ideas about Slavic unity continue to affect world events. Russia's support for Serbian interests during the conflicts of the 1990s is due partly to their shared Slavic heritage.

Activity: Have students conduct research on the degree to which Russia's support of Serbia contributed to the start of World War I. Have students prepare a newscast for the day Russia declared war on the Austro-Hungarian Empire.

◄ **Visual Record Answer**

The lower photo shows that the bridge and wall have been destroyed. The buildings have also been damaged.

Teaching Objective 2

ALL LEVELS: (Suggested time: 45 min.) Organize the class into small groups. Ask each group to plan a feast for a major Roman Catholic, Orthodox Christian, or Muslim holiday. They should include the history of the religion in southeastern Europe as background. Have students use the information in the text as a starting point for creating their menu and conduct additional research as necessary. Then have the groups present their menus and historical summaries to the class. **COOPERATIVE LEARNING**

Teaching Objective 3

ALL LEVELS: (Suggested time: 25 min.) Copy the following graphic organizer onto the chalkboard, omitting the italicized answers. Call on volunteers to fill in the charts. Then lead a discussion on how the various groups have affected changes in the region's governments during the 1980s and 1990s. **ENGLISH LANGUAGE LEARNERS**

Ethnic and Religious Groups of Southeastern Europe

	Major ethnic or religious group	Important minority group
Yugoslavia	*Orthodox Christians*	*Albanian Muslims*
Bosnia and Herzegovina	*Muslims*	*Catholic Croats, Orthodox Serbs*
Croatia	*Roman Catholics*	*Orthodox Christian Serbs*
Slovenia	*Roman Catholics*	
Macedonia	*Orthodox Christians*	*Albanian Muslims*
Romania	*Romanian*	*Roma, Hungarians*
Moldova	*(diverse)*	
Bulgaria	*Bulgarians*	*Turks, Macedonians*
Albania	*Muslims*	

HUMAN SYSTEMS

During the conflict in Kosovo in the late 1990s, American teenagers used e-mail to keep in touch with teens in the war-torn region. The Kosovar youths shared their concerns and fears. The American teens distributed those messages to the news media and to charity groups in order to spread the word about the war's effects.

Critical Thinking: How has access to e-mail affected the world?

Answer: It has improved communication around the world and has allowed people to communicate almost instantaneously.

▲ Ethnic Albanians worship at a mosque in Pristina, Serbia.

Culture

The Balkans are the most diverse region of Europe in terms of language, ethnicity, and religion. It is the largest European region to have once been ruled by a Muslim power. It has also been a zone of conflict between eastern and western Christianity. The three main Indo-European language branches—Romance (from Latin), Germanic, and Slavic—are all found here, as well as other branches like Albanian. Non-Indo-European languages like Hungarian and Turkish are also spoken here.

Balkan diets combine the foods of the Hungarians and the Slavs with those of the Mediterranean Greeks, Turks, and Italians. In Greek and Turkish cuisines, yogurt and soft cheeses are an important part of most meals, as are fresh fruits, nuts, and vegetables. Roast goat or lamb are the favorite meats for a celebration.

In the Balkans Bosnian and ethnic Albanian Muslims celebrate the feasts of Islam. Christian holidays—Christmas and Easter—are celebrated on one day by Catholics and on another by Orthodox Christians. Holidays in memory of ancient battles and modern liberation days are sources of conflict between ethnic groups.

✓ **READING CHECK:** *Places and Regions* Why is religion an important issue in southeastern Europe? source of conflict within and between countries

Our Amazing Planet

The Danube Delta, on the Romanian coast of the Black Sea, is part of a unique ecosystem. Most of the Romanian caviar-producing sturgeon are caught in these waters. Caviar is made from the salted eggs from three types of sturgeon fish. Caviar is considered a delicacy and can cost as much as $50 per ounce.

Southeastern Europe Today

Like other southeastern European countries, Yugoslavia was occupied by Germany in World War II. A Communist government under Josip Broz Tito took over after the war. Tito's strong central government prevented ethnic conflict. After Tito died in 1980, Yugoslavia's Communist government held the republics together. Then in 1991 the republics of Slovenia, Croatia, Bosnia and Herzegovina, and Macedonia began to break away. Years of bloody civil war followed. Today the region struggles with the violence and with rebuilding economies left weak by years of Communist-government control.

Yugoslavia The republics of Serbia and Montenegro remain united and have kept the name of Yugoslavia. Belgrade is the capital of both Serbia and Yugoslavia. It is located on the Danube River. The capital of Montenegro is Podgorica (PAWD-gawr-eet-sah). The Serbian government supported ethnic Serbs fighting in civil wars in Croatia and in Bosnia and Herzegovina in the early 1990s. Tensions between ethnic groups also have been a problem within Serbia. About 65 percent of the people in Serbia and Montenegro are Orthodox Christians. In the

➤ASSIGNMENT: Copy the following sentence onto the chalkboard: *Times are always changing in southeastern Europe.* Have students copy the sentence into their notebooks. Ask students to write a paragraph supporting or disputing this statement. *(To agree with the statement, students might cite the many changes in governmental control over the countries in the region. Those who dispute the statement might state that constant change and ethnic or religious conflict are so common in southeastern Europe that times do not really change.)*

Teaching Objectives 1–3

LEVEL 1: (Suggested time: 40 min.) Give each student two or three index cards. Assign or have each student select a country of southeastern Europe. Be sure that each country is selected by at least one student. Instruct students to provide basic information about the country (relative location, history, culture, economy, and government) on one side of the card and to draw a simple map of the country on the other side. Have students exchange cards to evaluate them.

southern Serbian province of Kosovo, the majority of people are ethnic Albanian and Muslim. Many of the Albanians want independence. Conflict between Serbs and Albanians led to civil war in the late 1990s. In 1999 the United States, other Western countries, and Russia sent troops to keep the peace.

Bosnia and Herzegovina Bosnia and Herzegovina generally are referred to as Bosnia. Some 40 percent of Bosnians are Muslims, but large numbers of Roman Catholic Croats and Orthodox Christian Serbs also live there. Following independence, a bloody civil war broke out between these groups as they struggled for control of territory. During the fighting the once beautiful capital of Sarajevo (sar-uh-YAY-voh) was heavily damaged.

These Muslim refugees are walking to Travnik, Bosnia, with the assistance of UN troops from Britain in the 1990s.

Interpreting the Visual Record What effect might the movement of refugees have on a region?

Croatia Croatia's capital is Zagreb (ZAH-greb). Most of the people of Croatia are Roman Catholic. In the early 1990s, Serbs made up about 12 percent of the population. In 1991 the ethnic Serbs living in Croatia claimed part of the country for Serbia. This resulted in heavy fighting. By the end of 1995 an agreement was reached and a sense of stability returned to the country. Many Serbs left the country.

Slovenia Slovenia is a former Austrian territory. It looks to Western European countries for much of its trade. Most people in Slovenia are Roman Catholic, and few ethnic minorities live there. Partly because of the small number of ethnic minorities, little fighting occurred after Slovenia declared independence from Yugoslavia. The major center of industry is Ljubljana (lee-oo-blee-AH-nuh), the country's capital.

Slovenia's capital, Ljubljana, lies on the Sava River.

National Geography Standard 10

Human Systems In the 1380s, Muslim Turkish armies began raiding the region that is now Bosnia and Herzegovina. The Turks had gained complete control of the area within about 100 years.

Unlike in most other parts of Europe conquered by the Ottoman Empire, a large percentage of Bosnians converted to Islam. There were several reasons for this. First, Muslims had a higher legal status than Christians. Second, Sarajevo and Mostar were mainly Muslim, and those who wanted to participate fully in city life had to convert. Also, the Bosnian Catholic Church was relatively weak.

Discussion: Lead a discussion about the goals, needs, and ideas that people of different religions might have in common.

Visual Record Answer

possible answer: ethnic conflict, overcrowding, political ramifications, new cultural traditions introduced in new countries

LEVELS 2 AND 3: (Suggested time: 45 min.) Have students use the cards they created for the Level 1 lesson to design a card trading game. The game should have a goal, rules, and a point system. For example, the goal might be to acquire wealth by collecting the countries with the strongest economies. As an example of the point system, countries with coastlines along the Mediterranean might be worth more points than landlocked countries. Have students play the game and summarize what they learned about the countries for which they did not create a card.

CLOSE

Read the Links to Our Lives feature at the beginning of the chapter to the class. Call on volunteers to add a detail or to explain each point further. Then ask students to propose a single sentence to answer the question "Why should we study Eastern Europe?" *(Possible answer: Events in Eastern Europe can affect people everywhere.)*

Section 3 Review

Answers

Define For definitions, see: Roma, p. 302

Working with Sketch Maps Maps will vary, but listed places should be labeled in their approximate locations. Slovenia, Croatia, Bosnia and Herzegovina, Macedonia, Serbia, and Montenegro

Reading for the Main Idea

1. Invasions and foreign control have led to multiple religious and ethnic groups living in the region. (NGS 9)
2. conflict among ethnic, religious groups (NGS 13)

Critical Thinking

3. created after World War I in response to Slav nationalism
4. independent countries without Communist governments; political and economic troubles; developing market economies

Organizing What You Know

5. languages—Romance, Germanic, Slavic, and Non-Indo-European languages; foods—yogurt, soft cheeses, fresh fruits, nuts, vegetables, goat, lamb; celebrations—Catholic and Orthodox Christian holidays, Islamic feasts, national holidays

Chart Answer

23 percent

Ethnic Groups in Macedonia

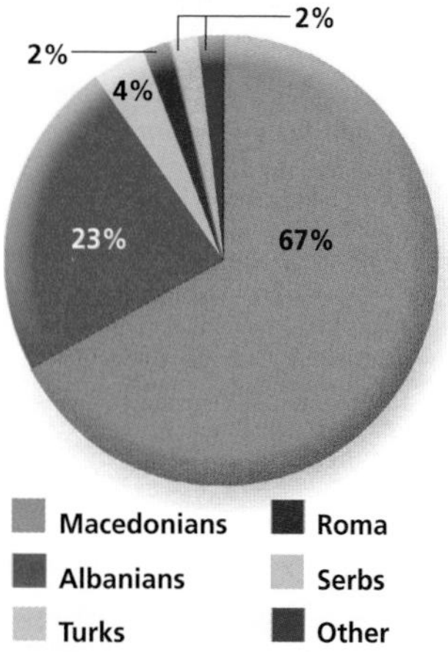

Source: Central Intelligence Agency, *The World Factbook 2001*

In 2001 ethnic Albanian rebels launched months of fighting in hopes of gaining more rights for Macedonia's Albanian minority.

Interpreting the Chart **What percentage of Macedonia's people are Albanian?**

Macedonia When Macedonia declared its independence from Yugoslavia, Greece immediately objected to the country's new name. Macedonia is also the name of a province in northern Greece that has historical ties to the republic. Greece feared that Macedonia might try to take over the province.

Greece responded by refusing to trade with Macedonia until the mid-1990s. This slowed Macedonia's movement from the command—or government-controlled—economy it had under Communist rule to a market economy in which consumers help to determine what is to be produced by buying or not buying certain goods and services. Despite its rocky start, in recent years Macedonia has made progress in establishing free markets.

Romania A Communist government took power in Romania at the end of World War II. Then in 1989 the Communist government was overthrown during bloody fighting. Change, however, has been slow. Bucharest, the capital, is the biggest industrial center. Today more people work in agriculture than in any other part of the economy. Nearly 90 percent of the country's population is ethnic Romanian. **Roma**, or Gypsies as they were once known, make up almost 2 percent of the population. They are descended from people who may have lived in northern India and began migrating centuries ago. Most of the rest of Romania's population are ethnic Hungarian.

Moldova Throughout history control of Moldova has shifted many times. It has been dominated by Turks, Polish princes, Austria, Hungary, Russia, and Romania. Not surprisingly, the country's population reflects this diverse past. Moldova declared its independence in

Turkish Roma girls

REVIEW AND ASSESS

Have students complete the Section Review. Then put the names of all the countries on slips of paper in a hat and have each student draw a name. (You will need to repeat country names.) Ask students to write a sentence that begins "If I lived in this country, I would . . . " and complete it with a detail from Section 3 about the country they have drawn. Have other students guess which country is being discussed. Then have students complete Daily Quiz 13.3.

RETEACH

Have students complete Main Idea Activity 13.3. Then, pair students and provide each pair with an outline map of the region. Have students take turns finding two facts about each country on the map.
ENGLISH LANGUAGE LEARNERS, COOPERATIVE LEARNING

EXTEND

Have interested students create maps of southeastern Europe to show how boundaries have changed, areas where different ethnic and religious groups have settled, and areas where conflicts have been most severe. Let the students who created the maps lead a discussion on conclusions that may be drawn from the maps. **BLOCK SCHEDULING**

1991 from the Soviet Union. However, the country suffers from difficult economic and political problems. About 40 percent of the country's labor force works in agriculture. Chişinău (kee-shee-NOW), the major industrial center of the country, is also Moldova's capital.

Bulgaria Mountainous Bulgaria has progressed slowly since the fall of communism. However, a market economy is growing gradually, and the people have more freedoms. Most industries are located near Sofia (SOH-fee-uh), the capital and largest city. About 9 percent of Bulgaria's people are ethnic Turks.

Albania Albania is one of Europe's poorest countries. The capital, Tiranë (ti-RAH-nuh), has a population of about 270,000. About 70 percent of Albanians are Muslim. Albania's Communist government feuded with the Communist governments in the Soviet Union and, later, in China. As a result, Albania became isolated. Since the fall of its harsh Communist government in the 1990s, the country has tried to move toward both democracy and a free market system.

✓ **READING CHECK:** *Human Systems* What problems does the region face, and how are they reflections of its Communist past? ethnic violence—Communist government prevented conflict; economic conditions—years of government control left economies weak; democratic reforms—Communist government limited freedoms

Southeastern Europe

Country	Population/ Growth Rate	Life Expectancy	Literacy Rate	Per Capita GDP
Albania	3,510,484 / 0.9%	69, male / 75, female	93%	$3,000
Bosnia and Herzegovina	3,922,205 / 1.4%	69, male / 74, female	not available	$1,700
Bulgaria	7,707,495 / −1.1%	68, male / 75, female	98%	$6,200
Croatia	4,334,142 / 1.5%	70, male / 78, female	97%	$5,800
Macedonia	2,046,209 / 0.4%	72, male / 76, female	not available	$4,400
Moldova	4,431,570 / 0.05%	60, male / 69, female	96%	$2,500
Romania	22,364,022 / −0.2%	66, male / 74, female	97%	$5,900
Yugoslavia (Serbia and Montenegro)	10,677,290 / −0.3%	70, male / 76, female	93%	$2,300
United States	281,421,906 / 0.9%	74, male / 80, female	97%	$36,200

Sources: Central Intelligence Agency, *The World Factbook 2001;* U.S. Census Bureau

Interpreting the Chart Based on the table, which southeastern European country has the highest level of economic development?

Section 3 Review

go.hrw.com Homework Practice Online
Keyword: SK3 HP13

Define and explain: Roma

Working with Sketch Maps On the map you drew in Section 2, label the region's countries and their capitals. They are listed at the beginning of the section. In a box in the margin, identify the countries that once made up Yugoslavia.

Reading for the Main Idea

1. *Human Systems* How has the region's history influenced its religious and ethnic makeup?
2. *Human Systems* What events and factors have contributed to problems in Bosnia and other countries in the region since independence?

Critical Thinking

3. **Summarizing** How was Yugoslavia created?
4. **Analyzing Information** How are the region's governments and economies changing?

Organizing What You Know

5. **Categorizing** Copy the following graphic organizer. Use it to identify languages, foods, and celebrations in the region.

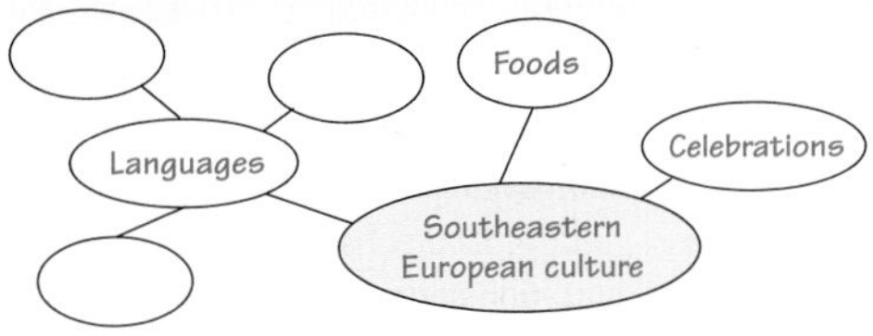

CHAPTER 13 Review ANSWERS

Building Vocabulary

For definitions, see: oil shale, p. 291; lignite, p. 291; amber, p. 291; Indo-European, p. 292; Roma, p. 303

Reviewing the Main Ideas

1. Northern European Plain, Great Hungarian Plain, Dinaric Alps, Balkan Mountains, Carpathian Mountains, Transylvanian Alps (NGS 4)
2. Balts, hunting people from the Ural Mountains, Huns, Slavs, Magyars, Mongols, Germans; language, customs (NGS 9)
3. air, soil, and water pollution; deforestation, destruction of natural resources (NGS 14)
4. communism; moving to democracy and a market economy; collapse of the Soviet Union in 1991 (NGS 9)
5. Slovenia, Croatia, Bosnia and Herzegovina, Macedonia; ethnic identity and religion (NGS 13)

ASSESS

Have students complete the Chapter 13 Test.

RETEACH

Supply pairs of students with poster board or butcher paper. Help students create time lines showing major events in Eastern Europe. Have students annotate and illustrate the time lines with details about how the events affected culture. Display the time lines around the classroom.
ENGLISH LANGUAGE LEARNERS, COOPERATIVE LEARNING

PORTFOLIO EXTENSIONS

1. Although cultural conflict is part of Eastern Europe's history, so is cooperation. Have students work in groups to write constitutions for "Cromavania"—an imaginary Eastern European country where people live together peacefully. When the groups have completed their constitutions, lead a discussion on the strong points of each. Have students place their constitutions and a summary of what they learned in their portfolios.
2. Have students work in groups to write a series of short skits with the republics that have been part of Yugoslavia as characters. Each skit should depict a different period in the region's history. The countries should express how they feel at each stage in Yugoslavia's history. Have each group perform its skit. Place scripts in student portfolios.

Chapter 13 Review ANSWERS

Understanding Environment and Society

Answers may vary according to the resources available to the student. Students may find that COMECON was founded in 1949. The purpose of the organization was to aid its members in economic integration. The Soviet Union provided fuel and raw materials. In turn, Eastern Europe supplied the Soviet Union with finished machinery and goods. Economies are moving from command to market economies. Use Rubric 29, Presentations, to evaluate student work.

Thinking Critically

1. Possible answer: Balkan diets combine the foods of the Hungarians and the Slavs with those of the Mediterranean Greeks, Turks, and Italians.
2. Danube River; there are several cities located along its route.
3. its location on the Vistula River in central Poland; perhaps less connected to the rest of Poland, more open to attack or Western influence
4. Yugoslavia—violent; Czechoslovakia—peaceful

Chapter 13 Reviewing What You Know

Building Vocabulary

On a separate sheet of paper, write sentences to define each of the following words.

1. oil shale
2. lignite
3. amber
4. Indo-European
5. Roma

Reviewing the Main Ideas

1. *Places and Regions* What are the major landforms of Eastern Europe?
2. *Human Systems* What groups influenced the culture of Eastern Europe? How can these influences be seen in modern society?
3. *Environment and Society* What were some environmental effects of Communist economic policies in Eastern Europe?
4. *Human Systems* What political and economic systems were most common in Eastern Europe before the 1990s? How and why have these systems changed?
5. *Human Systems* List the countries that broke away from Yugoslavia in the early 1990s. What are the greatest sources of tension in these countries?

Understanding Environment and Society

Economic Geography

During the Communist era, the Council for Economic Assistance, or COMECON, played an important role in planning the economies of countries in the region. Create a presentation comparing the practices of this organization to the practices now in place in the region. Consider the following:

- How COMECON organized the distribution of goods and services.
- How countries in the region now organize distribution.

Thinking Critically

1. **Drawing Inferences and Conclusions** How has Eastern Europe's location influenced the diets of the region's people?
2. **Analyzing Information** What is Eastern Europe's most important river for transportation and trade? How can you tell that the river is important to economic development?
3. **Identifying Cause and Effect** What geographic factors help make Warsaw the transportation and communication center of Poland? Imagine that Warsaw is located along the Baltic coast of Poland or near the German border. How might Warsaw have developed differently?
4. **Comparing/Contrasting** Compare and contrast the breakups of Yugoslavia and Czechoslovakia.
5. **Summarizing** How has political change affected the economies of Eastern European countries?

FOOD FESTIVAL

Kolaches are popular Czech or Polish pastries. For a shortcut version, use frozen bread dough. Or, make a sweetened yeast dough from scratch. After the dough has risen, roll it out, cut it into circles 2–3 inches across, and indent the center of each. Add a filling of fruit preserves, cottage cheese with egg and sugar, or a sweetened poppyseed paste. Let rise again. Sprinkle with a streusel topping of sugar, cinnamon, butter, and a little flour. Bake at 375° for 15–20 minutes. There are many *kolache* recipes on the Internet. *Kolacky* and *kolachke* are alternate spellings.

CHAPTER 13 REVIEW AND ASSESSMENT RESOURCES

Reproducible
- Readings in World Geography, History, and Culture 37, 38, and 39
- Critical Thinking Activity 13
- Vocabulary Activity 13

Technology
- Chapter 13 Test Generator (on the One-Stop Planner)
- HRW Go site
- Audio CD Program, Chapter 13

Reinforcement, Review, and Assessment
- Chapter 13 Review, pp. 304–05
- Chapter 13 Tutorial for Students, Parents, Mentors, and Peers
- Chapter 13 Test
- Chapter 13 Test for English Language Learners and Special-Needs Students
- Unit 3 Test
- Unit 3 Test for English Language Learners and Special-Needs Students

Building Social Studies Skills

Map ACTIVITY

On a separate sheet of paper, match the letters on the map with their correct labels.

Baltic Sea	**Balkan Mountains**
Adriatic Sea	**Carpathian Mountains**
Black Sea	
Danube River	
Dinaric Alps	

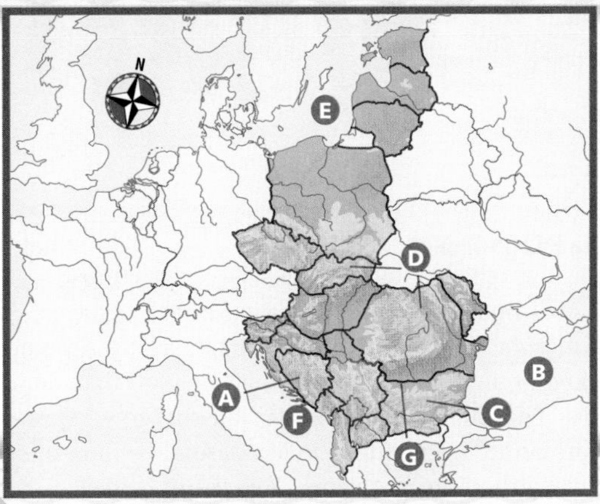

Mental Mapping Skills ACTIVITY

Using the chapter map or a globe as a guide, draw a freehand map of Eastern Europe and label the following:

Bosnia and Herzegovina	**Hungary**
Croatia	**Macedonia**
Czech Republic	**Poland**
Estonia	**Yugoslavia**

WRITING ACTIVITY

Imagine that you are a teenager living in Romania and want to write a family memoir of life in Romania. Include accounts of life for your grandparents under strict Soviet rule and life for your parents during the Soviet Union's breakup. Also describe your life in free Romania. Be sure to use standard grammar, spelling, sentence structure, and punctuation.

Alternative Assessment

Portfolio ACTIVITY

Learning About Your Local Geography

Cooperative Project Ask international agencies or search the Internet for help in contacting a boy or girl in Eastern Europe. As a group, write a letter to the teen, telling about your daily lives.

internet connect

Internet Activity: **go.hrw.com**
KEYWORD: **SK3 GT13**

Choose a topic to explore Eastern Europe:
- Investigate the conflicts in the Balkans.
- Take a virtual tour of Eastern Europe.
- Learn about Baltic amber.

5. Change to capitalism and market economies is happening at the same time as the change from communism to democracy.

Map Activity
A. Dinaric Alps
B. Black Sea
C. Balkan Mountains
D. Carpathian Mountains
E. Baltic Sea
F. Adriatic Sea
G. Danube River

Mental Mapping Skills Activity
Maps will vary, but listed places should be labeled in their approximate locations.

Writing Activity
Answers will vary, but students should discuss all three generations. Details should be consistent with text information. Use Rubric 40, Writing to Describe, to evaluate student work.

Portfolio Activity
Students might tell about their community, families, school, sports, or other topics. Use Rubric 25, Personal Letters, to evaluate student work.

internet connect

GO TO: **go.hrw.com**
KEYWORD: **SK3 Teacher**
FOR: a guide to using the Internet in your classroom

Focus On Government

Analyzing an Economic System

Ask students to list the 15 current member countries of the European Union *(Austria, Belgium, Denmark, Finland, France, Germany, Greece, Ireland, Italy, Luxembourg, the Netherlands, Portugal, Spain, Sweden, and the United Kingdom).* Point out that each country had to meet specific requirements before it was admitted into the union. Organize the class into pairs. Assign one EU country to each pair or call on some pairs to volunteer for more than one country. Have students conduct research on the requirements their assigned country had to meet to become a member of the EU and what changes the country had to make. Ask the pairs to compile lists of the requirements and changes.

Geography Sidelight

As it is working to unite the economic and political aspects of the member countries, the EU is also supporting the preservation of particular ethnic cultures. For example, almost 50 million Europeans speak minority languages such as Wendish, Frisian, or Basque. The European Union supports several organizations that try to ensure the survival of these languages.

The EU has an ambitious range of goals. If successful, it will have eased intercountry commerce among its member countries and strengthened the region's political and social systems while maintaining the unique culture of each member country.

Critical Thinking: How is the EU working to link its member countries?

Answer: by easing intercountry commerce, strengthening the region's political and social systems

➤ **This Focus On Government feature addresses National Geography Standards 6, 11, and 13.**

Focus On Government

The European Union

What if . . . ? Imagine you are traveling from Texas to Minnesota. You have to go through a border checkpoint in Oklahoma to prove your Texas identity. The guard charges a tax on the cookies you are bringing to a friend in Minnesota. Buying gas presents more problems. You try to pay with Texas dollars, but the attendant just looks at you. You discover that they speak "Kansonian" in Kansas and use Kansas coins. All this would make traveling from one place to another much more difficult.

The European Union Fortunately, that was just an imaginary situation. However, it is similar to what might happen while traveling across Europe. European countries have different languages, currencies, laws, and cultures. For example, someone from France has different customs than someone from Ireland.

However, many Europeans also share common interests. For example, they are interested in peace in the region. They also have a common interest in Europe's economic success.

A shared belief in economic and political cooperation has resulted in the creation of the European Union (EU). The EU has 15 countries that are members. They are: Austria, Belgium, Denmark, Finland, France, Germany, Great Britain, Greece, Ireland, Italy, Luxembourg, the Netherlands, Portugal, Spain, and Sweden.

The Eurostar train carries passengers from London to Paris. These two cities are only about 200 miles (322 km) apart. However, they have different cultures and ways of life.

The Beginnings of a Unified Europe Proposals for an economically integrated Europe first came about in the 1950s. After World War II, the countries of Europe had many economic problems. A plan was made to unify the coal and steel production of some countries. In 1957 France, Germany, Italy, Belgium, the Netherlands, and Luxembourg formed the European Economic Community (EEC). The name was later shortened to simply the European Community (EC). The goal of the EC was to combine each country's economy into a single market. Having one market would make trading among them easier. Eventually, more countries became interested in joining the EC. In 1973 Britain, Denmark, and Ireland joined. In the 1980s Greece, Portugal, and Spain joined.

Going Further: Thinking Critically

Have students identify the issues and benefits awaiting the member countries of the European Union during the early phase of unification. Then have students conduct research on the economic issues that faced the United States as the country was evolving from a group of colonies into a unified country. As a class, compare and contrast the U.S. experience with those arising now for the European countries. Have students prepare a report that warns European member countries of potential pitfalls, recommends how to overcome them, and lists likely benefits.

◀ The 15 EU countries produce a wide range of exports and are one of the world's richest markets.

▲ The flag of the EU features 12 gold stars on a blue background. The EU's currency, the euro, replaced the currencies of most EU countries.

In the early 1990s, a meeting was held in Maastricht, the Netherlands, to discuss the future of the EC. This resulted in the Maastricht Treaty. The treaty officially changed the EC's name from the European Community to the European Union (EU).

The Future Some people believe the EU is laying the foundation for a greater sense of European identity. A European Court of Justice has been set up to enforce EU rules. According to some experts, this is helping to build common European beliefs, responsibilities, and rights.

The introduction of a common currency, the euro, will continue in the future. Currently, each country has its own form of money. For example, France uses the Franc. Italy uses the lira. On January 1, 2002, the euro became a common form of money for most EU countries. With the exception of Denmark, Sweden, and the United Kingdom, the euro replaced the currencies of member countries. Another goal of the EU is to extend membership to other countries.

The EU has resulted in many important changes in Europe. Cooperation between member countries has increased. Trade has also increased. EU members have adopted a common currency and common economic laws. The EU is creating a more unified Europe. Some people even believe that the EU might someday lead to a "United States of Europe."

Understanding What You Read

1. What was the first step toward European economic unity?
2. What is the euro? How will it affect the other currencies of the EU countries?

Understanding What You Read

Answers

1. The first step toward European economic unity was a proposal that came about in 1950, following World War II.
2. The euro is the common currency of most of the European Union. It replaced the currencies of most of the EU countries.

GeoSkills

Going Further: Thinking Critically

Edge cities are defined as "nodal concentrations of retail and office space that are situated on the outer fringes of metropolitan areas, typically near major highway intersections." Essentially, edge cities are major commercial centers that grow up near major cities. They differ from residential suburbs in that they contain much more office, retail, and hotel space than the communities made up mainly of houses and apartments. Tysons Corner, outside Washington, D.C., in Fairfax County, Virginia, is a typical example. Edge cities are particularly common near the rapidly growing cities of the Sunbelt.

Have students research the development of an edge city in your state or region. Organize the class into groups and have them pursue specific aspects of the topic. Here are some suggestions for the different groups' investigations:

- Contact the administration of the nearby large city for information on how the edge city has affected it.
- Acquire old and recent maps of the area.
- Find old and new photos of the development.
- Find information on how the edge city has affected utility services.
- Ask highway department officials how traffic patterns have changed.
- Investigate changes in property values and taxes in the tax assessor's office.
- Search newspaper files for news stories about the development.
- Research the effect on employment and unemployment patterns.

When the research is complete, have the groups compile their information in a large flowchart.

Practicing the Skill

1. Students should compare their definitions with dictionary definitions. They should note that the differences among the three types of settlements is primarily size; a city is larger than a town, which is larger than a village. Students should note how their definitions are different from the dictionary's.
2. Students should note the apparent age of their settlement and about how many people live there. Students should describe some jobs in their settlement and how their settlement is connected to others, such as by highway or rail. Students might note unique features, such as buildings, parks, and cultural attractions.
3. Students might suggest capital cities, mountain cities, river cities, port cities, colonial cities, modern cities, tourist cities, or other types.

➤ **This GeoSkills feature addresses National Geography Standards 4, 6, 12, and 17.**

Geo Skills

Building Skills for Life: Analyzing Settlement Patterns

There are many different kinds of human settlements. Some people live in villages where they farm and raise animals. Others live in small towns or cities and work in factories or offices. Geographers analyze these settlement patterns. They are interested in how settlements affect people's lives.

All settlements are unique. Even neighboring villages are different. One village might have better soil than its neighbors. Another village might be closer to a main road or highway. Geographers are interested in the unique qualities of human settlements.

Geographers also study different types of settlements. For example, many European settlements could be considered medieval cities. Medieval cities are about 500–1,500 years old. They usually have walls around them and buildings made of stone and wood. Medieval cities also have tall churches and narrow, winding streets.

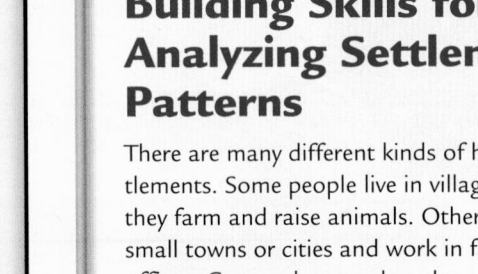

▲ This illustration shows a German medieval city in the 1400s.

Analyzing settlement patterns is important. It helps us learn about people and environments. For example, the architecture of a city might give us clues about the culture, history, and technology of the people who live there.

You can ask questions about individual villages, towns, and cities to learn about settlement patterns. What kinds of activities are going on? How are the streets arranged? What kinds of transportation do people use? You can also ask questions about groups of settlements. How are they connected? Do they trade with each other? Are some settlements bigger or older than others? Why is this so?

Practicing the Skill

1. How do you think a city, a town, and a village are different from each other? Write down your own definition of each word on a piece of paper. Then look them up in a dictionary and write down the dictionary's definition. Were your definitions different?
2. Analyze the settlement where you live. How old is it? How many people live there? What kinds of jobs do people have? How is it connected to other settlements? How is it unique?
3. Besides medieval cities, what other types of cities can you think of? Make a list of three other possible types.

HANDS ON GEOGRAPHY

One type of settlement is called a planned city. A planned city is carefully designed before it is built. Each part fits into an overall plan. For example, the size and arrangement of streets and buildings might be planned.

There are many planned cities in the world. Some examples are Brasília, Brazil; Chandigarh, India; and Washington, D.C. Many other cities have certain parts that are planned, such as individual neighborhoods. These neighborhoods are sometimes called planned communities.

Suppose you were asked to plan a city. How would you do it? On a separate sheet of paper, create your own planned city. These guidelines will help you get started.

1. First, decide what the physical environment will be like. Is the city on the coast, on a river, or somewhere else? Are there hills, lakes, or other physical features in the area?
2. Decide what to include in your city. Most cities have a downtown, different neighborhoods, and roads or highways that connect areas together. Many cities also have parks, museums, and an airport.
3. Plan the arrangement of your city. Where will the roads and highways go? Will the airport be close to downtown? Try to arrange the different parts of your city so that they fit together logically.
4. Draw a map of your planned city. Be sure to include a title, scale, and orientation.

Some people think the city plan for Brasília looks like a bird, a bow and arrow, or an airplane. **Interpreting the Visual Record What do you notice about the arrangement of Brasilia's streets?**

Lab Report

1. How was your plan influenced by the physical environment you chose?
2. How do you think planned cities are different from cities that are not planned?
3. What problems might people have when they try to plan an entire city?

Lab Report

Answers

1. Students might note that streets would have to fit around physical features such as rivers or mountains. They may also note that the environment affected the businesses in their cities. For example, a port city might have docks and businesses related to fishing, travel, or recreation located near the waterfront.
2. Students might suggest that planned cities are more efficient, have stronger economies, or have a higher standard of living. They might also argue that planned cities have less character or "atmosphere."
3. Students might suggest the difficulty of accommodating different populations, such as locating the airport near downtown for easy business travel or on the outskirts so that noise pollution bothers fewer people. Students may mention other problems, such as the difficulty of predicting growth.

Visual Record Answer

possible answers: wide, symmetrical arrangement; attractive pattern

UNIT 4

RUSSIA AND NORTHERN EURASIA

USING THE ILLUSTRATIONS

Direct students' attention to the photographs on these pages. Point out that the bear is a symbol of Russia. Ask what other animal symbols students have seen and for which countries they stand. *(Possible answers: bald eagle—United States; kangaroo, koala—Australia; panda, dragon, tiger—China)* Ask why animal symbols seem to be popular. *(Possible answers: portray strength and power, not associated with any one ethnic group or political party, refer to the country's natural heritage)*

Refer to the photo of the Buryat people. Ask how they compare to the Russian citizens and politicians that are typically seen on news broadcasts. *(Possible answers: more colorful clothing, facial features similar to those of Chinese people)* Point out that the Buryat and some other ethnic groups in Russia have more in common with the peoples of Mongolia and China than they do with Russians who live in cities thousands of miles to the west.

Point out the domes on St. Basil's Cathedral in Moscow. Ask students to suggest why they are shaped like onions. *(so snow does not accumulate)* The cathedral was built in the 1550s to commemorate a major military victory.

UNIT OBJECTIVES

1. Describe the physical features of Russia and northern Eurasia and link the region's natural environment with its economic development.
2. Examine the progression of the region's history, from early periods through communism to the breakup of the Soviet Union.
3. Analyze the relationships among the region's ethnic and culture groups.
4. Use special-purpose maps to analyze relationships among climate, population patterns, and economic activities in Russia and northern Eurasia.
5. Identify the challenges that lie ahead for the governments and peoples of Russia and northern Eurasia.

UNIT 4

Russia and Northern Eurasia

Dancers in Russian national dress

St. Basil's Cathedral, Moscow, Russia

Church overlooking the Black Sea, Ukraine

Chapters

Notes from the Field

Journalists in Russia

Journalists Gary Matoso and Lisa Dickey traveled more than 5,000 miles across Russia. They wrote this account of their visit with Buyanto Tsydypov. He is a Buryat farmer who lives in the Lake Baikal area. The Buryats are one of Russia's many minority ethnic groups. **WHAT DO YOU THINK?** ***If you visited a Buryat family, what would you like to see or ask?***

"You came to us like thunder out of the clear blue sky," said our host. The surprise of our visit did not, however, keep him from greeting us warmly.

Buyanto brought us to a special place of prayer. High on a hillside, a yellow wooden frame holds a row of tall, narrow sticks. On the end of each stick, Buddhist prayer cloths flutter in the biting autumn wind.

In times of trouble and thanks, Buryats come to tie their prayer cloths—called *khimorin*—to the sticks and make their offerings to the gods. Buyanto builds a small fire. He unfolds an aqua-blue *khimorin* to show the drawings.

"All around are the Buddhist gods," he says, "and at the bottom we have written our names and the names of others we are praying for."

He fans the flames slowly with the cloth, purifying it with sacred smoke. After a time he moves to the top of the hill where he ties the *khimorin* to one of the sticks.

Buryat people, Lake Baikal area, Russia

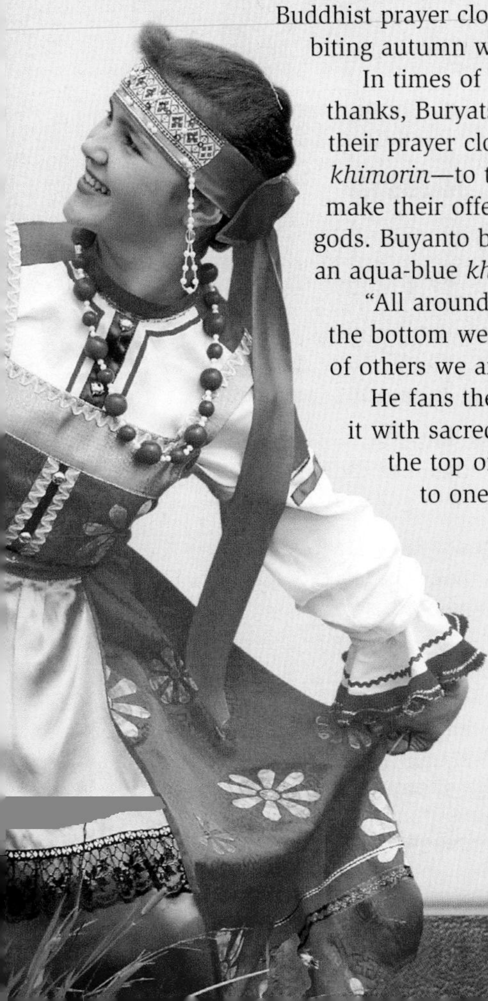

Brown bear

Understanding Primary Sources

1. How do you know that Buyanto Tsydypov was surprised to meet the two American journalists?
2. What is a *khimorin*?

More from the Field

In the early 1930s the Soviets began enforcing a policy of state ownership of farmland. Harsh measures were used—land confiscations, arrests, and deportations to prison camps. Soviet farm policies resulted in a famine during 1932–33 in which millions of people died.

Since the fall of the Soviet Union, the Russian government has given land grants and tax breaks to private farmers. They decide for themselves what crops to plant and are not required to sell to state agencies. Buyanto Tsydypov, the Buryat farmer described on this page, is one such private farmer. Many people in his village admire his independence and determination.

Critical Thinking: What type of economic system did the Soviet Union have?

Answer: command

Understanding Primary Sources

Answers

1. He compares their arrival to thunder out of a clear blue sky.
2. prayer cloth

OVERVIEW

In this unit, students will learn about the land and people of Russia and northern Eurasia—a vast region that is undergoing many political and economic changes.

Huge open plains and deep evergreen forests dominate Russia's landscape. The population is concentrated in the western plains, where the best farmland and the largest cities are located. Harsh climates keep Siberia and Russia's Far East thinly populated. Development of the region's rich resources may increase settlement there.

For most of the country's history, Russia's leaders have limited citizens' personal freedoms. During the more-than-70 years of communist rule, the Soviet Union became a superpower. Since the collapse of the Soviet Union, Russia has struggled with rising crime, corruption, and unemployment.

The other countries of northern Eurasia were also part of the Soviet Union. They are now independent republics. Their economies are emerging, with varying degrees of success. Distinctive religious and ethnic groups live in the area.

PEOPLE IN THE PROFILE

Note that the elevation profile crosses the West Siberian Plain. The Khanty are one of the many indigenous groups that live in this region of subarctic forest and bogs. They have adapted to the harsh environment. Winters are very cold, and the summer skies are filled with mosquitoes.

The Khanty, who settle in extended family groups, live mainly by fishing, hunting, and reindeer herding. They make almost everything they use, including their shelters, fishing boats, winter clothing, and containers. Arts and crafts include colorful birchbark baskets.

Russia's oil wealth poses problems for the Khanty and other Siberian peoples. Oil spills and pollution threaten the wetlands. Raised roads trap water, causing floods and ruining the land's ability to function as pasture for reindeer. Fires caused by worker carelessness and oil-soaked debris are common. Acid rain has damaged huge areas of land. The Khanty have responded to these threats by organizing to protect their land and way of life from further decline.

Critical Thinking: To what harsh conditions have the Khanty adapted?

Answer: very cold winters, summer skies filled with mosquitoes

UNIT 4 ATLAS The World in Spatial Terms

Russia and Northern Eurasia

Elevation Profile

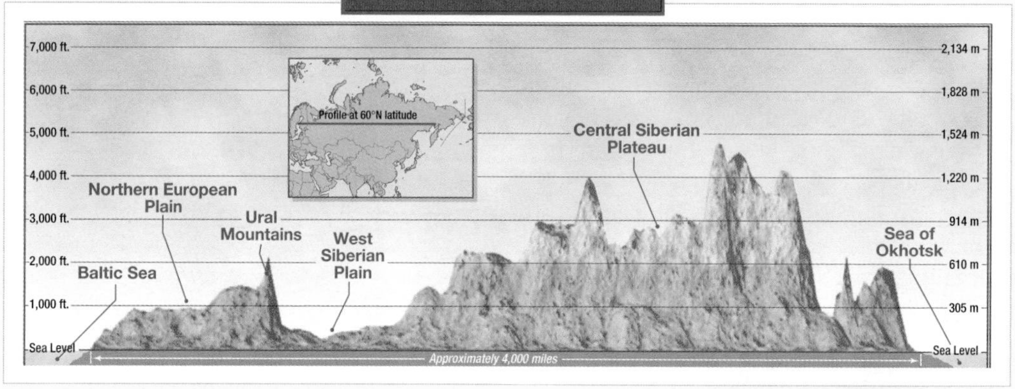

The United States and Russia and Northern Eurasia Comparing Sizes

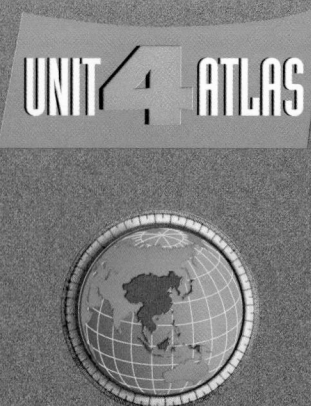

GEOSTATS:

Russia

- World's largest country in area: 6,659,328 sq. mi. (17,075,200 sq km)
- World's sixth-largest population: 145,470,197 (July 2001 estimate)
- World's largest lake: Caspian Sea–143,244 sq. mi. (371,002 sq km)
- World's deepest lake: Lake Baikal–5,715 ft. (1,742 m)
- Largest number of time zones: 11
- Highest mountain in Europe: Mount Elbrus–18,510 ft. (5,642 m)

Using the Physical Map

Focus students' attention on the **physical map** of Russia and northern Eurasia on this page. What is the first thing that students notice about Russia? *(Many students will note its large size.)* Challenge students to estimate the distance between Russia's eastern and western borders. *(about 6,000 miles or 9,650 km)*

Note that the Northern European Plain is a physical feature of western Russia and that the eastern portion of the country is called Siberia. Ask students what physical feature appears to separate European Russia from Siberia. *(Ural Mountains)*

Russia and Northern Eurasia: Physical

UNIT 4 ATLAS

internet connect

ONLINE ATLAS
GO TO: go.hrw.com
KEYWORD: SK3 MAPSU4
FOR: Web links to online maps of the region

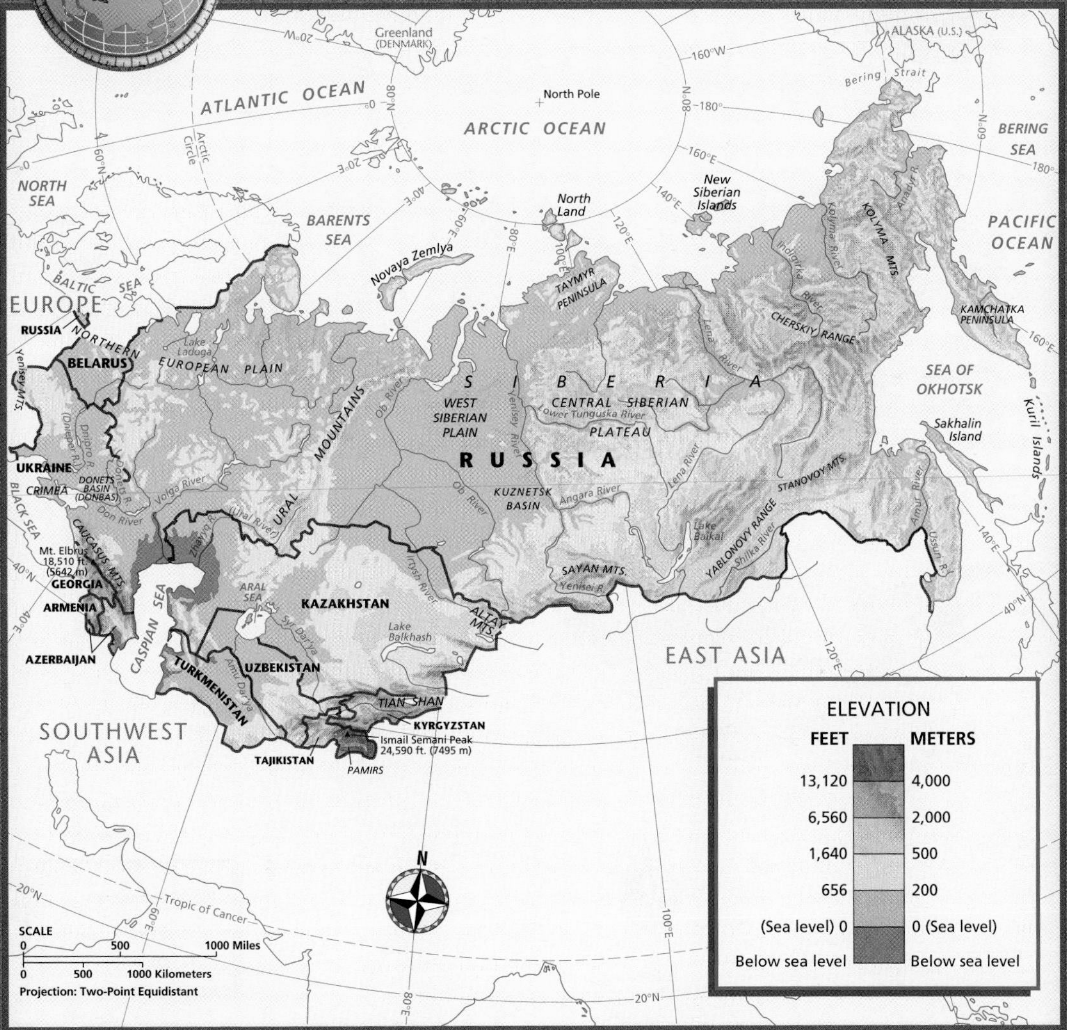

1. Places and Regions In what general direction do the great rivers of Siberia flow?
2. Places and Regions Which countries have areas that are below sea level?

Critical Thinking

3. **Drawing Inferences and Conclusions** Russia has often been invaded by other countries. Which part of Russia might be easy to invade? Why do you think this area would be a good invasion route?
4. **Comparing** Northern Russia appears to have many good harbors. Compare this map to the **climate map** of the region. Why have few harbors been developed on Russia's north coast?

Physical Map

Answers

1. north
2. Azerbaijan, Russia, Kazakhstan, and Turkmenistan

Critical Thinking

3. western Russia; because it is an open plain
4. because the water there is frozen much of the year

UNIT 4 ATLAS

USING THE POLITICAL MAP

Refer students to the **political map** of Russia and northern Eurasia on this page. Tell the class that all the countries represented in color on the map used to be part of the Soviet Union, but that the Soviet Union broke up in 1991. New independent countries emerged.

Ask students to use the **physical** and **land use and resources maps** to predict how the breakup of the Soviet Union may have affected Russia's economy and transportation links. *(lost access to valuable mineral and energy resources, reduced access to the Caspian and Black Seas)*

internet connect

ONLINE ATLAS
GO TO: **go.hrw.com**
KEYWORD: **SK3 MAPSU4**

Your Classroom Time Line

These are the major dates and time periods for this unit. Have students enter them on the time line you created earlier. You may want to watch for these dates as students progress through the unit.

600s B.C. Greeks establish trading colonies along the Black Sea.

500s B.C. The Persian Empire controls the Caucasus.

100s B.C. Chinese trade and military expeditions begin moving into Central Asia.

A.D. 400 The Georgian language has its own alphabet by this date.

Political Map

Answers

1. no

Critical Thinking

2. less than 100 mi. (161 km); about 780 mi. (1,255 km)
3. All of the irregular curved borders appear to follow natural features.
4. The boundary crosses a wide, flat desert. No physical features, such as rivers, that might define a natural boundary appear on the maps.

Russia and Northern Eurasia: Political

UNIT 4 ATLAS

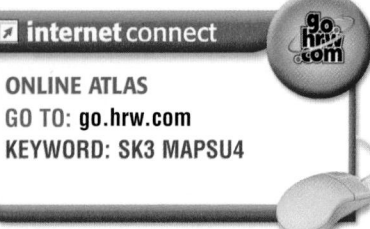

1. *Places and Regions* Compare this map to the **physical map** of the region. Do any physical features define a border between Russia and the countries of Europe?

Critical Thinking

2. **Analyzing Information** About how far apart are Russia and the Alaskan mainland? the Russian mainland and the North Pole?
3. **Analyzing Information** Which borders separating the countries south and southeast of Russia appear to follow natural features?
4. **Drawing Inferences and Conclusions** Compare this map to the **physical** and **climate maps** of the region. Why do you think the boundary between Kazakhstan and northwestern Uzbekistan is two straight lines?

Using the Climate Map

Have students examine the **climate map** of Russia and northern Eurasia on this page. Ask them to name the climates of the largest areas of Russia. *(subarctic, tundra)* Have them compare the climate map to the **population map** on the next page. Ask students to answer the following questions: In which climate region is the population density highest? *(humid continental)* In what three types of climate is the population density lowest? *(tundra, subarctic, desert)*

Russia and Northern Eurasia: Climate

UNIT 4 ATLAS

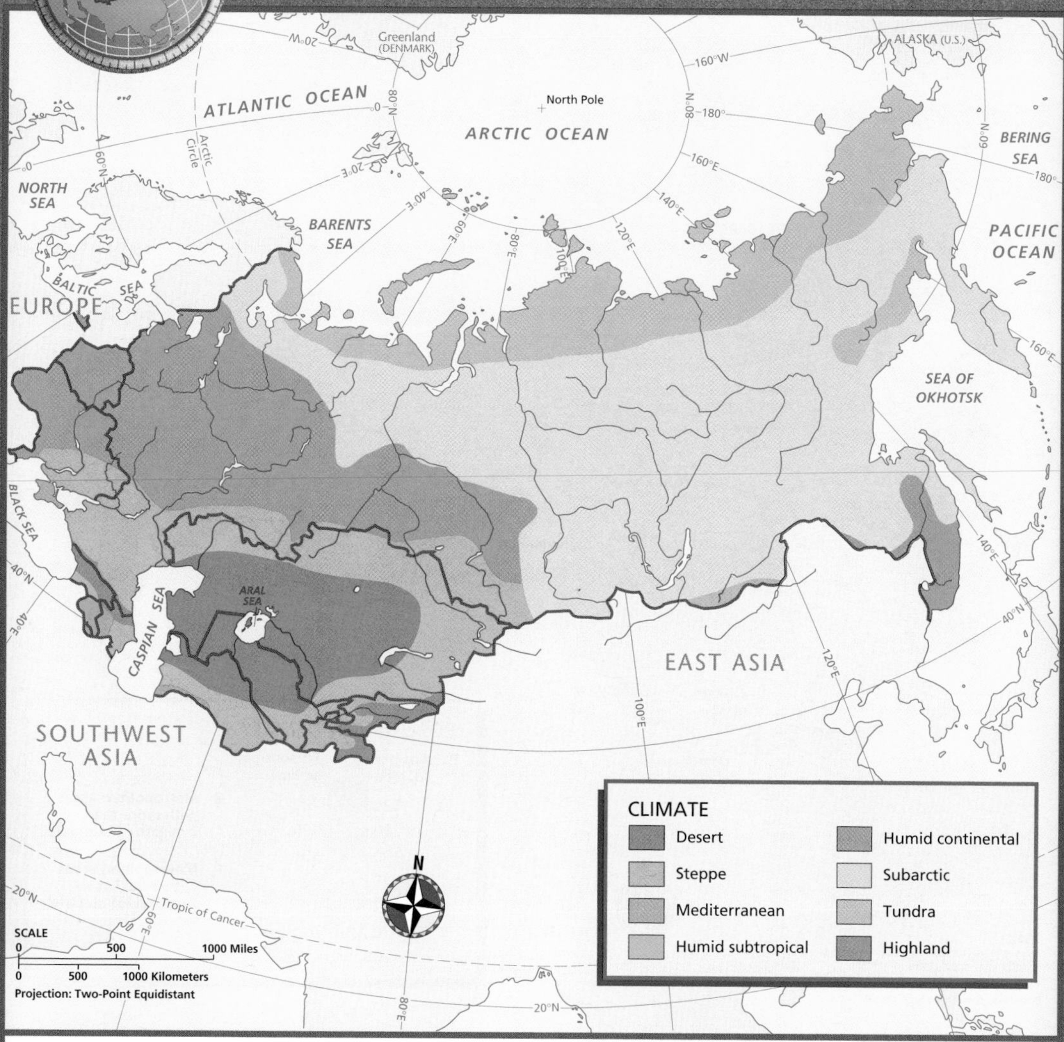

1. *Places and Regions* Compare this map to the **political map**. Which is the only country that has a humid subtropical climate?
2. *Physical Systems* Which climate types stretch across Russia from Europe to the Pacific?
3. *Places and Regions* Compare this map to the **political map**. Which country has only a humid continental climate?

Critical Thinking

4. **Drawing Inferences and Conclusions** Compare this map to the **land use and resources map**. Why do you think nomadic herding is common east of the Caspian Sea?
5. **Comparing** Compare this to the **political map**. In which countries would you expect to find the highest mountains? Why?

Your Classroom Time Line, (continued)

- **500s** Turkic-speaking nomads from northern Asia spread through Central Asia.
- **700s** Arab armies take over much of Central Asia.
- **800s** The Rus establish Kiev as their capital.
- **1200s** Tatars sweep across the steppes.
- **late 1400s** Muscovy wins control over parts of Russia from the Mongols.
- **late 1400s** The Kremlin's red brick walls and towers are built.
- **1547** Ivan IV crowns himself czar of all Russia.
- **early 1700s** Peter the Great expands the Russian Empire.

Climate Map

Answers

1. Georgia
2. tundra and subarctic
3. Belarus

Critical Thinking

4. The land is flat, and the climate is dry; herd animals can live on land too dry for farming.
5. Kyrgyzstan and Tajikistan; because they have large areas of highland climates

UNIT 4 ATLAS

USING THE POPULATION MAP

Direct your students' attention to the **population map** on this page. Ask them to write a question, with answer, about the population patterns they see on the map. *(Possible answer: Most people live in the western part of the region; the eastern part is sparsely populated.)* Have students exchange their questions with another student and then answer the exchanged questions. Then ask students to suggest reasons for the patterns of population density. They might consult the **climate map** for assistance.

internet connect

ONLINE ATLAS
GO TO: **go.hrw.com**
KEYWORD: **SK3 MAPSU4**

Your Classroom Time Line (continued)

- **1700s** Russian fur traders establish settlements along North America's Pacific coast.
- **late 1800s** The Russian Empire begins to decline.
- **1867** Russia sells Alaska to the United States.
- **1891** Construction of the Trans-Siberian Railroad begins.
- **1893** Composer Peter Tchaikovsky dies.
- **1917** Czar Nicholas II abdicates his throne.
- **1917** The Russian Revolution begins.
- **1918** Writer Aleksandr Solzhenitsyn is born.
- **1922** The Soviet Union is established.
- **1922** Georgia, Armenia, and Azerbaijan unite to oppose Soviet rule.

Population Map

Answers

1. Russia, Ukraine
2. Russia

Critical Thinking

3. that the land is fairly flat and that farming is common in western Russia
4. Charts, graphs, databases, and models should accurately reflect population figures for cities on the map.

Russia and Northern Eurasia: Population

UNIT 4 ATLAS

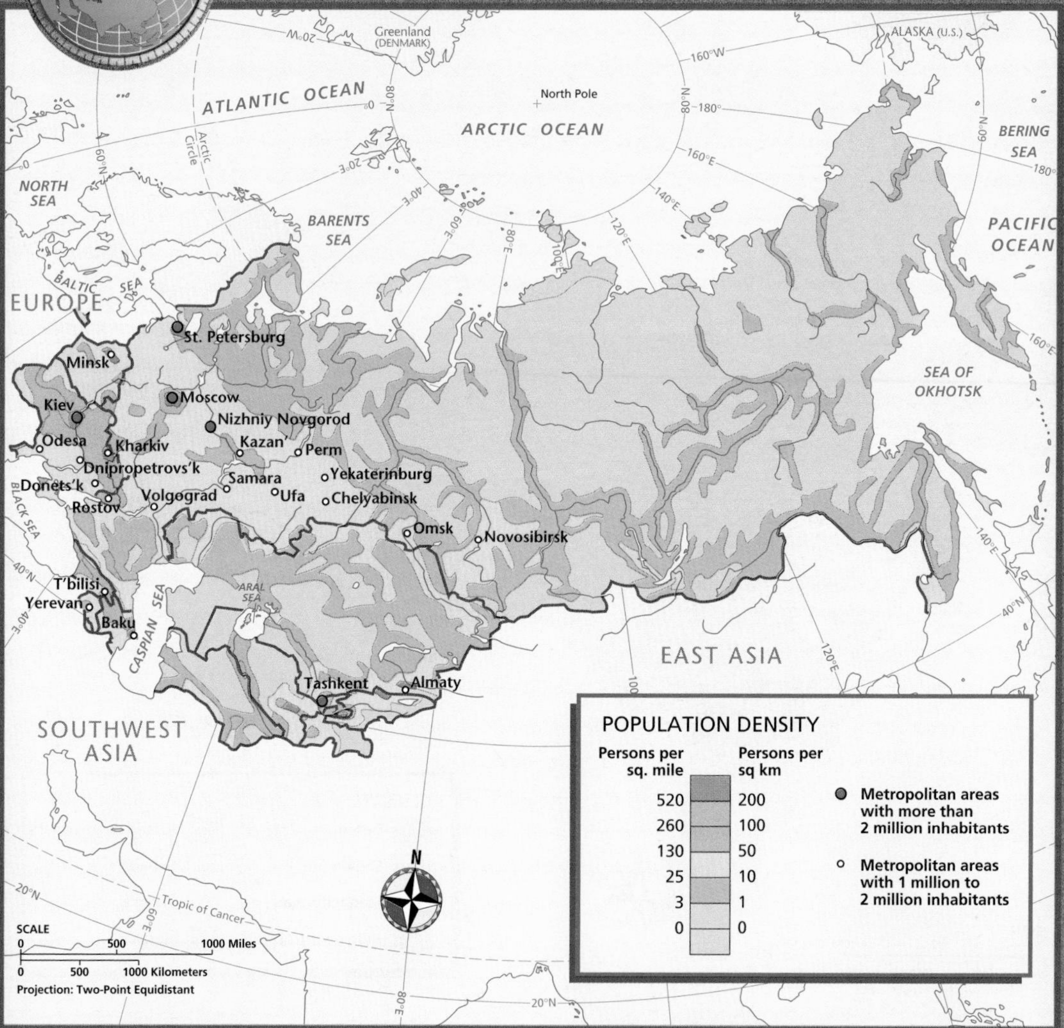

1. *Places and Regions* Which countries have a large area with more than 260 people per square mile and cities of more than 2 million people?
2. *Places and Regions* Which country has areas in the north where no one lives?

Critical Thinking

3. **Making Generalizations and Predictions** What can you assume about landforms and farming in western Russia just by looking at the **population map**? Check the **physical** and **land use and resources maps** to be sure.
4. **Analyzing Information** Use the map on this page to create a chart, graph, database, or model of population centers in Russia and northern Eurasia.

USING THE LAND USE AND RESOURCES MAP

Direct students' attention to the **land use and resources map** on this page. Then have them compare it to the **population map.** Ask: What area is rich in natural resources but sparsely populated? *(Siberia)* Challenge students to use information from this and the other maps to suggest why Russia may have trouble making use of Siberia's resources. *(Possible answers: vast distance from population centers, difficulties in transporting goods across the region)*

Russia and Northern Eurasia: Land Use and Resources

UNIT 4 ATLAS

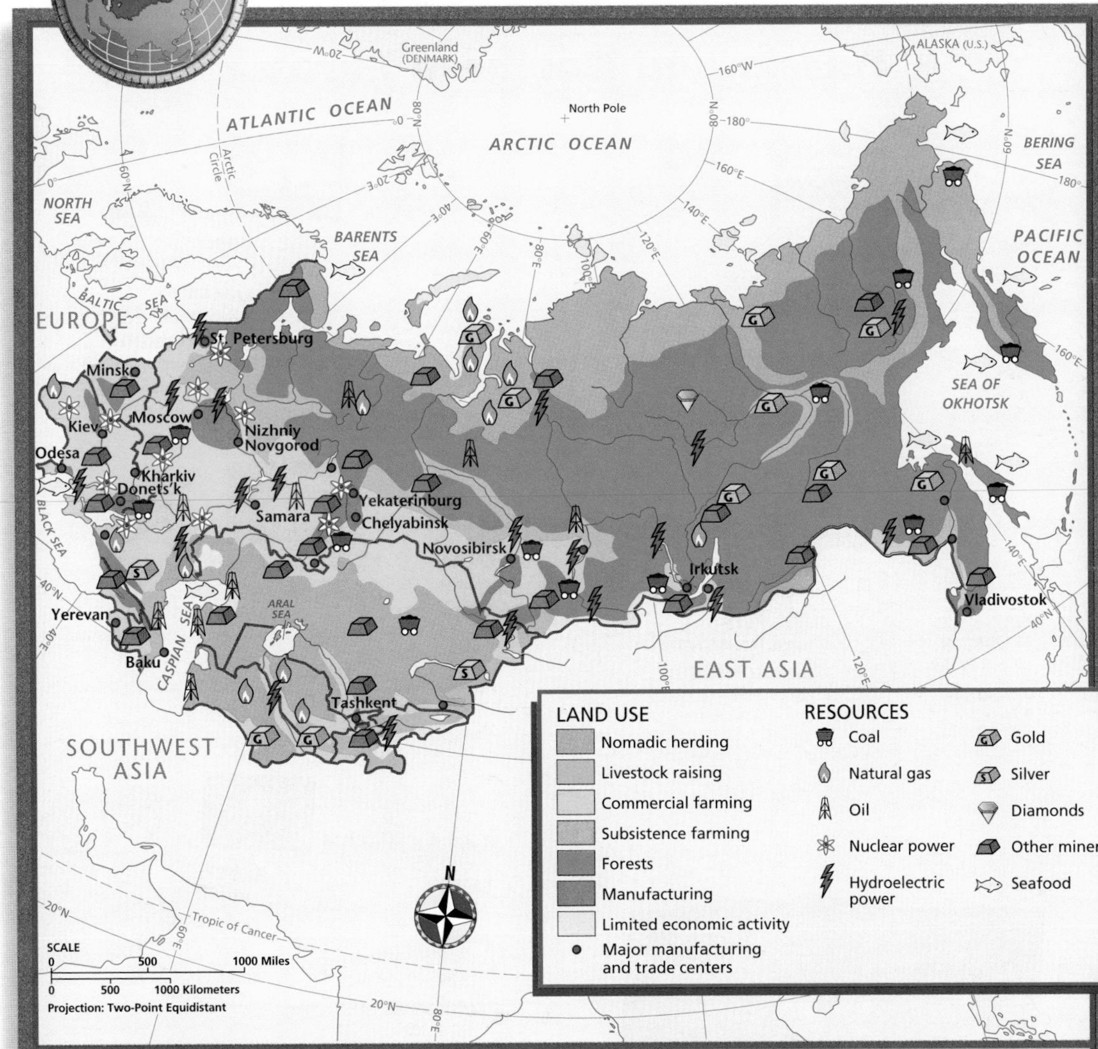

1. *Places and Regions* Where are most oil reserves in the region located?

2. *Places and Regions* Where are most gold mines in the region located?

3. *Places and Regions* Which country has diamonds? Which countries in the region have silver deposits?

Critical Thinking

4. **Comparing** Compare this to the **physical map**. Which waterways might be used to transport mineral resources mined near Irkutsk to manufacturing and trade centers?

5. **Analyzing Information** Use the map on this page to create a chart, graph, database, or model of economic activities in the region.

Your Classroom Time Line (continued)

- 1939–45 World War II is fought.
- late 1940s The Cold War begins.
- 1957 The Soviets launch Sputnik.
- 1989 The Baikal–Amur Mainline (BAM) railway is completed.
- 1986 The Chernobyl nuclear reactor disaster occurs.
- 1991 The Soviet Union collapses.
- 1990s Many civil wars erupt in the former Soviet republics.
- 1991 Russia and the other former Soviet republics form the Commonwealth of Independent States (CIS).
- 1995 In the Russian Far East, an earthquake causes severe damage on Sakhalin Island.

Land Use and Resources Map

Answers

1. in the Caspian Sea area and in Siberia
2. eastern Siberia
3. Russia; Russia and Kazakhstan

Critical Thinking

4. Angara River and Lake Baikal
5. Charts, graphs, databases, and models should accurately reflect information from the map.

RUSSIA AND NORTHERN EURASIA

FAST FACTS ACTIVITIES

LEVEL 1: (Suggested time: 25 min.) Explain that the unemployment rate shows the percentage of people who could work but cannot find a job. The U.S. unemployment rate of 4 percent meant that 4 percent of the people who were of age to work—that is, older than 16 and younger than 64—and could work did not have jobs when the data was collected.

Have students create a two-column chart with these headings: Less than 10 percent unemployment and More than 10 percent unemployment. Ask students to place each of the unit's countries except Turkmenistan in the correct column.

LEVEL 2: (Suggested time: 30 min.) After the collapse of the Soviet Union, many northern Eurasian countries were left with unstable economies and high unemployment and underemployment rates. To demonstrate a 20 percent unemployment rate, have students compute 20 percent of the class. *(Multiply the total number of students by 0.2.)* Have 20 percent of the class stand on one side of the room while the remaining students stand on the other.

Define the term "underemployment." *(Underemployed workers can only find jobs that require less skill than they have, or pay less money than their*

UNITED STATES OF AMERICA

CAPITAL: Washington, D.C.
AREA: 3,717,792 sq. mi. (9,629,091 sq km)
POPULATION: 281,421,906
MONEY: U.S. dollar (US$)
LANGUAGES: English, Spanish (spoken by a large minority)
UNEMPLOYMENT: 4 percent

Russia and Northern Eurasia

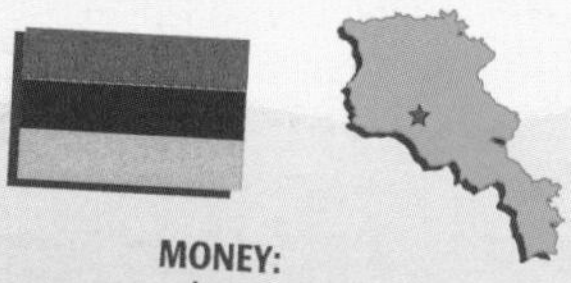

ARMENIA

CAPITAL: Yerevan
AREA: 11,506 sq. mi. (29,800 sq km)
POPULATION: 3,336,100
MONEY: dram (AMD)
LANGUAGES: Armenian, Russian
UNEMPLOYMENT: 20 percent

BELARUS

CAPITAL: Minsk
AREA: 80,154 sq. mi. (207,600 sq km)
POPULATION: 10,350,194
MONEY: Belarusian rubel (BYB/BYR)
LANGUAGES: Byelorussian, Russian
UNEMPLOYMENT: 2.1 percent (and many underemployed workers)

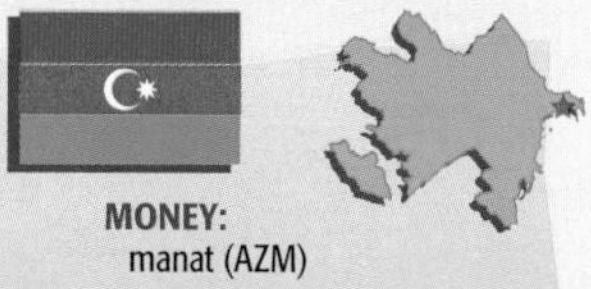

AZERBAIJAN

CAPITAL: Baku
AREA: 33,436 sq. mi. (86,600 sq km)
POPULATION: 7,771,092
MONEY: manat (AZM)
LANGUAGES: Azeri, Russian, Armenian
UNEMPLOYMENT: 20 percent

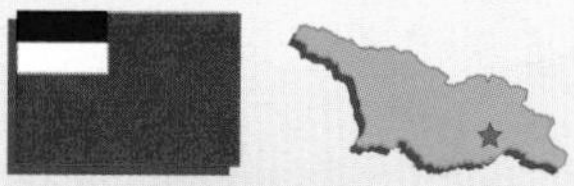

GEORGIA

CAPITAL: T'bilisi
AREA: 26,911 sq. mi. (69,700 sq km)
POPULATION: 4,989,285
MONEY: lari (GEL)
LANGUAGES: Georgian (official), Russian, Armenian, Azeri
UNEMPLOYMENT: 14.9 percent

Geese flock to this pasture in Ukraine

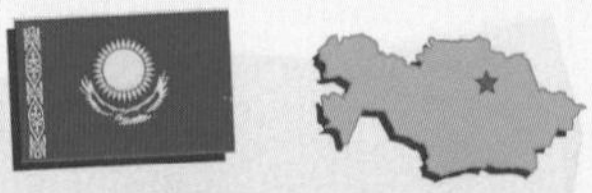

KAZAKHSTAN

CAPITAL: Astana
AREA: 1,049,150 sq. mi. (2,717,300 sq km)
POPULATION: 16,731,303
MONEY: tenge (KZT)
LANGUAGES: Kazakh, Russian
UNEMPLOYMENT: 13.7 percent

KYRGYZSTAN

CAPITAL: Bishkek
AREA: 76,641 sq. mi. (198,500 sq km)
POPULATION: 4,753,003
MONEY: Kyrgyzstani som (KGS)
LANGUAGES: Kirghiz, Russian
UNEMPLOYMENT: 6 percent

Countries not drawn to scale.

skills are generally worth, or both.) Separate out another 20 percent from the employed group of students to represent the underemployed. Ask: Why might a country have a high rate of underemployment? *(Workers are likely to accept any job available, whether or not it meets their skill level or salary needs.)* When students have sat down, have them summarize their observations.

LEVEL 3: (Suggested time: 45 min.) Explain to students that during the Soviet era, the economies of the areas under Soviet control were closely interconnected. When the Soviet Union broke up, many countries did not have enough resources to sustain their populations independently. Have students use the **land use and resources map** to make connections between resources and unemployment rates.

Organize the class into small groups. Have each group brainstorm and create a list of possible solutions to the problems of unemployment and underemployment in the independent republics. Then have each group present its list of solutions to the class.

UNIT 4 ASSESSMENT RESOURCES

Reproducible
- Unit 4 Test
- Unit 4 Test for English Language Learners and Special-Needs Students

internet connect
COUNTRY STATISTICS
GO TO: go.hrw.com
KEYWORD: SK3 FACTSU4

Highlights of Country Statistics
- *CIA World Factbook*
- Library of Congress country studies
- Flags of the world

UNIT 4 ATLAS

Mosque in Uzbekistan

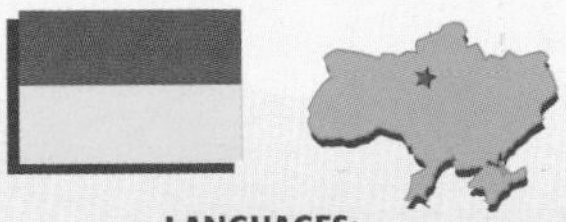

UKRAINE

CAPITAL: Kiev
AREA: 233,089 sq. mi. (603,700 sq km)
POPULATION: 48,760,474
MONEY: hryvna
LANGUAGES: Ukranian, Russian, Romanian, Polish, Hungarian
UNEMPLOYMENT: 4.3 percent officially registered (and many unregistered or underemployed)

RUSSIA

CAPITAL: Moscow
AREA: 6,592,735 sq. mi. (17,075,200 sq km)
POPULATION: 145,470,197
MONEY: Russian ruble (RUR)
LANGUAGES: Russian
UNEMPLOYMENT: 10.5 percent (and many underemployed workers)

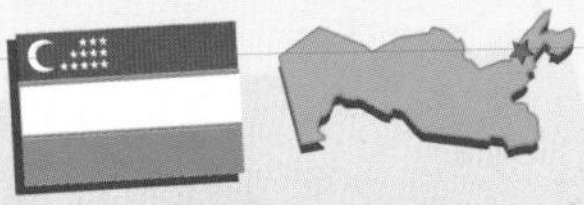

UZBEKISTAN

CAPITAL: Tashkent
AREA: 172,741 sq. mi. (447,400 sq km)
POPULATION: 25,155,064
MONEY: Uzbekistani sum (UZS)
LANGUAGES: Uzbek, Russian, Tajik
UNEMPLOYMENT: 10 percent (and many underemployed)

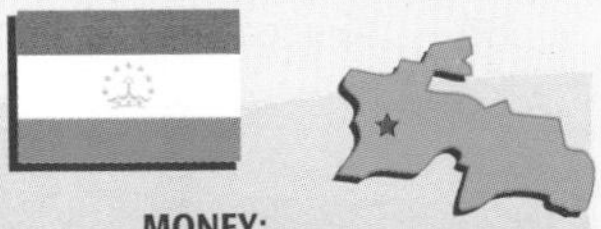

TAJIKISTAN

CAPITAL: Dushanbe
AREA: 55,251 sq. mi. (143,100 sq km)
POPULATION: 6,578,681
MONEY: somoni (SM)
LANGUAGES: Tajik (official), Russian
UNEMPLOYMENT: 5.7 percent (and many underemployed workers)

internet connect
COUNTRY STATISTICS
GO TO: go.hrw.com
KEYWORD: SK3 FACTSU4
FOR: more facts about Russia and northern Eurasia

TURKMENISTAN

CAPITAL: Ashgabat
AREA: 188,455 sq. mi. (488,100 sq km)
POPULATION: 4,603,244
MONEY: Turkmen manat (TMM)
LANGUAGES: Turkmen, Uzbek, Russian
UNEMPLOYMENT: data not available

Sources: Central Intelligence Agency, *The World Factbook 2001; The World Almanac and Book of Facts 2001;* pop. figures are 2001 estimates.

CHAPTER 14 Russia

Chapter Resource Manager

Objectives	Pacing Guide	Reproducible Resources
SECTION 1 **Physical Geography** (pp. 321–24) **1.** Identify the physical features of Russia. **2.** Identify the climates and vegetation found in Russia. **3.** Describe the natural resources of Russia.	**Regular** .5 day **Block Scheduling** .5 day *Block Scheduling Handbook, Chapter 14*	**RS** Guided Reading Strategy 14.1 **E** Creative Strategies for Teaching World Geography, Lessons 12 and 13
SECTION 2 **History and Culture** (pp. 325–30) **1.** Describe Russia's early history. **2.** Describe how the Russian empire grew and then fell. **3.** Describe the former Soviet Union. **4.** Describe what Russia is like today.	**Regular** 2 days **Block Scheduling** .5 day *Block Scheduling Handbook, Chapter 14*	**RS** Guided Reading Strategy 14.2 **RS** Graphic Organizer 14 **E** Cultures of the World Activity 4 **SM** Geography for Life Activity 14 **IC** Interdisciplinary Activities for Middle Grades 17, 18, 20 **SM** Map Activity 14
SECTION 3 **The Russian Heartland** (pp. 331–33) **1.** Describe why European Russia is considered the country's heartland. **2.** Identify the characteristics of the four major regions of European Russia.	**Regular** 1 day **Block Scheduling** .5 day *Block Scheduling Handbook, Chapter 14*	**RS** Guided Reading Strategy 14.3
SECTION 4 **Siberia** (pp. 334–36) **1.** Describe the human geography of Siberia. **2.** Identify the economic features of the region. **3.** Describe how Lake Baikal has been threatened by pollution.	**Regular** 1 day **Block Scheduling** .5 day *Block Scheduling Handbook, Chapter 14*	**RS** Guided Reading Strategy 14.4
SECTION 5 **The Russian Far East** (pp. 337–39) **1.** Describe how the Russian Far East's climate affects agriculture in the region. **2.** Identify the major resources and cities of the region. **3.** Identify the island regions that are part of the Russian Far East.	**Regular** 1 day **Block Scheduling** .5 day *Block Scheduling Handbook, Chapter 14*	**RS** Guided Reading Strategy 14.5

Chapter Resource Key

RS Reading Support
IC Interdisciplinary Connections
E Enrichment
SM Skills Mastery
A Assessment
REV Review
ELL Reinforcement and English Language Learners
Transparencies
CD–ROM
Music
Video
go.hrw.com Internet
Holt Presentation Maker Using Microsoft® Powerpoint®

One-Stop Planner CD-ROM

See the ***One-Stop Planner*** for a complete list of additional resources for students and teachers.

One-Stop Planner CD–ROM

It's easy to plan lessons, select resources, and print out materials for your students when you use the ***One-Stop Planner CD–ROM with Test Generator.***

Technology Resources	Review, Reinforcement, and Assessment Resources
One-Stop Planner CD–ROM, Lesson 14.1 Geography and Cultures Visual Resources with Teaching Activities 31–36 *ARGWorld* CD–ROM: Planning in Russia and Its Neighbors Homework Practice Online HRW Go site	**ELL** Main Idea Activity 14.1 **REV** Section 1 Review, p. 324 **A** Daily Quiz 14.1 **ELL** English Audio Summary 14.1 **ELL** Spanish Audio Summary 14.1
One-Stop Planner CD–ROM, Lesson 14.2 Homework Practice Online HRW Go site	**ELL** Main Idea Activity 14.2 **REV** Section 2 Review, p. 330 **A** Daily Quiz 14.2 **ELL** English Audio Summary 14.2 **ELL** Spanish Audio Summary 14.2
One-Stop Planner CD–ROM, Lesson 14.3 *ARGWorld* CD–ROM: Land Values in Post-Soviet Moscow Homework Practice Online HRW Go site	**ELL** Main Idea Activity 14.3 **REV** Section 3 Review, p. 333 **A** Daily Quiz 14.3 **ELL** English Audio Summary 14.3 **ELL** Spanish Audio Summary 14.3
One-Stop Planner CD–ROM, Lesson 14.4 Homework Practice Online HRW Go site	**ELL** Main Idea Activity 14.4 **REV** Section 4 Review, p. 336 **A** Daily Quiz 14.4 **ELL** English Audio Summary 14.4 **ELL** Spanish Audio Summary 14.4
One-Stop Planner CD–ROM, Lesson 14.5 Geography and Cultures Visual Resources with Teaching Activities 24–29 Homework Practice Online HRW Go site	**ELL** Main Idea Activity 14.5 **REV** Section 5 Review, p. 339 **A** Daily Quiz 14.5 **ELL** English Audio Summary 14.5 **ELL** Spanish Audio Summary 14.5

internet connect

HRW ONLINE RESOURCES

GO TO: go.hrw.com
Then type in a keyword.

TEACHER HOME PAGE
KEYWORD: SK3 TEACHER

CHAPTER INTERNET ACTIVITIES
KEYWORD: SK3 GT14

Choose an activity to:
- take a trip on the Trans-Siberian Railroad.
- examine the breakup of the Soviet Union.
- view the cultural treasures of Russia.

CHAPTER ENRICHMENT LINKS
KEYWORD: SK3 CH14

CHAPTER MAPS
KEYWORD: SK3 MAPS14

ONLINE ASSESSMENT
Homework Practice
KEYWORD: SK3 HP14
Standardized Test Prep Online
KEYWORD: SK3 STP14
Rubrics
KEYWORD: SS Rubrics

COUNTRY INFORMATION
KEYWORD: SK3 Almanac

CONTENT UPDATES
KEYWORD: SS Content Updates

HOLT PRESENTATION MAKER
KEYWORD: SK3 PPT14

ONLINE READING SUPPORT
KEYWORD: SS Strategies

CURRENT EVENTS
KEYWORD: S3 Current Events

Meeting Individual Needs

Ability Levels

Level 1 Basic-level activities designed for all students encountering new material

Level 2 Intermediate-level activities designed for average students

Level 3 Challenging activities designed for honors and gifted-and-talented students

English Language Learners Activities that address the needs of students with Limited English Proficiency

Chapter Review and Assessment

E Readings in World Geography, History, and Culture 40, 41, 42

SM Critical Thinking Activity 14

REV Chapter 14 Review, pp. 340–41

REV Chapter 14 Tutorial for Students, Parents, Mentors, and Peers

ELL Vocabulary Activity 14

A Chapter 14 Test

Chapter 14 Test Generator (on the One-Stop Planner)

Audio CD Program, Chapter 14

A Chapter 14 Test for English Language Learners and Special-Needs Students

CHAPTER 14

LAUNCH INTO LEARNING

Tell students that Russia occupies more land area than any other country in the world. Then ask them to consider the advantages and disadvantages of such enormous size. *(Students may mention as advantages the possibility of many natural resources and access to trade with many other countries. Disadvantages might include difficulties in communication, defense, distribution of goods and services, and maintaining a sense of unity.)*

Section 1

Objectives

1. Identify the physical features of Russia.
2. Name the climates and vegetation that are found in Russia.
3. List Russia's natural resources.

LINKS TO OUR LIVES

You may wish to highlight to students these reasons why we should know more about Russia:

- Russia occupies more land area than any other country on Earth. It also has one of the largest populations.
- For many years, the Soviet Union and the United States were enemies. Now economic and cultural connections between the countries are increasing.
- As one of the few countries with nuclear weapons, Russia could pose a threat to U.S. national security.
- Russia continues to struggle with economic, environmental, and political problems. Because of Russia's size and importance, these problems can affect people around the world.

CHAPTER 14

Russia

Now we will learn about Russia. First we will meet Polina, who lives in Moscow, the capital. She is in the eleventh grade at State School 637 and will graduate in the spring.

Privyet! (Hi!) My name is Polina and I am 17. I live in an apartment in Moscow with my mother and father. Our apartment has two rooms. Every day except Sunday I wake up at 7:00 A.M., have some bread and cheese with tea, and take the subway to school. At the end of eighth grade, we had to choose whether to study science or humanities. I chose humanities. My favorite subjects are history, literature, and English—my history teacher is great!

We have about five or six classes with a 15-minute break between each one. During the breaks, I often eat a snack like *pirozhki*, a small meat pie, at the school snack bar. I go home at 2:00 P.M. for lunch (meat, potatoes, and a salad of cooked vegetables and mayonnaise) and a nap. When I wake up, I go out with my friends to a park. Sometimes my parents and I join my uncle, aunt, and grandmother for Sunday dinner. My uncle makes my favorite dishes, like meat salad with mayonnaise. I love ice cream, too!

Translation: Hi! I live in Moscow.

Focus

LET'S GET STARTED

Copy the following instructions onto the chalkboard: *Look at the physical-political map of Russia in your textbook. Choose a place on the map and write a sentence describing what you think that place might look like.* Discuss student responses in terms of the landscapes found across Russia. Tell students that in Section 1 they will learn more about the physical geography of Russia.

Using the Physical-Political Map

Have students examine the map on this page. Call on students to describe the general physical characteristics of Russia. *(more mountains in eastern half, flat lowlands, long rivers, one large lake)* Ask if they are already familiar with any of the features on the map. *(Possible answers: Moscow, Siberia, Arctic Circle, Arctic Ocean)* Ask students to predict the general climate of Russia. *(cold)*

Building Vocabulary

Write the key terms on the chalkboard. Tell students that **taiga** and **steppe** come from Russian words. You may want to remind students that "steppe" is also a climate type. Call on volunteers to find the definitions of the Define terms in the section text and to write sentences using them.

Section 1 Physical Geography

Read to Discover

1. What are the physical features of Russia?
2. What climates and vegetation are found in Russia?
3. What natural resources does Russia have?

Define

taiga
steppe

Locate

Arctic Ocean
Caucasus Mountains
Caspian Sea
Ural Mountains
West Siberian Plain
Central Siberian Plateau
Kamchatka Peninsula
Kuril Islands
Volga River
Baltic Sea

WHY IT MATTERS

Like many nations, Russia is concerned about environmental issues. Use CNNfyi.com and other **current events** sources to investigate environmental concerns there. Record your findings in your journal.

A Siberian sable

Russia: Physical-Political

Section 1 RESOURCES

Reproducible
- Block Scheduling Handbook, Chapter 14
- Guided Reading Strategy 14.1
- Creative Strategies for Teaching World Geography, Lessons 12 and 13

Technology
- One-Stop Planner CD–ROM, Lesson 14.1
- Homework Practice Online
- Geography and Cultures Visual Resources with Teaching Activities 31–36
- *ARGWorld* CD–ROM: Planning in Russia and Its Neighbors
- HRW Go site

Reinforcement, Review, and Assessment
- Section 1 Review, p. 324
- Daily Quiz 14.1
- Main Idea Activity 14.1
- English Audio Summary 14.1
- Spanish Audio Summary 14.1

TEACH

Teaching Objectives 1–3

LEVEL 1: (Suggested time: 15 min.) Copy the following graphic organizer onto the chalkboard, omitting the italicized answers. Call on volunteers to fill in the circle's quarters with Russia's major landforms, rivers, climate and vegetation types, and natural resources.
ENGLISH LANGUAGE LEARNERS

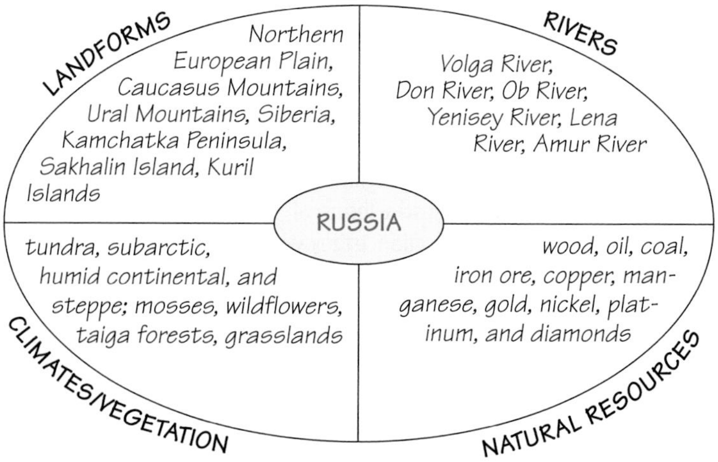

EYE ON EARTH

In addition to being the world's largest lake, the Caspian Sea is also one of the greatest salt lakes. Stretching 746 miles (1,200 km) long and 270 miles (434 km) wide, the Caspian Sea covers an area of about 143,550 square miles (371,795 sq km)—an area almost larger than Japan. Scientific studies have shown that until recent times, geologically speaking, the Caspian Sea was actually linked to the Atlantic Ocean through the Sea of Azov, the Black Sea, and the Mediterranean Sea.

Activity: Have students use the physical-political map to find which bodies of water once connected the Caspian Sea to the Atlantic Ocean.

internet connect
GO TO: **go.hrw.com**
KEYWORD: **SK3 CH14**
FOR: **Web sites about the Caspian Sea**

Visual Record Answer

that Siberia has a cold climate and barren, rugged mountains

internet connect
GO TO: **go.hrw.com**
KEYWORD: **SK3 CH14**
FOR: **Web sites about Russia**

▲ A train chugs through the cold Siberian countryside.

Interpreting the Visual Record **What does this photograph tell you about the physical features and climate of Siberia?**

Our Amazing Planet

The coldest temperature ever recorded outside of Antarctica in the last 100 years was noted on February 6, 1933, in eastern Siberia: –90° F (–68°C).

Physical Features

Russia was by far the largest republic of what was called the Union of Soviet Socialist Republics, or the Soviet Union. Russia is the largest country in the world. It stretches 6,000 miles (9,654 km), from Eastern Europe to the Bering Sea and Pacific Ocean.

The Land Much of western, or European, Russia is part of the Northern European Plain. This is the country's heartland, where most Russians live. To the north are the Barents Sea and the Arctic Ocean. Far to the south are the Caucasus (KAW-kuh-suhs) Mountains. There Europe's highest peak, Mount Elbrus, rises to 18,510 feet (5,642 m). The Caucasus Mountains stretch from the Black Sea to the Caspian (KAS-pee-uhn) Sea. The Caspian is the largest inland body of water in the world.

East of the Northern European Plain is a long range of eroded low mountains and hills. These are called the Ural (YOOHR-uhl) Mountains. The Urals divide Europe from Asia. They stretch from the Arctic coast in the north to Kazakhstan in the south. The highest peak in the Urals rises to just 6,214 feet (1,894 m).

East of the Urals lies a vast region known as Siberia. Much of Siberia is divided between the West Siberian Plain and the Central Siberian Plateau. The West Siberian Plain is a large, flat area with many marshes. The Central Siberian Plateau lies to the east. It is a land of elevated plains and valleys.

A series of high mountain ranges runs through southern and eastern Siberia. The Kamchatka (kuhm-CHAHT-kuh) Peninsula, Sakhalin (sah-kah-LEEN) Island, and the Kuril (KYOOHR-eel) Islands surround the Sea of Okhotsk (uh-KAWTSK). These are in the Russian Far East. The rugged Kamchatka Peninsula and the Kurils have active volcanoes. Earthquakes and volcanic eruptions are common. The Kurils separate the Sea of Okhotsk from the Pacific Ocean.

Rivers Some of the world's longest rivers flow through Russia. These include the Volga (VAHL-guh) and Don Rivers in European Russia. The Ob (AWB), Yenisey (yi-ni-SAY), Lena (LEE-nuh), and Amur (ah-MOOHR) Rivers are located in Siberia and the Russian Far East. The Amur forms part of Russia's border with China.

The Volga is Europe's longest river. Its course and length make it an important transportation route. It flows southward for 2,293 miles (3,689 km) across the Northern European Plain to the Caspian Sea. Barges can travel by canal from the Volga to the Don River. The Don empties into the Black Sea. Canals also connect the Volga to rivers that drain into the Baltic Sea far to the northwest.

In Siberia, the Ob, Yenisey, and Lena Rivers all flow thousands of miles northward. Eventually, they reach Russia's Arctic coast. These and other Siberian rivers that drain into the Arctic Ocean freeze in winter. In spring, these rivers thaw first in the south. Downstream in

LEVEL 2: (Suggested time: 20 min.) Ask students to imagine that they are taking part in a summer camp exchange program in Russia. First, students should choose a rural area where their camp is located. Then tell students to write a letter home to their families. They should describe the land, weather conditions, vegetation, and the outdoor recreational activities they are enjoying. Students should also mention some of the natural resources that are located near their camp.

LEVEL 3: (Suggested time: 20 min.) Have students write an outline for a documentary film script about the physical geography of Russia. Outlines should indicate the various physical features, climates, vegetation types, and resources found in Russia. Ask students to write descriptions of the scenes that the film would include.

the north, however, the rivers remain frozen much longer. As a result, ice jams there block water from the melting ice and snow. This causes annual floods in areas along the rivers.

✓ **READING CHECK:** *Places and Regions* What are the major physical features of Russia? Northern European Plain, Caucasus Mountains, Caspian Sea, Ural Mountains, Siberia, Kamchatka Peninsula, Sakhalin Island, Kuril Islands, several long rivers, including the Volga

Climate and Vegetation

Nearly all of Russia is located at high northern latitudes. The country has tundra, subarctic, humid continental, and steppe climates. Because there are no high mountain barriers, cold Arctic winds sweep across much of the country in winter. Winters are long and cold. Ice blocks most seaports until spring. However, the winters are surprisingly dry in much of Russia. This is because the interior is far from ocean moisture.

Winters are particularly severe throughout Siberia. Temperatures often drop below –40°F (–40°C). Although they are short, Siberian summers can be hot. Temperatures can rise to 100°F (38°C).

Vegetation varies with climates from north to south. Very cold temperatures and permafrost in the far north keep trees from taking root. Mosses, wildflowers, and other tundra vegetation grow there.

The vast **taiga** (TY-guh), a forest of mostly evergreen trees, grows south of the tundra. The trees there include spruce, fir, and pine. In European Russia and in the Far East are deciduous forests. Many temperate forests in European Russia have been cleared for farms and cities.

Wide grasslands known as the **steppe** (STEP) stretch from Ukraine across southern Russia to Kazakhstan. Much of the steppe is used for growing crops and grazing livestock.

✓ **READING CHECK:** *Physical Systems* How does Russia's location affect its climate? Arctic winds blow, creating winters that are very cold.

BUILD on WHAT You Know

Do you remember what you learned about tundra climates? See Chapter 3 to review.

Camels graze on open land in southern Siberia near Mongolia.

National Geography Standard 8

Physical Systems Tundra plants are well adapted to Russia's harsh climate. Vegetation in Russia's tundra climate is limited to lichens and mosses, small shrubs, and grasses. Most plants grow low to the ground to reduce the effects of cold and wind, but may have massive root systems to store nutrients. Other plants have waxy or hairy surfaces that protect them from freezing winds.

Critical Thinking: What adaptations do humans make to survive cold environments like the tundra? How do the adaptations made by humans compare to those made by plants?

Answer: Students might mention that like plants, humans cover themselves with insulating materials, try to keep abundant supplies of food, and try to minimize exposure to wind and cold.

CLOSE

Refer students to the chapter map and the time zone map in the textbook's Skills Handbook. Call out physical features and ask students what time it is where that feature is located when it is 3 P.M. in Moscow. *(Examples: western Siberian lowlands: 5 P.M.; White Sea: 3 P.M.; mouth of the Lena River: 9 P.M.)*

REVIEW AND ASSESS

Have students complete the Section Review. Then pair students. Tell each student to create five fill-in-the-blank questions based on the material in Section 1. Then have students take turns asking their questions. Students should copy the questions and answers into their notes. Then have students complete Daily Quiz 14.1. **COOPERATIVE LEARNING**

RETEACH

Have students complete Main Idea Activity 14.1. Then have students work in groups to create large sketch maps of Russia. Students should take turns labeling physical features, climates, or locations of natural resources on their maps. **ENGLISH LANGUAGE LEARNERS, COOPERATIVE LEARNING**

EXTEND

Have interested students conduct research on the Tunguska Event of 1908, when a huge explosion of mysterious origin leveled about 1,200 square miles of the Siberian wilderness. Ask students to identify when plant and animal life returned. **BLOCK SCHEDULING**

Section 1 Review

Answers

Define For definitions, see: taiga, p. 323; steppe, p. 323

Working with Sketch Maps Maps will vary, but listed places should be labeled in their approximate locations. The Volga River is Europe's longest river. It flows across the Northern European Plain to the Caspian.

Reading for the Main Idea

1. the Ural Mountains (NGS 4)
2. by canals to the Don and other rivers that empty into Baltic and Black Seas (NGS 4)
3. severe, with temperatures as low as –40° F (–40° C), and surprisingly dry; answers will vary, but students might mention various human adaptations to the cold, such as more indoor activities. (NGS 15)

Critical Thinking

4. by providing Russia with materials to export as well as limiting its need for imported goods

Organizing What You Know

5. climates—tundra, subarctic, humid continental, and steppe; vegetation—mosses, wildflowers, tundra vegetation, evergreen trees, deciduous forests, and grasslands; resources—oil, coal, iron ore, copper, manganese, gold, platinum, nickel, diamonds, forests

A blast furnace is used to process nickel in Siberia. Nickel is just one of Russia's many natural resources.

Resources

Russia has enormous energy, mineral, and forest resources. However, those resources have been poorly managed. For example, much of the forest west of the Urals has been cut down. Now wood products must be brought long distances from Siberia. Still, the taiga provides a vast supply of trees for wood and paper pulp.

Russia has long been a major oil producer. However, many of its oil deposits are far from cities, markets, and ports. Coal is also plentiful. More than a dozen metals are available in large quantities. Russia also is a major diamond producer. Many valuable mineral deposits in remote Siberia have not yet been mined.

✓ **READING CHECK:** *Places and Regions* How might the location of its oil deposits prevent Russia from taking full advantage of this resource? It makes transporting to other markets difficult.

go.hrw.com Homework Practice Online
Keyword: SK3 HP14

Section 1 Review

Define and explain: taiga, steppe

Working with Sketch Maps On a map of Russia that you sketch or that your teacher provides, label the following: Arctic Ocean, Caucasus Mountains, Caspian Sea, Ural Mountains, West Siberian Plain, Central Siberian Plateau, Kamchatka Peninsula, Kuril Islands, Volga River, and Baltic Sea.

Reading for the Main Idea

1. *Places and Regions* What low mountain range in central Russia divides Europe from Asia?
2. *Places and Regions* How is the Volga River linked to the Baltic and Black Seas?
3. *Environment and Society* What are winters like in much of Russia? How might they affect people?

Critical Thinking

4. **Making Generalizations and Predictions** How might Russia's natural resources make the country more prosperous?

Organizing What You Know

5. **Categorizing** Copy the following graphic organizer. Use it to list the climates, vegetation, and resources of Russia.

Climates	Vegetation	Resources

Section 2

Objectives

1. Outline the major events in Russia's early history.
2. Describe the growth and decline of the Russian Empire.
3. Report on the Soviet Union.
4. Identify characteristics of present-day Russia.

FOCUS

LET'S GET STARTED

Arrange the classroom so that the desks are in two large groups with one aisle down the middle. Ask a volunteer to walk from one corner of the room to the opposite corner by the easiest route. *(Most students will use the aisle.)* Then, explain to the class that the great Russian steppe, or plain, was like a flat aisle or hallway that created easy access to the Russian interior for both invaders and immigrants. Tell students that in Section 2 they will learn more about those groups as they study the history and culture of Russia.

Building Vocabulary

Write the Define terms on the board. Tell students that **czar** comes from a Latin word, *caesar*, which means "emperor." Point out that **abdicate** comes from two Latin word parts: *ab-* ("away") and *dicare* ("to proclaim"). Have students suggest other words related to **allies**. Call on volunteers to define **superpowers** and **consumer goods** based on the terms' component words. Ask students to contrast the **Cold War** with other wars they have studied. Check all suggestions against definitions.

Section 2 History and Culture

Read to Discover

1. What was Russia's early history like?
2. How did the Russian Empire grow and then fall?
3. What was the Soviet Union?
4. What is Russia like today?

Define

- czar
- abdicated
- allies
- superpowers
- Cold War
- consumer goods

Locate

- Moscow

Russian caviar, blini (pancakes), and smoked salmon

WHY IT MATTERS

Russia is well known for its ballet companies. Use CNNfyi.com or other **current events** sources to discover more about Russia's culture and its international recognition in the field of dance and other arts. Record your findings in your journal.

Early Russia

The roots of the Russian nation lie deep in the grassy plains of the steppe. For thousands of years, people moved across the steppe bringing new languages, religions, and ways of life.

Early Migrations Slavic peoples have lived in Russia for thousands of years. In the A.D. 800s, Viking traders from Scandinavia helped shape the first Russian state among the Slavs. These Vikings called themselves Rus (ROOS). The word *Russia* comes from their name. The state they created was centered on Kiev. Today Kiev is the capital of Ukraine.

In the following centuries, missionaries from southeastern Europe brought Orthodox Christianity and a form of the Greek alphabet to Russia. Today the Russian language is written in this Cyrillic alphabet.

Mongols After about 200 years, Kiev's power began to decline. In the 1200s, Mongol invaders called Tatars swept out of Central Asia across the steppe. The Mongols conquered Kiev and added much of the region to their vast empire.

The Mongols demanded taxes but ruled the region through local leaders. Over time, these local leaders established various states. The strongest of these was Muscovy, north of Kiev. Its chief city was Moscow.

✓ **READING CHECK:** *Human Systems* What was the effect of Viking traders on Russia? **They helped shape the first Russian state, centered on Kiev, now the capital of Ukraine.**

▲ This painting from the mid-1400s shows a battle between soldiers of two early Russian states.

Section 2 RESOURCES

Reproducible

- Guided Reading Strategy 14.2
- Graphic Organizer 14
- Geography for Life Activity 14
- Interdisciplinary Activity for the Middle Grades 17, 18, 20
- Map Activity 17
- Cultures of the World Activity 4

Technology

- One-Stop Planner CD–ROM, Lesson 14.2
- Homework Practice Online
- HRW Go site

Reinforcement, Review, and Assessment

- Section 2 Review, p. 330
- Daily Quiz 14.2
- Main Idea Activity 14.2
- English Audio Summary 14.2
- Spanish Audio Summary 14.2

TEACH

Teaching Objective 1

ALL LEVELS: (Suggested time: 20 min.) Have students draw a time line of the major events in Russian history up to the Russian Revolution in 1917. You may want to have students color code the entries based on the ethnic or religious group involved. Display the time lines around the classroom. **ENGLISH LANGUAGE LEARNERS, COOPERATIVE LEARNING**

Teaching Objective 2

ALL LEVELS: (Suggested time: 20 min.) Copy the following graphic organizer onto the chalkboard, omitting the italicized answers. Use it to help students understand the rise and fall of the Russian Empire. Call on volunteers to fill in details to connect the events. **ENGLISH LANGUAGE LEARNERS**

RISE AND FALL OF THE RUSSIAN EMPIRE

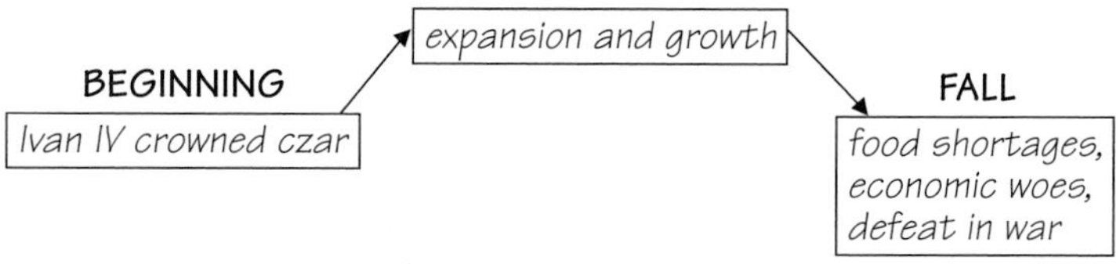

Linking Past to Present

Invasions of Russia Russia has been invaded several times during its long history. Each invasion eventually failed. In (1707–09) Sweden's king Charles XII invaded Russia, only to be defeated by brutal winter weather, burned land and crops, and the czar's army. In 1812, Napoléon I of France invaded Russia with about 400,000 men. Only 30,000 or so survived the battles, cold, disease, hunger, and attacks by Russian soldiers and citizens. In 1941 Nazi Germany invaded the Soviet Union. Again, cold was a major factor in the Nazis' defeat.

Critical Thinking: What roles might Russia's size and landforms have played in these defeats?

Answer: Students might mention that the size made supply lines extremely long and thus difficult to maintain. Armies had to cover such great distances that severe weather could trap unprepared troops before they could return home. The featureless steppe offered little shelter.

Map Answer ►

between 1801 and 1945

History of Russian Expansion

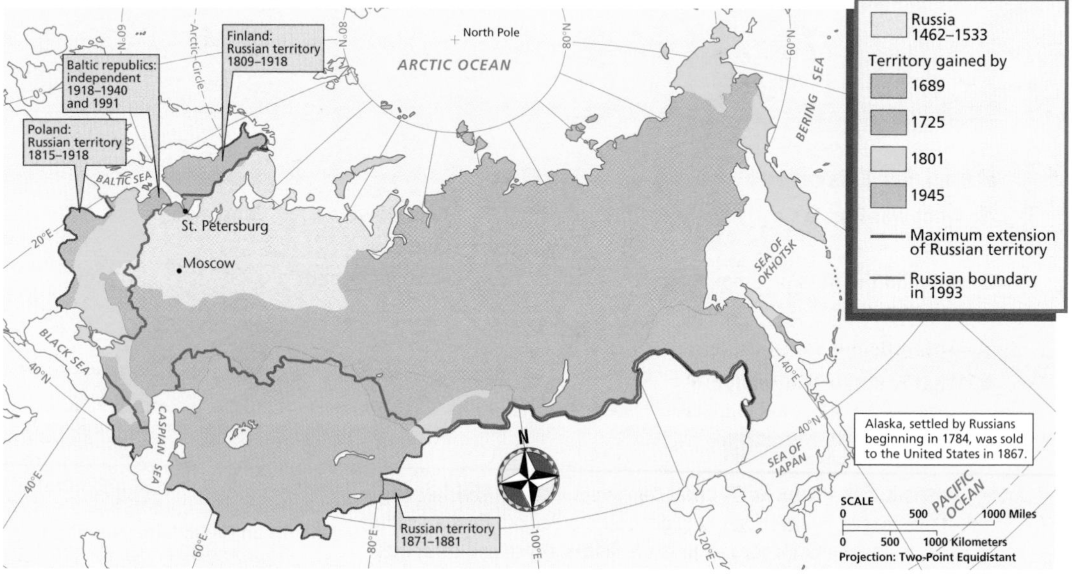

▲ The colors in this map show land taken by the Russian Empire and the Soviet Union over time.

Interpreting the Map **When was the period of Russia's greatest expansion?**

▲ Ivan the Terrible became grand prince of Moscow in 1533. He was just three years old. He ruled Russia from 1547 to his death in 1584.

The Russian Empire

In the 1400s Muscovy won control over parts of Russia from the Mongols. In 1547 Muscovy's ruler, Ivan IV—known as Ivan the Terrible—crowned himself **czar** (ZAHR) of all Russia. The word *czar* comes from the Latin word *Caesar* and means "emperor."

Expansion Over more than 300 years, czars like Peter the Great (1672–1725) expanded the Russian empire. By the early 1700s the empire stretched from the Baltic to the Pacific.

Russian fur traders crossed the Bering Strait in the 1700s and 1800s. They established colonies along the North American west coast. Those colonies stretched from coastal Alaska to California. Russia sold Alaska to the United States in 1867. Around the same time, Russia expanded into Central Asia.

Decline The Russian Empire's power began to decline in the late 1800s. Industry grew slowly, so Russia remained largely agricultural. Most people were poor farmers. Far fewer were the rich, factory workers, or craftspeople. Food shortages, economic problems, and defeat in war further weakened the empire in the early 1900s.

In 1917, during World War I, the czar **abdicated**, or gave up his throne. Later in 1917 the Bolshevik Party, led by Vladimir Lenin, overthrew the government. This event is known as the Russian Revolution.

✓ **READING CHECK:** *Human Systems* What conflict brought a change of government to Russia? the Russian Revolution in which the Bolshevik Party overthrew the government

Teaching Objective 3

LEVEL 1: (Suggested time: 20 min.) Pair students and have them design a political poster for the Soviet Union. Posters should include references to the Russian Revolution, past leaders, and Cold War elements concerning competition between the Soviet Union and the United States. **ENGLISH LANGUAGE LEARNERS, COOPERATIVE LEARNING**

LEVEL 2: (Suggested time: 30 min.) Have students create editorial cartoons referring to the Soviet Union during the Cold War from an American viewpoint. Ask them to explain how their cartoons show Americans' frame of reference regarding the Cold War. **ENGLISH LANGUAGE LEARNERS**

LEVEL 3: (Suggested time: 30 min.) Have students play the role of an American journalist living in the Soviet Union just after its collapse in 1991. Tell them to write a newspaper article describing the conditions and factors leading up to the collapse of the Soviet Union, doing additional research as needed. Have each student write an appropriate headline for his or her article and share it with the class.

The Soviet Union

The Bolsheviks, or Communists, established the Soviet Union in 1922. Most of the various territories of the Russian Empire became republics within the Soviet Union.

Under Lenin and his successor, Joseph Stalin, the Communists took over all industries and farms. Religious practices were discouraged. The Communists outlawed all other political parties. Many opponents were imprisoned, forced to leave the country, or even killed.

The Soviet leaders established a command economy, in which industries were controlled by the government. At first these industries grew dramatically. However, over time the lack of competition made them inefficient and wasteful. The quality of many products was poor. Government-run farms failed to produce enough food to feed the population. By the late 1950s the Soviet Union had to import large amounts of grain.

Cold War The Soviet Union in the 1950s was still recovering from World War II. The country had been a major battleground in the war. The United States and the Soviet Union had been **allies**, or friends, in the fight against Germany. After the war the two **superpowers**, or powerful countries, became rivals. This bitter rivalry became known as the **Cold War**. The Cold War lasted from the 1940s to the early 1990s. The Soviet Union and the United States built huge military forces, including nuclear weapons. The two countries never formally went to war with each other. However, they supported allies in small wars around the world.

Collapse of the Soviet Union The costs of the Cold War eventually became too much for the Soviet Union. The Soviet government spent more and more money on military goods. **Consumer goods** became expensive and in short supply. Consumer goods are products used at home and in everyday life. The last Soviet leader, Mikhail Gorbachev, tried to bring about changes to help the economy. He also promoted a policy allowing more open discussion of the country's problems. However, the various Soviet republics pushed for independence. Finally, in 1991 the Soviet Union collapsed. The huge country split into 15 republics.

In late 1991 Russia and most of the other former Soviet republics formed the Commonwealth of Independent States, or CIS. The CIS does not have a strong central government. Instead, it provides a way for the former Soviet republics to address shared problems.

✓ READING CHECK: *Human Systems* What was the Cold War, and how did it eventually cause the Soviet Union's collapse? **rivalry between Soviet Union, U.S.; military costs eventually ruined Soviet economy**

▲ Tourists can visit the czar's Summer Palace in St. Petersburg.

Interpreting the Visual Record **How do you think the rich lifestyle of the czars helped the Bolsheviks gain support?**

A man lights candles in front of portraits of the last czar, Nicholas II, and his wife. Russians are divided over what kind of government their country should have today.

Interpreting the Visual Record **Why do you think some Russians might wish to have a czar again?** ▼

Environment and Society

Russia is faced with many environmental problems that it has inherited from the era of Soviet rule. During the 1900s, the Soviet government expanded settlement into Siberia with little concern for environmental impacts. The size and richness of the land made it seem that there was no limit to the resources or to the land on which to dump wastes.

Now Russia is left with large areas that have polluted air, soil, or water. The Russian economy is troubled, and Russia has limited resources to devote to leftover environmental damage.

Discussion: Where should environmental issues, such as cleaning up pollution and toxic waste, rank in national priorities? Have students discuss the issue from different points of view. *(Students may suggest that such issues are important but that they decline in priority in the face of an economic crisis. Others may feel environmental issues should take first priority because life itself depends on a healthy environment.)*

▲ Visual Record Answers

by alienating the common people from the czar

◄ in hopes of regaining their cultural heritage

➤ASSIGNMENT: Have students design a tombstone or write an obituary for the Soviet Union. You may want to provide students with examples of obituaries from the daily newspaper to use as models for their work. Display tombstones or obituaries around the classroom.

Teaching Objective 4

LEVELS 1 AND 2: (Suggested time: 20 min.) Have students use their textbook to create a chart on modern-day Russia. The chart should include information on the country's ethnic make up, religions, foods and festivals, arts and sciences, and government. When all students have completed their charts, ask volunteers to share their charts with the class. **ENGLISH LANGUAGE LEARNERS**

LEVEL 3: (Suggested time: 30 min.) Pair students and instruct them to create a chart with two columns, one labeled "Russia Today" and the other "United States." Using their textbooks, students should select a fact about life in Russia and write it in the first column. Then they should compare or contrast that fact with life in the United States in the second column. Tell them to be sure to include government, economy, and culture in their charts. **COOPERATIVE LEARNING**

HUMAN SYSTEMS

The Tatars, also known as the Tartars, are an ethnic group living in various parts of Russia. One branch of this group is the Crimean Tatars. They have a unique history. The Crimean Tatars formed their own Soviet republic in 1921, but this republic was dissolved in 1945 when they were accused of helping the Germans in World War II. Most of the Crimean Tatars were deported to other parts of the Soviet Union and forbidden to speak their native language. In 1956 they regained their civil rights but were still refused the right to move back to the Crimea region. It was not until the breakup of the Soviet Union in 1991 that Crimean Tatars were allowed back into that region. Today there are about 270,000 Tatars in the Crimea.

Critical Thinking: Why might a government outlaw a specific language?

Answer: to try to break up a certain group of people

Connecting to Literature

Answers

1. She is living in a Soviet republic where there is little information about the United States.
2. a wonderful place with many consumer goods

CONNECTING TO *Literature*

AUNT RIMMA'S TREAT

The former Soviet Union was composed of many republics, which are now independent countries. Nina Gabrielyan's The Lilac Dressing Gown *is told from the point of view of an Armenian girl living in Moscow before the Soviet breakup.*

Aunt Rimma. . . came to visit and gave me a pink caramel which I, naturally, popped straight in my mouth. "Don't swallow it," Aunt Rimma says in an odd sort of voice. "You're not supposed to swallow it, only chew it." "Why," I ask, puzzled by her solemn tone. "It's chewing gum," she says with pride in her eyes. "Chewing gum?" I don't know what she means. "American chewing gum," Aunt Rimma explains. "Mentor's sister sent it to us from America." "Oh, from America? Is that where the capitalists are? What is she doing there?" "She's living there," says Aunt Rimma, condescending to my foolishness.

But I am not as foolish as I used to be. I know that Armenians live in Armenia. Our country is very big and includes many republics: Armenia, Georgia, Azerbaijan, Tajikistan, Uzbekistan, Ukraine, Belorussia [Belarus], the Caucasus and Transcaucasia. All this together is the Soviet Union. Americans . . . live in America. Clearly, Mentor's sister cannot possibly be American. . . . Rimma goes on boasting: "Oh! the underwear they have there! . . . And the children's clothes!"

I begin to feel a bit envious. . . . Nobody in our house has anyone living in America, but Aunt Rimma does. My envy becomes unbearable. So I decide to slay our boastful neighbor on the spot: "Well we have cockroaches! This big! Lots and lots of them!'"

Analyzing Primary Sources

1. How does the Armenian girl's frame of reference affect her view of the United States?
2. What does Aunt Rimma seem to think the United States is like?

Russia Today

Russia has been making a transition from communism to democracy and a free market economy since 1991. Change has been slow, and the country faces difficult challenges.

People and Religion More than 146 million people live in Russia today. More than 80 percent are ethnic Russians. The largest of Russia's many minority groups are Ukrainians and Tatars. These Tatars are the descendants of the early Mongol invaders of Russia.

TEACHER TO TEACHER

Teaching Objectives 1–4

ALL LEVELS: (Suggested time: 30 min.) Organize the class into groups. Have each group design a poster for a film about one of the major eras of Russian history. Each poster should include the film title, a phrase to describe the action, and an illustration. Ask students to present and explain their posters to the class. **ENGLISH LANGUAGE LEARNERS, COOPERATIVE LEARNING**

Marcia Clevenger of Charleston, West Virginia, suggests the following activity to help students learn about Russia: Have students create a pop-up book or picture book about Russia. Students should cut up old magazines, use Internet pictures, or draw illustrations to tell a story about life in Russia. Ask volunteers to present their books to the class.

In the past, the government encouraged ethnic Russians to settle in areas of Russia far from Moscow. They were encouraged to move to places where other ethnic groups were in the majority. Today, many non-Russian peoples in those areas resent the domination of ethnic Russians. Some non-Russians want independence from Moscow. At times this has led to violence and even war, as in Chechnya in southern Russia.

Russia

COUNTRY	POPULATION/ GROWTH RATE	LIFE EXPECTANCY	LITERACY RATE	PER CAPITA GDP
Russia	145,470,197 / –0.4%	62, male 73, female	98%	$7,700
United States	281,421,906 / 0.9%	74, male 80, female	97%	$36,200

Sources: Central Intelligence Agency, *The World Factbook 2001;* U.S. Census Bureau

Interpreting the Chart **How many times greater is the U.S. population than the Russian population?**

Since 1991 a greater degree of religious expression has been allowed in Russia. Russian Orthodox Christianity is becoming popular again. Cathedrals have been repaired, and their onion-shaped domes have been covered in gold leaf and brilliant colors. Muslims around the Caspian Sea and the southern Urals are reviving Islamic practices.

Food and Festivals Bread is an important part of the Russian diet. It is eaten with every meal. It may be a rich, dark bread made from rye and wheat flour or a firm white bread. As in other northern countries, the growing season is short and winter is long. Therefore, the diet includes many canned and preserved foods, such as sausages, smoked fish, cheese, and vegetable and fruit preserves.

Black caviar, one of the world's most expensive delicacies, comes from Russia. The fish eggs that make up black caviar come from sturgeon. Sturgeon are fish found in the Caspian Sea.

The anniversary of the 1917 Russian Revolution was an important holiday during the Soviet era. Today the Orthodox Christian holidays of Christmas and Easter are again becoming popular in Russia. Special holiday foods include milk puddings and cheesecakes.

The Arts and Sciences Russia has given the world great works of art, literature, and music. For example, you might know *The Nutcracker,* a ballet danced to music composed by Peter Tchaikovsky (1840–93). It is a popular production in many countries.

Ballet dancers perform Peter Tchaikovsky's *Swan Lake* at the Mariinsky Theater in St. Petersburg.

Across the Curriculum
ART

Fabergé Eggs For centuries Russians have exchanged eggs at Easter. In 1884 the czar gave his wife a special jeweled egg made by master craftsman Peter Carl Fabergé. It opened to reveal a smaller gold egg, which in turn contained a tiny golden chicken and glittering crown. The czarina was so delighted that the Fabergé egg became an annual tradition. Each was a small triumph of the jeweler's art.

Critical Thinking: How do Fabergé eggs reflect Russian traditions?

Answer: Their creation began with the czar's gift to his wife, reflecting a tradition of exchanging eggs at Easter.

Chart Answer

1.9 times greater

Close

Remind students that the United States and the Soviet Union were bitter enemies for many years. Ask students to suggest reasons why the Cold War never exploded into World War III. *(Students might mention the threat of mutual destruction by nuclear weapons.)*

Review and Assess

Have students complete the Section Review. Then pair students and have each student write five major events in Russian history, in random order. Students should exchange lists with their partners and place the events in chronological order. Then have students complete Daily Quiz 14.2.
COOPERATIVE LEARNING

Reteach

Have students complete Main Idea Activity 14.2. Then have students work in groups to create time lines of major events in Russian history.
ENGLISH LANGUAGE LEARNERS, COOPERATIVE LEARNING

Extend

Have interested students conduct research on the current Russian government and another government of their choice. Ask them to create a chart comparing the two governments. **BLOCK SCHEDULING**

Section Review 2

Answers

Define For definitions, see: czar, p. 326; abdicated, p. 326; allies, p. 327; superpowers, p. 327; Cold War, p. 327; consumer goods, p. 327

Working with Sketch Maps Moscow should be labeled in its approximate location. Kiev was the first Russian state among the Slavs.

Reading for the Main Idea

1. traders, who called themselves Rus and organized the first Russian state (NGS 4)
2. Its leaders overthrew the Russian government in 1917. (NGS 13)
3. struggling economy, corruption, competing political groups (NGS 4)

Critical Thinking

4. Russian Empire—food shortages, economic problems, and defeat in war; Soviet Union—lack of consumer goods, too much money spent on military goods

Organizing What You Know

5. Answers will vary but should include information from the section.

Many Russian writers are known for how they capture the emotions of characters in their works. Some writers, such as Aleksandr Solzhenitsyn (1918–), have written about Russia under communism.

Russian scientists also have made important contributions to their professions. For example, in 1957 the Soviet Union launched *Sputnik.* It was the first artificial satellite in space. Today U.S. and Russian engineers are working together on space projects. These include building a large space station and planning for a mission to Mars.

Government Like the U.S. government, the Russian Federation is governed by an elected president and a legislature called the Federal Assembly. The Federal Assembly includes representatives of regions and republics within the Federation. Non-Russians are numerous or in the majority in many of those regions and republics.

The government faces tough challenges. One is improving the country's struggling economy. Many government-owned companies have been sold to the private sector. However, financial problems and corruption have made people cautious about investing in those companies.

Corruption is a serious problem. A few people have used their connections with powerful government officials to become rich. In addition, many Russians avoid paying taxes. This means the government has less money for salaries and services. Agreement on solutions to these problems has been hard.

Republics of the Russian Federation

Adygea	Karachay-Cherkessia
Alania	Karelia
Bashkortostan	Khakassia
Buryatia	Komi
Chechnya	Mari El
Chuvashia	Mordvinia
Dagestan	Sakha
Gorno-Altay	Tatarstan
Ingushetia	Tuva
Kabardino-Balkaria	Udmurtia
Kalmykia	

✓ **READING CHECK:** *Human Systems* What are the people and culture of Russia like today? 80 percent ethnic Russian; free practice of religion; bread and preserved foods important; many great works of art, literature, and ballet, and scientific contributions

go.hrw.com Homework Practice Online
Keyword: SK3 HP14

Section Review 2

Define and explain: czar, abdicated, allies, superpowers, Cold War, consumer goods

Working with Sketch Maps On the map you created in Section 1, label Moscow. In the margin, explain the role Kiev played in Russia's early history.

Reading for the Main Idea

1. *Places and Regions* How did Russia get its name?
2. *Human Systems* What was the Bolshevik Party?
3. *Places and Regions* What are some of the challenges that Russia faces today?

Critical Thinking

4. **Comparing** Compare the factors that led to the decline of the Russian Empire and the Soviet Union. List the factors for each.

Organizing What You Know

5. **Summarizing** Copy the following graphic organizer. Use it to identify important features of Russia's ethnic population, religion, food, and arts and sciences.

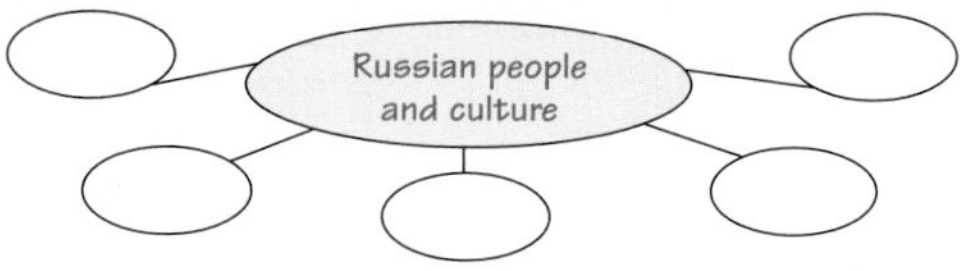

Section 3

Objectives

1. Explain why European Russia is considered the heartland of the country.
2. Describe the four major regions of European Russia.

FOCUS

LET'S GET STARTED

Copy the following "equation" and question onto the chalkboard: *St. Petersburg → Petrograd → Leningrad → St. Petersburg. What do you think this means?* Tell students that "-burg" and "-grad" mean "city" and that the city now called St. Petersburg has changed names several times. Ask students to speculate what might have prompted the name change. *(changes in government or rulers)* Tell students that in Section 3 they will learn more about St. Petersburg and other cities of Russia's heartland.

Building Vocabulary

Write the terms **light industry** and **heavy industry** on the chalkboard. Based on what students know about industry, ask them to speculate what the adjective before each of these terms does to the meaning of the term. *(Light and heavy generally reflect the weight of the product created by the industry.)* Have a student locate and read aloud a definition of **smelters**. Call on a volunteer to describe the relationship between smelters and heavy industry. *(Smelters process metal ores that are primarily used in heavy industry.)*

Section 3 The Russian Heartland

Read to Discover

1. Why is European Russia considered the country's heartland?
2. What are the characteristics of the four regions of European Russia?

Define

light industry
heavy industry
smelters

Locate

St. Petersburg
Nizhniy Novgorod
Astrakhan
Yekaterinburg
Chelyabinsk
Magnitogorsk

WHY IT MATTERS

Following the fall of Communism, some Russian cities' landscapes began to change. In the larger cities like Moscow there are newer buildings and more restaurants. Use CNNfyi.com or other **current events** sources to find information about Moscow and other large Russian cities. Record your findings in your journal.

Jeweled box made by Peter Carl Fabergé

Section 3 RESOURCES

Reproducible

- Guided Reading Strategy 14.3

Technology

- One-Stop Planner CD–ROM, Lesson 14.3
- Homework Practice Online
- *ARGWorld* CD–ROM: Land Values in Post-Soviet Moscow
- HRW Go site

Reinforcement, Review, and Assessment

- Section 3 Review, p. 333
- Daily Quiz 14.3
- Main Idea Activity 14.3
- English Audio Summary 14.3
- Spanish Audio Summary 14.3

The Heartland

The European section of Russia is the country's heartland. The Russian nation expanded outward from there. It is home to the bulk of the Russian population. The national capital and large industrial cities are also located there.

The plains of European Russia make up the country's most productive farming region. Farmers focus mainly on growing grains and raising livestock. Small gardens near cities provide fresh fruits and vegetables for summer markets.

The Russian heartland can be divided into four major regions. These four are the Moscow region, the St. Petersburg region, the Volga region, and the Urals region.

✓ **READING CHECK:** *Places and Regions* Why is European Russia the country's heartland? The bulk of the population, the national capital, and large industrial cities are there.

The Moscow Region

Moscow is Russia's capital and largest city. More than 9 million people live there. In addition to being Russia's political center, Moscow is the country's center for transportation and communication. Roads, railroads, and air routes link the capital to all points in Russia.

At Moscow's heart is the Kremlin. The Kremlin's red brick walls and towers were built in the late 1400s. The government offices, beautiful palaces, and gold-domed churches within its walls are popular tourist attractions.

Twenty towers, like the one in the lower left, are spaced along the Kremlin's walls.

Interpreting the Visual Record What was the advantage of locating government buildings and palaces within the walls of one central location?

◄ **Visual Record Answer**

for protection from attack

TEACH

Teaching Objectives 1–2

ALL LEVELS: (Suggested time: 15 min.) Copy the following graphic organizer onto the chalkboard, omitting the italicized answers. Call on students to fill in the chart with descriptive characteristics that make this region Russia's heartland. **ENGLISH LANGUAGE LEARNERS**

REGIONS OF THE RUSSIAN HEARTLAND

Moscow	St. Petersburg	Volga	Ural
• *capital and largest city* • *huge industrial area*	• *second-largest city* • *major seaport* • *important centers of learning*	• *major shipping route* • *many factories and industries*	• *many mineral resources*

Teaching Objective 2

ALL LEVELS: (Suggested time: 30 min.) Organize the class into four groups, assigning each group a region of the Russian heartland. Have students create a "Welcome to ____" billboard to be placed at the region's border. Students should emphasize positive aspects of their region. Display the billboards around the classroom. **ENGLISH LANGUAGE LEARNERS, COOPERATIVE LEARNING**

Across the Curriculum
SCIENCE

The Pulkovo Observatory Among the many learning institutions of the St. Petersburg region is the Pulkovo Observatory. The observatory's 15-inch (38 cm) refracting telescope was the world's largest when it was built in 1839. Known for its quality of observations, the observatory doubled the size of its refracting telescope to 30 inches (76 cm) in 1878. Although it was destroyed during World War II, the Pulkovo Observatory was rebuilt in 1951.

Critical Thinking: Why might an observatory be destroyed during a war?

Answer: The enemy might have targeted all scientific sites in an effort to destroy technological advancements and to reduce potential surveillance.

Visual Record Answer

It is practiced freely.

Vendors sell religious art and other crafts at a sidewalk market in Moscow.

Interpreting the Visual Record What do the items in this market suggest about the status of religion in Russia since the communist era?

Moscow is part of a huge industrial area. This area also includes the city of Nizhniy Novgorod, called Gorky during the communist era. About one third of Russia's population lives in this region.

The Soviet government encouraged the development of **light industry**, rather than **heavy industry**, around Moscow. Light industry focuses on the production of lightweight goods, such as clothing. Heavy industry usually involves manufacturing based on metals. It causes more pollution than light industry. The region also has advanced-technology and electronics industries.

The St. Petersburg Region

Northwest of Moscow is St. Petersburg, Russia's second-largest city and a major Baltic seaport. More than 5 million people live there. St. Petersburg was Russia's capital and home to the czars for more than 200 years. This changed in 1918. Palaces and other grand buildings constructed under the czars are tourist attractions today. St. Petersburg was known as Leningrad during the communist era. Much of the city was heavily damaged during World War II.

The surrounding area has few natural resources. Still, St. Petersburg's harbor, canals, and rail connections make the city a major center for trade. Important universities and research institutions are located there. The region also has important industries.

READING CHECK: *Human Systems* Why are Moscow and St. Petersburg such large cities? centers of politics, transportation, communication, industry, trade, education

The Mariinsky Theater of Opera and Ballet is one of St. Petersburg's most beautiful buildings. It was called the Kirov during the communist era.

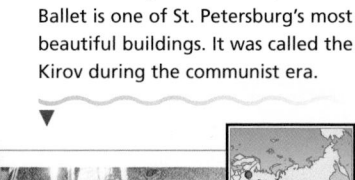

CLOSE

Invite students to write their names in Russian style. To create a middle name, boys add *-ovich*, meaning "son of," and girls add *-ovna*, meaning "daughter of," to the father's first name. Girls also add *-a* to the last name. If the last name ends in *-sky* or *-ski*, girls use *-skaya* instead of the *-a*. So Yuri and Helena, the son and daughter of Viktor Maksimov, would be named Yuri Viktorovich Maksimov and Helena Viktorovna Maksimova.

REVIEW AND ASSESS

Have students complete the Section Review. Then pair students. Ask each student to write five fill-in-the-blank questions on the section material and to exchange papers with his or her partner. Then have students complete Daily Quiz 14.3. **COOPERATIVE LEARNING**

RETEACH

Have students complete Main Idea Activity 14.3. Then organize the class into four groups, one for each of the regions in the Russian heartland. Have students draw pictures that show characteristics of their region and place them on a wall map or sketch map of the region. Discuss the picture maps. **ENGLISH LANGUAGE LEARNERS, COOPERATIVE LEARNING**

EXTEND

Have interested students conduct research and write a report on the founding of St. Petersburg by Peter the Great. Ask them to include the difficulties created by the region's physical geography. **BLOCK SCHEDULING**

The Volga Region

The Volga region stretches along the middle part of the Volga River. The Volga is often more like a chain of lakes. It is a major shipping route for goods produced in the region. Hydroelectric power plants and nearby deposits of coal and oil are important sources of energy.

During World War II, many factories were moved to the Volga region. This was done to keep them safe from German invaders. Today the region is famous for its factories that produce goods such as motor vehicles, chemicals, and food products. Russian caviar comes from a fishery based at the old city of Astrakhan on the Caspian Sea.

The Urals Region

Mining has long been important in the Ural Mountains region. Nearly every important mineral except oil has been discovered there. Copper and iron **smelters** are still important. Smelters are factories that process copper, iron, and other metal ores.

Many large cities in the Urals started as commercial centers for mining districts. The Soviet government also moved factories to the region during World War II. Important cities include Yekaterinburg (yi-kah-ti-reem-BOOHRK) (formerly Sverdlovsk), Chelyabinsk (chel-YAH-buhnsk), and Magnitogorsk (muhg-nee-tuh-GAWRSK). Now these cities manufacture machinery and metal goods.

✓ **READING CHECK:** *Places and Regions* What industries are important in the Volga and Urals regions? automobile, chemical, food, mining, machinery, metal goods

▲ A fisher gathers sturgeon in a small shipboard pool in the Volga region. The eggs for making caviar are taken from the female sturgeon. Then the fish is released back into the water.

Homework Practice Online
go.hrw.com
Keyword: SK3 HP14

Section 3 Review

Define and explain: light industry, heavy industry, smelters

Working with Sketch Maps On the map you created in Section 2, label St. Petersburg, Nizhniy Novgorod, Astrakhan, Yekaterinburg, Chelyabinsk, and Magnitogorsk. In the margin of your map, write a short caption explaining the significance of Moscow and St. Petersburg.

Reading for the Main Idea

1. *Places and Regions* Why might so many people settle in Russia's heartland?
2. *Places and Regions* Where did the Soviet government move factories during World War II?

Critical Thinking

3. **Drawing Inferences and Conclusions** Why do you think the Soviet government encouraged the development of light industry around Moscow?
4. **Finding the Main Idea** What role has the region's physical geography played in the development of European Russia's economy?

Organizing What You Know

5. **Contrasting** Use this graphic organizer to identify European Russia's four regions. Write one feature that makes each region different from the other three.

Section 3 Review

Answers

Define For definitions, see: light industry, p. 332; heavy industry, p. 332; smelters, p. 333

Working with Sketch Maps Maps will vary, but listed places should be labeled in their approximate locations. Moscow is Russia's capital and largest city. St. Petersburg is Russia's second-largest city and a major Baltic seaport.

Reading for the Main Idea

1. because it is home to the national capital and large industrial cities; contains the country's most productive farmland (NGS 4)
2. Volga and Urals (NGS 4)

Critical Thinking

3. because it causes less pollution
4. The region's natural resources allowed it to develop an industrial economy.

Organizing What You Know

5. Answers will vary but could include the following: Moscow—center of transportation and communication; St. Petersburg—major seaport; Volga—known for heavy-machine industries and giant factories; Urals—important mining area.

Section 4

Objectives

1. Describe the human geography of Siberia.
2. Identify the economic features of the region.
3. Analyze how Lake Baikal has been threatened by pollution.

Focus

LET'S GET STARTED

Copy the following question onto the chalkboard: *Where are the coldest places in the world?* If students do not name Siberia, refer them to the "Our Amazing Planet" feature in Section 1. You may want to point out Verkhoyansk, in Siberia, where the temperature was recorded, on a more detailed wall map of Russia. Tell students that in Section 4 they will learn more about Siberia.

Building Vocabulary

Write **habitation fog** on the chalkboard. Ask students to infer the meaning of the term based on what they know about the words "habitat" or "inhabit" and "fog." *(Habitation fogs occur over cities—places people inhabit.)* To confirm the meaning, call on a volunteer to find and read the term's definition aloud.

Section 4 Resources

Reproducible
- Guided Reading Strategy 14.4

Technology
- One-Stop Planner CD–ROM, Lesson 14.4
- Homework Practice Online
- HRW Go site

Reinforcement, Review, and Assessment
- Section 4 Review, p. 336
- Daily Quiz 14.4
- Main Idea Activity 14.4
- English Audio Summary 14.4
- Spanish Audio Summary 14.4

Visual Record Answer

subarctic

Section 4 Siberia

Read to Discover

1. What is the human geography of Siberia like?
2. What are the economic features of the region?
3. How has Lake Baikal been threatened by pollution?

Define

habitation fog

Locate

Siberia
Trans-Siberian Railroad
Baikal-Amur Mainline
Kuznetsk Basin
Ob River
Yenisey River
Novosibirsk
Lake Baikal

Why It Matters

It takes more than a week by train to cross Russia. The Trans-Siberian Railroad and the Baikal-Amur Mainline let travelers see more of the country. Use CNNfyi.com or other **current events** sources to learn more about these rail systems. Record your findings in your journal.

Russian matryoshka *nesting doll*

A Sleeping Land

East of European Russia, across the Ural Mountains, is Siberia. Siberia is enormous. It covers more than 5 million square miles (12.95 million sq. km) of northern Asia. It extends all the way to the Pacific Ocean. That is nearly 1.5 times the area of the United States! To the north of Siberia is the Arctic Ocean. To the south are the Central Asian countries, Mongolia, and China.

Many people think of Siberia as simply a vast, frozen wasteland. In fact, in the Tatar language, *Siberia* means "Sleeping Land." In many ways, this image is accurate. Siberian winters are long, dark, and severe. Often there is little snow, but the land is frozen for months. During winter, **habitation fog** hangs over cities. A habitation fog is a fog caused by fumes and smoke from cities. During the cold Siberian winter, this fog is trapped over cities.

Siberia has lured Russian adventurers for more than 400 years. It continues to do so today. This vast region has a great wealth of natural resources. Developing those resources may be a key to transforming Russia into an economic success.

Reindeer graze around a winter camp in northern Siberia.

Interpreting the Visual Record What climate does this area appear to have, tundra or subarctic?

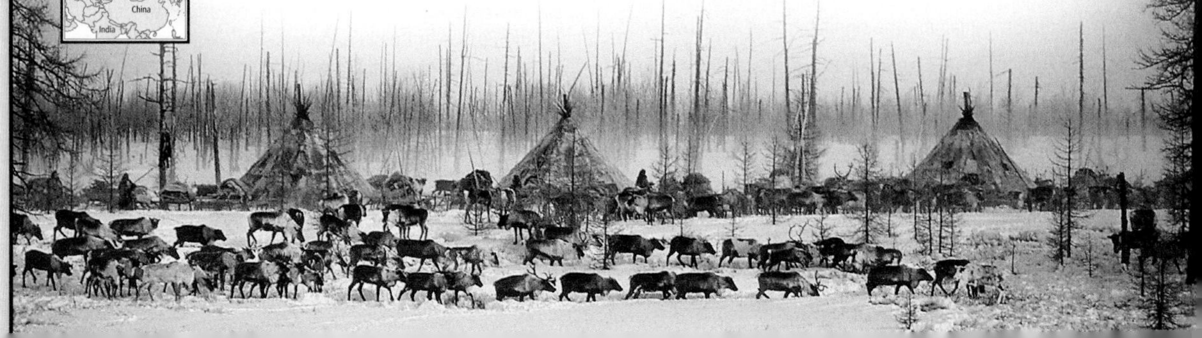

TEACH

Teaching Objectives 1–2

ALL LEVELS: (Suggested time: 20 min.) Copy the following graphic organizer onto the chalkboard, omitting the italicized answers. Have students list human geography and economic reasons for or against living in Siberia. **ENGLISH LANGAGE LEARNERS**

Living in Siberia	
Reasons for:	**Reasons against:**
great wealth and natural resources, Trans-Siberian Railroad and the BAM, higher wages	*severe climate, sparsely populated, little industry*

Teaching Objective 3

ALL LEVELS: (Suggested time: 30 min.) Pair students and have them write contracts for new commercial development at or near Lake Baikal. Contracts should start with a review of Lake Baikal's special qualities, include a review of past problems, and conclude with requirements for operating a business that will not harm the lake or its plants and animals. **ENGLISH LANGUAGE LEARNERS, COOPERATIVE LEARNING**

People Siberia is sparsely populated. In fact, large areas have no human population at all. Most of the people live in cities in western and southern parts of the region.

Ethnic Russians make up most of the population. However, minority groups have lived there since long before Russians began to expand into Siberia.

Settlements Russian settlement in Siberia generally follows the route of the Trans-Siberian Railroad. Construction of this railway started in 1891. When it was completed, it linked Moscow and Vladivostok, a port on the Sea of Japan.

Russia's Trans-Siberian Railroad is the longest single rail line in the world. It is more than 5,700 miles (9,171 km) long. For many Siberian towns, the railroad provides the only transportation link to the outside world. Another important railway is the Baikal-Amur Mainline (BAM), which crosses many mountain ranges and rivers in eastern Siberia.

The Omsk (AWMSK) Cathedral in Omsk, Siberia, provides an example of Russian architecture. Omsk was founded in the early 1700s.

✓ **READING CHECK:** *Places and Regions* Where is Russian settlement located in Siberia, and why do you think this is the case? along the Trans-Siberian Railroad; because it provides a means of transportation and communication in this vast land

Siberia's Economy

The Soviet government built the Baikal-Amur Mainline so that raw materials from Siberia could be easily transported to other places. Abundant natural resources form the foundation of Siberia's economy. They are also important to the development of Russia's struggling economy. Siberia's natural resources include timber, mineral ores, diamonds, and coal, oil, and natural gas deposits.

Although Siberia has rich natural resources, it contains a small percentage of Russia's industry. The harsh climate and difficult terrain have discouraged settlement. Many people would rather live in European Russia, even though wages may be higher in Siberia.

Lumbering and mining are the most important Siberian industries. Large coal deposits are mined in the Kuznetsk Basin, or the Kuzbas. The Kuzbas is located in southwestern Siberia between the Ob and Yenisey Rivers. It is one of Siberia's most important industrial regions.

Siberia's largest city, Novosibirsk, is located near the Kuznetsk Basin. The city's name means "New Siberia." About 1.5 million people live there. It is located about halfway between Moscow and Vladivostok on the Trans-Siberian Railroad. Novosibirsk is Siberia's manufacturing and transportation center.

✓ **READING CHECK:** *Environment and Society* How do Siberia's natural resources influence the economies of Siberia and Russia? They are the foundation of Siberia's economy and could help develop Russia's economy.

A worker repairs an oil rig in Siberia.

Interpreting the Visual Record How is this worker protected from the cold Siberian climate?

HUMAN SYSTEMS

The *Rossiya* train takes 153 hours and 49 minutes to travel from Moscow to Vladivostok on the Trans-Siberian Railroad. The famous railroad required 12 years and more than 70,000 workers to complete. It opened Siberia to settlement and provided access to the region's vast natural resources for development.

The number of Russians using the Trans-Siberian Railroad declined about 50 percent between 1991 and 1997. Even so, trains carry half of all passenger traffic in Russia, compared to less than 1 percent in the United States.

Critical Thinking: How did the Trans-Siberian Railroad make economic development of the region possible?

Answer: It opened the region to settlement and gave access to the natural resources there.

internet connect go.hrw.com

GO TO: go.hrw.com
KEYWORD: SK3 CH14
FOR: Web sites about the Trans-Siberian Railroad

◄ **Visual Record Answer**

with heavy clothing

Close

Remind students that Siberia alone is 1.5 times the size of the United States. Lead a discussion on the various problems Russia faces governing such a large area.

Review and Assess

Have students complete the Section Review. Then have each student create a web diagram with "Siberia" in the center. When they have finished, call on volunteers to reproduce their webs on the chalkboard. Then have students complete Daily Quiz 14.4.

Reteach

Have students complete Main Idea Activity 14.4. Then organize the class into small groups and have groups create a script for an educational television special on Siberia. The script should cover the physical, human, economic, and environmental geography of the region. Have students add pictures and other visual aids if possible. **ENGLISH LANGUAGE LEARNERS, COOPERATIVE LEARNING**

Extend

Have interested students conduct research on the rivers of Siberia, the problems caused by their freezing and thawing, and what life is like on board the rivers' freight barges. They may want to compile their findings into a job description and résumé for a barge worker. **BLOCK SCHEDULING**

Section 4 Review

Answers

Define For definition, see: habitation fog, p. 334

Working with Sketch Maps Maps will vary, but listed places should be labeled in their approximate locations.

Reading for the Main Idea

1. west—Ural Mountains; east—Pacific Ocean; south—Central Asian republics, Mongolia, and China; north—Arctic Ocean (NGS 4)
2. in cities in the western and southern parts along the Trans-Siberian Railroad; because it is a means of transportation and communication (NGS 9)
3. The harsh climate and difficult terrain have discouraged settlement there. (NGS 9)

Critical Thinking

4. Answers will vary, but students should include reasons to justify their responses.

Organizing What You Know

5. natural resources—timber, mineral ores, diamonds, coal, oil, and natural gas; major industries—lumbering and mining

Visual Record Answer ►

cause health problems for the plants and animals

The scenery around Lake Baikal is breathtaking. The lake is seven times as deep as the Grand Canyon.

Interpreting the Visual Record
How would pollution affect this lake and the plants and animals that live there?

Lake Baikal

Some people have worried that economic development in Siberia threatens the region's natural environment. One focus of concern has been Lake Baikal (by-KAHL), the "Jewel of Siberia."

Baikal is located north of Mongolia. It is the world's deepest lake. In fact, it holds as much water as all of North America's Great Lakes. The scenic lake and its surrounding area are home to many kinds of plants and animals. Some, such as the world's only freshwater seal, are endangered.

For decades people have worried about pollution from a nearby paper factory and other development. They feared that pollution threatened the species that live in and around the lake. In recent years scientists and others have proposed plans that allow some economic development while protecting the environment.

Our Amazing Planet

Lake Baikal covers less area than do three of the Great Lakes: Superior, Huron, and Michigan. Still, Baikal is so deep that it contains about one fifth of all the world's freshwater!

✓ **READING CHECK:** *Environment and Society* How has human activity affected Lake Baikal? **The paper factory and other developments have posed an environmental threat.**

Section 4 Review

go.hrw.com Homework Practice Online
Keyword: SK3 HP14

Define and explain: habitation fog

Working with Sketch Maps On the map you created in Section 3, label Siberia, Trans-Siberian Railroad, Baikal-Amur Mainline, Kuznetsk Basin, Ob River, Yenisey River, Novosibirsk, and Lake Baikal.

Reading for the Main Idea

1. *Places and Regions* What are the boundaries of Siberia?
2. *Human Systems* Where do most people in Siberia live? Why?
3. *Places and Regions* Why does this huge region with many natural resources have little industry?

Critical Thinking

4. **Making Generalizations and Predictions** Do you think Russians should be more concerned about rapid economic development or protecting the environment? Why?

Organizing What You Know

5. **Categorizing** Use this organizer to list the region's resources and industries that use them.

Natural Resources	⇨	Major Industries

Section 5

Objectives

1. Explain the relationship between climate and agriculture in the Russian Far East.
2. List the major resources and cities of the Russian Far East.
3. Identify the islands of the Russian Far East.

FOCUS

LET'S GET STARTED

Ask students to predict what the economy of the Russian Far East might be like based solely on its location. *(Students might suggest that shipping, fishing, and lumbering are important there.)* Tell students that in Section 5 they will learn more about the economy of the Russian Far East.

Building Vocabulary

Write **icebreakers** on the chalkboard. Tell students that when a group of people gets together for the first time, such as in a classroom or at a club meeting, the leader of the group often starts with an activity called an icebreaker. Ask students if they have ever been part of such an activity and to describe its purpose. *(Possible answer: to break up the cold social atmosphere so that people can better communicate and work together)* Ask students to relate that description to the ships called icebreakers.

Section 5 The Russian Far East

Read to Discover

1. How does the Russian Far East's climate affect agriculture in the region?
2. What are the major resources and cities of the region?
3. What island regions are part of the Russian Far East?

Define

icebreakers

Locate

Sea of Okhotsk
Sea of Japan
Amur River
Khabarovsk
Vladivostok
Sakhalin Island

WHY IT MATTERS

Because of conflict over the Kuril Islands, Russia and Japan did not sign a peace agreement to end World War II. Use CNNfyi.com or other **current events** sources to find information on this controversy and other political concerns. Record your findings in your journal.

A Russian figurine

Section 5 RESOURCES

Reproducible

- Guided Reading Strategy 14.5

Technology

- One-Stop Planner CD–ROM, Lesson 14.5
- Homework Practice Online
- HRW Go site

Reinforcement, Review, and Assessment

- Section 5 Review, p. 339
- Daily Quiz 14.5
- Main Idea Activity 14.5
- English Audio Summary 14.5
- Spanish Audio Summary 14.5

Agriculture

Off the eastern coast of Siberia are the Sea of Okhotsk and the Sea of Japan. Their coastal areas and islands make up a region known as the Russian Far East.

The Russian Far East has a less severe climate than the rest of Siberia. Summer weather is mild enough for some successful farming. Farms produce many goods, including wheat, sugar beets, sunflowers, meat, and dairy products. However, the region cannot produce enough food for itself. As a result, food must also be imported.

Fishing and hunting are important in the region. There are many kinds of animals, including deer, seals, rare Siberian tigers, and sables. Sable fur is used to make expensive clothing.

✓ **READING CHECK:** *Environment and Society* How does scarcity of food affect the Russian Far East? It forces people in the region to import food.

◄ The Siberian tiger is endangered. The few remaining of these large cats roam parts of the Russian Far East. They are also found in northern China and on the Korean Peninsula.

TEACH

Teaching Objective 1

ALL LEVELS: (Suggested time: 30 min.) Have students draw picture postcards detailing the relationship between agriculture and climate in the Russian Far East. **ENGLISH LANGUAGE LEARNERS**

Teaching Objectives 2–3

ALL LEVELS: (Suggested time: 15 min.) Copy the following graphic organizer onto the chalkboard, omitting the italicized answers. Call on volunteers to come fill in the diagram.
ENGLISH LANGUAGE LEARNERS

Section 5 Review

Answers

Define For definition, see: icebreakers, p. 338

Working with Sketch Maps Maps will vary, but listed places should be labeled in their approximate locations. Russia and Japan dispute possession of Sakhalin Island and the Kuril Islands.

Reading for the Main Idea

1. far less severe (NGS 4)
2. wheat, sugar beets, sunflowers, meat, dairy products; coal, oil, and geothermal energy (NGS 4)

Critical Thinking

3. It remains a major naval base and home port for a large fishing fleet.
4. The natural resources around these islands make them attractive to both countries.

Organizing What You Know

5. Answers will vary but could include the following: Khabarovsk—its location where the Trans-Siberian Railroad crosses the Amur River made it ideal for processing forest and mineral resources; Vladivostok—its location on the Sea of Japan made it ideal as a naval base and fishing port.

BUILD on WHAT You Know

Do you remember what you learned about plate tectonics? See Chapter 2 to review.

Economy

Like the rest of Siberia, the Russian Far East has a wealth of natural resources. These resources have supported the growth of industrial cities and ports in the region.

Resources Much of the Russian Far East remains forested. The region's minerals are only beginning to be developed. Lumbering, machine manufacturing, woodworking, and metalworking are the major industries there.

The region also has important energy resources, including coal and oil. Another resource is geothermal energy. This resource is available because of the region's tectonic activity. Two active volcanic mountain ranges run the length of the Kamchatka Peninsula. Russia's first geothermal electric-power station was built on this peninsula.

Cities Industry and the Trans-Siberian Railroad aided the growth of cities in the Russian Far East. Two of those cities are Khabarovsk (kuh-BAHR-uhfsk) and Vladivostok (vla-duh-vuh-STAHK).

More than 600,000 people live in Khabarovsk, which was founded in 1858. It is located where the Trans-Siberian Railroad crosses the Amur River. This location makes Khabarovsk ideal for processing forest and mineral resources from the region.

Vladivostok is slightly larger than Khabarovsk. *Vladivostok* means "Lord of the East" in Russian. The city was established in 1860 on the coast of the Sea of Japan. Today it lies at the eastern end of the Trans-Siberian Railroad.

Vladivostok is a major naval base and the home port for a large fishing fleet. **Icebreakers** must keep the city's harbor open in winter. An icebreaker is a ship that can break up the ice of frozen waterways. This allows other ships to pass through them.

Historical monuments and old architecture compete for attention in Vladivostok.

CLOSE

Ask students why residents of the Russian Far East might feel closer to Japan and other Pacific nations than to European Russia. *(Possible answers: physically closer to Japan, trade relationships with Pacific countries, isolation created by Siberia)* Ask students to predict if this isolation will increase or decrease. *(Students may predict that new communication technology will help decrease isolation.)*

REVIEW AND ASSESS

Have students complete the Section Review. Then have each student write two true-false questions about agriculture, resources, cities, and islands in the Russian Far East. Pair students and have them quiz each other. Then have students complete Daily Quiz 14.5. **COOPERATIVE LEARNING**

RETEACH

Have students complete Main Idea Activity 14.5. Then, organize the class into small groups, assigning resources, cities, agriculture, and islands to different groups. Ask each group to create a visual aid to illustrate the important facts about its topic. **ENGLISH LANGUAGE LEARNERS, COOPERATIVE LEARNING**

EXTEND

Have interested students conduct research on one of the non-Russian ethnic groups of the Russian Far East. Have them include information on the group's common occupations, history, language, and customs. **BLOCK SCHEDULING**

The Soviet Union considered Vladivostok very important for defense. The city was therefore closed to foreign contacts until the early 1990s. Today it is an important link with China, Japan, the United States, and the rest of the Pacific region.

✓ **READING CHECK:** *Environment and Society* How do the natural resources of the Russian Far East affect its economy? **They have supported mining and timber industries, as well as the growth of industrial cities and ports.**

Islands

The Russian Far East includes two island areas. Sakhalin is a large island that lies off the eastern coast of Siberia. To the south is the Japanese island of Hokkaido. The Kuril Islands are much smaller. They stretch in an arc from Hokkaido to the Kamchatka Peninsula.

Sakhalin has oil and mineral resources. The waters around the Kurils are important for commercial fishing.

Russia and Japan have argued over who owns these islands since the 1850s. At times they have been divided between Japan and Russia or the Soviet Union. The Soviet Union took control of the islands after World War II. Japan still claims rights to the southernmost islands.

Like other Pacific regions, Sakhalin and the Kurils sometimes experience earthquakes and volcanic eruptions. An earthquake in 1995 caused severe damage on Sakhalin Island, killing nearly 2,000 people.

✓ **READING CHECK:** *Environment and Society* How does the environment of the Kuril Islands and Sakhalin affect people? **Earthquakes, such as the one in 1995, and volcanoes pose a constant threat to lives and property.**

▲ An old volcano created Crater Bay in the Kuril Islands. The great beauty of the islands is matched by the terrible power of earthquakes and volcanic eruptions in the area.

Interpreting the Visual Record What do you think happened to the volcano that formed Crater Bay?

Section 5 Review

go.hrw.com Homework Practice Online
Keyword: SK3 HP14

Define and explain: icebreakers

Working with Sketch Maps On the map you created in Section 4, label the Sea of Okhotsk, the Sea of Japan, the Amur River, Khabarovsk, Vladivostok, and Sakhalin Island. In the margin, explain which countries dispute possession of Sakhalin Island and the Kuril Islands.

Reading for the Main Idea

1. *Places and Regions* How does the climate of the Russian Far East compare to the climate throughout the rest of Siberia?
2. *Places and Regions* What are the region's major crops and energy resources?

Critical Thinking

3. **Drawing Inferences and Conclusions** In what ways do you think Vladivostok is "Lord of the East" in Russia today?
4. **Drawing Inferences and Conclusions** Why do you think Sakhalin and the Kuril Islands have been the subject of dispute between Russia and Japan?

Organizing What You Know

5. **Finding the Main Idea** Copy the following graphic organizer. Use it to explain how the location of each city has played a role in its development.

Khabarovsk	Vladivostok

CHAPTER 14 Review ANSWERS

Building Vocabulary

For definitions, see: taiga, p. 323; steppe, p. 323; czar, p. 326, abdicated, p. 326; allies, p. 327, superpowers, p. 327; Cold War, p. 327; consumer goods, p. 327; light industry, p. 332; heavy industry, p. 332; smelters, p. 333; habitation fog, p. 334; icebreakers, p. 338

Reviewing the Main Ideas

1. Northern European Plain, Caucasus Mountains, Ural Mountains, Siberia, the Kamchatka Peninsula, Sakhalin Island, and the Kuril Islands; in the Kuznetsk Basin (NGS 4)
2. the Ural Mountains (NGS 4)
3. It has an elected executive and a legislative branch, like the United States. (NGS 12)
4. Moscow, St. Petersburg, Volga, and the Urals (NGS 4)
5. to the Volga and Urals regions; because of the threat of German invasion (NGS 4)

◄ **Visual Record Answer**

It became extinct and was filled with water.

ASSESS

Have students complete the Chapter 14 Test.

RETEACH

Organize the class into five groups and assign each group a different section from the chapter. Tell the students that each group will create one act for a play titled *Russia—Then, Now, and Forever.* Students should create one or more scenes for their act. The scene can relate to any topic, subject, or plot; however, students should weave facts from their section into the script. Have students perform their plays for the class.

ENGLISH LANGUAGE LEARNERS, COOPERATIVE LEARNING

PORTFOLIO EXTENSIONS

1. Have students create a "family tree" of important rulers and leaders of Russia and the Soviet Union. Some of the most significant are Rurik, Ivan I, Peter the Great, Catherine the Great, Nicholas II, Vladimir Lenin, Joseph Stalin, Nikita Khrushchev, Mikhail Gorbachev, and Boris Yeltsin. Students should use biographies to research each person then summarize how he or she affected Russian or Soviet history and culture.
2. Have students locate a copy of the Cyrillic alphabet. They should create a chart that compares the Cyrillic alphabet to the Latin alphabet. Ask students to report on how some Russian citizens of other ethnic groups are now using the Latin alphabet instead of the Cyrillic.

CHAPTER 14 Review ANSWERS

Understanding Environment and Society

Students' presentations should indicate an understanding of the Russian steppe as well as the agricultural resources of that region. Use Rubric 29, Presentations, to evaluate student work.

Thinking Critically

1. because it has natural resources for factories and industries
2. command; government-, not individual-, owned industries
3. because the size of the country and movement of goods and resources; built two transcontinental railroads
4. provides a trading port for Siberia and the Russian Far East
5. food shortages, economic problems, defeat in war, and lack of consumer goods; collapse of the Russian Empire and the Soviet Union (NGS 17)

Map Activity

A. Vladivostok
B. Kamchatka Peninsula
C. Moscow
D. Arctic Ocean

CHAPTER 14 Reviewing What You Know

Building Vocabulary

On a separate sheet of paper, write sentences to define each of the following words.

1. taiga
2. steppe
3. czar
4. abdicated
5. allies
6. superpower
7. Cold War
8. consumer goods
9. light industry
10. heavy industry
11. smelters
12. habitation fog
13. icebreakers

Reviewing the Main Ideas

1. *Places and Regions* What are the major physical features of Russia? Where in Siberia are large coal deposits?
2. *Places and Regions* What landform separates Europe from Asia?
3. *Human Systems* How is Russia's government organized, and how does it compare with that of the United States?
4. *Places and Regions* What four major regions make up European Russia?
5. *Places and Regions* Where were Russian factories relocated during World War II and why?

Understanding Environment and Society

Resource Use

The grasslands of the steppe are one of Russia's most valuable agricultural resources. Create a presentation on farming in the steppe. You may want to consider the following:

- The crops that are grown in the Russian steppe.
- The kinds of livestock raised in the region.
- How the climate limits agriculture in the steppe.

Thinking Critically

1. **Finding the Main Idea** In what ways might Siberia be important to making Russia an economic success?
2. **Contrasting** What kind of economic system did the Soviet Union have, and how did it differ from that of the United States?
3. **Drawing Inferences and Conclusions** Why is transportation an issue for Russia? What have Russians done to ease transportation between European Russia and the Russian Far East?
4. **Analyzing Information** How does Vladivostok's location make it an important link between Russia and the Pacific world?
5. **Identifying Cause and Effect** What problems existed in the Russian Empire and the Soviet Union in the 1900s, and what was their effect?

Food Festival

Here is a basic recipe for borscht, a traditional Russian soup. In a large pot, sauté a chopped onion in 2 tbsp. butter. Stir in 1½ lb. sliced raw red beets, ¼ c. red wine vinegar, 1 tsp. sugar, 2 chopped fresh tomatoes, 1 tsp. salt, and some black pepper. Pour in ½ c. beef stock, cover, and simmer one hour. Pour in 5 c. more of beef stock and ½ lb. shredded cabbage. Bring to a boil. Add ¼ lb. cubed ham, 1 lb. cooked sliced beef, ½ c. chopped parsley, and a bay leaf. Simmer for 30 minutes. Garnish with sour cream.

Chapter 14 Review and Assessment Resources

Reproducible
- Readings in World Geography, History, and Culture 41, 42, and 43
- Critical Thinking Activity 14
- Vocabulary Activity 14

Technology
- Chapter 14 Test Generator (on the One-Stop Planner)
- Audio CD Program, Chapter 14
- HRW Go site

Reinforcement, Review, and Assessment
- Chapter 14 Review pp. 340–41
- Chapter 14 Tutorial for Students, Parents, Mentors, and Peers
- Chapter 14 Test
- Chapter 14 Test for English Language Learners and Special-Needs Students

Building Social Studies Skills

Map Activity

On a separate sheet of paper, match the letters on the map with their correct labels.

Arctic Ocean
Caucasus Mountains
Caspian Sea
West Siberian Plain
Central Siberian Plateau
Kamchatka Peninsula
Volga River
Moscow
St. Petersburg
Vladivostok

Mental Mapping Skills Activity

On a separate sheet of paper, draw a freehand map of Russia and label the following:

Baltic Sea
Kuril Islands
Lake Baikal
Sakhalin Island
Siberia
Ural Mountains

Writing Activity

Imagine that you are a tour guide on a trip by train from St. Petersburg to Vladivostok. Use the chapter map or a classroom globe to write a one-page description of some of the places people would see along the train's route. How far would you travel? Be sure to use standard grammar, spelling, sentence structure, and punctuation.

E. West Siberian Plain
F. Volga River
G. Caucasus Mountains
H. Central Siberian Plateau
I. Caspian Sea
J. St. Petersburg

Mental Mapping Skills Activity

Maps will vary, but listed places should be labeled in their approximate locations.

Writing Activity

Students' descriptions should include the various physical features along the route and an accurate estimation of the distance of the trip. Use Rubric 40, Writing to Describe, to evaluate student work.

Portfolio Activity

Students should develop an accurate list of organizations and projects for their community. Ask students who volunteered to share their experiences with the class. Use Rubric 30, Research, to evaluate student work.

Alternative Assessment

Portfolio Activity

Learning About Your Local Geography

Youth Organizations The Baikal-Amur Mainline (BAM) was built partly by youth organizations. Make a list of some projects of youth organizations in your community.

internet connect

Internet Activity: go.hrw.com
KEYWORD: SK3 GT14

Choose a topic to explore about Russia:
- Take a trip on the Trans-Siberian Railroad.
- Examine the breakup of the Soviet Union.
- View the cultural treasures of Russia.

internet connect

GO TO: **go.hrw.com**
KEYWORD: **SK3 Teacher**
FOR: a guide to using the Internet in your classroom

CHAPTER 15 Ukraine, Belarus, and the Caucasus

Chapter Resource Manager

Objectives	Pacing Guide	Reproducible Resources
SECTION 1 **Physical Geography** (pp. 343–45) **1.** Identify the region's major physical features. **2.** Identify the climate types and natural resources found in the region.	**Regular** .5 day **Block Scheduling** .5 day *Block Scheduling Handbook, Chapter 15*	**RS** Guided Reading Strategy 15.1 **E** Creative Strategies for Teaching World Geography, Lesson 13
SECTION 2 **Ukraine and Belarus** (pp. 346–50) **1.** Identify the groups that have influenced the history of the Ukraine and Belarus. **2.** Identify some important economic features and environmental concerns of Ukraine. **3.** Describe how the economy of Belarus has developed.	**Regular** 1 day **Block Scheduling** .5 day *Block Scheduling Handbook, Chapter 15*	**RS** Guided Reading Strategy 15.2 **RS** Graphic Organizer 15 **E** Cultures of the World Activity 4
SECTION 3 **The Caucasus** (pp. 351–53) **1.** Identify the groups that have influenced the early history and culture of the Caucasus. **2.** Describe the economy of Georgia. **3.** Describe what Armenia is like today. **4.** Describe what Azerbaijan is like today.	**Regular** 1 day **Block Scheduling** .5 day *Block Scheduling Handbook, Chapter 15*	**RS** Guided Reading Strategy 15.3 **E** Cultures of the World Activity 4 **SM** Geography for Life Activity 15 **SM** Map Activity 15

Chapter Resource Key

RS Reading Support

IC Interdisciplinary Connections

E Enrichment

SM Skills Mastery

A Assessment

REV Review

ELL Reinforcement and English Language Learners

Transparencies

CD–ROM

Music

Video

go.hrw.com Internet

Holt Presentation Maker Using Microsoft® Powerpoint®

One-Stop Planner CD–ROM

See the ***One-Stop Planner*** for a complete list of additional resources for students and teachers.

One-Stop Planner CD–ROM

It's easy to plan lessons, select resources, and print out materials for your students when you use the ***One-Stop Planner CD–ROM with Test Generator.***

Technology Resources	Review, Reinforcement, and Assessment Resources
One-Stop Planner CD–ROM, Lesson 15.1 Geography and Cultures Visual Resources with Teaching Activities 31–35 Homework Practice Online HRW Go site	**ELL** Main Idea Activity 15.1 **REV** Section 1 Review, p. 345 **A** Daily Quiz 15.1 **ELL** English Audio Summary 15.1 **ELL** Spanish Audio Summary 15.1
One-Stop Planner CD–ROM, Lesson 15.2 Homework Practice Online HRW Go site	**ELL** Main Idea Activity 15.2 **REV** Section 2 Review, p. 350 **A** Daily Quiz 15.2 **ELL** English Audio Summary 15.2 **ELL** Spanish Audio Summary 15.2
One-Stop Planner CD–ROM, Lesson 15.3 Homework Practice Online HRW Go site	**ELL** Main Idea Activity 15.3 **REV** Section 3 Review, p. 353 **A** Daily Quiz 15.3 **ELL** English Audio Summary 15.3 **ELL** Spanish Audio Summary 15.3

internet connect

HRW ONLINE RESOURCES

GO TO: go.hrw.com
Then type in a keyword.

TEACHER HOME PAGE
KEYWORD: SK3 TEACHER

CHAPTER INTERNET ACTIVITIES
KEYWORD: SK3 GT15

Choose an activity to:
- trek through the Caucasus Mountains.
- design Ukrainian Easter eggs.
- investigate the Chernobyl disaster.

CHAPTER ENRICHMENT LINKS
KEYWORD: SK3 CH15

CHAPTER MAPS
KEYWORD: SK3 MAPS15

ONLINE ASSESSMENT
Homework Practice
KEYWORD: SK3 HP15
Standardized Test Prep Online
KEYWORD: SK3 STP15
Rubrics
KEYWORD: SS Rubrics

COUNTRY INFORMATION
KEYWORD: SK3 Almanac

CONTENT UPDATES
KEYWORD: SS Content Updates

HOLT PRESENTATION MAKER
KEYWORD: SK3 PPT15

ONLINE READING SUPPORT
KEYWORD: SS Strategies

CURRENT EVENTS
KEYWORD: S3 Current Events

Meeting Individual Needs

Ability Levels

Level 1 Basic-level activities designed for all students encountering new material

Level 2 Intermediate-level activities designed for average students

Level 3 Challenging activities designed for honors and gifted-and-talented students

English Language Learners Activities that address the needs of students with Limited English Proficiency

Chapter Review and Assessment

E Readings in World Geography, History, and Culture 43 and 44
SM Critical Thinking Activity 15
REV Chapter 15 Review, pp. 354–55
REV Chapter 15 Tutorial for Students, Parents, Mentors, and Peers
ELL Vocabulary Activity 15
A Chapter 15 Test
Chapter 15 Test Generator (on the One-Stop Planner)
Audio CD Program, Chapter 15
A Chapter 15 Test for English Language Learners and Special-Needs Students

CHAPTER 15

LAUNCH INTO LEARNING

Read the following quotation from Russian writer Nikolay Gogol's *Taras Bulba: "The farther they penetrated the steppe, the more beautiful it became. . . . Nothing in nature could be finer. The whole surface resembled a golden-green ocean, upon which were sprinkled millions of different flowers. . . . Oh, steppes, how beautiful you are!"* Ask what resources the region described might have. *(arable land)* Tell students that in this chapter they will learn more about the fertile area described, which extends into the countries of Belarus and Ukraine. The countries of the Caucasus Mountains, which have a very different landscape, are also covered in this chapter.

Section 1

Objectives

1. Identify the region's major physical features.
2. Describe the climate types and natural resources found in the region.

LINKS TO OUR LIVES

You might point out to students that there are many reasons why we should know more about Ukraine, Belarus, and the three nations of the Caucasus, these among them:

- The nations in this region were part of the Soviet Union. Now they are struggling to adopt freer economic and political practices. Americans are now free to travel to these countries and to invest in them.
- These countries possess many natural resources, including deposits of oil, natural gas, metals, and minerals. The United States may wish to establish stronger trade relationships with these countries.
- Ukraine's government has the task of ensuring that the damaged Chernobyl nuclear reactor is adequately contained. Inadequate safeguards that result from the country's economic hardships could have far-reaching effects.

CHAPTER 15

Ukraine, Belarus, and the Caucasus

This region consists of plains in the north and mountains in the south. Both of these physical features made this area important to ancient invaders. Before you learn the history of this region, you should meet Ana.

Hi! I am a senior in high school in the city of Tbilisi, Georgia. I live with my parents and my younger sister. I go to school from 9:00 A.M. to 2:00 P.M. and study foreign languages—English and Spanish. I hope to be a journalist. In school the teachers decide which classes everyone must take.

After school I do my homework as fast as possible and then get together with my friends. I come home in the early evening and listen to music, read, or watch television.

We also have great food. My favorite dish is baked chicken with nuts. If you came to Georgia, I would take you to the mountains, to the seaside, and to some hot springs. We might also go to a festival where you could see Georgians in the country's national dress. Women wear a long red or purple robe with a white head scarf. Men wear a black suit or robe with gold embroidery.

Translation: Hi! I am Ana.

FOCUS

LET'S GET STARTED

Copy the following passage and instructions onto the chalkboard: *Ukraine was considered the breadbasket of the Soviet Union, while Georgia was a popular vacation spot. Name a pair of states in the United States that could fit the same description.* Discuss student responses. *(Possible answers: Nebraska/Florida; Iowa/California; Kansas/Hawaii)* Tell students that in Section 1 they will learn more about the physical geography of Ukraine, Belarus, and the Caucasus nations.

Using the Physical-Political Map

Have students examine the map on this page. Call on individual students to locate the Dnieper River, the Donets River, and the Black Sea. Point out why navigable waterways are important in agricultural and industrial regions such as Ukraine *(for shipping agricultural products, manufactured goods, and mineral resources to markets or processing plants).*

Building Vocabulary

Write the Define term on the chalkboard. Have a student read aloud a dictionary definition of a **reserve**. *(The definition should pertain to the idea of preserving something.)* Ask volunteers to speculate how **nature reserves** might serve the public interest. You may want to ask students to describe nature reserves they have visited or have seen on television programs.

Section 1 Physical Geography

Read to Discover

1. What are the region's major physical features?
2. What climate types and natural resources are found in the region?

Define

nature reserves

Locate

Black Sea
Caucasus Mountains
Caspian Sea
Pripyat Marshes
Carpathian Mountains
Crimean Peninsula
Sea of Azov
Mount Elbrus
Dnieper River
Donets Basin

WHY IT MATTERS

Ukraine is trying to create a nature reserve to protect its natural environment. Use CNNfyi.com or other **current events** sources to find information about how other countries are trying to protect their environments. Record your findings in your journal.

A gold pig from Kiev

Section 1 RESOURCES

Reproducible
- Block Scheduling Handbook, Chapter 15
- Guided Reading Strategy 15.1
- Creative Strategies for Teaching World Geography, Lesson 13

Technology
- One-Stop Planner CD–ROM, Lesson 15.1
- Homework Practice Online
- Geography and Cultures Visual Resources with Teaching Activities 31–35
- HRW Go site

Reinforcement, Review, and Assessment
- Section 1 Review, p. 345
- Daily Quiz 15.1
- Main Idea Activity 15.1
- English Audio Summary 15.1
- Spanish Audio Summary 15.1

Ukraine, Belarus, and the Caucasus: Physical-Political

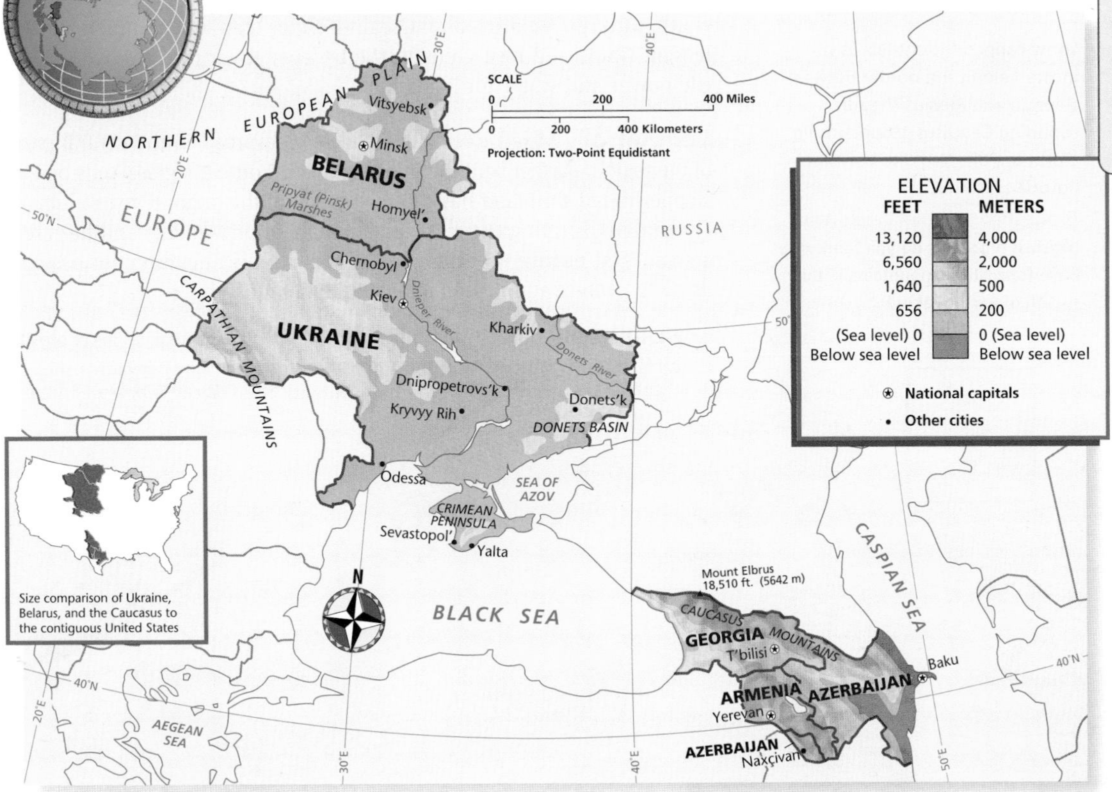

TEACH

Teaching Objective 1

ALL LEVELS: (Suggested time: 30 min.) Tell students to imagine that they are tourists visiting Ukraine, Belarus, and the Caucasus countries. Then have students write a postcard or letter that describes the region's physical features to a friend back home. Ask volunteers to read their postcards or letters to the class. **ENGLISH LANGUAGE LEARNERS**

Teaching Objective 2

ALL LEVELS: (Suggested time: 20 min.) Copy the following graphic organizer onto the chalkboard, omitting the italicized answers. Have students complete the organizer, placing the resource that all three have in common in the center. Then pair students and have each pair draw a map that depicts the region's various climates. Have students create keys for their maps. **ENGLISH LANGUAGE LEARNERS, COOPERATIVE LEARNING**

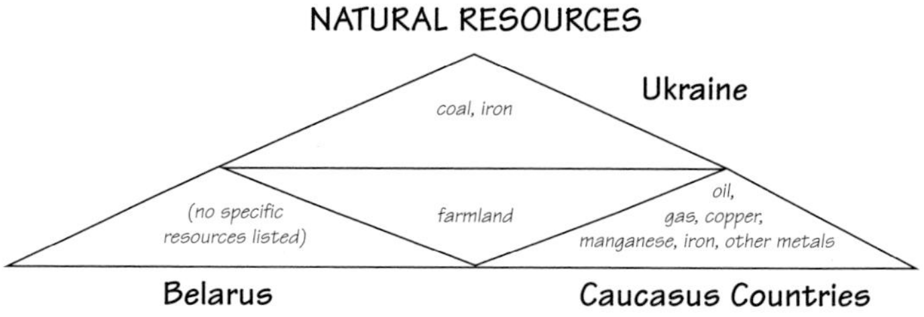

Linking Past to Present

Coal Mining in the Donets Basin Note in the section on resources that the Donets Basin is an important coal-mining area. Coal was first discovered there in 1721. However, coal mining became a significant industry only after 1869, when the first railway reached the region. Donets Basin coal mining reached its peak importance by 1913, when the region produced 87 percent of Russian coal. The main part of the coal field covers nearly 9,000 square miles (23,300 sq km) in Ukraine and southwestern Russia, an area slightly smaller than the state of Vermont.

Activity: Have students conduct further research on how the coal mines of the Donets Basin have affected the political and economic history of the region. Ask them to create a time line showing major developments.

Visual Record Answer

volcanic action

internet connect

GO TO: go.hrw.com
KEYWORD: SK3 CH15
FOR: Web sites about Ukraine, Belarus, and the Caucasus

Physical Features

The countries of Ukraine (yoo-KRAYN) and Belarus (byay-luh-ROOS) border western Russia. Belarus is landlocked. Ukraine lies on the Black Sea. Georgia, Armenia (ahr-MEE-nee-uh), and Azerbaijan (a-zuhr-by-JAHN) lie in a rugged region called the Caucasus (KAW-kuh-suhs). It is named for the area's Caucasus Mountains. The Caucasus region is located between the Black Sea and the Caspian Sea.

Landforms Most of Ukraine and Belarus lie in a region of plains. The Northern European Plain sweeps across northern Belarus. The Pripyat (PRI-pyuht) Marshes, also called the Pinsk Marshes, are found in the south. The Carpathian Mountains run through part of western Ukraine. The Crimean (kry-MEE-uhn) Peninsula lies in southern Ukraine. The southern Crimean is very rugged and has high mountains. It separates the Black Sea from the Sea of Azov (uh-ZAWF).

In the north along the Caucasus's border with Russia is a wide mountain range. The region's and Europe's highest peak, Mount Elbrus (el-BROOS), is located here. As you can see on the chapter map, the land drops below sea level along the shore of the Caspian Sea. South of the Caucasus is a rugged, mountainous plateau. Earthquakes often occur in this region.

Rivers One of Europe's major rivers, the Dnieper (NEE-puhr), flows south through Belarus and Ukraine. Ships can travel much of its length. Dams and reservoirs on the Dnieper River provide hydroelectric power and water for irrigation.

Vegetation Mixed forests were once widespread in the central part of the region. Farther south, the forests opened onto the grasslands of the steppe. Today, farmland has replaced much of the original vegetation.

Ukraine is trying to preserve its natural environments and has created several **nature reserves**. These are areas the government has set aside to protect animals, plants, soil, and water.

READING CHECK: *Places and Regions* What are the region's major physical features? **Caucasus Mountains, Northern European Plain, Pripyat Marshes, Carpathian Mountains, Crimean Peninsula, Black Sea, Caspian Sea, Sea of Azov, Dnieper River**

Snow-capped Mount Elbrus is located along the border between Georgia and Russia. The surrounding Caucasus Mountains lie along the dividing line between Europe and Asia.

Interpreting the Visual Record **What physical processes do you think may have formed the mountains in this region of earthquakes?**

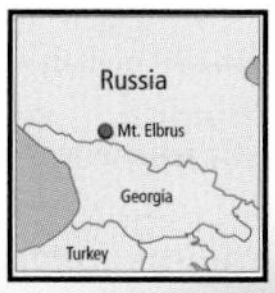

CLOSE

Ask students to describe some things they might do for fun if they visited this region. *(Possible answers include skiing Mount Elbrus, taking a boat ride on the Dnieper River, visiting a nature reserve in Ukraine, going to the beach in Georgia, and so on.)*

REVIEW AND ASSESS

Have students complete the Section Review. Then organize the class into triads and have each triad write eight quiz questions that match a country to its landform, climate, or natural features. Have groups exchange their quizzes and then solve them. Then have students complete Daily Quiz 15.1. **COOPERATIVE LEARNING**

RETEACH

Have students complete Main Idea Activity 15.1. Then have students illustrate the postcards or letters they created earlier with appropriate physical features, climate type(s), and natural resources. **ENGLISH LANGUAGE LEARNERS**

EXTEND

Have interested students conduct further research on the plants and animals of the Pripyat Marshes of Belarus. Then have students create a publicity campaign to raise awareness of the region's unique characteristics and the forces that endanger it. **BLOCK SCHEDULING**

Climate

Like much of western Russia, the northern two thirds of Ukraine and Belarus have a humid continental climate. Winters are cold. Summers are warm but short. Southern Ukraine has a steppe climate. Unlike the rest of the country, the Crimean Peninsula has a Mediterranean climate. There are several different climates in the Caucasus. Georgia's coast has a mild climate similar to the Carolinas in the United States. Azerbaijan contains mainly a steppe climate. Because it is so mountainous, Armenia's climate changes with elevation.

✓ **READING CHECK:** *Places and Regions* What climate types are found in this area? humid continental, steppe, Mediterranean

BUILD on WHAT You Know

Do you remember what you learned about steppe climates? See Chapter 3 to review.

Resources

Rich farmlands are Ukraine's greatest natural resource. Farming is also important in Belarus. Lowland areas of the Caucasus have rich soil and good conditions for farming.

The Donets (duh-NYETS) Basin in southeastern Ukraine is a rich coal-mining area. Kryvyy Rih (kri-VI RIK) is the site of a huge open-pit iron-ore mine. The region's most important mineral resources are Azerbaijan's large and valuable oil and gas deposits. These are found under the shallow Caspian Sea. Copper, manganese, iron, and other metals are also present in the Caucasus.

✓ **READING CHECK:** *Environment and Society* How have this region's natural resources affected economic development? Rich farmlands, minerals, metals, and oil and gas support the economy.

Section 1 Review

go.hrw.com Homework Practice Online
Keyword: SK3 HP15

Define and explain: nature reserves

Working with Sketch Maps On a map of Europe that you draw or that your teacher provides, label the following: Black Sea, Caucasus Mountains, Caspian Sea, Pripyat Marshes, Carpathian Mountains, Crimean Peninsula, Sea of Azov, Mount Elbrus, Dnieper River, and Donets Basin. Where in the region is a major coal-mining area?

Reading for the Main Idea

1. *Places and Regions* What three seas are found in this region?
2. *Places and Regions* What creates variation in Armenia's climate?

Critical Thinking

3. **Drawing Inferences and Conclusions** Why has so much farming developed in Ukraine, Belarus, and the Caucasus?
4. **Drawing Inferences and Conclusions** How do you think heavy mining in this region could create pollution?

Organizing What You Know

5. **Categorizing** Copy the following graphic organizer. Use it to describe the region's physical features, climates, and resources.

	Physical features	Climate	Resources
Belarus			
Caucasus			
Ukraine			

Section 1 Review

Answers

Define For definition, see: nature reserves, p. 344

Working with Sketch Maps Maps will vary, but listed places should be labeled in their approximate locations. The Donets Basin is a major coal-mining area.

Reading for the Main Idea

1. Black Sea, Sea of Azov, Caspian Sea (NGS 4)
2. elevation (NGS 4)

Critical Thinking

3. because of the areas' rich soil and good weather conditions
4. metals being washed into water supply

Organizing What You Know

5. Belarus—Northern European Plain, Pinsk Marshes, Dnieper River; humid continental; farmland; Caucasus region—Black Sea, Caspian Sea, Caucasus Mountains, Mount Elbrus; steppe climate, mild coastal climate; rich soil, copper, manganese, iron, oil, and gas; Ukraine—plains, steppe, Carpathian Mountains, Crimea, Black Sea, Sea of Azov, Dnieper River; humid continental, steppe, Mediterranean; farmland, coal, iron

Section 2

Objectives

1. Identify the groups that influenced the history of Ukraine and Belarus.
2. Discuss some important economic features and environmental concerns of Ukraine.
3. Trace the development of Belarus's economy.

Focus

LET'S GET STARTED

Copy the following instructions onto the chalkboard: *Compile a list of countries that formerly ruled different parts of what is now the United States.* Ask volunteers to share their lists with the class. *(Students might mention England, France, Spain, Mexico, and Russia.)* Tell students that in Section 2 they will learn more about the various groups that have influenced the culture and history of Ukraine and Belarus.

Building Vocabulary

Write the words **serfs**, **Cossacks**, and **soviet** on the chalkboard. Ask volunteers to look up the definitions and origins of the Define terms in the dictionary and to read the definitions and origins aloud. *(The word Serf is from the Latin servus, which means "slave"; Cossack, which has Polish, Ukrainian, and Turkish origins, means "pirate"; and soviet is a Russian word for a council.)* Tell students that the origins of these words indicate some of the historical influences on Ukraine and Belarus.

Section 2 Resources

Reproducible
- Guided Reading Strategy 15.2
- Graphic Organizer 15
- Cultures of the World Activity 4

Technology
- One-Stop Planner CD–ROM, Lesson 15.2
- Homework Practice Online
- HRW Go site

Reinforcement, Review, and Assessment
- Section 2 Review, p. 350
- Daily Quiz 15.2
- Main Idea Activity 15.2
- English Audio Summary 15.2
- Spanish Audio Summary 15.2

Section 2 Ukraine and Belarus

Read to Discover

1. Which groups have influenced the history of Ukraine and Belarus?
2. What are some important economic features and environmental concerns of Ukraine?
3. How has the economy of Belarus developed?

Define

serfs
Cossacks
soviet

Locate

Ukraine
Belarus
Kiev
Chernobyl
Minsk

Why It Matters

Energy created by nuclear power plants is important to the United States, Ukraine, and Belarus. Go to CNNfyi.com or other **current events** sources to find information about nuclear energy. Record your findings in your journal.

A hand-painted Ukrainian egg

гео·
гра·
фия

These are the syllables for the Russian word for geography, written in the Cyrillic alphabet.

History and Government

About 600 B.C. the Greeks established trading colonies along the coast of the Black Sea. Much later—during the A.D. 400s—the Slavs began to move into what is now Ukraine and Belarus. Today, most people in this region speak closely related Slavic languages.

Vikings and Christians In the 800s Vikings took the city of Kiev. Located on the Dnieper River, it became the capital of the Vikings' trading empire. Today, this old city is Ukraine's capital. In the 900s the Byzantine, or Greek Orthodox, Church sent missionaries to teach the Ukrainians and Belorussians about Christianity. These missionaries introduced the Cyrillic alphabet.

St. Sophia Cathedral in Kiev was built in the 1000s. It was one of the earliest Orthodox cathedrals in this area. Religious images decorate the dome's interior.

TEACH

Teaching Objective 1

LEVEL 1: (Suggested time: 20 min.) Pair students and have each pair create a collage that represents the various groups that have influenced the history of Ukraine and Belarus. Display collages around the classroom. **ENGLISH LANGUAGE LEARNERS, COOPERATIVE LEARNING**

LEVELS 2 AND 3: (Suggested time: 20 min.) Tell students to imagine that they are historians who specialize in the history of Ukraine and Belarus. Pair students and have each pair write an outline for a book on the various groups that have influenced the history of Ukraine and Belarus. **COOPERATIVE LEARNING**

Mongols and Cossacks A grandson of Genghis Khan led the Mongol horsemen that conquered Ukraine in the 1200s. They destroyed most of the towns and cities there, including Kiev.

Later, northern Ukraine and Belarus came under the control of Lithuanians and Poles. Under foreign rule, Ukrainian and Belorussian **serfs** suffered. Serfs were people who were bound to the land and worked for a lord. In return, the lords provided the serfs with military protection and other services. Some Russian and Ukrainian serfs left the farms and formed bands of nomadic horsemen. Known as **Cossacks**, they lived on the Ukrainian frontier.

▲ This watercolor on rice paper depicts Kublai Khan. He was the founder of the vast Mongol empire in the 1200s. The Mongols conquered large areas of Asia and Europe, including Ukraine.

The Russian Empire North and east of Belarus, a new state arose around Moscow. This Russian kingdom of Muscovy won independence from the Mongols in the late 1400s. The new state set out to expand its borders. By the 1800s all of modern Belarus and Ukraine were under Moscow's rule. Now the Cossacks served the armies of the Russian czar. However, conditions did not improve for the Ukrainian and Belorussian serfs and peasants.

Soviet Republics The Russian Revolution ended the rule of the czars in 1917. Ukraine and Belarus became republics of the Soviet Union in 1922. Although each had its own governing **soviet**, or council, Communist leaders in Moscow made all major decisions.

Ukraine was especially important as the Soviet Union's richest farming region. On the other hand, Belarus became a major industrial center. It produced heavy machinery for the Soviet Union. While Ukraine and Belarus were part of the Soviet Union, the Ukrainian and Belorussian languages were discouraged. Practicing a religion was also discouraged.

After World War II economic development continued in Ukraine and Belarus. Factories and power plants were built with little concern for the safety of nearby residents.

Kiev remained an important cultural and industrial center during the Soviet era. Parts of the city were destroyed during World War II and had to be rebuilt. Today tree-lined streets greet shoppers in the central city. ◄

ENVIRONMENT AND SOCIETY

Although the Soviet era left parts of Ukraine badly polluted, there are also large natural areas that are protected from industry.

Ukrainians have long supported conservation. Ukraine's first nature reserve began as a private wildlife refuge in 1875. This reserve covers about 27,400 acres (11,000 hectares) and protects a portion of virgin steppe. A successful breeding program for endangered species has been established there. Two of the species are onagers, or wild donkeys, and ostriches.

The Black Sea Nature Reserve, established in 1927, includes protected areas of the sea. The Danube Water Meadows was established for the scientific study and protection of the Danube River's tidewater plant life. The Ukrainian Steppe Reserve, consisting of three separate sections, preserves three special kinds of steppe: meadow steppe, black-earth steppe, and stony steppe. Other reserves protect forest-steppe woodlands, marshes, forests, and mountains.

Activity: Have students conduct additional research and create a collage of pictures of animals found in Ukraine's nature reserves.

Teaching Objective 2

LEVEL 1: (Suggested time: 30 min.) Tell students to imagine that they are compiling an almanac about contemporary Ukraine. Pair students and have each pair compile a list of important economic features and environmental concerns of Ukraine. *(Lists should mention Ukraine's rich soil and the importance of agriculture; it is the world's largest producer of sugar beets; it is a top steel producer; Ukraine experienced rapid industrial growth under the Soviets; the world's worst nuclear disaster occurred at Chernobyl in 1986.)* Ask volunteers to present their lists to the class. **ENGLISH LANGUAGE LEARNERS, COOPERATIVE LEARNING**

LEVELS 2 AND 3: (Suggested time: 30 min.) Have students use the lists they created for the *Level 1* activity as the basis for recommendations to Ukraine's government for developing the country's economy and protecting its environment.

Teacher to Teacher

Joanne Sadler of Buffalo, New York, suggests this activity to teach about the geography and culture of Ukraine and Belarus. Organize students into groups, and have groups study either Ukraine or Belarus. Have students write and illustrate another section for this textbook, to be titled Ukraine (or Belarus), Close-Up and in Detail.

Daily Life

The Ukrainian custom of the first haircutting, *postryzhyny,* marks a child's first birthday. The child's whole family gathers to participate in the event and to share a feast.

Guests contribute coins that are collected and placed in a soup bowl. Each of the child's godparents then takes a turn cutting the child's hair as the guests observe. First a lock is cut from the front of the child's head, then from the back, and then from each side. These cuttings are taken from four areas on the head, which represent the four directions of the compass.

After the haircutting, liquor is poured into the bowl to cover the coins. The baby's feet are then dipped into the bowl. This ritual symbolizes that the child will never be controlled by alcohol or money in years to come. The coins are then dried and saved for the child.

Activity: Have students recall rites of passage in their lives such as religious confirmations or graduations and explain the relationship between these rituals and their culture.

Chart Answer

the United States

Near the end of World War II, Soviet, American, and British leaders met at Livadia Palace in Yalta, Ukraine. There they planned the defeat and occupation of Germany.

Patients per Doctor

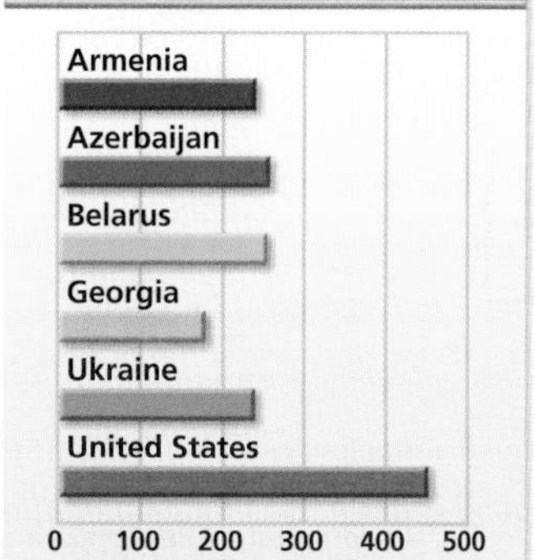

Source: *The World Book Encyclopedia of People and Places*

Interpreting the Graph **In what country is the number of patients per doctor greatest?**

End of Soviet Rule When the Soviet Union collapsed in 1991, Belarus and Ukraine declared independence. Each now has a president and a prime minister. Both countries still have economic problems. Ukraine has also had disagreements with Russia over control of the Crimean Peninsula and the Black Sea naval fleet.

✓ **READING CHECK:** *Human Systems* Which groups have influenced the history of Ukraine and Belarus? Greeks, Slavs, Vikings, Christians, Mongols, Lithuanians, Poles

Ukraine

Ethnic Ukrainians make up about 75 percent of Ukraine's population. The largest minority group in the country is Russian. There are other ties between Ukraine and Russia. For example, the Ukrainian and Russian languages are closely related. In addition, both countries use the Cyrillic alphabet.

Economy Ukraine has a good climate for growing crops and some of the world's richest soil. As a result, agriculture is important to its economy. Ukraine is the world's largest producer of sugar beets. Ukraine's food-processing industry makes sugar from the sugar beets. Farmers also grow fruits, potatoes, vegetables, and wheat. Grain is made into flour for baked goods and pasta. Livestock is also raised. Ukraine is one of the world's top steel producers. Ukrainian factories make automobiles, railroad cars, ships, and trucks.

Teaching Objective 3

ALL LEVELS: (Suggested time: 15 min.) Copy the graphic organizer onto the chalkboard, omitting the italicized answers. Use it to help students understand the development of Belarus's economy. Have each student complete the organizer. Ask volunteers to share their answers with the class. Conclude by leading a discussion about Belarus's culture, natural resources, and products. **ENGLISH LANGUAGE LEARNERS**

➤ASSIGNMENT: Distribute outline maps of Belarus to students. Have each student fill in the map with symbols that represent the various natural resources and products of Belarus. Have students create keys for their maps.

Economic Development in Belarus

• *WWII destroyed most of Belarus's agriculture and industry.* • *Belarus received the worst of the radiation fallout from Chernobyl.* • *The country has resisted economic reforms.* • *Belarus has limited mineral resources.*	⇨ ⇨ ⇨	*slow economic growth*

CONNECTING TO *Science*

A combine used during July harvest

Wheat: From Field to Consumer

Wheat is one of Ukraine's most important farm products. The illustration below shows how wheat is processed for use by consumers.

- The head of the wheat plant contains the wheat kernels, wrapped in husks. The kernel includes the bran or seed coat, the endosperm, and the germ from which new wheat plants grow.
- Whole wheat flour contains all the parts of the kernel. White flour is produced by grinding only the endosperm. Vitamins are added to some white flour to replace vitamins found in the bran and germ.
- People use wheat to make breads, pastas, and breakfast foods. Wheat by-products are used in many other foods.

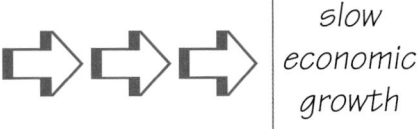

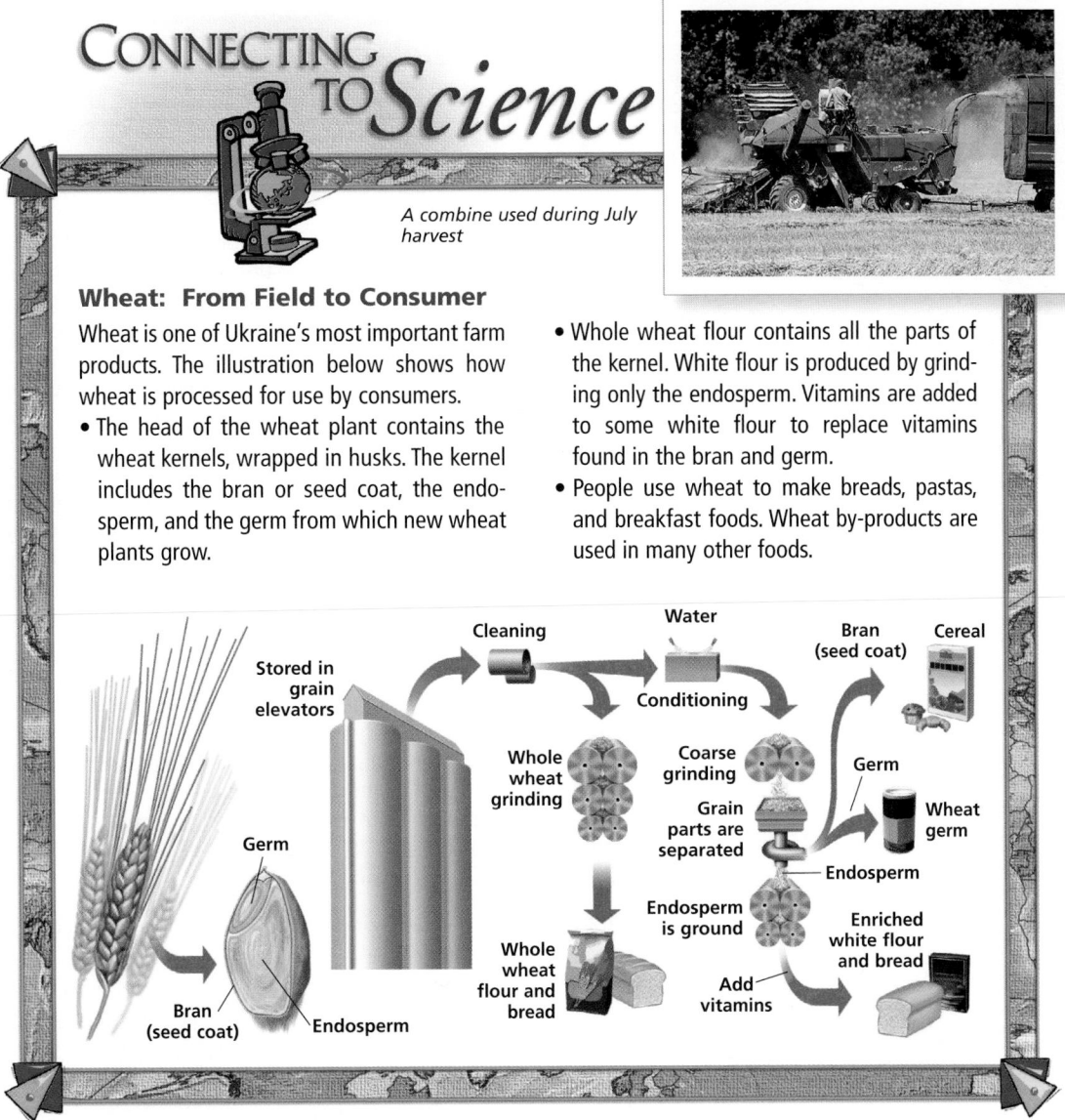

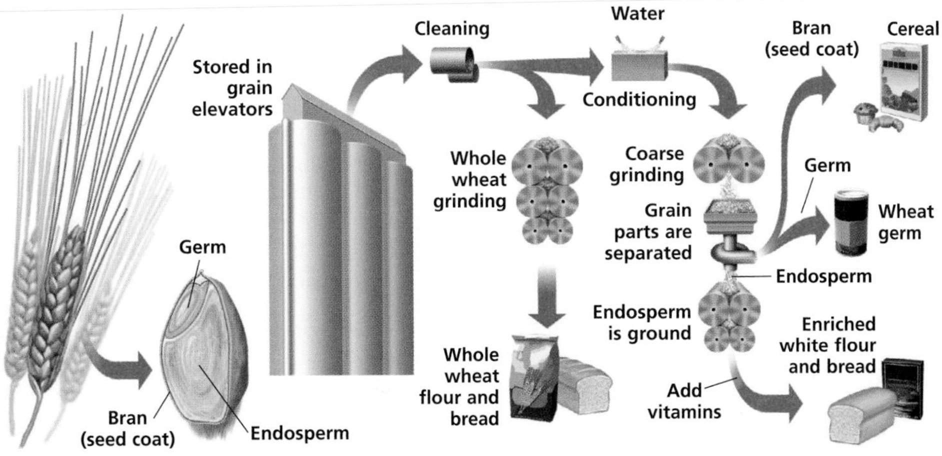

Environment During the Soviet period, Ukraine experienced rapid industrial growth. There were few pollution controls, however. In 1986 at the town of Chernobyl, the world's worst nuclear-reactor disaster occurred. Radiation spread across Ukraine and parts of northern Europe. People near the accident died. Others are still suffering from cancer. Many Ukrainians now want to reduce their country's dependence on nuclear power. This has been hard because the country has not developed enough alternative sources of power.

✓ **READING CHECK:** *Environment and Society* How has the scarcity of alternative sources of power affected Ukraine? It has forced a reliance upon nuclear power.

GLOBAL PERSPECTIVES

Some effects of the Chernobyl nuclear explosion only became apparent years after the accident. The resulting radioactive contamination of the region's food supply and environment has hurt human and animal health and the region's economies.

Some 10 years after the accident, children's cancer rates in parts of Belarus, Russia, and Ukraine were as much as 30 times higher than before the accident. Also, scientists have noted that genetic damage in humans and animals has been passed on to the next generation.

The contaminated part of Ukraine and Belarus was estimated to cover an area the size of England, Wales, and Northern Ireland. Moreover, Belarus has spent up to 20 percent of its annual budget to cope with Chernobyl's aftermath.

internet connect

GO TO: go.hrw.com
KEYWORD: SK3 CH15
FOR: Web sites about Chernobyl

Chart Answer (p. 350)

lower; answers will vary but might mention nuclear fallout, poverty, diet, or other causes.

CLOSE

Ask students to name some cultural and historical similarities between Ukraine and Belarus. *(Students should mention similarities in languages, use of a Cyrillic alphabet, and Soviet rule.)* Then ask students how the accident at Chernobyl has affected the region. *(The accident has affected people's health and has slowed down economic development.)*

REVIEW AND ASSESS

After students complete the Section Review, have them write four quiz questions related to the history and culture of Ukraine and Belarus. Collect the questions and use them to quiz the class before students complete Daily Quiz 15.2.

RETEACH

Have students complete Main Idea Activity 15.2. Then organize the class into small groups; have each group use the Define and Locate terms and a few other terms from this section to create a crossword puzzle with clues. Have groups exchange their puzzles and then solve them.
ENGLISH LANGUAGE LEARNERS, COOPERATIVE LEARNING

EXTEND

Have students research the Mongol conquest of the Ukraine or the role of the Cossacks in Russian and Ukrainian history. Ask students to prepare a five-minute oral report based on their research. **BLOCK SCHEDULING**

Section 2 Review

Answers

Define For definitions, see: serfs, p. 347; Cossacks, p. 347; soviet, p. 347

Working with Sketch Maps Maps will vary, but listed places should be labeled in their approximate locations.

Reading for the Main Idea

1. trading colonies, Slavic languages, city of Kiev, Christianity, Cyrillic alphabet (NGS 9)
2. Ukrainian, Belorussian, Russian (NGS 9)

Critical Thinking

3. Ukraine and Belarus declared independence but have economic problems. Ukraine and Russia have had disagreements about political and military issues.
4. contaminated region, killed people, led to cancer

Organizing What You Know

5. 600 B.C.—Greeks establish colonies; A.D. 400s—Slavs move into region; 800s—Kiev established; 900s—Christianity, Cyrillic alphabet brought in; 1200s—Mongols invade; late 1400s—Muscovy wins independence; 1800s—region ruled by Moscow; 1917—Russian Revolution, beginning of Soviet rule; 1991—Soviet Union collapses

Belarus and Ukraine

Country	Population/ Growth Rate	Life Expectancy	Literacy Rate	Per Capita GDP
Belarus	10,350,194 / −0.2%	62, male; 75, female	98%	$7,500
Ukraine	48,760,474 / −0.9%	61, male; 72, female	98%	$3,850
United States	281,421,906 / 0.9%	73, male; 80, female	97%	$36,200

Sources: Central Intelligence Agency, *The World Factbook 2001;* U.S. Census Bureau

Interpreting the Chart **How does life expectancy in the region compare to that of the United States? Why do you think this is the case?**

Belarus

The people of Belarus are known as Belorussians, which means "white Russians." Ethnically they are closely related to Russians. Their language is also very similar to Russian.

Culture Ethnic Belorussians make up about 75 percent of the country's population. Russians are the second-largest ethnic group. Both Belorussian and Russian are official languages. Belorussian also uses the Cyrillic alphabet. Minsk, the capital of Belarus, is the administrative center of the Commonwealth of Independent States.

Economy Belarus has faced many difficulties. Fighting in World War II destroyed most of the agriculture and industry in the country. Belarus also received the worst of the radiation fallout from the Chernobyl nuclear disaster, which contaminated the country's farm products and water. Many people developed health problems as a result. Another problem has been slow economic progress since the collapse of the Soviet Union. Belarus has resisted economic changes made by other former Soviet republics.

There are various resources in Belarus, however. The country has a large reserve of potash, which is used for fertilizer. Belarus leads the world in the production of peat, a source of fuel found in the damp marshes. Mining and manufacturing are important to the economy. Flax, one of the country's main crops, is grown for fiber and seed. Cattle and pigs are also raised. Nearly one third of Belarus is covered by forests that produce wood and paper products.

✓ **READING CHECK:** *Human Systems* How has the economy of Belarus developed? slowly, because of the collapse of the Soviet Union and resistance to economic change

Section 2 Review

go.hrw.com Homework Practice Online
Keyword: SK3 HP15

Define and explain: serfs, Cossacks, soviet

Working with Sketch Maps On your map from Section 1, label Ukraine, Belarus, Kiev, Chernobyl, and Minsk.

Reading for the Main Idea

1. *Human Systems* What contributions were made by early groups that settled in this region?
2. *Human Systems* What ethnic groups and languages are found in this region today?

Critical Thinking

3. **Finding the Main Idea** How did the end of Soviet rule affect Ukraine and Belarus?
4. **Summarizing** How has the nuclear disaster at Chernobyl affected the region?

Organizing What You Know

5. **Sequencing** Copy the time line below. Use it to trace the region's history from the A.D. 900s to today.

A.D. 900 ———————— Today

Section 3

Objectives

1. List the groups that influenced the early history and culture of the Caucasus.
2. Discuss Georgia's economy.
3. Describe today's Armenia.
4. Discuss what Azerbaijan is like today.

Focus

LET'S GET STARTED

Copy the following scenario onto the chalkboard: *Imagine that you live in a small country that has recently gained independence from a large country. Your country is struggling to change in ways that bring more freedom to your citizens. Independence has been difficult, however. Many people have concluded life was better under the large country's rule. Do you agree? Write down your response.* Discuss responses. Tell students that in Section 3 they will learn about three countries—Georgia, Armenia, and Azerbaijan—where this scenario is a reality.

Building Vocabulary

Have volunteers read aloud the definitions of **homogeneous** and **agrarian** from the glossary. Then ask students to apply these terms to describe various societies or economies they have already studied.

Section 3 The Caucasus

Read to Discover

1. What groups influenced the early history and culture of the Caucasus?
2. What is the economy of Georgia like?
3. What is Armenia like today?
4. What is Azerbaijan like today?

Define

homogeneous
agrarian

Locate

Georgia
Armenia
Azerbaijan

Cover of The Knight in Panther's Skin

Why It Matters

Each of the countries in this section has been involved in a war since the collapse of the Soviet Union in 1991. Use CNNfyi.com or other **current events** sources to find information about the reasons for this unrest. Record your findings in your journal.

Section 3 Resources

Reproducible

- Guided Reading Strategy 15.3
- Cultures of the World Activity 4
- Geography for Life Activity 15
- Map Activity 15

Technology

- One-Stop Planner CD–ROM, Lesson 15.3
- Homework Practice Online
- HRW Go site

Reinforcement, Review, and Assessment

- Section 3 Review, p. 353
- Daily Quiz 15.3
- Main Idea Activity 15.3
- English Audio Summary 15.3
- Spanish Audio Summary 15.3

History

In the 500s B.C. the Caucasus region was controlled by the Persian Empire. Later it was brought under the influence of the Byzantine Empire and was introduced to Christianity. About A.D. 650, Muslim invaders cut the region off from Christian Europe. By the late 1400s other Muslims, the Ottoman Turks, ruled a vast empire to the south and west. Much of Armenia eventually came under the rule of that empire.

Modern Era During the 1800s Russia took over eastern Armenia, much of Azerbaijan, and Georgia. The Ottoman Turks continued to rule western Armenia. Many Armenians spread throughout the Ottoman Empire. However, they were not treated well. Their desire for more independence led to the massacre of thousands of Armenians. Hundreds of thousands died while being forced to leave Turkey during World War I. Some fled to Russian Armenia.

After the war Armenia, Azerbaijan, and Georgia were briefly independent. By 1922 they had become part of the Soviet Union. They again became independent when the Soviet Union collapsed in 1991.

This wall painting is one of many at the ancient Erebuni Citadel in Yerevan, Armenia's capital. The fortress was probably built in the 800s B.C. by one of Armenia's earliest peoples, the Urartians.

TEACH

Teaching Objectives 1–4

ALL LEVELS: (Suggested time: 30 min.) Copy the following graphic organizer onto the chalkboard, omitting the italicized answers. Use it to help students understand the shared history and culture of the Caucasus countries and the economic and cultural differences between these countries. Have students work in pairs to complete the organizer. **ENGLISH LANGUAGE LEARNERS, COOPERATIVE LEARNING**

SHARED HISTORY AND CULTURE
• *ruled by ancient Persian Empire* • *under Byzantine influence* • *Caucasus isolated from Christian groups in Europe* • *united in 1922 to oppose Soviet rule; ruled by Soviet Union*

CURRENT CULTURAL AND ECONOMIC SITUATION		
GEORGIA	ARMENIA	AZERBAIJAN
• *Georgian language* • *shortage of good farmland* • *tea, citrus fruits, vineyards, and tourism important to economy* • *imports most of its energy*	• *Armenian language predominant* • *varied industry includes mining, carpets, clothes, and footwear* • *agriculture important*	• *Turkish language predominant* • *agrarian society* • *oil, natural gas, cotton, and fishing important to economy*

Section 3 Review

Answers

Define For definitions, see: homogeneous, p. 353; agrarian, p. 353

Working with Sketch Maps Maps will vary, but listed places should be labeled in their approximate locations. Proximity to Asia has led to Turkish, Persian, and Muslim influences. Geographic and cultural isolation from the rest of Europe is a result of the Caucasus Mountains.

Reading for the Main Idea

1. Persians, Byzantine Empire, Muslims, Ottoman Turks (NGS 9)
2. the Soviet Union (NGS 13)

Critical Thinking

3. The massive 1988 earthquake destroyed much of the country's industry.
4. around farming, oil

Organizing What You Know

5. Students should discuss physical, historical, and cultural differences. They might mention that the countries share a history of Soviet domination and reliance on agriculture.

Visual Record Answer ▲

cucumbers and other vegetables

►

possible answer: because the building is sacred to the people

▲ This Georgian family's breakfast includes local specialties such as *khachapuri*—bread made with goat cheese.

Interpreting the Visual Record

What other agricultural products do you see on the table?

Government Each country has an elected parliament, president, and prime minister. In the early 1990s there was civil war in Georgia. Armenia and Azerbaijan were also involved in a war during this time. Ethnic minorities in each country want independence. Disagreements about oil and gas rights may cause more regional conflicts in the future.

✓ **READING CHECK:** *Human Systems* How has conflict among cultures been a problem in this region? Ethnic minorities want independence; disagreements about oil and gas rights may cause problems.

Georgia

Georgia is a small country located between the high Caucasus Mountains and the Black Sea. It has a population of about 5 million. About 70 percent of the people are ethnic Georgians. The official language, Georgian, has its own alphabet. This alphabet was used as early as A.D. 400.

As in all the former Soviet republics, independence and economic reforms have been difficult. Georgia has also suffered from civil war. By the late 1990s the conflicts were fewer but not resolved.

Georgia has little good farmland. Tea and citrus fruits are the major crops. Vineyards are an important part of Georgian agriculture. Fish, livestock, and poultry contribute to the economy. Tourism on the Black Sea has also helped the economy. Because its only energy resource is hydropower, Georgia imports most of its energy supplies.

✓ **READING CHECK:** *Human Systems* In what way has scarcity of energy resources affected Georgia's economy? It has made Georgia dependent on other nations for energy imports.

Armenia

Armenia is a little smaller than Maryland. It lies just east of Turkey. It has fewer than 4 million people and is not as diverse as other countries

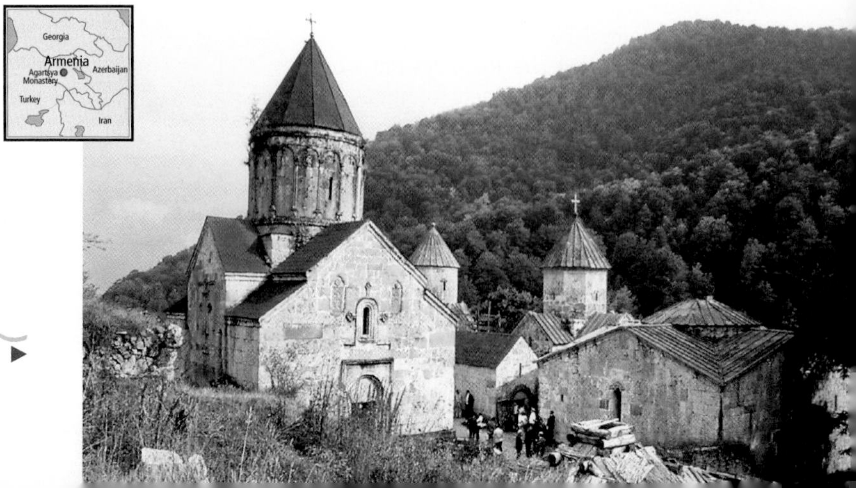

The Orthodox Christian Agartsya Monastery was built in Armenia in the 1200s.

Interpreting the Visual Record

Why do you think this building's exterior is so well preserved? ►

Close

Have students use the information in the graphic organizer to compose a few lines of rhyming verse about one of the countries.

Reteach

Have students complete Main Idea Activity 15.3. Then ask students to outline the section. ENGLISH LANGUAGE LEARNERS

Review and Assess

Have students complete the Section Review. Then organize students into groups. Have each group create eight flash cards using information in the section and exchange these flash cards with another group. Have students within each group use the flash cards to quiz each other. Then have students complete Daily Quiz 15.3. COOPERATIVE LEARNING

Extend

Have interested students conduct research on the cultural history of a resource or product of one of the Caucasus countries (for example, carpets in Armenia, vineyards in Georgia, or caviar in Azerbaijan). Then have each student deliver a short oral report to the class about his or her chosen subject. BLOCK SCHEDULING

in the Caucasus. Almost all the people are Armenian, belong to the Armenian Orthodox Church, and speak Armenian.

Armenia's progress toward economic reform has not been easy. In 1988 a massive earthquake destroyed nearly one third of its industry. Armenia's industry today is varied. It includes mining and the production of carpets, clothing, and footwear.

Agriculture accounts for about 40 percent of Armenia's gross domestic product. High-quality grapes and fruits are important. Beef and dairy cattle and sheep are raised on mountain pastures.

The Caucasus

Country	Population/ Growth Rate	Life Expectancy	Literacy Rate	Per Capita GDP
Armenia	3,336,100 / −0.2%	62, male / 71, female	99%	$3,000
Azerbaijan	7,771,092 / 0.3%	59, male / 68, female	97%	$3,000
Georgia	4,989,285 / −0.6%	61, male / 68, female	99%	$4,600
United States	281,421,906 / 0.9%	74, male / 80, female	97%	$36,200

Sources: Central Intelligence Agency, *The World Factbook 2001;* U.S. Census Bureau

Interpreting the Chart **Which country has the lowest per capita GDP in the region? Why might this be the case?**

✓ **READING CHECK:** *Environment and Society* How did the 1998 earthquake affect the people of Armenia? destroyed nearly one third of Armenian industry

Azerbaijan

Azerbaijan has nearly 8 million people. Its population is becoming ethnically more **homogeneous**, or the same. The Azeri, who speak a Turkic language, make up about 90 percent of the population.

Azerbaijan has few industries except for oil production. It is mostly an **agrarian** society. An agrarian society is organized around farming. The country's main resources are cotton, natural gas, and oil. Baku, the national capital, is the center of a large oil-refining industry. Oil is the most important part of Azerbaijan's economy. Fishing is also important because of the sturgeon of the Caspian Sea.

✓ **READING CHECK:** *Human Systems* What are some cultural traits of the people of Azerbaijan? They are mostly ethnically homogenous and speak a Turkic language.

Section 3 Review

go.hrw.com Homework Practice Online
Keyword: SK3 HP15

Define and explain: homogeneous, agrarian

Working with Sketch Maps On the map you created for Section 2, label Georgia, Armenia, and Azerbaijan. How has the location of this region helped and hindered its growth?

Reading for the Main Idea

1. *Human Systems* Which groups influenced the early history of the Caucasus?
2. *Human Systems* Which country controlled the Caucasus during most of the 1900s?

Critical Thinking

3. **Analyzing Information** Why has economic reform been difficult in Armenia?
4. **Finding the Main Idea** How is Azerbaijan's economy organized?

Organizing What You Know

5. **Comparing/Contrasting** Copy the following graphic organizer. Use it to show the similarities and differences among the countries of the Caucasus region.

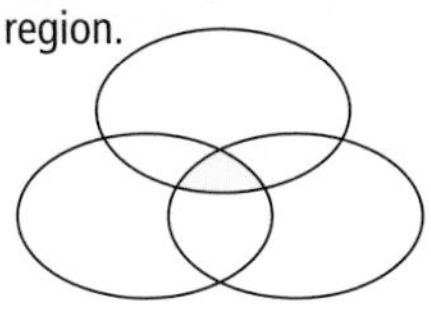

Chapter 15 Review Answers

Building Vocabulary

For definitions, see nature reserves, p. 344; serfs, p. 347; Cossacks, p. 347; soviet, p. 347; homogeneous, p. 353; agrarian, p. 353

Reviewing the Main Ideas

1. Ukraine—steel, coal, agriculture; Belarus—mining, manufacturing, wood and paper products; Georgia—fish, livestock, poultry, tourism; Armenia—agriculture, mining, carpets, clothes, footwear; Azerbaijan—cotton, oil, natural gas, agriculture, fishing (NGS 12)
2. on the Dnieper River; Viking traders (NGS 12)
3. the Greeks and Slavs (NGS 12)
4. Ukraine became rich farming region; Belarus became major industrial center; industrialization polluted some areas. (NGS 14)
5. destroyed much of Armenia's industry and economy (NGS 15)

▲ **Chart Answer**

Azerbaijan; answers will vary, but might mention because there is little industry.

Assess

Have students complete the Chapter 15 Test.

Reteach

Organize students into groups and have each group choose one of the five countries discussed in the chapter. Then have the group prepare a monologue that might be spoken by a senior citizen of their chosen country. The monologue should tell the story of the person's life, and selected facts about the person's country. Have one student from each group read or recite the group's monologue to the class. **COOPERATIVE LEARNING**

Portfolio Extensions

PORTFOLIO

1. Organize students into small groups. Have each group write a new national anthem for Ukraine or Belarus in which the lyrics pertain to their chosen country's geographical features and natural resources. Have volunteers perform their anthems for the class. Have students place their national anthems in their portfolios.

2. Tell students to imagine that they work for the tourist bureau of one of the Caucasus countries. Organize students into small groups and have each group choose a country. Then have each group prepare a three-minute radio commercial that advertises its chosen country's cultural and geographical features to tourists. Have groups perform their radio commercials for the class. Have students place their commercials in their portfolios.

Chapter 15 Review Answers

Understanding Environment and Society

Answers will vary, but the information should address oil and natural gas as it relates to the Caucasus region. Use Rubric 29, Presentations, to evaluate student work.

Thinking Critically

1. Students might suggest that ethnic rivalries can intensify conflicts.
2. Their status as Soviet Republics and proximity to Russia made them enemy targets.
3. Answers will vary. Students should use information from the chapter to justify their opinions.
4. Possible answer: They are physically isolated, have different physical features, and different ethnic groups.
5. Possible answer: Their climates and resources differ, therefore their economies differ.

Chapter 15 Reviewing What You Know

Building Vocabulary

On a separate sheet of paper, write sentences to define each of the following words.

1. nature reserves
2. serfs
3. Cossacks
4. soviet
5. homogeneous
6. agrarian

Reviewing the Main Ideas

1. *Human Systems* Which industries have traditionally been very important to the economies of the countries covered in this chapter?
2. *Human Systems* What is the relative location of Kiev? Which group of people founded Kiev?
3. *Human Systems* What peoples were apparently the first to settle in Belarus and Ukraine?
4. *Environment and Society* How did industrialization under Soviet rule affect these regions?
5. *Environment and Society* What effect did the earthquake in 1988 have on Armenia's economy?

Understanding Environment and Society

Resource Use

Prepare a chart and presentation on oil and natural gas production in the Caucasus. In preparing your presentation, consider the following:

- When and where oil and natural gas were discovered.
- The importance of oil and natural gas production to the former Soviet Union.
- How oil is transported from there to international markets today.

When you have finished your model and presentation, write a five-question quiz, with answers, about your chart to challenge fellow students.

Thinking Critically

1. **Drawing Inferences and Conclusions** How might ethnic diversity affect relations among countries?
2. **Analyzing Information** How did the location of Ukraine and Belarus contribute to their devastation during World War II?
3. **Analyzing Information** Of the countries covered in this chapter, which do you think was the most important to the former Soviet Union? Why do you think this was so?
4. **Summarizing** Why did the countries of the Caucasus develop so differently from Russia, Ukraine, and Belarus?
5. **Finding the Main Idea** Why are the economies of each of the Caucasus countries so different from one another?

Food Festival

Compote is a popular dessert in Ukraine. It is sweetened stewed fruit, cooked carefully to keep the fruit as whole as possible. This compote is a more modern version of compote as it does not use lemons, which are hard to get in many areas of Ukraine. To make the compote, wash and drain 1 pound of fresh strawberries or raspberries. Combine ¾ cup of sugar and 1 cup of water and bring it to a boil. Pour the boiling syrup over the fruit and let it stand for several hours before eating.

Chapter 15 Review and Assessment Resources

Reproducible
- Readings in World Geography, History, and Culture 43 and 44
- Vocabulary Activity 15

Technology
- Chapter 15 Test Generator (on the One-Stop Planner)
- HRW Go site
- Audio CD Program, Chapter 15

Reinforcement, Review, and Assessment
- Chapter 15 Review, pp. 354–55
- Chapter 15 Tutorial for Students, Parents, Mentors, and Peers
- Chapter 15 Test
- Chapter 15 Test for English Language Learners and Special-Needs Students

Building Social Studies Skills

Map Activity

On a separate sheet of paper, match the letters on the map with their correct labels.

Caucasus Mountains	**Crimean Peninsula**
Pripyat Marshes	**Mount Elbrus**
Carpathian Mountains	**Donets Basin**
	Chernobyl

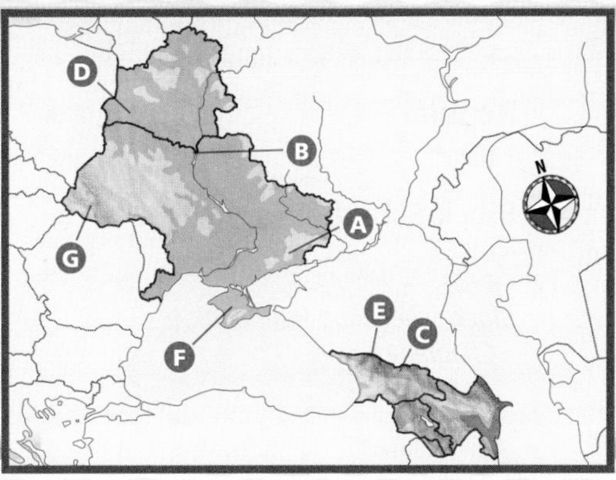

Mental Mapping Skills Activity

On a separate sheet of paper, draw a freehand map of Ukraine, Belarus, and the Caucasus. Include Russia and Turkey for location reference. Make a map key and label the following:

Armenia	**Dnieper River**
Azerbaijan	**Georgia**
Belarus	**Russia**
Black Sea	**Turkey**
Caspian Sea	**Ukraine**

Writing Activity

Choose one of the countries covered in this chapter to research. Write a report about your chosen country's struggle to establish stability since 1991. Include information about the country's government and economic reforms. Describe the social, political, and economic problems the country has faced. Be sure to use standard grammar, sentence structure, spelling, and punctuation.

Alternative Assessment

Portfolio Activity

Learning About Your Local Geography

Cooperative Project What is an important crop in your region of the United States? Work with a partner to create an illustrated flowchart that shows how this crop is made ready for consumers.

internet connect

Internet Activity: **go.hrw.com**
KEYWORD: **SK3 GT15**

Choose a topic to explore about Ukraine, Belarus, and the Caucasus.
- Trek through the Caucasus Mountains.
- Design Ukrainian Easter eggs.
- Investigate the Chernobyl disaster.

Map Activity
A. Donets Basin
B. Chernobyl
C. Caucasus Mountains
D. Pripyat Marshes
E. Mount Elbrus
F. Crimean Peninsula
G. Carpathian Mountains

Mental Mapping Skills Activity
Maps will vary, but listed places should be labeled in their approximate locations.

Writing Activity
Answers will vary but should consider the social, political, and economic problems faced by the country. Use Rubric 30, Research, to evaluate student work.

Portfolio Activity
Answers will vary based on the region. Students should include all facets of production of the crop, and display the information in a flowchart. Use Rubric 7, Charts, to evaluate student work.

internet connect

GO TO: **go.hrw.com**
KEYWORD: **SK3 Teacher**
FOR: **a guide to using the Internet in your classroom**

CHAPTER 16

Central Asia

Chapter Resource Manager

Objectives	Pacing Guide	Reproducible Resources
SECTION 1 **Physical Geography** (pp. 357–59) **1.** Identify the main landforms and climates of Central Asia. **2.** Identify the resources that are important to Central Asia.	**Regular** .5 day **Block Scheduling** .5 day *Block Scheduling Handbook, Chapter 16*	**RS** Guided Reading Strategy 16.1 **SM** Geography for Life Activity 16 **SM** Map Activity 16
SECTION 2 **History and Culture** (pp. 360–62) **1.** Describe how trade and invasions affected the history of Central Asia. **2.** Describe political and economic conditions in Central Asia today.	**Regular** 1 day **Block Scheduling** .5 day *Block Scheduling Handbook, Chapter 16*	**RS** Guided Reading Strategy 16.2 **RS** Graphic Organizer 16 **E** Cultures of the World Activity 4 **IC** Interdisciplinary Activity for the Middle Grades 19
SECTION 3 **The Countries of Central Asia** (pp. 363–65) **1.** Identify aspects of culture in Kazakhstan. **2.** Describe Kyrgyz culture. **3.** Explain politics in Tajikistan. **4.** Identify art forms in Turkmenistan. **5.** Descrube Uzbekistan's population.	**Regular** 1 day **Block Scheduling** .5 day *Block Scheduling Handbook, Chapter 16*	**RS** Guided Reading Strategy 16.3

Chapter Resource Key

- **RS** Reading Support
- **IC** Interdisciplinary Connections
- **E** Enrichment
- **SM** Skills Mastery
- **A** Assessment
- **REV** Review
- **ELL** Reinforcement and English Language Learners
- Transparencies
- CD–ROM
- Music
- Video
- go.hrw.com Internet
- Holt Presentation Maker Using Microsoft® Powerpoint®

One-Stop Planner CD–ROM

See the ***One-Stop Planner*** for a complete list of additional resources for students and teachers.

One-Stop Planner CD–ROM

It's easy to plan lessons, select resources, and print out materials for your students when you use the ***One-Stop Planner CD–ROM with Test Generator.***

Technology Resources	Review, Reinforcement, and Assessment Resources
One-Stop Planner CD–ROM, Lesson 16.1 *ARGWorld* CD–ROM: An Inland Sea in Central Asia Geography and Cultures Visual Resources with Teaching Activities 31–35 Homework Practice Online HRW Go site	**ELL** Main Idea Activity 16.1 **REV** Section 1 Review, p. 359 **A** Daily Quiz 16.1 **ELL** English Audio Summary 16.1 **ELL** Spanish Audio Summary 16.1
One-Stop Planner CD–ROM, Lesson 16.2 Homework Practice Online HRW Go site	**ELL** Main Idea Activity 16.2 **REV** Section 2 Review, p. 362 **A** Daily Quiz 16.2 **ELL** English Audio Summary 16.2 **ELL** Spanish Audio Summary 16.2
One-Stop Planner CD–ROM, Lesson 16.3 Homework Practice Online HRW Go site	**ELL** Main Idea Activity 16.3 **REV** Section 3 Review, p. 365 **A** Daily Quiz 16.3 **ELL** English Audio Summary 16.3 **ELL** Spanish Audio Summary 16.3

internet connect

HRW ONLINE RESOURCES

GO TO: go.hrw.com
Then type in a keyword.

TEACHER HOME PAGE
KEYWORD: SK3 TEACHER

CHAPTER INTERNET ACTIVITIES
KEYWORD: SK3 GT16

Choose an activity to:
- study the climate of Central Asia.
- travel along the historic Silk Road.
- learn about nomads and caravans.

CHAPTER ENRICHMENT LINKS
KEYWORD: SK3 CH16

CHAPTER MAPS
KEYWORD: SK3 MAPS16

ONLINE ASSESSMENT
Homework Practice
KEYWORD: SK3 HP16
Standardized Test Prep Online
KEYWORD: SK3 STP16
Rubrics
KEYWORD: SS Rubrics

COUNTRY INFORMATION
KEYWORD: SK3 Almanac

CONTENT UPDATES
KEYWORD: SS Content Updates

HOLT PRESENTATION MAKER
KEYWORD: SK3 PPT16

ONLINE READING SUPPORT
KEYWORD: SS Strategies

CURRENT EVENTS
KEYWORD: S3 Current Events

Meeting Individual Needs

Ability Levels

Level 1 Basic-level activities designed for all students encountering new material

Level 2 Intermediate-level activities designed for average students

Level 3 Challenging activities designed for honors and gifted-and-talented students

English Language Learners Activities that address the needs of students with Limited English Proficiency

Chapter Review and Assessment

IC Interdisciplinary Activities for the Middle Grades 13, 14, 15
E Readings in World Geography, History, and Culture 34, 35, 36
SM Critical Thinking Activity 16
REV Chapter 16 Review, pp. 398–99
REV Chapter 16 Tutorial for Students, Parents, Mentors, and Peers
ELL Vocabulary Activity 15
A Chapter 16 Test
Chapter 16 Test Generator (on the One-Stop Planner)
Audio CD Program, Chapter 16
A Chapter 16 Test for English Language Learners and Special-Needs Students

CHAPTER 16

Launch into Learning

Read to the class the following description by traveler Marco Polo: "For twelve days the course is along this elevated plain, which is named Pamer [Pamir Plateau]. . . . So great is the height of the mountains, that no birds are to be seen near their summits; and however extraordinary it may be thought, it was affirmed, that from the keenness of the air, fires when lighted do not give the same heat as in lower situations, nor produce the same effect in cooking food." Explain to students that Polo is describing the thinness of air at higher elevations. Ask students to name other areas where this effect might be observed. *(Possible answers: high in the Rockies, the Sierra Nevada, the Andes, or the Alps)* Tell students they will learn more about the physical and cultural geography of Central Asia in this chapter.

Section 1

Objectives

1. Identify the main landforms and climates of Central Asia.
2. Describe the resources that are important to Central Asia.

Links to Our Lives

You may want to remind students that even a region as remote as Central Asia connects with us in several ways:

- The region has large oil and gas deposits, which the United States may want to use.
- People in the region are struggling with the shift from communism to more open governments.
- Damage from Soviet nuclear tests in the area may last far into the future.
- The region is opening to tourism. It offers some of the most spectacular scenery in the world.

CHAPTER 16

Central Asia

Leila is a student from Central Asia, a region of grasslands, scorching deserts, and high mountains.

Salam! (Hi!)—we also say "privyet," which is "hi" in Russian. My name is Leila, and I am 16. I live in Turkmenabat, the "city of Turkmen," with my parents, my three older brothers, and my younger sister. My oldest brother's wife also lives with us. My father is a professor of British studies at Turkmenabat University and my mother teaches cooking and sewing at the high school. Our house is on a canal with lots of trees. It is surrounded by a high wall, and has a balcony and an open roof where we can go if we want to be outdoors in privacy.

I am in the ninth and last year at School Number Five. The school has about 2,000 students in nine grades from kindergarten to high school. We go to school six days a week and study 18 different subjects. There is no choice of courses. I could have gone to a Russian school, but my parents chose a Turkmen one. At school, we line up by class and do 5 minutes of exercises before classes start. Each class has 25 boys and girls who stay together from kindergarten through high school. We are like a second family.

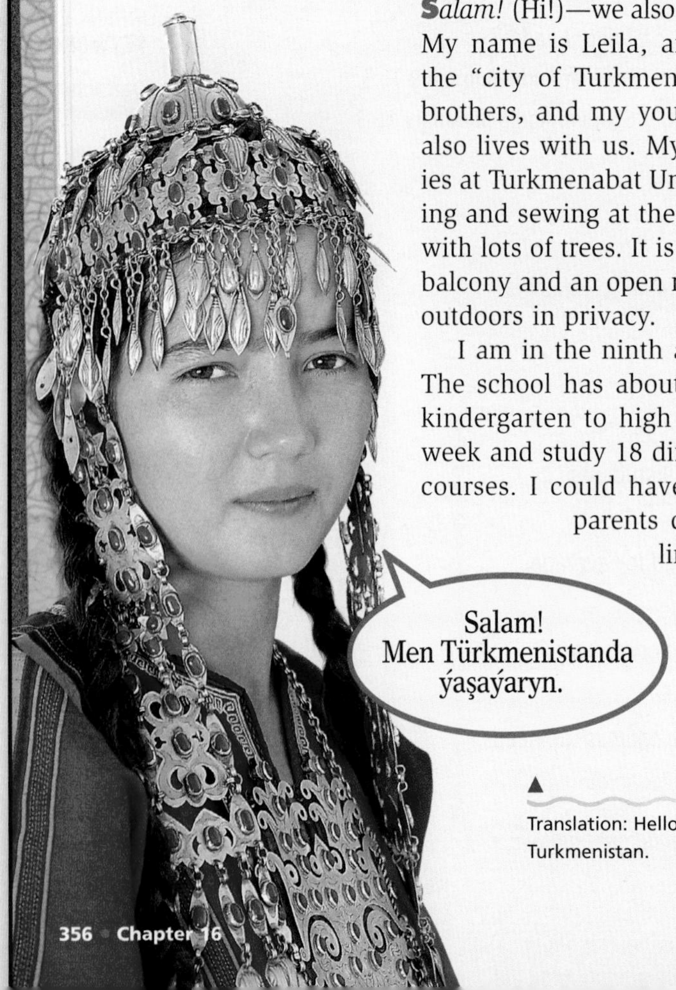
Salam!
Men Türkmenistanda ýaşaýaryn.

Translation: Hello, I live in Turkmenistan.

FOCUS

LET'S GET STARTED

Copy the following instructions onto the chalkboard: *Look at the map on this page. Estimate the distance between Semey and Tashkent. How many major cities are there between Semey and Mary?* Have students share their responses. *(The distance is about 750 miles. There is one city—Bukhara.)* Then ask what this information tells us about the urban density of Kazakhstan. *(Students might suggest that Kazakhstan has a low urban density.)* Tell students that in Section 1 they will learn more about the physical geography of Central Asia.

Using the Physical-Political Map

Have students examine the map on this page. Ask volunteers to describe the boundaries of the Central Asian countries and to name the nations that border the region. *(Afghanistan, China, Iran, and Russia)* Ask students why Russia might want to stay on good terms with the countries of Central Asia. *(Possible answer: to maintain a buffer zone between itself and other countries)*

Building Vocabulary

Copy the following terms onto the chalkboard: **landlocked** and **oasis**. Have volunteers read aloud the definitions from the text's glossary. Then ask students to explain what these terms might tell us about Central Asia's need for water.

Section 1 Physical Geography

Read to Discover

1. What are the main landforms and climates of Central Asia?
2. What resources are important to Central Asia?

Define

landlocked
oasis

Locate

Pamirs
Tian Shan
Aral Sea
Kara-Kum
Kyzyl Kum
Syr Dar'ya
Amu Dar'ya
Fergana Valley

WHY IT MATTERS

The countries of Uzbekistan, Kazakhstan, and Turkmenistan have large oil and natural gas reserves. Use CNNfyi.com or other **current events** sources to find examples of efforts to develop these resources. Record your findings in your journal.

A Bactrian camel and rider

Central Asia: Physical-Political

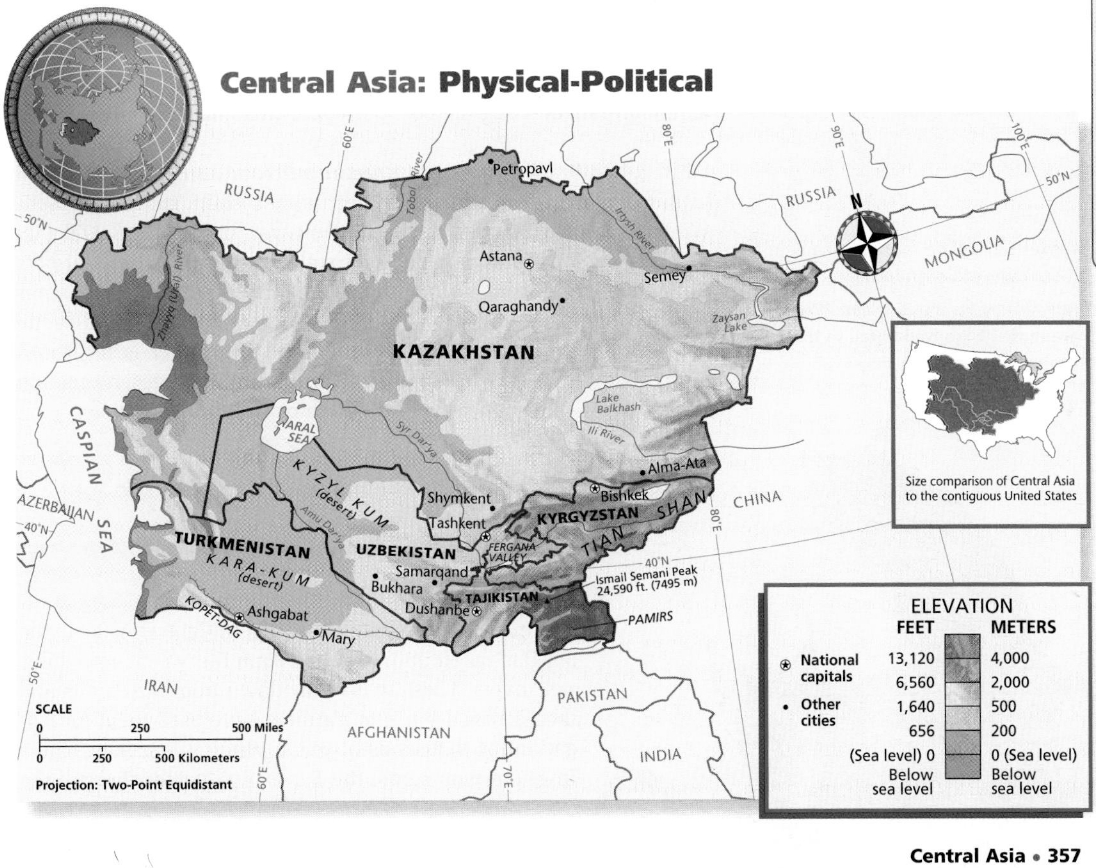

Section 1 RESOURCES

Reproducible

- Block Scheduling Handbook, Chapter 16
- Guided Reading Strategy 16.1
- Geography for Life Activity 16
- Map Activity 16

Technology

- One-Stop Planner CD–ROM, Lesson 16.1
- Homework Practice Online
- Geography and Cultures Visual Resources with Teaching Activities 31–35
- *ARGWorld* CD–ROM: An Inland Sea in Central Asia
- HRW Go site

Reinforcement, Review, and Assessment

- Section 1 Review, p. 359
- Daily Quiz 16.1
- Main Idea Activity 16.1
- English Audio Summary 16.1
- Spanish Audio Summary 16.1

TEACH

Teaching Objectives 1–2

ALL LEVELS: (Suggested time: 20 min.) Copy the following graphic organizer onto the chalkboard, omitting the italicized answers. Use it to help students identify the main landforms, climates, and resources of Central Asia. Pair students and have each pair complete the organizer. **ENGLISH LANGUAGE LEARNERS, COOPERATIVE LEARNING**

THE PHYSICAL GEOGRAPHY OF CENTRAL ASIA

Landforms	*mountains—the Pamirs, and Tian Shan; plains; low plateaus; deserts—Kara-Kum and Kyzyl Kum*
Climates	*steppe, desert, and highland*
Resources	*water—Syr Dar'ya and Amu Dar'ya; the Fergana Valley; oil, gas, gold, copper, uranium, zinc, lead, and coal*

➤ASSIGNMENT: Have students write a short story about an adventure in the mountains or deserts of Central Asia. Ask them to use elements in the Section 1 illustrations and Our Amazing Planet as ideas for the plot.

EYE ON EARTH

Lake Sarez in eastern Tajikistan was created in 1911 when a massive earthquake caused part of a mountain to collapse, blocking the Murgab River. Today the lake stretches about 37 miles (60 km) behind the dam created by the landslide. Downstream, waters of the Murgab flow into the Pyandzh River which becomes the Amu Dar'ya.

If another earthquake were to cause the dam to collapse, a huge wall of water would destroy entire villages and threaten the lives of thousands of people. The dam's isolation in the Pamirs and the poverty of Central Asian countries make it difficult to find a solution to this potentially disastrous problem.

Discussion: Ask students to consider how wealthy nations could help Central Asian and other poor countries meet the challenges posed by problems such as Lake Sarez.

internet connect

GO TO: go.hrw.com
KEYWORD: SK3 CH16
FOR: Web sites about the Pamirs

Visual Record Answer

Their heavy fur coats protect them from the cold.

Mountain climbers make camp before attempting to scale a peak in the Pamirs.

Our Amazing Planet

With temperatures above 122°F (50°C), it is not surprising that the creatures that live in the Kara-Kum are a tough group. Over a thousand species live there, including cobras, scorpions, tarantulas, and monitor lizards. These lizards can grow to more than 5 feet (1.5m) long.

Landforms and Climate

This huge, **landlocked** region is to the east of the Caspian Sea. Landlocked means the region does not border an ocean. The region lies north of the Pamirs (puh-MIRZ) and Tian Shan (TYEN SHAHN) mountain ranges.

Diverse Landforms As the name suggests, Central Asia lies in the middle of the largest continent. Plains and low plateaus cover much of this area. Around the Caspian Sea the land is as low as 95 feet (29 m) below sea level. However, the region includes high mountain ranges along the borders with China and Afghanistan.

Arid Lands Central Asia is a region of mainly steppe, desert, and highland climates. Summers are hot, with a short growing season. Winters are cold. Rainfall is sparse. However, north of the Aral (AR-uhl) Sea rainfall is heavy enough for steppe vegetation. Here farmers can grow crops using rain, rather than irrigation, as their water source. South and east of the Aral Sea lie two deserts. One is the Kara-Kum (kahr-uh-KOOM) in Turkmenistan. The other is the Kyzyl Kum (ki-ZIL KOOM) in Uzbekistan and Kazakhstan. Both deserts contain several **oasis** settlements where a spring or well provides water.

✓ **READING CHECK:** *Places and Regions* What are the landforms and climates of Central Asia? plains, low plateaus, high mountain ranges; steppe, desert, highland

Lynxes can still be found in the mountains of Central Asia.

Interpreting the Visual Record How are these lynxes well suited to the environment in which they live?

Resources

The main water sources in southern Central Asia are the Syr Dar'ya (sir duhr-YAH) and Amu Dar'ya (uh-MOO duhr-YAH) rivers. These rivers flow down from the Pamirs and then across dry plains. Farmers have used them for irrigation for thousands of years. When it first flows down from the mountains, the Syr Dar'ya passes through the

CLOSE

Tell students that the mountains in Kyrgyzstan, particularly the Tian Shan—also known as the Celestial Mountains—are considered by some to be as beautiful as the mountains in Nepal and Switzerland. Have students share memories and impressions of mountains they have visited.

REVIEW AND ASSESS

Have students complete the Section Review. Then organize students into triads. Have each triad write eight matching questions that link landforms, climate, or natural resources to a particular country in Central Asia. Have groups exchange their questions and answer them. Then have students complete Daily Quiz 16.1. **COOPERATIVE LEARNING**

RETEACH

Have students complete Main Idea Activity 16.1. Then have students draw storyboards for a television commercial attempting to attract tourists to Central Asia. Commercials should highlight the region's landscape, climate, and resources. **ENGLISH LANGUAGE LEARNERS**

EXTEND

Tell students about the Torugart Pass, an overland route from Bishkek, Kyrgyzstan, through the Tian Shan to China. The pass was until recently closed to international travelers. Have interested students conduct research on and map the overland trade routes to China.
BLOCK SCHEDULING

Fergana Valley. This large valley is divided among Uzbekistan, Kyrgyzstan, and Tajikistan. As the river flows toward the Aral Sea, irrigated fields line its banks.

During the Soviet period, the region's population grew rapidly. Also, the Soviets encouraged farmers to grow cotton. This crop grows well in Central Asia's sunny climate. However, growing cotton uses a lot of water. Increased use of water has caused the Aral Sea to shrink.

A Dying Sea Today, almost no water from the Syr Dar'ya or Amu Dar'ya reaches the Aral Sea. The rivers' waters are used up by human activity. The effect on the Aral Sea has been devastating. It has lost more than 60 percent of its water since 1960. Its level has dropped 50 feet (15 m) and is still dropping. Towns that were once fishing ports are now dozens of miles from the shore. Winds sweep the dry seafloor, blowing dust, salt, and pesticides hundreds of miles.

Mineral Resources The Central Asian countries' best economic opportunity is in their fossil fuels. Uzbekistan, Kazakhstan, and Turkmenistan all have huge oil and natural gas reserves. However, transporting the oil and gas to other countries is a problem. Economic and political turmoil in some surrounding countries has made it difficult to build pipelines.

Several Central Asian countries are also rich in other minerals. They have deposits of gold, copper, uranium, zinc, and lead. Kazakhstan has vast amounts of coal. Rivers in Kyrgyzstan and Tajikistan could be used to create hydroelectric power.

✓ **READING CHECK:** *Environment and Society* How has human activity affected the Aral Sea? **It has used up the water from the rivers that used to flow into the sea and has depleted the sea's water.**

internet connect
GO TO: go.hrw.com
KEYWORD: SK3 CH16
FOR: Web sites about Central Asia

This boat sits rusting on what was once part of the Aral Sea. The sea's once thriving fishing industry has been destroyed.

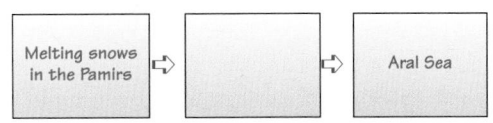

Homework Practice Online
Keyword: SK3 HP16

Section 1 Review

Define and explain: landlocked, oasis

Working with Sketch Maps On a map of Central Asia that you draw or that your teacher provides, label the following: Pamirs, Tian Shan, Aral Sea, Kara-Kum, Kyzyl Kum, Syr Dar'ya, Amu Dar'ya, and Fergana Valley.

Reading for the Main Idea

1. *Environment and Society* What has caused the drying up of the Aral Sea?
2. *Places and Regions* What mineral resources does Central Asia have?

Critical Thinking

3. **Analyzing Information** Why did the Soviets encourage Central Asian farmers to grow cotton?
4. **Finding the Main Idea** What factors make it hard for the Central Asian countries to export oil and gas?

Organizing What You Know

5. **Sequencing** Copy the following graphic organizer. Use it to describe the courses of the Syr Dar'ya and Amu Dar'ya, including human activities that use water.

Melting snows in the Pamirs	⇨		⇨	Aral Sea

Section 1 Review Answers

Define For definitions, see: landlocked, p. 358; oasis, p. 358

Working with Sketch Maps Maps will vary, but listed places should be labeled in their approximate location.

Reading for the Main Idea

1. The rivers' waters are used up by human activity upstream. (NGS 14)
2. oil, natural gas, gold, copper, uranium, zinc, lead, and coal (NGS 4)

Critical Thinking

3. probably because cotton grows well in Central Asia
4. Economic and political turmoil in some surrounding countries has made it difficult to build pipelines in the region.

Organizing What You Know

5. irrigation of crops

Section 2

Objectives

1. Explain how trade and invasions affected Central Asia's history.
2. Describe the political and economic conditions in Central Asia today.

Focus

LET'S GET STARTED

Copy the following instructions onto the chalkboard: *Write down three things about life as a nomad. Use knowledge you may have of the Plains Indians, touring performers, migrant workers, and so on.* Discuss responses. Tell students that in Section 2 they will learn more about the nomadic peoples of Central Asia.

Building Vocabulary

Write the Define terms **nomads** and **caravans** on the chalkboard. Have volunteers read aloud the definitions from the text's glossary. Tell students that the word *nomad* comes from the Greek language and that the word *caravan* is Persian in origin. Point out that Greece and Persia are two of the major cultural influences on Central Asia.

Section 2 RESOURCES

Reproducible
- Guided Reading Strategy 16.2
- Graphic Organizer 16
- Cultures of the World Activity 4

Technology
- One-Stop Planner CD–ROM, Lesson 16.2
- Homework Practice Online
- HRW Go site

Reinforcement, Review, and Assessment
- Section 2 Review, p. 362
- Daily Quiz 16.2
- Main Idea Activity 16.2
- English Audio Summary 16.2
- Spanish Audio Summary 16.2

Visual Record Answer ►

Students might mention the mosques and minarets.

Section 2 History and Culture

Read to Discover

1. How did trade and invasions affect the history of Central Asia?
2. What are political and economic conditions like in Central Asia today?

Define

nomads
caravans

An ancient Kyrgyz stone figure

WHY IT MATTERS

Even though the Soviet Union has collapsed, traces of its influence remain in Central Asia, particularly in government. Use CNNfyi.com or other **current events** sources to find examples of political events in these countries. Record your findings in your journal.

History

For centuries, Central Asians have made a living by raising horses, cattle, sheep, and goats. Many of these herders lived as **nomads**, people who often move from place to place. Other people became farmers around rivers and oases.

Trade At one time, the best land route between China and the eastern Mediterranean ran through Central Asia. Merchants traveled in large groups, called **caravans**, for protection. The goods they carried included silk and spices. As a result, this route came to be called the Silk Road. Cities along the road became centers of wealth and culture.

Central Asia's situation changed after Europeans discovered they could sail to East Asia through the Indian Ocean. As a result, trade through Central Asia declined. The region became isolated and poor.

Bukhara, in Uzbekistan, was once a powerful and wealthy trading center of Central Asia.

Interpreting the Visual Record
What architectural features can you see that distinguish Bukhara as an Islamic city?

▼

TEACH

Teaching Objectives 1–2

ALL LEVELS: (Suggested time: 20 min.) Copy the following graphic organizer onto the chalkboard, omitting the italicized answers. Use it to help students understand how trade and invasions affected Central Asia's history and its current political and economic conditions. Have volunteers fill in the organizer on the chalkboard. Tell students to copy the organizer into their notebooks. **ENGLISH LANGUAGE LEARNERS**

Central Asia's History and Current Conditions

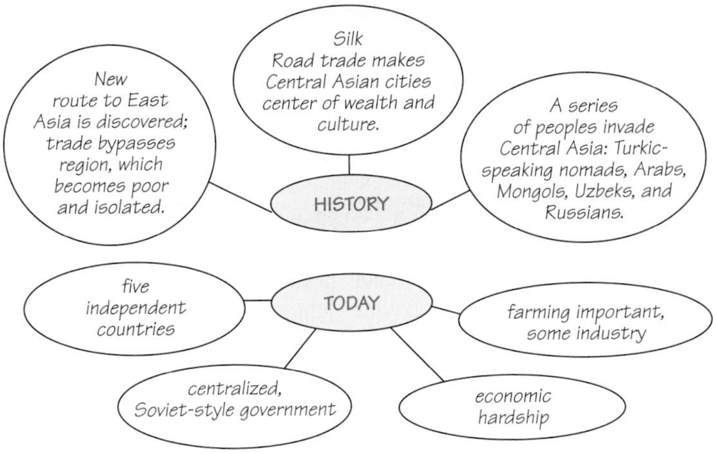

HISTORY
New route to East Asia is discovered; trade bypasses region, which becomes poor and isolated.
Silk Road trade makes Central Asian cities center of wealth and culture.
A series of peoples invade Central Asia: Turkic-speaking nomads, Arabs, Mongols, Uzbeks, and Russians.
TODAY
five independent countries
farming important, some industry
centralized, Soviet-style government
economic hardship

CONNECTING TO *History*

The Silk Road

Silk processing in modern Uzbekistan

The Silk Road stretched 5,000 miles (8,000 km) across Central Asia from China to the Mediterranean Sea. Along this route passed merchants, armies, and diplomats. These people forged links between East and West.

The facts of the Silk Road are still wrapped in mystery. Chinese trade and military expeditions probably began moving into Central Asia in the 100s B.C. Chinese trade goods soon were making their way to eastern Mediterranean ports.

Over the next several centuries, trade in silk, spices, jewels, and other luxury goods increased. Great caravans of camels and oxcarts traveled the Silk Road in both directions. They crossed the harsh deserts and mountains of Central Asia. Cities like Samarqand and Bukhara grew rich from the trade. In the process, ideas and technology also moved between Europe and Asia.

Travel along the Silk Road was hazardous. Bandits often robbed the caravans. Some travelers lost their way in the desert and died. In addition, religious and political turmoil occasionally disrupted travel.

Understanding What You Read

1. What was the Silk Road?
2. Why was the Silk Road important?

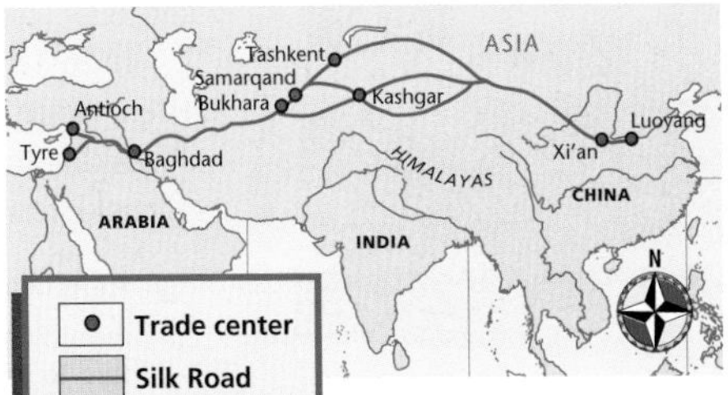

ASIA
Tashkent
Samarqand
Bukhara
Kashgar
Antioch
Tyre
Baghdad
Luoyang
Xi'an
HIMALAYAS
CHINA
ARABIA
INDIA
N
Trade center
Silk Road

Invasions and the Soviet Era About A.D. 500, Turkic-speaking nomads from northern Asia spread through Central Asia. In the 700s Arab armies took over much of the region, bringing Islam. In the 1200s the armies of Mongol leaders conquered Central Asia. Later, another Turkic people, the Uzbeks, took over parts of the region. In the 1800s the Russian Empire conquered Central Asia.

After the Russian Revolution, the Soviet government set up five republics in Central Asia. The Soviets encouraged ethnic Russians to move to this area and made the nomads settle on collective ranches or farms. Religion was discouraged. Russian became the language of government and business. The government set up schools and hospitals. Women were allowed to work outside the home.

✓ **READING CHECK:** *Human Systems* What type of government system did the five republics set up by the Soviet Union have? one in which the government controlled many aspects of life

GLOBAL PERSPECTIVES

Between 1949 and 1989 the Soviet government conducted more than 400 nuclear tests above, on, or below the surface of an area in Kazakhstan called the Polygon. Nearby residents—some of whom could see mushroom clouds from their homes—were told that the tests were important scientific research. They were not informed of the dangers.

Radiation has since devastated area residents. Thousands have died from cancer, including rare forms of the disease. As many as 100,000 people who received the heaviest doses of radiation have passed genetic defects on to their children and grandchildren. Many babies in the area are born with health problems.

Critical Thinking: What responsibility does Kazakhstan's new government have to its citizens who were hurt by Soviet nuclear testing?

Answer: Some students might mention that the country has limited responsibility but should nonetheless help its citizens.

Connecting to History

Answers

1. a trade route linking China to the Mediterranean Sea
2. linked East and West; helped ideas move between Europe and Asia

Close

Have students examine the section again to identify the different time periods discussed. Then have them determine when in Central Asia's history they would choose to live. Ask them to write a paragraph describing their daily lives.

Review and Assess

Have students complete the Section Review. Then organize them into groups of four. Have each group write eight true/false questions about the section. Have groups exchange their questions and answer them. Then have students complete Daily Quiz 16.2. **COOPERATIVE LEARNING**

Reteach

Have students complete Main idea Activity 16.2. Then have students suggest designs for commemorative stamps for each political, cultural, and economic period in Central Asia's history.
ENGLISH LANGUAGE LEARNERS

Extend

Have interested students create a plan for a family of four—two parents and two children—to live as nomads in the United States today. Remind them to consider how the family will make its living, how the children will be educated, what provisions the family will need, and where it will go.
BLOCK SCHEDULING

Section 2 Review

Answers

Define For definitions, see: nomads, p. 360; caravans, p. 360

Working with Sketch Maps Places should be labeled in their approximate locations.

Reading for the Main Idea

1. farming, herding, trading (NGS 14)
2. Possible answers: Turkic-speaking nomads, Arab armies, Mongols, Uzbeks, the Russian Empire (NGS 13)
3. Soviets set up five republics, made nomads settle on collective ranches or farms, discouraged religion, made Russian official language, established schools. (NGS 12)

Critical Thinking

4. reaction against long Soviet rule; desire to be more Western

Organizing What You Know

5. primary—growing crops, mining metals, raising livestock, drilling for oil; secondary—making cloth, food products, chemicals from oil; manufacturing tractors

Visual Record Answer ▶

Students might mention the desks or the students' uniforms.

Build on What You Know

Do you remember what you learned about acculturation? See Chapter 5 to review.

A Kyrgyz teacher conducts class.
Interpreting the Visual Record
How is this class similar to yours?

Central Asia Today

The five republics became independent countries when the Soviet Union broke up in 1991. All have strong economic ties to Russia. Ethnic Russians still live in every country in the region. However, all five countries are switching from the Cyrillic alphabet to the Latin alphabet. The Cyrillic alphabet had been imposed on them by the Soviet Union. The Latin alphabet is used in most Western European languages, including English, and in Turkey.

Government All of these new countries have declared themselves to be democracies. However, they are not very free or democratic. Each is ruled by a strong central government that limits opposition and criticism.

Economy Some of the Central Asian countries have oil and gas reserves that may someday make them rich. For now, though, all are suffering economic hardship. Causes of the hardships include outdated equipment, lack of funds, and poor transportation links.

Farming is important in the Central Asian economies. Crops include cotton, wheat, barley, fruits, vegetables, almonds, tobacco, and rice. Central Asians raise cattle, sheep, horses, goats, and camels. They also raise silkworms to make silk thread.

Industry in Central Asia includes food processing, wool textiles, mining, and oil drilling. Oil-rich Turkmenistan and Kazakhstan also process oil into other products. Kazakhstan and Uzbekistan make heavy equipment such as tractors.

✓ **READING CHECK:** *Human Systems* How do political freedoms in the region compare to those of the United States? Citizens' criticism of or opposition to the government is limited, whereas in the United States people can speak freely.

go.hrw.com Homework Practice Online
Keyword: SK3 HP16

Section 2 Review

Define and explain: nomads, caravans

Working with Sketch Maps On the map you created in Section 1, draw and label the five Central Asian countries.

Reading for the Main Idea

1. *Environment and Society* How have the people of Central Asia made a living over the centuries?
2. *Human Systems* What are four groups that invaded Central Asia?
3. *Human Systems* How did Soviet rule change Central Asia?

Critical Thinking

4. **Drawing Inferences and Conclusions** What does the switch to the Latin alphabet suggest about the Central Asian countries?

Organizing What You Know

5. **Categorizing** Copy the following graphic organizer. Use it to categorize economic activities in Central Asia. Place the following items in the chart: making cloth, growing crops, mining metals, making food products, raising livestock, making chemicals from oil, drilling for oil, and manufacturing tractors.

Primary industries	Secondary industries

Section 3

Objectives

1. Identify some important aspects of culture in Kazakhstan.
2. Explain how Kyrgyz culture reflects nomadic traditions.
3. Discuss the political violence of recent years in Tajikistan.
4. Describe two important art forms in Turkmenistan.
5. Analyze how Uzbekistan's population is significant.

Focus

LET'S GET STARTED

Copy the following instructions onto the chalkboard: *Imagine that a student from Central Asia has joined our class for the next year. Write down four customs or other facets of our lives you believe that he or she should see or experience.* Ask volunteers to share their lists with the class. Tell students that in Section 3 they will learn more about the individual countries of Central Asia.

Building Vocabulary

Write the terms **yurt** and **mosques** on the chalkboard. Have volunteers locate and read aloud the definitions from the text's glossary. Ask students to suggest structures they have seen or heard of with similar shapes or functions. *(Possible answers: church, temple, tepee, or trailer)*

Section 3 The Countries of Central Asia

Read to Discover

1. What are some important aspects of culture in Kazakhstan?
2. How does Kyrgyz culture reflect nomadic traditions?
3. Why have politics in Tajikistan in recent years been marked by violence?
4. What are two important art forms in Turkmenistan?
5. How is Uzbekistan's population significant?

Define

yurt
mosques

Locate

Tashkent
Samarqand

A warrior's armor from Kazakhstan

WHY IT MATTERS

Since the collapse of the Soviet Union, religious freedom is more common in Central Asia. Use CNNfyi.com or other **current events** sources to find examples of religious and ethnic differences in the countries of Central Asia. Record your findings in your journal.

Section 3 Resources

Reproducible

- Guided Reading Strategy 16.3

Technology

- One-Stop Planner CD–ROM, Lesson 16.3
- Homework Practice Online
- HRW Go site

Reinforcement, Review, and Assessment

- Section 3 Review, p. 365
- Daily Quiz 16.3
- Main Idea Activity 16.3
- English Audio Summary 16.3
- Spanish Audio Summary 16.3

internet connect

GO TO: **go.hrw.com**
KEYWORD: **SK3 CH16**
FOR: **Web sites about carpets**

Kazakhstan

Of the Central Asian nations, Kazakhstan was the first to be conquered by Russia. Russian influence remains strong there. About one third of Kazakhstan's people are ethnic Russians. Kazakh and Russian are both official languages. Many ethnic Kazakhs grow up speaking Russian at home and have to learn Kazakh in school.

Kazakhstanis celebrate the New Year twice—on January 1 and again on Nauruz, the start of the Persian calendar's year. Nauruz falls on the spring equinox.

Food in Central Asia combines influences from Southwest Asia and China. Rice, yogurt, and grilled meat are common ingredients. One Kazakh specialty is smoked horsemeat sausage with cold noodles.

✓ **READING CHECK:** *Human Systems* How has Kazakhstan been influenced by Russia? was first to be conquered by Russia; about one third of the people are ethnic Russians; Russian is commonly spoken

Kyrgyzstan

Kyrgyzstan has many mountains, and the people live mostly in valleys. People in the southern part of the country generally share cultural ties with Uzbekistan. People in northern areas are more linked to nomadic cultures and to Kazakhstan.

A woman in Uzbekistan grills meat on skewers.

TEACH

Teaching Objectives 1–5

ALL LEVELS: (Suggested time: 30 min.) Copy the following graphic organizer onto the chalkboard, omitting the italicized answers. Point out to students that each circle represents one of the Central Asian countries. The lines indicate that the countries have much in common. Refer students to the Read to Discover questions. Call on students to suggest phrases that answer the questions and that highlight the countries' distinct traits. Write the phrases in the circles. **ENGLISH LANGUAGE LEARNERS**

Distinctive Traits of Central Asian Countries

Kazakhstan
• *Russian influence strong*
• *Kazakh and Russian official languages*
• *combined influences in food*
Turkmenistan
• *English—second official language*
• *Islamic principles taught in schools*
• *important art forms—carpets, poetry*
Kyrgyzstan
• *in north—linked to nomadic cultures*
• *clan membership still important*
• *black and white felt hats to show class status*
• *yurts—movable round house*
Tajikistan
• *civil war in mid-1990s: Soviet-style government against reformers*
• *language related to Persian*
• *Persian literature part of heritage*
Uzbekistan
• *largest population in Central Asia*
• *Uzbek—official language*
• *Tashkent and Samarkand—Silk Road cities*
• *traditional art—embroidering with gold*

Section 3 Review

Answers

Define For definitions, see: yurt, p. 364; mosques, p. 364

Working with Sketch Maps Maps should correctly depict the locations of the two cities.

Reading for the Main Idea

1. Kazakhstan; about one third of people there are ethnic Russians (NGS 4)
2. type of government—Soviet-style or reform (NGS 13)
3. It has experienced a revival. (NGS 10)

Critical Thinking

4. Answers might include yurts, carpets.

Organizing What You Know

5. Soviet—communist; Russian; Cyrillic; Islam discouraged; today—so-called democracies but actually centralized, Soviet-style; in some places, Russian and local language, like Kazakh; Latin; government supports revival

Chart Answer ►

It is much greater.

Visual Record Answer ►

Students might say that carpets can be transported to any location.

Ethnic Makeup of Kazakhstan and Uzbekistan

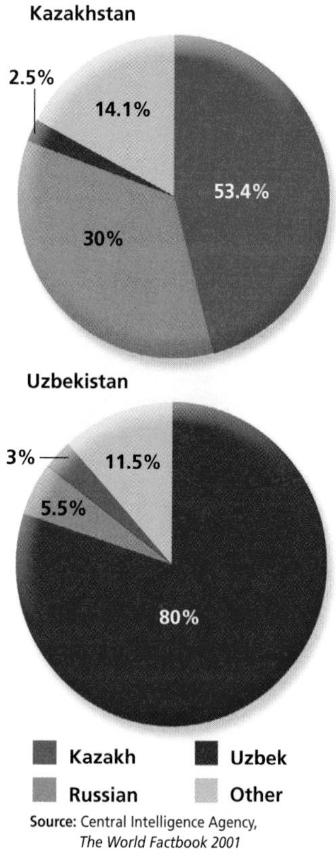

Source: Central Intelligence Agency, *The World Factbook 2001*

Interpreting the Chart **How does the number of Russians in Kazakhstan compare to that in Uzbekistan?**

The word *kyrgyz* means "forty clans." Clan membership is still important in Krygyz social, political, and economic life. Many Kyrgyz men wear black and white felt hats that show their clan status.

Nomadic traditions are still important to many Kyrgyz. The **yurt** is a movable round house of wool felt mats over a wood frame. Today the yurt is a symbol of the nomadic heritage. Even people who live in cities may put up yurts for weddings and funerals.

✓ **READING CHECK:** *Human Systems* In what ways do the Kyrgyz continue traditions of their past? **by emphasizing clan membership and constructing yurts**

Tajikistan

In the mid-1990s Tajikistan experienced a civil war. The Soviet-style government fought against a mixed group of reformers, some of whom demanded democracy. Others called for government by Islamic law. A peace agreement was signed in 1996, but tensions remain high.

The other major Central Asian languages are related to Turkish. However, the Tajik language is related to Persian. Tajiks consider the great literature written in Persian to be part of their cultural heritage.

✓ **READING CHECK:** *Human Systems* What has happened in politics in recent years in Tajikistan? **In the mid-1990s there was a civil war between reformers and Soviet-style government; tensions remain high.**

Turkmenistan

The major first language of Turkmenistan is Turkmen. In 1993 Turkmenistan adopted English, rather than Russian, as its second official language. However, some schools teach in Russian, Uzbek, or Kazakh.

Islam has experienced a revival in Central Asia since the breakup of the Soviet Union. Many new **mosques**, or Islamic houses of worship, are being built and old ones are being restored. Donations from other Islamic countries, such as Saudi Arabia and Iran, have helped

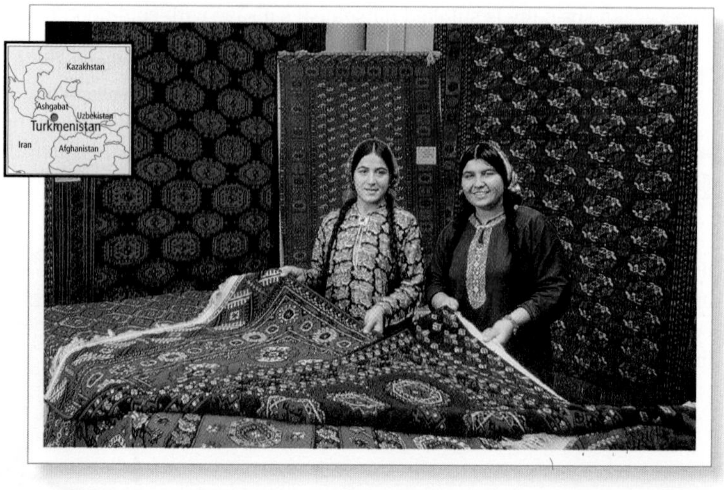

Turkoman women display carpets. Central Asian carpets are famous for their imaginative patterns, bright colors, and expert artistry.

Interpreting the Visual Record **Why were carpets suited to the nomadic way of life?**

CLOSE

Ask students whether they can note any similarities between cultural practices in the United States and those in Central Asia.

REVIEW AND ASSESS

Have students complete the Section Review. Then organize students into groups of four. Have each group write five multiple-choice questions about the section. Have groups trade questions and then answer them. Then have students complete Daily Quiz 16.3. **COOPERATIVE LEARNING**

RETEACH

Have students complete Main Idea Activity 16.3. Then have students design flags for each Central Asian nation, including symbols for peoples, customs, and religion. **ENGLISH LANGUAGE LEARNERS**

EXTEND

Have interested students conduct research on the resurgence of barter as a means of doing business in the Central Asian republics. Have students develop their own barter system for basic goods. **BLOCK SCHEDULING**

these efforts. The government of Turkmenistan supports this revival and has ordered schools to teach Islamic principles. However, like the other states in the region, Turkmenistan's government views Islam with some caution. It does not want Islam to become a political movement.

Historically, the nomadic life required that all possessions be portable. Decorative carpets were the essential furniture of a nomad's home. They are still perhaps the most famous artistic craft of Turkmenistan. Like others in Central Asia, the people of Turkmenistan also have an ancient tradition of poetry.

✓ **READING CHECK:** *Human Systems* What are two forms of art in Turkmenistan, and how do they reflect its cultural traditions? decorative carpets, poetry; ancient forms of artistic expression

Central Asia

COUNTRY	POPULATION/ GROWTH RATE	LIFE EXPECTANCY	LITERACY RATE	PER CAPITA GDP
Kazakhstan	16,731,303 / .03%	58, male / 69, female	98%	$5,000
Kyrgyzstan	4,753,003 / 1.4%	59, male / 68, female	97%	$2,700
Tajikistan	6,578,681 / 2.1%	61, male / 67, female	98%	$1,140
Turkmenistan	4,603,244 / 1.9%	57, male / 65, female	98%	$4,300
Uzbekistan	25,155,064 / 1.6%	60, male / 68, female	99%	$2,400
United States	281,421,906 / 0.9%	74, male / 80, female	97%	$36,200

Sources: Central Intelligence Agency, *The World Factbook 2001;* U.S. Census Bureau

Interpreting the Chart **Which country has the lowest per capita GDP in the region?**

Uzbekistan

Uzbekistan has the largest population of the Central Asian countries—about 24 million people. Uzbek is the official language. People are required to study Uzbek to be eligible for citizenship.

Tashkent and Samarqand are ancient Silk Road cities in Uzbekistan. They are famous for their mosques and Islamic monuments. Uzbeks are also known for their art of embroidering fabric with gold.

✓ **READING CHECK:** *Human Systems* What is one of an Uzbekistan citizen's responsibilities? to study the Uzbek language

go.hrw.com Homework Practice Online
Keyword: SK3 HP16

Section 3 Review

Define and explain: yurt, mosques

Working with Sketch Maps On the map you created in Section 2, label Tashkent and Samarqand.

Reading for the Main Idea

1. *Places and Regions* In which Central Asian nation is the influence of Russia strongest? Why is this true?
2. *Human Systems* What were the two sides in Tajikistan's civil war fighting for?
3. *Human Systems* What is the role of Islam in the region today?

Critical Thinking

4. **Finding the Main Idea** What are two customs or artistic crafts of modern Central Asia that are connected to the nomadic lifestyle?

Organizing What You Know

5. **Contrasting** Copy the following graphic organizer. Use it to describe the conditions in Central Asia during the Soviet era and today.

	Soviet era	Today
Type of government		
Official language		
Alphabet		
Government attitude toward Islam		

CHAPTER 16 Review ANSWERS

Building Vocabulary

For definitions, see: landlocked, p. 358; oasis, p. 358; nomads, p. 360; caravans, p. 360; yurt, p. 364; mosques, p. 364

Reviewing the Main Ideas

1. steppe, desert, and highland
2. Much of the fishing industry has been destroyed, and dust from the sea floor endangers people's health. (NGS 4)
3. Soviets required that nomads settle on collective farms and ranches, discouraged the practice of religion, and made Russian the official language. (NGS 15)
4. They have strong economic ties, and many ethnic Russians still live throughout the region. (NGS 13)
5. Russian, Kazakh, Tajik, English, Uzbek, and other local languages (NGS 4)

▲ **Chart Answer**

Tajikistan

Case Study

Setting the Scene

In areas like Central Asia where climate and soil are not favorable for agricultural development, pastoral nomadism is a lifestyle that allows people to feed themselves. The Kazakhs have traditionally supported themselves by raising camels, cattle, horses, and sheep. In order to feed their herds and to ensure that fresh pastures are always available, the Kazakhs have developed a complex seasonal migration system. This migratory lifestyle is reflected in the kinds of homes they build—yurts. During the Soviet rule of Kazakhstan, new systems of agriculture requiring fixed settlements were introduced into the area. These farms became obstacles to the seasonal movement of animals.

Building a Case

After students have read the case study, ask them the following questions: Who are pastoral nomads? *(people who move around during the year to maintain herds of livestock)* What are some of the animals that pastoral nomads herd? *(cattle, horses, sheep, goats, yaks, reindeer, and camels)* Why must animals in desert regions be kept moving throughout the year? *(to prevent overgrazing and provide fresh grass shoots)* How far do some Kazakh families move in a year? *(500 miles or 805 km)* How do the kinds of houses built by Kazahks reflect their nomadic lifestyle? *(Yurts are impermanent structures that are designed to be easily moved.)*

Historical Geography

An example of a current conflict between farmers and nomads is the fighting between the Hutu and Tutsi in eastern Africa. The Hutu first settled the area that is now Rwanda and Burundi from about 500 B.C. to the year 1000. Hutu life centered on small-scale agriculture. During the 1300s and 1400s the Tutsi entered the region. They were pastoral nomads who depended on large herds of big-horned cattle. The Tutsi soon gained control over the area, gave up their nomadic lifestyle and became "lords" over Hutu farmers. Despite their small numbers, the Tutsi dominated the Hutu economically and politically for almost 600 years. In the late 1990s violence erupted between the Hutu and the Tutsi. In 1994 the United Nations sent peacekeepers to stabilize the region.

Activity: Have students conduct research on the current relationship between the Hutu and the Tutsi. Ask students to find out how successful UN forces were at bringing peace to the region. Have students report their findings in the form of a newsmagazine article.

➤ **This Case Study feature addresses National Geography Standards 3, 4, 8, 9, and 15.**

CASE STUDY

KAZAKHS: PASTORAL NOMADS OF CENTRAL ASIA

Nomads are people who move around from place to place during the year. Nomads usually move when the seasons change so that they will have enough food to eat. Herding, hunting, gathering, and fishing are all ways that different nomadic groups get their food.

Nomads that herd animals are called pastoral nomads. Their way of life depends on the seasonal movement of their herds. Pastoral nomads may herd cattle, horses, sheep, goats, yaks, reindeer, camels, or other animals. Instead of keeping their animals inside fenced pastures, pastoral nomads let them graze on open fields. However, they must make sure the animals do not overgraze and damage the pastureland. To do this, they keep their animals moving throughout the year. Some pastoral nomads live in steppe or desert environments. These nomads often have to move their animals very long distances between winter and summer pastures.

The Kazakhs of Central Asia are an example of a pastoral nomadic group. They have herded horses, sheep, goats, and cattle for hundreds of years. Because they move so much, the Kazakhs do not have permanent homes. They bring their homes with them when they travel to new places.

A Kazakh nomad keeps a watchful eye over a herd of horses.

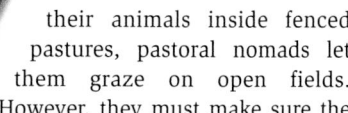

Drawing Conclusions

Focus discussion on the challenges confronted by the pastoral nomads in Kazakhstan. Ask students when Russians and eastern Europeans began settling the area. *(late 1800s)* How was their food production system different? *(They were farmers, not herders.)* How did this system conflict with the nomadic way of life? *(It blocked access to pasture lands.)* Here is some additional information to share with students: Between 1906 and 1912, Russian settlers established more than 500,000 farms. Many Kazakhs were displaced. In the 1950s and 1960s Soviet officials tried to settle the remaining nomads and expand cultivation. Some 60 percent of Kazakhstan's pasture lands were planted with crops. Today, 84 percent of pastures are devoted to cattle and sheep. Point out that this suggests the area may be more suited to herding than farming.

Going Further: Thinking Critically

The Kazakhs are only one example of nomads. Other examples include the Inuit of the Arctic, Plains Indians of North America, !Kung of the Kalahari Desert, Ona of Tierra del Fuego, Fulani of West Africa, Bedouins of the Arabian Desert, Masai of East Africa, Chukchi of Siberia, Lapps of Finland, and Tuareg of the Sahara. Have students work in small groups to conduct research on a nomadic group. Ask them to answer these questions: What is the people's main food source? What is the climate and soil like where they live? What environmental pressures do they experience? What kind of houses do they build? Have they had conflicts with settled farming peoples? If so, what were the issues? What was the outcome? You may want to have each group create a puppet show to present its findings.

◄ Yurts are carefully stitched together by hand. When it is time to move to new pastures, they are carried from place to place on horseback or on small wagons.

The Kazakhs live in tent-like structures called yurts. Yurts are circular structures made of bent poles covered with thick felt. Yurts can be easily taken apart and moved. They are perfect homes for the Kazakhs' nomadic lifestyle.

During the year, a Kazakh family may move its herds of sheep, horses, and cattle as far as 500 miles (805 km). For one Kazakh family, each year is divided into four different parts. The family spends the first part of the year in winter grazing areas. Then, in early spring, they move to areas with fresh grass shoots. When these spring grasses are gone, the family moves their animals to summer pastures. In the fall, the animals are kept for six weeks in autumn pastures. Finally, the herds are taken back to their winter pastures. Each year, the cycle is repeated.

The nomadic lifestyle of the Kazakhs has changed, however. In the early 1800s people from Russia and eastern Europe began to move into the region. These people were farmers. They started planting crops in areas that the Kazakhs used for pasture. This made it more difficult for the Kazakhs to move their animals during the year. Later, when Kazakhstan was part of the Soviet Union, government officials encouraged the Kazakhs to settle in villages and cities. Many Kazakhs still move their animals during the year. However, tending crops has also become an important way to get food.

Seasonal Movement of a Kazakh Family

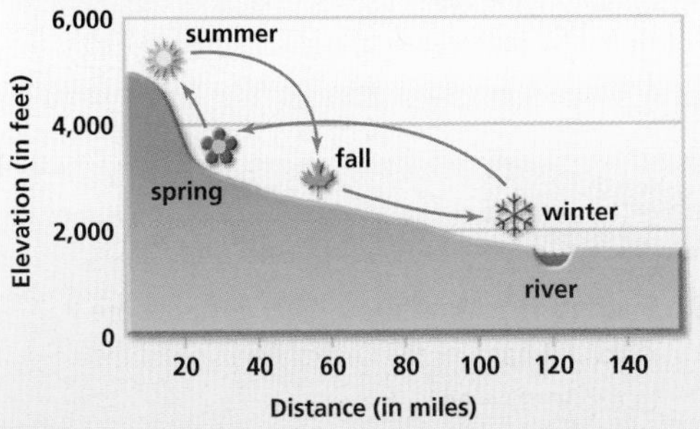

▲ During the year, a Kazakh family may move its herds to several different pasture areas as the seasons change.

Interpreting the Graph **Why do you think animals are moved to higher elevations during the summer and to lower elevations during the winter?**

Understanding What You Read

1. Why do some pastoral nomads have to travel such great distances?
2. How has the nomadic lifestyle of the Kazakhs changed during the last 100 years?

Understanding What You Read

1. Pastoral nomads rely on raising animals for their food supply. Animals must be constantly moved to prevent overgrazing and to access the best pasture grasses.
2. The Kazakhs have lost much of their pasturelands to Russian farmers. They are now more reliant on raising crops.

Graph Answer

Students might suggest to keep the animals cooler in the summer and warmer in the winter.

367

ASSESS

Have students complete the Chapter 16 Test.

RETEACH

Point out that making carpets and writing poetry are two forms of art highly valued in Central Asia, particularly in Turkmenistan. Organize the class into two groups, with one group creating a carpet design and the other writing a poem. Have students meet within their groups to decide on a overall concept and to volunteer for certain topics. Topics will include landforms, mineral resources, invasions, and other subjects listed in the chapter. Then let individual students work on their topics, either as designs or verses. Finally, have students combine their contributions into one carpet design and one poem. **ENGLISH LANGUAGE LEARNERS, COOPERATIVE LEARNING**

PORTFOLIO

PORTFOLIO EXTENSIONS

1. Have students build model yurts with sticks, felt, and cloth. Point out that to be useful as movable structures, yurts must be constructed so they can be folded and loaded onto a wagon or horse. Have students photograph their yurts for their portfolios.
2. Have students conduct research on calligraphy, which is important to Uzbekistan, and use it to render the Latin and Cyrillic alphabets in pen and ink. Ask students to write their names in both alphabets. You might also challenge students to find a sentence or phrase in the Uzbek language and write it in calligraphic script.

CHAPTER 16

Review ANSWERS

Understanding Environment and Society

Students may find that beginning in the 1960s the water level was reduced because water was being diverted for irrigation. Governmental agencies have suggested options for saving the sea, including reducing the amount of cotton being grown, more efficient irrigation methods, or shifting from an agricultural economy. Students may find that consequences such as habitat destruction, species loss, and negative health effects may worsen if the sea's water level continues to drop. Use Rubric 29, Presentations, to evaluate student work.

Thinking Critically

1. Soviets imposed Russian, Cyrillic alphabet on Central Asia; after Soviet collapse, some nations returned to local languages, adopted Latin alphabet.
2. Answers may vary, but carpets and yurts—two art forms—can be moved around.
3. because a growing population needed more cotton; caused the Aral Sea to shrink

CHAPTER 16

Reviewing What You Know

Building Vocabulary

On a separate sheet of paper, write sentences to define each of the following words.

1. landlocked
2. oasis
3. nomads
4. caravans
5. yurt
6. mosques

Reviewing the Main Ideas

1. *Places and Regions* What types of climates are most common in Central Asia?
2. *Environment and Society* What problems have resulted from the shrinking of the Aral Sea?
3. *Human Systems* How did Soviet rule change Central Asians' way of life?
4. *Human Systems* What kinds of ties do the Central Asian countries have to Russia today?
5. *Places and Regions* What are the various languages spoken in Central Asia?

Understanding Environment and Society

Aral Sea in Danger

The rapid disappearance of the Aral Sea is a serious concern in Central Asia. Research and create a presentation on the Aral Sea. You may want to think about the following:

- Actions that could be taken to preserve the Aral Sea.
- What could be done to help slow the dropping of the sea's water level.
- Possible consequences if the level of the Aral Sea continues to drop.

Include a bibliography of the sources you used.

Thinking Critically

1. **Summarizing** How have politics influenced language and the alphabet used in Central Asia?
2. **Finding the Main Idea** How do the artistic crafts of Central Asia reflect the nomadic lifestyle?
3. **Analyzing Information** Why did the Soviets encourage cotton farming in the region? What were the environmental consequences?
4. **Finding the Main Idea** What obstacles are making it hard for the Central Asian countries to export their oil?
5. **Summarizing** What are some reasons the Central Asian countries have experienced slow economic growth since independence? How are they trying to improve the situation?

FOOD FESTIVAL

Here is an easy recipe for Kazakh Rice Salad. You may want to ask students to contribute the ingredients. Then mix up the salad in class. This recipe serves six. Double or triple it to serve everyone. Combine these ingredients in a large bowl: 1 c. cooked brown rice, cold; 1 c. cooked buckwheat, cold; ½ c. dates, chopped; 3 cloves garlic, minced; 2 tbs. fresh ginger, minced; ¾ c. cashews or almonds, crushed; ½ c. rice wine vinegar; ½ c. extra virgin olive oil; ¾ c. fresh cilantro, chopped.

CHAPTER 16 REVIEW AND ASSESSMENT RESOURCES

Reproducible
- Readings in World Geography, History, and Culture 45 and 46
- Critical Thinking Activity 16
- Vocabulary Activity 16

Technology
- Chapter 16 Test Generator (on the One-Stop Planner)
- HRW Go site
- Audio CD Program, Chapter 16

Reinforcement, Review, and Assessment
- Chapter 16 Review, pp. 368–69
- Chapter 16 Tutorial for Students, Parents, Mentors, and Peers
- Chapter 16 Test
- Chapter 16 Test for English Language Learners and Special-Needs Students
- Unit 4 Test
- Unit 4 Test for English Language Learners and Special-Needs Students

Building Social Studies Skills

Map ACTIVITY

On a separate sheet of paper, match the letters on the map with their correct labels.

Caspian Sea	**Kara-Kum**
Pamirs	**Kyzyl Kum**
Tian Shan	**Tashkent**
Aral Sea	**Samarqand**

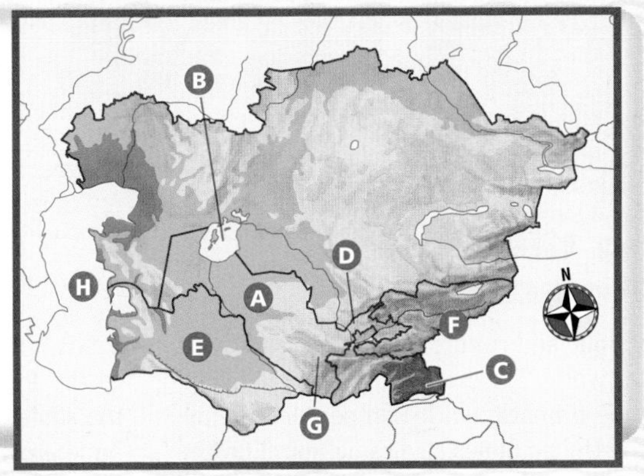

Mental Mapping Skills ACTIVITY

On a separate sheet of paper, draw a freehand map of Central Asia. Make a key for your map and label the following:

Amu Dar'ya	**Syr Dar'ya**
Aral Sea	**Tajikistan**
Kazakhstan	**Turkmenistan**
Kyrgyzstan	**Uzbekistan**

WRITING ACTIVITY

Imagine that you are a caravan trader traveling along the Silk Road during the 1200s. Write a journal entry describing your journey from the Mediterranean Sea through Central Asia. Be sure to use standard grammar, spelling, sentence structure, and punctuation.

Alternative Assessment

Portfolio ACTIVITY

Learning About Your Local Geography

History The Mongol conqueror Genghis Khan is a hero to many Central Asians. Use biographies or interviews with residents to find out about a person who is special to your area. Report your findings.

internet connect

Internet Activity: **go.hrw.com**
KEYWORD: SK3 GT16

Choose a topic to explore about Central Asia:
- Study the climate of Central Asia.
- Travel along the historic Silk Road.
- Learn about nomads and caravans.

4. economic, political turmoil has made it difficult to build pipelines
5. outdated equipment, lack of funds, poor transportation links; some nations are trying to diversify their economic products.

Map Activity

A. Kyzyl Kum	E. Kara-Kum
B. Aral Sea	F. Tian Shan
C. Pamirs	G. Samarqand
D. Tashkent	H. Caspian Sea

Mental Mapping Skills Activity

Maps will vary, but listed places should be labeled in their approximate locations.

Writing Activity

Information could include difficult travel due to unfamiliar terrain, varied climate, warring factions. Use Rubric 15, Journals, to evaluate work.

Portfolio Activity

Students' findings will vary, but they should include information about a local hero. Use Rubric 4, Biographies, to evaluate student work.

internet connect

GO TO: **go.hrw.com**
KEYWORD: SK3 Teacher
FOR: a guide to using the Internet in your classroom

Focus On Culture

Recalling Concepts

Review with students what they learned in the preceding unit about the political and economic relationships between the countries of Central Asia and the former Soviet Union. *(Students should mention that these countries were Soviet republics, that the economies of these countries were part of the Soviet economy, and that the Soviet Union set up schools and hospitals and made Russian the language of government and business.)* Point out that these countries were part of the Soviet Union for approximately 70 years.

Remind students that the cultural and ethnic differences between the people of these countries and of Russia contributed to the drive by Central Asians to gain independence from the Soviet Union. On a wall map show students the relative location of Kazakhstan, Uzbekistan, Turkmenistan, Azerbaijan, Kyrgyzstan, and Tajikistan between Russia and the countries of Southwest Asia.

Geography Sidelight

Many Central Asians have been working to restore the practice of Islam in their region. These efforts have been met with a certain amount of governmental resistance. In order to limit the power of Islamic political movements, some governments in the region have imposed restrictions on Muslims. For example, some Islamic-based political groups have been banned. Even the wearing of long beards, which can symbolize adherence to fundamentalist Islamic beliefs, has been outlawed in some areas.

Kazakhstan, Kyrgyzstan, and Turkmenistan are viewed as less politically vulnerable to Islamic fundamentalist movements. Russian cultural influences are more widespread in these three countries than in Uzbekistan and Tajikistan. In addition to being geographically close to Iran and Afghanistan—countries where Islamic political movements have been very powerful—Uzbekistan and Tajikistan have well-established Islamic institutions.

Critical Thinking: What is the relationship of Islam to Central Asian countries?

Answer: Many people have tried to increase its influence, while governments have tried to limit it.

➤ **This Focus On Culture feature addresses National Geography Standards 3, 5, 10, and 13.**

Facing the Past and Present

Patterns of trade and culture can change quickly in our modern world. For example, the United States used to trade primarily with Europe. Most immigrants to the United States also came from Europe. Today, the American connection to Europe has faded. The United States now trades more with Japan and other Pacific Rim countries than with Europe. New ideas, new technology, and immigrants to the United States come from all around the world.

Central Asia Since the breakup of the Soviet Union, similar changes have taken place in Central Asia. In the past, Central Asia had many ties to the Soviet Union. For example, the economies of the two regions were linked. Central Asia exported cotton and oil to Russia and to countries in Eastern Europe. In exchange, Central Asia received a variety of manufactured goods. The Soviet Union also heavily influenced the culture of Central Asia. Many Central Asians learned to speak Russian.

Looking South Today, Central Asia's links to the former Soviet Union have weakened. At the same time, its ancient ties to Southwest Asia have grown stronger. The Silk Road once linked Central Asian cities to Southwest Asian ports on the Mediterranean. Now the peoples of Central Asia are looking southward once again. New links are forming between Central Asia and Turkey. Many people in Central Asia are traditionally Turkic in culture and language. Turkey's business leaders are working to expand their industries in Central Asia. Also, regular air travel from Turkey to cities in Central Asia is now possible as well.

Religion also links both Central Asia and Southwest Asia. Islam was first introduced into Central Asia in the A.D. 700s. It eventually became the region's dominant religion. However, Islam declined during the Soviet era. Missionaries from Arab countries and Iran are now working to strengthen this connection. Iran is also spending millions of dollars to build roads and rail lines to Central Asia.

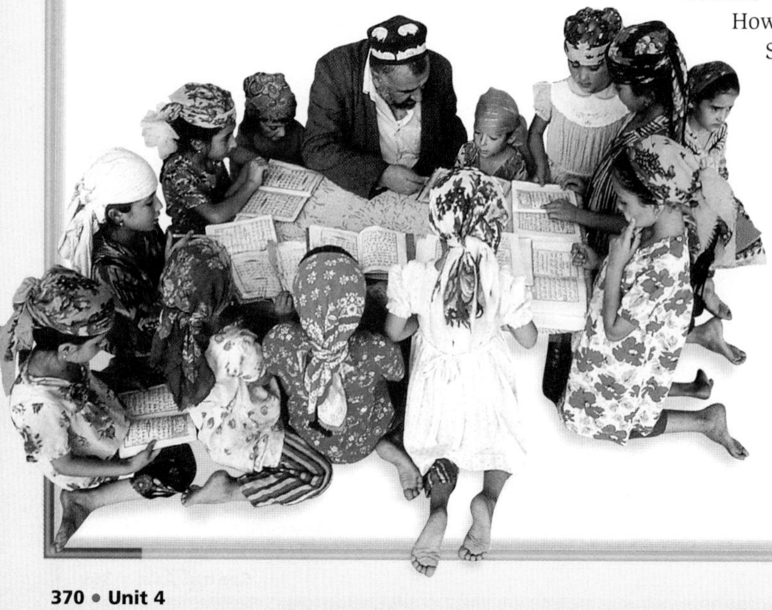

◀ These children are learning about Islam in Dushanbe, Tajikistan. Although the former Communist government discouraged the practice of religion, today Islam flourishes in the independent Central Asian republics.

Going Further: Thinking Critically

Direct students' attention to the map on this page. Have them name the countries that share each language group. Then ask the following questions: What language is common in a very small area of Southwest Asia and Central Asia? *(Greek)* What is the main language group of Central Asia? *(Turkic)* What language groups are most common in Southwest Asia? *(Iranic and Semitic)* Then organize the class into three groups and assign each group one of the three dominant language groups. Have each group conduct research on the history of the language group in Central Asia or Southwest Asia.

Language Groups of Southwest and Central Asia

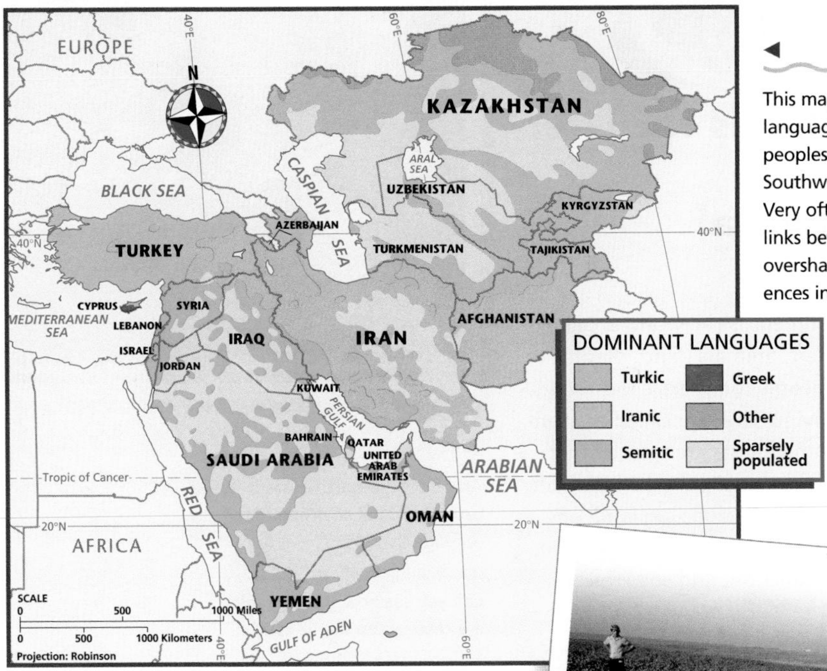

This map shows the major language groups that link peoples throughout Southwest and Central Asia. Very often, however, the links between peoples are overshadowed by differences in culture and history.

Many people in Central and Southwest Asia grow cotton, such as here in Uzbekistan.

Central Asia and Southwest Asia share a similar climate, environment, and way of life. Both regions are dry, and water conservation and irrigation are important. Many people in both regions grow cotton and herd animals. In addition, both Central Asia and Southwest Asia are dealing with changes caused by the growing influence of Western culture. Some people are worried that compact discs, videotapes, and satellite television from the West threaten traditional beliefs and ways of life. Shared fears of cultural loss may bring Central Asia and Southwest Asia closer together.

Defining the Region As the world changes, geographers must reexamine this and other regions of the world. Will geographers decide to include the countries of Central Asia in the region of Southwest Asia? Will Russia regain control of Central Asia? The geographers are watching and waiting.

Understanding What You Read

1. What ties did Central Asia have to the Soviet Union in the past?
2. Why are ties between Central Asia and Southwest Asia growing today?

Understanding What You Read

Answers

1. economic—Central Asia exported cotton and oil to Soviet territories and imported Soviet manufactured goods; cultural—Soviet Union strongly influenced culture of Central Asia, and many Central Asians learned how to speak Russian
2. because Turkey's business leaders are trying to expand their industries into Central Asia; because of expanded air travel, roads, and rail lines; because of the resurgence of Islam; and because of common reaction against Western culture

GeoSkills

Going Further: Thinking Critically

Have students investigate ways in which their peers are involved in environmental movements. Organize students into groups. Have each group find an example of young people's activism in environmental issues. Students may select an individual or an organization to profile. If a group has selected an individual, ask its members to prepare a résumé for that person. If a group selected an organization, ask its members to prepare a brochure about the organization. Have students include answers to as many of the following questions as possible:

- What is the individual's or organization's main goal?
- Did a certain specific incident or issue inspire the activist(s) to get involved?
- How might other students become involved in this effort?
- Are there aspects of the work done by the organization or individual that might appeal especially to young people?
- How is work done by the organization or individual publicized?
- What is the most important accomplishment achieved so far by the individual or organization?

Have each group present its work to the class.

Practicing the Skill

1. Students should describe a local environmental problem fully. They should relate the environmental problem to their community.
2. Students should present options for solving the problem.
3. Students should list advantages and disadvantages of options.
4. Students should choose an option, create a plan with the option, explain their plan, and present it.

Visual Record Answer

Possible answer: by containing the oil within a flexible barrier and removing it manually from the surface

➤ This GeoSkills feature addresses National Geography Standards 4 and 14.

Geo Skills

Building Skills for Life: Addressing Environmental Problems

The natural environment is the world around us. It includes the air, animals, land, plants, and water. Many people today are concerned about the environment. They are called environmentalists. Environmentalists are worried that human activities are damaging the environment. Environmental problems include air, land, and water pollution, global warming, deforestation, plant and animal extinction, and soil erosion.

People all over the world are working to solve these environmental problems. The governments of many countries are trying to work together to protect the environment. International organizations like the United Nations are also addressing environmental issues.

An oil spill in northwestern Russia caused serious environmental damage in 1995.

Interpreting the Visual Record

Can you see how these people are cleaning up the oil spill?

Practicing: The Skill

1. **Gather Information.** Create a plan to present to the city council for solving a local environmental problem. Select a problem and research it using databases or other reference materials. How does it affect people's lives and your community's culture or economy?
2. **List and Consider Options.** After reviewing the information, list and consider options for solving this environmental problem.
3. **Consider Advantages and Disadvantages.** Now consider the advantages and disadvantages of taking each option. Ask yourself questions like, "How will solving this environmental problem affect business in the area?" Record your answers.
4. **Choose, Implement, and Evaluate a Solution.** After considering the advantages and disadvantages, you should create your plan. Be sure to make your proposal clear. You will need to explain the reasoning behind the choices you made in your plan.

HANDS ON GEOGRAPHY

The countries of the former Soviet Union face some of the worst environmental problems in the world. For more than 50 years, the region's environment was polluted with nuclear waste and toxic chemicals. Today, environmental problems in this region include air, land, and water pollution.

One place that was seriously polluted was the Russian city of Chelyabinsk. Some people have called Chelyabinsk the most polluted place on Earth. The passage below describes some of the environmental problems in Chelyabinsk. Read the passage and then answer the Lab Report questions.

Chelyabinsk was one of the former Soviet Union's main military production centers. A factory near Chelyabinsk produced nuclear weapons. Over the years, nuclear waste from this factory polluted a very large area. A huge amount of nuclear waste was dumped into the Techa River. Many people in the region used this river as their main source of water. They also ate fish from the river.

In the 1950s many deaths and health problems resulted from pollution in the Techa River. Because it was so polluted, the Soviet government evacuated 22 villages along the river. In 1957 a nuclear accident in the region released twice as much radiation as the Chernobyl accident in 1986. However, the accident near Chelyabinsk was kept secret. About 10,000 people were evacuated. The severe environmental problems in the Chelyabinsk region led to dramatic increases in birth defects and cancer rates.

The village of Mitlino was evacuated after a nuclear accident in 1957.

Lab Report

1. How did environmental problems near Chelyabinsk affect people who lived in the region?
2. What might be done to address environmental problems in the Chelyabinsk region?
3. How can a geographical perspective help to solve these problems?

Lab Report

Answers

1. The environmental problems in Chelyabinsk caused many deaths and health problems, including birth defects and cancer.
2. Students might suggest cleaning up the nuclear waste in Chelyabinsk and the Techa River. Students may also suggest further evacuating the area until it is safe for people to live there.
3. Physical geographic studies can help locate sources of pollution and demonstrate the effects of pollution on air, land, soil, and water. Cultural geographic studies can show the effects of pollution on people. These studies can propose ways to prevent similar problems in the future.

internet connect

GO TO: **go.hrw.com**
KEYWORD: **SK3 CH14**
FOR: **Web sites about children and pollution**

UNIT 5

SOUTHWEST ASIA

USING THE ILLUSTRATIONS

Direct students' attention to the photographs on these pages. Point out the young Israeli girls in the center. If students in your school wear uniforms, you may want to ask students to compare theirs with the uniforms the Israeli girls are wearing.

Ask students how they think the bottle trees have adapted to Yemen's rocky terrain and dry climate. *(The trunks store water. Students may also suggest that because the trees have relatively few leaves they lose less water through evaporation.)*

Ask which photos relate to religion. *(Students should identify the scene in Jerusalem and the family preparing a feast.)* Tell students that Southwest Asia is the birthplace of three of the world's major religions—Judaism, Christianity, and Islam. Focus students' attention on the photo showing a family preparing a feast. Tell students similar photos could have been taken in any one of several countries, including the United States, because Islamic families around the world fast during the month of Ramadan. Feasts that mark the end of the fast are also common. You may want to read the More from the Field feature about Ramadan on the opposite page aloud to the students.

UNIT OBJECTIVES

1. Describe how the physical geography and economic geography of Southwest Asia affect ways of life there.
2. Trace the history and development of religion in the region, and analyze the role that religion has played in events in Southwest Asia.
3. Identify the political, social, and environmental challenges facing the countries of Southwest Asia.
4. Understand the significance of the region's strategic location and natural resources.
5. Use special-purpose maps to interpret the relationships among climate, population patterns, and economic resources in Southwest Asia.

UNIT 5

Southwest Asia

CHAPTER 17

The Arabian Peninsula, Iraq, Iran, and Afghanistan

CHAPTER 18

The Eastern Mediterranean

Western Wall and Dome of the Rock, Jerusalem, Israel

Bottle trees, Yemen

CLOSE

Ask students to describe the characteristics of monarchy as a form of gov-ernment. Elicit opinions as to which form students would expect to be more responsive to public opinion—democracy or monarchy?

REVIEW AND ASSESS

Have students complete the Section Review. Then have them work in groups to reread the section silently and then discuss it among themselves. Then have students complete Daily Quiz 17.2.
COOPERATIVE LEARNING

RETEACH

Have students complete Main
groups. Assign each group one of
each group devise symbols for its co
make a flag incorporating the symbols.
ENGLISH LANGUAGE LEARNERS, CO

EXTEND

Have interested students conduct research on the
1973. Have them use their findings to write a report
immediate and long-term economic effects of the embar
Asia and the United States. **BLOCK SCHEDULING**

Yemen Yemen is located on the southern corner of the Arabian Peninsula. It borders the Red Sea and the Gulf of Aden. Yemen was formed in 1990 by the joining of North Yemen and South Yemen. The country has an elected government and several political parties. However, political corruption and internal conflicts have threatened this young democracy.

In ancient times, farmers in this area used very advanced methods of irrigation and farming. Yemen was famous for its coffee. Today, Yemen is the poorest country on the Arabian Peninsula. Oil was not discovered there until the 1980s. It now generates a major part of the national income.

✓ **READING CHECK:** *Human Systems* How are the governments and economies of the countries of the Arabian Peninsula organized? All but Yemen are monarchies. Oil is a major part of the economies.

An important part of Yemen's culture is its distinctive architecture, which features tall buildings and carved wooden windows. ▶

Section 2 Review

Answers

Define For definitions, see: Muslims, p. 388; caliph, p. 388; Sunni, p. 388; Shia, p. 388; Qur'an, p. 389; OPEC, p. 389

Working with Sketch Maps Riyadh is more centrally located.

Reading for the Main Idea
1. monarchy (NGS 4)
2. most important export (NGS 15)

Critical Thinking
3. brings people together for a common reason
4. needed for work in oil industry

Organizing What You Know
5. S. Arabia—Arabian Peninsula, monarchy, oil; Kuwait—between S. Arabia and Iraq, monarchy, oil; Bahrain—Persian Gulf islands, monarchy, oil, banking, tourism; Qatar—peninsula on Persian Gulf, monarchy, oil and gas; UAE—on Persian Gulf, monarchy, oil and gas; Oman—on Arabian Sea, monarchy, oil; Yemen—on Arabian Peninsula, elected government, farming and oil

◀ **Visual Record Answer**
sunny; rocky and barren

Section 2 Review

go.hrw.com Homework Practice Online
Keyword: SK3 HP17

Define and explain: Muslims, caliph, Sunni, Shia, Qur'an, OPEC

Working with Sketch Maps On the map you created in Section 1, label Mecca and Riyadh. Why is Riyadh's location more suitable than Mecca's to be the capital city of Saudi Arabia?

Reading for the Main Idea

1. *Places and Regions* What kind of government does Saudi Arabia have?
2. *Environment and Society* What role does oil play in the economies of small countries on the Arabian Peninsula?

Critical Thinking

3. **Drawing Inferences and Conclusions** How might trips to Mecca help create a sense of community among Muslims?
4. **Drawing Inferences and Conclusions** Why would a country like Saudi Arabia or the United Arab Emirates bring in large numbers of foreign workers?

Organizing What You Know

5. **Comparing** Copy the following graphic organizer. Use it to list these countries' locations, governments, and economies. Place them in the appropriate part of the chart.

	Location	Government	Economy
Saudi Arabia			
Kuwait			
Bahrain			
Qatar			
UAE			
Oman			
Yemen			

...rtance of oil in the world today. ...of oil are shipped from producers to ...ankers and pipelines. The discovery and ...have a dramatic effect on a country's ...nal politics. A good example of such an economic transformation is Saudi Arabia. In less than 100 years, ...ing one of the world's poorest countries to one of ...direct result of the discovery and exploitation of its ...eum accounts for 90 percent of Saudi Arabia's exports. ...ealth has had far-reaching implications for Saudi Arabia's

Building a Case

After students have read "Saudi Arabia: How Oil Has Changed a Country," engage them in a discussion by asking the following questions. What was life like for most Saudis before the discovery of oil? *(They herded animals, farmed and fished.)* When was oil discovered? *(In the mid-1930s)* How did oil wealth affect education in Saudi Arabia? *(New schools and universities were built. More people had access to education, and the literacy rate improved.)* How did the oil industry change the population of Saudi Arabia? *(Resulted in the presence of foreign workers, including Yemeni, Egyptians, Palestinians, Syrians, and others)* How did oil wealth change Saudi Arabia's relations with other countries? *(OPEC gave it greater influence over international affairs. Its military power increased. Foreign aid donations increased.)*

HISTORICAL GEOGRAPHY

The Persian Gulf region is not the only place to have been transformed by oil wealth. At the beginning of the 1900s, prior to the discovery of oil in Saudi Arabia, Azerbaijan produced half of the world's oil. As a result, Baku, the capital of Azerbaijan, developed into a beautiful, prosperous city. Wealthy European industrialists left behind palatial mansions and lush parks.

Recent estimates suggest that more oil exists in the region. Some believe as much as 200 billion barrels worth of oil may still be found under the Caspian Sea near Baku. In 1994 several international oil companies signed an agreement to begin drilling in the area.

Activity: Have students conduct research on the history of the Baku oil fields. Then have students predict how Baku might be affected if indeed more oil is found under the Caspian Sea. How might the architecture, transportation, education, communications, and political influence of the region be changed?

➤ **This Case Study feature addresses National Geography Standards 9, 11, 12, and 16.**

CASE STUDY

SAUDI ARABIA: HOW OIL HAS CHANGED A COUNTRY

At the beginning of the 1900s, Saudi Arabia was a poor country. Most people followed traditional ways of life and lived by herding animals, farming, or fishing. There were few good roads or transportation systems, and health care was poor.

Major Oil Fields in Saudi Arabia

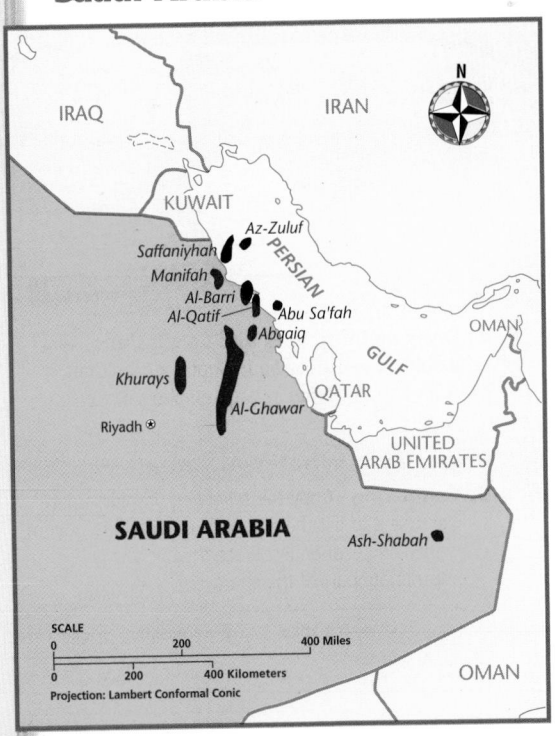

▲ Saudi Arabia's huge oil fields contain about one quarter of the world's known oil reserves.

Then, in the mid-1930s, oil was discovered in Saudi Arabia. Soon, even larger oil reserves were found. Eventually it was learned that Saudi Arabia had the largest oil reserves in the world. Most of the country's oil reserves are located in eastern Saudi Arabia along the Persian Gulf coast. The Al-Ghawar oil field, discovered in 1948, is the largest in the world. In addition to this field, Saudi Arabia has at least nine other major oil fields. The discovery of these huge oil reserves changed Saudi Arabia's economy and society.

Rising income from oil exports gave Saudi Arabia's government more and more money to invest. New airports, apartments, communications systems, oil pipelines, and roads were built. In 1960 Saudi Arabia had about 1,000 miles (1,600 km) of roads. By 1997 it had about 91,000 miles (145,600 km). Saudi Arabia's cities grew as people left small villages. These changes helped modernize Saudi Arabia's economy.

Saudi society was also affected. The standard of living rose, and people had more money to spend on goods. Foreign companies began to open stores and restaurants in Saudi Arabia. New schools were built throughout the country, and education became available to all citizens. The literacy rate increased from about 3 percent to about 63 percent. Thousands of Saudis traveled to other countries, and many new universities were opened. Health care also improved.

As Saudi Arabia's economy grew, many foreign workers came to the country to work. In the early 2000s foreign workers made up about 25 percent of Saudi Arabia's population. These workers included people from Yemen, Egypt, Palestine, Syria, Iraq, South Korea, and the Philippines. Americans and Europeans also went to Saudi Arabia to work in the oil industry.

The development of the oil industry greatly increased Saudi Arabia's importance in the world. In 1960 Saudi Arabia was a founding member of

Drawing Conclusions

Ask students to consider the future prospects of Saudi Arabia. Emphasize that oil is a limited resource and will not provide a source of wealth indefinitely. Some facts to consider: the world's estimated oil reserves are being used up. If current rates of consumption continue, shortages will occur in the mid-2000s. Point out that many of the investments the Saudis are making will have long term benefits—specifically improved education, communication, and transportation systems. Explain that agricultural development is very important to the Saudi government. Major irrigation projects financed by oil profits were constructed in the 1970s to decrease dependence on imported food. From 1976 to 1993, the amount of cultivated land increased so much that Saudi Arabia had become a food exporter by the mid-1990s.

Going Further: Thinking Critically

The discovery of precious metals can also have a major impact on a region. Gold, diamonds, and silver have all had significant impacts on different parts of the world. Provide students with a global map of mineral resources and one of the following: a set of encyclopedias, a computer with access to the Internet, or an almanac that includes brief country descriptions. Then organize them into groups and have them answer the following questions:

- What countries have deposits of diamonds, gold, and silver?
- When were these resources discovered?
- How did the discovery of resources affect settlement of the area?
- How much of the country's economy is based on these resources?

Oil is refined in eastern Saudi Arabia. Then it is pumped into ships and exported to countries around the world.

the Organization of Petroleum Exporting Countries (OPEC). Members of OPEC try to influence the price of oil on the world market. Saudi Arabia's huge oil reserves make it one of OPEC's most powerful member countries.

The Saudi government has also bought military equipment from the United States and other countries. This military strength has increased Saudi Arabia's importance in the Persian Gulf region. Since 1991, Saudi Arabia has given large sums to countries that sided with it against Iraq in the Persian Gulf War. Egypt, Turkey, and Syria have especially benefited from this aid.

Today, Saudi Arabia is a wealthy country. This wealth has come almost entirely from the sale of oil. Saudi Arabia is currently the world's leading oil exporter. Oil provides about 90 percent of the government's export earnings. Saudi Arabia exports oil to Japan, the United States, South Korea, and many other countries.

Since the mid-1960s Saudi Arabia has expanded its transportation network. A modern road system now connects many parts of the country.

Understanding What You Read

1. What was Saudi Arabia like before oil was discovered there?
2. How has oil changed Saudi Arabia's economy and society?

Understanding What You Read

Answers

1. The people of the region depended primarily on herding animals, farming, and fishing. Few roads existed. Health care was poor. Literacy rates were lower.
2. New infrastructure like roads, airports, schools, and communications facilities has been built. The standard of living rose and literacy rates have improved. Foreign workers have become a significant portion of the population.

Section 3

Objectives

1. Relate the key events in Iraq's history.
2. Examine Iraq's government and economy.
3. Describe the makeup of Iraq's population.

Focus

LET'S GET STARTED

Copy the following questions onto the chalkboard: *Do you know anyone who fought in the Persian Gulf War? What have you heard about their experiences there?* Discuss responses. Tell students that the United States and Iraq went to war in 1990 and that in Section 3 they will learn more about Iraq's history and culture.

Building Vocabulary

Write the key term **embargo** on the chalkboard. Ask a volunteer to read aloud the definition from the text or glossary. Point out the Spanish derivation of the word; it contains "bar" from *barra,* which means a bar that can be put across something. Ask students to discuss the relationship between the word history and the definition.

Section 3 **Resources**

Reproducible
- Guided Reading Strategy 17.3
- Environmental and Global Issues Activity 5

Technology
- One-Stop Planner CD–ROM, Lesson 17.3
- Homework Practice Online
- HRW Go site

Reinforcement, Review, and Assessment
- Section 3 Review, p. 398
- Daily Quiz 17.3
- Main Idea Activity 17.3
- English Audio Summary 17.3
- Spanish Audio Summary 17.3

Section 3 Iraq

Read to Discover

1. What were the key events in Iraq's history?
2. What are Iraq's government and economy like?
3. What is the makeup of Iraq's population?

Define

embargo

Locate

Baghdad

Why It Matters

The United States has used both economic and military action against Iraq in the 1990s and early 2000s. Use CNNfyi.com or other **current events** sources to learn about U.S. relations with Iraq today. Record your findings in your journal.

A Sumerian clay tablet

Murals and posters of Saddam Hussein can be found throughout Baghdad, Iraq's capital.

History

The history of Mesopotamia, an ancient region in Iraq, stretches back to some of the world's first civilizations. The Sumerian, Babylonian, and Assyrian cultures arose in this area. The Persians conquered Mesopotamia in the 500s B.C. Alexander the Great made it part of his empire in 331 B.C.

In the A.D. 600s the Arabs conquered Mesopotamia. The people of the area gradually converted to Islam. The city of Baghdad became one of the world's greatest centers of trade and culture. However, it was destroyed by the Mongols in 1258.

In the 1500s Mesopotamia became part of the Ottoman Empire. During World War I it was taken over by the British. In 1932 the British set up the kingdom of Iraq and placed a pro-British ruler in power. In the 1950s a group of Iraqi army officers overthrew this government. After several more changes in government, the Ba'ath Party took power in 1968.

A Ba'ath leader named Saddam Hussein slowly gained more power. In the late 1970s he became president of Iraq and leader of the armed forces.

In 1980 Iraq invaded Iran. Saddam Hussein hoped to take advantage of the confusion following the Iranian Revolution. The Iranians fought back, however, and the Iran-Iraq War dragged on until 1988. Hundreds of thousands of people were killed on both sides. Each side tried to attack the other's oil tankers and refineries. Both countries' economies were damaged.

In 1990, Iraq invaded the small, oil-rich country of Kuwait. Saddam Hussein used an old claim that Kuwait was part of Iraq to justify this

Chapters

17 The Arabian Peninsula, Iraq, Iran, and Afghanistan
explores the connections among natural resources, climate, and human activity. It concentrates on the religious heritage of interior Southwest Asia and the role of oil in the region.

18 The Eastern Mediterranean
describes the landforms, climates, economies, resources, and histories of Turkey, Israel and the Occupied Territories, Syria, Lebanon, and Jordan.

Notes from the Field

An Exchange Student in Turkey

Sara Lewis was an American exchange student in Turkey. Here she describes how teenagers live in Istanbul, Turkey's largest city, and the month-long fast of Ramadan. **WHAT DO YOU THINK?** *What would it be like to live in a place where you can see the remains of thousands of years of history?*

The people of Istanbul are very traditional and family-oriented, but today's Turk also has European-style tastes. Turkish teens go dancing and hang out in coffeehouses. All around them, though, are reminders of the past. There are many monuments left over from Greek, Roman, Byzantine, and Ottoman times.

Islam plays a big part in daily life. Five times every day I can hear the people being called to prayer. For more than a month, my host parents fasted during Ramadan. They didn't eat or drink anything while the sun was up. My host sister and I fasted for one day. By the time the sun went down we were starving! I'm glad we weren't expected to continue the fast. Then we shopped for new clothes. It is the custom to wear new clothes to the feast at the end of Ramadan. The fresh clothes seem to stand for the cleanliness one achieves during the month of fasting.

Family preparing food for the end of Ramadan

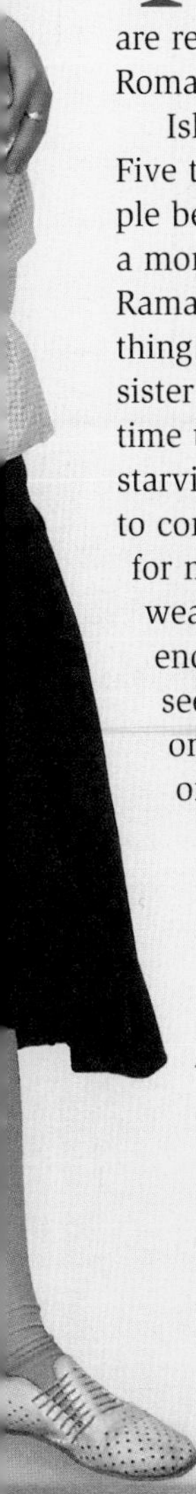

Jewish girls, Zefat, Israel

Understanding Primary Sources

1. What do modern Turkish teenagers do for fun?
2. How is Ramadan observed in Turkey?

Sooty falcon

More from the Field

Islam teaches that Muslims should fast during Ramadan, the ninth month of the Islamic calendar. Muslims fast during Ramadan for several reasons. Fasting is said to help people focus on their spiritual lives, to teach self-control, and to promote compassion for the poor.

When Ramadan ends, Muslims celebrate 'Eid-ul-Fitr, or the Festival of Fast-Breaking. People wear their best clothes and attend morning prayers. People wish each other *Eid mubarak*—"a blessed 'Eid". It is customary to give to charity. Often children receive gifts.

Activity: Ask students to find information about how Muslims in your community or state observe Ramadan. Have them write newspaper articles to describe local observances.

internet connect

GO TO: **go.hrw.com**
KEYWORD: **SK3 CH17**
FOR: **Web sites about Ramadan**

Understanding Primary Sources

Answers

1. go dancing and hang out in coffeehouses
2. through fasting, wearing new clothes to the feast

UNIT 5 ATLAS

OVERVIEW

In this unit, students will learn about the geography, ancient history, and rapid modernization of the region often referred to as the Middle East.

Some of the world's earliest civilizations developed in the Fertile Crescent, in what is now Iraq. Wave after wave of conquerors have invaded the region. In the A.D. 600s Islam spread from the Arabian Peninsula across much of the area. Tensions between major Islamic groups date back to the religion's early years and continue today.

Two other major faiths—Judaism and Christianity—also originated in Southwest Asia. Over the centuries there have been many conflicts among followers of the three religions.

Arid and semiarid climates dominate Southwest Asia. Farming is possible only in limited areas, so nomadic herding is common. Overall population density is low.

The discovery of immense oil reserves in the early 1900s changed the region's economy dramatically. Some countries, particularly those along the Persian Gulf, have become very wealthy. Oil wealth has brought political power to several Southwest Asian countries. However, many people still live in poverty.

PEOPLE IN THE PROFILE

The elevation profile crosses northern Saudi Arabia and southern Iraq, homeland of the nomadic Bedouins. A traditional Bedouin home is a long, low black tent made of woven goat hair. There are few household items besides carpets, utensils, and a portable stove. Long robes of thick material, a head covering, and sandals form the basic wardrobe. Milk products, dried fruit, and grains from villagers are staple foods.

In the 1900s life began to change for the Bedouin. The countries where they wandered exercised more control over their movement. The Bedouin had to reduce their raids and pursue peaceful commercial relationships. Many settled in urban areas. Airplanes, telephones, trucks, and other by-products of oil wealth continue to change Bedouin ways.

Critical Thinking: How did the Bedouin belief system affect their use of technology, and how has this changed as a result of their relationship with surrounding cultures?

Answer: Bedouins were nomads who did not use technology until countries where they lived began to control their movements and influence their use of technology.

UNIT 5 ATLAS The World in Spatial Terms

Southwest Asia

Elevation Profile

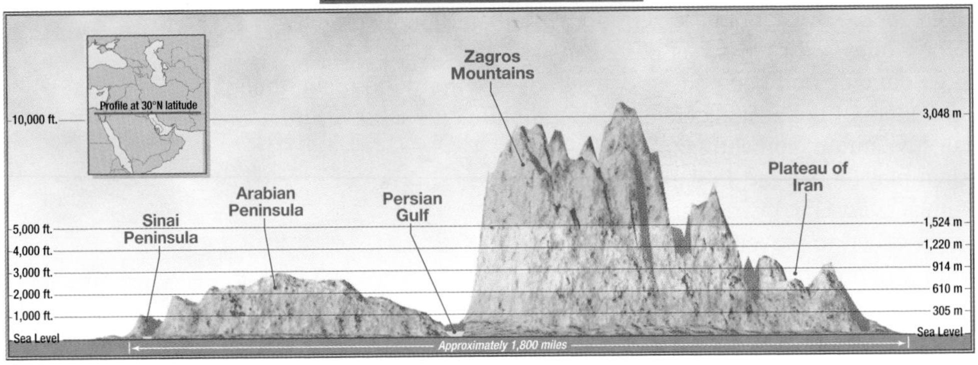

The United States and Southwest Asia: Comparing Sizes

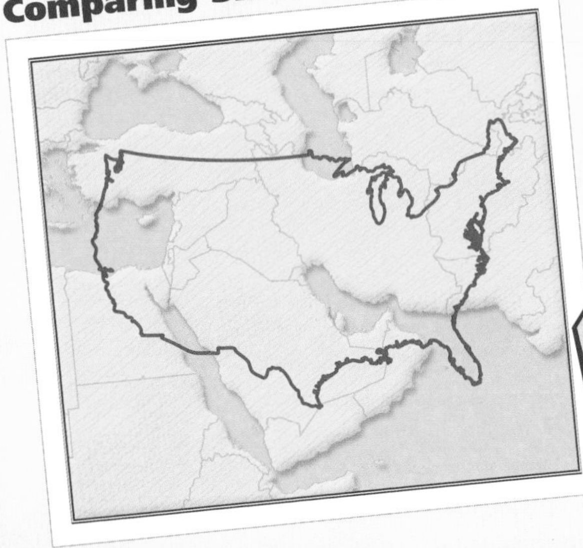

GEOSTATS:

- World's lowest point on land: the Dead Sea, in Israel and Jordan–1,312 feet (400 m) below sea level
- World's leading exporter of oil: Saudi Arabia
- Approximate amount of proven oil reserves in Saudi Arabia: 261 billion barrels
- Estimated number of barrels of oil that pass through the Strait of Hormuz every day: 15.4 million

USING THE PHYSICAL MAP

Focus students' attention on the **physical map** on this page. Most of Southwest Asia lacks certain physical features that are important to other places the class has studied. What are these features? *(major rivers and lakes)* Call on a volunteer to point out where in Southwest Asia the lack of water is most severe. *(Arabian Peninsula)* Ask students to suggest what residents of the region might do to increase their access to water. You may want to remind students about desalinization plants at this point.

Southwest Asia: Physical

UNIT 5 ATLAS

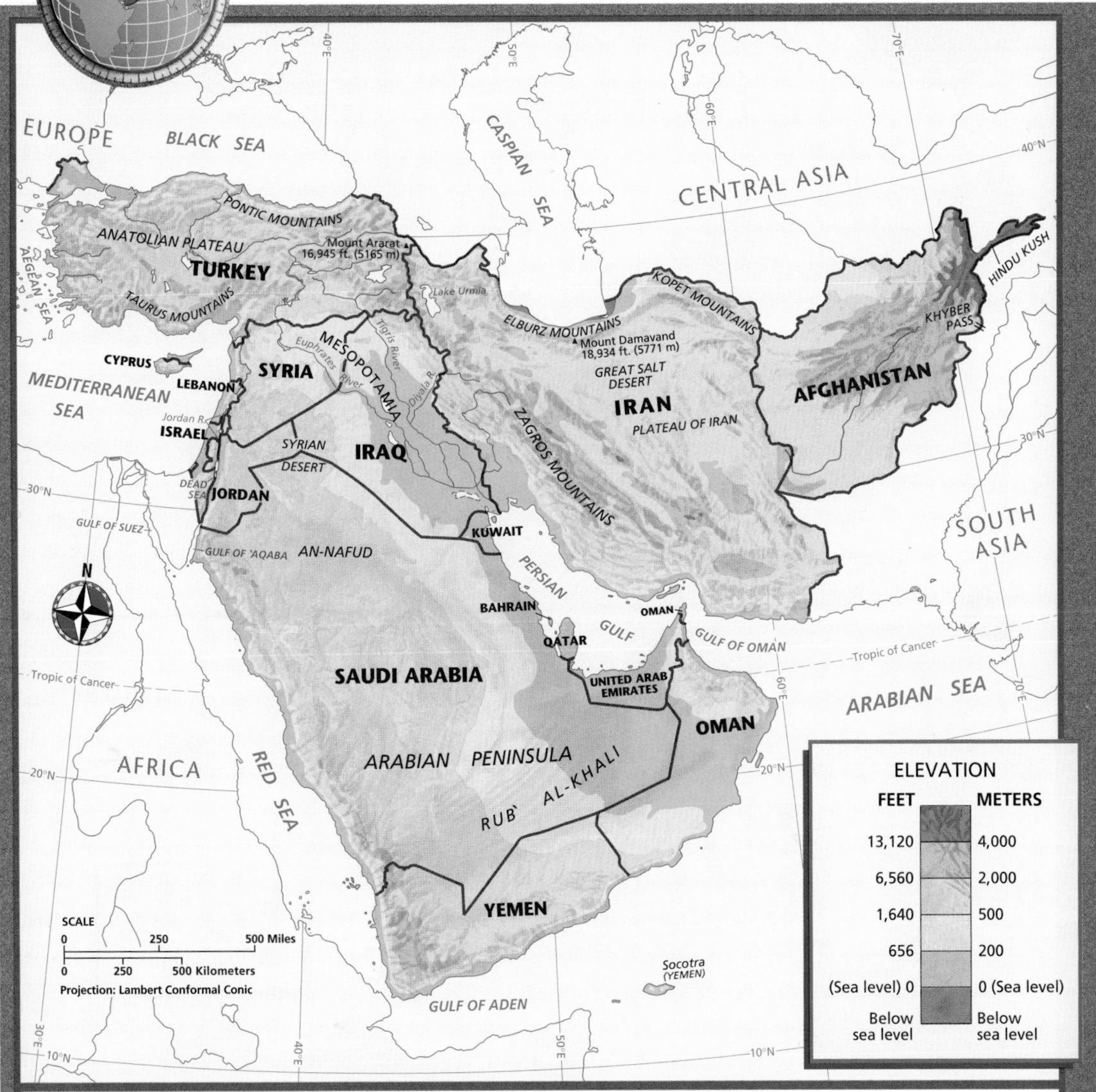

1. Environment and Society What landforms might make north-south travel difficult in Iran?
2. Places and Regions Where are the region's highest mountains? What are these mountains called?
3. Places and Regions Which country is partly in Europe and partly in Asia?

Critical Thinking

4. **Analyzing Information** How might one travel overland from Syria to Oman? Why?
5. **Comparing** Compare this map to the **population map**. What physical features contribute to Iraq's relatively high population density?

internet connect

ONLINE ATLAS
GO TO: go.hrw.com
KEYWORD: SK3 MapsU5
FOR: Web links to online maps of the region

Physical Map

Answers

1. Zagros Mountains, Great Salt Desert
2. Afghanistan; Hindu Kush
3. Turkey

Critical Thinking

4. down the Euphrates River and through the Persian Gulf and the Gulf of Oman; avoids deserts
5. the Tigris and Euphrates Rivers, a large area of open plains

UNIT 5 ATLAS

Using the Political Map

While students focus on the **political map** on this page, ask which three continents meet at Southwest Asia. *(Europe, Asia, Africa)* How might this location have affected the region's history and culture? *(invasions, trade, and ideas from many directions)* Which small countries border the Persian Gulf? *(Kuwait, Bahrain, Qatar, United Arab Emirates, Oman)* Which small countries border the Mediterranean Sea? *(Syria, Lebanon, Israel)* What nearby small country is landlocked? *(Jordan)* Why might all these countries want to get or keep access to the sea? *(for trade, travel, communication with other countries)*

internet connect
go.hrw.com
ONLINE ATLAS
GO TO: go.hrw.com
KEYWORD: SK3 MapsU5

Your Classroom Time Line

These are the major dates and time periods for this unit. Have students enter them on the time line you created earlier. You may want to watch for these dates as students progress through the unit.

500s B.C. The Persian Empire is established; Persians conquer Mesopotamia.

330s B.C. Alexander the Great conquers Asia Minor.

60s B.C. The Roman Empire conquers Palestine.

c.* 610 Muhammad begins to preach the religion of Islam.

*c. stands for *circa* and means "about."

Political Map

Answers

1. Saudi Arabia and Iran; Bahrain

Critical Thinking

2. because in the desert there are few natural features that would help define boundaries; because it is in the desert and may be unmarked or may be in dispute
3. Iran, Iraq, Kuwait, Saudi Arabia, Bahrain, Qatar, United Arab Emirates, and Oman; maintaining access to the Arabian Sea through these important waterways

Southwest Asia: Political

UNIT 5 ATLAS

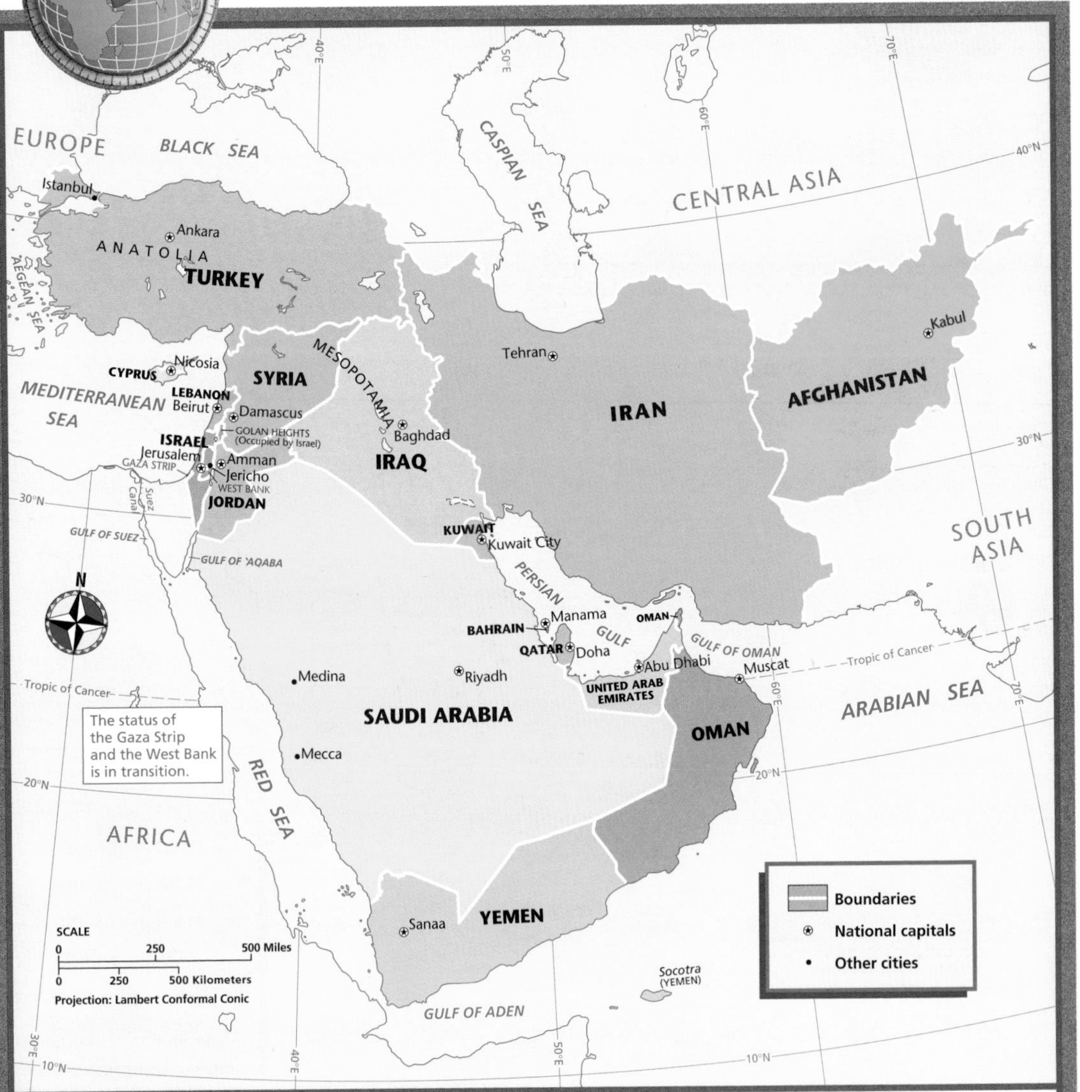

1. **Places and Regions** What are the region's two largest countries? the region's smallest?

Critical Thinking

2. **Comparing** Examine the **climate map**. Why do you think so many of the boundaries in this region are straight lines? Why might some maps show Saudi Arabia's southern border as a dotted line?
3. **Drawing Inferences and Conclusions** Which countries lie along the Persian Gulf and the Gulf of Oman? Why might conflicts occur among these countries?

Using the Climate Map

Have students look at the **climate map** of Southwest Asia on this page. Ask them to list the region's climate types in the order of the amount of area each one appears to cover *(desert, steppe, Mediterranean, highland, humid subtropical)*. Refer students to the **political map** on the preceding page and ask which countries they would choose to live in or visit based solely on climate.

Southwest Asia: Climate

UNIT 5 ATLAS

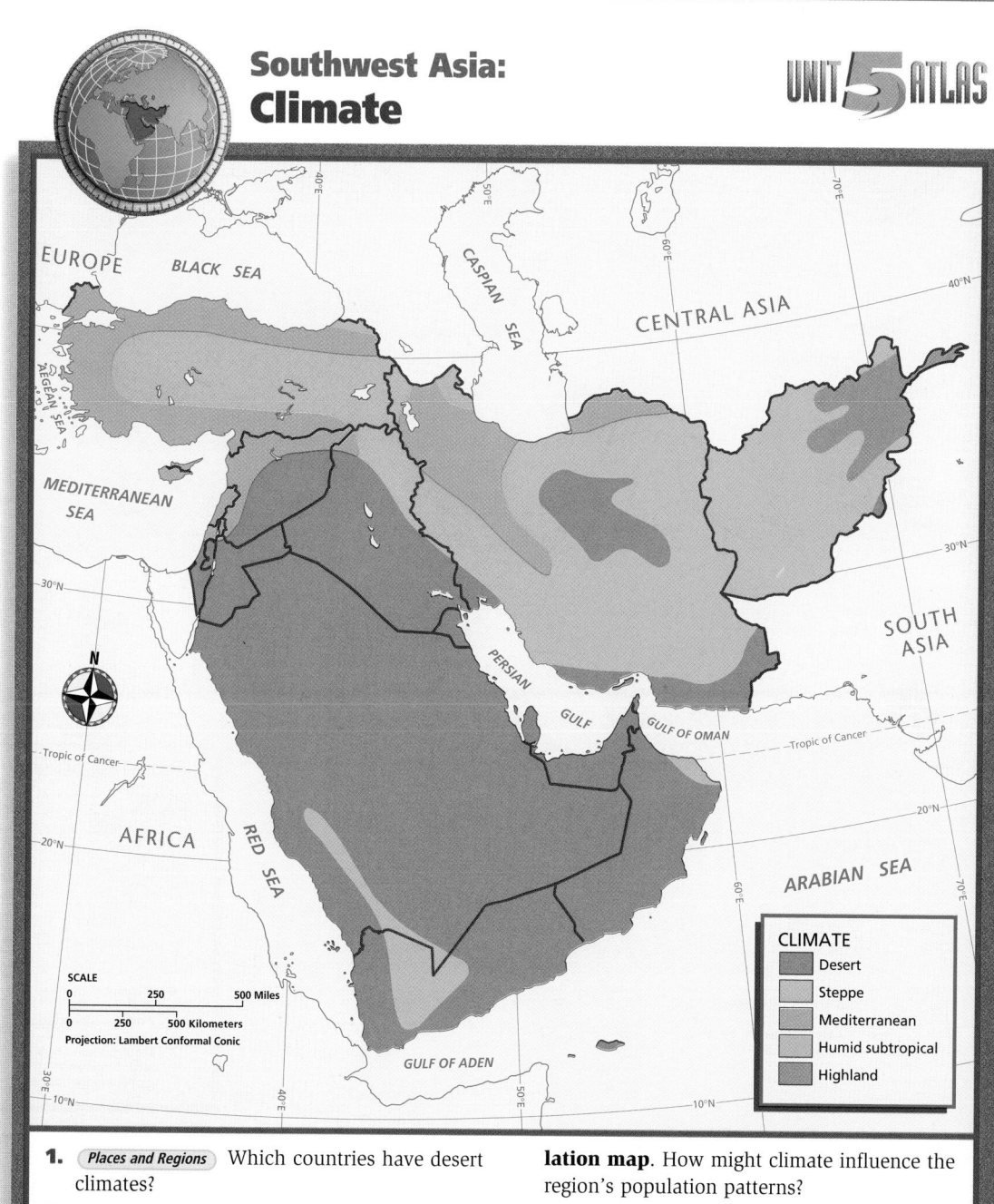

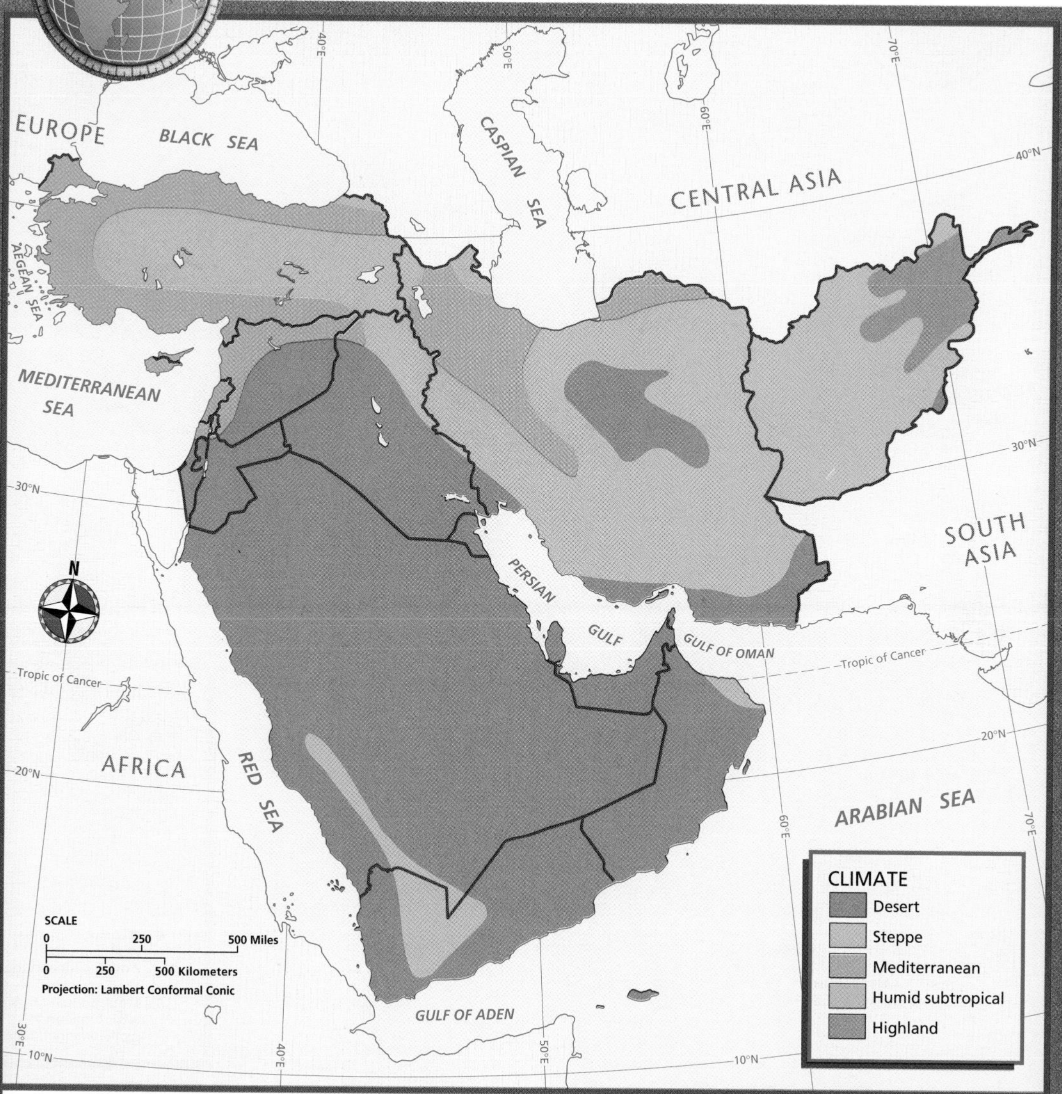

1. *Places and Regions* Which countries have desert climates?

2. *Places and Regions* Which two countries have only a Mediterranean climate?

Critical Thinking

3. **Comparing** Compare this map to the **population map**. How might climate influence the region's population patterns?

4. **Analyzing Information** Compare this map to the **physical** and **land use and resources maps**. Why is commercial farming possible in central Iraq?

Your Classroom Time Line, (continued)

- A.D. 600s Arabs conquer Mesopotamia and Palestine.
- 1000s The Seljuk Turks invade Asia Minor.
- 1000s–1200s The Crusades occur.
- 1099 Crusaders capture Jerusalem.
- 1258 Mongols destroy Baghdad.
- 1500s Mesopotamia becomes part of the Ottoman Empire.
- 1700s Kuwait is established.
- late 1800s The Zionist movement begins.
- 1925–1979 Shahs rule Iran.
- 1930 Constantinople is renamed Istanbul.
- 1930s Oil is discovered on the Arabian Peninsula.
- 1932 The kingdom of Saudi Arabia is founded.

Climate Map

Answers

1. All the countries have desert climates except Cyprus, Lebanon, and Turkey.
2. Cyprus, Lebanon

Critical Thinking

3. Desert areas have low population densities.
4. irrigation provided by rivers

☑ internet connect

ONLINE ATLAS
GO TO: go.hrw.com
KEYWORD: SK3 MapsU5

USING THE POPULATION MAP

Have students examine the **population map** of Southwest Asia on this page. Ask a volunteer to compare the map to the **physical map** of the region and to name the most densely populated country in the region. *(Israel)* Ask students to point out countries in the region with large areas that are sparsely populated. *(Saudi Arabia, Jordan, Syria, Iraq, Iran, Afghanistan, Yemen, Oman)*

Your Classroom Time Line, (continued)

1932 The kingdom of Iraq is founded.

1940s Syria, Lebanon, and Jordan gain independence.

1948 The State of Israel is created.

1948–74 Israel and its Arab neighbors fight several wars.

1950s Iraqi army officers overthrow the government.

1952–99 King Hussein rules Jordan.

1968 The Ba'ath Party gains power in Iraq.

1970s A civil war begins in Lebanon.

1979 The Iranian Revolution overthrows Iran's shah.

late 1970s Saddam Hussein becomes president of Iraq.

1979 The Soviet Union invades Afghanistan.

Population Map

Answers

1. Turkey
2. Rub` Al-Khali
3. Yemen

Critical Thinking

4. Charts, graphs, databases, and models should reflect population figures on the map.

5. Trade from the Black Sea and areas to the north has to pass through Istanbul on its way to the Mediterranean.

Southwest Asia: Population

POPULATION DENSITY

Persons per sq. mile	Persons per sq. km
520	200
260	100
130	50
25	10
3	1
0	0

● Metropolitan areas with more than 2 million inhabitants

○ Metropolitan areas with 1 million to 2 million inhabitants

SCALE
0 250 500 Miles
0 250 500 Kilometers
Projection: Lambert Conformal Conic

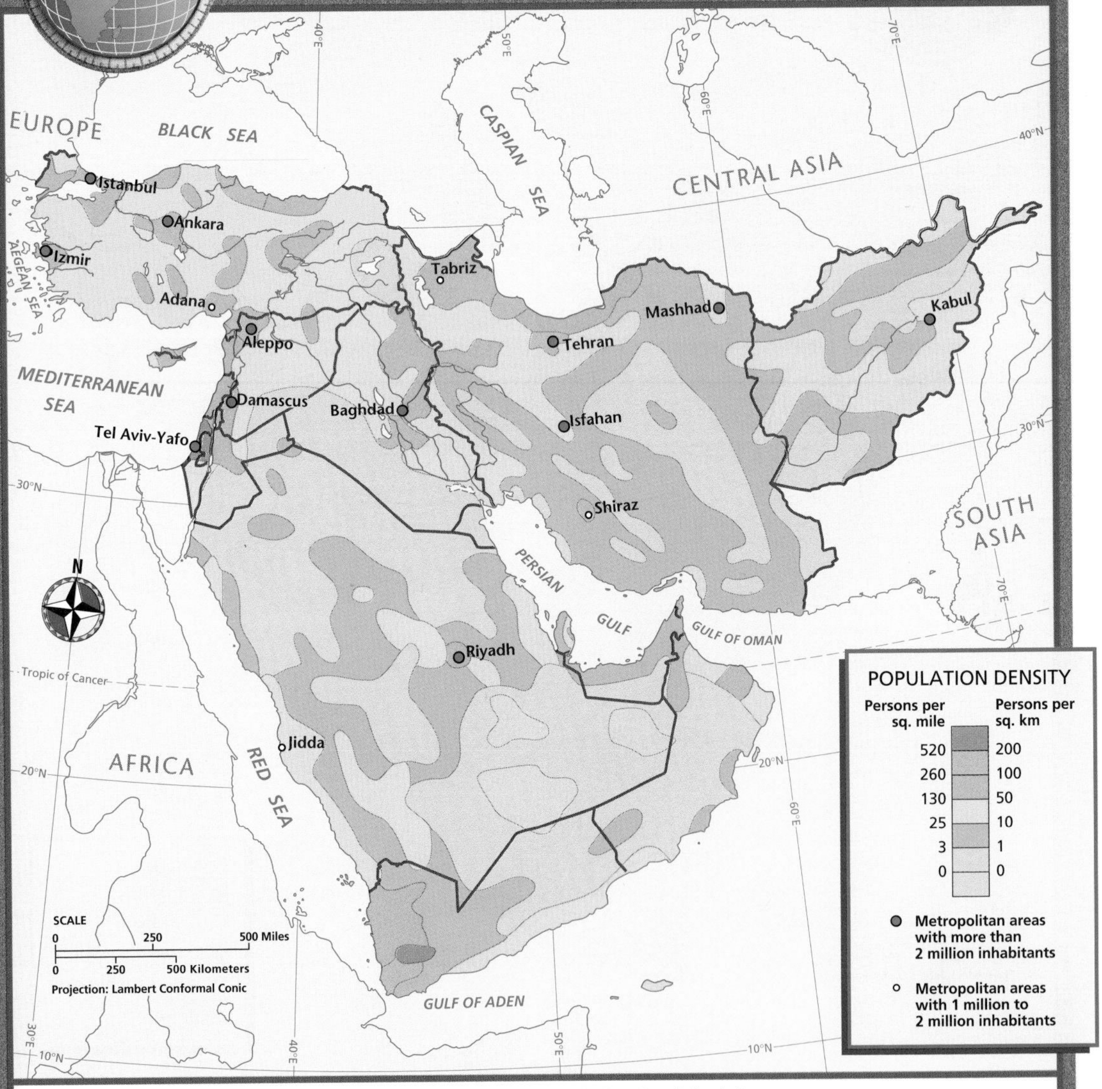

1. (Places and Regions) Which country has two cities of more than 2 million people?

2. (Places and Regions) Compare this map to the **physical map** of the region. Which desert area has almost no residents?

3. (Places and Regions) Which country has an area of high population density but no big cities?

Critical Thinking

4. Analyzing Information Use this map to create a chart, graph, database, or model of population centers in Southwest Asia.

5. Analyzing Information Look at Istanbul's location. Why is it good for a large city?

Using the Land Use and Resources Map

Ask students to look at the **land use and resources map** on this page. Have them identify the economic activities and resources that seem to dominate Southwest Asia. *(nomadic herding, oil)* Refer students to the **climate map** and ask how the region's harsh climate affects these two economic activities. *(limits nomadic herding to some extent, has no effect on oil industry)* Also, have students compare this map to the **population map** of the region and identify the type of farming that takes place in the areas with the greatest population density. *(commercial farming)*

Southwest Asia: Land Use and Resources

UNIT 5 ATLAS

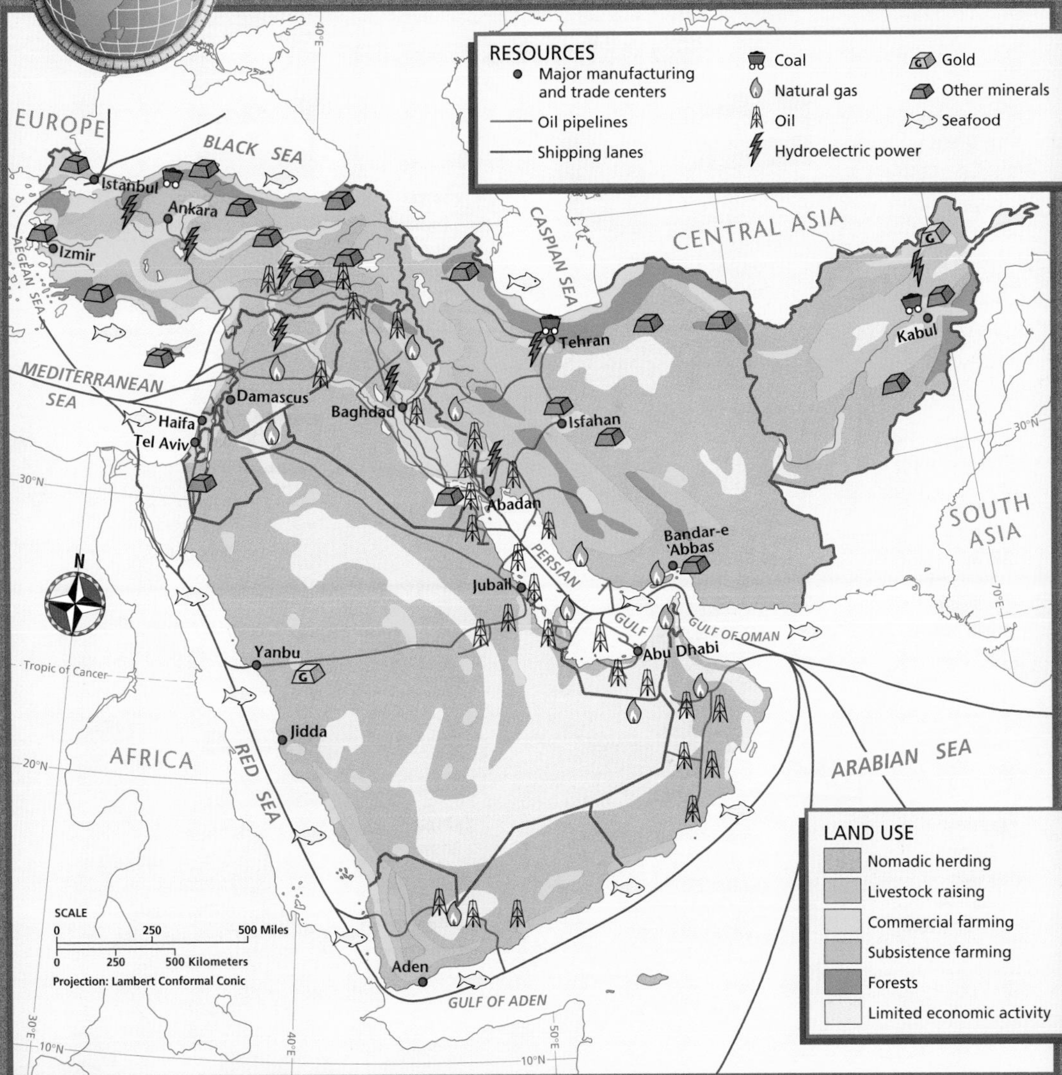

1. *Human Systems* By what route would an oil tanker travel from Abadan to Mediterranean ports?
2. *Places and Regions* What two energy resources are often found together in the region?
3. *Places and Regions* Which countries have gold? Which countries have deposits of coal?

Critical Thinking

4. **Analyzing Information** Use this map to create a chart, graph, database, or model of economic activities in Southwest Asia.
5. **Drawing Inferences and Conclusions** Why might vegetables be costly in Kuwait?

Your Classroom Time Line, (continued)

- 1980 Iraq invades Iran.
- 1989 Soviet troops withdraw from Afghanistan.
- 1990 Iraq invades Kuwait.
- 1990 Yemen is founded.
- 1991 Persian Gulf War ends.
- 1990s Israel agrees to turn over parts of the Occupied Territories.
- mid-1990s The Taliban rise to power in Afghanistan.
- 2001 United States attacks Taliban and terrorist targets in Afghanistan.

Land Use and Resources Map

Answers

1. southeast through Persian Gulf and Gulf of Oman to Arabian Sea, southwest to Gulf of Aden, northwest through Red Sea, and through Suez Canal to Mediterranean Sea
2. natural gas and oil
3. Afghanistan, Saudi Arabia; Afghanistan, Iran, Turkey

Critical Thinking

4. Charts, graphs, databases, and models should accurately reflect information on the map.
5. no farmland indicated on map; fruits, grains, vegetables would have to be imported, driving up prices

SOUTHWEST ASIA

FAST FACTS ACTIVITIES

LEVEL 1: (Suggested time: 25 min.) Have students calculate the number of telephones per person in two different countries in Southwest Asia. *(Divide the number of telephones by the number of people.)* Ask students to present each answer as a whole-number ratio of telephones to people in the form of a sentence. *(Example: In Cyprus there is one telephone for every two people.)*

Then challenge students to use the data in the tables on the following pages to identify the relationship between income and telephone ownership in their selected countries. *(Determine if countries with a higher per capita GDP tend to have more phones per person.)*

LEVEL 2: (Suggested time: 35 min.) Have the students refer to the **political map** and **land use and resources map** in this unit's atlas to identify two countries whose economies depend on oil production and two that depend more on other resources. Have students use the tables on the following pages to create a bar graph showing the per capita GDP of citizens in their selected countries. What can they conclude?

UNITED STATES OF AMERICA

CAPITAL:
Washington, D.C.
AREA:
3,717,792 sq. mi.
(9,629,091 sq km)
POPULATION:
281,421,906
MONEY:
U.S. dollar (US$)
LANGUAGES:
English, Spanish (spoken by a large minority)
TELEPHONES:
194,000,000 (1997 estimate)

Southwest Asia

AFGHANISTAN

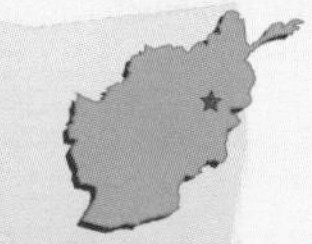

CAPITAL: Kabul
AREA: 250,000 sq. mi. (647,500 sq km)
POPULATION: 26,813,057
MONEY: afghani (AFA)
LANGUAGES: Pashtu, Afghan Persian (Dari), Turkic languages
TELEPHONE LINES: 29,000 (1996)

IRAQ

CAPITAL: Baghdad
AREA: 168,753 sq. mi. (437,072 sq km)
POPULATION: 23,331,985
MONEY: Iraqi dinar (IQD)
LANGUAGES: Arabic, Kurdish
TELEPHONE LINES: 675,000 (1997)

BAHRAIN

CAPITAL: Manama
AREA: 239 sq. mi. (620 sq km)
POPULATION: 645,361
MONEY: Bahraini dinar (BHD)
LANGUAGES: Arabic, English, Farsi, Urdu
TELEPHONE LINES: 152,000 (1997)

ISRAEL

CAPITAL: Jerusalem
AREA: 8,019 sq. mi. (20,770 sq km)
POPULATION: 5,938,093
MONEY: new Israeli shekel (ILS)
LANGUAGES: Hebrew (official), Arabic, English
TELEPHONE LINES: 2,800,000 (1999)

CYPRUS

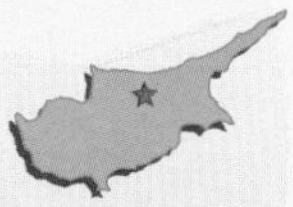

CAPITAL: Nicosia
AREA: 3,571 sq. mi. (9,250 sq km)
POPULATION: 762,887
MONEY: Cypriot pound (£C), Turkish lira (TRL)
LANGUAGES: Greek, Turkish, English
TELEPHONE LINES: Greek Cypriot area: 405,000 (1998); Turkish Cypriot area: 83,162 (1998)

JORDAN

CAPITAL: Amman
AREA: 35,637 sq. mi. (92,300 sq km)
POPULATION: 5,153,378
MONEY: Jordanian dinar (JOD)
LANGUAGES: Arabic, English
TELEPHONE LINES: 403,000 (1997)

IRAN

CAPITAL: Tehran
AREA: 636,293 sq. mi. (1,648,000 sq km)
POPULATION: 66,128,965
MONEY: Iranian rial (IRR)
LANGUAGES: Persian, Turkic, Kurdish
TELEPHONE LINES: 6,313,000 (1997)

KUWAIT

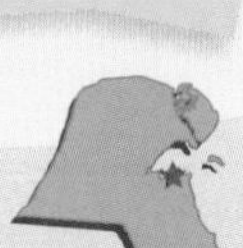

CAPITAL: Kuwait
AREA: 6,880 sq. mi. (17,820 sq km)
POPULATION: 2,041,961
MONEY: Kuwaiti dinar (KWD)
LANGUAGES: Arabic, English
TELEPHONE LINES: 412,000

Countries not drawn to scale.

Gross Domestic Product—per Capita	
Afghanistan	$800
Bahrain	$15,900
Cyprus (Greek)	$16,000
(Turkish)	$5,300
Iran	$6,300
Iraq	$2,500
Israel	$18,900
Jordan	$3,500
Kuwait	$15,700
Lebanon	$5,000
Oman	$7,700
Qatar	$20,300
Saudi Arabia	$10,500
Syria	$3,100
Turkey	$6,800
United Arab Emirates	$22,800
Yemen	$820

Source: CIA *World Factbook, 2001*

Fast Facts Activities

LEVEL 3: (Suggested time: 40 min.) Ask students to imagine that they live in Southwest Asia and have a pen pal in the United States. Have them write letters describing how modernization and Western culture have influenced their way of life. Students should indicate if they feel the effects on their region have been positive or negative. They should back up their opinions with examples from the unit.

Unit 5 Assessment Resources

Reproducible

- Unit 5 Test
- Unit 5 Test for English Language Learners and Special-Needs Students

internet connect

COUNTRY STATISTICS
GO TO: **go.hrw.com**
KEYWORD: **SK3 FactsU5**

Highlights of Country Statistics

Links to online country statistics for Southwest Asia include:

- *CIA World Factbook*
- Library of Congress country studies
- Flags of the world

Unit 5 Atlas

Lebanon

CAPITAL: Beirut
AREA: 4,015 sq. mi. (10,400 sq km)
POPULATION: 3,627,774
MONEY: Lebanese pound (LBP)
LANGUAGES: Arabic (official), French
TELEPHONE LINES: 700,000 (1999)

Syria

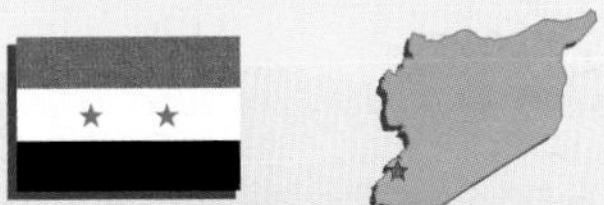

CAPITAL: Damascus
AREA: 71,498 sq. mi. (185,180 sq km)
POPULATION: 16,728,808
MONEY: Syrian pound (SYP)
LANGUAGES: Arabic (official), Kurdish
TELEPHONE LINES: 1,313,000 (1997)

Oman

CAPITAL: Muscat
AREA: 82,031 sq. mi. (212,460 sq km)
POPULATION: 2,662,198
MONEY: Omani rial (OMR)
LANGUAGES: Arabic (official), English
TELEPHONE LINES: 201,000 (1997)

Turkey

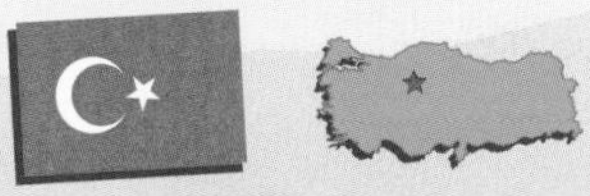

CAPITAL: Ankara
AREA: 301,382 sq. mi. (780,580 sq km)
POPULATION: 66,493,970
MONEY: Turkish lira (TRL)
LANGUAGES: Turkish (official), Kurdish
TELEPHONE LINES: 19,500,000 (1999)

Qatar

CAPITAL: Doha
AREA: 4,416 sq. mi. (11,437 sq km)
POPULATION: 769,152
MONEY: Qatari rial (QAR)
LANGUAGES: Arabic (official), English
TELEPHONE LINES: 142,000 (1997)

United Arab Emirates

CAPITAL: Abu Dhabi
AREA: 32,000 sq. mi. (82,880 sq km)
POPULATION: 2,407,460
MONEY: Emirati dirham (AED)
LANGUAGES: Arabic (official), Persian
TELEPHONE LINES: 915,223 (1998)

Saudi Arabia

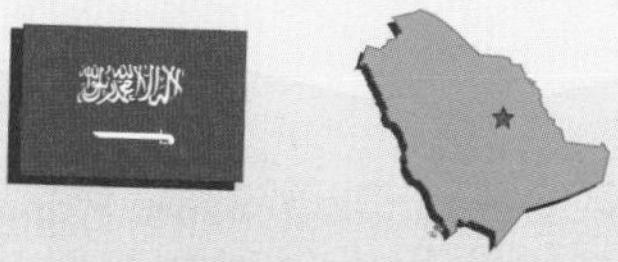

CAPITAL: Riyadh
AREA: 756,981 sq. mi. (1,960,582 sq km)
POPULATION: 22,757,092
MONEY: Saudi riyal (SAR)
LANGUAGES: Arabic
TELEPHONE LINES: 3,100,000

Yemen

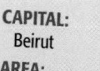

CAPITAL: Sanaa
AREA: 203,849 sq. mi. (527,970 sq km)
POPULATION: 18,078,035
MONEY: Yemeni rial (YRI)
LANGUAGES: Arabic
TELEPHONE LINES: 291,359 (1999)

internet connect

COUNTRY STATISTICS
GO TO: **go.hrw.com**
KEYWORD: **SK3 FactsU5**
FOR: **more facts about Southwest Asia**

Sources: Central Intelligence Agency, *The World Factbook 1999; The World Almanac and Book of Facts 1999*

CHAPTER 17 The Arabian Peninsula, Iraq, Iran, and Afghanistan

Chapter Resource Manager

Objectives	Pacing Guide	Reproducible Resources
SECTION 1 **Physical Geography** (pp. 385–87) 1. Identify the major physical features of the region. 2. Describe the climates found in this region. 3. Identify the region's important resources.	**Regular** .5 day **Block Scheduling** .5 day *Block Scheduling Handbook, Chapter 17*	**RS** Guided Reading Strategy 17.1 **IC** Interdisciplinary Activities for the Middle Grades 21, 23
SECTION 2 **The Arabian Peninsula** (pp. 388–93) 1. Describe what Saudi Arabia's history, government, and people are like. 2. Identify the kinds of government and economy the other countries of the Arabian Peninsula have.	**Regular** 2.5 days **Block Scheduling** .5 day *Block Scheduling Handbook, Chapter 17*	**RS** Guided Reading Strategy 17.2 **RS** Graphic Organizer 17 **E** Cultures of the World Activity 5 **E** Creative Strategies for Teaching World Geography, Lessons 14 and 15 **SM** Geography for Life Activity 17 **SM** Map Activity 17
SECTION 3 **Iraq** (pp. 396–98) 1. Identify the key events in Iraq's history. 2. Describe Iraq's government and economy. 3. Describe the makeup of Iraq's population.	**Regular** 1 day **Block Scheduling** .5 day *Block Scheduling Handbook, Chapter 17*	**RS** Guided Reading Strategy 17.3 **E** Environmental and Global Issues Activity 5
SECTION 4 **Iran and Afghanistan** (pp. 399–401) 1. Identify some important events in Iran's history. 2. Describe what Iran's government and people are like. 3. Describe the problems Afghanistan faces today.	**Regular** 1 day **Block Scheduling** .5 day *Block Scheduling Handbook, Chapter 17*	**RS** Guided Reading Strategy 17.4

Chapter Resource Key

- **RS** Reading Support
- **IC** Interdisciplinary Connections
- **E** Enrichment
- **SM** Skills Mastery
- **A** Assessment
- **REV** Review
- **ELL** Reinforcement and English Language Learners
- Transparencies
- CD–ROM
- Music
- Video
- go.hrw.com Internet
- Holt Presentation Maker Using Microsoft® Powerpoint®

One-Stop Planner CD–ROM

See the ***One-Stop Planner*** for a complete list of additional resources for students and teachers.

One-Stop Planner CD–ROM

It's easy to plan lessons, select resources, and print out materials for your students when you use the *One-Stop Planner CD–ROM with Test Generator.*

Technology Resources	Review, Reinforcement, and Assessment Resources
One-Stop Planner CD–ROM, Lesson 17.1 *ARGWorld* CD–ROM: Control of Water in Southwest Asia Geography and Cultures Visual Resources with Teaching Activities 37–41 Homework Practice Online HRW Go site	**ELL** Main Idea Activity 17.1 **REV** Section 1 Review, p. 387 **A** Daily Quiz 17.1 **ELL** English Audio Summary 17.1 **ELL** Spanish Audio Summary 17.1
One-Stop Planner CD–ROM, Lesson 17.2 *ARGWorld* CD–ROM: Mapping Oil Trade from the Persian Gulf Homework Practice Online HRW Go site	**ELL** Main Idea Activity 17.2 **REV** Section 2 Review, p. 393 **A** Daily Quiz 17.2 **ELL** English Audio Summary 17.2 **ELL** Spanish Audio Summary 17.2
One-Stop Planner CD–ROM, Lesson 17.3 Homework Practice Online HRW Go site	**ELL** Main Idea Activity 17.3 **REV** Section 3 Review, p. 398 **A** Daily Quiz 17.3 **ELL** English Audio Summary 17.3 **ELL** Spanish Audio Summary 17.3
One-Stop Planner CD–ROM, Lesson 17.4 Homework Practice Online HRW Go site	**ELL** Main Idea Activity 17.4 **REV** Section 4 Review, p. 401 **A** Daily Quiz 17.4 **ELL** English Audio Summary 17.4 **ELL** Spanish Audio Summary 17.4

internet connect

HRW ONLINE RESOURCES

GO TO: go.hrw.com
Then type in a keyword.

TEACHER HOME PAGE
KEYWORD: SK3 TEACHER

CHAPTER INTERNET ACTIVITIES
KEYWORD: SK3 GT17

Choose an activity to:
- visit Mesopotamia.
- discover the importance of camels.
- see Arabian arts and crafts.

CHAPTER ENRICHMENT LINKS
KEYWORD: SK3 CH17

CHAPTER MAPS
KEYWORD: SK3 MAPS17

ONLINE ASSESSMENT
Homework Practice
KEYWORD: SK3 HP17
Standardized Test Prep Online
KEYWORD: SK3 STP17
Rubrics
KEYWORD: SS Rubrics

COUNTRY INFORMATION
KEYWORD: SK3 Almanac

CONTENT UPDATES
KEYWORD: SS Content Updates

HOLT PRESENTATION MAKER
KEYWORD: SK3 PPT17

ONLINE READING SUPPORT
KEYWORD: SS Strategies

CURRENT EVENTS
KEYWORD: S3 Current Events

Meeting Individual Needs

Ability Levels

Level 1 Basic-level activities designed for all students encountering new material

Level 2 Intermediate-level activities designed for average students

Level 3 Challenging activities designed for honors and gifted-and-talented students

English Language Learners Activities that address the needs of students with Limited English Proficiency

Chapter Review and Assessment

- **IC** Interdisciplinary Activities for the Middle Grades 24
- **E** Readings in World Geography, History, and Culture 47, 48, and 49
- **SM** Critical Thinking Activity 17
- **REV** Chapter 17 Review, pp. 402–03
- **REV** Chapter 17 Tutorial for Students, Parents, Mentors, and Peers
- **ELL** Vocabulary Activity 17
- **A** Chapter 17 Test
- Chapter 17 Test Generator (on the One-Stop Planner)
- Audio CD Program, Chapter 17
- **A** Chapter 17 Test for English Language Learners and Special-Needs Students

CHAPTER 17

Launch into Learning

Copy the following headings onto the chalkboard: *What I know about Southwest Asia* and *What I want to know about Southwest Asia.* Ask volunteers to compile lists under the headings. If most of the items in the first column relate to oil or political unrest, point out that there is much more to learn about the region. Tell students that they will learn about the history, economy, and cultures of the Arabian Peninsula, Iraq, Iran, and Afghanistan in this chapter.

Section 1

Objectives

1. Name the major physical features of this region.
2. List the climates found in this region.
3. Identify the region's important resources.

Links to Our Lives

Tell students that there are many reasons why they should know more about the Arabian Peninsula, Iraq, Iran, and Afghanistan. Here are some of those reasons:

- The world's first urban civilization, Sumer, developed in what is now Iraq. The mighty Persian Empire ruled the area later. These civilizations have left behind great works of art, architecture, and literature.
- This region contains most of the world's known oil reserves.
- The U.S. military has been involved in the region. For example, U.S. personnel played a major role in Operation Desert Storm in 1990–91.
- This region is the cradle of Islam, and home to many of its followers. Islam is the fastest-growing religion in the United States.

CHAPTER 17

The Arabian Peninsula, Iraq, Iran, and Afghanistan

Let's meet Mitra, a girl who lives in Tehran, the capital of Iran.

I am Mitra and I am 13 years old. I live with my sister and my parents on the top floor of a house in Tehran, the capital of Iran. On the ground floor, my father has a business printing pictures on metal, and the lower two floors are rented to tenants. My dad is also the president of an architectural firm in Tehran, but he is always working at home when we are home from school. Our house has a wall all around, touching the wall of the houses on both sides. Behind the house is a huge garden with roses and fruit trees. My sister is seven years older and works with my dad making architectural drawings. She raises chickens in the garden and collects their eggs. Food is sometimes hard to get, so we are very glad to have the chickens. My dad built them a very nice house, and it is exciting to watch the baby chicks hatch. The roof of our house is flat, and in nice weather we sit up there and look out to the mountains around the city. In the summer we sleep there too.

▲ Translation: Greetings to you.

Focus

LET'S GET STARTED

Copy the following instructions onto the chalkboard: *Make a list of adjectives to describe sand—what it looks like, feels like, where you find it, what it does, and so on.* Discuss responses. Tell students that in Section 1 they will learn about a region with one of the great sand deserts of the world.

Using the Physical-Political Map

Have students examine the map on this page. Call on volunteers to name the major bodies of water in the region, and to point out the Strait of Hormuz and the Bab al-Mandab. Ask why some or all of the region's bodies of water might be of strategic or economic importance. Then ask what countries border these bodies of water, and how they might be able to use their locations to their advantage in the region or in the world.

Teach

Teaching Objectives 1–2

ALL LEVELS: (Suggested time: 20 min.) Copy the following graphic organizer onto the chalkboard, omitting the italicized answers. Have students complete it to learn about the physical features and climate of the area. Encourage students to use the map and the text to complete the organizer. **ENGLISH LANGUAGE LEARNERS**

Section 1 Physical Geography

Read to Discover

1. What are the major physical features of the region?
2. What climates are found in this region?
3. What are the region's important resources?

Define

exotic rivers
wadis
fossil water

Locate

Persian Gulf
Arabian Peninsula
Red Sea
Arabian Sea
Tigris River
Euphrates River
Elburz Mountains
Zagros Mountains
Hindu Kush
Rub' al-Khali

WHY IT MATTERS

This region holds about half of the world's known reserves of oil, or petroleum. Use CNNfyi.com or other **current events** sources to find out how and why oil is so important in the world today. Record your findings in your journal.

Golden lion from Iran

Section 1 Resources

Reproducible

- Block Scheduling Handbook, Chapter 17
- Guided Reading Strategy 17.1
- Interdisciplinary Activities for the Middle Grades 21, 23

Technology

- One-Stop Planner CD–ROM, Lesson 17.1
- Homework Practice Online
- Geography and Cultures Visual Resources with Teaching Activities 37–41
- *ARGWorld* CD–ROM: Control of Water in Southwest Asia
- HRW Go site

Reinforcement, Review, and Assessment

- Section 1 Review, p. 387
- Daily Quiz 17.1
- Main Idea Activity 17.1
- English Audio Summary 17.1
- Spanish Audio Summary 17.1

The Arabian Peninsula, Iraq, Iran, and Afghanistan: Physical-Political

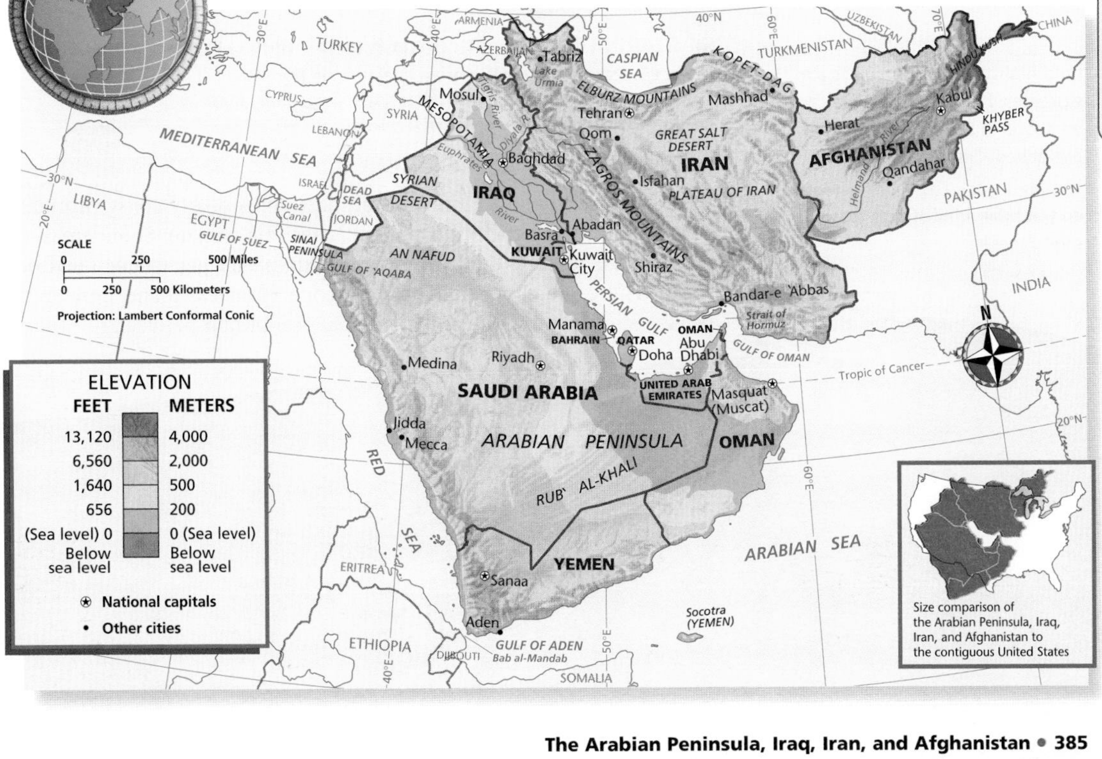

Physical Geography of the Region

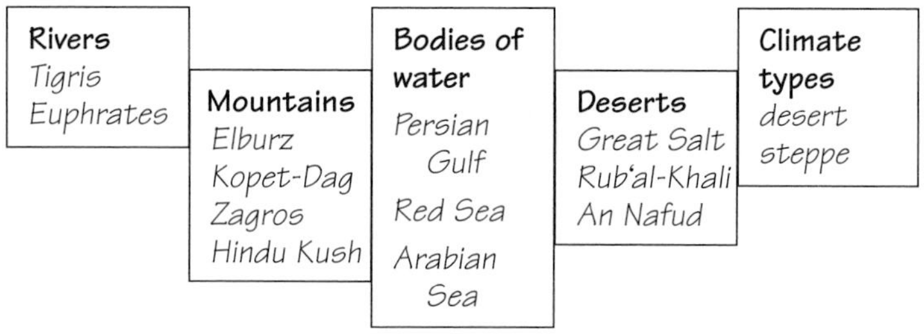
Rivers
Tigris
Euphrates
Mountains
Elburz
Kopet-Dag
Zagros
Hindu Kush
Bodies of water
Persian Gulf
Red Sea
Arabian Sea
Deserts
Great Salt
Rub'al-Khali
An Nafud
Climate types
desert
steppe

Teaching Objective 3

ALL LEVELS: (Suggested time: 45 min.) Organize the class into groups. Have each group create a visual presentation about the oil industry in Southwest Asia by collecting from magazines and other sources, pictures of oil rigs, supertankers, refineries, oil ministers at OPEC meetings, and so on. Then have a student from each group present the group's work to the class. Finally, refer students to the land use and resources map in the Unit Atlas. Have students use the map to compare land use and resources of the Arabian Peninsula with land use and resources elsewhere in the region. **ENGLISH LANGUAGE LEARNERS, COOPERATIVE LEARNING**

EYE ON EARTH

Satellite images have detected traces of an ancient river that apparently flowed from the Hejaz Mountains of western Saudi Arabia eastward across the desert, forming a large delta that once covered what is now Kuwait. The river, which may have been three miles wide in places, probably dates to a relatively wet period that occurred between 5,000 and 11,000 years ago. Similar discoveries in North Africa have yielded new sources of groundwater. That possibility exists for the ancient Saudi Arabian river also.

Critical Thinking: How might the use of satellite imagery to locate a new water source affect this region?

Answer: Answers will vary but could include new settlements or greater crop production.

internet connect

GO TO: go.hrw.com
KEYWORD: SK3 CH17
FOR: Web sites about the Arabian Peninsula, Iraq, Iran, and Afghanistan

BUILD on WHAT You Know

Do you remember what you learned about desert climates? See Chapter 3 to review.

Physical Features

The 10 countries of this region are laid out like a semicircle, with the Persian Gulf in the center. They are Saudi Arabia, Kuwait (koo-WAYT), Bahrain (bah-RAYN), Qatar (KAH-tuhr), the United Arab Emirates (E-muh-ruhts), Oman (oh-MAHN), Yemen (YE-muhn), Iraq (i-RAHK), Iran (i-RAN), and Afghanistan (af-GA-nuh-stan).

This area can be divided into three landform regions. The Arabian Peninsula is a large rectangular area. The Red Sea, Gulf of Aden, Arabian Sea, and Persian Gulf border the peninsula. North of the Arabian Peninsula is the plain of the Tigris (TY-gruhs) and Euphrates (yooh-FRAY-teez) Rivers. In ancient times, this area was called Mesopotamia (me-suh-puh-TAY-mee-uh), or the "land between the rivers." East of this plain is a region of mountains and plateaus. It stretches through Iran and Afghanistan.

The surface of the Arabian Peninsula rises gradually as one moves westward from the Persian Gulf. The highest point is in the southwest, in the mountains of Yemen.

North of the Arabian Peninsula is a low, flat plain. It runs from the Persian Gulf into northern Iraq. The Tigris and Euphrates Rivers flow across this plain. They are what are known as **exotic rivers**. Exotic rivers begin in humid regions and then flow through dry areas.

East of this low plain the land climbs sharply. Most of Iran is a plateau bordered by mountains. The Elburz (el-BOOHRZ) Mountains and Kopet-Dag range lie in the north. The Zagros (ZA-gruhs) Mountains lie in the southwest. Afghanistan includes many mountain ranges, such as the towering Hindu Kush range.

✓ **READING CHECK:** *Places and Regions* What are the major physical features of this area? Arabian Peninsula, plain of the Tigris and Euphrates Rivers, mountains of Yemen, Elburz Mountains, Kopet-Dag, Hindu Kush

Climate

Most of Southwest Asia has a desert climate. A nearly constant high-pressure system in the atmosphere causes this climate. Some areas—mostly high plateaus and the region's edges—do get winter rains or snow. These areas generally have steppe climates. Some mountain peaks receive more than 50 inches (130 cm) of rain per year.

Cold nighttime temperatures and extremely hot days help break rock into sand in Saudi Arabia's Rub' al-Khali, or "empty quarter."

Interpreting the Visual Record **How do you think wind affects these sand dunes?**

Visual Record Answer ▶

changes their shape and size

Close

Tell students that what we now call Southwest Asia has also been known as the Middle East. Ask students to speculate why that term is now used less often. Explain that *Middle East* described the region from a European perspective, in terms of the region's physical relationship to Europe.

Review and Assess

Have students complete the Section Review. Then have each student use the physical-political map of the region and the text to write two matching questions that connect countries to physical features, climates, or resources. Collect the questions and quiz the class. Then have students complete Daily Quiz 17.1.

Reteach

Have students complete Main Idea Activity 17.1. Then have students draw storyboards for a trailer advertising a documentary film about the region's physical geography. **ENGLISH LANGUAGE LEARNERS**

Extend

Have interested students conduct research on camels, including the physiological characteristics that suit them to the desert, the history of their use by humans, and their current role in Arab life. Then have students create an advertisement describing the benefits of owning a camel. **BLOCK SCHEDULING**

The desert can be both very hot and very cold. In summer, afternoon temperatures can reach 129°F (54°C). During the night, however, the temperature may drop quickly. Temperatures sometimes dip below freezing during winter nights.

The Rub' al-Khali (ROOB ahl-KAH-lee), or "empty quarter," of southern Saudi Arabia is the largest sand desert in the world. In northern Saudi Arabia is the An Nafud (ahn nah-FOOD), another desert.

✓ **READING CHECK:** *Places and Regions* What are the climates of this region? desert, steppe

Resources

Water is an important resource everywhere, but in this region it is crucial. Many desert regions are visited only by nomads and their animal herds. In many places, springs or wells provide water. Nomads sometimes get water by digging shallow wells into dry streambeds called **wadis**. Wells built with modern technology can reach water deep underground. The groundwater in these wells is often **fossil water**. Fossil water is water that is not being replaced by rainfall. Wells that pump fossil water will eventually run dry.

Other than water, oil is the region's most important mineral resource. Most of the oilfields are located near the shores of the Persian Gulf. The countries of the region are not rich in resources other than oil. Iran is an exception, with deposits of many metals.

✓ **READING CHECK:** *Places and Regions* What are the region's important resources? water, oil, metals in Iran

Beneath the Red Sea lie many valuable resources, such as oil, sulfur, and metal deposits.

Section 1 Review

go.hrw.com Homework Practice Online — Keyword: SK3 HP17

Define and explain: exotic rivers, wadis, fossil water

Working with Sketch Maps On a map of the Arabian Peninsula, Iraq, Iran, and Afghanistan that you draw or that your teacher provides, label the following: Persian Gulf, Arabian Peninsula, Red Sea, Arabian Sea, Tigris River, Euphrates River, Elburz Mountains, Zagros Mountains, Hindu Kush, and Rub' al-Khali.

Reading for the Main Idea

1. *Places and Regions* Why do you think Mesopotamia was important in ancient times?
2. *Places and Regions* What is the region's climate?

Critical Thinking

3. **Drawing Inferences and Conclusions** Why do you think the Persian Gulf is important to international trade?
4. **Drawing Inferences and Conclusions** What settlement pattern might you find in this region?

Organizing What You Know

5. **Summarizing** Copy the following graphic organizer. Use it to list as many details of landforms, resources, and climate as you can. Place them in the correct part of the diagram.

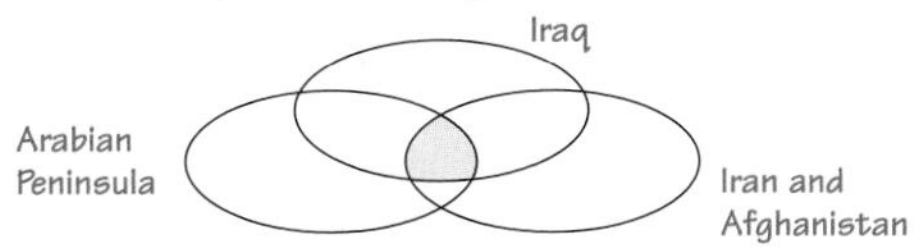

Section 1 Review — Answers

Define For definitions, see: exotic rivers, p. 386; wadis, p. 387; fossil water, p. 387

Working with Sketch Maps Maps will vary, but listed places should be labeled in their approximate locations.

Reading for the Main Idea

1. fertile, well-watered area (NGS 4)
2. desert or steppe (NGS 4)

Critical Thinking

3. large oilfields near its shores, used for transport of oil in tankers
4. widely scattered settlements, because they need to be near water sources

Organizing What You Know

5. Students' answers should describe landforms, resources, and climate. They may mention that all three regions have a desert climate, oil, and limited water supplies.

Section 2

Objectives

1. Describe Saudi Arabia's history, government, and people.
2. Review the kinds of government and economy found in other nations of the Arabian Peninsula.

FOCUS

LET'S GET STARTED

Copy the following instructions onto the chalkboard: *Look at the photo on this page. What do you think is happening in this city?* Discuss responses. Tell students that people go to Mecca on a pilgrimage, or a journey to a sacred place. Making the journey is an important duty for people who follow Islam, one of the world's major religions. Tell students that in Section 2 they will learn more about Islam and about the region where it began.

Building Vocabulary

Write the key terms **Muslims**, **caliph**, **Sunni**, **Shia**, **Qur'an**, and **OPEC** on the chalkboard. Have volunteers look up the definitions in the glossary and read them aloud, then use them in a sentence. Point out that these terms refer to two items that are important to societies of the Arabian Peninsula—Islam and oil.

Section 2 RESOURCES

Reproducible
- Guided Reading Strategy 17.2
- Graphic Organizer 17
- Geography for Life Activity 17
- Map Activity 17
- Cultures of the World Activity 5
- Creative Strategies for Teaching World Geography, Lessons 14 and 15

Technology
- One-Stop Planner CD–ROM, Lesson 17.2
- *ARGWorld* CD–ROM: Mapping Oil Trade from the Persian Gulf
- Homework Practice Online
- HRW Go site

Reinforcement, Review, and Assessment
- Section 2 Review, p. 393
- Daily Quiz 17.2
- Main Idea Activity 17.2
- English Audio Summary 17.2
- Spanish Audio Summary 17.2

Section 2 The Arabian Peninsula

Read to Discover

1. What are Saudi Arabia's history, government, and people like?
2. What kinds of government and economy do the other countries of the Arabian Peninsula have?

Define

Muslims
caliph
Sunni
Shia
Qur'an
OPEC

Locate

Mecca
Riyadh

WHY IT MATTERS

Saudi Arabia has long been an important U.S. ally in this region. Use CNNfyi.com or other **current events** sources to learn about current relations between the United States and Saudi Arabia. Record your findings in your journal.

Spicy Arabian food

Non-Muslims are not allowed to enter Mecca, Islam's holiest city.

Saudi Arabia

Saudi Arabia is by far the largest country of the Arabian Peninsula. Although the kingdom of Saudi Arabia was not created until the 1930s, the region has long been an important cultural center.

Islam The history of the Arabian Peninsula is closely linked to Islam. This religion was founded by Muhammad, a merchant from the Arabian town of Mecca. Around A.D. 610 he had a vision that he had been named a prophet by Allah, or God. Arab armies and merchants carried Muhammad's teachings to new areas. Islam spread quickly across North Africa, much of Asia, and parts of Europe. Followers of Islam are called **Muslims**. Islam provides a set of rules to guide human behavior.

The Islamic world was originally ruled by a religious and political leader called a **caliph**. Gradually this area broke up into several empires. There are also religious divisions within Islam. Followers of the largest branch of Islam are called **Sunni**. Followers of the second-largest branch of Islam are called **Shia**. In the late 600s

TEACH

Teaching Objective 1

LEVEL 1: (Suggested time: 25 min.) Copy the graphic organizer onto the chalkboard, omitting the italicized answers. Have students fill in the four headings with details about the topic. Students may provide more details than those supplied. Then lead a class discussion on how Islam and oil relate to various aspects of Saudi Arabian society today. **ENGLISH LANGUAGE LEARNERS**

Saudi Arabia's History, Government, and Economy

History of Islam	Political Developments
A.D. 610—Muhammad reports his vision. Islam spreads. late 600s—Sunni and Shia Muslims disagree about leadership of Islam.	*1920s—Local ruler conquers neighbors, creates kingdom of Saudi Arabia in 1932. King rules with advice of others. Saudi Arabia and United States establish close relationship.*
Traditional Economy	**Oil Industry**
Barley, dates, fruits, millet, vegetables, and wheat are traditional crops. Nomads kept sheep, goats, horses, and camels.	*1930s—oil discovered in Saudi Arabia. Country becomes world's leading oil exporter. Oil becomes most important part of economy and pays for desalination plants.*

these two groups disagreed over who should lead the Islamic world. There are many smaller groups within Islam as well.

Islamic culture helps to unite Muslims around the world. For example, all Muslims learn Arabic to read the **Qur'an**, the holy book of Islam. Muslims are also expected to visit the holy city of Mecca at least once. These practices and many others help make Muslims part of one community.

Government and Economy In the 1920s a local ruler from the Saud family of central Arabia conquered his neighbors and in 1932 created the kingdom of Saudi Arabia. Members of the Saud family have ruled the country ever since. Riyadh, a city near the center of the country, became the capital.

Saudi Arabia is a monarchy with no written constitution or elected legislature. Most government officials are relatives of the king. The king may ask members of his family, Islamic scholars, and tribal leaders for advice on important decisions.

In recent years, Saudi Arabia and the United States have established a close relationship. Both countries have strategic military and economic interests in the region. Saudi Arabia purchases U.S. weapons, such as fighter planes. In 1990, when Iraq invaded Kuwait, the Saudi government allowed U.S. military forces to operate from Saudi Arabia.

Oil and related industries are the most important part of the Saudi economy. Saudi Arabia has the world's largest reserves of oil. It is also the world's leading exporter of oil. Saudi Arabia is a leader of the Organization of Petroleum Exporting Countries, or **OPEC**. The members of OPEC try to influence the price of oil on world markets.

Estimated Oil Reserves, 1999

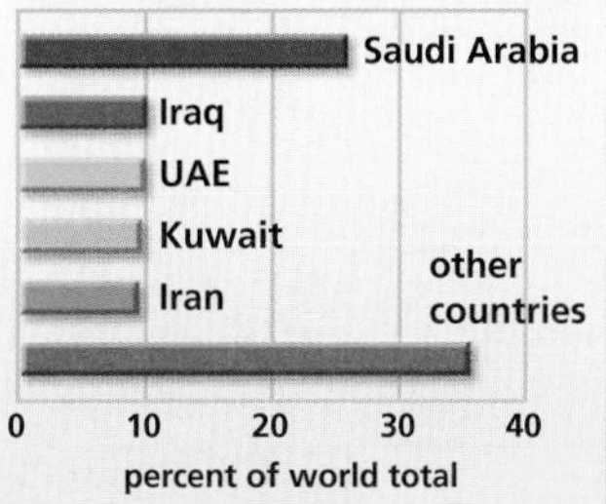

Source: *Encyclopaedia Britannica*

Interpreting the Graph **Which nation in the region has the largest oil reserves?**

Opened in 1986, this causeway links Saudi Arabia to Bahrain.

Interpreting the Visual Record **How important do you think modern transportation systems are to a country's economy?**

▼

DAILY LIFE

One of the duties of a Muslim is to make a hajj, or pilgrimage, to the holy city of Mecca at least once in his or her lifetime. To make the pilgrimage acceptable to Allah, the hajji, or pilgrim, must follow rules laid out in the Qur'an. For example, pilgrims must study Islam before going. They must be debt-free and leave behind enough money to take care of their families while they are gone. If the hajji is old or physically challenged, the hajji must be accompanied by an able-bodied person. During the pilgrimage, pilgrims are not allowed to cut their hair or fingernails or to shave.

Critical Thinking: Why do you think that the Qur'an describes such detailed guidelines for making a hajj?

Answer: Answers will vary, but students might suggest that following detailed guidelines may indicate a strong commitment to Islam.

▲ **Graph Answer**

Saudi Arabia

◄ **Visual Record Answer**

very important for moving raw materials and products

LEVELS 2 AND 3: (Suggested time: 30 min.) Tell students to imagine that they lived in Mecca in the A.D. 650s and that they found a time machine in the desert that brought them forward in time to the present. Have students write a monologue about their experiences in the present day in which they reflect on how much life in Saudi Arabia has changed. Ask them to record observations about history, the government, economy, and people.

Teaching Objective 2

LEVEL 1: (Suggested time: 20 min.) Pair students and have pairs list the six other countries of the Arabian Peninsula in their notebooks. Ask them to designate each country with an **M** if it is a monarchy, an **E** if the government is elected, an **O** if oil is an important part of the economy, and an **N** if new industries are starting to play a major role in the economy. Then have each pair list at least two other details about that country. Call on volunteers to discuss the designations they assigned and the details they recorded. **COOPERATIVE LEARNING**

LEVELS 2 AND 3: (Suggested time: 30 min.) Have students write letters to an advice column for a newspaper in one of the six smaller countries of the Arabian Peninsula. Tell students to use the issue of oil wealth, the presence of foreign workers, or limitation of political rights as the subject of their letters. Encourage them to use ideas inspired by the text's illustrations or material on the history of Islam on previous pages.

Eye on Earth

Sabchat is the Arabic word used to describe a slushy, salty, flat area that lies just above the water table. There are many in the desert areas of Qatar and Saudi Arabia.

A *sabchat* forms when the wind blows away the sand and soil down to the water table. Silt, clay, and salts accumulate. A crust forms on the surface. The crust may be a thick armorlike layer of salt or a thin layer over quicksand. A thin sheet of sand may drift over a *sabchat*, hiding it from view.

Critical Thinking: What physical processes create a *sabchat*?

Answer: Wind blows away the sand and soil down to the water table, which is then covered by a crust.

internet connect

GO TO: go.hrw.com
KEYWORD: SK3 CH17
FOR: Web sites about *sabchat* and deserts

Visual Record Answer

because the irrigation sprinklers rotate around the water source to form a circle

Fossil water has been used to convert desert land northwest of Riyadh into circular fields.

Interpreting the Visual Record **Why do you think these fields are circular?**

Our Amazing Planet

A *sabchat* is a thick, slushy deposit of sand, silt, mud, and salt found in coastal Saudi Arabia. A *sabchat* can trap people and animals who do not see it in time.

Oil was discovered in Saudi Arabia in the 1930s. Before this time farming and herding had been the main economic activities. Crops included barley, dates, fruits, millet, vegetables, and wheat. Nomads kept herds of sheep, goats, horses, and camels.

Like other oil-rich states in the region, Saudi Arabia has tried to increase its food production. Because freshwater is scarce, desalination plants are used to remove the salt from seawater. This water is then used in farming. Income from oil allows the Saudi government to pay for this expensive process. Even so, Saudi Arabia imports much of its food.

People and Customs Nearly all Saudis are ethnic Arabs and speak Arabic. About 85 percent are Sunni. The rest are Shia. Most Saudis now live in cities, and a sizable middle class has developed. The Saudi government provides free health care and education to its citizens.

More than 5 million foreigners live and work in Saudi Arabia. Most come from Turkey, Egypt, Jordan, Yemen, and countries in southern and eastern Asia. Many Americans and Europeans work in the country's oil industry.

Saudi laws and customs limit women's freedoms. For example, a woman rarely appears in public without her husband or a male relative. Women are also not allowed to drive cars. In 1999, women made up just 5 percent of Saudi Arabia's workforce.

Islamic practices are an important part of Saudi Arabia's culture. Muslims pray five times each day. Friday is their holy day. Because Islam encourages modesty, Saudi clothing keeps arms and legs covered. Men traditionally wear a loose, ankle-length shirt. They often wear a cotton headdress held in place with a cord. These are practical in the desert, giving protection from sun, wind, and dust. Saudi women usually wear a black cloak and veil when they are in public.

Two major celebrations mark the Islamic calendar. Each year, some 2 million Muslims travel to Mecca to worship. The journey ends with 'Id al-Adha, the Festival of Sacrifice. The year's second celebration is 'Id al-Fitr, a feast ending the month of Ramadan. During Ramadan, Muslims do not eat or drink anything between dawn and sunset.

✓ **READING CHECK:** *Human Systems* What is significant about Saudi Arabia's history, government, and people? history—linked to Islam; government—monarchy with no constitution; people—mostly Sunni, city dwellers, follow Islamic practices

Other Countries of the Arabian Peninsula

Six small coastal countries share the Arabian Peninsula with Saudi Arabia. All are heavily dependent on oil. All but Yemen are monarchies, and each is overwhelmingly Islamic. Oil is a major part of each country's economy. However, possession of differing amounts of oil has made some countries much richer than others.

Teaching Objectives 1–2

LEVEL 1: (Suggested time: 15 min.) Prepare in advance slips of paper on which you have written roles for various individuals of the Arabian Peninsula. Here are some examples: a caliph of the 600s, a member of Saudi Arabia's royal family, a Saudi Islamic scholar, a general in the United States army stationed in Saudi Arabia, a foreign worker, a nomadic herder, an emir of the UAE, or a member of Yemen's government. Prepare a role for each student. You may need to repeat some. Have students draw a slip from a hat and create a sentence that gives a clue to their identity. Call on volunteers to read their sentences. Ask other students to attempt to identify the role of each speaker and explain his or her significance.

LEVEL 2: (Suggested time: 30 min.) Organize students into groups. Have each group use the roles from the Level 1 activity to create a conversation that those people might have if they were all invited to the same dinner party. **COOPERATIVE LEARNING**

LEVEL 3: (Suggested time: 45 min.) Have students use the Level 2 activity and library resources to create a written plan for the entire dinner party, including a menu of local dishes, a program of entertainment, and proper etiquette for the occasion. You may need to schedule an extra class day for this activity.

Connecting to Math

Muslim Contributions to Mathematics

Muslim astronomers in the 1500s

During the period of Western history called the Dark Ages, European art, literature, and science declined. However, during this same period Islamic civilization, stretching from Spain to the borders of China, was flowering. Muslim scholars made important advances in art, literature, medicine, and mathematics.

The system of numerals we call Arabic, including the use of the zero, was first created in India. However, it was Muslim thinkers who introduced that system to Europe. They also developed algebra, geometry, and trigonometry. In fact, words like *algebra* and *algorithm* are translations of Arabic words.

Other Muslims advanced the study of astronomy and physics. Arab geographers calculated distances between cities, longitudes and latitudes, and the direction from one city to another. Muslims developed the first solution for cubic equations. They also defined ratios and used mathematics to explain the rainbow.

Understanding What You Read

1. Where did the Arabic system of numerals originate, and how did it get to Europe?
2. How did Muslim scholars contribute to our knowledge of geography?

Kuwait The country of Kuwait was established in the mid-1700s. Trade and fishing were once the main economic activities there. Oil, which was discovered in the 1930s, has made Kuwait very rich. The Iraqi invasion of 1990 caused massive destruction.

As in Saudi Arabia, a royal family dominates politics in Kuwait. In 1992, however, Kuwait held elections for a legislature. Less than 15 percent of Kuwait's population were given the right to vote. These people were all men from well-established families.

Bahrain and Qatar Bahrain is a group of small islands in the western Persian Gulf. It is a constitutional monarchy that is headed by a ruling royal family and a legislature. These islands have been a center of trade since ancient times. In 1986 a 15.5 mile (25 km) bridge was completed that connects Bahrain to Saudi Arabia, making movement between the two countries easier.

Across the Curriculum
TECHNOLOGY

Cell Phones in Southwest Asia Cellular phones are very popular in Southwest Asia. Kuwait, for example, has a ratio of more than one cellular phone for every 13 inhabitants. The phones are status symbols. They are also convenient.

The widespread use of cellular phones is also partly due to the religious conservatism of Southwest Asia. It is reported that in Kuwait, young men contrive to meet young women by driving around with two cellular phones. When they see a likely prospect, they toss one of the phones to her, then use the other to call her.

Discussion: Lead a discussion about cell phone use and Kuwaiti culture. Then compare cell phone use in the United States with usage in Kuwait.

Connecting to Math

Answers

1. India; brought to Europe by Muslim thinkers
2. figured distances between cities, longitudes and latitudes, and the direction from one city to another

➤ASSIGNMENT: Have students use library resources to conduct research on Islamic practices and holidays, including Ramadan and the annual arrival of Muslim pilgrims in Mecca. Tell students to examine and explain the significance of these practices.

TEACHER TO TEACHER

Sandra Rojas of Commerce City, Colorado, suggests the following activity to help students learn about the Arabian Peninsula: On a large sheet of paper draw a wall map of the Arabian Peninsula with the region's political boundaries marked. Organize the class into seven groups and assign each group one of the countries of the Arabian Peninsula. Have each group locate information on the economy, culture, and government of its country. Then have members from each group copy the information about their country onto the wall map. Use different colored markers for each country. Display the wall map while students complete the chapter.

HUMAN SYSTEMS

Note the distinctive architecture in the photo on the opposite page. In Sanaa, Yemen's capital, some houses still have windows made of alabaster. The Arabic word for alabaster translates into English as moonstone. Although you cannot see through it, alabaster allows a bright but soft light to enter a room.

To make the windows, alabaster was mined by hand and then cut into slabs 1/2 to a full centimeter thick. Although beautiful from inside and out, alabaster darkens and cracks with age. Because alabaster needs to be replaced as often as every five years, glass is much less expensive.

Activity: Have students conduct further research on Yemeni architecture and building techniques. Have students build scale models of particular examples.

internet connect

GO TO: **go.hrw.com**
KEYWORD: **SK3 CH17**
FOR: **Web sites about Yemen**

Chart Answer ▲

Yemen

Visual Record Answer ►

Answers may vary. Students may suggest that religious law requires the separation.

The Arabian Peninsula

COUNTRY	POPULATION/ GROWTH RATE	LIFE EXPECTANCY	LITERACY RATE	PER CAPITA GDP
Bahrain	645,361 (228,424 noncitizens) / 1.7%	71, male 76, female	85%	$15,900
Kuwait	2,041,961 (1,519,913 noncitizens) / 3.4%	75, male 77, female	79%	$15,000
Oman	2,622,198 / 3.4%	70, male 74, female	80%	$7,700
Qatar	769,152 / 3.2%	70, male 75, female	79%	$20,300
Saudi Arabia	22,757,092 (5,360,526 noncitizens) / 3.3%	66, male 70, female	63%	$10,500
United Arab Emirates	2,407,460 (1,576,472 noncitizens) / 1.6%	72, male 77, female	79%	$22,800
Yemen	18,078,035 / 3.4%	58, male 62, female	38%	$820
United States	281,421,906 / 0.9%	74, male 80, female	97%	$36,200

Sources: Central Intelligence Agency, *The World Factbook 2001;* U.S. Census Bureau

Interpreting the Chart **Which country has the lowest standard of living?**

Oil was discovered in Bahrain in the 1930s, creating wealth for the country. However, by the 1990s this oil was starting to run out, and banking and tourism are now becoming important to the economy. Bahrain also refines crude oil imported from nearby Saudi Arabia.

Qatar occupies a small peninsula in the Persian Gulf. Like Bahrain, Qatar is ruled by a powerful monarch. In the 1990s, Qatar's ruler announced a plan to make the country more democratic. He also ended censorship of Qatari newspapers and television.

Qatar has sizable oil reserves and even larger reserves of natural gas—some of the largest in the world.

The United Arab Emirates The United Arab Emirates, or UAE, consists of seven tiny kingdoms. They are ruled by emirs. This country also has great reserves of oil and natural gas. Profits from these resources have created a modern, comfortable lifestyle for the people of the UAE. The government has also worked to build up other industries.

Like Saudi Arabia and many of the small Persian Gulf countries, the UAE depends on foreign workers. In the UAE, foreign workers outnumber citizens.

Oman Oman is located just outside the mouth of the Persian Gulf. The country is slightly smaller than Kansas. In ancient times, Oman was a major center of trade for merchants traveling the Indian Ocean. Today, Oman's economy is heavily dependent on oil. However, Oman does not have the oil wealth of Kuwait or the UAE. Therefore, the government, ruled by the Al Bu Sa'id family, is attempting to create new industries.

◄ Kindergarten students in the United Arab Emirates kneel to pray.

Interpreting the Visual Record **For what cultural reason do you think these girls and boys are seated in separate rows?**

TEACH

Teaching Objective 1

ALL LEVELS: (Suggested time: 15 min.) Have students list the various cultures and key events that have shaped Iraq's history from the Sumerians to the Persian Gulf War. **ENGLISH LANGUAGE LEARNERS**

Teaching Objective 2

ALL LEVELS: (Suggested time: 15 min.) List *Government* and *Economy* as column heads on the chalkboard. Have volunteers list as many terms as they can under each column to describe Iraq. **ENGLISH LANGUAGE LEARNERS**

Teaching Objective 3

ALL LEVELS: (Suggested time: 20 min.) Copy the following graphic organizer. Have students complete and color it to show the makeup of Iraq's population. What religion do most Iraqis practice? *(Islam)* **ENGLISH LANGUAGE LEARNERS**

Iraq's Population

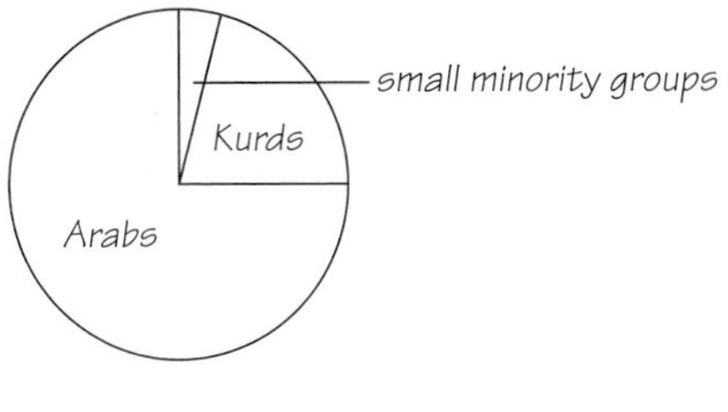

invasion. Many Western leaders believed that Iraq should not be allowed to conquer its neighbors. They were also concerned that Iraq would control such a large share of the world's oil. Also, Iraq had missiles, poison gas, and perhaps nuclear weapons. An alliance of countries, led by the United States and Great Britain, sent troops, tanks, and planes to Saudi Arabia. In the Persian Gulf War of 1991 this alliance forced the Iraqis out of Kuwait.

✓ **READING CHECK:** *Human Systems* What are some key events in Iraq's history? creation in 1932, overthrow of pro-British ruler in 1950s, Ba'ath to power in 1968, Saddam Hussein becomes president in late 1970s, invasion of Iran in 1980, of Kuwait in 1990, forced out of Kuwait in 1991

Iraq

Country	Population/ Growth Rate	Life Expectancy	Literacy Rate	Per Capita GDP
Iraq	23,332,985 2.8%	66, male 68, female	58%	$2,500
United States	281,421,906 0.9%	74, male 80, female	97%	$36,200

Sources: Central Intelligence Agency, *The World Factbook 2001;* U.S. Census Bureau

Interpreting the Chart **What percentage of the Iraqi population is literate?**

Government and Economy

The government of Iraq is still firmly controlled by the Ba'ath Party and Saddam Hussein. The Iraqi leader has built up a large army and a secret police force. He has also placed his relatives in many important government positions.

Iraq has the world's second-largest known oil reserves, but its economy has suffered recently. Before the war with Iran, Iraq was the world's second-largest oil exporter. That war and the Persian Gulf War damaged Iraq's oil industry. Furthermore, the United Nations placed an **embargo**, or limit on trade, on Iraq. Today, Iraq exports much less oil than it did in the 1980s.

Other industries include construction, mining, and manufacturing. Many factories produce weapons for Iraq's army. More than 3 million people live in Baghdad, and most of Iraq's industries are there.

Irrigation from the Tigris and Euphrates Rivers supports Iraq's farming sector. Chief crops include barley, dates and other fruit, cotton, rice, vegetables, and wheat.

✓ **READING CHECK:** *Human Systems* How is Iraq's government organized, and what is its economy like? dictatorship—Ba'ath Party and Saddam Hussein; economy is damaged from war but focuses on oil exports, construction, mining, manufacturing

Our Amazing Planet

Dust storms occur throughout Iraq and can rise to several thousand feet above the ground. These storms often happen during the summer. Five or six usually strike central Iraq in July, the worst month.

Many Iraqis shop for food in local markets.

Cultural Kaleidoscope
The Ma'dan

Disruptions of the 1900s threaten a unique culture. Iraq's wars and irrigation projects on the Tigris and Euphrates Rivers have affected the Ma'dan, also known as the Marsh Arabs.

The Ma'dan are an ancient people who occupy the marshy land just north of where the rivers meet. They do little farming. Instead, they fish and raise water buffalo. They cut river reeds to build their homes and furniture. The Ma'dan also harvest cereal grains and date palms from the marshes. To get around they use canoe-like boats.

The Iraqi government has built dams upriver. Because the rivers no longer flow freely, the marsh has been reduced to mud in some places. Fish and bird habitats have been destroyed.

Activity: Refer to the photo of the Ma'dan settlement on the following page. Have students point out ways the Ma'dan way of life appears to differ from that of city or desert dwellers.

internet connect

GO TO: **go.hrw.com**
KEYWORD: **SK3 CH17**
FOR: **Web sites about the Ma'dan**

▲ **Chart Answer**

58%

Close

Ask students to imagine what life would be like under a dictator like Saddam Hussein. Have them describe and explain how political life would be different if they lived in Iraq.

Review and Assess

Have students complete the Section Review. Then have them write 4 multiple-choice questions about the section. Collect the questions and quiz the class orally. Then have students complete Daily Quiz 17.3.

Reteach

Have students complete Main Idea Activity 17.3. Have students work in groups to create flow charts or other graphic displays that depict the answers to the Read to Discover questions.
ENGLISH LANGUAGE LEARNERS, COOPERATIVE LEARNING

Extend

Have interested students conduct research on the environmental devastation that resulted from the Persian Gulf War. Have them share their findings with the class in the form of posters or dioramas. **BLOCK SCHEDULING**

Section 3 Review

Answers

Define For definition, see: embargo, p. 397

Working with Sketch Maps Most of Iraq's population is concentrated around the plains of the Tigris and Euphrates River Valleys because of water resources.

Reading for the Main Idea

1. Sumerians, Babylonians, Assyrians, Arabs, the Ottoman Empire, and the British (NGS 13)
2. dictatorship—Saddam Hussein maintains a large army and secret police force (NGS 12)

Critical Thinking

3. to prevent Iraq from conquering its neighbors and to protect the region's oil resources
4. They have damaged the economy and resulted in limits being placed on the country's oil exports.

Organizing What You Know

5. Answers should include the Arab conquest of Mesopotamia (A.D. 600s), occupation by the Ottoman Empire (1500s) and the British (during World War I), the Ba'ath Party takeover (1968), the Iran war (1980–88), and the Persian Gulf War (1990–91).

▲ The Ma'dan, sometimes called the Marsh Arabs, are a minority group in Iraq. They live in Iraq's southern marshes.

People

Nearly all Iraqis are Muslim. More than 75 percent of the population are Arabs. Some 15 to 20 percent are Kurds, and the rest belong to small minority groups. Many Kurds living in Iraq's northern mountains have resisted control by the Baghdad government. After the Gulf War, Shia Arabs living in southern Iraq also rebelled against Saddam Hussein. The Iraqi government has used military force to crush these uprisings.

✓ **READING CHECK:** *Human Systems* What is the makeup of Iraq's population? 75 percent Arabs, 15–20 percent Kurds, remainder is small minority groups

go.hrw.com Homework Practice Online
Keyword: SK3 HP17

Section 3 Review

Define and explain: embargo

Working with Sketch Maps On the map you created in Section 2, label Baghdad. Where would you expect most of Iraq's population to be concentrated? Why?

Reading for the Main Idea

1. *Human Systems* What are some different groups that have controlled Mesopotamia?
2. *Human Systems* What is Iraq's government like?

Critical Thinking

3. **Finding the Main Idea** Why did several countries force Iraqi troops out of Kuwait?
4. **Analyzing Information** How have recent wars affected Iraq's economy?

Organizing What You Know

5. **Sequencing** Create a time line tracing the history of the area now known as Iraq. List events from the Arab conquest to 1991.

A.D. 600s ———————— 1991

Section 4

Objectives

1. List some of the important events in Iran's history.
2. Describe Iran's government and people.
3. Identify problems that Afghanistan faces today.

Focus

LET'S GET STARTED

Copy these instructions onto the chalkboard: *Look at the diagram on this page. How does this system seem to compare to the irrigation system indicated in the aerial photo in Section 2? Do you think one system may be more efficient than the other?* Disuss responses. Tell students that the *qanat* system is used in Iran and that in Section 4 they will learn more about Iran.

Building Vocabulary

Write the key terms **shah** and **theocracy** on the chalkboard. Call on volunteers to locate the word histories in a dictionary and read them to the class. Lead a discussion comparing theocracy and democracy.

Section 4 Iran and Afghanistan

Read to Discover

1. What were some important events in Iran's history?
2. What are Iran's government and people like?
3. What problems does Afghanistan face today?

Define

shah
theocracy

Locate

Tehran
Khyber Pass
Kabul

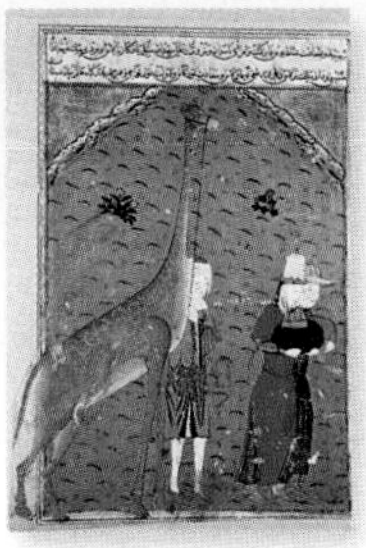

A Persian manuscript

WHY IT MATTERS

The nation of Afghanistan has been involved in many conflicts in recent history. Use CNNfyi.com or other **current events** sources to learn about the situation in Afghanistan today. Record your findings in your journal.

Iran

Iran is a large country with a large population, rich history, and valuable natural resources. A revolution in 1979 made Islam the guiding force in Iran's government.

History The Persian Empire, established in the 500s B.C., was centered in what is now Iran. It was the greatest empire of its time and was an important center of learning. In the 300s B.C. Alexander the Great conquered the Persian Empire. In the A.D. 600s Arabs invaded the region and established Islam. Persian cultural and scientific contributions became elements of Islamic civilization. Later, different peoples ruled the region, including the Mongols and the Safavids.

Qanat Irrigation System

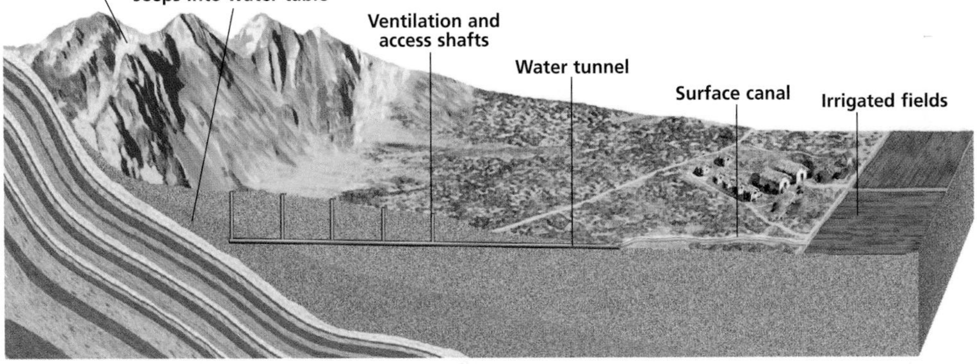

In the dry climates of Iran and Afghanistan, some farmers still use an ancient method of irrigating crops. Runoff from mountains is moved through underground tunnels, called *qanats* (kuh-NAHTS), to fields below the mountains.

Section 4 Resources

Reproducible
- Guided Reading Strategy 17.4

Technology
- One-Stop Planner CD–ROM, Lesson 17.4
- Homework Practice Online
- HRW Go site

Reinforcement, Review, and Assessment
- Section 4 Review, p. 401
- Daily Quiz 17.4
- Main Idea Activity 17.4
- English Audio Summary 17.4
- Spanish Audio Summary 17.4

TEACH

Teaching Objective 1

ALL LEVELS: (Suggested time: 30 min.) Copy the following graphic organizer onto the chalkboard, omitting the italicized answers. Have students add annotations to the time line to identify key events in Iran's history. **ENGLISH LANGUAGE LEARNERS**

Teaching Objective 2

ALL LEVELS: (Suggested time: 20 min.) Lead a discussion in which students compare Iran's current government and people to the government and people of other Southwest Asian countries. Ask: How are they different? *(Iran's government is a theocracy, while others are mainly monarchies.)* How are they similar? *(Islam is the main religion; population is diverse.)* Students may add other details.

Teaching Objective 3

ALL LEVELS: (Suggested time: 30 min.) Have students write a brief news report stating the problems that Afghanistan faces today.

Key Events in Iran's History

500s B.C.	300s B.C.	A.D. 600s	1921	1979	1980
Persian Empire established	*Persian Empire conquered by Alexander the Great*	*Arab invasion of the region and establishment of Islam*	*Shah in power*	*Shah overthrown by Islamic republic*	*Beginning of war with Iraq*

Section 4 Review

Answers

Define For definitions, see: shah, p. 400; theocracy, p. 400

Working with Sketch Maps Maps will vary, but listed places should be labeled in their approximate locations. Tehran and Kabul are both located in high desert plateaus.

Reading for the Main Idea

1. ethnic Persians, ethnic Azerbaijanis, Kurds, Arabs, Turkmans; Shia, Sunni Muslims, Christians, and Jews (NGS 4)
2. damaged industry, trade, and transportation (NGS 13)

Critical Thinking

3. Students should note possible internal problems between the government and religious leaders and problems with foreign countries because of Iran's support of terrorist organizations.
4. The country's geography makes it difficult to organize and rule people. (NGS 15)

Organizing What You Know

5. shah gains power; Islamic republic overthrows shah; U.S. Embassy attacked in Tehran; American hostages captured; war with Iraq

In 1921 an Iranian military officer took power and encouraged reform. He claimed the old Persian title of **shah**, or king. In 1941 the Shah's son took control. He was an ally of the United States and Britain and tried to modernize Iran. His programs were unpopular, however, and in 1979 he was overthrown.

Iran's new government set up an Islamic republic. Soon afterward, relations with the United States broke down. A mob of students attacked the U.S. Embassy in Iran's capital, Tehran, in November 1979. They took Americans hostage with the approval of Iran's government. More than 50 Americans were held by force for a year. The Iranian Revolution itself was soon followed by a long, destructive war with Iraq beginning in 1980.

Government and Economy Iran is a **theocracy**—a government ruled by religious leaders. The country has an elected president and legislature. An expert on Islamic law is the supreme leader, however.

Iran's government has supported many hard-line policies. For example, the country's government has called for the destruction of the state of Israel. Iran has also supported terrorist groups in other countries. However, in the late 1990s signs indicated that Iran was trying to make democratic reforms.

Iran has the fifth-largest oil reserves in the world. Oil is its main industry. Iran is a member of OPEC. Iran's other industries include construction, food processing, and the production of beautiful woven carpets. About one third of Iran's workforce is employed in agriculture.

During Iran's New Year celebrations grass is used to symbolize spring and life.

People and Customs Iran has a population of about 65 million—one of the largest in Southwest Asia. It is also quite diverse. Ethnic Persians make up a slight majority. Other groups include ethnic Azerbaijanis, Kurds, Arabs, and Turkmans. Persian is the official language. Almost all Iranians speak Persian, although some speak it as a second language. The region's other languages include several Kurdish dialects, some Turkic languages, and Arabic.

The Shia branch of Islam is Iran's official religion. About 90 percent of Iranians are Shia. About 10 percent of Iran's residents are Sunni Muslim. The rest are Christian, Jewish, or practice other religions.

In addition to the Islamic holy days, Iranians celebrate Nauruz—the Persian New Year. They also recognize the anniversary of the Iranian Revolution on February 11. Iranian food features rice, bread, vegetables, fruits, and lamb. Strong tea is a popular drink among many Iranians.

✓ **READING CHECK:** ***Human Systems*** What are Iran's government and people like? **theocracy with an elected president and legislature; supreme leader is Islamic law expert; diverse population; Persian most commonly spoken language; Shia branch of Islam is official religion**

CLOSE

Remind students that Iran and Afghanistan have both affected U.S. history. Ask students to identify events in U.S. history that were linked to these two countries. *(Iran hostage crisis of 1979–81, Iran-Contra affair of the 1980s, boycott of the 1980 Summer Olympics in Moscow in response to the Soviet Union's invasion of Afghanistan)*

REVIEW AND ASSESS

Have students complete the Section Review. Then pair students and have pairs quiz each other on terms, places, and dates from the section. Then have students complete Daily Quiz 17.4. **COOPERATIVE LEARNING**

RETEACH

Have students complete Main Idea Activity 17.4. Organize the class into groups. Have each group design official seals for Iran and Afghanistan. Tell students to include as many references to the information in the section as possible. **ENGLISH LANGUAGE LEARNERS, COOPERATIVE LEARNING**

EXTEND

Have interested students conduct research on Zoroastrianism and Baha'i, two religions that originated in Iran. Ask them to compare and contrast the basic beliefs of the two religions to other major religions. **BLOCK SCHEDULING**

Afghanistan

Afghanistan is a landlocked country of high mountains and fertile valleys. Merchants, warriors, and missionaries have long used the Khyber (KY-buhr) Pass to reach India. This narrow passage through the Hindu Kush lies between Afghanistan and Pakistan.

History and People In 1979 the Soviet Union sent troops into Afghanistan to help the communist government there in a civil war. This led to a long war between Soviet troops and Afghan rebels. The Soviets left in 1989, and an alliance of Afghan groups took power. Turmoil continued, and in the mid-1990s a radical Muslim group known as the Taliban arose. Its leaders took over most of the country, including the capital, Kabul. The Taliban ruled Afghanistan strictly. For example, they forced women to wear veils and to stop working outside the home.

In 2001 Taliban officials came into conflict with the United States. Investigation of terrorist attacks on September 11 on Washington, D.C., and New York City led to terrorist Osama bin Laden and his al Qaeda network, based in Afghanistan. The United States and Great Britain then attacked Taliban and al Qaeda targets.

The long period of war has damaged Afghanistan's industry, trade, and transportation systems. Farming and herding are the most important economic activities now.

Afghans belong to many different ethnic groups. The most numerous are the Pashtun, Tajik, Hazara, and Uzbek. Almost all Afghans are Muslims, and about 84 percent are Sunni.

✓ **READING CHECK:** *Human Systems* What are some of the challenges Afghanistan faces today? damaged economy, political conflict

Iran and Afghanistan

Country	Population/ Growth Rate	Life Expectancy	Literacy Rate	Per Capita GDP
Afghanistan	26,813,057 / 3.5%	47, male / 45, female	32% (1999)	$800
Iran	66,128,965 / 0.7%	69, male / 71, female	79% (1999)	$6,300
United States	278,058,881 / 0.9%	74, male / 80, female	97% (1994)	$36,200

Sources: Central Intelligence Agency, *The World Factbook 2001;* U.S. Census Bureau

Interpreting the Chart **How many times greater is the U.S. per capita GDP than Afghanistan's?**

Section 4 Review

go.hrw.com Homework Practice Online
Keyword: SK3 HP17

Define and explain: shah, theocracy

Working with Sketch Maps On the map you created in Section 3, label Tehran, Khyber Pass, and Kabul. What physical features do Tehran and Kabul have in common?

Reading for the Main Idea

1. *Places and Regions* What ethnic and religious groups live in Iran?
2. *Human Systems* How have Afghanistan's recent wars affected the country?

Critical Thinking

3. **Making Generalizations and Predictions** What challenges might the Iranian government face in the future?
4. **Drawing Inferences and Conclusions** How might Afghanistan's political problems be affected by its rugged physical geography?

Organizing What You Know

5. **Sequencing** Copy the following graphic organizer. Use it to show the main events in Iran in recent decades.

☐	⇨ ☐	⇨ ☐	⇨ ☐	⇨ ☐

Chapter 17 Review ANSWERS

Building Vocabulary

For definitions, see: exotic rivers, p. 386; wadis, p. 387; fossil water, p. 387; Muslims, p. 388; Sunni, p. 388; Shia, p. 388; Qur'an, p. 389; OPEC, p. 389; embargo, p. 397; theocracy, p. 400

Reviewing the Main Ideas

1. Arabian Peninsula, plains of Tigris and Euphrates Rivers, mountains and plateaus (NGS 4)
2. oil (NGS 4)
3. Saudi Arabia, Iraq, Iran, Kuwait, Qatar, UAE, Oman, Yemen (NGS 4)
4. Qatar (NGS 4)
5. tension between countries, disrupted governments, caused internal problems (NGS 4)

Understanding Environment and Society

Information included should be consistent with text material. Students should conclude that the economy would suffer greatly and there would be a decline in wealth. Use Rubric 29, Presentations, to evaluate student work.

▲ **Chart Answer**

around 46 times greater

ASSESS

Have students complete the Chapter 17 Test.

RETEACH

Have students work individually or in groups to create poster collages that portray each of the chapter's sections. Each poster should provide answers to the section's Read to Discover questions. Collages might include realistic drawings, symbols, student-generated newspaper headlines or poems, magazine and newspaper clippings, bumper stickers, and so on. Encourage students to be creative. **ENGLISH LANGUAGE LEARNERS**

PORTFOLIO EXTENSIONS

1. Have students conduct research on Arab civilization during the Dark Ages (roughly A.D. 467 to 1000) and write a report contrasting it with European life during the same time period.
2. Have students conduct research on the conventions of Islamic art. *(It cannot include depictions of animals or people.)* Then have them create a "name poem" of an important person, place, or term associated with Southwest Asia and decorate it in Islamic style, with plant forms and geometric patterns. A name poem arranges the letters of the name vertically, then for each letter provides a phrase that pertains to the poem's subject and that begins with the letter.

Chapter 17 Review ANSWERS

Thinking Critically

1. Economies are hurt when oil prices fall.
2. the region's desert climate and lack of water
3. Answers will vary but should indicate an understanding of the political and economic situation of Iran and the region.
4. to protect independent countries and oil supplies in the area
5. main religion of the area; daily prayer, modesty, pilgrimage to Mecca, month of fasting called Ramadan

Map Activity

A. Baghdad
B. Kabul
C. Riyadh
D. Tigris River
E. Euphrates River
F. Mecca
G. Hindu Kush
H. Khyber Pass
I. Tehran
J. Persian Gulf

Chapter 17 Reviewing What You Know

Building Vocabulary

On a separate sheet of paper, write sentences to define each of the following words.

1. exotic rivers
2. wadis
3. fossil water
4. Muslims
5. Sunni
6. Shia
7. Qur'an
8. OPEC
9. embargo
10. theocracy

Reviewing the Main Ideas

1. *Places and Regions* What are the three landform regions that make up this part of Southwest Asia?
2. *Places and Regions* What is the most important resource of this region?
3. *Places and Regions* Which countries in this area have large reserves of oil?
4. *Places and Regions* Which small country has some of the largest natural gas reserves in the world?
5. *Places and Regions* How have wars affected politics in this region?

Understanding Environment and Society

Resource Use

Use the graph in Section 2 and other print sources to create a presentation on the importance of the oil industry in one or more of the countries discussed in this chapter. As you prepare your presentation you may want to think about the following:

- How the oil industry affects people's daily lives in this region.
- What might happen to this region's economy if scientists discover an inexpensive alternative to oil.

Thinking Critically

1. Why might it be dangerous for these countries' economies to be almost entirely dependent on the sale of oil?
2. Some of the countries of Southwest Asia are trying to increase crop production. What environmental factors might make this difficult?
3. How might Iran's recent history have been different if the 1979 revolution had not taken place?
4. Why has the United States become involved in the politics of this region?
5. What is Islam, and what are some of its practices and holidays?

FOOD FESTIVAL

This Persian apple dessert uses rose water or orange-flower water—typical Southwest Asian ingredients—for flavoring. Both are available at gourmet and specialty food shops. Peel and cut up 3 medium apples. Place the apples in a blender with 2½ tbs sugar, 2 tbs lemon juice, 1 tbs rose or orange-flower water, and a pinch of salt. Process 20 to 30 seconds, until coarsely chopped.

CHAPTER 17 REVIEW AND ASSESSMENT RESOURCES

Reproducible
- Readings in World Geography, History, and Culture 47, 48, and 49
- Critical Thinking Activity 17
- Vocabulary Activity 17

Technology
- Chapter 17 Test Generator (on the One-Stop Planner)
- Audio CD Program, Chapter 17
- HRW Go site

Reinforcement, Review, and Assessment
- Chapter Review, pp. 402–03
- Chapter 17 Tutorial for Students, Parents, Mentors, and Peers
- Chapter 17 Test
- Chapter 17 Test for English Language Learners and Special-Needs Students

Building Social Studies Skills

Map ACTIVITY

On a separate sheet of paper, match the letters on the map with their correct labels.

Persian Gulf	**Riyadh**
Tigris River	**Baghdad**
Euphrates River	**Tehran**
Hindu Kush	**Khyber Pass**
Mecca	**Kabul**

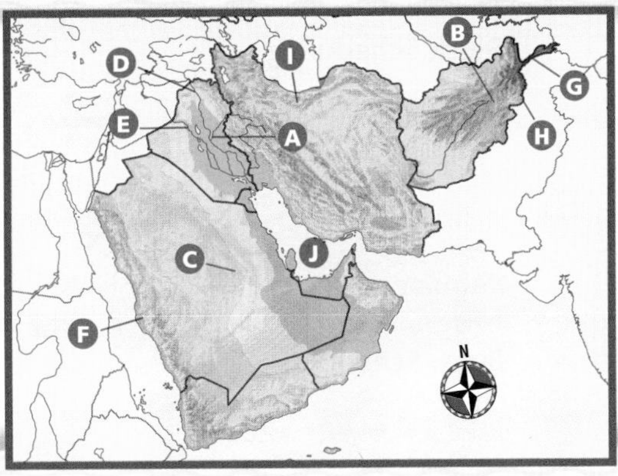

Mental Mapping Skills ACTIVITY

On a separate sheet of paper, draw a freehand map of the region. Make a key for your map and label the following:

Afghanistan	**Qatar**
Iran	**Saudi Arabia**
Iraq	**United Arab Emirates**
Kuwait	
Oman	**Yemen**

WRITING ACTIVITY

Imagine that you have been asked to write a travel brochure about one of the following countries: Afghanistan, Iran, Iraq, or Saudi Arabia. Use your textbook, the library, and the Internet to research your country. Then write a few paragraphs about its climate, attractions, and cultural events.

Alternative Assessment

Portfolio ACTIVITY

Learning About Your Local Geography

Individual Project Contact your local utility company for information on where your electricity is generated and what fuel or energy source is used. Draw a diagram of your local power supply system.

internet connect

Internet Activity: **go.hrw.com**
KEYWORD: **SK3 GT17**

Choose a topic to explore about the Arabian Peninsula, Iraq, Iran, and Afghanistan:
- Visit Mesopotamia.
- Discover the importance of camels.
- See Arabian arts and crafts.

Mental Mapping Skills Activity
Maps will vary, but listed places should be labeled in their approximate locations.

Writing Activity
Answers will vary depending on the country each student selects. Paragraphs should include information about climate, interesting sites, cultural events, and other information that might be useful for a travel brochure. Encourage students to add pictures or illustrations to their brochure. Use Rubric 42, Writing to Inform, to evaluate student work.

Portfolio Activity
Answers will vary, but paragraphs should indicate an understanding of the local power supply system. Use Rubric 13, Graphic Organizers, to evaluate student work.

internet connect

GO TO: **go.hrw.com**
KEYWORD: **SK3 Teacher**
FOR: a guide to using the Internet in your classroom

CHAPTER 18

The Eastern Peninsula

Chapter Resource Manager

Objectives	Pacing Guide	Reproducible Resources
SECTION 1 **Physical Geography** (pp. 405–07) **1.** Identify the main physical features of the eastern Mediterranean. **2.** Identify the climate types of the region. **3.** Identify the natural resources found in this area.	**Regular** .5 day **Block Scheduling** .5 day *Block Scheduling Handbook, Chapter 18*	**RS** Guided Reading Strategy 18.1 **RS** Graphic Organizer 18
SECTION 2 **Turkey** (pp. 408–11) **1.** Describe the history of the area that is now Turkey. **2.** Describe the government and economy of Turkey. **3.** Describe how Turkish society is divided.	**Regular** 1 day **Block Scheduling** .5 day *Block Scheduling Handbook, Chapter 18*	**RS** Guided Reading Strategy 18.2 **E** Cultures of the World Activity 5
SECTION 3 **Israel and the Occupied Territories** (pp. 412–15) **1.** Describe the early history of Israel. **2.** Describe modern Israel. **3.** Describe the conflict over the Occupied Territories.	**Regular** 1 day **Block Scheduling** .5 day *Block Scheduling Handbook, Chapter 18*	**RS** Guided Reading Strategy 18.3 **SM** Geography for Life Activity 18 **IC** Interdisciplinary Activities for the Middle Grades 22 **SM** Map Activity 18
SECTION 4 **Syria, Lebanon, and Jordan** (pp. 416–19) **1.** Describe the government and economy of Syria. **2.** Describe how Lebanese society is divided. **3.** Identify the events that have shaped the history of Jordan.	**Regular** 1 day **Block Scheduling** .5 day *Block Scheduling Handbook, Chapter 18*	**RS** Guided Reading Strategy 18.4

Chapter Resource Key

RS Reading Support

IC Interdisciplinary Connections

E Enrichment

SM Skills Mastery

A Assessment

REV Review

ELL Reinforcement and English Language Learners

Transparencies

CD–ROM

Music

Video

Internet

Holt Presentation Maker Using Microsoft® Powerpoint®

One-Stop Planner CD–ROM

See the ***One-Stop Planner*** for a complete list of additional resources for students and teachers.

One-Stop Planner CD–ROM

It's easy to plan lessons, select resources, and print out materials for your students when you use the ***One-Stop Planner CD–ROM with Test Generator.***

Technology Resources	Review, Reinforcement, and Assessment Resources
One-Stop Planner CD–ROM, Lesson 18.1 *ARGWorld* CD–ROM: Control of Water in Southwest Asia Geography and Cultures Visual Resources with Teaching Activities 37–41 Homework Practice Online HRW Go site	**ELL** Main Idea Activity 18.1 **REV** Section 1 Review, p. 407 **A** Daily Quiz 18.1 **ELL** English Audio Summary 18.1 **ELL** Spanish Audio Summary 18.1
One-Stop Planner CD–ROM, Lesson 18.2 Homework Practice Online HRW Go site	**ELL** Main Idea Activity 18.2 **REV** Section 2 Review, p. 411 **A** Daily Quiz 18.2 **ELL** English Audio Summary 18.2 **ELL** Spanish Audio Summary 18.2
One-Stop Planner CD–ROM, Lesson 18.3 Homework Practice Online HRW Go site	**ELL** Main Idea Activity 18.3 **REV** Section 3 Review, p. 415 **A** Daily Quiz 18.3 **ELL** English Audio Summary 18.3 **ELL** Spanish Audio Summary 18.3
One-Stop Planner CD–ROM, Lesson 18.4 Homework Practice Online HRW Go site	**ELL** Main Idea Activity 18.4 **REV** Section 4 Review, p. 419 **A** Daily Quiz 18.4 **ELL** English Audio Summary 18.4 **ELL** Spanish Audio Summary 18.4

internet connect

HRW ONLINE RESOURCES

GO TO: go.hrw.com
Then type in a keyword.

TEACHER HOME PAGE
KEYWORD: SK3 TEACHER

CHAPTER INTERNET ACTIVITIES
KEYWORD: SK3 GT18

Choose an activity to:
- visit the Dead Sea.
- compare Israeli and Arab foods.
- travel to historic Jerusalem.

CHAPTER ENRICHMENT LINKS
KEYWORD: SK3 CH18

CHAPTER MAPS
KEYWORD: SK3 MAPS18

ONLINE ASSESSMENT
Homework Practice
KEYWORD: SK3 HP18
Standardized Test Prep Online
KEYWORD: SK3 STP18
Rubrics
KEYWORD: SS Rubrics

COUNTRY INFORMATION
KEYWORD: SK3 Almanac

CONTENT UPDATES
KEYWORD: SS Content Updates

HOLT PRESENTATION MAKER
KEYWORD: SK3 PPT18

ONLINE READING SUPPORT
KEYWORD: SS Strategies

CURRENT EVENTS
KEYWORD: S3 Current Events

Meeting Individual Needs

Ability Levels

Level 1 Basic-level activities designed for all students encountering new material

Level 2 Intermediate-level activities designed for average students

Level 3 Challenging activities designed for honors and gifted-and-talented students

English Language Learners Activities that address the needs of students with Limited English Proficiency

Chapter Review and Assessment

- **IC** Interdisciplinary Activities for the Middle Grades 24
- **E** Readings in World Geography, History, and Culture 50, 51, and 52
- **SM** Critical Thinking Activity 18
- **REV** Chapter 18 Review, pp. 420–21
- **REV** Chapter 18 Tutorial for Students, Parents, Mentors, and Peers
- **ELL** Vocabulary Activity 18
- **A** Chapter 18 Test
- **A** Unit 5 Test
- Chapter 18 Test Generator (on the One-Stop Planner)
- Audio CD Program, Chapter 18
- **A** Chapter 18 Test for English Language Learners and Special-Needs Students
- **A** Unit 5 Test for English Language Learners and Special-Needs Students

CHAPTER 18

Launch into Learning

Copy the names of the countries of the eastern Mediterranean region—*Turkey, Lebanon, Syria, Jordan,* and *Israel*—onto the chalkboard, leaving several inches of space between each country's name. Ask students to respond by sharing facts that they know about a particular country. *(Students might mention that part of Turkey is in Asia and the other part of the country is in Europe. They might also mention that the modern state of Israel was created after World War II, or that Lebanon was torn apart by ethnic and religious conflicts.)* Write each response under the name of the appropriate country. Tell students that they will learn more about the eastern Mediterranean region in this chapter.

Section 1

Objectives

1. Identify the main physical features of the eastern Mediterranean.
2. Describe the climate types of this region.
3. List this region's natural resources.

Links to Our Lives

You might point out to students that there are many reasons why we should know more about the countries of the eastern Mediterranean, these among them:

- The eastern Mediterranean region, where Europe, Asia, and Africa meet, is a geographical, economic, and cultural crossroads. The region has made many contributions to world civilization. It has also been the scene of many conflicts over the years.
- Judaism and Christianity began in the region. Both faiths have affected American culture, no matter what religion one might follow.
- The United States has been a supporter of Israel since its creation. This policy has contributed to anti-American feelings and actions among some Arabs.

CHAPTER 18

The Eastern Mediterranean

The next student we will meet lives in Turkey, a country that lies partly in Europe, partly in Asia.

My name is Adalet, and I am in the tenth grade at Ted College, a private school in Ankara, the capital of Turkey. I live in an apartment a little outside the city with my mom and dad. We live on the twelfth floor, and have a view of the city, the distant mountains, and of course the parking lot. In the summers, my favorite time is when I can go to stay with my grandma and my grandpa in their summer house on the Aegean Sea, in Kusadasi near the ancient Greek city of Ephesus. I sleep until 11:00 A.M. or noon, then spend the day at the beach with my friends until the sun goes down.

On school days, from September to June, I get up at 8:00 A.M., put on my school uniform, and have breakfast of corn flakes or bread and cheese with milk or tea. At school we go directly to our classes. I am studying biology, physics, algebra, geometry, history, Turkish, and English. The English, science, and math classes are taught in English, the others in Turkish.

Türkiye'den selamlar!

Translation: Greetings from Turkey!

FOCUS

LET'S GET STARTED

Copy the following instructions onto the chalkboard: *Look at the map on this page. Write down all the intersections you see occurring in the eastern Mediterranean.* Discuss student responses. *(Students should note the intersection of continents as well as countries.)* Tell students that in Section 1 they will learn more about the eastern Mediterranean.

Using the Physical-Political Map

Have students examine the map on this page. Call attention to the shapes of the countries and the region's landforms. Point out that the region has few natural boundaries. Tell students that they will learn how the borders of the eastern Mediterranean countries changed several times during the 1900s as a result of political events.

Building Vocabulary

Write the terms **phosphates** and **asphalt** on the chalkboard. Have volunteers use a dictionary to search for the definitions and roots of the terms and read them aloud. Point out to students that both words are derived from Greek. The root word *phós* means light, and the root word *asphaltos* means to make firm or secure.

Section 1 Physical Geography

Read to Discover

1. What are the main physical features of the eastern Mediterranean?
2. What are the climate types of the region?
3. What natural resources are found in this area?

Define

phosphates
asphalt

Locate

Dardanelles
Bosporus
Sea of Marmara
Jordan River
Dead Sea
Syrian Desert
Negev

An ancient temple in Petra, Jordan

WHY IT MATTERS

The people who live in the Eastern Mediterranean region have needed irrigation and other technology to adapt to the land. Use CNNfyi.com or other **current events** sources to find information on ways that other people have modified their environment. Record your findings in your journal.

The Eastern Mediterranean: Physical-Political

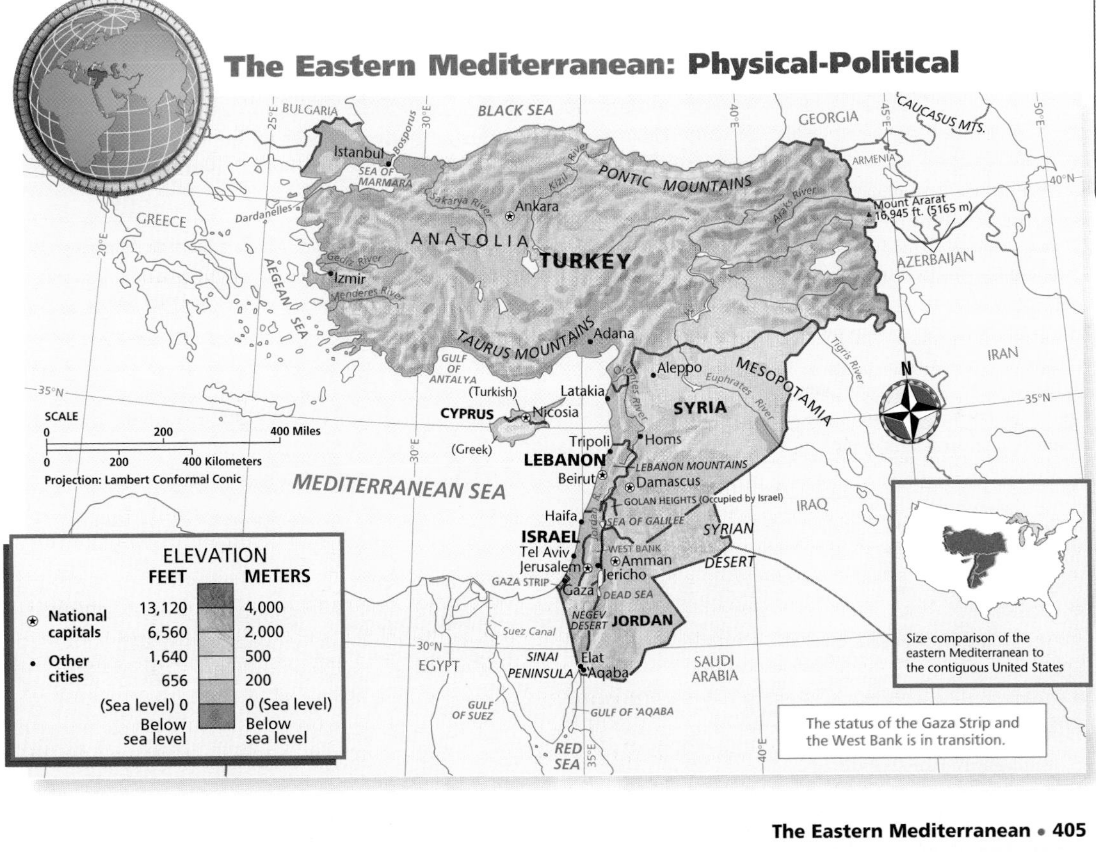

Section 1 RESOURCES

Reproducible

- Block Scheduling Handbook, Chapter 18
- Guided Reading Strategy 18.1
- Graphic Organizer 18

Technology

- One-Stop Planner CD–ROM, Lesson 18.1
- Homework Practice Online
- Geography and Cultures Visual Resources with Teaching Activities 37–41
- *ARGWorld* CD–ROM: Control of Water in Southeast Asia
- HRW Go site

Reinforcement, Review, and Assessment

- Section 1 Review, p. 407
- Daily Quiz 18.1
- Main Idea Activity 18.1
- English Audio Summary 18.1
- Spanish Audio Summary 18.1

TEACH

Teaching Objectives 1–3

ALL LEVELS: (Suggested time: 30 min.) Copy the graphic organizer onto the chalkboard, omitting the italicized answers. Use it to help students learn about the physical features, climate types, and resources of the eastern Mediterranean region. Have students copy the organizer into their notebooks and complete it. **ENGLISH LANGUAGE LEARNERS**

THE EASTERN MEDITERRANEAN

Physical Features	Climate Types	Resources
Balkan Peninsula, Dardanelles, Bosporus, Sea of Marmara, Euphrates River, Jordan River, Dead Sea, Syrian Desert, Negev	*Mediterranean, desert, humid subtropical*	*limited farmland and pastureland, sulfur, mercury, phosphates, asphalt*

EYE ON EARTH

One of the desert climate areas of this region is the Negev. By the 1960s many native species that roamed the Negev during biblical times had become endangered or extinct. Conservationists launched programs to protect and reintroduce wildlife.

Today, three fourths of the Negev have been designated as nature preserves. Visitors can see foxes, gazelles, hyenas, ostriches, wolves, and zebras. The onager, a type of donkey, and the oryx, a type of antelope, have also returned to the Negev. With more prey available, the leopard again prowls the Negev. Because it is at the top of the desert's food chain, the leopard's return shows that the full range of the area's wildlife is being restored.

Activity: Have students use pictures of the Negev's animals to create a travel poster for the region.

internet connect

GO TO: go.hrw.com
KEYWORD: SK3 CH18
FOR: Web sites about the Negev

Visual Record Answer

possible answer: connects the Black Sea area and Russia to the Mediterranean world

internet connect

GO TO: go.hrw.com
KEYWORD: SK3 CH18
FOR: Web sites about the eastern Mediterranean

Our Amazing Planet

The Dead Sea, which covers an area of just 394 square miles (1,020 sq km), contains approximately 12.7 billion tons of salt. Each year the Jordan River deposits 850,000 additional tons.

Physical Features

The countries of the eastern Mediterranean are Turkey, Lebanon, Syria, Jordan, and Israel. In addition to its own territory, Israel controls areas known as the Occupied Territories. These include the West Bank, the Gaza Strip, and the Golan Heights.

On Two Continents The eastern Mediterranean region straddles two continents. A small part of Turkey lies on Europe's Balkan Peninsula. This area consists of rolling plains and hills. A narrow waterway, made up of the Dardanelles (dahrd-uhn-ELZ), the Bosporus (BAHS-puh-ruhs), and the Sea of Marmara (MAHR-muh-ruh), separates Europe from Asia. The larger, Asian part of Turkey is mostly plateaus and highlands.

Hills, Valleys, and Plains Heading south from Turkey and into Syria, we cross a narrow plain. The Euphrates River, fed by precipitation in Turkey's eastern mountains, flows southeast through this plain. Farther south are more hills and plateaus. Two main ridges run north-south. One runs from southwestern Syria through western Jordan. The other, closer to the coast, runs through Lebanon, Israel, and the West Bank. The Jordan River valley separates these two ridges. A narrow coastal plain rims the region along its seacoasts. In western Turkey the coastal plain is wider.

River and Sea The Jordan River begins in Syria and flows south. Israel and the West Bank lie on the west side of the river. The country of Jordan lies on the east side. The Jordan River flows into the Dead Sea. This unusual body of water is the lowest point on any continent—1,312 feet (400 m) below sea level. It is so salty that swimmers cannot sink in it.

✓ **READING CHECK:** *Places and Regions* What are the region's main physical features? rolling plains; hills; narrow waterway of Dardanelles, Bosporus, and Sea of Marmara; plains; Euphrates River; Jordan River; Dead Sea

Climate

Dry climates are the rule in most of this region. However, there are important variations. Turkey's Black Sea coast and the Mediterranean coast all the way to Israel have a Mediterranean climate. Central Syria

Istanbul lies on both sides of the waterway known as the Bosporus. This waterway divides Europe from Asia.

Interpreting the Visual Record **Why do you think the Bosporus would be an important crossroads of trade?**

Close

Remind students that this region is at the intersection of three continents—Asia, Europe, and Africa. Ask students to describe the possible economic and cultural consequences of living in such a region. *(Students might mention trade and cultural exchanges. They might also mention geopolitical struggles for control over a strategically important region.)*

Reteach

Have students complete Main Idea Activity 18.1. Then organize students into small groups. Have each group create a map for one of the eastern Mediterranean countries. Each map should identify the main physical features, climate types, and resources of the assigned country. **COOPERATIVE LEARNING, ENGLISH LANGUAGE LEARNERS**

Review and Assess

Have students complete the Section Review. Then have students work in groups of three to create eight matching questions that relate each country to its corresponding landforms, climates, or natural features. Finally, have students complete Daily Quiz 18.1. **COOPERATIVE LEARNING**

Extend

Have interested students conduct research on the cedars of Lebanon and how deforestation changed the country's landscape. **BLOCK SCHEDULING**

---

and lands farther south have a desert climate. A small area of northeastern Turkey has a humid subtropical climate.

The Syrian Desert covers much of Syria and Jordan. It usually receives less than five inches (12.7 cm) of rainfall a year. Another desert, the Negev (NE-gev), lies in southern Israel.

✓ **READING CHECK:** *Places and Regions* What are the climates of the eastern Mediterranean? Mediterranean, desert, humid subtropical, desert

Resources

Unlike nearby countries in Southwest Asia, the countries of the eastern Mediterranean do not have large oil reserves. The people of this region make their living from the land in other ways.

Limited Farming Commercial farming is possible only where rain or irrigation provides enough water. Subsistence farming and livestock herding are common in drier areas. Desert areas support a few nomadic herders.

Mineral Resources Many minerals, including sulfur, mercury, and copper, are found in the region. **Phosphates**—mineral salts containing the element phosphorus—are produced in Syria, Jordan, and Israel. Phosphates are used to make fertilizers. The area also exports **asphalt**—the dark tar-like material used to pave streets. The Dead Sea is a source of mineral salts.

✓ **READING CHECK:** *Places and Regions* What natural resources are found in this area? minerals, including sulfur, mercury, copper, phosphates, asphalt

A shepherd in eastern Turkey watches his sheep.

Interpreting the Visual Record Why is livestock herding common in many parts of the eastern Mediterranean region?

▼

Homework Practice Online

Keyword: SK3 HP18

Section 1 Review

Define and explain: phosphates, asphalt

Working with Sketch Maps On a map of the eastern Mediterranean that you draw or that your teacher provides, label the following: Dardanelles, Bosporus, Sea of Marmara, Jordan River, Dead Sea, Syrian Desert, and the Negev.

Reading for the Main Idea

1. *Places and Regions* What country lies on two continents?
2. *Places and Regions* What are the most common climates of the region?
3. *Places and Regions* How do geographic factors affect the economic activities of the region?

Critical Thinking

4. **Drawing Inferences and Conclusions** How do you think the Dead Sea got its name?

Organizing What You Know

5. **Categorizing** Copy the following graphic organizer. Use it to list the major landforms and bodies of water of each region.

Country/ territory	Landforms and bodies of water

Section 1 Review

Answers

Define For definitions, see: phosphates, p. 407; asphalt, p. 407

Working with Sketch Maps Places should be labeled in their approximate locations.

Reading for the Main Idea

1. Turkey (NGS 4)
2. dry (NGS 4)
3. areas with rain or irrigation—commercial farming; drier areas—subsistence farming or herding; mineral resources support mining (NGS 4)

Critical Thinking

4. possible answer: from the lack of plant and animal life

Organizing What You Know

5. Turkey—hills, plains, plateaus, highlands, Dardanelles, Bosporus, Sea of Marmara, Euphrates River; Syria—mountain ridge, Jordan River, Syrian Desert; Lebanon—mountain ridge; Jordan—Jordan River, mountain ridge, Dead Sea, Syrian Desert; Israel and Occupied Territories—mountain ridge, Jordan River, Dead Sea, Negev

◄ **Visual Record Answer**

Climate is too dry for farming but supports enough vegetation for livestock.

Section 2

Objectives

1. Discuss the history of the area that is now Turkey.
2. Describe Turkey's government and economy.
3. Identify the divisions in Turkish society.

Focus

LET'S GET STARTED

Copy the following instructions onto the chalkboard: *Write down any images, words, or facts associated with the country of Turkey that come to mind.* Discuss student responses. *(Students might mention Turkish baths, the phrase "young Turks," or the fact that the Ottoman Empire was a world power.)* Tell students that in Section 2 they will learn more about Turkey.

Building Vocabulary

Write the term **secular** on the chalkboard. Have a volunteer read the definition aloud from the glossary. Ask students how this word might apply to American society. *(Students might mention that the separation of church and state is a principle of the U.S. Constitution.)*

Section 2 Resources

Reproducible
- Guided Reading Strategy 18.2
- Cultures of the World Activity 5

Technology
- One-Stop Planner CD–ROM, Lesson 18.2
- Homework Practice Online
- HRW Go site

Reinforcement, Review, and Assessment
- Section 2 Review, p. 411
- Daily Quiz 18.2
- Main Idea Activity 18.2

Section 2 Turkey

Read to Discover

1. What is the history of the area that is now Turkey?
2. What kind of government and economy does Turkey have?
3. How is Turkish society divided?

Define

secular

Locate

Ankara
Istanbul

A replica of the legendary Trojan Horse in Turkey

WHY IT MATTERS

Turkey built dams on the Tigris and Euphrates Rivers to provide electricity and water for its people. The dams created conflict with Turkey's neighbors, Syria and Iraq. Use CNNfyi.com or other **current events** sources to find information about water issues around the world. Record your findings in your journal.

▲ Ottoman monarchs lived in the Topkapi Palace built in 1462.

History

Turkey, except for the small part that lies in Europe, makes up a region called Asia Minor. In ancient times this area was part of the Hittite and Persian Empires. In the 330s B.C. Alexander the Great conquered Asia Minor. Later it became part of the Roman Empire. Byzantium, renamed Constantinople, was one of the most important cities of the empire. After the fall of Rome, Constantinople became the capital of the Byzantine Empire.

In the A.D. 1000s the Seljuk Turks invaded Asia Minor. The Seljuks were a nomadic people from Central Asia who had converted to Islam. In 1453 another Turkish people, the Ottoman Turks, captured the city of Constantinople. They made it the capital of their Islamic empire.

Ottoman Empire During the 1500s and 1600s the Ottoman Empire was very powerful. It controlled territory in North Africa, Southwest Asia, and southeastern Europe. In the 1700s and 1800s the empire gradually weakened.

In World War I the Ottoman Empire fought on the losing side. When the war ended, the Ottomans lost all their territory outside of what is now Turkey. Greece even invaded western Asia Minor in an attempt to take more land. However, the Turkish army pushed out the invaders. Military officers

TEACH

Teaching Objective 1

ALL LEVELS: (Suggested time: 20 min.) Copy the following graphic organizer, omitting the italicized answers. Use it to help students understand the significance of individuals or groups on the history of this area. **ENGLISH LANGUAGE LEARNERS**

EVENTS AND ERAS IN TURKISH HISTORY

Alexander the Great conquers the area ⇨	*Seljuk Turks invade* ⇨	*Ottoman Turks invade and establish an empire* ⇨	*Kemal Atatürk gains power and establishes modern Turkey*

Teaching Objective 2

ALL LEVELS: (Suggested time: 30 min.) Have students write a letter to Kemal Atatürk to tell him about the legacy of his work, as shown in Turkey's government and economy. *(Students' letters should mention that Turkey is a secular country with a democratic government; the military has taken control of Turkey's government three times but has restored civilian control each time; Turkey's economy has a mix of traditional and modern industries.)* Ask volunteers to read their letters to the class.

◄ A boy holds a Turkish national flag during celebrations on Republic Day. A banner behind him shows Atatürk, the founder of modern Turkey.

then took over the government. Their leader was a war hero, Mustafa Kemal. He later adopted the name Kemal Atatürk, which means "father of the Turks." He formally dissolved the Ottoman Empire and created the nation of Turkey. He made the new country a democracy and moved the capital to Ankara. Constantinople was renamed Istanbul in 1930.

Modern Turkey Atatürk wanted to modernize Turkey. He believed that to be strong Turkey had to westernize. He banned the fez, the traditional hat of Turkish men, and required that they wear European-style hats. The Latin alphabet replaced the Arabic one. The European calendar and metric system replaced Islamic ones. Women were encouraged to vote, work, and hold office. New laws made it illegal for women to wear veils.

✓ **READING CHECK:** *Human Systems* What is the history of what is now Turkey? Asia Minor conquered by Alexander the Great; later part of Roman, Byzantine Empires; invaded by Seljuk Turks, Ottoman Turks; empire lost; Mustafa Kemal took over government

Government and Economy

Today, Turkey has a legislature called the National Assembly. A president and a prime minister share executive power. The Turkish military has taken over the government three times. However, each time it has returned power to civilian hands.

Although most of its people are Muslim, Turkey is a **secular** state. This means that religion is kept separate from government. For example, the religion of Islam allows a man to have up to four wives. However, by Turkish law a man is permitted to have just one wife. In recent years Islamic political parties have attempted to increase Islam's role in Turkish society.

Turkish women harvest grapes.

Interpreting the Visual Record **What kind of climate do Turkey's coastal plains have?** ▼

National Geography Standard 13

Human Systems A major challenge facing the Turkish government is rebuilding areas hit by a devastating earthquake. On August 17, 1999, the worst earthquake in 60 years struck Turkey. The quake measured 7.4 on the Richter scale, and the trembling was felt 180 miles (340 km) away. The quake actually pushed parts of Turkey several feet westward.

The death toll had reached more than 18,000 within a week. Countries around the world contributed to the recovery effort. Greece, Turkey's old adversary, was the first country to fly food and medicine to the victims. The two countries had had their last military confrontation only three years earlier.

Activity: Have students conduct research and report on the effects of the 1999 earthquake on Turkish citizens and on the relief efforts provided by many countries.

internet connect go.hrw.com

GO TO: go.hrw.com
KEYWORD: SK3 CH18
FOR: Web sites about modern earthquakes

◄ **Visual Record Answer**

a Mediterranean climate

Teaching Objective 3

LEVELS 1 AND 2: (Suggested time: 25 min.) Tell students to imagine that they are foreign-exchange students living in Turkey. Then pair students and have each pair write a postcard to a friend or family member that describes the cultural and political divisions in Turkish society. *(Postcards should mention the Kurdish desire for independence, the split between the urban middle class and villagers with a more traditional outlook, and political divisions between secularists and Islamists.)* Ask volunteers to read their postcards to the class.

ENGLISH LANGUAGE LEARNERS, COOPERATIVE LEARNING

LEVEL 3: (Suggested time: 30 min.) Tell students to imagine that they are Turks. Then pair students and have each pair write a dialogue about the divisions in Turkish society. *(See the Levels 1 and 2 lesson for the correct divisions.)* Have volunteers perform their dialogues for the class.

➤ASSIGNMENT: To help students learn more about Turkey's history, have them conduct research on some aspect of Turkish economic, political, or cultural history. Then have students prepare short oral reports to present their findings to the class.

Global Perspectives

One of the many ethnic groups of Turkey are the Kurds. Kurds make up about 20 percent of Turkey's population. Their ancient homeland, often called Kurdistan, includes southeastern Turkey and parts of northeastern Syria, northern Iraq, and western Iran. The Kurds speak languages unrelated to Turkish. Their traditional way of life is nomadic. National borders that were drawn after World War I disrupted the nomadic routes of the Kurds.

Many Kurds in all four countries want an independent state. By the late 1990s, a Kurdish rebellion in Turkey had lasted for 15 years and cost approximately 30,000 lives. The leader of Turkey's Kurdish rebels, Abdullah Ocalan, was arrested in 1999.

Activity: Have students locate the area known as Kurdistan on an outline map of Southwest Asia.

Connecting to Literature Answers

1. because the large number of people made too much noise
2. for all the people who died in the flood

Connecting to Literature

Statue of Gilgamesh

Gilgamesh is the hero of this ancient story that was popular all over Southwest Asia. In this passage Utnapishtim (oot-nuh-peesh-tuhm), whom the gods have given everlasting life, tells Gilgamesh about surviving a great flood. Utnapishtim does as he is told. He builds a boat and fills it with supplies, his family, and animals. Then terrible rains come and flood Earth.

Epic of Gilgamesh

"In those days . . . the people multiplied, the world bellowed like a wild bull, and the great god was aroused by the clamor[1]. Enlil (en-LIL) heard the clamor and he said to the gods in council, 'The uproar of mankind is intolerable[2] and sleep is no longer possible by reason of the babel[3].' So the gods agreed to exterminate[4] mankind. Enlil did this, but Ea (AY-uh) because of his oath warned me in a dream. . . . 'Tear down your house, I say, and build a boat. . . . then take up into the boat the seed of all living creatures.'

"When the seventh day dawned the storm from the south subsided, the sea grew calm, the flood was stilled; I looked at the face of the world and there was silence, all mankind was turned to clay. . . . I opened a hatch and the light fell on my face. Then I bowed low, I sat down and I wept, . . . for on every side was the waste of water."

Analyzing Primary Sources

1. Why did the god bring the flood?
2. Why does Utnapishtim cry?

Vocabulary [1]clamor: noise [2]intolerable: not bearable [3]babel: confusing noise [4]exterminate: kill off

Turkey's economy includes modern factories as well as village farming and craft making. The most important industries are clothing, chemicals, and oil processing. About 43 percent of Turkey's labor force works in agriculture. Grains, cotton, sugar beets, and hazelnuts are major crops. The Turkish economy has grown rapidly in recent years, but inflation is a problem. Large numbers of Turks have left Turkey in search of better jobs. As of 1994 an estimated 1.5 million Turks were working abroad to earn higher wages.

In the 1990s Turkey began building dams on the Tigris and Euphrates Rivers. These will provide electricity and irrigation water. However, the dams have caused concern for Syria and Iraq. They are disturbed that another country controls the sources of their water.

✓ **READING CHECK:** *Human Systems* What kind of government and economy does Turkey have? secular state with legislature and a shared executive; modern industrial, agricultural

Close

Tell students that Turkey has become a major international tourist destination. Ask students to explain the physical or cultural attractions that might draw tourists to Turkey. *(Students might mention the Bosporus waterway or Byzantine architecture.)*

Review and Assess

Have students complete the Section Review. Then organize students into groups of four. Have each group write 10 multiple-choice questions about the section. Then have students complete Daily Quiz 18.2. **COOPERATIVE LEARNING**

Reteach

Have students complete Main Idea Activity 18.2. Then have students write newspaper headlines for stories about events in Turkish history and the country's current situation. **ENGLISH LANGUAGE LEARNERS**

Extend

Have interested students conduct research on the history of the Turkish-Greek conflict over Cyprus. Tell students to also consider the current state of relations between Greece and Turkey. Have students create an annotated time line to present their findings. **BLOCK SCHEDULING**

People and Culture

Turkey has more than 64 million people. Ethnic Turks make up 80 percent of the population. Kurds are the largest minority. They are about 20 percent of the population. Since ancient times the Kurds have lived in what is today southeastern Turkey. Kurds also live in nearby parts of Iran, Iraq, and Syria. In the 1980s and 1990s some Kurds fought for independence from Turkey. The Turkish government has used military force against this rebellion.

Kemal Atatürk's changes created a cultural split between Turkey's urban middle class and rural villagers. The lifestyle and attitudes of middle-class Turks have much in common with those of middle-class Europeans. Most Turks, though, are more traditional. Islam influences their attitudes on matters such as the role of women. This cultural division is a factor in Turkish politics.

Turkish cooking is much like that of the rest of the Mediterranean region. It features olives, vegetables, cheese, yogurt, and bread. Shish kebab—grilled meat on a skewer—is a favorite Turkish dish.

✓ **READING CHECK:** *Human Systems* What are the divisions in Turkish society? Kurds; cultural, between urban and rural people

Crowds pass through a square near the University of Istanbul. Different styles of dress reflect the diverse attitudes that exist in Turkey today.

Section 2 Review

go.hrw.com Homework Practice Online Keyword: SK3 HP18

Define and explain: secular

Working with Sketch Maps On the map you created in Section 1, label Ankara and Istanbul. What are the advantages of Istanbul's location?

Reading for the Main Idea

1. *Human Systems* How did Atatürk try to modernize Turkey?
2. *Human Systems* What foods are popular in Turkish cooking?

Critical Thinking

3. **Finding the Main Idea** Why were some Turks unhappy with changes brought about by Atatürk?
4. **Analyzing Information** Why is the Turkish government building dams on the Tigris and Euphrates Rivers? How will this affect countries downriver?

Organizing What You Know

5. **Sequencing** Create a timeline listing major events in Turkey's history. List major people, groups, invasions, empires, and changes in government.

400 B.C. — A.D. 2000

Section 2 Review Answers

Define For definition, see: secular, p. 409

Working with Sketch Maps Istanbul's location places it at a strategic crossroads.

Reading for the Main Idea

1. banning fez; introducing Latin alphabet, European calendar, metric system; increasing women's role
2. olives, vegetables, cheese, yogurt, bread, shish kebab (NGS 12)

Critical Thinking

3. Many Turks preferred a more traditional way of life.
4. to provide electricity and irrigation water; may reduce the amount of water available to countries downstream

Organizing What You Know

5. 330s B.C.—Alexander the Great conquers Asia Minor; post-Alexander—Asia Minor becomes part of Roman then Byzantine Empires; A.D. 1000s—Seljuk Turks invade Asia Minor; 1453—Ottoman Turks invade; World War I—Ottoman Empire is lost; post–World War I—Atatürk establishes modern Turkey

Section 3

Objectives

1. Recount the early history of Israel.
2. Describe what modern Israel is like.
3. Explain the conflict over the Occupied Territories.

Focus

LET'S GET STARTED

Copy the following instructions onto the chalkboard: *Write down three or four things that you have heard about the country of Israel.* Compile a list of students' responses on the chalkboard. Tell students that the United States has political, cultural, and religious ties with Israel. Inform them that in Section 3 they will learn more about Israel and the Occupied Territories.

Building Vocabulary

Write the terms **Diaspora** and **Zionism** on the chalkboard. Have volunteers read aloud the definitions from the glossary. Point out to students that the words *Diaspora* and *Zionism* refer to two movements that occurred during different periods in history—the first to the dispersal of the Jewish population of ancient Israel, the second to the creation of a Jewish homeland in modern Israel.

Section 3 Resources

Reproducible

- Guided Reading Strategy 18.3
- Geography for Life Activity 18
- Interdisciplinary Activity for the Middle Grades 22
- Map Activity 18

Technology

- One-Stop Planner CD–ROM, Lesson 18.3
- Homework Practice Online
- HRW Go site

Reinforcement, Review, and Assessment

- Section 3 Review, p. 415
- Daily Quiz 18.3
- Main Idea Activity 18.3
- English Audio Summary 18.3
- Spanish Audio Summary 18.3

Section 3 Israel and the Occupied Territories

Read to Discover

1. What was the early history of Israel like?
2. What is modern Israel like?
3. What is the conflict over the Occupied Territories?

Define

Diaspora
Zionism

Locate

Jerusalem
Gaza Strip
Golan Heights
West Bank
Tel Aviv

Why It Matters

For thousands of years there has been conflict surrounding present-day Israel and the Occupied Territories. Use CNNfyi.com or other **current events** sources to find information about the current state of affairs in this region. Record your findings in your journal.

A fragment of the Dead Sea Scrolls

Ancient Israel

The Hebrews, the ancestors of the Jews, first established the kingdom of Israel about 3,000 years ago. It covered roughly the same area as the modern State of Israel. In the 60s B.C. the Roman Empire conquered the region, which they called Palestine. After a series of Jewish revolts, the Romans forced most Jews to leave the region. This scattering of the Jewish population is known as the **Diaspora**.

During the era of Roman control, a Jewish man named Jesus began preaching a new religion. He said he was the son of God. Jesus taught that faith and love for others were more important than Judaism's many laws about daily life. His teachings particularly appealed to the poor and powerless. Both Roman and Jewish rulers saw Jesus's teachings as dangerous. Jesus was tried and executed. His followers believe he was resurrected. Christianity—Jesus's teachings and the belief in his resurrection—spread through the Roman Empire. In time, Christianity became the most common religion of the Mediterranean region.

The Arabs conquered Palestine in the mid-600s. From the 1000s to the 1200s, European armies launched a series of invasions called the Crusades. The Crusaders captured Jerusalem in 1099. In time the Crusaders were pushed out of the area altogether.

The ancient port of Caesarea lies on the coast of Israel. It was the regional capital during the time of Roman control. Today the harbor structure is partly underwater.

TEACH

Teaching Objectives 1–2

ALL LEVELS: (Suggested time: 40 min.) Lead a discussion about events in the early history of Israel. *(Students should mention the ancient kingdom of Israel, Roman rule and the Diaspora, the spread of Christianity, the Arab conquest of Palestine, the Crusades, and Ottoman rule in Palestine.)* Then copy the following graphic organizer onto the chalkboard, omitting the italicized answers. Use it to help students learn about modern Israel. **ENGLISH LANGUAGE LEARNERS**

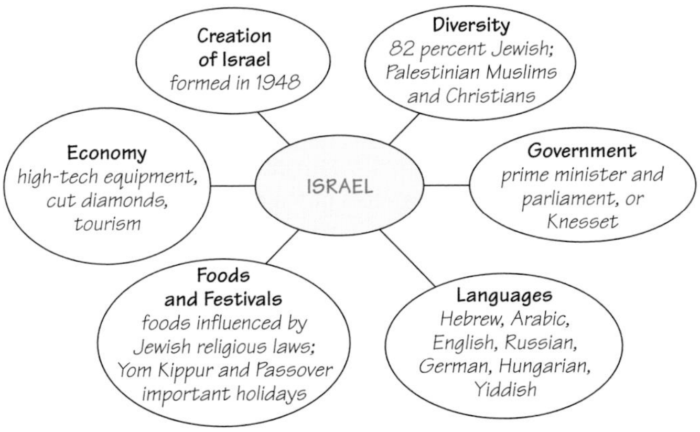

Israeli soldiers stand guard on a hill overlooking the Gaza Strip.

From the 1500s to World War I, Palestine was part of the Ottoman Empire. At the end of the war, it came under British control.

✓ **READING CHECK:** *Human Systems* What significant events occurred in the early history of the area of Israel? kingdom of Israel established 3,000 years ago; 60s B.C., conquered by Roman Empires; Christianity developed; mid-600s, conquered by Arabs, 1500s to World War I, Ottoman Empire control

Modern Israel

In the late 1800s a movement called **Zionism** began among European Jews. Zionism called for Jews to establish a country or community in Palestine. Tens of thousands of Jews moved to the area.

After World War II, the United Nations recommended dividing Palestine into Arab and Jewish states. This plan created conflict among the peoples there. Fighting broke out between Israel and surrounding Arab countries. The Israeli forces defeated the Arabs.

Many Palestinians fled to other Arab states, particularly to Jordan and Lebanon. Some used terrorist attacks to strike at Israel. Israel and its Arab neighbors also fought wars in 1956, 1967, and 1973.

Eastern Mediterranean

COUNTRY	POPULATION/ GROWTH RATE	LIFE EXPECTANCY	LITERACY RATE	PER CAPITA GDP
Israel*	5,938,093 / 1.6%	77, male 81, female	95%	$18,900
Jordan	5,153,378 / 3%	75, male 80, female	87%	$3,500
Lebanon	3,627,774 / 1.4%	69, male 74, female	86%	$5,000
Syria	16,728,808 / 2.5%	68, male 70, female	71%	$3,100
Turkey	66,493,970 / 1.2%	69, male 74, female	85%	$6,800
United States	281,421,906 / 0.9%	74, male 80, female	97%	$36,200

*Does not include the West Bank and Gaza Strip.
Sources: Central Intelligence Agency, *The World Factbook 2001;* U.S. Census Bureau

Interpreting the Chart **Which country in the region has the largest population?**

Government and Economy Israel has a prime minister and a parliament, called the Knesset. There are two major political parties and many smaller parties.

Israel has built a strong military for protection from the Arab countries around it. Terrorist attacks have also occurred. At age 18 most Israeli men and women must serve in the military.

DAILY LIFE

A unique element in Israel's economy is the use of kibbutzim. A kibbutz is a cooperative community. The country's 270 kibbutzim generate at least one third of Israel's farm output. In a kibbutz most property is owned jointly by the members, who share the work and decision making. Members live in small cabins, but meals are served in a communal dining hall. Clothing, laundry service, recreational facilities, medical care, and education are provided by the kibbutz. The living standard is relatively high. For example, some kibbutzim have cable television and swimming pools.

Critical Thinking: Why might people wish to live on a kibbutz?

Answer: Answers will vary, but students might mention a high living standard in a very small society that is essentially owned and run by its members.

Chart Answer

Turkey

Teacher to Teacher

Marcia Caldwell of Austin, Texas, suggests the following activity to help students learn about Israel and the Occupied Territories: Organize students into small groups and distribute one posterboard cube with eight-inch-wide faces to each group. Have students affix images and phrases that relate to the six essential elements of geography on the faces of the cube. You may wish to direct students to newspapers, magazines, and the Internet as they search for pertinent images and phrases. Hang the completed cubes from the ceiling around the classroom.

Teaching Objective 3

ALL LEVELS: (Suggested time: 30 min.) Pair students and have each pair draw and color a map showing Israel after the statehood declaration and today, with the Israeli-occupied territories. Then have each pair write a proposal to end the conflict over the Occupied Territories. The proposal should address the grievances of all countries involved and should attempt a compromise. *(Students should create compromise settlements to address the Israeli-Syrian dispute over Israel's annexation of the Golan Heights, Israel's occupation and settlement of the West Bank, and Israel's annexation of East Jerusalem.)*

ENGLISH LANGUAGE LEARNERS, COOPERATIVE LEARNING

Eye on Earth

Along with Jerusalem, the Dead Sea is a major tourist attraction for Israel. The Dead Sea is 51 miles (82 km) long. It is fed by the Jordan River, but it has no outlet to the sea. When the freshwater that comes in evaporates, it leaves behind a rich concentrate of minerals and salts. No wildlife can live in the salty water, which is so buoyant that swimmers cannot sink in it.

There are several resorts on the Dead Sea's shores. The sea's black mud is said to improve the complexion, and many people claim the water relieves their aching joints.

Activity: Have students research the mineral and salt content of the Dead Sea. As closely as possible, reproduce the sea's water in a glass.

The West Bank in Transition

The Gaza Strip is densely populated and has few natural resources. Israel captured this territory from Egypt in 1967.

▼

Israel has a modern, diverse economy. Items like high-technology equipment and cut diamonds are important exports. Tourism is a major industry. Israel's lack of water limits farming. However, using highly efficient irrigation, Israel has successfully increased food production. It imports grain but exports citrus fruit and eggs.

Languages and Diversity Israel's population includes Jews from all parts of the world. Both Hebrew and Arabic are official languages. When they arrive in Israel, many Jews speak English, Russian, German, Hungarian, Yiddish, or Arabic. The government provides classes to help them learn Hebrew.

About 82 percent of Israel's population is Jewish. The rest is mostly Palestinian. About three fourths of these are Muslim. The rest are Christian.

Food and Festivals Israeli food is influenced by Jewish religious laws. Jews are forbidden to eat pork and shellfish. They also cannot eat meat and milk products at the same meal. The country's food is as diverse as the population. Eastern European dishes are popular, as are Southwest Asian foods.

For Jews, Saturday is a holy day. Yom Kippur, the most important Jewish holiday, is celebrated in October. Passover, in the spring, celebrates the Hebrews' escape from captivity in Egypt. During Passover, people eat matzo (MAHT-suh), a special bread without yeast.

✓ **READING CHECK:** *Human Systems* What are modern Israel's government, economy, and culture like? gov.—prime minister, parliament; econ.—modern, diverse; culture—Jews, Palestinians, Muslims, Christians

The Occupied Territories

In 1967 Israel captured the Gaza Strip, the Golan Heights, and the West Bank. These are sometimes called the Occupied Territories.

Disputed Land The Gaza Strip is a small, crowded piece of coastal land. More than a million Palestinians live there. The area has almost no resources. The Golan Heights is a hilly area on the Syrian

CLOSE

Have students summarize the role Israel plays in world politics as well as its role as a religious center. *(Students might mention that Israel is the world's only Jewish state, that Israel has been at the center of many disputes, and that Israel draws many religious pilgrims to its holy sites.)* Ask them to speculate how Israel's role might change in the future.

REVIEW AND ASSESS

Have students complete the Section Review. Then pair students and have each pair create five true-or-false questions about Israel. Have pairs exchange questions with other pairs and have students within pairs quiz each other. Then have students complete Daily Quiz 18.3.
COOPERATIVE LEARNING

RETEACH

Have students complete Main Idea Activity 18.3. Then create an outline for the section on the chalkboard and have volunteers fill in facts under topics such as ancient and modern history, government and economy, people, and territories. **ENGLISH LANGUAGE LEARNERS**

EXTEND

Have interested students conduct research on the significance of certain Jewish religious holidays and observances, such as Yom Kippur and Rosh Hashanah. Then have students create posters showing how these holidays are observed. **BLOCK SCHEDULING**

border. In 1981 Israel formally declared the Golan Heights part of Israel. Syria still claims this territory.

The West Bank is the largest of the occupied areas, with a population of about 1.6 million. Since Israel took control of the West Bank, more than 100,000 Jews have moved into settlements there. The Palestinians consider this an invasion of their land. This has caused tension and violence between Arabs and Israelis.

Israel annexed East Jerusalem in 1980. Even before this, the Israeli government had moved the capital from Tel Aviv to Jerusalem. Most foreign countries have chosen not to recognize this transfer. The Palestinians still claim East Jerusalem as their rightful capital.

Control of Jerusalem is a difficult and often emotional question for Jews, Muslims, and Christians. The city contains sites that are holy to all three religions.

The Future of the Territories In the 1990s Israel agreed to turn over parts of the Occupied Territories to the Palestinians. In return, the Palestinian leadership—the Palestinian Authority—agreed to work for peace. Parts of the Gaza Strip and West Bank have been transferred to the Palestinian Authority. More areas of the West Bank are expected to be handed over in the future.

The future of the peace process is uncertain. Some Palestinian groups have continued to commit acts of terrorism. Some Jewish groups believe for religious reasons that Israel must not give up the West Bank. Other Israelis fear they would be open to attack if they withdrew from the territories.

✓ **READING CHECK:** *Human Systems* Why have the Occupied Territories been a source of conflict? Jews, Muslims, and Christians all have religious ties to Jerusalem.

Muslim women gather to pray at the Dome of the Rock in Jerusalem.

Section 3 Review

go.hrw.com Homework Practice Online
Keyword: SK3 HP18

Define and explain: Diaspora, Zionism

Working with Sketch Maps On the map you created in Section 2, label Jerusalem, Gaza Strip, Golan Heights, West Bank, and Tel Aviv.

Reading for the Main Idea

1. *Environment and Society* How has technology allowed Israel to increase its food production?
2. *Human Systems* What historical factors helped create a culturally diverse region in Israel?

Critical Thinking

3. **Finding the Main Idea** How do political boundaries in Israel create conflicts?
4. **Summarizing** What are some difficult issues involved in the Israeli-Palestinian peace process?

Organizing What You Know

5. **Sequencing** Copy the following graphic organizer. Use it to list the sequence of events that led to the formation of modern Israel.

[ ] ⇨ [ ] ⇨ [ ] ⇨ [ ] ⇨ [ ]

Section 3 Review

Answers

Define For definitions, see: Diaspora, p. 412; Zionism, p. 413

Working with Sketch Maps Maps will vary, but listed places should be labeled in their approximate locations.

Reading for the Main Idea

1. through water from irrigation (NGS 14)
2. When it was created, people moved to Israel from countries around the world. (NGS 9)

Critical Thinking

3. They cut across cultural lines, and each feels entitled to the land.
4. occupation of the Gaza Strip, Golan Heights, and West Bank; new settlements in the West Bank; annexation of East Jerusalem; Israeli government's moving to Jerusalem; control of Jerusalem

Organizing What You Know

5. Hebrews form ancient kingdom of Israel; Romans expel Jews in the Diaspora; Zionist movement begins; tens of thousands of Jews move to area; British control ends in 1948, and Jews establish State of Israel

Section 4

Objectives

1. Describe Syria's government and economy.
2. Discuss the divisions in Lebanese society.
3. Recount events that have shaped Jordan's history.

FOCUS

LET'S GET STARTED

Copy the following instructions onto the chalkboard: *Imagine that our school is made up of three distinct groups that have different languages, different religions, different loyalties, and different customs. What do you think our most difficult challenge would be?* Then have students discuss what links or separates cultures and societies. Tell them that this is the system that exists in Lebanon today. Tell students that in Section 4 they will learn more about Lebanon and its neighbors.

Building Vocabulary

Write the term **mandate** on the chalkboard. Ask students to suggest possible definitions. *(Some may suggest that a mandate is a law or guideline.)* Tell students that the word *mandate* also refers to a system of governing former colonies.

Section 4 RESOURCES

Reproducible
- Guided Reading Strategy 18.4

Technology
- One-Stop Planner CD–ROM, Lesson 18.4
- Homework Practice Online
- HRW Go site

Reinforcement, Review, and Assessment
- Section 4 Review, p. 419
- Daily Quiz 18.4
- Main Idea Activity 18.4
- English Audio Summary 18.4
- Spanish Audio Summary 18.4

Section 4 Syria, Lebanon, and Jordan

Read to Discover

1. What kind of government and economy does Syria have?
2. How is Lebanese society divided?
3. What events have shaped the history of Jordan?

Define

mandate

Locate

Damascus
Beirut
Amman

Spices in an outdoor market

WHY IT MATTERS

Syria, Jordan, and Lebanon have experienced conflicts due to religious or ethnic differences of people living there. Use CNNfyi.com or other **current events** sources to find information about other countries struggling with such conflicts. Record your findings in your journal.

Syria

The capital of Syria, Damascus, is believed to be the oldest continuously inhabited city in the world. For centuries it was a leading regional trade center. Syria became part of the Ottoman Empire in the 1500s. After World War I, France controlled Syria as a **mandate**. Mandates were former territories of the defeated nations of World War I. They were placed under the control of the winning countries after the war. Syria finally became independent in the 1940s.

Politics and Economy From 1971 to 2000, the Syrian government was led by Hafiz al-Assad. Assad increased the size of Syria's military. He wanted to match Israel's military strength and protect his rule from his enemies within Syria. Assad's son, Bashar, was elected president after his father's death in 2000.

Syria's government owns the country's oil refineries, larger electrical plants, railroads, and some factories. Syria's key manufactured goods are textiles, food products, and chemicals. Agriculture remains important.

Roman columns still stand in the ancient city of Apamea, Syria.

TEACH

Teaching Objective 1

ALL LEVELS: (Suggested time: 20 min.) Have each student write five headlines for articles discussing features of Syria's government and economy. *(Headlines should mention that Syria was ruled by one person—Hafiz al-Assad—until 2001; Assad increased the size of Syria's military; the government owns many of the country's utilities and industries; Syria's most important goods are textiles, food products and chemicals; agriculture remains a key part of the economy; or that Syria has limestone, basalt, and phosphates.)* **ENGLISH LANGUAGE LEARNERS**

Teaching Objective 2

LEVEL 1: (Suggested time: 40 min.) Pair students and have each pair create a graphic organizer to represent the divisions in Lebanese society. *(Organizers should show the religious division between Christians and Muslims; the division of Muslims into Sunni, Shia, and Druze; and the division of Christians into the Maronite majority and other smaller groups.)* To conclude, lead a discussion about possible ways in which such differences might be resolved through political means.
ENGLISH LANGUAGE LEARNERS, COOPERATIVE LEARNING

LEVELS 2 and 3: (Suggested time: 20 min.) Have students write a speech addressing the divisions in Lebanon's society. *(See the Level 1 lesson for the divisions.)* Speeches should offer possible political solutions.

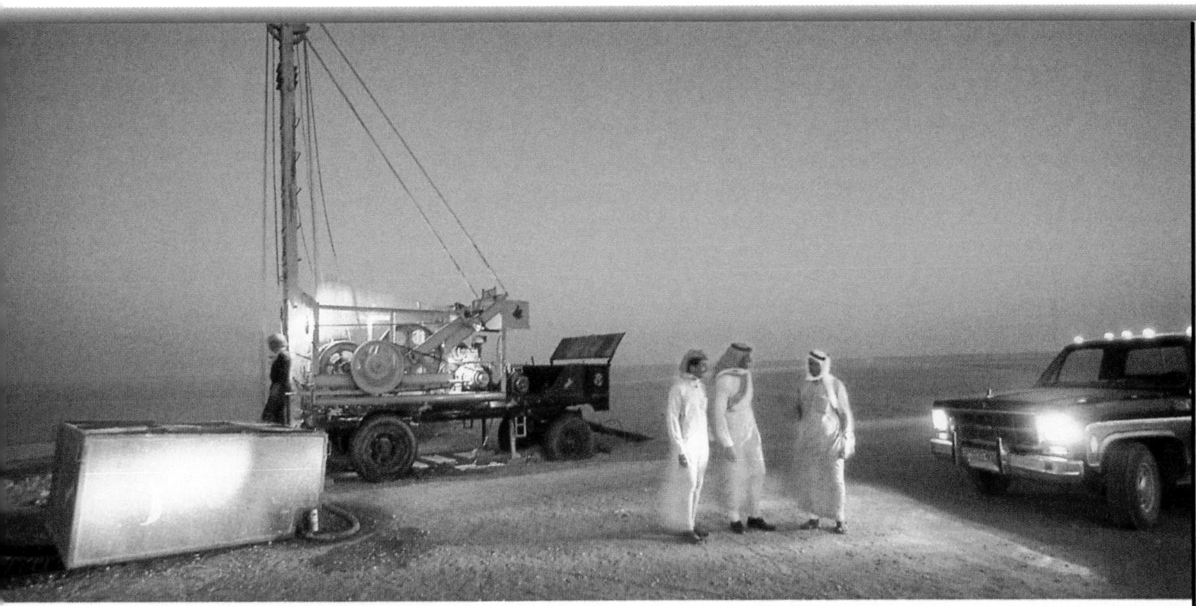

▲ People must drill for water in dry areas of Syria.

Interpreting the Visual Record
Judging from this photo, how has technology affected the lifestyle of people in desert areas?

Syria has only small deposits of oil and natural gas. It is rich in limestone, basalt, and phosphates.

People Syria's population of more than 17 million is about 90 percent Arab. The other 10 percent includes Kurds and Armenians. About 74 percent of Syrians are Sunni Muslim. Another 16 percent are Alawites and Druze, members of small branches of Islam. About 10 percent of Syrians are Christian. There are also small Jewish communities in some cities.

✓ **READING CHECK:** *Places and Regions* How is Syria's economy organized?
economy—mostly command, agricultural

Lebanon

Lebanon is a small, mountainous country on the Mediterranean coast. It is home to several different groups of people. At times these different groups have fought each other.

History and People During the Ottoman period many religious and ethnic minority groups settled in Lebanon. After World War I Lebanon, along with Syria, became a French mandate. Lebanon finally gained independence in the 1940s.

The Lebanese are overwhelmingly Arab, but they are divided by religion. Most Lebanese are either Muslim or Christian. Each of those groups is divided into several smaller groups. Muslims are divided into Sunni, Shia, and Druze. The Maronites are the largest of the Christian groups in Lebanon. At the time of independence, there were slightly more Christians than Muslims. Over time, however, Muslims became the majority.

▲ This photograph from the early 1900s shows a tall cedar tree in the mountains of northern Lebanon. Lebanon's cedars have long been a symbol of the country.

National Geography Standard 12

Human Systems Jubayl, formerly known as Byblos, is one of the oldest continuously inhabited cities in the world. The city lies just north of Beirut, Lebanon. It was a fishing village about 7,000 years ago. Later, the town became an important port from which cedar wood and oil were shipped to Egypt. During the 900s B.C. it was the main city in Phoenicia, a region that included present-day Lebanon.

Byblos was invaded several times. Today, the visitor can see many remnants of its long history, including Egyptian temples, a Roman amphitheater, and a castle built during the Crusades.

Activity: Have students find Jubayl on a large map or globe. Call on volunteers to point out possible routes that would have connected the city to Egypt and Rome. Lead a discussion about the advantages and disadvantages that the city's location presented.

▲ **Visual Record Answer**

provided modern transportation and mobile drilling equipment

Teaching Objective 3

ALL LEVELS: (Suggested time: 25 min.) Copy the graphic organizer at right onto the chalkboard, omitting the italicized answers. Use it to help students describe the history of Jordan. Organize the class into groups of three and have each group complete the organizer. In each group one student should identify important events in Jordan's history, another student should summarize these events, and the final student should comment on the significance and importance of each event. Have groups present their organizers to the class.

ENGLISH LANGUAGE LEARNERS, COOPERATIVE LEARNING

JORDAN'S HISTORY

Event	Summary	Significance
borders established	*Great Britain drew borders after WWI*	*foreign influence in creation of country*
monarchy established	*Abdullah named as monarch*	*British ally named to rule country*
independence	*full independence gained in 1940s*	*independence from foreign rule*
war of 1948	*war with Israel*	*annexed Arab lands of the West Bank*
Arab-Israeli wars	*wars between Arab nations and Israel*	*Palestinian refugees immigrate to Jordan*
rule of Hussein	*King Hussein rules from 1952 to 1999*	*considered one of the best rulers of region*

Section 4 Review

Answers

Define For definition, see: mandate, p. 416

Working with Sketch Maps Maps will vary, but listed places should be located accurately.

Reading for the Main Idea

1. ruled from 1971 to 2001 and increased the size of the military (NGS 12)
2. religious and political divisions between Christians and Muslims (NGS 13)
3. Jordan (NGS 4)

Critical Thinking

4. drew its borders, established its monarchy, immigrated there, and provided economic aid

Organizing What You Know

5. Answers should reflect information from the section.

Visual Record Answer

Beirut badly damaged, Lebanon's economy hurt

A vendor in Beirut sells postcards of what the city looked like before it was scarred by war.

Interpreting the Visual Record **What were the effects of the civil war in Lebanon?**

Civil War For some decades after independence, Christian and Muslim politicians managed to share power. A complex system assigned certain government positions to different religious groups. For example, the president was always a Maronite. However, over time this cooperation broke down. The poorest group, the Shia, grew rapidly but were not given additional power. Tensions mounted. Adding to the divisions between Lebanese was the presence of hundreds of thousands of Palestinian refugees living in Lebanon. Ethnic and religious groups armed themselves, and in the 1970s fighting broke out. Warfare between Lebanese groups lasted until 1990. Tens of thousands of people died, and the capital, Beirut, was badly damaged.

During the 1990s Lebanon's economy slowly recovered from the civil war. The refining of crude oil brought in by pipeline is a leading industry. Other industries include food processing, textiles, cement, chemicals, and jewelry making. Lebanese farmers produce tobacco, fruit, grains, and vegetables.

✓ **READING CHECK:** *Human Systems* What is causing divisions in Lebanese society? conflict between Christians, Muslims; Palestinian refugees

Jordan

Jordan's short history has been full of conflict. Great Britain drew its borders, and Jordan's royal family is actually from Arabia. The country has few resources and several powerful neighbors. In addition, most of its people think of another country as their homeland. Yet Jordan has survived.

History and Government The country of Jordan (called Transjordan until 1949) was created from Ottoman territory following World War I. The British controlled the area as a mandate. They established an Arabian prince named Abdullah as the monarch of the new country. Abdullah had helped the British in World War I, but he had been driven out of Saudi Arabia. In the 1940s the country became fully independent. After the creation of Israel and the war of 1948, Jordan annexed the Arab lands of the West Bank.

At the time of its independence, Jordan's population was small. Most Jordanians lived a nomadic or seminomadic life. After each of the Arab-Israeli wars of 1948 and 1967, hundreds of thousands of Palestinian Arab refugees came to live in Jordan. These immigrants strained Jordan's resources. In addition, a cultural division arose between the Palestinians and the "original" Jordanian Arabs. After 1967 Palestinians actually made up a majority of Jordan's people.

From 1952 to 1999 Jordan was ruled by King Hussein. Most observers, both inside and outside Jordan, considered him one of the

Jordanian children play with a camel. At one time a majority of Jordanians were nomads. Today most of Jordan's people live in urban areas.

CLOSE

Ask for volunteers to complete the following sentence: The most important challenge facing Syria, Jordan, or Lebanon is ______________. Encourage the class to discuss the answers. *(Students might mention relations with Israel, internal conflict, or improving the economy.)*

REVIEW AND ASSESS

Have students complete the Section Review. Then organize the class into three groups. Have each group represent one of the countries in this section and lead the class in a discussion on challenges facing its assigned country. Then have students complete Daily Quiz 18.4.

COOPERATIVE LEARNING

RETEACH

Have students complete Main Idea Activity 18.4. Have students use handwritten headlines, magazine pictures, newspaper articles, and their own drawings to create a collage that answers the Read to Discover questions.

ENGLISH LANGUAGE LEARNERS

EXTEND

Have interested students conduct research on King Hussein and Jordan's political role in this region. Have students present their findings in a report and speculate on how Jordan might be affected by his death.

BLOCK SCHEDULING

◄ A truck crosses the Jordan River near the Dead Sea. The river is a key source of water for agriculture and other uses.

Interpreting the Visual Record **What countries and territories border the Jordan River?**

best rulers in the region. Hussein's popularity allowed him to begin some democratic reforms in the 1980s and 1990s. Today, the division between Palestinian and Jordanian Arabs causes less conflict.

Economy and Resources Jordan is a poor country with limited resources. The country does produce phosphates, cement, and potash. Tourism and banking are becoming important industries. Jordan depends on economic aid from the oil-rich Arab nations and the United States. Amman, the capital, is Jordan's only large city.

Jordanian farmers raise fruits and vegetables in the Jordan River valley, using irrigation. Some highland areas receive enough winter rainfall to grow grains. Raising sheep and goats is an important source of income. However, overgrazing has caused soil erosion. A crucial resource issue for Jordan is its shortage of water.

✓ **READING CHECK:** *Human Systems* How did King Hussein affect Jordan's history? He was considered by many to be one of the best rulers in the region and enacted democratic reforms.

BUILD on WHAT You Know

Do you remember what you learned about uneven resource distribution? See Chapter 5 to review.

go.hrw.com **Homework Practice Online**
Keyword: SK3 HP18

Section 4 Review

Define and explain: mandate

Working with Sketch Maps On the map you created in Section 3, label Damascus, Beirut, and Amman.

Reading for the Main Idea

1. *Human Systems* How did Hafiz al-Assad affect Syria?
2. *Human Systems* What divisions led to conflict in Lebanon?
3. *Places and Regions* Which of the countries discussed in this section does not border the Mediterranean Sea?

Critical Thinking

4. **Finding the Main Idea** How have foreign countries influenced Jordan?

Organizing What You Know

5. **Categorizing** Use the graphic organizer to gather information about Syria, Lebanon, and Jordan.

	Syria	Lebanon	Jordan
Major religion(s)			
Type of government			
Major problem(s)			
Greatest strength(s)			

CHAPTER 18

Review ANSWERS

Building Vocabulary

For definitions, see: phosphates, p. 407; asphalt, p. 407; secular, p. 409; Diaspora, p. 412; Zionism, p. 413; mandate, p. 416

Reviewing the Main Ideas

1. Asia, Africa, and Europe (NGS 10)
2. Islam, Christianity, Judaism (NGS 10)
3. modernized factories, changed alphabet and clothing laws, changed calendar, encouraged women to work, created a cultural split (NGS 10)
4. Several countries believe that they should control these territories. (NGS 13)
5. by using irrigation for agriculture in dry areas, through animal herding and mining in areas with mineral resources (NGS 14)

▲ **Visual Record Answer**

Lebanon, Syria, Israel, Jordan, Golan Heights, and the West Bank

ASSESS

Have students complete the Chapter 18 Test.

RETEACH

Organize the class into five groups. Assign each group one of the countries of this region. Have each group write eight questions about its assigned country. When all groups have completed their questions, collect them and use them for a chapter-review game. Have groups compete against each other to see which group can answer the most questions correctly. **COOPERATIVE LEARNING**

PORTFOLIO EXTENSIONS

PORTFOLIO

1. Have students conduct research on the Crusades and write a story from the point of view of a young Muslim soldier fighting for Salāh ad-Din. Have each of your students include a discussion of military tactics, equipment, fighting techniques, and the motives of both sides.
2. Have students draw and embellish a map of Jerusalem showing the sacred sites of the three major religions. Students should write appropriate captions for each site that explain the relationship between the religious ideas about the site and the culture of the region.

CHAPTER 18

Review ANSWERS

Understanding Environment and Society

Information should be consistent with text. Presentations should focus on water sources and use in the area and include problems that could occur from a lack of freshwater. Use Rubric 29, Presentations, to evaluate student work.

Thinking Critically

1. to move it farther into Turkey and to separate it from the old Ottoman Empire's capital site
2. People different backgrounds have moved to Israel, making it culturally diverse.
3. high concentration of salt; mineral salts used in chemical products
4. Jordan's resources were strained and a cultural division arose.
5. The assignment of certain government positions to certain religious groups caused tension between the groups.

Map Activity

A. Istanbul
B. Damascus

CHAPTER 18

Reviewing What You Know

Building Vocabulary

On a separate sheet of paper, write sentences to define each of the following words.

1. phosphates
2. asphalt
3. secular
4. Diaspora
5. Zionism
6. mandate

Reviewing the Main Ideas

1. *Human Systems* People and customs from what three continents have influenced the eastern Mediterranean region?
2. *Human Systems* What three major religions have holy sites in Jerusalem?
3. *Human Systems* How has westernization changed Turkey's government, economy, and culture?
4. *Human Systems* What is the conflict over the Occupied Territories?
5. *Environment and Society* How do people in this area make their living from the land?

Understanding Environment and Society

Conflicts over Water

Freshwater is a scarce resource throughout much of the eastern Mediterranean. Create a presentation and model on issues of water supply and demand. Consider the following:

- Ways people in this region obtain freshwater.
- Human activities that consume water.
- Problems between countries when water supplies are strained.

Create a five-question quiz, with answers, about your model that you can use to challenge your classmates.

Thinking Critically

1. **Drawing Inferences and Conclusions** Why do you think Atatürk moved Turkey's capital from Istanbul to Ankara?
2. **Finding the Main Idea** How has human migration to Israel affected its population?
3. **Analyzing Information** How is the Dead Sea unusual? What are its commercial uses?
4. **Analyzing Information** How did refugees from Palestine affect Jordan?
5. **Drawing Inferences and Conclusions** Why did attempts to balance religious groups' participation in Lebanon's government fail?

FOOD FESTIVAL

Latkes are potato pancakes that are traditionally eaten during the Jewish holiday of Hanukkah. This recipe serves five to six people. First, grate three or four medium potatoes. Squeeze any excess water from the grated potatoes. Then grate one small onion and mix into the potatoes. Mix in one egg, 2 to 3 tbs. flour, and salt and pepper to taste. Form into patties and fry in a well-oiled skillet until both sides are golden-brown. Place the latkes on a paper towel to drain the excess oil. Serve them with apple sauce or sour cream.

CHAPTER 18 REVIEW AND ASSESSMENT RESOURCES

Reproducible
- Readings in World Geography, History, and Culture 50, 51, and 52
- Critical Thinking Activity 18
- Vocabulary Activity 18

Technology
- Chapter 18 Test Generator (on the One–Stop Planner)
- Audio CD Program, Chapter 18
- HRW Go site

Reinforcement, Review, and Assessment
- Chapter 18 Review, pp. 420–21
- Chapter 18 Tutorial for Students, Parents, Mentors, and Peers
- Chapter 18 Test
- Chapter 18 Test for English Language Learners and Special-Needs Students
- Unit 5 Test
- Unit 5 Test for English Language Learners and Special-Needs Students

Building Social Studies Skills

Map ACTIVITY

On a separate sheet of paper, match the letters on the map with their correct labels.

Dardanelles	**Negev**
Bosporus	**Istanbul**
Sea of Marmara	**Tel Aviv**
Jordan River	**Jerusalem**
Dead Sea	**Damascus**

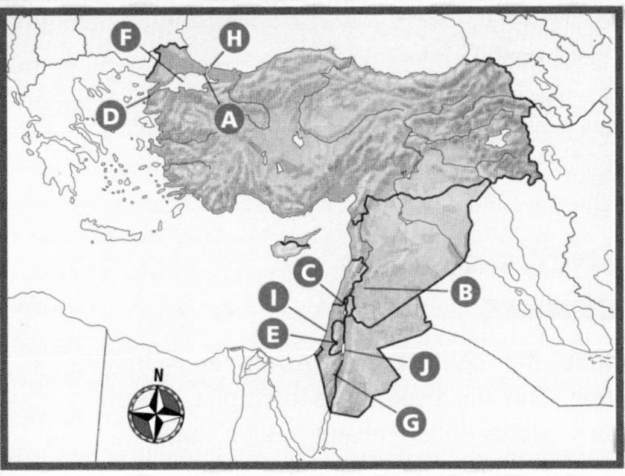

Mental Mapping Skills ACTIVITY

On a separate sheet of paper, draw a freehand map of the eastern Mediterranean and label the following:

Gaza Strip	**Lebanon**
Golan Heights	**Syria**
Israel	**Turkey**
Jordan	**West Bank**

WRITING ACTIVITY

Imagine that your family is about to travel to the eastern Mediterranean for a vacation. Your parents have asked you to help plan the trip. Write about the places you would like to visit and the reasons you would find them interesting. Be sure to use standard grammar, spelling, sentence structure, and punctuation.

Alternative Assessment

Portfolio ACTIVITY

Learning About Your Local Geography

Individual Project For thousands of years the eastern Mediterranean has been a crossroads between Europe, Asia, and Africa. Show how your community is linked to other towns and cities in a graphic organizer.

internet connect

Internet Activity: **go.hrw.com**
KEYWORD: **SK3 GT18**

Choose a topic to explore about the eastern Mediterranean:
- Visit the Dead Sea.
- Compare Israeli and Arab foods.
- Travel to historic Jerusalem.

C. Jordan River
D. Dardanelles
E. Jerusalem
F. Sea of Marmara
G. Negev
H. Bosporus
I. Tel Aviv
J. Dead Sea

Mental Mapping Skills Activity

Places should be labeled in their approximate locations.

Writing Activity

Answers will vary, but information included should be consistent with written material. Students should include sites from the area as well as reasons for visiting the sites. Use Rubric 37, Writing Assignments, to evaluate student work.

Portfolio Activity

Answers will vary, but graphic organizers should include all the required information and be presented in a clear, concise style. Use Rubric 13, Graphic Organizers, to evaluate student work.

internet connect

GO TO: **go.hrw.com**
KEYWORD: **SK3 Teacher**
FOR: a guide to using the Internet in your classroom

Focus On Regions

Mapping a Transition Zone

Have students work in pairs to draw maps of the transition zone between North Africa and the rest of the continent. Ask them to label the countries in the transition zone and indicate the dominant climate type and economic activities with shading and symbols as needed. Instruct students to include major cities on their maps. Have pairs write an extended caption explaining the characteristics that define this region of Africa as a transition zone. Call on volunteers to present and explain their maps to the class.

Geography Sidelight

Students may have heard the term *Middle East* used to describe a region that includes Southwest Asia and part of North Africa. However, many geographers believe *Middle East* is an inappropriate label, as the region is actually west and southwest of the rest of Asia.

Europeans had initially called this region the *Middle East* and Asia the *Far East*. Both terms may display a cultural bias because they describe these regions' locations as they relate to Europe. People in Asia and Southeast Asia may have different ideas about what is actually "middle-eastern" in relation to their location.

Critical Thinking: How might the term *Middle East* show cultural bias?

Answer: It describes the location of Southwest Asia and North Africa from the perspective of Europe.

➤ **This Focus On Regions feature addresses National Geography Standards 3, 5, and 10.**

Focus On Regions

Differences and Connections

In this textbook the countries of Africa are grouped together. This has been done to emphasize the region's connections. Yet any region as large as Africa has important differences from place to place. East Africa is different from West Africa, and West Africa is different from North Africa. This is also true of other large regions. In South America, for example, Brazil is different from Argentina in many important ways.

North Africa and Southwest Asia One subregion of Africa that geographers often include in another region is North Africa. These geographers see more connections between North Africa and Southwest Asia than between North Africa and the rest of the continent. For example, in both areas Arabic is the main language. The major religion is Islam. Political issues also tie North Africa to its eastern neighbors, as does physical geography. The countries of North Africa and Southwest Asia are part of a vast desert region. These countries face common issues such as water conservation and water management.

▲ The Muhammad Ali Mosque is one of many beautiful places of worship for Muslims in Cairo, Egypt. Although various religions, including Judaism and Christianity, are practiced in North Africa and Southwest Asia, most people in the two regions are Muslim.

The African Continent There are also many reasons for placing North Africa in a region with the rest of Africa. People in Africa share some important historical, cultural, and economic ties. For example, ancient Egyptians had close contact with other peoples in Africa. In turn, cultures south of the Sahara contributed to the development of Egypt's great Nile Valley civilizations. Farther west, Mediterranean peoples have long traded with West African kingdoms south of the Sahara. Such contact helped spread

◄ This sign in Morocco is in both Arabic and French. Arabic is spoken throughout North Africa and many parts of Southwest Asia.

Going Further: Thinking Critically

Organize the class into small groups in preparation for creating a time line about the historical, economic, and cultural connections between North Africa and Southwest Asia. Assign a section of the time line to each group. Have students use their textbook and other classroom and library resources to conduct research on events in the region's history. Students should include events such as Alexander the Great's conquests, the expansion of Rome into the region, the spread of Islamic merchants' trading networks, and the expansion of early African kingdoms. Then assemble the time line on a classroom wall. Have each group choose a spokesperson to present his or her group's section and to explain how the events affected the present-day characteristics of the transition region.

Regional Links

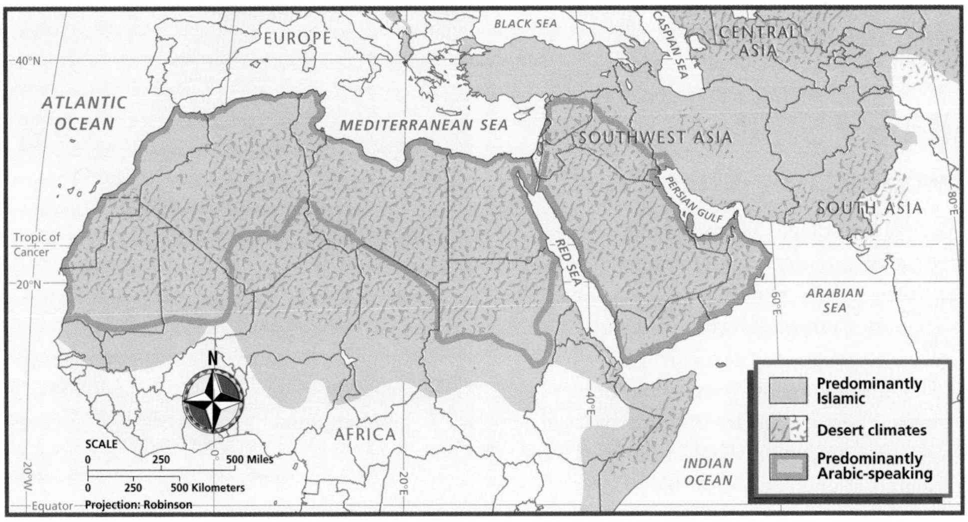

▲ Language, religion, and climate are some of the ties between North Africa and Southwest Asia. Arabic speakers, Muslims, and desert climates are dominant throughout much of the two regions and in parts of the surrounding transition zones.

Islam among the peoples of North Africa and the rest of the continent. Today, there are mosques as far south as Nigeria and Tanzania.

North Africa also has political connections to the rest of Africa. Many African countries face similar political and economic challenges. In part, these issues arise from their shared colonial history. To resolve some of these issues, the countries of North Africa are working with other African countries. They have formed associations such as the Organization of African Unity.

Transition Zones There are many differences and connections between North Africa and the rest of the continent. Perhaps they are most evident in a band of countries that lie just south of the Sahara. These countries stretch from Mauritania in the west to Ethiopia and Sudan in the east. They form a transition zone. In other words, they resemble both North Africa and Africa south of the Sahara. In Chad and Sudan, for example, strong political and cultural differences separate north and south. The northern regions of these countries are tied closely to North Africa. The southern regions of these countries are tied closely to African countries to the south.

North Africa and the countries in the transition zone have important connections to two major world regions. Which region they are placed in depends on the geographer's point of view. Understanding the differences among countries and their connections to other areas of the world is important. It is more important than deciding where on a map a region begins or ends.

Understanding What You Read

1. What ties are there between North Africa and Southwest Asia?
2. What ties are there between North Africa and the rest of the African continent?

Understanding What You Read

Answers

1. In both regions Arabic is the main language, and Islam is the main religion. Both regions are part of a vast desert area.
2. People of Africa share historical, cultural, and economic ties. Ancient Egyptians and Mediterranean peoples interacted with peoples south of the Sahara. North Africa also shares a colonial history with the rest of the continent.

GeoSkills

Going Further: Thinking Critically

To prepare, gather journal excerpts from the Lewis and Clark expedition of 1804–06. You might locate the excerpts in a library book or on the Internet. Locate a map of the expedition's route. Select several journal passages that describe the landscape. Organize the class into as many groups as there are selections from the journals. Provide each group member with a copy of the excerpt. Give a copy of the expedition map and a current map of the United States to each group. Ask students to read their selection carefully and then meet as a group to answer these questions:

- What place is being described in the journal excerpt?
- Are there cities and towns there now? if so, which?
- What landforms, bodies of water, plants, or animals are described? Which ones can still be found in that place today? Are the plants and animals described still common in the region, or have they been displaced by farms or buildings?
- Are there any obstacles, such as waterfalls, described in the journal selection that slowed the expedition? If so, do those physical features continue to slow or prevent transportation in some way? Can this information be obtained from the modern map?
- How does the weather as described in the excerpt compare to modern weather? For example, if the writer describes heavy snow, is heavy snow still common in the area?

Practicing the Skill

1. Students should speculate about the age of buildings and compare their apparent ages. Students might mention changes in materials, design, signage, or other features as a result of new technology, cultural influences, or what is popular at the time.
2. Students should describe the progression of events, the location, and how geography influenced the event. For example, a north Texas community may have been damaged by a tornado. Its location in "Tornado Alley" made the event more likely to occur.
3. Students should describe the settlement history of your state or community to the best of their knowledge. They may mention the area's physical geography—mild climate, good soil, location on a major river—as reasons why people were attracted to it.

internet connect go.hrw.com

CHAPTER WEB LINKS
GO TO: go.hrw.com
KEYWORD: SK3 Ch6
FOR: Web sites about Lewis and Clark

➤ **This GeoSkills feature addresses National Geography Standards 1, 4, and 17.**

Geo Skills

Building Skills for Life: Interpreting the Past

Both people and places are a product of the past. Therefore, understanding the geography of past times is important. It gives us a more complete understanding of the world today.

History and geography can hardly be separated. All historical events have to happen somewhere. These events are affected by local geography. For example, wheat and barley were domesticated in an area of Southwest Asia called the Fertile Crescent. This region received enough rainfall for these crops to grow.

Geographers who specialize in studying the past are called historical geographers. Historical geographers are interested in where things used to be and how they developed. They also try to understand how people's beliefs and values influenced historical events. In other words, why did people do what they did?

All geographers must be skilled at interpreting the past. Cultural geographers might study how old buildings and houses reflect earlier times. Physical geographers may need to reconstruct past landscapes. For example, many rivers have changed their course. Understanding where a river used to flow could help explain its present course.

In your study of geography, think about how a place's history has shaped the way it is today. Look for clues that can tell you about its past. Ask yourself how people may have thought about a place in earlier times. No matter where you are, the evidence of the past is all around!

▲ Ancient ruins in Petra, Jordan, can provide clues to the region's past.

Practicing the Skill

1. Look at some of the buildings in your community. How old do you think they are? Do some look older than others? Can you describe how building styles in your community have changed over time? Why do you think styles have changed?
2. Analyze an important historical event that occurred in your state. What happened? Where did it happen? How was the event influenced by local geography?
3. Interpret the settlement history of your state or community. When was it settled? What attracted people to it? How do you think they felt about the place at the time?

- What people lived in the area in the early 1800s? Are there many American Indians still living in the region?
- Are there physical features in the journal description that students cannot identify? If so, what clues can they gather from elsewhere in the excerpt?

Then have students create two posters—one showing how the place may have appeared to Lewis, Clark, and the expedition members, and another showing what it may look like now. You may want to have students communicate with students living in the chosen locale for better information on its modern-day appearance.

Lab Report

Answers

1. The stone was very large, black, and had "a glitter in it." Students may suggest that the stone was a meteorite.
2. Students may suggest that Ibn Battuta had never seen a stone that would not break.
3. Students might suggest that the story tells us that people from different regions came into contact with each other at this time or that people shared knowledge about discoveries and events.

HANDS ON GEOGRAPHY

One way to interpret the past is by studying old travel accounts. They often have detailed information about the people, places, and daily life of past times.

One famous travel account describes the journeys of Ibn Battuta. Ibn Battuta was one of the greatest travelers in history. During the mid-1300s he traveled about 75,000 miles (120,700 km) throughout parts of Asia and Africa. Near the end of his travels, he gave a long account of the many places he visited. This account is a valuable historical document today.

The following passage is taken from Ibn Battuta's travel account. In this passage, Ibn Battuta is visiting the Sultan of Birgi, a town in what is now western Turkey. Read the passage, and then answer the Lab Report questions.

"In the course of this audience the sultan asked me this question: "Have you ever seen a stone that fell from the sky?" I replied, "I have never seen one, nor ever heard tell of one." "Well," he said, "a stone did fall from the sky outside this town of ours," and then called some men and told them to bring the stone. They brought in a great black stone, very hard and with a glitter in it—I reckoned its weight to amount to a hundredweight. The sultan ordered the stonebreakers to be summoned, and four of them came and on his command to strike it they beat upon it as one man four times with iron hammers, but made no impression on it. I was astonished at this phenomenon, and he ordered it to be taken back to its place."

This painting shows Ibn Battuta during his travels in the mid-1300s.

Lab Report

1. What did the "stone that fell from the sky" look like? What do you think it was?
2. Why do you think Ibn Battuta was "astonished" by what he saw?
3. What can this story tell us about the historical geography of this region?

UNIT 6 AFRICA

Using the Illustrations

Direct students' attention to the photographs on these pages. Point out the Masai men in the center. The Masai are a nomadic people of East Africa. Their high-jumping dance may go on for hours. Masai dancing, game parks, and beautiful scenery attract tourists to Kenya. Ask students how the photo of the Masai dancers contrasts with the photo of Cape Town. *(Possible answer: contrast of traditional rural culture and modern urban culture)* Emphasize that Africa's cultures are very diverse, ranging from ways of life that are almost unchanged over many centuries to those that are influenced by the latest styles and technologies.

Focus attention on the photo at the top of this page. Tell students that the Sahara has spread over large tracts of land in northern and western Africa, turning grasslands into desert. Ask what the man in the photo is doing to stop or slow the process *(building fences to keep sand from covering the land).* Point out that low-technology efforts like this can be very effective and that one person can make a big difference in solving environmental problems.

Point out the woman using the microscope. Tell students that African medical personnel face many challenges, including a shortage of equipment.

Unit Objectives

1. Describe the various landforms, bodies of water, climates, and natural resources of Africa and relate resource distribution to economic development of the continent.
2. Identify ways that historical events have affected modern African societies.
3. Compare the cultures and ways of life of peoples in different regions of Africa.
4. Explain the economic, political, social, and environmental issues facing African nations.
5. Interpret maps, diagrams, charts, and photographs to better understand interrelationships of Africa's physical and human geography.

UNIT 6

Africa

Holding back the desert in Morocco

Cape Town, South Africa

CHAPTERS

NOTES FROM THE field

A Physician in Sierra Leone

Dr. James Li went to Africa fresh out of medical school to work as a general practice physician. Here he describes what it was like to work at a hospital in Sierra Leone. **WHAT DO YOU THINK?** ***What might Dr. Li's patients think about health care in the United States?***

Malaria, tuberculosis, and leprosy were widespread in our hospital population. We often saw patients who had walked over 100 miles for treatment. None of the patients in our area had access to clean water. Phone lines installed by the British had been cut down and the copper wires melted to make cooking pots. Our hospital was in the remote jungle, about six hours by dirt road from the nearest major city. It had 140 beds. In the clinic, we saw several hundred other patients each day. Food for the staff and patients was grown on the hospital grounds.

I came to love the people who were our patients, particularly the very young and the very old. In the midst of shortages, I met people who lived happily and suffered the tragedies of utter poverty with a dignity that would have failed me. These reflections fill me with a strange combination of awe, sadness, humility, and excitement.

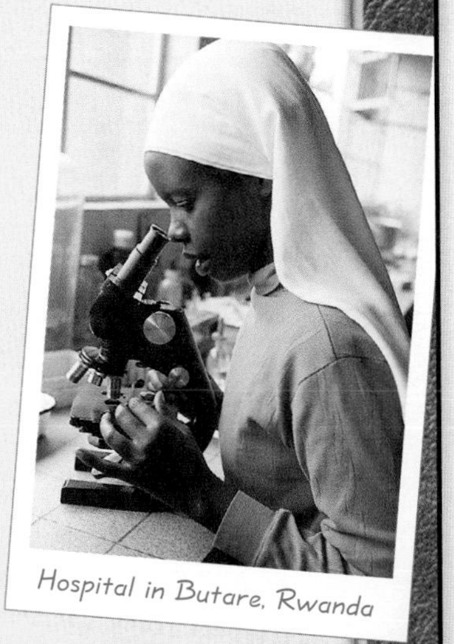
Hospital in Butare, Rwanda

Masai dancers, Kenya

Understanding Primary Sources

1. What were conditions like at Dr. Li's hospital?
2. How does Dr. Li describe the culture of Sierra Leone?

Giraffe

MORE FROM THE FIELD

Tuberculosis (TB) causes about 3 million deaths worldwide each year. It spreads through the air and infects the lungs. In recent years, doctors in the United States are seeing many TB cases that are resistant to drug therapy.

Malaria is primarily a tropical disease. It kills some 1.5 to 2.7 million people each year. Children under five are most vulnerable. More than 90 percent of malaria cases and the great majority of malaria deaths occur in tropical Africa. One basic method for preventing malaria is to eliminate the *Anopheles* mosquito, which transmits the disease. Draining marshes, swamps, and stagnant pools eliminates the mosquito's breeding places.

Activity: Have students create a world map showing areas with a high malaria risk. Discuss ways that individuals can eliminate standing water in their communities.

Understanding Primary Sources

Answers

1. busy, cut off from conveniences
2. resilient—they lived happily and suffered the tragedies of poverty with dignity

UNIT 6 ATLAS

Overview

In this unit, students will learn that Africa's physical and human geography are extremely diverse. The continent contains the world's largest desert (the Sahara) and the longest river (the Nile). There are snowcapped mountains that border vast plains. In the tropics are dense rain forests.

Africa's human mosaic is equally varied. Hundreds of different ethnic groups speak numerous languages. Archaeological digs in East Africa have unearthed remains indicating that the earliest humans lived there. Wealthy, sophisticated kingdoms developed in many regions. Later, European merchants arrived, and the international slave trade began. The slave trade left a tragic legacy. European colonizers were next to arrive. Europeans often ignored the ancient political and cultural divisions among the people. The negative effects of colonialism are still being felt.

Today, Africa faces the future with advantages and disadvantages. Desertification and high birthrates are widespread problems. Millions of Africans live in poverty. However, the continent has rich undeveloped resources. More African governments are responding to their citizens' needs.

People in the Profile

Note that the elevation profile crosses Africa just north of Mount Kenya. Known locally as Kirinyaga (meaning "Mountain of Whiteness"), Mount Kenya is the second-highest mountain in Africa. Mount Kenya is home to the Kikuyu people who farm the fertile lower slopes of the mountain.

The Kikuyu have played a significant role in Kenya's history. By the end of the 1800s, most of Africa had been colonized by European countries. The Kikuyu were the first ethnic group in Kenya to fight for independence from Britain. They started the Mau Mau rebellion, which in the 1950s advocated violent resistance to British rule. During the revolt, thousands of Kikuyu rebels were killed and many more were put into detention camps in an attempt by the British to stop the uprising. Despite these efforts, the Mau Mau rebellion sparked the Kenyan independence movement.

Jomo Kenyatta, the first prime minister and president of Kenya, was a member of the Kikuyu people. The Kikuyu continued to play a major role in Kenya's political system throughout Kenyatta's rule.

Critical Thinking: How did the Kikuyu fight British rule?

Answer: by starting the Mau Mau rebellion

UNIT 6 ATLAS The World in Spatial Terms

Africa

Elevation Profile

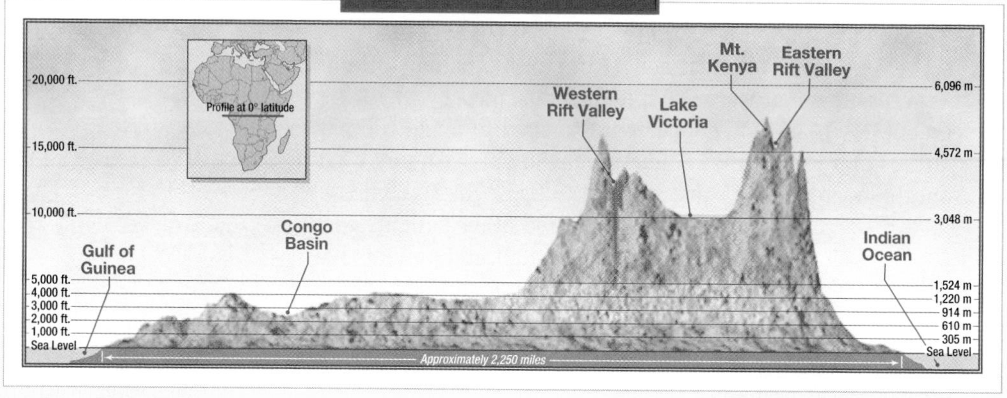

The United States and Africa: Comparing Sizes

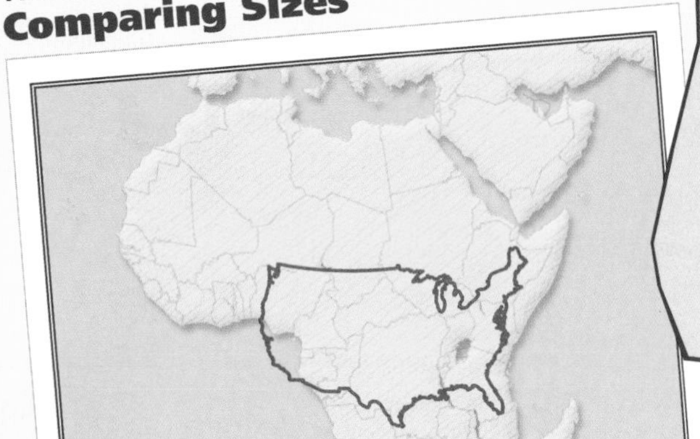

GEOSTATS:

- World's longest river: Nile River–4,160 miles (6,693 km)
- World's largest hot desert: Sahara–about 3,500,000 sq. mi. (9,065,000 sq km)
- World's highest recorded temperature: 136°F (58°C) in Al Azizyah, Libya on September 13, 1992

Using the Physical Map

As students examine the **physical map** on this page, tell the class that Africa is the second-largest continent. Then ask students what bodies of water surround Africa *(Mediterranean Sea, Atlantic Ocean, Gulf of Guinea, Indian Ocean, Gulf of Aden, Red Sea).* Africa has great rivers and deserts. Ask students to identify some of each *(deserts: Sahara, Kalahari, Namib; rivers: Nile, Niger, Congo, Zambezi, Orange).*

Point out that across most of Africa the major landforms are plateaus and basins. You may want to review the definitions of these terms. Call on a volunteer to identify the major basins *(Chad Basin, Sudan Basin, Congo Basin).* Then ask students where the region's largest plateau is located *(southern Africa).*

Africa: Physical

UNIT 6 ATLAS

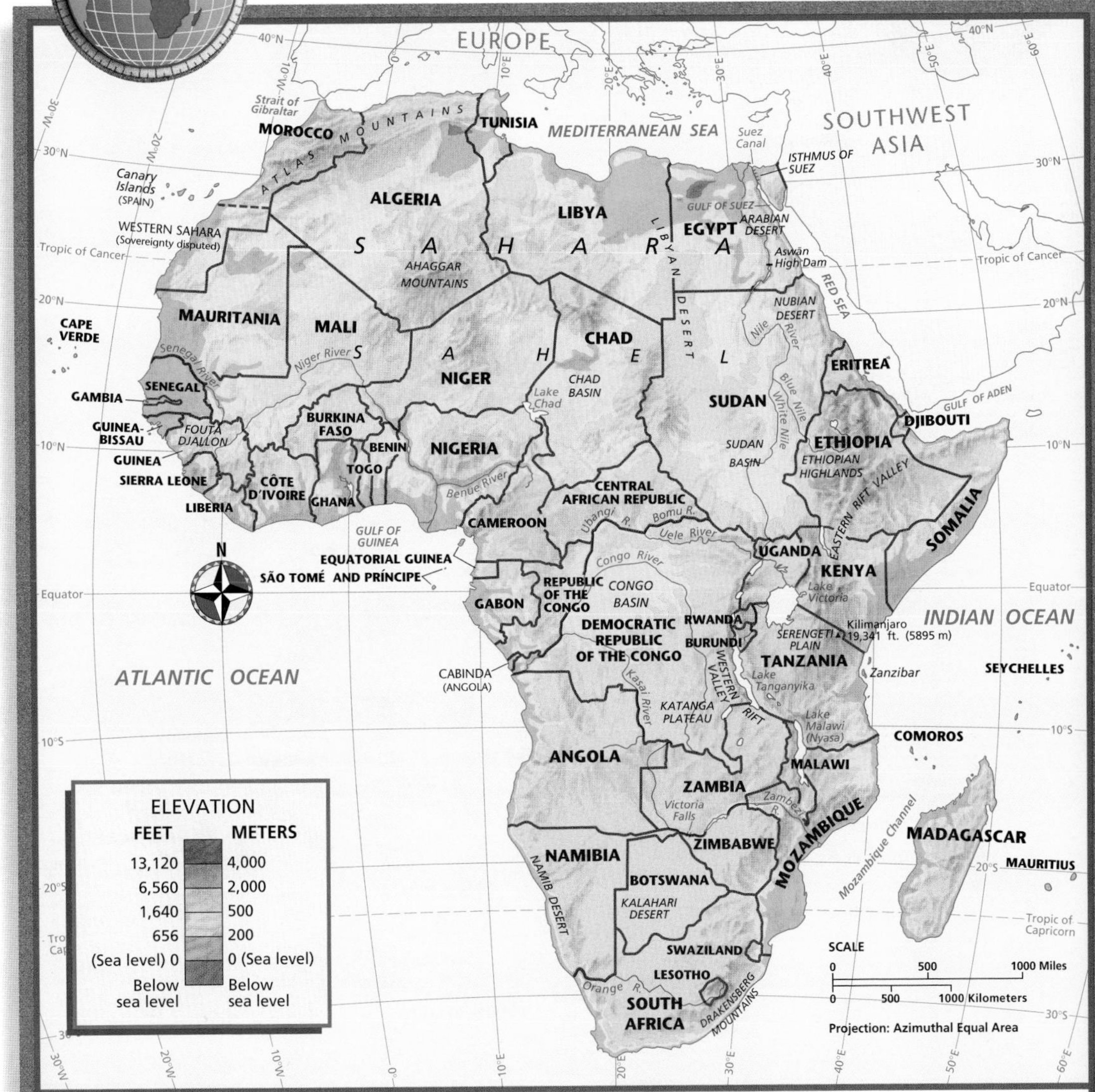

1. (Places and Regions) Which region of Africa has the highest mountains? Which country appears to have the largest number of high mountains?

2. (Places and Regions) Which countries have areas that lie below sea level? Where is the highest point in Africa?

Critical Thinking

3. **Analyzing Information** Which North African mountains might create a rain-shadow effect?

4. **Comparing** Compare this map to the **climate map** of the region. Which areas might have tropical rain forests?

internet connect go.hrw.com

ONLINE ATLAS
GO TO: **go.hrw.com**
KEYWORD: **SK3 Maps U6**
FOR: Web links to online maps of the region

Physical Map

Answers

1. eastern; Ethiopia
2. Algeria, Tunisia, Egypt, Ethiopia; Tanzania

Critical Thinking

3. Atlas Mountains
4. areas with humid tropical climates, such as the Congo Basin, the east coast of Madagascar, and the coasts of Sierra Leone and Liberia

UNIT 6 ATLAS

Using the Political Map

Focus students' attention on the **political map** of Africa. Call on a volunteer to identify all the mainland countries that have a seacoast while others name the landlocked countries they border. You may want to review the meaning of *landlocked*. Ask students to find the mainland country that is divided into two parts—one large and one very small *(Angola)*. Then ask them to name the largest country in Africa *(Sudan)*.

Have students compare this map to the **physical**, **climate**, and **population maps** of the continent. Ask students which of the region's boundaries appear to be unrelated to physical features *(straight lines in northern and southern Africa)*. Then ask students to identify one reason that these boundaries may have been drawn in this way. *(They are in dry, sparsely populated areas where there are no rivers or other major physical features to serve as natural boundaries.)*

internet connect

ONLINE ATLAS
GO TO: **go.hrw.com**
KEYWORD: **SK3 MapsU6**

Your Classroom Time Line

These are the major dates and time periods for this unit. Have students enter them on the time line you created earlier. You may want to watch for these dates as students progress through the unit.

- c.* 3200 B.C. The northern Nile River area unites into one Egyptian kingdom.
- c. 2000 B.C. The great pyramids at Giza are built for Egyptian rulers.
- 1500–1000 B.C. Egyptians carve tombs into the sides of cliffs.
- 332 B.C. Alexander the Great founds Alexandria, Egypt.
- A.D. 300s Christianity is adopted in Ethiopia.

*c. stands for *circa* and means "about."

Political Map

Answers

1. Lesotho
2. Equatorial Guinea
3. Madagascar; Cape Verde, São Tomé and Príncipe, Mauritius, Comoros, Seychelles

Critical Thinking

4. makes trade and communication easier; also, large settlements farther south limited by the Sahara

Africa: Political

UNIT 6 ATLAS

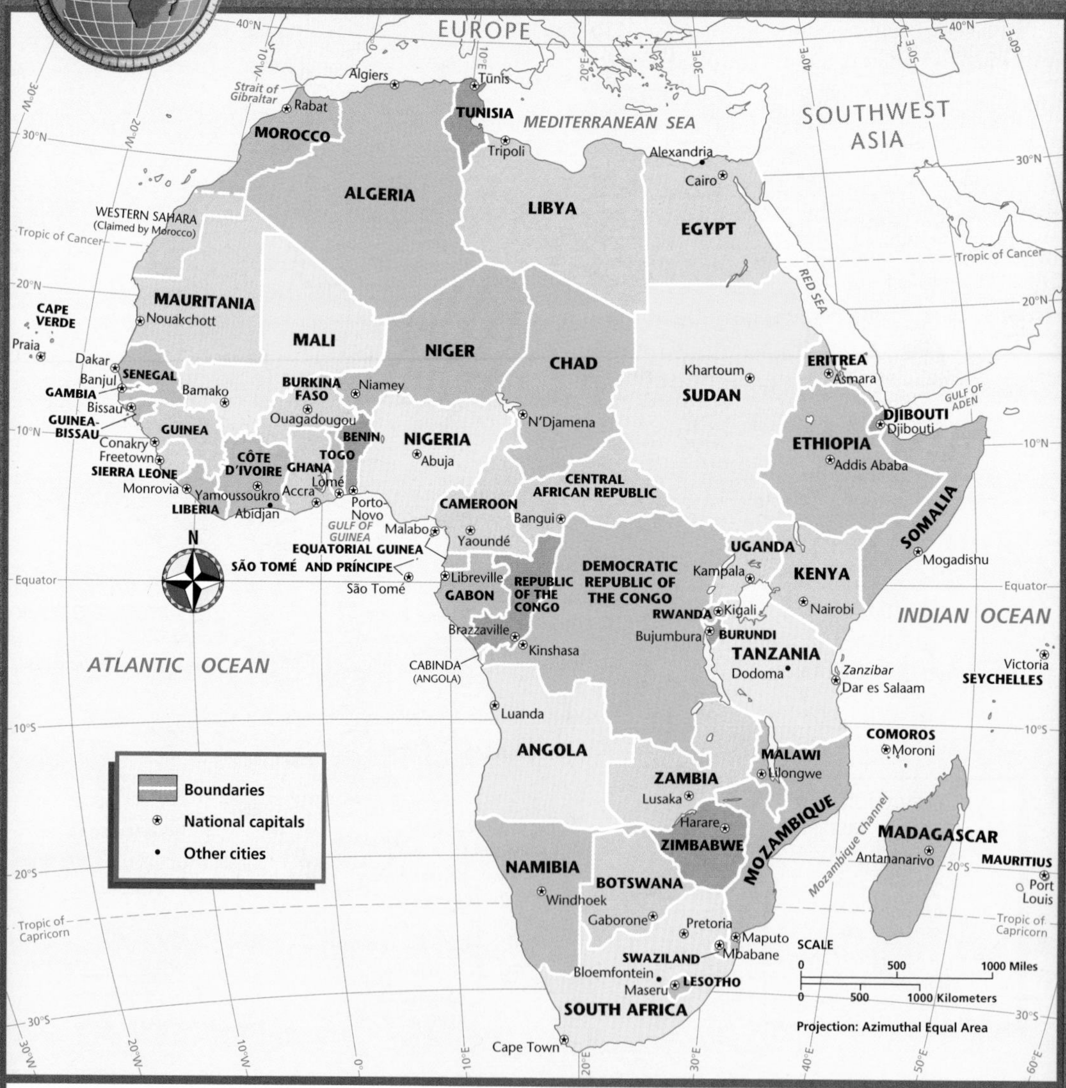

1. Places and Regions Which country is completely surrounded by another country?
2. Places and Regions Which country lies mostly on the mainland but has its capital on an island?
3. Places and Regions What is the largest African island country? What are the other island countries?

Critical Thinking

4. **Comparing** Compare this map to the **climate map** of the region. Why do you think the capitals of Algeria, Tunisia, and Libya all lie on the Mediterranean Sea?

Using the Climate Map

Have students review the **climate map** of Africa on this page and identify the climate region that most resembles their own local climate. Have students compare the latitude of their location with that of the corresponding African climate. Then ask students to write statements relating climate and latitude. *(Examples: Latitude affects climate. Places in low latitudes generally have a hot climate. Those in middle latitudes have a temperate climate. Places in high latitudes generally have a cold climate.)*

Then have students compare this map to the **physical map**. Ask them which countries have a highland climate region although they are on the equator or within 10 degrees of the equator *(Democratic Republic of the Congo, Uganda, Rwanda, Burundi, Kenya, Ethiopia)*. Then ask students how altitude affects climate. *(Higher altitudes create milder climates, even at very low latitudes.)*

Africa: Climate

UNIT 6 ATLAS

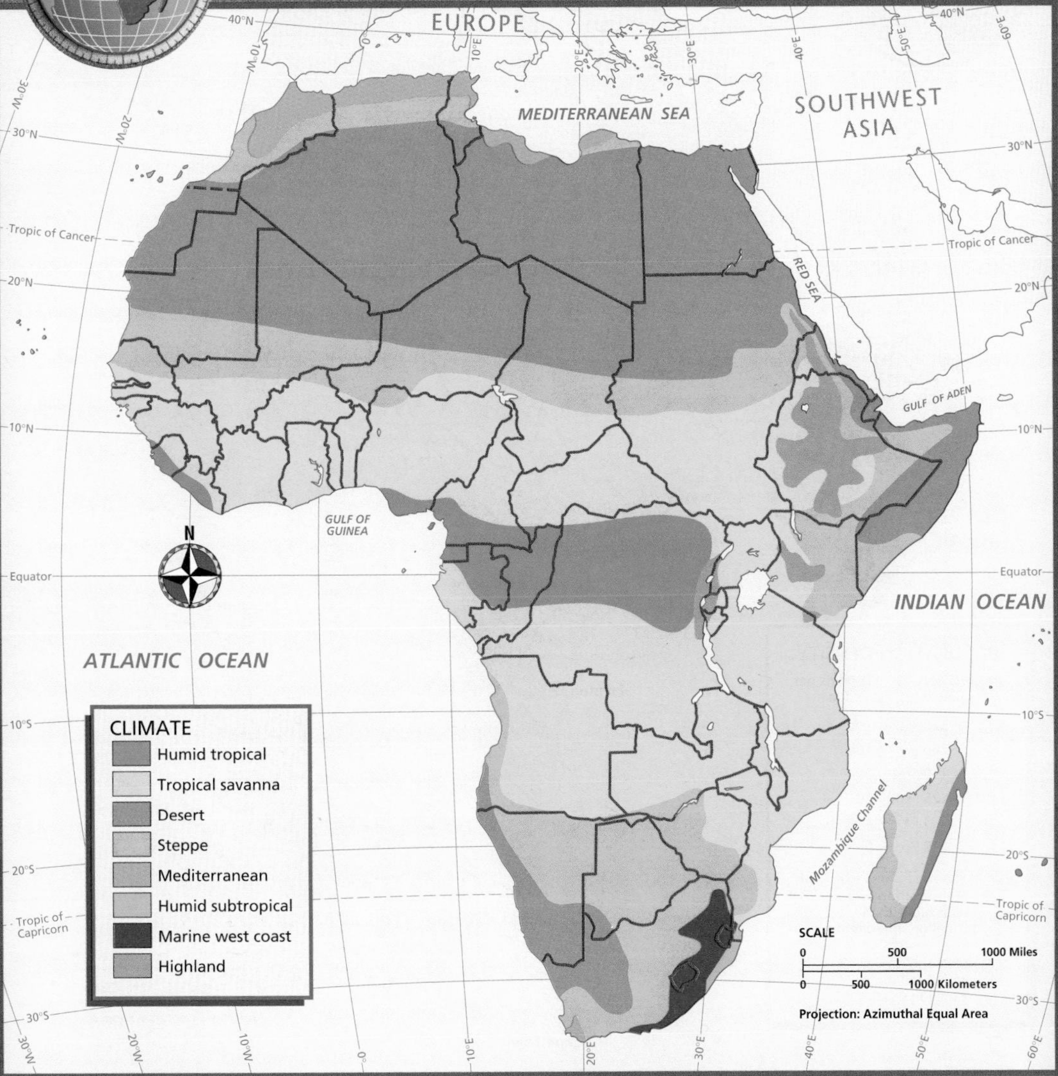

1. *Physical Systems* Where are humid tropical climates found in Africa?
2. *Places and Regions* Compare this map to the **land use and resources map** of the region. Which climate region in North Africa has limited economic activity?

Critical Thinking

3. **Comparing** Compare this map to the **land use and resources** and **population maps** of the region. Why might the eastern part of South Africa be more densely populated than the western part?

Your Classroom Time Line (continued)

- **late 600s–early 700s** Arab armies conquer Egypt and North Africa.
- **c. 800s** Kingdom of Ghana rises to power.
- **1300s** Mansa Mūsā, king of Mali, wins fame for his wealth and wise rule.
- **1500s** The Portuguese set up their first settlements on the East African coast.
- **1600s** The demand for slave labor in the American colonies increases.
- **1652** The Dutch set up a trade station at Cape Town.
- **1800s** Great Britain takes over southern Africa's Cape Colony.
- **1820s** Americans create the territory of Liberia.
- **1833** British ban slavery in their empire.

Climate Map

Answers

1. parts of central Africa, small part of coastal western Africa, east coast of Madagascar
2. desert

Critical Thinking

3. pleasant climate and plentiful resources

USING THE POPULATION MAP

Have students examine the **population map** of Africa on this page and identify relative locations of the areas with the highest population density. *(Examples: near the Gulf of Guinea, in the far northeast, near the lake in East Africa, in the far southeast)* List student responses on the chalkboard. Then ask students to compare this map to the **political map** of the continent. Ask volunteers to identify the countries that correspond to the areas of high population density and to write the countries' names next to the relative location descriptions on the chalkboard.

internet connect

ONLINE ATLAS
GO TO: go.hrw.com
KEYWORD: SK3 MapsU6

Your Classroom Time Line (continued)

- 1860s French build the Suez Canal.
- 1860s Dutch discover diamonds in the northern Cape Colony.
- 1886 Gold is discovered in Transvaal, southern Africa.
- 1899 The Boers and the British in southern Africa engage in warfare.
- 1912 Europeans control all of North Africa.
- 1912 The African National Congress (ANC) is formed.
- 1922 Egypt gains limited independence.
- 1936–41 Italy annexes Ethiopia.
- late 1940s South Africa establishes apartheid.

Population Map

Answers

1. Egypt; Nile River
2. Kinshasa, Johannesburg

Critical Thinking

3. Charts, graphs, databases, and models should accurately reflect population figures for cities on the map.

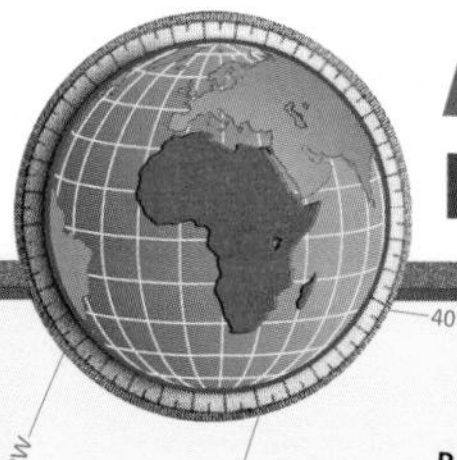

Africa: Population

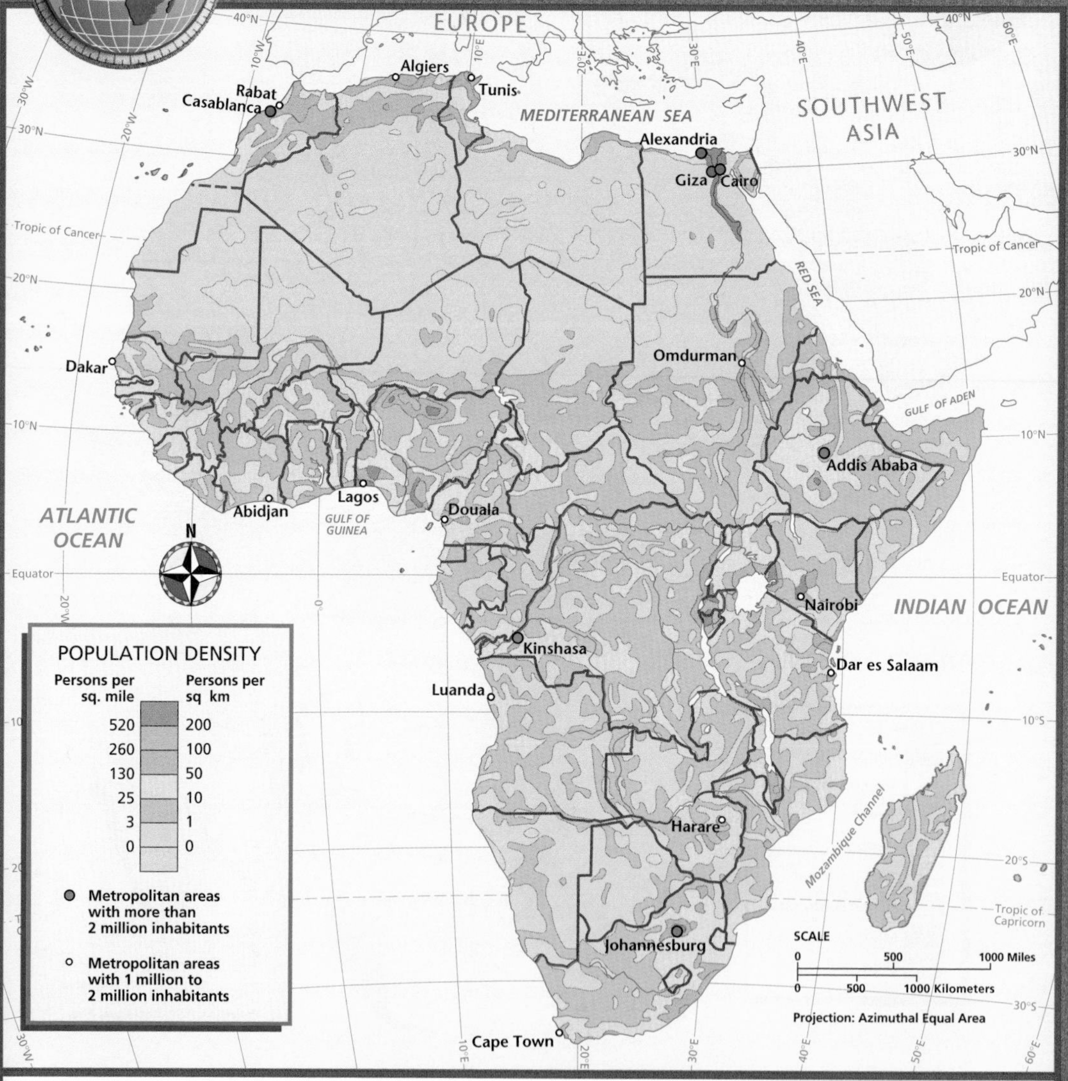

1. *Places and Regions* Which country has the most cities with more than 2 million people? Compare this map to the **physical map** of the region. Which river flows through or near these cities?
2. *Places and Regions* What are the two largest African cities shown south of the equator?

Critical Thinking

3. **Analyzing Information** Use the map on this page to create a chart, graph, database, or model of population centers in Africa.

Using the Land Use and Resources Map

Focus students' attention on the **land use and resources map** on this page. Call on volunteers for general statements about the distribution of resources. *(Examples: Resources are not distributed evenly. Most of the oil and gas industry is in northern Africa. Some countries have very few resources.)*

Have students compare this map to the **physical map**. Ask them which countries have no manufacturing centers, precious metals, or other minerals *(Mali, Chad, Somalia)* and what economic activities these countries depend on instead *(nomadic herding, subsistence farming, small amounts of commercial farming and livestock raising).*

Africa: Land Use and Resources

UNIT 6 ATLAS

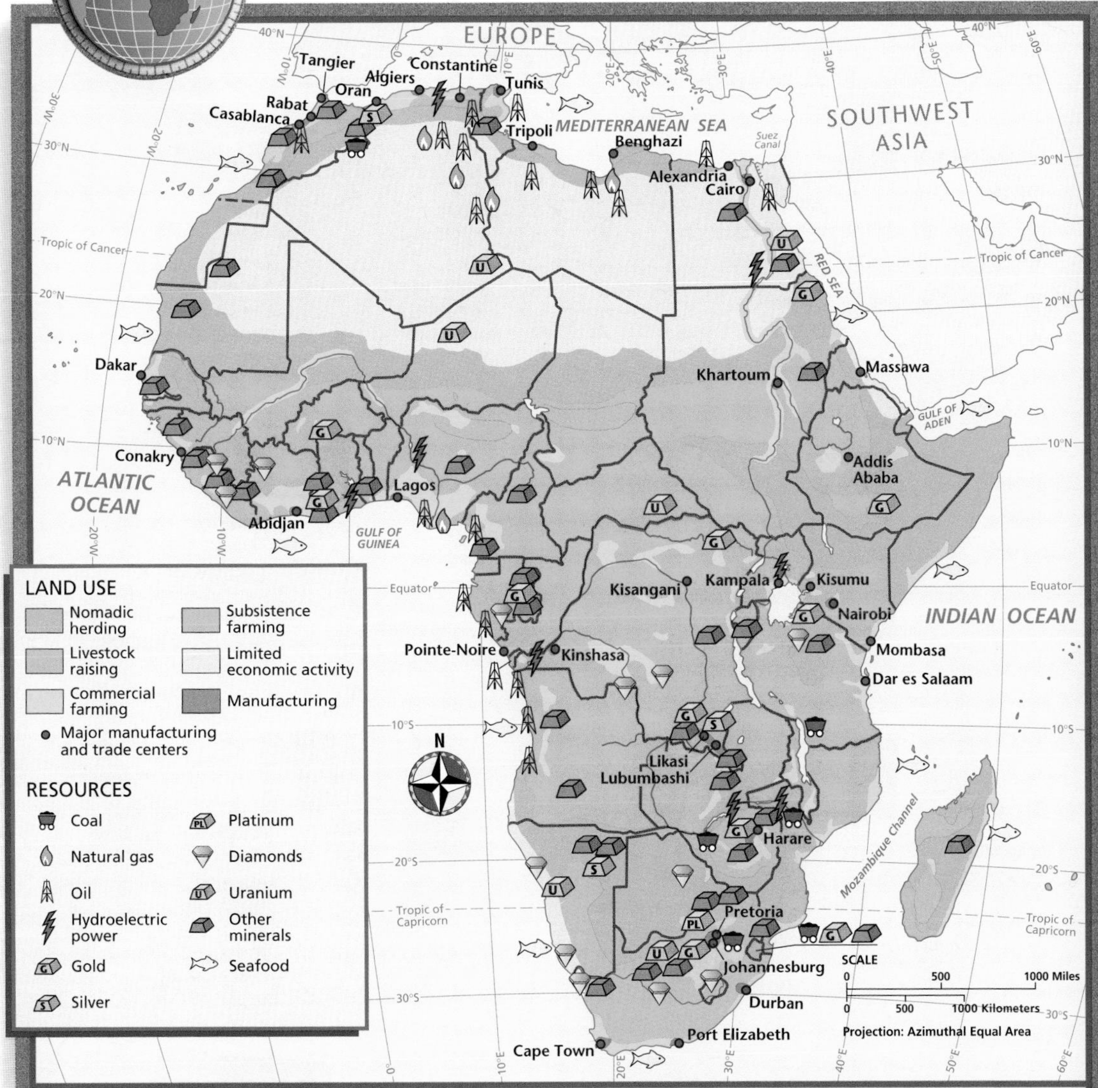

1. *Places and Regions* Which region of Africa seems to have the most mineral resources?
2. *Environment and Society* What type of farming is the most common in Africa?
3. *Environment and Society* Compare this map to the **physical map** of the region. Which rivers support commercial farming in dry areas? Which river in southeastern Africa provides hydroelectric power to the region?

Critical Thinking

4. **Analyzing Information** Use the map on this page to create a chart, graph, database, or model of economic activities in Africa.

Your Classroom Time Line (continued)

- **1950s–60s** Many African countries gain their independence.
- **1964** Nelson Mandela is sentenced to life in prison.
- **1969** Mu'ammar al-Gadhafi takes over Libya.
- **1979** Egypt signs a peace treaty with Israel.
- **1980s** Millions of Ethiopians suffer from famine.
- **late 1980s** South Africa begins to move away from the apartheid system.
- **1990s** Severe drought hits Somalia.
- **1990** Nelson Mandela is released from prison.
- **1994** Nelson Mandela is elected president of South Africa.

Land Use and Resources Map

Answers

1. southeastern
2. subsistence farming
3. Niger and Nile Rivers; Zambezi River

Critical Thinking

4. Charts, graphs, databases, and models should accurately reflect information from the map.

FAST FACTS ACTIVITIES

AFRICA

LEVEL 1: (Suggested time: 30 min.) Ask students to study the Fast Facts feature on African countries and to suggest some ways that European colonization has affected African countries. *(Possible answers: In many of the countries a European language is an official language. Many countries use money derived from European currency.)* Have students write a paragraph explaining their suggestions, citing facts to support them. In particular, students should note the widespread use of the Communaute Financiere Africaine franc (CFAF) and include their thoughts on what this indicates about colonial power.

LEVEL 2: (Suggested time: 40 min.) Direct students' attention to the figures that describe the number of people per doctor. Ask students to explain why they think this data is not available for many African countries *(Possible answers: Many of these countries are experiencing economic, political, or social tensions, and the governments are not stable enough to gather the information. Africa's physical geography—rivers, jungles, mountains, deserts—may make data collection difficult.)*

UNITED STATES OF AMERICA

CAPITAL: Washington, D.C.
AREA: 3,717,792 sq. mi. (9,629,091 sq km)
POPULATION: 281,421,906
MONEY: U.S. dollar (US$)
LANGUAGES: English, Spanish (spoken by a large minority)
PEOPLE PER DOCTOR: 365

Africa

ALGERIA

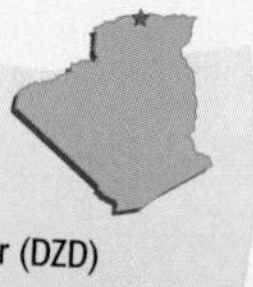

CAPITAL: Algiers
AREA: 919,590 sq. mi. (2,381,740 sq km)
POPULATION: 31,736,053
MONEY: Algerian dinar (DZD)
LANGUAGES: Arabic (official), French, Berber
PEOPLE PER DOCTOR: 1,066

BURKINA FASO

CAPITAL: Ouagadougou
AREA: 105,869 sq. mi. (274,200 sq km)
POPULATION: 12,272,289
MONEY: CFAF*
LANGUAGES: French (official), ethnic languages
PEOPLE PER DOCTOR: 27,158

ANGOLA

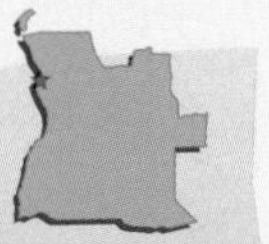

CAPITAL: Luanda
AREA: 481,351 sq. mi. (1,246,700 sq km)
POPULATION: 10,366,031
MONEY: kwanza (AOA)
LANGUAGES: Portuguese (official), Bantu languages
PEOPLE PER DOCTOR: data not available

BURUNDI

CAPITAL: Bujumbura
AREA: 10,745 sq. mi. (27,830 sq km)
POPULATION: 6,223,897
MONEY: Burundi franc (BIF)
LANGUAGES: Kirundi (official), French (official), Swahili
PEOPLE PER DOCTOR: 16,667

BENIN

CAPITAL: Porto-Novo
AREA: 43,483 sq. mi. (112,620 sq km)
POPULATION: 6,590,782
MONEY: CFAF*
LANGUAGES: French (official), Fon, Yoruba, other ethnic languages
PEOPLE PER DOCTOR: 14,216

CAMEROON

CAPITAL: Yaoundé
AREA: 183,567 sq. mi. (475,440 sq km)
POPULATION: 15,803,220
MONEY: CFAF*
LANGUAGES: English (official), French (official), 24 ethnic language groups
PEOPLE PER DOCTOR: 14,286

BOTSWANA

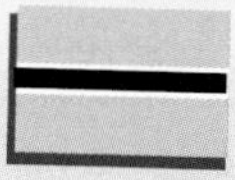

CAPITAL: Gaborone
AREA: 231,803 sq. mi. (600,370 sq km)
POPULATION: 1,586,119
MONEY: pula (BWP)
LANGUAGES: English (official), Setswana
PEOPLE PER DOCTOR: 4,395

CAPE VERDE

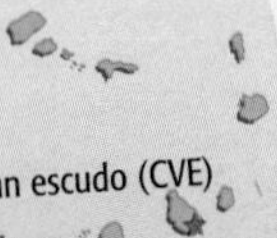

CAPITAL: Praia
AREA: 1,557 sq. mi. (4,033 sq km)
POPULATION: 405,163
MONEY: Cape Verdean escudo (CVE)
LANGUAGES: Portuguese, Crioulo
PEOPLE PER DOCTOR: 3,448

*Communaute Financiere Africaine franc

Countries not drawn to scale.

Fast Facts Activities

Create a bar graph on the chalkboard, listing figures in increments of 5,000 (starting at 0) through 50,000 along the y-axis (the vertical axis). These figures represent people. Title the graph People per Doctor in African Countries. Add a bar for the United States, which has about 365 people per doctor. (Point out that the figure for the United States is so low compared to many of the other countries that the representation of it is barely visible on the graph.) Then select students to come to the chalkboard and create and label a bar for each country. Continue until all the countries with available data, or a representative sample, have been placed on the graph.

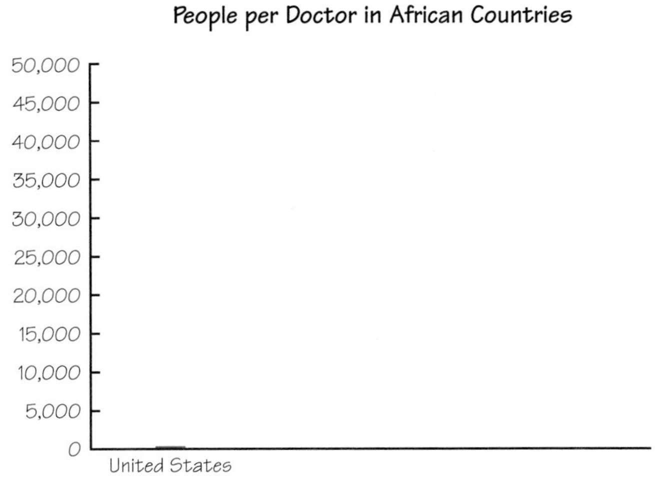

Unit 6 Assessment Resources

Reproducible

- Unit 6 Test
- Unit 6 Test for English Language Learners and Special-Needs Students

Unit 6 Atlas

Central African Republic

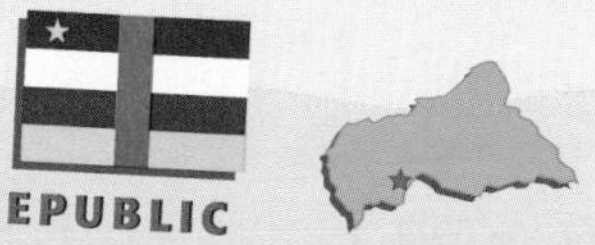

CAPITAL: Bangui
AREA: 240,534 sq. mi. (622,984 sq km)
POPULATION: 3,576,884
MONEY: CFAF*
LANGUAGES: French (official), Sangho (national language), Arabic, Hunsa, Swahili
PEOPLE PER DOCTOR: 18,660

Congo, Republic of the

CAPITAL: Brazzaville
AREA: 132,046 sq. mi. (342,000 sq km)
POPULATION: 2,894,336
MONEY: CFAF*
LANGUAGES: French (official), Lingala, Monokutuba
PEOPLE PER DOCTOR: 3,704

Chad

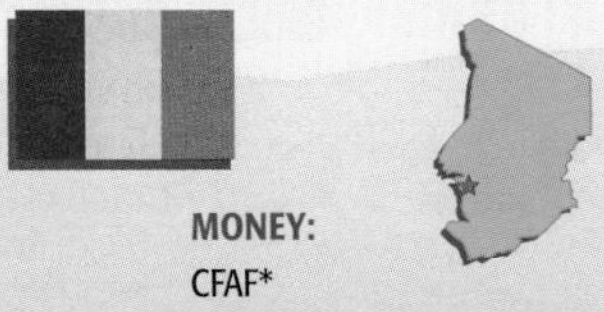

CAPITAL: N'Djamena
AREA: 495,752 sq. mi. (1,284,000 sq km)
POPULATION: 8,707,078
MONEY: CFAF*
LANGUAGES: French (official), Arabic (official), Sara and Sango
PEOPLE PER DOCTOR: 27,765

Côte d'Ivoire

CAPITAL: Yamoussoukro
AREA: 124,502 sq. mi. (322,460 sq km)
POPULATION: 16,393,057
MONEY: CFAF*
LANGUAGES: French (official), Dioula, ethnic languages
PEOPLE PER DOCTOR: data not available

Comoros

CAPITAL: Moroni
AREA: 838 sq. mi. (2,170 sq km)
POPULATION: 596,202
MONEY: Comoran franc (KMF)
LANGUAGES: Arabic (official), French (official), Comoran
PEOPLE PER DOCTOR: 10,000

Djibouti

CAPITAL: Djibouti
AREA: 8,494 sq. mi. (22,000 sq km)
POPULATION: 460,700
MONEY: Djiboutian franc (DJF)
LANGUAGES: French (official), Arabic (official), Somali, Afar
PEOPLE PER DOCTOR: 5,000

Congo, Democratic Republic of the

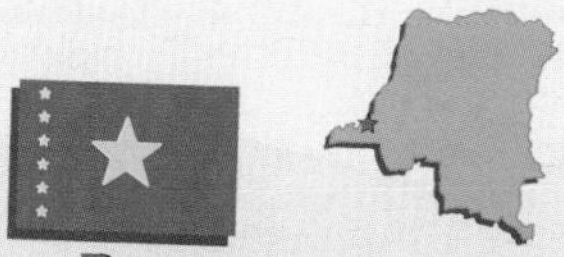

CAPITAL: Kinshasa
AREA: 905,563 sq. mi. (2,345,410 sq km)
POPULATION: 53,624,718
MONEY: Congolese franc (CDF)
LANGUAGES: French (official), Lingala, Kingwana, Kikongo, Tshiluba
PEOPLE PER DOCTOR: data not available

Egypt

CAPITAL: Cairo
AREA: 386,660 sq. mi. (1,001,450 sq km)
POPULATION: 69,536,644
MONEY: Egyptian pound (EGP)
LANGUAGES: Arabic (official), English, French
PEOPLE PER DOCTOR: 472

Sources: Central Intelligence Agency, *The World Factbook 2001; The World Almanac and Book of Facts 2001; United Nations Development Programme: Health Profile,* pop. figures are 2001 estimates.

internet connect

ONLINE ATLAS
GO TO: go.hrw.com
KEYWORD: SK3 FACTSU6

Highlights of Country Statistics

- *CIA World Factbook*
- Library of Congress country studies
- Flags of the world

Fast Facts Activities

Discuss the trends displayed in the graph. *(Sample questions: Which countries have the highest number of people per doctor? Which countries have the lowest number? Where do most of the countries fall on the graph?)* Invite students to offer possible explanations for the trends. Ask them how the number of doctors might affect public health in Africa.

LEVEL 3: (Suggested time: 45 min.) Have the students work in small groups to prepare for a debate. Have half of the groups prepare arguments to support using a single language throughout Africa. Ask the other groups to prepare arguments maintaining the wide variety of languages and dialects. Have students include a theory that explains how so many

Equatorial Guinea

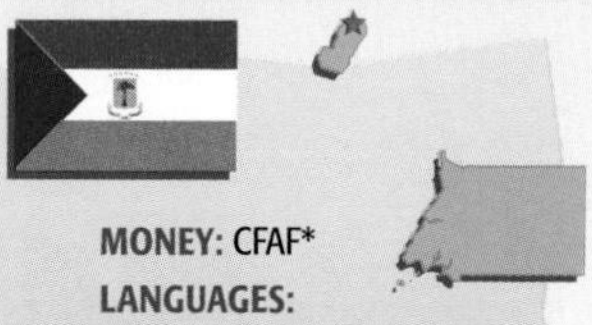

CAPITAL: Malabo
AREA: 10,830 sq. mi. (28,051 sq km)
POPULATION: 486,060
MONEY: CFAF*
LANGUAGES: Spanish (official), French (official), Fang, Bubi, Ibo
PEOPLE PER DOCTOR: 4,762

Eritrea

CAPITAL: Asmara
AREA: 46,842 sq. mi. (121,320 sq km)
POPULATION: 4,298,269
MONEY: nafka (EKN)
LANGUAGES: Afar, Amharic, Arabic, Tigre, Kunama, Tigrinya
PEOPLE PER DOCTOR: 36,000

Ethiopia

CAPITAL: Addis Ababa
AREA: 435,184 sq. mi. (1,127,127 sq km)
POPULATION: 65,891,874
MONEY: birr (ETB)
LANGUAGES: Amharic, Tigrinya, Orominga, Guaraginga, Somali, Arabic
PEOPLE PER DOCTOR: 25,000

Gabon

CAPITAL: Libreville
AREA: 103,346 sq. mi. (267,667 sq km)
POPULATION: 1,221,175
MONEY: CFAF*
LANGUAGES: French (official), Fang, Myene, Bateke, Bapounou/Eschira, Bandjabi
PEOPLE PER DOCTOR: 5,263

Gambia

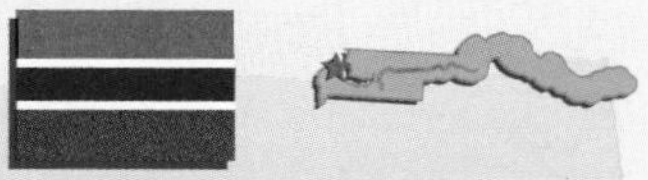

CAPITAL: Banjul
AREA: 4,363 sq. mi. (11,300 sq km)
POPULATION: 1,411,205
MONEY:dalasi (GMD)
LANGUAGES: English (official), Mandinka, Wolof, Fula
PEOPLE PER DOCTOR: 50,000

*Communaute Financiere Africaine franc

Ghana

CAPITAL: Accra
AREA: 92,100 sq. mi. (238,540 sq km)
POPULATION: 19,894,014
MONEY: new cedi (GHC)
LANGUAGES: English (official), Akan, Moshi-Dagomba, Ewe, Ga
PEOPLE PER DOCTOR: 22,970

Guinea

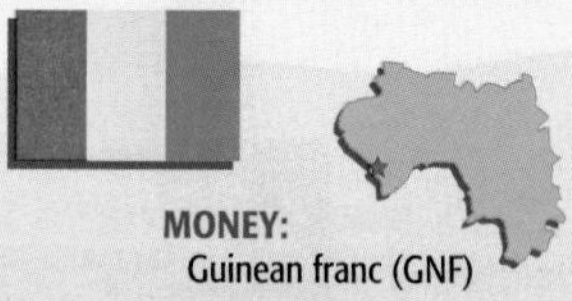

CAPITAL: Conakry
AREA: 94,925 sq. mi. (245,857 sq km)
POPULATION: 7,613,870
MONEY: Guinean franc (GNF)
LANGUAGES: French (official), ethnic languages
PEOPLE PER DOCTOR: 6,667

Guinea-Bissau

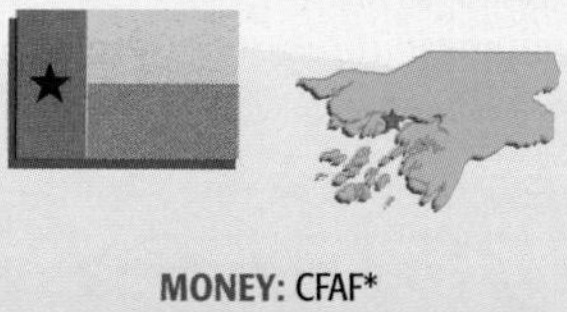

CAPITAL: Bissau
AREA: 13,946 sq. mi. (36,120 sq km)
POPULATION: 1,315,822
MONEY: CFAF*
LANGUAGES: Portuguese (official), Crioulo, ethnic languages
PEOPLE PER DOCTOR: 5,556

Kenya

CAPITAL: Nairobi
AREA: 224,961 sq. mi. (582,650 sq km)
POPULATION: 30,765,916
MONEY: Kenyan shilling (KES)
LANGUAGES: English (official), Kiwahili (official), ethnic languages
PEOPLE PER DOCTOR: 5,999

Lesotho

CAPITAL: Maseru
AREA: 11,720 sq. mi. (30,355 sq km)
POPULATION: 2,177,062
MONEY: loti (LSL), South African rand (ZAR)
LANGUAGES: English (official), Sesotho, Zulu, Xhosa
PEOPLE PER DOCTOR: 14,306

Countries not drawn to scale.

FAST FACTS ACTIVITIES

languages and dialects developed as well as the possible effects of linguistic diversity. You may want to stage the debate as if it took place at the United Nations.

UNIT 6 ATLAS

LIBERIA

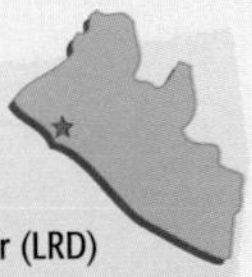

CAPITAL: Monrovia
AREA: 43,000 sq. mi. (111,370 sq km)
POPULATION: 3,225,837
MONEY: Liberian dollar (LRD)
LANGUAGES: English (official), many ethnic languages
PEOPLE PER DOCTOR: 8,333

LIBYA

CAPITAL: Tripoli
AREA: 679,358 sq. mi. (1,759,540 sq km)
POPULATION: 5,240,599
MONEY: Libyan dinar (LYD)
LANGUAGES: Arabic, Italian, English
PEOPLE PER DOCTOR: data not available

MADAGASCAR

CAPITAL: Antananarivo
AREA: 226,656 sq. mi. (587,040 sq km)
POPULATION: 15,982,563
MONEY: Malagasy franc (MGF)
LANGUAGES: French (official), Malagasy (official)
PEOPLE PER DOCTOR: 4,167

MALAWI

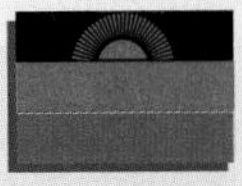

CAPITAL: Lilongwe
AREA: 45,745 sq. mi. (118,480 sq km)
POPULATION: 10,548,250
MONEY: Malawian kwacha (MWK)
LANGUAGES: English (official), Chichewa (official), many ethnic languages
PEOPLE PER DOCTOR: 50,000

MALI

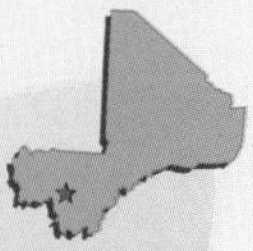

CAPITAL: Bamako
AREA: 478,764 sq. mi. (1,240,000 sq km)
POPULATION: 11,008,518
MONEY: CFAF*
LANGUAGES: French (official), Bambara, many ethnic languages
PEOPLE PER DOCTOR: 25,000

MAURITANIA

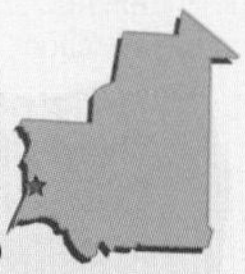

CAPITAL: Nouakchott
AREA: 397,953 sq. mi. (1,030,700 sq km)
POPULATION: 2,747,312
MONEY: ouguiya (MOR)
LANGUAGES: Hasaniya Arabic (official), Wolof (official), Pular, Soninke, French
PEOPLE PER DOCTOR: 11,085

MAURITIUS

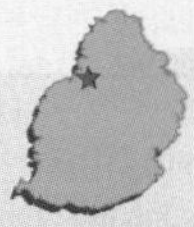

CAPITAL: Port Louis
AREA: 718 sq. mi. (1,860 sq km)
POPULATION: 1,189,825
MONEY: Mauritian rupee (MUR)
LANGUAGES: English (official), Creole, French, Hindi, Urdu, Hakka, Bojpoori
PEOPLE PER DOCTOR: 1,182

MOROCCO

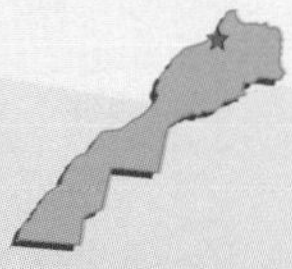

CAPITAL: Rabat
AREA: 172,413 sq. mi. (446,550 sq km)
POPULATION: 30,645,305
MONEY: Moroccan dirham (MAD)
LANGUAGES: Arabic (official), Berber, French
PEOPLE PER DOCTOR: 2,923

MOZAMBIQUE

CAPITAL: Maputo
AREA: 304,494 sq. mi. (801,590 sq km)
POPULATION: 19,371,057
MONEY: metical (MZM)
LANGUAGES: Portuguese (official), ethnic languages
PEOPLE PER DOCTOR: 131,991

Sources: Central Intelligence Agency, *The World Factbook 2001; The World Almanac and Book of Facts 2001; United Nations Development Programme: Health Profile;* pop. figures are 2001 estimates.

Namibia

CAPITAL: Windhoek

AREA: 318,694 sq. mi. (825,418 sq km)

POPULATION: 1,797,677

MONEY: Namibian dollar (NAD), South African rand

LANGUAGES: English (official), Afrikaans, German, Oshivambo, Herero, Nama

PEOPLE PER DOCTOR: 4,594

Niger

CAPITAL: Niamey

AREA: 489,189 sq. mi. (1,267,000 sq km)

POPULATION: 10,355,156

MONEY: CFAF*

LANGUAGES: French (official), Hausa, Djerma

PEOPLE PER DOCTOR: 35,141

Nigeria

CAPITAL: Abuja

AREA: 356,667 sq. mi. (923,768 sq km)

POPULATION: 126,635,626

MONEY: naira (NGN)

LANGUAGES: English (official), Hausa, Yoruba, Ibo, Fulani

PEOPLE PER DOCTOR: 4,496

Rwanda

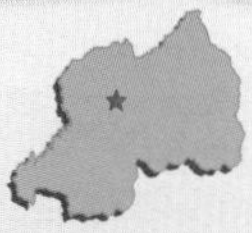

CAPITAL: Kigali

AREA: 10,169 sq. mi. (26,338 sq km)

POPULATION: 7,312,756

MONEY: Rwandan franc (RWF)

LANGUAGES: Kinyarwanda (official), French (official), English (official), Kiwahili

PEOPLE PER DOCTOR: 50,000

São Tomé and Príncipe

CAPITAL: São Tomé

AREA: 386 sq. mi. (1,001 sq km)

POPULATION: 165,034

MONEY: dobra (STD)

LANGUAGES: Portuguese (official)

PEOPLE PER DOCTOR: 3,125

Senegal

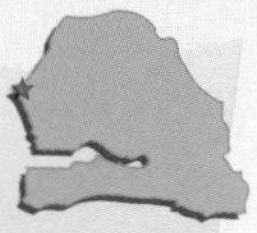

CAPITAL: Dakar

AREA: 75,749 sq. mi. (196,190 sq km)

POPULATION: 10,284,929

MONEY: CFAF*

LANGUAGES: French (official), Wolof, Pulaar, Diola, Mandinka

PEOPLE PER DOCTOR: 14,285

Seychelles

CAPITAL: Victoria

AREA: 176 sq. mi. (455 sq km)

POPULATION: 79,715

MONEY: Seychelles rupee (SCR)

LANGUAGES: English (official), French (official), Creole

PEOPLE PER DOCTOR: 906

Sierra Leone

CAPITAL: Freetown

AREA: 27,699 sq. mi. (71,740 sq km)

POPULATION: 5,426,618

MONEY: leone (SLL)

LANGUAGES: English (official), Mende, Temne, Krio

PEOPLE PER DOCTOR: 10,832

Somalia

CAPITAL: Mogadishu

AREA: 246,199 sq. mi. (637,657 sq km)

POPULATION: 7,488,773

MONEY: Somali shilling (SOS)

LANGUAGES: Somali (official), Arabic, Italian, English

PEOPLE PER DOCTOR: data not available

South Africa

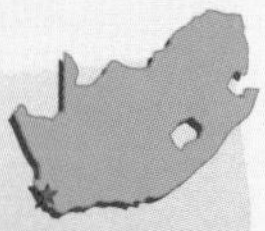

CAPITAL: Pretoria

AREA: 471,008 sq. mi. (1,219,912 sq km)

POPULATION: 43,586,097

MONEY: rand (ZAR)

LANGUAGES: 11 official languages, including Afrikaans, English, Ndebele, Pedi, Sotho, Swazi, Tsonga, Tswana, Venda, Xhosa, Zulu

PEOPLE PER DOCTOR: 1,742

*Communaute Financiere Africaine franc

Countries not drawn to scale.

UNIT 6 ATLAS

Sudan

CAPITAL: Khartoum
AREA: 967,493 sq. mi. (2,505,810 sq km)
POPULATION: 36,080,373
MONEY: Sudanese dinar (SDD)
LANGUAGES: Arabic (official), Nubian, Ta Bedawie, English, many ethnic languages
PEOPLE PER DOCTOR: 11,300

Swaziland

CAPITAL: Mbabane
AREA: 6,704 sq. mi. (17,363 sq km)
POPULATION: 1,104,343
MONEY: lilangeni (SZL)
LANGUAGES: English (official), siSwati
PEOPLE PER DOCTOR: data not available

Tanzania

CAPITAL: Dar es Salaam
AREA: 364,898 sq. mi. (945,087 sq km)
POPULATION: 36,232,074
MONEY: Tanzanian shilling (TZS)
LANGUAGES: Kiwahili (official), English (official), Arabic, many ethnic languages
PEOPLE PER DOCTOR: 20,511

Togo

CAPITAL: Lomé
AREA: 21,925 sq. mi. (56,785 sq km)
POPULATION: 5,153,088
MONEY: CFAF*
LANGUAGES: French (official), Ewe, Mina, Kabye, Dagomba
PEOPLE PER DOCTOR: 16,667

Tunisia

CAPITAL: Tunis
AREA: 63,170 sq. mi. (163,610 sq km)
POPULATION: 9,705,102
MONEY: Tunisian dinar (TND)
LANGUAGES: Arabic (official), French
PEOPLE PER DOCTOR: 1,640

Uganda

CAPITAL: Kampala
AREA: 91,135 sq. mi. (236,040 sq km)
POPULATION: 23,985,712
MONEY: Ugandan shilling (UGX)
LANGUAGES: English (official), Luganda, Swahili, Arabic
PEOPLE PER DOCTOR: 25,000

Zambia

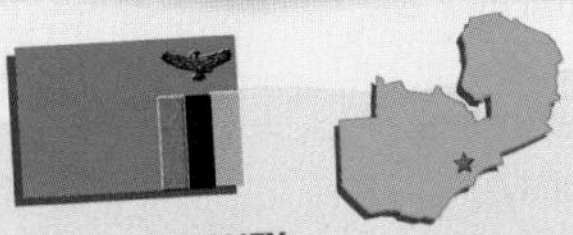

CAPITAL: Lusaka
AREA: 290,584 sq. mi. (752,614 sq km)
POPULATION: 9,770,199
MONEY: Zambian kwacha (ZMK)
LANGUAGES: English (official), Bemba, Kaonda, Lozi, Lunda, Luvale, many ethnic languages
PEOPLE PER DOCTOR: 10,917

Zimbabwe

CAPITAL: Harare
AREA: 150,803 sq. mi. (390,580 sq km)
POPULATION: 11,365,366
MONEY: Zimbabwean dollar (ZWD)
LANGUAGES: English (official), Shona, Sindebele
PEOPLE PER DOCTOR: 6,909

internet connect

COUNTRY STATISTICS
GO TO: **go.hrw.com**
KEYWORD: **SK3 FACTSU6**
FOR: **more facts about Africa**

Sources: Central Intelligence Agency, *The World Factbook 2001; The World Almanac and Book of Facts 2001; United Nations Development Programme: Health Profile;* pop. figures are 2001 estimates.

CHAPTER 19 North Africa

Chapter Resource Manager

Objectives	Pacing Guide	Reproducible Resources
SECTION 1 **Physical Geography** (pp. 441–44) **1.** Identify the major physical features of North Africa. **2.** Identify the climates, plants, and wildlife found in North Africa. **3.** Identify North Africa's major resources.	**Regular** 1 day **Block Scheduling** .5 day *Block Scheduling Handbook, Chapter 19*	**RS** Guided Reading Strategy 19.1 **SM** Geography for Life Activity 19 **E** Lab Activity for Geography and Earth Science, Hands-On 4
SECTION 2 **History and Culture** (pp. 445–49) **1.** Identify the major events in the history of North Africa. **2.** Identify some important facts about the people and culture of North Africa.	**Regular** 1 day **Block Scheduling** .5 day *Block Scheduling Handbook, Chapter 19*	**RS** Guided Reading Strategy 19.2 **RS** Graphic Organizer 19 **E** Cultures of the World Activity 6 **E** Creative Strategies for Teaching World Geography, Lessons 14, 15, and 16 **IC** Interdisciplinary Activity for the Middle Grades 26, 27 **SM** Map Activity 19
SECTION 3 **Egypt Today** (pp. 452–54) **1.** Describe what the people and cities of Egypt are like today. **2.** Identify Egypt's important economic activities. **3.** Identify the challenges Egypt faces today.	**Regular** .5 day **Block Scheduling** .5 day *Block Scheduling Handbook, Chapter 19*	**RS** Guided Reading Strategy 19.3
SECTION 4 **Syria, Lebanon, and Jordan** (pp. 455–57) **1.** Describe what the region's people and cities are like today. **2.** Identify the countries' important economic activities. **3.** Identify the challenges the countries face today.	**Regular** 1 day **Block Scheduling** .5 day *Block Scheduling Handbook, Chapter 19*	**RS** Guided Reading Strategy 19.4

Chapter Resource Key

RS Reading Support

IC Interdisciplinary Connections

E Enrichment

SM Skills Mastery

A Assessment

REV Review

ELL Reinforcement and English Language Learners

Transparencies

CD–ROM

Music

Video

go.hrw.com Internet

Holt Presentation Maker Using Microsoft® Powerpoint®

One-Stop Planner CD-ROM

See the ***One-Stop Planner*** for a complete list of additional resources for students and teachers.

One-Stop Planner CD–ROM

It's easy to plan lessons, select resources, and print out materials for your students when you use the ***One-Stop Planner CD–ROM with Test Generator.***

Technology Resources	Review, Reinforcement, and Assessment Resources
One-Stop Planner CD–ROM, Lesson 19.1 Geography and Cultures Visual Resources with Teaching Activities 43–41 Homework Practice Online HRW Go site	**ELL** Main Idea Activity 19.1 **REV** Section 1 Review, p. 414 **A** Daily Quiz 19.1 **ELL** English Audio Summary 19.1 **ELL** Spanish Audio Summary 19.1
One-Stop Planner CD–ROM, Lesson 19.2 Homework Practice Online HRW Go site	**ELL** Main Idea Activity 19.2 **REV** Section 2 Review, p. 419 **A** Daily Quiz 19.2 **ELL** English Audio Summary 19.2 **ELL** Spanish Audio Summary 19.2
One-Stop Planner CD–ROM, Lesson 19.3 Homework Practice Online HRW Go site	**ELL** Main Idea Activity 19.3 **REV** Section 3 Review, p. 484 **A** Daily Quiz 19.3 **ELL** English Audio Summary 19.3 **ELL** Spanish Audio Summary 19.3
One-Stop Planner CD–ROM, Lesson 19.4 Homework Practice Online HRW Go site	**ELL** Main Idea Activity 19.4 **REV** Section 4 Review, p. 487 **A** Daily Quiz 19.4 **ELL** English Audio Summary 19.4 **ELL** Spanish Audio Summary 19.4

internet connect

HRW ONLINE RESOURCES

GO TO: go.hrw.com
Then type in a keyword.

TEACHER HOME PAGE
KEYWORD: SK3 TEACHER

CHAPTER INTERNET ACTIVITIES
KEYWORD: SK3 GT19

Choose an activity to:
- journey through the Sahara.
- examine the rich history of North Africa.
- practice using Arabic calligraphy.

CHAPTER ENRICHMENT LINKS
KEYWORD: SK3 CH19

CHAPTER MAPS
KEYWORD: SK3 MAPS19

ONLINE ASSESSMENT
Homework Practice
KEYWORD: SK3 HP19
Standardized Test Prep Online
KEYWORD: SK3 STP19
Rubrics
KEYWORD: SS Rubrics

COUNTRY INFORMATION
KEYWORD: SK3 Almanac

CONTENT UPDATES
KEYWORD: SS Content Updates

HOLT PRESENTATION MAKER
KEYWORD: SK3 PPT19

ONLINE READING SUPPORT
KEYWORD: SS Strategies

CURRENT EVENTS
KEYWORD: S3 Current Events

Meeting Individual Needs

Ability Levels

Level 1 Basic-level activities designed for all students encountering new material

Level 2 Intermediate-level activities designed for average students

Level 3 Challenging activities designed for honors and gifted-and-talented students

English Language Learners Activities that address the needs of students with Limited English Proficiency

Chapter Review and Assessment

- **E** Readings in World Geography, History, and Culture 53 and 54
- **SM** Critical Thinking Activity 19
- **REV** Chapter 19 Review, pp. 458–59
- **REV** Chapter 19 Tutorial for Students, Parents, Mentors, and Peers
- **ELL** Vocabulary Activity 19
- **A** Chapter 19 Test
- Chapter 19 Test Generator (on the One-Stop Planner)
- Audio CD Program, Chapter 19
- **A** Chapter 19 Test for English Language Learners and Special-Needs Students

Launch into Learning

Tell students to look at the map on the following page, and direct their attention to the Sahara and North Africa's borders. Tell students that the Sahara has been compared to an ocean, and that camels have been referred to as "ships of the desert." Ask students to speculate why such comparisons might have been made. *(Students might mention that the Sahara is vast, as is an ocean; also, camels are a major mode of transportation in the desert, as ships are a major mode of transportation in the ocean.)* Then ask students to describe the southern borders of the North African countries. *(Students might mention that the borders are unusually straight.)* Tell students that borders in North Africa were drawn by Great Britain and France, which colonized the region. Tell students that they will learn more about the countries of North Africa in this chapter.

Section 1

Objectives

1. Identify the major physical features of North Africa.
2. Describe the climate, plants, and wildlife found in North Africa.
3. Identify the major resources of North Africa.

Links to Our Lives

You might want to emphasize the importance of learning about North Africa by sharing these points with your students:

- The United States has generally good relations with Egypt, Morocco, and Tunisia. These countries help maintain peace in the region.
- Events in the oil-producing region of North Africa can affect the price of gasoline and heating oil in the United States.
- North African countries have produced beautiful architecture, textiles, and other goods as well as music, literature, and foods that many Americans enjoy.
- The ancient Egyptians produced many magnificent monuments, cultural artifacts, and technological achievements. Many ancient Egyptian artifacts can be viewed in museums across the United States and in other countries.

Chapter 19

North Africa

The first region in Africa we will study is North Africa. Before we do that, we will meet Shaimaa. She and nearly all other Egyptians live along the Nile River or in the Nile Delta.

أنا طالبة في القاهرة،
في شمال افريقيا.
أهلاً وسهلاً بكم

Ahlan! (Hi!) My name is Shaimaa, and I am 18. I live with my mother and my little sister in an apartment. We live about an hour from downtown Cairo. My father lives in the United States but visits us every year.

Every day but Friday, I get up at 7:00 A.M., drink a glass of milk, and then meet my friends. School is about 15 minutes away on the metro (subway). We go to an all-girls school. We all have religious education in school. I study Islam with the other Muslim girls. The Christian girls meet with their religious teacher.

At about 3:00 P.M., I get home from school. I eat a big lunch of chicken and vegetables and sleep for a couple of hours. When I wake up, I have a lot of homework. At 10:00 P.M. we have a small meal of cheese, yogurt, or beans before bedtime.

On Fridays I usually go to movies with my girlfriends and walk along the Nile. Sometimes I stay home and listen to music.

Translation: I am a student in Cairo, in North Africa. Welcome to all of you.

FOCUS

LET'S GET STARTED

Copy the following instructions onto the chalkboard: *Recall the hottest summer temperatures you have experienced. How do you change your daily routine on hot summer days? Imagine what adjustments you would have to make in North Africa, where summer temperatures commonly rise higher than 120°F.* Tell students to respond in writing, and ask volunteers to share their responses with the class. Then tell students that in Section 1 they will learn more about the physical geography of North Africa.

Using the Physical-Political Map

Have students examine the map on this page and locate the five countries of the region *(Morocco, Algeria, Tunisia, Libya, Egypt)*. Have them name the mountain ranges *(Atlas, Ahaggar)*. Have them estimate the distance between Tripoli and the Ahaggar Mountains *(800 miles)*. Then have them find a similar distance on the map of their own region and compare the terrains.

Building Vocabulary

Write the key terms on the chalkboard. Show pictures of deserts with sand dunes and those with rocky plains. Explain that the term **ergs** refers to the first kind of desert and that the term **regs** refers to the second kind of desert. Then explain what **depressions** in a desert are and how **silt** carried by rivers helps produce fertile soil in areas flooded by those rivers.

Section 1 Physical Geography

Read to Discover

1. What are the major physical features of North Africa?
2. What climates, plants, and wildlife are found in North Africa?
3. What are North Africa's major resources?

Define

ergs
regs
depressions
silt

Locate

Red Sea
Mediterranean Sea
Sahara
Nile River
Sinai Peninsula
Ahaggar Mountains
Atlas Mountains
Qattara Depression
Nile Delta
Lake Nasser
Suez Canal

WHY IT MATTERS

Most Egyptians live along the banks of the Nile River. The river's flooding is both beneficial and destructive. Use CNNfyi.com or other **current events** sources to find out how Egyptians are affected by the river. Record your findings in your journal.

A tomb painting from ancient Egypt

North Africa: Physical-Political

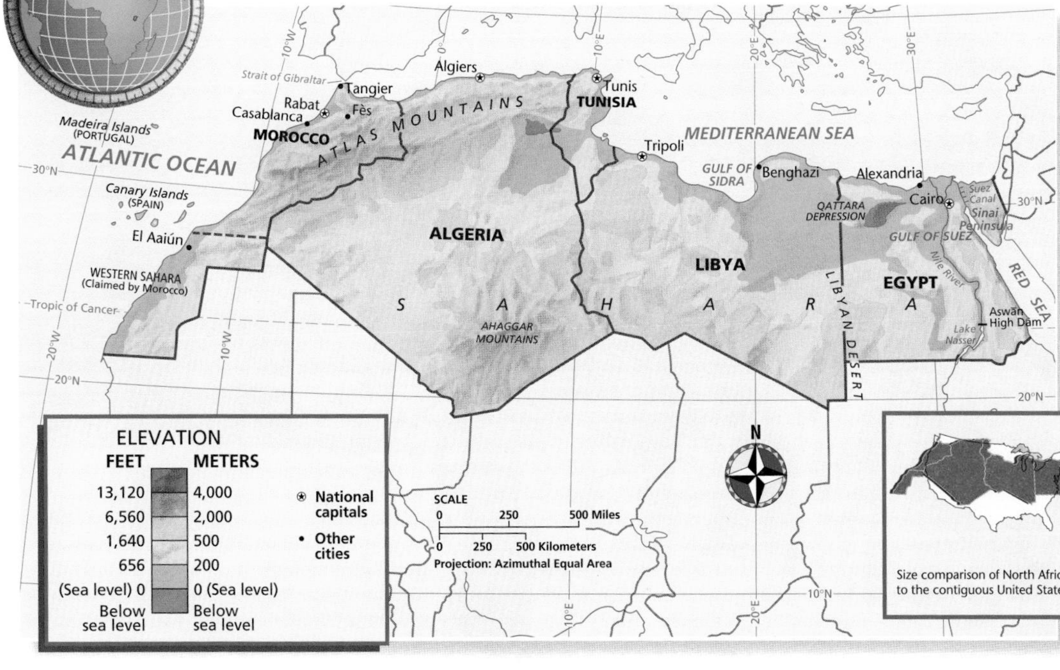

Section 1 RESOURCES

Reproducible
- Block Scheduling Handbook, Chapter 19
- Guided Reading Strategy 19.1
- Geography for Life Activity 19
- Lab Activity for Geography and Earth Science, Hands-On 4

Technology
- One-Stop Planner CD–ROM, Lesson 19.1
- Homework Practice Online
- Geography and Cultures Visual Resources with Teaching Activities 43–47
- HRW Go site

Reinforcement, Review, and Assessment
- Section 1 Review, p. 444
- Daily Quiz 19.1
- Main Idea Activity 19.1
- English Audio Summary 19.1
- Spanish Audio Summary 19.1

TEACH

Teaching Objective 1

ALL LEVELS: (Suggested time: 10 min.) Provide each student with an outline map of the region. Have students identify the locations of physical features and label them on their maps. Tell students to create a key explaining the meanings of any symbols.

ENGLISH LANGUAGE LEARNERS

Teaching Objective 2

ALL LEVELS: (Suggested time: 20 min.) Organize the class into three groups and assign each group a climate type found in North Africa. Give each group a sheet of butcher paper or poster board, an outline map of the region, colored pencils or markers, and a glue stick. Have each group use the materials to create a chart that shows the location of their climate type and the plants and animals found in the area. Instruct the groups to also describe their climate types in words or pictures. Have a representative from each group present the group's poster to the class.

ENGLISH LANGUAGE LEARNERS, COOPERATIVE LEARNING

Across the Curriculum

SCIENCE

The Sahara The Sahara has not always been as dry as it is in modern times. Neolithic petroglyphs show pictures of giant buffalo, elephant, rhinoceros, and hippopotamus in areas now too dry for such animals to exist. It is thought that the Sahara dried up around 4000 B.C.

Critical Thinking: What, in addition to climate changes, could account for the extinction of these animals from this region?

Answer: Student might suggest that overhunting or human destruction of the animals' habitat could have caused their extinction.

internet connect

GO TO: **go.hrw.com**
KEYWORD: **SK3 CH19**
FOR: Web sites about the Sahara

Visual Record Answer

Trees draw water from the oasis.

internet connect

GO TO: **go.hrw.com**
KEYWORD: **SK3 CH19**
FOR: Web sites about North Africa

Physical Features

North Africa includes Morocco, Algeria, Tunisia, Libya, and Egypt. The region stretches from the Atlantic Ocean to the Red Sea. Off the northern coast is the Mediterranean Sea. In the south is the Sahara (suh-HAR-uh), a vast desert. The region also has mountains, the northern Nile River valley, and the Sinai (SY-ny) Peninsula.

The Sahara The huge Sahara covers most of North Africa and stretches southward. The name *Sahara* comes from the Arabic word for "desert." The Sahara is the largest desert in the world. It is so big that nearly all of the United States would fit into it.

Large areas of this very dry region have few people or none at all. Great "seas" of sand dunes called **ergs** cover about a quarter of the desert. Much of the rest of the Sahara is made up of broad, windswept gravel plains. These gravel plains are called **regs**.

Mountains Do you think of deserts as flat regions? Well, the Sahara is not flat. Some sand dunes and ridges rise as high as 1,000 feet (305 m). The Sahara also has mountain ranges. For example, the Ahaggar (uh-HAH-guhr) Mountains are located in the central Sahara. Their highest peak is 9,842 feet (3,000 m). The Atlas Mountains on the northwestern side of the Sahara are higher. Mountains there rise to 13,671 feet (4,167 m).

There also are very low areas in the Sahara. These low areas are called **depressions**. The Qattara (kuh-TAHR-uh) Depression in western Egypt is 440 feet (134 m) below sea level. Other low areas often have large, dry lake beds. Water from rare rain storms collects there.

Oases like this one in Algeria are scattered throughout the vast Sahara.

Interpreting the Visual Record How do these trees survive in the desert's harsh, dry climate?

Teaching Objective 3

ALL LEVELS: (Suggested time: 15 min.) To help students learn about the resources of North Africa copy the following graphic organizer onto the chalkboard, omitting the connecting lines. Tell students to copy the organizer into their notebooks. Have each student complete the graphic organizer by connecting each box to the appropriate area or areas. Ask volunteers to share their answers with the class.
ENGLISH LANGUAGE LEARNERS

Major Resources of North Africa

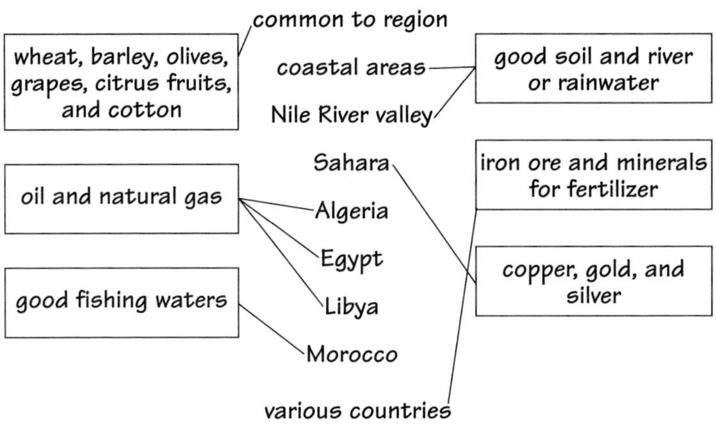

▲ Water from the Nile River irrigates rich farmlands along the river and in its delta. You can clearly see the Nile Delta in the satellite photograph. The irrigated farmlands of the delta are shown in red.

Interpreting the Visual Record
Why do you think most Egyptians live near the Nile River?

The Nile The world's longest river, the Nile, flows northward through the eastern Sahara. The Nile empties into the Mediterranean Sea. It is formed by the union of two rivers, the Blue Nile and the White Nile. They meet in Sudan, south of Egypt.

The Nile River valley is like a long oasis in the desert. Water from the Nile irrigates surrounding farmland. The Nile fans out near the Mediterranean Sea, forming a large river delta. About 99 percent of Egypt's population lives in the Nile River valley and the Nile Delta.

For centuries rain far to the south caused annual floods along the northern Nile that left rich **silt** in surrounding fields. Silt is finely ground soil good for growing crops. The Aswān High Dam, which was completed in 1971, was built to control flooding. Water trapped by the dam formed Lake Nasser in southern Egypt. However, the dam also traps silt, preventing it from being carried downriver. Today Egypt's farmers must use fertilizers to enrich the soil.

The Sinai and Suez Canal East of the Nile is the triangular Sinai Peninsula. Barren, rocky mountains and desert cover the Sinai. Between the Sinai and the rest of Egypt is the Suez Canal. The canal was built by the French in the 1860s. See the Case Study about the Suez Canal in this chapter.

✓ **READING CHECK:** *Places and Regions* What are the major physical features of North Africa? Sahara, mountains, Nile River valley, Sinai Peninsula

Climate, Vegetation, and Animals

There are three main climates in North Africa. A desert climate covers most of the region. Temperatures range from mild to very hot. How hot can it get? Temperatures as high as 136°F (58°C) have been recorded in Libya! However, the humidity is very low. As a result, temperatures can drop quickly after sunset.

Our Amazing Planet

The Arabian camel has long been used for transportation in the Sahara. It can store water in the fat of its hump. Camels have survived for more than two weeks without drinking.

National Geography Standard 14

Environment and Society
Although the Aswān High Dam controls the flooding of the Nile River, stores water for use, and provides hydroelectric power, it also has had a negative impact on the region. In addition to the loss of fertile soil, critics cite the incursion of saltwater from the Mediterranean Sea into the Nile Delta and a reduction in the number of fish along the delta shore as problems that have been caused by the dam.

Critical Thinking: In addition to the benefits of the dam that were mentioned in the passage, why might a dam-construction project be popular in a country?

Answer: Answers will vary, but students might mention the jobs that would be created by such a project or the excitement generated by a large public-works project.

◄ **Visual Record Answer**

access to fertile soil and water for irrigation, industry, and drinking

Close

Ask a student to point out one of the major landforms and a climate zone of North Africa on a wall map. Lead a discussion on how the landforms and climates of this region might affect the lives of people who live there.

Review and Assess

Have students complete the Section Review. Then organize students into groups and distribute five index cards to each group. Have each group create five where-am-I questions and swap its cards with the other groups. Then have students complete Daily Quiz 19.1.

COOPERATIVE LEARNING

Reteach

Have students complete Main Idea Activity 19.1. Then have each student create a collage to represent the climate types, natural resources, and physical features of North Africa. Have volunteers present their collages to the class. **ENGLISH LANGUAGE LEARNERS**

Extend

Have interested students conduct research on Nile River navigation throughout history. Ask them to find information on any or all of these topics: (1) ancient and modern boat construction, (2) prevailing winds, (3) effect of the rapids on navigation, and (4) income generated by modern tourist cruises. Have students present the results of their research as a written report or illustrated poster. **BLOCK SCHEDULING**

Section 1 Review

Answers

Define For definitions, see: ergs, p. 442; regs, p. 442; depressions, p. 442; silt, p. 443

Working with Sketch Maps Maps will vary, but listed places should be labeled in their approximate locations. Aswān High Dam created Lake Nasser.

Reading for the Main Idea

1. the Ahaggar and the Atlas Mountains (NGS 4)
2. the Sinai Peninsula (NGS 4)
3. It flooded and left deposits of fertile silt. (NGS 15)

Critical Thinking

4. likely to be near water sources

Organizing What You Know

5. climates—desert, Mediterranean, steppe; plants and animals—grasses, small shrubs, trees, insects, snails, small reptiles, gazelles, hyenas, baboons, foxes, weasels; resources—good soil and water sources in coastal areas, fishing off the Atlantic coast, oil and gas, iron ore and minerals for fertilizers, copper, gold, and silver

BUILD on WHAT You Know

Do you remember what you learned about climate regions? See Chapter 3 to review.

In some areas there has been no rain for many years. However, rare storms can cause flash floods. In places these floods as well as high winds have carved bare rock surfaces out of the land. Storms of sand and dust can also be severe.

Hardy plants and animals live in the desert. Grasses, small shrubs, and even trees grow where there is enough water. Usually this is in oases. Gazelles, hyenas, baboons, foxes, and weasels are among the region's mammals.

Much of the northern coast west of Egypt has a Mediterranean climate. Winters there are mild and moist. Summers are hot and dry. Plant life includes grasses, shrubs, and even a few forests in the Atlas Mountains. Areas between the Mediterranean climate and the Sahara have a steppe climate. Shrubs and grasses grow there.

✓ **READING CHECK:** *Physical Systems* How does climate affect the plants and wildlife of North Africa? It determines what lives in a region: desert—hardy plants and animals, more wildlife in oases; Mediterranean—grasses, shrubs, forests; steppe—shrubs, grasses.

Olives like these in Morocco are an important agricultural product in North Africa. Olives and olive oil are common ingredients in many foods around the Mediterranean.

Resources

Good soils and rain or river water aid farming in coastal areas and the Nile River valley. Common crops are wheat, barley, olives, grapes, citrus fruits, and cotton. The region also has good fishing waters.

Oil and gas are important resources, particularly for Libya, Algeria, and Egypt. Morocco mines iron ore and minerals used to make fertilizers. The Sahara has minerals such as copper, gold, and silver.

✓ **READING CHECK:** *Places and Regions* What are North Africa's major resources? good soils, river water, oil, gas, iron ore; minerals

go.hrw.com Homework Practice Online Keyword: SK3 HP19

Section 1 Review

Define and explain: ergs, regs, depressions, silt

Working with Sketch Maps On a map of North Africa that you draw or that your teacher provides, label the following: Red Sea, Mediterranean Sea, Sahara, Nile River, Sinai Peninsula, Ahaggar Mountains, Atlas Mountains, Qattara Depression, Nile Delta, Lake Nasser, and the Suez Canal. In a box in the margin, identify the dam that created Lake Nasser.

Reading for the Main Idea

1. *Places and Regions* What two mountain ranges are found in North Africa?
2. *Places and Regions* What part of Egypt is east of the Suez Canal?
3. *Environment and Society* How did the Nile affect farming in the river valley?

Critical Thinking

4. **Drawing Inferences and Conclusions** Where would you expect to find most of North Africa's major cities?

Organizing What You Know

5. **Summarizing** Use this graphic organizer to describe the physical features of North Africa.

Climates	Plants and Animals	Resources

Section 2

Objectives

1. Relate the major events in the history of North Africa.
2. Describe some important facts about the people and culture of North Africa.

Focus

LET'S GET STARTED

Copy the following instructions onto the chalkboard: *Write down what you already know about North Africa.* Have students write their responses, and ask volunteers to share their responses with the class. *(Students might mention that they have heard of the Sahara, the Sphinx, mummies, or the pyramids.)* Tell students that in Section 2 they will learn other interesting things about the history and culture of North Africa.

Building Vocabulary

Write the key terms on the chalkboard. Show pictures of **hieroglyphs** in the tombs of Egyptian **pharaohs**. Explain that hieroglyphs are a writing system that used pictures or symbols instead of an alphabet. Tell students that pharaohs were the kings of ancient Egypt. Then define **Bedouins** and ask students to describe what a Bedouin caravan might look like.

Section 2 History and Culture

Read to Discover

1. What are the major events in the history of North Africa?
2. What are some important facts about the people and culture of North Africa?

Define

pharaohs
hieroglyphs
Bedouins

Locate

Alexandria
Cairo
Fès
Western Sahara

Why It Matters

Egypt was home to one of the world's earliest and most complex civilizations. Use CNNfyi.com or other **current events** sources to find examples of new archaeological finds from Egypt. Record your findings in your journal.

Woven carpet from Morocco

History

The Nile River valley was home to some of the world's oldest civilizations. Sometime after 3200 B.C. lands along the northern Nile were united into one Egyptian kingdom.

The early Egyptians used water from the Nile to grow wheat, barley, and other crops. They also built great stone pyramids and other monuments. Egyptian **pharaohs**, or kings, were buried in the pyramids. How did the Egyptians build these huge monuments? See Connecting to Technology on the next page.

The Egyptians also traded with people from other places. To identify themselves and their goods, the Egyptians used **hieroglyphs** (HY-ruh-glifs). Hieroglyphs are pictures and symbols that stand for ideas or words. They were the basis for Egypt's first writing system.

West of Egypt were people who spoke what are called Berber languages. These people herded sheep and other livestock. They also grew wheat and barley in the Atlas Mountains and along the coast.

▲ The tombs of early Egyptian pharaohs were decorated with paintings, crafts, and treasures. You can see hieroglyphs across the top of this wall painting.

Invaders Because of North Africa's long coastline, the region was open to invaders over the centuries. Those invaders included people from the eastern Mediterranean, Greeks, and Romans. For example, one invader was the Macedonian king Alexander the Great. He founded the city of Alexandria in Egypt in 332 B.C. This city became an important seaport and trading center on the Mediterranean coast.

Beginning in the A.D. 600s, Arab armies from Southwest Asia swept across North Africa. They brought the Arabic language and Islam to the

Section 2 Resources

Reproducible

- Guided Reading Strategy 19.2
- Graphic Organizer 19
- Map Activity 19
- Cultures of the World Activity 6
- Creative Strategies for Teaching World Geography, Lessons 14, 15, and 16

Technology

- One-Stop Planner CD–ROM, Lesson 19.2
- Homework Practice Online
- HRW Go site

Reinforcement, Review, and Assessment

- Section 2 Review, p. 449
- Daily Quiz 19.2
- Main Idea Activity 19.2
- English Audio Summary 19.2
- Spanish Audio Summary 19.2

TEACH

Teaching Objective 1

LEVEL 1: (Suggested time: 20 min.) Copy the following chart onto the chalkboard, omitting the italicized answers. Organize the class into four groups. Assign one group Ancient North Africa, another group Invaders, the third European Control, and the last group Modern North Africa. Have groups compile information about significant groups in each category and put it in the chart.

ENGLISH LANGUAGE LEARNERS, COOPERATIVE LEARNING

History of North Africa

Ancient North Africa	Invaders	European Control	Modern North Africa
Egyptian kingdom united c. 3200 B.C.; Egyptians farmed, built pyramids, developed trade, used hieroglyphs; Berbers lived west of Egypt, herded livestock, grew wheat and barley	*Greeks, Romans; Alexander the Great—founded Alexandria; Arabs beginning in A.D. 600s—brought Arabic and Islam*	*control began in 1800s, completed by 1912; Italy—Libya; Spain—northern Morocco; France—rest of Morocco, Tunisia, Algeria; Great Britain—Egypt; all independent by 1962*	*building strong ties with other Arab countries; several wars against Israel; 1979, Egypt became first Arab nation to sign peace treaty with Israel; 1976, Morocco took over Western Sahara*

Linking Past to Present

The Necropolis at Bahariya In 1999, archaeologists announced the discovery three years earlier of the largest mummy burial site ever found in Egypt. Located at the Bahariya Oasis, 230 miles from the Giza pyramids, the site contained evidence dating it from the 300s B.C. to the A.D. 100s. This site was undisturbed by grave robbers and contained large numbers of gold artifacts.

Activity: Have students conduct research on and then present a news broadcast about one of the major archaeological discoveries such as the tomb of King Tutankhamen or the burial site at the Bahariya Oasis.

Connecting to Technology

Answers

1. pyramids, tombs, and stone temples; to show the power of Egyptian rulers
2. The Egyptians believed they needed to honor the dead rulers' power through massive monuments that required extensive technology to construct.

internet connect

GO TO: **go.hrw.com**
KEYWORD: **SK3 CH19**
FOR: **Web sites about Egyptian mummies**

CONNECTING TO *Technology*

Egyptian Monuments

The monuments of ancient Egypt are among the great wonders of the world. They are thousands of years old. These pyramids, temples, and other structures reflect the power of Egyptian rulers. They also show the skills of Egyptian engineers.

The most famous of Egypt's monuments are the huge stone pyramids at Giza. Giza is near Cairo. The pyramids there were built more than 4,000 years ago as tombs for Egyptian rulers. The largest structure is the Great Pyramid. At its base the pyramid's sides are each about 755 feet (230 m) long. The pyramid rises nearly 500 feet (152 m) above the desert floor.

Another famous set of monuments is found farther south at the Valley of the Kings. The Valley of the Kings lies along the Nile River. Between 1500 B.C. and 1000 B.C., the Egyptians built tombs into the sides of cliffs there. They also carved huge sculptures and columns on the cliff faces. Near the tombs they built giant stone temples.

How did the Egyptians build these giant monuments? Workers had to cut large blocks of stone far away and roll them on logs to the Nile. From there the blocks could be moved on barges. At the building site, the Egyptians finished carving the blocks with special tools. Then they built dirt and brick ramps on the sides of the structures. They hauled the blocks up the ramps.

The average weight of each of the 2.3 million blocks in the Great Pyramid is 2.5 tons (2.25 metric tons). It is estimated that many thousands of workers were needed to move the heavy blocks into place. Together, these workers and Egyptian engineers built amazing monuments that have stood for thousands of years.

Understanding What You Read

1. What kinds of monuments did the Egyptians build and why?
2. How did the Egyptians' belief systems affect their use of technology in building monuments?

The Sphinx and pyramids at Giza, Egypt

LEVELS 2 AND 3: (Suggested time: 45 min.) Organize students into four groups. Assign each group one of the following time periods: Ancient North Africa, Invaders, European Control, or Modern North Africa. Have each group prepare a time line of main events for its time period and write an essay assessing the significance of the individuals involved in the events. Once groups have finished their work, have one student from each group present his or her group's time lines to the class and discuss the significance of the events. **COOPERATIVE LEARNING**

Teaching Objective 2

LEVEL 1: (Suggested time: 20 min.) Draw two large circles on a piece of butcher paper. Divide one circle into six equal parts and cut it into six wedges. Organize the class into six groups and assign each group a topic: language, religion, food, festivals, art, and literature. Have each group write or draw pictures illustrating information about its topic on its wedge. Then have students glue their wedges onto the circle. **ENGLISH LANGUAGE LEARNERS, COOPERATIVE LEARNING**

LEVELS 2 AND 3: (Suggested time: 25 min.) Have each student complete the Level 1 lesson for this objective. Then have students select one of the wedges created in the Level 1 activity and write a short poem based on the images from the wedge.

region. Today most North Africans are Muslim and speak Arabic. Under Muslim rule, North African cities became major centers of learning, trade, and craft making. These cities included Cairo in Egypt and Fès in Morocco.

European Control In the 1800s European countries began to take over the region. By 1912 they controlled all of North Africa. In that year Italy captured Libya from the Ottoman Empire. Spain already controlled northern Morocco. France ruled the rest of Morocco as well as Tunisia and Algeria. Egypt was under British control.

Egypt gained limited independence in 1922. The British kept military bases there and maintained control of the Suez Canal until 1956. During World War II the region was a major battleground. After the war, North Africans pushed for independence. Libya, Morocco, and Tunisia each won independence in the 1950s.

Algeria was the last North African country to win independence. Many French citizens had moved to the country. They considered Algeria part of France. Algeria finally won independence in 1962, after a long, bitter war. Most French residents of Algeria then moved to France.

Modern North Africa Since independence, the countries of North Africa have tried to build stronger ties with other Arab countries. For example, Egypt led other Arab countries in several wars against Israel. However, in 1979 Egypt signed a peace treaty with Israel.

In 1976 Morocco took over the former Spanish colony of Western Sahara. Western Saharan rebels have been trying to win independence from Morocco since then.

✓ **READING CHECK:** *Places and Regions* What were some major events in the history of North Africa? **invasions by Mediterranean peoples, Greeks, Romans, Arab armies; control by European countries; independence**

▲ The ruins of an old Roman amphitheater remain in Alexandria, Egypt.

The inset photograph shows the beautiful doors of the royal palace in Fès, Morocco.

Interpreting the Visual Record What interesting architectural features do you see in the photograph?

▼

HUMAN SYSTEMS

France's colonial presence in Algeria began in 1830 when the French king invaded the country in hopes of gaining popularity at home. He had found his excuse to attack Algeria three years earlier when the country's ruler struck the French consul with a fly whisk during a meeting. Algeria remained under French control until the Algerian War brought independence in 1962.

Independence came at a heavy price. It is estimated that some 10,000 French soldiers and more than 100,000 Algerians were killed during the fighting. In addition, many villages were destroyed and some 2,000,000 peasants had to be relocated.

Activity: Have students create charts to categorize and list examples of French influence in Algeria.

◄ **Visual Record Answer**

Students' answers will vary, but they might mention Muslim influences.

Teacher to Teacher

Wendy Mason of San Marcos, Texas, suggests the following activity to help students learn about North African history and culture: Pair students and have them prepare a storyboard for a slide show presentation on North African history and culture. Have pairs draw their storyboards on large sheets of poster board and present them to the class. After all the pairs have presented their storyboards, lead a class discussion on combining ideas from all the storyboards to make one class storyboard.

➤ASSIGNMENT: Have students select a topic for a documentary film on North Africa and then prepare a movie poster to advertise the documentary. Encourage students to consider the region's history, languages, customs, interesting sites, and culture when choosing their documentary topics. Also encourage students to cut out pictures or to add illustrations to their posters for added interest.

Daily Life

Among the religious obligations of Muslims are five daily prayers. These prayers are to be offered at various times of the day, with the first being said before sunrise and the last before retiring to bed. Although tradition dictates that the prayers are to be offered in a mosque, Muslims who are unable to attend are permitted to offer individual prayers.

The muezzin is the person in a mosque who issues the daily calls to prayer. The muezzin traditionally stands in the door or tower of the mosque when proclaiming the call. However, many mosques today have replaced the muezzin with electronic recordings and amplifiers.

Activity: Have students conduct research on how Muslims incorporate the five daily prayers into their lives.

Visual Record Answers ▲

Students should suggest that Muslims need space to kneel and bow while praying.

►

The agricultural products are varied.

▲ The floor of the beautiful Muhammad Ali mosque in Cairo is covered with carpet. Muslim men kneel and bow with their faces to the ground to pray.

Interpreting the Visual Record **Why do you think there are no chairs in this picture?**

Culture

As you have read, the histories of the North African countries have much in common. You will also find many cultural similarities among those countries.

Language and Religion Egyptians, Berbers, and **Bedouins** make up nearly all of Egypt's population. Bedouins are nomadic herders who travel throughout deserts of Egypt and Southwest Asia. Most people in the countries to the west are of mixed Arab and Berber ancestry. Arabic is the major language. Some people speak Berber languages.

Most ethnic Europeans left North Africa after the region's countries became independent. However, French, Italian, and English still are spoken in many areas.

Most North Africans are Muslims. Of the region's countries, Egypt has the largest number of non-Muslims. About 6 percent of Egyptians are Christians or practice other religions.

A variety of vegetables and meat surround a large serving of couscous.

Interpreting the Visual Record **What does this food tell you about agricultural products of the region?** ▼

Food and Festivals What kind of food would you eat on a trip to North Africa? Grains, vegetables, fruits, and nuts are common there. Many meals include couscous (KOOS-koos). Couscous is made from wheat and looks like small pellets of pasta. It is steamed over boiling water or soup. Often it is served with vegetables or meat, butter, and olive oil. Some people mix their couscous with a fiery hot sauce called *harissa.*

A popular dish in Egypt is *fuul.* It is made from fava beans mashed with olive oil, salt, pepper, garlic, and lemons. The combination is then served with hard-boiled eggs and bread.

Important holidays in North Africa include the birthday of the prophet of Islam, Muhammad. The birthday is marked

Close

Ask students to give examples of the importance of the Nile River to Egyptians in the past and today. *(Students might note the location of temples and cities, transportation, source of drinking water and irrigation, and so on.)* Point out that the Nile made Egypt a magnet for traders, travelers, and invaders.

Reteach

Have students complete Main Idea Activity 19.2. Then organize students into three groups and have each group create a large outline map of North Africa on butcher paper. Using the outline map as a base, have the groups create collages from pictures that illustrate information about North Africa. **ENGLISH LANGUAGE LEARNERS, COOPERATIVE LEARNING**

Review and Assess

Have students complete the Section Review. Next, have students prepare crossword puzzles about North Africa. When the puzzles are completed, photocopy them and let each student work a puzzle created by another student. Then have students complete Daily Quiz 19.2.

Extend

Have students research and explain the significance of the major holidays of Islam, including Ramadan. Ask them to describe the origins of the holidays and how they are recognized in a written report. **COOPERATIVE LEARNING, BLOCK SCHEDULING**

with lights, parades, and special sweets of honey, nuts, and sugar. During the holy month of Ramadan, Muslims abstain from food and drink during the day.

Art and Literature North Africa has long been known for its beautiful architecture, wood carving, and other crafts. Women weave a variety of textiles. Among these are beautiful carpets that feature geometric designs and bright colors.

The region has also produced important writers and artists. For example, Egyptian poetry and other writing date back thousands of years. One of Egypt's most famous writers is Naguib Mahfouz. In 1988 he became the first Arab writer to win the Nobel Prize in literature. Egypt also has a growing movie industry. Egyptian films in Arabic have become popular throughout Southwest Asia and North Africa.

Many North Africans also enjoy popular music based on singing and poetry. The musical scale there has many more notes than are common in Western music. As a result, North African tunes seem to wail or waver. Musicians often use instruments such as the three-stringed *sintir* of Morocco.

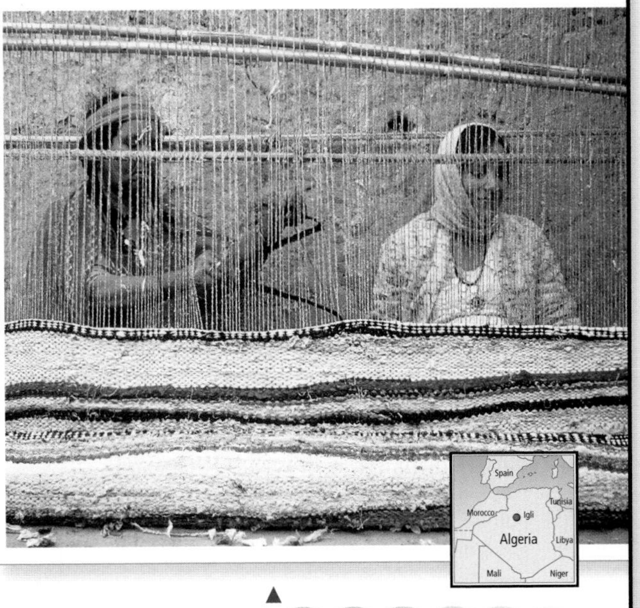

▲ This Algerian woman and her daughter are using a loom to weave a rug. The loom allows a rug maker to weave horizontal and vertical threads together.

✓ **READING CHECK:** *Human Systems* What are some important facts about the people and culture of North Africa? Arabic major language; Islam major religion; Ramadan major holiday; architecture, arts and crafts, literature major contributions to world

go.hrw.com **Homework Practice Online**
Keyword: SK3 HP19

Section 2 Review

Define and explain: pharaohs, hieroglyphs, Bedouins

Working with Sketch Maps On the map you drew in Section 1, label Alexandria, Cairo, Fès, and Western Sahara. In a box in the margin, identify the country that claims Western Sahara.

Reading for the Main Idea

1. *Places and Regions* What early civilization thrived along the northern Nile River about 3200 B.C.?
2. *Human Systems* How did Islam and the Arabic language come to North Africa?
3. *Human Systems* Which European countries controlled countries in North Africa by 1912?

Critical Thinking

4. **Drawing Inferences and Conclusions** Look at the world map in the textbook's Atlas. Why do you think the Suez Canal is an important waterway for world shipping?

Organizing What You Know

5. **Summarizing** Copy the following graphic organizer. Use it to identify at least six important facts about the languages, religions, food, festivals, art, and literature of North Africa.

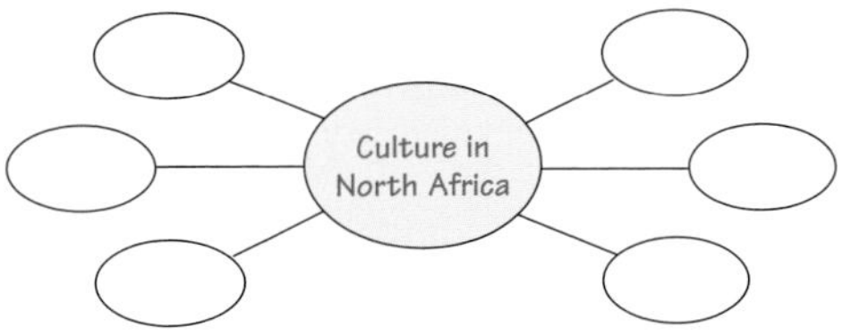

Section 2 Review

Answers

Define For definitions, see: pharaohs, p. 445; hieroglyphs, p. 445; Bedouins, p. 448

Working with Sketch Maps Alexandria, Cairo, Fès, and Western Sahara should be located accurately. Morocco claims Western Sahara.

Reading for the Main Idea

1. the Egyptians (NGS 4)
2. Arabic armies (NGS 9)
3. Italy—Libya; Spain—northern Morocco; France—rest of Morocco, Algeria, and Tunisia; Great Britain—Egypt (NGS 13)

Critical Thinking

4. provides a water route between the Indian Ocean and the Mediterranean Sea

Organizing What You Know

5. languages—Arabic, Berber, French, Italian, English; religions—Islam, Christianity, others; food—couscous and other grains, vegetables, fruit, nuts; festivals—Muhammad's birthday, Ramadan; art—architecture, woodcarvings, weaving, music; literature—long tradition of poetry

CASE STUDY

Setting the Scene

The Suez Canal was built by the French to span the Isthmus of Suez. Since it opened in 1869, the 101-mile-long canal has played an important role in international affairs because of its location. Soon after its completion, canal users recognized the importance of the channel. To protect this transportation corridor, a number of countries met in 1888 to sign an agreement that declared that the canal should be open to ships of all nations in times of war and peace. As the case study suggests, this agreement has not always been upheld.

Building a Case

Have students read "The Suez Canal: A Strategic Waterway." Ask them to explain the term *choke point. (narrow passage that restricts ship traffic)* Next, ask students to identify some other examples of choke points. *(Strait of Hormuz, Strait of Malacca, Panama Canal)* Conclude by asking the following questions: What bodies of water does the Suez Canal link? *(Mediterranean Sea and Red Sea)* How has the canal affected Europe and Asia? *(It has shortened the distance between the continents and made the movement of goods and people easier.)*

HISTORICAL GEOGRAPHY

Strategic choke points were also important in the ancient world. Some historians believe that the Trojan War was a battle over one such choke point. Greeks had set up colonies on the shores of the Black Sea starting in the 1200s B.C. All trade between Greece and these Black Sea colonies had to pass through the Dardanelles, a narrow strait connected to the Aegean Sea. At the entrance of the Dardanelles was the city of Troy. The Trojans demanded tolls from the Greeks to allow passage. Friction over the sea lane led to a war that lasted 10 years.

Activity: Have students write brief accounts of the Trojan War from the perspective of a Greek sailor, a Greek colonist, and a Trojan warrior.

➤ **This Case Study feature addresses National Geography Standards 1, 11, and 13.**

Map Answer ►

contributed to increased trade

CASE STUDY

THE SUEZ CANAL: A STRATEGIC WATERWAY

The movement of goods and people around the world has increased dramatically during the past 100 years. In the world's oceans, more ships are sailing from port to port through busy shipping lanes. Just as highways can get crowded with cars, shipping lanes can get crowded with ships. These crowded areas are called choke points. Choke points are narrow routes that ships must pass through to get to other areas. The Suez Canal, Panama Canal, Strait of Malacca, Strait of Hormuz, and Strait of Gibraltar are all examples of important choke points around the world.

Choke points often have great strategic importance. In times of war, they can be important for moving ships and troops to other parts of the world. During peacetime, ships must pass through them to transport goods to world markets.

The Suez Canal in Egypt is an important choke point in the world's shipping lanes. It is also a very strategic waterway. Opened in 1869, the Suez Canal connects the Mediterranean Sea with the Red Sea. This shortens the journey for ships sailing between Europe and Asia by about 6,000 miles (9,654 km).

The history of the Suez Canal is closely tied to the growth of world trade and ocean shipping. Before World War II, the canal linked Great Britain to its colonies in East Africa, Asia, and the Pacific. Beginning in the 1930s, the development of the oil industry in the Persian Gulf made the canal even more important. It became a major shipping route for tankers carrying oil to Europe and North America.

The Suez Canal was built mainly to serve international trade. However, it has also been used by military ships. In 1905 Russian navy ships used the canal during the Russo-Japanese War. In the 1930s Italy moved troops through the canal to eastern Africa before invading Ethiopia.

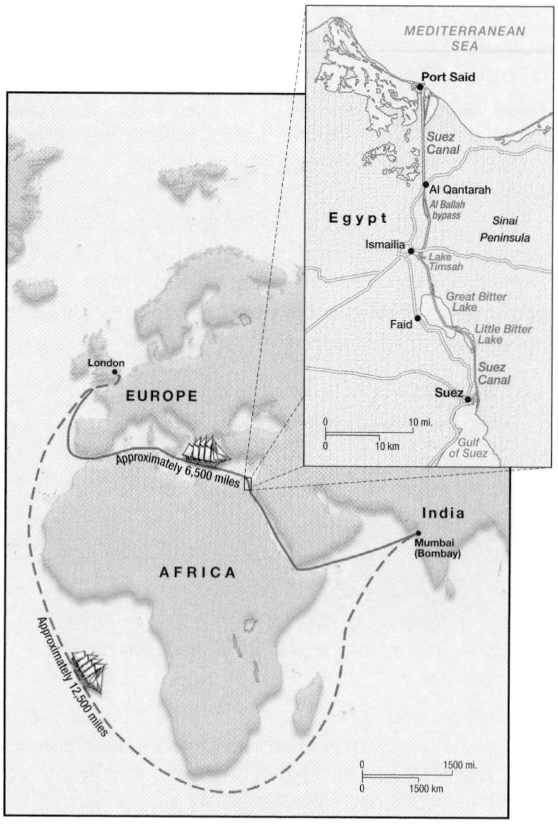

▲ A ship traveling from London to Mumbai has a much shorter journey if it travels through the Suez Canal.

Interpreting the Map How do you think the Suez Canal affected trade between Great Britain and India?

Drawing Conclusions

Lead a discussion about the importance of the Suez Canal by asking: How is the canal "strategic" with respect to trade? *(Goods traveling between Europe and Asia must pass through the canal.)* What are some examples of goods that are transported through the canal? *(oil, coal, wheat, corn, barley, metals)* How is the canal "strategic" in time of war? *(Troops can be moved to war zones more quickly.)* How might history have been altered if the canal had been blocked to military ships? *(Italy would have had difficulty invading Ethiopia. Russian ships would have had a longer voyage to fight in the Russo-Japanese War.)* How did the closure of the canal during the Israeli-Egyptian conflict undermine its strategic importance? *(created demand for larger ships that are too big to use the canal)*

Going Further: Thinking Critically

Provide detailed maps that depict the choke points listed in the case study. *(Suez Canal, Panama Canal, Strait of Malacca, Strait of Hormuz, Strait of Gibraltar)* Then have students work in small groups to determine what choke points they would pass through to make the shortest water voyages between the following destinations:

- New York to San Francisco *(Panama Canal)*
- Boston to Bombay *(Strait of Gibraltar, Suez Canal)*
- Taipei to Naples *(Strait of Malacca, Suez Canal)*

Ask selected groups to share their answers with the class.

An important test for the Suez Canal involved relations between Egypt and Israel. In the 1950s Egypt refused to allow ships from Israel to use the canal. Even ships from countries that traded with Israel were prevented from using it. In 1956–57 the Suez Canal was closed after Israeli and Egyptian forces began fighting. It was closed again during 1967–75 after more fighting between the two countries. For several years the canal was the front line between the Israeli and Egyptian armies. While it was closed, ships were forced to make the much longer journey around Africa. This created a demand for larger ships that could carry more cargo.

Israeli troops look across the Suez Canal in 1967.

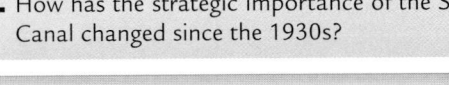

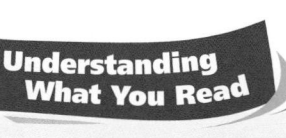

A French newspaper shows a warship crossing the Suez Canal in 1904 (left). Today, the canal is an important shipping lane (below).

Today, the Suez Canal is one of the busiest shipping lanes in the world. Each day, about 55 ships make the 15-hour journey through the canal. Ships heading north carry oil, coal, metals, and other goods to Europe and North America. Ships heading south carry wheat, corn, barley, manufactured goods, and other products to Asia.

Many modern ocean ships are now too big to use the Suez Canal. As a result, the canal's economic importance has declined in recent years. However, the Suez Canal is still an important waterway to many countries.

Understanding What You Read

1. What are choke points? Why are they important?
2. How has the strategic importance of the Suez Canal changed since the 1930s?

Understanding What You Read

1. Choke points are narrow, crowded routes that ships must pass through to get to other areas. They are important because they allow ships to transport goods to world markets and move troops during wars.
2. The Suez Canal increased in importance with the growth of the oil industry in the Persian Gulf. In recent years, however, its importance has declined somewhat because many newer ocean ships are too large to use it.

internet connect

GO TO: **go.hrw.com**
KEYWORD: **SK3 CH19**
FOR: **Web sites about the Suez Canal**

Section 3

Objectives

1. Describe the people and cities of Egypt today.
2. List Egypt's important economic activities.
3. Identify the challenges that Egypt faces today.

Focus

LET'S GET STARTED

Copy the following instructions onto the chalkboard: *Imagine what it would be like to live in a country with cities thousands of years old. What might be the benefits and drawbacks of living in such a country?* Tell students to respond in writing. Ask volunteers to share their responses with the class. *(Students might mention that such a country would have many antiquities and sites of historical interest or that streets and lanes might be narrow.)* Tell students that in Section 3 they will learn that Egypt is a modern nation with ties to the ancient world.

Building Vocabulary

Write the key term **fellahin** on the chalkboard. Ask a volunteer to find and read aloud the word's definition and origin *(Arabic)* from the dictionary. Then ask students to describe where they think most fellahin might live *(around the Nile River and the Nile Delta).*

Section 3 Resources

Reproducible
- Guided Reading Strategy 19.3

Technology
- One-Stop Planner CD–ROM, Lesson 19.3
- Homework Practice Online
- HRW Go site

Reinforcement, Review, and Assessment
- Section 3 Review, p. 454
- Daily Quiz 19.3
- Main Idea Activity 19.3
- English Audio Summary 19.3
- Spanish Audio Summary 19.3

internet connect

GO TO: go.hrw.com
KEYWORD: SK3 CH19
FOR: Web sites about Egypt today

Section 3 Egypt Today

Read to Discover

1. What are the people and cities of Egypt like today?
2. What are Egypt's important economic activities?
3. What challenges does Egypt face today?

Define

fellahin

Locate

Egypt

A gold mask from the tomb of King Tutankhamen

WHY IT MATTERS

An ongoing debate in Egypt is the place of Islam in society and the government. Use CNNfyi.com or other **current events** sources to find out how Egyptians view this debate. Record your findings in your journal.

People and Cities

Egypt is North Africa's most populous country. More than 67 million people live there.

Rural Egypt More than half of all Egyptians live in small villages and other rural areas. Most rural Egyptians are farmers called **fellahin** (fel-uh-HEEN). They own very small plots of land. Most fellahin also work large farms owned by powerful families. Many also depend on money sent home by family members working abroad. Many Egyptians work in Europe or oil-rich countries in Southwest Asia.

Cities Egypt's capital and largest city is Cairo. More than 10 million people live there. Millions more live in surrounding cities.

A muezzin (moo-E-zuhn) calls Muslims to prayer in Cairo. A muezzin often makes his calls from the door or the minaret, or tower, of a mosque. His calls—or recordings played through speakers—can be heard throughout Islamic communities five times daily.

TEACH

Teaching Objectives 1–3

ALL LEVELS: (Suggested time: 20 min.) Copy the following graphic organizer onto the chalkboard, omitting the italicized answers. Use the organizer to help students learn about Egypt's people, cities, economic activities, and challenges it faces today by having the class supply the missing information. **ENGLISH LANGUAGE LEARNERS**

Aspects of Modern Egypt

People	*More than half of all Egyptians live in small villages; most fellahin, or farmers, work large family-owned farms; many people work in Europe or Southwest Asia.*
Cities	*Cairo—largest city, problems include traffic, pollution, and crowding; Alexandria—second-largest city, major seaport*
Economic Activities	*textiles, tourism, oil, tolls from Suez Canal, farming (cotton, vegetables, grain, fruit)*
Today's Challenges	*need to fertilize heavily and import food, growing population, deciding role of Egypt in world affairs and role of Islam in government, health and literacy issues*

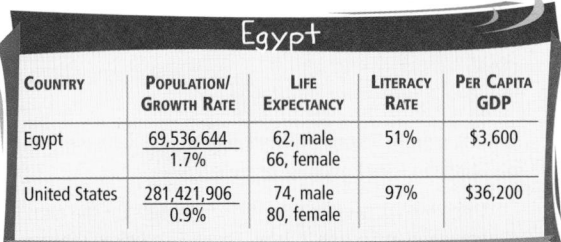

◄ Tourist ships sit on the Nile River in Luxor. Tourists visit the ruins of a beautiful temple built there more than 2,300 years ago. These and other historical sites make tourism an important part of Egypt's economy.

Cairo was founded more than 1,000 years ago along the Nile. Its location at the southern end of the delta helped it grow. The city lies along old trading routes between Asia and Europe. Later it was connected by railroad to Mediterranean ports and the Suez Canal.

Today Cairo is a mixture of modern buildings and small, mud-brick houses. People continue to move there from rural areas. Many live in makeshift housing. Traffic and pollution are serious problems.

Alexandria has more than 4 million people. It is located in the Nile Delta along the Mediterranean coast. The city is a major seaport and home to many industries.

✓ **READING CHECK:** *Places and Regions* What are Egypt's people and cities like today? **More than half of the people live in rural areas; cities are congested and heavily populated, with both modern and traditional architecture.**

Our Amazing Planet

Fish is an important food for people living along the Nile. One fish, the giant Nile perch, can grow to a weight of 300 pounds (136 kg).

Economy

To provide for its growing population, Egypt is working to expand its industries. Textiles, tourism, and oil are three of the most important industries. The Suez Canal is another source of income. Ships pay tolls to pass through it. Ships use the canal to avoid long trips around southern Africa. This makes the canal one of the world's busiest waterways.

About 40 percent of Egyptian workers are farmers. A warm, sunny climate and water for irrigation make the Nile Delta ideal for growing cotton. Farmlands along the Nile River are used for growing vegetables, grain, and fruit.

Egypt

Country	Population/ Growth Rate	Life Expectancy	Literacy Rate	Per Capita GDP
Egypt	69,536,644 / 1.7%	62, male / 66, female	51%	$3,600
United States	281,421,906 / 0.9%	74, male / 80, female	97%	$36,200

Sources: Central Intelligence Agency, *The World Factbook 2001;* U.S. Census Bureau

Interpreting the Chart **How does the growth rate of Egypt compare to that of the United States?**

✓ **READING CHECK:** *Places and Regions* How does the Suez Canal affect Egypt's economy? **provides income from tolls**

Linking Past to Present

Luxor The Temple of Luxor is made up of courtyards, halls, and chambers surrounded by rows of gigantic columns. The columns are shaped to resemble papyrus plants topped by buds.

Construction of the temple complex was begun by the pharaoh Amenhotep III, although there may have been an earlier temple on this spot. Several succeeding pharaohs added to the structure. Later still, a shrine to Alexander the Great was added. A Roman legion made its headquarters inside this temple, and both Christian churches and a mosque were built on different parts of the site.

Today Luxor is one of Egypt's major tourist attractions. In addition to the popular Nile River cruises, visitors can reach Luxor by train or airplane.

Activity: Have interested students research the Temple of Luxor and prepare maps of the complex as it might have looked during different periods of history.

◄ **Chart Answer**

It is higher.

CLOSE

Have students recall the challenges faced by Egypt today. Then have them compare the challenges facing Egypt with the challenges faced by the United States. Ask students if both countries can deal with similar challenges in the same way. Have volunteers justify and explain their answers.

REVIEW AND ASSESS

Have students complete the Section Review. Then have them work in pairs to write five questions about modern Egypt on index cards, placing the answers on the reverse side. When students have finished, have each pair exchange cards with another pair and answer the questions. Then have students complete Daily Quiz 19.3. **COOPERATIVE LEARNING**

RETEACH

Have students complete Main Idea Activity 19.3. Then organize students into four groups and assign each group Cities, People, Economy, or Challenges. Have groups use their textbooks to list information about their assigned topics. Then share the findings in class.
ENGLISH LANGUAGE LEARNERS, COOPERATIVE LEARNING

EXTEND

Have interested students conduct research on life in Egypt today and create a print or multimedia presentation outlining daily life in either a rural or urban setting. Have students share their presentations with the class.
BLOCK SCHEDULING

Section 3 Review

Answers

Define For definition, see: fellahin, p. 452

Working with Sketch Maps Egypt should be labeled in its approximate location. Cairo is the capital of Egypt.

Reading for the Main Idea

1. Some people want an Islamic government, which would mean fewer freedoms. (NGS 9)
2. textiles, tourism, oil, shipping through the Suez Canal, cotton, vegetables, grain, and fruit (NGS 4, 11)

Critical Thinking

3. for access to water for irrigation and travel
4. Its location at the southern end of the Nile delta helped it grow through trade.

Organizing What You Know

5. Answers should include the challenges of farming, crowded cities, health issues, literacy rate, and debates over Egypt's role in the world and the influence of Islam in the country's government.

▲ These Bedouin students attend a modern school in the Sinai. Improving education for all Egyptians is an important challenge facing the country. Among North African countries, only Morocco has a lower literacy rate than Egypt.

Challenges

Egypt faces important challenges today. For example, the country's farmland is limited to the Nile River valley and delta. To keep the land productive, farmers must use more and more fertilizer. This can be expensive. In addition, overwatering has been a problem. It has brought to the surface salts that are harmful to crops. These problems and a rapidly growing population have forced Egypt to import much of its food.

In addition, Egyptians are divided over their country's role in the world. Many want their country to remain a leader among Arab countries. However, others want their government to focus more on improving life for Egyptians at home.

Many Egyptians live in severe poverty. Many do not have clean water for cooking or washing. The spread of disease in crowded cities is also a problem. In addition, about half of Egyptians cannot read and write. Still, Egypt's government has made progress. Today Egyptians live longer and are much healthier than 50 years ago.

Another challenge facing Egyptians is the debate over the role of Islam in the country. Some Muslims want to shape the country's government and society along Islamic principles. However, some Egyptians worry that such a change would mean fewer personal freedoms. Some supporters of an Islamic government have turned to violence to advance their cause. Attacks on tourists in the 1990s were particularly worrisome. A loss of tourism would hurt Egypt's economy.

✓ **READING CHECK:** *Places and Regions* What are some of the challenges Egypt faces today? necessity of importing food due to the expense of fertilizer and from overwatering, division over country's role in the world; severe poverty, disease, illiteracy; role of Islam, violence

go.hrw.com Homework Practice Online
Keyword: SK3 HP19

Section 3 Review

Define and explain: fellahin

Working with Sketch Maps On the map you drew in Section 2, label Egypt. In a box in the margin, identify the capital of Egypt.

Reading for the Main Idea

1. *Human Systems* What is the relationship between religion and culture in Egypt today?
2. *Places and Regions* What industries and crops are important to Egypt's economy?

Critical Thinking

3. **Analyzing Information** Why do so many Egyptians live along the Nile and in the Nile Delta?
4. **Finding the Main Idea** How has Cairo's location shaped its development?

Organizing What You Know

5. **Summarizing** Copy the following graphic organizer. Use it to describe some of the challenges facing Egypt today.

Section 4

Objectives

1. Describe the region's people and cities today.
2. List the countries' important economic activities.
3. Identify the challenges the countries face today.

Focus

LET'S GET STARTED

Write the name of your town, city, or regional center on the chalkboard. Then copy the following instructions onto the chalkboard: *Identify old and new neighborhoods or districts in the community. What resources helped create the new features?* Tell students to respond in writing. Then tell students that in Section 4 they will learn how Libya, Tunisia, Algeria, and Morocco also display contrasts between old and new.

Building Vocabulary

Write **Casbah**, **souks**, and **free port** on the chalkboard. Explain that a Casbah is the old district of a city that served as a fortress, a souk is a marketplace usually found within the Casbah, and a free port is a city with few taxes placed on goods. Tell students that all of these terms are important to understanding the economy of this region. Then have a student look up **dictator** in the glossary. Mention that one of the countries of North Africa is ruled by a dictator.

Section 4 Libya, Tunisia, Algeria, and Morocco

Read to Discover

1. What are the region's people and cities like today?
2. What are the countries' important economic activities?
3. What challenges do the countries face today?

Define

Casbah
souks
free port
dictator

Locate

Libya
Tunisia
Algeria
Morocco
Tripoli
Benghazi
Algiers
Casablanca
Rabat
Tunis
Strait of Gibraltar

Why It Matters

Libya has been ruled by a dictator. Use CNNfyi.com or other **current events** sources to find out more about Libya's dictatorship. Record your findings in your journal.

Moroccan pottery

People and Cities

Western Libya, Tunisia, Algeria, and Morocco are often called the Maghreb (MUH-gruhb). This Arabic word means "west" or "the direction of the setting sun." Most of the Maghreb is covered by the Sahara. There you will find sandy plains and rocky uplands. Cities and farmland are located in narrow coastal strips of land. These strips lie between the Atlantic and Mediterranean coasts in the north and the Sahara and Atlas Mountains farther inland.

Libya is almost completely desert. Fertile land is limited to small areas along the coast. Cities and most of the population are found in those coastal areas. Libya is the most urbanized country in the region. More than 85 percent of Libya's more than 5 million people live in cities. The largest cities are Benghazi and the capital, Tripoli.

Algiers is a large city and is Algeria's capital. The central part of Algiers is a maze of winding alleys and tall walls. This old district is called the **Casbah**. The Casbah is basically an old fortress. **Souks**, or marketplaces, are found there today. The centers of other North African cities also have Casbahs.

Other large cities include Casablanca and Rabat in Morocco and Tunis in Tunisia. Another Moroccan city, Tangier, overlooks the Strait of Gibraltar. This beautiful city was once a Spanish territory. Today tourists can take a quick ferry ride from Spain across the strait to Tangier, a **free port**. A free port is a city in which almost no taxes are placed on goods sold there.

✓ **READING CHECK:** *Places and Regions* What geographic factors explain the region's population patterns? The region is mostly desert, and the cities and farmland are therefore located along the coast.

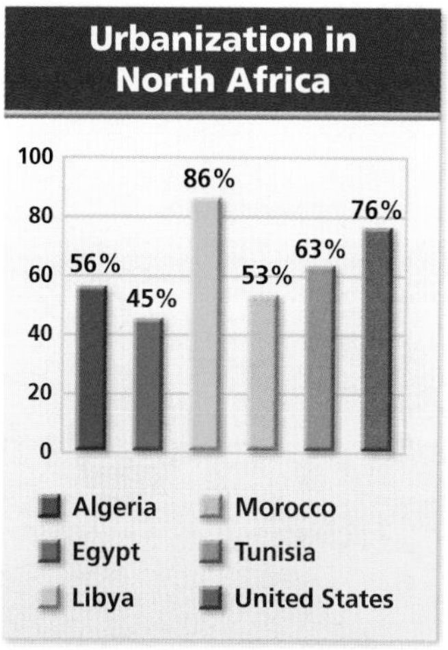

Source: *The World Almanac and Book of Facts 2001*
Percentages of population living in urban areas

Interpreting the Chart Which country is the most urbanized?

Section 4 Resources

Reproducible

- Guided Reading Strategy 19.4

Technology

- One-Stop Planner CD–ROM, Lesson 19.4
- Homework Practice Online
- HRW Go site

Reinforcement, Review, and Assessment

- Section 4 Review, p. 457
- Daily Quiz 19.4
- Main Idea Activity 19.4
- English Audio Summary 19.4
- Spanish Audio Summary 19.4

◄ **Chart Answer**

Libya

TEACH

Teaching Objectives 1–3

ALL LEVELS: (Suggested time: 30 min.) Copy the following graphic organizer onto the chalkboard, omitting the italicized answers. Use it to help students learn about the people, cities, economic activities, and today's challenges that characterize Libya, Tunisia, Algeria, and Morocco by having the class supply the missing information.

ENGLISH LANGUAGE LEARNERS

Features of the Maghreb

People	*most people live in the narrow coastal region between Atlantic and Mediterranean coasts in north and Sahara and Atlas Mountains farther inland*
Cities	*cities located in narrow coastal region; Libya most urbanized; largest cities are Algiers (Algeria), Tripoli and Benghazi (Libya), Casablanca and Rabat (Morocco), Tunis (Tunisia); Tangier (Morocco) is a free port*
Economic Activities	*oil, mining, tourism, natural gas, iron ore, lead, fertilizer, farming, trade*
Today's Challenges	*need for more economic and political freedom; dictatorship in Libya; deciding role of Islam in government*

Section 4 Review

Answers

Define For definitions, see: Casbah, p. 455; souks, p. 455; free port, p. 455; dictator, p. 457

Working with Sketch Maps The capitals are Libya—Tripoli; Algeria—Algiers; Morocco—Rabat; and Tunisia—Tunis.

Reading for the Main Idea

1. They include a Casbah and a souk. (NGS 4)
2. command; governments have moved to loosen their control (NGS 12)

Critical Thinking

3. along the coast; to be closer to water and away from the desert
4. Political rights of North Africans are limited, and people have little say in their governments, whereas in the United States people enjoy the right to vote and participate in the political process.

Organizing What You Know

5. industries—oil, mining, tourism, and fertilizer; resources—oil, natural gas, iron ore, and lead; farm products—wheat and other grains, olives, fruits, and nuts; trade partners—European countries

Chart Answer

Morocco

Libya, Tunisia, Algeria, and Morocco

Country	Population/ Growth Rate	Life Expectancy	Literacy Rate	Per Capita GDP
Algeria	31,736,053 1.7%	69, male 71, female	62%	$5,500
Libya	5,240,599 2.4%	74, male 78, female	76%	$8,900
Morocco	30,645,305 1.7%	67, male 72, female	44%	$3,500
Tunisia	9,705,102 1.7%	72, male 76, female	67%	$6,500
United States	281,421,906 0.9%	74, male 80, female	97%	$36,200

Sources: Central Intelligence Agency, *The World Factbook 2001;* U.S. Census Bureau

Interpreting the Chart **According to the data in the chart, which country is the least economically developed?**

Economy

Oil, mining, and tourism are important industries in these countries. Oil is the most important resource, particularly in Libya and Algeria. Money from oil pays for schools, health care, other social programs, and military equipment. The region's countries also have large deposits of natural gas, iron ore, and lead.

Morocco is the only North African country with little oil. However, the country is an important producer and exporter of fertilizer.

About 20 percent of the workers in Libya, Tunisia, and Algeria are farmers. In Morocco farmers make up about half of the labor force. North Africa's farmers grow and export wheat, other grains, olives, fruits, and nuts. However, the region's desert climate and poor soils limit farming, particularly in Libya. Libya imports most of its food.

The Maghreb countries have close economic relationships with European countries. This is partly because of old colonial ties between North Africa and Europe. In addition, European countries are located nearby, lying just across the Mediterranean Sea. Formal agreements between North African countries and the European Union (EU) also have helped trade. Today about 80 percent of Tunisia's trade is with EU countries. The largest trade partners of Algeria, Libya, and Morocco are also EU members. Many European tourists visit North Africa.

✓ **READING CHECK:** *Places and Regions* How does the oil industry affect Libya's and Algeria's schools, health care, and other social programs? provides money to fund them

Marrakech (muh-RAH-kish) is a popular tourist resort in central Morocco. It sits in the foothills of the Atlas Mountains.

CLOSE

Tell students that the Casbah of Algiers has been used as a metaphor for the Algerian nation. Then lead a class discussion about why certain sites or monuments might become symbols or metaphors for a nation. *(For example, the Statue of Liberty is sometimes viewed as a symbol of the American ideal of freedom.)*

REVIEW AND ASSESS

Have students complete the Section Review. Discuss the review in class. Then have students complete Daily Quiz 19.4.

RETEACH

Have students complete Main Idea Activity 19.4. Then organize students into four groups and assign each group one of the following topics: People, Places, Products, or Challenges in Libya, Tunisia, Algeria, and Morocco. Distribute poster board to the groups, and have each group create a graphic organizer to present information related to its topic.

ENGLISH LANGUAGE LEARNERS, COOPERATIVE LEARNING

EXTEND

Have interested students conduct research on the colonial history of Libya, Tunisia, Algeria, or Morocco. Have students use their findings to create an annotated time line for one of the four countries.

BLOCK SCHEDULING

Challenges

The countries of the Maghreb have made much progress in health and education. However, important challenges remain. Among these challenges is the need for more economic freedom. Each of these countries has had elements of a command economy, in which government owns and operates industry. However, in recent years the region's governments have moved to loosen that control. They have sold some government-owned businesses. They have also taken other steps to help private industry grow.

Political freedoms are limited for many North Africans. Many have little say in their governments. For example, since 1969 Libya has been ruled by a **dictator**, Mu'ammar al-Gadhafi. A dictator is someone who rules a country with complete power. Gadhafi has supported bombing, kidnapping, and other acts of violence against Israel and Israel's supporters. As a result, countries have limited their economic relationships with Libya.

As in Egypt, another challenge is conflict over the role of Islam in society. For example, in Algeria some groups want a government based on Islamic principles and laws. In 1992 the government canceled elections that many believed would be won by Islamic groups. Violence between Algeria's government and some Islamic groups has claimed thousands of lives since then.

✓ **READING CHECK:** *Places and Regions* What are some of the challenges the region's countries face today? economic and political freedom; conflict over Islam's role

These Islamic students at a religious school in Libya are reciting verses from the Qur'an. Religious education is important in this mostly Islamic region.

Section 4 Review

go.hrw.com Homework Practice Online
Keyword: SK3 HP19

Define and explain: Casbah, souks, free port, dictator

Working with Sketch Maps On the map you created in Section 3, label Libya, Tunisia, Algeria, Morocco, Tripoli, Benghazi, Algiers, Casablanca, Rabat, Tunis, and the Strait of Gibraltar. In a box in the margin, identify the national capitals of the region's countries.

Reading for the Main Idea

1. *Places and Regions* What are the old, central districts of many North African cities like?
2. *Places and Regions* What type of economy have these nations had, and how is this changing?

Critical Thinking

3. **Finding the Main Idea** Where are the region's farms and most of its people found today? Why?
4. **Analyzing Information** How are the political rights of North Africans different from those of people in the United States?

Organizing What You Know

5. **Summarizing** Copy the following graphic organizer. Use it to list industries, resources, farm products, and trade partners of the Maghreb countries.

Industries	
Resources	
Farm products	
Trade partners	

CHAPTER 19 Review ANSWERS

Building Vocabulary

For definitions, see: ergs, p. 442; regs, p. 442; depressions, p. 442; silt, p. 443; pharaohs, p. 445; hieroglyphs, p. 445; Bedouins, p. 448; fellahin, p. 452; Casbah, p. 455; souks, p. 455; free port, p. 455; dictator, p. 457

Reviewing the Main Ideas

1. The Sahara covers most of the region and settlement is limited to coastal areas. (NGS 4)
2. desert, Mediterranean, and steppe (NGS 4)
3. farming-related issues, rapidly growing population, health and literacy issues, its role as an Arab country, and the role of Islam in the country (NGS 4)
4. oil, mining, tourism, fertilizer, and textiles (NGS 4)
5. Libya (NGS 4)

Understanding Environment and Society

Answers will vary, but the information included should be consistent with text material. Use Rubric 29, Presentations, to evaluate student work.

ASSESS

Have students complete the Chapter 19 Test.

RETEACH

Create a large outline map of North Africa. Organize students into five groups and assign each group a topic: major cities, economic activities, unique characteristics, type of government, and challenges. Have each group find information on its topic for each of the five countries of North Africa and write that information on a small strip of construction paper. When finished, have students glue their strips onto the map.
ENGLISH LANGUAGE LEARNERS, COOPERATIVE LEARNING

PORTFOLIO

PORTFOLIO EXTENSIONS

1. Have five groups of students conduct research on one the five basic types of dunes. Provide students with sand and a large rectangular pan and have each group make a model of the dune. Groups should then write a summary of how their dune types are formed. Ask each group to present its model and summary to the other groups. Photograph the dune models and summaries and place them in students' portfolios.
2. Have students conduct research on foods common to North Africa. Then ask students to prepare a table of contents for a recipe book of North African dishes.

CHAPTER 19

Review ANSWERS

Thinking Critically

1. The harsh desert climate has caused settlements to be located near water sources.
2. to be near water sources
3. as a shipping route from the Mediterranean Sea to the Indian Ocean
4. Debate over the role of Islam in government has sometimes caused internal conflict; conflict with Israel has sometimes strained international relations.
5. political tensions, lack of adequate water resources, and problems related to population growth

Map Activity

A. Western Sahara
B. Ahaggar Mountains
C. Casablanca
D. Sinai Peninsula
E. Tripoli
F. Atlas Mountains
G. Strait of Gibraltar
H. Cairo
I. Algiers
J. Nile Delta

CHAPTER 19

Reviewing What You Know

Building Vocabulary

On a separate sheet of paper, write sentences to define each of the following words.

1. ergs
2. regs
3. depressions
4. silt
5. pharaohs
6. hieroglyphs
7. Bedouins
8. fellahin
9. Casbah
10. souks
11. free port
12. dictator

Reviewing the Main Ideas

1. *Places and Regions* How does geography affect settlement patterns in North Africa?
2. *Places and Regions* What are North Africa's main climates?
3. *Places and Regions* What are some important challenges facing Egypt today?
4. *Places and Regions* What are the most important industries in North Africa?
5. *Places and Regions* Which is the most urbanized country in North Africa?

Understanding Environment and Society

Farming and Fertilizer

Egyptian farmers must use large amounts of fertilizer. Create a presentation about the use of fertilizer. Consider the following:

- The kinds of chemicals, minerals, and other materials most commonly used as fertilizers.
- How fertilizers work.
- The positive and negative effects of using fertilizer.

Thinking Critically

1. **Finding the Main Idea** How do you think the Sahara has influenced settlement in the region?
2. **Finding the Main Idea** Why are most of North Africa's cities and population located in coastal areas?
3. **Analyzing Information** Why is the Suez Canal important to world trade?
4. **Finding the Main Idea** In what way has Islam influenced politics in North Africa?
5. **Making Generalizations and Predictions** Based on what you have read, what kinds of challenges do you think North Africans will face in coming decades? Explain your answer.

Food Festival

Explain to students that preparing couscous from scratch is very time consuming. Therefore, many people use packaged couscous. It can be prepared for serving in a variety of ways. For instance, steamed couscous can be served as a dessert when topped with raisins, sugar, and cinnamon. Purchase and prepare a package of couscous according to the directions. Prepare two separate batches of couscous, one as described above and one as described in Section 2.

Chapter 19 Review and Assessment Resources

Reproducible
- Readings in World Geography, History, and Culture 53 and 54
- Critical Thinking Activity 19
- Vocabulary Activity 19

Technology
- Chapter 19 Test Generator (on the One-Stop Planner)
- Audio CD Program, Chapter 19
- HRW Go site

Reinforcement, Review, and Assessment
- Chapter 19 Review, pp. 458–59
- Chapter 19 Tutorial for Students, Parents, Mentors, and Peers
- Chapter 19 Test
- Chapter 19 Test for English Language Learners and Special-Needs Students

Building Social Studies Skills

Map Activity

On a separate sheet of paper, match the letters on the map with their correct labels.

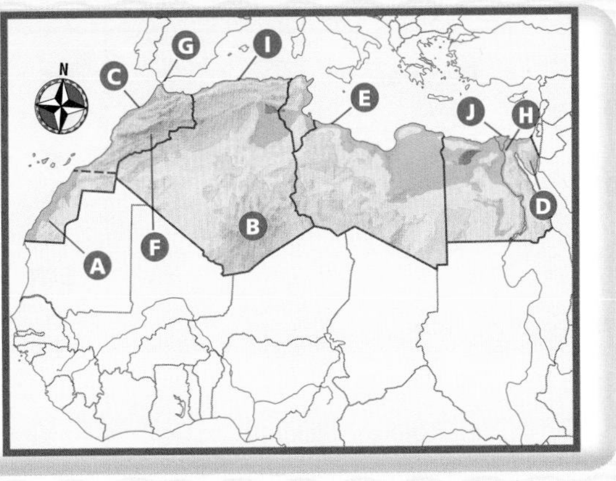

Sinai Peninsula	**Western Sahara**
Ahaggar Mountains	**Tripoli**
Atlas Mountains	**Algiers**
Nile Delta	**Casablanca**
Cairo	**Strait of Gibraltar**

Mental Mapping Skills Activity

On a separate sheet of paper, draw a freehand map of North Africa. Make a key for your map and label the following:

Algeria	**Red Sea**
Egypt	**Sahara**
Libya	**Tunisia**
Morocco	

Writing Activity

Imagine that you are a Bedouin teenager in the Sahara. Write a one-paragraph journal entry about a typical day in your life. How do you cope with the desert heat? What is it like living without a permanent home? What religion do you practice? What do you eat? Use the library and other resources to help you. Be sure to use standard grammar, spelling, sentence structure, and punctuation.

Alternative Assessment

Portfolio Activity

Learning About Your Local Geography

Research Project Use an almanac to identify your state's largest cities. Sketch a map of your state. Mark the three largest cities and major physical features such as rivers, mountains, and coastlines.

internet connect

Internet Activity: **go.hrw.com**
KEYWORD: **SK3 GT19**

Choose a topic to explore about North Africa:
- Journey through the Sahara.
- Examine the rich history of North Africa.
- Practice using Arabic calligraphy.

Mental Mapping Skills Activity

Maps will vary, but listed places should be in their approximate locations.

Writing Activity

Answers will vary, but the information included should be consistent with text material. Students should respond to the various questions as well as demonstrate an understanding of the Bedouin culture and lifestyle. Use Rubric 15, Journals, to evaluate student work.

Portfolio Activity

Maps should indicate the three largest cities and major physical features of your state. Use Rubric 20, Map Creation, to evaluate student work.

internet connect

GO TO: **go.hrw.com**
KEYWORD: **SK3 Teacher**
FOR: **a guide to using the Internet in your classroom**

CHAPTER 20

West Africa

Chapter Resource Manager

Objectives	Pacing Guide	Reproducible Resources
SECTION 1 **Physical Geography** (pp. 461–63) **1.** Identify the landforms and climates found in West Africa. **2.** Explain why the Niger River is important to the region. **3.** Identify the resources of West Africa.	**Regular** .5 day **Block Scheduling** .5 day *Block Scheduling Handbook, Chapter 20*	**RS** Guided Reading Strategy 20.1
SECTION 2 **History and Culture** (pp. 464–68) **1.** Identify the great African kingdoms that once ruled the region. **2.** Explain how contact with Europeans affected West Africa. **3.** Identify the challenges the governments of the region face. **4.** Identify some features of West African culture.	**Regular** 1 day **Block Scheduling** .5 day *Block Scheduling Handbook, Chapter 20*	**RS** Guided Reading Strategy 20.2 **RS** Graphic Organizer 20 **E** Cultures of the World Activity 6 **E** Creative Strategies for Teaching World Geography, Lesson 16 **SM** Geography for Life Activity 20 **IC** Interdisciplinary Activities for the Middle Grades 26, 27, 28
SECTION 3 **The Sahel Countries** (pp. 469–71) **1.** Describe what Mauritania, Mali, and Niger are like today. **2.** Describe the challenges Chad and Burkina Faso face.	**Regular** .5 day **Block Scheduling** .5 day *Block Scheduling Handbook, Chapter 20*	**RS** Guided Reading Strategy 20.3 **SM** Map Activity 20
SECTION 4 **The Coastal Countries** (pp. 472–75) **1.** Describe what life in Nigeria is like today. **2.** Identify some economic challenges the other countries of the region face.	**Regular** 1 day **Block Scheduling** .5 day *Block Scheduling Handbook, Chapter 20*	**RS** Guided Reading Strategy 20.4 **SM** Map Activity 20

Chapter Resource Key

RS Reading Support
IC Interdisciplinary Connections
E Enrichment
SM Skills Mastery
A Assessment
REV Review
ELL Reinforcement and English Language Learners
Transparencies
CD–ROM
Music
Video
Internet
Holt Presentation Maker Using Microsoft® Powerpoint®

One-Stop Planner CD–ROM

See the ***One-Stop Planner*** for a complete list of additional resources for students and teachers.

One-Stop Planner CD–ROM

It's easy to plan lessons, select resources, and print out materials for your students when you use the ***One-Stop Planner CD–ROM with Test Generator.***

Technology Resources	Review, Reinforcement, and Assessment Resources
One-Stop Planner CD–ROM, Lesson 20.1 Our Environment CD–ROM/ Seek and Tell/People Affecting Nature *ARGWorld* CD–ROM: Land-Use Decisions in West Africa Geography and Cultures Visual Resources with Teaching Activities 43–48 go.hrw.com: Homework Practice Online; HRW Go site	ELL Main Idea Activity 20.1 REV Section 1 Review, p. 463 A Daily Quiz 20.1 ELL English Audio Summary 20.1 ELL Spanish Audio Summary 20.1
One-Stop Planner CD–ROM, Lesson 20.2 go.hrw.com: Homework Practice Online; HRW Go site	ELL Main Idea Activity 20.2 REV Section 2 Review, p. 468 A Daily Quiz 20.2 ELL English Audio Summary 20.2 ELL Spanish Audio Summary 20.2
One-Stop Planner CD–ROM, Lesson 20.3 go.hrw.com: Homework Practice Online; HRW Go site	ELL Main Idea Activity 20.3 REV Section 3 Review, p. 471 A Daily Quiz 20.3 ELL English Audio Summary 20.3 ELL Spanish Audio Summary 20.3
One-Stop Planner CD–ROM, Lesson 20.4 go.hrw.com: Homework Practice Online; HRW Go site	ELL Main Idea Activity 20.4 REV Section 4 Review, p. 475 A Daily Quiz 20.4 ELL English Audio Summary 20.4 ELL Spanish Audio Summary 20.4

internet connect

HRW ONLINE RESOURCES

GO TO: go.hrw.com
Then type in a keyword.

TEACHER HOME PAGE
KEYWORD: SK3 TEACHER

CHAPTER INTERNET ACTIVITIES
KEYWORD: SK3 GT20

Choose an activity to:
- find out about giant baobab trees.
- meet the people of West Africa.
- learn about the history of the slave trade.

CHAPTER ENRICHMENT LINKS
KEYWORD: SK3 CH20

CHAPTER MAPS
KEYWORD: SK3 MAPS20

ONLINE ASSESSMENT
Homework Practice
KEYWORD: SK3 HP20
Standardized Test Prep Online
KEYWORD: SK3 STP20
Rubrics
KEYWORD: SS Rubrics

COUNTRY INFORMATION
KEYWORD: SK3 Almanac

CONTENT UPDATES
KEYWORD: SS Content Updates

HOLT PRESENTATION MAKER
KEYWORD: SK3 PPT20

ONLINE READING SUPPORT
KEYWORD: SS Strategies

CURRENT EVENTS
KEYWORD: S3 Current Events

Meeting Individual Needs

Ability Levels

Level 1 Basic-level activities designed for all students encountering new material

Level 2 Intermediate-level activities designed for average students

Level 3 Challenging activities designed for honors and gifted-and-talented students

English Language Learners Activities that address the needs of students with Limited English Proficiency

Chapter Review and Assessment

- **E** Readings in World Geography, History, and Culture 55, 56, and 57
- **SM** Critical Thinking Activity 20
- **REV** Chapter 20 Review, pp. 476–77
- **REV** Chapter 20 Tutorial for Students, Parents, Mentors, and Peers
- **ELL** Vocabulary Activity 20
- **A** Chapter 20 Test
- Chapter 20 Test Generator (on the One-Stop Planner)
- Audio CD Program, Chapter 20
- **A** Chapter 20 Test for English Language Learners and Special-Needs Students

CHAPTER 20

Launch into Learning

Hold up a jar of peanut butter. Ask students to name as many products as they can that are made from peanuts *(candy, cookies, oil for cooking).* Point out that peanuts originated in Africa, that they were brought from there to our continent, and that in the 1900s they replaced cotton as the main crop grown on many southern farms. Tell students that they will learn about West Africa, where peanuts are an important crop, in this chapter.

Section 1

Objectives

1. Describe the landforms and climates found in West Africa.
2. Explain the importance of the Niger River to the region.
3. Identify the resources found in West Africa.

Links to Our Lives

You may wish to tell students that there are many reasons for us to know more about West Africa:

- West Africa has a fascinating history. International relations are eased when we know the history of other regions.
- Many Americans are descended from Africans from this region.
- Americans established the West African country of Liberia and some freed African American slaves settled there.
- Products such as peanuts, coffee, and chocolate are exported from the region to the United States.

CHAPTER 20

West Africa

We move south now to West Africa. The land becomes much wetter as we move southward. Before we continue, we meet Ousseina from Niger.

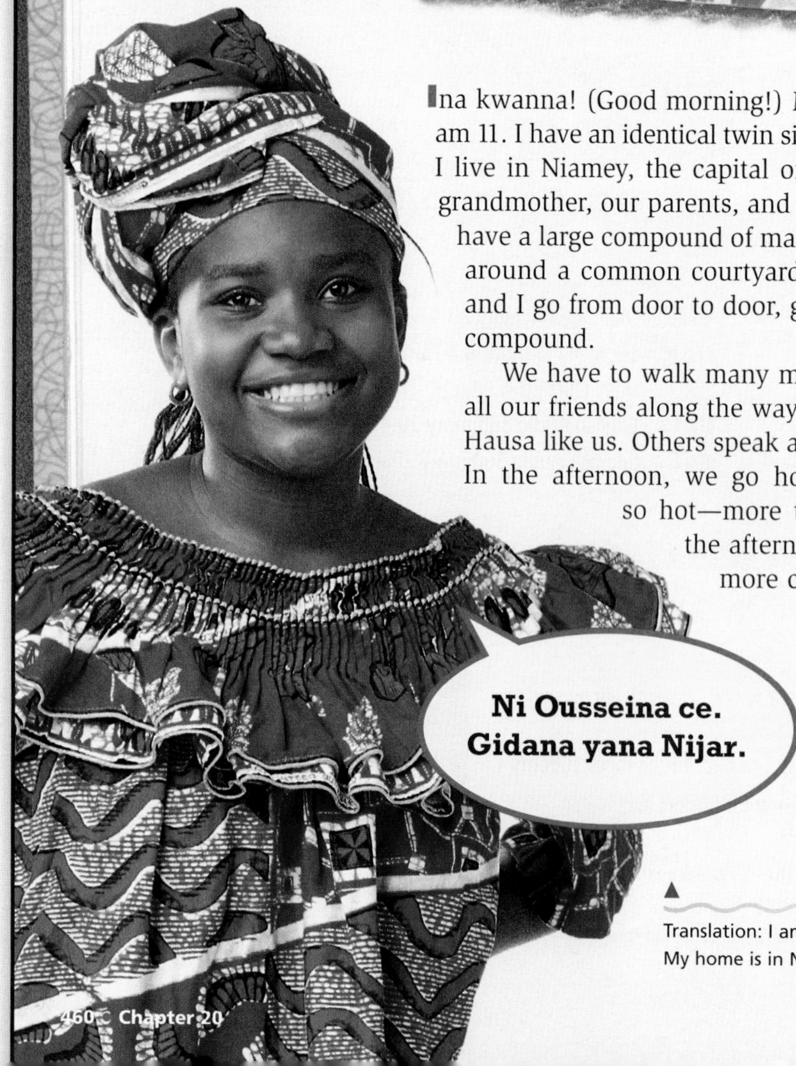

Ni Ousseina ce.
Gidana yana Nijar.

Ina kwanna! (Good morning!) My name is Ousseina, and I am 11. I have an identical twin sister whose name is Hassana. I live in Niamey, the capital of Niger, with my sister, our grandmother, our parents, and many aunts and uncles. We have a large compound of many one- or two-room houses around a common courtyard. After breakfast, my sister and I go from door to door, greeting all the elders in our compound.

We have to walk many miles to school, but we meet all our friends along the way. Some of our friends speak Hausa like us. Others speak a different language of Niger. In the afternoon, we go home to sleep because it is so hot—more than 110°F (43°C)! Late in the afternoon we return to school for more classes.

After school, we help grandma pound spices for dinner, sweep the room, and wash the dishes. At bedtime, grandma tells us stories and sings.

▲ Translation: I am Ousseina. My home is in Niger.

FOCUS

LET'S GET STARTED

Copy the following instructions onto the chalkboard: *List two ways that deserts might spread into nondesert areas. Discuss responses. (Possible answers: drought, overgrazing grasses, cutting trees, destroying plant life, changing climate).* Explain that plants and trees help hold soil in place. When vegetation is removed or dies and no rains fall, the topsoil blows away, causing deserts to spread. Tell students that in Section 1 they will learn more about the physical geography of West Africa, including how the desert is growing in the region.

Using the Physical-Political Map

Have students examine the map on this page. Call on volunteers to name the Sahel countries *(Mauritania, Mali, Niger, Burkina Faso, and Chad)* and the coastal countries *(Senegal, Gambia, Guinea-Bissau, Guinea, Sierra Leone, Liberia, Côte d'Ivoire, Ghana, Togo, Benin, and Nigeria).* Tell students that the Sahara is expanding into the Sahel countries. Have students discuss problems the encroaching desert may pose for the people of those countries and, in turn, for people in the coastal countries.

Section 1 Physical Geography

Read to Discover

1. What landforms and climates are found in West Africa?
2. Why is the Niger River important to the region?
3. What resources does West Africa have?

Define

zonal
Sahel
harmattan
tsetse fly
bauxite

Locate

Sahara
Niger River
Gulf of Guinea

Colorful Nigerian tapestry

WHY IT MATTERS

West African countries have often faced problems in managing their natural resources. Use CNNfyi.com or other **current events** sources to find problems faced by countries in the region regarding resources and the environment. Record your findings in your journal.

West Africa: Physical-Political

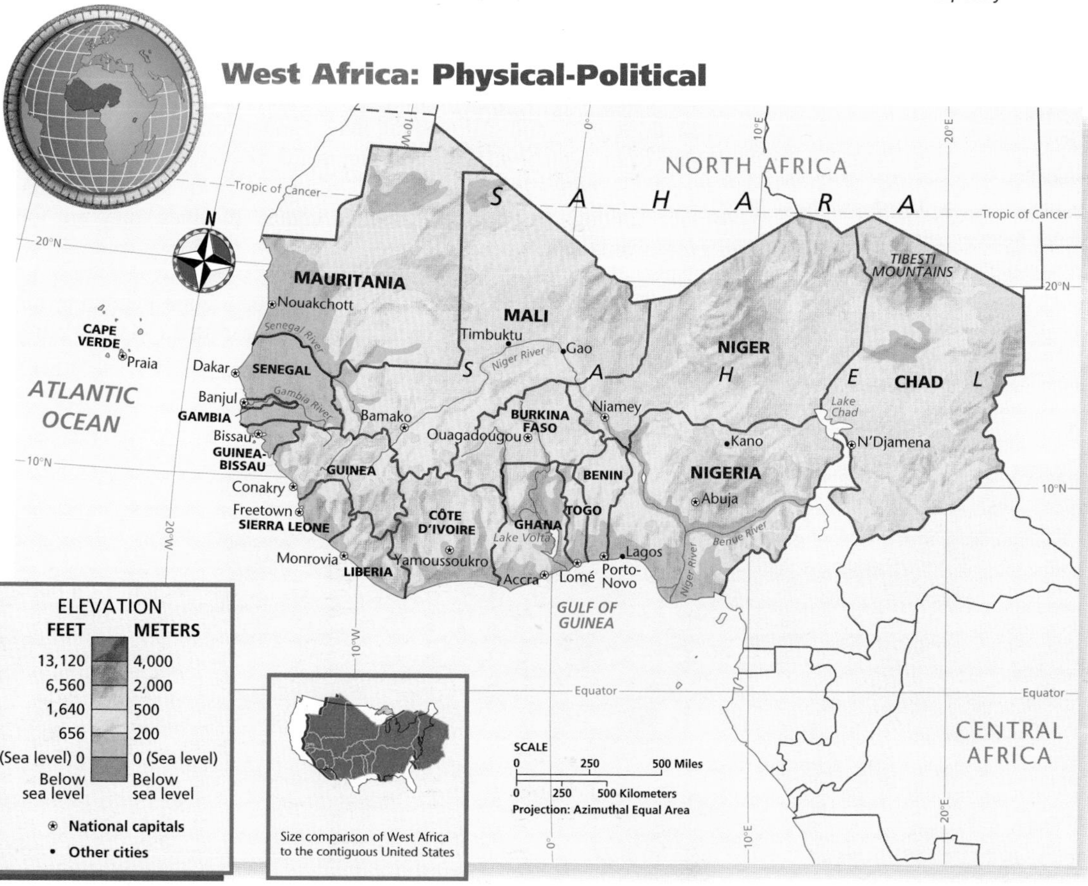

Section 1 RESOURCES

Reproducible
- Block Scheduling Handbook, Chapter 20
- Guided Reading Strategy 20.1

Technology
- One-Stop Planner CD–ROM, Lesson 20.1
- Homework Practice Online
- Geography and Cultures Visual Resources with Teaching Activity 43–48
- Our Environment CD–ROM/Seek and Tell/People Affecting Nature
- *ARGWorld* CD–ROM: Land-Use Decisions in West Africa
- HRW Go site

Reinforcement, Review, and Assessment
- Section 1 Review, p. 463
- Daily Quiz 20.1
- Main Idea Activity 20.1
- English Audio Summary 20.1
- Spanish Audio Summary 20.1

TEACH

Teaching Objective 1

ALL LEVELS: (Suggested time: 20 min.) Copy the following graphic organizer onto the chalkboard, omitting the italicized answers. Use it to help students describe West Africa's zones. Have students complete the organizer. **ENGLISH LANGUAGE LEARNERS**

Zones of West Africa

CLIMATE ZONE	DESCRIPTION
Sahara	*largest desert*
Sahel	*dry grasslands with steppe climate*
savanna	*good soil, thick grass, scattered trees*
coast and forest	*wet humid tropical climate, rain forests*

Teaching Objectives 2–3

ALL LEVELS: (Suggested time: 10 min.) Ask students to look at the chapter map and locate the cities that lie within Mali and Niger. Then ask students what the cities have in common. *(They are inland and on the Niger River.)* Then have students list the resources found in West Africa. Ask them to name the country whose resources became more valuable in the 1900s *(Nigeria).* **ENGLISH LANGUAGE LEARNERS**

ENVIRONMENT AND SOCIETY

Found only in Africa near the equator, the tsetse fly transmits a parasite to humans by biting them. The parasite causes the deadly illness trypanosomiasis (tri-pa-nuh-suh-MY-uh-suhs), also known as sleeping sickness. More than 66 million people suffer from the disease. Early symptoms include high fever, weakness, and headache. As the disease progresses, the parasite invades the central nervous system. Eventually, the patient lapses into a coma and dies.

Medical treatment for trypanosomiasis requires hospitalization. Protection can help prevent tsetse fly bites and includes wearing khaki or olive-colored long pants and long-sleeved shirts and using a bednet at night.

Activity: Have students conduct research on other insects that transmit disease in tropical areas. Then have them put on a health news show.

internet connect

GO TO: go.hrw.com
KEYWORD: SK3 CH20
FOR: Web sites about the tsetse fly

Visual Record Answer

It is carried in containers, rather than through an expensive irrigation system.

internet connect

GO TO: go.hrw.com
KEYWORD: SK3 CH20
FOR: Web sites about West Africa

BUILD ON WHAT You Know

Do you remember what you learned about desertification? See Chapter 4 to review.

Landforms and Climate

West Africa is largely a region of plains. There are low mountains in the southwest and high mountains in the northeast. Four major climate types stretch from east to west in bands or zones. Therefore, geographers say the region's climates are **zonal**.

The Sahara The northernmost parts of the region lie within the Sahara, the first climate zone. The Sahara is the world's largest desert. It stretches across northern Africa from the Atlantic Ocean to the Red Sea. Large areas of this dry climate zone have few or no people.

The Sahel South of the Sahara is a region of dry grasslands called the **Sahel** (sah-HEL). This second climate zone has a steppe climate. Rainfall varies greatly from year to year. In some years it never rains. During winter a dry, dusty wind called the **harmattan** (hahr-muh-TAN) blows south from the Sahara.

During the late 1960s a drought began in the Sahel. Crops failed for several years, and there was not enough grass for the herds. Animals overgrazed the land, and people cut the few large trees for firewood. Wind blew away fertile soil, and the Sahara expanded southward. Without plants for food, many animals and people died. Recent years have been rainier, and life has improved.

The Savanna Farther south is the savanna zone. It contains good soil, thick grass, and scattered tall trees. Farmers can do well when the rains come regularly. However, the region is home to a dangerous insect. The **tsetse** (TSET-see) **fly** carries sleeping sickness, a deadly disease. Although insecticides can help control the flies, they are too expensive for most people to buy.

The Coast and Forest The fourth climate zone lies along the Atlantic and Gulf of Guinea coasts. Many of West Africa's largest cities lie in this coastal zone. You will find a wet, humid tropical climate there. Plentiful rain supports tropical rain forests. However, many trees

Farmers use river water to irrigate onion fields in central Mali. Farming can be difficult in the dry Sahel.

Interpreting the Visual Record **How do you think economic factors have affected the use of technology in bringing water to the fields?**

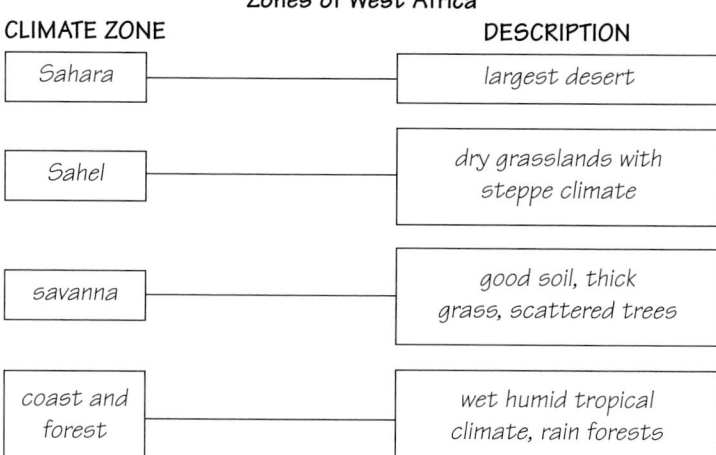

CLOSE

Ask students to describe the four environmental zones of West Africa and to identify their locations *(Sahara—largest desert, in the northern-most part; Sahel—dry grasslands, in the north; savanna—region of fertile grasslands and trees that receives seasonal rains, in the middle; coastal—tropical rain forests across the south and west coasts).* Then have students name the river that flows through the region *(Niger River).*

REVIEW AND ASSESS

Have students complete the Section Review. Then pair students and have them create flash cards. One side of each card should describe a climate zone, and the other name the zone. Have pairs exchange cards with other pairs and quiz their partners. Then have students complete Daily Quiz 20.1.
COOPERATIVE LEARNING

RETEACH

Have students complete Main Idea Activity 20.1. Then have students create a salt dough map of the region with labels for each climate type and the Niger River. Display and discuss the map.
ENGLISH LANGUAGE LEARNERS

EXTEND

Have interested students conduct research on the Niger River and write and illustrate a journal describing a trip from its source to its mouth.
BLOCK SCHEDULING

have been cut to make room for growing populations. As a result, environmental damage is a serious problem.

✓ **READING CHECK:** *Places and Regions* What are the region's landforms and climate zones? plains, mountains, desert, grasslands, forests; desert, steppe, savanna, humid tropical

The Niger River

The most important river in West Africa is the Niger (NY-juhr). The Niger River starts in low mountains just 150 miles (241 km) from the Atlantic Ocean. It flows eastward and southward for 2,600 miles (4,183 km) and empties into the Gulf of Guinea.

The Niger brings life-giving water to West Africa. In the Sahel it divides into a network of channels, swamps, and lakes. This network is known as the inland delta. The Niger's true delta on the Gulf of Guinea is very wide. Half of Nigeria's coastline consists of the delta.

✓ **READING CHECK:** *Environment and Society* Why is the Niger River important to West Africa? It sustains life in West Africa.

▲ The hippopotamus is just one of the many animal species living in the Niger region. Hippopotamuses are good swimmers and can stay underwater for as long as six minutes.

Resources

West Africa's mineral riches include diamonds, gold, iron ore, manganese, and **bauxite**. Bauxite is the main source of aluminum. Nigeria is a major exporter of oil. In fact, oil and related products make up about 95 percent of that country's exports.

✓ **READING CHECK:** *Places and Regions* What are some of the region's resources? minerals—diamonds, gold, iron ore, manganese, and bauxite; oil and oil products

Section 1 Review

Define and explain: zonal, Sahel, harmattan, tsetse fly, bauxite

Working with Sketch Maps On a map of West Africa, label the following: Sahara, Sahel, Niger River, and Gulf of Guinea.

Reading for the Main Idea

1. *Physical Systems* What effect has drought had on West Africa's vegetation?
2. *Places and Regions* What natural resources are found in West Africa?
3. *Places and Regions* What is West Africa's most important river? Describe its two delta regions.

go.hrw.com Homework Practice Online
Keyword: SK3 HP20

Critical Thinking

4. **Making Generalizations and Predictions** Where in the region would you expect to find the densest populations? Why?

Organizing What You Know

5. **Summarizing** Copy the following graphic organizer. Use it to list and describe the region's climate zones.

Zones	Characteristics

Section 1 Review Answers

Define For definitions, see: zonal, p. 462; Sahel, p. 462; harmattan, p. 462; tsetse fly, p. 462; bauxite, p. 463

Working with Sketch Maps The Sahara, the Sahel, the Niger River, and the Gulf of Guinea should be located accurately.

Reading for the Main Idea

1. caused crops to fail and vegetation to die (NGS 7)
2. diamonds, gold, iron ore, manganese, bauxite, and oil (NGS 4)
3. Niger River; inland delta—creeks, swamps, and lakes; true delta—half of Nigeria's coastline on the Gulf of Guinea (NGS 4)

Critical Thinking

4. along the coastal zone; because there is plenty of rain for growing crops

Organizing What You Know

5. Sahara—largest desert, dry; Sahel—dry grasslands, steppe climate; savanna— good soil, thick grass, scattered trees; coast and forest— wet humid tropical climate, plentiful rain

Section 2

Objectives

1. Describe the great kingdoms that once ruled the region.
2. Analyze how European contact affected West Africa.
3. Identify the challenges faced by the region's governments.
4. Describe some features of West African culture.

Focus

LET'S GET STARTED

Copy the following question onto the chalkboard: *What are three objects from our culture that might be found in later years by archaeologists?* Discuss responses. Then have volunteers state what those artifacts might tell future researchers about our culture. Tell students that in Section 2 they will learn that objects found in West Africa tell us much about the region's cultural history.

Building Vocabulary

Write the key terms on the chalkboard. Call on volunteers to read the definitions of **archaeology** and **oral history**. Point out that the suffix *-ism* usually refers to a system of beliefs and that **animism** refers to a system of beliefs based on the idea that natural objects, such as water, plants, or animals, have spirits.

Section 2 Resources

Reproducible

- Guided Reading Strategy 20.2
- Graphic Organizer 20
- Cultures of the World Activity 6
- Geography for Life Activity 20
- Creative Strategies for Teaching World Geography, Lesson 16
- Interdisciplinary Activities for the Middle Grades 26, 27, 28

Technology

- One-Stop Planner CD–ROM, Lesson 20.2
- Homework Practice Online
- HRW Go site

Reinforcement, Review, and Assessment

- Section 2 Review, p. 468
- Daily Quiz 20.2
- Main Idea Activity 20.2
- English Audio Summary 20.2
- Spanish Audio Summary 20.2

Visual Record Answer ►

human figures, animals, symbols, and abstract shapes

Section 2 History and Culture

Read to Discover

1. What great African kingdoms once ruled the region?
2. How did contact with Europeans affect West Africa?
3. What challenges do the region's governments face?
4. What are some features of West African culture?

Define

archaeology
oral history
animism

Locate

Timbuktu

Bronze figure of a hunter, lower Niger area

Why It Matters

Many West African countries struggled to win their independence from colonial powers and are still struggling to maintain stable governments. Use CNNfyi.com or other **current events** sources to find examples of political events in these countries. Record your findings in your journal.

West Africa's History

Much of what we know about West Africa's early history is based on **archaeology**. Archaeology is the study of the remains and ruins of past cultures. **Oral history**—spoken information passed down from person to person through generations—offers other clues.

Great Kingdoms Ancient artifacts suggest that the earliest trading towns developed in the Niger's inland delta. Traders brought dates and salt from the desert. People from the Sahel sold animals and hides. Other trade goods were grains, fish, kola and other tropical nuts, and metals, such as gold. (Much later, kola nuts provided the flavor for cola drinks.) This trade helped African kingdoms grow. One of the earliest West African kingdoms, Ghana (GAH-nuh), had become rich and powerful by about A.D. 800.

These early West African cliff paintings illustrate features from a ceremonial ritual for young people.

Interpreting the Visual Record **What kinds of images do you see on this cliff wall?**

▼

TEACH

Teaching Objectives 1–2

ALL LEVELS: (Suggested time: 30 min.) Copy the following graphic organizer onto the chalkboard, omitting the italicized answers. Use it to help students understand the region's history. For each footprint have students suggest a phrase to describe a step in West Africa's history. **ENGLISH LANGUAGE LEARNERS**

Major Steps in West Africa's History

Ghana comes to power. Merchants introduce Islam.
Kingdom of Mali replaces Ghana.
Kingdom of Songhay comes to power.
Slave trade begins.
European colonization begins.
Independence movements spread.

Teaching Objective 3

LEVEL 1: (Suggested time: 10 min.) Have students list as many challenges facing West African governments as possible. *(Students might mention civil wars and military rulers, high birthrates, crowded cities, unemployment, and lack of education.)* Then ask students to suggest possible solutions for each problem listed. **ENGLISH LANGUAGE LEARNERS**

Some 200 years later, North African merchants began crossing the Sahara to trade in Ghana. These merchants introduced Islam to West Africa. In time, Islam became the main religion practiced in the Sahel.

Later Ghana fell to Muslim warriors from Morocco. The Muslim empire of Mali (MAH-lee) replaced the kingdom of Ghana. Mali stretched from the Niger's inland delta to the Atlantic coast. Mansa Mūsā was king of Mali during the early 1300s. Famous for his wealth and wise rule, Mansa Mūsā supported artists, poets, and scholars.

The kingdom of Songhay (SAWNG-hy) came to power as Mali declined. With a university, mosques, and more than 100 schools, the Songhay city of Timbuktu was a cultural center. By about 1600, however, Moroccan invasions had weakened the kingdom.

Forested areas south of the Sahel were also home to great civilizations. In what is now Nigeria, wealthy kings were buried with brass sculptures and other treasures.

✓ **READING CHECK:** *Human Systems* What are some of the great African kingdoms that once ruled the region? **Ghana, Mali, Songhay**

The Slave Trade During the 1440s Portuguese explorers began sailing along the west coast of Africa. The Europeans called it the Gold Coast for the gold they bought there. Once they could buy gold where it was mined, the Europeans stopped buying it from Arab traders. As a result, the trans-Sahara trade and the great trade cities faded.

For a while, both Europeans and Africans profited from trade with each other. However, by the 1600s the demand for labor in Europe's American colonies changed everything. European traders met this demand by selling enslaved Africans to colonists. The slave trade was very profitable for these traders.

The slave trade had devastating effects on West African communities. Families were broken up when members were kidnapped and enslaved. Many Africans died on the voyage to the Americas. Most who survived were sent to the West Indies or Brazil. The slave trade finally ended in the 1800s. By then millions of Africans had been forced from their homes.

Colonial Era and Independence In the late 1800s, many European countries competed for colonies in West Africa. France claimed most of the region's northwest. Britain, Germany, and Portugal seized the rest.

In all of West Africa only tiny Liberia remained independent. Americans had founded it in the 1820s as a home for freed slaves. Sierra Leone, a British colony, also became a home for freed slaves.

Our Amazing Planet

We might think that salt is common and cheap, but it was precious to the traders of the Sahara. At one time it was worth its weight in gold.

▲ There are many reminders of the slave trade and its effects on West Africans. Performers here reenact the treatment of enslaved Africans in an old slave house in Dakar, Senegal.

Interpreting the Visual Record **How does this photograph show slavery's effect on creative expression?**

HUMAN SYSTEMS

Mansa Mūsā, emperor of Mali, made a pilgrimage to Mecca in 1324. It was this journey that displayed Mali's incredible wealth to other countries.

Some 60,000 retainers accompanied the emperor. Included were 12,000 slaves, all dressed in brocade and silk. Mansa Mūsā rode on horseback behind 500 slaves, each bearing a staff decorated with gold. Each of the 80 camels in the baggage train carried 300 pounds of gold.

Mansa Mūsā spent so much gold in Cairo that the market was flooded with it, causing a decline in gold prices that lasted at least 12 years.

Critical Thinking: Why might spending so much gold reduce its value?

Answer: Students might say that the more there is of a product, the lower its value.

◄ **Visual Record Answer**

Societal issues regarding slavery are reflected in a modern-day performance.

internet connect go.hrw.com

GO TO: go.hrw.com
KEYWORD: SK3 CH20
FOR: Web sites about Mansa Mūsā

LEVELS 2 AND 3: (Suggested time: 45 min.) Have students complete the Level 1 exercise for this objective. Then ask students to use their list and other resources to prepare a newspaper commentary on West Africa's challenges.

Teaching Objective 4

LEVEL 1: (Suggested time: 30 min.) Organize students into four groups. Assign each group one of the following culturally influenced characteristics of West African life: languages, religion, clothing, or housing. Have each group draw examples of those characteristics on slips of colored paper and create a collage by pasting their examples onto a piece of posterboard. **ENGLISH LANGUAGE LEARNERS, COOPERATIVE LEARNING**

LEVELS 2 AND 3: (Suggested time: 45 min.) Organize students into groups and have each group develop an illustrated cover for a magazine about West African culture. Then have all group members write one-paragraph summaries of articles that could be included in the magazine. Article subjects might include languages, religions, clothing and housing. **COOPERATIVE LEARNING**

➤ASSIGNMENT: Have students imagine they are exchange students living in West Africa. Ask students to write a letter to a friend comparing life in West Africa to life back home.

Linking Past to Present

Liberia As opposition to slavery took hold in the early 1800s, agents from the U.S. government and two officers from the American Colonization Society received permission from African chiefs to allow freed slaves from the United States to settle at the mouth of the Mesurado River. From 1822 to 1865, American freed slaves settled there.

Critical Thinking: Why might freed slaves settle in Liberia?

Answer: They had originally come from Africa and probably wished to return there.

Global Perspectives

The British Broadcasting Corporation (BBC) radio program *Focus on Africa* features African journalists reporting the region's latest news. This program broadcasts information that listeners are otherwise unable to get about their own countries. During the civil war in Liberia, it was said that the only safe time to walk in the street was during the BBC broadcast because all the fighters were listening.

Critical Thinking: Why might the BBC broadcast be so important to people of this region?

Answer: Students might say that war can prevent people from getting accurate news about their own countries.

Some Europeans moved to West Africa to run the colonies. They built roads, bridges, and railroads. Teachers and missionaries set up Christian churches and schools. After World War II, Africans increasingly worked for independence. Most of the colonies gained independence during the 1950s and 1960s. Portugal, the last European country to give up its West African colonies, did so in 1974.

✓ **READING CHECK:** *Human Systems* What impact did contact with Europeans have on West Africa? began the slave trade, which devastated West African communities

Challenges

Independence brought a new set of challenges to the region. The borders that the Europeans had drawn ignored human geography. Sometimes borders separated members of one ethnic group. Other borders grouped together peoples that did not get along. As a result, many West Africans were more loyal to their ethnic groups than to their new countries. In addition, too few people had been trained to run the new governments. Dictators took control in many countries. Civil wars and military rulers still trouble the region. Some countries have made progress, however. For example, in 1996 Chad created its first democratic constitution.

The governments of West African countries have several difficult problems in common. Birthrates are high. As a result, more and more people must make a living from the small amount of fertile land. In addition, many people are moving to already crowded cities even though urban jobs are few. These countries must also find ways to educate more of their people. Many families cannot afford to send their children to school.

✓ **READING CHECK:** *Places and Regions* What are three challenges the region faces? Political boundaries cut across culture regions, causing ethnic conflict and civil war; land cannot support large populations; lack of education is caused by poverty.

A roadside market provides a glimpse of crowded Lagos, Nigeria's largest city. More than 10 million people live in and around Lagos. Overcrowded cities are a problem throughout much of the region.

Teacher to Teacher

Cindy Herring of Round Rock, Texas, suggests the following activity to help students learn about the history of West Africa. Organize students into four groups and give each group a long sheet of butcher paper and art supplies. Have each group develop an illustrated time line for a designated period in West African history. Illustrations may be drawn or cut out of magazines. Combine all the groups' time lines to form one large time line. Display it on the wall of the classroom while students work on the chapter.
COOPERATIVE LEARNING

Close

Refer students to the feature on this page. Lead a discussion about how the father's ideas about marriage compare or contrast to most Americans' ideas.

Review and Assess

Have students complete the Section Review. Then organize students into four groups and give each group five index cards. Assign each group one of the following topics: great kingdoms; the slave trade; the colonial era and independence; and progress and challenges. Have each group write five questions about its topic on one side of each card and the answers on the other side. Have groups exchange cards and quiz each other. Then have students complete Daily Quiz 20.2. **COOPERATIVE LEARNING**

Connecting to Literature

Marriage Is a Private Affair

by Chinua Achebe

Chinua Achebe was born in an Ibo village in Nigeria in 1930. Many of his writings explore the changes colonialism brought to Africa. They also look at the conflict between old and new ways. In this story, Achebe looks at different views a father and son have about marriage.

A Nigerian church carving

"Father," began Nnaemeka suddenly, "I have come to ask for forgiveness."

"Forgiveness? For what, my son?" he asked in amazement.

"It's about this marriage question."

"Which marriage question?"

"I can't—we must—I mean it is impossible for me to marry Nweke's daughter."

"Impossible? Why?" asked his father.

"I don't love her."

"Nobody said you did. Why should you?" he asked.

"Marriage today is different . . ."

"Look here, my son," interrupted his father, "nothing is different. What one looks for in a wife are a good character and a Christian background."

Nnaemeka saw there was no hope along the present line of argument.

"Moreover," he said. "I am engaged to marry another girl who has all of Ugoye's good qualities, and who . . ."

His father did not believe his ears. "What did you say?" he asked slowly and disconcertingly[1]. . . .

"Whose daughter is she, anyway?"

"She is Nene Atang."

"What!" All the mildness was gone again. "Did you say Neneataga, what does that mean?"

"Nene Atang from Calabar. She is the only girl I can marry." This was a very rash reply and Nnaemeka expected the storm to burst.

Analyzing Primary Sources

1. What universal theme does the passage illustrate?
2. Why do you think the father and son disagree about marriage?

Vocabulary [1]disconcertingly: disturbingly

Across the Curriculum

MUSIC

Hubert Ogunde

Hubert Ogunde was an influential African writer. This Nigerian playwright, actor, theater manager, and musician pioneered Nigerian folk opera. His works often used satire to comment on colonization, political strife, and government corruption. Ogunde sketched the operas' plots and wrote down and rehearsed the songs but had the performers improvise the dialogue. His plays influenced urban pop culture in West Africa.

Activity: Have students compare this type of performance with television programs in the United States and explain the similarities and differences.

Connecting to Literature

Answers

1. the relationship between love and marriage
2. The father believes in traditional marriages, while the son believes in a modern approach to marriage.

RETEACH

Have students complete Main Idea Activity 20.2. Then have students draw an outline map of the region on butcher paper. Have them create a map key that depicts symbols for products traded in West Africa. Then have students draw or glue symbols on the map in the appropriate areas. Display and discuss the map. **ENGLISH LANGUAGE LEARNERS**

EXTEND

Have interested students conduct research on French influences that remain in West Africa's languages, architecture, art, music, government, or literature. Have students show and explain the influence in the form of posters, collages, or graphic organizers. **BLOCK SCHEDULING**

▲ These homes in Burkina Faso are made of a mixture of mud, water, and cow dung. Trees are scarce in the Sahel and savanna zones. As a result, there is little wood for construction. Women are responsible for painting and decorating the walls of the homes.

Interpreting the Visual Record

For what purpose are the roofs apparently used?

Culture

Hundreds of ethnic groups exist in West Africa today. Hundreds of languages are spoken in the region. In some areas, using the colonial languages of French or English helps people from different groups communicate. West African languages that many people share, such as Fula and Hausa, also aid communication.

Religion The traditional religions of West Africa have often been forms of **animism**. Animism is the belief that bodies of water, animals, trees, and other natural objects have spirits. Animists also honor the memories of ancestors. In some isolated areas animism still forms the basis of most religious practices. Today most people of the Sahel practice Islam. Farther south live many Christians.

Clothing and Homes Some West Africans, particularly in cities, wear Western-style clothing. Others wear traditional robes, pants, blouses, and skirts. These are often made from colorful patterned cotton fabric. Because of the warm climate, most clothing is loose and flowing. Many women wear beautiful folded and wrapped headdresses. In the desert men often wear turbans. Both men and women may use veils to protect their faces from blowing sand.

Rural homes are small and simple. Many homes in the Sahel and savanna zones are circular. Straw or tin roofs sit atop mud, mud-brick, or straw huts. However, in cities you will find some modern buildings.

✓ **READING CHECK:** *Human Systems* What are some features of West African culture? hundreds of ethnic groups and languages; animism; Western-style and traditional clothing; traditional and modern architecture

go.hrw.com **Homework Practice Online**
Keyword: SK3 HP20

Section 2 Review

Define and explain: archaeology, oral history, animism

Working with Sketch Maps On the map you created in Section 1, label Timbuktu. What made Timbuktu an important Songhay city?

Reading for the Main Idea

1. *Human Systems* How did European contact affect West Africa's people?
2. *Human Systems* Where in West Africa are Islam and Christianity practiced?

Critical Thinking

3. **Drawing Inferences and Conclusions** How might shared histories and challenges lead to more cooperation among the region's countries?
4. **Drawing Inferences and Conclusions** Why do you think mud bricks are used in West Africa?

Organizing What You Know

5. **Summarizing** Use the following graphic organizer to summarize challenges facing West Africa.

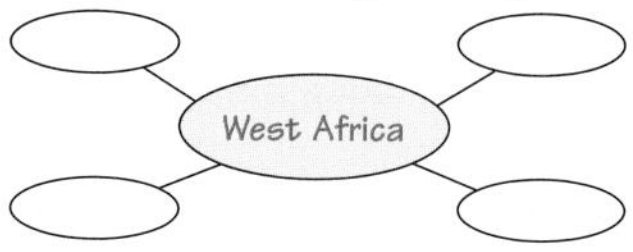

Section 2 Review

Answers

Define For definitions, see: archaeology, p. 464; oral history, p. 464; animism, p. 468

Working with Sketch Maps Timbuktu should be labeled in its approximate location. Timbuktu was important because it was a cultural center.

Reading for the Main Idea

1. Many West Africans died, and millions were forced from their homes as a result of the slave trade. (NGS 9)
2. the Sahel; in the southern region (NGS 10)

Critical Thinking

3. Students may suggest that people with shared histories and challenges are more likely to understand each others' point of view.
4. lack of wood or other building materials

Organizing What You Know

5. Possible answers include: military rulers, high birthrates, overcrowded cities, and lack of affordable education

Visual Record Answer ▲

to dry and store produce

Section 3

Objectives
1. Describe modern Mauritania, Mali, and Niger.
2. Analyze challenges facing Chad and Burkina Faso.

Focus

LET'S GET STARTED
Copy the following question onto the chalkboard: *What could you do to survive if all your community's sources of water—wells, lakes, rivers, aquifers, pipelines—were going dry?* Discuss responses. Point out that this is happening in parts of the Sahel countries. Explain to students that in Section 3 they will find out more about the causes and results of the region's water shortages.

Building Vocabulary
Write the key terms on the chalkboard and call on volunteers to read the definitions aloud. Tell students that **millet** and **sorghum** are examples of **staple** grain crops grown in the Sahel and that **malaria** is a disease spread by mosquitoes.

Section 3 The Sahel Countries

Read to Discover
1. What are Mauritania, Mali, and Niger like today?
2. What challenges do Chad and Burkina Faso face?

Define
millet
sorghum
malaria
staple

Locate
Nouakchott
Senegal River
Gao
Tibesti Mountains
Lake Chad
Ouagadougou

Why It Matters
The shrinking of Lake Chad over the last several decades has been a major concern for the people of Chad. Many blame a changing environment and poor managing of the lake's resources. Use CNNfyi.com or other **current events** sources to find information about Lake Chad. Record the findings in your journal.

Carved mask from Mauritania

Section 3 Resources

Reproducible
- Guided Reading Strategy 20.3
- Map Activity 20

Technology
- One-Stop Planner CD–ROM, Lesson 20.3
- Homework Practice Online
- HRW Go site

Reinforcement, Review, and Assessment
- Section 3 Review, p. 471
- Daily Quiz 20.3
- Main Idea Activity 20.3
- English Audio Summary 20.3
- Spanish Audio Summary 20.3

Mauritania, Mali, and Niger
Most of the people in these three large countries are Muslim. Mauritania, in fact, has laws based on Islam. These countries are also former French colonies, and French influence remains. In Mali and Niger, the official language is French. However, the people there speak more than 60 different local languages.

Today, drought and the expanding desert make feeding the people in these countries difficult. In the Sahel nomads depend on their herds of cattle, goats, and camels. In the savanna regions farmers grow **millet** and **sorghum**. These grain crops can usually survive drought.

Mauritania Many Mauritanians are Moors, people of mixed Arab and Berber origin. They speak Arabic. In the past, Moors enslaved some of the black Africans. Today, tension between the two groups continues.

Women carry goods for sale in a market in central Mali. Much of Mali's economic activity takes place in the Niger River's inland delta.
Interpreting the Visual Record **Why is Mali's economic activity centered around the inland delta?**

Visual Record Answer
because most of the country is desert with few resources

TEACH

Teaching Objectives 1–2

ALL LEVELS: (Suggested time: 35 min.) Copy the following graphic organizer onto the chalkboard, omitting the italicized answers. Use it to help students describe the Sahel countries. Have students complete the organizer. Then ask students to create their own graphic organizer to depict the challenges facing Chad and Burkina Faso. Lead a discussion in which students compare all five of the Sahel countries.

ENGLISH LANGUAGE LEARNERS

The Sahel Countries

Country	Climate	Crops	Economy	Challenges
Mauritania	*desert*	*(no specific crops listed)*	*farming, fishing, herding*	*poverty, tensions, expanding desert*
Mali	*desert*	*cotton*	*fishing, farming, tourism*	*malaria, poverty*
Niger	*desert*	*cotton, sorghum, peanuts, peas, rice, millet*	*farming, herding*	*poverty, expanding desert*

National Geography Standard 15

Environment and Society The economies of countries of the Sahel, such as Niger, fluctuate with the weather. When there are periods of ample rainfall, the land can support as many as 10 million cattle, sheep, and goats. However, the droughts of the 1970s and 1980s destroyed all but 1 million animals, bringing famine to the country.

Activity: Have students suggest other countries whose economies are linked to weather conditions.

Chart Answer

Mali, Nigeria

Mud and other local materials were used to build many mosques in the Sahel. This mosque is located in Djenné [je-NAY], Mali. The majority of people living in the Sahel are Muslims.

Our Amazing Planet

The Tuareg (TWAH-reg) people of the Sahara and Sahel pound powdered blue dye into their flowing robes. They do this rather than dip the fabric in precious water. The blue powder wears off onto the skin, where it may help hold in moisture.

Most Mauritanians were once nomadic herders. Today, the expanding Sahara has crowded more than half of the nomads into the cities. Just 40 years ago, Nouakchott (nooh-AHK-shaht), Mauritania's capital, was a small village. More than 700,000 people live there now. About half of the population lives in shacks at the city's edges.

Throughout the country, people are very poor. Only in the far south, near the Senegal River, can farmers raise crops. Fishing in the Atlantic Ocean is another source of income.

Mali To the east of Mauritania lies landlocked Mali. The Sahara covers much of northern Mali. In the south lies a wetter farming region. About 80 percent of Mali's people fish or farm along the Niger River. Cotton is the country's main export. Timbuktu and Gao (GOW), ancient trading cities, continue to attract tourists.

Health conditions in Mali are poor. **Malaria**, a disease spread by mosquitoes, is a major cause of death among children.

Niger The Niger River flows through just the southwestern corner of landlocked Niger. Only about 3 percent of Niger's land is good for farming. All of the country's farmland lies along the Niger River and near the Nigerian border. Much of the rest of Niger lies within the Sahara. Farmers raise cotton, peanuts, beans, peas, and rice. Millet and sorghum are two of the region's **staple**, or main, food crops. The grains are cooked like oatmeal. Nomads in the desert region depend on the dairy products they get from their herds for food.

The Sahel Countries

Country	Population/ Growth Rate	Life Expectancy	Literacy Rate	Per Capita GDP
Burkina Faso	12,272,289 / 2.7%	46, male 47, female	19%	$1,000
Chad	8,707,078 / 3.3%	49, male 53, female	48%	$1,000
Mali	11,008,518 / 3%	46, male 48, female	31%	$850
Mauritania	2,747,312 / 2.9%	49, male 53, female	47%	$2,000
Niger	10,355,156 / 2.7%	42, male 41, female	14%	$1,000
United States	281,421,906 / 0.9%	74, male 80, female	97%	$36,200

Sources: Central Intelligence Agency, *The World Factbook 2001;* U.S. Census Bureau

Interpreting the Chart **Based on the numbers in the chart, which two countries in the region are the least economically developed?**

✓ **READING CHECK:** *Places and Regions* What is it like to live in Mauritania, Mali, and Niger? People are generally poor farmers and herders; most are Muslim and speak French. Life is often difficult because of drought and the expanding desert.

CLOSE

Refer to the photo of the Djenné mosque on the previous page. Challenge students to suggest possible steps in the building's construction.

REVIEW AND ASSESS

Have students complete the Section Review. Then have students choose one of the section's countries and write three statements about their country. Have students play a where-am-I game using the statements. Then have students complete Daily Quiz 20.3.

RETEACH

Have students complete Main Idea Activity 20.3. Then organize the class into five groups and assign one of the Sahel countries to each group. Have groups create mobiles depicting characteristics of their assigned country. **ENGLISH LANGUAGE LEARNERS, COOPERATIVE LEARNING**

EXTEND

Have interested students conduct research on the Tuareg and other nomadic peoples of the Sahel countries and describe their cultures in a report. Tell them to include information about the relationships between these cultures and their artforms and/or music. **BLOCK SCHEDULING**

Chad and Burkina Faso

Drought has also affected the former French colonies of Chad and Burkina Faso (boohr-KEE-nuh FAH-soh). These countries are among the world's poorest and least developed. Most people farm or raise cattle.

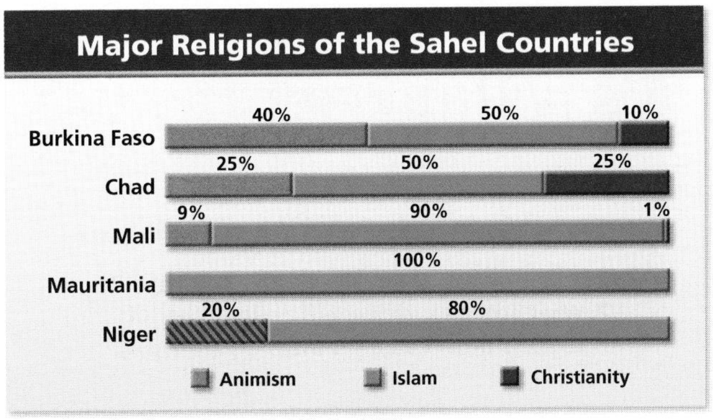

Source: Central Intelligence Agency, *World Factbook 2001*

▲ Although most people in the Sahel are Muslim, some practice forms of animism and Christianity.

Interpreting the Graph **Which country's population is entirely Muslim? Which countries have significant numbers of Christians?**

Chad Chad is located in the center of Africa. The Tibesti Mountains in northern Chad rise above the Sahara. Lake Chad is in the south. Not long ago, the lake had a healthy fishing industry. It even supplied water to several other countries. However, drought has evaporated much of the lake's water. At one time, Lake Chad had shrunk to just one third its size in 1950.

The future may be better for Chad. A civil war ended in the 1990s. Also, oil reserves now being explored may help the economy.

Burkina Faso This country's name means "land of the honest people." Most of its people follow traditional religions. The country has thin soil and few mineral resources. Few trees remain in or near the capital, Ouagadougou (wah-gah-DOO-goo). They have been cut for firewood and building material. Jobs in the city are also scarce. To support their families many young men work in other countries. However, foreign aid and investment are starting to help the economy.

✓ **READING CHECK:** *Places and Regions* What are the challenges facing Chad and Burkina Faso? drought, poverty, underdevelopment; scarcity of natural resources in Burkina Faso has led to economic problems

go.hrw.com Homework Practice Online
Keyword: SK3 HP20

Section 3 Review

Define and explain: millet, sorghum, malaria, staple

Working with Sketch Maps On the map you created in Section 2, draw the boundaries of Mauritania, Mali, Niger, Chad, and Burkina Faso. Then label Nouakchott, Senegal River, Gao, Tibesti Mountains, Lake Chad, and Ouagadougou. In a box in the margin, describe what has happened to Lake Chad in recent decades.

Reading for the Main Idea

1. *Places and Regions* Why has Nouakchott grown so rapidly?
2. *Places and Regions* Which European language is most common in the Sahel countries? Why?

Critical Thinking

3. **Drawing Inferences and Conclusions** How do the typical foods of Niger relate to the country's water resources?
4. **Comparing** What do Chad and Burkina Faso have in common with the other Sahel countries?

Organizing What You Know

5. **Comparing** Copy the following graphic organizer. Label each of the star's points with one of the Section 3 countries. In the center, list characteristics the countries share.

Section 3 Review

Answers

Define For definitions, see: millet, p. 469; sorghum, p. 469; malaria, p. 470; staple, p. 470

Working with Sketch Maps Places should be labeled in their approximate locations. Lake Chad has shrunk.

Reading for the Main Idea

1. The expanding Sahara has crowded nomads into Nouakchott. (NGS 4)
2. French; France controlled the Sahel countries as colonies. (NGS 4)

Critical Thinking

3. Farmers grow food crops in areas where there is enough water. Nomads in dry areas rely on products from their herds.
4. similar climate, colonial background, religions, economic activities

Organizing What You Know

5. five points—Burkina Faso, Chad, Mali, Mauritania, Niger; possible for center—Sahel climate zone, farming, low literacy rate, former French colonies, hurt by droughts

◄ **Graph Answer**

Mauritania; Burkina Faso and Chad

Section 4

Objectives

1. Describe life in Nigeria today.
2. Identify the economic challenges faced by other coastal West African countries.

FOCUS

LET'S GET STARTED

Copy the following instructions onto the chalkboard: *Look at the flags on the Fast Facts pages. Write down why you think Liberia's flag looks the way it does.* Discuss responses. Remind students that Liberia was founded as a home for freed slaves from the United States. Liberia's flag was based on the American flag. Tell students they will learn more about Liberia and the other coastal countries of West Africa in Section 4.

Building Vocabulary

Write the key terms on the chalkboard. Ask students if they know a person who tells really interesting stories. Point out that **griots** are storytellers who pass on the histories of peoples of Senegal and Gambia. Tell them that the seeds of the **cacao** tree are used to make chocolate and that to **secede** means to break away to form another country, an action that often leads to war.

Section 4 RESOURCES

Reproducible
- Guided Reading Strategy 20.4
- Map Activity 20

Technology
- One-Stop Planner CD–ROM, Lesson 20.4
- Homework Practice Online
- HRW Go site

Reinforcement, Review, and Assessment
- Section 4 Review, p. 475
- Daily Quiz 20.4
- Main Idea Activity 20.4
- English Audio Summary 20.4
- Spanish Audio Summary 20.4

Visual Record Answer

Answers will vary according to local styles.

Section 4 The Coastal Countries

Read to Discover

1. What is life in Nigeria like today?
2. What economic challenges do the region's other countries face?

Define

secede
griots
cacao

Locate

Abuja
Lagos
Dakar
Monrovia
Lake Volta

WHY IT MATTERS

Nigeria has made a great deal of economic progress but there are problems that still hurt the nation's economy. Use CNNfyi.com or other **current events** sources to find examples of Nigerian social, economic, and political affairs. Record your findings in your journal.

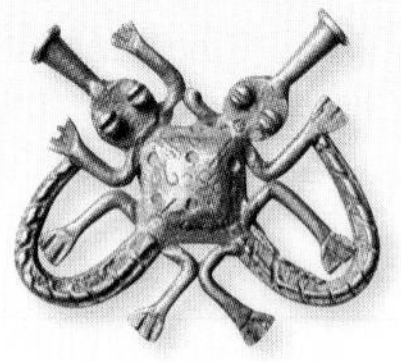

A gold ornament from Ghana

The faces of Nigeria are very young. About 45 percent of all Nigerians are younger than 15 years old. Only about 22 percent of all Americans are that young.

Interpreting the Visual Record

How is the clothing shown here similar to clothing worn by students in your school?

Nigeria

The largest country along West Africa's coast is Nigeria. With more than 113 million people, it has Africa's largest population.

Nigeria's People Nigeria was once an important British colony. Like many other colonies, Nigeria's borders included many ethnic groups. Today, a great variety of ethnic groups live in Nigeria. The Yoruba, Fula, Hausa, and Ibo are four of the largest ethnic groups. More than 200 languages are spoken there.

Nigeria's ethnic groups have not always gotten along. In the 1960s the Ibo tried to **secede**. That is, they tried to break away from Nigeria and form their own country. They called it Biafra (bee-AF-ruh). However, the Ibo lost the bloody war that followed.

Avoiding ethnic conflicts has continued to be an issue in Nigeria. It was important in choosing a site for a new Nigerian capital in the late 1970s. Leaders chose Abuja (ah-BOO-jah) because it was centrally located in an area of low population density.

Nigeria's Economy Nigeria has some of the continent's richest natural resources. Oil is the country's most important resource. Major oil fields are located in the Niger River delta and just off the coast. Oil accounts for 95 percent of the country's export earnings. Nigeria also

TEACH

Teaching Objective 1

ALL LEVELS: (Suggested time: 20 min.) Copy the following graphic organizer onto the chalkboard, omitting the italicized answers. Use it to help students describe modern Nigeria. Organize the class into two groups. Have one group find positive aspects of modern Nigeria. Have the second find challenges faced by modern Nigeria. Have volunteers from each group write their information on the organizer.
ENGLISH LANGUAGE LEARNERS, COOPERATIVE LEARNING

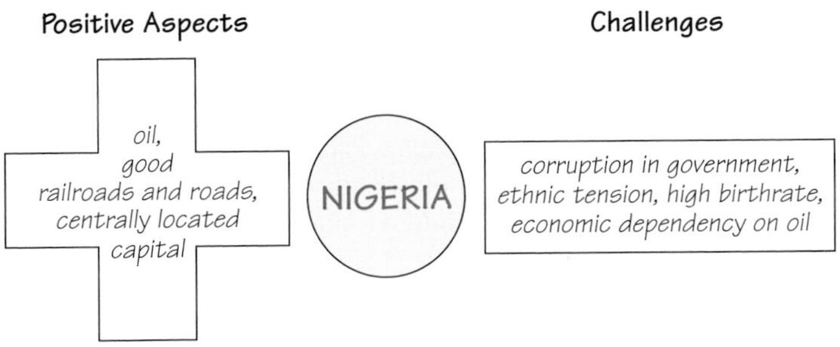

▲ Oil drilling rigs like this one are common in areas of southern Nigeria. Oil accounts for about 20 percent of Nigeria's GDP. Oil revenues pay for about 65 percent of the government's budget.

has good roads and railroads. Lagos (LAY-gahs), the former capital, is the country's largest city. The city is a busy seaport and trade center.

Although the country has rich resources, many Nigerians are poor. A major cause of the poverty there is a high birthrate. Nigeria can no longer feed its growing population without importing food. Another cause is the economy's dependence on oil. When prices are low, the whole country suffers. A third cause of Nigeria's poverty is a history of bad government. Corrupt government officials have used their positions to enrich themselves.

✓ **READING CHECK:** *Places and Regions* What are Nigeria's people and economy like today? many ethnic groups, languages; frequent ethnic conflict; economy based on oil; poverty is a problem

Other Coastal Countries

Several small West African countries lie along the Atlantic Ocean and the Gulf of Guinea. They are struggling to develop their economies.

Senegal and Gambia Senegal (se-ni-GAWL) wraps around Gambia (GAM-bee-uh). The odd border was created by French and British diplomats. Senegal, a former French colony, is larger and richer than Gambia, a former British colony. Dakar (dah-KAHR) is Senegal's capital and an important seaport and manufacturing center. Senegal and Gambia have many similarities. Peanuts are their most important crop. Common foods include chicken stew and fish with a peanut sauce. Tourism is growing slowly.

The headdresses and patterned clothing worn by these women in Dakar are common in Senegal. ▼

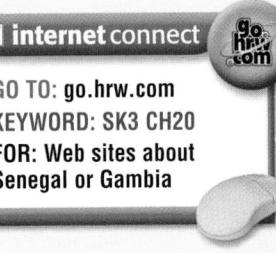

HUMAN SYSTEMS

Senegal and part of Gambia were once part of the Wolof Empire. They came under separate rule when Europeans colonized the area. Both the French and British established posts in the 1600s. The British and French fought at various times for control of the Gambia River and its trade. In the early 1800s the British gained control of the river basin, while the French maintained their hold on Senegal.

The two European powers ruled their colonies differently. The French colonizers of Senegal exerted direct control over the colony. They sought to assimilate the Senegalese into a French way of life. This practice proved effective only in the cities. In contrast, the British controlled their colony indirectly through Gambian rulers.

Discussion: Lead a class discussion about how differences in colonial rule might affect modern life in Senegal and Gambia.

internet connect
GO TO: **go.hrw.com**
KEYWORD: **SK3 CH20**
FOR: **Web sites about Senegal or Gambia**

Teaching Objective 2

LEVEL 1: (Suggested time: 40 min.) Organize the class into eleven groups and assign each group a country. Have each group fold a piece of paper lengthwise and crosswise, creating four boxes. Have groups fill in each box with the following information on economic development for their assigned country: per capita GDP, literacy rate, major economic resources, economic challenges, possible solutions to challenges. You may want to provide encyclopedias or other library resources so that students can add detail. **ENGLISH LANGUAGE LEARNERS, COOPERATIVE LEARNING**

LEVELS 2 AND 3: (Suggested time: 45 min.) Have students work in pairs to write stories about economic challenges in one of the coastal countries. Then ask them to convert their stories into rhyming poems, such as griots might perform. Have them recite their tales to the class. Allow students to use notes or cue cards. **COOPERATIVE LEARNING**

Section 4 Review

Answers

Define For definitions, see: secede, p. 472; griots, p. 474; cacao, p. 475

Working with Sketch Maps Maps will vary, but listed places should be labeled in their approximate locations. The largest and most populous country on West Africa's coast is Nigeria.

Reading for the Main Idea

1. West Africa's only island nation (NGS 4)
2. oil; makes up 95 percent of Nigeria's exports (NGS 4)

Critical Thinking

3. Civil wars disrupted both countries.
4. to help avoid further ethnic conflict

Organizing What You Know

5. Possible answers: high birthrate, economic dependence on oil, government corruption

Chart Answer ▲

Cape Verde

Visual Record Answer ►

shrubs, grasses, crops, flowering trees

The Coastal Countries

Country	Population/ Growth Rate	Life Expectancy	Literacy Rate	Per Capita GDP
Benin	6,590,782 3%	49, male 51, female	38%	$1,030
Cape Verde	405,163 0.9%	66, male 73, female	72%	$1,700
Côte d'Ivoire	16,393,221 2.5%	44, male 46, female	49%	$1,600
Gambia	1,411,205 3%	52, male 56, female	48%	$1,100
Ghana	19,894,014 1.8%	56, male 59, female	65%	$1,900
Guinea	7,613,870 2%	44, male 48, female	36%	$1,300
Guinea-Bissau	1,315,822 2.2%	47, male 52, female	54%	$850
Liberia	3,225,837 1.9%	50, male 53, female	38%	$1,100
Nigeria	126,635,626 2.6%	51, male 51, female	57%	$950
Senegal	10,284,929 2.9%	61, male 64, female	33%	$1,600
Sierra Leone	5,426,618 3.6%	43, male 49, female	31%	$510
Togo	5,153,088 2.6%	52, male 56, female	52%	$1,500
United States	281,421,906 0.9%	74, male 80, female	97%	$36,200

Sources: Central Intelligence Agency, *The World Factbook 2001;* U.S. Census Bureau

Interpreting the Chart Which country in the region has the highest life expectancy?

Many of the people speak a language called Wolof (WOH-lawf). **Griots** (GREE-ohz) are important to the Wolof-speakers and other West Africans. Griots are storytellers who pass on the oral histories of their tribes or peoples. Sometimes the griots combine music with their stories, which may take hours or days to tell. Wolof women wear complex hairstyles and gold jewelry.

Guinea, Guinea-Bissau, and Cape Verde Guinea's main natural resource is a huge supply of bauxite. Its small neighbor to the east, Guinea-Bissau (GI-nee bi-SOW), has undeveloped mineral resources. Cape Verde (KAYP VUHRD) is a group of volcanic islands in the Atlantic. It is West Africa's only island country. Farming and fishing bring in the most money there.

Liberia and Sierra Leone Liberia is Africa's oldest republic. Monrovia, Liberia's capital, was named for U.S. president James Monroe. The freed American slaves who settled Liberia and their descendants lived in coastal towns. They often clashed with the Africans already living there. Those Africans and their descendants were usually poorer and lived in rural areas. In the 1980s conflicts led to a bitter civil war. Sierra Leone (lee-OHN) has also experienced violent civil war. The fighting has wrecked the country's economy. Now, both Liberia and Sierra Leone must rebuild. They do have natural resources on which to build stronger economies. Liberia produces rubber and iron ore. Sierra Leone exports diamonds.

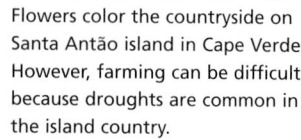

Flowers color the countryside on Santa Antão island in Cape Verde. However, farming can be difficult because droughts are common in the island country.

Interpreting the Visual Record What kinds of vegetation do you see on the island?

CLOSE

Focus students' attention on the photo of the Lagos roadside market in Section 2. Ask students to suggest why shopping at this market might be important for cultural as well as economic reasons. *(Possible answers: opportunity to socialize, meet friends, exchange news)*

REVIEW AND ASSESS

Have students complete the Section Review. Then organize students into groups and have each group create a crossword puzzle about the coastal countries of West Africa. Photocopy the puzzles and have groups exchange and complete them. Then have students complete Daily Quiz 20.4. **COOPERATIVE LEARNING**

RETEACH

Have students complete Main Idea Activity 20.4. Then have students outline the section. **ENGLISH LANGUAGE LEARNERS**

EXTEND

Have interested students conduct research and report on the artistic traditions of West Africa's coastal countries. Examples include terra cotta heads from the Nok civilization, bronze heads created in the Ife kingdom, and sculptures made by people of Benin in brass, bronze, and ivory. **BLOCK SCHEDULING**

Ghana and Côte d'Ivoire The countries of Ghana and Côte d'Ivoire (koht-dee-VWAHR) have rich natural resources. Those resources may help them build strong economies. Ghana is named for the ancient kingdom, although the kingdom was northwest of the modern country. Ghana has one of the largest human-made lakes in the world—Lake Volta. Gold, timber, and **cacao** (kuh-KOW) are major products. Cocoa and chocolate are made from the seeds of the cacao tree. The tree came originally from Mexico and Central America.

Côte d'Ivoire is a former French colony whose name means "Ivory Coast" in English. It is a world leader in cacao and coffee exports. Côte d'Ivoire also boasts Africa's largest Christian church building.

A storyteller passes on a legend to children in Côte d'Ivoire. Oral storytellers use facial expressions, gestures, and even music and dance. Those techniques help draw listeners into a story.

Interpreting the Visual Record **What role might oral storytellers like the one shown here play in a place where many people cannot read?**

Togo and Benin Unstable governments have troubled both Togo and Benin (buh-NEEN) since independence. Both have experienced periods of military rule. Their fragile economies have contributed to their unstable and sometimes violent politics. These long, narrow countries are poor. The people depend on farming and herding for income. Palm tree products, cacao, and coffee are the main crops in Togo and Benin.

✓ **READING CHECK:** *Places and Regions* What are characteristics of the economies of the coastal countries? Senegal, Gambia—some manufacturing, peanuts, tourism; Guinea, Guinea-Bissau—minerals; Cape Verde—farming, fishing; Liberia—rubber, iron ore; Sierra Leone—diamonds; Ghana—gold, timber, cacao; Côte d'Ivoire—cacao, coffee; Togo, Benin—farming, herding

go.hrw.com Homework Practice Online Keyword: SK3 HP20

Section 4 Review

Define and explain: secede, griots, cacao

Working with Sketch Maps On the map that you drew in Section 3, draw the boundaries for the coastal countries. Label Abuja, Lagos, Dakar, Monrovia, and Lake Volta. In a box in the margin, identify the largest and most populous country on West Africa's coast.

Reading for the Main Idea

1. *Places and Regions* How is Cape Verde different from the other countries in the region?
2. *Places and Regions* What is Nigeria's most important natural resource? Why?

Critical Thinking

3. **Finding the Main Idea** Why must Liberia and Sierra Leone rebuild their economies?
4. **Analyzing Information** Why was choosing a new capital important to Nigeria's future?

Organizing What You Know

5. **Summarizing** Copy the following graphic organizer. Use it to list three main causes of poverty in Nigeria today.

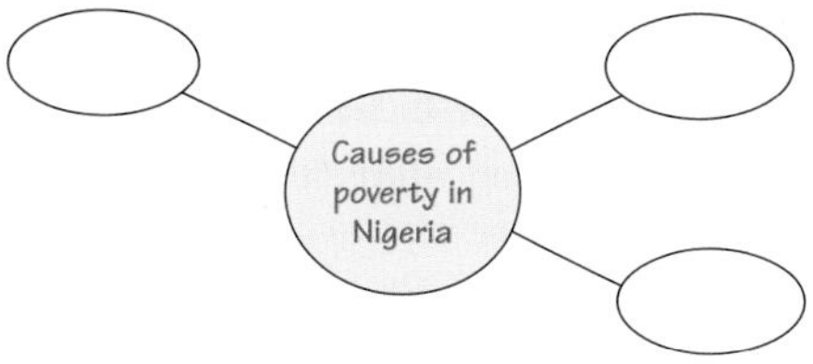

CHAPTER 20 Review ANSWERS

Building Vocabulary

For definitions, see: zonal, p. 462; Sahel, p. 462; harmattan, p. 462; tsetse fly, p. 462; bauxite, p. 463; archaeology, p. 464; oral history, p. 464; animism, p. 468; millet, p. 469; sorghum, p. 469; malaria, p. 470; staple, p. 470; secede, p. 472; griots, p. 474; cacao, p. 475

Reviewing the Main Ideas

1. Sahara, Sahel, savanna, and coast and forest (NGS 4)
2. The inland delta consists of swamps, creeks, and lakes, while the true delta is wide and located on the coast. (NGS 4)
3. with merchants from North Africa (NGS 9)
4. Possible answers: caused crops to fail, forced nomads into cities, dried up sources of water (NGS 15)
5. oil (NGS 15)

▲ Visual Record Answer

They can pass information and traditions on to later generations.

Assess

Have students complete the Chapter 20 Test.

Reteach

Draw a large outline map of West Africa. Organize students into four groups and assign each a climate zone. Give each group two different colors of paper. Have them cut the paper into strips. On one color they should write resources and on the other problems faced by countries in their region. Have them glue the strips onto the countries of their climate region. Display and discuss the map. **ENGLISH LANGUAGE LEARNERS, COOPERATIVE LEARNING**

Portfolio Extensions

1. Have students select a country of West Africa and use the six essential elements of geography to make an illustrated brochure about the country. Instruct students to create six panels for their brochures—one for each of the elements.
2. Have students examine pictures of various types of clothing worn in West Africa. Have them create an illustrated brochure describing the clothing, its origins, and the people who might wear it.

Chapter 20 Review ANSWERS

Understanding Environment and Society

Tsetse flies, mosquitoes, and other dangerous insects are common in the savanna zone and areas where there is standing water. They spread sleeping sickness and malaria, and they have limited the areas where people can live. Use Rubric 29, Presentations, to evaluate student work.

Thinking Critically

1. Possible answers: crowded with new arrivals, near water sources, and in better climate
2. Early trade brought Islam and increased the power and wealth of the kingdoms. European trade caused a decline in Saharan trade and introduced the slave trade.
3. Possible answers: language, religion, clothing, and buildings
4. European colonizers ignored ethnic rivalries when they drew Africa's borders
5. Liberia was founded for freed American slaves rather than as a European colony.

Chapter 20 Reviewing What You Know

Building Vocabulary

On a separate sheet of paper, write sentences to define each of the following words.

1. zonal
2. Sahel
3. harmattan
4. tsetse fly
5. bauxite
6. archaeology
7. oral history
8. animism
9. millet
10. sorghum
11. malaria
12. staple
13. secede
14. griots
15. cacao

Reviewing the Main Ideas

1. *Places and Regions* What are the four climate zones of West Africa?
2. *Places and Regions* How do the Niger River's two delta regions differ?
3. *Human Systems* How did Islam come to West Africa?
4. *Environment and Society* How has drought affected the countries of the Sahel?
5. *Environment and Society* On what natural resource does Nigeria's economy depend?

Understanding Environment and Society

Health

Prepare a presentation, including a chart, graph, database, or model, about disease in West Africa. Consider the following:

- The parts of West Africa in which the tsetse fly, mosquitoes, and other dangerous insects are common.
- What diseases the insects spread.
- How these insects have affected human settlement and activities.

Then create a five-question quiz, with answers, about your presentation to challenge fellow students.

Thinking Critically

1. **Drawing Inferences and Conclusions** Why are many of West Africa's largest cities located in the coastal and forest zone?
2. **Finding the Main Idea** What role did trade play in the early West African kingdoms and later European colonies in the region?
3. **Analyzing Information** What are three cultural features of West Africa influenced by Europeans?
4. **Drawing Inferences and Conclusions** How have borders set by European colonial powers led to conflicts such as Nigeria's war in Biafra in the late 1960s?
5. **Contrasting** How is Liberia's history different from that of other West African countries?

Food Festival

Explain that okra and yams are West African foods used in the United States. Okra can be served pickled, boiled, sautéed, or fried and is used to thicken and flavor gumbos and stews. Point out that yams are often confused with sweet potatoes, but the two are from different plant species. Sweet potatoes can be substituted in most recipes that call for yams. Have students research recipes for okra and yams and bring samples to class for everyone to taste.

Chapter 20 Review and Assessment Resources

Reproducible
- Readings in World Geography, History, and Culture 55, 56, and 57
- Critical Thinking Activity 20
- Vocabulary Activity 20

Technology
- Chapter 20 Test Generator (on the One-Stop Planner)
- Audio CD Program, Chapter 20
- HRW Go site

Reinforcement, Review, and Assessment
- Chapter 20 Review, pp. 476–77
- Chapter 20 Tutorial for Students, Parents, Mentors, and Peers
- Chapter 20 Test
- Chapter 20 Test for English Language Learners and Special-Needs Students

Building Social Studies Skills

Map Activity

On a separate sheet of paper, match the letters on the map with their correct labels.

Niger River	Tibesti Mountains
Gulf of Guinea	Lake Chad
Timbuktu	Abuja
Nouakchott	Lagos
Senegal River	Lake Volta

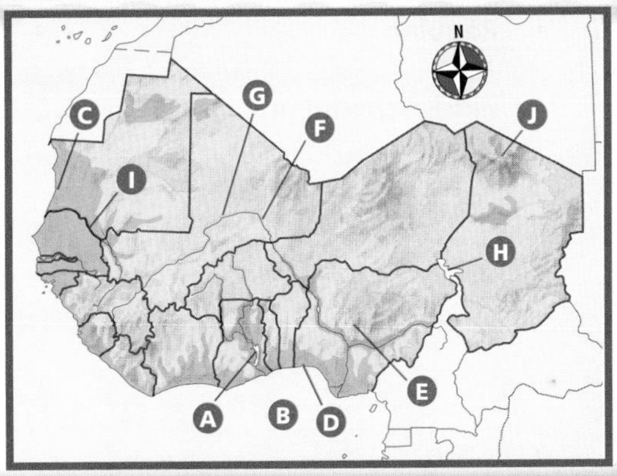

Mental Mapping Skills Activity

On a separate sheet of paper, draw a freehand map of West Africa. Label the following:

Burkina Faso	Mauritania
Chad	Niger
Liberia	Nigeria
Mali	Sahel

Writing Activity

Imagine that you are an economic adviser to a West African country of your choice. Use print resources to prepare a short economic report for your country's leader. Identify the country's important natural resources and what can be done with them. In addition, describe economic advantages or disadvantages of the country's climate, location, and physical features. Be sure to use standard grammar, spelling, sentence structure, and punctuation.

Alternative Assessment

Portfolio Activity

Learning About Your Local Geography

Group Project The borders of today's African countries were largely drawn by Europeans. With your group, research how the borders of your state were decided.

internet connect

Internet Activity: go.hrw.com
KEYWORD: SK3 GT20

Choose a topic to explore about West Africa:
- Find out about giant baobab trees.
- Meet the people of West Africa.
- Learn about the history of the slave trade.

Map Activity
A. Lake Volta
B. Gulf of Guinea
C. Nouakchott
D. Lagos
E. Abuja
F. Niger River
G. Timbuktu
H. Lake Chad
I. Senegal River
J. Tibesti Mountains

Mental Mapping Skills Activity
Maps will vary, but listed places should be labeled in their approximate locations.

Writing Activity
Information should be consistent with text material. Students should identify all the elements that relate to their country's economy. Use Rubric 42, Writing to Inform, to evaluate student work.

Portfolio Activity
Groups should note which political boundaries follow natural boundaries and note reasons for changed boundaries. Use Rubric 14, Group Activity, to evaluate student work.

internet connect

GO TO: go.hrw.com
KEYWORD: SK3 Teacher
FOR: a guide to using the Internet in your classroom

CHAPTER 21 East Africa

Chapter Resource Manager

Objectives	Pacing Guide	Reproducible Resources
SECTION 1 **Physical Geography** (pp. 479–81) **1.** Identify the major landforms of East Africa. **2.** Identify the rivers and lakes that are important in this region. **3.** Identify East Africa's climate types and natural resources.	**Regular** .5 day **Block Scheduling** .5 day *Block Scheduling Handbook, Chapter 21*	**RS** Guided Reading Strategy 21.1 **SM** Geography for Life Activity 21
SECTION 2 **History and Culture** (pp. 482–84) **1.** Identify the important events and developments that have influenced the history of East Africa. **2.** Describe the culture of East Africa.	**Regular** .5 day **Block Scheduling** .5 day *Block Scheduling Handbook, Chapter 21*	**RS** Guided Reading Strategy 21.2 **RS** Graphic Organizer 21 **E** Cultures of the World Activity 6 **E** Creative Strategies for Teaching World Geography, Lesson 16 **IC** Interdisciplinary Activities for the Middle Grades 26, 27, 28
SECTION 3 **The Countries of East Africa** (pp. 485–88) **1.** Explain why settlers come to Kenya. **2.** Describe how Tanzania was created. **3.** Describe Rwanda and Burundi. **4.** Describe Uganda. **5.** Identify the physical features of Sudan.	**Regular** 1 day **Block Scheduling** .5 day *Block Scheduling Handbook, Chapter 21*	**RS** Guided Reading Strategy 21.3 **E** Creative Strategies for Teaching World Geography, Lesson 17 **SM** Map Activity 21
SECTION 4 **The Horn of Africa** (pp. 489–91) **1.** Identify the main physical features of Ethiopia. **2.** Describe Eritrea. **3.** Identify the physical and cultural characteristics of Djibouti.	**Regular** .5 day **Block Scheduling** .5 day *Block Scheduling Handbook, Chapter 21*	**RS** Guided Reading Strategy 21.4

Chapter Resource Key

- **RS** Reading Support
- **IC** Interdisciplinary Connections
- **E** Enrichment
- **SM** Skills Mastery
- **A** Assessment
- **REV** Review
- **ELL** Reinforcement and English Language Learners
- Transparencies
- CD–ROM
- Music
- Video
- Internet
- Holt Presentation Maker Using Microsoft® Powerpoint®

One-Stop Planner CD–ROM

See the ***One-Stop Planner*** for a complete list of additional resources for students and teachers.

One-Stop Planner CD–ROM

It's easy to plan lessons, select resources, and print out materials for your students when you use the ***One-Stop Planner CD–ROM with Test Generator.***

Technology Resources	Review, Reinforcement, and Assessment Resources
One-Stop Planner CD–ROM, Lesson 21.1 Earth: Forces and Formations CD–ROM/Seek and Tell/Forces and Processes Geography and Cultures Visual Resources with Teaching Activities 43–47 Homework Practice Online HRW Go site	**ELL** Main Idea Activity 21.1 **REV** Section 1 Review, p. 481 **A** Daily Quiz 21.1 **ELL** English Audio Summary 21.1 **ELL** Spanish Audio Summary 21.1
One-Stop Planner CD–ROM, Lesson 21.2 Homework Practice Online HRW Go site	**ELL** Main Idea Activity 21.2 **REV** Section 2 Review, p. 484 **A** Daily Quiz 21.2 **ELL** English Audio Summary 21.2 **ELL** Spanish Audio Summary 21.2
One-Stop Planner CD–ROM, Lesson 21.3 Homework Practice Online HRW Go site	**ELL** Main Idea Activity 21.3 **REV** Section 3 Review, p. 488 **A** Daily Quiz 21.3 **ELL** English Audio Summary 21.3 **ELL** Spanish Audio Summary 21.3
One-Stop Planner CD–ROM, Lesson 21.4 Homework Practice Online HRW Go site	**ELL** Main Idea Activity 21.4 **REV** Section 4 Review, p. 491 **A** Daily Quiz 21.4 **ELL** English Audio Summary 21.4 **ELL** Spanish Audio Summary 21.4

internet connect

HRW ONLINE RESOURCES

GO TO: go.hrw.com
Then type in a keyword.

TEACHER HOME PAGE
KEYWORD: SK3 TEACHER

CHAPTER INTERNET ACTIVITIES
KEYWORD: SK3 GT21

Choose an activity to:
- hike Mount Kilimanjaro.
- learn about cultural groups in East Africa.
- travel back to ancient Nubian kingdoms.

CHAPTER ENRICHMENT LINKS
KEYWORD: SK3 CH21

CHAPTER MAPS
KEYWORD: SK3 MAPS21

ONLINE ASSESSMENT
Homework Practice
KEYWORD: SK3 HP21
Standardized Test Prep Online
KEYWORD: SK3 STP21
Rubrics
KEYWORD: SS Rubrics

COUNTRY INFORMATION
KEYWORD: SK3 Almanac

CONTENT UPDATES
KEYWORD: SS Content Updates

HOLT PRESENTATION MAKER
KEYWORD: SK3 PPT21

ONLINE READING SUPPORT
KEYWORD: SS Strategies

CURRENT EVENTS
KEYWORD: S3 Current Events

Meeting Individual Needs

Ability Levels

Level 1 Basic-level activities designed for all students encountering new material

Level 2 Intermediate-level activities designed for average students

Level 3 Challenging activities designed for honors and gifted-and-talented students

English Language Learners Activities that address the needs of students with Limited English Proficiency

Chapter Review and Assessment

E Readings in World Geography, History, and Culture 58, 59, and 60

SM Critical Thinking Activity 21

REV Chapter 21 Review, pp. 492–93

REV Chapter 21 Tutorial for Students, Parents, Mentors, and Peers

ELL Vocabulary Activity 21

A Chapter 21 Test

Chapter 21 Test Generator (on the One-Stop Planner)

Audio CD Program, Chapter 21

A Chapter 21 Test for English Language Learners and Special-Needs Students

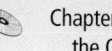

CHAPTER 21

LAUNCH INTO LEARNING

Have students imagine that they are going on safari in Africa. Ask them to imagine what kinds of animals they might see. *(Possible answers: elephants, lions, zebras, wildebeests, hippopotamuses, rhinoceroses)* What kinds of scenery might they see? *(Possible answers: grasslands, mountains in the distance, unusual trees, dry areas, villages with mud-walled houses with grass roofs)* Tell them that this is the view many people have of Africa, but it applies most appropriately to parts of East Africa. Tell students they will learn about East Africa in this chapter.

Section 1

Objectives

1. Describe the major landforms of East Africa.
2. Identify the important rivers and lakes in this region.
3. Describe the climate types and natural resources of East Africa.

LINKS TO OUR LIVES

Point out to students that there are many reasons why they should know more about East Africa. Here are some of them:

- Many scientists believe the human species has its origins in East Africa. Some of the most important fossil remains of humans come from the Olduvai Gorge in Tanzania.
- The United States imports agricultural products from the area.
- The region contains wildlife not found elsewhere. This wildlife is of interest to people around the world.
- Events in the region have influenced the foreign policy of the United States, including the use of armed forces and economic aid.

CHAPTER 21

East Africa

East Africa has been identified by historians as the cradle of the human race. It has made great contributions to the development of world civilization.

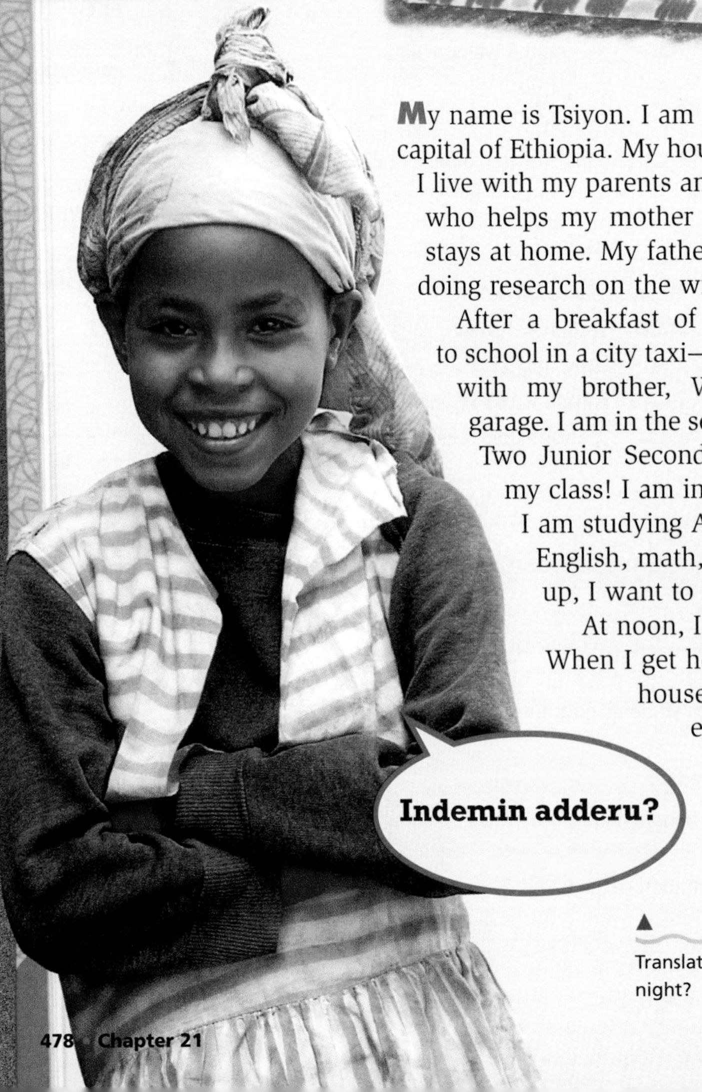

My name is Tsiyon. I am 13, and I live in Addis Ababa, the capital of Ethiopia. My house is very far from the city center. I live with my parents and my brothers. We have a servant who helps my mother with the housework. My mother stays at home. My father works for the game department, doing research on the wild animals of Ethiopia.

After a breakfast of bread and tea with milk, I ride to school in a city taxi—a small blue and white minibus—with my brother, Wondemagegn, who works in a garage. I am in the seventh grade at Freyhewat Number Two Junior Secondary School. There are 82 kids in my class! I am in school from 8:00 A.M. to 3:00 P.M. I am studying Amharic—an Ethiopian language—English, math, science, and sports. When I grow up, I want to be a doctor.

At noon, I eat a lunch I brought from home. When I get home, I help my mother sweep the house. Then I watch television. In the evenings, I do my homework.

Indemin adderu?

▲ Translation: How did you pass the night?

FOCUS

LET'S GET STARTED

Copy the following instructions onto the chalkboard: *Look at the map at the beginning of the chapter and notice the lakes. Why do you think the lakes form a chain?* Discuss responses. Tell students that they are looking at surface evidence of East Africa's rift system. Explain to students that in Section 1 they will learn more about the physical geography of East Africa.

Using the Physical-Political Map

Have students examine the map on this page. Then have them describe the location of the Great Rift Valley *(runs north through Tanzania up to Eritrea)*. Have students explain the effect East Africa's location might have on climate patterns in the region *(wet tropical on the coast, dry in the north and west away from the coast)*. How might highland areas affect climate? *(cooler temperatures, affecting rainfall)*

TEACH

Teaching Objectives 1–2

ALL LEVELS: (Suggested time: 20 min) Copy the following graphic organizer onto the chalkboard, omitting the italicized answers. Organize the class into four groups. Assign each group one of the following topics: the Great Rift Valley, mountains and plains, rivers, or lakes. Have each group use the text and the map to find names or characteristics of the topic. Ask members from each group to come up to the chalkboard and fill in the organizer. **ENGLISH LANGUAGE LEARNERS, COOPERATIVE LEARNING**

Section 1 Physical Geography

Read to Discover

1. What are the major landforms of East Africa?
2. Which rivers and lakes are important in this region?
3. What are East Africa's climate types and natural resources?

Define

rifts

Locate

Great Rift Valley
Mount Kilimanjaro
Lake Victoria
White Nile
Blue Nile

WHY IT MATTERS

Many scientists believe that the first humans on Earth lived on the land that we now call East Africa. Use CNNfyi.com or other **current events** sources to learn about recent discoveries related to the beginnings of humankind. Record your findings in your journal.

A bat-eared fox from Tanzania

East Africa: Physical-Political

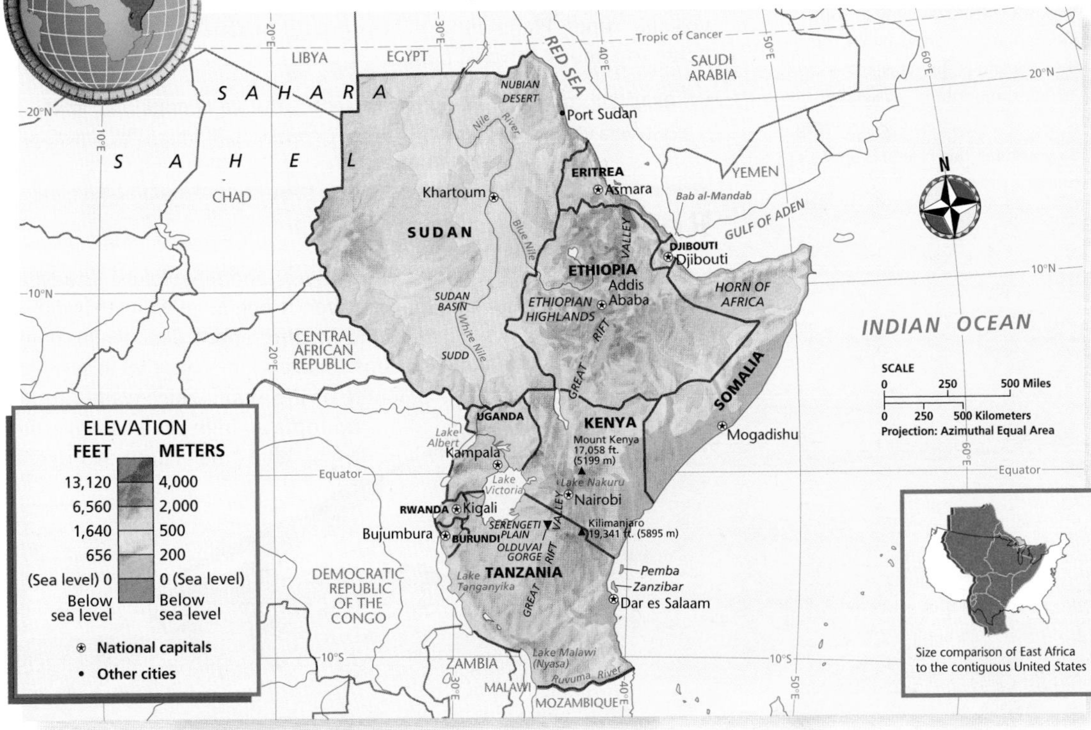

Section 1 RESOURCES

Reproducible
- Block Scheduling Handbook, Chapter 21
- Guided Reading Strategy 21.1
- Geography for Life Activity 21

Technology
- One-Stop Planner CD–ROM, Lesson 21.1
- Homework Practice Online
- Geography and Cultures Visual Resources with Teaching Activities 43–47
- Earth: Forces and Formations CD–ROM/Seek and Tell/Forces and Processes
- HRW Go site

Reinforcement, Review, and Assessment
- Section 1 Review, p. 481
- Daily Quiz 21.1
- Main Idea Activity 21.1
- English Audio Summary 21.1
- Spanish Audio Summary 21.1

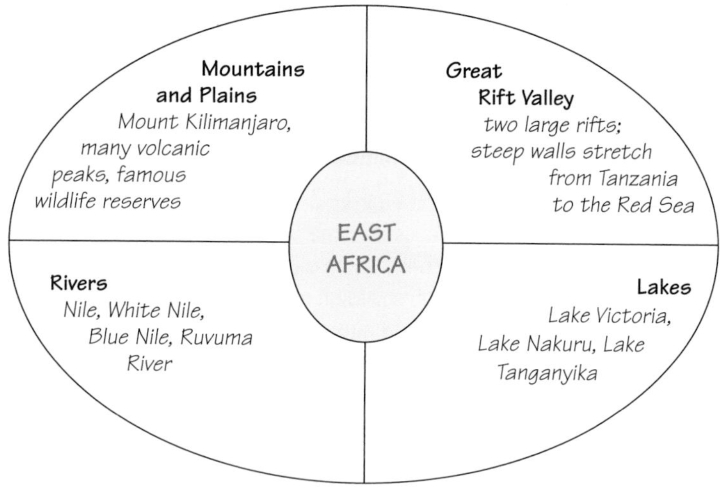

Teaching Objective 3

ALL LEVELS: (Suggested time: 30 min.) Have students sketch an outline map of East Africa. Then ask them to label the climate zones and shade them in different colors. Then have students list the natural resources found in the region. **ENGLISH LANGUAGE LEARNERS**

CLOSE

Point out the relationships between landforms and climates types found in East Africa. Describe how tectonic forces act upon landforms and bodies of water. Then have students recall other places in the world where tectonic forces act upon landforms. *(Possible answers: Iceland, Italy, places along the Pacific Rim, or other areas)*

EYE ON EARTH

Scientists studying Lake Victoria's origins have concluded that its 300 species of cichlids, which are brightly colored fish, have evolved just in the last 12,000 years.

However, the number of cichlids is dwindling. Because of overfishing, habitat loss, and the introduction to the lake of Nile perch, which feed on the cichlids, the number of cichlids has dropped. Because cichlids feed on algae, their destruction contributes to an increase in algae in the lake, which clouds the water and makes it difficult for the fish to find their brightly colored mates.

Critical Thinking: Why has the number of cichlids dwindled?
Answer: because of overfishing, habitat loss, and introduction of the Nile perch

internet connect
GO TO: go.hrw.com
KEYWORD: SK3 CH21
FOR: Web sites about Lake Victoria

Visual Record Answer

because of the many unique physical features and variety of wildlife

internet connect
GO TO: go.hrw.com
KEYWORD: SK3 CH21
FOR: Web sites about East Africa

Our Amazing Planet

Baobab (BOW-bab) trees are one of the few kinds of trees on the African savanna. It is also one of the largest trees in the world. It can grow as large as 30 feet (9 m) in diameter and to a height of 60 feet (18 m). The trunks are often used to store water or as temporary shelters.

The Land

East Africa is a land of high plains and plateaus. In the north, deserts and dry grasslands define the landscape. In the southwest, large lakes dot the plateaus. In the east, sandy beaches and beautiful coral reefs run along the coast. East Africa's most striking features are its great **rifts**. They cut from north to south across the region. Rifts are long, deep valleys with mountains or plateaus on either side. Rifts form when Earth's tectonic plates move away from each other.

The Rift Valleys Deep beneath East Africa's surface, Earth's mantle is churning. This movement causes the land to arch and split along the rift valleys. If you look at the Great Rift Valley from the air, it looks like a giant scar. The Great Rift Valley is made up of two rifts—the eastern rift and the western rift. The rift walls are usually a series of steep cliffs. These cliffs drop an average of about 9,000 feet (2,743 m) to the valley floor. The eastern rift begins north of the Red Sea. The rift continues south through Eritrea (er-uh-TREE-uh) and Ethiopia (ee-thee-OH-pee-uh) into southern Tanzania (tan-zuh-NEE-uh). The western rift extends from Lake Albert in the north to Lake Malawi (mah-LAH-wee), also known as Lake Nyasa, in the south.

Mountains and Plains East Africa also has many volcanic mountains. Mount Kilimanjaro (ki-luh-muhn-JAHR-oh), at 19,341 feet (5,895 m), is Africa's tallest mountain. Although this part of Africa is along the equator, the mountain is so high that snow covers its two volcanic cones. Plains along the eastern rift in Tanzania and Kenya are home to famous national parks.

✓ **READING CHECK:** *Places and Regions* What are the major landforms of East Africa? high plains, plateaus, deserts, grasslands, beaches, rifts, mountains

Rivers and Lakes

East Africa is the site of a number of rivers and large lakes. The Nile is the world's longest river. It begins in East Africa and flows north to the Mediterranean Sea. Water from small streams collects in Lake Victoria, the source of the White Nile. Waters from Ethiopia's highlands form the Blue Nile. These two rivers

This is a crater rim view of Mount Kilimanjaro. Another name for this volcano is Kilima Njaro—"shining mountain" in Swahili.
Interpreting the Visual Record **Why do you think this region is a major tourist attraction?**

Review and Assess

Have students complete the Section Review. Then pair students and distribute six index cards to each pair. Have them use the map and the text to draw a picture of a product, resource, or animal from East Africa on one side of each index card and the name of the region where it is found on the other. Have students exchange their completed cards and quiz themselves. Then have students complete Daily Quiz 21.1. **COOPERATIVE LEARNING**

Reteach

Have students complete Main Idea Activity 21.1. Then have students choose one of the lines of latitude marked on the physical-political map of East Africa. Ask students to write a description of the landforms, bodies of water, climates, and resources they would encounter if they made a journey from the coast of the Red Sea or Indian Ocean west along that line of latitude. Or, have students create a graphic depiction of the same information. **ENGLISH LANGUAGE LEARNERS**

Extend

Have students investigate the search for the origins of the Blue Nile and White Nile. Instruct them to pay particular attention to the difficulties imposed on the explorers by the region's physical geography. **BLOCK SCHEDULING**

meet at Khartoum, Sudan, to create the mighty Nile. The Nile provides a narrow, fertile lifeline through Sudan by providing irrigation in the desert.

Lake Victoria is Africa's largest lake in area, but it is shallow. Along the western rift is a chain of great lakes. Many of the lakes along the drier eastern rift are quite different. Heat from Earth's interior makes some of these eastern lakes so hot that no human can swim in them. Others, like Lake Nakuru, are too salty for most fish. However, algae in Lake Nakuru provides food for more than a million flamingos.

✓ **READING CHECK:** *Places and Regions* Which rivers and lakes are most important in this region? the Nile, Lake Victoria, Lake Nakuru

Lake Nakuru is known in part for the many flamingos that gather there.

Climate and Resources

Northern Sudan and the northeast coast have desert and steppe climates. The climate changes to tropical savanna as you travel south. However, the greatest climate changes occur along the sides of the rift valleys. The rift floors are dry, with grasslands and thorn shrubs. In contrast, the surrounding plateaus and mountains have a humid highland climate and dense forests. Rain falls at the high elevations, but the valleys are in rain shadows.

Most East Africans are farmers or herders. However, the region does have mineral resources such as coal, copper, diamonds, gold, iron ore, and lead.

✓ **READING CHECK:** *Places and Regions* What are East Africa's climate types and natural resources? desert, steppe, tropical savanna, humid highland; coal, copper, diamonds, gold, iron ore, lead

BUILD on WHAT You Know

Do you remember what you learned about rain shadows? See Chapter 3 to review.

go.hrw.com Homework Practice Online
Keyword: SK3 HP21

Section 1 Review

Define and explain: rifts

Working with Sketch Maps On a map of East Africa that you draw or that your teacher provides, label the following: Great Rift Valley, Mount Kilimanjaro, Lake Victoria, White Nile, and Blue Nile. How do the mountains help support the river systems?

Reading for the Main Idea

1. *Places and Regions* What are the major landforms of East Africa?
2. *Places and Regions* Which rivers and lakes are located in this part of Africa?
3. *Physical Systems* Why are volcanic mountains found in parts of East Africa?

Critical Thinking

4. **Drawing Inferences and Conclusions** How do you think the climate types found in East Africa influence what grows there?

Organizing What You Know

5. **Summarizing** Copy the following graphic organizer. Use it to describe what you know about East Africa's physical geography.

	Vegetation	Climates
Coasts		
Rift Valleys		
Plateaus/mountains		

Section 1 Review

Answers

Define For definition, see: rifts, p. 480

Working with Sketch Maps Maps will vary, but listed places should be labeled in their approximate locations. Snow falls in the mountains, melts, and flows into the river.

Reading for the Main Idea

1. high plains, plateaus, mountains, eastern and western rift, volcanic mountains (NGS 4)
2. Lake Victoria, Lake Nakuru, the White Nile, the Blue Nile, and the Nile River (NGS 4)
3. because the land is splitting along the rift valleys (NGS 7)

Critical Thinking

4. The mainly dry climate limits what grows in East Africa.

Organizing What You Know

5. Coasts—little to no vegetation, desert and steppe; rift valleys—grasslands and thorn shrub, dry; mountains/Plateaus—dense forests, humid highland climate

Section 2

Objectives

1. Describe the important events and developments that influenced the history of East Africa.
2. Analyze the culture of East Africa.

Focus

LET'S GET STARTED

Copy the following passage onto the chalkboard: *Karibu! Hujambo? is a common greeting in East Africa. What do you think it means?* Discuss responses. *(The greeting means Welcome! How are you?)* Tell students that in Section 2 they will learn more about the culture and history of East Africa.

Building Vocabulary

Write **Swahili** on the chalkboard. Explain to students that the phrase in the previous activity is in the Swahili language. Swahili is one of many features of the diverse culture of East Africa.

Section 2 RESOURCES

Reproducible

- Guided Reading Strategy 21.2
- Graphic Organizer 21
- Cultures of the World Activity 6
- Creative Strategies for Teaching World Geography, Lesson 16
- Interdisciplinary Activities for the Middle Grades 26, 27, 28

Technology

- One-Stop Planner CD–ROM, Lesson 21.2
- Homework Practice Online
- HRW Go site

Reinforcement, Review, and Assessment

- Section 2 Review, p. 484
- Daily Quiz 21.2
- Main Idea Activity 21.2
- English Audio Summary 21.2
- Spanish Audio Summary 21.2

Section 2 East Africa's History and Culture

Read to Discover

1. What important events and developments influenced the history of East Africa?
2. What is the culture of East Africa like?

Define

Swahili

WHY IT MATTERS

Recent conflicts in East Africa have taken millions of lives. Many of these conflicts are the result of the region's history. Use CNNfyi.com or other **current events** sources to learn more about conflicts in East Africa today. Record your findings in your journal.

A Turkana woman

History

Several early civilizations developed at the site known as Meroë, near where the branches of the Nile come together. These civilizations had their own forms of writing. Each controlled a major trade route. East Africans traded ivory and gold, among other things.

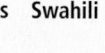

This statue depicts a prince from the early Nubian civilization of northern Sudan.

Christianity and Islam Like Egypt, Ethiopia was an early center of Christianity. In the A.D. 500s Christianity spread into neighboring Nubia, which is now part of Egypt and Sudan. In Nubia, Christian kingdoms lasted until about 1500. Ethiopia still has a large Christian population today.

Arab armies conquered Egypt and North Africa by about A.D. 700. However, these armies were not able to keep control of East Africa. Gradually, Arabic-speaking nomads spread into northern Sudan from Egypt. They brought their Islamic faith with them. At the same time, Islam spread to the coastal region of what is now Somalia. Christianity is believed to have been introduced in Ethiopia as early as the A.D. 300s. Christian kingdoms, particularly in Ethiopia, have fought wars with Muslim leaders. Religion continues to be a source of conflict in this East African region.

The Slave Trade The east coast slave trade dates back more than 1,000 years. Most slaves went to Islamic countries in Africa and Asia. The Portuguese had begun setting up forts and settlements on the East African coast by the early 1500s. At first, the Europeans made little

TEACH

Teaching Objective 1

ALL LEVELS: (Suggested time: 20 min) Copy the following graphic organizer onto the chalkboard, omitting the italicized answers. Organize students into three groups to find information about one of the following influences; religion, the slave trade, or colonization. Have members from each group fill in the appropriate part of the organizer. Then have students speculate on how these historical factors or events have influenced the region's culture today. **ENGLISH LANGUAGE LEARNERS, COOPERATIVE LEARNING**

EAST AFRICA'S HISTORY

Religion	Trade	Colonization
• *Arabic-speaking nomads spread Islamic faith* • *Christianity from Ethiopian kingdoms*	• *slaves traded to Arabia* • *Portuguese start European slave trade*	• *Great Britain* • *Germany* • *Italy annexed Ethiopia*

Teaching Objective 2

ALL LEVELS: (Suggested time: 20 min.) Pair students and have pairs compose a poem or riddle about the culture of East Africa. Call on volunteers to read their composition aloud. **COOPERATIVE LEARNING**

This sketch shows a slave market in Zanzibar. Arab traders took advantage of local rivalries and encouraged powerful African leaders to capture their enemies and sell them into slavery.

Interpreting the Visual Record **What does this piece of art tell us about the historical culture of this region?**

effort to move into the interior. However, in the late 1700s the East African island of Zanzibar became an international slave-trading center. Later, plantations like those of the Americas were set up with slave labor to grow cloves and sugarcane.

Africa Divided In the mid-1800s European adventurers traveled into the African interior searching for the source of the Nile. Here they found rich lands well suited for agriculture. In the 1880s the European powers divided up most of the continent. Most of Africa's modern borders resulted from this process. Control over much of East Africa went to the British. Germany colonized Tanzania, Rwanda, and Burundi. After World War I, with the defeat of Germany, the British took over Tanzania. Belgium gained control of Rwanda and Burundi.

Conflict Within East Africa, just Kenya was settled by large numbers of Europeans. The colonial rulers usually controlled their countries through African deputies. Many of these deputies were traditional chiefs, who often favored their own peoples. This tended to strengthen ethnic rivalries. These ethnic divisions have made it hard for governments to create feelings of national identity.

Independence Ethiopia was never colonized. Its mountains provided natural protection, and its peoples and emperors resisted colonization. It was, however, annexed by Italy from 1936 to 1941. Most East African countries were granted independence by European colonizers in the early 1960s. More recently, East Africa has become headquarters for some international companies and organizations.

✓ **READING CHECK:** *Human Systems* Which European countries influenced the history of East Africa? **Portugal, Britain, Germany, Belgium, Italy**

High Mountains of the World

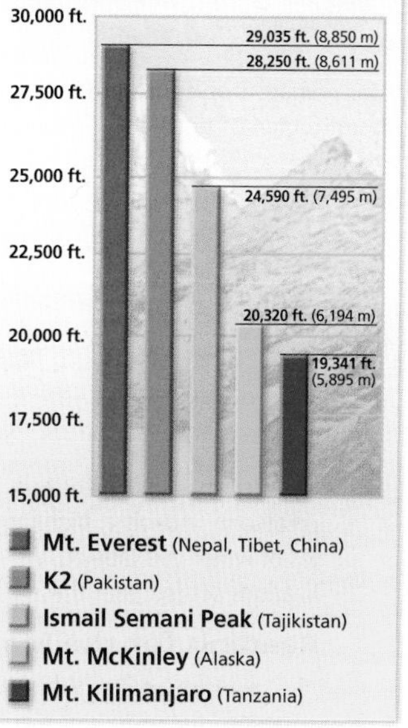

Interpreting the Graph **How much taller is Mt. Everest than K2?**

National Geography Standard 10

Human Systems Trade networks have connected East Africa to other regions for centuries. These trade links apparently extended as far as China. For example, an East African tomb from the 1700s includes Ming dynasty porcelain as part of its decoration. The island of Zanzibar, as a market for the trading of goods, was also a major point from which the Islamic religion and the Swahili language were introduced into the interior of Africa.

Swahili is a good example of the kind of cultural development that characterizes East Africa. The Swahili language is the result of the mixing of Arabic with African languages. Swahili also refers to the new culture that emerged in East Africa.

Critical Thinking: How might trade foster cultural change?

Answer: Students may say that along with goods, foreign merchants would introduce new ideas.

▲ Visual Record Answer

Possible answers: that Arabs and Africans came in contact with each other or that the slave trade existed there

◄ Graph Answer

785 feet (239 m)

Close

Remind students that the history of the region led to many of the conflicts in these countries today. Ask students to describe the history of these challenges and speculate what measures might be taken to solve them.

Review and Assess

Have students complete the Section Review. Then have pairs of students organize the challenges of East Africa into economic, political, religious, and ethnic areas. Have them identify a historical influence that led to each challenge. Then have students complete Daily Quiz 21.2.

Reteach

Have students complete Main Idea Activity 21.2. Then have them draw a time line on a piece of butcher paper. Assign groups of students periods in East African history and have them draw or find pictures to illustrate events. Have students glue their pictures to the time line. Display and discuss the time line. **ENGLISH LANGUAGE LEARNERS**

Extend

Have interested students conduct research on the history of Christianity in Ethiopia. Have them pay particular attention to the churches of Lalibela, which were hewn out of solid rock, starting at ground level. Allow students to present their findings in a written report. **BLOCK SCHEDULING**

Section 2 Review

Answers

Define For definition, see: Swahili, p. 484

Working with Sketch Maps Maps will vary, but bodies of water should be highlighted in their approximate locations. Early civilizations settled there because of the good climate and environment.

Reading for the Main Idea

1. Portugal, Great Britain, and Germany (NGS 9)
2. Christianity and Islam (NGS 10)

Critical Thinking

3. ethnic conflicts, such as that between the Tutsi and the Hutu in Rwanda; religious conflicts
4. Swahili may unify people who also speak other languages.

Organizing What You Know

5. A.D. 400—Sudan conquered and Christianity introduced; 700—Arab nomads introduce Islam; early 1500s—Portuguese establish settlements; 1700s—European slave trade begins; late 1800s—European colonization; early 1960s—independence; 1990s—ethnic conflicts

▲ Afar nomads stop for rest in a desert region of northern Ethiopia.

Culture

East Africa has the world's longest history of human settlement. Gradually, the region developed the continent's greatest diversity of people and ways of life. The **Swahili** language is widely spoken in East Africa. This Bantu language has been greatly influenced by Arabic. In fact, the word Swahili comes from the Arabic word meaning "on the coast."

East Africa shares many challenges with other African regions. One challenge is that populations are growing faster than the economies that support them. Many jobless people have crowded into the cities. Another challenge is religious and ethnic conflict. Such conflict and other political problems have slowed economic progress. Often a country's wealth is spent on weapons rather than helping people.

The ethnic conflicts have roots in the region's history. Colonial borders drawn by Europeans often lumped different ethnic groups into one country. Differences between groups have led to conflict in some countries since independence.

The worst ethnic conflict has been in Rwanda and Burundi. Thousands of Tutsi were killed by the Hutu in Rwanda in the 1990s. There also has been fighting between Muslims and Christians in the northern part of the region.

✓ **READING CHECK:** *Human Systems* What conflicts have occurred between groups in East Africa? ethnic conflicts, such as that between Tutsi and Hutu in Rwanda, and religious conflicts

go.hrw.com Homework Practice Online
Keyword: SK3 HP21

Section 2 Review

Define and explain: Swahili

Working with Sketch Maps On the map you created in Section 1, highlight the location of the bodies of water you labeled. Why do you think early civilizations settled near these bodies of water?

Reading for the Main Idea

1. *Human Systems* Which European countries influenced the history of this area?
2. *Human Systems* What are the main religions practiced in East Africa?

Critical Thinking

3. **Finding the Main Idea** What conflicts have occurred because of political boundary lines?
4. **Drawing Inferences and Conclusions** Why might Swahili be a particularly important language in East Africa?

Organizing What You Know

5. **Sequencing** Create a time line listing historical events in East Africa from the A.D. 400s to the 1990s. Be sure to include the adoption of Christianity, the slave trade, exploration, colonial rule, independence, and ethnic conflict.

A.D. 400 ——————————— 1990s

Section 3

Objectives

1. Explain why settlers came to Kenya.
2. Describe how Tanzania was created.
3. Explain what Rwanda and Burundi are like.
4. Explain what Uganda is like.
5. Describe the physical features of Sudan.

Focus

LET'S GET STARTED

Copy the following instructions onto the chalkboard: *Look at the chart of statistics on the next page. Write down one statement about the African countries listed that you can conclude from the chart.* Discuss responses. *(Possible answers: Burundi has the lowest literacy note. Kenya has the lowest population growth rate and the highest per capita GDP.)* Tell students that in Section 3 they will learn more about the countries of East Africa.

Building Vocabulary

Ask students to describe a canyon. Then have students visualize a narrow canyon with steep sides. Explain to students that they have just visualized a **gorge**. Tell students that a gorge is one of the major landforms of East Africa. Have volunteers explain what they know about the kinds of physical processes that create canyons and gorges.

Section 3 The Countries of East Africa

Read to Discover

1. Why did settlers come to Kenya?
2. How was Tanzania created?
3. What are Rwanda and Burundi like?
4. What is Uganda like?
5. What are the physical features of Sudan?

Define

gorge

Locate

Kenya
Tanzania
Rwanda
Burundi
Uganda
Sudan

Sudanese jar from the A.D. 100s

WHY IT MATTERS

Thousands of tourists enjoy the beautiful wildlife preserves and national parks of Kenya, Uganda, and Tanzania each year. Use CNNfyi.com or other **current events** sources to find out about travel and sightseeing opportunities in these countries. Record your findings in your journal.

Section 3 Resources

Reproducible

- Guided Reading Strategy 21.3
- Map Activity 21
- Creative Strategies for Teaching World Geography, Lesson 17

Technology

- One-Stop Planner CD–ROM, Lesson 21.3
- Homework Practice Online
- HRW Go site

Reinforcement, Review, and Assessment

- Section 3 Review, p. 488
- Daily Quiz 21.3
- Main Idea Activity 21.3
- English Audio Summary 21.3
- Spanish Audio Summary 21.3

Kenya

Kenya's first cities were founded along the coast of the Indian Ocean by Arab traders. Beginning in the 1500s Portugal controlled this coast for about 200 years. Arabs then recaptured it.

During the 1800s British merchants began trading on the coast. They built a railway from Mombasa to Lake Victoria. British settlers then came to take advantage of the fertile highlands. People from India and Pakistan also came to work on the Europeans' farms. Many of the local people, particularly the Kikuyu, moved out of their traditional areas. They became farmworkers or took jobs in the cities. After World War II many Africans protested British colonial rule. There were peaceful demonstrations as well as violent ones.

One conflict was over land. The British and the Kikuyu viewed land differently. The British considered land a sign of personal wealth, power, and property. The Kikuyu saw land as a source of food rather than something to be bought or sold. This caused conflict because the British wanted the land. Kenya gained independence from Britain in the 1960s, and its government has been quite stable ever since.

Kenya is a popular tourist destination. Tourism is a major source of income for the country. Kenya's greatest challenge is its rapidly increasing population. There is no empty farmland left in most areas. Much of Kenya has been set aside as national parkland.

As in the rest of Africa, Europeans once ruled colonies in East Africa. Fort Jesus, founded in 1593 by the Portuguese, is a national monument in Kenya.

TEACH

Teaching Objective 1

ALL LEVELS: (Suggested time: 20 min.) Have students design a mural that identifies the various groups that settled in Kenya and explains the reasons why they came. Ask students to refer to Section 2 for details. (Murals should depict these groups: Arabs—trade, slaves; Portuguese—trade, slaves; British—farming; Indians and Pakistanis—farm work) **ENGLISH LANGUAGE LEARNERS**

Teaching Objective 2

ALL LEVELS: (Suggested time: 10 min.) Have students write a brief news report on the creation of Tanzania. Students should mention how and when Tanzania was created as well as some of the physical features found in that country and how they have influenced its economic development. **ENGLISH LANGUAGE LEARNERS**

➤ASSIGNMENT: Ask students to imagine that they are explorers traveling through East Africa. Have them write three journal entries about the experience. Entries should include a description of the physical features as well as the people and customs of the area.

EYE ON EARTH

The Serengeti Plain provides a good example of a complex food chain. The plain's grasses and other plants are the chain's foundation. Termites, ants, and other insects eat the plants and serve as food for birds and small carnivores. Large herbivorous animals such as wildebeests, zebras, gazelles, and giraffes also feed on the plants. Big predators, including lions, leopards, and cheetahs, feed on the small and large plant-eating animals. Vultures, hyenas, and other scavengers eat what they leave. Beetles help scour the bones.

Wildebeests are a particularly interesting link in the Serengeti food chain. During the dry season they migrate north to greener fields. Thousands of animals go on the run, thundering across the plains. Lions and other predators follow, preying on slow or sick animals that cannot keep up with the herd.

Activity: Have students diagram the food chains described.

Chart Answer ▲

Burundi

East Africa

Country	Population/ Growth Rate	Life Expectancy	Literacy Rate	Per Capita GDP
Burundi	6,223,897 / 2.4%	45, male / 47, female	35%	$720
Kenya	30,765,916 / 1.3%	47, male / 48, female	78%	$1,500
Rwanda	7,312,756 / 1.2%	38, male / 40, female	48%	$900
Sudan	36,080,373 / 2.8%	56, male / 58, female	46%	$1,000
Tanzania	36,232,074 / 2.6%	51, male / 53, female	68%	$710
Uganda	23,985,712 / 2.9%	43, male / 44, female	62%	$1,100
United States	281,421,906 / 0.9%	74, male / 80, female	97%	$36,200

Sources: Central Intelligence Agency, *The World Factbook 2001;* U.S. Census Bureau

Interpreting the Chart **Which country has the lowest literacy rate?**

Many people would like to farm these lands. If the national parks are converted to farmland, however, African wildlife would be endangered. In addition, the tourism industry would likely suffer.

✓ **READING CHECK:** *Human Systems* Why did settlers come to Kenya? for trade, land

Tanzania

South of Kenya is the large country of Tanzania. It was created in the 1960s when Tanganyika and the island of Zanzibar united. Today many tourists come to explore numerous national parks and Mount Kilimanjaro. The mountain's southern slopes are a rich agricultural region that provides coffee and tea for exports. Also in Tanzania is the Serengeti Plain. On this plain, herds of antelope and zebras still migrate freely, following the rains. Nearby is a famous archaeological site, Olduvai Gorge. A **gorge** is a narrow, steep-walled canyon. Evidence of some of the earliest humanlike fossils have been found in Olduvai Gorge.

Tanzania is a country of mainly poor subsistence farmers. Poor soils and limited technology have restricted productivity. This country has minerals, particularly gold and diamonds. Although the Tanzanian government has tried to make the country more self-sufficient, it has not yet succeeded.

✓ **READING CHECK:** *Places and Regions* How was the country of Tanzania created? when Tanganyika and Zanzibar united

These giraffe feed on the tree-studded grasslands of the Serengeti. This reserve was opened in 1974 and is Kenya's most famous and popular animal reserve.

▼

Teaching Objectives 3–5

ALL LEVELS: (Suggested time: 15 min.) Copy the following graphic organizer onto the chalkboard, omitting the italicized answers, to help students compare Rwanda, Burundi, Uganda, and Sudan. Then lead a discussion about the causes of the challenges and possible solutions for them. **ENGLISH LANGUAGE LEARNERS**

	Physical Features	Challenges
Rwanda	*fertile highland*	*violence between ethnic groups*
Burundi	*fertile highland*	*violence between Tutsi and Hutu*
Uganda	*fertile plateau*	*violent dictatorship, collapsed economy*
Sudan	*Sahara, dry savannas, the Sudd*	*conflicts between Arab and traditionally African cultures*

TEACHER TO TEACHER

Susan Walker of Beaufort, South Carolina, suggests the following activity to help students learn about the discoveries at Olduvai Gorge: Have students conduct research on the discoveries and how they were found in the rock layers. Then have them construct a scale model of the site, using modeling clay or salt-flour dough tinted different shades to depict the seven major layers. Have students place small objects in the clay to stand for artifacts and fossils. Finally, demonstrate how tectonic action and erosion exposed the remains by cutting and lifting parts of the clay "landscape."

CONNECTING TO *Science*

Olduvai Gorge, Tanzania

OLDUVAI GORGE

Many scientists believe the human species has its origins in Africa. Archaeologists there have discovered fossil remains of humans and humanlike animals several million years old. Some of the most important finds have occurred at a site known as Olduvai Gorge.

Located in Tanzania, Olduvai is a steep-sided canyon some 30 miles (48 km) long. It is up to 300 feet (90 m) deep. The exposed sides of this gorge contain fossil deposits estimated to be more than 4 million years old. Along with the fossils, scientists found stone tools and the remains of numerous humanlike animals.

Archaeologists Louis and Mary Leakey played a key role in uncovering Olduvai's secrets. In 1931 Louis Leakey found remains of ancient tools and bones in the gorge. Then in 1959 Mary Leakey found the skeleton of an *Australopithecus*, the first humanlike creature to walk upright. Several years later, the Leakeys found the remains of a more advanced species. The new find was known as *Homo habilis*. The species could make stone tools.

These discoveries helped provide some of the missing links between humans and their ancestors. Today archaeological work at Olduvai Gorge continues to add to our understanding of human origins.

Understanding What You Read

1. What is Olduvai Gorge?
2. What part has Olduvai played in the search for human origins?

Rwanda and Burundi

These two countries in fertile highlands were once German colonies. After World War I the Belgians ruled them. In the 1960s, after they gained independence, they were divided into two countries. Both countries are mostly populated by two ethnic groups—the Tutsi and the Hutu. Violence between the groups has killed thousands. Rwanda and Burundi have the densest rural settlement in Africa. Foreign aid has helped improve farming and health care.

✓ **READING CHECK:** *Human Systems* Why did the creation of Rwanda and Burundi create conflict? **Political boundaries were drawn across cultural lines, leading to fighting between the Hutu and the Tutsi.**

Our Amazing Planet

Red colobus monkeys of Zanzibar eat charcoal, which absorbs poisons in the fruit-tree leaves the monkeys sometimes eat.

PHYSICAL SYSTEMS

Tectonic activity of the Olduvai Gorge area created conditions that were good for both the preservation of ancient remains and their eventual discovery.

The Olduvai fossil beds were laid down in a lake basin. Frequent changes in the lake's levels and ashfalls from volcanoes would bury remains quickly, preserving them. Volcanic action and sedimentation created seven major layers. Then relatively constant fault movements and volcanic action cut through the layers of rock and sediment that had accumulated over the millennia. Flowing water further exposed the layers and skeletal remains they contained.

Activity: Have students draw diagrams to depict how skeletal remains were preserved and then exposed in Olduvai Gorge. Call on students to explain their diagrams to the class.

Connecting to Science Answers

1. a steep-sided canyon in Tanzania
2. Remains of ancient tools and skeletons were found in the exposed sides of the canyon.

Close

Tell students that the locations and landforms of East African countries greatly affected their history and the challenges they face today. Have students summarize the region's challenges. *(Possible answers: lack of resources, political unrest, ethnic conflict, religious conflict, economic development)*

Review and Assess

Have students complete the Section Review. Then have each student write a short-answer question for each of the six countries in Section 3. Collect the questions and use them to quiz the class orally.

Reteach

Have students complete Main Idea Activity 21.3. Then draw an outline map of this region on butcher paper. Organize students into five groups and assign each group a country. Have students write the resources and challenges of their assigned country on small slips of paper. Have students label the countries and glue their strips of paper onto the map. **ENGLISH LANGUAGE LEARNERS, COOPERATIVE LEARNING**

Extend

Have interested students investigate wildlife conservation programs in East Africa. Ask them to concentrate on government efforts to balance economic progress, tourism, and conservation. **BLOCK SCHEDULING**

Section 3 Review

Answers

Define For definition, see: gorge, p. 486

Working with Sketch Maps Maps will vary, but listed places should be labeled in their approximate locations. The agriculturally fertile countries are located in different climate zones than the others.

Reading for the Main Idea

1. good farmland, refuge for wildlife, important tourist destination (NGS 4)
2. Kenya's wildlife refuges, Mount Kilimanjaro, the Serengeti Plain, and the Olduvai Gorge (NGS 4)

Critical Thinking

3. provides a water supply and makes it possible to farm
4. has taken thousands of lives, limited foreign investment, and slowed economic development

Organizing What You Know

5. Kenya—fertile farmland, growing population; Tanzania—gold and diamonds, limited farming productivity; Rwanda and Burundi—fertile land, ethnic conflicts; Uganda—fertile farmland, collapsed economy; Sudan—minerals and oil, civil wars and religious conflicts

Uganda

Uganda, another site of an ancient empire, is found on the plateau north and west of Lake Victoria. Economic progress has been slow. Foreign investment stopped as a result of a violent dictatorship. In the 1970s the country's economy collapsed. Limited peace and democracy were achieved in the late 1980s.

✓ **READING CHECK:** *Human Systems* What is Uganda like today? relatively peaceful and democratic

▲ Alfred Louis Sargent created this engraving of Khartoum, Sudan, in the 1800s.

Sudan

Sudan is Africa's largest country. It has three physical regions. The Sahara makes up the northern half of the country. Dry savannas extend across the country's center. Much of southern Sudan is taken up by a swamp called the Sudd. Sudan is mainly an agricultural country, but it is also developing some of its mineral resources. Oil reserves have not yet been developed.

Modern Sudanese culture shows influences of Arab and traditionally African cultures. Arab Muslims make up about 40 percent of the population and have political power. They dominate northern Sudan. Khartoum, the capital, is located in this area. During the last several decades there has been a civil war between northern Muslims and southerners who practice Christianity or traditional African religions.

✓ **READING CHECK:** *Human Systems* What conflict has been occurring in Sudan? civil war between different religious groups

go.hrw.com Homework Practice Online
Keyword: SK3 HP21

Section 3 Review

Define and explain: gorge

Working with Sketch Maps On the map you created in Section 2, label Kenya, Tanzania, Rwanda, Burundi, Uganda, and Sudan. Why are some of the countries of this region agriculturally fertile while others are not?

Reading for the Main Idea

1. *Places and Regions* What makes the highlands of Kenya important?
2. *Places and Regions* Which areas of East Africa are tourist attractions?

Critical Thinking

3. **Making Generalizations and Predictions** How might irrigation help a region's economic development?
4. **Finding the Main Idea** How has unrest hurt Rwanda and some other East African countries?

Organizing What You Know

5. **Summarizing** Copy the following graphic organizer. Use it to list each country in this section. In one column list resources important to its development, and in the next column list obstacles that could prevent the country's economic success.

Country	Resources	Obstacles

Section 4

Objectives

1. Describe the main physical features of Ethiopia.
2. Describe what Eritrea is like.
3. Explain what Somalia is like.
4. Explain the physical and cultural characteristics of Djibouti.

FOCUS

LET'S GET STARTED

Copy the following instructions onto the chalkboard: *Look at the photos in Section 4 and read the captions. Write down one question that you have about what you see in the photos.* Discuss responses. Ask students to keep their questions handy and try to answer them as they read Section 4. Tell students they will learn more about the countries that occupy the Horn of Africa in Section 4.

Building Vocabulary

Ask students what is the longest period without rain that they can remember. Discuss responses. Tell students a **drought** is a period of little rain. Mention that droughts have affected many people who live in the Horn of Africa.

Section 4 The Horn of Africa

Read to Discover

1. What are the main physical features of Ethiopia?
2. What is Eritrea like?
3. What is Somalia like?
4. What are the physical and cultural characteristics of Djibouti?

Define

droughts

Locate

Ethiopia
Eritrea
Somalia
Djibouti
Bab al-Mandab

WHY IT MATTERS

The United States sent food aid and 15,000 troops to Somalia in the 1990s to try to bring peace to the war-torn nation. Use CNNfyi.com or other **current events** sources to investigate current conditions in Somalia. Record your findings in your journal.

A classic Swahili sailing dhow

Section 4 RESOURCES

Reproducible

- Guided Reading Strategy 21.4

Technology

- One-Stop Planner CD–ROM, Lesson 21.4
- Homework Practice Online
- HRW Go site

Reinforcement, Review, and Assessment

- Section 4 Review, p. 491
- Daily Quiz 21.4
- Main Idea Activity 21.4
- English Audio Summary 21.4
- Spanish Audio Summary 21.4

Ethiopia

Ethiopia is one of the world's poorest countries. The rugged mountain slopes and upland plateaus have rich volcanic soil. Agriculture is Ethiopia's chief economic activity. It exports coffee, livestock, and oilseeds. However, during the last 30 years the region has experienced serious **droughts**. Droughts are periods when little rain falls and crops are damaged. Aid has been sent from around the world. Nevertheless, several million people starved during the 1980s.

Except for a time when Ethiopia was at war with Italy, the Ethiopian highlands have never been under foreign rule. The mountains protected the interior of the country from invasion. Most of the highland people are Christian, while most of the lowland people are Muslim.

READING CHECK: *Places and Regions* What physical features are found in Ethiopia? mountains, plateaus

As Afar nomads in Ethiopia move their encampment, they are continually challenged by the environment.
Interpreting the Visual Record How important are these camels for the Afar? What purpose do they serve?

◄ **Visual Record Answer**

probably very important; they make survival in the harsh environment possible

TEACH

Teaching Objectives 1–4

ALL LEVELS: (Suggested time: 25 min.) Copy the following graphic organizer onto the chalkboard, omitting the italicized answers. Have students copy and complete it. Then ask for volunteers to provide answers for the graphic on the chalkboard. Use the completed graphic organizer to discuss similarities and differences among these societies.
ENGLISH LANGUAGE LEARNERS, COOPERATIVE LEARNING

Country	Location or Physical Features	Key Characteristics
Ethiopia	*on Red Sea, mountains and plateaus*	*agricultural, poor, suffered droughts*
Eritrea	*on Red Sea*	*independent since 1993, improving economy*
Somalia	*deserts and dry savannas*	*mainly herders, less diverse culture, troubled by civil war and drought*
Djibouti	*on Bab al-Mandab, partly below sea level*	*independent since 1977, port provides income, depends on Ethiopian agriculture*

Section 4 Review

Answers

Define For definition, see: droughts, p. 489

Working with Sketch Maps Maps will vary, but listed places should be labeled in their approximate locations. The rich volcanic soil of Ethiopia and the proximity of the Indian Ocean help some of the economies.

Reading for the Main Idea

1. mountains (NGS 4)
2. Eritrea (NGS 4)

Critical Thinking

3. Djibouti's location on the Bab al-Mandab strait makes it important.
4. Massive droughts and civil wars have prompted foreign aid.

Organizing What You Know

5. Ethiopia—Christianity and Islam; Eritrea—Christianity and Islam; Somalia—Islam; Djibouti—Islam

Chart Answer

Djibouti; highest literacy rate and per capital GDP

Visual Record Answer

Possible answers: arches, domes, minarets

The Horn of Africa

Country	Population/ Growth Rate	Life Expectancy	Literacy Rate	Per Capita GDP
Djibouti	460,700 2.6%	49, male 53, female	46%	$1,300
Eritrea	4,298,269 3.8%	54, male 59, female	25%	$710
Ethiopia	65,891,874 2.7%	44, male 46, female	36%	$600
Somalia	7,448,773 3.5%	45, male 48, female	24%	$600
United States	281,421,906 0.9%	74, male 80, female	97%	$36,200

Sources: Central Intelligence Agency, *The World Factbook 2001;* U.S. Census Bureau

Interpreting the Chart **Which country in the region has the highest level of economic development, and why?**

Eritrea

Eritrea, located on the Red Sea, was once part of Ethiopia. In the late 1800s the Italians made this area a colony. In the 1960s it became an Ethiopian province. After years of war, Eritrea broke away from Ethiopia in 1993. The economy has slowly improved since then. The population is made up of Muslims and Christians.

✓ **READING CHECK:** *Places and Regions* What is Eritrea like today? improved economy; population made up of Muslims and Christians

Somalia

Somalia is a land of deserts and dry savannas. Most Somalis are nomadic herders. Livestock and bananas are the main exports. Somalia is less diverse than most other African countries. Most residents of Somalia are members of the Somali people. Most Somali share the same culture, religion (Islam), language (Somali), and way of life (herding). Somalia has been troubled by civil war. In the 1990s widespread starvation caused by the war and a severe drought attracted international attention. The United Nations sent aid and troops to the country. U.S. troops were sent to Somalia to assist with this operation.

✓ **READING CHECK:** *Environment and Society* How have drought and conflict affected Somalia? caused widespread starvation

The main mosque, Khulafa el Rashidin, was built in 1937 with Italian Carrara marble in Asmara, Eritrea.

Interpreting the Visual Record **What architectural elements of this building have you seen in other regions you have studied?**

CLOSE

Lead a discussion about the roles the United States has played or could play in this region. *(Possible answers: has provided economic and military aid; could continue to supply aid)*

REVIEW AND ASSESS

Have students complete the Section Review. Organize the class into small groups. Have each group create a 10-term crossword puzzle about the Horn of Africa. Photocopy the puzzles and have groups exchange and complete them. Then have students complete Daily Quiz 21.4. **COOPERATIVE LEARNING**

RETEACH

Have students complete Main Idea Activity 21.4. Then have students list five facts for each of the countries in Section 4. **ENGLISH LANGUAGE LEARNERS**

EXTEND

Have interested students examine the roles that strategic waterways, such as the Bab al-Mandab, have had on the history and economic development of countries where they are located. Tell them to compare the Bab al-Mandab, the Strait of Hormuz, and the Strait of Gibraltar and present their findings in chart form. **BLOCK SCHEDULING**

▲ Djibouti's Lake Assal has one of the lowest surface levels on the planet. It lies 515 feet (157 m) below sea level. The only way to reach this area is by use of a four-wheel drive vehicle.

Interpreting the Visual Record **What about the surrounding physical geography would appear to limit access to this lake?**

Djibouti

Djibouti is a small desert country. It lies on the Bab al-Mandab. This the narrow strait that connects the Red Sea and the Indian Ocean. The strait lies along a major shipping route. This has helped Djibouti's economy. In the 1860s Djibouti came under French control. It gained independence in 1977. The French government still contributes economic and military support to the country. Its port, which serves landlocked Ethiopia, is a major source of income. Djibouti is heavily dependent on food imports.

The people of Djibouti include the Issa and the Afar. The Issa are closely tied to the people of Somalia. The Afar are related to the people of Ethiopia. Members of both groups are Muslim. Somalia and Ethiopia have both wanted to control Djibouti. So far the country has maintained its independence.

✓ **READING CHECK:** *Places and Regions* What are Djibouti's physical and cultural features? desert country on a strait; Issa and Afar peoples

go.hrw.com Homework Practice Online
Keyword: SK3 HP21

Section 4 Review

Define and explain: droughts

Working with Sketch Maps On the map you created in Section 3, label Ethiopia, Eritrea, Somalia, Djibouti, and Bab al-Mandab. What physical features help the economies of the region?

Reading for the Main Idea

1. *Places and Regions* What physical features have helped protect Ethiopia from foreign invasion?
2. *Places and Regions* What country was part of Ethiopia until it broke away in 1993?

Critical Thinking

3. **Drawing Inferences and Conclusions** Why do you think France remains interested in Djibouti?
4. **Finding the Main Idea** Why have foreign aid agencies been involved in East Africa?

Organizing What You Know

5. **Analyzing Information** Copy the following graphic organizer. Use it to show the major religions of these countries. Add boxes as needed.

Country	Religion

CHAPTER 21 Review ANSWERS

Building Vocabulary

For definitions, see: rifts, p. 480; Swahili, p. 484; gorge, p. 486; droughts, p. 489

Reviewing the Main Ideas

1. fertile land, coal, copper, diamonds, and iron ore (NGS 4)
2. movement of two tectonic plates away from each other (NGS 7)
3. They all controlled it. (NGS 9)
4. poor soil in some areas and limited technology (NGS 15)
5. Droughts caused millions to starve, but aid from foreign countries has been received. (NGS 15)

Understanding Environment and Society

Answers will vary, but the information included should be consistent with text material. Students may find that there have been conflicts over preservation of wildlife refuges versus development.

◄ **Visual Record Answer**

mountains and rough desert

Assess

Have students complete the Chapter 21 Test.

Reteach

Organize the class into nine groups. Assign each group a country. Have them write three descriptive statements about their country on one piece of paper and draw its flag on another piece of paper. Distribute flags to different groups. Collect the statements and read samples aloud. As you read the statements, have the group with the flag of that country hold it up as their answer. **COOPERATIVE LEARNING**

PORTFOLIO

Portfolio Extensions

1. Have students create dioramas of East Africa's remarkable ecosystems, such as the Sudd, the Serengeti, Lake Nakuru, or Kilimanjaro. Have them depict physical features, vegetation, animal life, and whatever makes the place unique. Photograph dioramas for placing in student portfolios.
2. Have students conduct research on the ancient kingdoms of Ethiopia and Sudan. Instruct them to create posters that depict what structures remain from those cultures and what they may have looked like when first built.

Chapter 21

Review ANSWERS

Use Rubric 29, Presentations, to evaluate student work.

Thinking Critically

1. provided trade routes and the location of ancient civilizations
2. Unstable governments and constant fighting have kept many foreign investors away.
3. Answers will vary, but students should note the limiting factors of the region's physical features and the distribution of water and other natural resources.
4. ethnic conflict; students might mention either allowing ethnic groups to relocate to be united or changing political boundaries to unite ethnic groups.
5. The mountains protected the interior from invasion.

Map Activity

A. Blue Nile
B. Bab al-Mandab
C. Mount Kilimanjaro
D. Great Rift Valley
E. Lake Victoria
F. White Nile

Chapter 21

Reviewing What You Know

Building Vocabulary

On a separate sheet of paper, write sentences to define each of the following words.

1. rifts
2. Swahili
3. gorge
4. droughts

Reviewing the Main Ideas

1. *Places and Regions* What are East Africa's main natural resources?
2. *Physical Systems* What caused the formation of the Great Rift Valley of East Africa?
3. *Human Systems* How have Arabs, Portuguese, the British, and the Kikuyu affected Kenya?
4. *Environment and Society* What factors have slowed Tanzania's economic growth?
5. *Environment and Society* How have droughts in the Horn of Africa affected its people and their relationship with the rest of the world?

Understanding Environment and Society

Environmental Issues

Many East African countries have endured ethnic, religious, and political conflicts. Sometimes environmental issues can cause conflict. Prepare a presentation, including a graph, database, or model, comparing conflicts in East Africa over environmental issues. You may want to think about the following:

- Specific conflicts about environmental issues.
- Steps East Africans might take to eliminate these conflicts.

You may want to compare your data to facts you have learned about other regions in Africa for more ideas.

Thinking Critically

1. **Finding the Main Idea** What is the significance of the Nile River in this region's human history?
2. **Analyzing Information** Why are foreign investors hesitant to invest in many countries of this region?
3. **Summarizing** Why do most East Africans make their living by farming and herding?
4. **Making Generalizations and Predictions** What problems have resulted in part from the boundary lines drawn by European powers, and how might the countries of East Africa overcome these problems?
5. **Analyzing Information** How has Ethiopia avoided falling under foreign rule for most of its history?

Food Festival

The staple of Ethiopian cuisine is a sourdough pancake called *injera.* Ethiopians break off pieces of it with the right hand and use it to scoop up bits of spicy stew. Have students bring stew and use pita bread as the *injera.* Students can scoop up bits of stew with the pieces of bread. Or, challenge students to locate recipes for *injera,* make it at home, and bring samples to class.

Chapter 21 Review and Assessment Resources

Reproducible
- Readings in World Geography, History, and Culture 58, 59, and 60
- Critical Thinking Activity 21
- Vocabulary Activity 21

Technology
- Chapter 21 Test Generator (on the One-Stop Planner)
- HRW Go site
- Audio CD Program, Chapter 21

Reinforcement, Review, and Assessment
- Chapter 21 Review, pp. 492–93
- Chapter 21 Tutorial for Students, Parents, Mentors, and Peers
- Chapter 21 Test
- Chapter 21 Test for English Language Learners and Special-Needs Students

Building Social Studies Skills

Map Activity

On a separate sheet of paper, match the letters on the map with their correct labels.

Great Rift Valley
Mount Kilimanjaro
Lake Victoria
White Nile
Blue Nile
Bab al-Mandab

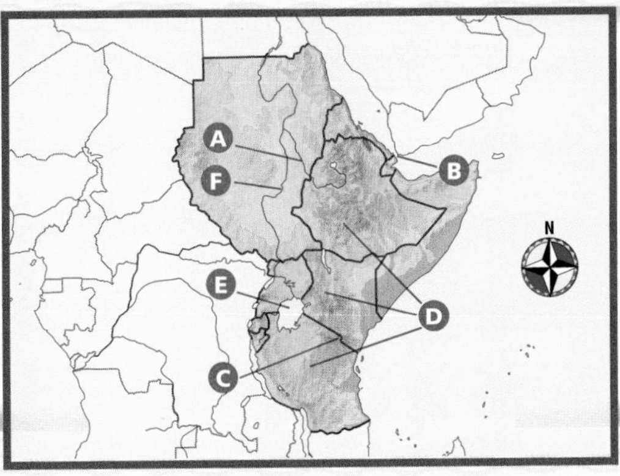

Mental Mapping Skills Activity

On a separate sheet of paper, draw a freehand map of East Africa. Make a key for your map and label the following:

Burundi
Djibouti
Eritrea
Ethiopia
Kenya
Rwanda
Somalia
Sudan
Tanzania
Uganda

Writing Activity

Imagine that you have been awarded an all-expenses-paid vacation to East Africa. Write a letter to your travel agent explaining what you want to do on your trip. List and describe physical features you want to see and African wildlife and historical sites you would like to explore. Be sure to use standard grammar, spelling, sentence structure, and punctuation.

Alternative Assessment

Portfolio Activity

Learning About Your Local Geography

Cooperative Project Much of Africa's precolonial history was memorized by elders, rather than written down. Prepare an oral history of how older people in your community recall a certain historical event.

internet connect

Internet Activity: go.hrw.com
KEYWORD: SK3 GT21

Choose a topic to explore about East Africa:
- Hike Mount Kilimanjaro.
- Learn about cultural groups in East Africa.
- Travel back to ancient Nubian kingdoms.

Mental Mapping Skills Activity

Maps will vary, but listed places should be labeled in their approximate locations.

Writing Activity

Answers will vary, but the information included should be consistent with text material. Students' itineraries should include physical features found in the region, historical sites, and African wildlife. Use Rubric 42, Writing to Inform, to evaluate student work.

Portfolio Activity

Answers will vary, but groups should develop a list of questions, and their oral histories should be about a historical event and be recorded from the original source. Use Rubric 14, Group Activity, to evaluate student work.

internet connect

GO TO: go.hrw.com
KEYWORD: SK3 Teacher
FOR: a guide to using the Internet in your classroom

CHAPTER 22 Central Africa
Chapter Resource Manager

Objectives	Pacing Guide	Reproducible Resources
SECTION 1 **Physical Geography** (pp. 495–97) **1.** Identify the major physical features of central Africa. **2.** Identify the climates, plants, and animals found in the region. **3.** Identify the major natural resources of central Africa.	**Regular** .5 day **Block Scheduling** .5 day *Block Scheduling Handbook, Chapter 22*	**RS** Guided Reading Strategy 22.1 **SM** Map Activity 22
SECTION 2 **History and Culture** (pp. 498–501) **1.** Describe the history of central Africa and the challenges people there face today. **2.** Describe the people and cultures of central Africa.	**Regular** 1 day **Block Scheduling** .5 day *Block Scheduling Handbook, Chapter 22*	**RS** Guided Reading Strategy 22.2 **RS** Graphic Organizer 22 **E** Cultures of the World Activity 6 **E** Creative Strategies for Teaching World Geography, Lesson 16 **IC** Interdisciplinary Activities for the Middle Grades 26, 27, 28
SECTION 3 **The Democratic Republic of the Congo** (pp. 502–04) **1.** Describe the history of the Democratic Republic of the Congo. **2.** Describe the people and culture of the country. **3.** Describe the economy of the country.	**Regular** 1 day **Block Scheduling** .5 day *Block Scheduling Handbook, Chapter 22*	**RS** Guided Reading Strategy 22.3 **SM** Geography for Life Activity 22
SECTION 4 **The Other Central African Countries** (pp. 505–07) **1.** Describe the people and economies of the northern central African countries. **2.** Describe the people and economies of the southern central African countries.	**Regular** .5 day **Block Scheduling** .5 day *Block Scheduling Handbook, Chapter 22*	**RS** Guided Reading Strategy 22.4

Chapter Resource Key

- **RS** Reading Support
- **IC** Interdisciplinary Connections
- **E** Enrichment
- **SM** Skills Mastery
- **A** Assessment
- **REV** Review
- **ELL** Reinforcement and English Language Learners
- Transparencies
- CD–ROM
- Music
- Video
- go.hrw.com Internet
- Holt Presentation Maker Using Microsoft® Powerpoint®

One-Stop Planner CD–ROM

See the *One-Stop Planner* for a complete list of additional resources for students and teachers.

One-Stop Planner CD–ROM

It's easy to plan lessons, select resources, and print out materials for your students when you use the ***One-Stop Planner CD–ROM with Test Generator.***

Technology Resources	Review, Reinforcement, and Assessment Resources
One-Stop Planner CD–ROM, Lesson 22.1 Geography and Cultures Visual Resources with Teaching Activities 43–47 Homework Practice Online HRW Go site	**ELL** Main Idea Activity 22.1 **REV** Section 1 Review, p. 497 **A** Daily Quiz 22.1 **ELL** English Audio Summary 22.1 **ELL** Spanish Audio Summary 22.1
One-Stop Planner CD–ROM, Lesson 22.2 Homework Practice Online HRW Go site	**ELL** Main Idea Activity 22.2 **REV** Section 2 Review, p. 501 **A** Daily Quiz 22.2 **ELL** English Audio Summary 22.2 **ELL** Spanish Audio Summary 22.2
One-Stop Planner CD–ROM, Lesson 22.3 Homework Practice Online HRW Go site	**ELL** Main Idea Activity 22.3 **REV** Section 3 Review, p. 504 **A** Daily Quiz 22.3 **ELL** English Audio Summary 22.3 **ELL** Spanish Audio Summary 22.3
One-Stop Planner CD–ROM, Lesson 22.4 Homework Practice Online HRW Go site	**ELL** Main Idea Activity 22.4 **REV** Section 4 Review, p. 507 **A** Daily Quiz 22.4 **ELL** English Audio Summary 22.4 **ELL** Spanish Audio Summary 22.4

internet connect

HRW ONLINE RESOURCES

GO TO: go.hrw.com
Then type in a keyword.

TEACHER HOME PAGE
KEYWORD: SK3 TEACHER

CHAPTER INTERNET ACTIVITIES
KEYWORD: SK3 GT22

Choose an activity to:
- identify threats to the Congo Basin.
- visit the people of central Africa.
- learn the importance of African cloth.

CHAPTER ENRICHMENT LINKS
KEYWORD: SK3 CH22

CHAPTER MAPS
KEYWORD: SK3 MAPS22

ONLINE ASSESSMENT
Homework Practice
KEYWORD: SK3 HP22
Standardized Test Prep Online
KEYWORD: SK3 STP22
Rubrics
KEYWORD: SS Rubrics

COUNTRY INFORMATION
KEYWORD: SK3 Almanac

CONTENT UPDATES
KEYWORD: SS Content Updates

HOLT PRESENTATION MAKER
KEYWORD: SK3 PPT22

ONLINE READING SUPPORT
KEYWORD: SS Strategies

CURRENT EVENTS
KEYWORD: S3 Current Events

Chapter Review and Assessment

E Readings in World Geography, History, and Culture 61, 62, and 63
SM Critical Thinking Activity 22
REV Chapter 22 Review, pp. 508–09
REV Chapter 22 Tutorial for Students, Parents, Mentors, and Peers
ELL Vocabulary Activity 22
A Chapter 22 Test
Chapter 22 Test Generator (on the One-Stop Planner)
Audio CD Program, Chapter 22
A Chapter 22 Test for English Language Learners and Special-Needs Students

Meeting Individual Needs

Ability Levels

Level 1 Basic-level activities designed for all students encountering new material

Level 2 Intermediate-level activities designed for average students

Level 3 Challenging activities designed for honors and gifted-and-talented students

English Language Learners Activities that address the needs of students with Limited English Proficiency

CHAPTER 22

Launch into Learning

Distribute green paper to students. Have them cut a heart shape out of the paper. Explain that they are going to study the heart of Africa. Ask students to speculate about why green was the color chosen. *(The region contains dense rain forests and jungles.)* Tell them that the region has other climate and vegetation types as well as tropical rain forests. Ask what colors they might choose for a tropical savanna climate and forested highlands. Instruct students to keep their green hearts for a later activity. Tell students they will learn more about central Africa in this chapter.

Section 1

Objectives

1. Identify the major physical features of central Africa.
2. Describe the climates, plants, and animals found in the region.
3. Identify the major natural resources of central Africa.

Links to Our Lives

There are many reasons why students in the United States should know about the physical and human geography of central Africa, these among them:

- Events in the region have affected U.S. foreign policy in recent decades.
- The region contains rich deposits of minerals that are useful to people throughout the world.
- Central Africa is home to many exotic plant and animal species not found elsewhere in the world.
- The tropical rain forest of central Africa plays a significant role in influencing the world's climate.

CHAPTER 22

Central Africa

The fourth region in Africa we will study includes the 10 countries of central Africa. Before we begin, meet Akalemwa Ngenda, a student living in Zambia.

Bwanji? (How are you?) This is how you say *hi* in Nyanja, a common language spoken by most people in my country, Zambia. At home, I would say "Mucwañi?" to my father in Lozi, my own language. My name is Akalemwa, and I am 18. *Akalemwa* means "the one you cannot outrun." I am in the twelfth grade at Choma Secondary School in Zambia's southern province. Choma is a small town. My family's house is near the hospital, about 2.5 kilometers (1.6 mi.) from town. My father is a medical assistant in charge of village health clinics for our region. My mother is a nurse. I have an older half sister, two older brothers, two younger sisters, and five younger brothers. Our house has three bedrooms and a vegetable garden. I attend a Protestant missionary school. We sleep in a dormitory at school. Our school year has three three-month terms, starting in January. We have a month vacation after each one. I want to be a doctor, so I am studying math, physics, chemistry, biology, English, and geography.

Lumela, ni muituti mwa naha Zambia.

▲ Translation: Hi, I am a student in Zambia.

FOCUS

LET'S GET STARTED

Copy the following instructions onto the chalkboard: *Look at the physical-political map of central Africa. Work with a partner to select a country, a landform, and a river on the map. Then describe the relative locations of the items you and your partner selected.* Allow students time to write their responses. Discuss the relative locations students provide. Then tell students that in Section 1 they will learn more about the importance of the features they described.

Using the Physical-Political Map

Have students examine the map on this page. Ask them to describe the main physical features of the region. Their descriptions should include rivers, plains, and mountains. Draw students' attention to the Congo River, which is a vital transportation route of central Africa.

Section 1 Physical Geography

Read to Discover

1. What are the major physical features of central Africa?
2. What climates, plants, and animals are found in the region?
3. What major natural resources does central Africa have?

Define

basin
canopy
copper belt
periodic markets

Locate

Congo Basin
Western Rift Valley
Lake Tanganyika
Lake Malawi
Congo River
Zambezi River

WHY IT MATTERS

The rain forests of Central Africa are home to many plant and animal species. Use CNNfyi.com or other **current events** sources to find information about rain forests around the world. Record your findings in your journal.

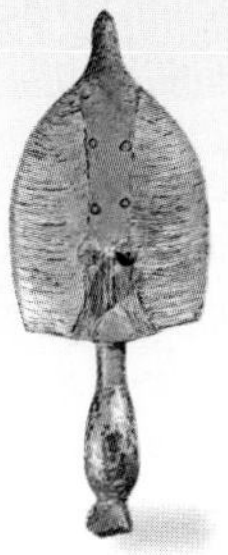

A copper and wood religious artifact from Gabon

Central Africa: Physical-Political

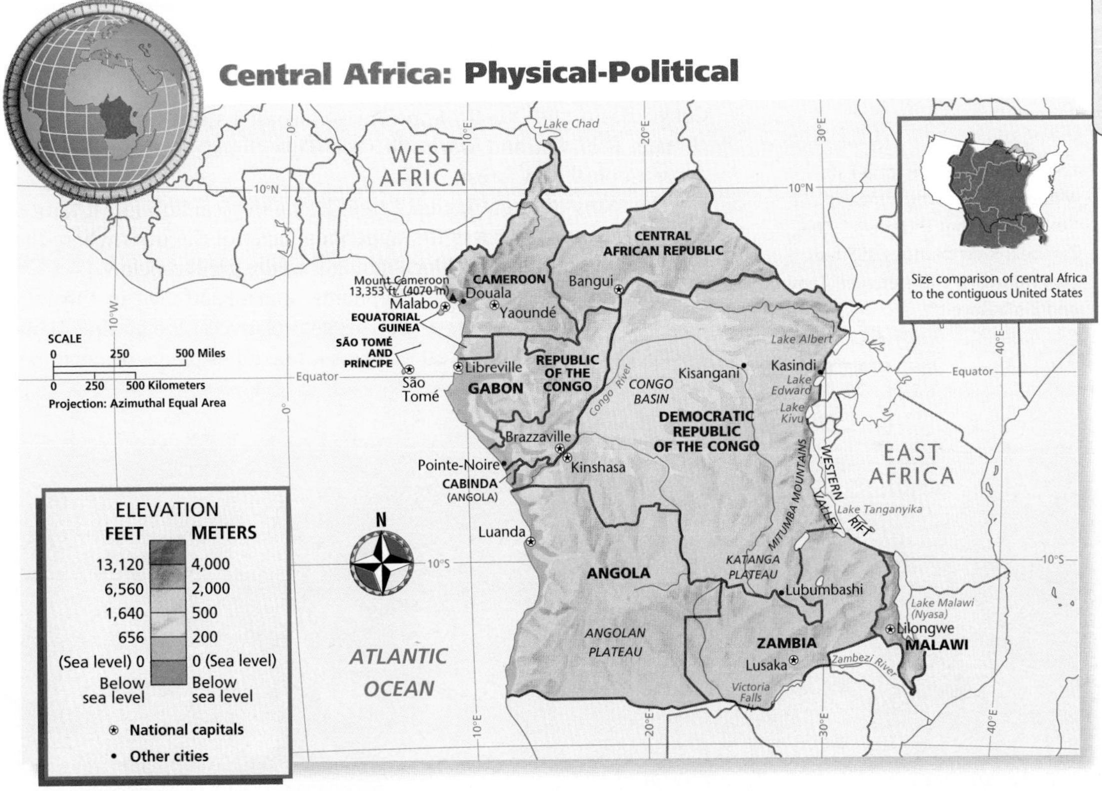

Section 1 RESOURCES

Reproducible

- Block Scheduling Handbook, Chapter 22
- Guided Reading Strategy 22.1
- Map Activity 22

Technology

- One-Stop Planner CD–ROM, Lesson 22.1
- Homework Practice Online
- Geography and Cultures Visual Resources with Teaching Activities 43–47
- HRW Go site

Reinforcement, Review, and Assessment

- Section 1 Review, p. 497
- Daily Quiz 22.1
- Main Idea Activity 22.1
- English Audio Summary 22.1
- Spanish Audio Summary 22.1

TEACH

Teaching Objectives 1–3

ALL LEVELS: (Suggested time: 20 min.) Copy the following graphic organizer onto the chalkboard, omitting the italicized answers. Use it to help students describe central Africa's physical geography. Call on volunteers to fill in the ovals with appropriate words and phrases for each category. Have students use this unit's atlas also.
ENGLISH LANGUAGE LEARNERS

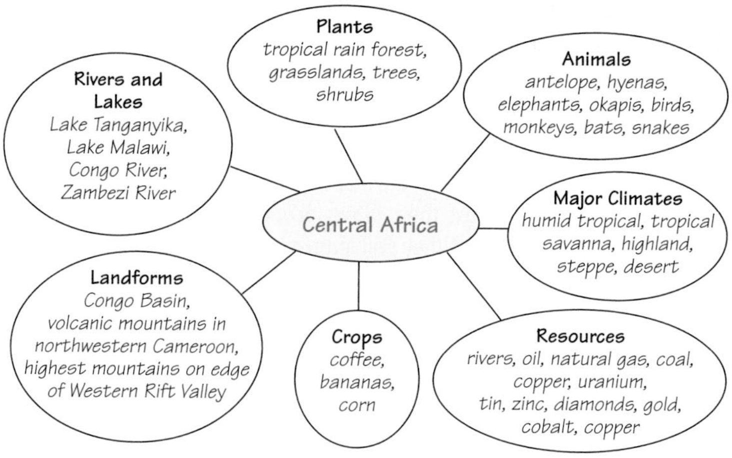

EYE ON EARTH

The climate of central Africa is affected by a belt of converging trade winds that encircle Earth near the equator. Known as the Intertropical Convergence Zone, this belt shifts north and south seasonally as our planet's position relative to the Sun changes during Earth's annual revolution.

Unstable moist tropical air is forced upward where it is cooled and condensed to produce heavy rainfall. In July and August this rainfall occurs mainly in the northern part of central Africa. Then the zone shifts southward, and as a result the southern regions receive the most rain in January and February. In March, it shifts northward again. Altitude, ocean currents, and proximity to the sea also influence the amount of rainfall specific regions receive.

Critical Thinking: Have students look at this chapter's physical-political map. Ask students where in Angola the Intertropical Convergence Zone would bring the least rain.

Answer: southwestern Angola

internet connect
GO TO: go.hrw.com
KEYWORD: SK3 CH22
FOR: Web sites about tropical climates

internet connect
GO TO: go.hrw.com
KEYWORD: SK3 CH22
FOR: Web sites about central Africa

Our Amazing Planet

One of the Congo River's most common animals is the crocodile. This sharp-toothed reptile can grow to about 20 feet (6 m) in length. It swims by sweeping its long tail from side to side.

Physical Features

Central Africa stretches southward from Cameroon and the Central African Republic to Angola and Zambia. The Atlantic Ocean lies off the western coast.

Think of the region as a big soup bowl with a wide rim. Near the middle of the bowl is the Congo Basin. A **basin** is a generally flat region surrounded by higher land such as mountains and plateaus.

In northwestern Cameroon are volcanic mountains. The highest is 13,353 feet (4,070 m). Central Africa's highest mountains lie along the Western Rift Valley. Some of these snow-capped mountains rise to more than 16,700 feet (5,090 m). The Western Rift Valley stretches southeastward from the Democratic Republic of the Congo. Lake Tanganyika (tan-guhn-YEE-kuh) and Lake Malawi are found there.

Two major river systems drain the region. In the north the Congo River flows westward to the Atlantic Ocean. Hundreds of smaller rivers flow into the Congo. In the south the Zambezi (zam-BEE-zee) River flows eastward to the Indian Ocean. The Zambezi is famous for its great falls, hydroelectric dams, and lakes.

✓ **READING CHECK:** *Places and Regions* What are the major physical features of central Africa? Congo Basin, volcanic mountains, Western Rift Valley, Lake Tanganyika, Lake Malawi, Congo River, Zambezi River

Climates, Plants, and Animals

Central Africa lies along the equator and in the low latitudes. The Congo Basin and much of the Atlantic coast have a humid tropical climate. It is wet and warm all year. This climate supports a large, dense tropical rain forest.

The many different kinds of trees in the tropical rain forest form a complete **canopy**. This is the uppermost layer of the trees where the limbs spread out. Leaves block sunlight to the ground below.

Small antelopes, hyenas, elephants, and okapis live in the rain forest region. The okapi is a short-legged relative of the giraffe. Many insects also live in the forest. However, few other plants or creatures

Local people call Victoria Falls *Mosi-oa-Tunya*, which means "the smoke that thunders." The Zambezi River plunges 355 feet (108 m) over a cliff between Zambia and Zimbabwe.

CLOSE

Refer to the photo of Victoria Falls on the previous page. Ask students what a local person might tell them about the falls. *(Possible answers: how far away the sound can be heard, stories about people who have fallen over the falls)*

REVIEW AND ASSESS

Have students complete the Section Review. Then pair students and have pairs draw symbols on their green hearts to represent vegetation types, landforms, climates, animals, and resources of central Africa. Have students create a key for their hearts. Then have students complete Daily Quiz 22.1. **COOPERATIVE LEARNING**

RETEACH

Have students complete Main Idea Activity 22.1. Then organize students into three groups. Have each group outline information on physical features; climates, plants, and animals; or resources. **ENGLISH LANGUAGE LEARNERS**

EXTEND

Have interested students conduct research on the Congo Basin and design covers for a book about the area. **BLOCK SCHEDULING**

live on the forest floor. This is because little sunlight shines through the canopy. Many animals live in the trees. They include birds, monkeys, bats, and snakes. Large areas of the tropical rain forest are being cleared rapidly for farming and timber. This threatens the plants, animals, and people who live there.

North and south of the Congo Basin are large areas with a tropical savanna climate. Those areas are warm all year, but they have distinct dry and wet seasons. There are grasslands, scattered trees, and shrubs. Only in the high eastern mountains is there a highland climate. Dry steppe and even desert climates are found in the far south.

✓ **READING CHECK:** *Places and Regions* What are the region's climates, plants, and animals? humid tropical, tropical savanna, highland, steppe, desert; tropical rain forest, grasslands, trees, shrubs; antelopes, hyenas, elephants, okapis, birds, monkeys, insects, bats, snakes

Resources

Central Africa's rivers are among the region's most important natural resources. They are used for travel, trade, and producing hydroelectricity. Other energy resources are oil, natural gas, and coal.

Central Africa has many minerals, including copper, uranium, tin, zinc, diamonds, gold, and cobalt. Most of Africa's copper is found in an area called the **copper belt**, which includes northern Zambia and the southern Democratic Republic of the Congo.

Central African countries have mostly traditional economies. Most people in central Africa are subsistence farmers. However, an increasing number of farmers are growing crops for sale. Common crops include coffee, bananas, and corn. In rural areas, people trade their products in **periodic markets**. A periodic market is an open-air trading market. It is set up regularly at a crossroads or in a town.

✓ **READING CHECK:** *Places and Regions* What is central Africa's most important natural resource, and why? rivers; support travel, trade, and produce hydroelectricity

▲ Elephants have created a network of trails and clearings throughout the tropical rain forest. Many animals will gather at forest clearings.

go.hrw.com Homework Practice Online
Keyword: SK3 HP22

Section 1 Review

Define and explain: basin, canopy, copper belt, periodic markets

Working with Sketch Maps On a map of central Africa that you draw or that your teacher provides, label the following: Congo Basin, Western Rift Valley, Lake Tanganyika, Lake Malawi, Congo River, and Zambezi River. In a box in the margin, describe the location of the highest mountains.

Reading for the Main Idea

1. *Places and Regions* What major landforms and rivers are found in the region?
2. *Physical Systems* Why do few plants and animals live on the floor of the tropical rain forest?
3. *Places and Regions* For the most part, what type of economy do Central African countries have?

Critical Thinking

4. **Making Generalizations and Predictions** How might central Africa become a rich region?

Organizing What You Know

5. **Summarizing** Copy the following graphic organizer. Use it to describe the region's climates, plants and animals, and major resources.

Climates	Plants and Animals	Resources

Section 1 Review

Answers

Define For definitions, see: basin, p. 496; canopy, p. 496; copper belt, p. 497; periodic markets, p. 497

Working with Sketch Maps Maps will vary, but listed places should be labeled in their approximate locations. The highest mountains lie along the edge of the Western Rift Valley.

Reading for the Main Idea

1. Congo Basin, volcanic mountains, edge of Western Rift Valley, Congo and Zambezi Rivers (NGS 4)
2. Sunlight does not get through the canopy. (NGS 8)
3. traditional (NGS 4)

Critical Thinking

4. It has rich resources, particularly metals, minerals, and oil.

Organizing What You Know

5. Climates—humid tropical, tropical savanna, highland, steppe, desert; Plants and Animals—antelope, hyenas, elephants, okapis, birds, monkeys, bats, snakes, insects; Resources—rivers, oil, natural gas, coal, copper, uranium, tin, zinc, diamonds, gold, cobalt

Section 2

Objectives

1. Analyze the history of central Africa, and discuss the challenges the people there face today.
2. Describe the people and cultures of central Africa.

Focus

LET'S GET STARTED

Copy this question onto the chalkboard: *What type of economy might you expect to find in a region with fertile farmland, rich mineral deposits, and a mighty river system?* Discuss responses. *(Possible answer: a prosperous region)* Tell students that although central Africa has many natural resources, most of its people are poor. Point out that the region's history may provide some clues to why the region is not wealthy today. Explain to students that in Section 2 they will learn more about central Africa's history and culture.

Building Vocabulary

Write the key terms on the chalkboard. Call on volunteers to find the definitions in a dictionary and read them aloud. Explain that elephant tusks are made of **ivory**. Because elephants are killed for their tusks, it is now illegal to import ivory into the United States. Point out that a **dialect** is usually associated with a certain geographic region. You may want to lead a discussion about how some people in different parts of the United States speak different dialects of English. However, simply using one or two words differently does not create a different dialect.

Section 2 Resources

Reproducible
- Guided Reading Strategy 22.2
- Graphic Organizer 22
- Cultures of the World Activity 6
- Creative Strategies for Teaching World Geography, Lesson 16
- Interdisciplinary Activities for the Middle Grades 26, 27, 28

Technology
- One-Stop Planner CD–ROM, Lesson 22.2
- Homework Practice Online
- HRW Go site

Reinforcement, Review, and Assessment
- Section 2 Review, p. 501
- Daily Quiz 22.2
- Main Idea Activity 22.2
- English Audio Summary 22.2
- Spanish Audio Summary 22.2

Section 2 History and Culture

Read to Discover

1. What is the history of central Africa, and what challenges do the people there face today?
2. What are the people and cultures of central Africa like?

Define

ivory
dialects

Locate

Brazzaville
Kinshasa

Why It Matters

One of the major challenges for central Africa is dealing with malaria. Use CNNfyi.com or other current events sources to find information about the disease and how to prevent it. Record your findings in your journal.

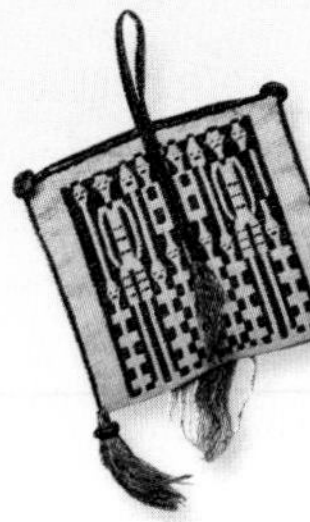

Bag made from raffia palm fibers, Cameroon

This carved mask was created by a Bantu-speaking people called the Fang. Most live in Cameroon, Equatorial Guinea, and Gabon. Their ancestors moved there in the 1800s.

History

Early humans lived in central Africa many thousands of years ago. They had different languages and cultures. About 2,000 years ago new peoples began to move into the region from western Africa. Those new peoples spoke what are called Bantu languages. Today, Bantu languages are common in most of the region.

Early History Several early Bantu-speaking kingdoms formed in central Africa. Among the most important was the Kongo Kingdom. It was located around the mouth of the Congo River. The Kongo and other central Africans traded with peoples in western and eastern Africa.

Some of the early kingdoms used slaves. In the late 1400s, Europeans came to the region. They began to trade with some African kingdoms for slaves. The Europeans took many enslaved Africans to the Americas. Europeans also wanted the region's forest products and other resources, such as **ivory**. Ivory is a cream-colored material that comes from elephant tusks. It is used in making fine furniture, jewelry, and crafts.

Some African kingdoms became richer by trading with Europeans. However, all were gradually weakened or destroyed by the Europeans. European countries divided all of central Africa into colonies in the late 1800s. The colonial powers were France, the United Kingdom, Belgium, Germany, Spain, and Portugal.

The Europeans drew colonial borders that ignored the homelands of central Africa's ethnic groups. Many groups were lumped together in colonies. These groups spoke different languages and had different

TEACH

Teaching Objective 1

LEVEL 1: (Suggested time: 20 min.) Copy the following graphic organizer onto the chalkboard, omitting the italicized answers. Use it to help students understand central Africa's history and the challenges that exist there today. Have students copy the organizer into their notebooks and break down the history of central Africa into nine steps. Ask them to concentrate on how past events have played a part in creating the challenges the region faces today. Then complete the organizer as a class. **ENGLISH LANGUAGE LEARNERS**

Major Steps in Central Africa's History

Step 1: *West Africans moved into central Africa 2,000 years ago.*
Step 2: *Bantu-speaking kingdoms were set up and began trading with western and eastern Africa.*
Step 3: *Early African kingdoms used slaves.*
Step 4: *Europeans arrived and traded for slaves.*
Step 5: *European countries ignored ethnic groups when dividing Africa into colonies.*
Step 6: *African colonies gained independence.*
Step 7: *Ethnic differences caused problems.*
Step 8: *The United States and Soviet Union supported African allies in wars.*
Step 9: *Wars prevented development of natural resources and cooperation to prevent diseases.*

ways of life. These differences resulted in conflicts, particularly after the colonies won independence.

Modern Central Africa African colonies did not gain independence until after World War II. The largest central African colony was the Belgian Congo. It is now the Democratic Republic of the Congo. That country won independence in 1960. Angola won independence from Portugal in 1975. It was the last European colony in central Africa.

Independence for some African countries came after bloody wars. After independence, fighting between some ethnic groups continued within the new countries. The region also became a battleground in the Cold War. The United States and the Soviet Union supported their allies in small wars throughout Africa. The region's wars killed many people and caused great damage. Some fighting continues off and on in the region.

Challenges Today Ending these wars is one of central Africa's many challenges today. The region's countries must also develop their natural resources more effectively. This would help the many poor people who live there. Another great challenge is stopping the spread of diseases such as malaria and AIDS. These diseases are killing millions of people and leaving many orphans.

✓ **READING CHECK:** *Human Systems* What role did Europeans play in central Africa's history? initially traded with kingdoms, then colonized central Africa, dividing it up without regard to ethnic boundaries

Culture

Today, about 100 million people live in central Africa. They belong to many different ethnic groups with varying customs.

Our Amazing Planet

In 1986 a cloud of carbon dioxide killed many people and animals near Cameroon's Lake Nyos. The lake is located in the center of a volcanic mountain. An earthquake may have allowed the gas to escape from deep in the lake.

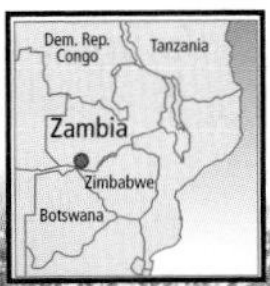

Zambian women are grinding wheat to prepare it for cooking. They are using rods called pestles. The container is called a mortar. This technique is used for grinding many ingredients.

Interpreting the Visual Record **How and why is the technology used to grind wheat different from what people in other places might use?**

GLOBAL PERSPECTIVES

During the Cold War, the Soviet Union and the United States sought to gain influence over newly emerging nations. Angola, located on the west coast of Africa, gained independence in 1974. Its strategic location near the oil-tanker routes around the Cape of Good Hope attracted attention.

Three ethnic groups sought to gain power in newly independent Angola. One group was supported by the Soviet Union and Cuba, another had ties to China, and the third was backed by Zaire. When 10,000 Cuban soldiers were airlifted to Angola by the Soviet Union in 1975, President Gerald Ford sought to assist noncommunist Angolans. However, many Americans feared that intervention could create a situation similar to the war in Vietnam. Congress prevented overt U.S. assistance.

Critical Thinking: Why was Angola's location considered strategic?

Answer: A strong naval presence in Angola could disrupt oil supply routes.

◄ **Visual Record Answer**

possible answers: They are using traditional tools for grinding because they live in a society where the machine technology other people use for grinding wheat is not available.

LEVELS 2 AND 3: (Suggested time: 45 min.) Have students use the steps from the Level 1 activity to write song lyrics describing the region's history.

TEACHER TO TEACHER

Jean Eldredge of Altamonte Springs, Florida, suggests this activity. Have students label lines of latitude and longitude—from 40°N to 40°S and 20°W to about 55°E—on a blank sheet of paper. Then have students draw an outline of Africa onto their grid. Then instruct them to fill in borders and physical features of the central African countries.

Teaching Objective 2

ALL LEVELS: (Suggested time: 30 min.) Draw a large circle on a piece of white butcher paper. Cut the circle into quarters. Organize the class into four groups. Assign each group one of the following topics: languages, religion, food, or arts. Give each group a quarter of the paper circle and colored markers. Have students write or draw information about their topic on their paper. Instruct them to use different colors to show how culture traits spread, if known. For example, they might use brown for Muslim influences, blue for French, and so on. Tape completed quarters together to create a cultural kaleidoscope of central Africa.

ENGLISH LANGUAGE LEARNERS, COOPERATIVE LEARNING

PLACES AND REGIONS

Although Spanish and French are now the official languages of Equatorial Guinea, Portugal was the first European country to control the area.

Equatorial Guinea lies chiefly on the African mainland and includes the island of Bioko, which was sighted by the Portuguese in about 1472. Portugal controlled the area until 1778, when it ceded the land to Spain. The Spanish withdrew after a yellow fever outbreak. The British, who outlawed the slave trade in 1807, took control. The Royal Navy used the island's strategic location in its efforts to stop slave traders.

Later, Spain regained control of Bioko. In 1959, Africans of Spanish Guinea gained the same rights as Spaniards. The country became independent in 1968.

Activity: Have students conduct research on the region's history and have them write ships' logs to describe the experiences of Portuguese, Spanish, and British sailors at Bioko.

Connecting to History

Answers

1. areas that are now part of Nigeria and Cameroon
2. brought new ways for growing food, used tools of iron, brought Bantu languages

CONNECTING TO History

BANTU LANGUAGES

A headrest from early Luanda, a Bantu kingdom

About 2,000 years ago, the movement of groups of people began to change Africa. Traders, farmers, and other people moved across the southern third of Africa. Experts believe these people came from areas that are now part of Nigeria and Cameroon. Their movement across Africa lasted many centuries. During this period new languages developed. We call them Bantu languages. The grammar and root sounds of the different Bantu languages remain similar.

Some Bantu speakers moved southward along the Atlantic coast. Others moved eastward to Kenya and then turned southward. Over time they reached the tip of southern Africa. The Bantu speakers mixed with peoples who already lived in these lands.

The migration of Bantu speakers had important effects on African life. They brought new ways for growing food. They used tools made of iron, which others also began to use. The Bantu-speakers of course brought their languages. Today, many Africans speak Swahili, Zulu, and other Bantu languages.

Understanding What You Read

1. From where did the Bantu peoples come?
2. How did their movement shape Africa?

BUILD ON WHAT You Know

Do you remember what you learned about world religions? See Chapter 5 to review.

Peoples and Languages As you have read, many central Africans speak Bantu languages. However, those languages can be very different from each other. In fact, hundreds of different languages and **dialects** are spoken in the region. A dialect is a variation of a language.

Many people in the region speak African languages in everyday life. However, the official languages of the central African countries are European. French is the official language of the former French colonies in the north. It is also the official language of the Democratic Republic of the Congo. English is the official language in Zambia and Malawi. Portuguese is spoken in Angola and the island country of São Tomé and Príncipe. Spanish and French are the official languages of tiny Equatorial Guinea.

Religion Many people in the former French, Spanish, and Portuguese colonies are Roman Catholic. Protestant Christianity is most common in former British colonies. Many people practice traditional African religions. In some cases Christian and African practices have been combined.

CLOSE

Ask students to suggest other ways the various invasions and occupations of the central African countries may have influenced the region's culture. *(Students might mention new forms of music and art or new foods that combine influences of various cultures.)*

REVIEW AND ASSESS

Have students complete the Section Review. Then pair students. Have each student use material in Section 2 to prepare a five-word crossword puzzle. Ask the pairs to exchange and solve each other's puzzles. Then have students complete Daily Quiz 22.2. **COOPERATIVE LEARNING**

RETEACH

Have students complete Main Idea Activity 22.2. Assign the following topics to groups: history, challenges, languages and religions, and food and the arts. Have each group draw information in appropriate places on an outline map of central Africa. Display and discuss the map. **ENGLISH LANGUAGE LEARNERS, COOPERATIVE LEARNING**

EXTEND

Have students conduct research on Jinga Mbande, a woman who fought against the Portuguese colonialists in Angola in the 1600s. A good source is *The Warrior Queens*, by Antonia Frasier. You may want to have students report their findings in the form of a eulogy—a speech given upon a famous person's death. **BLOCK SCHEDULING**

◄ These Cameroon juju dancers are calling attention to the destruction of the tropical rain forest. Note the headpieces that represent some of the forest's animals.

Many Muslims live near the mostly Muslim countries of the Sahel in the north. Zambia also has many Muslims as well as Hindus. The Hindus are the descendants of immigrants from southern Asia.

Food In most central African countries, corn, rice, grains, and fish are common foods. In the tropical rain forest, plantains, cassava, and various roots are important foods. For example, in Cameroon you might eat a dish called *fufu. Fufu,* a thick, pasty ball of mashed cassava, yams, or plantains, is served with chicken, fish, or a beef gravy.

The Arts *Makossa* dance music from Cameroon has become popular throughout Africa. It can be played with various instruments, including guitars and electric keyboards. The cities of Brazzaville and Kinshasa on the Congo River are the home of *soukous* music.

The region is also famous for carved masks, sculpture, and beautiful cotton gowns. The gowns are dyed in bright colors. They often show pictures that represent things important to the wearer.

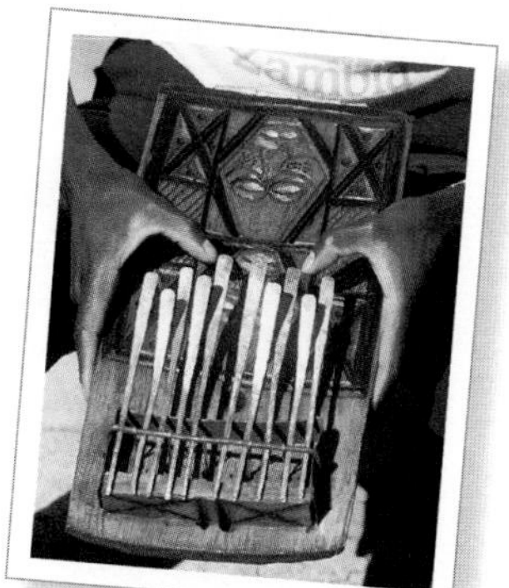

▲ The *likembe*, or thumb piano, was invented in the lower Congo region. Today its music is heard in many African countries.

Interpreting the Visual Record
How is the *likembe* played?

✓ **READING CHECK:** *Human Systems* What are some characteristics of the people and culture of central Africa? Bantu and European languages spoken; Christianity and traditional African religions practiced; traditional music, carved masks, sculpture, cotton gowns are important arts

go.hrw.com Homework Practice Online
Keyword: SK3 HP22

Section 2 Review

Define and explain: ivory, dialects

Working with Sketch Maps On the map you created in Section 1, label Brazzaville and Kinshasa.

Reading for the Main Idea

1. *Human Systems* Why are many languages that are spoken today in central Africa related?
2. *Human Systems* What have been some of the causes of wars in central Africa?
3. *Human Systems* What arts are popular in the region?

Critical Thinking

4. **Finding the Main Idea** How did Europeans influence the culture of the region?

Organizing What You Know

5. **Summarizing** Copy the following graphic organizer. Use it to identify central Africa's challenges.

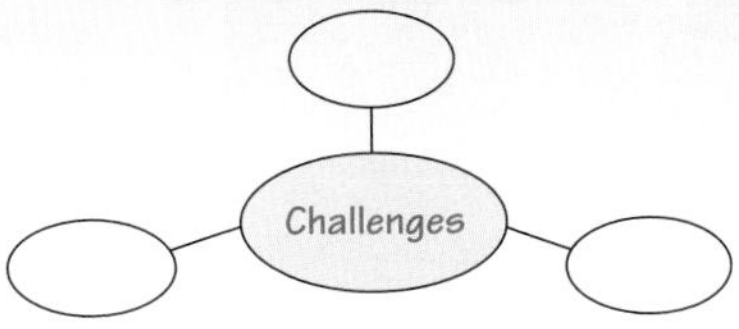

Section 2 Review

Answers

Define For definitions, see: ivory, p. 498; dialects, p. 500

Working with Sketch Maps Brazzaville and Kinshasa should be located accurately.

Reading for the Main Idea

1. related to languages spread by Bantu speakers long ago (NGS 9)
2. fight for independence, ethnic troubles, Cold War rivalries (NGS 13)
3. *makossa* music, *soukous* music, carved masks, sculpture, cotton gowns (NGS 10)

Critical Thinking

4. brought European languages, religions; separated some ethnic groups, lumped others together into colonies

Organizing What You Know

5. ending wars, developing resources more efficiently, stopping spread of diseases such as malaria and AIDS

▲ **Visual Record Answer**

held in front, gripped with both hands, thumbs used to press keys

Section 3

Objectives

1. Recount the history of the Democratic Republic of the Congo.
2. Describe the people and culture of the country.
3. Analyze the economy of the country.

Focus

LET'S GET STARTED

Copy this information onto the chalkboard: *All of these names are for the same country: Congo Free State, Belgian Congo, Republic of the Congo, Zaire, and the Democratic Republic of the Congo. What can you conclude from this country's many name changes?* Discuss students' responses. *(Possible answers: changes in government, instability)* Tell students the country has indeed had a troubled history. Explain to students that in Section 3 they will learn more about the history and culture of the Democratic Republic of the Congo.

Building Vocabulary

Write **civil war** on the chalkboard. Call on a volunteer to read the definition. Ask students why the phrase might be considered an oxymoron, or two words in a term that seem to contradict each other. *(The word* civil *can mean polite. Wars are not polite.)*

Section 3 Resources

Reproducible
- Guided Reading Strategy 22.3
- Geography for Life Activity 22

Technology
- One-Stop Planner CD–ROM, Lesson 22.3
- Homework Practice Online
- HRW Go site

Reinforcement, Review, and Assessment
- Section 3 Review, p. 504
- Daily Quiz 22.3
- Main Idea Activity 22.3
- English Audio Summary 22.3
- Spanish Audio Summary 22.3

Section 3 The Democratic Republic of the Congo

Read to Discover

1. What is the history of the Democratic Republic of the Congo?
2. What are the people and culture of the country like?
3. What is the economy of the country like?

Define

civil war

Locate

Lubumbashi

A wooden cup of the Congo region

WHY IT MATTERS

The Democratic Republic of the Congo faces many challenges. Use CNNfyi.com or other **current events** sources to find out more about some of the challenges facing the nation today. Record your findings in your journal.

A dancer performs in a coming-of-age ceremony for young people. Many such African traditions have survived the period of colonial rule by Europeans.

History

Portuguese sailors made contact with the Kongo Kingdom in 1482. Over time the slave trade and other problems weakened the Kongo and other African kingdoms.

A Belgian Colony In the 1870s King Leopold II of Belgium took control of the Congo Basin. The king ruled the Congo Free State as his personal colony until 1908. His soldiers treated the Africans harshly. They forced people to work in mines and on plantations. These policies brought international criticism.

The Belgian government took control of the colony from the king in 1908. Many Belgian businesses and people moved there. They mined copper and other resources. The giant colony won independence from tiny Belgium in 1960.

A New Country Many Belgians fled the country after 1960. There were few teachers, doctors, and other professionals left in the former colony. In addition, people from different areas and ethnic groups fought each other. These problems were partly to blame for keeping the country poor.

A dictator, who later changed his name to Mobutu Sese Seko, came to power in 1965. He was an ally of the United States during the Cold War. Mobutu changed the country's name to Zaire in 1971. During his rule, the country suffered from economic problems and government corruption.

502

TEACH

Teaching Objective 1

ALL LEVELS: (Suggested time: 20 min.) Copy the following graphic organizer onto the chalkboard, omitting the italicized answers. Use it to help students describe historical actions taken and the results of these actions in the history of the Democratic Republic of the Congo. Call on volunteers to describe results from the causes listed. Encourage students to refer to Section 2 for more information on regional history.

ENGLISH LANGUAGE LEARNERS

History of the Democratic Republic of the Congo

Portuguese contact → *Slave trade weakened kingdoms.*
Leopold II treated people harshly. → *international criticism*
Belgian government took control. → *Belgians moved there for business.*
Congo gained independence in 1960. → *Many Belgian professionals left.*
Few professionals, many problems → *Country stayed very poor.*
Mobutu Sese Seko in power → *economic problems, corruption*
Civil war erupted. → *New government took over.*

A new government took over in 1997 after a **civil war**. A civil war is a war between two or more groups within a country. The new government changed the country's name to the Democratic Republic of the Congo. However, fighting between ethnic groups has continued.

✓ **READING CHECK:** *Human Systems* What has the history of the Democratic Republic of the Congo been like? filled with numerous conflicts

The Democratic Republic of the Congo

COUNTRY	POPULATION/ GROWTH RATE	LIFE EXPECTANCY	LITERACY RATE	PER CAPITA GDP
Democratic Republic of the Congo	53,624,718 3.1%	47, male 51, female	77%	$600
United States	281,421,906 0.9%	74, male 80, female	97%	$36,200

Sources: Central Intelligence Agency, *The World Factbook 2001;* U.S. Census Bureau

Interpreting the Chart How does the literacy rate of the Democratic Republic of the Congo compare to that of the United States?

The People

More than 50 million people live in the Democratic Republic of the Congo today. The population is very diverse. It is divided among more than 200 ethnic groups. The Kongo people are among the largest groups. These groups speak many different languages, but the official language is French. About half of the country's people are Roman Catholic. Protestant Christians, Muslims, and people who practice traditional African religions also live in the country.

More than 4 million people live in Kinshasa, the capital and largest city. Kinshasa is a river port located along the Congo River near the Atlantic coast. The crowded city has some modern buildings. However, most of the city consists of poor slums.

✓ **READING CHECK:** *Human Systems* What are the people and culture of the Democratic Republic of the Congo like? diverse, many different languages spoken, half Roman Catholic

A family sells charcoal along a road in the northern Democratic Republic of the Congo. Charcoal is a major fuel source. In many rural areas there is no electricity.
▼

DAILY LIFE

Periodic markets are a common part of rural life in central Africa. On market days, sellers rent stalls or display tables where they place their wares for sale. Shoppers can buy a wide range of items, from cosmetics to audio cassettes. Shoppers compare the merchants' goods and bargain for the best deals.

Kasindi, in the eastern part of the Democratic Republic of the Congo, is an important market town. The periodic market there is held on Tuesdays and Fridays.

Critical Thinking: Have students locate Kasindi on this chapter's physical-political map and examine the location on this unit's atlas maps. Ask students why Kasindi is a good location for a periodic market.

Answer: Kasindi is located near an international border and at the edge of different climate zones, where products from other countries and ecosystems can be traded. It is also near a lake and river, which may serve as transportation routes for merchants and buyers.

▲ **Chart Answer**

It is 20 percent lower.

Teaching Objectives 2–3

ALL LEVELS: (Suggested time: 30 min.) Assign one group the topic People and the other Economy. Have each group create a mobile from cardboard, wire hangers, and string about its topic as it relates to the Democratic Republic of the Congo. **COOPERATIVE LEARNING**

CLOSE

Ask students to think of uses for the country's major resources. Provide encyclopedias for students to look up unfamiliar resources, such as cobalt.

REVIEW AND ASSESS

Have students complete the Section Review. Then have pairs of students outline the section. Then have students complete Daily Quiz 22.3. **COOPERATIVE LEARNING**

RETEACH

Have students complete Main Idea Activity 22.3. Then have students create illustrated travel guides for the Democratic Republic of the Congo, including information on history, the people, cities, and the economy. **ENGLISH LANGUAGE LEARNERS**

EXTEND

Ask students to prepare proposals for developing the country's natural resources. Have them present their proposals on storyboards for a slide show for government officials. **BLOCK SCHEDULING**

Section 3 Review

Answers

Define For definition, see: civil war, p. 503

Working with Sketch Maps Lubumbashi should be located accurately. Most of the area's copper is shipped through Lubumbashi.

Reading for the Main Idea

1. Belgium (NGS 4)
2. few professionals, ethnic fighting, poverty (NGS 4)

Critical Thinking

3. more than 200 ethnic groups, many different languages and religions; Kongo people
4. would allow resources to be shipped more efficiently; educated citizens might bring greater economic development

Organizing What You Know

5. 1482—Portuguese make contact with Kongo Kingdom; 1870s—Leopold II takes control of Congo Basin; 1908—Belgian government takes control; 1960—independence; 1965—Mobutu takes power; 1971—changed to Zaire; 1997—Mobutu loses power after civil war

Chart Answer

11 percent

The Congo River lies in the distance in this photograph of Kinshasa.

Major World Copper Producers

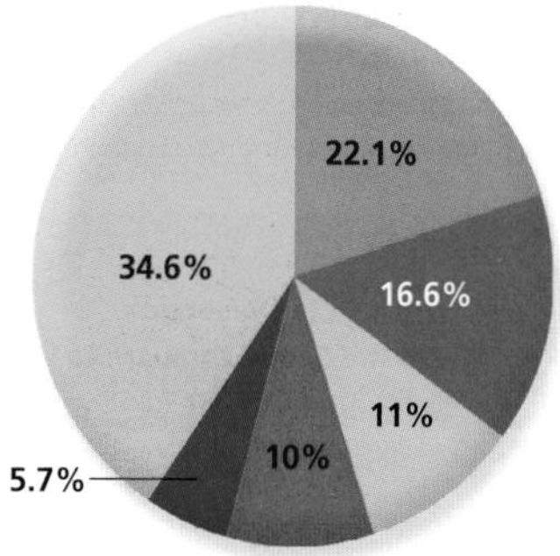

- Chile
- United States
- Zambia, Democratic Republic of the Congo
- Former Soviet republics
- Peru
- Rest of the world

Source: *Europa Book of the Year 1998*

Interpreting the Chart **How much of the world's copper do Zambia and the Democratic Republic of the Congo produce?**

The Economy

The Democratic Republic of the Congo is a treasure chest of minerals and tropical resources. For example, the south is part of central Africa's rich copper belt. Much of the copper from the area is shipped through the city of Lubumbashi (loo-boom-BAH-shee). The country also has gold, diamonds, and cobalt. The country's tropical rain forest also supplies wood, food, rubber, and other products.

However, the country's people are very poor. Most people live in rural areas. They must farm and trade for food. Civil war, bad government, and crime have scared many foreign businesses away. As a result, the country's rich resources have helped few of its people.

For the economy to improve, the country needs peace and a stable government. Schools must be improved and better health care provided. The country also must repair and expand roads and railways. If its challenges are met, the country's resources can make the future brighter.

✓ **READING CHECK:** *Environment and Society* How might the economy of the country improve? If government, schools, health care, and transportation networks improve, the country could take advantage of its rich resources.

go.hrw.com **Homework Practice Online** Keyword: SK3 HP22

Section 3 Review

Define and explain: civil war

Working with Sketch Maps On the map you created in Section 2, label the Democratic Republic of the Congo and Lubumbashi. Why is Lubumbashi economically important?

Reading for the Main Idea

1. *Places and Regions* What European country ruled what is now the Democratic Republic of the Congo as a colony until 1960?
2. *Places and Regions* What were some problems the country faced after independence?

Critical Thinking

3. **Finding the Main Idea** In what ways is the Democractic Republic of the Congo a culturally diverse country? What ethnic group is among the largest?
4. **Drawing Inferences and Conclusions** How might better schools and transportation help the country's economy?

Organizing What You Know

5. **Sequencing** Copy the following time line. Use it to identify important groups and individuals in the history of the present-day Democratic Republic of the Congo.

1482 —————————————— 2000

Section 4

Objectives

1. Describe the people and economies of the northern central African countries.
2. Describe the people and economies of the southern central African countries.

Focus

LET'S GET STARTED

Copy this question onto the chalkboard: *What are some physical challenges you might face if you lived close to the equator?* Allow students time to write their answers. Discuss responses. *(Possible answers: heat, humidity, insects)* Point out that central Africa lies near the equator. Tell students that in Section 4 they will learn more about how the location and other physical factors affect people's lives in the other countries of central Africa.

Building Vocabulary

Write **exclave** on the chalkboard. Tell students that the prefix *ex-* usually means "out of." The words *excavate* and *exclude* are examples. Explain that an exclave is a part of a country that is separated by territory of other countries. Ask students if the United States has an exclave *(Alaska).*

Section 4 The Other Central African Countries

Read to Discover

1. What are the people and economies of the northern central African countries like?
2. What are the people and economies of the southern central African countries like?

Define

exclave

Locate

Douala
Yaoundé
Luanda

Why It Matters

Poverty is a serious problem for the people of northern central Africa. Use CNNfyi.com or other **current events** sources to find information about poverty and other challenges in the region. Record your findings in your journal.

Fruit and nuts from the tropical rain forest

Northern Central Africa

Six countries make up northern central Africa. Four are Cameroon, the Central African Republic, Gabon, and the Republic of the Congo. They gained independence from France in 1960. Cameroon had been a German colony until after World War I. Tiny Equatorial Guinea gained independence from Spain in 1968. The island country of São Tomé and Príncipe won independence from Portugal in 1975.

The People Cameroon is by far the most populous country in this region. About 15.8 million people live there. The Central African Republic has the second-largest population, with 3.5 million. Tiny São Tomé and Príncipe has only about 165,000 people.

A large variety of ethnic groups are found in northern central Africa. Many people are moving from rural areas to cities to search for jobs. Governments are struggling to provide basic services in those crowded cities.

Douala (doo-AH-lah) and Yaoundé (yown-DAY) in Cameroon are the largest cities. Each has more than 1 million people. Douala is an important seaport on the Atlantic coast. Yaoundé is Cameroon's capital. Brazzaville, the capital of the Republic of the Congo, also has 1 million people. The city is a major Congo River port.

The Economy Most of the area's countries are very poor. Most people are farmers. Gabon has the strongest economy

Although the region's cities are growing rapidly, most people still live in rural areas. This small village is located in northern Cameroon. As in much of rural Africa, several buildings make up a family's home.

Interpreting the Visual Record What is similar about the construction of these houses?

Section 4 Resources

Reproducible

- Guided Reading Strategy 22.4

Technology

- One-Stop Planner CD–ROM, Lesson 22.4
- Homework Practice Online
- HRW Go site

Reinforcement, Review, and Assessment

- Section 4 Review, p. 507
- Daily Quiz 22.4
- Main Idea Activity 22.4
- English Audio Summary 22.4
- Spanish Audio Summary 22.4

◄ Visual Record Answer

possible answer: made from local materials, cone-shaped roofs made from thatch, some apparently in clusters

TEACH

Teaching Objectives 1–2

ALL LEVELS: (Suggested time: 20 min.) Copy the following graphic organizer onto the chalkboard, omitting the italicized answers. Use it to help students describe the region's people and economies. Ask students to copy it into their notebooks and to fill in the name of the country best described by the phrase. **ENGLISH LANGUAGE LEARNERS**

Central Africa's People and Economies

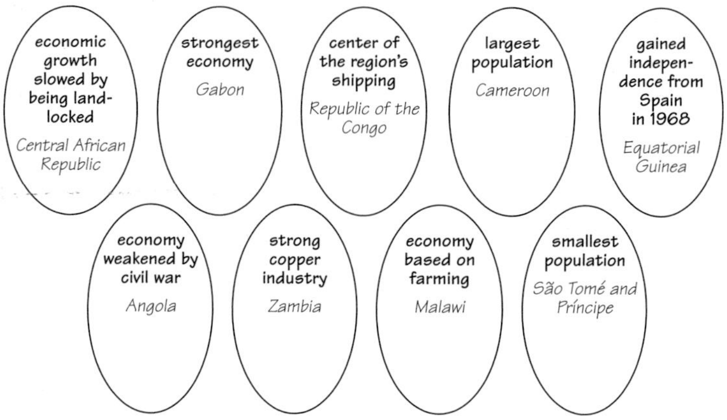

Section 4 Review

Answers

Define For definition, see: exclave, p. 507

Working with Sketch Maps Douala, Yaoundé, and Luanda should be located accurately. Luanda, a city of more than 2 million people, has many high-rise buildings and factories, but its people have suffered from poverty and years of war.

Reading for the Main Idea

1. Gabon (NGS 4)
2. most important resource (NGS 15)
3. civil war (NGS 13)

Critical Thinking

4. major shipping port along the Congo River

Organizing What You Know

5. Cameroon, Central African Republic, Gabon, Republic of the Congo—France, 1960; Equatorial Guinea—Spain, 1968; São Tomé and Príncipe, Angola—Portugal, 1975; Zambia, Malawi—Great Britain, 1964

Chart Answer ▲

São Tomé and Príncipe, Cameroon

Visual Record Answer ►

possible answers: dressed alike, bright plumes, decorated poles

Northern Central Africa

Country	Population/ Growth Rate	Life Expectancy	Literacy Rate	Per Capita GDP
Cameroon	15,803,220 / 2.4%	54, male 55, female	63%	$1,700
Central African Republic	3,576,884 / 1.9%	42, male 46, female	60%	$1,700
Congo, Republic of the	2,894,336 / 2.2%	44, male 51, female	75%	$1,100
Equatorial Guinea	486,060 / 2.5%	52, male 56, female	78%	$2,000
Gabon	1,221,175 / 1.0%	49, male 51, female	63%	$6,300
São Tomé and Príncipe	165,034 / 3.2%	64, male 67, female	73%	$1,100
United States	281,421,906 / 0.9%	74, male 80, female	97%	$36,200

Sources: Central Intelligence Agency, *The World Factbook 2001;* U.S. Census Bureau

Interpreting the Chart Which country in the region has the smallest population? the largest?

in central Africa. More than half of the value of its economy comes from the oil industry. Oil is important in Cameroon as well as in the Republic of the Congo.

The mighty Congo River is also a vital trade and transportation route. As a result, it plays a major role in the region's economy. Many of the region's goods and farm products are shipped down river to Brazzaville. Brazzaville lies across the Congo River from Kinshasa. From Brazzaville, goods are shipped by railroad to a port on the Atlantic coast.

✓ **READING CHECK:** *Human Systems* What are the people and economies of northern central Africa like? large variety of ethnic groups; mostly poor farming nations; oil important in Gabon, Cameroon, and Democratic Republic of the Congo

Southern Central Africa

Zambia, Malawi, and Angola make up the southern part of central Africa. The British gave Zambia and Malawi independence in 1964. Angola won independence from Portugal in 1975.

The People The populations of the three southern countries are nearly the same size. They range from about 9.7 million in Zambia to more than 11 million in Angola. Angola and Zambia are much larger in area than Malawi. Large parts of Angola and Zambia have few people. Most people in the region live in rural areas. They grow crops and herd cattle and goats.

Riverboats carry the king of Zambia's Lozi people to a dryland home during the flooding season. This event is an annual ritual for the Lozi, who live in western Zambia. As in other central African countries, many different ethnic groups live in Zambia.

Interpreting the Visual Record How can you tell these people are taking part in a formal tradition?

➤**ASSIGNMENT:** Have students create a new flag for one of the countries of central Africa, using text information to choose relevant colors and symbols. Ask students to write a paragraph explaining the colors and symbols.

CLOSE

Lead a class discussion on how living in the exclave of Cabinda might affect daily life.

REVIEW AND ASSESS

Have students complete the Section Review. Then have each student write a one-sentence description for one of the countries in Section 4 for a where-am-I game. Call on students to read their descriptions and ask others to identify the country. Then have students complete Daily Quiz 22.4.

RETEACH

Have students complete Main Idea Activity 22.4. Then have students create a chart that identifies the major challenges facing central Africa by country. **ENGLISH LANGUAGE LEARNERS**

EXTEND

Have students conduct research on the copper industry. Ask them to report their results on illustrated posters. **BLOCK SCHEDULING**

Angola's capital, Luanda, is the southern region's largest city. More than 2 million people live there. Seen from the sea, Luanda looks like a modern city. The city has many high-rise buildings and factories. Unfortunately, Luanda and its people have suffered from poverty and years of war. Rebels fought the Portuguese in the 1960s and early 1970s. After independence, the country plunged into civil war. Fighting has continued off and on since then. Many people have been killed or injured by land mines.

The Economy In peacetime, the future of Angola could be bright. There are many places with fertile soils. The country has large deposits of diamonds and oil. The oil is found offshore north of Luanda and in the **exclave** of Cabinda. An exclave is part of a country that is separated by territory of other countries. Cabinda is separated from the rest of Angola by the Democratic Republic of the Congo.

Much of Zambia's income comes from rich copper mines. However, 85 percent of Zambia's workers are farmers. Most of the country's energy comes from hydroelectric dams and power plants along rivers.

Almost 90 percent of Malawi's people live in rural areas. Nearly all of them are farmers. The building of factories and industries has been slow. Aid from other countries and religious missionaries has been important to the economy.

Southern Central Africa

Country	Population/ Growth Rate	Life Expectancy	Literacy Rate	Per Capita GDP
Angola	10,366,031 / 2.2%	37, male 40, female	42%	$1,000
Malawi	10,548,250 / 1.5%	37, male 38, female	56%	$900
Zambia	9,770,199 / 1.9%	37, male 38, female	78%	$880
United States	281,421,906 / 0.9%	74, male 80, female	97%	$36,200

Sources: Central Intelligence Agency, *The World Factbook 2001;* U.S. Census Bureau

Interpreting the Chart **Which country in the region has the highest literacy rate?**

✓ **READING CHECK:** *Human Systems* What are the people and economies of the southern central African countries like? mostly rural dwellers who farm and herd; Angola—economy hurt by conflict, large supply of diamonds and oil; Zambia—copper mining, farming; Malawi—farming, foreign aid

Section 4 Review

go.hrw.com Homework Practice Online Keyword: SK3 HP22

Define and explain: exclave

Working with Sketch Maps On the map you created in Section 3, label the countries of northern and southern central Africa, and Douala, Yaoundé, and Luanda. In a box in the margin, describe Luanda.

Reading for the Main Idea

1. *Places and Regions* Which country has the strongest economy in central Africa?
2. *Environment and Society* What significance does oil have in the economies of many countries in the region?
3. *Human Systems* What has happened in Angola since that country won its independence?

Critical Thinking

4. **Analyzing Information** Why is Brazzaville important to the region's economy?

Organizing What You Know

5. **Summarizing** Copy the following graphic organizer. Divide it into nine rows below the headings. Use it to list the countries discussed in this section. Then identify the European country that colonized each central African country. Finally, list the date each country won independence.

Country	European colonial ruler	Year of independence

CHAPTER 22 Review ANSWERS

Building Vocabulary

For definitions, see: basin, p. 496; canopy, p. 496; copper belt, p. 497; periodic markets, p. 497; ivory, p. 498; dialects, p. 500; civil war, p. 503; exclave, p. 507

Reviewing the Main Ideas

1. Congo, Zambezi; edge of Western Rift Valley (NGS 4)
2. migration of Bantu speakers from western Africa (NGS 9)
3. Roman Catholicism, Protestant Christianity, Islam, traditional African religions (NGS 10)
4. France, United Kingdom, Belgium, Portugal, Spain (NGS 13)
5. It has made it the strongest in the region. (NGS 15)

Understanding Environment and Society

Students should cover basic facts about malaria, how it is spread, where it is most common, and how it can be prevented. Use Rubric 29, Presentations, to evaluate student work.

▲ **Chart Answer**

Zambia

ASSESS

Have students complete the Chapter 22 Test.

RETEACH

Ask students to draw a wall map of the region and to include the boundaries between the countries. Have them label the countries, rivers, rifts, and other major physical features. Inside the outline of each country, have students list that country's European influences, resources, and challenges. **ENGLISH LANGUAGE LEARNERS**

PORTFOLIO EXTENSIONS

1. Have students plan a festival to highlight and celebrate the musical styles of central Africa, including *makossa* and *soukous*. Ask them to obtain recordings of the music and explain the music's importance in the region. Place programs, photographs, or recordings in portfolios.
2. Have students write stories based on how they think the history might have developed if: (a) the Portuguese or Belgians had never come to the Congo; (b) the Portuguese or Belgians had come as trading partners only and had never controlled the region; or (c) the region had a marine west coast climate instead of tropical climates.

CHAPTER 22 Review ANSWERS

Thinking Critically

1. Because little sunlight reaches the forest floor, few plants provide food or shelter for the animals.
2. ethnic groups with rivalries lumped together in colonies; other ethnic groups divided
3. made them important shipping ports for regional goods
4. common languages in countries where many ethnic languages are spoken; helpful in business and government as language bridge between groups
5. Cold War rivalries drew the United States and Soviet Union into the area's small regional wars and made them worse.

Map Activity

A. Luanda
B. Western Rift Valley
C. Congo River
D. Douala
E. Congo Basin
F. Lake Tanganyika
G. Zambezi River
H. Kinshasa
I. Lake Malawi
J. Lubumbashi

CHAPTER 22 Reviewing What You Know

Building Vocabulary

On a separate sheet of paper, write sentences to define each of the following words.

1. basin
2. canopy
3. copper belt
4. periodic markets
5. ivory
6. dialects
7. civil war
8. exclave

Reviewing the Main Ideas

1. *Places and Regions* What two river systems are most important in central Africa? Where are the region's highest mountains found?
2. *Human Systems* How were the Bantu languages introduced to central Africa?
3. *Human Systems* What religions are practiced by people in the Democratic Republic of the Congo?
4. *Human Systems* What European countries once colonized central Africa?
5. *Environment and Society* How important has oil been to Gabon's economy?

Understanding Environment and Society

Health

Malaria has been a problem for thousands of years. Create a presentation about malaria. In preparing your presentation, consider the following:
- What malaria is and how it is spread.
- Where malaria is most common.
- What health experts have done to fight malaria.

Thinking Critically

1. **Drawing Inferences and Conclusions** Why do few large animals live on the floor of central Africa's tropical rain forests?
2. **Finding the Main Idea** How have borders drawn by European colonial powers contributed to ethnic conflicts in central Africa?
3. **Analyzing Information** How have their locations on major rivers or near important natural resources aided the growth of cities such as Brazzaville and Lubumbashi?
4. **Drawing Inferences and Conclusions** Why do you think European languages are still the official languages in the countries of central Africa?
5. **Finding the Main Idea** How did the Cold War contribute to problems in central Africa?

FOOD FESTIVAL

Central African cooks serve *fufu* with meat dishes that have a gravy, like we might serve mashed potatoes. Here is an instant, Americanized version of *fufu*. Bring six cups of water to a boil in a large pot. Add two cups each of instant mashed potato flakes and tapioca. Using a strong wooden spoon, stir constantly for 10 to 15 minutes. If the *fufu* is thinner than mashed potatoes, add more of the dry ingredients. The *fufu* should be very thick. Shape into balls and serve immediately. *Fufu* is eaten by tearing off a small handful and using it to scoop up the meat and sauce.

CHAPTER 22 REVIEW AND ASSESSMENT RESOURCES

Reproducible
- Readings in World Geography, History, and Culture 61, 62, and 63
- Critical Thinking Activity 22
- Vocabulary Activity 22

Technology
- Chapter 22 Test Generator (on the One-Stop Planner)
- Audio CD Program, Chapter 22
- HRW Go site

Reinforcement, Review, and Assessment
- Chapter 22 Review, pp. 508–09
- Chapter 22 Tutorial for Students, Parents, Mentors, and Peers
- Chapter 22 Test
- Chapter 22 Test for English Language Learners and Special-Needs Students

Building Social Studies Skills

Map ACTIVITY

On a separate sheet of paper, match the letters on the map with their correct labels.

Congo Basin	**Congo River**
Western Rift Valley	**Zambezi River**
Lake Tanganyika	**Kinshasa**
Lake Malawi	**Lubumbashi**
	Douala
	Luanda

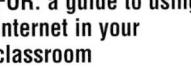

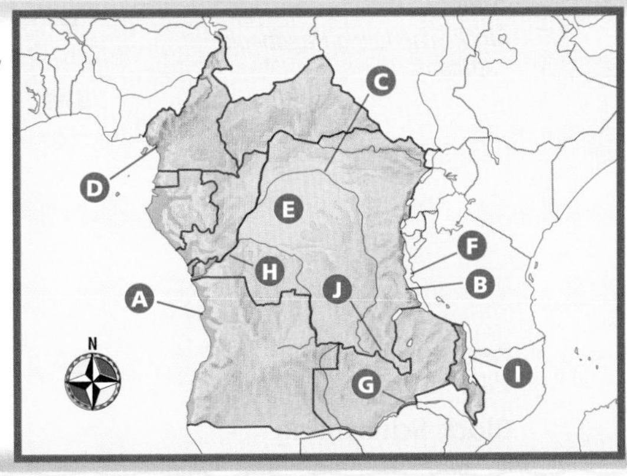

Mental Mapping Skills ACTIVITY

On a separate sheet of paper, draw a freehand map of central Africa. Make a key for your map and label the following:

Angola	**Malawi**
Democratic Republic of the Congo	**Republic of the Congo**
	Zambia

WRITING ACTIVITY

Imagine that you have been hired to write a magazine article describing challenges facing central Africans. Write a descriptive headline and a brief summary for your proposed article. The article should cover the region's economic, political, and health challenges. Be sure to use standard grammar, spelling, sentence structure, and punctuation.

Alternative Assessment

Portfolio ACTIVITY

Learning About Your Local Geography

Group Project Plan a periodic market. List items grown or produced locally. Then design an advertisement to persuade farmers or craft makers to sell at the market.

internet connect

Internet Activity: **go.hrw.com**
KEYWORD: **SK3 GT22**

Choose a topic to explore about central Africa:
- Identify threats to the Congo Basin.
- Visit the people of central Africa.
- Learn the importance of African cloth.

Mental Mapping Skills Activity

Maps will vary, but listed places should be labeled in their approximate locations.

Writing Activity

Paragraphs will vary, but should be consistent with text material. Students should include economic, political, and health challenges in their paragraphs. Use Rubric 42, Writing to Inform, to evaluate student work.

Portfolio Activity

Students should list local products, determine a location for the market, explain why it was chosen, and provide an advertisement for it. Use Rubric 2, Advertisements, to evaluate student work.

internet connect

GO TO: **go.hrw.com**
KEYWORD: **SK3 Teacher**
FOR: a guide to using the Internet in your classroom

CHAPTER 23 Southern Africa

Chapter Resource Manager

Objectives	Pacing Guide	Reproducible Resources
SECTION 1 **Physical Geography** (pp. 511–13) **1.** Identify the major physical features and climates of southern Africa. **2.** Identify the resources found in the region.	**Regular** .5 day **Block Scheduling** .5 day *Block Scheduling Handbook, Chapter 23*	**RS** Guided Reading Strategy 23.1
SECTION 2 **Southern Africa's History and Culture** (pp. 514–17) **1.** Describe what the early history of southern Africa was like. **2.** Describe how Europeans gained control of southern Africa, and what changes this caused.	**Regular** 1 day **Block Scheduling** .5 day *Block Scheduling Handbook, Chapter 23*	**RS** Guided Reading Strategy 23.2 **RS** Graphic Organizer 23 **E** Cultures of the World Activity 6 **E** Creative Strategies for Teaching World Geography, Lesson 16 **SM** Geography for Life Activity 23 **IC** Interdisciplinary Activities for the Middle Grades 26, 27, 28
SECTION 3 **South Africa Today** (pp. 518–21) **1.** Describe South Africa's policy of apartheid. **2.** Identify the factors that led to the end of apartheid. **3.** Describe what South Africa's economy is like. **4.** Describe South Africa's prospects for the future.	**Regular** 1 day **Block Scheduling** .5 day *Block Scheduling Handbook, Chapter 23*	**RS** Guided Reading Strategy 23.3
SECTION 4 **The Other Southern Africa Countries** (pp. 522–25) **1.** Identify the groups that have influenced Namibia's culture. **2.** Identify the factors that have helped Botswana's economy. **3.** Describe Zimbabwe's economy. **4.** Explain why Mozambique has remained so poor. **5.** Identify the events that have marked Madagascar's recent history.	**Regular** 1 day **Block Scheduling** .5 day *Block Scheduling Handbook, Chapter 23*	**RS** Guided Reading Strategy 23.4 **SM** Map Activity 23

Chapter Resource Key

RS	Reading Support	**REV**	Review		Video
IC	Interdisciplinary Connections	**ELL**	Reinforcement and English Language Learners	go.hrw.com	Internet
E	Enrichment		Transparencies		Holt Presentation Maker Using Microsoft® Powerpoint®
SM	Skills Mastery		CD–ROM		
A	Assessment		Music		

One-Stop Planner CD–ROM

See the ***One-Stop Planner*** for a complete list of additional resources for students and teachers.

One-Stop Planner CD–ROM

It's easy to plan lessons, select resources, and print out materials for your students when you use the ***One-Stop Planner CD–ROM with Test Generator.***

Technology Resources	Review, Reinforcement, and Assessment Resources
One-Stop Planner CD–ROM, Lesson 23.1 Geography and Cultures Visual Resources with Teaching Activities 43–47 go.hrw.com: Homework Practice Online; HRW Go site	**ELL** Main Idea Activity 23.1 **REV** Section 1 Review, p. 513 **A** Daily Quiz 23.1 **ELL** English Audio Summary 23.1 **ELL** Spanish Audio Summary 23.1
One-Stop Planner CD–ROM, Lesson 23.2 go.hrw.com: Homework Practice Online; HRW Go site	**ELL** Main Idea Activity 23.2 **REV** Section 2 Review, p. 517 **A** Daily Quiz 23.2 **ELL** English Audio Summary 23.2 **ELL** Spanish Audio Summary 23.2
One-Stop Planner CD–ROM, Lesson 23.3 *ARGWorld* CD–ROM: Malaria and Other Diseases in Africa go.hrw.com: Homework Practice Online; HRW Go site	**ELL** Main Idea Activity 23.3 **REV** Section 3 Review, p. 521 **A** Daily Quiz 23.3 **ELL** English Audio Summary 23.3 **ELL** Spanish Audio Summary 23.3
One-Stop Planner CD–ROM, Lesson 23.4 go.hrw.com: Homework Practice Online; HRW Go site	**ELL** Main Idea Activity 23.4 **REV** Section 4 Review, p. 525 **A** Daily Quiz 23.4 **ELL** English Audio Summary 23.4 **ELL** Spanish Audio Summary 23.4

internet connect

HRW ONLINE RESOURCES

GO TO: go.hrw.com
Then type in a keyword.

TEACHER HOME PAGE
KEYWORD: SK3 TEACHER

CHAPTER INTERNET ACTIVITIES
KEYWORD: SK3 GT23

Choose an activity to:
- explore the Namib Desert.
- go on a South African safari.
- investigate apartheid.

CHAPTER ENRICHMENT LINKS
KEYWORD: SK3 CH23

CHAPTER MAPS
KEYWORD: SK3 MAPS23

ONLINE ASSESSMENT
Homework Practice
KEYWORD: SK3 HP23
Standardized Test Prep Online
KEYWORD: SK3 STP23
Rubrics
KEYWORD: SS Rubrics

COUNTRY INFORMATION
KEYWORD: SK3 Almanac

CONTENT UPDATES
KEYWORD: SS Content Updates

HOLT PRESENTATION MAKER
KEYWORD: SK3 PPT23

ONLINE READING SUPPORT
KEYWORD: SS Strategies

CURRENT EVENTS
KEYWORD: S3 Current Events

Meeting Individual Needs

Ability Levels

Level 1 Basic-level activities designed for all students encountering new material

Level 2 Intermediate-level activities designed for average students

Level 3 Challenging activities designed for honors and gifted-and-talented students

English Language Learners Activities that address the needs of students with Limited English Proficiency

Chapter Review and Assessment

- **E** Readings in World Geography, History, and Culture 64, 65, and 66
- **SM** Critical Thinking Activity 23
- **REV** Chapter 23 Review, pp. 526–27
- **REV** Chapter 23 Tutorial for Students, Parents, Mentors, and Peers
- **ELL** Vocabulary Activity 23
- **A** Chapter 23 Test
- **A** Unit 6 Test
- Chapter 23 Test Generator (on the One-Stop Planner)
- Audio CD Program, Chapter 23
- **A** Chapter 23 Test for English Language Learners and Special-Needs Students
- **A** Unit 6 Test for English Language Learners and Special-Needs Students

CHAPTER 23

Launch into Learning

Write the following phrases on the chalkboard: *gold, diamonds, elephants, baboons, crocodiles, sailing ships, waterfalls, mountains, cattle, nomadic herders, modern cities, mountains, a mosque, a cross, and a cup of coffee.* Ask students what these items have in common. *(They are found in the region of southern Africa).* Tell students they will learn more about how these concepts relate to southern Africa in this chapter.

Section 1

Objectives

1. Identify the major physical features and climates of southern Africa.
2. Describe the resources found in the region.

Links to Our Lives

You may wish to point out that there are many reasons why we should know more about the countries of southern Africa, these among them:

- The region is rich in minerals, many of which are imported by the United States.
- Many Americans have ancestors from this region.
- Some of the countries of southern Africa are struggling to improve relationships among different racial and ethnic groups. Their efforts can have wide-ranging results.
- Many Americans travel to the region to see wildlife in their native habitat.
- Art of the region is unique and in demand throughout the world.

CHAPTER 23

Southern Africa

Southern Africa is a region going through many changes. Some people in the area have a comfortable lifestyle, but others live in severe poverty.

Abatsu! (Good morning!) My name is Kha//'an[1] and I am a San person from Namibia. I am in ninth grade at the Tsumkwe[2] Secondary School. We do not have a house, only a small shelter to store our things. We live near the gate in the fence at M'Kata. If you turn east when you arrive there, you will see our blankets under the tree. When we are seeking shade or shelter, we go and sit under the tree.

My mother and father passed away when I was five years old, and my elder brother took care of me. I live with my brother, and the government helps us survive by giving us food rations.

At school I live in a hostel. We wake up at five in the morning and have breakfast at six. Breakfast is only bread. We have maize meal, tea, and milk at noon. My favorite subjects are English, history, physical science, and mathematics. When I grow up, I want to be a doctor.

Translation: Greetings! I am Kha//'an.

[1] The "//" is a click made by clucking the tongue at the sides of the mouth.

[2] The "k" in Tsumkwe is another kind of click. It is made by placing the tongue on the roof of the mouth and bringing it down with a "pop."

[3] The = is a click made by placing the tip of the tongue on the ridge behind the upper front teeth and bringing it down with a "pop."

FOCUS

LET'S GET STARTED

Copy the following questions onto the chalkboard: *What are some possible advantages of living on a peninsula between two oceans? What are some possible disadvantages of such a location?* Have students respond in writing. *(Students might suggest that the location would facilitate trade but also make it easier for invaders to attack and take over the region.)* Discuss students' answers. Then tell students that for many centuries southern Africa's location and resources have made the region attractive to outsiders. Tell students that in Section 1 they will learn more about the physical geography of southern Africa.

Using the Physical-Political Map

Have students examine the map on this page. Then ask students to describe the relative location of southern Africa in terms of its natural boundaries. *(south of the Zambezi River, west of the Indian Ocean, east of the Atlantic Ocean)* Then have students use the map to identify the region's landforms. *(island, cape, deserts, escarpment, mountains, peninsula)*

Section 1 Physical Geography

Read to Discover

1. What are the major physical features and climates of southern Africa?
2. What resources are found in the region?

Define

enclaves
the veld
pans

Locate

Drakensberg
Inyanga Mountains
Cape of Good Hope
Kalahari Desert
Namib Desert
Orange River
Aughrabies Falls
Limpopo River

WHY IT MATTERS

The rich mineral resources of southern Africa, such as gold, diamonds, and platinum, are important to international economic stability. Use CNNfyi.com or other **current events** sources to find examples of how this region is using its resources. Record your findings in your journal.

Pygmy mouse lemur

Section 1 RESOURCES

Reproducible
- Block Scheduling Handbook, Chapter 23
- Guided Reading Strategy 23.1

Technology
- One-Stop Planner CD–ROM, Lesson 23.1
- Homework Practice Online
- Geography and Cultures Visual Resources with Teaching Activities 43–47
- HRW Go site

Reinforcement, Review, and Assessment
- Section 1 Review, p. 513
- Daily Quiz 23.1
- Main Idea Activity 23.1
- English Audio Summary 23.1
- Spanish Audio Summary 23.1

Southern Africa: Physical-Political

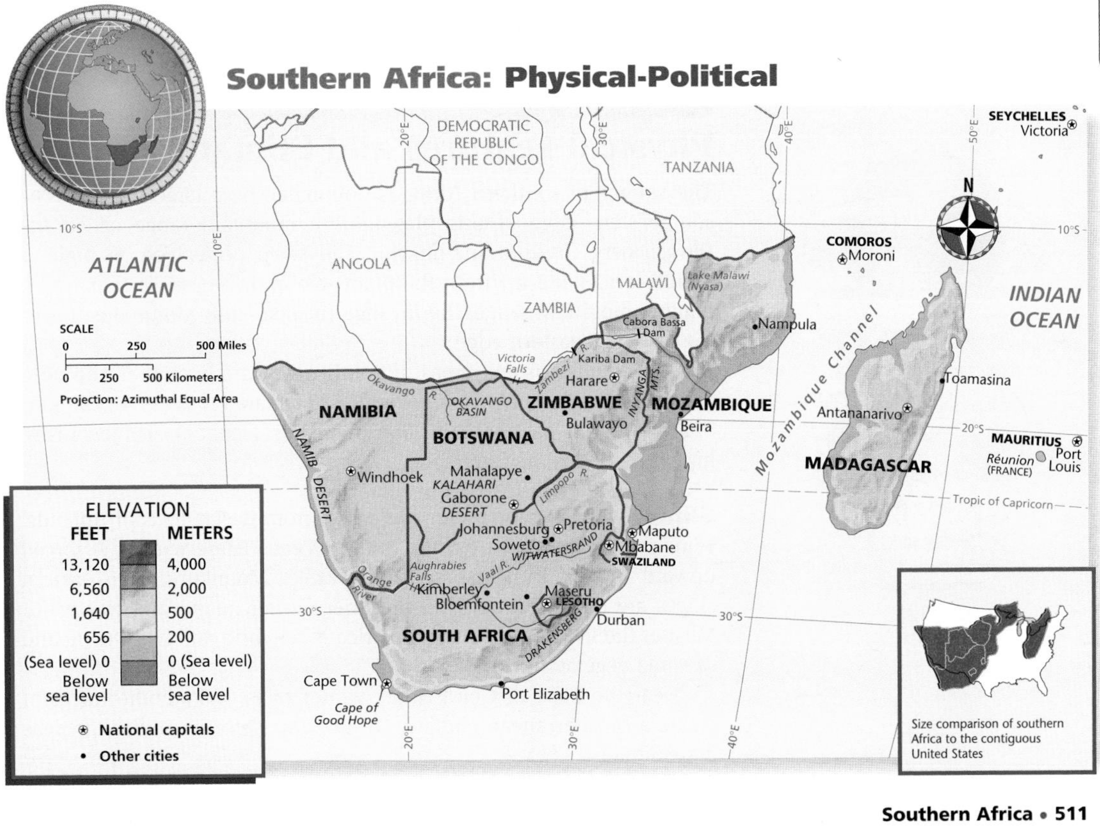

TEACH

Teaching Objectives 1–2

ALL LEVELS: (Suggested time: 30 min.) Copy the following graphic organizer, omitting the italicized answers. Organize the class into five groups. Assign one group Countries, the second Landforms, the third Climates, the fourth Bodies of Water, and the fifth Resources. Have each group list information about its topic. Then have students write their findings on the organizer. **ENGLISH LANGUAGE LEARNERS, COOPERATIVE LEARNING**

SOUTHERN AFRICA'S PHYSICAL GEOGRAPHY

COUNTRIES	LANDFORMS	CLIMATES	BODIES OF WATER	RESOURCES
South Africa, Namibia, Botswana, Zimbabwe, Mozambique, Madagascar, Swaziland, and Lesotho	*large plateau, the veld, the Drakensberg, Inyanga Mountains, Kalahari Desert, and Namib Desert*	*desert, semi-arid, steppe, savanna, Mediterranean*	*Orange River, Aughrabies Falls, Limpopo River*	*gold, diamonds, platinum, copper, uranium, coal, and iron ore*

EYE ON EARTH

The Namib Desert stretches 80 to 100 miles (129–161 km) eastward from Namibia's west coast toward the Great Escarpment. Its name means "an area where there is nothing," and except for a few towns this arid place is totally uninhabited. An unusual plant found here is the *tumboa,* also known as *welwitschia,* the only plant of its genus. It has a taproot about 25 to 50 inches (60–120 cm) in diameter that extends about 12 inches (30 cm) above the ground and resembles a giant radish. Two broad, flat leaves reaching 10 feet (3 m) from the crest of the huge root crown grow throughout the plant's 100-year lifespan.

Activity: Have students conduct research on the Namib Desert and create a diorama of the region.

internet connect

GO TO: **go.hrw.com**
KEYWORD: **SK3 CH23**
FOR: **Web sites about the Namib Desert**

Visual Record Answer

for its scenic beauty and wildlife

The Drakensberg range rises sharply in eastern South Africa.

Interpreting the Visual Record Why do you think tourists are attracted to this region?

Our Amazing Planet

Drakensberg means "dragon mountain" in the Afrikaans language. The Zulu, one of the peoples of the region, call it Kwathlamba, meaning "barrier of pointed spears" or "piled-up rocks."

Countries of the Region

Lining southern Africa's coasts are Namibia (nuh-MI-bee-uh), South Africa, and Mozambique (moh-zahm-BEEK). Botswana (bawt-SWAH-nah), Zimbabwe (zim-BAH-bway), and the two tiny countries of Lesotho (luh-SOH-toh) and Swaziland (SWAH-zee-land) are all landlocked. Lesotho and Swaziland are **enclaves**—countries surrounded or almost surrounded by another country. Madagascar (ma-duh-GAS-kuhr), off the east coast, is the world's fourth-largest island.

Physical Features and Climate

The surface of southern Africa is dominated by a large plateau. The southeastern edge of this plateau is a mountain range called the Drakensberg (DRAH-kuhnz-buhrk). The steep peaks rise as high as 11,425 feet (3,482 m) from the plains along the coast. Farther north, another mountain range, the Inyanga (in-YANG-guh) Mountains, forms the plateau's eastern edge.

The open grassland areas of South Africa are known as **the veld** (VELT). Kruger National Park covers 7,523 square miles (19,485 sq km) of the veld. The park contains lions, leopards, elephants, rhinoceroses, hippos, baboons, and antelope.

Climate The region's climates range from desert to cool uplands. Winds carry moisture from the Indian Ocean. These winds are forced upward by the Drakensberg and Inyanga Mountains. The eastern slopes are rainy, but climates are drier farther inland and westward. Most of the interior of southern Africa is semiarid and has steppe and savanna vegetation.

Near the Cape of Good Hope, winter rains and summer drought create a Mediterranean climate. Off the Cape, storms and rough seas are common.

CLOSE

Remind students that the location and resources of the region attracted people from other parts of the world and continue to affect the region's economy, politics, and culture. Ask students to identify some of the resources of the region and explain how location and resources affect the economy of the region.

REVIEW AND ASSESS

Have students complete the Section Review. Then tell each student to write four multiple-choice questions based on the information in this section. Pair students and have them take turns asking each other their questions. Then have students complete Daily Quiz 23.1. **COOPERATIVE LEARNING**

RETEACH

Have students complete Main Idea Activity 23.1. Then have groups create wall maps of the region with symbols for physical features, natural resources, and climate types. Display the maps. **ENGLISH LANGUAGE LEARNERS, COOPERATIVE LEARNING**

EXTEND

Until the 1800s, millions of quaggas—zebra-like animals—roamed the veld of South Africa. The last captive quagga died in 1883. Have interested students conduct research on efforts by South African scientists to bring back the quagga by tracing their DNA from zebras. Have students record their findings in a written report. **BLOCK SCHEDULING**

Deserts and Rivers In the central and western parts of the region, savanna and steppe give way to two major deserts. The Kalahari (ka-luh-hahr-ee) occupies most of Botswana. Here ancient streams have drained into low, flat areas, or **pans**. Minerals left behind when the water evaporated form a glittering white layer.

The Namib (NAH-mib) Desert lies along the Atlantic coast. Inland, the Namib blends into the Kalahari and steppe. Almost no rain falls, but at night fog rolls in from the ocean. Some plants and animals survive by using the fog as a source of water.

Southern Africa has some of the world's most spectacular rivers and waterfalls. The Orange River passes through the Aughrabies (oh-KRAH-bees) Falls as it flows to the Atlantic. When the water is highest, the Aughrabies Falls are several miles wide. The water tumbles down 19 separate waterfalls. The Limpopo (lim-POH-poh) River is the region's other major river. It flows into the Indian Ocean.

✓ **READING CHECK:** *Places and Regions* What are southern Africa's physical features and climate? plateau, Drakensberg mountain range, the veld, rivers, waterfalls; desert, cool uplands, semiarid with steppe and savanna vegetation, mediterranean

internet connect

GO TO: go.hrw.com
KEYWORD: SK3 CH23
FOR: Web sites about southern Africa

▲ The Orange River flows down the spectacular Aughrabies Falls. The falls are near the Namibian border in northwestern South Africa.

Resources

Southern Africa is very rich in mineral resources. Gold, diamonds, platinum, copper, uranium, coal, and iron ore are all found in the region. Where rain is plentiful or irrigation is possible, farmers can grow a wide range of crops. Ranchers raise livestock on the high plains. Some nomadic herders still live in desert areas.

✓ **READING CHECK:** *Places and Regions* What are the main resources of southern Africa? minerals—gold, diamonds, platinum, copper, uranium, coal, iron ore; farmland

go.hrw.com Homework Practice Online
Keyword: SK3 HP23

Section 1 Review

Define and explain: enclaves, the veld, pans

Working with Sketch Maps On a map of southern Africa that you draw or that your teacher provides, label the following: Drakensberg, Inyanga Mountains, Cape of Good Hope, Kalahari Desert, Namib Desert, Orange River, Aughrabies Falls, and Limpopo River. The Orange River forms part of the border between what two countries?

Reading for the Main Idea

1. *Places and Regions* What are the major landforms of southern Africa?
2. *Places and Regions* What are the main natural resources of the region?
3. *Physical Systems* How do physical processes affect southern Africa's climate?

Critical Thinking

4. **Drawing Inferences and Conclusions** How do you think the climate off the Cape of Good Hope has affected shipping in the area?

Organizing What You Know

5. **Summarizing** Copy the following graphic organizer. Use it to list the landforms, climates, and resources of southern Africa.

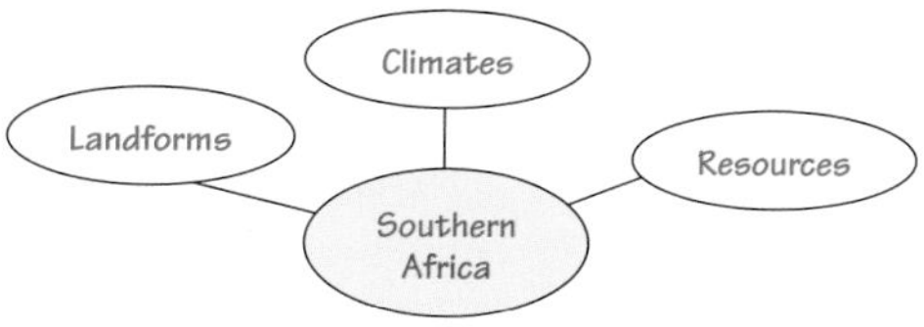

Section 1 Review Answers

Define For definitions, see: enclaves, p. 512; the veld, p. 512; pans, p. 513

Working with Sketch Maps Maps will vary, but listed places should be labeled in their approximate locations. The Orange River forms the border between Namibia and South Africa.

Reading for the Main Idea

1. large plateau, the Drakensberg, the Inyanga Mountains, and the veld (NGS 4)
2. gold, diamonds, platinum, copper, uranium, and iron ore (NGS 4)
3. The Drakensburg forces moisture from the Indian Ocean upward and causes the eastern slopes to be rainy, while climates are drier farther inland. (NGS 7)

Critical Thinking

4. limited shipping because of rough seas

Organizing What You Know

5. landforms—large plateau, the Drakensberg, the Inyanga Mountains, and the veld; climates—desert, semiarid, steppe, savanna, and Mediterranean; resources—gold, diamonds, platinum, copper, uranium, iron ore, farmland, and ranchland

Section 2

Objectives

1. Describe the early history of southern Africa.
2. Explain how Europeans gained control of the region, and identify the changes this caused.

FOCUS

LET'S GET STARTED

Copy the following question onto the chalkboard: *How might the physical geography of southern Africa have affected its history?* Allow for students to respond in writing. Discuss students' responses. Students might mention that the abundant resources and the location of southern Africa may have attracted foreigners. Tell students that in Section 2 they will learn more about the history and culture of southern Africa.

Building Vocabulary

Write the key term on the chalkboard. Explain that **Boers** is the name given to descendants of the original white Dutch colonial settlers of South Africa who first settled there in the 1600s.

Section 2 RESOURCES

Reproducible
- Guided Reading Strategy 23.2
- Graphic Organizer 23
- Geography for Life Activity 23
- Cultures of the World Activity 6
- Creative Strategies for Teaching World Geography, Lesson 16
- Interdisciplinary Activities for the Middle Grades 26, 27, 28

Technology
- One-Stop Planner CD–ROM, Lesson 23.2
- Homework Practice Online
- HRW Go site

Reinforcement, Review, and Assessment
- Section 2 Review, p. 517
- Daily Quiz 23.2
- Main Idea Activity 23.2
- English Audio Summary 23.2
- Spanish Audio Summary 23.2

Visual Record Answer

These images show scenes of daily life, which was probably filled with hunting for survival.

Section 2 Southern Africa's History and Culture

Read to Discover

1. What was the early history of southern Africa like?
2. How did Europeans gain control of southern Africa, and what changes did this cause?

Define

Boers

Locate

Cape Town

WHY IT MATTERS

Among the many problems facing southern Africa are frequent natural disasters. Use CNNfyi.com or other **current events** sources to find examples of these disasters and how officials work to deal with them. Record your findings in your journal.

Modern sculpture from Zimbabwe

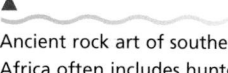

Ancient rock art of southern Africa often includes hunters and animals.

Interpreting the Visual Record **What do these images suggest about the people who made them?**

Early History

Southern Africa's landscape and climate have influenced the region's history. For example, monsoon winds blow from the Indian Ocean to southern Africa from November to February. From May to September the wind blows the other way, from Africa to Asia. Ancient ships used these winds to make regular trading voyages between the two continents.

The Khoisan Some of the oldest human fossils have been found in southern Africa. By about 18,000 B.C., groups of hunter-gatherers were living throughout the mainland region. They left distinctive paintings of people and animals on rock surfaces. Some descendants of these people still live in certain desert regions. They speak languages of the Khoisan language family, which share unusual "click" sounds. However, most Khoisan people were absorbed into groups that moved into the region later.

Bantu Migrations Some 1,500–2,000 years ago a different group of people spread from central Africa into southern Africa. They spoke another family of languages known as Bantu. Today, most southern Africans speak one of the more than 200 Bantu languages. Scholars believe the early Bantu people introduced the use of iron to make tools. The Bantu are also thought to have introduced cattle herding to the region.

TEACH

Teaching Objective 1

LEVELS 1 AND 2: (Suggested time: 30 min.) Copy the following graphic organizer onto the chalkboard, omitting the italicized answers. Organize the class into three groups. Assign the first group Khoisan and Bantu, the second Shona and Swahili, and the third Asians and Portuguese. Have each group find information about how, when, or where each group arrived, and the cultural changes each brought.
ENGLISH LANGUAGE LEARNERS, COOPERATIVE LEARNING

GROUP	HOW, WHEN, WHERE	CULTURAL CHANGES
Khoisan	*hunter-gatherers, 18000 B.C., throughout mainland*	*Khoisan language family, rock paintings*
Bantu	*spread from central Africa, 1,500–2,000 years ago, southern Africa*	*Bantu language family, use of iron, cattle herding*
Shona	*built an empire, A.D. 1000, Zimbabwe and Mozambique*	*farmed, raised cattle, traded gold*
Swahili	*sailors and traders, A.D. 1100, east coast*	*adopted Islam and Arab customs, traded with East Asia*
Asians	*settled on island, A.D. 700, Madagascar*	*Asian and African influences, language related to Indonesia*
Portuguese	*traders, early 1500s, Mozambique*	*set up forts, slave use and trade*

Shona and Swahili By about A.D. 1000 one Bantu group, the Shona, had built an empire. It included much of what is now Zimbabwe and Mozambique. They farmed, raised cattle, and traded gold with other groups on the coast. They also constructed stone-walled towns called *zimbabwe*. The largest town, now called Great Zimbabwe, may have had 10,000 to 20,000 residents. Great Zimbabwe was abandoned in the 1400s.

Among Great Zimbabwe's trading partners were the Swahili-speaking people of the east coast. These were Africans who had adopted Islam and many Arab customs by the A.D. 1100s. The Swahili-speakers were sailors and traders. Archaeologists have found Chinese porcelain at Great Zimbabwe. This suggests that Africa and East Asia were connected by an Indian Ocean trade network.

People fish with nets in the Indian Ocean. Fishing is an important industry in Mozambique and Madagascar.

Madagascar Madagascar's early history is quite different from the rest of southern Africa. Madagascar's first settlers came from Asia, rather than Africa, about A.D. 700. The island's culture shows the influence of both Africa and Asia. Malagasy, Madagascar's official language, is related to languages spoken in Indonesia. Malagasy also includes many words from the Bantu language family.

Mozambique In the early 1500s the Portuguese set up forts in Mozambique. They hoped to take over the region's gold trade from the Swahili-speakers and Arabs. The Portuguese also established large estates along the Zambezi River that used slave labor. In the 1700s and 1800s Mozambique became an important part of the slave trade. Africans were captured there and sent as slaves to Brazil and other parts of the world.

✓ **READING CHECK:** *Human Systems* What were some key events in the early history of southern Africa? 1,500–2,000 years ago—Bantu migrations; A.D. 1000—Shona empire; A.D. 700—Asian settlers reach Madagascar; 1500—Portuguese build forts in Mozambique; 1700s and 1800s—Mozambique important to slave trade

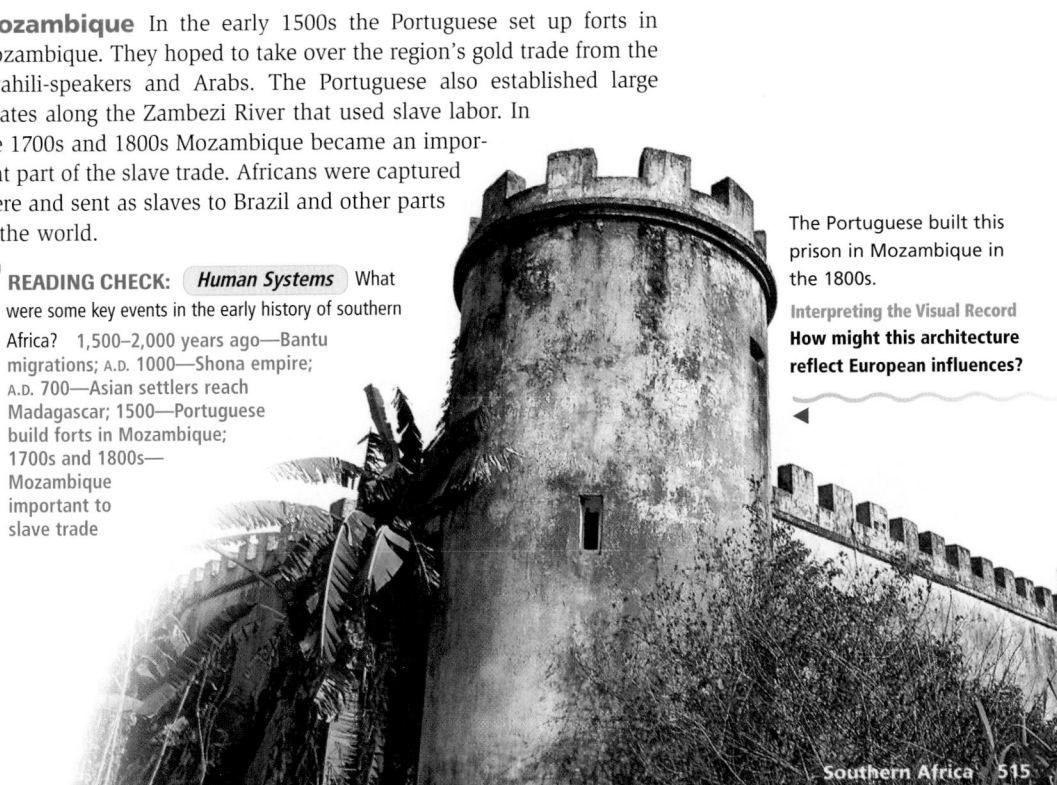

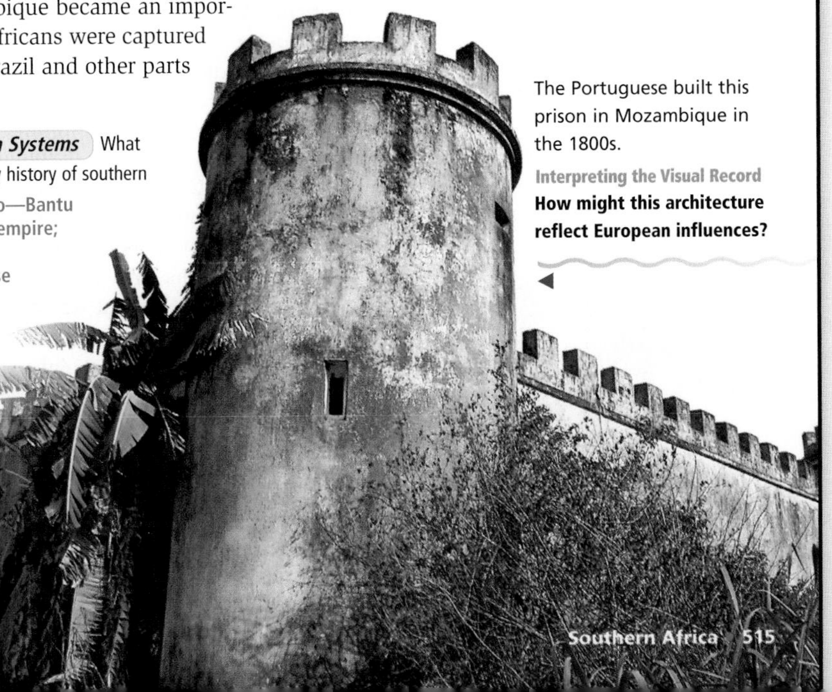

The Portuguese built this prison in Mozambique in the 1800s.

Interpreting the Visual Record
How might this architecture reflect European influences?

EYE ON EARTH

Along with a unique early history, Madagascar has unique wildlife. The island has an estimated 200,000 species of plants and animals, three fourths of which are not found anywhere else in the world. Among the animals unique to the island is the lemur, a suborder of primates that has both nocturnal and diurnal species. One lemur active during the day is the indri, which has rounded ears and soft black and white fur. Its appearance resembles a cross between a teddy bear and a panda. It has extraordinary gymnastic ability and is able to propel backwards nearly 30 feet (10 m), flipping in midair to land face forward. The indri and other species are endangered by deforestation, which has reduced the original forest by 90 percent. Species of plants that may hold cures for human diseases are also endangered.

Critical Thinking: How has human activity threatened plant and animal species in Madagascar?

Answer: Deforestation is threatening them.

Visual Record Answer

Students should mention the turret and other features that look like a European castle.

LEVEL 3: (Suggested time: 25 min.) Have students write a report on an archaeological study of southern Africa. Students should list artifacts and fossils that represent the various groups and how they influenced southern Africa's early history. Each artifact or fossil should also include where it was found and its approximate age. *(cave paintings dating back to 18,000 B.C. to represent the Khoisan; iron tools 1,500–2,000 years old to represent the Bantu; gold and Chinese porcelain from A.D. 1000 in Zimbabwe and Mozambique to represent the Shona; guns and European tools from the early 1500s in Mozambique to represent the Portuguese)*

Teaching Objective 2

ALL LEVELS: (Suggested time: 45 min.) Organize students into six groups. Assign each group one of Section 2's subsections, starting with "The Dutch, British, and Portuguese in South Africa." Give each group a large sheet of paper and have the group design a mural for its section. Combine each group's designs into one large mural and display it in the classroom. Discuss the finished product. **ENGLISH LANGUAGE LEARNERS, COOPERATIVE LEARNING**

Linking Past to Present
Shaka Zulu

Shaka (c. 1787–1828), the powerful founder of the Zulu Empire, was a legendary figure even in his own time. An imposing 6'3" warrior, Shaka gained his authority to rule mainly through the fear he inspired. He engaged the small Zulu clan—one of hundreds in southern Africa—in a series of tribal wars that resulted in the Zulu becoming the dominant group in the region.

Shaka Zulu has not been forgotten. He has become a symbol of Zulu strength and greatness. He also plays a role in popular culture. For example, the singing group Ladysmith Black Mambazo released an album under the name *Shaka Zulu.* Tourists can visit Shakaland, a cultural village in South Africa, to learn about Zulu traditions.

Critical Thinking: How is Shaka's popularity reflected in South African culture today?

Answer: Ladysmith Black Mambazo's album is named *Shaka Zulu;* Shakaland is a tourist site.

internet connect

GO TO: **go.hrw.com**
KEYWORD: **SK3 CH23**
FOR: **Web sites about Zulu**

Visual Record Answer

by wind

In this painting, British ships sail past the Cape of Good Hope. The distinctively shaped Table Mountain is visible in the distance.

Interpreting the Visual Record How are these ships propelled?

Our Amazing Planet

Much of southern Africa was a malaria zone. Quinine, a medicine made from a South American tree, could combat malaria. Using this medicine, Europeans were able to move into lowland areas where malaria was common.

The Dutch, British, and Portuguese in South Africa

The land around the Cape of Good Hope lacked the gold and copper of the Zambezi Valley. However, it had a Mediterranean climate. It also was free of the mosquitoes and tsetse flies that spread tropical diseases. In 1652 the Dutch set up a trade station at a natural harbor near the Cape. This small colony would eventually become known as Cape Town. It provided supplies to Dutch ships sailing between the Dutch East Indies and Europe. The Dutch brought in slaves to work in the colony. Some were Malays from Southeast Asia. Others were Africans bought at slave markets in other parts of Africa.

Afrikaners and Afrikaans The main language spoken in the colony was Dutch. Over time, Khoisan, Bantu, and Malay words were added, creating a new language called Afrikaans. White descendants of the original colonists are called Afrikaners. Some European men married Khoisan or Malay women. People descended from Malays, Khoisan, or a mixture of these with Europeans are called Coloureds.

A British Colony Afrikaner frontier farmers called **Boers** gradually spread out from the original colony. Then, in the early 1800s, Great Britain took over the area of the Cape. The Boers resisted the British colonial government. Many Boers packed all their belongings into wagons and moved farther east and north. This movement was called the Great Trek.

At about the same time, a Bantu-speaking group, the Zulu, became a powerful fighting force. They conquered the surrounding African peoples, creating their own empire. When the Boers moved into the northern plains, they entered Zulu territory. The two sides clashed over control of the land. Eventually the Zulu were challenged by the British and defeated after a series of battles.

In this 1935 photograph, elephant tusks and rhino horns are inspected in a London warehouse.

CLOSE

Call on volunteers to name all the outside groups that entered the area, how and why they arrived, the areas they controlled, and the cultural impact of their presence.

REVIEW AND ASSESS

Have students complete the Section Review. Have each student prepare three "Who Am I?" questions covering the groups discussed in the text. Collect questions and use them to review. Then have students complete Daily Quiz 23.2.

RETEACH

Have students complete Main Idea Activity 23.2. Create an outline of the important information presented in the section on the chalkboard. Leave some parts of the outline blank and have students provide the missing information. Then have all students copy the outline for use as a study reference. **ENGLISH LANGUAGE LEARNERS**

EXTEND

Have interested students conduct research on the Great Trek of the Boers. Ask students to use their research to write and illustrate descriptions of what the journey must have been like. Have them share their work with the class. **BLOCK SCHEDULING**

Trade in Slaves and Ivory The British banned slavery in their empire in 1833. The Portuguese colonies of Angola and Mozambique remained as Africa's main slave markets. The slave trade eventually ended in the late 1800s. African trade began to focus on ivory—the tusks of elephants. Hunters wiped out the entire elephant population in some parts of southern Africa.

Diamonds, Gold, and Colonies In the 1860s diamonds were discovered in the northern part of the Cape Colony. In 1886 gold was discovered in the Transvaal, an area controlled by the Boers. Thousands of British and others came to South Africa. Railroads were built to connect the interior with the coast.

As the British moved north from the Cape Colony, some Boers moved into what is now Botswana. Afraid that the Boers would take over his country, Botswana's ruler asked for British protection. In 1885 Botswana (then known as Bechuanaland) came under British control. What is now Namibia became German South-west Africa—a German colony. In 1889 what is now Zimbabwe came under control of the British South Africa Company as part of Rhodesia. It became a self-governing British colony in 1923.

▲ This old steam train still operates in South Africa. The British built this and other railroads during the colonial period.

Interpreting the Visual Record

Why did the Europeans connect the interior with the coast?

South Africa In 1899 tensions over land and mineral wealth led to war between the Boers and the British. The Boers were greatly outnumbered, but held off the British army for three years. In the end the Boers were defeated. Their territory was added to the British colony of South Africa. In 1920, following Germany's defeat in World War I, Namibia was placed under South Africa's control.

✓ **READING CHECK:** *Human Systems* How did Europeans gain control of southern Africa, and what changes did this cause? initially founded colonies, then spread out, conquering native peoples militarily; European groups struggled for control over area, established trade in slaves, ivory, and minerals

go.hrw.com Homework Practice Online
Keyword: SK3 HP23

Section 2 Review

Define and explain: Boers

Working with Sketch Maps On the map you created in Section 1, label Cape Town. What were the advantages of Cape Town's location?

Reading for the Main Idea

1. *Human Systems* How did the Bantu affect the history of southern Africa?
2. *Human Systems* How do archaeologists believe Chinese porcelain came to Great Zimbabwe?

Critical Thinking

3. **Finding the Main Idea** How did the slave trade affect southern Africa?
4. **Analyzing Information** What European groups settled in southern Africa? How did these groups interact with each other?

Organizing What You Know

5. **Sequencing** Copy the following time line. Use it to mark important events in the history of southern Africa from 18,000 B.C. to 1920.

18,000 B.C. ———————— 1920

Section 2 Review

Answers

Define For definition, see: Boers, p. 516

Working with Sketch Maps Cape Town should be labeled in its approximate location; its location is good for shipping.

Reading for the Main Idea

1. They established Bantu languages there and are thought to have introduced iron tools and cattle herding. (NGS 9)
2. It came through an Indian Ocean trade network. (NGS 11)

Critical Thinking

3. Many European countries started colonies in the area, and many people were captured as slaves.
4. Portuguese, Dutch, Germans, and British; established separate colonies, traded goods and slaves

Organizing What You Know

5. Time lines may vary but should include relevant events from the section.

◄ **Visual Record Answer**

for easier access to farmland and natural resources

Section 3

Objectives

1. Describe South Africa's policy of apartheid.
2. Identify the factors that led to the end of apartheid.
3. Describe South Africa's economy.
4. Analyze South Africa's prospects for the future.

Focus

LET'S GET STARTED

Copy the following questions onto the chalkboard: *Would you like to visit South Africa? Why or why not?* Have students respond in writing. Explain that in the mid- to late 1900s most U.S. citizens did not travel to South Africa because of the way the government treated a majority of its people. Now that the government has changed, tourists from the United States are enjoying South Africa's many attractions. Tell students that in Section 3 they will learn more about apartheid and South Africa.

Building Vocabulary

Have students look at the first half of the word **apartheid** *(apart)*. What can they conclude about the word's meaning? In Dutch, *-heid* is similar to the English suffix *-hood*, meaning a "condition" or "quality." Therefore, apartheid is the condition of being apart or separate. Tell students that **townships** and **sanctions** are terms that can be related to apartheid.

Section 3 Resources

Reproducible

- Guided Reading Strategy 23.3

Technology

- One-Stop Planner CD–ROM, Lesson 23.3
- Homework Practice Online
- *ARGWorld* CD–ROM: Malaria and Other Diseases in Africa
- HRW Go site

Reinforcement, Review, and Assessment

- Section 3 Review, p. 521
- Daily Quiz 23.3
- Main Idea Activity 23.3
- English Audio Summary 23.3
- Spanish Audio Summary 23.3

Section 3 South Africa Today

Read to Discover

1. What was South Africa's policy of apartheid?
2. What factors led to the end of apartheid?
3. What is South Africa's economy like?
4. What are South Africa's prospects for the future?

Define

apartheid
townships
sanctions

Locate

Witwatersrand
Johannesburg
Durban
Port Elizabeth

Zulu in traditional warrior's clothing

WHY IT MATTERS

South Africa is working to make sure all citizens are treated fairly today. Use CNNfyi.com to find examples of recent events in South Africa that reflect these efforts. Record your findings in your journal.

Nelson Mandela was sentenced to life imprisonment in 1964 but was released in 1990. In 1994 he was elected president of South Africa.

Racial Divisions

In the early 1900s South Africa's government, which was dominated by Afrikaners, became increasingly racist. Some black South Africans opposed the government. They formed the African National Congress (ANC) in 1912 to defend their rights. However, the trend toward racial division and inequality continued.

After World War II South Africa became an independent country. The South African government set up a policy of separation for its different peoples. This policy was called **apartheid**, meaning "apartness." The government divided people into three groups: whites, coloureds and Asians, and blacks—the overwhelming majority. Coloureds and Asians were only allowed to live in certain areas. Each African tribe or group was given its own rural "homeland."

The whites owned most of the good farmland. They also owned the mines and other natural resources. Black Africans had no rights in white areas. Blacks' land, housing, and health care were poor compared to those for whites. Education for blacks was limited, and classes were often taught in Afrikaans. Coloureds' facilities were poor but slightly better than those for blacks. People who protested these rules were sent to prison. One of those imprisoned was a lawyer named Nelson Mandela, a leader of the ANC.

Many blacks found work in white-owned industries, mines, shops, and farms. They had to live in separate areas called **townships**. These were often crowded clusters of tiny homes. They were far from the jobs in the cities and mines.

✓ **READING CHECK:** *Human Systems* What was apartheid? South African government policy of separating its different peoples

TEACH

Teaching Objective 1

ALL LEVELS: (Suggested time: 30 min.) Write *What Apartheid Was Like* on the chalkboard. Have students volunteer information from the text that describes the policy of apartheid. Write their descriptions on the board. Pair students and have each pair create a news interview about life for black South Africans under apartheid. Have students present their interviews to the class. **ENGLISH LANGUAGE LEARNERS, COOPERATIVE LEARNING**

Teaching Objective 2

ALL LEVELS: (Suggested time: 20 min.) Have students list the types of sanctions used against South Africa. Then have them describe types of protests used within the country. Have students write a paragraph that explains why these sanctions and protests might persuade a government to change its policy. **ENGLISH LANGUAGE LEARNERS**

TEACHER TO TEACHER

Jim Corley of Durant, Oklahoma, suggests the following activity to help students learn more about southern Africa: Organize students into groups of five. Assign each group a region or country of southern Africa and each group member an essential element of geography. After group members have researched the region focusing on an element, have them combine their information and present it to the class.

▲ In townships like Kayalitsha, black workers lived in crude shacks.

Interpreting the Visual Record **How did the apartheid system affect the roles and responsibilities of South Africans?**

Pressure against South Africa

Many people around the world objected to South Africa's apartheid laws. Some countries banned trade with South Africa. Some companies in the United States and Europe refused to invest their money in South Africa. Many international scientific and sports organizations refused to include South Africans in their meetings and competitions. These penalties, called **sanctions**, were intended to force South Africa to end apartheid.

South Africa and Its Neighbors During the 1960s, 1970s, and 1980s other countries in southern Africa gained their independence from colonial rule. British colonists in Rhodesia protested Britain's decision to grant independence. They declared their own white-dominated republic in 1970. This break resulted in years of violence and civil war. Finally the white government agreed to hold elections. They turned the country over to the black majority. The new government renamed the country Zimbabwe.

Mozambique was granted independence in 1975 after 10 years of war against Portuguese rule. However, rebels backed by Rhodesia and South Africa plunged Mozambique into another long war. Despite violent resistance, Namibia continued to be ruled by South Africa until independence in 1990.

The End of Apartheid As other countries in southern Africa gained independence, South Africa became more and more isolated. Protest within the country increased. The government outlawed the ANC. Many ANC members were jailed or forced to leave the country. Antiapartheid protesters turned increasingly to violence. South African

GLOBAL PERSPECTIVES

The economic sanctions that were imposed on the Republic of South Africa as a means of pressuring the country to abandon its apartheid policies were very effective. South Africa lost billions of dollars in trade and investment because of these sanctions. In September 1993 African National Congress leader Nelson Mandela made a historic plea before the antiapartheid committee of the UN General Assembly. He asked that remaining economic sanctions against South Africa be lifted. Mandela's speech came one day after South Africa's parliament voted to allow all races to participate in the country's politics for the first time.

Critical Thinking: Why might Mandela have asked for sanctions to be lifted so soon after the parliament's vote?

Answer: Students may suggest that Mandela hoped to become a political leader in South Africa and therefore wanted to resolve the country's economic problems.

◄ **Visual Record Answer**

It gave all political powers to whites and severely limited the role that black Africans and others could play in both politics and society as a whole.

Teaching Objective 3

ALL LEVELS: (Suggested time: 10 min.) Copy the following graphic organizer onto the chalkboard, omitting the italicized answers. Have students complete it to help them understand South Africa's economy. **ENGLISH LANGUAGE LEARNERS**

SOUTH AFRICA'S ECONOMY		
Resources	**Industries**	**Concerns**
coal, hydroelectric power, uranium, gold, diamonds, copper, platinum, iron ore, and chromium	*computers, cars, televisions, and other products needed for modern life*	*better working conditions for black workers and farmers, most mineral wealth and industries still owned by whites*

Teaching Objective 4

LEVEL 1: (Suggested time: 20 min.) Pair students and have each pair develop an agenda for a conference about South Africa's future. The agenda should address challenges facing South Africa as well as opportunities for the future. **ENGLISH LANGUAGE LEARNERS, COOPERATIVE LEARNING**

LEVELS 2 AND 3: (Suggested time: 25 min.) Have students draft a campaign speech for a presidential candidate in South Africa. Speeches should address challenges facing the country, possible solutions to those challenges, and the country's prospects for the future. Have volunteers read their speeches aloud to the class.

GLOBAL PERSPECTIVES

As South Africa works to improve its economy, one industry is making changes throughout Africa. A radio invented in Great Britain may provide people in countries with little electricity and no money for batteries the ability to receive information on everything from health to election news. Operated by windup springs that provide power, the radios are manufactured in South Africa and are sufficiently inexpensive that people in many developing countries can afford them. The company that manufactures the radio also makes windup flashlights. The company provides needed jobs and produces needed products for sale in developing countries.

Activity: Have students prepare a radio show about the daily life of American students that could be broadcast to students in a developing country.

Visual Record Answer ▶

Some diamonds are not as clear and beautiful as is necessary for jewelry.

Chart Answer (p. 521) ▶

$8,500

▲ Members of the African National Congress lead a rally celebrating Nelson Mandela's release from prison.

forces attacked suspected rebel bases in neighboring countries like Botswana and Zimbabwe.

Finally, in the late 1980s South Africa began to move away from the apartheid system. In 1990 the government released Nelson Mandela from prison. Mandela was elected president in 1994 after all South Africans were given the right to vote.

Today all races have equal rights in South Africa. Public schools and universities are open to all, as are hospitals and transportation. However, economic equality has been slow in coming. White South Africans are still wealthier than the vast majority of black South Africans. Also, divisions between different black ethnic groups have caused new tensions. Still, South Africans now hope for a better future.

✓ **READING CHECK:** *Human Systems* Why and how did South Africa do away with apartheid? sanctions, isolation brought pressure; released Mandela, allowed equal rights, multiracial elections

South Africa's Economy

South Africa's government is trying to create jobs and better conditions for black workers and farmers. However, South Africa's mineral wealth and industries are still mostly owned by white people. Even officials who favor reform are afraid that too-rapid change will weaken the economy. They fear it will drive educated and wealthy whites to leave the country. The government has avoided taking white-owned farmland to divide among black farmers.

South Africa's energy resources include coal and hydroelectric power. Rich uranium mines provide fuel for nuclear power plants. In addition to gold and diamonds, mineral resources include copper, platinum, iron ore, and chromium.

The Witwatersrand region around Johannesburg is the continent's largest industrial area. South African and foreign companies build computers, cars, televisions, and many other products needed for modern life. The major port is Durban on the Indian Ocean coast. Cape Town and Port Elizabeth are other important ports.

✓ **READING CHECK:** *Human Systems* How has apartheid continued to affect South Africa's economy today? Mineral wealth, industries, and farmland are mostly owned by whites.

These are uncut, or rough, diamonds. Diamonds are the hardest mineral. This makes them useful for certain types of cutting and drilling.

Interpreting the Visual Record
Why do you think some diamonds are more suitable for industrial uses than for jewelry? ▶

South Africa's Future

South Africa has more resources and industry than most African countries but faces severe problems. It must begin to deliver equal education and economic opportunities to the entire population.

CLOSE

Remind students that the end of apartheid in South Africa offers hope and challenges for the future. Ask students to identify some of the hopes for South Africa's future.

REVIEW AND ASSESS

Have students complete the Section Review. Have students create crossword puzzles about South Africa. Photocopy the puzzles and have students exchange and complete them. Then have students complete Daily Quiz 23.3.

RETEACH

Have students complete Main Idea Activity 23.3. Organize students into four groups. Assign the first group Apartheid, the second End of Apartheid, the third South Africa's Economy, and the fourth Challenges. Have each group use pictures and text from magazines to create a collage of information about its topic. Display and discuss the collages.
ENGLISH LANGUAGE LEARNERS, COOPERATIVE LEARNING

EXTEND

Have interested students conduct research on the African National Congress and make a scrapbook depicting important events in the organization's history. Display the scrapbooks in the classroom.
BLOCK SCHEDULING

Cape Town has grown into a large industrial city. In this photograph, the harbor is visible beyond the tall buildings of the business district.

New problems have arisen since the end of apartheid. Crime has increased in the large cities. Also, South Africa, like the rest of the region, is facing a terrible AIDS epidemic.

There are 11 official languages in South Africa, although English is used in most areas. South Africa produces fine wines from the region around Cape Town. It has a unique cooking style combining Dutch, Malay, and African foods. The country also has a lively tradition of literature and the arts. Today, traditional ethnic designs are used in clothing, lamps, linens, and other products. These are sold to tourists and locals alike.

South Africa

Country	Population/ Growth Rate	Life Expectancy	Literacy Rate	Per Capita GDP
South Africa	43,586,097 0.3%	48, male 49, female	82%	$8,500
United States	281,421,906 0.9%	74, male 80, female	97%	$36,200

Sources: Central Intelligence Agency, *The World Factbook 2001;* U.S. Census Bureau

Interpreting the Visual Record What is South Africa's per capita GDP?

✓ **READING CHECK:** *Places and Regions* What are the challenges faced by South Africa? unequal education and economic opportunities, crime, AIDS

go.hrw.com Homework Practice Online
Keyword: SK3 HP23

Section 3 Review

Define and explain: apartheid, townships, sanctions

Working with Sketch Maps On the map you created in Section 2, label Witwatersrand, Johannesburg, Durban, and Port Elizabeth.

Reading for the Main Idea

1. *Human Systems* What was the system of apartheid, and how did it affect the roles and responsibilities of South Africans?
2. *Human Systems* How did people around the world protest apartheid?
3. *Places and Regions* What challenges remain for South Africa?

Critical Thinking

4. **Analyzing Information** Why is South Africa the continent's most economically developed country?

Organizing What You Know

5. **Summarizing** Use this graphic organizer to list information about life in South Africa during apartheid and since the system ended.

During apartheid	Since apartheid ended

Section 3 Review

Answers

Define For definitions, see: apartheid, p. 518; townships, p. 518; sanctions, p. 519

Working with Sketch Maps Maps will vary, but listed places should be labeled in their approximate locations.

Reading for the Main Idea

1. a system of racial segregation; limited power to whites (NGS 13)
2. by banning trade with South Africa, refusing to invest money in South Africa, refusing South Africa's admission to meetings and competitions (NGS 12)
3. lack of equal education and economic opportunities for all races, increased crime rate, and an AIDS epidemic (NGS 4)

Critical Thinking

4. energy/mineral resources, developed industries, and important ports

Organizing What You Know

5. apartheid—whites owned most natural resources, Black Africans had no rights in white areas, were forced to live in townships with poor facilities; since the end of apartheid—all races have equal rights, political prisoners freed

Section 4

Objectives

1. Identify the groups that influenced Namibia's culture.
2. Describe the factors that have helped Botswana's economy.
3. Describe Zimbabwe's economy.
4. Explain why Mozambique remains poor.
5. Describe the events that marked Madagascar's recent history.

Focus

LET'S GET STARTED

Copy the following instruction and question onto the chalkboard: *Look at the physical-political map in Section 1. What physical features might affect the economic and political conditions in Namibia, Botswana, Zimbabwe, Mozambique, and Madagascar?* Have students respond in writing. *(possible answers: deserts, some are landlocked, isolated island)* Tell students that in Section 4 they will learn more about these southern African countries.

Building Vocabulary

Write the key term on the chalkboard. Ask students why a country might join the **Organization of African Unity**. Explain to students that this organization tries to promote cooperation between African countries.

Section 4 Resources

Reproducible
- Guided Reading Strategy 23.4
- Map Activity 23

Technology
- One-Stop Planner CD–ROM, Lesson 23.4
- Homework Practice Online
- HRW Go site

Reinforcement, Review, and Assessment
- Section 4 Review, p. 525
- Daily Quiz 23.4
- Main Idea Activity 23.4
- English Audio Summary 23.4
- Spanish Audio Summary 23.4

Visual Record Answer ►

lack of resources and the encroachment of modern society into their region

Section 4 The Other Southern African Countries

Read to Discover

1. What groups have influenced Namibia's culture?
2. What factors have helped Botswana's economy?
3. What is Zimbabwe's economy like?
4. Why has Mozambique remained so poor?
5. What events have marked Madagascar's recent history?

Define

Organization of African Unity (OAU)

Locate

Windhoek
Gaborone
Harare
Maputo

Grilled shrimp from Mozambique

Why It Matters

Like many African countries, Madagascar has suffered in recent years from environmental problems. Use CNNfyi.com or other **current events** sources to find examples of how Madagascar is coping with its environmental problems. Record your findings in your journal.

A San grandmother and child drink water from an ostrich egg. The San people traditionally lived by hunting and gathering. However, just a few live this way today.

Interpreting the Visual Record **What factors do you think might lead many San people to give up their traditional way of life?**

▼

Namibia

Most Namibians live in the savannas of the north or in the cooler central highlands. Windhoek, the capital, is located in these highlands. About 7 percent of the population is white, mainly of German descent. The rest of the population is divided among several different ethnic groups. Most Namibians are Christian. English is the official language. However, schooling was in Afrikaans until recently.

At independence in 1990, white farmers held most of the productive land. Most of Namibia's income comes from the mining of diamonds, copper, lead, zinc, and uranium. Fishing in the Atlantic Ocean and sheep ranching are also important sources of income.

Namibian culture shows many different influences. In many rocky areas, ancient rock engravings and paintings of the Khoisan are preserved. Beer and pastries reflect the period of German colonization.

✓ **READING CHECK:** *Human Systems* What are Namibia's cultural influences? white people of German descent; many other different cultures, including ancient Khoisan peoples

Botswana

Botswana is a large, landlocked, semiarid country. Thanks to mineral resources and stable political conditions, Botswana is one of Africa's success stories. Cattle ranching and mining of copper and diamonds are the principal economic activities. Recently, international companies have set up factories here. A new capital,

TEACH

Teaching Objective 1

ALL LEVELS: (Suggested time: 25 min.) Pair students and have each pair design a magazine cover for an issue featuring Namibian culture. Display covers around the room. **ENGLISH LANGUAGE LEARNERS, COOPERATIVE LEARNING**

Teaching Objectives 2–3

ALL LEVELS: (Suggested time: 30 min.) Pair students and have them write an essay that compares and contrasts the economies of Botswana and Zimbabwe. Essays should include what factors helped make the economies successful.
ENGLISH LANGUAGE LEARNERS, COOPERATIVE LEARNING

➤ASSIGNMENT: Tell students to imagine that they are traveling through southern Africa. Have them write journal entries describing each country they visit. Entries should mention such items as economy, culture, language, and challenges.

Gaborone, was built after independence. Like the other countries of the region, Botswana belongs to the **Organization of African Unity (OAU)**. The OAU, founded in 1963, tries to promote cooperation between African countries.

Botswana's population is less than 1.6 million. About 79 percent belong to a single ethnic group, the Tswana. Most of the population live in the savanna and steppe areas of the east and south. The San and other minority groups mostly live in the northern swamps and the Kalahari Desert. About half of Botswana's people are Christian. The rest follow traditional African religions.

Botswana's major river, the Okavango, flows from Angola into a huge basin. The swamps of this basin are home to elephants, crocodiles, antelope, lions, hyenas, and other animals. Many tourists travel to Botswana to see these wild animals in their habitat.

Traditional crafts of Botswana include ostrich-eggshell beadwork and woven baskets with complex designs. People there also produce colorful wool tapestries and rugs.

✓ **READING CHECK:** *Places and Regions* Why might Botswana be considered a success story? because it has had stable political conditions, has made good use of its resources, and has attracted foreign investment

▲ The Okavango River spreads out to form a large swampy area. Dense vegetation allows only small boats to move through the narrow channels.

Zimbabwe

Zimbabwe's capital is Harare. Zimbabwe gained independence in 1980. Since then, the country has struggled to create a more equal distribution of land and wealth. White residents make up less than 1 percent of the population. However, they still own most of the large farms and ranches.

Zimbabwe exports tobacco, corn, sugar, and beef. It now manufactures many everyday items, including shoes, batteries, and radios. Exports of gold, copper, chrome, nickel, and tin are also important to Zimbabwe's economy.

The AIDS epidemic threatens to kill hundreds of thousands of Zimbabwe's people, leaving many orphans behind. These effects will make economic growth harder. Other diseases such as malaria and tuberculosis are often deadly. There are also tensions between the majority Shona people and the minority Ndebele.

Artists in Zimbabwe have revived the tradition of stone sculpture found at Great Zimbabwe. Some of the larger pieces are among the most striking examples of modern art in the world.

✓ **READING CHECK:** *Places and Regions* What is Zimbabwe's economy like? exports agricultural products, minerals; manufactures everyday items

Other Southern African Countries

Country	Population/ Growth Rate	Life Expectancy	Literacy Rate	Per Capita GDP
Botswana	1,586,119 / 0.5%	37, male 38, female	70%	$6,600
Madagascar	15,982,563 / 3%	53, male 58, female	80%	$800
Mozambique	19,371,057 / 1.3%	37, male 36, female	42%	$1,000
Namibia	1,797,677 / 1.4%	42, male 39, female	38%	$4,300
Zimbabwe	11,365,366 / 0.2%	39, male 36, female	85%	$2,500
United States	281,421,906 / 0.9%	74, male 80, female	97%	$36,200

Sources: Central Intelligence Agency, *The World Factbook 2001;* U.S. Census Bureau

Interpreting the Chart **Which countries are the least economically developed in the region, according to the chart?**

Linking Past to Present

Equality in Zimbabwe In the years following Zimbabwe's war for independence from white colonialists, Zimbabwean women became accustomed to equality with men. The 1980 constitution prohibited discrimination. In 1982 a law was passed stating that women more than 18 years old could not be considered minors, despite traditions that deprived them of rights in marriage and divorce, adoption, and inheritance of property. Then, in 1999 Zimbabwe's Supreme Court decided that women are not equal to men because of African custom and that laws cannot supersede custom. A woman had filed suit against her brother-in-law after he evicted her from the house she believed she inherited from her father. The woman expected Zimbabwe's laws and the international human rights treaties signed by Zimbabwe to protect her rights, but the 5-to-0 decision against her was final.

Critical Thinking: How and why has Zimbabwe limited the roles and responsibilities of its female citizens?

Answer: It has declared women unequal to men because of African custom.

◄ **Chart Answer**

Madagascar, Mozambique

Teaching Objective 4

LEVEL 1: (Suggested time: 15 min.) Have students write a letter home as if they were exchange students in Mozambique. Letters should describe the causes of the country's poverty and explain what the country is doing to try to solve its economic challenges. **ENGLISH LANGUAGE LEARNERS**

LEVELS 2 AND 3: (Suggested time: 25 min.) Have students write scripts for an oral presentation on Mozambique. Presentations should state why Mozambique is a poor country as well as offer suggestions for rebuilding its economy. Have volunteers present their information to the class.

Teaching Objective 5

ALL LEVELS: (Suggested time: 20 min.) Copy the following graphic organizer onto the chalkboard, omitting the italicized answers. Have students complete the time line to document recent events in Madagascar's history. **ENGLISH LANGUAGE LEARNERS**

MADAGASCAR'S HISTORY

1800s	early 1990s		1996
French colony	*rule of socialist dictator ends*	*period of democracy*	*former dictator voted back into power*

Section 4 Review

Answers

Define For definition, see: Organization of African Unity, p. 523

Working with Sketch Maps Windhoek, Gaborone, and Harare are located on the plateau.

Reading for the Main Idea

1. threaten to kill thousands and leave many orphans, and make economic growth harder (NGS 9)
2. slowed economic development (NGS 13)
3. diamonds, copper, lead, zinc, uranium; provide money through mining and sale of the minerals

Critical Thinking

4. Students might suggest that Botswana has not experienced as much ethnic political conflict as other southern African countries.

Organizing What You Know

5. Charts should include information from the section.

Connecting to History Answers

1. by European participants at a conference in Berlin
2. divided some ethnic groups into separate colonies and combined other groups that were hostile to each other

Connecting to History

Africa's Borders

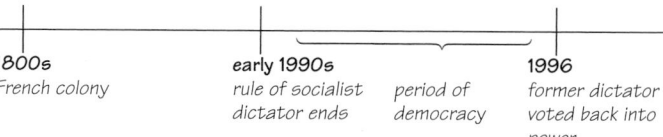

The Granger Collection, New York

Cartoon of France and Britain dividing Africa

In 1884–85, representatives of Europe's colonial countries met in Berlin. These countries included Belgium, France, Germany, Great Britain, Italy, Portugal, and Spain. Each was conquering areas in Africa. Their claims to territory were beginning to overlap. Leaders began to worry that a rivalry in Africa might trigger a war in Europe.

The Berlin conference was called to agree to boundaries for these African colonies. No Africans were invited to the conference. The European representatives divided up Africa among themselves. The new borders sometimes followed physical features, such as lakes and mountains. Many simply followed straight lines of latitude or longitude. Often people from the same ethnic group were separated by the new borders. In other places, ethnic groups hostile to each other were grouped together.

Most of the European colonies in Africa became independent after 1960. However, the leaders of the new African countries have mostly avoided drawing new boundary lines. So these countries still struggle with the borders they inherited from the Berlin Conference. These borders have made it hard for many African countries to build national loyalty among their citizens.

Understanding What You Read

1. How were Africa's borders established?
2. What consequences have these borders had for modern Africa?

Mozambique

Mozambique is one of the world's poorest countries. Its economy was badly damaged by civil war after independence from Portugal. Today, Mozambique's ports of Maputo—the capital—and Beira once again ship many products from interior Africa. The taxes collected on these shipments are an important source of revenue. Energy sources include coal and new hydroelectric dams on the Zambezi River. Plantations grow cotton, cashews, sugar, and tea.

Most of Mozambique's people belong to various Bantu ethnic groups. Each group has its own language. However, the country's official language is Portuguese. About 30 percent of the people are Christian, and 20 percent are Muslim.

BUILD on WHAT You Know

Do you remember what you learned about uneven resource distribution? See Chapter 5 to review.

CLOSE

Remind students that location and climate as well as political and ethnic issues have created challenges for many southern African countries. Have students identify how location and climate have created challenges for southern Africa and suggest possible solutions to these challenges.

REVIEW AND ASSESS

Have students complete the Section Review. Then organize students into five groups. Assign each group a southern African country and have its members write six questions and answers on index cards about their country. Have groups exchange cards and quiz each other. Then have students complete Daily Quiz 23.4. **COOPERATIVE LEARNING**

RETEACH

Have students complete Main Idea Activity 23.4. Pair students and have them develop a chart that compares the economies of all the countries of southern Africa. **ENGLISH LANGUAGE LEARNERS, COOPERATIVE LEARNING**

EXTEND

Have interested students create dioramas of the Namib Desert or the Kalahari Desert. They should display the landforms, plant and animal life, economic activities, and the presence of mineral resources. **BLOCK SCHEDULING**

Mozambique is famous for its fiery pepper or *peri-peri* sauces. They are often served on shrimp and rice.

✓ **READING CHECK:** *Places and Regions* What factor limited Mozambique's economic development? civil war

Madagascar

Madagascar is a former French colony. It was ruled by a socialist dictator until the early 1990s. At that point the people demanded a new political system. The optimism that came with democracy faded as the new leaders struggled with poverty. Surprisingly, in 1996 the people voted the former dictator back into power.

Nearly all of Madagascar's people are still very poor. There is little industry. Most of the country's income comes from exports of coffee, sugar, vanilla, and cloves. Most of the people depend on subsistence farming.

Madagascar has many animals found nowhere else. This is because the island has been separated from the African mainland for millions of years. Some 40 species of lemurs, relatives of apes, live only on this island. However, destruction of the rain forests threatens many of Madagascar's animals with extinction.

Malagasy and French are spoken throughout Madagascar. About 55 percent of the people follow traditional African religions. Some 40 percent are Christian, and about 5 percent are Muslim.

✓ **READING CHECK:** *Places and Regions* What has Madagascar's recent history been like? People voted former dictator into power, people are poor, little industry

Remnants of French culture can still be seen in Madagascar.

Interpreting the Visual Record **Which of these vendors' foods suggests a connection to French culture?**

Section 4 Review

go.hrw.com Homework Practice Online
Keyword: SK3 HP23

Define and explain: Organization of African Unity (OAU)

Working with Sketch Maps On the map you created in Section 3, label Windhoek, Gaborone, Harare, and Maputo. Which cities are located on the plateau of the southern African interior?

Reading for the Main Idea

1. *Human Systems* How have diseases affected Zimbabwe's people and economy?
2. *Human Systems* How did civil war affect Mozambique?
3. *Environment and Society* What minerals does Namibia have, and how do they affect its economy?

Critical Thinking

4. **Drawing Inferences and Conclusions** How do you think the fact that Botswana's population is almost entirely of a single ethnic group has affected its politics?

Organizing What You Know

5. **Summarizing** Copy the following graphic organizer. Use it to list information about southern African cultures.

Country	Culture
Namibia	
Botswana	
Zimbabwe	
Mozambique	
Madagascar	

CHAPTER 23 Review ANSWERS

Building Vocabulary

For definitions, see: enclaves, p. 512; the veld, p. 512; pans, p. 513; Boers, p. 516; apartheid, p. 518; townships, p. 518; sanctions, p. 519; Organization of African Unity, p. 523

Reviewing the Main Ideas

1. The Kalahari has pans and mineral deposits. The Namib Desert receives almost no rain, but a night fog provides water for some plants and animals. (NGS 4)
2. monsoon winds that blow across the Indian Ocean (NGS 15)
3. the Bantu language family, use of iron, and cattle herding (NGS 9)
4. Gold, diamonds, copper, platinum, iron ore, and chromium contribute to its wealth. (NGS 15)
5. whites, Coloureds and Asians, and Blacks; Blacks (NGS 9)

Visual Record Answer

the long loaves of bread

Assess

Have students complete the Chapter 23 Test.

Reteach

Attach a large piece of paper to the wall and draw an outline map of the region on it. Organize students into groups and assign each group one of the countries of the region. Have groups represent each country's physical features, economic activities, and cultural aspects on the map. Display and discuss the map. **ENGLISH LANGUAGE LEARNERS, COOPERATIVE LEARNING**

Portfolio Extensions

PORTFOLIO

1. Have students create a design for a commemorative stamp for a country of southern Africa. The stamp should honor or celebrate an important event or person. Place designs in students' portfolios.
2. Have students locate and listen to contemporary southern African music. Students might search for artists such as Ladysmith Black Mombazo, Dilika, Johnny Clegg, and The Boyoyo Boys. Then ask students to imagine they are music critics for a magazine and have them write reviews for the music they selected.

CHAPTER 23

Review ANSWERS

Understanding Environment and Society

Presentations should identify what physical features have contributed to cities' growth and what southern African cities owe their locations to mineral deposits. Use Rubric 29, Presentations, to evaluate student work.

Thinking Critically

1. on the eastern coast; good climate and soil for farming
2. economic isolation from other countries; increased protests
3. white descendants of original Dutch colonists who challenged British rule
4. unstable governments, lack of industry, ethnic conflicts, and disease
5. encouraged many Europeans to settle there and establish colonies

CHAPTER 23

Reviewing What You Know

Building Vocabulary

On a separate sheet of paper, write sentences to define each of the following words.

1. enclaves
2. the veld
3. pans
4. Boers
5. apartheid
6. townships
7. sanctions
8. Organization of African Unity (OAU)

Reviewing the Main Ideas

1. *Places and Regions* What are southern Africa's two deserts like?
2. *Environment and Society* What made it possible for early southern Africans to trade with Asians?
3. *Human Systems* What did the original Bantu migrants bring to southern Africa?
4. *Environment and Society* How have mineral resources affected South Africa?
5. *Human Systems* What three racial groups were defined and separated by apartheid? Which group made up the majority of the population?

Understanding Environment and Society

City Location

Johannesburg grew because of gold deposits nearby. Cape Town's location is excellent for a port. Create a presentation on why cities are located where they are. As you prepare your presentation you may want to think about the following:

- What physical features seem to contribute to the growth of cities around the world?
- Which major cities of southern Africa owe their locations to deposits of minerals?

Thinking Critically

1. **Drawing Inferences and Conclusions** In what parts of southern Africa do you think most farming takes place? Why?
2. **Analyzing Information** What finally motivated the South African government to end the apartheid system?
3. **Summarizing** Who were the Afrikaners, and what role did they play in the history of southern Africa?
4. **Summarizing** What factors have slowed the economic development of the countries in this region?
5. **Finding the Main Idea** How did the discovery of diamonds and gold affect the settlement of Europeans in southern Africa?

Food Festival

Have students prepare a southern Africa food safari. Serve fruit in a vanilla-sugar syrup topped with lichee nuts (Madagascar), Portuguese cheese and black olives (Mozambique), and Mealie Bread (South Africa). Here is a recipe for Mealie Bread: 2 c. biscuit mix, 1 c. canned creamed corn, 2 tbs sugar, 1 egg, ½ c. milk. Mix and spread in 9 in. baking pan. Spread 2 oz. melted butter over mixture. Bake in a 400° oven for 20 min.

Chapter 23 Review and Assessment Resources

Reproducible
- Readings in World Geography, History, and Culture 64, 65, and 66
- Critical Thinking Activity 23
- Vocabulary Activity 23

Technology
- Chapter 23 Test Generator (on the One-Stop Planner)
- HRW Go site
- Audio CD Program, Chapter 23

Reinforcement Review, and Assessment
- Chapter 23 Review, pp. 526–27
- Chapter 23 Tutorial for Students, Parents, Mentors, and Peers
- Chapter 23 Test
- Chapter 23 Test for English Language Learners and Special-Needs Students
- Unit 6 Test
- Unit 6 Test for English Language Learners and Special-Needs Students

Building Social Studies Skills

Map Activity

On a separate sheet of paper, match the letters on the map with their correct labels.

Drakensberg	**Kalahari Desert**
Inyanga Mountains	**Namib Desert**
Cape of Good Hope	**Orange River**
	Limpopo River
	Harare

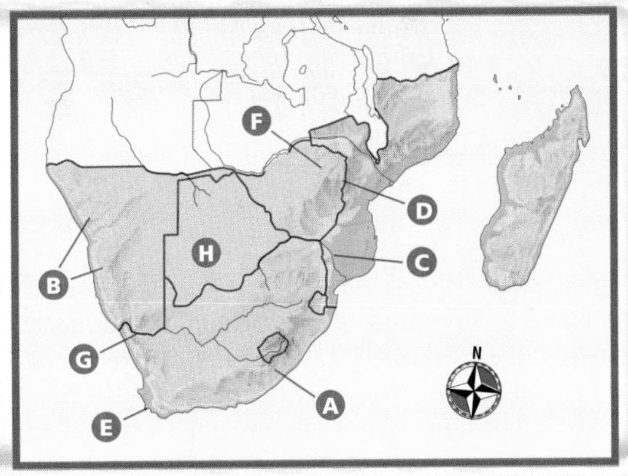

Mental Mapping Skills Activity

On a separate sheet of paper, draw a freehand map of southern Africa. Make a key for your map and label the following:

Botswana	**Namibia**
Madagascar	**South Africa**
Mozambique	**Zimbabwe**

Writing Activity

Write a brief essay in which you compare and contrast Botswana and Mozambique. How are the physical features of these two countries similar or different? How do these features affect the economies of Botswana and Mozambique? What are the major industries? Be sure to use standard grammar, spelling, sentence structure, and punctuation.

Alternative Assessment

Portfolio Activity

Learning About Your Local Geography

Mineral Resources Southern Africa is rich in minerals. Working in a group, list mineral resources found in your area. Then create a display about local mineral resources.

internet connect

Internet Activity: **go.hrw.com**
KEYWORD: **SK3 GT23**

Choose a topic to explore about southern Africa:
- Explore the Namib Desert.
- Go on a South African safari.
- Investigate apartheid.

Map Activity
A. Drakensberg
B. Namib Desert
C. Limpopo River
D. Inyanga Mountains
E. Cape of Good Hope
F. Harare
G. Orange River
H. Kalahari Desert

Mental Mapping Skills Activity
Maps will vary, but listed places should be labeled in their approximate locations.

Writing Activity
Student essays should describe the physical features, economies, and major industries of the two countries. Use Rubric 9, Comparing and Contrasting, to evaluate student work.

Portfolio Activity
Group displays should discuss several mineral resources of your area. Use Rubric 14, Group Activity, to evaluate student work.

internet connect

GO TO: **go.hrw.com**
KEYWORD: **SK3 Teacher**
FOR: **a guide to using the Internet in your classroom**

FOCUS ON ENVIRONMENT

Solving Local Problems

After students have read the feature, ask if they were inspired by Wangari Maathai's story. Begin a class discussion about ways that individuals can be effective agents for change in their communities. Encourage students to talk about people they may know who address local problems as Maathai has done. Ask students to consider why Maathai's program was successful. *(The people who were directly affected by the problem of soil erosion were encouraged to participate directly in the solution.)*

GEOGRAPHY SIDELIGHT

In a country where women often do not participate in public life, Wangari Maathai is an exception. She holds a doctorate from the University of Nairobi and has received international recognition for her environmental activism.

In the late 1980s Maathai again acted on behalf of the environment. The major political party had wanted to build a skyscraper in a popular open space in downtown Nairobi. Maathai protested that it is as important to leave some land undeveloped for people as it is to preserve land for animals. She also protested the high cost of developing the area. The money was to be borrowed from foreign banks. Maathai protested that any more borrowed money should go to feeding, educating, and helping Kenya's people.

In early 1999 Maathai and several supporters were severely beaten when they tried to plant trees in a forest on Nairobi's outskirts.

Critical Thinking: How has Wangari Maathai worked to protect the environment?

Answer: She protested the development of a space in downtown Nairobi, saying that the money should be used for the people.

➤ **This Focus On Environment feature addresses National Geography Standards 4, 13, 14, and 16.**

FOCUS ON ENVIRONMENT

Kenya's Tree Planters

Ten million young trees scattered throughout Kenya offer new hope to farmers and families. The trees were planted by women who farm these lands. "We are planting trees to ensure our own survival," says Wangari Maathai. She is the founder of Kenya's Green Belt Movement.

Deforestation

Like many other developing countries, Kenya is a country of farmers and herders. The most productive farmlands are in the highlands. This area was once green with trees and plants. Over the past century, however, much of this land has been cleared. Today, less than 10 percent of the original forests remain. Many trees were cut for firewood, which people use for cooking.

Without tree roots to hold the soil in place, the land eroded. The land was losing its fertility. Farmers moved to the savannas in search of better land.

Wangari Maathai recognized what was happening. "When I would visit the village where I was born," she says, "I saw whole forests had been cleared for cultivation and timber. People were moving onto hilly slopes and riverbeds and marginal areas that were only bush when I was a child." Maathai was shocked to find children suffering from malnutrition. People were no longer eating foods such as beans and corn. Instead, they were eating refined foods such as rice. They were doing this because these foods need less cooking—and thus less firewood.

The Green Belt Movement

On June 5, 1977, in honor of World Environment Day, Maathai and a few supporters planted seven trees in Nairobi, Kenya's capital. Thus began the Green Belt Movement. This movement then spread through the Kenyan highlands and captured people's attention around the world.

From the beginning, Maathai knew that the success of her efforts depended mainly on women. In Kenya, men tend cash crops such as coffee and cotton. Women collect firewood and grow corn, beans, and other food crops. It is the women who grow the food their families eat. The women were the first to see the connection between poor soil and famine.

◄ Many of the trees in Kenya's highlands have been cleared for crops.

Going Further: Thinking Critically

Point out that the feature emphasizes the geography element of environment and society. Ask students to draw a before-and-after picture that illustrates how the interaction between environment and society was both harmful and beneficial to Kenya *(harmful: people cleared the forests to obtain firewood, causing soil erosion; beneficial: people planted trees to restore the soil and water supply)*. Ask students if they think such problems can be avoided in the future. Why or why not? How might the problems be avoided?

The movement's workers encouraged women in Kenya to plant trees. They pointed out that women would not have to walk miles to collect firewood. They would have wood available nearby for fires, fences, and buildings. If the tree seedlings survived, the women would also be paid a small sum of money.

First, a few small nurseries were started to give out free seedlings. The nurseries are staffed by local women who are paid for their work. Nurseries also train and pay local people known as Green Belt Rangers. The Rangers visit farms, check on seedlings, and offer advice.

▲ Workers from the Green Belt Movement teach schoolchildren in Kenya how to care for tree seedlings.

Before long, nurseries were appearing in communities throughout the highlands. Kenyan women talked to friends and neighbors about the benefits of planting trees. Neighbors encouraged Esther Wairimu to plant seedlings. Five years later, her fields were surrounded by mango, blue gum, and other trees. "I have learned that a tree… is life," she says.

Today, the advantages of planting trees are clear. Farmers now have fuel and shade. Even more important, the soil is being protected from erosion. The number of Green Belt nurseries has grown to 3,000. Most are run by women.

Wangari Maathai (right) has encouraged Kenyan women to plant millions of trees. Women are paid for the seedlings they plant that survive. ▼

Expanding the Movement

Maathai believes that local people must work together to protect the environment. She stresses that the Green Belt Movement relies on farmers. The movement receives little government support. Most of the money comes from small personal donations.

Now Maathai dreams of spreading her movement to other African countries. "We must never lose hope…," she says. "One person *can* make a difference."

Your Turn

Working in small groups, think of an environmental issue facing your community or state.

1. Develop a plan that demonstrates how "one person can make a difference."
2. Present your group's plan to the rest of the class.

Your Turn

Answers

1. Students' plans will vary but should detail an important environmental issue and outline an effective and realistic solution.
2. Use Rubric 29, Presentations, to evaluate student work.

GeoSkills

Going Further: Thinking Critically

Ask students to identify an organism that is endangered or approaching extinction and to conduct research on the ecosystem in which their chosen organism lives. Then ask them to write an essay explaining the role of their organism in its ecosystem and how the disappearance of that organism would affect the ecosystem. Finally, have each student draw two diagrams—one that illustrates the organism's ecosystem and another that shows what the ecosystem would be like if the organism disappeared.

You may want to have students concentrate on plants or animals often viewed as unattractive or a nuisance. For example, some people do not like snakes or bats, and many species of both animals are endangered. They are as important to their ecosystems as koalas, tigers, or giant pandas are to theirs. If you choose this option, have students create a public awareness campaign that highlights the entire ecosystem's health. You might also ask students to create a mascot for the campaign to enhance the image of the species.

Practicing the Skill

1. Students should describe an ecosystem in their region and note how the plants and animals that live there depend on each other for survival.
2. Students might suggest that the disappearance of one group would seriously affect the well-being of other groups.
3. Students might suggest that construction or pollution have affected ecosystems in their community. Students will probably note that some plants and animals are disappearing as a result of pressures from human activities.

➤ **This GeoSkills feature addresses National Geography Standards 8, 14, and 18.**

Geo Skills

Building Skills for Life: Understanding Ecosystems

The plants and animals in an area, together with the nonliving parts of their environment, form an ecosystem. An important thing to remember about ecosystems is that each part is interconnected.

Life on Earth depends on the energy and nutrients flowing through ecosystems. Energy and nutrients move between plants, animals, and soils through food chains and food webs.

Most ecosystems involve three groups of organisms: producers, consumers, and decomposers. Producers make their own food. Plants are producers. They make food by combining carbon dioxide, nutrients from the soil, sunlight, and water. Consumers are unable to make food. They have to get food from producers or from other consumers. Humans are consumers. We eat plants (producers) and animals (consumers). Decomposers get food from dead organisms and wastes. Bacteria and fungi are decomposers.

Elephants have adapted to the ecosystem of Africa's Namib Desert.

Like all forms of life, humans depend on ecosystems for survival. Knowing how ecosystems work helps us understand how we are connected to both living and nonliving things. It can also help us manage, protect, and use our environments wisely.

Practicing the Skill

1. Describe an ecosystem in the region where you live. What plants and animals live there? How are they connected?
2. Make a table showing some of the producers, consumers, and decomposers in an ecosystem. What do you think would happen if one of these groups suddenly disappeared?
3. Try to identify some ways that human activities have changed ecosystems in your community. Have some plants or animals disappeared?

HANDS ON GEOGRAPHY

Like all organisms, lions are part of an ecosystem. Lions survive by killing and eating other animals, such as zebras and gazelles. After lions have killed an animal, they eat their fill. Then other animals such as vultures and hyenas eat what is left.

The lion itself is food for other organisms. Small animals like ticks, fleas, and mosquitoes drink the lion's blood. The lion's waste serves as food for organisms that live in the soil. When the lion dies, it will be eaten by other animals.

How does the lion fit into its ecosystem? You can answer this question by drawing a connections web. These guidelines will help you get started.

1. Draw a picture of a lion.
2. Think of interactions that lions have with other parts of their environment.
3. Include these interactions in your drawing. For example, lions eat zebras, so you could add a zebra to your drawing.
4. Draw lines connecting the lion to other parts of its environment. For example, you could draw a line connecting the lion and the zebra.
5. Be sure to extend the connections to include nonliving parts of the environment as well. For example, lions eat zebras, zebras eat grass, and grass depends on sunlight.
6. Continue making connections on your diagram. When you are done, answer the Lab Report questions.

Lab Report

1. Label the producers, consumers, and decomposers on your connections web. To which group do lions belong?
2. How are lions dependent on nonliving parts of their environment?
3. Imagine there are no more lions. How do you think the ecosystem shown in your connections web would be affected?

Lab Report

Answers

1. Students' webs should be clearly labeled and should identify lions as consumers.
2. Lions are indirectly dependent on nonliving parts of the environment, such as sunlight. Sunlight helps grass grow. Zebras eat the grass. Lions eat the zebras.
3. Students should demonstrate how the disappearance of lions could cause overpopulation and eventual starvation for the zebras. The vultures and hyenas may also be adversely affected. Also, insects and small organisms in the soil would have less food.

UNIT 7

EAST AND SOUTHEAST ASIA

Using the Illustrations

Direct students' attention to the photographs on these pages. Have them select the photo that may portray a nomadic way of life *(the Mongolian man outside his home).* Ask what led them to this conclusion. *(Possible answer: The house appears temporary or movable.)* Then inquire which photo shows a settled way of life and why students chose it *(rice farmer; shows agriculture).* Ask how the Chinese farmer's field compares to farmland students have seen in the United States. *(Possible answers: The plants are in standing water. There are no distinct rows.)* Ask what conclusions students can draw from this. *(Rice farming requires plenty of water. Much work is done by hand.)*

Point out the Shwedagon Pagoda, a Buddhist shrine in Myanmar. The solid brick structure is covered with gold. It is more than 300 feet (90 m) tall.

Finally, ask students to look at the people in the center photo. They are dressed for the Festival of the Ages parade held every year in October in Kyoto, Japan. The procession illustrates clothing styles from the 700s through the 1800s. Thousands of costumed people join the parade.

Unit Objectives

1. Describe how climate and landforms affect the way people live in East and Southeast Asia.
2. Identify the political, economic, social, and cultural characteristics of East and Southeast Asia.
3. Explain the importance of East and Southeast Asia as a major region of the world.
4. Understand links among the physical geography, historical geography, and cultural geography of East and Southeast Asia.
5. Interpret special-purpose maps to analyze relationships among climate, population patterns, and economic resources in East and Southeast Asia.

UNIT 7

East and Southeast Asia

CHAPTER 24
China, Mongolia, and Taiwan

CHAPTER 25
Japan and the Koreas

CHAPTER 26
Southeast Asia

Shwedagon Pagoda, Yangon, Myanmar

Rice paddies near Guilin, China

Chapters

24 China, Mongolia, and Taiwan
explains the physical geography, history, and culture of these three countries.

25 Japan and the Koreas
describes the geography and interrelated histories and cultures of Japan, North Korea, and South Korea.

26 Southeast Asia
examines the physical and human geography of the five mainland countries (Vietnam, Laos, Cambodia, Thailand, Myanmar) and the five island countries (Malaysia, Singapore, Brunei, Indonesia, Philippines).

Notes from the Field

A Peace Corps Volunteer in Mongolia

Matt Heller served as a Peace Corps volunteer in Mongolia. He worked as an English teacher and coordinated a greenhouse reconstruction project. **WHAT DO YOU THINK?** ***What kinds of changes would you have to make in your life to live in Mongolia?***

I've been a Peace Corps volunteer in Mongolia for eighteen months. I live in a *ger*, a tent with a small wood stove in the center. It is strong and practical, perfect for a nomadic herder. I, however, am an English teacher in a small school in rural Mongolia. *Ger* life is hard. It makes twenty-year-olds look thirty-five. It makes your soul hard.

Mongolians are very proud of their history. Once, while sitting on the train going from Ulaanbaatar to my own town, Bor-Undur, a Mongolian pointed to his arm and said, "In here is the blood of Genghis Khan. Beware." There is no argument to that statement. I responded, "Yes, older brother (a respectful title addressed to elders), your country is beautiful. Mongolians are lucky."

Along with many other things, I'm learning how Mongolians live. In the steppe there is very little snow, only biting wind and dust. It gets as cold as –50 degrees, not counting the wind chill factor. If I leave tea in a mug, it will freeze by morning. I've broken three mugs that way.

A Mongolian outside his home

Festival of Ages parade, Kyoto, Japan

Understanding Primary Sources

1. What conclusion does Matt Heller make about Mongolians and their past?
2. How does Matt Heller describe the climate in Mongolia?

Giant panda

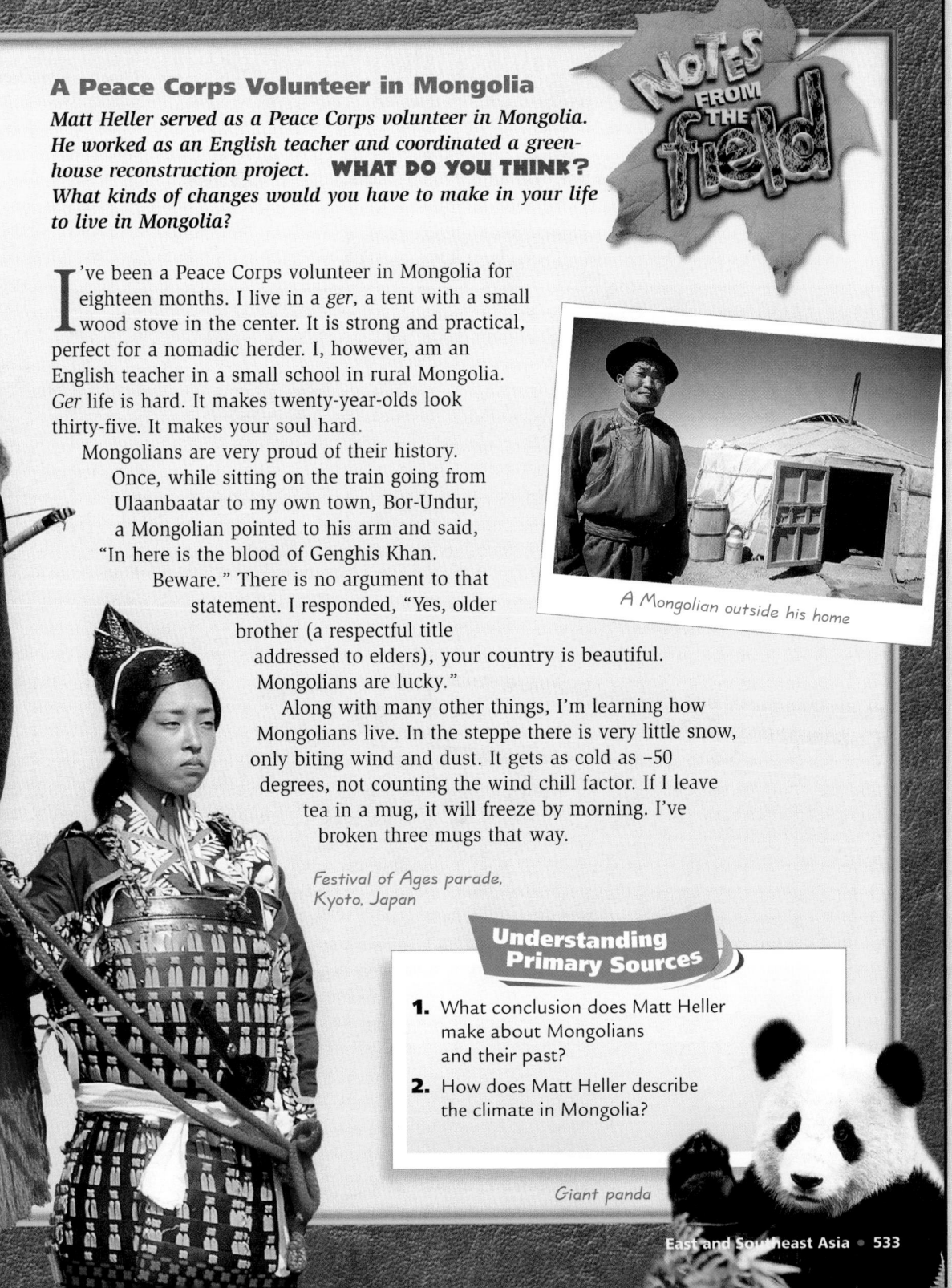

More from the Field

Genghis Khan was a conqueror who unified nomadic Mongolian tribes and then, in the early 1200s, extended his power across Asia all the way to the Black Sea. His fast-moving cavalry relied on hardy ponies that needed no feed other than the grass of the steppe. Mare's milk was a staple food. The Mongols' *gers*, or yurts, were made of horsehair. In fact, the Mongols were so dependent on their horses that their country was said to be "the back of a horse."

Activity: Have students conduct research on nomadic Mongolians of today. Ask them to compare the role of horses in modern Mongolian society to the role they played in Genghis Khan's day. Then have them work in groups to compose song lyrics that relate their research results.

Understanding Primary Sources

Answers

1. proud of and closely connected to their past
2. cold, windy, dry, dusty

UNIT 7 ATLAS

OVERVIEW

This unit introduces students to East and Southeast Asia. The region stretches from Myanmar (also called Burma) in the west to Japan in the northeast. Russia lies to the north and the Indian Subcontinent to the southwest. The Indian Ocean is at the southern edge, and the Pacific Ocean is at the eastern edge.

Because it includes such a vast area, the region exhibits an equally wide range of landforms, climate types, economies, and cultures. People who live much as their ancestors did thousands of years ago contrast with busy workers crowding the sidewalks of dynamic modern cities. In all, approximately one third of the world's population lives in this region.

Although only a small percentage of the region's huge land area is suitable for farming, most of the people there are involved in agriculture. Industry continues to gain in importance, however.

PEOPLE IN THE PROFILE

Note that the elevation profile crosses the Plateau of Tibet, which has an average elevation of over 15,000 feet (4,570 m). People who travel to Tibet from low-lying areas can suffer or even die from altitude sickness if they do not take time to become acclimated.

How are the Tibetans able to live at such high altitudes? Recent studies suggest that their metabolism has adapted in response to the lack of oxygen. It appears that the hearts and brains of those living in Tibet's mountains use glucose and oxygen differently than people who live at low elevations. Also, they have developed enzymes that process oxygen more efficiently. Some scientists who have compared the Quechua of the Andes to the Sherpa of Tibet found that the Sherpa had adapted more fully to high altitudes because, as a people, they have been in Tibet at least 10,000 years longer than the Quechua have been in Peru.

UNIT 7 ATLAS The World in Spatial Terms

East and Southeast Asia

Elevation Profile

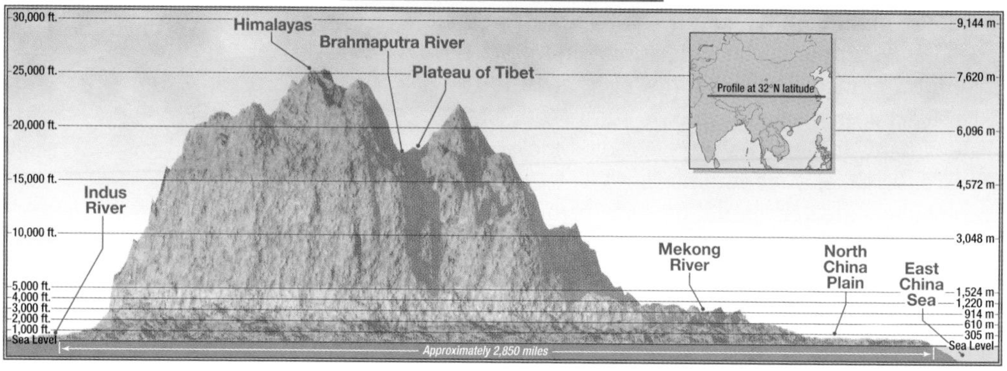

The United States and East and Southeast Asia: Comparing Sizes

GEOSTATS: China

World's largest population: 1,273,111,290 (July 2001 estimate)

Population growth rate: 0.9% (2001 estimate)

Highest point: Mount Everest–29,035 ft. (8,850 m)

Lowest point: Turpan Depression–505 ft. (154 m) below sea level

Using the Physical Map

Call on a volunteer to describe the landforms of China and Mongolia. Then have students examine the **political map** on the next page and draw conclusions about the connections between China's landforms and the building of the Great Wall. *(This area lacks natural barriers to movement from the north, and rulers wanted to protect the North China Plain from invasions by the people of the Mongolian Plateau.)* Point out to students that the Philippines, Japan, and Indonesia are on the Pacific Ring of Fire. Then remind them of the volcanic activity that occurs along the edges of the tectonic plates.

East and Southeast Asia: Physical

UNIT 7 ATLAS

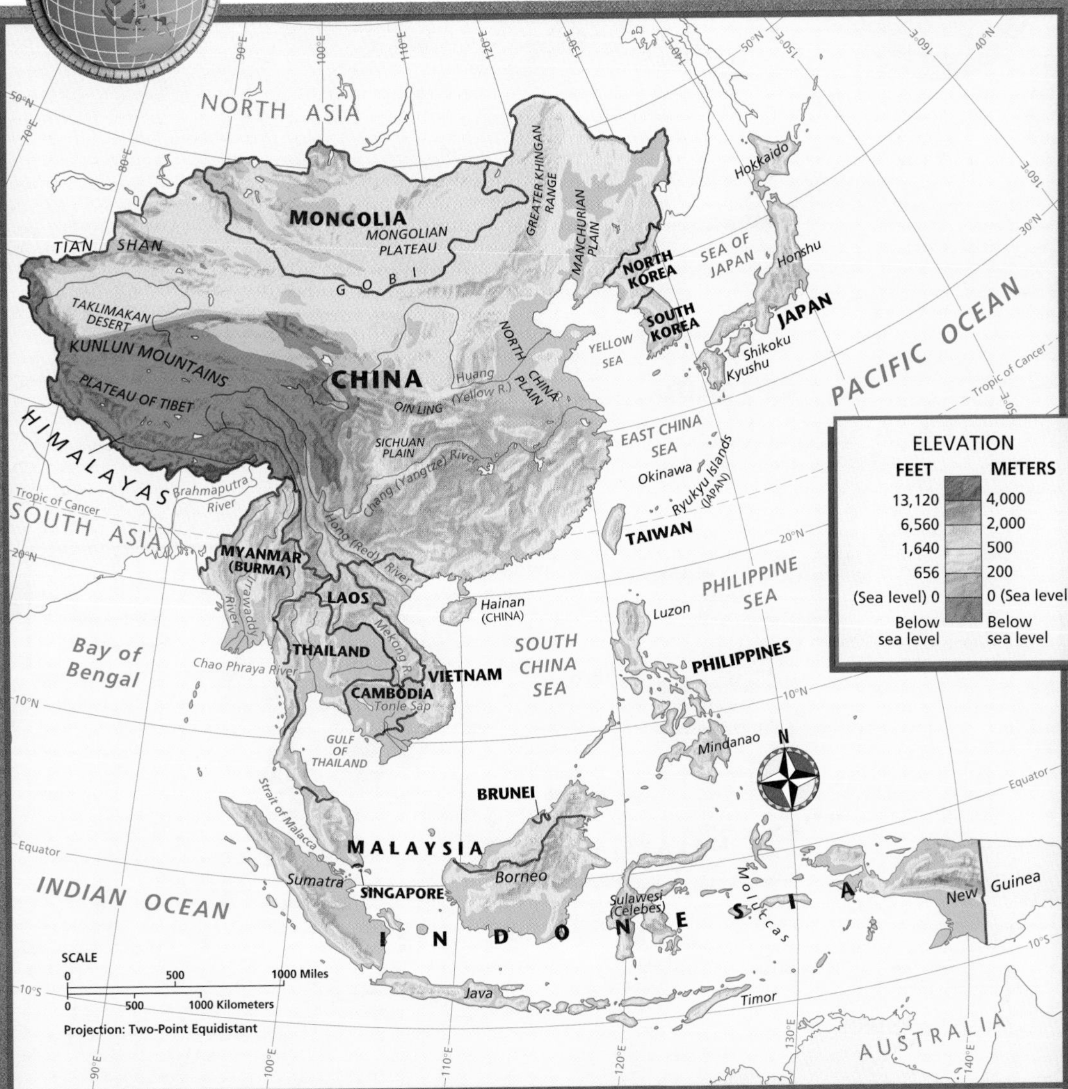

1. Places and Regions Which of the region's physical features could make travel and trade difficult?
2. Places and Regions What are the region's three largest island countries?
3. Places and Regions Which countries have territory on the mainland and on islands?

Critical Thinking

4. **Comparing** Compare this map to the **population map**. What physical features might prevent western China from becoming densely populated? Why might eastern China be so densely populated?

internet connect

ONLINE ATLAS
GO TO: **go.hrw.com**
KEYWORD: **SK3 MapsU7**
FOR: Web links to online maps of the region

Physical Map

Answers

1. large expanses of water, high mountains, vast deserts
2. Indonesia, Japan, Philippines
3. China, Malaysia

Critical Thinking

4. mountains and deserts; presence of rivers, wide plains

UNIT 7 ATLAS

Using the Political Map

As students examine the **political map** on this page, have them name the largest country on the map *(China)* and the smallest *(Singapore)*. Call on volunteers to name the mainland countries *(Mongolia, China, North Korea, South Korea, Myanmar, Thailand, Laos, Vietnam, Cambodia, part of Malaysia)* and the island countries *(Japan, Taiwan, Philippines, Indonesia, Singapore, Brunei, part of Malaysia)*. Then ask students which of the mainland countries has the largest number of international borders *(China)*.

internet connect

ONLINE ATLAS
GO TO: **go.hrw.com**
KEYWORD: SK3 MapsU7

Your Classroom Time Line

These are the major dates and time periods for this unit. Have students enter them on the time line you created earlier. You may want to watch for these dates as students progress through the unit.

c.* 5000 B.C. Rice farming develops near the Chang River.

c. 2000 B.C. Chinese living in the Huang valley form kingdoms.

551 B.C. Confucius is born.

200s B.C. The Great Wall of China is completed.

206 B.C.–A.D. 220 The Han dynasty rules China.

108 B.C. Chinese invade Korea.

by A.D. 100 Buddhism reaches China from India.

600s Tibet's Potala Palace is built.

early 900s The Koryo dynasty rises to power in Korea.

*c. stands for *circa* and means "about."

Political Map

Answers

1. Mongolia
2. Laos; Mongolia

Critical Thinking

3. Himalayas; Mekong River, river between North Korea and China

East and Southeast Asia: Political

UNIT 7 ATLAS

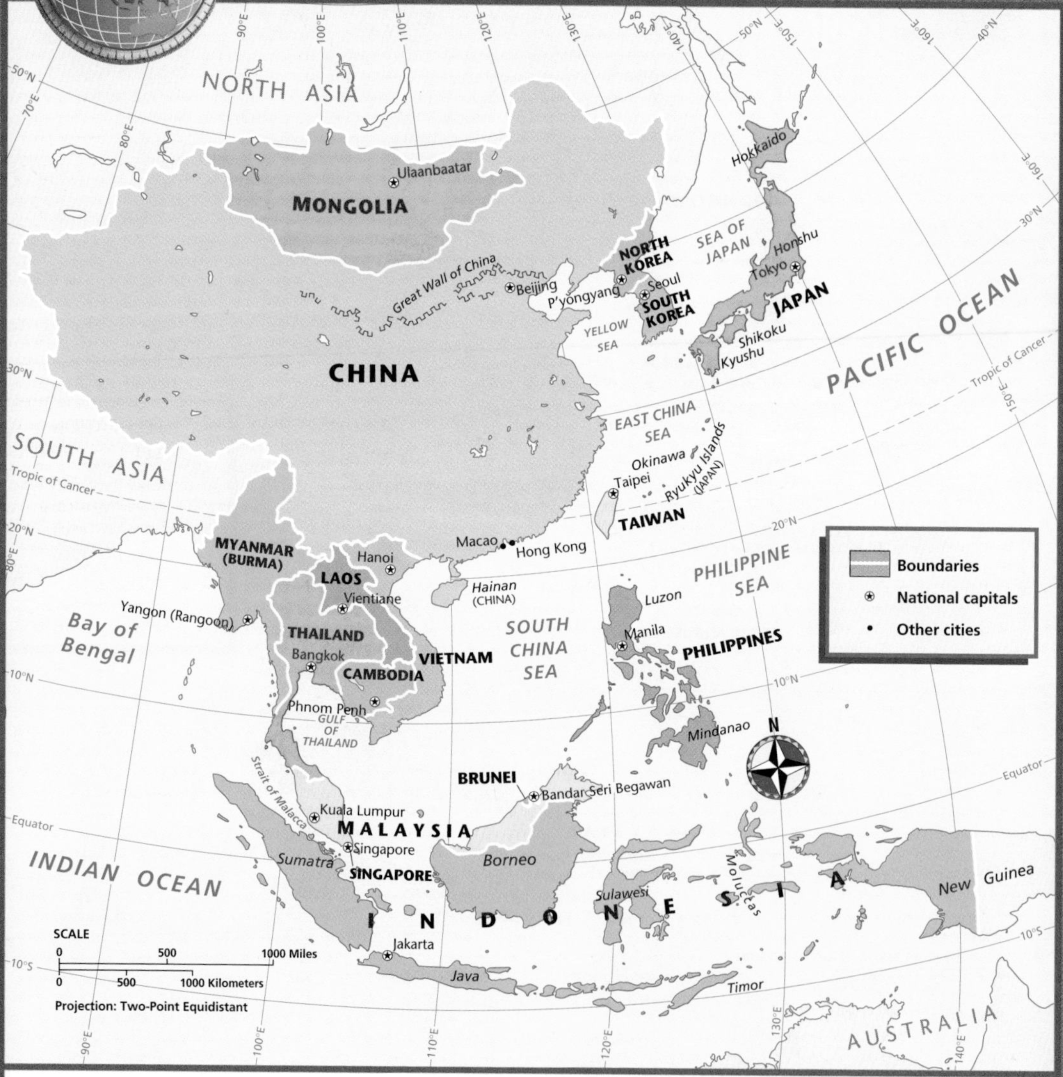

1. *Places and Regions* Which country entirely on the mainland shares a border with just one of the region's other countries?
2. *Places and Regions* What is the only landlocked country in Southeast Asia? Which country in East Asia is landlocked?

Critical Thinking

3. **Comparing** Compare this map to the **physical map** of the region. Which physical feature forms the border of southwestern China? Which other physical features form natural borders in the region?

USING THE CLIMATE MAP

As students examine the **climate map** on this page, point out that the region stretches from 50° north of the equator to 10° south. Ask students what effect this wide latitude range would be likely to have on the climate of the region *(a wide range of climate types)*. Have them discuss the extremes of the region's climates, which range from subarctic to humid tropical. Have students identify other countries that lie as far north as Mongolia *(such as Canada)* and as far south as Indonesia *(such as Brazil)*.

East and Southeast Asia: Climate

UNIT 7 ATLAS

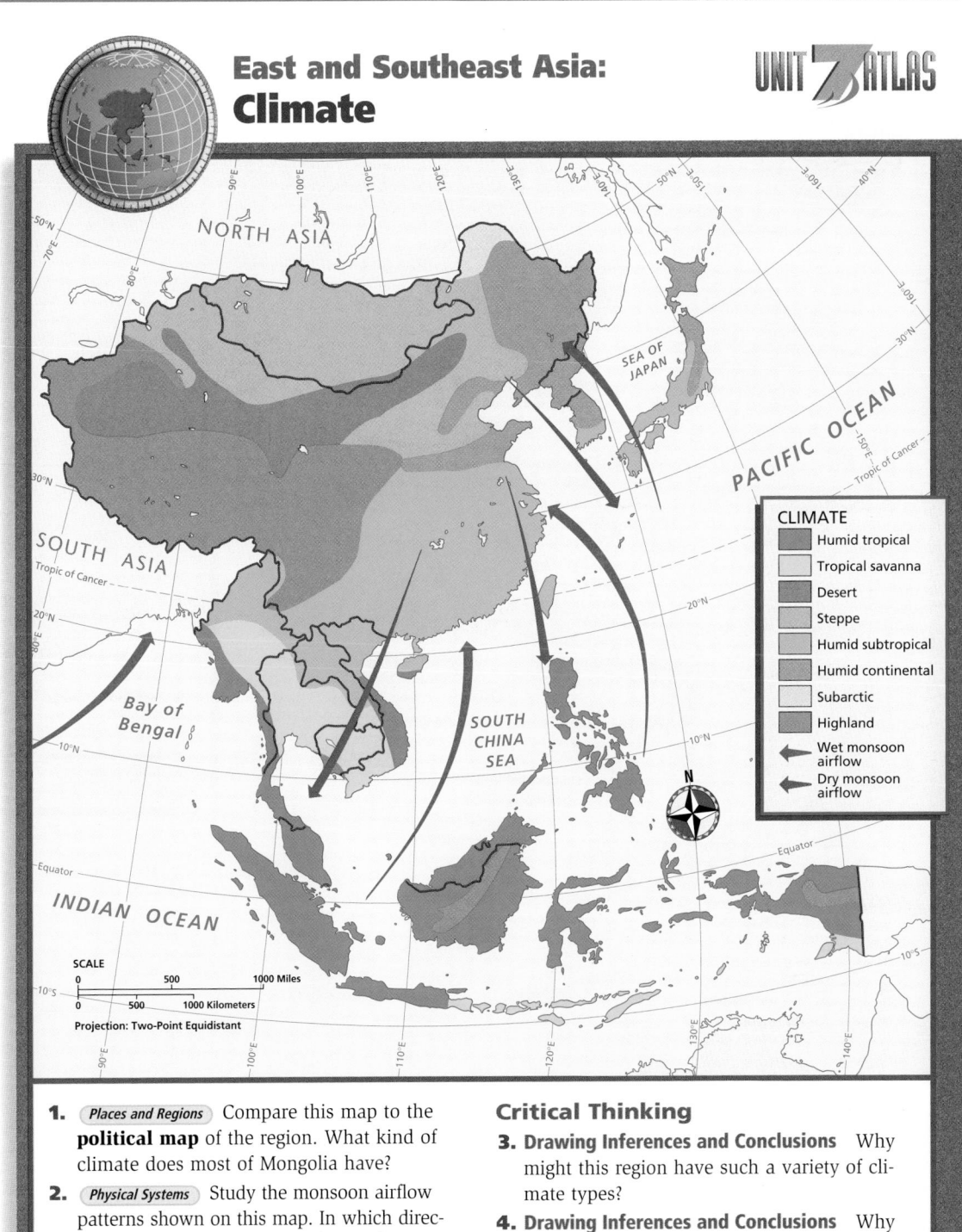

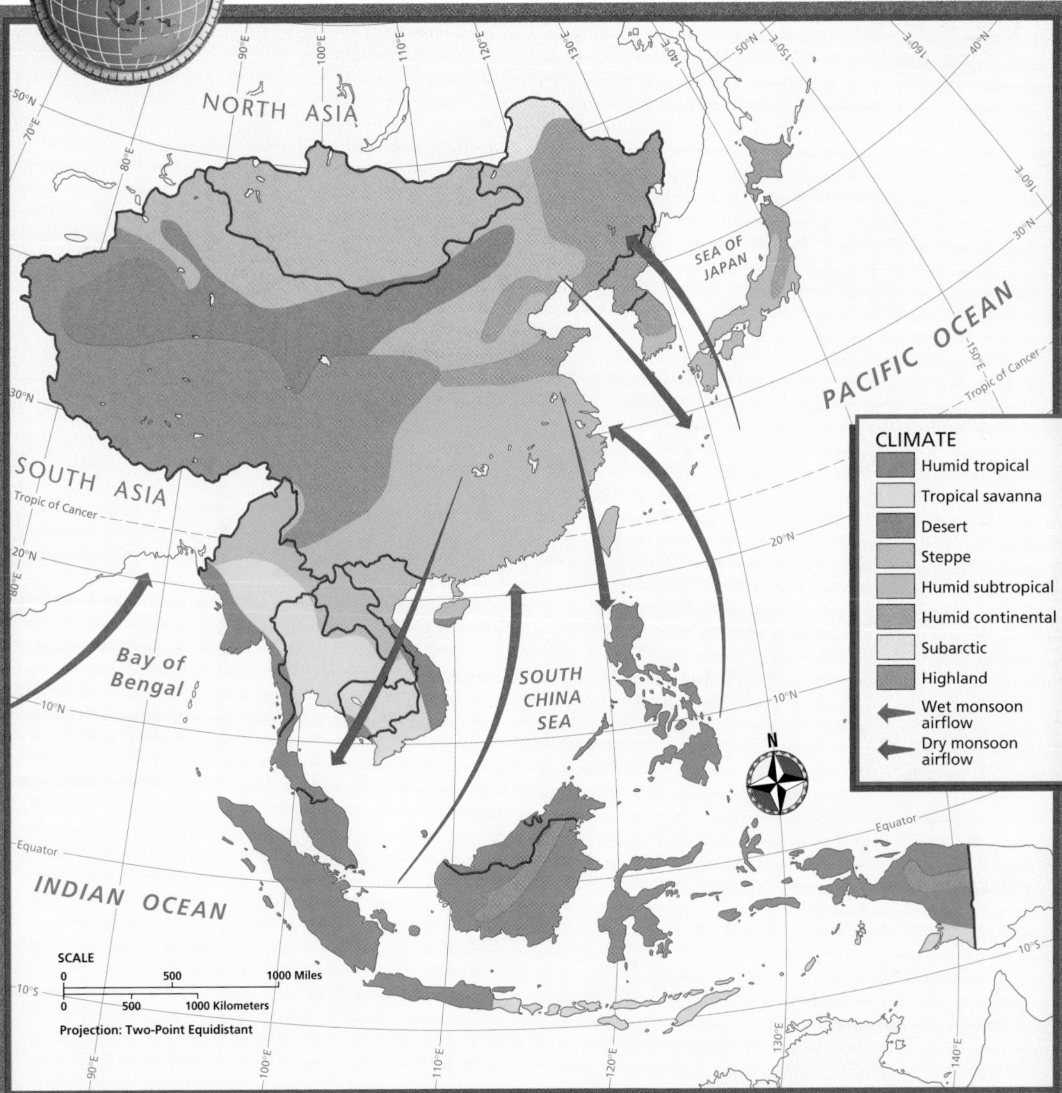

1. *Places and Regions* Compare this map to the **political map** of the region. What kind of climate does most of Mongolia have?
2. *Physical Systems* Study the monsoon airflow patterns shown on this map. In which directions do the wet and dry monsoons flow?

Critical Thinking

3. **Drawing Inferences and Conclusions** Why might this region have such a variety of climate types?
4. **Drawing Inferences and Conclusions** Why might the South China Sea be stormy?

Your Classroom Time Line (continued)

- **1279–1368** Mongols rule China.
- **late 1200s** Marco Polo visits China.
- **1368–1644** The Ming dynasty rules China.
- **1500s** Europeans begin establishing colonies in Southeast Asia.
- **1644–1912** The Manchu dynasty rules China.
- **1853** Commodore Matthew Perry's ships sail into Tokyo Bay.
- **1895** Japan defeats China in the Sino-Japanese War.
- **1898** The United States wins the Philippines from Spain.
- **1910** Japan annexes Korea.
- **1912** Sun Yat-sen establishes the Republic of China.
- **1930s** Japanese soldiers invade Manchuria and China.
- **1941** Japan attacks Pearl Harbor; United States declares war on Japan.

Climate Map

Answers

1. steppe
2. wet—south to north; dry—north to south

Critical Thinking

3. wide variety of latitudes, landforms, and elevations
4. warm climate, monsoon winds

UNIT 7 ATLAS

Using the Population Map

As students examine the **population map**, ask which countries do not have metropolitan areas with more than 2 million inhabitants *(Brunei, Cambodia, Laos, Mongolia).* Then ask which country has low population density throughout its entire area *(Mongolia).* Have students compare the **population map** with the **physical** and **political maps** on previous pages. Ask which sea might be the most at risk for pollution by agricultural and urban runoff *(Yellow Sea).*

internet connect

ONLINE ATLAS
GO TO: **go.hrw.com**
KEYWORD: **SK3 MapsU7**

Your Classroom Time Line (continued)

- 1945 World War II ends.
- 1946 Philippines gains its independence.
- 1949 Mao Zedong establishes the communist People's Republic of China.
- 1948 Korea divides into two countries.
- early 1950s Korean War is fought.
- 1960s–70s The United States fights in the Vietnam War.
- 1966–76 Mao Zedong implements the Cultural Revolution in China.
- 1989 Chinese troops kill protesters in Tiananmen Square.
- 1997 British return control of Hong Kong to China.
- 1999 Macao is returned to China.

Population Map

Answers

1. more than 520 persons per square mile (200 per sq km)
2. Taklimakan Desert

Critical Thinking

3. Charts, graphs, databases, and models should accurately reflect population figures on the map.

East and Southeast Asia: Population

UNIT 7 ATLAS

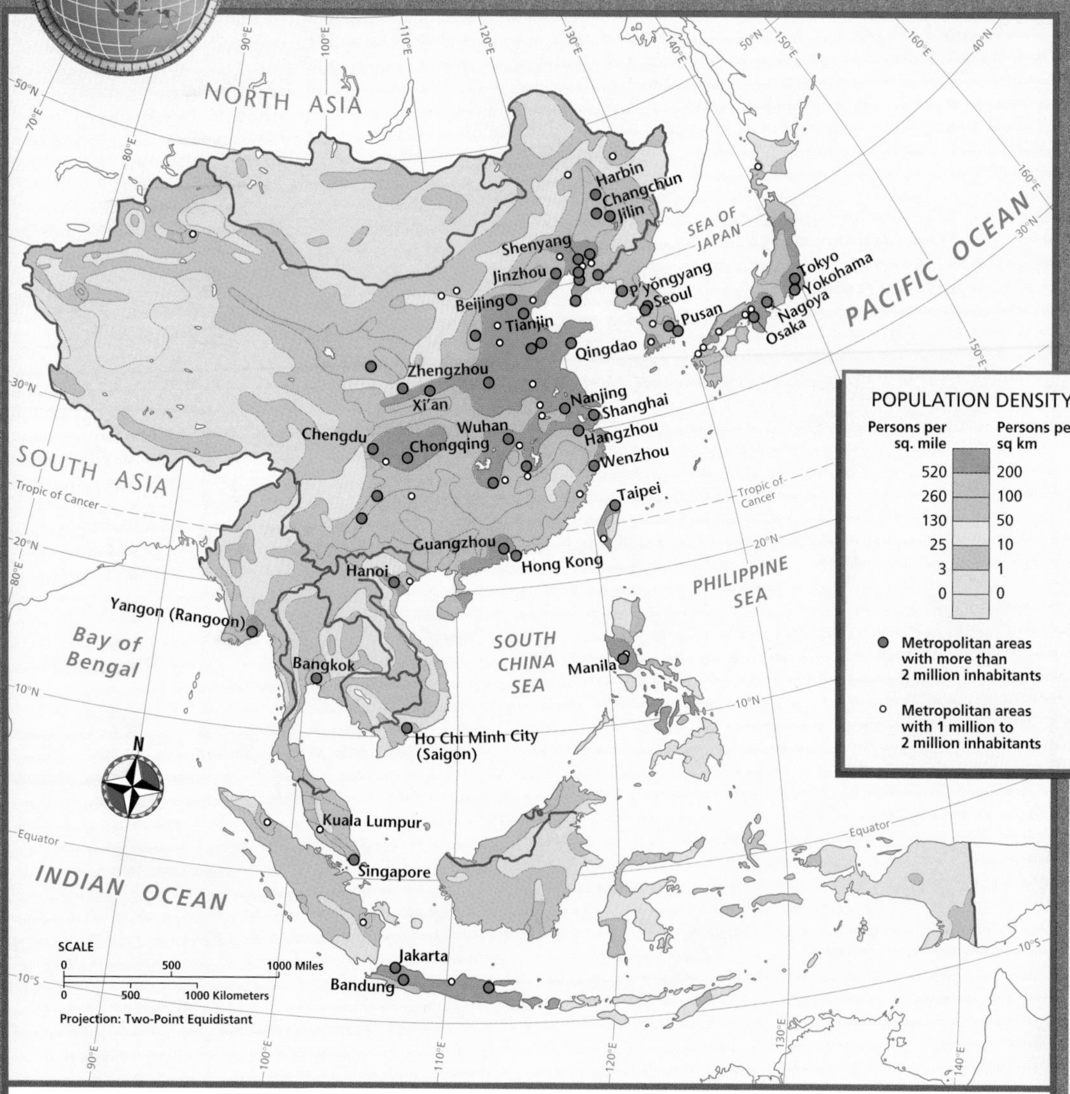

1. *Places and Regions* What is the population density in the area between Shanghai and Beijing?
2. *Environment and Society* Compare this map to the **physical map** of the region. Which physical features help explain the low population density in western China?

Critical Thinking

3. **Analyzing Information** Use the map on this page to create a chart, graph, database, or model of population centers in East and Southeast Asia.

Using the Land Use and Resources Map

As students examine the **land use and resources map** on this page, point out that most residents of the region are involved in agriculture. Ask students which type of agriculture is practiced in the most densely populated areas *(commercial farming)*. Then ask which types of economic activities are practiced in Mongolia *(nomadic herding)* and in the island countries along the equator *(forestry)*.

UNIT 7 ATLAS

East and Southeast Asia: Land Use and Resources

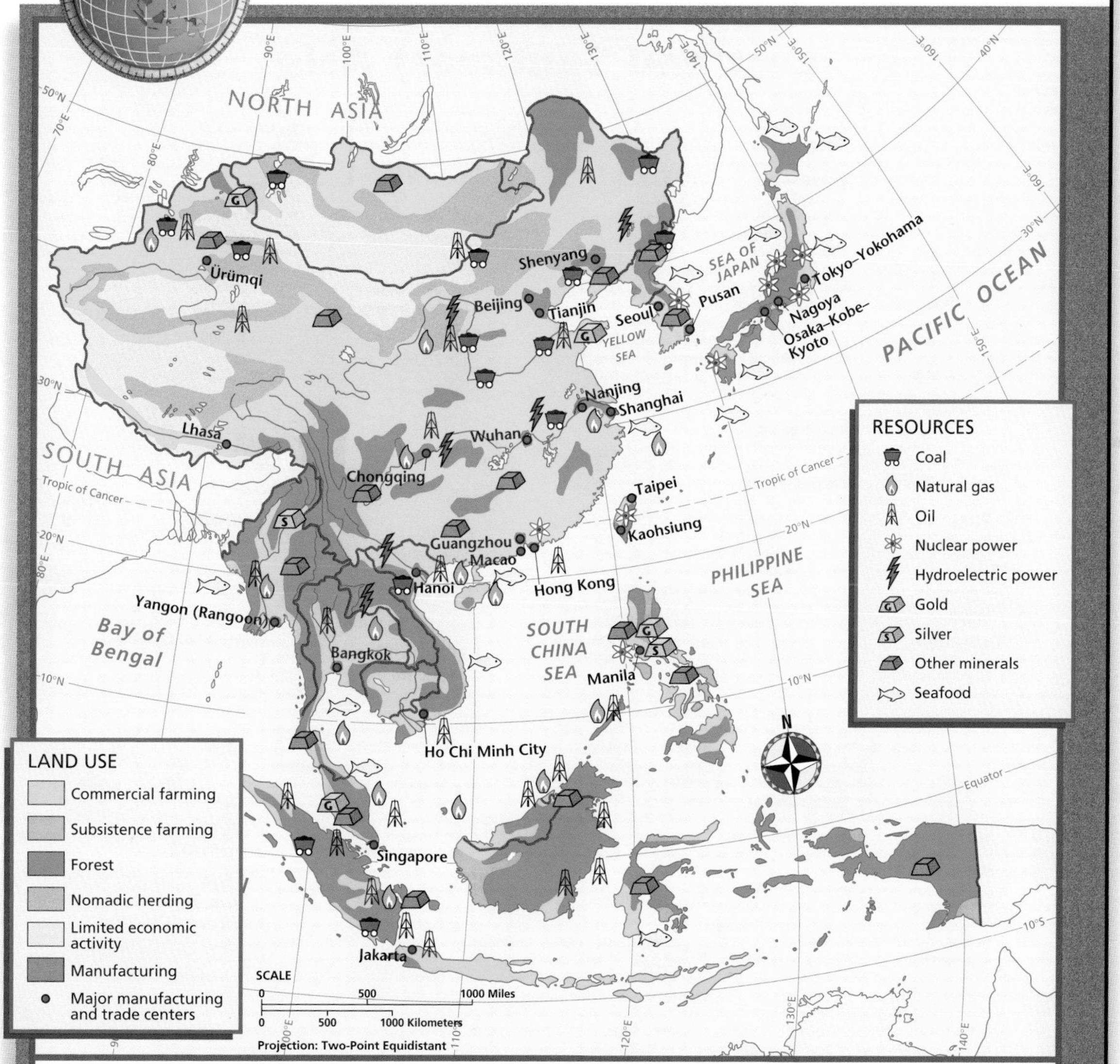

1. *Places and Regions* Which two countries have large areas where nomadic herding is common?
2. *Places and Regions* Where in China is most commercial farming found? In which countries is subsistence farming common?

Critical Thinking

3. **Analyzing Information** Use the map on this page to create a chart, graph, database, or model of economic activities in East and Southeast Asia.

Land Use and Resources Map

Answers

1. China and Mongolia
2. eastern and northeastern regions; China, all the other countries except Brunei, Singapore, and Japan

Critical Thinking

3. Charts, graphs, databases, and models should accurately reflect information on the map.

Fast Facts

EAST AND SOUTHEAST ASIA

FAST FACTS ACTIVITIES

LEVEL 1: (Suggested time: 20 min.) Have students calculate the number of televisions per person in an industrialized country such as Japan, Taiwan, or South Korea and a nonindustrialized country such as Myanmar, Cambodia, or Mongolia. *(Students should divide the number of televisions by the number of people.)* Ask students to write a sentence or two describing the relationship between industrialization and television ownership.

Then challenge students to interpret the meaning of a number smaller than one. *(Example: There are .008 televisions per person in Cambodia, which means that there are about 8 televisions for every 1,000 people.)*

LEVEL 2: (Suggested time: 45 min.) Have students use the accompanying table to calculate the actual number of people aged 14 and younger in the countries of East and Southeast Asia. Then have them create a chart that shows the number of people from 0 to 14 years old in the countries of East and Southeast Asia. Have students use the chart in the chapter on Japan and the Koreas comparing the populations of Japan and the United States as a model, but have students use one figure for every 500,000 people.

UNITED STATES OF AMERICA

CAPITAL: Washington, D.C.
AREA: 3,717,792 sq. mi. (9,629,091 sq km)
POPULATION: 281,421,906
MONEY: U.S. dollar (USD)
LANGUAGES: English, Spanish (spoken by a large minority)
TELEVISIONS: 215,000,000

Fast Facts

East and Southeast Asia

BRUNEI

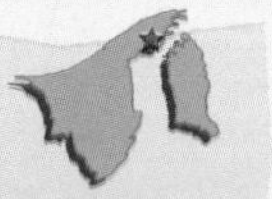

CAPITAL: Bandar Seri Begawan
AREA: 2,228 sq. mi. (5,770 sq km)
POPULATION: 343,653
MONEY: Bruneian dollar (BND)
LANGUAGES: Malay (official), English, Chinese
NUMBER OF TELEVISIONS: 201,900 (1998)

JAPAN

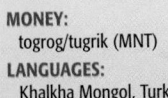

CAPITAL: Tokyo
AREA: 145,882 sq. mi. (377,835 sq km)
POPULATION: 126,771,662
MONEY: yen (JPY)
LANGUAGES: Japanese
NUMBER OF TELEVISIONS: 86,500,000 (1997)

CAMBODIA

CAPITAL: Phnom Penh
AREA: 69,900 sq. mi. (181,040 sq km)
POPULATION: 12,491,501
MONEY: riel (KHR)
LANGUAGES: Khmer (official), French
NUMBER OF TELEVISIONS: 94,000 (1997)

LAOS

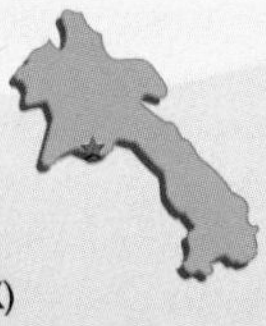

CAPITAL: Vientiane
AREA: 91,428 sq. mi. (236,800 sq km)
POPULATION: 5,635,967
MONEY: kip (LAK)
LANGUAGES: Lao (official), French, English, ethnic languages
NUMBER OF TELEVISIONS: 52,000

CHINA

CAPITAL: Beijing
AREA: 3,705,386 sq. mi. (9,596,960 sq km)
POPULATION: 1,273,111,290
MONEY: yuan (CNY)
LANGUAGES: Mandarin, Yue, other forms of Chinese
NUMBER OF TELEVISIONS: 400,000,000 (1997)

MALAYSIA

CAPITAL: Kuala Lumpur
AREA: 127,316 sq. mi. (329,750 sq km)
POPULATION: 22,229,040
MONEY: ringgit (MYR)
LANGUAGES: Bahasa Melayu (official), English, Chinese dialects, ethnic languages
NUMBER OF TELEVISIONS: 10,800,000 (1999)

INDONESIA

CAPITAL: Jakarta
AREA: 741,096 sq. mi. (1,919,440 sq km)
POPULATION: 228,437,870
MONEY: Indonesian rupiah (IDR)
LANGUAGES: Bahasa Indonesia (official), English, Dutch, Javanese
NUMBER OF TELEVISIONS: 13,750,000

MONGOLIA

CAPITAL: Ulaanbaatar
AREA: 604,247 sq. mi. (1,565,000 sq km)
POPULATION: 2,654,999
MONEY: togrog/tugrik (MNT)
LANGUAGES: Khalkha Mongol, Turkic, Russian
NUMBER OF TELEVISIONS: 168,800 (1999)

Countries not drawn to scale.

Percentage of Population Aged 0–14

Brunei	31	Myanmar (Burma)	29
Cambodia	41	North Korea	26
China	25	Philippines	37
Indonesia	30	Singapore	18
Japan	15	South Korea	22
Laos	43	Taiwan	21
Malaysia	35	Thailand	23
Mongolia	33	Vietnam	32

Source: Central Intelligence Agency, *The World Factbook 2001*

Fast Facts Activities

LEVEL 3: (Suggested time: 45 min.) Have students use the historical and economic information in this unit to write an essay that explains the population differences illustrated in the accompanying table. Encourage them to use supplemental resources.

Unit 7 Assessment Resources

Reproducible
- Unit 7 Test
- Unit 7 Test for English Language Learners and Special-Needs Students

Unit 7 Atlas

Myanmar (Burma)

CAPITAL: Yangon (Rangoon)
AREA: 261,969 sq. mi. (678,500 sq km)
POPULATION: 41,994,678
MONEY: kyat (MMK)
LANGUAGES: Burmese, many ethnic languages
NUMBER OF TELEVISIONS: 320,000 (2000)

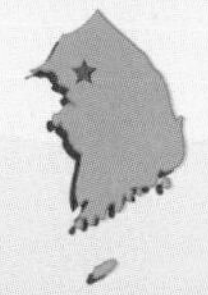

South Korea

CAPITAL: Seoul
AREA: 38,023 sq. mi. (98,480 sq km)
POPULATION: 47,904,370
MONEY: South Korean won (KRW)
LANGUAGES: Korean, English
NUMBER OF TELEVISIONS: 15,900,000 (1997)

North Korea

CAPITAL: P'yŏngyang
AREA: 46,540 sq. mi. (120,540 sq km)
POPULATION: 21,968,228
MONEY: North Korean won (KPW)
LANGUAGES: Korean
NUMBER OF TELEVISIONS: 1,200,000 (1997)

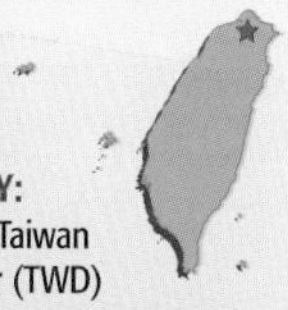

Taiwan

CAPITAL: Taipei
AREA: 13,892 sq. mi. (35,980 sq km)
POPULATION: 22,370,461
MONEY: New Taiwan dollar (TWD)
LANGUAGES: Mandarin, Taiwanese (Min), Hakka dialects
NUMBER OF TELEVISIONS: 8,800,000

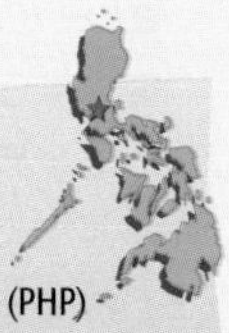

Philippines

CAPITAL: Manila
AREA: 115,830 sq. mi. (300,000 sq km)
POPULATION: 82,841,518
MONEY: Philippine peso (PHP)
LANGUAGES: Filipino, English
NUMBER OF TELEVISIONS: 3,700,000

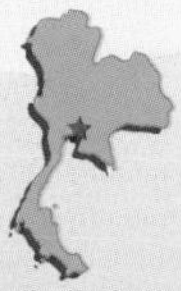

Thailand

CAPITAL: Bangkok
AREA: 198,455 sq. mi. (514,000 sq km)
POPULATION: 61,797,751
MONEY: baht (THB)
LANGUAGES: Thai, English, ethnic languages
NUMBER OF TELEVISIONS: 15,190,000

Singapore

CAPITAL: Singapore
AREA: 250 sq. mi. (648 sq km)
POPULATION: 4,300,419
MONEY: Singapore dollar (SGD)
LANGUAGES: Chinese, Malay, Tamil, English
NUMBER OF TELEVISIONS: 1,330,000

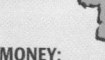

Vietnam

CAPITAL: Hanoi
AREA: 127,243 sq. mi. (329,560 sq km)
POPULATION: 79,939,014
MONEY: dong (VND)
LANGUAGES: Vietnamese (official), Chinese, English, French, Khmer, ethnic languages
NUMBER OF TELEVISIONS: 2,900,000

internet connect
COUNTRY STATISTICS
GO TO: **go.hrw.com**
KEYWORD: **SK3 FactsU7**
FOR: more facts about East and Southeast Asia

Sources: Central Intelligence Agency, *The World Factbook 2001; The World Almanac and Book of Facts 2001;* pop. figures are 2001 estimates.

internet connect
COUNTRY STATISTICS
GO TO: **go.hrw.com**
KEYWORD: **SK3 FactsU7**

Highlights of Country Statistics
- *CIA World Factbook*
- Library of Congress country studies
- Flags of the world

CHAPTER 24 China, Mongolia, and Taiwan

Chapter Resource Manager

Objectives	Pacing Guide	Reproducible Resources
SECTION 1 **Physical Geography** (pp. 543–46) **1.** Identify the physical features of China, Mongolia, and Taiwan. **2.** Identify the types of climate found in China, Mongolia, and Taiwan. **3.** Identify the natural resources of China, Mongolia, and Taiwan.	**Regular** 1 day **Block Scheduling** .5 day *Block Scheduling Handbook, Chapter 24*	**RS** Guided Reading Strategy 24.1 **E** Lab Activity for Geography and Earth Science, Hands-On 1
SECTION 2 **China's History and Culture** (pp. 547–53) **1.** Identify the major events in the history of China. **2.** Name some features of China's culture.	**Regular** 2 days **Block Scheduling** .5 day *Block Scheduling Handbook, Chapter 24*	**RS** Guided Reading Strategy 24.2 **RS** Graphic Organizer 24 **E** Cultures of the World Activity 7 **E** Creative Strategies for Teaching World Geography, Lesson 18 **SM** Geography for Life Activity 24 **IC** Interdisciplinary Activities for the Middle Grades 29, 30, 31 **SM** Map Activity 24
SECTION 3 **China Today** (pp. 554–57) **1.** Describe where most of China's people live. **2.** Identify the major cities in China and what they are like. **3.** Describe China's economy. **4.** Identify the challenges that China faces.	**Regular** 1 day **Block Scheduling** .5 day *Block Scheduling Handbook, Chapter 24*	**RS** Guided Reading Strategy 24.3
SECTION 4 **Mongolian and Taiwan** (pp. 558–61) **1.** Describe how Mongolia's culture developed. **2.** Describe Taiwan's culture.	**Regular** 1 day **Block Scheduling** .5 day *Block Scheduling Handbook, Chapter 24*	**RS** Guided Reading Strategy 24.4 **E** Cultures of the World Activity 7 **E** Creative Strategies for Teaching World Geography, Lesson 18

Chapter Resource Key

- **RS** Reading Support
- **IC** Interdisciplinary Connections
- **E** Enrichment
- **SM** Skills Mastery
- **A** Assessment
- **REV** Review
- **ELL** Reinforcement and English Language Learners
- Transparencies
- CD–ROM
- Music
- Video
- Internet
- Holt Presentation Maker Using Microsoft® Powerpoint®

One-Stop Planner CD–ROM

See the *One-Stop Planner* for a complete list of additional resources for students and teachers.

One-Stop Planner CD–ROM

It's easy to plan lessons, select resources, and print out materials for your students when you use the ***One-Stop Planner CD–ROM with Test Generator.***

Technology Resources	Review, Reinforcement, and Assessment Resources
One-Stop Planner CD–ROM, Lesson 24.1 Geography and Cultures Visual Resources with Teaching Activities 49–54 Homework Practice Online HRW Go site	**ELL** Main Idea Activity 24.1 **REV** Section 1 Review, p. 546 **A** Daily Quiz 24.1 **ELL** English Audio Summary 24.1 **ELL** Spanish Audio Summary 24.1
One-Stop Planner CD–ROM, Lesson 24.2 Homework Practice Online HRW Go site	**ELL** Main Idea Activity 24.2 **REV** Section 2 Review, p. 553 **A** Daily Quiz 24.2 **ELL** English Audio Summary 24.2 **ELL** Spanish Audio Summary 24.2
One-Stop Planner CD–ROM, Lesson 24.3 *ARGWorld* CD–ROM: Population and Development in China Homework Practice Online HRW Go site	**ELL** Main Idea Activity 24.3 **REV** Section 3 Review, p. 557 **A** Daily Quiz 24.3 **ELL** English Audio Summary 24.3 **ELL** Spanish Audio Summary 24.3
One-Stop Planner CD–ROM, Lesson 24.4 Homework Practice Online HRW Go site	**ELL** Main Idea Activity 24.4 **REV** Section 4 Review, p. 591 **A** Daily Quiz 24.4 **ELL** English Audio Summary 24.4 **ELL** Spanish Audio Summary 24.4

internet connect

HRW ONLINE RESOURCES

GO TO: go.hrw.com
Then type in a keyword.

TEACHER HOME PAGE
KEYWORD: SK3 TEACHER

CHAPTER INTERNET ACTIVITIES
KEYWORD: SK3 GT24

Choose an activity to:
- follow the Great Wall of China.
- visit the land of Genghis Khan.
- see the artistic treasures of China.

CHAPTER ENRICHMENT LINKS
KEYWORD: SK3 CH24

CHAPTER MAPS
KEYWORD: SK3 MAPS24

ONLINE ASSESSMENT
Homework Practice
KEYWORD: SK3 HP24
Standardized Test Prep Online
KEYWORD: SK3 STP24
Rubrics
KEYWORD: SS Rubrics

COUNTRY INFORMATION
KEYWORD: SK3 Almanac

CONTENT UPDATES
KEYWORD: SS Content Updates

HOLT PRESENTATION MAKER
KEYWORD: SK3 PPT24

ONLINE READING SUPPORT
KEYWORD: SS Strategies

CURRENT EVENTS
KEYWORD: S3 Current Events

Chapter Review and Assessment

- **E** Readings in World Geography, History, and Culture 67, 68, and 69
- **SM** Critical Thinking Activity 24
- **REV** Chapter 24 Review, pp. 562–63
- **REV** Chapter 24 Tutorial for Students, Parents, Mentors, and Peers
- **ELL** Vocabulary Activity 24
- **A** Chapter 24 Test
- Chapter 24 Test Generator (on the One-Stop Planner)
- Audio CD Program, Chapter 24
- **A** Chapter 24 Test for English Language Learners and Special-Needs Students

Meeting Individual Needs

Ability Levels

Level 1 Basic-level activities designed for all students encountering new material

Level 2 Intermediate-level activities designed for average students

Level 3 Challenging activities designed for honors and gifted-and-talented students

English Language Learners Activities that address the needs of students with Limited English Proficiency

CHAPTER 24

Launch into Learning

Hold up a compass, a porcelain object, and a piece of paper. Tell students to describe in writing the importance of each object for human civilization. *(Students might mention that the compass made possible wide-ranging exploration, that porcelain has functional and decorative value, and that paper has enabled people to communicate and reproduce their ideas.)* Ask students to speculate on the origins of each item. If students are unaware of the objects' origins, tell them that they are based on Chinese innovations. Tell students that in this chapter they will learn about other contributions to world culture made by the people of China, Mongolia, and Taiwan.

Section 1

Objectives

1. Identify the physical features of China, Mongolia, and Taiwan.
2. Describe the types of climate found in China, Mongolia, and Taiwan.
3. List the natural resources of China, Mongolia, and Taiwan.

Links to Our Lives

There are many reasons why American students should know more about China, Mongolia, and Taiwan. Here are a few of them:

- Americans buy many goods that were produced in China and Taiwan; these countries are important trading partners of the United States.
- China has the largest population of any country.
- China's strategic location and its status as a nuclear power influence the foreign policy of the United States.
- Mongolia and China have shaped world history, and China has made important cultural and technological contributions to the world.

CHAPTER 24

China, Mongolia, and Taiwan

This region of Asia with its varied landscape and cultures is home to one of the world's oldest living civilizations.

Hello! My name is Lu Hua. Lu is my family name, and Hua is my given name. In China the family name comes first. I am 16 and live in Jin Shan County, outside of Shanghai, with my parents and my brother. My father is a clerk in a Volkswagen factory. My ancestors have lived in this village for hundreds of years. All 200 people in this village are named Lu.

I am in my last year at Jin Shan County High School. To get into this school, which is the best in the county, I had to pass a very difficult exam when I was 11. I had the best score that year. The school goes from seventh to twelfth grade. Each grade has four classes with 50 kids in each class. Now I am hoping to go on to a university. In China only one or two out of a hundred kids can go to college.

Most of my friends want to be scientists. I think I would like to be a diplomat, to travel, and have adventures. My family are common people, though, not Chinese Communist Party members, so I may not get into the diplomatic college.

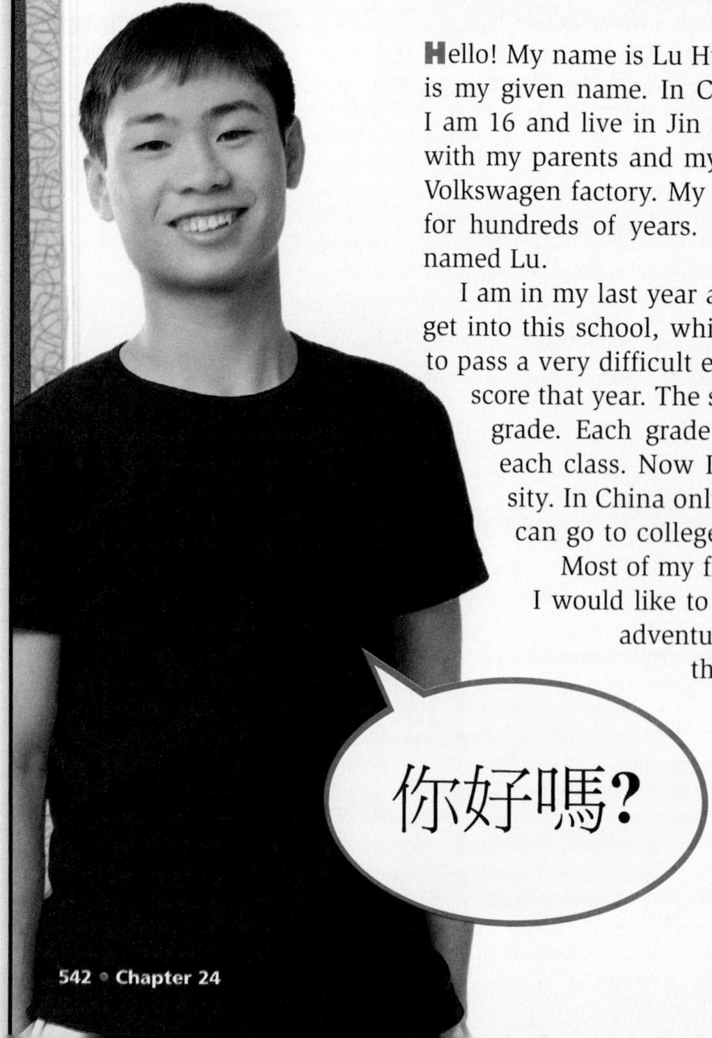

◀ Translation: How are you?

Focus

LET'S GET STARTED

Display examples of traditional Chinese ink-on-paper landscape painting. (Find examples in art history texts and books on China.) Ask students what they can learn about China's physical geography by looking at the paintings. *(Possible answers: mountainous, rainy, highly eroded, has places where bamboo grows)* Tell students that China has a wide range of landforms and climates, as the paintings indicate. Then ask students to describe the landscapes. *(Possible answers: peaceful, dreamy, fantastic)* If students say the landscapes are imaginary refer them to the photo of rice paddies near Guilin in the unit opener photos to show that the high, narrow landforms in many of the paintings are indeed real. Tell students they will learn more about the physical geography of China, Mongolia, and Taiwan in this chapter.

Using the Physical-Political Map

Have students examine the map on this page. Ask volunteers to locate the mountain ranges, plains, plateaus, and rivers. Then ask a volunteer to identify the physical features that one would encounter if traveling directly from Lhasa to Harbin and from Guangzhou to Ulaanbaatar.

Building Vocabulary

Copy the Define terms, **dikes** and **arable**, onto the chalkboard. Have volunteers read the definitions aloud. Ask students how dikes and arable land might be related. *(Dikes could protect arable land from floods.)* Ask students to review the textbook's discussion about land reclamation in the Netherlands in a previous chapter. Tell students that people in the Netherlands also build dikes to prevent floods.

Section 1 Physical Geography

Read to Discover

1. What are the physical features of China, Mongolia, and Taiwan?
2. What types of climate are found in China, Mongolia, and Taiwan?
3. What natural resources do China, Mongolia, and Taiwan have?

Define

dikes
arable

Locate

Himalayas
Mount Everest
Kunlun Mountains
Tian Shan
Plateau of Tibet
Taklimakan Desert
Tarim Basin
Gobi
North China Plain
Huang He
Chang River
Sichuan (Red) Basin
Xi River

Why It Matters

China was once one of the most isolated countries in the world. Today, however, it plays a major role in the world community. Use CNNfyi.com or other **current events** sources to find out more about China's role in world affairs. Record your findings in your journal.

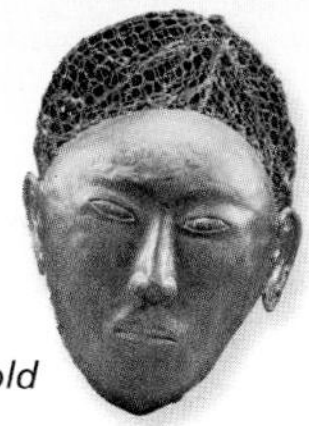

Mask made of gold

China, Mongolia, and Taiwan: Physical-Political

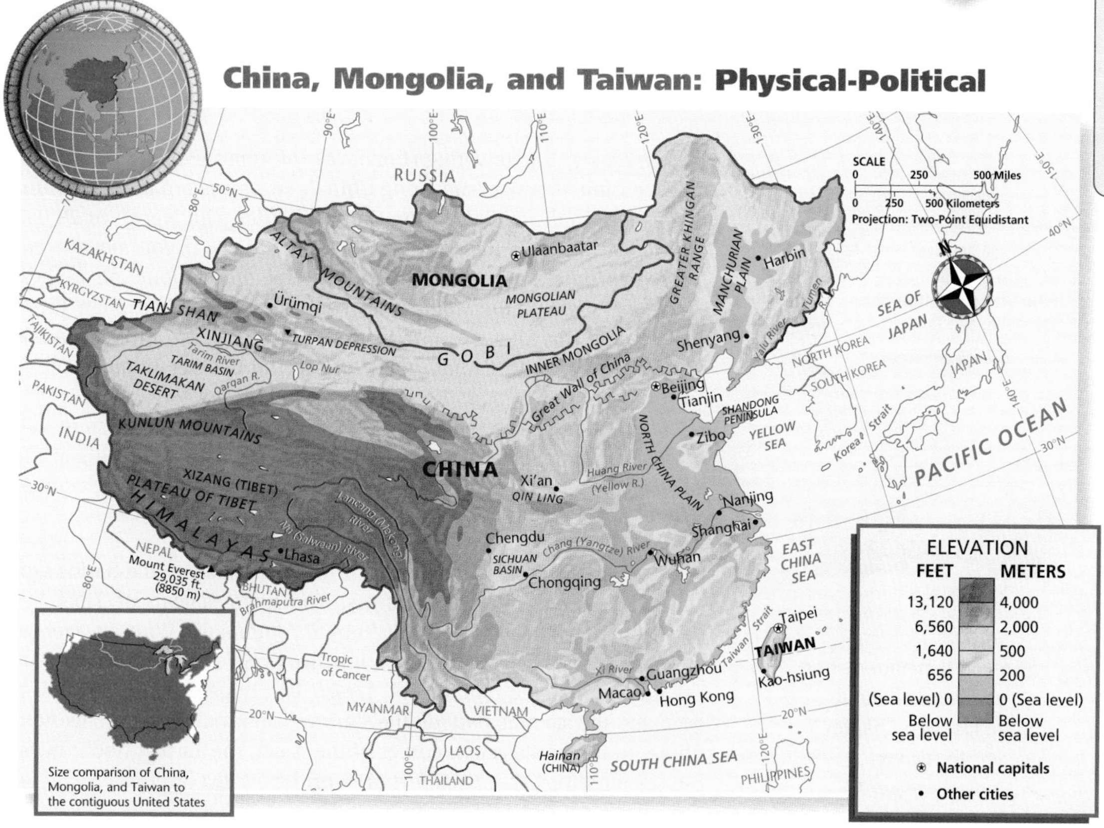

Section 1 Resources

Reproducible
- Block Scheduling Handbook, Chapter 24
- Guided Reading Strategy 24.1
- Lab Activity for Geography and Earth Science, Hands-On 1

Technology
- One-Stop Planner CD–ROM, Lesson 24.1
- Homework Practice Online
- Geography and Cultures Visual Resources with Teaching Activities 49–54
- HRW Go site

Reinforcement, Review, and Assessment
- Section 1 Review, p. 546
- Daily Quiz 24.1
- Main Idea Activity 24.1
- English Audio Summary 24.1
- Spanish Audio Summary 24.1

TEACH

Teaching Objective 1

LEVEL 1: (Suggested time: 45 min.) Organize the class into small groups. Provide students with a large piece of cardboard, construction paper, markers, modeling clay, glue, and scissors. Have each group prepare a large, three-dimensional outline map of the region that includes the major landforms and rivers. *(Maps should show the Himalayas and Kunlun and Altay Mountains, the Plateau of Tibet, the Mongolian Plateau, the Taklimakan Desert and the Gobi, the Tarim Basin, and the Chang, Huang, and Xi Rivers.)* Display maps around the classroom.
ENGLISH LANGUAGE LEARNERS, COOPERATIVE LEARNING

LEVELS 2 AND 3: (Suggested time: 15 min.) Tell students that some ancient Chinese maps have north and south reversed, with north at the bottom of the page. Call volunteers to the chalkboard to sketch outline maps of China, Taiwan, and Mongolia in this configuration. Ask other volunteers to draw in the major landforms and rivers. *(See the Level 1 lesson for the correct landforms and features.)* Ask other students to participate by providing any necessary hints and guidance to students at the chalkboard. You may wish to provide students with colored chalk.

PLACES AND REGIONS

Mongolia is located far from the ocean at high latitudes and high elevations. Mongolia's low southern deserts receive fewer than four inches (10 cm) of precipitation each year. Northern mountains receive some 14 inches (35.5 cm) of precipitation each year, most of which falls during summer months. The country has between 220 and 260 clear, sunny days each year, but its weather can be severe and unpredictable, with fierce blizzards, sand storms, and hailstorms.

Critical Thinking: What region of the United States might have a climate similar to that of Mongolia?

Answer: Students might suggest the northern Great Plains.

Across the Curriculum
LITERATURE

Legends In his writings, Marco Polo described mirages experienced by travelers crossing the Taklimakan Desert. *Taklimakan* is a Turkic word meaning "enter and you will not come out."

Activity: Have students conduct research on the Taklimakan Desert and write a folktale about how it received its name.

Visual Record Answer ►

highland climates

▲ These snow-covered mountains are in the Wolong Nature Reserve in central China.

Interpreting the Visual Record What climate types would you expect to find in this area of the reserve?

internet connect
GO TO: go.hrw.com
KEYWORD: SK3 CH24
FOR: Web sites about China, Mongolia, and Taiwan

BUILD ON WHAT You Know

Do you remember what you learned about plateaus? See Chapter 2 to review.

Physical Features

China has some of the world's tallest mountains, driest deserts, and longest rivers. Mongolia (mahn-GOHL-yuh) is China's neighbor to the north. It is a large, rugged, landlocked country. Burning hot summers and bitter cold winters are common there. In contrast, Taiwan (TY-WAHN) is a green tropical island just off the coast of mainland China.

Mountains The towering Himalayas (hi-muh-LAY-uhz), the world's tallest mountain range, run along China's southwestern border. Mount Everest lies in the Himalayas on China's border with Nepal. At 29,035 feet (8,850 m), it is the world's tallest mountain. If you move north from the Himalayas, you will find several other mountain ranges. These are the Kunlun Mountains (KOON-LOON), the Tian Shan (TYEN SHAHN), and the Altay Mountains (al-TY). To the east, on Mongolia's eastern border with China, you will see the Greater Khingan (KING-AHN) Range.

Mountains stretch the length of Taiwan and cover the eastern half of the island. In some places, the mountains end in steep cliffs at the edge of the Pacific Ocean. To the west of the mountains is a fertile coastal plain.

Plateaus, Basins, and Deserts Isolated plateaus and basins separate the region's mountain ranges. The huge Plateau of Tibet lies between the Himalayas and the Kunlun Mountains. With an average elevation of 16,000 feet (4,877 m), it is the world's highest plateau. The Taklimakan (tah-kluh-muh-KAHN) Desert is a huge expanse of sand. It occupies the central part of the Tarim (DAH-REEM) Basin in western China. In the northeastern corner of the basin, the Turpan (toohr-PAHN) Depression drops about 505 feet (154 m) below sea level.

Teaching Objective 2

ALL LEVELS: (Suggested time: 10 min.) Copy the following graphic organizer onto the chalkboard, omitting the italicized answers. Have each student complete the chart by filling in the climate regions of China, Mongolia, and Taiwan and their corresponding climate types. Ask volunteers to share their answers with the class. Then ask students where they think most of China's population might live *(eastern half)* and why *(enough rainfall for growing crops).*

Region	Climate Type
Southeastern coastal region	*Humid and wet*
Northwest	*Dry*
Extreme northwest	*Desert*

Teaching Objective 3

ALL LEVELS: (Suggested time: 20 min.) Organize the class into small groups. Provide each group with an outline map of the region. Have each group create a map key with symbols to represent the various resources found in each country. Tell students to draw the symbols on the maps in appropriate places. *(Maps should indicate that China has gold, iron ore, lead, salt, tungsten, zinc, uranium, coal and oil; Mongolia has coal, oil, iron ore, and copper; Taiwan has arable land.)* Display and discuss student maps. **ENGLISH LANGUAGE LEARNERS, COOPERATIVE LEARNING**

The Mongolian Plateau covers most of the country of Mongolia. The Gobi (GOH-bee) takes up much of the central and southeastern sections of the plateau. The Gobi is the coldest desert in the world and covers more than 500,000 square miles (1,295,000 sq km). Much of the Gobi is gravel and bare rock.

Plains China and Mongolia have few areas of lowlands made up of coastal and river floodplains. However, these fertile plains support the major population centers. Millions of people live in the North China Plain. It is the largest plain in China and is crossed by major rivers.

Rivers The river known as the Huang (HWAHNG) rises on the eastern edge of the Plateau of Tibet. It flows eastward through the North China Plain and empties into the Yellow Sea. It takes its name, which means "yellow river," from the yellowish mud it carries. Winds carry loess, a yellowish-brown soil, from the Gobi to northern China. The Huang picks up the loess as the river flows through the region. On its way to the sea the river dumps the loess, raising the river bottom. This can lead to flooding. Floodwaters deposit a layer of rich silt that is good for farming but also cause great damage and loss of life. As a result, the Huang has long been known as China's Sorrow. The Chinese have tried to control the Huang by building **dikes**. These high banks of earth or concrete help reduce flooding.

The Chang (CHAHNG), or Yangtze (YAHNG-TSE), River also rises in the Plateau of Tibet. It flows eastward for 3,434 miles (5,525 km) across central China through the fertile Sichuan (SEE-CHWAHN), or Red, Basin. The Chang is China's—and Asia's—longest river. In fact, its name means "long river." The Chang is one of China's most important transportation routes. It is connected to the Huang by the world's oldest and longest canal system, the Grand Canal. The Xi (SHEE) River is southern China's most important river and transportation route.

✓ **READING CHECK:** *Places and Regions* What are the major physical features of this region? mountains, plateaus, basins, deserts, plains, rivers

Climate

China, Mongolia, and Taiwan are part of a huge region with several different climates. China's precipitation varies. The southeastern coastal region is the country's most humid area. As you move northwestward the climate becomes steadily drier. The extreme northwest has a true desert climate.

Seasonal monsoon winds greatly affect the climate of the region's southern and eastern parts. In winter, winds from Central Asia bring dry, cool-to-cold weather to eastern

Our Amazing Planet

China has greater potential for hydroelectric power than any other country in the world. When completed in 2009, the Three Gorges Dam on the Chang River will be the world's largest dam.

The Gobi is the world's third-largest desert. Herders ride Bactrian camels.

Interpreting the Visual Record **What characteristics of a desert environment can be seen in this photo?**

Human Systems

The Grand Canal, which links Beijing and Hangzhou, is more than 1,000 miles (1,609 km) long. Construction on the canal may have begun as long ago as the 300s B.C. The Sui dynasty rebuilt this oldest section during the A.D. 600s, and each succeeding dynasty expanded the canal until the Mongols completed it in the 1200s. The canal provided an important transportation and communication link between cities and outlying lands of the empire. In 1958 the Chinese government began work to straighten, dredge, and widen the canal to allow larger vessels to pass through.

Critical Thinking: What influence might the Grand Canal have on China's economy?

Answer: good for shipping and trade

Section 1 Review

Answers

Define For definitions, see: dikes, p. 545; arable, p. 546

Working with Sketch Maps The North China Plain and coastal plains of Taiwan are the most fertile areas.

◄ **Visual Record Answer**

bare ground, no vegetation

CLOSE

Tell students that in 1999 the official figure for Mount Everest's elevation was revised. Scientists operated Global Positioning System satellite equipment at the mountain's top to determine that the peak is about seven feet higher than the previous official measurement, which had been made in 1954. The mountain's official elevation is now 29,035 feet (8,850 m).

REVIEW AND ASSESS

Have students complete the Section Review. Then pair students and have each pair create six flashcards for a where-am-I game. Students should describe a climate region or geographic feature on the front of each card and write the answer on the back. Have pairs exchange cards with other pairs and answer the other pairs' questions. Then have students complete Daily Quiz 24.1. **COOPERATIVE LEARNING**

RETEACH

Have students complete Main Idea Activity 24.1. Then organize students into small groups and have each group create an outline of the section. Have each group present its outline to the class.
ENGLISH LANGUAGE LEARNERS, COOPERATIVE LEARNING

EXTEND

Have interested students conduct research on the typhoons that hit China's coastal areas. Students should investigate how typhoons are formed and how they affect China. **BLOCK SCHEDULING**

Reading for the Main Idea

1. Plateau of Tibet between the Himalayas and the Kunlun Mountains; Mount Everest; on China's border with Nepal (NGS 4)
2. Huang, Chang River, Xi River; because it often floods (NGS 15)
3. Gobi and Taklimakan Deserts (NGS 4)

Critical Thinking

4. latitude, ocean currents, high mountains, large interior, and monsoons

Organizing What You Know

5. China—Himalayas, Kunlun Mountains, Tian Shan, Altay Mountains, Plateau of Tibet, Tarim Basin, Turpan Depression, North China Plain; from tropical to desert; gold, iron ore, lead, salt, zinc, uranium, coal, oil; Mongolia—Mongolian Plateau, Khingan Range; hot summers, cold winters, desert; coal, oil, iron ore, and copper; Taiwan—mountains and coastal plains; tropical; arable land

Chart Answer ▲

Mongolia

Visual Record Answer ►

the ears, snout, and scale-like pattern

China, Mongolia, and Taiwan

Country	Population/ Growth Rate	Life Expectancy	Literacy Rate	Per Capita GDP
China	1,273,111,290 / 0.9%	70, male / 74, female	82%	$3,600
Mongolia	2,654,999 / 1.5%	62, male / 67, female	83%	$1,780
Taiwan	22,370,461 / 0.8%	74, male / 80, female	94%	$17,400
United States	281,421,906 / 0.9%	74, male / 80, female	97%	$36,200

Sources: Central Intelligence Agency, *The World Factbook 2001;* U.S. Census Bureau

Interpreting the Chart **Which country has the fastest rate of population growth?**

This bronze vessel dates to the A.D. 1000s. Found in a tomb, it is just over a foot long (30 cm) and is covered with detailed animal designs.

Interpreting the Visual Record **What features of this object suggest animal characteristics?** ►

Asia. In summer, winds from the Pacific bring warm, wet air. This creates hot, rainy summers. Typhoons sometimes hit the coastal areas during the summer and fall. Typhoons are violent storms with high winds and heavy rains similar to hurricanes. They often bring flooding and cause a great deal of damage.

✓ **READING CHECK:** *Places and Regions* What are the climates of China, Mongolia, and Taiwan like? varied, with humid, dry, and desert regions

Resources

China has a wide range of mineral resources. These include gold, iron ore, lead, salt, uranium, and zinc, as well as energy resources such as coal and oil. China has greater coal reserves than any other country. At the present rate of use, these reserves will last another 1,000 years. China also produces enough oil to meet most of its own needs.

Mongolia has deposits of coal, copper, gold, iron ore, and oil. Taiwan's most important natural resource is its arable land, or land that is suitable for growing crops.

✓ **READING CHECK:** *Places and Regions* What are the region's resources? gold, iron ore, lead, salt, uranium, zinc, coal, copper, oil, arable land

Section 1 Review

go.hrw.com Homework Practice Online
Keyword: SK3 HP24

Define and explain: dikes, arable

Working with Sketch Maps On a map of China, Mongolia, and Taiwan that you draw or that your teacher provides, label the following: the Himalayas, Mount Everest, Kunlun Mountains, Tian Shan, Plateau of Tibet, Taklimakan Desert, Tarim Basin, Gobi, North China Plain, Huang He, Chang River, Sichuan Basin, and Xi River. Where do you think the most fertile areas of the region are located?

Reading for the Main Idea

1. *Places and Regions* What and where is the world's largest plateau? What is the world's tallest mountain? Where is it located?
2. *Environment and Society* What are three major rivers in eastern and southern China? How does the Huang affect China's people?
3. *Places and Regions* What is the region's driest area?

Critical Thinking

4. **Drawing Inferences and Conclusions** What do you think might be the major factors that influence this region's climate?

Organizing What You Know

5. **Summarizing** Summarize the natural environments of China, Mongolia, and Taiwan.

Country	Physical Features	Climates	Resources

Section 2

Objectives

1. Identify some of the major events in the history of China.
2. Discuss some features of China's culture.

FOCUS

LET'S GET STARTED

Copy the following passage onto the chalkboard: *Chinese artisans from various eras created impressive objects. Look at the photographs of the bronze flying horse, the clay soldiers, and the vase in this section. During which dynasty was each object created? What might have been the purpose of each object? Why might we value these objects today?* Discuss responses. Tell students that in Section 2 they will learn about China's history and culture.

Building Vocabulary

Copy the Define terms—**emperor, dynasty, porcelain, martial law,** and **pagodas**—onto the chalkboard. Have volunteers read the terms' definitions aloud. Point out that a dynasty can also be any family or group that maintains great power, wealth, or position for several generations. Ask students to give examples of real-life or fictional dynasties that fit either or both meanings. *(Example: a basketball team that wins the championship for several years in a row)*

Section 2 China's History and Culture

Read to Discover

1. What are some of the major events in the history of China?
2. What are some features of China's culture?

Define

emperor
dynasty
porcelain
martial law
pagodas

Locate

China
Great Wall

A bronze flying horse from the Han dynasty

WHY IT MATTERS

The Chinese government has been criticized by much of the world community for its political system. Use CNNfyi.com or other **current events** sources to find out more about efforts to encourage the Chinese government to enact democratic political reforms. Record your findings in your journal.

Section 2 RESOURCES

Reproducible

- Guided Reading Strategy 24.2
- Graphic Organizer 24
- Cultures of the World Activity 7
- Geography for Life Activity 24
- Creative Strategies for Teaching World Geography, Lesson 18
- Interdisciplinary Activities for the Middle Grades 29, 30, 31
- Map Activity 24

Technology

- One-Stop Planner CD–ROM, Lesson 24.2
- Homework Practice Online
- HRW Go site

Reinforcement, Review, and Assessment

- Section 2 Review, p. 553
- Daily Quiz 24.2
- Main Idea Activity 24.2
- English Audio Summary 24.2
- Spanish Audio Summary 24.2

History

Farmers have cultivated rice in southern China for some 7,000 years. Warm, wet weather made the region ideal for growing rice. Rice remains one of the region's main sources of food. Farmers in drier northern China grew a grain called millet and other crops. The early Chinese also grew hemp for fiber for clothing and spun silk from the cocoons of silkworms. Various cultures developed, particularly along the region's rivers.

The Qin Dynasty and the Great Wall Beginning about 2000 B.C. northern Chinese living in the Huang valley formed kingdoms. As Chinese civilization began to develop, peoples from various regions organized into large states. Each state was governed by an **emperor**—a ruler of a large empire. An emperor is often a member of a **dynasty**. A dynasty is a ruling family that passes power from one generation to the next. Beginning in about 500 B.C., the Chinese began building earthen

The Great Wall of China, including its branches and curves, stretches more than 2,000 miles (3,218 km).

TEACH

Teaching Objective 1

LEVEL 1: (Suggested time: 40 min.) Organize the class into eight groups. Assign each group one of the following topics: Ancient China, Qin dynasty, Han dynasty, Mongols, Ming dynasty and the Manchu, the Republics, Mao, and Post-Mao. Have groups illustrate on large, unlined notecards the events occurring during their assigned time period. Create a time line on the chalkboard or on butcher paper. Then have each group attach its notecard in the appropriate place to create a pictorial time line of events in Chinese history. *(The time line should include: Ancient China—rice, hemp, silk, and formation of kingdoms; Qin dynasty—Great Wall, written history, artifacts, and country's name; Han dynasty—expansion of kingdom; extension of Great Wall, compass, paper, and porcelain; Mongols—conquest of China; Ming dynasty and the Manchu—overthrow of Mongols, strengthened Great Wall, development of Chinese culture, control by the Qing dynasty; Republics—overthrow of Qing dynasty, Japanese invasion of China; Mao—People's Republic of China, Cultural Revolution; post-Mao—Tiananmen Square protest, free market reforms.)* **ENGLISH LANGUAGE LEARNERS, COOPERATIVE LEARNING**

DAILY LIFE

The Chinese civil service system, created during the Qin dynasty, helped maintain stability within the empire for more than 2,000 years. People who scored well on certain examinations were chosen for the civil service. These civil servants helped carry out imperial laws and keep records. Other countries used the Chinese civil service system as a model when creating their own civil services.

These government servants became known as mandarins. The word *mandarin* has Spanish, Portuguese, Malay, and Sanskrit roots. It eventually referred both to the officials and the dialect they spoke. Today the country's official language is Mandarin Chinese.

Critical Thinking: Why might mandarins have so greatly influenced language in China?

Answer: Students might suggest that because government business was conducted in Mandarin, it became important for people to know the dialect.

Visual Record Answer ►

Answers will vary according to which cultures the class has studied.

▲ Archaeologists have discovered 6,000 of these uniquely crafted soldiers near Xi'an. Each of these figures from the Qin dynasty has different facial features.

walls hundreds of miles long. These walls separated the kingdoms from the northern nomads and from each other. Records show that the first emperor of the Qin, also spelled Ch'in (CHIN), dynasty ordered the building of the Great Wall along China's northern border. People began to connect the sections of walls about 200 B.C.

The Qin dynasty is well known for its contributions to China's culture. It left behind many historical artifacts. For example, when the first emperor died, he was buried with thousands of life-sized warriors and horses made of clay. You might wonder why someone would want their tomb filled with clay figures. It was an ancient Chinese funeral tradition to bury masters with clay soldiers for protection. Since the Qin emperor had made many enemies during his life, he wanted protection after his death.

During the Qin dynasty the Chinese used a writing system to record their history. This system was similar to the one used in China today. China's name also dates from this time period. In Chinese, China means "Qin kingdom" or "middle kingdom." This name may refer to the Chinese belief that China was the center, or middle, of the world.

This vase from the Ming dynasty is an example of early Chinese art.

Interpreting the Visual Record How does this vase compare to artifacts left by other early cultures you have studied? ►

The Han Dynasty The Han dynasty came after the Qin dynasty. From the 200s B.C. to the A.D. 200s, the Han dynasty expanded its kingdom southward. The Han also extended the Great Wall westward to protect the Silk Road. This road was originally used by trading caravans taking silk and other Chinese goods to regions west of China. During the Han dynasty the Chinese invented the compass, which aided travel. The dynasties that followed the Han made China even more powerful. The Chinese continued to make important contributions to society. Later contributions include paper and **porcelain** (POHR-suh-luhn), a type of very fine pottery.

LEVEL 2: (Suggested time: 15 min.) Tell students to imagine that newspaper reporters have existed since ancient times. Pair students and assign each pair a period of Chinese history. Then have each pair use the time line in this section to write three or four headlines to describe the major events of its assigned period. *(See the Level 1 lesson for the correct periods and events. Example: For the Mongol period a headline might read Genghis Khan invades China! Use of horses proves military advantage.)* Ask volunteers to read their headlines to the class. **COOPERATIVE LEARNING**

LEVEL 3: (Suggested time: 45 min.) Pair students and assign each pair one period of Chinese history. Tell one student in each pair to imagine that he or she is a reporter, and tell the other student to imagine that he or she is a historian. Have each pair write a transcript of an interview, including questions and answers, about its assigned historical period. *(See the Level 1 lesson for the correct events.)* Have students conduct their interviews for the class. **COOPERATIVE LEARNING**

Mongols, Ming Dynasty, and the Manchu In the 1200s Mongol armies led by Genghis Khan conquered China. *Khan* is a title that means "ruler." The Mongols were feared and known for spreading terror throughout the region. Their use of horses added to their military advantage.

Within 100 years the Ming dynasty seized control of China. After several battles with the Mongols, the Ming emperors closed China to outsiders. These emperors strengthened the Great Wall and focused on the development of their own culture.

In the 1600s a group called the Manchu began expanding from their home in Manchuria. Manchuria is located in far northeastern China. The Manchus conquered Inner Mongolia, Korea, and all of northern China. Led by the Qing (CHING) dynasty, the Manchu controlled China for more than 260 years. The dynasty's strong government slowly weakened, however, and was overthrown in the early 1900s.

The Catalan Atlas from the 1300s shows Marco Polo's family traveling by camel caravan.

Interpreting the Visual Record What kind of information might this atlas provide?

Outside Influences Marco Polo was one of the few Europeans to visit China before the 1500s. Europeans reached China by following the Silk Road. None came by sea before the 1500s. In the 1500s Portuguese sailors established a trade colony at Macao (muh-KOW) in south China. French and British sailors and traders followed. The Chinese believed that foreigners had little to offer other than silver in return for Chinese porcelain, silk, and tea. Even so, Europeans introduced crops like corn, hot chili peppers, peanuts, potatoes, sweet potatoes, and tobacco. By the 1800s the European countries wanted to control China's trade. A series of conflicts caused China to lose some of its independence. For example, during this period the British acquired Hong Kong. The British, Germans, and French also forced China to open additional ports. China did not regain total independence until the mid-1900s.

The Republics of China In 1912 a revolutionary group led by Sun Yat-sen (SOOHN YAHT-SUHN) forced the last emperor to abdicate, or give up power. This group formed the first Republic of China. Mongolia and Tibet each declared their independence.

After Sun Yat-sen's death the revolutionaries split into two groups, the Nationalists and the Communists. A military leader named Chiang Kai-shek (chang ky-SHEK) united China under a Nationalist government. The Communists opposed him, and a civil war began. During

This time line reviews major events in China's rich history. The last Chinese dynasty was overthrown in 1912.

Interpreting the Time Line What events have shaped China's government in the 1900s?

Historic China: A Time Line

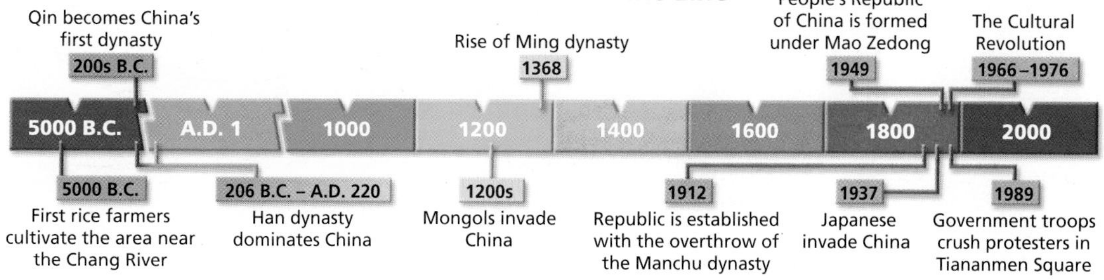
Qin becomes China's first dynasty
200s B.C.
Rise of Ming dynasty
1368
People's Republic of China is formed under Mao Zedong
1949
The Cultural Revolution
1966–1976
5000 B.C.
A.D. 1
1000
1200
1400
1600
1800
2000
5000 B.C.
First rice farmers cultivate the area near the Chang River
206 B.C. – A.D. 220
Han dynasty dominates China
1200s
Mongols invade China
1912
Republic is established with the overthrow of the Manchu dynasty
1937
Japanese invade China
1989
Government troops crush protesters in Tiananmen Square

Linking Past to Present

The Forbidden City The imperial palace complex in Beijing was completed in 1421 to house the Ming emperors. It became known as the Forbidden City because most imperial subjects were barred entry into the area. The gate to the formal entrance stands 125 feet high (138 m). Beyond the gate lies a large courtyard that contains a stream crossed by five marble bridges.

The seven-acre (three-hectare) area in the heart of the Forbidden City contains the Hall of Supreme Harmony, where the emperor held court. The area could hold tens of thousands of subjects who came to honor the emperor. In 1998 an international production of Puccini's opera *Turandot* was held there. The story of *Turandot* is set in the Forbidden City.

Critical Thinking: Why might the palace have been forbidden to most subjects?

Answer: It kept royalty and commoners in their distinct places in society.

▲ **Visual Record Answer**

imprecise locations and distances

◄ **Time Line Answer**

overthrow of the Manchu dynasty, Japanese invasion of China, formation of the People's Republic of China, the Cultural Revolution, crushing of Tiananmen Square protest

➤ASSIGNMENT: Have each student locate a news article about current events in China. Students' articles should pertain to the topics in this section. Ask each student to write a brief summary of his or her chosen article. Then have volunteers read their summaries to the class. Hold a question-and-answer session to review the material.

Teaching Objective 2

LEVEL 1: (Suggested time: 20 min.) Copy the following graphic organizer onto the chalkboard, omitting the italicized answers. Pair students and ask each pair to complete the organizer. Ask volunteers to share their answers with the class. **ENGLISH LANGUAGE LEARNERS, COOPERATIVE LEARNING**

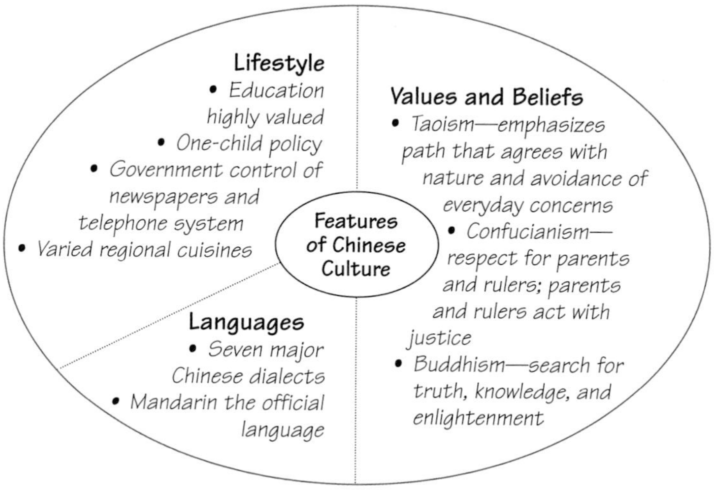

Features of Chinese Culture
Lifestyle
• Education highly valued
• One-child policy
• Government control of newspapers and telephone system
• Varied regional cuisines
Values and Beliefs
• Taoism—emphasizes path that agrees with nature and avoidance of everyday concerns
• Confucianism—respect for parents and rulers; parents and rulers act with justice
• Buddhism—search for truth, knowledge, and enlightenment
Languages
• Seven major Chinese dialects
• Mandarin the official language

Linking Past to Present

Famine in China

A great famine in China began in 1958 despite a record grain crop. There were many reasons for the famine. The government had set very high targets for grain production. Feeling pressured, government-run farms provided exaggerated estimates of their grain harvests. As a result, the government sent large amounts of grain to the cities. Some rural areas lacked enough grain to feed the people.

Also, the government had sent many farmers to work in factories. With fewer farmers working in the fields, grain production dropped during 1960 and 1961.

The Chinese famine ended in 1962. Today, the food supply in China is much larger than it was during the 1960s. Furthermore, because Chinese farmers hold much larger grain stocks than they once did, the chances of another rural famine have decreased greatly.

Activity: Have students write essays explaining how the famine illustrates the need for limiting the power of government.

World War II both groups fought the Japanese. The Communists finally defeated the Nationalists in 1949. Led by Mao Zedong (MOW ZUH-DOOHNG), the Communists set up the People's Republic of China. Mao's version of communism is known as Maoism. Only one political party—the Communist Party—was allowed.

Chiang Kai-shek and his Nationalists retreated to Taiwan. There they created a government for what they called the Republic of China. This government maintained its control through **martial law**, or military rule, for many years.

▲ This image of Mao Zedong is a recognizable symbol of the Cultural Revolution.

Mao's China Under Mao the government took over the country's economy. His government seized private land and organized it into large, government-run farms. Factories were also put under state control. The central government decided the amount and type of food grown on a farm. It also regulated the production of factory goods, owned all housing, and decided where people should live. Sometimes families were separated or forced to relocate. Women were given equal status and assigned equal work duties. To control China's huge population, the government tried to restrict couples to one child per family. Religious worship was prohibited. Despite the efforts to organize the economy, there were planning errors. In the 1960s a famine killed about 30 million people.

The Cultural Revolution In 1966 Mao began a movement called the Cultural Revolution. The Revolution was an attempt to make everyone live a peasant way of life. Followers of Mao were known as Red Guards. They closed schools and universities. Millions of people were sent to the countryside to work in the fields. Opponents were imprisoned or executed.

The Potala Palace in Tibet was built in the A.D. 600s. Today, it has more than 1,000 rooms and is used for religious and political events. ▼

550 Chapter 24

LEVEL 2: (Suggested time: 20 min.) Have each student create a collage that illustrates the impact of traditional Chinese culture, including languages, values and beliefs, on the modern Chinese lifestyle. *(See the graphic organizer from the Level 1 lesson for the correct features.)* Ask volunteers to present and explain their collages to the class.

LEVEL 3: (Suggested time: 20 min.) Tell students to imagine that they are foreign exchange students who are living in China. Have each student write a letter to a friend or family member back home to describe various features of Chinese culture, including languages, values and beliefs, and the modern Chinese lifestyle. *(See the graphic organizer from the Level 1 lesson for the correct features.)* Ask volunteers to read their letters to the class.

➤ASSIGNMENT: Have each student create a design for a commemorative postage stamp that pertains to some aspect of Chinese history that has had an impact on contemporary China. Display students' stamps around the classroom.

➤ASSIGNMENT: Tell students to imagine that they are Chinese university students living during the Cultural Revolution. Have each student write two or three journal entries to describe Mao's actions during the Revolution and the consequences for the Chinese people.

After Mao's death in 1976, the new Chinese communist leadership admitted some past mistakes. It tried to modernize the government. Today, farmers can grow and market their own crops on a portion of their rented land. Almost every inch of productive land is used. China is able to meet most of its food needs. The Chinese enjoy a more varied diet of chicken, fish, fruits, meat, and vegetables. Many of China's large, inefficient state-run factories are being closed or turned over to private industries. Millions of Chinese have started small businesses.

A few Chinese have become wealthy. Some business owners can afford to build private homes and to buy cars, computers, and televisions. However, most Chinese are poor. State employees are paid low wages, live in tiny apartments, and cannot afford cars.

Although the government has allowed individuals some economic freedom, it restricts freedom of speech and religion. In 1989 the Chinese army was called in to attack pro-democracy student demonstrators in Tiananmen Square in Beijing (BAY-JING). Many students were injured or killed. Other rebellions among China's ethnic minorities, particularly in Tibet, have been crushed.

✓ **READING CHECK:** ***Human Systems*** What are some major events in China's history? dynasties, republics, establishment of communist government, Cultural Revolution, death of Mao, Tiananmen Square

Culture

About 92 percent of China's population consider themselves Han Chinese. Almost everyone can speak one of the seven major Chinese dialects. Mandarin Chinese is the official language and the most common.

Values and Beliefs Several philosophies and religions began in China. Taoism (TOW-i-zuhm), or Daoism (DOW-i-zuhm), is an ancient Chinese religion. Taoists believe that humans should try to follow a path that agrees with nature and avoids everyday concerns. The word *dào* means "the path." Each object or natural feature is thought to have its own god or spirit that may reward good deeds or bring bad luck.

The teachings of Confucius also have been important to Chinese culture. Confucius was a philosopher who lived from 551 to 479 B.C. His teachings stressed the importance of family. Confucius believed that children should

▲ The Chinese New Year is also called the Lunar New Year. This is because the cycles of the moon are the basis for the Chinese calendar. The New Year is sometimes called the Spring Festival.

Interpreting the Visual Record **What object are parade participants carrying? What does it resemble?**

ENVIRONMENT AND SOCIETY

For many centuries, natural geographic barriers kept Tibet isolated from the rest of the world. Although the Chinese influenced Tibet's government, style of dress, and food, Tibetan culture remained unique. The practice of Tibetan Buddhism is an important part of Tibet's culture. Tibet's religious and political leader is known as the Dalai Lama.

In the 1800s Britain attempted but failed to gain control of Tibet. China, with British permission, moved to take over Tibet in 1904. However, after the Chinese Revolution eight years later, Tibetans expelled all the Chinese from their country and declared independence. China again invaded and assumed control of Tibet in 1950. The present Dalai Lama fled to India in 1959. Human rights groups have protested some of China's actions in Tibet. From his base in India, the Dalai Lama works to educate the world about Tibet's situation.

Activity: Have a student find and read aloud the definition of the word *theocracy* from the dictionary. Then ask what kind of government Tibet had under the Dalai Lama's leadership. *(a theocracy)*

◄ **Visual Record Answer**

a large puppet; a dragon

TEACHER TO TEACHER

Marcia Caldwell of Austin, Texas, suggests the following activity to help students understand Chinese philosophy: First, tell students that in ancient Chinese thought there are two complementary forces contained in all things—Yin and Yang. Yin is conceived of as Earth, female, dark, passive, and yielding. It is associated with the moon, winter, even numbers, valleys and streams, the tiger, the color orange, and a broken line. Yang is conceived of as the Sun, male, light, active, and forceful. It is associated with odd numbers, mountains, summer, the dragon, the color blue, and an unbroken line. According to the belief, as one of the principles increases the other decreases. This interaction is believed to describe the actual processes of the universe; their remaining in balance is essential to universal harmony.

Have students write down the qualities associated with each principle. Provide students with a large representation of the Yin-Yang symbol. See the example at right. Lead a discussion about how the two sides of the symbol balance and interconnect.

Tell students to use their lists as inspiration for a topic for a poem. Then have them brainstorm with a partner to choose words associated with both the Yin and Yang aspects of the topic. Instruct them to write the words on the correct sides of the symbol and incorporate the words into a brief poem, which they write below the symbol. Finally, invite students to illustrate their poems.

Across the Curriculum
LITERATURE

Nüshu Hundreds of years ago, women in southern Hunan province developed a form of writing known as *nüshu,* or "women's writing." Women invented *nüshu* as a secret script. It enabled women to communicate with each other. *Nüshu* gave women a way to share stories, poems, and letters about their lives. *Nüshu* also allowed women to express ideas that challenged male control of society. Most women who read and wrote *nüshu* were unable to read and write Chinese script.

The Chinese government suppressed the use of *nüshu* during the 1960s. Most women are now educated to read and write Chinese but not *nüshu.* Researchers believe that only two women in rural villages still use *nüshu.* They fear that the language will soon be lost forever.

Discussion: Lead a class discussion on how *nüshu* was related to male control of society.

Connecting to Literature
Answers

1. written by a follower of Confucius, whose teachings stressed the importance of family
2. to encourage children to take care of their parents

CONNECTING TO *Literature*

Confucius, a Chinese philosopher

This Chinese tale comes from the Hsiao Ching, *or* Book of Filial Piety*—which means "devotion to parents." A student of a follower of Confucius is believed to have written it about 400* B.C. *This story encourages children to protect their parents.*

A Loving Son

Wu Meng was eight years old and very dutiful to his parents. His family was so poor that they could not afford to furnish their beds with mosquito-curtains. Every summer night thousands of mosquitoes attacked them, feasting upon their flesh and blood.

Wu Meng looked at his tired parents asleep on their bed as thousands of mosquitoes fiercely attacked them. Wu saw them sucking his parents' blood, which caused his heart to grieve.

To protect his parents, Wu decided that he would not drive the mosquitoes away from himself. Lying on the bed, he threw off his clothes, and soon feeling the pain of the mosquito attacks, he cried: "I have no fear of you, nor have you any reason to fear me. Although I have a fan, I will not use it, nor will I strike you with my hand. I will lie very quietly and let you gorge to the full." Such was his love for his parents!

Analyzing Primary Sources

1. How does this tale reflect traditional Chinese philosophical beliefs?
2. Why do you think the author of the *Hsiao Ching* wrote this story?

respect their parents and subjects should respect their ruler. He believed people should treat those under their control justly and argued that state power should be used to improve people's lives.

The religion called Buddhism also has been important in China. It was founded by an Indian prince, Siddhartha Gautama. Gautama was born in Nepal about 563 B.C. He decided to search for truth and knowledge. Enlightenment—peace and a sense of being one with the universe—came to him while he was sitting under a Bo or Bodhi tree. As a result, he was given the name Buddha, which means "awakened or enlightened one." Buddhism reached China from India about A.D. 100. It became the country's main religion between the 300s and 500s. Indian architecture also became popular in China. Chinese **pagodas**, or Buddhist temples, are based on Indian designs. Pagodas have an upward-curving roof. Some pagodas are 15 stories tall.

CLOSE

Draw five or more large squares on the chalkboard. Ask volunteers to sketch or write in each box something they might see if they visited China today. Discuss the historical or cultural significance of each sketch or description.

REVIEW AND ASSESS

Have students complete the Section Review. Then have each student write two quiz questions pertaining to the information in Section 2. Collect the questions and use them to quiz the class orally. Then have students complete Daily Quiz 24.2.

RETEACH

Have students complete Main Idea Activity 24.2. Then organize students into four groups—one for each of these topics: Dynasties, Outside Influences, Republics of China, and Culture. Have each group create a poster about its assigned topic. Have students present their posters to the class. **ENGLISH LANGUAGE LEARNERS, COOPERATIVE LEARNING**

EXTEND

Have interested students conduct research on the Chinese origins of paper-making and wood-block printing. Have students write a report to discuss their findings. You may also want them to use potatoes to create their own "wood blocks." Ask volunteers to share their reports with the class. **BLOCK SCHEDULING**

Lifestyle Chinese culture highly values education. Chinese children are required to attend nine years of school. However, just 1 to 2 percent of students pass the difficult entrance exams to get into a university.

The Chinese government tries to control many aspects of everyday life. For example, parents are allowed to have just one child. This is because the government is trying to slow population growth.

The government also controls the newspapers and telephone system. This allows it to limit the flow of information and ideas. Satellite TV, the Internet, and e-mail are becoming more widespread, however. This makes it more difficult for the government to control communication between individuals.

Chinese food varies widely from region to region. Food in Beijing is heavily salted and flavored with garlic and cilantro. Sichuan-style cooking features hot pepper sauces. Cantonese cooking was introduced to the United States by immigrants from Guangzhou (GWAHNG-JOH).

Traditional Chinese medicine stresses herbal products and harmony with the universe. People around the world have used acupuncture. This therapy involves inserting fine needles into specific parts of the body for pain relief. Many Chinese herbal remedies have been used by American drug companies as the basis for modern medicines.

China has rich literary traditions. Painting, porcelain, sculpture, and carving of ivory, stone, and wood are also popular. Performing arts emphasize traditional folktales and stories shown in dances or operas with elaborate costumes.

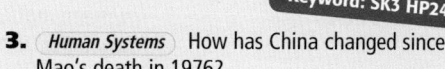

▲ For more than 200 years the Beijing Opera, or Peking Opera, has been recognized worldwide for its artistic contributions. Originally performed for the royal family, it is now viewed by the public and is aired on Chinese television and radio stations.

✓ **READING CHECK:** *Human Systems* What is China's culture like? beliefs include Taoism, Confucianism, Buddhism; education important; government-controlled society; rich traditions

go.hrw.com Homework Practice Online Keyword: SK3 HP24

Section 2 Review

Define and explain: emperor, dynasty, porcelain, martial law, pagodas

Working with Sketch Maps On the map you created in Section 1, label China and the Great Wall. Why do you think the Chinese chose to build the Great Wall where they did?

Reading for the Main Idea

1. *Human Systems* What contributions did the Qin and Han dynasties make to Chinese history?
2. *Human Systems* How did Mao's rule change China?
3. *Human Systems* How has China changed since Mao's death in 1976?

Critical Thinking

4. **Summarizing** What are three philosophies or religions that have been important in China? Describe them.

Organizing What You Know

5. **Summarizing** Copy the following graphic organizer. Use it to describe the Ming and the Manchu.

Ming Dynasty	Manchu

Section 2 Review

Answers

Define For definitions see: emperor, p. 547; dynasty, p. 547; porcelain, p. 548; martial law, p. 550; pagodas, p. 552

Working with Sketch Maps The north was more vulnerable.

Reading for the Main Idea

1. Qin—built Great Wall, kept history, and gave the country its name; Han—expanded kingdom; extended Great Wall; invented compass, paper, and porcelain (NGS 12)
2. government took control of property, housing, jobs; gave women equal status; one-child families; suppressed religion (NGS 12)
3. modernized government, turned over parts of the economy to private interests (NGS 12)

Critical Thinking

4. Taoism—follow a path that agrees with nature and avoids everday concerns; Confucianism—respect for elders and rulers; Buddhism—becoming one with the universe

Organizing What You Know

5. Ming—overthrew Mongols, strengthened Great Wall, focused on Chinese culture; Manchu—from Manchuria, conquered large area, strong government

Section 3

Objectives

1. Explain where most of China's people live.
2. Identify the major cities of China, and describe what they are like.
3. Discuss China's economy.
4. Describe the challenges China faces.

Focus

LET'S GET STARTED

Copy the following instructions onto the chalkboard: *Compile a list of three or four things that you know about China's population, cities, or economy.* Discuss responses. *(Students might mention that China is the world's most populous country, that China's cities are crowded and rapidly modernizing, or that China's economy is growing rapidly.)* Tell students that in Section 3 they will learn more about life in China today.

Building Vocabulary

Copy the following Define terms onto the chalkboard: **command economy**, **multiple cropping**, and **most-favored-nation status.** Ask students to examine the terms and then define any parts of the terms with which they are familiar. *(Students might know the meanings of* command, multiple, economy, *and* most-favored.*)* Have volunteers read the terms and their definitions aloud from the section's text or glossary.

Section 3 Resources

Reproducible
- Guided Reading Strategy 24.3

Technology
- One-Stop Planner CD–ROM, Lesson 24.3
- Homework Practice Online
- *ARGWorld* CD–ROM: Population and Development in China
- HRW Go site

Reinforcement, Review, and Assessment
- Section 3 Review, p. 557
- Daily Quiz 24.3
- Main Idea Activity 24.3
- English Audio Summary 24.3
- Spanish Audio Summary 24.3

internet connect

GO TO: go.hrw.com
KEYWORD: SK3 CH24
FOR: Web sites about China's changing demographics

Section 3 China Today

Read to Discover

1. Where do most of China's people live?
2. What are the major cities in China, and what are they like?
3. What is China's economy like?
4. What challenges does China face?

Define

command economy
multiple cropping
most-favored-nation status

Locate

Beijing
Shanghai
Nanjing
Wuhan
Chongqing
Hong Kong
Macao

The Bei Si pagoda

Why It Matters

In 1997 Hong Kong was returned to China after nearly 100 years of British rule. Use CNNfyi.com or other **current events** sources to find out more about developments in Hong Kong's free market economy since its return to Chinese rule. Record your findings in your journal.

China's Population

China has the largest population in the world—some 1.25 billion people. That number is equal to about 20 percent of the world's population. More people live in China than in all of Europe, Russia, and the United States combined. China's population is growing rapidly—by about 11 million each year. Some years ago, China's leaders took steps to bring the growth rate under control. They encouraged people to delay getting married and starting families. As you have read, they also have limited the size of families.

China's population is not evenly distributed across the land. The western half of the country, which is mostly desert and mountain ranges, is almost empty. Just 10 percent of China's people live there. The rest are crowded into the country's eastern half. In fact, more people live in the North China Plain than in the entire United States. However, this region is only about the size of Texas. Most Chinese live in the countryside. Even so, China has 40 cities with populations greater than 1 million.

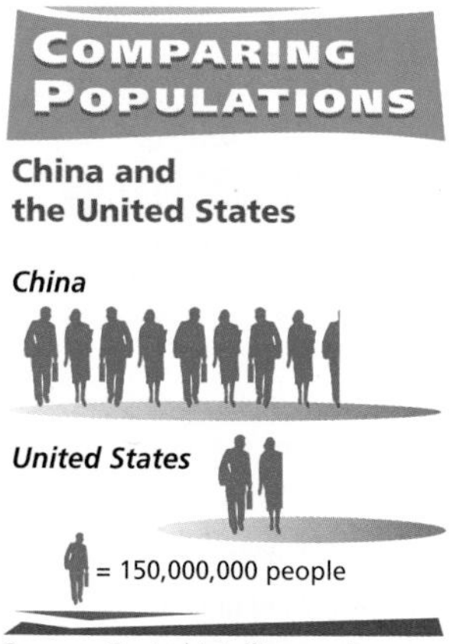

Source: Central Intelligence Agency, *The World Factbook 2001*

✓ **READING CHECK:** *Human Systems* Where do most of China's people live? in the eastern half of the country

China's Cities

Studying China's physical features helps explain why its residents live where they do. By locating rivers and river valleys, we can see where millions of people could best survive.

TEACH

Teaching Objectives 1–2

All LEVELS: (Suggested time: 20 min.) First, discuss how settlement patterns in China are related to geographic factors by reviewing the population map in this unit's atlas. *(The population is concentrated in China's eastern half because deserts and mountains dominate the western half.)* Then copy the following graphic organizer onto the chalkboard, omitting the italicized answers. Have students complete the organizer by identifying and describing China's major cities. Ask volunteers to share their answers with the class. Finally, point out that most of the cities are in eastern China. **ENGLISH LANGUAGE LEARNERS**

Eastern China
- *Beijing—China's capital, largest city in northern China*
- *Nanjing—iron-ore and coal mines*
- *Shanghai—China's largest city, leading industrial center and a major seaport, much new construction*
- *Wuhan—iron-ore and coal mines*
- *Guangzhou—southern China's largest city, major industrial city*
- *Hong Kong—former British colony that maintains free-market economy, densely populated, a major seaport, major banking center, commerce, and tourism*
- *Macao—former Portuguese colony, port*

Interior
- *Chongqing—one of few large cities in China's interior*

Several of China's most important cities are located on the Chang River. Shanghai, the country's largest city, lies on the Chang Delta. It serves as China's leading industrial center and is the major seaport. Shanghai's skyline is constantly changing. New hotels and office buildings seem to rise from the city center almost daily.

Using the map in Section 1, follow the course of the Chang River inland. Locate the cities of Nanjing and Wuhan (WOO-HAHN). These two industrial centers were built around iron-ore and coal mines. If you continue to follow the river upstream, you will reach Chongqing (CHOOHNG-CHING), located in the Sichuan Basin. It is one of the few large cities in China's interior. Guangzhou, located at the mouth of the Xi River, is southern China's largest city. Long famous as a trading center, it was known in the West as Canton. Today Guangzhou is one of China's major industrial cities.

Beijing, also known as Peking, is China's capital. It was established more than 3,000 years ago as a trading center. Beijing is the largest city in northern China and is well known for its cultural heritage. Beijing has famous tourist sites like the "Forbidden City," the great palace of the last emperor.

About 90 miles (145 km) southeast of Guangzhou is Hong Kong, a former British colony. With a population of 6.5 million, Hong Kong is one of the world's most densely populated places. It is only half as large as Rhode Island, but has more than seven times as many residents. Hong Kong is China's major southern seaport and is a center for banking and international trade. It is also a major tourist destination.

Hong Kong's dragon-boat races have changed over the past 2,000 years from a cultural festival to an international sport.

Towering skyscrapers mark Shanghai's skyline.

Interpreting the Visual Record How can you tell this city is growing rapidly?

EYE ON EARTH

Although it is a large, crowded city, Hong Kong is home to more than a bustling economy. On hiking trails through the many parks, visitors might spot monkeys, some of the many birds that live there, or some of the 390 native tree species.

Some 40 percent of Hong Kong's land is protected from development. It is too late for some animals, such as tigers, that have been crowded out. The South China red fox is among the few carnivores that still live in Hong Kong. Pollution from increased development poses a threat to the remaining wildlife.

Conservation efforts are underway in Hong Kong. Local environmental groups are working to survey the remaining wildlife and to promote environmental awareness.

Activity: Have students conduct research on how protecting the environment affects economic development. Have students present their findings to the class.

Visual Record Answer

several buildings under construction, buildings crowded together

Teaching Objective 3

ALL LEVELS: (Suggested time: 20 min.) Remind students of the concepts of command, traditional, and market economy. Then organize students into small groups. Have each group create a poster comparing the U.S. economic system to that of China. The posters should mention the types of economies and the role of government in each. *(Posters should note that China has a command economy with some elements of free enterprise, and that the United States has a mostly market, free-enterprise system.)* Ask groups to present their posters to the class and to explain the benefits of the U.S. free enterprise system. **COOPERATIVE LEARNING**

Teaching Objective 4

ALL LEVELS: (Suggested time: 20 min.) Organize the class into two large groups. Tell one group that it is responsible for reporting the economic progress that China has made in recent decades. Tell the other group to list the challenges that China still faces. *(One group should mention that China has progressed by increasing its standard of living, by expanding industrial production, and by instituting economic freedoms; the other group should mention that China is still relatively poor, that it faces environmental problems, that its most-favored-nation status is in jeopardy, and that it still has not instituted political reforms.)* **COOPERATIVE LEARNING**

Environment and Society

Despite the potential loss of fertile farmland, in the 1990s the Chinese government began construction of a gigantic dam on the Chang River. The Three Gorges Dam, scheduled for completion in 2009, will create a reservoir hundreds of miles long. It will flood two cities, more than 100 towns, and almost 2,000 villages. This project has forced more than 1 million people to relocate.

The plan is to control the flooding of the river and to produce needed electric power. The government hopes that the dam will attract commerce to China's interior and that it will help modernize the country's economy.

Opponents of the project fear the loss of farmland, the potential pollution, and the loss of hundreds of historical sites. Wildlife is also threatened. For example, the habitat of the river dolphin will be destroyed. Only about 100 individuals of this rare species are left.

Activity: Have students conduct further research on the Three Gorges Dam. Then have students discuss the different points of view regarding the dam's impact on the Chinese people.

Visual Record Answer ►

plentiful water, human labor, and animal power

▲ In 1997 Hong Kong was returned to China. The return put an end to more than 150 years of British control.

The British occupied the island of Hong Kong in the 1830s. In the late 1800s Hong Kong was leased to the British for 99 years. The lease ran out in 1997, and Hong Kong then became a special administrative region of China. The British left. Hong Kong has some political independence and is allowed to maintain its free-market economy. Macao, a nearby port city, was once a Portuguese colony. At the end of 1999, it was returned to China. It was the last foreign territory in China.

✓ **READING CHECK:** What are the major cities of China? Shanghai, Nanjing, Wuhan, Chongqing, Guangzhou, Beijing, Hong Kong

China's Economy

When the Chinese Communists took power in 1949 they set up a **command economy**. In this type of economy, the government owns most industries and makes most economic decisions. It set almost all production goals, prices, and wages. In the late 1970s, however, the government began to introduce some elements of free enterprise.

Agriculture Only about 10 percent of China's land is good for farming. Nevertheless, China is a world leader in the production of many crops. China's huge workforce makes this possible. More than 50 percent of Chinese workers earn a living from farming. Having many farmers means the land can be worked intensively to produce high yields. China's farmers have also increased production by cutting terraces into hillsides to create new farmland.

China is divided between rice-growing and wheat-growing regions. The divide lies midway between the Huang and the Chang River. To the south, rice is the main crop. Here **multiple cropping** is common because the weather is warm and wet. Multiple cropping means that two or three crops are raised each year on the same land. This practice makes southern China more prosperous than northern China. Wheat and sorghum are northern China's main crops.

Models of rice fields like these have been found in Han dynasty tombs.

Interpreting the Visual Record **What resources appear to be necessary for growing rice in this region?** ▼

CLOSE

Based on the information presented in this section, ask students to predict what news stories they might read about in the next few years related to China's population, cities, economy, and government.

REVIEW AND ASSESS

Have students complete the Section Review. Then pair students and have pairs reread the section and discuss it among themselves. Then have students complete Daily Quiz 24.3. **COOPERATIVE LEARNING**

RETEACH

Have students complete Main Idea Activity 24.3. Then ask them to create a map of China. Have groups cut out figures to represent population, major cities, the economy, and challenges facing China. Paste their information onto the map. **ENGLISH LANGUAGE LEARNERS, COOPERATIVE LEARNING**

EXTEND

Have students create an annotated time line of Hong Kong's history. Have students write a caption explaining how aspects of its history are reflected in modern Hong Kong society. **BLOCK SCHEDULING**

Industry When the Communists came to power, the Chinese economy was based almost entirely on farming. The Communist government introduced programs to build industry. Today, China is an industrial giant. It produces everything from satellites and rockets to toys and bicycles. Mining is also important. For example, China is a leading producer of iron ore.

✓ **READING CHECK:** *Human Systems* What kind of economy does China have? command, with some elements of free enterprise

Future Challenges

China has enjoyed remarkable economic success in recent years. As a result, the standard of living has improved in most of China. Most Chinese are much better off than they were just a few years ago. However, by global standards China remains a relatively poor country. In addition, China's drive to industrialize has caused major problems. The smoke and waste pumped out by industries have badly polluted the air and water.

Another challenge involves the government's unwillingness to match the new economic freedoms with political reforms. China's human rights record has affected its economic relations with other countries. The U.S. government has considered canceling China's **most-favored-nation status** several times. Countries with this status get special trade advantages from the United States. China's future economic health might depend on its government's willingness to accept political reforms and join the world economic community.

✓ **READING CHECK:** *Human Systems* What challenges does China face? relatively poor, polluted; poor human rights record threatens trade status

By the mid-1990s clothing, electrical equipment, footwear, textiles, and other consumer goods were among China's leading exports. Employees of the Bei Bei Shoe Factory glue soles on by hand on an assembly line in Shanghai.

Interpreting the Visual Record **Why might companies choose to use people rather than machines on this type of assembly line?**

▼

go.hrw.com **Homework Practice Online**
Keyword: SK3 HP24

Section 3 Review

Define and explain: command economy, multiple cropping, most-favored-nation status

Working with Sketch Maps On the map you created in Section 2, label Beijing, Shanghai, Nanjing, Wuhan, Chongqing, Hong Kong, and Macao. In the margin of your map, draw a box for each city. List the characteristics of each city in its box.

Reading for the Main Idea

1. *Places and Regions* Where do most people in China live? Why?
2. *Places and Regions* Along which rivers are several of China's most important cities located? Why?
3. *Environment and Society* What farming practices have allowed the Chinese to increase production?

Critical Thinking

4. **Finding the Main Idea** How has the Chinese government changed its economic policies since the late 1970s? What has been the impact of these changes?

Organizing What You Know

5. **Summarizing** Copy the following graphic organizer. Use it to describe the kinds of challenges facing China today. Identify specific environmental, political, and economic challenges.

Challenges

Section 3 Review

Answers

Define For definitions, see command economy, p. 556; multiple cropping, p. 556; most-favored-nation status, p. 557

Working with Sketch Maps Maps will vary, but listed places should be labeled in their approximate locations. Students may list any of several characteristics.

Reading for the Main Idea

1. in the eastern half; because the western half is mostly desert and mountain ranges (NGS 4)
2. Chang and Xi Rivers (NGS 4)
3. cutting terraces into hillsides; practicing multiple cropping (NGS 14)

Critical Thinking

4. introduced elements of free enterprise; improved standard of living and rise to status as industrial giant

Organizing What You Know

5. environmental—polluted air and water; political—poor human rights record; economic—relatively poor, most-favored-nation status in jeopardy

◄ **Visual Record Answer**

Human labor may be cheaper than machines or the work may require tasks that only people can perform.

Section 4

Objectives

1. Recount how Mongolia's culture has developed.
2. Describe Taiwan's culture.

Focus

LET'S GET STARTED

Copy the following instructions onto the chalkboard: *Look at the photos in Section 4 and read the captions. What can the photos tell you about differences in the physical and human geography of Mongolia and Taiwan?* Discuss responses. *(Possible answers: Mongolia is flatter and drier than Taiwan. Herding is common in Mongolia, but probably not common in Taiwan.)* Tell students that Mongolia and Taiwan are indeed very different and that they will learn about these places in Section 4.

Building Vocabulary

Draw students' attention to the photograph of ***gers*** in this section. Explain that they are Mongolian homes. Ask students what this photograph tells them about Mongolian life. *(Students might mention that* gers *are not fixed to the ground and can be moved fairly easily. Students might conclude that Mongolians shift locations quite often.)*

Section 4 RESOURCES

Reproducible
- Guided Reading Strategy 24.4
- Cultures of the World Activity 7
- Creative Strategies for Teaching World Geography, Lesson 18

Technology
- One-Stop Planner CD–ROM, Lesson 24.4
- Homework Practice Online
- HRW go site

Reinforcement, Review, and Assessment
- Section 4 Review, p. 561
- Daily Quiz 24.4
- Main Idea Activity 24.4
- English Audio Summary 24.4
- Spanish Audio Summary 24.4

Section 4 Mongolia and Taiwan

Read to Discover

1. How has Mongolia's culture developed?
2. What is Taiwan's culture like?

Define

gers

Locate

Mongolia
Ulaanbaatar
Taiwan
Kao-hsiung
Taipei

WHY IT MATTERS

Tensions between China and Taiwan have remained high since 1949. These tensions affect not only China and Taiwan but also their neighbors in the region. Use CNNfyi.com or other **current events** sources to find the latest information about relations between China and Taiwan. Record your findings in your journal.

Mythical animal statue at Taroka Gorge, Taiwan

Mongolia

Mongolia is home to the Mongol people and has a fascinating history. You will learn of invaders and conquests and a culture that prizes horses.

Mongolia's History Today when people discuss the world's leading countries, they do not mention Mongolia. However, 700 years ago Mongolia was perhaps the greatest power in the world. Led by Genghis Khan, the Mongols conquered much of Asia, including China. Later leaders continued the conquests, building the greatest empire the world had seen. The Mongol Empire reached its height in the late 1200s.

This engraving shows early Mongolian soldiers.

It stretched from Europe's Danube River in the west to the Pacific Ocean in the east. Over time, however, the empire declined. In the late 1600s Mongolia fell under the rule of China.

In 1911, with Russian support, Mongolia declared its independence from China. Communists took control of the country 13 years later and established the Mongolian People's Republic. The country then came under the influence of the Soviet Union. Mongolia became particularly dependent on the Soviet Union for economic aid. This aid ended when the Soviet Union collapsed

TEACH

Teaching Objective 1

LEVEL 1: (Suggested time: 30 min.) Pair students and have each pair create a collage to represent elements of Mongolia's culture. *(Collages should represent aspects of the nomadic lifestyle, including herding and* gers. *Collages should also represent the importance of horses.)* Display students' collages around the classroom. **ENGLISH LANGUAGE LEARNERS, COOPERATIVE LEARNING**

LEVELS 2 AND 3: (Suggested time: 20 min.) Have each student write a short poem about Mongolia's culture. *(See the Level 1 lesson for the correct aspects of Mongolia's culture.)* Ask volunteers to recite their poems to the class.

➤ASSIGNMENT: Tell students to imagine that for six months they will live with Mongolian nomads. Their hosts herd livestock from horseback and live in *gers*. Each student's luggage may weigh no more than 30 pounds. Have students gather information on the weight of each item they might take and then consider the consequences of choosing some items but not others. Encourage students to take Mongolia's climate and culture into account. Then have students create their lists and discuss which are the most important items.

Nomads of Mongolia live in *gers* like those shown here.

Interpreting the Visual Record What in this photo suggests this would not be a permanent settlement?

in the early 1990s. Since then, Mongolians have struggled to build a democratic government and a free-market economy.

Mongolia's Culture Despite years of Communist rule and recent Western influence, the Mongolian way of life remains quite traditional. Many people still follow a nomadic lifestyle. They live as herders, driving their animals across Mongolia's vast grasslands. They make their homes in ***gers*** (GUHRZ). These are large, circular felt tents that are easy to raise, dismantle, and move.

Since most people live as herders, horses play an important role in Mongolian life. Mongolian children learn to ride when they are very young—often before they are even five years old. In Mongolia, the most powerful piece in the game of chess is the horse, not the queen.

Mongolia Today Mongolia is a large country—slightly larger than Alaska. Its population numbers just over 2.5 million. Some 25 percent of Mongolians live in Ulaanbaatar (oo-lahn-BAH-tawr), the capital city. Ulaanbaatar is also Mongolia's main industrial and commercial center. Mongolia's other cities are quite small. Not one has a population greater than 100,000.

✓ **READING CHECK:** *Human Systems* What are some elements of Mongolian culture? traditional, nomadic lifestyle; people living in *gers* and herding animals

Taiwan

For many years the island of Taiwan was known in the West as Formosa. This name came from Portuguese sailors who visited the island in the late 1500s. They thought the island was so lovely that they called it *Ilha Formosa*, or "beautiful island."

In the Gobi, temperatures can range from -40°F (-40°C) in January to 113°F (45°C) in July. Some areas of this desert receive little more than 2 inches (5 cm) of rain each year.

EYE ON EARTH

Mongolia is the home of a much sought-after falcon, the saker. Valued by Genghis Khan for their hunting abilities, sakers are still prized for their speed and agility.

There is a limited legal trade in the birds for research purchases. There is also an illegal trade; falconers from the Arab states will pay up to $200,000 for a saker. As a result of its popularity, the falcon's survival is in doubt. Illegal trading of sakers has caused the bird to become endangered in the wild.

Activity: Have students conduct research on the illegal international trade in endangered species. Ask students to use their findings to create maps showing the sources, routes, and destinations of the trade. The maps should also show which species are targeted.

▲ **Visual Record Answer**

Gers are collapsible. There are no streets or permanent structures.

Teaching Objective 2

ALL LEVELS: (Suggested time: 15 min.) Copy the following graphic organizer onto the chalkboard, omitting the italicized answers. Have students complete the organizer by filling in facts about Taiwan's culture. Ask volunteers to share their answers with the class.

ENGLISH LANGUAGE LEARNERS

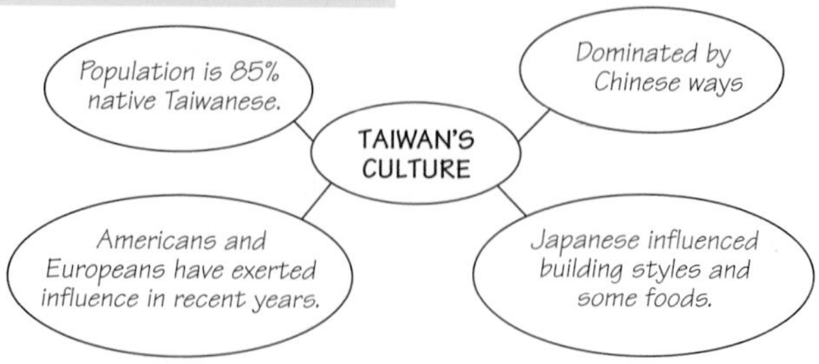

TAIWAN'S CULTURE
Population is 85% native Taiwanese.
Dominated by Chinese ways
Americans and Europeans have exerted influence in recent years.
Japanese influenced building styles and some foods.

LEVELS 2 AND 3: (Suggested time: 45 min.) Draw a series of seven footsteps on the chalkboard and label each foot with a date in Taiwan's history. Ask volunteers to recount what happened at each step in Taiwan's history and to write the correct responses into the footsteps. *(Students should mention that in the 600s, the Chinese first settle in Taiwan; in the 1100s, the Japanese take control of eastern Taiwan; in the 1500s and 1600s, Europeans try to set up bases and are driven out; in 1895 the Chinese give Taiwan to the Japanese; in 1945 China regains control; in 1949 Chiang Kai-shek and the nationalists establish a government in Taiwan; today Taiwanese enjoy expanded democratic rights, but the island is claimed by China.)* Have students copy the footsteps and events into their notes. Then have students consult library resources to find details about how these events affected Taiwan's culture.

Section 4 Review

Answers

Define For definition, see: *gers*, p. 559

Working with Sketch Maps Maps will vary, but listed places should be labeled in their approximate locations. Ulaanbaatar—capital of Mongolia, 25 percent of population, commerce and industry; Taipei—capital of Taiwan, overcrowded and polluted, on coastal plain

Reading for the Main Idea

1. by herding animals (NGS 12)
2. Chinese settle in 600s; Japanese control in 1100s; Chinese raiders drive out Europeans in 1600s; treaty gives Japanese control in 1895; China gets control after WWII; Chiang Kai-shek and nationalists flee to Taiwan (NGS 13)
3. changed from agriculture to industry (NGS 12)

Critical Thinking

4. overcrowding and pollution

Organizing What You Know

5. Postcards will vary but information should be consistent with text material.

Visual Record Answer

Answers will vary.

▲ Chiang Kai-shek served in the Japanese army before returning to China to help overthrow the Manchu dynasty. He was head of the Nationalist government in China for 20 years, then moved with his followers to Taiwan.

Taiwan's History The Chinese began settling Taiwan in the A.D. 600s. Some 600 years later the Japanese took control of eastern Taiwan. The search for spices brought European traders to Taiwan. The Dutch, Portuguese, and Spanish all tried to set up bases there. However, raiders from mainland China drove out these Europeans in the mid-1600s.

The struggle among the Chinese, Japanese, and Europeans for control of Taiwan continued until the late 1800s. In 1895 a treaty between the Chinese and the Japanese gave Taiwan to Japan. The Japanese then tried to force their way of life on the people of Taiwan. The Taiwanese rebelled against these efforts, but their revolts were crushed by the Japanese military.

After Japan's surrender at the end of World War II in 1945, China once again took command of Taiwan. In 1949 Mao Zedong established the People's Republic of China in mainland China. Chiang Kai-shek and the Nationalist Chinese government fled to Taiwan. The Nationalist government controlled Taiwan through martial law for decades. In recent years, however, the government has expanded democratic rights. China still claims that Taiwan is a province of China—not an independent country.

Taiwan's Culture Taiwan's history is reflected in its culture. Its population is about 85 percent native Taiwanese. They are descendants of people who migrated from China to Taiwan over hundreds of years. Chinese ways dominate Taiwan's culture. However, some building styles and certain foods reflect Japanese influences. European and American practices and customs have strongly influenced Taiwan's way of life in recent years. This is particularly true in the cities.

Taiwan has good farmlands, but the island's eastern half is mountainous.

Interpreting the Visual Record How do the trees shown here compare to trees common where you live? ▼

China
Mt. Ho-Huan
Taiwan
Philippines
Vietnam

CLOSE

Tell students to imagine that Mongolia and Taiwan are redesigning their flags. Ask students what symbols Mongolia and Taiwan might use on a newly designed flag. Have students explain their responses.

REVIEW AND ASSESS

Have students complete the Section Review. Then pair students and have each pair create a graphic organizer to present important facts about the culture and history of Mongolia and Taiwan. Have students present their organizers to the class. Then have students complete Daily Quiz 24.4. **COOPERATIVE LEARNING**

RETEACH

Have students complete Main Idea 24.4. Then ask each student to write down one fact about Mongolia and one fact about Taiwan. Collect the facts and review them with the class. **ENGLISH LANGUAGE LEARNERS**

EXTEND

Have interested students conduct research on the conquests of Genghis Khan. Have students use their findings to create a short oral report about the Mongolian Empire under Genghis Khan and a map showing the extent of the empire. **BLOCK SCHEDULING**

Taiwan Today Taiwan has a modern, industrial economy and a population of nearly 22 million. These people live on an island the size of Delaware and Maryland combined. Most people live on the western coastal plain of Taiwan. Population densities there can reach higher than 2,700 per square mile (1,042 per sq km). Taiwan's two largest cities, Kao-hsiung (KOW-SHYOOHNG) and Taipei (TY-PAY), are located on the coastal plain. Taipei is the capital city. It faces serious overcrowding and environmental problems. The thousands of cars, motorcycles, and trucks that clog Taipei's streets each day cause severe air pollution.

In the early 1950s Taiwan's economy was still largely based on agriculture. Today, however, only about 10 percent of workers make a living as farmers. Even so, Taiwan still produces enough rice—the country's chief food crop—to feed all of its people. Taiwan's farmers also grow fruits, sugarcane, tea, and vegetables.

Taiwan now has one of Asia's most successful economies. It is a world leader in the production and export of computers and sports equipment.

▲ Taipei, Taiwan, was founded in the 1700s and has developed into an important city for overseas trade.

Interpreting the Visual Record **What characteristics of Taipei shown here resemble other cities?**

✓ **READING CHECK:** *Human Systems* What are some elements of the culture of Taiwan? 85 percent native Taiwanese; Chinese influence dominant; Japanese influence in some building styles and foods; some European and American customs

go.hrw.com **Homework Practice Online**
Keyword: SK3 HP24

Section 4 Review

Define and explain: *gers*

Working with Sketch Maps On the map you created in Section 3, label Mongolia, Ulaanbaatar, Taiwan, Kao-hsiung, and Taipei. In the border of your map, draw a box for each city. Describe each city in its box. How has the history of each city played a part in its growth?

Reading for the Main Idea

1. *Human Systems* How do most people earn a living in Mongolia?
2. *Human Systems* Write a brief outline of the significant individuals or groups that have influenced Taiwan's history.
3. *Human Systems* How has Taiwan's economy changed since the early 1950s?

Critical Thinking

4. **Analyzing Information** What are some problems Taiwan faces today?

Organizing What You Know

5. **Summarizing** Copy the following graphic organizer. Use it to design and write two postcards to a friend describing life in Mongolia. In your postcards, note how life follows traditional patterns yet is also undergoing changes.

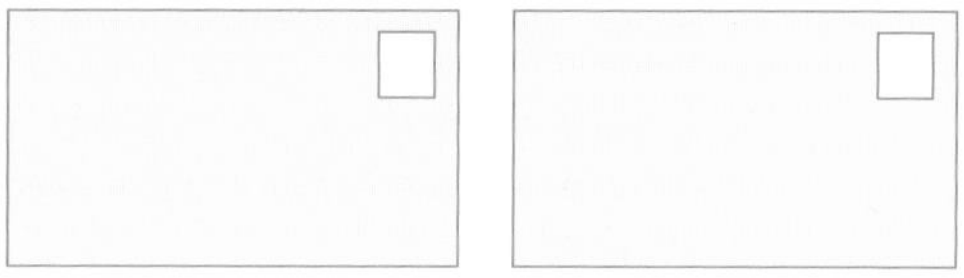

CHAPTER 24 Review ANSWERS

Building Vocabulary

For definitions, see: dikes, p. 545; arable, p. 546; emperor, p. 547; dynasty, p. 547; porcelain, p. 548; martial law, p. 550; pagodas, p. 552; command economy, p. 556; multiple cropping, p. 556; most-favored-nation status, p. 557; *gers*, p. 559

Reviewing the Main Ideas

1. isolated plateaus and basins (NGS 4)
2. carries loess, a yellowish-brown soil (NGS 4)
3. ordered construction of the Great Wall (NGS 14)
4. Mongolia (NGS 13)
5. satellites, rockets, toys, bicycles, and mining (NGS 11)

◄ **Visual Record Answer**

advertisements, bright lights, signs, traffic

Assess

Have students complete the Chapter 24 Test.

Reteach

Organize the class into three groups and assign each group China, Mongolia, or Taiwan. Have each group prepare an outline to present pertinent historical, cultural, and economic information about its country. Then have each group copy its outline onto the chalkboard. Ask the class to fill in any information that might be missing from the outlines.

ENGLISH LANGUAGE LEARNERS, COOPERATIVE LEARNING

PORTFOLIO

Portfolio Extensions

1. Have students construct Chinese lanterns from construction paper and write information about China on the lanterns. Fold the lanterns to insert into portfolios.
2. Obtain several abacuses from an Asian import shop. Invite an expert to demonstrate basic addition and subtraction on an abacus. On a sheet of paper, have students devise math problems that can be solved using an abacus. Have students write the correct answers on a separate sheet of paper. After students exchange problems and solve them, use the answer key to check students' solutions. Put papers and answer keys in portfolios.

CHAPTER 24

Review ANSWERS

Understanding Environment and Society

Presentations will vary, but the information included should be consistent with text material. Students should describe the one-child policy and environmental, economic, and social consequences of failure to maintain a balance. Use Rubric 29, Presentations, to evaluate student work.

Thinking Critically

1. They had to travel overland, which entailed a long, difficult journey.
2. Cultural isolation might have continued.
3. It imposed the peasant's lifestyle on people and resulted in the closing of schools, people being sent to the countryside to work, and the imprisonment of opponents.
4. Electronic media make information from other countries available to the Chinese people. This free flow of information could threaten communist control.

CHAPTER 24

Reviewing What You Know

Building Vocabulary

On a separate sheet of paper, write sentences to define each of the following words.

1. dikes
2. arable
3. emperor
4. dynasty
5. porcelain
6. martial law
7. pagodas
8. command economy
9. multiple cropping
10. most-favored-nation status
11. *gers*

Reviewing the Main Ideas

1. *Places and Regions* What physical features separate this region's mountain ranges?
2. *Places and Regions* Why is the Huang called the Yellow River?
3. *Environment and Society* What idea did the first emperor of the Qin dynasty have that greatly affected the landscape of China?
4. *Human Systems* What country discussed in this chapter once ruled a vast empire that stretched into Europe?
5. *Environment and Society* What are Taiwan's leading exports?

Understanding Environment and Society

Resource Use

Growing industries and cities are taking over farmland in China. Create a presentation about potential problems caused by this situation. Consider the following:

- Actions China's government has already taken to try to solve the problem of its growing population.
- Problem in balancing population and food supply.

Thinking Critically

1. **Drawing Inferences and Conclusions** Why do you think so few Europeans reached China before the 1500s?
2. **Drawing Inferences and Conclusions** How might Chinese history have been different if Europeans had not forced trade upon the Chinese?
3. **Finding the Main Idea** What was the Cultural Revolution and how did it affect life in China?
4. **Making Generalizations and Predictions** In what ways does modern technology threaten the Chinese government's ability to control the flow of information in the country? What changes might the free flow of information bring to China?
5. **Analyzing Information** How is Taiwan's history reflected in the island's culture today?

FOOD FESTIVAL

Chinese Almond Cakes Have students make these cakes for a New Year's celebration. Sift together 2½ c. all-purpose flour and ¾ tsp baking powder. Blend in 1 c. shortening and 1½ c. sugar. Stir in ¼ tsp almond extract, 2 Tbs beaten egg and 1 Tbs water. Knead dough and let stand for 5 minutes. Then form dough into 1½ inch balls. Flatten balls into cake-like shapes. Press an almond into each cake. (You will need about 30 almonds.) Bake on greased baking sheets in a 375° oven for 5 minutes. Reduce heat to 300° and bake for 8–10 minutes more, or until golden.

CHAPTER 24 REVIEW AND ASSESSMENT RESOURCES

Reproducible
- Readings in World Geography, History, and Culture 67, 68, and 69
- Critical Thinking Activity 24
- Vocabulary Activity 24

Technology
- Chapter 24 Test Generator (on the One-Stop Planner)
- Audio CD Program, Chapter 24
- HRW Go site

Reinforcement, Review, and Assessment
- Chapter 24 Review, pp. 562–63
- Chapter 24 Tutorial for Students, Parents, Mentors, and Peers
- Chapter 24 Test
- Chapter 24 Test for English Language Learners and Special-Needs Students

Building Social Studies Skills

Map ACTIVITY

On a separate sheet of paper, match the letters on the map with their correct labels.

Mount Everest	Chang River
Plateau of Tibet	Great Wall
North China Plain	Shanghai
Huang	Hong Kong
	Ulaanbaatar

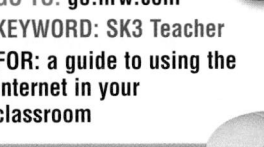

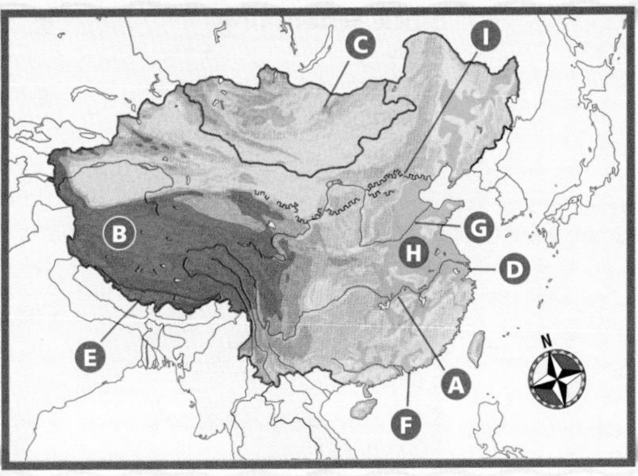

Mental Mapping Skills ACTIVITY

On a separate sheet of paper, draw a freehand map of China, Mongolia, and Taiwan. Make a key for your map and label the following:

China	Mongolia
Gobi	Pacific Ocean
Himalayas	Taiwan

WRITING ACTIVITY

Imagine that you are a Chinese university professor. Using the time line in section 2, the text, and other sources, write a brief lesson plan on China's history. You may want to include some visuals, such as photographs of artifacts, in your lesson plan. Be sure to use standard grammar, spelling, sentence structure, and punctuation.

Alternative Assessment

Portfolio ACTIVITY

Learning About Your Local Geography

Cooperative Project Fishing is an important source of food in China. Is fishing important in your community? Draw and label three types of fish available in your community.

internet connect

Internet Activity: **go.hrw.com**
KEYWORD: **SK3 GT24**

Choose a topic to explore about China, Mongolia, and Taiwan:
- Follow the Great Wall of China.
- Visit the land of Genghis Khan.
- See the artistic treasures of China.

5. Most people are descendants of Chinese immigrants, so Chinese is dominant. Other cultural features reflect Japanese, American, and European influences.

Map Activity

A. Chang River
B. Plateau of Tibet
C. Ulaanbaatar
D. Shanghai
E. Mount Everest
F. Hong Kong
G. Huang He
H. North China Plain
I. Great Wall

Mental Mapping Skills Activity

Maps will vary, but listed places should be labeled in their approximate locations.

Writing Activity

Lesson plans will vary but should be consistent with text material. Use Rubric 40, Writing to Describe, to evaluate student work.

Portfolio Activity

Drawings will vary, but the information included should be accurate according to local sources. Use Rubric 28, Posters, to evaluate student work.

internet connect

GO TO: **go.hrw.com**
KEYWORD: **SK3 Teacher**
FOR: **a guide to using the Internet in your classroom**

CHAPTER 25

Japan and the Koreas

Chapter Resource Manager

Objectives	Pacing Guide	Reproducible Resources
SECTION 1 **Physical Geography** (pp. 565–68) **1.** Identify the physical features of Japan and Korea. **2.** Identify the natural resources of the region. **3.** Identify the climate types of the region.	**Regular** 1 day **Block Scheduling** .5 day *Block Scheduling Handbook, Chapter 25*	**RS** Guided Reading Strategy 25.1 **E** Lab Activities for Geography and Earth Science, Demonstrations 5, 6, 7
SECTION 2 **The History and Culture of Japan** (pp. 569–71) **1.** Describe Japan's early history and culture. **2.** Describe the modernization of Japan.	**Regular** 1 day **Block Scheduling** .5 day *Block Scheduling Handbook, Chapter 25*	**RS** Guided Reading Strategy 25.2 **RS** Graphic Organizer 25 **E** Creative Strategies for Teaching World Geography, Lesson 18, 19 **E** Cultures of the World Activity 7
SECTION 3 **Japan Today** (pp. 572–76) **1.** Describe where most Japanese live. **2.** Describe what most Japanese cities are like. **3.** Describe life in Japan. **4.** Describe how the Japanese economy has developed.	**Regular** 1 day **Block Scheduling** .5 day *Block Scheduling Handbook, Chapter 25*	**RS** Guided Reading Strategy 25.3 **SM** Map Activity 25
SECTION 4 **The History and Culture of the Koreas** (pp. 577–79) **1.** Explain what Korea's ancient history was like. **2.** Identify the major events of Korea's early modern period. **3.** Describe the division of Korea and its effects.	**Regular** .5 day **Block Scheduling** .5 day *Block Scheduling Handbook, Chapter 25*	**RS** Guided Reading Strategy 25.4 **IC** Interdisciplinary Activities for the Middle Grades 29, 30, 31
SECTION 5 **South and North Korea Today** (pp. 580–83) **1.** Describe South Korea's government and society. **2.** Describe South Korea's economy. **3.** Describe North Korea. **4.** Describe how North Korea's government has affected the country's development.	**Regular** 1 day **Block Scheduling** .5 day *Block Scheduling Handbook, Chapter 25*	**RS** Guided Reading Strategy 25.5 **SM** Geography for Life Activity 25

Chapter Resource Key

RS Reading Support
IC Interdisciplinary Connections
E Enrichment
SM Skills Mastery
A Assessment
REV Review
ELL Reinforcement and English Language Learners
Transparencies
CD–ROM
Music
Video
Internet
Holt Presentation Maker Using Microsoft® Powerpoint®

One-Stop Planner CD–ROM

See the ***One-Stop Planner*** for a complete list of additional resources for students and teachers.

One-Stop Planner CD–ROM

It's easy to plan lessons, select resources, and print out materials for your students when you use the ***One-Stop Planner CD–ROM with Test Generator.***

Technology Resources	Review, Reinforcement, and Assessment Resources
One-Stop Planner CD–ROM, Lesson 25.1 *ARGWorld* CD–ROM: Naming Places in East Asia Geography and Cultures Visual Resources with Teaching Activities 49–52 Homework Practice Online HRW Go site	**ELL** Main Idea Activity 25.1 **REV** Section 1 Review, p. 568 **A** Daily Quiz 25.1 **ELL** English Audio Summary 25.1 **ELL** Spanish Audio Summary 25.1
One-Stop Planner CD–ROM, Lesson 25.2 Homework Practice Online HRW Go site	**ELL** Main Idea Activity 25.2 **REV** Section 2 Review, p. 571 **A** Daily Quiz 25.2 **ELL** English Audio Summary 25.2 **ELL** Spanish Audio Summary 25.2
One-Stop Planner CD–ROM, Lesson 25.3 Homework Practice Online HRW Go site	**ELL** Main Idea Activity 25.3 **REV** Section 3 Review, p. 576 **A** Daily Quiz 25.3 **ELL** English Audio Summary 25.3 **ELL** Spanish Audio Summary 25.3
One-Stop Planner CD–ROM, Lesson 25.4 Homework Practice Online HRW Go site	**ELL** Main Idea Activity 25.4 **REV** Section 4 Review, p. 579 **A** Daily Quiz 25.4 **ELL** English Audio Summary 25.4 **ELL** Spanish Audio Summary 25.4
One-Stop Planner CD–ROM, Lesson 25.5 Homework Practice Online HRW Go site	**ELL** Main Idea Activity 25.5 **REV** Section 5 Review, p. 583 **A** Daily Quiz 25.5 **ELL** English Audio Summary 25.5 **ELL** Spanish Audio Summary 25.5

internet connect

HRW ONLINE RESOURCES

GO TO: go.hrw.com
Then type in a keyword.

TEACHER HOME PAGE
KEYWORD: SK3 TEACHER

CHAPTER INTERNET ACTIVITIES
KEYWORD: SK3 GT25

Choose an activity to:
- investigate volcanoes.
- visit Japan and the Koreas.
- compare your school to a Japanese school.

CHAPTER ENRICHMENT LINKS
KEYWORD: SK3 CH25

CHAPTER MAPS
KEYWORD: SK3 MAPS25

ONLINE ASSESSMENT
Homework Practice
KEYWORD: SK3 HP25
Standardized Test Prep Online
KEYWORD: SK3 STP25
Rubrics
KEYWORD: SS Rubrics

COUNTRY INFORMATION
KEYWORD: SK3 Almanac

CONTENT UPDATES
KEYWORD: SS Content Updates

HOLT PRESENTATION MAKER
KEYWORD: SK3 PPT25

ONLINE READING SUPPORT
KEYWORD: SS Strategies

CURRENT EVENTS
KEYWORD: S3 Current Events

Meeting Individual Needs

Ability Levels

Level 1 Basic-level activities designed for all students encountering new material

Level 2 Intermediate-level activities designed for average students

Level 3 Challenging activities designed for honors and gifted-and-talented students

English Language Learners Activities that address the needs of students with Limited English Proficiency

Chapter Review and Assessment

E Readings in World Geography, History, and Culture 70 and 71

SM Critical Thinking Activity 25

REV Chapter 25 Review, pp. 584–85

REV Chapter 25 Tutorial for Students, Parents, Mentors, and Peers

ELL Vocabulary Activity 25

A Chapter 25 Test

Chapter 25 Test Generator (on the One-Stop Planner)

Audio CD Program, Chapter 25

A Chapter 25 Test for English Language Learners and Special-Needs Students

LAUNCH INTO LEARNING

Create two columns on the chalkboard. Label them *What you already know about Japan and the Koreas* and *What you want to know about Japan and the Koreas*. Call on volunteers to offer items for both categories. Tell students that while studying this chapter they will be able to check their knowledge and add to it by filling the gaps they have mentioned. Add that Japan, North Korea, and South Korea affect the economy and security of countries around the world, including the United States. Tell students they will learn more about the connections among Japan, the Koreas, and the United States in this chapter.

Section 1

Objectives

1. Describe the physical features of Japan and the Koreas.
2. Identify the region's natural resources.
3. Describe the climate types found in the region.

LINKS TO OUR LIVES

You may wish to point out to the students that there are many reasons why we should know more about Japan and the Koreas, these among them:

▶ The region's economies are directly linked with ours. An East Asian recession in the late 1990s threatened to harm the American economy.

▶ North Korea has developed nuclear power and perhaps even missiles that could strike its neighbors.

▶ During World War II the United States and Japan were enemies. Since then Japan has been an ally. Occasionally trade or environmental issues threaten the relationship.

▶ Japanese and Korean culture offer a wealth of literature, art, philosophy, music, crafts, foods, and other traditions that people around the world enjoy.

CHAPTER 25

Japan and the Koreas

Now we continue east to North and South Korea and the island nation of Japan. First we meet Akiko, a Japanese student whose school day may be very different from yours.

Konichiwa! (Good afternoon!) I'm Akiko, and I'm in the seventh grade at Yamate school. Every morning except Sunday I put on my school uniform and eat rice soup and pickles before I leave for school. The train I take is so crowded I can't move. At school, I study reading, math, English, science, and writing. I know 1,800 Japanese characters, but I need to know about 3,000 to pass the ninth grade exams. For lunch, I eat rice and cold fish my mom packed for me. Before we can go home, we clean the school floors, desks, and windows. My dad usually isn't home until after 11:00 P.M., so my mom helps me with my homework in the "big" (8 feet by 8 feet) room of our three-room apartment. In the evenings, I go to a *juku* school to study for the ninth grade exams. If I do not do well, I will not go to a good high school, and my whole family will be ashamed. On Sundays, I sometimes go with my parents to visit my grandparents, who are rice farmers. I like rock music a lot, especially U2.

こんにちは.
私は東京に
住んでいます.

Translation: Good afternoon. I live in Tokyo.

FOCUS

LET'S GET STARTED

Copy the following instructions onto the chalkboard: *Look at the photo of the bullet train and Mount Fuji on the next page. What do you think Mount Fuji's shape can tell us about how it was formed? Write down your ideas.* Allow students time to write down their answers. Discuss responses. Then tell the students that Fuji's conical shape indicates that it is a volcano and that volcanoes and earthquakes have affected life in Japan. Tectonic events are rare in the Koreas, however. Tell students that in Section 1 they will learn more about the region's physical geography.

Using the Physical-Political Map

Have students examine the map on this page. Ask volunteers to name the principal landforms that Japan, North Korea, and South Korea occupy and the bodies of water that surround them. Then, comparing the latitude of these countries to the East Coast of the United States, have students predict their climates. Point out the Oyashio and Japan Currents. Ask students to speculate on how these currents affect Japan's climate.

Section 1 Physical Geography

Read to Discover

1. What are the physical features of Japan and the Koreas?
2. What natural resources does the region have?
3. Which climate types are found in the region?

Define

tsunamis
Oyashio Current
Japan Current

Locate

Korean Peninsula
Sea of Japan
Hokkaido
Honshu
Shikoku
Kyushu

WHY IT MATTERS

Weather plays an important role in Japanese life. Use CNNfyi.com or other **current events** sources to find examples of how weather has affected Japanese history and society. Record the findings in your journal.

Japanese bonsai

Japan and the Koreas: Physical-Political

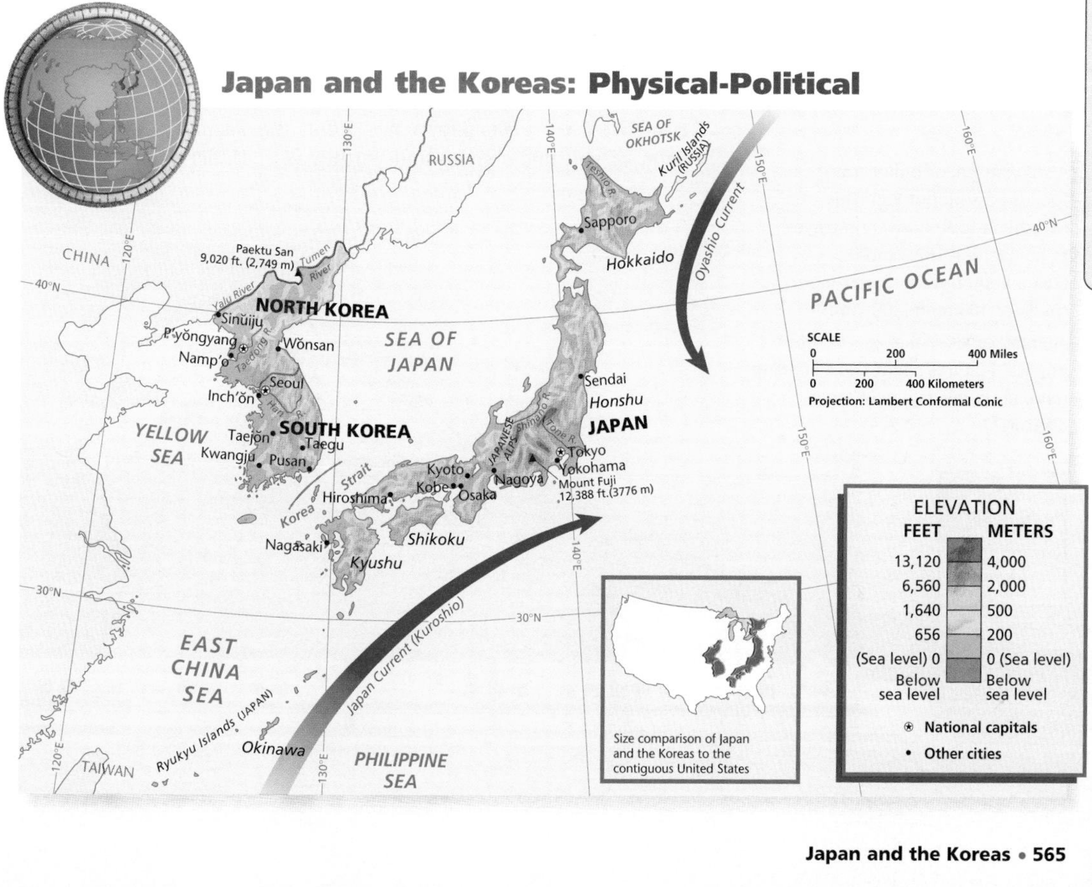

Section 1 RESOURCES

Reproducible

- Block Scheduling Handbook, Chapter 25
- Guided Reading Strategy 25.1
- Lab Activities for Geography and Earth Science, Demonstrations 5, 6, 7

Technology

- One-Stop Planner CD–ROM, Lesson 25.1
- Homework Practice Online
- Geography and Cultures Visual Resources with Teaching Activities 49-52
- *ARGWorld* CD–ROM: Naming Places in East Asia
- HRW Go site

Reinforcement, Review, and Assessment

- Section 1 Review, p. 568
- Daily Quiz 25.1
- Main Idea Activity 25.1
- English Audio Summary 25.1
- Spanish Audio Summary 25.1

TEACH

Teaching Objective 1

ALL LEVELS: (Suggested time: 15 min.) Copy the following graphic organizer onto the chalkboard, omitting the italicized answers. Use it to help students create lists of the physical features of Japan and the Koreas. Call on students to come to the chalkboard and fill in words and phrases to describe the features. In the center, have them list features the countries have in common. **ENGLISH LANGUAGE LEARNERS**

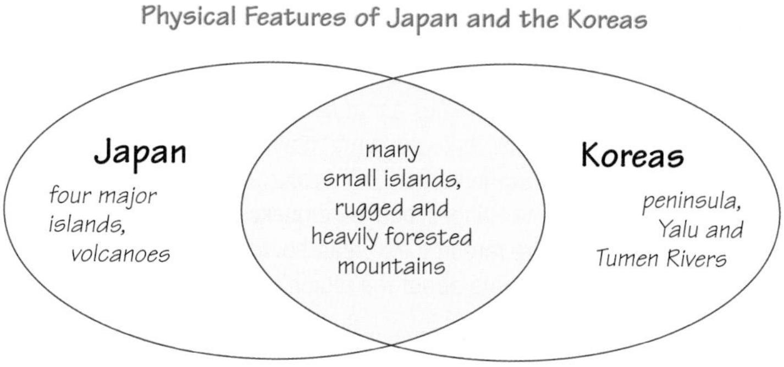

ENVIRONMENT AND SOCIETY

Climbing Mount Fuji is a popular activity in Japan. The ascent has long been a pilgrimage for Japanese Buddhists, who regard the peak as sacred. Now recreational climbers can drive halfway up the mountain and walk the rest of the way. The rising numbers of automobile tourists and climbers, however, are creating significant problems with erosion, garbage, pollution, and traffic.

Activity: Have students research U.S. national parks that are threatened by their own popularity. *(Examples: Yellowstone, Grand Canyon, Yosemite)* Ask them to compare those problems with Mount Fuji's. How are the issues similar or different?

Visual Record Answer

possible answer: mountains

internet connect

GO TO: **go.hrw.com**
KEYWORD: **SK3 CH25**
FOR: **Web sites about Japan and the Koreas**

Our Amazing Planet

The world's largest crab lives off the southeastern coast of Japan. The giant spider crab can grow larger than 12 feet (3.6 m) across (from claw to claw). It can also weigh more than 40 pounds (18 kg)!

Physical Features

The Korean Peninsula extends southward about 600 miles (965 km) from mainland Asia. The peninsula is relatively close to southern Japan.

The Korean Peninsula is about the same size as Utah. It contains two countries, North Korea and South Korea. The Yalu and Tumen Rivers separate North Korea from China. The Tumen River also forms a short border with Russia. Off the coast of the Korean Peninsula lie more than 3,500 islands.

The Sea of Japan separates Japan from the Eurasian mainland. The narrow Korea Strait lies between South Korea and the island country of Japan. No place in Japan is more than 90 miles (145 km) from the sea. Japan is about the size of California. It is made up of four large islands called the home islands. The country also includes more than 3,000 smaller islands. The home islands from north to south are Hokkaido (hoh-KY-doh), Honshu (HAWN-shoo), Shikoku (shee-KOH-koo), and Kyushu (KYOO-shoo). South of the home islands are Japan's Ryukyu (ree-YOO-kyoo) Islands. Okinawa is the largest of these islands. Fewer than half of the Ryukyus are inhabited.

Mountains Rugged and heavily forested mountains are a common sight in the landscape of this region. Mountains cover about 75 percent of Japan. Many of Japan's mountains were formed by volcanic activity. The country's longest mountain range, the Japanese Alps, forms a volcanic spine through Honshu. The small amount of plains in these countries is found along the coasts and river valleys.

The Ring of Fire Japan lies along the Pacific Ring of Fire—a region of volcanic activity and earthquakes. Under Japan the dense

A *shinkansen*, or bullet train, speeds past Mount Fuji. This Hiroshima-to-Kokura train travels at an average 162.3 mph (261.8 kmh) but has a maximum speed of 186 mph (300 kmh).

Interpreting the Visual Record What physical features might make building railroads in this region of Japan difficult?

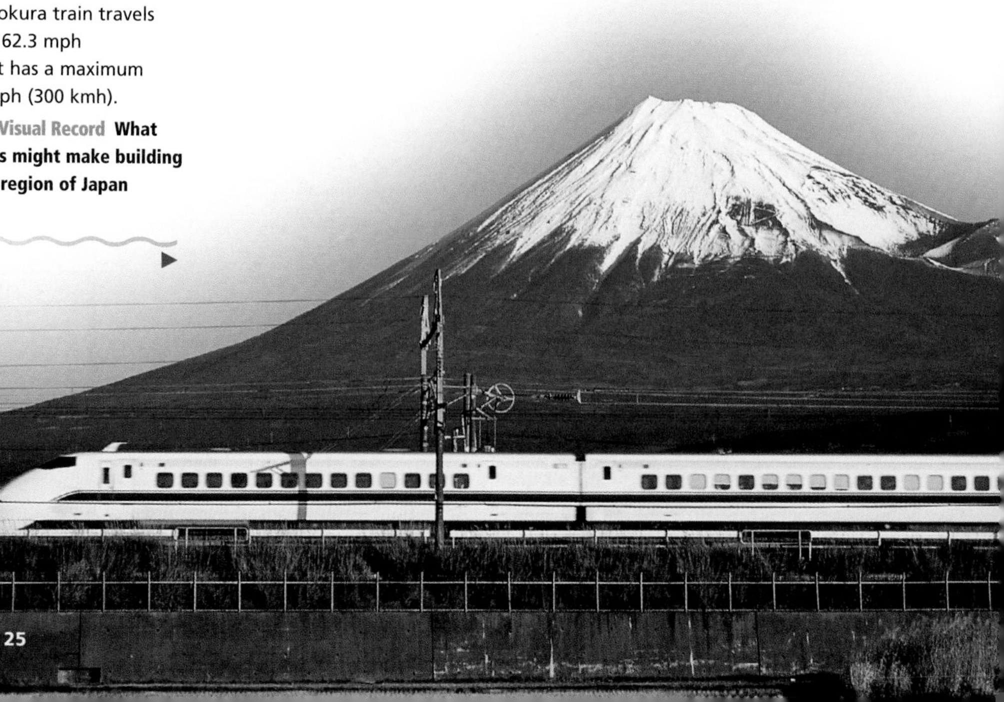

Teaching Objective 2

LEVEL 1: (Suggested time: 20 min.) Have students create an illustration for the cover of a book about Japan's natural resources. Instruct them to show what they know about those resources in the illustration. Display the book covers. **ENGLISH LANGUAGE LEARNERS**

LEVELS 2 AND 3: (Suggested time: 30 min.) Have students write a title and subtitle for their book from the Level 1 activity. Then have them write a summary paragraph that might appear in an advertisement for the book.

Teaching Objective 3

ALL LEVELS: (Suggested time: 30 min.) Assign each of five groups one of these locations: northern Hokkaido, mouth of the Yalu River, Okinawa, Seoul, or Tokyo. Have each group write weather reports for typical summer and winter days for their locales. **COOPERATIVE LEARNING**

CLOSE

Call on a volunteer to pick random spots on a wall map of Japan and the Koreas. Ask the class to suggest ways that physical features, volcanic activity, vegetation, or climate in those places would affect their daily lives if they lived there.

Pacific plate dives beneath the lighter Eurasian and Philippine plates. This subduction zone borders the Pacific side of the Japanese islands, forming the Japan Trench. This is one of the deepest places on the ocean floor. The movement of one tectonic plate below another builds up tension in Earth's crust. The Eurasian plate buckles and pushes up, creating mountains and fractures in the crust. Magma flows up through these fractures. Where magma rises to the surface, it forms volcanoes. Today, Japan has about 40 active volcanoes. Mount Fuji (FOO-jee), Japan's highest peak, is an inactive volcano.

Because Japan lies along a subduction zone, earthquakes are also common. As many as 1,500 occur every year. Most are minor quakes. In 1995, however, an earthquake killed more than 5,000 people in Kobe.

Underwater earthquakes sometimes create huge waves called **tsunamis** (tsooh-NAH-mees). These dangerous waves can travel hundreds of miles per hour. They can also be as tall as a 10-story building when they reach shore. In 1993 a tsunami caused terrible destruction when it struck the coast of Hokkaido.

Unlike Japan, the Korean Peninsula is not located in a subduction zone. As a result, it has no active volcanoes and is dominated by eroded mountains. Earthquakes are quite rare.

BUILD on WHAT You Know

Do you remember what you learned about plateaus? See Chapter 2 to review.

✓ **READING CHECK:** *Places and Regions* What are the physical features of Japan and the Koreas? Korean Peninsula, Yalu and Tumen Rivers, mountains, volcanoes, Japan Trench

Natural Resources

Except for North Korea, the region is not rich in natural resources. It has no oil or natural gas. The Korean Peninsula's mountainous terrain and rivers, however, are good for producing hydroelectric power. North Korea also has iron ore, copper, zinc, lead, and coal.

Japan lies near one of the world's best fisheries. East of Japan the cool **Oyashio** (oh-YAH-shee-oh) **Current** dives beneath the warm, less dense **Japan Current**. The cool water scours the bottom, bringing nutrients to the surface. Fish can find plentiful food to eat in this rich marine environment.

✓ **READING CHECK:** *Places and Regions* What are the region's natural resources? rivers for hydroelectric power; iron ore, copper, zinc, lead, coal in North Korea; fishing

▲ Every hour an artificial volcano erupts at Ocean Dome, the world's largest indoor beach, at Miyazaki, Japan.

Across the Curriculum

SCIENCE

The Great Kanto Earthquake Just as they were preparing for lunch on September 1, 1923, Yokohama residents were rocked by an earthquake that may have lasted 10 full minutes. More than 800 aftershocks would follow. The devastation was unimaginable—not just from the tremors but also from the fires and the thousands of landslides unleashed by the quake. In one enclosed location alone, about 40,000 people perished by fire. Not surprisingly, the quake also created a tsunami, but it caused relatively little damage. Some 143,000 people were killed. About 694,000 houses were damaged or destroyed.

Activity: Have students conduct research on how scientists analyzed the Kanto earthquake in an effort to save lives the next time the earth shook. Have them create models that show some of the innovations in building codes.

internet connect

GO TO: go.hrw.com
KEYWORD: SK3 CH25
FOR: Web sites about earthquakes

Review and Assess

Have students complete the Section Review. Then pair students. Have each student create four multiple-choice questions based on the information in Section 1. Have pairs exchange and complete each other's quizzes. Then have students complete Daily Quiz 25.1. **COOPERATIVE LEARNING**

Reteach

Have students complete Main Idea Activity 25.1. Then organize them into three groups. Have each group create a wall map of the region. Ask one group to create symbols for physical features, the second symbols for natural resources, and the third symbols for climate types. Have the groups draw or glue the symbols onto their maps in the correct areas. Display the maps. **ENGLISH LANGUAGE LEARNERS**

Extend

Have interested students conduct research on the connections between the physical geography of the region and how its landscape, plants, and animals are portrayed in traditional arts, such as pottery, paintings on silk, lacquerware, and embroidery on silk. Ask students to match natural design elements with the species or places they portray. **BLOCK SCHEDULING**

Section 1 Review

Answers

Define For definitions, see: tsunamis, p. 567; Oyashio Current, p. 567; Japan Current, p. 567

Working with Sketch Maps Maps will vary, but listed places should be labeled in their approximate locations. Korean rivers produce hydroelectric power. Japan relies on the sea for fish.

Reading for the Main Idea

1. Japan has volcanoes and frequent earthquakes. (NGS 7)
2. Oyashio Current—keeps northern coastal areas cool in summer; Japan Current—warms southern Japan (NGS 7)

Critical Thinking

3. people live on the coastal plains, use resources from the sea
4. rely on imports, trade

Organizing What You Know

5. Japan—mountains, volcanoes, coastal plains; few resources; humid continental, humid subtropical; South Korea—mountains, coastal plains; few resources; humid subtropical; North Korea—mountains, coastal plains; iron ore, copper, zinc, lead, coal; humid continental

Visual Record Answer

walls trap water in fields

Climate

The Koreas and Japan each have two climate regions. Hokkaido, northern Honshu, and the northern part of the Korean Peninsula have a humid continental climate. The Oyashio Current keeps areas near the coast of northern Japan cool in summer. Winters are long and cold, and the growing season is short. The rest of the Korean Peninsula and Japan have a humid subtropical climate. Winters are mild. Summers are hot and humid. The Japan Current, which flows northward from the tropical North Pacific, warms southern Japan. This part of the region experiences heavy rains and typhoons during the summer. Some areas receive up to 80 inches (203 cm) of rain each year.

✓ **READING CHECK:** *Places and Regions* What is the region's climate? humid continental, humid subtropical

◄ Terracing creates more arable land for some South Korean farmers. These fields overlook the Sea of Japan.

Interpreting the Visual Record **How do the terraces here hold water?**

go.hrw.com Homework Practice Online
Keyword: SK3 HP25

Section 1 Review

Define and explain: tsunamis, Oyashio Current, Japan Current

Working with Sketch Maps On a map of Japan and the Koreas that you draw or that your teacher provides, label the following: the Korean Peninsula, Sea of Japan, Hokkaido, Honshu, Shikoku, and Kyushu. On what physical features might Japan and the Koreas depend for their economies?

Reading for the Main Idea

1. *Physical Systems* How has Japan's location in a subduction zone made it different from the Koreas?
2. *Physical Systems* How do ocean currents affect the climates of Japan?

Critical Thinking

3. **Drawing Inferences and Conclusions** How do you think residents are affected by this region's mountainous terrain and the nearness of the sea?
4. **Making Generalizations and Predictions** What can you predict about South Korea's and Japan's economies?

Organizing What You Know

5. **Summarizing** List the landforms, natural resources, and climates of Japan, North Korea, and South Korea.

	Landforms	Resources	Climate
Japan			
South Korea			
North Korea			

Section 2

Objectives

1. Describe Japan's early history and culture.
2. Examine how the modernization of Japan took place.

Focus

LET'S GET STARTED

Copy the following question and instruction onto the chalkboard: *How could living on an island make life both easier and more difficult? Work with a partner to write down two reasons for each.* Give students time to record their ideas. Discuss responses. Point out that its island location has given Japan both advantages and disadvantages. Ask students to refer to their lists as they study the history and culture of Japan. Tell students that in Section 2 they will learn about early Japan and learn how modern Japan was created.

Building Vocabulary

Write the key terms on the chalkboard. Call on volunteers to find and read their definitions aloud. Point out that **Shintoism** is based on the Japanese word *Shinto*. **Samurai** and **shogun** are Japanese words. **Diet** comes from a Middle English word for "day's journey" or "day for meeting." (The Japanese word for the Diet is *Kokkai*.) Ask a volunteer to relate the meaning of *Diet* to the word's history. Ask a student to propose another context in which we might use the word **shamans**. *(Possible answer: Modern religious leaders might be labeled as shamans in news reports.)*

Section 2 The History and Culture of Japan

Read to Discover

1. What was Japan's early history and culture like?
2. How did the modernization of Japan take place?

Define

- Shintoism
- shamans
- samurai
- shogun
- Diet

Why It Matters

Japanese cultural traditions are important not only in Japan but also in other societies. Use CNNfyi.com or other **current events** sources to find information on Japanese cultural traditions. Record the findings in your journal.

A ukiyo-e, a Japanese woodblock print

Early Japan

Japan's first inhabitants came from central Asia thousands of years ago. Rice farming was introduced to Japan from China and Korea about 300 B.C. As the population increased, farmers irrigated new land for growing rice. They also built dikes and canals to channel water into the rice paddies. Local chieftains organized the workers and controlled the flow of water. This control allowed the chieftains to extend their political power over larger areas.

Religion The earliest known religion of Japan, **Shintoism**, centers around the *kami*. *Kami* are spirits of natural places, sacred animals, and ancestors. Many of Japan's mountains and rivers are sacred in Shintoism. **Shamans**, or priests who communicated with the spirits, made the *kami*'s wishes known.

Buddhism and Confucianism were later introduced from China. Buddhist shrines were often located next to older *kami* shrines. Today, as in the past, most Japanese practice Shintoism and Buddhism. As you learned in Chapter 27, the principles of Confucianism include respect for elders, parents, and rulers.

▲ Todaiji Temple in Nara, Japan, contains the largest wooden hall in the world and a statue of Buddha that is more than 48 feet (15 m) tall.

Interpreting the Visual Record **What interesting features do you see in this building's architecture?**

Section 2 Resources

Reproducible

- Guided Reading Strategy 25.2
- Graphic Organizer 25
- Creative Strategies for Teaching World Geography Lessons 18, 19
- Cultures of the World Activity 7

Technology

- One-Stop Planner CD–ROM, Lesson 25.2
- Homework Practice Online
- HRW Go site

Reinforcement, Review, and Assessment

- Section 2 Review, p. 571
- Daily Quiz 25.2
- Main Idea Activity 25.2
- English Audio Summary 25.2
- Spanish Audio Summary 25.2

◄ **Visual Record Answer**

possible answers: large roofs, wide doors, decorative designs

TEACH

Teaching Objectives 1–2

ALL LEVELS: (Suggested time: 20 min.) Copy the following graphic organizer onto the chalkboard, omitting the italicized answers. Use it to help students describe early and modern Japan. Ask students to copy the organizer into their notebooks and fill it in with the main events and features of Japan's early and modern periods. Discuss the organizers as a class. **ENGLISH LANGUAGE LEARNERS**

Early and Modern Japan

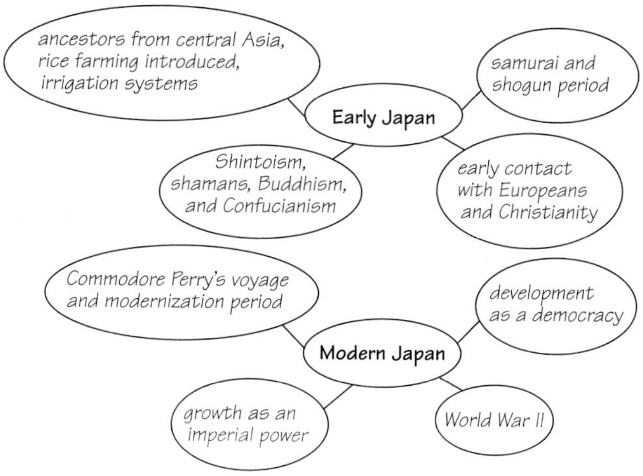

Across the Curriculum
ART

Japanese Influence In the late 1700s Chinese art was very popular with Europeans. As a result, people were prepared to appreciate new forms of Asian arts when Japanese wood-block prints started arriving in Europe. These colorful pictures of outdoor scenes and city life were packed around the china dishes exported to Europe. Hokusai (1760–1849) and Hiroshige (1797–1858) were two of the most successful printmakers.

The clean lines, balanced design, and subject matter appealed to many European artists, particularly in the late 1800s. Many of them either tried to copy the Japanese style or included aspects of it in their work. Vincent van Gogh and James A. McNeill Whistler are two notable examples.

Activity: Have students conduct research on Japanese printmakers and their work. Ask them to look for other examples of Western art that seem to be inspired by the Japanese prints. Ask them to compare the two groups.

Analyzing Primary Sources

Answers

1. They are based on the traditional belief that making a thousand cranes will please the gods.
2. She is trying to bring hope to her sick friend.

CONNECTING TO *Literature*

Sadako and the Thousand Paper Cranes by Eleanor Coerr

On August 6, 1945, the United States dropped an atomic bomb on Hiroshima, Japan, trying to bring an end to World War II. Sadako and the Thousand Paper Cranes *is based on the story of a real little girl who lived in Hiroshima at the time of the bombing. Ten years later, she died as a result of the radiation from the bomb.*

Sadako and the Thousand Paper Cranes

That afternoon Chizuko was Sadako's first visitor. She smiled mysteriously as she held something behind her back. "Shut your eyes," she said. . . . Chizuko put some pieces of paper and scissors on the bed. "Now you can look," she said.

"What is it?" Sadako asked, staring at the paper.

Chizuko was pleased with herself. "I've figured out a way for you to get well," she said proudly. "Watch!" She cut a piece of gold paper into a large square. In a short time she had folded it over and over into a beautiful crane.

Sadako was puzzled. "But how can that paper bird make me well?"

"Don't you remember that old story about the crane?" Chizuko asked. "It's supposed to live for a thousand years. If a sick person folds one thousand paper cranes, the gods will grant her wish and make her healthy again." She handed the crane to Sadako. "Here's your first one."

Analyzing Primary Sources

1. How do the cranes reflect traditional Japanese culture?
2. How does Chizuko's visit to Sadako reflect her feelings about her friend?

The Shoguns In the A.D. 700s Japan began to develop a political system of its own. Many small feudal domains were each ruled by a lord. **Samurai** (SA-muh-ry) were warriors who served the lords. Rivalries were put aside when a foreign threat appeared. For example, the feudal domains united against the Mongols in the 1200s. After a victory, the emperor sometimes named the warriors **shogun**. Shogun means "great general" and is the highest rank for a warrior.

In the mid-1500s, Portuguese traders arrived in Japan. Spanish missionaries followed, introducing Christianity to Japan. Later, Europeans were forced to leave. Japanese leaders feared that foreign ideas might undermine Japanese society. Japan remained cut off from the Western world until the mid-1850s.

✓ **READING CHECK:** *Human Systems* What was early Japan like? controlled by chieftains; Shintoism developed; Buddhism and Confucianism introduced from China; feudal domains developed; cut off from Western world until mid-1850s

CLOSE

Tell students that among the gifts that Commodore Perry brought with him to Japan were a scale-model train, telescope, clocks, telegraph transmitter, and three lifeboats. Ask them to consider what the Japanese may have thought of these items. Ask students what products from a previously unknown land would convince them to trade with that country and why.

REVIEW AND ASSESS

Have students complete the Section Review. Then organize the class into pairs. Have each student create five who-am-I questions based on the section's information about individuals and groups. Have students take turns asking each other their questions. Then have students complete Daily Quiz 25.2. **COOPERATIVE LEARNING**

RETEACH

Have students complete Main Idea Activity 25.2. Then have students use information from the text and class notes to create an annotated time line depicting the major events of Japanese history. **ENGLISH LANGUAGE LEARNERS**

EXTEND

Have interested students imagine they are foreign correspondents assigned to Japan in the 1860s. Ask them to write newspaper articles describing the conditions and experiences of the modernization period, conducting additional research as needed. Have students share their articles with the class. **BLOCK SCHEDULING**

Modern Japan

In 1853 U.S. commodore Matthew Perry's warships sailed into Tokyo Bay. Perry displayed U.S. naval power and brought gifts that showed the wonders of American technology. Perry's arrival convinced the Japanese that they needed to become as politically strong as the Americans and Europeans. In the 1860s Japan began to industrialize and modernize its educational, legal, and governmental systems.

Commodore Matthew Perry arrives in Japan in 1853.

An Imperial Power Japan needed resources in order to industrialize. As a result, it began to expand its empire around 1900. Japan annexed, or took control of, Korea in 1910. Japan also took over northeastern China and its supply of coal and iron ore. Japan continued to expand in Asia during the late 1930s.

World War II During World War II Japan was an ally of Germany and Italy. Japan brought the United States into the war in 1941 by attacking the U.S. naval base at Pearl Harbor, Hawaii. Japan conquered much of Southeast Asia and many Pacific islands before being defeated by U.S. and Allied forces in 1945. With the end of World War II Japan lost its empire.

Government After World War II the United States occupied Japan until 1952. With U.S. aid, Japan began to rebuild into a major world industrial power. Japan also established a democratic government. Today, Japan is a constitutional monarchy with several political parties. The government is made up of the **Diet** (DY-uht)—an elected legislature—and a prime minister. Japan's emperor remains a symbol of the nation, but he has no political power.

✓ **READING CHECK:** *Places and Regions* How did Japan modernize? As a result of American contact, Japan industrialized and modernized its institutions; expanded its empire.

go.hrw.com Homework Practice Online
Keyword: SK3 HP25

Section 2 Review

Define and explain: Shintoism, shamans, samurai, shogun, Diet

Working with Sketch Maps On the map you created in Section 1, label China and Korea. In a box in the margin, identify the body of water that separates Korea from Japan. Why do you think China and Korea had a strong influence on Japan's culture and history?

Reading for the Main Idea

1. *Human Systems* What religions have been practiced in Japan, and from where did they come?
2. *Places and Regions* Why did Japan decide to trade with the United States and Europe?

Critical Thinking

3. **Finding the Main Idea** How and why did rice farming develop in Japan?
4. **Analyzing Information** What influences do the principles of Confucianism have on the Japanese?

Organizing What You Know

5. **Sequencing** Copy the following graphic organizer. Use it to show important developments in modern Japanese history from the 1850s to today.

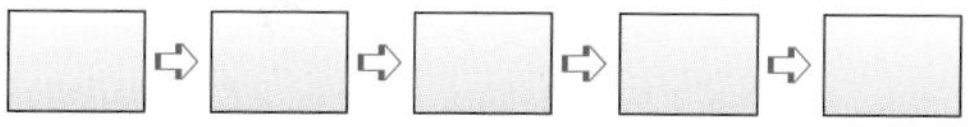

Section 2 Review Answers

Define For definitions, see: Shintoism, p. 569; shamans, p. 569; samurai, p. 570; shogun, p. 570; Diet, p. 571

Working with Sketch Maps China and the Sea of Japan should be located accurately. China and Korea were close to Japan and had well-developed cultures.

Reading for the Main Idea

1. Shintoism—early Japanese religion; Buddhism and Confucianism—introduced from China (NGS 9)
2. U.S. military power and industry convinced Japanese they needed to become strong (NGS 4)

Critical Thinking

3. introduced from China and Korea around 300 B.C.; irrigation of new land to feed increased population; dikes built to channel water
4. respect for elders, parents, and rulers

Organizing What You Know

5. Commodore Perry's visit convinced Japan to trade with the West; Japan began to industrialize and modernize; Japan expanded throughout Korea, northeastern China, and Asia; Japan fought and lost World War II; Japan rebuilt as an industrialized, democratic country.

Section 3

Objectives

1. Describe where most Japanese live.
2. Analyze the characteristics of Japan's major cities.
3. Describe what life is like in Japan.
4. Explain how the Japanese economy has developed.

Focus

LET'S GET STARTED

Copy this haiku and these instructions onto the chalkboard:

If I could bundle
Fuji's breezes back to town . . .
*What a souvenir!**

How do you think the poet feels about the city? Write down your ideas. Discuss responses. Tell students haiku are Japanese poems that contain 17 syllables. They are arranged in three lines with five syllables in the first and last lines and seven in the middle. Haiku do not rhyme. Many haiku are about nature. In this example from Matsuo Basho, the poet captures his feelings in just a few words. Tell students that they will try writing their own haiku about Japanese city life. Then tell students that in Section 3 they will learn more about life in Japan today.

**translated by Peter Beilenson and Harry Behn*

Section 3 Resources

Reproducible
- Guided Reading Strategy 25.3
- Map Activity 25

Technology
- One-Stop Planner CD–ROM, Lesson 25.3
- Homework Practice Online
- HRW Go site

Reinforcement, Review, and Assessment
- Section 3 Review, p. 576
- Daily Quiz 25.3
- Main Idea Activity 25.3
- English Audio Summary 25.3
- Spanish Audio Summary 25.3

Section 3 Japan Today

Read to Discover

1. Where do most Japanese live?
2. What are the major Japanese cities like?
3. What is life in Japan like?
4. How has the Japanese economy developed?

Define

megalopolis
kimonos
futon
intensive cultivation
work ethic
protectionism
trade surplus

Locate

Inland Sea
Osaka
Tokyo
Kobe
Kyoto

Why It Matters

The Japanese economy has struggled for the last several years due to a number of factors. The health of the Japanese economy is not only important to Japan but to much of the world. Use CNNfyi.com or other **current events** sources to find information on the Japanese economy. Record the findings in your journal.

Small child in a traditional kimono

▲ Beyond Tokyo's Nijubashi Bridge is the Imperial Palace, the home of the emperor.

Where People Live

Japan is one of the world's most densely populated countries. It is slightly smaller than California but has nearly four times as many people! There are an average of 863 people per square mile (333/sq km). However, Japan is very mountainous. Within its area of livable land, population density averages 7,680 people per square mile (2,964/sq km).

Most people live on the small coastal plains, particularly along the Pacific and the Inland Sea. Japan's major cities and farms compete for space on these narrow coastal plains. Only about 11 percent of Japan's land is arable, or fit for growing crops.

The Japanese have reclaimed land from the sea and rivers. In some places, they have built dikes to block off the water. They have drained the land behind the dikes so it could be used for farming or housing. They have even built artificial islands. The airport near Osaka, for example, was built on an artificial island in the early 1990s.

✓ **READING CHECK:** *Human Systems* Where do most people in Japan live? on the small coastal plains, particularly along the Pacific and the Inland Sea

TEACH

Building Vocabulary

Have students find the key terms and their definitions in the text. Then create two columns on the chalkboard, one titled Daily Life and the other titled Economy. Have students assign each of the terms to one of the columns. *(**Kimonos** and **futon** relate to daily life. The remaining terms relate to the economy.)*

Teaching Objective 1

ALL LEVELS: (Suggested time: 15 min.) Organize the class into five groups. Assign each group a different set of places to use as the end points of an elevation profile, such as Nagasaki to Hiroshima. (See the elevation profile at the beginning of this unit for an example.) Have the students draw the cross sections and label the locations of cities, mountains, and coastal plains. Ask them also to write a sentence to summarize where on the elevation profile the population is concentrated *(on the coastal plains).* **ENGLISH LANGUAGE LEARNERS, COOPERATIVE LEARNING**

Japan's Cities

Japan's cities, like major cities everywhere, are busy, noisy, and very crowded. Almost 30 million people live within 20 miles of the Imperial Palace in Tokyo. This densely populated area forms a **megalopolis**. A megalopolis is a giant urban area that often includes more than one city as well as the surrounding suburban areas. Yokohama is Japan's major seaport.

Most of Tokyo was built recently. An earthquake in 1923 and bombings during World War II destroyed most of the old buildings.

Tokyo is the capital and the center of government. Japan's banking, communications, and education are also centered here. Tokyo is densely populated, and land is scarce. As a result, Tokyo's real estate prices are among the world's highest.

Tokyo's Ginza shopping district is the largest in the world. Some department stores sell houses and cars and provide dental care. They also offer classes on how to properly wear **kimonos**—traditional robes—and to arrange flowers.

High rents in Ginza and elsewhere in Tokyo encourage the creative use of space. Tall buildings line the streets. However, shops are also found below the streets in the subway stations. Another way the Japanese have found to maximize space is the "capsule hotel." The guests in these hotels sleep in compartments too small to stand in upright. Businesspeople often stay in these hotels rather than commuting the long distances to their homes.

So many people commute to and from Tokyo that space on the trains must also be maximized. During peak travel periods, commuters are crammed into cars. They are helped by workers hired to push as many people into the trains as possible.

Another megalopolis in Japan is located in the Kansai region. It has three major cities: Osaka (oh-SAH-kah), Kobe (KOH-bay), and Kyoto (KYOH-toh). Industrial Osaka has been a trading center for centuries. Kobe is an important seaport. Kyoto was Japan's capital for more than 1,000 years.

READING CHECK: *Places and Regions* What is life like in Japan's major cities? busy, noisy, crowded, expensive; space is limited and thus is maximized

Japan

Country	Population/ Growth Rate	Life Expectancy	Literacy Rate	Per Capita GDP
Japan	126,771,662 0.2%	78, male 84, female	99%	$24,900
United States	281,421,906 0.9%	74, male 80, female	97%	$36,200

Sources: Central Intelligence Agency, *The World Factbook 2001;* U.S. Census Bureau

Interpreting the Chart How does life expectancy in Japan compare to that of the United States?

Japanese workers stay focused on their responsibilities on an electronics production line.

National Geography Standard 2

The World in Spatial Terms

In Tokyo only the major streets are named. Many city addresses are determined by dividing the city into wards and other districts. A ward is divided into sections, which in turn are further divided into subdivisions. Each block of each subdivision is numbered, as is each house on a block. The houses are numbered in the order they were built, not in consecutive order. Japan's residents identify an address by using a system that gives numbers for the block, house, and subdivision, and names for the section and ward.

Here is a fictional address: 3–1–2 Yamabuki, Chiyoda-ku. That address means "subdivision 3, block 1, house 2, in the Yamabuki section of Chiyoda ward."

Activity: Have students sketch maps to show how local addresses are created. Ask them to compare their maps to the Japanese model and describe the differences.

internet connect

GO TO: go.hrw.com
KEYWORD: SK3 CH25
FOR: Web sites about Tokyo

Chart Answer

It is higher.

Teaching Objective 2

LEVEL 1: (Suggested time: 15 min.) Have students create lists of phrases to describe the major Japanese cities.

LEVELS 2 AND 3: (Suggested time: 30 min.) Review the characteristics of haiku. Ask students to use the lists they created for the Level 1 activity to write a haiku that describes some aspect of life in a large Japanese city. Call on volunteers to read their poems aloud.

Teaching Objective 3

ALL LEVELS: (Suggested time: 30 min.) Have students work in pairs to conduct interviews for a television program to be called A Day in the Life of the Japanese People. **COOPERATIVE LEARNING**

TEACHER TO TEACHER

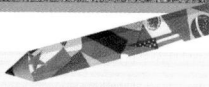

Celeste Smith of Austin, Texas, suggests the following activity to help students learn about life in Japan. Have students work in groups to create collages that summarize home life in Japanese cities. Collages may include pictures from magazines, newspapers, drawings, and pieces of fabric. Have students include pictures of Japanese homes, gardens, products, celebrations, families, and anything else students find relevant. Ask students to tell what the illustrations depict. Display the collages.

ENVIRONMENT AND SOCIETY

In mid-March the blooming of the cherry trees symbolically begins the Japanese year. At this time schools begin the new academic year and recently hired employees join their companies.

Many Japanese organize flower-viewing parties, called *ohanami.* When the cherry trees in Tokyo's Ueno Park are in full bloom, as many as 250,000 people come each day to look. They enjoy picnics, sing songs, and photograph the blossoms. About three days after the blossoms are at their best, the petals begin to fall. A breeze causes a *hana fubuki*—a storm of falling flower petals—that turns the pavement white.

Activity: Ask students what activities or celebrations related to plants occur in your community. *(Possible answers: celebrations related to wildflowers, autumn foliage, harvests, or gardens)* Have them gather information about how these events began, who participates, and what takes place during the observances. Ask students to create a mural to show their findings.

Visual Record Answer

It is in Japanese.

The port of Kobe is located in the central part of the Japanese islands.

Japanese and American all-star baseball teams compete in Yokohama.

Interpreting the Visual Record How is this scoreboard different from one in the United States?

Life in Japan

Japan is a very homogeneous nation. In other words, almost everyone—more than 99 percent of the population—is ethnically Japanese and shares a common language and culture. Japanese society has traditionally been dominated by men, but this is changing. More Japanese women have jobs today than in the past, but most women are still expected to be dutiful wives and mothers. Many quit their jobs when they marry.

Many Japanese families live in suburbs, where housing is cheaper. As a result, many Japanese spend as much as three hours commuting to and from work.

Because land is so scarce, most Japanese homes do not have large yards. Most homes are also smaller than typical American homes. Rooms are usually used for more than one purpose. For example, a living room may also serve as a bedroom. During the day, people sit on cushions at low tables. At night, they sleep on the floor on a **futon** (FOO-tahn)—a lightweight cotton mattress. In the morning they put the mattress away, and the bedroom becomes a living room again.

For most occasions, people wear Western-style clothing. Many Japanese wear kimonos at festivals and weddings. Listening to music and playing video games are popular leisure activities. Baseball, golf, and skiing are also popular. On festival days, families may gather to enjoy the cherry blossoms or visit a local shrine or temple. They might also watch the ancient sport of sumo wrestling or traditional dramas on television. Many Japanese enjoy traditional arts such as the tea ceremony, flower arranging, growing dwarf potted trees called bonsai, and kite flying.

✓ **READING CHECK:** *Human Systems* What is life like in Japan? more than 99 percent of the people are ethnically Japanese, share a common language and culture; many families live in suburbs; homes are smaller than American homes; people wear Western-style clothing, enjoy sports, traditional arts

➤**ASSIGNMENT:** Have students design flags or banners that might be displayed in a parade celebrating Japanese farming.

Teaching Objective 4

ALL LEVELS: (Suggested time: 15 min.) Copy the following graphic organizer onto the chalkboard, omitting the italicized answers. Use it to help students describe Japan's economy. Lead a discussion about Japan's dependence on imported raw materials, oil, and food and the importance of its fishing industry. Then call on students to fill in the bulleted lines with the main features of Japan's economy.

Japan's Economy

Agriculture	Industry	Global Market
• *most farms on Honshu* • *small farms* • *terraced hillsides an example of intensive cultivation* • *government price supports for rice, limits on rice imports*	• *automobiles, televisions, cameras, CD players* • *work ethic* • *good employer-employee relations* • *investment in other countries*	• *trade surplus created by protectionism* • *objections from other countries* • *competition from other Asian countries*

Japan's Economy

Japan has few natural resources. It therefore imports many of the raw materials it uses to run its industries. Oil is one key material that Japan must import. The country produces about one third of its energy through nuclear power.

The sea is an important source of food. Japan has the world's largest fishing industry. It also imports fish from all over the world. Fish is a major part of the Japanese diet. In fact, Tokyo's largest fish market sells about 5 million pounds of seafood each day. The average Japanese eats more than 100 pounds of fish each year. In contrast, the average American eats less than 5 pounds each year.

Agriculture Most Japanese farms are located on Honshu. Many farmers own their land and live in small villages. Farms in Japan are much smaller than those in the United States. The average Japanese farm is about 2.5 acres (1 hectare). Most American farms are about 150 times larger. Japan's shortage of land means that there is little pastureland for livestock. Meat is a luxury. The Japanese get most of their protein from fish and soybeans.

Farmers make the most of their land by terracing the hillsides. This means cutting the hillside into a series of small flat fields. The terraces look like broad stair steps. The terraces give farmers more room to grow crops. Japanese farmers use **intensive cultivation**—the practice of growing food on every bit of available land. Even so, Japan must import about two thirds of its food.

Farmers are encouraged to stay on the land and to grow as much rice as possible. However, many farms are too small to be profitable. To solve this problem, the Japanese government buys the rice crop. The price is set high enough to allow farmers to support their families. The government then resells it at the same price. Because this price is much higher than the world market price, the government restricts rice imports.

Seeds of the tea plant were first brought to Japan about A.D. 800. Tea is now an important product of southern Japan. Top-quality teas are harvested by hand only. Workers pick just the tender young leaves at the plant's tip.

Interpreting the Visual Record

How might using machines for harvesting leaves affect the quality of the tea?

▼

Environment and Society

Cultured pearls are among the many products from Japanese seas. The world's best-known pearl producer, Mikimoto, is located at Ago Bay in southern Japan. In 1893, Kokichi Mikimoto discovered that carefully tended oysters could grow nearly perfect pearls around an implanted nucleus. The process takes three to six years. Westerners at first believed that Mikimoto's cultured pearls were fakes, but the technique eventually gained widespread acceptance. Japanese underwater pearl farms now yield about 70 tons (63 metric tons) of pearls each year, despite pollution problems and competition from rapidly expanding Chinese producers.

Critical Thinking: How might people have first discovered the pearl creation process?

Answer: They may have found pearls in different stages in oysters they were eating. They may also have cut a pearl open and found the bit of sand or other nucleus at the center.

◄ **Visual Record Answer**

Machines are unable to select only the young, tender leaves.

CLOSE

Ask students to imagine that they have recently moved to Japan. Have them write letters to friends in the United States about Japan. Ask them to describe where in Japan they live, the local physical features, the local economy, and how Japanese daily life is different from American daily life.

REVIEW AND ASSESS

Have students complete the Section Review. Then organize them into groups. Assign each group one of the section's main topics. Ask them to write three sentences to summarize their topic's main points and read their sentences to the class. Then have students complete Daily Quiz 25.3. COOPERATIVE LEARNING

RETEACH

Have students complete Main Idea Activity 25.3. Then create an outline on the chalkboard of the most important information from the section. Leave some of the categories blank, and have students fill in the missing information. ENGLISH LANGUAGE LEARNERS

EXTEND

Organize students into groups. Assign one of Japan's political parties to each group. Have students use additional resources to write a brief history explaining the significance of the assigned political party. Then stage a debate between two "party leaders" that allows them to share the information with the class. COOPERATIVE LEARNING, BLOCK SCHEDULING

Section 3 Review

Answers

Define For definitions, see: megalopolis, p. 573; kimonos, p. 573; futon, p. 574; intensive cultivation, p. 575; work ethic, p. 576; protectionism, p. 576; trade surplus, p. 576

Working with Sketch Maps The Inland Sea, Osaka, Tokyo, Kobe, and Kyoto should be located accurately. The cities are large, busy, noisy, crowded, and part of a megalopolis.

Reading for the Main Idea

1. little land for farming, so dikes are built and land is reclaimed from the sea (NGS 15)
2. imports raw materials for industries (NGS 11)

Critical Thinking

3. Western clothing, music, video games, baseball, golf, skiing
4. because many countries have a trade deficit with Japan

Organizing What You Know

5. Imperial Palace, part of megalopolis, center of government, banking, communications, education, Ginza. Paragraphs will vary.

▲ A worker in this Japanese automobile factory does his job with the help of a robot.

Industry Japan imports most of its raw resources. These resources are then used to make goods to sell in other countries. For example, Japan is known for its high-quality automobiles. Japan also makes televisions, cameras, and compact disc players.

There are many reasons for Japan's economic success. Most Japanese have a strong **work ethic**. This is the belief that work in itself is worthwhile. Most Japanese work for large companies and respect their leaders. In return, employers look after workers' needs. They offer job security, exercise classes, and other benefits.

The Japanese have also benefited from investments in other countries. For example, some Japanese companies have built automobile factories in the United States. Other Japanese companies have invested in the American entertainment and real estate industries.

Japan and the Global Market Japan's economy depends on trade. In the past the government set up trade barriers to protect Japan's industries from foreign competition. This practice is called **protectionism**. This has helped Japan build up a huge **trade surplus**. A trade surplus means that a nation exports more than it imports. Other countries have objected to Japan's trade practices. Some countries have even set up barriers against Japanese goods. As a result, Japan has eased some trade barriers.

Japan has other economic problems, too. Some Asian countries that pay lower wages are able to produce goods more cheaply than Japan. The most important problem Japan—and Asia—faced in the 1990s was an economic slowdown. It threatened the country's prosperity. Japan is now in a recovery and slow-growth period.

✓ **READING CHECK:** *Human Systems* How have Japan's leaders tried to protect the nation's economy? by setting up trade barriers to protect Japan's industries and build up a trade surplus

go.hrw.com Homework Practice Online Keyword: SK3 HP25

Section 3 Review

Define and explain: megalopolis, kimonos, futon, intensive cultivation, work ethic, protectionism, trade surplus

Working with Sketch Maps On the map you created in Section 2, label Inland Sea, Osaka, Tokyo, Kobe, and Kyoto. Draw a box in the margin of your map. What do the cities have in common? Write your answer in the margin box.

Reading for the Main Idea

1. *Environment and Society* How does Japan's physical geography affect farming?
2. *Human Systems* How has Japan developed its industries without plentiful raw materials?

Critical Thinking

3. **Drawing Inferences and Conclusions** In what ways do the daily lives of Japanese reflect influences of Western culture?
4. **Drawing Inferences and Conclusions** Why do you suppose other countries might be concerned about Japan's surplus?

Organizing What You Know

5. **Summarizing** Copy the following graphic organizer. Use it to list the activities and services available in Tokyo.

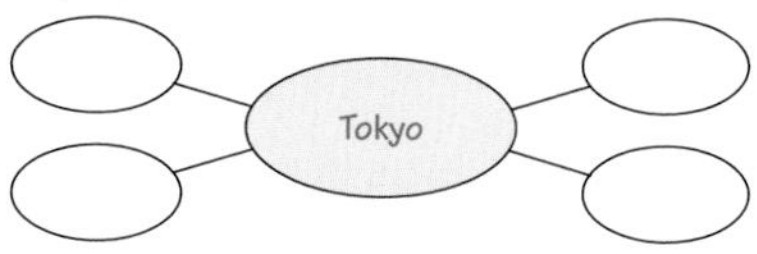

Section 4

Objectives

1. Describe Korea's ancient history.
2. Analyze the major events of Korea's early modern period.
3. Explain the history and results of Korea's division after World War II.

Focus

LET'S GET STARTED

Copy the following instructions onto the chalkboard: *Work with a partner to make a folding fan from a sheet of paper. Illustrate it with scenes reflecting natural environments.* Ask volunteers to display their fans. Then tell students that fans have long been part of Korean life in the summer. Artists painted views of waterfalls, mountains, and other scenes on them. Tell students that in Section 4 they will learn more about Korean culture and history.

Building Vocabulary

Write the term **demilitarized zone** on the board and circle the prefix *de-*. Explain that *de-* means "removal" or "reversal." When *de-* is added to *militarized*, *demilitarized* means "the removal of the military." Point out Korea's demilitarized zone on a map and explain that no troops are allowed in this area. Ask students what other words they know using *de-* that carry the meaning of removal. *(Possible answers: deice, deport, derail)*

Section 4 The History and Culture of the Koreas

Read to Discover

1. What was Korea's ancient history like?
2. What were the major events of Korea's early modern period?
3. Why was Korea divided after World War II, and what were the effects of the division?

Define

demilitarized zone

Locate

North Korea
South Korea

Why It Matters

The Korean War left many lingering problems for Koreans and Americans alike. Use CNNfyi.com or other **current events** sources to find examples of current relations between the United States and North and South Korea. Record the findings in your journal.

A tray decorated with mother-of-pearl inlay

Section 4 Resources

Reproducible
- Guided Reading Strategy 25.4
- Interdisciplinary Activities for the Middle Grades 29, 30, 31

Technology
- One-Stop Planner CD–ROM, Lesson 25.4
- Homework Practice Online
- HRW Go site

Reinforcement, Review, and Assessment
- Section 4 Review, p. 579
- Daily Quiz 25.4
- Main Idea Activity 25.4
- English Audio Summary 25.4
- Spanish Audio Summary 25.4

Ancient Korea

Korea's earliest inhabitants were nomadic hunters from north and central Asia. About 1500 B.C. they adopted rice farming, which had been introduced from China. Then, in 108 B.C., the Chinese invaded Korea. This invasion marked the beginning of a long period of Chinese influence on Korean culture. The Chinese introduced their system of writing and their system of examinations for government jobs. They also introduced Buddhism and Confucianism to Korea.

Korea's original religion—shamanism—continued to be practiced, along with the newer traditions introduced from China. According to shamanism, natural places and ancestors have spirits. Many mountains are particularly sacred to Koreans. Shamanism is still practiced in South Korea.

Over the centuries, Korean tribes eventually recaptured most of the peninsula. In the A.D. 600s the kingdom of Silla (SI-luh) united the peninsula. Korea's golden age began. Korea became known in Asia for its architecture, painting, ceramics, and fine jewelry.

▲ A weaver demonstrates his craft near Seoul.

◄ This celadon vase is from the A.D. 1000s.

TEACH

Teaching Objectives 1–2

ALL LEVELS: (Suggested time: 45 min.) Have the students create storyboards for key scenes in an imaginary television special entitled Korea through Good Times and Bad.

Teaching Objective 3

ALL LEVELS: (Suggested time: 15 min.) Copy the following graphic organizer onto the chalkboard, omitting the italicized answers. Use it to list five items for each category. **ENGLISH LANGUAGE LEARNERS**

Korea before World War II	Korea after World War II
1. many influences, invasion from China	*1. U.S. and Soviet troops remained after war*
2. shamanism, newer religious traditions from north	*2. Soviets helped communists take over north*
3. Silla, Koryo periods—many accomplishments	*3. invasion of South Korea by North Korea*
4. invasion by China, Hermit Kingdom period	*4. UN and Chinese troops sent in, Korean War*
5. invasion by Japan	*5. truce and creation of DMZ*

Across the Curriculum
LITERATURE

Pansori Many Koreans enjoy traditional folk tales. The most famous of these is a kind of ballad called a *pansori*, which is often chanted to the beat of drums. One of the best-known *pansori* is the Tale of Sim Chong, a story about a devoted daughter who was willing to sacrifice herself in order to restore her blind father's eyesight. This tale, like many other Korean folk tales, reinforces the Confucian value of duty to the family.

A film made recently in South Korea recounts an ancient tale of cruelty, true love, and perseverance from Korea's dynastic era. The story is told in *pansori* form so the audience is immersed in both Korean history and culture.

Discussion: Lead a discussion about which stories teach traditional values to American children. Students might mention stories like Aesop's fable about the tortoise and the hare (perseverance), the tale of George Washington and the cherry tree (honesty), or others.

Heavy fencing and explosives have kept people out of the demilitarized zone for decades. As a result, this land has provided a safe home for rare animal and plant species. Some scientists hope that in the future the DMZ can be set aside as a nature preserve, which would attract tourists while protecting the wildlife.

During the Silla period Korea began using the results of examinations to award government jobs. Generally only boys who were sons of noblemen could take the examinations. People from the lower classes could not rise to important positions by studying and passing the examinations, as they could in China.

By the early 900s a new kingdom had taken power. The modern name of Korea comes from this kingdom's name, Koryo. During the Koryo dynasty, Korean artisans invented the first movable metal type. During the following dynasty, scholars developed the Hangul (HAHN-gool) alphabet, which was officially adopted in 1446. Hangul was much easier to use with the Korean language than Chinese characters had been. Because Hangul had only 24 symbols, it was easier to learn. It had previously been necessary to memorize about 20,000 Chinese characters to read the Buddhist scriptures.

READING CHECK: *Human Systems* How did the Chinese influence Korea's ancient history? They invaded the region and influenced Korean culture.

Early Modern Korea

By the early 1600s China again controlled Korea. For 300 years Korea remained under Chinese control. Closed off to most other outsiders, Korea became known as the Hermit Kingdom. During this period, Catholicism was introduced into Korea through missionaries in China. Korea's Christian community was sometimes persecuted and remained small until the mid-1900s.

In the mid-1890s Japan defeated China in the Sino-Japanese War. This cleared the way for Japan to annex Korea in 1910. The Japanese ruled harshly. They took over the Korean government and many businesses and farms. Koreans were forced to take Japanese names, and

Close

Remind students that a Korean king directed the development of the Hangul alphabet. Lead a discussion on what steps students would take to design a new alphabet for the English language.

Review and Assess

Have students complete the Section Review. Then organize the class into pairs. Have each student write a fill-in-the-blank question from each of the text subsections. Ask students to trade papers and answer each other's questions. Then have students complete Daily Quiz 25.4.
COOPERATIVE LEARNING

Reteach

Have students complete Main Idea Activity 25.4. Then have them create a sequential-step diagram that describes Korea's history. Have students fill in the major events and add details about Korean culture.
ENGLISH LANGUAGE LEARNERS

Extend

Have interested students conduct research on the traditional dress, or *hanbok*, of Korea. You may want to have them create a poster describing *hanbok* and illustrating it using fabric scraps. More ambitious students may want to construct doll-sized examples. **BLOCK SCHEDULING**

Japanese was taught in the schools. Japan ruled Korea until the end of World War II.

✓ **READING CHECK:** *Places and Regions* What were some major events in Korea's early modern period? 1600s—controlled by China, Catholicism introduced; 1910—Japan annexed Korea and ruled it hardy until the end of World War II

A Divided Korea

At the end of World War II, U.S. and Soviet troops oversaw the Japanese departure from Korea. The Soviets helped Communist leaders take power in the north. The United States backed a democratic government in the south. The United Nations hoped that the Koreas would reunite. However, the United States and the Soviet Union could not agree on a plan. In 1948 South Korea officially became the Republic of Korea. North Korea became the Democratic People's Republic of Korea led by Korean Communist dictator Kim Il Sung.

In 1950 North Korea tried to unify the country by invading South Korea, resulting in the Korean War. The United Nations sent troops—mostly U.S.—to defend South Korea. Communist China sent forces to North Korea. A truce was declared in 1953, but Korea remains divided. The border between North Korea and South Korea is a strip of land roughly 2.4 miles (4 km) wide called the **demilitarized zone** (DMZ). This buffer zone separates the two countries. A total of about 1 million U.S., South Korean, and North Korean soldiers patrol the DMZ. It is the world's most heavily guarded border.

In the early 1970s and early 1990s North and South Korea tried to reach an agreement for reunification. The negotiations failed. Still, Koreans remain hopeful that their country will one day be reunited.

✓ **READING CHECK:** *Places and Regions* What were some events in Korea's history after World War II? U.S. and Soviet troops oversaw Japanese departure, influence caused the division of Korea; 1950—North invaded South, UN sent troops for the South's defense; truce declared in 1953, but country remains divided

Our Amazing Planet

Casting bronze bells is an ancient Korean craft. The largest bell in South Korea, completed in A.D. 771, is more than 12 feet (3.6 m) tall and weighs about 25 tons (110,000 kg). When struck, it is said that the bell's tone can be heard 40 miles (64 km) away.

go.hrw.com Homework Practice Online
Keyword: SK3 HP25

Section 4 Review

Define and explain: demilitarized zone

Working with Sketch Maps On the map you created in Section 3, label North Korea and South Korea. In the margin, draw a box and include in it information that explains the significance of the DMZ. When was it established?

Reading for the Main Idea

1. *Human Systems* What were some of the accomplishments of the early Koreans before about 1600? What dynasty gave Korea its name?
2. *Human Systems* What long-lasting effect did the Korean War have on the Korean Peninsula?

Critical Thinking

3. **Analyzing Information** How has the Korean Peninsula been influenced by other countries?
4. **Making Generalizations and Predictions** What might be preventing North Korea and South Korea from reuniting?

Organizing What You Know

5. **Sequencing** Copy the following time line. Use it to list the important events of Korean history from 1500 B.C. through the 1500s.

1500 B.C. ———————————————— 1500s

Section 4 Review

Answers

Define For definition, see: demilitarized zone, p. 579

Working with Sketch Maps North Korea and South Korea should be located accurately. The DMZ prevents people and goods from crossing the border between North and South Korea. It was established after the Korean War.

Reading for the Main Idea

1. architecture, painting, ceramics, jewelry, movable metal type, Hangul alphabet; Koryo (NGS 9)
2. physical, economic, and political division (NGS 13)

Critical Thinking

3. China—origin of writing and political systems, religion; Japan—conquered Korea, forced Koreans to take Japanese names and speak Japanese
4. possible answer: philosophical and political differences

Organizing What You Know

5. 1500 B.C.—adoption of rice farming; 108 B.C.—Chinese invade; A.D. 600—kingdom of Silla and golden age begins; 900—Koryo dynasty begins; 1446—Hangul adopted

Section 5

Objectives

1. Describe South Korea's government and society.
2. Analyze South Korea's economy.
3. Describe North Korea.
4. Explain how the government has affected North Korea's development.

Focus

LET'S GET STARTED

Copy this scenario onto the chalkboard: *Imagine that you live in what used to be one country, but it has been divided into two countries by war. Your friends and family live in the other section, but you cannot visit them. Write a paragraph that you might include in a diary if you lived in this place.* Allow time for students to write their paragraphs and discuss responses. Remind students that Korea is divided into two countries. Compare the diary entries to Korea's situation. Tell students that in Section 5 they will learn more about the impact of division on Korea.

Building Vocabulary

Write the key terms on the chalkboard. Have students read the definitions in the text. You may want to read aloud the recipe for **kimchi** from Food Festival or the sidebar feature on the next page that mentions the Kimchi Festival. If possible, bring a sample to class. Discuss if or how Korean ***chaebol*** compare to any American business relationships with which students are familiar. You may want to ask students about news reports they have seen about **famine** in other countries.

Section 5 Resources

Reproducible

- Guided Reading Strategy 25.5
- Geography for Life Activity 25

Technology

- One-Stop Planner CD–ROM, Lesson 25.5
- Homework Practice Online
- HRW Go site

Reinforcement, Review, and Assessment

- Section 5 Review, p. 583
- Daily Quiz 25.5
- Main Idea Activity 25.5
- English Audio Summary 25.5
- Spanish Audio Summary 25.5

Chart Answer

It is much lower.

Section 5 South and North Korea Today

Read to Discover

1. What are South Korea's government and society like?
2. What is South Korea's economy like?
3. What is North Korea like?
4. How has North Korea's government affected the country's development?

Define

entrepreneurs
chaebol
kimchi
famine

Locate

Seoul
Pusan
P'yŏngyang

Why It Matters

North and South Korea are still struggling to reach an understanding after some 50 years since the end of the Korean War. Use CNNfyi.com or other **current events** sources to find examples of the political, economic, and diplomatic relations between the two countries. Record the findings in your journal.

Korean-style vegetables

North Korea and South Korea

Country	Population/ Growth Rate	Life Expectancy	Literacy Rate	Per Capita GDP
North Korea	21,968,228 / 1.2%	68, male 74, female	99%	$1,000
South Korea	47,904,370 / 0.9%	71, male 79, female	98%	$16,100
United States	281,421,906 / 0.9%	74, male 80, female	97%	$36,200

Sources: Central Intelligence Agency, *The World Factbook 2001;* U.S. Census Bureau

Interpreting the Chart How does North Korea's per capita GDP compare to that of South Korea?

South Korea's People and Government

South Korea is densely populated. There are 1,197 people per square mile (462/sq km). Most people live in the narrow, fertile plain along the western coast of the Korean Peninsula. Travel in the peninsula's mountainous interior is difficult, so few people live there. South Korea's population is growing slowly, at about the same rate as in most industrialized countries.

South Korea's Cities Because South Korea is densely populated, space is a luxury—just as it is in Japan. Most South Koreans live in small apartments in crowded cities. Seoul (SOHL) is the country's capital and largest city. The government, the economy, and the educational system are centered there. After the Korean War, the population exploded because refugees flocked to Seoul seeking work and housing. By 1994 the city had nearly 11 million residents. Today, Seoul is one of the world's most densely populated cities. It has some 7,000 people per square mile (2,703/sq km).

South Korea's second-largest city is Pusan (POO-sahn). This major seaport and industrial center lies on the southern coast. Pusan also has an important fishing industry.

The rapid growth of South Korea's cities has brought problems. Housing is expensive. The many factories, cars, and coal-fired heating

TEACH

Teaching Objective 1

LEVEL 1: (Suggested time: 15 min.) Have students list the main points they would include in a newspaper entitled South Korea—Then and Now. **ENGLISH LANGUAGE LEARNERS**

LEVELS 2 AND 3: (Suggested time: 45 min.) Have students use the lists they created for the Level 1 activity to write and illustrate their story.

Teaching Objective 2

ALL LEVELS: (Suggested time: 15 min.) Copy the following graphic organizer onto the chalkboard, omitting the italicized answers. Use it to help students describe how agriculture and industry are part of South Korea's economy. **ENGLISH LANGUAGE LEARNERS**

Economy

Industry	Agriculture
many businesses run by families or chaebol	*less than 20 percent of land can be farmed*
nuclear power encouraged by government	*must import food*
high-technology industry, electronic goods, steel, shipbuilding, automobiles, textiles	*small farms along western and southern coasts*
economic slowdown and recovery	*much work done by hand*
	rice, Chinese cabbage, soybeans

systems sometimes cause dangerous levels of air pollution. Industrial waste has also polluted the water.

Postwar Government South Korea is technically a democracy, but it was run by military dictators until the late 1980s. More recently, South Korea introduced a multiparty democratic government. The government controls economic development but does not own businesses and property.

✓ **READING CHECK:** *Human Systems* What kind of government does South Korea have? a multiparty democracy

South Korean Society

Like Japan's, South Korea's population is homogeneous. Most Koreans complete high school. About half go on to some form of higher education. Women are beginning to hold important jobs.

Traditional Families Most Koreans marry someone they meet through their parents. Most families still value sons. This is because only a son can take over the family name. Only a son can lead the ceremonies to honor the family's ancestors. Some couples who do not have a son adopt a boy with the same family name. This is not too difficult because there are few family names in Korea.

Religion Today, Christianity is the most common religion, followed closely by Buddhism. Whatever their religion, most Koreans take part in ceremonies to honor their ancestors. Most also follow Confucian values. Many Koreans still ask shamans for personal advice.

✓ **READING CHECK:** *Human Systems* What is South Korea's society like? homogenous, well educated, male dominated

This view looks out over the busy harbor of Pusan. Travelers can take a ferry from Pusan across the Korean Strait to Japan.

Interpreting the Visual Record How does this photo show the importance of shipping to this region's economy?

National Geography Standard 10

Human Systems Visitors to South Korea can enjoy a wide range of festivals. Among the many activities featured at various festivals are snowboarding, ice sculpting, bull riding, making paper, catching fireflies, flying kites, designing jewelry, and guessing the weight of raw fish.

In October the city of Kwangju hosts the annual four-day Kimchi Festival. Festival organizers hope to attract tourists and educate the world about kimchi. The event begins with an opening ceremony and a parade. At the display area visitors can taste, buy, or just look at all kinds of kimchi dishes. There are several competitions, with a special category for foreigners.

Discussion: Ask students what food festivals are held in your area and how they reflect the region's culture. What other foods are typical of your region?

◄ **Visual Record Answer**

possible answer: many large ships are docked

Teaching Objectives 3–4

ALL LEVELS: (Suggested time: 45 min.) Have students present skits that portray life in North Korea and how the government's control of the economy affects the people. **ENGLISH LANGUAGE LEARNERS**

Teaching Objectives 1–4

LEVEL 1: (Suggested time: 30 min.) Have students create an outline that includes information for a Web site about North Korea or South Korea. Ask students to include a basic map and titles for the main links. **ENGLISH LANGUAGE LEARNERS**

LEVELS 2 AND 3: (Suggested time: 30 min.) Have students imagine that they are veterans of the Korean War living in either North Korea or South Korea. Then ask them to write letters that compare pre- and postwar history, politics, and living conditions. Ask volunteers from each "side" to read their letters to the class.

Section 5 Review

Answers

Define For definitions, see: *chaebol*, p. 582; kimchi, p. 582; famine, p. 583

Working with Sketch Maps They are located on waterways and are important for trade.

Reading for the Main Idea

1. rapid growth, pollution, housing shortage (NGS 4)
2. best farmland along west coast, only 14 percent can be farmed (NGS 15)
3. South Kores—market, industrialized, modern; North Korea—command, government-owned factories, old machinery (NGS 4)

Critical Thinking

4. use money and talents to start a business; businesses are sometimes linked in a *chaebol*

Organizing What You Know

5. North—communist, industry undeveloped, farming by cooperatives; South—democratic, industrialized, much farm work done by hand; both—farming on coasts, import food, nuclear power

Visual Record Answer ►

possible answer: it is much larger

This is one of many statues of Kim Il Sung in P'yŏngyang. He led North Korea from the end of World War II until his death in 1994. Although dead, Kim Il Sung was declared the "Eternal President" of North Korea in 1998.

Interpreting the Visual Record **How does this statue compare to monuments honoring important people in the history of the United States?**

South Korea's Economy

After the war, South Korea industrialized quickly, and its market economy grew. By the 1990s it had become one of the strongest economies in Asia.

Industry Koreans' strong sense of family often carries over into work. Large groups of relatives may become **entrepreneurs**. This means they use their money and talents to start and manage a business. Businesses are sometimes linked through family and personal ties into huge industrial groups called ***chaebol***.

The government has encouraged the use of nuclear power. It also has encouraged the growth of high-technology industries. These industries make electronic goods for export. Other important industries are steel, shipbuilding, automobiles, and textiles. In the late 1990s South Korea, like many other Asian countries, experienced an economic slowdown. It is now making a rapid recovery.

Agriculture South Korea has the peninsula's richest agricultural land. However, less than 20 percent of the land can be farmed. The shortage of land means that South Korea must import about half of its food.

Most South Korean farms are small and lie along the western and southern coasts. The rugged terrain makes using heavy machinery difficult. As a result, farmers must do much of their work by hand and with small tractors. Since the late 1980s there has been a shortage of farmworkers.

Farmers grow rice on about half their land. Other important crops are Chinese cabbage and soybeans. Soybeans are used to make soy sauce and tofu, or bean curd. Chinese cabbage that has been spiced and pickled is called **kimchi**. This is Korea's national dish.

✓ **READING CHECK:** *Places and Regions* What is South Korea's economy like? **industrialized, with high-technology industries, as well as steel, shipbuilding, automobiles, and textiles**

North Korea's People and Government

North Korea's Communist Party controls the government. For many years, North Korea had ties and traded mostly with other Communist countries. Since the Soviet Union's breakup in 1991, North Korea has been largely isolated from the rest of the world. In the late 1990s North Korea angered many countries by trying to develop nuclear weapons and missiles.

Population Like South Korea and Japan, North Korea has a homogeneous population. North Korea is not as densely populated as South Korea. In North Korea, there are 513 people per square mile (198/sq km).

582

CLOSE

Challenge students to imagine that they are members of a committee called New Ideas for a New Millennium. The committee was created to propose steps to move North and South Korea closer to reunification. Brainstorm suggestions.

REVIEW AND ASSESS

Have students complete the Section Review. Ask students to suggest phrases or sentences that describe either North or South Korea. Then have students complete Daily Quiz 25.5.

RETEACH

Have students complete Main Idea Activity for 25.5. Then organize the class into groups. Give each a sheet of newsprint with two columns headed North Korea and South Korea. Have each student add a fact to each column until all the basic information in the section has been covered. **ENGLISH LANGUAGE LEARNERS, COOPERATIVE LEARNING**

EXTEND

Have interested students conduct research on aspects of the Korean War, such as important personalities or battles. Some might research the U.S. Army's medical service during the war and compare their findings with how it was portrayed in the television program *M*A*S*H.*
BLOCK SCHEDULING

North Korea's Capital The capital of North Korea is P'yŏngyang (pyuhng-YANG). About 2.6 million people live there. North Korea's only university is in P'yŏngyang. Few private cars can be seen on the city streets. Most residents use buses or the subway system to get around. At night, many streets are dark because the city frequently experiences shortages of electricity.

✓ **READING CHECK:** *Places and Regions* What is North Korea like? homogenous population; Communist; shortages of electricity in cities

▲ Rice farming requires large amounts of human energy for transplanting, weeding, and harvesting. These farmers are planting seedlings.

Interpreting the Visual Record Why do you think it would be difficult to use machinery for transplanting here?

North Korea's Economy

North Korea has a command economy. This means that the central government plans the economy and controls what is produced. The government also owns all the land and housing and controls access to jobs.

North Korea's best farmland is along the west coast. Only 14 percent of North Korea's land can be farmed. Most of this land is owned by the state. It is farmed by cooperatives—groups of farmers who work the land together. Some people have small gardens to grow food for themselves or to sell at local markets.

North Korea does not produce enough food to feed its people. It lost its main source of food and fertilizer when the Soviet Union collapsed. Poor harvests in the mid-1990s made the situation worse. **Famine**, or severe food shortages, resulted. The government's hostility toward the West made getting aid difficult. Thousands starved.

North Korea is rich in mineral resources. It has also developed a nuclear power industry. North Korea makes machinery, iron, and steel. However, its factories use outdated technology.

✓ **READING CHECK:** *Human Systems* How has North Korea's government affected the country's economic development? It has limited its modernization because it owns all industry and controls access to jobs.

go.hrw.com Homework Practice Online
Keyword: SK3 HP25

Section 5 Review

Define and explain: entrepreneurs, *chaebol*, kimchi, famine

Working with Sketch Maps On the map you created in Section 4, label Seoul, Pusan, and P'yŏngyang. Why do you think these cities remain important? Write your answer in a box in the margin.

Reading for the Main Idea

1. *Places and Regions* How did South Korea's cities change after the Korean War?
2. *Environment and Society* How does North Korea's physical geography affect its farmers?
3. *Places and Regions* What kind of economies do North and South Korea have, and how are they different?

Critical Thinking

4. **Finding the Main Idea** How have entrepreneurs affected South Korea's economy?

Organizing What You Know

5. **Categorizing** List characteristics of the government, industry, and agriculture of North and South Korea since World War II. Where the circles overlap, list things the countries share.

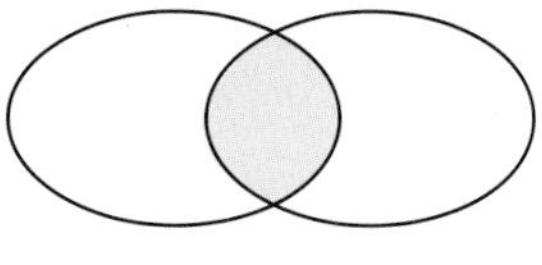

CHAPTER 25 Review ANSWERS

Building Vocabulary

For definitions, see: tsunamis, p. 567; shamans, p. 569; samurai, p. 570; shogun, p. 570; Diet, p. 571; kimonos, p. 573; futon, p. 574; intensive cultivation, p. 575; work ethic, p. 576; protectionism, p. 576; trade surplus, p. 576; demilitarized zone, p. 579; entrepreneurs, p. 582; kimchi, p. 582; famine, p. 583

Reviewing the Main Ideas

1. islands, mountains, volcanoes, coastal plain; peninsula, mountains, coastal plain (NGS 4)
2. China—rice farming, Buddhism, Confucianism; Korea—rice farming (NGS 9)
3. strong work ethic, good employer-employee relationships, active foreign trade (NGS 4)
4. because North Korea invaded South Korea in 1950 (NGS 13)
5. because it attracted refugees seeking jobs in industry; crowded, polluted, expensive real estate (NGS 4)

◄ **Visual Record Answer**

small space available for working

ASSESS

Have students complete the Chapter 25 Test.

RETEACH

Tell students that comic books are popular among Japanese adults and children. Have students create comic books that describe each of the chapter's sections. Each comic book should provide answers to the section's Read to Discover questions. Encourage students to be creative while maintaining accuracy. Once students have finished, have them take turns reading each other's comic books.

PORTFOLIO

PORTFOLIO EXTENSIONS

1. Have students make models or dioramas of a typical Japanese or Korean home, including features of everyday life, such as furniture. Take photographs of the dioramas for inclusion in student portfolios.

2. Organize the class into groups and ask them to write a brief one-act play that incorporates elements from some of the comic books they wrote. Then have them perform their plays for the class. Place a program and plot summary in student portfolios along with the comic books.

CHAPTER 25

Review ANSWERS

Understanding Environment and Society

Students may note freshwater and saltwater pearls of different colors, sizes, and shapes. Pearls are produced mainly in Japan, China, Australia, Indonesia, Tahiti and other South Pacific nations (black pearls), and the United States (freshwater pearls only). Students should discuss the process and threats to the industry, such as pollution, loss of markets, lack of cash for investment, or competition from countries that can produce pearls more cheaply. Use Rubric 29, Presentations, to evaluate student work.

Thinking Critically

1. Mountainous interiors forced settlement on coastal plains.
2. Japan—many islands, volcanoes and earthquakes are common; Koreas—peninsula, no volcanoes, earthquakes are rare
3. irrigated land to grow rice, built dikes and canals
4. Possible answer: It grew so large that it began to spread and join other large cities nearby.

CHAPTER 25

Reviewing What You Know

Building Vocabulary

On a separate sheet of paper, write sentences to define each of the following words.

1. tsunamis
2. shamans
3. samurai
4. shogun
5. Diet
6. kimonos
7. futon
8. intensive cultivation
9. work ethic
10. protectionism
11. trade surplus
12. demilitarized zone
13. entrepreneurs
14. kimchi
15. famine

Reviewing the Main Ideas

1. *Places and Regions* What are the geographic features of Japan? of Korea?
2. *Human Systems* Which other Asian countries have influenced the culture of Japan? What contributions did they make?
3. *Places and Regions* What factors have contributed to Japan's economic success?
4. *Human Systems* Why were there foreign troops on the Korean Peninsula in the 1950s?
5. *Places and Regions* Why has Seoul grown rapidly since the Korean War? What is life in the city like?

Understanding Environment and Society

Resource Use

Japan's cultured-pearl industry is an example of how the Japanese have adapted to their limited amount of land by making creative use of the sea. Create a presentation about the pearl industry. As you create your presentation, consider the following:

- The kinds of pearls that are found there.
- Where and how pearls are produced.
- Things that could hurt Japan's role in the pearl industry.

Thinking Critically

1. **Analyzing Information** How have geographic features affected where people live in Japan and the Korean Peninsula?
2. **Contrasting** What physical features make Japan different from the Koreas? Create a chart to organize your answer.
3. **Finding the Main Idea** How have the Japanese changed the physical landscape to meet their needs?
4. **Analyzing Information** How has Tokyo developed into a megalopolis?
5. **Contrasting** How do the economies of North and South Korea differ?

FOOD FESTIVAL

There are countless recipes for kimchi, a favorite Korean dish. Here is one version. First, wash and drain 2 pounds cabbage and cut into pieces. Sprinkle with 2 tbs salt and let stand for 4 hours. Press out liquid. Mix in 1½ tbs minced onion, 1 tsp minced garlic, ⅔ tsp minced ginger, and ½–1 tbs crushed red pepper. Place mixture in large glass jar with a tight-fitting lid. Refrigerate for 4 days. Koreans serve kimchi as a relish at almost every meal.

CHAPTER 25 REVIEW AND ASSESSMENT RESOURCES

Reproducible
- Readings in World Geography, History, and Culture 70 and 71
- Critical Thinking Activity 25
- Vocabulary Actitivy 25

Technology
- Chapter 25 Test Generator (on the One-Stop Planner)
- Audio CD Program, Chapter 25
- HRW Go site

Reinforcement, Review, and Assessment
- Chapter 25 Review, pp. 584–85
- Chapter 25 Tutorial for Students, Parents, Mentors, and Peers
- Chapter 25 Test
- Chapter 25 Test for English Language Learners and Special-Needs Students

Building Social Studies Skills

Map ACTIVITY

On a separate sheet of paper, match the letters on the map with their correct labels.

Hokkaido	**Tokyo**
Honshu	**Seoul**
Shikoku	**P'yŏngyang**
Kyushu	

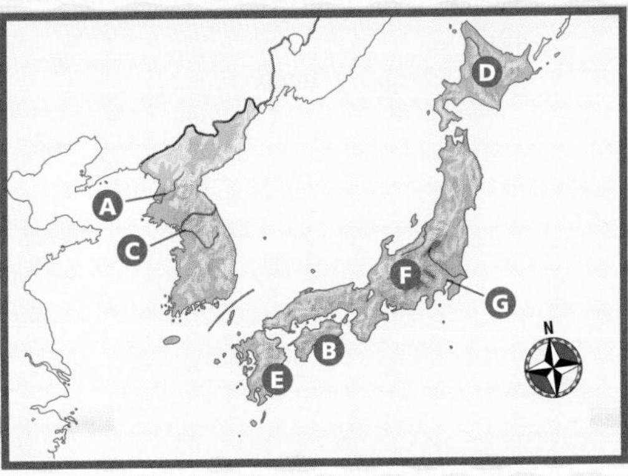

Mental Mapping Skills ACTIVITY

On a separate sheet of paper, draw a freehand map of Japan and the Koreas. Make a key for your map and label the following:

China	**North Korea**
Inland Sea	**Sea of Japan**
Japan	**South Korea**

WRITING ACTIVITY

Imagine that you are traveling in Japan, North Korea, or South Korea. Write a one-page letter to a friend describing the places you have visited and an adventure you have had during your stay. Be sure to use standard grammar, spelling, sentence structure, and punctuation.

Alternative Assessment

Portfolio ACTIVITY

Learning About Your Local Geography

Cooperative Project Japan and South Korea are very densely populated. How densely populated is your state? With your group, research the population densities of counties within your state. Create a population density map of the area you have researched.

internet connect

Internet Activity: **go.hrw.com**
KEYWORD: SK3 GT25

Choose a topic to explore Japan and the Koreas:
- Investigate volcanoes.
- Visit Japan and the Koreas.
- Compare your school to a Japanese school.

5. North Korea—command economy run by government, produces some machinery, iron, steel, but lags behind in high-quality goods; South Korea—highly industrialized market economy, produces quality export goods

Map Activity

A. P'yŏngyang
B. Shikoku
C. Seoul
D. Hokkaido
E. Kyushu
F. Honshu
G. Tokyo

Mental Mapping Skills Activity

Maps will vary, but listed places should be labeled in their approximate locations.

Writing Activity

Answers will vary, but the information included should be consistent with text material. Use Rubric 40, Writing to Describe, to evaluate student work.

Portfolio Activity

Counties chosen should accurately reflect different population densities. Use Rubric 20, Map Creation, to evaluate student work.

internet connect

GO TO: **go.hrw.com**
KEYWORD: SK3 Teacher
FOR: a guide to using the Internet in your classroom

CHAPTER 26 Southeast Asia
Chapter Resource Manager

Objectives	Pacing Guide	Reproducible Resources
SECTION 1 **Physical Geography** (pp. 587–89) **1.** Identify the major physical features of Southeast Asia. **2.** Identify the climates, vegetation, and wildlife found in the region. **3.** Identify the resources found in Southeast Asia.	**Regular** .5 day **Block Scheduling** .5 day *Block Scheduling Handbook, Chapter 26*	**RS** Guided Reading Strategy 26.1
SECTION 2 **History and Culture** (pp. 590–93) **1.** Identify some important events in the history of Southeast Asia. **2.** Describe what the people and culture of Southeast Asia are like today.	**Regular** 1 day **Block Scheduling** .5 day *Block Scheduling Handbook, Chapter 26*	**RS** Guided Reading Strategy 26.2 **RS** Graphic Organizer 26 **E** Cultures of the World Activity 7 **E** Creative Strategies for Teaching World Geography, Lesson 18 **SM** Geography for Life Activity 26 **IC** Interdisciplinary Activities for the Middle Grades 29, 30 **SM** Map Activity 26
SECTION 3 **Mainland Southeast Asia Today** (pp. 594–96) **1.** Identify where people in the mainland countries live today. **2.** Describe the economies of the main.land countries.	**Regular** .5 day **Block Scheduling** .5 day *Block Scheduling Handbook, Chapter 26*	**RS** Guided Reading Strategy 26.3 **IC** Interdisciplinary Activity for the Middle Grades 31
SECTION 4 **Island Southeast Asia Today** (pp. 597–601) **1.** Identify the major cities of the island countries. **2.** Describe the economies of island countries.	**Regular** 1 day **Block Scheduling** .5 day *Block Scheduling Handbook, Chapter 26*	**RS** Guided Reading Strategy 26.4 **IC** Interdisciplinary Activity for the Middle Grades 31

Chapter Resource Key

- **RS** Reading Support
- **IC** Interdisciplinary Connections
- **E** Enrichment
- **SM** Skills Mastery
- **A** Assessment
- **REV** Review
- **ELL** Reinforcement and English Language Learners
- Transparencies
- CD–ROM
- Music
- Video
- go.hrw.com Internet
- Holt Presentation Maker Using Microsoft® Powerpoint®

One-Stop Planner CD–ROM

See the *One-Stop Planner* for a complete list of additional resources for students and teachers.

One-Stop Planner CD–ROM

It's easy to plan lessons, select resources, and print out materials for your students when you use the ***One-Stop Planner CD–ROM with Test Generator.***

Technology Resources	Review, Reinforcement, and Assessment Resources
One-Stop Planner CD–ROM, Lesson 26.1 Geography and Cultures Visual Resources with Teaching Activities 49–53 Homework Practice Online HRW Go site	**ELL** Main Idea Activity 26.1 **REV** Section 1 Review, p. 589 **A** Daily Quiz 26.1 **ELL** English Audio Summary 26.1 **ELL** Spanish Audio Summary 26.1
One-Stop Planner CD–ROM, Lesson 26.2 Homework Practice Online HRW Go site	**ELL** Main Idea Activity 26.2 **REV** Section 2 Review, p. 593 **A** Daily Quiz 26.2 **ELL** English Audio Summary 26.2 **ELL** Spanish Audio Summary 26.2
One-Stop Planner CD–ROM, Lesson 26.3 Homework Practice Online HRW Go site	**ELL** Main Idea Activity 26.3 **REV** Section 3 Review, p. 596 **A** Daily Quiz 26.3 **ELL** English Audio Summary 26.3 **ELL** Spanish Audio Summary 26.3
One-Stop Planner CD–ROM, Lesson 26.4 *ARGWorld* CD–ROM: Comparing Regions in Indonesia *ARGWorld* CD–ROM: Export Processing Zones in Indonesia Homework Practice Online HRW Go site	**ELL** Main Idea Activity 26.4 **REV** Section 4 Review, p. 631 **A** Daily Quiz 26.4 **ELL** English Audio Summary 26.4 **ELL** Spanish Audio Summary 26.4

internet connect

HRW ONLINE RESOURCES

GO TO: go.hrw.com
Then type in a keyword.

TEACHER HOME PAGE
KEYWORD: SK3 TEACHER

CHAPTER INTERNET ACTIVITIES
KEYWORD: SK3 GT26

Choose an activity to:
- explore an Indonesian rain forest.
- learn about shadow puppets.
- see the buildings of Southeast Asia.

CHAPTER ENRICHMENT LINKS
KEYWORD: SK3 CH26

CHAPTER MAPS
KEYWORD: SK3 MAPS26

ONLINE ASSESSMENT
Homework Practice
KEYWORD: SK3 HP26
Standardized Test Prep Online
KEYWORD: SK3 STP26
Rubrics
KEYWORD: SS Rubrics

COUNTRY INFORMATION
KEYWORD: SK3 Almanac

CONTENT UPDATES
KEYWORD: SS Content Updates

HOLT PRESENTATION MAKER
KEYWORD: SK3 PPT26

ONLINE READING SUPPORT
KEYWORD: SS Strategies

CURRENT EVENTS
KEYWORD: S3 Current Events

Meeting Individual Needs

Ability Levels

Level 1 Basic-level activities designed for all students encountering new material

Level 2 Intermediate-level activities designed for average students

Level 3 Challenging activities designed for honors and gifted-and-talented students

English Language Learners Activities that address the needs of students with Limited English Proficiency

Chapter Review and Assessment

- **E** Readings in World Geography, History, and Culture 72, 73, and 74
- **SM** Critical Thinking Activity 26
- **REV** Chapter 26 Review, pp. 604–05
- **REV** Chapter 26 Tutorial for Students, Parents, Mentors, and Peers
- **ELL** Vocabulary Activity 26
- **A** Chapter 26 Test
- **A** Unit 7 Test
- Chapter 26 Test Generator (on the One-Stop Planner)
- Audio CD Program, Chapter 26
- **A** Chapter 26 Test for English Language Learners and Special-Needs Students
- **A** Unit 7 Test for English Language Learners and Special-Needs Students

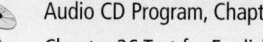

LAUNCH INTO LEARNING

Locate Southeast Asia's mainland and islands on a wall map. Point out that the region lies between India and China, on the Indochina Peninsula. Explain to students how this location has influenced Southeast Asia throughout history. *(colonization, the spread of various languages, religions, and customs, etc.)* Then call attention to the region's location between the Pacific and Indian Oceans. Use this information as a springboard for helping students focus on the physical, economic, and political characteristics of Southeast Asia.

Section 1

Objectives

1. Identify the major physical features of Southeast Asia.
2. Describe the climates, vegetation, and wildlife of Southeast Asia.
3. Name the resources found in Southeast Asia.

LINKS TO OUR LIVES

You may want to reinforce students' interest in Southeast Asia by pointing out the following concepts:

▶ The economies of some Pacific Rim countries are so fast-growing they have been called the tigers. The tigers have many trade links to the United States. Investments are so interwoven that business events there can affect our stock market.

▶ The Vietnam War has affected the American consciousness. Our experiences there continue to affect U.S. policies and actions, particularly regarding the military.

▶ The struggle of East Timor to win independence from Indonesia required the assistance of a UN peacekeeping force.

CHAPTER 26

Southeast Asia

Our study of the world now takes us to Southeast Asia. This vast region stretches from Myanmar eastward to New Guinea in the Pacific Ocean.

Sawaddee! (May you have good fortune!) I am Chosita, and I am 14 years old. I live in Bangkok with my parents and my older sister. We get up early for school because traffic in Bangkok is very heavy. By 6:00 A.M. we are on the road. My school has an eatery where street vendors sell all kinds of food—noodles in broth with beef, stir-fried noodles with meat and greens, dessert cakes of taro, pumpkin, and sticky rice, and fruits like rambutan, mangosteen, durian, and mango.

We go to school from June to September and from November to February. Our big vacation is March through May.

Our school has 38 students and two teachers in each class. I will not learn to use the computer until next year because we are the last class under the old school policy. The new policy has all students begin learning the computer in fourth grade.

สวัสดีค่ะ ดิฉันชื่อ โชสิตา ดิฉันอยู่ที่กรุงเทพฯ ซึ่งเป็นเมืองหลวงของประเทศไทยค่ะ

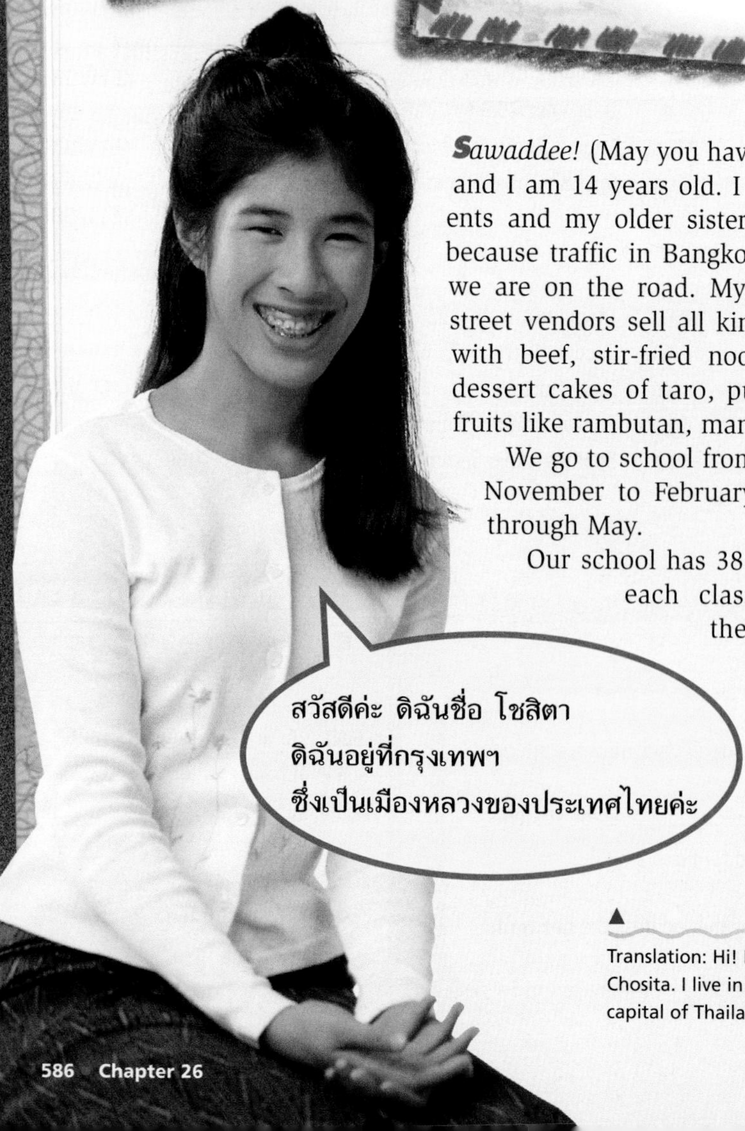

▲ Translation: Hi! My name is Chosita. I live in Bangkok, the capital of Thailand.

Focus

LET'S GET STARTED

Copy the following instructions onto the chalkboard: *Look at the photos and map in this section. What do they tell you about the physical geography of Southeast Asia?* Have students write their answers. Discuss their responses. Ask students to keep what they have written at hand so they can compare their original impressions with what they learn as they progress through the section. Tell students that in Section 1 they will learn more about the physical geography of Southeast Asia.

Using the Physical-Political Map

Have students examine the map on this page. Then ask students to locate the various countries of mainland and island Southeast Asia. Call on volunteers to identify a peninsula *(the Indochina or Malay Peninsula)* and a delta *(Irrawaddy, Chao Phraya, Mekong, or Hong).* Point out to students the difference in size between Singapore and Brunei and the other countries. *(Singapore and Brunei are much smaller.)*

Building Vocabulary

Write the terms **mainland** and **archipelagos** on the chalkboard. Have volunteers read aloud the definitions of the terms. Ask students if they have ever heard these terms used to describe other places in the world. On a world map, have students point out examples of a mainland and examples of archipelagos. *(Students might mention the continents as examples of a mainland and the East Indies, West Indies, British Isles, or Japan as examples of archipelagos.)*

Section 1 Physical Geography

Read to Discover

1. What are the major physical features of Southeast Asia?
2. What climates, vegetation, and wildlife are found in the region?
3. What resources does Southeast Asia have?

Define

mainland
archipelagos

Locate

Indochina Peninsula
Malay Peninsula
New Guinea
Malay Archipelago
Philippines
Irian Jaya
Borneo
Java
Sumatra
Mekong River

WHY IT MATTERS

Indonesia is made up of thousands of islands. Many have their own culture. There are ongoing conflicts between these groups. Use CNNfyi.com or other **current events** sources to find examples of this problem. Record your findings in your journal.

Golden statue from Thailand

Southeast Asia: Physical-Political

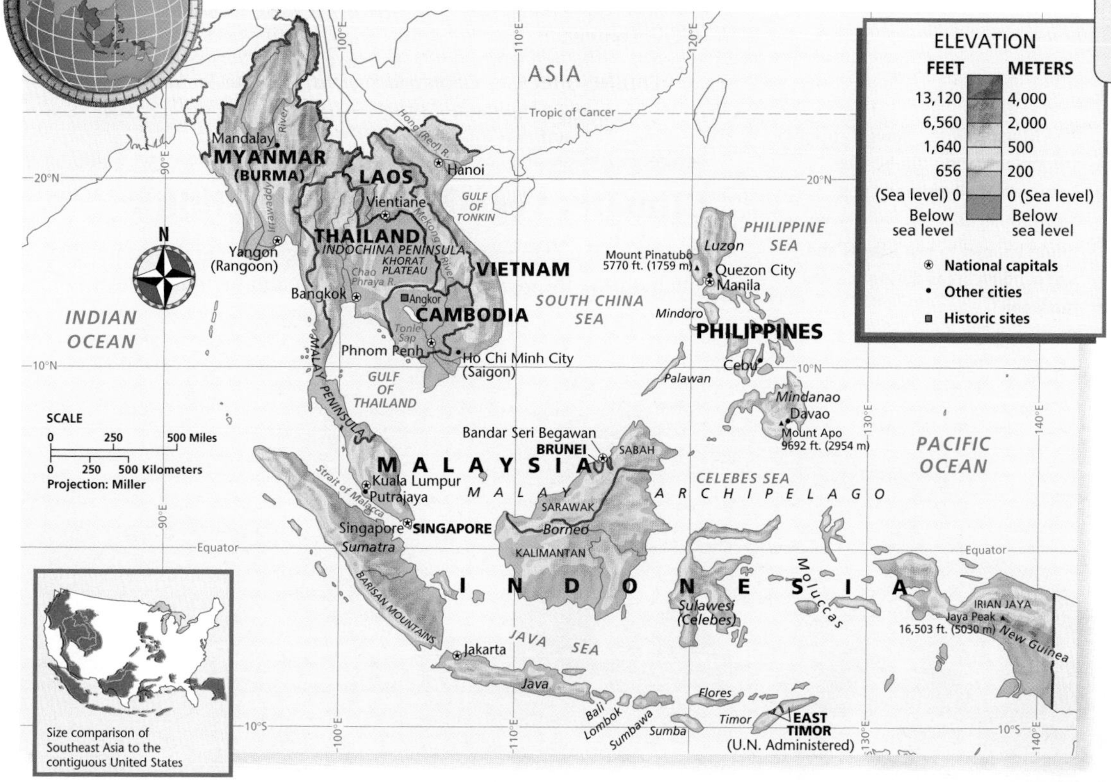

Section 1 Resources

Reproducible

- Block Scheduling Handbook, Chapter 26
- Guided Reading Strategy 26.1

Technology

- One-Stop Planner CD–ROM, Lesson 26.1
- Homework Practice Online
- Geography and Cultures Visual Resources with Teaching Activities 49–53
- HRW Go site

Reinforcement, Review, and Assessment

- Section 1 Review, p. 589
- Daily Quiz 26.1
- Main Idea Activity 26.1
- English Audio Summary 26.1
- Spanish Audio Summary 26.1

TEACH

Teaching Objectives 1–3

ALL LEVELS: (Suggested time: 20 min.) Copy the following graphic organizer onto the chalkboard, omitting the italicized answers. Call on students to provide words and phrases to describe the physical features, climates, vegetation and wildlife, and resources of the mainland and island regions of Southeast Asia. **ENGLISH LANGUAGE LEARNERS**

SOUTHEAST ASIA

Physical Features	Climates	Vegetation and Wildlife	Resources
Indochina and Malay Peninsulas, Malay Archipelago, the Philippine Islands, five major river systems, and some mountains	*tropical, with seasonal monsoons*	*rain forest, rhinoceroses, orangutans, tigers, and elephants*	*forests, rich farmland, tin, iron ore, oil, gas*

EYE ON EARTH

Climbing high mountains is often considered one of the greatest challenges to human adventurers. However, Low's Gully in Malaysia presents a different kind of challenge. This steep ravine on the island of Borneo is about 8,600 feet (2,600 m) deep. Above it rises Mount Kinabalu, the highest mountain in Southeast Asia.

Adventurers who want to descend to the floor of Low's Gully must first climb about three quarters of the way up Mount Kinabalu. The descent then takes them down narrow cracks and steep granite cliffs. Dense vegetation and frequent flash floods near the gully's floor provide more challenges.

Activity: Have students write a scene for the script of an action movie in which two characters descend into Low's Gully.

Visual Record Answer

They might get flooded and become submerged by rising river waters.

internet connect

GO TO: go.hrw.com
KEYWORD: SK3 CH26
FOR: Web sites about Southeast Asia

BUILD on WHAT You Know

Do you remember what you learned about the Pacific Ring of Fire? See Chapter 2 to review.

Physical Features

Southeast Asia is made up of two peninsulas and two large island groups. The Indochina and Malay (muh-LAY) Peninsulas lie on the Asian **mainland**. A mainland is a region's main landmass. The two large groups of islands, or **archipelagos** (ahr-kuh-PE-luh-gohs), lie between the mainland and New Guinea. They are the Malay Archipelago—made up mostly of Indonesia—and the Philippines. The Philippines are sometimes considered part of the Malay Archipelago. Western New Guinea is called Irian Jaya. It is part of Indonesia.

Landforms Southeast Asia's highest mountains are on the mainland in northern Myanmar (MYAHN-mahr). Mountain ranges fan out southward into Thailand (TY-land), Laos (LOWS), and Vietnam (vee-ET-NAHM). Between the mountains are low plateaus and river floodplains. The floodplains are rich farmlands.

Some of the large islands also have high mountains. Those islands include Borneo, Java, Sumatra, New Guinea, and some in the Philippines. They are part of the Pacific Ring of Fire. Earthquakes and volcanic eruptions often shake this part of the world.

Rivers Five major river systems drain the mainland. Many people and the largest cities are found near these rivers. The greatest river is the Mekong (MAY-KAWNG). The Mekong River flows southeast from China to southern Vietnam. You will read about the other rivers later in this chapter.

✓ **READING CHECK:** *Places and Regions* What are Southeast Asia's major physical features? **peninsulas, islands, mountain ranges, plateaus, river floodplains, river systems**

Climate, Vegetation, and Wildlife

The warm temperatures of this tropical region generally do not change much during the year. However, northern and mountain areas tend to be cooler.

The Mekong River flows through a floodplain along the border between Thailand and Laos.

Interpreting the Visual Record **What might happen to low islands and surrounding areas during the wet monsoon?**

CLOSE

Ask students to identify items they are familiar with that utilize Southeast Asian products such as coconuts, sugarcane, and rubber. *(Students might mention candy bars, soft drinks, or bicycle tires.)*

REVIEW AND ASSESS

Have students complete the Section Review and use the map and text to create four true-or-false questions for the region's physical features, climate, wildlife and vegetation, and resources. Have students quiz a partner using their questions. Then have students complete Daily Quiz 26.1. **COOPERATIVE LEARNING**

RETEACH

Have students complete Main Idea Activity 26.1. Then organize the class into groups. Have each group design graphics and text for a Web site providing information about the physical features, climate, wildlife, and resources of Southeast Asia. **ENGLISH LANGUAGE LEARNERS, COOPERATIVE LEARNING**

EXTEND

Have interested students conduct research on volcanic eruptions and earthquakes in the Pacific Ring of Fire during the 1900s. Have them mark affected areas on a map and use it as a visual aid in a short oral presentation to the class. **BLOCK SCHEDULING**

Much of the rainfall on the mainland is seasonal. Wet monsoon winds from nearby warm oceans bring heavy rains in the summer. Dry monsoons from the northeast bring drier weather in winter. Most of the islands are wet all year. Typhoons bring heavy rains and powerful winds to the island countries.

The region's tropical rain forests are home to many kinds of plants and animals. About 40,000 kinds of flowering plants grow in Indonesia alone. Rhinoceroses, orangutans, tigers, and elephants also live in the region. However, many of these plants and animals are endangered. Southeast Asia's rain forests are being cleared for farmland, tropical wood, and mining.

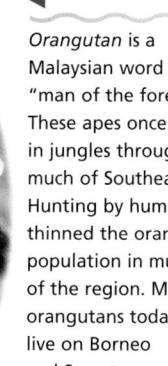

◄ *Orangutan* is a Malaysian word for "man of the forest." These apes once lived in jungles throughout much of Southeast Asia. Hunting by humans has thinned the orangutan population in much of the region. Most orangutans today live on Borneo and Sumatra.

✓ **READING CHECK:** *Places and Regions* What are the region's climates, vegetation, and wildlife like? **tropical; tropical rain forests; many different plants and animals, some endangered**

Resources

Southeast Asia's rain forests produce valuable wood and other products. Thailand, Indonesia, and Malaysia (muh-LAY-zhuh) are the world's largest producers of natural rubber. The rubber tree is native to South America. However, it grows well in Southeast Asia's tropical climates.

Rich volcanic soils, floodplains, and tropical climates are good for farming. Abundant water and good soils in river deltas are ideal for growing rice. Coconuts, palm oil, sugarcane, coffee, and spices are also key products. Countries here also mine tin, iron ore, oil, and gas.

✓ **READING CHECK:** *Places and Regions* What are the region's important resources? **rain forests, rich soils, abundant water, tin, iron ore, oil, gas**

Our Amazing Planet

A plant in Indonesia produces the world's largest flower—about 3 feet (1 m) across!

go.hrw.com Homework Practice Online
Keyword: SK3 HP26

Section 1 Review

Define and explain: mainland, archipelagos

Working with Sketch Maps On a map of Southeast Asia that your teacher provides or that you draw, label the following: Indochina Peninsula, Malay Peninsula, New Guinea, Malay Archipelago, Philippines, Irian Jaya, Borneo, Java, Sumatra, and the Mekong River. In a box in the margin, describe the Mekong River.

Reading for the Main Idea

1. *Places and Regions* Where are the region's highest mountains?
2. *Places and Regions* Where are large cities found?
3. *Places and Regions* Which countries are the world's largest producers of natural rubber?

Critical Thinking

4. **Making Generalizations and Predictions** What do you think might happen to the region's wildlife if much of the tropical rain forests continue to be destroyed?

Organizing What You Know

5. **Summarizing** Copy the following graphic organizer. Use it to describe the region's climates, vegetation and wildlife, and resources.

Climates	Vegetation and wildlife	Resources

Section 1 Review Answers

Define For definitions, see: mainland, p. 588; archipelagos, p. 588

Working with Sketch Maps Maps will vary, but the listed places should be labeled in their approximate locations.

Reading for the Main Idea

1. northern Myanmar (NGS 4)
2. on or near the region's five major rivers (NGS 4)
3. Thailand, Indonesia, Malaysia (NGS 4)

Critical Thinking

4. Much wildlife will die and disappear as habitat is destroyed.

Organizing What You Know

5. climates—tropical wet and dry seasons on mainland, most islands wet all year; vegetation and wildlife—tropical rain forest, many kinds of plants, trees, and animals, including rhinos, orangutans, tigers, elephants; resources—forests, good farmland, tin, iron ore, oil, gas

Section 2

Objectives

1. Analyze important events in the history of Southeast Asia.
2. Describe the people and culture of Southeast Asia today.

Focus

LET'S GET STARTED

Copy the following instructions onto the chalkboard: *Why might it be difficult for countries made up of many islands to maintain a peaceful union? Why might having many islands be beneficial? Write down two possible answers for each question.* Allow students time to write their responses. Discuss students' answers. Tell students that in Section 2 they will learn more about the people and cultures of Southeast Asia.

Building Vocabulary

Write the key term **refugees** on the chalkboard. Point out to students that *refuge,* the root of the word, means "a safe place." Select a volunteer to locate and read aloud the meaning of *refugees.* Point out the relationship between the meaning of *refugees* and the meaning of *refuge.*

Section 2 Resources

Reproducible

- Guided Reading Strategy 26.2
- Graphic Organizer 26
- Cultures of the World Activity 7
- Map Activity 26
- Geography for Life Activity 26
- Creative Strategies for Teaching Geography, Lesson 18
- Interdisciplinary Activities for the Middle Grades 26, 30

Technology

- One-Stop Planner CD–ROM Lesson 26.2
- Homework Practice Online
- HRW Go site

Reinforcement, Review, and Assessment

- Section 2 Review, p. 593
- Daily Quiz 26.2
- Main Idea Activity 26.2
- English Audio Summary 26.2
- Spanish Audio Summary 26.2

Visual Record Answer

Answers will vary according to students' experiences.

Section 2 History and Culture

Read to Discover

1. What are some important events in the history of Southeast Asia?
2. What are the people and culture of Southeast Asia like today?

Define

refugees

Locate

Angkor, Cambodia, Thailand, Vietnam, Laos, Indonesia, Malaysia, Timor, Myanmar, Singapore

Why It Matters

Sea trade routes have always been important in this region but have also created a piracy problem. Use CNNfyi.com or other **current events** sources to find examples of piracy in this region. Record your findings in your journal.

Assorted peppers from Myanmar

History

Southeast Asia was home to some of the world's earliest human settlements. Over time many peoples moved there from China and India. The Khmer (kuh-MER) developed the most advanced of the region's early societies. The Khmer Empire was based in Angkor in what is now Cambodia (kam-BOH-dee-uh). It controlled a large area from the early A.D. 800s to the mid-1200s.

Colonial Era Europeans began to establish colonies in Southeast Asia in the 1500s. By the end of the 1800s, the Portuguese, British, Dutch, French, and Spanish controlled most of the region. The United States won control of the Philippines from Spain after the Spanish-American War in 1898. Just Siam (sy-AM), now called Thailand, remained independent.

The Khmer built the beautiful Angkor Wat in the A.D. 1100s in Angkor. This vast temple in present-day Cambodia was dedicated to Vishnu, a Hindu god.

Interpreting the Visual Record **How does this temple's architecture compare to other religious buildings you have seen?**

TEACH

Teaching Objective 1

LEVEL 1: (Suggested time: 20 min.) Copy the following graphic organizer onto the chalkboard, omitting the italicized answers. Ask students to copy the organizer into their notebooks and fill it in to show the progression of influences on Southeast Asia. The first one is filled in for them. **ENGLISH LANGUAGE LEARNERS**

INFLUENCES ON SOUTHEAST ASIA

Settlers from China and India	→	*Khmer Empire*	→	*Europeans*	→	*Various governments (communist, democratic, etc.)*

LEVELS 2 AND 3: (Suggested time: 45 min.) Give students an outline map of Southeast Asia. Have students create a symbol for each of the influences identified in the graphic organizer in the previous activity. *(See the Level 1 lesson for the correct influences.)* Have students connect their symbols with arrows to demonstrate the directions from which each influence advanced. Then have students present a brief oral report to the class on the history of movements to Southeast Asia, using their map as a visual aid.

◄ A Filipino official releases a dove during his country's independence celebrations in Manila. The Philippines was the largest U.S. overseas possession from 1898 to 1946.

Japan invaded and occupied most of Southeast Asia during World War II. After Japan was defeated in 1945, the United States granted the Philippines independence a year later. European countries tried to regain control of their colonies in the region. Some Southeast Asians decided to fight for independence. One of the bloodiest wars was in French Indochina. The French finally left in 1954. Their former colonies of Vietnam, Laos, and Cambodia became independent. By the mid-1960s, European rule had ended in most of the region.

Modern Era Unfortunately, fighting did not end in some countries when the Europeans left. Vietnam split into two countries. In the 1960s the United States sent troops to defend South Vietnam against communist North Vietnam. Civil wars also raged in Laos and Cambodia. Communist forces took power in all the countries in 1975. North and South Vietnam were then united into one country.

The region's wars caused terrible destruction. Millions died, including more than 50,000 Americans. About 1 million Vietnamese **refugees** tried to escape the communist takeover in South Vietnam. Refugees are people who flee their own country, usually for economic or political reasons. Many refugees from the region came to the United States.

In Cambodia more than 1 million people died under a cruel communist government. That government ruled from 1975 to 1978. Then Vietnam invaded Cambodia in 1978, sparking another conflict. That war continued off and on until the mid-1990s. Many Cambodian refugees fled to Thailand.

Communists and other groups also fought against governments in the Philippines, Indonesia, and Malaysia. In 1975 Indonesia invaded the former Portuguese

Hanoi's Ho Chi Minh Mausoleum honors the communist leader who fought for Vietnam's independence. He led North Vietnam from 1954 to his death in 1969. Many U.S. soldiers and Vietnamese died during the war he led against South Vietnam. ▼

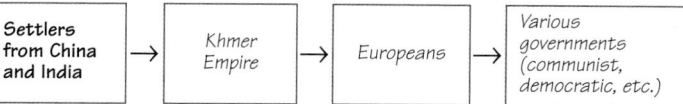

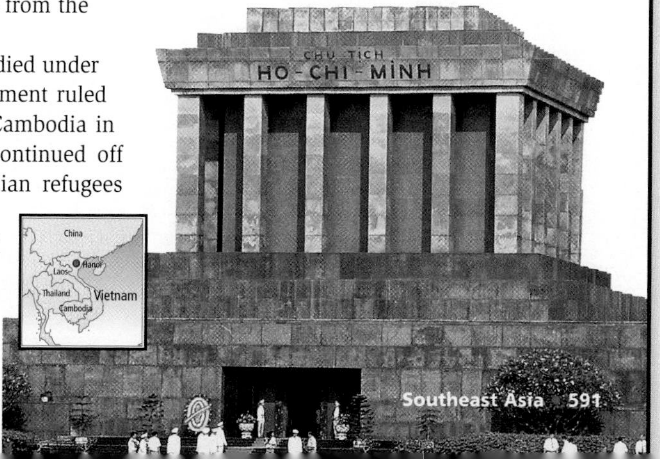

HUMAN SYSTEMS

In the 1970s the Khmer Rouge regime in Cambodia attempted to establish an agrarian society based on Marxism. It's leaders' strategy included the systematic destruction of both the traditional rural culture and the Western-influenced culture of the country's cities. More than 1 million Cambodians died during the four years of Khmer Rouge domination.

Thousands of people fled Cambodia attempting to escape the regime. Many eventually made it to the United States after spending between one and six years in Thailand's refugee camps.

Continued political unrest in the 1980s resulted in the migration of more than 500,000 additional Cambodians to Thailand. Many of these refugees eventually settled elsewhere, such as in France or in the United States.

Discussion: Lead a class discussion that addresses the following questions: Under what conditions would you seek refuge in another country? Do you feel other countries have a responsibility toward people seeking refuge? Why or why not? If so, to what extent?

Teaching Objective 2

LEVEL 1: (Suggested time: 30 min.) Call students' attention to the characteristics of the people and cultures of Southeast Asia that are discussed throughout this section. Call on students to take turns reading aloud the parts of the section that describe these attributes of human geography, while a volunteer lists them on the chalkboard. Have students copy the characteristics into their notebooks. **ENGLISH LANGUAGE LEARNERS**

LEVEL 2: (Suggested time: 25 min.) Have each student design a cover for a magazine entitled Southeast Asia Today. Students' covers should reflect the information covered in this section. Encourage students to use drawings and pictures as well as a variety of type sizes to distinguish items of greater and lesser interest.

LEVEL 3: (Suggested time: 30 min.) Have students write an article for a magazine entitled Southeast Asia Today. Encourage students to relate the region's history and culture to life there today.

GLOBAL PERSPECTIVES

Political unrest in East Timor was not the only challenge facing Indonesia in the late 1990s. In 1997 uncontrolled fires began burning in parts of Indonesia, destroying up to 17,600 square miles (45,580 sq km) of forest. Health-crippling smog settled over Borneo and Sumatra, in what is known as "the era of smoke."

Government regulations to prevent fires went unenforced. For example, exploitative logging practices resulted in dry and flammable tracts of forest, and plantation owners regularly burned land to clear it.

The problem has continued into the 2000s. Some environmental organizations have begun to pressure Indonesia to enforce the laws more strictly. The government has threatened to revoke logging licenses if the fires continue.

Activity: Have students conduct research on the recent fires in Indonesia.

internet connect

GO TO: go.hrw.com
KEYWORD: SK3 CH26
FOR: Web sites about current events in Indonesia

Visual Record Answer

towers, spires, columns, layered roofs, small windows

colony of East Timor. The East Timorese demanded independence. However, the Indonesian military kept a tight grip on the region. The people of East Timor voted for independence in 1999. East Timor then plunged again into violence. The United Nations sent troops to restore peace and manage the area before independence was achieved.

Governments The region's countries have had different kinds of governments. Many have been ruled by dictators. Some countries, such as the Philippines and Indonesia, now have governments elected by the people.

In other countries, the people still have little say in their government. For example, Myanmar is ruled by a military government. That government has jailed and even killed its opponents. Vietnam and Laos are still ruled by Communist governments. Only recently have Indonesians been allowed to vote in free elections. In some countries, such as Singapore, the same party always wins elections.

✓ READING CHECK: *Places and Regions* What were some key events in Southeast Asian history? 1500s—Europeans established colonies and began to control regions; World War II—Japan invaded; mid 1960s—European rule had mostly ended, Vietnam War, civil wars in Laos and Cambodia

Culture

The populations of most countries in Southeast Asia are very diverse. This is because many different peoples have moved to the area over time. Today, for example, nearly 70 percent of the people in Myanmar are Burmese. However, Chinese, Asian Indians, and many other ethnic groups also live there.

The Shwedagon Pagoda is a beautiful Buddhist shrine in Yangon, Myanmar. Pagodas are important parts of a Buddhist temple complex. **Interpreting the Visual Record** **What architectural features do you see in the photograph?**

CLOSE

Ask students to name important aspects of the history and culture of Southeast Asia. Write them on the chalkboard as students name them. Then have students vote on which are the most important.

REVIEW AND ASSESS

Have students complete the Section Review. Then organize them into groups. Provide each group with two sets of index cards. On one set, each group should describe a feature of the people or culture. On the other each group should illustrate each feature. Then have students complete Daily Quiz 26.2. **COOPERATIVE LEARNING**

RETEACH

Have students complete Main Idea Activity 26.2. Then have students design storyboards depicting key aspects of the region's history, culture, and current conditions. Finally, have students share their work with a partner. **ENGLISH LANGUAGE LEARNERS, COOPERATIVE LEARNING**

EXTEND

Have interested students conduct research on recent Southeast Asian immigrants in the United States. If possible, assign a separate immigrant group—Hmong, Vietnamese, Cambodian, Thai, Laotian, or Montagnard—to each student. Have them create a pamphlet providing information on U.S. culture and institutions and advice for new immigrants. **BLOCK SCHEDULING**

Many ethnic Chinese live in the largest cities of most Southeast Asian countries. In Singapore they are a majority of the population—more than 75 percent. Singapore is a tiny country at the tip of the Malay Peninsula.

Languages and Religions The peoples of Southeast Asia speak many different languages. For example, in the former Dutch colony of Indonesia, most people speak Bahasa Indonesia. However, Javanese, other local dialects, English, and Dutch are also spoken there. European and Chinese languages are spoken in many other countries.

In addition, Indians, Chinese, Arab traders, and Europeans brought different religions to the region. For example, Hinduism is practiced in the region's Indian communities. However, Buddhism is the most common religion in the mainland countries today. Islam is the major religion in Malaysia, Brunei, and Indonesia. In fact, Indonesia has the largest Islamic population in the world. Nearly 90 percent of its more than 228 million people are Muslim.

Christians are a minority in most of the former European colonies. However, more than 80 percent of people in the Philippines, a former Spanish colony, are Roman Catholic.

▲ Women in Laos sprinkle water on passing Buddhist monks during a New Year festival. This custom symbolizes the washing away of the old year. It is said to bring good luck to people in the year ahead.

Interpreting the Visual Record
What kinds of clothes are the monks wearing?

Food Southeast Asian foods have been influenced by Chinese, South Asian, and European cooking styles. There are many spicy, mild, and sweet varieties. Rice is the most important food in nearly all of the countries. It is served with many other foods and spices, such as curries and chili peppers. Coconut is also important. It is served as a separate dish or used as an ingredient in other foods.

✓ **READING CHECK:** *Human Systems* How have migration and cultural borrowing influenced the region's culture? great cultural diversity, many languages spoken; religions practiced; foods reflect influence of many different regions

go.hrw.com Homework Practice Online
Keyword: SK3 HP26

Section 2 Review

Define and explain: refugees

Working with Sketch Maps On the map you created in Section 1, label the region's countries, Angkor, and Timor. In a box in the margin, describe the recent history of East Timor.

Reading for the Main Idea

1. *Places and Regions* What was the Khmer Empire?
2. *Human Systems* How did Europeans influence the region's history and culture?

Critical Thinking

3. **Drawing Inferences and Conclusions** Why do you think European countries wanted to regain their Southeast Asian colonies following World War II?
4. **Finding the Main Idea** What religion is most common in the mainland countries? in the island countries?

Organizing What You Know

5. **Sequencing** Copy the following time line. Use it to identify important people, years, periods, and events in Southeast Asia's history.

A.D. 800 ———————————— 2000

Section 2 Review Answers

Define For definition, see: refugees, p. 591

Working with Sketch Maps East Timor voted for independence in 1999; plunged into violence; UN force helped restore peace and administer region.

Reading for the Main Idea

1. the most advanced of the region's early societies (NGS 4)
2. controlled the region and introduced new languages, religions, and foods (NGS 10, 13)

Critical Thinking

3. to restore their empires and to control the region's resources
4. most common: Buddhism; Islam, except for Christianity in the Philippines

Organizing What You Know

5. A.D. 800–1200s—Khmer; 1500s–1800s—Europeans; 1898—U.S. control of Philippines; 1946—Phil. independence; 1954—French leave Indochina; 1975—Vietnam War ends, Indonesia takes East Timor; 1978—Vietnam invades Cambodia; 1999—E. Timor votes for independence

◄ **Visual Record Answer**

orange robes and sandals

Section 3

Objectives

1. Identify the areas where mainland Southeast Asians live.
2. Describe the economies of the mainland countries.

Focus

LET'S GET STARTED

Copy the following question and instruction onto the chalkboard: *Would you rather live in a rural area or in the city? Write your answer and a few sentences explaining why in your notebook.* Allow students time to write their responses. Explain to students that most mainland Southeast Asians live in rural areas. Tell students that in Section 3 they will learn more about where people of mainland Southeast Asia live.

Building Vocabulary

Write the term **klongs** on the chalkboard. Ask a volunteer to locate the term and to read aloud its definition. Point out that *klongs* is a Thai word. Then ask students to suggest words we use in English to describe the same thing. *(Possible answers: canal, waterway, aqueduct, drainpipe, sewer tunnel)*

Section 3 Resources

Reproducible
- Guided Reading Strategy 26.3
- Interdisciplinary Activity for the Middle Grades 31

Technology
- One-Stop Planner CD–ROM Lesson 26.3
- Homework Practice Online
- HRW Go site

Reinforcement, Review, and Assessment
- Section 3 Review, p. 596
- Daily Quiz 26.3
- Main Idea Activity 26.3
- English Audio Summary 26.3
- Spanish Audio Summary 26.3

Visual Record Answer

to shade their head and face from the Sun

Section 3 Mainland Southeast Asia Today

Read to Discover

1. Where do people in the mainland countries live today?
2. What are the economies of the mainland countries like?

Define

klongs

Locate

Bangkok
Yangon
Hanoi
Ho Chi Minh City
Chao Phraya River
Irrawaddy River
Hong (Red) River
Vientiane
Phnom Penh

WHY IT MATTERS

The region's cities are growing rapidly. This can create problems with the city's everyday functioning. Use CNNfyi.com or other **current events** sources to find examples of urbanization problems in the region. Record your findings in your journal.

Golden door of a Buddhist temple

People and Cities

Most mainland Southeast Asians today live in rural areas. Many are farmers in fertile river valleys and deltas. Fewer people live in remote hill and mountain villages.

However, the region's cities have been growing rapidly. People are moving to urban areas to look for work. The cities have many businesses, services, and opportunities that are not found in rural areas. Many of the cities today are crowded, smoggy, and noisy.

Look at the chapter map. You will find that the largest cities are located along major rivers. Location near rivers places these cities near important rice-growing areas. Access to rivers also makes them key

A Vietnamese man sells incense sticks in Ho Chi Minh City.

Interpreting the Visual Record **Why might people in warm, sunny climates wear hats like the one in this photo?**

TEACH

Teaching Objective 1

ALL LEVELS: (Suggested time: 20 min.) Copy the following graphic organizer onto the chalkboard, omitting the italicized answers. Have students list descriptions of life in mainland Southeast Asia.
ENGLISH LANGUAGE LEARNERS

Teaching Objective 2

ALL LEVELS: (Suggested time: 30 min.) Before class, prepare large cutouts of Vietnam, Laos, Cambodia, Thailand, and Myanmar. Do not label them. Organize the class into five groups and give each group one of the cutouts. Have each group identify its country and list three facts about its economy on the cutout. Then allow groups to assemble their cutouts to configure a map of Southeast Asia. Each group should report its facts to the class while placing its cutout on the chalkboard.
ENGLISH LANGUAGE LEARNERS, COOPERATIVE LEARNING

Rural Life	Both	Urban Life
• *in remote hill and mountain villages* • *most people live here*	• *near major rivers* • *near rice-growing areas*	• *has businesses* • *crowded, smoggy, noisy* • *important shipping areas*

shipping centers for farm and factory products. The largest cities are Bangkok, Yangon, Hanoi, and Ho Chi Minh City.

Bangkok The mainland's largest city is Bangkok, Thailand's capital. Bangkok lies near the mouth of the Chao Phraya (chow PRY-uh) River. More than 7 million people live there. Much of Bangkok is connected by ***klongs***, or canals. The *klongs* are used for transportation and for selling and shipping goods. They also drain water from the city.

▲ Boat traffic can be heavy along Bangkok's *klongs*. These canals have been part of the city's original transportation system for centuries.

Other Cities The region's second-largest city is Yangon, formerly known as Rangoon. It is Myanmar's capital and major seaport. The city is located in the Irrawaddy River delta on the coast of the Andaman Sea. To the east, Vietnam's largest cities are also located in major river deltas. The capital, Hanoi (ha-NOY), is located in the Hong (Red) River delta in the north. Ho Chi Minh City is located in the Mekong River delta in the south. Ho Chi Minh City was once known as Saigon and was South Vietnam's capital. Today it is an important seaport and business center with more than 4.6 million people.

✓ **READING CHECK:** *Places and Regions* Where do most people in mainland Southeast Asia live today and why? mostly in rural areas; because many are farmers and need access to fertile river valleys and deltas

Economy

War, bad governments, and other problems have slowed progress in most of the mainland countries. However, rich resources could make the future brighter for the people there.

Vietnam Vietnam's economy has been slowly recovering since the end of the war in 1975. In recent years, the Communist government has begun moving from a command economy to a more market-oriented one. Some people are now allowed to own private businesses. Most Vietnamese remain farmers.

Most of Vietnam's factories, coal, oil, and other resources are in the north. The Hong and Mekong River deltas are major farming areas. Rice is the most important crop and food. In many places it is planted twice each year.

Laos This mountainous, landlocked country has few good roads, no railroads, and few telephones and televisions. Only some cities have

Mainland Southeast Asia

Country	Population/ Growth Rate	Life Expectancy	Literacy Rate	Per Capita GDP
Cambodia	12,491,501 2.3%	55, male 59, female	35%	$1,300
Laos	5,635,967 2.5%	52, male 55, female	57%	$1,700
Myanmar	41,994,678 0.6%	54, male 57, female	83%	$1,500
Thailand	61,797,751 0.9%	66, male 72, female	94%	$6,700
Vietnam	79,939,014 1.5%	67, male 72, female	94%	$1,950
United States	281,421,906 0.9%	74, male 80, female	97%	$36,200

Sources: Central Intelligence Agency, *The World Factbook 2001;* U.S. Census Bureau

Interpreting the Chart **According to the data in the chart, which country in the region is the most economically developed?**

EYE ON EARTH

Until recently, very little was known about Laos. The country's geography and politics largely isolated it from the outside world. However, in recent years the Communist government has been trying to attract foreign travelers to boost Laos's economy. As a result, more tourists are visiting the country.

Some visitors journey to a lovely valley hidden deep in the forest. The only way to reach the valley is by sailing along Hin Boun River, which flows into the valley through a cave about 6 miles (10 km) long. Other tourists prefer to visit cultural sites, such as the Pak Ou caves. Visitors to the caves find thousands of Buddha statues that were moved there more than 400 years ago to protect them from invaders.

Activity: Have students conduct research on the Internet for information on interesting sites in Laos. Then have students create travel fliers featuring those sites.

internet connect

GO TO: go.hrw.com
KEYWORD: SK3 CH26
FOR: Web sites about traveling to Laos

◄ **Chart Answer**

Thailand

CLOSE

Take a poll: Would students rather live in rural or urban areas of Southeast Asia? Why? Compare students' answers with the initial preferences indicated in the section opener.

REVIEW AND ASSESS

Have students complete the Section Review. Then have students imagine they are writing a letter to a pen pal in mainland Southeast Asia. Have them ask questions that address the main ideas of the section. Then have students trade their letters with a partner and answer their partner's questions. Then have students complete Daily Quiz 26.3. **COOPERATIVE LEARNING**

RETEACH

Have students complete Main Idea Activity 26.3. Then have students write three facts related to the Read to Discover questions at the beginning of the section. When complete, select volunteers to share their facts with the class. **ENGLISH LANGUAGE LEARNERS**

EXTEND

Have interested students conduct research on King Mongkut of Thailand who kept Thailand from being colonized. Then have them compare what they have learned to any of the movies, plays, or books about him and share their findings with the class. **BLOCK SCHEDULING**

Section 3 Review

Answers

Define For definition, see: *klongs*, p. 595

Working with Sketch Maps region's largest city, Bangkok, is Thailand's capital; lies near mouth of Chao Phraya River; connected by *klongs*

Reading for the Main Idea

1. rural areas, particularly in fertile river valleys and deltas; because most are farmers (NGS 4)
2. to look for work (NGS 9)
3. Thailand; because of its rich natural resources (NGS 4)

Critical Thinking

4. war, nondemocratic governments, and other political problems

Organizing What You Know

5. Bangkok—mainland's largest city, *klongs;* Yangon—Myanmar's capital, major seaport; Hanoi—Vietnam's capital, in Hong Delta; Ho Chi Minh City—formerly Saigon, important seaport and business center, Mekong Delta; Vientiane—Laos's capital; Phnom Penh—Cambodia's capital, in southern rice-growing area

Chart Answer ▲

Thailand

Major Producers of Natural Rubber

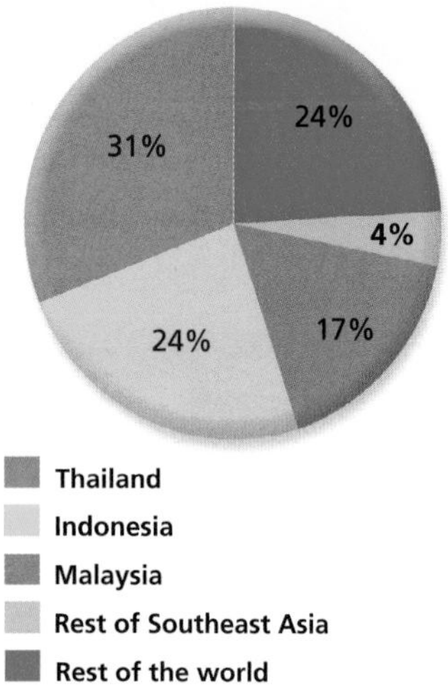

Source: United Nations, *Monthly Bulletin of Statistics*, July 1999

Interpreting the Chart **Which country is the largest producer of natural rubber?**

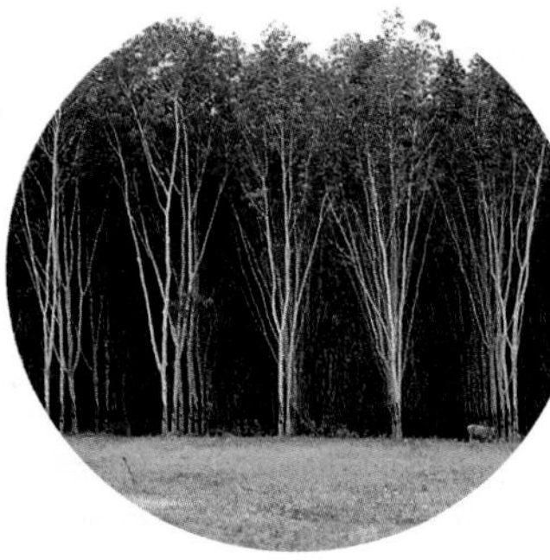

▶ Trees like these in Thailand make the region the world's largest producer of natural rubber.

electricity. The economy is mostly traditional—most people are subsistence farmers. They produce just enough food for themselves and their families. The Communist government in Vientiane (vyen-TYAHN), the capital, has also recently begun allowing more economic freedom.

Cambodia Economic progress in Cambodia has been particularly slow because of war and political problems. Agriculture is the most important part of the economy. The capital and largest city, Phnom Penh (puh-NAWM PEN), is located along the Mekong River. It lies in Cambodia's southern rice-growing area.

Thailand Thailand's economy has had problems but is the strongest of the mainland countries. This is partly because Thailand has rich resources. These resources include timber, natural rubber, seafood, rice, many minerals, and gems. Factories produce computers and electronics. Many Thai operate small businesses. Tourism is also important.

Myanmar This former British colony is also called Burma. It gained independence in 1948 and was officially renamed Myanmar in 1989. It has rich resources, including copper, tin, iron ore, timber, rubber, and oil. However, a harsh military government has limited political freedom. This has slowed economic progress.

✓ **READING CHECK:** *Places and Regions* What are the mainland economies like? some moving from traditional or command to market; all suffering from problems, including war, political problems

go.hrw.com Homework Practice Online
Keyword: SK3 HP26

Section 3 Review

Define and explain: *klongs*

Working with Sketch Maps On the map you created in Section 2, label Bangkok, Yangon, Hanoi, Ho Chi Minh City, Chao Phraya River, Irrawaddy River, Hong (Red) River, Vientiane, and Phnom Penh. Describe the mainland's largest city.

Reading for the Main Idea

1. *Places and Regions* Where do most mainland Southeast Asians live and why?
2. *Human Systems* Why are many people moving to cities?
3. *Places and Regions* Which country has the strongest economy and why?

Critical Thinking

4. **Finding the Main Idea** What kinds of problems appear to have slowed economic progress in the region?

Organizing What You Know

5. **Summarizing** Copy the following graphic organizer. Use it to describe the mainland's major cities. In each of its six circles, write the name of a city. In the circles, write important facts about the cities.

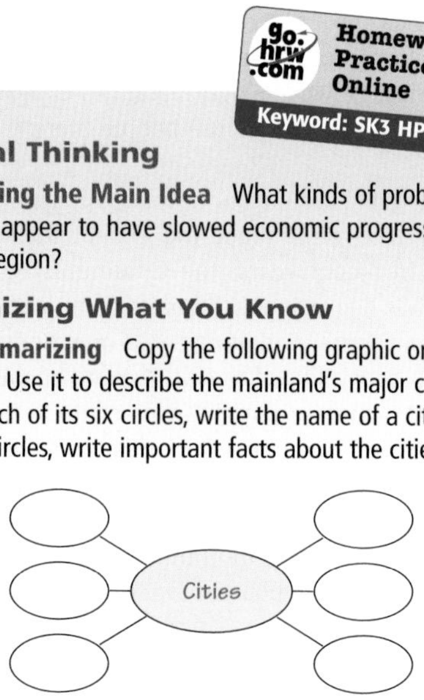

Section 4

Objectives

1. Name the major cities of the island countries of Southeast Asia.
2. Describe the economies of the island countries of Southeast Asia.

FOCUS

LET'S GET STARTED

Copy the following instructions onto the chalkboard: *Indonesia's national motto is "The many are one." Write the reasons you think this is appropriate for a country comprised of more than 7,000 individual islands.* Allow students to write their responses. Lead a discussion based on students' answers. Tell students that in Section 4 they will learn more about the people and culture of Indonesia and the other island countries of Southeast Asia.

Building Vocabulary

Write **kampongs** and **sultan** on the chalkboard and have volunteers locate and read the definitions. Then have students write sentences using the words. Select volunteers to share their sentences with the class.

Section 4 Island Southeast Asia Today

Read to Discover

1. What are the major cities of the island countries?
2. What are the economies of the island countries like?

Define

kampongs
sultan

Locate

Jakarta
Manila
Kuala Lumpur
Luzon
Bali

A decorative Indonesian fabric called a batik (buh-TEEK)

WHY IT MATTERS

Some countries in this region have higher per capita GDPs as a result of their growing economies. Use CNNfyi.com or other **current events** sources to check economic growth in this region. Record your findings in your journal.

People and Cities

Indonesia is the largest of the island countries and the world's fourth-most-populous country. The country's more than 17,000 islands were known as the Dutch East Indies until 1949. Malaysia, Singapore, and Brunei were British colonies. The British granted independence to Malaysia in 1963. Singapore split from Malaysia in 1965. In 1984 Brunei became the region's last European colony to gain independence. As you have read, the Philippines gained independence from the United States in 1946. More than 7,000 islands make up that country.

Many people live in rural areas in the island countries. However, the island countries are more urbanized than the mainland countries. As on the mainland, many people are moving to cities in search of jobs. One country, Singapore, is simply a large city on a small island.

Modern skyscrapers tower over colonial-era buildings in Singapore. The city has one of the world's busiest ports.

Interpreting the Visual Record What does Singapore's architecture tell you about its economy and culture?

Section 4 RESOURCES

Reproducible

- Guided Reading Strategy 26.4
- Interdisciplinary Activity for the Middle Grades 31

Technology

- One-Stop Planner CD–ROM Lesson 26.4
- Homework Practice Online
- *ARG World* CD–ROM: Comparing Regions in Indonesia
- *ARG World* CD–ROM: Export Processing Zones in Indonesia
- HRW Go site

Reinforcement, Review, and Assessment

- Section 4 Review, p. 601
- Daily Quiz 26.4
- Main Idea Activity 26.4
- English Audio Summary 26.4
- Spanish Audio Summary 26.4

Visual Record Answer

good economy, modern western influence

TEACH

Teaching Objective 1

LEVEL 1: (Suggested time: 20 min.) Copy the following graphic organizer onto the chalkboard, omitting the italicized answers. Use it to help students identify the major cities and list at least one important feature of each city. Ask for volunteers to fill in answers for the organizer on the chalkboard. **ENGLISH LANGUAGE LEARNERS**

MAJOR CITY	IMPORTANT FEATURE
Jakarta	*largest city, many kampongs*
Singapore	*modern, clean, low crime rate, strict laws*
Manila	*capital of Philippines, major seaport*
Kuala Lumpur	*Malaysia's capital, cultural and business center*

LEVELS 2 AND 3: (Suggested time: 45 min.) Singapore is a colonial city. The other major cities also have been affected by colonialism. Organize students into four groups. Assign each of the groups Singapore, Jakarta, Manila, or Kuala Lumpur. Have groups conduct research on how colonizers affected the assigned city's development. Then have groups conduct a panel discussion to compare their results. **COOPERATIVE LEARNING**

DAILY LIFE

Singapore's very strict government is devoted to preserving order. Visitors must follow the rules carefully, or they will be punished. Some punishable offenses include driving without a seatbelt, smuggling chewing gum into the country, and not flushing toilets in public buildings. In addition to paying a fine, people caught littering must undergo counseling.

Because Singapore's government operates much like an efficiently run corporation, the country has been referred to as a nation-corporation. For the tourist willing to follow the rules, Singapore offers a wide range of attractions, from lively festivals and parades to a nighttime cable-car ride down a mountain. There are also many dining options. The city's low crime rate helps make Singapore a popular tourist destination.

Critical Thinking: How does Singapore's government compare to that of the United States?

Answer: It is much stricter about law breaking.

Connecting to Art

Answers

1. in ancient societies, including Greece, Rome, China, and India
2. lasts all night, is accompanied by the traditional music of Java

CONNECTING TO Art

Shadow Puppets

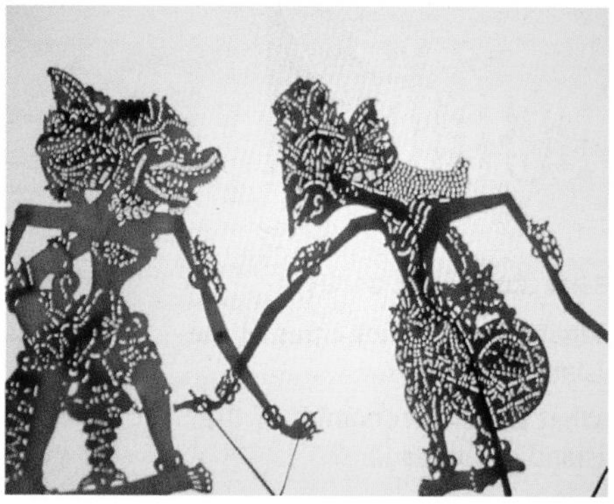

Wayang *puppets*

Puppetry is an art form with roots in ancient civilizations, including Greece, Rome, China, and India. On the Indonesian island of Java, one of the world's great puppet traditions is known as *wayang*. This shadow puppet theater still entertains audiences.

Wayang puppets are beautiful works of art. The puppets are made from thin sheets of painted leather. They are pierced with holes so that light can shine through them. Then they are mounted on sticks. The performance takes place behind a screen. A light source is placed behind the puppets. The puppet's shadows fall on the screen and are visible to the audience on the other side.

Wayang performances tell stories from the *Ramayana* and the *Mahabharata*. These are two long poems of the Hindu religion. Hinduism came to Java from India hundreds of years ago. The puppets play the parts of gods, heroes, and villains. A performance usually lasts all night and includes the traditional music of Java. The music is played by an orchestra that includes gongs and other traditional instruments.

Over the years, *wayang* artists have developed other types of puppets. Some puppets are wooden forms. A new generation of artists is even creating computerized stories for *wayang* theater.

Understanding What You Read

1. Where did puppetry originate?
2. What is a *wayang* performance like?

Our Amazing Planet

In 1883 a huge volcanic eruption on the Indonesian island of Krakatau killed thousands. Ash in the atmosphere colored sunsets around the world for months.

Jakarta The region's largest city is Jakarta, Indonesia's capital. More than 11 million people live there. It is located on Java, which is by far Indonesia's most populous island. Many Indonesians live in **kampongs** around Jakarta. A kampong is a traditional village. It has also become the term for the crowded slums around large cities.

Singapore If you traveled from Jakarta to Singapore, you would find two very different cities. Singapore is one of the most modern and cleanest cities in the world. Crime rates also are very low. How has Singapore accomplished this?

Its government is very strict. For example, fines for littering are stiff. People caught transporting illegal drugs can be executed. The government even bans chewing gum and certain movies and music.

Teaching Objective 2

LEVEL 1: (Suggested time: 30 min.) Organize the class into five groups and assign each group one of the following countries: Indonesia, the Philippines, Singapore, Malaysia, or Brunei. Have each group prepare a list of facts about the economy of its country. When complete, have students compare the economic organization of the nations.
ENGLISH LANGUAGE LEARNERS, COOPERATIVE LEARNING

LEVELS 2 AND 3: (Suggested time: 30 min.) Organize the class into groups of five and assign each group one of the countries in this section. Have each group categorize each fact compiled in the Level 1 lesson that pertains to its assigned country as positive or negative. Then have each group write a public service announcement accentuating positive aspects and downplaying negative circumstances of its country's economic situation. Have groups present their announcement to the class, then discuss the techniques each group used to make its country's economy seem bright.
COOPERATIVE LEARNING

Is the lack of some individual freedoms a good trade-off for less crime, a clean city, and a strong economy? Some people in Singapore say yes. Others believe Singapore can be just as successful with less government control.

Other Cities The region's other large cities include Manila and Kuala Lumpur. More than 10 million people live in Manila, the capital of the Philippines. The city is a major seaport and industrial center on Luzon. Luzon is the country's largest and most populated island.

Kuala Lumpur is Malaysia's capital as well as its cultural, business, and transportation center. It is a modern city with two of the world's tallest buildings, the twin Petronas Towers.

✓ **READING CHECK:** *Places and Regions* What are some of the major cities of the island countries?
Jakarta, Singapore, Manila, Kuala Lumpur

Island Southeast Asia

Country	Population/ Growth Rate	Life Expectancy	Literacy Rate	Per Capita GDP
Brunei	343,653 / 2.1%	71, male; 76, female	88%	$17,600
Indonesia	228,437,870 / 1.6%	66, male; 71, female	84%	$2,930
Malaysia	22,229,040 / 2.4%	68, male; 74, female	84%	$7,900
Philippines	82,841,518 / 1.7%	65, male; 71, female	95%	$3,600
Singapore	4,300,419 / 1.7%	77, male; 83, female	94%	$26,500
United States	281,421,906 / 0.9%	74, male; 80, female	97%	$36,200

Sources: Central Intelligence Agency, *The World Factbook 2001*; U.S. Census Bureau

Interpreting the Chart **According to the chart, which country's economic development is closest to that of the United States?**

Economy

The economies of the island countries grew rapidly until the mid-1990s. Then economic and political problems slowed growth for a while. However, rich resources are helping the economies there to recover. In addition, wages and labor costs are low in many of the countries. This means that companies there can manufacture many products more cheaply for export.

Indonesia Europeans once called Indonesia the Spice Islands because of its cinnamon, pepper, and nutmeg. Today, its rich resources include natural rubber, oil, natural gas, and timber. Indonesia also has good farmlands for rice and other crops. Busy factories turn out clothing, electronics, and furniture. Some islands, such as Bali, are popular with tourists.

Farming is an important economic activity in the Philippines. In mountain areas, farmers plant rice and other crops in terraced fields. These flat terraces hold water and slow erosion along mountainsides.
▼

Environment and Society

Found on the island of Luzon in the Philippines, the spectacular Banaue Rice Terraces have been called the eighth human-made wonder of the world. Begun an estimated 2,000 years ago by the Ifugao people, the terraces cover whole mountainsides, from riverbed to summit. The terraces make it possible to farm in an area where there is no level land.

Unfortunately, the rice terraces are deteriorating. They require regular maintenance, but because many people have moved from the area to the cities to look for jobs the terraces are being neglected. Also, some of the mountain springs that feed water to the terraces have run dry due to illegal logging. As a result, erosion has become a problem.

The Philippine government has declared Banaue a national park. Now the local people can apply for government funds to maintain the terraces.

Activity: Have students write a paragraph explaining how humans have modified the environment with the Banaue Rice Terraces.

▲ **Chart Answer**

Singapore

➤ASSIGNMENT: Have students prepare for an interview for the position of economic adviser to one of the countries in the Southeast Asian islands. Students should research and describe how the factors of production have influenced the country's economic history and recommend steps for improving the economy. Have students write notes to use for interviews in which they demonstrate their expertise.

TEACHER TO TEACHER

Sandra Rojas of Commerce City, Colorado, suggests this activity to help students learn about the islands of Southeast Asia. Have each student conduct research as if he or she were planning a trip around the Southeast Asian islands. Explain to students that they should plan as if they were going to stay with local people. They should, therefore, study local customs. Students should prepare a presentation for the class in which they show their intended route on a map and display drawings and pictures that show the sights they intend to see. They should also describe how they might adjust their ways of dressing, eating, or socializing in order to fit in with their host families.

Section 4 Review

Answers

Define For definitions, see: kampongs, p. 598; sultan, p. 601

Working with Sketch Maps Java and Luzon are the most heavily populated islands.

Reading for the Main Idea

1. Jakarta; capital cities, centers of industry, commerce, and transportation (NGS 4)
2. because of the many spices there; the Netherlands (NGS 4)

Critical Thinking

3. Answers will vary, but students should demonstrate knowledge of the government's strict policies.
4. Singapore—located on major shipping route, home to many foreign companies; Brunei—oil companies

Organizing What You Know

5. Cambodia, Laos, Vietnam—France; Brunei, Malaysia, Myanmar, Singapore—Britain; Indonesia—the Netherlands (East Timor—Portugal); Philippines—Spain, then United States; Thailand—independent

Visual Record Answer ▶

to prevent dust particles from contaminating machinery or products

▲ Bali dancers are popular tourist attractions in Indonesia, a country with many different dance styles. This *Barong* dancer uses her hands, arms, and eyes to help tell a traditional story.

Large areas of Indonesia's tropical rain forests are often burned for farming. Smoke from the fires can spread for hundreds of miles. The smoke sometimes blots out sunlight, smothering cities in haze.

The Philippines The Philippines is mostly an agricultural country today. A big problem is the gap between rich and poor Filipinos. A few very rich Filipinos control most of the land and industries. Most farmers are poor and own no land.

The economy has improved in recent years. Companies sell many electronics and clothing products to overseas customers, particularly in the United States. The country also has rich resources, including tropical forests, copper, gold, silver, and oil. Farmers grow sugarcane, rice, corn, coconuts, and tropical fruits.

Singapore Singapore is by far the most economically developed country in all of Southeast Asia. The British founded the city at the tip of the Malay Peninsula in 1819. This location along major shipping routes helped make Singapore rich. Goods are stored there before they are shipped to their final stop. In addition, many foreign companies have opened banks, offices, and high-technology industries there.

Malaysia Malaysia is made up of two parts. The largest part lies on the southern Malay Peninsula. The second part lies on the northern portion of Borneo. Well-educated workers and rich resources make Malaysia's future look bright. The country produces natural rubber, electronics, automobiles, oil, and timber. The government is trying to attract more high-technology companies to the country. Malaysia is also the world's leading producer of palm oil.

▶ Electronics and technology products are increasingly important to the region's economies. This Filipino is working at a semiconductor manufacturing plant.

Interpreting the Visual Record **Why do you think the person in the photograph is dressed this way?**

CLOSE

Ask each student to write down one reason why it is important to know about each of the island groups of Southeast Asia. Call on students to share their responses. Compile students' answers on the chalkboard.

REVIEW AND ASSESS

Have students complete the Section Review. Then have students examine the illustrations in the section. Ask students to write a question about each illustration. Collect the questions and use them to lead a class discussion about the section. Then have students complete Daily Quiz 26.4.

RETEACH

Have students complete Main Idea Activity 26.4. Then have students draw and label pictures to illustrate important facts about the people, cities, and economies of the Southeast Asian islands.
ENGLISH LANGUAGE LEARNERS

EXTEND

Focus students' attention on the photo of the *Barong* dancer in this section. Have interested students conduct research on traditional music and dance styles of Bali and how they relate to Balinese society. Ask students to learn a basic dance using hand, arm, and eye movements and demonstrate it to the class. **BLOCK SCHEDULING**

▲ This beautiful mosque in Brunei's capital reflects the country's oil wealth. Money from oil and natural gas production pays for many social services there.

Brunei Large deposits of oil have made Brunei rich. This small country on the island of Borneo is ruled by a **sultan**. A sultan is the supreme ruler of a Muslim country. Brunei shares the island of Borneo with Indonesia and Malaysia.

✓ **READING CHECK:** *Places and Regions* How do rich resources affect the economies of the island countries? They are helping countries regain the prosperity of the mid-1990s.

Section 4 Review

go.hrw.com Homework Practice Online
Keyword: SK3 HP26

Define and explain: kampongs, sultan

Working with Sketch Maps On the map you created in Section 3, label Jakarta, Manila, Kuala Lumpur, Luzon, and Bali. In a box in the margin, identify the most heavily populated islands in Indonesia and the Philippines.

Reading for the Main Idea

1. *Places and Regions* What is the region's largest city? What is the importance of Manila and Kuala Lumpur to their countries?
2. *Places and Regions* Why was Indonesia once called the Spice Islands? What European country once controlled nearly all of Indonesia?

Critical Thinking

3. **Analyzing Information** Some Singaporeans say that limiting some individual freedoms is a good trade-off for less crime and a better economy. Do you agree? Why or why not?
4. **Comparing/Contrasting** How have Singapore and Brunei become rich countries?

Organizing What You Know

5. **Summarizing** Copy the following graphic organizer. Use it to list the nine Southeast Asian countries that are former European colonies. Next to each country's name, write the name of the European country that once controlled it.

Southeast Asian country	European colonial power

CHAPTER 26
Review ANSWERS

Building Vocabulary

For definitions, see: mainland, p. 588; archipelagos, p. 588; refugees, p. 591; *klongs,* p. 595; kampongs, p. 598; sultan, p. 601

Reviewing the Main Ideas

1. Indochina and Malay Peninsulas, Malay Archipelago, and the Philippines (NGS 4)
2. northern Myanmar (NGS 4)
3. war, undemocratic governments, political problems (NGS 4)
4. forests, farmland, oil, gas, various minerals and metals (NGS 4)
5. through their colonies (NGS 9)

Understanding Environment and Society

Students' presentations will vary, but students should include information describing how rice is grown, what growing conditions are necessary for rice cultivation, and where rice is grown in Southeast Asia. Use Rubric 29, Presentations, to evaluate student work.

CASE STUDY

Setting the Scene

Indonesia is a good example of how physical and cultural factors can affect ethnic diversity. Because Indonesia is comprised of many different islands, physical isolation has resulted in many different ethnic groups. Of the more than 17,000 islands that make up Indonesia, at least 6,000 are inhabited. Cultural factors, such as language and a shared colonial history, have a unifying rather than a fragmenting effect. These factors are used by the central government to promote a national identity. The ongoing turmoil in East Timor, which erupted into violence in 1999, illustrates the difficulties the central government faces.

Building a Case

Have students read "Multiethnic Indonesia." Then conduct a classroom discussion about the multiethnic character of Indonesia. Use the following questions to guide the discussion: How many different ethnic groups are there in Indonesia? *(more than 300)* What are some of the larger ethnic groups? *(Javanese, Sundanese, Madurese, and Coastal Malays)* Now look at a detailed map of Indonesia. Show students that each of the ethnic groups listed in the text corresponds to an island *(e.g. Javanese: Java; Balinese: Bali, and so on)*. Point out that island isolation leads to the development of different ethnic groups over time. How many different islands are there in Indonesia? *(more than 17,000)* Suggest that many unrecognized groups might exist.

HISTORICAL GEOGRAPHY

As the colonial presence of the Dutch and Portuguese suggests, Indonesia has been a place where foreigners have mixed with the local people. Indonesians have terms to distinguish between indigenous people *(pribumi)* and outsiders *(keturunan)*. Before the Dutch and Portuguese arrived, Arab traders introduced Islam to the region. As a result, about 88 percent of the Indonesian people are Muslim. Hinduism and Buddhism were introduced even earlier.

More recently, an influx of Chinese immigrants has arrived, further complicating the ethnic mosaic of Indonesia. There may be as many as 6 million Chinese Indonesians. They are engaged heavily in commerce and are generally more prosperous than their fellow citizens. In 1998 this disparity in wealth led to some anti-Chinese demonstrations.

Activity: List in chronological order the arrival of each of the outside groups mentioned above. What enduring cultural elements did they contribute to Indonesia's culture?

➤ This Case Study Feature addresses National Geography Standards 5, 10, and 15.

CASE STUDY

MULTIETHNIC INDONESIA

Indonesia is a multiethnic country—a country with many different ethnic groups. The national motto of Indonesia is *Bhinneka Tunggal Ika*, which means "the many are one." This motto comes from the many different ethnic groups that live there.

More than 300 different ethnic groups live in Indonesia. Most of these groups speak their own language and have their own way of life. No single ethnic group holds a majority. The largest are the Javanese, Sundanese, Madurese, and Coastal Malays. The country also has many smaller ethnic groups, such as the Dayaks and the Balinese. Why does Indonesia have so many different ethnic groups? Part of the answer lies in the country's diverse physical geography.

Indonesia is a very large country. It is made up of more than 17,000 islands. About 228 million people live on these islands. Indonesia's islands, mountains, and dense rain forests have served as boundaries between different ethnic groups. Many small ethnic groups lived in isolation and had very little contact with other peoples. Over time, these groups developed their own cultures, languages, and ways of life.

The modern country of Indonesia has its roots in the early 1600s. About this time, Dutch traders built forts in the area. They wanted to protect the trade routes used by Dutch ships to transport spices and other goods. The Dutch remained an important force in the region until Indonesia became independent in 1949. The long history of

Most Indonesians are related to the peoples of East Asia. However, in the eastern islands, most people are of Melanesian origin. Over the centuries, many Arabs, Indians, and Europeans have added to the country's ethnic diversity.

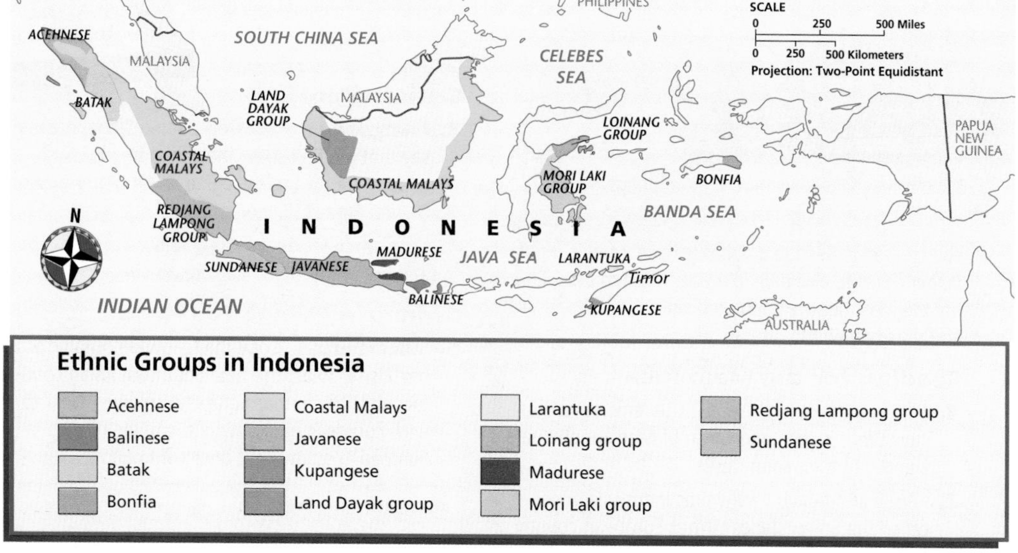

Drawing Conclusions

Shift the discussion to focus on how such a diverse nation binds itself together. Highlight the role of the government in uniting the mosaic of peoples. What has the government done to foster a national identity in Indonesia? *(It has promoted a common language, common education system, and national celebrations.)* Point out as well that it has attempted to block regional autonomy. How has a shared history helped create an Indonesian identity? *(There is a long history of Dutch control of Indonesia dating back to the 1600s).* Point out that the island of Timor was settled and controlled by the Portuguese—not the Dutch. Suggest that having had a different colonizer contributed to a desire on the part of the Timorese for independence.

Going Further: Thinking Critically

The Pacific has many islands spread out over a large area. Organize the students into small groups and provide each group with an atlas. Have students answer the following questions:

- How many islands belong to France? the United States? What other countries have possessions in the Pacific?
- How many independent countries can you find?
- If you were drawing national boundaries based only on the groupings of islands regardless of colonial history, how might the national boundaries be different?
- How might the colonizing nations and the physical geography have influenced the number of different ethnic groups in the Pacific?

The Balinese are one of Indonesia's many ethnic groups. Unlike most other Indonesians, Hinduism is their main religion.

Dutch control helped unify the islands into the modern country of Indonesia.

In addition to this shared past, several other factors have helped unify Indonesia. For example, Indonesia's government has promoted the country's official language, Bahasa Indonesia. Although most Indonesians speak more than one language, Bahasa Indonesia is used in schools and government. The use of this language has been an important force in uniting the country. The government has also tried to develop a common Indonesian culture. It has promoted national holiday celebrations, education, popular art, and television and radio programs.

A shared history, a common education system, and an official language help give isolated ethnic groups an Indonesian identity. However, the country's multiethnic society still faces some important challenges. In certain parts of Indonesia, people want independence. For example, in 1999 people in East Timor voted for independence from Indonesia. This caused a great deal of unrest. Most people there supported the vote, but others did not. When some groups rioted, many Timorese left the area for their own safety. The Acehnese, an ethnic group on the island of Sumatra, have also been seeking independence.

Ethnic Groups in Indonesia

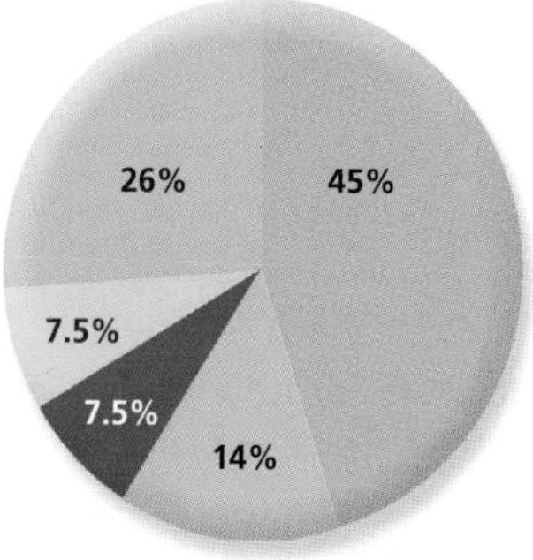

Source: Central Intelligence Agency, *The World Factbook 2001*

Understanding What You Read

1. How diverse is Indonesia's society? What has helped to promote cooperation among the different ethnic groups that live in the country?
2. What are some important challenges facing Indonesia today?

Understanding What You Read

1. Indonesia has more than 300 ethnic groups with different languages and ways of life; a shared past, an official language, a common culture promoted by the government
2. The government faces the challenge of uniting the country by promoting a common language and culture. Independence movements threaten to pull Indonesia apart.

ASSESS

Have students complete the Chapter 26 Test.

RETEACH

Have students work in pairs to write two- to three-sentence summaries of each section. Then have them compile the summaries to form a chapter summary. Encourage students to use their summaries to help them study the chapter. **ENGLISH LANGUAGE LEARNERS, COOPERATIVE LEARNING**

PORTFOLIO EXTENSIONS

1. Have students research a Buddhist myth, and then make a Thai mask to represent one of the myth's characters. Have students present their masks to the class. Photograph student work for portfolios.
2. In Indonesia, puppet theater is often used to convey ethical messages and commentary. Have students make puppets and present a skit about an ethical dilemma. Place a copy of the script and a photograph of the puppet in students' portfolios.

CHAPTER 26 Review ANSWERS

Thinking Critically

1. located near major rice-growing areas, good locations for shipping goods
2. European, Chinese, and Indian
3. Philippines—United States granted independence in 1946; Vietnam—long war that ended in 1954 pushed French out
4. mainland—wet and dry seasons; islands—dry throughout year; typhoons, earthquakes, volcanic eruptions
5. Labor costs are low, keeping manufacturing costs low and exporting less expensive.

Map Activity

A. Indochina Peninsula
B. Mekong River
C. Malay Peninsula
D. Borneo
E. Manila
F. Malay Archipelago and the Philippines
G. Timor
H. Bangkok
I. Jakarta
J. Irian Jaya

CHAPTER 26 Reviewing What You Know

Building Vocabulary

On a separate sheet of paper, write sentences to define each of the following words.

1. mainland
2. archipelagos
3. refugees
4. *klongs*
5. kampongs
6. sultan

Reviewing the Main Ideas

1. *Places and Regions* What peninsulas and archipelagos make up Southeast Asia?
2. *Places and Regions* Where will you find the highest mountains in Southeast Asia?
3. *Places and Regions* What factors have slowed economic progress in mainland Southeast Asia?
4. *Places and Regions* What resources are important to the economies of the region?
5. *Human Systems* How did European countries influence Southeast Asia?

Understanding Environment and Society

Rice Farming

Rice is the main crop and most common food in Southeast Asia. Develop a presentation about rice farming. In developing your presentation, consider the following:

- How rice is grown.
- What growing conditions are needed.
- Where rice is grown in Southeast Asia.

Thinking Critically

1. **Drawing Inferences and Conclusions** Why are some cities located near river deltas so large?
2. **Finding the Main Idea** What outside cultures have strongly influenced the development of Southeast Asian culture?
3. **Analyzing Information** How did the Philippines and Vietnam gain independence?
4. **Finding the Main Idea** How do the climates of mainland countries differ from those in the island countries? What natural disasters are a danger in the region?
5. **Analyzing Information** How do labor costs affect the island countries' economies?

Food Festival

Cha Thai (Thai Iced Tea)

Here is a refreshing version of iced tea from Thailand. Place 8 black decaffeinated tea bags in a teapot. Pour 4 cups boiling water over the tea. Let steep for 5 minutes. Fill four 10 oz. glasses with ice cubes; then fill two thirds full with tea. Into each glass, stir 2 tbs evaporated milk and a dash of ground cinnamon. Sweeten to taste. Note that this recipe serves four. Be prepared to increase the recipe's proportions to serve your entire class.

Chapter 26 Review and Assessment Resources

Reproducible
- Readings in World Geography, History, and Culture 72, 73, and 74
- Critical Thinking Activity 26
- Vocabulary Activity 26

Technology
- Chapter 26 Test Generator (on the One-Stop Planner)
- HRW Go site
- Audio CD Program, Chapter 26

Reinforcement, Review, and Assessment
- Chapter Review, pp. 604–05
- Chapter 26 Tutorial for Students, Parents, Mentors, and Peers
- Chapter 26 Test
- Chapter 26 Test for English Language Learners and Special-Needs Students
- Unit 7 Test
- Unit 7 Test for English Language Learners and Special-Needs Students

Building Social Studies Skills

Map Activity

On a separate sheet of paper, match the letters on the map with their correct labels.

Indochina Peninsula	Mekong River
Malay Peninsula	Timor
Malay Archipelago and the Philippines	Bangkok
Irian Jaya	Jakarta
Borneo	Manila

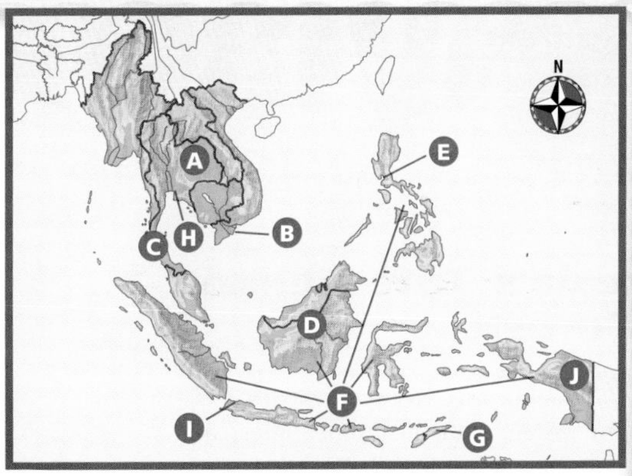

Mental Mapping Skills Activity

On a separate sheet of paper, draw a freehand map of Southeast Asia. Make a key for your map and label the following:

Andaman Sea	Philippines
Hanoi	Singapore
Java	Sumatra
Kuala Lumpur	Yangon

Writing Activity

Imagine that you are an economic adviser for a poor Southeast Asian country. Write a one-paragraph summary explaining how some countries in the region built stronger economies. Use the report to suggest policies your chosen country might adopt to build its economy. Be sure to use standard grammar, spelling, sentence structure, and punctuation.

Alternative Assessment

Portfolio Activity

Learning About Your Local Geography

Individual Project Find out how your local community was affected by the Vietnam War. Interview community members, locate and read old newspapers or firsthand accounts, and present an oral report.

internet connect

Internet Activity: go.hrw.com
KEYWORD: SK3 GT26

Choose a topic to explore about Southeast Asia:
- Explore an Indonesian rain forest.
- Learn about shadow puppets.
- See buildings of Southeast Asia.

Mental Mapping Skills Activity

Maps will vary, but the listed places should be labeled in their approximate locations.

Writing Activity

Check to see that students have included accurate information; most students will have focused on the strategies used to achieve Singapore's and Thailand's economic success. Details should be consistent with text information. Use Rubric 43, Writing to Persuade, to evaluate student work.

Portfolio Activity

Students' presentations will vary, but the information included should address the community's involvement in and feelings about the war. Use Rubric 24, Oral Presentations, to evaluate student work.

internet connect

GO TO: go.hrw.com
KEYWORD: SK3 Teacher
FOR: a guide to using the Internet in your classroom

Focus On Environment

Linking Local and Global Perspectives

Have students recall the problems caused by human interaction with the environment that were discussed in this Focus On Environment feature. Ask them to identify measures people in the region are taking to resolve these issues and reduce the effects of human activity on the environment. *(People in Indonesia are trying to protect the rain forests by setting up parks and nature reserves that are off-limits to logging companies. Some groups are finding ways to earn a living without cutting down trees.)* Then have volunteers identify international measures that are being taken to resolve these issues. *(Environmental organizations are pressuring wealthy nations to stop buying wood that comes from tropical rain forests.)* Lead a discussion to address the following questions: Is it important for issues such as deforestation to be resolved at both local and global levels? Why or why not? What types of issues are best addressed locally? Which are best addressed on the global level?

Geography Sidelight

The islands of Indonesia are home to a vast variety of wildlife. Many of these species are endemic, or found nowhere else on Earth. For example, Komodo dragons, which can reach about 12 feet (4 m) in length, are found only on the Indonesian islands of Komodo and Rinca. To help protect these rare creatures, Indonesia has established a system of nature preserves and parks. For example, the Kulon Peninsula National Park on Java is home to the endangered Sumatran rhinoceros.

Indonesia's Mount Leuser National Park and Tanjung Puting Park are home to another endangered animal, the orangutan. Once found throughout Southeast Asia, the orangutan is now found only on the islands of Borneo and Sumatra. Indonesia's efforts to establish even more nature preserves are important steps toward protecting the country's many rare and endangered species.

Critical Thinking: How has Indonesia worked to protect its endangered species?

Answer: by establishing national parks and preserves

➤ **This Focus On Environment feature addresses National Geography Standards 4, 8, 14, and 18.**

Focus On Environment

Indonesia's Threatened Rain Forests

Why Are Rain Forests Important? Do you know where bananas, pineapples, and oranges originally came from? Each of these plants first grew in a tropical rain forest.

Tropical rain forests are considered by many to be the most important forests in the world. It is estimated that about half of all species of plants, animals, and insects on Earth live in tropical rain forests. Rain forest trees and plants help maintain global temperatures. They also help hold rain-drenched soil in place. This prevents it from washing away and clogging rivers. About one fourth of all medicines currently found in drugstores come from tropical rain forests.

Tropical rain forests can be found in many countries along the equator and between about 20° north and south latitude. One of the largest rain forests is in Brazil's Amazon Basin. In Africa, rain forests are found in many countries, such as Gabon and the Democratic Republic of the Congo. In Southeast Asia, rain forests are found in countries such as Thailand, Vietnam, and Indonesia.

▲ Found in the rain forests of Sumatra, the *Rafflesia arnoldii* is the largest known flower in the world. It can weigh up to 24 pounds (11 kg) and can measure about 3 feet (1 meter) across.

Deforestation in Indonesia Indonesia has large areas of tropical rain forest in Borneo, Sumatra, and Irian Jaya. These areas are home to many unusual species of plants and animals. For example, the largest flower in the world, the *Rafflesia arnoldii*, is found there.

Indonesia's tropical rain forests are being rapidly cut down. About 4,700 square miles (12,170 sq km) of rain forest are lost each year. This rate of deforestation is second only to Brazil's. Some people have predicted that much of Indonesia's rain forests will be gone in just 10 years. When the rain forests are cleared, animals such as the endangered orangutan do not have a home.

There are many reasons that Indonesia's tropical rain forests are being cleared. Trees from tropical rain forests produce beautiful woods. They are used to make furniture, boats, and houses. The demand for special trees and wood has made logging a profitable business. Much of

Going Further: Thinking Critically

Have students conduct research on another Indonesian environmental story, this one with a more positive conclusion. Many Indonesian farmers have found that heavy pesticide use is much more harmful than helpful and can actually result in increased numbers of harmful insects. Through the Integrated Pest Management (IPM) program they have learned to inspect their fields carefully and use pesticides only when absolutely necessary. Crop yields have increased. Ask students to investigate Indonesia's IPM program. Then have them create publicity materials that IPM promoters could use to begin grassroots efforts to save the country's rain forests. Have students highlight techniques and arguments that have been effective against heavy pesticide use. Ask them to determine which ones might also be effective in the struggle to save the rain forests.

the logging is done by large corporations that do not replant the areas that are cut.

Deforestation in Indonesia also occurs because people need land to farm and raise animals. They also need wood for fuel. People clear the land using a method called slash-and-burn. Large trees are cut, or slashed, and left on the ground. Then the land is burned during the dry season. This clears the land of vegetation and prepares it for farming. In 1997 large areas of land in Indonesia were cleared. Huge fires burned out of control. The fires burned an area roughly the size of Denmark. Smoke filled the sky and caused some airplanes and ships to crash.

Protecting Indonesia's Rain Forests

Some people in Indonesia are trying to protect the rain forests. Parks and nature reserves have been set up that are off-limits to logging companies. Some groups are finding ways to earn money without cutting down trees. Selling fruits and nuts from the rain forest is one way. Also, international organizations such as the Rainforest Action Network are helping to protect the forests. Some environmental groups are even pressuring countries to stop buying trees that come from tropical rain forests.

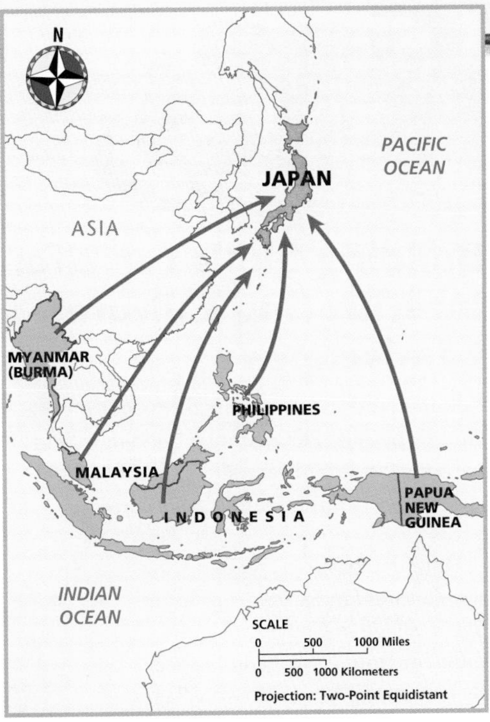

▲ Much of the timber cut in Indonesia and other Southeast Asian countries is exported to Japan. Indonesia exports about 2 million tons of plywood and 145,000 tons of lumber to Japan each year.

◀ Orangutans live in the tropical rain forests of Borneo and Sumatra. Deforestation has seriously reduced their habitat. The word *orangutan* means "man of the forest."

Understanding What You Read

1. Why are Indonesia's tropical rain forests being cut down?
2. What is being done to protect Indonesia's tropical rain forests?

Understanding What You Read

Answers

1. Indonesia's tropical rain forests are being cut down to bring profit to large logging corporations, to provide land to farm or raise cattle, and to provide wood for fuel.
2. People in Indonesia are trying to protect the rain forests by setting up parks and nature reserves that are off-limits to logging companies. Some groups are finding ways to earn a living without cutting down trees. Environmental organizations are pressuring wealthy nations to stop buying trees that come from tropical rain forests.

Going Further: Thinking Critically

Have students imagine that they are photojournalists assigned to portray their community's culture in a photo essay for a foreign magazine. Ask students to identify at least 10 elements in their community that they feel would accurately depict their culture to a person in another country. Have students write a few sentences about each element, explaining why it should be included in the photo essay. You may wish to extend the activity by having students create an actual photo essay with their own photographs or pictures from local publications.

PRACTICING THE SKILL

1. Students might suggest dams, highways, houses, movie theaters, office buildings, parking lots, shopping malls, streets, supermarkets, utility poles and wires, or other forms. Answers will vary according to the forms suggested.
2. Students should explain how the unique qualities of the television show or movie, such as architectural styles, are linked to the place it was filmed.
3. Students' answers should reflect the ability to read the cultural landscape in the pictures.

➤ **This GeoSkills feature addresses National Geography Standards 4, 6, and 10.**

GeoSKILLS

Building Skills for Life: Interpreting Cultural Landscapes

Cultural landscapes are the forms put on the land by people. For example, buildings, field patterns, and roads are all a part of cultural landscapes. Cultural landscapes show a people's way of life.

Different cultures create distinctive cultural landscapes. For example, a village in China looks very different from a village in France. By comparing how the two villages look, we can begin to see how their cultures are different.

Geographers interpret cultural landscapes. They observe a landscape, describe what they see, and try to explain how it reflects the culture of the place. This is called reading the cultural landscape.

You can read cultural landscapes too. To read a cultural landscape, start by describing what you see. What kinds of buildings are there? What are they used for? What kinds of clothing are people wearing? Then, think about how what you see relates to the place's culture. What would it be like to live there? What do people there do for fun?

Cultural landscapes tell a story. By reading and interpreting these stories, you can learn a lot about people and geography.

Architecture is an important part of the cultural landscape at the Black Dragon Pool in southern China.

PRACTICING THE SKILL

1. Try to read the cultural landscapes of your community. What forms have people put on the land? What do they tell you about the daily life of the people who live there?
2. Watch a television show or movie and interpret the cultural landscapes you see. What is distinctive about them? How are they different from the cultural landscapes you are used to? Can you guess where the program was filmed?
3. Look carefully at the pictures in a newspaper or magazine without reading the captions. Do the pictures tell a story? Is culture a part of this story?

HANDS ON GEOGRAPHY

The photographs below show two very different cultural landscapes. What can these photographs tell us about each place's culture and way of life? Look closely at each photograph and then answer the Lab Report questions.

A cultural landscape in East Asia

A cultural landscape in Southeast Asia

Lab Report

1. What do these two photographs show? On a separate sheet of paper, write a short description of each photograph.
2. Do these two photographs tell you something about each place's culture and way of life? On a separate sheet of paper, describe what you think the culture of these two places is like.
3. Are there some things about a place's culture that you cannot learn from just looking at a photograph? What are they? If you took a trip to these two places, what else could you learn about their cultures?

Lab Report

Answers

1. Students might note how the modern skyscrapers in the background contrast with the ancient style of the junk (boat) in the foreground. This is a photo of Hong Kong. Students should note the ornate architecture and the unusual roofline extensions in the photo from Southeast Asia. The photo depicts temporary housing for guests attending a funeral in a Toraja village in Indonesia.
2. Students might suggest that the East Asia photo depicts a modern urban culture, with some elements of older ways of life. The other photo depicts a more traditional culture.
3. Students might suggest that customs, foods, language, religion, or types of recreation are not apparent in the photos. They might learn about these things if they traveled to the places pictured.

UNIT 8

SOUTH ASIA

Using the Illustrations

Direct students' attention to the photos on these pages. Point out the Karakoram Range on a wall map. The Karakorams are part of a mountain system that includes the Himalayas.

Note the small photo of the Bengal tiger. You may want to direct students' attention to the Focus On Environment feature at the end of this unit. It describes efforts to save the dwindling tiger population of South Asia.

Point out the polo players at an elephant festival. Polo, a sport that involves hitting a ball with a long-handled mallet, is traditionally played on horseback. In South Asia there are indeed polo matches played while riding elephants.

The girls in the center photo are from Rajasthan. You may want to locate a map of India that shows the individual states and have students find Rajasthan in the northwestern part of the country.

Finally, point out Swayambhunath Stupa. One may avoid the tongue-twisting name by using its other name, the Monkey Temple, which refers to the many monkeys that live there.

Unit Objectives

1. Describe how landforms, climate, and water resources affect the way people live in South Asia.
2. Identify the natural resources of South Asia, and link them to the region's economic development.
3. Trace the history of South Asia, and identify the interactions through time of the region's cultural groups.
4. Analyze the social and environmental challenges facing the nations of South Asia.
5. Interpret special-purpose maps for insight into how the people of South Asia adapt to geographical differences within the region.

UNIT 8

South Asia

CHAPTER 27
India

CHAPTER 28
The Indian Perimeter

Swayambhunath Stupa, Kathmandu, Nepal

Karakoram Range, Pakistan

CHAPTERS

27 India
explores the geography and cultural diversity of the largest country in South Asia. It also describes the serious economic and environmental challenges facing India.

28 The Indian Perimeter
profiles the six countries that surround India: Pakistan, Nepal, Bhutan, Bangladesh, Sri Lanka, and the Maldives.

NOTES FROM THE field

A Scholar in India

Emily K. Bloch coordinates educational programs at the South Asia Outreach center at the University of Chicago. Her special area of interest is South Asian children's literature. Here she describes the many types of transportation she has used in India.* WHAT DO YOU THINK? *Which type of transportation would you enjoy the most?

In my travels throughout India, I've had the good fortune to ride on a variety of vehicles. I've ridden in buses and cycle-rickshaws, taxis and three-wheeled scooters, on a motorcycle through Calcutta and on the crossbar of a bicycle. I rode in a howdah [seat] on the back of an elephant in a wildlife sanctuary. I even had an uncomfortable, but very welcome, lift in a bullock cart.

But my favorite mode of travel is also one of the most popular in India—riding the great trains. The railway, with more than 1.5 million employees, is the largest employer in the world. It has nearly 40,000 miles of track. Eleven thousand trains, connecting more than 7,000 stations, carry about 12 million people daily. Though the luxury of the princely rail lines beckons, and the first-class cars provide food and bed-linens, I love the crowded, second-class, wooden-benches experience and look forward to my next journey side by side with my fellow Indian and foreign travelers.

Polo players at elephant festival, Jaipur, India

Women in festival dress, Rajasthan, India

Understanding Primary Sources

1. What do Emily Bloch's transportation experiences illustrate about the use of technology in India?
2. Which method of travel does Emily Bloch prefer? Why?

White Bengal tiger

MORE FROM THE FIELD

As a British colony, India was a land where more than 1,000 languages were spoken. Society was further divided by religion and the caste system.

Construction of the railroads, which began in the 1850s, helped unite India politically and economically. The railroads provided a way for India's agricultural products and natural resources to reach the port cities and thus to reach world markets.

Critical Thinking: How did railroads affect Indian economic development?

Answer: provided a way to get products to port cities and world markets

Understanding Primary Sources

Answers

1. The many types of transportation indicate that there is a wide range in the use of technology in India, and that economic factors probably determine what type of transportation people in India use.
2. riding the trains; seems to provide a closer relationship with the people of India

OVERVIEW

In this unit, students will learn about the geographic and cultural contrasts of South Asia, one of the most densely populated regions on Earth.

The Indian Subcontinent is separated from the rest of Asia by the highest mountains in the world. The Himalayas drop to the fertile Gangetic Plain. Plateau and desert lands lie to the south and west. Climate patterns are greatly influenced by monsoon winds that bring wet and dry seasons to much of the region.

India dominates the subcontinent. With more than 1 billion people, it is the world's most populous democracy. It is a land of stunning contrasts—from nuclear technology and a strong computer industry to subsistence farming and homeless city-dwellers. Ancient customs sometimes clash with modern trends. Conflicts between India and its neighbors over territory increase political tensions.

The surrounding countries also face the difficult tasks of feeding their growing populations and of developing their economies while protecting the environment.

PEOPLE IN THE PROFILE

Note that the elevation profile crosses the eastern part of India that is almost cut off from the rest of the country. Assam is one of the states in the area.

About two thirds of the state's working population is involved in agriculture. Festivals in Assam are important social and cultural events. Not surprisingly, the three chief nonreligious festivals, called Bihu, are related to agriculture. The Bohag Bihu—also known as the Rongali Bihu—takes place in April and celebrates the new year, the arrival of spring, and the beginning of the planting season. The Bohag Bihu lasts for several days. One of its main attractions is dancing.

The Magh Bihu in mid-January is a harvest festival. Feasting is its key activity. The third main festival is the Kangali Bihu, which is held in October. *Kangali* means "poor," and this festival occurs at a time of the year when most people have already consumed most of their stored food grains.

Critical Thinking: How does Assam's dependence on agriculture influence its celebrations?

Answer: Festivals are based on agriculture and celebrate it with dancing and feasting.

UNIT 8 ATLAS The World in Spatial Terms

South Asia

Elevation Profile

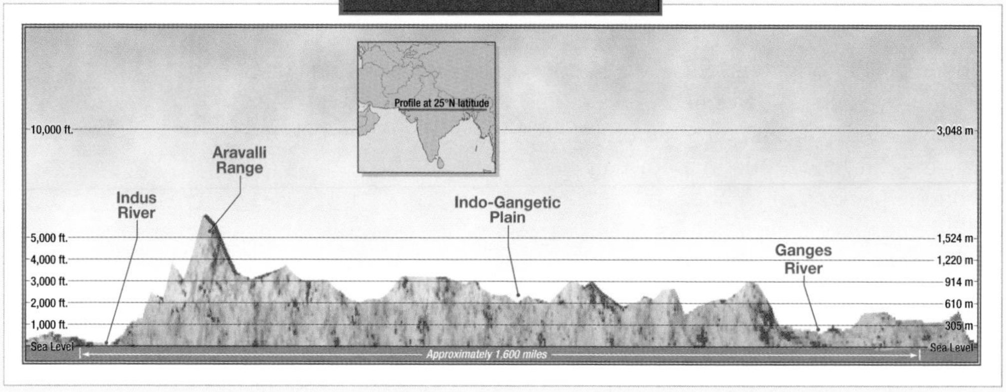

The United States and South Asia: Comparing Sizes

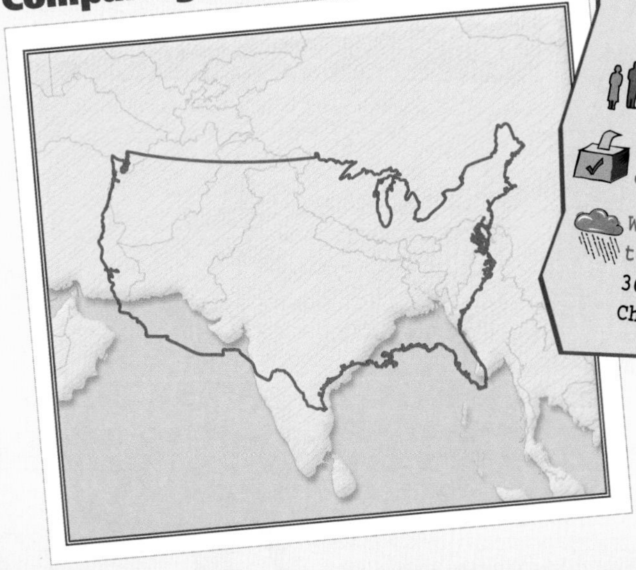

GEOSTATS:

India

World's second-largest population: 1,029,991,145 (July 2001 estimate)

Predicted population in 2025: 1,415,274,000

World's most populous democracy

World's greatest recorded total rainfall in one month: 366 in. (930 cm) in Cherrapunji, in July 1861

USING THE PHYSICAL MAP

Have students examine the **physical map** on this page and identify South Asia's three major landform regions. *(Students should name the Himalayas, the Gangetic Plain, and the Deccan Plateau.)* Then call on a student to describe the courses of the three main rivers that flow from the Himalayas—the Indus, the Ganges, and the Brahmaputra.

Call on a volunteer to identify the two mountain ranges that are part of the Himalayas *(Hindu Kush, Karakoram Range).*

South Asia: Physical

UNIT 8 ATLAS

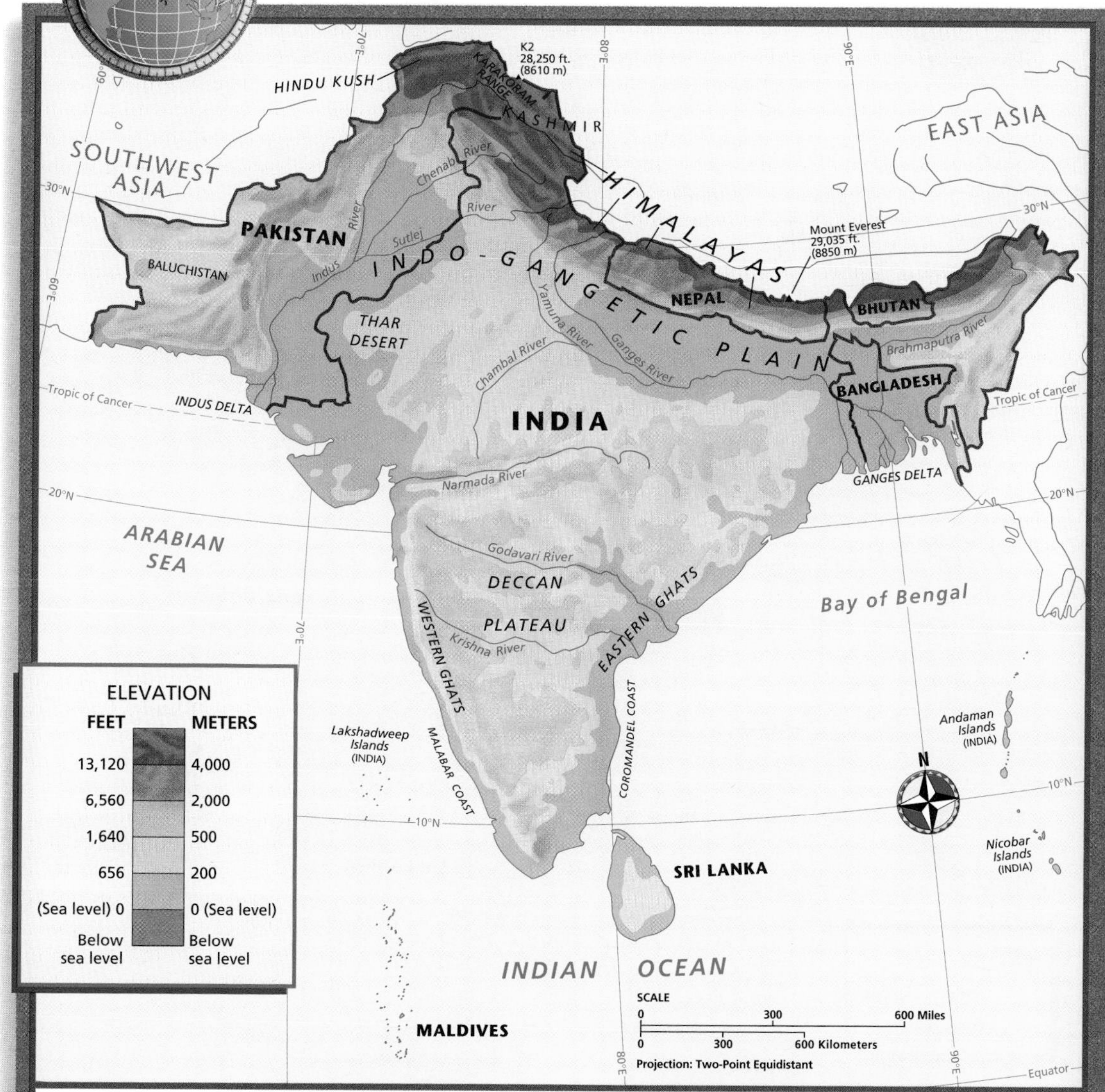

1. (Places and Regions) Which country has the lowest overall elevation? Which countries have mountains that stretch from their western to eastern borders?
2. (Places and Regions) Which two rivers that drain the Deccan Plateau flow directly into the Bay of Bengal?

Critical Thinking

3. **Drawing Inferences and Conclusions** Based on the map, do you think it would be easier for travelers or invaders to come to northern India by land or by sea? Why might this be the case?

internet connect

ONLINE ATLAS
GO TO: go.hrw.com
KEYWORD: SK3 MapsU8
FOR: Web links to online maps of the region

Physical Map

Answers

1. Bangladesh; Nepal, Bhutan
2. Godavari and Krishna Rivers

Critical Thinking

3. by sea; because they could more easily sail up the rivers than cross the mountain ranges

UNIT 8 ATLAS

Using the Political Map

While students examine the **political map** of South Asia on this page, ask what is unusual about the boundary of India. *(Students should note that a small part of the country connects with the remainder by a narrow strip of land.)* Call on a volunteer to suggest why part of the region known as Kashmir is in Pakistan and part is in India. *(The region is claimed by both countries.)*

internet connect
ONLINE ATLAS
GO TO: go.hrw.com
KEYWORD: SK3 MapsU8

Your Classroom Time Line

These are the major dates and time periods for this unit. Have students enter them on the time line you created earlier. You may want to watch for these dates as students progress through the unit.

- c. 2500 B.C. Civilization develops in the Indus River valley.
- c. 1750 B.C. Indo-Aryans move into northern India.
- c. 1500 B.C. The Indus civilization disappears.
- c. 500 B.C. Siddhartha Gautama establishes Buddhism.
- A.D. 1000 Muslim armies begin raiding northwestern India.

*c. stands for *circa* and means "about."

Political Map

Answers

1. Maldives and Sri Lanka; Sri Lanka
2. Bangladesh
3. Nepal and Bhutan

Critical Thinking

4. Students might suggest that the physical separation between the two territories may have fostered Bangladeshis' psychological sense of separation from Pakistan.

South Asia: Political

UNIT 8 ATLAS

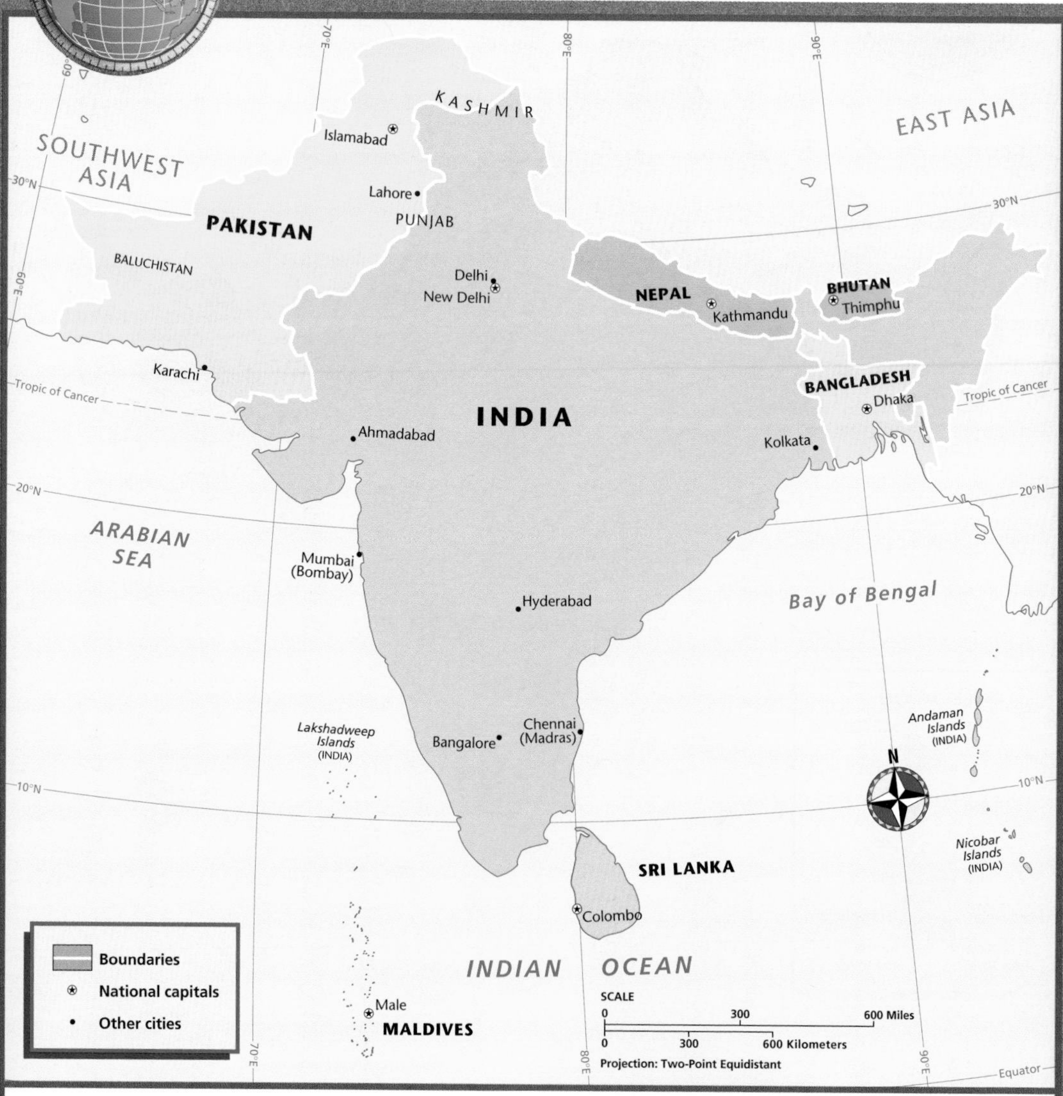

1. *Places and Regions* What are the region's two island countries? Which is larger?
2. *Places and Regions* Which country is almost completely surrounded by India?
3. *Places and Regions* Compare this map to the **physical map**. Which small countries might be called the "Mountain Kingdoms?"

Critical Thinking

4. **Drawing Inferences and Conclusions** Bangladesh was once part of Pakistan. What role do you think Bangladesh's location may have played in its drive for independence?

Using the Climate Map

Have students look closely at the **climate map** on this page and compare it to the **physical map** of the region. Ask students the following questions: What is the dominant climate in Pakistan? *(desert)* Which country in the region has mostly humid tropical and humid subtropical climates? *(Bangladesh)* Which island country has tropical savanna and humid tropical climates? *(Sri Lanka)* What low mountain range appears to catch much of the rain from the wet monsoon? *(Western Ghats)*

South Asia: Climate

UNIT 8 ATLAS

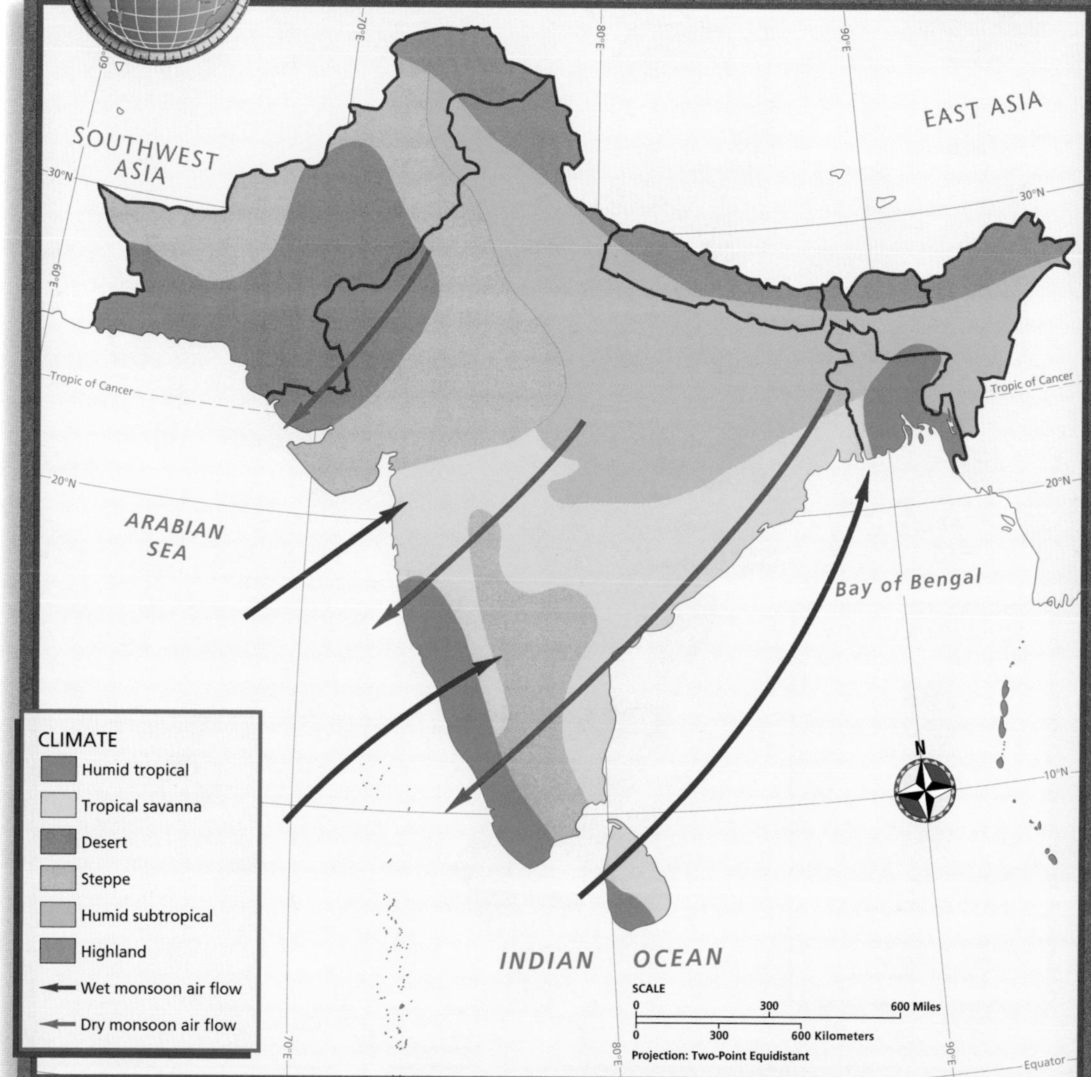

1. *Places and Regions* Compare this map to the **physical map.** Which landform in south central India has a steppe climate?
2. *Places and Regions* Compare this map to the **land use and resources map** of the region. Which climate regions have the least economic activity?

Critical Thinking

3. **Analyzing Information** Compare this map to the **political map** of the region. Other than those with mainly highland climates, which country appears to be least affected by monsoons?

Your Classroom Time Line, (continued)

- A.D. **1200s** A Muslim kingdom is established at Delhi.
- **1400s** Sikhism begins.
- **1505** The Portuguese establish a trading post in Sri Lanka.
- **early 1500s** The Mughal Empire comes to power.
- **1600s** The English East India Company wins rights to trade in the Mughal Empire.
- **1700s–1800s** The British slowly take control of India.
- **1700s** The British receive Sri Lanka from the Dutch.
- **1857** Indian sepoys revolt against the British.
- **1885** Nationalists create the Indian National Congress.
- **1947** The British divide their Indian colony into India and Pakistan.
- **1948** Sri Lanka gains its independence from Britain.
- **1953** Norkey Tenzing and Sir Edmund Hillary reach the summit of Mount Everest.

Climate Map

Answers

1. Deccan Plateau
2. highland and desert climates

Critical Thinking

3. Pakistan

UNIT 8 ATLAS

Using the Population Map

As students examine the **population map** on this page, ask them to write a general statement about the population pattern in South Asia. *(Example: South Asia is a densely populated region with many large cities.)*

Have students compare this map to the **physical map**. Call on a volunteer to identify the country with more than 520 people per square mile throughout most of its area. *(Bangladesh)* What desert area has more than 260 residents per square mile? *(a narrow strip of land in southern Pakistan)* What physical feature makes this dense population possible? *(the Indus River)*

internet connect

ONLINE ATLAS
GO TO: **go.hrw.com**
KEYWORD: **SK3 MapsU8**

Your Classroom Time Line, (continued)

- **1971** East Pakistan breaks away from West Pakistan, forming independent Bangladesh.
- **1984** The Indian government uses force to defeat Sikh rebels.

South Asia: Population

UNIT 8 ATLAS

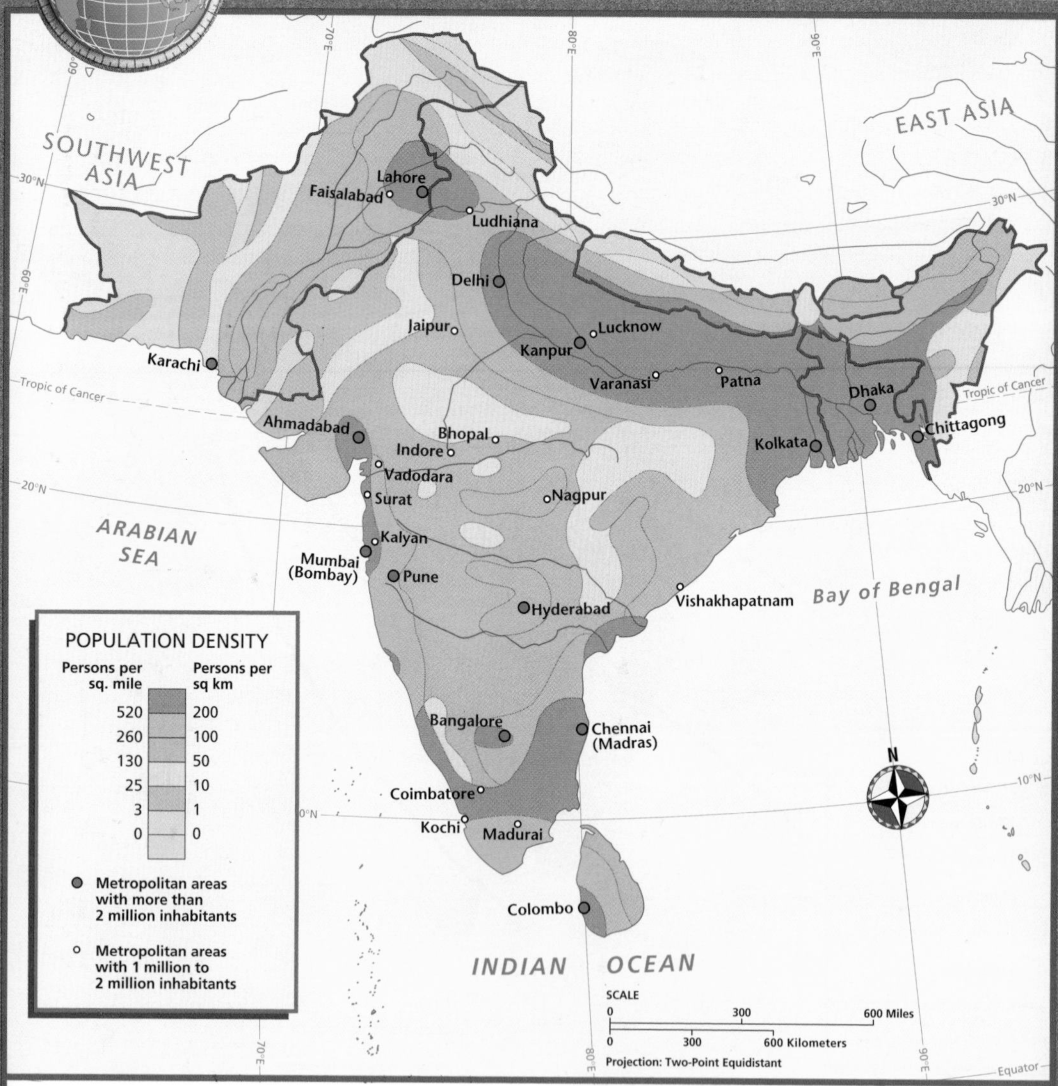

1. *Places and Regions* Compare this map to the **physical map**. Which large Indian city is located on the Ganges Delta?
2. *Places and Regions* Compare this map to the **physical map** of the region. What is the name of the large densely populated area in the northeastern part of the region?

Critical Thinking

3. **Analyzing Information** Use the map on this page to create a chart, graph, database, or model of population centers in South Asia.

Population Map

Answers

1. Kolkata
2. Indo-Gangetic Plain

Critical Thinking

3. Charts, graphs, databases, and models should accurately reflect population figures for cities on the map.

USING THE LAND USE AND RESOURCES MAP

Focus students' attention on the **land use and resources map** on this page. Then ask students to answer the following questions: Which types of land use are most widespread in the region? *(commercial farming, subsistence farming)* Where is India's largest manufacturing and trade region? *(on the Arabian Sea)* Which cities are in that region? *(Ahmadabad, Vadodara, Mumbai, and Pune)*

Have students compare this map to the **physical map** of South Asia. Call on a volunteer to identify where in India timber is harvested and which river may be used to transport the timber to manufacturing centers *(northern edge of the Deccan Plateau; Narmada River)*.

South Asia: Land Use and Resources

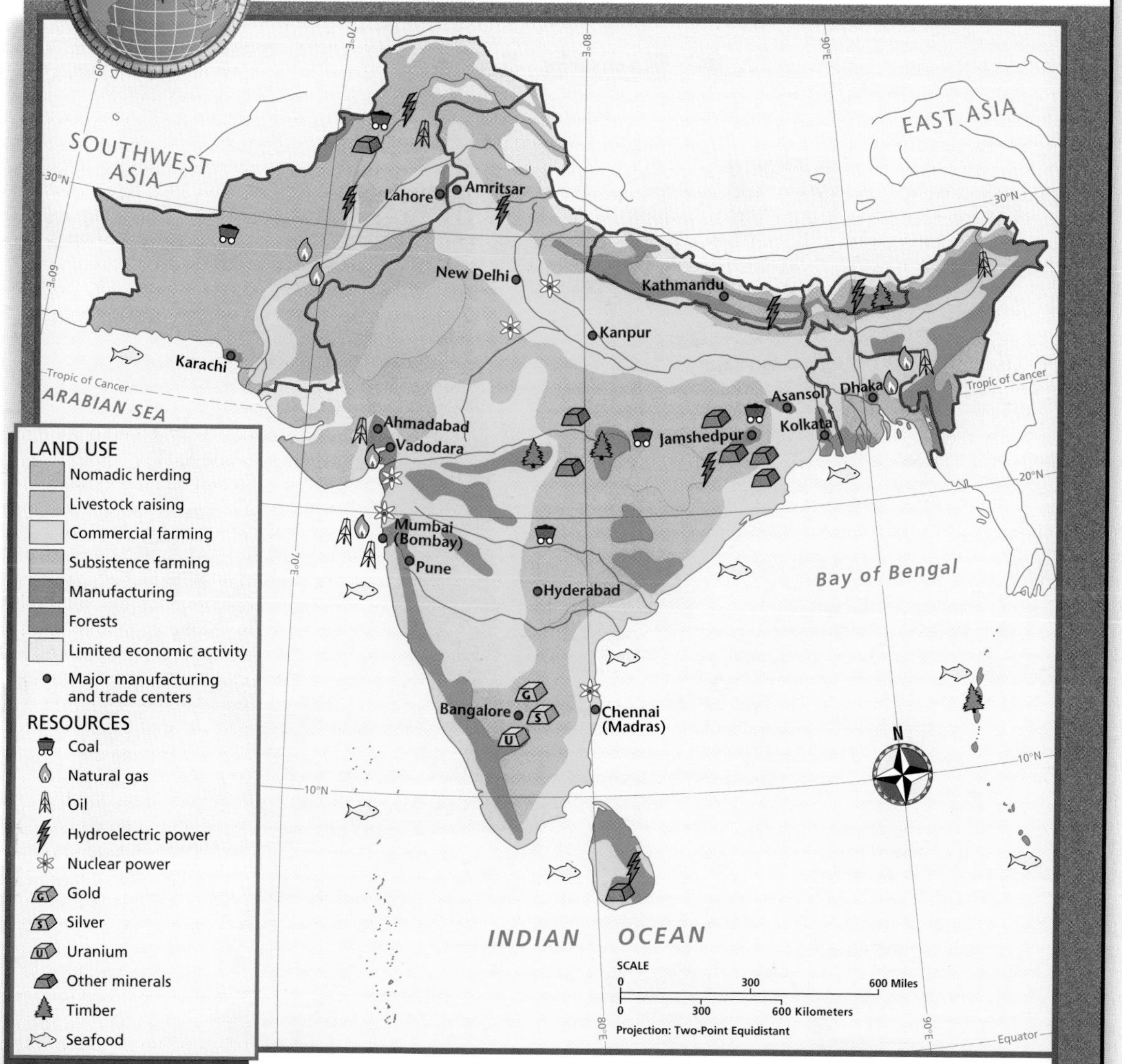

1. (Places and Regions) Which country has the largest area used for nomadic herding?
2. (Places and Regions) Which country has the largest area used for manufacturing?
3. (Places and Regions) Which city is near the region's gold, silver, and uranium mines?
4. (Places and Regions) Which country does not produce hydroelectric power?

Critical Thinking

5. **Analyzing Information** Use this map to create a chart, graph, database, or model of economic activities in South Asia.

Land Use and Resources Map

Answers

1. Pakistan
2. India
3. Bangalore
4. Bangladesh

Critical Thinking

5. Charts, graphs, databases, and models should accurately reflect information from the map.

SOUTH ASIA

FAST FACTS ACTIVITIES

LEVEL 1: (Suggested time: 30 min.) Provide students with a copy of the following table. Explain that imports are goods bought from other countries and that exports are goods sold to other countries. Ask students which two South Asian countries import more than twice as much as they export *(Maldives and Nepal)*. If students do not know the term, point out that this type of imbalance is called a trade deficit. Then ask students what effect a trade deficit might have on a country's economy. *(A trade deficit generally weakens a country's economy.)*

Imports and Exports of South Asia

	Imports	Exports
Bangladesh	$8.1 billion	$5.9 billion
Bhutan	$269 million	$154 million
India	$60.8 billion	$43.1 billion
Maldives	$372 million	$88 million
Nepal	$1.2 billion	$485 million
Pakistan	$9.6 billion	$8.6 billion
Sri Lanka	$6.1 billion	$5.2 billion
United States	$1.22 trillion	$776 billion

Source: *The World Almanac and Book of Facts, 2001*

UNITED STATES OF AMERICA

CAPITAL: Washington, D.C.
AREA: 3,717,792 sq. mi. (9,629,091 sq km)
POPULATION: 281,421,906
MONEY: U.S. dollar (USD)
LANGUAGES: English, Spanish (spoken by a large minority)
ARABLE LAND: 19 percent

South Asia

BANGLADESH

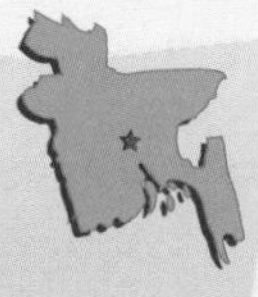

CAPITAL: Dhaka
AREA: 55,598 sq. mi. (144,000 sq km)
POPULATION: 131,269,860
MONEY: taka (BDT)
LANGUAGES: Bangla (official), English
ARABLE LAND: 73 percent

Crystal clear waters in the Maldives

BHUTAN

CAPITAL: Thimphu
AREA: 18,147 sq. mi. (47,000 sq km)
POPULATION: 2,049,412
MONEY: ngultrum (BTN), Indian rupee (INR)
LANGUAGES: Dzongkha
ARABLE LAND: 2 percent

INDIA

CAPITAL: New Delhi
AREA: 1,269,338 sq. mi. (3,287,590 sq km)
POPULATION: 1,029,991,145
MONEY: Indian rupee (INR)
LANGUAGES: Hindi (official), English (associate official), 14 other official languages, many ethnic languages
ARABLE LAND: 56 percent

MALDIVES

CAPITAL: Male
AREA: 116 sq. mi. (300 sq km)
POPULATION: 310,764
MONEY: rufiyaa (MVR)
LANGUAGES: Maldivian Divehi, English
ARABLE LAND: 10 percent

Countries not drawn to scale.

Fast Facts Activities

LEVEL 2: (Suggested time: 40 min.) Call attention to the percentage of arable land listed for each country in the Fast Facts feature. Remind students that arable land is land suitable for growing crops. Refer students to this unit's atlas, to note the landform and climatic conditions of the South Asian countries. Have students write a paragraph about the correlations between the type of landforms, climate, and the percent of arable land. Ask them to provide specific examples to support their reasoning.

LEVEL 3: (Suggested time: 45 min.) Have students work in groups to prepare news bulletins for a South Asian country. Coverage may include weather, economic outlook, and other stories. Encourage students to use the Fast Facts feature, this unit's atlas, and information throughout the unit in their reports. *(Example: "Welcome to Bangladesh Nightly News. Earlier today, the annual monsoon season kicked off to a roaring start, drenching the capital city of Dhaka. In economic news, economists predict that the taka's value will soon change as imports again surpass exports, climbing past the $8 billion mark.")*

UNIT 8 ATLAS

Buddhist monks in Kandy, Sri Lanka

Nepal

CAPITAL: Kathmandu
AREA: 54,363 sq. mi. (140,800 sq km)
POPULATION: 25,284,463
MONEY: Nepalese rupee (NPR)
LANGUAGES: Nepali (official), other ethnic languages
ARABLE LAND: 17 percent

Sri Lanka

CAPITAL: Colombo
AREA: 25,332 sq. mi. (65,610 sq km)
POPULATION: 19,408,635
MONEY: Sri Lankan rupee (LKR)
LANGUAGES: Sinhalese, Tamil, English
ARABLE LAND: 14 percent

Pakistan

CAPITAL: Islamabad
AREA: 310,401 sq. mi. (803,940 sq km)
POPULATION: 144,616,639
MONEY: Pakistani rupee (PKR)
LANGUAGES: Punjabi, Sindhi, Siraiki, Pashtu, Urdu (official), Balochi, Hindko, Brahui, English, many ethnic languages
ARABLE LAND: 27 percent

internet connect
COUNTRY STATISTICS
GO TO: **go.hrw.com**
KEYWORD: **SK3 FactsU8**
FOR: more facts about South Asia

Sources: Central Intelligence Agency, *The World Factbook 2001; The World Almanac and Book of Facts 2001;* pop. figures are 2001 estimates.

UNIT 8 Assessment Resources

Reproducible
- Unit 9 Test
- Unit 9 Test for English Language Learners and Special-Needs Students

internet connect
COUNTRY STATISTICS
GO TO: **go.hrw.com**
KEYWORD: **SK3 FactsU8**

Highlights of Country Statistics

Links to online country statistics for South Asia include:
- *CIA World Factbook*
- Library of Congress country studies
- Flags of the world

CHAPTER 27

India

Chapter Resource Manager

Objectives	Pacing Guide	Reproducible Resources
SECTION 1 **Physical Geography** (pp. 621–23) **1.** Identify the main landform regions of India. **2.** Identify the major rivers in India. **3.** Identify the climate types of India. **4.** Identify the natural resources of India.	**Regular** .5 day **Block Scheduling** .5 day *Block Scheduling Handbook, Chapter 27*	**RS** Guided Reading Strategy 27.1
SECTION 2 **India's History** (pp. 624–29) **1.** Identify the outside groups that affected India's history. **2.** Describe the Mughal Empire. **3.** Describe how Great Britain gained control of India. **4.** Describe why India was divided when it became independent.	**Regular** 2 days **Block Scheduling** .5 day *Block Scheduling Handbook, Chapter 27*	**RS** Guided Reading Strategy 27.2 **RS** Graphic Organizer 27 **E** Cultures of the World Activity 8 **E** Creative Strategies for Teaching World Geography, Lessons 20 and 21 **SM** Geography for Life Activity 27 **IC** Interdisciplinary Activities for the Middle Grades 33, 34, 35 **SM** Map Activity 27
SECTION 3 **India Today** (pp. 630–35) **1.** Identify the four major religions that originated in India. **2.** Explain the caste system. **3.** Identify the languages that are important in India. **4.** Describe the government and economy of India.	**Regular** 2 days **Block Scheduling** .5 day *Block Scheduling Handbook, Chapter 27*	**RS** Guided Reading Strategy 27.3

Chapter Resource Key

RS Reading Support
IC Interdisciplinary Connections
E Enrichment
SM Skills Mastery
A Assessment
REV Review
ELL Reinforcement and English Language Learners
Transparencies
CD–ROM
Music
Video
Internet
Holt Presentation Maker Using Microsoft® Powerpoint®

One-Stop Planner CD–ROM

See the *One-Stop Planner* for a complete list of additional resources for students and teachers.

One-Stop Planner CD–ROM

It's easy to plan lessons, select resources, and print out materials for your students when you use the *One-Stop Planner CD–ROM with Test Generator.*

Technology Resources	Review, Reinforcement, and Assessment Resources
One-Stop Planner CD–ROM, Lesson 27.1 Geography and Cultures Visual Resources with Teaching Activities 55–60 Homework Practice Online HRW Go site	**ELL** Main Idea Activity 27.1 **REV** Section 1 Review, p. 623 **A** Daily Quiz 27.1 **ELL** English Audio Summary 27.1 **ELL** Spanish Audio Summary 27.1
One-Stop Planner CD–ROM, Lesson 27.2 Homework Practice Online HRW Go site	**ELL** Main Idea Activity 27.2 **REV** Section 2 Review, p. 629 **A** Daily Quiz 27.2 **ELL** English Audio Summary 27.2 **ELL** Spanish Audio Summary 27.2
One-Stop Planner CD–ROM, Lesson 27.3 Homework Practice Online HRW Go site	**ELL** Main Idea Activity 27.3 **REV** Section 3 Review, p. 635 **A** Daily Quiz 27.3 **ELL** English Audio Summary 27.3 **ELL** Spanish Audio Summary 27.3

internet connect

HRW ONLINE RESOURCES

GO TO: go.hrw.com
Then type in a keyword.

TEACHER HOME PAGE
KEYWORD: SK3 TEACHER

CHAPTER INTERNET ACTIVITIES
KEYWORD: SK3 GT27

Choose an activity to:
- tour the regions of India.
- travel to ancient India.
- learn about Mohandas Gandhi.

CHAPTER ENRICHMENT LINKS
KEYWORD: SK3 CH27

CHAPTER MAPS
KEYWORD: SK3 MAPS27

ONLINE ASSESSMENT
Homework Practice
KEYWORD: SK3 HP27
Standardized Test Prep Online
KEYWORD: SK3 STP27
Rubrics
KEYWORD: SS Rubrics

COUNTRY INFORMATION
KEYWORD: SK3 Almanac

CONTENT UPDATES
KEYWORD: SS Content Updates

HOLT PRESENTATION MAKER
KEYWORD: SK3 PPT27

ONLINE READING SUPPORT
KEYWORD: SS Strategies

CURRENT EVENTS
KEYWORD: S3 Current Events

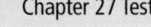

Meeting Individual Needs

Ability Levels

Level 1 Basic-level activities designed for all students encountering new material

Level 2 Intermediate-level activities designed for average students

Level 3 Challenging activities designed for honors and gifted-and-talented students

English Language Learners Activities that address the needs of students with Limited English Proficiency

Chapter Review and Assessment

E Readings in World Geography, History, and Culture 75 and 76

SM Critical Thinking Activity 27

REV Chapter 27 Review, pp. 636–37

REV Chapter 27 Tutorial for Students, Parents, Mentors, and Peers

ELL Vocabulary Activity 27

A Chapter 27 Test

Chapter 27 Test Generator (on the One-Stop Planner)

Audio CD Program, Chapter 27

A Chapter 27 Test for English Language Learners and Special-Needs Students

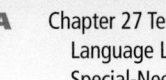

CHAPTER 27

Launch into Learning

Ask students to think of images, places, or scenes that they associate with India. *(Students might mention elephants, mountains, the Taj Mahal, cobras, or other images.)* Tell students that India is undergoing rapid economic changes and that it is the site of a thriving software industry. Tell students that India also makes more movies than any other country in the world. Tell students they will learn that India is a mix of ancient and modern, urban and rural, rich and poor in this chapter.

Section 1

Objectives

1. Identify the three main landform regions of India.
2. List the major rivers in India.
3. Describe India's climate types.
4. Discuss India's natural resources.

Links to Our Lives

There are many reasons why American students should know more about India. Here are a few of them:

- India is projected to become the world's most populous country within the next generation.
- India, like the United States, is one of the world's nuclear powers.
- It is essential to the future health of the world's economy that India, one of the world's poorest countries, continues to make economic progress.
- Many Indians and Indian-Americans now live and work in communities throughout the United States.

CHAPTER 27

India

India is a huge country with an ancient culture and a population of a billion people. However, you might find you have quite a bit in common with a student from India.

Hi! My name is Rojo, and I am 16. I live with my mother, father, and dog, Jacki. I live in Vaduthala, a small town outside the city of Cochin on the southern tip of India. I live in a one-floor house, with three bedrooms, two bathrooms, living area, dining area, kitchen, cooking terrace, front yard with lots of plants, and a road down to the lake. To cook, my mom usually goes outside to the cooking terrace and makes fish curry. That way, the whole house does not smell like the food.

I am a senior at the State Bank Officers Association High School. This term I am studying math, physics, chemistry, and computer science. My favorite subject is math, because I can use it in so many different ways. The computer language I am studying is based on math. We speak in English in school. At home or with friends, we speak Malayalam, the most common language of the state of Kerala.

Translation: What's new?

Focus

LET'S GET STARTED

Copy the following instruction and question onto the chalkboard: *Examine the physical-political map in Section 1. What natural boundaries separate India from other countries?* Allow students to respond in writing. Students might suggest the Himalayan Mountains, the Thar Desert, the Arabian Sea, and the Bay of Bengal. Tell students that in Section 1 they will learn more about the physical geography of India.

Using the Physical-Political Map

Have students examine the map on this page. Have students name the major bodies of water that border India. Then ask them to name and describe the location of major landforms. Note the variety of landforms and elevations in India. Ask students where they think population is most heavily concentrated and why. *(Possible answers: on the Gangetic Plain and along rivers, because of the availability of water)*

Building Vocabulary

Copy the key term **teak** onto the chalkboard. Ask students whether they can name any objects that might be made out of teak. Have a student find and read aloud the definition from the glossary. Then tell students that teak is one of India's important resources.

Section 1 Physical Geography

Read to Discover

1. What are the three main landform regions of India?
2. What are the major rivers in India?
3. What climate types does India have?
4. What natural resources does India have?

Define

teak

Locate

Gangetic Plain
Deccan
Eastern Ghats
Western Ghats
Ganges River
Bay of Bengal
Brahmaputra River
Thar Desert

An Indian black cobra

WHY IT MATTERS

India is the largest country of the Indian Subcontinent. Its climate helps to create its fertile soil. Use CNNfyi.com or other **current events** sources to find information about other ways that India's climate affects daily life. Record your findings in your journal.

India: Physical-Political

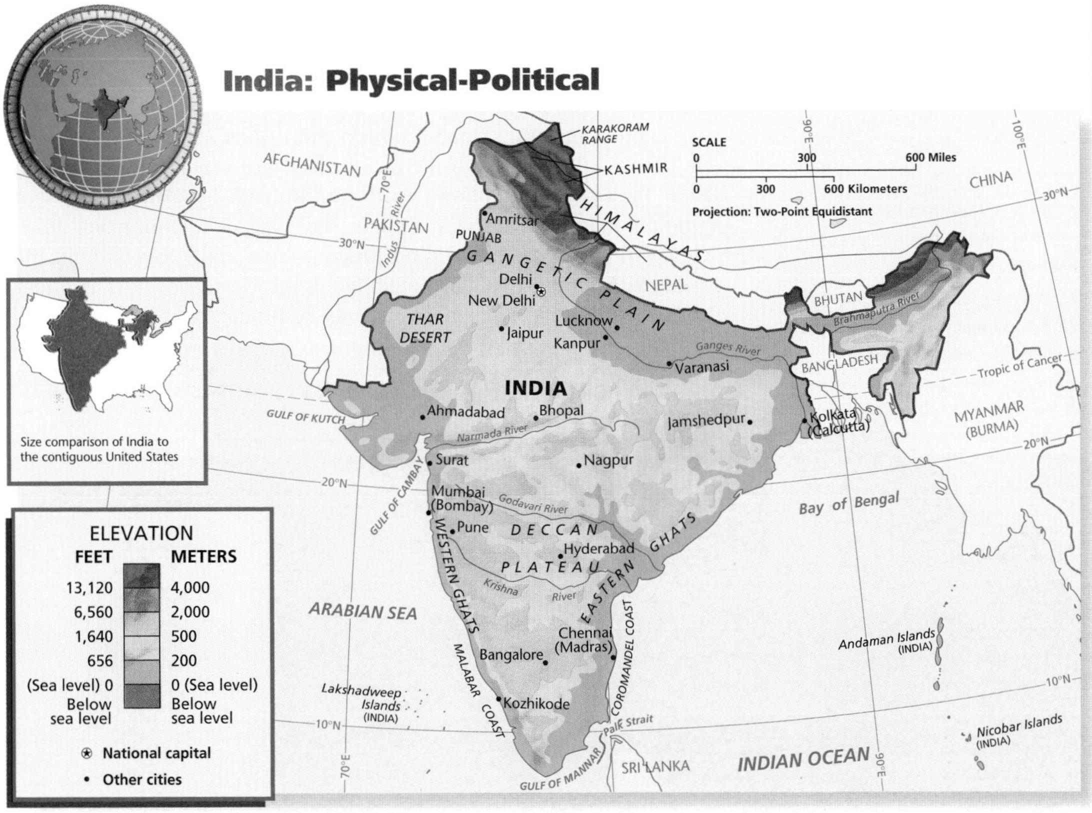

Section 1 Resources

Reproducible
- Block Scheduling Handbook, Chapter 27
- Guided Reading Strategy 27.1

Technology
- One-Stop Planner CD–ROM, Lesson 27.1
- Homework Practice Online
- Geography and Cultures Visual Resources with Teaching Activities 55–60
- HRW Go site

Reinforcement, Review, and Assessment
- Section 1 Review, p. 623
- Daily Quiz 27.1
- Main Idea Activity 27.1
- English Audio Summary 27.1
- Spanish Audio Summary 27.1

TEACH

Teaching Objectives 1–3

ALL LEVELS: (Suggested time: 30 min.) Have each student draw a simple map of India. Then have each student label and shade in different colors the three main landform regions of India, draw the four major rivers in blue, and then label the country's climate regions. *(Students should label the Gangetic Plain, the Deccan, and the Himalayas; they should draw and identify the Narmada, Ganges, Godavari, and Krishna Rivers; they should also label the highland, humid tropical, tropical savanna, and steppe climates in the appropriate places.)* Ask volunteers to share their maps with the class. **ENGLISH LANGUAGE LEARNERS**

Teaching Objective 4

ALL LEVELS: (Suggested time: 10 min.) Copy the following graphic organizer onto the chalkboard, omitting the italicized answers. Use it to help students identify India's natural resources. Ask volunteers to share their answers. **ENGLISH LANGUAGE LEARNERS**

India's Natural Resources

Mineral and Other Resources	Agricultural and Other Resources
iron ore, bauxite, uranium, coal, oil deposits, and gemstones	*fertile farmlands and forests*

DAILY LIFE

Many Hindus believe that in Varanasi, a Hindu holy city, the waters of the Ganges River can wash away sickness and sin. Religious pilgrims travel to Varanasi to bathe in the river. Also, some Hindus wish to be cremated on the banks of the Ganges. They believe that they will be freed from an endless cycle of rebirths.

Unfortunately, the Ganges River is highly polluted. Sewage and industrial waste contaminate the river. In addition, the dead bodies of people who cannot afford to be cremated are dumped into the river. However, local residents continue to use the river for doing laundry, bathing, or washing their hair. Some devoted Hindus refuse to believe that the holy river could be polluted. Since the mid-1980s environmental groups have worked to clean up the Ganges.

Critical Thinking: How has human activity changed the Ganges?

Answer: Sewage and industrial waste, dead bodies pollute the river.

Visual Record Answer ►

It is a very fertile area for farmland.

internet connect

GO TO: go.hrw.com
KEYWORD: SK3 CH27
FOR: Web sites about India

BUILD ON WHAT You Know

Do you remember what you learned about plate tectonics? See Chapter 2 to review.

Landforms

India has three main landform regions: the Himalayas, the Gangetic (gan-JE-tik) Plain, and the Deccan (DE-kuhn). The Himalayas run along the country's northern border and were created when two tectonic plates collided and pushed Earth's crust up.

The vast Gangetic Plain lies to the south of the Himalayas. It stretches about 1,500 miles (2,415 km) across northern India. About half of India's population lives there.

South of the Gangetic Plain is the triangular peninsula known as the Deccan. Most of its area is a plateau, which is divided by many hills and valleys. The plateau's edges are defined by the Eastern Ghats (GAWTS) and Western Ghats. These low mountain ranges separate the plateau's eastern and western edges from narrow coastal plains.

✓ **READING CHECK:** *Places and Regions* What are the three main landform regions of India? Himalayas, Gangetic Plain, Deccan

Rivers

India's most important river, the Ganges (GAN-jeez), begins on the southern slopes of the Himalayas. It then flows southeastward across northern India. It spreads into a huge delta before flowing into the Bay of Bengal. Hindus call the Ganges the "Mother River" and consider it sacred. Rich silt left by the Ganges has made the Gangetic Plain India's farming heartland.

The Brahmaputra (brahm-uh-POO-truh) River starts in the Plateau of Tibet. It flows through the far northeastern corner of India. From there the Brahmaputra flows southward through Bangladesh, where it empties into the Ganges Delta. The Narmada (nuhr-MUH-duh), Godavari (go-DAH-vuh-ree), and Krishna (KRISH-nuh) Rivers drain the Deccan. A large irrigation project along the Narmada River is under construction. It will include more than 40 branch canals.

✓ **READING CHECK:** *Places and Regions* What are India's most important rivers? Ganges, Brahmaputra, Narmada, Godavari, Krishna

A farmer plows rice fields in northern India. The Himalayas rise in the distance.

Interpreting the Visual Record

Why do so many of India's people live on the Gangetic Plain?

CLOSE

Locate weather information on your area in a local newspaper to find the average annual rainfall. Ask students to compare this amount with the world record recorded in Cherrapunji, India, listed on this page in the Our Amazing Planet feature. Lead a discussion on how life in your area would change if you received the same amount of rain as Cherrapunji.

REVIEW AND ASSESS

Have students complete the Section Review. Then pair students and have each pair write eight quiz questions about the landforms, bodies of water, climate types, and resources of India. Have each pair exchange its quiz with another pair and complete the quiz it receives. Then have students complete Daily Quiz 27.1. **COOPERATIVE LEARNING**

RETEACH

Have students complete Main Idea Activity 27.1. Then trace a wall map of India. Have each of three groups label one of the following: landform regions, climate types and regions, or natural resources. **ENGLISH LANGUAGE LEARNERS, COOPERATIVE LEARNING**

EXTEND

Have interested students conduct research on Indian mythology associated with either the Ganges or Brahmaputra Rivers. Have students present an oral report summarizing the myth as well as a visual illustration of some aspect of the myth. **BLOCK SCHEDULING**

Climate

India has a variety of climate types. Areas in the Himalayas have highland climates with snow and glaciers. The Thar Desert near the border with Pakistan is hot and dry year-round. The Gangetic Plain has a humid tropical climate. Farther south in the Deccan there are tropical savanna and steppe climates.

Seasonal winds—monsoons—bring moist air from the Indian Ocean in summer. In winter the wind brings dry air from the Asian interior. The timing of the monsoons is very important to farmers in India. If the summer rains come too soon or too late, food production suffers.

✓ **READING CHECK:** *Places and Regions* What are India's climates?
highland, desert, humid tropical, tropical savanna, steppe

Resources

India's fertile farmlands are important to its economy. Most of India's people work in agriculture. The country also produces cash crops for export. These include cashew nuts, cotton, jute, spices, sugarcane, tea, and tobacco.

Large deposits of iron ore, bauxite, uranium, and coal are among India's mineral resources. There are some oil reserves, but not enough to meet the country's needs. Gemstones are a valuable export.

India's forests are an important resource, as well as home to wildlife. **Teak**, one of the most valuable types of wood, grows in India and Southeast Asia. Teak is very strong and durable and is used to make ships and furniture.

✓ **READING CHECK:** *Places and Regions* What are India's natural resources?
fertile soil, minerals, some oil, forests

Our Amazing Planet

The city of Cherrapunji (cher-uh-POOHN-jee), in northeastern India, holds the world's record for rainfall in one year: almost 87 feet (26.5 m)!

Section 1 Review

go.hrw.com Homework Practice Online
Keyword: SK3 HP27

Define and explain: teak

Working with Sketch Maps On a map of India that you draw or that your teacher provides, label the following: Gangetic Plain, Deccan, Eastern Ghats, Western Ghats, Ganges River, Bay of Bengal, Brahmaputra River, and Thar Desert. How would you describe the relative location of India?

Reading for the Main Idea

1. *Places and Regions* What is the main mountain range of India?
2. *Places and Regions* What are some of India's most important cash crops? What type of wood is a valuable forest product?

Critical Thinking

3. **Making Generalizations and Predictions** What do you think happens to India's crops if the monsoon rains come too soon? too late?
4. **Summarizing** How does the Ganges affect economic activities in India?

Organizing What You Know

5. **Summarizing** Copy the following graphic organizer. Use it to list the landforms, climates, and resources of India.

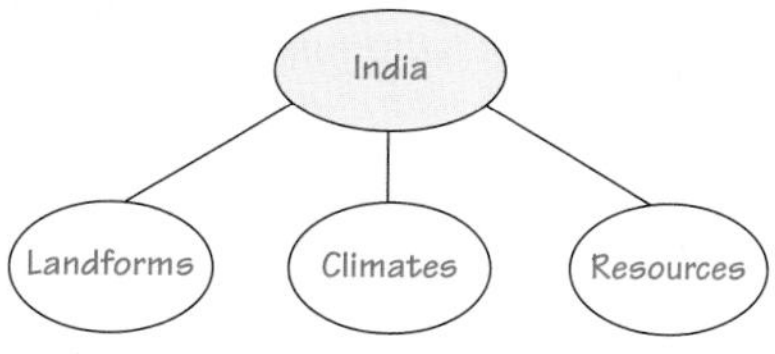

Section 1 Review

Answers

Define For definition, see: teak, p. 623

Working with Sketch Maps Maps will vary, but listed places should be labeled in their approximate locations. India's relative location is in southern Asia.

Reading for the Main Idea

1. the Himalayas (NGS 4)
2. cashew nuts, cotton, sugarcane, jute, spices, tea, and tobacco; teak (NGS 4)

Critical Thinking

3. Crops get rain too early and fail; crops dry up and fail.
4. enriches the Gangetic Plain, makes it India's farming heartland

Organizing What You Know

5. landforms—the Himalayas, the Gangetic Plain, and the Deccan; climates—highland, desert, humid tropical, savanna, and steppe; resources—farmland, iron ore, bauxite, uranium, coal, oil deposits, gemstones, forests, and teak

Section 2

Objectives

1. Identify the outside groups that have affected India's history.
2. Describe the Mughal Empire.
3. Explain how Great Britain gained control of India.
4. Discuss why India was divided when it became independent.

Focus

LET'S GET STARTED

Copy the following question onto the chalkboard: *How might the physical boundaries of India have affected its history?* Have students respond in writing and discuss answers. Students might suggest that before technological advances permitted long sea voyages, India would have been protected from invasion. Tell students that in Section 2 they will learn more about the history and culture of India.

Building Vocabulary

Copy the Define terms, **Sanskrit**, **sepoys**, and **boycott**, onto the chalkboard. Explain that Sanskrit is an ancient language that is used today for Hindu religious rituals. Sepoys were local Indian soldiers commanded by British officers. A boycott, or refusal to buy certain goods in protest, was an important tactic used by Indians against their British colonial rulers.

Section 2 Resources

Reproducible
- Guided Reading Strategy 27.2
- Graphic Organizer 27
- Geography for Life Activity 27
- Map Activity 27
- Cultures of the World Activity 8
- Creative Strategies for Teaching World Geography, Lessons 20 and 21
- Interdisciplinary Activities for the Middle Grades 33, 34, 35

Technology
- One-Stop Planner CD–ROM, Lesson 27.2
- HRW Go site

Reinforcement, Review, and Assessment
- Section 2 Review, p. 629
- Daily Quiz 27.2
- Main Idea Activity 27.2
- English Audio Summary 27.2
- Spanish Audio Summary 27.2

Visual Record Answer

by the uniform size of the bricks and the straight rows of buildings

Section 2 India's History

Read to Discover

1. What outside groups affected India's history?
2. What was the Mughal Empire like?
3. How did Great Britain gain control of India?
4. Why was India divided when it became independent?

Define

Sanskrit
sepoys
boycott

Locate

Delhi
Kolkata
Mumbai

Why It Matters

India's history has been shaped by invasions, conquests, and colonization. Use CNNfyi.com or other **current events** sources to find information about how daily life in India has been shaped by its history. Record your findings in your journal.

Coat of arms of the East India Company

Early Indian Civilizations

The first urban civilization on the Indian Subcontinent was centered around the Indus River valley. Its territory was mainly in present-day Pakistan but also extended into India. Scholars call this the Harappan civilization after one of its cities, Harappa. By about 2500 B.C. the people of this civilization were living in large, well-planned cities. Scholars believe the Harappans traded by sea with the peoples of Mesopotamia. The Harappans had a system of writing, but scholars have not been able to read it. As a result, very little is known about Harappan religion and customs.

Mohenjo Daro was one of the largest cities of the Harappan civilization.

Interpreting the Visual Record **How might you tell from this photo that Harappan cities were well planned?**

TEACH

Teaching Objective 1

ALL LEVELS: (Suggested time: 15 min.) Copy the following graphic organizer onto the chalkboard, omitting the italicized answers. Use it to help students identify various groups that affected India's history. Have each student complete the organizer by filling in the names of the outside and inside groups identified in the section. Have each student also provide a short description of how each outside group affected India's history. Ask volunteers to share their answers with the class.

ENGLISH LANGUAGE LEARNERS

Groups That Affected India's History

OUTSIDE GROUPS

- *Indo-Aryans had brought their language and religious beliefs by about 1500 B.C.*
- *First Muslim armies came about A.D. 1000 and later established Delhi sultanate.*
- *Invaders from central Asia came in early 1500s and established Mughal Empire.*
- *British started arriving in 1600s and colonized India.*

INSIDE GROUP

- *Harappans*

The Indo-Aryans By about 1500 B.C. a new group of people had come into northern India. Scholars call these people Indo-Aryans. Their language was an early form of **Sanskrit**. Sanskrit is still used in India in religious ceremonies.

The Indo-Aryans took control of northern India. These new arrivals mixed with Indian groups that were already living there. Their religious beliefs and customs mixed as well, forming the beginnings of the Hindu religion.

Hills and mountains prevented the Indo-Aryans from conquering southern India. However, Sanskrit and other Indo-Aryan cultural traits spread to the south.

This carved-lion pillar comes from Sarnath, an ancient city in northern India. The pillar was created in the 200s B.C. and is now the state emblem of India.

The Coming of Islam About A.D. 1000, Muslim armies began raiding northwestern India. In the early 1200s a Muslim kingdom was established at Delhi. Because the monarch was known as a sultan, this kingdom was called the Delhi sultanate. The Delhi sultanate eventually gained control over most of northern India. It also became a leading center of Islamic art, culture, and science. Most Indians, however, kept their own religions and did not convert to Islam.

Over the next two centuries the Delhi sultanate expanded into the Deccan. However, in the early 1500s a new invasion from Central Asia swept into India. This new conquest marked the beginning of the Mughal (MOO-guhl) Empire.

✓ **READING CHECK:** ***Human Systems*** How did outside groups affect early Indian history? Indo-Aryans—brought Sanskrit, formed beginnings of Hindu religion; Muslim armies—formed Delhi sultanate; Central Asians—conquered Delhi sultanate, established Mughal Empire

The Mughal Empire

The founder of the Mughal Empire was Babur, whose name meant "the Tiger." Babur was descended from Mongol emperor Genghis Khan. He was not only a brilliant general, but also a gifted poet. Babur defeated the last sultan of Delhi and took over most of northern India. After his death, however, Babur's lands were divided among his sons. They fought each other for years.

Babur's grandson, Akbar, finally emerged to reunite the Mughal Empire. He recaptured northern India and then expanded his empire into central India. Akbar was a good ruler as well as a successful conqueror. He reorganized the government and the tax system to make them more efficient. The fertile farmland and large population of the Gangetic Plain made the Mughal Empire rich. It quickly became one of the most powerful states in the world. The reign of Akbar and his successors was a golden age of architecture, painting, and poetry.

This illustration is from a book of the life of Babur, the first Mughal emperor. It shows Babur surrounded by servants and nobles.

Across the Curriculum
LITERATURE

Sanskrit Sanskrit is one of the two classical languages of India. The other classical language is Tamil, a South Indian language that belongs to the Dravidian family of languages. Many ancient works of literature are written in both languages; however, only Tamil continues to be widely spoken today. Sanskrit is used mainly for Hindu religious rituals and ceremonies.

Unlike many languages, Sanskrit was formally developed by grammarians. These scholars created a broad set of rules for Sanskrit's usage. The grammarian Panini, who lived about 400 B.C., wrote the most famous set of rules for Sanskrit.

Many works of Indian literature were written in Sanskrit. Famous Sanskrit works include the *Ramayana* and the *Mahabharatha,* which are known around the world. Both are epic poems that examine themes such as loyalty, duty, and love.

Activity: Have students find and read an epic poem. Students might choose the *Iliad, Song of Roland,* or *Beowulf,* for example. Then lead a class discussion on what information these epics provide about the cultures in which they were created.

Teaching Objective 2

LEVEL 1: (Suggested time: 20 min.) Pair students and have each pair create a bulleted list of the Mughal Empire's achievements. *(Lists should mention the efficient government and tax system; the wealth gained from farmland and the empire's large population; the flowering of architecture, painting, and poetry; and religious tolerance.)* Ask volunteers to share and discuss their lists with the class. Compile a master list of features on the chalkboard and have students copy the list into their notes.
ENGLISH LANGUAGE LEARNERS, COOPERATIVE LEARNING

LEVELS 2 AND 3: (Suggested time: 20 min.) Tell students to imagine that they are editors who must write a summary for a book jacket. Then have each student write a short summary to describe achievements of the Mughal Empire. *(See the Level 1 lesson for the correct achievements.)* Ask volunteers to read their summaries to the class.

➤**ASSIGNMENT:** Have each student create a collage to represent various achievements of the Mughal Empire. Display students' collages around the classroom.

Across the Curriculum
LITERATURE

Arundhati Roy Indian writers have gained attention recently. Many Indian novelists write in the English language, which allows a wide audience to read their work. The popularity of Indian novels may also be explained by the rise of a new, global culture.

One Indian writer who has gained fame is Arundhati Roy. Her novel, *The God of Small Things,* won England's top book prize in 1997. Among other issues, Roy's novel deals with violence and discrimination against lower-caste Indians. Despite the praise Roy's work received, some Indians were angry about things she wrote in her novel. Roy's book attracted a lawsuit in India.

Critical Thinking: What might the disputes involving Roy's book indicate about the importance people place on literature?

Answer: Students might observe that the disputes indicate that people took it very seriously.

Connecting to Literature
Answers

1. tree-shaded, steep
2. Manja's descriptions might have related to the journey's difficulty.

Connecting to Literature

Indian writer R.K. Narayan's My Days: A Memoir *describes the author's childhood in the early 1900s. In the following passage Narayan recalls traveling from boarding school to his parents' new home. At that time, travel was complicated and sometimes dangerous.*

My Days: A Memoir

At the proper time, I was awakened and put into a huge mat-covered wagon drawn by a pair of bullocks; I sat on a bed of straw covered over with a carpet; a stalwart[1] peon[2] from Hassan high school was seated beside the driver. Manja was his name . . . Part of the way as we traveled along, Manja got off and walked ahead of the caravan, carrying a staff menacingly. Some spots in that jungle and mountain country were well-known retreats of highway robbers; one form of protection was to travel in a closely moving caravan with Manja waving a staff at the head of the column, uttering blood-curdling challenges. That was enough to keep off robbers in those days.

We passed along miles and miles of tree-shaded highway, gigantic mango and blueberry trees and lantana[3] shrubs in multicolored bloom stretching away endlessly. A couple of times the bullocks were rested beside a pond or a well. The road wound up and down steep slopes—the sort of country I had never known before. . . . The overpowering smell of straw in the wagon and the slow pace of the bullocks with their bells jingling made me drowsy . . . After hours of tossing on straw, we came to a bungalow[4] set in a ten-acre field. [It was my parents' new home.] . . . The moment I was received into the fold at the trellised ivy-covered porch, I totally ignored Manja, and never looked in his direction, while he carried my baggage in.

A bullock cart and a double-decker bus in India

Analyzing Primary Sources

1. What are some of the words the author uses to describe the countryside?
2. How do you think Manja's description of this journey might be different from the author's?

Vocabulary [1]stalwart: strong and reliable [2]peon: a menial laborer [3]lantana: shrub with colorful flowers [4]bungalow: one-story house with low roof that originated in Bengal, India

Teaching Objective 3

ALL LEVELS: (Suggested time: 30 min.) Pair students and have each pair create a flowchart on posterboard to show how Great Britain gained control of India. *(Flowcharts should show that first they took control of small trading posts; then they gained larger pieces of territory; they used local Indian soldiers in their army; and they backed one Indian ruler against another; by the mid-1850s they controlled more than half of India.)* Ask volunteers to display and explain their flowcharts to the class.
ENGLISH LANGUAGE LEARNERS, COOPERATIVE LEARNING

TEACHER TO TEACHER

Jim Corley of Durant, Oklahoma, suggests the following activity to help students understand the history of India: Organize the class into four groups. Have one group conduct research on early civilizations that affected India. Have the second group conduct research on the Mughal Empire. Have the third group conduct research on Great Britain's control of India. The final group should conduct research on India's independence. Have each group use its information to create an illustrated presentation. Have each group present its illustrations, and lead a class discussion on each subject.

▲ The Taj Mahal is one of the most famous buildings in the world.

The ruling Mughals were Muslim, but Islam remained a minority religion in India. Most people continued to practice Hinduism. Akbar himself was tolerant and curious about other religions. He invited religious scholars and priests—including Christians, Hindus, Jains, and Muslims—to his court. He even watched them debate.

Akbar's grandson, Shah Jahan, is remembered for the impressive buildings and monuments he had built. These include the famous Taj Mahal. This grand building contains the tomb of Shah Jahan's beloved wife, Mumtaz Mahal.

In the 1600s and 1700s the Mughal Empire slowly grew weaker. Wars in the Deccan and revolts in many parts of the empire drained Mughal resources. At about this time, Europeans became an important force in Indian history.

✓ **READING CHECK:** *Human Systems* What was the Mughal Empire like? powerful, efficient government; many artistic achievements

Our Amazing Planet

Construction of the Taj Mahal began in 1631 and was not completed until 1653. Almost 20,000 people worked on the building.

The British

During the 1700s and 1800s the British slowly took control of India. At first this was done by the English East India Company. This company won rights to trade in the Mughal Empire in the 1600s. The East India Company first took control of small trading posts. Later the British gained more Indian territory.

Company Rule As the Mughal Empire grew weaker, the British East India Company expanded its political power. The company also built up its own military force. This army was made up mostly of

Cultural Kaleidoscope

The Sepoy Mutiny

The British rulers of India attempted to change traditional Hindu and Muslim social customs. These changes contributed to rising discontent before the Indian Mutiny.

The rebellion began when British officers ordered sepoys to load cartridges into new rifles. The sepoys were ordered to bite off the ends of cartridges rumored to be lubricated with cow and pig fat. Because Muslims are barred from eating pigs, and Hindus are barred from eating cows, many Indian soldiers refused to load their guns. As punishment, they were chained and put into prison. This provoked the rebellion's outbreak.

A large section of the civilian population supported the sepoys. Although the British put down the rebellion, they began to take Indian opinions into account when making laws for India.

Critical Thinking: What does the cartridge incident indicate about British attitudes toward Indian culture?

Answer: Students might say they appeared to be either ignorant of or insensitive to Indian cultures.

Teaching Objective 4

LEVEL 1: (Suggested time: 25 min.) Tell students that Muslims were a minority in India prior to Indian independence. Pair students and tell one student in each pair to imagine that he or she is a member of the Indian National Congress. Tell the other student to imagine that he or she is a Muslim who would like a separate Muslim state. Then have each pair write a short dialogue in which both partners explain their positions. *(Students should discuss Muslim fears of being a minority ruled by Hindus, the wish of the Congress Party to keep the country intact, and the rising tensions between Hindus and Muslims.)* Ask students to perform their dialogues for the class. **ENGLISH LANGUAGE LEARNERS, COOPERATIVE LEARNING**

LEVELS 2 AND 3: (Suggested time: 20 min.) Have each student write a short poem to explain the partition of India. *(See the All Levels lesson for the reasons why India was partitioned.)* Ask volunteers to recite their poems for the class.

CLOSE

Have students identify and summarize the cultural, political, and economic influences of outside groups on India.

Linking Past to Present

Gandhi Gandhi's political philosophy, based on truth, love, and nonviolence, was revolutionary. Dr. Martin Luther King Jr. adopted Gandhi's tactics of nonviolent protest as part of the U.S. civil rights movement. Today, organizers of a wide variety of movements continue to use Gandhi's methods.

Critical Thinking: How has Gandhi's political message remained important today?

Answer: His tactics of nonviolent protest are used around the world.

Section 2 Review

Answers

Define For definitions, see: Sanskrit, p. 625; sepoys, p. 628; boycott, p. 628

Working with Sketch Maps Maps will vary, but listed places should be labeled in their approximate locations. The Ganges River is important to Delhi, the Bay of Bengal and Ganges River are important to Kolkata, and the Arabian Sea is important to Mumbai.

Visual Record Answer ►

The British government began to rule India directly.

In September 1857, British and loyal Sikh troops stormed the gate of Delhi, defended by rebel sepoys. Bloody fighting continued until late 1858.

Interpreting the Visual Record

How did the Indian Mutiny lead to a change in the way India was governed?

Mohandas Gandhi was known to his followers as the Mahatma, or the "great soul."

sepoys, Indian troops commanded by British officers. The British used the strategy of backing one Indian ruler against another in exchange for cooperation. By the mid-1800s the company controlled more than half of India. The rest was divided into small states ruled by local princes.

The British changed the Indian economy to benefit British industry. India produced raw materials, including cotton, indigo—a natural dye—and jute. These materials were then shipped to Britain for use in British factories. Spices, sugar, tea, and wheat were also grown in India for export. Railroads were built to ship the raw materials to Calcutta (now Kolkata), Bombay (now Mumbai), and other port cities. India also became a market for British manufactured goods. Indians, who had woven cotton cloth for centuries, were now forced to buy British cloth.

The Indian Mutiny British rule angered and frightened many Indians. In 1857, the sepoy troops revolted. They killed their British officers and other British residents. The violence spread across northern India. Large numbers of British troops were rushed to India. In the end the British crushed the rebellion.

The Indian Mutiny convinced the British government to abolish the British East India Company. The British government began to rule India directly, and India became a British colony.

Anti-British Protest During the late 1800s Indian nationalism took a different form. Educated, middle-class Indians led this movement. In 1885 these Indian nationalists created the Indian National Congress to organize their protests. At first they did not demand independence. Instead, they asked only for fairer treatment, such as a greater share of government jobs. The British refused even these moderate demands.

After World War I more and more Indians began demanding the end of British rule. A lawyer named Mohandas K. Gandhi became the most important leader of this Indian independence movement.

Gandhi and Nonviolence Gandhi reached out to the millions of Indian peasants. He used a strategy of nonviolent mass protest. He called for Indians to peacefully refuse to cooperate with the British. Gandhi led protest marches and urged Indians to **boycott**, or refuse to buy, British goods. Many times the police used violence against marchers. When the British jailed Gandhi, he went on hunger strikes. Gandhi's determination and self-sacrifice attracted many followers. Pressure grew on Britain to leave India.

✓ **READING CHECK:** *Human Systems* What role did the British play in India? **They slowly took control in the 1700s and 1800s; after the Indian Mutiny the British government took complete control and ruled India as a colony.**

REVIEW AND ASSESS

Have students complete the Section Review. Then pair students. Have each student within the pair create a four-question multiple-choice quiz based on the information discussed in this section. Have students within the pair take turns quizzing each other. Then have students complete Daily Quiz 27.2. **COOPERATIVE LEARNING**

RETEACH

Have students complete Main Idea Activity 27.2. Then organize the class into four groups. Assign the first the Harappan civilization, the second the Indo-Aryans, the third the Muslims and Mughals, and the fourth the British. Have each group find a lasting influence or influences of each of the groups. Ask each group to share its ideas with the class. **ENGLISH LANGUAGE LEARNERS, COOPERATIVE LEARNING**

EXTEND

Encourage interested students to conduct research on tensions in India between Hindus and Muslims by reading newsmagazines and newspapers from recent years. Ask students to try to discover the sources of tension, the ways groups have expressed their frustrations, and the methods used by political leaders to handle these conflicts. Have them present their findings to the class. **BLOCK SCHEDULING**

▲ In the chaotic days of August 1947, millions of people left their homes to cross the new border between India and Pakistan. These Muslims are preparing to leave New Delhi by train.

Independence and Division

After World War II the British government decided to give India independence. The British government and the Indian National Congress wanted India to become one country. However, India's Muslims demanded a separate Muslim state. Anger and fear grew between Hindus and Muslims. India seemed on the verge of civil war.

Finally, in 1947 the British divided their Indian colony into two independent countries, India and Pakistan. India was mostly Hindu. Pakistan, which then included what is today Bangladesh, was mostly Muslim. However, the new boundary left millions of Hindus in Pakistan and millions of Muslims in India. Masses of people rushed to cross the border. Hundreds of thousands were killed in rioting and panic.

✓ **READING CHECK:** *Places and Regions* Why was India divided when it became independent? Muslims and Hindus were in conflict, and India seemed on the verge of civil war.

Section 2 Review

go.hrw.com Homework Practice Online Keyword: SK3 HP27

Define and explain: Sanskrit, sepoys, boycott

Working with Sketch Maps On the map you created in Section 1, label Delhi, Kolkata, and Mumbai. What bodies of water are important to each of these cities?

Reading for the Main Idea

1. *Human Systems* What made the Mughal Empire one of the most powerful states in the world?
2. *Human Systems* How did the British East India Company gain control of most of India?
3. *Human Systems* Who was the most important leader of the Indian independence movement, and what was his strategy?

Critical Thinking

4. **Finding the Main Idea** Why was the British colony of India divided into two countries?

Organizing What You Know

5. **Sequencing** Copy the following time line. Use it to mark important events in Indian history from 2500 B.C. to A.D. 1947.

2500 B.C. ———————— A.D. 1947

Reading for the Main Idea

1. reorganized, efficient government and tax system; the Gangetic Plain's fertile farmland and large population (NGS 12)
2. took control of small trading posts; gained larger pieces of territory; used Indian soldiers; and backed Indian rulers against each other (NGS 13)
3. Mohandas K. Gandhi; nonviolent protest (NGS 12)

Critical Thinking

4. Muslims demanded a separate state, and tensions between Hindus and Muslims grew.

Organizing What You Know

5. c. 2500 B.C.—Harappan civilization established; c. 1500 B.C.—Indo-Aryans in northern India; c. A.D. 1000—Muslim armies raiding northwest India; c. 1200—Delhi sultanate established; 1500s—Mughal Empire established; 1600s—English East India Company wins right to trade in Mughal Empire; mid-1800s—East India Company controls more than half of India; 1857—Indian Mutiny; 1885—Indian National Congress created; after WWI—nationalist movement gains momentum; 1947—independence and division

Section 3

Objectives

1. Identify the four major religions that originated in India.
2. Explain the caste system.
3. Identify the languages that are important in India.
4. Describe what kind of government India has, and discuss India's economy.

Focus

LET'S GET STARTED

Copy the following instructions onto the chalkboard: *India's population is predicted to reach 1.7 billion in 2050. Compile a list of problems facing a poor country with a large and growing population.* Discuss responses. *(Students might mention shortages of food, medical supplies, and housing, or the possibility of group conflicts over resources.)* Tell students that in Section 3 they will learn more about India today.

Building Vocabulary

Copy the following Define terms onto the chalkboard: **reincarnation**, **karma**, **nirvana**, **caste system**, **Dalits**, and **green revolution**. Ask students if they are familiar with any of the terms and if so, which ones. *(Students might mention reincarnation or karma.)* Then ask volunteers to read the terms' definitions aloud from the glossary.

Section 3 Resources

Reproducible
- Guided Reading Strategy 27.3

Technology
- One-Stop Planner CD–ROM, Lesson 27.3
- Homework Practice Online
- HRW Go site

Reinforcement, Review, and Assessment
- Section 3 Review, p. 635
- Daily Quiz 27.3
- Main Idea Activity 27.3
- English Audio Summary 27.3
- Spanish Audio Summary 27.3

Section 3 India Today

Read to Discover

1. What four major religions originated in India?
2. What is the caste system?
3. What languages are important in India?
4. What kind of government does India have, and what is India's economy like?

Define

reincarnation
karma
nirvana
caste system
Dalits
green revolution

Locate

Kashmir

WHY IT MATTERS

Religion has played a tremendous role in both the past and present culture of India. Use CNNfyi.com or other **current events** sources to find information about religion in India. Record your findings in your journal.

Masala dosa, *a rice and lentil dish from India*

This bronze statue depicts the Hindu god Siva.

Cows mingle with pedestrians and bicyclists in the street of an Indian town.

Religions of India

Religion is an important part of Indian culture. Four major religions—Hinduism, Buddhism, Jainism, and Sikhism—originated in India. Christianity and Islam, both of which originated elsewhere, also have millions of followers in India. About 81 percent of India's people are Hindu. About 12 percent of Indians are Muslim, and 2.3 percent are Christian. Around 2 percent are Sikh, and 2.5 percent are Buddhist, Jain, Parsi, or followers of another religion. Remember that 1 percent of India's population is about 10 million people!

Hinduism Hinduism is one of the oldest religions in the world. Hindus worship many gods. These include Brahma the Creator, Vishnu the Preserver, and Siva the Destroyer. Hinduism teaches that all gods and all living beings are part of a single spirit.

Two beliefs central to the Hindu religion are **reincarnation** and **karma**. Reincarnation is the belief that the soul is reborn again and again in different forms. Karma is the positive or negative force caused by a person's actions. Hindus believe that a person with good karma may be reborn as a person of higher status. A person with bad karma may be reborn with lower status, or as an animal or insect.

TEACH

Teaching Objective 1

ALL LEVELS: (Suggested time: 15 min.) Copy the following graphic organizer onto the chalkboard, omitting the italicized answers. Use it to help students identify four major religions that originated in India. Pair students and have each pair complete the organizer by filling in the names of the religions along with short descriptions of the religions' central beliefs or aspects of the religions' histories. Ask volunteers to share their answers with the class. **ENGLISH LANGUAGE LEARNERS, COOPERATIVE LEARNING**

Facts About Indian Religions

Buddhism—
- *Founded by Siddhartha Gautama, the Buddha*
- *By not lying, stealing, or being greedy, people can reach nirvana.*

Jainism—
- *All things in nature have souls.*
- *Reject all forms of violence against living things*

Sikhism—
- *Founded in the late 1400s*
- *Combines elements of Hinduism and Islam*

Hinduism—
- *Belief in reincarnation and karma*
- *Gods include Vishnu, Brahma, and Siva.*
- *All gods and living beings are part of one spirit.*

Hinduism also teaches a special respect for cows. Hindus do not eat beef. Even today, cows can be seen roaming cities and villages.

Buddhism Buddhism was founded in northern India in the 500s B.C. by a man named Siddhartha Gautama. Gautama became known as the Buddha, or "Enlightened One."

The Buddha taught his followers they could avoid sorrow if they followed certain rules. For example, he told them not to lie, steal, or be greedy. By following these rules, Buddhists believe they can escape from the suffering of life. This escape from suffering is known as **nirvana**.

Buddhism spread from India to other parts of Asia. Buddhism is no longer widely practiced in India. However, the religion is very important in Sri Lanka, China, Japan, and Southeast Asia.

The Buddha is often depicted in poses of meditation.

Interpreting the Visual Record

What elements of this statue could show examples of meditation?

Jainism Jainism was founded at about the same time as Buddhism. It teaches that all things in nature—animals, plants, and stones—have souls. Jains reject all forms of violence against any living thing. They are strict vegetarians. Some even cover their noses with cloth to avoid breathing in insects and thus killing them.

Jains make up only a small minority. However, they have made many contributions to Indian art, mathematics, and literature.

Sikhism Sikhism was founded in the late 1400s. It combines elements of Hinduism and Islam. Members of this religion are called Sikhs. Traditionally, many Sikh men have become soldiers. They continue to play an important role in India's army.

Jain women worship at a temple in Kolkata.

DAILY LIFE

India's diverse religious communities provide many opportunities for festivals. Muslims celebrate 'Id al-Fitr and Mohurrum. Sikhs celebrate Guru Nanak Jayanti, Jains celebrate Mahavir Jayanti, Buddhists celebrate Buddha's birthday, and Christians celebrate Christmas.

Some Hindu festivals revolve around birthday celebrations for Hindu gods. Hindu celebrations combine religious ceremonies, processions, music, dances, and feasting. One festival, Diwali, takes place in October. It honors the goddess Laksmi and is celebrated through feasting, worship, ceremonial lights, and fireworks. Tiny oil lamps are lit to outline every house and hut. Navaratri, a nine-day festival, is celebrated throughout India but activities differ according to the region.

Critical Thinking: What does Diwali celebrate?

Answer: honors Laksmi; celebrated with feasting, worship, ceremonial lights, fireworks

internet connect go.hrw.com

GO TO: go.hrw.com
KEYWORD: SK3 CH27
FOR: Web sites about festivals of India

Visual Record Answer

possible answer: serene expression and calm posture

Teaching Objective 2

LEVEL 1: (Suggested time: 20 min.) Tell students that although caste has diminished somewhat in importance, it is still a feature of Indian society. Call on volunteers to describe the caste system in their own words. *(Students should mention that castes are groups of people whose birth determines their social position; that castes are ranked in status, from highest to lowest; that people from a higher caste cannot marry or even touch people of lower castes; and that people at the bottom of the caste system are called Dalits.)* Write correct descriptions on the chalkboard. Conclude by leading a discussion about why abolishing a traditional social system such as the caste system is so difficult. **ENGLISH LANGUAGE LEARNERS**

LEVELS 2 AND 3: (Suggested time: 40 min.) Ask each student to write a short essay about the features of the caste system and why it might be so difficult to abolish a traditional social system like the caste system. *(See the Level 1 lesson for the correct features.)* Ask volunteers to read their essays to the class.

➤ASSIGNMENT: Ask students to conduct further research on the caste system in India today. Tell students to concentrate on specific examples of government programs to help lower-caste Indians, the progress that lower-caste Indians have made, and the challenges that they still face. Have students prepare oral reports to present their findings.

HUMAN SYSTEMS

The majority of Sikhs live in the state of Punjab, where their religion originated. The Harmandir Sahib, or Golden Temple as it is more commonly known, is the religion's holiest site. The temple was built in the early 1600s under the leadership of the Sikhs' fifth guru.

The Sikhs have been forced to defend their temple against various invaders. Although it has been destroyed many times, Sikhs have rebuilt the temple each time. The front of the temple is covered with a thin layer of gold. Visitors enter the temple by walking across a marble causeway, which crosses a lake.

Critical Thinking: Why might many Sikh men have become soldiers?

Answer: Students might suggest that Sikh men have had a history of defending the Golden Temple.

Visual Record Answers ▲

Answers will vary. Students might mention Hasidic Jews, the Amish, or some Muslim women.

►

Students might suggest poverty and continued caste prejudice.

A Sikh guards the Golden Temple, the center of the Sikh religion, in Amritsar, India. Sikh men can be recognized by their beards and special turbans.

Interpreting the Visual Record **Do you know of any other religious or cultural groups whose members have a distinctive way of dressing?**

Most Sikhs live in the state of Punjab in northern India. Some Sikhs want to break away from India and form an independent country. The Indian government has refused to allow this, and violent clashes have resulted.

✓ **READING CHECK:** *Human Systems* What four religions originated in India? **Hinduism, Buddhism, Jainism, Sikhism**

Castes

Another key feature of Indian society is the **caste system**. Castes are groups of people whose birth determines their position in society. The castes are ranked in status, from highest to lowest. People from a higher caste cannot marry or even touch people of lower castes.

Many Dalits, like this woman in Goa, still perform jobs that Indians consider dirty or impure. Despite government efforts, most Dalits still live in poverty.

Interpreting the Visual Record **What factors might keep most Dalits from getting better jobs?**

Teaching Objective 3

LEVEL 1: (Suggested time: 40 min.) Have the class brainstorm the challenges that linguistic diversity would present to a country. Then pair students and have each pair draw a comic strip to show the role of languages in Indian society. *(Comic strips should show the diversity of languages, that Hindi is the national language, that South Indians have resisted adopting Hindi, and that English is still commonly used in government, business, and higher education. Comic strips should also show that languages are a source of division in India.)* Ask students to display and explain their comic strips to the class. **ENGLISH LANGUAGE LEARNERS, COOPERATIVE LEARNING**

LEVEL 2: (Suggested time: 30 min.) Lead a class discussion about why linguistic diversity might present challenges to a country. Then have each student create a poster to show how language divides Indian society. *(See the Level 1 lesson for the correct answers.)* Ask volunteers to present and explain their posters to the class.

LEVEL 3: (Suggested time: 30 min.) Have each student write a short poem about the different languages that are important in Indian society, and the challenges that linguistic diversity might present in India. *(See the Level 1 lesson for the correct answers.)* Ask volunteers to recite their poems for the class.

The people at the bottom, called **Dalits**, do work that higher castes consider unclean. They wash and cremate dead bodies, process cow hides into leather goods, and sweep up trash.

Gandhi tried to improve people's attitudes toward the Dalits. He called them "Children of God." After independence, the Indian government officially ended the caste system. However, it is still a strong force in Indian society. The government also tried to improve economic conditions for the Dalits. Today some Dalits are educated and have good jobs. The majority, though, are still poor.

✓ **READING CHECK:** *Human Systems* How does the caste system create conflict in Indian society? separates society, keeps some people poor

India

Country	Population/ Growth Rate	Life Expectancy	Literacy Rate	Per Capita GDP
India	1,029,991,145 1.7%	62, male 64, female	52%	$2,200
United States	281,421,906 0.9%	74, male 80, female	97%	$36,200

Sources: Central Intelligence Agency, *The World Factbook 2001;* U.S. Census Bureau

Interpreting the Chart How does India's economic development compare with that of the United States?

Languages

India's people speak an amazing number of different languages. There are 24 languages with a million or more speakers, plus hundreds of other languages. Hindi is the main language of about 30 percent of the people, mostly in northern India. In 1965 it became the official national language. However, the states of southern India have resisted the push to adopt Hindi. English is still commonly used in government, business, and higher education throughout India.

✓ **READING CHECK:** *Human Systems* What languages are important in India? Hindi, English, and 22 others

Government and Economy

India has made a great deal of economic progress since gaining independence, but the country remains poor. Rapid population growth strains the country's resources, and the divisions among India's people make it difficult to govern.

Government India is ruled by a democratic government. With more than 1 billion people, the country is the world's largest democracy. The structure of the government is based on Britain's parliamentary system. However, as in the United States, India's central government shares power with state governments.

Indian politics have sometimes been marked by violence and assassinations. The government used force to defeat Sikh rebels in 1984. There have also been outbreaks of

Parliament House, New Delhi, is the home of India's legislative branch.

Interpreting the Visual Record What form of government does India have? What is one feature that India's government shares with the U.S. government?

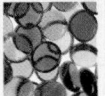

Cultural Kaleidoscope

Government and Diversity The Indian parliament consists of the president and two houses. The Lok Sabha is the lower house of parliament, and the Rajya Sabha is the upper house. The position of president in India is largely ceremonial—the prime minister wields the real power.

Much of India's constitution was built on laws enacted during British rule. However, the diversity of India's population has created unique challenges for the government. India's constitution is one of the longest in the world. Its 395 articles, numerous schedules, and more than 70 amendments affect the country's government all the way to the local level.

Critical Thinking: Why might India's constitution be so long and detailed?

Answer: Students might suggest that India's constitution must protect the country's many ethnic and religious minorities.

▲ **Chart Answer**

much lower

◄ **Visual Record Answer**

limited government—a parliamentary democracy; shares power with states

Teaching Objective 4

LEVELS 1 AND 2: (Suggested time: 30 min.) Have each student create a chart on India's government and economy. *(Charts should mention that India is a limited government—parliamentary democracy—is the world's largest democracy, and is based on a federal system. Charts should also mention that India's economy is a mixture of the traditional and market; that about 65 percent of India's workforce are farmers; that India is considered a developing country, but its economy ranks among the 10 largest in the world; that India has modern factories and high-tech service industries; and that India lacks good roads and good telecommunications.)* Ask students to discuss how India's government and economy compare to that of the United States. **ENGLISH LANGUAGE LEARNERS**

LEVEL 3: (Suggested time: 40 min.) Ask each student to create a collage to represent the important features of India's government and economy. *(See the Level 1 lesson for the correct features.)* Ask volunteers to display and explain the different elements of their collages to the class.

Section 3 Review

Answers

Define For definitions, see: reincarnation, p. 630; karma, p. 630; nirvana, p. 631; caste system, p. 632; Dalits, p. 633; green revolution, p. 635

Working with Sketch Maps Kashmir should be located accurately. Pakistan claims Kashmir.

Reading for the Main Idea

1. Hinduism, Islam, Christianity, Sikhism, Buddhism, Jainism; Hinduism (NGS 10)
2. by instituting practices including greater use of fertilizer and pesticides and encouraging farmers to grow new varieties of wheat and rice (NGS 14)

Critical Thinking

3. The majority are poor, and caste prejudice still exists in Indian society.
4. because millions of Indians cannot read

Organizing What You Know

5. crops—rice, wheat, cotton, tea, sugarcane, and jute; industries—moviemaking, textiles, jewelry, cars, bicycles, oil products, chemicals, food processing, and electronics

Visual Record Answer

horse-drawn carts

India has stationed large numbers of troops in Kashmir.

violence between Hindus and Muslims. In 1992 a mob of Hindus tore down a mosque that stood on a Hindu holy site. Riots broke out in many parts of India as a result.

India's border with Pakistan has been in dispute since 1947. Both countries claim a mountainous region called Kashmir. Before India gained independence, Kashmir was ruled by a Hindu prince. Most of its people were Muslim, however. India and Pakistan have fought over Kashmir several times. Today, both countries have nuclear weapons, making the prospect of a future war even more frightening.

Economy India's economy is a mixture of the traditional and the modern. In thousands of villages, farmers work the fields just as they have for centuries. At the same time, modern factories and high-tech service industries demonstrate India's potential for wealth. However, the country still does not have enough good roads and telecommunications systems.

Close to 60 percent of India's workforce are farmers. Farming makes up 25 percent of India's GDP. Most farmers work on small farms less than 2.5 acres (1 hectare) in size. Many grow barely enough to feed themselves and their families. In recent years, the government has worked to promote commercial farming.

India's leading crops include rice, wheat, cotton, tea, sugarcane, and jute. Cattle and water buffalo are raised to pull plows and to provide milk.

Peppers are harvested in northern India.

Interpreting the Visual Record **What is one way agricultural products are transported in India?**

CLOSE

Ask students to recount facts about India's religions, government, and economy.

REVIEW AND ASSESS

Have students complete the Section Review. Then have students write one question on each of the following topics: religion and caste system, languages, government, and economy. Collect students' questions and use them to quiz the class. Then have students complete Daily Quiz 27.3.

RETEACH

Have students complete Main Idea Activity 27.3. Then organize students into three groups and assign each a section of the chapter. Have each group prepare an outline of its section and present the outline to the class. **ENGLISH LANGUAGE LEARNERS, COOPERATIVE LEARNING**

EXTEND

Have interested students conduct research on Lakshmi Bai, the Rani of Jhansi, who led rebels against British troops during the Indian Mutiny, and on Phoolan Devi, whose life became the subject of the film *Bandit Queen.* They might compare and contrast the lives of these two women to more traditional Indian women's roles. **BLOCK SCHEDULING**

Beginning in the 1960s, the Indian government started agricultural programs known as the **green revolution**. This effort encouraged farmers to adopt more modern methods. It promoted greater use of fertilizers, pesticides, and new varieties of wheat and rice. Crop yields increased. In years with good weather, India is self-sufficient in food and can export farm products.

India is considered a developing country. However, its economy is large enough to rank among the world's top 10 industrial countries. India's industries include textiles, jewelry, cars, bicycles, oil products, chemicals, food processing, and electronics.

India's moviemaking industry is one of the world's largest. Mumbai is a major moviemaking center. Movies are an incredibly popular form of entertainment, as millions of Indians cannot read. Many Indian movie stars have gone into politics. Indian movies have a distinctive style. They usually feature music and dancing and often draw on themes from Indian myths. Indian movies are popular in many other countries as well.

India now has a large, well-educated middle class. These people have enough money for luxuries like cable television and personal computers. Some Indians are very rich. Yet the majority of Indians are still poor.

▲ Red-hot steel is poured into molds in a foundry near Kolkata.

Interpreting the Visual Record
Which of India's industries might use the steel produced here?

✓ **READING CHECK:** *Places and Regions* What are India's government and economy like? government—limited, parliamentary democracy; economy—mix of traditional and market, among the 10 largest in the world

Section 3 Review

go.hrw.com Homework Practice Online
Keyword: SK3 HP27

Define and explain: reincarnation, karma, nirvana, caste system, Dalits, green revolution

Working with Sketch Maps On the map that you created in Section 2, label the Kashmir region of India. What other country claims Kashmir?

Reading for the Main Idea

1. *Human Systems* What are the main religions practiced in India? Which religion has the largest number of followers?
2. *Environment and Society* How did India's government increase agricultural output in the 1960s?

Critical Thinking

3. **Drawing Inferences and Conclusions** Why might India's government find it difficult to improve the economic situation of the Dalits?
4. **Finding the Main Idea** Why are movies such a popular form of entertainment in India?

Organizing What You Know

5. **Categorizing** Copy the following graphic organizer. Use it to list the leading crops and industries of India.

Leading crops	Leading industries

CHAPTER 27

Review ANSWERS

Building Vocabulary

For definitions, see: teak, p. 623; Sanskrit, p. 625; sepoys, p. 628; boycott, p. 628; reincarnation, p. 630; karma, p. 630; nirvana, p. 631; caste system, p. 632; Dalits, p. 633; green revolution, p. 635

Reviewing the Main Ideas

1. Two tectonic plates collided and pushed Earth's crust up. (NGS 4)
2. the Harappan civilization (NGS 9)
3. Indo-Aryans (NGS 9)
4. reincarnation—human soul is reborn many times in different bodies; karma—a person's actions create a positive or negative force (NGS 10)
5. religion, caste, and language (NGS 10)

Understanding Environment and Society

Presentations will vary but should be consistent with text material. Use Rubric 29, Presentations, to evaluate student work.

◄ **Visual Record Answer**

manufacturing cars or bicycles

ASSESS

Have students complete the Chapter 27 Test.

RETEACH

Provide students with a large piece of posterboard. Have them create a map of India showing its physical features. Organize students into three large groups to create symbols of Indian history, culture, and economy. Students can glue or tape their symbols to the map. Display and discuss students' work. **ENGLISH LANGUAGE LEARNERS, COOPERATIVE LEARNING**

PORTFOLIO

PORTFOLIO EXTENSIONS

1. Have students use the six essential elements of geography to create clothes-hanger mobiles about India. You may wish to refresh students' memories about the six elements. Display students' mobiles around the classroom. Then remove the hangers to place mobiles in students' portfolios.
2. Organize students into groups of five or six and have each group write a play about India's independence movement. Have students conduct biographical research on important Indian nationalist or British figures. Have students perform their plays for the class. Make copies of the scripts and place them in students' portfolios.

CHAPTER 27

Review ANSWERS

Thinking Critically

1. Southern Indian states have resisted adopting Hindi. (NGS 9)
2. They are used for textiles, jewelry, cars, bicycles, oil products, chemicals, and food processing.
3. encouraged farmers to change their methods of farming through new technology and crops; increased crop yields and exports
4. The empire was very powerful, and Akbar and his successors sponsored architecture, painting, and poetry. They also promoted religious tolerance.
5. might bring more income to the country

Map Activity

A. Brahmaputra River
B. Himalayas
C. Deccan
D. Western Ghats
E. Eastern Ghats
F. Thar Desert
G. Gangetic Plain
H. Ganges River

CHAPTER 27

Reviewing What You Know

Building Vocabulary

On a separate sheet of paper, write sentences to define each of the following words.

1. teak
2. Sanskrit
3. sepoys
4. boycott
5. reincarnation
6. karma
7. nirvana
8. caste system
9. Dalits
10. green revolution

Reviewing the Main Ideas

1. *Places and Regions* How were the Himalayas formed?
2. *Human Systems* What was the first urban civilization on the Indian Subcontinent?
3. *Human Systems* Who brought Sanskrit to India?
4. *Human Systems* What are the Hindu ideas of reincarnation and karma?
5. *Human Systems* What factors divide Indian society?

Understanding Environment and Society

Cows in India

The Hindu respect for cows has influenced the way Indians interact with their environment. Create a presentation about cows in India. As you prepare your presentation, consider the following:

- How Hindus' respect for cows has affected cattle populations in India.
- What cows eat, and whether humans ever have to compete with cows for food in India.
- How Hindu ideas of reincarnation and karma affect Indians' treatment of cows.

Thinking Critically

1. **Drawing Inferences and Conclusions** Why do you think Hindi has not become the language of all of India?
2. **Summarizing** How do India's natural resources affect its economy?
3. **Finding the Main Idea** How did the Indian government's promotion of new farming technology affect India's farming culture?
4. **Drawing Inferences and Conclusions** Why do you think the Mughal Empire under Akbar and his successors was considered a "golden age"?
5. **Drawing Inferences and Conclusions** How might the growth of a well-educated, high-tech workforce affect India's economy?

FOOD FESTIVAL

A tandoor is a clay oven often used in India to cook meat that has been marinated. Traditional tandoori marinade includes a mixture of plain yogurt, lemon juice, ginger, garlic, cloves, cumin, coriander, hot peppers, and turmeric. The meat can be served with rice and chutney or sliced and served on bread with mayonnaise flavored with mint, cilantro, chili, onion, and vinegar. Many Indian dishes contain blends of herbs, spices, fruits, nuts, grains, and yogurt.

CHAPTER 27 REVIEW AND ASSESSMENT RESOURCES

Reproducible
- Readings in World Geography, History, and Culture 75 and 76
- Critical Thinking Activity 27
- Vocabulary Activity 27

Technology
- Chapter 27 Test Generator (on the One-Stop Planner)
- Audio CD Program, Chapter 27
- HRW Go site

Reinforcement, Review, and Assessment
- Chapter 27 Review, pp. 636–37
- Chapter 27 Tutorial for Students, Parents, Mentors, and Peers
- Chapter 27 Test
- Chapter 27 Test for English Language Learners and Special-Needs Students

Building Social Studies Skills

Map ACTIVITY

On a separate sheet of paper, match the letters on the map with their correct labels.

Himalayas	**Ganges River**
Gangetic Plain	**Brahmaputra River**
Deccan	**Thar Desert**
Eastern Ghats	
Western Ghats	

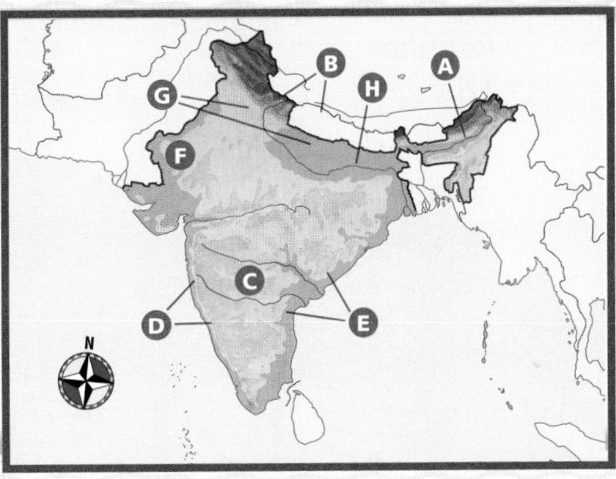

Mental Mapping Skills ACTIVITY

On a separate sheet of paper, draw a freehand map of the Indian Subcontinent. Make a key for your map and label the following:

Bangladesh	**Kashmir**
Bay of Bengal	**Nepal**
India	**Pakistan**
Indian Ocean	

WRITING ACTIVITY

Write a one- to two-page report about how religion influences life in India. List and describe India's major religions, then write about how they relate to politics, border disputes, social classes, daily life, languages, the economy, and so on. Be sure to use standard grammar, spelling, sentence structure, and punctuation.

Alternative Assessment

Portfolio ACTIVITY

Learning About Your Local Geography

The Seasons The change of the seasons is important to daily life in India. Create a poster on how temperature and precipitation of all four seasons affect daily life in your area.

internet connect

Internet Activity: **go.hrw.com**
KEYWORD: **SK3 GT27**

Choose a topic to explore about India:
- Tour the regions of India.
- Travel to ancient India.
- Learn about Mohandas Gandhi.

Mental Mapping Skills Activity

Maps will vary, but listed places should be labeled in their approximate locations.

Writing Activity

Reports will vary, but the information included should be consistent with text material. Use Rubric 42, Writing to Inform, to evaluate student work.

Portfolio Activity

Posters will vary, but the information included should be consistent with text material. Use Rubric 28, Posters, and 30, Research, to evaluate student work.

internet connect

GO TO: **go.hrw.com**
KEYWORD: **SK3 Teacher**
FOR: a guide to using the Internet in your classroom

CHAPTER 28

The Indian Perimeter

Chapter Resource Manager

Objectives	Pacing Guide	Reproducible Resources
SECTION 1 **Physical Geography** (pp. 639–41) **1.** Identify the major physical features located in the Indian Perimeter. **2.** Identify the climates and natural resources found in the region. **3.** Describe the physical features of the island countries.	**Regular** .5 day **Block Scheduling** .5 day *Block Scheduling Handbook, Chapter 28*	**RS** Guided Reading Strategy 28.1 **SM** Geography for Life Activity 28 **E** Lab Activity for Geography and Earth Science, Demonstration 10
SECTION 2 **Pakistan and Bangladesh** (pp. 642–45) **1.** Describe the history of Pakistan. **2.** Identify some features of Pakistan's culture. **3.** Describe the history of Bangladesh. **4.** Identify the challenges that face Bangladesh today.	**Regular** 1 day **Block Scheduling** .5 day *Block Scheduling Handbook, Chapter 28*	**RS** Guided Reading Strategy 28.2 **RS** Graphic Organizer 28 **E** Cultures of the World Activity 8 **IC** Interdisciplinary Activities for the Middle Grades 33, 34, 35
SECTION 3 **The Himalayan Countries** (pp. 648–53) **1.** Identify some important features of Nepal. **2.** Describe how Bhutan developed over time. **3.** Describe what Sri Lanka is like. **4.** Identify some economic activities of the Maldives.	**Regular** 2 days **Block Scheduling** .5 day *Block Scheduling Handbook, Chapter 28*	**RS** Guided Reading Strategy 28.3 **IC** Interdisciplinary Activities for the Middle Grades 33, 34, 35 **SM** Map Activity 28

Chapter Resource Key

RS Reading Support

IC Interdisciplinary Connections

E Enrichment

SM Skills Mastery

A Assessment

REV Review

ELL Reinforcement and English Language Learners

Transparencies

CD–ROM

Music

Video

go.hrw.com Internet

Holt Presentation Maker Using Microsoft® Powerpoint®

One-Stop Planner CD-ROM

See the ***One-Stop Planner*** for a complete list of additional resources for students and teachers.

One-Stop Planner CD–ROM

It's easy to plan lessons, select resources, and print out materials for your students when you use the ***One-Stop Planner CD–ROM with Test Generator.***

Technology Resources	Review, Reinforcement, and Assessment Resources
One-Stop Planner CD–ROM, Lesson 28.1 Geography and Cultures Visual Resources with Teaching Activities 55–59 Homework Practice Online HRW Go site	**ELL** Main Idea Activity 28.1 **REV** Section 1 Review, p. 641 **A** Daily Quiz 28.1 **ELL** English Audio Summary 28.1 **ELL** Spanish Audio Summary 28.1
One-Stop Planner CD–ROM, Lesson 28.2 Earth: Forces and Formations CD–ROM/Seek and Tell/Forces and Processes *ARGWorld* CD–ROM: Predicting Floods in Bangladesh Homework Practice Online HRW Go site	**ELL** Main Idea Activity 28.2 **REV** Section 2 Review, p. 645 **A** Daily Quiz 28.2 **ELL** English Audio Summary 28.2 **ELL** Spanish Audio Summary 28.2
One-Stop Planner CD–ROM, Lesson 28.3 Homework Practice Online HRW Go site	**ELL** Main Idea Activity 28.3 **REV** Section 3 Review, p. 653 **A** Daily Quiz 28.3 **ELL** English Audio Summary 28.3 **ELL** Spanish Audio Summary 28.3

internet connect

HRW ONLINE RESOURCES

GO TO: go.hrw.com
Then type in a keyword.

TEACHER HOME PAGE
KEYWORD: SK3 TEACHER

CHAPTER INTERNET ACTIVITIES
KEYWORD: SK3 GT28

Choose an activity to:
- climb the Himalayas.
- learn about Sri Lanka's history.
- visit the regions of Pakistan.

CHAPTER ENRICHMENT LINKS
KEYWORD: SK3 CH28

CHAPTER MAPS
KEYWORD: SK3 MAPS28

ONLINE ASSESSMENT
Homework Practice
KEYWORD: SK3 HP28
Standardized Test Prep Online
KEYWORD: SK3 STP28
Rubrics
KEYWORD: SS Rubrics

COUNTRY INFORMATION
KEYWORD: SK3 Almanac

CONTENT UPDATES
KEYWORD: SS Content Updates

HOLT PRESENTATION MAKER
KEYWORD: SK3 PPT28

ONLINE READING SUPPORT
KEYWORD: SS Strategies

CURRENT EVENTS
KEYWORD: S3 Current Events

Meeting Individual Needs

Ability Levels

Level 1 Basic-level activities designed for all students encountering new material

Level 2 Intermediate-level activities designed for average students

Level 3 Challenging activities designed for honors and gifted-and-talented students

English Language Learners Activities that address the needs of students with Limited English Proficiency

Chapter Review and Assessment

E Readings in World Geography, History, and Culture 77 and 78
SM Critical Thinking Activity 28
REV Chapter 28 Review, pp. 654–55
REV Chapter 28 Tutorial for Students, Parents, Mentors, and Peers
ELL Vocabulary Activity 28
A Chapter 28 Test
A Unit 8 Test
Chapter 28 Test Generator (on the One-Stop Planner)
Audio CD Program, Chapter 28
A Chapter 28 Test for English Language Learners and Special-Needs Students
A Unit 8 Test for English Language Learners and Special-Needs Students

CHAPTER 28

Launch into Learning

Provide small jars containing some of the aromatic spices used in the cuisines of the Indian Perimeter. Possibilities include asafetida, black pepper, cardamom, cloves, coriander, cumin, fennel, and tumeric. (Most of these are available from supermarkets.) Allow students to pass the jars around and smell the spices. Ask them what images the scents bring to mind. Discuss responses. *(Students might mention busy markets, caravans, or other images.)* Point out that cooks in the region grind their own spices and create their own unique spice mixtures. Tell students that they will learn more about the history, cultures, and foods of the Indian Perimeter countries in this chapter.

Section 1

Objectives

1. Identify the major physical features located in the Indian Perimeter.
2. Discuss the climates and natural resources found in this region.
3. Name the physical features of the island countries.

Links to Our Lives

You may wish to point out to students that there are many reasons why we should know more about the countries of the Indian Perimeter, these among them:

- These countries contain many sites of historical and cultural interest as well as spectacular landscapes that have been popular with adventurers and tourists.
- Pakistan and India have developed nuclear weapons. Consequently, political developments in the region might gain global significance.
- Some countries of the region are very poor and have rapidly growing populations. Developing the human potential of these populations and establishing economic stability in the region can benefit the world economy.

CHAPTER 28

The Indian Perimeter

The countries of this region, along with the Himalayas, help create India's border. After you meet Rehan you will learn that this land is one of majestic beauty with a rich heritage.

I am Rehan, and I am 14. I am an only child and live with my parents in Karachi, a big sprawling city like Los Angeles. On one side is the sea, on the other is the desert.

If you came to visit me in Pakistan, I would take you to the beach to watch the beautiful sunsets and ride on a camel. My parents used to take me there for camel rides when I was very little. Next I would take you to see the old colonial architecture in the city center. Then we would go and have a meal in a roadside cafe—grilled beef or lamb kabobs on a stick. I usually get up very early for school. By 7:10 A.M. I have breakfast—cereal and toast—and leave for school with my father. He is a doctor with an office near my school. I am in the second year (equivalent to grade 10) of a boys' private school styled after the British public school system.

Next year, I am going to America with my mother. My parents want me to have a chance to go to a world-class university.

▲ Translation: God's peace be upon you!

Focus

LET'S GET STARTED

Copy the following instructions onto the chalkboard: *In 1998 Bangladesh suffered from flooding so severe that two thirds of the land was under water. Use the physical and climate maps in this unit's atlas to list factors that might have contributed to the flooding.* Discuss responses. *(Students might mention low elevation, proximity to rivers, or heavy rains.)* Tell students that in Section 1 they will learn more about how the region's physical geography contributes to flooding in Bangladesh. They will also learn about the physical geography of the Indian Perimeter in Section 1.

Using the Physical-Political Map

Have students examine the map on this page. Then ask students to name all the mainland Indian Perimeter countries. *(Pakistan, Nepal, Bhutan, and Bangladesh)* Then ask students to locate the island nations that are part of the Indian Perimeter. *(Sri Lanka and the Maldives)* Conclude by discussing the ways in which India might dominate this region geographically, politically, strategically, and culturally.

Building Vocabulary

Write the terms **cyclones** and **storm surges** on the chalkboard. Have volunteers read the terms' definitions aloud from Section 1 or the glossary. Point out to students that these words refer to natural forces that have frequently caused disasters in this region, particularly in Bangladesh.

Section 1 Physical Geography

Read to Discover

1. What major physical features are located in the Indian Perimeter?
2. What climates and natural resources are found in this region?
3. What are the physical features of the island countries?

Define

cyclones
storm surges

Locate

Brahmaputra River
Ganges River
Himalayas
Mount Everest
Tarai
Karakoram Range
Hindu Kush
Khyber Pass
Indus River
Thar Desert

Why It Matters

In 2001 Pakistan played an important role in world events when the United States asked to use its airspace to conduct military actions against terrorist leader Osama bin Laden. Use CNNfyi.com or other **current events** sources to find the latest information on Pakistan's role in world affairs. Record your findings in your journal.

A relief sculpture of a Bodhi tree

The Indian Perimeter: Physical-Political

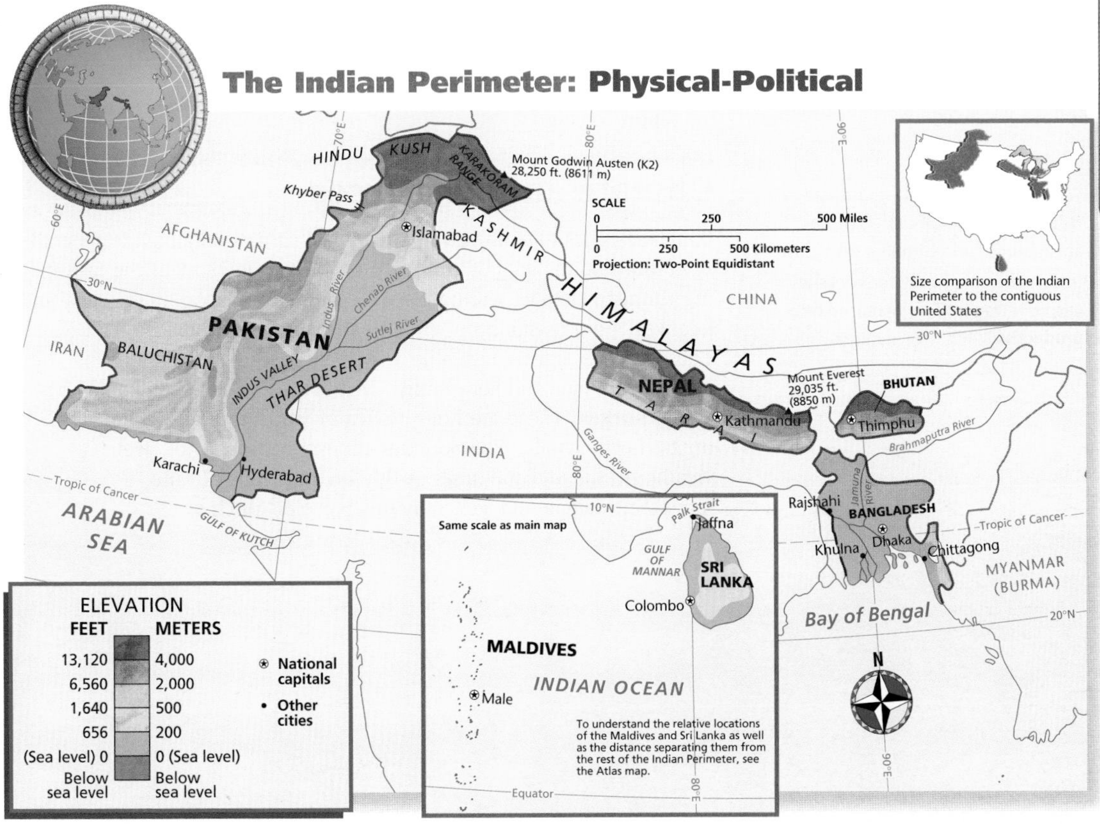

Section 1 Resources

Reproducible

- Block Scheduling Handbook, Chapter 28
- Guided Reading Strategy 28.1
- Geography for Life Activity 28
- Lab Activity for Geography and Earth Science, Demonstration 10

Technology

- One-Stop Planner CD–ROM, Lesson 28.1
- Homework Practice Online
- Geography and Cultures Visual Resources with Teaching Activities 55–59
- HRW Go site

Reinforcement, Review, and Assessment

- Section 1 Review, p. 641
- Daily Quiz 28.1
- Main Idea Activity 28.1
- English Audio Summary 28.1
- Spanish Audio Summary 28.1

TEACH

Teaching Objectives 1–2

ALL LEVELS: (Suggested time: 20 min.) Pair students and have each pair create a mock Web page that lists the physical features, climate types, and natural resources of the northern Indian Perimeter countries. *(Web pages should mention the Brahmaputra, Ganges, and Indus Rivers; Brahmaputra and Ganges Deltas; Himalayas and Mount Everest; Karakoram Range, Hindu Kush, Khyber Pass, Indus Valley, and Thar Desert; the wet climate of Bangladesh, lowland climate and cool mountain climates of Bhutan and Nepal, and the desert climate of Pakistan; farmland and timber in Bangladesh; farmland and minerals in Nepal and Bhutan; and natural gas, coal, limestone, and salt in Pakistan.)* Display Web pages around the classroom. **ENGLISH LANGUAGE LEARNERS, COOPERATIVE LEARNING**

Teaching Objective 3

ALL LEVELS: (Suggested time: 10 min.) Copy the following graphic organizer onto the chalkboard, omitting the italicized answers. Have each student complete the organizer. Ask volunteers to share their answers with the class. **ENGLISH LANGUAGE LEARNERS**

Physical Features of the Island Countries

SRI LANKA	MALDIVES
• *plains in northern half and coastal areas* • *mountains and hills in south-central part*	• *group of about 1,200 tiny islands* • *elevation of 6 feet or less above sea level*

National Geography Standard 13

Human Systems The Siachen Glacier in the Karakoram Range bears the distinction of being the world's highest battlefield. Since 1984, Indian and Pakistani troops have fought for control of the strip of inhospitable land 47 miles (76 km) long. Each country is afraid that the other will gain control of this previously unclaimed glacier.

Thus far, very few soldiers on either side have been killed by hostile fire. Temperatures on the glacier can drop to –40°C (–40°F), and some troops are stationed as high as 22,000 feet (6,706 m) above sea level. The punishing altitude, weather, and terrain have caused the majority of casualties.

Activity: Have students conduct additional research on military conflicts between India and Pakistan. Ask them to concentrate on the causes of the conflicts and possible solutions.

Visual Record Answer

by allowing farmers to grow more crops in a limited space

BUILD on WHAT You Know

Do you remember what you learned about deltas? See Chapter 2 to review.

internet connect

GO TO: **go.hrw.com**
KEYWORD: **SK3 CH28**
FOR: **Web sites about the Indian Perimeter**

Physical Features

The broad delta formed by the Brahmaputra (brahm-uh-POO-truh) and Ganges (GAN-jeez) Rivers covers most of Bangladesh. Some 200 rivers and streams crisscross this eastern part of the Indian Subcontinent. These numerous waterways, the low elevation of the land, and heavy monsoon rains combine to bring frequent floods to Bangladesh. Although these floods cause great damage, they leave behind a layer of fertile soil.

North of Bangladesh is Bhutan. This tiny country lies high in the mountain range known as the Himalayas (hi-muh-LAY-uhz). To the west is Nepal. The Himalayas occupy some 75 percent of Nepal's land area. Mount Everest, Earth's highest mountain, is located on Nepal's border with China. The Tarai (tuh-RY) is a low plain along Nepal's southern border. It is the country's main farming area. West of the Himalayas is the Karakoram (kah-rah-KOHR-oohm) Range. To the west the Karakorams merge into another mountain range, the Hindu Kush.

On Pakistan's western border is the Khyber (KY-buhr) Pass. For centuries, invaders and traders have traveled through this high mountain pass to India. East of the Khyber Pass is the Indus River. The Indus Valley lies mostly to the east of the river. This valley is Pakistan's main farming region and its most heavily populated area. East of these fertile lands is the Thar (TAHR) Desert, or the Great Indian Desert. A barren, hilly, and dry plateau in western Pakistan joins the plateaus of Iran.

✓ **READING CHECK:** *Places and Regions* What are the major physical features of this region? **delta, rivers, streams, Himalayas, Khyber Pass, Indus Valley, deserts**

Climate and Resources

Bangladesh has one of the world's wettest climates. Rainfall is generally more than 60 inches (127 to 152 cm) each year. Most of the rain falls from June to October, during the wet summer monsoon. In the early and late weeks of the monsoon, **cyclones** sweep in from the Bay of Bengal. These violent storms resemble the hurricanes of the Caribbean. They bring high winds and heavy rain. Cyclones are often accompanied by **storm surges**. These are huge waves of water that are whipped up by fierce winds. The summer monsoon brings hot, wet weather to the lowland areas of Bhutan and Nepal. In the mountains climates are generally much cooler. Much of

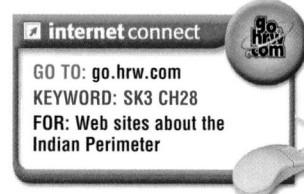

Terraced fields in the mountains of Nepal allow farmers to increase their production of millet and corn.

Interpreting the Visual Record **How can terracing lead to increased crop production?**

CLOSE

Read descriptions of or list geographic features of the countries mentioned in Section 1. Have students identify the country described.

REVIEW AND ASSESS

Have students complete the Section Review. Then organize students into small groups and have each group write five matching questions about each country's physical geography. Have groups quiz other groups with the questions they have prepared. Then have students complete Daily Quiz 28.1. **COOPERATIVE LEARNING**

RETEACH

Have students complete Main Idea Activity 28.1. Then have students design their own graphic organizers to depict answers to the Read to Discover questions. **ENGLISH LANGUAGE LEARNERS**

EXTEND

Have interested students conduct research on one of the invasions that occurred through the Khyber Pass. Have each student give a short presentation about his or her chosen event. Ask students to use maps or other visual resources in their presentations. **BLOCK SCHEDULING**

Pakistan has a desert climate, receiving less than 10 inches (25 cm) of rain each year. Summer temperatures can reach as high as 120°F (49°C).

Bangladesh's most important resource is its fertile farmland. About 15 percent of Bangladesh is forested, so it has some timber supplies. However, severe deforestation and soil erosion have plagued the region, particularly in Nepal. Bhutan and Nepal have farmland in lowland areas. Both countries have some minerals, but few are mined. Pakistan has large natural gas reserves but limited oil supplies. It has to import oil to meet its energy needs. Pakistan's other natural resources include coal, limestone, and salt.

✓ **READING CHECK:** *Places and Regions* What are the natural resources of the region? fertile farmland, timber, some minerals, natural gas

Our Amazing Planet

In May 1997 a cyclone and storm surges devastated Bangladesh. More than 1.5 million people were left homeless.

The Island Countries

The Indian Perimeter region also includes Sri Lanka and the Maldives. Sri Lanka is a large island located just off the southeastern tip of India. Plains cover most of the island's northern half and coastal areas. Mountains and hills rise in the south-central part of the island.

About 1,200 tiny tropical islands in the Indian Ocean make up the Maldives. The island group stretches from south of India to the equator. Only about 200 of the islands are inhabited. None rises more than 6 feet (1.8 m) above sea level.

✓ **READING CHECK:** *Places and Regions* What are the physical features of the island countries? plains, mountains, hills, low elevation

go.hrw.com Homework Practice Online
Keyword: SK3 HP28

Section 1 Review

Define and explain: cyclones, storm surges

Working with Sketch Maps On a map of India and the Indian Perimeter that you draw or that your teacher provides, label the following: Brahmaputra River, Ganges River, Himalayas, Mount Everest, Tarai, Karakoram Range, Hindu Kush, Khyber Pass, Indus River, and Thar Desert. Where is Earth's highest mountain located? What plains area is Nepal's main farming region?

Reading for the Main Idea

1. *Places and Regions* Why has the Khyber Pass been important in the history of Pakistan and India?
2. *Places and Regions* How have erosion and deforestation affected the region? Which country has been most affected by these problems?
3. *Places and Regions* What island countries are found in the region? How are they different from each other?

Critical Thinking

4. **Drawing Inferences and Conclusions** How do you think climate affects life in Pakistan and Bangladesh?

Organizing What You Know

5. **Summarizing** Copy the following graphic organizer. Use it to describe the landforms, climates, and resources of the region.

	Major Landforms	Climates	Resources
Pakistan			
Bangladesh			
Bhutan			
Nepal			

Section 1 Review Answers

Define For definitions, see: cyclones, p. 640; storm surges, p. 640

Working with Sketch Maps Places should be labeled in their approximate locations. The mountain is on Nepal's border with China; the Tarai is the farming area.

Reading for the Main Idea

1. used for centuries for access into Indian interior (NGS 4)
2. depleted available timber and fertile soil; Nepal (NGS 4)
3. Sri Lanka—large island; plains, mountains, hills; Maldives—about 1,200 tiny flat tropical islands (NGS 4)

Critical Thinking

4. Possible answer: by determining when crops can be planted

Organizing What You Know

5. Pakistan—Hindu Kush, Karakoram Range, Indus Valley, Thar Desert; desert; natural gas, coal, limestone, salt; Bangladesh—delta of Brahmaputra and Ganges Rivers; wet; farmland and forest; Bhutan—Himalayas; hot and wet, cool; farmland; Nepal—Himalayas, Mount Everest, Tarai; hot and wet, cool; farmland

Section 2

Objectives

1. Discuss Pakistan's history.
2. Describe some features of Pakistan's culture.
3. Discuss Bangladesh's history.
4. Identify challenges facing Bangladesh today.

FOCUS

LET'S GET STARTED

Copy the following passage onto the chalkboard: *How would you learn about the thoughts and emotions of people who lived long ago if you could not read their language? Think about what other things might serve as clues about a civilization.* Discuss responses. Tell students that there was an advanced civilization in the Indus Valley whose language we cannot read. Tell students that in Section 2 they will learn about the history and culture of Pakistan and Bangladesh.

Building Vocabulary

Write the term **cholera** on the chalkboard. Have a volunteer read aloud the term's definition from the dictionary as well as the definition of the term's root word, *choler.* Ask students what the term's root word might indicate about the nature of the disease. *(Students might respond that the root word suggests the violent nature of the disease.)*

Section 2 RESOURCES

Reproducible
- Guided Reading Strategy 28.2
- Graphic Organizer 28
- Cultures of the World Activity 8
- Interdisciplinary Activities for the Middle Grades 33, 34, 35

Technology
- One-Stop Planner CD–ROM, Lesson 28.2
- Homework Practice Online
- Earth: Forces and Formations CD–ROM/Seek and Tell/Forces and Processes
- *ARGWorld* CD–ROM: Predicting Floods in Bangladesh
- HRW Go site

Reinforcement, Review, and Assessment
- Section 2 Review, p. 645
- Daily Quiz 28.2
- Main Idea Activity 28.2
- English Audio Summary 28.2
- Spanish Audio Summary 28.2

Visual Record Answer

Answers will vary according to communities' traditions.

Section 2 Pakistan and Bangladesh

Read to Discover

1. What is the history of Pakistan?
2. What are some features of Pakistan's culture?
3. What is the history of Bangladesh?
4. What challenges face Bangladesh today?

Define

cholera

Locate

Pakistan, Bangladesh, Karachi, Lahore, Islamabad, Dhaka

WHY IT MATTERS

Since 1947, when Great Britain granted both nations independence, Pakistan and India have fought three wars. In recent years the countries have clashed over a revolt in the Indian state of Kashmir. Use CNNfyi.com or other **current events** sources to find out more about Pakistan's relations with India. Record your findings in your journal.

A floral-patterned container from Pakistan

These tombs in Pakistan were built in the 1500s–1700s. Women's graves had floral carvings. Horses and swords decorated men's tombs.

Interpreting the Visual Record **How do these tombs differ from modern-day grave sites in your community?**

Pakistan's History

An ancient civilization developed in the Indus River valley about 2500 B.C. Ruins of cities show that this was a large and well-organized society. The cause of the disappearance of this civilization about 1500 B.C. remains a mystery.

Over time the fertile Indus River valley was inhabited and conquered by many different groups. It has been part of the empires of Persia, Alexander the Great, and the Mughals. About A.D. 1000 Turkish invaders established Islam in the area. Islam has been the main religion there ever since.

In the early 1600s merchants from England formed the English East India Company to increase the spice trade. It later became known as the British East India Company. Although trade was only moderately successful, the company established England's power in India. It was not until 1947 that India gained independence.

Upon independence, India became two countries. The division into two countries was based on religion. India was mostly Hindu. East and West Pakistan were inhabited mainly by Muslims. Although East Pakistan and West Pakistan were mostly Muslim, they had other cultural differences. These differences led East Pakistan to break away in 1971. It became known as Bangladesh.

✓ **READING CHECK:** *Human Systems* What was Pakistan's early history like?

early civilizations in Indus valley; Turkish invaders established Islam; British dominated India; Indian independence and division in 1947

TEACH

Teaching Objectives 1 and 3

ALL LEVELS: (Suggested time: 20 min.) Pair students and have each pair create an annotated and forked time line that shows the common history of Pakistan and Bangladesh as well as their separate histories. *(Time lines should mention Indus civilization, invasion and establishment of Islam c. A.D. 1000, the Mughal Empire, British trade and rule of the region, establishment of Pakistan in 1947, and Bangladesh's secession from Pakistan in 1971.)* Ask volunteers to explain their time lines to the class.

ENGLISH LANGUAGE LEARNERS, COOPERATIVE LEARNING

Teaching Objective 2

LEVEL 1: (Suggested time: 20 min.) Pair students and have each pair create a collage to represent features of Pakistani culture. *(Collages should show the predominance of the Muslim religion, the variety of languages and ethnic groups, male domination of society, arranged marriages, and Pakistani cuisine's similarity to Iranian and Indian cuisine.)* Have volunteers discuss their collages with the class.

ENGLISH LANGUAGE LEARNERS, COOPERATIVE LEARNING

Pakistan Today

Pakistan's population is 97 percent Muslim. The country has many different languages and ethnic groups. A small number of Christians, Buddhists, and Hindus also live in Pakistan. Urdu is Pakistan's official language. However, less than 10 percent of the population speak it as their primary language. Many upper-class Pakistanis speak English.

Population Pakistan has the world's seventh largest population. The cities contain about a third of that population. The largest are the port city of Karachi (kuh-RAH-chee), Lahore (luh-HOHR), and the capital, Islamabad. Like the rest of the region, Pakistan is experiencing rapid population growth. In fact, the Indian Subcontinent accounted for about 30 percent of the world's population growth in the late 1990s. How does Pakistan support so many people? The Indus River valley has one of the world's largest irrigation systems. It allows the country to grow enough food for the large population. However, overall economic progress has been slow.

Culture In Pakistan a woman joins her husband's family at marriage. Marriages are usually arranged by parents. The young woman's parents often pay a large amount of money to the young man's family.

Pakistan celebrates many of the same Islamic festivals as other Southwest Asian countries. Festival meals are similar to those of Iran in their emphasis on rice, the region's staple food. Meals also include the grilled meat of chickens, goats, and sheep, as well as delicious breads. Pakistani foods feature strong spices and flavors.

✓ **READING CHECK:** *Human Systems* What is the culture of Pakistan like?
arranged marriages, Islamic traditions, spicy foods

Pakistan and Bangladesh

Country	Population/ Growth Rate	Life Expectancy	Literacy Rate	Per Capita GDP
Bangladesh	131,269,860 / 1.6%	61, male / 60, female	56%	$1,570
Pakistan	144,616,639 / 2.1%	61, male / 62, female	43%	$2,000
United States	281,421,906 / 0.9%	74, male / 80, female	97%	$36,200

Sources: Central Intelligence Agency, *The World Factbook 2001;* U.S. Census Bureau

Interpreting the Chart **Which country in the region has a larger population?**

Pakistan's flag symbolizes the country's commitment to Islam and the Islamic world. The crescent and star are symbols of Islam. The vertical white stripe represents religious minorities.

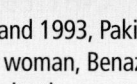

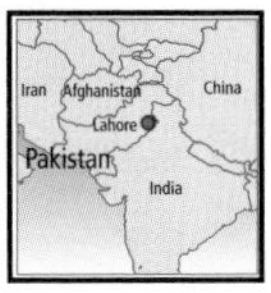

DAILY LIFE

In 1988 and 1993, Pakistanis elected a woman, Benazir Bhutto, to be the country's prime minister. Unlike the Oxford-educated Bhutto, however, most Pakistani women do not participate in public life.

The majority of women in Pakistan and other South Asian countries are poor. The roles of most of these women are limited by social customs, particularly in villages and remote areas. Such women can expect to be married at a young age and to have limited educational and economic opportunities.

In contrast, women who live in large cities and belong to the upper classes of Pakistani society have become doctors, politicians, accountants, and trade-union leaders. These well-educated and wealthy women encounter few of the social restrictions faced by Pakistani women who are poor.

Critical Thinking: What might the contrasting lifestyles of rich and poor Pakistani women suggest about the roles of women in Pakistan society?

Answer: Students might suggest that a woman's social role has a strong relationship to her economic level.

▲ **Chart Answer**

Pakistan

LEVELS 2 AND 3: (Suggested time: 30 min.) Tell students to imagine that they are social scientists who are writing a book about Pakistan's culture. Then have each student write an outline for the book. Ask volunteers to share their outlines with the class.

Teaching Objective 4

ALL LEVELS: (Suggested time: 15 min.) Copy the graphic organizer onto the chalkboard, omitting the italicized answers. Have students complete the organizer. Use it to help students learn about the challenges facing Bangladesh. Then challenge students to draw lines between those shown to depict how some of the challenges are connected. *(For example, flooding can spread disease.)* **ENGLISH LANGUAGE LEARNERS**

large population in a poor county

frequent flooding

Challenges Faced by Bangladesh

disease

limited educational opportunities for women

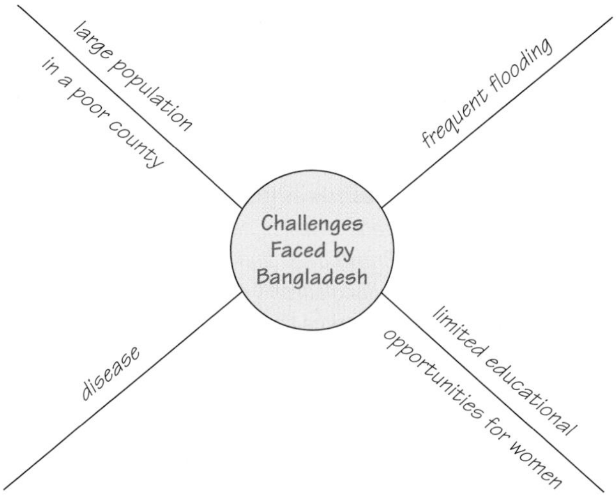

Across the Curriculum
TECHNOLOGY

Small Loans, Big Results
The Grameen Bank provides small loans to the poorest people in Bangladesh. Loan recipients use the money to purchase supplies or equipment needed for operating a small business. For example, one recipient might buy a sewing machine for a home tailoring business.

One Grameen program sells mobile telephones to individuals in small villages that have no other modern communications equipment. The owner then sells—for a small fee—access to the phone to other members of the community. The program's goal is to provide telephone service to 100 million people in nearly 70,000 villages.

Critical Thinking: How can improved communications link cultures and help spread cultural traits?

Answer: Answers will vary, but students might point out that improved communication can introduce modern culture to isolated traditional areas of the world.

Chart Answer ▶

88 percent

▲

Multan is a commercial and industrial center with several colleges and a university. The city came under Muslim control in the A.D. 700s.

Religions of Bangladesh

1%

11%

88%

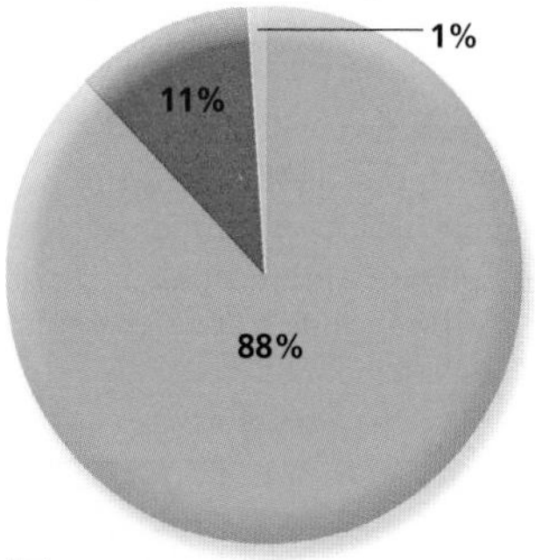

■ Muslim ■ Other
■ Hindu

Source: Central Intelligence Agency, *The World Factbook 1999*

Interpreting the Chart What percentage of the population of Bangladesh is Muslim?

Bangladesh's History

Bangladesh is part of a region known as Bengal. Bengal's many rivers are used for transporting goods. This is one reason why the area was important to Europeans in the 1500s. From the 1500s through the 1700s, the Mughal Empire combined Islam and regional Indian traditions to create a distinctive culture in this region. Eventually, the Mughal Empire weakened. The British East India Company then expanded its control over the area.

Because the British East India Company was so powerful, it could control what was traded. Therefore, British goods came into the region, but few goods other than rice and jute were exported. With a decline in trade, industry suffered. As a result, the region's economy became more agricultural in the 1800s. As you learned earlier, East Pakistan became the independent country of Bangladesh in 1971.

✓ **READING CHECK:** *Places and Regions* What has Bangladesh's history been like? **British East India Company controlled trade and limited exports, so industry suffered and made the country more agricultural.**

Bangladesh Today

Unlike other South Asian countries, Bangladesh has only one main ethnic group—the Bengalis. They make up 98 percent of the population. Social standing is based mostly on wealth and influence rather than on heredity or caste. Muslims may move up or down in status. Even among the country's Hindus, caste has much less importance than it does in India.

CLOSE

List six of the historical events discussed in Section 2 on the chalkboard, but organize the events out of chronological order. Call on volunteers to arrange the events in the correct order.

REVIEW AND ASSESS

Have students complete the Section Review. Then organize students into small groups and have each group write an outline of the section. Have groups share their outlines with the class. Then have students complete Daily Quiz 28.2.

RETEACH

Have students complete Main Idea Activity 28.2. Then organize students into small groups and have each group create an illustrated magazine cover, including story titles, for either Pakistan or Bangladesh. Have groups present their magazine covers to the class.
ENGLISH LANGUAGE LEARNERS, COOPERATIVE LEARNING

EXTEND

Have interested students conduct research on Mohenjo Daro or Harappa, two significant Indus civilization sites, and focus on the street plan, construction techniques and materials, or the social organization of either city. Students should also compare the ancient cities to modern cities.
BLOCK SCHEDULING

Population Bangladesh is about the same size as Wisconsin. However, it has a population nearly half the size of the entire U.S. population. Only 24 percent of Bangladesh's population lives in cities. The country's capital and largest city is Dhaka (DA-kuh). Cities and rural areas are densely populated. On average in Bangladesh there are 3,324 people per square mile (885/sq km).

Flooding and disease are two of the country's biggest challenges. Occasional violent tropical cyclones bring huge storm surges on shore. In addition, runoff from heavy rains in the distant Himalayas causes extensive flooding. When the land is flooded, sewage often washes into the water. As a result, Bangladesh has often suffered from epidemics of diseases like **cholera**. Cholera is a severe intestinal infection.

Culture Family life in Bangladesh is different from that in Pakistan. For example, many Bangladeshi women keep close ties to their own families after marriage. However, as in other South Asian countries, marriages are arranged by parents. Most couples typically do not know each other prior to their wedding.

More girls are going to secondary schools and universities than ever. Even so, the literacy rate for women remains just about half that of men. Bangladesh's official language, Bengali (Bangla), is the language spoken in schools. The country celebrates Islam's main festivals with feasts that feature fish and rice.

▲ Pedicabs crowd a Dhaka street. A pedicab is a ricksha pulled by a bicycle. It is a common form of transportation in parts of Asia.

✓ **READING CHECK:** *Places and Regions* What are some challenges that Bangladesh faces today? flooding, disease

go.hrw.com Homework Practice Online
Keyword: SK3 HP28

Section 2 Review

Define and explain: cholera

Working with Sketch Maps On the map you created in Section 1, label Pakistan, Bangladesh, Karachi, Lahore, Islamabad, and Dhaka. What does the name of Pakistan's capital indicate about the country's dominant religion?

Reading for the Main Idea

1. *Places and Regions* Why were East and West Pakistan separated from India in 1947? When did East and West Pakistan split into two independent countries?
2. *Environment and Society* What technology allows Pakistan to grow enough food for its large population?
3. *Environment and Society* What climatic condition contributes to the high disease rate in Bangladesh? How?

Critical Thinking

4. **Contrasting/Comparing** How are marriage and married life in Pakistan different from marriage and married life in Bangladesh? How are they similar?

Organizing What You Know

5. **Comparing/Contrasting** Copy the following graphic organizer. Use it to compare and contrast Pakistan and Bangladesh.

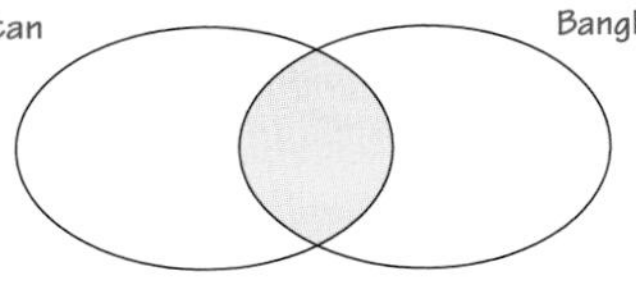

Section 2 Review Answers

Define For definition, see: cholera, p. 645

Working with Sketch Maps Maps will vary, but listed places should be labeled in their approximate locations. The name *Islamabad* indicates the importance of Islam.

Reading for the Main Idea

1. British division of India into predominantly Muslim and Hindu areas; 1971 (NGS 4)
2. irrigation of the Indus Valley NGS 14)
3. cyclones; result in flooding, washing sewage into water supply (NGS 15)

Critical Thinking

4. Pakistani women join their husbands' families at marriage. Bangladeshi women keep close ties to their own families. However, in both countries most marriages are arranged.

Organizing What You Know

5. Pakistan—many languages and ethnic groups, desert climate; Bangladesh—Bengali ethnic group, flooding; in common—predominantly Muslim countries with similar cultural heritage, large populations, high rates of illiteracy

CASE STUDY

Setting the Scene

Tourism is an important source of revenue for many developing countries. It can provide many job opportunities for local residents as well as help government modernize the country's infrastructure. The popularity of mountaineering has enabled Nepal to develop a tourism industry. The benefits of the new tourist economy, however, must be weighed against the negative effects it can have. As the number of tourists visiting the country has increased, Nepal has witnessed changes in traditional cultures and suffered from pressures on its environmental resources.

Building a Case

Have students read "Tourism in Nepal." Lead a class discussion by posing the following questions to students: How much money is generated worldwide each year by tourism? *($450 billion)* What is everyday life like for most Nepalese? *(Most are farmers living in small isolated villages.)* Who are the Sherpas? *(a group of people living near Mount Everest many of whom now work as guides)* Have students estimate the amount of income the government receives from climbing permits. *(Multiply 600 to 900 climbers at $50,000 each. The result is between $3 million and $4.5 million.)* Call on volunteers to answer the questions at the end of the selection.

HISTORICAL GEOGRAPHY

The Galápagos Islands of Ecuador provide another example of the effect that tourism can have in the developing world. The Galápagos, visited by Charles Darwin in 1835, are home to a unique assemblage of animals. Giant tortoises, iguanas, albatross, sea lions, cormorants, and fur seals are just a few of the animals that live there.

In 1968 the Galápagos National Park was created. Tourists soon overwhelmed the islands. Eight large towns soon eclipsed the two small original villages. Pizzerias, T-shirt shops, and nightclubs can now be found there. Airplanes bring tourists to the San Cristóbal airport, and cruise ships provide tours of the islands. Some tourists leave litter and graffiti.

Activity: Have students conduct research on Charles Darwin's voyage to the Galápagos Islands. Then organize students into groups of three. Have each group write a dialogue that might take place during a fictional encounter between Darwin, the Ecuadorian minister of tourism, and an American tourist, with each of the students assuming a role. Then have groups perform their dialogues for the class.

➤ This Case Study feature addresses National Geography Standards 4 and 14.

CASE STUDY

TOURISM IN NEPAL

Tourism is one of the largest industries in the world. It generates about $450 billion each year! In many developing countries, tourism has become an important source of income. By creating new and better-paying jobs, tourism helps raise the standard of living. For some developing countries, income from tourism might be the difference between poverty and having basic education or health care. However, tourism can also have negative effects. It can disrupt local cultures, change people's traditional way of life, and damage the environment.

Nepal has experienced both the advantages and disadvantages that tourism can have for developing countries. Nepal is a mountainous, landlocked country with a very low standard of living. Most people in Nepal are farmers who live in small, isolated villages. However, tourism is bringing an end to Nepal's isolation. Tourism is creating new economic opportunities for some of Nepal's residents.

One group in Nepal that is being affected by tourism is the Sherpa. The Sherpa are a group of people that live near Mount Everest. They are very skilled mountain climbers. As tourism to Nepal grows, people are traveling to this region to hike and climb. These adventure tourists hire Sherpa to help them climb Himalayan peaks like Mount Everest. As a result, some Sherpa communities have become much wealthier than their neighbors.

The growth of tourism has changed the Sherpa's way of life in several ways. With more men working for tourists as mountain guides, Sherpa women have become more involved in farming. Income from tourism has also created a gap between the people with money and people without money. At some local religious festivals, tourists now outnumber the local Sherpa. These changes and others are affecting the Sherpa's culture and daily life.

◄ Aluminum cans are crushed at Everest Base Camp and then taken to Nepal's capital to be recycled. A government ban on glass bottles has forced hikers and local residents to switch to aluminum cans.

Drawing Conclusions

Write the following question on the chalkboard: *How can we enjoy nature without disrupting it?* Have the class brainstorm and come up with a list of all of the services and facilities that tourists might expect. Compile a master list on the chalkboard. Examples might include roads, shelter, electricity, food, water, waste disposal, heat in the winter, air conditioning in the summer, souvenir shops, housing for workers, and telephone service. Ask students to think about how providing these facilities and services might disrupt the natural environment. *(For example, trees might be cut to build hotels and gift shops, to make way for roads, or to provide heating fuel. Dams built to provide electricity or drinking water might be harmful to streams and the wildlife that depend on them.)*

Going Further: Thinking Critically

The U.S. National Park Service tries to minimize the negative effects of tourism on the natural areas it oversees. Have students conduct research on a national park they have visited or would like to visit. You might suggest one of the following: Yosemite National Park, Yellowstone National Park, Petrified Forest National Park, Hawaii Volcanoes National Park, or Glacier National Park. Then tell students to imagine that they work for the Park Service and have each student create an informational brochure that answers the following questions: *How many people visit the site each year? How has the site been affected by tourism? What efforts has the Park Service made to limit the impact of tourists?* Ask volunteers to present their brochures to the class.

▲ Photographer Alison Wright stands next to a Sherpa guide near Solu Khumbu, Nepal. The Sherpa are famous for their mountain-climbing skills.

► Many Sherpa work as porters for mountain-climbing expeditions.

Tourism has brought more money into Nepal. Each tourist has to pay for a permit to climb Nepal's mountains. For example, it costs about $50,000 for a permit to climb Mount Everest. Some 50 teams of climbers attempted to reach the summit of Everest in 2000. Tourism has become an important and growing source of income for Nepal's government.

However, tourism has also damaged the country's environment. Hikers in Nepal leave behind an estimated 110,000 pounds (50,000 kg) of garbage each year! Garbage has become a serious concern in the Mount Everest region. In fact, some people have called Everest Base Camp the "world's highest garbage dump." Large garbage dumps have also appeared in other mountain-climbing areas of Nepal. Chopping down trees for firewood is another concern. Tourist lodges use large amounts of wood for cooking and for heating. In some areas, this demand for wood has caused deforestation and soil erosion.

Understanding What You Read

1. How has tourism changed the daily life of the Sherpa?
2. What are some of the advantages of tourism for Nepal? What are some disadvantages?

Understanding What You Read

1. Sherpas now have more income. Women's roles have changed as they have become more involved in farming. Income gaps have developed, and tourists now outnumber Sherpas at some local religious festivals.
2. Advantages include more job opportunities for the Nepalese and additional money for the government. Disadvantages include damage to the environment such as trash dumps, deforestation, and soil erosion.

internet connect

GO TO: go.hrw.com
KEYWORD: SK3 CH28
FOR: Web sites about tourism in Nepal

Section 3

Objectives

1. Describe some important features of Nepal.
2. Explain how Bhutan has developed.
3. Discuss what Sri Lanka is like.
4. Identify some economic activities of the Maldives.

FOCUS

LET'S GET STARTED

Copy the following questions onto the chalkboard: *What is the highest mountain in the world? What would it be like to climb that mountain?* Discuss responses. *(Mount Everest is the highest mountain. Students' descriptions will vary.)* Point out that many tourists come to Nepal to view Mount Everest. Tell students that in Section 3 they will learn more about the Himalayan countries, Nepal and Bhutan. They will also learn about the island countries, Sri Lanka and the Maldives.

Building Vocabulary

Write the terms **stupas, graphite,** and **atolls** on the chalkboard. Ask volunteers to look up the terms and their origins in a dictionary and to read them aloud to the class. Tell students that the terms *stupa* and *atoll* originated in the region and that they describe cultural and physical phenomena that are not found in Europe.

Section 3 RESOURCES

Reproducible
- Guided Reading Strategy 28.3
- Interdisciplinary Activities for the Middle Grades 33, 34, 35
- Map Activity 28

Technology
- One-Stop Planner CD–ROM, Lesson 28.3
- Homework Practice Online
- HRW Go site

Reinforcement, Review, and Assessment
- Section 3 Review, p. 653
- Daily Quiz 28.3
- Main Idea Activity 28.3
- English Audio Summary 28.3
- Spanish Audio Summary 28.3

Visual Record Answer

Indian architectural style

Section 3 The Himalayan and Island Countries

Read to Discover

1. What are some important features of Nepal?
2. How has Bhutan developed over time?
3. What is Sri Lanka like?
4. What are some economic activities of the Maldives?

Define

stupas
graphite
atolls

Locate

Nepal	Kathmandu
Bhutan	Thimphu
Sri Lanka	Colombo
Maldives	Male

A sculpture from Sri Lanka's Temple of the Tooth

WHY IT MATTERS

Nepal is the home of Mount Everest, the highest mountain in the world. Climbing Mount Everest has become increasingly popular in recent years. Use CNNfyi.com or other **current events** sources to find information on the environmental impact of climbing on Mount Everest. Record your findings in your journal.

A temple to the Hindu god Krishna can be seen from Durban Square in Patan, near Kathmandu, Nepal. The pillar in the square honors Krishna's companion, Garuda.

Interpreting the Visual Record **What architectural style is evident in these buildings?**

Nepal

Nepal's history is linked to the early civilizations of northern India. Nepal shared its languages, culture, and economic base with peoples to the south. It was probably in Nepal that Gautama, the Buddha, was born about 563 B.C. Some of the oldest monuments to the Buddha are **stupas** in Nepal. They date back to before 300 B.C. Stupas are mounds of earth or stones covering the ashes of the Buddha or relics of Buddhist saints. Like India, Nepal adopted Buddhism at first but was mostly Hindu by A.D. 1200. About 90 percent of Nepal's people are now Hindus.

The British never formally ruled Nepal. Their army did defeat the Nepalese army in the early 1800s. As a result, the British took away the territories claimed by Nepal and sent an official to maintain indirect control. The British later recognized Nepal's independence. Nepal now is a constitutional monarchy with a king and elected parliament.

One of Nepal's ethnic groups is the Sherpa. The Sherpa guide and serve as porters for Himalayan expeditions. Norkey Tenzing, a Sherpa, and Sir Edmund Hillary, an explorer from New Zealand, climbed Mount Everest in 1953. They were the first to reach the summit.

✓ **READING CHECK:** *Places and Regions* What has the history of Nepal been like? linked to early northern Indian civilizations; believed to be the birthplace of the Buddha; indirectly controlled by the British; gained independence and is now a constitutional monarchy

TEACH

Teaching Objective 1

ALL LEVELS: (Suggested time: 15 min.) Copy the following graphic organizer onto the chalkboard, omitting the italicized answers. Use it to help students identify the important features of Nepal. Have each student complete the organizer by filling in information about Nepal's culture, history, and economy. Ask volunteers to share their answers with the class. **ENGLISH LANGUAGE LEARNERS**

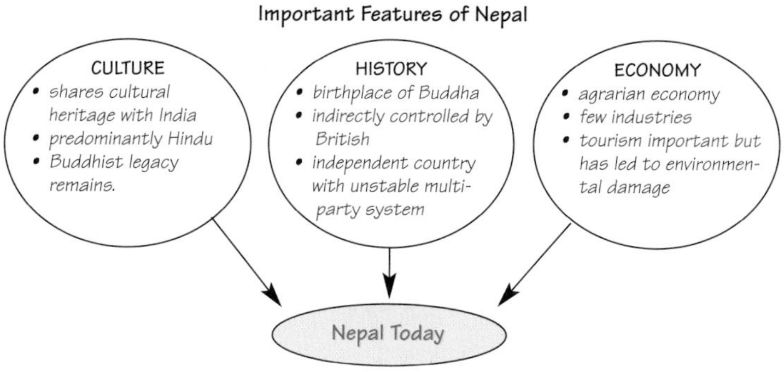

Important Features of Nepal
CULTURE
• shares cultural heritage with India
• predominantly Hindu
• Buddhist legacy remains.
HISTORY
• birthplace of Buddha
• indirectly controlled by British
• independent country with unstable multi-party system
ECONOMY
• agrarian economy
• few industries
• tourism important but has led to environmental damage
Nepal Today

CONNECTING TO *Technology*

CLIMBING MOUNT EVEREST

The world's highest peak, Mount Everest, sits on the border between China and Nepal. At 29,035 feet (8,850 m), it is one of the world's greatest mountaineering challenges.

After many failed attempts, two climbers—Edmund Hillary and Norkey Tenzing—finally reached the summit in 1953. One factor in their success was the use of advanced equipment. This equipment included special boots, oxygen bottles, and radio gear. Since then, technology has continued to play a key role in efforts to scale the Himalayas.

Today's climbers use a wide range of sophisticated gear. This gear includes everything from space-age mountaineering clothes to remote sensors and satellite-tracking devices. Even so, climbing Everest remains very dangerous. In the 1990s more than 50 climbers lost their lives trying to scale the peak. Now scientists are testing new medical technologies on Mount Everest in hopes of preventing such tragedies in the future.

A joint project of the National Aeronautics and Space Administration (NASA) and Yale University has been analyzing the effects of high-altitude climbing on the human body. In 1998 this project sent a team of climbers up Mount Everest wearing "bio-packs." These packs included high-tech devices to monitor heart rate, blood oxygen, and other vital signs. This data was transmitted instantly to a base camp on Everest. There doctors could analyze the results. Scientists believe that this kind of "telemedicine" will help save lives in other extreme environments, including outer space.

Understanding What You Read

1. How do climbers use technology on Mount Everest?
2. How might climbers benefit from the use of new medical technology?

Sir Edmund Hillary (left) and Norkey Tenzing (right)

Nepal Today

Like the other countries of the Indian Perimeter, Nepal's economy is based on agriculture. More than 80 percent of the people make a living from farming. Most are subsistence farmers. The best farmland is found on the Tarai—Nepal's "breadbasket." Farmers there grow such crops as rice, other grains, and sugarcane. In the hills north of the Tarai, farmers grow fruits, grains, and vegetables. Kathmandu (kat-man-DOO), Nepal's capital and largest city, is located in this central region. For the most part, the Himalayan region is not good for

ENVIRONMENT AND SOCIETY

Mountaineering on Mount Everest has come at a cost to the environment. Thousands of pounds of trash have been scattered across the world's highest mountain. This litter includes old tents, ropes, oxygen bottles, stoves, empty food containers, and human waste. An unknown number of corpses of unlucky climbers also remain on Everest.

Recently environmentalists have begun cleanup efforts. One group has removed 17,000 pounds of garbage since 1994. To prevent the future buildup of garbage, the group set up a system to pay Sherpas to bring used supplies back down.

Another expedition cleaned up a heavily used part of the mountain. Retrieved gas canisters and used batteries were carried by yak to a nearby city, transported by plane to Kathmandu, shipped to Thailand, and then to a hazardous waste site in California.

Discussion: Lead a discussion about ways students can limit the impact they have on the environment when visiting parks or other natural areas.

Connecting to Technology Answers

1. They use specialized equipment for mountaineering.
2. It might save the lives of injured climbers.

TEACHER TO TEACHER

Joanne Sadler of Buffalo, New York, suggests the following activity to help students learn about the countries of the Indian Perimeter: Tell students to imagine that they are foreign exchange students who have been living in Nepal, Bhutan, Sri Lanka, or the Maldives for the past year. Have each student write a short article for your newspaper's travel section that describes life in that country today as well as aspects of the country's culture and history. Ask volunteers to read their letters to the class.

Teaching Objective 2

LEVEL 1: (Suggested time: 15 min.) Pair students and have each pair compile a bulleted list that describes the history, culture, and economy of Bhutan. *(Lists should mention that little is known about Bhutan's early history; most Bhutanese are of Tibetan origin and are Buddhist; Bhutan is fairly isolated today; most people are subsistence farmers; timber and hydroelectric power are important resources; limited tourism provides some income for the country.)* Ask volunteers to read their lists to the class. **ENGLISH LANGUAGE LEARNERS, COOPERATIVE LEARNING**

LEVEL 2: (Suggested time: 30 min.) Have students use their lists from the Level 1 activity to write a short poem that describes Bhutan. Ask volunteers to recite their poems. **COOPERATIVE LEARNING**

National Geography Standard 6:

Places and Regions

Bhutan's natural boundaries—dense forests and mountains—have helped keep it isolated. Little is known about the early history of Bhutan. Bhutanese call their country Druk-yul, which means "the land of the thunder dragon." The name refers to the country's dominant Buddhist sect. The name *Bhutan* is used only in the country's English-language communication.

Critical Thinking: What might Bhutan's traditional name suggest about the way the Bhutanese view their country?

Answer: Students might suggest that the country's religious identity is important to the Bhutanese.

▲ Nepalese women are experts at winnowing grain. Winnowing is the process of separating chaff from grain by fanning.

farming. Farmers graze cattle, goats, sheep, and yaks on the lower slopes. Yaks are large, longhaired oxen. In addition, some farmers grow grain and fruit trees on terraces cut into mountainsides.

With few resources and limited access to world trade routes, Nepal has developed few industries. Most of its industries are based on farm products, such as jute. Jute is a plant fiber that is used in making twine. Tourism is one of the fastest-growing and most important industries in Nepal. Tourists like to hike and climb in the mountains.

Tourism brings in much-needed income to Nepal. However, the constant stream of hikers and climbers threatens the country's environment. How? Many visitors leave behind trash. Also, their need for firewood contributes to deforestation and soil erosion. This problem is worsened as Nepalese clear woodland for farming and for wood for fuel. In response, Nepal's government has restricted access by visitors to some of the most affected areas.

Folk dancers in traditional dress perform near Thimphu, Bhutan. ▼

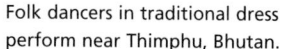

✓ **READING CHECK:** *Places and Regions* What is Nepal like? **agricultural; tourism an important industry but threatens environment, as do deforestation and erosion**

Bhutan

Little is known about the early history of Bhutan. Buddhism was practiced there as early as the A.D. 600s. Most of the people are of Tibetan origin and are Buddhist. In the 1600s a Tibetan monk helped organize Bhutan as a unified state. Then in the late 1700s the East India Company's business dealings increased British influence.

Bhutan, although very isolated, was under Indian control during British rule over that country.

Teaching Objective 3

LEVEL 1: (Suggested time: 20 min.) Pair students and have each student create a graphic organizer to present information about Sri Lanka's history, culture, and economy. *(Graphic organizers should mention that Buddhism spread to Sri Lanka from India; Sri Lanka has been influenced by the Portuguese, Dutch, and British; Sri Lanka gained independence in 1948; the Buddhist Sinhalese and Hindu Tamils are the main ethnic groups; the Sinhalese are the majority group; violence between Tamils and the government is a problem in Sri Lanka; Sri Lanka exports coconuts, tea, and rubber; and mining of graphite and precious gems and textile manufacturing are important to the economy.)* Ask volunteers to present their organizers to the class. **COOPERATIVE LEARNING**

LEVELS 2 AND 3: (Suggested time: 30 min.) Tell students to imagine that they must write an encyclopedia entry about Sri Lanka. Have each student write a short entry describing Sri Lanka's history, culture, and economy. Ask students to include information from the photos, photo captions, and the Case Study in their entries. Ask volunteers to read their entries to the class.

Bhutan's ruler, or maharaja, declared a constitutional monarchy in 1969.

READING CHECK: *Places and Regions* What has the history of Bhutan been like? isolated; under Indian control; constitutional monarchy established in 1969

Bhutan Today

Until the mid-1970s, the government of Bhutan followed a policy of near total isolation. Even today, the country's international ties are somewhat limited.

Much of Bhutan's economy is traditional. More than 90 percent of the people make a living as subsistence farmers. Most grow rice, wheat, corn, and potatoes in fertile valleys and on mountainside terraces. A few grow fruits and spices for export. On mountain pastures, farmers raise cattle, goats, sheep, and yaks. Thimphu (thim-POO) is Bhutan's capital.

Timber and hydroelectricity are among Bhutan's most important resources. The country sells both to India, its main trading partner. Tourism is also a source of income for Bhutan. However, the government limits the number of visitors to just 7,000 a year. By restricting tourism, the government hopes to protect Bhutan's way of life from outside influences and environmental damage.

READING CHECK: *Places and Regions* What is Bhutan like today? limited international ties, traditional economy, limited tourism

Himalayan and Island Countries

Country	Population/ Growth Rate	Life Expectancy	Literacy Rate	Per Capita GDP
Bhutan	2,049,412 / 2.1%	53, male; 52, female	42%	$1,100
Maldives	310,764 / 3.0%	61, male; 64, female	93%	$2,000
Nepal	25,284,463 / 2.3%	59, male; 58, female	28%	$1,360
Sri Lanka	19,408,635 / 0.9%	70, male; 75, female	90%	$3,250
United States	281,421,906 / 0.9%	74, male; 80, female	97%	$36,200

Sources: Central Intelligence Agency, *The World Factbook 2001;* U.S. Census Bureau

Interpreting the Chart **Which country has the lowest literacy rate?**

Sri Lanka

When Buddhism swept over India between the 400s and 200s B.C., Buddhist missionaries came to Sri Lanka. From Sri Lanka, traders spread Buddhism eastward to what are now Thailand and Indonesia. About 74 percent of Sri Lankans today belong to the Sinhalese ethnic group, and most are Buddhists. Sri Lanka has many dome-shaped stupas honoring the Buddha. Many are shaded by trees said to have been grown from cuttings of the original bo tree. It is believed the Buddha achieved enlightenment under this tree.

Throughout its history, Sri Lanka has been largely independent from India. In the 1500s the Portuguese established a trading post on the island. They were overthrown by the Dutch. The Dutch turned over Sri Lanka, then known as Ceylon, to the British in the late 1700s. The British granted independence in 1948. The new government made Sinhalese the official language and promoted Buddhism. An Indian and Hindu minority group, the Tamils, protested. Fighting between Tamils and the government has killed many people.

READING CHECK: *Places and Regions* How have Europeans played a role in Sri Lanka's history? Portuguese—established a trading post in the 1500s; Dutch—overthrew Portuguese and turned over Ceylon to British in the late 1700s; British—granted independence in 1948

▲ Buddhist monks prepare to pray at the top of Adam's Peak. The "yellow" robe is a symbol of Buddhism. The color of the robe varies according to the dye used. Some are bright orange, or saffron.

COOPERATIVE LEARNING

Bhutan and Nepal are known for their biological richness. The Himalayas have prevented plant species from spreading northward from India and southward from China. Botanists have estimated that there are at least 6,500 species of flowering plants in Nepal.

There may be even more species in Bhutan, partly because the country has cold and warm climate regions within a small area. Also, the country's isolation has saved it from mass deforestation. There are conifers, oaks, maples, laurels, and magnolias unique to Bhutan.

Organize students into six groups and assign each group one of the following topics: Nepal's climates, Nepal's plant life, Nepal's animal life, Bhutan's climates, Bhutan's plant life, or Bhutan's animal life. Have groups conduct research on their topic. Then have the groups use their findings and work together to create murals that depict the connections among climates, plant life, and animal life for the two countries. Have students create two murals—one for each country.

▲ **Chart Answer**

Nepal

Teaching Objective 4

LEVEL 1: (Suggested time: 20 min.) Pair students and have each pair create a collage to show the main economic activities in the Maldives. *(Collages should show fishing, tourism, clothing industry, boat-building, and boat repair.)* Ask each pair to present its collage.
ENGLISH LANGUAGE LEARNERS, COOPERATIVE LEARNING

LEVELS 2 AND 3: (Suggested time: 20 min.) Pair students and have each pair write a song about the main economic activities in the Maldives and their relations to the country's physical geography. *(See the Level 1 lesson for the correct economic activities.)* Ask volunteers to sing their songs for the class. **COOPERATIVE LEARNING**

➤ASSIGNMENT: Have each student create a map of the Maldives and use symbols to represent the main economic activities of the islands. Students should also create a key to explain their symbols. Display students' maps around the classroom.

Section Review 3

Define For definitions, see: stupas, p. 648; graphite, p. 652; atolls, p. 653

Working with Sketch Maps Maps will vary, but listed places should be labeled in their approximate locations.

Reading for the Main Idea

1. by grazing animals on the land and raising crops on terraces cut into hillsides (NGS 14)
2. keeps people from associating with those of lower castes (NGS 10)
3. fishing and the processing of fish (NGS 4)

Critical Thinking

4. Students' answers should discuss cultural, military, political, and religious links.

Organizing What You Know

5. Benefits—brings income; Drawbacks—results in environmental damage

Visual Record Answer ▲

Students may mention decorative walls, many-sided buildings, or other features.

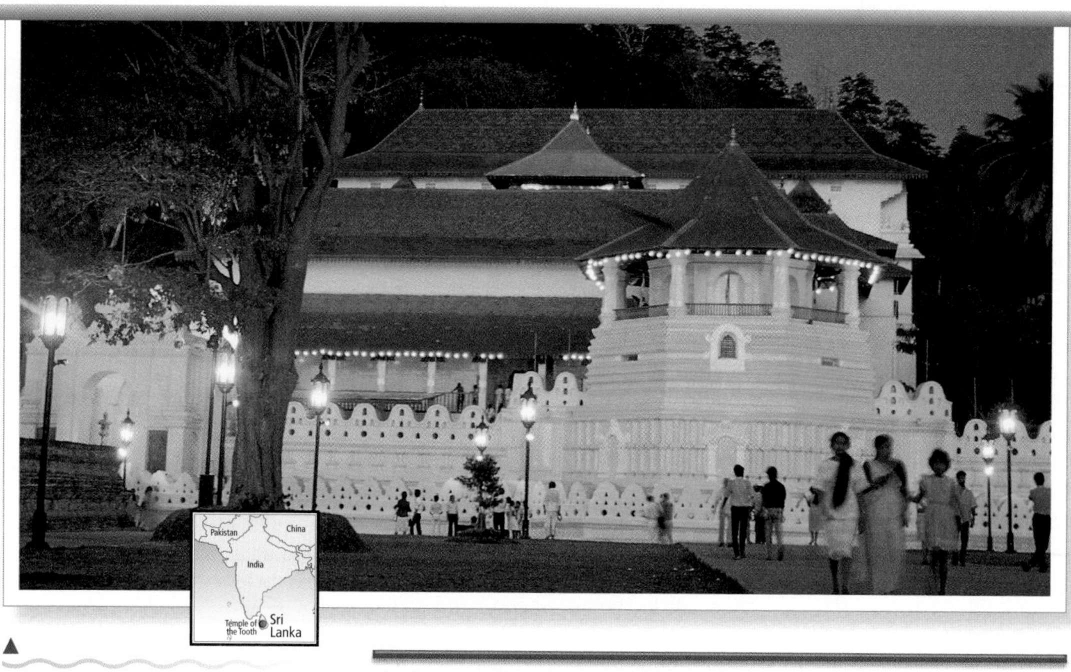

▲ The Dalada Maligawa (Temple of the Tooth) houses the sacred tooth of Buddha. Pilgrims come to pay respect to the Tooth Relic during daily ceremonies.

Interpreting the Visual Record What architectural features of this temple have you seen in other world regions?

Sri Lanka Today

Caste determines how Sri Lankans behave toward one another. As in India, many Sri Lankans believe associating with a member of a lower caste brings bad karma. Bad karma forces a person to undergo endless reincarnations, or rebirths. It also prevents him or her from finding nirvana, or spiritual release.

Sri Lanka has several features in common with the countries of the mainland Indian Perimeter. First, it is a mainly rural country. Less than 25 percent of Sri Lankans live in cities. Second, about 38 percent of Sri Lankans make a living from farming. They grow rice, fruits, and vegetables. Sri Lankan farmers also grow coconuts, rubber, and tea for export. Sri Lanka has a number of industries. They include food processing, textiles and apparel, telecommunications, insurance, and banking.

Mining is another important economic activity in Sri Lanka. The main resources mined are graphite and precious gems. Sri Lanka leads the world in **graphite** exports. Graphite is a form of carbon used in pencils and many other products. Recently, manufacturing—particularly of textiles and clothing—has grown in importance. Many textile and clothing factories are found in Colombo, the capital of Sri Lanka. Located on the west coast, Colombo is the country's largest city. It is also its most important commercial center and seaport. Sri Lanka's economy has grown in recent years. However, continuing ethnic conflicts between the Sinhalese majority and the Tamil minority threaten the country's progress.

Semiprecious stones such as the sapphire, ruby, cat's eye, topaz, and garnet are mined in Sri Lanka. ▼

✓ **READING CHECK:** *Places and Regions* What is Sri Lanka like today?
caste system; mainly rural; some industry, mining; ethnic conflict

CLOSE

Ask students to recall what they have learned about global warming. Then ask students how the Maldives might be affected by global warming. *(Students might mention that rising water levels could threaten the Maldives, which lie at a low elevation.)*

REVIEW AND ASSESS

Have students complete the Section Review. Then pair students and have each pair use the Section 3 Define and Locate terms to create a hidden-word puzzle. Have pairs solve and discuss other pairs' puzzles. Then have students complete Daily Quiz 28.3. **COOPERATIVE LEARNING**

RETEACH

Have students complete Main Idea Activity 28.3. Then have students design commemorative stamps for Nepal, Bhutan, Sri Lanka, or the Maldives. **ENGLISH LANGUAGE LEARNERS**

EXTEND

Have interested students conduct research on Anuradhapura, a town in Sri Lanka that for hundreds of years was the capital of the Sinhalese kings. It is sacred to Buddhists as the site where the ruler converted to Buddhism. Ask students to create a poster about Anuradhapura's history and cultural attractions. **BLOCK SCHEDULING**

The Maldives

The Maldives were originally settled by Buddhists from Sri Lanka. Today the main religion is Islam. Education includes traditional teachings of the Qur'an. However, there are no colleges or universities.

The Maldives consists of 19 **atolls**. An atoll is a ring of coral surrounding a body of water called a lagoon. These atolls sit atop an ancient submerged volcanic plateau. Altogether, the islands cover about 115 square miles (298 sq km). The largest island, Male (MAH-lay), is home to about one fourth of the population. A city by the same name serves as the country's capital. Only about 25 percent of the people live in cities.

Tourism is the Maldives largest industry. Fishing, particularly of tuna, is another of the country's chief economic activities. Exports of fresh, dried, frozen, and canned fish account for about half the country's income. Breadfruit—a fruit that resembles bread when baked—and coconuts are the Maldives' main food crops. Most other food must be imported. A small clothing industry and boat-building and repair are other sources of income for this island country.

▲ These fishers are working together to corral fish in large nets.

Interpreting the Visual Record **Why must so many people be involved in catching fish in this way?**

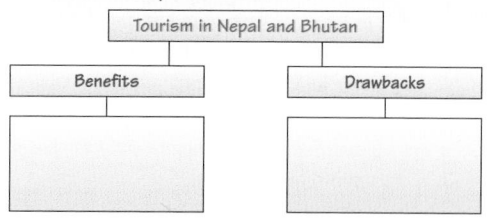

Coconuts are the fruit of the coconut palm. The inside of the fruit contains edible meat and juice. Fiber from the outer husk can be used to make brushes, matting, and fishnets. ◄

✓ **READING CHECK:** What geographic factors affect the economic activities of the Maldives? Good soil, ocean waters contribute to fishing and farming.

go.hrw.com Homework Practice Online
Keyword: SK3 HP28

Section 3 Review

Define and explain: stupas, graphite, atolls

Working with Sketch Maps On the map you created in Section 2, label Nepal, Bhutan, Sri Lanka, Maldives, Kathmandu, Thimphu, Colombo, and Male.

Reading for the Main Idea

1. *Environment and Society* How have the people of Nepal used the lower slopes of the Himalayas?
2. *Human Systems* How does the caste system affect social interaction in Sri Lanka?
3. *Places and Regions* What are the major economic activities of the Maldives?

Critical Thinking

4. **Comparing** In what ways have the cultures, people, and histories of India and the countries in this section been linked?

Organizing What You Know

5. **Categorizing** Copy the following graphic organizer. Use it to list the benefits and drawbacks of tourism in Nepal and Bhutan.

Tourism in Nepal and Bhutan	
Benefits	Drawbacks

CHAPTER 28 Review ANSWERS

Building Vocabulary

For definitions, see: cyclones, p. 640; storm surges, p. 640; cholera, p. 645; stupas, p. 648; graphite, p. 652; atolls, p. 653

Reviewing the Main Ideas

1. the Indus River; the Ganges and Jamuna Rivers (NGS 4)
2. Persians, Alexander the Great, Mughals, Arabs, British, Dutch (NGS 13)
3. Pakistan, Bangladesh, and the Maldives; Sri Lanka, Bhutan; Nepal (NGS 9)
4. seafood, textiles, timber, and hydroelectric power (NGS 4)
5. atolls (NGS 4)

Understanding Environment and Society

Answers will vary according to the country, but students may find that tourism, clearing land for farming, and cutting wood for fuel have caused deforestation in Nepal and other countries. Use Rubric 29, Presentations, to evaluate student work.

◄ **Visual Record Answer**

net's size and the absence of machinery

ASSESS

Have students complete the Chapter 28 Test.

RETEACH

Organize students into groups and have each group choose one of the countries discussed in the chapter. Then have each group prepare an annotated time line that presents the important events in its chosen country's history or write a national anthem describing the prominent features of its chosen country. Ask volunteers to present their time lines or sing their national anthems to the class. **ENGLISH LANGUAGE LEARNERS, COOPERATIVE LEARNING**

PORTFOLIO

PORTFOLIO EXTENSIONS

1. Organize students into two groups. Tell one group to prepare arguments in favor of tourism. Have the other group prepare arguments against tourism. Allow students to prepare. Then conduct the debate, acting as the moderator. Have students place their notes for the debate in their portfolios.
2. Tell students to imagine that they are economic advisers to the government of Bangladesh. Organize students into small groups and have each group create a program to help solve some of the challenges faced by Bangladesh. Have each group present its economic development plan to the class. Have students place their plans in their portfolios.

CHAPTER 28

Review ANSWERS

Thinking Critically

1. The pass has allowed several groups to invade and conquer it.
2. brought income; environmental damage
3. initiated British colonization of the region
4. rural; economies based on agriculture
5. Pakistan and Bangladesh were created originally as West and East Pakistan because the people there were mainly Muslims, in contrast to the Indian population, which was mainly Hindu.

Map Activity

A. Hindu Kush
B. Indus River
C. Himalayas
D. Tarai
E. Ganges River
F. Khyber Pass
G. Brahmaputra River
H. Thar Desert

CHAPTER 28

Reviewing What You Know

Building Vocabulary

On a separate sheet of paper, write sentences to define each of the following words.

1. cyclones
2. storm surges
3. cholera
4. stupas
5. graphite
6. atolls

Reviewing the Main Ideas

1. *Places and Regions* What is Pakistan's most important river? What two rivers are important to Bangladesh?
2. *Human Systems* What are some peoples who have conquered and ruled parts of the region over time?
3. *Human Systems* Which three countries of the Indian Perimeter are dominated by Islam? Which two are primarily Buddhist? Which one is populated mostly by Hindus?
4. *Places and Regions* What are some important resources of the Indian Perimeter?
5. *Places and Regions* What kinds of islands make up the Maldives?

Understanding Environment and Society

Deforestation

Nepal and other countries of the world are having problems with deforestation. How might the Nepalese government and others work to solve this problem? As you prepare your presentation, you may want to think about the following:

- The causes of deforestation in Nepal and other parts of the world.
- Examples of what other countries did in their efforts to slow deforestation.

Thinking Critically

1. **Drawing Inferences and Conclusions** What role do you think the Khyber Pass might have played in the history of this region?
2. **Finding the Main Idea** How has tourism benefited countries in the region? What problems has it caused?
3. **Analyzing Information** How did the British East India Company affect the history of this region?
4. **Analyzing Information** Is the population of the Indian Perimeter countries mostly rural or mostly urban? Explain your answer.
5. **Drawing Inferences and Conclusions** How has religion played a role in shaping the region's borders?

Food Festival

Banana ice cream is a popular dessert in Pakistan. To prepare ice cream for six, combine four ripe bananas and a 14-ounce can of sweetened condensed milk in a large bowl. Whisk in ¾ cup of whipping cream. Pour the banana mixture into a cake pan. Cover the pan and freeze until the mixture is softly set, stirring it occasionally. Transfer the mixture to a large bowl. Using an electric mixer, beat the ice cream until it is fluffy. Return the ice cream to the pan and cover and freeze it until firm, about six hours.

Chapter 28 Review and Assessment Resources

Reproducible
- Readings in World Geography, History, and Culture 77 and 78
- Critical Thinking Activity 28
- Vocabulary Activity 28

Technology
- Chapter 28 Test Generator (on the One-Stop Planner)
- Audio CD Program, Chapter 28
- HRW Go site

Reinforcement, Review, and Assessment
- Chapter Review, pp. 654–55
- Chapter 28 Tutorial for Students, Parents, Mentors, and Peers
- Chapter 28 Test
- Chapter 28 Test for English Language Learners and Special-Needs Students
- Unit 8 Test
- Unit 8 Test for English Language Learners and Special-Needs Students

Building Social Studies Skills

Map Activity

On a separate sheet of paper, match the letters on the map with their correct labels.

Brahmaputra River, Ganges River, Himalayas, Tarai, Hindu Kush, Khyber Pass, Indus River, Thar Desert

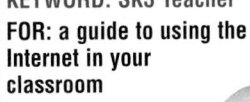

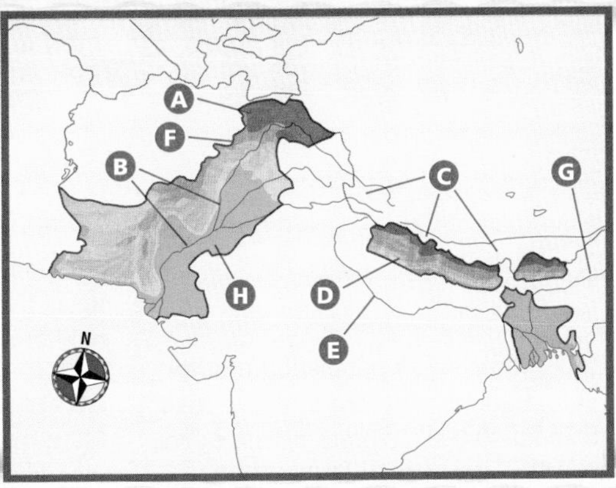

Mental Mapping Skills Activity

On a separate sheet of paper, draw a freehand map of India and the Indian Perimeter. Make a key for your map and label the following:

Bangladesh, Bhutan, India, Maldives, Mount Everest, Nepal, Pakistan, Sri Lanka

Writing Activity

Imagine that you are moving to one of the countries of the Indian Perimeter. In which country would you choose to live? State your choice and explain your reasons in two or three paragraphs. Explain how landforms, climate, culture, and economy influenced your decision. Be sure to use standard grammar, spelling, sentence structure, and punctuation.

Alternative Assessment

Portfolio Activity

Learning About Your Local Geography

Individual Project Floods often devastate Bangladesh. Present a newscast on your state's most recent flood or other natural disaster. What body of water flooded? What caused the flood?

internet connect

Internet Activity: go.hrw.com
KEYWORD: SK3 GT28

Choose a topic to explore about the Indian Perimeter.
- Climb the Himalayas.
- Learn about Sri Lanka's history.
- Visit the regions of Pakistan.

Mental Mapping Skills Activity

Maps will vary, but listed places should be labeled in their approximate locations.

Writing Activity

Answers will vary, but the information included should be consistent with text material. Each student should explain his or her choice by relating how landforms, climate, culture, and economy influenced the choice. Use Rubric 37, Writing Assignments, to evaluate student work.

Portfolio Activity

Newscasts will vary, but students should find information about the state's most recent flood, the body of water affected, and causes of the flood. They might also report on the amount of damage that occurred and emergency measures and flood prevention measures implemented by the state.

internet connect

GO TO: go.hrw.com
KEYWORD: SK3 Teacher
FOR: a guide to using the Internet in your classroom

Focus On Environment

Solving Problems

Tell students to imagine that they are Indian farmers living near a wildlife preserve. They are meeting to find a solution that will allow their villages to survive and grow while protecting the tigers and their habitat. Have students create a list of their specific concerns. *(Possible answers: the need for farmland and firewood, protection against attacks by tigers)* Then have them suggest ways of addressing these concerns. Write students' suggestions on the chalkboard. Then have students examine each suggestion's consequences for the tigers and reject any that do not also help protect the animals. Then have students work in pairs to refine the remaining solution(s) into a written proposal.

Geography Sidelight

The Sundarbans is an area of thick forest and swamps on the coast of the Indian Ocean. As part of the Ganges Delta, it spans the India-Bangladesh border. The Sundarbans is one of the most important protected tiger habitats in South Asia.

People who live in the Sundarbans area would prefer not to enter the forest, for fear of tiger attacks. They take the risk, however, to fish and gather wood or honey.

Several ingenious methods have been used to combat human-eating tigers. Life-sized dummies rigged with car batteries have been placed in the forest. The dummy emits a mild electric shock upon contact, thereby training tigers to fear humans. Local people also wear government-issued face masks on the backs of their heads when they enter the forest. Tigers usually attack from the back, so the mask confuses the animals. It appears, however, that the Sundarbans tigers may have figured out the trick.

Critical Thinking: How have people in the Sundarbans area tried to prevent tiger attacks?

Answer: by training the tigers through the use of dummies and face masks

➤ **This Focus On Environment feature addresses National Geography Standards 4, 14, 15, and 18.**

Focus On Environment

Saving Tigers

The Tiger

In the early 1800s India's tiger population was very large. By 1900, hunters had reduced the tiger population to about 40,000. To protect these tigers, India's British rulers placed strict limits on hunting them. Laws against illegal hunting were strictly enforced. After India won its independence in 1947, the hunting laws were often ignored, however. As a result, fewer than 2,000 tigers remained in India by 1972.

Why Are Tigers Endangered?

Unfortunately, laws have not kept people from hunting tigers. Tiger bones are used in traditional Chinese medicines. The bones are in high demand. International treaties now make it illegal to sell any part of an endangered species. However, the international trade in tiger bones continues.

Loss of habitat also threatens tiger survival. India's population has about doubled in the last 50 years. More land is now farmed to feed these additional people. As a result, many forests where tigers lived have been cleared for farming. As tiger habitat has decreased, so has the number of tigers.

International Efforts to Save the Tiger

Many people realized that tigers would not survive in the wild without protection. As a result, several international organizations and governments are working to ensure the tiger's future.

India's Project Tiger In 1973 the Indian government took steps to try to save its remaining tigers. With support from the World Wildlife Fund, nine tiger reserves were created as part of Project Tiger. By 1997, 23 reserves had been set up throughout India.

Each reserve has a core area. No one can enter this area without a park ranger. The core area is surrounded by a buffer zone. People can use land in these zones in ways that do not harm the environment.

Once on the verge of extinction, tigers have rebounded in India due largely to Project Tiger.

Going Further: Thinking Critically

There are many endangered species in the United States. Have students contact your state's wildlife agency or conservation organizations for information on endangered species in your area. Then have students examine how local attitudes toward the species may affect its prospects. Organize the class into two groups—one to propose an advertising campaign to influence attitudes and the other to plan on-site conservation efforts. Then have the groups work together to make certain the plans they have devised comprise a coordinated effort.

Realm of the Tiger

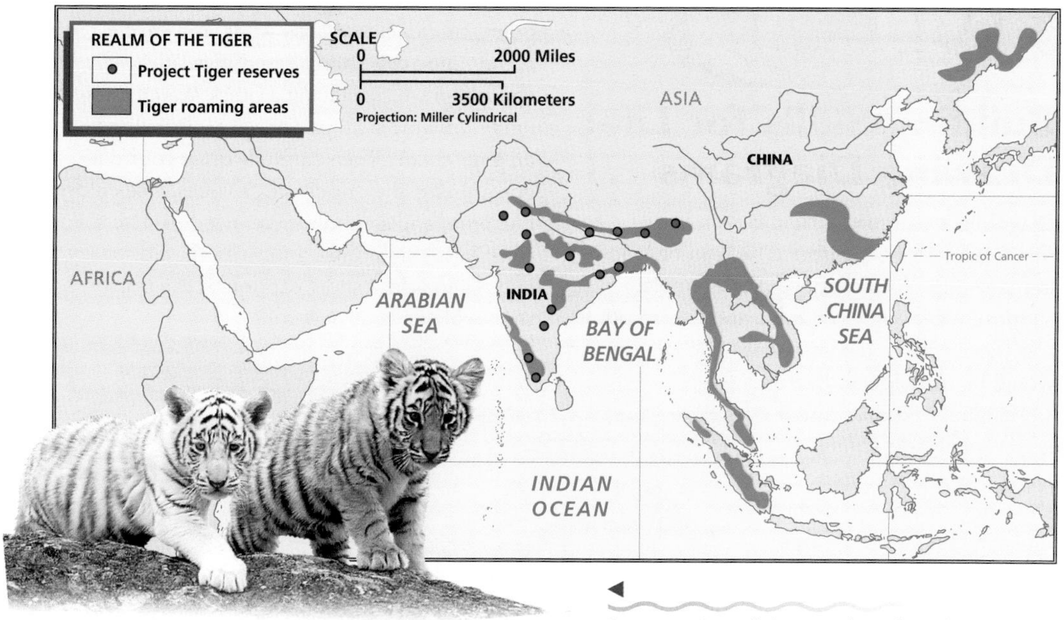

Some rare Bengal tigers, such as this cub, are white. In the last 100 years, only about a dozen white tigers have been seen in the wild.

Some conservationists are reaching out to nearby villagers. Villagers have often felt that tigers are being protected at their expense. The conservationists want to show that tiger reserves can also help villagers. One organization has introduced new types of water buffalo that need very little grazing land. To boost the local economy, it has helped women sell their handicrafts in urban areas. The Global Environment Facility and the World Bank also want to help. They plan to spend about $10 million to aid villages around India's Ranthambhore National Park.

China's Year of the Tiger Despite these efforts, the number of tigers remains low. In 1998, the Chinese Year of the Tiger, conservationists in Hong Kong tried a new approach. They started an advertising campaign to remind people how important tigers are in traditional Chinese culture. The advertisements showed children dressed in clothes with popular tiger designs. The advertisements read, "They protect your children; who will protect them?" Chinese conservationists hope their campaign will encourage people to stop using medicines made with tiger bones. They also hope it will encourage people to join the fight to save the world's tigers.

Your Turn

Imagine you are planning a worldwide campaign to save tigers in the wild. Write a short report that explains your plan and answers the following questions:

1. How would the Project Tiger reserves and the Chinese advertising campaign fit into your plan?
2. Do you think it is more important to save tiger habitat or to keep people from hunting them? why?

Your Turn

Answers

1. Students' reports should express a reasonable link between the Project Tiger reserves and the Chinese advertising campaign and the student's plan.
2. Students should support their answers.

internet connect

CHAPTER WEB LINKS
GO TO: **go.hrw.com**
KEYWORD: **SK3 Ch27**
FOR: Web sites about tigers

GeoSkills

Going Further: Thinking Critically

Organize the class into small groups. Prepare in advance slips of paper naming everyday skills that young children may not yet have mastered but at which your students are adept. Possibilities include riding a bicycle, programming a VCR, loading a dishwasher, changing a lightbulb, assembling a triple-decker sandwich, or scoring a soccer goal. Prepare one skill for each group. Ask each group to draw a skill from a hat. Allow groups to trade their assigned skills if they wish. Then have groups use the steps described in Building Skills for Life: Drawing Diagrams to draw a diagram that shows how to complete the task. Then have the groups exchange their diagrams and challenge them to improve another group's diagram. Finally, return the diagrams to the groups that created them and ask the students to summarize what they learned from the activity.

Practicing the Skill

1. Students' answers should describe their chosen diagram and discuss why it is difficult or easy to understand.
2. You may want to refer students to the photo of a sailing ship in Chapter 3. Students' answers should address each of the steps for drawing a diagram.
3. Students' answers should address each of the steps for drawing a diagram. Diagrams might show the rocks in place at the top of the mountain, breaking off, and creating the landslide.

➤ **This GeoSkills feature addresses National Geography Standards 1 and 7.**

Geo Skills

Building Skills for Life: Drawing Diagrams

Diagrams are drawings that explain how things work or fit together. They can be helpful in many different situations. For example, suppose you bought a bicycle and had to put it together. How would you do it? The easiest way would be to follow a diagram.

The goal of a diagram is to explain something. Diagrams are not intended to show exactly how something looks in real life. Diagrams should be easy to understand, neat, and simple.

In geography, diagrams are often used to show how different things are related. Physical geographers use diagrams to explain how sea and land breezes develop or how volcanoes form. Cultural geographers use diagrams to show settlement patterns and housing styles.

You can draw diagrams too. To draw a diagram, first decide exactly what you want it to show. Choose a short, descriptive title. Plan your diagram by making a quick sketch. How many drawings will you need? How will you arrange them? Then carefully draw your diagram. Label the important features. Use colors, patterns, and symbols if you need to. Use a ruler for straight lines and write the title at the top. Then add a short caption at the bottom that explains your diagram.

Sea and Land Breezes

Warm air
As warm air rises, it creates an area of low pressure over the land.
The cool air moves toward the land, producing a sea breeze.
Cool air
Air over the water is cooler and creates an area of high pressure.
Cool air
Air over land is cooler and creates an area of high pressure.
The cool air moves toward the water, producing a land breeze.
Warm air
Air over the water is warmer and creates an area of low pressure.

Breezes from the ocean cool the warmer land surface during the day. At night, cooler land surface breezes blow toward the ocean.

Practicing the Skill

1. Choose a diagram from a book, magazine, or newspaper. Write a short paragraph about the diagram. What does it show? Is it clear and easy to understand? why or why not?
2. Draw a diagram that shows how wind moves a sailboat through water.
3. Draw a diagram that shows how rocks can break off a mountain and cause a landslide.

HANDS ON GEOGRAPHY

The Himalayas are the highest mountains on Earth. Nine of the ten highest peaks in the world are located there. The Himalayas have 110 peaks that rise above 24,000 feet (7,315 m).

How did the Himalayas get to be so high? When were they formed? Drawing a diagram is one way to show how the Himalayas became the world's highest mountains.

Read the paragraph below. It describes how and when the Himalayas formed. Use this information to draw a diagram showing how the Himalayas became the world's highest mountains.

The Himalayas were formed long ago by the movement of Earth's tectonic plates. About 180 million years ago, the Indo-Australian plate broke off from the ancient supercontinent Gondwana and began moving north. This plate included what is now the Indian Subcontinent. About 50 million years ago, the Indo-Australian plate collided with the Eurasian plate and was forced under it. As the Indo-Australian plate was pulled under the Eurasian plate, the Himalayas began to rise. They grew slowly at first. Then, about 30 million years ago, the Himalayas began rising faster. However, they did not become the highest mountains in the world until about 500,000 years ago. The Himalayas are still rising today.

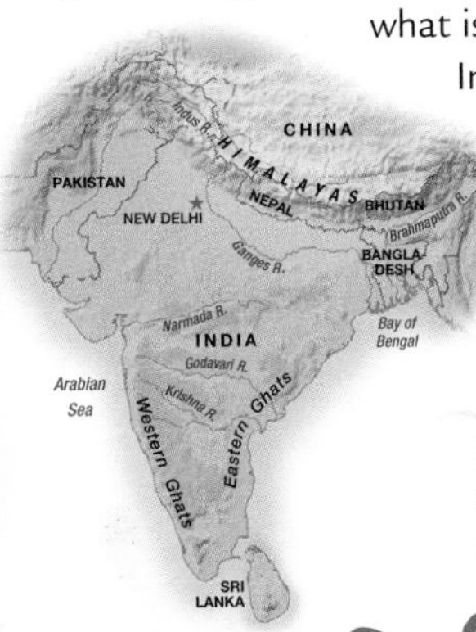

Mountain peaks rise above the clouds in the world's highest mountains.

Lab Report

1. Is your diagram clear and easy to understand? Show your diagram to another person to see if he or she can easily understand it.
2. Did you use colors or symbols in your diagram? why or why not?
3. By drawing a diagram of the Himalayas, did you learn something about diagrams in general? What did you learn?

Lab Report

Answers

1. Students should share their diagrams and discuss whether they are easy to understand.
2. Students should explain the use of colors and symbols in their diagrams.
3. Students might suggest that they learned more about the amount of skill and thought that a good diagram requires.

UNIT 9

THE PACIFIC WORLD AND ANTARCTICA

Using the Illustrations

Direct students' attention to the photos on these pages. Tell students that the young people in the center photo are from Melbourne, Australia. Have them locate Melbourne on a map. Ask students what they might conclude from the Australian teens' clothing and hair styles. *(Possible answer: Their culture has much in common with ours.)* Point out that Wellington, New Zealand, in the photo on the right-hand page, appears to have some features in common with American cities. Most New Zealanders also share many culture traits with Americans.

Ask what climate extremes are represented in the small photos *(very cold—Antarctica; tropical—Papua New Guinea)*. Ask students to suggest why the penguins do not appear to be frightened by the photographer's presence. *(Possible answer: Because they seldom see humans, they are not afraid of them.)* Refer back to the Papua New Guinea photo. Each village in the Sepik River area has such a house for each clan in the village. Ask students to speculate why the houses are built on stilts *(to keep them above the river's floodwaters)*.

Unit Objectives

1. Describe the landforms and climates in the Pacific world and Antarctica.
2. Identify the culture groups of the region and relate their experiences to historical events.
3. Explain the relationship of the region's resources to the economic development of Australia, New Zealand, and the Pacific Islands.
4. Identify the environmental challenges that confront the Pacific world and Antarctica.
5. Develop awareness of how events, immigration, politics, and shared goals create global connections among peoples and nations.

UNIT 9

The Pacific World and Antarctica

CHAPTER 29

Australia and New Zealand

CHAPTER 30

The Pacific Islands and Antarctica

King penguins watching photographer, Antarctica

House in Sepik River area, Papua New Guinea

Teenagers of Melbourne, Australia

Chapters

29 Australia and New Zealand
describes the countries' landforms, climates, water and mineral resources, unique plant and animal life, history, agriculture, industry, and population distribution.

30 The Pacific Islands and Antarctica
introduces the distinctive characteristics of the islands of the Pacific region and Antarctica.

A Film Critic in New Zealand

David Gerstner is a film critic and lecturer from New York City. In 1999 he moved to New Zealand. He found the country was very different than he thought it would be. **WHAT DO YOU THINK?** ***What do you know about New Zealand?***

First I was surprised that New Zealand has such lively cities. I had thought it was all mountains and bush. I was also surprised by the laid-back New Zealand approach to personal interaction. Celebrities, politicians, and the "common folk" have an easier time getting together. For example, I have become friends with some of New Zealand's well-known filmmakers, writers, and fashion designers—all within a year. The social playing field is more level than in the United States.

In fact, a friend and I literally ran into Jenny Shipley, who was then New Zealand's Prime Minister, on a busy street in the capital, Wellington. She was with her daughter. There were no secret service guards with them—like there would be with the U.S. president. We even had our picture taken with her! The easy-going Kiwi style of security is refreshing. However, for a New Yorker it is sometimes unsettling.

Wellington, New Zealand

Understanding Primary Sources

1. Why was David Gerstner surprised by New Zealand's cities?
2. What does this passage illustrate about New Zealand's culture?

Koalas

More from the Field

The Maori people first came to the islands of New Zealand about 1,000 years before Captain Cook landed there in 1769. During British rule (1840–1907) conflicts arose. European settlers took most of the best land. Many Maori were displaced to the cities.

Maori culture did not die. During the 1980s Maori activism became intense. Maori people wanted their language used in education and broadcasting. They also sought compensation for natural resources that they felt had been taken from them.

Today, almost 10 percent of New Zealand's population is Maori. By law, 6 of the 120 seats in New Zealand's parliament are reserved for Maori members.

Activity: Have students conduct research on Maori society in the 2000s. You may want to have them draw editorial cartoons that express Maori concerns.

Understanding Primary Sources

Answers

1. He had thought it was all mountains and bush.
2. easy-going, laid-back approach to personal interaction

UNIT 9 ATLAS

Overview

In this unit, students will learn about the people and geography of Australia, New Zealand, the Pacific Islands, and Antarctica.

Australia's unique ecosystems are the result of the continent's isolation. Geological activity makes New Zealand's landscape highly varied. Both countries have majority populations of British heritage and minority indigenous populations. The economies combine agriculture, manufacturing, and services. Australia and New Zealand have democratic governments. Most of the people enjoy a high standard of living.

The Pacific Islands' beauty disguises some economic difficulties. Many depend on aid from other countries.

Antarctica has too harsh an environment for permanent settlement. Several nations have established scientific research stations there.

People in the Profile

Note that the elevation profile follows the Tropic of Capricorn, which lies just north of Pitcairn Island. It is one of the most remote inhabited islands on Earth.

Although Polynesians lived on the island 600 years ago, it was abandoned until 1790. That year, men from the British ship *Bounty*—who had mutinied against their captain—settled on Pitcairn, along with several Tahitian men and women.

Within 10 years, all the mutineers but one, John Adams, had died. When British naval officers rediscovered Pitcairn in 1814 they were so impressed by Adams's virtues that they did not arrest him for the mutiny. Although Pitcairn's isolation was ended, the tiny population still struggled to survive. Twice the entire community moved to another island. Both times some of the settlers returned. In 1940 the sale of postage stamps began to bring in much-needed cash.

Today, Pitcairn has about 50 permanent residents. Government jobs provide the only steady employment.

Critical Thinking: How do Pitcairn's residents make a living?

Answer: mostly through government jobs

UNIT 9 ATLAS The World in Spatial Terms

The Pacific World and Antarctica

Elevation Profile

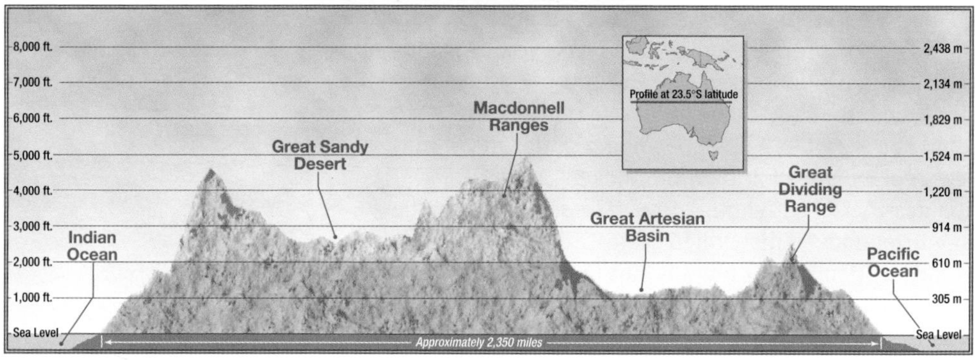

The United States and the Pacific World and Antarctica: Comparing Sizes

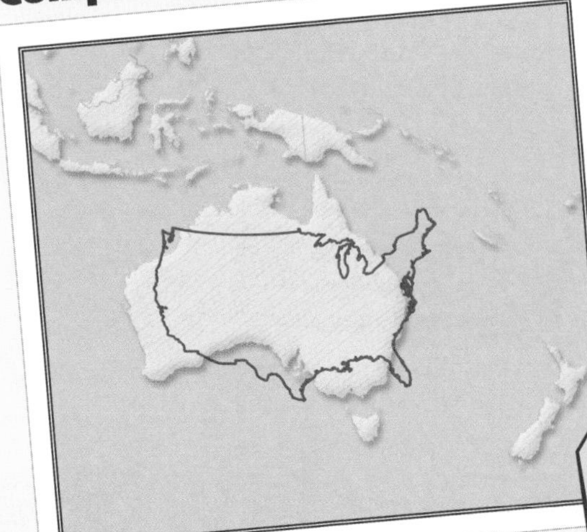

GEOSTATS: Antarctica

- World's highest, driest, coldest, and windiest continent
- World's lowest recorded temperature: -129°F (-89.2°C) on July 21, 1983
- Amount of the world's freshwater stored as ice in Antarctica: about 70 percent
- Average thickness of ice: over 1 mile (1.6 km)
- Highest mountain in Antarctica: Vinson Massif–16,066 ft. (4,897 m)

USING THE PHYSICAL MAP

Have students use the scales of the main map and the inset map on this page to draw a conclusion about the relative sizes of Australia and Antarctica. *(The scales are similar. Antarctica is larger than Australia.)* Then have students compare the physical features of the two continents. Ask why little information about Antarctica's landforms is available from the map. *(Antarctica is covered by an ice cap.)* Then ask students what physical feature the continents have in common *(a mountain range).*

The Pacific World and Antarctica: Physical

UNIT 9 ATLAS

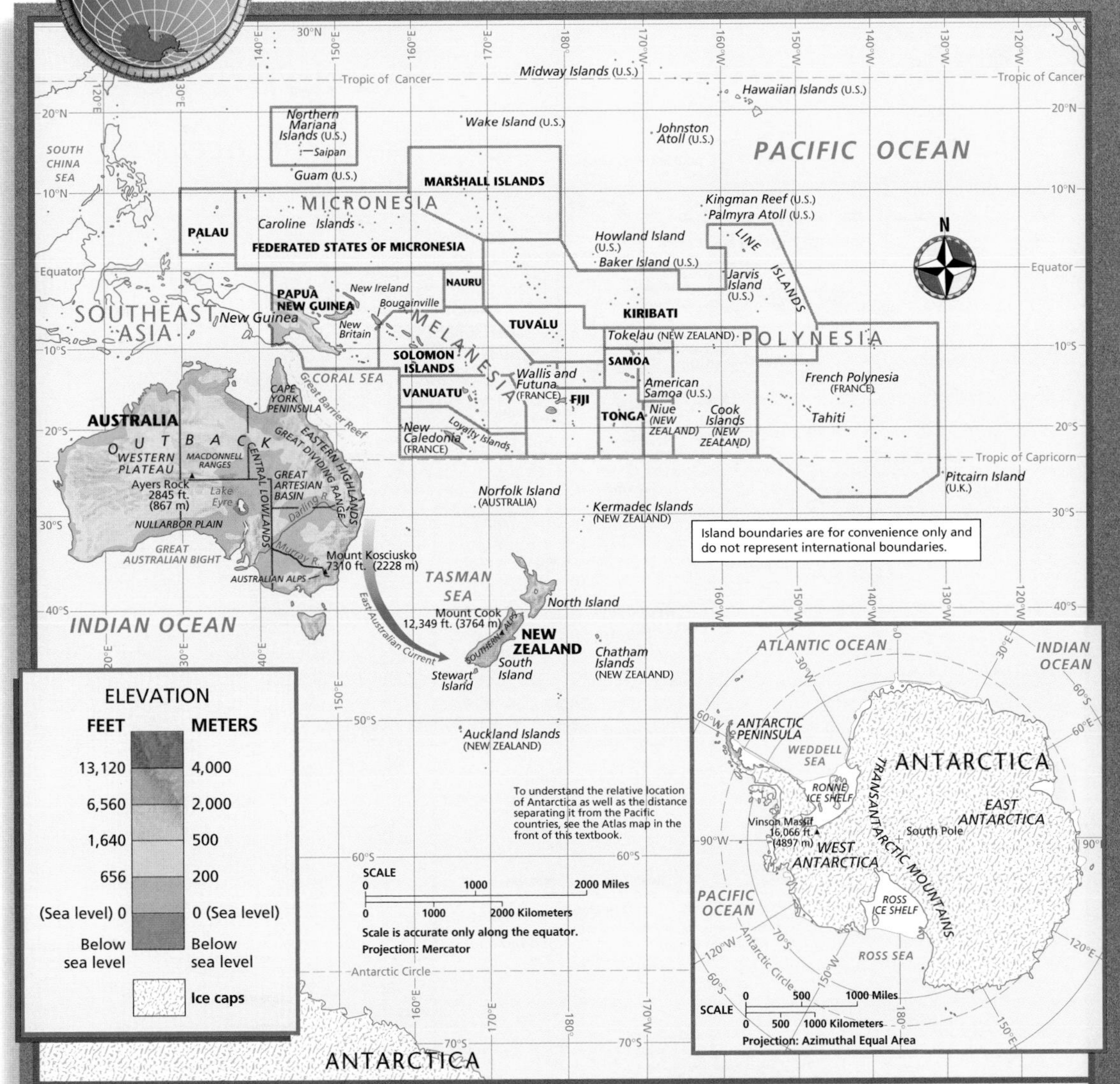

1. *Places and Regions* What is the highest point in Australia? in the region?
2. *Places and Regions* About how far apart are Guam and Tahiti? New Zealand and Australia?
3. *Places and Regions* What physical features cover most of Australia?

Critical Thinking

4. **Comparing** Which country is more mountainous, Australia or New Zealand?
5. **Analyzing Information** Compare this map to the **climate map** of the region. Which Australian mountain range seems to cause a rain-shadow effect?

internet connect

ONLINE ATLAS
GO TO: go.hrw.com
KEYWORD: SK3 MapsU9
FOR: Web links to online maps of the region

Physical Map
Answers

1. Mount Kosciusko; Vinson Massif, Antarctica
2. more than 5,000 miles (8,050 km); about 1,300 miles (2,092 km)
3. plains and plateaus

Critical Thinking

4. New Zealand
5. Great Dividing Range

USING THE POLITICAL MAP

Focus students' attention on the **political map** on this page. Point out that the islands of the southwestern Pacific Ocean are often grouped for convenience into three clusters. Ask students to name the groups *(Melanesia, Micronesia, Polynesia)*. Tell the class that the prefixes in these terms refer to the dark skin of many of the people *(mela-)*, the small size of the islands *(micro-)*, and the large number of islands *(poly-)*. Call on a volunteer to identify some of the islands within the three groups. Point out that the three terms do not relate to political boundaries.

Then ask students to name the capitals of Australia *(Canberra)* and New Zealand *(Wellington)*. Call on volunteers to identify the capitals of some of the island countries.

internet connect
ONLINE ATLAS
GO TO: go.hrw.com
KEYWORD: SK3 MapsU9

Your Classroom Time Line

These are the major dates and time periods for this unit. Have students enter them on the time line you created earlier. You may want to watch for these dates as students progress through the unit.

- c.* 38,000 B.C. Aborigines move into Australia.
- c. A.D. 1000 First settlers land in New Zealand.
- 1500s Ferdinand Magellan explores the Pacific Ocean.
- 1769 Captain James Cook visits New Zealand.
- 1770s Captain James Cook explores the Antarctic coast.
- 1788 British begin settling colonies in Australia.
- 1840 British sign a treaty with the Maori in New Zealand.

*c. stands for *circa* and means "about."

Political Map

Answers

1. Papua New Guinea; New Zealand
2. Papua New Guinea

Critical Thinking

3. Many of the island groups are not independent; they are controlled by other countries.
4. The islands of Fiji would be split.

The Pacific World and Antarctica: Political

UNIT 9 ATLAS

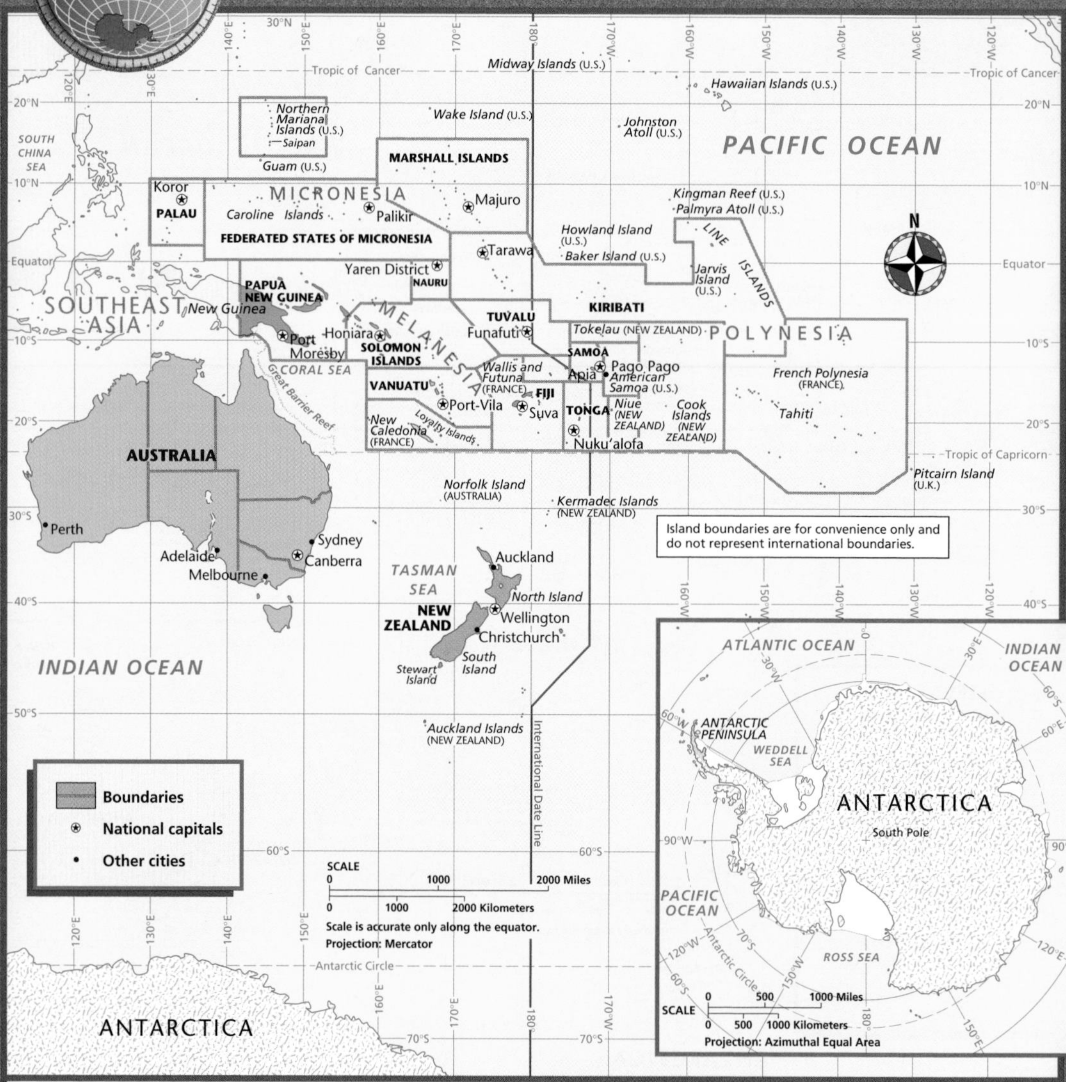

1. *Places and Regions* Which country occupies half of a large island? Which country occupies two large islands?
2. *Places and Regions* What is the only country in the region that has a land boundary with another country?

Critical Thinking

3. **Analyzing Information** What is one thing you can tell about governments in the region just by looking at the map?
4. **Drawing Inferences and Conclusions** Why do you think the international date line is not a straight line?

USING THE CLIMATE MAP

As students examine the **climate map** on this page, point out that all the climate types of the world are represented. Have students compare this map to the **physical map** of the region. For each climate type, call on a volunteer to describe the relative location of a place with that climate. *(Examples: humid tropical—northeast coast of Papua New Guinea; tropical savanna—north coast of Australia; desert—central and western Australia)* Ask which climate region includes the most land area *(ice cap).* Remind students that Antarctica is larger than Australia.

The Pacific World and Antarctica: Climate

UNIT 9 ATLAS

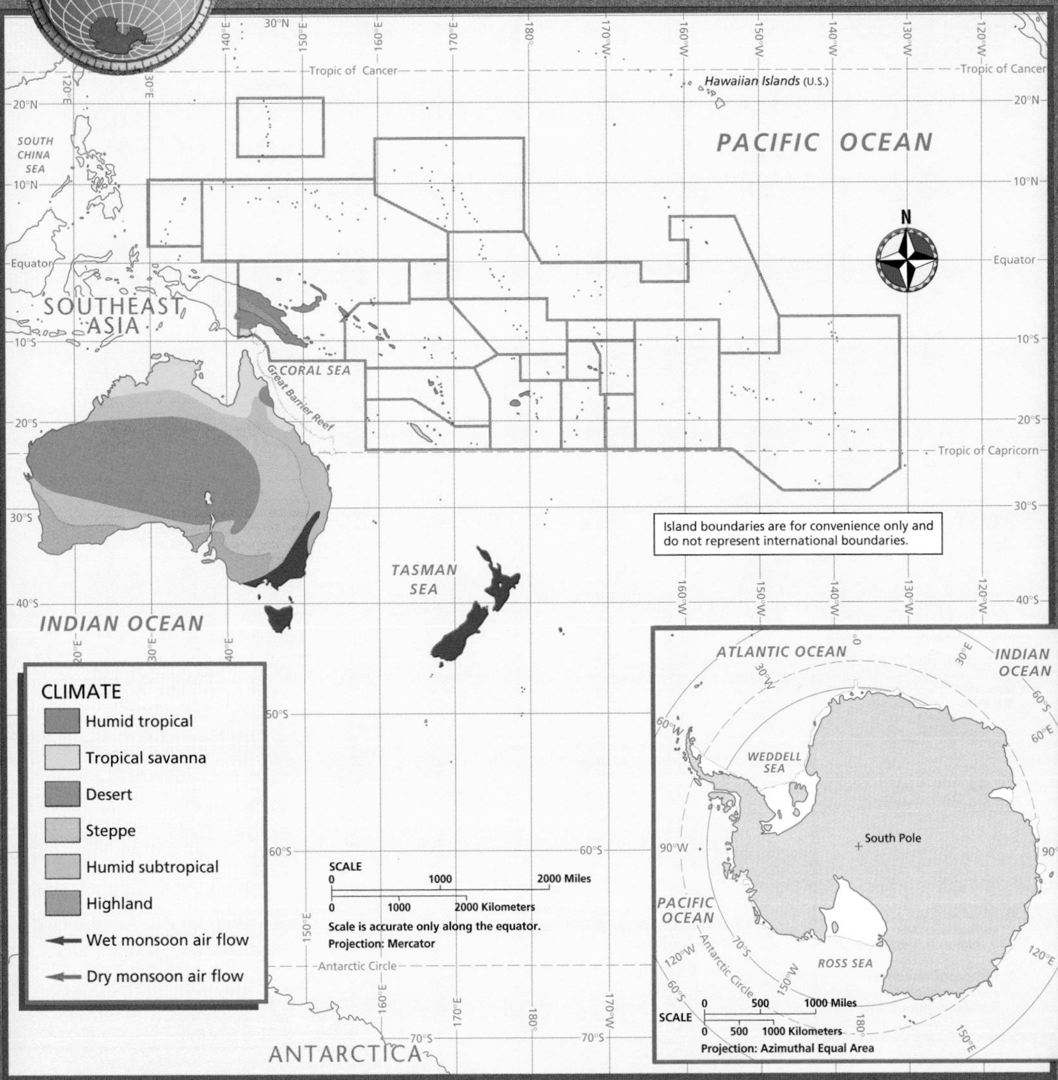

1. *Places and Regions* What type of climate does New Zealand have?
2. *Places and Regions* Which country has the greatest variety of climate types?
3. *Places and Regions* Compare this map to the **population map** of the region. What type of climate does Perth have?

Critical Thinking

4. **Making Generalizations and Predictions** Compare this map to the **physical map**. How might the East Australian Current affect New Zealand's weather?
5. **Comparing** How might climate affect Papua New Guinea's population patterns?

Your Classroom Time Line (continued)

1898 The United States wins Guam from Spain.

1901 The British grant Australia its independence.

1907 New Zealand gains its independence.

1911 First expedition reaches the South Pole.

1959 Many countries sign a treaty to preserve Antarctica.

1991 An international agreement forbids mining and drilling and limits tourism in Antarctica.

Climate Map

Answers

1. marine west coast
2. Australia
3. Mediterranean

Critical Thinking

4. brings warm water, helps keep New Zealand's climate mild
5. Population density is probably greater on coasts where there is a humid tropical climate.

UNIT 9 ATLAS

internet connect

ONLINE ATLAS
GO TO: **go.hrw.com**
KEYWORD: **SK3 MapsU9**

USING THE POPULATION MAP

Focus students' attention on the **population map** on this page. Ask them to write a statement to summarize the region's population density. *(Examples: Population density is low in the region and concentrated in a few cities. Population density is heaviest along the southeast coast of Australia.)* Call on volunteers to read their sentences.

Ask students which of New Zealand's islands appears to have more people *(North Island).* Have students compare this map to the **physical** and **climate maps**. Then ask them why Papua New Guinea has an east-west strip of sparse population density *(mountainous area with highland climate).*

The Pacific World and Antarctica: Population

UNIT 9 ATLAS

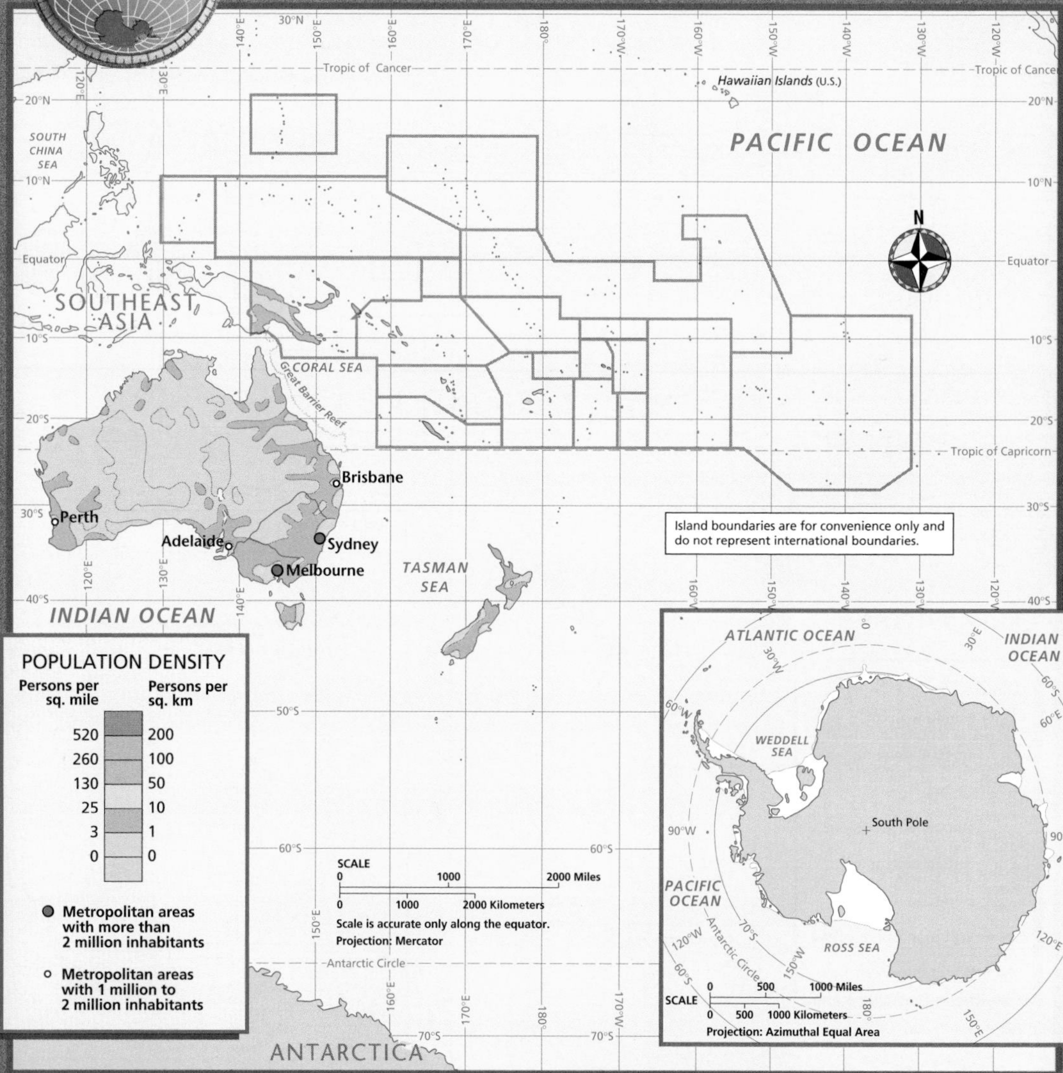

1. *Places and Regions* What is the only country with cities of over 2 million people?
2. *Places and Regions* What region of Australia has the highest population density?
3. *Places and Regions* Compare this map to the **physical map**. What is the densely populated island northwest of New Zealand?

Critical Thinking

4. **Analyzing Information** What large landmass has no permanent population? Why?
5. **Analyzing Information** Use the map on this page to create a chart, graph, database, or model of population centers in the Pacific world and Australia.

Population Map

Answers

1. Australia
2. southeast coast
3. New Caledonia

Critical Thinking

4. Charts, graphs, databases, and models should accurately reflect population figures for cities on the map.

USING THE LAND USE AND RESOURCES MAP

As students examine the **land use and resources map** explain that Australia's outback has been described as a storehouse of mineral wealth. Have students list the mineral resources found in Australia *(coal, gold, diamonds, uranium, and other minerals).* Then ask them where silver is mined *(Papua New Guinea).*

Have students identify the fossil fuel resources that are available in the region *(coal, oil, gas).* Finally, ask students which country has geothermal power *(New Zealand).*

The Pacific World and Antarctica: Land Use and Resources

UNIT 9 ATLAS

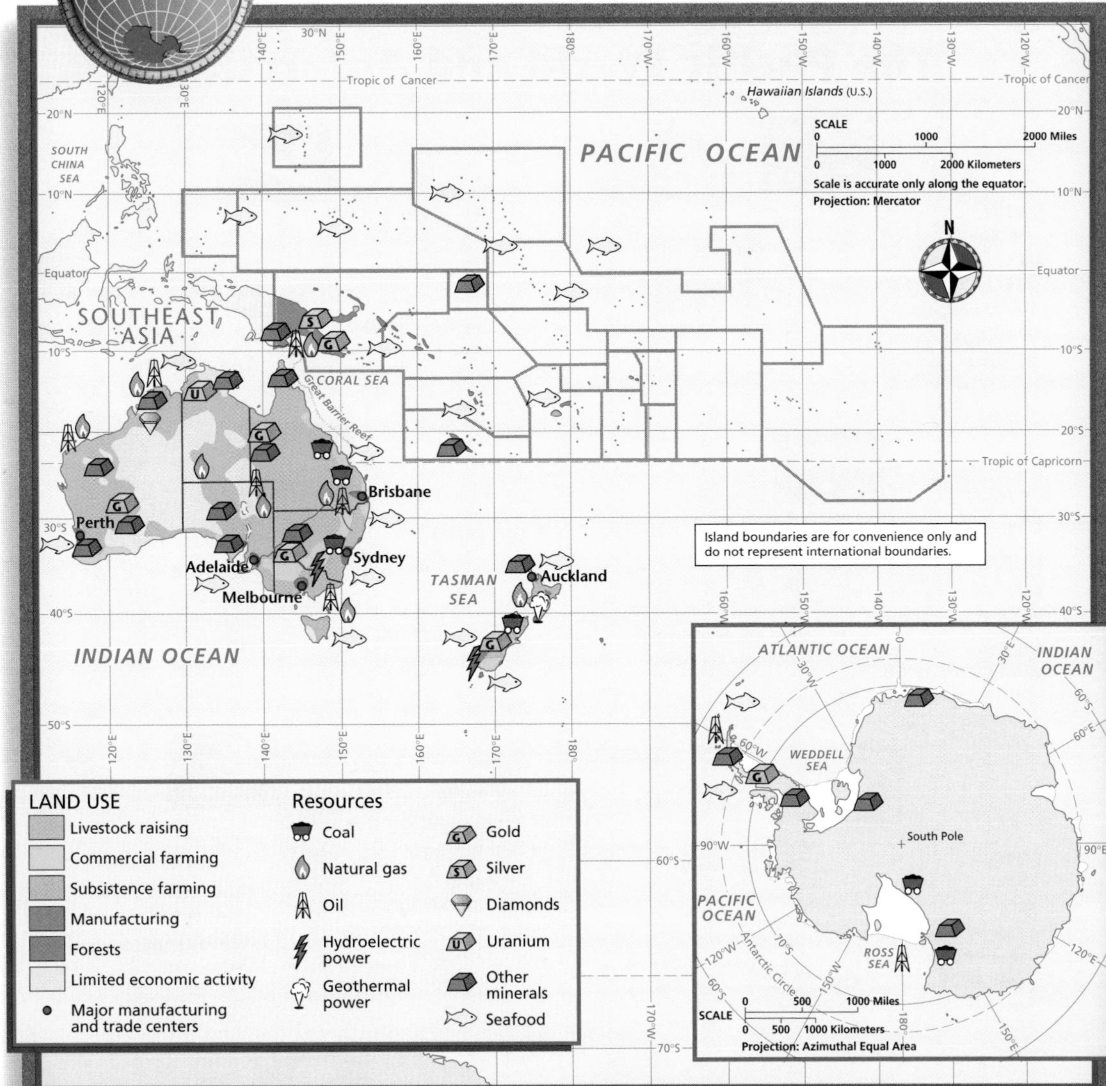

1. *Places and Regions* What cities are part of Australia's major manufacturing and trade centers?
2. *Places and Regions* What resources are available in Papua New Guinea but not in Australia?
3. *Places and Regions* What is the main resource of most of the small Pacific islands?

Critical Thinking

4. **Analyzing Information** Use the map on this page to create a chart, graph, database, or model of economic activities in the Pacific world and Australia.

Land Use and Resources Map

Answers

1. Perth, Adelaide, Melbourne, Sydney, Brisbane
2. forests and silver
3. seafood

Critical Thinking

4. Charts, graphs, databases, and models should accurately reflect information on the map.

Fast Facts

THE PACIFIC WORLD AND ANTARCTICA

FAST FACTS ACTIVITIES

LEVEL 1: (Suggested time: 30 min.) Refer students to the figures that show what percentage of the population is from 0 to 14 years old for countries in the Pacific world. Ask what is meant by 0 years old *(newborn to age 1).* Have students use these figures to create a bar graph for the countries of the Pacific world. Ask students to add the United States (21 percent) to their graphs.

Then have students write a few sentences recording their observations about the graphed figures. *(Examples: Every country's figures are higher than 20 percent. A third of countries have nearly half of their population in this age group.)*

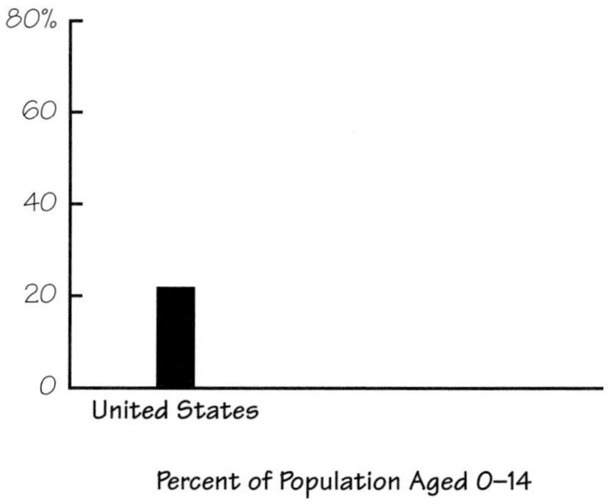

UNITED STATES OF AMERICA

CAPITAL: Washington, D.C.
AREA: 3,717,792 sq. mi. (9,629,091 sq km)
POPULATION: 281,421,906
MONEY: U.S. dollar (USD)
LANGUAGES: English, Spanish (spoken by a large minority)
POPULATION AGED 0–14: 21 percent

Fast Facts

The Pacific World and Antarctica

AUSTRALIA

CAPITAL: Canberra
AREA: 2,967,893 sq. mi. (7,686,850 sq km)
POPULATION: 19,357,594
MONEY: Australian dollar (AUD)
LANGUAGES: English, ethnic languages
POPULATION AGED 0-14 YEARS: 21 percent

MARSHALL ISLANDS

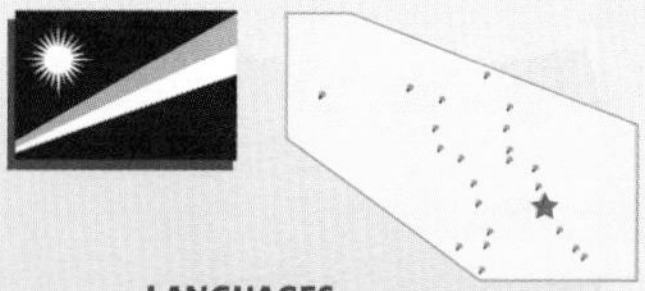

CAPITAL: Majuro
AREA: 70 sq. mi. (181.3 sq km)
POPULATION: 70,822
MONEY: United States dollar (USD)
LANGUAGES: English (official), Marshallese dialects, Japanese
POPULATION AGED 0-14 YEARS: 49 percent

FEDERATED STATES OF MICRONESIA

CAPITAL: Palikir
AREA: 271 sq. mi. (702 sq km)
POPULATION: 134,597
MONEY: United States dollar (USD)
LANGUAGES: English (official), Trukese, Pohnpeian, Yapese, Kosrean
POPULATION AGED 0-14 YEARS: data not available

NAURU

CAPITAL: no official capital
AREA: 8 sq. mi. (21 sq km)
POPULATION: 12,088
MONEY: Australian dollar (AUD)
LANGUAGES: Nauruan (official), English
POPULATION AGED 0-14 YEARS: 40 percent

FIJI

CAPITAL: Suva
AREA: 7,054 sq. mi. (18,270 sq km)
POPULATION: 844,330
MONEY: Fijian dollar (FJD)
LANGUAGES: English (official), Fijian, Hindustani
POPULATION AGED 0-14 YEARS: 33 percent

NEW ZEALAND

CAPITAL: Wellington
AREA: 103,737 sq. mi. (268,680 sq km)
POPULATION: 3,864,129
MONEY: New Zealand dollar (NZD)
LANGUAGES: English (official), Maori
POPULATION AGED 0-14 YEARS: 22 percent

KIRIBATI

CAPITAL: Tarawa
AREA: 277 sq. mi. (717 sq km)
POPULATION: 94,149
MONEY: Australian dollar (AUD)
LANGUAGES: English (official), I-Kiribati
POPULATION AGED 0-14 YEARS: 41 percent

PALAU

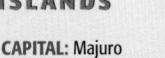

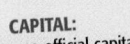

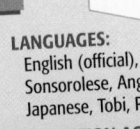

CAPITAL: Koror
AREA: 177 sq. mi. (458 sq km)
POPULATION: 19,092
MONEY: United States dollar (USD)
LANGUAGES: English (official), Sonsorolese, Angaur, Japanese, Tobi, Palauan
POPULATION AGED 0-14 YEARS: 27 percent

Countries not drawn to scale.

LEVEL 2: (Suggested time: 40 min.) Draw students' attention to the figures for the percentage of population that is from 0 to 14 years old. Have students write a paragraph describing the special needs of this population. *(Needs may include food, shelter, child care, health care, education, supervised leisure activities, or others.)*

Point out that people under 14 years old are only a few years away from childbearing and employment age. Have students write an additional paragraph explaining the possible effects that having a large segment of the population under 14 years may have on a country—particularly small island countries—such as those in the Pacific world. *(Possible answers: Unemployment may rise. Population may increase beyond available housing. Greater demands on food supplies may require increased imports.)*

LEVEL 3: (Suggested time: 45 min.) Ask students to imagine that they are young teens living in one of the Pacific world countries. Have them write a letter to their country's leader expressing their hopes and concerns for the upcoming decade. As young teens, issues may include job and educational opportunities and others as discussed in the Level 2 activity. Encourage students to draw information from the Fast Facts feature, this unit's atlas, and the chapters to support their statements.

UNIT 9 ATLAS

Papua New Guinea

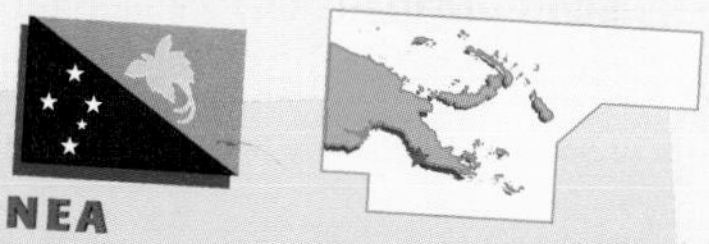

CAPITAL: Port Moresby

AREA: 178,703 sq. mi. (462,840 sq km)

POPULATION: 5,049,055

MONEY: kina (PGK)

LANGUAGES: 715 ethnic languages, pidgin English, English

POPULATION AGED 0-14 YEARS: 39 percent

Samoa

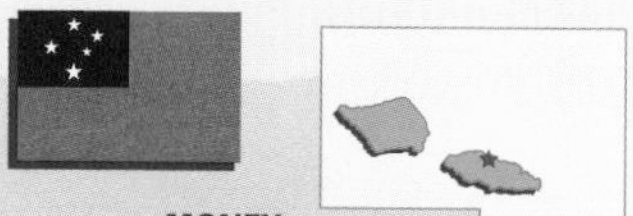

CAPITAL: Apia

AREA: 1,104 sq. mi. (2,860 sq km)

POPULATION: 179,058

MONEY: tala (WST)

LANGUAGES: Samoan (Polynesian), English

POPULATION AGED 0-14 YEARS: 32 percent

Solomon Islands

CAPITAL: Honiara

AREA: 10,985 sq. mi. (28,450 sq km)

POPULATION: 480,442

MONEY: Solomon Islands dollar (SBD)

LANGUAGES: Melanesian pidgin, 120 ethnic languages, English

POPULATION AGED 0-14 YEARS: 44 percent

Tonga

CAPITAL: Nuku'alofa

AREA: 289 sq. mi. (748 sq km)

POPULATION: 104,227

MONEY: pa'anga (TOP)

LANGUAGES: Tongan, English

POPULATION AGED 0-14 YEARS: 41 percent

Tuvalu

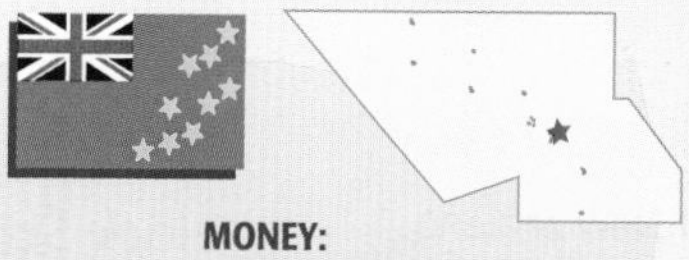

CAPITAL: Funafuti

AREA: 10 sq. mi. (26 sq km)

POPULATION: 10,991

MONEY: Tuvaluan dollar or Australian dollar (AUD)

LANGUAGES: Tuvaluan, English

POPULATION AGED 0-14 YEARS: 33 percent

Vanuatu

CAPITAL: Port-Vila

AREA: 4,710 sq. mi. (12,200 sq km)

POPULATION: 192,910

MONEY: vatu (VUV)

LANGUAGES: English (official), French (official), pidgin (known as Bislama or Bichelama)

POPULATION AGED 0-14 YEARS: 36 percent

internet connect

COUNTRY STATISTICS
GO TO: **go.hrw.com**
KEYWORD: **SK3 FactsU9**
FOR: more facts about the Pacific World and Antarctica

Sources: Central Intelligence Agency, *The World Factbook 2001; The World Almanac and Book of Facts 2001;* pop. figures are 2001 estimates.

UNIT 9 ASSESSMENT RESOURCES

Reproducible

- ◆ Unit 9 Test
- ◆ Unit 9 Test for English Language Learners and Special-Needs Students

internet connect

COUNTRY STATISTICS
GO TO: **go.hrw.com**
KEYWORD: **SK3 FactsU9**

Highlights of Country Statistics

- *CIA World Factbook*
- Library of Congress country studies
- Flags of the world

CHAPTER 29 Australia and New Zealand

Chapter Resource Manager

Objectives	Pacing Guide	Reproducible Resources
SECTION 1 **Australia** (pp. 671–77) **1.** Identify Australia's natural features and resources. **2.** Describe the history of Australia. **3.** Describe the people and culture of Australia. **4.** Describe what Australia is like today.	**Regular** 2.5 days **Block Scheduling** .5 day *Block Scheduling Handbook, Chapter 29*	**RS** Guided Reading Strategy 29.1 **E** Creative Strategies for Teaching World Geography, Lessons 22 and 23 **E** Environmental and Global Issues Activity 8 **SM** Geography for Life Activity 29 **IC** Interdisciplinary Activity for the Middle Grades 36 **SM** Map Activity 29
SECTION 2 **New Zealand** (pp. 680–83) **1.** Identify the natural features of New Zealand. **2.** Describe the history and culture of New Zealand. **3.** Describe New Zealand's cities and economy.	**Regular** 1 day **Block Scheduling** .5 day *Block Scheduling Handbook, Chapter 29*	**RS** Guided Reading Strategy 29.2 **E** Cultures of the World Activity 9

Chapter Resource Key

- **RS** Reading Support
- **IC** Interdisciplinary Connections
- **E** Enrichment
- **SM** Skills Mastery
- **A** Assessment
- **REV** Review
- **ELL** Reinforcement and English Language Learners
- Transparencies
- CD–ROM
- Music
- Video
- go.hrw.com Internet
- Holt Presentation Maker Using Microsoft® Powerpoint®

One-Stop Planner CD–ROM

See the ***One-Stop Planner*** for a complete list of additional resources for students and teachers.

One-Stop Planner CD–ROM

It's easy to plan lessons, select resources, and print out materials for your students when you use the ***One-Stop Planner CD–ROM with Test Generator.***

Technology Resources	Review, Reinforcement, and Assessment Resources
One-Stop Planner CD–ROM, Lesson 29.1 *ARGWorld* CD–ROM: Graphing In-migration to Australia Geography and Cultures Visual Resources with Teaching Activities 61–65 Homework Practice Online HRW Go site	**ELL** Main Idea Activity 29.1 **REV** Section 1 Review, p. 677 **A** Daily Quiz 29.1 **ELL** English Audio Summary 29.1 **ELL** Spanish Audio Summary 29.1
One-Stop Planner CD–ROM, Lesson 29.2 Homework Practice Online HRW Go site	**ELL** Main Idea Activity 29.2 **REV** Section 2 Review, p. 683 **A** Daily Quiz 29.2 **ELL** English Audio Summary 29.2 **ELL** Spanish Audio Summary 29.2

internet connect

HRW ONLINE RESOURCES

GO TO: go.hrw.com
Then type in a keyword.

TEACHER HOME PAGE
KEYWORD: SK3 TEACHER

CHAPTER INTERNET ACTIVITIES
KEYWORD: SK3 GT29

Choose an activity to:
- explore the Great Barrier Reef.
- learn about the Aborigines of Australia.
- tour New Zealand.

CHAPTER ENRICHMENT LINKS
KEYWORD: SK3 CH29

CHAPTER MAPS
KEYWORD: SK3 MAPS29

ONLINE ASSESSMENT
Homework Practice
KEYWORD: SK3 HP29
Standardized Test Prep Online
KEYWORD: SK3 STP29
Rubrics
KEYWORD: SS Rubrics

COUNTRY INFORMATION
KEYWORD: SK3 Almanac

CONTENT UPDATES
KEYWORD: SS Content Updates

HOLT PRESENTATION MAKER
KEYWORD: SK3 PPT29

ONLINE READING SUPPORT
KEYWORD: SS Strategies

CURRENT EVENTS
KEYWORD: S3 Current Events

Meeting Individual Needs

Ability Levels

Level 1 Basic-level activities designed for all students encountering new material

Level 2 Intermediate-level activities designed for average students

Level 3 Challenging activities designed for honors and gifted-and-talented students

English Language Learners Activities that address the needs of students with Limited English Proficiency

Chapter Review and Assessment

E Readings in World Geography, History, and Culture 79 and 80
SM Critical Thinking Activity 29
REV Chapter 29 Review, pp. 684–85
REV Chapter 29 Tutorial for Students, Parents, Mentors, and Peers
ELL Vocabulary Activity 29
A Chapter 29 Test
Chapter 29 Test Generator (on the One-Stop Planner)
Audio CD Program, Chapter 29
A Chapter 29 Test for English Language Learners and Special-Needs Students

CHAPTER 29

LAUNCH INTO LEARNING

Inform students that both Australia and New Zealand have many species unique to their countries. Ask students to look at the map to determine why this may be. *(The region is isolated. Both countries are islands.)* Ask students if they are already familiar with any animals from this region. *(Possible answers: koalas, kangaroos, kiwis)* Tell students they will learn more about the physical and human geography of Australia and New Zealand in this chapter.

Section 1

Objectives

1. Identify Australia's natural features and resources.
2. Examine the history of Australia.
3. Describe the people and culture of Australia.
4. Explain what Australia is like today.

LINKS TO OUR LIVES

You may want to reinforce interest in Australia and New Zealand by pointing out the following:

- Australia and New Zealand are both home to many rare species of plants and animals.
- Australia's Great Barrier Reef is the world's largest coral reef and is home to a complex ecosystem. It is threatened by many aspects of modern life. We can learn from Australians' efforts to preserve it.
- Most of both countries' residents share many cultural traits with U.S. citizens; clothing, food, homes, entertainment—all are similar to ours.
- Both Australia and New Zealand are attempting to redress injustices done to native peoples and are finding ways to recognize and respect minority cultures and religions.

CHAPTER 29

Australia and New Zealand

In this chapter we will study Australia and nearby New Zealand. First we meet Jared and Ashleigh. Their ancestors lived in Australia long before Europeans arrived.

Nyurramba garran, bulmba nganydjin ngunda!

Hi! My name is Jared. I have a twin sister, Ashleigh. We are 14 years old and live in Cooroy, Australia, which is about two hours north of Brisbane. We live with our mother in a three-bedroom house on a hill. Our land has a big creek with a dam on it. The dam forms a big pond, or a billabong. We can jump into it from the trees along the edge. Our father lives up north in the traditional lands of our people, the Djabugayndgi.

We used to speak our native language, but we've forgotten lots of it since we've been going to school. My totem, or special animal protector, is the *Ngumba*, or platypus. My sister's is the *Badgigal,* or freshwater turtle.

Since we were three years old, we have been singing and dancing in an Aboriginal dance group called *Bibayungen*. At Christmas, the whole family gathers with about 1,000 people in our traditional lands for a big dance and music festival (*warrima* in our language).

▲ Translation: Come and see our country!

FOCUS

LET'S GET STARTED

Copy the following instructions onto the chalkboard: *With a partner, compile a list of several things that come to mind when you think of Australia.* When students have completed their work, invite the pairs to write their answers on the chalkboard. Discuss responses. Have students save their lists to use in the Review and Assess activity at the end of the section. Tell students that in Section 1 they will learn more about Australia's physical geography, history, and culture.

Using the Physical-Political Map

Have students examine the map on this page. Then have students compare Australia's landforms with the landforms of the United States. *(Australia has more desert areas, fewer rivers, and lower mountains.)*

Section 1 Australia

Read to Discover

1. What are Australia's natural features and resources?
2. What is the history of Australia?
3. What is Australia like today?

Define

artesian wells
coral reef
endemic species
marsupials
outback
rugby
bush

Locate

Eastern Highlands
Central Lowlands
Western Plateau
Great Dividing Range
Tasmania
Murray-Darling Rivers
Great Artesian Basin
Great Barrier Reef
Sydney
Melbourne
Brisbane
Perth

Australian opal

WHY IT MATTERS

Use CNNfyi.com or other **current events** sources to find out about Aborigine culture in Australia. Record your findings in your journal.

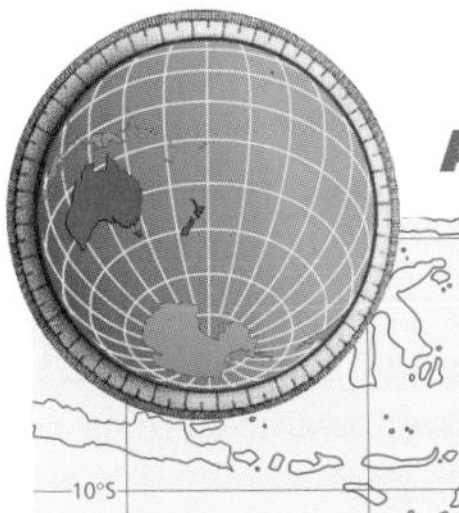

Australia and New Zealand: Physical-Political

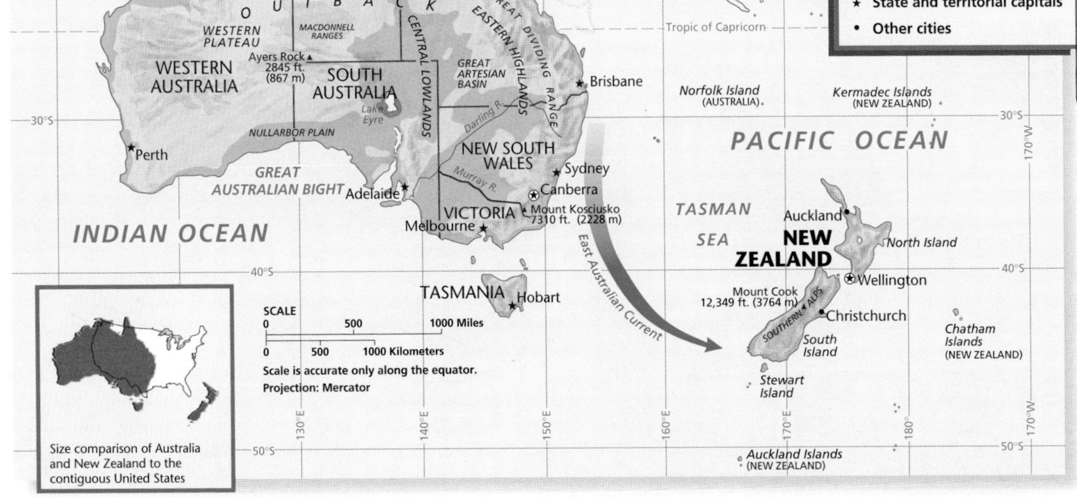

Section 1 RESOURCES

Reproducible

- Block Scheduling Handbook, Chapter 29
- Graphic Organizer 29
- Guided Reading Strategy 29.1
- Geography for Life Activity 29
- Map Activity 29
- Environmental and Global Issues Activity 8
- Creative Strategies for Teaching World Geography, Lessons 22 and 23
- Interdisciplinary Activity for Middle Grades 36

Technology

- One-Stop Planner CD–ROM, Lesson 29.1
- Homework Practice Online
- Geography and Cultures Visual Resources with Teaching Activities 61–65
- HRW Go site

Reinforcement, Review, and Assessment

- Section 1 Review, p. 677
- Daily Quiz 29.1
- Main Idea Activity 29.1
- English Audio Summary 29.1
- Spanish Audio Summary 29.1

TEACH

Teaching Objective 1

LEVEL 1: (Suggested time: 25 min.) Copy the following graphic organizer onto the chalkboard, omitting the italicized answers. Use it to help students describe Australia's physical features and climate types. Call on students to provide words and phrases that describe the landforms, bodies of water, climates, and resources of the country.
ENGLISH LANGUAGE LEARNERS

The Physical Geography of Australia

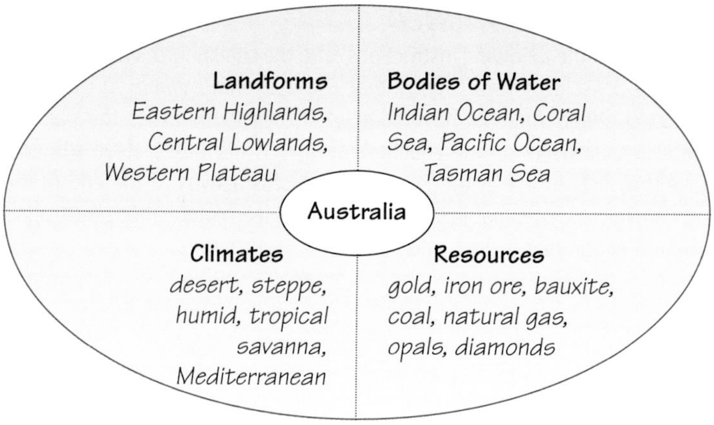

EYE ON EARTH

Uluru-Kata Tjuta National Park is home to Ayers Rock, a large monolith well known for its rich red color. The rusty color is caused in part by the chemical decay of minerals in the rock.

Ayers Rock has deep vertical crevices running down its sides and caves and inlets at its base. Like the red color of the rock, these formations were formed by chemical decay and sand-driven erosion.

In addition to Ayers Rock, the park has dunes reaching about 98 feet (30 m) in height and extensive sand plains. The land is valued for its rich landscape, geology, and desert ecosystems. The park management is committed to protecting its desert ecosystems.

Activity: Have students write a plan for a trip to Uluru-Kata Tjuta National Park. Tell students to plan how they would accommodate the need to preserve the park. For example, what would students do with their empty food containers?

Visual Record Answer

possible answers: crevices or grooves, smooth areas

internet connect

GO TO: go.hrw.com
KEYWORD: SK3 CH29
FOR: Web sites about Australia and New Zealand

Our Amazing Planet

Ayers Rock in central Australia is composed of sandstone with bits of reflective minerals. The sandstone changes color as the Sun moves across the sky. At sunset Ayers Rock turns a fiery orange-red.

Natural Features

Australia is the world's smallest, flattest, and lowest continent. In addition, it is the only country that is also a continent. The country sometimes is called the Land Down Under because it lies south of the equator. Seasons there are reversed from those north of the equator. For example, when it is winter in the United States, it is summer in Australia.

Land Australia has three main landform regions. These are the Eastern Highlands, the Central Lowlands, and the Western Plateau.

The Eastern Highlands are a system of ridges, plateaus, and valleys in the eastern part of Australia. They include the Great Dividing Range. This range stretches along the eastern coast and includes the island of Tasmania. It divides Australia's rivers into those that flow eastward and those that flow westward. Australia's one major river system, the Murray-Darling, flows westward from the range.

Australia's highest mountain is in the Great Dividing Range. It rises to just 7,310 feet (2,228 m). This tells us that Australia is very old. Erosion from wind and water has lowered the continent's mountains over millions of years.

The Central Lowlands are flatter and lower than the Eastern Highlands. The Central Lowlands include the Great Artesian Basin, Australia's largest source of underground water. **Artesian wells** dot this area. Artesian wells are those in which water rises toward the surface without being pumped. The groundwater comes from rain falling on nearby mountains. As the amount of groundwater increases, some is pushed to the surface. Much of the well water here is of poor quality, so it is used mostly for watering sheep.

Ayers Rock, called Uluru by Aborigines, rises more than 1,100 feet (335 m) above the surrounding Western Plateau. It is all that is left after erosion wore away an ancient mountain.

Interpreting the Visual Record How has erosion affected Ayers Rock?

LEVEL 2: (Suggested time: 35 min.) Write the words *Landforms, Bodies of Water, Climates, and Resources* along the top of a piece of butcher paper. Provide students with magazines and brochures about Australia. Have students cut out pictures of the features identified in the Level 1 lesson and tape them under the appropriate category on the paper. When complete, decide as a class if the pictures have been placed accurately and rearrange them as needed. Then glue them onto the paper to complete the display.

LEVEL 3: (Suggested time: 45 min.) Organize the class into groups. Have each group trace a map of Australia and divide it into regions as described in Section 1. Ask students to invent symbols to represent the natural features and mineral resources described in the section and to add those symbols to the appropriate regions on the map. Then have each group predict which part of the country is most important to Australia's economy. **COOPERATIVE LEARNING**

Teaching Objective 2

LEVEL 1: (Suggested time: 30 min.) Organize the class into groups. Have each group member select and illustrate an event in Australian history. Then have the group members work together to put the events in chronological order. Then have the groups check each other's work. **ENGLISH LANGUAGE LEARNERS, COOPERATIVE LEARNING**

The Great Barrier Reef is really thousands of small reefs and tiny islands. Some rise barely above water level when the tide is low.

Farther west is the Western Plateau, which covers more than half of the continent. The treeless Nullarbor Plain stretches along the southern edge of the plateau. This plain is the flattest large area on any continent.

The Great Barrier Reef Off the northeastern coast of Australia is the Great Barrier Reef. It is the world's largest **coral reef**, stretching more than 1,250 miles (2,000 km). A coral reef is a ridge found close to shore in warm, tropical waters. It is made of rocky limestone material formed by the skeletons of tiny sea animals. You will read more about coral reefs in the next chapter.

The Great Barrier Reef and its shallow waters are home to many kinds of marine animals. They include fish, shellfish, and sea birds. Australia's government has made most of the reef a national park.

Climate Australia has been described as a desert with green edges. Dry desert and steppe climates cover most of the country. The eastern and southeastern edges and Tasmania have humid climates. Two coastal areas in the south and southwest have a Mediterranean climate. Summers there are long, dry, and sunny. Winters are mild and wet. The north has a tropical savanna climate. There monsoons bring strong wet and dry seasons.

Plants and Animals Australia has many **endemic species**. Endemic species are plants and animals that developed in one particular region of the world. Why does Australia have many endemic species? Australia has been separated from other continents for millions of years. However, over time many nonnative plants and animals have been brought to Australia. See the Case Study in this chapter.

BUILD on WHAT You Know

Do you remember what you learned about monsoons? See Chapter 3 to review.

Our Amazing Planet

The shallow ocean waters along Australia's northern coast are home to the deadly sea wasp jellyfish. Each jellyfish has as many as 60 tentacles that hang down six and a half feet (2 m). The venom from its sting can kill a human in less than five minutes.

National Geography Standard 8

Physical Systems Australia's Great Barrier Reef, stretching more than 1,250 miles (2,000 km), is the largest structure in the world made by nonhuman living organisms. It is home to hundreds of species of fish, seaweed, birds, coral, mollusks, sea snakes, and sea turtles.

Like all coral reefs, it is made of the skeletons of hundreds of millions of coral polyps. New polyps grow only under very specific conditions. The water temperature must stay above about 68°F (20°C), the coral polyps must receive sufficient sunlight, and the water must be clear and salty.

The Great Barrier Reef is very fragile. Pollution, vibrations from boats, fuel spills, and even human sweat can damage it. Coral reefs can also be damaged if people walk on them or drag diving gear over them.

Discussion: Lead a discussion addressing these questions: Should people be allowed near the reef? Why or why not? What might be done to protect this delicate ecosystem?

internet connect

GO TO: go.hrw.com
KEYWORD: SK3 CH29
FOR: Web sites about the Great Barrier Reef

LEVEL 2: (Suggested time: 45 min.) Pair students and have pairs conduct research and write a short script for a radio show titled The Effects of European Settlement on Aboriginal Life. Tell students their shows should describe Aboriginal life before and after European settlement. Have each pair produce an audiotape of its program and play the tapes to the class. **COOPERATIVE LEARNING**

LEVEL 3: (Suggested time: 45 min.) Organize the class into two groups. Have one group prepare arguments for why Australia should leave the British Commonwealth, and the other prepare arguments for why the country should remain in the Commonwealth. Ask the students to take into account the country's history of settlement and independence. Encourage students to use encyclopedias and other resources as sources of background information. Call on volunteers to present their positions to the class. **COOPERATIVE LEARNING**

Teacher to Teacher

Jean Eldredge of Altamonte Springs, Florida, suggests the following activity to help students trace the importance of wool in Australia's economy. Have students conduct research on the wool industry, starting with the annual shearing at an Australian sheep station and concluding with the wool's shipment abroad. Have students apply the six essential elements of geography to steps in the process, as appropriate. For example, for Environment and Society have students examine how the ranchers adapt to the outback's arid environment. For Human Systems, ask them to analyze how the wool gets to market.

Environment and Society

Coober Pedy is a small town in the Australian outback. The name *Coober Pedy* comes from an Aboriginal phrase meaning "white man's burrow." In fact, many of the residents of the Australian outback town live and work underground to avoid the scorching heat. Homes, shops, and a church are all carved out of the rock. Most of the world's total production of opals comes from mines in the area.

Activity: Have students work in groups to conduct research on Australian opals, including the physical processes that create opals. Then ask them to create posters showing the different types. Students might use colored foil, cellophane, and tissue papers to depict the opals' colors. Display the posters around the classroom.

Connecting to Literature

Answers

1. far from other countries, winter in July
2. by calling the seas her neighbors and stating the long distances between Australia and other lands

Connecting to Literature

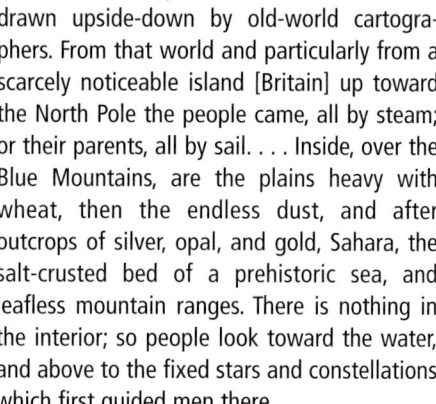

Christmas on an Australian beach

The Land Down Under

Europeans began settling in Australia in the late 1700s. The following selection describes what settlers found in the Land Down Under. The selection comes from Christina Stead's For Love Alone, *published in 1944.*

In the part of the world Teresa came from, winter is in July, spring brides marry in September, and Christmas is consummated[1] with roast beef [and] suckling[2] pig . . . at 100 degrees in the shade, near the tall pine tree loaded with gifts and tinsel as in the old country, and old carols have rung out all through the night.

This island continent lies in the water hemisphere. On the eastern coast, the neighboring nation is Chile, though it is far, far east, Valparaiso being more than six thousand miles away in a straight line; her northern neighbors are those of the Timor Sea, the Yellow Sea; to the south is that cold, stormy sea full of earth-wide rollers, which stretches from there without land, south to the Pole.

The other world—the old world, the land [Northern] hemisphere—is far above her as it is shown on maps drawn upside-down by old-world cartographers. From that world and particularly from a scarcely noticeable island [Britain] up toward the North Pole the people came, all by steam; or their parents, all by sail. . . . Inside, over the Blue Mountains, are the plains heavy with wheat, then the endless dust, and after outcrops of silver, opal, and gold, Sahara, the salt-crusted bed of a prehistoric sea, and leafless mountain ranges. There is nothing in the interior; so people look toward the water, and above to the fixed stars and constellations which first guided men there.

Analyzing Primary Sources

1. How does life differ from that in the "old world" in Teresa's land?
2. How does the writer give the reader a sense of Australia's isolation?

[1]consummated: completed, marked, or celebrated; [2]suckling: a young mammal that is still nursing

Two of Australia's most famous native animals are the kangaroo and the koala. Both animals are **marsupials** (mahr-soo-pee-uhls). Marsupials are animals that carry their young in pouches. Eucalyptus (yoo-kuh-LIP-tuhs) is Australia's most common tree.

Minerals Australia's **outback**, or inland region, has many mineral resources such as gold, iron ore, and bauxite. Australia also has coal, natural gas, oil, and gemstones such as opals and diamonds.

✓ **READING CHECK:** *Places and Regions* What are the natural features of Australia? three main landform regions; Murray-Darling river system; artesian wells; Great Barrier Reef; desert and steppe climates mostly, also humid, Mediterranean, and tropical savanna; many endemic species; mineral resources

Teaching Objective 3

ALL LEVELS: (Suggested time: 20 min.) Have students use the information presented in this section to create a pie chart depicting the ethnic diversity of the Australian population. *(British or other European background: 92 percent, Ethnic Asians: 7 percent, Aboriginal and other: 1 percent)* At the bottom of their pie charts, have students write a few sentences to provide details about the significance of each group's history or culture.

➤ASSIGNMENT: Have students write two paragraphs comparing and contrasting the people and culture of the United States and Australia, naming three similarities in the first paragraph and three differences in the second paragraph. Then have students write a third paragraph explaining which country would make a better home for an emigrant from a war-torn country. Students should support their answers with information presented in the first two paragraphs. Ask volunteers to read their papers to the class.

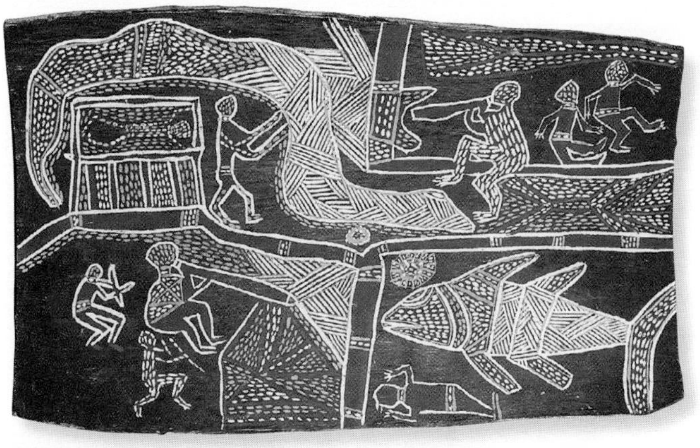

◄ A bark painting provides a glimpse of some Aborigines' spiritual beliefs. The painting shows the path taken by a soul on its journey to another world. Traditional Aboriginal culture almost disappeared after Europeans arrived in Australia.

History

The first humans to live in Australia were the Aborigines (a-buh-RIJ-uh-nees). They came from Southeast Asia at least 40,000 years ago. Early Aborigines hunted animals and gathered food from wild plants. They had many different languages, traditions, and customs. The arrival of Europeans changed life for the Aborigines.

European Settlers The British began settling colonies in Australia in 1788. Many of the first settlers were British prisoners, but other British settlers came, too. As the settlers built farms and ranches, they took over the Aborigines' lands. In addition, many Aborigines died of diseases brought unintentionally by the Europeans.

Independence The British granted independence to the Australian colonies in 1901. The colonies were united into one country within the British Commonwealth of Nations. Australia fought on the side of the British and other Allied forces in World Wars I and II.

Today Australia has six states and one large territory. The national capital is Canberra. Because Australia is part of the Commonwealth, the British monarch also is Australia's monarch. However, a prime minister and parliament make Australia's laws. Some Australians want their country to leave the Commonwealth and replace the monarch with an Australian president or other official.

States and Territories of Australia

State	Capital
New South Wales	Sydney
Victoria	Melbourne
Queensland	Brisbane
Western Australia	Perth
South Australia	Adelaide
Tasmania	Hobart
Territory	**Capital**
Northern Territory	Darwin
Australian Capital Territory	Canberra

Interpreting the Chart **What is the capital of Tasmania?**

✓ **READING CHECK:** *Human Systems* How has migration affected Australia's history? British settlers took Aborigine land and passed on diseases; led to membership in British Commonwealth

Culture

More than 90 percent of Australians today are of British or other European ancestry. Since the 1970s Asians have been moving to

Across the Curriculum
ART

Aboriginal Art The Aborigines of Australia have one of the oldest surviving cultures on Earth. Since they had no written language, they have traditionally used song, dance, and art to pass on knowledge and to celebrate their religion. Since ancient times the Aborigines have created rock engravings and mosaics painted on the ground. Aborigine artists now paint on canvas as well.

The artists' symbols have spiritual meanings and are handed down from one generation to the next. Wavy lines, dots, and rods all convey different ideas. Sometimes plain or concentric circles or animal tracks are also depicted.

Artists gravitate toward different styles, depending on where they live. For example, artists from Arnhem Land use many straight lines and cross-hatching. Dotted designs are common in art produced by the Aborigines of the central desert regions.

Activity: Have students try painting in the Aboriginal style using an American animal or legend as the subject matter.

◄ **Chart Answer**

Hobart

Teaching Objective 4

LEVEL 1: (Suggested time: 30 min.) Have students summarize Australia's strengths and the challenges facing the country today. Select volunteers to write items in a list on the chalkboard. Ask students if the United States has similar strengths or challenges. Discuss the comparisons with the class. **ENGLISH LANGUAGE LEARNERS**

LEVELS 2 AND 3: (Suggested time: 45 min.) Tell students to imagine that they are consultants who are applying for jobs to help Australia overcome one of its challenges. Students may select a challenge from the class list compiled in the Level 1 lesson. Have each student write a proposal listing several possible solutions to the selected issue. When students have finished their proposals, organize the class into groups according to the issues selected. Have members decide which of the proposed solutions would be most effective. Then have each group propose its solution to the class. **COOPERATIVE LEARNING**

National Geography Standard 10

Human Systems Australia's ethnic composition is changing rapidly. An increasing number of the country's immigrants are from areas close to Australia, such as Asia. Recent immigrants have come from China, Malaysia, the Philippines, and Vietnam.

In recent years, Australia's response to migration has been a policy of multiculturalism. The aim of this approach is to maintain the integrity of each individual culture, while promoting education and awareness among all of Australia's ethnic groups.

Australian schools have adopted the use of multicultural curricula. In particular, many schools teach Asian languages and customs to students beginning in the elementary grades.

Critical Thinking: How does Australia's current multicultural policy differ from the attitudes of its first European settlers?

Answer: Students may mention that, in contrast to Australia's current multicultural policy, the first settlers did not respect the Aborigines' culture.

Chart Answer ►

It is higher.

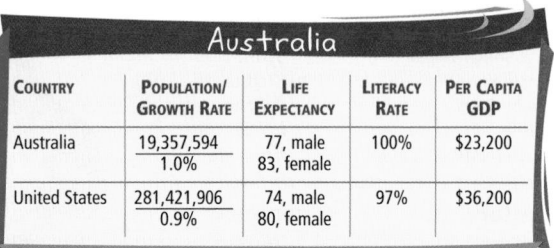

▲ The Sydney Opera House opened on the city's harbor front in 1973. It is one of Australia's most well-known buildings.

Australia in growing numbers. As a result, about 7 percent of Australia's population today is ethnic Asian. Only about 1 percent of the country's people are Aborigines. Nearly all Australians speak English, but some Aborigines also speak native languages.

Religion and Holidays Most Australians are Christians. They celebrate many of the same holidays as people in the United States. However, their traditions may be different. For example, December falls during summer in Australia. Christmas there is time for a beach party, picnic, or some other outdoor activity.

Outdoor activities are popular in sunny Australia. Many people enjoy picnics, swimming, and sports. Popular sports include the British game of cricket, sailing, horse racing, surfing, and Australian Rules football. Many Australians also enjoy **rugby**. Rugby is a British game similar to football and soccer.

The Arts Music and other arts are also popular in Australia. In fact, art by Aborigines has become popular around the world. The Australian government is an important source of money for the arts. For example, the government has built many public performance halls. One of the most famous is the Opera House in Australia's largest city, Sydney.

Australia

Country	Population/ Growth Rate	Life Expectancy	Literacy Rate	Per Capita GDP
Australia	19,357,594 1.0%	77, male 83, female	100%	$23,200
United States	281,421,906 0.9%	74, male 80, female	97%	$36,200

Sources: Central Intelligence Agency, *The World Factbook 2001;* U.S. Census Bureau

Interpreting the Chart **How does Australia's literacy rate compare to that of the United States?**

✓ **READING CHECK:** *Human Systems* What are some characteristics of Australia's people? 90 percent British or European ancestry, 7 percent ethnic Asian, 1 percent Aborigine; mostly Christian

CLOSE

Call on students to answer these questions: What is a key feature of Australia's physical geography? What is a key feature of the country's human geography? Discuss responses.

REVIEW AND ASSESS

Have students complete the Section Review. Then have students refer to the lists they made earlier of what they knew about Australia. Have students work with the same partners to correct inaccuracies and to add facts they learned in this section. Then have students complete Daily Quiz 29.1.

RETEACH

Have students complete Main Idea Activity 29.1. Then have students work in groups to prepare and perform a television commercial about Australia. Each commercial should include the information covered in the Read to Discover questions. **ENGLISH LANGUAGE LEARNERS, COOPERATIVE LEARNING**

EXTEND

Have interested students conduct research on the Tasmanian wolf (or Tasmanian tiger, *Thylacinus cynocephalus*), a marsupial believed to have become extinct in the 1930s. Have students focus on the causes of its extinction and reasons why some people think that the animal still survives. **BLOCK SCHEDULING**

Australia Today

About 85 percent of Australians live in and around cities. The rest live in the **bush**, or lightly populated wilderness areas. The largest cities are Sydney and Melbourne in the southeast. Brisbane, on the eastern coast, enjoys a warm, tropical climate that attracts tourists. Towns have grown around mining and ranching areas in parts of the dry, rugged outback. Large parts of the interior have few people. The seaport of Perth is the largest city in the west.

Economy Australia is a rich, economically developed country. It is a leading producer of agricultural goods such as wool, meat, and wheat. Australia supplies nearly half of the world's wool used in clothing. One of the country's most important industries is mining, particularly in the outback. Other industries include steel, heavy machines, and computers.

Challenges Australia faces important challenges. Among these is improving the economic and political status of the Aborigines. They have only recently gained back some of the rights they lost after Europeans colonized Australia. The country also must continue to build ties with Asian and Pacific countries. At one time Australia's strongest ties were with Britain. Today, its most important trading partners are much closer Asian countries and the United States. In addition, Australia must deal with the rapid growth of its cities. It also must protect native animals and the environment.

✓ **READING CHECK:** *Places and Regions* What is Australia's economy like? rich, economically developed

Major Producers of Wool

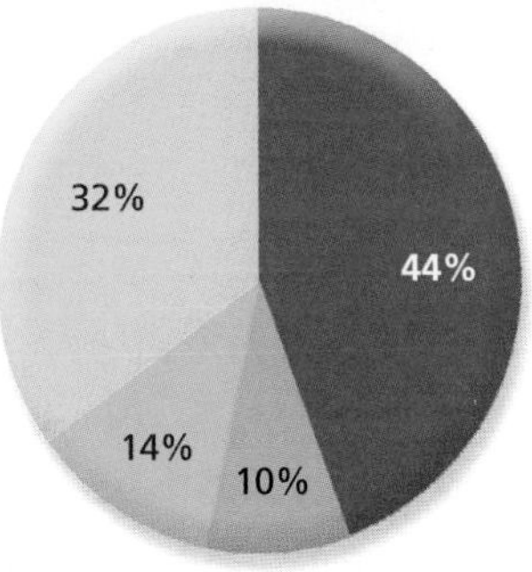

Source: Based on data from The Woolmark Company

▲ Wool is produced from sheep's hair. Australia has one of the world's largest flocks of sheep.

Interpreting the Chart **What nearby country is the world's second-largest wool producer?**

go.hrw.com **Homework Practice Online** Keyword: SK3 HP29

Section 1 Review

Define and explain: artesian wells, coral reef, endemic species, marsupials, outback, rugby, bush

Working with Sketch Maps On a map of Australia and New Zealand that you draw or that your teacher provides, label the following: Eastern Highlands, Central Lowlands, Western Plateau, Great Dividing Range, Tasmania, Murray River, Darling River, Great Artesian Basin, Great Barrier Reef, Sydney, Melbourne, Brisbane, and Perth. In a box in the margin, describe why the Great Artesian Basin is important to a dry country like Australia.

Reading for the Main Idea

1. *Human Systems* How has migration affected Australia's population? Which ethnic group is the largest?
2. *Places and Regions* What are two of Australia's most famous native marsupials?
3. *Human Systems* Who were the first humans to live in Australia? When did Europeans start settling there?

Critical Thinking

4. **Drawing Inferences and Conclusions** Why is building ties with Asian countries important for Australia today?

Organizing What You Know

5. **Categorizing** Copy the following graphic organizer. Use it to list Australia's climates, plants and animals, and mineral resources.

Climates	Plants and Animals	Resources

Section 1 Review

Answers

Define For definitions, see: artesian wells, p. 672; coral reef, p. 673; endemic species, p. 673; marsupials, p. 674; outback, p. 674; rugby, p. 676; bush, p. 677

Working with Sketch Maps Maps will vary, but listed places should be labeled in their approximate locations. It provides water for sheep and other livestock.

Reading for the Main Idea

1. people of British and other European ancestry and Asian ancestry have moved there, leading to decline of Aborigine population; British and other European ancestry (NGS 9)
2. koala bear, kangaroo (NGS 4)
3. Aborigines; 1788 (NGS 9)

Critical Thinking

4. Possible answer: Trade with Asian countries is good for Australia's economy.

Organizing What You Know

5. climates—desert, steppe, humid, Mediterranean, tropical savanna; plants and animals—eucalyptus, marsupials; resources—gold, iron ore, bauxite, coal, natural gas, opals, diamonds

▲ **Chart Answer**

New Zealand

CASE STUDY

Setting the Scene

Isolation can have a major effect on the distribution of plants and animals around the globe. Australian plant and animal life evolved to fit the continent's natural environment. Plants adapted to the low fertility of soils and gained the ability to live in a wide range of continental habitats. Marsupials filled the niches typically occupied by placental mammals on other continents. The arrival of Europeans ended the isolation of Australia's plants and animals. European settlers converted large areas of grassland into fields for agriculture and pasture. They also introduced animal species to the region that competed with Australia's native species for resources.

Building a Case

Have students read "Nonnative Species in Australia" and engage them in a discussion with the following questions: How long was Australia isolated from other continents? *(35 million years)* What kind of agriculture did the Europeans bring when they arrived in the late 1700s? *(cattle, sheep, and goat raising; wheat and sugarcane farming)* Have volunteers answer the questions at the end of the feature. Here are some additional facts to consider: About 10 percent of Australia's plants are nonnative. Introduced species threaten 64 percent of Australia's mammals, 27 percent of its birds, and 22 percent of its reptiles. More than half of Australia's endemic animal species are threatened by nonnative species.

HISTORICAL GEOGRAPHY

The European settlement of Australia resulted in an exchange of species between the two continents. European botanists were very interested in the potential economic value of the new plants and animals they found in Australia. In particular, the eucalyptus tree provided a number of uses, including oils for medicine and perfume, timber, windbreaks, and ornamentation.

British botanists collected seeds and brought them back to England for cultivation. Kew Gardens near London has an excellent collection of Australian plants. Today large stands of eucalyptus can be found worldwide, from California and the Congo to Spain and Portugal.

Activity: California has recently begun an effort to remove some of its eucalyptus trees and restore native vegetation. Have students conduct research on the distribution, use, and environmental impact of eucalyptus trees.

Visual Record Answer

possible answer: trampling of the river bank, which could lead to erosion

➤ **This Case Study feature addresses National Geography Standards 8, 12, and 14.**

CASE STUDY

NONNATIVE SPECIES IN AUSTRALIA

For about the last 35 million years, Australia has been separated from the other continents by oceans. During this time Australia's plants and animals developed in isolation from the rest of the world. Eventually, a unique Australian ecosystem developed. Many interesting species of plants and animals existed only in Australia. For example, kangaroos, koalas, and eucalyptus trees were all native to Australia. They were not found anywhere else in the world.

Europeans arrived in Australia in the late 1700s. They dramatically changed the natural balance that had developed over millions of years. Europeans brought with them many new species of plants and animals. These nonnative species have had a major impact on the environment in Australia.

European settlers turned large areas of grasslands into fields for agriculture. They set up huge wheat farms in the south and southwest. They planted sugarcane fields in the northeast. These crops had never been grown in Australia. Farmers also brought nonnative animals, such as sheep, cows, and goats, into Australia. In less than 100 years, Australia had more than 100 million sheep and 8 million cows.

Cattle are herded on the island of Tasmania in Australia.

Interpreting the Visual Record Can you see how these animals are changing the environment?

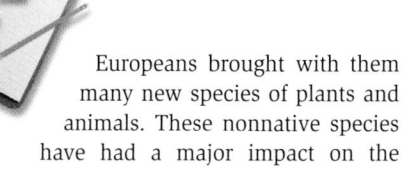

Introduced into Australia from Southeast Asia, water buffalo now eat food that Australia's native animals need to survive.

Drawing Conclusions

Lead a classroom discussion on the connection between soils, grasses, and animals in Australia. List each of these parts of the ecosystem on the chalkboard and have students answer the following questions:

- What was the impact of introduced species on soil? on grasses? on animals?
- How are these aspects of the natural world related to one another?
- How might changing one part of the system affect the other parts?
- Can you give examples from the text? from other sources?

Going Further: Thinking Critically

Islands are often home to rare species, which can be threatened by the introduction of new plants and animals. New Zealand, Hawaii, and Puerto Rico are just a few places where nonnative species have taken hold. Here is the percentage of nonnative species for each:

Location	Mammals	Plants	Birds	Fish
Australia	Unavailable	10%	Unavailable	13%
Hawaii	94%	47.4%	40%	76%
New Zealand	93%	46.7%	19%	53%
Puerto Rico	Unavailable	11.5%	23%	91%

Ask: How does the quantity of nonnative species in Australia compare to the other places? Why might this be so?

Understanding What You Read

1. Isolation led to the development of a unique ecosystem in Australia. For example, kangaroos, koalas, and eucalyptus are native to this Australian ecosystem.
2. Europeans brought cows, goats, sheep, cane toads, camels, rabbits, and prickly pear cactus. Many of these species increased rapidly and competed with native species for resources such as food and water. Some overgrazed and caused soil erosion.

When Europeans settled in Australia they brought with them many new plants and animals. Some of these plants and animals have become pests because they damage the environment.

Many of the nonnative species that were brought into Australia spread across the continent. Some are now considered pests. These plants and animals are seriously damaging Australia's environment. For example, camels were introduced into Australia as pack animals. Now they run wild in the desert interior. Water buffalo were introduced from Southeast Asia. They trample riverbanks, killing grasses and eroding the soil. The prickly pear cactus was introduced into Australia from the United States as a garden plant. It spread across much of the country and has been very hard to control. Cane toads were introduced in the 1930s to protect Australia's sugarcane fields from cane beetles. However, they release toxic chemicals when attacked and are now threatening native species and household pets.

Of the nonnative animals introduced into Australia, rabbits have probably been the most destructive. Brought in by hunters in the 1800s, rabbits multiplied and spread quickly. With no natural predators and lots of food, their numbers increased to about 500 million. Rabbits overgrazed Australia's grasslands and caused major soil erosion.

The introduction of nonnative species into Australia has damaged the environment in several ways. Some native species have become extinct because they could not compete with nonnative species. About 13 species of mammals and one species of bird have become extinct in Australia since the late 1700s. Many more are endangered. Nonnative species such as cows, sheep, and rabbits have caused soil erosion. These animals also consume food and water that Australia's native plants and animals need to survive. As the case of Australia shows, the introduction of nonnative species into a new environment can cause serious environmental problems.

Understanding What You Read

1. How did isolation affect Australia's ecosystem?
2. What are some of the species that Europeans brought with them to Australia? How does Australia's environment now reflect their influence?

internet connect

GO TO: go.hrw.com
KEYWORD: SK3 CH29
FOR: Web sites about endemic species

Section 2

Objectives

1. Identify the natural features of New Zealand.
2. Describe the history and culture of New Zealand.
3. Explain what New Zealand's cities and economy are like today.

Focus

LET'S GET STARTED

Copy the following instructions onto the chalkboard: *Look at the map at the beginning of the chapter and write a short answer to this question: What would it be like to live in a country like New Zealand that is separated from other countries by hundreds of miles of ocean?* When students have finished writing, discuss their answers. Tell students that in Section 2 they will learn more about life in New Zealand.

Building Vocabulary

Write the term **kiwi** on the chalkboard. Ask a volunteer to read aloud the definition from the glossary. Inform students that this flightless bird is so deeply identified with the country that New Zealanders are often referred to as "Kiwis." Ask volunteers to suggest nicknames for Americans using a similar logic.

Section 2 Resources

Reproducible
- Guided Reading Strategy 29.2
- Cultures of the World Activity 9

Technology
- One-Stop Planner CD–ROM, Lesson 29.2
- Homework Practice Online
- HRW Go site

Reinforcement, Review, and Assessment
- Section 2 Review, p. 683
- Daily Quiz 29.2
- Main Idea Activity 29.2
- English Audio Summary 29.2
- Spanish Audio Summary 29.2

internet connect

GO TO: **go.hrw.com**
KEYWORD: **SK3 CH29**
FOR: **Web sites about New Zealand's national parks**

Section 2 New Zealand

Read to Discover

1. What are the natural features of New Zealand?
2. What are the history and culture of New Zealand?
3. What are New Zealand's cities and economy like today?

Define

kiwi

Locate

Tasman Sea
South Island
North Island
Southern Alps
Wellington
Auckland
Christchurch

WHY IT MATTERS

Tourism is an important part of New Zealand's economy. Use CNNfyi.com or other **current events** sources to find out what attracts tourists to New Zealand. Record your findings in your journal.

A Maori pendant

Natural Features

New Zealand lies southeast of Australia across the Tasman Sea. It includes two large islands—South Island and North Island. Put together, the islands are about the size of the U.S. state of Colorado.

Land The highest mountains are the Southern Alps on South Island. These rugged, glacier-capped mountains cover the western half of the island. The highest peak reaches 12,349 feet (3,764 m). A narrow coastal plain stretches along South Island's eastern side.

North Island has three major volcanic peaks separated by a volcanic plateau. Volcanic eruptions are common. Most of the rest of North Island is covered by hills and coastal plains.

The Southern Alps stretch across much of New Zealand's South Island. New Zealand's government has established large national parks throughout the mountains.

TEACH

Teaching Objective 1

ALL LEVELS: (Suggested time: 25 min.) Copy the following graphic organizer onto the chalkboard, omitting the italicized answers. Use it to help students compare and contrast the physical geography of New Zealand's North Island and South Island. Call on students to provide words and phrases to complete the diagram. In the center, have students list features the islands have in common.

Comparison of the Physical Geography of New Zealand's North Island and South Island

North Island
- *three volcanic peaks separated by volcanic plateau*
- *hills and coastal plains*
- *northern area—warmer, more tropical climate*

Both Islands
- *marine west coast climate*
- *forests, pastures, farms*
- *endemic species*
- *fewer resources than Australia*

South Island
- *largest island*
- *Southern Alps—highest mountains in New Zealand*
- *narrow coastal plain along eastern side*
- *heavy rain and snow on western side*
- *grassy plains in rain shadow*

◄ Autumn turns South Island's countryside golden near the city of Christchurch. New Zealand has four distinct seasons.

Climate, Wildlife, and Resources Unlike mostly dry Australia, New Zealand is humid. It has a mild marine west coast climate. Rain falls throughout the year. Winds from the west bring very heavy rain or snow to South Island's western side. The grassy plains of the drier eastern side are in a rain shadow. The north of North Island, in the lower latitudes, has a warmer, more tropical climate.

Much of New Zealand was once forested. Today, more than half is covered by pastures and farms. Sheep and other livestock graze in New Zealand's pastures. Sheep and their wool have long been important to the country's economy.

As in Australia, many animals in New Zealand are endemic species. The country's endemic species include different kinds of bats and flightless birds such as the **kiwi** (KEE-wee). The kiwi has hairlike feathers and sharp senses of smell and hearing. This bird is so linked with the country that New Zealanders are sometimes called Kiwis.

Many other species have been brought to New Zealand over time. For example, deer and trout have been introduced to provide sport for hunters and fishers.

New Zealand has fewer mineral resources than Australia. Its natural resources include gold, iron ore, natural gas, and coal.

✓ **READING CHECK:** *Places and Regions* What are the natural features of New Zealand? two large islands; mountains, volcanic plateau, hills, coastal plains; many endemic species

BUILD ON WHAT YOU KNOW

Do you remember what you learned about the rain-shadow effect of mountains? See Chapter 3 to review.

History and Culture

New Zealand's first settlers came from other Pacific islands more than 1,000 years ago. Their descendants, the Maori (MOWR-ee), still live in New Zealand. The early settlers' main source of food was the moa—a giant, flightless bird. However, overhunting by New Zealand's early peoples wiped out the moa. The bird was extinct by the time Europeans arrived in New Zealand.

Maori traditional art, like this carved entrance arch, survived New Zealand's colonial period.

Interpreting the Visual Record What role does art play in Maori culture? ▼

Linking Past to Present

Sailing the Pacific In ancient times, Polynesian sailors roamed the Pacific in outrigger canoes. A Tahitian named Kupe is credited with discovering New Zealand in A.D. 950.

Some 400 years later, a large number of Tahitians used instructions left by Kupe and migrated to New Zealand. According to legend, the instructions were to "let the vessel's prow be directed to the left of the setting Sun."

Activity: Send students outside to write navigation instructions to items in the schoolyard. Have students use natural features, such as the Sun, like Kupe did.

◄ **Visual Record Answer**

Traditional art is important in Maori culture.

Teaching Objectives 1–3

LEVEL 1: (Suggested time: 45 min.) Organize the class into five groups and have them make posters of the following topics about New Zealand: history and government, culture, people, natural features, and life today. Display the posters around the classroom. **ENGLISH LANGUAGE LEARNERS, COOPERATIVE LEARNING**

LEVEL 2: (Suggested time: 30 min.) Ask students to imagine that they are students in New Zealand. Have them write a letter to a friend in the United States to persuade the friend to move to New Zealand. Instruct students to discuss the country's physical geography and its cultural history as well as its cities and economy. Have each student read his or her letter to a partner. Discuss which points in the letters were particularly persuasive. **COOPERATIVE LEARNING**

LEVEL 3: (Suggested time: 45 min.) Organize the class into groups. Have each group develop a front page for a New Zealand newspaper. Ask students to include information about the country's physical features and history as well as issues facing the country today using articles, cartoons, editorials, and other typical newspaper features. Then have a spokesperson from each group present his or her group's newspaper to the class. **COOPERATIVE LEARNING**

Section 2 Review

Answers

Define For definition, see: kiwi, p. 681

Working with Sketch Maps Maps will vary, but listed places should be labeled in their approximate locations. Wellington is the capital, and Auckland is the largest city.

Reading for the Main Idea

1. other Pacific islands (NGS 9)
2. market; agriculture is important; it has become more industrialized. (NGS 12)

Critical Thinking

3. North Island has volcanic peaks; South Island has glacier-capped mountains. The rest of North Island is covered by hills and coastal plains. The rest of South Island is a narrow coastal plain.
4. Possible answer: It has a more pleasant climate.

Organizing What You Know

5. climates—marine west coast, tropical; plants and animals—wheat, sheep, kiwi, deer, trout; resources—gold, iron ore, natural gas, coal

Visual Record Answer ▲

Possible answer: A Maori cultural tradition has been mixed with a game that is part of a culture of European heritage.

▲ New Zealand's All Black Rugby Union team performs a version of the Haka (HAH-kah) before international games. The Haka is a traditional Maori dance.

Interpreting the Visual Record
How has cultural borrowing affected this performance of the Haka?

European Settlers A Dutch explorer in 1642 was the first European to sight New Zealand. British explorer Captain James Cook visited New Zealand in 1769. Many British settlers started to arrive after the British signed a treaty with the Maori in 1840. However, fighting broke out as British settlers took over Maori lands.

Independence The British granted New Zealand independence in 1907. Like Australia, New Zealand became a member of the British Commonwealth of Nations. New Zealanders fought alongside the British in World Wars I and II.

The government of New Zealand is led by a prime minister. An elected parliament makes the country's laws. The capital, Wellington, is located at the southern tip of North Island.

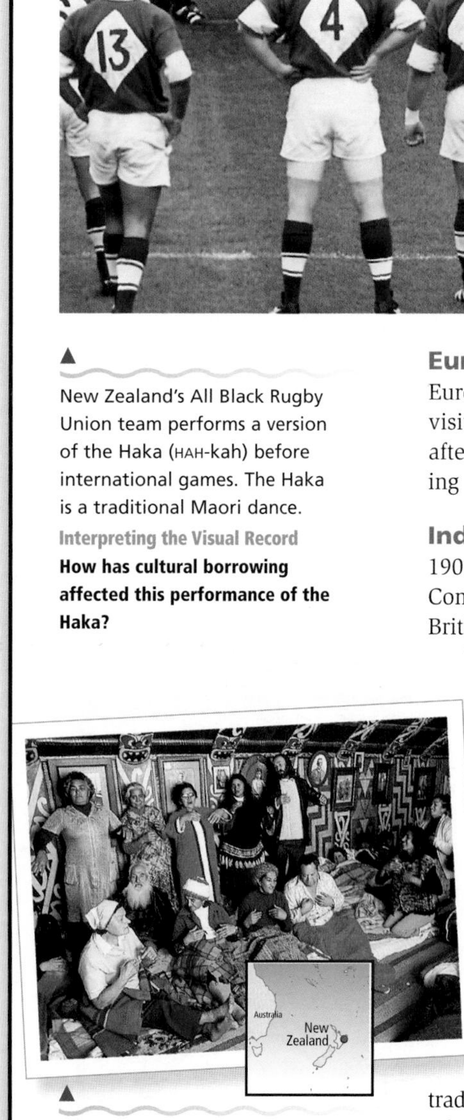

▲ Mourners attend a funeral in a Maori carved meeting house.

Culture Like Australia, most of New Zealand's 3.8 million people are of European ancestry. A growing number of ethnic Asians also live in the country. The Maori make up nearly 10 percent of the population. People from other Pacific islands have also come to New Zealand.

Most New Zealanders speak English. Some Maori also speak the Maori language. Most New Zealanders also are Christian. Sports, particularly rugby, are popular.

Maori culture has not disappeared. Traditional art, music, and dance remain important in Maori culture. Many Maori weddings, funerals, and other special events still are held in traditional carved houses. The Maori call these meetinghouses *whare whakairo*. Many meetinghouses are beautifully carved and decorated.

✓ **READING CHECK:** *Human Systems* How have conflict and cooperation affected the history of New Zealand? fighting between British and Maori initially; today, culture groups live together

CLOSE

Ask your students which country, Australia or New Zealand, they would prefer to visit. Have them state reasons for their preferences.

REVIEW AND ASSESS

Have students complete the Section Review. Then have them work in groups to write a six-question multiple-choice quiz about the section. Have groups trade and check each other's quizzes. Then have students complete Daily Quiz 29.2. COOPERATIVE LEARNING

RETEACH

Have students complete Main Idea Activity 29.2. Then have students compile fact sheets on New Zealand's physical features, history, cities, and economy. Students' facts should cover the main ideas of the section. Call on volunteers to read their fact sheets to the class.
ENGLISH LANGUAGE LEARNERS

EXTEND

Have interested students work in pairs to conduct research on the history of the Maori people. Instruct them to include the Maori's origins and customs. Ask them also to examine Maori political activism. Have each pair write a mock interview between an anthropologist and a Maori and present it to the class. COOPERATIVE LEARNING, BLOCK SCHEDULING

New Zealand Today

Today about 75 percent of New Zealanders, including most Maori, live on North Island. Most of the country's industries and agriculture are also located there.

Cities About 80 percent of New Zealanders live in urban areas. The country's largest city and seaport, Auckland, is located in the northern part of North Island. Christchurch is South Island's largest city.

New Zealand

COUNTRY	POPULATION/ GROWTH RATE	LIFE EXPECTANCY	LITERACY RATE	PER CAPITA GDP
New Zealand	3,864,129 / 1.1%	75, male / 81, female	99%	$17,700
United States	281,421,906 / 0.9%	74, male / 80, female	97%	$36,200

Sources: Central Intelligence Agency, *The World Factbook 2001;* U.S. Census Bureau

Interpreting the Chart How does New Zealand's life expectancy compare to that of the United States?

Economy New Zealand is a rich, modern country with a market economy. Its mild, moist climate helped make agriculture an important part of the economy. The country is a major producer of wool, meat, dairy products, wheat, kiwifruit, and apples.

New Zealand has become more industrialized in recent years. Factories turn out processed food, wood, paper products, clothing, and machinery. Banking, insurance, and tourism are also important industries. Australia, the United States, Japan, and the United Kingdom are the country's main trade partners.

✓ **READING CHECK:** *Places and Regions* What are some of New Zealand's cities, and what is its economy like?
Auckland, Christchurch; market; agricultural and industrial

Cable cars provide transportation for residents and visitors in Wellington.

go.hrw.com Homework Practice Online
Keyword: SK3 HP29

Section 2 Review

Define and explain: kiwi

Working with Sketch Maps On the map you created in Section 1, label the Tasman Sea, South Island, North Island, Southern Alps, Wellington, Auckland, and Christchurch. In a box in the margin, identify New Zealand's capital and the country's largest city.

Reading for the Main Idea

1. *Human Systems* From where did the ancestors of the Maori come?
2. *Human Systems* Describe New Zealand's economy. How has it changed in recent years?

Critical Thinking

3. **Contrasting** How is the physical geography of North Island different from that of South Island? Describe the physical geography of each island.
4. **Drawing Inferences and Conclusions** Why do you think most New Zealanders live on North Island?

Organizing What You Know

5. **Summarizing** Copy the following graphic organizer. Use it to list the climates, plants and animals, and resources of New Zealand.

Climates	Plants and Animals	Resources

CHAPTER 29 Review ANSWERS

Building Vocabulary

For definitions, see: artesian wells, p. 672; coral reef, p. 673; endemic species, p. 673; marsupials, p. 674; outback, p. 674; rugby, p. 676; bush, p. 677; kiwi, p. 681

Reviewing the Main Ideas

1. Eastern Highlands, Central Lowlands, Western Plateau (NGS 4)
2. desert, steppe, humid, Mediterranean, tropical savanna (NGS 4)
3. 1788; 1840 (NGS 9)
4. Aborigines; Maori (NGS 4)
5. wool (NGS 4)

Understanding Environment and Society

Answers will vary, but the information included should be consistent with text material. Check to see that students have included details on economic, geographic, environmental, and cultural issues. Use Rubric 29, Presentations, to evaluate student work.

▲ **Chart Answer**

It is higher.

Assess

Have students complete the Chapter 29 Test.

Reteach

Organize students into two groups. Assign Australia to one group and New Zealand to the other. Have each group write a basic plan or treatment for a documentary film about travel in its assigned country. When complete, have each group share its treatment with the class.

ENGLISH LANGUAGE LEARNERS, COOPERATIVE LEARNING

Portfolio Extensions

1. Have students conduct research on the daily lives of teens living at a sheep station in the outback. Students should focus on how the teens do schoolwork, what work they do at the station, and their entertainment options. Have students summarize their research in short reports. Then, in a class discussion, have students answer these questions: If I were moving to the Australian outback, what would I most like to take with me? What would I miss the most? Place the essays in the students' portfolios.
2. The Aborigines of Tasmania remained isolated from other cultures for about 10,000 years. Have students research this group of people and construct a diorama to depict the Tasmanian Aborigines' daily life.

Chapter 29 Review Answers

Thinking Critically

1. possible answer: opportunity to explore a rich ecosystem
2. because of its arid climate
3. They built farms, ranches, and cities and introduced nonnative species.
4. New Zealand is more humid. Australia: desert, steppe, humid, Mediterranean, tropical savanna; New Zealand: marine west coast
5. The arts have been important to Aboriginal culture; recently the Australian government has been an important source of support for the arts.

Map Activity

A. Perth
B. Tasmania
C. Auckland
D. Great Barrier Reef
E. Central Lowlands
F. Melbourne
G. Sydney
H. Western Plateau

Chapter 29 Reviewing What You Know

Building Vocabulary

On a separate sheet of paper, write sentences to define each of the following words.

1. artesian wells
2. coral reef
3. endemic species
4. marsupials
5. outback
6. rugby
7. bush
8. kiwi

Reviewing the Main Ideas

1. *Places and Regions* What are the three main landform regions of Australia?
2. *Places and Regions* What are the climates of the interior and the "edges" of Australia like?
3. *Human Systems* When did Europeans begin to settle in Australia and New Zealand?
4. *Places and Regions* What native peoples lived in Australia and New Zealand before the arrival of Europeans?
5. *Places and Regions* Australia and New Zealand are the world's two top producers of what agricultural product?

Understanding Environment and Society

Ranching

Create a presentation on ranching in Australia and New Zealand. In preparing your presentation, you may want to consider the following:

- How ranching contributes to the economies of the two countries.
- The geographic factors that have helped make ranching productive there.
- How ranching has affected the environment and culture of the two countries.

Thinking Critically

1. **Drawing Inferences and Conclusions** Why do you think the Great Barrier Reef would interest tourists?
2. **Drawing Inferences and Conclusions** Why do you think Australia's interior is lightly populated?
3. **Finding the Main Idea** How have humans changed the natural environments of Australia and New Zealand since the arrival of Europeans?
4. **Contrasting** How is New Zealand's climate different from Australia's? Identify the main climates in each country.
5. **Analyzing Information** What role have the arts played in Australian society?

Food Festival

For a New Zealand kiwi spritzer, combine 3 diced kiwifruit, 2 tbs sugar, and ½ c. orange juice in a blender. Blend until smooth. Pour ½ c. of mixture into a glass and stir in ¼ c. club soda. Note that this recipe should make approximately 4 servings. Be prepared to make more to serve your class.

Chapter 29 Review and Assessment Resources

Reproducible
- Readings in World Geography, History, and Culture 79 and 80
- Critical Thinking Activity 29
- Vocabulary Activity 29

Technology
- Chapter 29 Test Generator (on the One-Stop Planner)
- HRW Go site
- Audio CD Program, Chapter 29

Reinforcement, Review, and Assessment
- Chapter 29 Review, pp. 684–85
- Chapter 29 Tutorial for Students, Parents, Mentors, and Peers
- Chapter 29 Test
- Chapter 29 Test for English Language Learners and Special-Needs Students

Building Social Studies Skills

Map Activity

On a separate sheet of paper, match the letters on the map with their correct labels.

Central Lowlands	**Sydney**
Western Plateau	**Melbourne**
Tasmania	**Perth**
Great Barrier Reef	**Auckland**

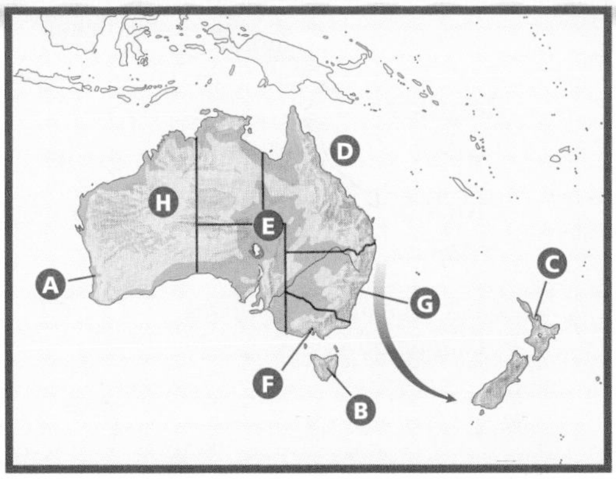

Mental Mapping Skills Activity

On a separate sheet of paper, draw a map of Australia and New Zealand. Make a key for your map and label the following:

Darling River	**Murray River**
Eastern Highlands	**North Island**
Great Dividing Range	**Southern Alps**
	South Island
	Tasman Sea

Writing Activity

Imagine that you are a European journalist in the late 1700s writing about the peoples and unique plants and animals in Australia and New Zealand. Write two paragraphs describing those peoples and endemic species. Include details that might surprise your European readers. Be sure to use standard grammar, spelling, sentence structure, and punctuation.

Alternative Assessment

Portfolio Activity

Learning About Your Local Geography

Individual Project Make a poster about plants and animals that are common in your area, including those that are endemic species and those that have been brought to the area from elsewhere.

internet connect

Internet Activity: **go.hrw.com**
KEYWORD: **SK3 GT29**

Choose a topic to explore about Australia and New Zealand:
- Explore the Great Barrier Reef.
- Learn about the Aborigines of Australia.
- Tour New Zealand.

Mental Mapping Skills Activity

Maps will vary, but listed places should be labeled in their approximate locations.

Writing Activity

Check to see that students have included accurate information about the region's plants and animals in the late 1700s. Details should be consistent with text information. Use Rubric 42, Writing to Inform, to evaluate student work.

Portfolio Activity

Students' posters will vary, but the information included should be accurate for your region. Use Rubric 28, Posters, to evaluate student work.

internet connect

GO TO: **go.hrw.com**
KEYWORD: **SK3 Teacher**
FOR: a guide to using the Internet in your classroom

CHAPTER 30 The Pacific Islands and Antarctica

Chapter Resource Manager

Objectives	Pacing Guide	Reproducible Resources
SECTION 1 **Physical Geography** (pp. 687–91) **1.** Identify the physical features and resources of the Pacific Islands. **2.** Identify the physical features and resources of Antarctica.	**Regular** 1.5 days **Block Scheduling** .5 day *Block Scheduling Handbook, Chapter 30*	**RS** Guided Reading Strategy 30.1 **SM** Geography for Life Activity 30 **E** Lab Activities for Geography and Earth Science, Demonstrations 6 and 7
SECTION 2 **The Pacific Islands** (pp. 692–96) **1.** Describe the history of the Pacific Islands. **2.** Describe what the people and culture of the Pacific Islands are like. **3.** Identify some challenges that Pacific Islanders face today.	**Regular** 1.5 days **Block Scheduling** .5 day *Block Scheduling Handbook, Chapter 30*	**RS** Guided Reading Strategy 30.2 **E** Cultures of the World Activity 9 **IC** Interdisciplinary Activities for the Middle Grades 37 **SM** Map Activity 30
SECTION 3 **Antarctica** (pp. 697–99) **1.** Describe how Antarctica was explored. **2.** Describe what research in Antarctica can tell us about our planet. **3.** Identify the problems that threaten Antarctica's environment.	**Regular** .5 day **Block Scheduling** .5 day *Block Scheduling Handbook, Chapter 30*	**RS** Guided Reading Strategy 30.3 **E** Environmental and Global Issues Activities 1 and 2 **IC** Interdisciplinary Activity for the Middle Grades 38

Chapter Resource Key

- **RS** Reading Support
- **IC** Interdisciplinary Connections
- **E** Enrichment
- **SM** Skills Mastery
- **A** Assessment
- **REV** Review
- **ELL** Reinforcement and English Language Learners
- Transparencies
- CD–ROM
- Music
- Video
- go.hrw.com Internet
- Holt Presentation Maker Using Microsoft® Powerpoint®

One-Stop Planner CD–ROM

See the *One-Stop Planner* for a complete list of additional resources for students and teachers.

One-Stop Planner CD–ROM

It's easy to plan lessons, select resources, and print out materials for your students when you use the ***One-Stop Planner CD–ROM with Test Generator.***

Technology Resources	Review, Reinforcement, and Assessment Resources
One-Stop Planner CD–ROM, Lesson 30.1 Earth: Forces and Formations CD–ROM/Seek and Tell/ Forces and Processes Geography and Cultures Visual Resources with Teaching Activities 61–66 HRW Go site	**ELL** Main Idea Activity 30.1 **REV** Section 1 Review, p. 691 **A** Daily Quiz 30.1
One-Stop Planner CD–ROM, Lesson 30.2 HRW Go site	**ELL** Main Idea Activity 30.2 **REV** Section 2 Review, p. 696 **A** Daily Quiz 30.2
One-Stop Planner CD–ROM, Lesson 30.3 HRW Go site	**ELL** Main Idea Activity 30.3 **REV** Section 3 Review, p. 699 **A** Daily Quiz 30.3

internet connect

HRW ONLINE RESOURCES

GO TO: go.hrw.com
Then type in a keyword.

TEACHER HOME PAGE
KEYWORD: SK3 TEACHER

CHAPTER INTERNET ACTIVITIES
KEYWORD: SK3 GT30

Choose an activity to:
- tour the Pacific Islands.
- explore Antarctica.
- learn about native traditions of the Pacific Islands.

CHAPTER ENRICHMENT LINKS
KEYWORD: SK3 CH30

CHAPTER MAPS
KEYWORD: SK3 MAPS30

ONLINE ASSESSMENT
Homework Practice
KEYWORD: SK3 HP30
Standardized Test Prep Online
KEYWORD: SK3 STP30
Rubrics
KEYWORD: SS Rubrics

COUNTRY INFORMATION
KEYWORD: SK3 Almanac

CONTENT UPDATES
KEYWORD: SS Content Updates

HOLT PRESENTATION MAKER
KEYWORD: SK3 PPT30

ONLINE READING SUPPORT
KEYWORD: SS Strategies

CURRENT EVENTS
KEYWORD: S3 Current Events

Meeting Individual Needs

Ability Levels

Level 1 Basic-level activities designed for all students encountering new material

Level 2 Intermediate-level activities designed for average students

Level 3 Challenging activities designed for honors and gifted-and-talented students

English Language Learners Activities that address the needs of students with Limited English Proficiency

Chapter Review and Assessment

E Readings in World Geography, History, and Culture 81 and 82
SM Critical Thinking Activity 30
REV Chapter 30 Review, pp. 700–01
REV Chapter 30 Tutorial for Students, Parents, Mentors, and Peers
ELL Vocabulary Activity 30
A Chapter 30 Test
A Unit 9 Test
Chapter 30 Test Generator (on the One-Stop Planner)
Audio CD Program, Chapter 30
A Chapter 30 Test for English Language Learners and Special-Needs Students
A Unit 9 Test for English Language Learners and Special-Needs Students

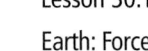

CHAPTER 30

LAUNCH INTO LEARNING

Have students examine the photographs in the chapter and describe what they see. *(Students are likely to describe both tropical and cold scenes.)* Point out that the Pacific Island and Antarctic regions are not only very different geographically but also have been treated very differently by other nations. While international leaders have tried to preserve Antarctica's environment, Pacific Island cultures and ecosystems have sometimes suffered from international involvement. Tell students that in this chapter they will study these two very different regions.

Section 1

Objectives

1. Describe the physical features and resources of the Pacific Islands.
2. Identify the physical features and resources of Antarctica.

LINKS TO OUR LIVES

You may wish to point out to students that there are many reasons to study the Pacific Islands and Antarctica. These reasons include the following:

- The Pacific Ocean covers about one third of Earth's surface. The Pacific Island region contains more water than land. Because of this, the region is sometimes called Oceania.
- Researchers in Antarctica have studied the rising levels of carbon dioxide in Earth's atmosphere and the thinning of the ozone layer over Antarctica. Rising levels of carbon dioxide and the thinning of the ozone layer may have adverse effects on Earth's climate.
- The natural beauty and pleasant climates of the Pacific Islands attract many tourists. Students may wish to visit the Pacific Islands someday.

CHAPTER 30

The Pacific Islands and Antarctica

In this chapter we will study the Pacific islands and Antarctica, two very different regions.

My name is Jean Vanessa, and I am 13 years old. I live in Vabukori, a village in Port Moresby, the capital of Papua New Guinea. My mother is a journalist. My grandmother bakes our bread in a drum oven. Sometimes we sell the bread.

My house is built on stilts. It is near the village square where we have meetings and play sports. After breakfast, I take a bus to Port Moresby Grammar School. I make my own sandwiches to take for lunch.

In my society, when you are an eldest child and a girl, and if you have younger brothers and sisters, you are your mother's helper. You have to learn to cook, clean, and take care of younger children.

My favorite holidays are Christmas and New Year's. We celebrate with feasting and singing. The villagers are divided into two groups. On Christmas Day one group cooks for the other group. They sing prophet songs about Bible stories from dawn to dusk. On New Year's Day the groups exchange roles.

Translation: Good morning.

FOCUS

LET'S GET STARTED

Copy the following instructions onto the chalkboard: *What mental images do you form when you think of the Pacific Islands and Antarctica? Write down what you have imagined.* Allow students time to complete their responses. Then have volunteers share their images with the class. *(Students might describe tropical scenes, ice, or penguins.)* Tell students that in Section 1 they will learn more about the physical geography of these regions.

Using the Physical-Political Map

Have students examine this unit's physical and political maps. Remind students that an atoll is a ring of small islands that have formed on a coral reef surrounding a shallow lagoon. Point out that there are more than 20,000 islands in the Pacific Ocean and that many of the smaller ones are atolls.

Building Vocabulary

Write the terms **ice shelf, icebergs, polar desert**, and **krill** on the chalkboard. Have volunteers read aloud the definitions from the glossary. Ask the students which region they think these terms describe *(Antarctica)*. Then have students write one sentence using each of the terms.

Section 1 Physical Geography

Read to Discover

1. What are the physical features and resources of the Pacific Islands?
2. What are the physical features and resources of Antarctica?

Define

ice shelf
icebergs
polar desert
krill

Locate

Melanesia
Micronesia
Polynesia
Tahiti
New Guinea
Papua New Guinea
Marshall Islands
New Caledonia
Transantarctic Mountains
Vinson Massif
Antarctic Peninsula

WHY IT MATTERS

Many Pacific Islands are formed by coral atolls—rings of small islands on a coral reef. Use CNNfyi.com or other **current events** sources to find out what environmental problems threaten coral atolls. Record your findings in your journal.

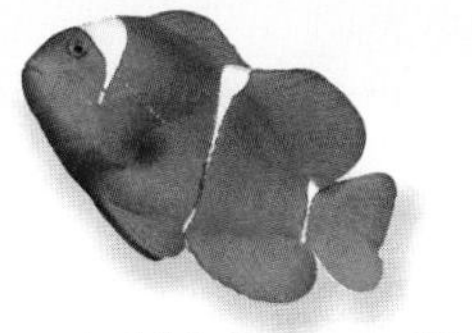

A tropical fish from near Fiji

The Formation of an Atoll

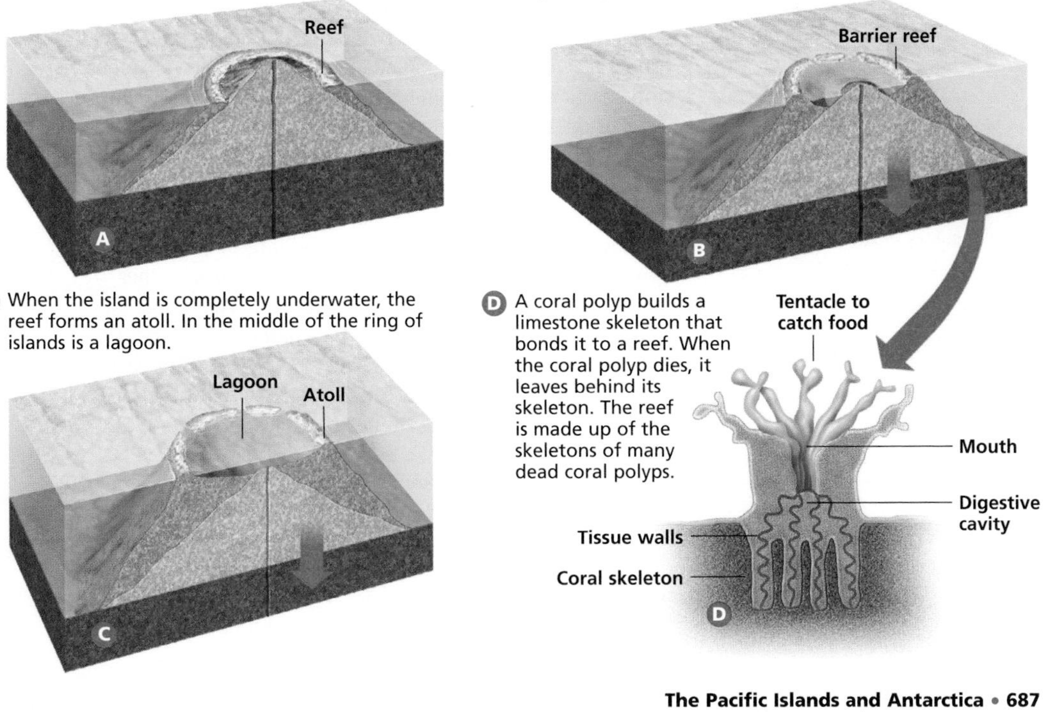

A A coral reef forms along the edges of a volcanic island.

B As the island sinks into the ocean, the reef continues to grow upward. It forms a barrier reef.

C When the island is completely underwater, the reef forms an atoll. In the middle of the ring of islands is a lagoon.

D A coral polyp builds a limestone skeleton that bonds it to a reef. When the coral polyp dies, it leaves behind its skeleton. The reef is made up of the skeletons of many dead coral polyps.

Section 1 RESOURCES

Reproducible
- Block Scheduling Handbook, Chapter 30
- Guided Reading Strategy 30.1
- Geography for Life Activity 30
- Lab Activities for Geography and Earth Science, Demonstrations 6 and 7

Technology
- One-Stop Planner CD–ROM, Lesson 30.1
- Homework Practice Online
- Geography and Cultures Visual Resources with Teaching Activities 61–66
- Earth: Forces and Formations CD–ROM/Seek and Tell/ Forces and Processes
- HRW Go site

Reinforcement, Review, and Assessment
- Section 1 Review, p. 691
- Daily Quiz 30.1
- Main Idea Activity 30.1
- English Audio Summary 30.1
- Spanish Audio Summary 30.1

TEACH

Teaching Objective 1

LEVEL 1: (Suggested time: 15 min.) Copy the following graphic organizer onto the chalkboard, omitting the italicized answers. Use it to help students describe similarities and differences between the high and low islands. Call on students to fill in the organizer.
ENGLISH LANGUAGE LEARNERS

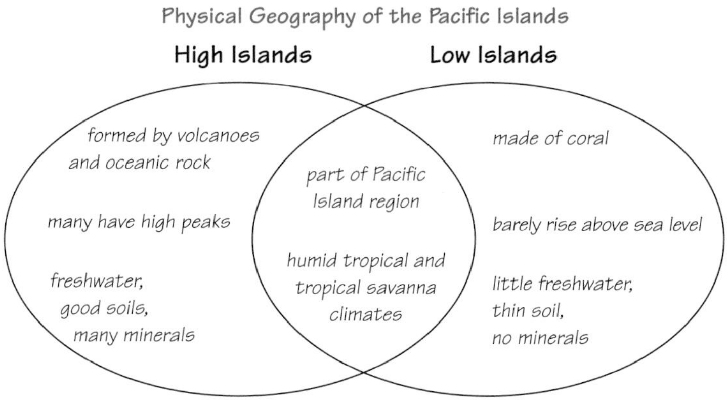

COOPERATIVE LEARNING

Organize students into small groups. Tell them to imagine that they are on the staff of a travel magazine and must prepare an issue on the island of New Guinea. Students should conduct research on the natural geography of New Guinea as well as its historical geography.

Have each group prepare one feature article and one sidebar about the island to be compiled into a class magazine. Within each group, individual roles may include artist, desktop publisher, researcher, and writer.

internet connect
GO TO: go.hrw.com
KEYWORD: SK3 CH30
FOR: Web sites about New Guinea

The airport on the left supports a healthy tourism industry in Bora-Bora, French Polynesia.

The Pacific Islands

The Pacific Ocean covers more than one third of Earth's surface. East of Indonesia and the Philippines are thousands of large and small islands. These islands are divided into three regions: Melanesia, Micronesia, and Polynesia. These regions have high and low islands. Now we will look at these islands, their climates, and their resources.

High Islands There are two kinds of high islands: oceanic and continental. Oceanic high islands were formed by volcanoes rising from the sea floor. The Polynesian islands of Tahiti and Hawaii are examples of oceanic high islands. Many continental high islands also have volcanoes. However, these large islands were formed from continental rock. They lie on Australia's continental shelf. New Guinea, which is in Melanesia, is a continental high island.

New Guinea is the world's second-largest island. Only Greenland is larger. A long mountain chain stretches across the central part of New Guinea. The range's highest mountain lies in the western half of the island. It reaches an elevation of 16,535 feet (5,040 m). The western part of New Guinea is called Irian Jaya and is part of Indonesia. Papua New Guinea occupies the eastern half of the island.

Low Islands Most of the low islands are made of coral. They barely rise above sea level. Many are atolls. For example, the Marshall Islands include two parallel chains of coral atolls. The two chains are about 800 miles (1,287 km) long. The highest point

A Forested Country: Land Use in Papua New Guinea

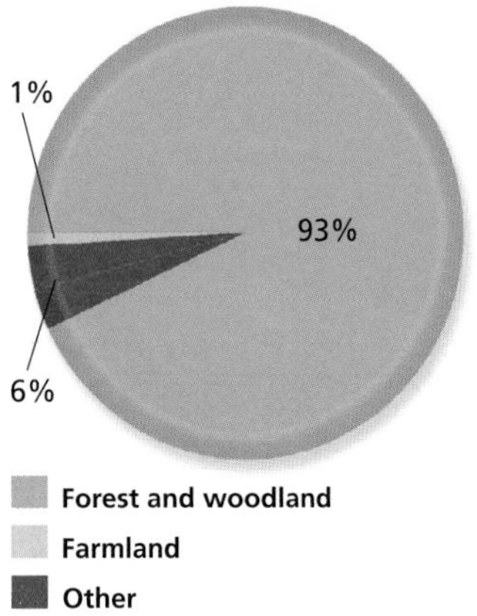

Source: Central Intelligence Agency, *The World Factbook 2001*

Interpreting the Chart **What percentage of Papua New Guinea is farmland?**

Chart Answer
1 percent

LEVEL 2: (Suggested time: 20 min.) Have students write paragraphs supporting the following statement: "The limitations of the low islands' physical geography have resulted in few people living there." *(See the graphic organizer from the Level 1 activity for the correct limitations.)* Call on volunteers to read their paragraphs aloud to the class.

LEVEL 3: (Suggested time: 45 min.) Pair students and have each pair conduct research on the forests of New Guinea from the perspective of a conservationist. Have students present oral reports describing the flora and fauna native to the forests and advocating the preservation of these areas. Students' reports should describe the benefits of preservation. **COOPERATIVE LEARNING**

Teaching Objective 2

LEVEL 1: (Suggested time: 25 min.) Pair students. Have one student in each pair use the textbook and classroom resources to list the land features and resources of Antarctica. The other student should list Antarctica's climatic and wildlife features. Then have students exchange lists with their partners and check each other's work. After discussing the lists, students can identify which of the continent's features they find most interesting and explain why to their partners. **ENGLISH LANGUAGE LEARNERS, COOPERATIVE LEARNING**

is just 33 feet (10 m) above sea level. Measured from the ground to their roofs, many buildings in your town are probably higher!

Climates and Plants Most of the Pacific Islands lie in the tropics and have a humid tropical climate. The temperatures are warm, and rain is common all year. New Guinea's central highland regions are cooler. Rainfall there is heavy, particularly on the southern slopes of the highlands. In some years more than 300 inches (762 cm) of rain fall there. Some islands, such as New Caledonia, have a tropical savanna climate. Rain there falls mostly in the summer.

The low islands have thin soils. These islands support few trees other than the coconut palm. However, the high islands have dense tropical rain forests. In fact, Papua New Guinea is one of the world's most densely forested countries. (See the pie graph on the previous page.) Many peoples living in the forests and rugged mountains of the central highlands were isolated for centuries.

Resources The islands' natural beauty and pleasant climates attract many tourists. In addition, the high islands have freshwater, good soils, and forest resources. Continental high islands also have many minerals. For example, New Guinea has copper, gold, silver, and oil. However, mining these resources is difficult in the rugged highlands and dense tropical forests.

Low islands have few resources. There is little freshwater, and thin soils limit farming. Partly because of this, low islands have smaller populations than high islands. Coconut palms and the sea are important sources of food.

✓ **READING CHECK:** *Places and Regions* What are the physical features and resources of the Pacific Islands? high—formed by volcanoes or from continental rock, sometimes mountainous; low—mostly made of coral, are atolls; resources—high islands: natural beauty, freshwater, good soils, forests, some minerals; low islands: few resources, including the sea

internet connect

GO TO: go.hrw.com
KEYWORD: SK3 CH30
FOR: Web sites about the Pacific Islands and Antarctica

Miners search for gold along a river in Papua New Guinea. They gently wash river water over pans to separate gold particles from less dense sand.

National Geography Standard 14

Environment and Society

In addition to having few resources, the low island of Guam is threatened by an invasion of brown tree snakes. In the late 1950s brown tree snakes—native to northeastern Australia, New Guinea, and some of the islands surrounding New Guinea—arrived in Guam. They were probably accidentally stowed away in ship cargoes from New Guinea. The snake can easily hide itself in small spaces.

The snake has thrived on Guam because of the availability of its natural prey and the lack of sufficient predators. By the late 1990s there were as many as 13,000 snakes per square mile in forested areas. The snake has virtually wiped out Guam's native forest birds. Several bird species—some found nowhere else in the world—have already disappeared, and many others are close to extinction.

The snake has also affected humans. Guam has experienced a major power outage once every three days since 1980 because the snakes cross power lines and cause short circuits.

Discussion: Ask students the following questions: What measures could have been taken to prevent the problems caused by the brown tree snakes? What can be done now to save Guam's wildlife?

LEVELS 2 AND 3: (Suggested time: 40 min.) Have students imagine that they are scientists conducting research on Antarctica. They have agreed to keep a journal of their discoveries and activities during their week-long stay. Students should write diary entries describing and explaining the formation of the physical features and resources of this continent and include illustrations of the features they have found particularly interesting. Have volunteers share their entries with the class.

➤ASSIGNMENT: Have students conduct research to create brochures highlighting the opportunities for scientists in Antarctica. Tell students that brochures should include drawings or pictures of Antarctica's unique features and information that explains how researchers have used technology to allow them to live in the continent's environment. Students might mention the continent's size, its icebergs, its complete darkness during winter, or its short summer when the Sun never sets. Display completed brochures around the classroom.

EYE ON EARTH

For many years scientists thought that icebergs breaking off the ice sheet of Antarctica were causing the world's ocean levels to rise. In the 1900s the sea level rose 7 inches (18 cm) and about 5.5 inches (14 cm) of that had been attributed to the breaking off of icebergs.

However, a recent study suggests that there might be other explanations for the rising ocean levels. Over a five-year period researchers used satellites to measure changes in the thickness of the Antarctic ice sheet. They determined that the thickness has dropped by less than half an inch per year and that melting within the ice sheet has caused ocean levels to rise less than one inch over the last 100 years.

Scientists suggest other possible reasons to explain the rising ocean levels. These reasons include global warming, which is melting mountain glaciers and contributing to rising sea levels.

Critical Thinking: Why might scientists have thought icebergs were causing sea levels to rise?

Answer: Students might suggest that the breakaway icebergs were more immediately apparent to researchers.

Visual Record Answer ►

wind, waves, and other water erosion including melting in warmer waters

▲ Huge icebergs like this one are found in the ocean waters around Antarctica. Icebergs are most numerous in the spring and summer. The warmer weather causes more ice to break away from ice sheets and glaciers.

Interpreting the Visual Record **What forces do you think shaped this iceberg?**

Antarctica

In the southernmost part of the world is the continent of Antarctica. This frozen land is very different from the tropical Pacific Islands.

The Land Antarctica is larger than the United States and Mexico combined. Ice covers about 98 percent of Antarctica's 5.4 million square miles (14 mil sq km). This ice sheet contains more than 90 percent of the world's ice. On average the ice sheet is more than one mile (1.6 km) thick.

The Transantarctic Mountains divide the continent into East Antarctica and West Antarctica. Antarctica's highest mountain peak is Vinson Massif. It rises to 16,864 feet (5,140 m). The continent also includes a few dry coastal valleys and the Antarctic Peninsula. (See the map of Antarctica in Focus On Environment at the end of this chapter.)

The weight of Antarctica's ice sheet causes the ice to flow slowly off the continent. When the ice reaches the coast, it forms a ledge over the water. This ledge is called an **ice shelf**. Sometimes huge chunks of ice, called **icebergs**, break away and drift into the ocean. Some of these icebergs are larger than the state of Rhode Island.

Climate and Wildlife Antarctica is the planet's coldest, driest, highest, and windiest continent. During the Southern Hemisphere's winter, the continent is hidden from the Sun. It remains in total darkness. During Antarctica's short summer, the Sun never sets.

CLOSE

Ask students the following questions: Given the opportunity, would you choose to travel through the Pacific Islands in an outrigger canoe or explore Antarctica with modern equipment? why?

REVIEW AND ASSESS

Have students complete the Section Review. Then ask each student to create three multiple-choice questions for this section. Pair students and have them complete and then grade their partners' quizzes. Then have students complete Daily Quiz 30.1. **COOPERATIVE LEARNING**

RETEACH

Have students complete Main Idea Activity 30.1. Then write *high islands*, *low islands*, and *Antarctica* on the chalkboard. Ask students to name terms that describe the physical features and climates of each. List them under the appropriate heading. **ENGLISH LANGUAGE LEARNERS**

EXTEND

Have interested students conduct research on the effect that global warming might be having on the low islands of the Pacific region and measures being taken to protect these islands. Have students present their findings to the class. **BLOCK SCHEDULING**

Antarctica's temperatures can drop below -120°F (-84°C). Less precipitation falls in this **polar desert** than in the Sahara in Africa. A polar desert is a high-latitude region that receives little precipitation. However, there is almost no evaporation or melting of ice. As a result, Antarctica's ice has built up over thousands of years.

The continent's warmest temperatures are found on the Antarctic Peninsula, which has a tundra climate. In January, during Antarctica's summer, temperatures on the coast average just below freezing.

Only tundra plant life survives in the rare ice-free areas. A few insects are the frozen land's only land animals. Antarctica has never had a permanent human population. Marine animals live in the icy waters around the continent. These animals include penguins, seals, and whales. They depend on tiny shrimplike creatures called **krill** for food.

The emperor penguin is one of the most common marine animals found in Antarctica. Penguins are flightless birds and are awkward on land. However, they are very good swimmers and are able to live in the icy climate.

Krill can be found in huge swarms in the waters around Antarctica.

Resources Antarctica has many mineral resources, including iron ore, gold, copper, and coal. However, there is debate over whether these resources should be mined. Some people worry that mining would harm the continent's environment. Others question whether mining in Antarctica would even be worthwhile for businesses.

✓ **READING CHECK:** *Places and Regions* What are the physical features and resources of Antarctica? ice sheet, mountains, limited tundra plant life, few insects, marine animals; mineral resources

go.hrw.com Homework Practice Online
Keyword: SK3 HP30

Section 1 Review

Define and explain: ice shelf, icebergs, polar desert, krill

Working with Sketch Maps On maps of the Pacific Islands and Antarctica that you draw or that your teacher provides, label the following: Melanesia, Micronesia, Polynesia, Tahiti, New Guinea, Papua New Guinea, Marshall Islands, New Caledonia, Transantarctic Mountains, Vinson Massif, and Antarctic Peninsula.

Reading for the Main Idea

1. *Places and Regions* What kinds of islands are found in the Pacific? How were they formed?
2. *Places and Regions* What animal life is found in Antarctica?
3. *Places and Regions* Which of the Pacific Islands is the second-largest island in the world?

Critical Thinking

4. **Making Generalizations and Predictions** What might happen if people were to begin mining Antarctica's mineral deposits on a large scale?

Organizing What You Know

5. **Summarizing** Copy the following graphic organizer. Use it to list physical features, climates, and resources of the Pacific Islands and Antarctica.

	Pacific Islands	Antarctica
Physical features		
Climates		
Resources		

Section 1 Review

Answers

Define For definitions, see: ice shelf, p. 690; icebergs, p. 690; polar desert, p. 691; krill, p. 691

Working with Sketch Maps Maps will vary, but listed places should be labeled in their approximate locations.

Reading for the Main Idea

1. high islands—formed by volcanoes or continental rock; low islands—made of coral (NGS 4)
2. a few insects, penguins, seals, whales (NGS 4)
3. New Guinea (NGS 4)

Critical Thinking

4. could disturb the environment and create pollution problems

Organizing What You Know

5. Pacific Islands—high and low islands; humid tropical, tropical savanna; pleasant climates, freshwater, good soils, forests, minerals; Antarctica—mountains, ice sheets, glaciers; very cold climates, tundra; iron ore, gold, copper, coal

Section 2

Objectives

1. Explain the history of the Pacific Islands.
2. Describe the people and culture of the Pacific Islands.
3. Identify some of the challenges facing Pacific Islanders today.

Focus

LET'S GET STARTED

Copy the following instructions onto the chalkboard: *What do you think might be a benefit of living in the Pacific Islands today? What might be a challenge? Write your answers in your notebook.* Have students complete their answers. Discuss student responses. Tell students that in Section 2 they will learn more about the history and challenges facing people of the Pacific Islands.

Building Vocabulary

Write the terms **trust territories** and **Exclusive Economic Zones** on the chalkboard. Have volunteers read the definitions aloud from the glossary. Then ask students to suggest possible reasons a country would want an EEZ *(to ensure profit from its coastal resources)* as well as how a country might enforce such a policy. *(Possible answer: patrolling its coasts)*

Section 2 Resources

Reproducible

- Guided Reading Strategy 30.2
- Graphic Organizer 30
- Cultures of the World Activity 9
- Interdisciplinary Activities for the Middle Grades 37
- Map Activity 30

Technology

- One-Stop Planner CD–ROM Lesson 30.2
- Homework Practice Online
- HRW Go site

Reinforcement, Review, and Assessment

- Section 2 Review, p. 696
- Daily Quiz 30.2
- Main Idea Activity 30.2
- English Audio Summary 30.2
- Spanish Audio Summary 30.2

internet connect

GO TO: **go.hrw.com**
KEYWORD: **SK3 CH30**
FOR: **Web sites about Tonga**

Visual Record Answer

fairly accurate

Section 2 The Pacific Islands

Read to Discover

1. What is the history of the Pacific Islands?
2. What are the people and culture of the Pacific Islands like?
3. What are some challenges that Pacific Islanders face today?

Define

trust territories
Exclusive Economic Zones

Locate

Guam
Northern Mariana Islands
Wake Island
French Polynesia

A tiki, a religious object from Polynesia

Why It Matters

The physical beauty of the Pacific Islands has made many into popular vacation destinations. However, many still struggle economically. Use CNNfyi.com or other **current events** sources to learn about the economic challenges facing Pacific Island countries. Record your findings in your journal.

This map of the South Pacific was made by a European cartographer in 1798.

Interpreting the Visual Record

Compare this map to the unit map. How accurate was this 1798 map?

History

Scholars believe that people began settling the Pacific Islands at least 40,000 years ago. Most early settlers came from Southeast Asia. The large islands of Melanesia were the first islands in the region settled. Over time, people moved to the islands of Micronesia and Polynesia.

Europeans in the Pacific In the early 1500s Ferdinand Magellan became the first European to explore the Pacific. In the late 1700s British captain James Cook explored the region. He visited all the main island regions of the Pacific. By the late 1800s European countries controlled most of the Pacific Islands. These European countries included France, Germany, the Netherlands, Spain, and the United Kingdom.

Modern History The Pacific Islands were battlegrounds during several wars in the colonial era. For example, the Spanish-American War of 1898 cost Spain the Philippines and Guam. They became U.S. territories after the war. Spain then sold other island territories to Germany. After World War I ended in 1918, Japan took over Germany's territories in the Pacific.

TEACH

Teaching Objective 1

ALL LEVELS: (Suggested time: 30 min.) Organize students into groups. Assign each group a significant event in the history of the Pacific Islands. Ask each group to illustrate the event and to write a caption for its illustration. Have groups display their completed pictures on the chalkboard. Ask volunteers to arrange all the events in chronological order. **ENGLISH LANGUAGE LEARNERS, COOPERATIVE LEARNING**

Teaching Objective 2

LEVEL 1: (Suggested time: 30 min.) Organize students into three groups. Give each group a sheet of paper and assign each group one of the regions in the Pacific: Melanesia, Micronesia, or Polynesia. Have group members list aspects of the people and culture of their assigned region. Have each group present its completed list to the class. **ENGLISH LANGUAGE LEARNERS, COOPERATIVE LEARNING**

Japan conquered many other islands in World War II. The United States and its allies eventually won them back and defeated Japan. The United Nations then made some islands **trust territories**. Trust territories are areas placed under the temporary control of another country. When the territory later sets up its own government, it gains independence. U.S. trust territories included much of Micronesia.

Most of the island countries won independence in the last half of the 1900s. Australia, France, New Zealand, the United Kingdom, and the United States still have Pacific territories. U.S. territories include the Northern Mariana Islands, Guam, and Wake Island.

✓ **READING CHECK:** *Places and Regions* What are some events in the history of the Pacific Islands? initial settlement 40,000 years ago; 1500s—Pacific explored by Ferdinand Magellan; 1700s—explored by Captain James Cook; 1800s—islands European-controlled; battleground during several wars; 1900s—independence

Culture

About 7 million people live in Melanesia, Micronesia, and Polynesia today. Check the unit map as we look more closely at each region and its people.

Melanesia Melanesia stretches from New Guinea to Fiji. It is the most populous of the three Pacific Island regions. Papua New Guinea and Fiji have the largest populations. Nearly two thirds of all Pacific Islanders live in Papua New Guinea.

Most Melanesians live in rural areas. Many homes in Melanesia and the other regions are made of timber and thatch, or straw. Papua New Guinea's capital, Port Moresby, is Melanesia's largest city. Nearly 200,000 people live there.

Melanesia's population includes ethnic Europeans and Asians, particularly Indians and Chinese. Many ethnic Asians are descended from people brought to the islands to work on colonial plantations. In Fiji, Indians make up nearly half of the population.

Either English or French is the official language on nearly all of the islands. This is a reflection of the region's colonial history. However, hundreds of local languages also are spoken there. In fact, about 700 languages are spoken in New Guinea alone. Many Papua New Guineans live in rugged, forested areas. They have had little contact with people from other areas. These isolated peoples developed their own languages.

Europeans brought Christianity to Melanesia and the other island regions. Today, most Pacific Islanders are Christian. However, some Melanesians still practice traditional local religions.

Micronesia and Polynesia Micronesia includes more than 2,000 tiny islands north of Melanesia. It stretches from Palau in the west to Kiribati in the east. Polynesia is the largest Pacific region. Its

▲ Students learn the day's lessons at a school in Papua New Guinea. Papua New Guinea's government has struggled to make education available for all students. Many countries in Micronesia and Polynesia offer schooling for students through high school.

HUMAN SYSTEMS

Although Papua New Guinea has slightly more than 5 million people, its residents speak more than 700 languages. There are some 6,000 languages spoken in the entire world. Most of the Papuan languages are complex and are spoken by very small groups of people.

To communicate with each other, the people of New Guinea use Tok Pisin, a language based on English. Tok Pisin is a pidgin language with a limited vocabulary. About 50,000 residents use Tok Pisin as their first language. Some 2 million more use it as their second language.

Tok Pisin plays a vital role in uniting Papuans. It is the language used most frequently in commerce and in the parliament.

Critical Thinking: Why do you think the people of Papua New Guinea speak so many different languages?

Answer: Students might suggest that the island's physical geography and the existence of many small, isolated groups have contributed to its linguistic diversity.

LEVELS 2 AND 3: (Suggested time: 45 min.) Organize students into groups of four. Have each group use drawings or pictures from magazines to write and illustrate a guidebook describing the people and culture of the Pacific Islands. Students should organize their work so that all group members have the opportunity to do both writing and illustrating.
COOPERATIVE LEARNING

Teaching Objective 3

LEVEL 1: (Suggested time: 20 min.) Copy the following graphic organizer onto the chalkboard, omitting the italicized answers. Use it to help students identify challenges facing the people of the Pacific Islands.
ENGLISH LANGUAGE LEARNERS

Challenges Pacific Islanders Face

Economy	Environment	Culture
• *Many countries import more than they export.* • *Many countries rely on Australia, Great Britain, Japan, the European Union, and the United Nations for economic aid.*	• *Nuclear tests were conducted in the region from the 1940s through the 1990s.*	• *Many residents fear loss of traditions and beliefs.* • *Modern travel and communications have introduced influences from other regions.* • *Residents worry about the effects of tourism, television, processed food, and alcohol.*

Linking Past to Present

Tapa Early Polynesians made innovative use of available materials. They wove mats and roofs from the leaves of the coconut palm and made meals from the coconut's meat and liquid.

Early Polynesians also used tree bark to make ornamental cloth known as tapa. They pounded the bark into a feltlike cloth and decorated it with geometric designs. Polynesians used tapa for clothing, interior decoration, and bedding, among other things. Many Polynesians continue the tradition of tapa making today, and tapa craft items are important to the tourist trade.

Activity: Have students conduct research on the types of tapa cloth designs used by Polynesians. Why do they use these designs? Then have students reproduce some of the designs on paper.

Connecting to Technology

Answers

1. to travel the long distances between islands
2. It is slowly disappearing from the region.

Connecting to Technology

Pacific Boatbuilders

Long before Europeans sailed into the region, Pacific Islanders were sailing over the open sea. Pacific peoples—particularly Polynesians—regularly sailed hundreds of miles from one island to the other. This seafaring tradition depended not just on sailing skills. It also depended on the craft of boatbuilding.

The best boats were built on the Fijian island of Kabara. This island has the finest forests of *vesi* trees in the South Pacific. The *vesi* tree grows tall and straight. Its trunk is heavy and tough enough to hold up to the stresses of sea travel.

Looking out to sea from Bora-Bora

The people of Kabara built different types of boats. Some were simple canoes. Some were large boats with two hulls, or frames. The largest were as long as 100 feet (30 m). These boats were fast and could carry as many as 300 people. They were highly prized for both trading and warfare.

In recent decades the ancient craft of boatbuilding has slowly disappeared from the region. A few older people remember the traditional skills—cutting and joining the wood and weaving the sails. Many islanders worry that unless more people learn the craft, traditional boatbuilding may be forgotten.

Understanding What You Read

1. Why were good boats needed by the Pacific Islanders?
2. What has happened to the craft of boatbuilding in recent years?

corners at New Zealand, Hawaii, and Easter Island form a huge triangle.

The populations of most Micronesian and Polynesian islands are much smaller than those of Melanesia. However, towns on these small islands can be very crowded. Micronesia is the most urban of the Pacific Island regions. Most Polynesians live in rural areas.

As in Melanesia, ethnic Europeans and Asians live in Micronesia and Polynesia. Most people in these regions are Christian and speak either English or French. Some speak local languages or Japanese.

✓ **READING CHECK:** *Human Systems* What are some characteristics of the people and cultures of Melanesia, Micronesia, and Polynesia? mostly rural dwellers, ethnic Europeans and Asians, mostly French and English-speaking, Christian

TEACHER TO TEACHER

LEVELS 2 AND 3: (Suggested time: 35 min.) Have students imagine that they are authors living 100 years in the future and that they have been asked to write a children's book about the challenges the Pacific Islanders faced in 2000. Students should select an issue from one of the three areas covered in the previous activity and write a paragraph summarizing the challenge and the efforts Pacific Islanders made to overcome it. Have volunteers share their paragraphs with the class.

Susan Walker of Beaufort, South Carolina, suggests the following activity to help students learn about the challenges facing the people of the Pacific Islands today. Have students work in groups or individually to draft a treaty for the Pacific Islands similar to the treaty drafted to protect the environment of Antarctica in 1959. Encourage students to use their textbooks as well as outside resources to complete this project.

The Pacific Islands Today

Many people imagine sunny beaches and tourists when they think of the Pacific Islands today. The islands do attract many tourists. For example, Tahiti, in French Polynesia, is a popular vacation spot. Many vacationers from South Korea and Japan enjoy visiting Guam, in Micronesia. Despite the region's healthy tourism industry, however, the Pacific countries face important challenges.

Economy The Pacific Islands are trying to build stronger economies. Tourism, agriculture, and fishing are already important there. Some countries, particularly Papua New Guinea, export valuable minerals and forest products.

Each Pacific country claims control of the fishing and seabed minerals around its islands. The 200-nautical-mile (370 km) zones they claim are called **Exclusive Economic Zones** (EEZs). Most of the world's countries also claim EEZs. A country must pay fees to fish or mine in another country's EEZ.

Natural resources should help the island economies grow. However, many Pacific countries import more products from abroad than they export. Many countries rely on the United Nations, the European Union, Great Britain, Australia, and Japan for economic aid.

Environment Many Pacific islanders are concerned about their region's environment. Many have been angered by nuclear weapons tests conducted in the region by other countries. The United States held such tests in the islands from the 1940s to the 1960s. Radiation left some islands unsafe for people for many years. France held nuclear tests in the region until the mid-1990s.

Pacific Islands

Country	Population/ Growth Rate	Life Expectancy	Literacy Rate	Per Capita GDP
Fiji	844,330 / 1.4%	66, male 71, female	92%	$7,300
Kiribati	94,149 / 2.3%	57, male 63, female	Not available	$850
Marshall Islands	70,822 / 3.9%	64, male 68, female	93%	$1,670
Micronesia, Federated States of	134,597 / Not available	Not available	89%	$2,000
Nauru	12,088 / 2.0%	58, male 65, female	Not available	$5,000
Palau	19,092 / 1.7%	66, male 72, female	92%	$7,100
Papua New Guinea	5,049,055 / 2.4%	61, male 66, female	72%	$2,500
Samoa	179,058 / –0.2%	67, male 72, female	97%	$3,200
Solomon Islands	480,442 / 3.0%	69, male 74, female	Not available	$2,000
Tonga	104,227 / 1.8%	66, male 71, female	99%	$2,200
Tuvalu	10,991 / 1.4%	65, male 69, female	Not available	$1,100
Vanuatu	192,910 / 1.7%	60, male 62, female	53%	$1,300
United States	281,421,906 / 0.9%	74, male 80, female	97%	$36,200

Sources: Central Intelligence Agency, *The World Factbook 2001*; U.S. Census Bureau

Interpreting the Chart **What is the literacy rate like in most of the Pacific Islands?**

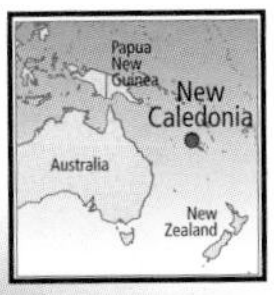

A fisher casts his net along the coast of New Caledonia.

GLOBAL PERSPECTIVES

After World War II, the United States received permission from the ruler of Bikini Atoll—part of the Marshall Islands—to perform atomic weapons tests there. Residents were evacuated.

The United States set off more than 60 blasts. Most of the radioactive fallout occurred over the testing area, but some radioactive particles fell on inhabited islands.

In 1969 the U.S. military told the island's former residents that it was safe to return to Bikini. Several families then returned only to discover that their food and water were not safe to consume. Today, Bikini Atoll has been declared safe for scuba diving and fishing, but many residents still face health concerns.

Activity: Have students imagine they are post–World War II Bikini residents. Have them write letters accepting or rejecting the U.S. request to test atomic weapons on their land.

▲ **Chart Answer**

very high

Section 2 Review

Answers

Define For definitions, see: trust territories, p. 693; Exclusive Economic Zones, p. 695

CLOSE

Ask students to answer the following question: How has colonization affected the people of the Pacific Islands? *(Students should consider the economic and cultural effects of colonization.)*

REVIEW AND ASSESS

Have students complete the Section Review. Then have them work in groups of four to write six multiple-choice questions about the information presented in this section. Have groups complete and check each other's quizzes. Then have students complete Daily Quiz 30.2.

RETEACH

Have students complete Main Idea Activity 30.2. Then organize the class into groups of four or five. Have each group compose a song about the people, history, and culture of the Pacific Islands.
ENGLISH LANGUAGE LEARNERS, COOPERATIVE LEARNING

EXTEND

Have interested students conduct research on the effects of nuclear testing on Bikini Atoll. Tell students to focus on the effects of tests on the environment, on people's health, and the efforts made to restore the environment. Have students prepare oral reports to present their findings to the class.
BLOCK SCHEDULING

Working with Sketch Maps Maps will vary, but listed places should be labeled in their approximate locations.

Reading for the Main Idea

1. Southeast Asia; Melanesia (NGS 9)
2. France, Germany, Japan, the Netherlands, Spain, the United Kingdom, and the United States (NGS 12)

Critical Thinking

3. Many Pacific Islanders speak European languages and practice Christianity.
4. developing economies, protecting the environment, and preserving cultures

Organizing What You Know

5. Time lines will vary but should include the following: 1500s—Magellan explores the Pacific; late 1700s—Cook visits the Pacific Islands; late 1800s—most islands become European colonies; 1898—the Philippines and Guam become U.S. territories; 1918—Japan given German colonies; 1945—the Allies retake islands Japan had captured in World War II; late 1900s—most Pacific Islands become independent

Visual Record Answer ►

clock, clothes, portable radio

▲ This Fijian chief's home has many traditional handicrafts. Fiji's Great Council of Chiefs has influence in the country's political system and culture.

Interpreting the Visual Record **What elements of this photograph show influences of outside cultures?**

Some people who live on islands still controlled by foreign countries want independence, which has led to outbreaks of violence. This happened in the 1980s in the French territory of New Caledonia. France has agreed to give New Caledonians more control over their local government.

Culture Many Pacific Islanders are also concerned about the loss of traditional customs and beliefs. Modern travel and communications have introduced influences from other regions. Islanders worry about the cultural effects of tourism, television, processed food, and alcohol.

✓ **READING CHECK:** *Human Systems* What are some challenges Pacific Islanders face today? need stronger economies; face environmental dangers from nuclear testing; some conflict regarding foreign control; loss of traditional customs

go.hrw.com **Homework Practice Online** Keyword: SK3 HP30

Section 2 Review

Define and explain: trust territories, Exclusive Economic Zones

Working with Sketch Maps On the map of the Pacific islands that you created in Section 1, label the region's countries. Then label the territories of Guam, Northern Mariana Islands, Wake Island, and French Polynesia.

Reading for the Main Idea

1. *Human Systems* From where did the Pacific Islands' first settlers come? Which Pacific Island region was settled first?
2. *Human Systems* What colonial powers once controlled most of the Pacific islands?

Critical Thinking

3. **Finding the Main Idea** In what ways did Europeans influence the islands' cultures?
4. **Finding the Main Idea** What are some of the challenges facing Pacific Islanders today?

Organizing What You Know

5. **Sequencing** Use this time line to explain important people and events in the region's history.

1500 ———————————— 2000

Section 3

Objectives

1. Describe how Antarctica was explored.
2. Explain what research in Antarctica can tell us about our planet.
3. Identify problems that threaten Antarctica's environment.

Focus

LET'S GET STARTED

Copy the following instructions onto the chalkboard: *Work with a partner to write five facts about Antarctica's physical geography. Try to write your facts without using your textbook.* Allow time for students to write their responses. Discuss students' answers. Then ask students to use their lists to predict some ways that Antarctica's physical geography may affect exploration and research on the continent. *(Possible answers: The conditions make it hard to explore; because of the continent's difficult conditions, there is much left to explore.)*

Tell students that in Section 3 they will learn more about exploration of and research in Antarctica.

Building Vocabulary

Write the term **antifreeze** on the chalkboard. Draw a vertical line between *anti* and *freeze*. Have students suggest meanings for the prefix. When the correct answer *(against)* is given, ask volunteers to suggest meanings for the entire word. Have students check their guesses against the definition in the book.

Section 3 Antarctica

Read to Discover

1. How was Antarctica explored?
2. What can research in Antarctica tell us about our planet?
3. What problems threaten Antarctica's environment?

Define

antifreeze

Locate

South Pole
Ross Ice Shelf

Why It Matters

Antarctica offers many opportunities for scientific research. Use CNNfyi.com or other **current events** sources to find out what current scientific investigations are taking place in Antarctica. Record your findings in your journal.

A sign near Scott Base in Antarctica

Section 3 Resources

Reproducible

- Guided Reading Strategy 30.3
- Environmental and Global Issues Activities 1 and 2
- Interdisciplinary Activities for the Middle Grades 38

Technology

- One-Stop Planner CD–ROM, Lesson 30.3
- Homework Practice Online
- HRW Go site

Reinforcement, Review, and Assessment

- Section 3 Review, p. 699
- Daily Quiz 30.3
- Main Idea Activity 30.3
- English Audio Summary 30.3
- Spanish Audio Summary 30.3

Early Explorers

Can you imagine a time when a large continent like Antarctica was a complete mystery? Today, orbiting satellites give us views of all of Earth's surface. Jet airliners and modern ships take people to all points on the globe. However, all of this has become possible only in the last century. For a long time, stormy ocean waters hid Antarctica from explorers. In the 1770s British explorer James Cook sighted icebergs in the waters around Antarctica. These icebergs suggested the existence of the vast, icy continent.

Other explorers followed. Some died in Antarctica's dangerous conditions while attempting to reach the South Pole and return. The huts of famous Antarctic explorers are still scattered across the continent. The first human expedition reached the South Pole in 1911.

Some countries have claimed parts of Antarctica. These countries and others agreed in 1959 to preserve the continent "for science and peace." The Antarctic Treaty of 1959 prevented more claims to the continent. It banned military activity there and set aside the whole continent for research.

✓ **READING CHECK:** *Environment and Society* How did the physical environment affect early efforts to explore Antarctica?

British explorer Robert Scott and his team reached the South Pole in January 1912. Norwegian Roald Amundsen's team had beaten them to the spot one month earlier. Scott and the rest of his team died trying to return from the Pole.

TEACH

Teaching Objectives 1–3

ALL LEVELS: (Suggested time: 20 min.) Copy the following graphic organizer onto the chalkboard, omitting the italicized answers. Use it to help students understand exploration, research, and environmental threats in Antarctica. Have them copy the organizer into their notebooks and complete it. **ENGLISH LANGUAGE LEARNERS**

Antarctica

Exploration	Research	Environmental Threats
• *James Cook sights icebergs in the waters surrounding Antarctica in the 1770s.* • *First expedition reaches the South Pole in 1911.* • *Antarctic Treaty of 1959 prevents more claims to the continent, bans military activity, and promotes research.*	• *Gases trapped in ice lead to conclusions that carbon dioxide levels have risen and may be responsible for global warming.* • *Fish with natural antifreeze help us understand how some animals adapt to harsh environments.*	• *There are current debates about allowing mining in the area; people fear mining will result in spills and other problems.* • *Researchers and tourists have left trash behind.* • *Oil spills have caused problems.*

➤ASSIGNMENT: Have students use magazines, newspapers, drawings, and objects to create collages illustrating problems threatening Antarctica's environment. Encourage students to refer to the chart created in the previous exercise for ideas.

Section 3 Review

Answers

Define For definition, see: antifreeze, p. 698

Working with Sketch Maps Places should be labeled in their approximate locations.

Reading for the Main Idea

1. Amundsen, 1911; Scott, 1912 (NGS 9)
2. preserved Antarctica for science and peace, prevented more claims to it, banned military activity there, and set aside the whole continent for research; answers will vary but students may say that it is easier to protect Antarctica because it is uninhabited (NGS 16)

Critical Thinking

3. They may have environmental consequences, such as how to protect Earth's environment. (NGS 15)
4. because of fears that mining, other development, and tourism would pollute and otherwise disturb the environment

Organizing What You Know

5. air pollution—gases trapped in ice show rising levels of carbon dioxide; air pollution—thinning of the ozone layer; life—how life adapts to harsh environments

Visual Record Answer ►

how these species survive in cold waters

BUILD on WHAT You Know

Do you remember what you learned about ozone? See Chapter 2 to review.

Research in Antarctica

Today, researchers are the only people who live in Antarctica. They live in a number of bases or stations. U.S. stations include Palmer, on the Antarctic Peninsula, and McMurdo, on the Ross Ice Shelf. The United States also maintains a base at the South Pole. Researchers at these bases are looking for clues to Earth's past and future.

Air Pollution The researchers have made important discoveries. For example, some have studied gases trapped in old Antarctic ice. They have compared these gases with gases in Earth's atmosphere today. Their studies have shown that carbon dioxide levels in the air have risen over time. Some scientists believe high levels of carbon dioxide are responsible for global warming.

Scientists are also looking for evidence that air pollution is damaging Earth's ozone layer. The ozone layer protects living things from the harmful effects of the Sun's ultraviolet rays. Scientists have found a thinning in the ozone layer above Antarctica.

Life Other research helps us understand mysteries of life on Earth. For example, researchers have studied a kind of fish that produces a natural **antifreeze**. Antifreeze is a substance added to liquid to keep the liquid from turning to ice. Natural antifreeze in their blood protects the fish in the icy waters around Antarctica. These fish may help us understand how some animals adapt to harsh environments.

✓ **READING CHECK:** *Environment and Society* What can research in Antarctica tell us about our planet? information about pollution and Earth's atmosphere and about how animals adapt to environments

Researchers are conducting experiments in icy waters off the coast of Antarctica.

Interpreting the Visual Record **What might these underwater researchers be trying to learn?**

CLOSE

Tell students that young people can go to Antarctica on contract jobs to clear snow, drive machines, and so on. Ask students if they would like to go and why.

REVIEW AND ASSESS

Have students complete the Section Review. Then organize students into three groups. Assign each group a section from the chapter. Have groups write quiz questions about the main ideas of their sections. When groups are finished have them take turns quizzing the class. Then have students complete Daily Quiz 30.3. **COOPERATIVE LEARNING**

RETEACH

Have students complete Main Idea Activity 30.3. Then have each student pretend to be a seal or a penguin. Have them write narratives from their animals' perspectives that address the Read to Discover questions. **ENGLISH LANGUAGE LEARNERS**

EXTEND

Have interested students read Richard Byrd's *Alone*, which tells of the year he spent living alone in an Antarctic base camp. Have students write reports on the book to share with the class. **BLOCK SCHEDULING**

Cruise ships allow tourists to witness the icy beauty of Antarctica. About 10,000 tourists visited Antarctica annually during summer months in the late 1990s. A typical tourist trip lasts about two weeks. The only other economic activity in the region involves offshore fishing.

Interpreting the Visual Record **What potential hazards do you see for ships?**

Environmental Threats

Antarctica's environment is an excellent place for research. This is because humans have disturbed little of the continent. That is changing, however. As you have read, there is already debate about whether to allow mining in Antarctica. In addition, tourists and even researchers have left behind trash, polluting the local environment. Oil spills have also caused problems.

Some people fear that mining Antarctica's resources will result in other spills and problems. To prevent this, a new international agreement was reached in 1991. This agreement forbids most activities in Antarctica that do not have a scientific purpose. It bans mining and drilling and limits tourism.

✓ **READING CHECK:** *Environment and Society* What are some of the problems threatening Antarctica's environment? mining, tourism, pollution, oil spills

Our Amazing Planet

Some of Antarctica's rocks contain bacteria and other tiny forms of life. They live in small spaces that can only be seen with a microscope. These tough survivors may be thousands of years old!

go.hrw.com **Homework Practice Online**
Keyword: SK3 HP30

Section 3 Review

Define and explain: antifreeze

Working with Sketch Maps On the map of Antarctica you created in Section 1, label the South Pole and Ross Ice Shelf.

Reading for the Main Idea

1. *Human Systems* Who were the first Antarctic explorers to reach the South Pole, and when did they do so?
2. *Environment and Society* What did the 1959 Antarctic Treaty do? Why do you think this kind of agreement has not been made about other places on Earth?

Critical Thinking

3. **Environment and Society** How might future scientific discoveries in Antarctica affect the world?
4. **Finding the Main Idea** Why did many countries in 1991 agree to ban mining and to limit tourism in Antarctica?

Organizing What You Know

5. **Summarizing** Copy the following graphic organizer. Use it to describe research in Antarctica.

Research	Purpose

CHAPTER 30 Review ANSWERS

Building Vocabulary

For definitions, see: ice shelf, p. 690; icebergs, p. 690; polar desert, p. 691; krill, p. 691; trust territories, p. 693; Exclusive Economic Zones, p. 695; antifreeze, p. 698

Reviewing the Main Ideas

1. Melanesia, Micronesia, Polynesia; United States took control of trust territories in Micronesia (NGS 4)
2. The 1959 Antarctic Treaty set aside the continent for peace and research and prevented more claims to Antarctica; a 1991 treaty banned mining and drilling and limits tourism. (NGS 16)
3. because there is almost no evaporation or melting of ice (NGS 7)
4. the high islands, particularly Papua New Guinea; coconut palms (NGS 4)
5. early 1500s; Ferdinand Magellan (NGS 9)

▲ **Visual Record Answer**

a rocky coast and floating ice

ASSESS

Have students complete the Chapter 30 Test.

RETEACH

Organize students into groups of five. Have each group write and illustrate a book for young children about the Pacific Islands and Antarctica. Have them refer to the Read to Discover questions in this chapter for topics to cover. When the books are complete, have students take turns reading them aloud to the class. **ENGLISH LANGUAGE LEARNERS, COOPERATIVE LEARNING**

PORTFOLIO EXTENSIONS

PORTFOLIO

1. Have students conduct research on the explorations of the Polynesian people and build model outrigger canoes. Students should also prepare maps depicting some of the major migrations that resulted in the settlement of the Pacific Islands. Place the maps and a photograph of the model in the students' portfolios.
2. Have students conduct research on a traditional religion of the Pacific Islands. Students should focus on one religious ceremony and create a presentation explaining its significance and identifying any items needed to perform the ceremony.

CHAPTER 30 Review ANSWERS

Understanding Environment and Society

Presentations will vary, but the information included should be consistent with text material. Students should mention that Antarctica has mineral resources, including iron ore, gold, copper, and coal. They should also mention that it would be profitable to mine these resources but that mining would disturb the environment and cause pollution. Use Rubric 29, Presentations, to evaluate student work.

Thinking Critically

1. to help them develop their own economies
2. English and French are many islands' official languages, and most people are Christian.
3. Answers may vary, but students might suggest teaching more about traditional cultures in schools or promoting traditional cultures in public events.
4. The locations provide researchers opportunities to study Antarctica in various locations and in varying conditions.
5. to help researchers get information on what conditions were like long ago

CHAPTER 30 Reviewing What You Know

Building Vocabulary

On a separate sheet of paper, write sentences to define each of the following words.

1. ice shelf
2. icebergs
3. polar desert
4. krill
5. trust territories
6. Exclusive Economic Zones
7. antifreeze

Reviewing the Main Ideas

1. *Places and Regions* What are the three Pacific Island regions? How was the United States linked to these regions after World War II?
2. *Environment and Society* What agreements have been made to protect Antarctica's environment?
3. *Physical Systems* Why is Antarctica covered in ice and snow even though it receives little precipitation?
4. *Places and Regions* Where are dense tropical forests found in the Pacific Islands? What trees grow on low islands?
5. *Human Systems* When did the first European explore the Pacific? Who was he?

Understanding Environment and Society

Resources

Create a presentation about natural resources in Antarctica. Include a thematic map showing Antarctica's resources. In preparing your presentation consider the following:

- Why some countries might want to mine them.
- What threats mining might create for the continent's environment.

Thinking Critically

1. **Drawing Inferences and Conclusions** Why do you think the Pacific Island countries want to keep their 200-nautical-mile Exclusive Economic Zones?
2. **Finding the Main Idea** How have Europeans influenced the languages and religions of the Pacific Islands?
3. **Making Generalizations and Predictions** How might Pacific Islanders work to preserve their traditional cultures?
4. **Drawing Inferences and Conclusions** Look at the map of Antarctica in the Focus On Environment feature. Why do you think researchers chose the specific locations of the U.S. stations?
5. **Making Generalizations and Predictions** Why do you think protecting Antarctica's environment is important for the research done there?

Food Festival

Although it is not feasible to have a complete luau buffet in the classroom, these ideas can set the tone as students learn about this Polynesian tradition. Decorate the room with flowers. If you can, play some Polynesian music. Dice pineapples and serve them with cherries for students to eat with toothpicks. Explain that while modern luaus serve contemporary foods such as hamburgers and hot dogs, traditional menus include dishes such as Kalua pork, Huli-Huli chicken, Lomi Lomi salmon, chicken long rice, haupia, and poi.

Chapter 30 Review and Assessment Resources

Reproducible
- Readings in World Geography, History, and Culture 81 and 82
- Critical Thinking Activity 30
- Vocabulary Activity 30

Technology
- Chapter 30 Test Generator (on the One-Stop Planner)
- HRW Go site
- Audio CD Program, Chapter 30

Reinforcement, Review, and Assessment
- Chapter 30 Review, pp. 700–01
- Chapter 30 Tutorial for Students, Parents, Mentors, and Peers
- Chapter 30 Test
- Unit 9 Test
- Chapter 30 Test for English Language Learners and Special-Needs Students
- Unit 9 Test for English Language Learners and Special-Needs Students

Building Social Studies Skills

Map Activity

On a separate sheet of paper, match the letters on the map with their correct labels.

Papua New Guinea	New Caledonia
Fiji	Guam
Marshall Islands	French Polynesia
	Palau
	Solomon Islands

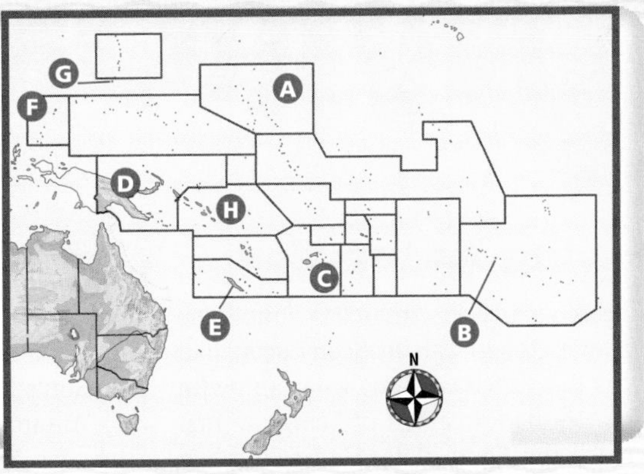

Mental Mapping Skills Activity

On two separate sheets of paper, draw freehand maps of the Pacific Islands and of Antarctica. Make keys for your maps and label the following:

Antarctica	Micronesia
Antarctic Peninsula	Polynesia
Melanesia	Transantarctic Mountains

Writing Activity

Imagine that you are one of the first explorers trying to reach the South Pole. Write journal entries in which you describe what you have seen. Describe the conditions you have endured and the strategies you have used to survive. Use your textbook, the library, and the Internet to help you write your journal entries. Be sure to use standard grammar, spelling, sentence structure, and punctuation.

Alternative Assessment

Portfolio Activity

Learning About Your Local Geography

Cooperative Project Tourism is important to many Pacific islands. What is a tourist attraction in your community or region? Working with a group, create a brochure to attract visitors to your area.

internet connect

Internet Activity: **go.hrw.com**
KEYWORD: **SK3 GT30**

Choose a topic to explore about the Pacific Islands and Antarctica.
- Tour the Pacific Islands.
- Explore Antarctica.
- Learn about native traditions.

Map Activity
A. Marshall Islands
B. French Polynesia
C. Fiji
D. Papua New Guinea
E. New Caledonia
F. Palau
G. Guam
H. Solomon Islands

Mental Mapping Skills Activity
Maps will vary, but listed places should be labeled in their approximate locations.

Writing Activity
Journal entries will vary, but the information included should be consistent with text material. Students should describe conditions and survival strategies. Use Rubric 15, Journals, to evaluate student work.

Portfolio Activity
Brochures will vary, but the information included should be consistent with text material. Brochures should describe a local tourist attraction. Use Rubric 43, Writing to Persuade, to evaluate student work.

internet connect

GO TO: **go.hrw.com**
KEYWORD: **SK3 Teacher**
FOR: a guide to using the Internet in your classroom

Focus On Environment

Expressing Points of View

Lead a class discussion to address the following questions: Should individual nations be allowed to claim and control Antarctica? How would these claims be verified and enforced? Should Antarctica be declared a world park on behalf of all Earth's people? How would Antarctica's neutrality be upheld? Require that students support their statements. Call on volunteers to summarize students' arguments for and against the various proposals on the chalkboard.

Geography Sidelight

Antarctica offers many opportunities for astronomical and space research. NASA sometimes uses the continent as a training ground for astronauts and as a laboratory for developing new technologies. One of NASA's latest projects in Antarctica is the testing of a four-wheeled robot named Nomad. This robot may someday serve as a prototype for future planetary exploration. It has successfully located meteorites in a remote area of eastern Antarctica.

This project will provide valuable information about the origins of life and help scientists fine-tune the robot for space exploration on planets with environments similar to Antarctica.

Critical Thinking: What effects might future scientific discoveries and technological innovations using the robot have?

Answer: might have both social and environmental consequences if it provides more information about the origins of life

internet connect

CHAPTER WEB LINKS
GO TO: go.hrw.com
KEYWORD: SK3 CH30
FOR: Web links about NASA in Antarctica

➤ **This Focus On Environment feature addresses National Geography Standards 1, 8, 13, 14, and 18.**

Focus On Environment

Preserving Antarctica

In 1912 a search party in Antarctica found the body of explorer Robert Falcon Scott. Scott had written in his diary, "Great God, what an awful place this is." He had dreamed of being the first person to reach the South Pole. However, when Scott's party arrived they found the tent and flag of Norwegian explorer Roald Amundsen. Amundsen had arrived five weeks earlier.

Almost a century later, Antarctica's modern-day explorers are research scientists. Antarctica has provided a location for international scientific research and cooperation.

A Scientific Laboratory

A wasteland to many, Antarctica is a scientific laboratory to some. Studying the continent helps scientists understand our planet. For example, the icy waters around Antarctica move north, cooling warmer waters. This movement affects ocean currents, clouds, and weather patterns. The world's climate is also affected by Antarctic sea ice. The ice acts as a shield. It keeps Earth cool by reflecting the Sun's heat energy.

Antarctic ice also provides information about the past. Buried deep within the ice are gas bubbles that are a record of Earth's air. Scientists have compared atmospheric gases trapped in Antarctic ice with atmospheric gases of today. They learned that the use of fossil fuels has raised the amount of carbon dioxide in the air. Carbon dioxide levels are now the highest in human history.

Eyes on Antarctica

Antarctica is not owned by any single country. Some countries claim parts of Antarctica, but these claims are not recognized. In 1959 the Antarctic Treaty established Antarctica as a continent for

How Antarctic Ice Affects Climate

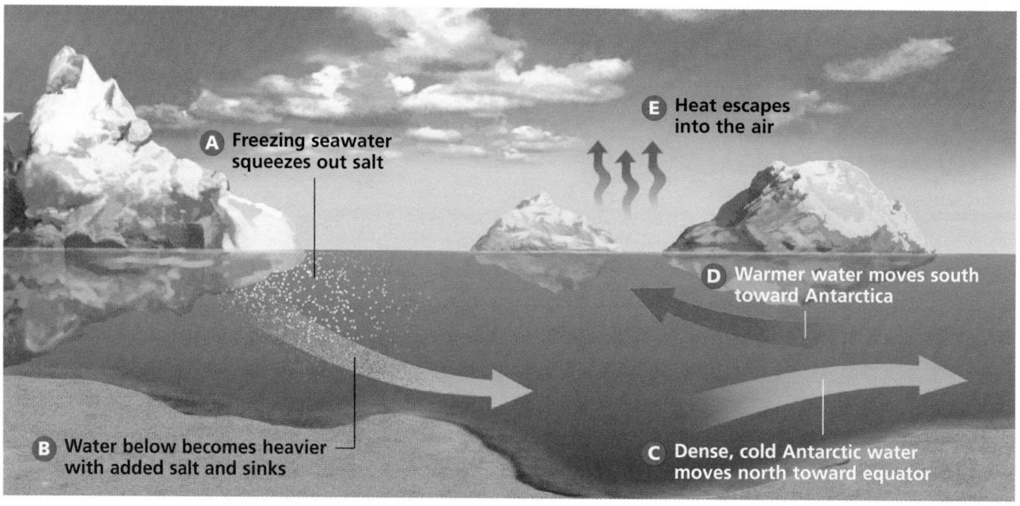
A Freezing seawater squeezes out salt
B Water below becomes heavier with added salt and sinks
C Dense, cold Antarctic water moves north toward equator
D Warmer water moves south toward Antarctica
E Heat escapes into the air

Going Further: Thinking Critically

To help reinforce students' understanding of the geographic concept of region, have students formulate plans for Antarctica. Organize the class into four groups: research scientists, environmentalists, a mining cartel, and an international tourism company. Have each group prepare a presentation outlining their group's goals for Antarctica, its response to the Madrid Protocol, and its vision for the continent in the year 2050. Encourage students to make sketches to show how their group envisions the Antarctic region's physical, economic, and population geography in the future. Have members from each group select a spokesperson to present a report to the class.

science and peace. The treaty was originally signed by 12 countries and was later agreed to by more than a dozen others. The Antarctic Treaty banned military activity in the region. It also made Antarctica a nuclear-free zone and encouraged scientific research.

However, the Antarctic Treaty did not cover mining rights on the continent. When Antarctica's mineral riches were discovered, some countries wanted rights to this new source of wealth. Geographers, scientists, and environmentalists also took notice. They feared that if mining took place, Antarctica's environment would suffer. In particular, they feared that a practical method of obtaining the offshore oil would be found. Antarctica's coastline and marine life could then be threatened by oil spills.

At the same time, evidence of environmental neglect at some of Antarctica's research stations appeared. Environmentalists voiced concern about scientists' careless disposal of trash and sewage. A U.S. Coast Guard captain who worked on icebreakers described pollution at McMurdo, a U.S. research station. "Trash was just rolled down the hill. . . . One of the jobs of the icebreakers was to break up the ice where the trash was and push it out to sea." In 1989, an Argentine ship ran aground in Antarctica, spilling thousands of gallons of oil. As a result, environmental concerns for the region increased.

Antarctica

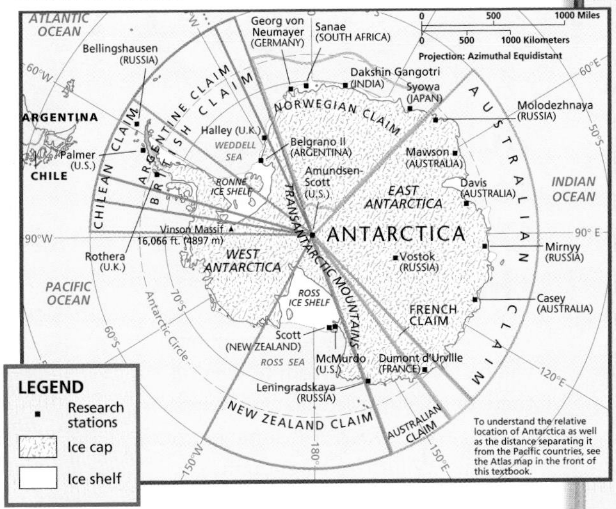

Protecting Antarctica

In 1991, the countries that had signed the original Antarctic Treaty signed a new agreement called the Madrid Protocol. It forbids most activities on Antarctica that do not have a scientific purpose. It also bans mining and drilling and sets limits on tourism. Pollution concerns are addressed specifically.

In 50 years the agreement can be changed if enough of the signing countries agree. Then it becomes the responsibility of another generation to preserve Antarctica.

▲ Many countries have set up research stations and bases in Antarctica, such as Argentina's Camara Base on Half Moon Island.

Your Turn

Imagine you are helping to develop the Madrid Protocol.

1. Write a short description of the provisions you want included in the treaty.
2. Be sure to explain why you think these provisions are important.

Your Turn

Answers

1. Students may suggest expanding the agreement to include tourism and development of the region. Some students may suggest further limiting access to Antarctica by reducing the number of research projects allowed each year. Others may suggest keeping the agreement as it is.
2. Students should use the information available in the feature to provide reasonable support for the provisions suggested.

Going Further: Thinking Critically

Tell students to imagine that the city council in your community has just received funding for a new park. The council must decide where the park should be located and will hear proposals from various interested parties. Have students work in groups to develop a proposal to the city council for the new park's location. Have each group create a list of the considerations they think are necessary to find a location, such as where land for sale is available, and where the most potential park users live. Ask the groups to use a map of the community and any other information that might be helpful. Students should examine positive aspects of the project, such as increased recreation facilities, as well as negative ones, such as the possible loss of tax revenue if the land is removed from private hands. After each group has decided on a location, have them write a proposal for the city council. Finally, have students present their proposals to the class.

Practicing THE SKILL

1. Students might suggest the need to know the physical landscape, nearby cities or towns, types of businesses, and the amount of traffic in the area.
2. Students should demonstrate knowledge of issues that affect highway construction.
3. Students should list options.
4. Students should predict consequences.
5. Students' proposals should be clearly thought out and should present different options, along with the chosen option. Presentations should include visual material.

➤ **This GeoSkills feature addresses National Geography Standards 2, 12, and 18.**

Geo Skills

Building Skills for Life: Making Decisions About the Future

What do you think your community will be like 10 years from now? Will there be new movie theaters, neighborhoods, and shops? Will some old ones be gone? Will there be enough schools? What will traffic be like? These are the kinds of questions to ask if you are planning for the future. People need to plan ahead. Often, this requires the collection and analysis of geographic information, such as population patterns, growth rates, income levels, and other data.

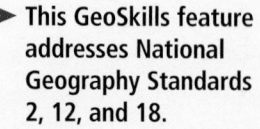

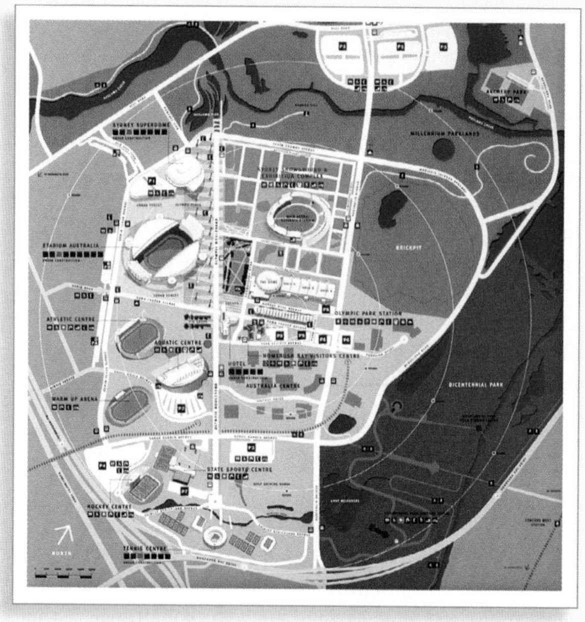

▲ Map used to plan for the 2000 Olympics in Australia

Practicing THE SKILL

1. **Identify a situation that requires a decision.** Suppose you were asked by your local planning commission to select the best route for a new highway. What information would you need?
2. **Gather information.** Use several resources to find information that might influence the building of a highway.
3. **Identify options.** Consider the many options you might give to a planning commission regarding a proposed highway. Be sure to record your possible options.
4. **Predict consequences.** Now take each option you came up with and consider what might be the outcome of each course of action. Ask yourself questions like, "How is this highway going to affect surrounding neighborhoods?"
5. **Take action to implement your decision.** Create a proposal for the planning commission. Explain why you chose your option and rejected others. Include maps, graphs, charts, models, or databases showing the information you collected.

HANDS ON GEOGRAPHY

With nearly 4 million residents, Sydney is the biggest city in Australia. Like other large cities, Sydney is dealing with some environmental problems. The number of people and cars in Sydney is growing. As a result, the city's air pollution and traffic problems are getting worse. To protect Sydney's environment and quality of life, some residents are planning for the future.

One environmental group in Sydney has suggested building a network of trails and parks called greenways. These greenways could help relieve the city's traffic problems and reduce air pollution. Can you think of another way people in Sydney might plan for the future? These guidelines will help you get started.

1. Think of one way that residents of Sydney could plan for the future. For example, they could build an underground subway system. This plan might help reduce air pollution.
2. Create a short proposal for your plan. In your proposal, list some of the benefits that your plan would offer.
3. Create a map or drawing for your proposal. Maps and drawings help people imagine what your plan would look like.
4. Include a timetable with your proposal. How long would it take for your plan to become reality?

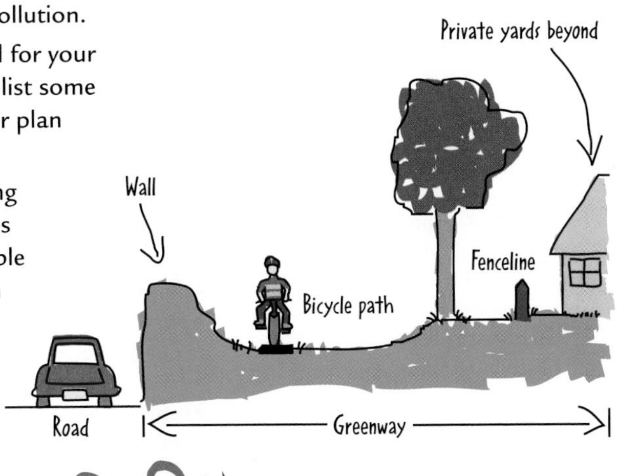

Lab Report

1. Imagine you went to Sydney to discuss your plan for the future. Who might be interested in it?
2. How important do you think it would be for your plan to have public support?
3. Do you think your proposal would be a good idea for other cities too? why or why not?

Lab Report

Answers

1. Students might suggest political leaders, transportation experts, and environmental experts and leaders.
2. Students should note that public support is typically necessary before a plan's proposals can be carried out.
3. Answers will vary. Students should support their arguments.

GAZETTEER

Phonetic Respelling and Pronunciation Guide

Many of the key terms in this textbook have been respelled to help you pronounce them. The letter combinations used in the respelling throughout the narrative are explained in this phonetic respelling and pronunciation guide. The guide is adapted from Webster's Tenth New College Dictionary, Merriam-Webster's New Geographical Dictionary, and Merriam-Webster's New Biographical Dictionary.

MARK	AS IN	RESPELLING	EXAMPLE
a	alphabet	a	*AL-fuh-bet
ā	Asia	ay	AY-zhuh
ä	cart, top	ah	KAHRT, TAHP
e	let, ten	e	LET, TEN
ē	even, leaf	ee	EE-vuhn, LEEF
i	it, tip, British	i	IT, TIP, BRIT-ish
ī	site, buy, Ohio	y	SYT, BY, oh-HY-oh
	iris	eye	EYE-ris
k	card	k	KAHRD
ō	over, rainbow	oh	OH-vuhr, RAYN-boh
u̇	book, wood	ooh	BOOHK, WOOHD
ȯ	all, orchid	aw	AWL, AWR-kid
ȯi	foil, coin	oy	FOYL, KOYN
au̇	out	ow	OWT
ə	cup, butter	uh	KUHP, BUHT-uhr
ü	rule, food	oo	ROOL, FOOD
yü	few	yoo	FYOO
zh	vision	zh	VIZH-uhn

A syllable printed in small capital letters receives heavier emphasis than the other syllable(s) in a word.

A

Abu Dhabi (24°N 54°E) capital of the United Arab Emirates, 385
Abuja (ah-BOO-jah) (9°N 7°E) capital of Nigeria, 461
Accra (6°N 0°W) capital of Ghana, 461
Addis Ababa (9°N 39°E) capital of Ethiopia, 479
Adriatic Sea sea between Italy and the Balkan Peninsula, 125, 289
Aegean (ee-JEE-uhn) **Sea** sea between Greece and Turkey, 122, 125, 231
Afghanistan landlocked country in central Asia, 385
Africa second-largest continent; surrounded by the Atlantic Ocean, Indian Ocean, and Mediterranean Sea, A2–A3
Ahaggar Mountains mountain range in southern Algeria, 441
Albania country in Eastern Europe on the Adriatic Sea, 289
Alexandria (31°N 30°E) city in northern Egypt, 441
Algeria country in North Africa located between Morocco and Libya, 441
Algiers (37°N 3°E) capital of Algeria, 441
Alps major mountain system in south-central Europe, 125, 249
Amman (32°N 36°E) capital of Jordan, 405
Amsterdam (52°N 5°E) capital of the Netherlands, 249
Amu Dar'ya (uh-MOO duhr-YAH) river in Central Asia that drains into the Aral Sea, 357
Amur (ah-MOOHR) **River** river in northeast Asia forming part of the border between Russia and China, 321
Andorra European microstate in the Pyrenees mountains, A15
Andorra la Vella (43°N 2°E) capital of Andorra, A15
Angkor ancient capital of the Khmer Empire in Cambodia, 587
Angola country in central Africa, 495
Ankara (40°N 33°E) capital of Turkey, 405
Antananarivo (19°S 48°E) capital of Madagascar, 511
Antarctica continent around the South Pole, A22, 663
Antarctic Circle line of latitude located at 66.5° south of the equator; parallel beyond which no sunlight shines on the June solstice (first day of winter in the Southern Hemisphere), A4–A5
Antarctic Peninsula peninsula stretching toward South America from Antarctica, A22
Antwerp (51°N 4°E) major port city in Belgium, 249
Apennines (A-puh-nynz) mountain range in Italy, 231
Apia (14°S 172°W) capital of Western Samoa, 664
Arabian Peninsula peninsula in Southwest Asia between the Red Sea and Persian Gulf, 107, 385
Arabian Sea sea between India and the Arabian Peninsula, 385
Aral (AR-uhl) **Sea** inland sea between Kazakhstan and Uzbekistan, 357
Arctic Circle line of latitude located at 66.5° north of the equator; the parallel beyond which no sunlight shines on the December solstice (first day of winter in the Northern Hemisphere), A4–A5
Arctic Ocean ocean north of the Arctic Circle; world's fourth-largest ocean, A2–A3
Armenia country in the Caucasus region of Asia; former Soviet republic, 343
Ashgabat (formerly Ashkhabad) (40°N 58°E) capital of Turkmenistan, 359
Asia world's largest continent; located between Europe and the Pacific Ocean, A3

Asmara (15°N 39°E) capital of Eritrea, 479
Astana (51°N 71°E) capital of Kazakhstan, 357
Astrakhan (46°N 48°E) old port city on the Volga River in Russia, 321
Athens (38°N 24°E) capital and largest city in Greece, 122, 125
Atlantic Ocean ocean between the continents of North and South America and the continents of Europe and Africa; world's second-largest ocean, A2
Atlas Mountains African mountain range north of the Sahara, 441
Auckland (37°S 175°E) New Zealand's largest city and main seaport, 671
Aughrabies (oh-KRAH-bees) **Falls** waterfalls on the Orange River in South Africa, 511
Australia only country occupying an entire continent (also called Australia); located between the Indian Ocean and the Pacific Ocean, A3, 150, 671
Austria country in west-central Europe south of Germany, 249
Azerbaijan country in the Caucasus region of Asia; former Soviet republic, 343

B

Bab al-Mandab narrow strait that connects the Red Sea with the Indian Ocean, 479
Baghdad (33°N 44°E) capital of Iraq, 385
Bahrain country on the Persian Gulf in Southwest Asia, 385
Baku (40°N 50°E) capital of Azerbaijan, 343
Bali island in Indonesia east of Java, 587
Balkan Mountains mountain range that rises in Bulgaria, 289
Baltic Sea body of water east of the North Sea and Scandinavia, 269
Bakamo (13°N 8°W) capital of Mali, 461
Bandar Seri Begawan (5°N 115°E) capital of Brunei, 587
Bangkok (14°N 100°E) capital and largest city of Thailand, 587
Bangladesh country in South Asia, 639
Bangui (4°N 19°E) capital of the Central African Republic, 495
Banjul (13°N 17°W) capital of Gambia, 461
Barcelona (41°N 2°E) Mediterranean port city and Spain's second-largest city, 231
Basel (48°N 8°E) city in northern Switzerland on the Rhine River, 249
Bay of Bengal body of water between India and the western coasts of Myanmar (Burma) and the Malay Peninsula, 621
Bay of Biscay body of water off the western coast of France and the northern coast of Spain, 249
Beijing (40°N 116°E) capital of China, 543
Beirut (34°N 36°E) capital of Lebanon, 405
Belarus country located north of Ukraine; former Soviet republic, 343
Belfast (55°N 6°W) capital and largest city of Northern Ireland, 269
Belgium country between France and Germany in west-central Europe, 249
Belgrade (45°N 21°E) capital of Serbia and Yugoslavia on the Danube River, 289
Benghazi (32°N 20°E) major coastal city in Libya, 441
Benin (buh-NEEN) country in West Africa between Togo and Nigeria, 461
Bergen (60°N 5°E) seaport city in southwestern Norway, 269
Berlin (53°N 13°E) capital of Germany, 249
Bern (47°N 7°E) capital of Switzerland, 249
Bhutan South Asian country in the Himalayas located north of India and Bangladesh, 639
Birmingham (52°N 2°W) major manufacturing center of south-central Great Britain, 269
Bishkek (43°N 75°E) capital of Kyrgyzstan, 357
Bissau (12°N 16°W) capital of Guinea-Bissau, 461
Black Sea sea between Europe and Asia, A14, 107, 122
Blue Nile East African river that flows into the Nile River in Sudan, 479
Bombay see Mumbai.
Bonn (51°N 7°E) city in western Germany; replaced by Berlin as the capital of reunified Germany, 249
Borneo island in the Malay Archipelago in Southeast Asia, 587
Bosnia and Herzegovina country in Eastern Europe between Serbia and Croatia, 289
Bosporus a narrow strait separating European and Asian Turkey, 405
Botswana country in southern Africa, 511
Brahmaputra River major river of South Asia that begins in the Himalayas of Tibet and merges with the Ganges River in Bangladesh, 621
Bratislava (48°N 17°E) capital of Slovakia, 289
Brazil largest country in South America, 150
Brazzaville (4°S 15°E) capital of the Republic of the Congo, 495
Brisbane (28°S 153°E) seaport and capital of Queensland, Australia, 671
British Isles island group consisting of Great Britain and Ireland, A15, 269
Brittany region in northwestern France, 249
Brunei (brooh-NY) country on the northern coast of Borneo in Southeast Asia, 587
Brussels (51°N 4°E) capital of Belgium, 249
Bucharest (44°N 26°E) capital of Romania, 289
Budapest (48°N 19°E) capital of Hungary, 289
Bujumbura (3°S 29°E) capital of Burundi, 479
Bulgaria country on the Balkan Peninsula in Eastern Europe, 289
Burkina Faso (boor-KEE-nuh FAH-soh) landlocked country in West Africa, 461
Burma see Myanmar.
Burundi landlocked country in East Africa, 479

C

Cairo (30°N 31°E) capital of Egypt, 441
Calcutta see Kolkata.
Cambodia country in Southeast Asia west of Vietnam, 587
Cameroon country in central Africa, 495
Canberra (35°S 149°E) capital of Australia, 671
Cantabrian (kan-TAY-bree-uhn) **Mountains** mountains in northwestern Spain, 231
Cape Horn (56°S 67°W) cape in southern Chile; southernmost point of South America, 150
Cape of Good Hope cape of the southwest coast of South Africa, 511
Cape Town (34°S 18°E) major seaport city and legislative capital of South Africa, 511
Cape Verde island country in the Atlantic Ocean off the coast of West Africa, 461
Cardiff (52°N 3°W) capital and largest city of Wales, 269
Caribbean Sea arm of the Atlantic Ocean between North and South America, A10, 150

Carpathian Mountains mountain system in Eastern Europe, 289

Casablanca (34°N 8°W) seaport city on the western coast of Morocco, 441

Caspian Sea large inland salt lake between Europe and Asia, A16

Caucasus Mountains mountain range between the Black Sea and the Caspian Sea, 321

Central African Republic landlocked country in central Africa located south of Chad, 495

Central America narrow southern portion of the North American continent, 150

Central Lowlands area of Australia between the Western Plateau and the Great Dividing Range, 671

Central Siberian Plateau upland plains and valleys between the Yenisey and Lena Rivers in Russia, 321

Chad landlocked country in northern Africa, 461

Chang River major river in Central China, 115, 543

Chao Phraya (chow-PRY-uh) **River** major river in Thailand, 587

Chelyabinsk (chel-YAH-buhnsk) (55°N 61°E) manufacturing city in the Urals region of Russia, 321

Chernobyl (51°N 30°E) city in north-central Ukraine; site of a major nuclear accident in 1986, 343

China country in East Asia; most populous country in the world, 151, 543

Chişinău (formerly Kishinev) (47°N 29°E) capital of Moldova, 289

Chongqing (30°N 108°E) city in southern China along the Chang River, 543

Christchurch (44°S 173°E) city on the eastern coast of South Island, New Zealand, 671

Cologne (51°N 7°E) manufacturing and commercial city along the Rhine River in Germany, 249

Colombo (7°N 80°E) capital city and important seaport of Sri Lanka, 639

Comoros island country in the Indian Ocean off the coast of Africa, 511

Conakry (10°N 14°W) capital of Guinea, 461

Congo Basin region in central Africa, 495

Congo, Democratic Republic of the largest and most populous country in central Africa, 495

Congo, Republic of the central African country located along the Congo River, 495

Congo River major navigable river in central Africa that flows into the Atlantic Ocean, 495

Constantinople (now called Istanbul) (41°N 29°E) former city in Turkey, 405

Copenhagen (56°N 12°E) seaport and capital of Denmark, 269

Cork (52°N 8°W) seaport city in southern Ireland, 269

Côte d'Ivoire (KOHT-dee-VWAHR) (Ivory Coast) country in West Africa, 461

Crete largest of the Greek islands, 122

Crimean Peninsula peninsula in Ukraine that juts southward into the Black Sea, 343

Croatia Eastern European country and former Yugoslav republic, 289

Cyprus island republic in the eastern Mediterranean Sea, 405

Czech Republic Eastern European country and the western part of the former country in Czechoslovakia, 289

D

Dakar (15°N 17°W) capital of Senegal, 461

Damascus (34°N 36°E) capital of Syria and one of the world's oldest cities, 405

Danube River major river in Europe that flows into the Black Sea in Romania, 249

Dardanelles narrow strait separating European and Asian Turkey, 405

Dar es Salaam (7°S 39°E) capital and major seaport of Tanzania, 479

Dead Sea salt lake on the boundary between Israel and Jordan in southwestern Asia, 405

Deccan the southern part of the Indian subcontinent, 621

Delhi (29°N 77°E) city in India, 621

Denmark country in northern Europe, 269

Dhaka (24°N 90°E) capital and largest city of Bangladesh, 639

Dinaric Alps mountains extending inland from the Adriatic coast to the Balkan Peninsula, 289

Djibouti country located in the Horn of Africa, 479

Djibouti (12°N 43°E) capital of Djibouti, 479

Dnieper River major river in Ukraine, 343

Doha (25°N 51°E) capital of Qatar, 385

Donets Basin industrial region in eastern Ukraine, 343

Douro River river on the Iberian Peninsula that flows into the Atlantic Ocean in Portugal, 231

Drakensberg mountain range in southern Africa, 511

Dublin (53°N 6°W) capital of the republic of Ireland, 269

Durban (30°S 31°E) port city in South Africa, 511

Dushanbe (39°N 69°E) capital of Tajikistan, 357

E

Eastern Ghats mountains on the eastern side of the Deccan Plateau in southern India, 621

Eastern Highlands mountain ranges in eastern Australia, 671

East Timor eastern section of Timor administered by the United Nations until formal independence, 587

Ebro River river in Spain that flows into the Mediterranean Sea, 231

Egypt country in North Africa located east of Libya, 107, 112, 441

Elburz Mountains mountain range in northern Iran, 385

England southern part of Great Britain and part of the United Kingdom in northern Europe, 269

English Channel channel separating Great Britain from the European continent, 249

equator the imaginary line of latitude that lies halfway between the North and South Poles and circles the globe, A4–A5

Equatorial Guinea central African country, 495

Eritrea (er-uh-TREE-uh) East African country located north of Ethiopia, 479

Essen (51°N 7°E) industrial city in western Germany, 249

Estonia country located on the Baltic Sea; former Soviet republic, 289

Ethiopia East African country in the Horn of Africa, 479

Euphrates River major river in Iraq in southwestern Asia, 107, 109, 111, 385

Europe continent between the Ural Mountains and the Atlantic Ocean, A3

F

Fergana Valley fertile valley in Uzbekistan, Kyrgyzstan, and Tajikistan, 357

Fertile Crescent rich farmland region in southwestern Asia, 109

Fès (34°N 5°W) city in north-central Morocco, 441
Fiji South Pacific island country in Melanesia, 663
Finland country in northern Europe located between Sweden, Norway, and Russia, 269
Flanders northern coastal part of Belgium where Flemish is the dominant language, 249
Florence (44°N 11°E) city on the Arno River in central Italy, 231
France country in west-central Europe, 249
Frankfurt (50°N 9°E) main city of Germany's Rhineland region, 249
Freetown (9°N 13°W) capital of Sierra Leone, 461
Funafuti (9°S 179°E) capital of Tuvalu, 664

G

Gabon country in central Africa located between Cameroon and the Republic of the Congo, 495
Gabarone (24°S 26°E) capital of Botswana, 511
Galway (53°N 9°W) city in western Ireland, 269
Gambia country along the Gambia River in West Africa, 461
Ganges River major river in India flowing from the Himalayas southeastward to the Bay of Bengal, 621
Gangetic (gan-JE-tik) **Plain** vast plain in northern India, 621
Gao (GOW) (16°N 0°) city in Mali on the Niger River, 461
Gaza Strip area occupied by Israel from 1967 to 1994; under Palestinian self-rule since 1994, 405
Geneva (46°N 6°E) city in southwestern Switzerland, 249
Genoa (44°N 10°E) seaport city in northwestern Italy, 231
Georgia (Eurasia) country in the Caucasus region; former Soviet republic, 343
Germany country in west-central Europe located between Poland and the Benelux countries, 249
Ghana country in West Africa, 461
Giza (30°N 31°E) Egyptian city on the west bank of the Nile, 112
Glasgow (56°N 4°W) city in Scotland, United Kingdom, 269
Gobi desert that makes up part of the Mongolian plateau in East Africa, 543
Golan Heights hilly region in southwestern Syria occupied by Israel, 405
Göteberg (58°N 12°E) seaport city in southwestern Sweden, 269
Great Artesian Basin Australia's largest source of underground well water; located in interior Queensland, 671
Great Barrier Reef world's largest coral reef; located off the northeastern coast of Australia, 671
Great Britain major island of the United Kingdom, 269
Great Dividing Range mountain range of the Eastern Highlands in Australia, 671
Great Rift Valley valley system extending from eastern Africa to Southwest Asia, 479
Greece country in southern Europe located at the southern end of the Balkan Peninsula, 122, 231
Greenland self-governing province of Denmark between the North Atlantic and Arctic Oceans, 269
Guadalquivir (gwah-thahl-kee-VEER) **River** important river in southern Spain, 231
Guam (14°N 143°E) South Pacific island and U.S. territory in Micronesia, 663
Guinea country in West Africa, 461
Guinea-Bissau (GI-nee bi-SOW) country in West Africa, 461
Gulf of Bothnia part of the Baltic Sea west of Finland, 269
Gulf of Finland part of the Baltic Sea south of Finland, 289
Gulf of Guinea part of the Atlantic Ocean south of the West African countries, 461

H

Hamburg (54°N 10°E) seaport on the Elbe River in northwestern Germany, 249
Hanoi (ha-NOY) (21°N 106°E) capital of Vietnam, 587
Harare (18°S 31°E) capital of Zimbabwe, 511
Harappa early city of the Indus River Valley, civilization of India and Pakistan, 113
Helsinki (60°N 25°E) capital of Finland, 269
Himalayas mountain system in Asia; world's highest mountains, 543
Hindu Kush high mountain range in northern Afghanistan, 385
Hispaniola large Caribbean island divided into the countries of Haiti and the Dominican Republic, 150
Ho Chi Minh City (formerly Saigon) (11°N 107°E) major city in southern Vietnam; former capital of South Vietnam, 587
Hokkaido (hoh-KY-doh) major island in northern Japan, 565
Hong Kong (22°N 115°E) former British colony in East Asia; now part of China, 543
Hong (Red) River major river that flows into the Gulf of Tonkin in Vietnam, 587
Honiara (9°S 160°E) capital of the Solomon Islands, 664
Honshu (HAWN-shoo) largest of the four major islands of Japan, 565
Huang River (Yellow River) one of the world's longest rivers; located in northern China, 115, 543
Hungary country in Eastern Europe between Romania and Austria, 289

I

Iberian Peninsula peninsula in southwestern Europe occupied by Spain and Portugal, 231
Iceland island country between the North Atlantic and Arctic Oceans, 269
India country in South Asia, 113, 151, 621
Indian Ocean world's third-largest ocean; located east of Africa, south of Asia, west of Australia, and north of Antarctica, A3
Indochina Peninsula peninsula in southeastern Asia that includes the region from Myanmar (Burma) to Vietnam, 587
Indonesia largest country in Southeast Asia; made up of more than 17,000 islands, 587
Indus River major river in Pakistan, 113, 639
Inland Sea body of water in southern Japan between Honshu, Shikoku, and Kyushu, 565
Inyanga Mountains mountain region of Zimbabwe and Mozambique, 511
Iran country in southwestern Asia; formerly called Persia, 385
Iraq (i-RAHK) country located between Iran and Saudi Arabia, 385
Ireland country west of Great Britain in the British Isles, 269
Irian Jaya western part of the island of New Guinea and part of Indonesia, 587
Irish Sea sea between Great Britain and Ireland, 269
Irrawaddy River important river in Myanmar (Burma), 587
Islamabad (34°N 73°E) capital of Pakistan, 639
Israel country in southwestern Asia, 405
Istanbul (formerly Constantinople) (41°N 29°E) largest city and leading seaport in Turkey, 405
Italy country in southern Europe, 231

J

Jakarta (6°S 107°E) capital of Indonesia, 587
Japan country in East Asia consisting of four major islands and more than 3,000 smaller islands, 191, 565
Java major island in Indonesia, 587
Jerusalem (32°N 35°E) capital of Israel, 405
Johannesburg (26°S 28°E) city in South Africa, 511
Jordan Southwest Asian country stretching east from the Dead Sea and Jordan River into the Arabian Desert, 405
Jordan River river in southwestern Asia that separates Israel from Syria and Jordan, 109, 405
Jutland Peninsula peninsula in northern Europe made up of Denmark and part of northern Germany, 269

K

Kabul (35°N 69°E) capital and largest city of Afghanistan, 385
Kalahari Desert dry plateau region in southern Africa, 511
Kamchatka Peninsula peninsula along Russia's northeastern coast, 321
Kampala (0° 32°E) capital of Uganda, 479
Kaohsiung (23°N 120°E) Taiwan's second-largest city and major seaport, 543
Karachi (25°N 69°E) Pakistan's largest city and major seaport, 639
Karakoram Range high mountain range in northern India and Pakistan, 639
Kara-kum (kahr-uh-KOOM) desert region in Turkmenistan, 357
Kashmir mountainous region in northern India and Pakistan, 621
Kathmandu (kat-man-DOO) (28°N 85°E) capital of Nepal, 639
Kazakhstan country in Central Asia; former Soviet republic, 357
Kenya country in East Africa south of Ethiopia, 479
Khabarovsk (kuh-BAHR-uhfsk) (49°N 135°E) city in southeastern Russia on the Amur River, 321
Khartoum (16°N 33°E) capital of Sudan, 479
Khyber Pass major mountain pass between Afghanistan and Pakistan, 113, 385
Kiev (50°N 31°E) capital of Ukraine, 343
Kigali (2°S 30°E) capital of Rwanda, 479
Kinshasa (4°S 15°E) capital of the Democratic Republic of the Congo, 495
Kiribati South Pacific country in Micronesia and Polynesia, 664
Kjølen (CHUHL-uhn) **Mountains** mountain range in the Scandinavian Peninsula, 269
Kobe (KOH-bay) (35°N 135°E) major port city in Japan, 565
Kolkata (Calcutta) (23°N 88°E) giant industrial and seaport city in eastern India, 621
Korea peninsula on the east coast of Asia, 565
Koror capital of Palau, 664
Kosovo province in southern Serbia, 289
Kuala Lumpur (3°N 102°E) capital of Malaysia, 587
Kuril (KYOOHR-eel) **Islands** Russian islands northeast of the island of Hokkaido, Japan, 321
Kuwait country on the Persian Gulf in southwestern Asia, 385
Kuwait City (29°N 48°E) capital of Kuwait, 385
Kuznetsk Basin (Kuzbas) industrial region in central Russia, 321
Kyoto (KYOH-toh) (35°N 136°E) city on the island of Honshu and the ancient capital of Japan, 565
Kyrgyzstan (kir-gi-STAN) country in Central Asia; former Soviet republic, 357
Kyushu (KYOO-shoo) southernmost of Japan's main islands, 565
Kyzyl Kum (ki-zil KOOM) desert region in Uzbekistan and Kazakhstan, 357

L

Lagos (LAY-gahs) (6°N 3°E) former capital of Nigeria and the country's largest city, 461
Lahore (32°N 74°E) industrial city in northeastern Pakistan, 614
Lake Baikal (by-KAHL) world's deepest freshwater lake; located north of the Gobi in Russia, 321
Lake Chad shallow lake between Nigeria and Chad in western Africa, 461
Lake Malawi (also called Lake Nyasa) lake in southeastern Africa, 495
Lake Nasser artificial lake in southern Egypt created in the 1960s by the construction of the Aswan High Dam, 441
Lake Tanganyika deep lake in the Western Rift Valley in Africa, 495
Lake Victoria large lake in East Africa surrounded by Uganda, Kenya, and Tanzania, 479
Lake Volta large artificial lake in Ghana, 461
Laos landlocked country in Southeast Asia, 587
Lapland region extending across northern Finland, Sweden, and Norway, 269
Latvia country on the Baltic Sea; former Soviet republic, 289
Lebanon country in Southwest Asia, 405
Lesotho country completely surrounded by South Africa, 511
Liberia country in West Africa, 461
Libreville (0° 9°E) capital of Gabon, 495
Libya country in North Africa located between Egypt and Algeria, 441
Liechtenstein microstate in west-central Europe located between Switzerland and Austria, 249
Lilongwe (14°S 34°E) capital of Malawi, 495
Limpopo River river in southern Africa forming the border between South Africa and Zimbabwe, 511
Lisbon (39°N 9°W) capital and largest city of Portugal, 231
Lithuania European country on the Baltic Sea; former Soviet republic, 289
Ljubljana (lee-oo-blee-AH-nuh) (46°N 14°E) capital of Slovenia, 289
Lomé (6°N 1°E) capital of Togo, 461
London (52°N 0°) capital of the United Kingdom, 269
Luanda (9°S 13°E) capital of Angola, 495
Lubumbashi (loo-boom-BAH-shee) (12°S 27°E) industrial city in the Democratic Republic of the Congo, 495
Luxembourg small European country bordered by France, Germany, and Belgium, 249
Luxembourg (50°N 7°E) capital of Luxembourg, 249
Luzon chief island of the Philippines, 587

M

Macao (22°N 113°E) former Portuguese territory in East Asia, now part of China, 543
Macedonia Balkan country; former Yugoslav republic, 289
Madagascar largest of the island countries off the eastern coast of Africa, 511
Madrid (40°N 4°W) capital of Spain, 231
Magnitogorsk (53°N 59°E) manufacturing city of the Urals region of Russia, 321

Majuro (7°N 171°E) capital of the Marshall Islands, 664
Malabo (4°N 9°E) capital of Equatorial Guinea, 495
Malawi (muh-LAH-wee) landlocked country in central Africa, 495
Malay Archipelago (ahr-kuh-PE-luh-goh) large island group off the southeastern coast of Asia including New Guinea and the islands of Malaysia, Indonesia, and the Philippines, 587
Malay Peninsula peninsula in Southeast Asia, 587
Malaysia country in Southeast Asia, 587
Maldives island country in the Indian Ocean south of India, 639
Male (5°N 72°E) capital of the Maldives, 639
Mali country in West Africa, 461
Malta island country in southern Europe located in the Mediterranean Sea between Sicily and North Africa, 222
Manama (26°N 51°E) capital of Bahrain, 385
Manchester (53°N 2°W) major commercial city in west-central Great Britain, 269
Manila (15°N 121°E) capital of the Philippines, 587
Maputo (27°S 33°E) capital of Mozambique, 511
Marseille (43°N 5°E) seaport in France on the Mediterranean Sea, 249
Marshall Islands Pacific island country in Indonesia, 664
Maseru (29°S 27°E) capital of Lesotho, 511
Masqat (Muscat) (23°N 59°E) capital of Oman, 385
Mauritania African country stretching east from the Atlantic coast into the Sahara, 461
Mauritius island country located off the coast of Africa in the Indian Ocean, 511
Mbabane (26°S 31°E) capital of Swaziland, 511
Mecca (21°N 40°E) important Islamic city in western Saudi Arabia, 385
Medina (Al Madinah) (24°N 39°E) important Islamic city north of Mecca, 385
Mediterranean Sea sea surrounded by Europe, Asia, and Africa, 122, 125
Mekong River important river in Southeast Asia, 587
Melanesia island region in the South Pacific that stretches from New Guinea to Fiji, 663
Melbourne (38°S 145°E) capital of Victoria, Australia, 671
Mexico country in North America, 150
Micronesia island region in the South Pacific that includes the Mariana, Caroline, Marshall, and Gilbert island groups, 687
Micronesia, Federated States of island country in the western Pacific, 663
Milan (45°N 9°E) city in northern Italy, 231
Minsk (54°N 28°E) capital of Belarus, 343
Mogadishu (2°N 45°E) capital and port city of Somalia, 479
Moldova Eastern European country located between Romania and Ukraine; former Soviet republic, 289
Monaco (44°N 8°E) European microstate bordered by France, 249
Mongolia landlocked country in East Asia, 543
Monrovia (6°N 11°W) capital of Liberia, 461
Montenegro See Yugoslavia.
Morocco country in North Africa south of Spain, 441
Moroni (12°S 43°E) capital of Comoros, 511
Moscow (56°N 38°E) capital of Russia, 321
Mount Elbrus (43°N 42°E) highest European peak (18,510 ft.; 5,642 m); located in the Caucasus Mountains, 343
Mount Everest (28°N 87°E) world's highest peak (29,035 ft.; 8,850 m); located in the Himalayas, 543
Mount Kilimanjaro (3°S 37°E) (ki-luh-muhn-JAHR-oh) highest point in Africa (19,341 ft.; 5,895 m); located in northeast Tanzania, near Kenya border, 479
Mozambique (moh-zahm-BEEK) country in southern Africa, 511
Mumbai (Bombay) (19°S 73°E) India's largest city, 621
Munich (MYOO-nik) (48°N 12°E) major city and manufacturing center in southern Germany, 249
Murray-Darling Rivers major river system in southeastern Australia, 671
Muscat See Masqat.
Myanmar (MYAHN-mahr) (Burma) country in Southeast Asia between India, China, and Thailand, 587

N

Nairobi (1°S 37°E) capital of Kenya, 479
Namib Desert Atlantic coast desert in southern Africa, 511
Namibia (nuh-MI-bee-uh) country on the Atlantic coast in southern Africa, 511
Nanjing (32°N 119°E) city along the upper Chang River in China, 543
Naples (41°N 14°E) major seaport in southern Italy, 231
Nauru South Pacific island country in Micronesia, 663
N'Djamena (12°N 15°E) capital of Chad, 461
Negev desert region in southern Israel, 405
Nepal South Asian country located in the Himalayas, 639
Netherlands country in west-central Europe, 249
New Caledonia French territory of Melanesia in the South Pacific Ocean east of Queensland, Australia, 663
New Delhi (29°N 77°E) capital of India, 621
Newfoundland eastern province in Canada including Labrador and the island of Newfoundland, 150
New Guinea large island in the South Pacific Ocean north of Australia, 587
New Zealand island country located southeast of Australia, 671
Niamey (14°N 2°E) capital of Niger, 461
Nice (44°N 7°E) city in the southeastern coast in France, 249
Nicosia (35°N 33°E) capital of Cyprus, 405
Niger (NY-juhr) country in West Africa, 461
Nigeria country in West Africa, 461
Niger River river in West Africa, 107, 461
Nile Delta region in northern Egypt where the Nile River flows into the Mediterranean Sea, 441
Nile River world's longest river (4,187 miles; 6,737 km); flows into the Mediterranean Sea in Egypt, 441
Nile Valley area around the Nile River where distinct cultures developed, 112
Nizhniy Novgorod (Gorky), Russia (56°N 44°E) city on the Volga River east of Moscow, 321
Normandy (France) region of northwestern France, 249
North America continent including Canada, the United States, Mexico, Central America, and the Caribbean Islands, A2
North China Plain region of northeastern China, 543
Northern European Plain broad coastal plain from the Atlantic coast of France into Russia, 249
Northern Ireland the six northern counties of Ireland that remain part of the United Kingdom; also called Ulster, 269
Northern Mariana Islands U.S. commonwealth in the South Pacific, 664
North Island northernmost of two large continental islands of New Zealand, 671
North Korea country on the northern part of the Korean Peninsula in East Asia, 565
North Pole the northern point of Earth's axis, A22
North Sea major sea between Great Britain, Denmark, and the Scandinavian Peninsula, 249

Northwest Highlands region of rugged hills and low mountains in Europe, including parts of the British Isles, northwestern France, the Iberian Peninsula, and the Scandinavian Peninsula,

Norway European country located on the Scandinavian Peninsula, 269

Nouakchott (nooh-AHK-shaht) (18°N 16°W) capital of Mauritania, 461

Nova Scotia province in eastern Canada, 150

Novosibirsk (55°N 83°E) industrial center in Siberia, Russia, 321

Nuku'alofa capital of Tonga, 664

Nuuk (Godthab) (64°N 52°W) capital of Greenland, 269

O

Ob River large river system that drains Russia and Siberia, 321

Oman country in the Arabian Peninsula; formerly known as Muscat, 385

Orange River river in southern Africa, 511

Osaka (oh-SAH-kuh) (35°N 135°E) major industrial center on Japan's southwestern Honshu island, 565

Oslo (60°N 11°E) capital of Norway, 269

Ouagadougou (wah-gah-DOO-GOO) (12°N 2°W) capital of Burkina Faso, 461

P

Pacific Ocean Earth's largest ocean; located between North and South America and Asia and Australia, A2–A3

Pakistan South Asian country located northwest of India, 639

Palau South Pacific island country in Micronesia, 664

Palikir capital of the Federated States of Micronesia, 664

Pamirs mountain area mainly in Tajikistan in Central Asia, 357

Papua New Guinea country on the eastern half of the island of New Guinea, 664

Paris (49°N 2°E) capital of France, 249

Peloponnesus (pe-luh-puh-NEE-suhs) peninsula forming the southern part of the mainland of Greece, 231

Persian Gulf body of water between Iran and the Arabian Peninsula, 385

Perth (32°S 116°E) capital of Western Australia, 671

Peru country in South America, 150

Philippines Southeast Asian island country located north of Indonesia, 587

Phnom Penh (12°N 105°E) capital of Cambodia, 587

Plateau of Tibet high, barren plateau of western China, 543

Podgorica (PAWD-gawr-ett-sah) capital of Montenegro, 289

Poland country in Eastern Europe located east of Germany, 289

Polynesia island region of the South Pacific that includes the Hawaiian and Line island groups, Samoa, French Polynesia, and Easter Island, 663

Po River river in northern Italy, 231

Port Elizabeth (34°S 26°E) seaport city in South Africa, 511

Port Louis (20°S 58°E) capital of Mauritius, 511

Port Moresby (10°S 147°E) seaport and capital of Papua New Guinea, 664

Porto-Novo (6°N 3°E) capital of Benin, 461

Portugal country in southern Europe located on the Iberian Peninsula, 231

Port-Vila (18°S 169°E) capital of Vanuatu, 664

Prague (50°N 14°E) capital of the Czech Republic, 289

Praia (PRIE-uh) (15°N 24°W) capital of Cape Verde, 461

Pretoria (26°S 28°E) administrative capital of South Africa, 511

Pripyat Marshes (PRI-pyuht) marshlands in southern Belarus and northwest Ukraine, 343

Pusan (35°N 129°E) major seaport city in southeastern South Korea, 565

P'yŏngyang (pyuhng-YANG) (39°N 126°E) capital of North Korea, 565

Pyrenees (PIR-uh-neez) mountain range along the border of France and Spain, 249

Q

Qatar Persian Gulf country located on the Arabian Peninsula, 385

Qattara Depression lowland region (436 feet below sea level; 133 m) in western Egypt, 441

R

Rabat (34°N 7°W) capital of Morocco, 441

Rangoon See Yangon.

Red Sea sea between the Arabian Peninsula and northeastern Africa, 107, 385

Reykjavik (RAYK-yuh-veek) (64°N 22°W) capital of Iceland, 269

Rhine River major river in Western Europe, 249

Riga (57°N 24°E) capital of Latvia, 289

Riyadh (25°N 47°E) capital of Saudi Arabia, 385

Romania country in Eastern Europe, 389

Rome (42°N 13°E) capital of Italy, 231

Ross Ice Shelf ice shelf of Antarctica, 663

Rub' al-Khali uninhabited desert area in southeastern Saudi Arabia, 385

Russia world's largest country, stretching from Europe and the Baltic Sea to eastern Asia and the coast of the Bering Sea, 321

Rwanda country in East Africa, 479

S

Sahara desert region in northern Africa, 441

St. Petersburg (formerly Leningrad; called Petrograd 1914 to 1924) (60°N 30°E) Russia's second largest city and former capital, 321

Sakhalin Island Russian island north of Japan, 321

Salzburg state of central Austria, 249

Samarqand (40°N 67°E) city in southeastern Uzbekistan, 357

Samoa South Pacific island country in Polynesia, 663

Sanaa (15°N 44°E) capital of Yemen, 385

San Marino microstate in southern Europe surrounded by Italy, 231

São Tomé (1°N 6°E) capital of São Tomé and Príncipe, 495

São Tomé and Príncipe island country located off the Atlantic coast of central Africa, 495

Sarajevo (sar-uh-YAY-voh) (44°N 18°E) capital of Bosnia and Herzegovina, 186, 289

Saudi Arabia country occupying much of the Arabian Peninsula in southwestern Asia, 385

Scandinavian Peninsula peninsula of northern Europe occupied by Norway and Sweden, 269

Scotland northern part of the island of Great Britain, 269

Sea of Azov sea in Ukraine connected to and north of the Black Sea, 343

Sea of Japan body of water separating Japan from mainland Asia, 321

Sea of Marmara sea between European Turkey and the Asian peninsula of Anatolia, 405

Sea of Okhotsk inlet of the Pacific Ocean on the eastern coast of Russia, 321

Seine River river that flows through Paris in northern France, 249

Senegal country in West Africa, 461

Senegal River river in West Africa, 461

Seoul (38°N 127°E) capital of South Korea, 565

Serbia Yugoslav republic, 185. *See also* Yugoslavia.

Seychelles island country located east of Africa in the Indian Ocean, 511

Shanghai (31°N 121°E) major seaport city in eastern China, 543

Shikoku (shee-KOH-koo) smallest of the four main islands of Japan, 565

Siberia vast region of Russia extending from the Ural Mountains to the Pacific Ocean, 321

Sichuan Basin rich agriculture and mining area along the Chang (Yangtze) River in central China, 543

Sicily island region of Italy located in the Mediterranean Sea, 231

Sierra Leone West African country located northwest of Liberia, 461

Sinai (SY-ny) **Peninsula** peninsula in northeastern Egypt, 441

Singapore island country located at the tip of the Malay Peninsula in Southeast Asia, 587

Skopje (SKAW-pye) (42°N 21°E) capital of Macedonia, 289

Slovakia country in Eastern Europe; formerly the eastern part of Czechoslovakia, 289

Slovenia country in Eastern Europe; former Yugoslav republic, 289

Sofia (43°N 23°E) capital of Bulgaria, 289

Solomon Islands South Pacific island country in Melanesia, 664

Somalia East African country located in the Horn of Africa, 479

South Africa country in southern Africa, 511

South America continent in the Pacific and Atlantic Oceans separated from North America by Middle America, A2

Southern Alps mountain range in South Island, New Zealand, 671

South Island the southern island of two main islands of New Zealand, 671

South Korea country occupying the southern half of the Korean Peninsula, 565

South Pole the southern point of Earth's axis, A22

Spain country in southern Europe occupying most of the Iberian Peninsula, 231

Sri Lanka island country located south of India; formerly known as Ceylon, 639

Stockholm (59°N 18°E) capital of Sweden, 269

Strait of Gibraltar (juh-BRAWL-tuhr) strait between the Iberian Peninsula and North Africa that links the Mediterranean Sea to the Atlantic Ocean, 231

Sudan East African country; largest country in Africa, 479

Suez Canal canal linking the Red Sea to the Mediterranean Sea in northeastern Egypt, 441

Sumatra large island of Indonesia, 587

Suva (19°S 178°E) capital of Fiji, 664

Swaziland country in southern Africa, 511

Sweden country in northern Europe, 269

Switzerland country in west-central Europe located between Germany, France, Austria, and Italy, 249

Sydney (34°S 151°E) largest urban area and leading seaport in Australia, 671

Syr Dar'ya (sir duhr-YAH) river draining the Pamirs in Central Asia, 357

Syria Southwest Asian country located between the Mediterranean Sea and Iraq, 405

Syrian Desert desert region covering parts of Syria, Jordan, Iraq, and northern Saudi Arabia, 405

T

Tagus River longest river on the Iberian Peninsula in southern Europe, 231

Tahiti French South Pacific island in Polynesia, 663

Taipei (25°N 122°E) capital of Taiwan, 543

Taiwan (TY-WAHN) island country located off the southeastern coast of China, 543

Tajikistan (tah-ji-ki-STAN) country in Central Asia; former Soviet republic, 357

Taklimakan Desert desert region in western China, 543

Tallinn (59°N 25°E) capital of Estonia, 289

Tanzania East African country located south of Kenya, 479

Tarai (tuh-RY) region in Nepal along the border with India, 639

Tarawa capital of Kiribati, 664

Tarim Basin arid region in western China, 543

Tashkent (41°N 69°E) capital of Uzbekistan, 357

Tasmania island state of Australia, 671

Tasman Sea part of South Pacific Ocean between Australia and New Zealand, 671

T'bilisi (42°N 45°E) capital of Georgia in the Caucasus region, 343

Tehran (36°N 52°E) capital of Iran, 385

Tel Aviv (tehl uh-VEEV) (32°N 35°E) largest city in Israel, 405

Thailand (TY-land) country in Southeast Asia, 587

Thar (TAHR) **Desert** sandy desert of northwestern India and eastern Pakistan; also called the Great Indian Desert, 621

Thessaloníki (41°N 23°E) city in Greece, 231

Thimphu (28°N 90°E) capital of Bhutan, 639

Tian Shan (TIEN SHAHN) high mountain range separating northwestern China from Russia and some Central Asian republics, 357

Tiber River river that flows through Rome in central Italy, 231

Tibesti Mountains mountain group in northwest Chad, 461

Tigris River major river in southwestern Asia, 107, 385

Timbuktu (17°N 3°W) city in Mali and an ancient trading center in West Africa, 461

Tiranë (ti-RAH-nuh) (42°N 20°E) capital of Albania, 289

Togo West African country located between Ghana and Benin, 461

Tokyo (36°N 140°E) capital of Japan, 565

Tonga South Pacific island country in Polynesia, 664

Transantarctic Mountains major mountain range that divides Antarctica into East and West, 664

Tripoli (33°N 13°E) capital of Libya, 441

Tropic of Cancer parallel 23.5° north of the equator; parallel on the globe at which the Sun's most direct rays strike the earth during the June solstice (first day of summer in the Northern Hemisphere), A4–A5

Tropic of Capricorn parallel 23.5° south of the equator; parallel on the globe at which the Sun's most direct rays strike the earth during the December solstice (first day of summer in the Southern Hemisphere), A4–A5

Tunis (37°N 10°E) capital of Tunisia, 441

Tunisia country in North Africa located on the Mediterranean coast between Algeria and Libya, 441

Turin (45°N 8°E) city in northern Italy, 231

Turkey country of the eastern Mediterranean occupying Anatolia and a corner of southeastern Europe, 405

Turkmenistan country in Central Asia; former Soviet republic, 357

Tuvalu South Pacific island country of Polynesia, 663

U

Uganda country in East Africa, 479

Ukraine country located between Russia and Eastern Europe; former Soviet republic, 343

Ulaanbaatar (oo-lahn-BAH-tawr) (48°N 107°E) capital of Mongolia, 543

United Arab Emirates country located on the Arabian Peninsula, 385

United Kingdom country in northern Europe occupying most of the British Isles; Great Britain and Northern Ireland, 269

Ural Mountains mountain range in west central Russia that divides Asia from Europe, 321

Uzbekistan country in Central Asia; former Soviet republic, 357

V

Vaduz (47°N 10°E) capital of Liechtenstein, 249

Valletta (36°N 14°E) capital of Malta, 222

Vanuatu South Pacific island country in Melanesia, 663

Vatican City (42°N 12°E) European microstate surrounded by Rome, Italy, 231

Victoria (1°S 33°E) capital of the Seychelles, 511

Vienna (48°N 16°E) capital of Austria, 249

Vientiane (18°N 103°E) capital of Laos, 587

Vietnam country in Southeast Asia, 587

Vilnius (55°N 25°E) capital of Lithuania, 289

Vinson Massif (78°S 87°W) highest mountain (16,066 ft.; 4,897 m) in Antarctica, 663

Vistula River river flowing through Warsaw, Poland, to the Baltic Sea, 289

Vladivostok (43°N 132°E) chief seaport of the Russian Far East, 321

Volga River Europe's longest river; located in west central Russia, 321

W

Wake Island (19°N 167°E) U.S. South Pacific island territory north of the Marshall Islands, 663

Wales part of the United Kingdom occupying a western portion of Great Britain, 269

Wallonia region in southern Belgium, 249

Warsaw (52°N 21°E) capital of Poland, 289

Wellington (41°S 175°E) capital of New Zealand, 671

West Bank area of Palestine west of the Jordan River; occupied by Israel in 1967; political status is in transition, 405

Western Ghats (GAWTS) hills facing the Arabian Sea on the western side of the Deccan Plateau, India, 621

Western Plateau large, flat plain covering more than half of Australia, 671

Western Rift Valley westernmost of two deep troughs beginning near Lake Malawi (Lake Nyasa) in eastern Africa and continuing north into the Red Sea and then into Syria, 495

Western Sahara disputed territory in northwestern Africa; claimed by Morocco, 441

West Siberian Plain region with many marshes east of the Urals in Russia, 321

White Nile part of the Nile River system in eastern Africa, 479

Windhoek (22°S 17°E) capital of Namibia, 511

Witwatersand (WIT-wawt-uhrz-rahnd) a range of low hills in north central South Africa, 511

Wuhan (31°N 114°E) city in south central China, 543

X

Xi River river in southeastern China, 543

Y

Yamoussoukro (7°N 5°W) capital of Côte d'Ivoire, 461

Yangon (Rangoon) (17°N 96°E) capital of Myanmar (Burma), 587

Yaoundé (4°N 12°E) capital of Cameroon, 495

Yekaterinburg (formerly Sverdlovsk) (57°N 61°E) city in the Urals region in Russia, 321

Yemen country located in the southwestern corner of the Arabian Peninsula, 385

Yenisey (yi-ni-SAY) major river in central Russia, 321

Yerevan (40°N 45°E) capital of Armenia, 343

Yugoslavia former Eastern European country of six republics; now including only the republics of Serbia and Montenegro, 289

Z

Zagreb (46°N 16°E) capital of Croatia, 289

Zagros Mountains mountain range of southwestern Iran, 385

Zambezi (zam-BEE-zee) **River** major river in central and southern Africa, 495

Zambia country in central Africa, 495

Zimbabwe (zim-BAH-bway) country in southern Africa, 511

Zurich (47°N 9°E) Switzerland's largest city, 249

GLOSSARY

Phonetic Respelling and Pronunciation Guide

Many of the key terms in this textbook have been respelled to help you pronounce them. The letter combinations used in the respelling throughout the narrative are explained in this phonetic respelling and pronunciation guide. The guide is adapted from *Webster's Tenth New College Dictionary*, *Merriam-Webster's New Geographical Dictionary*, and *Merriam-Webster's New Biographical Dictionary*.

MARK	AS IN	RESPELLING	EXAMPLE
a	alphabet	a	*AL-fuh-bet
ā	Asia	ay	AY-zhuh
ä	cart, top	ah	KAHRT, TAHP
e	let, ten	e	LET, TEN
ē	even, leaf	ee	EE-vuhn, LEEF
i	it, tip, British	i	IT, TIP, BRIT-ish
ī	site, buy, Ohio	y	SYT, BY, oh-HY-oh
	iris	eye	EYE-ris
k	card	k	KAHRD
ō	over, rainbow	oh	OH-vuhr, RAYN-boh
u̇	book, wood	ooh	BOOHK, WOOHD
ȯ	all, orchid	aw	AWL, AWR-kid
ȯi	foil, coin	oy	FOYL, KOYN
au̇	out	ow	OWT
ə	cup, butter	uh	KUHP, BUHT-uhr
ü	rule, food	oo	ROOL, FOOD
yü	few	yoo	FYOO
zh	vision	zh	VIZH-uhn

**A syllable printed in small capital letters receives heavier emphasis than the other syllable(s) in a word.*

A

abdicated Gave up the throne, **326**

absolute authority Monarch's power to make all the governing decisions, **153**

absolute location The exact spot on Earth where something is found, often stated in latitude and longitude, **7**

acculturation The process of cultural changes that result from long-term contact with another society, **77**

acid rain A type of polluted rain, produced when pollution from smokestacks combines with water vapor, **64**

Age of Exploration A period when Europeans were eager to find new and shorter sea routes so that they could trade with India and China, **150**

aggression Warlike action, such as an invasion or an attack, **199**

agrarian A society organized around farming, **353**

air pressure The weight of the air, measured by a barometer, **39**

alliance A formal agreement or treaty among nations formed to advance common interests or causes, **164**

allies Friendly countries that support one another against enemies, **327**

alluvial fan A fan-shaped landform created by deposits of sediment at the base of a mountain, **32**

amber Fossilized tree sap, **291**

animism A religious belief that bodies of water, animals, trees, and other natural objects have spirits, **468**

Antarctic Circle The line of latitude located at 66.5° south of the equator, **20**

antifreeze A substance added to liquid to keep the liquid from turning to ice, **698**

anti-Semitism Hatred of Jews, **202**

apartheid The South African government policy of separation of races, which began to disappear in the 1980s, **196, 518**

aqueduct A bridgelike structure that carries water from the mountains to the cities, **62, 128**

aquifers Underground, water-bearing layers of rock, sand, or gravel, **62**

arable Suitable for growing crops, **546**

archaeology The study of the remains and ruins of past cultures, **464**

archipelago (ahr-kuh-PE-luh-goh) A large group of islands, **588**

Arctic Circle The line of latitude located at 66.5° north of the equator, **20**
arid Dry with little rainfall, **46**
armistice An agreement to stop fighting, **188**
arms race Competition to create and have more advanced weapons, **206**
artesian wells Wells in which water rises toward the surface without being pumped, **672**
asphalt The tarlike material used to pave streets, **407**
atmosphere The layer of gases that surrounds Earth, **22**
atolls Rings of coral surrounding lagoons, **653**
axis An imaginary line that runs from the North Pole through Earth's center to the South Pole, **19**

B

balance of power A way of keeping peace when no one nation or group of nations is more powerful than the others, **169**
bankrupt Having no money, **191**
basins Regions surrounded by mountains or other higher land, **496**
bauxite The most important aluminum ore, **463**
Bedouins Nomadic herders in the deserts of Egypt and Southwest Asia, **448**
bloc A group of nations united under a common idea or for a common purpose, **205**
Boers Afrikaner frontier farmers, white descendents of South Africa's original European colonists, **516**
bog Soft ground that is soaked with water, **280**
boycotted Refused to buy, **195, 628**
bush Lightly populated wilderness areas, such as parts of Australia, **677**

C

cacao (kuh-KOW) A small tree on which cocoa beans grow, **475**
caliph A religious and political ruler in the Muslim world, a title which means "successor to the Prophet Muhammed," **134, 388**
canopy The uppermost layer of a forest's trees where limbs spread out and block out sunlight, **496**
cantons Political and administrative districts in Switzerland, **263**
capital The money and tools needed to make a product, **172**
capitalism Economic system in which private individuals control the factors of production, **175**
caravans Groups of people who travel together for protection, **360**
carrying capacity The maximum number of a species that can be supported by an area, **89**
cartography The art and science of mapmaking, **13**
Casbah The old fortress and central part of some North African cities, **455**
caste system A system in which people's position in society is determined by their birth into a particular caste or group, **632**
cathedrals Huge churches sometimes decorated with elaborate stained-glass windows, **140**
chaebol Huge industrial groups of South Korean companies, banks, and other businesses, **582**
chancellor Germany's head of government, or prime minister, **257**
chivalry A code of behavior including bravery, fairness, loyalty, and integrity, **139**
cholera A life-threatening intestinal infection, **645**
city-states Self-governing cities, such as those of ancient Greece, **124, 234**
civil war A conflict between two or more groups within a country, **503**
civilization A highly complex culture with growing cities and economic activity, **80**
clergy Officials of the church, including the priests, bishops, and pope, **140**
climate The weather conditions in an area over a long period of time, **37**
climatology The field of tracking Earth's larger atmospheric systems, **13**
coalition governments Governments in which several political parties join together to run a country, **239**
collective farms Large farms owned and controlled by the central government, **193**
colony Territory controlled by people from a foreign land, **151**
command economy An economy in which the government owns most of the industries and makes most of the economic decisions, **85**
commercial agriculture A type of farming in which farmers produce food for sale, **80**
communism An economic and political system in which the government owns or controls almost all of the means of production, industries, wages, and prices, **85, 204**
condensation The process by which water changes from a gas into tiny liquid droplets, **24**
constitution A document that outlines basic laws that govern a nation, **154**
constitutional monarchy A government with a monarch as head of state and a parliament or other legislature that makes the laws, **273**
consumer goods Products used at home and in everyday life, **327**
continental shelf The gently sloping underwater land surrounding each continent, **25**
continents Earth's large landmasses, **29**
copper belt A major copper-mining region of central Africa, **497**

coral reef A ridge made up of the skeletal remains of tiny sea animals and found close to shore in warm, tropical waters, **673**
core The inner, solid part of Earth, **28**
cork The bark stripped from a certain type of oak tree and often used as stoppers and insulation, **245**
cosmopolitan Having many foreign influences, **262**
Cossacks Nomadic horsemen who once lived on the Ukrainian frontier, **347**
Counter-Reformation Attempt by the Catholic Church, following the Reformation to return the Church to an emphasis on spiritual matters, **148**
couriers Messengers, **135**
crop rotation A system of growing different crops on the same land over a period of years, **59**
Crusades A long series of battles starting in 1096 between the Christians of Europe and the Muslims to gain control of Palestine, **140**
crust The outer, solid layer of Earth, **28**
culture A learned system of shared beliefs and ways of doing things that guide a person's daily behavior, **75**
culture region Area of the world in which people share certain culture traits, **75**
culture traits Elements of culture, **75**
currents Giant streams of ocean water that move from warm to cold or from cold to warm areas, **41**
cyclones Violent storms with high winds and heavy rain in South Asia, similar to hurricanes in the Caribbean, **640**
czar (ZAHR) Emperor of the Russian Empire, **326**

D

daimyo Local lords in Japan, **136**
Dalits People at the bottom of the Indian caste system who do the work that is considered unclean, **633**
deforestation The destruction or loss of forest area, **61**
deltas Landforms created by the deposits of sediment at the mouths of rivers, **32**
demilitarized zone A buffer zone that serves as a barrier separating two countries, such as North and South Korea, **579**
democracy A political system in which a country's people elect their leaders and rule by majority, **86**
depressions Low areas, **442**
desalinization The process in which the salt is taken out of seawater, **63**
desertification The long-term process of losing soil fertility and plant life, **60**
developed countries Industrialized countries that have strong secondary, tertiary, and quaternary industries, **84**
developing countries Countries in different stages of moving toward development, **84**
dialect A variation of a language, **115, 244, 500**
Diaspora The scattering of the Jewish population from Palestine under Roman rule, **412**
dictator One who rules a country with complete authority, **192, 457**
Diet (DY-uht) Japan's elected legislature, **571**
diffusion The movement of ideas or behaviors from one cultural region to another, **9**
dikes High banks of earth or concrete built along waterfronts to help reduce flooding, **545**
direct democracy Government in which citizens take part in making all decisions, **124**
division of labor Organization of society in which each person performs a specific job, **106**
domestication The growing of a plant or taming of an animal by a people for their own use, **79**
droughts Periods when little rain falls and crops are damaged, **489**
dynasty A ruling family that passes power from one generation to the next, **112, 547**

E

earthquakes Sudden, violent movements along a fracture in the Earth's crust, **29**
ecology The study of connections among different forms of life, **51**
ecosystem All of the plants and animals in an area together with the nonliving parts of their environment, **54**
embargo A limit on trade, **397**
emigrate Leave one's country to move to another, **179**
emperor A ruler of a large empire, **547**
enclaves Countries surrounded or almost surrounded by another country, **512**
endemic species Plants and animals that developed in one particular region of the world, **673**
Enlightenment An era of new ideas from the mid-1600s through the 1700s, **159**
entrepreneurs People who use their money and talents to start a business, **85, 582**
equinoxes The two days of the year when the Sun's rays strike the equator directly, **21**
ergs Great "seas" of sand dunes in the Sahara, **442**
erosion The movement by water, ice, or wind of rocky materials to another location, **32**
ethnic groups Cultural groups of people who share learned beliefs and practices, **75**
evaporation The process by which heated water becomes water vapor and rises into the air, **24**

exclave A part of a country that is separated by territory of other countries, **507**
Exclusive Economic Zones Areas off a country's coast within which the country claims and controls all resources, **695**
exotic rivers Rivers that begin in humid regions and then flow through dry areas, **386**
extinct Something that dies out completely; is no longer present, **51**

F

factors of production The natural resources, money, labor, and capital needed for business operation, **85, 172**
factory A building to house workers and their equipment, **172**
famine A great shortage of food, **278, 583**
fascism A political movement that puts the needs of the nation above the needs of the individual, **192**
fault A fractured surface in Earth's crust where a mass of rock is in motion, **31**
fellahin (fel-uh-HEEN) Egyptian farmers who own very small plots of land, **452**
feudalism A system after the 900s under which most of Europe was organized and governed by local leaders based on land and service, **139**
fief A grant of land, **139**
fjords (fee-AWRDS) Narrow, deep inlets of the sea set between high, rocky cliffs, **270**
floodplain A landform of level ground built by sediment deposited by a river or stream, **32**
food chain A series of organisms in which energy is passed along, **51**
fossil fuels Nonrenewable resources formed from the remains of ancient plants and animals, **68**
fossil water Water that is not being replaced by rainfall, **387**
free enterprise An economic system in which people, not government, decide what to make, sell, or buy, **85**
free port A city in which almost no taxes are placed on goods sold there, **455**
front The unstable weather where a large amount of warm air meets a large amount of cold air, **40**
futon (FOO-tahn) A lightweight cotton mattress, often used in Japan, **574**

G

genocide The planned killing of a race of people, **202**
geography The study of Earth's physical and cultural features, **3**
geothermal energy A renewable energy resource produced from the heat of Earth's interior, **70**
gers (GURHZ) Large, circular tents that are easy to raise, dismantle, and move; used by nomadic herders in Mongolia, **559**
geysers Hot springs that shoot hot water and steam into the air, **284**
glaciers Large, slow-moving sheets or rivers of ice, **33**
glen A Scottish term for a valley, **275**
global warming A slow increase in Earth's average temperature, **64**
globalization Process in which connections around the world increase and cultures around the world share similar practices, **211**
gorge A narrow, steep-walled canyon, **486**
graphite A form of carbon used in pencils and many other products, **652**
Great Depression Period after the Stock Market crashed in 1929 when worldwide business slowed down, banks closed, prices and wages dropped, and many people were out of work, **191**
greenhouse effect The process by which Earth's atmosphere traps heat, **38**
green revolution A program, begun by the Indian government in the 1960s, that encouraged farmers to modernize their methods to produce more food, **635**
griots (GREE-ohz) West African storytellers who pass on the oral histories of their tribes or people, **474**
gross domestic product The value of all goods and services produced within a country, **83**
gross national product The value of all goods and services that a country produces in one year within or outside the country, **83**
groundwater The water from rainfall, rivers, lakes, and melting snow that seeps into the ground, **25**

H

habitation fog A fog caused by fumes and smoke trapped over Siberian cities by very cold weather, **334**
harmattan (HAR-muh-TAN) A dry, dusty wind that blows south from the Sahara during winter, **462**
heavy industry Industry that usually involves manufacturing based on metals, **332**
hieroglyphics A writing system using pictures and symbols, **113**
hieroglyphs (HY-ruh-glifs) Pictures and symbols used to record information in ancient Egypt, **445**
history A written record of human civilization, **107**
Holocaust The mass murder of millions of Jews and other people by the Nazis in World War II, **202, 257**
hominid An early humanlike creature, **103**
homogeneous Sharing the same characteristics, such as ethnicity, **353**

human geography The study of people, past or present, **11**
humanists Scholars of the Renaissance who studied history, poetry, grammar, and other subjects taught in ancient Greece and Rome, **144**
humus Decayed plant and animal matter, **55**
hurricanes Tropical storms that bring violent winds, heavy rain, and high seas, **48**
hydroelectric power A renewable energy resource produced from dams that harness the energy of falling water to power generators, **70**

I

icebergs Large chunks of ice that break away from glaciers and ice shelves and drift into the ocean, **690**
icebreakers Ships that can break up the ice of frozen waterways, allowing other ships to pass through them, **338**
ice shelf A ledge of ice over coastal water, **690**
impressionism A form of art that developed in France in the late 1800s and early 1900s, **255**
individualism A belief in the political and economic independence of individuals, **160**
Indo-European A language family that includes many languages of Europe, such as Germanic, Baltic, and Slavic languages, **292**
Industrial Revolution Period that lasted through the 1700s and 1800s when advances in industry, business, transportation, and communications changed people's lives in almost every way, **171**
intensive cultivation The practice of growing crops on every bit of available land, **575**
irrigation A system of bringing water from rivers to fields through ditches and canals to water crops, **105**
isthmus A neck of land connecting two larger land areas, **28**
ivory A cream-colored material that comes from elephant tusks and is used in making fine jewelry and handicrafts, **498**

J

Japan Current A warm ocean current east of Japan, **567**

K

kampongs A traditional village in Indonesia; also the term for crowded slums around Indonesia's large cities, **598**
karma Among Hindus and Buddhists, the positive or negative force caused by a person's actions, **630**
kimchi Chinese cabbage that has been spiced and pickled; Korea's national dish, **582**
kimonos Traditional Japanese robes, **573**
kiwi (KEE-wee) A flightless bird in New Zealand; a name sometimes applied to New Zealanders, **681**
klongs Canals throughout Bangkok, Thailand, **595**
knight A nobleman who serves as a professional warrior, **139**
krill Tiny shrimplike marine animals that are an important food for larger Antarctic marine life, **691**

L

land bridges Strips of dry land between continents caused by sea levels dropping, **104**
landforms The shapes of land on Earth's surface, **28**
landlocked Completely surrounded by land, with no direct access to the ocean, **358**
lava Magma that has broken through the crust to Earth's surface, **29**
levees Large walls, usually made of dirt, built along rivers to prevent flooding, **10**
light industry Industry that focuses on the production of lightweight goods, such as clothing, **332**
lignite A soft form of coal, **291**
limited government Government in which government leaders are held accountable by citizens through their constitutions and democratic processes, **86**
limited monarchy Monarchy in which the powers of the king were limited by law, **153**
literacy The ability to read and write, **178**
lochs Scottish lakes located in valleys carved by glaciers, **270**
loess (LES) Fine, windblown soil that is good for farming, **251**
Loyalists Colonists loyal to Great Britain, **164**

M

magma Melted rock in the upper mantle of Earth, **28**
mainland A region or country's main landmass, **232, 588**
malaria A deadly disease spread by mosquitoes, **470**
mandate Former territories of defeated World War I countries that were placed under the control of winning countries after the war, **416**
mantle The liquid layer that surrounds Earth's core, **28**

market economy An economy in which consumers help determine what is to be produced by buying or not buying certain goods and services, **85**
marsupials (mahr-SOO-pee-uhls) Animals that carry their young in pouches, **674**
martial law Military rule, **550**
mass production System of producing large numbers of identical items, **175**
medieval Refers to the period from the collapse of the Roman Empire to about 1500, **253**
megalopolis A giant urban area that includes a string of cities that have grown together, **573**
mercantilism Economic theory using colonies to increase a nation's wealth by gaining access to labor and natural resources, **151**
metallic minerals Shiny minerals, like gold and iron, that can conduct heat and electricity, **65**
meteorology The field of forecasting and reporting rainfall, temperatures, and other atmospheric conditions, **13**
middle class Class of skilled workers between the upper class and poor and unskilled workers, **141, 176**
militarism Use of strong armies and the threat of force to gain power, **185**
millet A grain crop that can survive drought, **469**
minerals Inorganic substances that make up Earth's crust, occur naturally, are solids in crystalline form, and have a definite chemical composition, **65**
monarchy A territory ruled by a king who has total power to govern, **153**
monsoon The seasonal shift of air flow and rainfall, which brings alternating wet and dry seasons, **45**
Moors Muslim North Africans, **243**
mosaics (moh-ZAY-iks) Pictures created from tiny pieces of colored stone, **236**
mosques Islamic houses of worship, **134, 364**
most-favored-nation status A status that grants special trade advantages from the United States, **557**
multicultural A mixture of different cultures within the same country or community, **75**
multiple cropping A type of agriculture in which two or three crops are raised each year on the same land, **556**
Muslims Followers of Islam, **388**

N

nationalism The demand for self-rule and a strong feeling of loyalty to one's nation, **181, 264**
NATO North Atlantic Treaty Organization, a military alliance of various European countries, the United States, and Canada, **206, 253**
nature reserves Areas a government has set aside to protect animals, plants, soil, and water, **344**
navigable Water routes that are deep enough and wide enough to be used by ships, **250**
neutral Not taking a side in a dispute or conflict, **282**
New Deal Franklin D. Roosevelt's program to help end the Great Depression, **191**
nirvana Among Buddhists, the escape from the suffering of life, **631**
nobles People who were born into wealthy, powerful families, **139**
nomads People who often move from place to place, **360**
nonmetallic minerals Minerals that lack the characteristics of metal, **66**
nonrenewable resources Resources, such as coal and oil, that cannot be replaced by Earth's natural processes or that are replaced slowly, **65**
North Atlantic Drift A warm ocean current that brings mild temperatures and rain to parts of northern Europe, **271**
nutrients Substances promoting growth, **52**

O

oasis A place in the desert where a spring or well provides water, **358**
oil shale Layered rock that yields oil when heated, **291**
OPEC Organization of Petroleum Exporting Countries, which tries to influence the price of oil on world markets, **389**
oppression Cruel and unjust use of power against others, **166**
oral history Spoken information passed from one generation to the next, **117, 464**
orbit The path an object makes around a central object, such as a planet around the Sun, **17**
Organization of African Unity (OAU) An organization, founded in 1963, that tries to promote cooperation among African countries, **523**
outback Australia's inland region, **674**
Oyashio (oh-YAH-shee-oh) **Current** A cool ocean current east of Japan, **567**
ozone A form of oxygen in the atmosphere that helps protect Earth from harmful solar radiation, **22**

P

pagodas Buddhist temples, **552**
Pangaea (pan-GEE-uh) Earth's single, original super-continent from which today's continents were separated, **31**
pans Low, flat, desert areas of southern Africa into which ancient streams drained, **513**

Parliament An English assembly made up of nobles, clergy, and common people which had the power to pass and enforce laws, **153**
partition To divide a land area into smaller parts, **207**
Patriots Colonists who wanted independence from British rule, **164**
peat Matter made from dead plants, usually mosses, **280**
peninsula Land bordered by water on three sides, **28**
periodic markets Open-air trading markets in central Africa, **497**
permafrost The layer of soil that stays frozen all year in tundra climate regions, **49**
perspective Point of view based on a person's experience and personal understanding, **3**
petroleum An oily liquid that can be refined into gasoline and other fuels and oils, **68**
pharaohs Ancient Egyptian kings, **112, 445**
phosphates Mineral salts containing the element phosphorus; used to make fertilizers, **407**
photosynthesis The process by which plants convert sunlight into chemical energy, **51**
physical geography The study of Earth's natural landscape and physical systems, including the atmosphere, **11**
plain A nearly flat area on Earth's surface, **28**
plant communities Groups of plants that live in the same area, **53**
plant succession The gradual process by which one group of plants that replaces another, **54**
plateau An elevated flatland on Earth's surface, **28**
plate tectonics The theory that Earth's surface is divided into several major, slowly moving plates or pieces, **28**
polar desert A high-latitude region that receives little precipitation, **691**
police state A country in which the government has total control over the people using the police, **192**
pope The bishop of Rome and the head of the Roman Catholic Church, **239**
popular sovereignty Governmental principle based on just laws and on a government created by and subject to the will of the people, **161**
population density The average number of people living within a square mile or square kilometer, **81**
porcelain A type of very fine pottery, **548**
precipitation The process by which water falls back to Earth, **24**
prehistory A time before written records, **103**
primary industries Economic activities that directly involve natural resources or raw materials, such as farming and mining, **83**
protectionism The practice of setting up trade barriers to shield industries at home from foreign competition, **576**
Protestants Christians who broke away from the Catholic Church during the Reformation, **147**
Puritan Member of a group of Protestants who rebelled against the Church of England , **153**

Q

quaternary industries Economic activities that include specialized skills or knowledge and work mostly with information, **83**
Qur'an The holy book of Islam, **389**

R

race A group of people who share inherited physical or biological traits, **76**
rain shadow The dry area on the leeward side of a mountain or mountain range, **44**
reactionaries People who want to return to an earlier political system, **170**
reason Logical thinking, **160**
refineries Factories where crude oil is processed, **68**
reforestation The planting of trees in places where forests have been cut down, **61**
reform Making something better by removing its faults, **180**
Reformation A movement in Europe to reform Christianity in the 1500s, **147, 256**
refugees People who flee to another country, usually for economic or political reasons, **591**
regs Broad, windswept gravel plains in the Sahara, **442**
Reign of Terror Period from 1789 to 1794 when thousands of people died at the guillotine in France, **168**
reincarnation The belief that the human soul is reborn again and again in different bodies, **630**
relative location The position of a place in relation to another place, **7**
Renaissance (re-nuh-SAHNS) French word meaning "rebirth" and referring to a new era of learning that began in Europe in the 1300s, **143, 239**
renewable resources Resources, such as soils and forests, that can be replaced by Earth's natural processes, **59**
republic Government in which voters elect leaders to run the state, **126**
Restoration The reign of Charles II in English history during which the monarchy was restored to power, **154**
revolution One complete orbit around the Sun, **19**
rifts Long, deep valleys with mountains or plateaus on either side, **480**
Roma An ethnic group also known as Gypsies who are descended from people who may have migrated from India to Europe long ago, **302**

rotation One complete spin of Earth on its axis, **19**
rugby A game with British origins; similar to football and soccer, **676**
rural An area of open land that is often used for farming, **4**

S

Sahel (sah-HEL) A dry grasslands region with a steppe climate south of the Sahara, **462**
samurai (SA-muh-ry) Warriors who served Japanese lords, **136, 570**
sanctions An economic or political penalty, such as an embargo, used by one or more countries to force another country to cease an illegal or immoral act, **519**
Sanskrit An early language form in South Asia; used as a sacred language in India today, **625**
satellite A body that orbits a larger body, **18**
scarcity Situation that occurs when demand is greater than supply, **87**
Scientific Revolution The period during the 1500s and 1600s when mathematics and scientific instruments were used to learn more about the natural world, **149**
secede To break away from a country to form another, **472**
secondary industries Economic activities that change raw materials created by primary industries into finished products, **83**
secular Kept separate from religion, as in a secular government or state, **409**
secularism Playing down the importance of religion, **160**
semiarid Relatively dry with small amounts of rain, **62**
sepoys Indian troops commanded by British officers during the colonial era in India, **628**
serfs People who were bound to the land and worked for a lord, **140, 347**
shah An ancient Persian word for king, **134, 400**
shamans Shinto priests, **569**
Shia The second-largest branch of Islam, **388**
Shintoism The earliest known religion of Japan, **569**
shogun "Great General," the highest Japanese warrior rank, **136, 570**
silt Finely ground soil, **443**
sirocco (suh-RAH-koh) A hot, dry wind from North Africa that blows across the Mediterranean to Europe, **233**
smelters Factories that process metal ores, **333**
social contract Idea that government should be based on an agreement made by the people**, 161**
solar energy A renewable energy resource produced from the Sun's heat and light, **71**
solar system The Sun and the objects that move about it, including planets, moons, and asteroids, **17**
solstice The days when the Sun's vertical rays are farthest from the equator, **21**
sorghum A grain crop that can survive drought, **469**
souks Marketplaces in North Africa, **455**
soviet A council of Communists who governed republics and other places in the Soviet Union, **347**
spatial perspective Point of view based on looking at where something is and why it is there, **3**
staple A region or country's main food crop, **470**
steppe (STEP) A wide, flat grasslands region that stretches from Ukraine across southern Russia to Kazakhstan, **323**
steppe climate A dry climate type generally found between desert and wet climate regions, **46**
stock market An organization through which shares of stock in companies are bought and sold**, 190**
storm surges Huge waves of water that are whipped up by fierce winds, particularly from cyclones, hurricanes, and other tropical storms, **640**
stupas Mounds of earth or stones covering the ashes of the Buddha or relics of Buddhist saints, **648**
subcontinent A very large land mass that is smaller than a continent, **113**
subduction The movement of one of Earth's heavier tectonic plates underneath a lighter tectonic plate, **29**
subregions Small areas of a region, **8**
subsistence agriculture A type of farming in which farmers grow just enough food to provide for themselves and their own families, **80**
suburbs Areas just outside or near a city, **179**
suffragettes Women who fought for all women's right to vote**, 180**
sultan The supreme ruler of a Muslim country, **601**
Sunni The largest branch of Islam, **388**
superpowers Powerful countries, **327**
Swahili A Bantu language that is widely spoken in areas of Africa, **484**
symbol A word, shape, color, flag, or other sign that stands for something else, **77**

T

taiga (TY-guh) A forest of evergreen trees growing south of the tundra of Russia, **323**
teak A valuable type of wood; grown in India and Southeast Asia, **623**
terraces Horizontal ridges built into the slopes of steep hillsides to prevent soil loss and aid farming, **60**

tertiary industries Economic activities that handle goods that are ready to be sold to consumers, **83**
textiles Cloth products, **273**
theocracy A government ruled by religious leaders, **400**
the veld (VELT) Open grasslands areas of South Africa, **512**
townships Special areas of crowded clusters of tiny homes for black South Africans living outside cities, **518**
trade surplus The value of exports is greater than the value of imports, **576**
traditional economy Economy based on custom and tradition, **85**
tributary Any smaller stream or river that flows into a larger stream or river, **25**
Tropic of Cancer The line of latitude that is 23.5° north of the equator, **21**
Tropic of Capricorn The line of latitude that is 23.5° south of the equator, **21**
trust territories Areas placed under the temporary control of another country until they set up their own government, **693**
tsetse (TSET-see) **fly** A fly in Africa south of the Sahara that spreads sleeping sickness, a deadly disease, **462**
tsunamis (tsooh-NAH-mees) Huge waves created by undersea tectonic activity, such as earthquakes, **567**
tundra climate A cold region with low rainfall, lying generally between subarctic and polar climate regions, **49**
typhoons Tropical storms that bring violent winds, heavy rain, and high seas, **48**

U

U-boats German submarines during World War I, **187**
uninhabitable Not capable of supporting human settlement, **284**
unlimited governments Governments that have total control over their citizens, **86**
urban An area that contains a city, **4**

V

vassals People who held land from a feudal lord and received protection in return for service to the lord, especially in battle, **139**
vernacular Everyday speech which varies from place to place, **142**

W

wadis Dry steambeds in Southeast Asia, **387**
water cycle The circulation of water from Earth's surface to the atmosphere and back, **23**
water vapor The gaseous form of water, **23**
weather The condition of the atmosphere at a given place and time, **37**
weathering The process of breaking rocks into smaller pieces through heat, water, or other means, **31**
work ethic The belief that work in itself is worthwhile, **576**
working class Unskilled and semi-skilled workers with low-paying jobs, **177**

Y

yurt A movable round house of wool felt mats over a wood frame, **364**

Z

Zionism The movement to establish a Jewish country or community in Palestine, **413**
zonal The climates of West Africa, which stretch east to west in bands, **462**

SPANISH GLOSSARY

Phonetic Respelling and Pronunciation Guide

Many of the key terms in this textbook have been respelled to help you pronounce them. The letter combinations used in the respelling throughout the narrative are explained in this phonetic respelling and pronunciation guide. The guide is adapted from *Webster's Tenth New College Dictionary, Merriam-Webster's New Geographical Dictionary,* and *Merriam-Webster's New Biographical Dictionary.*

MARK	AS IN	RESPELLING	EXAMPLE
a	alphabet	a	*AL-fuh-bet
ā	Asia	ay	AY-zhuh
ä	cart, top	ah	KAHRT, TAHP
e	let, ten	e	LET, TEN
ē	even, leaf	ee	EE-vuhn, LEEF
i	it, tip, British	i	IT, TIP, BRIT-ish
ī	site, buy, Ohio	y	SYT, BY, oh-HY-oh
	iris	eye	EYE-ris
k	card	k	KAHRD
ō	over, rainbow	oh	OH-vuhr, RAYN-boh
u̇	book, wood	ooh	BOOHK, WOOHD
ȯ	all, orchid	aw	AWL, AWR-kid
ȯi	foil, coin	oy	FOYL, KOYN
au̇	out	ow	OWT
ə	cup, butter	uh	KUHP, BUHT-uhr
ü	rule, food	oo	ROOL, FOOD
yü	few	yoo	FYOO
zh	vision	zh	VIZH-uhn

**A syllable printed in small capital letters receives heavier emphasis than the other syllable(s) in a word.*

A

abdicated/abdicar Renunciar al trono, **326**
absolute authority/autoridad absoluta Poder de un rey o reina para tomar todas las decisiones de gobierno, **153**
absolute location/posición exacta Lugar exacto de la tierra donde se localiza un punto, por lo general definido en términos de latitud y longitud, **7**
acculturation/aculturación Proceso de asimilación de una cultura a largo plazo por el contacto con otra sociedad, **77**
acid rain/lluvia ácida Tipo de lluvia contaminada que se produce cuando partículas de contaminación del aire se combinan con el vapor de agua de la atmósfera, **64**
Age of Exploration/Edad de la exploración Periodo en que los europeos estaban ansiosos por hallar rutas nuevas y más cortas para comerciar con la India y China, **150**
aggression/agresión Acción militar, **199**
agrarian/agrario Sociedad basada en la agricultura, **353**
air pressure/presión atmosférica Peso del aire, se mide con un barómetro, **39**
alliance/alianza Acuerdo entre diferentes países para respaldarse los temas de interès y las causas, **164**
allies/aliados Países que se apoyan entre sí para defenderse de sus enemigos, **327**
alluvial fan/abanico aluvial Accidente geográfico en forma de abanico que se origina por la acumulación de sedimentos en la base de una montaña, **32**
amber/ámbar Savia de árbol fosilizada, **291**
animism/animismo Creencia religiosa que explica que los cuerpos de agua, los animales, los árboles y otros objetos de la naturaleza tienen un espíritu, **468**
Antarctic Circle/círculo antártico Meridiano localizado a 66.5° al sur del ecuador, **20**
antifreeze/anticongelante Sustancia que se agrega a un líquido para evitar que se congele, **698**
anti-Semitism/antisemitismo Hatred of Jews, **202**
apartheid The South African government policy of separation of races, which began to disappear in the 1980s, **196 518**
aqueduct/acueducto Canale artificiale usado para transportar agua, **62, 128**
aquifers/acuíferos Capas subterráneas de roca, arena y grava en las que se almacena el agua, **62**
arable/cultivable Tierra con características que favorecen el cultivo, **546**
archaeology/arqueología Estudio de los restos de culturas pasadas, **464**
archipelago/archpiélago Grupo grande de islas, **588**
Arctic Circle/círculo ártico Meridiano localizado a 66.5° al norte del ecuador, **20**
arid/árido Territorio done la lluvia es muy escasa, **46**
armistice/armisticio Tregua, **188**
arms race/carrera armamentista Competencia entre países para producir y tener armas más avanzadas, **206**
artesian wells/pozos artesianos Pozos en los que el agua sube a la superficie de la tierra sin ser impulsada por medios artificiales, **672**

asphalt/asfalto Material oscuro usado para pavimentar calles, **407**
atmosphere/atmósfera Capa de gases que rodea a la tierra, **22**
atolls/atolones Anillos de coral que se forman alrededor de las lagunas, **653**
axis/eje Línea imaginaria que corre del polo norte al polo sur, pasando por el centro de la Tierra, **19**

B

balance of power/equilibrio de poder Condición que surge cuando varios países o alianzas mantienen niveles tan similares de poder evitar guerras, **169**
bankrupt/bancarrota Sin dinero, **191**
basins/cuencas Regiones rodeadas por montañas u otras tierras altas, **496**
bauxite/bauxita El mineral con contenido de aluminio más importante, **463**
Bedouins/beduinos Ganaderos nómadas del desierto de Egipto y el sudoeste de Asia, **448**
bloc/bloque Grupo de naciones unidas por una idea o un propósito común, **205**
Boers/boers agricultores africanos de raza blanca, descendientes de los primeros colonizadores europeos de Sudáfrica, **516**
bog/ciénaga Tierra suave, humedecida por el agua, **280**
boycott/boicot Rechazo de compra, **195, 628**
bush Zonas salvajes de escasa población, como ciertas regiones de Australia, **677**

C

cacao/cacas Árbol pequeño que produce los granos de cacao, **475**
caliph/califa Líder político y religioso del mundo musulmán, **134, 388**
canopy/dosel Capa superior de un bosque espeso en el que las ramas se entrelazan, bloqueando el paso de la luz solar, **496**
cantons/cantones Distritos políticos y administrativos de Suiza, **263**
capital/capital Dinero ganado, ahorrado e invertido para conseguir ganancias, **172**
capitalism/capitalismo Sistema económico en el que los negocios, las industrias y los recursos son de propiedad privada, **175**
caravans/caravanas Grupos de personas que viajan juntas por razones de seguridad, **360**
carrying capacity/capacidad de carga Número máximo de especies que puede haber en una zona determinada, **89**
cartography/cartografía Arte y ciencia de la elaboración de mapas, **13**
Casbah/casbah Antigua fortaleza y centro de las ciudades del norte de África, **455**
caste system/sistema de castas Sistema en el que la posición de una persona en la sociedad es determinada por el nivel social del grupo en el que nace, **632**
cathedrals/catedrales Iglesias grandes, **140**
chaebol/chaebol Enormes grupos industriales formados por compañías, bancos y otros negocias en Corea del Sur, **582**
chancellor/canciller jefe de gobierno o primer ministro alemán, **257**
chivalry/caballerosidad Código o sistema medieval de cavellería, **139**
cholera/cólera Infección intestinal seria que puede provocar a muerta, **645**
city-states/ciudades estado Ciudades con un sistema de autogobierno, como en la antigua Grecia, **124, 234**
civil war/guerra civil conflicto entre dos o más grupos dentro de un país, **503**
civilization/civilización Cultura altamente compleja con grandes ciudades y abundante actividad económica, **80**
clergy/clérigo Oficiante de la Iglesia, **140**
climate/clima condiciones meteorológicas registradas en un periodo largo, **37**
climatology/climatología Registro de los sistemas atmosféricos de la Tierra, **13**
coalition governments/gobiernos de coalición Gobiernos en los que la administración del país es regida por varios partidos políticos a la vez, **239**
collective farms/fincas colectivas Tierras mancomunadas en grandes fincas, donde las personas trabajan juntas como un grupo, **193**
colony/colonia Territorio controlado por personas de otro país, **151**
command economy/economía autoritaria Economía en la que el gobierno es propietario de la mayor parte de las industrias y toma la mayoría de las decisiones en materia de economía, **85**
commercial agriculture/agricultura comercial Tipo de agricultura cuya producción es exclusiva para la venta, **80**
communism/comunismo Sistema politica y económico en el cual los gobiernos poseen los medios de producción y controlan el planeamiento de la economia, **85, 204**
condensation/condensación Proceso mediante el cual el agua cambia de estado gaseoso y forma pequeñas gotas, **24**
constitution/constitución Documento que contiene las leyes y principios básicos que gobiernan una nación, **154**
constitutional monarchy/monarquía constitucional Gobierno que cuenta con un monarca como jefe de estado y un parlamento o grupo legislador similar para la aprobación de leyes, **273**
consumer goods/blenes de consumo Productos usados en la vida cotidiana, **327**
continental shelf/plataforma continental zona costera de pendiente suave que bordea a todos los continentes, **25**
continents/continentes Grandes masas de territorio sobre la Tierra, **29**
Copper Belt/Región del cobre Importante región minera de producción de cobre localizada en la parte central de África, **497**
coral reef/arrecife coralino Formaciones creadas por la acumulación de los restos de animales marinos diminutos cerca de las costas en las aguas templadas de las regiones tropicales, **673**

core/núcleo Parte sólida del interior de la Tierra, **28**
cork/corcho Corteza extraída de cierto tipo de roble, usada principalmente como material de bloqueo y aislante, **245**
cosmopolitan/cosmopolita Que tiene influencia de muchas culturas, **262**
Cossacks/cosacos Arrieros nómadas que habitaban en la región fronteriza de Ucrania, **347**
Counter-Reformation/Contrareforma Intento de la Iglesia Católica, luego de la Reforma, por devolver a la Iglesia a un énfasis en asuntos espirituales, **148**
courier/correo Persona que tiene por oficio llevar cartas y mensajes de un lugar a otro, **135**
crop rotation/rotación de cultivos Sistema agrícola en el que se siembran productos differentes en periodos específicos, **59**
Crusades/Cruzadas Expediciones hecas por los cristianos para recuperar la Tierra Santa de los musulmanes, **140**
crust/coteza Capa sólida de la superficie de la Tierra, **28**
culture/cultura Sistema de creencias y costumbres comunes que guía la conducta cotidiana de las personas, **75**
culture region/región cultural Región del mundo en la que se comparten ciertos rasgos culturales, **75**
culture traits/rasgos culturales Características de una cultura, **75**
currents/corrientes Enormes corrientes del océano que transportan agua tibia a las regiones frías y viceversa, **41**
cyclones/ciclones Tormentas violentas con fuertes lluvias, paracidas a los huracanes del Caribe, comunes en el sur de Asia, **640**
czar/zar Emperador ruso, **326**

D

daimyo/daimyo Señores o amos feudales en japón, **136**
Dalits/dalits Personas de la clase social más baja de la India, cuyas actividades son consideradas poco salubres, **633**
deforestation/deforestación Destrucción o pérdida de un área boscosa, **61**
deltas/deltas Formaciones creadas por la acumulación de sedimentos en las desembocadura de los ríos, **32**
demilitarized zone/zona desmilitarizada Zona de protección que sirve como barrera entre dos países en conflicto, como Corea del Norte y Corea del Sur, **579**
democracy/democracia Sistema político en el que la población elige a sus líderes mediante el voto de mayoría, **86**
depressions/cavidads Zonas de elevación muy baja, **442**
desalinization/desalinización Proceso mediante el cual se extrae la sal del agua de mar, **63**
desertification/desertificación Proceso a largo plazo en el que el suelo pierde su fertilidad y vegetación, **60**
developed countries/países desarrollados Países industrializados que cuentan con industrias primarias, secundarias, terciarias y cuaternarias, **84**
developing countries/países en vaís de desarrollo países que se encuentran en alguna etapa de su proceso de desarrollo, **84**
dialect/dialecto Variación de un idioma, **115, 244, 500**
Diaspora/Diáspora Dispersión de la población judía que emigró de territorio palestino durante el imperio romano, **412**
dictator/dictadore Persona que ejercen total autoridad sobre un gobierno, **192, 457**
Diet/Dieta Legislatura electa de Japón, **571**
diffusion/difusión Extensión de ideas o conducta de una región cultura a otra, **9**
dikes/diques Grandes muros de tierra o concreto construidos para contener a un cuerpo de agua y evitar inundaciones, **545**
direct democracy/democracia directa Forma de democracia en la que todos los ciudadanos participan directamente en la toma de decisiones, **124**
division of labor/división de labores Característa de las civilizaciones en la cual diferentes personas realizan diferentes trabajos, **106**
diversify/diversificar En la agircultura, se refiere a la siembra de varios productos y uno solo, **242**
domestication/domesticación Cuidado de una planta o animal para uso personal, **79**
droughts/sequías Periodos en los que los cultivos sufren daños debido a la escasez de lluvia, **489**
dynasty/dinastía Familia que gobierna y hereda el podor de generación en generación, **112, 547**

E

earthquakes/terremotos Movimientos repentinos y fuertes que se producen en las fisuras de la superficie de la tierra, **29**
ecology/ecología Estudio de las relacions entre las diversas formas de vida del planeta, **51**
ecosystem/ecosistema Conjunto de plantas y animales que habitan una región, junto con los elementos no vivos de ese entorno, **54**
embargo/embargo Límite impuesto a las relaciones comerciales, **397**
emigrate/emigrar Dejar un país para ir a otro, **179**
emperor/emperador Gobernante supremo de vastos territorios, **547**
enclaves/enclaves Países rodeados en su mayor parte o en su totalidad por otro país, **512**
endemic species/especies endémicas Plantas y animales que se desarrollan en una región particular del planeta, **673**
Enlightenment/Ilustración Período en los años 1700, cuando los filósofos creían que podían aplicar el método cientifico y el uso de la razón para explicar de manera lógica la naturaleza humana, **159**
entrepeneurs/empresarios Personas que usan su dinero y su talento para iniciar un negocio, **85, 582**
equinoxes/equinoccios Los dos días del año en que los rayos de sol caen directamente sobre el ecuador, **21**
ergs/ergs Grandes "mares" de arena formados por las dunas del desierto del Sahara, **442**
erosion/erosión Desplazamiento de agua, hielo, viento o minerales a otro lugar, **32**
ethnic groups/grupos étnicos Grupos culturales que comparten creencias y prácticas comunes, **75**
evaporation/evaporación Proceso mediante el cual el agua se convierte en vapor y se eleva en el aire, **24**

R21

exclave/exclave Parte de un país separada por el territorio de uno o más países, **507**
Exclusive Economic Zones/zonas de exclusividad económica Zonas costeras de un país en las que éste tiene derecho a extraer y controlar los recursos existentes, **695**
exotic rivers/ríos exóticos Ríos originados en regiones húmedas que fluyen a zonas más secas, **386**
extinct/extinto Que deja de existir por completo, **51**

F

factors of production/factores de producción Los recursos naturales, el dinero, el trabajo y los empresarios que se necesitan para las operaciones de negocios, **85, 172**
factory/factoría Fábrica o conjunto de fábricas donde se elabora o produce algún producto, **172**
famine/hambruna Gran escasez de alimeno, **278, 583**
fascism/fascismo Teoría politica que demanda la creación de un gobierno fuerte encabezado por un solo individuo donde el estado sea más importante que el individuo, **192**
fault/falla Fractura de la superficie de la tierra que causa el movimiento de grandes masas de rocas, **31**
fellahin/fellahin agricultores egipcios dueños de pequeñas porciones de terreno, **452**
feudalism/feudalismo Sistema de gobierno local basado en la concesión de tierras como pago por lealtad, ayuda militar y otros servicios, **139**
fief/feudo Concesión de tierras de un amo a su vasayo, **139**
fjords/fiordos Grietas estrechas y profundas localizadas entre altos acantilados donde se acumula el agua de mar, **270**
floodplain/llanura aluvial Especie de plataforma a nivel de la tierra, formada por la acumulación de los sedimentos de una corriente de agua, **32**
food chain/cadena alimenticia Serie de organismos que proveen alimento y energía unos o otros, **51**
fossil fuels/combustibles fósiles Recursos no renovables formados por restos muy antiguos de plantas y animales, **68**
fossil water/aguas fósiles Agua que no es reemplazada por el agua de lluvia, **387**
free enterprise/libre empresa Sistema económico en el que las personas, y no el gobierno, deciden qué productos fabrican, venden y compran, **85**
free port/puerto libre Cuidad en la que casi no se aplican impuestos a los productos que allí se adquieren, **455**
front/frente Inestabilidad climatológica en la que una gran masa de aire tibio choca con una gran masa de aire frío, **40**
futon/futón Especie de sofá ligero, también usado como cama, muy común en Japón, **574**

G

gauchos/gauchos Arrieros argentinos, **351**
genocide/genocidio Aniquilamiento intencional de un pueblo, **202**
geography/geografía Estudio de las características físicas y culturales de la Tierra, **3**
geothermal energy/energía geotérmica Fuente energética no removable producida por el calor del interior de la tierra, **70**
gers/gers Grandes tiendas circulares que son fáciles de armar, desarmar y transportar; usadas por los ganaderos nómades de Mongolia, **559**
geysers/géiseres Manantiales que lanzan chorros de agua caliente y vapor a gran altura, **284**
glaciers/glaciares Grandes bloques de hielo que se desplazan con lenitud sobre el agua, **33**
glen/glen Término de origen escocés que se sinónimo de valle, **275**
global warming/calentamiento global Aumento lento y constante de la temperatura de la Tierra, **64**
globalization/globalización Proceso mediante el que as comunicaciones alrededor del mundo se han incrementado haciendo a las culturas más parecidas, **211**
gorge/garganta Cañón estrecho y muy profundo, **486**
graphite/grafito Tipo de carbón usado para fabricar puntas de lápices y muchos otros productos, **652**
Great Depression/Gran Depresión Depresión mundial a principios de los años 1930, cuando los salarios cayeron, la actividad comercial bajó y hubo mucho desempleo, **191**
greenhouse effect/efecto invernadero Proceso mediante el cual la atmósfera terrestre atrapa el calor de su superficie, **38**
green revolution/revolución verde Programa iniciado por el gobierno de la India en la década de 1960 para modernizar los métodos y producir mayor cantidad de alimento, **635**
griots/griots Narradores de historias de África Occidental que pasan sus tradiciones tribales de manera oral, **474**
gross domestic product/producto interno bruto Valor de todos los bienes y servicios producidos en un país, **83**
gross national product/producto nacional bruto Valor de todos los bienes y servicios producidos en un año por un país, dentro o fuera de sus límites, **83**
groundwater/agua subterránea Agua de lluva, ríos, lagos y nieve derretida que se filtra al subsuelo, **25**

H

habitation fog/humo residente Especie de niebla producida por el humo atrapado en la atmósfera de las cuidades siberianas debido al intenso frío, **334**
harmattan/harmattan Viento seco y polvoso que sopla con fuerza hacia el sur durante el envierno en el desierto del Sahara, **462**
heavy industry/industra pesada Industria basada en la manufactura de metales, **332**
hieroglyphs/jeroglificos Imágenes y simbolos usados para registrar información en el antiguo Egipto, **445**
hieroglyphics/jeroglificos Forma antigua de escritura con imágenes y símbolos usados para registrar información, **113**

history/historia Registro escrito de la civilización humana, **107**

Holocaust/haulocausto Asesinato masivo de millones de judíos y personas de otros grupos a manos. de los nazis durante la Segunda Guerra Mundial, **202, 257**

hominid/homínido Primera criatura similares al hombre, **103**

homogeneous/homogéneo Agrupamiento que comparte ciertas características, como el origin étnico, **353**

human geography/geografía humana estudio del pasado y presente de la humanidad, **11**

humanists/humanistas Filósofos del Renacimiento que hacían énfasis en la individualidad, los logros personales y la razón, **144**

humus/humus Materia vegetal o animal en descomposición, **55**

hurricanes/huracanes Tormentas tropicales con intensos vientos, fuertes lluvias y altas mareas, **48**

hydroelectric power/energía hidroeléctrica Fuente energética renovable producida en generadores impulsados por caídas de agua, **70**

I

icebergs/icebergs Grandes bloques de hielo que se separan de los glaciales y flotan a la deriva en el océano, **690**

icebreakers/rompehielos Barcos que rompen la capa de hielo que se forma en la superficie de algunos cuerpos de agua para permitir el paso de otras embarcaciones, **338**

ice shelf/capa de hielo Cubierta de hielo que se forma en aguas costeras, **690**

impressionism/impresionismo Forma de arte desarrollada en Francia a finales del siglo XIX y principios del siglo XX, **255**

individualism/individualismo Creencia en la independencia económica y política de los individuos, **160**

Indo-European/Indoeuropeo Familia que incluye muchos idiomas europeos como el germánico, el báltico y los dialectos eslavos, **292**

Industrial Revolution/Revolución Industrial Cambios producidos a principios de los años 1700, cuando las maquinarias empezaban a hacer mucho del trabajo que las personas tenían que hacer antes, **171**

intensive cultivation/cultivo intenso Cultivo de productos en cualquier terreno disponible, **575**

irrigation/riego Proceso mediante el cual el agua se hace llegar a los cultivos de manera artificial, **105**

isthmus/istmo Franja de tierra que conecta dos áreas de mayor tamaño, **28**

ivory/marfil Material de color crema extraído de los colmillos de los elefantes que se usa para fabricar joyería y artículos decorativos, **498**

J

Japan Current/Corriente de Japón corriente oceánica de aguas tibias que fluye al este de Japón, **567**

K

kampongs/kampongs Aldea tradicional de Indonesía; el término también se usa para referirse a las grandes poblaciones humanas establecidas en los alrededores de las ciudades de Indonesia, **598**

karma/karma Para los hinduistas y budistas, es la fuerza positiva o negativa generada por las acciones de una persona, **630**

kimchi/kimchi Especie de col china aderezada y avinagrada que se sirve como plato tradicional en Corea, **582**

kimonos/kimonos vestidos tradicionales japoneses, **573**

kiwi/kiwi Ave que no vuela, originaria de Nueva Zelanda; a veces, este término se usa para referirse a los neozelandeses, **681**

klongs/klongs Canales de Bangkok, una ciudad de Tailandia, **595**

knight/caballero Guerrero profesional noble, **139**

krill/krill Animales marinos diminutos que son una importante fuente alimenticia para otras especies marinas del Océano Atlántico, **691**

L

land bridges/puentes de terreno Franjas de terreno seco que conecta grandes masas de tierra, **104**

landforms/accidentes geográficos Forma de la tierra en differentes partes de la superficie, **28**

landlocked/sin salida al mar Zona rodeada de agua por completo y sin acceso directo al océano, **358**

lava/lava Magma que emerge del interior de la tierra por un orificio de la corteza, **29**

levees/diques Paredes altas, por lo general de tierra, que se construyen a la orilla de un río para prevenir inundaciones, **10**

light industry/indutria ligera Industria que se enfoca en la manufactura de objetos ligeros como la ropa, **332**

lignite/lignita Tipo de carbón suave, **291**

limited government/gobierno limitado Gobierno en el que los ciudadanos hacen responables a los dirigentes por medio de la constitución y los procesos democráticos, **86**

limited monarchy/monarquía limitada Sistema de gobierno dirigido por una reina o un rey que no tiene el control absoluto de un país, **153**

literacy/alfabetismo Capacidad de leer y escribir, **178**

lochs/lagos Lagos escoceses enclavados en valles labrados por los glaciales, **270**

loess/limo Suelo fino de arenisca, excelente para la agricultura, **251**

Loyalists/Loyalistas, Leales Colonistas americanos que se oponían a independización de Intalterra, **164**

M

magma/magma Roca fundida que se localiza en el manto superior de la tierra, **28**

mainland/región continental Región donde se localiza la mayor porción de terreno de un país, **232, 588**
malaria/malaria Enfermedad mortal que se difunde por medio de los mosquitos, **470**
mandate/mandato Territorios que formaban parte de los países derrotados en la Primera Guerra Mundial, y que pasaron a control de los países vencedores, **416**
mantle/manto Capa líquida que rodea al centro de la Tierra, **28**
market economy/economía de mercado Tipo de economía en la qué los consumidores ayudan a determinar qué productos se fabrican al comprar o rechazar ciertos bienes y servicios, **85**
marsupials/marsupiales Animales que transportan a sus crías en un saco, **674**
martial law/ley marcial Ley militar, **550**
mass production/producción en serie Sistema de producción de grandes cantidades de productos idénticos, **175**
medieval/medieval Periodo de colapso del imperio romano, aproximadamente en el año 1,500 de nuestra era, **253**
megalopolis/megalópolis Enorme zona urbana que abarca una serie de ciudades que se han desarrollado juntas, **573**
mercantilism/mercantilismo Creación y conservación de riquezas mediante un control minucioso de intercambios comerciales, **151**
metallic minerals/minerales metálicos Minerales brillantes, como el oro y el hierro, que conducen el calor y la electricidad, **65**
meteorology/meteorología Predicción y registro de lluvias, temperaturas y otras condiciones atmosféricas, **13**
middle class/clase media Clase formada por comerciantes, patronos de pequeña y mediana industria y profesiones liberales. Está entre la clase noble y la clase campesina en la Edad Media, **141, 176**
militarism/militarismo Uso de armamento pesado y amenazas para obtener poder, **185**
millet/mijo Tipo de cultivo resistente a las sequías, **469**
minerals/minerales Sustancias inorgánicas que conforman la corteza de la tierra; en su medio natural, aparecen en forma cristalina y tienen una composición química definida, **65**
monarchy/monarquía Sistema de gobierno dirigido por un rey o una reina, **153**
monsoon/monzón Cambio de corrientes aire y lluvias que produce temporadas alternadas de humedad y sequía, **45**
Moors/moros Musulmanes del norte de África, **243**
mosaics/mosaicos Imágenes creadas con pequeños fragmentos de piedras coloreadas, **236**
mosques/mezquitas Casas de adoración islámica, **134, 364**
most-favored-nation status/estatus de nación favorecida Estatus que otorga privilegios de intercambio comercial entre Estados Unidos y otros países, **557**
multicultural/multicultural Mezcla de culturas en un mismo país o comunidad, **75**
multiple cropping/cultivo múltiple Tipo de agricultura en la que se producen dos o tres cultivos cada año en las mismas tierras, **556**
Muslims/Musulmanes Seguidores del Islam, **388**

N

nationalism/nacionalismo Demanda de autogo-bierno y fuerte sentimiento de lealtad hacia una nación, **181, 264**
NATO/OTAN (Organización del Tratado del Atlántico Norte); alianza militar formada por varios países europeos, Estados Unidos y Canadá, **206, 253**
nature reserves/reservas naturales Zonas asignadas por el gobierno para la protección de animales, plantas, suelo y agua, **344**
navigable/navegable Rutas acuáticas de profunidad suficiente para la navegación de barcos, **250**
neutral/neutral Que no toma ningún partido en una disputa o conflicto, **282**
New Deal//New Deal Pragrama del presidente Franklin D. Roosevelt en el que el gobierno federal estableció un amplio programa de obras públicas para crear empleo y conceder dinero a cada estado para sus necesidades, **191**
nirvana/nirvana Para los budistas, es el escape de los sufrimientos de la vida, **631**
nobles/nobles Personas que nacen entre familias ricas y poderosas, **139**
nomads/nómadas Personas que se mudan frecuentemente de un lugar a otro, **360**
nonmetallic minerals/minerales no metálicos Minerales que no tienen las características de los metales, **66**
nonrenewable resources/recursos no renovables Recursos, como el carbón mineral y petróleo, que no pueden reemplazarse a corto plazo por medios naturales, **65**
North Atlantic Drift/Corriente del Atlántico Norte corriente de aguas tibias que aumenta la temperatura y genera lluvias en el norte de Europa, **271**
nutrients/nutrientes sustancias que favorecen el crecimiento, **52**

O

oasis/oasis Lugar del desierto donde un manantial proporciona una fuente natural de agua, **358**
oil shale/pizarra petrolífera Capa de roca que al calentarse produce petróleo, **291**
OPEC/OPEP Organización de países exportadores de petróleo; grupo formado para ejercer influencia en el precio de los mercados petroleros mundiales, **389**
oppression/opresión Uso de poder cruel e injusto contra otros, **166**
oral history/historia oral Información oral transmitida de una persona a otra y de generación en generación, **117, 464**
orbit/órbita Trayectoria que sigue un objeto que gira alrededor de otro objeto (como la Tierra alrededor del Sol), **17**
Organization of African Unity (OAU)/Organización Africa Unida (OAU) Grupo fundado en 1963 para promover la cooperación entre los países africanos, **523**
outback/*Outback* (Interior) Región interior de Australia, **674**

Oyashio Current/Corriente Oyashio Corriente de aguas frías que fluye al este de Japón, **567**

ozone/ozono Forma del oxígeno en la atmósfera que ayuda a proteger a la Tierra de los daños que produce la radiación solar, **22**

P

pagodas/pagodas Templos budistas, **552**

Pangaea/Pangaea Supercontinente original y único del que se separaron los continentes actuales, **31**

pans/pans Regiones desérticas del sur de África, de terreno bajo y plano, en las que desaguaban corrientes antiguas, **513**

Parliament/Parlamento Órgano británico de legislación, **153**

partition/repartir Separar un terreno formando grupos más pequeños, **195**

Patriots/Patriotas Colonistas americanos que favorecieron con la independencia de Inglaterra, **164**

peat/turba Sustancia formada por plantas muertas, por lo general musgos, **280**

peninsula/península Tierra rodeada de agua en tres lados, **28**

periodic markets/mercados periódicos Mercados al aire libre en el África central, **497**

permafrost/permafrost Capa de suelo que permanece congelada todo el año en las regiones con clima de la tundra, **49**

perspective/perspectiva Punto de vista basado en la experiencia y la comprensión de una persona, **3**

petroleum/petróleo Líquido graso que al refinarse produce gasolina y otros combustibles y aceites, **68**

pharaohs/faraones Reyes del antiguo Egipto, **112, 445**

phosphates/fosfatos Sales minerales que contienen el elemento fósforo; se usa para hacer fertilizantes, **407**

photosynthesis/fotosíntesis Proceso por el que las plantas convierten la luz solar en energía quimica, **51**

physical geography/geografía física Estudio del paisaje natural y los sistemas físicos de la Tierra, entre ellos la atmósfera, **11**

plain/planicie Área casi plana de la superficie terrestre, **28**

plant communities/comunidades de plantas Grupos de plantas que viven en la misma zona, **53**

plant succession/sucesión de plantas Proceso gradual por el que un grupo de plantas reemplaza a otro, **54**

plateau/meseta Terreno plano y elevado sobre la superficie terrestre, **28**

plate tectonics/tectónica de placas Teoría de que la superficie terrestre está dividida en varias placas enormes que se mueven lentamente, **28**

polar desert/desierto polar Región de latitudes altas donde cae muy poca precipitación, **691**

police state/estado totalitario País en que el gobierno tiene un control total sobre la vida de las personas **192**

pope/papa Obisco de Roma y líder de la Iglesia católica romana, **239**

popular sovereignty/Soberanía Popular Principio gubernamental basado en leyes justas, y en un gobierno creado y sujeto a la voluntad del pueblo, **161**

population density/densidad de población Número promedio de personas que viven en una milla cuadrada o un kilómetro cuadrado, **81**

porcelain/porcelana Tipo de vajilla muy fina, **548**

precipitation/precipitación Proceso por el que el agua vuelve de regreso a la Tierra, **24**

prehistory/prehistoria Tiempo antiguo del que no se conservan registros escritos, **103**

primary industries/industrias primarias Actividades económicas que involucran directamente recursos naturales o materia prima, tales como la agricultura y la minería, **83**

protectionism/proteccionismo Práctica de poner barreras comerciales para proteger a las industrias nacionales de la competencia de las industrias extranjeras, **576**

Protestants/protestantes Reformistas que protestaban por la realización de ciertas prácticas de la Iglesia Católica, **147**

Puritan/Puritano Persona que aspiraban a una doctrina más pura que la propuesta por la Iglesia Católica Inglesa, **153**

Q

quaternary industries/industrias cuaternarias Actividades económicas que abarca destrezas o conocimiento especializados, y trabajan generalmente con la información, **83**

Qur'an/Corán El libro sagra do del Islam, **389**

R

race/raza Grupo de personas que comparten características físicas o biológicas heredades, **76**

rain shadow/barrera montañosa Área seca en el sotavento de una montaña o de una cordillera, **44**

reactionaries/reaccionarios Extremistas que se oponen al cambio, y desean deshacer ciertos cambios, **170**

reason/razón Razonamiento lógico, **160**

refineries/refinerías Fábricas donde se procesa el petróleo crudo, **68**

reforestation/reforestación Plantación de árboles donde los bosques han sido talados, **61**

reform/reformar Mejorar algo reparando sus errores, **180**

Reformation/Reforma Movimiento europeo del siglo XVI para reformar el cristianismo, **147, 256**

refugees/refugiados Personas que han escapado a otro país, generalmente por razones económicas o políticas, **591**

regs/regs Extensas planicies de grava azotadas por el viento, en el Sahara, **442**

Reign of Terror/Régimen de Terror Período durante la Revolución Francesca cuando la Convención Nacional trabajó para suprimir toda oposición, **168**

reincarnation/reencarnación Creencia de que el alma humana vuelva a nacer una y otro vez en diferentes cuerpos, **630**

relative location/ubicación relativa Posición de un lugar en relación con otro, **7**

Renaissance/Renaissance Palabra francesa que significa "renacimiento" y se refiere a una nueva era de conocimiento que empieza en Europa en el 1300s, **143, 239**
renewable resources/recursos renovables Recursos, como el suelo y los bosques, que pueden reemplazarse por medio de procesos naturales de la Tierra, **59**
republic/república Forma de gobierno representativo en que la soberania reside en el pueblo, **126**
Restoration/Restauración Período en que reinó Carlos II de Inglaterra, cuando la monarquía fué restablecida, **154**
revolution/revolución Una vuelta completa alrededor del Sol, **19**
rifts/hendeduras Valles largos y profundos con montañas o mesetas a cada lado480
Roma/Roma Grupo étnico, también conocido como gitanos, que pudo haber migrado de la India a Europa hace mucho tiempo, **302**
rotation/rotación Una vuelta completa de la Tierra sobre su propio eje, **19**
rugby/rugby Juego de origen británico similar a fútbol, **676**
rural/rural Área de terreno abierto que se usa para la agricultura, **4**

S

Sahel/sahel Región de pastizales secos con clima estepario del Sur del Sahara, **462**
samurai/samurai Guerreros al servicio de señores japoneses, **136, 570**
sanctions/sanciones Penalidad económica o política, como un embargo, que uno o más países usan para obligar a otro país a dejar de cometer un acto ilegal o inmoral, **519**
Sanskrit/sánscrito Idioma antiguo del Sur de Asia; en la actualidad se usa en la India como lengua sagrada en la, **625**
satellite/satélite cuerpo que da vueltas al rededor de un cuerpo grande, **18**
scarcity/escasez Situación que resulta cuando la demanda es mayor que la oferta, **87**
Scientific Revolution/Revolución Cientifica Transformación de pensamiento ocurrido durante 1500 y 1600, causada por la observación científica, la experimentación, y el cuestionamiento de las opiniones tradicionales, **149**
secede/separar Dividir un país para formar otro, **472**
secondary industries/industrias secundarias Actividades económicas que convierten en productos terminados la materia prima que producen las industrias primarias, **83**
secular/seglar Que está separado de la religión, como un gobierno o estado secular, **409**
secularism/secularismo Énfasis en asuntos no religiosos, **160**
semiarid/semiárido Terreno relativamente seco, con lluvias escasas, **62**
sepoys/cipayos Tropas indias dirigidas por oficiales británicos durante el periodo colonial de la India, **628**
serfs/siervos Personas que, en Rusia, estaban atados a una tierra y trabajaban para un señor, **140, 347**
shah/sha Rey de Irán, **134, 400**
shamans/shaman Monje sintoísta, **569**
Shia/Shia La segunda más grande rama del Islam, **388**
Shintoism/shintoismo Religión más antigua conocida de Japón, **569**
shogun/shogún "Gran General" el más alto rango entre los guerreros japoneses, **136, 570**
silt/cieno Tierra de granos muy finos, **443**
sirocco/siroco Viento seco y caliente del norte de África que viaja por el mar Mediterráneo hacia Europa, **233**
slash-and-burn agriculture/agricultura de corte y quema Tipo de agricultura en que los bosques se talan y se queman para limpiar el terreno y plantarlo, **307**
smelters/fundidoras Fábricas que procesan menas de metal, **333**
social contract/contrato social Idea de que el gobierno se basa en un acuerdo entre las personas, **161**
solar energy/energía solar Recurso de energía renovable que produce a luz y el calo r del Sol, **71**
solar system/sistema solar El Sol y los cuerpos celestes que se mueven en trono a él, entre otros los planetas, las lunas y los asteroides, **17**
solstice/solsticio Días en que los rayos verticales del Sol están más lejos del ecuador, **21**
sorghum/sorgo Grano de cultivo que puede sobrevivir a las sequías, **469**
souks/souks Mercados del norte de África, **455**
soviet/soviet supremo Consejo comunista que gobernó la república y otras regiones de la Unión Soviética, **347**
spatial perspective/perspectiva espacial Punto de vista basado o visto en relación con el lugar en que se encuentra un objeto, así como la razón por la que está ahí, **3**
staple/producto básico Cultivo principal de una región o un país, **470**
steppe/estepa Gran llanura de pastos altos que se extiende desde Ucrania, pasa por el sur de Rusia y llega hasta Kazajistán, **323**
steppe climate/clima estepario Tipo de clima, generalmente seco, que se encuentra entre las regiones de climas desértico y húmedo, **46**
stock market/mercado de valores Organización mediante la cual se venden y se compran partes de compañías, **190**
storm surges/mareas de tormenta Grandes ondas de agua que se elevan por la fuerza del viento, en particular de los ciclones, huracanes y otras tormentas tropicales, **640**
stupas/stupas Montículos de tierra o piedras que cubren las cenizas del Buda o las reliquias de los santos budistas, **648**
subcontinent/subcontinente Grandes masas de tierra menores que un continente, como el subcontinente de la India, **113**
subduction/subducción Movimiento en el que una placa tectónica terrestre más gruesa se sumerge debajo de una más delgada, **29**
subregions/subregiones Áreas pequeñas de una región, **8**
subsistence agriculture/agricultura de subsistencia Tipo de agricultura en que los campesinos siembran sólo lo necesario para mantenerse a ellos mismos y a sus familias, **80**

suburbs/suburbios Áreas residenciales en las afueras de la ciudad, **179**
suffragettes/sufragistas Mujeres que lucharon por el derecho de las mujeres de votar, **180**
sultan/sultán Gobernante supremo de un país musulmán, **601**
Sunni/sunita La rama más grande del Islam, **388**
superpowers/superpotencias Países poderosos, **327**
Swahili/suahili Idioma bantú que se habla extensamente en África, **484**
symbol/símbolo Palabra, forma, color, estadarte o cualquier otra cosa que se use en representación de algo, **77**

T

taiga/taiga Bosque de árboles siempre verdes que existen en el sur de la tundra en Rusia, **323**
teak/teca Tipo de madera preciosa que crece en la India y al sudeste de Asia, **623**
terraces/terrazas Crestas horizontales que se construyen sobre las laderas de las colinas para prevenir la pérdida de suelo y favorecer la agricultura, **60**
tertiary industries/industrias terciarias Actividades económicas que trabajan con productos listos para vender a los consumidores, **83**
textiles/textiles Productos para fabricar ropa, **273**
theocracy/teocracia Gobierno regido por líderes religiosos, **400**
the veld/el veld regiones de pastos altos en el sur de África, **512**
townships/municipios Regiones de multitudes de pequeñas casas apiñadas que habitan los sudafricanos de raza negra en las afueras de las ciudades, **518**
trade surplus/excedente comercial Ocurre cuando el valor de las imporaciones es mayor que el de las importaciones, **576**
traditional economy/economía tradicional Economía basada en las costumbres y las tradiciones, **85**
tributary/tributario Cualquier corriente pequeña o río que fluye hacia un río o una corriente más grande, **25**
Tropic of Cancer/Trópico de Cáncer Línea de latitud que está a 23.5 grados al norte del ecuador, **21**
Tropic of Capricorn/Trópico de Capricornio Línea de latitud que está a 23.5 grados al sur del ecuador, **21**
trust territories/territorios bajo administración fiduciaria Regiones que están bajo el control temporal de otro país hasta que establezca su propio gobierno, **693**
tsetse fly/mosca tse tse Mosca africana del sur del Sahara que transmite el mal del sueño, una enfermedad mortal, **462**
tsunamis/tsunamis Olas muy grandes que se forman por la actividad submarina de las placas tectónicas, tales como los terremotos, **567**
tundra climate/clima de la tundra Región fría de lluvias escasas, que por lo genera se encuentra entre los climas de las regiones subártica y polar, **49**
typhoons/tifones Tormentas tropicales de vientos violentos, fuertes lluvias y altas marejadas, **48**

U

U-boats/U-boats Submarinos alemanes usados en la Primer Guerra Mundial, **187**
uninhabitable/inhabitable Que no es propicio para el establecimiento de seres humanos, **284**
unlimited governments/gobiernos ilimitados Gobiernos que tienen un control total sobre sus ciudadanos, **86**
urban/urbano Área en que se encuentra una ciudad, **4**

V

vassals/vassalos Personas a la que un amo le concedía tierras, como pago por sus servicios, **139**
vernacular/vernácular Lenguaje doméstico, natino, propio de un país, **142**

W

wadis/wadis Lechos secos de corrientes en el sudoeste de Asia, **387**
water cycle/ciclo del agua Circulación del agua del la superficie de la Tierra a la atmósfera y su regreso, **23**
water vapor/vapor de agua Estado gaseoso del agua, **23**
weather/tiempo Condiciones de la atmósfera en un tiempo y un lugar determinados, **37**
weathering/desgaste Proceso de desintegración de las rocas en pedazos pequeños por la acción del calor, el agua y otros medios, **31**
work ethic/ética laboral Creencia de que el trabajo es un mérito en sí mismo, **576**
working class/clase trabajadora Personas capacitadas y poco capacitadas con empleos de salarios bajos, **177**

Y

yurt/yurta Tienda redonda y portátil de lana tejida que se coloca sobre una armazón de madera, **364**

Z

Zionism/sionismo Movimiento que trata de establecer un país o comunidad judía en Palestina, **413**
zonal/zonal Clima del este de África que se extiende en franjas de este a oeste, **462**

INDEX

S *indicates Skills Handbook* g *indicates graphic* m *indicates map* p *indicates photograph*

A

F

G

H

INDEX

I

J

K

N

O

P

Q

R

S

V

W

X

Y

Z

ACKNOWLEDGMENTS

For permission to reprint copyrighted material, grateful acknowledgment is made to the following sources:

Casa Juan Diego: Adapted from *Immigrants Risk All: Cry for Argentina* by Ana Maria from *Houston Catholic Worker*, July/August 1997, accessed February 2, 2000, at http://www.cjd.org/stories/risk.html.

Doubleday, a division of Random House, Inc.: From "Marriage Is a Private Affair" from *Girls at War and Other Stories* by Chinua Achebe. Copyright © 1972, 1973 by Chinua Achebe.

FocalPoint f/8: From "October 5—Galtai" from "Daily Chronicles" and from "Buddhist Prayer Ceremony" from "Road Stories" by Gary Matoso and Lisa Dickey from *The Russian Chronicles* from *FocalPoint f/8*, accessed October 14, 1999, at http://www.f8.com/ FP/Russia.

Glas Publishers (Russia): From "The Lilac Dressing Gown" by Nina Gabrielyan, translated by Joanne Turnbull from *A Will and a Way: New Russian Writing*, edited by Natasha Perova and Arch Tait. Copyright © 1996 by Glas: New Russian Writing.

HarperCollins Publishers, Inc.: From *My Days* by R. K. Narayan. Copyright © 1973, 1974 by R. K. Narayan.

James Li, M.D.: From "Africa" by James Li, M.D from *eMedicine*, accessed October 11, 1999, at http://www.emedicine.com/emerg/topic726.htm. Copyright © 1999 by James Li.

Ms. Magazine: From "Foresters Without Diplomas" by Wangari Maathai from *Ms.*, vol. 1, no. 5, March/April 1991. Copyright © 1991 by *Ms.* Magazine.

Penguin Books Ltd.: From *The Epic of Gilgamesh*, translated by N. K. Sandars (Penguin Classics 1960, Third Edition, 1972). Copyright © 1960, 1964, 1972 by N. K. Sandars.

Puffin Books, a division of Penguin Putnam Inc.: From *Sadako and the Thousand Paper Cranes* by Eleanor Coerr. Copyright © 1977 by Elizabeth Coerr.

Sandy Wiseman: From "Photo-Journey Through a Costa Rican Rainforest" by Sandy Wiseman from *EcoFuture™ PlanetKeepers*, accessed October 14, 1999, at http://www.ecofuture.org/ecofuture/pk/ pkar9512.html. Copyright © 1999 by Sandy Wiseman.

Writer's House, Inc. c/o The Permissions Company: From *For Love Alone* by Christina Stead. Copyright © 1944 by Harcourt, Inc.; copyright renewed © 1972 by Christina Stead.

SOURCES CITED:

From "The Aztecs in 1519" from *The Discovery and Conquest of Mexico* by Bernal Díaz. Published by Routledge and Kegan Paul, London, 1938.

ART CREDITS

Abbreviated as follows: (t) top, (b) bottom, (l) left, (r) right, (c) center.

Unit flags of United States created by One Mile Up, Inc. Unit flags of Canadian provinces created by EyeWire, Inc. Other flags, country silhouettes, feature maps and atlas maps created by MapQuest.com., Inc. All other illustrations, unless otherwise noted, contributed by Holt, Rinehart and Winston.

Study Skills: Page S2, MapQuest.com, Inc.; S3, MapQuest.com, Inc.; S4 (cl, bl), MapQuest.com, Inc.; S5, MapQuest.com, Inc.; S6 (tr), MapQuest.com, Inc.; S7, MapQuest.com, Inc.; S8 (tr), MapQuest.com, Inc.; S9, MapQuest.com, Inc.; S10 (tc, br), Leslie Kell; S11 (cr), Leslie Kell; S11 (b), Ortelius Design; S12 (b), Leslie Kell; S12 (tr), Rosa + Wesley; S13, Uhl Studios, Inc.; S14 (br), MapQuest.com, Inc.; S15 (bl), MapQuest.com, Inc.

Chapter 1: Page 5 (cl), MapQuest.com, Inc.; 6 (bl), MapQuest.com, Inc.; 7 (tl, tr), MapQuest.com, Inc.; 8 (l, bc), MapQuest.com, Inc.; 9 (tr), MapQuest.com, Inc.; 10 (tc), MapQuest.com, Inc.; 11 (bc), MapQuest.com, Inc.; 15 (tr), MapQuest.com, Inc.

Chapter 2: Page 17 (b), Uhl Studios, Inc.; 18 (b), Uhl Studios, Inc.; 19 (bl), Uhl Studios, Inc.; 20 (b), Uhl Studios, Inc.; 22 (tl), Robert Hynes; 23 (cr), MapQuest.com, Inc.; 24 (tr), Uhl Studios, Inc.; 25 (tr, br), Uhl Studios, Inc.; 26 (b), MapQuest.com, Inc.; 28 (b), Uhl Studios, Inc.; 29 (tl), MapQuest.com, Inc.; 29 (c, br), Uhl Studios, Inc.; 31 (tr), MapQuest.com, Inc.; 32 (tr), Ortelius Design; 35 (tr), MapQuest.com, Inc.

Chapter 3: Page 38 (tl), Uhl Studios, Inc.; 38 (cl), MapQuest.com, Inc.; 39 (bl), MapQuest.com, Inc.; 40 (tr), Uhl Studios, Inc.; 42 (cr), Rosa + Wesley; 42 (b), MapQuest.com, Inc.; 43 (t), Rosa + Wesley; 45 (tl), Uhl Studios, Inc.; 46 (br), MapQuest.com, Inc.; 47 (all), MapQuest.com, Inc.; 48 (tl, bc), MapQuest.com, Inc.; 49 (tr), MapQuest.com, Inc.; 50 (tl), MapQuest.com, Inc.; 52 (tr), Robert Hynes; 55 (tr), Uhl Studios, Inc.; 57 (tr), MapQuest.com, Inc.

Chapter 4: Page 60 (tc), MapQuest.com, Inc.; 62 (bl), MapQuest.com, Inc.; 63 (tr, br), MapQuest.com, Inc.; 65 (br), Leslie Kell; 67 (t), MapQuest.com, Inc.; 69 (tl, br), MapQuest.com, Inc.; 70 (tl), MapQuest.com, Inc.; 73 (tr), MapQuest.com, Inc.

Chapter 5: Page 76 (t, bl), MapQuest.com, Inc.; 77 (tr), MapQuest.com, Inc.; 78 (bl), MapQuest.com, Inc.; 81 (cr), MapQuest.com, Inc.; 83 (br), MapQuest.com, Inc.; 86 (c), Rosa + Wesley; 88 (cl), MapQuest.com, Inc.; 89 (tc), MapQuest.com, Inc.; 91 (tr), MapQuest.com, Inc.; 93 (tl), MapQuest.com, Inc.; 98 (t, bl), Ortelius Design.

Chapter 6: Page 106, (t) MapQuest.com,Inc.; 107, (b) MapQuest.com,Inc.; 109, (bl) MapQuest.com,Inc.; 110, MapQuest.com,Inc.; 111, (t) MapQuest.com,Inc.; 112, (bl) MapQuest.com,Inc.; 114, (bl) MapQuest.com,Inc.; 115, (tl) MapQuest.com,Inc.; 119, (tr) MapQuest.com,Inc.; 120, (tl) MapQuest.com,Inc.; 122, (b) MapQuest.com,Inc.; 125, (br) MapQuest.com,Inc.; 127, (b) MapQuest.com,Inc.; 131, (tr) MapQuest.com,Inc.

Chapter 7: Page 138, (br) MapQuest.com,Inc.; 157, (tr) MapQuest.com,Inc.

Chapter 8: Page 153 MapQuest.com, Inc.; 171 MapQuest.com, Inc.; 171 MapQuest.com, Inc.

Chapter 9: Page 165, MapQuest.com,Inc.; 183, MapQuest.com,Inc.; 186, MapQuest.com,Inc.; 205, (bl) MapQuest.com,Inc.; 209, MapQuest.com,Inc.; 213, (tr) MapQuest.com,Inc.; 215, (tr) MapQuest.com,Inc.

Chapter 10: Page 231 (b), MapQuest.com, Inc.; 232 (bl), MapQuest.com, Inc.; 233 (tr), MapQuest.com, Inc.; 233 (cr), Leslie Kell; 234 (br), MapQuest.com, Inc.; 235 (tc), MapQuest.com, Inc.; 237 (tr), Rosa + Wesley; 238 (cl), MapQuest.com, Inc.; 239 (tc), MapQuest.com, Inc.; 240 (bl), MapQuest.com, Inc.; 241 (tr), Rosa + Wesley; 242 (br), MapQuest.com, Inc.; 243 (c), MapQuest.com, Inc.; 244 (tl), Rosa + Wesley; 245 (tl), MapQuest.com, Inc.; 247 (tr), MapQuest.com, Inc.

Chapter 14: Page 249 (b), MapQuest.com, Inc.; 250 (tl, bc), MapQuest.com, Inc.; 251 (tr), MapQuest.com, Inc.; 254 (cr), MapQuest.com, Inc.; 255 (cl), Rosa + Wesley; 256 (bl), MapQuest.com, Inc.; 257 (cr), MapQuest.com, Inc.; 258 (tl), Leslie Kell; 258 (c), MapQuest.com, Inc.; 259 (cr), Rosa + Wesley; 260 (cl), MapQuest.com, Inc.; 262 (cl), Rosa + Wesley; 264 (tl), MapQuest.com, Inc.; 265 (cr), Rosa + Wesley; 267 (tr), MapQuest.com, Inc.

Chapter 15: Page 269 (b), MapQuest.com, Inc.; 370 (bl), MapQuest.com, Inc.; 272 (cl), MapQuest.com, Inc.; 273 (c), MapQuest.com, Inc.; 274 (bl), Rosa + Wesley; 277, Nenad Jakesevic; 275 (cr), Leslie Kell; 278 (bl), MapQuest.com, Inc.; 280 (tl), Rosa + Wesley; 280 (cl), MapQuest.com, Inc.; 281 (br), Rosa + Wesley; 282 (tl, bc), MapQuest.com, Inc.; 284 (tl), MapQuest.com, Inc.; 285 (cr), MapQuest.com, Inc.; 287 (tr), MapQuest.com, Inc.

Chapter 16: Page 289 (b), MapQuest.com, Inc.; 290 (bl), MapQuest.com, Inc.; 291 (tr), MapQuest.com, Inc.; 294 (tl), Rosa + Wesley; 294 (cl), MapQuest.com, Inc.; 296 (bc), MapQuest.com, Inc.; 297 (tr), MapQuest.com, Inc.; 298 (bl), MapQuest.com, Inc.; 299 (cr), MapQuest.com, Inc.; 300 (tc), MapQuest.com, Inc.; 301 (tr, cr), MapQuest.com, Inc.; 302 (tl), Leslie Kell; 302 (bl), MapQuest.com, Inc.; 303 (tr), Rosa + Wesley; 305 (tr), MapQuest.com, Inc.; 307 (tl), MapQuest.com, Inc.; 312 (t, bl), Ortelius Design.

Chapter 17: Page 321 (b), MapQuest.com, Inc.; 323 (br), MapQuest.com, Inc.; 326 (t), MapQuest.com, Inc.; 357 (tr, bc), MapQuest.com, Inc.; 329 (tr, cr), Rosa + Wesley; 329 (br), MapQuest.com, Inc.; 321 (br), MapQuest.com, Inc.; 322 (tl, bl), MapQuest.com, Inc.; 333 (tr), MapQuest.com, Inc.; 334 (bl), MapQuest.com, Inc.; 335 (tl), MapQuest.com, Inc.; 336 (tl), MapQuest.com, Inc.; 338 (bl), MapQuest.com, Inc.; 339 (tr), MapQuest.com, Inc.; 371 (tr), MapQuest.com, Inc.

Chapter 18: Page 343 (b), MapQuest.com, Inc.; 344 (bl), MapQuest.com, Inc.; 347 (bl), MapQuest.com, Inc.; 348 (tl), MapQuest.com, Inc.; 349 (c), Uhl Studios, Inc.; 350 (tl), Leslie Kell; 350 (cl), Rosa + Wesley; 352 (bc), MapQuest.com, Inc.; 353 (tr), Rosa + Wesley; 355 (tr), MapQuest.com, Inc.

Chapter 19: Page 357 (b), MapQuest.com, Inc.; 359 (cr), MapQuest.com, Inc.; 340 (cl), MapQuest.com, Inc.; 341 (cr), MapQuest.com, Inc.; 354 (l), Leslie Kell; 364 (bc), MapQuest.com, Inc.; 365 (tr), Rosa + Wesley; 367 (cr), Leslie Kell; 369 (tr), MapQuest.com, Inc.; 371 (t), MapQuest.com, Inc.; 376 (t, bl), Ortelius Design.

Chapter 20: Page 385 (b), MapQuest.com, Inc.; 386 (bl), MapQuest.com, Inc.; 388 (cl), MapQuest.com, Inc.; 389 (tr), Leslie Kell; 389 (cr), MapQuest.com, Inc.; 390 (tl), MapQuest.com, Inc.; 392 (tl), Rosa + Wesley; 393 (tr), MapQuest.com, Inc.; 394, MapQuest.com, Inc.; 396 (bl), MapQuest.com, Inc.; 397 (tr), Rosa + Wesley; 397 (br), MapQuest.com, Inc.; 399 (b), Uhl Studios, Inc.; 390 (tl), Rosa + Wesley; 400 (bl), MapQuest.com, Inc.; 403 (tr), MapQuest.com, Inc.

Chapter 21: Page 405 (b), MapQuest.com, Inc.; 406 (bl), MapQuest.com, Inc.; 407 (cr), MapQuest.com, Inc.; 408 (bl), MapQuest.com, Inc.; 411 (r), MapQuest.com, Inc.; 412 (bl), MapQuest.com, Inc.; 413 (tc), MapQuest.com, Inc.; 413 (br), Rosa + Wesley; 414 (tl), MapQuest.com, Inc.; 416 (bc), MapQuest.com, Inc.; 419 (tc), MapQuest.com, Inc.; 421 (tr), MapQuest.com, Inc.; 423 (t), MapQuest.com, Inc.; 455, MapQuest.com, Inc.; 428 (t,bl), Ortelius Design.

Chapter 22: Page 441 (b), MapQuest.com, Inc.; 442 (cl), MapQuest.com, Inc.; 443 (tl), MapQuest.com, Inc.; 447 (cl,tr), MapQuest.com, Inc.; 448 (tl), MapQuest.com, Inc.; 449 (tr), MapQuest.com, Inc.; 450, Ortelius Design; 452 (bl), MapQuest.com, Inc.; 483 (tc), MapQuest.com, Inc.; 453 (br), Rosa + Wesley; 455 (br), Leslie Kell; 456 (tl), Rosa + Wesley; 456 (bl), MapQuest.com, Inc.; 459 (tr), MapQuest.com, Inc.

Chapter 23: Page 461 (b), MapQuest.com, Inc.; 462 (bl), MapQuest.com, Inc.; 464 (bl), MapQuest.com, Inc.; 465 (br), MapQuest.com, Inc.; 466 (cl), MapQuest.com, Inc.; 468 (tl), MapQuest.com, Inc.; 469 (bc), MapQuest.com, Inc.; 470 (tl), MapQuest.com, Inc.; 470 (bl), Rosa + Wesley; 471 (cr), Leslie Kell; 472 (cl), MapQuest.com, Inc.; 473 (tr), MapQuest.com, Inc.; 5474 (tc), Rosa + Wesley; 504 (bc), MapQuest.com, Inc.; 477 (tr), MapQuest.com, Inc.

Chapter 24: Page 479 (b), MapQuest.com, Inc.; 480 (cr), MapQuest.com, Inc.; 481 (tr), MapQuest.com, Inc.; 483 (br), MapQuest.com, Inc.; 485 (br), MapQuest.com, Inc.; 486 (cl), MapQuest.com, Inc.; 486 (tl), Rosa + Wesley; 490 (tl), Rosa + Wesley; 490 (cr), MapQuest.com, Inc.; 491 (tr), MapQuest.com, Inc.; 493 (tr), MapQuest.com, Inc.

Chapter 25: Page 495 (b), MapQuest.com, Inc.; 496 (bl), MapQuest.com, Inc.; 499 (cr), MapQuest.com, Inc.; 501 (tc), MapQuest.com, Inc.; 503 (tr), Rosa + Wesley; 504 (tr), MapQuest.com, Inc.; 504 (cl), Leslie Kell; 505 (br), MapQuest.com, Inc.; 506 (tl), Rosa + Wesley; 507 (tr), Rosa + Wesley; 509 (tr), MapQuest.com, Inc.

Chapter 26: Page 511 (b), MapQuest.com, Inc.; 512 (tl), MapQuest.com, Inc.; 513 (cr), MapQuest.com, Inc.; 516 (cl), MapQuest.com, Inc.; 519 (br), Leslie Kell; 520 (cl), MapQuest.com, Inc.; 521 (cr), Rosa + Wesley; 522 (cl), MapQuest.com, Inc.; 523 (br), Rosa + Wesley; 525 (cr), MapQuest.com, Inc.; 527 (tr), MapQuest.com, Inc.; 534 (t,bl), Ortelius Design.

Chapter 27: Page 543 (b), MapQuest.com, Inc.; 544 (cl), MapQuest.com, Inc.; 545 (bc), MapQuest.com, Inc.; 546 (tl), Rosa + Wesley; 548 (tr), MapQuest.com, Inc.; 549 (b), Leslie Kell; 551 (tr, br), MapQuest.com, Inc.; 553 (cr), MapQuest.com, Inc.; 554 (bl), Rosa + Wesley; 555 (tr, cr), MapQuest.com, Inc.; 556 (tl, bl), MapQuest.com, Inc.; 557 (cr), MapQuest.com, Inc.; 559 (tl), MapQuest.com, Inc.; 560 (cl), MapQuest.com, Inc.; 563 (tr), MapQuest.com, Inc.

Chapter 28: Page 565 (b), MapQuest.com, Inc.; 566 (bl), MapQuest.com, Inc.; 567 (br), MapQuest.com, Inc.; 569 (bc), MapQuest.com, Inc.; 571 (cr), MapQuest.com, Inc.; 573 (tr, cr), Rosa + Wesley; 574 (tl, cl), MapQuest.com, Inc.; 577 (br), MapQuest.com, Inc.; 578 (tl), MapQuest.com, Inc.; 580 (cl), Rosa + Wesley; 581 (cr), MapQuest.com, Inc.; 582 (bc), MapQuest.com, Inc.; 585 (tr), MapQuest.com, Inc.

Chapter 29: Page 587 (b), MapQuest.com, Inc.; 588 (bl), MapQuest.com, Inc.; 590 (bl), MapQuest.com, Inc.; 591 (tc, br), MapQuest.com, Inc.; 593 (tr), MapQuest.com, Inc.; 594 (bl), MapQuest.com, Inc.; 595 (tr), MapQuest.com, Inc.; 595 (br), Rosa + Wesley; 596 (tl), Leslie Kell; 597 (br), MapQuest.com, Inc.; 599 (tr), Rosa + Wesley; 599 (br), MapQuest.com, Inc.; 600 (tl), MapQuest.com, Inc.; 601 (tr), MapQuest.com, Inc.; 602 (b), MapQuest.com, Inc.; 603 (cl), Leslie Kell; 605 (tr), MapQuest.com, Inc.; 607 (tr), MapQuest.com, Inc.; 612 (t, bl), Ortelius Design.

Chapter 30: Page 621 (b), MapQuest.com, Inc.; 622 (bl), MapQuest.com, Inc.; 624 (bl), MapQuest.com, Inc.; 627 (tr), MapQuest.com, Inc.; 630 (bl), MapQuest.com, Inc.; 631 (br), MapQuest.com, Inc.; 632 (tl, bl), MapQuest.com, Inc.; 633 (tr), Rosa + Wesley; 633 (c), MapQuest.com, Inc.; 637 (tr), MapQuest.com, Inc.

Chapter 31: Page 639 (b), MapQuest.com, Inc.; 642 (bl), MapQuest.com, Inc.; 643 (tr), Rosa + Wesley; 643 (br), MapQuest.com, Inc.; 644 (tl), MapQuest.com, Inc.; 644 (bl), Leslie Kell; 645 (tr), MapQuest.com, Inc.; 648 (bl), MapQuest.com, Inc.; 650 (bl), MapQuest.com, Inc.; 651 (tr), Rosa + Wesley; 651 (br), MapQuest.com, Inc.; 652 (tl), MapQuest.com, Inc.; 653 (tr), MapQuest.com, Inc.; 655 (tr), MapQuest.com, Inc.; 657, MapQuest.com, Inc.; 658 (cr), Stephen Durke/Washington Artists; 659 (cl), Ortelius Design; 662 (t, bl), Ortelius Design.

Chapter 32: Page 671 (b), MapQuest.com, Inc.; 672 (bl), MapQuest.com, Inc.; 676 (cl), MapQuest.com, Inc.; 676 (bl), Rosa + Wesley; 677 (tr), Leslie Kell; 679 (t), Nenad Jakesevic; 680 (bl), MapQuest.com, Inc.; 681 (tr), MapQuest.com, Inc.; 682 (bl), MapQuest.com, Inc.; 683 (tr), Rosa + Wesley; 683 (cr), MapQuest.com, Inc.; 685 (tr), MapQuest.com, Inc.

Chapter 33: Page 687 Uhl Studios, Inc.; 688 (tl), MapQuest.com, Inc.; 689 (tr), Leslie Kell; 689 (cl), MapQuest.com, Inc.; 695 (tr), Rosa + Wesley; 695 (br), MapQuest.com, Inc.; 696 (tl), MapQuest.com, Inc.; 701 (tr), MapQuest.com, Inc.; 702 (b), Uhl Studio, Inc.; 703 (tr), MapQuest.com, Inc.; 705 (cr), Leslie Kell.

PHOTO CREDITS

Cover and Title Page: (child image) Steve Vidler/Nawrocki Stock Photo; (bkgd) Image Copyright ©2003 PhotoDisc, Inc./HRW

Table of Contents: iv, (bl) © SuperStock; ix, (tl) © Alain Le Garsml/Panos Pictures; ix, (b) © Marc Riboud / Magnum; v, (tl) © S. Sherbell/SAbA Press Photos, Inc.; vi, (c) © From the Collections of Henry Ford Museum and Greenfield Village; vii, (b) © Index Stock Photography; vii, (tl) © travelpix/FPG International LLC; viii, (b) © Steve Raymer/NGS Image Collection; viii, (bl) © the State Russian Museum/CORbIS; x, (tr) © William Maynard Owen/NGS Image Collection; x, (bl) © Alex Wasinski/FPG International LLC; xi, (tl) © FPG International LLC; xi, (b) © Daniel J. Cox/Liaison Agency; xii, (tr) © Richard bickel/CORbIS; xiii, (bl) © C. Rennie/ trIP Photo Library; xiii, (b) © Ric Ergenbright; xiii, (cl) © CORbIS; xiv, (br) © Digital Stock Corp./HRW; xiv, (cl) © Frans Lanting/Minden Pictures; xv, (br) © Jim Zuckerman/CORBIS; xxiii, CORBIS Images/HRW; xxiv (tl), David Young-Wolf/PNI; xxiv (b), Klaus Lahnstein/Getty Images; S20 (b), © Nik Wheeler/ CORBIS; S21 (cl), Jeffrey Aaronson/Network Aspen; S21 (c), Index Stock Imagery, Inc.; S21 (tr), Amit Bhargava/Newsmakers/Getty Images; S22 (l), Norman Owen Tomalin/Bruce Coleman, Inc.; S22 (cl), © Fritz Polking/Peter Arnold, Inc.; S22 (tl), Larry Kolvoord Photography; S22 (tr), Runk/ Schoenberger/Grant Heilman Photography; S22 (tr), Wolfgang Kaehler Photography; S27 (b), Rosenback/ ZEFA/Index Stock Imagery, Inc.; **Unit 1:** 1 (b), Photo © Transdia/Panoramic Images, Chicago 1998; 1 (cl), © Joe Viesti/The Viesti Collection; 1 (tr), Francois Gohier/Photo Researchers, Inc.; 1 (t), © Norbert Wu/www.norbertwu.com; 1 (br), Steven David Miller/Animals Animals/Earth Scenes; **Chapter 1**: 2 (cl), © Stone/Philip & Karen Smith; 2 (bl), Sam Dudgeon/HRW Photo; 2 (tr), CORBIS; 2 (c), © Joseph Sohm; ChromoSohm Inc./CORBIS; 3 (bl), © Stone/Ken McVey; 3 (tr), British Library, London, Great Britain/ Art Resource, NY; 4, Luca Turi/AP/Wide World Photos; 5 (cl), The Stock Market/Jose Fuste Raga; 5 (cr), NASA; 6 (cl), Robert Caputo/Aurora; 6 (tr), © Stone/Robert Frerck; 7-8 © Bob Daemmrich; 8 (b), K.D. Frankel/Bilderberg/Aurora; 9 (t,br), © Wolfgang Kaehler; 10, © Nik Wheeler/CORBIS; 11 (cr), © Ilene Perlman/Stock, Boston; 11 (tr), © Archivo Iconografico S.A./CORBIS; 13 (cr), © Chris Rainier/CORBIS; **Chapter 2**: 16 (c), © Roger Ressmeyer/CORBIS; 16 (tl), Digital Stock Corp./HRW; 16 (t), Steve Winter/ Black Star; 16 (cl), T.A. Wiewandt/DRK Photo; 16 (bl), Scala/Art Resource, NY; 17, Monticello. Photo: Edward Owen; 19, © Stone/Earth Imaging; 20 (tl), courtesy of NASA, Marshall Space Flight Center; 21 (t,b), Global Hydrology and Climate Center, NASA Marshall Space Flight Center.; 23 (t), Image copyright ©2002 PhotoDisc, Inc.; 23 (b), © Gerald French/FPG International LLC; 24, © Tom Bean/CORBIS; 26 (t), © Norbert Wu/www.norbertwu.com; 26 (b), Boll/ Liaison Agency; 27, © Ed Kashi; 28, © Gary Braasch/ CORBIS; 30, © Stone/Andrew Rafkind; 31, © Galen Rowell/CORBIS; 32, © Yann Arhtus-Bertrand/CORBIS; **Chapter 3**: 36 (c), © Stone/Glenn Christianson; 36 (tr), © Stone/A. Witte/C. Mahaney; 36 (br), © Stone/ Brian Stablyk; 36 (b), © 1999 Michael DeYoung/ AlaskaStock.com; 36 (tl), Corbis Images; 37, Thomas A. Wiewandt, Ph.D., Wild Horizons; 38, © Layne Kennedy/CORBIS; 39, © Chinch Gryniewicz; Ecoscene/CORBIS; 40, © SuperStock; 41, © Flip Nicklin/Minden Pictures; 43 (cr), NASA; 44 (bl), Image copyright ©1998 PhotoDisc, Inc./HRW; 44 (t), Paul Seheult; Eye Ubiquitous/CORBIS; 45 (br), © Stone/Martin Puddy; 46 © Laurence Parent; 48 (tl), © Gail Mooney/CORBIS; 48 (b), © DiMaggio/Kalish/ Peter Arnold, Inc.; 49, © Darrell Gulin/CORBIS; 50, © Galen Rowell/CORBIS; 51 (c), © Darrell Gulin/CORBIS; 51 (t), Image Copyright ©2002 PhotoDisc, Inc.; 53 (t), Robert Maier; 54 (tl), © E & P Bauer/Bruce Coleman, Inc.; 54 (tr), © Scott T. Smith/CORBIS; 54 (bl), © F. Krahmer/Bruce Coleman, Inc.; 54 (br), © Gary Braasch/CORBIS; 55, Photo by William E. Ferguson © William E. Ferguson; **Chapter 4**: 58 (cl), © Philip James Corwin/CORBIS; 58 (bl), Aaron Chang/ The Stock Market; 58 (cr), © Morton Beebe/Corbis; 58 (t), Mike Huser/DRK Photo; 59 (b), Larry Lefever/ Grant Heilman Photography; 59 (t), Coronado Rodney Jones/HRW Photo; 60, © Michael Busselle/CORBIS; 61, © Kevin Schafer; 62 (bl), © Tom Carroll/FPG International LLC; 62 (t), © Mark Reinstein/FPG International LLC; 63 (tr), Grant Heilman/Grant Heilman Photography; 63 (br), Steven Burr Williams/ Liaison International; 64 (all), NASA, Goddard Space Flight Center; 65, Image Copyright ©2002 PhotoDisc, Inc./HRW; 66 (t), Sam Dudgeon/HRW Photo; 66 (b), Dr. Bode; 68 (b), © Stone/Mike Abrahams; 68 (t), © Stone/Vince Streano; 69, © Ernest Manewal/FPG International LLC; 70 (b), © Telegraph Colour Library/ FPG International LLC; 71, Mark Burnett/Photo Researchers, Inc.; **Chapter 5**: 74 (t), © Gerald Brimacombe/International Stock Photography; 74 (cr), © Bob Firth/International Stock Photography; 74 (b), © Ahu Tongariki/Bruce Coleman, Inc.; 74 (c), © Wally McNamee/CORBIS; 75 (b), © Bob Daemmrich/Stock, Boston; 75 (t), © Galen Rowell/CORBIS; 76, © Bohdam Hrynewch/Stock, Boston/PNI; 77 (b), © Stone/Ron Sherman; 77 (t), © Stone/Rich La Salle; 78 (b), Bruno Barbey/Magnum Photos; 78 (t), S. Sherbell/SABA Press Photos, Inc.; 79, © Werner Forman/CORBIS; 80, © Eric and David Hosking/CORBIS; 81 (b), © Rich Iwasaki/AllStock/PNI; 81 (t), © Stone/Bob Thomas; 83 (t), © Digital Vision; 84 (tl), © Michelle Gabel/The Image Works, Woodstock, NY; 84 (tr), © Richard Hamilton Smith/CORBIS; 84 (bl), Henry Friedman; 84 br), © Stephen Frisch/Stock, Boston/PNI; 85, Carolyn Schaefer/SCHAE/Bruce Coleman, Inc.; 87 (b), © Joanna B. Pinneo/Aurora; 87 (t), © Joel W. Rogers/ CORBIS; 88, © Pramod Mistry/The Image Works, Woodstock, NY; 89 (t), © Ed Kashi; **Unit 2:** 98, (tl) © Pix 2000/FPG International; 98, (tr) © the british Museum/the Art Archive; 98, (c) © Private Collection, Milan/Canali Photobank, Milan/SuperStock; 98, (bl) © George Grigoriou/ Stone; 99, (tr) Eirik Irgens Johnsen/University Museum of National Antiquities, Oslo, Norway; 99, (tl) © Archivo Iconografico, S.A./CORBIS; 99, (bl) © the Pierpont Morgan Library/Art Resource, NY; 99, (br) © Nimatallah/Art Resource, NY; 99, (cr) © Dagli Orti/Russian Historical Museum, Moscow/the Art Archive; 100, (c) © Dagli Orti/Musée du Chateau de Versailles/the Art Archive; 100, (bl) © Robert Harding Picture Library; 100, (tl) © Victoria & Albert Museum, London/bridgeman Art Library; 100, (tr) © Museo del Prado, Madrid/the Art Archive; 100, (br) © Philip Mould, Historical Portraits Ltd, London, UK/bridgeman Art Library; 101, (tr) © Hulton Archive by Getty Images; 101, (bl) © Culver Pictures; 101, (tr) © Sovfoto/Eastfoto; 101, (br) © AFP/CORBIS; **Chapter 6:** 102, (cr) © Gianni Dagli Orti/CORBIS; 102, (b) © Hulton-Deutsch Collection/CORBIS; 102, (tr) © Sheridan/Ancient Art & Architecture Collection; 102, (cl) © Dagli Orti/Musée du Louvre, Paris/the Art Archive; 103, (b) © Dr. Owen Lovejoy and students, Kent State University. Photo © 1985 David L. brill; 103, (t) © Nik Wheeler/CORBIS; 104, (t) © Courtesy Mammoth Site Museum, Hot Springs, SD; 104, (b) © Gianni Dagli Orti/CORBIS; 105, © Paul Almasy/CORBIS; 106, (b) © Martha Avery/Asian Art & Archaeology, Inc./CORBIS; 108, © Zafer Kizilkaya/Atlas Geographic; 109, (tr) © the british Museum/Compass; 110, (cl) © Gianni Dagli Orti/CORBIS; 110, (br) © CORbIS; 111, (br) © Sheridan/Ancient Art & Architecture Collection; 112, (tl) © Erich Lessing/Art Resource, NY; 25 Erich Lessing/Art Resource, NY; 112, (br) © Hugh Sitton/Stone; 113, (tl) © Erich Lessing/Art Resource, NY; 113, (br) © Robert Harding/CORBIS; 114, (tl) © Luca I. tettoni/CORBIS; 115, (br) © O. Louis Mazzatenta/NGS Image Collection; 116, (tr) © Giraudon/Art Resource, NY; 116, (c) © Dennis Cox/ChinaStock; 117, (tr) © boltin Picture Library; 117, (bl) © Mike Yamashita/Woodfin Camp & Associates; 118, (tl) © trip/M Jelliffe/the Viesti Collection; 118, (bl) © Woodfin Camp & Associates; 119, (br) © Salim Amin/Camerapix Ltd, Kenya; 120, (bl) © MIt Collection/CORBIS; 121, (tr) © Giraudon/Art Resource, NY; 122, (t) © Ashmolean Museum, Oxford, UK/bridgeman Art Library; 123, (t) © National Archaeological Museum, Athens, Greece/bridgeman Art Library; 124, (t) © Gian berto Vanni/Art Resource; 124, (b) © Phyllis Picardi/Stock South/PictureQuest; 125, (tr) © Art Resource, NY; 126, (b) © VCG/FPG International; 128, (t) © Jose Fuste Raga/CORBIS Stock Market; 128, (b) © Gene Plaisted, OSC/the Crosiers; **Chapter 7:** 132, (tr) © Sheridan/Ancient Art & Architecture Collection; 132, (c) © Erich Lessing/Art Resource, NY; 132, (b) © Gail Mooney/CORBIS; 132, (cl) © bibliothèque Nationale, Paris/AKG London; 133, (tr) © Giraudon/Art Resource, NY; 133, (br) © Nabeel turner/Stone; 134, (tl) © Giraudon/Art Resource, NY; 134, (br) © Madar-i-Shah Madrasa, Isfahan, Iran/bridgeman Art Library; 135, (tr) © Musée Guimet, Paris, France/Lauros-Giraudon, Paris/SuperStock; 135, (br) © Ortelius Design; 136, (tl) © Werner Forman Archive/Victoria & Albert Museum, London/Art Resource, NY; 137, (t) © AKG London; 137, (b) © Robert W. Madden/Paris/AKG London; 137, (br) © bibliothèque Nationale, Paris/AKG London; 138, (tl) © AKG London; 139, (t) © bettmann/CORBIS; 140, (t) © bridgeman Art Library/SuperStock; 140, (br) © the Pierpont Morgan Library/Art Resource, NY; 141, (t) © Musée de la tapisserie, bayeux, France/bridgeman Art Library; 141, (b) © Giraudon/Art Resource, NY; 142, (tl) © bibliothèque Royale de belgique, brussels, belgium/bridgeman Art Library; 142, (b) © Archivo Iconografico, S.A./CORBIS; 143, (t) © british Library, London/bridgeman Art Library, London/SuperStock; 143, (b) © Dagli Orti/Galleria degli Uffizi, Florence/the Art Archive; 144, (c) © bettmann/CORBIS; 144, (t) © Ali Meyer/bridgeman Art Library; 144, (br) © Kunsthistorisches Museum, Vienna, Austria/bridgeman Art Library; 145, (c) © the Pierpont Morgan Library/Art Resource, NY; 145, (tr) © Erich Lessing/Art Resource, NY; 145, (br) © Art Resource, NY; 146, (b) © Gary Yeowell/Stone; 146, (tl) © Giraudon/Art Resource, NY; 147, (tr) © Scala/Art Resource, NY; 147, (br) © belvoir Castle, Leicestershire/bridgeman Art Library; 148, (tl) © bridgeman Art Library; 149, (t) © Scala/Art Resource, NY; 149, (b) © New York Historical Society, New York, USA/bridgeman Art Library; 150, (bl) © Réunion des Musées Nationaux/bridgeman Art Library; 151, (cr) © David Parker/Science Photo Library/Photo Researchers; 152, (cl) © Museum of the City of New York/CORBIS; 153, (t) © Palace of Versailles, France/Lauros-Giraudon, Paris/SuperStock; 153, (b) © National Portrait Gallery, London/SuperStock; 154, (t) © bridgeman Art Library; 154, (bl) © Dreweatt Neate Fine Art Auctioneers, Newbury/bridgeman Art Library; **Chapter 8:** 158, (tr) © Sheridan/Ancient Art & Architecture Collection; 158, (br) © Lee Snider/CORBIS; 158, (cl) © SuperStock ; 159, (t) © David Muench/CORBIS; 159, (b) © Stock Montage/SuperStock; 160, (b) © Louvre, bibliothèque, Paris, France/Erich Lessing/Art Resource, NY; 160, (tl) © tate Gallery, London/Art Resource, NY; 161, (cr) © Giraudon/bridgeman; 161, (bl) © Erich Lessing/Art Resource, NY; 163, (t) © Colonial Willimasburg Foundation; 163, (b) © Stock Montage; 164, (bl) © SuperStock ; 165, (bl) © Christie's Images/SuperStock; 166, (b) © AKG, berlin/SuperStock; 167, (tr) © Dagli Orti/Musée de l'Affiche, Paris/the Art Archive; 167, (b) © Dagli Orti/Musée Carnavalet, Paris/the Art Archiv; 168, (tl) © Dagli Orti/Musée Carnavalet, Paris/the Art Archive; 168, (bl) © Dagli Orti/Musée de Versailles/the Art Archive; 169, (tr) © Erich Lessing/Art Resource, NY; 170, (bl) © Stock Montage; 170, (tr) © CORbIS/Hulton-Deutsch Collection Limited; 172, (b) © Steidle Collection, College of Earth and Mineral Sciences, Pennsylvania State University/Steidle Art Collection/SuperStock; 173, (tr) © AKG London; 174, (bl) © the Art Archive; 174, (tl) © Palubniak Studios; 175, (t) © From the Collections of Henry Ford Museum and Greenfield Village; 176, (t) © Hulton Archive by Getty Images; 176, (bl) © Gernsheim Collection, Harry Ransom Humanities Research Center, the University of texas at Austin; 177, (bl) © John D. Cunningham/Visuals Unlimited; 177, (tr) © Culver Pictures; 178, (c) © New York Journal, Jan 5, 1896, Courtesy OldNews, Inc. & New York World, 1883, Courtesy OldNews, Inc.; 178, (bl) © Winslow Homer, American, 1836–1910, Croquet Scene, oil on canvas, 1866, 15 7/8 x 26 1/16in, Friends of American Art Collection, 1942, photograph ©1998, the Art Institute of Chicago, All Rights Reserved; 179, (tr) © Erich Lessing/Art Resource, NY, (frame) Image Farm; 180, (t) © Hulton-Deutsch Collection/CORBIS; 181, (tr) © AKG berlin/SuperStock; **Chapter 9:** 184, (tr) © Sheridan/Ancient Art & Architecture Collection; 184, (br) © SuperStock; 185, (t) © AKG London; 187, (br) © Culver Pictures; 188, (bl) © bettmann/CORBIS; 189, (cr) © brown brothers; 191, (b) © AP/Wide World Photos; 191, (t) © brown brothers; 192, (tl) © AKG London; 192, (bl) © AKG London; 193, (cr) © AKG London; 194, (c) © Hulton Archive by Getty Images; 194, (tr) © Paul Velasco/Gallo Images/Corbi; 195, (tr) © Courtesy Organization of African Unity; 195, (bl) © black Star; 196, (tl) © Sean Gallup/Getty News Services; 196, (bl) © A. Ramey/ Woodfin Camp & Associates; 197, (tr) © Hulton-Deutsch Collection/CORBIS; 197, (br) © Hulton Archive by Getty Images; 198, (tl) © Paul Almasy/CORBIS; 199, (t) © AKG London; 199, (bl) © Giraudon/Art Resource, NY; 200, (tl) © UPI/bettmann/CORBIS; 200, (br) © bilderdienst Suddeutscher Verlag; 201, (tl) © AKG London; 201, (br) © American Stock/Hulton Archive by Getty Images, Hulton Archive by Getty Images; 202, (tl) © Loomis Dean/timePix; 202, (bl) © bruno P. Zehnder/Peter Arnold, Inc; 204, (cl) © AP/Wide World Photos; 204, (t) © PhotoDisc, Inc.; 205, (tr) © AP/Wide World Photos; 206, (tl) © CORBIS; 831 AKG London; 206, (bl) © Getty News Services; 207, (cr) © Hanan Isachar/CORBIS; 207, (t) © bruno barbey/Magnum/PictureQuest; 208, (c) © AKG London; 209, (tr) © Magnus bartlet/Woodfin Camp & Associates; 209, (bl) © Mario Corvetto/Evergreen Photo Alliance; 210, (cl) © Novosti/Science Photo Library/Photo Researchers; 210, (br) © David Young-Wolff/PhotoEdit; 210, (tl) © bob Daemmrich/CORBIS Sygma; 211, (tr) © European Communities; 214, (bl) © Stuart Ramson/AP/Wide World Photos; 214, (tr) © Mark Wilson/Getty Images; 215, (tr) © Galen Rowell/Mountain Light Photography; 217, (cr) © Scala/Art Resource, NY; 217, (tr) © Francis G. Mayer/CORBIS; 217, (cl) © Royal Holloway and bedford New College, Surrey, UK/bridgeman Art Library; **Unit 3:** 218-239 (c), SuperStock; 218 (t), © E. Nagele/FPG International LLC; 218 (bl), Alfredo Venturi/Masterfile; 219 (t), FPG International LLC; 219 (br), Robert Maier/Animals Animals/Earth Scenes; **Chapter 10**: 230, Steve Ewert Photography ; 231, © Robert Frerck/Odyssey/Chicago; 232, Photo © Earl Bronsteen 1/Panoramic Images, Chicago 1998; 233, © Gary Braasch/CORBIS; 234 (b), © Travelpix/ FPG International LLC; 234 (t), Museo Archeologico Nazionale, Naples, Italy/Bridgeman Art Library; 235 (b), © Joe Viesti/The Viesti Collection; 235 (t), © Leo de Wys/Steve Vidler; 236 (bl), SEF/Art Resource, New York; 236 (br), © Araldo de Luca/CORBIS; 238 (l), © Archivo Iconografico, S.A./C; 238 (r), Christie's Images Ltd.; 239 (b), © Gianni Dagli Orti/CORBIS; 239 (t), © Louis Goldman/FPG International LLC; 240 (b), © Leo de Wys/Siegfried Tauquer; 240 (t), © Bill Staley/ FPG International LLC; 242 (t,b), © Index Stock Photography; 243, © Jean Kugler/FPG International LLC; 244 (b), © Robert Frerck/Odyssey/Chicago; 245 (t), © Sitki Tarlan/Panoramic Images, Chicago 1998; 245 (b), © Charles O'Rear/CORBIS ; **Chapter 11**: 248, Steve Ewert Photography; 249, © Stone/Hideo Kurihara; 250 (b), © Leo de Wys/John Miller; 250 (t), © Leo de Wys/J. Messerschmidt; 251, © The Stock Market/H. P. Merten; 252 (t), Image Copyright ©2002 PhotoDisc, Inc./HRW; 252 (l), Castres, Musee Goya /Art Resource, NY; 253 (b), Scala/Art Resource; 253 (t), The Art Archive; 254 (lc), © Joe Viesti/The Viesti Collection; 254 (r), © Leo de Wys/Mike Busselle; 254 (tl), Pierre Witt/Rapho/ Liaison Agency; 256 (b), © Wolfgang Kaehler; 256 (t), © 1999 Ron Kimball/Ron Kimball Photography; 257 (inset), Archive Photos; 257 (cr), AKG Photo, London; 257 (t), CORBIS-Bettmann; 258, © Leo de Wys/Fridmar Damm; 260 (l), © Leo de Wys/Siegfried Tauqueur; 260 (t), Image Copyright ©2002 PhotoDisc, Inc./HRW; 261 (t), © Wolfgang Kaehler; 261 (b), © Steve Vidler/Nawrocki Stock Photo; 263 (t) Image Copyright ©2002 PhotoDisc, Inc./HRW; 263 (b), © Ken Ross/FPG International LLC; 264 (b), © G. Wagner/The Viesti Collection; 264 (t), © Stone/Siegfried Layda; **Chapter 12**: 268, Steve Ewert Photography; 269, © Werner Forman/CORBIS; 270, © Walter Bibikow/The Viesti Collection; 272 (b), British Museum, London, UK/Bridgeman Art Library; 272 (t), Bridgeman Art Library; 272 (c), © Yann Arthus-Bertrand/CORBIS; 273 (b), © Stone/Ed Pritchard; 273 (t),

Private Collection/ Bridgeman Art Library; 274 (b), © Jim Richardson; 274 (t), Popperfoto/Archive Photos; 276 © Historical Picture Archive/CORBIS; 277 U.S. National Library of Medicine, National Institutes of Health; 278 (t), Image Copyright ©2002 PhotoDisc, Inc./HRW; 278-309, b Joe Englander/Viesti Collection; 279 (r), © Becky Luigart-Stayner/CORBIS; 281 (t), Susan Marie Anderson/ Foodpix ; 281 (b), © Index Stock Photography; 282 (b), © Walter Bibikow/FPG International LLC; 283 (t), © Index Stock Photography and/or Zefa 1999; 284, © The Stock Market/Tom Stewart; 285, © Bryan & Cherry Alexander; **Chapter 13**: 288, Steve Ewert Photography; 289 (t), © Dr. Paul A. Zahl, Nat'l Audubon Society Collection/Photo Researchers; 290 (b), Robert Tixador/Figaro/Liaison Agency; 291, V. Leloup/ Figaro/Liaison Agency; 292 (b), © Paul Almasy/ CORBIS; 292 (t), © Archivo Iconografico, S.A./ CORBIS; 293 (b), Aldo Pavan/Liaison Agency; 294, © David Bartruff/FPG International LLC; 295 (t), Image Copyright © 2002 PhotoDisc, Inc./HRW; 296 (b), © Fergus O'Brien/FPG International LLC; 296 (t), © Travelpix/FPG International LLC; 297, © Garbor Feher/CORBIS-Sygma; 298 (t), © Jonathan Blair/ CORBIS; 298 (l), © Francoise de Mulder/CORBIS; 329 (b), Krpan Jasmin/Liaison Agency; 299 (t), © Michael S. Yamashita/CORBIS; 300, Marleen Daniels/ Liaison Agency; 301 (b), Francis Li/Liaison Agency; 301 (t), © Patrick Chauvel/CORBIS-Sygma; 302, © Matt Glass/CORBIS-Sygma; 306, © Bill Ross/CORBIS; 307 (tr), © European Communities ; **Unit 4:** 310 (bl), © SuperStock; 310 (br), © Ed Kashi; 310-341 (c), Mark Wadlow/Russia and Eastern Images; 311 (t), Bruce Coleman Inc.; 311 (br), Gerard Lacz/Peter Arnold, Inc.; 318, Robert S. Semeniuk/Black Star; 319 (t), © Gérard Degeorge/CORBIS; 319, Steve Ewert Photography; **Chapter 14**: 320, Steve Ewert Photography; 321, © Itar-Tass/Sovfoto/Eastfoto; 322, © Tass/Sovfoto/Eastfoto; 323, © Bryan & Cherry Alexander; 324, © Hans J. Burkard/Bilderberg/Aurora; 326, CORBIS-Bettmann; 325 (t), © StockFood America/Eising; 325 (b), Battle of the Novgorodians with the Suzdalians, Novgorod School, mid 15th century (tempera and gold on panel)/Tretyakov Gallery, Moscow, Russia/Bridgeman Art Library, London/New York ; 327 (b), © Steve Raymer/CORBIS; 327 (t), © Steve Raymer/CORBIS; 328 (t), © Wally McNamee/ CORBIS; 329, © Steve Raymer/CORBIS; 331 (b), © Vladimir Pcholkin/FPG International LLC; 331 (t), © Adam Woolfitt/CORBIS; 332 (b), © Steve Raymer/ CORBIS; 332 (t), © Steve McCurry/Magnum; 333, © Claus Meyer/Black Star/PNI; 334 (t), Image Copyright ©2002 PhotoDisc, Inc./HRW ; 334 (b), 335 (b), © Bryan & Cherry Alexander; 335 (t), Sovfoto/Eastfoto; 336, © Dean Conger/CORBIS; 337 (b), © Planet Earth Pictures/FPG International LLC; 337 (t), © The State Russian Museum/CORBIS; 338, © Wolfgang Kaehler/ CORBIS; 339, © Michael S. Yamashita/CORBIS ; **Chapter 15**: 342, Steve Ewert Photography; 343, Sisse Brimberg/NGS Image Collection; 344, © Dean Conger/ CORBIS; 346 (t), Image Copyright © 2002 PhotoDisc, Inc./HRW; 346 (b), AKG Photo, London; 347 (b), Steve Raymer/NGS Image Collection; 347 (t), AKG Photo, London; 348, Steve Raymer/NGS Image Collection; 349 © Randall Hyman; 351 (b), © Charles Lenars/CORBIS; 351 (t), © Dean Conger/CORBIS; 352 (b), © Sovfoto/Eastfoto; 352 (t), Stephanie Maze, USA; **Chapter 16**: 356, Brian Vikander/Vikander Photography; 357, © Wolfgang Kaehler/CORBIS; 358 b), © Hans Reinhard/Bruce Coleman Inc.; 358 t), © Tass/Sovfoto/Eastfoto; 359, © Yann Arthus-Bertrand/ CORBIS; 360 (b), © K.M. Westermann/CORBIS; 360 (t), © David Samuel Robbins/CORBIS; 361, © Wolfgang Kaehler/CORBIS; 362, © Nevada Wier/CORBIS; 363 (b), © Wolfgang Kaehler/CORBIS; 363 (t), © David S. Robbins/CORBIS; 364, © Dean Conger/CORBIS; 366, Alain Le Garsml/Panos Pictures; 367, Chris Stowers/ Panos Pictures; 370, © 1990 Abbas/Magnum Photos; 371 (c), Marc Garanger/CORBIS; 373, Sovfoto/ Eastfoto; **Unit 5:** 374 (bl), Richard A. Lobell Photography; 374 (br), Kevin Rushby; 374-405 (c), © Beryl Goldberg; 375 (cr), Annie Griffiths Belt/National Geographic Society Image Collection; 375 (br), Hill, M. Osf/Animals Animals/Earth Scenes; **Chapter 17**: 384, Steve Ewert Photography; 385, © Gianni Dagli Orti/CORBIS; 386, © Marc Riboud / Magnum; 387, © Stephen Frink/CORBIS; 388 (t), Christine Osborne Pictures/ MEP; 388 (b), © Abbas/Magnum; 389, © Adam Woolfitt/CORBIS; 390, NASA Johnson Space Center; 391, University Library Istanbul/The Art Archive; 392, © TRIP Photo Library/H Rogers; 393, © AbbieEnock; Travel Ink/CORBIS; 395 (t), © Burnett H. Moody/Bruce Coleman, Inc.; 396 (b), © Stephen Wallace/Black Star; 396 (t), © Gianni Dagli Orti/CORBIS; 397, © Alexandra Avakian/Contact Press Images; 398, © Nik Wheeler/CORBIS; 399, © Burstein Collection/CORBIS; 400, © Alexandra Avakian/Contact Press Images; **Chapter 18**: 404, © Zafer KIZILKAYA; 405, © Wolfgang Kaehler/CORBIS; 406-439, © Robert Frerck/Odyssey/Chicago; 409 (t), © AFP (Staton R. Winter)/CORBIS; 410 (t), Colossus of Gilgamesh Gripping a Lion, relief from the Palace of Sargon II at Khorsabad, Iraq, Assyrian Period, c. 725 BC (gypsum)/Ruenion des Musees Nationaux/Bridgeman Art Library, London/New York; 411, © Robert Frerck/ Odyssey/Chicago; 412 (l), Bill Curtsinger/NGS Image Collection; 412 (r), Israel Antiquities Authority; 413, © Peter Stone/Black Star/PNI; 414, © Alexandria Avakian/Contact Press Images; 415, © Richard T. Nowitz/CORBIS; 416 (b), © Ed Kashi; 416 (t), © Dave Bartuff/CORBIS; 417 (t), © Ed Kashi ; 417 (b), William Maynard Owen/NGS Image Collection; 418 (b), © Richard T. Nowitz/CORBIS; 418 (t), © Tomas Muscionico/Contact Press Images; 419, © Ed Kashi; 422 (b), Roger Antrobus/CORBIS; 422 (t), Pictor Uniphoto; **Unit 6:** 426 (bl), © Stone/Theo Allofs; 426 (t), © Robert Frerck/Odyssey/Chicago; 426-457 (b), Herb Zulpier/Masterfile; 427 (cr), © Charles Henneghien/Bruce Coleman, Inc.; 427 (br), S. Michael Bisceglie/Animals Animals/Earth Scenes; 439, Steve Ewert Photography; **Chapter 19**: 440, Steve Ewert Photography; 441, © Robert Frerck/Odyssey/ Chicago; 442, Guiseppe Bizzarri/Panos Pictures; 443 (tl), © Staffan Widstrand/CORBIS; 443 (tr), Photo Researchers, Inc.; 444, © Roger Tidman/CORBIS; 445 (t), Jean Léo Dugast/Panos Pictures; 445 (cr), Kenneth Garrett/NGS Image Collection; 446, Image Copyright ©2002 PhotoDisc, Inc./HRW; 447 (t), © M. Timothy O'Keefe/Bruce Coleman, Inc.; 447 (b), © Carmen Redondo/CORBIS; 447 (c), © Sharon Smith/ Bruce Coleman, Inc.; 448 (t), Frank and Helen Schreider/Photo Researchers, Inc.; 448 (bl), Jean-Léo Dugast/Panos Pictures; 449, Kazuyoshi Nomachi/ Photo Researchers, Inc.; 451 (cl), Leonard de Selva/ CORBIS; 451 (bl), © SuperStock; 451 (t), Bettman/ CORBIS; 452 (t), Egyptian National Museum, Cairo, Egypt/SuperStock; 452 (b), © Christine Osborne/ CORBIS; 453, © Yann Arthus-Bertrand/CORBIS; 454, © Jeffrey L. Rotman/CORBIS; 455, Jean-Léo Dugast/ Panos Pictures; 456, J. PH. Charbonnier/Photo Researchers, Inc.; 457, James L. Stanfield/NGS Image Collection; **Chapter 20**: 460, Steve Ewert Photography; 461, Betty Press/Woodfin Camp & Associates; 462, © Wolfgang Kaehler; 463, © Gail Shumway/FPG International LLC; 464 (t) Michael Holford; 464 (b), 465, © Wolfgang Kaehler; 466, © Stone/Will Curtis; 467, M & M Bernheim/Woodfin Camp & Associates; 468, © Peter Guttman/LIFE Magazine; 469 (all), © Wolfgang Kaehler; 470, © John Elk/Bruce Coleman, Inc.; 472 (c), © Stone/Sally Mayman; 472 (t), Goldweight representing two crocodiles with a shared stomach, Asante, Ghana 18th-19th century (brass)/Heini Schneebeli/Bridgeman Art Library, London/New York; 473 (b), © Alex Wasinski/ FPG International LLC; 473 (t), © Stone/James Nelson; 474, © SuperStock; 475, Marc & Evelyn Bernheim/Woodfin Camp & Associates; **Chapter 21**: 478, Neil Cooper/Panos Pictures; 479, © Norman Owen Tomalin/Bruce Coleman, Inc.; 480, © Gerald Cubitt; 481, M. Denis-Huot/Liaison Agency; 482 (t), M. & E. Bernheim/ Woodfin camp & Associates; 482 (bl), Giraudon/Art Resource, NY; 483, New York Public Library; 484, Victor Englebert; 485 (br), Dave G. Houser; 485 (t), Jar decorated with lion masks and cobra goddesses on lotus flowers, from Tomb 1090, Faras, Sudan, 1st-2nd century AD (painted pottery)/ Ashmolean Museum, Oxford, UK/Bridgeman Art Library, London/New York; 486, Daniel J. Cox/Liaison Agency; 487 (tr), © Wolfgang Kaehler; 488, Khartoum, Sudan, in the 1860's, engraved by Alfred Louis Sargent (b.1828) (engraving)/Private Collection/ Bridgeman Art Library, London/New York; 489 (b), Victor Englebert; 489 (t), Dave G. Houser; 490, Betty Press/Woodfin Camp & Associates; 491, Maya Kardum/Panos Pictures; **Chapter 22**: 494, Steve Ewert Photography; 495, M & E. Bernheim/Woodfin Camp & Associates; 496, © Gerald Cubitt; 497, Michael Nichols/NGS Image Collection; 498 (bl), Christie's Images; 498 (tr), Schomburg Center, The New York Public Library/Art Resource, NY; 499, © Stone/Ian Murphy; 500, M. & E. Bernheim/Woodfin Camp & Associates; 501 (tl), M. Edwards/Still Pictures/Peter Arnold, Inc.; 501 (cr), © David Reed/CORBIS; 502 (bl), Jose Azel/Aurora; 502 (tr), Aldo Tutino/Art Resource, NY; 503, © SuperStock; 504, Robert Caputo/Aurora; 505 (br), M & E Bernheim/Woodfin Camp & Associates; 505 (tr), Nicolet Gilles/Bios/Peter Arnold, Inc.; 506, Jason Lauré/Lauré Communications; **Chapter 23**: 510, Trygve Bolstad/Panos Pictures; 511, © Gerald Cubitt; 512, © Stone/John Lamb; 513, © Gerald Cubitt; 514 (cl), © Gerald Cubitt; 514 (tr), Robert Holmes; 515 (br), Peggy Kelsey; 515 (tr), Jason Lauré/Lauré Communications; 516 (tr), The Rabler off Table Mountain, Cape Town, c 1865/Jersey Museums Service, UK/Bridgeman Art Library, London/New York; 516 (bl), ©Stone/Hulton Getty; 517, © Gerald Cubitt; 518 (cl), Archive Photos/Express Newspapers; 518-550 (all), Jason Lauré/Lauré Communications; 521, © Stone/Steve Vidler; 522 (bl), © Gerald Cubitt; 522 (cr), Jason Lauré/Lauré Communications; 523, © Gerald Cubitt; 524, The Granger Collection, New York; 525, © Gerald Cubitt; 528, © Fritz Polking/Peter Arnold, Inc.; 529 (b), Peter Arnold, Inc.; 529 (t), © William Campbell/Peter Arnold, Inc.; **Unit 7:** 532 (bl), Vision Photo/Pacific Stock; 532 (br), © Stone/Margaret Gowan; 532-563 c © Ken Ross/FPG International LLC; 533 (cr), © Dean Conger/CORBIS; 533 (br), © John Giustina/FPG International LLC; **Chapter 24**: 542, Steve Ewert Photography; 543, © Asian Art and Archaeology/CORBIS; 544, © Stone/Colin Prior; 545, © Brian Vikander; 546, Courtesy of Freer Gallery of Art, Smithsonian Institution, Washington, D.C.; 547 (b), © Stone/D. E. Cox; 547 (t), National Museum, Beijing, China/Photograph by Erich Lessing/Art Resource, NY; 548 (t), © James Montgomery/Bruce Coleman, Inc.; 548 (bl), Giraudon/Art Resource, NY; 548 (inset), Dennis Cox/China Stock; 549, Catalan Atlas, detail showing the family of Marco Polo (1254-1324) travelling by camel caravan/British Library, London, UK/Bridgeman Art Library, London/New York; 550 (t), Harry Redl/Black Star; 550-581 b © James Montgomery/Bruce Coleman, Inc.; 551 (t), Jeffrey Aaronson/Network Aspen; 552, © Keren Su/FPG International LLC; 553, © 1997 Michele Burgess; 554, A. Ramey/Woodfin Camp & Associates, Inc.; 555 (b), © Keren Su/FPG International LLC; 555 (t), Jeffrey Aaronson/Network Aspen; 556 (b), © Stone/Yann Layma; 556 (t), CORBIS; 557, Jeffrey Aaronson/ Network Aspen; 558 (b), Mongolian Eight flags soldiers from Ching's military forces, engraved by R.Rancati (colour engraving)/Private Collection/ Bridgeman Art Library, London/New York; 558 (t), © The Purcell Team/Corbis; 559 (t), © Stone/Michel Setboun; 559 (inset), © Wolfgang Kaehler; 560 (b), Orion Press/Pacific Stock; 560 (t), © Stone/Hulton Getty; 561, © Stone/Hugh Sitton; **Chapter 25**: 564, Steve Ewert Photography; 565, © M. Fairman/Trip Photo Library; 566, © Uniphoto Pictor ; 567, ©SuperStock; 568, © Toyohiro Yamada/FPG International LLC; 569 (br), © Michael S. Yamashita/ CORBIS; 569 (cr), © Asian Art and Archaeology, Inc./CORBIS; 570, Cover of < U > Sadako and the Thousand Paper Cranes < U > by Eleanor Coerr © 1999 Puffin Books, a member of Penguin Putnam Inc., Books for Young Readers. Cover illustration copyright © Kazuhiko Sano, 1999; 571 (tr), © Bettman/CORBIS; 572 (bl), © SuperStock; 572 (tr), © C. Rennie/ TRIP Photo Library; 573, © Michael S. Yamashita/CORBIS; 574 (t), Toyohiro Yamada/FPG International LLC; 574 (bl), © David Wade/FPG International LLC; 575, Orion Press/Pacific Stock; 576, 577 (cr), Kim Newton/ Woodfin Camp & Associates, Inc; 577 (b), City Art Gallery, Leeds/Bridgeman Art Library, London/ SuperStock; 577 (tl), Spink and Son Ltd., London/ Bridgeman Art Library, London/SuperStock; 578, © SuperStock; 580, © Michael Boys/CORBIS; 581 (b), Nathan Benn/Woodfin Camp & Associates, Inc.; 581 (tr), © SuperStock; 582 (l), 583, Jeffrey Aaronson/ Network Aspen; **Chapter 26**: 586, Steve Ewert Photography; 587, 588, © Gerald Cubitt; 589, Fred Hoogervorst/Panos Pictures; 590 (b), Brian Vikander; 590 (t), © John Elk III/Bruce Coleman, Inc.; 591 (t), Bullit Marquez/AP/Wide World Photos; 591 (b), Charlyn Zlotnik/Woodfin Camp & Associates, Inc.; 592, © Richard Bickel/CORBIS; 593, Chris Sattlberger/ Panos Pictures; 594 (b), © John Elk III/Bruce Coleman, Inc.; 594 (t), Brian Vikander; 595, © Stefano Amantini/Bruce Coleman, Inc.; 596, © Gerald Cubitt; 597 (b), R. Ian Lloyd/The Stock Market; 597 (t), Robert Fried Photography; 598, Vision Photo/Pacific Stock; 599, © John Elk III/Bruce Coleman, Inc.; 600 (b), Veronica Garbutt/Panos Pictures; 600 (t), © Wolfgang Kaehler; 601, © Chris Salvo/FPG International LLC; 603 (t), Michele Burgess/The Stock Market; 606, © Frans Lanting/Minden Pictures; 607, © Picture Press/CORBIS; **Unit 8:** 610 (t), © Stone/Hugh Sitton; 610 (b), © Stone; 610 (c), © Stone/Art Wolfe; 611 (c), © Stone; 611 (b), Robert Winslow/Animals Animals/Earth Scenes; 618, © Adam Woolfit/CORBIS; 619 (t), Index Stock Imagery, Inc.; 619 (b), Steve Ewert Photography; **Chapter 27**: 620, Steve Ewert Photography; 621, © Dinodia/TRIP Photo Library; 622, Ric Ergenbright; 624, © R Graham/TRIP Photo Library; 625 (c), © H. Rogers/TRIP Photo Library; 625 (cr), The Memoirs of Babur/Nat'l Museum of India, New Delhi, India/Bridgeman Art Library, London/ New York; 626, © Cheryl Sheridan/Odyssey/Chicago; 627, Craig Lovell/ Eagle Visions; 628 (bl), © Stone/ Hulton Getty ; 628 (tl), Atkinson, George Franklin (1822-59)/British Library, London/Bridgeman Art Library, London/New York; 629, © Stone/Hulton Getty; 630 (bl), Ric Ergenbright; 630 (cl), © H. Rogers/ TRIP Photo Library; 630 (cr), © Resource Foto/TRIP Photo Library; 631 (tr), © Rogers/TRIP Photo Library; 631 (bl), © B. Turner/TRIP Photo Library; 632 (tr), © Robert Frerck/Odyssey/Chicago; 632 (b), © Mary Altier; 633, © Robert Frerck/Odyssey/Chicago; 634 (tl), Martin Adler/Panos Pictures; 634 (b), Ric Ergenbright; 635, © Robert Frerck/Odyssey/Chicago; **Chapter 28**: 638, Steve Ewert Photography; 639, © Luca I. Tettoni/CORBIS; 640, © Galen Rowell/CORBIS; 642 (c), © Nik Wheeler/CORBIS; 642 (t), CORBIS; 643, © Ed Kashi; 644, © Mike Goldwater/Network/ SABA; 645, Kevin Bubriski, courtesy of Aramco World Magazine. Used with permission.; 646, Binod Joshi/ AP/Wide World Photos; 647 (t), © Alison Wright/ CORBIS; 647 (br), © Christine Kolisch/CORBIS; 648 (l), William Thompson; 648 (t), © Robert Frerck / Odyssey/Chicago; 649, CORBIS/Bettman; 650 (t), © Stone/Paula Bronstein; 650 (b), © Evelyn Scott/Bruce Coleman, Inc.; 651, © Robert Frerck / Odyssey / Chicago; 652 (b), © Robert Frerck / Odyssey / Chicago; 652 (t), © Masha Nordbye/Bruce Coleman, Inc.; 653 (t), E. Valentin/Liaison Agency; 653 (c), Image Copyright ©2002 PhotoDisc, Inc./HRW; 659, © Galen Rowell/CORBIS; **Unit 9:** 660 (cl), © Stone/Glen Allison; 660 (t), © Frans Lanting/Minden Pictures; 660-691 b © Bill Bachman; 661 (cr), © Stone/Greg Probst; 661 (br), © Picture Finders Ltd./Leo de Wys; 669, Steve Ewert Photography; **Chapter 29**: 670, Penny Tweedie/HRW Photo; 671, © Alfred Pasieka/ Peter Arnold, Inc.; 672 (b), Digital Stock Corp./HRW; 673 (t), Digital Stock Corp./HRW; 674, W. Robert Moore/NGS Image Collection; 675, Aboriginal Bark painting showing the path taken by the soul on its journey to the other world/Private Collection/ Bridgeman Art Library, London/New York; 676, Digital Stock Corp./HRW; 678 (c), Wayne Lawler/ Photo Researchers, Inc.; 678 (b), © Four by Five/ SuperStock; 680 (b), © Clyde H. Smith/Peter Arnold, Inc.; 680 (t), Pendant in the form of the god Hei-Tiki from the Maori, New Zealand, late 18th century (nephrite)./Bonhams, London/Bridgeman Art Library, London/New York; 681 (t), © Colin Monteath/ Auscape; 681 (b), SEF/Art Resource, NY; 682 (t), Michael Steele/Allsport; 682 (b), John Eastcott/NGS Image Collection; 683, © Lance Nelson/The Stock Market; **Chapter 30**: 686, © Zafer KIZILKAYA; 687, © Amos Nachoum/CORBIS; 688, © Stone/ Paul Chelsey; 689 (b), © Bojan Brecelj/CORBIS; 689 (inset), © Arne Hodalic/CORBIS; 690, 691 (t), © Frans Lanting/ Minden Pictures; 691 (inset), Maria Stenzel/NGS Image Collection; 692 (l), Liaison Agency; 692 (t), © Index Stock Photography and / or Horst Von Irmer; 693, © David Austen / Stock, Boston / PNI; 694, Picture Finders Ltd. /Leo de Wys; 695, © Jack Fields/ CORBIS; 696, © Craig Lovell/CORBIS; 697 (b), CORBIS; 697 (t), © Galen Rowell/CORBIS; 698, © Norbert Wu/www.norbertwu.com; 699, © Stone/Art Wolfe.